WATER QUALITY ENGINEERING

WATER QUALITY ENGINEERING

Physical/Chemical Treatment Processes

MARK M. BENJAMIN
DESMOND F. LAWLER

WILEY

Cover design: John Wiley & Sons, Inc.
Cover photograph: Courtesy of Brain Haws

Copyright © 2013 by John Wiley & Sons, Inc. All rights reserved.

Published by John Wiley & Sons, Inc., Hoboken, New Jersey.
Published simultaneously in Canada.

No part of this publication may be reproduced, stored in a retrieval system, or transmitted in any form or by any means, electronic, mechanical, photocopying, recording, scanning, or otherwise, except as permitted under Section 107 or 108 of the 1976 United States Copyright Act, without either the prior written permission of the Publisher or authorization through payment of the appropriate per-copy fee to the Copyright Clearance Center, Inc., 222 Rosewood Drive, Danvers, MA 01923, +(978) 750-8400, fax +(978) 750-4470, or on the web at www.copyright.com. Requests to the Publisher for permission should be addressed to the Permissions Department, John Wiley & Sons, Inc., 111 River Street, Hoboken, NJ 07030, +(201) 748-6011, fax +(201) 748-6008, or online at http://www.wiley.com/go/permission.

Limit of Liability/Disclaimer of Warranty: While the publisher and author have used their best efforts in preparing this book, they make no representations or warranties with respect to the accuracy or completeness of the contents of this book and specifically disclaim any implied warranties of merchantability or fitness for a particular purpose. No warranty may be created or extended by sales representatives or written sales materials. The advice and strategies contained herein may not be suitable for your situation. You should consult with a professional where appropriate. Neither the publisher nor author shall be liable for any loss of profit or any other commercial damages, including but not limited to special, incidental, consequential, or other damages.

For general information on our other products and services or for technical support, please contact our Customer Care Department within the United States at +(800) 762-2974, outside the United States at +(317) 572-3993 or fax +(317) 572-4002.

Wiley also publishes its books in a variety of electronic formats. Some content that appears in print may not be available in electronic formats. For more information about Wiley products, visit our web site at www.wiley.com.

Library of Congress Cataloging-in-Publication Data:

Benjamin, Mark M.
  Water quality engineering: physical/chemical treatment processes/Mark Benjamin, Desmond Lawler.
    pages cm
    Includes bibliographical references and index.
    ISBN 978-1-118-16965-0 (cloth)
1. Water—Purification. 2. Sewage—Purification. I. Lawler, Desmond F. II. Title.
  TD430.B386 2013
  628.1′66–dc23

2012023641

Printed in the United States of America

10 9 8 7 6 5 4 3 2 1

CONTENTS

PREFACE   xxi

ACKNOWLEDGMENTS   xxv

PART I   REACTORS AND REACTIONS IN WATER QUALITY ENGINEERING

1 Mass Balances   3
    1.1 Introduction: The Mass Balance Concept, 3
    1.2 The Mass Balance for a System with Unidirectional Flow and Concentration Gradient, 7
       The Storage Term, 8
       The Advective Term, 10
       The Diffusion and Dispersion Terms, 11
       The Chemical Reaction Term, 15
       Combining the Terms into the Overall Mass Balance, 17
       The Differential Form of the One-Dimensional Mass Balance, 18
    1.3 The Mass Balance for a System with Flow and Concentration Gradients in Arbitrary Directions, 20
       The Advection Term, 20
       The Diffusion and Dispersion Terms, 21
       The Storage and Reaction Terms, 23
       The Overall Mass Balance, 23
    1.4 The Differential Form of the Three-Dimensional Mass Balance, 24
    1.5 Summary, 25
       References, 26
       Problems, 27

2 Continuous Flow Reactors: Hydraulic Characteristics   29
    2.1 Introduction, 29
    2.2 Residence Time Distributions, 30
       Tracers, 31
       Pulse Input Response, 33

Step Input Response, 35
Statistics of Probability Distributions and the Mean Hydraulic
   Detention Time, 37
2.3 Ideal Reactors, 42
　　Plug Flow Reactors, 42
　　　　Pulse Input to a PFR: Fixed Frame of Reference (Eulerian View), 43
　　　　Pulse Input to a PFR: Moving Frame of Reference (Lagrangian View), 44
　　Continuous Flow Stirred Tank Reactors, 45
　　　　Pulse Input to a CFSTR, 45
　　　　Step input to a CFSTR, 47
2.4 Nonideal Reactors, 48
　　Tracer Output from Nonideal Reactors, 48
　　　　Relating Tracer Input and Output Curves via the Convolution Integral, 48
　　Modeling Residence Time Distributions of Nonideal Reactors, 50
　　　PFR with Dispersion, 50
　　　CFSTRs in Series, 55
　　　Modeling Short-Circuiting and Dead Space, 57
　　　PFRs in Parallel and Series: Segregated Flow and Early Versus Late
      Mixing, 59
　　　Nonequivalent CFSTRs in Series, 62
　　　Simple Indices of Hydraulic Behavior, 62
2.5 Equalization, 62
　　Flow Equalization, 63
　　Concentration Equalization, 66
　　Concurrent Flow and Concentration Equalization, 69
2.6 Summary, 70
　　Appendix 2A. Introduction to Laplace Transforms as a Method of Solving
   (Certain) Differential Equations, 71
　　　Examples of the Use of Laplace Transforms, 73
　　References, 73
　　Problems, 74

3 Reaction Kinetics　　　　　　　　　　　　　　　　　　　　　　　　　　　　81

3.1 Introduction, 81
3.2 Fundamentals, 82
　　Terminology, 82
　　The Kinetics of Elementary Reactions, 84
　　　Frequency of Molecular Collisions, 84
　　　Energetics of Molecular Collisions, 84
　　The Kinetics of Nonelementary Reactions, 87
　　Power Law and Other Rate Expressions for Nonelementary Reactions, 88
3.3 Kinetics of Irreversible Reactions, 88
　　The Mass Balance for Batch Reactors with Irreversible Reactions, 89
　　The Integral Method of Reaction Rate Analysis, 89
　　Analysis of Reaction Half-Times, 91
　　Kinetics Expressions Containing Terms for the Concentrations of More
      Than One Reactive Species, 93
　　The Differential Method of Reaction Rate Analysis, 96
　　Analysis of Nonpower-Law Rate Expressions, 97
　　Characteristic Reaction Times, 97
3.4 Kinetics of Reversible Reactions, 99
　　Reversible Reactions, 99

Characteristic Times and Limiting Cases for Reversible Reactions, 103
Simplification of Reaction Rate Expressions for Limiting Cases, 104
   Very Rapid and Very Slow Approach to Equilibrium as Limiting Cases, 104
   Reaction Quotients, Equilibrium, and the Assumption of Irreversibility, 105
   Nearly Complete Reaction as a Limiting Case, 106
   Summary of Limiting Cases, 107
3.5 Kinetics of Sequential Reactions, 107
   The Progress of Consecutive Reactions and the Rate-Controlling Step, 108
   The Thermodynamics of Sequential Reactions, 111
   Steady State: Definition and Comparison with Chemical Equilibrium, 112
3.6 The Temperature Dependence of the Rates of Nonelementary Reactions, 114
3.7 Summary, 115
References, 116
Problems, 116

4 Continuous Flow Reactors: Performance Characteristics with Reaction   121

4.1 Introduction, 121
4.2 Extent of Reaction in Single Ideal Reactors at Steady State, 121
   Extent of Reaction in a Continuous Flow Stirred Tank Reactor at Steady State, 121
      First-Order Irreversible Reactions, 122
      Non-First-Order Irreversible Reactions, 123
   Extent of Reaction in a Plug Flow Reactor at Steady State, 123
      Fixed Frame of Reference (Eulerian View), 124
      Moving Frame of Reference (Lagrangian View), 125
      Irreversible nth-Order Reactions, 125
   Comparison of CFSTRs and PFRs for Irreversible Reactions, 126
   Reversible Reactions, 129
4.3 Extent of Reaction in Systems Composed of Multiple Ideal Reactors at Steady State, 130
   PFRs in Series, 130
   CFSTRs in Series, 130
      Application to Chemical Disinfection, 133
   CFSTRs or PFRs in Parallel, 135
   Using Reactors with Flow to Derive Rate Expressions, 135
4.4 Extent of Reaction in Reactors with Nonideal Flow, 135
   Fraction Remaining Based on the Exit Age Distribution, 136
   Fraction Remaining Based on the Dispersion Model, 140
   Summary of Steady-State Performance in Nonideal Reactors, 141
4.5 Extent of Reaction Under Non-Steady-Conditions in Continuous Flow Reactors, 141
   Extent of Conversion in PFRs Under Non-Steady-State Conditions, 141
   Extent of Conversion in CFSTRs Under Non-Steady-State Conditions, 142
   Extent of Conversion in Nonideal Reactors Under Non-Steady-State Conditions, 144
4.6 Summary, 146
References, 147
Problems, 147

PART II REMOVAL OF DISSOLVED CONSTITUENTS FROM WATER

5 Gas Transfer Fundamentals   155

5.1 Introduction, 155
- Importance of Gas Transfer in Environmental Engineering, 155
- Overview of Gas/Liquid Equilibrium, 155
- Overview of Transport and Reaction Kinetics in Gas Transfer Processes, 157
- Incorporating Gas Transfer into Mass Balances, 157
- Chapter Overview, 158

5.2 Types of Engineered Gas Transfer Systems, 159

5.3 Henry's Law and Gas/Liquid Equilibrium, 162
- Volatilization and Dissolution as a Chemical Reaction, 162
- Partition Coefficients, Equilibrium Constants, and the Formal Definition of Henry's Law, 162
- Dimensions of c_L, c_G, and Henry's Law Constant, 164
- Factors Affecting Gas/Liquid Equilibrium, 167

5.4 Relating Changes in the Gas and Liquid Phases, 170

5.5 Mechanistic Models for Gas Transfer, 170
- Fluid Dynamics and Mass Transport in the Interfacial Region, 170
- The Mass Balance on a Volatile Species Near a Gas/Solution Interface, 171
  - Gas Transfer and Transport Through a Fluid Packet at the Interface, 171
  - Flux Under Limiting-Case Scenarios: Short and Long Packet Residence Times, 174
  - Accounting for the Packet Age and Packet Residence Time Distribution, 175
  - The Gas Transfer Coefficient and Its Interpretation, 175

5.6 The Overall Gas Transfer Rate Coefficient, K_L, 179
- The Combined Resistance of the Gas and Liquid Phases, 179
- Comparing Gas-Phase and Liquid-Phase Resistances, 181
- Coupled Transport and Reaction, 183

5.7 Evaluating k_L, k_G, K_L, and a: Effects of Hydrodynamic and Other Operating Conditions, 187
- Approaches for Estimating Gas Transfer Rate Coefficients, 188
  - Gas-in-Liquid Systems, 188
  - Liquid-in-Gas Systems, 192
- Effects of Other Parameters on Gas Transfer Rate Constants, 195
  - Temperature, 195
  - Solution Chemistry, 196

5.8 Summary, 196

Appendix 5A. Conventions Used for Concentrations and Activity Coefficients When Computing Henry's Constants, 197
- Overview, 197
- Conventions for the Physicochemical Environment in the Standard State, 198

Appendix 5B. Derivation of the Gas Transfer Rate Expression for Volatile Species That Undergo Rapid Acid/Base Reactions, 199

References, 202

Problems, 204

6 Gas Transfer: Reactor Design and Analysis   **207**

    6.1 Introduction, 207
    6.2 Case I: Gas Transfer in Systems with a Well-Mixed Liquid Phase, 207
        The Overall Gas Transfer Rate Expression for Case I Systems, 211
        Analysis of Case I Systems in Batch Liquid Reactors, 213
        Limiting Cases of the General Kinetic Expression, 216
           Overview, 216
           Macroscopic (Advective) Limitation on the Gas Transfer Rate, 217
           Microscopic (Interfacial) Limitation on the Gas Transfer Rate, 218
           Summary of Rate Limitations on Overall Gas Transfer Rate, 219
        Case I Systems with Continuous Liquid Flow at Steady State, 220
           Reactors with Plug Flow of Liquid, 220
           Reactors with Flow and a Uniform Liquid-Phase Composition
              (CFSTRs with Respect to Liquid), 220
        Case I CFSTRs in Series, 225
        Design Constraints and Choices for Case I Systems with Flow, 226
    6.3 Case II: Gas Transfer in Systems with Spatial Variations in the Concentrations
        of Both Solution and Gas, 226
        The Mass Balance Around a Section of a Gas Transfer Tower:
           The Operating Line, 226
        The Mass Balance Around a Differential Section of a Gas Transfer Tower:
           Development of the Design Equation for Case II Systems, 229
        Pressure Loss and Liquid Holdup, 233
        Use of the Design Equation for Case II Systems, 236
           Description of the Influent Stream, Treatment Objectives, and Design
              Assumptions, 236
           Exploration of Feasible Designs for Meeting the Treatment Criteria, 236
           Sensitivity of the Column Size to Design Choices and Uncertainty in
              Parameter Values, 240
        Case II Systems Other than Packed Columns, 240
    6.4 Summary, 241
        Appendix 6A. Evaluation of $K_L a$ in Gas-in-Liquid Systems for Biological
        Treatment, 243
        References, 246
        Problems, 246

7 Adsorption Processes: Fundamentals   **257**

    7.1 Introduction, 257
        Background and Chapter Overview, 257
        Terminology and Overview of Adsorption Phenomena, 259
    7.2 Examples of Adsorption in Natural and Engineered Aquatic Systems, 262
        Use of Activated Carbon for Water and Wastewater Treatment, 262
        Sorption of NOM During Coagulation of Drinking Water, 264
        Sorption of Cationic Metals onto Fe and Al Oxides, 265
        Reactors for Adsorption onto Metal Hydroxide Solids, 266
    7.3 Conceptual, Molecular-Scale Models for Adsorption, 266
        Two Views of the Interface and Adsorption Equilibrium, 267
           Adsorption as a Surface Complexation Reaction, 267
           Adsorption as a Phase Transfer Reaction, 267
           Adsorption of Ions as Electrically Induced Partitioning:
              Donnan Equilibrium, 268
           Which Model is Best?, 268
    7.4 Quantifying the Activity of Adsorbed Species and Adsorption Equilibrium
        Constants, 268

7.5 Quantitative Representations of Adsorption Equilibrium: The Adsorption Isotherm, 269
  Model Adsorption Isotherms According to the Site-Binding Paradigm, 270
    Characterizing the Adsorbent Sites: Surface Site Distribution Functions, 270
    The Single-Site Langmuir Isotherm, 271
    Possible Reasons for Non-Langmuir Behavior, 273
    The Multisite Langmuir Isotherm, 274
    Modeling Surfaces with a Semicontinuous Distribution of Site-Types: The Freundlich Isotherm, 276
    Comparison of Multisite Langmuir and Freundlich Isotherms, 281
  Bidentate Adsorption, 281
  The Adsorption Distribution or Partition Coefficient, 282
  Competitive Adsorption in the Context of the Site-Binding Model of Adsorption, 283
    Competitive Langmuir Adsorption, 283
    A Special Case of Competitive Langmuir Adsorption: Ion Exchange Equilibrium, 284
    Sorption onto Ion Exchange Resins, 285
    Homovalent Ion Exchange, 286
    Heterovalent Ion Exchange, 288
    Some Special Nomenclature and Conventions Used for Ion Exchange Reactions, 289
    Modeling Ion Exchange Based on Donnan Equilibrium, 292
  Competitive Adsorption in the Context of the Site-Binding Model for Adsorbates that Obey Freundlich Isotherms, 294
7.6 Modeling Adsorption Using Surface Pressure to Describe the Activity of Adsorbed Species, 296
  The Surface Pressure Concept, 296
  Computation of the Surface Pressure from Surface Tension or Isotherm Data, 297
  Competitive Adsorption and Surface Pressure: The Ideal Adsorbed Solution Model, 302
7.7 The Polanyi Adsorption Model and the Polanyi Isotherm, 306
  Description of the Polanyi Model, 306
  Comparison of Conceptual Models for Adsorption and Their Relationships to the Linear, Langmuir, and Freundlich Isotherms, 313
7.8 Modeling Other Interactions and Reactions at Surfaces, 314
  The Structure of Charged Interfaces and the Electrostatic Contribution to Sorption of Ions, 314
    Effects of Electrical Potential on Binding of Ions to Surfaces, 314
    The Profile of Adsorbates and Electrical Potential in the Interfacial Region, 315
    The Electrostatic Contribution to the Equilibrium Constants in Competitive Adsorption Reactions, 318
  Phase Transitions Involving Ionic Adsorbates: Pore Condensation and Surface Precipitation, 319
7.9 Summary, 320
  References, 321
  Problems, 323

8 Adsorption Processes: Reactor Design and Analysis   **327**

    8.1 Introduction, 327
    8.2 Systems with Rapid Attainment of Equilibrium, 328
        Batch Systems, 328
        Systems with Continuous Flow of Both Water and Adsorbent, 331
        Sequential Batch Reactors, 332
        Fixed Bed Adsorption Systems, 333
            Qualitative Description, 333
            The Mass Balance on a Fixed Bed Reactor with Rapid Equilibration, 335
            Systems with Rapid Equilibration and Plug Flow, 336
    8.3 Systems with a Slow Approach to Equilibrium, 340
        Pore Diffusion Versus Surface Diffusion in Porous Adsorbent Particles, 341
        Adsorption in Batch Systems with Transport-Limited Adsorption Rates, 343
        Adsorption in Fixed Bed Systems with Transport-Limited
            Adsorption Rates, 350
    8.4 The Movement of the Mass Transfer Zone Through Fixed Bed Adsorbers, 354
    8.5 Chemical Reactions in Fixed Bed Adsorption Systems, 356
    8.6 Estimating Long-Term, Full-Scale Performance of Fixed Beds from
        Short-Term, Bench-Scale Experimental Data, 357
    8.7 Competitive Adsorption in Column Operations: The Chromatographic
        Effect, 359
        Systems with Rapid Attainment of Adsorptive Equilibrium, 359
        Competitive Adsorption in Systems That Do Not Reach Equilibrium
            Rapidly, 364
    8.8 Adsorbent Regeneration, 365
    8.9 Design Options and Operating Strategies for Fixed Bed Reactors, 366
        The Minimum Rate of Adsorbent Regeneration or Replacement, 366
        Design Options for Fixed Bed Adsorption Systems, 367
            Single Bed Designs, 367
            Packed Adsorption Beds in Series: "Merry-Go-Round" Systems, 368
            Packed Adsorption Beds in Parallel, 369
    8.10 Summary, 369
    References, 371
    Problems, 371

9 Precipitation and Dissolution Processes   **379**

    9.1 Introduction, 379
    9.2 Fundamentals of Precipitation Processes, 380
        Formation and Growth of Particles, 380
        Solute Transport, Surface Reactions, and Reversibility, 381
        Fundamentals of Solid/Liquid Equilibrium, 382
            The Solubility Product, 382
            The Activity of Solid Phases and Solid Solutions, 383
            Thermodynamics of Precipitation Reactions, 383
    9.3 Precipitation Dynamics: Particle Nucleation and Growth, 384
        Thermodynamics of Nucleation, 385
        Particle Growth and Size Distributions in Precipitation Reactors, 389
    9.4 Modeling Solution Composition in Precipitation Reactions, 394
        Quantitative Significance of the Solubility Product, 395
        Accounting for Soluble Speciation of the Constituents of the Solid, 395
    9.5 Stoichiometric and Equilibrium Models for Precipitation Reactions, 397
        Precipitation of Hydroxide Solids, 404
            Metal Speciation and the Metal Hydroxide—pH Relationship, 404
            Acid–Base Requirements for Metal Hydroxide Precipitation, 404

Precipitation of Carbonate Solids, 409
   Precipitative Softening, 409
   The Stoichiometric Model of Precipitative Softening , 410
   The Equilibrium Model of Precipitative Softening, 414
   Recarbonation of Softened Water, 417
   Precipitation of Other Metal Carbonates and Hydroxy-Carbonates, 418
Other Solids with pH-Dependent Metal and Ligand Speciation, 420
Effects of Complexing Ligands on Metal Solubility, 421
Precipitation Resulting from Redox Reactions, 421
9.6 Solid Dissolution Reactions, 422
9.7 Reactors for Precipitation Reactions, 426
9.8 Summary, 428
References, 429
Problems, 431

10 Redox Processes and Disinfection   435

10.1 Introduction, 435
10.2 Basic Principles and Overview, 435
   Applications of Redox Processes in Water and Wastewater Treatment, 435
   Oxidation, 436
     Control of Iron and Manganese, 436
     Destruction of Tastes and Odors, 436
     Color Removal, 436
     Aid to Coagulation, 436
     Oxidation of Synthetic Organic Chemicals, 436
     Destruction of Complexing Agents in Industrial Wastes, 437
   Reduction, 437
   Thermodynamic Aspects, 437
   Terminology for Oxidant Concentrations, 441
   Kinetics of Redox Reactions, 441
10.3 Oxidative Processes Involving Common Oxidants, 441
   Oxygen, 441
   Chlorine, 444
   Reactions of Free Chlorine with Inorganic Compounds, 446
     Reactions with Iron and Manganese, 446
     Reaction with Reduced Sulfur Compounds, 447
     Reactions with Bromide, 448
     Reactions with Organic Compounds, 448
   Chloramines, 455
     Formation of Chloramines, 455
     Reactions of Chloramines with Inorganic Compounds, 458
   Chlorine Dioxide, 459
     Generation of Chlorine Dioxide, 460
     Reactions of Chlorine Dioxide with Inorganic Compounds, 460
     Reactions of Chlorine Dioxide with Organic Compounds, 461
   Ozone, 461
     Ozone Generation, 462
     Potassium Permanganate, 466
     Generation of Permanganate, 467
     Reactions of Permanganate with Ferrous and Manganous Species, 467
10.4 Advanced Oxidation Processes, 469
   Reactions of OH Radicals with Inorganics, 470
   Reactions of OH Radicals with Organics, 470

    Generation and Fate of OH Free Radicals in Ozonation and Some
    Specific AOPs, 476
        UV/Hydrogen Peroxide, 476
        Ozone, 477
        O_3/UV and O_3/H_2O_2, 480
        UV/Semiconductor, 481
        Wet Air Oxidation, 482
        Sonolysis, 483
    Fenton-Based Systems, 483
        Dark Fenton Process, 483
        Light-Mediated Fenton Processes, 485
        Heterogeneous Fenton Processes, 485
        Electrochemical Fenton Processes, 486
        Cathodic Fenton Processes, 486
        Anodic Fenton Processes, 486
        Full-Scale Applications, 486
10.5 Reductive Processes, 486
    Sulfur-Based Systems, 486
    Iron-Based Systems (Fe(II), Fe(s)), 487
10.6 Electrochemical Processes, 488
10.7 Disinfection, 488
    Modeling Disinfection, 489
    Design and Operational Considerations, 493
    Characteristic Performance of Specific Disinfectants, 494
        Chlorine, 494
        Chloramines, 495
        Chlorine Dioxide, 497
        Ozone, 498
        Ultraviolet Radiation, 500
        OH Free Radicals, 502
10.8 Summary, 502
    References, 503
    Problems, 509

PART III REMOVAL OF PARTICLES FROM WATER

11 Particle Treatment Processes: Common Elements   519

11.1 Introduction, 519
11.2 Particle Stability, 521
    Particle Charge, 522
        Isomorphic Substitution, 522
        Chemical Reactions at the Surface, 522
        Adsorption on the Particle Surface, 523
    Characteristics of the Diffuse Layer, 524
    Interaction of Charged Particles, 525
    Van der Waals Attraction, 526
    Interactions of a Particle and Flat Plate, 530
    Experimental Measurements Related to Charge and Potential, 531
11.3 Chemicals Commonly Used for Destabilization, 532
    Inorganic Species, 532
    Organic Polymers, 533
11.4 Particle Destabilization, 535
    Compression of the Diffuse Layer, 535
    Adsorption and Charge Neutralization, 536

xiv   CONTENTS

        Enmeshment in a Precipitate–Sweep Flocculation, 537
        Adsorption and Interparticle Bridging, 541
11.5 Interactions of Destabilizing Chemicals with Soluble Materials, 542
        Combinations of Additives, 543
11.6 Mixing of Chemicals into the Water Stream, 544
11.7 Particle Size Distributions, 546
        Experimental Measurements of Size Distributions, 550
11.8 Particle Shape, 551
11.9 Particle Density, 552
11.10 Fractal Nature of Flocs, 552
11.11 Summary, 553
      References, 555
      Problems, 557

12 Flocculation   563

12.1 Introduction, 563
12.2 Changes in Particle Size Distributions by Flocculation, 564
12.3 Flocculation Modeling, 565
        Rate Equation for Floc Formation, 567
        Interpretation of the Rate Constant, 567
        The Smoluchowski Equation, 568
        Characteristic Reaction Times in Flocculation, 570
        Design Implications of the Smoluchowski Equation, 571
12.4 Collision Frequency: Long-Range Force Model, 572
        Collisions by Fluid Shear, 572
        Collisions by Differential Sedimentation, 575
        Collisions by Brownian Motion, 577
        The Total Collision Frequency Function, 578
        Design Implications of the Long-Range Force Model, 580
12.5 Collision Efficiency: Short-Range Force Model, 581
        Design Implications of the Short-Range Model, 589
12.6 Turbulence and Turbulent Flocculation, 589
        Turbulence, 589
        Turbulent Flocculation, 589
12.7 Floc Breakup, 592
12.8 Modeling of Flocculation with Fractal Dimensions, 594
12.9 Summary, 596
      Appendix 12A. Calculation Equations for Collision Efficiency
      Functions in the Short-Range Force Model, 597
      References, 598
      Problems, 599

13 Gravity Separations   603

13.1 Introduction, 603
13.2 Engineered Systems for Gravity Separations, 605
13.3 Sedimentation of Individual Particles, 607
        Stokes' Law, 607
        Inertial Effects on Sedimentation, 610
13.4 Batch Sedimentation: Type I, 612
        Monodisperse Suspension, 612
        Bimodal Suspension, 614
        Heterodisperse Suspension, 614

13.5 Batch Sedimentation: Type II, 618
    Mathematical Analysis, 619
    Experimental Analysis, 620
13.6 Continuous Flow Ideal Settling, 622
    Separating Influences of Suspension and Reactor, 622
    Interpreting the Reactor Settling Potential Function, 623
    Upflow Reactor, 624
    Ideal Horizontal Flow Reactors, 626
        Rectangular Reactor, 626
        Circular Reactor, 629
        Type I versus Type II Suspensions in Horizontal Flow Reactors, 631
        Correspondence of Batch and Continuous Flow Reactors for Ideal Horizontal Flow, 633
    Tube Settlers, 634
    Summary of Sedimentation in Ideal Flow Reactors, 638
13.7 Effects of Nonideal Flow on Sedimentation Reactor Performance, 639
    Nonideal, Tiered Flow, 639
    Nonideal, Channeled Flow, 640
    Mixed Flow, 642
    Summary of Nonideal Flow Effects, 644
13.8 Thickening, 644
    Batch Thickening, 645
    Solids Flux, 647
    Continuous Flow Thickening, 650
    Design of Continuous Flow Gravity Thickeners, 655
13.9 Flotation, 655
    Flotation Sytems Overview, 657
    Saturator, 658
    Bubble Formation, 661
    Flotation for Low Concentration Suspensions, 663
    Contact Zone Modeling, 663
        Flocculation Model, 664
        Filtration Model, 665
        Comparison of the Contact Zone Models, 667
    Separation Zone Modeling, 667
    Sludge Thickening by Dissolved Air Flotation, 668
13.10 Summary, 669
    References, 670
    Problems, 671

14 Granular Media Filtration   677

14.1 Introduction, 677
    A Typical Filter, 679
14.2 A Typical Filter Run, 680
14.3 General Mathematical Description of Particle Removal: Iwasaki's Model, 683
14.4 Clean Bed Removal, 684
    The Single Spherical Isolated Collector Model, 684
    Removal Mechanisms and Transport Efficiencies for a Single Isolated Collector, 686
        Particle–Collector Collisions by Interception, 686
        Particle–Collector Collisions by Sedimentation, 687
        Particle–Collector Collisions by Brownian Motion, 688
        Overall Particle–Collector Collision Efficiency, 689

Single Spherical Collector in Packed Medium Model, 690
Updated Packed Medium Model, 692
Other Advanced Models, 693
14.5 Predicted Clean Bed Removal in Standard Water and Wastewater Treatment Filters, 694
Design Tradeoffs, 696
14.6 Head Loss in a Clean Filter Bed, 698
14.7 Filtration Dynamics: Experimental Findings of Changes with Time, 700
Immediately After Backwashing, 700
Ripening, 701
Breakthrough, 701
Head Loss, 703
Filtration Dynamics: Effects of Design and Operational Variables, 705
Bed Depth, 705
Media Size, 706
Filtration Velocity, 707
Depth, Media Size, and Velocity, 708
Influent Concentration, 709
Summary of Effects of Independent Variables, 709
14.8 Models of Filtration Dynamics, 709
Macroscopic Models, 710
Ripening Model of O'Melia and Ali, 710
Ripening Models of Tien and Coworkers, 712
Mackie et al. Ripening Model, 713
Use and Value of Dynamic Models, 713
14.9 Filter Cleaning, 714
Surface Wash, 716
Air Scour, 716
14.10 Summary, 717
References, 718
Problems, 720

PART IV   MEMBRANE-BASED WATER AND WASTEWATER TREATMENT

15 Membrane Processes   731

15.1 Introduction, 731
15.2 Overview of Membrane System Operation, 732
15.3 Membranes, Modules, and the Mechanics of Membrane Treatment, 734
Membrane Structure, Composition, and Interactions with Water, 734
Driving Forces for Membrane Processes, 736
Pressure-Driven Processes, 736
Processes Driven Primarily by Concentration Differences or Electrical Forces, 738
Configuration and Hydraulics of Membrane Systems, 738
Configuration of Membrane Elements, 738
Configuration of Membrane Arrays, 741
15.4 Parameters Used to Describe Membrane Systems, 742
Location, Concentration, and Pressure, 742

Transport of Fluid and Solutes, 743
    Recovery, 743
    Flux, 743
    Specific Flux and Permeance, 744
    Resistance, 744
    Permeability, 745
    Effect of Temperature on Water Permeation, 746
Membrane Selectivity: Rejection and Separation, 747
    Polarization, 747
    Rejection, 748
    Challenge Tests and MWCO, 748
    Separation, 749
15.5 Overview of Pressure-Driven Membrane Systems, 749
    Similarities and Differences Among MF, UF, NF, and RO, 749
    Operation and Trends in the Performance of Pressure-Based Membrane Systems, 750
15.6 Quantifying Driving Forces in Membrane Systems, 752
    Energy and Driving Force in Membrane Systems, 752
    The Osmotic Pressure, 754
    Relative Magnitudes of Different Driving Forces for Transport in Membrane Systems, 757
15.7 Quantitative Modeling of Pressure-Driven Membrane Systems, 759
    Conceptual Models for Transport in Pressure-Driven Membranes, 759
    The Pore-Flow Model, 760
        Changes in Solution Composition at the Concentrate/Membrane Interface, 760
        Permeation of Solution, 764
        Relating Permeation to Contaminant Rejection, 764
    The Solution–Diffusion Model, 766
        The Transmembrane Pressure Profile According to the Solution–Diffusion Model, 766
        Concentration Changes at the Membrane/Solution Interfaces, 766
        Permeation Through the Membrane and the Concentration Profile Across the Membrane, 767
    Comparison of Predicted Fluxes and Transmembrane Parameter Profiles in the Pore-Flow and Solution–Diffusion Models, 771
    Summary of Transport Through Membranes, 772
15.8 Modeling Transport of Water and Contaminants From Bulk Solution to the Surface of Pressure-Driven Membranes, 773
    Overview, 773
    Physicochemical State of Contaminants Near the Membrane, 773
    Transport Through the Boundary Layer and Concentration Polarization in Frontal Filtration, 774
    Relating Permeation to TMP in Systems with Frontal Filtration, 777
        The Coupling Force Exerted by Rejected Contaminants on the Permeate Flow, 778
        Effect of Rejected Particles on Flux, 778
        Effect of Rejected Solutes on Flux, 780
        Expressing the Effects of the CP Layer on Permeation in Terms of Resistance, 781
        Comparing the Effects of Particles and Solutes on Flux Through the CP Layer, 782
    The Formation of Cakes, Gels, or Scales, and the Limiting Flux, 782
        Formation of a Compact Layer and the Definition of c_{lim}, 782
        The Limiting Flux and the Film-Gel Model, 783

The Hydraulic Resistance of the Compact Layer, 783
Estimating c_{lim} and k_{mt} in Systems with Compact Layers and the Flux Paradox, 784
Concentration Polarization and Precipitative Fouling, 784
Relating Parameter Profiles in the CP Layer with Those in the Membrane, 786
Nonsteady-State Fouling During Frontal or Dead-End Filtration, 787
Empirical Measures of Fouling: The MFI and SDI Tests, 790
Summary: Modeling Membrane Performance in Frontal Filtration, 792

15.9 Effects of Crossflow on Permeation and Fouling, 792
General Considerations in Modeling Fluid Flow and Particle Transport in Crossflow Filtration, 792
Fluid Flow in Crossflow Filtration, 793
The Pressure Profile on the Concentrate Side of Crossflow Membrane Systems, 795
Contaminant Transport Mechanisms in Crossflow Filtration, 798
Brownian and Shear-Induced Diffusion, 798
Deterministic Transport Mechanisms in Crossflow Filtration, 800
Relative Importance of Different Back-Transport Mechanisms, 802
Modeling Contaminant Transport and Flux in Crossflow Filtration Systems, 802
Overview, 802
Assumptions Commonly Used in Modeling Crossflow Filtration, 803
The Mechanics of Crossflow Filtration Modeling, 803
Modeling the Thickness of the CP Layer and $k_{\text{mt,CP}}$ in Crossflow Filtration, 804
Systems with a Ubiquitous Compact Layer: Applying the Film-Gel Model in Systems with Crossflow Filtration, 805
Modified Versions of the File and Film-Gel Models for Systems with Crossflow, 807
Systems with no Compact Layer, 809
Systems in which a Compact Layer is Present along only a Portion of the Membrane Element, 811
Modeling Crossflow Filtration of Particles Subject to Significant Inertial Lift, 813
Summary of Equations for Modeling Crossflow Filtration, 814
Nonsteady-State Fouling Patterns in Crossflow Filtration, 814
Summary of Modeling Approaches and Results for Crossflow Filtration, 815

15.10 Electrodialysis, 816
Transport in Systems with a Gradient in Electrical Potential, 819
Transport due to Advection, 819
Transport due to Diffusion, 819
Transport in Solution due to an Electric Field, 820
Overall Transport and Current Densities in Systems with a Gradient in Electrical Potential, 821
Modeling Electrodialysis Systems, 822
Overview, 822
Transport in the x Direction in a Single Cell-Pair, 822
Relating Ion Fluxes, Electrical Current Density, and the Electrical Potential Difference, 825
Analysis of the Two-Dimensional ED System, 828
Macroscopic Mass Balance on ED Reactor, 828

Microscopic Mass Balance on a Differential Element in an ED Reactor, 829
Ramifications for Design of Electrodialysis Systems, 830
Complications of Real Systems, 832
Water Flow Through Ion-Exchange Membranes, 832
Nonideal Behavior of Membranes, 832
Multicomponent Systems, 832
Overlimiting Current, 833
Additional Sources of Potential Drop, 833
Summary, 834
15.11 Modeling Dense Membrane Systems Using Irreversible Thermodynamics, 834
15.12 Summary, 838
References, 839
Problems, 841

INDEX   **847**

PREFACE

PURPOSE

This book has been written primarily as a textbook for a graduate course in physical/chemical treatment processes for water and wastewater. However, it should also be useful to working environmental engineers in providing a thorough and cohesive understanding of processes that would not be easily achieved by reading journal articles. While some introductory material is given for each subject, this book has been written with the assumption that the reader has had previous exposure, by class work or experience, with standard water and wastewater treatment processes. To illustrate specific applications, examples are woven throughout this book and problems are given at the end of each chapter. This book is divided into four parts, as explained below.

ORGANIZATION

Part I: Chapters 1–4. This part of this book describes the fundamental tools for investigating and studying water and wastewater treatment processes. It sets the stage for the subsequent chapters by presenting the background that is common to the analysis and understanding of many treatment processes. At the end of this section, the reader should have an advanced understanding of how mass balances are used in continuous flow systems in which reactions occur, and thereby will be able to understand and predict the changes in water quality that occur in such reactors. Details of the construction of mass balances are presented in Chapter 1; descriptions of flow characteristics are given in Chapter 2; the study of reaction kinetics is introduced in Chapter 3; and the material in the first three chapters is synthesized in Chapter 4. These chapters are written in a completely generic manner (i.e., with little attention to the application of the material to specific processes of interest in water and wastewater treatment) to emphasize that the material is usable in a wide variety of situations.

Part II: Chapters 5–10. This part of the book describes processes for removing soluble contaminants from water (or, in a few cases, inserting chemicals of interest into water). The processes are used in various applications, from treatment of municipal and industrial wastes to the production of drinking water or high-purity industrial process water. All the major processes that are used broadly to remove soluble contaminants are covered in this section or in the membrane chapter that ends this book. In each case, the emphasis is on fundamental understanding of the process dynamics through an analysis of batch (no flow) systems, followed by the interplay between the reaction kinetics and the flow characteristics of systems in which these processes are often carried out. The effects of process variables (i.e., hydraulic, equilibrium, and kinetic) on process design and process performance are emphasized.

The order of these chapters is such that they build on each other, and on the earlier chapters. Gas transfer is covered in Chapters 5 and 6, because it is usually independent of other processes (even if they happen simultaneously) and it builds directly on the earlier fundamental chapters. Gas transfer is one of the few common processes for which design is always closely related to theory, and so the analysis can be quite fundamental and still provide a picture of how gas transfer is achieved in real-engineered systems. Adsorption (Chapters 7 and 8) has many similarities to gas transfer and is often considered as a possible alternative process to gas stripping for the removal of specific contaminants, so it is presented next. For both gas transfer and adsorption, two chapters are provided, with the first describing the underlying fundamental science and the second presenting the application in water treatment engineering.

Precipitation (Chapter 9) relies (in part) on adsorption and creates particles that must be removed in subsequent processes (that are covered in the third part of this book). Chemical oxidation (and, to a lesser extent, reduction) (Chapter 10) is widely used for the destruction of organic compounds and the transformation of objectionable inorganics to less toxic species; the use of so-called advanced oxidation processes is expanding rapidly in the field and will continue to do so in future years. Disinfection (the inactivation of microorganisms) relies directly on oxidation processes and is included in the chapter on oxidation and reduction.

Part III: Chapters 11–14. This part of the book focuses on processes for removing particulate materials from water. Many contaminants of water are particles to begin with, are made into particles by precipitation, or are associated with particles by adsorption; hence, particle removal processes are used to remove contaminants that came into the treatment system as soluble materials as well as to remove those that entered as particles. Chapter 11 describes the fundamentals that are common to all the particle removal processes, such as properties of particles and interactions of particle surfaces with chemicals in solution that determine much of particle behavior and particle–particle interactions. Particle removal processes are intrinsically physical/chemical processes—the chemistry of all these processes is essentially identical and is therefore covered in this chapter. This chapter also draws heavily on the chapters on adsorption and precipitation, as the chemistry of particle removal processes relies on these phenomena. Subsequently, the chapters on flocculation (Chapter 12), gravity separations (Chapter 13), and granular media filtration (Chapter 14) emphasize the physical aspects of these processes. These processes (or some subset of them) are often performed in series, and the order of the chapters reflects the order of their appearance in a treatment train.

Part IV: Chapter 15. An extensive chapter on membrane processes ends this book; membrane processes are used to remove both soluble and particulate materials. Rather than divide this material into separate chapters so that they could be incorporated into Parts II and III, we thought that this subject should stand alone and be treated in a unified way. The distinction between "what is a particle" and "what is a soluble entity" is blurry, and the continuum of membrane processes does not force one to make an arbitrary dividing line between them.

APPROACH

The approach throughout this book is to elucidate the fundamentals of physical/chemical treatment processes for water and wastewater, that is, to provide the basis for understanding that underlies the application of these processes in various treatment systems. Our belief is that fundamental understanding requires both a conceptual or qualitative picture of each process and a mathematical description of the process. Throughout this book, we have related the conceptual and mathematical descriptions. The level of mathematical treatment is, in a few cases, sophisticated but always accessible to beginning graduate students and to professionals with a Master's degree. Different people have different ways of learning with respect to the conceptual/mathematical continuum—some learn the concepts through the mathematics, others learn to appreciate the mathematical description only because they see the concepts first. It has been our goal to write this book in such a way that both these types of learners can reach a deep level of understanding of these processes. Detailed examples are provided throughout this book to illustrate the linkage between the mathematical and conceptual approaches to this understanding.

We also believe that the best practical knowledge is a good understanding of theory. When a process is not working properly, or when conditions have changed since the time of design, a good understanding of the fundamental theory is often the only guide for changing the operation to solve the problem. When faced with a new problem that is not envisioned today (and very few of the current hot topics in environmental engineering were envisioned 15–25 years ago when many professionals were graduate students), a good understanding of theory is the place to start developing a solution.

Throughout this book, we have used a consistent approach. After the generic description of mass balances in continuous flow systems in Part I, the processes described in every subsequent chapter are explained and analyzed by writing the relevant mass balance for the process, reactor geometry, and flow pattern under consideration. Our goal is to equip readers with this tool, so that as a question arises about how some process will work, they can generate and solve an appropriate mass balance that helps to answer that question. While the details of the mass balances and associated solutions vary for different unit processes, the approach is consistent. After describing the common ways in which each process is configured, the role of different variables in achieving the process objectives is discussed. Each chapter also gives the reader an overview of the ways in which the process is used in water and wastewater treatment.

In writing this book, one of us would write a first draft of a chapter or section and then the other would comment or edit; the process went back and forth until both of us were satisfied. That is, the tasks of writing and editing were mutually shared throughout the production of this book. In light of this process, the author order for this book was chosen as alphabetical.

HOW TO USE THIS BOOK?

This book is too long to complete in a one-semester (or one-quarter) course. Professors need to choose how to use this book in one or two courses, and how it will fit into a broader curriculum for environmental engineering students interested in treatment processes. The material in Part I is covered in greater depth than is done in most curricula, but still can be covered in depth within approximately four to five weeks. Part I can be combined with either Part II or with Parts III and IV to form a full semester course. Much of the material that is in Part II is closely related to what is taught in many Water Chemistry courses, and so it is best if students studying Part II have already taken such a course, or are taking it simultaneously. Although the order of the chapters is purposeful, Part III (and to a lesser extent, Part IV) can be studied independently of Part II without too much difficulty.

MARK M. BENJAMIN
University of Washington–Seattle

DESMOND F. LAWLER
University of Texas at Austin

ACKNOWLEDGMENTS

We have many people to thank; without them, this book would not exist. First, we offer sincere thanks to our three colleagues who made major contributions to specific chapters in this book: Paul Anderson of the Illinois Institute of Technology (Chapter 9), David Waite of the University of New South Wales (Chapter 10), and Mark Wiesner of Duke University (Chapter 15). Next, and equally important, we thank many teachers who gave us the understanding that is reflected in this book: these include our formal teachers, several colleagues, and many students, especially our PhD students. Our formal teachers include Jim Leckie, Phil Singer, and the late Charlie O'Melia; the latter two gave us many suggestions for various parts of this book throughout its production. Several colleagues at our own and other institutions have been helpful with suggestions, correcting errors, supplying problems, and in various other ways: Bill Ball (Johns Hopkins University), Shankar Chellam (University of Houston), David Dzombak (Carnegie-Mellon University), Marc Edwards (Virginia Tech), Jim Edzwald (formerly of the University of Massachusetts), Meny Elimelech (Yale University), John Ferguson (University of Washington), Lynn Katz (University of Texas), Gregory Korshin (University of Washington), Ismail Koyuncu (Duke University), Kara Nelson (University of California—Berkeley), Richard Palmer (University of Massachusetts), David Stensel (University of Washington), Scott Summers (University of Colorado), and John Tobiason (University of Massachusetts). Of these, Bill Ball was particularly helpful on several chapters of this book, consistently giving us corrections and excellent ideas for improvement. The contributions of several former students in aiding our understanding of much of this material (and making suggestions about the writing) cannot go unmentioned; at the risk of leaving some out, we note particularly Lee Blaney, Phil Brown, Leonard Casson, Bob Cushing, Jeannie Darby, Bruce Dvorak, Lauren Greenlee, Mooyoung Han, Younggy Kim, JiHyang Kweon, Jessie Li, Jeff Nason, Caroline Gerwe Russell, and Shane Walker who studied at the University of Texas, and Yujung Chang, Chiwang Li, and Zhenxiao Cai, who studied at the University of Washington. In addition, we note that many students gave us feedback that helped in increasing the clarity and correcting the errors as we went through drafts used in our classes.

Finally, we thank our wives, Judith Benjamin and Alice Lawler, and our families for their love, support, and patience over the lengthy period that this book was in preparation.

MARK M. BENJAMIN
DESMOND F. LAWLER

PART I

REACTORS AND REACTIONS IN WATER QUALITY ENGINEERING

1

MASS BALANCES

1.1 Introduction: The mass balance concept
1.2 The mass balance for a system with unidirectional flow and concentration gradient
1.3 The mass balance for a system with flow and concentration gradients in arbitrary directions
1.4 The differential form of the three-dimensional mass balance
1.5 Summary
References
Problems

1.1 INTRODUCTION: THE MASS BALANCE CONCEPT

This book deals with the analysis of physical and chemical processes that alter water quality in aquatic systems. The primary systems of interest are the reactors used in water and wastewater treatment operations, although the principles apply equally well to natural aquatic systems, such as rivers, lakes, and oceans.

The first part of this book (Chapters 1–4) deals with reaction engineering and reactor analysis. These terms refer to the overall process of characterizing how the physical features of a system (e.g., volume, mixing characteristics, layout) combine with chemical factors (reaction rates, equilibrium relationships) to control the extent to which a reaction proceeds. In this definition, the word *reactor* is not restricted to engineered systems, but rather refers to any region of space where a reaction is occurring. Thus, a reactor may be as small as a single water droplet or as large as an entire lake. Reactor analysis is an essential step in predicting how a system responds to varying input rates or concentrations, in deciding how big a reactor must be to accomplish a predefined treatment objective, or in comparing different design options. Whatever the specific goal, the analysis consists of writing one or more mass balance equations, determining the values of the parameters in these equations, and then solving the equations for the quantities of interest.

A *mass balance* is an accounting tool that is used to keep track of how much of a substance is present in a given region of space at a given time. In other words, it is a formal statement of the principle of conservation of mass. When analyzing water chemistry problems, mass balances are frequently written to apply the principle of conservation of mass to systems in which there is no gradient of concentration with respect to time or distance. In such cases, the mass balance states that all the components that were originally added to the system must still be there once equilibrium is attained, even though they might be present in a different form. By combining mass balances with the relationships expressed by equilibrium constants and the charge balance, we can determine the speciation of the chemical components of the system at equilibrium.

In this book, mass balances are used most often to characterize systems in which concentrations are a function of time or location, and the mass balance equations are most frequently written in terms of the time derivative of mass. That is, rather than equating the total *amount* of a species i in the system before and after equilibration, the mass balances describe the *rates* at which i enters, exits, reacts, and accumulates in the system. In words,

$$\begin{pmatrix} \text{Rate of change} \\ \text{of mass of } i \\ \text{stored in the} \\ \text{system} \end{pmatrix} = \begin{pmatrix} \text{Rate at which} \\ i \text{ enters the} \\ \text{system from} \\ \text{outside} \end{pmatrix} - \begin{pmatrix} \text{Rate at which} \\ i \text{ exits the} \\ \text{system to} \\ \text{the outside} \end{pmatrix} + \begin{pmatrix} \text{Rate at which} \\ i \text{ is generated} \\ \text{inside the} \\ \text{reactor} \end{pmatrix} - \begin{pmatrix} \text{Rate at which} \\ i \text{ is destroyed} \\ \text{inside the} \\ \text{reactor} \end{pmatrix}$$

This statement is a valid representation of the principle of conservation of mass and is applicable to any item in any reactor. The space defining the reactor (i.e., the space over which the mass balance is being written) is referred to as the *control volume* (CV). To convert the previous statement into

---

Water Quality Engineering: Physical/Chemical Treatment Processes, First Edition. By Mark M. Benjamin, Desmond F. Lawler.
© 2013 John Wiley & Sons, Inc. Published 2013 by John Wiley & Sons, Inc.

a mathematical form that is applicable to a specific system of interest, it is necessary to

(1) Identify the substance for which the balance is to be written (i.e., what is i?).
(2) Define the system boundaries so that it is clear whether a given molecule of i is inside or outside the CV.
(3) Establish all the ways that i can enter or exit the CV.
(4) Establish all the ways that i can be created or destroyed within the CV.

In carrying out the first two steps, the importance of being precise when identifying both the item of interest and the boundaries of the system deserves emphasis. For instance, if a mass balance is being written for copper, is it a mass balance on total copper, dissolved copper, or free, uncomplexed cupric ion? If the system is a lake, does it include only the aqueous phase above the bottom sediment, or all the material in the water column (solution and suspended solids)? Does it include a portion of the interstitial solution in the lake bed or any rooted aquatic plants? Choosing the system boundaries to include or exclude any of these items is acceptable, and we could write a valid mass balance for any of these choices. However, depending on what we hope to learn by solving the mass balance, one set of system boundaries might be more useful than another.

Three choices for the system boundaries for a mass balance characterizing a water treatment process are shown in Figure 1-1. The flow diagram shown is used in the conventional activated sludge process for biological treatment of wastewater, and also for certain precipitation processes in water and wastewater treatment operations (specifically, precipitation processes that incorporate sludge recycle). The boundaries shown in Figure 1-1a isolate the tank where the key biological or chemical conversions occur, those shown in Figure 1-1b isolate the portion of the system where solid/liquid separation is accomplished, and those shown in Figure 1-1c encompass the entire treatment system. As noted in the previous paragraph, a valid mass balance could be written using any of these choices for the system boundaries, but the type of information that could be derived by solving the mass balance would be different in each case. A numerical example demonstrating this point is presented at the end of this section.

Conceptually, the third and fourth steps needed to write a mass balance are straightforward. In most systems, the only way in which a substance can enter or leave a system is by physical movement (transport), and the only way in which it can be formed or destroyed inside the system is by chemical reactions (including biochemical and nuclear reactions as special types of chemical reactions).

Physical transport processes include advection, diffusion, and dispersion, all of which can move molecules across the system boundaries either into or out of the CV. *Advection*, or

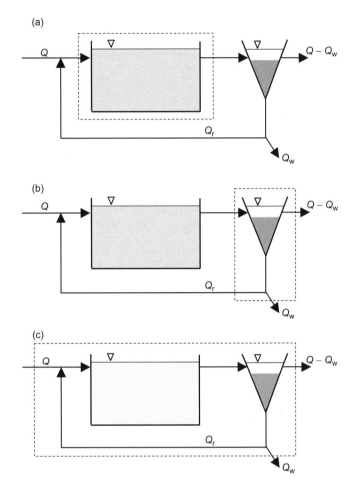

FIGURE 1-1. Three different choices (indicated by the broken lines) for the boundaries of the control volume for a mass balance in a system consisting of a reactor and a settling basin, with recycle of settled solids to the reactor influent. Q, Q_r, and Q_w are the influent flow rate, the flow rate of the recycle stream, and the flow rate of the waste (sludge) stream, respectively.

bulk flow, is flow in response to a driving force (e.g., a pressure gradient) that acts on all molecules in a region of space to move them in a given direction. This process contrasts with *diffusion*, which is the net effect of random molecular movement caused by the molecules' thermal-kinetic energy. Although the diffusive motion of any individual molecule is random, when the random motion of all the molecules is coupled with a concentration gradient, the net effect is transport in the direction that diminishes local concentration differences.

Dispersion might be considered as a cross between advection and diffusion. When a force is applied to a fluid, different portions of the fluid tend to move at different velocities. In laminar flow in a pipe, the parabolic distribution of velocities leads to some molecules of fluid (and any chemicals in that fluid) advancing through the pipe faster than others; in this case, the velocity is a well-defined function of distance from the wall. In laminar flow in porous media, packets of fluid pass through some pores quickly and other pores much more slowly, leading to differences in the rate of transport of chemicals by different packets. In contrast to the situation in pipes, the velocity profile in the latter situation does not bear any well-defined relationship to location in the flow pattern. The same statement is true in turbulent flow, in which eddies form and transport mass from one point to another; in this case, the velocity of each packet of fluid is characterized by fluctuations around a mean value, although the mean velocity of the bulk fluid might remain constant.

In all the systems described above, the portion of the transport of mass that is different from the average advective transport is attributed to *dispersion*. In cases where dispersive transport is dominated by random fluctuations in velocity (such as the latter two situations described in the preceding paragraph), it can be modeled and analyzed identically to diffusion, with packets of water in dispersive mixing being the analogs of molecules in diffusive mixing. A thorough description of dispersion, especially as it applies to porous media, is given by Bear (1979).

Chemical reactions can also alter the mass of i in the region of space enclosed by the system boundaries. However, in the case of chemical reactions, the key processes are creation of i from other molecules, or conversion of i into other molecules, within (as opposed to across) the system boundaries.

Based on the preceding discussion, the determination of whether or not a transport or reaction process should be included in a mass balance on i for a specific situation can be made by applying two simple rules:

(1) Terms describing physical transport are included in the mass balance if and only if they cause i to cross the system boundaries. If a molecule of i moves from one point to another strictly within or strictly outside the CV, such movement is not reflected in the mass balance.
(2) Terms describing chemical reactions in which i participates are included if the reactions occur within the system boundaries and are excluded if they occur outside the system boundaries.

An easy way to decide whether a physical transport process should be included in a mass balance expression is to imagine that the system is enclosed in a permeable balloon. If the process carries molecules of i across the surface of the balloon, it must be included in the mass balance; on the other hand, if it causes molecules of i to move around either inside or outside the balloon, but does not carry i across the surface, it should not be included.

To understand the basic principle and the essential simplicity of a mass balance, it might be useful to consider an analogy between a mass balance and the accounting that applies to a savings account. When balancing a savings account, we need to consider deposits and withdrawals, which might be considered as physical transfers of money from/to the external world. On the other hand, interest payments and fees paid for the bank's services might be considered as transfers that occur internally in the bank. While it may be convenient to separate these two groups of transactions when analyzing an account, it is also true that a dollar deposited and a dollar of interest are indistinguishable once they are in the account. Similarly, in a chemical mass balance, a molecule of i that enters the system by physically crossing a system boundary is indistinguishable from one that is generated internally in the system by a reaction. Thus, if a molecule of i is found in the system at some time, is it neither necessary, nor in general is it possible, to decide how the molecule got there. All that is known about the molecule, and all that matters, is that it is there.

Once the steps listed above are completed, the system is fully defined, and it should be possible to write a detailed mass balance (in words) that characterizes i in the system. Such a mass balance will have terms that fit into the categories shown in the following statement, but which are more detailed and system-specific.

$$\begin{array}{l} \text{Rate of change} \\ \text{of the amount of} \\ i \text{ stored within} \\ \text{the system} \end{array} = \begin{array}{l} \text{Net rate} \\ (\text{in}-\text{out}) \text{ at} \\ \text{which } i \text{ enters} \\ \text{by advection} \end{array}$$

$$+ \begin{array}{l} \text{Net rate} \\ (\text{in}-\text{out}) \text{ at which } i \\ \text{enters by diffusion} \\ \text{and dispersion} \end{array} + \begin{array}{l} \text{Net rate} \\ (\text{formation}-\text{destruction}) \\ \text{at which } i \text{ is created by} \\ \text{chemical reaction} \end{array}$$

For instance, the flow scheme shown in Figure 1-1a might be used in a water softening process, in which soluble calcium (all of which we assume to be Ca^{2+}) is removed from solution as a $CaCO_3(s)$ precipitate. A mass balance on Ca^{2+} in such a system, with system boundaries chosen to enclose only the precipitation reactor, might be expressed as follows:

$$\begin{array}{l} \text{Rate of} \\ \text{change of} \\ \text{mass of} \\ Ca^{2+} \text{ in the} \\ \text{reactor} \end{array} = \begin{array}{l} \text{Rate at which} \\ Ca^{2+} \text{ enters into} \\ \text{the mixed} \\ \text{influent/} \\ \text{recycle stream} \end{array} - \begin{array}{l} \text{Rate at which} \\ Ca^{2+} \text{ leaves in} \\ \text{the effluent} \\ \text{from the} \\ \text{reactor} \end{array}$$

$$+ \begin{array}{l} \text{Rate at which } Ca^{2+} \\ \text{enters the reactor by} \\ \text{diffusion and dispersion} \\ \text{at the points where the} \\ \text{inlet and outlet pipes} \\ \text{connect to the reactor} \end{array} - \begin{array}{l} \text{Rate at which} \\ Ca^{2+} \text{ precipitates} \\ \text{to form } CaCO_3(s) \\ \text{in the reactor} \end{array}$$

In some systems, it is useful to write two or more mass balances with different system boundaries. In many such cases, an input term for one set of boundaries is an output term for another. For instance, in a system where volatile organic compounds (VOCs) are being stripped from solution by a gas, the rate at which VOCs leave the liquid equals the rate at which they enter the gas, so terms representing this transfer have the same magnitude but different signs in mass balances on the two phases.

Although the qualitative mass balances described above are of some value, to be truly useful they must be converted into quantitative mathematical expressions and solved, so that we can compute the concentration of i explicitly as a function of time and/or location. The steps needed to carry out this conversion are detailed in the remainder of the chapter, where mathematical expressions are derived for each of the terms in the mass balance. With the aid of these expressions, it is possible to identify the physical/chemical factors that control a given process, evaluate these factors in an existing system, and/or use the information to design a process to accomplish a desired goal.

A Note on Dimensions. Since mass balances for the systems of interest describe rates at which material enters or leaves the system, the appropriate dimensions for all terms in the mass balance are mass/time. By using these dimensions, mass balances can be written even for systems in which the volume, pressure, temperature, or shape are variable. In some cases, it is convenient to divide all terms in the mass balance by the system volume, so that they have dimensions of mass per unit volume per time or, equivalently, concentration per time. In this book, we refer to the mass balances written in the latter way as *volume-normalized mass balances*.

In a few cases, an equation essentially identical to a mass balance is written on a basis other than mass. The most common such equations represent an accounting based on numbers, such as the number of particles in a suspension. The dimensions for these equations are specific to the item of interest; for example, for a particle number balance, the dimensions of each term are time^{-1}, or, for a volume-normalized number balance, (volume $\times$ time)$^{-1}$.

■ **EXAMPLE 1-1.** A common approach to soften water is to adjust the chemistry of the solution to cause $CaCO_3(s)$ to precipitate. As noted previously, such processes often utilize a flow scheme like that shown in Figure 1-1. New solids are generated in the main reaction tank, and then the solids are separated from solution in a settling tank. However, rather than discarding all of the solids, a fraction of the settled sludge is returned to the reaction tank to provide seed crystals onto which more solid can precipitate. The advantages of this operating scheme are discussed in Chapter 9. The relevant reaction is

$$Ca^{2+} + CO_3^{2-} \rightarrow CaCO_3(s)$$

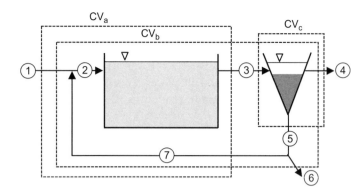

FIGURE 1-2. Various control volumes for mass balances that are useful for solving the example problem.

A proposed water softening process will treat a flow of $3.0\,m^3/s$. The objective is to induce precipitation of $CaCO_3(s)$ to reduce the dissolved Ca^{2+} concentration from 150 to 40 mg/L. The influent contains no $CaCO_3(s)$. The settling step is expected to be very efficient, so the $CaCO_3(s)$ concentration in the treated effluent is negligible. Figure 1-2, which is a specific version of the more general Figure 1-1, shows some flow streams labeled and some potentially useful control volume (CV) boundaries identified. The flow rates at the indicated points are as follows:

$Q_1 = 3000\,L/s, \quad Q_2 = 3240\,L/s, \quad Q_3 = 3240\,L/s,$
$Q_4 = 2916\,L/s, \quad Q_5 = 324\,L/s, \quad Q_6 = 84\,L/s, \quad Q_7 = 240\,L/s$

Assuming that the water volumes in the various tanks, the water flow rates, and the concentrations of all the chemicals in the system are steady over time, write mass balances using appropriate CVs to determine the concentration of $CaCO_3(s)$ solids (mg/L) in the pipe between the main reactor and the settling basin (pipe 3), the pipe leading to the sludge processing facility (pipe 6), and the pipe conveying recycled solids to the main reaction vessel (pipe 7). Assume also that the precipitation reaction occurs only in the reactor, and not in the settling tank.

Solution. The $CaCO_3(s)$ concentrations (which we represent as c_s) at locations 1 and 4 are given as zero. Given this information, the location of interest where it is easiest to calculate c_s is point 6, based on a mass balance on $CaCO_3(s)$ in control volume CV_b (because only one of the three streams crossing the boundary of CV_b contains $CaCO_3(s)$.

The mass balance states that the rate of change of the mass of $CaCO_3(s)$ stored in the CV equals the rate at which $CaCO_3(s)$ enters the CV, minus the rate at which it exits the CV, plus the rate at which it is generated inside the CV. In this case, the rate at which $CaCO_3(s)$ enters or leaves in any stream i can be expressed as $Q_i c_{s,i}$. Because the system is specified to be operating with steady volumes and concentrations throughout, the mass of $CaCO_3(s)$ in the CV is not changing, so the mass balance on $CaCO_3(s)$ in CV_b can be written as follows:

$$\begin{pmatrix} \text{Rate of change of mass} \\ \text{of } CaCO_3(s) \text{ stored in } CV_b \end{pmatrix}$$
$$= \cancel{Q_1 c_{s,1}} - \cancel{Q_4 c_{s,4}} - Q_6 c_{s,6} + \begin{pmatrix} \text{Rate of } CaCO_3(s) \\ \text{generation in } CV_b \end{pmatrix}$$
$$Q_6 c_{s,6} = \begin{pmatrix} \text{Rate of } CaCO_3(s) \\ \text{generation in } CV_b \end{pmatrix}$$

By our assumption, the only place where $CaCO_3(s)$ is generated by chemical precipitation in CV_b is the main reaction tank. Because we have information about the rates at which Ca^{2+} enters and leaves CV_b, and because we can relate the loss of Ca^{2+} to the production of $CaCO_3(s)$ via the reaction stoichiometry, it is convenient to proceed by writing a mass balance on Ca^{2+} in CV_b, as follows:

$$\begin{pmatrix} \text{Rate of change of} \\ \text{mass of } Ca^{2+} \text{ in } CV_b \end{pmatrix} = Q_1 c_{Ca^{2+},1} - Q_4 c_{Ca^{2+},4} - Q_6 c_{Ca^{2+},6}$$
$$+ \begin{pmatrix} \text{Rate of appearance of } Ca^{2+} \\ \text{in } CV_b \text{ by precipitation} \end{pmatrix}$$

The mass of Ca^{2+} stored in the reactor is not changing over time, so the term on the left is zero. Furthermore, all the Q and c terms on the right are given in the problem statement. (Because no chemical reaction occurs in the settling basin, the concentration of Ca^{2+} in both streams leaving that basin is the target concentration of 40 mg/L.) Also, because we know that Ca^{2+} is removed from solution by the precipitation reaction, it is convenient to write the final term in the previous equation as a subtraction of the rate of loss of Ca^{2+} in the CV. Making this change and substituting the known values into the previous equation, we obtain

$$0 = (3000 \, L/s)(150 \, mg/L) - (2916 \, L/s)(40 \, mg/L)$$
$$- (84 \, L/s)(40 \, mg/L) - \begin{pmatrix} \text{Rate of loss of } Ca^{2+} \text{ in} \\ CV_b \text{ by precipitation} \end{pmatrix}$$
$$\begin{pmatrix} \text{Rate of loss of } Ca^{2+} \text{ in} \\ CV_b \text{ by precipitation} \end{pmatrix} = 330{,}000 \, mg/s = 330 \, g/s$$

Noting that the molecular weights of Ca^{2+} and $CaCO_3(s)$ are 40 and 100, respectively, we compute the rate of $CaCO_3(s)$ generation as follows:

$$\text{Rate of } CaCO_3(s) \text{ generation} = (330 \, g/s \, Ca^{2+}) \frac{100 \, g \, CaCO_3(s)}{40 \, g \, Ca^{2+}}$$
$$= 825 \, g/s \, CaCO_3(s) \text{ generated}$$

Inserting this value into the mass balance on $CaCO_3(s)$ derived previously, we can find $c_{s,6}$ as

$$Q_6 c_{s,6} = \begin{pmatrix} \text{Rate of } CaCO_3(s) \\ \text{generation in } CV_b \end{pmatrix}$$
$$c_{s,6} = \frac{825 \, g/s}{84 \, L/s} = 9.82 \, g/L$$

We can find $c_{s,3}$ by a mass balance on $CaCO_3(s)$ using CV_c, as follows:

$$\begin{pmatrix} \text{Rate of change of mass} \\ \text{of } CaCO_3(s) \text{ in } CV_c \end{pmatrix} = Q_3 c_{s,3} - Q_4 c_{s,4} - Q_5 c_{s,5}$$
$$+ \begin{pmatrix} \text{Rate of generation of } CaCO_3(s) \\ \text{in } CV_c \text{ by precipitation} \end{pmatrix}$$

Because the splitting of the flow stream at the junction of points 5, 6, and 7 does not involve any redistribution of solids, the solid concentrations in all three of these streams are the same; that is, $c_{s,5} = c_{s,6} = c_{s,7} = 9.82 \, g/L$. Substituting these equalities into the mass balance, we find

$$0 = (3240 \, L/s) c_{s,3} - (2916 \, L/s)(0.0 \, g/L)$$
$$- (324 \, L/s)(9.82 \, g/L) + 0$$
$$c_{s,3} = 0.982 \, g/L = 982 \, mg/L \quad \blacksquare$$

1.2 THE MASS BALANCE FOR A SYSTEM WITH UNIDIRECTIONAL FLOW AND CONCENTRATION GRADIENT

In this section, we derive a mathematical expression for a mass balance on a substance i in a CV with only one inlet and one outlet through which water flows directly into or out of the system, as shown schematically in Figure 1-3.

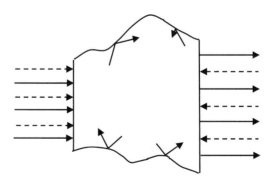

FIGURE 1-3. Schematic representation of the control volume, flow pattern (solid arrows), and concentration gradients (dashed arrows) in a hypothetical system in which the flow and the concentration gradients are perpendicular to the boundaries at the inlet and outlet.

Furthermore, the water is assumed to be well-mixed in the direction perpendicular to the flow as it crosses the system boundary; for example, if the flow is in the $+x$ direction, the concentration of i may vary in the x direction, but not in the $y-z$ plane at the inlet or outlet. These assumptions are reasonable for many real systems, such as a mixed-tank reactor with inflow and outflow through pipes, a tubular reactor with no internal axial mixing (a *plug flow reactor*), or a section of a river. No constraints are placed on the geometrical shape of the system or on the gradients that might exist inside it, as long as the gradients at the boundaries are exclusively in the direction of bulk flow.

The Storage Term

The mass of i present (stored) in any differentially small volume element dV ($=dxdydz$) within the system equals the product of the concentration of i in that volume element and the volume, or cdV.[1] To determine the total mass of i in the entire CV, it is necessary to sum the mass of i in all these differential volumes; that is, to integrate this term over the entire volume

$$\text{Mass of } i \text{ in volume } V = \int c\,dV \quad (1\text{-}1)$$

The derivative of this quantity with respect to time gives the rate of change of the amount of i stored in volume V. Thus, the storage term in the mass balance can be represented as

$$\text{Rate of change of mass of } i \text{ in volume } V = \frac{\partial}{\partial t}\int c\,dV \quad (1\text{-}2)$$

In both engineered reactors and natural systems, conditions are often encountered in which one of two simplifications applies to the storage term. First, in many systems, the concentration of the substance of interest is approximately uniform throughout the CV; such systems are sometimes referred to as *well-mixed*, although the condition can sometimes be met without physical mixing. Examples include tanks where concentrated chemical reagents (e.g., acids or bases, alum in coagulation processes, or hypochlorite species in disinfection processes) are added to and, ideally, dispersed rapidly throughout the solution by intensive, mechanically induced mixing. In other cases, the water in a tank is thoroughly mixed by rising gas bubbles when air or another gas (e.g., pure oxygen, ozone, or carbon dioxide) is injected to dissolve the injected gas or strip a volatile contaminant from the solution. In such systems, the right side of Equation 1-2 can be written as $V(dc/dt) + c(dV/dt)$. Furthermore, in many such cases, the volume of the system is constant, in which case the expression simplifies to $V(dc/dt)$. (If the system volume is constant but the concentration varies from one location to the next, the integral can be written as $\int (\partial c/\partial t)dV$, but this modification does not substantially ease the evaluation of the term.)

Second, many systems operate under conditions in which their composition and volume are approximately invariant over time (although the composition might vary as a function of location). In these systems, which are referred to as being at *steady state*, the storage term (i.e., the term describing the rate of change of the mass of i in the CV) is zero. The form of the storage term in the general case and for various simplifications is shown in Table 1-1.

■ **EXAMPLE 1-2.** Downstream of the point where wastewater is discharged into a small river, a contaminant decays in such a way that its concentration can be approximated by the equation $c(x) = c(0)/(1 + kc(0)x)$, where c is the contaminant concentration, x is the distance downstream, and k is a constant characterizing the reactivity of the contaminant. If the contaminant concentration in the river at the discharge point is 2 mg/L and $k = 3(\text{mg/L})^{-1}\text{km}^{-1}$, how much contaminant is stored in the 5-km segment of the river downstream of the plant? Over this stretch, the river has an approximately uniform cross-sectional area of $A = 24\,\text{m}^2$.

Solution. To determine the mass of contaminant stored in the river, we can define a CV that is bounded by the cross-section of the river and a distance dx in the direction of flow. The volume of water in this differential section is therefore $A\,dx$, and the mass of contaminant stored in the section is $c(x)A\,dx$. Designating the concentration at the upstream end of

TABLE 1-1. Expressions for the Storage Term in Mass Balance Equations

Rate of Change of Mass of i in Volume $V =$	
General case	$\dfrac{\partial}{\partial t}\int c\,dV$
Concentration independent of location (well-mixed)	$V\dfrac{dc}{dt} + c\dfrac{dV}{dt}$
Concentration independent of location, and system volume fixed	$V\dfrac{dc}{dt}$
Concentration and volume constant over time (steady state)	0

---

[1] Throughout this chapter, mass balances are written for an arbitrary constituent of interest, i. To minimize clutter, the concentration of this constituent is written simply as c, rather than c_i; a similar convention applies to other parameters characterizing i.

the region of interest (i.e., $c(0)$) as c_o, the mass of contaminant in the entire 5-km segment can be computed by integrating this function over the length of the segment; that is

$$M = \int_{x=0}^{5\,\text{km}} c(x) A \, dx$$

$$= A \int_{x=0}^{5\,\text{km}} \frac{c_o}{1 + k c_o x} \, dx$$

$$= \frac{A}{k} \ln[1 + k c_o x]_0^{5\,\text{km}}$$

$$= \frac{A}{k} \ln \frac{1 + k c_o (5\,\text{km})}{1 + k c_o (0\,\text{km})} = \frac{A}{k} \ln\{1 + k c_o (5\,\text{km})\}$$

Inserting the known values of A, k, and c_o, and several conversion factors

$$M = \frac{24\,\text{m}^2}{3(\text{mg/L})^{-1}\text{km}^{-1}} \frac{10^3\,\text{m}}{\text{km}} \frac{10^3\,\text{L}}{\text{m}^3} \frac{1\,\text{kg}}{10^6\,\text{mg}}$$

$$\ln\left[1 + \left(3(\text{mg/L})^{-1}\text{km}^{-1}\right)(2\,\text{mg/L})(5\,\text{km})\right] = 27.5\,\text{kg} \quad \blacksquare$$

■ **EXAMPLE 1-3.** The first step in many water and wastewater treatment processes often involves nothing more than mixing the incoming water thoroughly with water that arrived earlier in a large mixing tank. This process, known as *equalization*, mitigates temporal variations in the composition of the water that enters downstream reactors and thereby eases process control.

(a) Consider an equalization tank in an industrial wastewater treatment plant that contains 500 m³ of water with a Cu^{2+} concentration of 65 mg/L. One of the process lines feeding the tank is online, so that the incoming water contains 125 mg/L Cu^{2+}. If the flow rate into and out of the tank is 6 m³/min, what is the rate of change of the mass of Cu^{2+} stored in the tank; that is, what would be the value of the storage term in a mass balance on Cu^{2+} in the equalization tank at the given time?

(b) The tank is intended to prevent the Cu^{2+} concentration in the exiting water from fluctuating at a rate greater than 1 mg/L-min. Is it achieving this objective at the instant in question?

Solution.
(a) We can compute the desired values by writing a mass balance on Cu^{2+}, using the space occupied by all the water in the tank as the CV:

Rate of change of mass of Cu^{2+} in the tank	=	Rate at which Cu^{2+} enters in the influent to the tank	−	Rate at which Cu^{2+} leaves in the tank effluent
	+	Rate at which Cu^{2+} enters the reactor by diffusion and dispersion across the inlet and outlet pipes	+	Rate at which Cu^{2+} is generated by reactions in the tank

Because the flow rates into and out of the tank are equal and the Cu^{2+} concentration is uniform throughout the tank (because the water is thoroughly mixed), we can use the simplification shown in the third row of Table 1-1 to represent the storage term, as follows:

$$\text{Rate of change of mass of } Cu^{2+} \text{ in tank} = V \frac{dc_{Cu^{2+}}}{dt}$$

In the system of interest, Cu^{2+} does not participate in any chemical reaction, and the mass rates at which it enters and leaves the tank by advection are $Qc_{Cu^{2+},\text{in}}$ and $Qc_{Cu^{2+},\text{out}}$, respectively. The mass balance can therefore be written in a mathematical form as follows:

$$V \frac{dc_{Cu^{2+}}}{dt} = Qc_{Cu^{2+},\text{in}} - Qc_{Cu^{2+},\text{out}} = Q(c_{Cu^{2+},\text{in}} - c_{Cu^{2+},\text{out}})$$

$$= 6\,\text{m}^3/\text{min}(125\,\text{mg/L} - 65\,\text{mg/L})$$

$$\times \left(\frac{1000\,\text{L}}{\text{m}^3}\right)\left(\frac{1\,\text{kg}}{10^6\,\text{mg}}\right) = 0.36\,\text{kg/min}$$

The value of the storage term (i.e., the term on the left of the above equation) is therefore

$$\left(\begin{array}{c}\text{Rate of change of}\\ \text{mass of } Cu^{2+} \text{ in tank}\end{array}\right) = V \frac{dc_{Cu^{2+}}}{dt} = 0.36\,\text{kg/min}$$

(b) The rate of change of the Cu^{2+} concentration is computed by dividing the storage term in the mass balance by the liquid volume as

$$\frac{dc_{Cu^{2+}}}{dt} = \frac{V \dfrac{dc_{Cu^{2+}}}{dt}}{V}$$

$$= \frac{3.6 \times 10^5\,\text{mg/min}}{500\,\text{m}^3} \frac{1\,\text{m}^3}{1000\,\text{L}}$$

$$= 0.72\,\text{mg/L-min}$$

The result indicates that the equalization process is working well: even though the influent contains 60 mg/L more Cu^{2+} than the water in the tank, the Cu^{2+} concentration in the water leaving the tank and entering the downstream treatment processes is changing at a rate of <1 mg/L-min; that is, it is meeting the equalization objective. ∎

The Advective Term

Whereas the storage term in the mass balance describes changes inside the control volume, the terms for physical transport (advection, diffusion, and dispersion) describe processes at the boundaries. As noted previously, movement of fluid inside the CV does not transport i into or out of the CV and is therefore not reflected in the physical transport terms of the mass balance.

The volumetric rate (Q, volume/time) at which fluid crosses a boundary is given by the product of the velocity component perpendicular to the boundary ($v_\perp$) and the area (A) of the boundary. In the simplified system in which the flow is perpendicular to the boundary, the volumetric flow rates into and out of the system are $Q_{in} = v_{in}A_{in}$ and $Q_{out} = v_{out}A_{out}$, respectively. The rate at which a substance i that is dissolved or suspended in the water crosses the boundary by advection is the product of the volumetric flow rate and the concentration of i in the water; that is, $v_{in}A_{in}c_{in}$ or $Q_{in}c_{in}$ at the inlet, and $v_{out}A_{out}c_{out}$ or $Q_{out}c_{out}$ at the outlet. (Recall that, in the simplified case, c_{in} and c_{out} are assumed to be uniform across the areas A_{in} and A_{out}, respectively. This assumption is implicit in the preceding expressions.) Note that the products vAc and Qc have the desired dimensions of mass per time. The net advective term in the mass balance for the simplified case is therefore

Net rate of transport of i into CV by advection

$$= v_{in}A_{in}c_{in} - v_{out}A_{out}c_{out} \quad (1\text{-}3)$$

$$= Q_{in}c_{in} - Q_{out}c_{out} \quad (1\text{-}4)$$

The product cv is called the advective flux, often designated J_{adv}.[2] The advective flux represents the mass of i being advected in the direction of interest (in this case, the direction of bulk flow) per unit cross-sectional area per unit time. That is, if an imaginary flat surface of unit area were placed perpendicular to the direction of interest, the advective flux would indicate the mass of i passing through that surface per unit time as a result of the flow.

This result can be extended to a system containing multiple inlets and outlets by including an additional, similar term in the expression for each advective input or output.

Net rate of transport of i into CV by advection

$$= \sum_{\text{inlets}} Q_{in}c_{in} - \sum_{\text{outlets}} Q_{out}c_{out} \quad (1\text{-}5)$$

It is worth noting that advection can affect the amount of i stored in the CV only if the mass rates of inflow and outflow

---

[2] Formally, both the velocity and flux are vectors, related by $\mathbf{J}_{adv} = c\mathbf{v}$. However, since we are considering only flow in a single direction in this analysis, the direction is accounted for most simply by the signs on the terms (plus sign for input, and minus sign for output).

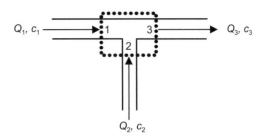

FIGURE 1-4. Schematic diagram for the junction of two streams.

of i differ. That is, substance i might be entering and leaving at very large rates, but unless these rates are different, there is no net effect on the amount of i in the system.

In many applications, advection is the only process that has a significant effect on the mass of a substance stored in the CV, such as at the junction of two pipes or channels, or when a single stream is divided into components. In such cases, the advective term is the only one that appears on the right side of the mass balance.

Consider, for example, a CV that encloses the junction of two pipes, as shown in Figure 1-4. The CV, indicated in the figure by the dotted box, includes the entire zone where the two streams mix, as well as a small portion of each pipe outside the mixing zone. We will evaluate the mass balances on water and on a conservative (nonreactive) substance i in the water, assuming that the system is at steady state over the time frame of the analysis.

The mass balance for water in this control region contains no dispersion or reaction terms, since these terms are unimportant for water.[3] By the assumption of steady state, the mass of water in the CV is constant, so the storage term in the mass balance is also zero, leaving only the advective terms in the mass balance. Formally, the mass balance for water is written as follows:

$$\begin{bmatrix} \text{Rate of change} \\ \text{of mass of water} \\ \text{stored in the} \\ \text{CV} \end{bmatrix} = \begin{bmatrix} \text{Rate at which water} \\ \text{enters the CV} \\ \text{by advection in} \\ \text{both influent streams} \end{bmatrix}$$

$$- \begin{bmatrix} \text{Rate at which water} \\ \text{leaves the CV} \\ \text{by advection in} \\ \text{the effluent stream} \end{bmatrix}$$

$$\int_V \frac{\partial \rho}{\partial t} dV = 0 = \rho Q_1 + \rho Q_2 - \rho Q_3 \quad (1\text{-}6)$$

---

[3] The reaction term is unimportant, because the water undergoes no reaction that significantly affects the mass of water in the system. Similarly, dispersion is unimportant, since no gradient exists in the concentration of water. Thus, even though dispersive mixing of the water might be substantial, it has no effect on the concentration of water anywhere in the system.

(This equation is based on the assumption that the volumetric flow rate of the entire solution can be equated with the volumetric flow rate of the water; that is, that water molecules account for essentially 100% of the volume of the solution. This assumption is virtually always valid for the systems of interest in environmental engineering.) Canceling the density of water from all terms, the equation reduces to the simple and fully expected result:

$$Q_3 = Q_1 + Q_2 \quad (1\text{-}7)$$

Consider next the mass balance on i. Because i is stipulated to be nonreactive, the reaction term in the mass balance is zero. While some dispersion certainly occurs at the junction, the net transport of i across the boundaries of the CV by dispersion is likely to be negligible in comparison to that by advection, so the dispersion term in the mass balance can be ignored. Finally, since the system is at steady state, the storage term is also zero. Under these conditions, the mass balance on i can be written as follows:

$$\begin{bmatrix} \text{Rate of change} \\ \text{of mass of } i \\ \text{stored in the} \\ \text{CV} \end{bmatrix} = \begin{bmatrix} \text{Rate at which } i \\ \text{enters the CV} \\ \text{by advection in} \\ \text{both influent streams} \end{bmatrix} - \begin{bmatrix} \text{Rate at which } i \\ \text{leaves the CV} \\ \text{by advection in} \\ \text{the effluent stream} \end{bmatrix}$$

$$\int_V \frac{\partial c}{\partial t} dV = 0 = Q_1 c_1 + Q_2 c_2 - Q_3 c_3 \quad (1\text{-}8)$$

$$c_3 = \frac{Q_1 c_1 + Q_2 c_2}{Q_3} = \frac{Q_1 c_1 + Q_2 c_2}{Q_1 + Q_2} \quad (1\text{-}9)$$

This result is likely to be familiar. The idea is, in fact, no different from what is used to describe mixing problems in elementary school mathematics, such as finding the cost of mixed nuts when the cost and proportion of each component are known. What is valuable here is to see that this well-known result can be obtained formally using the mass balance tool. The discussion also makes clear the assumptions that have to be met for the result to be valid, namely that the reaction, dispersion, and storage terms in the mass balance are all negligibly small in comparison to the advection term. These assumptions are not necessarily met in all cases of a junction of two streams, so we must be careful when applying Equation 1-9. For example, when free chlorine is used to disinfect a potable water or wastewater, a portion of the chlorine reacts almost instantaneously with some constituents of the water. As a result, the mass rate at which chlorine exits the mixing zone is less than the rate at which it is added, even if the water remains in the mixing zone for only an extremely short time.

Note that this analysis does not require the concentration to be the same everywhere in the CV for the result to be valid; indeed, the inclusion in the CV of a small portion of the water upstream of the mixing zone in pipes 1 and 2 ensures that this condition is not met. The only requirement is that the mixing of the streams be negligible before the water crosses the influent planes (points 1 and 2 in the figure), and complete before the water crosses the effluent plane (point 3).

■ **EXAMPLE 1-4.** A suspension flowing at $0.5\,\text{m}^3/\text{s}$ is to be dosed with a polymeric coagulant to induce large numbers of small, suspended particles to grow into fewer, larger ones. The coagulant is present in a stock solution at a concentration of 1000 mg/L, and it is to be dosed at a rate that yields a concentration of 0.3 mg/L in the mixed solution. What flow rate of stock solution is required?

Solution. The mixing process is, in effect, mixing at a junction, and hence can be analyzed using Equation 1-8. Referring to the water undergoing treatment as stream 1, the coagulant stock solution as stream 2, and the mixed solution as stream 3, the mass balance on coagulant is

$$0 = Q_1 c_1 + Q_2 c_2 - Q_3 c_3 = Q_1 c_1 + Q_2 c_2 - (Q_1 + Q_2) c_3$$

In this case, Q_2 is expected to be much less than Q_1; hence we can make the approximation that $Q_1 + Q_2 \approx Q_1$. Substituting the approximation and the given values and rearranging, we obtain

$$\begin{aligned} Q_2 &= \frac{Q_1(c_3 - c_1)}{c_2} \\ &= \frac{(0.5\,\text{m}^3/\text{s})((0.3 - 0.0)\,\text{mg/L})}{1000\,\text{mg/L}} \\ &= 0.15 \times 10^{-3}\,\text{m}^3/\text{s} = 0.15\,\text{L/s} \end{aligned}$$

Note that the assumption that Q_2 is negligible in comparison to Q_1 is valid. ■

The Diffusion and Dispersion Terms

Diffusion occurs because, even in a fluid that is stagnant at the macroscopic level, motion occurs at the molecular level. Molecules of water and all solutes in the water are in constant (small-scale) motion. The concentration of water molecules is essentially the same at all locations in the solution, so the small-scale motion of these molecules tends to be equal in all directions and to have no noticeable effect on the solution composition at any point. However, the concentration of a solute might be different at different locations. If the solute concentration differs in adjacent packets of fluid, random motion of these molecules is likely to cause more of them to move from the high-concentration to the low-concentration packet than vice versa, leading to a net movement of solute toward the low-concentration region. Over time, this motion tends to diminish differences in the solute concentration throughout the solution.

Similarly, in a flowing fluid, different molecules of fluid (and solute) flow at different rates, even if on a larger scale the fluid flow is uniform and steady. As noted previously, in laminar flow, fluid velocity tends to vary spatially across a plane perpendicular to the net flow, and this variation causes mass transport to be different from that described by the average advection. Turbulence also causes different molecules in the same solution to move at different velocities, in this case because of temporal variations of velocity around the mean at any point (or plane). In all these situations, the mass transport that is not captured by the average advective term is described by dispersion. Both time averaging (in the case of turbulence) and spatial averaging (in laminar flow) of the microscopic motions across a plane often lead to terms that are identical in form to those that describe molecular diffusion; here, we do not distinguish among the various phenomena that lead to this dispersion. The classic work of Taylor (1953, 1954) defined the dispersion in both laminar and turbulent flow. A succinct description of these phenomena is given by Holley[4] (1969), while a more thorough analysis and application to flow through porous media is given by Bear (1972, 1979).

Similar to advection, transport of i by diffusion and dispersion is described in terms of fluxes. According to *Fick's law*, the diffusive flux of a substance in any direction equals the negative product of the local concentration gradient of that substance and a *diffusion coefficient* or *diffusivity*, D.[5] By analogy, the dispersive flux in any given direction equals the negative product of the concentration gradient and a *dispersion coefficient* or *dispersivity*, ϵ. Both D and ϵ have dimensions of length squared per time. Except in very quiescent systems (e.g., the bottom of a quiet lake, the stagnant fluid in biofilms, or the pores inside granular activated carbon particles), the dispersion coefficient, ϵ, is generally much greater than the diffusion coefficient, D.

The diffusion coefficient depends on the particular substance being considered, and, in general, would be written as D_i. On the other hand, ϵ depends on the hydrodynamic characteristics of the system and applies equally to all substances in the water. However, because the intensity of dispersion might be different in different directions, ϵ depends on the direction of interest. For this simplified analysis, we are assuming that a concentration gradient exists only in the x direction, as shown schematically in Figure 1-5. For such a case, we can write[6]

$$\text{Flux in } +x \text{ direction due to diffusion} \equiv J_{\text{diff},x} = -D\frac{\partial c}{\partial x} \quad (1\text{-}10)$$

$$\text{Flux in } +x \text{ direction due to dispersion} \equiv J_{\text{disp},x} = -\epsilon_x\frac{\partial c}{\partial x} \quad (1\text{-}11)$$

Flux in $+x$ direction due to combined diffusion and

$$\text{dispersion} \equiv J_{\text{diff},x} + J_{\text{disp},x} \equiv J_{\mathbf{D},x} = -(D + \epsilon_x)\frac{\partial c}{\partial x} \quad (1\text{-}12)$$

where $J_{\mathbf{D},x}$ is the combined diffusive and dispersive flux in the x direction.

The concentration gradient $\partial c/\partial x$ is positive if the concentration increases in the $+x$ direction. Similarly, flux is defined as positive if material moves in the $+x$ direction. The negative sign in Equations 1-10–1-12 indicates that the direction of the diffusive and dispersive fluxes is opposite to the direction of the concentration gradient; that is, concentration increasing in the downstream ($+x$) direction leads to net diffusion and dispersion in the upstream ($-x$) direction.

The mass rate of input of i into the CV by diffusion and dispersion is the product of the flux and the appropriate area

Net rate of transport of i into CV via diffusion/dispersion

$$= [J_{\mathbf{D},x}A]_{\text{inlet}} - [J_{\mathbf{D},x}A]_{\text{outlet}} \quad (1\text{-}13\text{a})$$

$$= -\left\{\left((D+\epsilon_x)\frac{\partial c}{\partial x}A\right)_{\text{inlet}} - \left((D+\epsilon_x)\frac{\partial c}{\partial x}A\right)_{\text{outlet}}\right\} \quad (1\text{-}13\text{b})$$

$$= -(D+\epsilon_x)\left\{\left(\frac{\partial c}{\partial x}A\right)_{\text{inlet}} - \left(\frac{\partial c}{\partial x}A\right)_{\text{outlet}}\right\} \quad (1\text{-}13\text{c})$$

$$= (D+\epsilon_x)\left\{\left(\frac{\partial c}{\partial x}A\right)_{\text{outlet}} - \left(\frac{\partial c}{\partial x}A\right)_{\text{inlet}}\right\} \quad (1\text{-}13\text{d})$$

Note that the equality of Equations 1-13b and 1-13c relies on the assumption that the diffusion and dispersion coefficients are the same at the outlet as at the inlet. This assumption might be acceptable for flow through engineered systems with similar inlet and outlet structures, but many other situations exist in which the assumption does not apply.

---

[4] Holley reserves the term diffusion for time-averaged processes and the term dispersion for spatially averaged processes, but this distinction is not used universally. In this book, we reserve the word diffusion to refer to molecular diffusion of any chemical in water, and dispersion to refer to any process related to fluid motion that causes transport down a concentration gradient.

[5] When considering solutes, this statement assumes that the direction of decreasing concentration is also the direction of decreasing chemical activity, which is almost always true in aqueous systems.

[6] Although the dispersivity ϵ_x could vary with location (e.g., as a function of x in a reactor with flow in the x direction), we assume here that it is constant.

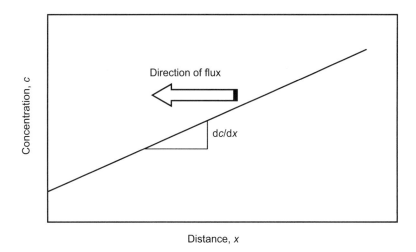

FIGURE 1-5. Schematic showing a (positive) concentration gradient ($dc/dx > 0$), leading to flux in the $-x$ direction.

Equation 1-13 indicates that the net transport of i into the CV by diffusion and dispersion equals the diffusive/dispersive transport in the $+x$ direction at the inlet minus that in the $+x$ direction at the outlet. Note that this statement does not imply that diffusion and dispersion must be into the CV at the inlet and out of it at the outlet. It is entirely possible for $\partial c/\partial x$ to be positive at the inlet point, in which case the diffusion/dispersion term would be negative; that is, out of the CV. Thus, the words inlet and outlet in Equation 1-13 describe locations where the bulk flow enters or leaves and where the area and concentration gradient must be evaluated to determine the diffusive/dispersive flux; they do not imply that diffusive/dispersive transport must be in any particular direction.

In many situations, we are interested in a CV that is differentially thick in the x direction and that has equal areas bounding it at its two ends (in the y–z plane). In such a case, Equation 1-13d can be rewritten as follows:

Net rate of transport of i into the CV via diffusion/

$$\text{dispersion} = (D + \epsilon_x)A\left\{\left(\frac{\partial c}{\partial x}\right)_{x+dx} - \left(\frac{\partial c}{\partial x}\right)_x\right\} \quad (1\text{-}14)$$

From the definition of the derivative, $(\partial c/\partial x)_{x+dx} - (\partial c/\partial x)_x = (\partial/\partial x)(\partial c/\partial x)dx = (\partial^2 c/\partial x^2)dx$, so that

Net rate of transport of i into the CV via diffusion/

$$\text{dispersion} = (D + \epsilon_x)A\frac{\partial^2 c}{\partial x^2}dx \quad (1\text{-}15)$$

■ **EXAMPLE 1-5.** Assume that, at the bottom of a deep lake, the sediment surface is flat and the water is quiescent. This region of the lake has become anaerobic, and phosphate species are being released from the sediment as a result of microbial activity. The pH of the lake is 8.8, causing almost all the phosphate to be present as the species HPO_4^{2-}. Assume that the concentration of this species at the sediment surface is approximately steady at 1.0 mg/L as P.[7] Under the ambient conditions, the diffusion coefficient of HPO_4^{2-} is 1.2×10^{-5} cm^2/s. Assuming that the HPO_4^{2-} does not participate in any chemical reactions in the water near the sediment, and that c_P in the water column is negligible at time $t = 0$, compute the concentration of HPO_4^{2-} and the flux of this species into the water thereafter, as a function of location and time. The water column can be considered deep enough that the phosphate species do not reach the top of the stagnant zone in the time frame of interest.

Solution. Choosing a CV that is differentially thick in the vertical direction (which we define as the x direction for this problem), a mass balance on the total dissolved phosphate includes only the diffusive term on the right, because the phosphate species are nonreactive and are not brought into or out of the CV by flow. Dropping the subscript on c, and noting that $Adx = dV$, the mass balance can be written and simplified as follows:

$$dV\frac{\partial c}{\partial t} = AD\frac{\partial^2 c}{\partial x^2}dx \quad (1\text{-}16)$$

$$\frac{\partial c}{\partial t} = D\frac{\partial^2 c}{\partial x^2} \quad (1\text{-}17)$$

Equation 1-17 describes how the changes in the concentration of a nonreactive material over distance and time are related in a stagnant solution. Because the differential equation is first order in time and second order in space, we need an initial condition and two boundary conditions to

---
[7] This terminology indicates that the concentration of HPO_4^{2-} molecules in the water is such that the water contains 1.0 mg/L of P.

solve it. For this problem, these equations are written as

Initial condition: $\quad c(x, 0) = 0$

Boundary conditions: $\quad c(0, t > 0) = c_o = 1.0\,\text{mg/L}$

$$c(\infty, t) = 0$$

Even with the simplifying assumptions that the water is stagnant and the phosphate is nonreactive, this is not an easy problem to solve. The solution, which is developed in many textbooks on differential equations (see, e.g., Kreyszig, 1993), can be expressed as follows:

$$c(x, t) = c_o \left(1 - \text{erf}\left\{ \frac{x}{\sqrt{4Dt}} \right\} \right) = c_o \,\text{erfc}\left\{ \frac{x}{\sqrt{4Dt}} \right\} \tag{1-18}$$

where $\text{erf}(z) \equiv (2/\sqrt{\pi}) \int_0^z e^{-u^2} du$ and $\text{erfc}(z) \equiv 1 - \text{erf}(z)$.

The function $\text{erf}(z)$ is known as the error function, the gaussian distribution, or the probability integral. The function $\text{erfc}(z)$ is known as the complementary error function and equals the same integral as $\text{erf}(z)$, except integrated from z to ∞, rather than from 0 to z. Values of these functions are widely tabulated and are available in many spreadsheet programs. Using such a program, we can determine for a given system (i.e., a given value of D) the concentration profile ($c(x)$) at a given time t, the concentration variation over time ($c(t)$) at a given location x, or $c(x, t)$ for selected values of x and t. Some results of such calculations are shown in Figure 1-6. The concentration gradient is steep at short times and distances, and it becomes flatter and extends farther into the water column at later times. Note that molecular diffusion is an extremely slow process for mixing

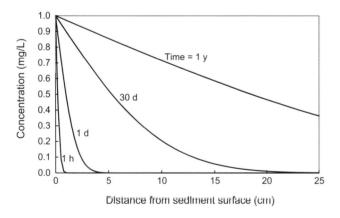

FIGURE 1-6. Concentration profile as a function of time and distance for unidirectional diffusion of a nonreactive contaminant (HPO_4^{2-}, in the current example) into stagnant water. The concentration of HPO_4^{2-} at the sediment surface is assumed to be constant at 1.0 mg/L.

a constituent into a solution. For example, even after 30 days, dissolved phosphate is barely detectable a mere 20 cm from its source.

The flux out of the sediment is given by the product of the diffusivity and the concentration gradient as

$$\text{Flux} = J_{\text{diff}} = -D \frac{dc}{dx} \tag{1-19}$$

$$= -Dc_o \frac{d}{dx}\left(1 - \text{erf}\left\{ \frac{x}{\sqrt{4Dt}} \right\} \right)$$

$$= Dc_o \frac{d}{dx}\left(\text{erf}\left\{ \frac{x}{\sqrt{4Dt}} \right\} \right) \tag{1-20a}$$

which is equivalent to

$$J_{\text{diff}} = Dc_o \frac{d}{dx}\left(\frac{2}{\sqrt{\pi}} \int_0^{x/\sqrt{4Dt}} \exp\{-u^2\} du \right) \tag{1-20b}$$

Using Leibniz's rule for differentiating an integral,[8] we can rewrite Equation 1-20b as

$$J_{\text{diff}} = Dc_0 \frac{2}{\sqrt{\pi}} \exp\left(-\frac{x^2}{4Dt} \right) \frac{d}{dx}\left(\frac{x}{\sqrt{4Dt}} \right) \tag{1-21}$$

and, at $x = 0$,

$$J_{\text{diff}, x=0} = Dc_0 \frac{2}{\sqrt{\pi}} (1.0) \left(\frac{1}{\sqrt{4Dt}} \right) = c_0 \sqrt{\frac{D}{\pi t}} \tag{1-22}$$

Inserting values into this expression, we find the following fluxes at the interface at times corresponding to the four curves in the plot:

Time	Flux at $x = 0$ (mg/m^2 d)
1 h	28.14
1 d	5.74
30 d	1.05
1 y	0.30

Diffusion and dispersion are extremely important in many natural and engineered environmental processes. In particular, diffusion controls the behavior of solutes in systems, or parts of systems, where advection is minimal, such as in the

---

[8] This rule is given by Beyer (1987) as $\partial \int_{a(y)}^{b(y)} (f(z, y) dz)/\partial y = \int_{a(y)}^{b(y)} (\partial f(z, y)/\partial y) dz + f(b(y), y)(\partial a/\partial y) - f(a(y), y)(\partial b/\partial y)$. For the application here, we define $\eta \equiv x/\sqrt{4Dt}$, and use Liebniz's rule with $y = \eta$, $z = u$, and the limits of integration $a(y) = 0$ and $b(y) = \eta$.

deep sediments of lakes and in the interior of porous particles (e.g., granules of activated carbon, which is used to collect many contaminants from solution via adsorption). Similarly, dispersion can control the spreading of solutes upstream and downstream along the length of a river and in reactors that have nearly uniform flow in one direction. Reactors that are designed to be *plug flow reactors* (with exactly uniform flow in the axial direction) often do not achieve that ideal, in which case dispersion can control the longitudinal mixing in the reactor. Such reactors include packed beds that are used to carry out adsorption of solutes and filtration of particles, and tanks used for water disinfection.

In many other commonly encountered situations, on the other hand, diffusion and dispersion are negligible. In particular, in CVs that are large and that connect with regions outside the system across relatively little surface area, the diffusion/dispersion term can often be ignored. The most common such systems are large tanks receiving influent and discharging effluent through pipes in which the water velocity is rapid. In some systems, these flows are not even continuous with the water in the tank (e.g., if the influent enters by being discharged from a free flowing pipe above the water level). In such cases, advection carries molecules across the system boundaries much more rapidly than diffusion and dispersion do; as a result, diffusion and dispersion can be ignored.

The Chemical Reaction Term

The final term in the mass balance represents the net rate of change of i due to chemical reactions in the CV. Contrary to advection, diffusion, and dispersion, chemical reactions affect the mass balance only if they occur inside the CV, rather than across its boundaries.

Conventionally, the rate of a reaction in a homogeneous phase is normalized to the volume and is given as r_V, the amount of i being formed per unit volume per time. Thus, in any differentially small region dV of the CV, the rate of formation of i by reaction is $r_V dV$. If i is being destroyed by chemical reaction, then r is negative. The rate at which the mass of i changes in the entire CV as a result of homogeneous chemical reactions is obtained by summing the rates applicable in all the dV-sized regions within the volume; that is, by integrating over the entire CV

Net rate of change of mass of i in CV by homogeneous

$$\text{reactions} = \int r_V dV \quad (1\text{-}23)$$

Many reactions that are important for water quality occur at the interface between two phases. Examples of such reactions include the transfer of gases into and out of solution, precipitation and dissolution of solid phases, corrosion of water distribution pipes, and the adsorption from solution of trace compounds onto the surfaces of activated carbon or minerals. Depending on the geometry and other characteristics of the system, the interfacial surfaces at which reactions occur might all be at the boundary of the CV or distributed throughout the CV. The former case would apply, for instance, if we chose a single droplet of water as the CV for the analysis of gas transfer in a spray aeration system (a system in which water droplets are sprayed into the air to enhance transfer of oxygen into the water). On the other hand, when analyzing the adsorption of pesticides onto powdered activated carbon in a well-mixed, batch system, the most appropriate CV might be a portion of fluid with a representative amount of activated carbon dispersed in it, in which case the interfacial area is distributed throughout the CV.

When interfacial reactions are important in a mass balance analysis, it is often convenient to normalize their rates to the amount of surface area in the system, rather than to the system volume. That is, the rate of a reaction occurring at a surface might be quantified by a term $r_{i,\sigma}$, indicating the rate at which i is being created per unit amount of surface area present. Thus, $r_{i,\sigma}$ would have dimensions of mass per area per time. The corresponding term in the mass balance would then be $r_{i,\sigma} a dV$, where a is the surface area concentration in the system (surface area per unit volume).

Writing the terms for homogeneous and interfacial reactions individually, the net rate of formation of i by chemical reactions in the CV is

Net rate of change of mass of i in CV by chemical

$$\text{reactions} = \int (r_V + r_\sigma a) dV \quad (1\text{-}24)$$

Recall that a system in which the composition is uniform throughout the CV is referred to as being well-mixed. Since the rates of chemical reactions depend on the system composition, the reaction rate is also spatially uniform in well-mixed systems; that is, it does not depend on the location. In such cases, the reaction terms can be taken out of the integral, leading to the following simpler form:

Net rate of change of mass of i in a well-mixed CV by

$$\text{chemical reactions} = (r_V + r_\sigma a)V \quad (1\text{-}25)$$

Each term in Equations 1-24 and 1-25 might be a composite of several others, each of which represents the rate of a single reaction by which i is generated or destroyed within the CV. For example, in a given CV, oxygen might be consumed by a chemical reaction with ferrous iron and by a biochemical reaction with degradable organic matter (mediated by organisms), and it might simultaneously enter the solution by transfer from injected air bubbles. Each of these reactions would be included individually in the mass balance on oxygen, as a component of the term shown generically above as the (single) reaction term.

The symbol r is used as a generic descriptor of the overall reaction rate of i in mass balances. However, the rate of each reaction that is occurring in the system is characterized by a distinct dependence on the concentrations of the reactants, solution temperature, pH, and other parameters. An example of such a relationship, and how it can be used in a mass balance, is provided below. Reaction rate expressions are the major focus of Chapter 3.

■ **EXAMPLE 1-6.** When drinking water is disinfected with chlorine-containing compounds (especially hypochlorous acid and hypochlorite ion [HOCl and OCl$^-$, respectively]), chlorinated organic compounds such as chloroform (CHCl$_3$) and chloro-acetic acid (CH$_x$Cl$_{3-x}$COOH) are generated. The organic reactants in this process comprise a diverse collection of molecules referred to as natural organic matter (NOM), and the chlorinated products of the reaction are referred to as disinfection byproducts (DBPs).

The kinetics of DBP formation is complex, both because of the diversity of the NOM molecules and because the disinfectant can simultaneously participate in many other reactions that do not form DBPs. Nevertheless, a number of mathematical models have been proposed to describe the kinetics, especially for the formation of CHCl$_3$. One such model suggests that the rate of CHCl$_3$ formation can be approximated by a function of the form $r_{CHCl_3} = k(NOM)(HOCl)^3$, where (NOM) and (HOCl) are the concentrations of NOM and HOCl, respectively, and k is an empirical constant. Chloroform is essentially inert in water.

The following data have been collected in a batch treatment test, under conditions where k is estimated to have a value of $3.0 \times 10^{-6}\,(mg/L)^{-3}min^{-1}$. Estimate the CHCl$_3$ concentration in the solution at $t = 1$ h.

t (min)	0	5	10	20	30	40	50	60
NOM (mg/L)	5.2	5.0	4.9	4.9	4.9	4.9	4.9	4.9
HOCl (mg/L)	7.0	5.1	4.8	4.4	4.0	3.7	3.4	3.2

Solution. The data for the NOM and HOCl concentrations are plotted in Figure 1-7a, and the corresponding rate of formation of CHCl$_3$ at each time (computed from the given expression for r_{CHCl_3}) is plotted in Figure 1-7b.

Since the only process that is affecting the CHCl$_3$ concentration in solution is the reaction, and since that

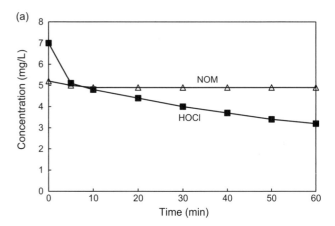

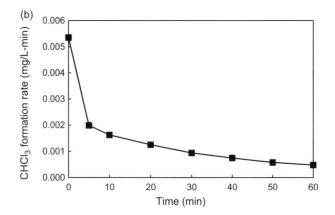

FIGURE 1-7. Conditions in an example, batch system undergoing chlorination. (a) Reactant concentrations; (b) the rate of chloroform formation, calculated as $k(NOM)(HOCl)^3$.

concentration is uniform throughout the solution, the mass balance on CHCl$_3$ can be expressed as follows:

$$V\frac{dc_{CHCl_3}}{dt} = Vr_{CHCl_3}$$

$$\int_{c_0}^{c_{60\,min}} dc_{CHCl_3} = \int_0^{60\,min} r_{CHCl_3}dt$$

$$c_{CHCl_3,\,60\,min} = \int_0^{60\,min} r_{CHCl_3}dt$$

Thus, the concentration of CHCl$_3$ in the solution at $t = 60$ min can be computed as the time integral of the formation rate over the period from 0 to 60 min. Although this integral cannot be evaluated analytically, it can be evaluated as the area under the curve in Figure 1-7b. The result is that, at $t = 60$ min, $c_{CHCl_3} = 0.073$ mg/L $= 73\,\mu g/L$. ■

Combining the Terms into the Overall Mass Balance

The overall mass balance for a system with unidirectional flow, a concentration gradient only in the direction of flow, and only one inlet and one outlet can be written by combining the terms described in the preceding sections for the individual processes as

$$\frac{\partial}{\partial t}\int c\,dV = Q_{in}c_{in} - Q_{out}c_{out} + (D+\epsilon_x)\left\{\left[A\frac{\partial c}{\partial x}\right]_{outlet} - \left[A\frac{\partial c}{\partial x}\right]_{inlet}\right\} + \int (r_V + r_\sigma a)\,dV \quad (1\text{-}26)$$

Although Equation 1-26 is applicable to any system with unidirectional flow, it is often quite daunting to use. Fortunately, the criteria for applying several of the simplifications described in the discussion of the individual terms are often met in systems of interest. Some mass balances in these cases are summarized in Table 1-2. (Note that, although the CVs for these equations are written as V, the equations might be applied to differential-sized CVs, dV.)

■ **EXAMPLE 1-7.** The amount of dissolved oxygen (DO) that would be consumed by microbial oxidation of the biodegradable organic matter in a solution is referred to as the biochemical oxygen demand of the solution and is commonly symbolized by the letter L. Note that, although L is a measure of the concentration of organics in the water, it is expressed in terms of the concentration of oxygen that would be consumed if the organics decayed. Over time, if the organics in solution are actually degraded, and if no biodegradable inputs enter the solution, the biochemical oxygen demand remaining in the solution declines; that is, L decreases over time. The rate of this decline is commonly represented as follows:

$$r_L = -k_d L$$

Consider a well-mixed lake (no concentration gradients in the lake) with volume 5×10^8 L that is fed by a stream flowing at 2.4×10^7 L/d. The stream contains 8 mg/L of DO and an amount of biodegradable organic matter that corresponds to $L = 10$ mg/L. Waste from a small municipality ($L = 95$ mg/L, DO $= 0$ mg/L) flows into the lake at a rate of 4.8×10^6 L/d. The value of k_d in the lake is $0.10\,\text{d}^{-1}$.

In addition to the input of oxygen via the stream, oxygen can enter the lake from the atmosphere at a rate given by the expression $r_{O_2} = k_r(\text{DO}^* - \text{DO})$, where k_r is a constant known as the reaeration constant, and DO* is the saturation (equilibrium) value of DO in the water. For the given system, $k_r = 0.05\,\text{d}^{-1}$ and DO$^* = 11.2$ mg/L. The values of DO and L in the lake are approximately constant over time, and diffusion and dispersion contribute negligibly to the transport of either substance into or out of the lake. The system is shown schematically in Figure 1-8.

(a) Write mass balances on biochemical oxygen demand and oxygen in the lake and determine the values of L and DO.
(b) Compute the rate (kg/d) at which each individual process (advection, reaeration, biological reaction) increases or decreases DO and L in the lake.

Known: $Q_1 = 2.4 \times 10^7$ L/d $\quad Q_3 = Q_1 + Q_2 = 2.88 \times 10^7$ L/d
$\quad\quad\quad\;$ DO$_1 = 8$ mg/L $\quad\quad\;$ DO$_3 = ?$
$\quad\quad\quad\;$ $L_1 = 10$ mg/L $\quad\quad\quad$ $L_3 = ?$
$\quad\quad\quad\;$ $Q_2 = 4.8 \times 10^6$ L/d $\quad$ $V = 5 \times 10^8$ L
$\quad\quad\quad\;$ DO$_2 = 0$ mg/L $\quad\quad\;$ $k_d = 0.10\,\text{d}^{-1}$
$\quad\quad\quad\;$ $L_2 = 95$ mg/L $\quad\quad\;$ DO$^* = 11.2$ mg/L
$\quad\quad\quad\quad\quad\quad\quad\quad\quad\quad\;\;$ $k_r = 0.05\,\text{d}^{-1}$
$\quad\quad\quad\;$ $L_{\text{in lake}} = ?$
$\quad\quad\quad\;$ DO$_{\text{in lake}} = ?$

Solution.

(a) This system is at steady state and is well-mixed, so it meets the criteria for using Equation 1-28. Substituting the appropriate expression for the reaction term, the mass balance on biochemical oxygen demand is

$$0 = Q_1 L_1 + Q_2 L_2 - Q_3 L_3 - k_d V L_{\text{in lake}}$$

Note that the reaction is taking place *in* the lake, so the L that is used to compute r_L is the concentration

TABLE 1-2. Simplified Mass Balances Applicable to Many Systems of Interest

Well-mixed reactor with fixed volume
$$V\frac{dc}{dt} = Q(c_{in} - c_{out}) + (r_V + r_\sigma a)V \quad (1\text{-}27)$$

Well-mixed reactor with fixed volume, at steady state
$$0 = Q(c_{in} - c_{out}) + (r_V + r_\sigma a)V \quad (1\text{-}28)$$

Plug flow reactor (negligible axial diffusion), at steady state
$$0 = Q(c_{in} - c_{out}) + \int(r_V + r_\sigma a)\,dV \quad (1\text{-}29)$$

Well-mixed batch reactor (no flow) with fixed volume
$$V\frac{dc}{dt} = (r_V + r_\sigma a)V \quad (1\text{-}30)$$

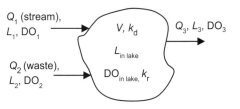

FIGURE 1-8. Schematic showing the inflows and outflow from the lake in the example system.

in the lake. On the other hand, the advective terms describe processes at the inlets and outlet, so the concentrations used to compute these terms are the concentrations at the system boundaries. However, because the lake is well-mixed, L in the outlet stream (L_3) equals that in the lake ($L_{\text{in lake}}$). Using L_3 to represent both these quantities, we obtain

$$0 = (2.4 \times 10^7 \text{ L/d})(10 \text{ mg/L}) + (4.8 \times 10^6 \text{ L/d})(95 \text{ mg/L}) \\ - (2.88 \times 10^7 \text{ L/d})L_3 - (0.10 \text{ d}^{-1})(5 \times 10^8 \text{ L})L_3$$

$$L_3 = \frac{(2.4 \times 10^7 \text{ L/d})(10 \text{ mg/L}) + (4.8 \times 10^6 \text{ L/d})(95 \text{ mg/L})}{(2.88 \times 10^7 \text{ L/d}) + (0.10 \text{ d}^{-1})(5 \times 10^8 \text{ L})}$$

$$L_3 = L_{\text{in lake}} = 8.83 \text{ mg/L}$$

The mass balance on oxygen is almost identical in form to the mass balance on L, except that the reaeration reaction provides a pathway for oxygen to enter the system (and therefore an additional term in the mass balance) that does not apply to L. Because of the definition of the oxygen demand, the term in the mass balance on DO that accounts for DO consumption by the biodegradation reaction is identical to the term in the mass balance on L for consumption of L. The mass balance on DO is therefore

Rate of change of mass of DO stored in the lake = Net rate (in−out) at which oxygen enters by advection

+ Rate at which oxygen is generated by biochemical reactions (< 0) + Rate at which oxygen transfers into lake water from the air (> 0)

$$\frac{d}{dt}[(\text{DO}_{\text{in lake}})V] = Q_1(\text{DO}_1) + Q_2(\text{DO}_2) - Q_3(\text{DO}_3) \\ - k_d(L_{\text{in lake}})V + k_r(\text{DO}^* - \text{DO}_{\text{in lake}})V$$

Applying the assumptions of complete mixing ($\text{DO}_{\text{in lake}} = \text{DO}_3$) and steady state, the equation becomes

$$0 = Q_1(\text{DO}_1) + Q_2(\text{DO}_2) - Q_3(\text{DO}_3) - k_d(L_3)V \\ + k_r(\text{DO}^* - \text{DO}_3)V$$

All the values in the above equation except DO_3 are known. Substituting these values and solving for DO_3, we obtain

$$0 = (2.4 \times 10^7 \text{ L/d})(8 \text{ mg/L}) + (4.8 \times 10^6 \text{ L/d})(0 \text{ mg/L}) \\ - (2.88 \times 10^7 \text{ L/d})(\text{DO}_3) \\ - (0.1 \text{ d}^{-1})(8.83 \text{ mg/L})(5 \times 10^8 \text{ L}) \\ + (0.05 \text{ d}^{-1})(11.2 \text{ mg/L} - \text{DO}_3)(5 \times 10^8 \text{ L})$$

$$\text{DO}_3 = \text{DO}_{\text{in lake}} = 0.57 \text{ mg/L}$$

(b) The rates of the various terms in the mass balance are computed as follows:

Advective inflow of biochemical oxygen demand:

$$Q_1(L_1) + Q_2(L_2) = [(2.4 \times 10^7 \text{ L/d})(10 \text{ mg/L}) \\ + (4.8 \times 10^6 \text{ L/d})(95 \text{ mg/L})][10^{-6} \text{ kg/mg}] \\ = 240 \text{ kg/d} + 456 \text{ kg/d} = 696 \text{ kg/d}$$

Advective outflow of biochemical oxygen demand:

$$Q_3(L_3) = (2.88 \times 10^7 \text{ L/d})(8.83 \text{ mg/L})(10^{-6} \text{ kg/mg}) \\ = 254 \text{ kg/d}$$

Rate of L utilization (i.e., the rate of DO utilization by biochemical reactions). (Note that, because r_L is defined to be positive when L increases, the rate of utilization of biochemical oxygen demand is defined as $-r_L V$.):

$$-r_L V = k_d(L_{\text{in lake}})V \\ = (0.10 \text{ d}^{-1})(8.83 \text{ mg/L})(5 \times 10^8 \text{ L}) \\ = 4.42 \times 10^8 \text{ mg/d} = 442 \text{ kg/d}$$

Advective inflow of O_2:

$$Q_1(\text{DO}_1) + Q_2(\cancel{\text{DO}_2}) = (2.4 \times 10^7 \text{ L/d})(8 \text{ mg/L}) \\ (10^{-6} \text{ kg/mg}) = 192 \text{ kg/d}$$

Advective outflow of O_2:

$$Q_3(\text{DO}_3) = (2.88 \times 10^7 \text{ L/d})(0.57 \text{ mg/L})(10^{-6} \text{ kg/mg}) \\ = 16 \text{ kg/d}$$

Rate of reaeration:

$$r_r V = k_r(\text{DO}^* - \text{DO})V \\ = (0.05 \text{ d}^{-1})(11.2 - 0.57 \text{ mg/L})(5 \times 10^8 \text{ L}) \\ = 2.66 \times 10^8 \text{ mg/d} = 266 \text{ kg/d}$$

The terms in the mass balance equations for L and DO, in kg/d, are summarized schematically in Figure 1-9. ∎

The Differential Form of the One-Dimensional Mass Balance

The overall mass balance derived in the preceding sections is conceptually satisfying and is useful in many applications. However, in some systems, we are more interested in the rate of change of the concentration of i at a single location than in the rate of change of the mass of i stored in a specified volume. We can convert Equation 1-26 into a form that is applicable at a unique location by writing it for a CV that is of differential length, dx. In that case, the cross-sectional

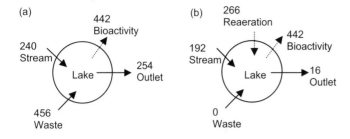

FIGURE 1-9. Contributions of various terms in the mass balance on (a) biochemical oxygen demand and (b) dissolved oxygen for the lake described in Example 1-7. Solid lines represent advective processes, and broken lines represent reactions. All numbers represent mass rates in kg/d.

areas at the inlet and outlet can be assumed to equal one another ($A_{\text{inlet}} = A_{\text{outlet}} = A$). Also, Q_{in} must equal Q_{out}, and we can write both as $A v_x$. Making these substitutions, writing dV as $A dx$, and dividing by A yields

$$\frac{\partial}{\partial t} \int c \, dx = v_x c_{\text{in}} - v_x c_{\text{out}} + (D + \epsilon_x) \left\{ \left[\frac{\partial c}{\partial x}\right]_{\text{outlet}} - \left[\frac{\partial c}{\partial x}\right]_{\text{inlet}} \right\}$$
$$+ \int (r_V + r_\sigma a) dx \quad (1\text{-}31)$$

Then, since the distance from the inlet to the outlet is dx, we can write $c_{\text{in}} - c_{\text{out}}$ as $-dc$ or, equivalently, $-(\partial c/\partial x) dx$. Similarly, $\{[\partial c/\partial c]_{\text{outlet}} - [\partial c/\partial c]_{\text{inlet}}\}$ can be written as $(\partial/\partial x)[\partial c/\partial x] dx$, yielding

$$\frac{\partial}{\partial t} \int c \, dx = -v_x \frac{\partial c}{\partial x} dx + (D + \epsilon_x) \left\{ \frac{\partial}{\partial x}\left[\frac{\partial c}{\partial x}\right] \right\} dx + \int (r_V + r_\sigma a) dx$$

$$\frac{\partial}{\partial t} \int c \, dx = -v_x \frac{\partial c}{\partial x} dx + (D + \epsilon_x) \frac{\partial^2 c}{\partial x^2} dx + \int (r_V + r_\sigma a) dx \quad (1\text{-}32)$$

Finally, differentiating both sides with respect to x, we obtain:

$$\frac{d}{dx}\left[\frac{\partial}{\partial t} \int c \, dx\right] = -v_x \frac{\partial c}{\partial x} + (D + \epsilon_x) \frac{\partial^2 c}{\partial x^2} + \frac{d}{dx} \int (r_V + r_\sigma a) dx \quad (1\text{-}33)$$

$$\frac{\partial c}{\partial t} = -v_x \frac{\partial c}{\partial x} + (D + \epsilon) \frac{\partial^2 c}{\partial x^2} + (r_V + r_\sigma a) \quad (1\text{-}34)$$

Equation 1-34 is known as the *one-dimensional advective diffusion equation*. As suggested above, it is the mass balance on a differential volume in a system in which concentration varies in only one dimension, and for this reason it is sometimes referred to as the mass balance "at a point." The equation is often the most convenient starting point for analyzing systems with unidirectional flow. If the initial and boundary conditions of the system are known, the equation can be integrated over the length of the reactor (x values from 0 to L) to determine the concentration profile as a function of location in the system.

■ **EXAMPLE 1-8.** Consider a plug flow reactor with uniform cross-section (i.e., one with flow in only one direction and negligible mixing in that direction), so that the solution has a uniform concentration perpendicular to the flow direction. A reaction is occurring in the solution at a rate given by $r_V = -kc$, and no surface reactions are occurring. Determine the relationship between the influent and effluent concentrations, if the system is at steady state.

Solution. To solve this problem, it is convenient to use Equation 1-34.

$$\frac{\partial c}{\partial t} = -v \frac{\partial c}{\partial x} + (D + \epsilon) \frac{\partial^2 c}{\partial x^2} + (r_V + r_\sigma a) \quad (1\text{-}34)$$

where x is the direction of flow. Steady state indicates that $\partial c/\partial t = 0$, and the assumptions of plug flow and negligible axial diffusion indicate that the middle term on the right side of the equation is zero. Therefore, for the given situation, the equation can be simplified as follows:

$$0 = -v \frac{\partial c}{\partial x} - kc \quad (1\text{-}35)$$

$$\frac{dc}{dx} = -\frac{k}{v} c \quad (1\text{-}36)$$

The partial differential operator in Equation 1-35 has been converted into an ordinary differential in Equation 1-36, because c_i is a function of only one variable (x). Equation 1-36 can be rearranged and integrated between the inlet ($x = 0$), where the concentration is c_{in}, and the outlet ($x = L$), where the concentration is c_{out}

$$\int_{c_{\text{in}}}^{c_{\text{out}}} \frac{dc}{c} = -\frac{k}{v} \int_0^L dx \quad (1\text{-}37)$$

$$\ln \frac{c_{\text{out}}}{c_{\text{in}}} = -\frac{k_1 L}{v} \quad (1\text{-}38)$$

$$c_{\text{out}} = c_{\text{in}} \exp\left(-\frac{k_1 L}{v}\right) \quad (1\text{-}39)$$

Multiplying the numerator and denominator of the exponential argument in Equation 1-39 by the cross-sectional area of the reactor, A, and noting that LA is the volume of the reactor (V), and vA is the volumetric flow rate (Q), we obtain

$$c_{\text{out}} = c_{\text{in}} \exp\left(-\frac{k_1 L}{v}\right) = c_{\text{in}} \exp\left(-\frac{kLA}{vA}\right) = c_{\text{in}} \exp\left(-\frac{kV}{Q}\right)$$

$$c_{\text{out}} = c_{\text{in}} \exp(-k\tau) \quad (1\text{-}40)$$

where τ is the hydraulic detention time, V/Q, equal to the amount of time that the fluid resides in the reactor. The significance of this term is discussed in detail in Chapter 2. ∎

1.3 THE MASS BALANCE FOR A SYSTEM WITH FLOW AND CONCENTRATION GRADIENTS IN ARBITRARY DIRECTIONS

The Advection Term

We next consider how we can extend the preceding analysis to a system in which the flow can approach the surface at an arbitrary angle, and the concentration gradients causing diffusion and dispersive flux are not necessarily perpendicular to the boundary of the CV. Such a system is shown schematically in Figure 1-10. In this system, mass is capable of entering and leaving the CV across any point on its surface, and the bulk water flow is allowed to approach the CV from any angle. The analysis of such a system is broadly applicable, but the generality of the analysis unavoidably introduces greater mathematical complexity than is present in the case of unidirectional flow.

Consider first the advective flux in such a system. As in the simpler system, that flux is given by the product $v_\perp c$. However, in this system, it is possible that only a portion of the flow will be perpendicular to the system boundary; that is, $v_\perp$ might be less than v. To quantify the relationship between v and $v_\perp$, consider a differentially small region dA on the surface of the CV, and define the angle between the direction of flow (advection) and a line perpendicular to and directed into the surface as θ_{SAd}, as shown in Figure 1-11.

The velocity vector near dA can be divided into three orthogonal components, two of which are parallel to the boundary, while the third is perpendicular to it. The components of flow parallel to the boundary cannot cross it, so the only component that is relevant for the mass balance is the perpendicular component. Mathematically, the magnitude of

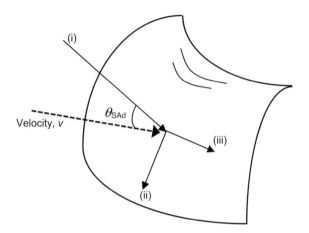

FIGURE 1-11. Definition diagram for a system in which flow can approach the boundary of the control volume from arbitrary directions. The flow toward the surface (broken line) can be represented as a combination of a component perpendicular to the surface (i) and two components parallel to the surface (ii and iii).

this component equals the product of the magnitude of v and the cosine of the angle θ_{SAd}

$$v_\perp = v \cos \theta_{SAd} \qquad (1\text{-}41)$$

The volumetric flow rate crossing dA and entering the CV is the product of the velocity component perpendicular to and into the CV and the area dA; that is,

$$dQ_{in} = v \cos \theta_{SAd}\, dA \qquad (1\text{-}42)$$

Note that, if the actual bulk flow is perpendicular to dA, the angle θ_{SAd} is 0. In this case, $\cos \theta_{SAd}$ is 1.0, and the entire flow enters the CV. This situation applies to the simplified one-dimensional model discussed in Section 1-2. Similarly, if $\theta_{SAd} = 180°$, $\cos \theta_{SAd}$ equals -1.0; in that case, the magnitude of $v \cos \theta_{SAd}$ equals that of the bulk velocity, but the sign is negative, indicating that flow is directly out of the CV (this applies to the outflow in the simplified model). For any other value of θ_{SAd}, the component of the flow perpendicular to and therefore crossing the system boundary is less than the entire flow. Thus, the product $v \cos \theta_{SAd} dA$ yields a value that indicates both the magnitude of the flow that is crossing the boundary of the CV and the direction in which that fluid transfer is occurring.

As noted in the prior discussion of the simplified system, the rate at which a substance i enters the CV by advection across a patch dA is the product of the concentration of i in the water and the volumetric flow rate across that area. Thus, for the more general case under consideration here, the rate at which i enters the CV by advection is given by $cv \cos \theta_{SAd} dA$. Since c can never be negative, the direction of solute advection is always the same as the direction of

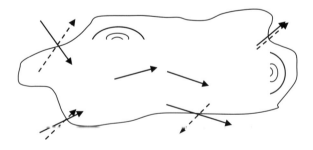

FIGURE 1-10. Schematic representation of a control volume, flow pattern (solid arrows), and concentration gradients (dashed arrows) in a system with an arbitrary boundary shape and with the flow direction and the direction of the concentration gradient bearing no simple relationship with the shape of the surface.

bulk fluid advection and is incorporated into the result by virtue of the sign of the computed value (positive for transfer into the CV, negative for transfer out of it).

To compute the overall rate of advective transport of i into the CV, the value of $cv \cos\theta_{SAd} dA$ must be computed for all patches of surface enclosing that volume, and summed (i.e., integrated), as follows:

Net rate of transport of i into CV by advection

$$= \oint cv \cos\theta_{SAd} dA \quad (1\text{-}43)$$

$$= \oint J_{adv} \cos\theta_{SAd} dA \quad (1\text{-}44)$$

The circle on the integral sign indicates that the integration is carried out around a surface that completely encloses the CV. The dimensions of c, v, and A are mass per volume, length per time, and area, respectively, giving a product with dimensions of mass per time, as required.

Equation 1-43 is often written in a slightly different form, using vector notation. To understand the alternative form, recall that a spatial vector $\boldsymbol{A}$ can be decomposed into the sum of components in the x, y, and z directions as: $\boldsymbol{A} \equiv a_x \boldsymbol{i} + a_y \boldsymbol{j} + a_z \boldsymbol{k}$, where a_x, a_y, and a_z, are the magnitudes of the vector in the three directions, and $\boldsymbol{i}, \boldsymbol{j}$, and $\boldsymbol{k}$ are unit vectors in these directions. The magnitude of $\boldsymbol{A}$ can be computed as $A \equiv \sqrt{a_x^2 + a_y^2 + a_z^2}$.[9]

Recall also that the dot product of two vectors is a scalar whose value equals the product of the magnitudes of the two vectors and the cosine of the angle between them

$$\boldsymbol{A} \cdot \boldsymbol{B} = AB \cos\alpha \quad (1\text{-}45)$$

If a vector $\boldsymbol{m}$ is defined whose magnitude is 1.0 and whose direction is perpendicular to and directed into the surface of the CV, the dot product of the velocity vector with $\boldsymbol{m}$ is

$$\boldsymbol{v} \cdot \boldsymbol{m} = v(1.0) \cos\theta_{SAd} = v \cos\theta_{SAd} \quad (1\text{-}46)$$

Thus, Equation 1-43 could be written in vector form as

Net rate of transport of i into CV by advection $= \oint c\boldsymbol{v} \cdot \boldsymbol{m} dA$

$$(1\text{-}47)$$

By convention, however, the vector defining the perpendicular to the surface is assigned a magnitude of 1.0 and a direction *out* of the surface. It is called the *outwardly directed unit normal* and is designated $\boldsymbol{n}$. Since $\boldsymbol{m} = -\boldsymbol{n}$, $\boldsymbol{v} \cdot \boldsymbol{m} = -\boldsymbol{v} \cdot \boldsymbol{n}$, and the conventional way of writing the advective term in the mass balance using vector notation is

Net rate of transport of i into CV by advection $= \oint c\boldsymbol{v} \cdot \boldsymbol{n} dA$

$$(1\text{-}48)$$

The product $c\boldsymbol{v}$ is the vector representation of the advective flux of i, $\boldsymbol{J}_{adv}$. $\boldsymbol{J}_{adv}$ characterizes the mass of i transported across a unit area per unit time in the direction of flow (i.e., the direction of $\boldsymbol{v}$). The product $\boldsymbol{n} dA$ is sometimes written as $d\boldsymbol{A}$, defined as a vector with magnitude equal to dA and direction perpendicular to and directed out of the surface. Thus, a few alternative forms of Equation 1-48 are as follows:

Net rate of transport of i into CV by advection

$$= \oint c\boldsymbol{v} \cdot d\boldsymbol{A} \quad (1\text{-}49a)$$

$$= -\oint \boldsymbol{J}_{adv} \cdot \boldsymbol{n} dA \quad (1\text{-}49b)$$

$$= -\oint \boldsymbol{J}_{adv} \cdot d\boldsymbol{A} \quad (1\text{-}49c)$$

The various forms for representing the advective term in mass balances, both for the general case and the simplified system with unidirectional flow, are summarized in Table 1-3.

The Diffusion and Dispersion Terms

Consider now the diffusive and dispersive fluxes for the general case represented in Figure 1-10. In this case, the concentration of i may vary in all directions, rather than just in the direction perpendicular to the boundary, so diffusion and dispersion can occur in all directions as well. Furthermore, whereas the diffusion coefficient is the same in all

TABLE 1-3. Summary of Equations Describing Advective Transport Into a Control Volume

Net rate of transport of i into control volume by advection =	
Arbitrary surface shape and flow pattern	Flow directly into and out of control volume across a flat boundary
Geometric or scalar representation	
$\oint cv \cos\theta_{SAd} dA$ (1-43)	$\sum_{inlets} Q_{in} c_{in} - \sum_{outlets} Q_{out} c_{out}$ (1-5)
$\oint J_{adv} \cos\theta_{SAd} dA$ (1-44)	
Vector representation	
$-\oint c\boldsymbol{v} \cdot \boldsymbol{n} dA$ (1-48)	
$-\oint c\boldsymbol{v} \cdot d\boldsymbol{A}$ (1-49a)	
$-\oint \boldsymbol{J}_{adv} \cdot \boldsymbol{n} dA$ (1-49b)	
$-\oint \boldsymbol{J}_{adv} \cdot d\boldsymbol{A}$ (1-49c)	

---
[9] In this book, vectors are shown in bold, and the magnitude of a vector is shown as the same symbol in normal text.

directions, the dispersion coefficient is often different in different directions. As a result, the dispersive flux in a given direction depends on both the concentration gradient and the dispersion coefficient in that direction, and the maximum dispersive flux might not be in the direction of either the maximum concentration gradient or the maximum dispersivity; rather, it is always in the direction of the maximum product of these two terms.

The variation in ϵ with direction complicates the mathematics beyond the scope of this discussion. This mathematics is presented in many advanced fluid dynamics textbooks (e.g., Bear, 1972; Domenico and Schwartz, 1990). Here, we focus on two limiting cases: one in which dispersion is significant in only one direction (the situation analyzed previously, yielding the various forms of Equation 1-13), and the other for so-called *isotropic turbulence*, in which the dispersion coefficient is identical in all directions, and the directions of maximum diffusion and maximum dispersion are identical. In the latter case, the diffusive, dispersive, and combined fluxes in any given direction can be computed as follows:

$$J_{\text{diff},l} = -D\frac{\partial c}{\partial l} \tag{1-50}$$

$$J_{\text{disp},l} = -\epsilon\frac{\partial c}{\partial l} \tag{1-51}$$

$$J_{\mathbf{D},l} = -(D+\epsilon)\frac{\partial c}{\partial l} \tag{1-52}$$

where l is a measure of distance in the direction of interest. As is the case for the advective term, only the components of these fluxes that are perpendicular to the surface of the CV and therefore causing i to enter or leave the CV should be considered in the mass balance. That is, we are interested in the value of $\partial c/\partial l$ along a line normal to the surface of the CV, which we will refer to as $\partial c/\partial l_n$.

If we knew the concentration profile in all directions at a given location on the CV boundary, and if we knew the orientation of the surface at that location, we could compute $\partial c/\partial l_n$ and the corresponding flux across a differential patch of surface at the location. By carrying out this process repeatedly until we have considered the boundary of the entire CV, we could compute the transport of i into the CV as follows:

Net rate of transport of i into CV by diffusion and

$$\text{dispersion} = -\oint (D+\epsilon)\frac{\partial c}{\partial l_n}dA \tag{1-53}$$

In some cases, the known information is the magnitude and direction of the maximum concentration gradient at various locations in the system, rather than the concentration gradient in the direction normal to the surface. In such a case, we can represent the maximum concentration gradient as the vector sum of three components, much as we did for the velocity vector: two components parallel to the surface, and one perpendicular to it. The component of the concentration gradient that is perpendicular to the surface, and that therefore is responsible for all the diffusive and dispersive transport across the CV boundary, is

$$\frac{\partial c}{\partial l_n} = \frac{\partial c}{\partial l'}\cos\theta_{nl'} \tag{1-54}$$

where $\partial c/\partial l_n$ is the gradient at the point of interest in the direction perpendicular to the surface, $\partial c/\partial l'$ is the largest concentration gradient at that location, and $\theta_{nl'}$ is the angle between these two gradients.

Then, substituting Equation 1-54 into Equation 1-53:

Net rate of transport of *i* into CV by diffusion and

$$\text{dispersion} = -\oint (D+\epsilon)\frac{\partial c}{\partial l'}\cos\theta_{nl'}dA \tag{1-55}$$

For consistency with the notation used in the equations derived earlier for advective flux, we subsequently refer to the angle between the normal to the surface and the steepest concentration gradient ($\theta_{nl'}$) as $\theta_{S\mathbf{D}}$; that is

Net rate of transport of *i* into CV by diffusion and

$$\text{dispersion} = -\oint (D+\epsilon)\frac{\partial c}{\partial l'}\cos\theta_{S\mathbf{D}}dA \tag{1-56}$$

$$= \oint J_{\mathbf{D},\max}\cos\theta_{S\mathbf{D}}dA \tag{1-57}$$

where $J_{\mathbf{D},\max}$ is the maximum diffusive and dispersive flux in any direction at the location where the integral is being evaluated.

The steepest gradient in some property (e.g., concentration, temperature, or electrical potential) is of interest in a number of engineering contexts and can be represented conveniently using vector notation. Specifically, for a property P, the vector characterizing the magnitude and direction of the steepest gradient in P is given the symbol ∇P and can be computed as

$$\nabla P = \frac{\partial P}{\partial x}\mathbf{i} + \frac{\partial P}{\partial y}\mathbf{j} + \frac{\partial P}{\partial z}\mathbf{k} \tag{1-58}$$

In this context, the vector of interest is ∇c, which has the magnitude and direction of $\partial c/\partial l'$. The vector $\mathbf{n}$ has a magnitude of unity and a direction perpendicular to the surface. Therefore, the dot product of ∇c and $-\mathbf{n}$ is

$$\nabla c \cdot (-\mathbf{n}) = \frac{\partial c}{\partial l'} \cos\theta_{SD} \quad (1\text{-}59)$$

Substituting this equality into Equation 1-56, we obtain

Net rate of transport of i into CV by diffusion and

$$\text{dispersion} = \oint (D+\epsilon)\nabla c \cdot \mathbf{n}\,dA \quad (1\text{-}60)$$

The product $(D+\epsilon)\nabla c$ is the flux in the direction of the steepest gradient in concentration. Representing this term as $J_{D,\max}$, we can write Equation 1-60 in the following alternative form:

Net rate of transport of i into CV by diffusion

$$= -\oint J_{D,\max} \cdot \mathbf{n}\,dA \quad (1\text{-}61)$$

Several equations derived above for the diffusive and dispersive terms in the mass balance equation are summarized in Table 1-4.

The Storage and Reaction Terms

Because the storage and reaction terms in the mass balance describe processes that occur inside the CV rather than at its boundary, these terms are identical regardless of whether the direction of flow and the direction of the concentration gradient are perpendicular to the boundary or not; that is, they have the same form for the systems as shown in both Figure 1-3a and 1-3b.

The Overall Mass Balance

The overall mass balance for the system with arbitrary flow and boundary characteristics, but constant volume and isotropic turbulence, can therefore be written by combining the expressions for advective and diffusive/dispersive flux with those derived earlier for storage and reaction. Two equivalent forms of the resulting equation are as follows:

$$\int \frac{\partial c}{\partial t}\,dV = \oint (-c\mathbf{v} + (D+\epsilon)\nabla c) \cdot \mathbf{n}\,dA + \int_V (r_V + r_\sigma a)\,dV \quad (1\text{-}66)$$

$$\int \frac{\partial c}{\partial t}\,dV = -\oint \{J_{adv} + J_{D,\max}\} \cdot \mathbf{n}\,dA + \int_V (r_V + r_\sigma a)\,dV \quad (1\text{-}67)$$

The two types of integrals in Equations 1-66 and 1-67 reflect the distinction between the terms in the mass balance that account for changes *within* the CV (the storage and reaction terms) and those that account for transport of molecules *across the boundaries* of the CV (the advection and dispersion terms). The idea of the volume integrals (for the storage and reaction terms) is that one moves through the CV and, in each differential element of volume evaluates $(\partial c/\partial t)\,dV$ or $r\,dV$. All these differential terms are then summed to get the values of the corresponding integrals. The idea of the surface integrals (for the advective and dispersion terms) is that one moves around the surface that encloses the CV, considering one differential area after another, and at each area evaluates the local concentration of i, the gradient of i, and the magnitude and direction of flow. The appropriate differential terms are then computed and summed to determine the value of the corresponding integrated terms.

The equations for the advective and dispersive fluxes of i in the simplified case can all be derived from these for the more general case by making appropriate substitutions for θ_{SAd} and θ_{SD} and assuming that the velocity and concentration are uniform across the cross-sections of the inlet and outlet; that is, in the y and z directions.

TABLE 1-4. Summary of Equations Describing Diffusive and Dispersive Transport Into a Control Volume

Net rate of transport of i into control volume by diffusion and dispersion =

Arbitrary surface shape and flow pattern, but isotropic turbulence		*Flow directly into and out of control volume across a flat boundary*	
Geometric or scalar representation			
$-\oint (D+\epsilon)\dfrac{\partial c}{\partial l'}\cos\theta_{SD}\,dA$	(1-62)	$[J_D A]_{\text{inlet}} - [J_D A]_{\text{outlet}}$	(1-13)
$\oint J_{D,\max}\cos\theta_{SD}\,dA$	(1-63)	$-(D+\epsilon_x)\left\{\left[\dfrac{\partial c}{\partial x}A\right]_{\text{inlet}} - \left[\dfrac{\partial c}{\partial x}A\right]_{\text{outlet}}\right\}$	(1-13)[a]
Vector representation			
$\oint (D+\epsilon)\nabla c \cdot \mathbf{n}\,dA$	(1-64)		
$-\oint J_{D,\max} \cdot \mathbf{n}\,dA$	(1-65)		

[a]Assumes $(D+\epsilon)_{\text{inlet}} = (D+\epsilon)_{\text{outlet}}$

1.4 THE DIFFERENTIAL FORM OF THE THREE-DIMENSIONAL MASS BALANCE

As is the case for the one-dimensional mass balance, it is sometimes of interest to shrink the CV to differential dimensions for mass balances in three-dimensional space. Equations 1-66 and 1-67 can be converted into such a form by using a mathematical equality known as the divergence theorem, which provides a means for relating phenomena at the boundaries of a surface to quantities in the space that the surface encloses. This theorem states that, for any vector X

$$\oint X \cdot n \, dA = \int \boldsymbol{\nabla} \cdot X \, dV \qquad (1\text{-}68)$$

The symbol $\nabla \cdot X$ is a scalar referred to as the divergence of X and defined as

$$\nabla \cdot X = i \cdot \frac{\partial X}{\partial x} + j \cdot \frac{\partial X}{\partial y} + k \cdot \frac{\partial X}{\partial z} \qquad (1\text{-}69a)$$

$$= \frac{\partial X_x}{\partial x} + \frac{\partial X_y}{\partial y} + \frac{\partial X_z}{\partial z} \qquad (1\text{-}69b)$$

where X_x is the x-component of X, and so on. The divergence operation is, in a way, analogous to the gradient operation: the gradient operation is applied to a scalar and yields a vector with the magnitude and direction of the steepest change in that scalar, whereas the divergence operation is applied to a vector and yields a scalar with the magnitude of the steepest change in that vector (and, of course, no direction, since it is a scalar).[10]

In essence, the divergence theorem states that any material that crosses the boundary enclosing a space (the area integral on the left side of the equation) can be expressed in terms of the increase in the amount of material inside the boundary (the volume integral on the right side). Applying this theorem to the right side of Equation 1-66, the mass balance can be rewritten as follows:

$$\int \frac{\partial c}{\partial t} dV = \oint (-c\boldsymbol{v} + (D+\epsilon)\boldsymbol{\nabla} c) \cdot n \, dA + \int (r_V + r_\sigma \sigma) dV \qquad (1\text{-}66)$$

$$\int \frac{\partial c}{\partial t} dV = \int [\boldsymbol{\nabla} \cdot \{-c\boldsymbol{v} + (D+\epsilon)\boldsymbol{\nabla} c\} + (r_V + r_\sigma a)] dV \qquad (1\text{-}70)$$

The terms on the right side of Equation 1-70 are the same terms that appear in the mass balances derived previously, except that now the advective and diffusive/dispersive fluxes have been converted into a volume basis.

---
[10] See Schey (1973) for a thorough and readable discussion of these mathematical operations.

The first term on the right side of Equation 1-70 can be expanded based on the chain rule, as follows:

$$\nabla \cdot (-c\boldsymbol{v}) = -\left(\frac{\partial(cv_x)}{\partial x} + \frac{\partial(cv_y)}{\partial y} + \frac{\partial(cv_z)}{\partial z}\right) \qquad (1\text{-}71)$$

$$= -c\left(\frac{\partial v_x}{\partial x} + \frac{\partial v_y}{\partial y} + \frac{\partial v_z}{\partial z}\right) - \left(v_x \frac{\partial c}{\partial x} + v_y \frac{\partial c}{\partial y} + v_z \frac{\partial c}{\partial z}\right) \qquad (1\text{-}72)$$

$$= -c(\boldsymbol{\nabla} \cdot \boldsymbol{v}) - \boldsymbol{v}(\boldsymbol{\nabla} c) \qquad (1\text{-}73)$$

According to the continuity equation, $\boldsymbol{\nabla} \cdot \boldsymbol{v}$ is zero for any incompressible fluid, such as water, so Equation 1-73 indicates that, for any constituent in an aqueous system[11]

$$\nabla \cdot (-c\boldsymbol{v}) = -\boldsymbol{v} \cdot \nabla v \qquad (1\text{-}74)$$

Furthermore, since we are assuming that ϵ is independent of direction (i.e., we are assuming isotropic dispersion), the term accounting for transport by diffusion and dispersion in Equation 1-70 can be written as follows:

$$\nabla \cdot \{(D+\epsilon)\nabla c\} = (D+\epsilon)\nabla \cdot \nabla c \qquad (1\text{-}75)$$

$$= (D+\epsilon)\nabla^2 c \qquad (1\text{-}76)$$

where $\nabla^2 c$ is a scalar known as the Laplacian of c. This symbol is a short-hand expression for the following summation:

$$\nabla^2 c \equiv \frac{\partial^2 c}{\partial x^2} + \frac{\partial^2 c}{\partial y^2} + \frac{\partial^2 c}{\partial z^2} \qquad (1\text{-}77)$$

Substituting Equations 1-74 and 1-76 into Equation 1-70, we obtain

$$\int \frac{\partial c}{\partial t} dV = \int \left(-\boldsymbol{v} \cdot \nabla c + (D+\epsilon)\nabla^2 c + (r_V + r_\sigma a)\right) dV \qquad (1\text{-}78)$$

By differentiating both sides of Equation 1-78 with respect to volume, the equation becomes a mass balance on a differential volume

$$\frac{\partial c}{\partial t} dV = \left(-\boldsymbol{v} \cdot \nabla c + (D+\epsilon)\nabla^2 c + (r_V + r_\sigma a)\right) dV \qquad (1\text{-}79)$$

---
[11] The continuity equation is a mass balance applied to the fluid (in our case, water) in a differential control volume of fixed dimensions. The equation indicates that, since water is incompressible, it cannot accumulate in the control volume, so it must enter and exit at the same rate. Any decrease in its velocity in one direction must therefore be compensated by an increase in its velocity in a different direction. The equation is derived formally in most texts dealing with fluid mechanics and/or environmental transport processes; see, for example, Clark (2009).

Finally, dividing by dV, we obtain a version of the volume-normalized mass balance (i.e., an equation for the rate of change of concentration in a differential CV) that is commonly referred to as the mass balance at a point.

$$\frac{\partial c}{\partial t} = -\boldsymbol{v} \cdot \boldsymbol{\nabla} c + (D + \epsilon)\nabla^2 c + (r_V + r_\sigma a) \qquad (1\text{-}80)$$

As shown in Equation 1-72, the term $\boldsymbol{v} \cdot \boldsymbol{\nabla} c$ in Equation 1-80 is a scalar equal to $v_x(\partial c/\partial x) + v_y(\partial c/\partial y) + v_z(\partial c/\partial z)$. This term is very frequently represented as $v\nabla c$, with the implicit understanding that v is a vector and that the product is a dot product. Equation 1-80 is the advective diffusion equation written for a system in which parameters can vary in all three spatial dimensions. Similar to Equation 1-34, Equation 1-80 is sometimes referred to as a mass balance at a point.

■ **EXAMPLE 1-9.** The HOCl concentration in drinking water decays as the water flows through the distribution system, due to reactions with NOM (as described in Example 1-6) and with the pipe walls (HOCl can react with both the metal and attached biomass). Assume that, in a given system, the overall decay rate accounting for reactions both in solution and on the surface (i.e., $r_V + r_\sigma a$) can be characterized by the expression $r_{HOCl} = -kc_{HOCl}$, where k has a value of $0.1\,h^{-1}$.

Compare the magnitudes of the net advection and reaction terms in the mass balance for a 5-km long section of 1-m diameter pipe in which the flow rate is $1.5\,m^3/s$. The concentration of HOCl entering this section of pipe is 3 mg/L. Assume that the pipe section can be characterized by plug flow and that the system is operating at steady state.

Solution. The most general form of the mass balance on HOCl for this system is given by Equation 1-26

$$\frac{\partial}{\partial t}\int_V c\,dV = Q_{in}c_{in} - Q_{out}c_{out} + (D+\epsilon_x)\left\{\left[A\frac{\partial c}{\partial x}\right]_{outlet} - \left[A\frac{\partial c}{\partial x}\right]_{inlet}\right\} + \int_V (r_V + r_\sigma a)dV$$

However, since the system is at steady state, the rate of change of HOCl stored in the system (the term on the left hand side of the mass balance) is zero, and since the system is characterized by plug flow, the middle term on the right is zero as well. Noting that $Q_{in} = Q_{out} = Q$, the mass balance can therefore be simplified as follows:

$$0 = Qc_{in} - Qc_{out} + \int_V (r_V + r_\sigma a)dV \qquad (1\text{-}81)$$

The difference $Qc_{in} - Qc_{out}$ is the net advective term in the mass balance, and the volume integral is the net reaction term. Since the pipe behaves as a plug flow reactor and the reaction rate is directly proportional to the HOCl concentration, we can use the result from Example 1-8 (specifically, Equation 1-40) to compute c_{out}. To do so, we need to first find the hydraulic residence time, τ. This value is

$$\tau = \frac{V}{Q} = \frac{(\pi d^2/4)L}{Q} = \frac{(\pi(1\,m)^2/4)(5000\,m)}{1.5\,m^3/s}\left(\frac{1\,h}{3600\,s}\right) = 0.73\,h$$

Therefore, c_{out} can be computed as follows:

$$c_{out} = c_{in}\exp(-k\tau)$$
$$= (3.0\,mg/L)\exp\left[-(0.1\,h^{-1})(0.73\,h)\right] = 2.79\,mg/L$$

The net advective term in the mass balance is therefore

$$\text{Net advective term} = Q(c_{in} - c_{out})$$
$$= (1.5\,m^3/s)(10^3\,L/m^3)((3.0 - 2.79)\,mg/L)$$
$$= 316\,mg/s$$

The net reaction term could be computed formally by substituting the reaction rate expression ($-kc_{HOCl}$) for $r_V + r_\sigma a$ and then integrating throughout the pipe volume. However, Equation 1-81 indicates that the reaction term is just the opposite of the advection term. This must be the case, since the system is at steady state (so no HOCl is accumulating in the CV), and the only factors that affect the HOCl concentration in the CV are advection and reaction. Therefore, the net reaction term in the CV must be $-316\,mg/s$. The negative sign indicates that the reaction removes HOCl from the system. ■

1.5 SUMMARY

Mass balances are accounting tools that can be used to identify the relative importance of different processes occurring in a reactor and to predict how the system behaves under any conditions. The critical steps in writing a mass balance are identifying the substance for which the balance is to be written, defining the system boundaries, and establishing all the ways that the substance can enter or exit the CV or be created or destroyed within it.

The overall mass balance states that the rate of change of the mass of i stored in the CV is a function of the flow rates into and out of the CV, the concentration gradient at the boundary of the CV, the reaction rate within the CV, and several physical variables describing the system (D, ϵ, V). In most cases, the initial condition of the system is known, and many of the parameters in the equation relating to the physical properties of the system are known. Techniques for estimating some of the physical parameters that are not known in advance are presented in Chapter 2. Similarly, explicit expressions describing the reaction rate (r) as a function of concentration are presented and discussed in Chapter 3.

TABLE 1-5. Summary of Mass Balance Equations Derived in this Chapter

Storage term	=	Net advection term	+	Net diffusion/dispersion term	+	Net chemical reaction term	Eq. no.

Equation describing the rate of change of mass of i in a control volume for one-dimensional flow:

$$\frac{\partial}{\partial t}\int c\,dV = Q_{in}c_{in} - Q_{out}c_{out} + (D+\epsilon_x)\left\{\left[A\frac{\partial c}{\partial x}\right]_{outlet} - \left[A\frac{\partial c}{\partial x}\right]_{inlet}\right\} + \int (r_V + r_\sigma a)dV \quad (1\text{-}26)$$

Equation describing the rate of change of concentration of i (i.e., the volume-normalized mass balance) at a point for one-dimensional flow:

$$\frac{\partial c}{\partial t} = -v_x\frac{\partial c}{\partial x} + (D+\epsilon_x)\frac{\partial^2 c}{\partial x^2} + (r_V + r_\sigma a) \quad (1\text{-}34)$$

Equations describing the rate of change of mass of i in a control volume for an arbitrary flow pattern:

$$\frac{\partial}{\partial t}\int c\,dV = \oint \{-c\mathbf{v} + (D+\epsilon)\nabla c\}\cdot \mathbf{n}\,dA + \int (r_V + r_\sigma a)dV \quad (1\text{-}66)$$

$$\frac{\partial}{\partial t}\int c\,dV = -\oint (\mathbf{J}_{adv} + \mathbf{J}_{D,max})\cdot \mathbf{n}\,dA + \int (r_V + r_\sigma a)dV \quad (1\text{-}67)$$

Equation describing the rate of change of concentration of i (i.e., the volume-normalized mass balance) at a point for an arbitrary flow pattern:

$$\frac{\partial c}{\partial t} = -\mathbf{v}\cdot\nabla c + (D+\epsilon)\nabla^2 c + (r_V + r_\sigma a) \quad (1\text{-}80)$$

Equation describing the rate of change of mass of i in a well-mixed reactor with continuous flow, but with fixed volume:

$$V\frac{dc}{dt} = Q(c_{in} - c_{out}) + (r_V + r_\sigma a)V \quad (1\text{-}27)$$

Equation describing the rate of change of concentration of i in a well-mixed batch (no flow) reactor:

$$\frac{dc}{dt} = r_V + r_\sigma \sigma \quad (1\text{-}30)$$

Equation describing the rate of change of mass of i in a well-mixed reactor with continuous flow at steady state:

$$0 = Q(c_{in} - c_{out}) + (r_V + r_\sigma a)V \quad (1\text{-}28)$$

In theory, if all the physical parameters and the reaction rates are characterized, the mass balance equation can be solved for the concentration of i at any given time and location. However, even for relatively simple geometries and reaction rates, solving the mass balance equation can be formidable, especially if all the terms are significant. The equation becomes much easier to solve if additional simplifying assumptions can be made. For instance, in many cases the flows into and out of the reactor are equal, or the system composition is approximately invariant over time ($\partial c/\partial t = 0$). Sometimes, systems are set up in which several of the terms in the mass balance are designed to be negligible, to make it easier to evaluate those that remain. For example, reactors are conveniently evaluated by passing nonreactive constituents (tracers) through them, so that the chemical reaction term is zero ($r_V = r_\sigma = 0$). In other cases, chemical reactions are conveniently studied in well-mixed batch systems, where there is no flow and no concentration gradient, so the net advective, diffusive, and dispersive terms are all zero. These types of systems are discussed in Chapters 2 and 3, after which the problem of predicting the behavior of systems in which all the processes are occurring is addressed in Chapter 4.

The various forms of the mass balance equation derived throughout the chapter are summarized in Table 1-5.

REFERENCES

Bear, J. (1972) *Dynamics of Fluids in Porous Media*. American Elsevier Publishers, New York. (Also reprinted by Dover, Mineola, NY.)

Bear, J. (1979) *Hydraulics of Groundwater*. McGraw Hill, New York.

Beyer, W. H. (1987) *CRC Standard Mathematical Tables*. 28th ed., CRC Press, Boca Raton, FL, p. 232.

Clark, M. M. (2009) *Transport Modeling for Environmental Engineers and Scientists*. 2nd ed., Wiley Interscience, New York.

Domenico, P. A., and Schwartz, F. W. (1990) *Physical and Chemical Hydrogeology*. John Wiley and Sons, New York.

Holley, E. R. (1969) Unified view of diffusion and dispersion. *J. Hydraulics Div. (ASCE)*, 95 (HY2), 621–631.

Kreyszig, E. (1993) *Advanced Engineering Mathematics*. 7th ed., John Wiley and Sons, New York.

Schey, H. M. (1997) *Div, Grad, Curl, and All That: An Informal Text on Vector Calculus.* W.W. Norton & Co., Inc., New York, NY.

Taylor, G. (1953) Dispersion of soluble matter in solvent flowing slowly through a tube. *Proc. R. Soc. Lond. A*, 219, 186–203.

Taylor, G. (1954) The dispersion of matter in turbulent flow through a pipe. *Proc. R. Soc. Lond. A*, 223, 446–468.

PROBLEMS

1-1. The raw water supply for a community contains 18 mg/L total particulate matter. It is to be treated by addition of 60 mg alum ($Al_2(SO_4)_3 \cdot 14H_2O$) per liter of water treated. Essentially, all the added alum precipitates represented by the following reaction:

$$Al_2(SO_4)_3 \cdot 14H_2O \rightarrow 2\,Al(OH)_3(s) + 3SO_4^{2-} + 6H^+ + 8H_2O$$

For a total flow of 7500 m³/d, compute the daily alum requirement, the total concentration of suspended solids in the water following alum addition, and the daily load of particulate solids requiring disposal (including both those initially present and those formed during treatment).

1-2. Manufacturing of electronic chips or wafers requires ultrapure water (UPW) in substantial quantities; the primary use of this water is to clean the wafers after various manufacturing steps. As a result, wafer fabrication plants have a UPW production plant to supply this water to the manufacturing plant. Typically, the UPW plant takes municipal tap water and treats it further to remove a substantial fraction of the solutes. The UPW then flows into the manufacturing plant, where it is contaminated as it is used to clean the wafers. The typical situation, in which the contaminated water is collected and discharged without any reuse, is shown schematically in Figure 1-Pr2a. If the manufacturing process continuously adds a specific contaminant into the UPW at a rate $\dot{X}$ (mass per time), the key flows can be represented as in Figure 1-Pr2b.

It is very expensive to make UPW, and the contamination levels in the manufacturing process are quite low. Therefore, it makes sense to consider recycling some of the wastewater back to the start of the UPW production plant. Under these circumstances, the schematic would be as shown in Figure 1-Pr2c. Note that the flow rate of water supplied to the manufacturing plant (Q) remains the same, but the use of municipal water is decreased (to $Q - Q_R$).

(a) Unfortunately, some contaminants that are added to the water in the chip manufacturing process are

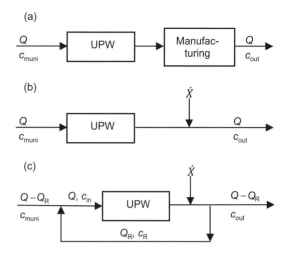

FIGURE 1-Pr2. Water and contaminant flows in a hypothetical electronic chip production plant with internal production of ultrapure water. (a) Generic flow diagram; (b) flow diagram assuming a continuous input of contaminant from the chip manufacturing process; (c) flow diagram from (b), modified to account for recycle of a portion of the contaminated water back to the inlet of the UPW production process.

not removed at all in the UPW production plant. As a result, they accumulate in the system with recycle until a steady-state condition is achieved. If the particular chemical that is added at the rate $\dot{X}$ is not removed at all in the UPW treatment, and if it is also not present at all in the influent (municipal water), what is the steady-state concentration in the effluent (c_{out}) in terms of $\dot{X}$ and the flow rates indicated? (Note: We are *not* concerned in this problem with the gradual rise of concentration until this concentration is reached, only with the ultimate steady-state condition.)

(b) Problems arise if the contaminant concentration in the influent to the manufacturing step rises above a critical acceptable level (c_{acc}). Again, assuming that the contaminant concentration in the municipal water is zero and that the UPW system does not remove any of this contaminant, develop an expression relating the recycle ratio ($R = Q_R/Q$) to the known quantities Q, c_{acc}, and $\dot{X}$. (Hint: Develop an expression relating Q_R, Q, c_{acc}, and $\dot{X}$ first, and then change it to include R.)

(c) If the water used in the manufacturing process is 2500 L/min, $\dot{X}$ is 300 mg/min, and c_{acc} is 0.2 mg/L, what flow rate of municipal water would have to be supplied?

1-3. A settling tank has a single influent flow and two effluent flows, as indicated in Figure 1-Pr3. The

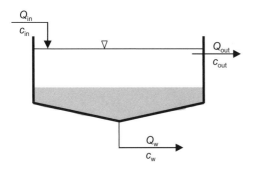

FIGURE 1-Pr3. A settling tank for separating and concentrating solids, and generating an effluent liquid stream with low solids concentration.

objective of sedimentation is to make c_{out} be much less than c_{in}, while making c_w much greater than c_{in}. A primary settling tank in a wastewater treatment plant is operating under steady-state conditions and has an influent concentration of 220 mg/L of suspended solids, an effluent concentration (c_{out}) of 45 mg/L, and a waste concentration (c_w) of 10,000 mg/L. Assume that no reactions involving these solids occur in the reactor.

(a) If the influent flow rate is 450 m³/h, what is the waste flow rate (Q_w)?

(b) What percentage of the influent flow is the waste flow?

1-4. Water flows into and out of a well-mixed, 1.6×10^5-L tank at a rate of 1000 L/min. The dissolved oxygen (DO) concentration in the influent is 1.0 mg/L. In the tank, bacteria use the oxygen for metabolism in proportion to the amount of oxygen present; that is, they use it at a rate $k_1(DO)$, where $k_1 = 2 \times 10^{-3}$ min^{-1}. Oxygen is dissolving into the water at a rate proportional to the degree of undersaturation; that is, at a rate given by $k_2(DO^* - DO)$, where DO^* is 10 mg/L and $k_2 = 4 \times 10^{-3}$ min^{-1}. Write a mass balance for DO in the solution, and find the steady-state concentration of DO.

1-5. A pond of volume 10^6 L receives influent stream water containing 75 mg/L soluble organic carbon and 6 mg/L inorganic carbon (H_2CO_3, HCO_3^-, and CO_3^{2-}) at a rate of 0.01 m³/s. Advection is much more important than diffusion or dispersion as a mechanism for transporting material into and out of the pond. In the pond, 75% of the incoming organic matter is consumed by microorganisms, which convert 30% of the ingested organic carbon into new cellular material and release 70% as CO_2 after respiration. The CO_2 combines immediately with water to form H_2CO_3 and then rapidly becomes distributed among the three inorganic carbon species listed above, with the distribution dependent on solution pH. Assume that 50% of the cellular mass is carbon. The entire system is at steady state.

(a) Write mass balances on total organic carbon (TOC), dissolved organic carbon (DOC), and total inorganic carbon (TIC), and compute the steady-state concentrations of these species.

(b) How much biomass is created per liter of influent?

1-6. In a biological waste treatment system, solids may enter in the influent, be created or destroyed by biological growth or decay, and may leave in the effluent. Consider a waste treatment plant treating 1 m³/s of an influent containing 55 mg/L of degradable organic solids. The wastewater also contains 180 mg/L of dissolved organic carbon (DOC), which may be degraded. For every gram of dissolved organic matter that is degraded, the overall reaction converts a portion of the material into CO_2 and H_2O, and another portion into 0.4 g of new biomass. Analytically, this biomass and the degradable solids in the influent are both quantified as volatile suspended solids (VSS); that is, the degradable solids in the influent and the new biomass that grows in the reactor are indistinguishable from one another.

(a) In a particular system, the dissolved organic matter (i.e., the substrate, S) is removed from solution at an overall rate (including both the conversion into CO_2 and into new cells) given by $r_S = -k_1 S^2 X/(S^2 + K_s)$, where X is the VSS concentration in the reactor. The values of k_1, K_s, and X in the reactor of interest are 8 mg DOC/mg VSS d, 110 (mg DOC/L)², and 120 mg VSS/L, respectively. Write a mass balance and compute the hydraulic detention time, τ (i.e., the average amount of time that water resides in the reactor, equal to V_L/Q_L) necessary to reduce the concentration of substrate to 3 mg DOC/L in a complete mix reactor operating at steady state.

(b) Solids decay in the reactor at a rate equal to $k_d X$, where k_d is the decay rate constant. Write a mass balance on VSS and compute the value of k_d.

2

CONTINUOUS FLOW REACTORS: HYDRAULIC CHARACTERISTICS

2.1 Introduction
2.2 Residence time distributions
2.3 Ideal reactors
2.4 Nonideal reactors
2.5 Equalization
2.6 Summary
    Appendix 2A. Introduction to Laplace transforms as a method of solving (certain) differential equations
    References
    Problems

2.1 INTRODUCTION

Most natural and engineered systems in which reactions occur have a continuous flow of water through them. The mixing patterns in a reactor can make an important difference in the amount of reaction that occurs and in the composition of the effluent from the system. The extent of mixing in continuous flow reactors spans a spectrum. Reactors with absolutely zero mixing lie at one extreme, and reactors with complete and instantaneous mixing between the influent and all of the water already in the reactor lie at the other. The extremes are commonly called *ideal* cases, and the range between the two ideal cases defines *nonideal flow*.

In this chapter, we focus strictly on the hydraulic and mixing characteristics of continuous flow reactors; that is, no reactions are considered. With respect to the mass balances considered in Chapter 1, the concern is only with the terms for the mass crossing the boundary of the control volume and the consequent rate of change of mass within that control volume. In subsequent chapters, reaction kinetics are considered, first alone and then in combination with the concepts developed in this chapter, to analyze systems in which both flow and reaction occur.

A useful way of considering the hydraulic characteristics (or flow regime) of a continuous flow reactor is to imagine that we could keep track of all the water that came into the reactor over an infinitesimally brief period and determine the time periods for which parts of that aliquot of influent stayed in the reactor. We might find, for example, that 5% of the molecules stay in the reactor for less than 1 min, 15% stay for between 1 and 2 min, 35% stay for between 2 and 3 min, and so forth until the full 100% was accounted for. The same information can be expressed in a cumulative fashion; that is, 5% of the molecules that enter at some instant stay in the reactor for less than 1 min, 20% stay for less than 2 min, 55% stay less than 3 min, and so forth. Such descriptions are called *residence time distributions* (RTDs). Much of this chapter is devoted to the subject of RTDs and what they tell us about the hydraulic or mixing characteristics of a reactor. An extensive presentation of RTDs is given by Nauman and Buffham (1983).

In reality, once an aliquot of influent enters the reactor, it is not possible to distinguish the water molecules in that aliquot from water molecules that enter earlier or later; that is, it is not possible to label and keep track of water molecules to carry out the conceptual experiment just described. However, the same type of information about the RTD of water can be obtained by adding tracers into the influent.

Tracer studies are used in water and wastewater treatment systems primarily to elucidate a reactor's hydraulic behavior. Understanding this behavior is important because, as shown in Chapter 4, the hydraulic characteristics of a reactor can dramatically influence the treatment efficiency that the reactor achieves. Thus, for instance, a tracer study might be carried out to evaluate alternative models for the mixing that occurs in a reactor. The RTD derived from that study might then be used to predict the conversion (removal) efficiency that would be achieved for a particular contaminant in the reactor. Such an exercise might be carried out, for example, to predict the maximum contaminant concentration that would occur in an industrial wastewater treatment plant following an upset that suddenly dumped the contaminant into the plant influent.

Tracer studies can also be used as diagnostic tools, for example, to explore whether the reason that a reactor is not

Water Quality Engineering: Physical/Chemical Treatment Processes, First Edition. By Mark M. Benjamin, Desmond F. Lawler.
© 2013 John Wiley & Sons, Inc. Published 2013 by John Wiley & Sons, Inc.

operating as efficiently as expected might be that the reactor has an undesirable mixing pattern. Finally, while most reactors used in water and wastewater treatment are constructed with the intention of facilitating a chemical reaction, some reactors (equalization reactors) are used not to carry out a chemical reaction, but only to mitigate fluctuations in the influent flow rate or solution composition. Again, tracer tests can be used to assess how well that design goal has been met.

Tracer studies require that a reactor already exists, but the hydraulic characteristics of a reactor are determined by the design, and therefore are best considered at the time of design (i.e., before the reactor exists). Until recently, the design of reactors was based on the experience gathered from existing reactors; rules of thumb for designing influent and effluent structures and for placement of baffles were developed over time. Now, however, the field of computational fluid dynamics (CFD) has developed to the point that it is becoming common for the hydraulic characteristics of reactors, and the sensitivity of those characteristics to possible design alternatives, to be studied at the time of design. Sophisticated software to accomplish this task is commercially available. Using that software, one can generate mathematically the kind of data that heretofore could only be obtained in a tracer study.

Computational fluid dynamics represents a major advance in our ability to understand the hydraulic characteristics of reactors. The concepts of CFD are not developed in this chapter; however, the concepts in this chapter can be equally well applied to results obtained from mathematical analysis (CFD) as from experimental analysis (tracer studies). An example of how tracer studies and CFD can be used together has been provided by Grayman et al. (1996) in a study of a storage reservoir that also served as a disinfection reactor.

The first part of this chapter introduces concepts about RTDs, mathematical representations of those distributions, and common ways of introducing tracers into reactors. Next, the RTDs for the extreme (ideal) conditions of zero and complete mixing are described, followed by those for intermediate degrees of mixing (i.e., mixing in nonideal reactors). Subsequently, the response of nonideal reactors in tracer studies is investigated, and the techniques for converting tracer responses to RTDs are presented. Mathematical models that characterize the nonideal flow reactors with a few parameters are then described. The design and evaluation of equalization reactors are discussed in the final section of the chapter.

2.2 RESIDENCE TIME DISTRIBUTIONS

Cumulative and differential RTDs can be given in discrete terms, as described previously. However, it is usually preferable to define continuous functions that accomplish the same task; that is, that describe the fraction of the fluid in the effluent that has been in the reactor less than a certain time (the cumulative distribution) or the fraction of the fluid that has stayed in the reactor between any two times, t_1 and t_2. To see how such functions can be developed, imagine a system (a lake, an engineered reactor, or a segment of a river) with water flowing in and out at the same rate (so that the volume of water in the system is constant). Assume that the flow is steady; that is, that it has been flowing in the same way for a long period and continues flowing in that way for at least the duration of the test. Imagine further that somehow we could label all the molecules that enter the system in an infinitesimal time period centered around the time defined as time zero; call the number of such labeled molecules N_∞. Finally, imagine that we could easily detect the labeled molecules as they left the reactor.

One possible result of such an experiment is shown in Figure 2-1a, in which the cumulative number of labeled molecules that have been counted in the effluent, $N(t)$, is shown as a function of time. The results shown in the figure suggest that, for this hypothetical system, not many labeled molecules come out in the early time period, then they come out more rapidly for a while, and then their appearance at the detection point slows down once again; eventually, all of them have come out so that $N(t)$ reaches N_∞.

The rate at which the labeled molecules come out can be found as the derivative of the curve shown in Figure 2-1a, that is, as the limit of $\Delta N(t)/\Delta t$ as Δt approaches zero. This derivative, evaluated at each time, is shown in Figure 2-1b and indicates directly that, in the example system, the rate at which the labeled molecules leave the system is small at the earliest times, higher for intermediate times, and approaches zero at later times. Since the function described in Figure 2-2b is the derivative of the function shown in Figure 2-2a, the integral of the function in Figure 2-2b from time zero to any time t yields $N(t)$. Taken over all time (from time zero to time infinity), that integral yields N_∞.

Normalizing the derivative function shown in Figure 2-2b by dividing by N_∞ makes the result independent of the absolute number of influent molecules that were labeled. The normalized result is called the *exit age distribution* of the system and is given the symbol $E(t)$; that is,

$$E(t) = \frac{dN/dt}{N_\infty} \qquad (2\text{-}1)$$

$$E(t)\,dt = \frac{dN}{N_\infty} \qquad (2\text{-}2)$$

This exit age distribution is shown in Figure 2-1c. $E(t)$ has dimensions of inverse time. Since dN is the *number* of labeled molecules that exit the reactor in the period dt (i.e., between t and $t + dt$), dN/N_∞ is the *fraction* of all labeled molecules that exit the reactor during this period. Therefore, since all the labeled molecules entered the reactor at $t = 0$, $E(t)\,dt$ can be identified as the fraction of the fluid that has residence time between t and $t + dt$. Correspondingly, the fraction of the fluid

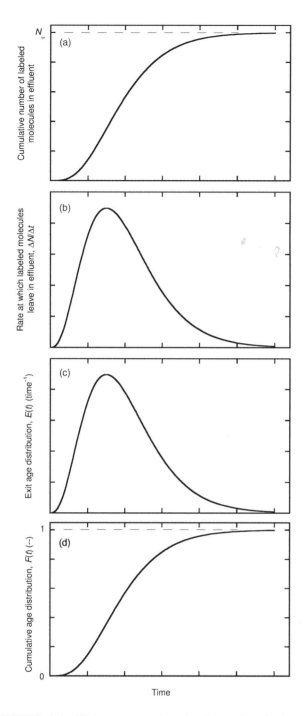

FIGURE 2-1. Various presentations for the results of a hypothetical experiment to evaluate the passage of labeled water molecules through a reactor.

with residence times between two discrete times is the time integral of $E(t)\,dt$ over that time period:

Fraction of fluid in the system with residence time

$$\text{greater than } t_1 \text{ and less than } t_2 = \int_{t_1}^{t_2} E(t)\,dt \qquad (2\text{-}3)$$

Note that, since 100% of the fluid must have a residence time less than infinity, the integral of $E(t)$ from $t=0$ to $t=\infty$ must be one; that is,

$$\int_0^\infty E(t)\,dt = 1 \qquad (2\text{-}4)$$

Finally, it is useful to describe the *cumulative age distribution*; that is, the fraction of the fluid with residence time less than or equal to a given value. This function can be derived by normalizing the data in Figure 2-1a by dividing by N_∞. The result (shown in Figure 2-1d) has the same shape as that in Figure 2-1a but now has a range from 0 to 1; this cumulative age distribution is normally given the symbol $F(t)$ and is dimensionless. From the integral/derivative relationship described above between the functions shown in Figure 2-1a and 2-1b, it should be clear that $F(t)$ is the integral of $E(t)$ from zero to time t; that is,

$$F(t) = \int_0^t E(t)\,dt \qquad (2\text{-}5)$$

The differential and cumulative age distributions; that is, $E(t)$ and $F(t)$, are useful descriptors of the flow characteristics (under steady flow conditions) of any system, as shown subsequently. Although we have not used statistical terminology in developing these functions, those familiar with statistical analysis will recognize $E(t)$ as a standard *probability density function*, and $F(t)$ as a *cumulative probability function*. These properties enable one to describe the flow characteristics in probabilistic terms. For example, we can say that the probability that a molecule will stay in a reactor less than time t is $F(t)$, and the probability that a molecule will stay in the reactor for longer than t_1 but less than t_2 is the integral expressed in Equation 2-3.

Tracers

The labeling of the influent water molecules at some instant and the subsequent measurement over time of those molecules in the effluent are, of course, not possible by any reasonable means. However, the same type of information about the RTD can be obtained by introducing a tracer into the influent water and measuring its concentration in the effluent over time. Ideally, tracers used to analyze the RTD of a reactor meet four criteria: their hydraulic behavior is identical to that of water (e.g., they are neutrally buoyant); their addition does not change the properties of the solution (e.g., the viscosity) appreciably; they are nonreactive; and their concentration in water can be measured well, preferably at low concentrations. In essence, these characteristics assure that the molecules of tracer behave identically to molecules of water in the reactor. However, because the tracer molecules can be differentiated easily from the water molecules, they serve the same function as the labeled molecules in the conceptual experiment. They therefore enable us to determine the differential and cumulative RTDs, $E(t)$ and $F(t)$, respectively.

Fluoride is a common choice as a tracer in water treatment plants, because it is often added as a treatment chemical to protect dental health (and is therefore readily available), and it is easy to measure accurately. However, it is also reactive with aluminum, and so is not an appropriate tracer in coagulation and flocculation facilities where aluminum is present. Rhodamine, a red dye, is a common choice as a tracer in wastewater plants, because it is nonreactive, measurable at low concentrations, and not otherwise present in the water. However, it cannot be used in drinking water treatment plants, because the color it imparts is unacceptable in the finished water.

Tracer studies should be done under steady flow conditions, because the various analyses described in the following sections assume that the flow pattern remains the same throughout the duration of the test. Also, it is desirable to carry out the test under more than one flow condition, for example, at an average flow and at a nearly maximum flow, and at various temperatures (times of the year) to see whether the hydraulic characteristics are substantially affected by these variables.

In real systems, it is often not possible to meet all of the various constraints specified in the preceding paragraph, so the conclusions reached from the studies must be tempered to recognize the imperfections. In this chapter, however, we assume that the conditions of the tracer test are ideal. Tracer studies are usually done by injecting the tracer into the reactor influent either as a pulse or as a step input. In an ideal *pulse input*, a known amount of tracer is put in instantaneously; that is, within an infinitesimally short time. The time at which the tracer is injected is usually defined as time zero. This input is similar to the labeling of all the molecules that come into the reactor in an infinitesimally short time as described earlier. In a *step input*, the tracer is put into the reactor at a constant concentration beginning at a defined instant (usually designated time zero) and continuing (ideally) forever; in reality, "forever" means for a sufficiently long time that the effluent concentration reaches essentially the same level as it would if the test were continued indefinitely.

For quantitative analysis of tracer experiments, it is necessary to describe these two different types of inputs mathematically. A step input function, starting at time zero and continuing forever, is shown in Figure 2-2a. Time zero is shown away from the vertical axis to emphasize that the tracer input is at some constant value (often zero) for a long time before $t = 0$ and at a different concentration, c_{in}, for all times thereafter.

A step function whose value jumps from 0 to 1 when the value of an independent variable z equals z^* is called the *Heaviside function* of z, applied at z^*:

$$H_{z^*}(z) = \begin{cases} 0 & \text{at } z < z^* \\ 1 & \text{at all other } z \end{cases} \quad (2\text{-}6)$$

This function is shown in Figure 2-2b. With this terminology, the most common type of step input tracer test (one with $c_{in} = 0$ at $t < 0$) can be described as the product of c_{in} and the Heaviside function applied at time zero; that is, $c_{in}(t) = c_{in} H_0(t)$.

The mathematical definition of the pulse input is a bit more obscure, as it includes the use of the *Dirac delta function*, $\delta_{z^*}(z)$, where z is some variable of interest. This function, shown graphically in Figure 2-3a, is the derivative of the Heaviside function. Its value can therefore be equated to the slope of Figure 2-2b; that is, its value is infinite when $z = z^*$ and zero at all other values of z. The dimensions of δ_{z^*} are the inverse dimensions of z; for example, if z is time, $\delta_{z^*}(z)$ has dimensions of 1/time. The integral of the Dirac function over any interval z_1 to z_2 is $H_{z^*}(z_2) - H_{z^*}(z_1)$. This difference is 1.0 over any interval that includes z^* and is zero over any interval that does not include it. In symbols, the

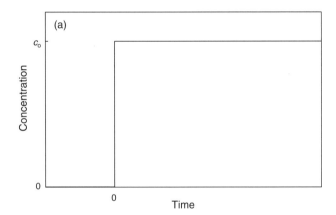

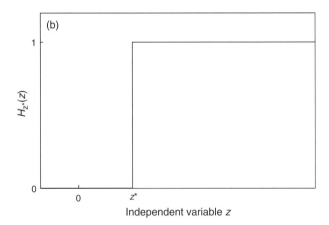

FIGURE 2-2. (a) The input tracer concentration for a step input test. (b) The Heaviside function of a variable z, applied at z^*. The function in (a) can be described as $c_{in} H_0(t)$.

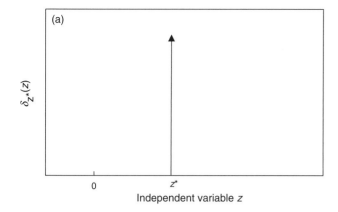

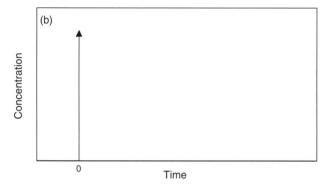

FIGURE 2-3. (a) The Dirac delta function of a variable z; (b) The Dirac function characterizing the influent in a pulse input tracer study.

Dirac delta function can be described as follows:

$$\delta_{z^*}(z) = \frac{dH_{z^*}(z)}{dz} = \begin{cases} \infty & \text{at } z = z^* \\ 0 & \text{at all other } z \end{cases} \quad (2\text{-}7a)$$

$$\int_{z_1}^{z_2} \delta_{z^*}(z)\,dz = H_{z^*}(z_2) - H_{z^*}(z_1) = \begin{cases} 1.0 & \text{if } z_1 < z^* < z_2 \\ 0 & \text{over any other interval} \end{cases}$$
$$(2\text{-}7b)$$

The Dirac function can also be defined as a limiting condition: if we imagine a continuous distribution $f(z)$ defined from $z = -\infty$ to $z = +\infty$ with an integral over that range of 1.0, and then imagine gradually shrinking the range of z values over which $f(z)$ is nonzero (while the area under the curve remains equal to 1.0), the limit as that range goes to zero is the delta function, and the value of z where $f(z) \neq 0$ is z^*.

As noted previously, the purpose of tracer tests is to gain information about the RTDs $E(t)$ and $F(t)$. We next consider how the concentration profile of tracer in the effluent of a reactor during a tracer test, whether that input is a pulse or a step, can be used to determine these RTDs.

Pulse Input Response

Consider the concentration of tracer in the effluent of some arbitrary reactor that has received a pulse input of mass $M_{p,in}$, so that the profile of the input concentration is as shown in Figure 2-3b. Assume that the reactor contains a volume V of water and has been receiving a constant flow Q for sufficient time that steady flow conditions exist in the reactor. If we imagine, for example, that this arbitrary reactor is a large open tank in which the water comes in at one end and flows out the other, it is reasonable to think that the influent over some small time period will mix somewhat with the water that is in the vicinity of the influent port at that time, but that no tracer will be transported immediately to the other end of the reactor. A person taking samples at the effluent will detect no tracer immediately after the pulse injection; as time proceeds, the concentration of tracer will rise and subsequently fall. Eventually all of the tracer injected at time zero will have exited the reactor and the concentration of tracer will again be zero. This response of the reactor ($c_p(t)$, the effluent concentration as a function of time)[1] is depicted in Figure 2-4a. The exact shape of the curve depends on the amount of mixing in the reactor, but that is not our concern now; it is sufficient to understand that a reactor could reasonably give the response shown.

We noted previously that $E(t)$ equals the rate at which labeled molecules of water exit the reactor, normalized to the total number of labeled molecules injected. The analog of the number of labeled molecules in the conceptual experiment is the mass of tracer in the real experiment. The exit age distribution, $E(t)$, can thus be evaluated as the rate at which tracer mass exits the reactor (the product of the flow and effluent concentration) divided by the total mass of tracer put into the reactor, $M_{p,in}$; that is, in symbols[2,3]

$$E(t) = \frac{Q}{M_{p,in}} c_p(t) \quad (2\text{-}8)$$

---

[1] The expression $c(t)$ is used in this text to denote the concentration of a substance as a function of time. In the current context, $c(t)$ is used for the time-varying concentration of tracer in the effluent from any reactor. $c_p(t)$ is used to indicate the effluent tracer profile for a pulse input tracer test, and $c_s(t)$ is used to indicate the corresponding profile for a step input tracer test.
[2] The normalization can also be based on the total mass of tracer that is detected over all time in the effluent from the reactor, M_{out}. The following derivations are carried out using $M_{p,in}$ as the normalizing factor. A discussion about potential differences between $M_{p,in}$ and $M_{p,out}$ is provided subsequently.
[3] This equation applies if the pulse is input at the time defined as $t = 0$. In rare cases, $t = 0$ is defined differently, in which case $c_p(t)$ in Equation 2-8 must be replaced by $c_p(t - t_p)$, where t_p is the time as which the pulse was input, so $t - t_p$ is the elapsed time since the pulse was injected. In the remainder of the chapter, unless otherwise noted, the assumption is made that the pulse is input at $t = 0$.

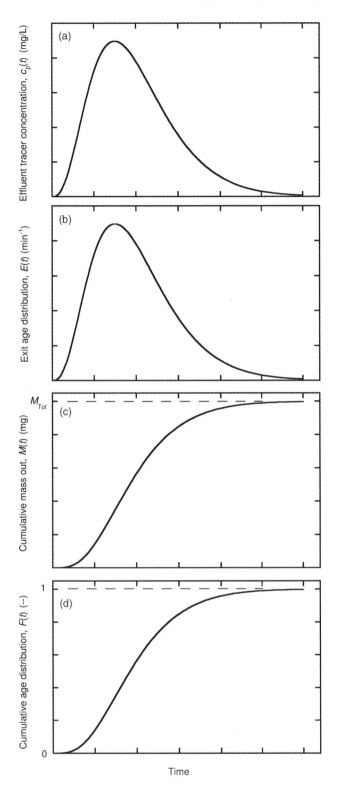

FIGURE 2-4. Experimental results for the tracer effluent concentration for a pulse input test, and the corresponding residence time distributions, in an arbitrary reactor.

The exit age distribution, $E(t)$, is shown for the reactor under consideration in Figure 2-4b; the shape is identical to that of $c_p(t)$, but the dimensions and values are different.

Equation 2-8 makes clear that $E(t)$ can be thought of as a normalized response to a pulse input tracer test; the normalization is accomplished by multiplying the (time-varying) direct response, $c_p(t)$, by the (constant) factor $Q/M_{p,in}$. By normalizing the data in this way, we assure that the $E(t)$ function derived from pulse input tracer tests will be the same, regardless of the amount of tracer injected, even though the $c_p(t)$ data would differ in the different experiments. Such a result is essential, since $E(t)$ describes how water flows through the reactor (at a particular flowrate Q) and is not intrinsically related to the tracer; tracers are simply a useful way of determining the flow characteristics of the reactor.

Equation 2-8 is often written in an alternative form by dividing the numerator and denominator of the fraction by the reactor volume, V, and rearranging, as follows:

$$E(t) = \frac{Q/V}{M_{p,in}/V} c_p(t) = \frac{1}{V/Q} \frac{c_p(t)}{M_{p,in}/V} = \frac{1}{V/Q} \frac{c_p(t)}{c_o} \quad (2\text{-}9)$$

where c_o is the nominal tracer concentration that would exist if all the mass of tracer injected were dispersed throughout the whole reactor volume. Note that c_o is a hypothetical concentration that, except in one special case described later in the chapter, does not ever actually exist in the reactor.

Based on the various descriptions of $E(t)$ to this point, we can think of the $E(t)$ function in three equivalent ways. First, for a sample of the influent, $E(t)$ expresses the differing amounts of time that molecules of water will stay in the reactor. If, for example, $\int_{6\,\text{min}}^{10\,\text{min}} E(t)dt = 0.15$, then 15% of the molecules in any sample of influent will stay in the reactor between 6 and 10 min. Second, for an aliquot of the effluent, $E(t)$ characterizes the differing amount of time that the molecules were in the reactor. So, using the same example, 15% of the molecules in any sample of effluent were in the reactor between 6 and 10 min. Finally, $E(t)$ is a probability density function so, for the same example, for any molecule that enters (or is in, or leaves) the reactor, the probability that the molecule will (or did) stay in the reactor between 6 and 10 min is 15%.

The cumulative age distribution, $F(t)$, can also be developed from a pulse input tracer test. The cumulative mass of tracer that has come out of the reactor up to any time t can be determined by integrating, from time zero to time t, the rate at which the mass is leaving. The rate at which tracer leaves the reactor is the product of flow and concentration, so assigning the symbol $M_p(t)$ to the mass of tracer that has exited the reactor up until time t, this quantity can be computed as

$$M_p(t) = Q \int_0^t c_p(t)\, dt \quad (2\text{-}10)$$

Equation 2-10 is a general expression for any reactor receiving a pulse input of tracer; for the arbitrary reactor under consideration, $M_p(t)$ is shown graphically in Figure 2-4c. Again exploiting the analogy between hypothetical, labeled water molecules and the pulse input of tracer, we note that $F(t)$ is the fraction of the pulse input mass ($M_{p,in}$) that has exited the reactor up until time t. Thus, $F(t)$ can be computed as the ratio of $M_p(t)$ to $M_{p,in}$; that is,

$$F(t) = \frac{M_p(t)}{M_{p,in}} = \frac{Q}{M_{p,in}} \int_0^t c_p(t)\,dt = \int_0^t E(t)\,dt \qquad (2\text{-}11)$$

where the first two equalities are specific for the pulse input case, and the equality of $F(t)$ with the final expression on the right is always true. For the case under consideration, $F(t)$ is shown in Figure 2-4d; $F(t)$ has the same shape as $M_p(t)$ but is dimensionless and has a maximum value (asymptote) of 1.0.

For a discrete data set, Equation 2-11 can be approximated using the trapezoidal rule as follows:

$$F(t_i) = \sum_{\text{all } j \leq i} \left(\frac{E(t_j) + E(t_{j-1})}{2}\right)(t_j - t_{j-1}) = \sum_{\text{all } j \leq i} E_{\text{ave}}(t_j)\Delta t_j \qquad (2\text{-}12)$$

where $E_{\text{ave}}(t_j)$ is the average value of $E(t)$ over the time interval Δt_j.

Step Input Response

The exit age distribution, $E(t)$, and cumulative age distribution, $F(t)$, can also be developed using a step input tracer test. Recall that, in such a test, the input concentration of a nonreactive tracer is steady at one value prior to time zero and at a different value at all times after time zero; here, for convenience, we assume that the change is from $c = 0$ to $c = c_{in}$. (Although the tracer is added beginning at time zero, the water flow is assumed to have been established previously and to be steady throughout the test period.) Again, our question is: What is the response of the reactor to this test; that is, how does the effluent concentration vary with time after the initiation of the tracer injection into the influent? For the example reactor whose response to a pulse input of tracer was described in the previous section, our intuition leads us to think that no tracer will be found in the effluent immediately, but the effluent concentration will gradually rise thereafter from zero to the constant influent concentration. Such a result is shown graphically in Figure 2-5a.

Any sample of effluent taken after the initiation of the tracer test can be thought of as a mixture of two parts: one fraction is water that came in before time zero and contains no tracer, and the other is water that came in after time zero and contains tracer at concentration c_{in}. Consider an effluent

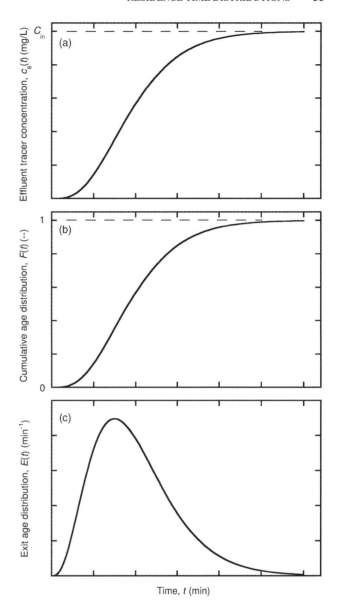

FIGURE 2-5. Experimental results for the tracer effluent concentration for a step input test, and the corresponding residence time distributions, in an arbitrary reactor.

sample collected at some time t^*. The fraction of the sample that entered the reactor less than time t^* ago (between $t = 0$ and $t = t^*$) is, by definition, the value of the cumulative age distribution for t^*; that is, it is $F(t^*)$. This fraction of the sample contained tracer molecules at concentration c_{in} when it entered the reactor. The other fraction, equal to $1 - F(t^*)$, came in longer than t^* ago (i.e., before $t = 0$), and contained no tracer when it entered. The tracer concentration in the sample is the weighted average of the concentrations in the two fractions:

$$c(t^*) = F(t^*)c_{in} + (1 - F(t^*))(0) = F(t^*)\,c_{in}$$

so,

$$F(t^*) = \frac{c(t^*)}{c_{in}} \quad (2\text{-}13a)$$

The preceding analysis can, of course, be applied at any time t, so the relationship between the time record of the effluent concentration in a 0-to-c_{in} step input tracer test, $c_s(t)$, and the cumulative age distribution, $F(t)$, can be generalized as

$$F(t) = \frac{c_s(t)}{c_{in}} \quad (2\text{-}13b)$$

Hence, the $F(t)$ curve can be obtained by normalizing the effluent concentration curve from a step input tracer, where the normalization involves simply dividing by the influent concentration. For the reactor under consideration, $F(t)$ is shown in Figure 2-5b. Further, from the earlier discussion, it is clear that $E(t)$ can be obtained as the derivative of $F(t)$, and that function is shown in Figure 2-5c.

The cumulative age distribution, $F(t)$, has three equivalent interpretations, analogous to those described previously for the exit age distribution, $E(t)$. First, $F(t)$ expresses the anticipated cumulative RTD of the influent water molecules that enter the reactor. If $F(t_1) = 0.6$, for example, 60% of the water molecules in any sample of influent will stay in the reactor for a period less than t_1. Second, $F(t)$ expresses the cumulative RTD of the molecules in the effluent. That is, if $F(t_1) = 0.6$, 60% of the water molecules in any sample of effluent spent a time less than t_1 in the reactor, while the rest $(1 - F(t_1))$, or 40% stayed in the reactor longer than t_1. Finally, $F(t)$ expresses the probability that a single water (or tracer) molecule stays in the reactor less than a certain time; for example, if $F(t_1) = 0.6$, the probability that a water molecule stays in the reactor less than t_1 is 60%.

We have now seen that both the exit age distribution, $E(t)$, and the cumulative age distribution, $F(t)$, can be determined from either pulse or step tracer tests. The $E(t)$ and $F(t)$ functions describe how water flows through a reactor and that result is independent of the way in which the tracer test is conducted. Every sample of water, over all time in a reactor with steady flow, behaves identically to samples collected during the tracer test; the tracer test simply allows us to see and quantify that behavior.

■ **EXAMPLE 2-1.** A pulse input tracer test is conducted on a reactor with a water volume of 240 m³ and a flow rate of 4 m³/min. A mass of 10 kg of tracer is added to the reactor at time zero (pulse input), and the effluent concentration is recorded every 10 min as shown in the following table.

TABLE 2-1. Spreadsheet Analysis to Convert Pulse Tracer Data into $E(t)$ and $F(t)$

	A Time (min)	B Concentration (mg/L)	C $E(t)$ (min^{-1})	D $E_{ave}(t)\Delta t$ (–)	E $F(t)$ (–)
1	0	0	0		0
2	10	16	0.0064	0.032	0.032
3	20	34	0.0136	0.100	0.132
4	30	40	0.0160	0.148	0.280
5	40	35	0.0140	0.150	0.430
6	50	31	0.0124	0.132	0.562
7	60	26	0.0104	0.114	0.676
8	70	21	0.0084	0.094	0.770
9	80	16	0.0064	0.074	0.844
10	90	12	0.0048	0.056	0.900
11	100	8	0.0032	0.040	0.940
12	110	6	0.0024	0.028	0.968
13	120	3	0.0012	0.018	0.986
14	130	1	0.0004	0.008	0.994
15	140	0	0	0.002	0.996
16	150	0	0	0.000	0.996
	SUM			0.996	

Determine and plot the differential and cumulative age distributions, $E(t)$ and $F(t)$.

Solution. The raw data are plotted in Figure 2-6, and the calculations needed to compute $E(t)$ and $F(t)$ are summarized in Table 2-1. $E(t)$ is computed by the substitution of the appropriate values into Equation 2-8. For example, the value in Row 3 associated with $t = 20$ min is

$$E(20\,\text{min}) = \frac{Q}{M_{p,in}} c_p(t)$$

$$= \left(\frac{4\,\text{m}^3/\text{min}}{10\,\text{kg}}\right)(34\,\text{mg/L})\left(\frac{1\,\text{kg}}{10^6\,\text{mg}}\right)\left(\frac{10^3\,\text{L}}{\text{m}^3}\right)$$

$$= 0.0136\,\text{min}^{-1}$$

Column D has no entry in Row 1 because the values in this column are associated with *intervals* of time and are shown in the table at the end of the interval. For example, the entry of 0.100 shown for $E(t)\Delta t$ at 20 min (in cell D3) is the product of the average value of $E(t)$ during the preceding interval $= (0.0064 + 0.0136)/2\,\text{min}^{-1}$ and the duration of the time interval $(20 - 10\,\text{min})$. This value is the approximate area under the $E(t)$ versus t curve for the interval from 10 to 20 min.

$F(t)$ is computed as the integral of $E(t)$ from time 0 to t, which is approximated numerically as shown in Equation 2-12. $F(t)$ values in column E are the running

Time (min)	0	10	20	30	40	50	60	70	80	90	100	110	120	130	140	150
Concentration (mg/L)	0	16	34	40	35	31	26	21	16	12	8	6	3	1	0	0

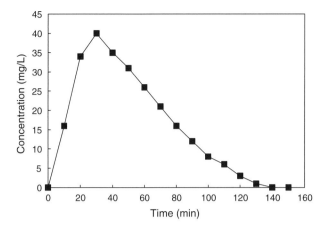

FIGURE 2-6. Tracer output curve for Example 2-1.

sum of the values in column D. The value of $F(t)$ in Row 3, for example, is the value of $F(t)$ in Row 2 plus the value of $E_{ave}(t)\Delta t$ in cell D3. Note that these data do not yield a final value of $F(t) = 1.0$, although the final value is close to 1.0. We consider the implications of this outcome subsequently. $E(t)$ and $F(t)$ are plotted against t in Figure 2-7.

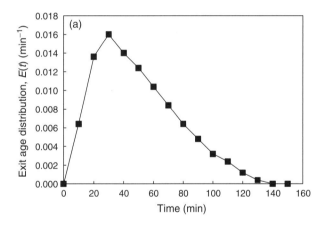

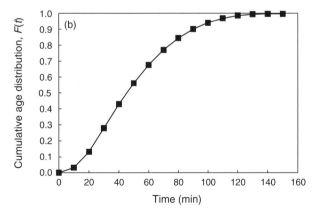

FIGURE 2-7. (a) Differential and (b) cumulative residence time distribution functions for the example data set.

Statistics of Probability Distributions and the Mean Hydraulic Detention Time

One important characteristic of the RTD is the mean hydraulic detention time; that is, the average amount of time that influent molecules spend in the reactor. In this section, we consider how to determine the mean hydraulic residence time from RTDs or from tracer tests.

In any system at steady state, each substance within the system is being replaced at an average rate given by the amount of the substance in the system divided by the rate at which it is entering or leaving the system. This is equally true if the substance of interest is water, another chemical species, or even people. For instance, if 120 people are in a department store, and 10 people leave while 10 others enter every minute, then the number in the store stays steady at 120, and the average length of time that a person stays in the store is 120 people/(10 people/minute) = 12 min. Some may stay longer and some shorter, but 12 min is the average. Similarly, for a fluid moving through a control volume, the mass of fluid in the system is the product of the density and the volume (ρV), and the rate at which fluid enters or leaves the system is ρQ, so the average hydraulic residence time is $\rho V / \rho Q = V/Q$. As in the earlier example, some of the fluid may spend more time than V/Q in the reactor, and some less, but the average is given by this value.

While the computation of the hydraulic residence time for a given reactor might appear to be a trivial matter, since V and Q are usually easy to measure, the evaluation is not as simple as it might seem. If, for instance, there is *dead space* in the corners of the reactor, the volume through which water actually flows (the effective volume of the reactor, $V_{eff,r}$, equal to the total volume minus the volume of the dead space) would be less than the full, geometric volume of the reactor. Correspondingly, the mean hydraulic residence time ($=V_{eff}/Q$) would be less than the value computed as V/Q.

Differences between the geometric volume of a reactor and the volume that is accessible to the fluid can also arise in other, more subtle ways. One such scenario can be understood most easily by considering the two limiting cases for diffusive and dispersive transport across the reactor boundaries. At one extreme lie the reactors with so-called *closed boundaries*, across which no diffusive or dispersive transport occurs. Such a condition characterizes most engineered reactors, at least to a close approximation. For instance, if the reactor of interest is an open tank, and the influent to the reactor enters via a pipe or through a gate in which the fluid velocity is relatively high, the possibility that a molecule that has entered the reactor will recross the reactor boundary and migrate upstream into the influent solution is virtually nil. Similarly, the effluent from many reactors, especially in environmental engineering, flows by gravity, irreversibly, over a weir, after which it travels to the next treatment process. In such reactors, the geometric volume is the maximum volume accessible to the fluid, and V/Q is the maximum mean hydraulic residence time.

At the other extreme, we sometimes define a region of space as a conceptual reactor, even though no physical boundary separates the region from the material outside the reactor. Examples of such reactors include a stretch of a river (e.g., downstream of a waste discharge point), a biological treatment process designed with a "racetrack" geometry, and a segment of an aquifer. In these reactors with *open boundaries*, diffusion or dispersion might transport a constituent that is initially at the upstream end of the reactor back upstream and out of the reactor, before advection carries it back into and through the reactor. Similarly, even after it leaves the region defined as the reactor, a constituent might be transported back upstream to spend more time in that space, before, eventually, leaving it permanently.

Reactors with open boundaries can be characterized by two different residence times—one describing how long, on average, the constituents are actually inside the region defined as the reactor, and another describing the average amount of time required for a constituent that is initially at the upstream end of the reactor to leave the reactor for good. The former value equals V_{eff}/Q, where V_{eff} is again defined as the volume that is accessible to the fluid within the reactor boundaries. If all of the geometric volume in the reactor is accessible, $V_{eff} \approx V$.

On the other hand, the residence time defined in the latter way given earlier is greater than V_{eff}/Q, since it includes some time that the fluid spends outside the reactor boundaries (either upstream or downstream) before it exits the reactor permanently. This residence time is important not only conceptually, but also for a very practical reason because we cannot prevent molecules from spending time outside the reactor boundaries as they move from the upstream to the downstream end of the reactor, this residence time is the only one that can be determined experimentally in a system with open boundaries.[4]

The preceding discussion indicates that the mean hydraulic residence time determined experimentally could be either smaller (if the reactor has dead space) or larger (if the reactor boundaries are open) than the ratio V/Q. In this text, we use the Greek letter τ to represent the theoretical residence time of water inside a reactor based on its geometry (V/Q), and $\bar{t}$ to represent the experimentally measured mean residence time. If water flows through the entire volume of the reactor, if the tracer has been well chosen (i.e., if it really does have the same mixing characteristics in the reactor as the water does), if the reactor boundaries are closed, and if the experimental data are complete and accurate, then $\bar{t} = \tau$.

[4] Keep in mind that the closed and open boundary systems described here represent limiting cases. In real systems, the upstream and downstream boundaries need not be identical, and either might be characterized not simply as open or closed, but by some condition that is between these two limits.

We next consider how tracer output can be used to compute $\bar{t}$ in a reactor receiving steady flow.

The average value of a variable is of interest in statistical analysis of many types of data, and some of the terms that appear in the analysis have been given special names. In particular, if events with various values of x can occur, and if $G(x)$ is the probability density function describing the likelihood that a single event will occur with value between x and $x + dx$, the mean value of x is given by

$$\bar{x} = \frac{\int_0^\infty x G(x) dx}{\int_0^\infty G(x) dx} \quad (2\text{-}14)$$

The term $\int_0^\infty x^n G(x) dx$ is called the nth moment of G, so, according to Equation 2-14, the mean value of x can be defined as the first moment of G divided by the zeroth moment.

The probability density function of residence times in a reactor was identified earlier as the $E(t)$ function, so the mean hydraulic detention time in a reactor can be computed from $E(t)$ or $F(t)$ as follows:

$$\bar{t} = \frac{\int_0^\infty t E(t) dt}{\int_0^\infty E(t) dt} \quad (2\text{-}15)$$

As shown in Equation 2-8, $E(t)$ can be equated with the normalized tracer effluent concentration from a pulse input test. By substituting Equation 2-8 into Equation 2-15, we can obtain an expression for $\bar{t}$ in terms of directly measurable experimental values:

$$\bar{t} = \frac{\int_0^\infty t(Q/M_{p,in}) c_p(t) dt}{\int_0^\infty (Q/M_{p,in}) c_p(t) dt} = \frac{\int_0^\infty t c_p(t) dt}{\int_0^\infty c_p(t) dt} \quad (2\text{-}16)$$

For a discrete data set, Equation 2-16 can be approximated as follows:

$$\bar{t} \approx \frac{\sum_{\text{all } i} \{(t_i + t_{i-1})/2\} \{(c_{p,i} + c_{p,i-1})/2\}(t_i - t_{i-1})}{\sum_{\text{all } i} \{(c_{p,i} + c_{p,i-1})/2\}(t_i - t_{i-1})}$$

$$= \frac{\sum_{\text{all } i} t_{\text{ave},i} c_{p,\text{ave},i} \Delta t_i}{\sum_{\text{all } i} c_{p,\text{ave},i} \Delta t_i} \quad (2\text{-}17)$$

An alternative way of manipulating Equation 2-15 is to recall that, since the probability of a water molecule having a residence time between zero and infinity is 100%, $\int_0^\infty E(t) dt = 1.0$. Therefore, Equation 2-15 can be simplified to

$$\bar{t} = \int_0^\infty t E(t) dt = \frac{Q}{M_{p,in}} \int_0^\infty t c_p(t) dt = \frac{\int_0^\infty t c_p(t) dt}{M_{p,in}/Q} \quad (2\text{-}18)$$

or, for the analysis of a discrete data set:

$$\bar{t} \approx \frac{\sum_{\text{all } i}\{(t_i + t_{i-1})/2\}\{(c_{p,i} + c_{p,i-1})/2\}(t_i - t_{i-1})}{M_{p,\text{in}}/Q}$$

$$= \frac{\sum_{\text{all } i} t_{\text{ave},i} c_{p,i,\text{ave}} \Delta t_i}{M_{p,\text{in}}/Q} \quad (2\text{-}19)$$

By inspection, Equations 2-16 and 2-18 give the same result if and only if,

$$\int_0^\infty c_p(t)\,dt = \frac{M_{p,\text{in}}}{Q} \quad (2\text{-}20)$$

$$Q\int_0^\infty c_p(t)\,dt = M_{p,\text{in}} \quad (2\text{-}21)$$

The left side of Equation 2-21 is the total mass of tracer that has exited the reactor over all time, which we designate $M_{p,out}$, and the right side is the total mass of tracer input, so it is clear that this equality should hold. However, as a practical matter, the mass of tracer detected in the effluent is often different from the amount injected. The discrepancy might arise because sampling was terminated while some tracer remained in the reactor, because flow and tracer data were inaccurate or incomplete, or a combination of those and other reasons. Regardless of the source of the discrepancy, the point is that the computed value of $\bar{t}$ can depend on whether $M_{p,\text{in}}$ (Equation 2-18) or $M_{p,\text{out}}$ (Equation 2-16) is used in the computation.

The computed values of $E(t)$ and $F(t)$ according to Equations 2-8 and 2-11 also depend on whether $M_{p,\text{in}}$ or $M_{p,\text{out}}$ is treated as the total mass of tracer used in the test. For example, if a tracer test was cut off too early, but all of the data were accurate, the correct value of $F(t)$ for the last sample would be less than 1.0. Normalizing $M_p(t)$ by $M_{p,\text{in}}$ would give the correct values for $F(t)$ at all times and cause the final, computed value of $F(t)$ to be less than 1.0, and so would be the logical choice in this case. On the other hand, if a calibration error caused the measured values of $c_p(t)$ to be in error by a consistent percentage, the evaluation of $E(t)$ or $F(t)$ using $M_{p,\text{out}}(t)$ for the normalization would correct for the error and hence would be the best choice. Of course, we normally do not know whether these types of errors have occurred, or their magnitude. Therefore, in general, it is a good idea to carry out the computation in both ways and, if the results differ by a significant amount, attempting to determine and correct for the cause of the discrepancy.

All of the expressions for $\bar{t}$ shown in Equations 2-16 to 2-19 can also be obtained using $F(t)$ instead of $E(t)$, by making the following substitution in Equation 2-15:

$$\bar{t} = \frac{\int_0^\infty tE(t)\,dt}{\int_0^\infty E(t)\,dt} = \frac{\int_0^\infty t(dF(t)/dt)\,dt}{\int_0^\infty (dF(t)/dt)\,dt} = \frac{\int_{F(0)}^{F(\infty)} t\,dF(t)}{\int_{F(0)}^{F(\infty)} dF(t)}$$

$$= \frac{\int_{F(0)}^{F(\infty)} t\,dF(t)}{F(\infty) - F(0)} \quad (2\text{-}22)$$

Since $F(\infty) = 1$ and $F(0) = 0$, Equation 2-22 can be simplified for a continuous or a discrete data set, respectively, as follows[5]:

$$\bar{t} = \int_{F(0)}^{F(\infty)} t\,dF(t) = \int_0^1 t\,dF(t) \quad (2\text{-}23)$$

$$\bar{t} \approx \sum_{\text{all } i}\left(\frac{t_i + t_{i-1}}{2}\right)(F_i(t) - F_{i-1}(t)) = \sum_{\text{all } i} t_{i,\text{ave}}\Delta F_i(t) \quad (2\text{-}24)$$

In addition to the mean value of a parameter x, it is useful to have a measure of the spread of values around the mean. In statistics, this spread is characterized by the *variance* of x values, σ_x^2, which is defined mathematically as the second moment around the mean divided by the zeroth moment:

$$\sigma_x^2 = \frac{\int_0^\infty (x - \bar{x})^2 G(x)\,dx}{\int_0^\infty G(x)\,dx} \quad (2\text{-}25a)$$

Variance has the dimensions of x^2. The variance can also be calculated as the difference between the ratio of the second moment around the axis to the zeroth moment, and the square of the mean:

$$\sigma_x^2 = \frac{\int_0^\infty x^2 G(x)\,dx}{\int_0^\infty G(x)\,dx} - \bar{x}^2 \quad (2\text{-}25b)$$

Two other functions related to the variance are often reported as indicators of the spread of a data set. These functions are the standard deviation (σ_x) of x, which is simply the square root of the variance, and the coefficient of variation (CV) of x, which is the standard deviation divided by the mean:

$$CV = \frac{\sigma_x}{\bar{x}} \quad (2\text{-}26)$$

---

[5] As noted earlier, the value of $F(t)$ depends on whether $M_{p,\text{in}}$ or $M_{p,\text{out}}$ is used in the calculation. If $M_{p,\text{in}}$ is used, the computed value of $F(\infty)$ might not be 1.0.

The variance of the RTD in a reactor can be obtained by substituting directly into either form of Equation 2-25, with t replacing x and E replacing G:

$$\sigma_{\text{RTD}}^2 = \frac{\int_0^\infty (t-\bar{t})^2 E(t)\,\mathrm{d}t}{\int_0^\infty E(t)\,\mathrm{d}t} = \frac{\int_0^\infty t^2 E(t)\,\mathrm{d}t}{\int_0^\infty E(t)\,\mathrm{d}t} - \bar{t}^2 \quad (2\text{-}27)$$

(The subscript on σ^2 has been written as RTD to emphasize that the term being computed is the variance of the RTD and not the variance of time *per se*.)

As when we analyzed $\bar{t}$, we can substitute $(Q/M_{\text{p,in}})c_{\text{p}}(t)$ for $E(t)$ to express the variance directly in terms of experimental values:

$$\sigma_{\text{RTD}}^2 = \frac{\int_0^\infty (t-\bar{t})^2 (Q/M_{\text{p,in}}) c_{\text{p}}(t)\,\mathrm{d}t}{\int_0^\infty (Q/M_{\text{p,in}}) c_{\text{p}}(t)\,\mathrm{d}t} = \frac{\int_0^\infty (t-\bar{t})^2 c_{\text{p}}(t)\,\mathrm{d}t}{\int_0^\infty c_{\text{p}}(t)\,\mathrm{d}t}$$

$$= \frac{\int_0^\infty t^2 c_{\text{p}}(t)\,\mathrm{d}t}{\int_0^\infty c_{\text{p}}(t)\,\mathrm{d}t} - \bar{t}^2 \quad (2\text{-}28)$$

Alternatively, we can take advantage of the fact that $\int_0^\infty E(t)\,\mathrm{d}t = 1$ to rewrite Equation 2-27 as[6]

$$\sigma_{\text{RTD}}^2 = \int_0^\infty (t-\bar{t})^2 E(t)\,\mathrm{d}t = \int_0^\infty t^2 E(t)\,\mathrm{d}t - \bar{t}^2 \quad (2\text{-}29)$$

Finally, we can write approximate forms of Equations 2-27 and 2-28, based on discrete numerical data, as follows:

$$\sigma_{\text{RTD}}^2 \approx \frac{\sum_{\text{all }i}\{(t_i + t_{i-1})/2\}^2\{(E_i + E_{i-1})/2\}(t_i - t_{i-1})}{\sum_{\text{all }i}\{(E_i + E_{i-1})/2\}(t_i - t_{i-1})} - \bar{t}^2$$

$$= \frac{\sum_{\text{all }i} t_{i,\text{ave}}^2 E_{i,\text{ave}}(t_i)\Delta t_i}{\sum_{\text{all }i} E_{i,\text{ave}}(t_i)\Delta t_i} - \bar{t}^2 \quad (2\text{-}30a)$$

$$\approx \left(\sum_{\text{all }i} t_{i,\text{ave}}^2 E_{i,\text{ave}}(t_i)\Delta t_i\right) - \bar{t}^2 \quad (2\text{-}30b)$$

$$\sigma_{\text{RTD}}^2 \approx \frac{\sum_{\text{all }i}\{(t_i + t_{i-1})/2\}^2\{(c_i + c_{i-1})/2\}(t_i - t_{i-1})}{\sum_{\text{all }i}\{(c_i + c_{i-1})/2\}(t_i - t_{i-1})} - \bar{t}^2 \quad (2\text{-}31a)$$

$$\approx \frac{\sum_{\text{all }i} t_{i,\text{ave}}^2 c_{i,\text{ave}}\Delta t_i}{\sum_{\text{all }i} c_{i,\text{ave}}\Delta t_i} - \bar{t}^2 \quad (2\text{-}31b)$$

---

[6] The denominator equals $F(\infty)$. As noted earlier, this value might not equal 1.0 if $M_{\text{p,in}}$ is used in the calculation of $E(t)$ and $F(t)$ values.

The variance of t has units of time squared. For certain calculations that are described later in the chapter, it is convenient to normalize the variance by dividing it by $\bar{t}^2$, to yield a nondimensional variance:

$$\tilde{\sigma}_{\text{RTD}}^2 \equiv \frac{\sigma_{\text{RTD}}^2}{\bar{t}^2} \quad (2\text{-}32)$$

■ **EXAMPLE 2-2.** For the reactor characterized by the tracer output in Example 2-1:

(a) Determine the fraction of the tracer that is recovered during the course of the experiment.
(b) Determine the mean detention time based on the concentration in the stream exiting the reactor, and the variance of the RTD.
(c) Compare the computed mean residence time to the theoretical mean detention time.

Solution. Information from Example 2-1 is repeated in Columns A through D in Table 2-2, and Columns E through J show the calculations needed to determine the requested parameters. Once again, values that are associated with the intervals of time are shown at the end of the interval, so they have no entry in Row 1.

(a) The mass of tracer exiting the reactor in any time period Δt is $Qc\Delta t$, and the total mass of tracer recovered during the test is the summation of the $Qc\Delta t$ values over all Δt (see Equation 2-21). Thus, the fraction of the tracer that was recovered can be computed as

$$\frac{M_{\text{p,out}}}{M_{\text{p,in}}} = \frac{\sum_{\text{all }i} Q c_{\text{p},i,\text{ave}}\Delta t_i}{M_{\text{p,in}}} = \frac{Q\sum_{\text{all }i} c_{\text{p},i,\text{ave}}\Delta t_i}{M_{\text{p,in}}}$$

$$= \frac{4\,\text{m}^3/\text{min}\,(2490\,\text{mg-min/L})\,10^3\,\text{L/m}^3}{(10\,\text{kg})\,10^6\,\text{mg/kg}}$$

$$= 0.996$$

This value is necessarily the same as the final value of $F(t)$, because $E(t)$ and $F(t)$ were calculated based on $M_{\text{p,in}}$; that is, $E(t_i)$ was calculated as $(Q/M_{\text{p,in}})c_{\text{p}}(t_i)$, and $F(t)$ was computed from those calculated $E(t)$ values. The fact that the fraction of mass recovered is close to unity suggests (but does not prove) that the tracer study was well done.

TABLE 2-2. Analysis Spreadsheet to Determine the Mean and Standard Deviation of RTDs

	A	B	C	D	E	F	G	H	I
	t (min)	$c(t)$ (mg/L)	$E(t)$ (min^{-1})	$F(t)$ (–)	$c_{\text{ave}}\Delta t$ (mg-min/L)	$t_{\text{ave}}c_{\text{ave}}\Delta t$ (mg-min^2/L)	$t_{\text{ave}}\Delta F(t)$ (min)	$\frac{(t_{\text{ave}})^2 c_{\text{ave}}\Delta t}{\sum_{\text{all }i} c_{\text{ave}}\Delta t}$ (min^2)	$\frac{(t_{\text{ave}})^2 E_{\text{ave}}(t)\Delta t}{\sum_{\text{all }i} E_{\text{ave}}(t)\Delta t}$ (min^2)
1	0	0	0	0					
2	10	16	0.0064	0.032	80	400	0.16	0.8	0.8
3	20	34	0.0136	0.132	250	3,750	1.50	22.6	22.6
4	30	40	0.0160	0.280	370	9,250	3.70	92.9	92.9
5	40	35	0.0140	0.430	375	13,125	5.25	184.5	184.5
6	50	31	0.0124	0.562	330	14,850	5.94	268.4	268.4
7	60	26	0.0104	0.676	285	15,675	6.27	346.2	346.2
8	70	21	0.0084	0.770	235	15,275	6.11	398.7	398.7
9	80	16	0.0064	0.844	185	13,875	5.55	417.9	417.9
10	90	12	0.0048	0.900	140	11,900	4.76	406.2	406.2
11	100	8	0.0032	0.940	100	9,500	3.80	362.4	362.4
12	110	6	0.0024	0.968	70	7,350	2.94	309.9	309.9
13	120	3	0.0012	0.986	45	5,175	2.07	239.0	239.0
14	130	1	0.0004	0.994	20	2,500	1.00	125.5	125.5
15	140	0	0	0.996	5	675	0.27	36.6	36.6
16	150	0	0	0.996	0	0	0	0.0	0.0
17	SUM				2490	123,300	49.32	3211.7	3211.7

(b) The mean hydraulic residence time can be computed from either Equation 2-17 or 2-24. Column E provides incremental values for the denominator of Equation 2-17, and column F provides the corresponding values for the numerator. The sum of the values in column E (cell E17) is the zeroth moment of $c_p(t)$, and the sum of the values in column F (cell F17) is the first moment. According to Equation 2-17, the ratio of these two values equals the mean hydraulic residence time:

$$\bar{t} = \frac{\sum_{\text{all }t} t_{\text{ave}} c_{\text{ave}} \Delta t}{\sum_{\text{all }t} c_{\text{ave}} \Delta t} = \frac{123{,}300\,(\text{mg-min}^2/\text{L})}{2{,}490\,(\text{mg-min/L})} = 49.5\,\text{min}$$

According to Equation 2-24, the mean hydraulic residence time can also be computed as the summation, over all time intervals, of the product of the average value of t during an interval and the increment in $F(t)$ during that interval. Values of $t_{\text{ave}}\Delta F$ are shown in column G of Table 2-2; the summation, shown in cell G17, is 49.3 min, which is close to the value of $\bar{t}$ computed by the first method. The slight discrepancy occurs because Equation 2-24 is based on the assumption that $F(t)$ reaches 1.0, but the maximum value of $F(t)$ in this data set is 0.996. If this fact is accounted for by using Equation 2-22 to calculate $\bar{t}$, and a value of 0.996 is inserted for $F(\infty)$, the calculated value of $\bar{t}$ becomes 49.3 min/ 0.996 = 49.5 min and the two values agree. If the tracer analysis were perfect and all of the input tracer were detected in the effluent, the original two estimates of $\bar{t}$ would be identical. Note that the computed value of $\bar{t}$ indicates the ratio of the accessible or effective reactor volume to the flow rate (i.e., $\bar{t} = V_{\text{eff}}/Q$) if the reactor has closed boundaries, but $\bar{t}$ is greater than that ratio if the boundaries are open. At this point, we make no assumption about which, if either, of those boundary conditions applies.

The variance of the RTD can be calculated as the second moment around the mean divided by the zeroth moment (Equations 2-31 and 2-30). The data needed to evaluate these terms include $\bar{t}$, which was calculated in part (a), and the summation shown in Row 17 of either column H or I. The calculation using the data in column I and Equation 2-30b is

$$\sigma_{\text{RTD}}^2 \approx \left(\sum_{\text{all }i} t_{i,\text{ave}}^2 E_{i,\text{ave}}(t_i) \Delta t_i\right) - \bar{t}^2$$

$$= 3211.7\,\text{min}^2 - (49.5\,\text{min})^2 = 760\,\text{min}^2$$

(c) The theoretical mean detention time is V/Q, so,

$$\tau = \frac{240\,\text{m}^3}{4\,\text{m}^3/\text{min}} = 60\,\text{min}$$

Thus, in this (made-up) example, the mean detention time calculated from the tracer results ($\bar{t}$) is only ~83% of the theoretical detention time τ, suggesting that the effective volume of the tank is less than its geometrical volume. Since essentially all of the injected tracer was recovered, the discrepancy between τ and $\bar{t}$ is due to nonideality in the flow pattern and not to a failure to carry out the test over a long enough period to allow all the tracer to exit. Therefore, if the reactor had closed boundaries, we could infer that 17% of the reactor volume was dead space. ∎

Having established the conceptual basis for RTDs and explored some experimental approaches for determining RTDs and representing them mathematically, we next consider what those RTDs can tell us about the flow patterns in reactors.

2.3 IDEAL REACTORS

The flow regime of a reactor refers to the amount and characteristics of internal mixing in the reactor. The two extremes of mixing intensity are no mixing whatsoever and complete mixing; continuous flow reactors with these characteristics are known as *plug flow reactors* (PFRs) and *continuous flow stirred tank reactors* (CFSTRs), respectively. CFSTRs are also referred to as completely mixed reactors (CMRs) or continuous stirred tank reactors (CSTRs).

In CFSTRs, fluid mixing is so intense that, as soon as a parcel of water enters the system, it mixes with and dilutes into the water in the entire reactor volume. At the other extreme of the mixing spectrum are PFRs, in which fluid moves steadily through the reactor along a defined path, without mixing at all with parcels of fluid ahead of or behind it in the axial direction. Mixing in the transverse direction is often considered complete, in which case there is no concentration gradient in that direction; however, that is not a core feature of the model. Axial diffusion and dispersion are both negligible in a PFR. Although ideal plug flow might never be achieved, some reactors and natural systems approach this behavior. Because no mixing occurs in the direction of flow, it is reasonable to imagine that each parcel of fluid in the reactor is physically isolated and behaves as if it were in a container moving through space without interaction with parcels upstream or downstream of it (as on a conveyor belt). Furthermore, any small parcel of fluid in the reactor contains molecules that all entered the reactor together at a specific earlier time. Thus, for instance, pesticides spilled into a river that behaves as an ideal plug flow system would move through the river system without spreading out into or being diluted by water that passed the point of the spill before or after the spill occurred.

In water and wastewater treatment systems, some reactors behave closely enough to one of these limiting cases that they can be modeled as ideal reactors. For instance, some reactors, such as rapid mix tanks in water treatment plants, are designed to have intense mixing so that the concentrated chemicals that are added to the water (e.g., coagulants, acids, bases, or disinfectant chemicals) are rapidly and uniformly distributed throughout the tank. Also, the contents of aeration tanks in activated sludge systems are often thoroughly mixed by bubbles that are injected through diffusers along the bottom of the tank and/or by mechanical mixers. These types of reactors can often be reasonably modeled as CFSTRs. At the other extreme, packed bed reactors (e.g., activated carbon adsorption columns, ion exchange columns, gas transfer towers for stripping or absorption) can often be treated as PFRs.

The CFSTR model is commonly invoked to describe the hydraulics of intensely mixed tanks, where the primary objectives are usually to disperse influent chemicals and to maintain a nearly uniform solution composition throughout the reactor.[7] In such tanks, turbulence is generated in various ways (e.g., mechanical mixers or intense aeration), and the contents of the reactors are mixed as the turbulence dissipates. More detailed discussion of the dynamics of turbulent mixing in tanks is provided in Chapter 11, where the importance of mixing for simultaneously dispersing chemicals and inducing collisions among particles in coagulation/flocculation reactors is explored.

In the following subsections, we determine the RTD for the two ideal reactor types by determining the expected result of tracer studies and converting that information into the RTDs.

Plug Flow Reactors

Consider a PFR of volume V that has operated with a steady flow Q for a substantial period. What would be the response of the reactor to a pulse input tracer test in which a mass M_{in} of tracer is put into the influent all at once at time zero?

From the conveyor belt description of a PFR given earlier, it should be obvious that all of the tracer would come out together after a time equal to the hydraulic residence time, τ. Whereas the average hydraulic residence time τ is V/Q for *any* reactor, τ is the exact residence time of every parcel of fluid (including the parcel that contains the tracer) for a PFR. For the described conditions, the output of the reactor contains the entire mass M_{in} at $t = \tau$ and contains no tracer at all other times. Thus, in mathematical terms, the mass of tracer in the output is $M_{\text{in}}\,\delta_\tau(t)$.

The $E(t)$ and $F(t)$ curves can also be determined easily based on this discussion. Because no tracer comes out before τ and all of it is out of the reactor after $t = \tau$, the cumulative age distribution $F(t)$ is the Heaviside function applied at $t = \tau$. Since $E(t)$ is the derivative of $F(t)$, $E(t)$ must be the Dirac delta applied at $t = \tau$. Thus,

$$F(t)_{\text{PFR}} = H_\tau(t) \qquad (2\text{-}33\text{a})$$

$$E(t)_{\text{PFR}} = \delta_\tau(t) \qquad (2\text{-}33\text{b})$$

These $E(t)$ and $F(t)$ functions are shown in Figure 2-8.

---
[7] Formally, the CFSTR model is also employed to describe mixing at a junction, where solutions of disparate compositions mix and generate a different, uniform composition in the fluid that leaves the junction point.

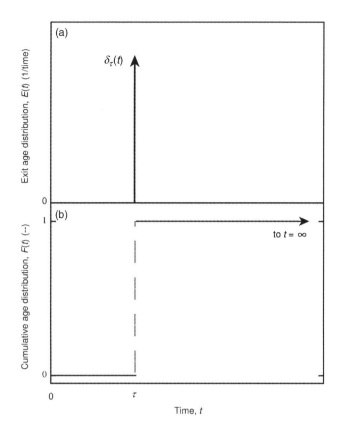

FIGURE 2-8. Residence time distributions of a plug flow reactor.

Although the $E(t)$ and $F(t)$ functions for a PFR can be inferred based on conceptual arguments alone, it is useful to consider their derivation mathematically. The derivation is based on the analysis of a differential control volume whose area is the cross-sectional area of the reactor, and whose length in the direction of flow is dx. Two choices for the boundaries of the control volume are reasonable, differing based on the frame of reference of the observer. In one choice, the control volume is a region of space that is fixed relative to an observer standing outside the reactor. This frame of reference is referred to as an Eulerian view. The other choice, called the Lagrangian view, is a region of space that is at the entrance of the reactor initially and that moves through the reactor at the average velocity of the fluid. The control volume in a Lagrangian view can be thought of as a region of space that is fixed relative to an observer who is traveling with the fluid. The PFR system is analyzed in the following subsections using both approaches.

Pulse Input to a PFR: Fixed Frame of Reference (Eulerian View) For a control volume fixed in space, as shown in Figure 2-9, the general mass balance for a tracer passing through a PFR can be written in words as follows:

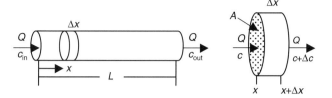

FIGURE 2-9. Definition diagram for mass balance on a PFR: Eulerian view.

Rate of change of tracer mass stored in the differential control volume	=	Rate of input of tracer mass into control volume by advection	−	Rate of output of tracer mass from control volume by advection

The mass balance contains no terms for input and output by dispersion/diffusion because, by the definition of a PFR, these terms are zero. The mass balance also lacks a reaction term, because the substance of interest is a nonreactive tracer. The volume of the small element of the reactor shown in Figure 2-9 is $A\Delta x$, where A is the cross-sectional area. Translating this general mass balance into symbols yields,

$$\frac{\Delta(cA\Delta x)}{\Delta t} = Qc - Q(c + \Delta c) \quad (2\text{-}34a)$$

$$\frac{\Delta(cA\Delta x)}{\Delta t} = -Q\Delta c \quad (2\text{-}34b)$$

If we divide through by $A\Delta x$ and replace Q/A by the fluid velocity v_x, we can then take the limit as Δx and Δt approach zero to obtain the following partial differential equation:

$$\frac{\partial c}{\partial t} = -v_x \frac{\partial c}{\partial x} \quad (2\text{-}35)$$

This differential equation is first order in both time and space and therefore requires one initial and one boundary condition. Both of these are contained in the description of the pulse input as a Dirac delta function at time zero: the tracer is injected all at once at $(x, t) = (0, 0)$ so the concentration at that instant at the influent end is infinite, and zero at all other locations in the reactor (initial condition); and, the tracer concentration at the influent end is zero at all other times (boundary condition).

To use the initial condition, we must define a concentration associated with the tracer input. In a real test, the tracer concentration in the input during a small interval around time zero would be large but finite, since all the tracer would be injected in a small volume of influent liquid. In the limit, this concentration is infinite, because the volume of solution containing the tracer is assumed to be infinitesimally small. In that case, the corresponding concentration can be defined as the mass M of the pulse input of tracer divided by the volume of an arbitrarily small aliquot of water. We identify this concentration as c''. The characteristics of the Dirac delta function make the calculus unusual, but the solution is that the tracer (described by a Dirac delta function) travels down the length of the reactor at the velocity v_x and therefore reaches the end of the reactor $(x=L)$ at $t = L/v_x = V/Q = \tau$. The tracer concentration at any time and location in the reactor output can be described by the following equation:

$$c_p^{PFR}(x,t) = c'' \delta_{x/v_x}(t) \qquad (2\text{-}36)$$

When this equation is applied to the effluent $(x=L)$, we find

$$c_{p,out}^{PFR}(t) = c'' \delta_\tau(t) \qquad (2\text{-}37)$$

The magnitude of the arbitrarily small aliquot of water turns out to be irrelevant to the result, because the Dirac function causes $c_p^{PFR}(x,t)$ to be infinitely large if $t = x/v_x$ and zero if t is any other value.

No molecule of water (or tracer) spends either more or less time in a plug flow reactor than τ. In probabilistic terms, we can say that the likelihood is 100% that a molecule will spend an amount of time equal to V/Q in the reactor, and the likelihood of it spending an amount of time other than τ is nil. That statement completely defines the RTD of a PFR.

Pulse Input to a PFR: Moving Frame of Reference (Lagrangian View) In the Lagrangian view, the frame of reference is not a fixed point in space but a certain parcel of the fluid. As noted earlier, we can think of a PFR something like a conveyor belt; that is, every parcel of fluid travels through the reactor as if it were a package on a conveyor belt with no interaction (dispersion/diffusion) with parcels upstream or downstream. The Lagrangian view of the system is based on a mass balance in which the control volume is a differential slice of space that travels from one end of the reactor to the other at the average fluid velocity.

Mathematically, the Lagrangian viewpoint can be represented by defining a coordinate system that moves with the water at velocity v_x. Axial distance on this coordinate system is defined as x^*. If the origin of the coordinate system $(x^*=0)$ is defined to coincide with the point $x=0$ at time 0, the translation from values of x (on the fixed axis) to those of x^* at any future time can be made using the equation: $x^* = x - v_x t$ (Figure 2-10). The coordinates of the two ends of the control volume are therefore always

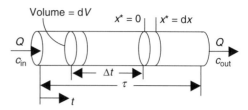

FIGURE 2-10. Definition diagram for a mass balance on a PFR: Lagrangian view. The coordinate system moves at the average velocity of the fluid, so $x^*=0$ is at the upstream end of the control volume at all times.

$x^*=0$ and $x^*=dx^*$, no matter how far the control volume has traveled through the reactor.

The general word mass balance for the moving control volume, again recognizing the lack of dispersion in a PFR and the lack of reactivity of the tracer, would be the same as that for the Eulerian analysis; that is,

Rate of change of tracer mass stored in the differential control volume	=	Rate of input of tracer mass into control volume by advection	−	Rate of output of tracer mass from control volume by advection

However, the Lagrangian view involves converting the word equation into a mathematical expression using x^*, rather than x, as the distance variable. Since the control volume is traveling with the fluid, no mass enters or leaves the control volume by advection, and the entire right side of mass balance is zero. The mass balance can therefore be written as follows:

$$dV \frac{\partial c(x^*,t)}{\partial t} = 0 \qquad (2\text{-}38)$$

Because the control volume is always finite, dV is nonzero, so this equation can be further simplified as follows:

$$\frac{\partial c(x^*,t)}{\partial t} = 0 \qquad (2\text{-}39)$$

The solution to Equation 2-39 is that c is a function of x^* only, independent of time. The profile at time zero is given by $c(x^*,0) = c'' \delta_0(x^*)$; that is, it is a Dirac delta function applied at $x^*=0$. Based on Equation 2-39, though, $c(x^*,t)$ does not change over time, so the concentration profile is given by $c'' \delta_0(x^*)$ at *all* times; that is,

$$c_p^{PFR}(x^*,t) = c'' \delta_0(x^*) \qquad (2\text{-}40)$$

The result is that $c_p^{PFR}(x^*,t)$ equals c'' times the Dirac spike whenever $x^*=0$, and zero when x^* has any nonzero value. Converting this result back to a function of x (rather than x^*) also converts it to a function of time. Since $x^* = x - v_x t$, $c_p^{PFR}(x, t)$ equals the product of c'' with the Dirac spike whenever $x - v_x t = 0$ (i.e., whenever $t = x/v_x$), and zero at all other times and locations, so

$$c_p^{PFR}(x, t) = c'' \delta_{x/v_x}(t) \tag{2-36}$$

As expected and necessary, the result based on the Lagrangian view (moving reference) is the same as that obtained previously (Equation 2-36) with the Eulerian view (fixed reference).

As noted previously, the response of a PFR to a tracer input can be deduced easily from a qualitative understanding of the flow pattern in such reactors; the derivations provided here are useful as formal proofs of the validity of the qualitative analysis. More importantly, they serve as a simple introduction to the Eulerian and Lagrangian viewpoints and to the methodology used later in this chapter and in Chapter 4 to analyze more complex systems, including reactors with nonnegligible diffusion/dispersion and in which chemical reactions are taking place.

Continuous Flow Stirred Tank Reactors

Pulse Input to a CFSTR Consider next the response, $c_p(t)$, of a CFSTR to a pulse input of tracer. The reactor is assumed to have closed boundaries and steady flow. The situation is described pictorially in Figure 2-11. We focus first on what happens at time zero. By definition, the concentration in a CFSTR is the same at all locations, and the influent is mixed instantaneously with all the fluid in the reactor. Thus, at time zero, the mass M of tracer is mixed instantly into the entire volume V, so that the concentration at any point in the reactor is M/V; as indicated in Equation 2-9, we designate this concentration as c_o. Because the effluent is taken from some point in the reactor, the concentration of the effluent at that instant (time 0) is also c_o; that is, $c_p(0) = c_o$.

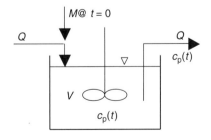

FIGURE 2-11. Definition diagram for a pulse input tracer test in a continuous flow stirred tank reactor.

To determine what happens at all later times, we write a mass balance on the tracer in the system, defining the system boundaries to include all the fluid in the tank. The input and output are the influent and effluent, respectively, and, because the tracer is nonreactive, the mass balance contains no reaction term. In any mass balance, diffusion and dispersion are relevant only at the boundaries of the control volume. We assume that tracer transport across the boundaries of the control volume by these mechanisms is negligible compared to transport by advection, so these terms are omitted in the mass balance.[8] The mass balance can therefore be described in words as follows:

$$\begin{matrix} \text{Rate of change} \\ \text{of mass} \\ \text{of tracer} \\ \text{stored in} \\ \text{reactor} \end{matrix} = \begin{matrix} \text{Rate of input} \\ \text{of tracer} \\ \text{mass into} \\ \text{reactor by} \\ \text{advection} \end{matrix} - \begin{matrix} \text{Rate of output} \\ \text{of tracer} \\ \text{mass from} \\ \text{reactor by} \\ \text{advection} \end{matrix} \tag{2-41}$$

As noted previously, the concentration is the same everywhere in the system, including in the effluent, so we can equate the concentration inside the reactor at time t with $c_p(t)$, the effluent concentration at that time. The mass stored in the reactor at any time is $Vc_p(t)$. Also, after the initial dumping of tracer into the reactor at time zero, no further additions of tracer are made into the tank, so the input term (the first term on the right) is zero. Therefore, in symbols

$$\frac{d(Vc_p(t))}{dt} = 0 - Qc_p(t) \tag{2-42}$$

Because V is a constant, it can be taken outside the derivative. Separating the variables, dividing by V, and recognizing that $\tau = V/Q$ yields

$$\frac{dc_p(t)}{c_p(t)} = -\frac{1}{\tau} dt \tag{2-43}$$

Integrating Equation 2-43 from zero to an indefinite time t, we obtain

$$\int_{c_p(0)}^{c_p(t)} \frac{dc_p(t)}{c_p(t)} = -\frac{1}{\tau} \int_0^t dt \tag{2-44}$$

---

[8] As noted previously, a system that has negligible diffusion and dispersion across the inlet and outlet is said to have *closed boundaries*. Such a situation can be assured in a CFSTR by a discontinuity in flow at each location, for example, by having the inlet and outlet flows discharge freely above the surface of the reactor and over a weir, respectively.

With the application of the initial condition that, at time zero, the (effluent) concentration is c_o, the solution is as follows:

$$\ln \frac{c_p^{CFSTR}(t)}{c_o} = -\frac{1}{\tau}t \quad (2\text{-}45a)$$

or

$$c_p^{CFSTR}(t) = c_o \exp\left(-\frac{t}{\tau}\right) \quad (2\text{-}45b)$$

where the superscript has been added to emphasize that the result applies only for an ideal CFSTR. The results are shown graphically according to both the arithmetic (Equation 2-45b) and logarithmic (Equation 2-45a) forms of the equation in Figure 2-12.

Immediately after the tracer is introduced and mixed throughout the reactor, the concentration is the highest it will ever be, because no more tracer is added after that time. Thereafter, the tracer is continually diluted by the (clean) influent, leading to a continual decline in the concentration in the reactor (and in its effluent). After one detention time ($t = \tau$), the concentration is still ~37% ($e^{-1} = 0.368$) of what it was immediately after the input of tracer. Even after three detention times ($t = 3\tau$), it is ~5% ($e^{-3} \cong 0.050$) of the initial value. In fact, theoretically, an infinite time is required to get the last molecules of tracer out of the reactor. This result contrasts with that for a PFR, for which none of the tracer exits before time τ, and none remains in the reactor after time τ.

Because it predicts the response of a CFSTR to a pulse input of tracer, Equation 2-45a provides an approach for testing whether a given reactor actually behaves as a CFSTR. That is, if experimental data for $\ln(c_p/c_o)$ are plotted against t for an existing reactor, the linearity of the relationship can serve as an indicator of how closely the reactor conforms to the CFSTR model—if the relationship is linear, the model applies, and the negative slope of the line is the mean hydraulic residence ($\bar{t}$), and if the relationship is not linear, then the model does not apply.

According to Equation 2-9, we can use the results from a pulse input tracer test to compute $E(t)$ as the product $(1/(V/Q))(c_p(t)/c_o)$. Recognizing now that V/Q is the theoretical hydraulic residence time τ, Equation 2-9 can be rewritten as

$$E(t) = \frac{1}{\tau}\frac{c_p(t)}{c_o} \quad (2\text{-}46)$$

For the particular case of a CFSTR, inserting Equation 2-45b into Equation 2-46 yields

$$E(t)_{CFSTR} = \frac{1}{\tau}e^{-t/\tau} \quad (2\text{-}47)$$

The mass of tracer that has come out of any reactor from time zero to any time t (i.e., $M_p(t)$) is given by Equation 2-10. For a CFSTR, inserting Equation 2-45b into Equation 2-10 and integrating the resulting expression yields the following relationship for $M_p(t)$:

$$M_p(t)_{CFSTR} = M_{in}(1 - e^{-t/\tau}) \quad (2\text{-}48)$$

Finally, we can find the cumulative age distribution, $F(t)$, as the integral to time t of $E(t)$; for a CFSTR, the result is as follows:

$$F(t) = \int_0^t E(t)dt = \int_0^t \frac{1}{\tau}\exp\left(-\frac{t}{\tau}\right)dt = -e^{-t/\tau}\Big|_0^t = 1 - e^{-t/\tau}$$

$$(2\text{-}49)$$

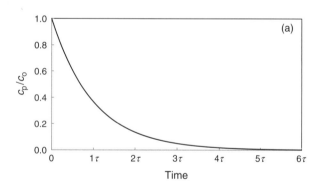

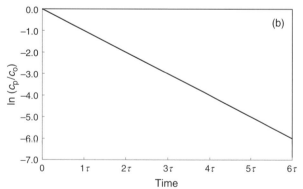

FIGURE 2-12. Response of a CFSTR to a pulse input tracer test. (a) Arithmetic representation; (b) logarithmic representation.

The four functions described in Equations 2-46 to 2-49 for the CFSTR are plotted in Figure 2-13. Two of the functions, $c_p(t)$ and $M_p(t)$, have values that depend on the mass of tracer input, the volume of the reactor, and the mean hydraulic residence time (τ), while the other two, $E(t)$ and $F(t)$, depend only on τ.

Summarizing these results, the molecules of water (or any nonreactive substance in the water) that enter a CFSTR are dispersed immediately throughout the reactor. All of those molecules are in the reactor in the instant following their entry. This means that, for tracers, the concentration is highest at $t = 0$. Molecules (of either water or tracer) that enter at any one time stay in the reactor for varying amounts of time. Many of the molecules that enter in one instant stay in the reactor for only a short time, and steadily smaller numbers of molecules stay for progressively longer times, with a very small number staying for extremely long periods. Correspondingly, the probability of the molecules leaving is highest immediately after their introduction. As a result, in any sample of the effluent, a large fraction of the molecules came in a short period ago, while a much smaller fraction spent a much longer time in the reactor.

Step Input to a CFSTR We now consider the response of a CFSTR to a step input of tracer, i.e., $c_s(t)$. As always, we can carry out the analysis by writing a mass balance to describe the situation for any time after time zero. The system is again all of the fluid in the reactor, and the constituent of interest is again the tracer. Recognizing that the tracer is nonreactive, so that no reaction term is necessary, the mass balance can be expressed in words identically as in Equation 2-41:

$$\begin{matrix}\text{Rate of change} \\ \text{of mass} \\ \text{of tracer} \\ \text{stored in} \\ \text{reactor}\end{matrix} = \begin{matrix}\text{Rate of input} \\ \text{of tracer} \\ \text{mass into} \\ \text{reactor by} \\ \text{advection}\end{matrix} - \begin{matrix}\text{Rate of output} \\ \text{of tracer} \\ \text{mass from} \\ \text{reactor by} \\ \text{advection}\end{matrix}$$

(2-41)

To convert this word equation into symbols, we note again that the mass of tracer in the CFSTR at any time is the product of the volume (V) and the concentration (c), and the rate of change of that mass with time is the time derivative of Vc. The mass input and output are the products of the constant flow rate (Q) and the influent and effluent concentrations (c_{in} and $c_s(t)$), respectively. The concentration in the effluent is the same as the concentration in the reactor so we can use $c_s(t)$ for both; this situation is a consequence of the thorough mixing and is unique to a CFSTR. In symbols

$$\frac{d(Vc_s(t))}{dt} = Qc_{in} - Qc_s(t) \quad (2\text{-}50)$$

Separating the variables, dividing by V, and recognizing that $Q/V = 1/\tau$, we obtain

$$\int_{c_s(0)}^{c_s(t)} \frac{dc_s(t)}{c_{in} - c_s(t)} = \frac{1}{\tau}\int_0^t dt \quad (2\text{-}51)$$

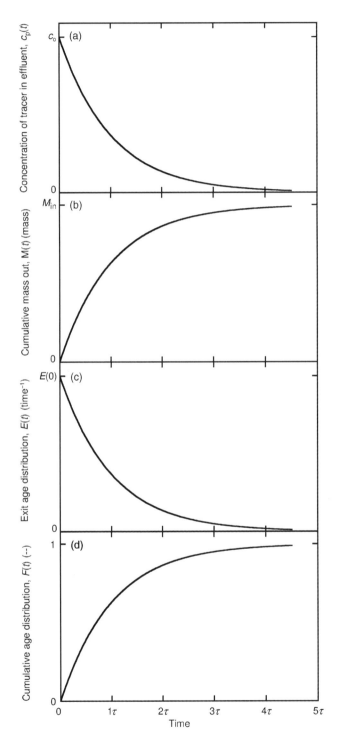

FIGURE 2-13. CFSTR response to pulse input tracer and exit age distributions.

To solve Equation 2-51, we need to know the initial condition. This condition can be slightly confusing, since the input concentration undergoes a step change at that time. The task is to find the concentration in the reactor (or in the effluent) at that time. As in the case of a pulse input, the approach is to consider the mass of tracer that enters the reactor between $t=0$ and an infinitesimal time δt later. In the step input case, this mass of tracer equals $c_{in} Q\, \delta t$, so the concentration of tracer in the reactor at that time (ignoring mass released in the effluent) is $c_{in} Q\, \delta t / V$. In the limit of δt approaching 0, this expression for the concentration goes to zero. Hence, the initial condition is $c_s(0) = 0$. The evaluation of the integrals in Equation 2-51 is then straightforward. The result, after algebraic manipulation, is as follows:

$$c_s^{\text{CFSTR}}(t) = c_{in}(1 - e^{-t/\tau}) \qquad (2\text{-}52)$$

The concentration profile indicated by the equation is as one might anticipate. At time zero, the concentration is zero. After very long periods (high values of t/τ), the concentration approaches the concentration of the influent. In the meanwhile, the concentration rises at an ever-decreasing rate toward the asymptote, c_{in}.

The cumulative age distribution follows directly from the equation describing the effluent concentration. According to Equation 2-13, the cumulative age distribution, $F(t)$, is obtained by dividing $c_s(t)$ by c_{in}. Carrying out this normalization yields the same result as for the pulse input test (Equation 2-49):

$$F(t)_{\text{CFSTR}} = \frac{c_s(t)}{c_{in}} = 1 - e^{-t/\tau} \qquad (2\text{-}49)$$

Finally, the exit age distribution, $E(t)$, can be obtained as the derivative of $F(t)$, as follows:

$$E(t)_{\text{CFSTR}} = \frac{1}{\tau} e^{-t/\tau} \qquad (2\text{-}47)$$

where, again, the result is the same as that obtained from the pulse input, as it must be. Plots of $F(t)$ and $E(t)$ for a CFSTR have already been presented (Figure 2-13c and 2-13d).

2.4 NONIDEAL REACTORS

Tracer Output from Nonideal Reactors

The two types of ideal reactors that we have described represent the two ends of the spectrum of mixing/no mixing. While these ideals are often approached in practice, most real reactors behave in a way that falls between the extreme cases; that is, they are *nonideal*.

As is the case for ideal reactors, many important features of the hydraulic behavior of nonideal reactors can be described by their $E(t)$ or $F(t)$ functions. The determination of these functions as well as other parameters of interest (e.g., the mean and variance of the RTD) from tracer test data was demonstrated in Examples 2-1 and 2-2. In the preceding section, we determined the expected response of ideal reactors to specific patterns of tracer input. We next address the more general question: what output profile can be expected for any arbitrary input of a nonreactive substance into a reactor? The analysis presented here is adapted from Levenspiel (1999).

Relating Tracer Input and Output Curves via the Convolution Integral In all cases other than PFRs, flow through a reactor causes the concentration profile of a conservative tracer to spread out and flatten, as shown in Figure 2-14a. To predict the output concentration profile from knowledge of the input profile, $c_{in}(t)$, for an arbitrary reactor with any given RTD, it is useful to plot both the input and output concentration profiles on the same axes, as shown in Figure 2-14b. Consider the effluent during a small Δt around time t, as shown. Each molecule of tracer that is in the output came into the reactor at some time earlier, but later than time zero. For instance, some part of the tracer that comes out in the time interval labeled Δt came into the reactor within a small interval of time ($\Delta t'$) that was earlier by an amount t'; that is, it entered around time $t - t'$. Stated another way, the small shaded section of the output in the time interval Δt represents *the same molecules of tracer* that are indicated by the small shaded section of the input in the time interval $\Delta t'$.

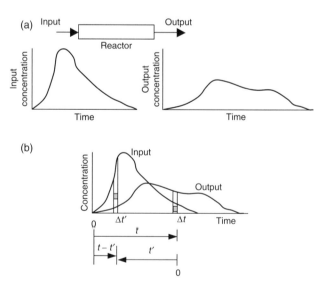

FIGURE 2-14. Transformation of input of conservative substance as it travels through a reactor system.

Extending the analysis, we see that the entire output in the small increment of time Δt is composed of parts of the input that came in at various times between 0 and t. Therefore, the total tracer output during time Δt can be obtained by considering each prior period $\Delta t'$, multiplying the concentration of tracer in the input at time $t - t'$ by the fraction of that input that stays in the reactor for time t', and then summing up the contributions for all possible $\Delta t'$ periods. Translating that sentence to a word equation yields the following:

$$\begin{pmatrix} \text{Mass of tracer} \\ \text{that comes out} \\ \text{in interval } \Delta t \end{pmatrix} = \sum_{\text{all intervals } \Delta t'} \begin{pmatrix} \text{Mass of tracer} \\ \text{that enters during} \\ \text{interval } \Delta t' \end{pmatrix}$$
$$\times \begin{pmatrix} \text{Fraction of input that} \\ \text{stays in the reactor for a} \\ \text{time between } t' \text{ and } t' + \Delta t' \end{pmatrix} \quad (2\text{-}53)$$

The term on the left side of Equation 2-53 is the product of the flow rate, Q, the (average) effluent concentration in the time interval Δt, and the time increment Δt; that is, $Q\, c_{\text{out}}(t)\, \Delta t$. Similarly, the first term within the summation on the right side is $Q c_{\text{in}}(t - t')\Delta t'$. The second term within the summation is $E(t')\Delta t'$. Thus, the equation can be written in symbols as follows:

$$Q\, c_{\text{out}}(t)\Delta t = \sum_{\substack{\text{all intervals} \\ \Delta t' \text{ for } t' < t}} \{Q c_{\text{in}}(t - t')\Delta t'\, E(t')\Delta t'\} \quad (2\text{-}54)$$

Assuming that the entire distribution is broken into equivalent increments, $\Delta t'$ is constant and can be taken out of the summation on the right. The flow rate is also constant and so can come out of the summation. The equation can therefore be simplified as follows:

$$Q\, c(t)\Delta t = Q \Delta t' \sum_{\substack{\text{all intervals} \\ \Delta t' \text{ for } t' < t}} \{c_{\text{in}}(t - t')E(t')\Delta t'\} \quad (2\text{-}55)$$

Although the actual times t and t' are offset from one another, the durations Δt and $\Delta t'$ can be chosen to be identical, in which case we can divide by that quantity and by Q and write the equation as

$$c(t) = \sum_{\substack{\text{all intervals} \\ \Delta t' \text{ for } t' < t}} \{c_{\text{in}}(t - t')E(t')\Delta t'\} \quad (2\text{-}56)$$

When the increment is allowed to shrink to infinitesimal size, the summation can be replaced by an integral, with the following result:

$$c(t) = \int_{t'=0}^{t'=t} c_{\text{in}}(t - t')\, E(t')\mathrm{d}t' \quad (2\text{-}57)$$

This integration is a special mathematical operation known as convolution; in words, we say that $c(t)$ is the convolution of $c_{\text{in}}(t)$ and $E(t)$. The symbol $*$ is used to represent the convolution process, so, equivalently to Equation 2-57, we can write,

$$c(t) = c_{\text{in}}(t) * E(t) \quad (2\text{-}58)$$

If $E(t)$ is described only at discrete times, as is usually the case in tracer tests, a numerical convolution can be performed as described by Equation 2-56. On the other hand, when the input and RTDs can be described by mathematical functions, Equation 2-58 can sometimes be solved using Laplace transforms because of the following property of the convolution integral[9]:

$$\mathbf{L}\{F_1(t) * F_2(t)\} = \mathbf{L}\{F_1(t)\} \times \mathbf{L}\{F_2(t)\} \quad (2\text{-}59)$$

where $\mathbf{L}\{F(t)\}$ is the Laplace transform of the function $F(t)$.[10] According to Equation 2-59, the Laplace transform of the convolution integral is the product of the Laplace transforms of the two functions being convolved. This property means that the integration can be avoided; to find $c(t)$, one simply looks up the Laplace transforms of $c_{\text{in}}(t)$ and of $E(t)$, multiplies them, and uses Laplace transform tables to find the inverse transform of that product. An example of this procedure follows.

■ **EXAMPLE 2-3.** The response of a CFSTR to a step tracer input was determined earlier in this chapter by using the appropriate mass balance and initial condition. Derive the same result using a convolution integral.

Solution. According to Equation 2-58, the effluent concentration as a function of time can be found as the convolution of the input (c_{in}) and the $E(t)$ function for a CFSTR, $E(t)_{\text{CFSTR}} = 1/\tau \exp(-t/\tau)$; that is,

$$c_{\text{s,out}}^{\text{CFSTR}}(t) = c_{\text{in}} * \frac{1}{\tau}\exp\left(-\frac{t}{\tau}\right) \quad (2\text{-}60)$$

---

[9] A brief explanation of Laplace transforms and some examples of their use are given in the appendix.

[10] By convention, Laplace transform functions are almost always represented by the symbol $f(s)$, and the function being transformed is written as $F(t)$. Do not confuse the generic $F(t)$ used here with the cumulative distribution function, which is also given the symbol $F(t)$.

Taking the Laplace transform of both sides of this equation, we obtain,

$$\mathbf{L}\{c_{\text{out}}(t)\} = \{\mathbf{L}(c_{\text{in}})\}\left[\mathbf{L}\left\{\frac{1}{\tau}\exp\left(-\frac{t}{\tau}\right)\right\}\right] \quad (2\text{-}61)$$

From Laplace transform tables, we find the three properties or formulas needed here: (1) the Laplace transform of a constant equals that constant divided by s, (2) the Laplace transform of the product of a constant and a function equals the product of the constant and the Laplace transform of the function, and (3) the Laplace transform of $\exp(at)$ is $1/(s-a)$. Using these properties yields

$$\mathbf{L}\{c_{\text{out}}(t)\} = \frac{c_{\text{in}}}{s}\frac{1}{\tau}\frac{1}{s-\left(-\frac{1}{\tau}\right)} = \frac{c_{\text{in}}}{\tau}\frac{1}{s\left(s+\frac{1}{\tau}\right)} \quad (2\text{-}62)$$

The inverse transformation requires the use of the second property noted in the preceding paragraph and the following inverse transform:

Laplace transform	Inverse transform
$\mathbf{L}\{F(t)\} = \dfrac{1}{(s-a)(s-b)}$	$F(t) = \dfrac{1}{a-b}\left(e^{at} - e^{bt}\right)$

With $a=0$ and $b=-1/\tau$, the inverse transformation yields

$$c_{\text{out}}(t) = \frac{c_{\text{in}}}{\tau}\frac{1}{(1/\tau)}\left(e^0 - e^{-t/\tau}\right) \quad (2\text{-}63\text{a})$$

$$= c_{\text{in}}\left(1 - e^{-t/\tau}\right) \quad (2\text{-}63\text{b})$$

This result, of course, agrees with that found earlier by using the mass balance approach. ∎

Modeling Residence Time Distributions of Nonideal Reactors

The RTDs $E(t)$ and $F(t)$ characterize the net hydraulic behavior of a reactor. (One could go into greater detail to describe the hydraulic behavior in a smaller section inside a reactor, but the concern here is only for the overall hydraulic behavior.) Although the numerical results from a tracer test capture all of the available information about these functions, using the results as a collection of individual data points is often inconvenient and, more importantly, can obscure important trends and similarities among different reactors. To overcome these drawbacks, it is useful to fit the data to a mathematical model that allows the functions to be compressed into a few parameters. Because such models rarely fit the data exactly, modeling usually includes some loss of information (or loss of precision), but the added convenience is often deemed to be worth the sacrifice. Assessing the loss of information by comparing the model with the data is an important part of the modeling process.

In this section, reactor models are presented in three categories, although the distinctions among the categories are somewhat arbitrary. In the first category, models that consider reactors to have hydraulic characteristics that are closely related to the ideal reactors (PFRs and CFSTRs) are presented; these models use the average hydraulic residence time and one additional parameter to fit the data. In these models, the idea is to describe the mixing in the reactor relative to the complete mixing of a CFSTR or the complete absence of mixing of a PFR as a base case, with the added fitting parameter describing the extent of deviation from the idealized conditions. In the second category, models that include the possibility of dead space in the reactor (i.e., space in which the water is stagnant so that the reactor behaves as if that space were not in the reactor) are presented. These models are combined with those in the first category and require two parameters beyond the hydraulic residence time: one characterizing the amount of dead space in the reactor and one characterizing the nonideal mixing. In the third category, more complex models with several parameters are presented; these models can be used to model almost any RTD with some stated accuracy. The greatest attention is on the first category, because these are the most commonly used models and they satisfy the goal of simplifying the use of $E(t)$ or $F(t)$ most completely. As the models become more complex, using all of the data for $E(t)$ or $F(t)$ might become simpler than finding the parameter values and using the model.

PFR with Dispersion The first model we consider assumes that the nonideal behavior of a reactor is caused by (or at least can be described by) some dispersion in a flow pattern that is nearly that of an ideal PFR. In this model, the constraint of zero mixing in the axial direction within a PFR is relaxed by considering dispersion (or, more properly, the sum of diffusion and dispersion) to be nonzero. To determine the RTD of such a reactor, one could perform a tracer study and analyze the results in terms of the Eulerian (fixed frame of reference) mass balance for the system, as developed in Chapter 1:

$$\frac{\partial c(x,t)}{\partial x} = -v_x\frac{\partial c(x,t)}{\partial t} + \mathbf{D}\frac{\partial^2 c(x,t)}{\partial x^2} \quad (2\text{-}64)$$

where v_x is the (average) velocity in the axial (x) direction and $\mathbf{D}$ is the sum of the dispersion and diffusion coefficients (recall that in the vast majority of engineered systems, the dispersivity is much larger than the diffusivity).

Equation 2-64 applies to any profile for a tracer input into a PFR with dispersion; the details of the input function for a given situation are included in the boundary and initial conditions. Although we could write those boundary and initial conditions to solve Equation 2-64 for a particular situation, the mass balance turns out to be simpler to analyze if written from a Lagrangian point of view; that is, if written for a control volume that is moving with the fluid at velocity v_x. Recall that, on the Lagrangian axis, distance is defined by a variable x^*, which is related to x by the equation: $x^* = x - v_x t$. Recall also that, from the Lagrangian viewpoint, fluid is not entering or leaving the control volume by advection, so the advective term can be set to zero. As a result, the mass balance can be written in terms of x^* as follows:

$$\frac{\partial c(x^*, t)}{\partial t} = \mathbf{D} \frac{\partial^2 c(x^*, t)}{(\partial x^*)^2} \quad (2\text{-}65)$$

which is the equation describing one-dimensional diffusion and dispersion from a planar source.

It is convenient to transform Equations 2-64 and 2-65 into dimensionless forms, so that the results can be understood most generally and are more broadly applicable. To do so, we can define dimensionless distance, time, and concentration as follows:

Dimensionless distance $\tilde{x}$: $\quad \tilde{x} = \dfrac{x}{L} \quad (2\text{-}66a)$

Dimensionless time $\tilde{t}$: $\quad \tilde{t} = \dfrac{t}{\tau_{\text{eff}}} = \dfrac{t}{V_{\text{eff}}/Q}$
$\qquad\qquad = \dfrac{t}{L/v_{x,\text{eff}}} \quad (2\text{-}66b)$

Dimensionless concentration $\tilde{c}$: $\quad \tilde{c} = \dfrac{c}{M/V_{\text{eff}}} = \dfrac{c}{c_o}$
$\qquad\qquad\qquad\qquad (2\text{-}66c)$

where τ_{eff}, V_{eff}, and $v_{x,\text{eff}}$ are the effective mean residence time, effective volume, and effective velocity, respectively. These "effective" values have identical meanings to τ, V, and v_x, respectively, except that they are computed considering only the volume or cross-sectional area that is accessible to the flow, rather than the volume or cross-section computed based on geometry. Differences between the geometrically based value and the effective value of each parameter can arise if some of the reactor volume is dead space, in which case τ_{eff} is less than τ, V_{eff} is less than V, and $v_{x,\text{eff}}$ is greater than v_x; dead space is discussed in more detail later in this chapter. In the remainder of this section, the subscript eff is omitted from all these variables for simplicity, but the variables should all be understood to represent effective values, rather than values based strictly on geometry; if the reactor has no dead space, then the geometric and effective values are identical. Substituting these dimensionless parameters into Equations 2-64 and 2-65 yields

$$\frac{\partial \tilde{c}(\tilde{x}, \tilde{t})}{\partial \tilde{t}} = -\frac{\partial \tilde{c}(\tilde{x}, \tilde{t})}{\partial \tilde{x}} + \frac{\mathbf{D}}{v_x L} \frac{\partial^2 \tilde{c}(\tilde{x}, \tilde{t})}{\partial \tilde{x}^2} \quad (2\text{-}67)$$

$$\frac{\partial \tilde{c}(\tilde{x}^*, \tilde{t})}{\partial \tilde{t}} = \frac{\mathbf{D}}{v_x L} \frac{\partial^2 \tilde{c}(\tilde{x}^*, \tilde{t})}{(\partial \tilde{x}^*)^2} \quad (2\text{-}68)$$

where $\tilde{x}^* \equiv \tilde{x} - \tilde{t}$. The dimensionless term $\mathbf{D}/v_x L$ appears in both Equations 2-67 and 2-68 and represents the single parameter of the model. Following the suggestion of Levenspiel (1999), $\mathbf{D}/v_x L$ is referred to here as the dispersion number.[11] The dispersion number gives a relative measure of the importance of dispersion and advection in the mass balance equation—the larger the dispersion number, the more significant dispersion is relative to advection.

Whether written in dimensional or dimensionless form, the defining equation is first order in time and second order in space. Therefore, to solve for the concentration profile, one initial and two boundary conditions are required. For a pulse input tracer test, the initial condition is described by the pulse input at time zero at the influent boundary and the absence of tracer elsewhere in the reactor. The boundary conditions are much more difficult to specify, and even subtle changes in those conditions can have a large effect on the predicted profile. Boundary conditions that have been proposed for different types of reactor inlet and outlet arrangements include specification of the flux at the boundaries or of the concentration either at the physical boundaries (the inlet and outlet of the reactor) or at the mathematical boundaries ($\tilde{x}^* = \pm\infty$).

As noted previously, reactor boundaries can be open (i.e., allow diffusion and dispersion to transport material into or out of the reactor), closed (do not allow such transport), or intermediate between these two extremes. For small values of $\mathbf{D}/v_x L$ (less than ~ 0.01), the solutions to the mass balance equation for different types of boundaries are very similar, but for larger values of $\mathbf{D}/v_x L$, the profiles can depend strongly on the boundary conditions. Possible boundary conditions for several types of reactors have been discussed by Kreft and Zuber (1978), Nauman (1981), and Levenspiel (1999).

To date, the mass balances have been solved analytically for only a few cases. For instance, for open boundaries at the physical limits of the control volume and boundary conditions of zero concentration at $\tilde{x}^* = \pm\infty$, the tracer output concentration following a pulse input can be expressed for

---

[11] In much of the literature on groundwater flow and the hydraulics of engineered reactors, this idea is expressed in terms of the Peclet number, Pe. Pe is defined as the inverse of the dispersion number; that is, $\text{Pe} \equiv v_x L/\mathbf{D}$, so the importance of dispersion *decreases* as Pe increases.

the dimensionless and dimensional formulations as follows (after Bear, 1972):

$$\tilde{c}(\tilde{x},\tilde{t}) = \frac{1}{\sqrt{4\pi(\mathbf{D}/v_x L)\tilde{t}}} \exp\left\{-\frac{(\tilde{x}-\tilde{t})^2}{4(\mathbf{D}/v_x L)\tilde{t}}\right\} \quad (2\text{-}69)$$

$$c(x,t) = \frac{c_o}{\sqrt{4\pi(\mathbf{D}/v_x L)t/\tau}} \exp\left\{-\frac{(x-v_x t/L)^2}{4(\mathbf{D}/v_x L)(t/\tau)}\right\} \quad (2\text{-}70)$$

These equations give the concentration distribution over time and space for a pulse input at time zero. They define a profile that spreads out with a diminishing peak as the tracer moves through the reactor. For any fixed value of t or $\tilde{t}$, the profile is a gaussian distribution along the distance axis. A family of curves from this equation, depicting the spread of tracer as it travels through the reactor, is shown in Figure 2-15a. The curves represent the tracer concentration throughout the reactor at different times, so the set of curves can be thought of as a time series of snapshots of the tracer profile.

Figure 2-15a describes the tracer concentration as a function of distance at various times; that is, each curve is represents a fixed value of time and various values of distance. Note that, when plotted in this way, the profile is always symmetric around its peak value. This property is emphasized by plotting the profile using a Lagrangian distance scale (Figure 2-15b). However, in an actual tracer study, the variable that is almost always measured is the tracer concentration as a function of time at a fixed location (i.e., in the reactor effluent, at $x = L$). An observer standing at this point first sees tracer that has been in the system for a short time and has dispersed downstream to reach the reactor effluent. Later, the observer sees a higher concentration of tracer, but this tracer has been in the system longer and has therefore been subject to more dispersion than the tracer seen earlier. Still later, after the peak has passed, the observer sees tracer that has been in the reactor and has been affected by dispersion for quite a long time. The fact that the parcels of fluid being observed at different times have been affected by dispersion to differing extents leads to a tracer profile that is skewed, with the tracer concentration increasing more rapidly at early times than it decreases at later times. A family of curves showing this trend for different values of $\mathbf{D}/v_x L$ for a reactor with closed boundaries is presented in Figure 2-16. At low values of $\mathbf{D}/v_x L$, the curves are approximately symmetric and gaussian, but as $\mathbf{D}/v_x L$ increases, they become more skewed.

The abscissa of Figure 2-16 is t/τ, which can be thought of as nondimensionalized time, and the ordinate is the product of $\tau E(t)$. As indicated by Equation 2-46, $\tau E(t)$ equals the ratio c/c_o for a pulse input tracer test, and hence can be

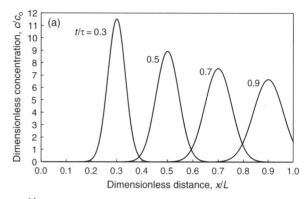

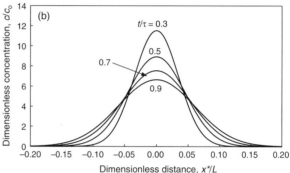

FIGURE 2-15. Snapshots of the tracer profile in a pulse input test, according to the dispersion model for $\mathbf{D}/v_x L = 0.002$. (a) Concentration as a function of distance using an Eulerian distance axis. (b) Concentration as a function of distance using a Lagrangian distance axis. Although the curves are computed using the equation for open boundaries, the dispersion number is small enough that the curves are good approximations for systems with closed boundaries as well.

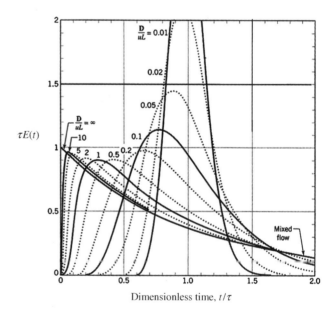

FIGURE 2-16. Model exit age distribution for a plug flow reactor with various amounts of dispersion, for a reactor with closed boundaries. *Source*: From Levenspiel (1999); reprinted with permission.

thought of as a nondimensionalized tracer concentration exiting the reactor. Plotting the data in this way (i.e., with dimensionless terms on both axes) allows us to characterize the response of reactors with any hydraulic residence time—from very short to extremely long—on a single, concise graph.

As is clear from the preceding discussion, the dispersion model uses the dispersion number as the single parameter that specifies where a system lies on the continuum between plug flow and complete mixing. The larger the dispersion number, the sooner a tracer can be detected downstream of its injection point, and the longer it takes for the total mass of tracer injected to be completely flushed out of the system. In the limit of extremely large dispersion, the system behaves very much like a CFSTR. Thus, a true PFR has zero dispersion, a CFSTR has infinite dispersion, and in the PFR-with-dispersion model, nonideal reactors can be characterized by various, intermediate amounts of dispersion.

The predicted $E(t)$ function for a PFR with dispersion and with open boundaries can be determined by specifying that $x = L$ in Equation 2-70 and combining that equation with Equation 2-9 to obtain the following:

$$E(t)_{\text{open boundaries}} = \frac{1}{\tau\sqrt{4\pi(\mathbf{D}/v_x L)(t/\tau)}} \exp\left\{-\frac{(1-(t/\tau))^2}{4(\mathbf{D}/v_x L)(t/\tau)}\right\} \quad (2\text{-}71)$$

As noted previously, $E(t)$ for a system with closed boundaries is almost identical to that for a system with open boundaries for small values of $\mathbf{D}/v_x L$, so Equation 2-71 is applicable to those systems as well. However, if $\mathbf{D}/v_x L$ is larger than ~ 0.01 and the system of interest has closed boundaries, the $E(t)$ function cannot be expressed analytically and must be computed numerically.

Using Equation 2-71 or a numerical solution to the mass balance equation, we can predict the $E(t)$ curve for a reactor that behaves similar to a PFR with dispersion. Of course, the conceptual image of a reactor is still highly idealized, albeit not as idealized as a PFR with no dispersion. In reality, even reactors in which the flow pattern seems to approximate plug flow with dispersion have peculiarities that cause localized mixing beyond that induced by dispersion, or regions where the dispersion coefficient differs from elsewhere in the reactor. Therefore, real reactors rarely fit the PFR-with-dispersion model exactly. As a result, we often face the task of identifying the dispersion number that provides the best fit to experimental data and assessing how closely that best-fit model reproduces the observed behavior.

Traditionally, the approach used to estimate the dispersion number that characterizes a reactor's hydraulics relies on matching the normalized variance of the RTD based on experimental (i.e., tracer) data with the normalized variance based on model calculations, where the normalized variance, $\tilde{\sigma}^2_{\text{RTD}}$, is defined as the ratio of the variance to the square of the mean residence time. That is, the criterion for choosing a value of $\mathbf{D}/v_x L$ to model the reactor's hydraulics is that

$$\tilde{\sigma}^2_{\text{RTD,exp't}} = \tilde{\sigma}^2_{\text{RTD,model}} \quad (2\text{-}72)$$

where, for either experimental or model data

$$\tilde{\sigma}^2_{\text{RTD}} \equiv \frac{\sigma^2_{\text{RTD}}}{\tau^2} \quad (2\text{-}73)$$

(Keep in mind that the value of τ in Equation 2-73 is the *effective* residence time, defined as the ratio of the flow rate to the effective volume, not the geometric volume.)

To use this criterion, we first determine the experimental values of $\bar{t}$ and $\sigma^2_{\text{RTD,expt}}$ directly from tracer data or from the $E(t)$ curve derived from those data. To obtain the normalized value of the experimental variance, we need to divide $\sigma^2_{\text{RTD,expt}}$ by τ^2. If the reactor of interest has closed boundaries, then τ can be equated to the experimentally determined $\bar{t}$:

$$\tau\bigg|_{\substack{\text{any reactor with}\\\text{closed boundaries}}} = \bar{t} \quad (2\text{-}74)$$

On the other hand, if the reactor has open boundaries, τ is less than $\bar{t}$, with the exact relationship depending on how the mixing across the boundaries occurs. The only open-boundary case we will consider here is for a PFR with dispersion, for which the relationship between τ and $\bar{t}$ is

$$\tau\bigg|_{\substack{\text{PFR + disp.model,}\\\text{open boundaries}}} = \frac{V}{Q} = \{1 + 2(\mathbf{D}/v_x L)\}^{-1}\bar{t} \quad (2\text{-}75)$$

Combining Equations 2-74 and 2-75 with the definition of the normalized variance, the criterion for choosing $\mathbf{D}/v_x L$ in PFRs with dispersion and with closed or open boundaries becomes

PFR + dispersion, closed boundaries:

$$\frac{\sigma^2_{\text{RTD,exp't}}}{\bar{t}^2} = \tilde{\sigma}^2_{\substack{\text{RTD,PFR+disp. model,}\\\text{closed boundaries}}} \quad (2\text{-}76)$$

PFR + dispersion, open boundaries:

$$\left(1 + 2\frac{\mathbf{D}}{v_x L}\right)^2 \frac{\sigma^2_{\text{RTD,exp't}}}{\bar{t}^2} = \tilde{\sigma}^2_{\substack{\text{RTD,PFR+disp. model,}\\\text{open boundaries}}} \quad (2\text{-}77)$$

Statistical analysis of the model $E(t)$ curves for a PFR with dispersion yields the following equations for the normalized variance in reactors with closed and open boundaries, respectively:

$$\tilde{\sigma}^2_{\substack{\text{RTD,PFR+disp. model,}\\\text{closed boundaries}}} = 2\frac{\mathbf{D}}{v_x L} - 2\left(\frac{\mathbf{D}}{v_x L}\right)^2 \left\{1 - \exp\left(-\frac{v_x L}{\mathbf{D}}\right)\right\} \quad (2\text{-}78)$$

$$\tilde{\sigma}^2_{\substack{\text{RTD,PFR+disp. model,}\\\text{open boundaries}}} = 2\frac{\mathbf{D}}{v_x L} + 8\left(\frac{\mathbf{D}}{v_x L}\right)^2 \quad (2\text{-}79)$$

Finally, combining Equations 2-76 to 2-79, we obtain

PFR + dispersion, closed boundaries:

$$\frac{\sigma^2_{\text{RTD,exp't}}}{\bar{t}^2} = 2\frac{\mathbf{D}}{v_x L} - 2\left(\frac{\mathbf{D}}{v_x L}\right)^2 \left\{1 - \exp\left(-\frac{v_x L}{\mathbf{D}}\right)\right\} \quad (2\text{-}80)$$

PFR + dispersion, open boundaries:

$$\frac{\sigma^2_{\text{RTD,exp't}}}{\bar{t}^2} = \frac{2(\mathbf{D}/v_x L) + 8(\mathbf{D}/v_x L)^2}{\{1 + 2(\mathbf{D}/v_x L)\}^2} \quad (2\text{-}81)$$

Thus, once the experimental data are analyzed to determine the left side of Equation 2-80 or 2-81, the value of $\mathbf{D}/v_x L$ that satisfies the equality can be determined and used when modeling the reactor as a PFR with dispersion. To test the goodness-of-fit of the model to the data, a model $E(t)$ curve for the system can be developed using this dispersion number, and the curve can be compared with the experimental $E(t)$ curve. An example of this process for a closed reactor is provided in the following section; the corresponding steps for an open system are described in Problem 2-10.

One consequence of the fact that the open and closed-boundary cases yield different estimates for the effective residence time, τ, is that they also lead to different estimates of the volume that is accessible to the flowing water, and hence different values of c_o (the mass of tracer added divided by the accessible volume). Nevertheless, both models yield the same result when tracer data are converted to an $E(t)$ curve, based on the following equalities:

$$E(t) = \frac{c(t)}{\tau c_o} = \frac{c(t)}{(V/Q)(M/V)} = \frac{Q}{M}c(t) \quad (2\text{-}82)$$

Since the ratio Q/M is model-independent, Equation 2-82 demonstrates that a given experimental data set of $c(t)$ versus t corresponds to a unique $E(t)$ versus t curve, regardless of which model is being tested. The implication is that, even though the values of τ and c_o depend on whether one is modeling the system boundaries as open or closed, the product τc_o is the same in either case. The value of this product can be calculated as

$$c_o \tau = \frac{M}{Q} = \frac{\int_0^\infty Qc\,dt}{Q} = \int_0^\infty c\,dt \approx \sum_{\text{all } t_i}\left(\frac{c_i + c_{i-1}}{2}\right)(t_i - t_{i-1}) \quad (2\text{-}83)$$

The $E(t)$ curve predicted using the best-fit value of $\mathbf{D}/v_x L$ often replicates the experimental $E(t)$ curve reasonably well. Several other approaches for estimating $\mathbf{D}/v_x L$ are also available, some of which might generate a better match to the $E(t)$ data. For example, Haas et al. (1997) have argued that using nonlinear regression to estimate the best-fit values of τ and $\mathbf{D}/v_x L$ is better than matching the experimental and theoretical variance, as described previously. Regardless of which approach is used, it is important to understand that, in some cases, no value of $\mathbf{D}/v_x L$ will provide a good fit to the tracer data; in such cases, we must conclude that the reactor hydraulics do not conform to the PFR-with-dispersion model at all.

In cases where the dispersion number $\mathbf{D}/v_x L$ is small (say, <0.01), the underlying concept on which the model is based—small parcels of water mixing small distances upstream and downstream in a completely random manner—might be fairly realistic, and the dispersion coefficient $\mathbf{D}$ can be interpreted as an indicator of the intensity of that mixing. These cases include most packed bed reactors (including groundwater systems), pipes flowing full under turbulent conditions, and open channels with high length/width and length/depth ratios. Other systems (other open channels, pipes with laminar flow or not flowing full, and tanks with or without mechanical agitation) are characterized by larger dispersion numbers. In these systems, mixing is likely to be dominated by large-scale phenomena, and the dispersion coefficient has little or no mechanistic significance. In these cases, the dispersion model is still frequently applied, but the dispersion number should be viewed as an empirical fitting parameter that is the only parameter of significance in the model; a computed diffusion coefficient has no independent meaning.

■ **EXAMPLE 2-4.** Assuming that the reactor characterized by the tracer data in Example 2-1 has closed boundaries, estimate the dispersion number for the reactor, and compare the experimental $E(t)$ curve with that predicted by the PFR-with-dispersion model.

Solution. The variance of t was estimated in Example 2-2 as 760 min^2, based on the experimental data. Dividing this value by $\bar{t}^2$ (i.e., 49.5^2 min^2) gives a value for the normalized variance of 0.31. Solving Equation 2-78 by trial and error, the dispersion number corresponding to this value of the normalized variance is 0.19.

The model $E(t)$ curve for a reactor with closed boundaries and a dispersion number of 0.19 can be developed by interpolation of the curves for dispersion numbers of 0.1 and 0.2 in Figure 2-16, but for our purposes, comparison with the curve for $\mathbf{D}/v_xL = 0.2$ is adequate. Using that curve and substituting $\tau = 49.5$ min to transform the values on the axes allows us to plot the $E(t)$ versus t curve that we would obtain for the example system, if the system followed the dispersion model exactly. Such a curve is compared with the experimental $E(t)$ data in Figure 2-17. The model prediction follows the general pattern of the data set, but the match is not excellent. Depending on how we intend to use the model, the absence of an excellent fit might or might not be of concern.

To the extent that the model fails to fit the data, we can infer that processes other than diffusion and random, eddy dispersion contribute to mixing of the fluid in the reactor. Depending on the intended use of the model, the discrepancy between the model and the data might or might not be of concern. ■

CFSTRs in Series A second model that is sometimes used to describe nonideal flow in a reactor represents the reactor as a series of N equal-sized smaller CFSTRs. To envision the model, one can imagine installing baffles in a reactor so that the water follows a path from one part of the reactor to the next with no possibility of returning to an earlier section. A schematic representation of the model is shown in Figure 2-18. With no baffles (Figure 2-18a), the reactor is a single CFSTR; as already discussed, some molecules of water appear in the effluent of such a reactor immediately after they enter. With one baffle, the reactor has two sections. In such a case, even if each section is well mixed, there is less overall mixing than in the system without the baffle. With three baffles, the reactor has four sections (as in

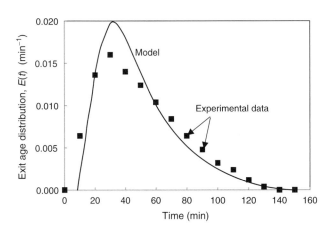

FIGURE 2-17. Experimental $E(t)$ values for the example data set, and predicted $E(t)$ curve based on the PFR-with-dispersion, closed boundary model, with $\mathbf{D}/v_xL = 0.2$.

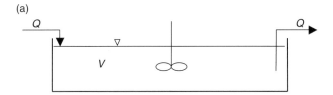

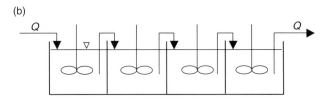

FIGURE 2-18. (a) A single CFSTR with volume V, and (b) the same tank divided into four equal-sized ($V/4$) CFSTRs in series.

Figure 2-18b), and still less mixing. In such a case, molecules that enter the first section at some instant are unlikely to mix with those that have been in the system for a long time, as the latter molecules are predominantly in downstream sections.

As the number of sections, N, increases, the mixing that takes place in each CFSTR affects smaller and smaller volumes of water. At very large N, a parcel of liquid moves through the series of reactors, mixing minimally with liquid ahead of or behind it. That is, as N increases, the reactor behaves increasingly like a PFR. This model is, logically enough, excellent at fitting tracer data for reactors that have real baffles (as, for example, disinfection reactors frequently do), but it is also useful for fitting tracer data for reactors with no physical baffles. In the latter case, the hydraulic equivalent of physical baffles might be provided by the combined effects of the arrangement of entry ports, the momentum of fluid entering the reactor, and the details of the effluent collection system.

A guide to the derivation of the expressions for $E(t)$ and $F(t)$ for this model is provided in Problem 2-4 at the end of this chapter. The results for any integer value of N are as follows:

$$E(t) = \frac{1}{\tau} \frac{N^N}{(N-1)!} \left(\frac{t}{\tau}\right)^{N-1} \exp\left(-\frac{Nt}{\tau}\right) \quad (2\text{-}84)$$

$$F(t) = 1 - \exp\left(-\frac{Nt}{\tau}\right) \sum_{i=1}^{N} \frac{1}{(i-1)!} \left(\frac{Nt}{\tau}\right)^{i-1} \quad (2\text{-}85a)$$

or

$$F(t) = 1 - \exp\left(-\frac{Nt}{\tau}\right) - \frac{Nt}{\tau}\exp\left(-\frac{Nt}{\tau}\right) - \frac{1}{2!}\left(\frac{Nt}{\tau}\right)^2$$
$$\times \exp\left(-\frac{Nt}{\tau}\right) - \cdots - \frac{1}{(N-1)!}\left(\frac{Nt}{\tau}\right)^{N-1} \exp\left(-\frac{Nt}{\tau}\right) \quad (2\text{-}85b)$$

where τ is the hydraulic detention time of the complete tank and N is the (conceptual) number of tanks into which the complete tank is divided.

Families of curves for both the exit age distribution, $E(t)$, and the cumulative age distribution, $F(t)$, are shown for different values of N in Figure 2-19. For generality (i.e., to make the curves independent of τ), the dimensionless product $\tau E(t)$ is used as the ordinate in Figure 2-19a; according to Equation 2-9, this ordinate is equivalent to $c_p(t)/c_o$, where $c_o = M_{p,\text{in}}/V$. Hence this ordinate can be thought of as a dimensionless concentration for a pulse input tracer test. As N increases, both distributions shift from the RTD characterizing a CFSTR ($N=1$) to that characterizing a PFR ($N=\infty$).

To use the CFSTRs-in-series model, the parameter N must be determined. Haas et al. (1997) suggested that the best means to fit the model is to use nonlinear regression directly with the data to find the best values of $\bar{t}$ and N. However, as in the case of the dispersion model, it is more common to use the normalized variance of the data from a tracer test to find a value of N. The variance of the residence time in a system consisting of N CFSTRs in series can be computed by substituting Equation 2-84 into Equation 2-27. Integration of the resulting equation yields the following theoretical relationship between N and the normalized variance of t:

$$N = \frac{\bar{t}^2}{\sigma_{\text{RTD}}^2} = \frac{1}{\tilde{\sigma}_{\text{RTD}}^2} \quad (2\text{-}86)$$

CFSTRs have closed boundaries by definition, since, if they did not, the complete mixing would extend indefinitely in the direction(s) where the boundary was open. Therefore, according to Equation 2-76, $\tau = \bar{t}$ for such reactors, and the normalized variance can be computed by dividing the absolute variance by $\bar{t}^2$. The use of this approach for estimating N is demonstrated in the following example.

■ **EXAMPLE 2-5.** Estimate the best-fit number of equal-sized CFSTRs in series to model the data in Example 2-1, and compare the model-predicted $E(t)$ curve with the experimental $E(t)$ curve, as was done in Example 2-4 for the PFR-with-dispersion model.

Solution. The normalized variance of the experimental $E(t)$ curve was determined to be 0.31 in Example 2-4. The best-fit number of equal-sized CFSTRs in series is the inverse of this number, or 3.22. While the number of CFSTRs in a real system would obviously have to be an integer, this is not necessarily a restriction for a hypothetical system. That is, if our goal is simply to use the CFSTRs-in-series model to describe the tracer pattern from the real reactor as best, we can in a compact form, treating N as a continuous variable might be acceptable.

Equation 2-84 gives the value of $E(t)$ for N ideal CFSTRs in series; for comparison with experimental data, however, $\bar{t}$ should be substituted for τ. Thus, the $E(t)$ curve for 3.22 CFSTRs can be estimated by the following modification of Equation 2-84, with $N = 3.22$:

$$E(t) = \frac{1}{\bar{t}} \frac{N^N}{(N-1)!} \left(\frac{t}{\bar{t}}\right)^{N-1} e^{-Nt/\bar{t}} \quad (2\text{-}87)$$

The value of $(N-1)!$ is defined only for integral values of N. However, a continuous function that is closely related to $(N-1)!$ has been identified that allows one, in essence, to estimate values of $(N-1)!$ when N is not an integer. This function, called the gamma function, $\Gamma(x)$, has the property that, if x is an integer $\Gamma(x) = (N-1)!$. Thus, for instance, $\Gamma(5) = 4!$. The value of $\Gamma(3.22)$, therefore, is a good approximation of 2.22!. Values of $\Gamma(x)$ are available in most spreadsheet programs, from which we can find

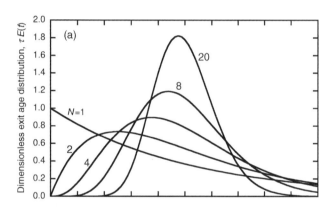

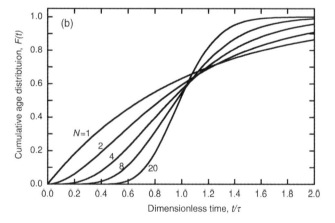

FIGURE 2-19. Residence time distribution for the CFSTRs-in-series model, for various values of N. As noted previously, t/τ can be thought of as dimensionless time, and $\tau E(t)$ can be thought of as the dimensionless concentration exiting a reactor in a pulse tracer test.

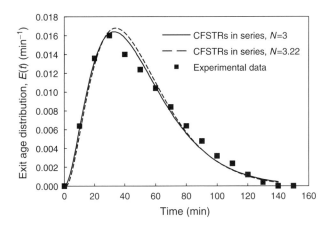

FIGURE 2-20. Experimental $E(t)$ curve and model curves for the same data using the CFSTRs-in-series model.

$\Gamma(3.22) = 2.47$. By inserting this value into Equation 2-87 for $(N-1)!$, along with the value of $\bar{t}$ computed in Example 2-1, we can calculate the expected $E(t)$ curve for $N = 3.22$. The $E(t)$ curves for N values of 3.00 and 3.22 are compared with the experimentally determined $E(t)$ values in Figure 2-20. The curves for $N = 3$ and $N = 3.22$ are not substantially different from one another, and both match the experimental data quite well. Thus, for this particular (hypothetical) data set, the CFSTRs-in-series model appears to match the experimental data significantly better than the PFR-with-dispersion model, when both models are constrained to match the experimental mean hydraulic detention time. ■

Modeling Short-Circuiting and Dead Space The preceding discussion describes how various degrees of mixing in a system might be represented, based on the implicit assumptions that all the fluid participates in the same flow pattern and that the entire volume of the reactor is involved in that flow pattern. However, either of these assumptions might not be valid in a given reactor. Two specific cases can be identified in which these assumptions are violated, and the terms *dead space* and *short-circuiting* are used to describe these cases.

Dead space is the simpler of the two cases to describe. As the name implies, this condition means that some part of the available space in the reactor is not used. The water in that portion of the reactor is essentially stagnant, and that portion of the reactor might as well be filled with concrete as with water. The effective volume of a reactor with dead space is less than the actual volume, and the entire flow passes through the effective volume. When this problem occurs, it usually is associated with corners or other specific regions of tanks where mixing is very poor. It can also occur because of differences in the density between the influent and water already in the tank, a situation that is most likely to occur in tanks with no mechanical mixing, such as tanks designed for sedimentation. If the influent water is at a different temperature and therefore has a different density than the water in the reactor, it can sink to the bottom or float across the top of the water in the tank, reducing the effective volume of the reactor.

In reactors with closed boundaries, dead space is directly quantifiable from tracer studies. In any such reactor, if all of the volume is used in the flow pattern (regardless of that flow pattern), the mean detention time of the tracer, $\bar{t}$, should equal the theoretical detention time, $\tau (= V_{tot}/Q)$. In a reactor with dead space, the entire flow Q passes through the effective reactor volume, V_{eff}, where $V_{eff} < V_{tot}$. The dead space, V_{dead}, is the difference between the total and effective volumes; that is,

$$V_{dead} = V_{total} - V_{eff} \quad (2\text{-}88)$$

Then, since $\bar{t}$ equals V_{eff}/Q, by definition

$$\frac{\bar{t}}{\tau} = \frac{V_{eff}/Q}{V_{tot}/Q} = \frac{V_{eff}}{V_{tot}} \quad (2\text{-}89)$$

The fractional dead space, m, is defined as V_{dead}/V_{tot} and can thus be determined from tracer information as follows:

$$m = 1 - \frac{V_{eff}}{V_{tot}} = 1 - \frac{\bar{t}}{\tau} \quad (2\text{-}90)$$

■ **EXAMPLE 2-6.** The results of a pulse input tracer test on a reactor with closed boundaries, using a dye as the tracer, are provided in the following table. The volume of water in the reactor is 50 m³, and the influent and effluent flow rates during the test were both 0.1 m³/min.

Time (h)	0	1	2	3	4
Dye concentration (mg/L)	35	31.6	24.1	21.6	17.9

Time (h)	5	6	7	8	9	10
Dye concentration (mg/L)	17.2	13.8	11.9	11.1	9.1	7.6

(a) Determine whether the reactor behaves as a CFSTR.
(b) Estimate the mean hydraulic residence time in the reactor based on: (i) the volume and flow rate, (ii) Equation 2-45b, and (iii) Equation 2-17. Discuss the different estimates and indicate the one that you find most trustworthy.

Solution.

(a) Whether a reactor behaves as a CFSTR can be determined by comparing the outcome of a tracer test with the expected outcome for an ideal CFSTR. According to Equation 2-45a, a plot of $\ln c(t)$ versus t for an ideal CFSTR should be linear with a slope of $-1/\bar{t}$ and an intercept of $\ln c_o$. The data are plotted using semilogarithmic coordinates in Figure 2-21 and are seen to fit the straight line expectation quite well. The best-fit slope is $-0.149/h$, corresponding to a hydraulic residence time of 6.7 h.

(b) The theoretical mean hydraulic residence time of a reactor is V/Q, which for this reactor is 500 min, or 8.33 h. The residence time computed according to Equation 2-45 is the result obtained in part (a); that is, 6.7 h. A numerical estimate of the residence time, based on Equation 2-17, suggests that the residence time is smaller still: only 3.8 h. These values are very different from one another, so it is important to determine which one, if any, is right.

A reasonable hypothesis for the difference between the first estimate (based on geometry) and the other two is that the effective volume of the reactor is significantly less than the total volume; that is, that some portion of the reactor is not participating in the mixing regime. As a result, the total flow is passing through only a portion of the geometric volume, causing the actual residence time to be less than that calculated based on the system geometry.

The explanation for the difference between the estimates of $\bar{t}$ using Equations 2-45 and 2-17 is a bit more subtle, and is based on the fact that the graphical approach is based on the *change* in tracer concentration during the period of analysis, whereas the numerical method represents an attempt to integrate the tracer data over the entire time period. That is, the numerical approach involves approximating the values of $\int_0^\infty ct\,dt$ and $\int_0^\infty c\,dt$ as $\sum_0^{t_{max}} ct\,dt$ and $\sum_0^{t_{max}} c\,dt$, respectively. The approximation of the integrals as summations turns out to be acceptable; the problem arises from the fact that the summation must extend to $t=\infty$, or at least to the time at which (nearly) all the tracer has exited the reactor. From the fact that the tracer concentration in the final sample is $>20\%$ of c_o, it is clear that this requirement is not met. The data at long residence times have a disproportionate effect on the estimation of $\bar{t}$, so that computing $\bar{t}$ without sufficient data at long times can lead to a severe underestimate.

In contrast, the graphical approach to estimating $\bar{t}$ (or a corresponding numerical approach) relies on the existing data rather than a simulated integration to long times, and therefore is not subject to the same bias. The result highlights the greater reliability of this latter approach. It also suggests that, if we have to use a numerical integration to estimate $\bar{t}$, it is probably worthwhile to extrapolate the data to an estimated time when all the tracer has exited the reactor; such an extrapolation certainly introduces some error, but it is still likely to improve the estimate of $\bar{t}$. ∎

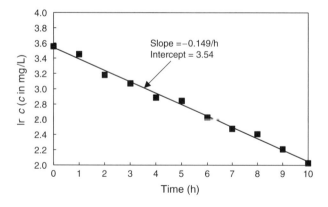

FIGURE 2-21. Data from example reactor, plotted as $\ln c$ versus time.

Short-circuiting is a term that has been used with less precision than dead space. Generally, it means that at least some of the fluid follows a different path through the reactor than the intended one and exits after a shorter-than-desired residence time. Here, we use the term to describe a situation in which the system behaves as though the effective volume of the reactor is divided into two or more sections in parallel. Short-circuiting can occur, for instance, in packed bed reactors when some portion of the flow travels along the walls of the reactor while the rest travels through the packed bed. In open reactors, the influent distribution system might direct a substantial portion of the flow through a small segment of the reactor, and a smaller portion through a larger segment; the former portion would be the short-circuiting portion of the flow. Similarly, the weir system that controls the effluent might draw a substantial portion of the flow through a small portion of the reactor, creating an imbalance in the spread of the flow throughout the reactor volume.

If flow through a reactor can, in fact, be represented as a composite of two or more distinct portions, then at least one portion of the flow will almost certainly have a mean hydraulic residence longer than the value characterizing the overall reactor ($\tau_{overall}$), and another portion will have a mean residence time shorter than $\tau_{overall}$. The portion(s) with mean residence time shorter than $\tau_{overall}$ is said to be short-circuiting.

Note that, in a CFSTR, a portion of the influent exits after a very short time due to random mixing processes. However, according to the definition used here, this scenario would not be described as short-circuiting, because that term is used only to describe situations in which a portion of the influent is envisioned to follow a defined, nonrandom path through the reactor to the effluent port.

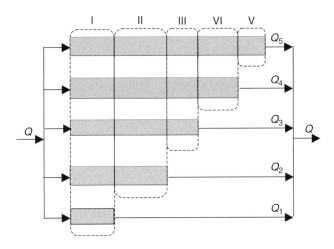

FIGURE 2-22. Schematic of PFRs-in-parallel model. The meaning of the dashed lines is discussed in the text.

PFRs in Parallel and Series: Segregated Flow and Early Versus Late Mixing Often, mixing patterns are more complex than those represented by any of the models described above. In such cases, a more complex network of ideal reactors must be used to model mixing in the real reactor. One conceptual approach for modeling complex mixing is to restrict the model network to several PFRs in parallel. Theoretically, one could define a different PFR for each parcel of water entering, and thereby accurately model the exit age distribution of virtually any real system.[12] A schematic showing a reactor that is represented by five PFRs in parallel, with various amounts of flow entering the reactors is shown in Figure 2-22.

The $E(t)$ curve for the model system represented in Figure 2-22 is a linear summation of the $E(t)$ curves for the individual reactors, weighted according to the flows through those reactors. That is, if 30% of the flow into the overall reactor could be modeled as though it passed through a PFR with a residence time of 10 min, then in a pulse input tracer test, 30% of the tracer input would enter that portion of the reactor and exit as a spike 10 min later. Others of the hypothetical reactors would contribute to the $E(t)$ curve at times corresponding to their respective residence times, with the contributions reflecting the fraction of the flow passing through each segment. The result is an $E(t)$ function for the model system consisting of five spikes, as shown in Figure 2-23a.

It is not possible to identify a set of Q_i and τ_i values for the hypothetical reactors that would cause the $E(t)$ function for the model system to match that of the experimental system, because the model $E(t)$ curve comprises a series of infinitely high and thin spikes. However, an appropriate set of Q_i and τ_i values can be chosen so that the model $F(t)$ curve is well matched to the experimental $F(t)$ curve. As can be seen in

[12] This statement is valid only if parcels of fluid that enter the reactor remain as coherent, distinct parcels until they leave it. This situation is known as *segregated* flow, and is discussed in more detail later in this section.

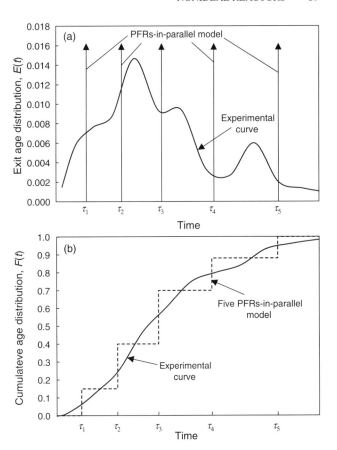

FIGURE 2-23. A residence time distribution modeled as five PFRs in parallel. (a) $E(t)$ curves; (b) $F(t)$ curves. The values of τ_i correspond to the residence times in the model reactors. The spikes in (a) are infinitely thin and high. The area under each spike in (a) and the height of each step in (b) equal the fraction of the total flow that passes through the corresponding reactor.

Figure 2-23b, the $F(t)$ curve for the model system consists of a stair-step plot; the height of each step corresponds to Q_i/Q_{tot}, and the location of the step on the abscissa corresponds to τ_i. It should be apparent that we could model any empirical $F(t)$ curve to any chosen degree of accuracy by using enough model PFRs in parallel. Thus, the PFRs-in-parallel model can represent either of the ideal extremes of mixing. A true PFR is represented as a single PFR that has residence time τ and receives all the flow. On the other hand, an ideal CFSTR is represented as an infinite number of PFRs; if the PFRs were assigned equally spaced values of τ_i, the fraction of the flow through the reactors would decrease exponentially with increasing τ_i. The model can also represent degrees of mixing that are intermediate between these two extremes or much more complex mixing patterns.

While the PFRs-in-parallel model is more flexible than the models discussed previously in terms of its ability to model a wide range of tracer output curves, it also has significant drawbacks. The two models of ideal reactors (ideal PFRs or CFSTRs) are each characterized by a single parameter, τ. The PFR-with-dispersion and CFSTRs-in-series models each add an adjustable parameter (dispersion number or number of equal-sized CFSTRs in series) to incorporate and characterize the degree of nonideality. In contrast, the PFRs-in-parallel model contains an arbitrarily large number of fitting parameters: the volume and flow rate through each of the hypothetical reactors. Because of this, the PFRs-in-parallel model has the advantage that it is capable of modeling the RTD in any reactor and the disadvantage that it is more complex and has more parameters to be evaluated than the other models presented here.

It is instructive to redraw Figure 2-22 so that the various sections of reactors that operate in parallel are merged; that is, each group of reactor sections enclosed in dotted lines in Figure 2-22 is treated as a single PFR. In such a case, the conceptual picture of the system is converted to a group of PFRs in series (Figure 2-24a). However, only a portion of the flow exiting each reactor enters the subsequent reactor;

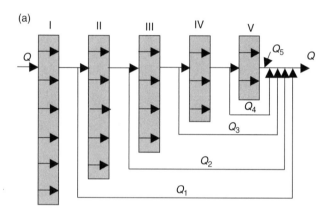

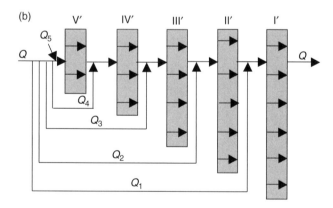

FIGURE 2-24. PFRs in series with multiple (a) output or (b) input points. The Roman numerals refer to the segments enclosed in dashed lines in Figure 2-22; each combination of segments enclosed by a dashed line in Figure 2-22 is represented as an individual PFR in Figure 2-24a. The roman numerals in (b) indicate reactors that have the same volumes and flow rates as the corresponding reactors in (a), for example, reactors I and I′ have the same volume and flow rate.

the remainder of the flow is shunted directly to the effluent of the entire reactor network, where it mixes with the flow that has passed through the other reactors. The RTDs for the reactors shown in Figures 2-22 and 2-24a are identical: in each case, the influent water enters the entire reactor system, portions of the water exit the system at various times thereafter, and all the portions are mixed at the most downstream point in the system.

PFRs can be combined in yet another way to yield the same tracer response as the systems characterized in Figures 2-22 and 2-24a. Specifically, Figure 2-24b describes a reactor system that is very similar to that of Figure 2-24a, but differs in a subtle way. In Figure 2-24a, all of the flow *enters* the system together, and portions of the flow are then steadily withdrawn and shunted past the *downstream* part of the system to the outlet. In contrast, in Figure 2-24b, portions of the influent are shunted past the *upstream* part of the system, and all the flow *exits* the system together. In fact, the system described by Figure 2-24b is simply the system described by Figure 2-24a run in reverse. If we eliminate the space between the various reactors, we see that Figure 2-24a can be described as a single PFR with multiple outlets placed along the length of the reactor, and Figure 2-24b can be described as a single PFR with multiple inlets (Figure 2-25). Once again, either purely conceptual arguments or a mathematical analysis can be applied to predict the tracer response of these systems. The conclusion is that the $E(t)$ functions for all the systems in Figures 2-22, 2-24, and 2-25 are identical.

In the system with multiple outlets along the reactor length (which is equivalent to a system with several independent PFRs in parallel), the separate flow streams are mixed at the latest possible point (the final effluent point). This model is sometimes referred to as the *complete segregation* or *segregated flow* model. In contrast, in the system with multiple inlets (Figure 2-24b or 2-25b), the streams are

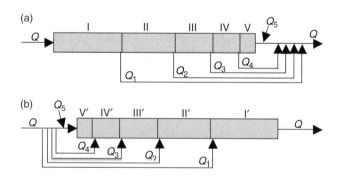

FIGURE 2-25. Representations of Figure 2-24a and 2-24b, respectively, as single reactors with multiple outlets or inlets. Subparts a and b represent complete segregation and maximum mixedness, respectively. The broken lines inside the reactor indicate the ends of the (conceptual) subreactor sections.

mixed at the earliest possible point that is consistent with the tracer data; this system is sometimes referred to as the *maximum mixedness* model. As shown in Chapter 4, despite responding identically in tracer tests, these systems sometimes behave quite differently if the input is a reactive substance. The conceptual description of a CFSTR, in which the influent is instantly and completely mixed with all the fluid in the reactor, is a maximum mixedness model.

■ **EXAMPLE 2-7.** Describe the parameters of a model reactor system containing five PFRs in parallel that might be a reasonable representation of the reactor described in Example 2-1. Compare the model-predicted $E(t)$ and $F(t)$ curves with the corresponding experimental curves, as was done earlier for the PFR-with-dispersion and CFSTRs-in-series models.

Solution. As in the previous approaches for modeling the experimental data, the goal is to define a model system that has RTD functions similar to the experimental ones. However, unlike those approaches, the use of the PFRs in parallel model does not involve use of the variance of $E(t)$ curve. The expected tracer output from a group of five PFRs in parallel is simply five spikes, each infinitely high and infinitely thin. Clearly, no matter how we configure the model system, an $E(t)$ curve that consists of five such spikes and values of zero at all other times will not match the experimental $E(t)$ curve.

The best approach for carrying out the modeling exercise is to try to match the predicted and experimental $F(t)$ curves, rather than the $E(t)$ curves. The $F(t)$ curve for the model system will be a stair–step function, because each model PFR contributes tracer to the effluent only at its residence time. The fraction of the total tracer that enters a given reactor i in the model system is Q_i/Q_{tot}. Therefore, $F(t)$ of the model system will be zero at all times less than τ_1, where τ_1 is the shortest hydraulic residence time of any reactor in the system. At $t = \tau_1$, $F(t)$ will jump to a value equal to Q_1/Q_{tot}, where it will remain until $t = \tau_2$, the second shortest residence time of the five reactors. At that time, $F(t)$ will instantly increase to $(Q_1 + Q_2)/Q_{tot}$. This pattern will continue until t equals the longest residence time of the five PFRs, at which point $F(t)$ becomes 1.0.

The modeling approach is therefore to draw the experimental $F(t)$ curve and to overlay a second curve that consists of five steps. The time and height of these steps can be chosen to provide the best match to the experimental data; often an eyeball fit is satisfactory. Such a plot is shown in Figure 2-26.

The times of the five vertical increases in $F(t)$ correspond to the residence times of the five model reactors, and the increase in $F(t)$ at each step indicates the fraction of the total flow that passes through the corresponding conceptual

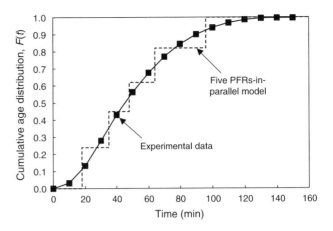

FIGURE 2-26. Experimental and model $F(t)$ curves. The model curve is for a system of five PFRs in parallel.

reactor. The model parameters can therefore be summarized as follows:

Reactor No.	τ (min)	Q_i/Q_{tot}
1	18	0.24
2	35	0.21
3	48	0.17
4	64	0.20
5	96	0.18

The model $E(t)$ curve is shown in Figure 2-27. Although the $E(t)$ curve for the model system does not match the experimental curve well, the $F(t)$ curves are reasonably well matched. It should be apparent that by increasing the number of hypothetical reactors being considered, we could improve the fit of the model $F(t)$ curve to the point where it provided an extremely good fit to the data, but the $E(t)$ curves could never be made to match. Note, however, that the inclusion of each additional reactor requires the assignment of two additional fitting parameters. ■

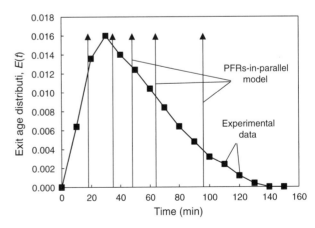

FIGURE 2-27. Experimental and model $E(t)$ curves. The five spikes are for a system of five PFRs in parallel.

Nonequivalent CFSTRs in Series Complex mixing can be modeled in numerous other ways. For instance, one might add flexibility to the CFSTR-in-series model by allowing the individual reactors to have different volumes. In such a model, the volume of each reactor would have to be specified, rather than just the total number of hypothetical reactors. Ultimately, no matter how complex the hydraulic flow pattern in a reactor, it can always be represented by some combination of CFSTRs and PFRs, in some combination of parallel and series connections. However, identifying a network of such reactors that is both physically reasonable and leads to a mathematically tractable mass balance can be difficult. Furthermore, with each additional component in the model system, additional fitting parameters need to be used to describe it, and the use of more than three or four fitting parameters is rarely justified by available data, unless the physical structure of the reactor provides support for the proposed model.

Simple Indices of Hydraulic Behavior As an alternative to characterizing the complete RTD of a reactor by the $E(t)$ and $F(t)$ functions, simpler approaches are sometimes used, in cases where detailed information about the RTD is not essential. For instance, in some circumstances, the times associated with specific values of the cumulative age distribution are adequate to convey the information of interest about the hydraulic behavior of the reactor. The most common of these indicators are the times associated with $F(t)$ equaling 0.1, 0.5, and 0.9; these times are often designated as T_{10}, T_{50}, and T_{90}, respectively, where the fractional values of $F(t)$ are converted to percentages when presented as the subscripts of the times, T. Various ratios of these values, to each other or to τ, are also sometimes used. For example, the Morrill index (Morrill, 1932) is the ratio T_{90}/T_{10}. This parameter was first used in a study of sedimentation tanks, but it is useful as an indicator of the dispersion characteristics of any reactor. The minimum value of 1.0 represents a reactor with nearly plug flow, and, in general, higher values suggest a reactor with more mixing. Some investigators have also used the inverse of this ratio.

Currently, the most common simple indicator of the hydraulic characteristics of a reactor is the value of T_{10} and the associated ratio T_{10}/τ. This ratio indicates how quickly (relative to the theoretical detention time) 10% of the water that enters at some instant leaves the reactor. Values of this ratio that are less than, say, 0.3 indicate a large amount of mixing and/or short-circuiting, whereas values greater than, say, 0.7 indicate a reactor in which little or no water makes it quickly from the influent to the effluent. The theoretically maximum value of the ratio is 1.0, which corresponds to a PFR. Hence, this ratio can give one a rough sense of the entire RTD.

In current U.S. drinking water regulations (Federal Register, 1989), the value of T_{10} is used as part of a calculation to determine whether a treatment plant is meeting the regulatory requirement for disinfection. The relationship to disinfection is discussed in Chapters 4 and 10, but the regulation, in essence, considers any disinfection reactor as if it were a PFR with detention time equal to T_{10}. Such a characterization is an oversimplification of the reality, but it is very convenient for purposes of enforcement. The *Guidance Manual* (USEPA, 1991) that accompanies the regulation includes estimates for the T_{10}/τ ratio for various types of reactors.

TABLE 2-3. Values for Indicators of Nonideal Flow for Specific Reactors

Reactor Type	$\dfrac{T_{10}}{\tau}$	$\dfrac{T_{50}}{\tau}$	$\dfrac{T_{90}}{\tau}$	$\dfrac{T_{90}}{T_{10}}$ (Morrill Index)
Plug flow	1	1	1	1
CFSTR	0.105	0.693	2.30	21.9
One reactor as three CFSTRs in series	0.367	0.891	1.77	4.82

■ **EXAMPLE 2-8.** Find the values of T_{10}/τ, T_{50}/τ, and the Morrill Index, T_{90}/T_{10}, for a PFR, a single CFSTR, and a single tank that behaves as three CFSTRs in series.

Solution. For a PFR, $T_{10} = T_{50} = T_{90} = \tau$, so all of the ratios equal 1.0. For a single CFSTR, according to Equation 2-49,

$$F(t) = 1 - \exp\left(-\frac{t}{\tau}\right)$$

and therefore $t/\tau = -\ln(1 - F(t))$. For $F(t)$ equal to 0.1, 0.5, and 0.9, the resulting values for t/τ are shown in Table 2-3.

For the reactor that behaves as three CFSTRs in series, according to Equation 2-85,

$$F(t) = 1 - e^{-3t/\tau} - \frac{3t}{\tau}e^{-3t/\tau} - \frac{9t^2}{2\tau^2}e^{-3t/\tau}$$

$$1 - F(t) = \left(e^{-3t/\tau}\right)\left(1 + \frac{3t}{\tau} + \frac{9t^2}{2\tau^2}\right)$$

Trial and error solution of this equation for the required values of $F(t)$ yields the values shown in Table 2-3. The results indicate that the 3-CFSTR reactor is intermediary between the two ideal reactors, as expected. ■

2.5 EQUALIZATION

The principles described in this chapter are applied directly in equalization, a pretreatment process sometimes used in water and wastewater treatment plants. Equalization is the process of reducing or eliminating temporal fluctuations in the water flow rate and/or the concentration of some constituent. For flow equalization, a tank in which the fluid volume can vary with time is used to dampen the flow fluctuations. The most common application is to equalize the influent variations in wastewater flow[13]; ideally, the effluent flow from this tank is the average flow. When the influent flow is greater than the average flow, the amount of water stored in the tank increases and, when the influent flow is less than average, the water level decreases. Similarly, for concentration equalization of a conservative chemical, a CFSTR with constant volume can dampen the concentration variations. When the influent concentration is greater than the concentration in the tank, the concentration in the tank (and its effluent) increases, but it remains less than that of the influent. When the influent concentration is less than that in the tank, the concentration in the tank decreases, while remaining greater than that of the influent. When equalization of both flow and concentration is required, a CFSTR with varying water volume is used.

Equalization is a pretreatment process; that is, it accomplishes no treatment (removal of pollutants) itself. As a result, its use can only be justified based on the effect of variation of flow or influent concentration on downstream treatment processes. For example, fluctuations in the flow rate will lead to fluctuation in the mean detention time (and the entire RTD) and therefore might adversely affect the removal efficiency; such a possibility is considered in Chapter 4. Also, changes in the influent concentration will change the effluent concentration in most treatment processes, as is demonstrated for specific processes in subsequent chapters. Equalization can dampen (or, ideally, eliminate) these effects. Equalization can also make operation of treatment processes simpler, particularly when a chemical must be fed to a process in stoichiometric proportion to the mass flux of a pollutant in the influent. The design question is: How large must the equalization volume be to reduce the fluctuations in the flow or concentration to acceptable levels. To address this issue, the effect of time-variant influent conditions on subsequent treatment processes must be assessed.

Flow Equalization

When considering flow equalization, the composition of the influent is irrelevant; the mass balance of interest is for water alone. The mass balance is written using the entire reactor as the control volume. In words, the mass balance is as follows:

Rate of change of mass of water in the reactor = Mass rate at which water in enters the reactor − Mass rate at which water exits the reactor

Translating into symbols

$$\frac{d\rho_L V(t)}{dt} = \rho_L Q_{in}(t) - \rho_L Q_{out}(t) \quad (2\text{-}91)$$

where ρ_L is the density of water. Recognizing that this density is a constant converts Equation 2-91 into a volume balance for water in the tank:

$$\frac{dV(t)}{dt} = Q_{in}(t) - Q_{out}(t) \quad (2\text{-}92)$$

As written, V, Q_{in}, and Q_{out} are all functions of time. The goal of flow equalization is to fix Q_{out} at a constant value, namely its average value over some time period, t'. The choice of that time period depends on the situation. For an industry with large flow variations over short cycles, a several hour period might be a satisfactory choice for t'. For plants with a diurnal cycle, a 24-h period is likely the best choice. For industries with negligible flow on weekends, storing water for the weekend (especially, for example, to keep a microorganism population alive in a biological treatment facility) might necessitate a choice of several days or a week as the averaging period. Once that choice is made (and assuming data are available), the determination of the average flow is straightforward.

Assuming that equalization works perfectly, Q_{out} can be set at the (constant) average flow (Q_{ave}) over the selected period. With that substitution, Equation 2-92 can be rearranged as follows:

$$\int_{V(0)}^{V(t)} dV = \int_0^t Q_{in}(t)dt - Q_{ave}\int_0^t dt \quad (2\text{-}93)$$

Upon integration, the result is

$$V(t) - V(0) = \int_0^t Q_{in}(t)dt - Q_{ave}t \quad (2\text{-}94a)$$

$$V(t) = V(0) + \int_0^t Q_{in}(t)dt - Q_{ave}t \quad (2\text{-}94b)$$

In words, Equation 2-94b states that the volume of water in the reactor at time t is the volume present at time zero, plus the cumulative volume of water input to the reactor since time zero, minus the cumulative volume of water taken out of the reactor since time zero. The design problem is to find the maximum value of $V(t)$ and provide that much storage capacity. Because the choice of time zero is arbitrary

---

[13] Flow equalization is also common in drinking water plants. In that case, the equalization is generally placed *after* the treatment train, meaning that the influent flow (from the treatment processes) is constant and the effluent flow varies in response to demand of customers in the distribution system. These post-treatment equalization tanks are both at the plant and in the distribution system. Elevated storage tanks are an example of such equalization facilities. The wastewater problem (varying influent flow, constant effluent flow) is used throughout the following analysis, but the end result (with a change in subscripts for in and out) is equally valid for the drinking water situation (constant inflow, varying outflow).

in a continuous flow reactor, $V(0)$ is also unknown and reflects the history prior to that time.

Both sides of Equation 2-94a represent the cumulative difference between the input and output from (the arbitrary) time zero to time t. If a set of flow values at various times (i.e., a data set describing $Q_{in}(t)$ vs. t) is available, the cumulative difference can be calculated using the right side of the equation. When the cumulative difference is positive, the basin contains more water than at time zero, and when the cumulative difference is negative, the basin has less water than at time zero.

Equalization can only be accomplished if the basin had enough water at time zero to offset the largest cumulative water deficit during the time period under consideration; otherwise the basin would be dry at some time, and the outflow could only equal the inflow. Hence, the minimum acceptable value of $V(0)$ equals the maximum cumulative deficit during the period under consideration. This value can be computed as

$$V(0)_{min} = (\text{Cumulative outflow} - \text{Cumulative inflow})_{max}$$
$$= \left(Q_{ave}t - \int_0^t Q_{in}(t)dt \right)_{max} \quad (2\text{-}95)$$

With $V(0)_{min}$ determined as in Equation 2-95, $V(t)$ at all other times can be calculated from Equation 2-94b and will always be ≥ 0. The maximum value of $V(t)$ from this calculation represents the required capacity of the basin to meet the equalization objective. This value, which we designate as $V_{eq'n,min}$, is the design value that we are seeking:

$$V_{eq'n,min} = (V(t))_{max} = V(0)_{min} + \left(\int_0^t Q_{in}(t)dt - Q_{ave}t \right)_{max}$$
$$(2\text{-}96)$$

■ **EXAMPLE 2-9.** Consider a small industry with a varying production of wastewater. To discharge to the local publicly owned treatment works (POTW), the flow must be equalized before entrance to the sewer. A typical set of hourly flow rates for 1 day (midnight to midnight) is shown in columns A and B of the spreadsheet shown in Table 2-4. What is the minimum volume of the equalization tank needed to allow the industry to discharge its waste at a steady rate on a day with typical flows?

TABLE 2-4. Spreadsheet for Design of Flow Equalization Basin in Example 2-9

A	B	C	D	E	F	G
		Average Flow	Cumulative	Cumulative		
	Instantaneous	During Previous	Volume Input	Volume Output	Cumulative Input	
Time	Flow Rate	Hour	Since $t=0$	Since $t=0$	Minus Output	$V(t)$
(h)	(m³/h)	(m³/h)	(m³)	(m³)	(m³)	(m³)
0	8		0.0	0.0	0.0	42.5
1	8	8.0	8.0	14.0	−6.0	36.5
2	5	6.5	14.5	28.0	−13.5	29.0
3	7	6.0	20.5	42.0	−21.5	21.0
4	8	7.5	28.0	56.0	−28.0	14.5
5	9	8.5	36.5	70.0	−33.5	9.0
6	10	9.5	46.0	84.0	−38.0	4.5
7	11	10.5	56.5	98.0	−41.5	1.0
8	15	13.0	69.5	112.0	−42.5	0.0
9	16	15.5	85.0	126.0	−41.0	1.5
10	19	17.5	102.5	140.0	−37.5	5.0
11	19	19.0	121.5	154.0	−32.5	10.0
12	19	19.0	140.5	168.0	−27.5	15.0
13	20	19.5	160.0	182.0	−22.0	20.5
14	20	20.0	180.0	196.0	−16.0	26.5
15	20	20.0	200.0	210.0	−10.0	32.5
16	20	20.0	220.0	224.0	−4.0	38.5
17	22	21.0	241.0	238.0	3.0	45.5
18	18	20.0	261.0	252.0	9.0	51.0
19	17	17.5	278.5	266.0	12.5	55.0
20	13	15.0	293.5	280.0	13.5	56.0
21	13	13.0	306.5	294.0	12.5	55.0
22	11	12.0	318.5	308.0	10.5	53.0
23	8	9.5	328.0	322.0	6.0	48.5
24	8	8.0	336.0	336.0	0.0	42.5

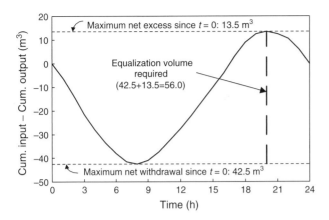

FIGURE 2-28. Variation in inflow for example system.

Solution. The cumulative input $\left(\sum_{j=1}^{i} Q_{in,j} \Delta t_j\right)$ of the flows is shown in column D of Table 2-4. The final value, 336 m^3, can be used to calculate the average flow, which is 14.0 m^3/h.

Since the goal is to maintain a steady output rate, the effluent flow is fixed at Q_{ave}. The cumulative output volume is therefore $Q_{ave}t$, which is shown in column E. The differences between the cumulative input and output volumes; that is, the values associated with the right side of Equation 2-94a, are shown in column F. These values are negative for the first several hours because the influent flow is less than the average (i.e., effluent) flow during these times. The maximum negative and maximum positive values in column F are -42.5 m^3 at $t = 8$ h and 13.5 m^3 at $t = 20$ h, respectively. Thus, at least 42.5 m^3 of water had to be present at time zero to allow the desired flow rate to be maintained through the whole period, and after $t = 20$ h, there will be 13.5 m^3 more water in the tank than at time zero. The sum of these volumes is the required design volume (56.0 m^3). A tank with this volume would be able to maintain a constant flow rate of 14.0 m^3/h despite the fluctuating influent flows, if this pattern of the flow variation was repeated daily. The same result can be obtained from a plot of the data in column F, as shown in Figure 2-28.

If, at time zero, the tank contained the minimum acceptable value of 42.5 m^3 of water, the volume of water in the tank at subsequent times could be computed according to Equation 2-94b. This information is shown in column G of the spreadsheet. Note that the required equalization volume is the maximum value in this column. ∎

As noted earlier, in most situations, flow patterns are somewhat cyclical, although the period of the cycle might be anywhere from hours to many days. The flow patterns will not be the same in every cycle, and therefore the required volume to equalize the flows will also vary. In such a case (again assuming data are available), one can carry out the analysis for a large number (N) of time periods. For each time period, a different value for the required volume is obtained. These values of required volumes can be rank-ordered from lowest to highest. Thus, if 25 cases have been analyzed, 25 values of the required volume (V) would have been computed, associated with k values of 1 to 25; the smallest value of V is assigned a k value of 1, the next smallest value of V is assigned a k value of 2, and so forth.

The rank number (k) is then converted to a fraction f by dividing it by $N+1$ (26 in the example system), and that fraction is interpreted as the probability that a given equalization volume will be required. For example, the third lowest value of V (821 m^3) has a k of 3, so the corresponding f value ($= k/(N+1)$) is 0.115. The interpretation is that an equalization volume of 821 m^3 will provide the desired amount of equalization 11.5% of the time. (The addition of one to N in the denominator is a common technique used in statistics to account for the fact that the available data might not include the most extreme condition that will ever occur. That is, using a value of $N+1$ in the denominator assures that f never reaches 100%, thereby allowing for the possibility that the required volume might exceed the largest value required in the past.)

A graph is then prepared showing f versus the required volume, V.[14] Using such a graph, the design can be statistically based. For example, if the objective is to be able to equalize the flow 90% of the time, the required volume is the value of V corresponding to $f = 0.9$. One might also use statistical techniques to attempt to predict the likelihood of extreme flows that are even larger or smaller than those during the period for which data are available.

■ **EXAMPLE 2-10.** The wastewater flows entering a municipal treatment plant were recorded at 15-min intervals. Twenty-eight days of data were analyzed to find the required volume necessary to completely equalize the flow for each day, using the methodology shown in the previous example. The 28 values of V required are shown in Table 2-5, rank ordered from lowest to highest. (Note that this ranking is unlikely to correspond to the temporal sequence.) Find the volumes that would be required to equalize the flow for 80, 85, 90, and 95% of the cases.

Solution. The rank number for each case needs to be converted into a fraction (or percentage) of cases with

---

[14] In preparing such graphs, a probability axis is sometimes used for the ordinate. Such a scale can linearize the data, if the data can be described by a standard probability function.

TABLE 2-5. Rank-ordered Required Volumes for Flow Equalization in Example 2-10

Rank	Required Volume (m³)	Rank	Required Volume (m³)	Rank	Required Volume (m³)	Rank	Required Volume (m³)
1	746	8	908	15	976	22	1193
2	777	9	962	16	977	23	1243
3	821	10	963	17	993	24	1280
4	869	11	966	18	1006	25	1287
5	872	12	969	19	1026	26	1348
6	874	13	972	20	1035	27	1438
7	878	14	974	21	1178	28	3292

required volume less than the value for that case, using the convention that the fraction is computed as the rank divided by $N + 1$. For example, since the value of 878 m³ is ranked seventh of the 28 cases, the associated fraction is 7/29 or 0.241. This means that 24.1% of the cases have a required volume less than or equal to 878 m³, so that providing this much equalization volume will allow the flow to be equalized 24.1% of all days. The data analyzed in this way are plotted in Figure 2-29a; because the data of interest are those associated with the upper portion of the data, those data are replotted in Figure 2-29b.

To find the volumes required for each specified fraction (or percentage), horizontal lines are drawn from given fractions on the ordinate to the graphed data, and then vertical lines are drawn from that point to the abscissa to find the associated volumes. The four required values are shown on the figure. Note that, in these real data from a wastewater plant in Texas, the highest value for required volume in this 28-day period (3292 m³, the last value in the table) is more than twice as high as the second highest value. In the analysis, the effect is a dramatic difference between the volumes required to equalize for 90% versus 95% of the cases. This result points out the problem of using a small data set; data for several months rather than 4 weeks should be used. ■

Concentration Equalization

Many wastewater treatment plants experience broad variations in the influent concentration. Because any CFSTR contains a mixture of the water and contaminants that came into the tank at all previous times (in proportions described by the exponentially decaying $E(t)$ distribution), CFSTRs tend to equalize the concentration; that is, the variation in concentration is less in the effluent than in the influent. This fact, intrinsically true for any CFSTR, has led to the use of CFSTRs as designed concentration equalization facilities (Novotny and Englande, 1974). In the analysis that follows, the constituent of interest is considered nonreactive in the equalization facility. That same constituent is, generally, reactive in some subsequent treatment process, in which the chemical, physical, or biological conditions might be different from those in the equalization tank. If the species is in fact reactive in the equalization facility, ignoring the reactivity should lead to a somewhat conservative design, which is usually an acceptable result.

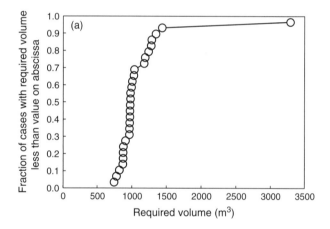

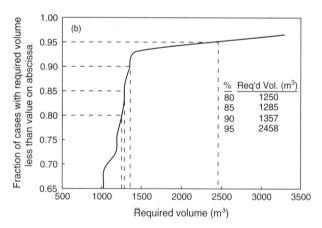

FIGURE 2-29. Probabilistic analysis of the equalization volume requirements. (a) A cumulative frequency plot for the required equalization volume for a wastewater treatment plant; (b) The upper portion of the curve in (a), and lines showing the equalization volume requirement for four specific probabilities.

The extent of concentration equalization that is accomplished in a CFSTR can be determined using the approach described in the preceding section for the response of CFSTRs to tracer inputs. For instance, consider a situation in which the flow is reasonably constant and the influent concentration to a CFSTR suddenly jumps to a new value. Imagine, for now, that the new influent concentration (c_2) remains constant for some time period. If the instant of the change is defined as time zero, the concentration in the tank at the instant the influent changes is $c(0)$, and this value reflects the previous history of the influent to the tank. The mass balance on the CFSTR for any time after time zero is

$$\frac{dVc_{\text{out}}}{dt} = Qc_2 - Qc_{\text{out}} \tag{2-97}$$

Dividing by V and noting that $Q/V = 1/\tau$, this equation yields

$$\frac{dc_{\text{out}}}{dt} = \frac{1}{\tau}c_2 - \frac{1}{\tau}c_{\text{out}} \tag{2-98a}$$

$$\frac{dc_{\text{out}}}{dt} + \frac{1}{\tau}c_{\text{out}} = \frac{1}{\tau}c_2 \tag{2-98b}$$

Rearranging and taking the Laplace transform of both sides (from the appendix), we obtain

$$sf(s) - [c_{\text{out}}(0)] + \frac{1}{\tau}f(s) = \frac{1}{\tau}c_2\frac{1}{s} \tag{2-99}$$

where $f(s)$ is the Laplace transform of $c_{\text{out}}(t)$. Rearranging again, $f(s)$ is found to be

$$f(s) = c_{\text{out}}(0)\frac{1}{s + \frac{1}{\tau}} + \frac{1}{\tau}c_2\frac{1}{s\left(s + \frac{1}{\tau}\right)} \tag{2-100}$$

Taking the inverse Laplace transform yields

$$c(t) = c(0)\exp\left(-\frac{t}{\tau}\right) + c_2\left[1 - \exp\left(-\frac{t}{\tau}\right)\right] \tag{2-101}$$

Equation 2-101 describes the change in the concentration over time of the effluent concentration from the reactor. Examining the two terms on the right side is instructive. The first term is identical to the response of a CFSTR to a pulse input of a tracer (Equation 2-45b) with the mass of tracer sufficient to yield the concentration $c(0)$ throughout the reactor at time zero. That term alone would characterize a CFSTR that contained tracer at a concentration $c(0)$ at time zero, and that received tracer-free influent from that time forward. The second term on the right side is identical to the response of a CFSTR that contained no tracer at time zero and that is then subjected to a step input of tracer (Equation 2-52) at concentration c_2. Overall, the response of the CFSTR shown in Equation 2-101 is the superposition of these two responses—the exponential die-off of the existing concentration and the rise toward the asymptote defined by the new influent concentration.

Equation 2-101 can be used, with minor alteration, to predict the changes in concentration in the effluent of a CFSTR that is receiving a time-varying input of a conservative substance. If input data are available at discrete times, the average input concentration for an interval Δt can be calculated and represented as $c_{\text{in},\Delta t}$. The concentration in the CFSTR at the end of that interval (time $t + \Delta t$) will reflect the decline of the concentration present at the start of the interval (time t) and the increase associated with the influent during the interval; that is,

$$c(t + \Delta t) = c(t)\exp\left(-\frac{\Delta t}{\tau}\right) + c_{\text{in},\Delta t}\left[1 - \exp\left(-\frac{\Delta t}{\tau}\right)\right] \tag{2-102}$$

Equation 2-102 is a recursive equation; that is, the concentration at one time is calculated based on the conditions in the tank at the previous discrete time and the influent during the interval. The equation is used successively to step through time.

To begin the analysis of concentration equalization in a CFSTR receiving a time-varying influent, we choose an arbitrary concentration that is assumed to exist in the tank at time zero ($c(0)$); a long-term average might be a reasonable choice, but any value that is in the range of possible influent concentrations is acceptable. Since this value is an arbitrary guess, though, we would want to discard the concentrations computed for the first few reactor hydraulic detention times when making a design decision, to avoid any significant influence of that choice.

For a given situation, the key design decision is the choice of τ. The analysis involves straightforward spreadsheet work. First, the influent concentrations and associated times are entered, and the time intervals of interest (Δt values) are selected. The average influent concentration during each interval is then calculated. Finally, an initial concentration is guessed, a trial hydraulic detention time (τ) is chosen, and Equation 2-102 is used recursively to calculate the expected effluent concentrations over time. Different values of τ are used until a predetermined criterion for the variation of the effluent concentration is attained. Possible criteria include not exceeding a maximum difference between the high and low values in the entire data set or in any chosen time period, a maximum standard deviation, or a maximum coefficient of variation (standard deviation divided by the mean).

■ **EXAMPLE 2-11.** The influent concentration of a contaminant during a 24-h period varies as shown in Table 2-6; the data are instantaneous values taken every hour. Assuming the flow rates into and out of the reactor are constant, and that

TABLE 2-6. Time-Varying Influent Concentrations for Example 2-11

Time (h)	Concentration (mg/L)	Time (h)	Concentration (mg/L)	Time (h)	Concentration (mg/L)
0	155	9	322	17	285
1	150	10	180	18	145
2	225	11	95	19	105
3	315	12	160	20	190
4	210	13	255	21	280
5	155	14	280	22	195
6	95	15	210	23	110
7	200	16	130	24	155
8	275				

TABLE 2-7. Summary of the Effects of Different Detention Times in Equalization Reactor

	Influent	$\tau=1$ h	$\tau=2$ h	$\tau=2.65$ h	$\tau=3$ h	$\tau=5$ h
Mean, $\bar{c}$ (mg/L)	196.8	196.8	196.8	196.8	196.8	196.8
SD, σ (mg/L)	54.4	37.9	24.3	19.6	17.7	14.0
CV (—)	0.28	0.19	0.12	0.10	0.09	0.07

the data shown in the table are typical of every day, determine the detention time required for a completely mixed equalization tank to reduce the coefficient of variation of the effluent to less than 10%.

Solution. Because the flow rates in and out are constant, the volume of water in the reactor is constant. The influent concentration during each hour can be approximated as the average of the values at the beginning and end of the hour. So, for example, the influent throughout the first hour is taken to be 152.5 mg/L. (Note, this step would not be necessary if the data were taken as composite samples.)

To initiate the calculations, assumptions about the concentration in the reactor at time zero and the hydraulic detention time are required. Any choices are okay; for the calculations here, values of 180 mg/L and 2 h have been chosen. Since data are available at 1-h intervals, we use this value as Δt. Then, we can use Equation 2-102 to find the concentration at the end of any time interval based on the concentration at the beginning of the interval. For instance, using $t = 0$ as the beginning of an interval, we find the concentration after the first hour as follows:

$$c(t + \Delta t) = c(t)\exp\left(-\frac{\Delta t}{\tau}\right) + c_{\text{in},\Delta t}\left[1 - \exp\left(-\frac{\Delta t}{\tau}\right)\right]$$

$$c(1\,\text{h}) = (180\,\text{mg/L})\exp\left(-\frac{1\,\text{h}}{2\,\text{h}}\right) + (152.5\,\text{mg/L})\left(1 - \exp\left(-\frac{1\,\text{h}}{2\,\text{h}}\right)\right)$$

$$= 169.2\,\text{mg/L}$$

The effluent concentration for each successive time is found using the same equation iteratively. Because the data are considered to be representative of every day, the spreadsheet can be continued for 48 or 72 h, repeating the data for each 24-h period. In that way, the choice of the starting concentration becomes irrelevant in the second or third 24-h period. Using one of these later 24-h periods, the mean and standard deviation of the effluent concentrations can be found, and then the coefficient of variation (CV, the ratio of the standard deviation to the mean).

The results of a spreadsheet analysis to determine the CV for various hydraulic detention times are summarized in Table 2-7 and Figure 2-30. The data in the table indicate that the minimum detention time to accomplish the goal is 2.65 h. The figure shows the temporal variations of the influent and effluent concentrations for various detention times; the heavy line is for the final result of $\tau = 2.65$ h. The influent data cycle several times in the 24-h period, leading to this small requirement for the detention time; a much longer detention time would be required if the 24-h period exhibited just one diurnal cycle. ■

The approach to concentration equalization given here is completely general and therefore can be applied in any situation. A more restrictive situation, but one that is often reasonably approximated in wastewater flows, is also instructive. Consider a situation where the flow is reasonably constant and the variation in concentration over time can be adequately described as a sine function. The concentration in such an influent is expressed mathematically as follows:

$$c_{\text{in}}(t) = c_{\text{ave}} + c_{\text{amp}}\sin(\omega t) \quad (2\text{-}103)$$

where c_{ave} is the average concentration in the influent, c_{amp} is the amplitude of the variation, and ω is the frequency of

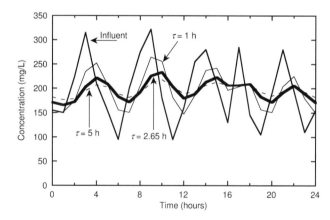

FIGURE 2-30. Concentration equalization for the example data set, using a CFSTR with a few detention times.

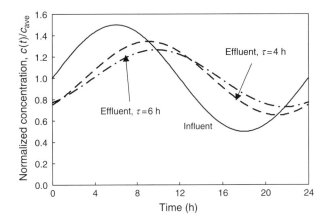

FIGURE 2-31. Concentration equalization of a sinusoidal influent.

the sine function. The frequency, ω, is defined as $2\pi/T$, where T is the period. If such an influent enters a CFSTR, it can be shown[15] that the effluent concentration varies as follows:

$$c_{out}(t) = c_{ave} + \frac{c_{amp}}{\sqrt{1+\omega^2\tau^2}} \sin(\omega t - \arctan(\omega\tau)) \quad (2\text{-}104)$$

Equation 2-104 indicates that the effluent concentration has the same long-term average concentration as the influent (because no reactions are considered to occur) and that it also has a sine variation like the influent. However, the amplitude of the sine curve is reduced by a factor of $\sqrt{1+\omega^2\tau^2}$, and the entire sine function is offset in time from the influent's sine function by the arctangent term. An example of these effects is shown in Figure 2-31. Results for two choices of the design detention time, τ, are shown. Generally, the offset of the effluent sine curve is unimportant; what is valuable is the reduction of the amplitude. For the examples shown in the figure, the amplitudes after equalization are 69 and 54% of the original amplitude with the 4- and 6-h detention times, respectively.

If the desired reduction in the amplitude is known, the required detention time can be calculated directly from

$$\frac{c_{amp,new}}{c_{amp,orig}} = \frac{1}{\sqrt{1+\omega^2\tau^2}} \quad (2\text{-}105)$$

---

[15] The solution is reasonably straightforward using Laplace transforms. The mass balance, as before, yields: $dc_{out}/dt = (1/\tau)c_{in} - (1/\tau)c_{out}$. To solve the equation, one needs to choose an initial condition. With the choice that $c(0) = c_{ave}$, the solution contains a third term not shown in Equation 2-104. That extra term is $c(0)\omega\tau \exp(-t/\tau)/(1+\omega^2\tau^2)$. The exponential decay with time makes this term infinitesimally small after a few periods and therefore irrelevant from the point of view of design, which is concerned with what happens over the long term.

If almost complete equalization is necessary, the time lag between the sine functions for the influent and effluent can be made to be half the period, so that blending a fraction of the unequalized influent with the equalization tank effluent results in a further reduction of the variation. This option, however, requires a large equalization volume. In a more general sense, the results in the figure demonstrate that the concentration in the effluent increases when the concentration of the influent is greater than that in the tank, and vice versa; that is, the maxima and minima of the effluent curves occur at the intersections with the influent curve.

Concurrent Flow and Concentration Equalization

Some situations, especially those encountered in industrial wastewater treatment, require that both flow and concentration be equalized; that is, the variation in each is unacceptably high. The approach to equalization analysis and design in such situations is a combination of the methods described for each separately. Concentration equalization occurs naturally in any CFSTR. Therefore, if a tank designed for flow equalization is provided with mixing, concentration equalization will occur to some degree at all times except when the tank is (nearly) empty. How much equalization occurs can be predicted by Equation 2-102, except in this case, the detention time is no longer a constant; that is $\tau = \tau(t)$. At any instant (or small interval), the mean hydraulic detention time is the ratio of the volume of water in the tank ($V(t)$, from Equation 2-94b) and the effluent flow rate (Q_{ave}).

If a preliminary design volume is determined based on flow equalization alone, $V(t)$ can be calculated as shown in Example 2-9. The time-varying value of the hydraulic residence time, $\tau(t)$, can then be computed as $V(t)/Q_{ave}$, and Equation 2-106 (which is just Equation 2-102 modified to include the possibility that τ values depend on t can be used recursively to compute $c(t)$ throughout the period of flow and concentration variation:

$$c(t+\Delta t) = c(t)\exp\left(-\frac{\Delta t}{\tau(t)}\right) + c_{in,\Delta t}\left\{1 - \exp\left(-\frac{\Delta t}{\tau(t)}\right)\right\} \quad (2\text{-}106)$$

If the concentration equalization that results naturally from flow equalization is insufficient, then a greater equalization volume must be allowed for in design, such that the tank is never completely emptied. The value of $V(t)$ for such a tank equals the value for the minimum-size flow equalization tank (i.e., a tank that is designed for flow equalization alone) plus the incremental volume to improve the concentration equalization. That is, $V(t)$ in the larger tank equals $V(t)$ as computed in column F of the spreadsheet in Example 2-9, plus a constant value V_{xc}, the extra volume added for concentration equalization. Various values of V_{xc} can be considered, and the concentration versus time profile for each can be computed according to Equation 2-106 until an acceptable degree of concentration equalization is achieved.

■ **EXAMPLE 2-12.** An industry produces a waste stream that has a flow variation similar to that described in Example 2-9, and a concentration variation similar to that described in Example 2-11. Assuming an initial concentration in the system of 100 mg/L, compare the fluctuations in the influent concentration with those in the effluent for basin volumes of 56.0 m^3 (the minimum for flow equalization alone, as per Example 2-9), 75 m^3, and 150 m^3.

Solution. The three scenarios are characterized in Figure 2-32, based on the application of Equation 2-106 to the given data. The results shown are for the expected concentrations during day 2, for an assumed concentration of 155 mg/L at the beginning of day 1. By focusing on the second day of operation, the effect of the arbitrary assumption regarding the concentration at the beginning of day 1 is virtually eliminated. Larger equalization volume clearly leads to more damping of the concentration. In the scenarios shown, the ratios of the maximum to the minimum concentration are 3.4, 2.2, 1.6, and 1.1 for equalization volumes 0, 56.5, 75, and 150 m^3, respectively. ■

When designing for either flow or concentration equalization, it is unwise to assume that a short period of observation will include the most extreme conditions ever experienced, so the preliminary data collection should be conducted over a large enough time span that these extremes are likely to be approached. If the flow and concentration vary in a somewhat regular pattern, the analysis can be conducted for each cycle to yield a value of the corresponding required volume for that cycle. These values can then be rank-ordered and converted to a cumulative probability distribution, as described previously for flow equalization alone, and the design value can be chosen on the basis of a predetermined objective stated in statistical terms.

When designing for simultaneous equalization of flow and concentration, the required volume depends on variations in both flow and concentration, and the risk that use of a small amount of data in the analysis will lead to an underestimate of the required equalization volume might be even higher than when designing for only one of these objectives. If the data for flow and concentration are negatively correlated (one increases when the other decreases), the design for flow equalization is likely to be sufficient for concentration equalization. If the data for these two variables are positively correlated, additional volume for concentration equalization is likely to be necessary, but the risk in the analysis is not much higher than in the flow equalization case. That is, the range of required volumes will be equally well characterized with the same number of data points. However, if the variations in flow and concentration are not correlated, the risk that a small number of data points will not characterize the variation in required volume rises—the data are less likely to include the extreme cases in which concentration is high when the flow is high and/or the volume in the tank is low. In this case, the designer needs to use a much greater number of data points in the analysis (if available), include a larger safety factor, design for a stricter objective, or use some combination of these approaches to ensure that the design meets the objectives.

2.6 SUMMARY

Various ways to evaluate and model the hydraulic characteristics (or mixing intensity) of reactors with continuous, steady flow have been presented in this chapter. The extremes of mixing intensity are represented as ideal reactor types: in an ideal plug flow reactor (PFR), there is zero mixing in the axial direction, and in an ideal continuous flow stirred tank reactor (CFSTR), the mixing is infinitely intense. Both ideal and nonideal reactors can be characterized by their RTDs (the exit age distribution, $E(t)$, and/or the cumulative age distribution, $F(t)$).

Nonideal reactors can sometimes be modeled quite well either by modifying the assumptions applicable to the ideal reactors or by representing the nonideal reactor as a linked network of ideal reactors. The models used most often to represent nonideal reactors include

1. A PFR with dispersion
2. A number of equal-sized CFSTRs in series
3. A combination of a few CFSTRs and PFRs connected in series and/or parallel, perhaps with dead space included
4. A network of several of PFRs in parallel or in series with multiple inlets and outlets.

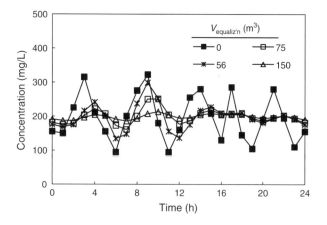

FIGURE 2-32. Concentration equalization in the system described in Examples 2-9 and 2-10.

The advantage of mathematical modeling is that it can reduce a large data set to a function that can be characterized using only a few parameters. For ideal reactors (CFSTRs and PFRs), a single parameter (τ) is sufficient to describe the reactor hydraulics completely. For the first two models for nonideality listed above, one additional parameter is needed. In the latter two models, the number of parameters needed increases by two for each additional model reactor used to represent the system. At some level of complexity, the advantage of modeling becomes small in comparison to the empirical $E(t)$ or $F(t)$ data (obtained from tracer tests).

The predicted tracer output curve using the best-fit values of the defining parameters for each model can be compared with the actual output curve. If a model gives a good fit, then one might infer that the hydraulics of the reactor have been characterized reasonably well. If the chosen models do not give a good match to the experimental output, the conclusion is that the actual flow pattern in the reactor is not similar to the model flow patterns investigated, and others may have to be postulated.

The decision regarding the number of fitting parameters to be used is a judgment which must be made based on the reliability of the RTD data, the need for an accurate representation, and the cost of improving the fit. However, in carrying out the analysis and interpreting the result, one must keep in mind that each of the mathematical models is based on a physical model of the flow pattern. If the physical model is correct, the results predicted by the mathematical model will fit the experimental ones. There may also be incorrect models of the physical process which nevertheless fit the tracer data quite well. No one model is universally better than the others; the best model is the one that gives a sufficiently accurate representation of the reactor behavior for the intended use.

Some real reactors often closely approximate one of the ideal cases. Rapid mix basins and flocculators are usually small open reactors with sufficient mixing to approximate a CFSTR. Packed beds (adsorption or ion exchange columns, deep bed filters, and perhaps even stripping towers) are close to being PFRs, although they are likely to have a small amount of dispersion. On the other hand, many reactors are nonideal. Aeration basins have intense mixing induced by the air movement or by mechanical aerators, but this mixing is usually relatively local, and such basins might be modeled as a small number of CFSTRs in series. Baffled reactors, such as disinfection chambers, might also act as a number of CFSTRs in series. Reactors (or pipes or channels) with a high ratio of length to width (or depth) will approximate plug flow, although some dispersion is likely. For accurate description, tracer studies and the types of analyses presented in this chapter are necessary.

Analysis of reactor hydraulics is directly applicable to the design of reactors for flow and/or concentration equalization. In such situations, the varying water volume stored in the equalization basin can allow a steady flow of water to downstream processes, even if the influent flow rate varies significantly. Similarly, the dilution of influent into a substantial volume of water collected during earlier times dampens the fluctuations in the composition of the feed to the downstream processes. Although equalization accomplishes no pollutant removal by itself, it can greatly facilitate smooth operation of the pollutant removal steps that follow it.

APPENDIX 2A. INTRODUCTION TO LAPLACE TRANSFORMS AS A METHOD OF SOLVING (CERTAIN) DIFFERENTIAL EQUATIONS

The process of using Laplace transforms for solving differential equations is similar to the well-known process of using logarithms to solve exponential problems. (Unfortunately, in this age of spreadsheets, some readers might be a bit rusty on the logarithm transform process, but the analogy is still useful.) Consider the following exponential equation:

$$x^a = b$$

where a and b are known and x is the unknown. To solve, we take the logarithmic transform of both sides of the equation to obtain

$$\log x^a = \log b$$

With this equation, we can say that we have transformed the problem onto the logarithmic plane. We then use the properties of logarithms to find

$$a \log x = \log b$$

and do algebra to find the solution on this logarithmic plane:

$$\log x = \frac{\log b}{a}$$

Finally, we do the inverse transformation (i.e., take the antilog) to find the solution on the normal plane:

$$x = \text{antilog} \frac{\log b}{a}$$

For example, following each of these steps to find the value of x in the equation:

$$x^{3.7} = 196.4$$

leads to the result $x = 4.17$.

The Laplace process for solving differential equations is similar to this logarithm process. We start with a differential equation in $F(t)$; $F(t)$ here is any (qualifying) function of time and not specifically the cumulative age distribution. The choice of this symbol for the function is perhaps unfortunate in this chapter, but its use by mathematicians is almost universal, and so it is used here for consistency with published mathematical tables. After taking the Laplace transform of both sides of the equation, we have an algebraic equation in the Laplace transform of $F(t)$, usually written alternately as either $\mathbf{L}\{F(t)\}$ or $f(s)$; the meaning of s is explained shortly. We then do algebra on the Laplace plane to solve for $f(s)$, and finally perform the inverse Laplace transformation, often written as $\mathbf{L}^{-1}$, to find $F(t)$.

The Laplace transform is defined as follows:

$$\mathbf{L}\{F(t)\} \equiv f(s) \equiv \int_{t=0}^{\infty} e^{-st} F(t)\, dt$$

where s is a *number* sufficiently large to make the integral converge (i.e., s is not a variable), $F(t)$ must be continuous (or piecewise continuous) for all $t > 0$, and $F(t)$ must be bounded near $t = 0$. This transformation lends itself to situations in which $F(t)$ is defined for $t \geq 0$ (but not for $t < 0$).

Fortunately, it is rarely necessary to use the definition and carry out the integration to apply Laplace transforms. Operations with Laplace transforms and transforms of many common functions have been tabulated; a convenient (although small) listing is given in the CRC Standard Mathematical Tables (Beyer, 1987), and several other listings are available in mathematical reference books.

The three most common operations (at least for the use associated with tracers in environmental systems) are easily derived from the definition. They are as follows:

$$\mathbf{L}\{A\, F(t)\} = A f(s) \qquad (2\text{A-}1)$$

That is, the Laplace transform of the product of a constant (A) and a function ($F(t)$) is the product of that constant and the Laplace transform of the function.

$$\mathbf{L}\{F_1(t) + F_2(t)\} = f_1(s) + f_2(s) \qquad (2\text{A-}2)$$

That is, the Laplace transform of a sum is the sum of the associated Laplace transforms.

$$\mathbf{L}\{F'(t)\} = \mathbf{L}\left\{\frac{dF(t)}{dt}\right\} = s f(s) - F(+0) \qquad (2\text{A-}3)$$

That is, the Laplace transform of the derivative of a function is the product of s and the Laplace transform of the function, minus the value of the function at time zero. (The $+0$ means that, if there is a discontinuity at zero, the limit as t approaches zero from positive t is used.) It is this operation that allows one to solve differential equations, because the transform of the time derivative of $F(t)$ is a function of $F(t)$ and not of the differential of $F(t)$.

A fourth operation is also often useful and was noted in the text, namely

$$\mathbf{L}\{F_1(t) * F_2(t)\} = \mathbf{L}\left\{\int_0^t F_1(t-\tau) F_2(\tau)\, d\tau\right\} = f_1(s) f_2(s) \qquad (2\text{A-}4)$$

That is, the Laplace transform of the convolution of two functions is the product of the Laplace transforms of the two functions.

A short listing of Laplace transforms of common functions is given in Table 2-A1; again, the reader is referred to the CRC Standard Mathematical Tables or other reference books for additional listings.

Notes: $\Gamma(k)$, which appears in transforms 5 and 9, is an integral function that, for positive values of k, is closely related

TABLE 2A-1. Laplace Transform of Common Functions

Number	Function ($F(t)$)	Laplace Transform of Function ($f(s)$)
1	1	$\dfrac{1}{s}$
2	t	$\dfrac{1}{s^2}$
3	$\dfrac{t^{n-1}}{(n-1)!} \quad (n = 1, 2, \ldots)$	$\dfrac{1}{s^n}$
4	$\dfrac{1}{\sqrt{\pi t}}$	$\dfrac{1}{\sqrt{s}}$
5	$t^{k-1} \quad (k > 0)$	$\dfrac{\Gamma(k)}{s^k}$
6	e^{at}	$\dfrac{1}{s-a}$
7	te^{at}	$\dfrac{1}{(s-a)^2}$
8	$\dfrac{1}{(n-1)!} t^{n-1} e^{at} \quad (n = 1, 2, \ldots)$	$\dfrac{1}{(s-a)^n}$
9	$t^{k-1} e^{at} \quad (k > 0)$	$\dfrac{\Gamma(k)}{(s-a)^k}$
10	$\dfrac{1}{a-b}\left(e^{at} - e^{bt}\right)$	$\dfrac{1}{(s-a)(s-b)}$
11	$\dfrac{1}{a-b}\left(ae^{at} - be^{bt}\right)$	$\dfrac{s}{(s-a)(s-b)}$
12	$\dfrac{(b-c)e^{at} + (c-a)e^{bt} + (a-b)e^{ct}}{(a-b)(b-c)(c-a)}$	$\dfrac{1}{(s-a)(s-b)(s-c)}$
13	$\dfrac{1}{a}\sin(at)$	$\dfrac{1}{s^2 + a^2}$

to the factorial function. The formal definition and values of this function can be found in mathematical handbooks.

Examples of the Use of Laplace Transforms

■ **EXAMPLE 2A-1.** *CFSTR with Pulse Input of Conservative Tracer at $t = 0$* The mass balance for this situation has been described in this chapter as follows:

$$V\frac{dc}{dt} = Q(0) - Qc$$

The reaction term is zero (the tracer is chosen to be nonreactive) and c is a function of time. As indicated, the input concentration is zero after time zero. Dividing by the volume of the reactor, V, yields

$$\frac{dc}{dt} = -\frac{1}{\tau}c$$

Taking the Laplace transform of each side gives

$$sf(s) - c(0) = \frac{1}{\tau}f(s)$$

where the concentration in the reactor at time zero, c_o, equals M_{in}/V. Rearranging and factoring yields

$$f(s)\left(s + \frac{1}{\tau}\right) = c_o$$

Solving for $f(s)$, the solution on the Laplace plane, gives

$$f(s) = c_o \frac{1}{s + (1/\tau)}$$

Using transform 6 from Table 2-A1 with the value of $a = -1/\tau$, and recognizing that the constant c_o will carry through into the equation during the inverse transformation (operation 1 shown earlier) yields the following:

$$c(t) = c_o e^{-t/\tau}$$

This result is, of course, the same as that found by straightforward integration and shown in the chapter. The value here is in recognizing the utility and simplicity of the Laplace transform method. ■

■ **EXAMPLE 2A-2.** *CFSTR with Step Input of Conservative Tracer at Time Zero* The mass balance for this situation (also given in this chapter) is

$$V\frac{dc}{dt} = Qc_{in} - Qc$$

where again c is a function of time, defined for all positive values of t, so the Laplace transform method is useful.

After dividing by V, recognizing that $Q/V = 1/\tau$, and rearranging, this equation becomes

$$\frac{dc}{dt} + \frac{1}{\tau}c = \frac{c_{in}}{\tau}$$

Taking the Laplace transform of both sides of this equation yields

$$sf(s) - c(0) + \frac{1}{\tau}f(s) = \frac{c_{in}}{\tau}\frac{1}{s}$$

Recognizing that the concentration of tracer in the reactor at time zero is zero (i.e., $c(0) = 0$) and rearranging yields the solution on the Laplace plane:

$$f(s) = \frac{c_{in}}{\tau}\frac{1}{s\left(s + \frac{1}{\tau}\right)}$$

Finally, applying transform 10 with $a = 0$ and $b = -1/\tau$ gives the solution as

$$c(t) = \frac{c_{in}}{\tau}\frac{1}{0 + \left(\frac{1}{\tau}\right)}\left(e^{0t} - e^{-t/\tau}\right) = c_{in}\left(1 - e^{-t/\tau}\right)$$

Of course, the solution is the same as derived earlier in two different ways (direct integration of the differential equation for the mass balance and by convolution of the $E(t)$ function for the CFSTR and the input function). In comparison with the integration method, the Laplace transform method is easier primarily because, after the transformation, the solution is found via algebra rather than calculus. For more complex problems, this benefit is even more apparent than in the two examples given here. The CFSTRs-in-series problem (Problem 2-4) or the sinusoidal concentration equalization problem are examples where the solution methodology using Laplace transforms is considerably easier than other methods (e.g., the integrating factor method). ■

REFERENCES

Bear, J. (1972) *Dynamics of Fluids in Porous Media*. American Elsevier Publishers, New York. (Also reprinted by Dover, Mineola, NY).

Beyer, W. H. (1987) *CRC Standard Mathematical Tables*, 28th edn. CRC Press, Boca Raton, FL, p. 232.

Federal Register (1989) National Primary Drinking Water Regulations: Filtration, Disinfection, Turbidity, *Giardia lamblia*, Viruses, *Legionella*, and Heterotrophic Bacteria. Fed. Reg. 54, 123, 27486 (June 29, 1989).

Grayman, W. M., Deininger, R. A., Green, A., Boulos, P. F., Bowcock, R. W., and Godwin, C. C. (1996) Water quality and mixing models for tanks and reservoirs. *JAWWA*, 88 (7), 60–73.

Haas, C. N., Joffe, J., Heath, M. S., Jacangelo, J. (1997) Continuous flow residence time distribution function characterization. *J. Environ. Eng. (ASCE)*, 123 (2), 107–114.

Kreft, A., and Zuber, A. (1978) On the physical meaning of the dispersion equation and its solutions for different initial and boundary conditions. *Chem. Eng. Sci.*, 33, 1471–1480.

Levenspiel, O. (1999) *Chemical Reaction Engineering*, 3rd edn. John Wiley & Sons, New York, NY.

Monteith, H. D., and Stephenson, J. P (1981) Mixing efficiencies in full-scale anaerobic digesters by tracer methods. *Journal of Water Pollution Control Federation*, 53 (1), 78–85.

Morrill, A. B. (1932) Sedimentation basin research and design, *JAWWA*, 24 (9), 1442–1463.

Nauman, E. B., and Buffham, B. A. (1983) *Mixing in Continuous Flow Systems*. John Wiley & Sons, Inc., New York, NY.

Nauman, E. B. (1981) Residence time distributions in systems governed by the dispersion equation. *Chem. Eng. Sci.*, 36 (6A), 957–966.

Novotny, V., and Englande, A. J., Jr. (1974) Equalization design techniques for conservative substances in wastewater treatment systems. *Water Research*, 8, 6, 325–332.

Rebhun, M., and Argaman, Y. (1965) Evaluation of hydraulic efficiency of sedimentation basins. *J. San. Eng. Div. (ASCE)*, 91 (SA5), 37–45.

U.S. Environmental Protection Agency (1991) Guidance Manual for Compliance with the Surface Water Treatment Requirements for Public Water Systems.

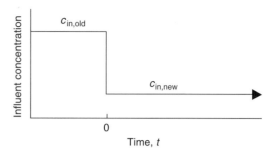

FIGURE 2-Pr3. Influent tracer concentration profile for a downward step change.

PROBLEMS

2-1. Show that the average detention time that tracer stays in a continuous flow stirred tank reactor (CFSTR) is the same as the hydraulic detention time; that is, that $\bar{t}_{CFSTR} = \tau$. Use either the response of a CFSTR to a pulse input tracer test, $c_p(t)$, or the exit age distribution, $E(t)$, as the starting point for this derivation.

2-2. (a) For a single CFSTR, at what value of t/τ is the concentration of the effluent 85% of the initial concentration in a pulse input tracer test?

(b) For the same system as in part (a), at what value of t/τ is the total mass of tracer that has left the reactor 85% of the mass initially injected?

(c) For a single CFSTR, at what value of t/τ is the concentration of the effluent 85% of the influent concentration in a step input tracer test?

2-3. The discussion of step-change tracer studies in this chapter focuses on systems in which the reactor is initially free of tracer, and a finite concentration of tracer is injected starting at time zero (and continuing indefinitely). However, the same principles apply if the concentration throughout the reactor before the step change is any finite value (designated here as $c_{in,old}$), and the step change in the influent concentration is to any other value (greater than or less than $c_{in,old}$). A graph describing such a situation for a case is shown in Figure 2-Pr3.

(a) For an arbitrary reactor with some intermediate amount of mixing (not a CFSTR and not a PFR), sketch a plausible graph showing the tracer effluent concentration as a function of time (i.e., $c_{out}(t)$), for a tracer input pattern like that shown in the graph. Show the influent tracer profile on the graph as well, for reference.

(b) For the arbitrary reactor considered in part (a), develop an expression for the cumulative age distribution, $F(t)$, in terms of $c_{in,old}$, $c_{in,new}$, and $c_{out}(t)$. Note that this expression must reflect the characteristics of any cumulative distribution; that is, $F(t)$ must increase monotonically from 0 to 1 as t goes from 0 to ∞. Make sure your answer to part (b) is consistent with your sketch in part (a).

(c) Derive an expression for $c_{out}(t)$ for a single CFSTR under the conditions of a "step down" tracer study as depicted in the figure.

(d) Use the expression developed in part (b) to convert your answer from part (c) into the cumulative age distribution, $F(t)$, for a CFSTR. Compare your result with Equation 2-49 in the text, and explain the observation.

(e) Use the expression developed in part (d) to derive the exit age distribution, $E(t)$, for a CFSTR.

2-4. You are the owner of a company that makes a chemical product in a large CFSTR with a theoretical detention time τ. Production has not been as good as you had hoped. An engineer hired to help solve problems in the plant suggests installing baffles in the reactor, effectively dividing it into a series of CFSTRs.

(a) Write out the mass balance and the resulting equation describing the effluent concentration as a function of time for a pulse input tracer test before any baffles are installed.

(b) Repeat part (a) for the concentration of the effluent from the tank once one baffle has been installed, dividing the tank into two identical CFSTRs. (Hints: The response of the first compartment to the pulse input is similar to that obtained in part (a), but you must revise your answer to account for the smaller volume. The revised result for the output from the first compartment constitutes the input to the second compartment. The solution for the concentration in the second compartment is most easily obtained using the convolution process described in the chapter and represents the response of the full, two-compartment reactor to the pulse input.)

(c) Repeat part (a) for the concentration of the effluent from the tank once two baffles have been installed, dividing the tank into three identical CFSTRs.

(d) Generalize your answers from parts (b) and (c) to develop the equation for a case where the tank is divided into N tanks.

(e) Assume that the detention time in the tank is 60 min, and that the amount of tracer injected is enough to cause the initial concentration in the unbaffled tank to be 100 mg/L. Plot, on a single graph, the effluent tracer concentration as a function of time for the cases of zero, one, and four baffles (corresponding to one, two, and five tanks in series), for $t=0$ to 200 min.

2-5. A reactor has an overall hydraulic residence time (τ) of 45 min, based on the total volume of water in the reactor and the total flow through it. The reactor can be reasonably represented as three ideal reactors in parallel. One of the conceptual reactors, accounting for 10% of the tank volume, is "nearly dead" space; this portion of the tank is completely mixed and has a residence time of 6τ. Another portion of the tank is occupied by short-circuiting flow; 30% of the influent flow passes through this unmixed (plug flow) section, with a residence time of 0.1τ. The remaining flow passes through the rest of the reactor volume and is well mixed.

(a) Draw a flowsheet to model the overall reactor, showing the volume, residence time, and fraction of the flow through each portion of the reactor.

(b) Write the equation and draw a curve showing $E(t)$ versus t for the reactor.

(c) What fraction of the water that enters the reactor stays there for a time $t < \tau$?

2-6. Determine and plot the $E(t)$ curve for a reactor system consisting of two CFSTRs in series, with hydraulic detention times of 1 h and 2 h, respectively. Compare the result with that for the same reactors in the reverse order. (Hint: write the mass balances for two CFSTRs in series with arbitrary mean residence times τ_1 and τ_2, to obtain a differential equation for the concentration exiting the second reactor. Solve the differential equation either analytically using Laplace transforms or numerically, and then convert the calculated c_{final} versus t data to $E(t)$ versus t.)

2-7. A hydraulic study of the flow-through characteristics of a sedimentation basin was made using NaCl as a tracer by injecting a pulse input of salt at the inlet and measuring the salt concentration at the outlet. The results of the study are shown in the following table:

Time (min)	Concentration of NaCl (mg/L at Outlet)
0	0
10	Trace
20	40
30	130
45	100
60	70
75	54
90	44
105	37
120	30
135	22
150	14
165	9
180	5

Note: The performance of this system is investigated further in the problems at the end of Chapter 4.

(a) Plot $c(t)$, $E(t)$, and $F(t)$ versus time. What is the value of $\bar{t}$ based on these data?

(b) Determine appropriate values of the adjustable parameters for each of the following four physical models of the system, and prepare a summary table showing your results. Note that, for the last model, a wide variety of answers could be considered acceptable, so briefly justify the choices that you make.

   (i) PFR with dispersion, for open boundaries.
   (ii) PFR with dispersion, for closed boundaries.
   (iii) N equal-sized CFSTRs in series.
   (iv) Seven ideal PFRs in parallel.

(c) On a single graph, plot $E(t)$ versus t curves for the model systems characterized in parts (i), (iii), and (iv) in part (b). On the same graph, sketch a corresponding curve for the model in part (ii), based on Figure 2-16. Add the experimental $E(t)$ data to the graph. (Hint: c_o can be found by writing an expression for M_{in} as a function of c_o, writing another expression for M_{out} in terms of the (known) effluent tracer concentrations, and assuming that $M_{in} = M_{out}$.) Prepare a second graph showing $F(t)$ versus t for each of the models, along with the experimental $F(t)$ data. Based on these graphs, would you choose one of these models to represent the data? If so, which one? If not, what other model might you test?

2-8. A stretch of a river can be modeled as a PFR with dispersion and open boundaries, with a dispersion number of 0.01. The river discharges into a lake that can be considered to be well-mixed, with a residence time of 1 d.

(a) A nondegradable contaminant is released into the river for 2 h, causing the concentration in the river water at the discharge point, after mixing across the river cross-section, to be 200 μg/L during the discharge period. The theoretical mean hydraulic residence time in the river between the release point and the lake (τ_{river}) is 6 h. As a first approximation, you decide to model the input as a pulse that contains all the mass that was actually discharged over the full release period, with the pulse entering the river at a time corresponding to the middle of the actual discharge. Estimate the time and magnitude of the peak in the contaminant concentrations entering and exiting the lake for this modeling approach, defining $t=0$ as the time at the beginning of the contaminant discharge. (Note that the time of the model, input pulse is not necessarily $t=0$. Consider how footnote 3 applies in this case.)

(b) What are the normalized variances of the concentration profiles entering and leaving the lake for the scenario described in part (a), based on your simulation? What is the theoretical value of the normalized variance at the entrance to the lake based on the PFR-with-dispersion model?

(c) According to your model, how long after the beginning of the contaminant release will it be before 95% of the contaminant has passed through the whole system?

(d) To test whether your simplification of the input function is acceptable, conduct a sensitivity analysis by modeling the input as 12 sequential pulses at 10-min intervals, rather than as a single pulse. In your model, assume the pulses are uniformly distributed at $t = 5, 15, 25, \ldots$, 115 min, and that each pulse contains one-twelfth of the total contaminant mass. The total concentration at any time at the point where the river discharges into the lake can be computed as the summation of the concentrations associated with the 12 simulated pulses. Briefly compare your results for parts (a) and (d).

2-9. In a number of situations, a real reactor might be modeled as a combination of two, linked reactors. For instance, in a stratified lake, the epilimnion and hypolimnion might have quite different compositions and reaction characteristics, yet have some continuous, relatively low level of water exchange between them across the thermocline. Similarly, engineered reactors often have zones in corners or along edges that are largely, but not completely, isolated from the core.

Consider such a reactor whose total volume is 100 m^3. Of this volume, 80 m^3 is a well-mixed core zone, and 20 m^3 is in a separate zone that is well-mixed itself but is not thoroughly mixed with the core. The flow pattern is such that the flow of 5 m^3/min into and out of the reactor enters and leaves through pipes that are in the core zone. The exchange of fluid between the core and secondary zones is at a steady rate of about 0.1 m^3/min. (Note: the performance of this system is investigated further in the problems at the end of Chapter 4.)

Ten grams of a tracer is injected as a pulse into the inlet pipe at $t=0$. Write the mass balance equations describing the concentration of tracer in each part of the reactor as a function of time, and calculate and plot those concentrations from $t=0$ until the time when 9.9 g of tracer has exited the overall reactor. Show the results for both reactors on the same graph. Discuss briefly the significance of the point of intersection of the two lines. (Hint: the mass balances yield two simultaneous differential equations that can be solved simultaneously, once the initial conditions are established. Once the equations are identified, they can be solved in several ways, including numerically, direct integration, or integration in Laplace space. To solve the problems by either of the latter two approaches, it is easiest to combine the two first-order differential equations to obtain a single, second-order differential equation.)

2-10. Assume that you obtained the tracer output data shown in Example 2-1, but you thought the reactor conformed more closely to the open boundary than the closed boundary case.

(a) What values would you estimate for τ and $\mathbf{D}/v_x L$ for the reactor in that case?

(b) Generate a predicted tracer output curve for the pulse input that was injected, using the model parameter values determined in part (a). Note that the appropriate value of c_o to use in the model calculations is the mass of tracer added divided by the effective volume of the reactor (not the geometric volume). How well does the PFR-with-dispersion model reproduce the experimental?

2-11. (a) Derive the $E(t)$ function for a set of two reactors in series: a CFSTR with mean hydraulic residence time τ_1 followed by a PFR with residence time τ_2.

(b) Repeat part (a) for the same reactors, but in the reverse order.

2-12. Rebhun and Argaman (1965) considered a reactor that exhibits nonideal flow to be conceptually divided into a dead space fraction (m) and an effective fraction ($1-m$), and further that the effective fraction consists of a plug flow fraction (p) and a completely mixed (CFSTR) fraction ($1-p$). The cumulative age distribution resulting from this model is as follows:

$$F(t) = 1 - \exp\left[-\frac{1}{(1-p)(1-m)}\left\{\frac{t}{\tau} - p(1-m)\right\}\right]$$

(a) Consider a continuous flow stirred tank reactor with a volume V_{CFSTR} and a steady flow Q. A step input tracer test is carried out on this reactor. Sketch a graph and give an equation to describe the response (effluent concentration as a function of time) of this reactor.

(b) Now consider a PFR with volume V_{PFR} that is added in series after the CFSTR described in part (a). Sketch a graph and give an equation to describe the response of this second reactor when the step input of tracer is added to the first reactor. Account properly for the definition of time zero.

(c) With the proper normalization, convert your answer to part (b) into the cumulative age distribution, $F(t)$.

(d) Consider the two reactors in series to have a total volume V. A fraction (p) of that volume is in the PFR, and the rest is in the CFSTR. Use the symbol τ for the theoretical detention time of the two-reactor system. Make the necessary adjustments to your equation to express $F(t)$ in terms of t, τ, and p.

(e) Finally, assume that, in addition to the volume V, there is space in the reactor system where water sits undisturbed by the flow; that is, dead space. Of the total volume V (dead space volume, plug flow volume, and CFSTR volume), the dead space fraction is called m. Considering this total volume, make the necessary adjustments to your equation to arrive at the equation shown by Rebhun and Argaman.

(f) To fit this model to data from a tracer test, one can plot $\ln(1 - F(t))$ versus t/τ. Show how the slope and abscissa intercept (as distinct from the usual ordinate intercept) can be used to evaluate the model parameters.

2-13. In tracer tests on any reactor other than an ideal PFR, some of the tracer will remain in the reactor indefinitely, at least in theory. In such cases, it is often unclear exactly how long one should collect and analyze the data to get a good estimate of the true residence time.

Consider a tracer test on an ideal CFSTR with an average hydraulic residence time of 60 min.

(a) Determine the average hydraulic residence time that you would compute based on the tracer output, if you collected perfect data starting at $t=0$, but stopped collecting data at various times up to 10τ. Recalling that $\bar{t}$ can be computed from the tracer data based on either M_{in} or M_{out} as the measure of total tracer used in the test, carry out the calculations twice—once using M_{in} and once using M_{out}.

(b) How long would you have to collect data in order to compute a value of $\bar{t}$ that is within 10% of τ, if the analytical data were perfect?

(c) If a systematic error occurred in the measurements of tracer concentrations such that all the measured values were 5% lower than the true values, how would the calculated values of $\bar{t}$ differ from those computed in part (a)?

2-14. The data shown in the following table are from a pulse input tracer study of an aerobic sludge digester for which the design mean hydraulic residence time is several weeks. The work was conducted by Monteith and Stephens (1981). A schematic showing the conceptual model they used to represent the digester is provided in Figure 2-Pr14. The total reactor volume (V) is assumed to be distributed between an active portion through which the flow passes, and an inactive portion (dead space). Furthermore, the active portion of the reactor comprises two parts: some of the water passes through a well-mixed (CFSTR) part of the reactor, while the rest is short-circuiting fluid that passes through the reactor as a plug and does not mix with the other fluid. The fractions of the total flow Q that are well mixed and that short-circuit through the reactor are designated

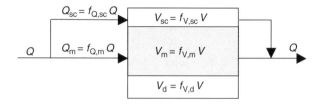

FIGURE 2-Pr14. Model representation of flow in a reactor with a well-mixed zone, short-circuiting, and dead space.

$f_{Q,m}$ and $f_{Q,sc}$, respectively. Correspondingly, the fractions of the total volume that are occupied by the mixed fluid, the short-circuiting fluid, and the dead space are designated $f_{V,m}$, $f_{V,sc}$, and $f_{V,d}$, respectively. Recall that $c_o = M_{in}/V$.

t/τ	c/c_o	t/τ	c/c_o
0.05^a	1.78	0.38	0.33
0.05^a	1.02	0.44	0.32
0.05^a	0.79	0.47	0.29
0.05^a	0.53	0.63	0.26
0.05^a	0.53	0.78	0.15
0.05^a	0.53	0.84	0.14
0.17	0.47	1.00	0.12
0.26	0.40	1.05	0.13

[a] These samples were the first ones collected. They were collected in the order shown (top to bottom) at various times during the first sampling date, at t/τ near 0.05.

(a) Why do you think such vastly different tracer output concentrations were observed over a relatively short sampling period after the tracer was injected (the first six data points)?

(b) Recall that, for any tracer output curve, $\int_{t_1}^{t_2} Qc\,dt$ is the total mass of tracer exiting the reactor from time t_1 to time t_2. Manipulate that integral to obtain an expression for $\int_{t_1}^{t_2}(c/c_o)(dt/\tau)$, which is the expression that the authors used in their analysis. What value would you expect to obtain for $\int_0^\infty (c/c_o)\,dt/\tau$?

(c) Write a mass balance around the well-mixed portion of the reactor, and develop an expression for the tracer concentration exiting that portion of the reactor (i.e., $c_m(t)$) in terms of τ, c_o, $f_{V,m}$, and $f_{Q,m}$, where τ is the overall mean hydraulic residence time in the reactor.

(d) Write another mass balance, this time around the point where the short-circuiting fluid joins the fluid that has passed through the mixed portion of the reactor, and develop an expression for the concentration of tracer exiting the entire reactor as a function of time $(c(t))$, in terms of the parameters used in part (c), plus $f_{V,sc}$ and $f_{V,d}$. Note that this is the concentration that would be detected by the investigators, who sampled a mixture of whatever water exited the reactor.

(e) Estimate the value of the integral $\int_0^\infty (c/c_o)\,dt/\tau$ for the experimental data. To do this, you will need to extrapolate the given data back to time zero and forward to times well beyond when the investigators stopped collecting data; one way to do this is to use the equation you developed in part (d). You will also need to estimate the (finite) area under the curve at short times, when the data indicate a very large rate of change in c. Why is the experimental value of $\int_0^\infty (c/c_o)\,dt/\tau$ different from the expected value determined in part (b)?

(f) Use the experimental data in conjunction with the equation you developed in part (d) to estimate the values of $f_{Q,m}$, $f_{V,m}$, and $f_{Q,sc}$ in the digester.

(g) It is not clear whether the investigators captured the peak of the tracer output as the short-circuiting fluid exited the system. Using your results from parts (e) and (f), and assuming that the peak can be approximated as a square wave with a width of 0.05 (i.e., a fixed value of c/c_o over a duration corresponding to $\Delta t/\tau = 0.05$), evaluate whether they did.

2-15. During an 8-h testing period, an industrial plant produces wastewater at a rate and with a contaminant concentration that varies as shown in the following table.

Time (h)	Average Flow (m³/h)	Average Concentration (mg/L)
0–1	11	109
1–2	12	121
2–3	14	140
3–4	18	164
4–5	20	202
5–6	25	201
6–7	50	124
7–8	10	112

(a) How large a tank is required to achieve complete flow equalization? Plot the volume of water in such tank as a function of time, using 5-min intervals.

(b) What is the long-term average contaminant concentration in the wastewater?

(c) The treatment process works best if the maximum concentration entering the process is no more than 50% greater than the minimum concentration (i.e., $c_{max}/c_{min} \leq 1.5$). If the flow equalization tank is a CFSTR, will it achieve this degree of concentration equalization? If not, approximately how large must the tank be to do so?

(d) What is the key assumption that was implicit in the analysis carried out in parts (a) and (c), regarding the flow rates and concentrations during the periods other than the test period? How might you modify your design decision to take this fact into account?

2-16. Water treatment plants usually include a storage tank or reservoir (called a clearwell) onsite, where fully treated water is stored before being pumped into the distribution system. The clearwell provides equalization capacity so that the plant can treat water at a steady rate and distribute it in response to unsteady demand.

Design a clearwell for a water treatment plant that has to meet a daily demand cycle similar to that shown in Figure 2-Pr16. The demand is low in the late evening and overnight, increases linearly during the morning, remains constant through the work day, and then decreases linearly in the evening. Find the

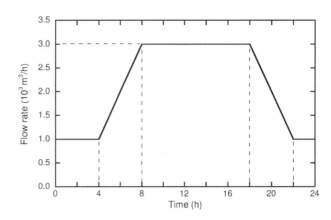

FIGURE 2-Pr16. Daily water demand pattern to be considered in the design of a clearwell.

minimum storage capacity that will allow the plant to be operated with a steady flow.

2-17. In Example 2-11, the analysis was carried out by assuming that the initial concentration in the tank was 180 mg/L. Using the detention time determined in that example, compare the predicted concentrations in the tank during a 24-h period, using 100, 180, and 400 mg/L as assumed initial concentrations. What do you conclude from the comparison?

3

REACTION KINETICS

3.1 Introduction
3.2 Fundamentals
3.3 Kinetics of irreversible reactions
3.4 Kinetics of reversible reactions
3.5 Kinetics of sequential reactions
3.6 The temperature dependence of the rates of nonelementary reactions
3.7 Summary
References
Problems

3.1 INTRODUCTION

The goal of a mass balance analysis is to describe the mass or concentration of a substance as a function of time and location in a given physical system. Ultimately, for the mass balance equation to be solvable, each of its terms must be reduced to a form in which time and location are the only independent variables, and mass or concentration is the only dependent variable.

In Chapter 1, the various terms that appear in mass balance equations were introduced, and some approaches for simplifying the mass balance were presented. It was shown that the storage term $(\partial(cV)/\partial t)$ can be simplified to $V dc/dt$ if the system is well-mixed (so that c is independent of location) and the volume is constant, and that it can be assigned a value of zero if the system is at steady state. Also, the advective and diffusive/dispersive terms in the mass balance were each segregated into two parts—one describing the concentration gradient (∇c) and the other characterizing the physical/chemical properties of the system, utilizing parameters such as Q, V, D, and ε.

In this chapter, we focus on the reaction term in the mass balance equation, exploring ways in which that term can be evaluated from experimental data and models for how the term depends on system composition and temperature.

In the mass balances shown in Chapter 1, the reaction term was usually written in a generic way as $V r_A$, with r_A representing the rate of formation of species A per unit volume of solution per unit time. Note that, because the constituents participating in any reaction must react in proportions defined by the reaction stoichiometry, the reaction rate of any other constituent in the reaction can be related to r_A. That is, for a reaction $a\text{A} + b\text{B} \rightarrow p\text{P}$, if r_A, r_B, and r_P are expressed on a molar basis (e.g., mol/L min), they must be related as follows:

$$-\frac{1}{a}r_A = -\frac{1}{b}r_B = \frac{1}{p}r_P \quad (3\text{-}1a)$$

$$r_A = \frac{a}{b}r_B = -\frac{a}{p}r_P \quad (3\text{-}1b)$$

Thus, by characterizing r_A as a function of system composition and temperature, we are implicitly characterizing r_B and r_P as well.

The chapter begins with an introduction to fundamental concepts and terminology that are used to describe reaction kinetics, followed by a discussion of the approaches that are commonly used to evaluate reaction kinetics empirically. The presentation focuses on the conversion of raw experimental data into mathematical expressions that can be used to model reaction kinetics, thereby allowing the rate of reaction to be predicted for conditions that have not been studied directly. The types of reactions that are explored in this section gradually increase in complexity, from single, irreversible reactions to overall reactions that might include multiple and/or reversible steps. In the process of describing these reactions, the idea of a characteristic reaction time is introduced, and the linkage between the kinetics of reversible reactions and the equilibrium constant for those reactions is explained. In the final section, the dependence of reaction rates on temperature is described.

Many textbooks have been written devoted to reaction kinetics and reactor engineering. Excellent texts that consider reaction kinetics primarily in the context of basic chemical and chemical engineering systems include those by Levenspiel (1999) and Hill (1977). Applications of

---

Water Quality Engineering: Physical/Chemical Treatment Processes, First Edition. By Mark M. Benjamin, Desmond F. Lawler.
© 2013 John Wiley & Sons, Inc. Published 2013 by John Wiley & Sons, Inc.

kinetics to environmental systems are introduced in many water chemistry texts (e.g., Stumm and Morgan, 1996; Morel and Hering, 1993; Benjamin, 2010), and are covered at a more advanced level in texts devoted specifically to that topic (e.g., Brezonik, 1994; Stumm, 1990).

Whereas Chapter 2 emphasized the hydrodynamics of reactors without regard to any reaction that occurs in them, this chapter emphasizes the kinetics of chemical reactions without regard to the hydrodynamics of the reactor in which the reaction is occurring. (The mixing pattern in the reactor is important for evaluating the kinetics, but the resulting rate expression applies to reactors with any hydrodynamic characteristics.) This emphasis is possible because reactions are inherently molecular-scale processes. That is, when molecules collide and react, their direct interaction is independent of the macroscale mixing that might be occurring in the reactor.

While it is convenient to study reactor hydrodynamics and reaction kinetics independently, ultimately both factors are critical for determining the extent to which a given reaction proceeds in a given reactor. We explore the interactions between these factors in subsequent chapters, where mass balances that apply to specific types of water and wastewater treatment processes (in which flow, mixing, and chemical reactions occur simultaneously) are explored.

■ **EXAMPLE 3-1.** Hexavalent chromium (i.e., Cr in the +6 oxidation state, commonly represented as Cr(VI)) is found in many industrial wastes at substantial concentrations, and is a trace contaminant in some groundwater. The treatment of these wastes often involves reduction of the Cr(VI) to trivalent Cr (i.e., Cr(III)), in which state the metal is very likely to precipitate as chromium hydroxide ($Cr(OH)_3(s)$). This treatment approach is attractive both because Cr(III) is much less of a health hazard than Cr(VI) and because the precipitated $Cr(OH)_3(s)$ can be removed from the system by settling and/or filtration.

The dominant form of Cr(VI) in mildly acidic, dilute wastes is as the bichromate ion, $HCrO_4^-$. Reaction of this ion with Fe^{2+} (ferrous) ions can generate $Cr(OH)_3(s)$ and $Fe(OH)_3(s)$ via the reaction

$$HCrO_4^- + 3\,Fe^{2+} + 8\,H_2O \rightarrow 3\,Fe(OH)_3(s) + Cr(OH)_3(s) + 5\,H^+$$

(a) How are the rates of reaction of $HCrO_4^-$, Fe^{2+}, and H^+ related to one another?

(b) If $r_{HCrO_4^-}$ by this reaction is -10^{-5} mol/L s, what is the rate at which alkalinity is being destroyed by the reaction, in equiv/L? (Note: each mole of H^+ produced destroys one equivalent of alkalinity.)

Solution.

(a) The rates of reaction of $HCrO_4^-$, Fe^{2+}, and H^+ are linked by the reaction stoichiometry. Applying Equation 3-1b to the reaction of interest, with $HCrO_4^-$, Fe^{2+}, and H^+ identified as A, B, and P, respectively, we can write

$$r_{HCrO_4^-} = \tfrac{1}{3} r_{Fe^{2+}} = -\tfrac{1}{5} r_{H^+}$$

(b) If $r_{HCrO_4^-} = -10^{-5}$ (mol/L s), r_{H^+} is $+5 \times 10^{-5}$ (mol/L s). Since one equivalent of alkalinity is destroyed per equivalent of H^+ produced, $r_{Alk,equiv/L} = -r_{H^+,mol/L}(1\,equiv/mol\,H^+) = -5 \times 10^{-5}$ (equiv/L s). ■

3.2 FUNDAMENTALS

Terminology

The step-by-step sequence of molecular collisions that converts reactants to products in a chemical reaction is called the *reaction mechanism* or *reaction pathway*, and each step in that sequence is called an *elementary reaction*. Thus, if a reaction A + B → P occurs by a collision between one molecule of A and one molecule of B to form one molecule of P, that reaction is elementary. At first glance, it might seem that all reactions of interest would be elementary. However, to the contrary, most overall reactions of interest are *nonelementary*. That is, the reaction mechanism consists of two or more elementary reactions, and the products that can be detected and measured at the macroscopic scale are not formed directly from the reactants in the manner expressed by the overall reaction stoichiometry. Nonelementary reactions are discussed in detail shortly.

For elementary reactions, the reaction rate expression is related directly to the stoichiometry. Specifically, if the reaction A + B → P is elementary, the reaction rate is proportional to the chemical activities and/or concentrations of the reactants, as given by the following, equivalent expressions:

$$r_A = r_B = -r_P = -k' a_A a_B \quad (3\text{-}2a)$$

$$= -k' \left(\gamma_A \frac{c_A}{c_{A,std.state}} \right) \left(\gamma_B \frac{c_B}{c_{B,std.state}} \right) \quad (3\text{-}2b)$$

where k' is a constant that depends on solution composition and temperature, with dimensions of mass/volume-time (e.g., mol/L min) a_i is the chemical activity of species i, dimensionless γ_i is the activity coefficient of i, dimensionless $c_{i,std.state}$ is the concentration of i in the standard state, mass/volume (e.g., mol/L)

The concentration terms in Equation 3-2b are often isolated from the rest of the terms on the right side by combining the latter terms into a composite parameter, k, as follows:

$$r_A = r_B = -r_P = -k c_A c_B \quad (3\text{-}3)$$

where $k \equiv \gamma_A \gamma_B / (c_{A,std.state} c_{B,std.state}) k'$, with dimensions of time-volume/mass (e.g., min L/mol). γ_A and γ_B depend primarily on the bulk composition of the solution (particularly, its ionic strength), so if that composition is approximately constant, k is constant as well.

Both k' and k are called *reaction rate constants*. The choice of representing the reaction rate expression in terms of chemical activities (Equation 3-2a) or concentrations (Equation 3-2b) (or, for that matter, as a mixture of concentrations and activities) is somewhat arbitrary. Use of concentration-based values is more common, but the activity-based values are more convenient in a few applications. In particular, if water is a reactant in an elementary reaction, it is common to write the rate expression in terms of the activity of water (a_{H_2O}, which has a value of ~ 1.0 under virtually all conditions of interest) rather than its concentration (c_{H_2O}, which has a value of ~ 55.6 mol/L).

The standard state concentrations $c_{A,std.state}$ and $c_{B,std.state}$ always have values of 1.0; they are included in Equation 3-2b (and therefore appear in the definition of k) only for consistency with thermodynamic conventions, which require that the activity of any species be dimensionless. For reactions involving only dissolved species in ideal solutions (in which case $\gamma_i = 1.0$ for all constituents), the numerical values of the concentration and activity are identical, so the distinction between k and k' becomes unimportant; in nonideal solutions, the values of both k and k' depend on the ionic strength of the solution. Unless otherwise specified, the ideal solution assumption is made throughout this text.[1,2]

The reaction $2A \rightarrow P$ can be written equivalently as $A + A \rightarrow P$. According to Equation 3-2, then, if this reaction were elementary, the reaction rate of A would be

$$r_A = -kc_A c_A = -kc_A^2 \quad (3\text{-}4)$$

Extending this reasoning, we can write the following rate expression for any elementary reaction with the generic stoichiometry $aA + bB + cC \rightarrow pP$

$$r_A = -kc_A^a c_B^b c_C^c \quad (3\text{-}5)$$

Although Equation 3-5 is universally accepted as the general form of the rate expression for an elementary reaction, theoretical calculations suggest that simultaneous collisions of three or more molecules are exceedingly rare in aqueous solutions under normal environmental conditions. Thus, as a practical matter, all elementary reactions in aqueous systems are assumed to involve two molecules; that is, to be *bimolecular*.

---
[1] A more detailed discussion of the relationship between concentration and activity is provided in Chapter 5 and in many water chemistry texts (see, for example, *Water Chemistry*, by Benjamin (2010)).
[2] The dependence of the rate constants can be explained and modeled, in large part, using the activated complex theory of reaction kinetics. This theory is outlined later in the chapter and is discussed in detail in most physical chemistry textbooks.

Empirically, the rates of many elementary reactions in solution can be represented by expressions of the form: $r_A = -kc_A$. At first glance, such an expression would appear to suggest that the reaction proceeds in the absence of molecular collisions (because it depends on the concentration of only a single species raised to the first power). A few reactions do indeed proceed at a rate that is governed by the speed of processes occurring within individual molecules, independent of their collisions with other molecules in the system. An important example of such a reaction is radioactive decay, a process that proceeds at the same rate regardless of intermolecular collision frequencies. However, in aquatic systems, elementary reactions that have a rate expression of the form $r_A = -k_A c_A$ usually occur via collisions with water molecules, and the rate expression shown actually represents a simplification of the true rate expression, which includes the activity of water. Thus, for instance, the true reaction rate of an acid dissociation process ($HA + H_2O \rightarrow H_3O^+ + A^-$) might be $r_{HA} = -k_{HA} c_{HA} a_{H_2O}$, but when this expression is combined with the assumption that $a_{H_2O} = 1.0$, the result is $r_{HA} = -k_{HA} c_{HA}$.

■ **EXAMPLE 3-2.** The hydration of $CO_2(aq)$ to form H_2CO_3, as shown in the following reaction, is elementary.

$$CO_2(aq) + H_2O \rightarrow H_2CO_3$$

The reaction rate expression can be represented as $r_{H_2CO_3} = k_{hyd} c_{CO_2(aq)}$, with k_{hyd} equal to approximately $0.03\,s^{-1}$. (This rate expression is a simplification of the more complete expression $r_{H_2CO_3} = k_{hyd} c_{CO_2(aq)} a_{H_2O}$, based on the assumption that $a_{H_2O} = 1.0$.) Determine the instantaneous rate of CO_2 hydration in an ideal solution containing 10^{-5} mol/L $CO_2(aq)$. Give the result as the fraction of the $CO_2(aq)$ that is being hydrated per second.

Solution. Substituting the initial values into the rate expression, the rate of hydration is

$$\begin{aligned} r_{H_2CO_3} &= k_{hyd} c_{CO_2(aq)} \\ &= (0.03\,s^{-1})(10^{-5}\,mol/L) = 3.00 \times 10^{-7}\,mol/L\,s \end{aligned}$$

Since the initial concentration of $CO_2(aq)$ is 10^{-5} mol/L, the fractional rate of hydration is

$$\frac{3 \times 10^{-7}\,(mol/L\,s)}{10^{-5}\,(mol/L)} = 0.03\,s^{-1} = 3\%\,s^{-1}$$

Note that, for this simple reaction rate expression (with the rate directly proportional to the concentration of a single constituent), the reaction rate constant can be interpreted as the fractional rate at which the constituent is consumed. ■

The Kinetics of Elementary Reactions

Frequency of Molecular Collisions The idea that the rate of an elementary reaction is proportional to the concentrations of the reacting species, as indicated by Equation 3-3, should make some intuitive sense. Reactions occur as a result of molecular collisions, and the rate of collisions between any two types of molecules is proportional to their respective concentrations. As a result, the rate of A–B collisions represents the maximum possible rate at which A and B can react with one another; in reality, not every collision leads to a reaction, so the reaction rate is less than the collision rate.

A theoretical analysis of the rate of molecular collisions in solution, based on work originally done by Smoluchowski (1917) to model the interactions of colloids, yields the following equation for the rate of "first encounters" of uncharged molecules:[3]

$$\text{Rate of encounters} = 4\pi(\lambda_A + \lambda_B)(D_A + D_B)N_{Av}c_Ac_B/1000 \quad (3\text{-}6)$$

where N_{Av} is Avogadro's number, λ_i and D_i are the radius and diffusion coefficient of molecule i in cm and cm^2/s, respectively, c_i is the concentration of i in mol/L, the factor of 1000 is the conversion between cm^3 and liters, and the rate of encounters is in mol/L s.[4]

If the molecules are assumed to behave as hard spheres moving through a continuum, the diffusion coefficients D_A and D_B can be related to the radii of the molecules, via a relationship known as the Stokes–Einstein equation,

$$D_i = \frac{RT}{6\pi N_{Av}\mu}\frac{1}{\lambda_i} \quad (3\text{-}7)$$

where R is the universal gas constant, T is absolute temperature, and μ is the solution viscosity.[5,6] Substituting Equation 3-7 into Equation 3-6, the rate of encounters becomes

$$\text{Rate of encounters} = \frac{2RT}{3\mu}\left(2 + \frac{\lambda_A^2 + \lambda_B^2}{\lambda_A\lambda_B}\right)\frac{c_Ac_B}{1000} \quad (3\text{-}8)$$

Equation 3-8 expresses the maximum rate of an elementary reaction between two uncharged species. According to Equation 3-3, however, the rate of such a reaction can also be expressed as the product of a rate constant and the concentrations of the reacting species. Thus, the collection of terms other than c_Ac_B on the right side of Equation 3-8 can be interpreted as the maximum possible value of the rate constant (k_{max}) for a reaction between two uncharged species; that is,

$$k_{max} = \frac{2RT}{3\mu}\left(2 + \frac{\lambda_A^2 + \lambda_B^2}{\lambda_A\lambda_B}\right)\frac{1}{1000} \quad (3\text{-}9)$$

For moderate sized, uncharged solutes in aqueous solutions at room temperature, the value of k_{max} is on the order of 10^{10} (mol/L)$^{-1}$s^{-1}. If the reacting species are oppositely charged ions, their attraction for one another can increase k_{max} by up to about an order of magnitude, and if they are ions with like charges, their mutual repulsion can decrease k_{max} by as much as two orders of magnitude. Reactions that proceed at the rates close to these maximum predicted values are said to be *diffusion-controlled*.

■ **EXAMPLE 3-3.** Two constituents of a solution, A and B, participate in a bimolecular, elementary reaction at diffusion controlled rates, with a rate constant of $k = 3 \times 10^{10}$ (mol/L)$^{-1}$s^{-1}. In a solution that initially contains 10^{-3} M A and 10^{-4} M B, what is the fractional rate of destruction of species B?

Solution. Since the reaction is elementary, the rate expression is

$$r_A = r_B = -kc_Ac_B$$
$$= -3 \times 10^{10} \,(\text{mol/L})^{-1}\,\text{s}^{-1}\,(10^{-3}\,\text{mol/L})(10^{-4}\,\text{mol/L})$$
$$= -3 \times 10^3 \,\text{mol/L s}$$

The rate of destruction of species B is $-r_B$, or 3×10^3 mol/L s. Dividing this value by the initial concentration of B yields the fractional rate of destruction as

$$\frac{r_B}{c_B} = \frac{3 \times 10^3 \,(\text{mol/L s})}{10^{-4}\,(\text{mol/L})} = 3 \times 10^7 \,\text{s}^{-1} = 3 \times 10^9 \%\,\text{s}^{-1}$$

The rate is fantastically rapid, many orders of magnitude faster than the rate of hydration of CO_2 found in Example 3-2. ■

Energetics of Molecular Collisions Although the measured reaction rates of some proton-transfer (acid–base) reactions approach those estimated based on diffusion control, rate constants for the vast majority of reactions of interest in water and wastewater treatment are less than k_{max} by several orders of magnitude. In such cases, the rate of the reaction is controlled not only by the frequency of

---

[3] An encounter is an instance in which the two molecules approach one another closely enough that they can subsequently collide. They then might remain close together and collide several times, but this is counted as only a single encounter. It is estimated that, each time dissolved molecules encounter one another in water at room temperature, they collide ~150 times before separating (Brezonik, 1994).

[4] See *Chemical Kinetics and Process Dynamics in Aquatic Systems*, by Brezonik (1994) for a discussion of the derivation of this equation.

[5] The Stokes-Einstein is derived in many textbooks. See, for example, *Diffusion: Mass Transfer in Fluid Systems*, by Cussler (1984) or *Water Chemistry*, by Benjamin (2010).

[6] R has dimensions of energy per mole per degree (temperature). Values of R in some common units are 1.987 cal/mol K, 8.314 J/mol K and 0.0821 L-atm/mol K.

collisions, but also by the energy of those collisions. That is, even though many collisions might occur, relatively few of them are sufficiently energetic to lead to the rearrangement of chemical bonds needed to form the product. The rates of such reactions are referred to as being *chemically controlled*, and they are modeled based on the energy transformations that occur as the reactant molecules approach one another and interact.

When reactant molecules approach one another, interactions between their electric fields cause the bonds in each molecule to become strained. This process increases the chemical potential energy of the molecules. The strain increases dramatically with decreasing separation. Since energy is conserved during the interaction, the increase in chemical potential energy must be balanced by a corresponding decrease in the energy elsewhere in the system. In this case, the dominant process is the conversion of molecular kinetic energy into chemical potential energy: the molecules slow down as the distance between them decreases and the bond strain increases. In the absence of other factors, the molecules would eventually stop their mutual approach and begin moving away from one another, thereby relieving the strain and reconverting the chemical energy into kinetic energy. From an energy perspective, the process is identical to that of a ball rolling up an incline, stopping, and reversing itself.

However, it is possible that, as the original molecular structures adjust in response to the strain, some bonds will weaken and others will begin to form. At some critical point, the original bonds become sufficiently distorted, and the new bonds form to a sufficient extent, that the strain can be more easily relieved by rearrangements that form product molecules rather than the original reactants. The amount of energy necessary to bring molecules from far apart (no interaction) to the critical point (the point where the strain is equally likely to be relieved by the formation of products or reformation of reactants) is called the *activation energy*, E^*. Similarly, the process of reaching the critical condition is commonly referred to as "overcoming the activation energy barrier." Molecules that have participated in reactions with enough energy to overcome the activation energy barrier are not identifiable as either the reactants or the products, but are intermediate species of negligible stability known as *activated complexes*. The overall model is referred to as the *activated complex model* or *transition state theory*.

In addition to the energy of the colliding molecules, the orientation of the molecules when they collide might be important in determining whether a reaction occurs. If the molecules are not spherically symmetric, only a fraction of all collisions can cause the bonds to distort in a way that leads to the formation of product, even if the collisions involve an amount of energy greater than the activation energy. Thus, the overall rate of reaction depends on the frequency with which reactant molecules interact, the likelihood that the colliding molecules are properly oriented and have sufficient kinetic energy to form the activated complex, and the rate at which the activated species is converted into products.

The mathematical formulation of the model involves representing the energy path that molecules follow as they are converted to products. This path is sometimes compared to a landscape, in which energy is initially required to bring the reactants to the top of a hill (the height of which represents the activation energy) before they can progress down the other side to form the product. For a generic bimolecular reaction between A and B, the activated complexes are commonly represented as AB^*, so the overall reaction can be represented as $A + B \rightarrow AB^* \rightarrow P$. A schematic of the energy relationships in the transition from reactants through the high-energy state to products is shown in Figure 3-1.

In any system, the energy associated with a given chemical species i is distributed among all the molecules of that species, with many molecules having a low energy, and progressively smaller numbers of molecules having progressively higher energies. A key assumption of the activated complex model is that, even though the energy of any given molecule of i changes from one instant to the next, the distribution of the available energy among molecules of i is constant (as long as the temperature of the bulk phase remains constant). That is, the fraction of all molecules of i that have molar energy greater than some

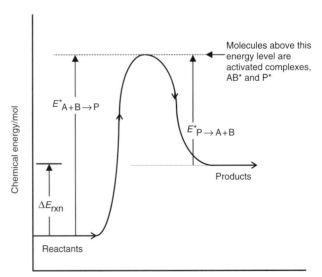

FIGURE 3-1. Schematic representation of the transition from reactants to products according to the activated complex model. The activated complexes are designated AB^* for the forward reaction and P^* for the reverse reaction, but are functionally indistinguishable.

specified value $\overline{E}_1$ is the same at any instant, even though the specific molecules that meet that criterion changes from one instant to the next. Furthermore, this energy distribution is assumed to be consistent with the *Boltzmann distribution*, which is characterized mathematically as follows:

$$\begin{pmatrix} \text{Probability that a molecule will have} \\ \text{molar energy between } \overline{E} \text{ and} \\ \overline{E} + d\overline{E}, \text{ when the temperature is } T \end{pmatrix}$$
$$= A \left(\frac{MW}{T}\right)^{1/2} \left(\frac{\overline{E}}{RT}\right) \exp\left(-\frac{\overline{E}}{RT}\right) \quad (3\text{-}10)$$

where A is a normalization factor that assures that the sum of all probabilities equals 1.0. At a given temperature, the pre-exponential term in Equation 3-10 increases rapidly with increasing $\overline{E}$, whereas the exponential term decreases. The former trend dominates at low values of $\overline{E}$, whereas the trend in $\exp(-\overline{E}/RT)$ dominates at higher values, leading to characteristic energy distributions among molecules like those shown in Figure 3-2.

Assuming that molecular energy of each reactant in a system obeys the Boltzmann distribution, then some fraction of the collisions between reactant molecules will have sufficient energy $(\overline{E}^*)$ to form activated complexes. This fraction is strongly dependent on the value of $\overline{E}^*$ and the temperature: lower values of $\overline{E}^*$ and higher temperatures allow more collisions to generate activated complexes. If, in addition to having sufficient energy, the colliding molecules have the proper orientation, then activated complexes form and decay to reaction products at a substantial rate.

Based on theoretical considerations that are beyond the scope of the current discussion, the rate of conversion of activated complexes to products is believed to be the same for all activated complexes, regardless of the reaction that is occurring.[7] Therefore, the dependence of the rate constant on temperature is controlled entirely by $\overline{E}^*$ and T, as follows:[8]

$$\frac{d \ln k}{dT} = \frac{\overline{E}^*}{RT^2} \quad (3\text{-}11)$$

$$\ln k_{T_2} = \ln k_{T_1} - \frac{\overline{E}^*}{R}\left(\frac{1}{T_2} - \frac{1}{T_1}\right) \quad (3\text{-}12)$$

Thus, if the rate constant of an elementary reaction is evaluated at several temperatures, a plot of $\ln k$ versus $1/T$ is expected to have a slope of $\overline{E}^*/R$, from which $\overline{E}^*$ can be determined. The rate constant at any other temperature can then be computed using Equation 3-12.

Catalysts operate by providing an alternative mechanism by which a reaction can occur. In particular, they allow reactants to be converted to products via a route that has a lower activation energy than the route that is taken in their absence. In terms of the energy landscape, a catalyst can be viewed as making an alternative path available that does not require quite so much of a climb before arriving at a point on the downhill slope. In some cases, catalysts can increase the rate of an overall reaction by many orders of magnitude. Note, however, that because the reactants and products of a reaction are the same in the presence or absence of catalysts, catalysts cannot alter the overall energy change associated with the reaction. Consequently, they have no effect on the equilibrium constant for the reaction and do not change the distribution of reactants and products once equilibrium is attained. Also, according to Equations 3-11 and 3-12, because catalysts lower $\overline{E}^*$, they decrease the dependence of the reaction rate on temperature.

Sometimes, the products of the reaction itself can act as catalysts. For instance, during the oxidation of ferrous ion (Fe^{2+}), ferric hydroxide solids ($Fe(OH)_3(s)$) are often formed. The surfaces of such oxides can catalyze the oxidation of the ferrous ions remaining in solution, so the effective rate constant increases as the reaction proceeds. Such reactions are called *autocatalytic*. In other cases, the products of the reaction may combine with a reactant to form

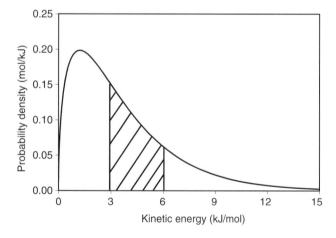

FIGURE 3-2. The Boltzmann distribution of kinetic energy among molecules at 20°C for a chemical with MW = 150. The cross-hatched area equals the fraction of the molecules that have kinetic energy between 3 and 6 kJ/mol (for reference, RT at this temperature is 2.44 kJ/mol.) With increasing MW, the distribution broadens somewhat, but the mean value of molar kinetic energy remains constant. With increasing temperature, the curve broadens and the peak shifts to higher energy, but the change is small over the range of temperatures in typical environmental systems.

---

[7] Additional information on both the conceptual underpinnings of the activated complex model and its implications are available in a number of water chemistry books as well as in virtually all physical chemistry textbooks.

[8] The given equation is an approximation that is applicable if $\overline{E}^* \gg RT$, which is valid for virtually any reaction of interest.

a species that does not participate in the original reaction. In those cases, the net result is that as product forms, the overall (apparent) rate constant decreases, and the reaction is said to be *autoretardant*.

As noted previously, most overall reactions of interest are not elementary. As a result, Equations 3-11 and 3-12 are not directly applicable to many reactions whose temperature dependence we might wish to characterize. However, the general form of those equations often applies nevertheless. We consider the effects of temperature on the rate constants of overall reactions in more detail later in this chapter.

The Kinetics of Nonelementary Reactions

As has been noted several times, most observable reactions in water are the net result of several elementary reactions proceeding in parallel and/or series. In fact, a completely rigorous analysis would indicate that the net rate of a reaction can never be fully described by a single, elementary reaction, since any reaction that is proceeding forward ($A \rightarrow B$) is, in theory, simultaneously proceeding in the reverse direction ($A \leftarrow B$).[9] In addition to participating in both the forward and reverse directions of a given reaction, a chemical might participate in two or more completely independent reactions simultaneously.

If the dominant reactions affecting a particular species cause it to be produced by one reaction and consumed by another, the reactions are said to be *sequential*, *consecutive*, or *series* reactions. For instance, when trisodium phosphate (Na_3PO_4) is added to an acidic solution, the salt dissolves to release Na^+ and PO_4^{3-} ions. Sequential protonation reactions can then generate HPO_4^{2-}, $H_2PO_4^-$, and (if the solution is sufficiently acidic) H_3PO_4. In this case, the reactions are so rapid that we do not normally bother analyzing their kinetics; rather, we assume that they reach equilibrium instantly. Nevertheless, it is clear that the reactions must proceed sequentially, so each step in the reaction could, in theory, be described by a kinetic expression.

In many sequential reactions, intermediate species are formed and then destroyed so rapidly that they are never present at a detectable level. For instance, in the generic reaction sequence $A \rightarrow B \rightarrow P$, the intermediate B might be so unstable that its existence is never noticed. In such a case, the overall reaction $A \rightarrow P$ would appear to be the only reaction occurring, and the rate expressions could be developed to describe the forward and reverse steps in that overall reaction. (The expressions would not be rigorously correct, but the errors might be undetectably small, being related to small changes in the concentration of B as the reaction proceeds.) The rate expressions might be adequate for characterizing the rate of the overall reaction under a wide variety of circumstances, but they would not be related to the reaction stoichiometry in the simple way that applies to elementary reactions (Equations 3-2, 3-4, and 3-5).

If the dominant reactions affecting a species cause it to be consumed in more than one reaction simultaneously, the reactions are said to operate in *parallel* and to be *competitive* for that reactant. In the same example system discussed earlier (addition of Na_3PO_4 to a solution), if the solution contained Ca^{2+} and Fe^{3+} ions, some PO_4^{3-} ions entering the solution would react with those ions, while others reacted with H^+. Thus, three reactions that convert PO_4^{3-} to other species would proceed in parallel, and the three cations might be described as competing with one another for the PO_4^{3-} ions.

Both sequential and parallel reactions abound in environmental engineering. Examples of the former include many biochemical transformations, such as the oxidation of ammonia to nitrite and then nitrate (Figure 3-3a), as well as many abiotic reactions, such as the precipitation of metals and oxidation of sulfides. Similarly, many chemicals undergo biologically mediated and abiotic reactions in parallel. For example, dissolved oxygen might participate simultaneously in biochemical oxidation reactions and the abiotic oxidation of ferrous and sulfide species (Figure 3-3b).

In many cases, a group of reactions forms a network that involves both sequential and parallel reactions. One particularly important example of such a network is the chlorination of natural organic matter (NOM) to form disinfection byproducts (DBPs), a process which was described in Example 1-6. In this process, some of the added hypochlorous acid (HOCl) reacts with NOM molecules in the system, and other HOCl molecules oxidize some inorganic species in solution. If any bromide (Br^-) is present, it is oxidized to HOBr, which can

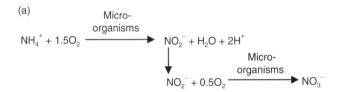

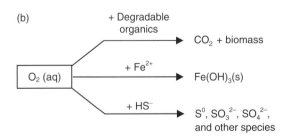

FIGURE 3-3. Examples of sequential and parallel reactions. (a) Oxidation of ammonia to nitrate in two sequential steps. (b) Competition for dissolved oxygen among three different reactions.

---

[9] The conditions under which an assumption of irreversibility is reasonable are discussed later in this chapter.

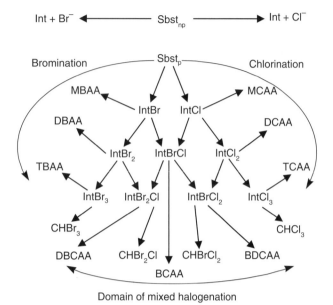

FIGURE 3-4. Schematic representation of the complexity of the reactions of HOCl and HOBr with NOM to form halogenated disinfection byproducts. Sbst$_p$ and Sbst$_{np}$ represent precursor and nonprecursor NOM molecules, respectively (those that generate DBPs, and those that generate nonhalogenated products, respectively). Unidentified intermediate species are labeled IntBr$_x$Cl$_y$, and some identifiable DBPs are shown as final products. Species shown as CHBr$_x$Cl$_{3-x}$ are trihalomethanes (like chloroform), and those that have AA as the last two letters are halogenated acetic acids (e.g., BDCAA is bromo-dichloro-acetic acid). *Source*: After Korshin et al. (2004).

react with NOM molecules in much the same way that HOCl does, forming brominated DBPs. Thus, initially, the different types of NOM molecules and Br$^-$ compete for the available HOCl. Then, once essentially all of the Br$^-$ has been oxidized, HOBr and the remaining HOCl compete for the available NOM. In addition to these competitive reactions, both the HOBr and HOCl participate in sequential reactions with NOM, forming mono-, then di-, and finally tri-halogenated species. A schematic of a portion of this reaction network is shown in Figure 3-4.

Power Law and Other Rate Expressions for Nonelementary Reactions

Often, even if a reaction $aA + bB \to pP$ is not elementary, the reaction rate of A can be modeled quite well as the product of a rate constant and the activities or concentrations of the reactants raised to some power. That is,

$$r_A = -k' a_A^\alpha a_B^\beta \tag{3-13a}$$

$$r_A = -k c_A^\alpha c_B^\beta \tag{3-13b}$$

where α and β are empirical constants, and $k \equiv k'\left(\gamma_A^\alpha \gamma_B^\beta / c_{A,\text{std.state}}^\alpha c_{B,\text{std.state}}^\beta\right)$.

When the rate expression takes this simple form, the sum of the exponents $(\alpha + \beta)$ is called the overall *order* of the reaction, and the reaction is said to be α-order with respect to A and β-order with respect to B. α and β are not necessarily integers and are not necessarily the stoichiometric coefficients of A and B in the reaction. Although many reactions have rate expressions that take the form of Equation 3-13, others do not; in the latter cases, one cannot speak of the reaction order.

It is important to understand the distinction between the exponents α and β in Equation 3-13, and a and b in Equation 3-5. Whereas the parameters a and b are the stoichiometric coefficients for the reactants in the elementary chemical reaction $aA + bB \to pP$, α and β are simply empirical constants that allow the experimental rate data to be fit by a power law equation. If the fit between the data and a power law expression is satisfactory, then the equation provides a concise and convenient way to summarize those data, and we might use it to predict the system behavior under conditions that have not been studied experimentally. If the fit is not good, then we have to search for another mathematical formulation that fits the data better. However, regardless of the outcome, the overall equation and the individual exponents have no direct, fundamental interpretation. In particular, if α and β happen to correspond to the stoichiometric coefficients of the reactants A and B, that result might mean that the reaction is elementary, or it might be purely coincidental. On the other hand, if the reaction of interest is elementary, the exponents in Equations 3-13a and 3-13b *must* be the stoichiometric coefficients of the reactants in that reaction. As noted earlier, this situation is the exception rather than the rule for reactions of interest in environmental engineering.

3.3 KINETICS OF IRREVERSIBLE REACTIONS

The preceding sections define terms and establish the framework for analysis of reaction kinetics. However, in practice, the central component in most engineering studies of reaction kinetics is the determination of the rate expression as a function of the concentrations of the reactants. The term reactant is used broadly here to refer to all species whose concentrations affect the rate of the overall reaction (i.e., all species that participate in the elementary reactions leading to the overall reaction), even if those species do not appear in the net reaction stoichiometry. In this stage of the analysis, empirical data describing the reaction rate under various conditions are collected, and an overall rate expression is proposed and used to predict the reaction kinetics under different conditions. Those predictions are then tested, and the proposed rate expression is accepted or modified, as appropriate. In some cases, though not always, an attempt is made to link the rate expression to the reaction mechanism (i.e., the elementary reactions that combine to yield the overall reaction). Ideally, the cycling between experimental and modeling efforts continues until the rate expression successfully describes the kinetics over a wide range of conditions.

Once the dependence of the reaction rate on reactant concentrations is established, the effects of other factors on the reaction rates might be explored. The most important of these factors is usually temperature. The rest of this chapter describes approaches for carrying out each of the analytical components described earlier. The discussion is presented in four sections. The first three sections focus on the effects of concentration on reaction progress for irreversible, reversible, and sequential reactions, respectively; the final section then focuses on the effects of temperature on reaction rates.

The Mass Balance for Batch Reactors with Irreversible Reactions

Generally, the information available for analysis of a rate expression includes the stoichiometry of the overall reaction and some experimental data describing the concentration versus time profiles for one or more species that participate in the reaction. The experiments used to generate the data may be conducted in any type of reactor. However, the simplest and most common approach is to characterize the rate at which the substances of interest are generated or destroyed in a well-mixed, batch (no flow) reactor. The analysis of such a reactor involves writing one or more mass balances in which the control volume includes all the fluid in the system, such as the following.

$$\begin{pmatrix} \text{Rate of change} \\ \text{of the mass} \\ \text{of } i \text{ stored in the} \\ \text{system} \end{pmatrix} = \begin{pmatrix} \text{Net rate} \\ (\text{in}-\text{out}) \text{ at} \\ \text{which } i \text{ enters} \\ \text{by advection} \end{pmatrix}$$
$$+ \begin{pmatrix} \text{Net rate} \\ (\text{in}-\text{out}) \text{ at which } i \\ \text{enters by diffusion} \\ \text{by advection and/or dispersion} \end{pmatrix} + \begin{pmatrix} \text{Net rate} \\ (\text{formation}-\text{destruction}) \\ \text{at which } i \text{ is created by} \\ \text{chemical reaction} \end{pmatrix}$$

Since there is no flow into or out of the reactor, the advective term is zero, and since the aqueous phase ends at the boundaries of the control volume, the diffusive term is zero as well. Applying these ideas and dividing through by V, the mass balance can be simplified to the following expression:

$$V\frac{\partial c_i}{\partial t} = \cancel{Q(c_{i,\text{in}} - c_{i,\text{out}})} + \cancel{(D_i + \epsilon)\frac{\partial^2 c_i}{\partial x^2}}V + (r_{i,V} + r_{i,\sigma}a)V$$

$$\frac{dc_i}{dt} = r_i \qquad (3\text{-}14)$$

where $r_i = r_{i,V} + r_{i,\sigma}a$.[10]

The overall reaction rate, r_i, has dimensions of mass per volume per time and may be either positive (if the reaction forms the substance under consideration) or negative (if the reaction destroys it). Unfortunately, the simplicity of Equation 3-14 leads to a common error whereby the term dc_i/dt is substituted for the reaction rate (r_i) in other mass balances (i.e., mass balances for nonbatch systems). While the equality is valid for a batch system, it is not valid for other systems.

Defining c_i as the concentration of i in moles per liter, the volume-normalized mass balances on A, B, and P in a batch system where the reaction $aA + bB \rightarrow pP$ is the only one affecting the concentrations of those constituents can be expressed as follows:

$$-r_A = -\frac{dc_A}{dt} = \text{rate of destruction of A} \qquad (3\text{-}15a)$$

$$-r_B = -\frac{dc_B}{dt} = \text{rate of destruction of B} \qquad (3\text{-}15b)$$

$$r_P = \frac{dc_P}{dt} = \text{rate of appearance of P} \qquad (3\text{-}15c)$$

In the following analysis, it is assumed that the reaction of interest occurs in a well-mixed batch reactor. It is also assumed for simplicity that the reaction is irreversible, so that Equation 3-15 applies. Data from experimental investigations of irreversible reactions are often analyzed by two techniques, referred to as the integral and differential methods. These approaches for analyzing kinetics data are described in the following sections.

The Integral Method of Reaction Rate Analysis

The integral method consists of postulating a rate expression, substituting it into Equation 3-14, integrating, and comparing the predicted c versus t response to the experimental one. For instance, if we postulated that a reaction $A \rightarrow B$ is first order with respect to A and independent of the concentration of B (i.e., $r_A = -k_1 c_A$), the differential and integrated forms of Equation 3-14 would be as follows:

$$\frac{dc_A}{dt} = -k_1 c_A \qquad (3\text{-}16)$$

$$\int_{c_A(0)}^{c_A(t)} \frac{dc_A}{c_A} = -k_1 \int_0^t dt \qquad (3\text{-}17)$$

$$\ln c_A(t) - \ln c_A(0) = \ln \frac{c_A(t)}{c_A(0)} = -k_1 t \qquad (3\text{-}18a)$$

$$c_A(t) = c_A(0)\exp(-k_1 t) \qquad (3\text{-}18b)$$

---

[10] For simplicity, unless otherwise noted, the discussion in this chapter assumes that the substances of interest do not participate in surface reactions ($r_{i,\sigma} = 0$), so $r_{i,V} = r_i$. The extension to include interfacial reactions is straightforward. Examples including such reactions are provided in subsequent chapters.

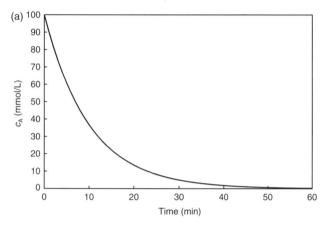

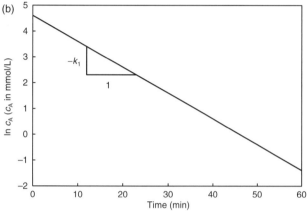

FIGURE 3-5. Two plots of the concentration versus time profile for a first order, irreversible reaction with $c(0) = 100\,\mu\text{mol/L}$ and $k_1 = 0.1\,\text{min}^{-1}$. (a) Unmodified data ($c_A$ vs t); (b) linearized data ($\ln c_A(t)$ vs t).

where $c_A(t)$ is the concentration of A remaining at time t, and $c_A(0)$ is the initial concentration. The result is that the concentration of A decays exponentially over time, as shown in Figure 3-5a for a reactant that is present in the initial solution at a concentration of $100\,\mu\text{mol/L}$. Figure 3-5b shows the same data plotted as $\ln c_A$ versus time; according to Equation 3-18a, such a plot must be linear with a slope of $-k_1$ and an intercept on the ordinate at $\ln c(0)$. Thus, if we suspected that a reaction was first order with respect to A, we could plot the experimental data as $\ln c_A(t)$ versus t. If the plot was linear (as in Figure 3-5b), the slope would be identified as $-k_1$, and the reaction could be treated as first order. If the plot were not linear, we would conclude that the postulated reaction rate expression was incorrect, and other rate expressions might be tried.

If the rate expression being tested is of the form $r_A = k_n c_A^n$, the prediction of the concentration versus time profile follows the earlier derivation and is relatively straightforward. The differential and integrated equations for such rate expressions are shown in Table 3-1. Algebraic manipulation of those equations converts them to a form that expresses the concentration at any time as a function of initial concentration and reaction time. These expressions are also shown in Table 3-1, and their use is illustrated in Example 3-4.

■ **EXAMPLE 3-4.** The oxidation of nitrite ion by monochloramine (NH_2Cl) can proceed via the following reaction:

$$NO_2^- + NH_2Cl + H_2O \rightarrow NO_3^- + NH_4^+ + Cl^-$$

TABLE 3-1. Differential and Integral Forms of the Reaction Rate Expression for an nth Order Reaction Occurring in a Batch Reactor

Reaction Order	Rate Expression — Differential Form	Rate Expression — Integral Form			Two Forms of the Solution to the Rate Expression[a]	
Zero	$r_A = \dfrac{dc_A}{dt} = -k_0$	$\int dc_A = -k_0 \int dt$	$c_A(t) - c_A(0) = -k_0 t$	(3-19a)	$c_A(t) = c_A(0) - k_0 t$	(3-19b)
One	$r_A = \dfrac{dc_A}{dt} = -k_1 c_A$	$\int \dfrac{dc_A}{c_A} = -k_1 \int dt$	$\ln \dfrac{c_A(t)}{c_A(0)} = -k_1 t$	(3-18a)	$c_A(t) = c_A(0)\exp(-k_1 t)$	(3-18b)
Two	$r_A = \dfrac{dc_A}{dt} = -k_2 c_A^2$	$\int \dfrac{dc_A}{c_A^2} = -k_2 \int dt$	$\dfrac{1}{c_A(t)} - \dfrac{1}{c_A(0)} = k_2 t$	(3-20a)	$c_A(t) = \left(\dfrac{1}{c_A(0)} + k_2 t\right)^{-1}$	(3-20b)
Any $n \neq 1$	$r_A = \dfrac{dc_A}{dt} = -k_n c_A^n$	$\int \dfrac{dc_A}{c_A^n} = -k_n \int dt$	$\dfrac{1}{n-1}\left(c_A(t)^{1-n} - c_A(0)^{1-n}\right) = k_n t$	(3-21a)	$c_A(t) = \left[\{c_A(0)\}^{1-n} + (n-1)k_n t\right]^{1/(1-n)}$	(3-21b)

[a]In the final row, the expressions shown are valid for any t if $n > 1$. If $n < 1$, they are valid only up to $t = \{c_A(0)\}^{1-n}/((1-n)k_n)$, at which time the computed value of c_A is zero. In reality, for reactions with apparent n values <1, n increases and approaches a value of one as the reactant concentration approaches zero.
Source: Levenspiel (1999).

In a wastewater solution at pH 7.5 that has been dosed with excess monochloramine, the rate expression for this reaction can be approximated as $r_{NO_2^-} = -kc_{NO_2^-}^{1.7}$, with $k = 2.3 (L/mol)^{0.7} \, min^{-1}$.

(a) If the wastewater initially contains 3.0×10^{-5} mol/L NO_2^-, how much NO_2^- would remain after 30 min of reaction in a well-mixed batch reactor?

(b) How much time would be required for 90% of the NO_2^- to be oxidized in the solution described in part (a), if all other conditions were held constant?

Solution.

(a) Substituting values into Equation 3-21b, the expression for the concentration of NO_2^- remaining after 30 min is

$$c_{NO_2^-}(t) = \left[\{c_{NO_2^-}(0)\}^{1-n} + (n-1)k_n t \right]^{1/(1-n)}$$

$$c_{NO_2^-}(30 \, min) = \left[(3 \times 10^{-5} \, mol/L)^{1-1.7} + (1.7-1) \right.$$

$$\left. \times \left(2.3 \frac{L^{0.7}}{mol^{0.7} \, min} \right)(30 \, min) \right]^{1/(1-1.7)}$$

$$= 2.86 \times 10^{-5} \, mol/L$$

Thus, less than 5% of the NO_2^- has been oxidized in 30 min.

(b) When 90% of the NO_2^- has been oxidized, only 10% (or $3 \times 10^{-6} \, mol/L$) remains. Using Equation 3-21a, the time required to achieve this amount of conversion is

$$t = \frac{1}{k_n(n-1)} \left[\{c_{NO_2^-}(t)\}^{1-n} - \{c_{NO_2^-}(0)\}^{1-n} \right]$$

$$= \frac{1}{(2.3 \, L^{0.7}/mol^{0.7} \, min)(1.7-1)}$$

$$\times \left\{ (3 \times 10^{-6} \, mol/L)^{1-1.7} - (3 \times 10^{-5} \, mol/L)^{1-1.7} \right\}$$

$$= 3652 \, min \approx 2.5 d \qquad \blacksquare$$

Analysis of Reaction Half-Times

Based on the equations in Table 3-1, we could test whether a reaction was zero order or second order with respect to A by plotting $c_A(t)$ or $1/c_A(t)$ versus t, respectively. If either plot was linear, that would confirm our hypothesis regarding the order of the reaction, and the slope of the plot would indicate the rate constant. Similarly, as shown earlier, we could test the linearity of a plot of $\ln c_A(t)$ versus t to determine whether a reaction was first order. However, in many cases n is not an integer. In such cases, using Equation 3-21 in conjunction with a series of guesses for the value of n is tedious, at best, and might not succeed at all.

An alternative approach for determining n that is often successful is based on the time required for the concentration of A to be reduced to one-half of its initial value; that is, the value of t when $c_A(t) = 0.5 c_A(0)$. Defining that time as the half-time, $t_{1/2}$, and substituting the expression $c_A(t_{1/2}) = 0.5 c_A(0)$ into Equations 3-18 to 3-21, we obtain the relationships shown in Table 3-2.

Taking the logarithm of both sides of Equation 3-25, we obtain

$$\log t_{1/2} = \log \left\{ \frac{2^{n-1} - 1}{k_n(n-1)} \left(c_A(0)^{1-n} \right) \right\} \qquad (3\text{-}26)$$

$$= \log k^* + (1-n) \log(c_A(0)) \qquad (3\text{-}27)$$

TABLE 3-2. Expressions for the Time Required for the Concentration of a Substance to Decrease to Half of its Initial Value for nth Order Reactions Occurring in a Batch Reactor

Reaction Order	Integrated Rate Expression	Condition at $t_{1/2}$	Value of $t_{1/2}$	
0	$c_A(t) - c_A(0) = -k_0 t$	$0.5 c_A(0) = k_0 t_{1/2}$	$t_{1/2} = \frac{0.5}{k_0} c_A(0)$	(3-22)
1	$\ln \frac{c_A(t)}{c_A(0)} = -k_1 t$	$\ln 0.5 = -k_1 t_{1/2}$	$t_{1/2} = \frac{\ln 2}{k_1}$	(3-23)
2	$\frac{1}{c_A(t)} - \frac{1}{c_A(0)} = k_2 t$	$\frac{1}{c_A(0)} = k_2 t_{1/2}$	$t_{1/2} = \frac{1}{k_2 c_A(0)}$	(3-24)
Any $n \neq 1$	$\frac{1}{n-1} \left(\{c_A(t)\}^{1-n} - \{c_A(0)\}^{1-n} \right) = k_n t$	$\frac{1}{n-1} \left(\{0.5 c_A(0)\}^{1-n} - \{c_A(0)\}^{1-n} \right) = k_n t_{1/2}$	$t_{1/2} = \frac{2^{n-1} - 1}{k_n(n-1)} \{c_A(0)\}^{1-n}$	(3-25)

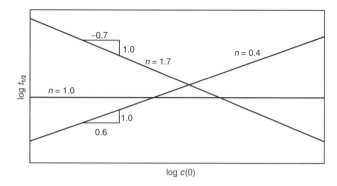

FIGURE 3-6. Logarithm of the half-time as a function of the initial concentration for irreversible reactions with rate expressions of the type $r_A = -k_n c_A^n$, in batch reactors.

Equation 3-27 has been derived for and is directly applicable only in cases where $n \neq 1$. However, comparison of Equation 3-27 with Equation 3-23 indicates that the former equation is also applicable to first-order reactions, if k^* is defined differently depending on the order of the reaction. Specifically, Equation 3-27 is applicable to any reaction that has power law kinetics if k^* is defined as follows:

$$k^* \equiv \begin{cases} \dfrac{2^{n-1}-1}{k_n(n-1)} & \text{for all } n \neq 1 \\ \dfrac{\ln 2}{k} & n = 1 \end{cases} \quad (3\text{-}28)$$

Therefore, we can test the hypothesis that a reaction rate is nth order with respect to a single species by plotting log $t_{1/2}$ versus log $c_A(0)$, for a series of values of $c_A(0)$. If the plot is linear, then the hypothesis is valid, and the slope equals $1 - n$; three examples of such a situation are shown in Figure 3-6. Furthermore, since the concentration defined as $c_A(0)$ need not truly be the concentration at the beginning of the experiment, several data points for a plot of log $t_{1/2}$ versus log $c_A(0)$ can be obtained from a single experiment. That is, we could choose several times during a test, define each time as $t = 0$ and the corresponding concentration as $c_A(0)$, and then determine how long it takes from that time for the concentration of A to decrease by one-half. These data could then be used to evaluate n. (Also, the choice of $c_A(t)/c_A(0) = 0.5$ is arbitrary; in theory, one could derive equally valid results and evaluate n for any choice of $c_A(t)/c_A(0)$, although the value of k^* would be different from that shown earlier.)

■ **EXAMPLE 3-5.** The following data were collected in a batch system in which the reaction A → P was proceeding.

t (min)	0	2	4	6	8	10	15	20
c_A, 10^{-4} M	25	12.9	7.9	5.2	3.6	2.8	1.6	1.0

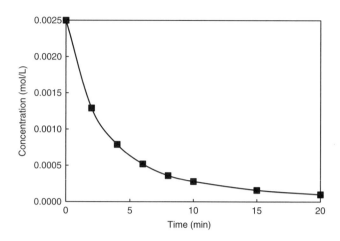

FIGURE 3-7. Plot of unmodified data for Example 3-5.

(a) Assuming that the rate expression is $r = k c_A^n$, find the value of n.
(b) What is the reaction rate constant?
(c) What concentration of A will be present in solution after 60 min?

Solution.

(a) A plot of c_A versus t is shown in Figure 3-7.
According to Equations 3-27 and 3-28, we can test whether the reaction kinetics are nth order, and if so determine the value of n, by plotting log $t_{1/2}$ versus log $c(0)$. To do so, we pick several times during the experiment, define each as $t = 0$, and determine how long it takes for the concentration of A to decrease by 50%. This approach works fine for $t < 10$ min, but for later times, it is difficult to determine the half-time because of the flatness of the curve. If the data at $t > 10$ min are trustworthy, then we should make an effort to include them in the analysis. That can be done by plotting the data as log c_A versus t, and noting that a 50% decrease in c_A corresponds to a decrease of 0.3 units in the value of log c_A (because log $0.5 = -0.3$). The data for such a plot are shown in the following table, and the plot is shown in Figure 3-8, from which it is clear that the logarithmic transformation makes the identification of $t_{1/2}$ at $t > 10$ min much easier.

t (min)	0	2	4	6	8	10	15	20
log c_A	−2.60	−2.89	−3.10	−3.28	−3.44	−3.55	−3.80	−4.00

Drawing a smooth curve through the data in Figure 3-8 and choosing several times throughout the experiment (not necessarily the times at which

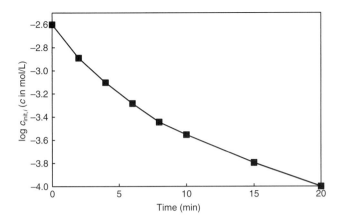

FIGURE 3-8. Data from Figure 3-7 replotted as $\log c$ versus t.

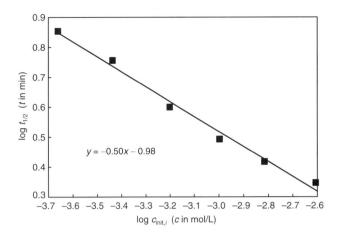

FIGURE 3-9. Log of half-time versus log of initial concentration for various starting points ($t_{\text{init},i}$).

the raw data were collected) as starting points, we can develop Table 3-3.

The values in columns A through C of the table identify six "initial" conditions. Column A shows the times associated with these conditions; we designate these times as $t_{\text{init},i}$ (i.e., $t_{\text{init},1}$, $t_{\text{init},2}$, and so forth.). Columns B and C show the concentrations ($c_{\text{init},i}$) and $\log c_{\text{init},i}$ values present in the solution at each $t_{\text{init},i}$. (Because all concentrations refer to the species A, the subscript A has been dropped from those terms.)

The time at which the concentration of A has declined to 0.5 $c(t_{\text{init},i})$ is designated as $t_{\text{fin},i}$. Values of $t_{\text{fin},i}$ can be estimated from Figure 3-8 by noting that, at time $t_{\text{fin},i}$, $\log c(t_{\text{fin},i}) = \log 0.5 + \log c(t_{\text{init}}) = -0.3 + \log c_A(t_{\text{init}})$. That is, at $t = t_{\text{fin},i}$, $\log c(t_{\text{fin},i})$ is 0.3 units less than $\log c(t_{\text{init},i})$. Values of $t_{\text{fin},i}$ are shown in column D of the table. Finally, the various $t_{1/2}$ values are computed as $t_{\text{fin},i} - t_{\text{init},i}$ and are shown in column E.

A plot of $\log t_{1/2}$ versus $\log c(t_{\text{init}})$ is shown in Figure 3-9. The data can be reasonably approximated by a straight line with a slope of -0.5. Because the slope equals $1 - n$, the reaction can be modeled as an nth order reaction with $n = 1.5$.

TABLE 3-3. Estimation of $t_{1/2}$ for Various Initial Concentrations in the Example Data Set

A	B	C	D	E	F
$t_{\text{init},i}$	$\log c(t_{\text{init},i})$	$\log c(t_{\text{fin},i})$	$t_{\text{fin},i}$	$t_{1/2}$ ($=t_{\text{fin},i} - t_{\text{init},i}$)	$\log (t_{1/2})$
0	−2.60	−2.90	2.2	2.2	0.35
1.5	−2.82	−3.12	4.1	2.6	0.42
3.0	−3.00	−3.30	6.1	3.1	0.50
5.0	−3.20	−3.50	9.0	4.0	0.60
8.0	−3.44	−3.74	13.7	5.7	0.76
12.0	−3.66	−3.96	19.1	7.1	0.85

(b) The $\log t_{1/2}$ versus $\log c(0)$ plot developed in part a is, in essence, a plot of Equation 3-27 for the example system. Therefore, the y-intercept of the plot can be equated with $\log k^*$. (Note that the intercept is at the point where $\log c(0) = 0$, which is far to the right of the data shown in Figure 3-9.) In the current case, that value is $\log k^* = -0.98$, so $k^* = 10^{-0.98} = 0.105$. The value of k_n can then be found from Equation 3-28

$$k_n = \frac{2^{n-1} - 1}{k^*(n-1)} = \frac{2^{0.5} - 1}{0.105(0.5)} = 7.91 \ (\text{L/mol})^{0.5} \ \text{min}^{-1}$$

(c) According to Equation 3-21b, we can find the concentration of A remaining at any time by

$$c_A(t) = \left[\{c_A(0)\}^{1-n} + (n-1)k_n t\right]^{1/(1-n)}$$

For the given conditions, $c_A(0)$ is 2.5×10^{-3} mol/L, n is 1.5, and k_n is 7.91 (L/mol)$^{0.5}$ min^{-1}, yielding a value of $c_A(60 \ \text{min}) = 1.51 \times 10^{-5}$ mol/L. ∎

Kinetics Expressions Containing Terms for the Concentrations of More Than One Reactive Species

Although direct integration of the rate expression and analysis of the reaction half-time can be used to determine the reaction order and the rate constant for rate expressions of the form $r_A = k_n c_A^n$, many rate expressions include the concentrations of more than one reactant. In those cases, the methods described in the preceding section are not directly applicable, and alternative approaches must be employed.

If we suspect that a rate expression depends on the concentrations of two or more species, we can still hypothesize a rate expression and test the fit of experimental data to the integrated equation, but the analysis is a bit more complex. For instance, if we are interested in the rate of a reaction with stoichiometry A + 2B → 2P, and we hypothesize that the rate expression has the form $r_A = -k_n c_A^\alpha c_B^\beta c_P^\gamma$, the mass balance on A in a batch reactor would be as follows:

$$V \frac{dc_A}{dt} = V r_A \quad (3\text{-}29)$$

$$\frac{dc_A}{dt} = -k_n c_A^\alpha c_B^\beta c_P^\gamma \quad (3\text{-}30)$$

Assuming that the stoichiometry of the overall reaction is known, we can use the stoichiometric relationships to express c_B and c_P in terms of c_A and thereby eliminate c_B and c_P from the equation. That is, because two moles of B are destroyed and two moles of P are generated for each mole of A that reacts, if all concentrations were expressed in moles per liter, we could write

$$c_B(t) - c_B(0) = 2(c_A(t) - c_A(0)) = 2\Delta c_A \quad (3\text{-}31a)$$
$$c_P(t) - c_P(0) = -2(c_A(t) - c_A(0)) = -2\Delta c_A \quad (3\text{-}31b)$$

Assuming the initial concentrations of all three species are known, Equations 3-31a and 3-31b can be solved for $c_B(t)$ and $c_P(t)$, respectively, and substituted into Equation 3-30 to yield

$$\frac{dc_A}{dt} = -k_n c_A^\alpha (c_B(0) + 2\Delta c_A)^\beta (c_P(0) - 2\Delta c_A)^\gamma \quad (3\text{-}32)$$

The resulting equation contains c_A as the only time-varying parameter and could be integrated numerically to predict $c_A(t)$ versus t for any assumed values of α, β, and γ. The predictions could then be compared with experimental data to see whether the assumed values were acceptable. Nevertheless, unless we have independent ways of estimating α, β, and γ, the chances of this approach being successful are small.

One approach for reducing the number of parameters that we have to guess in this exercise is to carry out the experiments with such high concentrations of species B and P that the changes in their concentrations are negligible compared to their initial concentrations. In this approach, B and P are referred to as being present in "great excess," meaning that the concentration initially present greatly exceeds the concentration that is destroyed or generated by the reaction. In that case, the following approximations apply:

$$\frac{dc_A}{dt} \approx k_n c_A^\alpha (c_B(0))^\beta (c_P(0))^\gamma \quad (3\text{-}33)$$

$$\frac{dc_A}{dt} \approx -k_n^* c_A^\alpha \quad (3\text{-}34)$$

where $k_n^* = k_n (c_B(0))^\beta (c_P(0))^\gamma$. Although the values of β and γ in Equation 3-33 are not necessarily known, they are constant, and, since $c_B(0)$ and $c_P(0)$ are also constant, the product $(c_B(0))^\beta (c_P(0))^\gamma$ can be incorporated into k_n^*, as shown. In such a case, the reaction behaves as if it were α order overall, so it can be integrated and analyzed using the equations in Table 3-1. Such reactions are sometimes referred to as being *pseudo-α-order*.

Once the reaction order with respect to A is determined, similar experiments can be conducted using great excesses of A and B, or A and P, to determine the reaction orders with respect to P and B, respectively. Note that, even if the dependence on c_B and c_P is not of the power law form, this technique can still be used to isolate and evaluate the dependence of the reaction rate on c_A, if that dependence is consistent with a power law function.

■ **EXAMPLE 3-6.** Oxidation of sulfide in natural waters is a complex process, due in part to the large number of intermediate products and side reactions. For instance, in addition to the dominant product (sulfate, SO_4^{2-}), the products of the reaction might include elemental sulfur (SO), sulfite (SO_3^{2-}), thiosulfate ($S_2O_3^{2-}$), and several other sulfur species. Nevertheless, Chen and Morris (1972) reported good agreement of oxidation rates with the following empirical equation:

$$r_{S(II)} = -k\, c_S^{1.34}\, c_{O_2}^{0.56}$$

where S(II) is the total sulfur in the −II oxidation state; that is, $c_S = c_{H_2S} + c_{HS^-} + c_{S^{2-}}$. For a solution at pH 7.5, k was reported to be 11.97 $(\text{mol/L})^{-0.9}\,\text{h}^{-1}$.

A batch system contains 2.0×10^{-4} mol/L total dissolved sulfide and is buffered at pH 7.5. Assuming the overall reaction is $HS^- + 2O_2 \to SO_4^{2-} + H^+$, determine the percentage of the initial total sulfide that reacts over a 400-h reaction period for the following initial concentrations of dissolved oxygen $[c_{O_2}(0)]$:

(a) 4.0×10^{-3} mol/L (much more O_2 than is required to oxidize all the sulfide)

(b) 4.0×10^{-4} mol/L (exactly the concentration of O_2 that is required to oxidize all the sulfide)

(c) 3.0×10^{-4} mol/L (less O_2 than is required to oxidize all the sulfide)

Solution.

(a) For the given initial sulfide concentration ($c_S(0) = 2.0 \times 10^{-4}$ mol/L), the maximum amount of O_2 that could react (based on the stoichiometry) is 4.0×10^{-4} mol/L, or 10% of $c_{O_2}(0)$. Under these circumstances, it is reasonable to assume that c_{O_2} is approximately constant throughout the experiment at the value $c_{O_2}(0)$. Because the reaction is taking place in a batch reactor, the mass balance on sulfide reduces to an equation indicating that the rate of reaction equals the rate of change of c_S:

$$\mathcal{V} r_{S(II)} = \mathcal{V}\frac{dc_S}{dt} = -\mathcal{V} k\, c_S^{1.34}\, c_{O_2}^{0.56} \quad (3\text{-}35)$$

$$\frac{dc_S}{dt} = -k^* c_S^{1.34} \quad (3\text{-}36)$$

where, for the given conditions, $k^* = k(c_{O_2}(0))^{0.56}$. Inserting the known values,

$$k^* = \left(11.97\,(\text{mol/L})^{-0.9}\,\text{h}^{-1}\right) \left(4 \times 10^{-3}\,\text{mol/L}\right)^{0.56}$$
$$= 0.543\,(\text{mol/L})^{-0.34}/\text{h}$$

Equation 3-36 has the form of a volume-normalized mass balance for a pseudo-nth-order reaction occurring in a well-mixed batch system, with $n = 1.34$. The solution to the equation, after integration, is given by Equation 3-21b in Table 3-1. Using the appropriate values for the current problem, we obtain

$$c_A(t) = \left[\{c_A(0)\}^{1-n} + (n-1)k_n t\right]^{1/(1-n)} \quad (3\text{-}21b)$$

$$c_S(400) = \left[(c_S(0))^{-0.34} + 0.34\left(0.543\,\frac{(\text{mol/L})^{-0.34}}{\text{h}}\right)(400\,\text{h})\right]^{1/-0.34}$$
$$= 1.67 \times 10^{-6}\,\text{mol/L}$$

$$\%\text{ reacted} = \frac{2.0 \times 10^{-4}\,(\text{mol/L}) - 1.67 \times 10^{-6}\,(\text{mol/L})}{2.0 \times 10^{-4}\,(\text{mol/L})} 100\%$$
$$= 99.2\%$$

Two moles of O_2 are consumed for each mole of S oxidized, so the amount of O_2 reacted is $(0.992)(2.0 \times 10^{-4}\,\text{mol/L})(2)$, or $3.97 \times 10^{-4}\,\text{mol/L}$. This is approximately 10% of the initial O_2 concentration, so the assumption of constant O_2 concentration is reasonable. If we wanted to be more accurate, we could resolve the problem using an average, but still constant O_2 concentration equal to 95% of $c_{O_2}(0)$, or to be as accurate as possible, we could consider the O_2 concentration to vary over time, as described in part c.

(b) The key to solving the problem for the case where $c_{O_2}(0) = 2c_S(0)$ is to note that, by stoichiometry, the amount of O_2 consumed $(-\Delta c_{O_2})$, is twice the amount of S consumed $(-\Delta c_S)$. Thus,

$$\Delta c_{O_2} = 2\Delta c_S$$
$$c_{O_2}(t) - c_{O_2}(0) = 2(c_S(t) - c_S(0))$$

Substituting the initial values, we obtain

$$c_{O_2}(t) - 4 \times 10^{-4}\,\text{mol/L} = 2(c_S(t) - 2 \times 10^{-4}\,\text{mol/L})$$
$$c_{O_2}(t) = 2c_S(t)$$

That is, since the initial O_2 and S concentrations are in the stoichiometric ratio, their concentrations remain in that ratio throughout the course of the reaction. Substituting this relationship into the mass balance (Equation 3-35), we obtain

$$-\frac{dc_S}{dt} = k\, c_S^{1.34}\, c_{O_2}^{0.56}$$
$$= k\, c_S^{1.34}\, (2c_S)^{0.56}$$
$$= 2^{0.56} k c_S^{1.90} = k'' c_{S(II)}^{1.90}$$

where $k'' = 2^{0.56}\,11.97\,(\text{mol/L})^{-0.9}\text{h}^{-1} = 17.6\,(\text{mol/L})^{-0.9}\text{h}^{-1}$.

Once again, we have a pseudo-nth-order reaction, in this case with $n = 1.90$, so we can again invoke Equation 3-21b in Table 3-1. The result is

$$c_S(400) = \left[(c_S(0))^{-0.90} + 0.90\left(17.6\,(\text{mol/L})^{-0.90}/\text{h}\right)(400\,\text{h})\right]^{1/-0.90}$$
$$= 4.31 \times 10^{-5}\,\text{mol/L}$$

$$\%\text{ reacted} = \frac{(2.0 \times 10^{-4} - 4.31 \times 10^{-5})\,\text{mol/L}}{2.0 \times 10^{-4}} * 100\% = 78\%$$

Note that the extent of oxidation is considerably less than in part (a) because the concentration of O_2 throughout the reaction period is less.

(c) For the conditions specified in this system, the reactants are not present in their stoichiometric ratio, and neither reactant can be approximated to be constant during the course of the reaction. Once again, the approach is to write and solve the mass balance on sulfide. However, in this case, the differential equation cannot be solved analytically. Nevertheless, a numerical solution is possible. Taking small time steps, the amount of S consumed during each step can be approximated from the rate expression, as follows:

$$-\frac{dc_S}{dt} = k c_s^{1.34} c_{O_2}^{0.56}$$
$$\Delta c_S \approx -k c_s^{1.34} c_{O_2}^{0.56} \Delta t$$

Also, by stoichiometry

$$\Delta c_{O_2} = 2\Delta c_S$$

Based on a spreadsheet analysis using one hour time steps and the recursive equation $c_S(t + \Delta t) =$

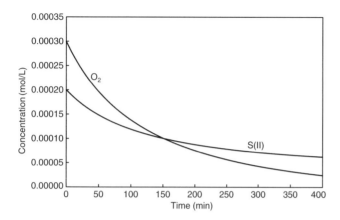

FIGURE 3-10. Concentrations of total sulfide species and dissolved oxygen as a function of time for Example 3-6c.

$c_S(t) + \Delta c_S$, the course of the reaction is followed. The concentrations of S and O_2 as a function of time are shown in Figure 3-10. After 400 h, c_S and c_{O_2} are 6.21×10^{-5} mol/L and 2.42×10^{-5} mol/L, respectively, and 69% of the sulfide has reacted. ∎

The Differential Method of Reaction Rate Analysis

The differential method for analyzing reaction rate expressions is very similar to the integral method, except that the differential method relies on analysis of dc_A/dt, rather than c_A. Differential analysis is generally somewhat less reliable than integral analysis, because the rate of change of a constituent is usually more difficult to measure than its absolute concentration, although modern instrumentation comes close to eliminating this drawback for many constituents. In any case, differential analysis is sometimes the only acceptable choice. For instance, if the products of a reaction alter its rate, it is usually much easier to determine the dependence of the rate on the reactant concentration in systems that contain none of the product than in systems where the product is constantly being generated. To carry out such an analysis, we can evaluate the initial reaction rate in experiments at several different initial reactant concentrations. To minimize the effect of product on the reaction rate, each experiment might have to be terminated shortly after it began, so the only data available would be the initial rate of reaction as a function of initial reactant concentration. Hypothesized rate expressions could be tested by comparing the data with the appropriate differential form of the rate expression.

For example, if the hypothesis is that the rate expression is of the form $r_A = -k_n c_A^n$, and the experiment is conducted in a batch system (so $r_A = dc_A/dt$), the logarithm of the absolute value of the experimental reaction rate could be plotted against the logarithm of the initial concentration. If the hypothesized rate expression were correct, the result would be a straight line of slope n and intercept $\ln k_n$

$$\frac{dc_A}{dt} = -k_n c_A^n \qquad (3\text{-}37)$$

$$\ln\left(-\frac{dc_A}{dt}\right) = \ln k_n + n \ln c_A \qquad (3\text{-}38)$$

∎ **EXAMPLE 3-7.** Luthy and Bruce (1979) studied the interactions of cyanide, sulfide, and thiocyanate as part of an investigation of the treatability of wastewater from steel manufacturing. In their study of the reaction between polysulfide ($S_x S^{2-}$) and cyanide (CN^-) to form a smaller polysulfide molecule ($S_{x-1} S^{2-}$) and thiocyanide (SCN^-), they used the initial rate method to minimize the effects of the reaction products on reaction kinetics. An idealized, hypothetical reaction of this sort is as follows:

$$S_x S^{2-} + CN^- \rightarrow S_{x-1} S^{2-} + SCN^-$$

Both polysulfide and cyanide ions undergo acid/base reactions in solution. The total dissolved concentrations of these species (considering all of their acidic and basic forms) can be written as $c_{(S_x S)_{tot}}$ and $c_{(CN)_{tot}}$, respectively. The researchers chose to write the reaction rate expression in terms of these total concentrations, rather than the concentration of the individual clients.

The data from an experiment at pH 10 are shown in Figure 3-11, with rates computed from the thiocyanide production during a 5-min period. In each experiment, one of the reactant concentrations was fixed and the other was varied (as indicated in the figure legend), and the rate of SCN^- formation was monitored. Assuming that the reaction rate is a power law function of the concentrations of the reactants (i.e., $r_{SCN^-} = k c_{(CN)_{tot}}^a c_{(S_x S)_{tot}}^b$), estimate the reaction order with respect to each reactant and the reaction rate constant.

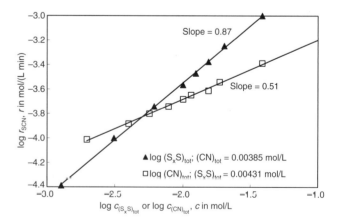

FIGURE 3-11. Initial rate of SCN^- formation at pH 10, as a function of polysulfide or cyanide concentration. *Source*: After Luthy and Bruce (1979).

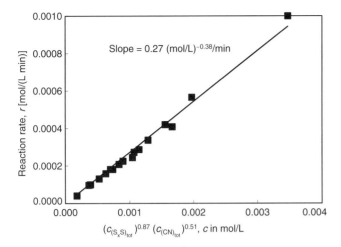

FIGURE 3-12. The reaction rate versus the product $c_{(CN)_{tot}}^{a} c_{(S_xS)_{tot}}^{b}$ for the example data set.

Solution. The two lines shown in Figure 3-11 indicate the reaction rate as a function of the polysulfide concentration for a fixed total cyanide concentration (triangles) and as a function of the total cyanide concentration for a fixed polysulfide concentration (squares). The fact that both data sets can be fit with straight lines on a log–log plot confirms that the rate expression is of the form hypothesized. The slopes of the two lines in the graph indicate that the reaction is approximately 0.51 order with respect to total cyanide and 0.87 order with respect to polysulfide. That is, the data indicate that the reaction rate can be modeled by the expression

$$r_{SCN^-} = k c_{(CN)_{tot}}^{0.51} c_{(S_xS)_{tot}}^{0.87} \tag{3-39}$$

Each data point shown provides an independent set of $c_{(CN)_{tot}}$, $c_{(S_xS)_{tot}}$, and r_{SCN} values that can be used in conjunction with Equation 3-39 to evaluate k. A plot of r_{SCN^-} versus $c_{(CN)_{tot}}^{0.51} c_{(S_xS)_{tot}}^{0.87}$ is shown in Figure 3-12. The rate constant k can be estimated as the best-fit value of the slope, equal in this case to $0.27 \, (mol/L)^{-0.38}/min$. Thus, the rate of formation of polysulfide can be written as

$$r_{SCN^-} = 0.27 \frac{(mol/L)^{-0.38}}{min} c_{(CN)_{tot}}^{0.51} c_{(S_xS)_{tot}}^{0.87}$$

Note that the net rate of formation of SCN^- in a batch system would be determined by combining the preceding expression with additional ones accounting for the effects of the reaction products on the rate. ∎

Analysis of Nonpower-Law Rate Expressions

The preceding discussion and examples suggest various ways in which the dependence of the rate expression on individual reactant concentrations might be assessed. Most of these methods succeed only in cases where the dependence of the reaction rate on the reactant of interest is of the power law form. The effect of other non power-law constituents can be eliminated by causing them to be present in great excess, but the rate dependence on those reactants cannot be assessed directly by the methods presented. The dependence of the reaction rate on the concentrations of such constituents can only be determined by trial and error.

Often, even if the dependence of the rate expression on a particular reactant does not follow a power law expression over the entire concentration range of interest, a power law expression applies over a limited concentration range. For instance, the rate expression $r_A = k_1 c_A c_B / (k_2 + c_A)$ can be approximated as first order with respect to species A if $c_A \ll k_2$, and zero order with respect to A if $c_A \gg k_2$.[11] In such cases, the methods outlined in the preceding sections can be used to evaluate the rate expression in each concentration range of interest.

Characteristic Reaction Times

In some situations, a rough assessment of the time frame over which a reaction proceeds might be adequate to meet our needs. Often, the half-time of a reaction ($t_{1/2}$) can provide such an indication. For instance, if a reaction has a half-time of one day, and the time available for the reaction to proceed is only a few minutes, then we can reasonably presume that the extent of reaction will be very small. On the other hand, if the time available for the same reaction to proceed is several weeks, it will probably progress nearly to its endpoint (either complete disappearance of reactant or equilibrium between reactants and products) during that period.

The time required for a reaction to proceed a significant extent toward its final endpoint is called the *characteristic reaction time*, t_{char}. Because "significant" is an imprecise term, the quantitative definition of the characteristic time is somewhat arbitrary; however, the imprecision is acceptable, because t_{char} is used only to give a rough idea of how far the reaction proceeds in a given situation. Although the half-time of a reaction would be an acceptable definition of the characteristic time, the most commonly used definitions are based on slightly different criteria. Defining $c(\infty)$ as the concentration of the reactive substance that would be present after an infinite amount of time passed, the most common definitions of the characteristic time are as follows:

- *Definition 1*: The time required for the difference between the instantaneous concentration ($c(t)$) and the ultimate concentration ($c(\infty)$) to be reduced to $1/e$ of its initial value in a batch reactor; that is, the time at which $|c(t) - c(\infty)|$ equals $(1/e)(|c(0) - c(\infty)|)$.

---

[11] This rate expression, commonly referred to as the Michaelis–Menten expression, has special significance in biological waste treatment processes and is derived later in this chapter.

TABLE 3-4. Characteristic Times of Irreversible nth Order Reactions

Value of n in $r = k_n c^n$	t_{char} (Definition 1)	t_{char} (Definition 2)
$n = 0$	$\dfrac{c_A(0)}{k_0}\left(1 - \dfrac{1}{e}\right)$	$\dfrac{c_A(0)}{k_0}$
$n = 1$	k_1^{-1}	k_1^{-1}
$n = 2$	$\dfrac{e - 1}{k_2 c_A(0)}$	$\dfrac{1}{k_2 c_A(0)}$
$n \neq 1$	$\dfrac{e^{n-1} - 1}{n - 1}\dfrac{1}{k_n[c_A(0)]^{n-1}}$	$\dfrac{1}{k_n[c_A(0)]^{n-1}}$

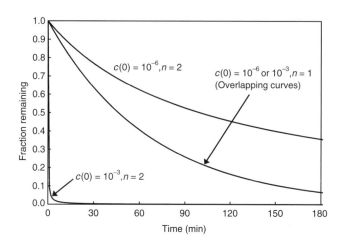

FIGURE 3-13. Fraction of initial reactant remaining as a function of time for first and second-order irreversible reactions taking place in a batch reactor. $k_1 = 0.015\,\text{min}^{-1}$; $k_2 = 10^{4.0}\,\text{L/mol min}$.

- *Definition 2*: The time that would be required for the concentration of the reactive substance to reach its ultimate value ($c(\infty)$) in a batch reactor if the reaction rate stayed at its initial value continuously.

Characteristic times based on the first definition can be computed for irreversible decay reactions (for which $c(\infty)$ is approximately zero) using the equations in Table 3-1. Since the initial rate of a reaction can be determined from the rate expression and the initial reactant concentrations, the calculation of the characteristic time according to the second definition is also straightforward. The resulting equations for computing the characteristic time of irreversible reactions with power law rate expressions are shown in Table 3-4, and an illustration of how these equations might be used is provided in the following example.

■ EXAMPLE 3-8.

(a) Plot the fraction of the initial concentration remaining as a function of time in a well-mixed batch reactor for irreversible reactions with the following rate expressions: $r_1 = -k_1 c$ and $r_2 = -k_2 c^2$, with $k_1 = 0.015/\text{min}$ and $k_2 = 10^{4.0}\,\text{L/mol min}$, respectively. For each reaction, consider $c(0)$ values of 10^{-3} and $10^{-6}\,\text{mol/L}$.

(b) Compute characteristic reaction times for the reactions described in part *a*, using each of the definitions of t_{char} and each initial condition specified. Also, determine the fraction of the constituent that remains after 1, 2, and 10 characteristic times have elapsed.

Solution.

(a) The fraction remaining versus time profiles for the various systems can be computed by the following modifications of Equations 3-18b and 3-20b for the first and second-order reaction, respectively. The results are shown in Figure 3-13.

$$\text{1st order:} \quad \frac{c(t)}{c(0)} = \exp(-k_1 t)$$

$$\text{2nd order:} \quad \frac{c(t)}{c(0)} = \frac{1}{1 + k_2 t c(0)}$$

The characteristic time for each case of interest can be computed using the equations in Table 3-4; the results are shown in Table 3-5. Several points are worth noting about these results. First, the characteristic time for the first-order reaction is independent of $c(0)$ and is the same when computed according to either definition. In contrast, the characteristic time for the second-order reaction (and, in fact, for all reaction orders other than one) depends strongly on $c(0)$ and also differs when calculated according to the two definitions of t_{char}. In the current example, t_{char} becomes dramatically shorter when $c(0)$ increases. On the other hand, although the absolute value of t_{char} for the second-order reaction changes dramatically when $c(0)$ changes, the fractional conversion of the reactant after 1, 2, or 10 characteristic times have elapsed is independent of $c(0)$.

Finally, although the fractional conversion of the reactant after a given number of characteristic reaction times does depend on the details of the rate expression, it is clear that a significant fraction of the original concentration remains after 1 or 2 characteristic times have elapsed, whereas the reactant is substantially depleted after 10 characteristic times have elapsed. ■

The utility of the t_{char} concept is that it provides a simple, unifying framework for comparing and predicting the progress of vastly different reactions (reactions with different

TABLE 3-5. Characteristic Times and the Extent of Reaction for System in Example 3-8

Reaction Order, n	$c(0)$ (mol/L)	$-r(0)$ (mol/L min)	Characteristic Time (min)		$\dfrac{c(t)}{c(0)}$ at $t = t_{\text{char}}$[a]		$\dfrac{c(t)}{c(0)}$ at $t = 2\,t_{\text{char}}$[a]		$\dfrac{c(t)}{c(0)}$ at $t = 10\,t_{\text{char}}$[a]	
			Definition 1	Definition 2	Definition 1	Definition 2	Definition 1	Definition 2	Definition 1	Definition 2
1	10^{-3}	1.5×10^{-5}	66.7	66.7	0.368	0.368	0.135	0.135	0.007	0.007
1	10^{-6}	1.5×10^{-8}	66.7	66.7	0.368	0.368	0.135	0.135	0.007	0.007
2	10^{-3}	10^{-2}	0.172	0.100	0.368	0.500	0.225	0.333	0.055	0.091
2	10^{-6}	10^{-8}	172	100	0.368	0.500	0.225	0.333	0.055	0.091

[a]In a well-mixed batch reactor

reaction orders, rate constants, rate expressions, and initial conditions) and for understanding the time frames in which a reaction can be considered to have proceeded either a negligible amount or nearly to completion. A reaction time equal to $3t_{\text{char}}$ is sometimes used as a criterion for making the latter assumption. However, as shown in Table 3-4 and Example 3-8, substantially longer times than $3t_{\text{char}}$ might be required for an assumption of nearly complete reaction to be valid in batch systems if the reaction order is greater than one.

Characteristic times can also be defined for physical systems such as reactors, and for transport processes such as diffusion and dispersion. Comparison of the characteristic times of various processes affecting a constituent is a powerful tool for understanding its behavior, and such comparisons are made for increasingly complex systems in subsequent chapters.

■ **EXAMPLE 3-9.** Assume the reactions described in Example 3-8 are occurring in a continuous flow system with a mean hydraulic residence time, τ, of 3.0 min. Evaluate qualitatively which of the reactions would proceed to a significant extent, based on their characteristic reaction times.

Solution. The characteristic times of the various reactions are given in Table 3-4. The characteristic reaction times for the first-order reaction with either value of $c(0)$ and for the second-order reaction with $c(0) = 10^{-6}$ mol/L are >60 min. Since the average amount of time that aliquots of influent spend in the reactor (i.e., τ) is less than 0.05 t_{char}, the reactions would probably proceed to a negligible extent in the reactor. In contrast, for the second-order reaction with $c(0) = 10^{-3}$ mol/L, t_{char} is less than 0.2 min, so the mean amount of time that water spends in the reactor is >15 t_{char}. Therefore, we expect the reactant to be very substantially depleted in that system. Techniques for computing the exact extent of conversion in these systems are presented in Chapter 4. ■

3.4 KINETICS OF REVERSIBLE REACTIONS

To this point, we have focused on approaches for determining the rate expression for an overall, irreversible reaction. Such a reaction might be elementary, or it might represent a network of linked elementary reactions that together can be described by a single overall rate expression. In this and the following section, two types of linked reactions—reversible reactions and sequential reactions—are analyzed. In addition to introducing a key concept related to sequential reactions (the rate-limiting step), the analysis illustrates the conceptual basis for complex rate expressions and the wide range of overall expressions that can result from combinations of relatively simple reactions.

Reversible Reactions

A reaction that can proceed along a given pathway can, in all cases, follow the identical path in the opposite direction. As a result, all chemical reactions are reversible, at least to some extent. If a reaction were truly irreversible, it would proceed until at least one of the reactants was completely depleted. No such reaction is known, although some reactions do approach this limit. Rather, reactions proceed until they reach an equilibrium condition, wherein all reactants and products are present, but there is no net driving force for the reaction to proceed in either direction. Thus, when a chemical reaction is at equilibrium, there is no net generation of the corresponding reactants or products *by that reaction*.

The quantitative representation of chemical equilibrium is the equilibrium constant, K_{eq}. This constant is defined as the ratio, in a system that has attained equilibrium, of the chemical activities of the products divided by those of the reactants, with each term raised to a power corresponding to its stoichiometric coefficient. The calculation of equilibrium constants from thermodynamic data and the determination of the equilibrium condition of a system are central components of water chemistry courses and are described in many textbooks. Note that the term equilibrium and the corresponding constant can be applied only to a reaction and not to an individual species.

The concept of chemical equilibrium is connected to that of reaction reversibility, because equilibrium is attained when the forward and reverse reactions proceed at equal rates, thereby preventing any net change in the concentrations of reactants or products as a result of the reaction. Consider, for example, the generic reaction shown in Equation 3-40,

$$aA + bB \underset{k_{PA}}{\overset{k_{AP}}{\rightleftharpoons}} pP + rR \qquad (3\text{-}40)$$

If the forward and reverse reactions are elementary, then the net reaction rate of A is given by Equation 3-41,[12]

$$r_A = -k'_{AP}(a_A)^a(a_B)^b + k'_{PA}(a_P)^p(a_R)^r \qquad (3\text{-}41)$$

At equilibrium, the magnitudes of the forward and reverse reaction rates are equal, so

$$k'_{AP}(a_{A,eq})^a(a_{B,eq})^b = k'_{PA}(a_{P,eq})^p(a_{R,eq})^r \qquad (3\text{-}42)$$

$$\frac{k'_{AP}}{k'_{PA}} = \frac{(a_{P,eq})^p(a_{R,eq})^r}{(a_{A,eq})^a(a_{B,eq})^b} \qquad (3\text{-}43)$$

where the subscript *eq* is included in the activity terms to indicate that Equations 3-42 and 3-43 are valid only at equilibrium. Since the right side of Equation 3-43 is defined as the equilibrium constant for the reaction, the equation can be written as

$$\frac{k'_{AP}}{k'_{PA}} = K_{eq} \qquad (3\text{-}44)$$

Equation 3-44 indicates that the equilibrium constant for any elementary reaction equals the ratio of the activity-based forward and reverse rate constants. Since equilibrium constants are defined in terms of chemical activities, this is one situation in which the distinction between concentration-based and activity-based rate constants is important. As noted earlier, if all the reactants are ideal solutes, the activity-based and concentration-based rate constants are identical; in such a case, the equilibrium constant would also equal the ratio of forward and reverse concentration-based rate constants (k_{AP}/k_{PA}).

■ **EXAMPLE 3-10.** Hydrofluoric acid (HF) is able to dissolve oxide layers that develop on aluminum surfaces, and it is therefore used industrially to clean aluminum parts before they are painted. The waste from such a process might be mixed with other streams containing dissolved metals, and the F^- ions can bind with those metals to form metal–ligand complexes. In doing so, the F^- replaces a molecule of water that is bound to the metal, although it is common to write the reaction without showing the H_2O molecules explicitly, as follows:

$$Ni^{2+} + F^- \leftrightarrow NiF^+$$

The activity-based forward rate constant (k'_f) for the formation of the complex in this reaction is 8.4×10^3 mol/L s, and the equilibrium constant for the reaction is $K_{eq} = 20$.

Assuming that the reaction is elementary in each direction, determine the activity-based rate constant for the reverse reaction.

In an ideal solution, initially containing 10^{-3} mol/L F^-, and 10^{-5} mol/L of both Ni^{2+} and NiF^+, what would be the rates of the forward, reverse, and overall reaction?

Solution.

(a) The activity-based rate constant for the reverse reaction (k'_r) can be determined by applying Equation 3-44 to the given reaction.

$$k'_r = \frac{k'_f}{K_{eq}} = \frac{8.4 \times 10^3 \text{ mol/L s}}{20} = 420 \text{ mol/L s}$$

(b) In an ideal solution, the activity of each solute has the same numerical value as its molar concentration, and that of water is 1.0. The rates of the forward and reverse reactions for the specified conditions would therefore be as follows:

$$r_f = k'_f a_{Ni^{2+}} a_{F^-}$$
$$= (8.4 \times 10^3 \text{ mol/L s})(10^{-3})(10^{-5})$$
$$= 8.4 \times 10^{-5} \text{ mol/L s}$$
$$r_r = k'_r a_{NiF^+}$$
$$= (420 \text{ mol/L s})(10^{-5}) = 4.2 \times 10^{-3} \text{ mol/L s}$$

Thus, for the given conditions, the reverse reaction would be proceeding 50 times as fast as the forward reaction, and the net reaction would be dissociation of NiF^+ complexes. Because one NiF^+ complex is formed for each Ni^{2+} that reacts, the rate of the forward reaction can be equated with the rate at which NiF^+ is generated by that reaction. The net rate of the overall reaction is therefore

$$r_{NiF^+,net} = r_f - r_r$$
$$= (8.4 \times 10^{-5} - 4.2 \times 10^{-3}) \text{ mol/L s}$$
$$= -4.1 \times 10^{-3} \text{ mol/L s} \qquad ■$$

The relationship shown in Equation 3-44 between the rate constants and the equilibrium constant applies only for

---

[12] In this text, rate constants are, in general, subscripted to indicate direction and reversibility of the reaction. For instance, if the reaction is considered irreversible (A→P), the rate constant is written as k_A. If it is reversible (A↔P), the rate constants for the forward and back reactions are written as k_{AP} and k_{PA}, respectively.

elementary reactions. For a nonelementary reaction, the equilibrium constant can still be derived by equating the forward and reverse reaction rates for the overall reaction at equilibrium. However, in that case, the relationship between K_{eq} and the rate constants of the elementary reactions that lead to the overall reaction is more complex than Equation 3-44.

The approach of a reversible reaction to equilibrium in a batch reactor can be analyzed following the same procedures as are used to analyze irreversible reactions in such reactors. That is, the rate expression for the overall reaction (in this case, including both the forward and reverse reactions) can be written and integrated. For instance, for a reaction $A \leftrightarrow B$ in which the forward and reverse reactions are both elementary, the net rate of reaction of A is

$$r_A = -k_{AB}c_A + k_{BA}c_B \qquad (3\text{-}45)$$

If the reaction is taking place in a batch reactor, and it is the only reaction affecting the concentration of A, the mass balance on A in the reactor indicates that dc_A/dt equals r_A, as is the case for irreversible reactions. Therefore, the mass balance on A in the system is

$$\frac{dc_A}{dt} = -k_{AB}c_A + k_{BA}c_B \qquad (3\text{-}46)$$

According to Equation 3-44, the equilibrium constant K_{eq} is the ratio of the forward and reverse activity-based rate constants, so Equation 3-45 can be rewritten as follows:

$$r_A = -k_{AB}c_A + \frac{k_{AB}}{K_{eq}}c_B \qquad (3\text{-}47)$$

$$r_A = -k_{AB}c_A + k_{AB}c_A^* = k_{AB}(c_A^* - c_A) \qquad (3\text{-}48)$$

where $c_A^* = c_B/K_{eq}$. By its definition, c_A^* is the concentration of A that *would be* in equilibrium with c_B. Thus, c_A^* is a hypothetical concentration, and it is a variable; that is, it changes as the concentration of B changes. The expression in parentheses in Equation 3-48 is the instantaneous difference between the actual concentration of A and the concentration of A that would be in equilibrium with B at that instant; that is, it is the instantaneous extent of disequilibrium of the reaction $A \leftrightarrow B$. Thus, the equation indicates that the net rate of the reaction is directly proportional to the gap between the current conditions and a corresponding, hypothetical equilibrium condition.

Equation 3-48 bears a strong resemblance to Equation 3-16, the mass balance for an irreversible first-order reaction in a batch reactor. The only difference between the equations is that the driving force in the case of the reversible reaction is the extent of disequilibrium, rather than the concentration of A. In fact, since c_A^* is zero for an irreversible reaction, the rate expression for the irreversible reaction (Equation 3-16) is seen to be a limiting case of the more general expression for the reversible system (Equation 3-48).

Defining the extent of disequilibrium as $\hat{c}_A (\hat{c}_A \equiv c_A^* - c_A)$, we can write

$$d\hat{c}_A \equiv dc_A^* - dc_A$$
$$= d\left(\frac{c_B}{K_{eq}}\right) - dc_A$$

K_{eq} is a constant and can be taken out of the differential. Also, the reaction stoichiometry requires that $dc_A = -dc_B$. Making that substitution, and then writing the equilibrium constant as the ratio of the forward and reverse rate constants, we obtain

$$d\hat{c}_A = -\frac{1}{K_{eq}}dc_A - dc_A$$
$$= -\left(\frac{1}{K_{eq}} + 1\right)dc_A = -\left(\frac{k_{BA}}{k_{AB}} + 1\right)dc_A \qquad (3\text{-}49)$$

By rearranging Equation 3-49, we obtain an expression for dc_A in terms of $d\hat{c}_A$

$$dc_A = -\left(\frac{k_{BA}}{k_{AB}} + 1\right)^{-1} d\hat{c}_A = -\frac{k_{AB}}{k_{AB} + k_{BA}} d\hat{c}_A \qquad (3\text{-}50)$$

Finally, by substituting Equation 3-50 into the left side of Equation 3-46 and the definition of $\hat{c}_A$ into the right side, and integrating, we obtain an expression for the rate at which a reversible reaction approaches equilibrium in a batch system:

$$-\frac{k_{AB}}{k_{AB} + k_{BA}} \frac{d\hat{c}_A}{dt} = k_{AB}\hat{c}_A \qquad (3\text{-}51)$$

$$\int_{\hat{c}_A(0)}^{\hat{c}_A(t)} \frac{d\hat{c}_A}{\hat{c}_A} = -(k_{AB} + k_{BA}) \int_0^t dt \qquad (3\text{-}52)$$

$$\hat{c}_A(t) = \hat{c}_A(0) \exp\{-(k_{AB} + k_{BA})t\} \qquad (3\text{-}53)$$

Equation 3-53, which applies to first-order reversible reactions, is analogous to Equation 3-18b for first-order irreversible reactions. In particular, for irreversible reactions, the concentration of the reactant undergoes exponential decay, and for reversible reactions, the extent of disequilibrium undergoes exponential decay. Noting that, for an irreversible reaction, the concentration of the reactant is the same as (the absolute value of) the extent of disequilibrium, and the rate of the reverse reaction (k_{BA}) is zero. We see once again that the equations derived in preceding sections for irreversible reactions can be viewed as limiting cases of the more general equations for reversible reactions.

To compute the actual concentrations of A and B at any time, we can substitute back into Equation 3-53 to replace $\hat{c}_A$ with $(c_B/K_{eq}) - c_A$ and k_{BA} with k_{AB}/K_{eq}. Then, using a mass balance on A to replace $c_B(t)$ with $c_A(0) + c_B(0) - c_A(t)$, and carrying out some algebra, we obtain the following result:

$$c_A(t) = \frac{c_A(0) + c_B(0)}{K_{eq} + 1} + \frac{K_{eq}c_A(0) - c_B(0)}{K_{eq} + 1}\exp\left(-\frac{K_{eq} + 1}{K_{eq}}k_{AB}t\right) \quad (3\text{-}54)$$

The first term on the right of Equation 3-54 can also be expressed as $c_{A,eq}$, the ultimate, equilibrium concentration of A in the system. The concentration of B at any time can be found by utilizing the stoichiometry of the reaction in combination with mass balances on A and B:

$$c_A(0) - c_A(t) = c_B(t) - c_B(0)$$
$$c_B(t) = c_B(0) + c_A(0) - c_A(t) \quad (3\text{-}55)$$

Although the exact form of the equations for reversible reactions with other stoichiometries and other rate expressions is more complex, the rate of approach to equilibrium is, in all cases, directly dependent on the extent of disequilibrium. It is important to recognize, however, that this conclusion applies when comparing the rate of a given reaction under different conditions. It does not imply that, when comparing two different reactions, the one farther from equilibrium will proceed more quickly.

■ **EXAMPLE 3-11.** For the solution described in Example 3-10b, plot the extent of disequilibrium of the complex formation reaction (i.e., $\hat{c}_{NiF^+}$) and the concentrations of Ni^{2+} and NiF^+ versus time, assuming that the reaction is occurring in a batch reactor. How close is the reaction to equilibrium after 1.0 s?

Solution. Formally, Equation 3-53 does not apply to this situation, because it describes the approach to equilibrium of a reaction that is first order in both directions, and the reaction of interest (even in its simpler form) is second order in at least one direction. However, F^- is present in great excess in the example system—its initial concentration is 10^{-3} mol/L, and the concentrations of Ni^{2+} and NiF^+ are only 10^{-5} mol/L, so the F^- concentration can change by, at most, $\pm 1\%$ as the reaction proceeds. As a result, the reaction can be considered pseudo-first-order in both directions, with the following rate expressions:

	True Rate Expression	Simplification/ Approximation	Pseudo-Rate Expression	Pseudo-Rate Constant
Forward reaction	$r_f = k'_f a_{Ni^{2+}} a_{F^-}$	$a_{F^-} \approx 10^{-3}$	$r^*_f = k''_f a_{Ni^{2+}}$	$k''_f = k'_f a_{F^-}$ $= (8400 \text{ mol/L s})10^{-3}$ $= 8.4 \text{ mol/L s}$
Reverse reaction	$r_r = k'_r a_{NiF^+}$			

If we use the pseudo-first-order rate expression for the forward reaction in conjunction with the true first-order rate expression for the reverse reaction, we can apply Equation 3-53 to solve for the extent of disequilibrium over time. To do so, we first compute the activity of NiF^+ that would be in equilibrium with Ni^{2+} and F^- at $t = 0$ as

$$a^*_{NiF^+}(0) = K_{eq}[a_{Ni^{2+}}(0)][a_{F^-}(0)]$$
$$= 20(10^{-5})(10^{-3}) = 2.0 \times 10^{-7}$$

By the assumption of ideality, the molar concentration of NiF^+ that would be in equilibrium at $t = 0$ is the same as its activity; that is, $c^*_{NiF^+}(0) = 2.0 \times 10^{-7}$ (mol/L). Therefore, the initial extent of disequilibrium of NiF^+ is

$$\hat{c}_{NiF^+}(0) = c^*_{NiF^+}(0) - c_{NiF^+}(0)$$
$$= 2.0 \times 10^{-7} - 1.0 \times 10^{-5}$$
$$= -9.8 \times 10^{-6} \text{ mol/L}$$

The negative value indicates that the equilibrium concentration of NiF^+ is smaller than the initial concentration, so the NiF^+ concentration must decline for the reaction to reach equilibrium.

The values of $\hat{c}_{NiF^+}(0)$, k''_f, and k'_r can then be inserted into Equation 3-53 to determine the extent of disequilibrium at any future time:

$$\hat{c}_{NiF^+}(t) = \hat{c}_{NiF^+}(0)\exp\{-(k''_f + k'_r)t\}$$

The results are plotted in Figure 3-14.

The concentration of Ni^{2+} as a function of time can be computed using Equation 3-54, except that, similar to the situation with the rate constants, we must convert the true equilibrium constant into a pseudo-equilibrium constant K''_{eq} that would apply if the reaction were simply $A \leftrightarrow B$. We can

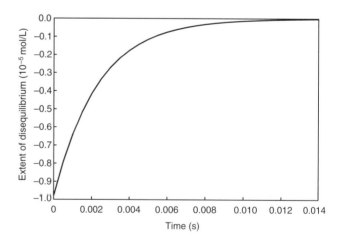

FIGURE 3-14. The extent of disequilibrium of NiF^+ with respect to its formation from Ni^{2+} and F^-, and its dissipation over time, for the conditions specified in Example 3-11.

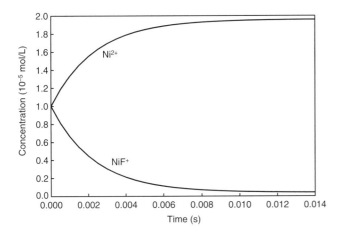

FIGURE 3-15. The approach to equilibrium of a system initially containing 10^{-5} mol/L each of Ni^{2+} and NiF^+, and 10^{-3} mol/L F^-.

accomplish that as follows.

$$K_{eq} = \frac{a_{NiF^+}}{a_{Ni^{2+}} a_{F^-}} = 20$$

$$a_{F^-} \approx 10^{-3}$$

$$K''_{eq} = \frac{a_{NiF^+}}{a_{Ni^{2+}}} = K_{eq} a_{F^-} = 0.020$$

Equation 3-54, modified to apply to the example system, is shown. We can use that equation to compute $c_{Ni^{2+}}$, and then compute c_{NiF^+} by knowing that one NiF^+ ion is lost for every Ni^{2+} ion that is generated. The results of these calculations are shown in Figure 3-15.

$$c_{Ni^{2+}}(t) = \frac{c_{Ni^{2+}}(0) + c_{NiF^+}(0)}{K''_{eq} + 1}$$
$$+ \frac{K''_{eq} c_{Ni^{2+}}(0) - c_{NiF^+}(0)}{K''_{eq} + 1} \exp\left\{ -\frac{K''_{eq} + 1}{K''_{eq}} k''_f t \right\}$$

Figures 3-14 and 3-15 indicate that the system reaches equilibrium very quickly, consistent with the usual observations for the formation of simple, inorganic metal–ligand complexes. At $t = 1.0$ s, the conditions are indistinguishable from equilibrium, with $c_{Ni^{2+}} = 1.96 \times 10^{-5}$ mol/L, $c_{NiF^+} = 3.95 \times 10^{-7}$ mol/L, and c_{F^-} still equal to 10^{-3} mol/L (by assumption). ∎

Characteristic Times and Limiting Cases for Reversible Reactions

The characteristic time of either the forward or reverse direction of a reversible reaction can be defined and quantified using the approach described previously for irreversible reactions. However, in the case of a reversible reaction, we are more often interested in the characteristic time for the composite of the forward and reverse reactions. In other words, we are interested in t_{char} for the approach of $\hat{c}_A(t)$ to zero; that is, the approach of the system to equilibrium. Applying either of the definitions of t_{char} given previously to a reaction that is first order in both the forward and reverse directions, we can use Equation 3-53 to compute t_{char} for the overall reaction. The result is compared with t_{char} for an irreversible first-order reaction in Table 3-6.

The result indicates that t_{char} for the overall reaction is shorter than that for either the forward or reverse reaction in isolation ($1/k_{AB}$ or $1/k_{BA}$, respectively). In many cases, one of the rate constants is so much larger than the other that the smaller rate constant can be ignored in the denominator of the fraction $1/(k_{AB} + k_{BA})$. In that case, the characteristic time for the overall reaction is approximately equal to the shorter of the characteristic times of the two individual reactions.

TABLE 3-6. Characteristic Times for Unidirectional (Irreversible) and Bidirectional (Reversible) Reactions

	Integrated Rate Expression[a]	Equation	t_{char}
Irreversible	$c_A(t) = c_A(0) \exp(-k_1 t)$	(3-18b)	$\dfrac{1}{k_1}$
Reversible	$\hat{c}_A(t) = \hat{c}_A(0)$ $\times \exp\{-(k_{AB} + k_{BA})t\}$	(3-53)	$\dfrac{1}{k_{AB} + k_{BA}}$

[a]For a batch system.

■ **EXAMPLE 3-12.** Consider the progress of a reversible reaction A ↔ B taking place in a batch reactor with two different initial conditions: one in which $c_A(0) = 100 \,\mu$mol/L and $c_B(0) = 0 \,\mu$mol/L, and another in which these two concentrations are reversed. Assume that both the forward and reverse reactions are first order, with $k_{AB} = 1$ min^{-1} and $k_{BA} = 0.1$ min^{-1}, and that the solutes behave ideally. How rapidly would these two systems approach equilibrium, and how does the rate of approach to equilibrium in each system compare to the rates of the individual reactions?

Solution. The changes in the concentrations of the two reactants over time in each system can be computed using Equation 3-54. The equilibrium constant for the reaction, equal to the ratio of the forward and reverse rate constants, is $K_{eq} = 10$. For the system with $c_A(0) = 100 \,\mu$mol/L, the conversion of A to B is rapid initially. However, within a few minutes, the equilibrium condition ($c_A = 9.1 \,\mu$mol/L, $c_B = 90.9 \,\mu$mol/L) is approached, and the reaction slows down to a negligible net rate (Figure 3-16a). The characteristic time for conversion of A to B is 1 min ($= 1/k_{AB}$), so the fact that the overall reaction is substantially complete in a few minutes is not surprising.

When the initial conditions are changed to $c_A(0) = 0 \,\mu$mol/L, $c_B(0) = 100 \,\mu$mol/L, the dominant reaction that proceeds as the system approaches equilibrium is B→A,

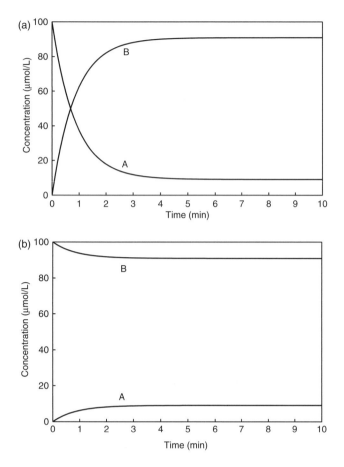

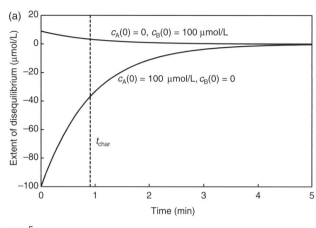

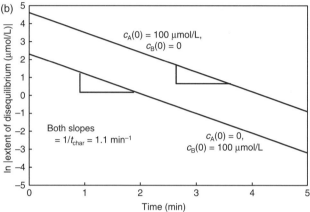

FIGURE 3-16. Concentration versus time profiles for A and B in a batch reactor, for the reversible reaction A ↔ B, with $k_A = 1.0$ /min, $k_B = 0.1$/min. (a) $c_A(0) = 100 \, \mu\text{mol/L}$, and $c_B(0) = 0$ μmol/L; (b) $c_A(0) = 0 \, \mu$mol/L, and $c_B(0) = 100 \, \mu$mol/L.

FIGURE 3-17. The decline in the extent of disequilibrium in the hypothetical system described in the text, plotted as (a) arithmetic and (b) logarithmic functions. (The absolute value of the extent of disequilibrium is shown in part b, since the logarithm of a negative number is undefined.)

which has a characteristic time of 10 min. Nevertheless, the approach to equilibrium is just as rapid in this case as in the previous one (Figure 3-16b). This result is consistent with the prediction that the characteristic time of the overall reaction is close to the shorter of the characteristic times for the forward and reverse reactions, regardless of which direction the net overall reaction is proceeding.

The results are shown in a slightly different way in Figure 3-17, in which the extent of disequilibrium ($\hat{c}_A$) and the logarithm of the absolute value of $\hat{c}_A$ are plotted against time. Consistent with Equation 3-53, the semi-logarithmic plot demonstrates that the extent of disequilibrium decreases by one natural log unit with each increment of time equal to t_{char}.

Simplification of Reaction Rate Expressions for Limiting Cases

Very Rapid and Very Slow Approach to Equilibrium as Limiting Cases Based on Equation 3-53, if either the forward or reverse rate of an equilibrium reaction is very fast, then it is reasonable to assume that the reaction reaches equilibrium quickly, perhaps even so quickly that the approach to equilibrium can be ignored. The most important reaction in environmental engineering for which this assumption applies is the dissociation and formation of water. The elementary forward and reverse reactions for this overall reaction are both second order: $2H_2O \leftrightarrow H_3O^+ + OH^-$, with forward and reverse concentration-based rate constants of approximately $4.5 \times 10^{-7} \, (\text{mol/L})^{-1} \, \text{s}^{-1}$ and $1.4 \times 10^{11} \, (\text{mol/L})^{-1} \, \text{s}^{-1}$, respectively.[13] Using Definition 1 to compute the characteristic time of a second-order reaction, the characteristic times of the forward and reverse reactions at pH 7.0 are about 19 h and 10^{-4} s, respectively. Although the characteristic time of

---

[13] The ratio $k_{forward}/k_{reverse}$ yields the equilibrium constant based on the concentration of all species. Defining the activity of water to be 1.0 when its concentration is 55.5 mol/L converts this equilibrium constant into the conventional value of $10^{-14.0}$.

the reaction forming H_2O depends on the exact conditions in the system, it is always orders of magnitude shorter than the characteristic time of the dissociation reaction, and therefore equilibrium is always approached at a rate corresponding to the characteristic time of the water-forming reaction.

The very short characteristic time of the water-forming reaction means that H^+, OH^-, and H_2O reach equilibrium with one another very shortly after the concentration of any of those species changes. Thus, for instance, when a strong acid is added to an aqueous solution, the pH of the solution decreases, and the relationship $a_{OH^-} = (10^{-14}/a_{H^+})$ applies almost instantly; after the acid increases the H^+ concentration, it takes almost no time at all for the OH^- concentration to adjust to the new situation. This assumption allows us to calculate the OH^- concentration in any solution where the pH is known, without conducting a kinetics analysis. The same assumption applies to most acid/base reactions, as well as many other reactions of importance in water and wastewater treatment systems.

At the other extreme, if the characteristic times for both the forward and reverse directions of a reaction are much longer than the time available, then the reaction proceeds negligibly toward equilibrium during that time. We need not conduct a kinetics analysis in this situation either, because the reacting species can be treated as nonreactive; that is, inert.

Reaction Quotients, Equilibrium, and the Assumption of Irreversibility The assumption that a reaction is irreversible can simplify kinetics analysis significantly, and such an assumption is made implicitly for many important reactions in environmental engineering. For instance, when hypochlorous acid (HOCl) is added to water as a disinfectant, it can react with contaminants in the water and be converted to chloride ion (Cl^-). Although, in theory, some chloride ions could revert to HOCl molecules, the rate of such reversion is negligibly small, and it is never considered in an analysis of the system. Similarly, it is never considered necessary to account for the reversion of corroded pipe materials back to their metallic form, or the spontaneous conversion of carbon dioxide and water to organic matter (in the absence of photosynthesis). The dissociation of many strong acids, bases, and salts (e.g., H_2SO_4, NaOH, and $Al_2(SO_4)_3 \cdot xH_2O$ (alum)), is also commonly treated as being irreversible. In this section, features of reaction rate expressions and equilibrium constants that can support an assumption of irreversibility are explored.

Consider the following generic elementary reaction and corresponding activity-based rate expression:

$$aA + bB \underset{k_{PA}}{\overset{k_{AP}}{\rightleftharpoons}} pP + rR \quad (3\text{-}56)$$

$$r_A = -k'_{AP} a_A^a a_B^b + k'_{PA} a_P^p a_R^r$$

If the reaction is not at equilibrium, the net rate of formation of A is nonzero. Multiplying and dividing the right side of Equation 3-56 by $k'_{AP} a_A^a a_B^b$, and then rearranging, we obtain

$$r_A = \frac{-k'_{AP} a_A^a a_B^b + k'_{PA} a_P^p a_R^r}{k'_{AP} a_A^a a_B^b} k'_{AP} a_A^a a_B^b \quad (3\text{-}57)$$

$$r_A = \left(-1 + \frac{(a_P^p a_R^r / a_A^a a_B^b)}{(k'_{AP}/k'_{PA})}\right) k'_{AP} a_A^a a_B^b \quad (3\text{-}58)$$

$$r_A = \left(-1 + \frac{Q}{K_{eq}}\right) k'_{AP} a_A^a a_B^b \quad (3\text{-}59)$$

where Q is the called the *reaction quotient* and is a ratio of activities analogous to the equilibrium constant, but not restricted to equilibrium conditions; that is,

$$Q = \frac{a_P^p a_R^r}{a_A^a a_B^b} \quad (3\text{-}60)$$

If Q is either much larger or much smaller than K_{eq}, the system is far from equilibrium; if Q is close to K_{eq}, the system is near equilibrium; and if Q equals K_{eq}, the system is at equilibrium.[14] Here, we define the ratio Q/K as the *degree of disequilibrium*, thereby distinguishing it from the difference $c^* - c$, or $\hat{c}$, which was defined previously as the *extent of disequilibrium*. Both Q/K and $\hat{c}$ are measures of how far the system is from an equilibrium condition, and both are useful in different contexts.

According to Equation 3-59, if Q is either much larger or much smaller than K_{eq}, the approximations shown in Table 3-7 apply.

Equations 3-61c and 3-62 are the rates of the reverse reaction alone and the forward reaction alone, respectively (the negative sign in Equation 3-62 indicates that the net change is disappearance of A). Thus, when the system is far from equilibrium, the overall rate of reaction can be approximated by the reaction

TABLE 3-7. Approximate Rate Equations for Systems Far from Equilibrium[a]

$Q \gg K_{eq}$		$Q \ll K_{eq}$	
$r_A \approx \dfrac{Q}{K_{eq}} k'_{AP} a_A^a a_B^b$	(3-61a)	$r_A \approx -k'_{AP} a_A^a a_B^b$	(3-62)
$\approx \dfrac{a_P^p a_R^r / a_A^a a_B^b}{k'_{AP}/k'_{PA}} k'_{AP} a_A^a a_B^b$	(3-61b)		
$\approx k'_{PA} a_P^p a_R^r$	(3-61c)		

[a] For the overall rate expression $r_A = -k'_{AP} a_A^a a_B^b + k'_{PA} a_P^p a_R^r$.

---

[14] The conventional thermodynamic measure of chemical disequilibrium is the molar Gibbs energy change associated with the reaction ($\Delta \overline{G}_r$), which is directly related to the ratio of Q to K_{eq}: $\Delta \overline{G}_r = RT \ln(Q/K_{eq})$.

in only one direction. In other words, a reaction that is approximately irreversible is one that is far from equilibrium. The same result can be seen for the specific case of first-order forward and reverse reactions by letting c_A^* be either much larger or much smaller than c_A in Equation 3-48.

Note that neither the equilibrium constant (K_{eq}) nor the instantaneous ratio of reactant activities to product activities (Q) can, by itself, indicate whether a reaction may be considered irreversible. Rather, the issue of reversibility hinges on the ratio of these quantities. Furthermore, because of certain universal aspects of the relationship between rate constants and the equilibrium constant, the term in parentheses in Equation 3-59 appears in the net rate expression not just for elementary reactions, but for all nonelementary reactions as well. As a result, the conclusion that reactions are approximately irreversible if Q is much larger or much smaller than K_{eq} is also universal; that is, it applies even to nonelementary reactions.

The earlier discussion indicates that *any* reaction might be reversible in some circumstances and approximately irreversible in others. Although the determination of whether a reaction can be treated as irreversible depends on the particular goals of the analysis, in most cases it would seem reasonable to assume that a reaction is irreversible if, throughout the time frame of interest, the rate in one direction is at least one to two orders of magnitude faster than the rate in the other direction.

■ **EXAMPLE 3-13.** Could the NiF^+ dissociation reaction be treated as irreversible under the conditions evaluated in Example 3-10b?

Solution. The analysis in Example 3-10b indicates that NiF^+ dissociation initially proceeds 50 times as fast as NiF^+ formation. Therefore, it might be reasonable to ignore the NiF^+ formation reaction initially and treat the reaction at that instant as being irreversible. However, as shown in Example 3-11, the extent of disequilibrium diminishes to near zero very shortly thereafter. Within 5 ms, the reaction is much closer to equilibrium, and an assumption of irreversibility would be inappropriate. ■

Nearly Complete Reaction as a Limiting Case The equilibrium condition for some reactions lies so far to one extreme that they can be considered irreversible under virtually all conditions of interest in environmental engineering. A few examples of such reactions were cited in the previous subsection. In such cases, given enough time, the reaction would proceed until one of the reactants was almost completely depleted. Therefore, if the equilibrium condition for a reaction lies far toward the product side, and if the time frame of interest is much longer than the characteristic reaction time, the approximation of *complete* or *stoichiometric reaction* applies. This approximation implies that 100% of the added reactant is converted into product (unless, of course, one of the other reactants runs out first).

The assumption of complete, stoichiometric reaction is often made when estimating the amounts of solids that form when precipitation reactions are induced in water or wastewater treatment systems. For instance, for the purpose of estimating sludge production, it is often assumed that all the metal (e.g., copper, zinc, chromium) that is in an industrial wastewater stream precipitates when the pH is increased to the range 10–12. All the aluminum or iron that is added to assist in coagulating the precipitated solids in such systems, or to facilitate coagulation in drinking water treatment, is also usually assumed to precipitate, as $Al(OH)_3(s)$ or $Fe(OH)_3(s)$. Similarly, many analytical tests are based on the assumption that the constituent being analyzed reacts completely with the added reagents.

■ **EXAMPLE 3-14.** Alum is frequently added to water and wastewater to increase the size of suspended particulate matter and thereby facilitate its removal by settling or filtration. Often, much of the aluminum added with the alum precipitates as $Al(OH)_3(s)$ by the reaction shown.

$$Al^{3+} + 3\,OH^- \leftrightarrow Al(OH)_3(s)$$

The equilibrium constant for this reaction written with the reactants and products reversed (i.e., for the dissolution of $Al(OH)_3(s)$) is referred to as the solubility product of the solid and is designated K_{s0}. For $Al(OH)_3(s)$, $K_{s0} = 10^{-31.62}$.

If the solution pH is known, and if the $Al(OH)_3(s)$ that precipitates is a pure solid (and therefore can be assigned an activity of 1.0), the solubility product can be used to compute the activity of free aquo Al^{3+} ions; that is, $a_{Al^{3+}}$. The equilibrium constants for formation of Al–OH complexes (i.e., soluble species with formulas of $Al_y(OH)_x^{3y-x}$) can then be used to calculate the activity of these species. At pH 6.0, the activity of all soluble Al species is approximately 200 times that of Al^{3+}.

Consider a system in which enough alum is added to a wastewater to provide a total aluminum concentration ($c_{Al,tot}$) of 10 mg/L Al to the solution. The alum dissolves completely, but $Al(OH)_3(s)$ then precipitates. The pH of the wastewater throughout the process is 6.0.

(a) Considering only formation of the hydroxo complexes (i.e., ignoring other inorganic and all organic complexes), compute Q/K_{eq} for the precipitation reaction shortly after the alum is first added. Assume that, at that time, all of the alum has dissolved, the Al–OH complexes have formed, and a very small amount of solid has precipitated. Do you think that precipitation of Al(OH)$_3$(s) can be treated as irreversible at this time? Assume ideal behavior of the solutes.

(b) Precipitation of Al(OH)$_3$(s) is rapid compared to the hydraulic residence time in the reactor under consideration, so it might be reasonable to assume that the reaction reaches equilibrium. In that case, do you think that the precipitation reaction could be treated as going to completion for the purposes of computing sludge production? Could the precipitation process be treated as going to completion for the purposes of estimating the amount of Al remains in solution?

Solution.

(a) The total molar concentration of Al added to the solution is

$$c_{\text{Al}_{\text{tot}}} = (10\,\text{mol/L})\left(\frac{1\,\text{mol Al}}{27000\,\text{mg}}\right) = 3.7 \times 10^{-4}\,\text{mol/L}$$

According to the problem statement, shortly after the alum addition, all of Al added is present as soluble species, and one two-hundredth of that Al is present as free Al^{3+}. Thus, at that time, the concentration of free aquo Al^{3+} is $c_{\text{Al}_{\text{tot}}}/200$, or 1.9×10^{-6} mol/L.

We are interested in the precipitation reaction, as shown in the problem statement. The value of Q for this reaction equals $\left(a_{\text{Al(OH)}_3(s)}/a_{\text{Al}^{3+}}a^3_{\text{OH}^-}\right) = \left(1/a_{\text{Al}^{3+}}a^3_{\text{OH}^-}\right) = \left(a_{\text{Al}^{3+}}a^3_{\text{OH}^-}\right)^{-1}$, the value of K_{eq} is K^{-1}_{s0}, and, at pH 6.0, $a_{\text{OH}^-} = K_w/a_{\text{H}^+} = 10^{-14.0}/10^{-6.0} = 10^{-8.0}$. Thus, Q/K_{eq} for the reaction is

$$\frac{Q}{K_{\text{eq}}} = \frac{\left(a_{\text{Al}^{3+}}a^3_{\text{OH}^-}\right)^{-1}}{K^{-1}_{s0}} = \frac{\left[(1.9 \times 10^{-6})(10^{-8.0})^3\right]^{-1}}{10^{+31.62}} = 0.013$$

Since Q is less than K_{eq}, the reaction is not at equilibrium and will proceed to the right (precipitation of Al(OH)$_3$(s)). The ratio of Q to K_{eq} is almost two orders of magnitude different from 1.0, meaning that the degree of disequilibrium is substantial, so it would be reasonable to treat the initial reaction as irreversible.

(b) The final equilibrium activity of Al^{3+} is given by the solubility product. Because the solution pH is assumed to remain at 6.0 throughout the process, $a_{\text{Al}^{3+}}$ at equilibrium can be computed as follows:

$$K_{\text{eq}} = \frac{1}{a_{\text{Al}^{3+}}a^3_{\text{OH}^-}}$$

$$a_{\text{Al}^{3+}} = \frac{1}{K_{\text{eq}}a^3_{\text{OH}^-}} = \frac{1}{10^{+31.62}(10^{-8.0})^3} = 10^{-7.62}$$

Again equating the numerical values of $a_{\text{Al}^{3+}}$ and $c_{\text{Al}^{3+}}$, we conclude that $c_{\text{Al}^{3+}}$ at equilibrium is $10^{-7.62}$ mol/L. The final concentration of total dissolved aluminum is $200\,c_{\text{Al}^{3+}}$, or 4.8×10^{-6} mol/L (0.13 mg/L).

The total dissolved Al concentration at equilibrium is only 1.3% of the total Al added, so approximately 99% of the Al that was added precipitates. Thus, the assumption of stoichiometric precipitation would be a good one, if the reaction really did reach equilibrium. Keep in mind though, that in a real system, some of the Al might form soluble complexes with ligands other than OH$^-$, and some colloidal Al(OH)$_3$(s) might remain suspended in the effluent. As a result, the concentration of total Al in the effluent could be substantially greater than the value computed using the preceding equations.

If the precipitation process were treated as going to completion to estimate the soluble Al in the effluent, the estimate of that concentration would be zero. Whether or not ignoring Al in the effluent is an acceptable assumption depends on the ultimate use of the information. If we were interested in potential effects of dissolved Al on the receiving water, for instance, the assumption might not be justified, since relatively small additions of Al to the water might have a significant effect. On the other hand, when computing the total dissolved solids (TDS) of the effluent, the contribution of the Al species would almost certainly be negligible. Thus, in this example, as always, the validity of an assumption depends on the context in which the assumption is made and in which the data are being interpreted. ∎

Summary of Limiting Cases The various limiting cases used to simplify reaction rate expressions are summarized and compared in Table 3-8.

3.5 KINETICS OF SEQUENTIAL REACTIONS

The progress of an overall reaction that consists of two or more reactions in series is presented in this section. The analysis begins with a few numerical simulations, the results of which allow us to generalize about the behavior of the reactants at various points in the sequence and the characteristic times of the overall reaction. The results are also used to identify limiting conditions that might be applicable in certain situations and that can simplify the mathematical analysis of the system. Finally, an historically important overall reaction rate expression is developed based on the presumed reaction mechanism.

TABLE 3-8. Summary of Limiting Cases Under Which the Reaction Rate Expression can be Simplified

Approximation	Criteria	Effect
Irreversible reaction	Forward rate $\gg$ reverse rate (or vice versa); Reaction far from equilibrium	Simplifies the rate expression in the mass balance
Reaction at equilibrium	Forward rate = reverse rate; $t_{char} \ll$ time available for reaction	Eliminates the forward and reverse rate terms for a given reaction from the mass balance, and substitutes the equilibrium relationship between reactants and products
Stoichiometric or complete reaction	$t_{char} \ll$ time available for reaction, and equilibrium lies far to product side	Eliminates the forward and reverse rate terms for a given reaction from mass balance, and substitutes a relationship based on stoichiometry
Negligible reaction	$t_{char} \gg$ time available for reaction	Eliminates the forward and reverse rate terms for a given reaction from mass balance, since neither reaction proceeds to a significant extent

The Progress of Consecutive Reactions and the Rate-Controlling Step

As material progresses through a reaction sequence, we can imagine that, at each step, it encounters a resistance that prevents it from being converted instantly into the next species. When chemicals encounter such a sequence of resistances, they tend to build up immediately upstream of high-resistance transition points. In this aspect, the process is identical to water flowing through a series of tanks connected by pipes with partially open valves (Figure 3-18), or traffic on a road with tollbooths located every few miles. In each of these systems, material (chemical species, water, or cars) builds up behind the points where the resistance to progress is greatest. If the resistance at one point in the sequence is much greater than that at any other point, that point represents a bottleneck that limits the rate of the overall process. In the case of sequential reactions, the reaction imposing the greatest resistance is called the *rate-determining*, *rate-limiting*, or *rate-controlling* step. In these cases, greater resistance is synonymous with increased characteristic time, so the rate-controlling step can also be defined as the step with the longest characteristic time.

Consider, for example, the progress of the reaction sequence $A \xrightarrow{k_A} B \xrightarrow{k_B} P$ in a batch reactor, if each reaction is first order. The rate of change of the concentrations of A, B, and P in the system can be derived from the corresponding mass balances. In each case, the mass balance reduces to an equality between the storage and reaction terms. After dividing by V, these equations can be written as follows:

$$\frac{dc_A}{dt} = -k_A c_A \quad (3\text{-}63)$$

$$\frac{dc_B}{dt} = k_A c_A - k_B c_B \quad (3\text{-}64)$$

$$\frac{dc_P}{dt} = k_B c_B \quad (3\text{-}65)$$

Equation 3-63 is identical to the mass balance on A that would apply if the reaction $A \to B$ were taking place in isolation. The implication is that the reaction of B to form P, after B has been formed from A, has no effect on the reaction rate of A. As a result, species A disappears at the rate derived previously for first order, irreversible decay; that is, $c_A(t) = c_A(0)\exp(-k_A t)$.

As A is depleted, B forms, so its concentration initially increases. However, the rate of formation of B is $k_A c_A$, and since c_A is steadily decreasing, the rate at which B forms decreases steadily as well. At the same time, B is being converted to P at a rate proportional to c_B; that is, the rate of disappearance of B increases as c_B increases. Thus, the overall pattern that develops is that B accumulates initially, until its rate of formation ($k_A c_A$) equals its rate of disappearance ($k_B c_B$). From that time on, B disappears more quickly than it is formed, and c_B decreases. Since P is formed at a rate proportional to c_B, P accumulates steadily, and its rate of accumulation is largest when c_B is largest.

Analytical solutions to the mass balances on B and P for $c_B(0) = c_P(0) = 0$ can be derived by integration of the corresponding rate expressions, yielding:[15]

$$c_B(t) = \frac{k_A c_A(0)}{k_B - k_A}(\exp(-k_A t) - \exp(-k_B t)) \quad (3\text{-}66)$$

$$c_P(t) = \frac{c_A(0)}{k_B - k_A}(k_B\{1 - \exp(-k_A t)\} - k_A\{1 - \exp(-k_B t)\}) \quad (3\text{-}67)$$

The concentration profiles for all three species in a system with $c_A(0) = 100\ \mu\text{mol/L}$ and $c_B(0) = c_P(0) = 0\ \mu\text{mol/L}$, and with $k_A = k_B = 0.1\ \text{min}^{-1}$ are shown in Figure 3-19. The results are consistent with the earlier discussion. In particular, B accumulates in the system until approximately

---

[15] The equations shown do not apply if $k_A = k_B$. In that case, the expressions of the concentrations of B and P are: $c_B = c_A(0) k_A t \exp(-k_A t)$ and $c_P = c_A(0)[1 - \exp(-k_A t) - k_A t \exp(-k_A t)]$

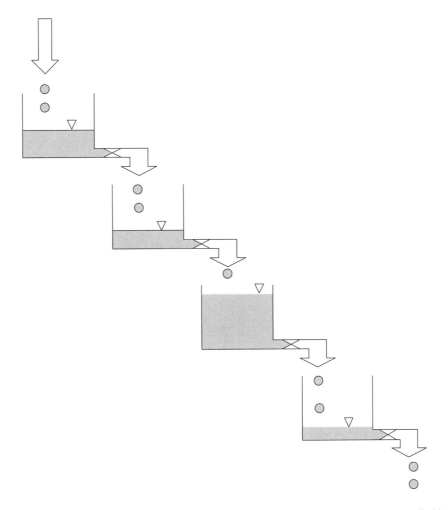

FIGURE 3-18. A sequence of tanks, with the rate at which water leaves each tank controlled by a valve. The rate at which water exits the overall system is controlled by the tightest valve (on the pipe leaving the third tank in the sequence), behind which water accumulates.

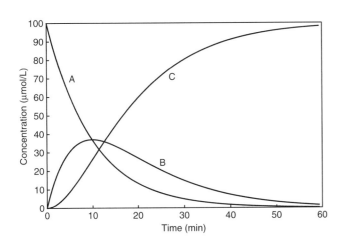

FIGURE 3-19. Concentration versus time profiles for a sequential irreversible reaction A → B → P in a batch reactor, with $c_A(0) = 100\,\mu\text{mol/L}$, and $c_B(0) = c_P(0) = 0\,\mu\text{mol/L}$, and with $k_A = k_B = 0.1/\text{min}$.

$t = 10$ min, at which time the rates of formation and destruction of B are equal, and $c_B = 36\,\mu\text{mol/L}$. Continuously thereafter, B reacts more rapidly than it forms, so its concentration decreases. The overall reaction is substantially complete ($c_P = 90\,\mu\text{mol/L}$) in about 40 min. This result is consistent with what we might expect based on the characteristic reaction times, which are 10 min for each of the constituent reactions. That is, we might expect each reaction to be substantially complete in a time equal to $\sim 3\,t_{\text{char}}$ (i.e., 30 min), with the beginning of that time for the second reaction being delayed until a substantial amount of B has formed ($\sim 1\,t_{\text{char}}$). In this case, because the characteristic time is identical for the two reactions, neither reaction can be said to be rate limiting.

Figures 3-20a and 3-20b show the concentration profiles for systems with the same initial composition as in Figure 3-19, but with one of the rate constants increased by a factor of 10, and the other rate constant decreased by the same factor. Thus, in these systems, the characteristic times are 1 min for

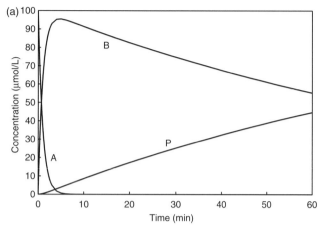

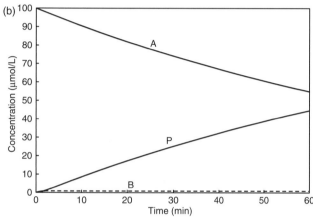

FIGURE 3-20. Concentration profiles for species A, B, and P for the same reaction sequence as characterized in Figure 3-19, except with different rate constants: (a) $k_A = 1.0/\text{min}$, $k_B = 0.01/\text{min}$; (b) $k_A = 0.01/\text{min}$, $k_B = 1.0/\text{min}$.

the faster reaction and 100 min for the slower one. The concentrations are still changing significantly at the end of the time period shown in Figures 3-20a and 3-20b but, as in Figure 3-19, they approach values of $c_A = c_B = 0$ and $c_P = 100\,\mu\text{mol/L}$ asymptotically at large t.

In the system in which k_A is increased, so that $k_A \gg k_B$, almost all the A is converted to B before significant amounts of B are converted into P (Figure 3-20a). As a result, c_B increases rapidly to a value near $c_A(0)$, after which the only relevant reaction is decay of B to P. That is, after a relatively short reaction time ($t < 5$ min, representing several characteristic reaction times for the reaction A $\to$ B, but much less than one characteristic reaction time for the reaction B $\to$ P), the composition of the system can be approximated as $c_A \approx 0\,\mu\text{mol/L}$, $c_B \approx c_A(0)$, $c_P \approx 0\,\mu\text{mol/L}$. Thereafter, the first reaction can be ignored, and the only reaction of relevance is the first-order decay of B to P.

The same conclusions can be reached by evaluating Equations 3-66 and 3-67 with $k_A \gg k_B$. In such a case, $\exp(-k_A t) \ll \exp(-k_B t)$, and the expressions simplify to

$$c_B(t) = c_A(0)\exp(-k_B t) \quad (3\text{-}68)$$

$$c_P(t) = c_A(0)\{1 - \exp(-k_B t)\} \quad (3\text{-}69)$$

Equations 3-68 and 3-69 are those that would apply for a first order, irreversible reaction in which P is formed from B, if the initial concentration of B were $c_A(0)$. In this case, conversion of B to P is the rate-limiting step, and the characteristic time for the overall reaction (i.e., the characteristic time for formation of P) is k_B^{-1}.

On the other hand, if $k_B \gg k_A$ (Figure 3-20b), then A decays quite slowly to form B, and B decays almost immediately to form P. As a result, the qualitative changes in c_B observed in the previous cases still occur (i.e., c_B increases, passes through a maximum when $k_A c_A = k_B c_B$, and then decreases), but the absolute value of the concentration of B at any time is low. The equations characterizing the concentrations of B and P in this case are as shown below. In this case, the conversion of A to B is the rate-limiting step in the overall reaction, so the characteristic time for formation of P is k_A^{-1}.

$$c_B = \frac{k_A}{k_B} c_A(0)\exp(-k_A t) \quad (3\text{-}70)$$

$$c_P = c_A(0)\{1 - \exp(-k_A t)\} \quad (3\text{-}71)$$

Generalizing the preceding result, when any irreversible reaction sequence is initiated, the first reaction in the sequence proceeds at the same rate as if it were the only reaction occurring. Material then begins accumulating upstream of each point of resistance in the rest of the sequence; that is, some of each intermediate species accumulates, providing the driving force for the next reaction in the sequence. If the reaction is occurring in a batch reactor, the concentration of each intermediate increases steadily until the driving force is sufficient to push material through that step as rapidly as it is arriving from the upstream step. Thereafter, the rate of arrival declines (because the initial supply of reactant becomes depleted), and the concentration of the intermediate decreases, decreasing the rate at which the subsequent reaction proceeds. This process continues until essentially all the material has passed through the whole sequence, and only the ultimate product of the reaction sequence is present.

The characteristic time of the overall sequence is at least as long as that of the rate-limiting step. If the characteristic time of the rate-limiting step is significantly greater than that of any other step in the sequence, then the characteristic time of the overall reaction is approximately equal to that of the rate-limiting step. (Although this point is made by considering only first-order reactions in the example system, it applies to other rate expressions as well.)

If the same reaction sequence occurred in a reactor system with a continuous input of the initial reactant (e.g., a CFSTR with the reactant present in the influent at some steady concentration), each intermediate would be generated continuously, and the whole system would reach a steady state in which material was flowing through each step at the same rate. To achieve that steady-state condition, the concentration of each intermediate would be large if it preceded a high-resistance reaction (one with a long t_{char}), and small if it preceded a low-resistance (short t_{char}) reaction. Again, the analogy to water flow through a series of tanks is apt.

Regardless of whether the reaction sequence occurs in a batch reactor or a reactor with flow, the greatest resistance, the longest characteristic time, and the greatest accumulation of material are all associated with the rate-limiting step. In many reaction sequences, the resistance associated with the rate-limiting step is so much greater than that of any other step in the sequence that it is reasonable to treat the system as though none of the other reactions impose any resistance at all. In such a case, intermediate species that precede the reactants for the rate-limiting step are formed and then depleted relatively rapidly, so they are all present at low concentrations. On the other hand, the reactants that participate in the rate-limiting step accumulate and then move through that step gradually, until they are eventually depleted.

Species that follow the rate-controlling step react relatively quickly, but they are only supplied at a slow, nearly steady rate from upstream. As a result, the concentrations of these species (except the final product in the sequence) remain low, only building up enough to maintain a reaction rate approximately equal to that of the rate-controlling step. Correspondingly, the ultimate product is also formed at that rate (i.e., the rate of the controlling step). Note that, because material moves through the rate-controlling step and all subsequent steps at approximately the same rate, it is not correct to state that the rate-controlling step is the slowest one in the sequence. What *is* true is that the rate-controlling step is the major impediment to speeding up the reaction; that is, material *could* move through the rest of the system faster, if the rate-limiting step were not holding everything up.

■ **EXAMPLE 3-15.** Two reactions proceed in sequence. In the first, reactants A and B are converted to a product C and an intermediate D. In the second, D reacts with water to form product E. Both reactions can be treated as irreversible. The reaction between species A and B is first order with respect to each of those species and second order overall, with a rate constant of $k_{AB} = 10^{-3}$ L/mol s. The subsequent reaction forming species E is pseudo-first-order, with $k = 10^{-2}$ s^{-1}.

$$A + B \rightarrow C + D \quad r_A = -(10^{-3}\,\text{L/mol s})c_A c_B$$
$$D + H_2O \rightarrow E \quad r_E = (10^{-2}/\text{s})c_D$$

(a) Determine the rate-limiting step for this reaction sequence in a system in which the initial concentrations of A and B are both 10^{-2} mol/L, and the system initially contains no C, D, or E.

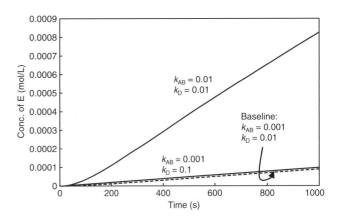

FIGURE 3-21. Concentration of species E as a function of time in the two-step, irreversible reaction described in the example problem statement, for various values of the forward rate constants k_{AB} and k_D. Values shown for k_{AB} are in L/mol s, and those for k_D are in s^{-1}.

(b) Verify your conclusion by comparing the production over time of species E in the initial system and in systems in which either k_{AB} or k_D is increased by a factor of ten. (The hypothetical cases where the rate constant is multiplied by ten might be accomplished by adding catalysts to the system.)

Solution.

(a) The characteristic time of the first reaction can be computed using the expressions in Table 3-3, since the initial concentrations of A and B are identical. Using the first definition for t_{char}, we find

$$t_{char} = \frac{e^{n-1}-1}{n-1}\frac{1}{k_n c_A^{n-1}(0)} = \frac{e-1}{1}\frac{1}{k_n c_A(0)} = 172{,}000\,\text{s} = 47.8\,\text{h}$$

Because the conversion of D to E is pseudo-first-order, its characteristic time is the inverse of the rate constant, or 100 s. The first reaction is therefore expected to be rate limiting and to provide almost all the resistance to formation of product.

(b) The conclusion reached in part (a) is confirmed by the simulations shown in Figure 3-21, which shows the results of a numerical integration of the rate expressions. Increasing the rate constant of the initial reaction increases the rate of formation of E dramatically, whereas increasing the rate constant of the second reaction has a much smaller effect. ■

The Thermodynamics of Sequential Reactions

As noted previously, the degree of disequilibrium of a reaction can be quantified by the ratio of the activity quotient to the equilibrium constant (Q/K_{eq}). This ratio is related to the molar Gibbs energy of the reaction, $\Delta \overline{G}_r$ (kJ/mol of reaction) by

$$\Delta \overline{G}_r = RT \ln(Q/K_{eq}) \qquad (3\text{-}72)$$

Thus, the farther a reaction is from equilibrium (i.e., the closer it is to being irreversible), the larger is the magnitude of its negative molar Gibbs energy of reaction.[16]

In a sequence of reactions, the Gibbs energies of the individual reactions are additive, so $\Delta \overline{G}_r$ for the overall reaction is the sum of the $\Delta \overline{G}_r$ values of the individual reactions. Thus, one can say that, in a reaction sequence, the Gibbs energy of the overall reaction is expended little by little in pushing material through the various steps. Extending the metaphor, other things being equal, the larger the resistance with which an individual step opposes the overall reaction, the larger the expenditure of Gibbs energy required to push material through that step. Thus, the reaction farthest from equilibrium provides the greatest resistance to the overall reaction.

Since reactions always proceed toward equilibrium, they always proceed in the direction that causes $\Delta \overline{G}_r$ to approach zero (and Q/K_{eq} to approach 1.0). This statement applies not only to the overall sequential reaction, but also to each individual reaction in the sequence. Thus, it is not possible for the Gibbs energy released from one step in a reaction sequence to allow another step to proceed "uphill" or to "overcome a Gibbs energy barrier." Such a statement implies that a chemical reaction proceeds in a nonthermodynamic direction (i.e., away from equilibrium) because the energy cost of doing so is being paid by another reaction. What is possible is that a product from one reaction can be depleted so rapidly by a subsequent reaction that its concentration is maintained at a very low level, much lower than it would be if the downstream reaction were not occurring. In this way, the Gibbs energy change for each step can be held at a negative value, so that the driving force for every reaction is in the forward direction. Such relationships are the core feature of sequential reactions.

Steady State: Definition and Comparison with Chemical Equilibrium

A species whose concentration at a given location is not changing over time is said to be at *steady state*, and, if the concentrations of all chemical species in a reactor are unchanging over time, then the reactor is said to be at steady state. (Note that the concentrations might change from point to point in the reactor; the definition of steady state requires only that they be constant at each point over time.) In reactors with flow, steady-state conditions can be established and maintained indefinitely. In a batch reactor, the concentrations of all reactants, intermediates, and products in a reaction sequence change over time, so a true steady state never develops. Nevertheless, the concentrations of the intermediates downstream of the rate-controlling step change relatively slowly. In such cases, an approximation is often made that those intermediates are at steady state during most of the time the reaction is proceeding. A classic example of the use of this approximation is provided in the following section of this chapter.

In many ways, steady state describes a condition for a particular species that is analogous to the condition that chemical equilibrium describes for a reaction: in both cases, an overall process is poised at a stable condition as a result of ongoing, balanced subprocesses. Because both concepts are central to the analysis of physical/chemical treatment systems, and because the distinctions between them are sometimes subtle, it is worth considering a few situations in which each concept is applied independently.

It was noted previously that the equilibration of water molecules with H^+ and OH^- is so rapid at near-neutral pH that it can be considered instantaneous. However, this statement does not imply that the H^+ and OH^- concentrations in a system are always at steady state. For instance, biological processes might consume or generate acidity continuously in a batch waste treatment process, so that the concentrations of H^+ and OH^- would be constantly changing; that is, neither species would be at steady state. The two species would nevertheless be in continuous equilibrium with one another and with H_2O via the association/dissociation reaction for water.

On the other hand, as noted earlier, a species can be at steady state in a reactor if it is formed and destroyed at equal rates by different chemical reactions, or if the summation of the rates at which it is formed by all reactions and the rate at which it enters the reactor by advection, diffusion, and dispersion equals the summation of the rates at which it is destroyed and/or removed from the reactor by the corresponding processes. In many systems, even though one or more individual species are at steady state, the chemical reactions in which those species participate are not at equilibrium.

Thus, a reaction can be at equilibrium even if the concentrations of the reactants and products in the system are changing, and a species can be at steady state even if all the reactions in which it participates are far from equilibrium. Additional important examples of steady-state systems are presented in Chapter 4, where the effects of fluid flow and dispersion on the concentration are considered in addition to those of chemical reaction.

The usefulness of the steady-state approximation can be illustrated by the development of the classical Michaelis–Menten relationship describing the enzyme-catalyzed

---

[16] Q might be either larger or smaller than K_{eq} in a system that is not at equilibrium. In either case, the reaction proceeds in the direction that causes Q to approach K_{eq}, which is the direction in which $\Delta \overline{G}_r < 0$. Q might be either larger or smaller than K_{eq} in a system that is not at equilibrium. In either case, the reaction proceeds in the direction that causes Q to approach K_{eq}, which is the direction in which $\Delta \overline{G}_r < 0$.

conversion of organic substrates to cell parts. This mechanism was postulated by Michaelis and Menten in 1913 and is still used as the basis for most mathematical modeling of biological reactions in wastewater treatment. The analysis would be quite complex if the variation in the concentrations of intermediates over time had to be considered, but it is greatly simplified if the steady-state assumption is made.

Michaelis and Menten hypothesized the following set of elementary equations:

$$E + S \underset{k_{ES^* \to S}}{\overset{k_{S \to ES^*}}{\rightleftarrows}} ES^* \quad (3\text{-}73)$$

$$ES^* \underset{k_{P \to ES^*}}{\overset{k_{ES^* \to P}}{\rightleftarrows}} P + E \quad (3\text{-}74)$$

where E is an enzyme, S is a biodegradable substrate, P is a product of the reaction, and ES^* is a reaction intermediate. Key assumptions of the model are that

(i) The concentration of ES^* reaches an approximate steady-state value, so ($dc_{ES^*}/dt \approx 0$).

(ii) For the given steady-state concentration of ES^*, $k_{ES^* \to P} c_{ES^*} \gg k_{P \to ES^*} c_P$; that is, the reaction $ES^* \to P$ is far from equilibrium, and so is essentially irreversible.

Assumption (ii) is usually considered applicable for any steady-state concentration of ES^*, implying that the reaction $ES^* \to P$ goes to completion. Note that, since the reaction forming ES^* is assumed to have a significant reverse rate while the reaction forming P is approximately irreversible, Q/K_{eq} is much smaller and $-\Delta \overline{G}_r$ is much larger for the second step; that is, most of the energy driving the reaction is released in the second step.

In accordance with these assumptions, the expressions for the net formation rates of the enzyme–substrate complex (ES^*) and the product in a batch system are

$$r_{ES^*} = \frac{dc_{ES^*}}{dt} = 0 = k_{S \to ES^*} c_E c_S - k_{ES^* \to S} c_{ES^*} - k_{ES^* \to P} c_{ES^*}$$
(3-75)

$$r_P = \frac{dc_P}{dt} = k_{ES^* \to P} c_{ES^*} \quad (3\text{-}76)$$

The feature of the biological system that is not typical of other reactions is that one of the reactants (E) is consumed in the first reaction step but is then regenerated in the second step. Therefore, in a batch reactor, the total amount of E in the system (the sum of the concentrations of ES^* and unbound E) is constant, regardless of how much S reacts. The enzyme is thus a catalyst; that is, a substance that participates in a reaction and affects the overall reaction rate, but is neither generated nor consumed by the overall reaction. Defining $c_{E,tot}$ as the total concentration of enzyme, we can write

$$c_{E,tot} = c_E + c_{ES^*} \quad (3\text{-}77)$$

Substituting Equation 3-77 into Equation 3-75 and then solving for the concentration of ES^* yields

$$0 = k_{S \to ES^*}(c_{E,tot} - c_{ES^*}) c_S - (k_{ES^* \to S} + k_{ES^* \to P}) c_{ES^*}$$
(3-78)

$$c_{ES^*} = \frac{k_{S \to ES^*} c_{E,tot} c_S}{k_{S \to ES^*} c_S + k_{ES^* \to S} + k_{ES^* \to P}} \quad (3\text{-}79)$$

Finally, inserting Equation 3-79 into Equation 3-76 yields, after some manipulation

$$r_P = \frac{k_{ES^* \to P} c_{E,tot} c_S}{\dfrac{k_{ES^* \to S} + k_{ES^* \to P}}{k_{S \to ES^*}} + c_S} = \frac{k_{ES^* \to P} c_{E,tot} c_S}{K_m + c_S} \quad (3\text{-}80)$$

where K_m is defined as the ratio $(k_{ES^* \to S} + k_{ES^* \to P}/k_{S \to ES^*})$ and has units of concentration and is known as the Michaelis constant or the half-velocity constant. Assuming the total enzyme concentration is proportional to the microorganism concentration c_X (i.e., $c_{E,tot} = k_X c_X$), and defining the product $k_X k_{ES^* \to P}$ as k_{max}, Equation 3-80 can be rewritten as follows:

$$r_P = \frac{k_{max} c_S c_X}{K_m + c_S} \quad (3\text{-}81)$$

Given that the concentration of ES^* is always small and changes very slowly, the rate of product formation, r_P, can be equated with the rate of substrate utilization, $-r_S$. This rate is commonly normalized to the concentration of organisms in the system and referred to as the specific rate of substrate utilization, computed as $-r_S/c_X$. Thus, Equation 3-81 can be rewritten to express the specific rate of substrate utilization as

$$-\frac{r_S}{c_X} = \frac{k_{max} c_S}{K_m + c_S} \quad (3\text{-}82)$$

Equation 3-82 is commonly evaluated in terms of the relative values of K_m and c_S. If $c_S \ll K_m$, the rate expression reduces to

$$-\frac{r_S}{c_X} = \frac{k_{max}}{K_m} c_S \quad (3\text{-}83)$$

According to Equation 3-83, under conditions where $c_S \ll K_m$, the specific rate of substrate utilization is proportional to the concentration of substrate (c_S); in such cases, the reaction is said to be *substrate-limited*. Under substrate-limited conditions, a significant amount of free enzyme is available in the system. As a result, the steady-state concentration of ES^*, and hence the rate at which the product is formed, can be increased by increasing the substrate concentration.

At the opposite extreme, if $c_S \gg K_m$, the Equation 3-82 reduces to

$$-\frac{r_S}{c_X} = k_{max} \quad (3\text{-}84)$$

Under these conditions, the reaction is said to be *enzyme-limited*. The system contains a negligible steady-state

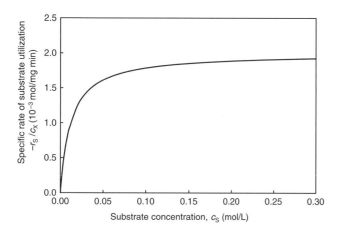

FIGURE 3-22. Production rate of P as a function of substrate (S) concentration for a Michaelis–Menten type reaction. $k_{max} = 2 \times 10^{-3}$ mol/mg min, $K_m = 0.012$ mol/L. c_X is assumed to be given in units of mg/L.

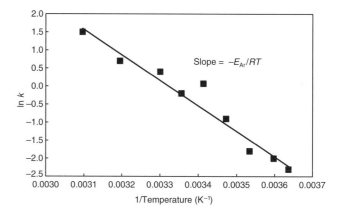

FIGURE 3-23. Characteristic plot of the rate constant versus inverse absolute temperature, from which the values of k_{Ar} and E_{Ar} can be computed. ($\ln k_{Ar}$) is found by extrapolating the straight line to the hypothetical condition $1/T = 0$.)

concentration of free enzyme, so adding more substrate cannot increase the concentration of ES* or the rate of product formation. Using the terms defined earlier in this chapter, the first reaction is poised at a condition of near completion, in which almost all the enzyme is in the form of product (ES*).

The terms substrate-limited and enzyme-limited can be somewhat misleading, since Equations 3-83 and 3-84 indicate that the rate of product formation is proportional to the total enzyme concentration in both cases. More descriptive terms would be substrate-and-enzyme-limited and enzyme-only-limited, respectively, but the shorter terms have been universally adopted.

The specific rate of substrate utilization is shown for a wide range of substrate concentrations in Figure 3-22. At low concentrations, the graph approximates a straight line with slope k_{max}/K_m (consistent with Equation 3-83), and at high concentrations, it approaches a value of k_{max} asymptotically (as per Equation 3-84); at intermediate concentrations, the curve can only be described by Equation 3-82.

Derivations such as the one presented earlier played an important role in the historical development of kinetics modeling. The need to make simplifying assumptions to analyze the kinetics of complex networks has diminished as high-speed computing equipment has been used to model the progress of all the steps in such reactions. Nevertheless, the steady-state assumption is still useful for developing simplified conceptual models of these systems and for interpreting the results of the computer simulations.

3.6 THE TEMPERATURE DEPENDENCE OF THE RATES OF NONELEMENTARY REACTIONS

The effects of temperature on reaction rates are important in environmental engineering because many industrial wastes are generated at elevated temperatures, and in some cases, the flow rate of a waste stream is small enough that increasing its temperature in an effort to increase the reaction rate (or, in the case of biological processes, to provide a selective advantage for growth of certain microbes) is an economically and technically attractive option. In addition, temperature is an important parameter to consider when comparing treatment processes in different climates, and seasonal changes in temperature can have a large effect on reaction rates in many natural and engineered aquatic systems. Investigations of the effect of temperature on reaction rates can also provide insight into the mechanisms controlling those rates.

The temperature dependence of elementary reactions was described in the context of the activated complex model for reaction kinetics earlier in this chapter. Prior to the development of that model, however, a similar relationship for the effect of temperature on rate constants had been proposed by Arrhenius, based on empirical studies of both elementary and nonelementary reactions. The relationship that he proposed is

$$k = k_{Ar} \exp\left(-\frac{E_{Ar}}{RT}\right) \qquad (3\text{-}85)$$

where E_{Ar} is an empirical constant, unique to a particular reaction, with units of energy per mole, R is the universal gas constant, and T is the absolute temperature.

Taking the logarithm of both sides of Equation 3-85, it can be written as follows:

$$\ln k = \ln k_{Ar} - \frac{E_{Ar}}{R}\frac{1}{T} \qquad (3\text{-}86)$$

Defining k_{T_1} and k_{T_2} as the reaction rate constants at temperatures T_1 and T_2, Equation 3-86 can be manipulated as follows:

$$\ln k_{T_2} - \ln k_{T_1} = \ln \frac{k_{T_2}}{k_{T_1}} = -\frac{E_{Ar}}{R}\left(\frac{1}{T_2} - \frac{1}{T_1}\right) \quad (3\text{-}87)$$

Equation 3-86 indicates that a plot of $\ln k$ versus $1/T$ should be a straight line with slope $-E_{Ar}/R$ and intercept $\ln k_{Ar}$, as shown in Figure 3-23. By preparing such a plot, we can determine these values and use them to compute the rate constant at other temperatures, either graphically or by inserting the constants into Equation 3-87.

3.7 SUMMARY

This chapter describes the techniques commonly used to analyze rate data from batch, well-mixed, homogenous (one-phase) systems. Such systems are often used for the study of reaction kinetics, because the mass balance reduces to a very simple form. Reaction kinetics is also sometimes studied in reactors with flow, either as a matter of preference or for more fundamental reasons. Such systems are discussed and the relevant mass balances are solved in Chapter 4. Subsequent chapters extend the analysis to multiphase systems, for example, systems where transfer of a substance into a gas or onto a solid adsorbent is an important process.

Chemical reactions generally occur via collisions among molecules. Reactions that result from a single collision are called elementary reactions, and their rates are given by the product of a temperature-dependent rate constant and the concentrations of the colliding species. In aqueous solutions, the maximum rate at which reactions can proceed is limited by the rates of diffusion of the reacting molecules. Although some reactions proceed at nearly the maximum rate, most proceed more slowly, indicating that they are controlled by the energetics of the reacting species.

Attempts to predict the rates of elementary reactions that are controlled by energetics led to the development of the activated complex model of reaction progress. This model represents reactions as requiring collisions among molecules that have enough energy and are properly oriented to overcome an activation energy barrier, after which conversion into product molecules is spontaneous. The model predicts that the reaction rate constant is strongly dependent on temperature and the activation energy of the reaction (i.e., the energy required to form activated complexes from the reactants).

Many overall reactions reflect the result of two or more elementary reactions. In such cases, the rate expression can take many mathematical forms, sometimes containing more than one constant and the concentrations of the reacting species raised to various powers.

Elucidation of rate expressions generally involves collection of experimental data and attempts to fit the data to potentially appropriate equations. Both integral and differential methods are used to test hypothesized rate expressions. The identification of an appropriate rate expression is reasonably straightforward in cases where the data can be fit with a power law expression. If the rate expression is not of the power law type, its identification is more problematic. If a reaction rate depends on the concentrations of several different species, the effect of individual species can be isolated by adding a relatively great excess of all species except one. By repeating the process and changing the species that is not added in excess, information about the rate dependence on each species can be obtained.

All reactions are reversible, at least in theory, and are therefore characterized by equilibrium constants. For elementary reactions, the equilibrium constant can be identified as the ratio of the forward and reverse rate constants. The degree of disequilibrium of a reaction is quantified by the ratio Q/K_{eq}, a quantity that is also related to the amount of Gibbs free energy that is released as a reaction proceeds. Reactions that are far from equilibrium can be approximated as being irreversible.

The characteristic reaction time provides a rough idea of the time frame over which a reaction proceeds. If the time available for reaction is far less than the characteristic time, negligible reaction occurs, and if the time available is far greater, the reaction proceeds almost to its endpoint (completion or equilibrium). The characteristic reaction time is also a qualitative indicator of how much a particular reaction resists conversion of reactants into products—the longer the characteristic time, the greater is the resistance.

Many overall reactions of interest consist of a sequence of approximately irreversible reactions. If one of the reactions in the sequence has a much longer characteristic time than the other, then it generates almost all the resistance to the progress of the reaction. Such a reaction is called rate-limiting, rate-determining, or rate-controlling step. This step represents a bottleneck in the overall process, and it limits the overall rate of product formation. Reactants upstream of the rate-controlling step are depleted relatively quickly, and those downstream (other than the ultimate product) attain gradually changing concentrations that allow them to proceed at approximately the same rate as the rate-controlling step.

Mathematical analysis of reaction networks can be complex even when the networks include only a few reactions. Approximations that some constituents are at steady state, or that some reactions are irreversible, rapidly equilibrated, or complete are sometimes made to reduce the mathematical and conceptual complexity of such systems. The Michaelis–Menten expression is a well-known and historically important example of the application of such simplifications. The proposed mechanism associated with the expression also shows how catalysts can participate in an overall reaction, even though they do not appear in the overall reaction stoichiometry.

No matter how persistent or determined the investigator is, some reaction rate expressions seem to defy efforts to represent them in reasonably tractable mathematical forms. In such cases, it must be remembered that the study of kinetics is inherently empirical, and that, in the absence of simple mathematical relationships describing the rate expressions, empirical data describing the reaction rate as a function of the concentrations of various constituents of the system can serve the same function.

REFERENCES

Benjamin, M. M. (2010) *Water Chemistry*. Waveland Press, Long Grove, IL.

Brezonik, P. L. (1994) *Chemical Kinetics and Process Dynamics in Aquatic Systems*. Lewis Publishers, Boca Raton, FL.

Chen, K. Y., and Morris, J. C. (1972) Kinetics of oxidation of aqueous sulfide by oxygen. *Environ. Sci. Technol.*, 6 (6), 529–537.

Curtis, G. P., and Reinhard, M. (1994) Reductive dehalogenation of hexachloroethane, carbon tetrachloride, and bromoform by anthrahydroquinone disulfonate and humic acid. *Environ. Sci. Technol.*, 28 (13), 2393–2401.

Cussler, E. L. (1984) *Diffusion: Mass Transfer in Fluid Systems*. Cambridge University Press, Cambridge, England.

Hill, C. G. (1977) *An Introduction to Chemical Engineering Kinetics and Reactor Design*. Wiley, New York.

Korshin, G. V., Benjamin, M. M., and Chang, H-S. (2004) Modeling DBP Formation Kinetics: Mechanistic and Spectroscopic Approaches, AwwaRF Report #91000F. American Water Works Research Foundation, Denver, CO.

Lee, D. S. (1990) Supercritical Water Oxidation of Acetamide and Acetic Acid. PhD dissertation, Dept. of Civil Engineering, Univ. of Texas at Austin, Austin, TX.

Levenspiel, O. (1999) *Chemical Reaction Engineering*, 3rd edn. Wiley, New York.

Luthy, R. G., and Bruce, S. B., Jr. (1979) Kinetics of reaction of cyanide and reduced sulfur species in aqueous solution. *Environ. Sci. Technol.*, 13 (12), 1481–1487.

Mitch, W. A., and Sedlak, D. L. (2002) Formation of N-nitrosodimethylamine (NDMA) from dimethylamine during chlorination. *Environ. Sci. Technol.*, 36 (4), 588–595.

Morel, F. M. M., and Hering, J. G. (1993) *Principles and Applications of Aquatic Chemistry*. Wiley Interscience, New York.

Morgan, J. J. (2005) Kinetics of reaction between O_2 and Mn(II) species in aqueous solutions. *Geochim, Cosmochim. Acta*, 69 (1), 35–48.

Oliver, B. G., and Schindler, D. B. (1980) Trihalomethanes from the chlorination of aquatic algae. *Environ. Sci. Technol.*, 14 (12), 1502–1505.

Piché, S., and Larachi F. (2007) Hydrosulfide oxidation pathways in oxic solutions containing iron(III) chelates. *Environ. Sci. Technol.*, 41 (4), 1206–1211.

Roberts, A. L., and Gschwend, P. M. (1991) Mechanism of pentachloroethane dehydrochlorination to tetrachloroethylene. *Environ. Sci. Technol.*, 25 (1), 76–86.

Rule, K. L., Ebbett, V. R., and Vikesland, P. J. (2005) Formation of chloroform and chlorinated organics by free-chlorine-mediated oxidation of triclosan. *Environ. Sci. Technol.*, 39 (9), 3176–3185.

Smoluchowski, M. (1917) Versuch einer mathematischen Theorie der koagulationskinetik Kolloider Lösungen. *Z. Physik. Chem.*, 92, 129–168.

Stumm, W. (1990) *Aquatic Chemical Kinetics*, Wiley Interscience, New York.

Stumm, W., and Morgan, J. J. (1996) *Aquatic Chemistry*, 3rd edn. Wiley Interscience, New York.

PROBLEMS

3-1. (a) Identify the overall order of the reaction and the order with respect to individual reactants in the rate expression: $r_A = -kc_A^{0.5} c_B$.

(b) The reaction whose kinetics is expressed in part (a) has the stoichiometry

$$\tfrac{1}{2} A + B \rightarrow P$$

What is the overall order of the reaction and the order with respect to individual reactants if the stoichiometry is rewritten as

$$A + 2B \rightarrow 2P$$

(c) For biological degradation of substrate S by microorganisms at concentration X, the rate of degradation is often expressed as follows: $r_S = -kXS/(K_S + S)$. Identify the order of the reaction with respect to the microorganism concentration, the substrate concentration, and the overall order.

(d) In some engineered systems for biological degradation, the substrate concentration is very low, so that $S \ll K_s$. In that circumstance, find the order of the reaction with respect to the microorganism concentration, the substrate concentration, and the overall order.

3-2. Oliver and Schindler (1980) presented the data shown in the following table for chloroform ($CHCl_3$) production from the chlorination of the algal species *Anabaena oscillarioides* at pH 7.0 and 20°C. The first table shows the $CHCl_3$ concentration over time in a batch system dosed with 10.3 mg/L chlorine (as Cl_2), and the second shows the $CHCl_3$ concentration after 24 h of reaction in systems with various chlorine doses. The algal concentration was 3.6 mg/L as dry weight in all experiments.

Time (h)	CHCl$_3$ (µg/L)
4	4
22	16
47	35
76	48
95	65

Chlorine Dose (mg/L)	CHCl$_3$ (µg/L)
1	2
2	6
5	11
10	20
20	23
40	30
75	24
150	30

(a) Assuming the reaction that generates chloroform is first order with respect to algal dry weight concentration, derive a conditional rate expression for the reaction at pH 7.0 and 20°C. Note: You will have to develop the expression based on your own interpretation of the data; there is no single correct rate expression that can be identified based on the given information. However, you should be able to identify an expression that is at least reasonably consistent with the trends shown.

(b) According to your expression in part a, how much chloroform would be generated in a batch system containing 8 mg/L dry weight algae exposed to 5.0 mg/L Cl$_2$ for 2 h at pH 7?

3-3. Piché and Larachi (2007) proposed the following reaction for the oxidation of polysulfide ions (S$_n^{2-}$, where n can be any value ≥ 2) to form thiosulfate ions (S$_2$O$_3^{2-}$) and colloidal sulfur (S$_8$):

$$S_n^{2-} + \frac{3}{2}O_2 \rightarrow S_2O_3^{2-} + \frac{n-2}{8}S_8$$

(a) For $n = 6$, write expressions for $r_{S_2O_3^{2-}}$ and r_{S_8} in terms of $r_{S_n^{2-}}$, if all the rates are given in units of mol/L h.

(b) Repeat part a, if the rates are in mg/L-h of the respective chemicals; designate these rates as r'.

(c) Repeat part a, if the rates are in mg S/L h; designate these rates as r''.

3-4. The following table gives the instantaneous rates of the disappearance of A in a reaction between A and B, at various concentrations of both reactants. Deduce the values of x and y and of the rate constant k, if the reaction rate expression is known to be $r_A = -kc_A^x c_B^y$.

c_A (mol/L)	c_B (mol/L)	$-r_A$ (mol/L s)
2.3×10^{-4}	3.1×10^{-5}	5.2×10^{-4}
4.6×10^{-4}	6.2×10^{-5}	4.16×10^{-3}
9.2×10^{-4}	6.2×10^{-5}	1.66×10^{-2}

3-5. An enzyme-mediated reaction is described by the Michaelis–Menten equation with K_m of 3×10^{-3} mol/L, $c_{E,tot}$ of 10^{-4} mol/L, and $k_{ES^* \rightarrow P}$ of 10 min^{-1}. How long would it take into convert 99% of the initial substrate S into product P in a batch process if the initial concentration of S is 10^{-2} mol/L? Compare the result based on an analytical solution to the problem (by integrating the relevant equation) with that obtained if the assumption is made that $c_S \gg K_m$.

3-6. Lee (1990) reported the following data for the oxidation of acetic acid by hydrogen peroxide under supercritical conditions ($T > 374$°C, P > 218 bar) in a batch system. The first table contains data collected at T close to 500°C, and the second table contains data over a range of temperatures.

Experimental results for variable reaction times at (approximately) the same temperature

Temperature (°C)	$c(0)$ (mmol/L)	$c(t)$ (mmol/L)	Time (s)
502.7	2.925	0.313	13.7
501.0	2.822	0.468	14.1
502.3	2.815	0.152	20.2
503.7	2.804	0.056	26.0

Experimental results for systems at different temperatures

Temperature (°C)	$c(0)$ (mmol/L)	$c(t)$ (mmol/L)	Time (s)
450.9	3.877	1.935	29.3
449.0	5.369	3.425	20.7
461.7	3.509	1.323	27.6
462.7	3.239	1.189	26.0
473.9	3.250	0.546	32.1
474.9	3.242	0.486	32.2
502.3	2.815	0.152	20.2
503.7	2.804	0.056	26.0
523.9	1.555	0.023	14.1
525.5	1.528	0.015	20.6

(a) Use integral methods to determine whether the data are best described by zero, first, or second-order reaction kinetics with respect to acetic acid, and estimate the rate constant for the chosen rate expression. (Note: these data are messy. Consider the implicit data point at time zero in addition to the reported data to aid in the analysis.)

(b) Based on the reaction order determined in part *a*, estimate the value of the rate constant k for each experiment in the second table, and estimate the value of E_{Ar} for the reaction.

3-7. The disinfection of microorganisms is often described by the "Chick–Watson law," which, in its simplest form, can be written as follows:

$$r_X = -kc_X c_D$$

where c_X is the concentration of viable microorganisms, c_D is concentration of disinfectant (e.g., chlorine), and k is a constant.

(a) In a batch reactor with a constant disinfectant concentration of 1.5 mg/L, 99% of the microorganisms are inactivated (killed) after 15 min. Find the value of k.

(b) Manipulation of your work in part (a) should convince you that the same degree of disinfection is achieved as long as the product of the concentration and time (the so-called *ct* product) is the same. Show a linear plot that illustrates this trade-off for the 99% inactivation of the microorganisms. (That is, manipulate the algebraic answer you developed to solve part (a) in such a way that you can draw a straight line on a plot, where every point on the line represents a combination of c and t that will give 99% inactivation.)

(c) On the same plot as you developed in part (b), draw a second line that describes 99.9% inactivation.

3-8. Many reactions between two species are described by second-order kinetics, first order with respect to each species. For a generic reaction $aA + bB \rightarrow pP$ that can be described by such kinetics, the rate expression for the disappearance of A would be

$$r_A = -k c_A c_B$$

In a batch reactor with initial concentrations c_{A_o} and c_{B_o} (that are *not* present in the stoichiometric ratio b/a), the concentration of A can be found by the following expression:

$$c_A(t) = \frac{c_{A_o} - \frac{a}{b} c_{B_o}}{1 - \frac{a\, c_{B_o}}{b\, c_{A_o}} \left\{ \exp\left[-\left(\frac{b c_{A_o}}{a c_{B_o}} - 1\right) c_{B_o} k t\right] \right\}}$$

(a) The homogeneous oxidation of manganese by oxygen can be described according to the stoichiometry

$$4\,Mn^{2+} + O_2 + 6\,H_2O \rightarrow 4\,MnOOH(s) + 8\,H^+$$

Morgan (2005) reported that, under a particular set of solution conditions, the kinetics of the reaction could be described as

$$r_{Mn^{2+}} = -k c_{Mn^{2+}} c_{O_2}$$

where $k = 1.22 \times 10^{-2}\,M^{-1}s^{-1}$. For a solution with initial concentrations of 2 mg/L O_2 and 500 µg/L Mn(II), prepare a graph depicting the decay of the Mn^{2+} concentration over time. Use a time scale of days on the figure, since the reaction proceeds slowly. Also, for solution conditions such that the given rate constant applies, find

(i) The time required to reduce the Mn^{2+} concentration to 20 µg/L.

(ii) The characteristic reaction time according to the first definition given in this chapter.

(iii) The characteristic reaction time according to the second definition in the chapter.

(b) Resolve part (a) under the assumption that oxygen is in great excess, and compare the results for the characteristic times and the time to achieve 20 µg/L to those found in part (a). Comment on whether the assumption of great excess gets better or worse as the reaction proceeds.

(c) Derive the expression given in the problem statement for $c_A(t)$. To do so, it is useful to define the initial concentration ratio c_{B_o}/c_{A_o} as M and the extent of the reaction of A at any time as $X_A = (c_{A_o} - c_A(t))/c_{A_o}$. The concentration of B at any time can also be expressed in these terms, accounting for the stoichiometry of the two reactants. The resulting differential equation in X_A can then be integrated using partial fractions to yield (after algebraic manipulation) the expression shown.

3-9. Mitch and Sedlak (2002) studied the formation of N-nitrosodimethylamine (NDMA) from dimethylamine ($NH(CH_3)_2$, or DMA) in response to the addition of chloramines to water as disinfectants. They proposed that the rate-limiting step in the reaction sequence is the formation of unsymmetric dimethylhydrazine (UDMH) according to the following elementary reaction of monochloramine (NH_2Cl, or MCA) with DMA:

$$NH_2Cl + DMA \rightarrow UDMH$$

The rate constant for the formation of UDMH via this reaction was reported to be $0.081\,M^{-1}s^{-1}$.

(a) What is the characteristic reaction time (according to the second definition) for the loss of DMA by this reaction under the conditions of the authors' experiments, which included an initial concentration of 1 mM of each of the reactants?

(b) What is the characteristic reaction time under more realistic conditions that could occur in water distribution or wastewater treatment applications, namely 3×10^{-5} M of each species?

3-10. Curtis and Reinhard (1994) researched the kinetics of the reductive dehalogenation of hexachloroethane. A figure from that article summarizing some possible pathways (set of elementary reactions) for the overall reaction is shown in Figure 3-Pr10a, and some of the results presented in the article are shown in Figure 3-Pr10b.

(a) Write the complete set of reaction rate expressions (i.e., one for each of the five constituents on the pathway to C_2Cl_4) for the degradation depicted in Figure 3-Pr10a.

(b) The complete reaction pathway in Figure 3-Pr10a is complex, but the authors note that reactions 1, 5, and 8 constitute the likely major pathway. Assuming that the other reactions are negligible, write rate expressions for the four constituents in this simplified reaction sequence.

(c) The authors note that k_5 is larger than k_1, and they claim that they can further simplify the reaction scheme as follows:

$$C_2Cl_6 \xrightarrow{k_1} C_2HCl_5 \xrightarrow{k_8} C_2Cl_4$$

Show that this simplification of the pathway noted in part (b) is correct under the stated circumstances, and write the rate expressions for this reaction scheme.

(d) The authors studied the kinetics of this reaction in batch systems, in which C_2Cl_6 was added, but no other constituents shown were added. Develop an expression for the concentration of C_2Cl_6 as a function of time (i.e., find $c_{C_2Cl_6}(t)$). Designate the initial C_2Cl_6 concentration, $c_{C_2Cl_6}(0)$, as A_o.

(e) Using the reaction sequence described in part (c), derive expressions for the concentrations of C_2HCl_5 and C_2Cl_4 as a function to time in a batch reactor, again designating A_o as the initial concentration of C_2Cl_6.

(f) Based on your answers to parts (d), (e), and (f) and the data given in Figure 3-Pr10b, estimate the rate constants k_1 and k_8 in the experiments performed by the authors.

3-11. Triclosan (structure shown in Figure 3-Pr11) is an antimicrobial agent used widely in hand soaps, toothpastes, deodorants, and other consumer products.

In recent years, a good deal of attention has been paid to the reaction of triclosan and other personal care products with chlorine during drinking water or

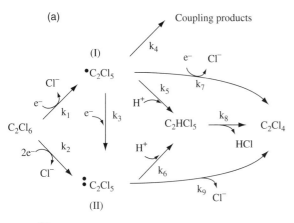

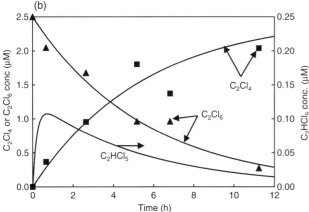

FIGURE 3-Pr10. Kinetics of the reductive dehalogenation of hexachloroethane (C_2Cl_6) to tetrachloroethene (C_2Cl_4). (a) Potential pathways, as shown by Curtis and Reinhard, based on work by Roberts and Gschwend (1991); (b) Experimental results for a system at pH 7.2 and 50°C, initially containing 2.5 µM C_2Cl_6 and 500 µM of the reducing agent AHQDS (2,6-anthrahydroquinone disulfonate). The curves are based on model calculations. *Source:* Part a reprinted and part b adapted with permission from Curtis and Reinhard (1994), copyright 1994 American Chemical Society.

FIGURE 3-Pr11. Chemical structure of triclosan.

wastewater disinfection processes. Many of these reactions depend strongly on solution pH, because both the compound of interest and the primary disinfectant compound (hypochlorous acid, HOCl) are weak acids. In the case of triclosan, the proton on the hydroxyl group can be released. The acid dissociation reactions and constants for HOCl and triclosan (TricH) are as follows:

$$HOCl \leftrightarrow OCl^- + H^+ \quad \log K_a = -7.55$$
$$TricH \leftrightarrow Tric^- + H^+ \quad \log K_a = -7.9$$

Rule et al. (2005) reported that, under typical water treatment conditions, the reaction between HOCl and $Tric^-$ proceeds much more rapidly than that between HOCl and TricH, or between OCl^- and either TricH or $Tric^-$. The reaction produces a chlorinated triclosan intermediate that then undergoes additional reactions with HOCl. They found that the dominant reaction is first order in both HOCl and $Tric^-$, with a rate constant of $k = 5.4 \times 10^3 \, (mol/L)^{-1} s^{-1}$. Ignoring the nondominant reactions, the rate of disappearance of total triclosan is therefore

$$r_{Tric_{tot}} = -\left[5.4 \times 10^3 \, (mol/L)^{-1} s^{-1}\right] c_{Tric^-} c_{HOCl}$$

(a) For a solution containing 0.5 μmol/L, total triclosan and 14 μmol/L total OCl (1 mg/L as Cl_2), plot the initial rate of triclosan reaction (μmol/L s) over the pH range 4–11. Use a logarithmic scale for the rate.

(b) For the initial conditions specified in part (a), plot the concentrations of TricH and $Tric^-$ over time until 99% of the total triclosan has reacted in a batch reactor, if the solution pH is constant at 7.5. Assume that the acid/base reactions proceed instantaneously (so that they are always at equilibrium), that the reaction stoichiometry is such that one molecule of HOCl reacts with one molecule of triclosan, and that the test is carried out in an idealized solution containing only the two reactants. Justify any other simplifying assumptions you make. Keep in mind that the reaction rate given in the problem statement is for total triclosan, but that the reactant is $Tric^-$, which represents only a fraction of $Tric_{tot}$.

(c) Repeat part (b), but for a test involving a natural water in which reactions of HOCl and OCl^- with other species in the system cause the total OCl concentration to decay according to a pseudo-first-order reaction, so that $c_{OCl_{tot}}(t) = c_{OCl_{tot}}(0) \exp[(-0.1/min)t]$.

4

CONTINUOUS FLOW REACTORS: PERFORMANCE CHARACTERISTICS WITH REACTION

4.1 Introduction
4.2 Extent of reaction in single ideal reactors at steady state
4.3 Extent of reaction in systems composed of multiple ideal reactors at steady state
4.4 Extent of reaction in reactors with nonideal flow
4.5 Extent of reaction under non-steady-state conditions in continuous flow reactors
4.6 Summary
References
Problems

4.1 INTRODUCTION

In the preceding chapters, we established the conceptual basis and mathematical form of the mass balance equation, and we explored models for the physical (diffusive, dispersive, advective) and chemical (reaction) terms in that equation. In Chapter 2, we found that it is convenient to use nonreactive tracers to investigate the hydraulic characteristics of reactors, thereby making the reaction term in the mass balance equation equal to zero. Correspondingly, in Chapter 3, we saw the convenience of using batch systems to investigate reaction kinetics, because in such systems the transport terms in the mass balance equation are zero. In this chapter, the material of the previous chapters is synthesized to characterize and predict behavior in systems where all the physical and chemical processes discussed previously can occur simultaneously. This situation is both the most realistic and the one that is most often of interest in the analysis of water and wastewater treatment systems. In most of this chapter, we retain one major simplification—that the system is at steady state. Similar to the simplifications mentioned earlier, this condition leads to one term in the mass balance equation—the storage term—being zero.

Systems with ideal, limiting-case flow characteristics (CFSTRs and PFRs) and a wide variety of reaction rate expressions are considered first. Subsequently, networks of such reactors are analyzed, followed by reactors with nonideal flow characteristics that can be described by the mathematical models presented in Chapter 2. Then, we consider systems with nonideal flow that cannot be represented adequately by those simplified models. Finally, the behavior of a few ideal reactor systems under non-steady-state conditions is considered.

4.2 EXTENT OF REACTION IN SINGLE IDEAL REACTORS AT STEADY STATE

The term *steady state* refers to a situation in which all parameters of interest in the system, including the chemical composition, are constant with time. In a steady-state system, a constituent may be transferred across the system boundaries and may react within the boundaries, but the concentration of the constituent at any point within the boundaries is unchanged over time. An assumption of steady-state conditions is often made for the analysis of water and wastewater treatment processes. In these cases, the system is rarely truly at steady state, but the changes in the input composition, temperature, flow rate, and so on, are typically gradual. If these changes occur gradually over several hours or days and the hydraulic detention time is only an hour or two, then the assumption of steady state is usually considered acceptable.

Extent of Reaction in a Continuous Flow Stirred Tank Reactor at Steady State

Consider the mass balance on a species i in a CFSTR in which a reaction is occurring, as depicted in Figure 4-1. We assume that the volume of water in the reactor is constant, that the system has only one inlet and one outlet, and also

---

Water Quality Engineering: Physical/Chemical Treatment Processes, First Edition. By Mark M. Benjamin, Desmond F. Lawler.
© 2013 John Wiley & Sons, Inc. Published 2013 by John Wiley & Sons, Inc.

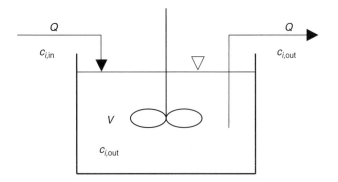

FIGURE 4-1. Definition diagram for a CFSTR with reaction.

that diffusion and dispersion across the boundaries are negligible. The entire water volume in the reactor is chosen as the control volume. Because the system is completely mixed, the concentration in the effluent is the same as that throughout the tank, as indicated in the figure by representing both concentrations as $c_{i,\text{out}}$.

The mass balance on i can be written as follows:

Rate of change of the mass of i stored within the system = Net rate (in − out) at which i enters by advection

+ Net rate (in − out) at which i enters by diffusion and dispersion + Net rate (formation − destruction) at which i is created by chemical reaction

Translating into an equation, for constant volume, it can be written as

$$V \frac{dc_{i,\text{out}}}{dt} = Q(c_{i,\text{in}} - c_{i,\text{out}}) + 0 + Vr_i \quad (4\text{-}1)$$

Since the concentration of i in the control volume is the same as that in the effluent stream, r_i is the rate of reaction associated with the effluent concentration, $c_{i,\text{out}}$. Equation 4-1 is valid for any constituent in a CFSTR with a constant volume (the usual case). If the reactor is at steady state, the left side of the equation is zero. Therefore, removing the subscript i, for steady-state conditions we can write

$$0 = Q(c_{\text{in}} - c_{\text{out}}) + Vr_{c_{\text{out}}} \quad (4\text{-}2)$$

Dividing by Q and substituting the hydraulic detention time τ for V/Q, we obtain

$$\tau_{\text{CFSTR}} = -\frac{c_{\text{in}} - c_{\text{out}}}{r_{c_{\text{out}}}} \quad (4\text{-}3)$$

First-Order Irreversible Reactions   Equation 4-3 relates the influent and effluent concentrations of a reactant passing through a CFSTR at steady state with the hydraulic residence time, for any reaction rate expression. For instance, if the reaction is a first-order irreversible decay, we can substitute the corresponding rate expression ($r_{c_{\text{out}}} = -k_1 c_{\text{out}}$), into Equation 4-3 and rearrange the equation as follows:

$$\frac{c_{\text{out}}}{c_{\text{in}}} = \frac{1}{1 + k_1 \tau} \quad (4\text{-}4)$$

Plots of $c_{\text{out}}/c_{\text{in}}$ versus τ for a few values of k_1 are shown in Figure 4-2a. Note, however, that Equation 4-4 indicates that $c_{\text{out}}/c_{\text{in}}$ depends solely on the value of the product $k_1 \tau$ in this type of system. Therefore, all the plots in Figure 4-2a collapse to a single curve when $c_{\text{out}}/c_{\text{in}}$ is plotted against this dimensionless product, as shown in Figure 4-2b.

A few rearrangements of Equation 4-4 are often useful, as follows. The effluent concentration, found by multiplying both the sides of Equation 4-4 by c_{in}, is a function of the

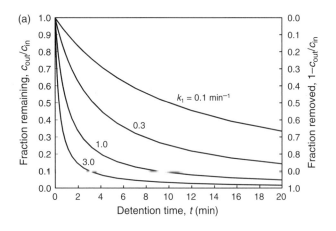 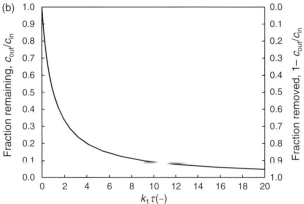

FIGURE 4-2. (a) Extent of reaction for a reactant undergoing a first-order irreversible decay reaction in a CFSTR for various values of k_1. (b) Same as (a), but with the abscissa shown as the product $k_1 \tau$. The data from all four lines in (a) collapse to the single curve shown in this figure.

influent concentration, the reaction rate constant, and the hydraulic detention time. The removal efficiency (η), a common parameter used to describe the effectiveness of a process in water and wastewater treatment, is

$$\eta = \frac{c_{in} - c_{out}}{c_{in}} = 1 - \frac{c_{out}}{c_{in}} = \frac{k_1 \tau}{1 + k_1 \tau} \quad (4\text{-}5)$$

Finally, the hydraulic detention time (which, for a given flow rate, establishes the reactor volume) required to achieve a desired fraction remaining is

$$\tau = \frac{1}{k_1}\left(\frac{c_{in}}{c_{out}} - 1\right) \quad (4\text{-}6a)$$

$$= \frac{c_{in} - c_{out}}{k_1 c_{out}} \quad (4\text{-}6b)$$

$$= \frac{1}{k}\frac{\eta}{1 - \eta} \quad (4\text{-}6c)$$

Recall from Chapter 3 that the characteristic reaction time for first-order irreversible reactions is $1/k_1$. In a continuous flow reactor of any type, we can define a *characteristic flow time* as the hydraulic residence time, τ. Therefore, the dimensionless product $k_1\tau$ can be thought of as the ratio of the characteristic time for flow to the characteristic time for reaction ($k_1\tau = \tau/(1/k_1)$), and Equation 4-4 indicates that this ratio controls the extent of reaction in the reactor. In other words, $k_1\tau$ is the number of characteristic reaction times provided by the hydraulic residence time. Thus, for example, when these two characteristic times equal one another, the product $k_1\tau$ equals 1.0, and the concentration of the reactant in the effluent from a CFSTR is one-half of the influent concentration.

A large value of $k_1\tau$ indicates that the hydraulic residence time equals many characteristic reaction times, so the reaction can proceed substantially toward completion before the water exits the reactor. (Note, however, that even if the hydraulic residence time is equivalent to several characteristic reaction times, a significant amount of reactant remains. For example, at $k_1\tau = 10$, c_{out} is still 9% of c_{in}.) Conversely, a value of $k_1\tau$ much less than 1.0 indicates that the water, on average, exits the reactor long before a single characteristic reaction time has passed; as a result, the extent of conversion is quite small.

■ **EXAMPLE 4-1.** Disinfection, or the inactivation (killing) of microorganisms, is sometimes considered to be a first-order reaction when a chemical disinfectant (e.g., chlorine) is used. For a given drinking water supply and a certain test organism, the first-order rate constant is found to be $1.38\,\text{min}^{-1}$. If 99% inactivation is desired, what detention time must be provided if the disinfection is to be accomplished in a CFSTR?

Solution. 99% inactivation means that only 1% of the test microorganisms are viable in the effluent from the CFSTR; that is, that $c_{out}/c_{in} = 0.01$, or that $c_{in}/c_{out} = 100$. Applying Equation 4-6a with the given rate constant yields

$$\tau = \frac{1}{1.38\,\text{min}^{-1}}(100 - 1) = 71.7\,\text{min}$$

Thus, a CFSTR with a hydraulic residence time of 71.7 min will accomplish the treatment objective. ■

Non-First-Order Irreversible Reactions Following the same steps as for the first-order reaction, a comparable analysis can be carried out for zero-, second-, and nth-order decay reactions occurring in CFSTRs under steady-state conditions. The mass balance equations for these systems are identical in form to that for a first-order reaction and are given as Equation 4-2. Similarly, all the mass balance equations can be manipulated to yield Equation 4-3 to characterize the required detention time to achieve a given degree of reaction. The equations for the different reaction orders differ only with respect to the rate expression (i.e., the expression for $r_{c_{out}}$) that is substituted when they are solved. The results (including those previously derived earlier for first-order reactions) are summarized in Table 4-1.

Note that, in all cases, the effluent concentration depends on three factors: the influent concentration, the reaction rate constant, and the hydraulic detention time. For fractional reaction orders or any integral order other than 0, 1, or 2, the equation for c_{out} can be solved numerically, but no closed-form solution exists. The fraction remaining, c_{out}/c_{in} (or, alternately, the removal efficiency $\eta = 1-(c_{out}/c_{in})$), depends on the same three parameters as c_{out}, except in the case of first-order reactions, where the result is independent of the influent concentration.

The fraction remaining for the general case of an nth-order irreversible reaction is $(1 + k_n \tau c_{out}^{n-1})^{-1}$. Furthermore, the product $k_n c_{out}^{n-1}$ is the inverse of the characteristic reaction time in the system based on Definition 2 in Table 3-3.[1] Thus, for any nth-order irreversible reaction occurring in a CFSTR, we can represent the fraction remaining as

$$\frac{c_{out}}{c_{in}} = \left(1 + \frac{\text{characteristic flow time}}{\text{characteristic reaction time}}\right)^{-1}$$

Extent of Reaction in a Plug Flow Reactor at Steady State

Consider next the performance of ideal plug flow reactors (PFRs) under steady-state conditions. Contrary to the situation in a CFSTR, the concentration of reactant varies from

---

[1] The characteristic reaction time as defined in Chapter 3 is based on a batch reaction and, for nonfirst-order reactions, depends on the initial concentration, $c(0)$. For a CFSTR at steady state, we are interested in the characteristic reaction time for the solution *in* the reactor (where the reaction is proceeding), so the appropriate concentration to use is $c_{reactor} = c_{out}$.

TABLE 4-1. Behavior of CFSTRs for nth-Order Reactions at Steady State

Reaction Order, n ($r = -k_n c_{out}^n$)	c_{out}	$\dfrac{c_{out}}{c_{in}}$	τ_{CFSTR}[a]
0	$c_{in} - k_0 \tau$	$1 - \dfrac{k_0 \tau}{c_{in}}$	$\dfrac{c_{out}}{k_0}\left(\dfrac{c_{in}}{c_{out}} - 1\right)$
1	$\dfrac{c_{in}}{1 + k_1 \tau}$	$\dfrac{1}{1 + k_1 \tau}$	$\dfrac{1}{k_1}\left(\dfrac{c_{in}}{c_{out}} - 1\right)$
2	$\dfrac{-1 + \sqrt{1 + 4k_2 \tau c_{in}}}{2 k_2 \tau}$	$\dfrac{1}{1 + k_2 \tau c_{out}}$	$\dfrac{1}{k_2 c_{out}}\left(\dfrac{c_{in}}{c_{out}} - 1\right)$
n	Numerical solution	$\dfrac{1}{1 + k_n \tau c_{out}^{n-1}}$	$\dfrac{1}{k_n c_{out}^{n-1}}\left(\dfrac{c_{in}}{c_{out}} - 1\right)$

[a]In all the expressions for τ, $(c_{in}/c_{out}) - 1$ can be rewritten as $\eta/(1 - \eta)$.

location to location within a PFR, making it impossible to characterize the entire reactor by a single value of concentration. Rather, to choose boundaries within which the concentration of a reactive species i is single-valued, the control volume must be differentially small.

As in Chapter 2, here we represent a PFR as a tube, with flow only in the axial dimension and with no axial dispersion. Recall that, in the previous analysis of PFRs, we identified two reasonable choices for the boundaries of the differential control volume, differing based on the frame of reference of the observer: in the Eulerian view, the control volume is a fixed region of space viewed from a stationary observation point outside the PFR, whereas in the Lagrangian view, the control volume is a fixed region of space viewed from a platform that moves at the velocity of the fluid. PFRs in which reactions are taking place are analyzed below using both approaches.

Fixed Frame of Reference (Eulerian View) The system of interest, defined according to the Eulerian view, is depicted as Figure 4-3a, and a differential section of the reactor around which we can write a mass balance is shown in 4-3b.

The control volume for the mass balance is $\Delta V = A \Delta x$, where A is the cross-sectional area, and, by the definition of a PFR, the terms for input and output by dispersion/diffusion are zero. Taking these two factors into account,

the mass balance can be written in words and symbols as follows:

Rate of change of the mass of i stored within the system = Net rate (in − out) at which i enters by advection + Net rate (in − out) at which i enters by diffusion and dispersion + Net rate (formation − destruction) at which i is created by chemical reaction

$$\frac{\Delta(cA\Delta x)}{\Delta t} = Qc - Q(c + \Delta c) + 0 + rA\Delta x \quad (4\text{-}7a)$$

$$= -Q\Delta c + rA\Delta x \quad (4\text{-}7b)$$

Recalling that A is a constant, dividing by Δx, and taking the limit as Δx and Δt approach zero yields the following partial differential equation:

$$A\frac{\partial c}{\partial t} = -Q\frac{\partial c}{\partial x} + rA \quad (4\text{-}8)$$

Applying the specification that the system is at steady state reduces the left side of the equation to zero. As a result, the concentration is a function of position only, so the partial differentials can be converted into ordinary differentials, with the following result:

$$0 = -Q\,dc + rA\,dx \quad (4\text{-}9)$$

$$\frac{1}{r}dc = \frac{A}{Q}dx \quad (4\text{-}10)$$

Equation 4-10 is the general mass balance for a PFR operating under steady-state conditions, developed from the Eulerian (fixed reference) point of view. To use this equation for any particular reaction, we can substitute the relevant expression for r and then integrate. Recalling that r is a

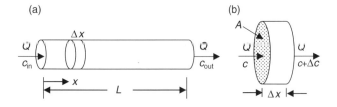

FIGURE 4-3. Definition diagram for mass balance on PFR in Eulerian view: (a) full reactor, (b) differential section.

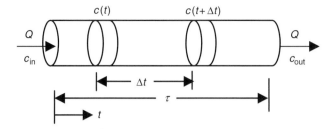

FIGURE 4-4. Definition diagram for mass balance on PFR: Lagrangian view. During time Δt, the parcel moves in the direction shown, but no fluid enters or leaves the parcel.

function of c, that A and Q are constants, and that $c|_{x=0} = c_{in}$ and $c|_{x=L} = c_{out}$, the integration yields

$$\int_{c_{in}}^{c_{out}} \frac{1}{r} dc = \frac{A}{Q} \int_0^L dx \qquad (4\text{-}11a)$$

$$= \frac{A}{Q} L = \frac{V}{Q} = \tau_{PFR} \qquad (4\text{-}11b)$$

Before proceeding with the integration of the left side of this equation for various rate expressions, we develop the same equation using the Lagrangian frame of reference.

Moving Frame of Reference (Lagrangian View) As described in Chapter 2, we consider a PFR as a conveyor belt; that is, every parcel of fluid travels through the reactor as if it were a package on a conveyor belt with no interaction (dispersion/diffusion) with upstream or downstream parcels. The definition diagram for this system is shown in Figure 4-4.

The general mass balance for such a differentially small parcel would be the same as that given earlier for the Eulerian view; that is,

Rate of change of the mass of i stored within the system = Net rate (in − out) at which i enters by advection

+ Net rate (in − out) at which i enters by diffusion and dispersion + Net rate (formation − destruction) at which i is created by chemical reaction

In the Lagrangian view, however, the relevant terms of the equation are different from those in the Eulerian view. First, no mass enters or leaves the control volume by either advection or diffusion in the Lagrangian view, so these terms in the mass balance are zero. Second, although the reaction term in the Lagrangian view is $r dV$, similar to the Eulerian view, in the Lagrangian view r is considered to be a function of time rather than location.[2] Finally, contrary to the case in the Eulerian view, the composition of the moving parcel that is the system of interest in the Lagrangian view varies with time (as the parcel travels from one end of the reactor to the other), so the left side of the equation is not zero. Thus, in the Lagrangian view, the storage term is nonzero and the net advective term is zero, whereas in the Eulerian view, the opposite is true.

Writing in terms of equations, applying the earlier reasoning, and recalling that dV is constant, the result is as follows:

$$\frac{d(cdV)}{dt} = dV \frac{dc}{dt} = r\, dV \qquad (4\text{-}12)$$

$$\frac{dc}{dt} = r \qquad (4\text{-}13)$$

Equation 4-13 is identical to the equation developed in Chapter 3 for the change in concentration over time of a reactive substance in a batch reactor. This fact is not coincidental; in essence, the Lagrangian view considers a parcel of fluid in a PFR as a tiny batch reactor that traveled from one end of the full-scale reactor to the other. The time of interest equals the travel time from one end of the reactor to the other, or the hydraulic detention time, τ. That is, when Equation 4-13 is rearranged and integrated, the limits of integration for time are from 0 to τ, and their associated concentrations are the influent and effluent concentrations

$$\int_{c_{in}}^{c_{out}} \frac{1}{r} dc = \int_0^\tau dt = \tau_{PFR} \qquad (4\text{-}14)$$

As expected and necessary, the result with the Lagrangian view is the same as that obtained earlier (Equation 4-11) with the Eulerian view. The insight that is gained with the Lagrangian view is that a PFR can be thought of as a moving batch reactor with a reaction time equivalent to the detention time. Thus, *the extent of reaction in a PFR with a given detention time is identical to the amount of reaction that occurs over an equivalent time period in a batch reactor.* This conclusion applies regardless of the reaction that is occurring.

Irreversible nth-Order Reactions To investigate the behavior of PFRs when treating species whose reaction rate expressions are of nth order (i.e., $r = -k_n c^n$), we can substitute the appropriate rate expression into Equation 4-14 and perform the integration. Because of the equivalence of PFRs and batch reactors, we can use the results for batch reactors (see Table 3-1) with appropriate substitutions of symbols: c_{in} for $c(0)$, c_{out} for $c(t)$, and τ for t. The resulting expressions for the effluent concentration, the fraction remaining, and the hydraulic detention time required to

---

[2] Recall that, in the Lagrangian view, the distance coordinate is x^*, defined as $x - v_x t$, so that the packet of interest is always at $x^* = 0$.

TABLE 4-2. Behavior of PFRs for nth-Order Reactions at Steady State

Reaction Order, n ($r = -k_n c^n$)	c_{out}	$\dfrac{c_{\text{out}}}{c_{\text{in}}}$	τ_{PFR}
0	$c_{\text{in}} - k_0 \tau$	$1 - \dfrac{k_0 \tau}{c_{\text{in}}}$	$\dfrac{1}{k_0}(c_{\text{in}} - c_{\text{out}})$
1	$c_{\text{in}} \exp(-k_1 \tau)$	$\exp(-k_1 \tau)$	$\dfrac{1}{k_1} \ln \dfrac{c_{\text{in}}}{c_{\text{out}}}$
2	$\dfrac{c_{\text{in}}}{1 + k_2 \tau c_{\text{in}}}$	$\dfrac{1}{1 + k_2 \tau c_{\text{in}}}$	$\dfrac{1}{k_2 c_{\text{in}}}\left(\dfrac{c_{\text{in}}}{c_{\text{out}}} - 1\right)$
Any $n \neq 1$[a]	$\left[(n-1)k_n \tau + c_{\text{in}}^{1-n}\right]^{1/(1-n)}$	$\left(1 + (n-1)k_n \tau c_{\text{in}}^{n-1}\right)^{1/(1-n)}$	$\dfrac{1}{(n-1)k_n c_{\text{in}}^{n-1}}\left[\left(\dfrac{c_{\text{in}}}{c_{\text{out}}}\right)^{n-1} - 1\right]$

[a] The expressions shown for c_{out} and $c_{\text{out}}/c_{\text{in}}$ are valid for any τ if $n > 1$. If $n < 1$, they are valid only up to $\tau = c_{\text{in}}^{1-n}/((1-n)k_n)$, at which τ the computed c_{out} is zero. In reality, for reactions with apparent n values <1, n increases and approaches a value of 1 as the reactant concentration approaches zero.
Source: Levenspiel (1999).

achieve a certain fraction remaining are shown for zero, first, second, and a general nth-order reaction in Table 4-2.

As in CFSTRs, the effluent concentration, the fraction remaining, and the detention time required to achieve a certain fraction remaining in PFRs all depend on the order of the reaction, the reaction rate constant, and the influent concentration. And also as with CFSTRs, an exception is that the fraction remaining, $c_{\text{out}}/c_{\text{in}}$ (or, alternately, the fraction removed, $1 - (c_{\text{out}}/c_{\text{in}})$) for first-order reactions is independent of the influent concentration.

■ **EXAMPLE 4-2.** Consider the same conditions cited in Example 4-1, but now consider that the disinfection is to be accomplished in a PFR rather than a CFSTR. Recall that disinfection is considered as a first-order reaction with a rate constant (for the given chemical, water, and test microorganism) of 1.38 min^{-1}, and that 99% inactivation is desired. What detention time must be provided if the disinfection is to be accomplished in a PFR?

Solution. For 99% inactivation, $c_{\text{out}}/c_{\text{in}} = 0.01$, or $c_{\text{in}}/c_{\text{out}} = 100$. Applying the equation from Table 4-2 for the detention time required in a PFR for first-order reactions yields

$$\tau = \frac{1}{1.38 \text{ min}^{-1}} \ln(100) = 3.34 \text{ min}$$

The required detention time in a PFR is far less than that in a CFSTR (71.7 min) for this first-order reaction. The reasons for this difference are discussed in the following section. ■

Comparison of CFSTRs and PFRs for Irreversible Reactions

The preceding examples demonstrate that the hydraulic characteristics of a reactor can have a dramatic effect on the detention time required to achieve a given amount of reaction, even if the influent composition and flow rate are unchanged. In the specific case of these examples, the results suggested that by eliminating the mixing (i.e., by using a PFR instead of a CFSTR), we could dramatically decrease the time required to achieve the treatment objective. We next address the questions: what is the conceptual basis for the effect of hydraulic behavior on conversion efficiency, and to what extent can the results from the examples be extrapolated to other reaction orders, influent compositions, and extents of reaction?

These issues can be approached in two ways. First, for a given design situation (Q, c_{in}, and desired c_{out}), we can compare the required detention times (or volumes) for the two types of reactors, as was done in Examples 4-1 and 4-2 for a first-order reaction. Alternatively, for a reactor receiving a certain influent (Q, c_{in}) and having a given size (and therefore τ), we can compare the effluent concentrations. Using the first approach, we can compute the value of τ required for a given degree of conversion using the equations given in Tables 4-1 and 4-2 for the two different ideal reactor types and then take their ratios to obtain the results shown in Table 4-3.

The results indicate that, for first- and second-order reactions, more detention time is required in a CFSTR than a PFR ($\tau_{\text{CFSTR}}/\tau_{\text{PFR}} > 1$) to accomplish the same amount of removal. Furthermore, the advantage of using a PFR increases dramatically with increasing removal efficiency (η) and is greater for second order than for first order reactions. Investigation of other reaction orders shows that the trend of increasing $\tau_{\text{CFSTR}}/\tau_{\text{PFR}}$ with increasing removal efficiency applies to all reactions with $n > 0$. In addition, the advantage of using PFRs increases dramatically with increasing n or η. For a zero-order reaction, on the other hand, the two reactors perform identically, regardless of the removal efficiency achieved. Although reactions with $n < 0$ are rare, they do

TABLE 4-3. Ratio of Sizes of CFSTRs and PFRs to Accomplish the Same Removal

Removal efficiency, η	$\dfrac{c_{in}}{c_{out}}$	$\dfrac{\tau_{CFSTR}}{\tau_{PFR}}$		
		$n=0$	$n=1$	$n=2$
General expression:	$\dfrac{1}{1-\eta}$	1	$\dfrac{(c_{in}/c_{out})-1}{\ln(c_{in}/c_{out})}$	$\dfrac{c_{in}}{c_{out}}$
0.80	5	1	2.49	5
0.90	10	1	3.91	10
0.95	20	1	6.34	20
0.99	100	1	21.5	100

exist (at least over certain concentration ranges), and in those cases, CFSTRs achieve greater removal efficiency than PFRs.[3]

A corollary of the results shown in Table 4-3 is that a PFR will achieve a greater removal efficiency than a CFSTR, if they have identical residence times and are treating the same influent, and if $n > 0$. This point is illustrated for a second-order reaction in Example 4-3.

■ **EXAMPLE 4-3.** Find the effluent concentrations from a PFR and a CFSTR, each treating the same pollutant and each with a detention time of one hour. The pollutant undergoes a second-order decay reaction with $k_2 = 0.1\,(\text{mg/L})^{-1}\,\text{min}^{-1}$, and the influent concentration, c_{in}, is 100 mg/L.

Solution. Using the equations given in Tables 4-1 and 4-2, the effluent concentrations are found as follows:

PFR:

$$c_{out} = \frac{c_{in}}{1 + k_2 \tau c_{in}}$$

$$= \frac{100\,\text{mg/L}}{1 + (0.1\,\text{L/mg min})(60\,\text{min})(100\,\text{mg/L})}$$

$$= 0.166\,\text{mg/L}$$

CFSTR:

$$c_{out} = \frac{-1 + \sqrt{1 + 4 k_2 \tau c_{in}}}{2 k_2 \tau}$$

$$= \frac{-1 + \sqrt{1 + 4(0.1\,\text{L/mg min})(60\,\text{min})(100\,\text{mg/L})}}{2(0.1\,\text{L/mg min})(60\,\text{min})}$$

$$= 4.0\,\text{mg/L}$$ ■

---

[3] In reactions with $n < 0$, the reaction rate increases as the concentration decreases. Such situations might occur with some biological reactions, if the substance being degraded was also an inhibitor. They can also arise if one of the products of the reaction catalyzes the reaction. In the latter case, the reaction rate would not increase if c was decreased by diluting the initial solution, but the rate might nevertheless increase as c declines in a given system, due to the accumulation of the catalytic product.

We can gain insight into *why* the generalizations developed earlier apply by considering how the rate of the reaction depends on the reactant concentration. Specifically, for any reaction whose rate increases with increasing reactant concentration (i.e., any reaction with $n > 0$), the reaction rate is greater at points where the reactant concentration is greater.

Now consider the range of concentrations that exists in each type of reactor. In a PFR, the reactant concentration is high at the inlet and gradually decreases as the fluid moves through the reactor. Thus, the reaction rates in the reactor range from that corresponding to c_{in} to that corresponding to c_{out}; for a reaction with $n > 0$, this means that the absolute value of the reaction rate declines steadily from the influent to the effluent end of the reactor. In a CFSTR, on the other hand, the concentration is the same everywhere, equal to c_{out}; the reaction rate in a CFSTR is therefore the rate that corresponds to c_{out} throughout the reactor. Thus, if the two reactors are used to carry out the same reaction between the same two limits (i.e., from the same c_{in} to the same c_{out}), the reaction rate throughout the CFSTR will be identical to that at the effluent port of the PFR. However, at all other (upstream) points in the PFR, the reaction will proceed more rapidly than at the effluent port, and hence the detention time required to accomplish the given degree of reaction will be less. On the other hand, if $n = 0$, the rates will be the identical ($r = -k_0$) at all points in both reactors, so the time required for the desired conversion will be the same.

This argument can be illustrated via a graphical interpretation of the relevant equations. The equations for the detention time required to achieve a given degree of contaminant destruction in a CFSTR (Equation 4-3) and a PFR (Equation 4-14) are repeated below, in a slightly modified form

$$\tau_{CFSTR} = \frac{1}{-r_{c_{out}}}(c_{in} - c_{out}) \tag{4-15}$$

$$\tau_{PFR} = \int_{c_{out}}^{c_{in}} -\frac{1}{r}\,dc \tag{4-16}$$

Both these expressions include the term $-1/r$, but this term is evaluated at different concentrations in the two equations: for CFSTRs (Equation 4-15), it is evaluated only at c_{out}, whereas for PFRs (Equation 4-16), it is evaluated over the entire range between c_{in} and c_{out} (i.e., at an infinite number of c values separated by intervals of dc). The significance of this difference is apparent in a plot of $-1/r$ versus c, as suggested by Levenspiel (1999). Such a plot is shown in Figure 4-5 for a hypothetical decay reaction in which the rate increases with concentration. Note that the curve shown describes only the reaction kinetics and hence is independent of the reactor type in which the reaction

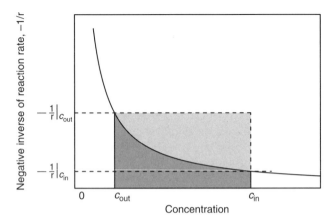

FIGURE 4-5. A graphical approach to evaluating τ as a function of influent and effluent concentrations in a CFSTR and a PFR. The curve is a plot of $-1/r$ for a hypothetical reaction. The shaded (both light and dark) rectangular area and the dark area alone indicate the required τ_{CFSTR} and τ_{PFR}, respectively, to reduce the concentration of the constituent of interest from c_{in} to c_{out}.

might be occurring. An arbitrary set of c_{in} and c_{out} values is also shown in the figure.

According to Equation 4-15, the required detention time to reduce the pollutant concentration from c_{in} to c_{out} in a CFSTR is equal to the product of $-1/r|_{c_{out}}$ and the difference $c_{in} - c_{out}$. In the graph, this product corresponds to the area of the shaded rectangle. On the other hand, Equation 4-16 indicates that the required detention time to accomplish the same removal in a PFR is equal to the integral of $-(1/r)dc$ between c_{out} and c_{in}, which corresponds to the dark area under the curve. Clearly, consistent with our earlier conclusion, $\tau_{PFR} < \tau_{CFSTR}$. Note that, if the reactions were of zero order, the line representing $-1/r$ would be horizontal, so the two areas (τ values) would be identical.

The graphical analysis reinforces the explanation given earlier for the difference between the required detention times in the two reactors: the reaction rate in a PFR changes throughout the reactor from that associated with c_{in} at the influent end to that associated with c_{out} at the effluent end, whereas the reaction rate everywhere in the CFSTR is the (low) rate associated with the effluent concentration. This difference causes PFRs to be more efficient than CFSTRs for any reaction in which $|r|$ increases with increasing c (causing $|1/r|$ to decrease with increasing c); that is, the detention time required to accomplish a certain degree of removal in a PFR will be less than that required to accomplish the same removal in a CFSTR. The requirement that $|r|$ increase with c is equivalent to a requirement that the reaction be of order > 0.

This discussion may lead us to question why CFSTRs are ever used, given that the vast majority of reactions of interest are of order greater than zero and that PFRs are always more efficient than CFSTRs for such reactions. One reason that is particularly applicable to water and wastewater treatment systems is that, as shown in Chapter 2 for conservative substances, completely mixed reactors can serve as concentration equalization basins. As shown at the end of this chapter, they can serve the same role for reactive substances. The damping of fluctuations in the influent to a CFSTR is achieved by the mixing the influent with the water already present in the reactor, whereas a PFR (which has no mixing) provides no such effects.

The influent to water and wastewater treatment plants is, to a large extent, beyond the control of the operators. Changes in the composition of the water in a treatment process can sometimes dramatically influence the process performance. In such cases, it may be advantageous to reduce the magnitude of these changes by taking advantage of the concentration equalization provided by CFSTRs. The benefits of equalization can be particularly important in systems where biological reactions occur, since organisms that are essential to the success of the process might be inhibited or killed by rapid changes in the solution composition or by a spiked dose of a toxic substance.

In addition to affecting the rate of response to transients in influent concentration, CFSTRs respond less severely than PFRs to long-term change in the influent flow rate. That is, for any reaction whose rate increases with concentration, the new steady-state condition that is ultimately achieved in a CFSTR in response to, say, a step increase in flow is not as different from the old condition as in a PFR. Inother words, the CFSTR is more forgiving than a PFR to such changes.

Also, mixing is desirable when chemicals must be added to the water to cause a reaction to occur. Typically, when analyzing the performance of a PFR, an assumption is made that the reactants are completely mixed along the cross-section of flow, but not mixed at all in the direction of flow. When dealing with the large flows typical of water and wastewater treatment plants, it is often extremely important to disperse any added reactants rapidly into the water so that they can react at the desired concentration with contaminants throughout the influent. If mixing is inadequate, the reactant might be present in some parts of the reactor in concentrated packets and be virtually absent from other portions of the system. As a practical matter, it is often easier to mix the entire volume of the reactor than to achieve complete mixing perpendicular to the direction of flow and no mixing parallel to it.

Finally, although the ideal extremes of a CFSTR and a PFR are both difficult to achieve in large systems, dead space and short-circuiting are far more likely in reactors where efforts are made to avoid mixing (PFRs) than in reactors where efforts are made to maximize mixing (CFSTRs). Water flow itself creates substantial mixing unless significant effort is made to prevent it. As a result, it is easier to achieve the ideal of a CFSTR than that of a PFR, and a design based on the assumption that ideal PFR conditions will be achieved might not meet the treatment objective, whereas designing for a CFSTR is much more likely to do so.

Reversible Reactions

The analysis in the preceding section applies to irreversible reactions. However, as noted in Chapter 3, not all reactions meet that criterion. In this section, we analyze the extent to which reversible reactions proceed in ideal CFSTRs and PFRs.

For the reversible reaction A ↔ B in which the forward and reverse reactions are both elementary, the rate expression for species A can be written in one of the two following forms (see Equations 3-45 and 3-48):

$$r_A = -k_{AB}c_A + k_{BA}c_B \quad (4\text{-}17a)$$

$$= k_{AB}(c_A^* - c_A) \quad (4\text{-}17b)$$

where c_A is the concentration of A in solution and c_A^* is the concentration of A that would be present if A were at equilibrium with the existing concentration of B. The value of c_A^* is c_B/K_{AB}, where K_{AB} is the equilibrium constant for the reaction (equal to k_{AB}/k_{BA}). Many nonelementary reactions that are important in environmental engineering also have rate expressions like Equation 4-17b.

If the reaction rate of A is characterized by Equation 4-17a, the mass balance on A in a CFSTR at steady state can be written as follows:

$$0 = Qc_{A,in} - Qc_{A,out} - Vk_{AB}c_{A,out} + Vk_{BA}c_{B,out} \quad (4\text{-}18)$$

The changes from influent to effluent in the concentrations of reactants and products are related by the reaction stoichiometry. For the one-to-one stoichiometry of this example, the relationship is[4]

$$c_{B,out} - c_{B,in} = c_{A,in} - c_{A,out} \quad (4\text{-}19)$$

Solving Equation 4-19 for $c_{A,out}$ or $c_{B,out}$, substituting these expressions into Equation 4-18, dividing by Q, and rearranging yield the following equations for the effluent concentrations of A and B as a function of the influent concentrations, reaction rate constant, and hydraulic residence time:

$$c_{A,out} = \frac{c_{A,in}(1 + k_{BA}\tau) + c_{B,in}k_{BA}\tau}{1 + k_{AB}\tau + k_{BA}\tau} \quad (4\text{-}20)$$

---

[4] In multiphase reactions such as precipitation, dissolution, and gas transfer, this relationship does not necessarily apply, since the reactant or product might transfer out of solution. In such cases, the mass balances on the two constituents in the two phases must be linked. These situations are analyzed in the chapters that deal with these specific processes.

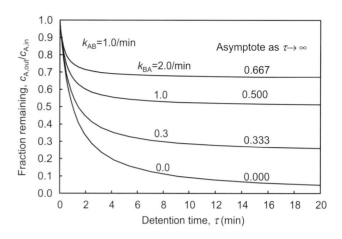

FIGURE 4-6. Extent of conversion of A into B via a reversible, elementary reaction in a CFSTR, with $c_{B,in} = 0$. The values of the asymptotes (i.e., $c_{A,out}/c_{A,in}$ at equilibrium) can be computed by rearranging Equation 4-22, with $c_{B,in} = 0$.

$$c_{B,out} = \frac{c_{A,in}k_{AB}\tau + c_{B,in}(1 + k_{AB}\tau)}{1 + k_{AB}\tau + k_{BA}\tau} \quad (4\text{-}21)$$

Note that, for an irreversible reaction, k_{BA} is zero, and Equation 4-20 reduces to Equation 4-4. On the other hand, if the reverse reaction is significant, the steady-state concentration of A depends on the influent concentrations of both A and B and also on both the forward and reverse rate constants, in addition to the detention time, τ.

The effects of variations in detention time and the reverse rate constant can be seen in Figure 4-6 for a system in which $c_{B,in} = 0$. For each value of the reverse rate constant (k_{BA}), the fraction of A remaining decreases dramatically at small values of τ, more or less as it would if the reaction were irreversible (corresponding to the curve shown for $k_{BA} = 0$). However, at larger values of τ, the fraction of A remaining in the reversible cases declines less than in the irreversible case and approaches a nonzero, asymptotic value at high values of τ. For the limiting case of very long detention time (τ approaching infinity), the products $k_{AB}\tau$ and $k_{BA}\tau$ are both $\gg 1$, and Equations 4-20 and 4-21 yield

$$c_{A,out}\big|_{\tau\to\infty} = \frac{k_{BA}(c_{A,in} + c_{B,in})}{k_{AB} + k_{BA}} \quad (4\text{-}22)$$

$$c_{B,out}\big|_{\tau\to\infty} = \frac{k_{AB}(c_{A,in} + c_{B,in})}{k_{AB} + k_{BA}} \quad (4\text{-}23)$$

The values of $c_{A,out}$ and $c_{B,out}$ computed using Equations 4-22 and 4-23 are, not surprisingly, those that correspond to equilibrium, a result that can be confirmed by taking their ratio

$$\frac{c_{B,out}}{c_{A,out}}\bigg|_{\tau\to\infty} = \frac{k_{AB}}{k_{BA}} = K_{AB} \quad (4\text{-}24)$$

That is, at long detention times in a CFSTR, an irreversible reaction goes to completion ($c_A = 0$), but a reversible reaction goes to equilibrium ($c_A = c_A^*$).

Recall from Chapter 3 that the characteristic reaction time for a reversible reaction that is first order in both directions ($t_{char} = 1/(k_{AB} + k_{BA})$) is shorter than for either the forward or reverse reaction, so that, in Figure 4-6, a given value of τ represents a greater number of characteristic reaction times as k_{BA} increases. As a result, equilibrium is approached more rapidly as either k_{AB} or k_{BA} increases. This result is apparent in the figure as k_{BA} increases from 0 to $2.0\,\text{min}^{-1}$. (As a point of reference, $\tau = 4$ min corresponds to 4 and 12 characteristic reaction times for k_{BA} values of 0 and $2.0\,\text{min}^{-1}$, respectively.)

The simultaneous changes in the steady-state values of c_A, c_B, and c_A^* for a particular set of k_{AB} and k_{BA} values are shown in Figure 4-7. The gap between c_A and c_A^* is the extent of disequilibrium in the reactor (and in the effluent) and is the driving force for the reaction. Consistent with the earlier discussion, as τ becomes very large, this driving force diminishes to nearly zero, and the system approaches equilibrium.

The extent to which a reversible reaction proceeds in a PFR can be inferred directly from the similarity between a PFR and a batch reactor. Specifically, if a reversible reaction with a rate expression $r_A = -k_{AB}c_A + k_{BA}c_B$ occurs in a PFR, the effluent concentration can be computed by using the result obtained in Chapter 3 for the same reaction in a batch reactor (Equation 3-54), but replacing $c(0)$, $c(t)$, and t in the equation for the batch system by $c_{A,in}$, $c_{A,out}$, and τ for the PFR. The result is

$$c_{A,out} = \frac{c_{A,in} + c_{B,in}}{K_{eq} + 1} + \frac{K_{eq}c_{A,in} - c_{B,in}}{K_{eq} + 1} \exp\left(-\frac{K_{eq} + 1}{K_{eq}} k_{AB}\tau\right) \quad (4\text{-}25)$$

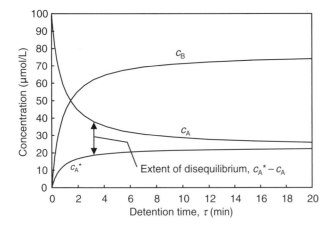

FIGURE 4-7. Concentrations of reactant (A) and product (B), and the extent of disequilibrium ($c_A^* - c_A$) as a function of τ in a CFSTR. The influent concentrations of A and B are 100 and 0 μM, respectively, and the rate constants are $k_{AB} = 1.0\,\text{min}^{-1}$ and $k_{BA} = 0.3\,\text{min}^{-1}$.

4.3 EXTENT OF REACTION IN SYSTEMS COMPOSED OF MULTIPLE IDEAL REACTORS AT STEADY STATE

As discussed in Chapter 2, real reactors with nonideal flow patterns can sometimes be represented as combinations of ideal reactors connected in series or in parallel. In addition, it is sometimes advantageous to design systems as combinations of reactors, rather than as a single reactor. In this section, the extent of reaction that can be expected in such reactors and reactor systems at steady state is derived.

In dealing with multiple reactor systems, problems to define parameters such as volume and detention time can easily arise. Here, the terms τ_{tot} and V_{tot} are used to represent the total detention time and volume, respectively, both for a single reactor that is conceptually (but not really) divided into multiple reactors, and for a true multiple reactor system. Detention times and volumes of parts of the total system are referred to with subscripts ($1, 2, \ldots i, \ldots, N$).

Based on the discussion in the previous section, we can conclude that zero-order reactions behave the same regardless of the flow pattern. Not only is the behavior of a zero-order reaction the same in a CFSTR and a PFR, but it is also the same in any combination of ideal reactors, whether in series or parallel. That is, the result of a zero-order reaction in a continuous flow reactor is determined solely by the hydraulic detention time (τ_{tot}) and not at all by the mixing pattern in the reactor. Hence, zero-order reactions are ignored in the ensuing discussion. The primary focus is on first-order reactions, although some attention is also given to other reaction rate expressions.

PFRs in Series

Because a PFR operates like a conveyor belt, placing several PFRs in series is equivalent to increasing the residence time in a single PFR. That is, the removal efficiency in a series of PFRs with the total residence time τ_{tot} is identical to that in a single PFR with the residence time τ_{tot}. This result is true for any reaction rate expression.

CFSTRs in Series

If two CFSTRs are connected in series, the effluent from the first is the influent to the second. Thus, for a first-order decay reaction ($r = -k_1 c$), the following results apply:

First reactor: $\quad c_{out,1} = \dfrac{c_{in}}{1 + k_1 \tau_1} \quad (4\text{-}26)$

Second reactor: $\quad c_{out,2} = \dfrac{c_{out,1}}{1 + k_1 \tau_2} \quad (4\text{-}27)$

Overall: $\quad c_{out,2} = \dfrac{c_{in}/(1 + k_1 \tau_1)}{1 + k_1 \tau_2}$

$\quad = \dfrac{c_{in}}{(1 + k_1 \tau_1)(1 + k_1 \tau_2)} \quad (4\text{-}28)$

TABLE 4-4. Analysis of Several Reactor Configurations with the Same Total Detention Time: First-Order Reaction Under Steady-State Conditions

Mode	τ_1 (min)	τ_2 (min)	$\dfrac{c_{\text{out}}}{c_{\text{in}}}$	c_{out} (mg/L)
PFR	60	0	$\exp(-60k_1)$	0.25
1 CFSTR	60	0	$\dfrac{1}{1+60k_1}$	14.3
2 CFSTRs	45	15	$\dfrac{1}{(1+45k_1)(1+15k_1)}$	7.3
2 CFSTRs	30	30	$\dfrac{1}{(1+30k_1)(1+30k_1)}$	6.3
2 CFSTRs	15	45	$\dfrac{1}{(1+15k_1)(1+45k_1)}$	7.3

This result can be extended to N CFSTRs in series in which a first-order decay reaction occurs, as follows:

$$c_{\text{out},N} = \frac{c_{\text{in}}}{(1+k_1\tau_1)(1+k_1\tau_2)\cdots(1+k_1\tau_N)} \quad (4\text{-}29a)$$

$$= \frac{c_{\text{in}}}{\prod_{i=1}^{N}(1+k_1\tau_i)} \quad (4\text{-}29b)$$

If all reactors are of the same size, Equation 4-29b simplifies to the following form:

$$c_{\text{out},N} = \frac{c_{\text{in}}}{\left(1+k_1\dfrac{\tau_{\text{tot}}}{N}\right)^N} \quad (4\text{-}30)$$

Equation 4-30 can be applied to model a single real reactor with nonideal flow that is conceptually divided into N equal-sized CFSTRs in series. The approach used to make the best estimate of the number N for the CFSTRs in series model for nonideal flow is presented in Chapter 2. In some cases, a single real reactor is physically divided into separate subreactors by the installation of baffles; in such situations, the subreactors might or might not have equal sizes.

■ **EXAMPLE 4-4.** Consider a first-order decay reaction occurring in a single CFSTR that is then divided into two smaller sequential reactors (not necessarily with equal volumes) by the installation of a baffle, with each smaller reactor behaving as a CFSTR. Because such a modification makes the flow more PFR-like (as shown in Chapter 2), and because reaction efficiency for any reaction of order $n > 0$ is greater in a PFR than a CFSTR, we might expect that dividing the reactor would improve the efficiency. To test this hypothesis for a particular reactor, consider a system receiving an influent with $c_A = 100$ mg/L of a reactant that undergoes a first-order decay reaction with $k_1 = 0.1$ min^{-1}. Assume that the system has a total hydraulic detention time of 1 h, but that it can be operated in any of five different modes: as a PFR, as a CFSTR, or as two CFSTRs in series, with the detention time split between the two CFSTRs in three different proportions. The effluent concentrations for the five possible operating modes, computed using the appropriate equations derived earlier in this chapter, are shown in Table 4-4.

As expected, the effluent concentrations for the CFSTRs in series are intermediate between those for a single CFSTR and a single PFR of the same total residence time (volume). Furthermore, at least in this example, the way the reactor is divided (e.g., 15 and 45 min, vs. 30 and 30 min) affects the fraction remaining, but, for a given method of dividing the reactor, the sequence of the two reactors (i.e., $\tau = 15$ min followed by $\tau = 45$ min, or vice versa) does not affect the result.

We can identify the optimal method of dividing the reactor by writing the general expression for c_{out} from the second reactor (Equation 4-28), substituting $\tau_{\text{tot}} - \tau_1$ for τ_2, and differentiating with respect to τ_1 as follows:

$$c_2 = \frac{c_{\text{in}}}{(1+k_1\tau_1)(1+k_1\tau_2)} \quad (4\text{-}31)$$

$$= \frac{c_{\text{in}}}{(1+k_1\tau_1)(1+k_1(\tau_{\text{tot}}-\tau_1))} \quad (4\text{-}32)$$

$$= \frac{c_{\text{in}}}{1+k_1\tau_{\text{tot}}+k_1^2\tau_1\tau_{\text{tot}}-k_1^2\tau_1^2} \quad (4\text{-}33)$$

$$\frac{dc_2}{d\tau_1} = \frac{-c_{\text{in}}\left(k_1^2\tau_{\text{tot}}-2k_1^2\tau_1\right)}{\left(1+k_1\tau_{\text{tot}}+k_1^2\tau_1\tau_{\text{tot}}-k_1^2\tau_1^2\right)^2} \quad (4\text{-}34)$$

The minimum value of c_2 can be found by setting $dc_2/d\tau_1$ to zero and solving for τ_1

$$0 = \frac{c_{\text{in}}\left(k_1^2\tau_{\text{tot}}-2k_1^2\tau_1\right)}{\left(1+k_1\tau_{\text{tot}}+k_1^2\tau_1\tau_{\text{tot}}-k_1^2\tau_1^2\right)^2} \quad (4\text{-}35)$$

$$0 = k_1^2 \tau_{\text{tot}} - 2k_1^2 \tau_1 \quad (4\text{-}36)$$

$$\tau_{\text{tot}} = 2\tau_1 \quad (4\text{-}37)$$

The result is that the optimal way to divide the reactor is into two equal portions, so that the residence time in each portion is one-half of the total residence time in the system. ■

Recalling the earlier discussion, it is clear why multiple CFSTRs lead to a better overall removal than a single CFSTR with the same τ_{tot} for reactions of order greater than zero. The fluid in the first reactor of the sequence has a higher reactant concentration than that in the subsequent reactors, so the reaction rate in that reactor is faster. The greater the number of reactors in the sequence, the greater number of steps down in the reaction rate between the first and last reactors, the larger the volume in which the reactant concentration is higher than in the final effluent (and final reactor), and hence the higher the overall average reaction rate (considering the entire reactor system). If the number of reactors in the system were increased indefinitely, keeping the overall residence time constant, the flow pattern and the removal efficiency would approach those in a PFR.

■ **EXAMPLE 4-5.** For a second-order reaction with $k_2 = 0.1$ (mg/L)$^{-1}$ min^{-1} and $c_{\text{in}} = 100$ mg/L (the same reaction as considered in Example 4-3), determine the extent of conversion in the same five reactor arrangements as in Example 4-4.

Solution. The analysis follows the same procedure as in Example 4-4, except that the expressions for $c_{\text{out}}/c_{\text{in}}$ for a second-order reaction are used. The results are shown in the first five rows of Table 4-5.

The results for the second-order reaction differ from those for the first-order reaction in several ways. Although the effluent concentration from the CFSTRs in series is again an intermediate between that of a single CFSTR and a single PFR, the sequence of the reactors does make a difference, as is evident from a comparison of the 15–45 min and the 45–15 min sequences. Further investigation identifies a 22–38 min arrangement (characterized in the last row of the table) as the optimal one, although the removal efficiency is not very sensitive to the design arrangement over a wide range of possibilities. ■

Similar calculations for other system conditions indicate that, for a reaction with $n > 0$, the benefits that can be obtained by dividing a single CFSTR into N equal-sized CFSTRs in series increase as the removal efficiency, the order of the reaction, or the number of reactors (N) increases. As N becomes very large, the performance of the CFSTR system approaches that of a single PFR. Also, as might be expected, the marginal benefit of an additional reactor is greatest for the initial division of the single reactor into two, and diminishes thereafter.

The two previous examples stress the improved performance when a single CFSTR with a certain detention time is replaced by a series of smaller CFSTRs with the same total detention time. Such a situation occurs when baffles are installed in a reactor to divide it physically into smaller segments, each of which has sufficient mixing to be considered as a CFSTR. A similar analysis could be performed to investigate the differences in the required total detention times to achieve a specified effluent concentration in a single CFSTR or N CFSTRs in series. In this case, the total required detention time diminishes with each successive addition of a reactor, although again the greatest savings is in going from one to two reactors. The savings is apparent in Figure 4-8,

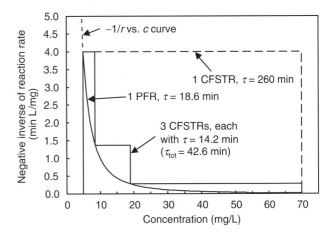

FIGURE 4-8. Graphical comparison of the hydraulic residence time required to reduce the concentration of a contaminant from 70 to 5 mg/L in three different reactor setups, for a second-order irreversible reaction with $k_2 = 0.01$ L/mg min. For the CFSTRs, the value on the ordinate corresponds to $-1/r$ in the reactor and therefore is computed for the concentration inside the reactor (equal to c_{out} for that reactor); this value of $-1/r$ is shown for the whole range from c_{in} to c_{out}, even though the concentration actually changes discontinuously. The residence time in each reactor is equal to the area under the corresponding curve, between the limits of c_{in} and c_{out} on the abscissa.

TABLE 4-5. Analysis of Several Reactor Configurations with the Same Total Detention Time: Second-Order Reaction Under Steady-State Conditions ($c_{\text{in}} = 100$ mg/L)

Mode	τ_1 (min)	τ_2 (min)	c_{out} (mg/L)
PFR	60	0	0.166
1 CFSTR	60	0	4.0
2 CFSTRs	45	15	1.450
2 CFSTRs	30	30	1.211
2 CFSTRs	15	45	1.213
2 CFSTRs	22	38	1.185

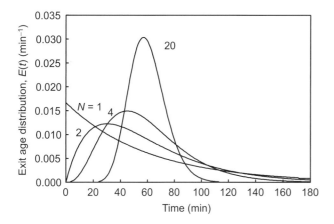

FIGURE 4-9. Residence time distributions for N CFSTRs in series, with a total residence time of 60 min.

which shows the residence times required for the same removal efficiency in one CFSTR, one PFR, or three CFSTRs in series, using the graphical presentation introduced in Figure 4-5. It should also be apparent from this figure that, if N were increased indefinitely, the total detention time in the series of CFSTRs would approach that in the single PFR. In practice, the use of two or more CFSTRs in series often represents a compromise between the benefits of equalization (provided by increased mixing) and those of increased reaction rate (provided by avoiding mixing).

Figure 4-9 shows exit age distributions ($E(t)$ functions), derived using Equation 2-84, for a system consisting of N equal-sized CFSTRs in series, with an overall mean hydraulic residence time of 60 min. Recalling that the fraction of the flow with residence times between t_1 and t_2 can be computed as the area under the $E(t)$ curve between t_1 and t_2, it is clear that the net effect of increasing N is to reduce the fractions of the flow with very long and very short residence times, while increasing the fraction with intermediate residence times. For this analysis, the most significant shift is the reduction in the fraction of the flow that resides for a very short time in the reactor, a reduction that is most dramatic as N increases from 1 to 2.

The longer a parcel of fluid resides in the reactor, the greater is the extent of reaction in that parcel. Thus, the narrowing of the $E(t)$ curve with increasing N corresponds to reducing the fractions of the fluid in which the reaction has proceeded only to a small extent (very short residence times) and to a large extent (long residence times), and simultaneously increasing the fraction in which the reaction has proceeded to an intermediate extent (residence times closer to τ). The observation that such a shift increases the overall extent of reaction for reactions with $n > 0$ indicates, in essence, that the benefit associated with decreasing fraction of the flow that resides for a short amount of time in the reactor outweighs the penalty associated with decreasing the fraction of the flow that resides for a long time in the reactor,

if these changes are carried out while keeping the mean residence time constant. A corollary is that a disproportionately large fraction of the reactant that is found in the overall effluent stream is associated with the parcels that have resided for a very short time in the reactor. This point is illustrated quantitatively in Example 4-7.

Application to Chemical Disinfection An interesting application of the CFSTRs-in-series model was presented by Lawler and Singer (1993) to demonstrate the effects of the hydraulic flow pattern on disinfection of drinking water, and how well these effects were considered by the existing regulations. Achieving adequate disinfection of drinking water is essential for the protection of the public health. Although more complex models for the kinetics of disinfection are sometimes used, the most common description is the Chick–Watson law

$$r_X = -k' c_D^n c_X \quad (4\text{-}38)$$

where c_X is the concentration of viable microorganisms (of a specific type), c_D the concentration of a particular disinfectant, n the order of the disinfection reaction with respect to the disinfectant, and k' the rate constant.

If it is assumed that $n = 1$ (a common assumption that is reasonably consistent with the experimental results for many organisms and disinfectants), the rate of disinfection in a batch reactor can be described by the following mass balance:

$$\frac{dc_X}{dt} = -k' c_D c_X \quad (4\text{-}39)$$

If we assume that c_D is constant, integration of Equation 4-39 yields

$$\ln \frac{c_X(t)}{c_X(0)} = -k' c_D t \quad (4\text{-}40)$$

It is difficult and time-consuming to measure microorganism concentrations, and therefore it is difficult to evaluate the left side of Equation 4-40 directly. For simplicity, therefore, the U.S. drinking water regulations (Federal Register, 1989) assume that the equation accurately describes disinfection kinetics and use the right side of the equation to compute the amount of disinfection that is achieved in a given reactor. The regulations specify the value of k' for a given situation, based on batch studies that were conducted at the time the regulations were promulgated. These batch studies used various target microorganisms, all the common chemical disinfectants, and water with various physicochemical characteristics (e.g., temperature, pH). Utilities then use the appropriate value of k along with the product of c_D and t that characterizes their system to compute the amount of disinfection (i.e., the value of the left side of Equation 4-40) that they can expect to achieve; the plant receives regulatory "credit" for achieving that amount of disinfection, without actually measuring the microorganism concentrations in the influent and effluent.[5]

The regulations recognize that, in truth, the value of c_D might vary in the reactor, so they require the concentration measured at the outlet of the reactor be used in the analysis. Also, as noted in Chapter 2, the value of t to be used is T_{10}, the time associated with a value of 0.10 for the cumulative age distribution, $F(t)$; that is, T_{10} is the time such that $F(T_{10}) = 0.10$. This choice indicates that the regulation treats all disinfection reactors as achieving the same amount of disinfection as a PFR with a detention time equivalent to T_{10}. Substituting these concepts into Equation 4-40 and also changing from the expression for the batch reactor to its equivalent PFR, we find that the U.S. drinking water regulations give credit for disinfection in any continuous flow reactor as follows:

$$\ln \frac{c_{X,\text{out}}}{c_{X,\text{in}}} = -k' c_{D,\text{out}} T_{10} \qquad (4\text{-}41a)$$

$$\log \frac{c_{X,\text{in}}}{c_{X,\text{out}}} = \frac{k'}{\ln 10} c_{D,\text{out}} T_{10} \qquad (4\text{-}41b)$$

Equation 4-41b is written to reflect the fact that disinfection is often described in terms of "log inactivation" values, wherein, for example, a 99% inactivation $(c_{X,\text{out}}/c_{X,\text{in}} = 0.01)$ is described as a 2-log inactivation $(\log(c_{X,\text{in}}/c_{X,\text{out}}) = 2)$. According to this equation, the amount of disinfection achieved for a given microorganism and a given disinfectant is determined by the CT product (or, in our terms, the product $c_{D,\text{out}} T_{10}$).

Lawler and Singer (1993) analyzed the difference between the amount of disinfection that would be credited according to the Surface Water Treatment Rule and the amount that would really occur in a nonideal reactor that fit the CFSTRs-in-series model, assuming that the model for disinfection kinetics used in the regulations was valid. A figure adapted from this work is shown in Figure 4-10, showing the amount of disinfection expected if the disinfectant concentration is constant in the reactor. The abscissa is the dimensionless product $k' c_D \tau$, and the ordinate in Figure 4-10a describes the log inactivation achieved. Four pairs of lines are shown, with the curved lines in Figure 4-10a indicating the degree of inactivation predicted from Equation 4-30 (with the substitution that $k_1 = k' c_D$) for a single reactor behaving as N CFSTRs in series; the straight lines in

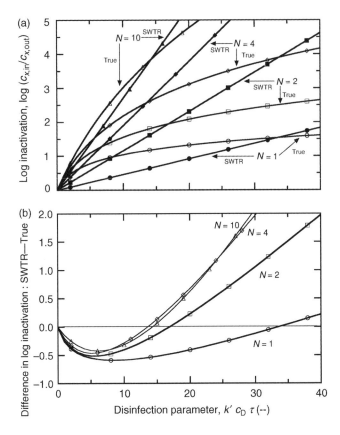

FIGURE 4-10. Comparison of disinfection expected in nonideal reactors and that credited by U.S. drinking water regulations. (Nonideal flow represented as N CFSTRs in series; SWTR refers to the USEPA's Surface Water Treatment Rule, which established these regulations.)

the figure reflect the disinfection credit that would be given according to the regulation. The T_{10} values to determine these straight lines can be found for each type of reactor by trial-and-error solution of Equation 2-85, with $F(t) = 0.1$.

The differences between the disinfection expected and that credited by the regulation are shown in Figure 4-10b. In practice, most reactors are likely to have values of the parameter $k' c_D \tau$ less than 15 for disinfectants and microorganisms of interest, suggesting that most plants are given credit for less disinfection than is likely to actually occur. Of most importance here, however, is the set of curved lines in Figure 4-10a; they indicate that the amount of disinfection that occurs in a reactor is dramatically affected by the value of N and by the number of conceptual CFSTRs that describes the hydraulic characteristics of a single real nonideal reactor. Inserting baffles to increase the N value is a highly efficient means to provide greater disinfection, or to improve the removal efficiency achieved by any other first-order reaction. Further discussion about the interaction of hydraulic design and disinfection kinetics is given by Ducoste et al. (2001).

---

[5] In actual practice, the value of k' is generally unknown to the user. The U.S. EPA published the values of the product of c_D and t that are associated with specific values of the amount of disinfection; these values reflect the value of k', as can be seen from the equation. Utilities can then use their value of the product of c_D and t to compute the "credit" that they receive for disinfection.

CFSTRs or PFRs in Parallel

The preceding sections demonstrate that the extent of reaction in a CFSTR or PFR depends on the reaction rate expression, the type of reactor, and the hydraulic residence time. If a single CFSTR is split into two or more CFSTRs in parallel and the influent flow is split in the same proportions as the volume of the reactors, the detention time in all the reactors is the same as in the original, larger reactor. Since splitting the flow has no effect on the reaction rate expression, the extent of conversion in each of the reactors would also be identical and equal to that of the original reactor. An analogous statement can be made about a single PFR that is split into several parallel PFRs with identical values of τ. Thus, splitting a reactor in this manner has no effect on the overall system performance.

If the flows are distributed differently than described earlier, so that different reactors have different detention times, but the total flow and total reactor volume remain constant, the overall amount of conversion accomplished does depend on the details of how the flow and volume are allocated. However, for reactions whose rates increase as the reactant concentration increases (any reaction with a positive reaction order with respect to the reactant of interest), the result is always that the extent of reaction is less if the detention times in the various reactors are unequal than if the flow and volume are allocated such that $\tau_1 = \tau_2 = \tau_i = \cdots = \tau_N$.

Thus, in most cases, no advantage in terms of removal efficiency is gained by using two reactors in parallel rather than a single reactor of the same total volume. The reason that systems are often designed with parallel reactors reflects a much more practical concern: if one reactor needs to be shut down for maintenance, the others can continue to operate and thereby mitigate the effects of the partial shutdown.

Using Reactors with Flow to Derive Rate Expressions

In addition to being useful for determining the expected conversion in ideal reactors when the rate expressions of interest are known, ideal reactors can be used to gather data from which previously undetermined rate expressions can be derived. For instance, we might use a laboratory-scale CFSTR operated at steady state to gather information about the removal efficiency as a function of detention time and influent concentration, and then use this information to determine the reaction order and the rate constant for the reaction. Similarly, it is common to use long, narrow tubular reactors as PFRs to study reaction kinetics; with the similarities observed between batch and PFRs, most of the techniques for analysis of reaction kinetics in batch reactors presented in Chapter 3 can be adapted to the continuous flow, steady-state case. To ensure (nearly) plug flow conditions in such reactors, turbulent flows (high Reynolds numbers) are required. For either type of ideal reactor, the assumption of ideal flow should be tested with tracer studies.

Choosing to study reaction kinetics in ideal continuous flow reactors can have two advantages. First, steady state is the conceptual equivalent of stopping the reaction at a certain point, and operating a system at steady state allows multiple measurements of the concentration to be obtained. In cases where accurate instantaneous measurements might be difficult, this advantage can be critical. Second, the use of CFSTRs might help to distinguish different reaction orders better or yield more accurate rate constants than batch (or PFR) reactors because of the reduced removal efficiency that is achieved in a CFSTR. That is, for some reactions, batch reactors or PFRs might give such high removal efficiencies that the error in concentration measurements makes it impossible to distinguish whether one reaction rate model is better than another; CFSTRs might be more sensitive (within a measurable range) to changes in operating conditions.

4.4 EXTENT OF REACTION IN REACTORS WITH NONIDEAL FLOW

The ultimate objective of reactor and reaction analysis is to be able to predict and interpret removal efficiencies in reactors with any flow pattern and in which reactions characterized by any rate expressions are occurring. The preceding sections of this chapter have shown how this can be done for single ideal reactors or combinations of such reactors, in which reactions characterized by some simple rate expressions are occurring. The approach that was developed is useful if a real reactor system can be well represented as a relatively simple combination of ideal reactors. For instance, if a system is designed to operate as two CFSTRs in series (of a known size and sequence), and tracer experiments confirm that the real flow pattern is consistent with this design, then the approach described earlier should yield reliable predictions for the removal efficiency when a reaction with a known rate expression is occurring in the system.

In cases where the details of the hydraulic behavior in a reactor are not known, we found in Chapter 2 how some information about that behavior can be acquired from tracer studies and represented in the form of the exit age distribution, $E(t)$, or the cumulative age distribution, $F(t)$, for the reactor. However, while each reactor has a specific $E(t)$ or $F(t)$ function, the converse is not true: a given $E(t)$ or $F(t)$ function does not uniquely identify a particular mixing regime. That is, the same $E(t)$ curve characterizes several (in fact, an infinite number of) reactor combinations and flow patterns. For instance, a PFR with a single inlet and multiple outlets (completely segregated flow, Figure 2-25a) can have an $E(t)$ curve that is identical to that of a PFR with multiple inlets along its length and a single outlet (maximum mixedness, Figure 2-25b). Recall also that any model reactor

TABLE 4-6. Effects of Amount and Timing of Mixing on Extent of Reaction[a]

	Increase Mixing	Delay Mixing
	Effect on RTD Parameters	
Any n	Broadens $E(t)$ curve, increases σ^2_{RTD}	No effect on $E(t)$ curve or σ^2_{RTD}
	Effect on Extent of Reaction	
$n < 0$	Increase	Decrease
$n = 0$	No effect	No effect
$0 < n < 1$	Decrease	Increase or decrease; depends on details
$n = 1$	Decrease	No effect
$n > 1$	Decrease	Increase

[a] For reactions with rate expressions of the form $r = \pm k_n c^n$.

system with multiple reactors in series has the same $E(t)$ function regardless of the sequence of the reactors.

The question we must now answer is: do reactors that have identical $E(t)$ curves necessarily also behave identically (i.e., do they accomplish the same degree of conversion) if a reactive substance is input? In other words, if we know the $E(t)$ function describing a reactor, the reaction rate expression for a reaction of interest, and the influent concentration, is this information sufficient to predict the effluent concentration?

The answer to this question was hinted at in previous examples, in which the overall removal efficiency in a system of two different sized CFSTRs in series was shown to be independent of their placement (whether the larger or smaller was first) for first-order reactions. In contrast, for second-order reactions, the extent of reaction was greater when the smaller CFSTR was placed upstream of the larger CFSTR, compared with the reverse order. Placing the smaller reactor upstream means that the inflow is initially mixed into less water; that is, it causes the majority of the mixing to be delayed and therefore corresponds to later mixing than when the larger reactor is placed upstream. Further analysis shows that later mixing leads to greater conversion for all nth-order reactions with $n > 1$, whereas earlier mixing leads to greater conversion for $n < 0$. For $0 < n < 1$, earlier mixing can either increase or decrease the extent of reaction, depending on the details of the system. For $n = 1$, the timing of the mixing is irrelevant, and for $n = 0$, both the timing and the extent of mixing are irrelevant (since in this case, the reaction rate does not depend on the reactant concentration).

As noted in Chapter 2, for a given *amount* of mixing (i.e., a given $E(t)$ curve), the limiting case of early mixing is known as maximum mixedness, and the limiting case of late mixing is known as segregated flow. Correspondingly, in a reactor with a given amount of mixing, segregated flow yields the greatest amount of conversion that can be achieved for a reaction with $n > 1$, while maximum mixedness yields the least amount of reaction; for reactions orders between 0 and 1, the reverse is true; and for reaction orders of 0 or 1, the timing of mixing has no effect on the extent of reaction. The effects of both the extent and timing of mixing are summarized in Table 4-6.

The noneffect of the timing of mixing on zero- or first-order reactions can be extended to reactions whose rates can be characterized by a combination of zero- and first-order expressions, which are known as *linear reactions*. For such reactions, knowledge of the reaction rate expression, $E(t)$ for the reactor of interest, and the influent composition is sufficient to compute the effluent composition. In contrast, for nonlinear reactions, the extent of conversion depends on the details of the mixing pattern in the reactor, so knowledge of the $E(t)$ curve is helpful, but not sufficient, to predict the extent of conversion. In these cases, we can predict the maximum and minimum extents of reaction that might occur, but we cannot predict the effluent composition unless we have additional information (or make additional assumptions) about the specific mixing pattern that generated this observed $E(t)$ function; in most real reactors, mixing probably occurs at various times after the feed has entered, so the observed extent of reaction is between the two limiting values. We next consider how to calculate the expected extent of reaction for linear reactions, and the maximum and minimum possible extents of reaction for nonlinear reactions, in reactors with any arbitrary RTD.

Fraction Remaining Based on the Exit Age Distribution

As noted earlier, the $E(t)$ function describes the age distribution of fluid in the exit stream from a reactor, but it does not indicate anything about the detailed mixing of the fluid. For instance, the effluent from a reactor might be a mixture of equal volumes of fluid that entered 10 min earlier and 20 min earlier. If we carry out a tracer test, we would be able to identify what proportion of the fluid in the effluent entered at each of the two prior times, but we would not be able to decide whether the mixing took place shortly after the "10-minute-old" parcel entered the reactor (10 min ago, corresponding to the maximum mixedness model) or just a few seconds before the two parcels of fluid arrived at the effluent port (corresponding to the segregated flow model).

If the reactor hydraulics corresponds to segregated flow, the reactor effluent is composed of many distinct parcels with different residence times. The concentration in the mixed effluent is the flow-weighted average concentration in the parcels, given as follows:

$$c_{\text{out}} = \sum_{\text{all } i} \frac{Q_i}{Q_{\text{tot}}} c_i \qquad (4\text{-}42)$$

where i is an index that refers to different parcels of water that reside in the reactor for different time periods. Q_i/Q_{out} is the fraction of the exiting fluid that is associated with parcel i, c_i is the concentration of reactant in this fluid, and the summation is over all parcels i that contribute to the effluent stream. For instance, if the effluent is composed of just three parcels that account for 25, 35, and 40% of the flow, and if these parcels contain, respectively, 1.0, 2.0, and 0.1 mg/L of the contaminant, the concentration in the effluent would be

$$c_{\text{out}} = 0.25(1.0\,\text{mg/L}) + 0.35(2.0\,\text{mg/L})$$
$$+ 0.40(0.1\,\text{mg/L}) = 0.99\,\text{mg/L}$$

If the parcels become differentially small, then each parcel is associated with a differential flow rate, and Equation 4-42 can be rewritten as an integral, as follows:

$$c_{\text{out}} = \int_{\text{all } i} c_i \frac{dQ_i}{Q_{\text{tot}}} \qquad (4\text{-}43)$$

Equation 4-43 applies to any reactive substance in any reactor with segregated flow. In such a reactor, each differential-sized parcel of water in the effluent can be associated with a specific residence time between t and $t + dt$, and we can reinterpret dQ_i/Q_{out} as the fraction of the exiting fluid that has resided between t and $t + dt$ in the reactor; we recognize this fraction as $E(t)dt$.

Furthermore, if each parcel of fluid in the effluent has remained segregated from all others during its flow through the reactor, it can be viewed as a small batch reactor that has reacted for time t. Then, from the influent concentration and the reaction rate expression, we can compute the value of c_i for a parcel that has resided between t and $t + dt$ in the reactor as $c(t)$, equal to the concentration that would be present in a batch reactor after time t (or in the effluent of a PFR with $\tau = t$).

Making these two substitutions into Equation 4-43, and recognizing that the integration over all parcels is equivalent to integrating it over all possible parcel residence times, we obtain

$$c_{\text{out}} = \int_0^\infty c(t) E(t)\, dt \qquad (4\text{-}44)$$

If the $E(t)$ function is known only at discrete times, Equation 4-44 can be approximated as follows:

$$c_{\text{out}} = \sum_{\text{all } i} c_{\text{ave}}(t_i) E_{\text{ave}}(t_i) \Delta t_i \qquad (4\text{-}45)$$

where the subscript "ave" indicates the average value of the function during the time interval, Δt_i. Note that the integral in Equation 4-44 and the summation in Equation 4-45 are not over time *per se*, but rather over the range of possible residence times experienced by different parcels of fluid exiting the reactor at any instant.

If $c(t)$ were known as a function of t for a batch system, and $E(t)$ were known from a tracer experiment, we could compute the product $c(t) E(t)$ for various values of t, and plot that product as a function of t. According to Equation 4-44, the area under the curve from $t = 0$ to ∞ would then be the overall effluent concentration expected if the given reactions were to occur in the given reactor, and if the reactors were characterized by segregated flow.[6]

If $c(t)$ is an explicit function of time, it might be possible to manipulate the earlier equations to obtain other useful relationships. For example, assume that we determined the $E(t)$ function for a reactor, and we are interested in predicting the fraction of a contaminant that will remain in the effluent. If the reactant undergoes a first-order decay reaction, $c(t)$ and $c_{\text{ave}}(t_i)$ can be expressed as $c_{\text{in}} \exp(-k_1 t)$ and $c_{\text{in}} \exp(-k_1 t_{\text{ave}})$, respectively, so Equations 4-44 and 4-45 can be rewritten as follows:

$$\frac{c_{\text{out}}}{c_{\text{in}}} = \int_0^\infty \exp(-k_1 t) E(t)\, dt \qquad (4\text{-}46)$$

$$\frac{c_{\text{out}}}{c_{\text{in}}} \approx \sum_{\text{all } i} \exp(-k_1 t_{\text{ave}}) E_{\text{ave}}(t_i) \Delta t_i \qquad (4\text{-}47)$$

Equations 4-46 and 4-47 were derived for a reactor with segregated flow, a condition that is probably rarely met. Recall, however, that reactions that can be characterized by linear rate expressions progress to the same extent (i.e., to the same effluent concentration) in any reactor with a given $E(t)$ function. Therefore, if a reactor has a given $E(t)$ function, a linear reaction will proceed in this reactor to the same extent as it would in a reactor with segregated flow, even if the flow pattern is not segregated in the real reactor. Therefore, we can use Equations 4-44 and 4-45 to predict the extent of a linear reaction in any reactor, and to predict the extent of any

---

[6] Nauman and Buffham (1983) provide an elegant description of the limitations of Equation 4-44, partially based on the work of Zwietering (1959); the reader is referred to their book for a more complete understanding.

reaction in a reactor with segregated flow. Conversely, it is not possible to predict the extent of a nonlinear reaction in a reactor, even with knowledge of the $E(t)$ curve, unless the flow in the reactor is segregated. These points are illustrated by the following example.

■ **EXAMPLE 4-6.** Consider a reactor with the tracer response curve that is shown in Figure 4-11a. This response suggests that the reactor has two distinct flow paths: 20% of the flow behaves as though it was in a PFR with a residence time of 10 min, and 80% of the flow behaves as though it was in a PFR with a residence time of 15 min. However, it does not provide sufficient information about the details of the mixing pattern.

(a) Compute the expected effluent concentration from this reactor for a substance that is present at a concentration of 20 mg/L in the influent and undergoes a linear reaction characterized by the rate expression $r_i = -0.2\,\text{min}^{-1} c_i - 0.04\,\text{mg/L min}$. Carry out calculations for the two extremes of segregated flow (shown schematically in Figure 4-11b) and maximum mixedness (shown in Figure 4-11c).

(b) Repeat the calculations for another reactant with the same influent concentration that undergoes a second-order reaction with k_2 equal to $0.03\,(\text{mg/L})^{-1}/\text{min}$.

Solution.

(a) We can compute what the concentration of the reactant undergoing the linear reaction would be as a function of the hydraulic residence time in a PFR by evaluating the mass balance for such a reactant in a batch reactor. Substituting the rate expression into the differential forms of the mass balance for a PFR (Equation 4-13), and integrating, we find

$$r_i = \frac{dc_i}{dt} = -0.2\,\text{min}^{-1} c_i - 0.04\,\text{mg/L min} \int_{c_{i,\text{in}}}^{c_{i,\text{out}}} \frac{dc_i}{-0.2\,\text{min}^{-1} c_i - 0.04\,\text{mg/L min}}$$

$$= \int_0^\tau dt - \frac{1}{0.2\,\text{min}^{-1}} \ln \frac{0.2\,\text{min}^{-1} c_{i,\text{out}} + 0.04\,\text{mg/L min}}{0.2\,\text{min}^{-1} c_{i,\text{in}} + 0.04\,\text{mg/L min}} = \tau$$

$$c_{i,\text{out}} = \frac{\left(0.2\,\text{min}^{-1} c_{i,\text{in}} + 0.04\,\text{mg/L min}\right) \exp\left(-0.2\,\text{min}^{-1} \tau\right) - 0.04\,\text{mg/L min}}{0.2\,\text{min}^{-1}}$$

The complete segregation model indicates that 20% of the flow would react for 10 min and then appear in the effluent. The concentration in this portion of the flow can be determined by substituting values of 20 mg/L for $c_{i,\text{in}}$ and 10 min for τ in the earlier expression, yielding $c_{i,\text{out}} = 2.53$ mg/L for this portion of flow. A similar calculation for the other 80% of the flow, which reacts for 15 min, indicates that it would

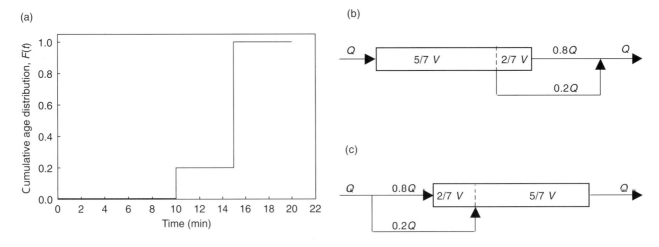

FIGURE 4-11. (a) Tracer output curve for a hypothetical system that might have a mixing pattern of (b) segregated flow or (c) maximum mixedness. Each portion of the reactor in (b) and (c) behaves as a PFR.

exit with a concentration of 0.81 mg/L. Given the mixing proportions in the effluent, the overall effluent concentration would be $0.2(2.53\,\text{mg/L}) + 0.8(0.81\,\text{mg/L})$, or 1.15 mg/L.

In the maximum mixedness model for the same reactor (i.e., the same $E(t)$ curve), 80% of the flow reacts for 5 min, at which point it contains a concentration of 7.23 mg/L. It then mixes with the remaining 20% of the influent to give a mixed concentration of $(0.8)(7.23\,\text{mg/L}) + (0.2)(20\,\text{mg/L})$, or 9.78 mg/L. This mixture then reacts for 10 more minutes before exiting the reactor at a concentration of 1.15 mg/L. Thus, the predicted effluent concentration is identical for the segregated and maximum mixedness cases, consistent with our expectation for a linear reaction.

(b) If the reaction is of second order with the specified rate constant, we can use the following equation (from Table 4-2) to compute c_{out} in each portion of the flow for the case of segregated flow:

$$c_{\text{out}} = \frac{c_{\text{in}}}{1 + k_2 \tau c_{\text{in}}}$$

The result is that 20% of the effluent has reacted for 10 min and has a concentration of 2.86 mg/L, and the remaining 80% had reacted for 15 min and has a concentration of 2.00 mg/L. The mixed effluent therefore has a concentration of 2.17 mg/L.

In the maximum mixedness case, the 80% of the influent that reacts for 5 min has a concentration of 5.00 mg/L. When this fluid mixes with the other 20% of the influent, the mixed concentration is 8.00 mg/L. This mixture then reacts for 10 min before exiting the reactor with a concentration of 2.35 mg/L. Thus, for the reactant that undergoes a second-order reaction, the details of the mixing do affect the extent of reaction, with earlier mixing (represented, in the limit, by maximum mixedness) achieving less removal than later mixing (represented, in the limit, by segregated flow). ∎

EXAMPLE 4-7. Consider again the reactor that was described in Example 2-1. There, a tracer study was described and the residence time distribution, $E(t)$, was calculated from the results. Consider now that a contaminant undergoes a first-order reaction with a rate constant $k_1 = 0.1\,\text{min}^{-1}$ in that reactor. Determine the expected fraction of the contaminant remaining in the effluent and the removal efficiency.

Solution. Because the reaction rate expression is linear, the destruction of the contaminant will be identical in any reactor that has the given $E(t)$ function. Therefore, we can carry out the analysis for a reactor with segregated flow and be confident that our result will apply to the real reactor, even if that reactor does not have segregated flow.

The analysis is conducted via spreadsheet calculations, as illustrated below. The first three columns of the spreadsheet are taken from the results shown in Chapter 2 for the calculation of the $E(t)$ function. The data in column C indicate the area under the $E(t)$ curve during the preceding interval, as found via the trapezoidal rule; that is, it is the product of the average value of $E(t)$ during the preceding interval and the length of that interval.

Column D shows the average value of t during the preceding time interval, and Column E shows the expected fraction of the contaminant remaining in parcels that have resided for that amount of time in the reactor $[c(t_{\text{ave}})/c_{\text{in}}]$; since the reaction is a first-order decay, this fraction is $\exp(-k_1 t_{\text{ave}})$.

Column F is the product of the values in Columns C and E and represents the contribution of parcels having residence times in the given range to the overall fraction of contaminant remaining in the effluent. The sum of all values in Column F, shown in the final row, represents the effluent value of $c_{\text{out}}/c_{\text{in}}$ for this reaction in this reactor. For the conditions in this example, the fraction remaining is expected to be 0.06, so the removal efficiency is $(1 - 0.06)\,100\%$ or 94%.

A	B	C	D	E	F
Time (min)	$E(t)$ (min^{-1})	$E_{\text{ave}}(t)\,\Delta t$ (–)	t_{ave} (min)	$c(t_{\text{ave}})/c_{\text{in}}{}^a$ (–)	$E_{\text{ave}}(t) \times \exp(-k_1 t_{\text{ave}})\,\Delta t^b$ (–)
0	0.0000	—	—	—	—
10	0.0064	0.032	5	0.607	1.94×10^{-2}
20	0.0136	0.100	15	0.223	2.23×10^{-2}
30	0.0160	0.148	25	8.21×10^{-2}	1.21×10^{-2}
40	0.0140	0.150	35	3.02×10^{-2}	4.53×10^{-3}
50	0.0124	0.132	45	1.11×10^{-2}	1.47×10^{-3}
60	0.0104	0.114	55	4.09×10^{-3}	4.66×10^{-4}
70	0.0084	0.094	65	1.50×10^{-3}	1.41×10^{-4}
80	0.0064	0.074	75	5.53×10^{-4}	4.09×10^{-5}
90	0.0048	0.056	85	2.03×10^{-4}	1.14×10^{-5}
100	0.0032	0.040	95	7.49×10^{-5}	2.99×10^{-6}
110	0.0024	0.028	105	2.75×10^{-5}	7.71×10^{-7}
120	0.0012	0.018	115	1.01×10^{-5}	1.82×10^{-7}
130	0.0004	0.008	125	3.73×10^{-6}	2.98×10^{-8}
140	0.0000	0.002	135	1.37×10^{-6}	2.74×10^{-9}
150	0.0000	0.000	145	5.04×10^{-7}	0
				Sum of column F values ($= c_{\text{out}}/c_{\text{in}}$):	0.0605

^{a} Computed as $\exp(-k_1 t_{\text{ave}})$.
^{b} Determined by the trapezoid rule; that is, taking the average value of $E(t)\exp(-k_1 t)$ in the preceding interval and multiplying by the width of the interval. This calculation is accomplished in the spreadsheet by multiplying Columns C and E. ∎

It is instructive to study the values in Column F of Example 4-7 to see how much of the contaminant in the composite effluent is contributed by the different parcels of fluid with different residence times. For example, the first three intervals represent only 28% of the flow (the sum of

$E\Delta t$), but they account for 89% of the unreacted pollutant (the sum of the first three values in Column F divided by the final sum of Column F). The fact that some parts of the fluid stay in the reactor for a long time and achieve a very high degree of removal efficiency cannot compensate for the low efficiency in those packets that exit quickly.

In general, the parcels in the earliest few time intervals account for most of the effluent concentration. Thus, this example emphasizes that *it is critically important to minimize short-circuiting to achieve high treatment efficiency*. It also emphasizes the importance of characterizing the early part of the residence time distribution function carefully— an error in $E(t)$ at small values of t has a much greater effect on the computed reaction efficiency than does an error at large values of t.

Fraction Remaining Based on the Dispersion Model

In most cases, Equation 4-44 (or, for first-order reactions, Equation 4-46) has no explicit solution, and a graphical or numerical integration is required. However, if $E(t)$ can be approximated by some simple mathematical function rather than as a series of discrete data points, and if $c(t)$ is a relatively simple function, it may be possible to solve the integral analytically. In such cases, we can predict the system output without resorting to numerical or graphical integration.

For instance, if $E(t)$ can be represented adequately by the PFR-with-dispersion model, and if the reactant undergoes a first-order irreversible decay, so that $c(t) = c(0)\exp(-k_1 t)$, the solution of Equation 4-46 in a reactor with either open or closed boundaries is[7,8]

$$\frac{c_{\text{out}}}{c_{\text{in}}} = \frac{4a\exp(v_x L/2\mathbf{D})}{(1+a)^2\exp(av_x L/2\mathbf{D}) - (1-a)^2\exp(-av_x L/2\mathbf{D})}$$

(4-48)

where $a = [1 + 4k_1\tau(\mathbf{D}/v_x L)]^{1/2}$. Note that the expected fraction remaining in the effluent is a function of two dimensionless groups, the dispersion number and the product of the reaction rate constant and the detention time. As

---

[7] This result can be derived by writing the mass balance equation for steady-state conditions, substituting the appropriate expression for the reaction rate, and retaining the dispersion term

$$0 = -v_x \frac{dc}{dx} + \mathbf{D}\frac{d^2 c}{dx^2} - k_1 c$$

Solving this equation in conjunction with the boundary conditions yields the result shown in Equation 4-48.

[8] Remember that Equation 4-44 is strictly valid only for reactors with segregated flow, a constraint that does not generally apply to reactors with dispersion. However, if we are interested in a linear reaction, the extent of reaction is identical for a reator with a given $E(t)$, regardless of the timing of mixing. As a result, for that scenario, the equation for the extent of reaction in a reactor with segregated flow applies to a reactor with dispersion.

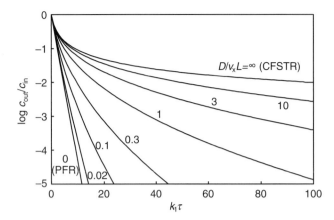

FIGURE 4-12. The effects of dispersion on the fraction remaining, for a reactant undergoing a first-order reaction in a reactor at steady state.

noted earlier, the latter can be considered as the ratio of the hydraulic residence time to the characteristic reaction time.

For a given reactor (detention time and dispersion number) and reaction (value of the first-order rate constant), the expected fraction remaining can be found by directly applying Equation 4-48. A few results of such calculations are shown in Figure 4-12. The steady and dramatic trend of increasing fraction remaining with increasing dispersion for a given value of $k_1\tau$ is evident.

Equation 4-48 indicates that any given, desired fraction remaining could be reached using many different combinations of detention time and dispersion number. In other words, the equation characterizes the acceptable trade-offs between investing resources in the reactor volume (increasing τ) versus investing them in measures that can reduce dispersion, while still meeting the design objective. The decision about which combination to select for the design can then be based on an economic analysis of all the technically feasible options, in conjunction with other site-specific considerations.

■ **EXAMPLE 4-8.** Consider the example reactor analyzed in Example 2-4, for which the hydraulic detention time and dispersion number ($\mathbf{D}/v_x L$) were estimated via tracer studies to be 49.5 min and 0.19, respectively. If the first-order reaction described in the preceding example (rate constant, $k_1 = 0.1\,\text{min}^{-1}$) occurs in that reactor, what is the predicted fraction of the influent concentration remaining in the effluent according to the dispersion model, and what is the removal efficiency?

Solution. The value of the dimensionless product $k_1\bar{t}$ is 4.95 and the value of $(\mathbf{D}/v_x L)$ is given as 0.19. With these values, the parameter a is $[1 + 4k\bar{t}(\mathbf{D}/v_x L)]^{1/2} = [1 + (4)(4.95)(0.19)]^{1/2} = 2.18$. (Note that $\bar{t}$, rather than τ, is used to calculate the parameter a, because $\bar{t}$ is the value of the hydraulic residence time that provides the best fit between the dispersion model and the experimental data.) Substituting these values into Equation 4-48 yields the fraction remaining, $c_{\text{out}}/c_{\text{in}} = 0.038$, so the removal efficiency is $(1 - 0.038)(100\%)$, or 96.2%. For comparison, the fraction remaining for the same reaction in a PFR with the same detention time is $(c_{\text{out}}/c_{\text{in}}) = \exp(-k\bar{t}) = \exp[(-0.1\,\text{min}^{-1})(49.5\,\text{min})] = 0.0071$, corresponding to a removal efficiency of $(1 - 0.0071)100\%$, or 99.29%. The dispersion lowers the removal efficiency significantly. ■

Summary of Steady-State Performance in Nonideal Reactors

To summarize, the expected extent of conversion of a reactant undergoing a linear reaction in a reactor with any known RTD can be computed by evaluating the integral in Equation 4-44. Although the derivation leading to that equation relied on an assumption of segregated flow, the result is applicable regardless of whether or not the reactor is actually characterized by segregated flow. Explicit expressions exist for the extent of reaction that will be achieved if a linear reaction proceeds in a reactor operating at steady state with certain, idealized RTDs (e.g., the RTD of a CFSTR or a PFR with dispersion), and thus it avoids evaluating the integral in Equation 4-44 in these cases.

On the other hand, if the reaction is nonlinear, then the extent of reaction can be predicted only for two limiting cases of the timing of mixing: segregated flow and maximum mixedness. If the reactor has segregated flow (i.e., mixing at the latest possible time in the reactor), Equation 4-44 applies. This scenario yields the maximum extent of reaction achievable (for the given RTD) if the reaction order is >1, and the minimum extent of reaction if the order is <1. If the reactor has maximum mixedness, the analysis is often complicated and requires step-by-step assessment of the amount of reaction that occurs between sequential mixing locations. However, for the special case of a CFSTR (which has maximum mixedness, by definition), the mass balance can be solved analytically or numerically to obtain the extent of reaction directly. Maximum mixedness maximizes the extent of reaction for a reaction with order <1 and minimizes it for reaction orders >1. Practically, most reactors probably have mixing that occurs neither instantly nor at the latest possible time, and they therefore produce extents of reaction that are intermediate between those calculated based on segregated flow and maximum mixedness for nonlinear reactions.

4.5 EXTENT OF REACTION UNDER NON-STEADY-CONDITIONS IN CONTINUOUS FLOW REACTORS

Until now in this chapter, we have considered only steady-state conditions in continuous flow reactors. Although an assumption of steady state is often satisfactory for the analysis of engineered systems, it is sometimes necessary to consider the transient response of a system when a change is imposed. It is also instructive to see how rapidly a system can be expected to respond to a changed condition, since such an analysis gives some insight into when an assumption of steady-state conditions might be reasonable even if that condition is not met absolutely. The analysis of nonsteady conditions is intrinsically more complex than that of steady conditions, because the mass balance contains another non-zero term; that is, the storage term, which is zero in steady-state conditions (at least in the common Eulerian view). In this section, we consider changes in either influent concentration or flow and, for simplicity, imagine that the change in conditions goes from one condition to another at a single instant and remains there indefinitely. Other types of changes can be considered using this scenario as a baseline.

Extent of Conversion in PFRs Under Non-Steady-State Conditions

Consider first the case of a PFR whose influent concentration abruptly changes from one value to another at a time defined to be $t = 0$. Because the incoming fluid does not mix with water already present in the reactor, the new influent acts independent of the old. The result is a step change in the effluent, at time $t = \tau_{\text{PFR}}$, corresponding to the expected extent of reaction under the new influent conditions.

A step change in the flow rate into a PFR provides a different, more complicated response. Assuming that the step change creates no hydraulic perturbations or mixing, each parcel or plug can again be considered separately. Any parcel that exited the reactor prior to the step change resided for a time equal to the original detention time, say τ_1, in the reactor. Similarly, any parcel that enters after the step change in flow rate will reside for a time equal to the new detention time, τ_2, in the reactor. However, parcels that were in the reactor at the time of the step change will reside for some time in the reactor traveling at one velocity and some other time traveling at another, and therefore are in the reactor for times between τ_1 and τ_2.

Considering a single parcel (i.e., in a Lagrangian view) that was somewhere in the reactor at the time of the step change, and then generalizing to any such parcel, allows us to determine that the actual residence time resided in the PFR changes linearly from τ_1 to τ_2 over a time period equal to τ_2. During this time, each exiting parcel has a residence time that we refer to as $\tau(t)$. The residence times of parcels that exit at any time can be described by the following equation:

$$\tau(t) = \begin{cases} \tau_1, & t \leq 0 \\ \tau_1 + \left(\dfrac{\tau_2 - \tau_1}{\tau_2}\right)t, & 0 < t \leq \tau_2 \\ \tau_2, & t > \tau_2 \end{cases} \quad (4\text{-}49)$$

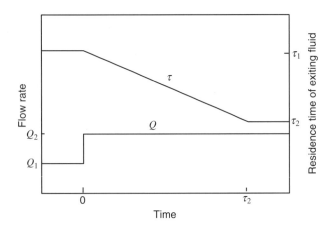

FIGURE 4-13. Response of a PFR to a step change in the flow rate.

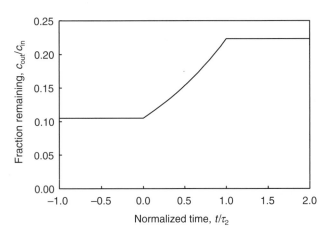

FIGURE 4-14. Response of a PFR with a first-order reaction to a step increase in flow rate. (Conditions specified in text.)

The situation is depicted for a step increase in flow at time zero in Figure 4-13.

■ **EXAMPLE 4-9.** Consider a PFR in which a first-order decay reaction with a rate constant $k = 0.05 \text{ min}^{-1}$ is occurring, and which is operating at steady state. At a time $t = 0$, the flow rate abruptly changes to decrease the detention time from 45 min (τ_1) to 30 min (τ_2). Determine the response of the PFR (i.e., effluent concentration as a function of time).

Solution. The effluent concentration at $t = 0$ can be calculated from the steady-state equation (PFR, first order) as follows:

$$c_{\text{out}} = c_{\text{in}} \exp(-k_1 \tau_1) \quad (t < 0)$$
$$= (100 \text{ mg/L}) \exp(-(0.05 \text{ min}^{-1})(45 \text{ min}))$$
$$= 10.5 \text{ mg/L} \quad (4\text{-}50)$$

At times greater than $t = \tau_2$ (i.e., after all the water that was originally in the reactor has exited), a new steady-state effluent concentration will be reached, as follows:

$$c_{\text{out}} = c_{\text{in}} \exp(-k_1 \tau_2) \quad (t > \tau_2)$$
$$= (100 \text{ mg/L}) \exp(-(0.05 \text{ min}^{-1})(30 \text{ min}))$$
$$= 22.3 \text{ mg/L} \quad (4\text{-}51)$$

At intermediate times, the fraction remaining gradually changes from the first of these conditions to the second; the fraction remaining at any time during this period is calculated by substituting the expression for $\tau(t)$ into the integrated batch rate expression to yield

$$c_{\text{out}} = c_{\text{in}} \exp(-k_1 \tau(t))$$
$$= \exp\left\{-k_1 \left[\tau_1 + \left(\frac{\tau_2 - \tau_1}{\tau_2}\right)t\right]\right\} \quad (0 < t < \tau_2)$$
$$(4\text{-}52)$$

For the conditions specified, the results are shown graphically in Figure 4-14. Note that the change in effluent concentration is not linear with time, and that the derivative of the response changes abruptly both when the flow changes initially and again at $t = \tau_2$. ■

The same type of analysis as demonstrated in Example 4-9 applies for any reaction rate expression: the integrated form of the expression can be used with the varying actual residence time to yield the non-steady-state response of the PFR. A zero-order reaction gives a linear change with time from the old steady state to the new steady state. Higher order reactions show increasing curvature.

Extent of Conversion in CFSTRs Under Non-Steady-State Conditions

Consider next the nonsteady-state response of a CFSTR in which a first-order reaction is occurring, by considering that the influent concentration undergoes a step change from $c_{\text{in},1}$ to $c_{\text{in},2}$ at $t = 0$. Assume that the flow rate (and therefore detention time) remains constant. Further assume (although it is not essential, as shown below) that the conditions prior to the change had existed long enough so that the reactor was at steady state at $t = 0$. Applying the appropriate equation from Table 4-1, the effluent concentration at $t \leq 0$ is

$$c_{\text{out}}(t) = \frac{c_{\text{in},1}}{1 + k_1 \tau} \quad \text{for} \quad t \leq 0 \quad (4\text{-}53)$$

The mass balance for the CFSTR for any time after the change in influent is

$$V \frac{dc(t)}{dt} = Q c_{\text{in},2} - Q c(t) - V k_1 c(t) \quad (4\text{-}54)$$

where $c(t)$ is the concentration in both the reactor and the effluent, and is changing with time. Dividing Equation 4-54 by V yields

$$\frac{dc}{dt} = \frac{1}{\tau}c_{in,2} - \frac{1}{\tau}c(t) - k_1 c(t) \quad (4\text{-}55a)$$

$$= \frac{c_{in,2}}{\tau} - \left(\frac{1}{\tau} + k_1\right)c(t) \quad (4\text{-}55b)$$

We can solve this differential equation by taking its Laplace transform, solving for $\mathbf{L}(c(t))$, and then taking the inverse transform to determine $c(t)$. Following convention, we represent the Laplace transform of the function of interest as $f(s)$; that is, $f(s) \equiv \mathbf{L}(c(t))$. The Laplace transforms of a variable $c(t)$, the time derivative of this variable (dc/dt), and a constant A are given in Appendix 2A of Chapter 2 as $f(s)$, $sf(s) - c(0)$, and A/s, respectively. Therefore, taking the Laplace transform of each side of Equation 4-55b, we obtain

$$sf(s) - c(0) = \frac{c_{in,2}}{\tau}\frac{1}{s} - \left(\frac{1}{\tau} + k_1\right)f(s) \quad (4\text{-}56)$$

After substituting from Equation 4-53 for $c(0)$, solving for $f(s)$ yields

$$f(s) = \frac{c_{in,1}}{1 + k_1\tau}\frac{1}{s + \left(\frac{1}{\tau} + k_1\right)} + \frac{c_{in,2}}{\tau}\frac{1}{s}\frac{1}{s + \left(\frac{1}{\tau} + k_1\right)} \quad (4\text{-}57)$$

Taking the inverse Laplace transform yields the final result, as follows:

$$c(t) = \frac{c_{in,1}}{1 + k_1\tau}\exp\left[-\left(\frac{1}{\tau} + k_1\right)t\right]$$

$$+ \frac{c_{in,2}}{1 + k_1\tau}\left\{1 - \exp\left[-\left(\frac{1}{\tau} + k_1\right)t\right]\right\} \quad (4\text{-}58a)$$

$$= \frac{c_{in,1}}{1 + k_1\tau}\exp\left[-(1 + k_1\tau)\frac{t}{\tau}\right]$$

$$+ \frac{c_{in,2}}{1 + k_1\tau}\left\{1 - \exp\left[-(1 + k_1\tau)\frac{t}{\tau}\right]\right\} \quad (4\text{-}58b)$$

Equations 4-58a and 4-58b look complex, but considering each of their components allows us to interpret them relatively easily. First, note that if the rate constant, k_1, were zero, the substance would be a conservative (nonreactive) substance; the result would then simplify to the same expression as we derived in Chapter 2 in the section on equalization (Equation 2-101).

Now, consider the exponential terms. In the analysis of nonreactive substances in a CFSTR under non-steady-state conditions (Chapter 2, tracers and equalization), we obtained an expression for $c(t)$ that contained exponential terms of the form $\exp(-t/\tau)$. Identifying τ as the characteristic time of the reactor, we found that, after one characteristic time in a CFSTR, only $1/e$ of the fluid (and tracer) present at time zero remains in the reactor. Similarly, in the analysis of first-order reactions in batch reactors in Chapter 3, we derived an expression for $c(t)$ that contained terms of the form $\exp(-k_1 t)$, and we identified $1/k_1$ as the characteristic reaction time. In that case, after one characteristic reaction time, only $1/e$ of the reactant that was present at time zero remains in the reactor. In the system under consideration in this section, reactant is being flushed out of the system by flow and is simultaneously being destroyed by reaction. Both processes are driving the reactant to its new, steady-state value, so the approach is faster than it would be if only one process were operative.

The overall characteristic time for the process is given by

$$\frac{1}{t_{char,overall}} = \frac{1}{t_{char,reaction}} + \frac{1}{t_{char,reactor}} \quad (4\text{-}59a)$$

$$= k_1 + \frac{1}{\tau} \quad (4\text{-}59b)$$

so that

$$t_{char,overall} = \frac{1}{1 + k_1\tau}\tau \quad (4\text{-}60)$$

Equation 4-59b shows that the overall characteristic reaction time is shorter than the characteristic time of either the reaction or the reactor. One consequence of this observation is that, if a system is designed to achieve high removal efficiency (so that τ is several times $t_{char,reaction}$), it is bound to have a hydraulic residence time that is several times $t_{char,overall}$. In other words, $t_{char,overall}$ is far less than one hydraulic residence time, and the system responds quite rapidly to a change in input conditions.

Third, consider what happens at long times; that is, when t/τ (or, more precisely, $(1 + k_1\tau)t/\tau$) is large. Under these circumstances, the exponential terms can be considered as zero and either form of Equation 4-58 reduces to $c(t) = c_{in,2}/(1 + k_1\tau)$. We recognize this fraction as the steady-state concentration expected under the new conditions.

Finally, if we consider the meaning of the two terms on the right side of Equation 4-58 individually, we note that the first term accounts for the decay of the original, and the second accounts for the rise to the ultimate concentration, with each term incorporating the effects of both advection and reaction.

■ **EXAMPLE 4-10.** A CFSTR is being used to treat an influent containing 100 mg/L of a reactant that undergoes a first-order decay reaction with a rate constant of 0.1 min^{-1}.

The hydraulic detention time in the reactor is 60 min, and the system has operated under the given conditions for long enough that it is at steady state. Then, at a time $t=0$, the influent concentration increases to 150 mg/L.

(a) Determine the response of the reactor. In particular, how much time is required before the ultimate change in the effluent concentration is 90% complete?

(b) What are the characteristic times for the reaction, the reactor, and the overall process?

Solution.

(a) The steady-state effluent concentration of a contaminant undergoing a first-order decay reaction in a CFSTR is given as Equation 4-4. After rearranging and substituting the given values, the concentration of the contaminant in the reactor effluent at $t=0$ is

$$c_{\text{out}} = \frac{c_{\text{in}}}{1+k_1\tau} = \frac{100\,\text{mg/L}}{1+(0.1\,\text{min}^{-1})(60\,\text{min})}$$
$$= 14.3\,\text{mg/L}$$

The response of the system to the increase in influent concentration is given by Equation 4-58 and is plotted in Figure 4-15. Both the decay and rise terms in Equation 4-58 are shown. The new steady-state effluent condition, computed again using Equation 4-4, is 21.4 mg/L, an increase of 7.1 mg/L over the original concentration. The response is therefore 90% complete when the concentration has increased by 0.9 (7.1 mg/L), or 6.4 mg/L, to a concentration of 20.7 mg/L. The effluent concentration reaches this value after only approximately 20 min has elapsed, corresponding to only one-third of one hydraulic detention time.

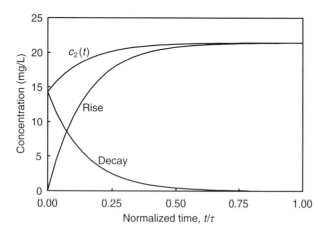

FIGURE 4-15. Non-steady-state response of a CFSTR to a step increase in influent concentration for a reactant undergoing a first-order reaction. $c_{\text{inf}} = 100\,\text{mg/L}$ at $t/\tau \leq 0$, and $c_{\text{inf}} = 150\,\text{mg/L}$ at $t/\tau > 0$; $\tau = 60\,\text{min}$; $k_1 = 0.1\,\text{min}^{-1}$.

(b) The characteristic times for the reaction and the reactor are k_1^{-1} and τ, corresponding to 10 and 60 min, respectively. According to Equation 4-59, the characteristic time for the overall reactive system is

$$\frac{1}{t_{\text{char,overall}}} = \frac{1}{t_{\text{char,reaction}}} + \frac{1}{t_{\text{char,reactor}}}$$

$$\frac{1}{t_{\text{char,overall}}} = \frac{1}{k_1^{-1}} + \frac{1}{\tau} = \frac{1}{10\,\text{min}} + \frac{1}{60\,\text{min}} = 0.117\,\text{min}^{-1}$$

$$t_{\text{char,overall}} = 8.57\,\text{min}$$

Consistent with the preceding discussion, the characteristic time for the reactive system is shorter than that of the reaction alone and is substantially shorter than that of the reactor alone. As a result, a reactor that would require several hours to respond fully to a change in the concentration of a tracer ($t_{\text{char}} = 60\,\text{min}$) is essentially at its new steady-state condition in less than half an hour when the concentration of the reactive substance changes.

∎

The fact that the reactivity of the substance accelerates the approach to steady state allows us to ignore non-steady-state conditions in many reactors, even if the influent conditions change fairly frequently. Unless the details of the rapid change are important to an investigator, treating the system as though the new steady state were reached instantly is often satisfactory.

Extent of Conversion in Nonideal Reactors Under Non-Steady-State Conditions

To extend the preceding analysis to a nonideal reactor that is modeled as N CFSTRs in series, we return back to the result for the single CFSTR case, as characterized in Equation 4-58b. Define $c_{\text{ss},j}$ as the steady-state concentration in the reactor under condition j, where $j=1$ and $j=2$ characterize the conditions before and after the step change, respectively. Based on Equation 4-4, $c_{\text{ss},j}$ is given as

$$c_{\text{ss},j} = \frac{c_{\text{in},j}}{1+k_1\tau} \quad (4\text{-}61)$$

Substituting the appropriate form of Equation 4-61 into the preexponential terms in Equation 4-58b, subtracting $c_{\text{ss},1}$ from both sides, and rearranging, we obtain

$$c(t) = c_{\text{ss},1}\exp\left[-(1+k_1\tau)\frac{t}{\tau}\right]$$
$$+ c_{\text{ss},2}\left\{1-\exp\left[-(1+k_1\tau)\frac{t}{\tau}\right]\right\} \quad (4\text{-}62)$$

$$c(t) - c_{\text{ss},1} = c_{\text{ss},1}\exp\left[-(1+k_1\tau)\frac{t}{\tau}\right] - c_{\text{ss},1}$$
$$+ c_{\text{ss},2}\left\{1-\exp\left[-(1+k_1\tau)\frac{t}{\tau}\right]\right\} \quad (4\text{-}63)$$

$$c(t) - c_{\text{ss},1} = -c_{\text{ss},1}\left\{1 - \exp\left[-(1 + k_1\tau)\frac{t}{\tau}\right]\right\}$$
$$+ c_{\text{ss},2}\left\{1 - \exp\left[-(1 + k_1\tau)\frac{t}{\tau}\right]\right\} \quad (4\text{-}64)$$

$$\frac{c(t) - c_{\text{ss},1}}{c_{\text{ss},2} - c_{\text{ss},1}} = 1 - \exp\left[-(1 + k_1\tau)\frac{t}{\tau}\right] \quad (4\text{-}65)$$

The left side of Equation 4-65 represents the fraction of the ultimate change that has occurred at time t; that is, it is a normalized response function for this reactive system with flow. Since the steady-state concentrations under the two influent conditions can be calculated from Equation 4-61, this equation gives a direct expression of the progress of the system, on a scale from zero to one, in response to the step change. In Chapter 2, we considered the normalized response to a step change for a nonreactive substance, and we denote this function as $F(t)$. Here, we denote the normalized response to a step change in influent concentration in the reactive system the symbol, $F_R(t)$; that is,

$$F_R(t) \equiv \frac{c(t) - c_{\text{ss},1}}{c_{\text{ss},2} - c_{\text{ss},1}} \quad (4\text{-}66)$$

Using this terminology, the results for the single CFSTR with a first-order reaction can be expressed as

$$F_{R,\text{CFSTR}}(t) = 1 - \exp\left[-(1 + k_1\tau)\frac{t}{\tau}\right] \quad (4\text{-}67)$$

and the results for the PFR (with any reaction) can be expressed as

$$F_{R,\text{PFR}}(t) = \begin{cases} 0 & \text{for} \quad t < \tau_{\text{PFR}} \\ 1 & \text{for} \quad t \geq \tau_{\text{PFR}} \end{cases} \quad (4\text{-}68a)$$

or, more succinctly

$$F_{R,\text{PFR}}(t) = H_{\tau_{\text{PFR}}}(t) \quad (4\text{-}68b)$$

Note, that for both the CFSTR and the PFR, $F_R(t) = F(t)$ if the influent substance is nonreactive ($k_1 = 0$), as we would expect.

In Chapter 2, nonideal flow was described as any hydraulic behavior between the two ideal extremes of a CFSTR and a PFR. It seems reasonable to expect, therefore, that the response to a step change in influent concentration in a nonideal reactor with a first-order reaction occurring would be between the results given in Equations 4-67 and 4-68.

If the nonideal flow is modeled as N CFSTRs in series, the results can be obtained via Laplace transforms of the appropriate mass balances (i.e., a mass balance at nonsteady state for each of the reactors). Only the results are shown below, but the mathematics is essentially identical to that used for the nonsteady response to a step input of tracer. The only difference in the two derivations is in a constant that reflects the hydraulic detention time alone for the tracer system and both the reaction rate constant and the detention time for the reactive system; in other words, the difference is only in the characteristic times for the two systems.

For a reactor with total detention time, τ, modeled as two CFSTRs in series each with detention time $\tau/2$, the result is as follows:

$$F_{R,N=2}(t) = \frac{c(t) - c_{\text{ss},1}}{c_{\text{ss},2} - c_{\text{ss},1}}$$
$$= 1 - \exp\left[-\left(1 + \frac{k_1\tau}{2}\right)\frac{2t}{\tau}\right]$$
$$- \left(1 + \frac{k_1\tau}{2}\right)\frac{2t}{\tau}\exp\left[-\left(1 + \frac{k_1\tau}{2}\right)\frac{2t}{\tau}\right] \quad (4\text{-}69)$$

For a single reactor modeled as N CFSTRs in series, the result can be extended as follows:

$$\frac{c(t) - c_{\text{ss},1}}{c_{\text{ss},2} - c_{\text{ss},1}} = 1 - \exp\left[-\left(1 + k_1\frac{\tau}{N}\right)\frac{Nt}{\tau}\right]$$
$$\times \sum_{i=1}^{N}\left\{\frac{1}{(i-1)!}\left[\left(1 + \frac{k_1\tau}{N}\right)\frac{Nt}{\tau}\right]^{i-1}\right\} \quad (4\text{-}70)$$

where, according to Equation 4-30, $c_{\text{ss},j}$ in both Equations 4-69 and 4-70 can be computed as $c_{\text{ss},j} = c_{\text{in},j}/(1 + (k_1\tau/N))^N$.

An example showing the difference between a single CFSTR and a reactor that can be modeled as two CFSTRs in series is shown in Figure 4-16. For the single CFSTR ($N = 1$), consistent with the results obtained in the preceding section, the reactive system responds to a change in the input concentration much more rapidly than how the tracer does. It is interesting to note, however, that when $N = 2$, the system does not react as quickly as a single CFSTR. Further analysis indicates that this trend continues if N is increased further (not shown). At short times after the step change in influent concentration, the response of the reactive system slows as N increases. The opposite is true at long times; that is, as for the tracer curves, the lines for the single CFSTR and the N CFSTRs in series cross in the upper part of the curve. In the limit of $N = \infty$, the reactor behaves like a PFR, just as for the tracer; as indicated earlier, the characteristic time for a PFR is unaffected by the reaction because no parcel of fluid interacts with any other.

Consider next a CFSTR in which an abrupt change in the influent flow rate occurs. We begin with the assumption that this change is immediately mirrored by a comparable change in the effluent flow rate, so the detention time changes

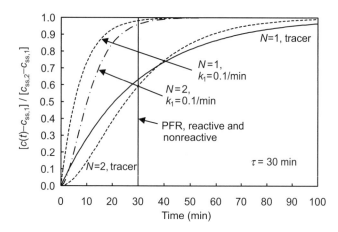

FIGURE 4-16. Normalized response of reactors to a step change in input concentration of a nonreactive (tracer) and a reactive (first-order) species. Reactors shown are a single CFSTR ($N=1$), a PFR, and a single tank with nonideal flow modeled as two CFSTRs in series ($N=2$). (Recall that, for a tracer, $F_R(t) = F(t)$.)

immediately to the value associated with the new flow rate.[9] The mass balance after the change in flow rate is identical to that given in Equation 4-54, with the provision that the value of Q to be used in the mass balance is the flow rate after the step change. The solution is therefore the same as in Equation 4-58.

If it is specified that the effluent value at the time of the change was the previous steady-state value, $c/(1+k_1\tau_1)$ can be substituted for $c(0)$, but it is necessary to differentiate between the old and new detention times (τ_1 and τ_2, respectively), as follows:

$$c_2(t) = \frac{c}{1+k_1\tau_1}\exp\left[-(1+k_1\tau_2)\frac{t}{\tau_2}\right]$$
$$+ \frac{c}{1+k_1\tau_2}\left\{1 - \exp\left[-(1+k_1\tau_2)\frac{t}{\tau_2}\right]\right\} \quad (4\text{-}71)$$

4.6 SUMMARY

This chapter completes the first section of this book, and it is useful to consider where we have been and where we are going at this juncture. In Chapter 1, the fundamental mass balance equation was developed in detail. The mass balance accounts for the fact that material can be transported into or out of a control volume by advection, diffusion, or dispersion, and that it can be generated or destroyed within the control volume by reaction. In Chapter 2, the hydraulic characteristics of continuous flow reactors were investigated by considering nonreactive substances. The results for ideal and nonideal reactors were described in terms of the residence time distributions. In this analysis, the overall mass balance was simplified because the reactive terms were considered zero. In Chapter 3, we considered reactions in batch reactors, in which case the mass balance is simplified because the transport terms are zero. The rate expressions for several types of reactions commonly encountered in environmental engineering were integrated, and approaches for analyzing experimental data to determine the form of the rate expression and the effects of temperature on reaction rates were elucidated. Finally, in this fourth chapter, both flow and reactions were considered simultaneously.

For most of the analyses presented in this chapter, the mass balance was simplified by considering steady-state conditions, again making one term (the storage term) in the mass balance equation zero. A wide variety of reaction rate expressions were considered in conjunction with both ideal and nonideal flow patterns to explore the expected extent of reaction for many situations that occur in environmental engineering. In addition to the results for specific reactor/reaction combinations, the presentation describes various approaches that might be used to predict the extent of reaction for other scenarios.

The key conclusion of these explorations is that accurate prediction of the extent of conversion requires an integrated analysis of the hydraulic characteristics of the reactor with the reaction rate expression. In the case of ideal PFRs or CFSTRs, the effluent concentration can often be predicted by an analytical solution for the corresponding mass balance, incorporating the appropriate reaction rate expression. For the most common types of reaction—those for which the rate of reaction increases with reactant concentration—PFRs are always more efficient than CFSTRs. In this case, better efficiency is evidenced by a greater extent of reaction for a given hydraulic residence time or, equivalently, the same extent of reaction being achieved with a shorter hydraulic residence time. Conceptually, this result can be explained by the fact that the concentration in a PFR changes gradually from c_{in} to c_{out}, whereas in a CFSTR it changes abruptly from c_{in} to c_{out} as soon as the influent enters the reactor. This result can be extended to nonideal reactors; the more intense and sooner the mixing, the less efficient the reactor is, for a reaction whose rate increases with concentration.

If the reactor is not an ideal PFR or CFSTR, but its hydraulic characteristics have been characterized in tracer tests, then the effluent concentration for a given influent concentration can be predicted if certain constraints are met. In particular, if the reaction rate expression is linear or if

---

[9] Note that this assumption is not always valid. In particular, in some reactors, the effluent flow rate is proportional to the volume of water above the weir level. In such cases, the volume of water in the reactor changes in response to a change in the influent flow rate, and a lag ensues before the new steady-state volume is attained.

mixing in the reactor can be represented as segregated flow, then the effluent concentration can be evaluated using Equation 4-44. In that case, the $E(t)$ function and the rate expression can be determined using the approaches described in Chapters 2 and 3, respectively, and the integration can be carried out either analytically or numerically. If the reactor hydraulics does not conform to segregated flow, the extent of reaction might be greater than, less than, or the same as for segregated flow, depending on the reaction order. Early mixing increases the extent of reaction for reaction orders less than zero, decreases the extent of reaction for reaction orders greater than one, and has no effect on the extent of reaction if the reaction order is zero or one; for reaction orders between zero and one, early mixing could either increase or decrease the extent of reaction, depending on system-specific conditions. Regardless of the timing of the mixing, a large proportion of the total reactant concentration in the effluent is often attributable to a fairly small proportion of the flow – the flow that exits after residing only a short time in the reactor. This result emphasizes the importance of minimizing short-circuiting in the reactor.

This chapter also characterizes the responses of a few reactive systems under nonsteady conditions; in these cases all the terms of the mass balance are relevant and included. The analysis of these systems is mathematically complex, so only a small number of cases involving idealized reactor flow patterns and simple reaction rate expressions were explored, but the results have some ramifications that go beyond these cases. The key result is that the characteristic time for the overall reactive system is shorter than either that of the reactor alone or the reaction alone. As a result, in reactive systems in which high degrees of removal are obtained, steady state is achieved quite quickly, thereby diminishing the need to consider unsteady conditions in many cases.

In subsequent chapters, many of the principles and results that are considered in this first part of the book are applied to specific physical and chemical processes that are used to treat water and wastewater.

REFERENCES

Barbash, J. E., and Reinhard, M. (1989) Abiotic dehalogenation of 1,2-dichloroethane and 1,2-dibromomethane in aqueous solution containing hydrogen sulfide, *Env. Sci. Technol*, 23 (11), 1349–1358.

Ducoste, K. H., Carlson, K. H., and Bellamy, W. D. (2001) The integrated disinfection design framework approach to reactor hydraulics characterization. *J. Water Supply: Res. Technol.*, AQUA, 50 (4), 245.

Federal Register (1989) National Primary Drinking Water Regulations: Filtration, Disinfection, Turbidity, *Giardia lamblia*, Viruses, *Legionella*, and Heterotrophic Bacteria. *Fed. Reg.* 54, 124, 27486 (June 29, 1989).

Gould, J. P. (1982) The kinetics of hexavalent chromium reduction by metallic iron, *Water Res.*, 16 (6), 871–877.

Lawler, D. F., and Singer, P. C. (1993) Analyzing disinfection kinetics and reactor design: a conceptual approach versus the SWTR. *JAWWA*, 85 (11), 67–76.

Levenspiel, O. (1999) *Chemical Reaction Engineering*, 3rd edn., John Wiley and Sons, New York.

Nauman, E. B., and Buffham, B. A. (1983) *Mixing in Continuous Flow Systems*, John Wiley and Sons, New York.

Pankow, J. F., and Morgan, J. F. (1981) Kinetics for the aquatic environment, *Env. Sci. Technol.*, 15 (10), 1155–1164.

Zwietering, T. N. (1959) The degree of mixing in continuous flow systems. *Chem. Eng. Sci.*, 11 (1), 1–15.

PROBLEMS

4-1. A pollutant has been spilled during a transfer operation and is slowly migrating through the soil toward a nearby stream. The flow rate toward the stream is 15 L/h, and the time of travel from the point of the spill to the stream is 1 day. The concentration of pollutant in the spilled water was 50 mg/L.

(a) The pollutant decays at a rate given by the expression: $r = -(0.3\,\text{mg/L h}) - (0.02\,c/\text{h})$, where c is the concentration in mg/L. Assuming that the flow from the point of the spill to the stream can be modeled as plug flow, determine the pollutant's concentration at the point where the spill enters the stream.

(b) A sample of the spill was collected immediately after the spill occurred and taken to a laboratory for analysis. In one batch test, the solution was put into a well-mixed beaker for one day and then analyzed. Would you expect the pollutant concentration in the beaker at the end of one day to be more, less, or the same as that entering the stream? Explain in one or two sentences.

4-2. A solution containing the pollutant described in Problem 4-1 has been discharged into a well-mixed lake for a long enough time that the lake has reached steady state. If the concentration of the pollutant exiting the lake is 0.1 mg/L and the residence time in the lake is 336 h (2 weeks), what is the concentration of the pollutant entering the lake?

4-3. A reaction A→B is known to proceed according to the rate expression

$$r_A = -\frac{k_1 c_A}{k_2 + c_A}$$

A pilot-scale CFSTR has been operated at steady state at several different flow rates to determine the coefficients k_1 and k_2. The reactor has a volume of 2.5 L, and all experiments were performed with an influent concentration $c_{A,in}$ of 100 mg/L. The effluent values of c_A are shown in the following table.

Flow Rate, L/h	Effluent Concentration c_A, mg/L
0.39	5.8
0.78	17.9
1.56	46.0
3.13	71.9
6.25	85.7
12.5	92.8
25	96.2
100	99.0

(a) Determine the reaction rate constants k_1 and k_2.

(b) Find the volume of a CFSTR that would be required to treat a flow rate of $10 \, m^3/h$ with an influent concentration, $c_{A,in}$, of 100 mg/L, if the steady-state effluent concentration is to be 1.0 mg/L.

(c) Find the volume of a PFR that would accomplish the same objective as the CFSTR does in (b).

4-4. A reaction $2A \rightarrow B$ is second order with respect to A. Batch experiments have shown that, for an initial solution containing $c_A = 3$ mg/L, the concentration of A in a system can be reduced by 99% in 25 min.

(a) What is the rate constant k_2?

(b) What would be the volume of a PFR to treat 100 L/min of this solution while meeting an effluent standard of 0.1 mg/L?

(c) What would be the volume of a CFSTR to meet the same effluent standard as in (b)?

4-5. It has been reported that the hydrolysis of iron can be modeled by a first-order rate equation of the type $r_{Fe^{3+}} = -kc_{Fe^{3+}}$. In a batch experiment at pH 3.65, 20% of the initial Fe^{3+} is hydrolyzed after 15 s. You may assume that the back reaction is negligible.

(a) How long would it be before 99% of the Fe was hydrolyzed in the batch system? (Note: it is not reasonable to assume a value of $c_{Fe^{3+}}$ at $t = 0$ in the absence of data, unless you can show that your result would be valid for other initial values of $c_{Fe^{3+}}$ as well.)

(b) What residence time would be required to achieve 99% hydrolysis in a CFSTR at steady state?

(c) What about in two equal-sized CFSTRs in series?

4-6. A first-order reaction causes cyanide (CN) to be destroyed at a rate given by $r_{CN} = -(0.8/h)\, c_{CN}$. The reaction is occurring in a reactor with a theoretical mean hydraulic residence time of 8.33 h, but because of some dead space in the reactor, the effective mean residence time is only 6.7 h. The active portion of the reactor is acting as a CFSTR. For an influent cyanide concentration of 18 mg/L, compute the expected steady-state effluent concentrations for the following scenarios:

(a) The reactor as it is now operating.

(b) The reactor if additional mixers are used so that it operates as a CFSTR with no dead space.

(c) The reactor described in (b) if a baffle is inserted in the middle, and the flow enters first one side and then the other, so that the overall reactor operates as two ideal CFSTRs in series.

(d) The reactor modified as in (c), but with the influent flow split into two equal portions, so that the reactor behaves as two ideal CFSTRs in parallel.

(e) The reactor modified so that it behaves as an ideal PFR.

4-7. Low-molecular-weight halogenated organic compounds are being found in trace concentrations in many water systems due to the uncontrolled release of these substances from industrial processes. Barbash and Reinhard (1989) studied the abiotic dehalogenation of some of these compounds and found that, under some circumstances, the abiotic reaction can be significant. They report that the debromination of 1,2 dibromoethane (EDB) is of first order.

An industry has a small process line (2.5 L/min) which contains 30 mg/L EDB and 3 mg/L sodium (which is nonreactive). The stream is treated in a 3000-L CFSTR under conditions where the first-order debromination rate constant is $7 \, d^{-1}$. The reactor has been operating at steady state, but a process upset has occurred, causing the influent concentrations to jump to 100 (mg/L) EDB and 200 (mg/L) sodium, where they remain for several days.

(a) What is the effluent EDB concentration prior to the upset?

(b) Write the mass balance for sodium and determine the effluent concentration 6 h after the upset.

(c) Repeat (b) for the EDB.

(d) Does one component respond more quickly than the other to the change in influent? Address this question by determining the ratio of the change in concentration that has occurred in 6 h to the expected, ultimate change (once the system has reached a new steady state) for each chemical.

4-8. A reactor has been constructed and is intended to operate as a PFR with a residence time of 4 h. It has been treating 10^{-3} M of a contaminant A with a steady-state conversion efficiency of 98% by addition of 0.03M of a second compound B. The reaction has stoichiometry $A + B \rightarrow C$ and is first order with respect to each reactant (second order overall), with a rate constant of 60 $(mol/L)^{-1}h^{-1}$.

(a) Assuming that, in reality, the reactor hydraulics includes some dispersion, estimate the dispersion number.

(b) What would be the conversion efficiency if the reactor were an ideal PFR?

4-9. Gould (1982) reported that Cr(VI), a carcinogenic form of chromium often found in industrial wastes, can be converted into its nontoxic form, Cr(III), by reaction with metallic iron, $Fe^\circ(s)$. The iron is oxidized to Fe(III). He found good agreement with the rate expression

$$r_{Cr(VI)} = -k c_{Cr(VI)}^{0.5} c_{H^+}^{0.5} A_{Fe(s)}$$

where $A_{Fe(s)}$ is the surface area of the iron in cm^2/L. The reported value of k is 5.45×10^{-5} L cm^{-2} min^{-1}. The experiments described in the paper were conducted with 10^{-4} to 4×10^{-3} M Cr(VI) initially in solution, and pH between 1.0 and 5.0.

(a) A plating waste containing 1.5×10^{-3} M Cr(VI)$_{tot}$ is to be treated by this process in a CFSTR with a detention time of 2 hours. The pH in the tank is to be held at 2.0. Assume that iron can be added as strands with a 100-μm diameter, such as steel wool, and that the density of the iron is 7.8 g/cm^3. Determine whether this tank could accomplish 99% Cr(VI) removal with a reasonable amount of $Fe^\circ(s)$ added.

(b) In a discussion of alternative designs, an engineer suggests doubling the iron dosage to reduce the required residence time. However, another one objects, arguing that, in light of the costs of both the iron filings and their subsequent disposal, the one-time cost of constructing a larger reactor is worthwhile. Comment on this discussion and, if you can, suggest a design that satisfies both people.

4-10. Tripolyphosphate ($P_3O_{10}^{5-}$) molecules hydrolyze in a two-step process, generating pyrophosphate ($P_2O_7^{4-}$) and orthophosphate (PO_4^{3-}) according to the following reactions:

$$P_3O_{10}^{5-} + H_2O \leftrightarrow P_2O_7^{4-} + PO_4^{3-} + 2H^+ \quad \text{Reaction (1)}$$

$$P_2O_7^{4-} + H_2O \leftrightarrow PO_4^{3-} + 2H^+ \quad \text{Reaction (2)}$$

The first-order rate constants for the above two irreversible reactions are $2.3 \times 10^{-6} s^{-1}$ and $1.0 \times 10^{-6} s^{-1}$, respectively. In a batch system, the functions which describe the molar concentrations of the three P species over time are (Pankow and Morgan, 1981)

$$c_{P_3O_{10}}(t) = c_{P_3O_{10}}(0)\exp(-k_1 t)$$

$$c_{P_2O_7}(t) = \frac{k_1 c_{P_3O_{10}}(0)}{k_2 - k_1}[\exp(-k_1 t) - \exp(-k_2 t)]$$

$$c_{PO_4}(t) = -c_{P_3O_{10}}(0)\left[\exp(-k_1 t) \right.$$
$$\left. + \frac{2}{k_2 - k_1}(k_2 \exp(-k_1 t) + M k_1 \exp(-k_2 t))\right]$$
$$+ c_{P_3O_{10}}(0)\left[1 + \frac{2}{k_2 - k_1}(k_2 + M k_1)\right] + c_{PO_4}(0)$$

where $M = \dfrac{(k_2 - k_1)c_{P_2O_7}(0)}{k_1 c_{P_3O_{10}}(0)} - 1$.

The coefficient 2 in the equation for c_{PO_4} reflects the formation of two PO_4 molecules for each P_2O_7 molecule that hydrolyzes.

(a) What would be the steady-state effluent concentrations of the three phosphate species in an ideal CFSTR treatment system with a hydraulic residence time of 15 d, if the influent contained P_3O_{10} at a concentration of 15 mg/L as P, and no P_2O_7 or PO_4? Recall that, although one mole of P_3O_{10} decays to form one mole each of P_2O_7 and PO_4, these 1:1 molar ratios do not apply if the mass balance is written in terms of the number of moles of P being converted from one species into another in the reaction.

(b) A tracer study suggests that the entire reactor in (a) does not behave as a CFSTR. Rather, a portion of the reactor behaves as a CFSTR, but 20% of the fluid short-circuits through the reactor with a hydraulic residence time of 1 day, and 15% of the reactor volume is dead space. The short-circuiting fluid behaves as though it passes through a PFR. What are the expected, steady-state concentrations of the P-containing species in the reactor effluent?

(c) Determine the steady-state concentrations (mg/L) of the three P-containing species in the effluent from two, equal-sized CFSTRs in series, each with $\tau = 7.5$ days.

(d) Plot the concentrations of the three P species as a function of residence time in a PFR for $0 < \tau < 30$ d. Compare the result for a PFR with $\tau = 15$ d with the results of (a) and (c).

4-11. Consider the two-compartment reactor described in Problem 2-9. If a reactant that undergoes a second-order reaction with a rate constant $k_2 = 1000 \text{ L/mol min}$ enters this reactor at an influent concentration of 10^{-4} M, compute the expected steady-state concentration in the effluent from the real reactor. Predict also what the concentration would be if additional mixing was induced in the system so that the reactor behaved as a single, ideal CFSTR.

4-12. The following reaction rate expression was given in Chapter 3 for the oxidation of sulfide:

$$-r_{\text{TOTS}} = k(\text{TOTS})^{1.34} c_{O_2}^{0.56}$$

where TOTS is the total sulfur in the $-$II oxidation state; that is, $\text{TOTS} = c_{H_2S} + c_{HS^-} + c_{S^{2-}}$. At pH 7.5, k is 11.97 $(\text{L/mol})^{-0.9} \text{ h}^{-1}$. Assume that this reaction is occurring in the reactor that was described in several examples in Chapters 2 and 4 (e.g., Example 2-1), but a catalyst has been added so that the rate constant is increased by a factor of 75. The influent concentrations of S(II) and O_2 are 2×10^{-4} M and 1.5×10^{-4} M, respectively, and the water is well-buffered at pH 7.5. Assuming that the reactor has segregated flow, what would be the expected effluent concentration of TOTS(II)?

4-13. As indicated in Problem 3-7, the disinfection of microorganisms is often described by the "Chick–Watson law." This law, in its simplest form, can be written as follows:

$$r_X = -k c_X c_D$$

where c_X is the concentration of viable microorganisms, c_D is the concentration of disinfectant (e.g., chlorine), and k is a constant.

(a) In a batch reactor with a constant disinfectant concentration of 1.5 mg/L, 99% of the microorganisms are inactivated (killed) after 15 min. Find the value of k. (Note: this is the same question as (a) of Problem 3-7. That problem dealt with a batch system; the remaining questions here ask about the same reaction in systems with continuous flow.)

(b) Find the detention time required in the following reactors to achieve 99% inactivation of the microorganisms at steady state.
   (i) PFR, if the chlorine concentration is 1.5 mg/L throughout the reactor.
   (ii) CFSTR in which the chlorine concentration is 1.5 mg/L.
   (iii) PFR, if the chlorine concentration is 1.0 mg/L throughout the reactor.
   (iv) CFSTR in which the chlorine concentration is 1.0 mg/L.

(c) Recall that, for disinfection, the EPA treats all reactors as if they are PFRs with a detention time equal to T_{10} (the time when $F(t)$ is 0.10). How much disinfection credit would EPA give for the reactor described in b(i)?

(d) How much credit would EPA give for the reactor described in b(ii)?

4-14. A disinfection reaction characterized by the rate expression $r_N = -k_1 c_X$ is proceeding in the reactor described in Problem 2-7, where c_X is the concentration of viable microorganisms remaining at time t, and $k_1 = 0.2 \text{ min}^{-1}$.

(a) For a system receiving an influent containing 10^6 organisms per 100 mL, determine the expected concentration of viable organisms exiting from the reactor according to each model representation of the hydraulic behavior in the reactor (i.e., for the dispersion model with closed boundaries, the CFSTRs-in-series model, and the seven-PFRs-in-parallel model), as well as for a single, ideal PFR and a single, ideal CFSTR of comparable size. Also, make an estimate of percent survival based on the $E(t)$ curve.

(b) Based on the estimate of percent survival from the $E(t)$ curve, develop a plot showing time on the abscissa and, on the ordinate, the fraction of organisms in the effluent that has resided for an amount of time $<t$ in the reactor. On the same graph, plot $F(t)$ versus t. Comment on the difference between the two curves.

(c) Assume that you had been hired to conduct this study (both the tracer tests and the prediction/analysis of disinfection efficiency). Discuss in a few sentences which model you would use in presenting the results to your client and any recommendations you would make. Can you recommend a different, but still simple, reactor model that might be better than the ones tested?

4-15. An engineer has determined that the flow patterns in a continuous flow reactor at a water treatment plant can be reasonably represented by a CFSTR with a 15-min hydraulic detention time, followed (in series) by a PFR with a 5-min detention time. This interpretation is based on the results of a step input tracer test. Hypochlorous acid (HOCl) is being dosed steadily into the influent; by convention, the disinfectant dose and the concentration of disinfectant remaining at any time thereafter are represented "as Cl_2". In these units, the dose is 8 mg Cl_2/L.

When the influent is chlorinated in a separate, batch test, 1.5 mg/L Cl_2 is consumed before the first data point is analyzed, which is at $t = 30$ s. For the next 30 min, the Cl_2 concentration in solution decays approximately according to the rate expression: $r_{Cl_2} = -0.02 - 0.15 c_{Cl_2}^{0.5}$, where r_{Cl_2} is in mg Cl_2/(L min) and c_{Cl_2} is in mg Cl_2/L.

(a) Sketch the effluent tracer curve that you think led the engineer to infer the CFSTR–PFR reactor arrangement.

(b) Suggest a reason for the rapid disappearance of the first 1.5 mg Cl/L in the batch test.

(c) What is the value of T_{10} that would be used to compute the CT product for this system?

(d) What is the value of C that would be used to compute the CT product for this system?

(e) The plant owners are considering installing a baffle that would convert the CFSTR portion of the system into two equal-sized CFSTRs in series. Would such a change increase, decrease, or have no effect on the values of C and T that would be used to compute the CT product? Explain your reasoning in a few sentences.

4-16. One of the advantages of CFSTRs over PFRs is the ability of the former to smooth out fluctuations in influent quality. A CFSTR and a PFR are being considered as potential reactors to reduce the concentration of a pollutant X from 10^{-3} M to 5×10^{-6} M. The pollutant undergoes a first-order reaction with rate constant 0.04 min^{-1}. The waste stream also contains a nonreactive tracer Y at concentration 5×10^{-3} M. Predict the effect on the effluent concentrations of both X and Y of the following two types of upset in each type of reactor:

(a) The influent concentration triples for 1 h and then returns to its normal value,

(b) The flow rate of the waste stream triples for half an hour and then returns to its normal value.

Assume the reactor is at steady state prior to each upset and that the volume of water in the reactor is constant. Determine the expected effluent concentrations of both X and Y from the time of the upset until the system has returned to a near-steady-state condition.

4-17. Answer Problem 4-16 for an upset in which the influent concentration of X triples and the flow rate of the waste stream is simultaneously reduced to one-third of its initial value. The upset lasts for half an hour, after which both parameters return to their pre-upset values. (This situation simulates a transient in which an input stream that represents two-third of the flow to the reactor but contains none of the contaminant is briefly shut off.) Assume that the reactor is at steady state prior to the upset and that the reactor volume is fixed. Show the expected effluent concentration of X from the time of the upset until the system has returned to a near-steady-state condition in each reactor.

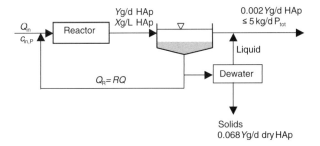

FIGURE 4-Pr18. Hypothetical flow diagram for a phosphorus precipitation and solid separation process.

Is the total mass of reactant that exits the reactors greater than, less than, or the same as would have exited in the absence of the upset?

4-18. The solid recycling reactor system shown schematically in Figure 4-Pr18 is to be used to remove phosphate from an industrial waste stream by precipitation of hydroxyapatite, $Ca_5(PO_4)_3OH(s)$ (designated HAp in the figure). As shown in the diagram, the effluent from the reactor enters a sedimentation basin, from which most of the sludge is returned to the reactor. The remaining sludge is dewatered and landfilled. A small dose of HAp was added to the reactor to initiate the reaction, but the system currently operates at steady state without HAp addition.

The influent contains 16.7 mg/L of soluble PO_4-P and no hydroxyapatite, and the flow rate is 3×10^6 L/d. The discharge permit requires that the *total* P in the effluent be not more than 5 kg/d. Of the solids entering the settling basin, 0.2% escape with the effluent, 6.8% are separated, dried, and sent to a landfill, and the remainder is recycled to the main reactor. The recycle ratio, R, is 0.2. The system is operating at steady state.

For the particular Ca^{2+} concentration, alkalinity, and pH in the reactor, the rate expression for removal of phosphate from solution is given as

$$r_{PO_4\text{-}P} = -7.75 \times 10^{-4} c_P^{1.7} c_{HAp}$$

where c_P and c_{HAp} are the concentrations of soluble phosphorus and solid hydroxyapatite in mg/L, respectively, and r is in mg P/L h.

(a) Determine the conversion efficiency of soluble P to HAp that must be achieved to meet the discharge permit requirement. Do not assume any particular mixing pattern in the reactor.

(b) What are the concentrations of PO_4-P and HAp in each flow stream?

(c) Determine the CFSTR volume that would be required to meet the treatment objective. Assume that all reactions occur in the reactor, and not in the settling basin.

(d) Repeat (d) if the reactor is a PFR.

PART II

REMOVAL OF DISSOLVED CONSTITUENTS FROM WATER

5

GAS TRANSFER FUNDAMENTALS

5.1 Introduction
5.2 Types of engineered gas transfer systems
5.3 Henry's law and gas/liquid equilibrium
5.4 Relating changes in the gas and liquid phases
5.5 Mechanistic models for gas transfer
5.6 The overall gas transfer rate coefficient, K_L
5.7 Evaluating k_L, k_G, K_L, and a: Effects of hydrodynamic and other operating conditions
5.8 Summary
   Appendix 5A. Conventions used for concentrations and activity coefficients when computing Henry's constants
   Appendix 5B. Derivation of the gas transfer rate expression for volatile species that undergo rapid acid/base reactions
   References
   Problems

5.1 INTRODUCTION

Importance of Gas Transfer in Environmental Engineering

Transfer of molecules between gaseous and aqueous phases is important in numerous environmental systems, both natural and engineered. Gas transfer into solution is commonly referred to as *gas absorption*, and transfer out of solution is referred to as *gas stripping*; however, the fundamental principles controlling both processes are identical.

Engineered gas transfer processes include absorption of oxygen and/or ozone into solutions to facilitate biological or chemical oxidation, absorption of sulfur dioxide and other pollutants from the gases generated in fossil fuel combustion processes, and stripping of volatile organic compounds (VOCs) from drinking water, wastewater, or contaminated groundwater by exposing the solutions to air.[1] In biological treatment systems, although absorption of oxygen into the water is usually the primary gas transfer goal, stripping of carbon dioxide, ammonia, and/or volatile trace organic compounds occurs simultaneously. And, in natural systems, gas transfer plays a central role in the global carbon and nitrogen cycles, helps determine the chemical composition of cloud droplets and hence of precipitation, and determines the rate and extent of reaeration of (i.e., absorption of oxygen into) surface waters that have been contaminated by oxygen-depleting wastes.

Overview of Gas/Liquid Equilibrium

Gas transfer processes combine a chemical reaction (the conversion of a gaseous species into a dissolved species, or vice versa, which occurs precisely at the interface) with physical transport steps (movement of the species between the interface and the bulk fluid phases).[2] The chemical reaction, like all chemical reactions, approaches an equilibrium condition that reflects the relative stabilities of the

---

[1] Any species that can exist in the gas phase is considered volatile, and increasing tendency to enter the gas phase corresponds to increasing volatility. The solubility of a gas refers to its tendency to enter solution. Thus, the phrases *increasing volatility* and *decreasing solubility* are synonymous.

[2] Some authors argue that, because the volatile molecule has the same chemical composition in the gaseous and aqueous phases, gas transfer should be considered as a strictly physical process. The counter-argument is that the molecule bonds, albeit weakly, to water molecules in solution, so the process of transferring between phases does involve a rearrangement of chemical bonds and hence can be reasonably viewed as a chemical reaction. Overall, it is probably best to view this process, and most others described in this text, as physicochemical and not attempt to identify them as strictly physical or chemical in nature. Nevertheless, for the sake of brevity, and to link gas transfer with other processes where the breaking of chemical bonds is more explicit, we refer to it as a chemical reaction throughout this chapter.

Water Quality Engineering: Physical/Chemical Treatment Processes, First Edition. By Mark M. Benjamin, Desmond F. Lawler.
© 2013 John Wiley & Sons, Inc. Published 2013 by John Wiley & Sons, Inc.

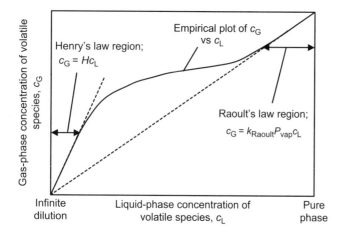

FIGURE 5-1. Gas–liquid equilibrium, demonstrating the Henry's law and Raoult's law regions. The designations of the regions are based on the assumption that the mixture contains only water and the volatile species.

reactants and products. Typically, this equilibrium condition is characterized by a relationship between the liquid- and gas-phase concentrations of the volatile species (c_L and c_G, respectively) that is qualitatively similar to that shown in Figure 5-1.[3]

The linear portions of the plot at the two extremes of Figure 5-1 are of particular importance. These sections indicate that c_G is directly proportional to c_L in two limiting cases: one that applies when the volatile species is very dilute, and the other that applies when it is the main constituent in the liquid phase. As indicated, these regions of the plot are referred to as the *Henry's law* and *Raoult's law* regions, respectively. The proportionality constant in the Henry's law region is usually designated as H and is referred to as the Henry's law constant (or as Henry's constant), whereas that in the Raoult's law region is the vapor pressure of a pure liquid of the volatile species (P_{vap}); that is

$$\text{Henry's law:} \quad c_G = H c_L \tag{5-1}$$

$$\text{Raoult's law:} \quad c_G = k_{Raoult} P_{vap} c_L \tag{5-2a}$$

$$P_G = P_{vap} x_L \tag{5-2b}$$

where k_{Raoult} is included in Equation 5-2a to ensure consistency between the units on the left and right sides of the equation. (As explained shortly, the units of H assure such consistency in Equation 5-1.) The most common expression for Raoult's law, shown in Equation 5-2b, uses the partial pressure of the volatile constituent in the gas phase as the measure of c_G, in which case the product $k_{Raoult} c_L$ equals the (dimensionless) mole fraction of the constituent in the liquid phase (x_L).

Although the figure shows a continuum between the two limiting cases, the volatile species and water might not be able to form a homogeneous solution at all mixing ratios. That is, at certain mixing ratios, the constituents might split into two different liquid phases—one consisting of water in which a small concentration of the volatile species is dissolved, and the other consisting of the volatile species in which a small concentration of water is dissolved. This scenario is commonly observed in mixtures of water with VOCs such as benzene, toluene, and trichloroethylene. In such cases, a portion of the middle of the curve in Figure 5-1 would not represent attainable conditions.

In the vast majority of environmental engineering applications, the interest is in the Henry's law region of the graph. Indeed, some volatile solutes (e.g., O_2, CO_2, H_2S) never exist as pure liquids at normal temperatures and pressures, in which case the Raoult's law portion of the graph is never relevant. On the other hand, some systems of interest are characterized by the Raoult's law region, such as soils where a nearly pure, nonaqueous-phase liquid (NAPL) might be present as the result of a spill or leakage from a storage tank. In addition, Raoult's law is sometimes used to define "ideal" solute behavior (i.e., the behavior that causes the activity coefficient of the solute to be 1.0) in dilute aqueous solutions of organic solutes.

Equilibrium between the gas and liquid phases is a limiting condition under which no net gas transfer occurs. However, many systems of interest in environmental engineering are not characterized by such equilibrium. If the dissolved concentration of the volatile species, c_L, is larger than the equilibrium value (for the extant value of the gas-phase concentration, c_G), then that species will tend to transfer out of the solution and into the gas. In such a case, the aqueous solution is said to be *supersaturated* with the gas. Correspondingly, if c_L is less than the equilibrium value, the solution is said to be *undersaturated* with respect to the gas, and the volatile species will tend to transfer from the gas into solution. Finally, if c_L is equilibrated with c_G, the solution is said to be *saturated* with the volatile species.

These conditions are shown graphically in Figure 5-2 for a system in the Henry's law region of Figure 5-1. If the conditions in a given system (i.e., the values of c_L and c_G in the system) correspond to a point to the right of the equilibrium line, then $c_L > c_G/H$, and the solution is supersaturated. On the other hand, if the conditions correspond to a point to the left of the line, then $c_L < c_G/H$, and the solution is undersaturated.

---

[3] Note that, in the figure and throughout this chapter, the subscript L is used to identify the liquid phase, but this phase is considered to be an aqueous solution in all cases.

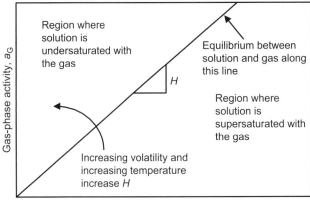

FIGURE 5-2. Schematic illustration of the relationship described by Henry's law, including regions where the solution is under and supersaturated with respect to the gas.

Overview of Transport and Reaction Kinetics in Gas Transfer Processes

The transfer of a molecule between a solution and a gas phase involves several steps. In any real system, these steps proceed simultaneously and, to some extent, in both directions. Nevertheless, it is common to analyze the system in terms of the net, unidirectional movement of the species of interest through a series of three to five conceptually distinct spatial zones. Listed in the order that a molecule traverses them if it is being stripped from solution, these zones include the bulk solution, the interfacial liquid region, the interface itself, the interfacial gas-phase region, and the bulk gas phase. These regions are shown schematically in Figure 5-3, along with a hypothetical concentration profile for the volatile constituent. The interfacial regions are defined as the regions of finite thickness near the interface in which the fluid dynamics are affected significantly by the interface and are therefore different from those in the bulk phases. The interface is defined as the infinitely thin boundary separating the two phases; when a molecule crosses this boundary, it is considered to have been transferred to the other phase.[4]

As we did when considering sequential chemical reactions, we can represent each step in the gas transfer process as contributing some resistance to the overall process, and we can equate the total resistance with the sum of the individual terms. The step with the greatest resistance can be identified as the rate-limiting step, providing us with both an understanding of how the process is operating and some clues about the most effective strategies to alter the transfer rate.

Empirically, the rate of gas transfer in a given system is almost always strongly correlated with the molecular diffusivity. That is, if several volatile chemicals are transferring between a gas phase and a solution in a given system (so that the same hydrodynamics applies to all the transferring species), the relative rates of transfer increase with increasing diffusivity. This result suggests that diffusive transport through one of the interfacial regions is the rate-limiting step. In such a case, transport through the bulk phases and the phase transfer reaction itself (i.e., crossing the infinitely thin boundary between the solution and gas phases) must contribute negligible resistance to the overall process. The implications of these observations are that (1) the concentration gradient in the bulk phases is negligible in most systems of interest and (2) at the interface, the gas transfer reaction proceeds so rapidly that we consider the two phases to be in equilibrium, as characterized by Henry's law. The first implication can be confirmed empirically, and, although the second one cannot be validated directly, it is widely accepted and is incorporated into all modern models of gas transfer. Hence, studies of gas transfer kinetics typically focus on the interfacial layers in both phases.

Incorporating Gas Transfer into Mass Balances

The approach for evaluating the rate of transfer between a solution and a gas is (as always) to write and solve one or more mass balances on the substance of interest. In Chapters 1–4, the right side of the mass balance on a substance i is represented as a sum of three terms: one each for appearance of i in (or its disappearance from) the control volume via advection, via the combination of diffusion and dispersion, and via chemical reaction. The fundamental requirement for the right side of the mass balance equation is that it take into account every process that affects the mass of i stored in the control volume. The separation and grouping of processes

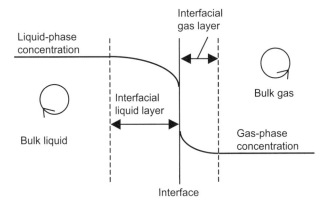

FIGURE 5-3. Schematic representation of the zones through which molecules must pass to transfer between the aqueous and gaseous phases.

---

[4] Schematics such as Figure 5-3 are often shown with linear concentration gradients in both interfacial regions. As shown later in this chapter, linear gradients can develop under certain limiting conditions, but they are not expected to be the norm in most gas transfer systems.

used in Chapters 1–4 is convenient conceptually as well as mathematically in many cases, as demonstrated by numerous examples in those chapters.

However, in systems where gas transfer is important, it is convenient to treat the gas transfer contribution as an individual term on the right side of the mass balance, rather than including it with the terms that we considered in previous chapters. That is, mass balances in such systems are commonly written with four terms on the right side of the equation: one for advection; one for diffusion and dispersion other than that associated with gas transfer; one for chemical reactions other than gas transfer; and one for the combined process of reaction and diffusion/dispersion associated with gas transfer. In the form of a word equation, such a mass balance would be

gas transfer can proceed at vastly different rates in systems with identical chemical composition and temperature but different physical characteristics; that is, $r_{L,gt}$ depends on several physicochemical parameters that can vary widely among systems of interest. This chapter is devoted to developing an understanding of the factors that affect $r_{L,gt}$ and approaches for evaluating this term.

Chapter Overview

In this chapter, each portion of the overview provided earlier is expanded. The major types of engineered gas transfer systems that are in common use are described first, followed by a formal representation of the equilibrium relationship between a given species in the gas phase and the same

$$\begin{pmatrix}\text{Rate of change of}\\ \text{the mass of } i \text{ stored}\\ \text{in the control}\\ \text{volume}\end{pmatrix} = \begin{pmatrix}\text{Net rate at which } i \text{ enters}\\ \text{the control volume by}\\ \text{advection}\end{pmatrix} + \begin{pmatrix}\text{Net rate at which } i \text{ enters the}\\ \text{control volume by diffusion}\\ \text{and/or dispersion, excluding}\\ \text{diffusion and dispersion}\\ \text{associated with gas transfer}\end{pmatrix}$$
$$+ \begin{pmatrix}\text{Net rate at which } i \text{ is}\\ \text{created within the control}\\ \text{volume by chemical}\\ \text{reaction, excluding the gas}\\ \text{transfer reaction at the}\\ \text{interface}\end{pmatrix} + \begin{pmatrix}\text{Net rate at which } i \text{ enters the}\\ \text{control volume by gas transfer}\end{pmatrix} \quad (5\text{-}3)$$

As we will see, when a mass balance is written for the volatile species in the liquid phase, the term characterizing the overall gas transfer process bears a strong functional resemblance to the reaction term. In particular, just as the reaction term can be written as the product $r_L V_L$, the gas transfer term can be usefully represented as the product of the rate of gas transfer per unit volume of solution, $r_{L,gt}$ (mass of i entering the (liquid) control volume by gas transfer per unit volume of solution per unit time), and the volume of the solution, V_L. Furthermore, $r_{L,gt}$ can be expressed as a function that bears many similarities to r_L for a first-order reversible reaction, so that the mass balance can, in theory, be evaluated just like the mass balances for systems where only homogeneous (i.e., single-phase) reactions are occurring.

Although analyzing gas transfer identically to homogeneous reactions is appealing in concept, it is difficult to implement in practice. The difficulty arises because the presence of a second phase makes the physics of gas transfer processes considerably more complex than homogeneous reactions. In particular, whereas the rate constant for homogeneous, first-order reactions depends only on the temperature, gas transfer reactions are characterized by apparent rate constants that depend strongly on the amount of gas/liquid interfacial area and the patterns and intensity of mixing in the system (in both the solution and gas phases). As a result,

species in the liquid phase in dilute aqueous solutions (i.e., Henry's law). The microscopic, interfacial phenomena involved in gas transfer (phenomena occurring at length scales up to a few thousand times the size of the relevant molecules) are then described, leading to the development of a rate expression for gas transfer. This expression is developed first for transport between the interface and the bulk fluid in a single phase, and is then combined with the Henry's law relationship to derive the form of the overall term $r_{L,gt}$. In the final section of the chapter, approaches are presented for calculating parameters that can be used to estimate the gas transfer rate constant for different types of engineered systems. The expression for $r_{L,gt}$ is subsequently used in our exploration of gas transfer at the macroscopic scale (i.e., at the scale of full-sized gas transfer reactors) in Chapter 6.[5]

---

[5] The solutions to mass balance equations for gas transfer systems, applicable to various control volumes and system designs, are presented in a number of textbooks. Most textbooks on environmental engineering processes (e.g., MWH, 2005 and AWWA, 2011) include a discussion of gas transfer explicitly, and many more textbooks deal with mass transport phenomena in general. For example, textbooks by Clark (2009) and Logan (1999) deal specifically with mass transport in environmental systems, whereas those by Cussler (1997) and Sherwood et al. (1975) take a more generic, chemical engineering approach.

5.2 TYPES OF ENGINEERED GAS TRANSFER SYSTEMS

Engineered gas transfer systems can be broadly classified as either gas-in-liquid or liquid-in-gas. The former group includes systems in which discrete bubbles are distributed in a continuous solution phase, whereas in the latter group, droplets of water are exposed to a continuous gas phase. While many systems fit neatly into one of these two categories, others represent a hybrid of the two limiting cases.

The most common gas-in-liquid systems of interest in environmental (water) engineering are those in which gas (usually air, but sometimes pure oxygen, carbon dioxide, or ozone) enters the reactor through diffusers. The diffusers have tiny holes through which the gas is forced (Figure 5-4a), generating small bubbles that rise through the water column and exit when they break through the water surface. As a bubble rises, some molecules are transferred from the bubble to the liquid (absorption) and/or from the liquid to the bubble (stripping). The aeration devices used in household aquaria provide a familiar example of a diffuser system; there, air is pumped through a diffuser stone and, as the bubbles rise, oxygen is transferred from each bubble to the water, maintaining a dissolved oxygen (DO) level high enough to keep the fish alive.

Diffuser systems are used very commonly in conjunction with activated sludge treatment of wastewaters. In this case, the very high rate of oxygen consumption by the aerobic microorganisms necessitates a high flow rate of air (or oxygen), typically supplied by a large number of diffusers distributed over the bottom of the aeration tanks. Pictorial representations of both an empty activated sludge aeration tank, showing the arrangement of diffusers, and a similar tank during operation are shown in Figures 5-4b and 5-4c.

Most air-in-water reactors have well-mixed aqueous phases. However, some (including, for example, many reactors used to transfer ozone into water) utilize counter-current flow, in which the two phases travel in opposing directions (gas up and liquid down), and a significant concentration gradient might exist in the solution phase. Other examples of gas-in-liquid systems encountered in water engineering are recarbonation reactors (which increase the total dissolved carbonate concentration and hence reduce the pH after precipitative water softening) and anaerobic digesters (in which the methane and carbon dioxide that are produced by biological degradation of sludges are partially removed from solution by the formation of gas bubbles). The defining characteristic of gas-in-liquid systems is that the volume of the reactor is taken up almost completely (certainly more than 90% and usually closer to 99%) by the liquid.

Liquid-in-gas systems are also commonly used for both stripping and absorption in environmental engineering practice. The most common example is a tower packed with a media designed to create a large interfacial area between the liquid and gas phases. The space inside such towers is filled with air, liquid, and the packing. The packing pieces can be small (2–8 cm) units that are dumped into the column ("random packing") or larger pieces (1–2 m on a side) that are placed in the column in a more structured manner. On a volumetric basis, the packing accounts for only approximately 5% of the space, and water for even less, so that more than 90% of the space is filled with air. In such towers, water is sprayed by nozzles or allowed to drip through numerous small holes at the top and falls by gravity from one piece of packing to the next, until it reaches a reservoir below the packing and is removed. Gas (invariably air) is blown into the bottom (usually) and travels upward through the packing and out the top.

Groundwater used as a source of drinking water is frequently aerated in this way to increase the oxygen content (and thereby oxidize reduced inorganic compounds such as Fe^{2+} and HS^-) and/or to strip low concentrations of volatile organics out of the water. Similarly designed columns can be used to absorb unwanted components of a gas stream (such as SO_2 from the burning of fossil fuels) into water. Another common liquid-in-gas system is a trickling filter, which (at least in terms of gas transfer) is similar to the packed towers just described but is used for biological treatment of wastewater. Images of a full-scale gas transfer tower and of some plastic packing media are shown in Figure 5-5.

Some gas transfer systems have elements of both gas-in-liquid and liquid-in-gas systems. For example, surface aerators (either spray or brush aerators) that are used in many activated sludge systems operate by spraying droplets of water into the air or by carrying a thin film of water into the air attached to the bristles of rotating brushes. This part of the system can be characterized as liquid-in-gas. On the other hand, as droplets fall into the water or when the brushes re-enter the bulk liquid phase, they entrain air bubbles that comprise a gas-in-liquid system. In these systems, a significant amount of transfer can also occur across the bulk water surface, a process that does not conform closely to either the gas-in-liquid or liquid-in-gas system. Images of some surface aerators are shown in Figure 5-6.

Tray towers, in which water falls from one tray to another, while each tray retains a pool of water a few centimeters deep, are approximately half filled with water and half filled with gas. They also have discontinuous liquid and gas phases and do not fall into either of the earlier categories. However, the flow patterns in both phases are similar to those in packed towers, and they can be analyzed similarly to such liquid-in-gas systems.

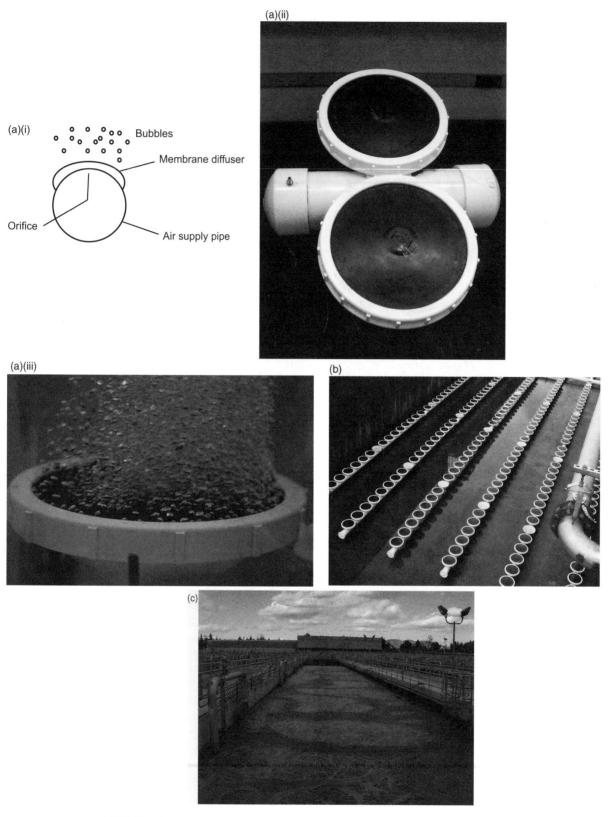

FIGURE 5-4. Components and result of diffused aeration in environmental engineering systems. (a) Schematic and photographs of a type of diffuser commonly used in aeration tanks of activated sludge treatment processes. (b) An array of diffusers at the bottom of an empty aeration tank, seated on the pipe supplying the air. (c) A full-scale aeration tank in operation, with the bubbles reaching the top forming a thin foam layer. *Source*: Photo credits: (a, c) D. Rosso; (b) J. Wilson.

FIGURE 5-5. Components of packed towers for gas transfer in environmental engineering. (a) A packed column used for scrubbing of acidic gases by absorption into water. (b) (i) A piece of random packing and (ii) molded packing of the type used in liquid-in-gas, gas transfer columns. (c) A typical water distribution system for a packed tower (in this case, a trickling filter). *Source*: Photo credits: (a) Envitech, Inc.; (b) Courtesy Lantec Products Inc.; (c) D. Rosso.

FIGURE 5-6. Surface aerators used in environmental engineering systems. (a, b) Two types of rotating brush aerators; (c) a floating, high-speed surface aerator. Note the intense mixing that such systems can generate. *Source*: (a) House Industries, Inc.; (b) J. Wilson; (c) Wikipedia (http://en.wikipedia.org/wiki/File:Surface_Aerator.jpg, accessed 4/26/2012).

5.3 HENRY'S LAW AND GAS/LIQUID EQUILIBRIUM

Volatilization and Dissolution as a Chemical Reaction

The transfer of a species A from an aqueous solution into a gas phase can be represented by the following reaction:

$$A(\text{aq}) \leftrightarrow A(\text{g}) \tag{5-4}$$

where $A(\text{aq})$ represents molecules of A dissolved in the aqueous solution; that is, surrounded by water molecules, and $A(\text{g})$ represents molecules of A in the gas phase.

In some cases, dissolved molecules form strong enough bonds with water that a water molecule is actually considered to be a part of the dissolved species. For instance, $CO_2(\text{aq})$ and $SO_2(\text{aq})$ are often written as H_2CO_3 and H_2SO_3, respectively, so that the reactions characterizing volatilization of these species are as follows:

$$H_2CO_3(\text{aq}) \leftrightarrow CO_2(\text{g}) + H_2O \tag{5-5a}$$

$$H_2SO_3(\text{aq}) \leftrightarrow SO_2(\text{g}) + H_2O \tag{5-5b}$$

At times, it might be important to know whether a dissolved gas molecule is chemically bonded or just adjacent to a water molecule. However, this distinction is usually unimportant for our purposes, so we can consider the corresponding pairs of species (e.g., $CO_2(\text{aq})$ and $H_2CO_3(\text{aq})$) to be identical.

Partition Coefficients, Equilibrium Constants, and the Formal Definition of Henry's Law

Based on Figure 5-1, Henry's constant can be defined as the ratio of the gas-phase concentration to the liquid-phase concentration of a volatile species i, when the system is at equilibrium and the species is present at an infinitely dilute concentration in a solution containing only the species and water. The figure also implies that the ratio $c_{G,i}/c_{L,i}$ is approximately constant and equal to H not only at infinite dilution, but also over an extended range of concentrations, designated as the Henry's law region. As a result, we can write

$$H \equiv \left.\frac{c_{G,i}}{c_{L,i}}\right|_{\text{eq,infinite dilution}} = \left.\frac{c_{G,i}}{c_{L,i}}\right|_{\text{eq,Henry's law region}} \tag{5-6}$$

where the subscript eq emphasizes that the ratio equals H only if the system is at equilibrium.

The vast majority of applications of gas transfer in environmental engineering (including all the applications that are discussed in this book) fall into the Henry's law region.[6] Therefore, the constraint that Equation 5-6 applies only in that region is rarely stated explicitly, and the equation is usually written as follows:

$$H = \left.\frac{c_{G,i}}{c_{L,i}}\right|_{eq} \quad (5\text{-}7)$$

The ratio of a constituent's concentrations in two phases at equilibrium is often referred to as a *distribution coefficient* or *partition coefficient*. Thus, Henry's constant as defined in Equations 5-6 and 5-7 can be described as the partition coefficient for species i between the gas and aqueous phases.

A related, but subtly different, definition of Henry's constant that is also commonly used defines this value as the equilibrium constant for the volatilization reaction shown in Equation 5-4. That is, using the symbol H' for this definition, we can write

$$H' \equiv \left.\frac{a_{G,i}}{a_{L,i}}\right|_{eq} \quad (5\text{-}8)$$

where $a_{G,i}$ and $a_{L,i}$ are the chemical activities of i in the gas and liquid phases, respectively. Comparing Equations 5-7 and 5-8, and recognizing H' as a true equilibrium constant, we treat H as a pseudo-equilibrium constant, computed using concentrations instead of activities in the calculation.

To quantify H', we need to establish conventions for computing the activity of the volatile species in both the gas phase and the solution. To do that, recall that the chemical activity of any species i in any phase equals the product of the activity coefficient of i (γ_i) and the species' concentration (c_i) in that phase, normalized to its concentration in some arbitrary, but clearly defined, standard state ($c_{i,\text{std.state}}$); that is

$$a_i = \gamma_i \frac{c_i}{c_{i,\text{std.state}}} \quad (5\text{-}9)$$

The (dimensionless) activity coefficient is an indicator of how the environment surrounding molecules of i compares with a clearly defined reference environment; if the actual environment is very similar to the reference environment, γ_i is close to 1.0, and if the two environments are very different, γ_i is far from 1.0. The reference environment for gaseous species is always chosen to be an ideal gas, but two different choices are commonly made for the reference environment of volatile species in solution: infinite dilution of the species

[6] The applicability of Henry's law implies that not only the species of interest, but all solutes, are so dilute that the volatile species behaves as though it were the only solute present. A corollary is that, if multiple volatile species are present in a single solution, and if they all are in the Henry's law region of the plot, then they all behave independently; that is, the behavior of each species can be assessed without considering the presence of the others.

in an aqueous solution (the Henry's law reference state), and a pure phase of the species, present as a liquid (the Raoult's law reference state).[7]

Combining Equations 5-8 and 5-9, we find

$$H' = \frac{\gamma_{G,i}}{\gamma_{L,i}} \left.\frac{(c_{G,i}/c_{G,i,\text{std.state}})}{(c_{L,i}/c_{L,i,\text{std.state}})}\right|_{eq} \quad (5\text{-}10)$$

Equation 5-10 makes clear that H', like all true equilibrium constants, is dimensionless. However, the numerical value of H' depends on the choices that are made for $c_{L,i,\text{std.state}}$ and $c_{G,i,\text{std.state}}$ and on the environmental conditions in the two reference states.

To understand the relationship between H' and H, it is useful to regroup the terms in Equation 5-10 as follows:

$$H' = \frac{1}{\gamma_{L,i}} \left(\gamma_{G,i} \frac{c_{L,i,\text{std.state}}}{c_{G,i,\text{std.state}}} \right) \left.\frac{c_{G,i}}{c_{L,i}}\right|_{eq}$$

$$H' = \frac{1}{\gamma_{L,i}} \left(\gamma_{G,i} \frac{c_{L,i,\text{std.state}}}{c_{G,i,\text{std.state}}} \right) H \quad (5\text{-}11)$$

The standard state concentrations in both phases ($c_{G,i,\text{std.state}}$ and $c_{L,i,\text{std.state}}$) are virtually always assigned values of 1.0, with whatever units are being used to express the concentrations. Also, the activity coefficients of gas-phase species, $\gamma_{G,i}$, are always assumed to be 1.0, because, at normal temperatures and pressures, the behavior of any real gas conforms closely to that of an ideal gas. Therefore, the term in parentheses in Equation 5-11 has a magnitude of 1.0 and units corresponding to those of $c_{L,i}/c_{G,i}$, and we can write:

$$H' = \frac{1}{\gamma_{L,i}} \left(1.0 \frac{\text{units of } c_{L,i}}{\text{units of } c_{G,i}} \right) H \quad (5\text{-}12)$$

Equation 5-12 establishes the relationship between the two common ways of defining Henry's constant: as an equilibrium constant, H', and as a pseudo-equilibrium constant, H. In addition to the fact that H has units that correspond to the ratio of gas-phase to liquid-phase concentrations, whereas H' is dimensionless, the two terms differ numerically by a factor equal to the activity coefficient of the dissolved, volatile species.

As noted previously, the value of this activity coefficient depends on how closely the actual environment of interest conforms to the reference state. Most solutions of interest in environmental engineering are relatively dilute. As a result,

[7] A detailed explanation of the meaning of the standard state and the activity coefficient, and a summary of the most common conventions for quantifying c_i and γ_i in gases and aqueous solutions, are provided in Appendix 5A.

if the infinite dilution reference state is chosen, then $\gamma_{L,i}$ is likely to be very close to 1.0, and the values of H' and H will be almost identical. If, on the other hand, the infinite dilution reference state is used and the actual solution of interest is not dilute (e.g., if it is seawater), then $\gamma_{L,i}$ can be very different from 1.0, leading to a large difference between the values of H' and H. For example, if the infinite dilution reference state is used, the activity coefficient for DO in seawater at 20°C is 1.24, indicating that oxygen is only 81% (= 1/1.24) as soluble in seawater as in very clean fresh water at this temperature. A similar difference exists between H' and H if the selected reference state is different from infinite dilution (e.g., if the Raoult's law reference state is used), but the actual solution of interest is dilute. As a more extreme case, if the Raoult's law reference state is adopted, the activity coefficient of chloroform in a dilute aqueous solution at 20°C is 830. (The basis for this latter calculation is shown in Appendix 5A.)

Any true equilibrium constant, such as H', has the same value in all systems at a given temperature. Thus, the value of H' for oxygen at 20°C is the same in fresh water and seawater, and the difference in the solubility of oxygen in the two solutions is attributed to a change in its activity coefficient, as explained earlier. On the other hand, H is a ratio of concentrations at equilibrium. Changes in the environmental conditions (such as the overall solution composition) can affect this ratio, so H can vary from one system to the next, even among systems at the same temperature.

Throughout this text, unless otherwise stated, we adopt the definition of Henry's constant as a pseudo-equilibrium constant (H), so that it has units that correspond to those of $c_{G,i}/c_{L,i}$. However (also, unless otherwise stated), we focus on dilute solutions, and we use the infinite dilution convention for the reference state. As a result, the solute is always assumed to behave ideally (i.e., $\gamma_{L,i} = 1.0$).

One other commonly used convention defines Henry's constant based on the reverse of reaction 5-4, so that the expressions for the partition and equilibrium constants are inverted.[8] In this case, the constant is often (but not always) written with K_H replacing H; that is

$$K_H = \frac{1}{H} = \left.\frac{c_{L,i}}{c_{G,i}}\right|_{eq} \quad (5\text{-}13\text{a})$$

$$K'_H = \gamma_{L,i}\left(1.0\,\frac{\text{units of } c_{G,i}}{\text{units of } c_{L,i}}\right)\frac{1}{H} \quad (5\text{-}13\text{b})$$

Dimensions of c_L, c_G, and Henry's Law Constant

The ideal gas law establishes the following relationship between concentration and pressure for any ideal gas:

$$c_{G,i,\text{mol}} = \frac{n_i}{V_G} = \frac{P_i}{RT} \quad (5\text{-}14)$$

where $c_{G,i,\text{mol}}$ is the gas-phase concentration of i in moles per unit volume, n_i is the number of moles of i in the gas phase, V_G is the gas volume, P_i is the pressure exerted by i in the gas phase (the *partial pressure* of i), T is the absolute temperature, and R is the universal gas constant. At standard temperature and pressure (STP) (i.e., 25°C [298.15K] and 1 atm), RT is 24.47 atm L/mol or 2479 kPa L/mol.[9]

According to Equation 5-14, the partial pressure exerted by any ideal gas i, regardless of its chemical identity, is given by

$$P_i = RT\,c_{G,i,\text{mol}} \quad (5\text{-}15)$$

Thus, the partial pressure exerted by any constituent of a gas is directly proportional to its concentration in the gas phase, and, at a given temperature, the partial pressure of a gaseous species is a direct measure of its molar concentration in the gas phase.

The total pressure exerted by all the N gases in a system is the sum of the partial pressures exerted by the individual gases, so

$$P_{\text{tot}} = \sum_{j=1}^{N} P_j = RT \sum_{j=1}^{N} c_{G,j,\text{mol}} \quad (5\text{-}16)$$

Thus,

$$\frac{P_i}{P_{\text{tot}}} = \frac{RT\,c_{G,i,\text{mol}}}{RT\sum_{j=1}^{N} c_{G,j,\text{mol}}} = \frac{c_{G,i,\text{mol}}}{\sum_{j=1}^{N} c_{G,j,\text{mol}}} \equiv y_i \quad (5\text{-}17)$$

where y_i is the dimensionless mole fraction of i in the gas phase, defined as

$$y_i \equiv \frac{\text{moles of } i \text{ in the gas phase}}{\text{total moles of all species in the gas phase}} \quad (5\text{-}18)$$

The mass concentration of i in the gas phase, which we represent as $c_{G,i,\text{mass}}$ (mass/volume), is the product of the molar concentration of i and its molecular weight (MW_i). Thus, the relationships among partial pressure (P_i), mole fraction (y_i), molar concentration ($c_{G,i,\text{mol}}$), and mass

---

[8] Historically, this form of the equilibrium constant was known as the Bunsen coefficient. However, at present, it is widely referred to as a Henry's constant, just like the form shown in Equation 5-3.

[9] SI units for expressing pressure are Pascals (1 Pa = 1 N/m²) or bars (1 bar = 10⁵ Pa = 100 kPa). However, units of atmospheres have long been conventional for expressing gas-phase pressures in environmental engineering, so these units are used in this discussion. 1.00 atm = 1.013 bar.

concentration ($c_{G,i,\text{mass}}$) in the gas phase can be summarized as follows:

$$P_i = y_i P_{\text{tot}} = RT c_{G,i,\text{mol}} = \frac{RT c_{G,i,\text{mass}}}{\text{MW}_i} \quad (5\text{-}19)$$

Dimensions that are typically used to report values of $c_{G,i}$, $c_{L,i}$, and H in the environmental engineering literature are shown in Table 5-1, and values of H for some environmentally significant gases are listed in Table 5-2. In this text, values of H expressed in different units are indicated by different subscripts, as shown in the table. Given the various forms in which Henry's constant is computed, we must be especially careful when interpreting and using tabulated data for its value; that is, it is essential that the units used to express the concentrations in both phases, the convention adopted to quantify the activity coefficient, and the direction of the reaction be specified to avoid errors of interpretation.

In environmental engineering applications involving volatile organic compounds (VOCs), it is common to express

TABLE 5-1. Units Commonly Used for Henry's Law Constants[a]

Dimensions Used for Gas-Phase Concentration	Dimensions Used for Aqueous-Phase Concentration	Symbol	Units of H
Mass concentration (μg i/L of gas)	Mass concentration (μg i/L of solution)	H_{cc}[b]	$\dfrac{\text{L liquid}}{\text{L gas}}\left(\dfrac{L_L}{L_G}\right)$
Molar concentration (mol i/L of gas)	Molar concentration (mol i/L of solution)	H_{mm}[b]	$\dfrac{\text{L liquid}}{\text{L gas}}\left(\dfrac{L_L}{L_G}\right)$
Partial pressure (atm)	Mass concentration (mg i/L of solution)	H_{pc}	$\dfrac{\text{atm-L}_L}{\text{mg } i}$
Partial pressure (atm)	Molar concentration (mol i/L of solution)	H_{pm}	$\dfrac{\text{atm-L}_L}{\text{mol } i}$
Partial pressure (atm)	Mole fraction (mol i/total moles of solution)	H_{px}	atm
Mole fraction[c] (moles of i/total moles of gas)	Molar concentration (mol i/L of solution)	H_{ym}	$\dfrac{L_L}{\text{mol of gas}}$
Mole fraction[c] (moles of i/total moles of gas)	Mole fraction (mol i/total moles of solution)	H_{yx}	$\dfrac{\text{mol of solution}}{\text{mol of gas}}$

[a]For Henry's law constant written as a partition coefficient for the volatilization reaction; that is, as $c_{G,i} = H c_{L,i}$.
[b]Values of H_{cc} and H_{mm} are always identical.
[c]Note that, when the units used for the gas-phase concentration are mole fractions, the value of H depends on the total pressure in the gas phase. Tabulated values of H_{ym} and H_{yx} invariably assume $P_{\text{tot}} = 1.0$ atm. Values applicable for other pressures can be computed as: $H_{\text{system}} = H_{\text{tabulated}}(P_{\text{tot,system}}/P_{\text{tot,tabulated}}(= 1.0 \text{ atm}))$.

TABLE 5-2. Henry's Constants of Some Environmentally Important Gases

	H[a]	
Compound	H_{pm} (atm/(mol/L$_L$))	H_{cc} or H_{mm} (L$_L$/L$_G$)[b]
Nitrogen	1590	65.0
Hydrogen	1280	52.3
Carbon monoxide	1050	42.9
Oxygen	769	31.4
Methane	769	31.4
Cyclohexane	182	7.44
Radon	108	4.41
Ozone	90.9	3.72
Carbon dioxide	29.4	1.20
Carbon tetrachloride	29.4	1.20
Tetrachloroethylene (PCE)	17.5	0.715
Chlorine	10.8	0.441
Hydrogen sulfide	10.0	0.409
Trichloroethylene (TCE)	10.0	0.409
Ethylbenzene	8.33	0.340
Toluene	6.67	0.273
Benzene	5.56	0.227
o-Xylene	4.55	0.186
Chloroform	4.00	0.163
Chlorine dioxide	1.00	0.0409
Sulfur dioxide	0.833	0.0340
Bromoform	0.556	0.0227
2,2′,5,5′-Tetrachlorobiphenyl	0.250	0.0102
Ethyl acetate	0.143	5.84×10^{-3}
Hydrogen cyanide	0.100	4.09×10^{-3}
Acetone	0.040	1.64×10^{-3}
Ammonia	0.017	6.95×10^{-4}
Methanol	4.6×10^{-3}	1.88×10^{-4}
Phenol	5.0×10^{-4}	2.04×10^{-5}
Acetic acid	2.0×10^{-4}	8.17×10^{-6}

[a]Values are for 25°C and are based on compilation by Sander (1999).
[b]Values in this column are valid for any units of concentration, provided that the same units are used for both the liquid and gas phases.

Henry's constants using mass per volume dimensions for the concentrations in both phases (i.e., as H_{cc}). This form of Henry's constant is used most extensively in the remainder of the text. Because the same concentration units are used in both phases, H_{cc} is often referred to as being dimensionless. However, the concentration in the numerator is the mass of the volatile species per unit volume of gas, and that in the denominator is the mass per unit volume of solution. As a result, the ratio has units of volume of liquid per volume of gas (e.g., L$_L$/L$_G$).

■ **EXAMPLE 5-1.** Henry's constant for oxygen at 25°C, in the form of H_{pm}, is given in Table 5-2 as 769 atm/(mol/L$_L$). Express this constant (a) in atm/(mol/m$_L^3$), (b) as H_{pc}, and (c) as the "dimensionless" Henry's constant, H_{cc}.

Solution. The differences among the various forms of Henry's constant are directly related to the different ways of expressing the concentrations in the gas and liquid phases. The constant given in the problem statement, H_{pm}, is based on values of $c_{G,i}$ and $c_{L,i}$ in units of atm and mol/L$_L$, respectively.

(a) To convert from units of atm/(mol/L$_L$) to atm/(mol/m$_L^3$), we need to convert the liquid-phase concentration from mol/L$_L$ to mol/m$_L^3$, as follows:

$$H_{pm} = (769\ \text{atm L}_L/\text{mol})(10^{-3}\ \text{m}_L^3/\text{L}_L)$$
$$= 0.769\ \text{atm m}_L^3/\text{mol}$$

(b) To convert from H_{pm} to H_{pc}, we need to convert the liquid-phase concentration from molar to mass units, utilizing the MW of O$_2$. The conversion is therefore

$$H_{pc} = \frac{H_{pm}}{\text{MW}_{O_2}} = \frac{769\ \text{atm}/(\text{mol}/\text{L}_L)}{32{,}000\ \text{mg/mol}}$$
$$= 0.0240\ \text{atm}/(\text{mg/L}_L)$$

(c) The required constant is the ratio of the gas- and aqueous-phase concentrations when both are expressed as mass per unit volume. The value of 769 atm/(mol/L$_L$) given in the problem statement already has the aqueous-phase concentration in mol/L$_L$, so the easiest approach is to convert the gas-phase concentration from atm to mol/L$_G$. To do so, we can use the ideal gas law (Equation 5-14) to relate partial pressure with the gas-phase molar concentration: $c_G = n/V_G = P/RT$. This equality indicates that a gas-phase concentration expressed as a pressure can be converted to one expressed in moles per unit volume by dividing by RT. Thus, H_{cc} is

$$H_{cc} = \frac{H_{pm}}{RT} = \frac{769\ \text{atm}/(\text{mol}/\text{L}_L)}{(0.082\ \text{atm L}_G/\text{mol K})298\ \text{K}}$$
$$= 31.5\ \text{L}_L/\text{L}_G \qquad \blacksquare$$

The value of Henry's constant expressed as H_{cc} is useful for acquiring an intuitive feel for the solubility or volatility of a chemical as follows. Consider a closed vial containing volumes V_L of liquid and V_G of gas. A volatile species is present in the two phases at concentrations c_L and c_G, respectively. If these two phases are at equilibrium, then $H_{cc} = c_G/c_L = (m_G/V_G)/(m_L/V_L)$. If we specify further that $m_G = m_L$ (i.e., that the same mass of the constituent of interest is in each phase), then $H_{cc} = V_L/V_G$. That is, Henry's constant expressed as H_{cc} equals the liquid-to-gas volume ratio that causes the same mass of the constituent to be in each of the two phases at equilibrium.

For example, H_{cc} for benzene at 25°C is 0.227 L$_L$/L$_G$. One interpretation of this value is that, at equilibrium, the mass of benzene in 0.227 L of solution is the same as that in 1.0 L of gas. On the other hand, H_{cc} for trichloroethylene (TCE, a common groundwater contaminant) is 0.409 L$_L$/L$_G$, indicating that (again, at equilibrium) 0.409 L of solution is required to hold the same mass of TCE as is present in 1.0 L of gas.

A similar comparison can be made for closed systems that have a fixed ratio of $V_L/V_G = 1.0\ \text{L}_L/\text{L}_G$; that is, equal volumes of the two phases (rather than a fixed ratio of $m_L/m_G = 1$). In this case, at equilibrium, $H_{cc} = c_G/c_L = m_G/m_L$. That is, for a system with equal volumes of gas and liquid, H_{cc} equals the ratio of the species' mass in the gas phase to that in the solution. This way of looking at H_{cc} leads to the conclusion that, in an equilibrium system with equal volumes of gas and liquid, benzene will be distributed in a ratio of 0.227 grams in the gas phase per gram in solution, whereas TCE will be distributed in a ratio of 0.409 grams in the gas per gram in solution.

Based on either of these two ways of comparing the two species, we conclude that TCE is more volatile (i.e., less soluble) than benzene, and that a higher Henry's constant corresponds to greater volatility. Both ways of thinking about volatility and Henry's constant are shown schematically in Figure 5-7 for TCE and benzene, and also for oxygen, which is much more volatile than either of the organics.

The equilibrium between the two phases in a closed vial is taken advantage of when headspace analysis is used in gas chromatography. In this application, the gas-phase concentration is measured, and the liquid-phase concentration is inferred, assuming gas/liquid equilibrium and taking into account the ratio of volumes of liquid and gas in the bottle.

■ **EXAMPLE 5-2.** You have received a sealed vial containing a water sample to be analyzed for cyclohexane. The volume of the vial is 10 mL, and the temperature is 25°C. You notice that the vial contains an air bubble, whose volume is 0.25 mL.

(a) If the sample originally contained 10^{-7} mol/L of the analyte, what concentration will you measure in the solution, assuming that it has equilibrated with the air bubble?

(b) What are the concentrations of cyclohexane (MW = 84) in the gas and liquid phases in the bottle, in μg/L?

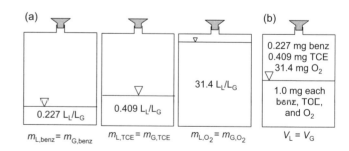

FIGURE 5-7. Various ways of interpreting H. (a) The solution-to-gas volume ratios required to have an equal mass of a given species in each phase. (b) Distribution of three species between the gas and solution in an equilibrium system with $V_L = V_G$.

Solution.

(a) The total number of moles of cyclohexane ($n_{\text{tot},i}$) in the vial is $(0.00975 \text{ L})(10^{-7.0} \text{ mol/L})$, or 9.75×10^{-10} mol. This mass will be distributed between the liquid and bubble at equilibrium in accordance with Henry's law, as follows:

$$n_{\text{tot}} = n_{\text{G}} + n_{\text{L}} = c_{\text{G}}V_{\text{G}} + c_{\text{L}}V_{\text{L}} = \frac{P_i}{RT}V_{\text{G}} + \frac{P_i}{H_{\text{pm}}}V_{\text{L}}$$

where the relationship $c_{\text{G},i} = P_i/RT$ is from the ideal gas law (Equation 5-14). The partial pressure of cyclohexane in the gas phase can therefore be computed as

$$P_i = \frac{n_{\text{tot},i}}{(V_{\text{G}}/RT) + (V_{\text{L}}/H_{\text{pm}})}$$

$$= \frac{9.75 \times 10^{-10} \text{ mol}}{0.00025 \text{ L}_{\text{G}}/((0.082 \text{ atm L}_{\text{G}}/\text{mol K})(298 \text{ K})) + 0.00975 \text{ L}_{\text{L}}/(182 \text{ atm L}_{\text{L}}/\text{mol})} = 1.53 \times 10^{-5} \text{ atm}$$

The equilibrium concentration in solution can be found by combining this result with H_{pm}:

$$c_{\text{L}} = \frac{P_i}{H_{\text{pm}}} = \frac{1.53 \times 10^{-5} \text{ atm}}{182 \text{ atm}/(\text{mol/L}_{\text{L}})} = 8.4 \times 10^{-8} \text{ mol/L}_{\text{L}}$$

The solution originally contained 10^{-7} mol/L of the analyte, so the calculation indicates that approximately 16% of the analyte volatilized when the solution equilibrated with the air bubble. This loss of the analyte would cause a corresponding 16% error in the concentration that would be detected if the analysis did not take the volatilization process into account. Note that the error caused by the presence of the bubble decreases with decreasing volatility (i.e., decreasing H).

(b) The value of c_{L} in mg/L can be computed as the product of the molar concentration and the MW, and c_{G} can then be computed using H_{cc} as

$$c_{\text{L}} = (8.4 \times 10^{-8} \text{ mol/L}_{\text{L}})(84 \times 10^6 \text{ µg/mol})$$

$$= 7.0 \text{ µg/L}_{\text{L}}$$

$$c_{\text{G}} = H_{\text{cc}} c_{\text{L}} = \left(7.44 \frac{\text{L}_{\text{L}}}{\text{L}_{\text{G}}}\right)(7.0 \text{ µg/L}_{\text{L}}) = 52 \text{ µg/L}_{\text{G}}$$

∎

Factors Affecting Gas/Liquid Equilibrium

The major factors controlling Henry's constant of a compound are its chemical structure and the temperature of the system. As noted previously, the solubility of a volatile species is also affected by the presence of other constituents in the solution, and this change in solubility is often reported as a change in Henry's constant. The effects of chemical structure, temperature, and solution composition on solubility of volatile compounds are reviewed briefly in this section.

In the gas phase, the space separating molecules is so large (on average) that attractive or repulsive interactions among the molecules are negligible, regardless of their identity. As a result, the volatility of a dissolved species is dictated primarily by its interactions with the aqueous phase: the more favorable its interactions with the water (i.e., the more hydrophilic the molecule), the less volatile it is. Therefore, increasing hydrophilicity (or, equivalently, decreasing hydrophobicity) is associated with decreasing values of H.

The hydrophilicity of a solute is controlled primarily by two factors: bonding between the solute and water molecules, and the mobility of the water molecules surrounding the solute. Whenever a volatile molecule transfers from the gas phase to an aqueous solution, some bonds between adjacent water molecules (hydrogen bonds) must break to create space for the volatile molecule to occupy, and new bonds form between the solute and water. The net effect of these changes is quantified by the enthalpy of dissolution, which is negative (favorable) for dissolution of any volatile species. In other words, although the need to break water–water bonds opposes dissolution, the new solute–water bonds that form are favorable, and their effect always dominates the former one (Israelachvili, 1985). The reason for this somewhat surprising outcome is that the water molecules can rearrange themselves to create the hole with almost no breakage of bonds.

If bond formation and breakage were the only factors controlling dissolution, then no species would be volatile, because dissolution would be favored under all circumstances. However, when solutes dissolve, they also orient the nearby water molecules in a way that reduces the freedom of movement of those molecules. If the solute is ionic or highly polar, then the adjacent water molecules are held in an orientation that maximizes favorable electrical interactions with the solute, and if it is nonpolar, the surrounding water molecules arrange themselves to form a cage-like structure around the solute (Figure 5-8). This decreased freedom of motion of the water molecules represents a decrease in entropy, which opposes dissolution. The balance between this decreased freedom of motion and the formation of solute–water bonds (which favors dissolution) determines the volatility of any given compound.

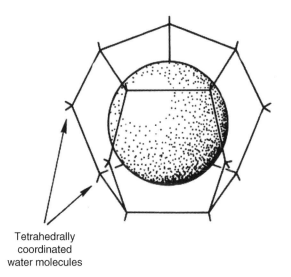

FIGURE 5-8. Schematic of hydrogen-bonded water molecules surrounding a hydrophobic solute. *Source:* From Israelachvili (1985).

Bonds between charged solutes (ions) and water are particularly strong, making all ions very hydrophilic and nonvolatile; that is, for all practical purposes, ionized species do not exist in the gas phase. Bonds between neutral molecules and water are weaker than those between ions and water, with the strength of the interaction decreasing with decreasing polarity of the solute.[10] As a result, Henry's constants of neutral molecules increase with decreasing polarity, other factors (e.g., molecular size) being equal.

On the other hand, for a given molecular structure (e.g., a series of straight-chain hydrocarbons), increasing molecular size steadily increases both the favorable solute–water interactions and the unfavorable immobilization of the water. Invariably, increasing molecular size magnifies the favorable interactions more than the unfavorable ones (Israelachvili, 1985), so that an increase in molecular size increases the net attractiveness of the dissolution reaction and decreases Henry's constant.

The preceding discussion focuses on the transfer of volatile molecules directly between an aqueous solution and the gas phase. An alternative and useful interpretation of the effects of molecular structure on volatility considers the pure compound in a liquid state (not an aqueous solution of the compound) as the starting point; that is, it considers volatility from a Raoult's law perspective.

The volatility of pure compounds is quantified by their vapor pressure. If the interactions of molecules of these compounds with water molecules and with one another were identical, then the trend in Henry's constants for a range of compounds would be the same as the trend in the vapor pressures of the pure compounds. However, most neutral compounds interact less favorably with water than with other like molecules. As a result, the compounds are more likely to volatilize from an aqueous solution than from the pure liquid; this fact is reflected in the greater slope in the Henry's law region than the Raoult's law region in Figure 5-1.

Thus, from this perspective, a Henry's constant can be viewed as reflecting two factors—the inherent volatility of the pure compound, as indicated by its vapor pressure at the temperature of interest, and the activity coefficient of the compound when it is dissolved in water (using the pure liquid as the standard state), which indicates the stability of the molecule when surrounded by water molecules. Some compounds, such as methanol, are quite stable when surrounded by water, but nevertheless have relatively high Henry's constants because of their inherent volatility. Compounds such as polychlorinated biphenyls (PCBs), on the other hand, have very low inherent volatility as pure compounds, but their volatility when dissolved in water is nevertheless significant, because their interactions with water molecules are so unfavorable. The decline in stability when a molecule moves from a pure liquid to an aqueous solution (which is related to the decrease in entropy of the water molecules, as discussed earlier) is often referred to as the *hydrophobic effect*.

In recent years, a number of attempts have been made to predict several important molecular properties, including Henry's constants, based strictly on molecular structure. These so-called quantitative structure–activity relationships (QSARs) have been quite successful in some cases and less so in others. The predictions of QSARs for Henry's constants of organic compounds are often quite good. Compilations of these predictions have been provided by Nirmalakhandan et al. (1997) and Schwarzenbach et al. (2002).

As suggested in Figure 5-2, Henry's constant also depends on the temperature. This effect is typically modeled using the van't Hoff equation, which describes the effect of temperature on any equilibrium constant:

$$\ln \frac{H_{T_2}}{H_{T_1}} = \frac{\Delta \overline{H}^\circ}{R} \left(\frac{1}{T_1} - \frac{1}{T_2} \right) \quad (5\text{-}20)$$

where H_{T_1} and H_{T_2} are Henry's constants at temperatures T_1 and T_2, respectively, $\Delta \overline{H}^\circ$ is the enthalpy change per mole of stoichiometric reaction under standard state conditions, and R is the universal gas constant.[11] As noted previously,

---

[10] These bonds arise from so-called London–van der Waals interactions, in which resonance between the electronic vibrations in adjacent molecules generates attractive forces between the molecules.

[11] Equation 5-20 is based on the assumption that the molar enthalpy of reaction under the conditions of interest equals the molar enthalpy under standard state conditions. This assumption is generally acceptable for environmental systems. Also, note that $\overline{H}^\circ$ is the universally accepted symbol for standard molar enthalpy; it is not a Henry's constant.

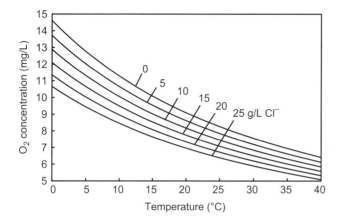

FIGURE 5-9. Temperature- and salt-dependence of the solubility of oxygen in water equilibrated with the atmosphere. The number shown for each curve indicates the chloride concentration in the water in grams per kilogram of solution. For reference, the Cl⁻ concentration in seawater is 19.3 g/kg. The data are for solutions that contain the indicated concentration of Cl⁻ and other salts in the same ratio to Cl⁻ as their ratio in seawater. *Source*: After APHA (2005).

transferring a molecule from solution to the gas phase always requires energy to break the bonds between the solute and neighboring (mostly water) molecules, so the enthalpy of volatilization is always positive (i.e., the reaction is endothermic). As a result, volatility (and therefore Henry's constant) always increases with increasing temperature. This trend is demonstrated in terms of solubility (the opposite of volatility) for oxygen in Figure 5-9 and in terms of Henry's constant for several VOCs in Figure 5-10. Because of the overwhelming importance of DO as a water quality parameter, its solubility as a function of temperature and salt content

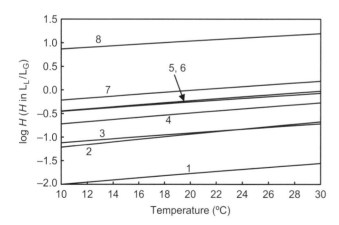

FIGURE 5-10. The temperature dependence of Henry's constant for several halogenated VOCs. 1: Bromoform; 2: Hexachloroethane; 3: Chloroform; 4: Trichloroethylene; 5: 1,1,1-Trichloroethane; 6: Tetrachloroethylene; 7: Carbon tetrachloride; 8: Dichlorodifluoromethane. *Source*: Based on data in Munz and Roberts (1987).

has been studied more so than that of any other gaseous species. The curves in Figure 5-9 are obtained from the following empirical correlation that was derived from such studies (APHA, 2005):

$$\ln c_{O_2}^* = -139.344 + \frac{1.576 \times 10^6}{T} - \frac{6.642 \times 10^7}{T^2}$$
$$+ \frac{1.244 \times 10^{10}}{T^3} - \frac{8.622 \times 10^{11}}{T^4}$$
$$- \text{Chl}\left[3.193 \times 10^{-2} - \frac{19.43}{T} + \frac{3.867 \times 10^3}{T^2}\right]$$
(5-21)

where $c_{O_2}^*$ is the equilibrium concentration of DO at a partial pressure of 0.21 atm, T is the temperature, and Chl is the chlorinity of the solution, defined as the chloride concentration in g/kg. The overall salt content of the solution is assumed to include all the salt ions that are in seawater that has been diluted or concentrated until it has the specified chlorinity.

Finally, the volatility of a molecule is affected not only by its own properties, but also by the presence of other species in solution, which might either increase or decrease the solubility of the volatile species. Most often, solubility decreases with increasing salt concentration (i.e., with increasing ionic strength), as shown for oxygen in Figure 5-9. Because other solutes can also affect the solubility of gases in ways that are difficult to predict, the saturation concentration of gases should be determined experimentally, rather than relying on the literature values, if highly accurate values are required.

In the waste treatment literature, it is conventional to represent the ratio of the oxygen saturation concentration in a particular water to the corresponding concentration in clean water at the same temperature by the symbol beta (β):

$$c_{O_2,\text{sat,actual}} = \beta c_{O_2,\text{sat,clean}} \quad (5\text{-}22)$$

In this case, the effect on Henry's constant is given by

$$H_{O_2,\text{actual}} = \frac{H_{O_2,\text{clean}}}{\beta} \quad (5\text{-}23)$$

Except in highly saline waters, β tends to be a fairly minor correction. For instance, in activated sludge systems treating domestic wastewater, β values are usually between 0.7 and 0.98, and a value of 0.95 is frequently assumed (Metcalf and Eddy Inc, 2003).[12]

---

[12] Note that, if Henry's constant is defined as the equilibrium constant for the reaction (i.e., as H'), then the effect of other solutes on the solubility of the volatile species would be interpreted as an effect on γ_i, while H' remained constant.

5.4 RELATING CHANGES IN THE GAS AND LIQUID PHASES

The stoichiometry of a reaction establishes that, if a moles of reactant A are consumed, p moles of product P are formed. In homogeneous reactions taking place in solution, both A and P are dissolved in the same volume of solution, so the stoichiometric mole ratios can be converted directly into concentration ratios by dividing by the volume; that is, reaction of a mol/L of A yields p mol/L of product P. A similar relationship exists with respect to the rate of reaction (conversion of a mol/L of A per minute generates p mol/L of P per minute).

In contrast, in a gas transfer reaction in which dissolved A is the reactant and gas phase A is the product, the mass balance requires that for every mole of A leaving the aqueous phase, one mole of A enter the gas phase. Since the volumes and molar densities of the two phases differ, it is incorrect to equate the number of moles per liter lost from the liquid with the number of moles per liter gained by the gas. Nevertheless, it is convenient to have a single expression that describes gas transfer and that applies identically to both phases. Such an expression can be written by normalizing the transfer to the interfacial area, rather than to the volume. When this is done, the rate of transfer is given as a flux J, with dimensions such as moles or milligrams of the volatile species transferred per unit area (A) of interface per unit time. Fluxes have directionality as well as magnitude; in this text, we define J_L and J_G as the fluxes *into* the liquid and gas phases, respectively. Thus, in a given reactor, $J_L = -J_G$.

As noted earlier, the term accounting for gas transfer in a mass balance on the liquid phase can be expressed as the product $r_{L,gt}V_L$, where $r_{L,gt}$ is the rate (mass or moles per unit volume of solution per unit time) at which the species of interest enters the solution. At times, it is useful to normalize $r_{L,gt}$ to the volume of the whole reactor (including both the solution and gas phases) rather than to the solution alone. This rate, which we designate as $r_{R,gt}$, can be expressed as follows:

$$r_{R,gt} = J_L \frac{A}{V_R} \quad (5\text{-}24)$$

Note that, because the reactor volume includes the space occupied by both the liquid and gas phases, in general, $V_R \neq V_L \neq V_G$.

Designating the amount of interfacial area per unit volume of reactor (i.e., the concentration of interfacial area in the reactor) as a_R, the products $J_L a_R$ and $J_G a_R$ give the rates of gas transfer into the liquid and gas phases, respectively, per unit volume of reactor (e.g., in mol/L min). Similarly, the ratio of interfacial area to the volume of liquid in the reactor (A/V_L) can be designated as a_L. Substituting these definitions into Equation 5-24, we obtain

$$r_{R,gt} = J_L \frac{A}{V_R} = J_L a_R = J_L a_L \frac{V_L}{V_R} \quad (5\text{-}25)$$

Correspondingly, the rate of gas transfer per unit volume of solution ($r_{L,gt}$) is obtained by an expression analogous to Equation 5-24, with V_L replacing V_R as follows:

$$r_{L,gt} = J_L \frac{A}{V_L} = J_L a_L \quad (5\text{-}26)$$

In the gas transfer literature, an unsubscripted a is often used to represent both a_R and a_L, in which case the meaning of the parameter must be inferred from context. In systems in which bubbles of gas are dispersed in a bulk liquid, the volume occupied by the gas phase is often much smaller than that occupied by the liquid, so $a_R \approx a_L$, $r_{R,gt} \approx J_L a_L$, and the distinction between a_R and a_L becomes unimportant. However, both for mathematical formality and because the values of a_R and a_L differ significantly in some systems, it is important to be clear about which means of normalizing the interfacial area is being used in a given analysis.

The flux describes the gas transfer rate across a given patch of surface and, as is shown subsequently, can be related to the chemical and physical properties in the immediate vicinity of the interface. It is thus a highly localized, or microscopic, characteristic. The term a (either a_R or a_L), on the other hand, characterizes macroscopic properties of the system. Even if the microenvironments near the gas–liquid interface in a stripping tower, an activated sludge aeration tank, a flowing river, and a quiescent pond are similar, so that they have comparable fluxes across each unit area of interface (equal J values), the very different macroscopic geometries (different a values) of these systems could lead to vastly different overall gas transfer rates.

5.5 MECHANISTIC MODELS FOR GAS TRANSFER

Fluid Dynamics and Mass Transport in the Interfacial Region

As noted previously, empirical evidence suggests that transport through the liquid- and/or gas-phase interfacial regions (as opposed to transport through the bulk fluid phases or the kinetics of the phase transfer reaction at the interface itself) is the rate-limiting step in most gas transfer processes. The reason that transport through the interfacial regions is impeded more than in the bulk phases has to do with the unusual environment near the interface. Whereas the forces operating on water and solute molecules are identical in all directions in the bulk solution, this symmetry does not extend all the way to the gas/liquid interface, because the gas does not exert as strong an attraction on the interfacial molecules as the solution does. As a result of this asymmetry, the motion of molecules near the interface is slightly constrained, making it more difficult to mix surface water with bulk water than it is to mix packets of bulk water with each other.

The resistance to mixing between interfacial water and bulk solution can be envisioned by assuming that tiny packets of bulk solution move to the interface, remain there for some period of time, and are then swept back into the bulk solution as they are replaced by other packets; the time that a packet spends at the interface is then an indicator of the resistance to packet exchange. If the interfacial region is assumed to be one packet thick, each packet is bounded on one side by the interface and on the other by bulk solution. While at the interface, the packets are assumed to have no internal mixing, so transport through them occurs solely via molecular diffusion. A similar process can be envisioned to occur on the gas-phase side of the phase boundary, although the dimensions of the packets and the frequency with which packets are exchanged might be different in the two phases.

All models of gas transfer used in environmental engineering incorporate the assumption that the resistance to transport through one or both of the fluid boundary layers dominates the resistance imposed by other steps in the gas transfer process, and all represent the fluid dynamics of the boundary layers by some version of the "exchanging packets" model described earlier. Here, we refer to all such models as two-resistance models. The absence of resistance in the bulk phases is incorporated into the models by assigning a uniform concentration throughout each phase (with the exception of the interfacial region), and the absence of resistance at the interface is incorporated by assuming that, right at the interface, the two phases are so close to equilibrium that their concentrations can be related by Henry's law.

In the following sections, the mathematics associated with some of these models is presented. Then, approaches are presented for evaluating whether the gas or liquid interfacial region contributes more resistance to the overall process, and ways in which this information can be used to characterize some important gas transfer processes in environmental engineering are explored. As in previous chapters, the analysis is carried out by writing mass balances around the key regions of interest, in this case the interfacial regions. For simplicity, the interface is assumed to be flat.

The Mass Balance on a Volatile Species Near a Gas/Solution Interface

Gas Transfer and Transport Through a Fluid Packet at the Interface The earliest two-resistance model for gas transfer (Whitman, 1923) was based on the assumption that the water and gas in the interfacial region remained there permanently; this model is referred to as the *two-film*, *stagnant-film*, or *fixed-film* model. The model has many useful features, but it predicts a larger variation in gas transfer rates among different species than is observed experimentally. To address this inconsistency, the possibility that packets of fluid from the bulk phases could exchange with packets at the interface was incorporated into later models, making the residence time distribution (RTD) of the packets a critical model parameter.

The mathematics of gas transfer under two limiting-case scenarios for the RTDs of the packets was then developed. One limiting case assumed that all packets had identical residence time at the interface, whereas the other assumed a distribution that decayed exponentially with increasing time. This latter assumption corresponds to a dynamic in which all packets that are at the interface at any instant have equal probability of being stripped away and returned to bulk solution in the next instant (regardless of how long they have already been at the interface). These RTDs correspond closely to the RTDs of water packets in a continuous flow reactor with plug flow and complete mix hydraulics, respectively. The model with uniform packet residence times was developed by Higbie (1935) and is called the *penetration model*, and that with an exponential RTD was developed by Danckwerts (1951) and is called the *surface renewal model*. In the context of these models, the stagnant-film model can be described as assuming an RTD in which all the packets spend so much time at the interface (approaching an infinite time) that the details of any differences among packet residence times are irrelevant. These different RTDs are shown schematically in Figure 5-11.

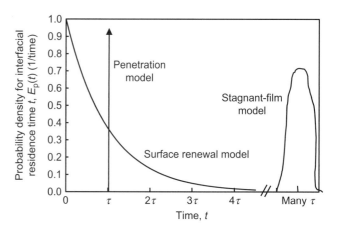

FIGURE 5-11. Packet residence time distributions according to the two-film, penetration, and surface renewal models. $E_p(t)$ is the probability density associated with t; that is, the probability that a packet will spend between t and $t + dt$ at the interface, divided by dt. In the example curves, the mean packet residence time is the same for the penetration and surface renewal models (τ_{Pen} and τ_{SR}, respectively). For the stagnant-film model, no explicit RTD is postulated, but all packets are assumed to have very long residence times near the interface.

Consider the scenario that ensues when a packet of bulk solution containing a volatile substance *i* arrives at the interface. Assume that, for the given gas-phase composition, the bulk solution is undersaturated with respect to species *i*. Some *i* molecules will transfer quickly from the gas into solution, establishing equilibrium across the interface and a concentration gradient in the liquid packet. In response, molecules of *i* begin to diffuse through the packet, away from the interface and toward the bulk solution, so the concentration of *i* throughout the packet increases.

Initially, the concentration gradient and resultant flux near the interface are large, and the corresponding values near the boundary of the packet with the bulk solution are very small.

The concentration profile in the packet at any time can be computed by solving the nonsteady-state mass balance written around the packet. A definition diagram for the system, representing the gas/liquid interface and the boundary of the interfacial layer with the bulk solution as planes and the packet as occupying the space between them, is provided in Figure 5-13.

The cross-sectional area of the packet and the thickness of the interfacial region are designated as A and δ_L, respectively, and we assume that a concentration gradient exists only perpendicular to the interface (in the x direction). Using a thin layer of the packet as the control volume, the mass balance can be written as follows:

$$\begin{pmatrix}\text{Rate of change}\\ \text{of mass of } i\\ \text{stored in the}\\ \text{control volume}\end{pmatrix} = \begin{pmatrix}\text{Net change in}\\ \text{mass of } i \text{ in the}\\ \text{control volume}\\ \text{due to advection}\end{pmatrix} + \begin{pmatrix}\text{Net change in}\\ \text{mass of } i \text{ in the}\\ \text{control volume}\\ \text{due to diffusion}\\ \text{and dispersion}\end{pmatrix} + \begin{pmatrix}\text{Net change in}\\ \text{mass of } i \text{ in}\\ \text{the control}\\ \text{volume due to}\\ \text{reaction}\end{pmatrix} \quad (5\text{-}27)$$

$$\frac{\partial c_L}{\partial t} A\,dx = -A v \frac{\partial c_L}{\partial x} dx + A(D_L + \varepsilon_L)\frac{\partial^2 c_L}{\partial x^2} dx + r_L A\,dx$$

If the packet stays at the interface for only a short time (compared with the characteristic time for diffusion through the packet), then the gradient near the bulk solution will remain small, and the packet will remain in an unsteady-state situation throughout its time at the interface. On the other hand, if the packet stays at the interface for a time that is long compared with the characteristic time for diffusion (but still short enough that the concentration in bulk solution does not change appreciably), then the concentration gradient will diminish near the interface and becomes significant near the bulk solution. These expectations are shown schematically in Figure 5-12.

Note that, because the phase transfer takes place outside the control volume, it does not appear in the mass balance. Both advection and dispersion across the boundaries of the control volume are zero because, by definition, no fluid enters or leaves the packet and no mixing occurs within the packet while it is at the interface. Thus, the first term on the right side of the mass balance has a value of zero

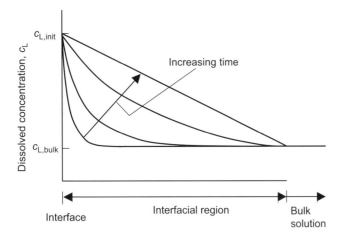

FIGURE 5-12. Schematic representation of the concentration profile in a packet of water after it reaches the interface, in a system where the solution is undersaturated with the volatile constituent.

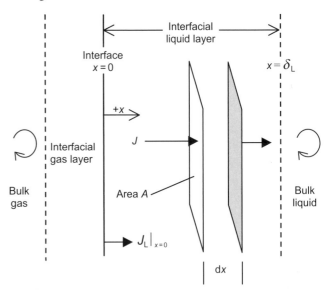

FIGURE 5-13. Definition diagram for the mass balance around a packet of liquid arriving at a planar interface. The system of interest is a packet of cross-sectional area A that extends from $x = 0^+$ to $x = \delta_L$. The mass balance (Equation 5-27) is written over a differential volume with area A and thickness dx (the area between the two planes in the middle of the interfacial layer).

(because $v = 0$), and we can assign $\varepsilon = 0$ in the second term. Assuming that i does not undergo any reaction in the interfacial region, r_L is zero as well. As a result, the molecular diffusion term is the only significant term on the right side, and Equation 5-27 simplifies to

$$\frac{\partial c_L}{\partial t} = D_L \frac{\partial^2 c_L}{\partial x^2} \qquad (5\text{-}28)$$

The concentration profile in the packet is found by integrating Equation 5-28 from a plane adjacent to the interface on the solution side (so that the gas transfer reaction remains outside the system boundaries) to $x = \delta_L$. Solution of the equation requires one initial condition and two boundary conditions. The initial condition is that, when the packet first arrives at the interface, the concentration throughout the packet equals that in the bulk solution ($c_{L,b}$). The boundary conditions apply at the interface ($x = 0$), where the concentration is $c_{L,int}$, and at the boundary between the interfacial region and the bulk solution ($x = \delta_L$), where it is $c_{L,b}$.

The values of $c_{L,int}$ and $c_{L,b}$ are steadily changing while the packet is at the interface. However, incorporating these changes into the boundary conditions makes the problem very difficult to solve. This difficulty can be circumvented by assuming that both $c_{L,int}$ and $c_{L,b}$ are constant during the short period that the packet stays at the interface. The former assumption reflects the idea that the actual gas transfer reaction proceeds very quickly, so that the concentration right at the interface changes rapidly when the packet first arrives and then only slightly during the remaining time that it resides there. We leave the identification of the exact value of $c_{L,int}$ for later; all that matters now is the assumption that it is a constant during the residence time of a packet of solution at the interface. The assumption that $c_{L,b}$ is constant during the same time period is considered reasonable and realistic, because the bulk-phase composition changes slowly, and the packet is assumed to stay at the interface for a relatively short time.

The first integration of Equation 5-28 yields an expression for the spatial gradient of c_L (i.e., $\partial c_L / \partial x$) as a function of location and time, and the second integration yields the concentration profile in the packet. The result for the second integration can be expressed as the following infinite series[13] and is shown for an example system in Figure 5-14a.

$$c_L(x,t) = c_{L,b} + (c_{L,int} - c_{L,b})$$
$$\times \sum_{i=0}^{\infty} \left\{ \operatorname{erfc} \frac{(2i+1)\delta_L - x}{2\sqrt{D_L t}} - \operatorname{erfc} \frac{(2i+1)\delta_L + x}{2\sqrt{D_L t}} \right\} \quad (5\text{-}29)$$

where erfc is the complementary error function.

[13] This equation is based on a solution given by Carslaw and Jaeger (1959) for diffusive heat transfer through a slab, a problem that has the same set of governing equations as the situation being analyzed here.

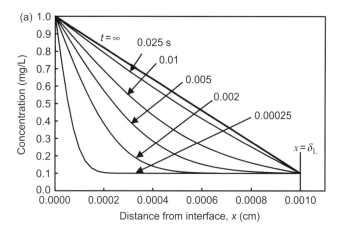

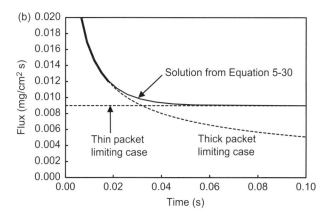

FIGURE 5-14. Characteristics of gas transfer into and through a stagnant packet of liquid at a gas/solution interface at various times after the packet arrives at the interface. (a) Concentration profiles through the packet. (b) Flux into solution. Assumed conditions: $c_{L,b} = 0.1$ mg/L, $c_{L,int} = 1.0$ mg/L, $\delta_L = 0.001$ cm = 10 μm, and $D_L = 1 \times 10^{-5}$ cm^2/s.

According to Fick's law (see Chapter 1), the negative product of the concentration gradient with the diffusion coefficient yields the flux. Thus, the flux of i into solution can be computed by evaluating this product immediately adjacent to the interface (i.e., at $x = 0$) as follows:

$$J_L = -D_L \left. \frac{\partial c_L}{\partial x} \right|_{x=0} \qquad (5\text{-}30)$$

Toor and Marcello (1958) combined Equations 5-29 and 5-30 to develop the following equation for the flux into solution:

$$J_L = \sqrt{\frac{D_L}{\pi t}} (c_{L,int} - c_{L,b}) \left[1 + 2 \sum_{i=1}^{\infty} \exp\left\{ -\frac{i^2 \delta_L^2}{D_L t} \right\} \right]$$
$$(5\text{-}31)$$

The flux into solution for the example system, computed according to Equation 5-31, is plotted as the solid line in

Figure 5-14b. (The significance of the broken lines labeled as limiting cases is explained shortly.)

Both the thickness of the interfacial layer and the average time that packets spend in this layer depend on the mixing intensity in the bulk fluid. For example, the interfacial thickness in the liquid phase is thought to range from a few micrometers for vigorously mixed aerated tanks to a few tens of micrometers for thin sheets of falling water (such as those moving over the surface of packing material in packed columns) and up to a few hundred micrometer for relatively stagnant water surfaces. The corresponding packet residence times range from approximately one-tenth to a few tenths of a second in vigorously mixed tanks to 10 seconds or more in nearly stagnant water. The conditions shown in Figure 5-14 fall in the middle of these ranges.

Taken together, Figures 5-14a and 5-14b indicate that, initially, the concentration gradient near the interface is steep and the flux into solution is correspondingly large. Over time, the gradient distributes itself across the entire packet, eventually becoming linear when the system reaches steady state, so the flux decreases and ultimately stabilizes at a fixed, nonzero value.

Flux Under Limiting-Case Scenarios: Short and Long Packet Residence Times Two limiting conditions for the flux and concentration profile are of special interest. The first is the limiting condition at short times, when c_L is depleted significantly near the interface, but is barely changed from its initial value of $c_{L,b}$ farther from the interface. The second is at very long times, when the profile becomes nearly linear, and the concentration in the packet differs from $c_{L,b}$ even a short distance from $x = \delta_L$.

In the first case, because the concentration perturbation has not fully penetrated the packet, the concentration profile in the packet is essentially the same as if the boundary condition was applied at $x = \infty$ instead of $x = \delta_L$. If this replacement is made in the boundary condition, the equations for the mass balance, the concentration profile, and the flux into solution have the closed-form solutions shown in the following equations:

$$\frac{\partial c_L}{\partial t} = D_L \frac{\partial^2 c_L}{\partial x^2} \tag{5-32}$$

$$c_L(x,t) = c_{L,\text{int}} + (c_{L,b} - c_{L,\text{int}})\operatorname{erf}\left(\frac{x}{\sqrt{4D_L t}}\right) \tag{5-33}$$

$$J_L = \sqrt{\frac{D_L}{\pi t}}(c_{L,\text{int}} - c_{L,b}) \tag{5-34}$$

Conceptually, the constraint that the concentration perturbation has not penetrated to $x = \delta_L$ must be met at very short times in packets of any size, and it will be met for longer periods in thick packets than in thin ones. Thus, Equations 5-33 and 5-34 are considered to be applicable to the limiting cases of either short packet residence times or thick packets. The computed flux for this limiting case is shown as the curved broken line in Figure 5-14b, which indicates that the approximation is very good up to times of ~0.02 s in the example system. As shown in Figure 5-14a, the concentration profile changes substantially during this time period.

The other limiting case applies if a packet spends a long time at the interface or if it is very thin. In this case, the results of the more general analysis indicate that the packet reaches a steady-state condition in which the concentration profile through the interfacial region is linear (shown by the line for $t = \infty$ in Figure 5-14a). The details of this steady state can be analyzed by setting the left side of Equation 5-28 to zero. A single integration of this equation yields the concentration gradient, and a second integration yields the concentration profile, as follows:

Mass balance equation:
$$0 = D_L \frac{d^2 c_L}{dx^2} \tag{5-35}$$

First integration:

$$\frac{dc_L}{dx} = \text{constant} = \frac{\Delta c_L}{\Delta x} \tag{5-36}$$

$$= \frac{c_{L,b} - c_{L,\text{int}}}{\delta_L - 0} = \frac{c_{L,b} - c_{L,\text{int}}}{\delta_L} \tag{5-37}$$

Second integration:

$$c_L(x) = c_{L,\text{int}} + \frac{x}{\delta_L}(c_{L,b} - c_{L,\text{int}}) \tag{5-38}$$

The diffusive flux into the solution can again be determined by applying Fick's law at $x = 0$, in conjunction with the computed concentration gradient (Equation 5-37):

$$J_L = -D_L \left.\frac{dc_L}{dx}\right|_{x=0} = -D_L \frac{c_{L,b} - c_{L,\text{int}}}{\delta_L} = \frac{D_L}{\delta_L}(c_{L,\text{int}} - c_{L,b}) \tag{5-39}$$

The results are consistent with the profile and flux shown for long residence times in Figure 5-14: the concentration profile is linear, with a slope equal to the ratio of the concentration difference across the region to the thickness of the packet. This linear concentration gradient leads to a constant flux through the packet and into solution, as shown by the horizontal broken line in Figure 5-14b. In the example system, the thin packet, long residence time limiting case applies at times greater than approximately 0.25 s. Note that, for this limiting case, the flux at the gas/liquid interface, computed according to Equation 5-39, is also the flux into the bulk solution (at $x = \delta_L$). In contrast, for the thick packet limiting case, almost none of the volatile species enters the bulk solution while the packet is at the interface; rather, the change in bulk concentration occurs only when the packets are swept away from the interface and mix with the rest of the solution.

Based on Equation 5-31, Toor and Marcello (1958) reported that the parameter that identifies whether either of the limiting cases applies is the dimensionless ratio $t/(\delta_L^2/D_L)$. If this ratio is small (less than approximately 0.2), the thick packet, short exposure time approximation leading to Equations 5-33 and 5-34 applies, and if it is large (greater than approximately 0.6), the thin packet, long exposure time approximation applies, and Equations 5-38 and 5-39 can be used; between these two extremes, neither approximation is justifiable, and the system condition must be evaluated using Equation 5-31.

The term δ_L^2/D_L has dimensions of time and can be considered as the characteristic time for diffusion through the liquid interfacial region. Thus, $t/(\delta_L^2/D_L)$ is the ratio of the actual time that a packet has been at the interface to the characteristic time for diffusion through the packet. A small value of this ratio indicates that the packet has been present for much less than one characteristic time for diffusion, consistent with the idea that the effects of diffusion have not penetrated through the entire packet. On the other hand, if the ratio is large, the packet has been at the interface for many characteristic diffusion times, so the diffusive flux approximates its ultimate (steady-state) value. For the example system shown in Figure 5-14, the characteristic time for diffusion, δ_L^2/D_L, is 0.10 s, and the curves for the five different values of t correspond to $t/(\delta_L^2/D_L)$ ratios ranging from 0.0025 to 0.25.[14] Note that, even if a packet spent 0.5 s in the interfacial region, so that the concentration profile was approximately linear when the packet re-entered bulk solution, the average flux during the packet's time at the interface would have been significantly greater than the steady-state flux, because of the higher fluxes that were achieved when it first arrived.

Accounting for the Packet Age and Packet Residence Time Distribution The equations developed in the preceding section describe the flux into a single packet of fluid at the interface. However, a real interface will have some packets that arrived recently and others that have been present for a long time, all exchanging the volatile species with the gas phase at different rates. In most cases, our interest is in the average flux, considering all the packets present at the interface at any given time. As noted in the introduction to this section, this average flux has been evaluated for the two model packet RTDs shown in Figure 5-11. Designating the mean packet residence time at the interface as $\tau_{p,L}$, the difference between the models is that, in the penetration model, the residence time of every packet equals $\tau_{p,L}$, whereas in the surface renewal model, different packets have different residence times, even though the mean value is still $\tau_{p,L}$. (Note that, even in the penetration model, at any instant the interfacial layer is occupied by packets that have been present for a range of times from 0 to $\tau_{p,L}$, so the flux differs from one packet to the next.)

When the RTDs of the two models are combined with the equations for flux from individual packets, equations for the average flux considering all the packets are obtained. These equations are provided in Table 5-3. Although the equations for the general case are complex, those for the limiting cases of small and large $t/(\delta_L^2/D_L)$ are fairly simple and bear strong similarities to those for flux through an isolated packet. Note also that the flux for the thin packet, long residence time scenario is independent of the packet RTD, because, as explained previously, this RTD presumes that all the packets have been at the interface long enough to have reached a steady-state concentration profile and flux.

Although the preceding discussion focuses on the liquid interfacial region, identical processes and equations characterize the gas side of the interface. The concentration profiles on both sides of the interface for the thin packet, long residence time limiting-case scenario (i.e., Whitman's two-film model) are shown Figure 5-15. This view of the interface is still widely used, probably more because of its conceptual simplicity than its applicability in real systems; for any detailed modeling, the likelihood that packets fail to reach steady state while at the interface should be taken into account.

It is worth noting that the assumption that the interface itself is infinitely thin (i.e., a plane) requires that the flux out of one phase equal the flux into the other. Combining the requirement for equality of the fluxes through the two boundary layers with the fact that D_G is typically about four orders of magnitude greater than D_L for the same substance (because there is so much less resistance to molecular movement in the gas phase than in solution) leads to the conclusion that the concentration gradient immediately adjacent to the interface in the gas boundary layer must be very much smaller than that adjacent to the interface in the liquid boundary layer. This statement applies regardless of any assumption being made about the packet RTD.

The Gas Transfer Coefficient and Its Interpretation The preceding discussion introduces various conceptual and mathematical models for the gas transfer process. From a

---

[14] Based on the interpretation of this ratio as the number of characteristic diffusion times that a packet stays in the interfacial region, we might expect that the criterion for the thin packet approximation to apply would be that $t/(\delta_L^2/D_L)$ be $\gg 1$, rather than >0.6 as suggested by Toor and Marcello (1958). However, the definitions of the characteristic times for the two processes (diffusion and packet replacement) are not completely consistent with one another, so the value of the ratio when the component processes are equally important does not necessarily equal 1.0. This situation is similar to the case in fluid mechanics, where the choice for a characteristic length to use in computing the Reynolds number (Re) causes the transition between laminar and turbulent conditions to occur at Re other than 1.0.

176 GAS TRANSFER FUNDAMENTALS

TABLE 5-3. Average Flux into Solution for Various Packet Properties and for Two Packet Interfacial RTDs

Any value of $\dfrac{\tau_{p,L}}{\delta_L^2/D_L}$ (most general case)

PFR-type packet RTD: $\quad J_L = 2\sqrt{\dfrac{D_L}{\pi \tau_{p,L}}}(c_{L,\text{int}} - c_{L,b})\left[1 + 2\sqrt{\pi}\sum_{i=1}^{\infty}\text{ierfc}\left\{\dfrac{i\delta_L}{\sqrt{D_L \tau_{p,L}}}\right\}\right]$ (5-40)

CFSTR-type packet RTD: $\quad J_L = \sqrt{\dfrac{D_L}{\tau_{p,L}}}(c_{L,\text{int}} - c_{L,b})\left[1 + 2\sum_{i=1}^{\infty}\exp\left\{-2i\delta_L\dfrac{1}{\sqrt{D_L \tau_{p,L}}}\right\}\right]$ (5-41)

$\dfrac{\tau_{p,L}}{\delta_L^2/D_L}$ small (thick packet, short times)

PFR-type packet RTD: $\quad J_L = 2\sqrt{\dfrac{D_L}{\pi \tau_{p,L}}}(c_{L,\text{int}} - c_{L,b})$ (5-42)

CFSTR-type packet RTD: $\quad J_L = \sqrt{\dfrac{D_L}{\tau_{p,L}}}(c_{L,\text{int}} - c_{L,b})$ (5-43)

$\dfrac{\tau_{p,L}}{\delta_L^2/D_L}$ large (thin packet, long times)

Any RTD: $\quad J_L = \dfrac{D_L}{\delta_L}(c_{L,\text{int}} - c_{L,b})$ (5-39)

---

practical perspective, the most important result of the analysis is that all the equations for the average flux into solution (i.e., all those in Table 5-3) can be written in the following form:

$$J_L = k_L(c_{L,\text{int}} - c_{L,b}) \quad (5\text{-}44)$$

where k_L is a constant known as the *liquid-phase gas transfer coefficient*, with units of length per time. Equation 5-44 indicates that the gas transfer coefficient equals the flux into solution per unit concentration difference across the packet. However, even though we derived the equation to characterize transport of a volatile species, a review of the derivation shows that k_L is not related fundamentally to gas transfer; in fact, it could be applied to any constituent that has a concentration gradient between the bulk solution and a phase boundary, if transport occurs by exchange of packets between the two locations. Therefore, k_L is often referred to more generically as a *liquid-phase mass transfer coefficient*. An equation analogous to Equation 5-44 can be written for the gas phase, in which case the coefficient is written as k_G and is referred to as a *gas-phase mass transfer* (or *gas transfer*) *coefficient*; that is,

$$J_G = k_G(c_{G,\text{int}} - c_{G,b}) \quad (5\text{-}45)$$

Like k_L, k_G has dimensions of length per time.[15]

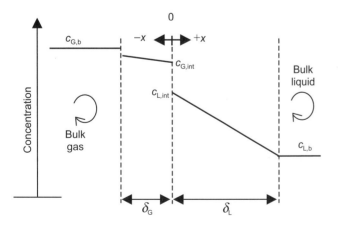

FIGURE 5-15. Schematic representation of the two-film model for gas transfer.

[15] Since c_L is based on a volume of solution, and since the volume of the interfacial region equals the product of the interfacial area and the thickness of the layer, the dimensions of k_L are volume *in the solution phase* per unit interfacial area per unit time. At times, it is useful to write the units of k_L as $cm_L^3/cm^2 s$, where the area term in the denominator is not subscripted because the interfacial area is not in either phase. Correspondingly, the units of k_G are volume in the gas phase per unit interfacial area per unit time (e.g., $cm_G^3/cm^2 s$). Nevertheless, the dimensions of both k_L and k_G are usually reported simply as length per time.

The liquid- and gas-phase mass transfer coefficients are measures of the ease with which a species can be transported between the interface and the corresponding bulk fluid. These coefficients reflect characteristics of both the physical system (the packet exchange frequency and the thickness of the interfacial layer in each phase) and the constituent of interest (diffusivity in each phase). Other factors being equal, we expect k_L for all species in a system to increase with increasing mixing intensity in the liquid, because this change would increase the packet exchange frequency (lower $\tau_{p,L}$) and decrease the thickness of the interfacial layer (lower δ_L). The same effects would occur for the gas-phase interfacial layer by increasing the mixing of the gas phase, when this change is possible. These effects have been confirmed experimentally.

Similarly, in a given system (with given values of $\tau_{p,L}$ and δ_L that are applicable to all solutes), we expect k_L for different species to increase in the same order as their diffusivities. This dependence is predicted by Equations 5-39, 5-42, and 5-43. However, the exact nature of the dependence is predicted to be different depending on what stage of the process an average packet is in. That is, if most of the packets are at a stage of the process such that the thick packet, short residence time approximation applies (i.e., if the concentration gradient has not yet penetrated completely through them), k_L is predicted to be proportional to the square root of diffusivity; on the other hand, if most packets are at a stage where the thin packet, long residence time approximation applies (i.e., if they have been at the surface long enough to reach steady state), k_L is predicted to be proportional to the molecule's diffusivity to the first power.

The evidence from many empirical studies of gas transfer coefficients suggests that k_L and k_G do indeed vary with the diffusivity raised to some power m; that is

$$\frac{(k_L a_L)_2}{(k_L a_L)_1} = \left(\frac{D_{L,2}}{D_{L,1}}\right)^m \quad (5\text{-}46)$$

Reported values of m are rarely less than 0.5 or greater than about 0.75 (Munz and Roberts, 1984; Versteeg et al., 1987). These results suggest that the stagnant-film model (which predicts that $m = 1.0$) is rarely a realistic representation of the system dynamics. Rather, most systems seem to be characterized by packets that are swept away from the interface during the early or intermediate stages of the gas exchange process, well before they have reached steady state.

While the generic packet exchange model that is used as the framework for the preceding discussion has proven quite useful and versatile, it is, of course, still an idealized representation of a complex process. Other models have been developed that make different assumptions and are, arguably, more realistic. For instance, Kishinevskii and coworkers (Kishinevskii and Pamfilev, 1949; Kishinevskii and Serebrianskii, 1956), Harriott (1962), and King (1966) all have proposed models in which interfacial packets are not replaced in their entirety, but only partially, when a turbulent eddy approaches the surface. In these models, the outer part of a packet (the part nearest the bulk solution) is predicted to be relatively easily replaced, whereas the portions nearer the interface are replaced much less frequently. With an appropriate function relating turbulence with distance from the interface, this model predicts that k_L is proportional to D_L^m, with the typical range of m being $0.5 \leq m \leq 0.75$. Versteeg et al. (1987) provided details on this and related models; several more recent modifications and refinements of the models have been published in the chemical engineering and fluid mechanics literature (e.g., Fan et al., 1993; Munsterer and Jahne, 1998; Jahne and Haussecker, 1998; Moog and Jirka, 1999).

Values of D_L for several volatile species of interest are provided in Table 5-4; larger compilations have been provided by Hayduk and Laudie (1974) and by Cussler (1997). In addition, a number of correlations have been developed to estimate diffusion coefficients based on the molecular properties of the species of interest and the bulk properties of the fluid (see, e.g., Cussler, 1997; Poling et al., 2001; Green and Perry, 2008). One of the earliest and still most widely used predictive equations is that of Wilke and Chang (1955). This equation was originally presented in a

TABLE 5-4. Diffusivities of Some Volatile Compounds in Water and Air at 20°C[a]

	D_L (10^{-5} cm²/s)	D_G (cm²/s)	D_G/D_L (—)
Ethyl benzene	0.81	0.088	1.09×10^4
Tetrachloroethylene (PCE)	0.85	0.092	1.08×10^4
Cyclohexane	0.85	0.080	1.06×10^4
m- and p-Xylenes	0.87	0.088	1.00×10^4
1,1,1-Trichloroethane	0.90	0.092	1.03×10^4
Toluene	0.90	0.097	1.09×10^4
Monochlorobenzene	0.90	0.092	1.03×10^4
Carbon tetrachloride	0.92	0.084	0.91×10^4
Dichlorobromomethane	0.92	0.110	1.19×10^4
Bromoform	0.92	0.088	0.96×10^4
Trichloroethylene (TCE)	0.94	0.097	1.04×10^4
Ethylene dibromide	0.94	0.103	1.10×10^4
Chlorodibromomethane	0.96	0.097	1.01×10^4
Chloroform	1.03	0.117	1.13×10^4
Acetone	1.10	0.103	0.94×10^4
Methylene chloride	1.28	0.125	0.98×10^4
Formaldehyde	1.56	0.136	0.87×10^4
Oxygen	2.05	0.200	0.98×10^4
Ammonia	2.15	0.274	1.27×10^4

[a]Data based on values given by Selleck et al. (1988), Cussler (1997), and Howard and Corsi (1996).

general form that could be applied to diffusion through any liquid. A slightly modified version (Hayduk and Laudie, 1974) that is applicable specifically to diffusion through aqueous solutions is (Green and Perry, 2008)

$$D_L = \frac{1.316 \times 10^{-4}}{\mu_L^{1.14} \overline{V}_i^{0.589}} \qquad (5\text{-}47)$$

where D_L is in cm²/s, μ_L is the solution viscosity in centipoise (10^{-2} g/cm s), and $\overline{V}_i$ is the molar volume of i at the boiling point (cm³/mol). Because, in liquids, viscosity decreases with an increase in temperature, diffusivity increases with increasing temperature.

Predictive equations for diffusion coefficients in gases indicate that D_G varies with T raised to a positive exponent (typically, between 1.5 and 2.0). For instance, the following expression has been suggested for diffusivities in air (Fuller et al., 1966):

$$D_G = 10^{-3} \frac{T^{1.75}[(1/28.8) + (1/\text{MW}_i)]^{1/2}}{P\left[(20.1)^{1/3} + (V_{D,i})^{1/3}\right]^2} \qquad (5\text{-}48)$$

where MW is the molecular weight, $V_{D,i}$ is the "diffusion volume" of i, which can be estimated based on its molecular structure (Fuller et al., 1966; Cussler, 1997; Green and Perry, 2008), and P is the total pressure in atmospheres; the values 28.8 and 20.1 are the average MW and diffusion volume, respectively, of the constituents of air. In general, diffusivities in air are much less sensitive to temperature than they are in water.

■ **EXAMPLE 5-3.** The diffusivity of chloroform at 20°C is given in Table 5-4 as 1.03×10^{-5} cm²/s in an aqueous solution and 0.117 cm²/s in a gas phase. The viscosity of water at temperatures from 0°C to 30°C is provided in the following table. Estimate the diffusivity of chloroform in both phases as a function of temperature over this range.

T (°C)	0	5	10	15	20	25	30
μ_L (10^{-3} N s/m²)	1.781	1.518	1.307	1.139	1.002	0.890	0.798

Solution. The diffusivity at each temperature can be estimated based on the given values at 20°C and Equations 5-47 and 5-48, which indicate that the ratios of diffusivities at two temperatures are approximately

$$\frac{D_{L,T_2}}{D_{L,T_1}} = \left(\frac{\mu_{L,T_1}}{\mu_{L,T_2}}\right)^{1.14} \quad \text{and} \quad \frac{D_{G,T_2}}{D_{G,T_1}} = \left(\frac{T_2}{T_1}\right)^{1.75}$$

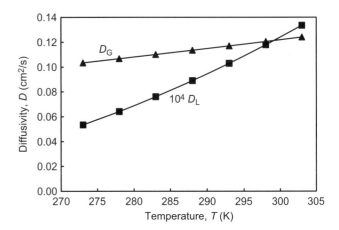

FIGURE 5-16. Temperature-dependence of gas- and solution-phase diffusivities of chloroform.

where T is the absolute temperature. Using $T_1 = 20°C = 293$ K, values of $D_\text{chloroform}$ in each phase at each of the other temperatures of interest have been computed and are plotted in Figure 5-16. Consistent with the generalization in this text, the diffusivity in solution is much more sensitive to temperature than that in air. ■

■ **EXAMPLE 5-4.** The aqueous-phase diffusivities of oxygen and trichloroethylene (TCE) at 20°C are 2.05×10^{-5} and 0.94×10^{-5} cm²/s, respectively. The results obtained by Hsieh et al. (1993) for k_L for these two species in a surface aeration system operated under various conditions yielded a best-fit ratio of 0.68 for $k_{L,\text{TCE}}/k_{L,O_2}$. Assuming that the k_L values are proportional to D_L^m, compute the best-fit value of m for this correlation. Are the results more consistent with the thick packet model, the thin packet model, or neither?

Solution. The value of m can be determined by substitution into a slightly modified form of Equation 5-46, as follows:

$$\frac{k_{L,\text{TCE}}}{k_{L,O_2}} = \left(\frac{D_\text{TCE}}{D_{O_2}}\right)^m$$

$$\log \frac{k_{L,\text{TCE}}}{k_{L,O_2}} = m \log \frac{D_\text{TCE}}{D_{O_2}}$$

$$m = \frac{\log(k_{L,\text{TCE}}/k_{L,O_2})}{\log(D_\text{TCE}/D_{O_2})}$$

$$= \frac{\log 0.68}{\log(0.94 \times 10^{-5}/2.05 \times 10^{-5})} = 0.49$$

Thus, the experimental results are consistent with the surface renewal and penetration models (i.e., the thick packet models) of gas transfer ($m = 0.5$) and are inconsistent with the stagnant-film (thin packet) model ($m = 1.0$). ∎

5.6 THE OVERALL GAS TRANSFER RATE COEFFICIENT, K_L

The Combined Resistance of the Gas and Liquid Phases

While Equations 5-44 and 5-45 are useful as a quantitative representation of our conceptual model of transport between the interface and the bulk phases, they are not particularly useful for practical design or analysis, since they contain terms (the interfacial concentrations) that are impossible to measure. In this section, we circumvent the problem by utilizing the assumption that exchange of volatile species across the interfacial plane is extremely rapid, so that Henry's law applies at this location at all times. As noted earlier, this assumption is equivalent to assuming that the transfer of molecules across the interface contributes negligibly to the total resistance of the system to gas transfer.

We begin the analysis by equating the fluxes into solution and out of the gas:

$$J_L = -J_G \tag{5-49a}$$

$$k_L(c_{L,\text{int}} - c_{L,b}) = k_G(c_{G,b} - c_{G,\text{int}}) \tag{5-49b}$$

For this analysis and many other calculations related to gas transfer processes, it is useful to define hypothetical concentrations that would be in equilibrium with the bulk gas or solution phase, even if the system is not, in reality, equilibrated. The hypothetical gas- and solution-phase concentrations that would be in equilibrium with a given solution and gas, respectively, are designated here as c_G^* and c_L^*, and are defined mathematically as follows:

$$c_G^* = Hc_L \tag{5-50}$$

$$c_L^* = \frac{c_G}{H} \tag{5-51}$$

The significance of c_G^* and c_L^* with respect to a plot of c_G versus c_L are shown in Figure 5-17.

The concentration of the volatile species in the bulk gas phase can be written as $c_{G,b} = Hc_L^*$. Making this substitution in Equation 5-49b and applying the assumption of equilibrium at the interface (i.e., $c_{G,\text{int}} = Hc_{L,\text{int}}$), the equation can be rewritten and solved for $c_{L,\text{int}}$ as follows:

$$k_L(c_{L,\text{int}} - c_{L,b}) = k_G(Hc_L^* - Hc_{L,\text{int}}) \tag{5-52}$$

$$c_{L,\text{int}} = \frac{k_G Hc_L^* + k_L c_{L,b}}{k_L + k_G H} \tag{5-53}$$

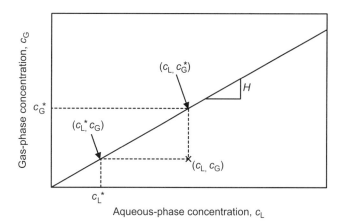

FIGURE 5-17. Aqueous and gas-phase concentrations in an example system (indicated by the ×), and the corresponding values of c_G^* and c_L^*.

The flux from the gas into the liquid film can then be written as follows:

$$J_L = k_L(c_{L,\text{int}} - c_{L,b}) \tag{5-44}$$

$$= k_L\left(\frac{k_G Hc_L^* + k_L c_{L,b}}{k_L + k_G H} - c_{L,b}\right)$$

$$= k_L\left(\frac{k_G Hc_L^* + k_L c_{L,b}}{k_L + k_G H} - \frac{k_L c_{L,b} + k_G Hc_{L,b}}{k_L + k_G H}\right)$$

$$= k_L\left(\frac{k_G Hc_L^* - k_G Hc_{L,b}}{k_L + k_G H}\right)$$

$$= \frac{k_L k_G H}{k_L + k_G H}(c_L^* - c_{L,b}) \tag{5-54}$$

Equation 5-54 is most often written in the following form:

$$J_L = K_L(c_L^* - c_{L,b}) \tag{5-55}$$

where K_L is the *overall gas transfer coefficient*, defined as

$$K_L \equiv \frac{k_L k_G H}{k_L + k_G H} \tag{5-56}$$

Equation 5-55 gives the flux into solution as the product of a constant and the difference between two macroscopic, measurable variables. Although it has essentially the same form as the equations describing gas transfer rates based on the (unmeasurable) interfacial concentrations (e.g., Equations 5-44 and 5-45), it differs from these equations in that K_L is an overall gas transfer rate coefficient, reflecting the composite effect of the individual gas transfer rate constants in both interfacial zones and of Henry's law constant.

As noted earlier (Equation 5-25), the rate of gas transfer per unit volume of reactor is the product of the flux and the concentration of surface area in the reactor (a_R). Thus, the rate of gas transfer into the liquid phase per unit volume of reactor is

$$r_{R,gt} = J_L a_R = K_L a_R (c_L^* - c_{L,b}) \quad (5\text{-}57)$$

Note that $r_{R,gt}$ is the rate of appearance of i in the aqueous phase, but the units of $r_{R,gt}$ are mass per unit time per unit volume of reactor (not per unit volume of solution).

The corresponding value for the rate of gas transfer per unit volume of solution can be computed by multiplying $r_{R,gt}$ by the ratio of the reactor volume to the aqueous volume, or, equivalently, by replacing a_R by a_L in Equation 5-57:

$$r_{L,gt} = K_L a_R \frac{V_R}{V_L}(c_L^* - c_{L,b}) \quad (5\text{-}58a)$$

$$= K_L a_L (c_L^* - c_{L,b}) \quad (5\text{-}58b)$$

The overall rate constant $K_L a$ (either $K_L a_R$ or $K_L a_L$) combines a transport-related term (K_L) that describes the ease with which the volatile species passes through the two interfacial layers, with a geometric term (a_R or a_L) that describes the concentration of surface area in the system. In most engineered gas transfer systems, the total interfacial area in the system depends on system-specific mixing and geometric considerations and is not easily measured. Therefore, when gas transfer is evaluated experimentally in such systems, the lumped parameter $K_L a_R$ or $K_L a_L$ is more frequently reported than are individual values of k_L, k_G, K_L, a_R or a_L. Defining equations for $K_L a_R$ or $K_L a_L$ can be obtained by multiplying both sides of Equation 5-56 by a_L or a_R:

$$K_L a_L = \frac{k_L k_G a_L H}{k_L + k_G H} \quad (5\text{-}59a)$$

$$K_L a_R = \frac{k_L k_G a_R H}{k_L + k_G H} \quad (5\text{-}59b)$$

$K_L a_R$ and $K_L a_L$ have the conventional dimensions for a first-order rate constant (time^{-1}).

The description of gas transfer earlier in this chapter emphasizes that the transfer of a constituent between a well mixed gas and a well mixed aqueous phase involves several transport steps as well as a chemical reaction. Nevertheless, Equations 5-57 and 5-58 indicate that this process can be modeled quantitatively by a rate expression that is very similar to that for a first-order chemical reaction alone, with a driving force equal to the extent of disequilibrium between the two bulk phases, and an overall rate constant $K_L a$. The expression for $r_{L,gt}$ or $r_{R,gt}$ can be substituted directly into a mass balance equation, just as the expression $-k_1 c$ is substituted if the constituent undergoes a first-order decay reaction.

■ **EXAMPLE 5-5.** Droplets of groundwater 0.5 mm in diameter and containing 3×10^{-5} mol/L H_2CO_3 and 4 mg/L O_2 are sprayed into the air. The value of K_L for both constituents is 0.05 cm/min, and the temperature of both air and water is 25°C. The partial pressures of O_2 and CO_2 in air are 0.21 and 0.00031 atm, respectively. What are the initial rates (mol/L min) and directions (into or out of the droplet) of CO_2 and O_2 transfer? Assume that the solution pH is low enough that ionization of H_2CO_3 is negligible.

Solution. Henry's constants for O_2 and CO_2 at 25°C are given in Table 5-2, from which the equilibrium concentrations of DO and carbon dioxide can be computed as follows:

$$c_{O_2}^* = \frac{P_{O_2}}{H_{O_2}} = \frac{0.21 \text{ atm}}{769 \text{ atm}/(\text{mol/L})} = 2.73 \times 10^{-4} \text{mol/L}$$

$$c_{CO_2}^* = \frac{P_{CO_2}}{H_{CO_2}} = \frac{3.1 \times 10^{-4} \text{ atm}}{29.4 \text{ atm}/(\text{mol/L})} = 1.05 \times 10^{-5} \text{mol/L}$$

The initial DO concentration in the water in moles per liter is

$$4 \text{ mg/L } O_2 \left(\frac{1 \text{ mol } O_2}{32,000 \text{ mg } O_2}\right) = 1.25 \times 10^{-4} \text{ mol/L}$$

Since $c_{L,O_2} < c_{L,O_2}^*$ and $c_{L,CO_2} > c_{L,CO_2}^*$ the water is initially undersaturated with respect to O_2 and supersaturated with respect to CO_2. As a result, the direction of gas transfer will be into the droplet for O_2 and out of it for CO_2.

Assuming that the droplets are spherical, the value of a_L can be computed from the geometry, as follows:

$$a_L = \frac{\text{surface area of droplet}}{\text{volume of droplet}} = \frac{\pi d^2}{\pi (d^3/6)} = \frac{6}{d}$$

$$a_L = \frac{6}{0.5 \text{ mm}} = 12 \text{ mm}^{-1} = 120 \text{ cm}^{-1}$$

We can compute the initial rate of gas transfer using Equation 5-58 as

$$r_{L,gt} = K_L a_L (c_L^* - c_{L,b})$$

The transfer rates are therefore

$$r_{L,gt,O_2} = (0.05 \text{ cm/min})(120 \text{ cm}^{-1})$$
$$\times (2.73 \times 10^{-4} \text{ mol/L} - 1.25 \times 10^{-4} \text{ mol/L})$$
$$= (8.88 \times 10^{-4} \text{ mol } O_2/\text{L min})$$
$$= (8.88 \times 10^{-4} \text{ mol } O_2/\text{L min})\left(\frac{32{,}000 \text{ mg}}{1 \text{ mol}}\right)$$
$$= 28.4 \text{ mg/L min}$$

$$r_{L,gt,CO_2} = (0.05 \text{ cm/min})(120 \text{ cm}^{-1})$$
$$\times (1.05 \times 10^{-5} \text{mol/L} - 3 \times 10^{-5} \text{mol/L})$$
$$= (-1.20 \times 10^{-4} \text{ mol } CO_2/\text{L min})$$
$$= (-1.20 \times 10^{-4} \text{ mol } CO_2/\text{L min})$$
$$\times \left(\frac{44{,}000 \text{ mg}}{1 \text{ mg}}\right) = -5.1 \text{ mg/L min}$$

Note that the signs on the rate terms are positive for oxygen transfer (O_2 transfers from the gas phase into solution) and negative for CO_2 transfer (from the solution into the gas), consistent with the conclusions from the first part of the problem. ∎

Comparing Gas-Phase and Liquid-Phase Resistances

As discussed in Chapter 3, the rate of any process can be thought of as the ratio of the driving force for the process to the resistance. Applying this concept to gas transfer reactions, and recognizing the product $Ja_L V_L$ as the overall reaction rate (mass transferred per unit time), we can identify the term $(K_L a_L V_L)^{-1}$ as the resistance to gas transfer as follows:

$$\text{Rate} = \frac{\text{Driving force}}{\text{Resistance}}$$

$$J_L a_L V_L = (c_L^* - c_{L,b}) K_L a_L V_L = \frac{c_L^* - c_{L,b}}{(K_L a_L V_L)^{-1}} \quad (5\text{-}60)$$

If we invert the equation defining K_L (Equation 5-56) and divide both sides by $a_L V_L$, we can express the overall resistance as the sum of two resistances in series, each reflecting the resistance to transport on one side of the interface:

$$K_L^{-1} = (k_G H)^{-1} + k_L^{-1} \quad (5\text{-}61a)$$

$$(K_L a_L V_L)^{-1} = (k_G H a_L V_L)^{-1} + (k_L a_L V_L)^{-1} \quad (5\text{-}61b)$$

$$R_{tot} = R_G + R_L \quad (5\text{-}62)$$

where R_{tot}, R_G, and R_L are the total resistance to gas transfer and the portions of that resistance residing in the gas and liquid phases, respectively.[16]

The ratio of R_L to R_G provides a criterion for deciding which of the two resistances governs the overall rate

$$\frac{R_L}{R_G} = \frac{k_G}{k_L} H \quad (5\text{-}63)$$

Correspondingly, the fraction of the total resistance that is attributable to the liquid interfacial region is given by

$$\frac{R_L}{R_{tot}} = \frac{R_L}{R_G + R_L} = \frac{1}{(R_G/R_L) + 1}$$
$$= \frac{1}{(k_L/k_G H) + 1} \quad (5\text{-}64)$$

Equations 5-63 and 5-64 indicate that, for given values of k_G and k_L (i.e., given hydrodynamics), the fraction of the resistance that resides in the liquid phase increases with increasing H. That is, the liquid-phase resistance becomes progressively more important (as a fraction of the total resistance) with decreasing solubility (increasing volatility), other factors being equal. This relationship is shown graphically in Figure 5-18. Typical values of k_G/k_L for gas transfer systems used in environmental engineering are also shown in this figure. These values make it clear that the vast majority of the resistance to transfer of oxygen (log $H = 1.50$) will be in the liquid phase in any system of interest; however, a substantial portion, and perhaps the majority, of the resistance to transfer of a number of VOCs, such as TCE (log $H = -0.35$), chloroform (log $H = -0.80$), and benzene (log $H = -2.05$), typically resides in the gas phase. We return to this issue shortly.

■ **EXAMPLE 5-6.** Consider a system at 25°C, in which the concentration of ammonia in bulk solution is 5×10^{-3} mol/L, and its partial pressure in the bulk gas is zero. Assume that the solution pH is >11, so that reactions of NH_3 with H^+ in solution can be ignored.

(a) For k_L and k_G values of 0.002 and 0.08 cm/s, respectively, compute the following:
  (i) The relative contributions of the aqueous and gaseous boundary layers to the overall resistance to gas transfer.

---

[16] The overall resistance to gas transfer and the contributions from the gas and liquid phases are often defined without the V_L term, as $(K_L a_L)^{-1}$, $(k_G H a_L)^{-1}$, and $(k_L a_L)^{-1}$, respectively, or without the $a_L V_L$ term, as K_L^{-1}, $(k_G H)^{-1}$, and k_L^{-1}. The definitions given here are more consistent with the definition of resistance in other systems (e.g., electrical systems). However, in the end, whether or not the V_L term is included is unimportant, since it affects all the calculated resistances equally, and we are only interested in the relative values of these quantities.

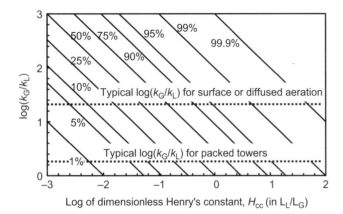

FIGURE 5-18. Relationships among the ratio of the mass transfer coefficients in the gas and liquid phases, Henry's constant, and the fraction of the overall resistance located in the liquid phase. *Source*: Modified from Munz and Roberts (1984); typical k_G/k_L values based on data of Munz and Roberts (1984); Roberts et al. (1985); and Parker et al. (1996).

(ii) The value of K_L.
(iii) The flux of ammonia from solution to the gas phase.
(iv) The concentration of ammonia at the interface, in both the solution and gas phases.
(v) The effect on the flux of ammonia of doubling either of the gas transfer coefficients, while the other remains constant (e.g., by increasing the mixing intensity in one phase or the other).

(b) Repeat the calculations in (a) for transfer of carbon tetrachloride (CCl_4), a much less soluble gas.

Solution.

(a) (i) We can use Equation 5-63 to compute R_L/R_G. To keep the units straight, it is helpful to represent k_L and k_G as described in footnote 15.

$$\frac{R_L}{R_G} = \frac{k_G}{K_L} H$$

$$= \frac{0.08 \text{ cm}_G^3/\text{cm}^2 \text{ s}}{0.002 \text{ cm}_L^3/\text{cm}^2 \text{ s}} (6.95 \times 10^{-4} \text{L}_L/\text{L}_G)$$

$$\times \left(\frac{1 \text{ L}_G}{1000 \text{ cm}_G^3}\right)(1000 \text{ cm}_L^3/\text{L}_L)$$

$$= 0.028$$

$$\frac{R_L}{R_{tot}} = \frac{1}{(R_G/R_L) + 1}$$

$$= \frac{1}{(1/0.028) + 1} = 0.027$$

Thus, the liquid phase contributes only 2.7% of the total resistance to gas transfer, and the gas phase contributes the remaining 97.3%.

(ii) The overall gas transfer coefficient based on the liquid phase, K_L, can be computed using Equation 5-61a. The result is

$$K_L = \left(k_L^{-1} + (k_G H)^{-1}\right)^{-1}$$

$$= \left\{(0.002 \text{ cm}_L^3/\text{cm}^2 \text{ s})^{-1} \right.$$

$$+ \left[0.08 \text{ cm}_G^3/\text{cm}^2 \text{ s}(6.95 \times 10^{-4} \text{L}_L/\text{L}_G)\right.$$

$$\left.\left.\times (1 \text{ L}_G/1000 \text{ cm}_G^3)(1000 \text{ cm}_L^3/\text{L}_L)\right]^{-1}\right\}^{-1}$$

$$= (500 \text{ s/cm}_L + 17,986 \text{ s/cm}_L)^{-1}$$

$$= 5.41 \times 10^{-5} \text{ cm}_L/\text{s}$$

In the previous summation, the value of 500 s/cm_L corresponds to k_L^{-1}, and the value of $17,986 \text{ s/cm}_L$ corresponds to $(k_G H)^{-1}$. The resulting value of K_L is close to that of $k_G H$ ($= 1/17,986$ or $5.56 \times 10^{-5} \text{ cm}_L/\text{s}$) and is much less than k_L ($1/500$ or $0.002 \text{ cm}_L/\text{s}$), reinforcing the concept that gas transfer in the system is controlled by the resistance of the gas-phase interfacial region.

(iii) The flux of ammonia into solution for the given conditions is, according to Equation 5-55

$$J_L = K_L(c_L^* - c_{L,b}) \quad (5\text{-}55)$$

where $c_{L,b}$ is the concentration of dissolved ammonia in bulk solution, which is given as 5×10^{-3} mol/L, and c_L^* is the concentration of dissolved ammonia that would be in equilibrium with the bulk gas phase. Formally, c_L^* can be computed as $c_{G,NH_3}/H_{NH_3}$, but in this case the calculation is trivial because the gas is devoid of ammonia; thus c_L^* is zero. The flux is therefore:

$$J_L = (5.41 \times 10^{-5} \text{cm/s})([0 - 5 \times 10^{-3}]\text{mol/L})$$

$$\times \left(\frac{1 \text{ L}}{10^3 \text{ cm}^3}\right) = -2.70 \times 10^{-10} \text{mol/cm}^2 \text{ s}$$

Multiplying the computed value by the atomic weight of nitrogen converts the result to more conventional mass units and indicates that -3.79×10^{-9} g of NH$_3$–N is transferred per second across each square centimeter of surface.[17] The fact that the flux is negative indicates that the direction of net transfer is out of solution and into the gas phase.

(iv) The concentration of ammonia in solution immediately adjacent to the interface can be computed by equating the overall flux to the flux through the aqueous interfacial zone, as follows:

$$J_L = K_L(c_L^* - c_{L,b}) = k_L(c_{L,int} - c_{L,b})$$

$$c_{L,int} = c_{L,b} + \frac{J_L}{k_L}$$

$$= 5 \times 10^{-3} \text{ mol/L}$$

$$+ \frac{-2.70 \times 10^{-10} \text{ mol/cm}^2 \text{ s}}{0.002 \text{ cm}_L/s}(10^3 \text{ cm}_L^3/L)$$

$$= 4.86 \times 10^{-3} \text{ mol/L}_L$$

Combining this result with Henry's law gives the gas-phase interfacial ammonia concentration as follows:

$$c_{G,int} = Hc_{L,int}$$

$$= (6.95 \times 10^{-4} \text{ L}_L/\text{L}_G)(4.86 \times 10^{-3} \text{ mol/L}_L)$$

$$= 3.38 \times 10^{-6} \text{ mol/L}_G$$

(v) If the same calculations are performed for a system in which the liquid-phase transfer coefficient (k_L) is twice as large, the overall gas transfer coefficient (K_L) and the flux increase by only about 1.4% each. This result is logical, since increasing the value of k_L makes it easier for the material to pass through a region (the solution near the interface) that offers relatively little resistance anyway. On the other hand, if the liquid-phase transfer coefficient remains at its original value but the gas-phase coefficient doubles, K_L and J_L increase by more than 94%, since this change reduces the resistance at the bottleneck in the system. In other words, diffusive transport through the interfacial region on the gas side of the interface is the rate-limiting step in the overall process.

---

[17] As is common in the environmental engineering literature, we express the concentration of ammonia here in terms of the concentration of N; that is, 17 mg/L NH$_3$ would be expressed as 14 mg/L NH$_3$-N.

TABLE 5-5. Summary of Calculations for Example Problem

Parameter	NH$_3$	CCl$_4$	Units
H	6.95×10^{-4}	1.20	L$_L$/L$_G$
$c_{L,b}$	5.0×10^{-3}	5.0×10^{-3}	mol/L
c_L^*	0	0	mol/L
k_L	0.002	0.002	cm$_L$/s
k_G	0.08	0.08	cm$_G$/s
$k_G H$	5.56×10^{-5}	9.60×10^{-2}	cm$_L$/s
K_L	5.41×10^{-5}	1.96×10^{-3}	cm$_L$/s
R_L/R_{tot}	0.027	0.980	–
$c_{L,int}$	4.86×10^{-3}	1.02×10^{-4}	mol/L
$c_{G,int}$	3.38×10^{-6}	1.22×10^{-4}	mol/L
J_L	-2.70×10^{-10}	-9.80×10^{-9}	mol/cm^2 s
J_L if k_L doubles	-2.74×10^{-10}	-1.92×10^{-8}	mol/cm^2 s
J_L if k_G doubles	-5.27×10^{-10}	-9.90×10^{-9}	mol/cm^2 s

(b) The calculations for CCl$_4$ are essentially identical to those for ammonia and are left as an exercise. The results are that the resistance of the gas-phase interfacial region is only 2.3% of the resistance in the liquid phase. Thus, in this case, the rate-limiting step is transport through the liquid-phase interfacial region. The results for all parts of the problem are summarized in Table 5-5. ∎

Coupled Transport and Reaction

The discussion and analysis to this point have been based on an assumption that the volatile species being transferred between the gas and the solution does not participate in a chemical or biochemical reaction in the solution phase. Such reactions can occur, of course, and they affect the mass balance on the volatile substance in various ways.

Consider first a species that can be destroyed by a chemical reaction in bulk solution, but at a rate that is too slow to affect the constituent significantly during the time it is in the interfacial region (i.e., during a typical packet residence time). For such a species, the gas transfer term in a mass balance written around the liquid phase would be written as earlier; that is, as $r_{L,gt}V_L$ or $r_{R,gt}V_R$, and the rate expression for the solution-phase reaction would be included as a separate term in the mass balance equation. The behavior of oxygen in activated sludge aeration basins is an example of such a system—oxygen is consumed in the bulk solution at a rate comparable to its rate of absorption from the gas, but it is assumed not to be depleted significantly by reactions in the interfacial region. In such a case, the net effect of the reaction in the bulk phase is to deplete the dissolved concentration of the volatile species, and thereby increase the driving force for gas absorption or decrease it for gas stripping.

In other systems, the volatile species might react so quickly that it is consumed to a significant extent during the residence time of a typical packet of fluid at the interface. In this case, the same trend as described in the preceding paragraph would apply, but the concentration of the volatile species in bulk solution (where the residence time is far greater than the interfacial packet residence time) would be near zero if the reaction was irreversible, and it would be the equilibrium concentration if the reaction was reversible. The destruction of the species within the interfacial region increases the gradient for absorption and therefore increases the flux into solution, as shown schematically in Figure 5-19.

The magnitude of the change in flux can be determined by writing a mass balance around a differentially thick control volume in a packet in the liquid interfacial region (Equation 5-27), with the advective term and eddy diffusivity set to zero as follows:

$$\frac{\partial c_L}{\partial t} A dx = -A v \frac{\partial c_L}{\partial x} dx + A(D_L + \varepsilon_L) \frac{\partial^2 c_L}{\partial x^2} dx + r_L A dx \quad (5-65)$$

Substituting a first-order decay expression for r_L and canceling $A dx$ from all the terms, we obtain

$$\frac{\partial c_L}{\partial t} = D_L \frac{\partial^2 c_L}{\partial x^2} - k_1 c_L \quad (5-66)$$

This equation can be solved analytically for the case of the thin packet, long residence time limiting case. The resulting expression for flux is identical in form to that for a system

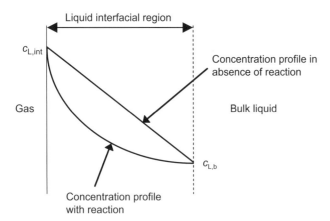

FIGURE 5-19. Schematic representation of the effect of a reaction in the interfacial region on the concentration profile for a species being absorbed from the gas phase. The linear profile in the absence of reaction would apply for a system with thin packets or long packet residence times. The destruction of the species depresses its concentration in the packet, increasing the gradient at the interface and therefore the flux into solution. A similar trend applies for thick packets and short residence times.

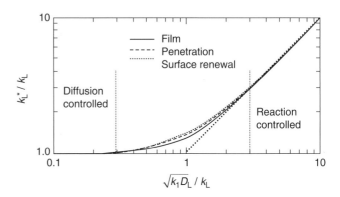

FIGURE 5-20. The enhancement of the liquid-phase mass transfer coefficient in systems where the volatile species undergoes a first-order, irreversible reaction in the liquid interfacial region. *Source*: After Sherwood et al. (1975).

with no reaction, but the value of the mass transfer coefficient is different. Specifically, we find

$$J_L = k_L^* (c_{L,int} - c_{L,b})$$

$$k_L^* = \sqrt{D_L k_1} \coth \frac{\sqrt{D_L k_1}}{k_L} \quad (5-67)$$

where k_L^* and k_L are the mass transfer coefficients in the presence and absence of the reaction, respectively. The ratio k_L^*/k_L is sometimes referred to as an *enhancement factor*, and its dependence on the diffusion coefficient and rate constant is shown in Figure 5-20. While Equation 5-67 was developed for the limiting case of thin packets and/or long packet residence times (i.e., the fixed-film case), it is also a close approximation to numerical solutions of Equation 5-66 for the penetration and surface renewal models, as shown in the figure.

The lines in the figure suggest that the relationship can be simplified at values of $\sqrt{D_L k_1}/k_L$ significantly less than or greater than 1.0: the enhancement factor is ~ 1.0 at low values of $\sqrt{D_L k_1}/k_L$ and approximately equal to $\sqrt{D_L k_1}/k_L$ at high values. These results are obtained because of the properties of the coth function in Equation 5-55. Specifically, for values of $x < 0.3$, coth (x) is approximately $1/x$, and for values of $x > 3$, coth (x) is very close to 1. The corresponding expressions for k_L^* are as follows:

$$\frac{\sqrt{D_L k_1}}{k_L} < 0.3 \quad k_L^* \approx k_L \quad (5\text{-}68a)$$

$$\frac{\sqrt{D_L k_1}}{k_L} > 3 \quad k_L^* \approx \sqrt{D_L k_1} \quad (5\text{-}68b)$$

At values of $\sqrt{D_L k_1}/k_L$ between 0.3 and 3, Equation 5-67 must be used.

In cases where Equation 5-68a applies, the effective mass transfer coefficient is the same as in the case without reaction; that is, the reaction is slow enough relative to the diffusion and packet transfer that the gradient near the interface is unaffected by the reaction. On the other hand, in cases where Equation 5-68b applies, the effective mass transfer coefficient is independent of k_L; that is, the gradient near the interface is controlled by the rate at which the volatile species disappears from solution by chemical reaction, rather than the rate at which it migrates toward the bulk solution. Under these conditions, when considering a control volume near the interface, the loss of the species by reaction is much greater than by diffusive migration.

■ **EXAMPLE 5-7.** (Adapted from Cussler, 1997) Chlorine gas is often injected into water to act as a disinfectant. The gas undergoes the following reaction upon entering solution:

$$Cl_2(aq) + H_2O \rightarrow HOCl + H^+ + Cl^-$$

The reaction is very rapid and essentially irreversible at near-neutral pH, with a pseudo-first-order rate constant of $25\,s^{-1}$.

A system for injecting $Cl_2(g)$ into a drinking water source has been tested using oxygen as a model nonreactive gas. Based on these tests and the relative diffusivities of $O_2(aq)$ and $Cl_2(aq)$, the value of k_L for $Cl_2(aq)$ in the absence of chemical reaction is expected to be 1.1×10^{-3} cm/s, and D_L for $Cl_2(aq)$ is $1.25 \times 10^{-5}\,cm^2/s$. Estimate k_L^* for $Cl_2(aq)$ in the reactor.

Solution. The ratio $\sqrt{D_L k_1}/k_L$ for $Cl_2(aq)$ in this system is

$$\frac{\sqrt{D_L k_1}}{k_L} = \frac{\sqrt{(1.25 \times 10^{-5}\,cm^2/s)(25\,s^{-1})}}{1.1 \times 10^{-3}\,cm/s} = 16.1$$

The reaction is sufficiently rapid that Equation 5-68b applies. We can therefore compute k_L^* as

$$k_L^* = \sqrt{D_L k_1} = \sqrt{(1.25 \times 10^{-5}\,cm^2/s)(25\,s^{-1})}$$
$$= 0.0177\,cm/s$$

Thus, the reaction enhances the liquid-phase transfer coefficient for $Cl_2(aq)$ approximately by a factor of 16. ■

Reactions in solution can alter the gas transfer coefficient in the liquid phase, but they have no effect on the coefficient in the gas phase. Therefore, as long as the liquid-phase resistance is the dominant resistance to gas transfer, the effect of the reaction is to increase the flux into solution in proportion to the k_L^*/k_L ratio. However, if this ratio becomes very large, the overall resistance becomes dominated by the resistance in the gas phase, and no further increase in flux can be achieved.

The effects of aqueous-phase reactions on gas transfer rates have been studied extensively in chemical engineering, and many more cases are described in the literature and in books on mass transfer in environmental systems (see, for example, Sherwood et al., 1975; Logan, 1999; Clark, 2009). Here, we consider only one other scenario that is of particular importance in environmental engineering applications—transfer of volatile species that undergo very rapid and reversible acid/base reactions in solution.

Volatile acids and bases that are important in environmental engineering include ammonia, carbon dioxide, hydrogen sulfide, and sulfur dioxide. The protonation and deprotonation reactions of these species are rapid, so that the characteristic time of the reaction is much shorter than that for transport through a packet of fluid. As a result, the reaction can be considered to be at equilibrium everywhere in the packet.[18]

The key difference between gas transfer of a species that undergoes acid/base (or other rapidly equilibrating) reactions and transfer of nonreactive species is that, in the former case, the species that is transported through the gas boundary layer and exists in the gas phase is not the only one (and often not even the dominant one) that is transported through and exists in the liquid. As a consequence, transport through the liquid boundary layer and across the interface is facilitated for the reactive species compared with a nonreactive species in the same system and with the same values of c_G and c_L.

Consider, for example, a system in which H_2S is being stripped out of a solution at pH 7.0 ($= pK_{a1}$). At this pH, the total sulfide in solution is distributed approximately evenly between $H_2S(aq)$ and HS^-. If a small amount of $H_2S(aq)$ leaves solution and enters the gas phase, the dissolved concentration of H_2S near the interface declines, thereby diminishing the driving force for additional stripping. However, if the acid/base reaction proceeds very rapidly, then some HS^- immediately becomes protonated to form $H_2S(aq)$. In the given scenario (assuming that the pH is well-buffered), one HS^- ion would be converted into $H_2S(aq)$ for every two $H_2S(aq)$ molecules that had been stripped, so the decline in the $H_2S(aq)$ concentration near the interface would be only one-half as great as in the case with no reaction, and the rate of H_2S transfer across the interface would be correspondingly higher.

Similarly, if $CO_2(g)$ were being absorbed into a solution at pH 9.0, hydration and deprotonation or reaction with a

---

[18] This inference assumes that, in cases where the gaseous molecule is hydrated in solution (e.g., as H_2CO_3 or H_2SO_3), the hydration and dehydration reactions are also very fast. In the case of H_2CO_3, the dehydration reaction is fast, but the hydration reaction is not (compared with the expected packet residence times), so this analysis and that in Appendix 5B overestimate the effect of pH on absorption of CO_2 into solution. Nevertheless, the trends indicated by the equations are applicable in all cases.

hydroxyl (OH$^-$) ion would convert most of the absorbed CO_2 into HCO_3^-. These reactions would reduce the concentration of CO_2(aq) near the interface and thereby maintain a greater driving force for further gas absorption compared with a case where no reaction occurred (e.g., at low pH). Note that, even though the acid/base reaction would diminish the gradient of CO_2(aq) from the interface into bulk solution, so that the rate of diffusion of that species would be less than in the absence of reaction, a corresponding gradient of HCO_3^- would be established. As a result, the overall rate of diffusion of the absorbed species into the bulk solution (the sum of the diffusive fluxes of CO_2(aq) and HCO_3^-) would not decline.

A quantitative analysis of gas transfer in systems in which rapid acid/base reactions occur is provided in detail in Appendix 5B. The key result is the following equation for the flux of the neutral species, i_{neut}, as a function of the extent of disequilibrium of that species between the two bulk phases:

$$J_{L,neut} = K_{L,i_{neut}}(c^*_{L,i_{neut}} - c_{L,i_{neut}}) \quad (5\text{-}69)$$

where

$$K_{L,i_{neut}} \equiv \frac{k_L k_G H}{k_L + \alpha_{neut} k_G H} \quad (5\text{-}70)$$

$$\alpha_{neut} \equiv \frac{c_{i,\text{neutral species}}}{\sum \text{all acid/base forms of } i} \quad (5\text{-}71)$$

Thus, for example, when considering gas transfer of sulfide species, α_{neut} would be defined as: $\alpha_{neut} = (c_{H_2S}/\text{TOTS}) = (c_{H_2S}/(c_{H_2S} + c_{HS^-} + c_{S^{2-}}))$. Using the conventions that are common in the water chemistry literature, α_{neut} corresponds to α_0 if the volatile species is the most acidic species of the acid/base group, whereas it corresponds to a different α value if the volatile species is a base (e.g., it is α_0 for the $H_2CO_3/HCO_3^-/CO_3^{2-}$ group, and α_1 for the NH_4^+/NH_3 group).

If dissolved i is present only as the acid/base species and not in other forms, α_{neut} is strictly a function of pH. Values of α_{neut} as a function of pH are shown for a few environmentally important acid/base groups in Figure 5-21. Note that α_{neut} can be exceedingly small for some of these groups at pH values that might be encountered in natural systems or engineered reactors.

Although the neutral species is the only one that actually crosses the air/water interface, our interest is often in the total dissolved concentration of the acid/base species, $c_{L,\text{TOT}i}$, so it is convenient to write Equation 5-69 in terms of this parameter. Because only the neutral species crosses the interface, the flux of TOTi equals the flux of i_{neut}. As a result, we can express the left side of Equation 5-69 in terms of TOTi by replacing $J_{L,neut}$ with $J_{L,\text{TOT}i}$. In addition, $c^*_{L,i_{neut}}$ and $c_{L,i_{neut}}$ can be expressed as $\alpha_{neut} c^*_{L,\text{TOT}i}$ and

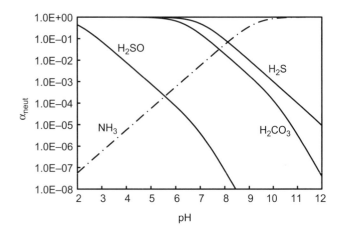

FIGURE 5-21. Fraction of the total dissolved acid/base group that is present as the neutral (volatile) species for four acid/base groups. (Recall that, in the gas phase, H_2CO_3 and H_2SO_3 exist as CO_2(g) and SO_2(g), respectively.)

$\alpha_{neut} c_{L,\text{TOT}i}$, respectively. Making these substitutions in Equation 5-69, we obtain

$$\begin{aligned} J_{L,\text{TOT}i} &= K_{L,i_{neut}}\left(\alpha_{neut} c^*_{L,\text{TOT}i} - \alpha_{neut} c_{L,\text{TOT}i}\right) \\ &= \alpha_{neut} K_{L,i_{neut}}\left(c^*_{L,\text{TOT}i} - c_{L,\text{TOT}i}\right) \quad (5\text{-}72) \\ &= K_{L,\text{TOT}i}\left(c^*_{L,\text{TOT}i} - c_{L,\text{TOT}i}\right) \end{aligned}$$

where

$$K_{L,\text{TOT}i} = \alpha_{neut} K_{L,i_{neut}} = \frac{k_L \alpha_{neut} k_G H}{k_L + \alpha_{neut} k_G H} \quad (5\text{-}73)$$

Equation 5-72 indicates that the flux of TOTi into or out of solution is proportional to a driving force defined as the difference between the instantaneous value of $c_{L,\text{TOT}i}$ and the corresponding value that would be in equilibrium with the gas phase. Note, however, that when the flux is expressed in this way, the mass transfer coefficient must be computed differently from when the driving force is written in terms of the extent of disequilibrium of the neutral species (as shown in Equation 5-73).

Equations 5-69 and 5-72 contain equivalent information, and both reduce to the equation for flux of a nonreactive species (Equation 5-11) if $\alpha_{neut} = 1$. They also demonstrate that if α_{neut} is substantially <1, the effects on gas transfer can be very dramatic. These effects are described in greater detail in Appendix 5B.

■ **EXAMPLE 5-8.** Compute $K_{L,i_{neut}}$, $K_{L,\text{TOT}i}$, and the flux of H_2S into a solution containing $10^{-5.0}$ M TOTS in contact with a gas phase in which $P_{H_2S} = 10^{-7.0}$ atm, if k_L and k_G for H_2S are 0.004 and 0.08 cm/s, respectively. Consider the pH range from 4 to 12. pK_{a1} for H_2S is 7.0.

Solution. Application of basic principles of water chemistry leads to the result that α_{neut} for H_2S (i.e., $\alpha_{H_2S}(aq)$) at any pH in the range of interest can be computed as follows:

$$\alpha_{H_2S(aq)} = \frac{c_{H^+}}{c_{H^+} + K_{a1}} = \frac{10^{-pH}}{10^{-pH} + 10^{-pK_{a1}}}$$

(The previous expression is based on the assumption that the fully dissociated species S^{2-} represents a negligible fraction of TOTS, which is acceptable because pK_{a2} is >13.0, and we are interested only in pHs up to 12.)

Once $\alpha_{H_2S}(aq)$ is known, $K_{L,i_{neut}}$ and $K_{L,TOTi}$ can be computed using Equations 5-70 and 5-73, respectively. The flux into solution can then be computed using Equations 5-69 and 5-72, respectively. The results of these calculations are shown in Figure 5-22.

Figure 5-22a shows the $K_{L,i_{neut}}$ approaches k_L ($= 0.004$ cm/s) at low pH (which corresponds to $\alpha_{neut} = 1.0$ in this system), suggesting that the resistance to gas transfer rate is primarily in the liquid phase under these conditions. At higher pH, $K_{L,i_{neut}}$ begins to increase, approaching a maximum value of $10^{-1.49}$ cm/s ($= 0.08$ cm/s), or k_G; thus, the transfer is limited by the gas-phase resistance under these conditions.[19] The flux of H_2S is out of the solution over most of the pH range of interest, but it declines with increasing pH, and reverses sign at pH 10.0 (i.e., H_2S absorption is favored at pH > 10.0)

The decrease in the overall resistance to gas transfer (leading to the increase in $K_{L,i_{neut}}$) with increasing pH is caused by a decline in the resistance to transport through the liquid boundary layer that accompanies the deprotonation of H_2S. When H_2S transfers into the gas phase at low pH, the H_2S concentration near the interface declines, and more H_2S can appear at the interface only by diffusion of this species through the boundary layer. In contrast, at pH values where H_2S is substantially deprotonated, the H_2S concentration at the interface can be replenished both by diffusive transport of H_2S through the boundary layer and by the protonation of HS^- that is already at the interface. As a result, the reliance on diffusive transport decreases, and the resistance to the overall process imposed by the liquid boundary layer diminishes. The total resistance to gas transfer, representing the sum of the resistances of the two boundary layers, therefore diminishes as well, until most of the resistance resides in the gas phase. At that point, increasing the pH continues to decrease the resistance contributed by the liquid boundary layer, but it has little effect on the overall resistance.

If k_G/k_L is substantially smaller (as it would typically be in a packed column), the variation in $K_{L,i_{neut}}$ with pH is much smaller (see Problem 5-15).

5.7 EVALUATING $k_L, k_G, K_L,$ AND a: EFFECTS OF HYDRODYNAMIC AND OTHER OPERATING CONDITIONS

To this point, we have explored some conceptual and mechanistic models for gas transfer, and used these models to characterize the overall rate of gas transfer in terms of the concentrations in the two bulk phases, Henry's constant, and the system-specific parameters k_L, k_G, and a (or the combination of these parameters that yields the overall rate constant $K_L a_L$ or $K_L a_R$). In the final section of this chapter, some theoretical considerations and empirical observations that are useful for estimating these system-specific parameters are presented. The discussion focuses on pure gas-in-liquid and liquid-in-gas systems (e.g., diffused aeration, bubble columns, and packed columns), for which correlations for the rate constants have been developed from first principles. No such correlations are available for

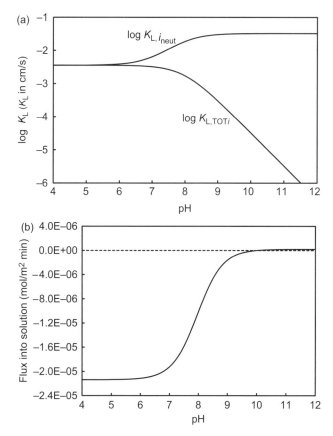

FIGURE 5-22. Aspects of H_2S transfer between the gas and solution in the example system. (a) Overall gas transfer coefficients in the system, applicable for two different ways of defining the gas transfer driving force. (b) H_2S flux in the system.

[19] The fraction of the resistance that resides in each phase can be computed using equations derived in Appendix 5B. The result is that the liquid contributes $>90\%$ of the overall resistance at pH < 7.4, 50% at pH 8.5, and $<10\%$ at pH > 9.5. Note that the gas-phase resistance becomes limiting at high pH because the liquid-phase resistance declines; the gas-phase resistance itself is unaffected by pH.

mechanically assisted surface aeration systems (e.g., brush aeration) for reasons noted earlier—such systems usually have both gas-in-water and water-in-gas components that are highly system-specific, and hence cannot be modeled by any well-defined, idealized limiting case. Some reference is also made to surface aeration across unbroken water surfaces (e.g., the surface of a lake or river), because of the extensive research that has been performed to study transfer of oxygen across such surfaces.

Approaches for Estimating Gas Transfer Rate Coefficients

Gas-in-Liquid Systems Theoretical analyses of gas-in-water systems typically begin by modeling the behavior of an isolated bubble rising through a quiescent solution. Superficially, such a system appears to be quite simple. However, this apparent simplicity is deceptive, as is evident from consideration of just one of the many variables that affects gas transfer in the system: bubble size. As a point of reference, diffused aeration systems used in environmental engineering processes typically have bubbles with equivalent diameters in the range of 0.2–2.5 cm.

Bubbles rising through a solution take various shapes, related primarily to their size, as shown schematically in Figure 5-23. These changes in shape are associated with corresponding changes in the rise velocity and the fluid dynamics in the two phases, which in turn help control the values of $k_L a$ and $k_G a$. Some model and experimental data for the correlations of bubble velocity and k_L with bubble size are shown in Figure 5-24.

Very small bubbles tend to behave as hard spheres (i.e., air movement inside the bubbles is negligible), in which case the rise velocity can be computed based solely on the water viscosity and the density difference between the gas and solution. As the bubble size increases, circulation cells develop in the bubble and reduce the friction at the interface, allowing the bubble to rise more rapidly than it would in the absence of the circulation. The increase in the rise velocity causes the bubble to have a larger effect on the solution, changing k_L, and the circulation of air in the bubble affects k_G. While these effects complicate the analysis, they can still be modeled reasonably well.

As the bubble size increases further, the bubble acquires a teardrop shape that becomes progressively flatter as the

FIGURE 5-23. Shapes of gas bubbles in aqueous solutions. The progression from left to right corresponds to increasing equivalent diameter (i.e., the diameter of a spherical bubble that has the same volume as the actual bubble) from roughly 0.2 to 2 cm. *Source*: After Calderbank et al. (1970).

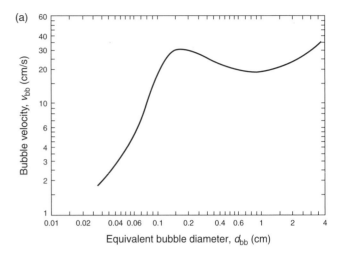

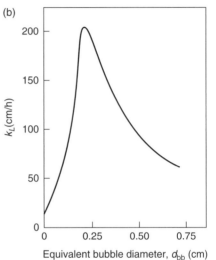

FIGURE 5-24. Important parameters in bubble aeration systems. (a) The rise velocity of air bubbles in water at 20°C (after Haberman and Morton, 1956); (b) k_L values for oxygen transfer from bubbles of various equivalent diameters rising in water at 20°C. *Source*: Modified from Barnhart (1969).

bubble size increases, and then rounds into a cap as the bubble becomes quite large (~2 cm). Over part of this range, the effect of increased cross-sectional area of the bubble in the horizontal plane dominates over the effect of increased buoyancy, so the rise velocity actually decreases with increasing bubble size (Figure 5-24a). When the bubble starts flattening, its trajectory becomes less linear, and it may jiggle as it rises. The combined effects of the changes in bubble size, shape, rise velocity, and trajectory cause both K_L and the products $K_L a_L$ and $K_L a_R$ to pass through a maximum at an equivalent spherical diameter near 0.2 cm (Figure 5-24b).

Gas transfer in these systems becomes progressively more difficult to model as bubble size increases. A similar statement applies to systems with gradually increasing bubble concentration (number per unit volume). In this case, the complications arise from the fact that, at high concentration, bubbles can coalesce, and even if they do not, the hydrodynamic disturbances caused by individual bubbles can overlap, leading to localized turbulence and correspondingly complicated patterns of fluid motion.

Values of k_L for even some of the more complicated situations described in the preceding paragraphs have been estimated from a fundamental, theoretical starting point. The results have then been compared with the behavior of experimental systems, and empirical modifications to the predictive equations have been made when necessary. An extensive compilation of these model equations is provided in *Perry's Chemical Engineers' Handbook* (Green and Perry, 2008). As is common in the chemical engineering literature, the equations are presented using dimensionless parameters for both the independent and dependent variables. Mathematical definitions and qualitative descriptions of these parameters are provided in Table 5-6, and a few of the predictive equations are presented in Table 5-7.

■ **EXAMPLE 5-9.** Water is to be aerated in a well-mixed tank by injecting bubbles that, at mid-depth, have an average diameter of 0.8 mm. The water temperature is 20°C and its viscosity is 10^{-2} g/cm s. Estimate k_L based on the correlation given in Table 5-7 for noninteracting bubbles, and estimate the thickness of the liquid boundary layer based on the stagnant-film model. At 20°C, the diffusion coefficient of O_2 in solution is 2.05×10^{-5} cm^2/s.

Solution. The rise velocity of the bubbles can be estimated from Figure 5-24 as 14 cm/s. The Reynolds, Schmidt, and Sherwood numbers for the bubbles are therefore

TABLE 5-6. Definitions and Descriptions of Dimensionless Numbers for Bubbles in Water, as Used in the Mass Transfer Correlations

Name	Symbol	Definition[a]	Parameter is Related to the Ratio of:
Sherwood number	Sh	$\dfrac{k_L d_{bb}}{D_{i,L}}$	Total mass transfer rate to diffusive mass transfer
Reynolds number	Re	$\dfrac{d_{bb} v_{bb} \rho_L}{\mu_L}$	Inertial to viscous forces acting on a bubble
Schmidt number	Sc	$\dfrac{\mu_L}{\rho_L D_{i,L}}$	Momentum diffusivity to mass diffusivity into a bubble
Grashof number	Gr	$\dfrac{d_{bb}^3 \rho_L g (\rho_L - \rho_{bb})}{\mu_L^2}$	Gravitational or buoyant forces to viscous forces acting on a bubble

[a] d, diameter; v, velocity; ρ, density; μ, viscosity; D, diffusivity; g, gravitational constant; k, gas transfer rate coefficient; bb, bubble; i, species being transferred; L, liquid phase.

TABLE 5-7. Some Mass Transfer Correlations for Gas Transfer Systems[a]

Equation	Comments
$Sh = Re^{1/3} Sc^{1/3}$	Noninteracting bubbles, $d_{bb} < 0.1$ cm
$Sh = 1.13 Re^{1/2} Sc^{1/2} \dfrac{d_{bb}}{0.45 + 0.2 d_{bb}}$	Noninteracting bubbles, $d_{bb} > 0.5$ cm, $500 < Re < 8000$
$Sh = 2 + 0.31 Sc^{1/3} Gr^{1/3}$	Noninteracting bubbles or bubble swarms, $d_{bb} < 0.25$ cm
$Sh = 0.42 Sc^{1/2} Gr^{1/3}$	Noninteracting bubbles or bubble swarms, $d_{bb} > 0.25$ cm

[a] From Green and Perry (2008) and the references therein.

$$Re = \frac{d_{bb} v_{bb} \rho_L}{\mu_L} = \frac{(0.08 \,\text{cm})(14 \,\text{cm/s})(1.0 \,\text{g/cm}^3)}{10^{-2} (\text{g/cm s})} = 112$$

$$Sc = \frac{\mu_L}{\rho_L D_{i,L}} = \frac{10^{-2} (\text{g/cm-s})}{(1.0 \,\text{g/cm}^3)(2.05 \times 10^{-5} \,\text{cm}^2/\text{s})} = 488$$

$$Sh = Re^{1/3} Sc^{1/3} = (112)^{1/3} (488)^{1/3} = 37.9$$

The estimated value of k_L can then be computed from the Sherwood number as

$$k_L = \frac{(Sh) D_{O_2,L}}{d_{bb}} = \frac{(37.9)(2.05 \times 10^{-5} \,\text{cm}^2/\text{s})}{0.08 \,\text{cm}} = 0.0097 \,\text{cm/s}$$

According to the two-film model, the mass transfer coefficient in the liquid film is $D_{L,O_2}/\delta_L$, so in this case, we could estimate δ_L as

$$\delta_L = \frac{D_{L,O_2}}{k_L} = \frac{2.05 \times 10^{-5} \,\text{cm}^2/\text{s}}{0.0097 \,\text{cm/s}} = 2.11 \times 10^{-3} \,\text{cm}$$

$$= 21.1 \,\mu\text{m} \quad ■$$

While the semitheoretical approaches described earlier are useful for gaining an appreciation of the factors that control k_L, they are often not practical to use for design calculations because of system-specific factors related with mixing patterns and water chemistry (affecting, e.g., the extent of bubble coalescence). In such cases, an approach that is commonly employed is to measure $K_L a_L$ for oxygen transfer under well-defined operating conditions, and then use more theoretical approaches to predict how $K_L a_L$ for oxygen transfer will change in response to different operating conditions or to predict $K_L a_L$ of other volatile species.

Because the primary goal of diffused aeration and surface aeration is so often oxygen transfer (e.g., for biological waste treatment processes), manufacturers of aeration equipment often provide estimates of $K_L a_L$ for oxygen (or other parameters, such as oxygen transfer efficiencies, that can be converted into $K_L a_L$ values) under specified conditions. Alternatively, pilot tests can be performed with the given equipment and feed water to develop a system-specific $K_L a_L$ value for oxygen. Standard procedures for conducting such tests have been developed, a few of which are described in Appendix 6A in Chapter 6.

Once a $K_L a_L$ value for oxygen has been obtained, corresponding values for many other volatile species can generally be estimated, as follows. First, recall that in virtually any oxygen gas transfer system, almost all the resistance resides in the aqueous phase (because H_{O_2} is large). With this assumption, we find

$$R_{tot,O_2} \approx \cancel{R_{G,O_2}} + R_{L,O_2} \quad (5\text{-}62)$$

$$(K_L a_L V_L)^{-1}_{O_2} \approx \cancel{(k_G H a_L V_L)^{-1}_{O_2}} + (k_L a_L V_L)^{-1}_{O_2} \quad (5\text{-}61b)$$

$$(K_L a_L)_{O_2} \approx (k_L a_L)_{O_2} \quad (5\text{-}74)$$

Recall also that the k_L values of different species in a given system are related with one another by Equation 5-46 (repeated in the following equation), with m usually between 0.5 and 0.75

$$\frac{(k_L a_L)_2}{(k_L a_L)_1} = \left(\frac{D_{L,2}}{D_{L,1}}\right)^m \quad (5\text{-}46)$$

Considering oxygen as compound 1 and the species of interest i as compound 2, the value of $(k_L a_L)_i$ can then computed from

$$(k_L a_L)_i = \left(\frac{D_{L,i}}{D_{L,O_2}}\right)^m (k_L a_L)_{O_2} \approx \left(\frac{D_{L,i}}{D_{L,O_2}}\right)^m (K_L a_L)_{O_2} \quad (5\text{-}75)$$

If the resistance to transfer of species i resides almost entirely in the aqueous phase (i.e., if $(k_L a_L)_i = (K_L a_L)_i$), then the following approximation holds, from which $K_L a_L$ for the species of interest can be computed directly as

$$\text{If } R_{G,i} \ll R_{L,i}: \quad (K_L a_L)_i = \left(\frac{D_{L,i}}{D_{L,O_2}}\right)^m (K_L a_L)_{O_2} \quad (5\text{-}76)$$

The ratio of $K_L a_L$ for a species i to $K_L a_L$ for oxygen is frequently designated by the symbol Ψ_i.[20]

$$\Psi_i \equiv \frac{(K_L a_L)_i}{(K_L a_L)_{O_2}} \quad (5\text{-}77)$$

---

[20] The definition of Ψ_i in the literature is not always consistent. Some authors define it as the ratio $(k_L a_L)_i / (k_L a_L)_{O_2}$, rather than as $(K_L a_L)_i / (K_L a_L)_{O_2}$. Although the denominators of these fractions are approximately equal to one another, the numerators (and therefore the values of the fractions) differ if gas-phase resistance is important for species i.

Comparison of Equations 5-75 and 5-77 indicates that, if $R_{G,i} \ll R_{L,i}, \Psi_i = \left(\frac{D_{L,i}}{D_{L,O_2}}\right)^m$.

■ **EXAMPLE 5-10.** The value of $K_L a_L$ for oxygen in a wastewater treatment process operating at 25°C and using mechanical surface aeration is $3.10\,h^{-1}$. Estimate $K_L a_L$ for carbon tetrachloride (CCl_4) in the system, if the majority of the resistance to transfer of both O_2 and CCl_4 is in the liquid phase, and the exponent m relating $k_L a_L$ values to diffusivities is 0.5. Assume that the ratio of the diffusivities of O_2 and CCl_4 is the same at 20°C and 25°C.

Solution. The calculation is a direct application of Equation 5-46. Rearranging this equation and substituting diffusivities from Table 5-4 and the given values of m and $K_L a_L$ for oxygen, we find

$$(K_L a_L)_{CCl_4} \approx (k_L a_L)_{CCl_4} = \left(\frac{D_{L,CCl_4}}{D_{L,O_2}}\right)^m (k_L a_L)_{O_2}$$

$$= \left(\frac{9.2 \times 10^{-6}\,cm^2/s}{2.05 \times 10^{-5}\,cm^2/s}\right)^{0.5} (3.10\,h^{-1}) = 2.08\,h^{-1}$$

■

While the approach described previously is adequate for estimating $K_L a_L$ values of volatile species for which the resistance to gas transfer is almost entirely in the liquid phase, this criterion is often not met for transfer of VOCs and many other compounds of interest. In such cases, to estimate the overall rate constant $K_L a_L$, it is necessary to estimate $k_G a_L$ in addition to $k_L a_L$.

Estimating $k_G a_L$ is more problematic than estimating $k_L a_L$, since no approach is available to estimate $k_G a_L$ directly for any single species (in a manner analogous to the estimation of $k_L a_L$ for oxygen), and because the relationships among $k_G a_L$ values for different species in a given system are not clear. With respect to the latter issue, an assumption is frequently made that these relationships are analogous to these among $k_L a_L$ values; that is

$$\frac{(k_G a_L)_2}{(k_G a_L)_1} = \left(\frac{D_{G,2}}{D_{G,1}}\right)^n \quad (5\text{-}78)$$

Although some evidence suggests that the value of n is different from that of m (Munz and Roberts, 1984), the value of n is not known with any confidence, so the assumption that $n = m$ is often made. In that case, dividing Equation 5-78 by Equation 5-46 yields

$$\frac{(k_G a_L)_2/(k_L a_L)_2}{(k_G a_L)_1/(k_L a_L)_1} = \left(\frac{D_{G,2}/D_{L,2}}{D_{G,1}/D_{L,1}}\right)^m \quad (5\text{-}79)$$

As suggested by the values in the final column of Table 5-4, the ratio of the gas-phase diffusivity to the liquid-phase diffusivity is close to 1.0×10^4 for most volatile compounds, so that the term in parentheses on the right side of Equation 5-79 is approximately 1.0. Making that substitution, the equation can be rewritten to show that the ratio $k_G a_L / k_L a_L$ (and therefore k_G / k_L) is approximately the same for all volatile species in a given system

$$\frac{(k_G a_L)_2 / (k_L a_L)_2}{(k_G a_L)_1 / (k_L a_L)_1} \approx 1.0$$

$$\frac{(k_G a_L)_2}{(k_L a_L)_2} = \frac{(k_G a_L)_1}{(k_L a_L)_1} \quad (5\text{-}80)$$

$$\left(\frac{k_G}{k_L}\right)_2 = \left(\frac{k_G}{k_L}\right)_1 \quad (5\text{-}81)$$

This result can be rationalized based on the idea that, for a given gas transfer system, the k_G / k_L ratio depends primarily on physical parameters (the interfacial thicknesses and packet RTDs) and hence is insensitive to the species being transferred. Thus, although the k_G / k_L ratio is expected to vary substantially from one gas transfer system to the next, its value might reasonably be approximately constant for all the volatile species in a given system.

Even if we accept the approximations leading to Equations 5-80 and 5-81, those equations alone cannot provide the information required to estimate $K_L a_L$ for a species for which gas-phase resistance is significant; to accomplish that, the value of the k_G / k_L ratio in the system must be known. While these ratios cannot be predicted from first principles, typical ranges for certain types of gas transfer systems have been established. These typical ranges are identified in Figure 5-18. Based on these values, we can decide whether, in fact, gas-phase resistance is likely to be important, and hence whether $K_L a_L$ can be approximated by the value of $k_L a_L$ computed using Equation 5-75. For instance, the figure indicates that 91–97% of the resistance to stripping of CCl_4 (log $H = 0.04$) in a diffused aeration system is expected to reside in the liquid phase; as a result, if we wish to estimate $K_L a_L$ for this compound in a diffused aeration system, the gas-phase resistance can be ignored, and $K_L a_L$ can be estimated using Equation 5-75. In contrast, for stripping of benzene (log $H = -2.05$) in the same system, only 8–30% of the resistance is expected to be in the liquid phase. Thus, while Equation 5-75 could be used to estimate the value of $k_L a_L$ for benzene in the system, that value would not be a good approximation of $K_L a_L$.

■ **EXAMPLE 5-11.** Estimate values for $k_L a_L$ and $K_L a_L$ for bromoform ($CHBr_3$) in the system described in Example 5-10, if the k_G / k_L ratio in the system is 20. As in Example 5-10, assume that $m = 0.5$ and the diffusivity ratios are the same at 20°C and 25°C.

Solution. Henry's constant for bromoform is given in Table 5-2 as $0.0227\, L_L / L_G$. For a system with this value of H and $k_G / k_L = 20$, Figure 5-18 indicates that the majority of the resistance to gas transfer for bromoform would be in the gas phase, so Equation 5-74 cannot be used to estimate $(K_L a_L)_{CHBr_3}$. However, because $(K_L a_L)_{O_2} \approx (k_L a_L)_{O_2}$, we can still use Equation 5-75 to estimate $(k_L a_L)_{CHBr_3}$:

$$(k_L a_L)_{CHBr_3} = \left(\frac{D_{L,CHBr_3}}{D_{L,O_2}}\right)^{0.5} (K_L a_L)_{O_2}$$

$$= \left(\frac{9.2 \times 10^{-6}\, cm^2/s}{2.05 \times 10^{-5}\, cm^2/s}\right)^{0.5} (3.10\, h^{-1}) = 2.08\, h^{-1}$$

The value of $(k_G a_L)_{CHBr_3}$ can then be computed from the given information about the k_G / k_L ratio:

$$(k_G a_L)_{CHBr_3} = \frac{k_G}{k_L} (k_L a_L)_{CHBr_3}$$

$$= 20(2.08\, h^{-1}) = 41.6\, h^{-1}$$

The known values of $(k_L a_L)_{CHBr_3}$ and $(k_G a_L)_{CHBr_3}$ can then be used to compute $(K_L a_L)_{CHBr_3}$ via Equation 5-59a, as follows:

$$K_L a_L = \frac{(k_L a_L)(k_G a_L) H}{k_L a_L + k_G a_L H}$$

$$= \frac{(2.08\, h^{-1})(41.6\, h^{-1})(0.0227\, L_L / L_G)}{2.08\, h^{-1} + (41.6\, h^{-1})(0.0227\, L_L / L_G)} = 0.649\, h^{-1}$$

The value of $K_L a_L$ for bromoform in this system is considerably less than would be computed using Equation 5-74, because of the influence of the gas-phase resistance. ■

The preceding discussion focuses on the relative $K_L a_L$ values for different species in a system. In a given system operated in a given way, these ratios are established by the system physics and chemistry. However, even after a system is installed, it might be possible to alter the absolute values of $K_L a_L$ for various species by altering the system operation. For instance, for both diffused aeration and surface aeration systems, $K_L a_L$ can often be increased by increasing the mixing intensity, as quantified by the power input per unit volume of water (the *specific power input*, P/V) (Roberts and Levy, 1985; Newbry, 1998). The increased mixing of the aqueous phase is thought to increase the $k_L a_L$ values for all volatile species in the system, but to have almost no effect on $k_G a_L$ (Libra, 1993). As a result, increasing P/V is expected to increase $K_L a_L$ for species whose mass transfer resistance is predominantly in the liquid phase. Specific power inputs to surface aeration systems in wastewater treatment plants

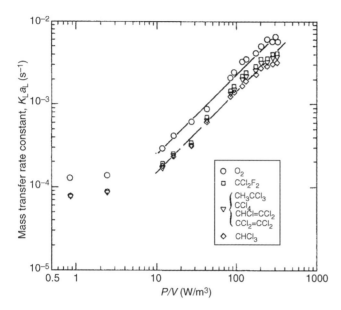

FIGURE 5-25. Dependence of $K_L a_L$ on the specific power input for a variety of volatile compounds in a surface aeration system. *Source:* Reprinted with permission from Roberts and Dändliker, (1983). Copyright 1983 American Chemical Society.

range from approximately 10 to 60 W/m³, with a typical value of 25 W/m³ (Paulson, 1979; WEF, 1988).

Such behavior is demonstrated in Figure 5-25 for absorption of oxygen and stripping of several volatile organic contaminants in a stirred, surface aeration system. In this study, the $K_L a_L$ values for all the species investigated were directly proportional to the specific power input over a very wide range of P/V (as indicated by the lines in the figure with slope nearly equal to 1.0), suggesting that the liquid-phase resistance dominated the gas-phase resistance for all the species investigated. Similar results were obtained in other laboratory tests by Hsieh et al. (1993) and in a field-scale study by Parker et al. (1996).

In all the systems characterized in Figure 5-25, the intense stirring of the solution was reported to generate a continuous spray of water and to entrain substantial amounts of air into the solution. When surface aeration occurs across an unbroken interface, such as in natural water bodies, a correlation can be developed between specific power dissipation (related to the velocity of the water) and $K_L a_L$; in these cases, both a theoretical analysis and experimental data suggest that the correlation is of the form $K_L a_L = k(P/V)^b$, where k and b are constants, and b is in the range from 0.3 to 0.5 (Kozinski and King, 1966; Hsieh et al., 1993).

Although increasing the specific power input can reduce the liquid-phase resistance (and hence increase $K_L a_L$) at any value of P/V, the unchanging gas-phase resistance might place an upper bound on $K_L a_L$ when P/V becomes large. Libra (1993) observed such a trend in experiments investigating the stripping of toluene, dichloromethane, and 1,2-dichlorobenzene in a stirred, diffused aeration system. In these systems, $K_L a_L$ for oxygen increased steadily as P/V was increased from 10 to 1000 W/m³ (Figure 5-26), indicating that $k_L a_L$ for oxygen was increasing; presumably, $k_L a_L$ for all three organic species increased as well. Nevertheless, $K_L a_L$ for the organic species did not increase, indicating that the rate of gas transfer of those species was being limited by the gas-phase resistance, which is unaffected by P/V.

Another parameter whose effect on $k_L a_L$ in diffused aeration systems is worth noting is the gas flow rate, Q_G. If the bubble size and the dimensions of the system are kept constant, then the number of bubbles present in the tank at any given time is expected to be directly proportional to Q_G. In that case, the concentration of surface area in the tank (a_L) would also be proportional to Q_G. If the mixing of the liquid was not dramatically affected by the change in Q_G, k_L would be approximately constant, and $k_L a_L$ would be directly proportional to Q_G. Such a result was obtained by Hsieh et al. (1993) in a bench-scale study of 20 VOCs. A similar result was obtained by Libra (1993) in a system with both diffused aeration and mechanical mixing, when the specific power input was large. However, at lower specific power inputs, $k_L a_L$ was less than proportional to Q_G in Libra's study and others cited therein. Perhaps because of the effects of specific power input and gas flow rate on the bubble size, the relationship was similar to that cited earlier for surface aeration in systems where the water surface was not broken; that is

$$k_L \propto Q_G^b \qquad (5\text{-}82)$$

The reported values of b ranged from approximately 0.3 to 0.6 at low P/V, and increased to close to 1.0 at higher P/V.

Liquid-in-Gas Systems The idealized representation of the contact between water and air in a packed tower is

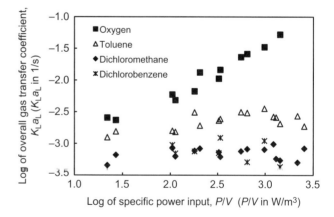

FIGURE 5-26. Dependence of overall gas transfer coefficients on specific power input for four species in a laboratory-scale, stirred diffused aeration system. *Source:* After Libra (1993).

based on a thin film of water flowing along a planar surface. Theoretical calculations for both heat and mass transfer in such systems have been performed, taking into consideration the thickness of the film, the affinity of the fluid for the solid (quantified via the surface tension), and the effect of counter-current air flow. In actual systems, of course, the flow over the packing is not uniform with respect to thickness, velocity, or surface smoothness (i.e., ripples might appear and disappear; water droplets form, coalesce, and then are converted into a film; and air flow might not be uniform across the cross-section of the column). Therefore, as is the case for diffused aeration, predictive equations for the gas transfer rate coefficients are based on a synthesis of fundamental models with empirical modifications that improve the correlation between the models and the experimental data.

A number of correlations that predict the values of one or both mass transfer coefficients (k_L and k_G) in packed towers have been proposed. Comparative studies of the accuracy of these models for application to stripping of VOCs (Roberts et al. 1985; Lamarche and Droste, 1989; Staudinger et al., 1990; Dvorak et al., 1996) have generally reached the conclusion that the correlations proposed by Onda et al. (1968) provide the best fit to the experimental data. The correlation is generally reported to be accurate in predicting $K_L a_L$ values to within ±30%, although larger deviations are sometimes reported, especially in cases where gas-phase resistance is important (Dvorak et al., 1996; Little and Mariñas, 1997).

Use of the Onda approach involves calculation of k_L, k_G, and a_R from three separate correlations, each of which includes some dimensionless parameters and some empirical fitting constants. The correlations, which have been presented by different researchers in a number of equivalent formats, are summarized in Table 5-8 as relationships among dimensionless terms.

The Onda correlations provide an approach for estimating the gas transfer rate coefficients in the packed portion of a bed. In many cases, a significant amount of gas transfer can occur near the top and (especially) the bottom of the tower, outside the packed region. The contributions of these zones to gas transfer can be assessed experimentally by studying the amount of gas transfer that is accomplished in the column in the absence of packing (Roberts et al., 1985).

■ **EXAMPLE 5-12.** Calculate the values of $k_L a_R$, $k_G a_R$, and $K_L a_R$ using the Onda correlations for stripping of chloroform from an aqueous solution at 20°C in a 1.5-m diameter column packed with 2-in (=0.051 m) polyethylene Berl saddles. The column is to be operated with a liquid flow rate of 17.7 L/s (0.0177 m³/s) and an air:water flow ratio, Q_G/Q_L, of 25 L$_G$/L$_L$. Relevant properties of the solution, the gas, and the packing are provided in

TABLE 5-8. Onda Correlations for Mass Transfer in Packed Towers

Correlations for computing k_L, k_G, and a^a

$$k_L \left(\frac{\rho_L}{\mu_L g}\right)^{1/3} = 0.0051 \left(\frac{Q_L \rho_L}{A_R a_R \mu_L}\right)^{2/3} \left(\frac{\mu_L}{\rho_L D_L}\right)^{-1/2} (a_T d_P)^{0.4}$$
(5-83a)

$$= 0.0051 \left(\frac{L_M}{a_R \mu_L}\right)^{2/3} \left(\frac{\mu_L}{\rho_L D_L}\right)^{-1/2} (a_T d_P)^{0.4}$$
(5-83b)

$$\left(\frac{k_G}{a_T D_G}\right) = 5.23 \left(\frac{Q_G \rho_G}{A_R a_T \mu_G}\right)^{0.7} \left(\frac{\mu_G}{\rho_G D_G}\right)^{1/3} (a_T d_P)^{-2} \quad (5\text{-}84\text{a})$$

$$= 5.23 \left(\frac{G_M}{a_T \mu_G}\right)^{0.7} \left(\frac{\mu_G}{\rho_G D_G}\right)^{1/3} (a_T d_P)^{-2} \quad (5\text{-}84\text{b})$$

$$\frac{a_R}{a_T} = 1 - \exp\left[-1.45 \left(\frac{\sigma_c}{\sigma_L}\right)^{0.75} \text{Re}^{0.1} \text{Fr}^{-0.05} \text{We}^{0.2}\right]$$
(5-85)

Definitions of dimensionless groups used to calculate a_R/a_T^b

Reynolds number: $\text{Re} = \dfrac{Q_L \rho_L}{A_R a_T \mu_L} = \dfrac{L_M}{a_T \mu_L}$ (5-86)

Froude number: $\text{Fr} = \dfrac{Q_L^2 a_T}{A_R^2 g} = \dfrac{L_M^2 a_T}{\rho_L^2 g}$ (5-87)

Weber number: $\text{We} = \dfrac{Q_L^2 \rho_L}{A_R^2 \sigma_L a_T} = \dfrac{L_M^2}{\rho_L \sigma_L a_T}$ (5-88)

Definitions of individual parameters
A_R = cross-sectional area of packed tower (m²)
a_T = surface area of packing per unit volume of reactor (m²/m³)
a_R = interfacial area per unit volume of reactor (m²/m³)
d_p = characteristic length (e.g., diameter) of packing (m)
D_L, D_G = diffusivity of solute in liquid and gas phases, respectively (m²/s)
g = gravitational constant (9.81 m/s²)
L_M, G_M = liquid and gas areal mass loading rates, equal to $Q\rho/A$ in each phase (kg/m² s)
Q_L, Q_G = liquid and gas volumetric flow rates (m³/s)
μ_L, μ_G = viscosity of liquid and gas phases, (N s/m²)
ρ_L, ρ_G = density of liquid and gas phases (kg/m³)
σ_c = critical surface tension of packing material (N/m)
σ_L = surface tension of liquid (N/m)

[a] Onda suggested using 2.00 as the coefficient in the expression for $(k_G/a_T D_G)$ if the characteristic size of the packing was <15 mm.
[b] The expressions after the second equal sign in Equations 5-86, 5-87, and 5-88 allow the calculations to be performed using mass flow rates instead of volumetric flow rates. This approach is valuable in systems where the temperature changes significantly in the column, because such a change alters the volumetric, but not the mass, flow rate of gas. Such situations arise only rarely in environmental applications.

the following table. What fraction of the overall mass transfer resistance resides in the gas phase?

$a_T = 105 \text{ m}^{-1}$	$H_{cc} = 0.16 \text{ m}_L^3/\text{m}_G^3$
$D_L = 1.03 \times 10^{-9} \text{ m}^2/\text{s}$	$D_G = 1.17 \times 10^{-5} \text{ m}^2/\text{s}$
$\mu_L = 1.0 \times 10^{-3} \text{ Pa s}$	$\mu_G = 1.82 \times 10^{-5} \text{ Pa s}$
$\rho_L = 998 \text{ kg/m}^3$	$\rho_G = 1.20 \text{ kg/m}^3$
$\sigma_L = 0.073 \text{ N/m}$	$\sigma_c = 0.033 \text{ N/m}$
$d_p = 0.051 \text{ m}$	

Solution. The calculation involves inserting the given information into the expressions in Table 5-8. We first compute the cross-sectional area of the column and the gas flow rate

$$A_R = \frac{\pi d_{column}^2}{4} = \frac{\pi (1.5 \text{ m})^2}{4} = 1.77 \text{ m}^2$$

$$Q_G = \frac{Q_G}{Q_L} Q_L = 25(0.0177 \text{ m}^3/\text{s}) = 0.443 \text{ m}^3/\text{s}$$

The Reynolds, Froude, and Weber numbers can then be computed as follows:

$$\text{Re} = \frac{Q_L \rho_L}{A_R a_T \mu_L} = \frac{(0.0177 \text{ m}^3/\text{s})(998 \text{ kg/m}^3)}{(1.77 \text{ m}^2)(105 \text{ m}^{-1})(1.0 \times 10^{-3} \text{ Pa s})}$$

$$\times \left(1 \frac{\text{Pa}}{\text{kg/m s}^2}\right) = 95.2$$

$$\text{Fr} = \frac{Q_L^2 a_T}{A_R^2 g} = \frac{(0.0177 \text{ m}^3/\text{s})(105 \text{ m}^{-1})}{(1.77 \text{ m}^2)^2 (9.81 \text{ m/s}^2)} = 1.07 \times 10^{-3}$$

$$\text{We} = \frac{Q_L^2 \rho_L}{A_R^2 \sigma_L a_T}$$

$$= \frac{(0.0177 \text{ m}^3/\text{s})^2 (998 \text{ kg/m}^3)}{(1.77 \text{ m}^2)^2 (0.073 \text{ N/m})(105 \text{ m}^{-1})} = 1.31 \times 10^{-2}$$

The value of a_R can now be computed, using the values of Re, Fr, and We, after which k_L and k_G can be determined as

$$a_R = \left\{1 - \exp\left[-1.45 \left(\frac{\sigma_c}{\sigma_L}\right)^{0.75} \text{Re}^{0.1} \text{Fr}^{-0.05} \text{We}^{0.2}\right]\right\} a_T$$

$$= \left\{1 - \exp\left[-1.45 \left(\frac{0.033}{0.073}\right)^{0.75} (95.2)^{0.1}\right.\right.$$

$$\left.\left. \times (1.07 \times 10^{-3})^{-0.05} (1.31 \times 10^{-2})^{0.2}\right]\right\} 105 \text{ m}^{-1}$$

$$= 55.2 \text{ m}^{-1}$$

$$k_L = 0.0051 \left(\frac{Q_L \rho_L}{A_R a_R \mu_L}\right)^{2/3} \left(\frac{\mu_L}{\rho_L D_L}\right)^{-1/2} (a_T d_p)^{0.4} \left(\frac{\rho_L}{\mu_L g}\right)^{-1/3}$$

$$= 0.0051 \left(\frac{(0.0177 \text{ m}^3/\text{s})(998 \text{ kg/m}^3)}{(1.77 \text{ m}^2)(55.2 \text{ m}^{-1})(1.0 \times 10^{-3} \text{ Pa s})}\right)^{2/3}$$

$$\times \left(\frac{1.0 \times 10^{-3} \text{ Pa s}}{(998 \text{ kg/m}^3)(1.03 \times 10^{-9} \text{ m}^2/\text{s})}\right)^{-1/2}$$

$$\times \left((105 \text{ m}^{-1})(0.051 \text{ m})\right)^{0.4}$$

$$\times \left(\frac{998 \text{ kg/m}^3}{(1.0 \times 10^{-3} \text{ Pa s})(9.81 \text{ m/s}^2)}\right)^{-1/3}$$

$$= 2.18 \times 10^{-4} \text{ m/s}$$

$$k_G = 5.23 \left(\frac{Q_G \rho_G}{A_R a_T \mu_G}\right)^{0.7} \left(\frac{\mu_G}{\rho_G D_G}\right)^{1/3} (a_T d_p)^{-2} (a_T D_G)$$

$$= 5.23 \left(\frac{(0.443 \text{ m}^3/\text{s})(1.20 \text{ kg/m}^3)}{(1.77 \text{ m}^2)(105 \text{ m}^{-1})(1.82 \times 10^{-5} \text{ Pa s})}\right)^{0.7}$$

$$\times \left(\frac{1.82 \times 10^{-5} \text{ Pa s}}{(1.2 \text{ kg/m}_G^3)(1.17 \times 10^{-5} \text{ m}^2/\text{s})}\right)^{1/3}$$

$$\times \left((105 \text{ m}^{-1})(0.051 \text{ m})\right)^{-2} \left((105 \text{ m}^{-1})\right)$$

$$\times (1.17 \times 10^{-5} \text{ m}^2/\text{s})$$

$$= 8.41 \times 10^{-3} \text{ m/s}$$

The values of $k_L a_R$, $k_G a_R$, and $K_L a_R$ are as follows:

$$k_L a_R = (2.18 \times 10^{-4} \text{ m/s})(55.2 \text{ m}^{-1}) = 1.20 \times 10^{-2} \text{ s}^{-1}$$

$$k_G a_R = (8.41 \times 10^{-3} \text{ m/s})(55.2 \text{ m}^{-1}) = 0.464 \text{ s}^{-1}$$

$$K_L a_R = \frac{H k_L k_G}{k_L + H k_G} a_R$$

$$= \frac{(0.16)(2.18 \times 10^{-4} \text{ m/s})(8.41 \times 10^{-3} \text{ m/s})}{(2.18 \times 10^{-4} \text{ m/s}) + (0.16)(8.41 \times 10^{-3} \text{ m/s})} \times 55.2 \text{ m}^{-1}$$

$$= 0.010 \text{ s}^{-1}$$

Finally, we can compute the fraction of the total resistance that is in the gas phase using Equation 5-64:

$$\frac{R_L}{R_{tot}} = \frac{1}{(k_L/k_G H) + 1}$$

$$= \frac{1}{(2.18 \times 10^{-4} \text{ m/s})/((8.41 \times 10^{-3} \text{ m/s})(0.16)) + 1}$$

$$= 0.861$$

86.1% of the resistance is in the liquid phase, so 13.9% is in the gas. ∎

Effects of Other Parameters on Gas Transfer Rate Constants

Temperature Changes in temperature might affect the value of $K_L a_L$ in a number of ways. First, the diffusivity of the species, which incorporates factors related to both the kinetic energy of the molecules and viscosity of the fluid, increases with temperature in both phases. In addition, increasing temperature leads to an increase in Henry's constant, altering the driving force for the transfer and shifting more of the resistance into the liquid phase. Finally, changes in temperature can alter the fluid dynamics in complex, system-specific ways that might, for example, change the packet exchange frequency or the thickness of the interfacial regions. Some of these effects are well established and can be modeled mathematically, whereas others are much less well understood.

Often, all the effects of temperature on gas transfer are lumped together and modeled empirically, using an approach that is based on that used to model the kinetics of homogeneous reactions. Specifically, since $K_L a_L$ is an overall rate constant, the effect of temperature on $K_L a_L$ is sometimes modeled using the Arrhenius expression. This expression, which applies to many chemical reaction rate constants, represents the logarithm of the rate constant as a linear function of the inverse absolute temperature (Equation 3-87).

If it is applied over a limited range of temperatures, the relationship expressed by the Arrhenius equation is also reasonably consistent with an equation of the following form:

$$k_{T_2} = k_{T_1} \theta^{T_2 - T_1} \qquad (5\text{-}89)$$

where θ is an empirical constant; reported values of θ range from approximately 1.01 to 1.04, with a value of 1.024 being common (Metcalf and Eddy Inc, 2003). In the waste treatment and stream reaeration literature, the dependence of $K_L a_L$ for oxygen on the temperature has often been modeled by such an equation, with T_1 frequently taken as 20°C and T_2 being in the range from 0°C to 35°C. The magnitude of the effects of temperature on $K_L a_L$ for this range of θ values is shown in Figure 5-27.

∎ **EXAMPLE 5-13.** In a test performed in the summer, with water at 28°C, the $K_L a_L$ value for oxygen transfer in a bubble aeration system is determined to be 0.12 min^{-1}. If all system operating parameters other than temperature remain the same, predict the difference in r_{L,O_2} for oxygen transfer into a solution with $c_{L,O_2} = 2.0 \text{ mg/L}$ between summer and winter, when the solution temperature is expected to be 5°C. Assume that the water contains negligible solutes and that the

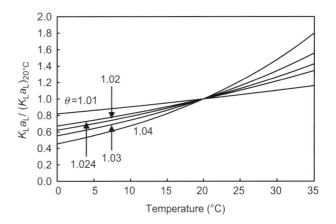

FIGURE 5-27. Effect of temperature on the overall gas transfer rate constant for oxygen, based on Equation 5-89 (T_2 in °C).

effect of temperature on $K_L a_L$ can be modeled using the Arrhenius equation given in the preceding section with $\theta = 1.024$.

Solution. The overall gas transfer rate constant in the winter can be determined by inserting the given information into the equation describing the effect of temperature on $K_L a_L$

$$(K_L a_L)_{T_2} = (K_L a_L)_{T_1} \theta^{T_2 - T_1}$$
$$(K_L a_L)_{5°C} = (K_L a_L)_{28°C} (1.024)^{5-28}$$
$$= (0.12 \text{ min}^{-1}) 1.024^{-23} = 0.070 \text{ min}^{-1}$$

The rate of gas transfer into solution in the given systems, r_{L,O_2}, is the product of $K_L a_L$ and the extent of disequilibrium. According to Figure 5-9, the solubility of oxygen in a solution containing no other solutes is 7.8 mg/L at 28°C, and 12.8 mg/L at 5°C. Therefore, the rates of O_2 transfer in the two systems are

$$r_{L,O_2,5°C} = \left[K_L a_L \left(c^*_{L,O_2} - c_{L,O_2} \right) \right]_{5°C}$$
$$= 0.070 \text{ min}^{-1} [(12.8 - 2.0) \text{ mg/L}] = 0.75 \text{ mg/L min}$$
$$r_{L,O_2,28°C} = \left[K_L a_L \left(c^*_{L,O_2} - c_{L,O_2} \right) \right]_{28°C}$$
$$= 0.12 \text{ min}^{-1} [(7.8 - 2.0) \text{ mg/L}] = 0.70 \text{ mg/L min}$$

Thus, the rate of oxygen transfer is expected to be approximately 7% higher in winter than in summer. That is, the decline in $K_L a$ at the lower temperature is more than compensated by the increased driving force (due to the increase in O_2 solubility at lower temperature), so that the rate of O_2 transfer into the solution is predicted to be greater in winter than in summer. ∎

Solution Chemistry Solution-phase chemistry can also have a large effect on gas transfer rates, both by modifying the solubility of the compound and by altering the structure of the interfacial region. In particular, the presence of surface-active agents (surfactants) can affect both k_L and the surface tension, which in turn affect the ease with which new interface is created and the value of a_L. Surface-active agents are chemicals that, for reasons described in Chapter 7, tend to accumulate at the boundary between water and another phase.

Detergents, oils and greases, and many other compounds that have at least some hydrophobic character, are surface active and tend to accumulate at the water/air interface. Surface accumulations of these substances provide an additional resistance through which gas molecules must pass to transfer between the phases, so the substances tend to lower k_L directly. They also tend to dampen disturbances at the water surface, thereby reducing the mixing of interfacial water with bulk water and increasing the resistance of the liquid interfacial layer. Some surfactants may form chemical associations with gas molecules as the latter move through the interfacial region. Surfactants are always present in domestic wastewater.

Because the identity and concentration of surfactants and other chemicals that affect $K_L a_L$ change over time in a given system and differ from one system to the next, it has not been possible to develop a useful approach for predicting their effect. Rather, all effects of water composition on $K_L a_L$ are typically lumped into an overall term, usually designated α, that accounts for differences in the values of $K_L a_L$ when a given system is operated with clean water and with the water of interest:

$$\alpha \equiv \frac{K_L a_L|_{\text{solution of interest}}}{K_L a_L|_{\text{clean water}}} \quad (5\text{-}90)$$

Thus, α expresses the same type of relationship for the gas transfer rate constant that β does for the saturation concentration of gas (as indicated in Equation 5-22). Both α and β, as well as the temperature correction factor θ, often appear in gas transfer rate expressions in the literature dealing with oxygen transfer into aerated, biological waste treatment systems. In such systems, if $K_L a_L$ and c_L^* are evaluated for systems with clean water at 20°C, the overall rate expression for oxygen transfer at other temperatures can be represented as follows:

$$r_{L,\text{gt},O_2} = \alpha K_L a_L \theta^{T-20} \left(\beta c_L^* - c_{L,b} \right) \quad (5\text{-}91)$$

The concepts of the α, β, and θ factors and the effects of water quality on the values of these parameters have been reviewed by Stenstrom and Gilbert (1981). These researchers reported typical ranges for the parameters in domestic wastewater (the application in which the α, β, and θ parameters are most often considered) as follows: $0.3 < \alpha < 0.8$, typically ~ 0.5; $0.7 < \beta < 1.0$, typically ~ 0.95; and $1.01 < \theta < 1.05$, typically ~ 1.02.

5.8 SUMMARY

In this chapter, the fundamentals of gas transfer processes are introduced and discussed. The gas transfer reaction takes place at the gas/liquid interface and is driven by the tendency of the system to reach chemical equilibrium at that location. The overall gas transfer process couples the reaction at the interface with transport between the interface and the two bulk phases. The process can be envisioned as proceeding via a series of steps, with each step imposing some resistance. In general, transfer through the liquid- or gas-phase boundary layer immediately adjacent to the interface imposes the greatest resistance. In many environmental engineering systems, the resistance to gas transfer is greater in the liquid-phase boundary layer than in the gas-phase boundary layer, but this generalization is not always true, particularly for species with low Henry's constants or those that undergo rapid reactions in the aqueous phase.

Molecular diffusion is the dominant transport mode through the boundary layers. The conditions in each boundary layer might be characterized by steady state and a linear concentration gradient (if the boundary layer is thin and packets of fluid spend a relatively long time in the layer) or by an unsteady state and a nonlinear gradient (if the boundary layer is thick and packets of fluid spend a relatively short time there). Empirical evidence suggests that the latter scenario is far more common than the former.

The flux of a volatile species across a gas/liquid interface depends on the ratio of the driving force (which can be quantified as the extent of disequilibrium) to the overall resistance. The overall rate of gas transfer is the product of the flux and the concentration of interfacial area in the system. Mathematically, these concepts lead to rate expressions for gas transfer of the form $r_{L,\text{gt}} = K_L a_L (c_L^* - c_{L,b})$, where $r_{L,\text{gt}}$ is the rate at which the volatile species enters solution from the gas phase, K_L is an overall gas transfer rate constant, a_L is the amount of interfacial area per unit volume of liquid in the system, c_L^* is the hypothetical aqueous concentration that would be in equilibrium with the bulk gas phase, and $c_{L,b}$ is the concentration of the species in the bulk solution. A similar expression can be written that normalizes the interfacial area and the gas transfer rate to the volume of the entire reactor, rather than only to the liquid in the reactor: $r_{R,\text{gt}} = K_L a_R (c_L^* - c_{L,b})$.

K_L is predicted to vary with the molecular diffusivity (D_L) of the species according to $K_L \propto D_L^m$, where m is a constant that is between 0.5 and 1.0. Mathematical models of the process suggest that the value of m depends on the thickness of the packets and the dynamics of fluid exchange between the interface and bulk phases.

Equations 5-57 and 5-58, are expressions for the rate of gas transfer that can be easily substituted into mass balance equations.

$$r_{R,gt} = J_L a_R = K_L a_R \left(c_L^* - c_{L,b} \right) \quad (5\text{-}57)$$

$$r_{L,gt} = K_L a_R \frac{V_R}{V_L} \left(c_L^* - c_{L,b} \right) \quad (5\text{-}58a)$$

$$= K_L a_L \left(c_L^* - c_{L,b} \right) \quad (5\text{-}58b)$$

In the preceding equations, the product $K_L a_L$ or $K_L a_R$ can be thought of as the rate constant for an overall gas transfer reaction that is first order with respect to the extent of disequilibrium between the bulk gas and bulk solution. Therefore, if we wish to increase the rate of approach to equilibrium, we could attempt to increase either K_L or a_L. K_L might be increased by increasing the temperature of the system or the mixing intensity in the phase that provides the greatest resistance.

Alternatively, if approaches for increasing K_L are impossible or impractical to implement, the rate constant can be increased by increasing the value of a_L. This might be accomplished by decreasing the size of the dispersed phase material (i.e., using smaller bubbles in a gas-in-water system or smaller water droplets in a water-in-gas system). The same effect could be achieved by altering the hydrodynamics or aerodynamics in the system in such a way that the dispersed phase spent more time in contact with the continuous phase, for instance by packing a spray tower with material that interrupts the downward path of the droplets; this alteration would increase the hold-up time of the liquid, thereby increasing the interfacial area per unit volume of reactor, without requiring a change in the overall flow rate of either liquid or gas.

If a volatile species undergoes a rapid reaction in solution, its transfer into solution is enhanced. This principle is particularly important when analyzing the transfer of a volatile species that is either an acid or a base. Defining α_{neut} as the fraction of the total dissolved concentration that is in the neutral (volatile) form, then, if a system initially favors gas stripping, decreasing α_{neut} will decrease the rate of stripping and, if α_{neut} is low enough, convert the system into one that favors gas absorption. If the initial system is one that favors absorption, decreasing α_{neut} will increase the rate of gas absorption. The conversion of the dissolved volatile species into another form increases the relative importance of gas-phase resistance to the overall resistance.

Values of the overall, first-order rate constant for gas transfer ($K_L a_L$ or $K_L a_R$) can be determined by combining experimental data with a mass balance on the volatile species, with Equation 5-57 or 5-58 substituted for $r_{L,gt}$. On the other hand, values of their corresponding rate constants for transfer through the interfacial region in either phase individually ($k_L a_L$, $k_L a_R$, $k_G a_L$, and $k_G a_R$) are more difficult to obtain experimentally, especially for full-scale systems. The latter values can often be estimated from correlations that have been developed for specific types of gas transfer systems.

In Chapter 6, the concepts developed in this chapter are combined with information about the macroscopic properties of liquid/gas contacting systems to develop design equations for analyzing full-scale systems and predicting their behavior under different operating conditions.

APPENDIX 5A. CONVENTIONS USED FOR CONCENTRATIONS AND ACTIVITY COEFFICIENTS WHEN COMPUTING HENRY'S CONSTANTS

Overview

The activity and concentration of volatile species can be expressed using a variety of conventions. Here, some of those conventions are described and compared.

Equation 5-8 shows that the version of Henry's constant that corresponds to an equilibrium constant can be expressed as the ratio of the gas-phase to the liquid-phase activity of the volatile species:

$$H' \equiv \left. \frac{a_{G,i}}{a_{L,i}} \right|_{eq} \quad (5\text{-}8)$$

Furthermore, according to Equation 5-9, the activity of i in either phase can be expressed as the product of its activity coefficient (γ_i) and its concentration normalized to the concentration in the standard state:

$$a_i = \gamma_i \frac{c_i}{c_{i,\text{std.state}}} \quad (5\text{-}9)$$

The activity coefficient indicates how differently the species of interest behaves in the real environment compared with its behavior in the standard state or reference environment. If the behavior of the molecules in the real system is identical to that in the reference environment, then γ_i is 1.0, and the molecules are said to behave *ideally*. Correspondingly, if the molecules behave very differently in the two environments, the activity coefficient is very different from

1.0: if the molecules are less "active" in the real environment than in the reference environment, γ_i is <1.0, and if they are more active, it is >1.0.

By definition, in the standard state, c_i equals $c_{i,\text{std.state}}$ and γ_i equals 1.0, so a_i is also 1.0. The activity of the species in a system of interest can then be interpreted with reference to the standard state. For instance, if a constituent has an activity of 10^{-3}, then it is only one one-thousandth as "active" as in the standard state.[21]

The choice of the standard state conditions is arbitrary; any system, with any concentration and any physicochemical environment, could be chosen as the standard state. Fortunately, only a few choices for the standard state are commonly employed; perhaps unfortunately, though, no single convention has been adopted universally. Hence, it is necessary to understand what these different standard states are, and which one is being used in a given situation, to interpret the available data.

Conventions for the Physicochemical Environment in the Standard State

A temperature of 25 °C and a total pressure of 1.0 atm are fairly universal choices for standard-state conditions in environmental engineering and science. Also, for gases, the molecular interactions in the standard-state environment are specified to be those that occur in an ideal gas. Gases at normal pressures behave very much like ideal gases (i.e., as if they were in the reference environment), so activity coefficients of gaseous species are almost always assumed to be 1.0 in environmental engineering.

In contrast, for solutes, different reference environments have been chosen by different groups of workers. For instance, chemists dealing with freshwater systems typically choose infinite dilution in pure water as the reference environment for solutes. That is, in the reference environment, a solute is envisioned to interact only with water molecules.[22] On the other hand, in chemical engineering (where much of the gas transfer literature evolved) the reference environment is often chosen as the pure chemical constituent, at least when that constituent is a liquid or solid at room temperature. Using this convention, the reference environment for benzene would be pure liquid benzene, where each benzene molecule interacts solely with other benzene molecules, as opposed to interacting solely with water molecules in the reference environment defined by infinite dilution. Neither of these choices is inherently better than the other; however, the different choices for the reference environment can have a large effect on the values of the activity coefficient and the activity of dissolved species.

If the reference environment chosen for a volatile species dissolved in water is infinite dilution, then its activity coefficient in real solutions is generally close to 1.0, for two reasons. First, the solutions usually *are* fairly dilute, so most of the molecules with which a dissolved species interacts are water molecules, as is the case in the reference environment. Second, the other constituents likely to be present in solution are mostly salt ions. These ions can have strong electrostatic interactions with one another. However, only uncharged species can exist in the gas phase in significant concentrations, and these species interact weakly with dissolved ions. As a result, in most solutions, volatile solutes behave similarly to how they would behave in pure water and therefore have activity coefficients very close to 1.0 when the reference state is infinite dilution.

If, however, the reference state is chosen to be a pure phase of the volatile species, the situation can be very different from that described in the preceding paragraph. In this case, "ideal behavior" is defined as a state in which each molecule of the volatile species interacts solely with other molecules that are identical to it (not with water). These interactions are likely to be much more favorable than those with water molecules, so the volatile species are more stable and less 'active' than when they are dissolved in an aqueous solution. By definition, the activity coefficient for the species is 1.0 in the reference state (the pure liquid), so γ_i is >1.0 (and often $\gg 1.0$) when the species is dissolved in water.

Consider a volatile species that is dissolved in an aqueous solution at a mole fraction x_i and is in equilibrium with the corresponding gaseous species at a partial pressure P_i (atm). If the reference state for the solute is chosen to be the pure liquid, and the standard state concentrations are $x_i^{\circ} = 1.0$ and $P_i^{\circ} = 1.0$ atm for the liquid and gas phases, respectively, the equilibrium constant for volatilization of i would be given by

$$^{\text{pure }i}H'_{\text{px}} = \frac{a_{G,i}}{a_{L,i}} = \frac{\gamma_{G,i}P_i/P_{i,\text{std.state}}}{\gamma_{L,i}^{\text{pure }i}x_i/x_{i,\text{std.state}}} = \frac{P_i}{\gamma_{L,i}^{\text{pure }i}x_i} \quad (5A\text{-}1)$$

where the superscript "pure i" indicates the reference state for dissolved i, and the final equality incorporates an assumption that $\gamma_{G,i}$ equals 1.0.

If the liquid is pure i ($x_i = 1.0$), then i in that liquid is in its reference state, and $\gamma_{L,i}^{\text{pure }i} = 1.0$. In this case, the

---

[21] The idea of a molecule being more active in one environment than another is not as vague as this discussion might imply. The activity is a quantitative reflection of the available energy stored in molecules, as indicated by the relationship between molar Gibbs free energy and chemical activity: $\overline{G}_i = G_i^{\circ} + RT \ln a_i$. According to this equation, for a system at 25 °C, whenever the activity of the constituent doubles, the Gibbs free energy of that constituent increases by 1.72 kJ/mol. The relationship between chemical activity, Gibbs free energy, and reactivity is discussed in all chemical thermodynamics texts and also in many environmental chemistry texts. See, for instance, the texts by Benjamin (2002) and Schwarzenbach et al. (2002), or the classic text of Lewis and Randall (1961).

[22] This choice of a reference state is sometimes referred to as the *Lewis and Randall convention*.

equilibrium gas-phase pressure of i is, by definition, the vapor pressure of i ($P_{\text{vap},i}$), and Equation 5A-1 simplifies to

$$^{\text{pure }i}H'_{\text{px}} = \frac{P_i}{\gamma_{\text{L},i}^{\text{pure }i}x_i} = \frac{(1.0)(P_{\text{vap},i})}{(1.0)(1.0)} = P_{\text{vap},i} \quad (5\text{A-2})$$

Although Equation 5A-2 was derived for a scenario in which the liquid phase was pure i, the same equilibrium constant must apply to gas/liquid partitioning of i for any other liquid phase containing i, as long as the same reference state is used for the liquid phase. Therefore, again utilizing the assumption that $\gamma_{\text{G},i} = 1.0$, but recognizing that $\gamma_{\text{L},i}^{\text{pure }i}$ and x_i might be different from 1.0, Equation 5–94 can be applied to a dilute aqueous solutions of i as

$$P_{\text{vap},i} = \frac{P_i}{\gamma_{\text{L},i}^{\text{pure }i}x_i} \quad (5\text{A-3})$$

Now consider a scenario in which beakers of pure liquid i and pure water are both in contact with the same gas phase, in a closed system. As the system equilibrates, molecules of i will evaporate from the pure liquid, and some of those gas-phase molecules will dissolve into the water. Assuming that the initial volume of liquid i is sufficient, the ultimate equilibrium system will include some remaining liquid i, a partial pressure of $P_{\text{vap},i}$ in the gas phase, and a mole fraction x_i in the aqueous solution.[23]

Because both liquids in this system are equilibrated with the same gas phase, they must be in equilibrium with one another. That is, the concentration of i dissolved in the water is equilibrated with pure liquid i, just as it would be if the two liquids were in direct contact with one another. We can therefore interpret x_i as the equilibrium solubility of pure i in water, expressed as a mole fraction; we designate this concentration $x_{\text{sat'n},i}$. To apply Equation 5A-3 to the aqueous solution, we substitute $P_{\text{vap},i}$ for P_i and $x_{\text{sat'n}}$ for x_i to obtain

$$P_{\text{vap},i} = \frac{P_{\text{vap},i}}{\gamma_{\text{L},i}^{\text{pure }i}x_{\text{sat'n},i}} \quad (5\text{A-4})$$

$$\gamma_{\text{L},i}^{\text{pure }i} = \frac{1}{x_{\text{sat'n},i}} \quad (5\text{A-5})$$

This result indicates that, when pure i is the reference state, the activity coefficient for a volatile species dissolved in an aqueous solution equals the inverse of the aqueous solubility of that species, expressed as a mole fraction.

Note that this result applies to any aqueous solution containing i, not just one that is saturated with i. Also, although Equation 5A-5 was derived for a species that forms a liquid when it is present as a pure compound, the same derivation applies if pure i forms a solid phase.

APPENDIX 5B. DERIVATION OF THE GAS TRANSFER RATE EXPRESSION FOR VOLATILE SPECIES THAT UNDERGO RAPID ACID/BASE REACTIONS

As noted in the body of this chapter, only neutral species can transfer into or out of the gas phase. However, when considering the composition of a solution, our interest is often in the total dissolved concentration of an acid/base group, rather than that of the neutral species alone. That is, for instance, we are often more interested in the total concentration of dissolved sulfide species ($c_{\text{L,TOTS}} = (c_{\text{H}_2\text{S(aq)}} + c_{\text{HS}^-} + c_{\text{S}^{2-}})$) than in $c_{\text{H}_2\text{S(aq)}}$, c_{HS^-}, or $c_{\text{S}^{2-}}$ individually. The relative contributions of the three dissolved sulfide species to TOTS$_\text{L}$ depend only on the solution pH and are commonly represented as so-called α values, defined as $\alpha_0 \equiv c_{\text{H}_2\text{S(aq)}}/c_{\text{L,TOTS}}$, $\alpha_1 \equiv c_{\text{HS}^-}/c_{\text{L,TOTS}}$, and $\alpha_2 \equiv c_{\text{S}^{2-}}/c_{\text{L,TOTS}}$, respectively. In the case of H$_2$S, pK_{a2} is >13, so the second deprotonation can be ignored in most solutions of interest. In the following discussion, we assume that α_2 is negligible, so $c_{\text{L,TOTS}} \approx c_{\text{H}_2\text{S(aq)}} + c_{\text{HS}^-}$.

More generically, for any acid or base group formed by protonation of a core molecule A, α_n is defined as $(c_{\text{A}_n}/c_{\text{L,TOTA}})$, where A$_n$ is the species that is formed when n protons are lost from the most protonated species in the group. For instance, for the carbonate group, the most protonated species is H$_2$CO$_3$. Therefore, α_2 is the ratio of the concentration of the species that has two fewer protons than H$_2$CO$_3$ (i.e., CO$_3^{2-}$) to $c_{\text{L,TOTCO}_3}$ ($= c_{\text{H}_2\text{CO}_3} + c_{\text{HCO}_3^-} + c_{\text{CO}_3^{2-}}$): $\alpha_2 \equiv (c_{\text{CO}_3^{2-}}/c_{\text{L,TOTCO}_3})$.

Consider now how the rapid protonation or deprotonation of a neutral molecule at the interface affects the mass balances that we write to derive $r_{\text{L,gt}}$. If the volatile species does not undergo any reaction in solution, then the mass balances need consider only a single species in each phase, so the rate at which the species enters solution can be equated with the rate at which it is lost from the gas. If, however, the volatile species undergoes rapid protonation or deprotonation, that equality no longer applies. For example, transfer of 1.0 µmol/s of H$_2$S into a pH 8.0 solution increases $c_{\text{L,TOTS}}$ at a rate of 1.0 µmol/s, but, because $\alpha_{\text{H}_2\text{S(aq)}} = 0.09$ at this pH, the number of moles of H$_2$S(aq) in solution increases at a rate of only 0.09 µmol/s. This complication has significant implications for both the driving force for gas transfer and the overall mass transfer coefficient.

---

[23] In reality, some water would evaporate dissolve into the liquid i, so that the liquid would no longer be pure. However, for simplicity, we assume here that the solubility of water in liquid i is negligible.

The easiest way to demonstrate these effects is by writing a liquid-phase mass balance on $c_{L,TOTS}$, and not solely on dissolved H_2S. Assuming that the only rapid reaction in which the sulfide species participate is protonation/deprotonation, a volume-normalized mass balance around the liquid interfacial zone can be written for each sulfide species. Each such mass balance includes only two terms: one for diffusive flux, and another for production of that species by chemical reaction. Assuming that $H_2S(aq)$ and HS^- have the same liquid-phase diffusivity,[24] these mass balances are

$$\frac{\partial c_{H_2S(aq)}}{\partial t} = D_{L,H_2S}\frac{\partial^2 c_{H_2S(aq)}}{\partial x^2} + r_{L,H_2S(aq)} \quad (5B\text{-}1a)$$

$$\frac{\partial c_{HS^-}}{\partial t} = D_{L,H_2S}\frac{\partial^2 c_{HS^-}}{\partial x^2} + r_{L,HS^-} \quad (5B\text{-}1b)$$

In these equations, $r_{L,H_2S(aq)}$ is, formally, the rate of generation of $H_2S(aq)$ by the reaction occurring in solution (in this case, deprotonation of $H_2S(aq)$), and r_{HS^-} is the rate of generation of HS^- by this same reaction. Thus, in a given system, $r_{L,HS^-} = -r_{L,H_2S(aq)}$. The mass balance on the sum of $H_2S(aq)$ and HS^- (i.e., on $TOTS_L$) can then be written as the sum of the right sides of Equations 5B-1a and 5B-1b, in which case the reaction terms cancel one another. Thus, we can write that mass balance as follows:

$$\frac{\partial c_{L,TOTS}}{\partial t} = D_{L,H_2S}\frac{\partial^2 c_{L,TOTS}}{\partial x^2} \quad (5B\text{-}2)$$

The initial and boundary conditions for Equation 5B-2 are essentially identical to those for a system where only a single species is diffusing (uniform concentration throughout the packet at $t=0$, and known concentrations at the boundaries of the packet at all times), except that they relate to $c_{L,TOTS}$ instead of to the single species. As a result, the integration yields a very similar result as that obtained earlier for the flux of a nonreactive volatile species (Equation 5-44):

$$J_{L,TOTS} = k_{L,TOTS}(c_{L,int,TOTS} - c_{L,TOTS}) \quad (5B\text{-}3)$$

A similar mass balance can be written around the gas-phase interfacial region, yielding

$$J_{G,TOTS} = k_{G,TOTS}(c_{G,int,TOTS} - c_{G,TOTS}) \quad (5B\text{-}4)$$

A key distinction between Equations 5B-3 and 5B-4 is that $c_{G,TOTS}$ equals the concentration of $H_2S(g)$ alone, whereas $c_{L,TOTS}$ includes contributions from both $H_2S(aq)$

---

[24] This assumption is not necessary, but it simplifies the algebra.

and HS^-. Nevertheless, it is useful to write the equations as shown, to emphasize their core similarity.

Although Henry's law relates only the concentrations of the volatile species (in this case, H_2S) in the two phases, a pH-dependent, pseudo-Henry's constant can be written to relate the concentrations of $c_{L,TOTS}$ in the two phases:

$$\hat{H}_{H_2S} \equiv \left.\frac{c_{G,TOTS}}{c_{L,TOTS}}\right|_{eq} = \alpha_{neut}H_{H_2S} \quad (5B\text{-}5)$$

Assuming that the gas and solution are equilibrated at the interface, the same algebra can be performed on Equations 5B-3 to 5B-5 as is done in the main body of this chapter for nonreacting species to yield an overall expression for gas transfer, but in which all the parameters refer to $c_{L,TOTS}$ instead of a single species:

$$J_{L,TOTS} = K_{L,TOTS}\left(c^*_{L,TOTS} - c_{L,TOTS}\right) \quad (5B\text{-}6)$$

$$K_{L,TOTS} = \frac{\hat{H}_{H_2S}k_{L,TOTS}k_{G,TOTS}}{k_{L,TOTS} + \hat{H}_{H_2S}k_{G,TOTS}} \quad (5B\text{-}7)$$

Equations 5B-6 and 5B-7 can be manipulated in a few ways to facilitate comparison with the rate of gas transfer in nonreactive systems. First, we note that k_L and k_G for TOTS are expected to be the same as the corresponding values for the H_2S species alone (based on the assumptions that $D_{L,H_2S(aq)}$ and D_{L,HS^-} equal one another and that no HS^- exists in the gas phase), so we can write these parameters as k_{L,H_2S} and k_{G,H_2S}, respectively. We can also use Equation 5B-5 to substitute for $\hat{H}_{H_2S}$ in Equation 5B-7. Finally, we note that, in many systems of interest, the concentration of the volatile species in the gas phase and the total concentration of the dissolved species are known, and the pH is adjustable. Therefore, it is convenient to isolate the effect of pH from that of the other parameters. To do that, we can express $c^*_{L,TOTS}$ as $(c^*_{L,H_2S(aq)}/\alpha_{neut})$, thereby separating the pH-dependent term (α_{neut}) from the partial pressure-dependent term ($c^*_{L,H_2S(aq)}$, which can be computed as $(c_{G,H_2S}/H_{H_2S})$). When these substitutions are made, Equations 5B-6 and 5B-7 can be rewritten as follows:

$$J_{L,TOTS} = K_{L,TOTS}\left(\frac{c^*_{L,H_2S(aq)}}{\alpha_{neut}} - c_{L,TOTS}\right) \quad (5B\text{-}8)$$

$$K_{L,TOTS} = \frac{\alpha_{neut}k_{L,H_2S}k_{G,H_2S}H_{H_2S}}{k_{L,H_2S} + \alpha_{neut}k_{G,H_2S}H_{H_2S}} \quad (5B\text{-}9)$$

Equations 5B-8 and 5B-9 indicate that, for a given gas-phase concentration of $H_2S(g)$ and a given total dissolved sulfide concentration, both the driving force for gas transfer and the overall rate constant depend on α_{neut}, and therefore on the solution pH. Under conditions where α_{neut} is approximately 1.0 (at pH less than approximately 6 in the case of H_2S), these equations simplify to the equations for nonreactive gases, as would be expected.

The effects of changes in α_{neut} on the gas transfer rate can be divided into two categories. First, reducing α_{neut} always increases the effective solubility of the species; that is, the ratio of $c_{L,\text{TOT}i}$ to $c_{G,i}$ at equilibrium. As a result, if the flux under some baseline condition is out of solution, a change in pH that decreases α_{neut} will certainly decrease the driving force for that process, and it might even reverse the direction of the gas transfer, so that gas absorption is favored instead. The magnitude of this effect for a hypothetical system is shown in Figure 5B-1. In the system shown, the two phases are in equilibrium at pH 10.0, at which $\alpha_{i,\text{neut}}$ (in this case, $\alpha_{H_2S(aq)}$) is $10^{-3.0}$. Thus, if the pH is increased from some low starting value, the driving force for H_2S dissolution steadily diminishes up to pH 10.0, above which the direction of transfer switches to H_2S absorption.

In addition to affecting the driving force for gas transfer, α_{neut} alters the gas transfer rate constant, such that a reduction in α_{neut} diminishes the importance of the liquid-phase resistance relative to the gas-phase resistance. Specifically, by following a derivation parallel to the one that led to Equation 5-64 for nonreactive species, we find the fraction of the overall resistance to gas transfer that resides in the liquid phase for an acidic or basic species to be

$$\frac{R_L}{R_{\text{tot}}} = \frac{1}{(k_L/\alpha_{\text{neut}}k_GH) + 1} = \frac{k_GH}{(k_L/\alpha_{\text{neut}}) + k_GH} \quad (5B\text{-}10)$$

The reason for this shift is that the acid/base reactions tend to replenish a species that is being stripped from solution, or consume a species that is entering solution, thereby doing the same thing that diffusion through the liquid boundary layer would do. To the extent that these changes occur by acid/base reactions, the "work" that has to be done by diffusion through the liquid boundary layer diminishes, thus diminishing the likelihood that diffusion through the liquid-phase boundary layer will be the rate-controlling step for gas transfer.

The qualitative trend that a decrease in α_{neut} decreases the fraction of the overall resistance attributable to the liquid boundary layer is always predicted. However, the extent to which this effect alters the gas transfer rate constant depends on the details of the system, as shown generically in Figure 5B-2 for a system with the (typical) characteristic that the overall resistance to gas transfer resides mostly in the liquid boundary layer when $\alpha_{\text{neut}} = 1$.

Consider first the trend shown for $K_{L,i_{\text{neut}}}$. If $\alpha_{\text{neut}} = 1$, the value of $K_{L,i_{\text{neut}}}$ is $k_{L,i}$, just as it is in a system with no acid/base reactions. However, as α_{neut} decreases, $K_{L,i_{\text{neut}}}$ increases proportionately (as indicated by the 1:1 slope); thus, under these conditions, if α_{neut} decreases to 0.1, $K_{L,i_{\text{neut}}}$ increases to $(k_{L,i}/\alpha_{\text{neut}}) = 10\,k_{L,i}$. These opposing changes in α_{neut} and $K_{L,i_{\text{neut}}}$ reflect the fact that, even though a decline in α_{neut} reduces the transport of the neutral species through the liquid boundary layer, this decline is exactly compensated by the transport of ionized forms of i.

The trend described in the preceding paragraph continues to a point, after which the change in $K_{L,i_{\text{neut}}}$ per unit change in α_{neut} steadily decreases, and $K_{L,i_{\text{neut}}}$ asymptotically approaches a maximum value of $Hk_{G,i}$. As is clear from the form of the limiting value, this transition reflects a shift

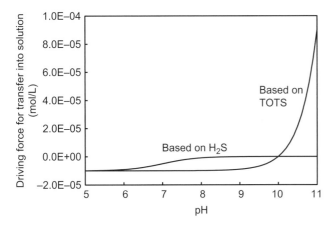

FIGURE 5B-1. The driving force for dissolution of H_2S into a solution containing $10^{-5.0}$ mol/L TOTS from a gas with $P_{H_2S} = 10^{-7.0}$ atm. The driving force based on TOTS is $c^*_{L,\text{TOTS}} - c_{L,\text{TOTS}}$, and that based on $H_2S(aq)$ is $c^*_{L,H_2S} - c_{L,H_2S}$.

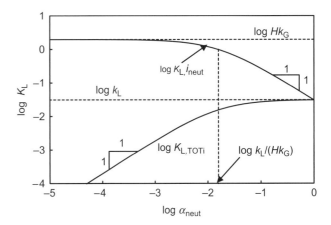

FIGURE 5B-2. The dependence of the overall gas transfer coefficient of a volatile acid or base on the extent of its ionization in solution. The figure is drawn for assumed values of $\log(k_L/k_GH) = -1.8$ and $\log k_L = -1.50$. The graph applies for any set of units, as long as the units of k_L, k_GH, and K_L are all the same.

from liquid-phase to gas-phase control of the gas transfer kinetics; once this shift has occurred, the fact that ionized forms of i can be transported through the liquid boundary layer no longer affects $K_{L,i_{neut}}$, because transport though that layer is no longer the bottleneck in the process.

As shown in the figure, the middle of the transition from liquid- to gas-phase control is at $\alpha_{neut} = k_{L,i}/k_{G,i}H$. As a consequence, the shift between these two limiting conditions occurs at different α_{neut} values for different volatile species in the same solution (because of their different H values), and different α_{neut} values for a given species in different solutions (because of the different $k_{L,i}/k_{G,i}$ ratios). In some cases, the value of $(K_{L,i}/Hk_{G,i})$ might be >1, so that transport through the gas phase controls gas transfer kinetics at all pH values, and $K_{L,i_{neut}}$ is independent of α_{neut} (and pH); in others, $(K_{L,i}/Hk_{G,i})$ might be so small that $K_{L,i_{neut}}$ increases with decreasing α_{neut} under all reasonable conditions.

Figure 5B-2 also shows the trend in $K_{L,TOTi}$ as α_{neut} changes. Although $K_{L,TOTi}$ is identical to $K_{L,i_{neut}}$ (and equal to $k_{L,i}$) at $\alpha_{neut} = 1$, the shape of the $K_{L,TOTi}$ versus α_{neut} curve is very different from that of $K_{L,i_{neut}}$ versus α_{neut}. The reason is that, in calculations using $K_{L,TOTi}$, the driving force is considered to be $c^*_{L,TOTS} - c_{L,TOTS}$; that is, the equations are written as though all the species that contribute to TOTi can exist in both the gas and liquid phases. As a result, $K_{L,TOTi}$ must decrease with decreasing α_{neut} to account for the fact that ionized species *cannot* contribute to transport through the gas-phase boundary layer (whereas $K_{L,i_{neut}}$ increases with decreasing α_{neut} to account for the fact that ionized species *can* contribute to transport through the liquid-phase boundary layer). The key point is that, as long as the expressions for the gas transfer coefficient and the driving force are consistent with one another, the correct value of flux can be computed using either approach; that is

$$J_{L,i} = K_{L,i_{neut}}\left(c^*_{L,i_{neut}} - c_{L,i_{neut}}\right) = K_{L,TOTi}\left(c^*_{L,TOTi} - c_{L,TOTi}\right) \quad (5B\text{-}11)$$

REFERENCES

APHA (2005) *Standard Methods for Examination of Water and Wastewater*, 20th edn., prepared and published jointly by APHA, AWWA, and WEF, Washington, DC.

AWWA (2011) In Edzwald, J. K. (ed.), *Water Quality and Treatment*, 6th ed., McGraw Hill, New York.

Barnhart, E. L. (1969) Transfer of oxygen in aqueous solutions. *J. Sanitary Eng. Div. (ASCE)*, 95 (SA3) 645–661.

Calderbank, P. S., Johnson, D. S. L., and Loudon, J. (1970) Mechanics and mass transfer of single bubbles in free rise through some Newtonian and non-Newtonian liquids. *Chem. Eng. Sci.*, 25, 235–256.

Carslaw, H. S., and Jaeger, J. C. (1959) *Conduction of Heat in Solids*. Oxford University Press, London, England.

Clark, M. M. (2009) *Transport Modeling for Environmental Engineers and Scientists*, 2nd ed., Wiley Interscience, New York, NY.

Cussler, E. L. (1997) *Diffusion: Mass Transfer in Fluid Systems*, Cambridge University Press, Cambridge, England.

Danckwerts, P. V. (1951) Significance of liquid-film coefficients in gas adsorption. *Ind Eng. Chem.*, 43(6), 1460–1467.

Dvorak, B. I., Lawler, D. F., Fair, J. R., and Handler, N. E. (1996) Evaluation of the Onda correlations for mass transfer with large random packings. *Environ. Sci. Technol.*, 30, 945–953.

Fan, L. T., Shen, B. C., and Chow, S. T. (1993) The surface-renewal theory of interphase transport – a stochastic treatment. *Chem. Eng. Sci.*, 48(23), 3971–3982.

Fuller, E. N., Schettler, P. D., and Giddings, J. C. (1966) A new method for prediction of binary gas-phase diffusion coefficients. *Ind. Eng. Chem.*, 58(5), 19–27.

Green, D. W., and Perry, R. H. (2008) *Perry's Chemical Engineers' Handbook*, 8th ed., McGraw Hill, New York.

Haberman, W. L., and Morton, R. K. (1956) An experimental study of bubbles moving in liquids. *ASCE Trans.*, 121, 227–250.

Harriott, P. (1962) A random eddy modification of the penetration theory. *Chem. Eng. Sci.*, 17, 149–154.

Hayduk, W., and Laudie, H. (1974) Prediction of diffusion coefficients for nonelectrolytes in dilution solutions. *AIChE J.*, 20, 611–615.

Higbie, R. (1935) The rate of absorption of a pure gas into a still liquid during short periods of exposure. *Trans. AIChE*, 31, 365–388.

Howard, C., and Corsi, R. L. (1996) Volatilization of chemicals from drinking water to indoor air: Role of the kitchen sink. *J. Air Waste Manag. Assoc.*, 46(9), 830–837.

Howe, K., and Lawler, D. F. (1989) Acid–base reactions in gas transfer: a mathematical approach. *JAWWA*, 81(1), 61–66.

Hsieh, C. C., Ro, K. S., and Stenstrom, M. K. (1993) Estimating emissions of 20 VOCs. 1. Surface aeration. *J. Environ. Eng. (ASCE)*, 119(6), 1077–1098.

Israelachvili, J. (1985) *Intermolecular and Surface Forces*, Academic Press, London.

Jahne, B., and Haussecker, H. (1998) Air–water gas exchange. *Ann. Rev. Fluid Mech.*, 30, 443–468.

Kishinevskii, M. K., and Pamfilev, A. V. (1949) The kinetics of absorption. *J. Appl. Chem. USSR*, 22, 117–126.

Kishinevskii, M. K., and Serebrianskii, V. T. (1956). Regarding the mechanism of gas transfer at the gas–liquid interface during intensive stirring. *J. Appl. Chem. USSR*, 29, 29–34.

King, C. J. (1966) Turbulent liquid phase mass transfer at a free gas–liquid interface. *Ind. Eng. Chem. Fund.*, 5, 1–8.

Kozinski, A. A., and King, C. J. (1966) The influence of diffusivity on liquid phase mass transfer to the free surface in a stirred vessel. *AIChE J.*, 12, 109–116.

Lamarche, P., and Droste, R. (1989) Air stripping mass transfer correlations for volatile organics. *JAWWA*, 81(1), 78–89.

Langmuir, D. (1997) *Aqueous Environmental Geochemistry*, Prentice Hall, Upper Saddle River, NJ.

Lewis, G., and Randall, M. (1961) revised by Pitzer K. S. and Brewer. L. (eds.), *Thermodynamics*, McGraw Hill, New York, NY.

Libra, J. (1993) *Stripping of Organic Compounds in an Aerated Stirred Tank Reactor*, VDI-Verlag, Düsseldorf.

Little, J. C., and Mariñas, B. J. (1997) Cross-flow versus countercurrent air-stripping towers. *J. Environ. Eng. (ASCE)*, 123, 668–674.

Logan, B. E. (1999) *Environmental Transport Processes*, Wiley Interscience. New York, NY.

Mackay, D. (1979) Finding fugacity feasible. *Environ. Sci. Technol.*, 13, 1218–1223.

Mackay, D., and Paterson, S. (1981) Calculating fugacity. *Environ. Sci. Technol.*, 15, 1006–1014.

Mackay, D. (1991) *Multimedia Environmental Models: The Fugacity Approach*. Lewis Publishers, Chelsea, MI.

Mackay, D., and Fraser, A. (2000) Bioaccumulation of persistent organic chemicals: mechanisms and models. *Environ. Pollut.*, 100(3), 375–391.

Metcalf and Eddy, Inc. (2003) In Tchobanoglous, G., Burton, F. L. and Stensel, H. D. (eds.), *Wastewater Engineering*, 4th ed., McGraw Hill, New York.

MWH (2005) revised by In Crittenden, J. C., Trussell, R. R., Hand, D. W., Howe, K. J., and Tchobanoglous, G. (eds.), *Water Treatment Principles and Design*. Wiley Interscience. New York, NY.

Moog, D. B., and Jirka, G. H. (1999) Air–water gas transfer in uniform channel flow. *J. Hydraulic Eng. (ASCE)*, 125(1), 3–10.

Munsterer, T., and Jahne, B. (1998) LIF measurements of concentration profiles in the aqueous mass boundary layer. *Exp. Fluids*, 25 (3), 190–196.

Munz, C., and Roberts, P. V. (1984) The ratio of gas-phase to liquid-phase mass transfer coefficients in gas–liquid contacting processes. In Brutsaert, W., and Jirka, G. H. (eds.), *Gas Transfer at Water Surfaces*, D. Reidel Publishing, Co., Boston, MA.

Munz, C. and Roberts, P. V. (1987) Air–water phase equilibria of volatile organic solutes. *JAWWA*, 79(5), 62–69.

Narbaitz, R. M., Mayorga, W. J., Torres, P., Greenfield, J. H., Amy, G. L., and Minear, R. A. (2002) Evaluating aeration-stripping media. *JAWWA*, 94(9), 97–111.

Newbry, B. (1998) Oxygen-transfer efficiency of fine-pore diffused aeration systems: energy intensity as a unifying evaluation parameter. *Water Environ. Res.*, 70(3), 323–333.

Nirmalakhandan, N., Brennan, R. A., and Speece, R. E. (1997) Predicting Henry's law constant and the effect of temperature on Henry's law constant. *Water Res.*, 31(6), 1471–1481.

Onda, K., Takeuchi, H., and Okumoto, Y. (1968) Mass transfer coefficients between gas and liquid phases in packed columns. *J. Chem. Eng. Japan*, 1(1), 56–61.

Parker, W. J., Monteith, H. D., Bell, J. P., and Melcer, H. (1996) A field scale evaluation of the airstripping of volatile organic compounds by surface aerators. *Water Environ. Res.*, 68(7), 1132–1139.

Paulson, W. C. (1979) Review of test procedures. In Boyle, W. C., (ed.), *Proceedings: Workshop Towards an Oxygen Transfer Standard*, EPA-600/9-78-021, Municipal Environmental Research Laboratory, USEPA, Cincinnati, OH.

Poling, B. E., Prausnitz, J. M., and O'Connell, J. P. (2001) *Properties of Gases and Liquids*, 5th ed., New York, McGraw Hill.

Roberts, P. V., Hopkins, G. D., Munz, C., and Riojas, A. H. (1985) Evaluating two-resistance models for air-stripping of volatile organic contaminants in a countercurrent, packed column. *Environ. Sci. Technol.*, 19, 164–173.

Roberts, P. V., and Dändliker, P. G. (1983) Mass transfer of volatile organic contaminants from aqueous solution to the atmosphere during surface aeration. *Environ. Sci. Technol.*, 17, 484–489.

Roberts, P. V., and Levy, J. A. (1985) Energy requirements for air stripping trihalomethanes. *JAWWA*, 77(4), 138–146.

Roberts, P. V., Munz, C., and Dändliker, P. (1984) Modeling volatile organic solute removal by surface and bubble aeration. *JWPCF*, 56(2), 157–163.

Sander, R. (1999) *Compilation of Henry's Law Constants for Inorganic and Organic Species of Potential Importance in Environmental Chemistry (Version 3)*. http://www.henrys-law.org. Accessed June 19, 2006.

Schwarzenbach, R. P., Gschwend, P. M., and Imboden, D. M. (2002) *Environmental Organic Chemistry*, 2nd ed., Wiley Interscience, New York, NY.

Selleck, R. E., Mariñas, B. J., and Diyamandoglu, V. (1988) *Treatment of Water Contaminants with Aeration in Counterflow Packed Towers: Theory, Practice, and Economics*. UCB/SEEHRL Report 88-3/1, University of California, Berkeley, CA.

Sherwood, T. K, Pigford, R. L., and Wilke, C. R. (1975) *Mass Transfer*, McGraw Hill, New York.

Staudinger, J., Knocke, W. R., and Randall, C. W. (1990) Evaluating the Onda mass transfer correlation for the design of packed-column air stripping. *JAWWA*, 82(1), 73–79.

Staudinger, J., and Roberts, P. V. (2001) A critical compilation of Henry's law constant temperature dependence relationships for organic compounds in dilute aqueous solutions. *Chemosphere*, 44, 561–576.

Stenstrom, M. K., and Gilbert, R. G. (1981) Effects of alpha, beta, and theta factor upon the design, specification and operation of aeration systems. *Water Res.*, 15, 643–654.

Stumm, W., and Morgan, J. J. (1996) *Aquatic Chemistry*, Wiley Interscience, New York, NY.

Toor, H. L., and Marcello, J. M. (1958) Film-penetration model for mass and heat transfer. *AIChE J.*, 4(1), 97–101.

Versteeg, G. F., Blauwhoff, M. M., and van Swaaij. W. P. M. (1987) The effect of diffusivity and gas–liquid mass transfer in stirred vessels. Experiments at atmospheric and elevated pressures. *Chem. Eng. Sci.*, 42(5), 1103–1119.

WEF (1988) *Aeration – MOP FD-13*. Joint American Society of Civil Engineers/WEF Publication, Water Environment Federation, Alexandria, VA.

Whitman, W. G. (1923) Preliminary experimental confirmation of the two-film theory of gas absorption. *Chem. Met. Eng.*, 29, 146–148.

Wilke, C. R., and Chang, P. (1955) Correlation of diffusion coefficients in dilute solutions. *AIChE J.*, 1, 264–270.

PROBLEMS

5-1. The vapor pressure of pure liquid benzene at 25°C is 0.13 atm.

(a) If the volatility of benzene followed Raoult's law under all conditions, what would be the partial pressure of an aqueous solution containing 0.1 mg/L benzene?

(b) What is the actual partial of benzene in such a solution, based on Henry's constant given in Table 5-2? Explain the difference between this value and the value computed in (a), based on the interactions of benzene molecules with one another versus these between benzene and water molecules.

(c) Explain qualitatively how the difference in the partial pressures computed in (a) and (b) is related to the aqueous-phase activity coefficient for benzene, if the reference state is defined as pure benzene at 25°C.

5-2. A solution at 25°C that contains 4 mg/L DO is being bubbled with air. Instantly, the gas supply is switched from air to pure oxygen. All other system parameters (the rate of gas injection, bubble size, etc.) remain the same. Assume $P_{tot} = 1.0$ atm throughout the process. To what extent do you expect the rate of oxygen dissolution into the water to change at that instant?

5-3. A water supply reservoir impounds water principally from melting snow (and therefore at 0°C) is at elevation 1000 m. The water flows by gravity in a closed conduit to a second reservoir at elevation 300 m, where it warms to 10°C. Assume that barometric pressure is 1.0 atm at sea level and decreases by 1% for every 270 m increase in elevation. How much oxygen will enter or leave each liter of water as it equilibrates in the lower reservoir? The molar standard (25°C) enthalpy of volatilization for oxygen is 13.3 kJ/mol. (Note: 1 kJ = 9.87 L atm.)

5-4. A research submarine suffers a catastrophic failure and sinks to the ocean floor at a depth of 1000 m. The gas phase in the submarine was originally normal air at 1.0 atm total pressure. A 1.0-L pocket of this gas is in contact with the ocean water, which is at 4°C. The average density of the seawater is 1030 kg/m³.

(a) What was the volume of this gas at the surface, where the temperature was 15°C?

(b) What is the oxygen concentration in the gas, in mg/L?

(c) What concentration of DO would be in equilibrium with the gas under the ambient (ocean bottom) conditions, if Henry's law applies?

5-5. Each of three closed (gas-tight) systems at 25°C contains 1.0 L of a 10 μmol/L chloroform solution, but they in equilibrium with different gas phases. System I contains 1.0 L of gas with $P_{tot} = 1.0$ atm, System II contains 2.0 L of gas with $P_{tot} = 1.0$ atm, and System III contains 1.0 L of gas with $P_{tot} = 2.0$ atm.

(a) What is the gas-phase concentration in each system?

(b) What is the total mass that was added to (and is still in) each system?

5-6. A drum that has been used to collect hazardous waste contains an organic phase in which the mole fractions of benzene, toluene, and *o*-xylene are 0.32, 0.27, and 0.21, respectively. At the ambient temperature of 25°C, the vapor pressures of these three chemicals are 0.126, 0.038, and 0.012 atm.

(a) If the organic phase behaves ideally with respect to Raoult's law, estimate the partial pressures of the three major constituents in the head space in the drum.

(b) Some rain water that has leaked into the drum and is present as a separate phase. If Henry's law applies to this phase, what are the mole fractions of the three constituents in the aqueous solution?

5-7. A biological treatment experiment is being conducted to assess the treatability of a well-buffered solution containing 50 mg/L each of acetone and phenol at pH 7.0 and 25°C. To test whether a significant fraction of the substrates might be lost due to volatilization, a control experiment is conducted in which a substance expected to completely inhibit biological activity is added to the solution. In this control test, 3 L of the solution is bubbled with air for 1 day, at a gas flow rate of 1.0 L/min.

(a) If as they rise, the injected air bubbles reach equilibrium with the solution, what fraction of each substrate will be stripped from the solution during the first minute of aeration? Assume initially that the solution composition is approximately constant during the time that a bubble remains in the tank, and then use your result to check that assumption.

(b) After some time, 50% of the acetone has been stripped from the solution. At that time, what fraction of the dissolved acetone will be stripped in a 1-min period?

(c) What do you expect the acetone and phenol concentrations to be after 1 day of aeration?

(d) Repeat (c), for a test being performed under thermophilic conditions, with the solution at 45°C. The standard enthalpies for the volatilization of acetone and phenol are 38 and 61 kJ/mol, respectively.

5-8. Figure 5-14b shows that the computed flux of a volatile species into solution in the thin packet limiting case is constant from the instant that the liquid and gas phases contact one another.

(a) Explain this result.

(b) In the same figure, the flux actually expected (shown by the thick line) is largest immediately upon contact of the two phases and then gradually declines until it reaches the value corresponding to the thin packet limiting case. However, the shape of the concentration profiles in Figure 5-14a suggests that the flux into the bulk solution (i.e., at $x = 10\,\mu\text{m}$ in the example system) would be zero initially and gradually *increase* to the thin packet limiting value. Explain this apparent anomaly.

5-9. A lake water at 15°C and containing 1.5 mg/L dissolved oxygen (DO) is being aerated in a fountain aerator. Droplets averaging 0.8 cm in diameter are released. For a mass transfer coefficient in the liquid phase (k_L) of 120 cm/h, and assuming that almost all of the resistance to oxygen transfer is in the liquid phase, how long would a droplet have to be in the air for the DO concentration to increase to 7.25 mg/L?

5-10. The value of k_L for oxygen in a particular gas transfer system that will be used to inject ozone (O_3) into a water supply has been estimated to be 0.015 cm/s. The ozone generator upstream of the gas transfer process converts 10% of the O_2 in the incoming air into O_3.

(a) If the water to be treated is at 25°C and contains 5 mg/L $O_2(aq)$ and no ozone, estimate the relative fluxes (on a molar basis) of O_2 and O_3 into the solution when the gas and liquid first come into contact. Assume that $m = 0.5$ and that the vast majority of the resistance to gas transfer is in the liquid phase for both species. Also assume $P_{tot} = 1.0$ atm. The diffusivity of O_3 in water at 25°C is $1.30 \times 10^{-5}\,\text{cm}^2/\text{s}$.

(b) If k_G/k_L is 20 for both O_3 and O_2, what fraction of the resistance to gas transfer resides in the liquid for each species?

(c) Do you think that the transfer of ozone would be enhanced, diminished, or not altered if the water were deaerated (so that the DO concentration were essentially zero) prior to the treatment process?

5-11. A spray aeration system is being considered to aerate pond water and oxidize the 40 mg/L of Fe^{2+} that is dissolved in it. The water is at 10°C and contains a very low concentration of dissolved salts; under these conditions, the saturation concentration of DO is 11.2 mg/L. For this system, the value of K_L for oxygen transfer into the droplets is 150 cm/h. Initially, there is no DO in the water. The droplets are 1.0 mm in diameter, and they are in contact with the air for 1 s.

(a) Compute the DO concentration in the droplets as they re-enter the pond, assuming that none of the oxygen reacts with the iron during the short time the droplets are airborne.

(b) The stoichiometry and rate of the reaction of Fe^{2+} with DO are as follows:

$$Fe^{2+} + \tfrac{1}{4}O_2 + \tfrac{5}{2}H_2O \rightarrow Fe(OH)_3(s) + 2H^+$$

$$r_{Fe^{2+}} = -k c_{Fe^{2+}} c_{O_2} c_{OH^-}^2$$

where k has a value at 10°C of 1.6×10^{12} (L/mg O_2)(L/mol OH^-)^2 min^{-1}. Convert the rate expression into a pseudo-first-order reaction with respect to DO for a scenario in which the pH and Fe^{2+} concentrations are approximately constant at their initial values (Hint: be careful about units). Based on Figure 5-20, for a solution at pH 7.0, would consideration of the reaction of DO with Fe^{2+} in the droplet significantly alter the answer to (a)? The diffusivity of oxygen in water at 10°C is $1.46 \times 10^{-5}\,\text{cm}^2/\text{s}$.

(c) Confirm your result from (b) by computing the maximum concentration of Fe^{2+} that could react with DO under the limiting-case condition that the DO concentration computed in (a) is attained instantaneously and is maintained throughout the time that the droplet is in the air.

5-12. (a) For chloroform, bromoform, and oxygen at 25°C, calculate the fraction of total resistance that is associated with the liquid phase in a systems with $k_G/k_L = 150$ (e.g., a large lake).

(b) Repeat the calculation (a) for bromoform in a system with $k_G/k_L = 20$ (e.g., a surface aeration system).

5-13. Howard and Corsi (1996) studied the volatilization of three organic contaminants from drinking water in systems intended to simulate the situation at kitchen sinks. In one test, they reported that $K_L a_L$ for toluene was $0.027\,\text{min}^{-1}$.

(a) Estimate the $K_L a_L$ value for acetone in the same system, assuming that the resistance to volatilization is essentially all in the liquid phase for both compounds, and that the temperature is 25°C. Assume that the exponent m relating $k_L a_L$ values to diffusivities is 0.6 and that the ratio of diffusivities of acetone and toluene is the same at 25°C as at 20°C.

(b) The authors estimated k_G/k_L to be 30 for this system. Make a new estimate for $K_L a_L$ of acetone, taking into account the fact that the k_G/k_L ratio applies to both volatile species.

(c) Do you think that gas-phase resistance is significant for transfer of either compound?

5-14. Compare the predicted flux computed in Example 5-8 with the value you would obtain if you ignored the effect of the acid/base reactions on K_L; that is, if you assumed that the value of K_L at $\alpha_{\text{neut}} = 1.0$ applied at all pH values. What fraction of the resistance is computed to reside in the gas phase if the acid/base reactions are ignored?

5-15. Example 5-8 shows the effect of pH on two forms of K_L for a volatile species (H_2S) that undergoes acid/base reactions in solution. That example used a ratio of $k_G/k_L = 20$, which is typical for surface aeration systems (see Figure 5-18). Repeat the calculations for a k_G/k_L ratio of 5 (by using a value of 0.02 cm/s for k_g), which is typical for packed towers designed to accomplish gas stripping or absorption.

5-16. Just like those whose focus is primarily on water quality, researchers and practitioners who focus on air quality are often interested in gas/liquid exchange reactions. In the air quality literature, it is common to write the expression for flux in terms of a driving force based on gas-phase composition. That is, the flux of a volatile species into the gas phase is written as $J_G = K_G a(c_G^* - c_G)$, where c_G^* is the gas-phase concentration that would be in equilibrium with the bulk liquid, and a might be either a_L or a_R.

A solution at 25°C that contains 4.5 mg/L DO is in contact with the normal atmosphere, and is being aerated by a system with $K_L a = 0.2\,\text{min}^{-1}$. Compute c_G, c_G^*, c_L^*, J_L, J_G and $K_G a$ in the system.

5-17. For the column described in Example 5-12, prepare plots showing $k_L a_R$, $k_G a_R$, $K_L a_R$, and the fraction of the total resistance in the gas phase for the following scenarios:

(a) Varying the liquid flow rate, Q_G, from 5 to 80 L/s while maintaining a fixed air:water ratio of 25 L_G/L_L (i.e., show each of the parameters as a function of Q_L over the specified range).

(b) Varying the air:water flow ratio from 10 to 60 L_G/L_L while maintaining a fixed liquid flow rate of 17.7 L/s.

5-18. Recall the definition of the residence time of a substance as the mass of the substance in the region of interest divided by the rate (mass/time) at which it moves through this region. Based on this definition, compute the average residence time of the volatile species in the liquid interfacial region in the example system shown in Figure 5-14, once a steady-state gradient has developed in the packet. Comment on the result in light of the limiting cases shown in Figure 5-14.

5-19. In the simplified two-film model developed for gas transfer, the assumption is commonly made that both boundary layers are flat. This assumption is often reasonable even for curved boundaries such as the surfaces of bubbles or droplets, since the radius of curvature is usually much larger than the thickness of the boundary layer. Nevertheless, there are cases of very small droplets where the boundary layer may be significantly curved, causing the surface area of the inner boundary to be less than that of the outer boundary.

Write a mass balance and derive the equation for steady-state gas transport through a spherical water droplet under conditions where the spherical geometry of the boundary layer must be considered. If a small spherical water droplet was falling through the air and the water in the droplet was not mixing internally at all (the hard sphere model), what boundary conditions would you write that would, in conjunction with your differential equation, allow you to solve for the concentration of a gaseous air pollutant as a function of location in the droplet?

6

GAS TRANSFER: REACTOR DESIGN AND ANALYSIS

6.1 Introduction
6.2 Case I: Gas transfer in systems with a well-mixed liquid phase
6.3 Case II: Gas transfer in systems with spatial variations in the concentrations of both solution and gas
6.4 Summary
    Appendix 6A. Evaluation of $K_L a$ in gas-in-liquid systems for biological treatment
    References
    Problems

6.1 INTRODUCTION

As discussed in the preceding chapter, transfer of molecules across gas/liquid interfaces occurs in a wide variety of physical systems. These systems might have continuous flow of both, one, or neither phase, and might be operated under steady or unsteady state conditions. In all these systems, the rate of transfer at any location, expressed as the mass of i entering the solution per unit volume of solution per unit time is given by Equation 6-1:

$$r_{L,gt} = K_L a_L (c_L^* - c_L) \quad (6\text{-}1)$$

where the parameter meanings are as described in Chapter 5.

Equation 6-1 characterizes each phase by a single concentration value, so the equation is meaningful only over a time frame and region of space in which the concentration in each bulk phase is uniform. In many systems of interest, however, the concentrations c_L and c_L^* change as a function of time and/or location, and the equation can be applied directly only over a differential time period or to a differential control volume. In such cases, the expression on the right side of Equation 6-1 can be inserted into a mass balance, but the mass balance must then be integrated to obtain a result applicable to longer times or larger volumes. The integrated equation can be used to analyze or predict the behavior of full-scale systems in terms of the major parameters of interest—the liquid and gas flow rates; the reactor dimensions; and the influent and effluent concentrations for systems with flow, or the initial and final concentrations for systems without flow.

In this chapter, we derive and analyze design equations for gas transfer in systems with a few common types of gas/liquid contacting arrangements. While we identify the key physical attributes of each system and take note of the situations in which each is commonly encountered, we distinguish among the systems primarily on the basis of the uniformity or nonuniformity of the composition of the two phases, and on the corresponding consequences for the mass balances. Specifically, the analysis is presented separately for two broad and important classes of reactors. One group (identified below as Case I) includes reactors in which the entire liquid phase has a uniform composition, and in which the composition of that phase is approximately constant over time periods comparable to the average gas residence time (e.g., the rise time of a bubble or the time that a droplet is airborne); many gas-in-liquid reactors, including well-mixed diffused aeration systems, fall into this category. The second group (Case II) includes reactors in which the liquid phase composition varies continuously in space, such as spray towers designed for gas stripping or absorption.

6.2 CASE I: GAS TRANSFER IN SYSTEMS WITH A WELL-MIXED LIQUID PHASE

In this section, we analyze gas transfer in systems containing a spatially uniform solution whose composition is approximately invariant over periods comparable to the average residence time of the gas. The systems might be either batch or continuous flow with respect to the solution. The uniformity of the solution phase might arise either from complete mixing of that phase or because the amount of gas transfer is insufficient to change the solution composition significantly over the time frame of interest. The composition of

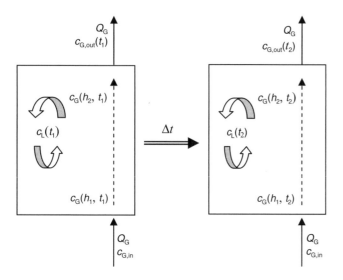

FIGURE 6-1. Schematic of a Case I gas transfer reactor: a diffused aeration system with a well-mixed batch liquid phase and steady flow of gas. c_L is the same throughout the control volume, but c_G varies from one location to the next.

the gas phase in these systems might vary with location in the reactor. The analysis is based largely on the work of Matter-Müller et al. (1980, 1981).

If the composition of the gas phase is nonuniform in the reactor, the value of c_L^* and, therefore, the rate of gas transfer will vary as a function of location (e.g., as a function of depth in a reactor with diffused aeration). The essential features of a batch liquid system with continuous flow of gas that meets these criteria are shown schematically in Figure 6-1.

In Figure 6-1, the volumetric flow rate of gas is shown as Q_G both entering and leaving the reactor. Two factors can cause a system to violate this equality. First, the amount of gas transferred might be large enough to change the mass flow rate of gas significantly during the gas residence time in the reactor. Such a situation might arise in, for example, recarbonation reactors using pure CO_2 as the injected gas phase, or aeration of a reservoir (where the air might be injected at depths of several tens of meters); this latter situation has been discussed in some detail by McGinnis and Little (2002). However, in most environmental engineering applications, an assumption that the mass flow rate of total gas is constant throughout the reactor is satisfactory, and we make that assumption throughout the chapter.[1] Second, the pressure differential between the gas inlet and outlet points might be large enough that the density of the gas phase differs significantly at those two points, so that $Q_{G,in}$ does not equal $Q_{G,out}$ even if the mass flow rates of gas are almost identical. This situation arises quite frequently in environmental engineering, particularly in deep tanks with diffused aeration.

Pressure, temperature, and bubble size (both diameter and volume) at different depths in gas-in-liquid system are all related via the ideal gas law and the pressure–depth relationship in water. These relationships are, respectively:

$$c_{G,mol} = \frac{n_G}{V_G} = \frac{1}{\overline{V}_G} = \frac{P}{RT} \quad (6\text{-}2)$$

where n_G is the number of moles in the gas phase, and $\overline{V}_G$ is the gas-phase molar volume (volume per mole); and

$$\frac{dP}{dz} = \rho_L g \quad (6\text{-}3)$$

where ρ_L is the liquid density and g is the gravitational constant. By combining Equations 6-2 and 6-3 with various geometric considerations, we can obtain the relationships shown in Table 6-1.

Although the volumetric gas flow rate at the system temperature and a characteristic system pressure (at either Z or mid-depth) is the value of most interest, Q_G is often reported in terms of the equivalent mass flow rate at a nominal temperature (typically 20°C) and pressure (invariably, 1.0 atm). That is, the reported value of Q_G is the volumetric flow rate that the gas would have if it were at those nominal conditions. When this convention is used, the units of the flow rate are often designated as a "standard" volume per unit time (e.g., as *standard cubic meters per hour* [SCM/h or SCMH]).

If the volumetric flow rate of a gas is given under standard conditions, the actual flow rate at any absolute temperature T and pressure P is

$$Q_G = Q_{G,std} \frac{P_{std}}{P} \frac{T}{T_{std}} \quad (6\text{-}4)$$

Since P_{std} is 1 atm and that is the condition at the surface, the volumetric gas flow rate at depth z and temperature T is

$$Q_G = Q_{G,std} \frac{P_{surf}}{P_{surf} + \rho_L g z} \frac{T}{T_{std}} \quad (6\text{-}5)$$

■ **EXAMPLE 6-1.** A diffuser injects air bubbles that are 1.2 mm in diameter into water at a depth of 4.5 m. The gas flow rate is reported as 120 SCM of air per square meter of diffuser area per hour. The water and air are at 5°C. Each diffuser has an area of 0.025 m^2, and the diffusers are installed at a spacing corresponding to three diffusers per square meter of floor space.

(a) What is the true air injection rate per unit floor area (meter cube of air per minute per square meter of floor area)?

---

[1] Cases where this assumption does not apply are analyzed in many chemical engineering texts; see, for example, the texts by Treybal (1980) and Sherwood et al. (1975).

TABLE 6-1. Relationships Among Depth (z), Pressure, and Bubble Properties in a Column of Liquid of Total Depth Z

Pressure versus depth[a]	$P_z = P_{surf} + \rho_L g z$	(6-6)
Bubble volume versus depth[a]	$V_{bb,z} = V_{bb,Z} \dfrac{P_Z}{P_z} = V_{bb,Z} \dfrac{P_{surf} + \rho_L g Z}{P_{surf} + \rho_L g z}$	(6-7)
Volumetric gas flow rate versus depth	$Q_{G,z} = Q_{G,Z} \dfrac{P_Z}{P_z} = Q_{G,Z} \dfrac{P_{surf} + \rho_L g Z}{P_{surf} + \rho_L g z}$	(6-8)
Bubble diameter versus depth	$d_{bb,z} = d_{bb,Z} \left(\dfrac{P_Z}{P_z}\right)^{1/3} = d_{bb,Z} \left(\dfrac{P_{surf} + \rho_L g Z}{P_{surf} + \rho_L g z}\right)^{1/3}$	(6-9)
Bubble area versus depth	$A_{bb,z} = A_{bb,Z} \left(\dfrac{P_Z}{P_z}\right)^{2/3} = A_{bb,Z} \left(\dfrac{P_{surf} + \rho_L g Z}{P_{surf} + \rho_L g z}\right)^{2/3} = \pi d_{bb,Z}^2 \left(\dfrac{P_{surf} + \rho_L g Z}{P_{surf} + \rho_L g z}\right)^{2/3}$	(6-10)

[a] In equations, "surf" refers to the water surface; that is, $z = 0$; bb = bubble.

(b) What are the average diameter, surface area, and volume of the bubbles in the tank?

(c) Estimate a_L in the tank.

Solution.

(a) Nominal airflow rates are given at the standard conditions of 1 atm pressure (which occurs at the water surface) and 20°C, so the given flow rate must be corrected for depth and temperature according to Equation 6-5. The product $\rho_L g$ for water at 5°C is

$$\rho_L g = \left(\dfrac{1000\ \text{kg}}{\text{m}^3}\right)(9.81\ \text{m/s}^2)\left(\dfrac{1\ \text{N}}{\text{kg m/s}^2}\right)\left(\dfrac{1\ \text{Pa}}{\text{N/m}^2}\right)\left(\dfrac{10^{-5}\ \text{atm}}{\text{Pa}}\right)$$

$$= 0.098\ \text{atm/m}$$

Therefore, the actual flow rate of the air at the depth of injection is

$$\dfrac{Q_G}{\text{m}^2\ \text{of diffuser}} = \dfrac{Q_{G,std}}{\text{m}^2\ \text{of diffuser}} \dfrac{P_{surf}}{P_{surf} + \rho g Z} \dfrac{T}{T_{std}} = \left(120\ \dfrac{\text{m}^3\ \text{air}}{\text{h m}^2\ \text{diffuser}}\right)$$

$$\times \dfrac{1\ \text{atm}}{1\ \text{atm} + 0.098\ (\text{atm/m})(4.5\ \text{m})} \dfrac{278\ \text{K}}{293\ \text{K}}$$

$$= 79.0\ (\text{m}^3\ \text{air})/(\text{h m}^2\ \text{diffuser})$$

Because the spacing is three diffusers per square meter of floor area, and each diffuser has an area of $0.025\ \text{m}^2$, we find

$$\dfrac{Q_G}{\text{m}^2\ \text{of floor}} = \left(79.0\ \dfrac{\text{m}^3\ \text{air}}{\text{h m}^2\ \text{diffuser}}\right)\left(3\ \dfrac{\text{diffusers}}{\text{m}^2\ \text{of floor}}\right)$$

$$\times \left(0.025\ \dfrac{\text{m}^2\ \text{diffuser}}{\text{diffuser}}\right) = 5.92\ (\text{m}^3\ \text{air})/(\text{h m}^2\ \text{floor})$$

(b) The surface area and volume of the bubbles when they are injected are

$$A_{bb,Z} = \pi d_{bb,Z}^2 = \pi (1.2\ \text{mm})^2 = 4.52\ \text{mm}^2$$

$$V_{bb,Z} = \dfrac{\pi d_{bb,Z}^3}{6} = \dfrac{\pi (1.20\ \text{mm})^3}{6} = 0.905\ \text{mm}^3$$

For systems with $Z \leq 10\ \text{m}$, the approximation that the average bubble size is the size at mid-depth ($z = Z/2$) is considered acceptable. The average diameter and surface area of the bubbles in the tank are therefore given by

$$d_{bb,avg} = d_{bb,Z} \left(\dfrac{P_{surf} + \rho_L g Z}{P_{surf} + \rho_L g Z/2}\right)^{1/3}$$

$$= (1.20\ \text{mm}) \left(\dfrac{1\ \text{atm} + (0.098\ \text{atm/m})(4.5\ \text{m})}{1\ \text{atm} + (0.098\ \text{atm/m})(2.25\ \text{m})}\right)^{1/3}$$

$$= 1.27\ \text{mm}$$

$$A_{bb,avg} = \pi d_{bb,avg}^2 = \pi (1.27\ \text{mm})^2 = 5.07\ \text{mm}^2$$

By geometry, the volume of an average bubble is therefore

$$V_{bb,avg} = \dfrac{\pi d_{bb,avg}^3}{6} = \dfrac{\pi (1.27\ \text{mm})^3}{6} = 1.07\ \text{mm}^3$$

Note that the bubble volume increases more than the bubble area as the bubbles rise, and that the area increases more than the diameter.

(c) The value of a_L is the gas/liquid interfacial area per unit volume of liquid. The rate of bubble generation ($\dot{N}_{bb}$) per unit floor area can be computed from the gas flow rate per unit floor area and the volume of individual bubbles:

$$\frac{\dot{N}_{bb}}{m^2 \text{ of floor}} = \frac{Q_G/(m^2 \text{ of floor})}{V_{bb,Z}}$$

$$= \frac{5.92(m^3 \text{ air}/h \, m^2 \text{ floor})}{(0.905(mm^3 \text{ air/bubble}))(10^{-9} \, m^3/mm^3)}$$

$$= 6.54 \times 10^9 \text{ bubble}/h \, m^2 \text{ floor}$$

The average bubble diameter was found in part b as 1.27 mm. From Figure 5-24, we can estimate the rise velocity of bubbles with this diameter to be 30 cm/s. Therefore, the average residence time of bubbles in the reactor is 450 cm/(30 cm/s), or 15 s, and the number of bubbles in the tank at any instant (N_{bb}) per square meter of floor area is

$$\frac{N_{bb}}{m^2 \text{ floor}} = \left(6.54 \times 10^9 \frac{\text{bubble}}{h \, m^2 \text{ floor}}\right)(15 \text{ s})\left(\frac{1 \text{ h}}{3600 \text{ s}}\right)$$

$$= 2.72 \times 10^7 \text{ bubbles}/m^2 \text{ floor}$$

Since the tank depth is 4.5 m, the water volume can be expressed as 4.5 m³/m² of floor, so,

$$a_L = \frac{A_{tot}}{V_L} = \frac{N_{bb}A_{bb,avg}}{V_L} = \frac{(N_{bb}/\text{floor area})A_{bb,avg}}{V_L/\text{floor area}}$$

$$= \frac{(2.72 \times 10^7 (\text{bubbles}/m^2 \text{ floor}))(5.07 \, mm^2/\text{bubble})(10^{-6} \, m^2/mm^2)}{4.5 (m^3 \text{ water})/(m^2 \text{ floor})}$$

$$= 30.6 \, m^{-1} \qquad \blacksquare$$

Systems that fit the conceptual model of Case I abound in environmental engineering. The most important such systems have continuous flow of water as well as gas, such as well-mixed activated sludge aeration basins, other aerated tanks or ponds, and so-called bubble columns, in which a gas such as ozone or carbon dioxide is injected at the bottom of a column of water that is typically flowing counter-current to the gas bubbles (i.e., from the top to the bottom of the column). In all these systems, the liquid phase is mixed by the same process that provides the gas/liquid contact—the injection and rising of bubbles in diffused aeration and bubble column systems, and the mechanical agitation that draws gas into the bulk liquid and throws water droplets into the bulk gas phase in surface aeration systems (photographs of a few such systems were shown in Figure 6.4). The mixing of the solution virtually eliminates concentration gradients in that phase. On the other hand, at least in diffused aeration and bubble column systems, the composition of the gas phase might change from one location to the next in the reactor (e.g., the gas exiting the top of the water column is likely to have a different composition from the injected gas). Thus, in such systems, the bubbles can be viewed as passing through a plug flow reactor (from bottom to top), while the water passes through the same physical reactor (from the liquid inlet to the outlet), but is completely mixed.

A thorough mathematical analysis of the functioning of systems such as the one depicted in Figure 6-1, as well as similar systems with flow of both liquid and gas, is provided in subsequent sections of this chapter. Before developing that analysis, however, it is worthwhile to consider the qualitative trends that we might expect to see in c_L and c_G over time. Imagine, for example, the changes that would occur over time if air bubbles were injected into a batch reactor that contained an (initially) anoxic solution. Assume that no reaction consuming or generating oxygen occurs in solution, and that the depth of the tank is small enough that we can ignore pressure differences between the bottom and the top.

The entering bubbles would always have the oxygen concentration in the atmosphere, so $c_{G,in}$ would be constant throughout the period of aeration. Initially, the dissolved oxygen concentration (c_L) would be zero, generating the maximum possible driving force for transfer of oxygen into solution. As a result, the rate of oxygen dissolution ($r_{O_2,gt}$) would be maximum at $t = 0$, and the concentration of oxygen remaining in the bubbles that exited the reactor at that time ($c_{G,out}$ at $t = 0$) would be the lowest of any time during the test. If the water/bubble contact time were long enough and the resistance to gas transfer small enough, then almost all of the oxygen might transfer out of the bubbles during their rise. On the other hand, if the rise time were relatively short and/or if significant resistance to gas transfer were present, then only a small fraction of the oxygen might transfer. In either case, however, $r_{O_2,gt}$ would have its maximum value at $t = 0$.

Over time, c_L would increase, decreasing the driving force for transfer and therefore the transfer rate, so that $c_{G,out}$ would increase. Eventually, c_L would reach the value that is in equilibrium with the incoming bubbles, so no more gas transfer would occur. At this time, $c_{G,out}$ would be the same as $c_{G,in}$. These trends are shown qualitatively in Figure 6-2.

All the trends described above are observed in real (batch) systems that meet the specified (Case I) criteria. In the following sections of the chapter, we derive equations that quantify the changes in the composition of both phases in Type I gas transfer reactors with diffused aeration, both for batch systems and those with flow. In such reactors, as in any system where gas transfer is occurring, Equation 6-1 applies locally over every region where the liquid- and gas-phase compositions are uniform. In Case I diffused aeration systems, such regions are of only differential dimension in the vertical direction, because c_L^* varies with height. However, by making certain reasonable assumptions and manipulating the equations appropriately, an algebraic expression for $r_{L,gt}$ can be found that characterizes the overall average gas transfer rate, considering the entire liquid phase.

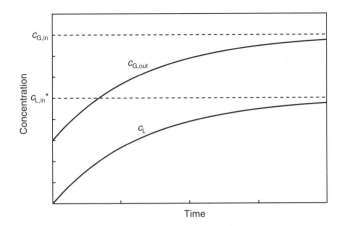

FIGURE 6-2. Expected trends in oxygen concentration in a batch reactor with diffused aeration, for a solution that is initially undersaturated with respect to the gas phase. c_L^* is the liquid phase concentration that would be in equilibrium with the incoming gas; that is, with $c_{G,in}$.

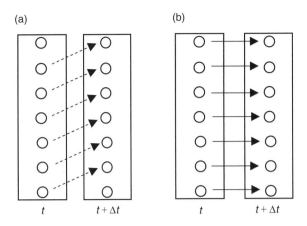

FIGURE 6-3. Schematic showing the bubbles in a Case I reactor at two times separated by a small Δt. (a) The arrows show the change in the positions of individual bubbles. (b) The reactor is shown at the same two times as in (a), but the arrows emphasize that the spatial distribution of bubbles is essentially identical at the two times.

The Overall Gas Transfer Rate Expression for Case I Systems

We begin the analysis by writing a mass balance on the gas phase in the reactor:

$$\begin{pmatrix} \text{Rate of change} \\ \text{of mass of } i \\ \text{stored in gas} \\ \text{bubbles} \end{pmatrix} = \begin{pmatrix} \text{Net} \\ \text{advective} \\ \text{input of } i \end{pmatrix} + \begin{pmatrix} \text{Net diffusive/} \\ \text{dispersive} \\ \text{input of } i \end{pmatrix}$$

$$+ \begin{pmatrix} \text{Rate at which } i \\ \text{is formed in gas} \\ \text{bubbles by} \\ \text{reactions other} \\ \text{than gas} \\ \text{transfer} \end{pmatrix} + \begin{pmatrix} \text{Rate at which } i \\ \text{enters gas} \\ \text{bubbles via gas} \\ \text{transfer} \end{pmatrix}$$

$$\int_{V_G} \frac{\partial c_G}{\partial t} dV_G = Q_G(c_{G,in} - c_{G,out}) + 0 + 0 + \int_A J_G dA \quad (6\text{-}11)$$

Note that, because the gas phase concentration varies as a function of height, the storage term must be computed (conceptually) by considering the rate of change of c_G in differential units of gas volume in the reactor and adding up (integrating) all of those differential terms. Similarly, the rate at which the volatile species enters the bubbles must be computed by considering differential patches of interfacial area and integrating over all of those patches in the reactor. We can equate the flux into the gas phase with that out of the liquid ($J_G = -J_L$), so Equation 6-11 can be rewritten and rearranged as follows:

$$\int_{V_G} \frac{\partial c_G}{\partial t} dV_G + \int_A J_L dA = Q_G(c_{G,in} - c_{G,out}) \quad (6\text{-}12)$$

The term $\int_A J_L dA$ is the overall rate of gas transfer into solution in the reactor, so it can be written as $\int_{V_L} r_{L,gt} dV_L$. Although $r_{L,gt}$ varies as a function of location in the reactor, we can define an overall, average rate of gas transfer into solution, $r_{L,gt\text{-}avg}$, such that $V_L r_{L,gt\text{-}avg} = \int_{V_L} r_{L,gt} dV_L$. Making these substitutions, Equation 6-12 becomes

$$\int_{V_G} \frac{\partial c_G}{\partial t} dV_G + V_L r_{L,gt\text{-}avg} = Q_G(c_{G,in} - c_{G,out}) \quad (6\text{-}13)$$

Equation 6-13 states, in words, that if the concentration of the volatile substance is less in the gas that exits the reactor than in the entering gas (i.e., if the right side of the equation is positive), the material that is "left behind" in the reactor must show up either as an increase in the total mass of the volatile species stored in all of the bubbles in the reactor $\left(\int_{V_G} (\partial c_G/\partial t) dV_G\right)$ or as material that is transferred to the solution phase $(V_L r_{L,gt\text{-}avg})$.

The two terms on the left side of Equation 6-13 can be visualized with the help of Figure 6-3, which shows two successive snapshots of the reactor taken a small Δt apart. In Figure 6-3a, the focus remains on the same bubbles as they move upward to different heights. Some mass of the volatile species i enters or leaves each bubble, and thereby leaves or enters the solution, during the period Δt. The total mass of i that enters or leaves the solution, Δm_L, is the sum of the changes in the mass of i in the individual bubbles, including bubbles at the bottom and top of the water column that have been in the system for only a portion of Δt. The rate of change of the mass of i stored in the liquid phase over the period Δt due to this phenomenon is $\Delta m_L/\Delta t$, or, over a differentially small time increment, dm_L/dt. By the definition of $r_{L,\text{gt-avg}}$, that same quantity (i.e., dm_L/dt) can be represented as the instantaneous value of $V_L r_{L,\text{gt-avg}}$ at that time.

Figure 6-3b, on the other hand, represents the change in the mass of i stored in the gas phase (m_G) over the time Δt, by considering the (different) bubbles that are present at the same heights at the two times. If we think of each bubble as an infinitesimally small fraction of the gas phase in the water column, the total mass of i in all the bubbles in the reactor at any instant can be represented as $\int_{V_G} c_G dV_G$. The shift from the left to the right portion of this figure (i.e., from t to $t + \Delta t$) represents the differential change in this summation; that is, $\partial(\int c_G dV_G)/\partial t$. Thus, Figures 6-3a and 6-3b provide a visual representation of the two terms on the left side of Equation 6-13.

In Case I systems with injected gas, it is commonly assumed that most of the difference between the rate at which a volatile species i enters the system in the injected bubbles and the rate at which it leaves in the exiting bubbles is accounted for by change in the mass of i in solution, rather than a change in the mass of i stored in the gas phase within the reactor. That is, $\partial(\int c_G dV_G)/\partial t \ll V_L r_{L,\text{gt-avg}}$. This assumption can be justified as follows. During a time on the order of the bubble rise time, enough volatile species might transfer into or out of a bubble to change c_G significantly, but given the much larger volume of the liquid, that same mass of i induces a much smaller change in c_L. As a result, a new bubble that is injected at the same time as a previous one exits contacts a solution that also has virtually the same composition as the previous bubble encountered. The bubble entering at the later time therefore behaves (in terms of its change in composition as it rises) almost identically to the bubble that entered earlier. The implication is that the gas-phase composition in the water column is almost the same from one instant to the next, so the mass of i stored in that gas phase changes very slowly. Compared to that rate, the mass of i stored in the liquid changes relatively rapidly, even though the *concentration* of i in the liquid changes relatively slowly. These ideas are sometimes expressed by stating that the gas phase in the system is at a quasi-steady state. Applying this assumption, Equation 6-13 can be simplified as follows:

$$V_L r_{L,\text{gt-avg}} = Q_G (c_{G,\text{in}} - c_{G,\text{out}}) \quad (6\text{-}14)$$

$$r_{L,\text{gt-avg}} = \frac{Q_G}{V_L} (c_{G,\text{in}} - c_{G,\text{out}}) \quad (6\text{-}15)$$

The liquid volume in the reactor, the gas flow rate, and the composition of the injected gas are typically all known or subject to operator control, so Equation 6-15 effectively gives $r_{L,\text{gt-avg}}$ as a function of $c_{G,\text{out}}$. That in itself is a major accomplishment, since it relates the overall rate of gas transfer to experimentally accessible parameters, whereas use of Equation 6-12 would require knowledge of the gas composition as a function of location in the reactor. Still, because our ultimate intent is to insert an expression for $r_{L,\text{gt-avg}}$ into a mass balance on the liquid phase, it would be useful to recast Equation 6-15 in such a way that $r_{L,\text{gt-avg}}$ is expressed strictly as a function of parameters in the solution phase.

To relate $c_{G,\text{out}}$ to these other parameters, we can write a mass balance on a single bubble as it rises through the water column. Such a mass balance has only the gas transfer term on the right side and is shown in words and symbols below:

Rate of change of mass stored in bubble as a result of gas transfer = (Flux into gas bubble) × (Surface area of bubble)

$$V_{\text{bb}} \frac{dc_G}{dt_{\text{bb}}} = -K_L (c_L^* - c_L) A_{\text{bb}} \quad (6\text{-}16)$$

where V_{bb} and A_{bb} are the volume and surface area of a single bubble, respectively.[2] The time variable in Equation 6-16 is written as t_{bb} to indicate that it refers to the time since the bubble was injected. In this way, we distinguish it from the variable t, which refers (in Equations 6-11 to 6-13) to the total time that the solution and gas phase have been in contact with one another.

Typically, the concentration of the bulk solution changes negligibly during the residence time of a single bubble, so we assume that c_L is not a function of t_{bb} in Equation 6-16; by contrast, c_L^* will depend on t_{bb} if the composition of the bubble changes significantly during its passage through the water.

The surface area and volume of a bubble can be represented as $V_L a_L / N_{\text{bb}}$ and V_G / N_{bb}, respectively, where N_{bb} is the number of bubbles in the reactor and V_G is the total volume of gas bubbles in the reactor at any instant. Furthermore, c_L^* can be written as c_G/H. Substituting these expressions into Equation 6-16, the mass balance can be rewritten as follows:

$$\frac{V_G}{N_{\text{bb}}} \frac{dc_G}{dt_{\text{bb}}} = -K_L \left(\frac{c_G}{H} - c_L \right) \frac{V_L a_L}{N_{\text{bb}}} \quad (6\text{-}17\text{a})$$

$$\frac{dc_G}{(c_G/H) - c_L} = -\frac{V_L K_L a_L}{V_G} dt_{\text{bb}} \quad (6\text{-}17\text{b})$$

$$\frac{dc_G}{c_G - H c_L} = -\frac{V_L K_L a_L}{H V_G} dt_{\text{bb}} \quad (6\text{-}17\text{c})$$

Equation 6-17c can be integrated between times 0 and t_{bb}, utilizing the assumption that the liquid phase concentration (c_L) is constant over the time frame of the integration, to

---

[2] As has been noted, the diameter of a bubble (and hence its volume and surface area) can sometimes change significantly as it rises through a water column due to changes in hydrostatic pressure, we ignore those changes here for simplicity.

yield the following expressions:

$$\ln \frac{c_G(t_{bb}) - Hc_L}{c_G(0) - Hc_L} = -\frac{V_L K_L a_L}{HV_G} t_{bb} \quad (6\text{-}18a)$$

$$c_G(t_{bb}) - Hc_L = (c_G(0) - Hc_L)\exp\left(-\frac{V_L K_L a_L}{HV_G} t_{bb}\right) \quad (6\text{-}18b)$$

Equation 6-18b indicates that the extent of disequilibrium between a bubble and the solution $(c_G - Hc_L)$ declines exponentially as the bubble rises in a reactor that meets the Case I assumptions. Because the aqueous phase concentration is assumed to remain constant during this short time, the approach to equilibrium reflects only changes in concentration in the gas phase.

To consider the change in the composition of a bubble over the total time that it is in the water column, we can insert the average rise time of a bubble, τ_{bb}, for t_{bb}. In addition, we can write $c_G(0)$ as $c_{G,\text{in}}$ and $c_G(\tau_{bb})$ as $c_{G,\text{out}}$, yielding:

$$c_{G,\text{out}} - Hc_L = (c_{G,\text{in}} - Hc_L)\exp\left(-\frac{V_L K_L a_L}{HV_G} \tau_{bb}\right) \quad (6\text{-}19)$$

Recognizing that V_G/τ_{bb} equals Q_G, rearranging the terms in the exponential argument, and solving for $c_{G,\text{out}}$, we obtain

$$c_{G,\text{out}} = Hc_L + (c_{G,\text{in}} - Hc_L)\exp\left(-\frac{K_L a_L}{Q_G H/V_L}\right) \quad (6\text{-}20)$$

Equation 6-20 expresses $c_{G,\text{out}}$ as a function of parameters that are typically known and/or controllable. Substituting this expression into Equation 6-15 and expressing $c_{G,\text{in}}$ as $Hc_{L,\text{in}}^*$, yields (after a few algebraic steps) the following expression for $r_{L,\text{gt-avg}}$:[3]

$$r_{L,\text{gt-avg}} = \frac{Q_G H}{V_L}\left[1 - \exp\left(-\frac{K_L a_L}{Q_G H/V_L}\right)\right]\left(c_{L,\text{in}}^* - c_L\right) \quad (6\text{-}21)$$

In most systems, all of the terms in Equation 6-21 preceding the driving force $\left(c_{L,\text{in}}^* - c_L\right)$ are constant, so the expression can be re-written to look much like a first-order reaction rate expression:

$$r_{L,\text{gt-avg}} = k_I\left(c_{L,\text{in}}^* - c_L\right) \quad (6\text{-}22)$$

[3] Note that the subscript "in" in the term $c_{L,\text{in}}^*$ refers to the incoming gas phase, not the incoming solution (Case I systems might or might not have flow of liquid).

where k_I is defined as shown below. The subscript I is used to emphasize that the result applies to Case I systems.

$$k_I \equiv \frac{Q_G H}{V_L}\left[1 - \exp\left(-\frac{K_L a_L}{Q_G H/V_L}\right)\right] \quad (6\text{-}23)$$

Equation 6-21, or, equivalently, Equations 6-22 and 6-23, represents an extremely important result—the rate expression for gas transfer in systems that fit the Case I assumptions. Insertion of that result into mass balances that have appropriate terms for other processes affecting the species of interest (e.g., advection, dispersion, chemical reactions other than gas transfer) allows us to explore alternative designs for gas transfer reactors, analyze data from existing reactors, or predict how a change in operating conditions will alter the performance of a reactor, as long as the reactor conforms to the assumptions of Case I. Many systems do conform to the Case I assumptions, making this result one of great practical as well as theoretical importance.

At the beginning of Chapter 5, several factors were listed that make gas transfer a more complicated process than the nth-order homogeneous reactions that were emphasized in Chapters 1–4. These factors include the independence of the flow rates and the mixing patterns in the two phases. Equation 6-22 suggests that, while it is true that those complications are present in gas transfer reactors, gas transfer can nevertheless be represented by a simple rate expression. Furthermore, the result suggests that the complicating factors primarily affect the rate constant, and that many of those effects can be anticipated and modeled.

Analysis of Case I Systems in Batch Liquid Reactors

We now turn our attention to the use of Equation 6-22, in conjunction with mass balances for various Case I reactors, to assess how gas transfer efficiency varies as a function of design choices and operating parameters. We start with the analysis of a batch, diffused gas transfer system in which no chemical reactions that involve the volatile species are occurring. The mass balance for the solution in such a system has only the storage term on the left and the gas transfer term on the right:

$$V_L \frac{dc_L}{dt} = V_L r_{\text{gt-avg}} \quad (6\text{-}24)$$

Dividing through by V_L and substituting the expression for $r_{L,\text{gt-avg}}$ from Equation 6-21, we obtain

$$\frac{dc_L}{dt} = \frac{Q_G H}{V_L}\left[1 - \exp\left(-\frac{K_L a_L}{Q_G H/V_L}\right)\right]\left(c_{L,\text{in}}^* - c_L\right) \quad (6\text{-}25a)$$

$$= k_I\left(c_{L,\text{in}}^* - c_L\right) \quad (6\text{-}25b)$$

In situations in which $c^*_{L,in}$ is known, an integral analysis can be carried out to yield an explicit expression for c_L as a function of time, as follows:

$$\int_{c_L(0)}^{c_L(t)} \frac{dc_L}{c^*_{L,in} - c_L} = k_I \int_0^t dt \qquad (6\text{-}26)$$

$$\ln\left(\frac{c^*_{L,in} - c_L(t)}{c^*_{L,in} - c_L(0)}\right) = -k_I t \qquad (6\text{-}27)$$

$$c^*_{L,in} - c_L(t) = \left(c^*_{L,in} - c_L(0)\right)\exp(-k_I t) \qquad (6\text{-}28)$$

Equation 6-28 indicates that the extent of disequilibrium between injected gas bubbles and a batch solution decays exponentially with time for as long as bubbles are being injected. As we would expect, the extent of disequilibrium diminishes to zero at very long times, meaning that the solution ultimately reaches equilibrium with the incoming gas.

Equation 6-28 can be rearranged in two ways to show the relationship between c_L and t for given values of the input and operational parameters, yielding design equations for the process being analyzed, as follows:

$$c_L(t) = c^*_{L,in} - \left(c^*_{L,in} - c_L(0)\right)\exp(-k_I t) \qquad (6\text{-}29)$$

$$t = -\frac{1}{k_I} \ln\left\{\frac{c^*_{L,in} - c_L(t)}{c^*_{L,in} - c_L(0)}\right\} \qquad (6\text{-}30)$$

Using these equations along with knowledge of the system characteristics and inputs (V_L, Q_G, $c_{L,in}$, $c_{G,in}$, $c_L(0)$, H, and $K_L a_L$), we can predict the dissolved concentration of a volatile species after any given aeration time (Equation 6-29) or the time required to achieve any desired final concentration (Equation 6-30). In a slightly modified form, the same equations can be used to evaluate $K_L a_L$ from experimental data:

$$\ln\left|c^*_{L,in} - c_L(t)\right| = -k_I t + \ln\left|c^*_{L,in} - c_L(0)\right| \qquad (6\text{-}31)$$

According to Equation 6-31, if the logarithm of the extent of disequilibrium between the solution and the incoming gas is plotted against the amount of time that bubbles have been injected into the solution, a straight line of slope $-k_I$ should be obtained (Figure 6-4). Since all the terms in the definition of k_I other than the $K_L a_L$ are independently measurable, the value of k_I can be used to compute $K_L a_L$.

In some situations, $c^*_{L,in}$ is not known. In those cases, a differential method of analysis can be used to estimate both k_I and $c^*_{L,in}$ from experimental data of c_L as a function of

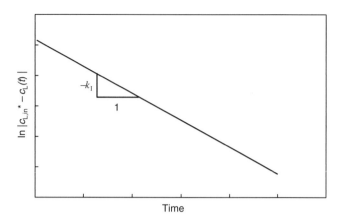

FIGURE 6-4. Generic plot of $\ln\left|c^*_{L,in} - c_L(t)\right|$ versus time for diffused aeration in a batch reactor with a well-mixed liquid phase.

time. In this approach, we approximate Equation 6-25b numerically as

$$\frac{\Delta c_L}{\Delta t} = k_I\left(c^*_{L,in} - \bar{c}_L\right) = k_I c^*_{L,in} - k_I \bar{c}_L \qquad (6\text{-}32)$$

where $\bar{c}_L$ is the average concentration during the time interval Δt. This equation suggests that a plot of $\Delta c_L/\Delta t$ versus $\bar{c}_L$ should yield a straight line with a slope equal to k_I and x-intercept equal to $c^*_{L,in}$.

■ **EXAMPLE 6-2.** A 1000-m³ batch of contaminated groundwater is to be aerated to increase the dissolved oxygen concentration and strip out trichloroethylene (TCE). The $K_L a_L$ value of the aerators for oxygen has been determined to be 14 h⁻¹. Because liquid-phase resistance almost always dominates gas-phase resistance for transfer of a highly volatile species like oxygen, assume that $K_L a_L \approx k_L a_L$ for O_2. For the purposes of a preliminary analysis, assume also that $k_L a_L$ for TCE is 68% of that for O_2, as was reported for the system analyzed in Example 5-4; for the system under consideration, this means that $(k_L a_L)_{TCE}$ is 9.52 h⁻¹. Assume that the k_G/k_L ratio for the system is 20.

The water is initially devoid of oxygen and contains 500 µg/L TCE. It is to be aerated at a gas flow rate of 25 m³/min until it contains at least 9 mg/L oxygen and less than 100 µg/L TCE. The oxygen concentration in the air is 310 mg/L, and the air contains no TCE. Henry's constants for oxygen and TCE at the ambient temperature of 20°C are 32.3 L_L/L_G and 0.45 L_L/L_G, respectively. (These values differ from those in from Table 5-2 because the values in the table are for $T = 25°C$.)

(a) Compute the concentrations of dissolved O_2 and TCE as a function of time, and estimate how long aeration will have to proceed to meet the treatment goals. Which species presents the more difficult treatment problem?

(b) Compute the following values at $t = 0$ and $t = 1\,\text{h}$.

   (i) Concentrations of O_2 and TCE in the exiting gas.

   (ii) Efficiency of the process with respect to the approach to equilibrium with individual bubbles (i.e., the percentage decrease in the extent of disequilibrium of each species as the bubbles pass through the water column).

   (iii) Efficiency of the aeration process with respect to oxygen dissolution (i.e., the fraction of the injected oxygen that enters the solution).

Solution.

(a) First, we must estimate $K_L a_L$ for TCE. To do this, we apply Equation 6.51, multiply both sides by a_L, and insert the information given in the problem statement:

$$K_L a_L = \frac{k_L k_G H}{k_L + k_G H} a_L$$

$$\left(\frac{K_L a_L}{k_L a_L}\right)_{\text{TCE}} = \frac{k_G H}{k_L + k_G H} = \frac{(k_G/k_L)H}{1 + (k_G/k_L)H}$$

$$= \frac{(20)0.45}{1 + (20)0.45} = 0.90$$

$$(K_L a_L)_{\text{TCE}} = 0.90(k_L a_L)_{\text{TCE}}$$

$$= 0.90(9.52\,\text{h}^{-1}) = 8.57\,\text{h}^{-1}$$

This result indicates that gas-phase resistance makes a small, but not negligible contribution to the overall resistance to transfer of TCE, causing $K_L a_L$ to be only 90% of $k_L a_L$ for TCE, whereas $K_L a_L$ is essentially identical to $k_L a_L$ for O_2.

The concentration profiles over time for oxygen and TCE can be computed using Equation 6-29 if $c^*_{L,\text{in}}$ and k_I for the two gases are known. These latter values can be computed using the given Henry's constants and Equation 6-23, respectively. The values are shown below, and the time profiles of c_L for both gases are shown in Figure 6-5. Note that k_I for O_2 equals 87% of $K_L a_L$ for O_2, whereas for TCE k_I is only 7.1% of $K_L a_L$.

	$c^*_{L,\text{in}}$ (mg/L)	k_I (h^{-1})
O_2	9.60	12.16
TCE	0	0.675

Consistent with the much larger value of k_I for oxygen than TCE, the dissolved oxygen concentration approaches equilibrium with the injected gas

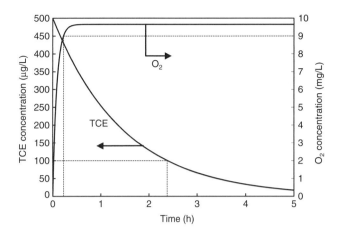

FIGURE 6-5. Approach to equilibrium of dissolved oxygen and TCE in the example system. The dashed lines indicate the times required to meet the treatment goals of 9 mg/L for O_2 and 100 μg/L for TCE.

much more quickly than TCE does. With respect to the treatment goals, the oxygen concentration exceeds 9 mg/L after approximately 13 min, when the TCE concentration has declined by less than 15%. Approximately 2.4 h would be required to meet the goal for TCE removal. Thus, it is clear that TCE presents the greater treatment challenge.

(b) (i) The values of c_L at $t = 0$ are given, and the values at $t = 1\,\text{h}$ can be computed as described in part (a). The concentrations in the exiting gas can then be computed by inserting the known information into Equation 6-20, yielding the following results.

	O_2 (mg/L)	TCE (μg/L)
$c_L, t = 0$	0	500
$c_L, t = 1\,\text{h}$	9.60	255
$c_{G,\text{in}}$	310	0
$c_{G,\text{out}}, t = 0$	232	225
$c_{G,\text{out}}, t = 1\,\text{h}$	310	115

(ii) The efficiency of the process with respect to the approach to equilibrium with individual bubbles can be defined as $\eta_{bb} \equiv 1 - ((c^*_{L,\text{out}} - c_L)/(c^*_{L,\text{in}} - c_L))$, where $c^*_{L,\text{in}}$ and $c^*_{L,\text{out}}$ refer to the liquid phase concentrations that would be in equilibrium with the influent and exiting gas bubbles, respectively. The values of c^*_L at each time can be computed using the results from part b(i) in conjunction with the relationship $c^*_L = c_G/H$. All the calculations are summarized in Table 6-2.

TABLE 6-2. Various Parameter Values Initially and After 1 h of Aeration in the Example System

	O_2 (mg/L)		TCE (μg/L)	
	$t=0$	$t=1$ h	$t=0$	$t=1$ h
c_L	0	9.60	500	255
$c^*_{L,in}$	9.60	9.60	0	0
$c^*_{L,out}$	7.18	9.60	500	255
$c^*_{L,in} - c_L$	9.60	0	−500	−255
$c^*_{L,out} - c_L$	7.18	0	0	0
η_{bb} (%)	25	25	100	100

η_{bb} is 100% for TCE transfer both at the beginning and the end of the aeration, meaning that, at both times, the bubbles reach equilibrium with the solution as they pass through it. For oxygen, the extent of disequilibrium decreases by 25% as the bubbles rise at the beginning of the hour. By the end of the hour, the solution has acquired enough dissolved O_2 that it is essentially equilibrated with the injected air, so the idea of a decrease in the extent of disequilibrium is almost meaningless. However, if one carries out the calculations to an unrealistic number of significant figures just to determine the theoretical decrease in disequilibrium, the result is that the extent of disequilibrium declines by the same percentage (25%) at the end of the hour as at the beginning. Thus, as each individual bubble rises, equilibrium is achieved more quickly with respect to TCE than O_2. However, as shown in part (a), the solution approaches equilibrium with the *incoming* gas more quickly for O_2 than TCE.

The fact that each bubble equilibrates with solution with respect to TCE before the bubble exits the system means that TCE is transferring into the bubble over only a portion of its rise time. By contrast, the same bubble remains disequilibrated with O_2 during the complete rise time, and O_2 transfers from the bubble into the water during this entire time. As a result, the disequilibrium between the incoming gas and solution dissipates faster for O_2 than for TCE. This point is discussed in greater detail in the following section.

(iii) Comparing the concentration of oxygen in the injected gas bubbles with that in the exiting gas, we find that the O_2 transfer efficiency (i.e., the fraction of the O_2 entering the reactor in the bubbles that is transferred to the solution) is 25% at the beginning of the hour and essentially zero at the end.

Note that the extent of disequilibrium between the solution and the incoming gas phase decays exponentially over time for both TCE and O_2, similar to the behavior of a reactant undergoing a first-order reaction in a batch reactor. However, even though the overall gas transfer rate constants ($K_L a_L$) for the two species differ by only 32%, their k_I values differ by much more than that, so the two species approach equilibrium at dramatically different rates. These different rates of equilibration can be expressed conveniently in terms of the characteristic times for the two processes. Recalling that the characteristic time of a first-order reaction is the inverse of the rate constant, the characteristic times for equilibration of TCE and O_2 with the incoming gas can be computed as k_I^{-1}, corresponding to 1.48 and 0.082 h (88.9 and 4.9 min), respectively. ∎

Limiting Cases of the General Kinetic Expression

Overview Equation 6-21 has two limiting values of interest, corresponding to large and small values of the exponential argument in the expression for k_I (Equation 6-23); that is, large and small values of the ratio of $K_L a_L$ to $Q_G H/V_L$. As shown in Table 6-3 and Figure 6-6, k_I

TABLE 6-3. Values of Terms in Equations 6-21 and 6-23 for Limiting Relative Values of $K_L a_L$ and $Q_G H/V_L$

	Value of Exponential Term	k_I	$r_{L,gt\text{-}avg}$	Equation Number for $r_{L,gt\text{-}avg}$
General case	$\exp\left(-\dfrac{K_L a_L}{Q_G H/V_L}\right)$	$\dfrac{Q_G H}{V_L}\left[1 - \exp\left(-\dfrac{K_L a_L}{Q_G H/V_L}\right)\right]$	$\dfrac{Q_G H}{V_L}\left[1 - \exp\left(-\dfrac{K_L a_L}{Q_G H/V_L}\right)\right](c^*_{L,in} - c_L)$	(6-21)
$\dfrac{K_L a_L}{Q_G H/V_L} \gg 1^a$	0	$\dfrac{Q_G H}{V_L}$	$\dfrac{Q_G H}{V_L}(c^*_{L,in} - c_L)$	(6-33)
$\dfrac{K_L a_L}{Q_G H/V_L} \ll 1^a$	$1 - \dfrac{K_L a_L}{(Q_G H/V_L)}{}^b$	$K_L a_L$	$K_L a_L(c^*_{L,in} - c_L)$	(6-34)

[a]The cases where $K_L a_L/(Q_G H/V_L)$ is much greater and much less than 1 correspond to situations where the gas transfer rate is limited by advective transport and by interfacial transport, respectively, as explained later in this chapter. Rate limitation by advective transport is also referred to as limitation by equilibrium.
[b]For small values of x, $\exp(-x)$ approaches $1 - x$.

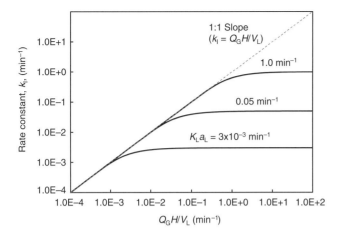

FIGURE 6-6. k_I as a function of the value of $Q_G H/V_L$ for three different values of $K_L a_L$ (the three different curves). The line with a 1:1 slope defines points where $k_I = Q_G H/V_L$. For any given value of $K_L a_L$, k_I is approximately equal to either $K_L a_L$ or $Q_G H/V_L$, except for a small region where it shifts from one of these limiting values to the other.

approaches the value of each of these parameter groups under conditions where that group is much smaller than the other. For example, the curve for $K_L a_L = 0.05\,\text{min}^{-1}$ in Figure 6-6 indicates that k_I is approximately equal to $Q_G H/V_L$ when $Q_G H/V_L$ is less than $0.02\,\text{min}^{-1}$, and approximately equal to $K_L a_L$ ($0.05\,\text{min}^{-1}$) when $Q_G H/V_L$ is greater than $0.2\,\text{min}^{-1}$.

k_I is a pseudo-first-order rate constant, so, as shown in Chapter 3, k_I^{-1} is the characteristic time for the overall gas transfer process in Case I systems. Correspondingly, the relationships in Table 6-3 and Figure 6-6 indicate that the characteristic time for gas transfer in Case I systems is usually close to, but slightly longer than, the larger of $(K_L a_L)^{-1}$ or $(Q_G H/V_L)^{-1}$. This result is reminiscent of that derived in Chapter 3 for an overall reaction that comprises two reactions in series; in that case, we identified the reaction with the longest characteristic time as the rate-limiting step.

In a Case I reactor, the overall gas transfer process can be thought of as comprising advective and interfacial steps in series. If the constituent is being stripped from the solution, it first passes through the interfacial region and is then advected out of the reactor with the carrier gas; if gas absorption is occurring, the steps occur in the reverse order: the species is first advected into the reactor and then must cross the interfacial region to enter the solution. We might suspect, then, that the limiting cases identified above correspond to situations in which one or the other of these steps limits the overall transfer rate. This inference is supported by the fact that the characteristic time for one of the limiting cases $\left((Q_G H/V_L)^{-1}\right)$ contains only terms that describe macroscopic and advective components of the system, while that for the other limiting case $\left((K_L a_L)^{-1}\right)$ contains only terms that relate to the microscopic, interfacial transfer process. We next explore this way of interpreting the overall rate expression.

Macroscopic (Advective) Limitation on the Gas Transfer Rate Consider first the limiting case in which $K_L a_L \gg Q_G H/V_L$; that is, the case in which the characteristic time for the macroscopic, advective transport process is much longer than that for the interfacial process, and hence the advective step is rate limiting. By manipulating the expression for $r_{L,\text{gt-avg}}$ given in Table 6-3 for these conditions, we obtain the following expression, which is also given in the table:

For $K_L a_L \gg \dfrac{Q_G H}{V_L}$:

$$r_{L,\text{gt-avg}} = \frac{Q_G H}{V_L}\left(c^*_{L,\text{in}} - c_L\right) = \frac{Q_G\left(c_{G,\text{in}} - c^*_G\right)}{V_L} \quad (6\text{-}33)$$

$c_{G,\text{in}}$ is the concentration in the incoming gas phase, and c^*_G is the gas-phase concentration that would be in equilibrium with the solution in the reactor at any instant. Therefore, the difference $c_{G,\text{in}} - c^*_G$ is the maximum change in the gas phase concentration that could possibly occur as the gas passes through the reactor, even if there were no interfacial resistance to gas transfer at all. Correspondingly, the product $Q_G\left(c_{G,\text{in}} - c^*_G\right)$ is the maximum rate (mass/time) at which transfer of the given species could occur in the reactor. Finally, dividing this product by V_L (to obtain the expression on the right side of Equation 6-33) yields the maximum rate of gas transfer that can occur, expressed as the mass of gas transferred per unit volume of solution per unit time. Note that these statements apply regardless of whether the process under consideration is gas stripping ($r_{L,\text{gt-avg}} < 0$) or absorption ($r_{L,\text{gt-avg}} > 0$).

The limiting case characterized by Equation 6-33 describes a situation in which the gas exiting the reactor is equilibrated with the solution, so gas transfer into or out of each bubble has ceased by the time the bubble exits the reactor. As a result, for a given volatile species, a given liquid volume, and a given composition of the incoming gas, the only way to increase the maximum transfer rate is to provide more carrier gas; that is, to increase Q_G. Breaking the bubbles into smaller ones to increase the interfacial area available for gas transfer, modifying the mixing pattern to keep bubbles in the system longer, or mixing the solution more intensely to reduce the interfacial resistance will have no effect whatsoever, since none of these modifications addresses the overriding constraint that the driving force for gas transfer has dissipated. Based on this understanding of the situation, this case can be described variously as rate limitation by the macroscopic transport step, by equilibrium between the gas and solution, or by (gas-phase) advective transport. Roberts et al. (1984) suggested that this limiting case applies to stripping of several VOCs (e.g., chloroform and TCE) in activated sludge aeration basins. They predicted that, for typical aeration rates, the exiting gas would be in equilibrium with the VOCs in the solution, so that only an increase in the gas flow rate could improve the stripping efficiency.

Microscopic (Interfacial) Limitation on the Gas Transfer Rate Next, we consider the other limiting condition ($K_L a_L \ll Q_G H/V_L$), in which the characteristic time for the microscopic process is much longer than that for the macroscopic process. When this case applies, the overall gas transfer rate is

For $K_L a_L \ll \dfrac{Q_G H}{V_L}$:

$$r_{L,\text{gt-avg}} = K_L a_L \left(c^*_{L,\text{in}} - c_L\right) \quad (6\text{-}34)$$

$$= \dfrac{K_L a_L}{H}\left(c_{G,\text{in}} - c^*_G\right) \quad (6\text{-}35)$$

The expression $r_{L,\text{gt}} = K_L a_L\left(c^*_L - c_L\right)$ (Equation 5-58b) characterizes the microscopic gas transfer rate across any interface. Comparing this expression with Equation 6-34, we see that the two expressions would be identical if c^*_L equaled $c^*_{L,\text{in}}$ everywhere in the reactor. Since $c^*_L = c_G/H$, this equality would apply if c_G throughout the reactor were the same as $c_{G,\text{in}}$; that is, if the change in the gas phase concentration were negligible as the gas passed through the reactor. Thus, this situation is at the opposite extreme from the scenario for rate limitation by the equilibrium constraint: in that limiting case, the gas phase composition changes as much as it possibly can (limited only by reaching equilibrium with the solution), and the gas exiting the system is virtually equilibrated with the solution; by contrast, in this case, the gas phase concentration changes negligibly, and the exiting gas is almost as far from equilibrium with the solution as the incoming gas was.

For a system characterized by this limiting case, for given values of c_L and $c_{G,\text{in}}$, Equations 6-34 and 6-35 indicate that the gas transfer rate can be changed only by changing $K_L a_L$. In this case, modifying the reactor to increase the interfacial area available for gas transfer or to reduce the interfacial resistance to gas transfer would increase the rate of gas transfer significantly, even if Q_G were not changed at all; increasing the time that the gas and liquid are in contact with one another would also increase the overall amount of gas transfer. We therefore refer to this case as rate limitation by microscopic processes, by interfacial transport, or by interfacial kinetics.

Although the preceding analysis has focused on systems in which the gas phase consists of bubbles rising through a water column, recall that well-mixed surface aeration systems (in which the gas phase is the bulk atmosphere) also meet the Case I criteria. In such systems, the gas phase has a constant composition regardless of how much of the volatile species enters or leaves the solution, so the systems also invariably meet the criterion that $K_L a_L \ll Q_G H/V_L$. Transfer of some species in diffused aeration systems (specifically, those in which the amount of the volatile substance that enters or exits the gas phase is insufficient to change the gas phase composition substantially) also meet the criterion. For instance, in the same study as cited above, Roberts et al. (1984) predicted that stripping of CCl_2F_2 (which is much more volatile than chloroform and TCE) would be limited by interfacial transfer in most activated sludge aeration basins. Typically, oxygen absorption in such systems is also limited by the interfacial transfer step, as the oxygen concentration in the bubbles changes by only a few percent as they pass through the water column. However, in recent years, the efficiency of oxygen transfer in such systems has improved dramatically, so the assumption of a nearly constant gas phase composition does not apply as commonly as it did in the past.

■ **EXAMPLE 6-3.** A spray aerator throws droplets 0.2 cm in diameter 2 m upward into the air. What will the concentration of dissolved oxygen be in the droplets as they re-enter the bulk solution, if the initial concentration is 3 mg/L and the overall gas transfer coefficient $K_L a_L$ is $0.13\,\text{s}^{-1}$? Assume that the fluid in each droplet is well mixed and that, under the ambient conditions, the saturation (equilibrium) concentration of dissolved oxygen is 9.2 mg/L.

Solution. The system can be characterized as a Case I gas transfer system with constant c_G, so Equation 6-22 applies with the simplification that $k_I \approx K_L a_L$. Values of $K_L a_L$, c^*_L, and $c_L(0)$ are given, and t can be computed from basic physics, as follows. The height z to which the droplet rises is $z = 1/2\,gt^2$, so the time for the droplet to reach a height of 2 m is given by

$$t = \sqrt{\dfrac{2z}{g}} = \sqrt{\dfrac{(2)(2\,\text{m})}{9.8\,\text{m/s}^2}} = 0.64\,\text{s}$$

The time that the droplet spends in the air is twice the time needed for it to reach its maximum height, so $t_{\text{tot}} = 1.28$ s. In this system, the gas "entering" the reactor is the gas that is in contact with the droplet when it is first formed; that is, it is the bulk atmosphere. Therefore, $c^*_{L,\text{in}}$ is the aqueous-phase concentration that would be in equilibrium with atmospheric O_2; we represent that concentration as $c^*_{L,\text{atm}}$. Plugging appropriate values into Equation 6-29 (and using the approximation that $k_I \approx K_L a_L$), we find the oxygen concentration in the droplet as it hits the water to be

$$c_L(t) = c^*_{L,atm} - \{c^*_{L,atm} - c_L(0)\}\exp(-K_La_Lt)$$

$$= (9.2\,\text{mg/L}) - ((9.2 - 3.0)\,\text{mg/L})\exp(-(0.13\,\text{s}^{-1})(1.28\,\text{s}))$$

$$c_L(t) = 3.95\,\text{mg/L}$$

Thus, the oxygen content of individual droplets increases by approximately 1 mg/L during their short period of contact with the atmosphere. The rate of change of dissolved oxygen in the bulk solution could be determined by writing a mass balance on the bulk phase, in which case the rate at which droplets bring oxygen into the solution would be included as an advective input. ■

■ **EXAMPLE 6-4.** Determine whether phase equilibrium (i.e., macroscopic, advective transport) or interfacial processes (microscopic transport) limit the rates of transfer of oxygen and TCE in Example 6-2, or whether both steps contribute significantly to the rate limitation. What operational changes might increase the overall rate of transfer of each species?

Solution. The values of K_La_L, Q_GH/V_L, the ratio of those terms, and k_I are shown in the following table.

	Rate Constant for Interfacial Transport $K_La_L\,(\text{h}^{-1})$	Rate Constant for Advective Transport $Q_GH/V_L\,(\text{h}^{-1})$	$\dfrac{K_La_L}{Q_GH/V_L}$	$k_I\,(\text{h}^{-1})$
O_2	14.0	48.2	0.29	12.1
TCE	8.57	0.675	12.7	0.675

The rate constants for interfacial and advective transport of oxygen are of comparable magnitude. As a result, the overall rate constant for oxygen (k_I) is less than the rate constant for either of the individual steps, and neither step can be said to be rate-limiting (although microscopic transport is somewhat more limiting than macroscopic transport). By contrast, the rate constant for microscopic transport of TCE is much greater than that for macroscopic transport, so macroscopic transport is rate limiting, and the overall rate constant is approximately equal to the rate constant for that step alone. As noted previously, these conclusions are independent of time and of the concentrations of O_2 or TCE in either phase.

Increasing the gas flow rate in this system would decrease the macroscopic resistance to gas transfer but (assuming that the changed flow rate did not affect K_La_L significantly) would not reduce the microscopic transport resistance. Because the capacity for macroscopic transport is the main factor limiting TCE transfer, an increase in Q_G would increase the value of k_I almost proportionately, and would have a correspondingly large effect on the overall TCE transfer. By contrast, for oxygen, both macroscopic and microscopic transport contribute significantly to the resistance to gas transfer, so increasing the gas flow rate (which affects the former but not the latter resistance) would result in a less-than-proportionate increase in k_I.

On the other hand, if the system were changed in a way that facilitated only microscopic transport (e.g., by injecting smaller bubbles or mixing the solution mechanically to increase K_La_L, while keeping Q_G constant), TCE transfer would be virtually unaffected, and the O_2 transfer rate would again increase, but by a factor that was less than proportionate to the increase in K_La_L. The most effective way to increase O_2 transfer in the system would be to decrease the resistance from both steps, for example, by injecting smaller bubbles and simultaneously increasing Q_G. ■

Summary of Rate Limitations on Overall Gas Transfer Rate To summarize, the overall gas transfer process in Case I reactors (those with a uniform value of c_L throughout the reactor) can be viewed as consisting of sequential interfacial transport and advective transport steps, either of which might be rate-limiting. The overall rate constant k_I for either limiting case, or for a case where both resistances are significant, is given by Equation 6-23; for a reactor that is batch with respect to the solution (no flow of solution through the reactor), k_I equals the negative slope of a plot like that shown in Figure 6-4.

The gas transfer rate in Case I systems is often limited primarily by one of the two transfer steps, in which case the rate constant contains only terms characterizing that step. Advective transport limitation is a consequence of the limited carrying capacity of the gas, whereas interfacial transport limitation is a consequence of the resistance to gas transfer through the liquid and gas interfacial regions. A characteristic of advective limitation is that the composition of an aliquot of gas (e.g., a bubble) changes enough as the gas passes through the system that it approaches equilibrium with the solution, whereas under interfacial limitation, the exiting gas phase is almost as far from equilibrium with the solution as the incoming gas is.

Whereas the interfacial resistance is exerted at a microscopic scale, the resistance associated with gas advection is macroscopic, in that it is related to the total flow of gas through the whole reactor. A gas transfer process limited by microscopic processes or interfacial transport is sometimes referred to as being kinetically limited, and one limited by macroscopic processes or advective transport gas is sometimes referred to as being equilibrium limited. The likelihood of equilibrium limiting the gas transfer rate increases with decreasing Henry's constant.

TABLE 6-4. Equations Describing the Aqueous Phase Concentration in Batch Gas Transfer Reactors with Spatially Uniform Solution Composition and Plug Flow of Gas: General and Limiting Cases

Qualitative and Quantitative Conditions for Limiting Cases		Equations Describing Dissolved Concentration of Volatile Species in Well-Mixed Batch Systems	
General case	—	$c_L(t) = c_{L,in}^* - \left(c_{L,in}^* - c_L(0)\right)\exp\left\{-\dfrac{Q_G H}{V_L}\left[1 - \exp\left(-\dfrac{K_L a_L}{Q_G H/V_L}\right)\right]t\right\}$	(6-36)
Rate limited by advective transport	$\dfrac{K_L a_L}{Q_G H/V_L} \gg 1$	$c_L(t) = c_{L,in}^* - \left(c_{L,in}^* - c_L(0)\right)\exp\left\{-\dfrac{Q_G H}{V_L}t\right\}$	(6-37)
Rate limited by interfacial transport	$\dfrac{K_L a_L}{Q_G H/V_L} \ll 1$	$c_L(t) = c_{L,in}^* - \left(c_{L,in}^* - c_L(0)\right)\exp\{-K_L a_L t\}$	(6-38)

The determination of whether one of the limiting cases applies, and if so, which one, depends solely on the relative values of $K_L a_L$ and $Q_G H/V_L$, and not on the value of either c_L or $c_{G,in}$. Therefore, if the transfer rate for a given species in a given reactor is limited by advective transport, that limitation will apply for as long as the reactor is operated, no matter what the values of c_L or $c_{G,in}$ are. However, the actual value of the maximum transfer rate does depend on c_L and $c_{G,in}$. As a result, if the gas transfer rate in a reactor was equilibrium-limited, and if the value of c_L changed over time (as would occur in a batch solution that was continuously sparged with gas), the rate of gas transfer would also change over time, albeit always being the maximum rate possible (limited only by equilibrium with the solution) for the current value of c_L. The equations for $c_L(t)$ in batch liquid systems where gas transfer is the only process affecting c_L, for the general case and each of the limiting cases, can be derived by combining the expressions for $k_{L,gtavg}$ in Table 6-3 with Equation 6-23; these equations are collected in Table 6-4.

Case I Systems with Continuous Liquid Flow at Steady State

Analysis of a gas transfer system with uniform solution composition and continuous flow of liquid is identical in most ways to that for a batch liquid system: an appropriate mass balance is written, the reaction rate expression for Case I conditions is inserted, and the equation is solved. Because the assumptions used to develop the Case I rate expression ($r_{L,gt-avg}$) apply equally to a system with liquid flow, the resulting expression can be used without alteration. Furthermore, the rate constant is subject to the same two rate-limiting conditions. However, in systems with continuous liquid flow, it is sometimes convenient to express the limiting conditions in terms of different parameters (e.g., τ_L) than in batch systems.

Reactors with Plug Flow of Liquid Recall that, for systems in which only homogeneous (single-phase) reactions are occurring, the analysis of batch reactors applies with minor modifications to plug flow reactors, because a PFR can be treated simply as a batch reactor on a conveyor belt. Essentially the same analogy and conclusion apply to systems in which heterogeneous reactions (i.e., multi-phase reactions, such as gas transfer) are occurring. In particular, many systems of interest can be approximated as having plug flow of a liquid that is continuously exposed to a gas phase. Examples of such systems include rivers (with continuous exposure to the atmosphere) and engineered reactors in which liquid flows horizontally and with minimal axial mixing while gas bubbles are injected uniformly at the bottom of the water column. In each of these example systems, the water can be imagined to be in a batch reactor that moves along a conveyor belt while being exposed to a gas phase. Thus, the equations in Tables 6-3 and 6-4 apply to such systems, with the modifications that t is replaced by the hydraulic residence time in the reactor, τ_L, and the concentrations at times 0 and t are replaced by those in the reactor influent and effluent, respectively.

Reactors with Flow and a Uniform Liquid-Phase Composition (CFSTRs with Respect to Liquid) In some continuous-flow gas transfer reactors, a bulk liquid phase is mixed with sufficient intensity, either by an external input of energy (e.g., by mechanical mixers) or by gas bubbles rising through it, that the liquid-phase composition is approximately uniform. Thus, it is reasonable to treat the liquid as passing through a CFSTR. By contrast, the gas bubbles in such a system might rise through the water column, all spending approximately the same time in the system and not coalescing with one another, so that bubbles that have spent a long time in the reactor do not mix with those that have spent a short time. That is, the gas phase behaves as it would in a PFR. The mass balance on a volatile species in the liquid phase of such a reactor is

Rate of change of mass of i stored in control volume = Net advective input of i + Net diffusive/dispersive input of i

+ Rate at which i enters by gas transfer + Rate at which i is formed by other reactions

$$V_L \frac{dc_L}{dt} = Q_L(c_{L,in} - c_L) + 0 + V_L r_{L,gt\text{-}avg} + 0 \quad (6\text{-}39)$$

Assuming steady-state, the equation can be solved to yield an expression for $c_{L,out}$:

$$0 = Q_L(c_{L,in} - c_{L,out})$$

$$+ Q_G H \left[1 - \exp\left(-\frac{K_L a_L}{Q_G H/V_L}\right)\right](c^*_{L,in} - c_{L,out}) \quad (6\text{-}40)$$

$$\frac{c_{L,in} - c_{L,out}}{c_{L,out} - c^*_{L,in}} = \frac{Q_G H}{Q_L}\left[1 - \exp\left(-\frac{K_L a_L}{Q_G H/V_L}\right)\right] \quad (6\text{-}41)$$

$$c_{L,in} - c_{L,out} = c_{L,out}\frac{Q_G H}{Q_L}\left[1 - \exp\left(-\frac{K_L a_L}{Q_G H/V_L}\right)\right]$$

$$- c^*_{L,in}\frac{Q_G H}{Q_L}\left[1 - \exp\left(-\frac{K_L a_L}{Q_G H/V_L}\right)\right] \quad (6\text{-}42)$$

$$c_{L,out} = \frac{c_{L,in} + \dfrac{Q_G H c^*_{L,in}}{Q_L}\left[1 - \exp\left(-\dfrac{K_L a_L}{Q_G H/V_L}\right)\right]}{1 + \dfrac{Q_G H}{Q_L}\left[1 - \exp\left(-\dfrac{K_L a_L}{Q_G H/V_L}\right)\right]} \quad (6\text{-}43)$$

Substituting V_L/τ_L for Q_L, we can write the pre-exponential terms and the exponential argument in Equations 6-41 and 6-43 in terms of k_I to derive the following relationships:

$$\frac{c_{L,in} - c_{L,out}}{c_{L,out} - c^*_{L,in}} = k_I \tau_L \quad (6\text{-}44)$$

$$\tau_L = \frac{1}{k_I}\frac{c_{L,in} - c_{L,out}}{c_{L,out} - c^*_{L,in}} \quad (6\text{-}45a)$$

$$k_I = \frac{1}{\tau_L}\frac{c_{L,in} - c_{L,out}}{c_{L,out} - c^*_{L,in}} \quad (6\text{-}45b)$$

$$c_{L,out} = \frac{c_{L,in} + c^*_{L,in} k_I \tau_L}{1 + k_I \tau_L} \quad (6\text{-}45c)$$

The various forms of Equation 6-45 all represent the same relationship among the design and operational parameters for gas transfer in a reactor that meets the assumptions of the derivation (continuous flow of liquid, steady-state, uniform solution composition, and plug flow for the gas phase). Equation 6-45a is useful for computing the hydraulic residence time required to achieve a desired change in the concentration of a volatile species if k_I is known; Equation 6-45b is useful for computing k_I (and thereby inferring $K_L a_L$) if operational data are available; and Equation 6-45c is useful for predicting the steady state concentration of the volatile species for a given set of operating conditions. The equations indicate that, as is true for batch aeration, the solution approaches equilibrium with the injected gas at long liquid residence times (τ_L in the reactor with flow, t in the batch reactor). Like the equations characterizing gas transfer in well-mixed batch systems, the expression for k_I in all the above equations reduces to simpler expressions in two limiting cases: $Q_G H/V_L$, if gas transfer is limited primarily by the carrying capacity of the gas phase, and $K_L a_L$, if transfer is limited by the rate of interfacial transport.

■ **EXAMPLE 6-5.** Assume that the system described in Example 6-2 has been modified so that it operates as a CFSTR. The water flow rate varies substantially, so the hydraulic residence time in the system might be as small as 10 min or as long as 5 h. As in the prior example, the system will be operated with a fixed liquid volume of 1000 m³ and a fixed gas flow rate of 25 m³/min. Develop a plot showing the steady state concentrations of dissolved oxygen and TCE as a function of the hydraulic residence time in the reactor. Does the dominant factor limiting gas transfer (interfacial transfer or equilibrium capacity) depend on τ_L?

Solution. The values of k_I are identical to those computed in Example 6-4; that is, $12.1\,\text{h}^{-1}$ for O_2 and $0.675\,\text{h}^{-1}$ for TCE. Those values can be inserted into Equation 6-45c to determine the effluent concentrations as a function of the hydraulic residence time; the results are plotted in Figure 6-7.

The results are qualitatively similar to those for the batch system: the solution approaches equilibrium with the incoming gas phase, with the approach getting ever closer as the solution residence time increases. The approach to equilibrium between the solution and the incoming gas is more rapid for oxygen than TCE, primarily because of the advective transport limitation on the latter. That is, because TCE is much less volatile than O_2 (the Henry's constant for TCE is almost two orders of magnitude less than that of O_2), the injected gas has a much lower carrying capacity for TCE than O_2.

By definition, increasing τ_L allows the water to stay in the reactor longer. When the system is operated with a fixed gas

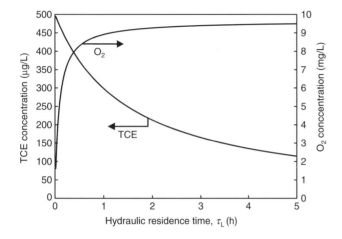

FIGURE 6-7. Steady-state dissolved oxygen and TCE concentrations as a function of hydraulic residence time in the example reactor.

flow rate, the rate constants associated with both advective transport and interfacial transport (the kinetic limitation) are fixed, independent of τ_L. Therefore, a change in τ_L has no effect on the relative importance of these two limitations to overall transport. As a result, as in the batch system, interfacial processes are more important than advective transport in limiting the oxygen transfer rate, and advective transport is much more important than interfacial processes in limiting transport of TCE, independent of τ_L.

In many reactors of interest, the liquid flow rate and/or the need for gas transfer varies considerably over time (e.g., in activated sludge systems, in which both the rate of inflow and the oxygen demand per liter of influent can change substantially over the course of a day as well as seasonally). In such cases, it is common to adjust the air flow rate in parallel with the liquid flow rate, so that a constant Q_G/Q_L ratio is maintained, rather than maintaining a fixed Q_G regardless of how Q_L changes. When analyzing or designing these types of systems, it is useful to express k_I and the steady-state aqueous phase concentration $c_{L,ss}$ as a function of Q_G/Q_L and τ_L. This way of representing k_I can be derived from Equation 6-23 as follows:

$$k_I = \frac{Q_G H}{V_L}\left[1 - \exp\left(-\frac{K_L a_L}{Q_G H/V_L}\right)\right] \quad (6\text{-}23)$$

$$= \frac{Q_G H}{Q_L \tau_L}\left[1 - \exp\left(-\frac{K_L a_L}{Q_G H/(Q_L \tau_L)}\right)\right] \quad (6\text{-}46a)$$

$$= \frac{(Q_G/Q_L) H}{\tau_L}\left[1 - \exp\left(-\frac{\tau_L K_L a_L}{(Q_G/Q_L) H}\right)\right] \quad (6\text{-}46b)$$

The corresponding expressions for $c_{L,out}$ for the two limiting cases are

Macroscopic (i.e., advective transport or equilibrium) limitation

$$c_{L,out} = \frac{c_{L,in} + c_{L,in}^* H Q_G/Q_L}{1 + H Q_G/Q_L} \quad (6\text{-}47)$$

Microscopic (i.e., interfacial transport or kinetic) limitation

$$c_{L,out} = \frac{c_{L,in} + c_{L,in}^* K_L a_L \tau_L}{1 + K_L a_L \tau_L} \quad (6\text{-}48)$$

Equation 6-47 indicates that gas transfer in a system limited by the carrying capacity is insensitive to τ_L, if the reactor is operated at constant Q_G/Q_L.[4] This result makes intuitive sense, because if carrying capacity limits gas transfer, the gas exiting the reactor is nearly equilibrated with the solution. In such a case, increasing τ_L, which increases the gas/liquid contact time, will not be able to overcome the limitation; only increasing the Q_G/Q_L ratio can do that. On the other hand, if gas transfer is limited by interfacial transport (Equation 6-48), increasing the liquid detention time allows more transfer to occur and causes $c_{L,out}$ to approach $c_{L,in}^*$ more closely.

The dimensionless parameter grouping $(Q_G/Q_L)H_{cc}$ (or, equivalently, $(Q_G/Q_L)H_{mm}$) is referred to as the stripping factor (S) in the literature on contaminant stripping. However, because Equation 6-46b is applicable to continuous flow Case I systems regardless of whether the process of interest is gas stripping or absorption, we refer to it here as the *gas transfer capacity factor*, R:

$$R \equiv \frac{Q_G H_{cc}}{Q_L} \quad (6\text{-}49)$$

Substituting this expression into Equation 6-47, that equation can be written as

$$c_{L,out} = \frac{c_{L,in} + c_{L,in}^* R}{1 + R} \quad (6\text{-}50)$$

We will see in the next section that the same term appears in the equations characterizing systems in which neither the gas nor the solution phase composition is uniform throughout the reactor (such as packed or spray towers). In those systems, R turns out to be a very useful design parameter.

---

[4] If Q_G and Q_L change significantly, $K_L a_L$ would almost certainly be affected as well, so k_I would change in a more complex way than suggested here. In this analysis, we consider only the direct effect of the flow rates on k_I.

CASE I: GAS TRANSFER IN SYSTEMS WITH A WELL-MIXED LIQUID PHASE   223

TABLE 6-5. Equations Describing the Aqueous Phase Concentration in Gas Transfer Reactors with Spatially Uniform Solution Composition and Plug Flow of Gas: General and Limiting Cases

	Qualitative and Quantitative Conditions for Limiting Cases	Equations Describing Dissolved Concentration of Volatile Species in Steady State Operation of Systems with Continuous Liquid Flow	
General case	—	$c_{L,out} = \dfrac{c_{L,in} + \dfrac{Q_G H c^*_{L,in}}{Q_L}\left[1 - \exp\left(-\dfrac{K_L a_L}{Q_G H/V_L}\right)\right]}{1 + \dfrac{Q_G H}{Q_L}\left[1 - \exp\left(-\dfrac{K_L a_L}{Q_G H/V_L}\right)\right]}$	(6-43)
Rate limited by advective transport	$\dfrac{K_L a_L}{Q_G H/V_L} \gg 1$	$c_{L,out} = \dfrac{c_{L,in} + c^*_{L,in}(Q_G H/V_L)\tau_L}{1 + (Q_G H/V_L)\tau_L}$	(6-47a)
Rate limited by interfacial transport	$\dfrac{K_L a_L}{Q_G H/V_L} \ll 1$	$c_{L,out} = \dfrac{c_{L,in} + c^*_{L,in} K_L a_L \tau_L}{1 + K_L a_L \tau_L}$	(6-47b)

The equations derived in this section showing the aqueous phase concentration for gas transfer in steady-state, Case I systems with flow of both gas and solution are collected in Table 6-5.

■ **EXAMPLE 6-6.** A preliminary design is being prepared for a steady-state activated sludge treatment process. The wastewater flow rate is 1.0 m³/s, the anticipated mean hydraulic residence time is 6 h, and it is estimated that 135 mg of oxygen will be consumed per liter of influent by the organisms carrying out the biodegradation process. The influent is at 20°C and is anoxic ($c_{L,O_2,in} = 0$ mg/L), and the water in the aeration basin must contain at least 2 mg/L dissolved oxygen to allow the organisms to function optimally. H_{cc} for oxygen at 20°C is 29.2 L_L/L_G, and the β factor for oxygen, which accounts for the effect of solution composition on H, is 0.95. K_L for oxygen was estimated to be 0.015 cm/s in tests with clean water, but is expected to be only 0.010 cm/s in the aeration basin, due to the presence of surfactants in the water.

In the proposed operation, the diffusers will each inject 0.06 SCM/min of air as 0.4-cm-diameter bubbles, where the standard conditions are defined to be 25°C and 1.0 atm. Estimate the density of diffusers on the reactor floor (number of diffusers per square meter of floor area) that would be required to treat the waste in an aeration tank that is 7 m deep.

Solution. We begin by writing a mass balance on oxygen in the aeration tank. Oxygen is entering the water in the tank by gas transfer, is leaving via advection, and is also being consumed by biological activity (the advective input is zero, since the incoming water is anoxic). The system will operate at steady-state, and the volume of the tank can be computed as $Q_L\tau_L$, or 3600 m³. The rate at which oxygen is being consumed by microorganisms is (135 mg/L)/6 h, or 0.00625 mg/L s. Writing the mass balance, applying the assumption of steady state, and inserting known values or expressions, we obtain

$$V_L \frac{dc_L}{dt} = Q_L(c_{L,in} - c_L) + V_L r_{gt,avg} + V_L r_{biol}$$

$$0 = (1.0\,\text{m}^3/\text{s})(0 - 2\,\text{mg/L}) + (3600\,\text{m}^3) r_{gt,avg}$$
$$+ (3600\,\text{m}^3)(-0.00625\,(\text{mg/L})/\text{s})$$

$$r_{gt,avg} = \frac{(1.0\,\text{m}^3/\text{s})(2\,\text{mg/L}) + (3600\,\text{m}^3)(0.00625\,(\text{mg/L})/\text{s})}{3600\,\text{m}^3}$$

$$= 0.00681\,(\text{mg/L})/\text{s}$$

The rate of oxygen transfer into the solution from the gas phase can also be computed using Equation 6-21.

$$r_{L,gt\text{-}avg} V_L = Q_G H \left[1 - \exp\left(-\frac{K_L a_L}{Q_G H/V_L}\right)\right]\left(c^*_{L,in} - c_L\right)$$

Rearranging to solve for Q_G, we obtain

$$Q_G = \frac{r_{L,gt\text{-}avg} V_L}{H\left[1 - \exp\left(-\dfrac{K_L a_L}{Q_G H/V_L}\right)\right]\left(c^*_{L,in} - c_L\right)} \quad (6\text{-}51)$$

The values of several parameters on the right side of Equation 6-51 are known, ($r_{L,gt\text{-}avg}$, V_L, K_L, c_L), but others (H, $c^*_{L,in}$, and the ratio $a_L/(Q_G/V_L)$) must be calculated.

According to Equation 5-23, the effective value of Henry's constant in the wastewater equals the Henry's constant in clean water divided by β, so,

$$H_{cc,\,\text{aeration basin}} = \frac{H_{cc,\,\text{clean H}_2\text{O}}}{\beta} = \frac{29.2\,\frac{L_L}{L_G}}{0.95} = 30.7\,\frac{L_L}{L_G}$$

$$H_{pm,\,\text{aeration basin}} = H_{cc,\,\text{aeration basin}}\,RT = \left(30.7\,\frac{L_L}{L_G}\right)$$

$$\left(0.082\,\frac{\text{atm-L}_G}{\text{mol}\,°\text{C}}\right)(293\,°\text{C}) = 738\,(\text{atm-L}_L)/\text{mol}$$

$c^*_{L,\text{in}}$ is the solution-phase oxygen concentration that would be in equilibrium with the composition of the injected gas. This value depends on the local pressure and therefore is different at different depths. An average value of $c^*_{L,\text{in}}$, which we represent as $\overline{c}^*_{L,\text{in}}$, can be computed based on the pressure at mid-depth (3.5 m). We can use Equation 6-6 to compute this pressure and then determine the concentration of interest:

$$P = P_{\text{surf}} + \rho_L g z$$

$$= 1.0\,\text{atm} + (998\,\text{kg/m}^3)(9.81\,\text{m/s}^2)(3.5\,\text{m})$$

$$\left(\frac{1\,\text{N}}{\text{kg m/s}^2}\right)\left(\frac{1\,\text{Pa}}{\text{N/m}^2}\right)\left(\frac{1\,\text{atm}}{10^5\,\text{Pa}}\right) = 1.34\,\text{atm}$$

$$P_{O_2,3.5\,\text{m}} = P_{\text{tot},3.5\,\text{m}}\,y_{O_2,\text{in}} = (1.34\,\text{atm})(0.21) = 0.28\,\text{atm}$$

$$\overline{c}^*_{L,\text{in}} = \frac{P_{O_2,3.5\,\text{m}}}{H_{O_2}} = \frac{0.28\,\text{atm}}{738\,\frac{\text{atm-L}_L}{\text{mol}}} = 3.79 \times 10^{-4}\,\frac{\text{mol}}{L_L}$$

$$= 3.79 \times 10^{-4}\,(\text{mol/L})(32000\,(\text{mg/mol})) = 12.1\,\text{mg/L}$$

The remaining parameter that we need to evaluate is the ratio $a_L/(Q_G/V_L)$. Noting that a_L equals A_{tot}/V_L, where A_{tot} is the total interfacial area in the tank at any instant, the ratio $a_L/(Q_G/V_L)$ can be written as follows:

$$\frac{a_L}{Q_G/V_L} = \frac{A_{\text{tot}}/V_L}{Q_G/V_L} = \frac{A_{\text{tot}}}{Q_G} = \frac{N_{bb}\,A_{bb}}{\dot{N}_{bb}\,V_{bb}} = \tau_{bb}\,\frac{6}{d_{bb}} \quad (6\text{-}52)$$

where N_{bb} is the number of bubbles in the tank at any instant, $\dot{N}_{bb}$ is rate at which bubbles are being injected, and τ_{bb} is bubble residence time. The average diameter of the bubbles (i.e., the diameter at mid-depth) in the aeration basin can then be computed from Equation 6-9:

$$d_{3.5\,\text{m}} = d_{7.0\,\text{m}}\left(\frac{P_{7.0\,\text{m}}}{P_{3.5\,\text{m}}}\right)^{1/3} = (0.40\,\text{cm})\left(\frac{1.70\,\text{atm}}{1.35\,\text{atm}}\right)^{1/3}$$

$$= 0.43\,\text{cm}$$

According to Figure 5-24, the velocity of 0.43-cm-diameter bubbles is ~22 cm/s, so the average residence time of bubbles in the water column is

$$\tau_{bb} = \frac{Z}{v_{bb}} = \frac{700\,\text{cm}}{22\,\text{cm/s}} = 32\,\text{s}$$

The ratio $a_L/(Q_G/V_L)$ can now be evaluated. Using the bubble diameter at mid-depth ($d_{3.5\,\text{m}}$) in conjunction with Equation 6-52, we find

$$\left(\frac{a_L}{Q_G/V_L}\right)_{3.5\,\text{m}} = \tau_{bb}\,\frac{6}{d_{bb,3.5\,\text{m}}} = (32\,\text{s})\frac{6}{0.43\,\text{cm}} = 447\,\text{s/cm}$$

Inserting all of the above values into Equation 6-51, we obtain

$$Q_{G,3.5\,\text{m}} = \frac{r_{L,\text{gt-avg}}\,V_L}{H\left[1 - \exp\left(-\frac{K_L a_L}{Q_G H/V_L}\right)_{3.5\,\text{m}}\right](\overline{c}^*_{L,\text{in}} - c_L)} = \frac{r_{L,\text{gt-avg}}\,V_L}{H\left[1 - \exp\left(-\frac{K_L}{H}\left(\frac{a_L}{Q_G/V_L}\right)_{3.5\,\text{m}}\right)\right](\overline{c}^*_{L,\text{in}} - c_L)}$$

$$= \frac{\left(0.00681\,\frac{\text{mg/L}_L}{\text{s}}\right)(3600\,\text{m}^3_L)}{30.7\,\frac{L_L}{L_G}\left[1 - \exp\left(-\frac{0.01\,\text{cm/s}}{30.7\,\frac{L_L}{L_G}}(447\,\text{s/cm})\right)\right]\left[(12.1 - 2.0)\,\frac{\text{mg}}{L_L}\right]\left(\frac{1\,\text{m}^3_L}{1000\,L_L}\right)\left(\frac{1000\,L_G}{1\,\text{m}^3_G}\right)}$$

$$= 0.58\,\text{m}^3_G/\text{s}$$

We can convert this volumetric flow at 3.5 m to the equivalent flow rate in SCM/s using a slightly modified form of Equation 6-5. Noting that we found the pressure at mid-depth in an earlier calculation to be 1.34 atm, we find

$$Q_{G,std} = Q_G \frac{P_{surf} + \rho_L g z}{P_{surf}} \frac{T_{std}}{T}$$

$$= \left(0.58 \frac{m_G^3}{s}\right)\left(\frac{1.34 \text{ atm}}{1.0 \text{ atm}}\right)\left(\frac{293°C}{293°C}\right) = 0.78 \text{ (SCM/s)}$$

For the specified gas flow rate per diffuser of 0.06 SCM/min, the required number of diffusers is

$$\text{Number of diffusers} = \frac{Q_{G,std}}{Q_{G,std} \text{ per diffuser}}$$

$$= \frac{(0.78 \text{ m}_G^3/\text{s})(60 \text{ s/min})}{0.06 \text{ m}_G^3/\text{min}} = 780$$

Thus, 780 diffusers would be required. Given the tank volume of 3600 m³ and depth of 7.0 m, the plan area of the tank is 514 m². This means that the spacing of the diffusers would be approximately one per 0.66 m². ∎

Case I CFSTRs in Series

Just like reactors in which homogeneous chemical reactions are occurring, the size of gas/liquid contactors required to achieve a given change in solution composition can sometimes be reduced dramatically by dividing the operation into two or more stages. If each reactor in the gas transfer system is consistent with the Case I assumptions, then the overall amount of transfer can be computed by applying Equation 6-44 to each reactor in the sequence. If the gas transfer objective is stripping of a contaminant into a gas that contains none of the contaminant initially, then Equation 6-44 describing the change in each reactor can be simplified as follows:

$$\frac{c_{L,in} - c_{L,out}}{c_{L,out} - c^*_{L,in}} = k_I \tau_L \qquad (6\text{-}53)$$

$$\frac{c_{L,in}}{c_{L,out}} - 1 = k_I \tau_L \qquad (6\text{-}54)$$

$$\frac{c_{L,out}}{c_{L,in}} = \frac{1}{1 + k_I \tau_L} \qquad (6\text{-}55)$$

The value of k_I in Equation 6-55 can be computed using the expressions in Table 6-3; if all the reactors are of equal size and are supplied with gas at equal rates, the values of k_I and τ_L are the same in each reactor, and the overall change in composition as water flows through the whole sequence of N reactors is given by an equation virtually identical to that for a sequence of CFSTRs in which a homogeneous reaction is occurring:

$$\frac{c_{L,out,N}}{c_{L,in,1}} = \left(1 + k_I \tau_{L,i}\right)^{-N} \qquad (6\text{-}56)$$

where $\tau_{L,i}$ is the hydraulic residence time in each reactor.

■ **EXAMPLE 6-7.** Compare the total hydraulic residence time that would be required to reduce the TCE concentration from 500 to 50 μg/L in a single CFSTR and in four equisized CFSTRs for the liquid and flow rates specified in Example 6-5. For the CFSTRs-in-series arrangement, assume that the gas flow is split equally among the four reactors.

Solution. The residence time required to accomplish the specified level of stripping in a single CFSTR can be computed with a modified form of Equation 6-55. The overall rate constant was computed in Example 6-4 to be 0.675 h⁻¹, so,

$$\tau_L = \frac{1}{k_I}\left(\frac{c_{L,in}}{c_{L,out}} - 1\right) = \frac{1}{0.675 \text{ h}^{-1}}\left(\frac{500 \text{ μg/L}}{50 \text{ μg/L}} - 1\right) = 13.3 \text{ h}$$

Because the value of Q_G/V_L is the same in each of the smaller reactors as in the overall reactor evaluated in Example 6-5, the value of k_I applicable to the overall reactor is also applicable to each of the smaller ones. We can therefore use Equation 6-56 to compute the TCE concentration exiting the sequence of smaller reactors, as follows:

$$\frac{c_{L,out,N}}{c_{L,in,1}} = \left(1 + k_I \tau_{L,i}\right)^{-N}$$

$$\left(\frac{c_{L,out,N}}{c_{L,in,1}}\right)^{-1/N} = 1 + k_I \tau_{L,i}$$

$$\tau_{L,i} = \frac{\left(\frac{c_{L,in,N}}{c_{L,out,1}}\right)^{1/N} - 1}{k_I}$$

$$= \frac{\left(\frac{500 \text{ μg/L}}{50 \text{ μg/L}}\right)^{1/4} - 1}{0.675 \text{ h}^{-1}} = 1.15 \text{ h}$$

The residence time in the sequence of four reactors is therefore 4 (1.15 h), or 4.6 h, which is only approximately one-third as long as in the single reactor that accomplishes the same amount of TCE stripping. ∎

Design Constraints and Choices for Case I Systems with Flow

To complete the analysis of Case I systems with flow, it is worthwhile to explore how the various concepts and equations derived in the preceding sections relate to design and analysis of specific types of systems. One approach for doing this is to list the parameters that might be specified arbitrarily for a given system, so that we can determine how many degrees of freedom we have in the design process. For instance, consider a Case I gas transfer system with continuous liquid flow designed to treat a given species (i.e., a species with a known Henry's constant). In most cases, the concentration of the volatile species in the liquid and gaseous influent streams would be known. In that case, the design and operating parameters that one could specify include the liquid and gas flow rates, the $K_L a_L$ value (by specifying the gas/liquid contacting scheme), and the volume of liquid in the system. Practical considerations might constrain these parameter values (especially $K_L a_L$), but in theory, any of them could take on any value without violating a fundamental equation (e.g., a mass balance or rate expression).

Once the six parameters identified above ($c_{L,in}$, $c_{G,in}$, Q_G, Q_L, $K_L a_L$, and V_L) were specified, the performance of the system with respect to gas transfer would be completely defined; that is, $c_{L,out}$, $c_{G,out}$, τ_L, $r_{L,gt\text{-}avg}$, and all other operating characteristics of the system would be established. Of course, we could just as well specify one or more of these latter four parameters in the design process, but for each one that was specified, one of the original six parameters would have to be eliminated from the specifications. The same could be said about specifying some parameter that is a combination of others (e.g., $Q_G H / V_L$). The same number of parameters can and must be specified to fully characterize a Case I system without liquid flow, except in that case τ_L, $c_{L,in}$ and $c_{L,out}$ would be replaced by the duration of gas/liquid contact (t) and the initial and final dissolved concentrations ($c_L(0)$ and $c_L(t)$), respectively.

Thus, a Case I gas transfer system is characterized by six degrees of freedom. Typically, in carrying out a design for a new treatment facility, we would know the liquid flow rate to be treated and the composition of the influent liquid and influent gas phases. We also would know the treatment goal, usually in terms of the required value of $c_{L,out}$ (or $c_L(t)$ for a batch system). In other cases, the treatment goal might be specified in terms of $r_{L,gt\text{-}avg}$, for example, in the case of a biological treatment process requiring a certain rate of oxygen transfer into solution. In any case, four of the six degrees of freedom are commonly established based on what the treatment process is supposed to accomplish. This leaves two degrees of freedom that are entirely under the designer's control. The choices available typically include specifying the aerator system to be used ($K_L a_L$), the hydraulic residence time (τ_L, which establishes the required liquid volume), and/or the gas flow rate; specifying any two of these three parameters establishes the value of the third one and constitutes a feasible design.

An important conclusion from this analysis is that, regardless of which parameters the designer chooses to specify, an infinite number of designs would, in theory, meet the treatment goals for a given water. The decision about which of these designs is best must therefore be made on the basis of criteria other than water quality. In almost all cases, the primary such criterion is cost. That is, the designer compares a variety of options that all meet the water quality goals and then chooses the one that is least costly. Several computer programs are available to assist in carrying out the relevant calculations.[5]

In making such a design decision, it is important to recognize that the design is likely to have impacts on downstream processes. For instance, use of a high gas flow rate might appear to be the least expensive way to strip a certain contaminant out of water. However, if the gas stream then has to be treated, the cost of that treatment is likely to increase as the gas flow rate increases. In such a case, the design that is best when considering the gas transfer process in isolation might not be best when the complete system is analyzed.

The linkages among treatment processes, not only in terms of cost but also water and air quality, sludge production, hazards to workers, and so on, and the impacts of design choices on those linkages, are critical considerations in the design process. They tend not to be emphasized in most textbooks (including this one), because the issues that arise are often site-specific, but their importance should not be overlooked.

6.3 CASE II: GAS TRANSFER IN SYSTEMS WITH SPATIAL VARIATIONS IN THE CONCENTRATIONS OF BOTH SOLUTION AND GAS

The Mass Balance Around a Section of a Gas Transfer Tower: The Operating Line

In many gas transfer systems, the concentrations of the volatile species in both the liquid and gas phases change significantly with location in the system. Here, we classify these systems as Case II systems. The most important examples of such systems are column contactors, which are typically operated with counter-current gas and liquid flow, although cocurrent flow and crosscurrent flow are occasionally used as well. In this section, we consider counter-current gas transfer in a column in which liquid

---

[5] See, for example, the ASAP® program (CenCITT, 1999).

enters at the top and gas is introduced at the bottom. The analysis applies equally regardless of whether water or gas is the continuous phase (so-called bubble columns or spray towers, respectively). In the latter case, the columns typically are packed with plastic media that water droplets strike and fall along in a thin film before forming new droplets and falling again.

The fundamental approach for designing or analyzing a gas transfer column is the same as for any other process: we write a generic mass balance and then incorporate constants or functions that specify the physical and chemical attributes of the particular system being considered. As is the case for the systems analyzed in Chapter 5 or in the preceding sections of this chapter, at any point in the system, the driving force for gas transfer equals $c_L^* - c_L$. However, when analyzing column contactors, the analysis must incorporate the fact that both c_L^* and c_L vary with location.

We begin the analysis by writing a mass balance on the constituent in both phases together, choosing a control volume that includes all the space in the reactor between the top and some arbitrary height z, as shown in Figure 6-8. We assume that the system is at steady state, that the constituent does not participate in any chemical reaction in the bulk liquid or bulk gas, and that both the liquid and gas flows can be approximated as being plug flow. Given these assumptions, the terms in the mass balance representing storage, diffusion, and dispersion, and reactions are all zero, so the only nonzero terms are those for advection into and out of the control volume.

Since the contaminant can enter and leave the control volume by advection in either phase, four advective terms must be included in the analysis, yielding the following mass balance:

$$\begin{bmatrix} \text{Rate of change of} \\ \text{mass of } i \text{ stored} \\ \text{in control volume} \end{bmatrix} = \begin{bmatrix} \text{Net (in - out)} \\ \text{liquid-phase advection} \\ \text{of } i \text{ into control volume} \end{bmatrix}$$
$$+ \begin{bmatrix} \text{Net (in - out)} \\ \text{gas-phase advection} \\ \text{of } i \text{ into control volume} \end{bmatrix}$$

$$0 = Q_L(c_L(Z) - c_L(z)) + Q_G(c_G(z) - c_G(Z)) \quad (6\text{-}57)$$

Solving the equation for $c_G(z)$, we obtain

$$c_G(z) = \frac{Q_L}{Q_G} c_L(z) + c_G(Z) - \frac{Q_L c_L(Z)}{Q_G} \quad (6\text{-}58)$$

On a plot with the liquid- and gas-phase concentrations as the abscissa and ordinate, respectively, Equation 6-58 corresponds to a straight line with slope Q_L/Q_G (Figure 6-9). This line is referred to as the *operating line* for the system.[6] At any height z in the column, the point $(c_L(z), c_G(z))$ alls on the operating line. The figure also includes a line characterizing the equilibrium relationship between the gas and liquid phases; that is, the Henry's law relationship, $c_G = Hc_L$.

In a system where gas is being stripped from solution, the operating line is always below and to the right of the equilibrium line, indicating that, at any height in the column, the aqueous phase concentration ($c_L(z_1)$) is greater than the value that would be in equilibrium with the gas at that point (i.e., $c_L(z_1) > c_L^*(z_1)$). Therefore, throughout the column, the volatile species transfers from solution to the gas phase. By similar reasoning, the operating line for a gas absorption system is always above and to the left of the equilibrium line, reflecting the fact that $c_L^*(z_1) > c_L(z_1)$, so the volatile species transfers into solution. The broken, horizontal lines show the difference between the actual and equilibrium concentrations of the dissolved volatile species at heights

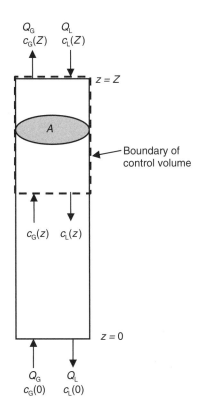

FIGURE 6-8. Definition diagram for a mass balance on the total volatile species (dissolved and gaseous) in the column between $z = Z$ and some arbitrary height z.

---

[6] In Equation 6.57, we have implicitly assumed that the volumetric flow rate of gas does not change with height, meaning that temperatures changes are small and that the molar rate of gas transfer is small compared to the molar flow rate of the carrier gas. This assumption is almost always valid for applications in environmental engineering. If it does not hold, the operating line is nonlinear.

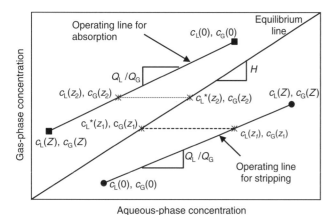

FIGURE 6-9. Equilibrium line and operating lines for gas transfer in a stripping and an absorption column. The broken lines at heights z_1 and z_2 indicate the driving force for stripping and absorption, respectively. The variable z represents the height measured from the bottom of the packing.

z_1 and z_2 and can therefore be identified as the driving force for gas transfer at those locations.[7]

For a stripping process, the concentrations characterizing the gas and the solution in the column move along the operating line from upper right to lower left as z decreases (moving down the column); in a gas absorption process, they move from lower left to upper right. Under most circumstances, the slope of the operating line, Q_L/Q_G, is subject to manipulation by the designer.

Note that the operating and equilibrium lines can never cross, because that would indicate that the two phases reached equilibrium (at the point of intersection), and that gas transfer continued as characterized by movement along the operating line. Such a situation is impossible, because once equilibrium is reached, the driving force for transfer disappears, and no further exchange can occur.

The ratio of the slope of the equilibrium line to the slope of the operating line is HQ_G/Q_L, the parameter combination defined earlier as the gas transfer capacity factor, R. Multiplying the numerator and denominator of this ratio by $c_{L,in}$, we obtain

$$R \equiv \frac{HQ_G}{Q_L} = \frac{HQ_G c_{L,in}}{Q_L c_{L,in}} = \frac{Q_G c^*_{G,out}}{Q_L c_{L,in}} \quad (6\text{-}59)$$

---

[7] It is also true that, at any height in the column, the extent of disequilibrium between the two phases could be expressed in terms of gas phase concentrations; that is, as $c^*_G(z_1) - c_G(z_1)$. When expressed in that way, the extent of disequilibrium can be computed as the vertical gap between the equilibrium line and the operating line. In that case, the overall gas transfer rate can be computed as $K_G a_R (c^*_G(z_1) - c_G(z_1))$, where K_G is the overall gas transfer rate constant based on gas phase concentrations. A derivation similar to that given for K_L in Chapter 5 shows that $K_G^{-1} = k_G^{-1} + (k_L/H)^{-1}$. This form of the rate expression is not emphasized here because our primary interest is in the solution phase.

(Note that $c_{L,in}$ refers to the top of the column, where the liquid enters. As a result, c^*_G at the top of the column equals $Hc_{L,in}$. Because gas exits at the top, this value of c^*_G is designated $c^*_{G,out}$.)

The denominator of the expression on the far right of Equation 6-59 ($Q_L c_{L,in}$) equals the mass flow rate of i into the column in the liquid phase, and the numerator ($Q_G c^*_{G,out}$) equals a hypothetical mass flow rate out of the column in the gas phase that would occur if the gas reached equilibrium with the incoming solution. Thus, for a stripping operation, the numerator is the maximum rate at which the gas phase can carry contaminant out of the column, for the given liquid-phase composition. Assuming that the incoming gas contains none of the contaminant, $Q_G c^*_{G,out}$ is also the maximum rate at which the gas phase can acquire contaminant from the liquid. If R is less than 1, then the gas is incapable of transporting all the contaminant out of the reactor, even if the gas acquires the maximum amount of contaminant it can (limited only by equilibrium). In that case, the maximum contaminant removal efficiency that could be achieved (i.e., the efficiency corresponding to equilibrium between the exiting gas and the incoming solution) would equal R.

If $R \geq 1$, then the gas has the potential to transport all the contaminant out of the reactor, and it might do so (or at least approach that limit) if the interfacial transport resistance were overcome. Put another way, for a stripping operation, if $R \geq 1$ and if gas transfer kinetic limitations were overcome, all of the contaminant would enter and leave the column at the top; if $R < 1$, some of the contaminant would leave the column at the bottom (in the solution), no matter how small the kinetic limitations to gas transfer became. Values of R greater than 1 are virtually always used in design.

By carrying out a similar manipulation as in Equation 6-59, but multiplying numerator and denominator by $c^*_{L,out}$, an analogous argument can be made for a gas absorption process. In this case, the interpretation of the conditions when $R < 1$ is that the incoming gas would be unable to supply enough of the volatile constituent to solution for the solution to equilibrate with $c_{G,in}$, even in the absence of interfacial resistance; that is, even if all of the volatile constituent were transferred from the gas to solution before the gas exited the reactor. Under these conditions, however, the liquid would be able to absorb essentially 100% of the volatile species from the incoming gas. As a result, if kinetic limitations to gas transfer were overcome, all of the transferable species would enter and leave the column at the bottom.

On the other hand, in an absorption column with $R \geq 1$, the liquid exiting the bottom of the column could (if interfacial resistance were small enough) reach equilibrium with the incoming gas, and some of the volatile species would exit with the gas at the top of the column. If the goal of the absorption process is primarily to increase the concentration of the volatile species in solution (e.g., oxygen transfer), then R values > 1

should be used, but if the goal is to transfer essentially all of the volatile species from the gas phase to solution (e.g., ozone transfer), then R values <1 are more appropriate.[8]

Thus, the gas transfer capacity factor provides a reference point for assessing the sufficiency of the gas flow rate for accomplishing the treatment goal. Although the parameter combination corresponding to R also appears in the equations characterizing Case I systems with continuous flow (Equation 6-50), in that case the term has only an abstract design application because of the uniformity of the liquid phase composition throughout the reactor. That is, since Case I systems have the same value of c_L throughout the reactor, it is not possible to strip almost all of a contaminant from the solution (so that $c_L \approx 0$) and simultaneously have the exiting gas approach equilibrium with the incoming solution (so that $c_L \approx c_{L,in}$). By contrast, in a counter-current column contactor used for stripping, the liquid phase composition truly can vary from $c_{L,in}$ to nearly zero in the reactor, and the liquid enters and leaves at opposite locations from where the gas does. These characteristics of counter-current Case II reactors allow the possibility that the incoming gas will approach equilibrium with the exiting solution, while the exiting gas is near equilibrium with the incoming solution. As a result, R has more practical significance and is a more central design parameter in Case II systems.

Recall that, for Case I systems with continuous liquid flow, the relative importance of microscopic (interfacial or kinetic) versus macroscopic (advective or equilibrium) limitations to gas transfer is indicated by the ratio $\tau_L K_L a_L/((Q_G/Q_L)H)$. The larger this ratio is, the more strongly equilibrium limits the rate of gas transfer into solution.

In Case II systems (which, by definition, always have steady liquid flow), the same ratio could be used to evaluate the relative importance of the two potentially rate-limiting factors. The denominator of this ratio is R, so the ratio can be expressed as $\tau_L K_L a_L/R$. The value of τ_L is often difficult to measure in Case II systems where the liquid is not the continuous phase, making it difficult to quantify the term $\tau_L K_L a_L/R$. Nevertheless, it is clear that, other factors being equal, an increase in R lessens the likelihood that gas transfer will be equilibrium-limited.

More importantly, the value of R places a bound on the maximum change in the liquid phase concentration. As noted previously, in a stripping process, the limit is that,

if $R < 1$, then R is the maximum fractional removal of the contaminant from the solution, and in an absorption process, if $R \geq 1$, then $1/R$ is the maximum fractional absorption of the species from the gas. How closely the change in the liquid composition approaches this maximum value depends on $K_L a_L \tau_L$; if $K_L a_L \tau_L$ is large enough, the change in the liquid composition will approximate the maximum possible change for the given R, whereas if $K_L a_L \tau_L$ is small, the amount of gas transfer will be less than that.

The Mass Balance Around a Differential Section of a Gas Transfer Tower: Development of the Design Equation for Case II Systems

Although the operating line for a gas transfer column contactor describes the overall changes in the composition of the phases as they pass through the system, we have not yet connected this information with the physical characteristics of the system. That is, for a given Q_L and Q_G, and given influent and effluent concentrations $c_L(Z)$, $c_G(Z)$, $c_L(0)$, and $c_G(0)$, we can determine the intermediate concentrations that the system must pass through at different points in the column, but we cannot yet answer key design questions, such as: How tall will the column have to be to achieve the desired amount of gas transfer? What will the required hydraulic and gas residence times be? And, how much will the gas transfer efficiency change for a given change in the gas flow rate? To answer these questions, we need to write another mass balance, using a control volume that includes only the solution phase in a region defined by the cross-section of the column and a differential length dz (Figure 6-10).

The mass balance on the volatile species dissolved in the liquid phase in this control volume can be written as follows, where dV_L is the differential volume of water in the control volume:

$$\begin{pmatrix}\text{Rate of change of the}\\ \text{mass of } i \text{ stored in the}\\ \text{solution phase in the}\\ \text{volume } A_R\, \mathrm{d}z\end{pmatrix} = \begin{pmatrix}\text{Net (in} - \text{out)}\\ \text{liquid-phase}\\ \text{advection of } i \text{ into}\\ \text{control volume}\end{pmatrix}$$
$$+ \begin{pmatrix}\text{Amount of } i \text{ entering}\\ \text{solution phase within}\\ \text{the control volume by}\\ \text{gas transfer}\end{pmatrix} \quad (6\text{-}60)$$

$$\frac{\partial}{\partial t}(c_L \mathrm{d}V_L) = Q_L[(c_L + \mathrm{d}c_L) - c_L] + r_{L,gt}\mathrm{d}V_L$$
$$= Q_L[(c_L + \mathrm{d}c_L) - c_L] + K_L a_L(c_L^* - c_L)\mathrm{d}V_L \quad (6\text{-}61)$$

The substitution of $K_L a_L(c_L^* - c_L)$ for $r_{L,gt}$ is justified by the fact that the liquid volume under consideration is differentially small, so the approximation that gas transfer

---

[8] The discussion here assumes that the incoming solution contains none of the constituent of interest. If the incoming solution already contains some of the volatile species, then less of the species would have to be transferred to reach equilibrium with the influent gas, and it would be possible for the exiting liquid to reach equilibrium with the incoming gas at R values <1. However, in this case, the absorption efficiency from the gas phase would always be $<100\%$, no matter how small R became. The analogous situation for stripping is that the incoming gas already contains some of the volatile species. In that case, an R value <1 would allow equilibrium between the incoming gas and exiting liquid, but some contaminant would remain in the effluent solution, no matter how large R became.

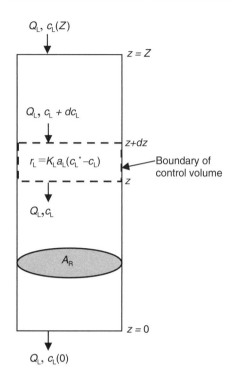

FIGURE 6-10. Definition diagram for a mass balance on the constituent of interest in the aqueous phase located in a zone of height dz in the reactor.

is limited by microscopic (interfacial) processes must apply (recall that the criterion for this limitation is that $K_L a \ll (Q_G H/V_L)$).

The interfacial area in the control volume is dispersed among many droplets and thin films if the reactor is a packed column, and among many bubbles if the reactor is a bubble column. In addition, the volume of water in the control volume is less (in the case of a packed column, much less) than the total volume of the reactor. Under the circumstances, it is convenient to normalize the interfacial area to the volume of the reactor rather than the liquid alone. To do that, we substitute $a_R V_R/V_L$ for a_L in Equation 6-61, with the implicit assumption that the liquid occupies a constant fraction of the reactor volume throughout the column (i.e., $dV_L/dV_R = V_L/V_R$). Then, noting that $(V_R/V_L)dV_L$ equals dV_R, we find.

$$\frac{\partial}{\partial t}(c_L dV_L) = Q_L[(c_L + dc_L) - c_L] + K_L a_R \left(c_L^* - c_L\right) \frac{V_R}{V_L} dV_L$$

$$= Q_L dc_L + K_L a_R \left(c_L^* - c_L\right) dV_R$$

$$= Q_L dc_L + K_L a_R \left(c_L^* - c_L\right) A_R dz \quad (6\text{-}62)$$

If the column is operating at steady state, the left side of Equation 6-62 is zero. Making that substitution and carrying out some algebra to isolate the concentration terms on the left side of the equation, we obtain

$$0 = Q_L dc_L + K_L a_R \left(c_L^* - c_L\right)(A_R dz) \quad (6\text{-}63)$$

$$\int_{c_L(0)}^{c_L(z)} \frac{dc_L}{c_L^* - c_L} = -\frac{K_L a_R}{Q_L/A_R} \int_0^z dz \quad (6\text{-}64)$$

$$\int_{c_L(0)}^{c_L(z)} \frac{dc_L}{c_L^* - c_L} = -\frac{K_L a_R}{Q_L/A_R} z \quad (6\text{-}65)$$

Equation 6-65 is one form of the design equation we sought, albeit not the most convenient form. More useful design equations can be developed by integrating the left side of the equation, as shown shortly. Nevertheless, the equation does relate factors that characterize the physical system (Q_L, A_R, z, a_R), chemical equilibrium (c_L^*), and gas transfer kinetics (K_L) to the change in the concentration of the dissolved species in the reactor, and therefore is an important result.[9,10]

---

[9] In addition to gas transfer, many other environmental engineering processes are carried out in columns (e.g., many types of filtration, adsorption, and biological treatment processes), and an equation similar to Equation 6.65 applies for any of these. That is, a mass balance similar to Equation 6.59 written for any kinetic process occurring in a column at steady state would yield:

$$(\text{Flow rate})\begin{pmatrix} \text{Change in} \\ \text{concentration} \end{pmatrix} = \begin{pmatrix} \text{Rate of removal} \\ \text{by reaction} \end{pmatrix} \begin{pmatrix} \text{Volume of} \\ \text{control section} \end{pmatrix}$$

$$(Q_L)(dc_L) = (r_L)(A_R dz)$$

Combining the mass balance with a rate expression and integrating would yield:Combining the mass balance with a rate expression and integrating would yield:

$$(\text{constant})(Z) = \int \frac{dc_L}{\text{driving force for reaction}}$$

where the constant includes the area of the column and constant terms that appear in the reaction rate expression. The driving force for the reaction is often, but not necessarily, the difference between an instantaneous and an equilibrium condition. However, regardless of whether this is the case, if we know the form of the function in the denominator, we can evaluate the integral numerically and solve for Z.

[10] It might seem odd that the gas flow rate, Q_G, does not appear in the design equation explicitly. However, that term is present implicitly, since it affects the slope of the operating line, and therefore the values of c_L and c_L^* as a function of z. As a result, changing Q_G changes the value of the integral on the left side of the equation where the constant includes the area of the column and constant terms that appear in the reaction rate expression. The driving force for the reaction is often, but not necessarily, the difference between an instantaneous and an equilibrium condition. However, regardless of whether this is the case, if we know the form of the function in the denominator, we can evaluate the integral numerically and solve for Z. It might seem odd that the gas flow rate, Q_G, does not appear in the design equation explicitly. However, that term is present implicitly, since it affects the slope of the operating line, and therefore the values of c_L and c_L^* as a function of z. As a result, changing Q_G changes the value of the integral on the left side of the equation.

CASE II: GAS TRANSFER IN SYSTEMS WITH SPATIAL VARIATIONS IN THE CONCENTRATIONS OF BOTH SOLUTION AND GAS

Because both c_L and c_L^* are variables that change with location in the column, the integration of an equation such as Equation 6-65 is, in general, nontrivial. However, the fact that the operating and equilibrium lines are linear in the present case means that the difference $c_L^* - c_L$ is a linear function of c_L, making it possible to solve the integral on the left side of the equation explicitly.

To carry out the integration, it is convenient to write the extent of disequilibrium, $c_L^* - c_L$, as $\hat{c}_L$. Making that substitution in Equation 6-65 and dividing and multiplying the integrand by $d\hat{c}_L$ yields

$$\frac{K_L a_R}{Q_L/A_R} z = -\int_{c_L(0)}^{c_L(z)} \frac{dc_L}{d\hat{c}_L} \frac{d\hat{c}_L}{\hat{c}_L} \tag{6-66}$$

We can relate the fraction $dc_L/d\hat{c}_L$ to the gas transfer capacity factor, R, by recalling that R is the ratio of the slopes of the equilibrium and operating lines for the process. The slope of the operating line is dc_G/dc_L, and the slope of the equilibrium line is H. For the current derivation, it is convenient to express H in terms of differentials, as follows:

$$\begin{aligned} c_G &= H c_L^* \\ dc_G &= H dc_L^* \\ H &= \frac{dc_G}{dc_L^*} \end{aligned} \tag{6-67}$$

Thus,

$$R = \frac{H}{\text{Slope of operating line}} = \frac{dc_G/dc_L^*}{dc_G/dc_L} = \frac{dc_L}{dc_L^*} \tag{6-68}$$

$$\frac{R}{R-1} = \frac{dc_L/dc_L^*}{dc_L/dc_L^* - 1} = \frac{dc_L}{dc_L - dc_L^*} = -\frac{dc_L}{d\hat{c}_L} \tag{6-69}$$

Substituting this expression into Equation 6-66, we obtain

$$\begin{aligned} \frac{K_L a_R}{Q_L/A_R} z &= -\int_{\hat{c}_L(0)}^{\hat{c}_L(z)} -\frac{R}{R-1} \frac{d\hat{c}_L}{\hat{c}_L} \\ &= \int_{\hat{c}_L(0)}^{\hat{c}_L(z)} \frac{R}{R-1} \frac{d\hat{c}_L}{\hat{c}_L} \end{aligned} \tag{6-70}$$

Note that the limits of integration have been rewritten in terms of $\hat{c}_L$, since the integration is now being carried out with respect to that variable instead of c_L.

For a given system (given Q_L and Q_G), the gas transfer capacity factor is constant, so the first term can be taken out of the integral. The integration can then be carried out to yield the following expressions describing the extent of disequilibrium as a function of height in the column and other system variables.[11]

$$\frac{K_L a_R}{Q_L/A_R} z = \frac{R}{R-1} \ln \frac{\hat{c}_L(z)}{\hat{c}_L(0)} \tag{6-71}$$

$$z = \frac{Q_L/A_R}{K_L a_R} \frac{R}{R-1} \ln \frac{\hat{c}_L(z)}{\hat{c}_L(0)} \tag{6-72}$$

$$\hat{c}_L(z) = \hat{c}_L(0) \exp\left(\frac{K_L a_R}{Q_L/A_R} \frac{R-1}{R} z\right) \tag{6-73a}$$

$$\hat{c}_L(0) = \hat{c}_L(z) \exp\left(-\frac{K_L a_R}{Q_L/A_R} \frac{R-1}{R} z\right) \tag{6-73b}$$

Equation 6-73b indicates that the extent of disequilibrium decays exponentially from the top to the bottom of the column; note that this decay reflects changes in both c_L and c_G, so it is generally *not* the case that the liquid phase concentration declines exponentially as the water falls through the packed bed. Evaluating Equations 6-72 and 6-73b at $z = Z$, substituting the designations bot for $z = 0$ and top for $z = Z$, and replacing the $\hat{c}_L$ terms with $c_L^* - c_L$, we obtain:

$$Z = \frac{Q_L/A_R}{K_L a_R} \frac{R}{R-1} \ln \frac{c_{L,\text{top}}^* - c_{L,\text{top}}}{c_{L,\text{bot}}^* - c_{L,\text{bot}}} \tag{6-74a}$$

$$= \frac{Q_L/A_R}{K_L a_R} \frac{R}{R-1} \ln \frac{\dfrac{c_{G,\text{top}}}{H} - c_{L,\text{top}}}{\dfrac{c_{G,\text{bot}}}{H} - c_{L,\text{bot}}} \tag{6-74b}$$

$$c_{L,\text{bot}}^* - c_{L,\text{bot}} = \left(c_{L,\text{top}}^* - c_{L,\text{top}}\right) \exp\left(-\frac{K_L a_R}{Q_L/A_R} \frac{R-1}{R} Z\right) \tag{6-75}$$

Using Equation 6-58 (the mass balance that was used to generate the operating line) to substitute for $c_{G,\text{top}}$ in the logarithmic argument in Equation 6-74b, we obtain:

$$\frac{\dfrac{c_{G,\text{top}}}{H} - c_{L,\text{top}}}{\dfrac{c_{G,\text{bot}}}{H} - c_{L,\text{bot}}} = \frac{\dfrac{c_{G,\text{bot}} + \dfrac{Q_L}{Q_G}\left(c_{L,\text{top}} - c_{L,\text{bot}}\right)}{H} - c_{L,\text{top}}}{\dfrac{c_{G,\text{bot}}}{H} - c_{L,\text{bot}}}$$

$$= \frac{\dfrac{c_{G,\text{bot}}}{H} + \dfrac{1}{R}\left(c_{L,\text{top}} - c_{L,\text{bot}}\right) - c_{L,\text{top}}}{\dfrac{c_{G,\text{bot}}}{H} - c_{L,\text{bot}}}$$

---

[11] Formally, the logarithmic arguments in Equation 6.71 should be written as the ratio of the absolute values of the gas transfer driving force at 0 and z, to avoid the impression that one might need to take the logarithm of a negative number. However, since $c_L^* - c_L$ always has the same sign throughout the column, the ratio is always positive (even if the driving force is negative), and the absolute value operation can be ignored.

Subtracting and adding $c_{L,bot}$ in the numerator and carrying out some algebra, we find

$$\frac{\frac{c_{G,top}}{H} - c_{L,top}}{\frac{c_{G,bot}}{H} - c_{L,bot}} = \frac{\frac{c_{G,bot}}{H} - c_{L,bot} + \frac{1}{R}(c_{L,top} - c_{L,bot}) - c_{L,top} + c_{L,bot}}{\frac{c_{G,bot}}{H} - c_{L,bot}}$$

$$= 1 + \frac{\frac{1}{R}(c_{L,top} - c_{L,bot}) - (c_{L,top} - c_{L,bot})}{\frac{c_{G,bot}}{H} - c_{L,bot}}$$

$$= 1 + \left(\frac{1}{R} - 1\right) \frac{c_{L,top} - c_{L,bot}}{\frac{c_{G,bot}}{H} - c_{L,bot}}$$

Finally, because both $1/R - 1$ and $c_{G,bot}/H - c_{L,bot}$ are typically less than zero in stripping columns, it is convenient to multiply both terms by -1, yielding:

$$\frac{\frac{c_{G,top}}{H} - c_{L,top}}{\frac{c_{G,bot}}{H} - c_{L,bot}} = 1 + \frac{R-1}{R} \frac{c_{L,top} - c_{L,bot}}{c_{L,bot} - \frac{c_{G,bot}}{H}} \quad (6\text{-}76)$$

Substituting Equation 6-76 into Equation 6-74b, we obtain the following design equation for the required height of a column to achieve a desired change in liquid phase concentration, for a known influent gas phase composition and known or adjustable operating parameters:

$$Z = \frac{Q_L/A_R}{K_L a_R} \left\{ \frac{R}{R-1} \ln\left(1 + \frac{R-1}{R} \frac{c_{L,top} - c_{L,bot}}{c_{L,bot} - \frac{c_{G,bot}}{H}}\right) \right\} \quad (6\text{-}77)$$

An alternative form of Equation 6-77 that is often cited is

$$Z = \frac{Q_L/A_R}{K_L a_R} \left\{ \frac{R}{R-1} \ln\left(\frac{1}{R} + \frac{R-1}{R} \frac{c_{L,top} - \frac{c_{G,bot}}{H}}{c_{L,bot} - \frac{c_{G,bot}}{H}}\right) \right\} \quad (6\text{-}78)$$

The right side of Equation 6-77 is sometimes split into two terms, as indicated by the expressions preceding and inside the brackets. The first term $((Q_L/A_R)/K_L a_R)$ has dimensions of length and is called the *height of a (gas) transfer unit* (HTU). The HTU can be thought of as a characteristic length associated with gas transfer. The remainder of the right side is dimensionless and is referred to as the *number of (gas) transfer units* (NTU). Since NTU depends explicitly on the equilibrium constant for gas transfer, but not on kinetic terms, it is thought of as incorporating the equilibrium limitations on gas transfer. The larger the value of NTU, the more significant is the equilibrium limitation on gas transfer. On the other hand, HTU depends on $K_L a_R$ but not directly on H. (If the gas phase resistance is significant, then HTU depends indirectly on H, since in that case, K_L depends on H.) Hence, if the resistance to gas transfer resides predominantly in the liquid phase, HTU can be thought of as an indicator of the kinetic resistance to gas transfer: for a given liquid loading rate, a larger value of HTU indicates a larger kinetic limitation to gas transfer.

Equation 6-75 can be manipulated in a similar way to yield the following equation for $c_{L,bot}$ as a function of the influent concentrations and Z:

$$c_{L,bot} = \frac{\left(c_{L,top} - \frac{c_{G,bot}}{H}\right)\frac{R-1}{R}}{\exp\left\{\frac{K_L a_R}{Q_L/A_R} \frac{R-1}{R} Z\right\} - \frac{1}{R}} + \frac{c_{G,bot}}{H} \quad (6\text{-}79)$$

Equations 6-77 and 6-79 are alternative versions of the design equation for Case II systems, in forms that are often more directly useful than Equations 6-74 and 6-75.[12] In cases where the influent gas contains none of the contaminant being stripped, Equation 6-79 can be simplified to yield the following expressions for the concentration of the dissolved volatile species at the bottom of the column, the removal efficiency (η) accomplished in the column, and the column height required to achieve a given fractional removal by stripping:

$$c_{L,bot} = \frac{c_{L,top} \frac{R-1}{R}}{\exp\left\{\frac{K_L a_R}{Q_L/A_R} \frac{R-1}{R} Z\right\} - \frac{1}{R}} \quad (6\text{-}80)$$

$$\eta \equiv 1 - \frac{c_{L,bot}}{c_{L,top}} = \frac{\exp\left\{\frac{K_L a_R}{Q_L/A_R}\left(\frac{R-1}{R}\right)Z\right\} - 1}{\exp\left\{\frac{K_L a_R}{Q_L/A_R}\left(\frac{R-1}{R}\right)Z\right\} - \frac{1}{R}}$$

$$= 1 - \frac{\frac{R-1}{R}}{\exp\left\{\frac{K_L a_R}{Q_L/A_R}\left(\frac{R-1}{R}\right)Z\right\} - \frac{1}{R}} \quad (6\text{-}81)$$

$$Z = \frac{Q_L/A_R}{K_L a_R} \frac{R}{R-1} \ln \frac{\frac{c_{L,top}}{c_{L,bot}}(R-1) + 1}{R} \quad (6\text{-}82)$$

---

[12] Several of the equations in the preceding derivation cannot be solved if $R = 1$, because the right hand sides of the equations reduce to an undefined value (0/0). In that case, the relationship between the amount of gas transfer and column height can be derived by recognizing that $R = 1$ means that the operating and equilibrium lines are parallel. As a result, the difference $c_L^* - c_L$ is the same everywhere in the column and can be taken out of the integral in Equation 6.70. When that is done, the integral can be solved directly, yielding the result:

$$Z = \frac{Q_L/A}{K_L a_R} \frac{c_{L,top} - c_{L,bot}}{c_{L,bot} - \frac{c_{G,bot}}{H}}$$

Note that, according to Equation 6-81, if the influent gas is devoid of the volatile species, the fraction of the contaminant stripped from solution is independent of the influent concentration.

■ **EXAMPLE 6-8.** Narbaitz et al. (2002) evaluated the stripping efficiency for several different contaminants in pilot-scale towers packed with various media. For 2-in. Nor-Pac media, in columns receiving a liquid areal loading rate of 5.6 L/m² s, they reported the $K_L a_R$ values shown in the following table for three contaminants at three different ratios of the air/water flow rates. The Henry's constants for the contaminants at the temperature of the experiments (~25°C) are also shown in the table.

(a) What depth of packing is required to achieve at least 98% removal of all three contaminants in a single column contactor at each air/water flow ratio?

(b) The input table indicates that $K_L a_R$ for chlorobenzene is insensitive to the Q_G/Q_L ratio. For a value of $K_L a_R = 18.5\,h^{-1}$, predict the stripping efficiency for chlorobenzene for a fixed gas transfer capacity factor (R) of 2.3 for the range of column heights $2.5 < Z < 10$ m, and also for a fixed column height of 3.5 m over the range $0.2 < R < 10$.

		$K_L a_R$ (h^{-1})		
		Volumetric Air/Water Ratio (Q_G/Q_L)		
Contaminant	H_{cc} (L$_L$/L$_G$)	25	50	100
Chlorobenzene	0.133	18.1	18.9	18.0
m-Xylene	0.282	19.0	19.2	21.1
Toluene	0.261	13.3	15.4	17.6

Solution.

(a) The required depth of packing for each contaminant can be computed using Equation 6-82, once the value of the gas transfer capacity factor R is evaluated for each condition of interest. The values of R are computed as the product of the air/water ratio (Q_G/Q_L) and the Henry's constant, yielding the following:

	R		
	Air/Water Ratio (Q_G/Q_L)		
Contaminant	25	50	100
Chlorobenzene	3.33	6.65	13.3
m-Xylene	7.05	14.1	28.2
Toluene	6.52	13.0	26.1

The required, 98% removal corresponds to a $c_{L,top}/c_{L,bot}$ ratio of 100/2 = 50. Therefore, for example, the packing depth required for chlorobenzene removal at an air/water ratio of 25 is

$$Z = \frac{Q_L/A_R}{K_L a_R} \frac{R}{R-1} \ln \frac{\frac{c_{L,top}}{c_{L,bot}}(R-1)+1}{R}$$

$$= \left(\frac{5.6\,\text{L/m}^2\,\text{s}}{18.1/\text{h}}\right)\left(\frac{3600\,\text{s}}{\text{h}}\right)\left(\frac{1\,\text{m}^3}{1000\,\text{L}}\right)\left(\frac{3.33}{3.33-1}\right)$$

$$\times \ln \frac{50(3.33-1)+1}{3.33} = 5.67\,\text{m}$$

Similar calculations for the other contaminants and other air/water ratios of interest yield the following values. (Note that the calculations can be carried out independently, since the constituents are assumed to have no interactions with one another.) The results indicate that the packing depth is controlled by the requirement for toluene stripping.

	Required Z (m)		
	Air/Water Ratio		
Contaminant	25	50	100
Chlorobenzene	5.67	5.72	6.14
m-Xylene	5.71	5.76	5.29
Toluene	8.12	7.17	6.34

(b) Both parts of this question can be answered by inserting appropriate values into Equation 6-81. At a fixed value of $R = 2.3$, the stripping efficiency is relatively high (>82%), even for a very short column (2.5 m); as the column height increases, the stripping efficiency does as well, exceeding 98% at $Z > 6.5$ m (Figure 6-11a). For a fixed height of 3.5 m, the efficiency is quite sensitive to R at low values but insensitive at high values. At $R < 1$, the maximum achievable efficiency equals R, and this efficiency is reached at R values up to approximately 0.6. At higher R, the efficiency continues to increase, but less rapidly, reaching a plateau of 93–95% at $R > 4$ (Figure 6-11b). The shape of this curve is typical, which is why R values greater than ~5 are rarely used. ■

Pressure Loss and Liquid Holdup

To this point, the discussion of gas transfer in towers has focused exclusively on the transfer of the volatile species between the solution and gas phases. However, in practice, the energy requirements for generating the flows of the two fluids represent an important component of the design. Furthermore, the energy requirements depend on several of the same parameters that control the gas transfer

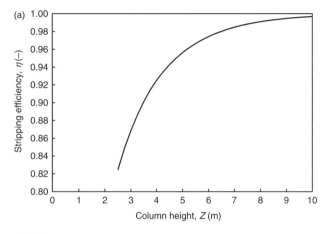

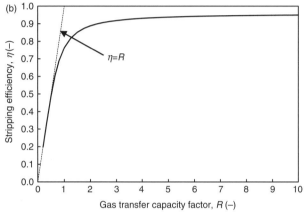

FIGURE 6-11. Response of the chloroform stripping efficiency to changes in (a) column height and (b) the gas transfer capacity factor, for the conditions specified in the example system.

efficiency, so the design decisions for these two aspects of process performance are linked. In this section, we consider the factors that control the pressure loss associated with gas flow through a column and explore the linkage between those factors and treatment efficiency.

The pressure gradient in a packed column (i.e., the pressure drop per unit column length) depends on the dimensions of the column, the liquid and gas flow rates, and the packing material. Historically, this gradient has been computed according to empirical correlations developed by Eckert (1961, 1975) and later modified and presented by Treybal (1980) in the form of a graph like that shown in Figure 6-12. A numerical approximation that mimics Eckert's correlations has been reported by Hand et al. (2011).

The value of C_f that appears in the ordinate of the graph is called the column packing factor and is a property of the type of packing used. Values of C_f for many types of packing are available in the literature (see, for example, Treybal (1980) or Perry and Green (1984)) or from the manufacturer,

and are generally in the range 10–50 for the plastic media used most often in environmental engineering applications. Although these values are typically reported as pure numbers, C_f actually has whatever units are required to make the fraction shown as the ordinate dimensionless; as a result, the value of C_f depends on the units used to express the other parameters.

As indicated in the figure, an upper limit for the pressure drop per unit column length is imposed by the need to avoid excessive hold-up of water in the column, which ultimately leads to column flooding. However, practical designs for stripping columns used in water and wastewater treatment invariably have pressure gradients that are far below 100 Pa/m, so the potential for flooding is not a concern. Flooding is more relevant in gas absorption processes, because in those systems, it is often desirable to maximize Q_G/A_R.

■ **EXAMPLE 6-9.** A maximum pressure gradient of 50 Pa/m has been set as a criterion for the treatment system described in Example 6-8. Which, if any, of the three air/liquid flow ratios explored in that example meet the criterion? The packing factor for the media is 21 when all the other parameters are expressed in SI units. The densities of air and water and the viscosity of water at 25°C are: $\rho_L = 997$ kg/m³; $\rho_G = 1.18$ kg/m³; and $\mu_L = 9.07 \times 10^{-4}$ kg/m s.

Solution. By substituting the three values of interest for Q_G/Q_L, we can compute the value of the abscissa of Figure 6-12 for each condition. Thus, for example, the abscissa for an air/water ratio of 25 is

$$\frac{Q_L}{Q_G}\frac{\rho_L}{\rho_G}\left(\frac{\rho_G}{\rho_L-\rho_G}\right)^{0.5} = \left(\frac{1}{25}\right)\left(\frac{997}{1.18}\right)\left(\frac{1.18}{997-1.18}\right)^{0.5} = 1.16$$

We can do the same to compute the ordinate, except that we first need to develop an expression for the gas-phase, areal loading rate based on other, known values. We can do that as follows:

$$\frac{Q_G}{A_R} = \frac{Q_G}{Q_L}\frac{Q_L}{A_R} = \frac{Q_G}{Q_L}\left(5.6\frac{L}{m^2\,s}\right) = \frac{Q_G}{Q_L}\left(0.0056\frac{m^3}{m^2\,s}\right)$$

Substituting the above expression into the expression for the ordinate, we calculate the ordinate for an air/water ratio of 25 as

$$\frac{(Q_G/A_R)^2 \rho_G C_f \mu_L^{0.1}}{\rho_L - \rho_G} = \frac{[(25)(0.0056)]^2(1.18)(21)(9.07\times 10^{-4})^{0.1}}{997-1.18}$$

$$= 2.42 \times 10^{-4}$$

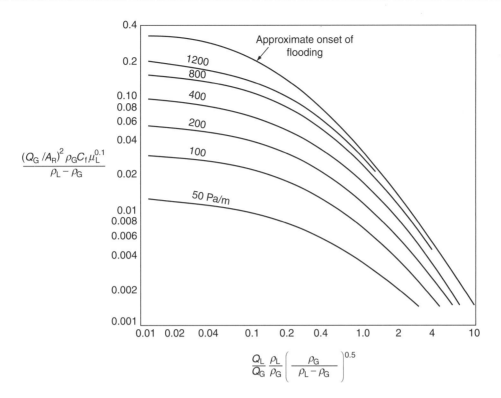

FIGURE 6-12. Pressure drop correlations for gas flow through packed columns. The label on each curve indicates the pressure gradient in the column. *Source*: After Treybal (1980).

The calculations for all three air/water ratios are summarized below:

Air/Water Ratio	Abscissa	Ordinate
25	1.16	2.42×10^{-4}
50	0.58	9.68×10^{-4}
100	0.29	3.87×10^{-3}

Locating these values in Figure 6-12, we see that the pressure gradients in the column would be well below 50 Pa/m at all the air/water ratios investigated. ∎

The pressure gradient in the column is potentially important for two aspects of gas transfer column design. First, it is relevant with respect to the technical feasibility of the design, in that it must be kept below the flooding threshold. As noted above, this consideration might be important for gas absorption systems, but is almost never relevant for gas stripping. Second, the product of the pressure gradient and the column height yields the overall pressure loss experienced by the gas as it passes through the column, which, although it does not bear on the technical ability to achieve the treatment goal, does bear on the economics of the process.

Although the correlations shown in Figure 6-12 are very commonly cited in discussions of stripping column design in the environmental engineering literature, the figure should not be relied upon to estimate the pressure gradient to closer than a factor of ~2. This caveat is especially relevant if the correlations have to be extrapolated to pressure gradients significantly less than 50 Pa/m, as is often the case in the systems of interest. In such cases, correlations provided by the manufacturer of the specific packing material of interest, or better yet, from pilot or full-scale systems under similar operating conditions, should be used.

Despite the limitations cited in the preceding paragraph, Figure 6-12 continues to be used to estimate pressure gradients in stripping columns, at least in preliminary design efforts. The reason seems to be that, at the pressure gradients of 50 Pa/m or less that typically characterize such systems, the energy cost of pumping water to the top of the column usually dominates over that for supplying the specified gas flow. As a result, even significant deviations from the predicted pressure gradient tend to have relatively little impact on the overall operational cost of the system. Thus, the main purpose of calculating the expected pressure gradient in a column is usually not to use that value in critical calculations related to the performance or cost of the system, but rather to assure that the design choices fall in a range where the pressure differential is not particularly important. This point is expanded upon in the following section, which outlines the overall design procedure for a packed-media stripping column.

Use of the Design Equation for Case II Systems

As for Case I systems, the design process for Case II systems allows a number of parameters to be chosen more or less arbitrarily while still achieving the design goals; that is, the systems have several degrees of freedom. However, because of the different physical attributes of the two types of systems, the parameters that are most convenient to specify and the details of the design approach are slightly different for the two cases.

The geometry of well-mixed (Case I) reactors for gas transfer is dictated in substantial measure by land costs and other externalities related to the reactors' footprint, and by the cost and capabilities of mixing devices. By contrast, the diameter of a packed column is rarely constrained by land costs, and it has a direct and easily computed effect on the areal liquid and gas loading rates (Q_L/A_R and Q_G/A_R). The effects of these latter parameters on $K_L a_R$ are predictable via reasonably well-established correlations (e.g., the Onda correlations given in Chapter 5, or similar ones), once the type of packing to be placed in the column has been chosen. As a result, column height (Z) and cross-sectional area (A_R) are usually considered to be independent parameters in design of towers, whereas only the overall liquid volume (i.e., the product of Z and A_R) is considered in design of tanks in which diffused or surface aeration is carried out (with the assumption that the gas volume is negligible, so that $V_L \approx V_R$).

Because Z and A_R are viewed as independent parameters in Case II systems, these systems can be considered to have one more degree of freedom (for a total of seven) than Case I systems in the design process. In addition to this extra degree of freedom, Case II systems also have an additional parameter of interest—the pressure loss across the column—that is not usually considered explicitly in the design of Case I systems.

Thus, one way to view the design of Case II systems is to define seven parameters ($c_{L,in}$, $c_{G,in}$, Q_G, Q_L, $K_L a_L$, Z, and A_R) as adjustable inputs that, once specified, fully define the performance of the system with respect to gas transfer; that is, they establish $c_{L,out}$, $c_{G,out}$, $r_{L,gt}$, ΔP, and all other operating characteristics of the system.[13] As in Case I systems, we could specify one or more of these dependent parameters (most likely, the contaminant concentration in the treated water [$c_{L,out}$]), or some parameter that was a combination of the other parameters (e.g., the air/water flow rate ratio, the gas transfer capacity factor [R], or the pressure gradient in the column [$\Delta P/Z$]) in lieu of an equivalent number of input parameters, as long as we ultimately specified seven independent terms.

In this final section of the chapter, we simulate the steps in a typical design exercise, identifying the key decisions and exploring the sensitivity of the outcome to those decisions. At least three papers in the drinking water literature of the 1980s contain examples of such a process for stripping of trace organic compounds from water (Kavanaugh and Trussell, 1980; Ball et al., 1984; Hand et al., 1986). In the first of these papers, a single, feasible design was developed to meet the treatment objectives. In the latter two, a similar procedure was used to develop several alternative designs that all met the treatment objective (with each paper using a different example influent water and treatment goal), and the costs of the feasible designs were estimated to demonstrate how an optimum (lowest cost) design would be selected. Subsequently, computer programs were developed to facilitate exploration of a wider range of designs (Dzombak et al., 1993; CinCett, 2002), and to assess the sensitivity of the design to uncertainty in the input parameters (Freiburger et al., 1993). The example system used in the discussion here is the one described by Hand et al. (1986), to which the optimization program of Dzombak et al. (1993) was subsequently applied.

Description of the Influent Stream, Treatment Objectives, and Design Assumptions The influent to the system described by Hand et al. (1986) was a drinking water supply that had been contaminated with several organic contaminants, including chlorinated organic solvents and gasoline additives commonly referred to as BTX compounds (for benzene, toluene, and xylene), all at trace levels ($<100\,\mu g/L$). The treatment process was designed to treat a flow of 0.095 m³/s of water at 10°C. Different treatment criteria were established for the different compounds. Based on an analysis like that shown in Example 6-8, the compound expected to place the greatest demands on the treatment process was TCE, which was present at a concentration of $72\,\mu g/L$, and for which a target of 95% removal was established. The Henry's constant for TCE in the raw water at the temperature of interest was determined to be 0.116. The design calculations were carried out using the Onda correlations to compute $k_L a_R$ and $k_G a_R$ for several types of packing media, and assuming that the incoming air contained none of the contaminants.

Exploration of Feasible Designs for Meeting the Treatment Criteria In the case under consideration, four parameters ($c_{L,in}$, $c_{L,out}$, $c_{G,in}$, and Q_L) are not subject to modification by the designer, so three degrees of freedom remain. In the original paper by Hand et al., these degrees of freedom were utilized to select a type of packing (and thereby specifying the values necessary to use the Onda correlations), a pressure gradient corresponding to one of the isopleths in Figure 6-12, and a volumetric air/liquid flow

---

[13] It would probably be more correct to specify the type of packing as an independent parameter, and the value of $K_L a_R$ as dependent, because the identification of the packing in conjunction with the other parameters determines the value of $K_L a_R$. However, since $K_L a_R$ is quantifiable and the type of packing is not, we represent $K_L a_R$ as an independent parameter here.

rate ratio (Q_G/Q_L). Various combinations of these parameters were selected, all of which yielded designs that met the treatment objective, and the dependence of the overall system cost on the design parameters was estimated, so that the least costly of the designs that were considered could be identified.

Procedures similar to those used by Hand et al. were used by Ball et al. (1984) and are incorporated into computer programs for column design (e.g., Dzombak et al., 1993; CenCitt, 1999). However, while the specification of a pressure gradient at the beginning of the calculations is acceptable, it is not particularly convenient. Not only is the pressure gradient less directly controllable than other parameters (such as the gas flow rate or column dimensions), but it also has no effect on the technical feasibility of the design, and it is likely not to even affect the economics very strongly. For those reasons, the three parameters that are specified in the approach presented here are the packing media, the column cross-sectional area (A_R), and the gas transfer capacity factor (R). The sequence of calculations can be summarized as follows:

- Choose one or more packing media to investigate. The packing media evaluated most thoroughly by Hand et al. was 0.0762-m (3-in) plastic saddles, so the calculations for that media are demonstrated here. The value of C_f reported for this media is 16.
- Choose a range of reasonable values for A_R, to yield areal liquid loading rates from $\sim$1–50 L/m^2 s. The values of A_R correspond to the total cross-sectional area available for flow, and might be distributed among several columns. For the example system here, we will assume that all the area is provided in a single column, so each choice of A_R corresponds to a specific column diameter.
- For each value of A_R, choose several reasonable values of R (e.g., from 1.5 to 10). Because each value of R corresponds to a specific gas flow rate (since H and Q_L are known), selecting reasonable values of R is equivalent to selecting reasonable values of Q_G for the given design goal. Similarly, since the values of A_R to be explored are also specified, the selection of a range of R values simultaneously specifies the range of areal gas loading rates (Q_G/A_R) that will be investigated.
- For each value of A_R and R, compute $K_L a_R$ from the Onda correlations, and then compute the packing depth Z required to achieve the treatment objective, using Equation 6-82.
- Plot Z versus R for fixed values of A_R. This plot shows numerous combinations of column sizes and gas flow rates that all meet the technical objective of the design; interpolation between the isopleths of fixed A_R identifies the universe of feasible designs within the ranges of R and A_R values explored.

- Estimate the pressure gradient ($\Delta P/Z$) and total pressure loss (($\Delta P/Z)Z$) associated with the feasible designs.
- Estimate the capital and operational costs associated with each feasible design, and select the design with the lowest overall cost. (This step would be part of a real design process, but will not be demonstrated here.)

Consider, for example, the situation for a column diameter of 2 m, corresponding to $A_R = 3.14$ m^2. Since the liquid flow rate is 0.095 m^3/s, the areal liquid loading rate is

$$\frac{Q_L}{A_R} = \frac{(0.095 \text{ m}^3/\text{s})(1000 \text{ L/m}^3)}{3.14 \text{ m}^2} = 30.2 \, (\text{L/m}^2\text{ s})$$

The gas flow rate corresponding to any value of the gas transfer capacity factor, and the corresponding areal gas loading rate can then be computed. For instance, for $R=3$, the values of these parameters are

$$Q_G = \frac{Q_L R}{H} = \frac{(0.095 \text{ m}^3/\text{s})3}{0.116} = 2.457 \text{ m}^3/\text{s}$$

$$\frac{Q_G}{A_R} = \frac{(2.457 \text{ m}^3/\text{s})(1000 \text{ L/m}^3)}{3.14 \text{ m}^2} = 782 \, (\text{L/m}^2\text{ s})$$

Based on the liquid and gas areal loading rates, the properties of the fluids at 10°C, and the identity of the packing, the Onda correlations provided in Chapter 5 can be used to calculate k_L, k_G, and a_R values for the given value of R as 3.13×10^{-4} m/s, 8.48×10^{-3} m/s, and 61.49 m^2/m^3, respectively. The value of $K_L a_R$ can then be computed with Equation 5-56 as

$$\begin{aligned} K_L a_R &= \frac{H k_L k_G}{k_L + H k_G} a_R \\ &= \frac{(0.116)(3.13 \times 10^{-4} \text{ m/s})(8.48 \times 10^{-3} \text{ m/s})}{(3.13 \times 10^{-4} \text{ m/s}) + (0.116)(8.48 \times 10^{-3} \text{ m/s})} \\ &\quad \times 61.49 \text{ m}^2/\text{m}^3 \\ &= 1.46 \times 10^{-2} \text{ s}^{-1} \end{aligned}$$

Equation 6-82 can now be used to determine the height of packing required to achieve the desired TCE removal efficiency. Noting that 95% TCE removal corresponds to a ratio of $c_{L,\text{top}}/c_{L,\text{bot}}$ equal to 20, we find:

$$\begin{aligned} Z &= \frac{Q_L/A_R}{K_L a_R} \frac{R}{R-1} \ln \frac{\frac{c_{L,\text{top}}}{c_{L,\text{bot}}}(R-1)+1}{R} \\ &= \frac{(0.0302 \text{ m}^3/\text{m}^2 \text{ s})}{1.46 \times 10^{-2} \text{ s}^{-1}} \frac{3}{3-1} \ln \frac{20(3-1)+1}{3} \\ &= 8.11 \text{ m} \end{aligned} \qquad (6\text{-}77)$$

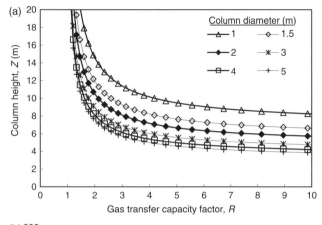

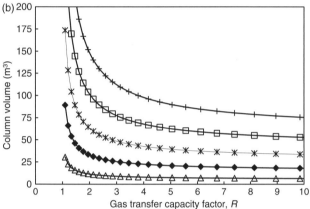

FIGURE 6-13. Sensitivity of (a) the height and (b) volume of a packed column to the gas transfer capacity factor, for several column diameters. See text for the system description. All points on the curves achieve the design goal of 95% stripping of TCE.

The above calculations can be repeated for several values of R for the given column diameter (d_R), and then the whole process can be repeated for several different values of d_R. Some results of such calculations are shown in Figure 6-13a. The dominant trends shown in the figure are that, for a given column diameter, the required packing depth decreases with increasing R value, and, for a given R value, the depth increases with decreasing column diameter.

Consider the former trend first. At values of R only slightly greater than 1 (i.e., small values of Q_G, for the given values of H and Q_L), the carrying capacity of the gas is sufficient to achieve the treatment goal. Nevertheless, the gas flow rate is low enough, and transfer of TCE into the gas is rapid enough, that the gas phase concentration at each height in the column is a significant fraction of the equilibrium gas phase concentration at that location. As a result, the driving force for TCE transfer ($c_L - c_L^*$) is significantly less than the driving force into clean, TCE-free air (for which c_L^* is zero). The rate of TCE transfer at any location in the column is therefore relatively small, and a tall tower that provides a correspondingly large amount of contact time for gas transfer to occur is needed.

As R increases, the concentration of TCE in the gas phase decreases, lowering c_L^* and increasing the driving force throughout the column, thereby allowing the same stripping efficiency to be achieved with shorter contact times (i.e., using a shorter column). However, as R increases even more, a point of diminishing returns is reached, because the TCE concentration in the gas phase becomes so low throughout the column that decreasing it even more (i.e., diluting the gas-phase TCE with more air) has a negligible effect on the driving force.

In addition to the effect on the driving force for stripping, increasing R by increasing Q_G in a column with a given diameter also increases $K_L a_R$ by decreasing the resistance of the two fluid boundary layers. However, this effect is much less important than effect on the driving force.

The effect of increasing column diameter on the required column height primarily reflects the fact that, when the areal liquid loading rate decreases, the water film flows along the surface of the packing media more slowly, so the average hydraulic residence time per unit length of column increases. This effect is counteracted partially by the lower value of $K_L a_R$ associated with the decreased areal loading rates, but the effect of the longer residence time per unit length of packed bed dominates, so an increase in diameter leads to a decrease in required depth.[14]

Hand et al. suggested that the total volume of the packed columns for different design options is a good indicator of the relative capital costs of those designs (assuming that, as is the case here, the volume is all in a single column). This parameter is plotted against the gas transfer capacity factor in Figure 6-13b. The figure demonstrates that, for a given column diameter, the required volume decreases with increasing value of R; this trend could be inferred directly from Figure 6-13a, which showed that, for a given column diameter, the column height decreases with increasing R. In addition, however, Figure 6-13b shows that, for a given value of R, the total column volume required to meet the treatment objective increases steadily with increasing column diameter. In other words, at a given R, increasing the cross-sectional area of the column leads to a less-than-proportional decrease in column height, so the column volume (and capital cost) tends to increase with increasing column cross-section.

The pressure gradient needed to provide the specified gas flow rate for each of the designs shown in Figure 6-13 can be estimated from Figure 6-12 (or the numerical correlations that simulate the plot), and the pressure loss across the

---

[14] The total height of the column must provide for water distribution and demisting near the top, and collection of water at the bottom. While these components add to the total column height, they also provide opportunities for additional gas transfer that can be significant, especially for low volatility compounds (Roberts et al., 1985).

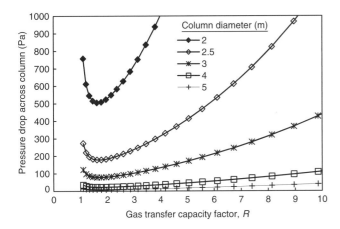

FIGURE 6-14. Pressure loss across the column for the design options shown in Figure 6-13. The pressure gradient was estimated using Figure 6-12, with $C_f = 16$.

columns can be computed as the product of this value with the column height. The results of such calculations are shown in Figure 6-14.

The major energy costs for operating these systems are associated with the fluid flows—pumping the water to the top of the column, and blowing the air through the column. The energy required for pumping the water is directly proportional to the column height, which is shown in Figure 6-13a, and the energy required for the air supply is proportional to the product of the pressure drop and the gas flow rate, which is shown in Figure 6-15.

Comparing Figures 6-13a and 6-15, we see that the energy required for both fluids decreases with increasing column diameter (i.e., cross-sectional area) at any given value of R. Furthermore, in both cases, the energy savings associated

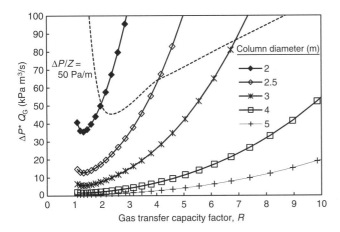

FIGURE 6-15. The product of the pressure loss across the column and the gas flow rate (an indicator of the energy costs for providing air to the system) for various designs of the example system. The broken line shows the isopleth along which the pressure gradient in the column is 50 Pa/m.

with increasing the column diameter are much larger when the diameter is small than when it is already relatively large (e.g., compare the benefits of increasing the diameter from 2 to 3 m with those of increasing it from 4 to 5 m). On the other hand, the capital cost increases with increasing column diameter, and this increase (at least in terms of column volume) is relatively insensitive to the column diameter (the incremental volume associated with increasing the diameter from 2 to 3 m is similar to that for increasing it from 4 to 5 m). The net result of these comparisons is that, for any given value of R, an optimal cross-sectional area exists that minimizes the total system cost. Once the optimum cross-sectional area has been identified for each value of R, the total cost of each system can be computed, and a global optimum can be identified.

A similar argument can be made by considering different values of R for any given column diameter. In this case, Figures 6-13a and 6-14 demonstrate that, at low values of R, the treatment objective can be reached, but only by paying a large penalty in terms of column height, with associated high costs for column volume and liquid pumping. At larger values of R, these penalties still apply, but the incremental cost of increasing R diminishes dramatically. On the other hand, at low values of R, the energy cost for blowing air through the system increases only slowly with increasing R, but at high values of R, this energy cost increases very rapidly. Thus, just as an optimum column diameter exists for any given value of R, an optimum value of R exists for any given value of column diameter. In this case, the next step would be to compare the overall cost of the system at the optimum value of R for each column diameter, and then identify the global optimum design.

The two approaches described above lead to the same optimal design, so the approach that is used is arbitrary. As noted previously, computer software exists that can carry out all the calculations described above, given the relevant technical and cost inputs.

The tower that was actually constructed to treat the water described in these calculations had a diameter of $d_R = 2.44$ m (corresponding to $Q_L/A_R = 20.3$ L/m^2 s) and a gas transfer capacity factor of $R = 3.5$. The theoretical depth of packing required to meet the design objective was $Z = 6.93$ m, and the tower that was actually built had $Z = 7.47$ m, to provide a safety factor. The pressure gradient for these conditions, computed using the Cummins and Westrick approximation of the Eckert curves, was 44 Pa/m.

Although the optimal design is, of course, specific to the inputs and treatment objectives for the particular system of interest, the general trends displayed in Figures 6-13 to 6-15 apply for high efficiency stripping of virtually any contaminant in drinking water or wastewater treatment systems. Typically, the gas transfer stripping factor that generates the least-cost design is in the range 2–10, and the corresponding column cross-section yields an areal liquid loading rate of

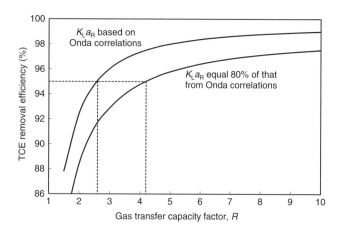

FIGURE 6-16. Effect of the gas transfer capacity factor on TCE removal efficiency in the example system, for $Z = 8.0$ m, and $d_R = 2.5$ m, and for two approaches for estimating $K_L a_R$.

1–50 L/m² s; hence, the focus on these ranges in the design calculations and the statements earlier in the chapter regarding the typical values used in design.[15]

Sensitivity of the Column Size to Design Choices and Uncertainty in Parameter Values The procedure described in the preceding section provides an approach not only for developing specific design recommendations, but also for evaluating how sensitive the design is to various parameters. That is, although the Henry's constant and the value of $K_L a_R$ for the critical contaminant are treated above as being well known, those values are in fact estimates based on the available experimental data. Furthermore, the values depend on the system temperature, which often fluctuates, and on the uncertain and perhaps variable composition of the water.

The effects of uncertainties in Henry's constant can be predicted in the example treatment system by exploring the effect of R on the TCE removal efficiency. That is, since Q_L, Q_G, and the tower dimensions are, presumably, well known, the only significant effect of a difference between the Henry's constant used in the calculations and the actual Henry's constant is on R.[16] The predicted TCE removal percentage is shown as a function of R in Figure 6-16. The plot suggests that an overestimate of Henry's constant by, say, 20% (leading to an equivalent overestimate of R) would decrease the TCE removal efficiency to approximately 93%. While that decline might not seem large, it represents a 40% increase in the TCE concentration remaining in solution, and therefore could be significant. In addition, the curvature of the correlation indicates that the removal efficiency could decline very dramatically if the error in estimating the Henry's constant were larger than 20%.

Figure 6-16 also shows a plot of TCE removal efficiency against R for a case where the Onda correlations overestimate the value of $K_L a_R$ by 25% (the correlations are reported to be accurate to within 20–30%). In this case, the removal efficiency at the design value of R is only 91.3% (corresponding to a 74% increase in the effluent TCE concentration over the target value). To achieve the target removal in this situation, the gas transfer capacity factor would have to be increased from 2.59 to 4.02, presumably by increasing the gas flow rate.

In both of the above scenarios, a packed tower designed to achieve 95% TCE removal could meet the treatment target, even if the value of Henry's constant or $K_L a_R$ used in the design process was in error by the amount discussed. However, in each case, the correction would require an increase in the gas flow rate, which would increase both the pressure drop across the column and the amount of gas that had to overcome that pressure drop. If the design is in the range where energy costs increase rapidly with Q_G, then this postdesign modification could lead to a substantial increase in operating costs.

The cost of constructing a column with a diameter slightly larger than the optimum is often not large. Therefore, such an approach is frequently employed, so that a margin of error exists that allows the gas flow rate to be increased without increasing the operational costs excessively. If the estimates of Henry's constant and $K_L a_R$ are in fact not seriously in error, this overdesign of the column has the effect of shifting the pressure drop to a range where it is a relatively small fraction of the overall operating cost, leading to the situation described previously: the pressure gradient in the column is less than 50 Pa/m, and the marginal cost of increasing the gas flow rate is not large. These issues are discussed in a more formal way and in greater detail by Freiburger et al. (1993).

Case II Systems Other than Packed Columns

The preceding discussion of Case II systems focuses almost exclusively on stripping in packed columns, which is indeed the primary application of such systems in environmental engineering. However, the usefulness of those systems for stripping of low volatility (i.e., low Henry's constant) compounds is limited, because of the large gas flow rate that is needed to achieve a reasonable gas transfer capacity factor. That is, when a low volatility compound must be treated, the gas phase approaches equilibrium with the solution when the gas-phase concentration is quite low, so a very large gas flow rate is required to achieve good treatment. Under these circumstances, the large gas flow rate can increase the cost of operating a packed column system to the point where

---

[15] For the liquid flow rate specified in the example system, column diameters of 1.0 and 5.0 m yield liquid-loading rates of 121 and 4.8 L/m² s, respectively.

[16] The Henry's constant also affects $K_L a_R$, but that effect is likely to be small compared to the effect on R.

alternative treatment processes are more attractive (Dvorak et al., 1994, 1996).

Alternative designs have been proposed for stripping of such compounds that provide opportunities for large amounts of air to contact the liquid without the large pressure losses associated with conventional packed columns. The key component of such designs is that the air flows horizontally across the column, while the water flows downward, so that the flow path for the air is much shorter than the whole depth of the column. Both packed columns with air blown across the column (referred to as *crossflow air stripping*) and systems in which the water falls onto numerous trays spaced at intervals throughout the column (*cascade air stripping*) have been used in these applications (Little and Mariñas, 1997; Mertooetomo et al., 1993; Nirmalakhandan et al., 1990, 1992).

6.4 SUMMARY

This chapter demonstrates the application of the gas transfer principles developed in Chapter 5 to large-scale systems. Whereas the material in Chapter 5 focused on factors that control the flux of a volatile species across a microscopic section of interface, this chapter shows how full-scale treatment systems, in which the flux can vary over time and space, can be designed to accomplish a desired objective. That objective can be either gas absorption or stripping, and the same equations apply to both objectives. A key result is that the rate of the overall process can be limited by either microscopic, interfacial processes, or macroscopic, advective transport processes, and that the identity of the rate-limiting step depends on design decisions. Generally, designs should be made so that the macroscopic processes do not impose the dominant limitation.

The development of the design equations for gas transfer in large scale systems involves the same general procedure—combining the microscopic mass balance for flux with a macroscopic mass balance that describes the advective flows of water and gas through the system—regardless of the details of the reactor configuration and regardless of whether the gas transfer process is absorption or stripping. However, it is convenient to develop those equations separately for systems with and without a spatial gradient in the liquid phase composition; here, we refer to systems with a spatially uniform liquid phase concentration as Case I systems, and those with a nonuniform concentration as Case II systems. Typical Case I systems are well-mixed tanks that are aerated by rising bubbles and/or surface mixing. The most common Case II system is a packed column with counter-current flow; other Case II systems include bubble aeration columns with co-current flow and cascade gas transfer systems.

For Case I systems with diffused aeration, the analysis involves consideration of two time frames that are usually very different from one another: the time that an individual rising bubble stays in the tank (typically a few to several seconds) and the characteristic time for the change in the dissolved concentration of the volatile species (typically several minutes or longer). During the rise time of an individual bubble, the solution composition is very nearly constant, and the disequilibrium between the bubble and the solution decays exponentially, due entirely to changes in the gas phase composition. In some cases, the extent of disequilibrium dissipates quickly, so that a near-equilibrium condition is attained while the bubble is still in the water column, and gas transfer is negligible for the rest of the bubble's rise time. However, in other cases, the disequilibrium dissipates minimally, and a substantial driving force for gas transfer exists throughout the water column; in the limit, the disequilibrium is barely diminished at all as the bubble rises, and the flux of the volatile constituent is approximately constant during the entire bubble residence time.

Whether one or the other of these limiting (Case I) scenarios applies is determined by the ratio $K_L a_L/(Q_G H/V_L)$ in batch liquid systems and either that ratio or, equivalently, $\tau_L K_L a_L/((Q_G/Q_L)H)$ in systems with continuous liquid flow. If this ratio is large (say, >5), most of the disequilibrium between the bubble and solution is dissipated as the bubble rises, and the only practical way to increase overall gas transfer is to increase the gas flow rate. In systems characterized by this condition, gas transfer is said to be limited by equilibrium, by macroscopic processes, or by advective transport. On the other hand, if the ratio is small (<0.2), only a small fraction of the disequilibrium dissipates as the bubble rises, and gas transfer can be enhanced by steps that increase $K_L a$, such as increased mixing intensity or increased gas/liquid surface area. In these cases, gas transfer is said to be limited by kinetics, by microscopic processes, or by interfacial transport.

Case I systems are often used to evaluate $K_L a$ experimentally for various types of gas transfer equipment. An overview of several approaches that are used to evaluate $K_L a$ for aerators used in biological treatment systems is provided in Appendix 6A.

Case II systems with counter-current flow can lead to very efficient gas transfer, because they allow the solution, as it moves through the reactor, to contact gas that is progressively cleaner (in a stripping reactor) or progressively more concentrated (in an absorption reactor) in the volatile constituent. Indeed, in a typical stripping reactor, the final gas that the solution contacts is devoid of the contaminant being stripped, so that if the gas flow is adequate and equilibrium is approached quickly, almost all of the contaminant will be transferred to the gas phase.

Distinguishing between microscopic and macroscopic rate limitation is less informative in Case II than in Case I systems. However, in Case II systems, the value of the denominator in the ratio $\tau_L K_L a_L/((Q_G/Q_L)H)$ can serve a related purpose. That term is referred to as the gas transfer capacity factor R (or, in the stripping literature, the stripping factor S).

In a stripping system, $R = 1$ corresponds to a condition where the gas flow rate is barely sufficient to acquire all of the contaminant from the incoming solution, if the exiting gas reaches equilibrium with that solution (and assuming that the incoming gas does not contain any of the contaminant). R values less than 1 correspond to designs in which it would be impossible to strip all the contaminant from the incoming solution, no matter how low an interfacial resistance is achieved and no matter what measures are taken to overcome the kinetic resistance to gas transfer (e.g., increasing the liquid/gas surface area, or increasing the height or diameter of the contactor to allow longer gas/liquid contact times). Under these conditions, the maximum stripping efficiency equals R. On the other hand, R values greater than or equal to 1 correspond to designs in which essentially all the contaminant could be stripped from the incoming solution if the kinetic resistance to gas transfer were sufficiently small. The constraints on gas transfer in a stripping or absorption column imposed by kinetic factors and by the approach to equilibrium between the gas and liquid phases are frequently separated into terms referred to as the height and number of transfer units (HTU and NTU), respectively, for the system.

For a stripping operation, if $R \geq 1$ and the conditions are such that gas transfer kinetic limitations are overcome (high $K_L a$ and/or long contact time generated by a large value of Z or a small value of Q_L/A), almost all of the contaminant would both enter and leave the column at the top (in the liquid and gas, respectively); if $R < 1$, some of the contaminant would leave the column at the bottom (in the solution), no matter what design choices are made to overcome the kinetic limitations to gas transfer. In typical designs of stripping towers, values of R between 2 and 5 are chosen, and other conditions are chosen to reduce the kinetic limitations on gas transfer, so that very efficient gas transfer can be achieved.

In absorption systems, an R value of 1 corresponds to the minimum gas flow needed to transport the volatile species into the reactor at a rate that would allow the exiting solution to equilibrate with the incoming gas. In this case, if $R \leq 1$, and if kinetic limitations to gas transfer were largely overcome, essentially all of the transferable species would enter (in the gas) and leave (in the liquid) the column at the bottom, and if $R > 1$, some of that species would exit out the top of the column (in the gas). If the goal is primarily to increase the concentration of the volatile species in solution (with little concern about some leaving in the gas phase), then R values >1 should be used, whereas if the goal is to transfer essentially all of the volatile species from the gas phase to solution, R values <1 are more appropriate.

In both Case I and II systems, many different system designs can lead to equivalent water quality outcomes. Under the circumstances, the goal is typically to identify the design that meets the water quality goal at the lowest cost, while meeting any other site-specific constraints. In general, Case I systems have six degrees of freedom. However, as a practical matter, one of these degrees of freedom is usually pre-specified (the target value of c_L in the treated water), and three others are not under the designer's control (c_L in the untreated water, $c_{G,in}$, and the volume or flow rate of water to be treated). Therefore, for many Case I design applications, the designer must specify the values of two other parameters to fully define the system characteristics and performance. These two parameters can be chosen from among $K_L a_L$, Q_G, and t for batch systems, and $K_L a_L$, Q_G, and τ_L for continuous flow systems; in each case, specification of any two of these three parameters fixes the value that the third one must have to meet the treatment objective. In the design process, two of the three parameters can be varied over reasonable ranges, and the cost of each combination of choices can be assessed and compared to help identify the most attractive design option.

In Case II systems, the choices of column height and cross-sectional area are considered independent in the design process, so these systems are considered to have seven degrees of freedom. Also, in Case II systems, the pressure drop from the bottom to the top of the column is an important consideration that can be incorporated into the analysis either in terms of the pressure gradient or the total pressure drop. As in Case I systems, four design variables are typically prespecified or not under the designer's control, so the design process involves assigning values to three other parameters. These parameters can be expressed in a variety of ways. Most commonly, three of the following parameters are specified in the design: the packing type, Q_L/A, R, d_R, $\Delta P/Z$, and Z. Once any three of these parameters are chosen, the others are fixed, and the design is complete. The task of comparing the universe of feasible designs for such systems is formidable, but can be accomplished with the help of available software and other resources.

Gas transfer is the environmental engineering process for which design of full-scale treatment systems can be accomplished most completely based on fundamental theory. While pilot scale testing is strongly recommended before a system is constructed, the required size of a gas transfer system to accomplish a given treatment objective can be calculated to a good approximation from a combination of first principles and well-established empirical correlations. As such, gas transfer provides an ideal transition between the introduction of mass balances and reactor theory in Section 6-1 of the text, and the treatment processes discussed in the remainder of the text. The treatment processes to be discussed in upcoming chapters can be interpreted using the same fundamentals as are applied to understand gas transfer. However, as we will see, the uncertainty about molecular behavior in those systems (i.e., the fact that the behavior of the molecules cannot be described by any models as simple and reliable as the ideal gas law and Henry's law) forces us to rely more strongly on empiricism in the design of those treatment processes than is the case for gas transfer systems.

APPENDIX 6A. EVALUATION OF K_La IN GAS-IN-LIQUID SYSTEMS FOR BIOLOGICAL TREATMENT

Measurement of K_La (either K_La_L or K_La_R) in full-scale wastewater treatment facilities using diffused aeration can pose special challenges, because of the size of the systems being characterized, the fact that the diffusers do not necessarily provide air uniformly to all portions of the solution, and the difficulty of determining the rate of biological oxygen consumption by the microorganisms in solution. The American Society of Civil Engineers (ASCE) (1997) has recommended three approaches for assessing oxygen transfer in such processes, and these approaches are described below. All of these approaches assume that interfacial resistance is the rate-limiting factor for the overall gas transfer process.

Approach 1. In this approach, the dissolved oxygen concentration in the solution is perturbed from its steady-state value, and K_La is evaluated based on the rate at which the system returns to the steady-state condition. The initial change in the oxygen concentration is induced quickly, and it is assumed that the biological population in the reactor does not adjust to the new conditions during the transient. The change in dissolved oxygen concentration can be induced by increasing the power input to the aerators, waiting long enough for the higher dissolved oxygen concentration to become uniformly distributed through the solution, and then returning the power to its normal operating value. Alternatively, the dissolved oxygen concentration can be altered by injecting a spike input of hydrogen peroxide (H_2O_2) into the solution. In the presence of reducing agents such as biodegradable organic compounds, H_2O_2 quickly releases oxygen by a reaction of the type:

$$2\,H_2O_2 \xrightarrow{\text{Organics}} 2\,H_2O + O_2$$

A mass balance on O_2(aq) in the well-mixed aqueous phase from that time forward is as follows:

$$V_L \frac{dc_L}{dt} = Q_L(c_{L,\text{in}} - c_L) + V_L K_L a_L(c_L^* - c_L) + V_L r_{\text{biochem}} \quad (6A\text{-}1)$$

where r_{biochem} is the rate of oxygen generation (i.e., $-r_{\text{biochem}}$ is the rate of oxygen utilization) by biological and chemical reactions in the reactor.

Eventually, the system returns to steady-state, at which time the same mass balance applies with $dc_L/dt = 0$. Designating steady-state values by the subscript *ss*, and dropping the subscript L on all terms except K_L and a_L (since all terms of interest refer to the liquid phase we have

$$0 = Q(c_{\text{in}} - c_{ss}) + V K_L a_L(c^* - c_{ss}) + V r_{\text{biochem,ss}} \quad (6A\text{-}2)$$

Although $r_{\text{biochem,ss}}$ is not directly measurable, Equation 6-84) can be rearranged to yield an expression for the product $V r_{\text{biochem,ss}}$ in terms of the other variables:

$$V r_{\text{biochem,ss}} = -Q(c_{\text{in}} - c_{ss}) - V K_L a_L(c^* - c_{ss}) \quad (6A\text{-}3)$$

Presuming that the short-term change in dissolved oxygen does not affect the rate of the oxygen-consuming reactions, this expression for $V r_{\text{biochem,ss}}$ can be substituted into Equation 6A-1 to yield

$$V \frac{dc}{dt} = Q(c_{\text{in}} - c) + V K_L a_L(c^* - c) - Q(c_{\text{in}} - c_{ss})$$
$$- V K_L a_L(c - c_{ss}) \quad (6A\text{-}4a)$$

$$V \frac{dc}{dt} = Q(c_{ss} - c) + V K_L a_L(c_{ss} - c) \quad (6A\text{-}4b)$$

$$\frac{dc}{c_{ss} - c} = \left(\frac{1}{\tau} + K_L a_L\right) dt \quad (6A\text{-}4c)$$

Integrating Equation 6A-4c between $t = 0$ (defined as the time immediately after the perturbation in dissolved oxygen concentration) and some later time, we obtain an expression for the approach of the dissolved oxygen concentration to its steady-state value:

$$\ln \frac{c_{ss} - c(t)}{c_{ss} - c(0)} = -\left(\frac{1}{\tau} + K_L a_L\right) t \quad (6A\text{-}5a)$$

$$c(t) = c_{ss} - (c_{ss} - c(0)) \exp\left[-\left(\frac{1}{\tau} + K_L a_L\right) t\right] \quad (6A\text{-}5b)$$

According to Equation 6A-5a, a plot of $\ln[(c_{ss} - c(t))/(c_{ss} - c(0))]$ versus time should be linear with a slope of $-(1/\tau + K_L a_L)$. Since the value of the logarithmic term is known from the experimental results and τ can be determined from theory (V/Q) or from a tracer test, the value of $K_L a_L$ can be computed from the slope of such a plot.

Approach 2. An alternative recommended approach for analyzing $K_L a_L$ in biologically active systems uses the inert gas krypton. Extensive experimental evidence suggests that the overall gas transfer coefficient for Kr is $83 \pm 4\%$ that of oxygen. Because this gas undergoes no reaction in solution and because the background concentration of krypton in both the aqueous and gas phases is negligible, the nonsteady state mass balance on dissolved krypton in a well-mixed aeration system following a spike input of the element is given by Equation 6A-5b with $c_{ss} = 0$; that is,

$$c_{Kr}(t) = c_{Kr}(0) \exp\left[-\left(\frac{1}{\tau} + K_{L,Kr}a_L\right)t\right] \quad (6A\text{-}6)$$

If the reactor is well-mixed, $K_{L,Kr}a_L$ can be evaluated by the following rearrangement of Equation 6A-6:

$$c_{Kr}(t) = c_{Kr}(0) \exp\left(-\frac{t}{\tau}\right) \exp(-K_{L,Kr}a_L t) \quad (6A\text{-}7a)$$

$$\ln \frac{c_{Kr}(t)}{c_{Kr}(0)} = -\frac{t}{\tau} - K_{L,Kr}a_L t \quad (6A\text{-}7b)$$

$$\ln \frac{c_{Kr}(t)}{c_{Kr}(0)} + \frac{t}{\tau} = -K_{L,Kr}a_L t \quad (6A\text{-}7c)$$

According to Equation 6A-7c, $K_{L,Kr}a_L$ can be evaluated directly from the time profile of krypton in the reactor, as the negative slope of a plot of the left side of the equation versus time. The computed value can then be divided by 0.83 to estimate $K_{L,O_2}a_L$.

Use of Equation 6A-7c is straightforward and leads to a correct result if the reactor is a CFSTR, but the equation does not apply for other reactor mixing patterns. To deal with such situations, the ASCE Guidelines recommend a slightly more complicated procedure involving injection of a second, nonvolatile compound (which we refer to as Cpd2) as a spike at the same time as the krypton is added. Because Cpd2 is nonvolatile, $K_{L,Cpd2}$ is zero, and the concentration of Cpd2 in the reactor effluent depends only on the initial concentration (after the spike has been mixed into the reactor contents) and the reactor RTD.

Recall that the integral of $E(t)dt$ between $t = 0$ and some later time t is the fraction of the fluid in the reactor effluent that has spent an amount of time less than t in the reactor. If we define $t = 0$ is the time of the spike input, then at some later time t_1, all of the fluid that has spent less than t_1 in the reactor contains no Kr or Cpd2. On the other hand, all of the fluid that has spent longer than t_1 in the reactor was present when the spikes were added and therefore contains some of those species. Because Cpd2 is nonvolatile, it is present at a concentration Cpd2(0) in all of this "older" fluid, whereas Kr is present at a concentration less than Kr(0), because some of it has volatilized. Correspondingly, the concentrations of the spiked species in the effluent at any time t can be found by an integral that considers only the fluid packets in that stream that have spent longer than t in the reactor. Specifically,

$$c_{Kr}(t) = c_{Kr}(0) \int_t^\infty \exp(-K_{L,Kr}a_L t) E(t) dt \quad (6A\text{-}8a)$$

$$c_{Cpd2}(t) = c_{Cpd2}(0) \int_t^\infty \exp(-K_{L,Kr}a_L t) E(t) dt$$

$$= c_{Cpd2}(0) \int_t^\infty E(t) dt \quad (6A\text{-}8b)$$

Defining the ratio of these concentrations as $R(t)$, we obtain,

$$R(t) \equiv \frac{c_{Kr}(t)}{c_{Cpd2}(t)} = \frac{c_{Kr}(0) \int_t^\infty \exp(-K_{L,Kr}a_L t) E(t) dt}{c_{Cpd2}(0) \int_t^\infty E(t) dt}$$

$$= R(0) \frac{\int_t^\infty \exp(-K_{L,Kr}a_L t) E(t) dt}{\int_t^\infty E(t) dt} \quad (6A\text{-}9)$$

Equation 6A-9 indicates that the concentration ratio at any time equals the product of the concentration ratio immediately after the spike and a ratio that accounts for the fractional loss of Kr from solution (by both volatilization and flow) to the fractional loss of Cpd2 (by flow only). Next, the equation is rearranged and approximated as follows:

$$\frac{R(t)}{R(0)} = \frac{\int_t^\infty \exp(-K_{L,Kr}a_L t) E(t) dt}{\int_t^\infty E(t) dt} \quad (6A\text{-}10)$$

$$\ln \frac{R(t)}{R(0)} = \ln \frac{\int_t^\infty \exp(-K_{L,Kr}a_L t) E(t) dt}{\int_t^\infty E(t) dt} \approx -K_{L,Kr}a_L t$$

$$(6A\text{-}11)$$

While the approximation shown in the second equality in Equation 6A-11 is not strictly justifiable mathematically, it simplifies the analysis considerably and has been found to provide a satisfactory estimate of $K_{L,Kr}a_L$ (which, according to equation, can be determined as the negative slope of a plot of $R(0)$ vs t).

The idea underlying the dual-spike approach is that by normalizing the decline in the Kr concentration to that of a nonvolatile species added at the same time, the loss of Kr from the reactor by flow can be accounted for. Those losses must be taken into account in any analysis; in the analysis leading to Equation 6A-7c, they were implicitly computed by assuming that the reactor was well-mixed, so that its $E(t)$ function was known. If we acknowledge the more realistic possibility that mixing is not ideal, the analysis leads to Equation 6A-10, which is a more general result, but also more difficult to use. Equation 6A-11 has, in essence, been developed as a compromise that takes nonideal mixing into account, but also includes a simplification to make the math more tractable.

The ASCE Guidelines indicate that tritiated water (^3H_2O) has often been used as the nonvolatile substance and that the test has often been carried out using ^{85}Kr as the krypton source. In such cases, very low levels of the compounds can be used, because their natural background concentrations are essentially nil and because extremely sensitive analytical instrumentation is available to detect them. However, non-radioactive Kr can also be used, along with any nonvolatile tracer of the hydrodynamics. Note that, in this case, the total dissolved concentration of the tracer is analyzed, so its reactivity in solution is unimportant; all that matters is that it not exit the aqueous phase. Thus, tritium (T^+) from the tritiated water might exchange with $^1H^+$ to form species such as TAc or NH_3T^+ (tritiated acetic acid or ammonium ion), but as long as these species do not volatilize or form solids, the tritium would be an acceptable tracer for the water in the system.

Approach 3. The approach described above for evaluating K_La_L in full-scale aeration basins circumvents the complexities associated with the reactivity of oxygen in biologically active systems by using a nonreactive gas as a surrogate for oxygen. The major complexity is that some of the oxygen that enters solution, and therefore should be included in the K_La_L calculation, is consumed in biochemical reactions and is not measured as part of the dissolved oxygen. In other words, the calculation needs to consider all of the oxygen that enters solution, but some of that oxygen does not show up in an analysis of the aqueous phase, and its rate of disappearance from the aqueous phase via biological activity is difficult to quantify accurately.

The final alternative discussed in the ASCE Guidelines for such systems circumvents the problem by analyzing the oxygen *lost* from the gas phase rather than the oxygen taken up by the solution. This approach succeeds because the oxygen loss from the gas bubbles as they transit through the solution includes both that oxygen that remains in solution and that which is consumed biologically.

If the test is carried out in a batch liquid system, the gas-phase mass balance on oxygen is given by Equation 6A-12:

$$\int_{V_G} \frac{\partial c_G}{\partial t} dV_G = Q_G(c_{G,in} - c_{G,out}) + \int_A J_G dA \quad (6A-12)$$

Furthermore, the flux out of the gas phase equals that into the solution, and, since gas transfer is assumed to be rate-limited by interfacial transport, we can express the gas transfer rate as $K_La_L(c_{L,in}^* - c_L)$. Applying these two ideas to Equation 6A-12, we obtain

$$\int_{V_G} \frac{\partial c_G}{\partial t} dV_G = Q_G(c_{G,in} - c_{G,out}) - \int_A J_L dA \quad (6A-13)$$

$$\int_{V_G} \frac{\partial c_G}{\partial t} dV_G = Q_G(c_{G,in} - c_{G,out}) - K_La_LV_L(c_{L,in}^* - c_L) \quad (6A-14)$$

Finally, utilizing the assumption that the rate of change of the mass of the volatile constituent stored in the gas bubbles in the reactor (the left side of Equation 6A-14) is much smaller than the overall rate of gas transfer into solution, $(K_La_LV_L(c_L^* - c_{L,in}))$[17] and rearranging, we obtain:

$$K_La_L = \frac{Q_G(c_{G,in} - c_{G,out})}{V_L(c_L^* - c_{L,in})} \quad (6A-15)$$

The major practical problem associated with use of Equation 6A-15 is the difficulty of obtaining representative samples of gas to analyze for oxygen. Some approaches for collecting the samples are described in the ASCE Guidelines. As one would expect, the larger the area over which the gas can be sampled, the more accurate is the resulting value of K_La_L, assuming that the conditions in the basin below the gas collection system are uniform.

The off-gas sampling approach is particularly useful for systems in which the oxygen concentration is known to vary as a function of location, such as in plug-flow activated sludge reactors. In such systems, it might be difficult to evaluate K_La_L based on analyses of the solution phase alone or based on the response of the reactor to a step change in inputs, because it is difficult to isolate the solution in a particular portion of the tank without changing the mixing properties. On the other hand, it is relatively easy to locate a gas collection system over a small portion of the tank, analyze the solution composition of the water directly below the collection system, and compute K_La_L of that

---

[17] Note: this is the same assumption as we used to obtain Equation 6-14.

section. If the gas injection is uniform throughout the tank, this value of $K_L a_L$ might be assumed to apply everywhere in the system, even though the dissolved oxygen concentration and biological population vary substantially from one point to the next. Alternatively, one could collect off-gas from various points in the reactor and evaluate $K_L a_L$ as a function of location.

REFERENCES

ASCE (1997) *Standard Guidelines for In-Process Oxygen Transfer Testing.* ASCE-18-96, Amer. Soc. Civil Eng., New York.

Ball, W. P., Jones, M. D., Kavanaugh, M. C., (1984) Mass transfer of volatile organic compounds in packed tower aeration. *J. Water Poll. Control Assn.*, 56 (2), 127–136.

CenCITT (1999) *Aeration System Analysis Program (ASAP)*, prepared by the National Center for Clean Industrial Treatment Technologies, Michigan Tech. Univ., Houghton, MI.

Dvorak, B. I., Lawler, D. F., Speitel, G. E., Jones, D., and Boadway, D. A. (1994) Selecting among physical/chemical processes for removing synthetic organics from water. *Water Environ. Res.*, 65 (7), 827–838.

Dvorak, B. I., Herbeck, C. J., Meurer, C. P., Lawler, D. F., and Speitel G. E., Jr. (1996) Selection among aqueous and off-gas treatment technologies for synthetic organic chemicals. *J. Environ. Eng. (ASCE)* 122 (7), 571–580.

Dzombak, D., Roy, S.B., and Fang, H.-J. (1993) Air-stripper design and costing computer program. *JAWWA*, 85 (10), 63–72.

Eckert, J. S. (1961) Design techniques for sizing packed towers. *Chem. Eng. Prog.*, 57 (9), 54–58.

Eckert, J. S. (1975) How tower packings behave. *Chem. Eng.*, 82, 70–76.

Freiburger, E. J., Jacobs, T. L., and Ball, W. P. (1993) Probabilistic evaluation of packed-tower aeration designs for VOC removal. *JAWWA*, 85 (10), 73–86.

Hand, D. W., Crittenden, J. C., Gehin, J. L., and Lykins, B. W. (1986) Design and evaluation of an air-stripping tower for removing VOCs from groundwater. *JAWWA*, 78 (9), 87–97.

Hand, D. W., Hokanson, D. R., and Crittenden, J. C. (2011) Gas-liquid processes: principles and applications. In Edzwald, J.K. (ed.), *Water Quality and Treatment*, 6th ed., Chapter 6, McGraw Hill, New York.

Little, J. C., and Mariñas, B. J. (1997) Crossflow versus counterflow air-stripping towers. *J. Environ. Eng. (ASCE)*, 122 (7), 571–580.

Matter-Müller, C., Gujer, W., Giger, W., and Stumm, W. (1980) Non-biological elimination mechanisms in a biological sewage-treatment plant. *Prog. Water Technol.*, 12, 299–314.

Matter-Müller, C., Gujer, W., and Giger, W. (1981) Transfer of volatile substances from water to the atmosphere. *Water Res.*, 15, 1271–1279.

McGinnis, D. F., and Little, J. C. (2002) Predicting diffused-bubble oxygen transfer rate using the discrete-bubble model. *Water Res.*, 36, 4627–4635.

Mertooetomo, E., Valsaraj, K. T., Wetzel, D. M., and Harrison, D. P. (1993) Cascade crossflow air stripping of moderately volatile compounds using high air-to-water ratios. *Water Res.*, 27 (7), 1139–1144.

Narbaitz, R. M., Mayorga, W. J., Torres, P., Greenfield, J. H., Amy, G. L., and Minear, R. A. (2002) Evaluating aeration-stripping media on the pilot scale. *JAWWA*, 94 (9), 87–97.

Nirmalakhandan, N., Jang, W., and Speece, R. E. (1990) Counter-current cascade air stripping for removal of low volatile organic contaminants. *Water Res.*, 24 (5), 615–623.

Nirmalakhandan, N., Jang, W., and Speece, R. E. (1992) Removal of 1,2-dibromo-3-chloropropane by countercurrent cascade air stripping. *J. Environ. Eng. (ASCE)*, 118 (2), 226–237.

Roberts, P. V., Munz, C., and Dändliker, P. (1984) Modeling volatile organic solute removal by surface and bubble aeration. *JWPCF*, 56 (2), 157–163.

Schulz, C. R., Schafran, G. C., Garrett, L. B., and Hawkins, R. A. (1995) Evaluating a high-efficiency ozone injection contactor. *JAWWA*, 87 (8), 85–99.

Sherwood, T. K., Pigford, R. L., and Wilke, C. R. (1975) *Mass Transfer*. McGraw-Hill, New York, NY.

Treybal, R. E. (1980) *Mass-Transfer Operations*. McGraw-Hill, New York, NY.

PROBLEMS

6-1. Air is injected through fine-bubble diffusers into a basin 3.5 m deep in a well-mixed, batch (no flow) reactor at 25°C. With sufficient time, the water will reach equilibrium with the air and no mass transfer is occurring. Under these conditions, answer the following:

(a) What is the pressure on the bubbles (in atm) at the bottom, middle, and top of the tank?

(b) What is the concentration of oxygen (in milligram per liter) in the bubbles at each of the three points of interest in part (a).

(c) What is the final (equilibrium) concentration of oxygen in the water (c_L^*), in milligram per liter? Because the basin is completely mixed, assume that the water is, on average, at equilibrium with bubbles at the mid-depth pressure.

(d) What would the value of c_L^* be if the injected gas were pure oxygen instead of air?

6-2. A gently mixed pond has an approximately rectangular surface and trapezoidal cross-section, with dimensions shown in Figure 6-Pr2. The hydraulic residence time in the pond is 6 h, and the dissolved

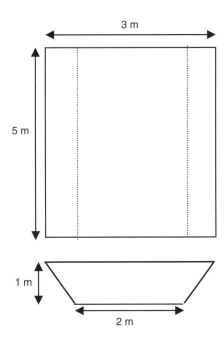

FIGURE 6-Pr2. Dimensions of a gently mixed pond.

oxygen concentration in the influent is approximately steady at 3 mg/L. During the day, algal growth in the pond causes the dissolved oxygen concentration to increase to 16 mg/L at sundown, even though the concentration in equilibrium with the air is only 9 mg/L under local conditions. Assume that photosynthetic activity stops abruptly when the sun sets at 6 pm, but that respiration consumes oxygen at a steady rate of 2.0 mg/L h continuously. If the overall gas transfer coefficient is 24 cm/h, compute the dissolved oxygen concentration in the pond at 6 AM.

6-3. As noted in Problem 5-11, the oxidation of Fe^{2+} by oxygen can be described by a rate expression of the form:

$$r_{Fe^{2+}} = -k \, c_{Fe^{2+}} \, c_{O_2} \, c_{OH^-}^2$$

where k has a value of 1.6×10^{12} (L/mg O_2)(L/mol OH^-)^2 min^{-1} at 10°C. The characteristics of a landfill leachate stream are presented below. In a proposed treatment system, the pH will be fixed at 7.5. Under these conditions, the rate expression describing BOD usage can be represented as follows:

$$r_{BOD} = -(0.2 \, d^{-1}) \frac{c_{BOD} c_{O_2}}{1.0 + c_{O_2}}$$

where c_{BOD} and c_{O_2} are both in milligram per liter. Henry's constant (H_{pm}) for O_2 in the leachate at the temperature of interest is 625 atm/(mol/L).

$$Q_L = 1150 \, m^3/d; \quad c_{BOD} = 12{,}000 \, mg/L;$$
$$c_{Fe^{2+}} = 435 \, mg/L; \quad c_{O_2} = 0 \, mg/L$$

(a) Determine the fraction of the Fe^{2+} that would be oxidized if the leachate were treated by diffused aeration in a CFSTR with a 1-h hydraulic detention time, if $K_L a_L = 10 \, h^{-1}$. Compare the magnitudes of the various terms in the mass balance on O_2 in the reactor.

(b) Determine the detention time that would be required to achieve 99% oxidation of the iron. Does the presence of the degradable organic matter have a large effect on the result?

(c) Determine the detention time that would be required to achieve 50% removal of the BOD.

6-4. A diffused-bubble gas transfer system is not operating as well as the designers had hoped for stripping two different volatile organic compounds from water. When the diameter of the injected bubbles is decreased by 50% while maintaining the same liquid and gas flow rates, the transfer efficiency for one of the VOCs improves dramatically, but that for the other VOC is barely affected. Explain these differing results, and suggest another change that might improve the stripping efficiency for the second gas.

6-5. A gas stream containing 3% ozone by volume is bubbled into a water column as part of a disinfection strategy. $K_L a_L$ for ozone in the system is $18 \, h^{-1}$, and the gas transfer process is limited by interfacial transport (i.e., kinetically or microscopically limited).

Ozone dissolved in water is unstable. The kinetics of decomposition are complex, but under some circumstances, the rate can be approximated as first-order with respect to dissolved ozone. The first-order rate constant is pH-dependent and, at pH 8.0 and 25°C, is $12 \, h^{-1}$ in the system of interest.

(a) Find the steady-state concentration of dissolved ozone in the reactor if it is operated with no flow of liquid.

(b) Find the steady-state concentration of dissolved ozone in the effluent if the reactor is operated as a CFSTR with a 15-min hydraulic retention time.

6-6. A diffused aeration unit is being tested in a batch liquid reactor consisting of a 3-m deep tank filled with 100 m^3 of water at 20°C. The gas being injected is pure O_2, and the diameter of the injected bubbles at the diffuser surface is 3.0 mm. For the experimental conditions, H_{pc} for O_2 is 0.019 atm L/mg. The gas flow rate is 120 SCM/h. Assume that the pressure throughout the tank can be approximated as the pressure at mid-depth, and that the pressure at the water surface is 1.0 atm.

(a) Assuming that the bubbles rise at an average velocity determined by their diameter at mid-depth, and that a negligible fraction of the total O_2 transfers out of the bubbles, estimate the total interfacial area in the tank at any instant.

(b) If the water in the tank is initially in equilibrium with air, and the concentration of dissolved O_2 increases to 31.0 mg/L after 10 min of aeration, what is K_L?

6-7. Biological wastewater treatment is being carried out in a CFSTR with steady flow. The basin depth is 3.5 m, the water temperature is 20°C, the liquid flow rate is 0.438 m³/s, and the hydraulic detention time is 4 h. The biological process is destroying 3000 kg/d of BOD_{ult}, and the steady-state oxygen concentration in the water (c_L) is maintained at 2.0 mg/L. The influent is anoxic (i.e., it contains negligible dissolved oxygen). The basin is aerated by fine-bubble diffusers, generating bubbles that have an average diameter of 2 mm. The air-flow rate through the reactor is 0.83 SCM/s, and the gas-phase volume fraction in the basin is estimated to be 0.005. For this wastewater, $\beta = 0.9$. H_{pc} and H_{cc} for oxygen in clean water at 20°C are 0.0231 atm L/mg and 30.77, respectively.

(a) What is the value of k_I in the system?
(b) What is the oxygen transfer efficiency, η_{Ox}, defined as $\eta_{Ox} = 1 - (c_{G,out}/c_{G,in})$?
(c) Estimate the values of the interfacial mass transfer parameters, k_L and a_L.

6-8. The data shown below were obtained for bubbling air into a 5-m deep activated sludge system operated as a CFSTR at steady state and 20°C. The saturation concentration of oxygen in clean water at this temperature is 9.3 mg/L, and the β value characterizing the effects of other components dissolved in the water on oxygen solubility is 0.85.

$Q_L = 2$ m³/min	Biological O_2 usage rate: 0.007 mg/L s
$V_L = 500$ m³	Average bubble diameter (at mid-depth): 0.6 cm
$c_{in} = 0$ mg O_2/L	Average bubble rise velocity: 20 cm/s
$c_{out} = 4$ mg O_2/L	$Q_G = 10$ m³/min at 20°C, 1 atm

(a) Find the overall gas transfer coefficient (K_L) and the product $K_L a_L$ for the system. Assume that the average properties of all the bubbles in the system can be approximated as their properties at mid-depth. Do not assume that interfacial resistance necessarily limits the rate of gas transfer.

(b) Compare the values of $K_L a_L$, $Q_G H/V_L$, and k_I for oxygen transfer in the system, and comment on the relative values.

(c) What fraction of the injected oxygen is actually transferred to the liquid phase?

(d) What fraction of the oxygen that enters solution is consumed by bioreactions?

6-9. The wastewater to be treated in the activated sludge system for which the preliminary design was carried out in Example 6-6 is expected to contain small amounts of contaminants from gasoline (benzene, toluene, and xylene, often abbreviated as BTX). Estimate the maximum fraction of each of these compounds that might be stripped from the water for the given air and liquid flow rates. For the given operational conditions, do you expect the actual removal fractions to be close to the maximum values? Assume that the α and β values for the BTX compounds are the same as those for oxygen.

6-10. Estimate the oxygen concentration that will be obtained in the effluent from a diffused aeration system in which 0.75 m³ of air at 25°C and 1 atm is injected per cubic meter of incoming water. The aeration tank is 3 m deep and behaves as a plug flow reactor operating at steady state, with a hydraulic residence time of 1 h. The bubbles are generated uniformly across the bottom of the tank. The diameter of the injected bubbles at mid-depth is 0.19 cm, and they rise at a uniform velocity of 22 cm/s. The bubbles cause the tank to be well mixed in the plane perpendicular to the water flow, although there is a gradient in the direction of flow. The initial dissolved oxygen concentration in the water is 2.5 mg/L, H_{pc} for oxygen in the system is 2.27×10^{-2} atm/(mg/L), and K_L is 10 cm/h. Treat the average total pressure in the tank as the pressure at mid-depth. State any assumptions you make, and check them if possible.

Hint: Find the total amount of interfacial area in the tank at any given time by multiplying the number of bubbles in the tank at any time by the surface area per bubble.

6-11. A surface aeration system is rated to provide a standard oxygen transfer efficiency (SOTE) of 1.04 kg O_2/kWh at a power-to-volume (P/V) ratio of 50 W/m³. The SOTE is defined as the standard oxygen transfer rate (SOTR, the transfer rate from air at 20°C and 1 atm total pressure to clean water maintained at zero dissolved oxygen content) divided by the P/V ratio.

(a) What value of $K_L a_L$ would the manufacturer report for the aerator under standard conditions, based on this test and an assumption that the oxygen transfer rate is limited by interfacial transport? The saturation value of dissolved oxygen under the test conditions is 9.06 mg/L.

(b) What oxygen transfer rate (OTR, in kg O_2/m^3 min) would the aerator achieve if it was operated with the same P/V ratio in an activated sludge aeration basin (a CFSTR) with a steady-state dissolved oxygen level of 2.0 mg/L, and a temperature of 30°C? Assume $\alpha = 0.85$, $\beta = 0.95$, and that the temperature correction factor for the gas transfer coefficient is $(K_L a_L)_{T_2} = (K_L a_L)_{T_1} \theta^{T_2 - T_1}$, where $\theta = 1.024$ and T is in °C. At 30°C, the oxygen solubility in clean water (in equilibrium with 0.21 atm O_2) is 7.56 mg/L.

(c) Now imagine water with the same characteristics as the wastewater but with no on-going oxygen consumption (no microbial activity).

  (i) If the wastewater enters the basin devoid of oxygen, what detention time is required in a CFSTR for the effluent (still at 30°C) to have a dissolved oxygen concentration of 6 mg/L?

  (ii) If a plug flow reactor were used, what detention time would be needed?

6-12. Small amounts of cyclohexane in an industrial wastewater are being stripped quite effectively in a CFSTR, gas-in-liquid (diffuser) treatment system. Recently, the industry has begun to use ethyl acetate in its work, and some of that gets into the wastewater as well. Information about the two chemicals is provided in Table 6-Pr12A, information about the CFSTR is provided in Table 6-Pr12B, and information about

TABLE 6-Pr12A. Chemical Information

Chemical	H_{cc} (L_L/L_G)	D_L (cm^2/s)	D_G (cm^2/s)
Cyclohexane	7.167	9.06E–6	0.088
Ethyl acetate	0.00532	9.66E–6	0.091

TABLE 6-Pr12B. Reactor Information

Parameter	Symbol	Units	Value
Liquid flow rate	Q_L	m^3/s	0.06
Gas flow rate	Q_G	m^3/s	5.51
Liquid volume	V_L	m^3	108

TABLE 6-Pr12C. Reactor Performance

Parameter	Symbol	Units	Value for Cyclohexane	Value for Ethyl Acetate
Overall mass transfer coefficient	$K_L a_L$	s^{-1}	0.050	
Fraction of gas transfer resistance in gas phase	–	–	0.02	No value needed
Liquid phase mass transfer coefficient	$k_L a_L$	s^{-1}		
Gas phase mass transfer coefficient	$k_G a_L$	s^{-1}		
Fraction remaining in CFSTR effluent	$c_{L,out}/c_{L,in}$	–		

the current behavior of cyclohexane in the CFSTR is provided in Table 6-Pr12C. Fill in the missing information for ethyl acetate in Table 6-Pr12C. (Note: it might be necessary to also calculate values that are not shown in the table.)

Comment on whether the CFSTR is effective at stripping ethyl acetate, and if not, how it can be improved. Assume that:

- Dependence of the mass transfer coefficients on the diffusion coefficients is the same as it is in the Onda correlations in both phases.
- Gas phase (air) has no cyclohexane and no ethyl acetate in it as it enters the diffusers.
- Values less than 0.2 are much less than 1, and values greater than 5 are much greater than 1.

6-13. As explained in the chapter, the kinetics of gas transfer in gas-in-liquid systems can be limited by phenomena occurring at the macroscopic or of microscopic scale. Alternatively, a system might not be characterized by either of these limiting case conditions.

(a) A diffused aeration CFSTR has been designed to strip 90% of a contaminant from the influent under steady state conditions, based on an assumption that the microscopic limitation would apply. The system operates with a liquid flow rate of 20 m^3/min and a detention time of 60 min. The injected air does not contain any of the contaminant. What value was used for the gas transfer rate constant?

(b) When the reactor was built and operated, the removal achieved was only 80%, because the assumption of microscopically limited kinetics was incorrect. What was the actual value of the rate constant for gas transfer?

(c) The gas flow rate in the system is 60 m^3/min, and H_{cc} for the target compound is 1.56 m$_L^3$/m$_G^3$.

Is this information consistent with your results for part (b)?

(d) Because the actual removal efficiency is less than desired, some changes to the reactor are being considered. One possibility is to install a baffle that would split the tank in half to create two CFSTRs in series. All other conditions will remain the same.

   (i) Will the value of the rate constant change in this case? Justify your response with words or calculations.

   (ii) What percentage removal of the contaminant will be achieved in this two-reactor system?

6-14. In practice, design calculations for diffuser systems are often carried out without taking the pressure variation in the tank into account at all. To see how much effect such variations might have on the design, carry out all the calculations in Example 6-6 without correcting for depth.

6-15. The results of four bench-scale oxygen absorption experiments using diffused aeration are shown in Figure 6-Pr15. In all cases, the water was initially bubbled with $N_2(g)$, so that the initial dissolved oxygen (DO) concentration was <0.5 mg/L. In the experiments, air at 1.0 atm P_{tot} was bubbled into the beakers at a flow rate of 1.0 L/min, and the DO concentration (in milligrams per liter) was measured as a function of time. The value of H_{cc} for oxygen at the temperature of the experiments was $30\,m_L^3/m_G^3$.

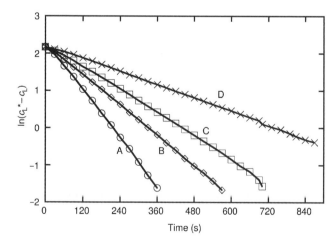

FIGURE 6-Pr15. Results of various laboratory experiments investigating gas transfer, as described in the problem statement.

The differences among different experiments were the volume of water (1.0 or 2.0 L) and the type of aerator (either a stone diffuser like those used in fish tanks or a tube; the stone would create much smaller bubbles). The diameter of the 2-L beaker was greater than the diameter of the 1-L beaker, so that the depth of water in the 2-L experiments was only slightly greater than that in the 1-L experiments, and the depth of water was small enough that corrections are unnecessary.

(a) Identify the conditions of Experiment A (i.e., 1.0 or 2.0 L, stone or tube aeration).

(b) Identify the conditions of Experiment D.

(c) Estimate the rate constant for gas transfer for Experiment A.

(d) For the conditions of Experiment A, do you think the gas transfer kinetics is limited by advective transport (macroscopic limitations) or by interfacial transport (microscopic limitations)? Carry out the appropriate calculations to justify your answer, even if you think it is obvious.

(e) On the basis of the information given, it is not possible to determine the conditions of Experiments B and C. However, if the experiments had all been done in beakers with equal diameters, so that the height of the water doubled when the volume doubled, the relationships among the lines would be different. How would they be different (from what they are now) and why?

(f) Why did bubbling pure nitrogen through the water reduce the oxygen concentration to (near) zero? Why this was a good thing to do?

6-16. (a) Calculate $K_L a_L$ for oxygen in an aeration system, based on the following dissolved oxygen (DO) concentrations during a test involving batch aeration of tap water.

Time (min)	Concentration (mg/L)
0	2.00
1	3.24
2	4.16
4	5.68
6	6.63
8	7.24
10	7.69
20	8.47
30	8.58
40	8.60
50	8.60

(b) What DO concentration would be expected in a batch experiment after 10 min of aeration if the same aerator is used on wastewater at the same temperature? For this wastewater, the DO concentration at time zero is 1.2 mg/L, $\alpha = 0.95$, and $\beta = 0.86$.

6-17. "Postaeration" tanks are CFSTRs that are designed to increase the oxygen concentration of a wastewater after biological treatment and before discharge to a receiving water. In a typical design situation, the oxygen concentration in the influent to the postaeration system and the required effluent concentration would both be known, along with the liquid flow rate. Assume that this oxygen transfer is accomplished by an air diffuser system.

(a) Develop an expression for the minimum volume of a CFSTR that could achieve a specified postaeration goal (i.e., a specified dissolved oxygen concentration), using an aeration system with known $K_L a_L$. Your result should be an algebraic expression for the volume in terms of known quantities. Indicate how each of those quantities is known (i.e., what in the description above allows you to know each quantity of interest?). Assume that all oxygen demands (i.e., reactions that consume oxygen) have been exerted in treatment prior to the postaeration tank, so that the only reaction that occurs is gas (oxygen) transfer. Assume also that gas transfer is kinetically limited.

(b) What is the required volume if two equi-sized CFSTRs in series are used, each equipped with the same type of diffuser system as considered in part (a)? What is the ratio of the volume required for the two-reactor system to that required for the one reactor system?

6-18. One special case of interest involving gas transfer in conjunction with another reaction in solution involves bubbling $O_2(g)$ into a solution containing Co(II) plus excess sulfite (SO_3^{2-}). Batch systems set up in this way are commonly used to determine $K_L a_L$ values for O_2. The cobalt acts as a catalyst for the oxidation of SO_3^{2-} so that the reaction of SO_3^{2-} with dissolved O_2 is essentially instantaneous and complete. As a result, the dissolved O_2 concentration is steadily zero even though oxygen is constantly entering solution. The stoichiometry is such that two moles of SO_3^{2-} react with one mol of O_2.

Although the assumption is rarely stated, the analysis of such systems implicitly assumes that the amount of oxygen transferred from the gas bubbles is insufficient to deplete the gaseous concentration of oxygen significantly; that is, the analysis assumes that the system meets the criteria of Case Ib. For a system in which this assumption is valid, develop an equation to determine $K_L a_L$ from batch aeration data based on the decline in the SO_3^{2-} concentration over a specified time interval.

6-19. Chloroform ($CHCl_3$) is present in the finished water of a drinking water treatment plant at a concentration of 150 μg/L. The water is at 25°C. The drinking water standards dictate that the total trihalomethane (chloroform plus other related compounds) concentration must be less than 80 μg/L. It is proposed to lower the chloroform concentration by bubbling the water with air (which contains no $CHCl_3$).

(a) What is the gas-phase concentration in equilibrium with the finished water?

(b) The stripping process will be carried out in a CFSTR. The plant personnel perform a batch stripping test in the tank and obtain the following results for the dissolved chloroform concentration as a function of time. Determine the mass transfer coefficient, $K_L a_L$, assuming that the transfer is limited by interfacial transport.

Time (min)	$CHCl_3$ Concentration (μg/L)
0	150.0
1	124.4
3	99.9
6	69.5
10	37.2
15	19.2

(c) Reanalyze the data from the stripping test, but this time assume that the transfer is limited by advective transport. What parameter can you determine from such an analysis, and what is its value?

(d) Determine the detention time in a steady-state CFSTR to achieve an effluent chloroform concentration of 40 μg/L, for the assumption in part (b). Would the computed detention time be shorter, longer, or the same if the assumption in part (c) applied?

(e) Describe how you could determine which, if either, of the assumptions in parts (b) and (c) were valid.

6-20. The diffused aeration system in a new activated sludge plant is being tested in batch experiments with tap water at 25°C, yielding the following data.

Time (min)	O$_2$ Concentration (mg/L)
0	2.10
1	2.50
2	2.88
3	3.22
4	3.57
5	3.90
6	4.20
7	4.50
8	4.75
10	5.25
12	5.69
14	6.06
16	6.40
18	6.70
20	6.98

(a) Estimate k_I and the saturation oxygen concentration in the water. Ignore the effect of pressure on c_L^* (i.e., assume that c_L^* at the bottom of the reactor is the same as it would be if P_{tot} at that location were 1.0 atm).

(b) In the test described in part (a), the liquid volume in the tank was 300 m^3, and the air flow rate at 25°C and 1 atm total pressure was 3 m^3/min. Estimate $K_L a_L$ in the reactor. Use the Henry's constant for O$_2$ at 25°C given in Table 5-2; that is, 31.4 L$_L$/L$_G$.

(c) Estimate the maximum oxygen transfer rate (g/min) that this aerator could achieve in a system with these values of V_L and Q_G. Do not restrict your consideration to the range of DO concentrations in the test.

(d) Although the test was carried out using normal air (21% O$_2$), the atmospheric pressure was not recorded, so the oxygen concentration in the incoming bubbles ($c_{G,in}$) is not known. Estimate the value of $c_{G,in}$ in milligrams per liter, and the rate at which gaseous oxygen enters the reactor in grams per minute.

(e) Use the results from parts (c) and (d) to calculate the maximum oxygen transfer efficiency; that is, the maximum fraction of the gas-phase oxygen entering the tank that can be transferred to the liquid phase.

6-21. A volatile solvent with $H_{cc} = 0.067$ is stripped from water in a well-mixed batch reactor by means of bubble aeration using clean air. The concentration of the solvent in solution decreases by 50% every 10 h. The volume of the reactor is 10 m^3 and the gas flow rate is 15 m^3/h.

(a) What is the first-order rate constant for gas transfer, k_I (in units of h^{-1}) for mass transfer in this reactor?

(b) What is $K_L a_L$ (in units of h^{-1}) in the system? Comment on the difference between this value and the value of k_I determined in part (a).

6-22. Although the derivation for the rate of gas transfer in gas-in-liquid systems in the text focuses on gas stripping systems, the analysis is equally valid for gas-in-liquid absorption systems.

(a) Schulz et al. (1995) discussed the design of high efficiency ozone contactors (reactors in which ozone is dissolved into the liquid phase through diffusers). Ozone is expensive to produce, so it is important that most of the gaseous ozone that is produced is transferred into the water. Typically, in ozone contactors, ozone makes up 4–12% of the gas phase, with the rest being normal air.

To maximize transfer from the bubbles to the solution, ozone contactors are usually quite deep and use diffusers that give very fine bubbles. The authors reported that ozone transfer efficiencies up to 98% could be achieved in these systems. Under these circumstances, is gas transfer likely to be kinetically limited, equilibrium limited, or neither, or is it impossible to know what limits the transfer efficiency? Justify your answer.

(b) After transferring into solution, ozone can undergo reactions in the liquid phase in which it self-decomposes and in which it reacts with solutes or microorganisms; we designate the rates of these groups of reactions as r_d and r_r, respectively. Assume that the rates of both groups of reactions are first order with respect to the ozone concentration and that no other species are included in the rate expressions. Derive an expression for the steady-state ozone concentration in the effluent of an ozone contactor that operates as a CFSTR, using kinetic expression for gas transfer that you argued for in part (a).

6-23. (a) A diffused aeration system has $K_L a_L = 0.150$ min^{-1} for oxygen transfer. Assuming that the ratio k_G/k_L for this system is 3.0, compute the gas- and liquid-phase mass transfer coefficients, $k_G a_L$ and $k_L a_L$ for oxygen, as well as the fraction of the total gas-transfer resistance that resides in each phase. For the given solution at the temperature of the test, H_{cc} for oxygen is 30 L$_L$/L$_G$.

(b) As in the Onda correlations, assume that $k_L a_L$ is proportional to $D_L^{0.5}$, and that $k_G a_L$ is proportional to $D_G^{0.667}$. Using the following information, estimate:

  (i) the liquid- and gas-phase mass transfer coefficients for TCE,

  (ii) the overall mass transfer coefficient for TCE,

  (iii) the fraction of the total resistance in each phase for TCE.

	D_L (cm^2/s)	D_G (cm^2/s)	H_{cc} (V_L/V_G)
Oxygen	2.03×10^{-5}	0.219	30
TCE	1.07×10^{-5}	0.088	0.32

(c) The gas transfer equipment characterized in parts (a) and (b) is installed in an activated sludge reactor that operates as a steady-state CFSTR with a detention time of 2 h. What fractional removal of TCE is expected to occur by stripping, assuming that the biological activity has no effect on TCE or gas transfer, and that TCE transfer is limited by interfacial transport (i.e., it is kinetically or microscopically limited).

(d) Air is supplied to the activated sludge at a rate of $0.05 \, m_G^3/m_L^3$ min. Using this value in the full expression for the rate of gas transfer, find the value of k_I that, in essence, replaces $K_L a_L$ in the overall mass balance. Re-do part (c) with this new kinetic expression, and compare your answers.

6-24. A manufacturer of printed circuit boards utilizes a batch process that takes place in a tank containing an aqueous solution of formaldehyde. The tank has a liquid volume of 8 m^3 and is bubbled with 20 m^3/h of air, both to provide mixing and to prevent an unwanted side reaction.

(a) One common approach for finding $K_L a_L$ for oxygen in this type of system is to deoxygenate the water (make the concentration of dissolved oxygen zero) and then measure the dissolved oxygen (DO) concentration over time as air is bubbled into the tank. In one such test, the DO level goes from zero to approximately 95% of the saturation (equilibrium) value in 30 min. The saturation value for oxygen at room temperature is 9.0 mg/L. Estimate $K_L a_L$ for oxygen in the tank, assuming that oxygen transfer is kinetically limited.

(b) Assuming that the ratio k_G/k_L for oxygen is 3.5, estimate $K_L a_L$ for formaldehyde in the system, based on the information given below. Assume that k_G varies as $D_G^{0.667}$ and k_L varies as $D_L^{0.5}$.

	D_L (cm^2/s)	D_G (cm^2/s)	H_{cc} (L_L/L_G)
Oxygen	2.05×10^{-5}	0.200	30
Formaldehyde	1.56×10^{-5}	0.136	2.8×10^{-5}

(c) The concentration of formaldehyde in the bath is approximately 19 g/L. Determine whether the stripping of formaldehyde is kinetically limited, equilibrium limited, or somewhere in between. If it is limited in either way, write out the operative, simplified rate expression for stripping.

(d) Based on all of the above information, estimate the mass emission rate of formaldehyde caused by the bubbling of air into the tank.

6-25. A completely mixed basin is being used for biological wastewater treatment at 20°C. The system is operating at steady-state with a dissolved oxygen concentration of 2 mg/L. The basin is aerated by air injected through fine-bubble diffusers, for which the efficiency of aeration ($\eta = 1 - c_{G,out}/c_{G,in}$) is reported to be 17%. The basin is 3.5 m deep, the liquid inflow rate is 31 m^3/min, and the residence time 4 h. The gas flow rate is 1.63 SCM/s, the gas-phase volume fraction in the basin is estimated to be 0.005, and the saturation concentration of dissolved O$_2$ in the water is 7.1 mg/L.

(a) Estimate $K_L a_L$ for oxygen in this system, as implied by the efficiency, η.

(b) Estimate the bubble residence time.

(c) Make an independent estimate of $K_L a_L$, based on boundary layer theory. Use the average bubble rise velocity implied by the bubble residence time and assume a bubble diameter of 2 mm.

6-26. The city of Cincinnati, OH, is located on the Ohio River and draws its drinking water from the river. Imagine a scenario in which a major spill of carbon tetrachloride (CCl$_4$) has occurred on the Kanawha River, which is an upstream tributary to the Ohio River. City officials wish to predict whether the concentration of CCl$_4$ at their intake will ever exceed 5 μg/L. The hypothetical situation is as follows. The chemical spill enters the Kanawha R. as it passes Charleston, WV, 100 km upstream of the river's confluence with the Ohio R. at Point Pleasant, WV. Cincinnati is on the Ohio R., 300 km downstream from Point Pleasant.

- The maximum CCl$_4$ concentration at the point of the spill is 10 mg/L.

- The average river velocity is 0.17 m/s in the Kanawha R. and 0.15 m/s in the Ohio R.
- The average reaeration coefficient (i.e., $K_L a_L$ for O_2) at 20°C is estimated to be $2.3\,d^{-1}$ in the Kanawha River between Cincinnati and Point Pleasant and $0.5\,d^{-1}$ in the Ohio River stretch between Point Pleasant and Cincinnati. (These estimates reflect the steeper slope and shallower depth of the Kanawha.)
- Average Kanawha River flow at Charleston is 303 m^3/s
- Average Ohio River flow at Cincinnati is 3144 m^3/s.

(a) Under these conditions, and with other assumptions of your own choosing (but which you should specify), estimate the maximum CCl_4 concentration at Cincinnati if the spill occurs in the summer, with $T = 20°C$.

(b) Make the same calculations under winter conditions, with 60% ice cover and a water temperature of 0°C.

(c) Discuss quantitatively the magnitude of the error in your calculation that would be associated with each of the following:
- Not accounting for differences in diffusivity between CCl_4 and O_2.
- 50% overestimation of both values of $K_L a_L$.
- 50% underestimation of total travel time.
- For part (b), failure to account for the ice cover.
- For part (b), failure to account for temperature effects on diffusivity (and therefore $K_L a_L$).

6-27. Water percolating through a landfill collects the by-products of microbial degradation of the buried wastes. The microorganisms rapidly deplete the dissolved oxygen, establishing conditions under which steel and other iron-containing substances can release Fe^{2+} into the water. The water that has passed through the waste, referred to as leachate, must be collected and treated. Often, one step in the treatment process is aeration, which causes oxygen to enters the water so that it can react with Fe^{2+} to form Fe^{3+} according to the following reaction:

$$Fe^{2+} + \tfrac{1}{4}O_2 + H^+ \leftrightarrow Fe^{3+} + \tfrac{1}{2}H_2O.$$

The rate of this reaction is highly pH dependent, as indicated by the following rate equation:

$$r_{Fe^{2+}} = -k\, c_{Fe^{2+}}\, c_{O_2}\, c_{OH^-}^2$$

$$k = 1.6 \times 10^{12} \left(\frac{L}{mg\, O_2}\right)\left(\frac{L}{mol\, OH^-}\right)^2 \left(\frac{1}{min}\right)$$

where all the concentration terms refer to the solution phase.

(a) It is proposed to aerate a leachate at pH 7.7 that contains 200 mg/L Fe^{2+} and no dissolved oxygen. The operation is to be carried out in a sequenced batch reactor by filling the reactor, injecting diffused air for 1.0 hour, and then emptying and refilling the reaction vessel with a new batch of leachate. The contents will be well mixed during the aeration process. The $K_L a_L$ value for the system under the proposed conditions is $5\,h^{-1}$. Assume the gas transfer is limited by interfacial transport, the pH stays constant throughout the reaction, Fe^{2+} is the only form of Fe(II) in the solution, and the saturation concentration of oxygen in the solution is 10.2 mg/L. Compute the concentrations of dissolved Fe^{2+} and oxygen at the end of the aeration step. (Hint: write mass balances on dissolved oxygen and Fe^{2+}, and solve them numerically to determine the concentrations of Fe^{2+} and dissolved oxygen as a function of time in the reactor.)

(b) The leachate also contains 450 mg/L inorganic carbon, generated as a byproduct of the oxidation of organic matter. Assuming that all the Fe^{2+} that oxidizes precipitates as $Fe(OH)_3(s)$ and that the carbonate and iron species control the solution acid/base balance, test the assumption of no pH change by computing what the final solution pH would be if the reactions proceeded to the extent found in part (a), and if the aeration step caused the solution to reach equilibrium with atmospheric CO_2. (Hint: the precipitation reaction consumes alkalinity, but the exchange of CO_2 between a gas and solution does not affect the alkalinity. Compute the change in alkalinity due to the precipitation reaction, then determine the pH from the alkalinity and the assumption of equilibrium with the atmosphere.) What are the implications of this calculation for the results of part (a)?

6-28. Consider a water treatment plant that includes a stripping tower designed to remove chloroform ($CHCl_3$) from the treated water. Some specifications relating to the physical features and the operation of the stripping tower are as follows:

- Height of packing, $Z = 4$ m,
- Volumetric liquid loading rate $= Q_L/A = 0.4$ m/min $= 0.00667$ m/s,
- Gas transfer capacity factor for chloroform, $R_{chloro} = 5$.

Recently, it has been discovered that the influent to the tower also contains bromoform (CHBr$_3$), another disinfection byproduct. Fill in the missing information in the table below, ultimately determining the removal efficiency expected for bromoform in the tower if it continues to be operated under the current conditions. Begin by determining the removal efficiency for chloroform, and then proceed by filling in the column for bromoform from top to bottom. The air supplied to the tower has no chloroform and no bromoform in it. All of the data shown in the table are for 20°C. Assume that the Onda correlations apply.

Name	Symbol	Units	CHCl$_3$	CHBr$_3$
Liquid diffusivity	D_L	m^2/s	1.03×10^{-9}	9.2×10^{-10}
Gas diffusivity	D_G	m^2/s	1.17×10^{-5}	8.8×10^{-6}
Henry's constant	H_{cc}	m$_L^3$/m$_G^3$	0.124	0.017
Gas transfer capacity factor	R	—	5	
Wetted area per volume of reactor	a_R	m^2/m^3	56.0	
Liquid mass transfer coefficient	k_L	m/s	0.000170	
Gas mass transfer coefficient	k_G	m/s	0.00570	
Overall mass transfer rate constant	$K_L a_R$	1/s	0.00768	
Removal efficiency	η	%		

6-29. Bromoform is being stripped from a solution at 25°C that is being fed at a rate of 500 L/min to a counter-current packed tower that is 2 m high and 1 m in diameter. The gas transfer capacity factor is 2, and the Henry's constant (H_{cc}) for bromoform in the system is 0.023. The system conditions are such that $k_L = 1.0 \times 10^{-4}$ m/s; $k_G = 2.5 \times 10^{-3}$ m/s, and $a_T = 105$ m^2/m^3.

(a) What are the liquid and gas loading rates, in units of kg/(m^2 s)?

(b) What is the height of a transfer unit (HTU)?

(c) What removal efficiency would you expect?

(d) Suppose that gas and liquid flow rates are held constant but that the tower diameter was decreased by a factor of two. In what direction would u_L (the superficial velocity, equal to Q_L/A_R), $K_L a_R$ and HTU vary (increase, decrease, or no change)? Briefly explain your answers.

6-30. A counter-current air stripping tower is intended to remove 95% of the bromoform from a liquid stream flowing at 400 L/min at 25°C. The packing is 1-in. (0.0254 m) polyethylene Intalox saddles ($a_T = 256$ m^2/m^3, packing factor, $C_f = 33$; $\sigma_c = 0.033$ for polyethylene). The tower is to be designed to operate with a pressure drop per unit length of tower height of 100 (N/m^2)/m, and a gas transfer capacity factor of 3. The surface tension of water at 25°C is 0.072 N/m, and the diffusivities of bromoform in water and in air at 25°C are 1.04×10^{-9} and 8.8×10^{-6} m^2/s, respectively.

(a) What airflow rate is needed and what tower diameter is required to give the cited level of pressure loss? (Do not be surprised by the very high air:water ratio required for removal of this hard-to-strip compound.)

(b) Compute the height of a transfer unit, the number of transfer units, and the height of the column required to achieve the specified removal efficiency.

6-31. Tetrachloroethylene (PCE) is being stripped from solution in a column contactor, under the following liquid and gas flux conditions: 2-in. (0.0508 m) plastic Berl saddles, liquid flux at 10 L/s m^2; gas flux at 200 L/s m^2; 20°C.

(a) Estimate the overall mass transfer coefficient ($K_L a_R$) using the Onda correlations.

(b) Calculate the PCE removal in a packed bed 3 m high, assuming no PCE in the influent air.

(c) How large a column diameter would be required to treat 10,000 m^3/day?

Useful information about 2-in. Berl saddles was provided in Example 5-11 and is repeated below, along with information about the properties of the fluid and PCE at 20°C.

$a_T = 105$ m^{-1} $H_{cc} = 0.53$ L$_L$/L$_G$
$D_L = 8.5 \times 10^{-10}$ m^2/s $D_G = 9.2 \times 10^{-6}$ m^2/s
$\mu_L = 1.0 \times 10^{-3}$ Pa s $\mu_G = 1.82 \times 10^{-5}$ Pa s
$\rho_L = 998$ kg/m^3 $\rho_G = 1.205$ kg/m^3
$\sigma_L = 0.073$ N/m $\sigma_c = 0.033$ N/m

7

ADSORPTION PROCESSES: FUNDAMENTALS

7.1 Introduction
7.2 Examples of adsorption in natural and engineered aquatic systems
7.3 Conceptual, molecular-scale models for adsorption
7.4 Quantifying the activity of adsorbed species and adsorption equilibrium constants
7.5 Quantitative representations of adsorption equilibrium: the adsorption isotherm
7.6 Modeling adsorption using surface pressure to describe the activity of adsorbed species
7.7 The Polanyi adsorption model and the Polanyi isotherm
7.8 Modeling other interactions and reactions at surfaces
7.9 Summary
References
Problems

7.1 INTRODUCTION

Background and Chapter Overview

Adsorption, or the selective accumulation of molecules at the interface between two phases, plays a critical role in the transport and fate of contaminants in both engineered and natural aquatic systems. Adsorption can cause a contaminant to accumulate in soil, in the sludge of a treatment plant, or in the silt that settles in a river delta, rather than remain in the aqueous solution that passes through such systems. In engineered treatment systems, adsorption often provides the most cost-effective means of reducing the concentrations of dissolved contaminants to extremely low levels.

Several different types of reactors are used to carry out adsorption processes. Generally, these reactors are designed to have a high concentration of solid surface, although with an appropriate choice of solids, that objective can often be met with a relatively low mass concentration. The two types of reactors that are used most often are continuous flow reactors in which the solid and soluble constituents move together (and then are separated in a subsequent step), and packed columns in which the solution passes through a stationary solid phase. Batch systems, in which a mixing step for contact between the two phases is followed by a separation step, are also used occasionally.

The solids used most widely to adsorb contaminants in water and wastewater treatment systems are activated carbon, metal oxides (in particular, oxides of aluminum and iron), and ion exchange resins. Activated carbon is used primarily to adsorb organics, while metal oxides and ion exchange resins are used primarily to adsorb inorganics.

The usefulness of activated carbon as an adsorbent derives from its extremely large surface area per unit mass (which allows relatively small volumes of the adsorbent to adsorb large amounts of contaminants), its strong affinity for a wide range of hydrophobic molecules (e.g., solvents and pesticides, humic substances that serve as precursors for disinfection by-products, and algal-generated compounds that impart tastes and odors to drinking water), and its relative inertness with respect to interactions with water and most hydrophilic solutes. It is available and used in both granular and powdered forms (often abbreviated as GAC and PAC, respectively). Granular activated carbon (GAC) is generally used in packed columns, whereas powdered activated carbon (PAC) is usually used in well-mixed reactors that are followed by solid/liquid separation.

Amorphous aluminum and ferric hydroxide solids $(Al(OH)_3(s)$ and $Fe(OH)_3(s)$, respectively) form when alum (hydrated $Al_2(SO_4)_3$), ferric chloride ($FeCl_3$), or other salts containing Al^{3+} or Fe^{3+} are added to water as coagulants. These solids have poorly crystallized, gelatinous structures with very large surface areas. They are typically capable of removing 40–80% of the natural organic matter (NOM) from drinking water, thereby reducing the tendency for unwanted by-products to form when the water is subsequently disinfected. The solids also can adsorb both cationic and anionic metal species (e.g., Cu^{2+}, Pb^{2+}, CrO_4^{2-}, $HAsO_4^{2-}$) and are used in treatment of both drinking water and industrial wastewater for this purpose.

Water Quality Engineering: Physical/Chemical Treatment Processes, First Edition. By Mark M. Benjamin, Desmond F. Lawler.
© 2013 John Wiley & Sons, Inc. Published 2013 by John Wiley & Sons, Inc.

Synthetic ion exchange adsorption media typically consist of small beads of a resin whose structure comprises a cross-linked, polymeric skeleton to which charged functional groups are covalently bound. The charges are neutralized by oppositely charged ions that are only loosely bound to the solid. When the resin is exposed to an aqueous solution, ions from the solution can adsorb, causing some of the loosely bound ions to be released and enter the solution; hence the name ion exchange. Some natural materials also have ion exchange properties. Ion exchange on clays can affect the composition of natural solutions that pass through soils and thereby play an important role in controlling nutrient availability to plants. Zeolites (aluminosilicate minerals with cavities large enough to accommodate relatively large ions) are sometimes mined and used in engineered ion exchange treatment processes.

Ion exchange is used for in-home water softening (adsorption of Ca^{2+} accompanied by desorption of Na^+ or K^+), for producing deionized water in laboratories or industries where ultrapure process water is required (via a two-stage process in which the resins release H^+ and OH^-, which then combine to form H_2O), and for recovering valuable metals from dilute process streams. Recently, such resins have received a good deal of attention as potential point-of-entry (POE) or point-of-use (POU) devices for removing arsenic (as the $H_2AsO_4^-$ or $HAsO_4^{2-}$ ion) from drinking water. Ion exchange reactions are more easily reversed than most other adsorption reactions. As a result, a batch of resin can be used to collect contaminants, and the contaminants can then be released by exposing the resin to a different solution (a process referred to as *regeneration* of the resin), so that the resin can be reused repeatedly.

Although the focus of this chapter is on the use of adsorption to remove soluble species from solution, the interaction is not one-way: Adsorption affects the properties and behavior of particles as well as solutes. In particular, adsorbed species can alter the tendency for particles to collide and stick to one another, so adsorbable chemicals are frequently added to suspensions to facilitate removal of the suspended particles. This process is widely used in drinking water and sludge treatment and is discussed in detail in Chapter 11.

Adsorption has many similarities to gas transfer processes, especially stripping. Both processes remove soluble contaminants from water by transferring them to a different phase and, in both cases, the removal efficiency that can be achieved is constrained by equilibrium between the two phases and, by the kinetics of mass transport. On the other hand, the two processes also differ in important ways. For instance, at least according to one widely used paradigm, solids have fixed, maximum capacities for adsorbed molecules. No such limit exists in gas transfer, because the gas phase volume and pressure are, in theory, capable of changing to accommodate any mass of the volatile species. A second paradigm for adsorption is more analogous to that for gas transfer in that it does not presuppose any limitation on the adsorptive capacity of the solid; both paradigms are presented in this chapter.

Another difference between adsorption and gas transfer is that, whereas a gas phase is envisioned to be uniform throughout, solid adsorptive surfaces are not homogeneous—they have different properties at different locations, such as on different crystal faces. Both the limited capacity of the adsorptive surface and its nonuniformity can complicate the equilibrium relationship between dissolved and adsorbed molecules, so that it differs from the simple Henry's law relationship that characterizes gas/liquid equilibrium. Several mathematical models for characterizing adsorptive equilibrium are in common use, and the most common of these are presented in this chapter.

Finally, whereas different chemical species behave independently in a gas transfer system (based on the assumption that the gas is ideal), adsorbed solutes can significantly influence the affinity of the surface for other solutes (or more of the same solute). For instance, when charged solutes (ions) adsorb, the surface can acquire an electrical charge that enhances the adsorption of oppositely charged ions and opposes the adsorption of like-charged ions.[1] Neutral adsorbed species can also affect the surface environment, increasing or decreasing its attractiveness to other solutes in various ways. These interactions can be important on surfaces, where the molecules might be closely packed, whereas they are never important in a gas phase, where the molecules are widely spaced.

In this text, two chapters are devoted to adsorption processes. This chapter presents the underlying principles of adsorptive equilibrium. Models for the kinetics of adsorption reactions, and approaches for applying the concepts developed in this chapter to the design and analysis of full-scale adsorption reactors are presented in Chapter 8.

In this chapter, we begin with an introduction to the terminology used to describe adsorption reactions and an overview of several specific applications of adsorption in environmental engineering. The most common approaches for modeling adsorption equilibrium, both conceptually and quantitatively, are then presented. A wide variety of such models have developed over the years, with different models preferred by researchers and practitioners dealing with different adsorptive systems. For example, models that are used extensively to describe adsorption of metals onto aluminum and iron oxides in industrial waste treatment processes are rarely used to describe adsorption of trace organic compounds (e.g., pesticides) onto activated carbon; yet other models are used to characterize the adsorption of

---

[1] The fact that ions adsorb at all provides one more distinction between adsorption and gas transfer. Recall that ions are extremely unstable in the gas phase, so we can ignore the possibility that they participate in any gas transfer reaction.

calcium or arsenate ions onto ion exchange resins in drinking water treatment.

Often, adsorption processes can be understood and modeled as reactions between an isolated, dissolved species and a surface. However, in many other cases, it is necessary to consider other reactions that occur in parallel or series with the adsorption reaction. For instance, adsorption of other species from solution is likely to compete with, and therefore reduce, adsorption of a target species, whereas interactions among species that have already adsorbed (e.g., formation of a new phase on the solid surface) might either facilitate or interfere with additional adsorption of the target.

In this chapter, each of these situations is discussed, and the models most commonly used in each application are presented. At the same time, however, an attempt is made to emphasize the unity of the underlying phenomena that control the partitioning of the contaminants in all applications.

Terminology and Overview of Adsorption Phenomena

Given the wide variety of applications in which adsorption is important, it is not surprising that several conceptual models have been developed to describe adsorption reactions. Unfortunately, no unified terminology for describing the process has been adopted. Some terms that have a well-accepted meaning are defined below; others that are associated with a specific model or application are described later in the chapter.

Adsorption is the selective accumulation of a chemical at the interface between two phases. The focus of this chapter is on adsorption at the interface between an aqueous solution and a solid, although much of the discussion also applies to adsorption at solution/gas and solid/gas interfaces. The reverse of the adsorption reaction (i.e., the release of an adsorbed substance to the bulk solution) is called *desorption*. The substance that adsorbs is called the *adsorbate* and, if it binds at a solid/liquid interface, the solid is called the *adsorbent*. In principle, adsorption—a surface or interfacial process—can be differentiated from *absorption*, in which a substance is transferred from one phase into the interior of another. However, it is often almost impossible to distinguish between adsorption and absorption, so the two processes are sometimes treated jointly. In such cases, the preceding terms are sometimes written as *sorption*, *sorbate*, and *sorbent*, respectively. Also, experimentally, any adsorbate that is associated with the solid is considered to be adsorbed. Therefore, if a solid is porous, adsorbate that is present in the solution inside the pores is considered to be adsorbed, even if it is not bound directly to the surface of the pore.

The surfaces of suspended solids can sometimes acquire an electrical charge that can enhance or impede sorption of ions from solution. If the tendency for a species to adsorb depends primarily on charge-based or other physical interactions between the surface and the adsorbate, the adsorbate is referred to as being *physically adsorbed* or *nonspecifically adsorbed*. On the other hand, if the tendency to adsorb depends strongly on the chemical identity of the adsorbate, the primary driving force for the reaction is attributed to specific chemical interactions between the adsorbate and the adsorbent. In such cases, the adsorbate is said to be *specifically* or *chemically adsorbed* (or *chemisorbed*). Chemically adsorbing molecules might bind to a surface even under conditions where electrostatic interactions oppose adsorption; that is, they can allow a cationic adsorbate to bind to a positively charged surface. In fact, however, most molecules interact with the surface in various ways that might be considered either physical or chemical, so these distinctions (and the associated terms) should be used and interpreted cautiously.

The amount of material adsorbed per unit amount of adsorbent is called the *adsorption density* and is commonly represented in equations as either q or Γ. The adsorption density can be quantified as the adsorbed mass per unit surface area (with units such as mg or μmoles of sorbate per m^2 of sorbent) or per unit mass of sorbent (e.g., mg or μmoles sorbate per gram of sorbent). It is common, though far from universal, to represent the adsorption density by Γ in the former case, and by q in the latter, and that convention is adopted here. The surface area per unit mass of the solid (meter square per gram) is called the *specific surface area* (SSA), so:

$$q_i = \Gamma_i(\text{SSA}) \qquad (7\text{-}1)$$

In the environmental engineering literature, adsorption densities are most often reported based on the mass of adsorbent (i.e., they are reported as q values). Therefore, throughout the chapter, generic equations that include the adsorption density are presented using q to represent that parameter. However, it should be understood that the same equations could be expressed in terms of Γ by applying the relationship shown in Equation 7-1.

When adsorption occurs in an aqueous system, the amount of a species that is adsorbed per unit volume of solution is also often of interest. This quantity can be expressed as a conventional concentration, for example, as mg of adsorbed i per liter of solution. Such terms are generally represented in equations like any other concentration; that is, as c_i, and can be computed as the product of the adsorption density and the concentration of adsorbent in the system:

$$c_{i,\text{adsorbed}}(\text{mg } i/\text{L}) = q_i(\text{mg } i/\text{g solid})c_{\text{solid}}(\text{g solid}/\text{L})$$
$$(7\text{-}2)$$

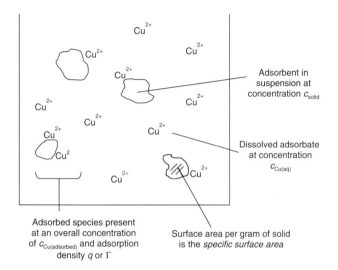

FIGURE 7-1. Schematic showing various components and descriptive terms for an adsorptive system.

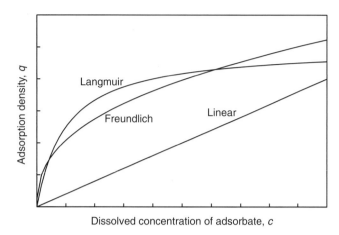

FIGURE 7-2. Generic adsorption isotherms. The mathematical basis for each isotherm shown is developed later in the chapter.

Several of the preceding definitions are illustrated schematically in Figure 7-1.

For a given system composition, only one distribution of adsorbate between the dissolved and the adsorbed states is consistent with chemical equilibrium. In the absence of other forces, the system would ultimately reach this condition. Relationships expressing the equilibrium adsorption density as a function of the dissolved adsorbate concentration are called *adsorption isotherms*.[2] The term "isotherm" emphasizes the fact that the relationship (like all equilibrium constants) is sensitive to temperature; adsorption reactions are invariably exothermic, so the tendency to adsorb declines with increasing temperature. Unfortunately, the term is used in two slightly different ways in the literature: in some usages, it represents the general relationship between adsorption density and dissolved concentration that is associated with a certain conceptual model of adsorption, and in others, it represents the specific mathematical relationships applicable to particular adsorbate–adsorbent pairs.

The isotherms (in the latter sense of the term) that are used most frequently to characterize adsorption in environmental systems are known as the *linear*, *Langmuir*, and *Freundlich* isotherms. The mathematical forms of these isotherms are shown in the following equations, and the isotherms are presented graphically in Figure 7-2.

$$\text{Linear:} \quad q = k_{\text{lin}} c \quad (7\text{-}3)$$

$$\text{Langmuir:} \quad q = \frac{q_{\max} K_{\text{Lang}} c}{1 + K_{\text{Lang}} c} \quad (7\text{-}4)$$

$$\text{Freundlich:} \quad q = k_{\text{f}} c^n \quad (7\text{-}5)$$

where k_{lin}, K_{Lang}, $q_{\max}$, k_{f}, and n are all empirical constants.[3]

The linear isotherm (Equation 7-3) describes systems in which the equilibrium adsorption density is proportional to the dissolved adsorbate concentration. Such an isotherm is often reported to characterize the sorption of trace compounds such as pesticides in systems, where the solids are abundant (e.g., in soils). The equation defining this isotherm is exactly analogous to Henry's law in systems, where gas/liquid equilibrium is being characterized.

Under conditions where $K_{\text{Lang}} c \ll 1$, the Langmuir isotherm (Equation 7-4) also describes a linear relationship between adsorption density and dissolved concentration. However, at larger values of $K_{\text{Lang}} c$, the relationship becomes curvilinear, and when $K_{\text{Lang}} c \gg 1$, the adsorption density approaches a constant, maximum value ($q_{\max}$), which is not exceeded no matter how large the dissolved adsorbate concentration becomes. For a given adsorbate/adsorbent pair and given background solution composition, K_{Lang} is constant, so the three regions of the isotherm are often described just in terms of the adsorbate concentration. That is, the isotherm is described as being linear at low c,

---

[2] More generally, an adsorption isotherm is a relationship between the chemical activities, at equilibrium, of a dissolved adsorbate and the same species when it is adsorbed to a surface. In most environmental systems, these activities are assumed to be proportional to the dissolved concentration of the adsorbate (c) and its adsorption density (q or Γ). However, in some important cases that are discussed later in this chapter, parameters other than Γ or q are used as indicators of the chemical activity of adsorbed species. Relationships between those other parameters and c are also called adsorption isotherms, as are q versus c or Γ versus c relationships.

[3] In the literature, the Freundlich isotherm is defined in two different ways: either as shown in Equation 7-5, or with the exponent shown as $1/n$, rather than as n. Because values of n are frequently reported without showing the isotherm equation explicitly, it is critical to establish which convention an author is using so that the value can be interpreted properly.

curvilinear at intermediate values of c, and independent of c at high values of c. Langmuir isotherms describe many systems in which all the adsorptive sites are nearly uniform, such as on ion exchange resins.

The Freundlich isotherm (Equation 7-5) is curvilinear at all values of c, with the intensity of the curvature depending on the value of n. Such isotherms describe sorption of most adsorbates onto activated carbon, and many adsorbates onto metal oxides, especially if the range of dissolved adsorbate concentrations considered is large (e.g., two or more orders of magnitude).

Typically, isotherms are derived by collecting experimental data for q and c in well-mixed, batch systems containing a range of adsorbate and/or adsorbent concentrations. The suspensions are mixed long enough for adsorptive equilibrium to be achieved, and the data are then fit to the isotherm equation. A conceptual basis and mathematical derivation is provided for each of the isotherms cited earlier, and a few others, later in this chapter.

Once an isotherm has been established for a given adsorbent–adsorbate pair, it can be used to predict the distribution of adsorbate between the surface and the solution for other conditions. For instance, assuming the system reaches equilibrium, the adsorbent dose needed to reduce the dissolved adsorbate concentration from some initial value to a lower target value in a batch system can be computed from the following mass balance analysis. Assuming that the adsorbate is neither created nor destroyed in the system, but simply transfers between the solution and the adsorbent, we can write

$$\begin{pmatrix} \text{Mass of adsorbate} \\ \text{in system initially} \end{pmatrix} = \begin{pmatrix} \text{Mass of adsorbate} \\ \text{in system at equilibrium} \end{pmatrix}$$

$$c_{init}V_L + q_{init}c_{solid}V_L = c_{fin}V_L + q_{fin}c_{solid}V_L \quad (7\text{-}6)$$

where c_{solid} is expressed as the mass of solid per unit volume of solution. (If c_{solid} is expressed as the mass of solid per unit volume of reactor, the same equation applies are with the exception that the terms containing c_{solid} multiplied by V_R instead of V_L.) Dividing through by V_L and rearranging yields

$$c_{init} - c_{fin} = (q_{fin} - q_{init})c_{solid} \quad (7\text{-}7)$$

$$c_{solid} = \frac{c_{init} - c_{fin}}{q_{fin} - q_{init}} \quad (7\text{-}8)$$

In the commonly encountered case in which fresh adsorbent is used (i.e., $q_{init} = 0$), this result simplifies to:

$$c_{solid} = \frac{c_{init} - c_{fin}}{q_{fin}} \quad (7\text{-}9)$$

If the adsorption isotherm characterizing the system has been established, it can then be used to substitute for q_{fin} in

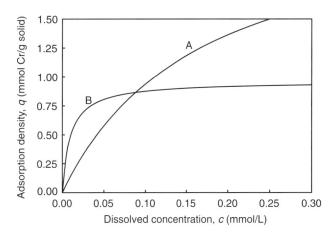

FIGURE 7-3. Hypothetical isotherms for adsorption of CrO_4^{2-} onto two adsorbents.

terms of c_{fin} in Equation 7-8 or 7-9, allowing c_{solid} to be calculated for any given values of c_{init} and c_{fin}.

■ **EXAMPLE 7-1.** A study of adsorption of CrO_4^{2-} onto two different adsorbents (A and B) yields the isotherms shown in Figure 7-3. You wish to reduce the concentration of CrO_4^{2-} in a wastewater from 0.2 to 0.02 mmol/L (roughly 10 to 1 mg Cr/L) by sorption in a batch treatment process, using the minimum dose (g/L) of solid.

(a) Which adsorbent would you use, and why?
(b) What adsorbent dose is required?

Solution.

(a) The initial CrO_4^{2-} concentration is 0.20 mmol/L, so it is necessary to remove 0.18 mmol/L CrO_4^{2-} from solution. Also, once the solution and solid have equilibrated, $c_{CrO_4^{2-}}$ will be 0.02 mmol/L.

For a given concentration of contaminant removed from solution, the larger the adsorption density, the less solid is required. Therefore, adsorbent B, which has the larger adsorption density in equilibrium with the final, dissolved CrO_4^{2-} concentration will be the preferred one. Note that adsorbent A has a higher adsorption density in equilibrium with the concentration in the raw water, but this fact is irrelevant for answering the question, since the ultimate adsorption density is in equilibrium with the treated water, not the initial solution.

(b) The adsorption density on adsorbent B when it is in equilibrium with a dissolved concentration of 0.02 mmol/L is ~0.65 mmol CrO_4^{2-}/g solid.

The required dose of the adsorbent can therefore be computed from Equations 7-7 to 7-9 as:

$$c_{\text{solid}} = \frac{c_{\text{init}} - c_{\text{fin}}}{q_{\text{fin}}}$$

$$= \frac{(0.20 - 0.02) \text{ mmol CrO}_4^{2-}/\text{L}}{0.65 \text{ mmol CrO}_4^{2-}/\text{g solid}}$$

$$= 0.277 \text{ g solid/L} = 277 \text{ mg solid/L} \quad \blacksquare$$

7.2 EXAMPLES OF ADSORPTION IN NATURAL AND ENGINEERED AQUATIC SYSTEMS

In this section, a brief description is provided of some of the most important applications of adsorption in environmental systems.

Use of Activated Carbon for Water and Wastewater Treatment

Activated carbon is the most widely used adsorbent in water and wastewater treatment systems and, as noted previously, is used in both granular and powdered form (GAC and PAC, respectively). Contaminants that can be removed from water by adsorption onto activated carbon include both naturally occurring and synthetic organic compounds (e.g., humic substances, compounds that impart tastes and odors to drinking water, and solvents and pesticides). Many of these contaminants are so strongly attracted to the activated carbon surface that their dissolved concentrations can be reduced to submicrogram per liter levels using dosages of carbon that are economically reasonable. As an example, Figure 7-4 shows the results of a study in which 20 mg/L PAC was able to remove 85–95% of the algal-generated, odor-producing compound geosmin from solutions initially containing <50 ng/L. Activated carbon is also used to sorb volatile organic compounds from gas phases, such as the gaseous emissions from stripping of contaminated groundwater and soils.

The ability of charcoal to reduce odors was recognized in ancient Greece, and its use to remove contaminants from both air and water dates to the late 1700s (Jankowska et al., 1991). An engineering process for inducing the formation of large numbers of tiny pores in charcoal, thereby dramatically increasing the surface area available for adsorption and "activating" the solid, was patented in 1900, and the first commercial plant for production of activated carbon was established in 1911. The production and uses of the material have increased steadily since that time.

Activated carbon has been prepared from a wide variety of raw materials, including coal, peat, the hard parts of fruits and nuts (shells, pits), and waste products with high carbon content (plastics). The essential steps in its production are *pyrolysis* or *carbonization* and *activation*. In the first step, the raw material is heated to a temperature of 500–800°C in

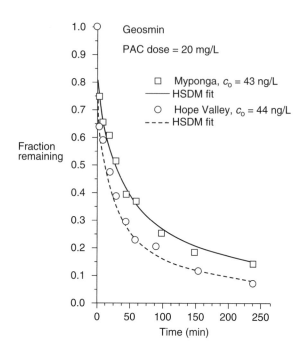

FIGURE 7-4. Adsorption kinetics of geosmin from two drinking water sources onto powdered activated carbon. The initial geosmin concentration in both waters was 44 ng/L. "HSDM" is a mathematical model for adsorption kinetics that is described later in this chapter. *Source*: Modified from Cook et al. (2001).

the absence of oxygen. This process rearranges the carbon atoms into a more ordered, graphitic structure, and drives off volatile molecules, leaving defects in the form of pores throughout the solid. Some amorphous (noncrystalline) carbon also remains in the solid. In the activation step, an oxidizing gas, generally steam or carbon dioxide, is contacted with the pyrolyzed solid at 800–1000°C. The gas diffuses into the pores, where it oxidizes and volatilizes virtually all of the amorphous carbon and some of the graphitic carbon, thereby unblocking some of the pre-existing pores and increasing the total pore volume. The resulting product typically has a specific surface area in the range 500–1500 m^2/g, and an internal pore volume of 0.5–1.0 cm^3/g. In most cases, at least 80% (and often more than 90%) of the surface area is associated with so-called *primary* and *secondary micropores* (pores with diameters <0.8 and 0.8–2 nm, respectively), and almost all of the remainder is attributed to *mesopores* (diameters 2–5 nm). Most of the pore volume is also usually associated with those sizes of pores, although in some cases a substantial fraction of the volume (but not the surface area) is in *macropores* (diameters >5 nm).

Individual particles of powdered activated carbon (PAC) typically have dimensions on the order of a few to a few 100 μm. PAC is convenient for intermittent use since it can be added to water in an existing mixing basin and then settled and/or filtered out downstream. Because the particles have a large amount of external surface area per unit mass, and the diffusion path to internal pores is short, the approach of the system to sorption equilibrium is relatively rapid. This characteristic is critical for applications of PAC, since the residence time of the adsorbent in the system is typically comparable to that of the water—on the order of minutes to hours.

Because PAC is typically applied in a well-mixed system, the maximum adsorption density on the PAC as it exits the system is governed by equilibrium between the PAC and the treated water. Since the contaminant concentration in that water is expected to be small, the adsorption density is also small (or at least, not as large as it would be if the PAC had equilibrated with a larger concentration in solution). As a result, the adsorption capacity of the PAC is often not used very efficiently in these types of applications. This idea is developed and demonstrated quantitatively in Chapter 8.

In contrast to PAC, GAC is generally used in packed bed systems—the carbon is placed in a column through which the water passes. A large fraction of the adsorptive surface in GAC particles is deep in the particles' interior, so a lengthy process of molecular diffusion into the pores is required for efficient utilization of the adsorptive capacity. On the other hand, plenty of time is generally available for this process to occur, since the GAC grains might remain in place for months to years before being replaced.[4] Also, the packed bed arrangement allows the GAC near the influent and throughout much of the column to achieve equilibrium with the influent contaminant concentration, while still producing very clean effluent (because the water comes into contact with nearly fresh GAC before it exits the column). As a result, the GAC in much of the column reaches an adsorption density that is much greater than is reached by PAC in a system with a completely mixed aqueous phase; that is, the GAC adsorption capacity is more fully utilized than that of the PAC. This point is also demonstrated quantitatively in Chapter 8.

Once contaminants begin appearing in the effluent from a GAC bed in unacceptable concentrations, the treatment process is halted, and the adsorbent is removed from the bed. The GAC might then be regenerated by a process similar to the initial carbonization and activation steps, or it might be replaced by fresh GAC. With each regeneration, some of the activated carbon is lost by oxidation, and some of the sorption capacity is lost due to failure to remove all of the adsorbed material. Eventually, regeneration is no longer justified, and the carbon is disposed of. The capital investment in a packed column is substantial, and the investment is generally justifiable only if the contaminants are present in the influent at unacceptable concentrations for a substantial fraction of the service time.

In full-scale GAC systems, adsorbates often behave very differently from their behavior in short-term tests using simplified solutions. For instance, when GAC is used to remove trace contaminants from drinking water, the contaminants often appear in the effluent much more quickly than would be predicted based on small-scale batch experiments. This behavior is often attributed to the fact that all potable waters contain molecules of natural organic matter (NOM) that can adsorb to the GAC and that are usually larger than the target adsorbates. The NOM molecules are generated primarily by the decay of plant material and have molecular weights typically ranging from a few hundred to a few thousand. Their molecular structure consists of a skeleton of aliphatic chains and aromatic rings to which various functional groups are attached. The dominant functional groups are carboxylic acids, which are mostly deprotonated at the pH of natural waters, so most NOM molecules are poly-anionic (i.e., they carry multiple negative charges) in most systems of interest.

Typically, the trace contaminants that are the targets of the treatment process adsorb more strongly to activated carbon than NOM molecules do, so they are adsorbed preferentially near the influent end of the column. As a result, when a treatment run begins, essentially all of the trace contaminants are removed near the influent end of the column for a fairly long period (often several months). During that time, the downstream portion of the column becomes extensively *preloaded* with NOM, meaning that the GAC is adsorbing NOM molecules, but no trace constituents. Later, the upstream GAC reaches equilibrium with the influent concentration of the trace contaminants, and those compounds appear for the first time in the downstream portion of the bed. These molecules have a strong affinity for the GAC, and substantial adsorptive capacity for the molecules might remain inside the GAC particles. However, if desorption of the previously bound NOM is kinetically inhibited, or if the adsorbed NOM blocks access of the trace compounds to the underlying pores, adsorption of the trace compounds is impeded. In such a case, the trace compounds can appear in the effluent long before they are predicted to do so based on short-term batch tests (in which preloading is not an issue).

Other differences between controlled laboratory tests and full-scale applications can arise because of reactions at the GAC surface. For instance, microorganisms tend to colonize the surface of the adsorbent, blocking access to part of it but also enhancing the apparent adsorptive capacity in some cases, by consuming the adsorbed molecules and releasing nonsorbable by-products. This latter process is sometimes referred to as bio-regeneration of the surface. Abiotic reactions at the surface, such as the polymerization of adsorbed phenolic species in the presence of adsorbed oxygen, can also alter system behavior (Vidic et al.,

---

[4] Note that the water spends only a short time in the reactor (typically 3–10 min). During that time, the adsorbate binds to the exposed, exterior surface of the GAC. After that binding occurs, the time available for migration of the adsorbate into the pores of the GAC is very long—essentially, until the GAC is removed from the column for regeneration or disposal.

1993). A comprehensive summary of the production, properties, and uses of activated carbon in drinking water treatment has been compiled by Sontheimer et al. (1988).

Sorption of NOM During Coagulation of Drinking Water

Sorption onto metal oxides and other inorganic solids plays a central role in controlling the behavior of NOM and many trace inorganic elements in both conventional water treatment operations and natural systems. The main adsorbents in these systems typically contain iron, aluminum, manganese, and/or calcium as cations, and oxide, hydroxide, and/or carbonate as anions. The following discussion focuses on a hydrated metal oxide as a model adsorbent, but the general principles apply to sorption onto most inorganic solids.

When a metal oxide is in contact with water, the surface oxide groups can react with water to become hydrated by reactions such as the following:

$$\equiv SOS\equiv \; + \; H_2O \rightarrow 2\equiv SOH$$

where $\equiv S$ represents a surface metal ion bound on one side to the bulk solid. The $\equiv SOH$ groups can undergo acid/base reactions with the solution to form two other surface species: $\equiv SOH_2^+$ and $\equiv SO^-$. Similar to the situation for acid/base groups in solution, the relative concentrations of the surface species depend on the solution pH and the inherent acid/base properties of the solid.

The sorption of NOM onto such hydrated oxide surfaces is commonly represented as a reaction between the carboxylic functional groups of the organic molecules and one or more surface sites, with the carboxyl groups replacing hydroxyl groups at the surface. Such a reaction is represented in Equation 7-10, with the reactants shown in the forms in which they might reasonably be present at near-neutral pH.

$$\equiv S\text{-OH} \; + \; \underset{^-O}{\overset{O}{\underset{\|}{C}}}R \; \leftrightarrow \; \equiv S\text{-O}-\overset{O}{\underset{\|}{C}}R + OH^- \quad (7\text{-}10)$$

Because the sorption reaction releases OH^- ions, it is favored by lowering the OH^- activity in solution; that is, by lowering the pH. On the other hand, at sufficiently low pH, both the surface sites and the carboxylic acid groups become protonated, and the dominant reaction becomes the following:

$$\equiv S\text{-OH}_2^+ \; + \; \underset{HO}{\overset{O}{\underset{\|}{C}}}R \; \leftrightarrow \; \equiv S\text{-O}-\overset{O}{\underset{\|}{C}}R + H_3O^+ \quad (7\text{-}11)$$

Under these conditions, the reaction releases hydronium ions to solution, and sorption becomes less favorable as the solution becomes more acidic. This combination of acid/base reactions affecting both the surface and the NOM molecules leads to a characteristic pattern in which sorption goes through a maximum at some pH. The exact pH of the maximum depends on the solids and NOM molecules involved, but for typical water treatment conditions, in which the adsorbent is a freshly precipitated hydrous oxide of either aluminum or iron, the pH of maximum adsorption is usually in the range from 4.5 to 6.0. This trend is shown for removal of NOM by alum in Figure 7-5.

The model reactions shown in Equations 7-10 and 7-11 are written to emphasize the interaction between the surface and a single functional group on an NOM molecule, and do not consider explicitly the fact that most NOM molecules have more than one carboxylic (or other) acid group. This property of the NOM might allow different parts of a single molecule to bind to the surface at two or more different sites, increasing the overall binding strength and making it quite unlikely that the molecule will desorb unless solution conditions are changed (i.e., individual bonds might break and re-form, but the chance of all the bonds breaking at once is small).

The maximum adsorption density of NOM on hydrous oxides is controlled by a combination of the total number of surface sites available, the build-up of negative charge at the surface that opposes adsorption of additional anionic molecules, and factors that control the size of NOM molecules. The importance of surface charge is suggested by the fact that increasing the concentration of dissolved divalent cations (particularly Ca^{2+}) usually increases the tendency of NOM to sorb. This observation is usually attributed to the formation of soluble complexes of Ca^{2+} ions with both soluble and adsorbed NOM molecules, lowering their net negative charge. The formation of such complexes might

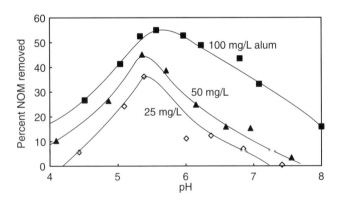

FIGURE 7-5. Removal of natural organic matter (quantified as TOC) by addition of aluminum sulfate (alum) as a function of pH. *Source*: From Semmens and Field (1980).

also decrease the space occupied by individual NOM molecules by neutralizing part of the negative charge on those molecules and allowing parts of the molecule to coil up (Cornel et al., 1986; Tipping and Ohnstad, 1984).

As noted previously, in many drinking water treatment systems, salts containing Al^{3+} and Fe^{3+} ions are added to the water, generating hydrous oxide solids (e.g., $Al(OH)_3(s)$ and $Fe(OH)_3(s)$). These freshly precipitated solids have large specific surface areas and are very effective adsorbents for NOM. The Al^{3+} and Fe^{3+} ions can also react directly with NOM molecules to form soluble complexes and solids, so that NOM removal in such systems is probably the result of both the formation of metal-NOM precipitates and sorption of NOM onto the hydrous metal oxides.

In NOM-free water, the surfaces of $Al(OH)_3(s)$ and $Fe(OH)_3(s)$ particles usually carry a slight positive charge at near-neutral pH. However, when low doses of the salts are added to a natural water, the NOM-to-metal ratio in the suspension is large, and NOM adsorption causes the solids' surfaces to have a net negative charge. This surface charge impedes collisions that are necessary for the particles to grow into larger flocs.[5] As the coagulant dose (and therefore adsorbent concentration) increases, the NOM adsorption density on the particles decreases, and the specific surface charge (the surface charge per unit area) becomes less negative. At sufficiently large coagulant doses, the electrostatic repulsion between particles becomes small enough to allow the particles to collide and grow, as desired. These phenomena are addressed in greater detail in Chapter 11.

Sorption of Cationic Metals onto Fe and Al Oxides

The same Fe- and Al-based salts that are used as coagulants in drinking water treatment are often added in industrial wastewater treatment processes, particularly those designed to remove metals from solution. For instance, in systems where pH adjustment is used to precipitate metal hydroxides, ferric chloride ($FeCl_3$) or alum might be added to coagulate the colloidal metal hydroxide particles that form. However, metal sorption onto freshly precipitated $Fe(OH)_3(s)$ (commonly called *ferrihydrite*) or $Al(OH)_3(s)$ is frequently an important component of the overall metal removal that occurs in such systems. Iron, aluminum, and other oxide minerals are also important adsorbents for metals in river systems, the oceans, sediments, and soils.

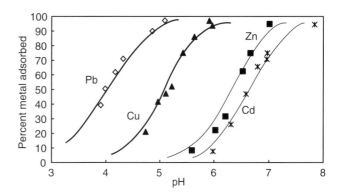

FIGURE 7-6. pH-adsorption edges for sorption of Pb^{2+}, Cu^{2+}, Zn^{2+}, and Cd^{2+} onto 10^{-3} M $Fe(OH)_3(s)$. All systems contained 10^{-3} M $Fe(OH)_3(s)$ and one metal adsorbate at a total concentration of 5×10^{-7} M. Ionic strength 0.1 M ($NaNO_3$). *Source*: After Benjamin and Leckie (1981).

Metal sorption onto hydrous oxides is typically strongly dependent on solution pH, as is shown for four metals in Figure 7-6. Similar diagrams apply for most cationic metals (e.g., Mn^{2+}, Co^{2+}, Co^{3+}, Cr^{3+}, Ni^{2+}, Zn^{2+}, Hg^{2+}, and Pb^{2+}). Typically, for a given metal concentration, sorption increases dramatically with increasing pH over a narrow pH range that is often referred to as a *pH-adsorption edge*.

Removal of metal ions from solution via precipitation (discussed in Chapter 9) is characterized by a pH dependence similar to that shown in Figure 7-6. However, typically, metal removal in the presence of adsorbents begins at a pH that is one to several units below the pH at which precipitation occurs in the absence of the adsorbent. This consistent observation has been taken as strong support for the idea that the removal mechanism is adsorption and not precipitation.

The location of the pH-adsorption edge for any given metal depends on the identity of the solid adsorbent, but the relative tendency of different metal cations to sorb to oxide surfaces is quite consistent from one solid to the next, typically following the sequence

Tendency to sorb to oxides:

$$Ca^{2+} < Cd^{2+}, Ni^{2+} < Zn^{2+}, Co^{2+}, Cu^+ < Cu^{2+}$$
$$< Pb^{2+} < Cr^{3+}, Fe^{3+}, Hg^{2+}$$

Many industrial wastewaters contain complexing agents (ligands) that bind to dissolved metal ions. These complexing agents are often added to the industrial process specifically to prevent precipitation of the metals, and, in most cases, the complexing agents also reduce the likelihood that the metal will adsorb. The simplest explanation for this observation is that the surface and the dissolved ligand compete for the metal ion. Therefore, the more metal that is bound by the complexing agent, and the more strongly that metal is bound,

---

[5] This situation is typical for suspended solids in many natural waters, where the combination of low solids concentrations and moderate to high NOM concentrations leads to large adsorption densities of the NOM. As a result, virtually all the suspended particles behave as though they had chemically similar, negatively charged, organic surfaces. The negative charge on the solids impedes collisions that might otherwise cause the particles to grow into larger flocs and settle.

the less that binds to the surface. In such cases, the presence of the complexing agent tends to shift the pH-adsorption edge to a higher pH range (Benjamin and Leckie, 1982).

The preceding scenario applies if the binding of metal–ligand complexes to the surface is weak, so that their contribution to the overall metal adsorption density is negligible. On the other hand, if the complexes adsorb to a significant extent, the overall sorption of metal reflects the combined sorption of free metal ions and complexes. This situation is particularly interesting in cases where the complexing agent can serve as a bridge between the surface and the metal. Such a scenario has been discussed by several researchers (e.g., Davis and Leckie (1978), Schindler (1990)).

A schematic depicting both possibilities is shown in Figure 7-7. In the figure, free Cd^{2+} ions are shown as potential adsorbates, and formation of chloro- or sulfato-complexes of Cd is shown as reducing Cd sorption since those complexes adsorb weakly or not at all. However, formation of a complex between Cd^{2+} and NOM is shown as potentially enhancing Cd sorption, because the NOM can sorb, forming a bridge between the surface and the metal ion.

Many anions, especially oxyanions such as phosphate (PO_4^{3-}), chromate (CrO_4^{2-}), arsenite and arsenate (AsO_3^{3-} and AsO_4^{3-}), and silicate (SiO_4^{4-}) can react with oxide surfaces in much the same way that NOM anions do (Figure 7-8), with minor differences attributable to the fact that the simpler anions are smaller and bind to fewer surface sites. As is the case with NOM sorption, the reactions often release OH^- to solution and cause the surface to acquire negative charge, and sorption often passes through a maximum value at a mildly acidic pH.

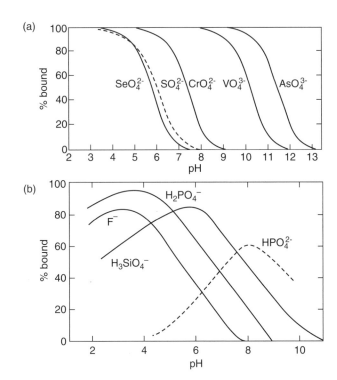

FIGURE 7-8. pH-adsorption edges for various anions. *Source*: After Stumm (1992).

Reactors for Adsorption onto Metal Hydroxide Solids

The $Fe(OH)_3(s)$ and $Al(OH)_3(s)$ solids that form in conventional coagulation processes are gelatinous solids that cannot be used in packed bed systems, because at high concentrations they form thick sludges that do not allow water to pass. However, granular forms of Fe and Al oxides are available that can be used in packed beds. The Al oxides that are used in this way are commonly referred to as activated alumina, with the activation step being a heating process that partially dehydrates the solid and increases its available surface area. Similarly, granular iron oxide adsorbents are available that consist of pure iron oxides that have been partially dehydrated and either pelletized or coated onto some other granular medium such as sand. In addition to being more space-efficient than complete-mix reactors, beds packed with oxide adsorbents tend to be easier to operate than coagulation systems (which require reagent dosing, pH adjustment, mixing, and solid/liquid separation).

7.3 CONCEPTUAL, MOLECULAR-SCALE MODELS FOR ADSORPTION

We now turn our focus to the conceptual and mathematical modeling of adsorption equilibrium. The models that are presented necessarily simplify the underlying phenomena

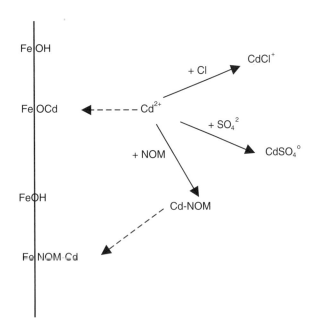

FIGURE 7-7. Schematic representation of the effects of complexation on Cd sorption onto an $Fe(OH)_3(s)$ particle.

to make them mathematically tractable, but they are thought to include the most salient features of adsorption processes.

Two Views of the Interface and Adsorption Equilibrium

Near the boundary between any two phases, a microscopic region exists wherein the physicochemical environment is different from that in either bulk phase. Adsorbed molecules reside in this region and, colloquially, can be considered to be "half in solution and half out." Two broad conceptual models have been developed to describe equilibrium between adsorbed species and those in bulk solution. In simple terms, the key difference between these two models is which "half" of the adsorbed species they focus on. That is, one model treats adsorbed molecules like dissolved species and makes adjustments to account for the fact that these species happen to be attached to solids, whereas the other model treats the adsorbed species as though they have been removed from solution, forming a separate phase as different from the dissolved phase as a solid or gas.

Adsorption as a Surface Complexation Reaction In adsorption models that focus on the similarity between adsorbed species and dissolved species, binding of the adsorbate to the adsorbent is viewed as being similar to the binding of two soluble species to form a new soluble species (e.g., $H^+ + HCO_3^- \leftrightarrow H_2CO_3$ or $Zn^{2+} + EDTA^{4-} \leftrightarrow ZnEDTA^{2-}$). The analogy that is most often made is to the combination of a metal ion and a dissolved ligand to form a metal/ligand complex. Based on this analogy, the adsorbent is sometimes described as a collection of *surface ligands*, and the adsorption reaction as a *surface complexation* reaction. This conceptual view, which was first popularized in the 1970s as an approach for describing adsorption of metals onto oxide minerals, is shown schematically in Figure 7-9. The surface ligands are presumed to correspond to specific sites on the adsorbent surface, so the model is sometimes referred to as a *site-binding* model. A key feature of site-binding models is that the total number of sites on the surface is limited, placing an upper limit on the adsorption density. Further background on this model is available in the original work of Huang and Stumm (1973), Yates et al. (1974), and Schindler et al. (1976), and in the more recent reviews by Dzombak and Morel (1990) and Davis and Kent (1990).

Ion exchange resins are good examples of adsorbents for which the site-binding model might be appropriate. Such resins have a well-defined number of binding sites per unit mass of solid; if that value and the mass concentration of the resin in a system are known, then the concentration of adsorption sites in the system (in, for instance, mol/L) can be computed and used in calculations in more or less the

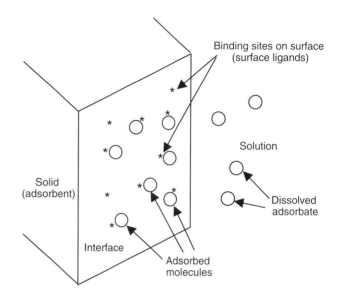

FIGURE 7-9. Schematic representation of an adsorptive solid/solution interface according to the site-binding model. The adsorbed molecules are located in the interfacial plane. The asterisks represent preferred binding sites on the surface (see text).

same manner as the concentration of dissolved species. In particular, chemical equilibrium expressions involving the site concentration can be written, (mostly) ignoring the fact that the sites happen to be part of a solid. The mathematics of such expressions are presented shortly.

Adsorption as a Phase Transfer Reaction An alternative model of the interface envisions adsorbed molecules not to be bound to specific sites on the solid surface, but to reside in a separate phase that is at or near the surface, but not really part of the solid. Adsorbed molecules are assumed to move around freely in the surface phase, as though it were a two-dimensional gas (Figure 7-10). As a result, even if only a small amount of material is adsorbed, it is considered to occupy the entire interfacial surface area, in the same sense that a bulk gas phase occupies all the volume available to it. In contrast to site-binding models, phase transfer models of adsorption place no absolute upper limit on the adsorption density; in theory, additional molecules could always be forced into the surface phase, if the driving force were large enough. The phase transfer model is used extensively to describe adsorption of organic contaminants onto soils (or components thereof) and onto suspended organic solids in natural water or wastewaters. The model has been reviewed in detail by Karickhoff (1984), Chiou (1998), Schwarzenbach et al. (2002), and Allen-King et al. (2002).

One version of the phase transfer model envisions a true phase transition to occur in pores of the adsorbent. This

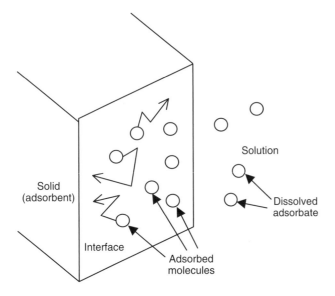

FIGURE 7-10. Schematic representation of an adsorptive solid/solution interface according to the phase transfer model. Adsorbed molecules are envisioned to move freely in the two-dimensional interphase region near the surface.

model, known as the *Polanyi model* (after the researcher who first developed it in the 1910s to describe adsorption from the gas phase), is used in environmental engineering primarily to characterize adsorption of hydrophobic adsorbates onto activated carbon. The model is based on the presumption that the adsorbates become so concentrated in the pores that they either condense to a nonaqueous liquid or precipitate as a solid. Thus, most of the adsorbed mass is envisioned to be present as tiny volumes of a bulk organic liquid, so that most of the adsorbed molecules are not even in direct contact with the surface. Nevertheless, the model is considered to describe adsorption, because it attempts to explain the behavior of species whose removal from solution is caused by the presence of a surface that selectively attracts the species (compared to water). This model and its application to adsorption of hydrophobic adsorbates onto activated carbon have been described in detail by Manes (1998).

Adsorption of Ions as Electrically Induced Partitioning: Donnan Equilibrium A third conceptual model for adsorption that is sometimes used is applicable only when the adsorbates are ions, such as in ion exchange reactions. According to this model, the adsorbed ions reside in an aqueous solution, but one that is in a different electrical environment from the bulk solution. In the specific case of ion exchange equilibrium, the solution inside the ion exchange resin is envisioned to be enriched in ions with positive or negative charge, and depleted in ions of the opposite charge, due to the high concentration of charged sites on the polymer strands that form the resin's structure. This model is less general than the two described previously and is discussed later in the chapter specifically in the context of ion exchange equilibrium.

Which Model is Best? Both the site-binding and phase transfer models of adsorption represent limiting cases; the real situation is not likely to conform to either model completely nor to be identical from one adsorptive system to the next. Furthermore, since adsorption of any molecule could be attributed to either attraction of the molecule to specific surface sites (according to the site-binding model) or a combination of exclusion of the molecule from the solution phase and non-site-specific attraction to the surface region (according to the phase transfer model), any given system could be represented by either model. Nevertheless, strong preferences have developed for representing certain adsorptive systems by one model or the other.

For ionic and highly polar adsorbates, the majority of the driving force for adsorption is usually thought to be the formation of strong surface-adsorbate bonds, and the site-binding model is preferred. Adsorption of cationic metals and oxyanions (e.g., cadmium, zinc, chromate and arsenate) onto minerals, precipitated solids (e.g., $Fe(OH)_3(s)$, $Al(OH)_3(s)$, or $CaCO_3(s)$), or ion exchange resins falls into this category. Conversely, the driving force for sorption of nonionic and nonpolar adsorbates is thought to be dominated by van der Waals attraction between the adsorbate and the undifferentiated surface (i.e., not with individual surface sites), and by the increase in entropy of the water when the hydrophobic constituent leaves the bulk solution. In these systems, adsorption can be highly favorable even if bonding of the adsorbate to identifiable surface sites is weak, and the phase transfer model is preferred. Such systems include, for example, adsorption of organic pesticides or solvents onto microbial surfaces or activated carbon.

Both the site-binding and phase transfer models of adsorption are widely used, so both models are developed in the following sections. Although the models differ in some important respects, they also share many fundamental similarities, and both the similarities and the differences are emphasized in the discussion.

7.4 QUANTIFYING THE ACTIVITY OF ADSORBED SPECIES AND ADSORPTION EQUILIBRIUM CONSTANTS

Adsorption reactions, like all chemical reactions, tend toward an equilibrium condition that can be characterized by an equilibrium constant. However, the form of the

TABLE 7-1. Adsorption Reactions According to the Site-Binding and Phase Transfer Adsorption Models

	Reaction	Equilibrium Constant	
Site-binding model[a]	$\equiv S + A(aq) \leftrightarrow \equiv SA$	$K_{ads,s\text{-}b} = \dfrac{a_{\equiv SA}}{a_{\equiv S} a_{A(aq)}}$	(7-12)
Phase transfer model	$A(aq) \leftrightarrow A(ads)$	$K_{ads,pt} = \dfrac{a_{A(ads)}}{a_{A(aq)}}$	(7-13)

[a] Unoccupied and occupied sites are represented by $\equiv S$ and $\equiv SA$, respectively. Unoccupied sites could also be represented as being occupied by a water molecule which is released when A adsorbs; that is, as $\equiv SH_2O$. Since the activity of H_2O is conventionally defined to be 1.0 in aqueous solutions, leaving the H_2O molecule out of the reaction has no effect on the value of K_{ads}.

equilibrium constant expression depends on which model of adsorption is adopted. The chemical reactions envisioned to occur and the corresponding forms of the equilibrium constant in the site-binding and phase transfer models of adsorption are shown in Table 7-1.

In the site-binding model, the reaction includes the participation of well-defined surface sites, and a term for the activity of those sites appears in the equilibrium constant expression. On the other hand, in the phase transfer model, the equilibrium constant contains only two terms—one for the activity of the adsorbate in each of the two phases where it is envisioned to reside. In either case, K_{ads} is a measure of the binding strength of the adsorbate to the adsorbent.

The chemical activities that appear in the adsorption equilibrium constant expressions (Equations 7-12 and 7-13) can be quantified in several different ways, yielding several different values for each equilibrium constant. This situation is analogous to that for gas transfer, in which the activity of volatile species can be defined in various different ways, leading to different numerical values for Henry's constant.

For adsorption reactions, the activities of adsorbed species and unoccupied adsorption sites are always presumed to be proportional to the concentrations of those species on the surface; that is, to their adsorption densities. When this convention is combined with assumptions about the activity coefficients in the system, pseudo-equilibrium constants for sorption according to the site-binding (sb) and phase transfer (pt) models can be written as follows:

$$K''_{ads,sb} = \frac{\Gamma_{\equiv SA}}{c_{A(aq)} \Gamma_{\equiv S}} = \frac{q_{\equiv SA}}{c_{A(aq)} q_{\equiv S}} \quad (7\text{-}14)$$

$$K''_{ads,pt} = \frac{\Gamma_{A(ads)}}{c_{A(aq)}} \quad (7\text{-}15a)$$

or

$$K'''_{ads,pt} = \frac{q_{A(ads)}}{c_{A(aq)}} \quad (7\text{-}15b)$$

Because $q_i = \Gamma_i$ (SSA), the conversion of $K''_{ads,sb}$ from the form based on Γ to the form based on q involves multiplying the numerator and denominator by the same term and hence has no effect on the numerical value. On the other hand, since Γ appears in the numerator but not the denominator of the expression for $K''_{ads,pt}$, the value of the pseudo-equilibrium constant for the phase transfer reaction does depend on the units used to express adsorption density, with $K'''_{ads,pt} = K''_{ads,pt}$(SSA). The ratio of the concentration of a given constituent in two phases is commonly referred to as a *partition coefficient* or *distribution coefficient*; $K''_{ads,pt}$ and $K'''_{ads,pt}$ can therefore be viewed as such coefficients.

As is true for gas transfer reactions, the distinctions between thermodynamic (activity-based, dimensionless) equilibrium constants defined in Equations 7-12 and 7-13, and the pseudo-equilibrium constants (concentration-based, dimensional) defined in Equations 7-14 and 7-15 are often ignored in the literature, so that K_{ads}, K''_{ads}, and K'''_{ads} are all referred to as equilibrium constants and are commonly represented as K_{ads} or simply K.

When the adsorption equilibrium constants defined above are quantified, $c_{A(aq)}$, $q_{\equiv S}$, $q_{\equiv SA}$, and $q_{A(ads)}$ can all be expressed using various units. Several commonly used sets of units are collected in Table 7-2.

7.5 QUANTITATIVE REPRESENTATIONS OF ADSORPTION EQUILIBRIUM: THE ADSORPTION ISOTHERM

In this section, a few isotherm equations that have been found useful for describing adsorption in aqueous systems are presented, along with conceptual models that suggest

TABLE 7-2. Units Used to Express Concentrations in Adsorptive Systems

	Comments
Aqueous Phase Concentration	
Mass concentration (mg i/L of solution)	Common
Molar concentration (mol i/L of solution)	Used commonly with site-binding model
Charge-equivalent concentration (normality) (equiv i/L of solution)	Used only to describe ion exchange equilibrium
Charge-equivalent fraction in solution (equiv i/total equiv of adsorbates in solution)	Used only to describe ion exchange equilibrium
Surface Concentration	
Mass or moles per unit mass of solid (mg, μg or mol i/g of adsorbent)	Common
Surface concentration on a real basis (mg, μg or mol i/m² of adsorbent)	Common
Surface mole fraction (moles of i/total moles of surface sites)	Fairly common, for site-binding model
Surface pressure (kPa m)	Used only for phase transfer model, usually in conjunction with IAST[a]
Equivalent-based adsorption density (equiv i/g of adsorbent)	Used only to describe ion exchange equilibrium
Surface equivalent fraction (equiv i/total equiv of all surface sites)	Used only to describe ion exchange equilibrium

[a]IAST is the ideal adsorbed solution theory. It is used to model competition among adsorbates for an adsorbent surface and is described in detail later in this chapter.

a theoretical basis for the equations. Both the equations and the models are presented first in the context of the site-binding paradigm, and subsequently in the context of the phase transfer paradigm.

Model Adsorption Isotherms According to the Site-Binding Paradigm

Characterizing the Adsorbent Sites: Surface Site Distribution Functions To represent the behavior of an adsorbate over a wide range of conditions, an adsorption isotherm must take into account characteristics of the adsorbent (e.g., the surface concentration of adsorptive sites), the solution (e.g., the concentrations of the target adsorbate and of other adsorbates that might compete with it for surface sites), and their interactions (e.g., the relative affinities of the different adsorbates for the sites). The approach we use here to characterize the surface and its affinity for adsorbates is to define a function $\Phi_{cum}(K)$ as the surface density of sites (e.g., moles of sites per square meter) having adsorption equilibrium constants less than or equal to K for a given adsorbate. We call this function the *cumulative surface site distribution function*, and we define the derivative of Φ_{cum} with respect to K; that is, $d\Phi_{cum}/dK$, as the *differential site distribution function*, $\Phi_{diff}(K)$.

The meanings of Φ_{cum} and Φ_{diff} can be illustrated by considering plots of those functions against K for an adsorbent on which all the sites are equivalent (Figure 7-11). Consider first the plot of Φ_{cum} for an adsorbate A (i.e., $\Phi_{cum,A}$). The equilibrium constant for sorption of A onto any surface site, under any conditions, is K_1. Because all of the sites have the same equilibrium constant for binding of A, the graph of $\Phi_{cum,A}$ jumps as a step function from zero to $\Phi_{cum,A}^{max}$ at $K = K_1$. The slope of $\Phi_{cum,A}$ versus K is zero at

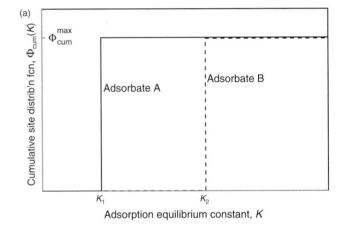

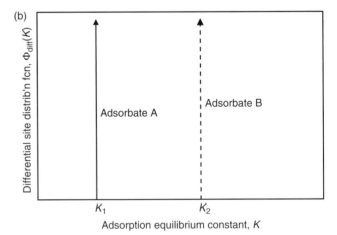

FIGURE 7-11. Cumulative and differential surface site distribution functions. (a) Φ_{cum} as a function of K for two adsorbates, for each of which the binding strength is independent of adsorption density. (b) Φ_{diff} for the systems shown in part *a*.

all values other than K_1 and infinite at $K = K_1$, so the plot of $\Phi_{\text{diff,A}}$ versus K consists of only an infinitely thin and infinitely tall spike at $K = K_1$. Thus, $\Phi_{\text{cum,A}}$ and $\Phi_{\text{diff,A}}$ can be represented as the product of $\Phi_{\text{cum,A}}^{\max}$ with the Heaviside function and the Dirac delta function, respectively:[6]

$$\Phi_{\text{cum,A}} = \Phi_{\text{cum,A}}^{\max} H_{K_1}(K) \qquad (7\text{-}16)$$

$$\Phi_{\text{diff,A}} = \Phi_{\text{cum,A}}^{\max} \delta_{K_1}(K) \qquad (7\text{-}17)$$

Figure 7-11 also shows plots of Φ_{cum} and Φ_{diff} versus K for a second adsorbate, B. The fact that the step increase in Φ_{cum} (and, correspondingly, the spike in Φ_{diff}) occurs at a value of K that is larger for adsorbate B than adsorbate A indicates that the affinity of the surface sites is greater for adsorbate B. Other than that, the functions are identical for the two adsorbates; in particular, they have identical values of $\Phi_{\text{cum}}^{\max}$, because the total population of sites is presumed to be the same regardless of which adsorbate is considered.

If the adsorbent characterized in Figure 7-11 were contacted with a solution containing dissolved A and B, anywhere from a negligible fraction to almost all of the potential binding sites could become occupied, depending on the details of the system composition. It is this relationship between system composition and surface occupation that we characterize when we develop the adsorption isotherm for a particular adsorbate–adsorbent pair. Thus, according to the site-binding model of adsorption, an isotherm can be viewed as representing the synthesis of a site distribution function for an adsorbent with information about the solution composition. Correspondingly, different site distribution functions lead to different isotherm equations, as is shown next.

The Single-Site Langmuir Isotherm The curves in Figure 7-11 characterize a surface whose affinity for a given adsorbate can be described by a single value (e.g., K_1 for adsorbate A) under all conditions. In the context of the site-binding paradigm, this situation can be interpreted as one in which all of the surface sites are identical, and in which binding of an adsorbate molecule to any given site has no effect on the equilibrium constant for binding of other molecules to the surface. Adsorption onto such a surface can be represented by the following chemical reaction and pseudo-equilibrium constant, where the fact that unbound A is in aqueous solution is implicit (so the designation (aq) is left off):

$$\equiv\!S + A \leftrightarrow \equiv\!SA \qquad (7\text{-}18)$$

$$K = \frac{q_{\equiv\text{SA}}}{q_{\equiv\text{S}} c_{\text{A}}} \qquad (7\text{-}19)$$

In Equation 7-19, $q_{\equiv\text{SA}}$ is the adsorption density of A in moles of A per unit mass of adsorbent, and $q_{\equiv\text{S}}$ can be considered the equivalent adsorption density of unoccupied sites; that is, the number of moles of unoccupied sites per unit mass of adsorbent. Note that the values of q in Equation 7-19 must be expressed in mole/mass units (as opposed to mass/mass), because the mass of unoccupied sites is undefined.

We can write a mole balance on surface sites to express $q_{\equiv\text{S}}$ in terms of the total surface concentration of adsorptive sites ($q_{(\equiv\text{S})_{\text{tot}}}$) and $q_{\equiv\text{SA}}$, as follows:

$$q_{\equiv\text{S}} = q_{(\equiv\text{S})_{\text{tot}}} - q_{\equiv\text{SA}} \qquad (7\text{-}20)$$

Substituting this expression into Equation 7-19 and carrying out some algebra, we obtain

$$K = \frac{q_{\equiv\text{SA}}}{\left(q_{(\equiv\text{S})_{\text{tot}}} - q_{\equiv\text{SA}}\right) c_{\text{A}}} \qquad (7\text{-}21)$$

$$K\left(q_{(\equiv\text{S})_{\text{tot}}} - q_{\equiv\text{SA}}\right) c_{\text{A}} = q_{\equiv\text{SA}} \qquad (7\text{-}22)$$

$$q_{(\equiv\text{S})_{\text{tot}}} K c_{\text{A}} = q_{\equiv\text{SA}} (1 + K c_{\text{A}}) \qquad (7\text{-}23)$$

$$q_{\equiv\text{SA}} = \frac{q_{(\equiv\text{S})_{\text{tot}}} K c_{\text{A}}}{1 + K c_{\text{A}}} \qquad (7\text{-}24)$$

$q_{\equiv\text{SA}}$ and $q_{(\equiv\text{S})_{\text{tot}}}$ are the adsorption density of A in the given system and the maximum possible adsorption density of A, respectively. These terms are commonly represented as q_{A} and $q_{\max}$, respectively, so Equation 7-24 can be written in the conventional form of an adsorption isotherm as:

$$q_{\text{A}} = \frac{q_{\max} K c_{\text{A}}}{1 + K c_{\text{A}}} \qquad (7\text{-}25)$$

Equation 7-25 is the *Langmuir isotherm*, as defined previously (Equation 7-4). The isotherm incorporates two constants—one ($q_{\max}$) establishing the maximum possible adsorption density and the other (K) the affinity of the adsorbent for the adsorbate. Thus, the Langmuir isotherm is a specific instance (corresponding to a specific pattern for the distribution of binding energies) of the more general site-binding model isotherm (i.e., Φ vs. K relationship).

The fractional occupation of surface sites is often of interest in adsorptive systems, and this parameter can be evaluated for systems that obey the Langmuir isotherm by dividing both sides of Equation 7-25 by $q_{\max}$:

$$\text{Fractional occupation of surface sites} = \frac{q_{\equiv\text{SA}}}{q_{(\equiv\text{S})_{\text{tot}}}} = \frac{q_{\text{A}}}{q_{\max}} = \frac{K c_{\text{A}}}{1 + K c_{\text{A}}} \qquad (7\text{-}26)$$

---

[6] The Heaviside and Dirac functions correspond, respectively, to a unit step-change and an infinitely high and narrow spike with unit area and are described in detail in Chapter 2.

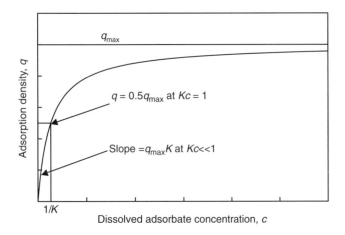

FIGURE 7-12. Graphical representation of the Langmuir isotherm.

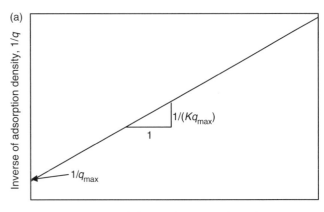

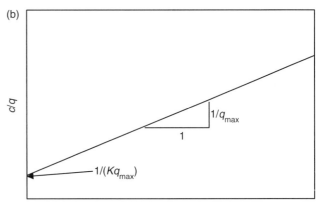

FIGURE 7-13. Graphical representations of the linearized forms of the Langmuir isotherm, according to (a) Equation 7-27 and (b) Equation 7-28.

It should be clear that q_{max} is essentially the same quantity as $\Phi_{cum,A}^{max}$—the maximum amount of A that can adsorb on a unit amount of the surface. Similarly, K in Equation 7-25 represents the same quantity as K_1 in Figure 7-11—the affinity of all the adsorbent sites for A. Thus, the Langmuir isotherm is the unique mathematical representation that characterizes systems with surface site distribution functions like those shown in Figure 7-11.

The Langmuir isotherm equation is plotted in Figure 7-12. Under conditions where $Kc_A \ll 1$, the denominator in Equation 7-25 is approximately equal to 1, and the isotherm becomes linear: $q_A \approx q_{max} K c_A$. On the other hand, if $Kc_A \gg 1$, the denominator is approximately equal to Kc_A, and $q_A \approx q_{max}$. Other factors being equal, an increase in the adsorption equilibrium constant increases the slope of the isotherm curve at low concentrations of dissolved adsorbate, and an increase in the value of q_{max} increases both the slope of the isotherm at low c_A and the adsorption density at high c_A.

To test whether experimental adsorption data fit the Langmuir isotherm, the data are often manipulated so that they can be plotted according to either of the following linearized versions of the isotherm, both of which can be derived by inverting Equation 7-25 and carrying out some algebra:

$$\frac{1}{q_A} = \frac{1}{q_{max} K}\frac{1}{c_A} + \frac{1}{q_{max}} \quad (7\text{-}27)$$

$$\frac{c_A}{q_A} = \frac{1}{q_{max}} c_A + \frac{1}{q_{max} K} \quad (7\text{-}28)$$

Based on Equations 7-27 and 7-28, if a plot of $1/q_A$ versus $1/c_A$ or a plot of c_A/q_A versus c_A is linear (Figure 7-13), the data are consistent with the Langmuir isotherm. These linearized equations have been used for many years to test the acceptability of the Langmuir isotherm to describe adsorption data. However, a plot based on Equation 7-27 tends to spread out the data at low values of c_A (i.e., high values of $1/c_A$), so that those data tend to have a greater influence in a regression analysis than the data at high c_A. Similarly, a plot based on Equation 7-28 spreads out the data at high values of c_A, giving more influence to those data in a regression analysis. Unless the data fit the Langmuir isotherm equation almost perfectly, the two methods yield different values for the parameters q_{max} and K.

With modern statistical packages for microcomputers, it is relatively easy to use nonlinear regression to find the best-fit values of the parameters by fitting the data directly to the Langmuir isotherm (Equation 7-25), without first transforming them. In most cases, this approach is preferable to testing the fit using either linearized equation.

■ **EXAMPLE 7-2.** The following data were obtained for the adsorption of toluene onto activated carbon in systems that have reached equilibrium.

c (mg/L)	q (mg/g)	c (mg/L)	q (mg/g)
0.3	27.6	5.6	210.8
0.4	36.8	8.1	199.0
0.6	46.2	11.7	234.2
0.9	65.1	16.9	206.1
1.3	79.9	24.4	215.7
1.9	121.7	35.3	293.2
2.7	127.2	50.9	247.0
3.9	179.7	–	–

(a) Determine the best-fit parameters for modeling the data according to the Langmuir isotherm.

(b) A batch treatment process has been designed to reduce the toluene concentration in a wastewater from 20.0 to 0.1 mg/L by adding an appropriate dose of fresh activated carbon ($q_{init} = 0$) and mixing the suspension long enough for equilibrium to be reached. What is the minimum dose of activated carbon needed to meet the treatment goal? What fraction of the activated carbon surface sites will be occupied?

(c) As a safety measure, the carbon dose added to each batch is 20% larger than the requirement calculated in part *b*. If a batch of water containing 30 mg/L toluene is received, will the treatment goal of 0.1 mg/L be reached?

Solution.

(a) The data are plotted in Figure 7-14, along with a smooth curve representing the fit to a Langmuir isotherm. The parameters for the model isotherm were determined using a nonlinear fitting program and have the values $K = 0.36$ L/mg, $q_{max} = 280$ mg/g.

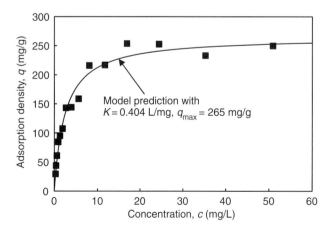

FIGURE 7-14. Langmuir isotherm plot fitting the data for adsorption of toluene onto activated carbon.

(b) The predicted adsorption density in equilibrium with the final solute concentration can be computed from the model parameters as follows:

$$q_{fin} = \frac{q_{max} K c_{fin}}{1 + K c_{fin}}$$
$$= \frac{(280\,\text{mg/g})(0.36\,\text{L/mg})(0.1\,\text{mg/L})}{1 + (0.36\,\text{L/mg})(0.1\,\text{mg/L})}$$
$$= 9.7\,\text{mg/g}$$

Then, using Equation 7-9, we can compute the required activated carbon dose:

$$c_{solid} = \frac{c_{init} - c_{fin}}{q_{fin}} = \frac{(20.0 - 0.1)\,\text{mg/L}}{9.7\,\text{mg/g}} = 2.1\,\text{g/L}$$

The fraction of the adsorption sites that is utilized can be estimated as the fraction of the maximum adsorption density that is attained, as indicated by Equation 7-26:

$$\frac{q_A}{q_{A,max}} = \frac{K c_A}{1 + K c_A}$$
$$= \frac{(0.36\,\text{L/mg})(0.1\,\text{mg/L})}{1 + (0.36\,\text{L/mg})(0.1\,\text{mg/L})} = 0.035$$

Thus, only 3.5% of the capacity of the activated carbon is being utilized.

(c) Since the activated carbon dose and initial toluene concentration are known, and the initial adsorption density is zero, the final toluene concentration can be computed by rearranging Equations 7-7 to 7-9 and substituting for q_{fin} based on the isotherm expression:

$$c_{fin} = c_{init} - q_{fin} c_{solid}$$
$$= c_{init} - \frac{q_{max} K c_{fin}}{1 + K c_{fin}} c_{solid}$$

Carrying out some algebra converts the above expression into the following quadratic equation:

$$K c_{fin}^2 + (c_{solid} q_{max} K + 1 - K c_{init}) c_{fin} - c_{init} = 0$$

Inserting a value of 120% × 2.1 g/L (=2.5 g/L) for c_{solid} and 30 mg/L for c_{init}, the result is that $c_{fin} = 0.12$ mg/L. Thus, the safety factor in the activated carbon dose would not be sufficient to meet the treatment objective in the scenario described. ∎

Possible Reasons for Non-Langmuir Behavior Although the Langmuir isotherm is adequate for describing adsorption in many systems, it fails in many others. Presumably, these deviations from the predicted behavior indicate that one or more of the assumptions used to derive the isotherm fail. The key assumption made in the derivation is that all surface sites are equivalent, expressed mathematically by representing Φ_{cum} as a step function. However, most adsorptive surfaces are neither uniformly flat nor infinite in extent, so some sites are in the middle of a crystal face of the solid, others are on edges, in corners, and so on. Sites having these different bonding structures are likely to have different affinities for adsorbate molecules.

In addition, the assumption that Φ_{cum} is independent of the adsorption density implies that binding of adsorbates to the surface has no effect on the value of K for other molecules. However, adsorbent sites are not necessarily independent, and a reaction at one point on the surface *can* affect the binding of other molecules to the surface, in at least two ways. First, sorption of ions can lead to a redistribution of electrical charge on the surface, altering the electrical interaction between the surface and other ions (both those already adsorbed and those that might bind subsequently). In addition, adsorbed molecules might interact directly with one another on the surface, either favorably or unfavorably.

Thus, the assumptions that underlie the Langmuir model are likely to be violated in many systems. In the following sections, we consider two plausible alternative surface site distribution functions and the isotherms that they lead to.

The Multisite Langmuir Isotherm The simplest approach for modeling a surface that contains groups of distinct sites is to maintain most of the assumptions of the single-site Langmuir model, but to apply those assumptions independently to each group of sites. That is, the surface can be represented as a mixture of several noninteracting sites (similar to a solution containing a mixture of metal-binding ligands), rather than as a single group of identical sites. The cumulative and differential surface site distributions for a hypothetical adsorbent containing three site-types are shown in Figure 7-15.

Note that, although the hypothetical surface has three different types of sites, each of those site types is assumed to have a single binding constant for the adsorbate, independent of how much adsorbate is bound to it or the other sites. In other words, each type of site is assumed to have the characteristics of a single-site system and to behave as it would if the other sites did not exist. In such a system, the various site-types would be occupied such that the Langmuir isotherm was satisfied for each type independently. The only linkage among the three groups of sites is indirect, in that they are all equilibrated with the same aqueous solution (i.e., with the same value of c_A). Complex natural adsorbent mixtures such as soils are often represented this way, and in some cases attempts are made to associate the different sites with different mineral phases in the mixtures.

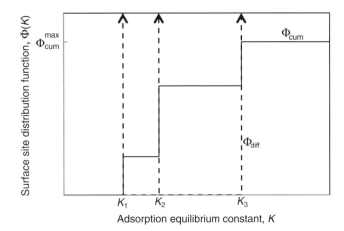

FIGURE 7-15. Cumulative and differential surface site distribution functions (solid and broken lines, respectively) for a surface with three types of sites, each of which has an affinity for the adsorbate that is independent of solution conditions. The Φ_{diff} curve includes all three spikes (when K equals K_1, K_2, and K_3) and has a value of zero at all other values of K.

Although the number of different types of sites which could be present on a given solid surface is, in theory, unlimited, adsorption data can often be fit by considering just two or three types of sites, implying that just a few site-types dominate adsorption in these systems. The most common multisite Langmuir models represent the surface as having two groups of sites: a large number of sites that bind the adsorbate weakly (low K_1), and a much smaller number of sites where the adsorptive bond is strong (high K_2). The surface site distribution functions and the overall isotherm for a single adsorbate in such a system are summations of the corresponding functions for the two sites; that is:

$$\Phi_{cum,tot} = \Phi_{cum,1} + \Phi_{cum,2} \quad (7\text{-}29)$$

$$\Phi_{diff,tot} = \Phi_{diff,1} + \Phi_{diff,2} \quad (7\text{-}30)$$

$$q_{A,tot} = q_{A,1} + q_{A,2} = \frac{q_{max,1} K_{ads,A,1} c_A}{1 + K_{ads,A,1} c_A} + \frac{q_{max,2} K_{ads,A,2} c_A}{1 + K_{ads,A,2} c_A} \quad (7\text{-}31)$$

Equation 7-31 is easily generalized to any arbitrary number N of surface sites as follows:

$$q_{A,tot} = \sum_{j=1}^{N} q_{A,j} = \sum_{j=1}^{N} \frac{q_{max,j} K_{ads,A,j} c_A}{1 + K_{ads,A,j} c_A} \quad (7\text{-}32)$$

The adsorption density on each type of site and the overall adsorption density on a hypothetical, two-site surface are shown in Figure 7-16.

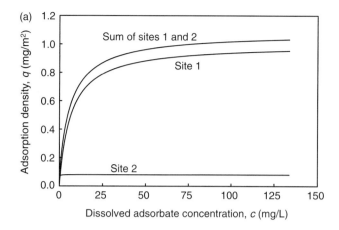

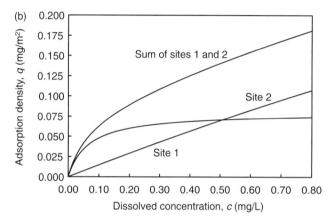

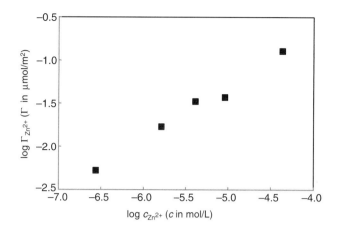

FIGURE 7-16. Adsorption onto a surface with two types of sites, each of which is occupied in accordance with the Langmuir isotherm. (a) Data over a wide range of q and c. (b) Data at very low q and c. The parameters used to generate the graphs were $q_{max,1} = 1$ mg/m^2, $K_1 = 0.15$ L/mg; $q_{max,2} = 0.08$ mg/m^2, $K_2 = 15$ L/mg.

FIGURE 7-17. Adsorption isotherm for binding of Zn^{2+} onto Cape Cod sediment. *Source*: After Davis et al. (1998).

■ **EXAMPLE 7-3.** The fit of a two-site Langmuir isotherm to data for adsorption of Zn^{2+} onto quartz at pH 5.0 is shown in Figure 7-17. The reactions and equilibrium constants used to model the data are shown below. The adsorption reaction in this system is envisioned as exchange of a Zn^{2+} ion for an H^+ ion bound to a surface oxide site, so the surface species are represented as $\equiv$SOZn$^+$ and $\equiv$SOH, respectively. The total surface concentrations of strong (St) and weak (Wk) binding sites on the adsorbent were estimated to be 0.033 and 3.84 μmol/m^2, respectively.

$$\equiv S_{St}OH + Zn^{2+} \leftrightarrow \equiv S_{St}OZn^+ + H^+$$

$$K_{St} = \frac{\Gamma_{\equiv S_{St}OZn^+} c_{H^+}}{\Gamma_{\equiv S_{St}OH} c_{Zn^{2+}}} = 10^{0.85}$$

$$\equiv S_{Wk}OH + Zn^{2+} \leftrightarrow \equiv S_{Wk}OZn^+ + H^+$$

$$K_{Wk} = \frac{\Gamma_{\equiv S_{Wk}OZn^+} c_{H^+}}{\Gamma_{\equiv S_{Wk}OH} c_{Zn^{2+}}} = 10^{-2.40}$$

(a) Derive Langmuir isotherms for sorption to each of the two sites in the given system (i.e., at pH 5.0).

(b) According to the model, what fraction of the total adsorbed zinc is bound to the strong binding sites at equilibrium Zn^{2+} concentrations of 10^{-7}, 10^{-6}, 10^{-5}, and 10^{-4} mol/L?

Solution.

(a) We can convert the given equilibrium constants into conditional constants applicable at a specific pH by dividing through by the value of c_{H^+} corresponding to that pH. When we do so, we obtain equations with the form of Equation 7-19. That is, designating the conditional equilibrium constants as K_{pH}:

$$K_{pH} = \frac{K}{c_{H^+}|_{pH}} = \frac{\Gamma_{\equiv SOZn^+}}{\Gamma_{\equiv SOH} c_{Zn^{2+}}}$$

Substituting the given K values and the known pH of 5 for the system of interest, we find

Strong binding sites:

$$K_{St,pH\,5} = \frac{\Gamma_{\equiv S_{St}OZn^+}}{\Gamma_{\equiv S_{St}OH} c_{Zn^{2+}}} = \frac{K_{St}}{c_{H^+}|_{pH\,5}} = \frac{10^{0.85}}{10^{-5.0}} = 10^{5.85}$$

Weak binding sites:

$$K_{Wk,pH\,5} = \frac{\Gamma_{\equiv S_{Wk}OZn^+}}{\Gamma_{\equiv S_{Wk}OH} c_{Zn^{2+}}} = \frac{K_{Wk}}{c_{H^+}|_{pH\,5}} = \frac{10^{-2.40}}{10^{-5.0}} = 10^{2.6}$$

The total surface concentrations of strong and weak sites correspond to $\Gamma_{\max}$ values, so the Langmuir isotherms can be written as follows:

$$\Gamma_{St} = \Gamma_{St,\max} \frac{K_{St,pH5} c_{Zn^{2+}}}{1 + K_{St,pH5} c_{Zn^{2+}}}$$

$$= (0.033 \, \mu mol/m^2) \frac{10^{5.85} c_{Zn^{2+}}}{1 + 10^{5.85} c_{Zn^{2+}}}$$

$$\Gamma_{Wk} = \Gamma_{Wk,\max} \frac{K_{Wk,pH5} c_{Zn^{2+}}}{1 + K_{Wk,pH5} c_{Zn^{2+}}}$$

$$= (3.84 \, \mu mol/m^2) \frac{10^{2.60} c_{Zn^{2+}}}{1 + 10^{2.60} c_{Zn^{2+}}}$$

(b) Substituting $c_{Zn^{2+}}$ values of 10^{-7}, 10^{-6}, 10^{-5}, and 10^{-4} mol/L, the adsorption densities on the two types of sites, and the fraction of the adsorbed zinc that is bound to strong sites, areas follows:

$c_{Zn^{2+}}$ (mol/L)	Γ_{St} ($\mu mol/m^2$)	Γ_{Wk} ($\mu mol/m^2$)	Γ_{tot} ($\mu mol/m^2$)	Γ_{St}/Γ_{tot}	Γ_{Wk}/Γ_{tot}
10^{-7}	0.0022	0.0002	0.0024	0.93	0.07
10^{-6}	0.0137	0.0015	0.0152	0.90	0.10
10^{-5}	0.0289	0.0152	0.0491	0.66	0.34
10^{-4}	0.0325	0.1470	0.1795	0.18	0.82

The speciation of adsorbed Zn^{2+} shifts from being dominated by strongly bound to weakly bound Zn^{2+} as the equilibrium dissolved Zn^{2+} concentration and the overall adsorption density increase. This result is obtained because, at relatively high Zn^{2+} concentration, the strong sites are nearly saturated (compare Γ_{St} with $\Gamma_{St,\max}$ at $c_{Zn^{2+}} = 10^{-4}$), but Zn^{2+} continues to bind to the weaker sites, which are present in higher numbers. ∎

If a surface contains several different types of sites, it might be tempting to make the approximation that the sites are occupied sequentially from strongest to weakest. However, as demonstrated in the preceding example, different groups of sites can contribute significantly to the total adsorption density over a wide range in c_{eq} values. Therefore, in general, it is better to recognize that the different types of sites are occupied simultaneously than to rely on the simplifying assumption that they are occupied sequentially.

Evidence for the presence of more than one type of surface site on an adsorbent often derives from a comparison of adsorption of different adsorbates on the same solid. If all the sites on the surface are identical, and if each adsorbate occupies a single site when it adsorbs, then the values of Φ_{cum}^{max} in the site distribution function and q_{max} in the Langmuir adsorption isotherm must be identical for all adsorbates.[7] Therefore, if the empirical data indicate that different adsorbates have different values of Φ_{cum}^{max} or q_{max}, then at least some of the sites must be available to some adsorbates but not others. In such cases, the system is sometimes represented as having a few distinct populations of sites that are not all accessible to all adsorbates, for example, one group available only to adsorbate A, one available only to adsorbate B, and one available to both A and B.

Modeling Surfaces with a Semicontinuous Distribution of Site-Types: The Freundlich Isotherm In many systems, it is not possible to represent experimental adsorption data satisfactorily even if the surface is treated as a collection of several site-types. Rather, the data suggest that the surfaces comprise a collection of sites with a virtually continuous distribution of binding energies. Some useful relationships that describe specific portions of such distributions are summarized in Table 7-3.

Plots of Φ_{cum} and Φ_{diff} versus K are shown in Figure 7-18 for two hypothetical surfaces with a wide variety of site-types. The adsorbent characterized in Figure 7-18a has sites with an irregular distribution of K values, which would be difficult to model by any simple equation. On the other hand, even though the range of K values for the adsorbent characterized in Figure 7-18b is large, Φ_{cum} and Φ_{diff} for this solid might be described fairly accurately as some simple function of K.

A heterogeneous surface might have sites with any distribution of adsorptive equilibrium constants, so there is no theoretical or fundamental reason to expect the surface site distribution function to have a particular form. However, empirically, many systems behave more or less as shown in Figure 7-18b. That is, they appear to comprise a very small number of sites with very strong affinity for the adsorbate (large equilibrium constant), and steadily increasing numbers of sites with steadily decreasing affinities. We will develop the isotherm corresponding to one particular such distribution shortly. First, though, we consider a general approach for developing the isotherm for complex surfaces such as those characterized in Figure 7-18.

Consider an adsorbent with a semicontinuous range of site-types that has equilibrated with a solution containing an adsorbate A at concentration c_A. We can represent the adsorption density of A on a group of sites that have K values within a given small range by defining functions analogous to Φ_{cum} and Φ_{diff}, except that they refer only to the sites that are occupied, rather than all the sites in that group. That is, we define the *cumulative site occupation function*, $\hat{q}_{cum}(K, c_A)$, as the adsorption density of A on sites that have adsorption equilibrium constants less than or equal to K, when the

---

[7] Note that the comparison must be on a molar rather than a mass basis.

TABLE 7-3. Various Subsets of Surface Binding Sites, as Expressed by the Surface Site Distribution Functions

$$\text{Concentration of sites with } K < K_1 = \Phi_{\text{cum}}(K_1) = \int_0^{K_1} \Phi_{\text{diff}}\, dK \qquad (7\text{-}33)$$

$$\text{Concentration of sites with } K > K_1 = (\equiv S)_{\text{tot}} - \Phi_{\text{cum}}(K_1) = \int_{K_1}^{\infty} \Phi_{\text{diff}}\, dK \qquad (7\text{-}34)$$

$$\text{Concentration of sites with } K_1 < K < K_2 = \Phi_{\text{cum}}(K_2) - \Phi_{\text{cum}}(K_1) = \int_{K_1}^{K_2} \Phi_{\text{diff}}\, dK \qquad (7\text{-}35)$$

$$\text{Total concentration of all sites} = (\equiv S)_{\text{tot}} = \Phi_{\text{cum}}(\infty) = \int_0^{\infty} \Phi_{\text{diff}}\, dK \qquad (7\text{-}36)$$

concentration of A in solution is c_A. Correspondingly, the *differential site occupation function*, $\hat{q}_{\text{diff}}(K, c_A)$, is the derivative of $\hat{q}_{\text{cum}}(K, c_A)$ with respect to K. Note that, unlike Φ_{cum} and Φ_{diff}, $\hat{q}_{\text{cum}}$ and $\hat{q}_{\text{diff}}$ depend on c_A. That is, the adsorption density on a specified subset of surface sites depends not only on the affinity of those sites for the adsorbate but also on the adsorbate concentration in solution.

Conceptually, the site distribution function, the site occupation function, and the equilibrium concentration of dissolved A can be related to one another by the following word equation.

$$\begin{pmatrix} \text{Adsorption density} \\ \text{on sites with binding} \\ \text{constants between } K \\ \text{and } K + \Delta K, \text{ for the given } c_A \end{pmatrix} = \begin{pmatrix} \text{Surface concentration} \\ \text{of all sites with binding} \\ \text{constants between } K \\ \text{and } K + \Delta K \end{pmatrix}$$
$$\times \begin{pmatrix} \text{Fractional coverage of} \\ \text{sites based on a } K \text{ value} \\ \text{in the middle of the} \\ K \text{ to } K + \Delta K \text{ range,} \\ \text{for the given } c_A \end{pmatrix} \qquad (7\text{-}37)$$

We can convert the word expressions in Equation 7-37 into mathematical terms as follows. The term on the left is the increment in total adsorption density attributable to the group of sites being considered and can be represented as $\hat{q}_{\text{cum}}(K + \Delta K, c_A) - \hat{q}_{\text{cum}}(K, c_A)$. If we multiply and divide this expression by ΔK, and let ΔK become differentially small, the expression can be simplified as follows:

$$\frac{\hat{q}_{\text{cum}}(K + \Delta K, c_A) - \hat{q}_{\text{cum}}(K, c_A)}{\Delta K} \Delta K = \frac{d\hat{q}_{\text{cum}}}{dK} dK = \hat{q}_{\text{diff}}\, dK \qquad (7\text{-}38)$$

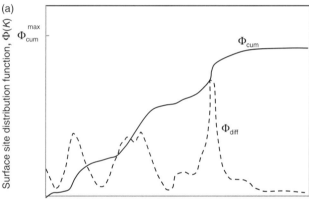

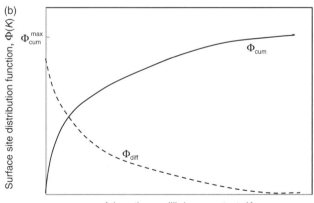

FIGURE 7-18. Examples of complex surface site distributions. (a) A surface with sites whose affinity for the adsorbate follows no obvious pattern. (b) A surface with a large concentration of sites with low affinity for the adsorbate and steadily smaller concentrations of sites with steadily higher K values.

By analogous reasoning, the first term on the right in Equation 7-37 can be expressed as $\Phi_{cum}(K + \Delta K) - \Phi_{cum}(K)$ and manipulated as follows:

$$\frac{\Phi_{cum}(K + \Delta K) - \Phi_{cum}(K)}{\Delta K}\Delta K = \frac{d\Phi_{cum}}{dK}dK = \Phi_{diff}dK \quad (7\text{-}39)$$

Finally, assuming that adsorption onto this small, nearly uniform group of sites can be characterized by the Langmuir isotherm, we can use Equation 7-26 to express the second term on the right in Equation 7-37 as $Kc_A/(Kc_A + 1)$. Substituting these three expressions into Equation 7-37 and rearranging slightly, we obtain

$$\hat{q}_{diff}\,dK = \Phi_{diff}\frac{Kc_A}{Kc_A + 1}dK \quad (7\text{-}40)$$

The overall adsorption density for the given value of c_A is obtained by integrating the adsorption densities over all sites; that is, considering all K values between 0 and ∞:

$$q_{tot} = \int_0^\infty \hat{q}_{diff}\,dK = \int_0^\infty \Phi_{diff}\frac{Kc_A}{Kc_A + 1}dK \quad (7\text{-}41)$$

Equation 7-41 expresses the equilibrium adsorption density of A in terms of its concentration in solution and Φ_{diff}. If Φ_{diff} is known as a function of K, the expression on the right can be integrated to yield an equation for q_{tot} for any specified value of c_A. By considering a range of c_A values, then, we can develop a plot of q_A versus c_A.

Consider, for instance, the limiting case of a surface on which all sites are identical, the total surface site concentration is Φ_{cum}^{max}, and the equilibrium constant for adsorption of species A on the sites is K_1. As indicated in Equation 7-17, such a surface has a differential site distribution function given by $\Phi_{cum,A}^{max}\delta_{K_1}(K)$. Substituting this expression into Equation 7-41 yields:

$$q_{tot} = \int_0^\infty \Phi_{cum,A}^{max}\delta_{K_1}(K)\frac{Kc_A}{Kc_A + 1}dK \quad (7\text{-}42)$$

The properties of the Dirac delta function cause the integrand to have a value of zero at all values of K other than K_1, leading to the following result for the integration:

$$q_{tot} = \Phi_{cum,A}^{max}\frac{K_1 c_A}{K_1 c_A + 1} \quad (7\text{-}43)$$

For this scenario, the maximum value of the cumulative site distribution can be equated with the maximum possible adsorption density of A, q_{max}. Making that substitution, and writing the total adsorption density of A as q_A rather than q_{tot}, we obtain the expected result that the adsorption density of A on the given surface is characterized by the Langmuir isotherm:

$$q_A = q_{max}\frac{K_1 c_A}{K_1 c_A + 1} \quad (7\text{-}44)$$

As noted earlier, site distribution functions are typically more complex than the limiting case that leads to the Langmuir isotherm, but they might nevertheless be characterized by fairly simple mathematical functions. For instance, site distribution functions like that shown in Figure 7-18b can often be modeled quite well by representing Φ_{diff} as a geometrically decreasing function of K; that is, $\Phi_{diff} = \alpha K^{-n}$, where α and n are constants and $0 \leq n \leq 1$.[8] Substitution of this function into Equation 7-41 yields the following expression for the adsorption density of A as a function of c_A:

$$q_A = \int_0^\infty \alpha K^{-n}\frac{Kc_A}{Kc_A + 1}dK = \int_0^\infty \frac{\alpha K^{1-n}c_A}{Kc_A + 1}dK \quad (7\text{-}45)$$

The integrals in Equation 7-45 are nonconvergent, because they are based on a distribution function that suggests that the site concentration becomes infinitely large as K becomes very small. However, if the distribution is specified to apply only down to some finite minimum value of K (i.e., for $K \geq K_{min}$, $\Phi_{diff} = \alpha K^{-n}$, but for $K < K_{min}$, $\Phi_{diff} = 0$), then the integral can be evaluated, yielding the following isotherm equation (Halsey and Taylor, 1947):

$$q_A = k_f c_A^n \quad (7\text{-}46)$$

where k_f is a constant related to α, n, and the thermodynamics of the adsorption reaction. The units of k_f

---

[8] The adsorption equilibrium constant can be expressed as $K_{ads} = \exp-(\Delta\overline{G}_{ads}^o/RT)$, where $\Delta\overline{G}_{ads}^o$ is the standard molar Gibbs energy of the adsorption reaction. Therefore, another way to describe the given distribution of site-types is

$$\Phi_{diff} = \alpha K_{ads}^{-n} = \alpha\left(\exp\frac{-\Delta\overline{G}_{ads}^o}{RT}\right)^{-n} = \alpha\exp\left(-\frac{n(-\Delta\overline{G}_{ads}^o)}{RT}\right)$$

This equation shows that the assumption about the site distribution is equivalent to an assumption that the site concentrations decrease exponentially with increasingly negative standard Gibbs energy of adsorption; that is, increasingly favorable adsorption energy.

correspond to those of the ratio q_A/c_A^n. For example, if q_A is in milligrams per gram, c_A is in milligrams per liter, and $n = 0.3$, k_f has the following units:

$$\text{Units of } k_f = \text{Units of } \frac{q_A}{c_A^n} = \frac{\text{mg/g}}{(\text{mg/L})^{0.3}} = \frac{\text{mg}^{0.7} \text{L}^{0.3}}{\text{g}} \quad (7\text{-}47)$$

The relationship in Equation 7-46, which was identified earlier (Equation 7-5) as the *Freundlich isotherm*, has been found to fit many experimental data sets quite well. Like the single-site Langmuir isotherm, a Freundlich isotherm is fully defined by two fitting parameters. k_f describes the adsorption density under standard conditions ($q = k_f$ when $c_A = 1$), and therefore can be interpreted as an indicator of the average, overall binding strength under those conditions. The parameter n, on the other hand, reflects the distribution of binding energies on the surface—the more dramatically the binding strength changes as the adsorption density changes, the smaller n becomes. As noted, the analysis leading to Equation 7-46 is valid only for $0 \leq n \leq 1$. Throughout this range, the average binding strength decreases with increasing surface coverage, and the lower the value of n, the more dramatic the decrease. Note that, if $n = 1$, the equation reduces to a linear isotherm ($q_A = kc_A$).

The Freundlich isotherm can be linearized by taking the logarithm of both sides of the equation, as shown in Equation 7-48; graphical representations of the Freundlich isotherm for a few values of k_f and n are shown in Figure 7-19.

$$\log q_A = \log k_f + n \log c_A \quad (7\text{-}48)$$

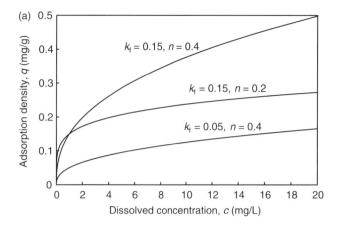
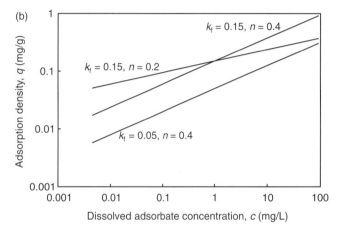

FIGURE 7-19. Graphical representation of the Freundlich isotherm for a hypothetical adsorbate. The units of k_f are $(\text{mg/g})/(\text{mg/L})^n$. The same data are shown on (a) linear and (b) logarithmic coordinates.

Freundlich isotherms characterize the cumulative adsorption onto all the sites on a surface. The total adsorption density on such a surface is envisioned as the summation of the adsorption densities on a wide variety of sites (i.e., sites with a wide variety of binding strengths), with the Langmuir isotherm applying to each group of sites individually. At any given value of c_A, some sites of all binding strengths are occupied, with the high-affinity sites being closer to saturation (i.e., complete coverage) than the low energy sites.

Sorption of organics onto activated carbon from dilute aqueous solutions ($c_{\text{fin}} < 500\ \mu\text{g/L}$) is sometimes characterized by a Langmuir isotherm, but more often by a Freundlich isotherm with $0.2 < n < 0.7$. Freundlich isotherms with values of n between 0.4 and 0.7 are also commonly reported for sorption of metals onto mineral adsorbents. Occasionally, data are reported that conform to a Freundlich isotherm with a value of $n > 1$ over a limited range of c_A values. Such data imply that, as the adsorption density increases, the affinity of the surface for the adsorbate also increases. Such a result is sometimes used as evidence of multilayer adsorption or conversion of adsorbed species into a separate, precipitated solid or condensed liquid phase at the surface.

Even if data fit a Freundlich isotherm over wide ranges of c and q, the fit would be expected to deteriorate at extremes of very small and very large adsorption densities. At extremely low q, even the strongest binding sites on the surface would be minimally occupied, so that the binding onto all site-types on the surface would be in the linear range of the Langmuir isotherm, and the overall adsorption would follow a linear isotherm ($n = 1.0$). The Freundlich equation fails under these conditions, because it is based on a site distribution that causes some sites to be filled almost to capacity for any concentration of dissolved adsorbate; that is, it treats the strongest binding sites as having $K = \infty$. The transition from a Freundlich isotherm with $n < 1$ to a linear isotherm at very low values of c_A has been demonstrated experimentally for metal adsorption onto ferrihydrite by Benjamin and Leckie (1981).

Similarly, at extremely high q, the surface is expected to become saturated, and a maximum adsorption density would be approached; no such maximum adsorption density is incorporated into the isotherm equation.

■ **EXAMPLE 7-4.** Churn and Chien (2002) reported that sorption of para-nitrophenol (PNP) onto granular activated carbon could be modeled by the Freundlich isotherm: $q = \left(3.69 (\text{mol/g})/(\text{mol/L})^{0.167}\right) c^{0.167}$. A system containing 250 mg/L of GAC is titrated with PNP, over a range of total PNP from 0 to 10 mmol/L. Plot the fraction of the PNP that is adsorbed as a function of c_{PNP} (the PNP concentration remaining in solution once equilibrium has been attained).

Solution. The distribution of PNP for any value of the total concentration of PNP in the system can be determined by solving the mass balance equation in conjunction with the isotherm equation:

$$c_{\text{PNP,tot}} = c_{\text{PNP,dissolved}} + q_{\text{PNP}} c_{\text{GAC}}$$

$$= c_{\text{PNP,dissolved}}$$

$$+ \left(3.69 \frac{\text{mol/g GAC}}{(\text{mol/L})^{0.167}} c_{\text{PNP,dissolved}}^{0.167}\right) (0.25 \, \text{g GAC/L})$$

Inserting various values of $c_{\text{PNP,tot}}$ and solving the equation for $c_{\text{PNP,dissolved}}$ yields the desired information, which is plotted in Figure 7-20. Both adsorbed and dissolved PNP increase as PNP is added to the system, but the dissolved concentration increases more rapidly than the adsorbed concentration, so the fraction sorbed decreases. ■

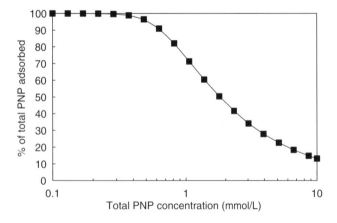

FIGURE 7-20. Fractional adsorption of PNP for a wide range of total PNP concentrations, in the example system.

If k_f is expressed in a set of units that is not consistent with the units in which c and q are given or desired, the unit conversion can be slightly confusing. For example, as noted previously, if the units of c and q are mg/L and mg/g, respectively, the units of k_f are $(\text{mg/g})(\text{L/mg})^n$. Sometimes, k_f is given in these units, but we wish to use the isotherm equation in conjunction with values of c in mol/L and q in mol/g, in which case k_f has units of $(\text{mol/g})(\text{L/mol})^n$. Designating the parameter with these latter units as k_f', the conversion can be carried out as follows:

$$k_f' (\text{mol/g})(\text{L/mol})^n$$

$$= \{k_f (\text{mg/g})(\text{L/mg})^n\}[\text{mol/mg}][\text{mg/mol}]^n$$

$$= k_f \times (\text{mol/mg})^{1-n}$$

$$= k_f \times \left(\frac{1 \, \text{mol}}{(1000 \times \text{MW}) \, \text{mg}}\right)^{1-n} \quad (7\text{-}49)$$

Any set of units for concentrations in both solid and aqueous phase can be converted into another set of units using an analogous procedure.

■ **EXAMPLE 7-5.** An adsorption reaction is characterized by the Freundlich isotherm $q = k_f c^{0.2}$, with $k_f = 100 \, (\text{mg/g})(\text{L/mg})^{0.2}$. The molecular weight of the adsorbate is 50.

(a) Compute the Freundlich constant k_f', for use in the isotherm equation when c is given in mol/L and q is in mol/g.

(b) Compute the value of q in equilibrium with a dissolved adsorbate concentration of 10 mg/L, using both k_f and k_f', and show that the results are equivalent.

Solution.

(a) According to Equation 7-49, the unit conversion is

$$k_f' = k_f \left(\frac{1 \, \text{mol}}{50,000 \, \text{mg}}\right)^{1-0.2} = 0.0174 (\text{mol/g})(\text{L/mol})^{0.2}$$

(b) If $c = 10 \, \text{mg/L} = (10 \, \text{mg/L})/(50,000 \, \text{mg/mol}) = 0.0002 \, \text{mol/L}$, the equilibrium value of q can be computed using k_f and k_f' as follows:

$$q = k_f c^{0.2} = \left[100 \, (\text{mg/g})(\text{L/mg})^{0.2}\right] (10 \, \text{mg/L})^{0.2}$$

$$= 158.5 \, \text{mg/g}$$

$$q = k'_f c^{0.2} = \left[0.0174 (\text{mol/g})(\text{L/mol})^{0.2}\right](0.0002\,\text{mol/L})^{0.2}$$
$$= 0.0032\,\text{mol/g}$$

Converting the value of q in moles per gram into the equivalent value in milligrams per gram, we find

$$q = (0.0032\,\text{mol/g})(50{,}000\,\text{mg/mol}) = 158.5\,\text{mg/g}$$

Thus, as we would anticipate, the same value of q is obtained using k_f and k'_f, proving that the conversion was done correctly. ∎

TABLE 7-4. Selected Adsorption Isotherms

Name	Equation	Name	Equation
Linear	$q = kc$	Tóth	$q = \dfrac{kq_{\max}c}{[1+(kc)^n]^{1/n}}$
Langmuir	$q = \dfrac{q_{\max}Kc}{1+Kc}$	Radke–Prausnitz	$q = \dfrac{k_1 k_2 c}{1+k_2 c^n}$
Freundlich	$q = kc^n$	Fritz–Schlunder	$q = \dfrac{k_1 c^{n_1}}{1+k_2 c^{n_2}}$
Langmuir–Freundlich	$q = \dfrac{q_{\max}(kc)^n}{1+(kc)^n}$		

Comparison of Multisite Langmuir and Freundlich Isotherms The preceding discussion describes the two most common approaches for modeling surfaces that have groups of sites with different affinities for an adsorbate. The multisite Langmuir model allows one to model any distribution of surface site concentrations and adsorption binding constants, but it suffers from the fact that each postulated type of site increases the number of adjustable parameters in the model by two. It is rarely possible to resolve experimental data well enough to justify proposing more than two types of sites on a given surface, and even then the evaluation of the four constants must be viewed somewhat skeptically.

On the other hand, the Freundlich model presupposes a great deal of variability in adsorption binding constants, but postulates a specific form for the site distribution function; that is, a specific relationship between the binding constants and the site concentrations. Because the Freundlich isotherm is a two-parameter model, it does not possess quite as much flexibility as multisite Langmuir models. Nevertheless, it often provides a remarkably good fit to experimental data over a range of several orders of magnitude in adsorbate concentration, and it is therefore frequently used to describe equilibrium adsorption relationships.

It should be emphasized that the isotherms that have been presented here have found popular acceptance because they are mathematically tractable and because they seem to fit experimental data, not because there is some fundamental physical or chemical reason to expect them to apply. Unfortunately, a false distinction between these isotherms has become widespread, whereby the Langmuir isotherm is viewed as theoretically based and the Freundlich isotherm as strictly empirical. Such a distinction is incorrect on both counts: as shown earlier, both the Langmuir and the Freundlich isotherms can be derived starting from a conceptual (theoretical) view of the surface and the adsorption process (i.e., the site-binding model, represented by a generic Φ vs. K distribution), but in both cases assumptions must be made about the surface site distribution function to derive the specific isotherm equation of interest. On the other hand, neither isotherm equation can be reliably predicted to apply in a given situation; use of either isotherm can be justified only by showing that empirical data fit the equation.

Other functional relationships could be (and have been) postulated between surface site densities and binding constants, and any such relationship would be perfectly acceptable as the basis of an isotherm, if the experimental data supported it. A few such isotherms are shown in Table 7-4. Often, data can be fit almost equally well by several different isotherms, so the fact that a data set fits a given equation should not be taken as proof that the conceptual model for that isotherm is correct.

Bidentate Adsorption

The discussion to this point has assumed implicitly that adsorbate molecules occupy only one surface site when they adsorb. Based on the analogy between adsorption and formation of metal/ligand complexes, such reactions are sometimes referred to as *monodentate adsorption*. However, in many systems of interest, an adsorbate can bind to more than one surface site, or can bind to a single site but cover adjacent sites, which has the same effect. The most common such situation involves elimination of two surface sites by adsorption of a single molecule. Such a process, referred to as *bidentate adsorption*, can be characterized by the following, generic reaction and pseudo-equilibrium constant:

$$2{\equiv}S + A \leftrightarrow {\equiv}S_2 A \qquad (7\text{-}50)$$

$$K = \frac{q_{\equiv S_2 A}}{q_{\equiv S}^2 \, c_A} \qquad (7\text{-}51)$$

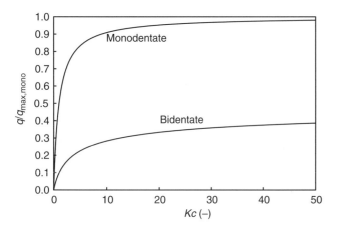

FIGURE 7-21. Comparison of the Langmuir isotherm for mono- and bi-dentate adsorbates.

A conventional isotherm equation (expressing q as a function of c) for bidentate adsorption can be obtained by combining Equation 7-51 (to characterize adsorptive equilibrium) with the following, mole-based site balance, which reflects the fact that two monodentate sites are eliminated for each bidentate surface complex that forms:

$$q_{\equiv S_{tot}} = q_{\equiv S} + 2q_{\equiv S_2 A} \qquad (7\text{-}52)$$

A derivation very similar to the one that led to the Langmuir isotherm for monodentate adsorption then leads to the following isotherm expressions, where q_{bi} is the adsorption density of the bidentate adsorbate, and $q_{max,mono}$ and $q_{max,bi}$ are the maximum, mole-based adsorption densities for monodentate and bidentate adsorption on the surface, respectively (so that $q_{max,bi} = q_{max,mono}/2$).

$$q_{bi} = \left(0.5 - \frac{\sqrt{1+8Kc}-1}{8Kc}\right) q_{max,mono} \qquad (7\text{-}53a)$$

$$= \left(1 - \frac{\sqrt{1+8Kc}-1}{4Kc}\right) q_{max,bi} \qquad (7\text{-}53b)$$

The isotherm described by Equations 7-53a and 7-53b is shown graphically, in nondimensionalized form, in Figure 7-21; a Langmuir isotherm for monodentate adsorption is also shown, for comparison. Although it is not immediately apparent from the equations, the relationship for bidentate adsorption approaches $q_{bi} = Kcq_{max,mono} = 2Kcq_{max,bi}$ at very low values of Kc, and $q_{bi} = 0.5q_{max,mono} = q_{max,bi}$ at very high values of Kc. Thus, the isotherm for bidentate adsorption has the same general shape as the Langmuir isotherm for monodentate adsorption; the conditions characterizing the limiting cases can be seen clearly in Figure 7-21.[9,10]

Bidentate adsorption has been shown to be important for the binding of some multivalent metals onto oxides and other inorganic minerals (Waychunas et al., 1995; Bargar et al., 1997; Hohl and Stumm, 1976). However, by far the most attention paid to the topic has been in the analysis of ion exchange processes, where bidentate adsorption is assumed to occur whenever a divalent ion adsorbs. Therefore, further discussion of bidentate adsorption is deferred to the subsequent section of this chapter where ion exchange reactions are discussed.

The Adsorption Distribution or Partition Coefficient

As noted previously and in Chapter 5, the ratio of the adsorption density of a species to its dissolved concentration (q_A/c_A) in a system at equilibrium is sometimes referred to as a distribution coefficient or partition coefficient (K_D):

$$K_{D,A} = \frac{q_A}{c_A} \qquad (7\text{-}54)$$

If adsorption is viewed as a phase transfer process, the partition coefficient can be interpreted as the adsorption pseudo-equilibrium constant (see Equation 7-14). Partition coefficients have no similar fundamental interpretation in the context of site-binding models, but they are nevertheless

---

[9] Some authors argue that adjacent, unoccupied sites in a bidentate reaction should be represented as a single, composite entity ($\equiv S_2$), since two monodentate sites must be adjacent to one another to constitute a bidentate site. When that convention is adopted, the adsorption equilibrium constant is written as follows:

$$K'' = \frac{q_{\equiv S_2 A}}{q_{\equiv S_2} c_A}$$

Benjamin (2002) showed that, for a surface where the sites are uniformly distributed spatially, the predicted partitioning of the adsorbate is essentially identical regardless of whether the reactant is written as $2\equiv S$ or $\equiv S_2$ up to approximately 85% site occupation by the bidentate adsorbate. The most widely adopted convention is to write the bidentate adsorption reaction as though it involves two independent sites (as in Equation 7-82) and to compute the equilibrium constant for the reaction as in Equation 7-51, as is done here.

[10] In some adsorption literature, especially that dealing with the surface complexation model, the equilibrium constant for bidentate adsorption is written with the activity of adsorbed species expressed in moles of adsorbed A and moles of unoccupied sites per unit volume of solution ($c_{\equiv SA}$ and $c_{\equiv S}$, respectively). Unfortunately, the conversion from Equation 7-51 to the corresponding equation for these units has frequently been done incorrectly. The correct expression is

$$K = \frac{c_{\equiv S_2 A}}{c_{\equiv S}^2 c_A} c_{(\equiv S)_{tot}}$$

where $c_{\equiv S_{tot}}$ is the total concentration of adsorptive sites in the system, in moles per liter.

sometimes reported for such systems, for reasons that are explained subsequently.

The partition coefficient for a Langmuir or Freundlich isotherm can be calculated from the respective isotherm equations as follows:

$$\text{Langmuir:} \quad K_{D,\text{Lang}} = \frac{q}{c} = \frac{q_{\max}K}{1 + Kc} \quad (7\text{-}55)$$

$$\text{Freundlich:} \quad K_{D,f} = \frac{q}{c} = k_f c^{n-1} \quad (7\text{-}56)$$

Systems characterized by linear isotherms have constant values of K_D, independent of c_A. However, for any isotherm that is concave downward when plotted as q versus c (i.e., the isotherm curve bends toward the x-axis as c increases), K_D decreases with increasing dissolved adsorbate concentration. This category includes all Langmuir isotherms, most Freundlich isotherms (those with $n < 1$), and many others. The dependence of K_D on c is shown for a few isotherms in Figure 7-22. For Langmuir isotherms, K_D is characterized by a maximum value of $Kq_{\max}$ at low c, whereas for Freundlich isotherms, K_D has no absolute upper limit.

K_D values can be useful because, in a system with a given adsorbent concentration (c_{solid}), they provide a direct measure of the relative amounts of the adsorbate that are adsorbed and in solution:

$$K_D c_{\text{solid}} = \frac{q_A c_{\text{solid}}}{c_A} \frac{V_L}{V_L} = \frac{\text{mass of A adsorbed}}{\text{mass of A dissolved}} \quad (7\text{-}57)$$

As a result, the partition coefficient can be interpreted as an indicator of the tendency for the adsorbate to bind to the adsorbent under the given conditions—the larger K_D, the larger the fraction of the adsorbate that is associated with the solid (again, for a fixed value of c_{solid}).

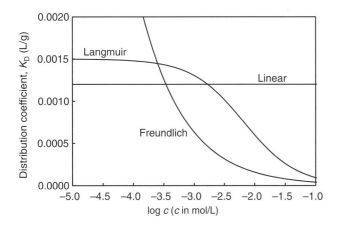

FIGURE 7-22. Distribution coefficients for a few types of isotherm equations. The isotherm parameter values used for the calculations are $K_{\text{lin}} = 0.0012\,\text{L/g}$, $K_{\text{Lang}} = 150\,\text{L/mol}$; $q_{\max,\text{Lang}} = 10^{-5}\,\text{mol/g}$; $k_f = 10^{-5}\,(\text{mol/g})/(\text{mol/L})^{0.4}$; $n = 0.4$.

If K_D increases with decreasing concentration, as is commonly the case, the fractional removal of adsorbate by a given concentration of adsorbent increases as the solution concentration decreases. This result is of particular interest in engineered processes where adsorption is carried out in packed beds. In such systems, the solution can be envisioned to move through sequential layers of the bed, each containing an identical concentration of adsorbent. If K_D increases with decreasing c_A, the fractional removal of adsorbate from solution in each layer of such a bed will be greater than in the preceding layer. Such a scenario can lead to extremely low dissolved adsorbate concentrations in relatively short beds. As a result, isotherms in which K_D increases with decreasing c_A are sometimes referred to as *favorable* isotherms. In those rarer cases in which the curvature in the isotherm is in the opposite direction, the isotherms are referred to as being *unfavorable*.

The partition coefficient provides a convenient and direct way to compare adsorption of different species in a given system, or of a given species in different systems. However, K_D values must be interpreted cautiously, because they can change dramatically as solution conditions change (as shown in Figure 7-22), even for a given adsorbate/adsorbent pair.

Competitive Adsorption in the Context of the Site-Binding Model of Adsorption

The isotherm equations derived in the preceding sections consider adsorption of only a single species. However, in many systems, multiple potential adsorbates are present simultaneously in solution. Adsorption of any of these adsorbates decreases the number of binding sites available to the others and hence affects adsorption of all the adsorbates. In such a case, the adsorbates are said to be competing for the surface sites. In this section, we consider how such competition affects the isotherms of the individual species.

Competitive Langmuir Adsorption As has been noted, the Langmuir adsorption isotherm for a single adsorbate represents the simplest of all possible cases of interest. Consider now a situation in which two monodentate adsorbates, each of which follows a Langmuir isotherm, compete for a single population of surface sites. In that case, an expression like Equation 7-19 applies to each adsorbate. Writing those expressions and then taking their ratio, we obtain

$$q_{\equiv SA} = K_A q_{\equiv S} c_A \quad (7\text{-}58a)$$

$$q_{\equiv SB} = K_B q_{\equiv S} c_B \quad (7\text{-}58b)$$

$$\frac{q_{\equiv SB}}{q_{\equiv SA}} = \frac{K_B c_B}{K_A c_A} \quad (7\text{-}59)$$

The mole-based site balance for this case is

$$q_{(\equiv S)_{tot}} = q_{\equiv S} + q_{\equiv SA} + q_{\equiv SB} \quad (7\text{-}60)$$

Using Equation 7-59 to substitute for $q_{\equiv SB}$ in the site balance, and solving the resulting equation for $q_{\equiv S}$, we obtain

$$q_{(\equiv S)_{tot}} = q_{\equiv S} + q_{\equiv SA} + \frac{K_B c_B}{K_A c_A} q_{\equiv SA} \quad (7\text{-}61)$$

$$q_{\equiv S} = q_{(\equiv S)_{tot}} - q_{\equiv SA} - \frac{K_B c_B}{K_A c_A} q_{\equiv SA} \quad (7\text{-}62)$$

Then, substituting Equation 7-62 into the equilibrium constant expression for adsorption of A (Equation 7-58a), collecting the terms containing $q_{\equiv SA}$, and rearranging

$$q_{\equiv SA} = K_A \left(q_{(\equiv S)_{tot}} - q_{\equiv SA} - \frac{K_B c_B}{K_A c_A} q_{\equiv SA} \right) c_A \quad (7\text{-}63)$$

$$(1 + K_A c_A + K_B c_B) q_{\equiv SA} = K_A q_{(\equiv S)_{tot}} c_A \quad (7\text{-}64)$$

$$q_{\equiv SA} = \frac{K_A c_A}{1 + K_A c_A + K_B c_B} q_{(\equiv S)_{tot}} \quad (7\text{-}65)$$

Writing $q_{\equiv SA}$ as q_A, and $q_{(\equiv S)_{tot}}$ as q_{max}, we obtain

$$q_A = \frac{K_A c_A}{1 + K_A c_A + K_B c_B} q_{max} \quad (7\text{-}66)$$

The analogous expression for q_B is

$$q_B = \frac{K_B c_B}{1 + K_A c_A + K_B c_B} q_{max} \quad (7\text{-}67)$$

Generalizing the above result to the case where j monodentate adsorbates compete for the sites leads to the following adsorption isotherm for any species i:

$$q_i = \frac{K_i c_i}{1 + \sum_{all\, j} K_j c_j} q_{max} \quad (7\text{-}68)$$

The numerator of Equation 7-68 is the same as in the equation for noncompetitive adsorption (Equation 7-25), and the denominator differs only in that it contains a term of the form $K_j c_j$ for each adsorbate in the system instead of just for the target adsorbate. Thus, each additional adsorbate in the system increases the denominator of Equation 7-68 and decreases the amount of i adsorbed. As would be expected, the larger the concentration of the competing adsorbates (c_j) and the greater their tendency to adsorb (K_j), the less i that adsorbs. For systems that include bidentate adsorption, the mathematics are more involved, but the general modeling approach and the qualitative conclusions are the same.

■ **EXAMPLE 7-6.** The Langmuir isotherm parameters for sorption of chromate onto adsorbent B in Example 7-1 are $q_{max} = 0.96$ mmol CrO_4^{2-}/g solid and $K = 104$ L/mmol CrO_4^{2-}. In that example, we computed that an adsorbent dose of 0.277 g/L was required to reduce the CrO_4^{2-} concentration from 0.2 to 0.02 mmol/L.

The values of q_{max} and K for sorption of sulfate onto the same solid are 0.96 mmol SO_4^{2-}/g solid and 20 L/mmol SO_4^{2-}, respectively. (Note that q_{max} is the same for the two adsorbates, implying that they have access to all the same sites.) If the solution to be treated contains 2.0 mmol/L sulfate in addition to the chromate, how large an adsorbent dose is needed to meet the CrO_4^{2-} removal target?

Solution. The isotherms for the two adsorbates are

$$q_{CrO_4^{2-}} = \frac{(0.96\,\text{mmol/g})(104\,\text{L/mmol})c_{CrO_4^{2-}}}{1 + (104\,\text{L/mmol})c_{CrO_4^{2-}} + (20\,\text{L/mmol})c_{SO_4^{2-}}}$$

$$q_{SO_4^{2-}} = \frac{(0.96\,\text{mmol/g})(20\,\text{L/mmol})c_{SO_4^{2-}}}{1 + (104\,\text{L/mmol})c_{CrO_4^{2-}} + (20\,\text{L/mmol})c_{SO_4^{2-}}}$$

In addition to the adsorption isotherms, we know that the mass balance equations on CrO_4^{2-} and SO_4^{2-} must be satisfied:

$$c_{CrO_4,tot} = c_{CrO_4^{2-}} + q_{CrO_4^{2-}} c_{solid}$$

$$c_{SO_4,tot} = c_{SO_4^{2-}} + q_{SO_4^{2-}} c_{solid}$$

Since the equilibrium concentration of CrO_4^{2-} is specified, we have four equations in four unknowns (the concentration of dissolved SO_4^{2-}, the adsorption densities of CrO_4^{2-} and SO_4^{2-}, and the adsorbent dose). When these equations are solved simultaneously, we obtain

$$c_{solid} = 1.59\,\text{g/L} \qquad q_{CrO_4^{2-}} = 0.11\,\text{mmol/g}$$
$$c_{SO_4^{2-}} = 0.74\,\text{mmol/L} \qquad q_{SO_4^{2-}} = 0.79\,\text{mmol/g}$$

Approximately two-thirds of the SO_4^{2-} adsorbs, occupying a significant fraction of the surface sites. As a result, more than five times as much adsorbent is required to achieve the treatment goal as in the noncompetitive system. ■

A Special Case of Competitive Langmuir Adsorption: Ion Exchange Equilibrium

Competitive adsorption is often viewed as a complication that can be added, when necessary, to the analysis of adsorption in single-adsorbate systems. However, in the case of ion exchange, competition between ions for the available charged sites is the essential feature of the process, and a slightly different way of describing that process has evolved. That description is presented in the following section, after a brief overview of the mechanics and applications of ion exchange processes.

Sorption onto Ion Exchange Resins Ion exchange is a label given to adsorption reactions in which one type of ion adsorbs and another desorbs, such that the overall reaction has no net effect on the amount of charge adsorbed. Ion exchange takes place on some clays and other minerals in natural systems, but it is most important in environmental engineering applications in systems where the adsorbent is a synthetic resin produced specifically for the purpose. As noted previously, such resins typically consist of a cross-linked, polymeric skeleton that is designed to be relatively inert, although in some cases exposure to strong oxidants or high or low pH can compromise its integrity. Charged functional groups are bound to the skeleton at fixed intervals, and the charges at those sites are neutralized by oppositely charged ions that are loosely bound to the solid. When the resin is exposed to an aqueous solution, ions from the solution exchange with some of the loosely bound ions at the surface.

The fixed functional groups can be either positively or negatively charged, making the resins anion or cation exchangers, respectively. A schematic representation of an ion exchange resin is shown in Figure 7-23.

The fixed negative charge on cation exchange resins is typically generated by either sulfonate or carboxylate functional groups, and the resin backbone is commonly represented simply as R, so the reactive groups in cation exchange resins are usually shown as R-SO$_3^-$ or R-COO$^-$. Sulfonates are extremely weak bases (i.e., they have very low affinity for H$^+$), so they virtually never combine with H$^+$. As a result, they can serve as ion exchange sites for cations even in the presence of very high H$^+$ concentrations (low pH). Carboxylate ions are also weak bases, but they are not as weak as sulfonates and do have significant affinity for H$^+$. As a result, when the pH is less than approximately 4, H$^+$ outcompetes most other ions for these groups (i.e., the groups are predominantly converted to the R-COOH form), and the resin no longer serves as an effective cation exchanger. Historically, even though the resins are useful as ion exchangers only when the binding sites are *not* protonated, the two types of resins have always been identified in terms of the acid strength of the protonated sites (R-SO$_3$H and R-COOH). Thus, resins with R-SO$_3^-$ and R-COO$^-$ groups are referred to as strong acid and weak acid cation exchangers, respectively.

The charge on anion exchange resins is almost always generated by quaternary amine sites; that is, sites where a nitrogen atom is attached to four organic groups (including the resin backbone), as shown in Figure 7-23c. This type of functional group can be thought of as similar to an ammonium ion (NH$_4^+$) in solution. However, the replacement of the H$^+$ ions surrounding the N atom by organic groups

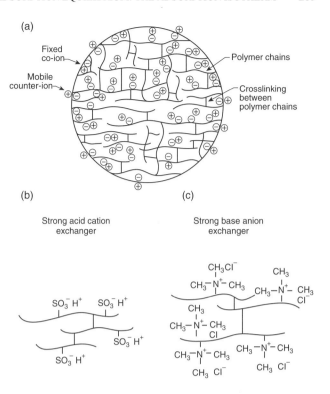

FIGURE 7-23. Schematic representation of ion exchange resins. (a) A cation exchange resin, with negative charges fixed to a polymeric structure, and positive charges loosely associated with those negative charges. (b, c) Expanded schematic representation of the structure of (b) a strong acid cation exchange resin and (c) a strong base anion exchange resin. *Source*: From Clifford et al. (2011). Reprinted from J. Edzwald, ed. *Water Quality and Treatment*, 6th edition by permission. Copyright 2011 by American Water Works Association.

makes the R$_4$N$^+$ group a much weaker acid than NH$_4^+$, so that the quaternary amine is very unlikely to combine with OH$^-$. Put another way, OH$^-$ groups in solution are poor competitors for the resin sites, so those sites are available to bind other types of anions.

Analogous to the situation with cation exchange resins, anion exchange resins with quaternary amine groups are characterized in the literature as strong base anion exchangers, because R$_4$N-OH groups are very strong bases (and hence never exist as important species in the resins). Weak base anion exchange resins can be prepared using tertiary amine sites (a nitrogen atom attached to three organic groups); in this case, the fourth bonding orbital on the nitrogen is occupied by a pair of unshared electrons.

Typical ion exchange reactions for strong acid, weak acid, strong base, and weak base resins are as follows:

$$2(\text{RSO}_3\text{-Na}) + \text{Ca}^{2+} \rightarrow (\text{RSO}_3)_2\text{-Ca} + 2\text{Na}^+ \quad (7\text{-}69)$$

$$\text{RCOO-H} + \text{Na}^+ \rightarrow \text{RCOO-Na} + \text{H}^+ \quad (7\text{-}70)$$

$$2(R_4N\text{-}Cl) + SO_4^{2-} \rightarrow (R_4N)_2\text{-}SO_4 + 2Cl^- \qquad (7\text{-}71)$$

$$R_3N\text{-}OH + Cl^- \rightarrow R_3N\text{-}Cl + OH^- \qquad (7\text{-}72)$$

Reactions 7-69 and 7-71 are referred to as *heterovalent ion exchange*, because two monovalent ions are released when a single divalent ion is adsorbed. Reactions 7-70 and 7-72, on the other hand, represent *homovalent ion exchange* (equal charge on the adsorbing and desorbing ions). Regardless of whether the exchange is homovalent or heterovalent, though, electroneutrality of the resin phase is preserved.

By passing water through both a strong acid cation exchanger and a strong base anion exchanger, cations and anions can be removed from solution and replaced, respectively, by H^+ and OH^-. These released ions combine with one another to form H_2O, so that the overall process removes ions from solution and replaces them with water. Such resins are sometimes combined in a single reactor (a *mixed ion exchange system*) to deionize water for laboratory and some commercial applications (e.g., to produce ultrapure water for semiconductor manufacturing, pharmaceutical production, and other industries).

In a typical engineered ion exchange process, at the beginning of a treatment cycle, the resin is loaded with the ions that are intended to replace the target ions in solution. Although batch systems are occasionally used, the resin is usually packed in a bed through which the influent is passed. The exchange of ions takes place as the water flows through the bed, ideally producing an effluent containing a very low concentration of the target ions. As upstream portions of the resin become progressively loaded with those ions, the driving force for their adsorption in that part of the column declines, and the zone of active adsorption moves downstream as a wave, as shown schematically in Figure 7-24.

For instance, at $t = t_3$, the concentration is equal to the influent concentration (c_{in}) in approximately the upper half of the column and close to zero in the lower half, with a small transition region between these two zones. The concentration of the ion being released from the resin (exchanged) would follow a profile that is the mirror image of that shown.

The hydraulics in the column often conform closely to plug flow. In such cases, the wave front is predicted to be virtually a square wave, so any curvature in the actual wave front is attributed to dispersion in the liquid phase. These characteristics of the wave front are developed quantitatively in Chapter 8.

Eventually, the concentration of the target ions in the column effluent exceeds the criterion for acceptable treatment, and the treatment cycle is terminated. A solution that is highly concentrated in the nontarget ion is then contacted with the resin, reversing the driving force and returning the resin approximately to its original condition; this step is referred to as *regeneration*. The entire treatment/regeneration process

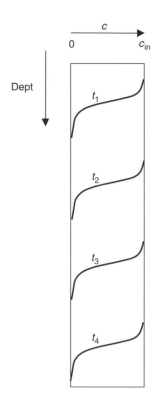

FIGURE 7-24. Schematic showing the concentration of the target adsorbate in solution as a function of location in the column at four different times during the operation of an ion exchange column.

is described in greater detail in Chapter 8. An extensive discussion of the applications of ion exchange to drinking water treatment is provided by Clifford et al. (2011).

Homovalent Ion Exchange The major conceptual difference between the conventional analysis of competitive adsorption (as described in the preceding section) and the conventional analysis of ion exchange is that, in ion exchange, an assumption is made that all the surface sites are occupied under all conditions; that is, $q_{\equiv S}$ is assumed to be zero always.[11] To incorporate this assumption into the analysis, we set $q_{\equiv S}$ to zero in the site balance. For example, for a system containing only two, monodentate (i.e., monovalent) adsorbates, the mole-based site balance equation is $q_{\equiv S_{tot}} = q_{\equiv SA} + q_{\equiv SB}$. The rest of the derivation follows the

---

[11] This assumption can be partially justified by the idea that the resin phase, as a whole, must be electrically neutral. Therefore, for every charged group that is part of the resin skeleton, an equal and opposite charge must reside nearby. In the model for ion exchange presented here, these neutralizing ions are assumed to bind to surface sites, with multivalent ions with charge n^+ or n^- binding to n surface sites. The net result of these assumptions is that every site is always occupied. An alternative model, which relaxes the assumption that all sites are occupied and takes into account the neutralizing ions in the nearby solution, has been presented by Dzombak and Hudson (1995).

same steps as shown in the preceding section, yielding the following isotherms for ion exchange equilibrium in such a system:

$$q_A = \frac{K_A c_A}{K_A c_A + K_B c_B} q_{max} \tag{7-73}$$

$$q_B = \frac{K_B c_B}{K_A c_A + K_B c_B} q_{max} \tag{7-74}$$

The corresponding isotherm equation for system containing j ions competing for exchange sites is

$$q_i = \frac{K_i c_i}{\sum_{all\,j} K_j c_j} q_{max} \tag{7-75}$$

Comparison of Equations 7-73 to 7-75 with those derived previously (Equations 7-66 to 7-68) indicates that, mathematically, the effect of assuming that all the ion exchange sites are occupied is to eliminate a term equal to 1 from the summation in the denominator of the isotherm expression. The reason for this is that the value of 1 in the denominator of Equations 7-66 to 7-68 accounts for the contribution of unoccupied sites to the total site concentration; the assumption that all sites are occupied changes this value to zero.

If we could evaluate K for the binding of each adsorbate to open sites on an ion exchange resin, we could use Equation 7-75 in conjunction with mass balances on the adsorbates to calculate the adsorption densities of all ions in a system with any given overall composition. However, that approach is not available to us because the fact that $q_{\equiv S}$ is assumed to be zero in ion exchange systems makes it impossible to evaluate K for an isolated ion. To overcome this problem, equilibrium constants for ion exchange reactions are never written solely for the adsorption part of an ion exchange reaction, but rather are written for the complete (adsorption and desorption) reaction.

For instance, consider an example system in which silver is being recovered from an industrial waste by exchange with Na^+. Representing negatively charged cation exchange resin sites as $\equiv R$, the overall reaction and equilibrium constant would be as follows:

$$\equiv RNa + Ag^+ \leftrightarrow \equiv RAg + Na^+ \tag{7-76}$$

$$K_{Ag/Na} = \frac{q_{Ag} c_{Na}}{q_{Na} c_{Ag}} = \frac{q_{Ag}/c_{Ag}}{q_{Na}/c_{Na}} \tag{7-77}$$

The equilibrium constant for the overall exchange reaction ($K_{Ag/Na}$) is referred to as the *selectivity coefficient* for Ag over Na.[12] The reaction would proceed to the right during the treatment step and to the left when the resin is regenerated.

All the terms on the right side of Equation 7-77 are experimentally accessible, so $K_{Ag/Na}$ is a measurable quantity. If Equation 7-77 is generalized to species A and B and rearranged to show the ratio of adsorbed B to adsorbed A, we obtain

$$\frac{q_B}{q_A} = K_{B/A} \frac{c_B}{c_A} \tag{7-78}$$

This expression for the ratio of adsorbed B to adsorbed A in ion exchange is identical to that for conventional, competitive Langmuir adsorption (Equation 7-59), with $K_{B/A} \equiv K_B/K_A$. Thus, the selectivity coefficient for an ion exchange reaction can be interpreted as the ratio of the pseudo-equilibrium constant for adsorption of the target ion (K_B) to the pseudo-equilibrium constant for adsorption of the released ion (K_A), even though neither of those equilibrium constants can be evaluated individually. Being ratios of adsorptive binding constants, selectivity coefficients are, in theory, also constants. In particular, they are expected to be independent of the absolute concentrations of the ions in the system. (In practice, selectivity coefficients depend on the system composition because the assumption that the activities of the various species can be represented by their concentrations is often not valid.)

The expressions for q_A and q_B in a binary ion exchange process (Equations 7-73 and 7-74) can be converted into expressions that contain the (measurable) selectivity coefficient $K_{B/A}$ rather than the (unmeasurable) adsorption constants K_A and K_B by dividing the numerator and denominator of those expressions by K_A, yielding

$$q_A = \frac{c_A}{c_A + (K_B/K_A)c_B} q_{max} = \frac{c_A}{c_A + K_{B/A} c_B} q_{max} \tag{7-79}$$

$$q_B = \frac{(K_B/K_A)c_B}{c_A + (K_B/K_A)c_B} q_{max} = \frac{K_{B/A} c_B}{c_A + K_{B/A} c_B} q_{max} \tag{7-80}$$

The corresponding adsorption isotherm equation for an ion exchange system containing any number of adsorbates, if all the adsorbates are monovalent, is

$$q_i = \frac{K_{i/k} c_i}{\sum_{all\,j} K_{j/k} c_j} q_{max} \tag{7-81}$$

where i is a target adsorbate species, k is the species used as the basis for comparison for reporting selectivity coefficients, the summation is over all adsorbates in the system (including k, if it is present), and $K_{k/k} \equiv 1$. Note that, if we remove the constraint that $q_{\equiv S} = 0$ and define k to be an empty site (or, equivalently, one occupied by water),

---

[12] Note that, for ion exchange reactions, q values are sometimes reported in equivalents (rather than moles) per unit amount of adsorbent. In the case of homovalent exchange, this convention has no effect on the computed value of $K_{A/B}$, but for heterovalent exchange (discussed subsequently), it does.

Equation 7-81 becomes the isotherm for competitive Langmuir adsorption (Equation 7-68), with $K_{i/k}$ and $K_{j/k}$ being the Langmuir adsorption constants for i and j, respectively.

■ **EXAMPLE 7-7.** Styrene-based ion exchange resins have been found to have very high affinities for perchlorate (ClO_4^-), an ion that has contaminated groundwater near industrial sites where rocket fuel is manufactured. The selectivity coefficients for nitrate (NO_3^-) and perchlorate over chloride (Cl^-) on a given resin are 2 and 125, respectively. Compute the adsorption densities of nitrate and perchlorate in equivalents per gram on a resin with a total site density of 1.7 meq/g, when the resin reaches equilibrium with a groundwater containing 31 mg/L NO_3^-, 199 µg/L ClO_4^-, and negligible Cl^-.

Solution. This problem can be solved by application of Equation 7-81. However, to use that equation, we need to convert the nitrate and perchlorate concentrations into units of mol/L or equiv/L. Since both ions are monovalent, the values using either set of units are identical. The molecular weights (and equivalent weights) of NO_3^- and ClO_4^- are 62 and 99.5, respectively, so the conversions are

$$c_{NO_3^-} = \frac{31\,\text{mg/L}}{62\,\text{mg/mmol}} = 0.5\,\text{mmol/L} = 0.5\,\text{meq/L}$$

$$c_{ClO_4^-} = \frac{199\,\mu\text{g/L}}{99,500\,\mu\text{g/mmol}} = 2.0 \times 10^{-3}\,\text{mmol/L}$$

$$= 2.0 \times 10^{-3}\,\text{meq/L}$$

Inserting appropriate values into Equation 7-81:

$$q_{NO_3^-} = \frac{K_{NO_3/Cl}\,c_{NO_3}}{K_{NO_3/Cl}\,c_{NO_3} + K_{ClO_4/Cl}\,c_{ClO_4}}\,q_{max}$$

$$= \frac{2(0.5\,\text{mmol/L})}{2(0.5\,\text{mmol/L}) + 125(2 \times 10^{-3}\,\text{mmol/L})}\,1.70\,\text{meq/g}$$

$$= 1.36\,\text{meq/g}$$

$$q_{ClO_4^-} = \frac{K_{ClO_4/Cl}\,c_{ClO_4}}{K_{NO_3/Cl}\,c_{NO_3} + K_{ClO_4/Cl}\,c_{ClO_4}}\,q_{max}$$

$$= \frac{125(2 \times 10^{-3}\,\text{mmol/L})}{2(0.5\,\text{mmol/L}) + 125(2 \times 10^{-3}\,\text{mmol/L})}\,1.70\,\text{meq/g}$$

$$= 0.34\,\text{meq/g}$$

Consistent with their relative affinities for the resin as indicated by their selectivity coefficients, the partition coefficient (q/c) for perchlorate ($0.34/0.002 = 170$) is much larger than that for nitrate ($1.36/0.5 = 2.72$). Note that the selectivity coefficients for the two ions relative to chloride are used in the calculations, even though no chloride is present in the solution. ■

Heterovalent Ion Exchange Consider next a heterovalent ion exchange reaction, for example, one describing adsorption of sulfate and release of chloride. This reaction and the corresponding selectivity coefficient are most often written as follows:[13]

$$2{\equiv}RCl + SO_4^{2-} \leftrightarrow {\equiv}R_2SO_4 + 2Cl^- \qquad (7\text{-}82)$$

$$K_{SO_4/Cl} = \frac{q_{{\equiv}R_2SO_4}\,c_{Cl}^2}{q_{{\equiv}RCl}^2\,c_{SO_4}} \qquad (7\text{-}83)$$

Equation 7-83 shows that the selectivity coefficient for the heterovalent reaction can be represented by the following relationship between the adsorptive binding constants:

$$K_{SO_4/Cl} = K_{SO_4}/K_{Cl}^2 \qquad (7\text{-}84)$$

Unlike the case for homovalent exchange reactions, however, $K_{SO_4/Cl}$ is not a direct indicator of the relative tendencies for the two ions to bind to the resin, because of the squared terms in the expression for $K_{SO_4/Cl}$. The use of selectivity coefficients to predict the ion distribution in a heterovalent ion exchange system is illustrated in the following example.

■ **EXAMPLE 7-8.** The selectivity coefficient of a cation exchange resin for Cu^{2+} over Na^+ is 4.5, when q and c are expressed in meq/g and meq/L, respectively. In a batch treatment process, an industrial waste containing 70 mg/L Cu^{2+} and no Na^+ is contacted with 5.0 g/L of an ion exchange resin that has been preloaded with Na^+. The exchange capacity of the resin is 1.5 meq/g.

What fraction of the copper in the original solution adsorbs, and what fraction of the sodium originally on the resin desorbs?

Solution. Applying Equation 7-83, the selectivity coefficient for this system is

$$K_{Cu/Na} = 4.5 = \frac{q_{Cu}\,c_{Na}^2}{q_{Na}^2\,c_{Cu}}$$

---

[13] As is the case for bidentate adsorption onto other kinds of solids, an alternative approach for writing the heterovalent ion exchange reaction is to represent the adjacent Cl-occupied resin sites as a single composite site:

$${\equiv}R_2Cl_2 + SO_4^{2-} \leftrightarrow {\equiv}R_2SO_4 + 2Cl^-$$

$$K''_{SO_4/Cl} = \frac{q_{{\equiv}R_2SO_4}\,c_{Cl}^2}{q_{{\equiv}R_2Cl_2}\,c_{SO_4}}$$

As indicated in footnote 9, under most circumstances, the predicted partitioning of adsorbate is almost the same regardless of which convention is adopted to represent the adjacent sites. In ion exchange, as in other adsorption reactions, it is most common to write the heterovalent ion exchange reaction as though it involves two independent monovalent sites (as in Equation 7-82), and to compute the selectivity coefficient for the reaction as in Equation 7-83.

A mass balance on this batch system reflects the fact that the total concentrations (dissolved plus adsorbed) of Na^+ and Cu^{2+} are not affected by the treatment step. These concentrations, which we designate as c_{tot}, can be computed based on the conditions in the waste and on the resin before they are mixed, as follows. (Note that the concentrations and adsorption densities are all based on equivalents, not moles.)

$$c_{i,tot} = c_{i,dissolved} + q_i c_{solid}$$

$$c_{Na,tot} = 0 + (1.5\,meq/g)(5.0\,g/L) = 7.5\,meq/L$$

$$c_{Cu,tot} = (70\,mg/L)\left(\frac{1\,mmol}{63.5\,mg}\right)(2\,meq/mmol)$$

$$+(0.0\,meq/g)(5.0\,g/L) = 2.2\,meq/L$$

The mass balances once the system reaches equilibrium are, therefore, as follows:

$$7.5\,meq/L = c_{Na,dissolved} + q_{Na} c_{solid}$$

$$2.2\,meq/L = c_{Cu,dissolved} + q_{Cu} c_{solid}$$

Given that the sites in the ion exchange resin are assumed to be always fully occupied by Cu^{2+} or Na^+, we can write the following equivalent-based site balance:

$$q_{tot} = 1.5\,meq/g = q_{Cu^{2+}(equiv)} + q_{Na^+(equiv)}$$

The balances on total Na^+, total Cu^{2+}, and total sites, plus the selectivity coefficient equation, provide four independent equations in four unknowns, and therefore can be solved simultaneously to yield a unique result. That result is

$$c_{Cu^{2+}} = 0.26\,meq/L = 0.13\,mmol/L = 8.26\,mg/L$$

$$c_{Na^+} = 1.94\,meq/L = 1.94\,mmol/L = 44.62\,mg/L$$

$$q_{Cu^{2+}} = 0.388\,meq/g = 0.194\,mmol/g = 12.32\,mg/g$$

$$q_{Na^+} = 1.112\,meq/g = 1.112\,mmol/g = 25.53\,mg/g$$

Based on these results, 8.26 mg/L Cu^{2+} remains in the treated solution, compared to 70 mg/L in the untreated solution, corresponding to 88.2% removal. The adsorption density of sodium on the original resin was 1.50 meq/g, of which 1.11 meq/g remains after the treatment step. Thus, 26% of the Na^+ is released. ∎

Some Special Nomenclature and Conventions Used for Ion Exchange Reactions As noted previously, the main substantive difference between the conventional modeling of ion exchange and other adsorption reactions is that, in ion exchange, the surface sites are assumed to be fully occupied under all circumstances; that is, $q_{\equiv S} = 0$. Additional practical differences between the two groups of reactions include the types of adsorbents of interest, the centrality of charge, and the equal focus on the adsorbing and desorbing species in ion exchange.

By solving the isotherm equations in conjunction with the mass balance equations on the various adsorbates in an ion exchange system, we can determine the concentrations and adsorption densities of all the adsorbates at equilibrium. However, it is often desirable to characterize the relative affinities of a resin for different adsorbates in a more generic way, without reference to a particular set of conditions. The parameter that is most often used for this purpose is known as the *separation factor* for species B over species A, represented as $\alpha_{B/A}$ and computed as follows:

$$\alpha_{B/A} \equiv \frac{q_B/c_B}{q_A/c_A} \quad (7\text{-}85)$$

Recall that the ratio q_i/c_i is the partition or distribution coefficient (K_D) for i, so separation factors can be viewed as ratios of the partition coefficients of a pair of adsorbates:

$$\alpha_{B/A} = \frac{K_{D,B}}{K_{D,A}} \quad (7\text{-}86)$$

Comparison of Equation 7-85 with Equation 7-77 shows that, for exchange of two monovalent ions, the separation factor is identical to the selectivity coefficient ($\alpha_{A/B} = K_{A/B}$); the same is true for other homovalent exchange reactions. Therefore, for such reactions, either the separation factor or the selectivity coefficient could be used as the generic indicator of resin preference. However, these two parameters are not equivalent for heterovalent exchange reactions. For example, for a divalent (B)/monovalent (A) ion exchange reaction (such as binding of SO_4^{2-} and release of Cl^-, as characterized by Reaction 7-82 and Equation 7-83), the relationships among the separation factor, the selectivity coefficient, and the distribution coefficients are

$$\alpha_{B^{2-}/A^-} \equiv \frac{q_{\equiv R_2B}/c_B}{q_{\equiv RA}/c_A} = K_{B/A}\frac{q_{\equiv RA}}{c_A} = K_{B/A}K_{D,A} \quad (7\text{-}87)$$

$$K_{B/A} = \frac{K_{D,B}}{K_{D,A}^2} \quad (7\text{-}88)$$

According to Equation 7-87, the separation factor for divalent–monovalent exchange equals the product of the selectivity coefficient for the divalent over the monovalent ion and the partition coefficient for the monovalent ion. Thus, when the composition of the system changes, the separation factor changes in the same direction as the partition coefficient of the monovalent ion; for example, in the example system, if $q_{\equiv RCl}/c_{Cl}$ increases, $\alpha_{SO_4/Cl}$ increases, and vice versa.[14]

The usefulness of separation factors becomes apparent if we manipulate Equation 7-85 as follows:

$$\alpha_{B/A} = \frac{q_B c_{resin} V_L / c_B V_L}{q_A c_{resin} V_L / c_A V_L}$$

$$= \frac{\text{amount of B adsorbed/amount of B dissolved}}{\text{amount of A adsorbed/amount of A dissolved}}$$

where the amounts shown in the final expression could be given as moles or equivalents. The ratio of the amount of a species that is adsorbed to the amount that is dissolved is a direct indication of the tendency of the solid to remove the adsorbate from the solution in the given system. This ratio implicitly accounts for the concentrations of adsorbent, adsorbate, and competing adsorbates in the system, as well as all the interactions among them. The separation factor $\alpha_{B/A}$ therefore characterizes the relative tendency of the solid to remove species B from solution compared to species A.

The effect of the separation factor on the removal of a target ion from solution in a simple system is illustrated in Figure 7-25. The plot shows the distribution of ions in systems with various separation factors, for exchange of two monovalent ions (Na^+ and Ag^+) in systems where they are the only two adsorbable ions. The axes of the plot show the *equivalent fraction* of each ion in each state (dissolved [x_i] or bound to the resin [y_i]), defined as:

$$\text{Equivalent fraction of A} = \frac{\text{equivalents of A}}{\text{equivalents of A} + \text{B}} \quad (7\text{-}89)$$

For the given example system, the equivalent fractions can be computed as follows:

$$x_{Ag^+} = \frac{c_{Ag^+}}{c_{Ag^+} + c_{Na^+}} \qquad x_{Na^+} = \frac{c_{Na^+}}{c_{Ag^+} + c_{Na^+}} = 1 - x_{Ag^+}$$

$$y_{Ag^+} = \frac{q_{Ag^+}}{q_{Ag^+} + q_{Na^+}} = \frac{q_{Ag^+}}{q_{tot}} \qquad y_{Na^+} = \frac{q_{Na^+}}{q_{tot}} = 1 - y_{Ag^+}$$

[14] Note that, if c and q are, respectively, the mole-based dissolved concentration and adsorption density of a multivalent ion, then the corresponding values based on equivalents are nc and nq, where n is the absolute value of the ionic charge. Values of q and c appear only as the ratio q/c in the calculation of selectivity coefficients and separation factors, so if both terms are expressed on a basis of equivalents, the n cancels. As a result, values of $K_{A/B}$ are the same regardless of whether they are computed using mole-based or equivalent-based units; the same statement applies to values of $\alpha_{A/B}$.

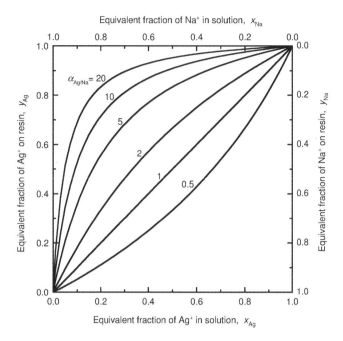

FIGURE 7-25. Distribution of Ag^+ between solution and resin for a binary ion exchange system with Na^+, for various Ag/Na separation factors.

Substituting these expressions into the definition of the separation factor, we find

$$\alpha_{Ag^+/Na^+} = \frac{q_{Ag^+}/c_{Ag^+}}{q_{Na^+}/c_{Na^+}} = \frac{y_{Ag^+}/x_{Ag^+}}{y_{Na^+}/x_{Na^+}}$$

$$= \frac{y_{Ag^+}/x_{Ag^+}}{(1 - y_{Ag^+})/(1 - x_{Ag^+})} \quad (7\text{-}90)$$

$$y_{Ag^+} = \frac{\alpha_{Ag^+/Na^+} x_{Ag^+}}{1 + \alpha_{Ag^+/Na^+} x_{Ag^+} - x_{Ag^+}} \quad (7\text{-}91)$$

Equation 7-91 is the basis for the curves plotted in Figure 7-25.

For homovalent ion exchange, the separation factor of a given resin for a given pair of ions is independent of the solution composition, but this constancy does not extend to heterovalent exchange. Consider, for example, the changes in partitioning of SO_4 and Cl that would occur in a batch ion exchange system containing a fixed concentration of total sulfate but a variable total concentration of chloride. Assume that the isotherm for binding of chloride is favorable; that is, that the isotherm is steeper at low values of c_{Cl} and q_{Cl} than at higher values. (Recall that Langmuir isotherms always fit this pattern, and Freundlich isotherms usually do.) In such a case, Equation 7-87 indicates that $\alpha_{SO_4/Cl}$ will decrease when the dissolved chloride concentration increases, as shown for a model system in Figure 7-26.

The trend shown in the figure has important implications for two practical situations. First, when ion exchange is

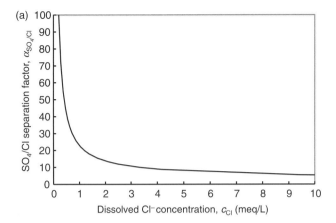

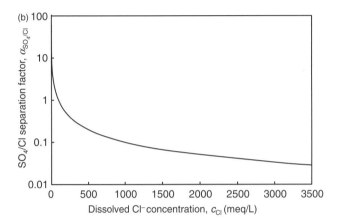

FIGURE 7-26. The decrease in the SO_4^{2-}/Cl^- separation factor with increasing $c_{Cl,tot}$ in a model ion exchange system. (a) For a fixed total concentration of 4 meq/L total SO_4 and a range of dissolved Cl concentrations that might be encountered in potable water sources. (b) For values of c_{Cl} applicable to resin regeneration conditions. The assumed conditions are $K_{SO_4/Cl} = 20$ (based on c in meq/L and q in meq/g), 5 meq/L total ion exchange sites, $c_{SO_4,tot} = 4$ meq/L, and $c_{Cl,tot}$ ranging from less than 1 to 3500 meq/L.

carried out in a packed bed, divalent ions often adsorb and are replaced in solution by monovalent ions as the water flows through the column. Thus, the total concentration of monovalent ions is higher at downstream than at upstream locations. This higher concentration of monovalent ions tends to decrease the divalent/monovalent separation factor. Fortunately, the affinity of ion exchange resins for divalent ions is usually sufficiently higher than that for monovalent ions that the ion exchange reaction is not severely impeded. For example, in the hypothetical system characterized in Figure 7-26a, $\alpha_{SO_4/Cl}$ remains >5 for c_{Cl} values up to 10 meq/L (350 mg/L Cl^-).

On the other hand, when the monovalent ion concentration is increased to several equivalents per liter, the separation factor might decline so much that the monovalent ion is favored by the resin over the divalent ion. Such a change can occur when a resin that has been used for adsorption of a multivalent ion is regenerated with a concentrated solution of the monovalent ion, for example, when a resin that is loaded with SO_4^{2-} is regenerated with a brine containing a high concentration of Cl^-. In that case, the decline in $\alpha_{SO_4/Cl}$ is desirable, because it encourages release of SO_4^{2-} and adsorption of Cl^-, which is exactly what we are trying to accomplish in the regeneration step. This situation is demonstrated in Figure 7-26b, which indicates that Cl^- is favored over SO_4^{2-} ($\alpha_{SO_4/Cl} < 1$) in the example system when $c_{Cl,tot}$ exceeds ~100 meq/L; in typical regenerant solutions, $c_{Cl,tot}$ might range from 1000 to 3000 meq/L.

■ **EXAMPLE 7-9.** In Example 7-8, the equilibrium composition of an ion exchange system for removing Cu^{2+} from solution and replacing it with Na^+ was determined. After the system has equilibrated, the solution is drained, and a small volume of regenerant containing 2.5 mol/L Na^+ is contacted with the resin. The resin concentration during this step is 500 g/L. Determine the equilibrium composition of the regenerant solution after it equilibrates with the resin and the regeneration efficiency (i.e., the fraction of the bound Cu that is released during the regeneration step), and compare $\alpha_{Cu/Na}$ in the two systems.

Solution. The Cu^{2+}/Na^+ selectivity coefficient is a pseudo-equilibrium constant for the exchange reaction and is the same (4.5) in the two systems. Ignoring (somewhat unrealistically) the changes in the activity coefficients of Cu^{2+} and Na^+ that accompany the change in ionic strength, we can treat K_{Cu^{2+}/Na^+} as a true equilibrium constant, in which case it would be the same (4.5) in the treatment and regeneration steps. With that assumption, after the resin equilibrates with the regenerant solution, we can write:

$$K_{Cu/Na} = 4.5 = \frac{q_{Cu} c_{Na}^2}{q_{Na}^2 c_{Cu}}$$

where, as before, the units of q and c are meq/g and meq/L, respectively. We can also write the following mass balances on total Na^+ and Cu^{2+} in the regenerant system:

$$c_{i,tot} = c_{i,\text{fresh regenerant}} + q_{i,\text{used resin}} c_{solid}$$

$$c_{Na,tot} = (2.5 \, \text{mol/L})(1.0 \, \text{equiv/mol}) + (1.112 \, \text{meq/g})$$

$$\times (500 \, \text{g/L}) \left(\frac{1 \, \text{equiv}}{1000 \, \text{meq}}\right) = 3.056 \, \text{equiv/L}$$

$$c_{Cu,tot} = 0.0 \, \text{eq/L} + (0.388 \, \text{meq/g})(500 \, \text{g/L}) \left(\frac{1 \, \text{equiv}}{1000 \, \text{meq}}\right)$$

$$= 0.194 \, \text{equiv/L}$$

Also as in the previous example, the following site balance applies:

$$q_{tot} = 1.5 \frac{\text{meq}}{\text{g}} = q_{Cu^{2+}} + q_{Na^+}$$

Solving the four equations in four unknowns that apply to this system, we find

$c_{Cu^{2+}} = 194 \text{ meq/L} = 97 \text{ mmol/L} = 6.15 \text{ g/L}$

$c_{Na^+} = 2.31 \text{ eq/L} = 2.31 \text{ mol/L} = 53.0 \text{ g/L}$

$q_{Cu^{2+}} = 3.69 \times 10^{-4} \text{ meq/g} = 1.84 \times 10^{-4} \text{ mmol/g}$

$\phantom{q_{Cu^{2+}}} = 0.012 \text{ mg/g}$

$q_{Na^+} = 1.50 \text{ meq/g} = 1.50 \text{ meq/g} = 34.5 \text{ mg/g}$

The regeneration is extremely effective, with 99.9% of the Cu^{2+} being released. The Cu^{2+} concentration in the regenerant solution is 88 times that in the original waste and, after regeneration, the resin contains virtually no Cu^{2+}, so it can be used as effectively in the next cycle as in the previous one.

The separation factor in the two solutions can be computed as:

$$\alpha_{Cu/Na} = \frac{q_{Cu}/c_{Cu}}{q_{Na}/c_{Na}} = K_{Cu/Na} \frac{q_{Na}}{c_{Na}}$$

The value of $\alpha_{Cu/Na}$ is 2.58 for the treatment step, but only 0.0029 for regeneration. Thus, the increase in the total concentration of monovalent species shifts the separation factor strongly to favor binding of that species. ∎

In systems where the separation factors are known, the surface speciation can be computed from an equation analogous to Equation 7-81, with the separation factors replacing the selectivity coefficients:

$$q_i = \frac{\alpha_{i/k} c_i}{\sum_{\text{all } j} \alpha_{j/k} c_j} q_{max} \quad (7\text{-}92)$$

Equation 7-92 can be applied even if heterovalent ion exchange is occurring. An example of the use of this equation to analyze ion exchange in a column system is given in Chapter 8.

Modeling Ion Exchange Based on Donnan Equilibrium When ion exchange is represented according to the site-binding model for adsorption, the ion exchange reactions the ions are envisioned to bind to individual sites in the ion exchange matrix. The bound ions are thereby removed from solution and exist as adsorbed species. This model implicitly assumes that the solution inside the ion exchange matrix is identical to the bulk solution, so that the interior ion exchange sites are exposed to and equilibrated with the same solution as the exterior sites.

An alternative model treats the counter-ions inside the ion exchange matrix not as specifically adsorbed to individual sites, but as *dissolved* ions in a region that has a nonzero electrical potential that is established by the charges on the resin structure. These charges are assumed to be sufficiently closely spaced that their electrical fields overlap extensively, causing the electrical potential to be uniform throughout that volume; this potential is termed the *Donnan potential*, ψ_{Donnan}. As a result, in this model, the concentration of dissolved ions inside the matrix at equilibrium equals the concentration in the bulk solution adjusted by a factor that accounts for the enhancement or diminishment caused by the electrical potential.

Although, in reality, the potential changes smoothly from ψ_{Donnan} to zero from slightly inside to slightly outside the resin, the changes is expected to be steep and is often idealized as a step change at the solid/water interface. The distribution of ions and the profile of electrical potential according to this model are shown schematically in Figure 7-27.

Equilibrium in any chemical system is defined as a condition in which each species has the same total potential (i.e., potential energy per mole) everywhere. In most

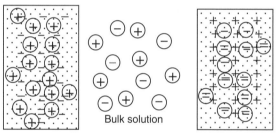

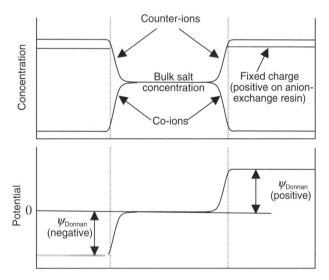

FIGURE 7-27. Profiles of concentration and electrical potential near ion exchange resins in equilibrium with the adjacent solution. Schematics and profiles for both a cation exchange resin and an anion exchange resin are shown (on the left and right, respectively) In the top figure, the mobile charges are circled and the fixed charges are not. *Source*: Adapted from Strathmann (2004).

systems of interest, all forms of potential other than the chemical potential ($\overline{G}_i$) are uniform, and equilibrium is achieved when $\overline{G}_i$ for each species is uniform. However, in the ion exchange model that includes the Donnan potential, the condition for equilibrium must take that potential into account. The electrical potential, ψ, is conventionally expressed in volts; to convert that term into conventional units for chemical potential (kJ/mol or kcal/mol), we multiply by $z_i\mathcal{F}$, where z_i is the charge on the species of interest (equiv/mol) and $\mathcal{F}$ is the Faraday constant (96.48 kJ/eq-V). The criterion for equilibrium of an ion i in an ion exchange system is therefore as follows:

$$\overline{G}^o_{i,\text{bulk}} + RT \ln a_{i,\text{bulk}} = \overline{G}^o_{i,\text{IX}} + RT \ln a_{i,\text{IX}} + z_i\mathcal{F}\psi_{\text{Donnan}} \quad (7\text{-}93)$$

where $\overline{G}^o_i$ is the standard chemical potential of i, and bulk and IX refer to the bulk solution and the solution in the ion exchange resin, respectively. Assuming that the same standard state is chosen for the solutions inside the resin and in the bulk solution, $\overline{G}^o_{i,\text{bulk}} = \overline{G}^o_{i,\text{IX}}$, and:

$$a_{i,\text{IX}} = a_{i,\text{bulk}} \exp\left(-\frac{z_i\mathcal{F}\psi_{\text{Donnan}}}{RT}\right) \quad (7\text{-}94)$$

Writing a_i as $\gamma_i c_i/c_{i,\text{std.state}}$, and solving for $c_{i,\text{IX}}$, we find

$$c_{i,\text{IX}} = \frac{\gamma_{i,\text{bulk}}}{\gamma_{i,\text{IX}}} \frac{c_{i,\text{IX,std.state}}}{c_{i,\text{b,std.state}}} c_{b,i} \exp\left(-\frac{z_i\mathcal{F}\psi_{\text{Donnan}}}{RT}\right) \quad (7\text{-}95)$$

By convention, the concentrations in the standard state, both in bulk solution and the inside the membrane, are assigned values of 1.0 mol/L. In addition, the activity coefficient ratio for each ion, $\gamma_{i,\text{bulk}}/\gamma_{i,\text{IX}}$, is usually assumed to be constant and is then incorporated into the Donnan potential.[15] Making these substitutions, Equation 7-95 simplifies to:

$$c_{i,\text{IX}} = c_{i,\text{bulk}} \exp\left(-\frac{z_i\mathcal{F}\psi_{\text{Donnan,eff}}}{RT}\right) \quad (7\text{-}96)$$

where the effective Donnan potential, $\psi_{\text{Donnan,eff}}$, equals $\psi_{\text{Donnan}} + (RT/z_i\mathcal{F})\ln(\gamma_{i,\text{IX}}/\gamma_{i,\text{bulk}})$. The distinction between ψ_{Donnan} and $\psi_{\text{Donnan,eff}}$ is usually ignored, so Equation 7-96 is typically written as:

$$c_{i,\text{IX}} = c_{i,\text{bulk}} \exp\left(-\frac{z_i\mathcal{F}\psi_{\text{Donnan}}}{RT}\right) \quad (7\text{-}97)$$

The net charge of all dissolved ions in the interior of the matrix must equal the total fixed charge associated with the ion exchange matrix. The void volume inside an ion exchange matrix (i.e., the volume occupied by solution) is difficult to measure, so concentrations inside the resin are typically normalized to the volume of the whole resin, rather than just the volume of the water. The fixed charge is normalized in the same way and is represented here as $z_{\text{fixed}} c_{\text{IX,fixed}}$, where z_{fixed} is the charge on the fixed sites ($+1$ or -1), and $c_{\text{IX,fixed}}$ is the concentration of the fixed sites inside the resin, in moles per liter. A charge balance on the resin and the interior solution is therefore as follows:

$$z_{\text{fixed}} c_{\text{IX,fixed}} = -(z_{\text{cat}} c_{\text{cat,IX}} + z_{\text{an}} c_{\text{an,IX}}) \quad (7\text{-}98)$$

where the subscripts cat and an refer to cations and anions, respectively. Similarly, a charge balance on bulk solution yields:

$$z_{\text{cat}} c_{\text{cat,bulk}} = -z_{\text{an}} c_{\text{an,bulk}} \quad (7\text{-}99)$$

Using Equation 7-97 to substitute for both $c_{\text{cat,IX}}$ and $c_{\text{an,IX}}$ in Equation 7-98, and then using Equation 7-99 to simplify the resulting expression, we find

$$\begin{aligned} z_{\text{fixed}} c_{\text{IX,fixed}} &= -\left(z_{\text{cat}} c_{\text{cat,bulk}} \exp\left(-\frac{z_{\text{cat}}\mathcal{F}\psi_{\text{Donnan}}}{RT}\right)\right. \\ &\quad \left. + z_{\text{an}} c_{\text{an,bulk}} \exp\left(-\frac{z_{\text{an}}\mathcal{F}\psi_{\text{Donnan}}}{RT}\right)\right) \\ &= -z_{\text{cat}} c_{\text{cat,bulk}} \left(\exp\left(-\frac{z_{\text{cat}}\mathcal{F}\psi_{\text{Donnan}}}{RT}\right)\right. \\ &\quad \left. - \exp\left(-\frac{z_{\text{an}}\mathcal{F}\psi_{\text{Donnan}}}{RT}\right)\right) \quad (7\text{-}100) \end{aligned}$$

In the case where the dissolved salts are all monovalent ($z_{\text{cat}} = -z_{\text{an}} = 1$), Equations 7-97 to 7-99 can be simplified and manipulated as follows:

$$c_{\text{IX,fixed}} = -c_{\text{cat,IX}} + c_{\text{an,IX}} \quad (7\text{-}101)$$

$$c_{\text{cat,bulk}} = c_{\text{an,bulk}} \equiv c_{\text{bulk}} \quad (7\text{-}102)$$

$$c_{\text{an,IX}} c_{\text{cat,IX}} = c_{\text{an,bulk}} \exp\left(\frac{\mathcal{F}\psi_{\text{Donnan}}}{RT}\right) c_{\text{cat,bulk}} \exp\left(-\frac{\mathcal{F}\psi_{\text{Donnan}}}{RT}\right)$$

$$= c_{\text{an,bulk}} c_{\text{cat,bulk}} = c_{\text{bulk}}^2 \quad (7\text{-}103)$$

Multiplying Equation 7-101 through by $c_{\text{cat,IX}}$ or $c_{\text{an,IX}}$ and substituting into the resulting equation based on Equation 7-103 yields quadratic equations that can be solved to obtain the following explicit expressions for the concentrations in the resin:

$$c_{\text{an,IX}} = \frac{c_{\text{IX,fixed}}}{2} + \sqrt{\frac{c_{\text{IX,fixed}}^2}{4} + c_{\text{bulk}}^2} \quad (7\text{-}104\text{a})$$

---

[15] Alternatively, only the activity coefficients inside the membrane might be assumed to be constant, and those for the dissolved ions can be calculated based on the solution composition.

$$c_{\text{cat,IX}} = -\frac{c_{\text{IX,fixed}}}{2} + \sqrt{\frac{c_{\text{IX,fixed}}^2}{4} + c_{\text{bulk}}^2} \quad (7\text{-}104b)$$

If the dissolved salts are not all monovalent, the concentration of each type of ion in the resin can be found by a numerical solution of Equations 7-97 to 7-99.

Solution of these equations indicates that the ratio of co-ions to counter-ions decreases dramatically as one moves from the bulk solution to the membrane, a phenomenon that is referred to as *Donnan exclusion*. The magnitude of the Donnan exclusion effect increases with decreasing concentration of i in solution.

■ **EXAMPLE 7-10.** An anion exchange resin with a fixed charge of 2.0 eq/L is equilibrated with a solution which, once equilibrium is achieved, contains 1000 mg/L NaCl. The exchange sites are always assumed to be available, so $c_{\text{IX,fixed}}$ is 2.0 mol/L.

(a) Assuming that both the bulk and internal solutions behave ideally, calculate the cation (Na^+) and anion (Cl^-) concentrations in the resin.

(b) Solve part *a* again, but with the solution containing 100 mg/L NaCl.

Solution.

(a) NaCl dissociates completely to yield Na^+ and Cl^- ions. Since the ions are monovalent, Equation 7-104a applies. The concentration of the ions in the bulk solution is

$$c_{Na^+} = c_{Cl^-} = \frac{1\,\text{g/L}}{58.5\,\text{g/mol}} = 0.017\,\text{mol/L}$$

Substituting these values into Equation 7-104a yields:

$$c_{\text{an,IX}} = \frac{c_{\text{IX,fixed}}}{2} + \sqrt{\frac{c_{\text{IX,fixed}}^2}{4} + c_{\text{bulk}}^2}$$

$$c_{Cl^-,\text{IX}} = \frac{2\,\text{mol/L}}{2} + \sqrt{\frac{(2\,\text{mol/L})^2}{4} + (0.017\,\text{mol/L})^2}$$

$$= 1 + 1.00014 = 2.00014\,\text{mol/L}$$

Similarly, using Equation 7-104b, we find that $c_{Na^+,\text{IX}} = -1 + 1.00014 = 0.00014\,\text{mol/L}$. The ratio of co-ions to counter-ions in the bulk solution is 1.0, whereas in the resin it is

$$\frac{c_{Na^+,\text{IX}}}{c_{Cl^-,\text{IX}}} = \frac{0.00014\,\text{eq/L}}{2.00014\,\text{eq/L}} = 7.0 \times 10^{-5}$$

Also, the ratio of counter-ions in the resin to that in the bulk solution is:

$$\frac{c_{Na^+,\text{IX}}}{c_{Na^+,\text{bulk}}} = \frac{0.00014\,\text{mol/L}}{0.017\,\text{mol/L}} = 8.2 \times 10^{-3}$$

(b) For the case that $c_{\text{NaCl}} = 100\,\text{mg/L} = 0.0017\,\text{mol/L}$, we find

$$c_{Cl^-,\text{IX}} = \frac{2\,\text{mol/L}}{2} + \sqrt{\frac{(2\,\text{mol/L})^2}{4} + (0.0017\,\text{mol/L})^2}$$

$$= 1 + 1.0000014 = 2.0000014\,\text{mol/L and}$$

$$c_{Na^+,\text{IX}} = -1 + 1.0000014 = 0.0000014\,\text{mol/L}.$$

In this case

$$\frac{c_{Na^+,\text{IX}}}{c_{Cl^-,\text{IX}}} = (0.0000014\,\text{mol/L})/(2.0000014\,\text{mol/L})$$

$$= 7.0 \times 10^{-7},$$

and

$$c_{\text{IX}}/c_{\text{bulk}} = (0.0000014\,\text{mol/L})/(0.0017\,\text{mol/L})$$

$$= 8.2 \times 10^{-4}.$$

Thus, the separation of the ions by the resin and the exclusion of the cations from the resin interior are substantially greater when the solution concentration is lower. ■

The model for ion exchange equilibrium presented above accounts for the different compositions inside and outside the resin but, because it distinguishes among ions based only on their charge, it does not explain selectivity of the resin for different ions of equal charge (i.e., Ag^+ vs. Na^+, or SO_4^{2-} vs. CO_3^{2-}). To do so, an additional, ion-specific term must be added to Equation 7-97 to account for the added or diminished attractiveness of the environment inside the resin for the particular ion; that is, the equation can be modified as follows:

$$c_{i,\text{IX}} = K_{\text{IX},i} c_{i,\text{bulk}} \exp\left(-\frac{z_i \mathcal{F} \psi_{\text{Donnan}}}{RT}\right) \quad (7\text{-}105)$$

where $K_{\text{IX},i}$ is the equilibrium constant for adsorption of i to the resin, and can be thought of as the ratio of the concentration of i in the resin to that in the bulk in the hypothetical situation where the Donnan potential was zero. The primary factor that is thought to affect $K_{\text{IX},i}$ is the ion size (increasing size decreases $K_{\text{IX},i}$), though other, species-specific factors related to chemical structure are undoubtedly also important.

Competitive Adsorption in the Context of the Site-Binding Model for Adsorbates that Obey Freundlich Isotherms

In systems where adsorption is characterized by the Freundlich isotherm, competitive effects can be significant if even a small fraction of the surface is occupied, since the sites that are preferentially occupied by the competing adsorbate might be the ones that bind the target adsorbate most strongly. That is, if a surface has only a small number of strong binding sites, a relatively small concentration of a strongly competing adsorbate might occupy almost all of those sites. In that case, even though the competing adsorbate has a negligible effect on the total number of available surface sites, the competitive effect could be significant, because the strongest sites would be largely unavailable to the target adsorbate.

To model such a situation, a logical conceptual approach would be to apply the competitive Langmuir model to each group of sites. Use of such an approach is reasonably straightforward if the adsorbates are characterized by Freundlich isotherms with the same n value. In such a case, the k_f values provide a direct measure of the relative affinity of the surface for the adsorbates. That is, if $n_A = n_B$ and $k_{f,A}/k_{f,B} = 2$, we can infer that every site on the surface has twice the attraction for species A as for species B. If, however, the adsorbates are characterized by different values of n, then A might outcompete B for sorption on some sites, and B might outcompete A for sorption on others. In such a case, additional assumptions or measurements relating the two isotherms are required to model competitive interactions.

Approaches have been proposed for analyzing competitive adsorption in such systems,[16] but the approaches that are based on site binding adsorption models tend to be cumbersome and to require extensive amounts of data to estimate the adjustable parameters. As a result, such approaches have not been widely adopted. On the other hand, the ability to model competitive adsorption of adsorbates that obey Freundlich isotherms represents a straightforward application of the phase transfer model for adsorption, which is discussed next.

Competitive Freundlich adsorption can lead to an interesting and somewhat surprising outcome if two adsorbates compete for the same sites, but the weaker-binding adsorbate is present at a much lower total concentration than the more strongly binding adsorbate. Such a situation might arise, for example, in the adsorption of synthetic organic compounds (SOCs) onto activated carbon from solutions that contain NOM (which is typically present at a concentration three or more orders of magnitude larger than the SOCs). Such SOCs almost invariably obey Freundlich isotherms with the Freundlich exponent n substantially less than 1 when they are the only adsorbate in the system. However, they frequently obey linear isotherms in the presence of NOM, at least over the range of concentrations where the SOC accounts for a small fraction of the total adsorption density (Matsui et al., 2002; Graham et al., 2000). Put another way, when an SOC adsorbs from clean water, the effective binding strength and the fraction of the SOC that

[16] For an example, see Sheindorf et al. (1981).

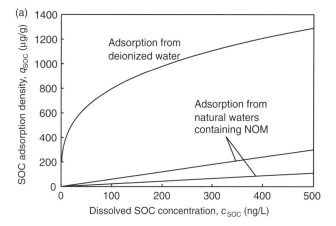

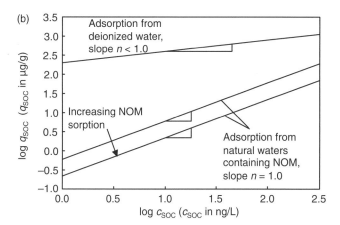

FIGURE 7-28. Adsorption of a hypothetical SOC from a solution in which the SOC is the only adsorbate and in two solutions in which NOM is also present. (a) Isotherm plots with arithmetic scales on the axes; (b) same data as in a, but with logarithmic scales.

adsorbs decrease as the total concentration of SOC in the system increases. However, when the same SOC adsorbs from a solution containing NOM, the binding strength and the fractional adsorption remain constant as the SOC concentration increases. Isotherms for a hypothetical SOC that adsorbs according to this pattern are shown in Figure 7-28.

This result can be rationalized by recalling that the Freundlich isotherm is consistent with a site distribution function in which the surface site concentrations increase as the binding strength decreases (as was shown in Figure 7-18b). Consider, for example, two systems with the same concentration of adsorbent but two different total concentrations of an SOC, such that the SOC adsorption densities in the two systems are 10^{-4} and 10^{-3} mol SOC/mol of sites, respectively. If no other adsorbates were present, and if SOC adsorption obeyed a Freundlich isotherm with n of, say, 0.5, then the sites occupied by the first 10% of the SOC molecules that adsorbed in the high-concentration system would be equivalent to those occupied by all the molecules that adsorbed in the low-concentration system. However, the remaining 90% of the molecules that adsorbed in the high-concentration system would bind to sites with lower affinities for the SOC, so the average binding strength characterizing the high-concentration system would be substantially less than in the low-concentration system.

Now consider the situation for the same two SOC adsorption densities, but in systems where NOM occupies 10% of all the adsorbent sites. If the SOC strongly outcompeted the NOM for the preferred sites, the NOM would have a negligible effect on the SOC isotherm. However, if the NOM outcompeted the SOC, the NOM would occupy at least some of the sites with the greatest affinity for the SOC, and the SOC would bind to sites that were weaker than in the NOM-free system. In the limit, the NOM might completely prevent the SOC from binding to the most preferred sites, so that the most attractive 10% of the total sites would have become unavailable to the SOC. In that case, the average binding strength of the adsorbed SOC would be much lower than in the NOM-free systems. In addition, however, because of the nature of the site distribution function for Freundlich isotherms, the binding strength of the (weaker) sites to which the SOC binds in the competitive system would vary much more gradually as a function of adsorption density than the (stronger) sites in the noncompetitive system. As a result, in the system with NOM, the SOC molecules would bind to sites with a smaller range of binding strengths, and the isotherm would be characterized by an n value closer to 1. In the limit, the affinities of the sites occupied by the SOC in this latter case would be almost identical, and the SOC adsorption isotherm would be linear (i.e., $n = 1.0$), as is in fact observed.

7.6 MODELING ADSORPTION USING SURFACE PRESSURE TO DESCRIBE THE ACTIVITY OF ADSORBED SPECIES

As noted previously, the phase transfer model for adsorption envisions adsorbed molecules to reside in a two-dimensional gas-like phase on the adsorbent surface. Correspondingly, Table 7-2 indicates that the activity or concentration of adsorbed species can be represented not only in terms of adsorption density, but alternatively as a *surface pressure*. In this section, the conceptual basis of the surface pressure is presented, and the linear, Langmuir, and Freundlich isotherms are developed in terms of the surface pressure. The major application of the surface pressure concept is in the modeling of competitive adsorption, and the key equations for this application are developed in the final portion of this section.

The Surface Pressure Concept

Water molecules have significant attraction for one another due to favorable electrostatic interactions between the hydrogen atoms on one molecule and the oxygen atoms on others (hydrogen bonds). Water molecules at the interface with another phase have opportunities for these types of favorable interactions in the direction of the bulk solution, but not in the direction of the other phase. (Some favorable interactions with molecules of the nonaqueous phase might exist, but they tend to be weaker than those with other water molecules.) As a result, water molecules at an interface experience a small force pulling them into the bulk phase.

The weaker the bonds between water molecules and those of the nonaqueous phase, the greater the force on the molecules at the interface pulling them toward the bulk solution, or, equivalently, the greater the force opposing movement of water molecules from the bulk to the interface. This force causes water (and, in fact, all liquids) to be cohesive and to resist changes that increase the interfacial area; one familiar consequence of this phenomenon is that adjacent water droplets tend to coalesce into larger ones (thereby reducing the total number of water molecules at the air/water interface), rather than spontaneously disperse into smaller ones.

The cohesiveness of a phase is quantified as the *surface tension*, conventionally represented as γ (not to be confused with an activity coefficient). Formally, surface tension is defined as the increase in the chemical potential energy (the Gibbs free energy) of the system per unit increase in surface area, while the temperature, pressure, and composition are held constant. Thus:

$$\gamma \equiv \left.\frac{\partial G_{sys}}{\partial A}\right|_{P,T,n_i} \qquad (7\text{-}106)$$

where G_{sys} is the Gibbs free energy of the system (with units such as kilojoules), A is interfacial area (e.g., square meters), and n_i is the number of moles of any species i in the system. Surface tension has units of energy per unit area or, equivalently, force per unit length. Because of the extensive hydrogen bonding among water molecules, the surface tension of pure water is significantly larger than that of most other liquids.[17]

The effects of surface tension on a solution and on adsorption at its boundary with another phase can be profound. To explain some of these effects, an analogy is sometimes made between the layer of molecules right at the interface and a thin, elastic membrane that has been slightly stretched. The molecules that make up the membrane have some optimal orientation and separation that

---

[17] Formally, the surface tension is a property of the interface and therefore depends on the identities of both phases that form the interface. However, surface tension of an individual phase can be defined as the hypothetical value of $\partial G/\partial A$ when that phase is in contact with an ideal gas with which it has no chemical interactions. The surface tension at the interface between two real materials can then be computed as the difference in the hypothetical surface tensions of the isolated phases.

maximizes their favorable interactions in the absence of external stresses, just as molecules of water do. This optimum molecular geometry is the norm in the unstretched membrane and the bulk water.

When the membrane is stretched, the average separation between the molecules increases. The membrane then exerts an opposing, contraction force, reflecting the preference of the molecules to return to their optimal separation. Similarly, the separation between water molecules at an interface is greater than optimal, albeit for a different reason; in this case, the extra separation arises because some surface molecules have been pulled into the bulk solution in response to the greater hydrogen bonding in that direction. Although the cause of the molecular separation is different in these two situations, the result is the same: the molecules that have greater than the optimal separation between them exert an opposing contraction force as they attempt to return to their optimal geometry. In the case of the interfacial layer, this force is manifested as the surface tension. Thus, the interfacial layer behaves like a membrane, made up primarily of water molecules, that is stretched across the surface of the bulk solution; the higher the surface tension, the tighter this imaginary membrane is stretched.

If one stretches a membrane over a flat surface (e.g., think of stretching a piece of plastic wrap tightly over a bowl), the tension in the membrane can be quantified as the energy required to stretch it divided by the area covered; this energy per unit area is exactly analogous to the surface tension defined above as $\partial G/\partial A$. Also, because it is stretched, the membrane pulls along the perimeter of the surface (the plastic wrap pulls the edge of the bowl toward the center) with a force that is proportional to the tension in the membrane. Therefore, the force per unit length around the perimeter is another, equivalent measure of the tension in the membrane. A similar idea—of surface tension as a force per unit length—applies to water at an interface. In this case, the analogy can be visualized as an elastic band stretched across the interface that opposes expansion of the surface.

Any process that lowers the surface tension (i.e., allows the water molecules to return to an average separation closer to the optimal one) relieves some of the strain in the system and is favored. Thus, if the surface tension at an adsorbent/solution interface can be reduced by placing a molecule of some solute in that layer (i.e., by adsorption of the solute), that reaction will be favored. The weaker the bonds between the solute and the bulk water, and the stronger the bonds between the solute and the adsorbent, the more favorable the adsorption reaction becomes. Opposing this tendency is thermal molecular motion, which tends to cause all molecules to distribute themselves randomly throughout the solution. Therefore, according to this view, adsorptive equilibrium can be interpreted as a balance between the tendencies of the system to minimize surface tension and to maximize the random distribution of molecules throughout the solution.

Computation of the Surface Pressure from Surface Tension or Isotherm Data

The pressure in a true, three-dimensional gas phase can be quantified as the force that the gas exerts on a unit area of any (two-dimensional) surface in contact with the gas. Similarly, a hypothetical, two-dimensional gas-like phase at a solid-/liquid interface can be considered to exert a surface pressure on any (one-dimensional) object in the layer, quantified as the force on the object per unit length; that is, the same units as surface tension. This surface pressure is normally represented as π.

One way to evaluate the pressure contributed by a particular constituent of a bulk gas phase is to put the gas both inside and outside a container, evacuate the constituent from the gas on the inside, and measure the force per unit area on the container's walls. By analogy, one can imagine evaluating the surface pressure contributed by an adsorbate by evacuating the adsorbate from a small region of the surface, thereby lowering the surface pressure in that region by an amount π_i. Because the pressure outside the evacuated area would be greater than that inside, a force would exist along the perimeter, directed toward its center (Figure 7-29). The force per unit length along the boundary could then be identified as the surface pressure attributable to the adsorbate in the unperturbed portion of the surface; in essence, this value is the partial surface pressure of i in the system.

The same scenario can be interpreted in the context of surface tension by considering that a pseudo-membrane

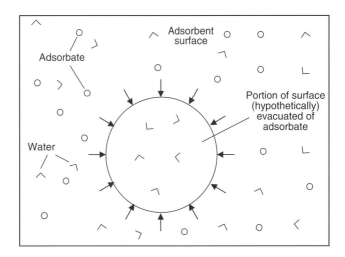

FIGURE 7-29. Schematic representation of the surface pressure. The arrows represent the surface pressure, which pushes toward the center of the area from which adsorbate molecules have been evacuated. The force per unit length is π, equal to the surface tension (γ) inside minus that outside the evacuated area.

inside the evacuated area is pulling the boundary toward the center with a force per unit length equal to $\gamma|_{\Gamma=0}$, while a different pseudo-membrane is pulling the boundary in the opposite direction with a (lesser) force per unit length equal to $\gamma|_{\text{actual }\Gamma}$.[18] Mathematically, for a system containing only one adsorbate, these concepts can be represented as follows:

Net surface pressure (force/length) pushing boundary toward evacuated area	=	Force/length pulling boundary toward evacuated area	−	Force/length pulling boundary toward nonevacuated area
π	=	$\gamma\|_{\Gamma=0}$	−	$\gamma\|_{\text{actual }\Gamma}$

(7-107)

Thus, $\pi = 0$ at an interface where no adsorption has occurred.

While the preceding conceptual approach for estimating π_i is useful for understanding the idea of surface pressure, a more convenient and practical approach for quantifying π can be derived based on a relationship known as the *Gibbs adsorption isotherm*, which relates the surface tension, the dissolved adsorbate concentration, and the adsorption density in any system at equilibrium, as follows:[19]

$$\pi = RT \int_0^{c_i} \frac{\Gamma}{c} dc \qquad (7\text{-}108a)$$

$$= RT \int_{-\infty}^{\ln c_i} \Gamma d\ln c = 2.303 RT \int_{-\infty}^{\log c_i} \Gamma d\log c \qquad (7\text{-}108b)$$

$$= RT \int_{-\infty}^{\ln c_i} \frac{d\ln c}{d\ln \Gamma} d\Gamma \qquad (7\text{-}108c)$$

where π is the surface pressure (Pa m), when the equilibrium dissolved adsorbate concentration is c_{eq}; R is the universal gas constant, 8.314 m^3-Pa/mol K; T is the temperature (K); c is the dissolved concentration of the adsorbate (moles per liter); Γ is the adsorption density in equilibrium with c (moles per meter square).

---

[18] Because the analysis of adsorption in terms of surface pressure is based on a view of the interfacial region as a separate phase from the liquid or solid, it is most convenient to carry out the analysis in terms of the amount of that phase that is present (*i.e.*, the surface area), rather than the amount of solid that is present. Therefore, the following discussion is presented using Γ rather than q as the measure of adsorption density. If desired, any of the equations can be recast in terms of q based on the relationship $\Gamma = q/\text{SSA}$.
[19] The Gibbs adsorption isotherm is derived in many texts dealing with colloid or interfacial phenomena. See, for instance, the texts by Adamson (1976), Hunter (1987), or Everett (1988).

The Gibbs isotherm is to the gas transfer model of adsorption what the Φ versus K function is to the site-binding model: a completely generic way to describe the relationship between the activities of a species in solution and at the surface. These generic isotherms posit no particular relationship between adsorption density and dissolved concentration, but they do establish a framework in which any such Γ versus c or q versus c relationship can be interpreted. For example, particular Γ versus c relationships (such as the Langmuir or Freundlich isotherm) can be interpreted in the context of the gas transfer model of adsorption, just as they can in the context of the site-binding model. These points are elucidated in the following section, which closely follows the presentation of Benjamin (2009).

Any of the forms of Equation 7-108 can be used to determine the surface pressure from experimental Γ versus c data. For example, if we prepared a plot of Γ/c versus c or Γ versus $\ln c$, the product of RT with the area under the curve between $c = 0$ ($\ln c = -\infty$) and the concentration corresponding to the adsorption density of interest would be the surface pressure associated with that adsorption density. This relationship is illustrated in Figure 7-30.

By carrying out the preceding steps for a range of c and Γ values, we can prepare an isotherm plot using surface pressure rather than adsorption density to quantify the activity of adsorbed species; that is, a plot of π versus c_{eq}. Alternatively, if the data conform to a linear, Langmuir, or Freundlich isotherm, we can use the isotherm equation to substitute for c or Γ in Equation 7-108 and integrate the resulting equation analytically. For instance, the surface pressure in a system that conforms to the Langmuir isotherm

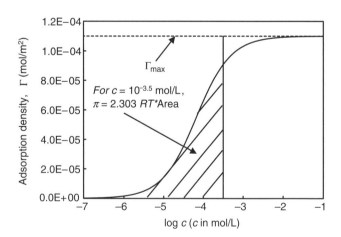

FIGURE 7-30. Computation of the surface pressure from a plot of Γ versus $\ln c$. The plot shown is for a Langmuir isotherm with $K = 15,000$ L/mol and $\Gamma_{max} = 1.1 \times 10^{-4}$ mol/m^2. The surface pressure could also be computed from the area under a plot of Γ/c versus c. *Source*: From Benjamin (2009).

can be computed as follows:

$$\pi = RT \int_0^c \frac{\frac{\Gamma_{max} Kc}{1+Kc}}{c} dc = RTK\Gamma_{max} \int_0^c \frac{dc}{1+Kc}$$

$$= RT\Gamma_{max} \ln(1+Kc) = 2.303 RT\Gamma_{max} \log(1+Kc) \tag{7-109}$$

Algebraic manipulation of the Langmuir isotherm shows that $1 + Kc = \Gamma_{max}/(\Gamma_{max} - \Gamma) = (1 - (\Gamma/\Gamma_{max}))^{-1}$, so the surface pressure can be expressed as a function of the adsorption density as:

$$\pi = RT\,\Gamma_{max} \ln\left(1 - \frac{\Gamma}{\Gamma_{max}}\right)^{-1}$$

$$= 2.303 RT\,\Gamma_{max} \log\left(1 - \frac{\Gamma}{\Gamma_{max}}\right)^{-1} \tag{7-110}$$

Similar derivations for systems that obey linear or Freundlich isotherms also lead to fairly simple expressions for surface pressure as a function of the dissolved adsorbate concentration or the adsorption density. The relationships for all three isotherms are collected in Table 7-5, and corresponding plots of π versus c are shown in Figure 7-31.

Equation 7-111 indicates that, if the isotherm is linear, the relationships between surface pressure and both the dissolved adsorbate concentration and the adsorption density are linear as well. By substituting N_{ads}/A for Γ in the expression on the right side of Equation 7-111, where N_{ads} is the number of moles adsorbed and A is the surface area of the adsorbent, we obtain an equation that is an exact analog of the ideal gas law, but for a two-dimensional system; that is, with surface pressure and area replacing three-dimensional pressure and volume:

Ideal gas law:

$$PV = RTN_{gas} \tag{7-113}$$

Linear adsorption isotherm expressed in terms of surface pressure:

$$\pi A = RTN_{ads} \tag{7-114}$$

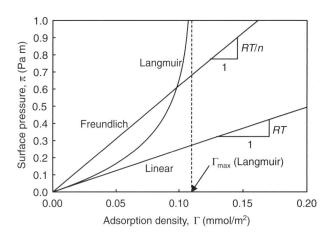

FIGURE 7-31. The relationship between surface pressure and adsorption density for various isotherms. (For the curve corresponding to the Langmuir isotherm, Γ_{max} was assigned a value of 1.1×10^{-4} mol/m^2, and for that corresponding to the Freundlich isotherm, n was assigned a value of 0.4.) *Source*: From Benjamin (2009).

This analogy suggests that, in an adsorptive system characterized by a linear isotherm, the adsorbed species can be thought of as forming an ideal two-dimensional gas. The characteristics of an ideal adsorbed phase, like those of an ideal gas, include (1) the pressure exerted by the molecules in the phase is directly proportional to the concentration of those molecules and does not depend on the identity of the molecules; and (2) the molecules interact by perfectly elastic collisions, but otherwise have no effect on one another; that is, each molecule in the real mixture behaves exactly as it would if the gas were made up entirely of that type of molecule. Based on these considerations, the plot of π versus Γ for the linear isotherm shown in Figure 7-31 applies for *any* linear isotherm, regardless of the identity of the adsorbate and adsorbent.

Since, according to the gas transfer model, an ideal adsorbed gas is reflected in a linear Γ versus c relationship, the extent to which that the Γ versus c relationship is nonlinear can be taken as an indicator of the nonideality of the surface phase. Such deviations from ideal behavior are much more common in adsorptive systems than in gaseous systems, both

TABLE 7-5. Surface Pressure Relationships for Various Isotherm Equations

Isotherm Type	Isotherm Equation	Expressions for Surface Pressure	
Linear	$\Gamma = kc$	$\pi = RTkc = RT\,\Gamma$	(7-111)
Langmuir	$\Gamma = \dfrac{\Gamma_{max} Kc}{1+Kc}$	$\pi = RT\Gamma_{max} \ln(1+Kc) = RT\,\Gamma_{max} \ln\left(1 - \dfrac{\Gamma}{\Gamma_{max}}\right)^{-1}$	(7-110)
Freundlich	$\Gamma = k_f c^n$	$\pi = \dfrac{RT}{n} k_f c^n = \dfrac{RT}{n}\Gamma$	(7-112)

because the underlying surface is nonuniform (as opposed to the empty and therefore uniform space in which gas molecules are dispersed) and because adsorbed molecules are packed much more densely than gas molecules, leading to the potential for significant intermolecular interactions. Nevertheless, linear adsorption isotherms do characterize many real systems, and they represent a useful limiting case with which other experimental isotherms can be compared.

In systems that obey the Langmuir isotherm, under conditions where $\Gamma \ll \Gamma_{max}$ (corresponding to $Kc \ll 1$), the surface pressure at a given adsorption density is identical to that for a system that has a linear isotherm. This result should not be surprising, since the Langmuir isotherm is itself linear under such conditions. Put another way, the surface pressure follows the ideal adsorbed gas law (Equation 7-114) under conditions where the Langmuir isotherm reduces to a linear isotherm. This result is also consistent with the conceptual model for the system, in that low adsorption densities correspond to large distances between adsorbed molecules and therefore minimal interactions among them. On the other hand, when the adsorption density becomes a significant fraction of Γ_{max}, the surface pressure increases at a dramatically greater rate, thereby opposing additional adsorption ever more forcefully. In the limit, π becomes infinitely large as the adsorption density asymptotically approaches Γ_{max}. In other words, whereas Γ_{max} is interpreted in the site-binding model as a limit on adsorption imposed by the finite number of sites on the surface, it is interpreted in the gas transfer model as a limit imposed by the buildup of an immensely large surface pressure, opposing transfer of more adsorbate molecules into the surface phase. Note that, when plotted as a function of Γ, the surface pressure depends on the proximity to Γ_{max}, but not on K; thus, a single plot of π versus Γ/Γ_{max} (Figure 7-32) characterizes all systems that follow Langmuir isotherms.

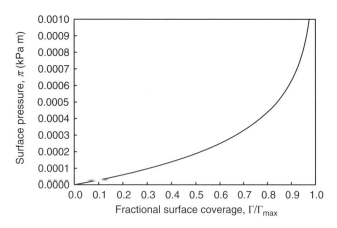

FIGURE 7-32. Universal relationship between surface pressure and fractional surface coverage for all systems that obey the Langmuir isotherm.

According to Equation 7-112, in systems that follow the Freundlich isotherm, π is directly proportional to Γ, just as it is for linear isotherms; however, the proportionality constant is RT/n, rather than RT. Thus, in such systems, the extent of nonideality is indicated by the value of n: the closer n is to 1.0, the closer the system conforms to the ideal adsorbed gas model.

For any Langmuir isotherm, and for Freundlich isotherms in which $n < 1$, each increment in adsorption density causes the surface pressure to increase more than would be predicted by the ideal gas model. These deviations from ideal behavior can be quantified by incorporating activity coefficients for the adsorbed species ($\gamma_{surf,i}$) into Equation 7-114:

$$\pi_i = \gamma_{surf} RT \frac{N_{surf}}{A} = \gamma_{surf} RT \, \Gamma \qquad (7\text{-}115)$$

Comparison of Equation 7-115 with the equations in Table 7-5 indicates that, in the case of Langmuir adsorption, the activity coefficient of the adsorbed species is ~ 1.0 at low adsorption densities and surface pressures, but increases steadily with increasing Γ and π, and gets very large as Γ approaches Γ_{max}. In the case of the Freundlich isotherm, the activity coefficient is $1/n$ (and therefore, typically, greater than 1.0) at all adsorption densities and surface pressures. The implication is that, for example, if a system was characterized by a Freundlich isotherm with $n = 0.5$, adsorption of a single molecule in the real system would increase the surface pressure by the same amount as would be expected for adsorption of two (1/0.5) molecules in an ideal system.

■ **EXAMPLE 7-11.** Adsorption of benzene and toluene onto activated carbon has been reported by Speth and Miltner (1990) to obey the following Freundlich isotherm equations, where c is in mg/L and q is in mg/g:

Benzene: $\quad q_{benz} = 50.1 \, c_{benz}^{0.533}$

Toluene: $\quad q_{tol} = 76.4 \, c_{tol}^{0.365}$

A solution at 25°C containing 0.50 mg/L benzene and 0.82 mg/L toluene is to be treated in a batch process to reduce the concentration of each contaminant to less than 0.01 mg/L. The adsorbent is activated carbon with a SSA of 650 m²/g.

(a) Develop curves of dissolved benzene and toluene concentration versus adsorbent dose, considering each contaminant in isolation. Plot the corresponding curves of surface pressure versus adsorbent dose.

(b) What is the required adsorbent dose to reach the treatment goal for each contaminant in isolation? What are the surface pressure and surface activity coefficient of each adsorbate when this adsorbent dose is added?

Solution.

(a) The dissolved concentration of each contaminant at equilibrium can be computed as a function of the activated carbon dose by combining the isotherm expression with the corresponding mass balance. For instance, for benzene, we can pick a value for c_{benz}, compute q_{benz} from the isotherm equation, and then compute the corresponding activated carbon dose from the mass balance, as follows:

$$q_{benz} = 50.1\, c_{benz}^{0.533}$$

$$c_{AC} = \frac{c_{tot,benz} - c_{benz}}{q_{benz}} = \frac{0.50 - c_{benz}}{50.1\, c_{benz}^{0.533}}$$

A parallel calculation for toluene yields:

$$c_{AC} = \frac{c_{tot,tol} - c_{tol}}{q_{tol}} = \frac{0.82 - c_{tol}}{76.4\, c_{tol}^{0.365}}$$

Substituting a range of values for c_{benz} and c_{tol} yields the activated carbon dose required to reach those concentrations. The resulting plots of for c_{benz} and c_{tol} versus c_{AC} are shown in Figure 7-33a.

Because both adsorbates follow Freundlich isotherms, the surface pressure associated with each condition is given by Equation 7-112 as $\pi = (RT/n)k_f c^n$, where π is in kPa m, R is in kPa m^3/mol K, and c is the molar concentration of the adsorbate in solution. For each dose of activated carbon, the dissolved concentrations of benzene and toluene can be converted from mg/L to mol/L units based on their molecular weights of 78 and 92, respectively. The corresponding plots of π versus c_{AC} are shown in Figure 7-33b.

(b) At the target concentration of 0.01 mg/L, the adsorption densities and activated carbon doses, calculated as described in part a, are

Benzene: $q_{benz} = 4.30$ mg/g $c_{AC} = 0.114$ g/L $= 114$ mg/L
Toluene: $q_{tol} = 14.23$ mg/g $c_{AC} = 0.057$ g/L $= 57$ mg/L

To compute the surface pressure, we need to express these adsorption densities in mol/m^2. This unit conversion can be achieved by dividing the given value of q by the molecular weight and the specific surface area, yielding:

$$\Gamma_{benz} = (4.30\, \text{mg/g AC})\left(\frac{1\,\text{mol}}{78{,}000\,\text{mg}}\right)\left(\frac{1\,\text{g AC}}{650\,\text{m}^2}\right)$$

$$= 8.49 \times 10^{-8}\,\text{mol/m}^2$$

$$\Gamma_{tol} = (14.23\,\text{mg/g AC})\left(\frac{1\,\text{mol}}{92{,}000\,\text{mg}}\right)\left(\frac{1\,\text{g AC}}{650\,\text{m}^2}\right)$$

$$= 2.38 \times 10^{-7}\,\text{mol/m}^2$$

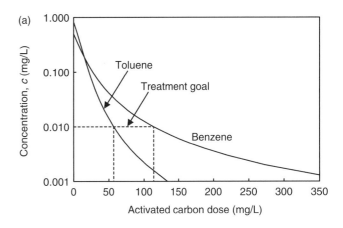

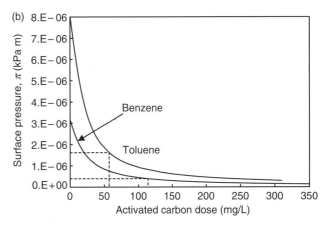

FIGURE 7-33. Conditions characterizing the example systems when dosed with various concentrations of activated carbon. (a) Dissolved benzene and toluene concentrations. (b) Surface pressures.

Substituting these values into Equation 7-112, we find

$$\pi_{benz} = \frac{RT}{n}\Gamma_{benz}$$

$$= \frac{(8.314 \times 10^{-3}\,\text{m}^3\,\text{kPa/mol K})(298\,\text{K})}{0.533}$$

$$\times (8.49 \times 10^{-8}\,\text{mol/m}^2)$$

$$= 3.95 \times 10^{-7}\,\text{kPa m}$$

$$\pi_{tol} = \frac{RT}{n}\Gamma_{tol}$$

$$= \frac{(8.314 \times 10^{-3}\,\text{m}^3\,\text{kPa/mol K})(298\,\text{K})}{0.365}$$

$$\times (2.38 \times 10^{-7}\,\text{mol/m}^2)$$

$$= 1.62 \times 10^{-6}\,\text{kPa m}$$

Since both adsorbates obey Freundlich isotherms, the surface activity coefficients are simply $1/n$. Thus, γ_{benz} is $1/0.533$, or 1.88, and γ_{tol} is $1.0/0.365$, or 2.74.

For the given solution and treatment target, benzene requires twice as large a dose of activated carbon to achieve the treatment target and, at equilibrium, has a lower adsorption density than toluene. The higher adsorption density of toluene, combined with its lower n value (greater nonideality), causes the surface pressure to be approximately four times as large for toluene as for benzene, for the conditions specified. ∎

Competitive Adsorption and Surface Pressure: The Ideal Adsorbed Solution Model

The main application of the surface pressure concept is in the modeling of competitive adsorption of hydrophobic molecules that obey nonlinear isotherms. The model that is most commonly used in these situations is known as the *ideal adsorbed solution theory* (IAST) or the *ideal adsorbed solution* (IAS) model. According to this model, the total surface pressure in an adsorbed phase, π_{tot}, can be considered to comprise independent contributions from each adsorbed species. By analogy with three-dimensional gas mixtures, we refer to each such contribution as the *partial surface pressure* of that species, π_i, with π_{tot} equal to the sum of all the π_i values:

$$\pi_{\text{tot}} = \sum_{j=1}^{m} \pi_i \qquad (7\text{-}116)$$

The key assumptions of the IAST, and the basis for the idea that the adsorbates form an "ideal" mixture, are that the partial surface pressure exerted by any adsorbate can be computed just like in single-adsorbate systems (i.e., using Equation 7-115), and that the activity coefficient $\gamma_{\text{surf},i}$ depends only on the total surface pressure, not on the identity of the adsorbates that generate that pressure (Benjamin, 2009). Note that each species in the system might behave nonideally as an individual adsorbate (i.e., it might have a nonlinear isotherm), but the nature of that nonideality is assumed to be identical in a mixed adsorbed phase as in a (hypothetical) pure phase at the same total surface pressure. Designating the conditions in this hypothetical, single-adsorbate system by an asterisk, we can write Equation 7-115 for the hypothetical system as follows:

$$\gamma_{\text{surf},i} = \frac{\pi_{\text{tot}}}{RT\Gamma_i^*} \qquad (7\text{-}117)$$

Then, combining Equations 7-115 and 7-117, we find

$$\pi_i = \frac{\pi_{\text{tot}}}{RT\Gamma_i^*}RT\,\Gamma_i = \frac{\Gamma_i}{\Gamma_i^*}\pi_{\text{tot}} \qquad (7\text{-}118)$$

While Equations 7-115 to 7-118 are applicable regardless of the adsorption characteristics of species i, recall that, for species that obey Freundlich isotherms (the most common case of interest), $\gamma_{\text{surf},i}$ equals $1/n_i$, independent of π_{tot}. Thus, for those adsorbates, $1/n_i$ can be substituted for $\gamma_{\text{surf},i}$ in Equation 7-115, and Γ_i^* can be eliminated from Equation 7-118, yielding:

$$\pi_i = \frac{RT}{n_i}\Gamma_i \qquad (7\text{-}119)$$

The implication of Equation 7-119 is that, if an adsorbate obeys a Freundlich isotherm when binding to a particular adsorbent in a noncompetitive system, it will have the same π_i versus Γ_i relationship when sorbing to that adsorbent in any other system, regardless of the presence or adsorption density of competing adsorbates. However, whereas each Γ_i is associated with a unique value of π_i in such systems, each (Γ_i, π_i) pair might be associated with many different values of c_i, depending on the identity and concentration of the competing adsorbates.

Returning to the derivation of the IAST equations for the more general case of systems where adsorbates might obey any isotherm equation, we note that, in any system, the activity of dissolved i can be related to the activity of adsorbed i by an equilibrium constant, K_{eq}:

$$a_{i,\text{sol'n}} = K_{\text{eq},i}a_{i,\text{surf}} \qquad (7\text{-}120)$$

The activity of dissolved i can be computed from its concentration and activity coefficient:

$$a_{i,\text{sol'n}} = \gamma_i \frac{c_i}{c_{i,\text{std.state}}} \qquad (7\text{-}121)$$

The activity of a species i in a three-dimensional gas is given by:

$$a_{i,\text{gas}} = \frac{P_i}{P_{i,\text{std.state}}} = \frac{y_i P_{\text{tot}}}{P_{i,\text{std.state}}} \qquad (7\text{-}122)$$

where P_i and P_{tot} are, respectively, the partial pressure of i and the total pressure in the gas phase; $P_{i,\text{std.state}}$ is the pressure in the standard state (typically, 1 atm); and y_i is the mole fraction of i in the gas phase. The analogous expression in the adsorbed phase is

$$a_{i,\text{surf}} = \frac{\pi_i}{\pi_{i,\text{std.state}}} = \frac{z_i \pi_{\text{tot}}}{\pi_{i,\text{std.state}}} \qquad (7\text{-}123)$$

in which z_i is the mole fraction in the surface phase. Substituting Equations 7-121 and 7-123 into Equation 7-120, and then rearranging, yields:

$$\gamma_i \frac{c_i}{c_{i,\text{std.state}}} = K_{\text{eq},i} \frac{z_i \pi_{\text{tot}}}{\pi_{i,\text{std.state}}} \quad (7\text{-}124)$$

$$c_i = \frac{c_{i,\text{std.state}} K_{\text{eq},i}}{\gamma_i \pi_{i,\text{std.state}}} z_i \pi_{\text{tot}} = K'_i \frac{\pi_{\text{tot}}}{\gamma_i} z_i = K_{\text{eff},i,\pi_{\text{tot}}} z_i \quad (7\text{-}125)$$

where K'_i is a composite constant equal to $c_{i,\text{std.state}} K_{\text{eq},i} / \pi_{i,\text{std.state}}$, and $K_{\text{eff},i,\pi_{\text{tot}}}$ is an effective or conditional equilibrium constant that relates c_i to z_i at the given π_{tot} and for the given dissolved-phase activity coefficient (γ_i).

The condition $z_i = 1.0$ corresponds to a one-adsorbate (i.e., noncompetitive) system. In such a system, Equation 7-125 simplifies to:

$$c_i = K_{\text{eff},i,\pi_{\text{tot}}} \quad (7\text{-}126)$$

which can be interpreted as indicating that $K_{\text{eff},i,\pi_{\text{tot}}}$ equals the concentration of i that, in a single-adsorbate system with the given γ_i, generates a surface pressure of π_{tot}. Per the convention defined earlier, this (hypothetical) concentration is designated c_i^*, so Equation 7-125 can be written as:

$$c_i = c_i^* z_i \quad (7\text{-}127)$$

Equations 7-116, 7-118, and 7-127, along with the following equations defining Γ_{tot} and quantifying the mass balances on the individual adsorbates, comprise the complete set of equations needed to solve for the equilibrium composition of a system with competitive adsorption among m adsorbates, according to the IAST.

$$\Gamma_{\text{tot}} = \sum_{j=1}^{m} \Gamma_j \quad (7\text{-}128)$$

$$c_{i,\text{tot}} = c_i + c_{\text{solid}} q_i = c_i + c_{\text{solid}}(\text{SSA})\Gamma_i \quad (7\text{-}129)$$

One additional equation, obtained by computing the ratio c_i/c_j according to Equation 7-127 and then solving for c_j, is also useful:

$$\frac{c_i}{c_j} = \frac{c_i^*(\Gamma_i/\Gamma_{\text{tot}})}{c_j^*(\Gamma_j/\Gamma_{\text{tot}})} = \frac{c_i^* \Gamma_i}{c_j^* \Gamma_j} = \frac{c_i^*(c_{i,\text{tot}} - c_i)/(c_{\text{solid}}[\text{SSA}])}{c_j^*(c_{j,\text{tot}} - c_j)/(c_{\text{solid}}[\text{SSA}])}$$

$$= \frac{c_i^*}{c_j^*} \frac{c_{i,\text{tot}} - c_i}{c_{j,\text{tot}} - c_j} \quad (7\text{-}130)$$

$$c_j = \frac{c_{j,\text{tot}}}{1 + \left((c_{i,\text{tot}}/c_i) - 1\right)\left(c_i^*/c_j^*\right)} \quad (7\text{-}131)$$

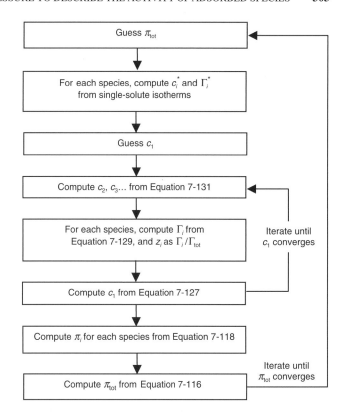

FIGURE 7-34. Flow chart for solving the IAST equations. *Source*: After Benjamin (2009).

A number of approaches have been suggested for solving the preceding equations (or equivalent ones based on slightly different derivations[20]) to find the equilibrium composition of the system (see, e.g., Crittenden et al., 1985; Tien, 1994; Moon and Tien, 1987). One relatively straightforward approach involves two nested loops of calculations, shown schematically in Figure 7-34. The steps shown in the flow chart can easily be programmed into a spreadsheet or numerical processing program such as Matlab® (Benjamin, 2009) and are illustrated in the following example.

■ **EXAMPLE 7-12.** Develop plots of c_{benz} and c_{tol} versus activated carbon dose (c_{AC}) and of π_{tot} versus c_{AC} for the solution described in Example 7-11, assuming that competitive adsorption can be modeled by the IAST. What activated carbon dose would be needed to meet the treatment objective with respect to both contaminants, when competitive adsorption is taken into account?

[20] In the literature, Equation 7-118 is frequently replaced by an equation derived by assuming that the surface area occupied by each adsorbate depends only on π_{tot}, and the sum of the surface areas occupied by all the adsorbates is the total adsorbent surface area (Crittenden et al., 1985). This equation is $1/\Gamma_{\text{tot}} = \sum_{\text{all } i}(x_i/\Gamma_i^*)$. The same result is obtained for the equilibrium distribution of adsorbates in the system whether one uses this equation or Equation 7-118.

Solution. This problem requires that we solve a set of simultaneous equations for various values of c_{AC}, until we identify the value of c_{AC} that just meets the treatment objective. For each dose, the system can be characterized by six unknowns:

$$c_{benz}, c_{tol}, \Gamma_{benz}, \Gamma_{tol}, \pi_{benz}, \text{ and } \pi_{tol}$$

To use the IAST, the isotherm parameters must be expressed in mole-based units. The total concentrations of benzene and toluene in the system can be converted into those units based on their respective molecular weights, and the corresponding k values can be converted into the appropriate units based on Equation 7-49.

Benzene: $c_{benz,tot} = \dfrac{0.50\,\text{mg/L}}{78{,}000\,\text{mg/mol}} = 6.41 \times 10^{-6}\,\text{mol/L}$

$$k'_f = k_f \times \left(\dfrac{1\,\text{mol}}{(1{,}000 \times \text{MW})\text{mg}}\right)^{1-n}$$

$$= 50.1 \dfrac{\text{mg/g}}{(\text{mg/L})^{0.533}} \left(\dfrac{1\,\text{mol}}{78{,}000\,\text{mg}}\right)^{1-0.533}$$

$$= 0.260 \dfrac{\text{mol/g}}{(\text{mol/L})^{0.533}}$$

Toluene: $c_{tol,tot} = \dfrac{0.82\,\text{mg/L}}{92{,}000\,\text{mg/mol}} = 8.91 \times 10^{-6}\,\text{mol/L}$

$$k'_f = k_f \times \left(\dfrac{1\,\text{mol}}{(1{,}000 \times \text{MW})\text{mg}}\right)^{1-n}$$

$$= \left(76.4\,(\text{mg/g})/(\text{mg/L})^{0.365}\right) \left(\dfrac{1\,\text{mol}}{92{,}000\,\text{mg}}\right)^{1-0.365}$$

$$= 0.0538\,(\text{mol/g})/(\text{mol/L})^{0.365}$$

According to Figure 7-33b, the surface pressures at an activated carbon dose of 50 mg/L are on the order of 1×10^{-6} and 2×10^{-6} kPa m, respectively, in the systems with benzene and toluene as the only adsorbate. In the system with both adsorbates, we expect the surface pressure to be higher than in either of the single-adsorbate systems, but less than the sum of the pressures in those two systems; a reasonable guess for π_{tot} might therefore be 2.3×10^{-6} kPa m. Both adsorbates obey Freundlich isotherms when they are the sole adsorbate present, so the value of Γ^* can be found from by applying Equation 7-112 to the hypothetical, single-adsorbate systems with the guessed value of π_{tot}:

$$\pi_{tot} = \dfrac{RT}{n}\Gamma^*$$

$$\Gamma^*_{benz} = \dfrac{n_{benz}\pi_{tot}}{RT} = \dfrac{(0.533)(2.3 \times 10^{-6}\,\text{kPa m})}{(8.314 \times 10^{-3}\,\text{m}^3\,\text{kPa/mol K})(298\,\text{K})}$$

$$= 4.95 \times 10^{-7}\,\text{mol/m}^2$$

$$\Gamma^*_{tol} = \dfrac{n_{tol}\pi_{tot}}{RT} = \dfrac{(0.365)(2.3 \times 10^{-6}\,\text{kPa m})}{(8.314 \times 10^{-3}\,\text{m}^3\,\text{kPa/mol K})(298\,\text{K})}$$

$$= 3.39 \times 10^{-7}\,\text{mol/m}^2$$

The corresponding values of $c*$ can then be computed from the isotherm expressions. Since the Γ^*: values are in mol/m^2, and k'_f is based on adsorption densities in mol/g, we must multiply Γ^* by the SSA to obtain c^* in mol/L. The calculation is as follows:

$$c^*_{benz} = \left(\dfrac{\Gamma^*_{benz}[\text{SSA}]}{k'_{f,benz}}\right)^{1/n_{benz}}$$

$$= \left(\dfrac{[4.95 \times 10^{-7}\,\text{mol/m}^2][650\,\text{m}^2/\text{g}]}{0.260\,(\text{mol/g})/(\text{mol/L})^{0.533}}\right)^{1/0.533}$$

$$= 3.50 \times 10^{-6}\,\text{mol/L}$$

$$c^*_{tol} = \left(\dfrac{\Gamma^*_{tol}[\text{SSA}]}{k'_{f,tol}}\right)^{1/n_{tol}}$$

$$= \left(\dfrac{[3.39 \times 10^{-7}\,\text{mol/m}^2][650\,\text{m}^2/\text{g}]}{0.0538\,(\text{mol/g})/(\text{mol/L})^{0.365}}\right)^{1/0.365}$$

$$= 2.87 \times 10^{-7}\,\text{mol/L}$$

We also compute the surface activity coefficients. In general, the calculation requires the use of Equation 7-117, but since both adsorbates obey Freundlich isotherms, the $\gamma_{surf,i}$ values are simply $1/n_i$:

$$\gamma_{surf,benz} = \dfrac{1}{n_{benz}} = \dfrac{1}{0.533} = 1.88$$

$$\gamma_{surf,tol} = \dfrac{1}{n_{tol}} = \dfrac{1}{0.365} = 2.74$$

We next make a preliminary guess for c_{benz} at equilibrium. We know from Figure 7-33a that, in a single-adsorbate system, this value would be ~ 0.03 mg/L, or 3.8×10^{-7} mol/L, and we expect the concentration to be higher in the competitive system (but, of course, less than $c_{benz,tot}$, which is 6.4×10^{-6} mol/L). A reasonable guess might therefore be 1.0×10^{-6} mol/L. Based on this guess, we can compute $c_{tol,tot}$ from Equation 7-131:

$$c_{tol} = \dfrac{c_{tol,tot}}{1 + \left((c_{benz,tot}/c_{benz}) - 1\right)\left(c^*_{benz}/c^*_{tol}\right)}$$

$$= \dfrac{8.91 \times 10^{-6}}{1 + ((6.41 \times 10^{-6}/1.00 \times 10^{-6}) - 1)(3.50 \times 10^{-6}/2.87 \times 10^{-7})}$$

$$= 1.33 \times 10^{-7} \tag{7-132}$$

where all the concentrations are in mol/L. Then, we use the mass balances (Equation 7-129) to compute the corresponding values of Γ_i. Each mass balance is of the form $c_{i,\text{tot}} = c_i + c_{\text{AC}}\Gamma_i(\text{SSA})$. Inserting the known values of $c_{i,\text{tot}}$, c_{AC}, and SSA and the current estimates of c_i, we find that $\Gamma_{\text{benz}} = 1.66 \times 10^{-7}$ mol/m^2, and $\Gamma_{\text{tol}} = 2.70 \times 10^{-7}$ mol/m^2. The corresponding mole fractions are $z_{\text{benz}} = 0.38$ and $z_{\text{tol}} = 0.62$. Inserting current values into Equation 7-127, we then find

$$c_{\text{benz}} = c^*_{\text{benz}} z_{\text{benz}} = (3.50 \times 10^{-6} \text{ mol/L})(0.38)$$
$$= 1.34 \times 10^{-6} \text{ mol/L}$$

This value is larger than our original guess (1.0×10^{-6} mol/L), so we make another guess for c_{benz} that is closer to the computed value, and then repeat the calculations of c_{tol}, the two Γ_i, the two z_i, and c_{benz}. This process is repeated until the computed c_{benz} equals the guessed value; for the given guess of π_{tot}, this occurs when $c_{\text{benz}} = 1.29 \times 10^{-6}$ mol/L. The corresponding value of c_{tol}, again using Equation 7-127, is 1.81×10^{-7} mol/L. The partial surface pressures of the two adsorbates under these conditions are found from either Equation 7-118 or 7-119 as $\pi_{\text{benz}} = 7.32 \times 10^{-7}$ kPa m and $\pi_{\text{tol}} = 1.82 \times 10^{-6}$ kPa m, and π_{tot} is found as their sum, or 2.56×10^{-6} kPa m. This value is larger than the original guess of 2.3×10^{-6} kPa m, so a smaller value is guessed, and the entire process is repeated. Using a spreadsheet, the values converge rapidly to yield, for an activated carbon dose of 50 mg/L:

$$\Gamma_{\text{benz}} = 1.51 \times 10^{-7} \text{ mol/m}^2 \quad \Gamma_{\text{tol}} = 2.67 \times 10^{-7} \text{ mol/m}^2$$

$$c_{\text{benz}} = 1.50 \times 10^{-6} \text{ mol/L} \quad c_{\text{tol}} = 2.34 \times 10^{-7} \text{ mol/L}$$

$$\pi_{\text{benz}} = 7.03 \times 10^{-7} \text{ Pa m} \quad \pi_{\text{tol}} = 1.81 \times 10^{-6} \text{ Pa m}$$

$$\pi_{\text{tot}} = 2.52 \times 10^{-6} \text{ Pa m}$$

The results for several other adsorbent doses, with concentrations converted to mol/L, are summarized in Table 7-6, and they are plotted in Figure 7-35, along with

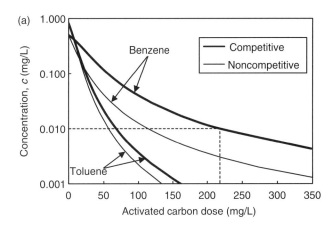
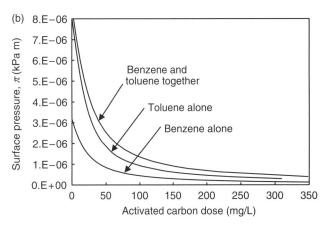

FIGURE 7-35. Comparison of conditions in the example system, with and without consideration of competitive adsorption according to the IAST. (a) equilibrium dissolved adsorbate concentrations; (b) surface pressure.

the results obtained previously for the noncompetitive systems. As we would expect from the prior analysis of the systems with only a single adsorbate, the required activated carbon dose is controlled by the need to meet the treatment goal for benzene. However, competitive adsorption causes that dose to almost double, from 114 to 218 mg/L. The surface pressure when the treatment goal

TABLE 7-6. Adsorption Characteristics of the Example System for Various Activated Carbon Doses

c_{AC} (mg/L)	c_{benz} (mg/L)	c_{tol} (mg/L)	q_{benz} (mg/g)	q_{tol} (mg/g)	Γ_{benz} (mol/m^2)	Γ_{tol} (mol/m^2)	π_{tot} (kPa m)
2.5	0.473	0.665	10.75	61.76	2.12×10^{-7}	1.03×10^{-6}	
10	0.391	0.344	10.94	47.60	2.16×10^{-7}	7.96×10^{-7}	
25	0.245	9.78×10^{-2}	10.20	28.88	2.01×10^{-7}	4.83×10^{-7}	
50	0.115	2.15×10^{-2}	7.70	15.96	1.52×10^{-7}	2.67×10^{-7}	
100	0.039	3.66×10^{-3}	4.61	8.16	9.09×10^{-8}	1.36×10^{-7}	1.33×10^{-6}
200	1.14×10^{-2}	5.68×10^{-4}	2.44	4.10	4.82×10^{-8}	6.85×10^{-8}	
300	5.41×10^{-3}	1.88×10^{-4}	1.65	2.34	3.25×10^{-8}	4.57×10^{-8}	
218	**0.010**	$\mathbf{4.49 \times 10^{-4}}$	**2.24**	**3.76**	$\mathbf{4.42 \times 10^{-8}}$	$\mathbf{6.28 \times 10^{-8}}$	$\mathbf{6.32 \times 10^{-7}}$

is achieved is also significantly higher in the competitive system than in the system with benzene alone (6.32×10^{-7} vs. 3.95×10^{-7} kPa m, respectively), and the concentration of dissolved toluene is 4.49×10^{-4} mg/L, far below the target value. ∎

One of the successes of the surface pressure concept and the IAST relates to the situation that was characterized in Figure 7-28, in which the adsorption isotherm of a trace adsorbate (e.g., an SOC) shifts from nonlinear (Freundlich, with $n < 1$) to linear when another adsorbate (e.g., NOM) is added to the system at a much higher concentration. This shift is predicted by the IAST, as follows. In a system in which the trace constituent is the only adsorbate, the total surface pressure equals the surface pressure exerted by that compound alone and varies with its adsorption density (Γ_{trace}) according to any of the forms of Equation 7-108. On the other hand, if NOM is present, the total surface pressure is dominated by the adsorbed NOM and hence does not change significantly when the trace compound adsorbs. As a result, π_{tot} is approximately independent of Γ_{trace}. Thus, for a given dose of adsorbent and a given initial concentration of NOM, π_{tot} is fixed, regardless of the total concentration of the trace compound. When this approximate constancy of π_{tot} is combined with Equation 7-125 and applied to the trace compound, that equation indicates that $K_{eff,trace,\pi_{tot}}$ is constant, and the equilibrium dissolved concentration of the trace compound is directly proportional to its mole fraction:

$$c_{trace} = K'_{trace} \frac{\pi_{tot}}{\gamma_{trace}} z_{trace} = K_{eff,trace,\pi_{tot}} z_{trace} \quad (7\text{-}133)$$

Because the trace compound accounts for a small fraction of the total adsorption density, its mole fraction on the surface is directly proportional to its adsorption density. As a result, Equation 7-133 implies that c_{trace} is directly proportional to Γ_{trace}, and adsorption of that compound is characterized by a linear isotherm.

7.7 THE POLANYI ADSORPTION MODEL AND THE POLANYI ISOTHERM

Description of the Polanyi Model

As indicated in the introduction to this chapter, one model of adsorption goes beyond making a conceptual analogy to phase transfer reactions and posits that a true phase change occurs in certain sorption reactions. This model, known as the *Polanyi adsorption model*, was first proposed in 1916 to describe adsorption of gases and has gained popularity in recent years in environmental engineering for modeling adsorption of hydrophobic adsorbates onto activated carbon (Manes, 1998; Crittenden et al., 1999; Xia and Ball, 2000).

The mathematics of the Polanyi model are based on the idea that a potential energy field exists adjacent to adsorptive surfaces, analogous to the gravitational field near a massive object or the electrical field near a charged object. When another object is exposed to the gravitational field, or an oppositely charged object is exposed to the electrical field, it experiences an attractive force toward the first object. The thermodynamic interpretation of this attraction is that the field causes the potential energy of the system to decline as the separation between the objects decreases.

In a similar way, the empirical observation that adsorbates are attracted to certain surfaces implies that the potential energy of those species is lower when they are near the surface than in the bulk phase. In the case of adsorption from solution, both the adsorbate and water might be attracted to the surface; the selective attraction of the adsorbate relative to water implies that the potential energy of the adsorbate is lowered more than that of water molecules. If we designate the total adsorptive potential energy of a system as PE_{ads}, and the *molar adsorption potential* of species i as ε_i, the value of ε_i at a location x can be defined as:

$$\varepsilon_i(x) = -\left.\frac{\partial PE_{ads}}{\partial n_i(x)}\right|_{P,T} \quad (7\text{-}134)$$

where $n_i(x)$ is the number of moles of i at location x.

The meaning of $\varepsilon_i(x)$ is best understood in terms of a hypothetical process in which a differential amount of i is brought to location x from a location where $\varepsilon_i = 0$, while all other system conditions (including the distribution of all other species, n_j) remain constant. Equation 7-134 indicates that $\varepsilon_i(x)$ could be computed as the decline in the total adsorptive potential energy of the system per mole of i transferred in such a process. By convention, the datum location for adsorptive potential for all species (i.e., the location where ε_i is defined to be zero) is the bulk fluid

---

[21] The definition of adsorption potential is very analogous to that of the chemical potential, in which case the energy of interest is the chemical potential energy, PE_{chem}. The common name for the chemical potential of a species i is the molar Gibbs free energy, $\overline{G}_i$, which can be defined as:

$$\overline{G}_i = \left.\frac{\partial PE_{chem}}{\partial n_i}\right|_{n_j,P,T} \quad (21)$$

Other than the type of potential energy being considered, the only difference in the definitions of ε_i and $\overline{G}_i$ is that ε_i is defined as an energy release, whereas $\overline{G}_i$ is defined as an energy input, so the defining equations have different signs.

THE POLANYI ADSORPTION MODEL AND THE POLANYI ISOTHERM

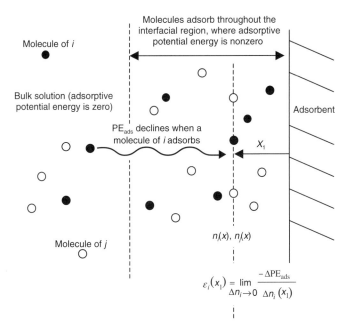

FIGURE 7-36. Schematic explanation of the definition of $\varepsilon_i(x)$. In the process, a differential amount of i moves from bulk solution to a distance x from the adsorptive surface.

phase, far from any adsorptive surface.[21] The hypothetical process is shown schematically in Figure 7-36.

In the adsorption processes of interest to us, when adsorbate i moves from bulk solution to a location near a surface (e.g., inside a pore of a particle of activated carbon), some water must be displaced from the pore to the bulk solution. Therefore, the overall change in PE$_{ads}$ accompanying adsorption of a small amount of adsorbate can be expressed as follows:

$$\mathrm{dPE}_{ads} = \mathrm{d}n_i(-\varepsilon_i(x) + \varepsilon_{i,\mathrm{bulk}}) + \mathrm{d}n_w(\varepsilon_w(x) - \varepsilon_{w,\mathrm{bulk}})$$
$$= \varepsilon_w(x)\mathrm{d}n_w - \varepsilon_i(x)\mathrm{d}n_i \quad (7\text{-}135)$$

where $\mathrm{d}n_i$ and $\mathrm{d}n_w$ are the number of moles of adsorbate and water involved in the process, and the second equality takes into account the fact that both $\varepsilon_{i,\mathrm{bulk}}$ and $\varepsilon_{w,\mathrm{bulk}}$ are zero.

Assuming that the pore always remains filled and that the solution is incompressible, the volume occupied by the adsorbate molecules must equal the volume of water displaced. Therefore, designating the molar volumes of the condensed adsorbate and water as $\overline{V}_i$ and $\overline{V}_w$, respectively, and the volume exchanged as $\mathrm{d}V$, we have

$$\mathrm{d}V = \overline{V}_w \mathrm{d}n_w = \overline{V}_i \mathrm{d}n_i \quad (7\text{-}136)$$

Using Equation 7-136 to substitute for $\mathrm{d}n_w$ and $\mathrm{d}n_i$ in Equation 7-135, we find

$$\mathrm{dPE}_{ads} = \frac{\varepsilon_{w,\mathrm{ads}}}{\overline{V}_w}\mathrm{d}V - \frac{\varepsilon_{i,\mathrm{ads}}}{\overline{V}_i}\mathrm{d}V$$
$$\frac{\mathrm{dPE}_{ads}}{\mathrm{d}V} = \frac{\varepsilon_w(x)}{\overline{V}_w} - \frac{\varepsilon_i(x)}{\overline{V}_i} \quad (7\text{-}137)$$

The ratio $\varepsilon_j(x)/\overline{V}_j$, where j might refer to either water or an adsorbate, can be interpreted as the adsorptive potential energy per unit volume of j at location x. This parameter, which we refer to as the *volumetric adsorption potential* at x and represent by the symbol $\hat{\varepsilon}_j(x)$, turns out to be very useful, because, empirically, its value has been found to be approximately the same for whole classes of organic compounds (Dubinen, 1960).[22] For example, at a given location in a pore, the value of $\hat{\varepsilon}_j(x)$ is approximately the same when j is any alkane, even though the values of $\varepsilon_j(x)$ for these same compounds span a wide range.

In a porous solid, both $\varepsilon_w(x)$ and $\varepsilon_i(x)$ (and, therefore, $\hat{\varepsilon}_w(x)$ and $\hat{\varepsilon}_i(x)$) are presumed to increase monotonically as one moves deeper into individual pores, because the pore diameter is presumed to decrease with depth, causing the amount of nearby solid surface per unit volume to increase. Furthermore, although the values of these four parameters might be very different at a particular location in a given adsorbent, the ratios $\varepsilon_w(x)/\varepsilon_i(x)$ and $\hat{\varepsilon}_w(x)/\hat{\varepsilon}_i(x)$ are usually approximately constant throughout the adsorbent. A schematic showing a plausible distribution of $\varepsilon_w(x)$ and $\varepsilon_i(x)$ values in a pore is shown in Figure 7-37.

A relationship among $\varepsilon_i(x)$, $\varepsilon_w(x)$, and the concentration of i can be derived from the principle that, at equilibrium, the total potential energy (PE$_{tot}$) of an isolated system (one that exchanges no mass or energy with the space outside its boundaries) must be at a minimum. A corollary is that, at equilibrium, any differential change within the system has no effect on PE$_{tot}$; that is:

$$\text{At equilibrium:} \quad \frac{\mathrm{dPE}_{tot}}{\mathrm{d}\xi} = 0 \quad (7\text{-}138)$$

where $\mathrm{d}\xi$ represents any differential, feasible change in the state of the system, such as a small amount of chemical reaction, or a small amount of adsorbate replacing water near the surface.

---

[22] In the literature, $\hat{\varepsilon}_j$ is sometimes referred to as the *adsorption potential* of j. The name *volumetric adsorption potential* is used here, to highlight both the similarity and difference between this parameter and ε_j, the *molar adsorption potential*.

308 ADSORPTION PROCESSES: FUNDAMENTALS

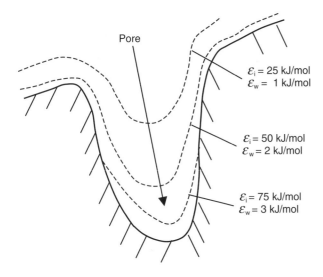

FIGURE 7-37. Schematic representation of the adsorption potential of a hydrophobic adsorbate i and of water in a pore of an activated carbon particle.

The total potential energy per mole of both water and adsorbate i in the system includes contributions from many sources (gravity, pressure, concentration, and so on). However, most of those contributions are, to a good approximation, constant regardless of where the species is located in the system. In the Polanyi model, an assumption is made that the only types of potential energy that depend on location are the adsorptive potential energy (PE_{ads}) and the chemical potential energy ($PE_{chem.}$); PE_{chem} is more commonly known as the Gibbs free energy, G. The change in the adsorptive potential energy per unit volume exchanged in the adsorption reaction is given by Equation 7-135; a similar derivation leads to the analogous change in chemical potential energy, yielding:

$$dPE_{chem} = dG = dn_i\left(\overline{G}_i(x) - \overline{G}_{i,bulk}\right) + dn_w\left(-\overline{G}_w(x) + \overline{G}_{w,bulk}\right) \quad (7\text{-}139)$$

where $\overline{G}_i$ and $\overline{G}_w$ are the chemical potential energy per mole of i and water, respectively. As before, we can substitute for dn_i and dn_w from Equation 7-136. Making that substitution and rearranging, we obtain

$$dG = \left(\frac{\overline{G}_i(x) - \overline{G}_{i,bulk}}{\overline{V}_i}\right)dV + \left(\frac{\overline{G}_{w,bulk} - \overline{G}_w(x)}{\overline{V}_w}\right)dV$$

$$\frac{dG}{dV} = \frac{\overline{G}_i(x) - \overline{G}_{i,bulk}}{\overline{V}_i} + \frac{\overline{G}_{w,bulk} - \overline{G}_w(x)}{\overline{V}_w} \quad (7\text{-}140)$$

Each $\overline{G}$ term in Equation 7-140 can be expanded as $\overline{G}^o + RT \ln a$, where $\overline{G}^o$ is the standard molar Gibbs energy of the species of interest, and a is its activity. Furthermore, for each species, $\overline{G}^o$ is the same regardless of whether the species is in bulk solution or in a pore. Applying these ideas, we find

$$\frac{dG}{dV} = \frac{\overline{G}_i^o(x) + RT \ln a_i(x) - \overline{G}_{i,bulk}^o - RT \ln a_{i,bulk}}{\overline{V}_i}$$
$$+ \frac{\overline{G}_{w,bulk}^o + RT \ln a_{w,bulk} - \overline{G}_w^o(x) - RT \ln a_w(x)}{\overline{V}_w}$$
$$= \frac{RT}{\overline{V}_i} \ln \frac{a_i(x)}{a_{i,bulk}} + \frac{RT}{\overline{V}_w} \ln \frac{a_{w,bulk}}{a_w(x)}$$

Finally, if we assume that the activity of water is close to unity both in the bulk solution and near the surface and that the activity coefficient of the adsorbate is approximately the same in both locations, we obtain

$$\frac{dG}{dV} = \frac{RT}{\overline{V}_i} \ln \frac{\gamma_i(x) c_i(x)/c_{i,std.state}}{\gamma_{i,bulk} c_{i,bulk}/c_{i,std.state}} + \frac{RT}{\overline{V}_w} \ln \frac{1.0}{1.0}$$
$$= \frac{RT}{\overline{V}_i} \ln \frac{c_i(x)}{c_{i,bulk}} \quad (7\text{-}141)$$

We can now combine Equations 7-137 and 7-141 to find the change in total potential energy of the system when a differential volume dV of adsorbate from bulk solution moves to the surface, and an equal volume of water moves in the opposite direction:

$$\frac{dPE_{tot}}{dV} = \frac{dPE_{ads}}{dV} + \frac{dG}{dV}$$
$$= \frac{\varepsilon_w(x)}{\overline{V}_w} - \frac{\varepsilon_i(x)}{\overline{V}_i} + \frac{RT}{\overline{V}_i} \ln \frac{c_i(x)}{c_{i,bulk}} \quad (7\text{-}142)$$

Invoking the requirement that, at equilibrium, dPE_{tot}/dV must be zero, we obtain

$$\frac{\varepsilon_i(x)}{\overline{V}_i} - \frac{\varepsilon_w(x)}{\overline{V}_w} = \frac{RT}{\overline{V}_i} \ln \frac{c_i(x)}{c_{i,bulk}} \quad (7\text{-}143)$$

$$\hat{\varepsilon}_i(x) - \hat{\varepsilon}_w(x) = \frac{RT}{\overline{V}_i} \ln \frac{c_i(x)}{c_{i,bulk}} \quad (7\text{-}144)$$

An alternate, convenient form of Equation 7-143 can be obtained by multiplying through by $\overline{V}_i$ to obtain:

$$\varepsilon_i(x) - \frac{\overline{V}_i}{\overline{V}_w} \varepsilon_w(x) = RT \ln \frac{c_i(x)}{c_{i,bulk}} \quad (7\text{-}145)$$

■ **EXAMPLE 7-13.** A solution at 25°C contains 50 μg/L of tetrachloroethene (C_2Cl_4, commonly referred to as perchlorethylene, PCE) and is to be treated by adsorption onto powdered activated carbon (PAC). What is the expected PCE concentration at a location in a pore where ε_{PCE} is 22.4 kJ/mol, and ε_w is 0.99 kJ/mol? The molar volume of PCE is 102 mL/mol, and that of water is 18 mL/mol.

Solution. Assuming that the activity coefficient of PCE is the same in bulk solution and in the pore, we can apply Equation 7-145. With PCE as i, we can solve for c_{PCE} and then substitute values into the equation to find

$$\varepsilon_{PCE}(x) - \frac{\overline{V}_{PCE}}{\overline{V}_w}\varepsilon_w(x) = RT \ln \frac{c_{PCE}(x)}{c_{PCE,bulk}}$$

$$c_{PCE}(x) = c_{PCE,bulk} \exp \frac{\varepsilon_{PCE}(x) - (\overline{V}_{PCE}/\overline{V}_w)\varepsilon_w(x)}{RT}$$

$$= 50\,\mu g/L \exp \frac{22.4\,kJ/mol - ((102\,mL/mol)/(18\,mL/mol))(0.99\,kJ/mol)}{(8.314 \times 10^{-3}\,kJ/mol\,K)298\,K}$$

$$= 43,863\,\mu g/L = 43.9\,mg/L$$

The result indicates that the organic is highly concentrated (by a factor of ~880) at the specified location in the pore, compared to the bulk solution. ∎

To this point, the argument x has been included in the parameters $\varepsilon(x)$, $\hat{\varepsilon}(x)$, and $c_i(x)$ to emphasize that the values of these parameters depend on location within a pore (in particular, the distance from the pore walls). However, for conciseness, these parameters will be shown without the argument in the remainder of the discussion.

If the adsorbate becomes sufficiently concentrated in the pore, its aqueous solubility can be exceeded, in which case it will condense as a pure liquid or precipitate as a pure solid. The concentration at which this transition occurs is the saturation concentration, c_{sat}. By inserting this value for c_i into Equation 7-145, we can compute the critical value of $\varepsilon_i - (\overline{V}/\overline{V}_w)\varepsilon_w$ necessary for condensation of the adsorbate, for any given bulk concentration, $c_{i,bulk}$:

$$\left(\varepsilon_i - \frac{\overline{V}_i}{\overline{V}_w}\varepsilon_w\right)_{crit} = RT \ln \frac{c_{sat}}{c_{i,bulk}} \qquad (7\text{-}146)$$

At locations deeper in the pore than the critical depth (i.e., the depth where Equation 7-146 is satisfied), the difference $\varepsilon_i - (\overline{V}/\overline{V}_w)\varepsilon_w$ is larger than the critical value, and the adsorbate will be present as a condensed phase; at shallower locations, the liquid will remain as an aqueous solution, albeit with a larger adsorbate concentration than in the bulk solution. This situation is shown schematically in Figure 7-38. Accordingly, the volume of condensed, pure adsorbate at any given value of $c_{i,bulk}$ equals the pore volume in which $\varepsilon_i - (\overline{V}_i/\overline{V}_w)\varepsilon_w$ is greater than $(\varepsilon_i - (\overline{V}_i/\overline{V}_w)\varepsilon_w)_{crit}$.

■ **EXAMPLE 7-14.** What is the critical value of ε_{PCE} at which PCE would be expected to condense as a pure liquid in the system described in Example 7-13? Assume that the ratio $\varepsilon_w/\varepsilon_i$ is constant throughout the pore space. The solubility of PCE in water at 25°C is 150 mg/L.

Solution. The ratio $\overline{V}_i/\overline{V}_w$ is $102/18 = 5.67$, and, based on the calculations in Example 7-13, $\varepsilon_w/\varepsilon_i$ is 0.044. Substituting those and other known values into Equation 7-146, we find

$$\left(\varepsilon_{PCE} - \frac{\overline{V}_i}{\overline{V}_w}\varepsilon_w\right)_{crit} = RT \ln \frac{c_{sat}}{c_{i,bulk}}$$

$$\times (\varepsilon_{PCE} - 5.67(0.044\varepsilon_{PCE}))_{crit} = RT \ln \frac{c_{PCE,sat}}{c_{PCE,bulk}}$$

$$0.75\varepsilon_{PCE,crit} = 8.314 \times 10^{-3}\,kJ/mol\,K(298\,K)$$

$$\times \ln \frac{150\,mg/L}{50\,\mu g/L(1\,mg/1000\,\mu g)} = 19.8\,kJ/mol$$

$$\varepsilon_{PCE,crit} = \frac{19.8\,kJ/mol}{0.75} = 26.4\,kJ/mol$$

Thus, at all locations in the pore where $\varepsilon_{PCE} > 26.4$ kJ/mol, PCE is expected to be present as a pure phase. ∎

The preceding discussion suggests that one useful way to characterize an adsorbent would be to establish the volumetric distribution of ε_i values in its pores; that is, to quantify the

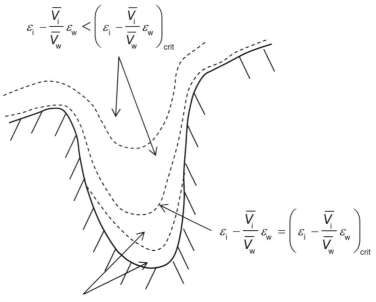

FIGURE 7-38. Schematic of the adsorption and condensation of a hydrophobic adsorbate in the pore of an activated carbon particle, as envisioned in the Polanyi model. The adsorbate is present in a condensed phase in the shaded region.

pore volume per unit mass of adsorbent wherein ε_i exceeds any particular value. To accomplish that, an assumption is made that, when hydrophobic adsorbates bind to activated carbon, the mass of adsorbate that condenses in the pores far exceeds the amount that is present in the aqueous solution in the pores. That is, the total mass adsorbed is approximated as the mass of the condensed phase in the pores. Since the density of the condensed organic phase can be determined independently, the mass of the condensed phase can be converted in to an equivalent volume. As a result, experiments conducted at various values of c_{bulk} can be used to infer the volume of pore space in a given sample of adsorbent in which ε_i exceeds the value necessary to induce condensation; that is, the value at which $c_i = c_{\text{sat}}$.

■ **EXAMPLE 7-15.** The following data were obtained for the PCE adsorption density as a function of the dissolved PCE concentration, for the same PAC as was characterized in Examples 7-13 and 7-14.

c (mg/L)	0.01	0.03	0.06	0.10	0.15
q (mg/g)	28	51	65	85	107

For each data point (i.e., each c, q pair), determine

(a) The corresponding critical molar adsorption potential.
(b) The volume of pore space in which ε_{PCE} exceeds $\varepsilon_{\text{PCE,crit}}$.

Solution.

(a) Assuming that $\varepsilon_w/\varepsilon_{\text{PCE}} = 0.044$, the critical value of ε_{PCE} for any bulk concentration can be computed as in Example 7-14. The results are summarized as follows:

c (mg/L)	0.01	0.03	0.06	0.10	0.15
$\varepsilon_{\text{PCE,crit}}$ (kJ/mol)	31.8	28.1	25.8	24.2	22.84

(b) The adsorption densities can be converted to apparent volumes of condensed PCE based on the molar volume of PCE (given in Example 7-13 as 102 mL/mol) and the molecular weight of PCE, which is 166. For example, for the data point considered in part *a* ($c_{PCE} = 0.01$ mg/L), the reported value of q is 28 mg/g, and the corresponding apparent volume of condensed PCE is

$$\begin{pmatrix} \text{Volume of PCE condensed} \\ \text{in pores, per gram of} \\ \text{PAC, when } q = 28 \text{ mg/g} \end{pmatrix}$$

$$= \frac{q_{PCE}}{MW_{PCE}} \overline{V}_{PCE}$$

$$= \frac{28 \text{ mg/g AC}}{166 \text{ mg/mmol}} 0.102 \text{ mL/mmol}$$

$$= 0.017 \text{ mL/g AC}$$

The interpretation is that, when the bulk concentration of PCE is 0.01 mg/L, 0.017 mL of PCE condenses in the pores in each gram of GAC. Thus, under these conditions, 0.017 mL of pore space per gram of GAC has a value of ε_{PCE} greater than the critical value necessary for condensation of PCE. As indicated in part *a*, this value is 31.8 kJ/mol. Thus, the overall interpretation is that, in the GAC, 0.017 mL of pore per gram has $\varepsilon_{PCE} > 31.8$ kJ/mol. The results for all the given data are summarized as follows.

c (mg/L)	0.01	0.03	0.06	0.10	0.15
q (mg/g)	28	51	65	85	107
$\varepsilon_{PCE,crit}$ (kJ/mol)	31.8	28.1	25.8	24.2	22.84
GAC pore volume (mL/g) with $\varepsilon_{PCE} > \varepsilon_{PCE,crit}$	0.017	0.031	0.040	0.052	0.066

∎

Example 7-15 demonstrates how the volume of pore space in which ε_i exceeds specific values can be determined from experimental q_i versus c_i data, if the (assumed constant) ratio $\varepsilon_w/\varepsilon_i$ is known. The question that remains, therefore, is: How are values for ε_w and ε_i, and thus their ratio, determined in the first place?

Since both water and the adsorbate are present in any experiment investigating adsorption from aqueous solution, ε_w and ε_i are not separable based on such an experiment. Therefore, the approach that has been used to evaluate these terms independently involves studying adsorption from the gas phase. (As noted previously, the Polanyi model was originally developed for this situation.) Such systems can be analyzed more simply than aqueous systems, because adsorption from the gas phase can be studied in single-component systems; that is, adsorption of organic compounds and water can be evaluated in separate, independent experiments.

For adsorption from the gas phase, the relationships corresponding to Equations 7-143 and 7-146 are

$$\frac{\varepsilon_i}{\overline{V}_i} = \hat{\varepsilon}_i = \frac{RT}{\overline{V}_i} \ln \frac{a_i}{a_{i,bulk}} = \frac{RT}{\overline{V}_i} \ln \frac{P_i}{P_{i,bulk}} \quad (7\text{-}147)$$

$$\varepsilon_{i,crit} = RT \ln \frac{P_{sat}}{P_i} \quad (7\text{-}148)$$

where P_i is the partial pressure of species i in the gas phase, P_{sat} is the partial pressure of i that is in equilibrium with the pure liquid (the vapor pressure of i at the temperature of the experiment), and all the other terms are as defined previously. Equations 7-147 and 7-148 apply regardless of whether i is an organic adsorbate or water. By assuming, in either case, that virtually all of the mass adsorbed from the gas phase is condensed, the distributions of pore volume as a function of ε_i and ε_w can be assessed. By dividing the pore volume corresponding to each value of ε by the mass of adsorbent, a curve can be generated showing the pore volume per unit mass in which ε exceeds a certain value. A typical such plot is shown in Figure 7-39.

Plots such as the one shown in Figure 7-39 provide essentially the same information as a plot of $\Phi_{cum}\{K\}$ versus K, albeit in a different conceptual framework. That is, each of those plots describes how much of the adsorbent has a given affinity for the adsorbate. In the case of the $\Phi_{cum}\{K\}$ versus K plot, the affinity is characterized in terms of site binding constants, whereas in Figure 7-39, it is characterized in terms of an adsorption potential, but the underlying information is essentially identical. Also, like the case for the site-binding model, specific forms of the relationships shown in Figure 7-39 lead to specific, widely recognized isotherms. For instance, if a plot of the pore volume per unit mass of adsorbent decays exponentially as a function of the adsorption potential, the corresponding q versus c data conform to the Freundlich isotherm (Dubinen, 1960).

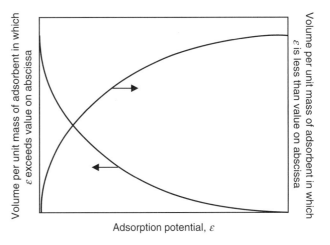

FIGURE 7-39. A generic curve showing the distribution of adsorption potential in the pores of an adsorbent.

EXAMPLE 7-16. A plot characterizing the pore volume with various values of ε for adsorption of PCE onto GAC is shown in Figure 7-40. The figure is consistent with the data in the preceding examples, and the extrapolation of the data at low adsorption potentials (indicated by the broken line) is based on an independent estimate that the GAC has 0.50 mL of total pore volume per gram of solid. What is the expected adsorption density, in mg PCE/g GAC, when the equilibrium, bulk PCE concentration is 400 μg/L?

Solution. The value of ε_{sat} for the given bulk concentration can be computed from Equation 7-146 as in Example 7-14. All the parameter values are the same as in that example, except that c_{PCE} is 400 μg/L instead of 50 μg/L. The result is that $\varepsilon_{PCE,crit} = 19.6$ kJ/mol. Based on the figure in the problem statement, the GAC has 0.10 mL of pore volume per gram in which ε_{PCE} is greater than this critical value, so that is the volume of PCE that will be adsorbed on (i.e., condensed in) the GAC at equilibrium. This value can be converted into a conventional adsorption density, in mg PCE/g GAC, using the molar volume and molecular weight of PCE, or, more directly, using the density of pure liquid PCE:

$$\rho_{pure\ PCE} = \frac{MW_{PCE}}{\overline{V}_{PCE}} = \frac{166\ g/mol}{102\ mL/mol} = 1.63\ g/mL$$

$$q_{PCE} = (0.10\ mL\ PCE/g\ GAC)(1.63\ g\ PCE/mL\ PCE)$$
$$\times (1000\ mg/g)$$
$$= 163\ mg\ PCE/g\ GAC$$

Plots of pore volume versus ε_i depend strongly on the identity of the adsorbate, because different adsorbates have different intensities of interaction with the surface. Fortunately, however, the empirical evidence is that the relationships among ε_i values of different species are quite consistent (i.e., the ratios $\varepsilon_i/\varepsilon_j$ and $\hat{\varepsilon}_i/\hat{\varepsilon}_j$ for any two species i and j tends to depend in a consistent way on the properties of the activated carbon, even for vastly different compounds). For instance, Wohleber and Manes (1971) found that $\hat{\varepsilon}_w/\hat{\varepsilon}_{hexane}$ was close to 0.28 on several activated carbons they studied. Later, Li et al. (2005) suggested that the value of this ratio was usually larger than that reported by Wohleber and Manes, and that it increased approximately linearly from 0.4 to 0.6 as the oxygen content of the activated carbon increased from 0 to 10 mmol O per gram of dry, ash-free adsorbent. As noted previously, Dubinen and co-workers had found that, within a given class of compounds, $\hat{\varepsilon}_i$ values for adsorption onto a given activated carbon are all approximately equal, even though the corresponding ε_i values differ (Dubinen, 1960). Thus, as a first approximation, we can expect $\hat{\varepsilon}_w/\hat{\varepsilon}_i$ to be close to the range reported by Wohleber and Manes or Li et al. if i is any normal alkane. This consistency of $\varepsilon_w/\varepsilon_i$ ratios and $\hat{\varepsilon}_i$ values makes the Polanyi model especially useful for predicting the adsorption of many different species on a given type of activated carbon, once the adsorption of a standard species i on the same solid has been characterized, as demonstrated in the following example. Crittenden et al. (1999) have expanded this idea to develop correlations that, in essence, predict the characteristics of curves like that in Figure 7-39 for a wide range of organic compounds and activated carbon adsorbents.

EXAMPLE 7-17. Assuming that trichloroethylene (TCE) has the same value of $\hat{\varepsilon}$ as PCE, use the data for PCE adsorption in Example 7-15 to develop a predicted adsorption isotherm for TCE on the same batch of activated carbon. The molar volume of pure TCE is 87.7 mL/mol, corresponding to a density of 1.5 g/mL, and its saturation concentration in water is $c_{sat} = 1000$ mg/L.

Solution. In Example 7-15, the GAC pore volume in which ε_{PCE} was greater than five specified values was computed. These ε_{PCE} values, and the corresponding pore volumes, are repeated in the first two rows of the following table.

ε_{PCE} (kJ/mol)	31.8	28.1	25.8	24.2	22.84
GAC pore volume (mL/g) with $\varepsilon_{PCE} > \varepsilon_{PCE}$	0.017	0.031	0.040	0.052	0.066
$\hat{\varepsilon}_{PCE}$ and $\hat{\varepsilon}_{TCE}$ (J/mL)	312	275	253	237	224
ε_{TCE} (kJ/mol)	27.3	24.2	22.2	20.8	19.6
ε_w (kJ/mol)	1.40	1.24	1.14	1.06	1.00
q_{TCE} (mg/g GAC)	25.5	46.5	60.0	78.0	99.0
c_{TCE} (mg/L)	0.26	0.66	1.21	1.83	2.61

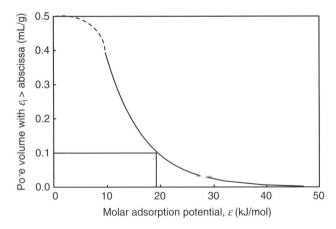

FIGURE 7-40. Cumulative distribution function showing the pore volume with various values of ε, for adsorption of PCE onto GAC in the example system.

Each value of ε_{PCE} is associated with a corresponding value of $\hat{\varepsilon}_{PCE}$, computed as $\hat{\varepsilon}_{PCE} = \varepsilon_{PCE}/\overline{V}_{PCE}$ and shown in the third row of the table; as indicated, these values are assumed to also equal $\hat{\varepsilon}_{TCE}$. Then, the value of ε_{TCE} associated with each pore volume can be computed as $\varepsilon_{TCE} = \hat{\varepsilon}_{TCE}\overline{V}_{TCE}$, yielding the results shown in the fourth row. Also, the value of ε_w associated with each pore volume is computed as 0.044 ε_{PCE} and is shown in the fifth row.

The mass of condensed TCE that would occupy a given volume is the product of that volume and the density of TCE. This mass, normalized to one gram of PAC, is shown in the sixth row for each corresponding pore volume (i.e., each volume shown in the second row). Thus, the TCE adsorption density shown is the one that would be present on (or in) the GAC, if the corresponding value of ε_{TCE} were the critical value that caused TCE to condense.

Finally, we calculate the bulk concentration of TCE that would cause each specified value of ε_{TCE} to be the critical value for condensation. That calculation is carried out by rearranging Equation 7-146, as follows:

$$\left(\varepsilon_i - \frac{\overline{V}_i}{\overline{V}_w}\varepsilon_w\right)_{crit} = RT \ln \frac{c_{sat}}{c_{i,bulk}}$$

$$c_{i,bulk} = c_{sat} \exp\left(-\frac{\varepsilon_i - (\overline{V}_i/\overline{V}_w)\varepsilon_w}{RT}\right)_{crit} \quad (7\text{-}149)$$

By substituting the values of ε_{TCE} and ε_w from each column in the table into Equation 7-149, we can compute the value of $c_{TCE,bulk}$ that would cause the corresponding pore volume to be filled with condensed TCE and would therefore lead to the value of q_{TCE} shown in the column. For example, for the values in the first column, we find

$$c_{i,bulk} = (1000\,\text{mg/L})$$

$$\times \exp\left(-\frac{27.3\,\text{kJ/mol} - ((87.7\,\text{mL/mol TCE})/(18.0\,\text{mL/mol}\,H_2O))(1.40\,\text{kJ/mol})}{(8.314 \times 10^{-3}\,\text{kJ/mol K})298\,\text{K}}\right)$$

$$= 0.257\,\text{mg/L}$$

The values of $c_{i,bulk}$ applicable for the other columns are shown in the table. The isotherm for TCE is shown in Figure 7-41, along with that for PCE, for comparison. ■

Comparison of Conceptual Models for Adsorption and Their Relationships to the Linear, Langmuir, and Freundlich Isotherms

A relationship between the volume adsorbed and the dissolved adsorbate concentration derived using the Polanyi model is sometimes referred to as the *Polanyi isotherm*. That designation uses the term "isotherm" in its generic sense, meaning a mathematical construct in which any relationship between the adsorption density and the equilibrium

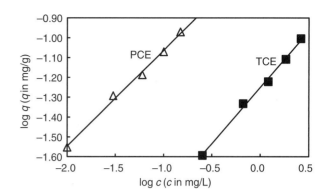

FIGURE 7-41. Experimental adsorption isotherm for PCE, and predicted isotherm for TCE, based on the Polanyi model.

dissolved adsorbate concentration can be analyzed. This is the sense in which the term is used when referring to the *Gibbs isotherm* as the mathematical underpinning for the gas transfer model of adsorption, or when a generic Φ versus K curve is referred to as the *site-binding isotherm*. None of these "isotherms" specifies a particular functional form for the relationship between adsorption density and dissolved adsorbate concentration, but each can be used to interpret or model any such relationship that is determined experimentally. For example, any of these conceptual models can be used to characterize data that conforms to a linear, Langmuir, or Freundlich isotherm, by making appropriate assignments for the site energy distribution function ($\Phi_{cum}(K)$ or $\Phi_{diff}(K)$) in the case of the site-binding model, the surface pressure versus adsorption density relationship in the case of the phase transfer model, and the adsorption potential versus volumetric adsorption density in the case of the Polanyi model.

As a practical matter, the main difference between the Polanyi model and the other generic models is in how they are used. The site-binding and gas transfer models have been used historically to provide a conceptual basis for understanding the behavior of individual adsorbates or competing adsorbates in the same system, but no significant effort has been made to predict adsorption of one adsorbate based on empirical data for adsorption of a different adsorbate. By contrast, such predictions have been the primary focus of research using the Polanyi model, with well-established physical/chemical properties of the adsorbates (e.g., their molar volume in pure liquid form) as the key parameters in the prediction methodology. While the predictive strength of these approaches requires improvement, they have shown

314 ADSORPTION PROCESSES: FUNDAMENTALS

significant potential already. Given the large number of organic contaminants for which adsorption is a potentially important factor for both contaminant transport in nature and removal in water and wastewater treatment processes, the importance of such predictive tools is bound to increase in the future.

7.8 MODELING OTHER INTERACTIONS AND REACTIONS AT SURFACES

In this final section of the chapter, we consider the effects of surface charge on adsorption of charged species and the effects of surface precipitation on adsorption isotherms. Surface charge can have a significant effect on adsorption of ionic species by hydrophilic solids (e.g., sorption of metals onto iron or aluminum hydroxides) and in controlling the movement of such solids in natural environments. Surface precipitation can be important in chemical coagulation processes (in which Al(III) or Fe(III) species might first adsorb on existing particles and later precipitate) or precipitation of other species such as Ca^{2+} in water softening operations or metals in some industrial waste treatment processes.

The Structure of Charged Interfaces and the Electrostatic Contribution to Sorption of Ions

In most cases, the contributions of individual factors to the overall binding energy of an adsorbate cannot be computed *a priori*. As a result, the determination of isotherm parameters is essentially empirical; it is not possible to predict either the absolute binding energy for a given adsorbate/adsorbent pair or the variation in that energy as a function of adsorption density from any theoretical basis. However, one factor that might contribute to variations in the binding energy of an ion—the variable electrical charge on the adsorbent's surface—has been modeled in a fundamental way.

It is clear that particles suspended in water can be electrically charged, since, when an aqueous suspension is placed in an electric field, the particles often migrate toward one or the other electrode. Regardless of whether the charge originates in the interior of the particle or at the surface, the particle behaves as though all the charge is distributed on its surface, and this charge affects the affinity of the surface for ionic adsorbates. Neutralization of the charge is a key step in inducing particle collisions and growth and is discussed in Chapter 11. An approach for modeling the effects of surface charge on adsorption of ions is presented in this section.

The key parameter that characterizes the electrical environment at any given location is the electrical potential, ψ. The modeling approach presented here involves characterizing ψ throughout the system and incorporating the corresponding attraction or repulsion of ions near the surface into the adsorptive driving force. The value of ψ in bulk solution is defined to be zero; therefore, the challenge is to estimate its value at the surface of a charged particle and to model the gradient of potential between the surface and the bulk solution.

Effects of Electrical Potential on Binding of Ions to Surfaces Factors leading to the development of surface charge are discussed in Chapter 11; in this section, we simply presume that such a charge can exist and investigate its effects on the nearby solution and on sorption reactions. Consider, for example, the electrical potential ψ near a particle that has some surface charge. We designate the value of ψ right at the surface as ψ_o and assume that ψ varies as a function of the distance x from the surface, decaying to zero in the bulk solution. Now consider an adsorbable cation A^{z+}, and assume that all the sites on the particle surface have equal affinity for A^{z+}, so that, in the absence of electrostatic interactions, the surface could be characterized by site distribution functions ($\Phi_{\text{cum},A}$ and $\Phi_{\text{diff},A}$) like those shown in Figure 7-11.

In the absence of surface charge ($\psi_o = 0$), $\Phi_{\text{cum},A}$ would increase from zero to $\Phi_{\text{cum},A}^{\max}$ at K_1. When ψ_o is nonzero, the plot of $\Phi_{\text{cum},A}$ versus K retains the shape of a Heaviside function, but the step change occurs at a K value that depends on ψ_o, shifting to higher or lower values if ψ_o is negative or positive, respectively. This shift can be rationalized by recognizing that, as ψ_o is increased (becoming progressively more positive), the electrostatic contribution to adsorption of a cation becomes progressively less favorable. Assuming that the contribution of purely chemical (covalent) bonding between the adsorbate and the surface sites is independent of ψ_o, the overall affinity of the surface for the cationic adsorbate decreases as ψ_o increases, leading to steadily lower values of K for the reaction.

If the adsorption of A^{z+} has a significant effect on the surface charge (increasing ψ_o as more A^{z+} adsorbs), the overall surface binding constant for A^{z+} will decrease as its adsorption density increases, other factors being equal. In such a case, if the surface is initially uncharged, sorption of the first molecule of A^{z+} will be characterized by a binding constant of K_1, but once a substantial amount of A^{z+} has bound, the binding constant will be smaller. Note that, unlike the case of a surface with nonuniform sites (on which the binding constant for the original molecule remains at K_1 when subsequent molecules adsorb), in this case the binding constants for *all* adsorbed A^{z+} molecules decline as more A^{z+} adsorbs.

Although the preceding discussion assumes that adsorbed molecules reside right at the surface and therefore experience a potential ψ_o, some adsorption models postulate that different types of adsorbates reside at different distances from the surface and hence might experience a potential anywhere between ψ_o and zero. The effect of this potential on the overall adsorption equilibrium constant is most easily explained in terms of binding energies.

The chemical (free) energy change accompanying a reaction is linked to the equilibrium constant for the reaction by the following relationship:

$$K_{eq} = \exp\left(-\frac{\Delta \overline{G}_r^o}{RT}\right) \quad (7\text{-}150)$$

where K_{eq} is the equilibrium constant, and $\Delta \overline{G}_r^o$ is the standard molar Gibbs free energy of reaction. Applied to an adsorption reaction, Equation 7-150 describes the chemical component of the affinity between the adsorbate and the adsorbent. The value of K_{eq} shown would characterize the species distribution at equilibrium if no other factors (such as electrostatics) affected the binding strength of the adsorbate to the adsorbent; in the adsorption literature, this value of K_{eq} is often referred to as the *intrinsic binding constant*, K_{intr}, for the reaction.

Electrostatic attraction or repulsion alters the net energy change associated with the adsorption reaction and thereby alters the overall binding constant. Specifically, the electrostatic energy required to move an ion from bulk solution (where the electrical potential is zero) to a location where ψ is finite is $z\psi$, where z is the ionic charge (formally, with units of equiv/mol). This energy is additive with the chemical energy of adsorption to yield an overall energy change ($\Delta \overline{G}_r^o + z\psi$) associated with the adsorption of an ion to a location where the electrical potential is ψ.

Conventionally, ψ is expressed in electrical terms with units of volts, causing the product $z\psi$ to have units of equiv-V/mol.[23] By contrast, $\Delta \overline{G}_r^o$ is traditionally reported in units more aligned with chemical processes, such as kJ/mol. The electrical energy can be expressed in these latter units by applying the Faraday constant, $\mathcal{F}(= 96.485 \text{ kJ/equiv V})$, as a conversion factor, so that the overall adsorptive binding constant, taking into account both chemical and electrical interactions, is

$$K_{overall} = \exp\left(-\frac{\Delta \overline{G}_r^o + z\mathcal{F}\psi}{RT}\right) \quad (7\text{-}151)$$

The two components of the overall constant are sometimes shown separately as follows:

$$K_{overall} = \exp\left(-\frac{\Delta \overline{G}_r^o}{RT}\right)\exp\left(-\frac{z\mathcal{F}\psi}{RT}\right) \quad (7\text{-}152)$$

$$K_{overall} = K_{intr}K_{elec} \quad (7\text{-}153)$$

where $K_{elec} \equiv \exp(-z\mathcal{F}\psi/RT)$.

[23] Equiv V is normally written as eV and referred to as electron volts.

Note that, if the electrical potential at the location of adsorption has the same sign as the charge on the ion, then $K_{elec} < 1$, so electrostatic interactions decrease the overall adsorption equilibrium constant. Conversely, if the charge on the adsorbate and the electrical potential have opposite signs, then $K_{elec} > 1$, and electrostatic interactions enhance adsorption. Note also that, in Equation 7-153, an analogy is clearly being made between K_{intr} and K_{elec}. Nevertheless, it is important to recognize that, whereas K_{intr} is a constant for a given adsorbate and adsorbent (or at least a given site-type), K_{elec} is *not* a constant; it depends on the electrical potential near the surface and therefore is sensitive to the details of the entire system, not just the identities of the reacting species.

The Profile of Adsorbates and Electrical Potential in the Interfacial Region Having established a quantitative relationship for the effect of ψ on the binding constant for adsorption, the key challenge remaining is to calculate ψ at the location(s) (i.e., the distance(s) from the surface) where adsorbed molecules reside. These calculations are complicated both by the uncertainty about exactly where those locations are and by the fact that ion adsorption and the electrical potential are implicitly linked to one another: the extent of ion adsorption affects the electrical potential at the interface, and vice versa.

The details of various models for calculating the profile of ψ as a function of distance from the surface are available in several texts and review articles (e.g., Davis and Kent, 1990; Westall, 1986; Dzombak and Morel, 1990) and are not presented here. However, because the ψ versus distance profile is important both in adsorption and in the evaluation of particle–particle interactions (described in Chapter 11), a generic description of the models used to calculate that profile is provided next.

The basic model for the interfacial structure that is currently used almost universally to account for electrostatics was first proposed in the 1910s and 1920s. This model represents the interfacial region as comprising one or two layers of molecules that are covalently bound to individual surface sites, and an adjacent layer of aqueous solution wherein the ionic concentrations are affected by the electrical potential, even though the ions are not bound directly to surface sites.

The structure of an iron oxide adsorbent according to one version of the basic model (the triple-layer model, TLM) is shown in Figure 7-42. The key features of this structure are the planes labeled as the surface (o), beta (β), and d planes in the figure. Some molecules are presumed to bind directly to a surface oxide ion (if the adsorbate is a cation) or by replacing a surface oxide ion and binding directly to the underlying Fe (if the adsorbate is an anion). Molecules that adsorb in this manner must lose at least the water of hydration on the side nearest the

316 ADSORPTION PROCESSES: FUNDAMENTALS

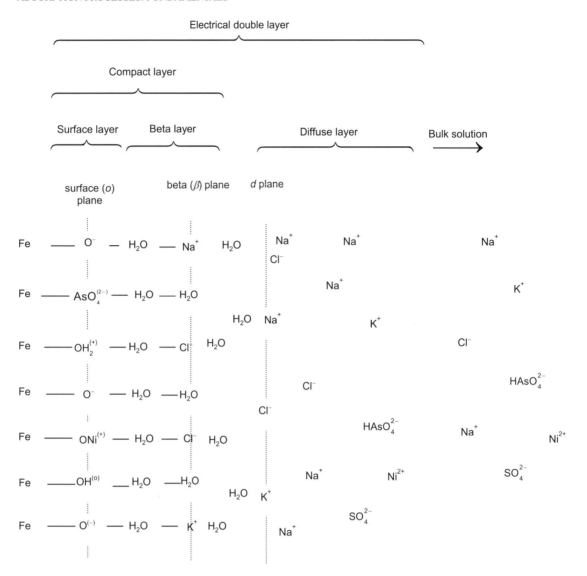

FIGURE 7-42. Schematic of the interfacial structure near an iron oxide surface, as envisioned in the triple-layer model. The charges shown on species in the surface (o) plane represent the sum of the charge from that species plus the Fe atom on the surface; that is, they represent the total charge expressed at the site.

solid. The surface complexes formed by such reactions are relatively strong and are sometimes referred to as *inner sphere complexes*. H^+ and OH^- are always presumed to bind in this way. The surface plane (sometimes called the naught plane) runs through the centers of the surface oxide ions and any adsorbates that are bound either directly to them or to surface Fe ions.

Other ions might also bind to the surface via chemical bonds, but without losing any waters of hydration. Such ions are therefore separated from the surface by a water molecule and form weaker complexes (sometimes called *ion pair* or *outer sphere complexes*) with the surface. The beta plane is defined to pass through these ions.

At distances farther from the surface than the beta plane, ions are assumed to be attracted to or repelled from the surface solely by electrostatic (i.e., not covalent) interactions. The electrical interactions increase the concentration of species that are oppositely charged from the surface (*counter-ions*) and decrease the concentrations of like-charged species (*co-ions*), compared to their concentrations in bulk solution. Over a distance of several molecular diameters, the net charge of these counter-ions and co-ions completely neutralizes the charge in the surface and beta layers, and the concentrations of the ions approach their values in bulk solution.

The region in which the concentrations of the nonspecifically adsorbed ions differ from their values in bulk solution is called the *diffuse layer*. The diffuse layer starts a short distance farther from the surface than the β layer, and the beginning of this region is defined as the d plane. The space between the β plane and the beginning of the diffuse layer is occupied by hydration waters of the ions in the β and d planes. Some models do not consider a β layer, in which case the d plane is separated from the surface layer only by the waters of hydration between the surface plane and the d plane. The region containing the surface and beta layers (and therefore containing all specifically adsorbed ions) is sometimes referred to as one or two *compact layer(s)*, and the combination of the compact layer(s) and the diffuse layer is called the *electrical double layer* (EDL).

The charge–potential relationships in the various layers are central to the quantitative analysis of adsorption. The relationship between the charge in the diffuse layer and the potential at its inner boundary (the d plane) was solved in 1910 for an idealized system that treated the ions as point charges, and this relationship is used by all current models. Except for geometric considerations, the analysis is just like that for the distribution of ions surrounding a central ion; that is, the analysis that yields a prediction for the ion activity coefficient according to the Debye–Huckel equation. The result for a solution of symmetric electrolyte (one in which the magnitude of the charge on the cations, $|z^+|$, is the same as that on the anions, $|z^-|$) for a system at 25°C is

$$\sigma_d = -0.1174 c^{0.5} \sinh\left(\frac{z\mathcal{F}\psi_d}{2RT}\right) \quad (7\text{-}154)$$

where σ_d is the equivalent charge density of the d layer, in coulombs per square meter (C/m^2), c is the electrolyte concentration in the bulk solution in mol/L, z is the absolute value of the ionic charge number of the electrolyte ions, $\mathcal{F}$ is the Faraday constant, ψ_d is the electrical potential at the d plane, and the product RT has its usual meaning. σ_d is referred to above as an equivalent charge density because it does not refer to charges that reside *in* the d plane, but rather to the net charge throughout the diffuse layer, treated *as though* it were all in the d plane.

When combined with a model for nonspecific adsorption of ions as a function of ψ (developed shortly), Equation 7-154 can be converted into an expression for ψ versus distance from the d plane. For most situations of interest in natural aquatic systems ψ_d is less than 25 mV, and ψ decays approximately exponentially through the diffuse layer. The characteristic distance for this decay (the distance needed for ψ to decay to ψ_d/e, sometimes called the Debye length and usually designated κ^{-1}) varies with the inverse square root of the ionic strength. Typical values of κ^{-1} in natural waters range from $\sim$10 nm in fresh water with an ionic strength of 10^{-3} mol/L NaCl to $\sim$0.4 nm in seawater. Further details about the decay of ψ with distance from the surface are given in Chapter 11.

The charge–potential relationship in the compact layer is less well agreed upon than that in the diffuse layer. The most common modeling approach is to assume that the potential changes linearly in this region, with one slope between the surface and beta planes and another between the beta and the d planes. The (assumed) fixed ratio of the change in potential between two layers to the charge is referred to as the *capacitance*, making an analogy between the parallel layers of adsorbed charge and a parallel plate capacitor (or, in models incorporating a beta layer, two capacitors in series). Thus, the relationships are

$$C_1 = \frac{\psi_o - \psi_\beta}{\sigma_o} \quad (7\text{-}155)$$

$$C_2 = \frac{\psi_\beta - \psi_d}{-\sigma_d} = \frac{\psi_d - \psi_\beta}{\sigma_d} \quad (7\text{-}156)$$

where C_1 and C_2 are the capacitances of the inner (o-to-β) and outer (β-to-d) layers, respectively, typically expressed in units of coulombs per volt (farads).[24]

The capacitances of the compact layers of adsorbents are not well known, and in any case the idea that all the adsorbates are aligned in a plane and that the charge is distributed uniformly on that plane is clearly an idealization. Therefore, C_1 and C_2 are usually used as fitting parameters to improve the match of model calculations to experimental data.

The triple-layer model is an elaboration of a two-layer model originally proposed by Stern (1924). That model considered only one compact layer adjacent to the surface, with the diffuse layer starting immediately outside of that layer. This layer (subsequently named the *Stern layer*) was presumed to contain hydrated ions. Two widely used simplifications of this model are based on opposite, limiting case assumptions about the relative amounts of charge in the Stern layer and the diffuse layer. In the *constant capacitance model*, all the ions that neutralize the surface charge are assumed to reside in the d plane, so that the diffuse layer is, in fact, not diffuse at all; this limiting case applies only in solutions with ionic strengths that are higher than most solutions encountered in environmental engineering. By contrast, in the *diffuse layer model*, no charge resides in the d plane, and all the neutralizing charge is spread out in the diffuse layer, so that $\psi_o = \psi_d$. A schematic of the potential–distance relationship through the interfacial region according to these models is shown in Figure 7-43, and the key characteristics of the models are compared in Table 7-7. Because both the constant capacitance and diffuse layer models can be derived as limiting cases of the triple-layer

---

[24] Parallel plate capacitors have equal and opposite charges on the two plates. When two capacitors are in series, the middle plate has a charge that is the opposite of the sum of the charges on the outer plates. Thus, in the current case, the model consists of a capacitor near the surface with a charge density of σ_o and one farther from the surface with a charge density of $-\sigma_d$.

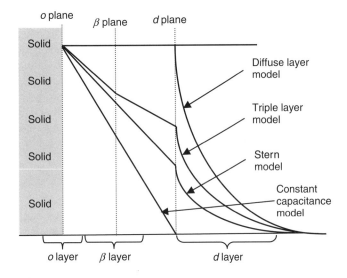

FIGURE 7-43. Profiles of ψ versus distance from the surface, as envisioned in three surface complexation adsorption models. Note that the length scales in the three regions are different—the distance from the surface to the d plane is thought to be on the order of a few nm, whereas the d layer is often several tens to $\sim$100 nm thick. *Source*: After Westall and Hohl (1980).

TABLE 7-7. Comparison of the Charge Distribution and Electrical Potential Profile Near the Surface in Four Surface Complexation Adsorption Models

Model	Charge Distribution
Stern model	• Only one compact layer (the Stern layer), containing specifically adsorbed species (o plane runs through the center of the Stern layer) • Diffuse layer starts outside Stern layer
Constant capacitance	• Only one compact layer (the Stern layer), containing specifically adsorbed species • Charge equal and opposite to that in the o plane provided by nonspecifically sorbed ions, all in d plane • Applicable in high ionic strength solutions
Diffuse layer	• No compact layer of charge • Charge equal and opposite to that in the o plane provided by nonspecifically sorbed ions distributed throughout d layer
Triple layer	• Dehydrated, specifically adsorbed species in o plane • Hydrated, specifically adsorbed species in o plane • Charge equal and opposite to that in the combined o and β planes provided by nonspecifically sorbed ions, starting in d plane and distributed throughout d layer

model (Westall, 1987), the following discussion is presented in the context of the latter, more general model.

The Electrostatic Contribution to the Equilibrium Constants in Competitive Adsorption Reactions

At equilibrium, the overall charge around an adsorbent particle, including the fixed charge, the charge of specifically adsorbed ions, and the charge in the diffuse layer must be zero.[25] Therefore, if an ion adsorbs, other processes must combine to expel some like-charged species from the interfacial region and/or bring oppositely charged species into the interfacial region, so that the overall electroneutrality of the particle and surrounding fluid is maintained. These processes need not occur at the point where the adsorbing ion resides, but they must take place in the interfacial region. Thus, sorption of an ion is *always* associated with either sorption of an oppositely charged ion or desorption of a like-charged ion; it never involves exchange exclusively with water molecules.

The fact that sorption of an ion always requires the simultaneous sorption or desorption of another ion is of central importance when evaluating the electrostatic contribution to such reactions. Consider, for example, a homovalent overall reaction composed of one adsorption and one desorption reaction at a site where the potential is ψ_x. According to Equations 7-152 and 7-153, the constituent adsorption and desorption reactions, as well as the overall reaction, can be represented as follows:

Adsorption	$\equiv S^y + A^+ \leftrightarrow \equiv SA^{y+1}$
	$K_{ads,A} = K_{intr,A} \exp\left(-\dfrac{(1)\mathcal{F}\psi_x}{RT}\right)$ (7-157)
Desorption	$\equiv SB^{y+1} \leftrightarrow \equiv S^y + B^+$
	$K_{des,B} = \left[K_{intr,B} \exp\left(-\dfrac{(1)\mathcal{F}\psi_x}{RT}\right)\right]^{-1}$ (7-158)
Overall	$\equiv SB^{y+1} + A^+ \leftrightarrow \equiv SA^{y+1} + B^+$
	$K_{overall} = K_{ads,A} K_{des,B} = K_{intr,A}/K_{intr,B}$ (7-159)

y is the charge on the unoccupied (or hydrated) surface site, and might be positive, negative, or zero.

Since ψ_x does not appear in the equilibrium constant expression for the overall reaction (Equation 7-159), the electrical potential at the location of the adsorbed species (or at the surface of the solid, or anywhere else in the system) has no effect on adsorption. This result is a consequence of

---

[25] This statement might seem to be in conflict with the fact that particles migrate under the influence of an electrical field. However, that situation is a nonequilibrium condition, in which the electrical force driving the particle in one direction and the counter-ions in the other direction is temporarily able to overcome the mutual attraction of the particles and counter-ions for one another. As soon as the electrical field is removed, the system returns to an equilibrium condition.

the fact that the overall reaction involves an equal exchange of charge at a fixed distance from the surface. In such a situation, the electrical potential at the point of adsorption enhances the adsorption reaction exactly as much as it impedes the desorption reaction.

The preceding scenario is, in essence, the classic model of an ion exchange reaction, and it points out an implicit assumption that is made in all ion exchange literature. Specifically, if the equilibrium constant for the exchange reaction (i.e., the selectivity coefficient) is indeed to be constant, then the exchanging species must bind at the same distance from the surface, or at least at locations where the electrical potential is identical. If they do, then the potential (ψ_x) that each ion experiences when it sorbs is the same, and that potential is the same regardless of the ratio of bound A to bound B. In such a case, ion exchange equilibria, which clearly involve species that have electrostatic interactions with the surface, can be modeled without taking those interactions into account explicitly.

If the ions involved in the adsorption and desorption reactions that comprise an overall reaction bind at different distances from the surface (for instance, if the adsorbing ion formed an inner sphere complex and the desorbing ion an outer sphere complex), the electrical contributions to the adsorption and desorption reactions would not cancel, and the surface potential would have a net effect on the reaction. This scenario is typically thought to be the case when strongly binding metal ions adsorb. The metal ions are usually assumed to bind in the surface or the beta plane, and the positive charge that they bring to the surface region is neutralized by release of a co-ion from, or adsorption of a counter-ion into, the diffuse layer. In such a case, if the metal is a cation, the overall reaction is opposed by a positive surface charge, even though the reaction has no effect on the total charge in the interfacial region. (Keep in mind, though, that even if the electrostatics are unfavorable, the overall reaction could be highly favorable, if the intrinsic chemical attraction of the metal for the surface site is strong.)

To summarize, electrical interactions between charged surfaces and ionic adsorbates increase the effective adsorption equilibrium constant for ions that are oppositely charged from the surface, and decrease the effective constant for ions that have the same sign charge as the surface. The modifications apply regardless of whether the sites on the surface are all identical to one another or differ from one surface location to another. The magnitude of these effects depends not only on the identity of the adsorbate and the surface, but also on the composition of the system as a whole, and in particular, on the locations where ions bind, or which they vacate, in the overall sorption reaction.

The general approach for modeling the electrostatic effect involves splitting the overall adsorption equilibrium constant into an intrinsic (chemical) and a variable (electrostatic) contribution, as described by Equation 7-152. The differences among the models currently in use reflect different assumptions about the structure of the near surface region (i.e., how closely different types of ions approach the surface). All such models are, of course, idealizations; real surfaces are likely to have microscale physical/chemical features and, perhaps, interactions among adsorbates that are not captured in the models. Nevertheless, several of the models do a good job of describing the overall effects of surface charge on adsorption reactions. As spectroscopic and other molecular-scale techniques for probing the surface environment improve, our ability to model electrostatic effects on adsorption is likely to improve in parallel.

Phase Transitions Involving Ionic Adsorbates: Pore Condensation and Surface Precipitation

One of the assumptions of the Langmuir model is the existence of an upper limit to the adsorption density. Although the validity of the assumption might seem unassailable, it can be violated in some situations. One example of such a situation—the condensation or precipitation of an organic adsorbate in the pores of activated carbon—was described previously in the context of the Polanyi adsorption model.

The process envisioned in the Polanyi model is usually referred to as pore condensation, although there is no fundamental reason why a similar process could not occur on an external surface if the driving forces for sorption and condensation were sufficiently strong. Subsequent adsorption then becomes a process of adding layers into the condensed material or dissolving molecules into the organic phase, and such a process would not necessarily have an upper limit on the amount of material that could sorb.

Condensation is the most likely phase transition for most organic adsorbates. When the adsorbate is inorganic (particularly if it is a metal), the corresponding reaction is surface precipitation; that is, the formation of an inorganic solid containing the adsorbate on the surface of the adsorbent.[26] At low adsorption densities of metal ions, each ion is likely to behave as though it were isolated from the others. However, as the adsorption density increases, adjacent metal ions might link to one another via adsorbed anions, eventually forming a two-dimensional and then three-dimensional lattice. Even if the anions that originally entered the solution with the metals are highly soluble, the formation of metal oxides or hydroxides at the surface is always possible, since OH^- ions can be provided by dissociation of water. Eventually, the original surface is completely covered, and the system behaves in many ways like a suspension of the pure precipitated adsorbate (Figure 7-44).

---

[26] Some condensed organics are also more stable as solids than liquids, and attempts have been made to apply the Polanyi theory to those organics as well as organics that condense as liquids.

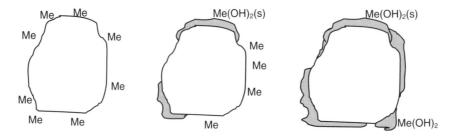

FIGURE 7-44. Schematic showing a metal ion adsorbate that is first bound to discrete sites on the adsorbent, but that gradually forms a surface precipitate covering much of the solid, so that the surface of the final product behaves almost like that of a particle of the pure metal hydroxide. *Source*: From Benjamin (2010).

Empirically, the manifestation of this process is that the adsorption density gradually increases in response to increasing dissolved adsorbate concentration but then, rather than approaching some maximum value, the apparent adsorption density increases without bound. In fact, the latter observation does not reflect continuous adsorption, but rather the formation of the precipitate.

Because the solubility product imposes a limit on the amount of the adsorbate that can be present in solution, addition of an increment of adsorbate after surface precipitation has been initiated leads to precipitation of virtually all of the added material. If adsorption is quantified simply by analyzing the amount of adsorbate that is removed from solution, the precipitation reaction can easily be mistaken for adsorption, causing the apparent adsorption density to increase indefinitely. Although the endpoints of this process (adsorption at very low dissolved adsorbate concentrations, and precipitation at high concentrations) are reasonably well defined, the transition between the two states is not. An approach for modeling this transition was proposed by Farley et al. (1985).

7.9 SUMMARY

Adsorption is the accumulation of selected molecules at the boundary between two phases. In aquatic systems, adsorption can occur at the interface of the solution with solids or gases, and can substantially alter the transport and fate of solutes, especially those that are present at trace concentrations. In engineered water treatment systems, adsorption is used to remove NOM, metals, synthetic organic compounds, and compounds that impart taste and odor from solution, employing adsorbents that include iron and aluminum oxides, activated carbon, and ion exchange resins. Adsorption is also a critical process in the transport of metals, pesticides, and volatile organic compounds that have entered surface and groundwater systems, helping to control the movement of those species through the environment and/or their removal from the aqueous phase.

Adsorption of an aqueous solute onto a solid can be modeled as a process in which solute molecules bind to specific sites on the solid surface, similar to the formation of complexes in solution, or as one in which the adsorbing molecules enter a surface phase in which they are highly mobile and are not bound to individual sites. Historically, the former model has been used primarily for adsorption of hydrophilic compounds onto inorganic minerals and ion exchange resins, while the latter has been used for adsorption of hydrophobic compounds onto organic solids (or organic coatings on inorganic solids) and activated carbon.

Adsorption bears many similarities to gas transfer, but it also differs in several ways: surfaces are not necessarily uniform; the number of surface sites available in a system (and hence the maximum amount of adsorption that can occur) is often limited; the surface sites are often close enough to one another that significant interactions can occur between adsorbed molecules; and the possibility that ions can adsorb means that electrostatic interactions between the surface and the adsorbates can be an important contributor or impediment to adsorption. These differences make the characterization of adsorptive equilibrium more complex than the characterization of gas/liquid equilibrium.

Adsorptive equilibrium is commonly characterized by adsorption isotherms, which are equations relating the adsorption density to the concentration of dissolved adsorbate, once equilibrium has been attained. In systems where none of the features that distinguish adsorption from gas transfer are significant, adsorptive equilibrium is characterized by a linear isotherm and a well-defined equilibrium constant that is analogous to a Henry's constant. However, if any of those distinguishing factors are important, the corresponding isotherm is generally curvilinear, and the apparent equilibrium constant might depend on the overall composition of the solution, rather than have a well-defined value that depends only on the identities of the adsorbate and adsorbent.

The Langmuir and Freundlich isotherms represent attempts to model adsorption in systems with a limited number of identical adsorptive sites and with certain types of site nonuniformity, respectively. These isotherms are used widely to describe experimental data, but other isotherms are also used and are equally acceptable, if they fit the observations. Other isotherms and adsorption models have been proposed to account for competitive adsorption (e.g., the competitive Langmuir and Freundlich isotherms, and isotherms developed based on the IAST), for condensation reactions at the surface and in pores (the Polanyi model), and for the effects of electrostatics on ionic adsorption (the triple layer, diffuse layer, and constant capacitance models).

Ion exchange reactions can be viewed as a subset of adsorption reactions, in which an assumption is made that all the surface sites are charged and are occupied by oppositely charged adsorbates under all circumstances. Because of this assumption, classic adsorption equilibrium constants that characterize the strength of binding of an isolated adsorbate to unoccupied sites cannot be evaluated for ion exchange reactions. Rather, the affinities of these adsorbents for various ions are reported in terms of selectivity coefficients or separation factors, indicating the relative binding strength of the adsorbents for different ions. In environmental engineering, most ion exchange processes use manufactured, synthetic resins as the adsorbent media. These resins have the advantage that they are relatively easily regenerated by exposure to a solution containing a high concentration of a nontarget ion. As a result, the resins can be reused repeatedly, and their use can be economical even though the media itself is usually considerably more expensive than other adsorbents on a mass or volume basis.

Electrostatic interactions between ions and solid surfaces can affect both entities; that is, the charge on the surface can affect the tendency for ions to adsorb, and the adsorption of ions can affect the charge on the surface. The surface charge on particles can also have a large effect on their interactions with one another; these effects are discussed in Chapter 11. Interactions among adsorbed species can lead to condensation or precipitation of the adsorbate on the surface or in the pores of the adsorbent. Models for these types of interactions have been proposed and applied, achieving varying levels of success at describing experimental data.

An understanding of the molecular-scale processes that control adsorption is invaluable for mechanistic interpretation of adsorption phenomena. Fortunately, however, complete knowledge of these phenomena is not necessary for designing adsorption-based water treatment processes. The key information needed for such efforts includes the adsorption isotherms for the various constituents of the solutions, the hydraulic characteristics of the reactor where adsorption will occur, and, possibly, knowledge of the adsorption kinetics. The synthesis of these factors, leading to development of design equations for adsorption-based treatment processes, is presented in Chapter 8.

REFERENCES

Adamson, A. W. (1976) *Physical Chemistry of Surfaces*. Wiley-Interscience, New York, NY.

Allen-King, R. M., Grathwohl, P., and Ball, W. P. (2002) New modeling paradigms for the sorption of hydrophobic organic chemicals to heterogeneous carbonaceous matter in soils, sediments, and rocks. *Adv. Water Res.*, 25, 985–1016.

Bargar, J. R., Brown, G. E., and Parks, G. A. (1997) Surface complexation of Pb(II) at oxide-water interfaces: II. XAFS and bond-valence determination of mononuclear Pb(II) sorption products and surface functional groups on iron oxides. *Geochim. Cosmochim. Acta*, 61 (13), 2639–2652.

Benjamin, M. M. (2010) *Water Chemistry*. Waveland Press, Long Grove, IL.

Benjamin, M. M. (2009) New conceptualization and solution approach for the ideal adsorbed solution theory (IAST). *Environ. Sci. Technol.*, 43, 2530–2536.

Benjamin, M. M. (2002) Modeling the mass-action expression for bidentate adsorption. *Environ. Sci. Technol.*, 36, 307–313.

Benjamin, M. M., and Leckie, J. O. (1981) Multiple-site adsorption of Cd, Cu, Zn, and Pb on amorphous iron oxyhydroxide. *J. Colloid Interface Sci.*, 79, 209–221.

Benjamin, M. M., and Leckie, J. O. (1982) Effects of complexation by Cl, SO_4 and S_2O_3 on adsorption behavior of Cd on oxide surfaces. *Environ. Sci. Technol.*, 16, 162–170.

Chiou, C. T. (1998) Soil sorption of organic pollutants and pesticides. In Meyers, R. A. (ed.), *Encyclopedia of Environmental Analysis and Remediation*, John Wiley and Sons, New York, NY.

Churn, J-M., and Chien, Y-W. (2002) Adsorption of nitrophenol onto activated carbon: Isotherms and breakthrough curves. *Water Res.*, 36 (3), 647–655.

Clifford, D., Sorg, T. J., and Ghury, G. L. (2011) Ionic exchange and adsorption of inorganic contaminants, In Edzwald, J. K. (ed.), *Water Quality and Treatment*, 6th ed. McGraw Hill, New York.

Cook, D., Newcombe, G., and Sztajnbok, P. (2001) The application of powdered activated carbon for MIB and geosmin removal: Predicting PAC doses in four raw waters. *Water Res.*, 35, 1325–1333.

Cornel, P. K., Summers, R. S., and Roberts, P. V. (1986) Diffusion of humic acid in dilute aqueous solution. *J. Colloid Interface Sci.*, 110, 149–164.

Crittenden, J. C., Luff, P., Hand, D. W., Oravitz, J. L., Loper, S. W., and Arl, M. (1985) Prediction of multicomponent adsorption equilibria using ideal adsorbed solution theory. *Environ. Sci. Technol.*, 19, 1037–1043.

Crittenden, J. C., Sanongraj, S., Bulloch, J. L., Hand, D. W., Rogers, T. N., Speth, T. F., and Ulmer, M. (1999) Correlation of aqueous-phase adsorption isotherms. *Environ. Sci. Technol.*, 33, 2926–2933.

Davis, J. A., Coston, J. A., Kent, D. B, and Fuller, C. P. (1998) Application of the surface complexation concept to complex mineral assemblages. *Environ. Sci. Technol.*, 32, 2820–2828.

Davis, J. A., and Kent, D. B. (1990) Surface complexation modeling in aqueous geochemistry. In Hochella, Jr., M. F., and White, A. F. (eds.), *Mineral-Water Interface Geochemistry*, Reviews in Mineralogy, Vol 23, 177–260.

Davis, J. A., and Leckie, J. O. (1978) Effect of adsorbed complexing ligands on trace metal uptake by hydrous oxides. *Environ. Sci. Technol.*, 12, 1309–1315.

Dubinen, M. M. (1960) The potential theory of adsorption of gases and vapors for adsorbents with energetically non-uniform surfaces. *Chem. Rev.*, 60 (2), 235–241.

Dzombak, D. A., and Hudson, R. J. M. (1995) Ion exchange: The contributions of diffuse layer sorption and surface complexation. In Huang, C. P., O'Melia, C. R., and Morgan, J.J. (eds.), *Aquatic Chemistry: Interfacial and Interspecies Processes*, Advances in Chemistry Series #244, American Chemical Society, Washington, DC

Dzombak, D. A., and Morel, F. M. M. (1990) *Surface Complex Modeling*. Wiley Interscience, New York, NY.

Everett, D. H. (1988) *Basic Principles of Colloid Science*. Royal Society of Chemistry, London.

Farley, K. J., Dzombak, D. A., and Morel, F. M. M. (1985) A surface precipitation model for the sorption of cations on metal oxides. *J. Colloid Interface Sci.*, 106, 226–242.

Graham, M. R., Summers, R. S., Simpson, M. R., and Macleod, B. W. (2000) Modeling equilibrium adsorption of 2-methylisoborneol and geosmin in natural waters. *Water Res.*, 34, 2291–2300.

Halsey, G., and Taylor, H. S. (1947) The adsorption of hydrogen on tungsten powders. *J. Phys. Chem.*, 15, 624–630.

Hohl, H., and Stumm, W. (1976) Interactions of Pb^{2+} with hydrous α-Al_2O_3. *J. Colloid Interface Sci.*, 55, 281–288.

Huang, C. P., and Stumm, W. (1973) Specific adsorption of cations on hydrous γ-Al_2O_3. *J. Colloid Interface Sci.*, 43, 409–420.

Hunter, R. J. (1987) *Foundations of Colloid Science*. Oxford University Press, Oxford, UK.

Jankowska, H., Swiatkowski, A., Choma, J., and Kemp, T. J. (1991) *Activated Carbon*. E. Horwood, New York, NY.

Karickhoff, S. W. (1984) Organic pollutant adsorption in aquatic systems. *J. Hydraulic Eng. (ASCE)*, 110, 707–735.

Li, L., Quinliven, P. A., and Knappe, D. R. U. (2005) Predicting adsorption isotherms for aqueous organic micropollutants from activated carbon and pollutant properties. *Environ. Sci. Technol.*, 39, 3393–3400.

Manes, M. (1998) Activated carbon adsorption fundamentals. In Meyers, R. A. (ed.), *Encyclopedia of Environmental Analysis and Remediation*, John Wiley and Sons, New York, NY.

Matsui, Y., Knappe, D. R. U., and Takagi, R. (2002) Pesticide adsorption by granular activated carbon adsorbers. 1. Effect of natural organic matter preloading on removal rates and model simplification. *Environ. Sci. Technol.*, 36, 3426–3431.

Moon, H., and Tien, C. (1987) Further work on multicomponent adsorption equilibria calculations based on the ideal adsorbed solution theory. *Ind. Eng. Chem. Res.*, 26, 2042–2047.

Schindler, P. W. (1990) Coadsorption of metal-ions and organic-ligands – formation of ternary surface complexes. *Rev. Mineral.*, 23, 281–307.

Schindler, P. W., Fürst, B., Dick, R., and Wolf, P. U. (1976) Ligand properties of surface silanol groups. 1. Surface complexation formation with Fe^{3+}, Cu^{2+}, Cd^{2+}, and Pb^{2+}. *J. Colloid Interface Sci.*, 55, 469–475.

Schwarzenbach, R. P., Gschwend, P. M., and Imboden, D. M. (2002) *Environmental Organic Chemistry*. Wiley Interscience, New York, NY.

Semmens, M. J., and Field, T. K. (1980) Coagulation: Experiences in organics removal. *J. AWWA*, 72 (8), 476–483.

Sheindorf, Ch., Rebhun, M., and Sheintuch, M. (1981) A Freundlich-type multicomponent isotherm. *J. Colloid Interface Sci.*, 79, 136–142.

Sontheimer, H., Crittenden, J. C., and Summers, R. S. (1988) *Activated Carbon for Water Treatment*, DVGW-Forschungsstelle (Germany), distributed in the USA by AWWA Research Foundation, Denver, CO.

Speth, T. F., and Miltner, R. J. (1990) Adsorption capacity of GAC for synthetic organics. *J. AWWA*, 82 (2), 72–75.

Stern, O. (1924) The theory of the electrical double layer (in German). *Z. Elektrochem.*, 30, 508–516.

Strathmann, H. (2004) *Ion-Exchange Membrane Separation Processes*. Science and Technology Series, Elsevier. Boston, MA.

Stumm, W. (1992) *Chemistry of the Solid-Water Interface*. Wiley-Interscience, New York, NY.

Tien, C. (1994) *Adsorption Calculations and Modeling*. Butterworth-Heinemann, Boston, MA.

Tipping, E., and Ohnstad, M. (1984) Aggregation of humic substances. *Chemical Geology*, 44, 349–357.

Vidic, R. D., Suidan, M. T., and Brenner, R. C. (1993) Oxidative coupling of phenols on activated carbon - impact on adsorption equilibrium. *Environ. Sci. Technol.*, 27, 2079–2085.

Waychunas, G. A., Davis, J. A., and Fuller, C. C. (1995) Geometry of sorbed arsenate on ferrihydrite and crystalline FeOOH: Re-evaluation of EXAFS results and topological factors in predicting sorbate geometry, and evidence for monodentate complexes. *Geochim. Cosmochim. Acta*, 59 (17), 3655–3661.

Westall, J. C. (1986) Chemical and electrostatic models for reactions at the oxide/solution interface. In Davis, J. A., and Hayes, K. F. (eds.), *Geochemical Processes at Mineral Surfaces*, ACS Symp. Ser. 323, Amer. Chem. Soc., Washington, DC.

Westall, J. C. (1987) Adsorption mechanisms in aquatic surface chemistry. In Stumm, W. (ed.), *Aquatic Surface Chemistry: Chemical Processes at the Particle-Water Interface*, John Wiley and Sons, New York, NY.

Westall, J. C., and Hohl, H. (1980) A comparison of electrostatic models for the oxide/solution interface. *Adv. Colloid Interface Sci.*, 12, 265–294.

Wohleber, D. A., and Manes, M. (1971) Application of the Polanyi adsorption potential to adsorption from solution on activated carbon. III. Adsorption of miscible liquids from water solution. *J. Phys. Chem.*, 75, 3720–3723.

Xia, G. S., and Ball, W. P. (2000) Polanyi-based models for the competitive sorption of low-polarity organic contaminants on a natural sorbent. *Environ. Sci. Technol.*, 34, 1246–1253.

Yates, D. E., Levine, S., and Healy, T. W. (1974) Site binding model of the electrical double layer at the oxide/water interface. *J. Chem. Soc. Faraday Trans. 1*, 70, 1807–1818.

PROBLEMS

TABLE 7-Pr2. Carbon Tetrachloride Adsorption

Bottle ID	Mass of CCl_4 (μg)	Water Volume (mL)	Mass of GAC (mg)	Final CCl_4 Concentration (mg/L)
A	50	100	20	0.12
B	10	100	20	0.019

7-1. Trichloroethylene (TCE) is an organic solvent that is present as a trace contaminant in many groundwaters and is quite strongly adsorbed onto activated carbon. The maximum adsorption capacity of a particular activated carbon for TCE is 65 mg TCE per gram carbon. When 1.0 L of a solution that initially contains 200 μg of TCE is dosed with 10 mg of the carbon, 15 μg/L TCE remains in solution at equilibrium.

(a) Assuming the Langmuir adsorption isotherm applies, find the adsorption equilibrium constant. Give appropriate units.

(b) At the end of the test, another 10 mg of activated carbon and 200 μg of TCE are added to the solution; that is, the total adsorbent and adsorbate concentrations are both doubled. Once the system re-equilibrates, will the concentration of dissolved TCE be larger, smaller, or the same as in part *a*?

(c) A water supply is to be treated to reduce the TCE concentration from 85 to 5 μg/L in a batch process. What is the required carbon dose (grams activated carbon per liter of water)? What fraction of the adsorption capacity of the activated carbon will be used?

7-2. The adsorption of carbon tetrachloride has been studied in a batch adsorption test using pulverized GAC. In the tests, a small amount of CCl_4 was added to bottles containing water and adsorbent, and the system was allowed to equilibrate under well-mixed conditions for 7 days, in sealed bottles that were free of gas headspace. Blanks (without GAC) were run to verify that no CCl_4 was lost from the system. At the end of the 7 days, the bottles were centrifuged and the equilibrium aqueous concentration of CCl_4 was measured. Table 7-Pr2 shows the results obtained for two representative samples.

(a) Use this "two-point isotherm" to estimate k_f and n for the Freundlich isotherm equation that describes CCl_4 adsorption onto this carbon. Include units where appropriate.

(b) Suppose that 30 μg of CCl_4 were added to a third bottle, also containing 100 mL of water and 20 mg of GAC. Estimate the expected final aqueous concentration of CCl_4.

7-3. Activated carbon is being used in a batch process to remove pesticides from a rural water supply. After each batch of water is treated, the carbon particles are allowed to settle, the water is drained off, and a new batch of water is mixed with the carbon. The adsorption is known to follow a Langmuir isotherm. After a number of batches of water have been treated in this way, the adsorption density on a sample of the activated carbon is given by: $q = (q_{max} K_{ads} c_{in})/(1 + K_{ads} c_{in})$, where c_{in} is the pesticide concentration in the untreated influent water. The value of q under these circumstances is much less than q_{max}. Do you think that the removal efficiency of pesticide in the next batch of water to be treated will be good or poor? Explain briefly.

7-4. A utility draws 50,000 m³/d of water from its water supply reservoir. After the local corn-growing season, the herbicide Alachlor ($C_{14}H_{20}ClNO_2$) is discovered in the reservoir at a concentration of 10 μg/L. The utility decides to add PAC to remove the Alachlor. The Freundlich isotherm parameters for Alachlor adsorption onto the selected PAC are: $k_f = 479$ (mg/g)/(mg/L)$^{0.26}$ and $n = 0.26$. If the goal is to bring Alachlor to below the MCL in drinking water (2 μg/L), and the system reaches equilibrium, how much PAC must be added to the water each day (kg/day)?

7-5. Determine whether the Langmuir or Freundlich isotherm provides a better fit to the data in Table 7-Pr5 for the uptake of the dye carrier methylnaphthalene by activated carbon. The data for c_{fin} were collected after equilibrium had been reached in batch adsorption experiments.

TABLE 7-Pr5. Methylnaphthalene Adsorption

PAC (g/L)	0.50	0.50	0.50	0.50	0.50	0.50	0.50	0.74	1.00	1.00
c_{init} (mmol/L)	0.10	0.23	0.35	0.505	0.67	0.91	1.36	1.59	1.71	0.100
c_{fin} (mmol/L)	0.02	0.03	0.05	0.065	0.13	0.22	0.61	0.44	0.32	0.008

7-6. Two equilibrium adsorption tests are carried out to characterize binding of chlorophenol to an activated carbon. Each test solution contained 1.0 L of water and 0.5 mg of chlorophenol. One gram of activated carbon was added to one solution (A), and 5.0 g was added to the other solution (B). After equilibration, the adsorption densities in these two systems were 0.45 and 0.0988 mg/g, respectively.

(a) Assuming the Freundlich isotherm applies to this system, evaluate the isotherm constants and write the isotherm equation.

(b) Based on the data given, can you decide whether the Freundlich equation is in fact a good model to use to represent adsorption in this system? That is, how confident can you be that the isotherm equation you developed in part *a* will apply to systems containing different amounts of chlorophenol and activated carbon than those investigated experimentally?

7-7. A researcher is studying desorption of various organic compounds from contaminated soils. In one set of experiments, various masses, M_s, of the same soil are dispersed into several jars, each containing a volume V_L of clean water. After mixing for a time thought to be long enough for equilibrium to be attained, the solids are separated from the liquid, and the liquid-phase concentration, c_i, is measured. A plot of $1/c_i$ versus the inverse of the adsorbent concentration (V_L/M_s) turns out to be linear.

Show that this result is consistent with a linear adsorption isotherm, as follows. Draw a linear isotherm, and write out the isotherm equation. Indicate two points on the graph, one representing the initial condition for both the liquid and solid phases, and the other representing the final, equilibrium condition for both phases. Then, write a mass balance that describes the changes in c_i and q_i between the initial and final conditions. Finally, substitute the isotherm equation into the mass balance, and manipulate the equation to develop a (linear) relationship between $1/c_i$ and V_L/M_s. Identify the meaning of the slope and intercept of the line represented by this relationship, in terms of the isotherm parameters.

7-8. The following data characterize the adsorption of a dimethylphenol (MP) on activated carbon.

c (mg/L)	q (mg/mg)	c (mg/L)	q (mg/mg)
0.05	0.007	3.1	0.263
0.12	0.018	4.4	0.347
0.21	0.030	7.2	0.322
0.63	0.088	14.4	0.426
1.5	0.163	25.1	0.499

(a) Compute the best-fit Langmuir constants for describing adsorption in the system.

(b) A waste solution containing 2.0 mg/L MP is generated at a rate of 2×10^6 L/d. You wish to treat the solution to attain a final concentration of 0.1 mg/L. If you chose to treat the waste in a batch system, how much carbon would be required per day, assuming equilibrium is attained? What is the adsorption density on the activated carbon at equilibrium?

(c) Consider the effect on treatment efficiency if the waste also contained phenol (Ph). When both q_{max} and K are expressed using moles as the unit of adsorbate mass, q_{max} is identical for the two adsorbates, but K_{Ph} is only one-half as great as K_{MP}. If the same amount of carbon were used as you computed in part *b*, but the waste contained 100 mg/L phenol in addition to the MP, how severely would the adsorption density of MP be decreased?

(d) As is noted in this chapter and explained in detail in Chapter 8, packed column adsorption systems are often operated until the adsorbent is essentially in equilibrium with the influent solution composition. What adsorption densities would be achieved in such a system treating the two influent solutions (2.0 mg/L MP, in the presence and absence of 100 mg/L phenol)?

7-9. The selectivity coefficient of an ion exchange resin for Cu^{2+} over Zn^{2+} is $K_{Cu/Zn} = 3$. A solution initially containing 6 mg/L Cu^{2+} and 30 mg/L Zn^{2+} is to be treated by addition of the resin to the waste solution in a batch process. The resin has a capacity of 3.5 meq/g, and is initially in the H^+ form, but at the pH of treatment, almost none of resin sites are expected to be occupied by H^+. Develop plots of Cu^{2+} and Zn^{2+} concentrations remaining in solution, the adsorption densities of the two metals, and the percentage removal of each metal from solution, as a function of the resin dose, for doses from 0 to 300 mg/L.

7-10. A home water softening unit contains 10 kg of ion exchange resin in the Na^+ form, with an exchange capacity of 3.0 meq/g. The water supplied to the home contains 20 mg/L Na^+, 70 mg/L Ca^{2+}, and 25 mg/L Mg^{2+}.

(a) What is the distribution of cations on the resin once the system has operated until the resin is no longer accomplishing any treatment (because it is in equilibrium with the influent composition)? The separation factors under these conditions are $\alpha_{Ca/Na} = 2.4$ and $\alpha_{Mg/Na} = 1.9$.

(b) Compute the selectivity coefficients for Ca^{2+} and Mg^{2+} over Na^+.

(c) If, once the system reaches the condition characterized in part a, the resin is equilibrated with 30 L of 2 M NaCl, what fractional regeneration of the Ca^{2+} and Mg^{2+} can be expected? What are the separation factors in the regeneration step? Assume that the selectivity coefficients are independent of ionic strength.

7-11. Two droplets of water, 2 mm in diameter, are gently released—one onto a glass slide, and the other onto a slide that has been coated with wax. The droplets begin to spread out, but the one on the glass spreads into a wide, thin layer, whereas the one on the waxed surface spreads only a little and then stabilizes as a bead that expands no more.

In each system, the surface tensions at the water/air and solid/air interfaces are much less than that at the water/solid interface. Describe the energy changes accompanying spreading of each droplet, and explain the result in terms of energy differences between the two systems. Based on the observation described, do you think the surface tension of the water/glass interface is larger or smaller than that of the water/wax interface?

7-12. Some results of batch tests performed to evaluate the adsorption isotherm for binding of phenol to an activated carbon are shown in following table. Each bottle contained 100 mL of a solution that initially contained 600 mg/L of dissolved phenol. The systems were then equilibrated at 20°C.

Bottle No.	Mass of Carbon (g)	Final Conc. (mg/L)
1	0.758	2.8
2	0.558	4.2
3	0.478	18.0
4	0.402	30.9
5	0.332	55.9
6	0.323	69.7
7	0.297	85.3
8	0.258	121
9	0.211	178
10	0.176	231
11	0.121	325
12	0.075	424

(a) Plot the data as a conventional isotherm (i.e., as q vs. c).

(b) Calculate the best-fit parameters for the Freundlich isotherm, using a graphical approach. Compare the model predictions with the raw data on the plot used to determine the model parameters.

(c) Calculate the best-fit parameters for the Langmuir isotherm, using both linearization methods, and using a nonlinear fitting program (e.g., the Solver® function in MS Excel®). Again, show the model predictions on the plots used to determine the model parameters.

(d) Plot the raw data and the calculated values from all four isotherms (determined in parts a–c) on a single plot of q versus c. Which isotherm yields the best fit?

(e) Plot surface pressure as a function of liquid-phase phenol concentration and as a function of phenol adsorption density, according to both the Langmuir and the Freundlich isotherms.

7-13. Freundlich isotherms for the adsorption of p-nitrophenol (PNP) and benzoic acid (BA) on activated carbon are as follows, for concentrations given in mol/L and adsorption densities in mg/g: $q_{PNP} = 89 c_{PNP}^{0.13}$; $q_{BA} = 140 c_{BA}^{0.22}$. The SSA of the activated carbon is 600 m²/g.

(a) You wish to treat a waste solution that contains 10 mg/L PNP to attain a final concentration of 0.1 mg/L. How much carbon would be required per liter of waste, assuming equilibrium is attained? What are the adsorption density and the surface pressure on the activated carbon at equilibrium?

(b) A solution contains 15 mg/L BA. If the dose of activated carbon computed in part a were added to this solution, what would the equilibrium concentration, the adsorption density, and the surface pressure of BA be?

(c) If the waste contained both PNP and BA at the concentrations specified and activated carbon were added at the dose computed in part a, what would the adsorption density of each species be at equilibrium if competition were consistent with the IAST model? What would the total surface pressure and the partial surface pressure of each contaminant be?

8

ADSORPTION PROCESSES: REACTOR DESIGN AND ANALYSIS

8.1 Introduction
8.2 Systems with rapid attainment of equilibrium
8.3 Systems with a slow approach to equilibrium
8.4 The movement of the mass transfer zone through fixed bed adsorbers
8.5 Chemical reactions in fixed bed adsorption systems
8.6 Estimating long-term, full-scale performance of fixed beds from short-term, bench-scale experimental data
8.7 Competitive adsorption in column operations: The chromatographic effect
8.8 Adsorbent regeneration
8.9 Design options and operating strategies for fixed bed reactors
8.10 Summary
References
Problems

8.1 INTRODUCTION

In Chapter 7, we explored the fundamentals of adsorption phenomena, considering the molecular–scale interactions that lead to adsorption, the factors that control adsorptive equilibrium, and the mathematical models that are commonly used to describe that equilibrium. In this chapter, approaches for the design and analysis of the adsorption-based treatment systems used most commonly in environmental engineering are presented. Knowledge of the adsorption isotherm relationship is essential for developing efficient designs for such systems, and, in cases where adsorptive equilibrium is closely approached in the treatment system, the isotherm equation is *all* that is needed for design calculations. In other cases, however, particularly those where a porous, granular adsorbent is used, equilibrium is not achieved quickly, and adsorption kinetics must also be considered.

Although several models for the kinetics of adsorption have been developed, quantitative predictions based on these models are, in general, less reliable than those from equilibrium models. As a result, greater caution (often manifested as larger safety factors) must be exercised when these models are used in design. Nevertheless, an understanding of the conceptual basis of the models, coupled with empirical data, often allows us to design new systems that work well, and to recommend reasonable approaches to improve the performance of systems that are not performing as desired.

Even though some steps in adsorption are fairly slow (e.g., diffusion of adsorbate into the interior of porous adsorbents), the design and operation of adsorption-based treatment processes are often dictated more by equilibrium constraints than by kinetics. This situation arises, in part, because the hydraulic detention time in adsorption reactors is usually longer than the characteristic time for the transport and binding of adsorbate molecules to the exterior of adsorbent particles. After that step is complete, even though the subsequent diffusion of the adsorbate into the interior of the particle might have a long characteristic time, the time available for that step to proceed (the residence time of the adsorbent in the system) is often even longer. As a result, calculations based on the attainment of adsorptive equilibrium are often useful for estimating performance factors such as the long-term adsorptive capacity of the media. Also, as a practical matter, the use of equilibrium-based equations for design calculations is often a tacit acknowledgment of deficiencies in our understanding of the adsorption kinetics for a particular application.

A completely general approach for analyzing adsorbate concentration and adsorption density as a function of time

---

Water Quality Engineering: Physical/Chemical Treatment Processes, First Edition. By Mark M. Benjamin, Desmond F. Lawler.
© 2013 John Wiley & Sons, Inc. Published 2013 by John Wiley & Sons, Inc.

and location in a given system involves writing three mass balances on the adsorbate, using three different sets of system boundaries. The control volume for one of those mass balances is the bulk solution phase (either all of that phase, or a differential portion of it), where advective transport and/or turbulent mixing are the dominant transport processes. A second mass balance can be written for a control volume that includes the liquid boundary layers around the adsorbent particles, where transport is dominated by exchange of small packets of fluid with the bulk solution and by diffusive transport into or out of those packets. The third mass balance is written around the adsorbent particles, in which transport occurs via diffusion, either through the liquid in the pores (*pore diffusion*) or along the interior surfaces of the adsorbent particles (*surface diffusion*).

Simultaneous solution of these three mass balances completely characterizes the behavior of an adsorptive system. Although no generic solution to the equations is available, in many cases we can make simplifying assumptions that allow us to obtain a closed form or numerical solution that is applicable to special cases. For instance, in systems that reach adsorptive equilibrium rapidly, the adsorbate concentration is uniform throughout the solution phase, and the adsorption density is uniform throughout the adsorbent. In other systems, the assumption of a uniform, well-mixed solution phase or of plug flow hydraulics is justified. In such situations, the assumptions reduce the mathematical complexity of the mass balance equations and simplify the analysis.

In this chapter, we develop the mass balances that characterize both the general case and several simplified cases. We start with systems that can be designed and analyzed based solely on equilibrium considerations, and then move to those in which the kinetics of adsorption must be considered. In each case, we consider both batch systems and fixed bed systems. The initial sections of the chapter focus on systems with a single target adsorbate; competitive adsorption is discussed in subsequent sections.

8.2 SYSTEMS WITH RAPID ATTAINMENT OF EQUILIBRIUM

Batch Systems

The simplest adsorptive system to analyze is a batch system in which the adsorbate is neither generated nor destroyed by chemical reactions. In such a system, the total mass of adsorbate in the system at any time equals the mass that was present initially; that is, when the solution and adsorbent were first mixed. Therefore, a mass balance on total adsorbate in the system at any time can be written as a relationship between the changes in mass in the two compartments of the system, as follows:

Change in total adsorbate mass in system due to adsorption between time 0 and t = Change in dissolved adsorbate mass + Change in adsorbed adsorbate mass

$$0 = V_L(c(0) - c(t)) + W(q(0) - q(t)) \tag{8-1}$$

where V_L is the volume of bulk solution in the reactor; W is the mass of adsorbent in the reactor; c is the dissolved concentration of adsorbate; and q is the adsorption density (mass of adsorbate/mass of adsorbent).

Rearranging Equation 8-1, we have

$$V_L(c(0) - c(t)) = -W(q(0) - q(t)) \tag{8-2}$$

If $c(t)$ and $q(t)$ are interpreted as the average concentration and average adsorption density in the system, respectively, then Equation 8-2 applies regardless of whether or not the system reaches equilibrium. That is, even if the adsorption density varies from one adsorbent particle to the next, or as a function of location in any given particle, or if the dissolved adsorbate concentration varies spatially in solution, the loss of adsorbate mass from solution can be equated with the increase of adsorbed adsorbate mass. However, if we restrict our consideration to systems that reach equilibrium, $c(t)$ and $q(t)$ must each be uniform throughout the system and must be related by the adsorption isotherm; that is,

$$V_L(c_{init} - c_{fin}) = -W(q_{init} - q_{fin}) = -W\left(q_{init} - q_{eq}\big|_{c_{fin}}\right) \tag{8-3}$$

where the subscripts init and fin refer to the initial (non-equilibrium) and final (equilibrium) conditions, respectively, and $q_{eq}\big|_{c_{fin}}$ is the equilibrium adsorption density associated with c_{fin}.

If adsorptive equilibrium is characterized by a Langmuir or Freundlich isotherm, Equation 8-3 leads to the following expressions:

Langmuir: $V_L(c_{init} - c_{fin}) = -W\left(q_{init} - q_{max}\dfrac{K_{ads}c_{fin}}{1 + K_{ads}c_{fin}}\right)$
(8-4a)

Freundlich: $V_L(c_{init} - c_{fin}) = -W\left(q_{init} - k_f c_{fin}^n\right)$ (8-4b)

In a typical application, we might know the initial dissolved adsorbate concentration, the target concentration after treatment, and the volume of solution to be treated. If fresh adsorbent were being added, q_{init} would be zero.

Assuming that the isotherm had been characterized in advance, $q_{eq}|_{c_{fin}}$ would also be known, so we could solve Equation 8-3 for the required adsorbent dose (W). Alternatively, if the initial dissolved adsorbate concentration and the adsorbent dose were fixed, the equation could be solved for the concentration of adsorbate remaining in solution once equilibrium had been attained.

■ **EXAMPLE 8-1.** Adsorption of phenol onto an activated carbon (AC) follows the Freundlich isotherm $q = 0.97((\text{mmol/g AC})/(\text{mmol/L})^{0.27})c^{0.27}$ in the range $0 < c < 5\,\text{mmol/L}$. A batch process is to be used to reduce the phenol concentration in a waste from 1.0 to 0.05 mmol/L. If the solution and adsorbent reach equilibrium in the system, what is the required dose of activated carbon?

Solution. Since adsorption is characterized by a Freundlich isotherm, we can use Equation 8-4b to compute the required adsorbent dose, as follows:

$$V_L(c_{init} - c_{fin}) = -W(q_{init} - k_f c_{fin}^n)$$

$$\frac{W}{V_L} = -\frac{c_{init} - c_{fin}}{q_{init} - k_f c_{fin}^n}$$

$$= -\frac{(1.0 - 0.05)\,\text{mmol/L}}{0\,\text{mmol/g} - 0.97((\text{mmol/g AC})/(\text{mmol/L})^{0.27})(0.05\,\text{mmol/L})^{0.27}}$$

$$= 2.2\,\text{g AC/L}$$

■

Equation 8-3 can also be used in conjunction with a graphical representation of the isotherm. For instance, say that the empirically determined isotherm was as shown in Figure 8-1. If the system reached equilibrium, the final condition would be characterized by a point (c, q) that falls somewhere on the isotherm curve. The initial condition would be characterized by a different point, (c_{init}, q_{init}),

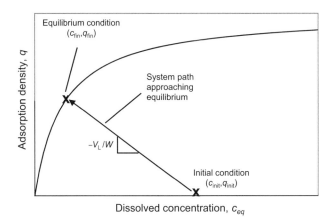

FIGURE 8-1. Graphical representation of the adsorption isotherm and mass balance for a system in which adsorbate must bind to the adsorbent to approach equilibrium.

not on the isotherm. In the typical case, the point (c_{init}, q_{init}) would be below the curve, indicating that material would have to adsorb for equilibrium to be attained; if the initial adsorption density was zero, the point would be on the abscissa; that is, at (c_{init}, 0).

The mass balance shown in Equation 8-2 indicates that, as the system approaches equilibrium, the ratio of the change in adsorption density, Δq, to the change in dissolved adsorbate concentration, Δc, equals $-V_L/W$:

$$q(t) - q(0) = -\frac{V_L}{W}(c(t) - c(0))$$
$$\Delta q(t) = -\frac{V_L}{W}\Delta c(t)$$
(8-5)

Or, in terms of the initial and final conditions:

$$q_{fin} - q_{init} = -\frac{V_L}{W}(c_{fin} - c_{init})$$
$$\Delta q = -\frac{V_L}{W}\Delta c$$
(8-6)

Thus, the approach to equilibrium can be depicted in Figure 8-1 as movement along a straight line from (c_{init}, q_{init}) toward, and ultimately reaching, the point (c_{fin}, q_{fin}) on the isotherm curve. The slope of this line is $-V_L/W$ and can be interpreted as the negative inverse of the adsorbent dose. This line is analogous to the operating line derived in the analysis of gas transfer processes. Note that, if we wish to achieve a very low value of c_{fin}, the isotherm equation will require that q_{fin} be small, which will force Δq to be small as well; correspondingly, for a given volume, V_L, of liquid to be treated, the required adsorbent dose, W, increases dramatically as the target value of c_{fin} gets small.

Adsorbents are often expensive to purchase or regenerate, so it makes sense to use as much of their capacity as possible in a given run. One approach for improving the efficiency with which the available capacity is utilized is demonstrated in the Example 8-2.

■ **EXAMPLE 8-2.** Determine the activated carbon dose needed to treat the solution described in Example 8-1, if the solution is treated in two, identical batch reactors in series. Assume that half of the activated carbon is added in each reactor, and that the adsorbent in the first reactor remains there when the water is transferred to the second. Answer the question using both (a) an analytical approach and (b) a graphical approach.

Solution.

(a) The mass balances in the two reactors are linked because the treated water from the first reactor is the input to the second, and because the problem states that W/V_L must be the same in both. We can therefore apply Equation 8-3 to obtain an expression for W/V_L in each reactor, and equate the resulting expressions to solve for the concentrations and adsorption densities of interest.

Defining the concentration and adsorption density of phenol at equilibrium in reactor i as c_i and q_i, respectively, and using the isotherm equation to compute q in terms of c in each reactor, we find (leaving out units for simplicity):

For Reactor 1: $\left(\dfrac{W}{V_L}\right)_1 = -\dfrac{c_{\text{init},1} - c_1}{q_{\text{init},1} - 0.97 c_1^{0.27}}$

For Reactor 2: $\left(\dfrac{W}{V_L}\right)_2 = -\dfrac{c_{\text{init},2} - c_2}{q_{\text{init},2} - 0.97 c_2^{0.27}}$

$= -\dfrac{c_1 - c_2}{q_{\text{init},2} - 0.97 c_2^{0.27}}$

The adsorbent is initially free of adsorbate, so $q_{\text{init},1} = q_{\text{init},2} = 0$. Making this substitution and equating the two expressions for W/V_L, we obtain

$$-\dfrac{c_{\text{init},1} - c_1}{0 - 0.97 c_1^{0.27}} = -\dfrac{c_1 - c_2}{0 - 0.97 c_2^{0.27}}$$

$$c_2^{0.27}\left(c_{\text{init},1} - c_1\right) = c_1^{0.27}(c_1 - c_2)$$

The values of $c_{\text{init},1}$ and c_2 are the concentrations of phenol in the untreated and fully treated water, respectively, corresponding to c_{init} and c_{fin} in the single-reactor system analyzed in Example 8-1. The value of c_1 is therefore the only unknown in the preceding equation. Solving the equation numerically, we find $c_1 = 0.40$ mmol/L, and inserting that value into the isotherm yields $q_1 = (0.97)(0.40)^{0.27}$, or 0.76 mmol/g. Finally, inserting the value of c_1 into the expression for W/V_L, we find that the required carbon dose in each reactor is 0.80 g/L, so the total dose is 1.60 g/L. This dose is ~73% of the dose required to treat the water in a single batch, as computed in Example 8-1.

(b) The problem can be solved graphically using Figure 8-2, which shows the isotherm and the conditions in both the one- and two-reactor systems. In the one-reactor system, the system path is a straight line from $(c_{\text{init}}, q_{\text{init}})$ to $(c_{\text{fin}}, q_{\text{fin}})$ (the unbroken straight line in the figure). All the adsorbent equilibrates with water containing the final dissolved adsorbate concentration (0.05 mmol/L), so the adsorption density in this system is the value of q in equilibrium with that concentration ($q = 0.97(0.05)^{0.27} = 0.43$ mmol/g).

In the two-reactor system, on the other hand, the changes caused by adsorption in the first reactor are characterized by

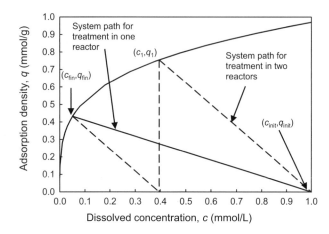

FIGURE 8-2. Graphical comparison of a one-reactor and a two-reactor system for achieving the same overall removal of contaminant by adsorption.

a line of slope $-V_L/W$, which intersects the isotherm at c_1. The treated water with concentration c_1 is then contacted with a second batch of fresh adsorbent ($q_{\text{init}} = 0$), so the initial condition in the second reactor is at the point $(c_1, 0)$. The fact that the adsorbent concentration (W/V_L) in the second reactor is identical to that in the first is reflected in the fact that the line from $(c_1, 0)$ to $(c_{\text{fin}}, q_{\text{fin}})$ has the same slope as that from $(c_{\text{init}}, 0)$ to (c_1, q_1). A trial and error approach that includes two, parallel line segments with the sawtooth pattern shown in the figure leads to the conclusion that $q_1 = 0.76$ mmol/g, and $W/V_L = 0.80$ g/L in each reactor. These results are, of course, identical to those obtained using the analytical approach. ∎

The preceding example indicates that less adsorbent is required in a system with two reactors in series than in a single reactor, for the same initial and final adsorbate concentrations. The reason for this difference is that, whereas the adsorbate concentration and adsorption density in the downstream reactor of the reactors-in-series system are identical to the corresponding parameters in the one-reactor system, the concentration and adsorption density in the upstream reactor are larger. That is, in the two-reactor system, the adsorption density on one-half of the adsorbent is identical to that in the one-reactor system, and the adsorption density on the other half of the adsorbent is larger.

The preceding explanation is essentially the same as the explanation for why a given treatment objective can be achieved more efficiently (i.e., with less reactor volume and a lower hydraulic residence time) in two CFSTRs in series than in a single CFSTR (for a reaction whose rate increases with concentration). In such a system, the reaction rate in the first reactor is greater than in the second reactor (or in a single reactor that achieves the same treatment goal), because the reactant concentration is larger in the first reactor.

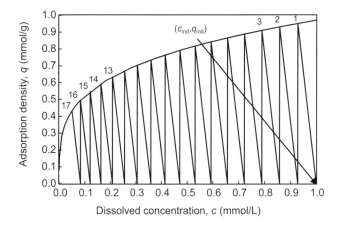

FIGURE 8-3. System path (zigzag line) describing equilibrium conditions in a series of 17 batch adsorption reactors. The isotherm curve and initial and final concentrations are the same as in Example 8-2.

Taken to its limit, the latter argument led us to conclude that, for a kinetically limited reaction, the most efficient type of reactor is a PFR (which can be viewed as an infinite number of differentially small CFSTRs in series). By analogy, we might expect that the most efficient adsorptive reactor would consist of a very large number of differentially small batch adsorption reactors in series (assuming that, as in the example, the adsorbent remains in each reactor as the water flows from one to the next). Such a system takes advantage of the largest available driving force for adsorption at each stage of treatment and thereby minimizes the required overall adsorbent dose. The conditions in a system that approaches this limit, for the influent and effluent described in Example 8-2, are shown in Figure 8-3. The total requirement for activated carbon in this system, consisting of 17 equi-sized adsorption reactors in series, would be 1.3 g/L, or approximately 60% of that required for treatment in a single batch. The benefits of using multiple reactors in series become considerably greater as the target concentration is lowered. Note that, if the adsorbent concentration in each reactor is allowed to become very large, then the limiting case becomes equivalent to a packed bed reactor.

Systems with Continuous Flow of Both Water and Adsorbent

Although the preceding discussion is presented in the context of batch systems, the analysis of systems with steady flow of both solution and adsorbent is essentially identical. A schematic of such a system is shown in Figure 8-4. In the figure, X is the mass rate at which adsorbent enters (and leaves) the reactor, with units such as mg/min, W is the mass of adsorbent in the reactor, and the other parameters have their usual meanings.

The mass balance on total adsorbate in such a system (the sum of the masses in solution and on the adsorbent) for a

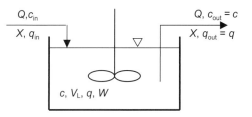

FIGURE 8-4. Schematic of a well-mixed reactor with suspended adsorbent.

nonreactive adsorbate, with a control volume defined by the boundaries of the reactor, is

$$\begin{array}{c}\text{Rate of change}\\\text{of mass of}\\\text{adsorbate stored}\\\text{in reactor}\end{array} = \begin{array}{c}\text{Net input}\\\text{(inflow} - \text{outflow)}\\\text{of dissolved}\\\text{adsorbate}\end{array} + \begin{array}{c}\text{Net input}\\\text{(inflow} - \text{outflow)}\\\text{of adsorbed}\\\text{adsorbate}\end{array}$$

$$V_L \frac{dc}{dt} + W \frac{dq}{dt} = Q(c_{in} - c_{out}) + X(q_{in} - q_{out}) \quad (8\text{-}7)$$

If the system is at steady state, the mass balance simplifies to

$$0 = Q(c_{in} - c_{out}) + X(q_{in} - q_{out}) \quad (8\text{-}8)$$

$$Q(c_{in} - c_{out}) = -X(q_{in} - q_{out}) \quad (8\text{-}9)$$

Equation 8-9 is an exact analog of Equation 8-2, with the influent and effluent concentrations in the flow-through system corresponding to the initial and final concentrations in the batch system, respectively, and the rates of entry of solution and adsorbent corresponding to the volume of solution and mass of adsorbent in the batch system. The equation applies to any control volume at steady state if c and q are interpreted as average values, and it applies to systems that reach equilibrium if q_{out} and c_{out} are assumed to be spatially uniform and related via the adsorption isotherm. Thus, the same mathematical and graphical approaches as described previously can be used to model the system. Specifically, the change from the influent to the effluent could be characterized graphically as movement from a point (c_{in}, q_{in}) that is below the isotherm curve, along a line of slope $-Q/X$, toward the isotherm, reaching the isotherm if the system reaches equilibrium. Note that nothing in the derivation specifies or constrains the mixing pattern inside the reactor. Thus, the result is valid for any arbitrary mixing pattern, including complete mixing, plug flow, or any other hydrodynamic regime.

Continuous flow systems with continuous addition of adsorbent are relatively easy to operate, and in many cases they can reduce dissolved contaminant concentrations to very low levels. Such systems are particularly convenient in situations where adsorbent is required only intermittently, for example, to remove taste- and odor-generating compounds that appear seasonally in a water supply system. However, they also have significant drawbacks, including the need to separate the adsorbent from the solution downstream of the adsorption step, and to process and, in most cases, dispose of the adsorbent sludge. In addition, such systems achieve equilibrium at a dissolved concentration equal to the low concentration in the effluent stream, so the adsorbent typically exits the system with a relatively low adsorption density and with much of its adsorptive capacity unutilized.

For situations in which the adsorbent is expensive and in which sorption is a permanent part of the treatment process, it is often more cost effective to carry out the process in a fixed bed adsorber. Such systems concentrate a large mass of adsorbent in a small volume (usually a column) through which the water passes. In many cases, once the adsorption capacity is exhausted, the adsorbent can be regenerated and reused. The dynamics of adsorption in such systems are analyzed next, and processes for regenerating various adsorbents are discussed later in this chapter.

Sequential Batch Reactors

As noted previously, one way to model a fixed bed adsorption system is as a group of small batch reactors in series. In such a model, during any given time step, the solution and adsorbent in each hypothetical reactor approach equilibrium, after which the solution moves to the next reactor, while the adsorbent remains in place. If equilibrium is attained rapidly, then Equation 8-3 can be used in conjunction with the adsorption isotherm to characterize the changes in adsorbate concentration and adsorption density in each reactor during a single equilibration step.

■ **EXAMPLE 8-3.** Consider an adsorption treatment system similar to the one described in Example 8-2, except containing three reactors in series. Each reactor has a volume of 10 L and contains 8.0 g of activated carbon, so that W/V_L is 0.8 g/L, as in that example. Initially, all the reactors contain clean water and virgin activated carbon. For the same influent as in Example 8-2, compute the concentration of phenol in each reactor and at each stage of treatment if the system is used in a way that approximates a fixed bed. Assume that 40 L of influent is treated in four sequential 10-L batches. Each batch of water equilibrates with the adsorbent in the first batch reactor and is then transferred to the second reactor, where it equilibrates with the activated carbon in that reactor. The activated carbon initially present in each reactor remains in place throughout the process. A schematic of the process is shown in Figure 8-5.

Solution. Define $c_{i,j}$ and $q_{i,j}$ as the adsorbate concentration and adsorption density in reactor i at the end of the jth step. The influent concentration of $c_{in} = 1.0$ mmol/L can be represented using this terminology as $c_{0,j}$ (the concentration in the "zeroth" reactor in all steps).

The conditions after treatment of the first batch of water are those determined in Example 8-2; that is, the dissolved

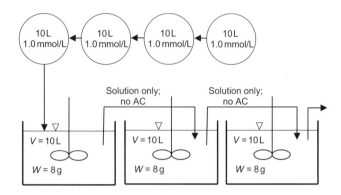

FIGURE 8-5. Schematic illustration of a hypothetical process in which sequential batches of water are treated in three well-mixed reactors in series, with the solution but not the adsorbent is transferred after each treatment step.

phenol concentrations in the first reactor at the end of the first step is 0.40 mmol/L, and the adsorption density in that reactor is 0.76 mmol/g. The concentration and adsorption density in the second reactor at the end of the first step are both zero, since no contaminant has reached that reactor yet. However, at the end of the second step, the conditions in the second reactor are the same as those computed in the second reactor in Example 8-2; that is, $c_{2,2} = 0.05$ mmol/L and $q_{2,2} = 0.43$ mmol/g.

When this same batch of water reaches the third reactor (at the beginning of the third step), it encounters more fresh adsorbent. At the end of that step, the solution and AC have equilibrated, leading to the following set of conditions:

$$\frac{W}{V_L} = -\frac{c_{2,2} - c_{3,3}}{q_{3,2} - q_{eq}|_{c_{3,3}}}$$

$$0.80 = -\frac{0.05 - c_{3,3}}{0.00 - 0.97\, c_{3,3}^{0.27}}$$

$$c_{3,3} = 3.87 \times 10^{-5}\, \text{mmol/L}$$

$$q_{3,3} = 0.97(3.87 \times 10^{-5})^{0.27} = 6.25 \times 10^{-2}\, \text{mmol/g}$$

When the second batch of influent enters the first reactor, it mixes with activated carbon that has already equilibrated with a previous batch of influent and therefore has some phenol adsorbed (specifically, $q_{1,1} = 0.76$ mmol/g). Nevertheless, we can apply Equation 8-3 to determine the conditions after the solution and adsorbent equilibrate, yielding:

$$\frac{W}{V_L} = -\frac{c_{0,1} - c_{1,2}}{q_{1,1} - q_{eq}|_{c_{1,2}}}$$

$$0.80 = -\frac{1.0 - c_{1,2}}{0.76 - 0.97\, c_{1,2}^{0.27}}$$

$$c_{1,2} = 0.86\, \text{mmol/L}$$

$$q_{1,2} = 0.97(0.86)^{0.27} = 0.93\, \text{mmol/g}$$

The solution with concentration $c_{1,2}$ enters the second reactor at the beginning of the third step. The adsorption density on the activated carbon in that reactor at that time is $q_{2,1}$. The conditions after equilibrium is reached can therefore be calculated as follows:

$$\frac{W}{V_L} = -\frac{c_{1,2} - c_{2,3}}{q_{2,2} - q_{eq}|_{c_{2,3}}}$$

$$0.80 = -\frac{0.86 - c_{2,3}}{0.43 - 0.97\, c_{2,3}^{0.27}}$$

$$c_{2,3} = 0.55\,\text{mmol/L}$$

$$q_{2,3} = 0.97\,(0.55)^{0.27} = 0.82\,\text{mmol/g}$$

Carrying out similar calculations until all four 10-L batches have passed through all three reactors yields the following results:

Step, j	$c_{1,j}$ (mmol/L)	$q_{1,j}$ (mmol/g)	$c_{2,j}$ (mmol/L)	$q_{2,j}$ (mmol/g)	$c_{3,j}$ (mmol/L)	$q_{3,j}$ (mmol/g)
0	0	0	0	0	0	0
1	0.40	0.76	0	0	0	0
2	0.86	0.93	0.05	0.43	0	0
3	0.97	0.96	0.55	0.82	3.9×10^{-5}	6.2×10^{-2}
4	1.00	0.97	0.98	0.94	0.14	0.57
5	–	–	1.00	0.96	0.73	0.89
6	–	–	–	–	0.95	0.96

The results are plotted in Figure 8-6. Figure 8-6a and 8-6b demonstrate the changes in the dissolved phenol concentration and the adsorption density, respectively, at given locations in the system as a function of time (i.e., step). Figure 8-6c, on the other hand, shows the changes experienced by a given batch of water as it moves through the system. In Figure 8-6c, the conditions at the bottom of each vertical line segment apply when a batch of water enters a reactor, and those at the intersections of the sloped line segments with the isotherm curve apply after the water has equilibrated with the activated carbon in the reactor. To minimize clutter, only the first two batches of water are shown.

The activated carbon in the first reactor is nearly equilibrated with the influent solution after the third batch of influent has contacted it (as indicated by the values for $c_{1,3}$ and $q_{1,3}$ in the preceding table), so almost no phenol is removed in that reactor from the fourth batch of water treated. Once three batches of water have passed through all three reactors (end of step 5), the adsorbent in the second and third reactors is also nearly equilibrated with the influent, so the overall treatment efficiency for the fourth batch of water is poor ($c_{3,6}$ is 0.95 mmol/L, corresponding to only 5% phenol removal). ■

Fixed Bed Adsorption Systems

Qualitative Description If the analysis of batch adsorption reactors in series is extended to the limit of an infinite number of differentially small reactors, it characterizes an ideal plug flow, fixed bed reactor. We analyze such systems

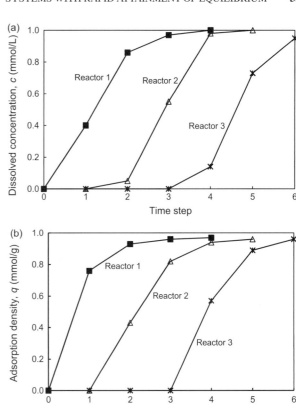
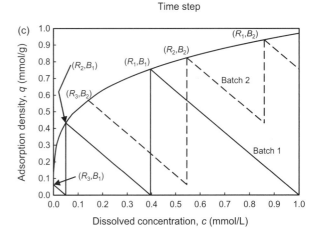

FIGURE 8-6. Adsorptive treatment of a solution in three sequential reactors containing adsorbent that is retained in each reactor. (a,b) Dissolved concentration, c and adsorption density, q in each reactor after equilibrating with successive batches of influent. (Lines are shown for clarity, but the data points are actually discontinuous.) (c) Changes in the conditions encountered by a given batch of water as it moves through the system. The point (R_i, B_j) characterizes the conditions in reactor i after batch j has reached equilibrium in that reactor.

next. However, before proceeding with a mathematical analysis of those systems, we consider the expected behavior qualitatively, to provide a conceptual framework into which the mathematical analysis can fit.

Consider a solution containing a single adsorbate that is being fed to a fixed bed of fresh adsorbent at a steady rate. Initially, most of the adsorbate binds near the top of the column, and the downstream sections are not exposed to any significant concentration of the adsorbate. Within the section where a significant amount of adsorption is occurring, each thin layer of the bed contains a lower adsorbate concentration and a lower adsorption density than the layer immediately upstream, much as is the case in the sequential batch reactors discussed in the preceding section.

Over time, the adsorbent in the upstream section accumulates enough adsorbate to equilibrate with the influent composition, after which it collects no additional adsorbate. It is important to recognize that this condition is reached when the adsorption density is the value of q that is in equilibrium with c_{in}, not the maximum value of q that could be achieved if the adsorbent were equilibrated with an arbitrarily high value of c. For example, if the system is characterized by a Langmuir isotherm, the maximum value of q in the column will correspond to $q_{max}K_{ads}c_{in}/(1+K_{ads}c_{in})$, not q_{max}. We designate this maximum attainable adsorption density as q_{in}^*.

Once the adsorbent nearest the inlet has equilibrated with the influent, the influent solution passes through that portion of the bed unaltered, and no adsorption occurs until it reaches a location at which $q < q_{in}^*$. Downstream of that point, the conditions are virtually identical to those that existed in the upstream section before it became saturated with adsorbate. As a result, a snapshot of the bed at that time would show the upstream portion characterized by c_{in} and q_{in}^*, a section immediately downstream with steadily decreasing c and q, and a section farther downstream where both c and q are near zero. This scenario is shown schematically in Figure 8-7 for several different times after treatment is initiated.

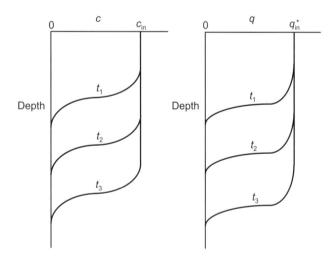

FIGURE 8-7. Qualitative patterns of adsorbate concentration and adsorption density in a fixed bed adsorption column at three different times. Formally, the vertical axis is the distance from the influent point; in the diagram, the flow is assumed to be downward, but essentially the same diagram would apply if the flow were upward or horizontal.

The portion of the bed in which the dissolved adsorbate concentration and the adsorption density are changing dramatically is commonly referred to as the *active zone* of the reactor or the *mass transfer zone* (MTZ). The shape of the MTZ is determined by the isotherm equation, the axial dispersion in the reactor, and, in cases where adsorptive equilibrium is not reached instantaneously, the kinetics of adsorption. That shape is derived mathematically in the following sections. For now, the important point is that, as the run proceeds, the MTZ moves through the bed maintaining a shape (once it is fully developed) that is approximately invariant.

Eventually, the leading edge of the MTZ reaches the outlet, and the concentration of adsorbate in the effluent begins to increase noticeably. If the run continues, the effluent concentration eventually increases to the influent value, at which point the bed is no longer achieving any removal at all. The period when the effluent concentration is rising is called *breakthrough*, and a plot of effluent concentration versus either time or volume of water processed is called a *breakthrough curve*.

As an alternative to using absolute concentration to characterize breakthrough, the effluent concentration is often expressed as the fractional breakthrough, c_{out}/c_{in}. The volume of water treated is also frequently expressed in nondimensional terms by normalizing it to either the total volume of the packed section of the reactor or the void volume in that space. The total volume of the packed section is commonly referred to as the *bed volume* (BV), and the cumulative volume of water treated divided by the BV is referred to as the *number of BVs treated*. In some cases, the cumulative volume of water treated is normalized by the void volume in the bed, rather than the total BV, in which case the ratio is referred to as the *number of void volumes* or the *number of pore volumes treated*. Formally, for a constant flow Q, these terms are computed as

Number of bed volumes treated

$$= \frac{\text{Cumulative volume treated}}{\text{Total reactor volume}} = \frac{Qt}{V_{reactor}} \quad (8\text{-}10)$$

Number of void volumes treated

$$= \frac{\text{Cumulative volume treated}}{\text{Reactor void volume}} = \frac{Qt}{\varepsilon V_{reactor}} \quad (8\text{-}11)$$

Simulated breakthrough curves for two systems (one with a short MTZ, and one with a long MTZ), with the abscissa and ordinate expressed in various ways, are shown in Figure 8-8.

When a breakthrough curve is shown as concentration versus volume treated, the area under the curve is the cumulative mass of adsorbate that has exited the bed, and the area above the curve and below c_{in} is the cumulative mass of adsorbate that has been removed from the influent. These areas are shown in Figure 8-8 at run times

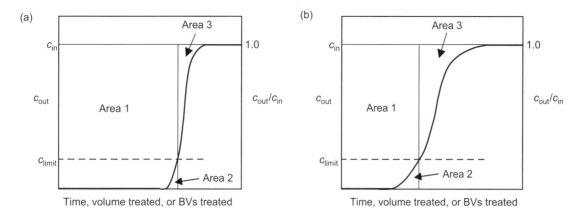

FIGURE 8-8. Plausible adsorption breakthrough curves for a system with (a) a short MTZ and (b) a long MTZ. The vertical lines indicate the conditions at 20% breakthrough.

corresponding to 20% breakthrough: Area 1 indicates the mass of adsorbate removed up to that point, and Area 2 indicates the mass of adsorbate that has exited.

Assume that the systems characterized in Figures 8-8a and b contain identical masses of the same adsorbent, but that the adsorbent particles in Figure 8-8b are larger than in Figure 8-8a. Assume also that the majority of the adsorbent surface area is in internal pores, so that the total surface area is the same in both beds despite the difference in particle sizes. Finally, assume that the two beds receive the same influent at the same loading rate. In such a case, over the course of a run that continued until 100% breakthrough, the systems would adsorb the same mass of contaminant. Nevertheless, comparison of the abscissas in the two figures at the point where 20% breakthrough occurs makes clear that the system with the steeper breakthrough curve performs better (i.e., treats more water and removes more contaminant) up to that point. Correspondingly, it has much less remaining adsorption capacity (Area 3), for the given influent composition, than the system with earlier and more gradual breakthrough.

Usually, the maximum allowable contaminant concentration in the effluent is well below the influent value, so that the run must be terminated before complete breakthrough occurs. In such cases, a sharp breakthrough curve (ideally, a square-wave) is preferred, because such a curve allows the largest possible volume of water to be processed and the greatest amount of adsorbate to be removed before the run is terminated. On the other hand, if the breakthrough curve is very steep, frequent and accurate monitoring of the effluent quality (and perhaps of the water quality slightly upstream) is needed to assure that a high concentration of the contaminant is not allowed to exit the column before the run is terminated. In the latter sections of this chapter, some approaches are presented for partially overcoming the problem of short runs caused by gradual breakthrough curves.

The Mass Balance on a Fixed Bed Reactor with Rapid Equilibration Having established some qualitative expectations for the behavior of fixed bed adsorbers, we now turn to the mathematical description of such systems. The definition diagram for such a system is shown in Figure 8-9.

The mass balance on the adsorbate in the liquid phase in a control volume bounded by the cross-section of the column and a length dz in the direction of flow is as follows:

Rate of change of mass of dissolved adsorbate stored in section	=	Net input of dissolved adsorbate into section via advection	+	Net input of dissolved adsorbate into section via diffusion and dispersion	−	Net rate of loss of dissolved adsorbate via adsorption	+	Net rate of loss of adsorbate by chemical reaction in solution
$\varepsilon A_R dz \dfrac{\partial c}{\partial t}$	=	$-Q\dfrac{\partial c}{\partial z}dz$	+	$\mathbf{D}_L \varepsilon A_R \dfrac{\partial^2 c}{\partial z^2}dz$	−	$dW\dfrac{\partial q}{\partial t}$	+	$\varepsilon A_R (dz) r_L$

$$(8\text{-}12)$$

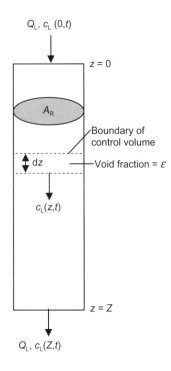

FIGURE 8-9. Definition diagram for a mass balance on the adsorbate in a control region consisting of the liquid in a volume defined by the cross-section of the column and height dz.

where ε is the void fraction in the packed section of the bed, A_R is the cross-sectional area of the reactor, $\mathbf{D}_L$ is the sum of the diffusion and dispersion coefficients for the adsorbate, and dW is the mass of adsorbent in the (differential) control volume. (Note that ε quantifies only the void space between the adsorbent particles, not the internal porosity of those particles.) Dividing through by $\varepsilon A_R dz$, and noting that Q/A_R is the average approach velocity v of the water upstream of the packed section (sometimes called the *superficial velocity*) yields the following expression for the volume-normalized mass balance:

$$\frac{\partial c}{\partial t} = -\frac{v}{\varepsilon}\frac{\partial c}{\partial z} + \mathbf{D}_L \frac{\partial^2 c}{\partial z^2} - \frac{1}{\varepsilon}\frac{dW}{A_R dz}\frac{\partial q}{\partial t} + r_L \quad (8\text{-}13)$$

The product $A_R dz$ equals the volume of the control space, dV_R, so the ratio $dW/A_R dz$ is the mass of adsorbent per unit volume of the control section. This ratio is the same throughout the reactor and is called the *packing density*, the *bulk density*, or the *bed density*, ρ_b.[1] Note that the

[1] This definition, in effect, simply gives a different name and symbol to the parameter that would be identified as the adsorbent concentration, c_{ads}, in batch adsorption systems. The subscript *b* is usually interpreted as referring to the bulk adsorbent; that is, the mass of a representative quantity of the adsorbent per unit volume it occupies, including the interstitial spaces between the particles. However, the bulk density of the adsorbent depends on how it is packed; for example, the bulk density as shipped might differ from that in a fixed bed. Therefore, we associate the subscript b with the word *bed*, thereby linking the parameter ρ_b directly to the physical system, which it is describing.

packing density is not the density of the adsorbent particles themselves, but rather the mass of adsorbent per L of reactor, including the volume of the interparticle void spaces. The density of the individual adsorbent particles, with air in any pore spaces, is commonly referred to as the *apparent particle density*, $\rho_{p,app}$. These two density-related parameters are related by

$$\rho_b = \rho_{p,app}(1 - \varepsilon) \quad (8\text{-}14)$$

Although both ρ_b and $\rho_{p,app}$ are encountered in the literature, ρ_b is used much more commonly in engineering practice.

Substituting for $dW/A_R dz$ in terms of ρ_b or $\rho_{p,app}$, we can write Equation 8-13 in either of the following two equivalent forms:

$$\frac{\partial c}{\partial t} = -\frac{v}{\varepsilon}\frac{\partial c}{\partial z} + \mathbf{D}_L \frac{\partial^2 c}{\partial z^2} - \frac{\rho_b}{\varepsilon}\frac{\partial q}{\partial t} + r_L \quad (8\text{-}15a)$$

$$\frac{\partial c}{\partial t} = -\frac{v}{\varepsilon}\frac{\partial c}{\partial z} + \mathbf{D}_L \frac{\partial^2 c}{\partial z^2} - \frac{\rho_{p,app}(1-\varepsilon)}{\varepsilon}\frac{\partial q}{\partial t} + r_L \quad (8\text{-}15b)$$

Systems with Rapid Equilibration and Plug Flow If the adsorbate is nonreactive in solution and if dispersion is negligible (i.e., if an assumption of plug flow is acceptable), Equations 8-15a and 8-15b can be simplified as follows:

$$\frac{\partial c}{\partial t} = -\frac{v}{\varepsilon}\frac{\partial c}{\partial z} - \frac{\rho_b}{\varepsilon}\frac{\partial q}{\partial t} \quad (8\text{-}16a)$$

$$\frac{\partial c}{\partial t} = -\frac{v}{\varepsilon}\frac{\partial c}{\partial z} - \frac{\rho_{p,app}(1-\varepsilon)}{\varepsilon}\frac{\partial q}{\partial t} \quad (8\text{-}16b)$$

In words, Equations 8-16a and 8-16b state that the rate of change of the dissolved adsorbate concentration in the control volume (the term on the left) equals the net rate of advection of dissolved adsorbate into that volume (the first term on the right) minus the rate at which adsorbate is removed from solution by adsorption (the second term on the right). In systems that reach equilibrium, q and c at each value of z are related by the adsorption isotherm (though different locations in the reactor might be characterized by different points on the isotherm).

Consider, for example, a system in which the adsorbate is nonreactive and the hydraulics conform to ideal plug flow, so that Equation 8-16a applies. Assume also that adsorptive equilibrium is reached rapidly. In that case, it is convenient to write $\partial q/\partial t$ as $(\partial q \partial c)/(\partial c \partial t)$ and to recognize that, since the system is always at adsorptive equilibrium, $\partial q/\partial c$ is the slope of the isotherm curve. Writing that slope as $(\partial q/\partial c)_{eq}$

and carrying out some algebra, we can manipulate Equation 8-16a as follows:

$$\frac{\partial c}{\partial t} = -\frac{v}{\varepsilon}\frac{\partial c}{\partial z} - \frac{\rho_b}{\varepsilon}\left(\frac{\partial q}{\partial c}\right)_{eq}\frac{\partial c}{\partial t}$$

$$\times \left(1 + \frac{\rho_b}{\varepsilon}\left(\frac{\partial q}{\partial c}\right)_{eq}\right)\frac{\partial c}{\partial t} = -\frac{v}{\varepsilon}\frac{\partial c}{\partial z} \quad (8\text{-}17)$$

$$\frac{\partial c}{\partial t} = -\frac{v}{\varepsilon + \rho_b(\partial q/\partial c)_{eq}}\frac{\partial c}{\partial z}$$

Equation 8-17 is a key result, expressing the rate of change of the concentration of dissolved adsorbate at a given location as a function of known system parameters and the gradient of the concentration at that location, for a system that meets the assumptions of the derivation (adsorptive equilibrium everywhere in the bed, nonreactive adsorbate, and plug flow). When this equation is solved in conjunction with appropriate initial and boundary conditions, the concentration versus time profile at any height in the column can be determined. Typically, these conditions would be that the adsorbate concentration is known and constant in the influent[2] and that it is zero throughout the column at time zero; that is,

$$c|_{z=0} = c_{in} \quad \text{for all } t \quad (8\text{-}18a)$$

$$c|_{t=0} = 0 \quad \text{for all } z \quad (8\text{-}18b)$$

Note that, since the system is assumed to be always in equilibrium at all heights z, the assumption that $c|_{t=0} = 0$ implies that $q|_{t=0} = 0$ as well; that is, the initial condition is that the bed is packed with fresh adsorbent.

■ **EXAMPLE 8-4.** The treatment objective described in Example 8-1 is to be accomplished by packing the activated carbon in a column through which the water containing 1.0 mmol/L phenol will be passed. The depth of the packed layer is 2.0 m, the apparent density of the adsorbent particles ($\rho_{p,app}$) is 1.4 g/cm^3, the void fraction ε is 0.40, and the superficial velocity of the feed is 9.0 m/h (15 cm/min). Assume that the solution and adsorbent equilibrate with one another almost instantly, and that the hydraulics approximate plug flow.

(a) How long can the treatment process operate before the effluent concentration exceeds the treatment target of 0.05 mmol/L? How long could it operate before 50% or 95% breakthrough occurred?

[2] This boundary condition is based on the assumption that transport into the reactor by diffusion and dispersion is negligible compared to that by advection. If this assumption does not hold (as might be the case when analyzing adsorption in slowly moving groundwater), the boundary condition is usually modified to indicate that the total flux into the system equals the sum of the advective and diffusive/dispersive fluxes: $vc_{in} = vc(0,t) - \mathbf{D}_L(\partial c(0,t)/\partial z)$.

(b) What is the average effluent concentration up to the time of 50% breakthrough identified in part (a)?

Solution.

(a) The system meets the requirements for Equation 8-17 to apply, along with the initial and boundary conditions specified in Equations 8-18a and 8-18b. Furthermore, values of v and ε are given, so the only additional term we need to specify before the equations can be solved is the product $\rho_b(\partial q/\partial c)_{eq}$. To minimize clutter, units are not shown in most of the following calculations; in all cases, the applicable units are g/cm^3 for ρ, mmol/L for c, and mmol/g for q.

The value of ρ_b can be computed as $(1-\varepsilon)\rho_{p,app}$, equal in this case to $(0.60)(1.4)$, or 0.84 g/cm^3. The isotherm equation is $q = 0.97\,c^{0.27}$, so the slope of the isotherm curve is given by

$$\left(\frac{\partial q}{\partial c}\right)_{eq} = (0.97)(0.27)c^{-0.73} = 0.26c^{-0.73}$$

The known or derived values can then be substituted into Equation 8-17 to obtain

$$\frac{\partial c}{\partial t} = -\frac{v}{\varepsilon + \rho_b(\partial q/\partial c)_{eq}}\frac{\partial c}{\partial z}$$

$$\frac{\partial c}{\partial t} = -\frac{0.15\,\text{m/min}}{0.40 + (0.84)(0.26c^{-0.73})(1000)}\frac{\partial c}{\partial z}$$

where the value of 1000 in the denominator is a unit conversion factor (from L to cm^3). This equation can be integrated in conjunction with the initial condition that the adsorbent contains no phenol at $t = 0$ and the boundary condition that the phenol concentration at the top of the column (i.e., in the influent) is always 1.0 mmol/L.

The integration has been carried out numerically to yield the results shown in Figure 8-10, which indicates that the breakthrough curve is virtually a step function. In the simulation shown, the column was divided into 100 segments, and the times corresponding to 5, 50, and 95% breakthrough (i.e., c_{out} values of 0.05, 0.50, and 0.95 mmol/L) were all within a 7-min span between 180 and 181 h.

(b) The breakthrough curve is very steep, with the effluent during the vast majority of the run containing essentially no phenol. The fraction of the total influent phenol that has appeared in the effluent during the run could, in theory, be computed as the ratio of the area under the effluent curve to the area under the influent curve, which is simply a horizontal line at $c = 1.0$ mmol/L. However, for the given results, the calculation need not be carried out, since the resulting

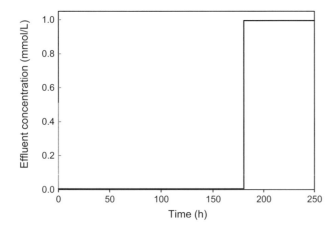

FIGURE 8-10. Simulation of phenol breakthrough in a model fixed bed adsorber.

value is essentially zero. In other words, the effluent during virtually the entire run is phenol-free. ∎

By manipulating Equation 8-17 to eliminate references to the system-specific details of geometry, velocity, and time, we can convert it into a universal expression for the behavior of fixed bed adsorption systems. To accomplish this revision, we rewrite Equation 8-17 in terms of the number of BVs of water treated, rather than time, as follows. We first define the *empty bed contact time* (EBCT) as the hydraulic residence time of the empty reactor; that is, the time that would be required for one BV of influent to pass through the empty (unpacked) reactor:

$$\text{EBCT} = \tau_{\text{empty bed}} = \frac{V_R}{Q} \tag{8-19}$$

Then, representing the number of empty BVs treated from time 0 to time t as the variable N_{BV} yields:

$$N_{BV} \equiv \frac{\text{Volume treated}}{\text{Volume of bed}} = \frac{tQ}{V_R} = \frac{t}{\tau_{\text{empty bed}}} \tag{8-20}$$

Substituting from Equation 8-20 into Equation 8-17 and rearranging, we obtain

$$\frac{\partial c}{\partial (N_{BV}(V_R/Q))} = -\frac{v}{\varepsilon + \rho_b (\partial q/\partial c)_{eq}} \frac{\partial c}{\partial z} \tag{8-21}$$

$$\frac{\partial c}{\partial N_{BV}} = -\frac{v(V_R/Q)}{\varepsilon + \rho_b (\partial q/\partial c)_{eq}} \frac{\partial c}{\partial z} \tag{8-22}$$

The approach velocity v divided by the flow rate Q is $1/A_R$. Therefore, vV_R/Q equals V_R/A_R, which equals the height Z of the packed bed. Defining a dimensionless parameter $\tilde{z}$ as z/Z (i.e., the distance from the influent end of the packing as a fraction of the total length of the packed section), Equation 8-22 can be rewritten as follows:

$$\frac{\partial c}{\partial N_{BV}} = -\frac{Z}{\varepsilon + \rho_b (\partial q/\partial c)_{eq}} \frac{\partial c}{\partial z}$$

$$= -\frac{1}{\varepsilon + \rho_b (\partial q/\partial c)_{eq}} \frac{\partial c}{\partial (z/Z)}$$

$$= -\frac{1}{\varepsilon + \rho_b (\partial q/\partial c)_{eq}} \frac{\partial c}{\partial \tilde{z}} \tag{8-23}$$

Finally, dividing both sides of Equation 8-23 by c_{in} and defining the fractional breakthrough of adsorbate (c/c_{in}) as $\tilde{c}$, we obtain

$$\frac{\partial \tilde{c}}{\partial N_{BV}} = -\frac{1}{\varepsilon + \rho_b (\partial q/\partial c)_{eq}} \frac{\partial \tilde{c}}{\partial \tilde{z}} \tag{8-24}$$

Equation 8-24 describes the fractional breakthrough at any location in the reactor as a function of the number of BVs treated, for a system that reaches equilibrium quickly and is characterized by plug flow. The parameters in the equation relate to the physical properties of the packed adsorbent (its mass per unit volume of reactor, and the void fraction) and the isotherm. Thus, assuming that the packing density is fairly consistent for a given adsorbent from one reactor to the next, a single curve showing $\tilde{c}$ versus $\tilde{t}$ at $\tilde{z} = 1$ could describe the breakthrough of a particular adsorbate in any plug-flow reactor packed with that adsorbent.

∎ **EXAMPLE 8-5.**

(a) Compute the number of empty BVs of water treated corresponding to 5% fractional breakthrough in the system analyzed in Example 8-4.

(b) Compare the activated carbon requirement per liter of water treated in the fixed bed system with that in the batch system in Example 8-1.

Solution.

(a) The empty bed detention time in the reactor can be computed by manipulating Equation 8-19 as follows, where A is the cross-sectional area of the bed:

$$\tau_{\text{empty bed}} = \frac{V_R}{Q} = \frac{V_R/A}{Q/A} = \frac{Z}{v} = \frac{2.0 \text{ m}}{9.0 \text{ m/h}} = 0.222 \text{ h}$$

The time of treatment can then be converted into an expression for the number of BVs treated (N_{BV}) using Equation 8-20. The time corresponding to 5% breakthrough is 180.7 h, so the number of BVs treated up to that point is

$$N_{BV,5\%} = \frac{t_{5\%}}{\tau_{\text{empty bed}}} = \frac{180.7 \text{ h}}{0.222 \text{ h}} = 813 \text{ BV}$$

(b) The fact that the run lasts 813 BVs before the target concentration is exceeded means that 813 L of water are treated per liter of the fixed bed. The bulk, packed density in the bed, ρ_b, was computed in Example 8-4 as 840 g/L. Therefore, the activated carbon requirement is

$$\frac{840\,\text{g AC/L of bed}}{813\,\text{L of influent/L of bed}} = 1.03\,\text{g AC/L of influent}$$

The corresponding value for the batch treatment system is 2.2 g activated carbon per L of influent. Thus, the adsorbent requirement is more than twice as large in the batch system as in the column system. This result is a direct consequence of the fact that, in the batch system, all the adsorbent is in equilibrium with $c = 0.05$ mmol/L, whereas in the fixed bed, most of the adsorbent is in equilibrium with 1.0 mmol/L (i.e., c_{in}) when c_{out} is 0.05 mmol/L. ∎

Although the extremely sharp breakthrough curve in the preceding examples is developed only for a specific set of operating conditions, the same qualitative result is obtained for virtually all analyses of fixed bed reactors that meet the conditions of rapid equilibration and plug flow. The reason for this outcome can be understood most easily if we modify the coefficient on the right side of Equation 8-24 as follows:

$$\frac{1}{\varepsilon + \rho_b(\partial q/\partial c)_{eq}} = \frac{1}{\varepsilon + \varepsilon(\partial(\rho_b q)/\partial(\varepsilon c))_{eq}}$$
$$= \frac{1}{\varepsilon\left[1 + (\partial(\rho_b q)/\partial(\varepsilon c))_{eq}\right]} \quad (8\text{-}25)$$

The product $\rho_b q$ is the mass of adsorbed adsorbate per unit volume of reactor, and εc is the mass of dissolved adsorbate per unit volume of reactor. Therefore, the term $(\partial(\rho_b q)/\partial(\varepsilon c))_{eq}$ represents the increase in the mass of adsorbed contaminant in a layer of the reactor per unit increase in the mass of dissolved contaminant in that same layer, if the adsorbent and solution reach equilibrium. Adsorption reactors are, of course, designed using adsorbents that have a large capacity to collect the target contaminant(s); if they did not have such a capacity, there would be little point in utilizing them. As a result, the ratio $(\partial(\rho_b q)/\partial(\varepsilon c))_{eq}$ is invariably in the range of at least a few hundred, and is often on the order of many thousands to hundreds of thousands.

Values of ε are typically 0.3–0.5, so the whole fraction shown in Equation 8-25 is typically on the order of 10^{-2} to 10^{-5}. As a result, the defining equation for the breakthrough curve (Equation 8-24) for most adsorbates in engineered, fixed bed adsorption reactors can be written as follows:

$$\frac{\partial \tilde{c}}{\partial N_{BV}} = -\left(10^{-2}\,\text{to}\,10^{-5}\right)\frac{\partial \tilde{c}}{\partial \tilde{z}} \quad (8\text{-}26)$$

Thus, although the exact value of the coefficient on the right side of Equation 8-26 depends on the details of the isotherm and the adsorbent packing in the bed, it is $\ll 1$ for all realistic engineered adsorption systems. This result leads to a virtually identical (nearly square-wave) predicted breakthrough profile for any adsorbate that reaches equilibrium in a system with plug flow.

Because of the sharpness of the breakthrough profile, essentially no adsorbate appears in the effluent until the mass input of adsorbate is sufficient to bring the adsorption density on all the adsorbent to the value that is in equilibrium with c_{in}; that is, to q_{in}^*. Thus, for example, if the bed contains a mass of adsorbent, M_{ads}, the mass of adsorbate bound to the solid at breakthrough will be $q_{in}^* M_{ads}$, and breakthrough will occur when that mass of adsorbate has been supplied in the influent. Noting that the cumulative volume input can be expressed as the product of the reactor volume and the number of BVs fed, expressions for the mass supplied and mass adsorbed at breakthrough can be written as follows:

Mass adsorbed at breakthrough $= M_{ads} q_{in}^* = \rho_b V_R q_{in}^*$

Mass supplied at breakthrough $= V_R N_{BV,bt} c_{in}$

where $N_{BV,bt}$ is the number of BVs treated at breakthrough. Equating the above two expressions, we find

$$\rho_b V_R q_{in}^* = V_R N_{BV,bt} c_{in} \quad (8\text{-}27)$$

$$N_{BV,bt} = \frac{\rho_b V_R q_{in}^*}{V_R c_{in}} = \frac{\rho_b q_{in}^*}{c_{in}} \quad (8\text{-}28)$$

The time to breakthrough can then be computed as the product of $N_{BV,bt}$ and the EBCT.

∎ **EXAMPLE 8-6.** Estimate the time and number of BVs treated at breakthrough for the system described in Example 8-5, assuming that the breakthrough curve can be approximated as a square wave.

Solution. If the breakthrough curve is a square wave, we can apply Equation 8-28. Substituting in the known information, we find

$$N_{BV,bt} = \frac{\rho_b q_{in}^*}{c_{in}}$$

$$= \frac{(840\,\text{g AC/L})\left(0.97(\text{mmol/g AC})/(\text{mmol/L})^{0.27}\right)c_{in}^{0.27}}{c_{in}}$$

$$= \frac{815(\text{mmol/L})^{0.73}}{c_{in}^{0.73}} = \frac{815(\text{mmol/L})^{0.73}}{(1.0\,\text{mmol/L})^{0.73}} = 815$$

The time to breakthrough is

$$t_{bt} = N_{BV,bt}(\text{EBCT}) = 815(0.222\,\text{h}) = 181\,\text{h}$$

TABLE 8-1. Summary of Some Key Equations Describing Behavior of Fixed Bed Adsorptive Systems

Equations applicable regardless of whether adsorptive equilibrium is reached[a]

Volume-normalized mass balance on adsorbate for a fixed bed	$\dfrac{\partial c}{\partial t} = -\dfrac{v}{\varepsilon}\dfrac{\partial c}{\partial z} + \mathbf{D}_L \dfrac{\partial^2 c}{\partial z^2} - \dfrac{\rho_b}{\varepsilon}\dfrac{\partial q}{\partial t} + r_L$	(8-15a)
	$\dfrac{\partial c}{\partial t} = -\dfrac{v}{\varepsilon}\dfrac{\partial c}{\partial z} + \mathbf{D}_L \dfrac{\partial^2 c}{\partial z^2} - \dfrac{\rho_{p,\text{app}}(1-\varepsilon)}{\varepsilon}\dfrac{\partial q}{\partial t} + r_L$	(8-15b)
Volume-normalized mass balance on adsorbate in a fixed bed with plug flow and no chemical reaction	$\dfrac{\partial c}{\partial t} = -\dfrac{v}{\varepsilon}\dfrac{\partial c}{\partial z} - \dfrac{\rho_b}{\varepsilon}\dfrac{\partial q}{\partial t}$	(8-16a)
	$\dfrac{\partial c}{\partial t} = -\dfrac{v}{\varepsilon}\dfrac{\partial c}{\partial z} - \dfrac{\rho_{p,\text{app}}(1-\varepsilon)}{\varepsilon}\dfrac{\partial q}{\partial t}$	(8-16b)

Equations applicable to systems in which adsorptive equilibrium is reached

Volume-normalized mass balance on adsorbate in a fixed bed with plug flow and no chemical reaction	$\dfrac{\partial c}{\partial t} = -\dfrac{v}{\varepsilon + \rho_b(\partial q/\partial c)_{\text{eq}}}\dfrac{\partial c}{\partial z}$	(8-17)
Typical boundary condition	$c\|_{z=0} = c_{\text{in}}\quad \text{for all } t$	(8-18a)
Typical initial condition	$c\|_{t=0} = 0\quad \text{for all } z$	(8-18b)
Mass balance for a fixed bed with plug flow and no chemical reaction, normalized to eliminate several system-specific parameters	$\dfrac{\partial \tilde{c}}{\partial N_{\text{BV}}} = -\dfrac{1}{\varepsilon + \rho_b(\partial q/\partial c)_{\text{eq}}}\dfrac{\partial \tilde{c}}{\partial \tilde{z}}$	(8-24)

[a] If the system does not reach equilibrium, q is the average value of adsorption density in the control volume.

The result is essentially identical to that found in Example 8-5, in which a more sophisticated (and complicated) analysis was carried out. ■

The key equations derived in this section for adsorption in fixed beds with plug flow and rapid equilibration of the solution with the adsorbent are collected in Table 8-1. The assumption of nearly plug flow hydraulics can be validated using tracer tests, and it is almost always found to be acceptable. Therefore, any deviations from the behavior described above are likely to be caused by failure of the system to reach equilibrium rapidly. In the following section, we consider how the equations can be modified to account for such a possibility, and how those modifications alter our predictions for the system behavior.

8.3 SYSTEMS WITH A SLOW APPROACH TO EQUILIBRIUM

To this point, we have considered only systems in which equilibrium between an adsorbent and the solution surrounding it (i.e., in the control volume) is reached instantaneously. For such a scenario to apply, the resistance to transport of adsorbate throughout the control volume must be negligible; that is, an adsorbate molecule must be able to reach all parts of the adsorbent surface in the control volume in a time that is short compared to the liquid residence time in that volume. As noted previously, many systems of interest do not fit this limiting case scenario (e.g., when the adsorbent is granular activated carbon (GAC) or other manufactured adsorbents with significant internal porosity). In this section, we modify the previous model to account for such a possibility.

Conceptually, adsorption from aqueous solution onto a solid adsorbent can be broken into the following sequence of steps:

1. Transport of adsorbate through bulk solution and into the boundary layer surrounding the adsorbent particle.
2. Transport through the boundary layer to the external surface of the adsorbent.

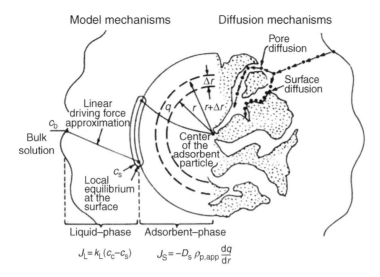

FIGURE 8-11. Schematic of the transport processes accompanying adsorption into a porous solid, and of some of the corresponding model representation. *Source:* From Hand et al. (1983). Reprinted with permission from ASCE.

3. The initial adsorption reaction, in which the adsorbate binds to the exterior of the adsorbent particle.
4. Migration of adsorbate into the interior of the adsorbent by surface diffusion, pore diffusion, or a combination of those processes.

Several of these steps are illustrated schematically in Figure 8-11.

Depending on the specific adsorbent, adsorbate, and contacting system of interest, any or all of the above steps might contribute a significant resistance to the overall process. For instance, in a quiescent system with very small and relatively nonporous adsorbent particles, the resistance associated with the first two steps is likely to be larger than that associated with the fourth step, while in a well-mixed system containing large adsorbent particles with many small pores, the opposite is likely to be the case. The resistance associated with the adsorption reaction (step 3) might be large if, for instance, the adsorbate must change configuration or undergo a dehydration reaction before it can adsorb.

The resistance associated with transport through bulk solution (step 1) can be estimated for many systems and can be partially controlled by altering the system hydrodynamics. As a practical matter, it appears to provide negligible resistance to the overall process in most systems of interest. Usually, the adsorption reaction (step 3) is also assumed to provide negligible resistance. This assumption is somewhat less well founded, but empirical evidence does seem to support it; that is, the approach to equilibrium can usually be modeled successfully without considering the kinetics of the sorption reaction itself. Thus, transport of the adsorbate either through the external boundary layer or through internal pores is assumed to be the rate-limiting step in most sorption processes.

In systems that reach equilibrium rapidly, the adsorbate concentrations in the boundary layer and the internal pores of the adsorbent are essentially identical to that in bulk solution, and the adsorption density is uniform throughout the particle. As a result, it is possible to analyze those systems without even acknowledging the existence of the boundary layer; we simply treat the solid as though it was in contact with the bulk solution. By contrast, in systems where adsorptive equilibrium is not necessarily attained, transport through the boundary layer must be considered (via a mass balance on that region), and the gradients of adsorbate concentration and adsorption density in the interior of the adsorbent particles must be evaluated explicitly.

In this section, we consider the kinetics of equilibration in typical adsorptive contacting systems, so that we can predict how closely a given system will approach the limiting, equilibrium condition. First, though, a brief discussion is provided of the relative importance of pore versus surface diffusion as modes for transporting adsorbate from the exterior surface of a particle to its interior.

Pore Diffusion Versus Surface Diffusion in Porous Adsorbent Particles

The issue of what portion of the transport of adsorbate within porous adsorbent particles is attributable to pore diffusion, and what part to surface diffusion, has been debated for years and remains unresolved. The flux associated with diffusion through the liquid filling the pore (i.e., pore diffusion) is given by Fick's law. However, the pores in a typical porous adsorbent are not aligned in the radial direction, but rather wind through the particle in a semi-random manner. As a result, when adsorbate molecules diffuse a distance dl in the pore, they are assumed to move closer to the center of the particle, but by a shorter distance dr. The average value of the ratio dl/dr in a particle is called the tortuosity, ξ. For activated carbon, ξ is typically estimated to be in the range from 2 to 10. Therefore, assuming that all pores eventually reach the center of the particle (an unrealistic, but necessary, simplification), the net radial flux through the pores can be expressed by a modified form of Fick's law that accounts for the tortuous path that the molecules must traverse, as follows:

$$J_{L,\text{pore}} = -\frac{D_L}{\xi}\frac{dc}{dr} \quad (8\text{-}29)$$

where D_L is the diffusivity of the contaminant in the liquid phase, r is the radial distance from the center of the particle, and $J_{L,\text{pore}}$ is the diffusive flux of adsorbate through the liquid filling the pore, in the $+r$ direction. Note that, because the $+r$ direction is from the center to the periphery of the particle, the normal situation of interest (transport from bulk solution into the particle) corresponds to $dc/dr > 0$ and $J_{L,\text{pore}} < 0$.

An expression analogs to Fick's law is also assumed to describe the flux along the surface of the pores (i.e., the flux due to surface diffusion), with the gradient expressed in terms of adsorption density.

$$J_s = -D'_s \frac{dq}{dr} \quad (8\text{-}30)$$

In Equation 8-30, D'_s is the surface diffusivity, with dimensions that cause J_s to have dimensions of mass/area-time for the given dimensions of q; for example, if q is in mg/g, r is in cm, and J_s is in mg/cm^2 s, then D'_s is in g/cm s. As is the case for pore diffusion, surface diffusion toward the center of the particle corresponds to a negative flux (i.e., $J_s < 0$).

It is common to define a parameter D_s as $D'_s/\rho_{p,\text{app}}$, and to write Equation 8-30 as follows:

$$J_s = -D_s \rho_{p,\text{app}} \frac{dq}{dr} \quad (8\text{-}31)$$

The product $\rho_{p,\text{app}}(dq/dr)$ is the gradient of adsorbed adsorbate, expressed in units that are comparable to those used for gradients in solution; that is, it is the change in the volumetric concentration of adsorbate (mass of adsorbate per unit volume of adsorbent) per unit distance in the radial direction. As a result, D_s has the same units (e.g., cm^2/s) and a similar meaning as a liquid-phase diffusivity, and values of D_s and D_L can be compared directly. (Keep in mind, though, that even if $D_L > D_s$, surface diffusion might be more important than pore diffusion, because the concentration gradient on the surface could be much greater than in the pore solution.)

The diffusive path along the surface is, presumably, just as tortuous as that through the pore solution, so it might seem appropriate to include a term for tortuosity in Equation 8-31. However, tortuosity is typically embedded in the value of D_s, rather than separated, since, as a practical matter, the two terms can never be evaluated independently. By contrast, D_L can be evaluated in bulk solution, allowing an independent estimate to be made for ξ based on the difference between D_L and the apparent diffusivity in the pore solution.

Taking into account both pore and surface diffusion, the overall flux of adsorbate inside the particle in the $+r$ direction can be expressed as follows:

$$J_{\text{overall}} = J_{L,\text{pore}} + J_s = -\frac{D_L}{\xi}\frac{dc}{dr} - D_s \rho_{p,\text{app}}\frac{dq}{dr} \quad (8\text{-}32)$$

If we assume that local equilibrium applies (i.e., that the solution and surface are in equilibrium in each microscopic section of the pore), then Equations 8-29 and 8-31 can be manipulated as follows, leading to the conclusion that the ratio of surface flux to pore flux is related to the slope of the adsorption isotherm:

$$\frac{J_s}{J_{L,\text{pore}}} = \frac{\rho_{p,\text{app}} D_s (dq/dr)}{D_L/\xi (dc/dr)} = \frac{\xi \rho_{p,\text{app}} D_s}{D_L}\left(\frac{dq}{dc}\right)_{eq} \quad (8\text{-}33)$$

Equation 8-33 suggests that, other factors being equal, surface diffusion becomes increasingly important as the slope of the isotherm increases. Thus, for a system that is characterized by a favorable isotherm, surface diffusion accounts for a larger fraction of the overall flux at low c and q than at higher values.[3]

While the validity of Equation 8-33 is widely accepted, the question of which transport mode dominates under given conditions in a given system remains open. The difficulty in resolving this issue arises from the uncertain geometry of the pores (including any interconnectedness) and the impossibility of sampling the fluid in the pores directly. Therefore, the arguments for dominance of one or the other transport mode rest on how well various models are able to reproduce experimental data for the overall rate of adsorption from bulk solution.

When the equations for the various models are solved and the predictions are compared with experimental data, the results are mixed. In some cases, an assumption that adsorbate diffuses into and through the particles strictly in the solution phase can account for the overall, observed rate of adsorption. However, in most cases, the predicted overall rate of adsorption based on pore diffusion alone is

---

[3] Recall from Chapter 7 that a favorable isotherm is one that, when plotted as q versus c is concave toward the abscissa (the c axis). All Langmuir isotherms and the vast majority of Freundlich isotherms are favorable.

significantly lower than the experimentally observed rate, suggesting that surface diffusion is important.

In such situations, the governing equations can be solved by taking both pore and surface diffusion into account, or, in limiting cases, by assuming that surface diffusion accounts for essentially all the internal transport. In either case, D_s is used as an adjustable parameter whose value is chosen to provide the best fit between the experimental and computed adsorption rates. When only surface diffusion is considered, the model is referred to as the homogeneous surface diffusion model (HSDM), and when both transport modes are considered, it is referred to as the pore/surface diffusion model (PSDM).

Both the HSDM and the PSDM are widely used to model environmental engineering adsorption systems. In either case, the value of D_s that provides the best fit to a given data set should be interpreted cautiously. While that value might be related to the true surface diffusivity, it also incorporates a host of uncertainties and assumptions. As a result, comparison of D_s values for different adsorbates and a given adsorbent might be a reasonable indicator of the relative rates of internal transport of those adsorbates, but any interpretation of the absolute value of D_s is dubious.

One interesting and important outcome of experiments to evaluate D_s for adsorption onto activated carbon is that D_s is often found to increase with increasing particle size, especially for larger adsorbates like natural organic matter (NOM). This trend has not been adequately explained from a theoretical perspective, but it is widely observed. Crittenden et al. (1991) proposed that the trend can be represented as follows:

$$D_s = k r_p^m \quad (8\text{-}34)$$

where k is an empirical constant that depends on the adsorbate and adsorbent, r_p is the particle radius, and m is a constant that varies between 0 and 1 and is often assumed to have one of those limiting values. If $m=0$, the surface diffusivity is independent of the particle size, and the system is referred to as having constant diffusivity (CD); if $m=1$, the system is referred to as having proportional diffusivity (PD).

To determine the value of m for a given adsorbate/adsorbent pair, D_s can be evaluated using two different sizes of particles. The results from those experiments can be used in conjunction with Equation 8-34 to estimate m, and that value of m can be used to predict D_s of the particles that might be used in a full-scale system.

The practical importance of Equation 8-34 arises from two implications. First, the advantages of using powdered-activated carbon (PAC) instead of granular-activated carbon (GAC) as a way of achieving equilibrium within a short contact time are often much less than might be anticipated based solely on consideration of the relative diffusion lengths in the two types of particles. And, second, the value of D_s determined using small adsorbent particles (whose use

is attractive because the complete transition from $q=0$ to $q = q_{eq}|_{c_{in}}$ throughout the particle can be studied over a short time period) might not be applicable to systems in which larger particles are used. This issue has particular importance in the use of small-scale packed beds to collect data for the design full-scale systems, which is addressed subsequently in this chapter.

The derivations presented in the following sections are based on the HSDM, because, even though pore diffusion is conceptually straightforward, its inclusion adds substantial mathematical complexity to the modeling. Additional details for use of both models are available in the literature (e.g., Crittenden and Weber, 1978; Hand et al., 1983, 1984; Crittenden et al., 1986), and computer software is available for applying both models either in design calculations or to estimate system parameters from experimental data (CenCITT, 1999).

Adsorption in Batch Systems with Transport-Limited Adsorption Rates

Consider first a batch adsorption system in which equilibrium is not necessarily reached in the time that the solid and solution are in contact with one another. A schematic defining several parameters in such a system is shown in Figure 8-12. The particles are idealized as being spherical

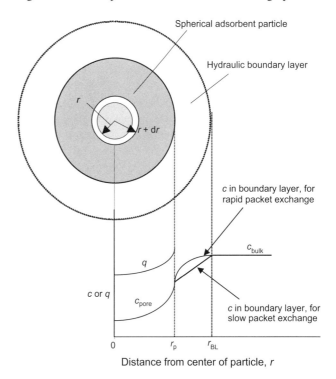

FIGURE 8-12. Schematic of the contaminant concentration profile in a porous adsorbent particle and surrounding fluid. Concentration profiles in the boundary layer are shown for both rapid and slow exchange of packets between that layer and bulk solution.

with a radius r_p, and having their internal surface area distributed uniformly throughout the particle. The solution is assumed to be well mixed, except in the boundary layer that surrounds the particles and extends from r_p to r_{BL}. As in gas transfer systems, transport into and out of this boundary layer is assumed to be by exchange of fluid packets with bulk solution on the fluid side, and by passive diffusion on the side in contact with the solid. The concentration gradient in the boundary layer is not specified, and might be either linear or nonlinear, depending on the hydrodynamics of packet exchange.

Based on the assumption that the adsorption reaction itself proceeds much more rapidly than the rate-limiting transport step(s), the interface between the water and particle is assumed to be characterized by local equilibrium (i.e., at any time, the values of q and c at r_p correspond to a point on the isotherm). This equilibrium would halt further adsorption, if not for the migration of adsorbed molecules from the surface into the interior of the particle, thereby re-establishing a driving force for more molecules to adsorb.

The only difference between the process described above and gas transfer into a bubble is that, in the adsorptive system, the concentration gradient in the solid is presumed to extend throughout the particle, whereas in gas transfer, the gradient is presumed to terminate at some short distance from the interface, where it merges with a well-mixed bulk phase. That is, modeling of the adsorptive system reflects the absence of mixing anywhere in the adsorbent particle.

The analysis proceeds by writing mass balances on adsorbate around two control volumes—the solution and the particles—and linking them via some assumptions regarding the boundary layer. Consider first the mass balance on adsorbate in bulk solution. Assuming that the adsorbate is neither created nor destroyed by chemical reactions, the only process that affects its mass in solution is transport into the boundary layer surrounding the particles. The mass balance can therefore be written as follows:

Rate of change of mass of adsorbate stored in solution = Net transport rate of adsorbate into bulk solution from the boundary layer

$$V_L \frac{dc}{dt} = J_L|_{r_{BL}} A_{r_{BL}} N_p \quad (8\text{-}35)$$

where c is the concentration in bulk solution, $J_L|_{r_{BL}}$ is the liquid-phase flux of contaminant in the $+r$ direction (from the boundary layer into solution) at r_{BL}, $A_{r_{BL}}$ is the surface area of a sphere at the outer periphery of the boundary layer, and N_p is the number of adsorbent particles in the system. In a system where adsorption is occurring, $J_L|_{r_{BL}}$ and dc/dt are both negative.

Because the number of adsorbent particles in a system is an awkward parameter to work with, Equation 8-35 is usually modified as follows. First, dividing through by V_L yields an expression for the volume-normalized mass balance:

$$\frac{dc}{dt} = J_L|_{r_{BL}} \frac{A_{r_{BL}} N_p}{V_L} \quad (8\text{-}36)$$

Assuming that the width of the boundary layer is small compared to the radius of the particle, $r_{BL} \approx r_p$, and we can substitute A_p, the external area of a single particle, for $A_{r_{BL}}$. Next, we substitute εV_R for V_L, where ε is the volume fraction of the system occupied by solution.[4] Finally, we multiply the right side of Equation 8-36 by the fraction $(1-\varepsilon)V_R/(V_p N_p)$, where V_p is the volume of a single particle. This fraction has the total volume of particles in the system as both the numerator and denominator, albeit expressed in two different ways. Making all of these substitutions, writing out the expressions for the surface area and volume of a sphere, and carrying out some algebra, we obtain

$$\frac{dc}{dt} = J_L|_{r_{BL}} \frac{A_p N_p}{\varepsilon V_R} \frac{(1-\varepsilon)V_R}{V_p N_p} = J_L|_{r_{BL}} \frac{A_p}{V_p} \frac{1-\varepsilon}{\varepsilon}$$

$$= J_L|_{r_{BL}} \frac{4\pi r_p^2}{(4/3)\pi r_p^3} \frac{1-\varepsilon}{\varepsilon} \quad (8\text{-}37)$$

$$\frac{dc}{dt} = J_L|_{r_{BL}} \frac{3}{r_p} \frac{1-\varepsilon}{\varepsilon}$$

Equation 8-37 is a first-order ordinary differential in time. The corresponding initial condition is the known contaminant concentration in solution at $t=0$:

$$c|_{t=0} = c_{init} \quad (8\text{-}38)$$

However, even with the initial condition specified, Equation 8-37 cannot be solved directly, because it contains $J_L|_{r_{BL}}$, which varies over time in an as yet unspecified manner. Under the circumstances, to solve for the solution composition over time, we need an independent equation relating $J_L|_{r_{BL}}$ to c and/or t. We develop that equation by considering the other mass balance characterizing the system.

---

[4] The definition of ε is consistent with its definition as the void fraction in a fixed bed reactor. In the current context, the void fraction is close to 1.00.

To write the mass balance on adsorbate in the particles, we choose as the control volume a thin, annular layer of adsorbent bounded by concentric spheres at radii of r and $r + dr$, as shown schematically in Figure 8-12. Transport of adsorbate into and out of that layer occurs only by surface diffusion (or, in the PSDM, by a combination of surface and pore diffusion), so the mass balance characterizing the control volume in a single particle can be written as follows:

Rate of change of mass of adsorbate stored in the annulus between the two concentric spheres	=	Rate (mass/time) at which adsorbate diffuses into the annulus (toward the center of the particle) at $r + dr$	−	Rate (mass/time) at which adsorbate diffuses out of the annulus (toward the center of the particle) at r

For convenience, this equation is written in terms of adsorbate transport toward the center of the particle (i.e., in the $-r$ direction), since that is the direction in which net diffusion would occur in an adsorption process. Since J_s is defined as the flux in the $+r$ direction, the flux in the $-r$ direction is $-J_s$. The two terms on the right can therefore be written as the product $-J_s A$ evaluated at $r + dr$ and at r, respectively, and the equation can be written in mathematical form as follows:

$$\frac{\partial}{\partial t}\left[q\rho_{p,app}(4\pi r^2 dr)\right] = (-J_s A)|_{r+dr} - (-J_s A)|_r \quad (8\text{-}39\text{a})$$

$$= (J_s A)|_r - (J_s A)|_{r+dr} \quad (8\text{-}39\text{b})$$

The product $r^2 dr$ is independent of t and can therefore be taken out of the differential on the left side, along with the constant $4\pi\rho_{p,app}$. Then, writing $J_s A|_{r+dr}$ as $J_s A|_r + (\partial J_s A/\partial r)|_r dr$, Equation 8-39b can be rewritten as follows:

$$\rho_{p,app}(4\pi r^2 dr)\frac{\partial q}{\partial t} = (J_s A)|_r - \left((J_s A)|_r + \frac{\partial(J_s A)}{\partial r}dr\right) \quad (8\text{-}40\text{a})$$

$$= -\frac{\partial(J_s A)}{\partial r}dr \quad (8\text{-}40\text{b})$$

$$= -\left(A\frac{\partial J_s}{\partial r} + J_s \frac{\partial A}{\partial r}\right)dr \quad (8\text{-}40\text{c})$$

Expressing A as $4\pi r^2$, substituting the pseudo-Fick's law expression (Equation 8-31) for J_s, and rearranging, we obtain

$$\rho_{p,app}(4\pi r^2 dr)\frac{\partial q}{\partial t}$$

$$= -\left(4\pi r^2 \frac{\partial}{\partial r}\left[-D_s \rho_{p,app}\frac{\partial q}{\partial r}\right] - D_s \rho_{p,app}\frac{\partial q}{\partial r}\frac{\partial(4\pi r^2)}{\partial r}\right)dr \quad (8\text{-}41\text{a})$$

$$= -\left(-4\pi r^2 D_s \rho_{p,app}\frac{\partial^2 q}{\partial r^2} - 4\pi D_s \rho_{p,app}\frac{\partial q}{\partial r}\frac{\partial(r^2)}{\partial r}\right)dr \quad (8\text{-}41\text{b})$$

$$= 4\pi D_s \rho_{p,app} dr \left(r^2 \frac{\partial^2 q}{\partial r^2} + \frac{\partial q}{\partial r}\frac{\partial(r^2)}{\partial r}\right) \quad (8\text{-}41\text{c})$$

Finally, writing $\partial(r^2)$ as $2r\partial r$, and dividing through by $4\pi r^2 \rho_{p,app} dr$:

$$\rho_{p,app}(4\pi r^2 dr)\frac{\partial q}{\partial t} = 4\pi D_s \rho_{p,app} dr\left(r^2\frac{\partial^2 q}{\partial r^2} + \frac{\partial q}{\partial r}\frac{2r \partial r}{\partial r}\right) \quad (8\text{-}41\text{d})$$

$$\frac{\partial q}{\partial t} = D_s \left(\frac{\partial^2 q}{\partial r^2} + \frac{2}{r}\frac{\partial q}{\partial r}\right) \quad (8\text{-}42)$$

Equation 8-42 is first order in time and second order in space, so it requires one initial condition and two boundary conditions. The initial condition specifies the adsorption density at time zero. In most cases the adsorbent is assumed to be free of adsorbate initially, in which case the initial condition is

$$q|_{t=0} = 0 \quad \text{for all } r \quad (8\text{-}43)$$

One boundary condition applies at the center of the particle ($r = 0$), where symmetry requires that the $q(r)$ (and $c(r)$) profiles be identical in all directions (i.e., along any radius). This condition applies because the penetration of the adsorbate (both on the pore surfaces and in the pore solution) proceeds identically from all points on the external surface of the particle. Since q and c are continuous, smooth functions along any path from one point on the surface, through the center, and back to another point on the surface, the gradient of adsorption density must be zero at the center (as shown in Figure 8-12). Thus, we can write

$$\left.\frac{\partial q}{\partial r}\right|_{r=0} = 0 \quad \text{for all } t \quad (8\text{-}44)$$

The second boundary condition applies at $r = r_p$ and reflects the assumption that equilibrium applies at all times at that point; that is, that $q(r_p, t)$ is related to $c(r_p, t)$ by the isotherm equation:

$$q_{r_p} = q_{eq}(c_{r_p}) \quad \text{for all } t \qquad (8\text{-}45)$$

where $q_{eq}(c_{r_p})$ is the function describing the applicable adsorption isotherm.

Like the mass balance on dissolved adsorbate (Equation 8-37), the mass balance on adsorbed adsorbate (Equation 8-45) cannot be solved directly, in this case because one of the boundary conditions (the isotherm equation) contains a parameter that is, as yet, unspecified (c_{r_p}). The two mass balances can be linked and solved by making two assumptions about the conditions in the boundary layer. The first assumption is that, even though the mass of adsorbate stored in the boundary layer changes over time, this change accounts for a negligible fraction of the adsorbate that leaves solution or enters the solid (or, in the case of desorption, that moves in the opposite direction). As a result, the flux out of solution at r_{BL} can be equated with the flux into the solid at r_p. Second, we assume that, at any instant, the concentration profile in the boundary layer is the steady-state profile for the current concentrations at r_p and r_{BL} and the dynamics of packet exchange in the system. Based on this assumption, we can equate the flux through the boundary layer with the product of the concentration difference across the layer and a mass transfer coefficient. In mathematical form, these assumptions can be expressed as follows:

$$J_L|_{r_p} = J_L|_{r_{BL}} = -J_s|_{r_p} = -k_L(c - c_{r_p}) \qquad (8\text{-}46)$$

where k_L is the mass transfer rate coefficient for contaminant transport through the boundary layer. Note that the dissolved adsorbate concentration at r_{BL} has been written as c, since the concentration at the periphery of the boundary layer is the bulk concentration.

By substituting the expression on the far right of Equation 8-46 for $J_L|_{r_{BL}}$ in Equation 8-37, we obtain

$$\frac{dc}{dt} = -k_L(c - c_{r_p}) \frac{3}{r_p} \frac{1-\varepsilon}{\varepsilon} \qquad (8\text{-}47)$$

We can also express the flux into the solid in terms of the gradient in adsorption density at the particle surface via Equation 8-31. Equating that expression with the flux through the boundary layer yields

$$k_L(c - c_{r_p}) = D_s \rho_{p,app} \left.\frac{\partial q}{\partial r}\right|_{r_p} \qquad (8\text{-}48)$$

Equation 8-48 provides the additional information needed to solve for the adsorbate distribution throughout the system over time. As an alternative to that equation, rather than equating the fluxes out of bulk solution and into the solid, we could equate the rate at which adsorbate passes through the boundary layer to the rate of accumulation of adsorbate in the entire adsorbent, as follows:

Rate at which adsorbate moves from the boundary layer into the solid at r_p

$$= -J_L|_{r_p} A_{r_p}$$
$$= k_L(c - c_{r_p})(4\pi r_p^2) \qquad (8\text{-}49)$$

Rate at which adsorbate accumulates in solid:

$$= \frac{\partial}{\partial t} \int_0^{r_p} q\rho_{p,app}(4\pi r^2 dr) \qquad (8\text{-}50)$$

Equating Equations (8-49) and (8-50):

$$k_L r_p^2 (c - c_{r_p}) = \frac{\partial}{\partial t} \int_0^{r_p} q\rho_{p,app} r^2 dr \qquad (8\text{-}51)$$

Equation 8-51 is a slightly different mathematical statement of the same point expressed by Equation 8-48—that essentially all of the adsorbate that leaves solution passes through the boundary layer and enters the solid. Thus, either Equation 8-48 or Equation 8-51 can be used to link the mass balance for the bulk solution with that for the solid.

To summarize the overall process, mass balances on adsorbate in the bulk solution phase and on the adsorbent must be solved simultaneously, in conjunction with appropriate initial and boundary conditions, to predict the behavior of adsorbate in a well-mixed batch suspension in which equilibrium is not attained instantly. These mass balances (and the physical regions they represent) are linked via the boundary layer surrounding the adsorbent particles. By making reasonable assumptions about the behavior of adsorbate in that layer, the equations can be solved. The key equations applicable to this scenario are collected in Table 8-2.

Numerical techniques are available to solve the equations in Table 8-2, and the results of such calculations for an example system are shown in Figure 8-13. The input data for the simulation are provided in the figure caption. Figure 8-13a shows profiles of adsorption density through an adsorbent particle after contact times up to 1 h. The simulation indicates that adsorbate penetrates approximately half way to the center of the particle within a few minutes, and the entire particle reaches equilibrium with the bulk solution ($q_{fin} = 17.5$ mg/g) at some time between 30 min and 1 h.

TABLE 8-2. Summary of Equations for Nonsteady-State Adsorption in a Well-Mixed Suspension of Spherical, Porous Adsorbent Particles

Equations characterizing the solution phase ($r > r_{BL}$)

Solution phase mass balance combined with assumptions about liquid boundary layer	$\dfrac{dc}{dt} = -k_L(c - c_{r_p})\dfrac{3}{r_p}\dfrac{1-\varepsilon}{\varepsilon}$	(8-47)
Initial condition	$c\vert_{t=0} = c_{init}$	(8-38)

Equations characterizing the solid phase ($r < r_p$)

Solid phase mass balance	$\dfrac{\partial q}{\partial t} = D_s\left(\dfrac{\partial^2 q}{\partial r^2} + \dfrac{2}{r}\dfrac{\partial q}{\partial r}\right)$	(8-42)
Initial condition	$q\vert_{t=0} = 0 \quad \text{for all } r$	(8-43)
Boundary conditions		
Symmetry condition: no gradient at center of particle	$\left.\dfrac{\partial q}{\partial r}\right\vert_{r=0} = 0 \quad \text{for all } t$	(8-44)
Equilibrium with solution at solid/water interface	$q_{r_p} = q_{eq}(c_{r_p}) \quad \text{for all } t$	(8-45)

Additional relationship linking $q(r_p, t)$ to $c(t)$, based on assumptions about the boundary layer

Either: flux out of solution equals flux into solid	$k_L(c - c_{r_p}) = D_s \rho_{p,app}\left.\dfrac{\partial q}{\partial r}\right\vert_{r_p}$	(8-48)
Or: rate of loss of contaminant mass from solution equals rate of gain of contaminant mass by adsorbent	$k_L r_p^2 (c - c_{r_p}) = \dfrac{\partial}{\partial t}\displaystyle\int_{r=0}^{r=r_p} q\rho_{p,app} r^2 dr$	(8-51)

Figure 8-13b shows the change in adsorption density at several different radii as the system approaches equilibrium. Interestingly, at the surface of the adsorbent particles ($r = 20\,\mu m$), the change in q is not monotonic: q is initially zero, then increases rapidly and overshoots the ultimate, equilibrium value, and finally decreases to reach that value. This behavior can be understood when the figure is considered in conjunction with Figure 8-13c, which shows the changes in the dissolved adsorbate concentration both in the bulk solution and at the particle surface. Whereas c_{bulk} declines steadily from its initial to its ultimate, equilibrium value (from 1.0 to 0.124 mg/L), c_{surf} follows a trend similar to that of q_{surf}: it starts near zero, overshoots the ultimate, equilibrium value, and then returns to the equilibrium value by the end of the simulation.

The changes in q_{surf} and c_{surf} reflect the assumption that local adsorptive equilibrium is reached instantaneously, combined with the fact that transport through the liquid boundary layer is much more facile than transport into the particle. When the adsorbent and solution are first brought into contact, the instantaneous equilibration at the interface increases the adsorption density and simultaneously depletes the local, solution-phase concentration, so that the surface is equilibrated with a dissolved concentration that is substantially less than the bulk concentration. Over the course of the next minute or so, adsorbate migrates from the bulk solution to the interface more rapidly than it is lost via adsorption, so c_{surf} increases (as does q_{surf}, in accord with the assumption of local equilibrium). The increase in c_{surf} and simultaneous decrease in the bulk concentration decrease the driving force for migration through the boundary layer, to the point where, after ~ 1 min, the rate of transport from the bulk to the surface is less than the rate of adsorption. As a result, c_{surf} and q_{surf} both begin to decrease. This situation remains in effect until the system reaches equilibrium, at which point the gradients in both c and q disappear and all transport ceases.

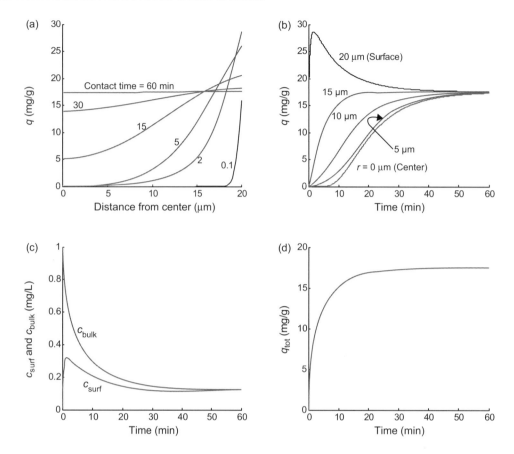

FIGURE 8-13. Modeling results for the approach to adsorptive equilibrium in a well mixed, batch system containing PAC and an adsorbate that obeys the isotherm $q = 50\, c^{0.5}$ (c in mg/L, q in mg/g). System conditions: $c(0) = 1.0$ mg/L; $q(0) = 0.0$ mg/g; $c_{PAC} = 50$ mg/L; $r_p = 20\,\mu$m; $D_s = 3.75 \times 10^{-10}$ cm^2/s; $k_L = 0.13$ cm/s; $\rho_{p,app} = 1.25$ g/cm^3.

Finally, Figure 8-13d shows the overall (average) adsorption density over time, computed as the ratio of the mass of adsorbate removed from solution to the mass of adsorbent in suspension. Although ultimate equilibrium is not attained until close to 1 h has elapsed, the vast majority of the ultimate adsorption density on the overall particle is achieved within 10–15 min. The temporal trend in the average adsorption density does not conform exactly to any simple kinetic expression. However, for systems in which diffusion through the liquid film contributes negligible resistance to the overall process, and in which equilibrium adsorption is characterized by a Freundlich isotherm, the curve in Figure 8-13d can be closely approximated by the following expression (Najm, 1996):

$$q(t) = q_{eq}[1 - 0.15 \exp(-143\lambda) - 0.16 \exp(-40\lambda) - 0.61 \exp(-10\lambda)] \quad (8\text{-}52)$$

where q_{eq} is the ultimate, equilibrium adsorption density, and λ is a dimensionless group defined as

$$\lambda \equiv \frac{D_s t}{r_p^2} \quad (8\text{-}53)$$

Figure 8-14 shows the sensitivity of the kinetics of the overall adsorption process to various system parameters. Tripling the film diffusion coefficient (k_L) and doubling the

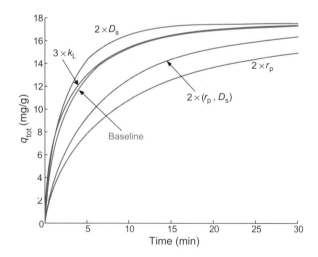

FIGURE 8-14. Sensitivity of the adsorption rate to various system parameters. Baseline is the system characterized in Figure 8-13.

surface diffusion coefficient (D_s) each increases the rate at which equilibrium is approached, but not dramatically. Doubling the particle radius, on the other hand, has a much more dramatic effect, delaying the time when any given adsorption density is reached by a factor of two to three. However, recall that the apparent surface diffusion coefficient tends to change along with the particle radius, at least for activated carbon particles, and that, in the limit, the changes in the two parameters are directly proportional to one another. The figure indicates that, in this limiting case (D_s doubles simultaneously with r_p), the effect on adsorption kinetics is only approximately half as great as when r_p is doubled and D_s remains constant. That is, if the surface diffusivity changes in conjunction with the particle radius, the effect on adsorption kinetics is considerably less dramatic than might otherwise be expected.

The simulations in Figures 8-13 and 8-14 characterize the behavior of PAC particles in a batch system and are therefore also applicable to adsorption onto PAC in a plug flow reactor. That is, if adsorption were occurring in a PFR under the conditions specified as the baseline case, and if τ_{PFR} were 20 min, every PAC particle exiting the reactor would be characterized by the values shown at $t = 20$ min in Figure 8-13, and the average adsorption density on every particle would be 17 mg/g. However, treatment with PAC is much more often carried out in a mixed reactor, in which case different exiting particles have spent different amounts of time in the reactor. In that case, each particle in the effluent has an adsorption density, $q(t)$, that corresponds to exposure to a solution with a fixed bulk concentration for the amount of time t that the particle spent in the reactor. The overall average adsorption density considering all the particles in the effluent (q_{out}) could then be computed using an expression analogous to Equation 4-44:

$$q_{out} = \int_0^\infty q(t) E_{PAC}(t) \, dt \tag{8-54}$$

where $E_{PAC}(t)$ is the exit age distribution of the particles in the effluent.

If the reactor is a CFSTR, then the residence time distribution of the exiting particles is identical to that of a tracer ($E_{PAC}(t) = (1/\tau_{CFSTR})\exp(-t/\tau_{CFSTR})$). When this RTD is combined with the $q(t)$ function from the analysis of adsorption in a system with a constant bulk concentration, the resulting integral in Equation 8-54 has no analytical solution. However, it can be approximated numerically if the resistance contributed by the liquid film is negligible and if adsorption follows a Freundlich isotherm. In that case, the average adsorption density on the effluent particles is given by Najm (1996):

$$q_{CFSTR} = \left[1 - \frac{6}{\pi^2} \sum_{i=1}^\infty \frac{1}{i^2(1+\pi^2 i^2 \lambda)}\right] q_{eq} \tag{8-55}$$

where all the variables are as defined previously (except that τ replaces t in the definition of λ).[5]

The preceding discussion describes the calculation of system dynamics in a case where all the critical parameters are known. It should be clear that an alternative use of the model equations is to estimate those parameters (in particular, D_s) from an experimental data set of c_{bulk} versus t in batch or continuous flow systems. Typical values for D_s determined in this way for sorption of synthetic organic compounds of common interest onto GAC are about two to three orders of magnitude smaller than those for the same molecules in solution (D_L); values for adsorption onto PAC are often still smaller, by one to two orders of magnitude. Nevertheless, the gradient in adsorption density is typically much greater than the gradient in the dissolved adsorbate concentration in the pores, so, as noted previously, the majority of the internal transport of adsorbate is usually calculated to be associated with surface diffusion.

■ **EXAMPLE 8-7.** Estimate the average adsorption density on PAC particles for continuous flow adsorption systems in (a) a CFSTR and (b) a PFR, both with $\tau = 20$ min. Assume that the same system conditions apply as were used to develop Figure 8-13, with the additional assumption that film diffusion can be ignored as a source of transport resistance.

Solution. Regardless of which type of reactor is used, the equilibrium value of the adsorption density can be computed based on a mass balance, as in several previous examples. With c in mg/L and q in mg/g, we can write:

$$c_{tot} = c_{eq} + q_{eq} c_{AC}$$
$$= c_{eq} + 50 c_{eq}^{0.5} c_{AC}$$
$$1.0 \, \text{mg/L} = c_{eq} + (50)\left(c_{eq}^{0.5}\right)(0.05 \, \text{g PAC/L})$$

Solving, we find $c_{eq} = 0.123$ mg/L. The corresponding value of q is 17.5 mg/g, as was indicated previously in the text. The other parameter whose value we need is λ, which can be computed as

$$\lambda = \frac{D_s \tau}{r_p^2} = \frac{(3.75 \times 10^{-10} \, \text{cm}^2/\text{s})(20 \, \text{min})(60 \, \text{s/min})}{(20 \times 10^{-4} \, \text{cm})^2} = 0.112$$

We can substitute the known values into Equation 8-55 to estimate the adsorption in a CFSTR. The summation in that equation converges very quickly, so that the contributions of the terms corresponding to i values of 4 and greater are

---

[5] Equations 8.52 and 8.55 are both based on the assumption that $q = 0$ on the influent PAC particles. If that is not the case, q_{out} and q_{eq} must be replaced by ($q_{out} - q_{in}$) and ($q_{eq} - q_{in}$), respectively.

negligible. (More terms are needed for smaller values of τ.) The result is that $q_{\text{out}} = 11.8 \text{ mg/g}$, or approximately 67% of q_{eq}.

If the adsorption process is carried out in a PFR, the dynamics are the same as in a batch reactor. Assuming that the liquid film resistance is negligible, the average adsorption density on the particles is given by Equation 8-52, with q_{PFR} replacing $q(t)$ and τ_{PFR} replacing t.

$$q_{\text{PFR}} = q_{\text{eq}}[1 - 0.15\exp(-143\lambda) - 0.16\exp(-40\lambda) - 0.61\exp(-10\lambda)]$$
$$= 17.5 \text{ mg/g}$$
$$\times \begin{bmatrix} 1 - 0.15\exp\{(-143)(0.112)\} - 0.16\exp\{(-40)(0.112)\} \\ -0.61\exp\{(-10)(0.112)\} \end{bmatrix}$$
$$= 14.0 \text{ mg/g}$$

(*Note*: In theory, this result should be identical to that in a batch system, as shown in Figure 8-13d. The difference in the two values is due to the use of the approximate Equation 8-52.) ∎

Adsorption in Fixed Bed Systems with Transport-Limited Adsorption Rates

Although packed beds in which adsorptive equilibrium is reached slowly compared to the solid/solution contact time are more complex than any of the systems that we have analyzed to this point, all the key elements needed to model the more complex systems have already been introduced. For the analysis, we use the same definition diagram and control volume (a section of the bed with length dz) as we used previously in the analysis of a rapidly equilibrating system (Figure 8-9). The mass balance on dissolved adsorbate has two terms that contribute to the change in adsorbate mass stored in this control volume: one accounting for net advective transport (into the volume at z and out of it at $z + dz$), and one for loss of adsorbate via transport into the boundary layers surrounding the particles.

The advective term describes the rates at which adsorbate is carried into and out of the control volume by the bulk flow, so it has the same form (as given in Equation 8-16a or 8-16b) regardless of whether the system reaches equilibrium rapidly or slowly. The typical boundary and initial conditions are also the same in the two types of systems; that is, $c = c_{\text{in}}$ at $z = 0$ at all times, and $c = 0$ throughout the column at $t = 0$. On the other hand, during the differential time period dt that the water spends in the control volume, the concentration in solution is approximately uniform and constant. Therefore, transport of adsorbate out of solution and through the boundary layer can be modeled identically to that in a batch reactor in which a slowly equilibrating adsorption reaction is proceeding, that is, as in Equation 8-47. Combining these ideas, we can write the solution phase mass balance for a slowly equilibrating fixed bed system as follows:

$$\frac{\partial c}{\partial t} = -\frac{v}{\varepsilon}\frac{\partial c}{\partial z} - k_{\text{L}}(c - c_{r_p})\frac{3}{r_p}\frac{1-\varepsilon}{\varepsilon} \quad (8\text{-}56)$$

By the same reasoning, the solid-phase mass balance in the differential control volume is essentially the same as that in a batch reactor with slow equilibration (Equation 8-42). The initial condition, the boundary conditions at the center of the particle and its exterior surface, and the additional equation linking transport into the solid with that through the boundary layer are also identical. Thus, the only differences between the equations characterizing adsorbate behavior in this system and in a batch system are the appearance of the advective term in the solution-phase mass balance (Equation 8-56) and the recognition that, in the current case, we need to express q as a function of z in addition to r and t.

The complete set of equations necessary to solve for the performance characteristics and breakthrough behavior in a fixed bed adsorption column in which the adsorbent does not equilibrate rapidly with the local solution are summarized in Table 8-3.

As in the case of slowly equilibrating batch systems, numerical methods have been developed to solve the set of equations shown in Table 8-3. The shape of the MTZ as it migrates through the bed is shown in Figure 8-15 for a simulated system in which chloroform adsorbs onto GAC in a fixed bed. The baseline conditions for these simulations are shown in Table 8-4.

The shape of the MTZ remains essentially constant as it propagates through the bed. For this example system, the zone from the location where c equals 5% of c_{in} to that where it equals 95% of c_{in} spans approximately 25% of the bed length. The details of this pattern are determined by the kinetic factors controlling adsorption—primarily the resistance to mass transport associated with adsorbate migration through the liquid boundary layer surrounding the adsorbent particles and through the interior of the particles via pore and surface diffusion. The greater these resistances are, the more the adsorbate profile spreads out in the bed. (Recall that, in the absence of kinetic limitations, the profile is almost a square wave.) The rate of movement of the MTZ through the column, on the other hand, is controlled primarily by equilibrium considerations; that is, by the rate at which incremental amounts of adsorbent equilibrate with the influent solution, given the rate at which adsorbate is supplied in the influent. This point is demonstrated quantitatively in the subsequent section.

Breakthrough curves for the baseline conditions and various modifications to those conditions are shown as a function of time and the number of BVs treated in Figures 8-16a and 8-16b, respectively. Consider first the effects of decreasing D_s (which might result, for example, from a decrease in temperature) or increasing r_p. Either of these changes increases the total resistance that adsorbates

TABLE 8-3. Summary of Equations Applicable to Adsorption onto Spherical Porous Adsorbent Particles in a Fixed Bed System with Plug Flow

Equations characterizing c(z, t) in the solution phase (i.e., at $r > r_{BL}$)

Solution phase mass balance combined with assumptions about liquid boundary layer	$\dfrac{\partial c}{\partial t} = -\dfrac{v}{\varepsilon}\dfrac{\partial c}{\partial z} - k_L\left(c - c_{r_p}\right)\dfrac{3}{r_p}\dfrac{1-\varepsilon}{\varepsilon}$	(8-56)
Initial condition	$c\big|_{t=0} = 0$ for all z	(8-38)

Equations characterizing q(z, r, t) in the solid phase

Solid phase mass balance	$\dfrac{\partial q}{\partial t} = D_s\left(\dfrac{\partial^2 q}{\partial r^2} + \dfrac{2}{r}\dfrac{\partial q}{\partial r}\right)$	(8-42)
Initial condition	$q\big|_{t=0} = 0$ for all r and z	(8-43)
Boundary conditions		
Symmetry condition: no gradient at center of particle	$\dfrac{\partial q}{\partial r}\bigg|_{r=0} = 0$ for all t and z	(8-44)
Equilibrium with solution at solid/water interface	$q_{r_p} = q_{eq}\left(c_{r_p}\right)$ for all t and z	(8-45)

Additional relationship linking q(z, r_p, t) to c(z, t)

Either: flux out of solution equals flux into solid	$k_L\left(c - c_{r_p}\right) = D_s \rho_{p,app} \dfrac{\partial q}{\partial r}\bigg|_{r_p}$	(8-48)
Or: rate of loss of contaminant mass from solution equals rate of gain of contaminant mass by adsorbent	$k_L r_p^2\left(c - c_{r_p}\right) = \dfrac{\partial}{\partial t}\displaystyle\int_{r=0}^{r=r_p} q\rho_{p,app} r^2\, dr$	(8-51)

encounter as they migrate to the center of the adsorbent particles, but neither change affects the ultimate equilibrium condition. Because of the increased resistance, less adsorbate enters the particles per unit time or per BV treated early in the run. As a result, breakthrough begins sooner (Figure 8-16a) and after fewer BVs have been treated (Figure 8-16b) than in the baseline system. The fact that the equilibrium adsorption density is identical in the three systems means that the decline in adsorbate removal from solution early in the run is exactly compensated by additional adsorption later in the run (i.e., the systems with modified conditions continue to adsorb material after complete breakthrough has occurred in the baseline system), so that the areas above the three curves are equal once complete breakthrough has occurred in all the systems.

Although it is not apparent in the figures, the curves for the modified conditions intersect the baseline curve at slightly greater than 50% breakthrough. In the limit, if D_s became very small or r_p became very large, some breakthrough would occur almost immediately after the run started, and the breakthrough curves for the modified conditions would intersect that for the baseline condition at a point well above 50% breakthrough. However, for most modifications that are of practical interest, the predicted behavior is more like that shown in the figures, with the breakthrough curves for the modified conditions appearing to pivot around the 50% breakthrough point on the baseline curve. As we would expect, increasing D_s or decreasing r_p would cause the curves to pivot in the opposite direction; that is, those changes would increase the slope of the breakthrough curve.

Next, consider the effect of a change in flow rate on the breakthrough curve. As shown in Figure 8-16a, when the influent flow rate is doubled, the entire curve shifts to an earlier time, but the shape of the curve is minimally affected. In fact, doubling the flow rate decreases the

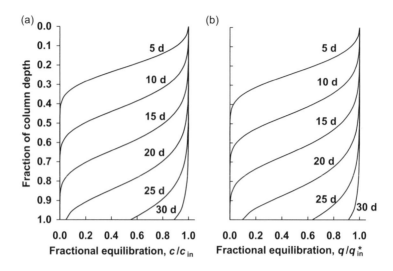

FIGURE 8-15. Simulated profiles of (a) concentration and (b) adsorption density for a fixed bed adsorption system. (Conditions listed in Table 8-4.)

time to breakthrough almost proportionately (50% breakthrough occurs at $t = 12.2$ min, as opposed to 24.6 min in the baseline case). As a result, the volume of influent treated at 50% breakthrough (equal to the product $Qt_{50\%}$) is almost exactly equal in the two systems. This latter fact is more apparent in Figure 8-16b, where the abscissa is a direct indicator of the volume of influent treated. Figure 8-16b also shows that, when breakthrough is plotted as a function of the number of BVs treated, the breakthrough curve is less steep in the system with higher flow rate than in the baseline system.

The implication of these observations is that, for the example system, the dominant control on the timing of breakthrough is the rate at which adsorbate is applied to the bed. The slight deviation from exact equality of $Qt_{50\%}$ in the two systems with different flow rates and the differences in the steepness of the breakthrough curves reflect changes in the kinetics of the adsorption process. When the flow rate doubles, the hydraulic residence time decreases by 50%, and the increased water velocity increases the mass transfer coefficient (k_L) by approximately 38% over the baseline value. However, separate simulations (not shown) indicate that a 38% increase in k_L without any change in Q has almost no effect on the breakthrough curve. We can therefore infer that, for the baseline system, transport through the liquid boundary layer imposes negligible resistance on the overall adsorption process. Correspondingly, the change in the shape of the breakthrough curve when the flow rate is doubled is caused primarily by the decline in the hydraulic residence time.

Ultimately, for a given c_{in}, the same mass of adsorbate is bound to the solid regardless of the flow rate, but the

TABLE 8-4. Baseline Conditions for the Simulations Shown in Figure 8-15

Influent and isotherm data
 Adsorbate: Chloroform ($CHCl_3$)
 D_L: 8.13×10^{-6} cm^2/s
 Isotherm equation: $q = 7.3c^{0.756}$ (q in mg/g GAC, c in mg/L)
 Q_L: 3.975 L/s
 EBCT: 10.0 min
 c_{in}: 0.500 mg/L
 T: 15°C

Adsorbent data
 Type: Calgon F400
 ρ_b: 0.803 g/cm^3
 r_p: 0.513 mm
 Internal porosity: 0.641

Bed data
 L: 2.765 m
 Column diameter: 1.048 m
 M_{ads} (mass of adsorbent in bed): 1000 kg
 ε: 0.478

Other kinetic data
 k_L: 3.28×10^{-3} cm/s
 D_s: 3.75×10^{-9} cm^2/s

Model used for calculations
 Pore and surface diffusion model (PSDM)[a]
 ETDOT™ Software v.1 (©CenCITT, 2002)

[a]Although the simulations were run using the PSDM, pore diffusion accounts for a negligible portion of the chloroform transport within the GAC particles, so the results would be essentially identical if the HSDM model were used.

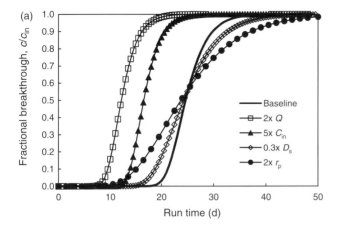

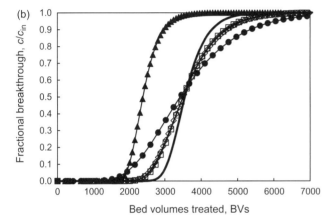

FIGURE 8-16. (a) Sensitivity of the breakthrough curve to various operational parameters; (b) same data as in part (a), plotted against number of bed volumes treated instead of time.

breakthrough curve begins after fewer BVs have been treated and ends after more BVs have been treated in a system with higher flow rate. As a result, and as is the case for changes in D_s and r_p, the area above the breakthrough curve and below a value of 1.0 is the same in the system with the higher flow rate as in the baseline system, when the data are plotted as in Figure 8-15b. Also, similar to the case for D_s and r_p, increasing Q indefinitely would lead to almost immediate breakthrough and an intersection of the breakthrough curve with that for the baseline system at considerably greater than 50% breakthrough; however, for normally encountered conditions, increasing Q has the effect of causing the breakthrough curve to pivot around that for the baseline system at the point of ~50% breakthrough, when the data are plotted as in Figure 8-16b.

Like increasing Q at constant c_{in}, increasing c_{in} at fixed Q increases the mass loading rate to the bed. However, the analysis is different in the two cases, because an increase in c_{in} leads to an increase in the equilibrium adsorption capacity of the column. For the example system, the effect of a five-fold increase in c_{in} is considered; for the given adsorption isotherm ($q = kc^{0.756}$), the corresponding increase in the equilibrium adsorption density in the column is a factor of $5^{0.756}$, or 3.38. Increasing c_{in} has no effect on the hydraulic residence time or k_L, so the response of the system to an increase in c_{in} reflects changes in adsorption capacity only. Therefore, since increasing c_{in} by a factor of five increases the rate at which adsorbate is supplied to the bed five-fold, but increases the adsorption capacity by a factor of only 3.38, breakthrough occurs after an amount of time or BVs treated that is only ~3.38/5.0, or 67%, as large as in the baseline system (50% breakthrough at $t = 16.6$ min).

The decrease in the time and number of BVs treated before breakthrough when c_{in} is increased is a direct consequence of the nonlinearity of the isotherm; in systems with linear adsorption isotherms, a change in c_{in} is predicted to have no effect on fractional breakthrough (i.e., c/c_{in}) as a function of time or the number of BVs treated. In that case, doubling c_{in} causes adsorbate to be provided to the bed at twice the baseline rate, but it also doubles the amount of adsorbate that the bed can acquire. The net result is a doubling of the rate of adsorption, leading to the same fractional removal from solution, and no net effect on fractional breakthrough as a function of either time or the number of BVs treated. This result would be obtained regardless of whether transport through the boundary layer or the solid was rate limiting, since the rates of diffusion in both those regions are first order in concentration, and hence would double.

Adsorption in engineered treatment systems is more often characterized by Freundlich isotherm with $n < 1$ than by linear isotherms, so the elapsed time or number of BVs treated before a given fractional breakthrough occurs is generally expected to decrease with an increase in c_{in}. For adsorptive systems that follow Langmuir isotherms, the effect of c_{in} on the timing of breakthrough depends on the value of c_{in} relative to $1/K_{ads}$: if $c_{in} \ll 1/K_{ads}$, the isotherm is approximately linear, and the timing of breakthrough is minimally affected by a change in c_{in}; on the other hand, if c_{in} is of the same order as or much greater than $1/K_{ads}$, an increase in c_{in} decreases the time to breakthrough substantially.

Although the shape of the MTZ in systems with transport-limited adsorption is predicted to change when c_{in} changes, the predicted change is usually small in systems with engineered adsorbents, compared to the change in the timing of the breakthrough curve (in terms of either absolute time or number of BVs treated). As explained previously, this result is obtained because the adsorptive capacity is always equivalent to the mass of adsorbate in a very large number of BVs of influent. In the example system considered here, the number of BVs treated between 10% and 90% breakthrough decreases from 1250 to 950 BV when c_{in} changes from the baseline value to five times that value. By contrast, for the same change in c_{in}, the center point of the breakthrough curve (50% breakthrough) decreases from 3690 to 2480 BV. This latter change in the timing (i.e., the center point) of breakthrough is generally of much greater significance than the change in slope; in fact, in Figure 8-16 (either part *a* or *b*), the change in the slope of the breakthrough curve is barely noticeable, whereas the change in the center point is dramatic.

8.4 THE MOVEMENT OF THE MASS TRANSFER ZONE THROUGH FIXED BED ADSORBERS

As demonstrated for the example system in Figure 8-15, the shape of the MTZ in a fixed bed adsorber is nearly invariant as the MTZ moves through the reactor. Thus, one can speak of the MTZ as an approximately constant band that is established shortly after the influent is introduced and that moves through the reactor at a velocity v_{MTZ}. In this section, we explore the factors that control that velocity.

Consider two times t_1 and t_2, during which the MTZ moves a distance Δz along the reactor without changing shape. During this time, the total mass of adsorbate entering the reactor is $Qc_{\text{in}}(t_2 - t_1)$, and the total mass exiting is essentially zero, so all of the entering adsorbate must adsorb. In reality, the recently adsorbed material is spread throughout the reactor, with the majority of it in the final MTZ and the region just upstream of the MTZ. However, a comparison of the profiles within the reactor at times t_1 and t_2 makes it clear that the increment in adsorbed material during the time period of interest equals the amount of adsorbate that is present in a volume $A_R \Delta z$ of packed bed, after it has equilibrated with the influent. This idea is shown schematically in Figure 8-17.

The mass of adsorbent in a reactor section of length Δz is $\rho_b A_R \Delta z$, so the mass of adsorbate bound to the solid in such a section once the adsorbent has equilibrated with the influent solution is $q_{\text{in}}^* \rho_b A_R \Delta z$, where, as before, q_{in}^* is the adsorption density in equilibrium with c_{in}. The mass of adsorbate in the solution pore spaces in the same section of the reactor is $c_{\text{in}} \varepsilon A_R \Delta z$. Equating the total adsorbate in the volume $A_R \Delta z$ with the amount of adsorbate entering the reactor between t_1 and t_2, we find

Total adsorbate mass in solution and on solid in section of volume $A_R \Delta z$	=	Total adsorbate mass entering reactor between t_1 and t_2

$$(c_{\text{in}} \varepsilon + q_{\text{in}}^* \rho_b) A_R \Delta z = Qc_{\text{in}}(t_2 - t_1) \quad (8\text{-}57)$$

The velocity of the MTZ is the ratio of Δz to the time interval, so

$$v_{\text{MTZ}} = \frac{\Delta z}{t_2 - t_1} = \frac{Qc_{\text{in}}}{c_{\text{in}} \varepsilon A_R + q_{\text{in}}^* \rho_b A_R} \quad (8\text{-}58)$$

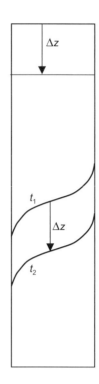

FIGURE 8-17. An approximation for the mass of adsorbate entering a fixed bed and being removed from solution between times t_1 and t_2. The adsorbate that enters between t_1 and t_2 is removed mostly in the region between the two curves in the lower part of the diagram, representing profiles of concentration or adsorption density at the two times. These changes are almost the same as would occur if a layer with thickness Δz were added to the top of the bed, containing concentration c_{in} in solution and adsorption density q_{in}^* on the solid.

By inverting both sides of Equation 8-58, the velocity of the MTZ can be divided into two components, one reflecting the contribution of the fluid velocity, and the other reflecting the extent to which the adsorbent slows down the movement of the adsorbate compared to that of bulk solution:

$$\frac{1}{v_{\text{MTZ}}} = \frac{c_{\text{in}} \varepsilon A_R + q_{\text{in}}^* \rho_b A_R}{Qc_{\text{in}}} = \frac{\varepsilon}{v} + \frac{\rho_b}{v} \frac{q_{\text{in}}^*}{c_{\text{in}}} \quad (8\text{-}59)$$

Since $v (= Q/A_R)$ is the approach velocity of the water upstream of the packed section of the bed, v/ε is the average velocity of the water in the direction of flow in the packed portion of the bed. Thus, Equation 8-59 indicates that the wave front of a nonadsorbing constituent (for which $q_{\text{in}}^* = 0$) moves through the reactor at the same velocity as the water moves through the pore spaces, as would be expected. The second term on the right side of Equation 8-59 represents the effect of sorption on the velocity of the MTZ. The larger this term is, the slower is the movement of the MTZ through the bed. Multiplying all terms by Z makes this point even more explicitly:

$$\frac{Z}{v_{MTZ}} = \frac{Z\varepsilon}{v} + \frac{\rho_b Z}{v}\frac{q^*_{in}}{c_{in}} \quad (8\text{-}60)$$

$$\tau_{MTZ} = \tau_w + \tau_{ads} \quad (8\text{-}61)$$

In Equation 8-61, τ_{MTZ} is the time required for the MTZ to move a distance equal to the length of the fixed bed, τ_w is the time required for water or a nonsorbing tracer to move through the bed, and τ_{ads} is the incremental increase in the "residence time" of the MTZ in the bed compared to that of water (or a tracer), due to the sorption process. As we would expect, the greater the partitioning of the adsorbate onto the solid (larger q^*_{in}/c_{in}), the larger are τ_{ads} and τ_{MTZ}; that is, the slower the movement of the adsorbate's wave front. The ratio of the residence times of the MTZ and water is sometimes called the retardation factor, R; that is,

$$R = \frac{\tau_{MTZ}}{\tau_w} = 1 + \frac{\tau_{ads}}{\tau_w} \quad (8\text{-}62)$$

In natural systems, one might be interested in the transport of weakly adsorbing species, for which τ_{ads}/τ_w might be on the order of 1.0 or less. However, in most engineered systems, the point of the adsorption process is to delay the breakthrough of contaminant for a very long time; that is, to have very large retardation factors. For instance, in fixed bed ion exchange columns and GAC columns, τ_w is typically less than 10 min, whereas τ_{ads} is usually at least several hours in the former systems and several weeks in the latter. In such cases, $R \approx \tau_{ads}/\tau_w$, the time to ~50% breakthrough (i.e., τ_{MTZ}) is essentially identical to τ_{ads}, and the velocity of the MTZ is simply:

$$v_{MTZ} \approx \frac{v}{\rho_b}\frac{c_{in}}{q^*_{in}} = \frac{c_{in}}{q^*_{in}\rho_b}v \quad (8\text{-}63)$$

By combining the mathematical result shown in Equation 8-63 with the analysis of the example system characterized in Figure 8-16, we can draw some useful conclusions about the effects of various parameters on the velocity and shape of the MTZ in a given system. Several such inferences are summarized in Table 8-5.

EXAMPLE 8-8. A solution at pH 7.0 that contains pentachlorophenol (PCP) is to be treated in the GAC adsorption column characterized below. At pH 7.0, adsorption of PCP onto this particular type of GAC can be described by the Freundlich isotherm $q = 150c^{0.42}$, where q is in mg PCP/g GAC, and c is in mg/L. Calculate the velocity at which the MTZ moves through the column for influent concentrations of 2 and 10 mg/L, once the shape of the MTZ has been fully developed. If the MTZ is very thin, how long can the system be operated before complete breakthrough when treating these two influents?

Column information

Liquid flow rate = 589 L/min
EBCT = 12 min
Mass of carbon in bed = 3386 kg
Column diameter = 1.5 m
Column length = 4.0 m
$\varepsilon = 0.35$
$\rho_b = 479$ kg/m³

Solution. For the given conditions, the superficial velocity, v (m/min), and the water residence time, τ_w (min), can be calculated as follows:

$$v = \frac{Z}{EBCT} = \frac{4.0\text{ m}}{12.0\text{ min}} = 0.33\text{ m/min}$$

$$\tau_w = \frac{Z\varepsilon}{v} = \frac{(4.0\text{ m})(0.35)}{0.33\text{ m/min}} = 4.24\text{ min} = 0.00293\text{ d}$$

The remaining parameters of interest depend on the influent PCP concentration. The equilibrium adsorption density can be computed from the isotherm equation, and τ_{ads} can be computed as the final term in Equation 8-60. For instance, for the 2 mg/L influent:

$$q^*_{in}(\text{mg/g}) = 150(c(\text{mg/L}))^{0.42} = 150(2.0)^{0.42} = 201(\text{mg/g})$$

$$\tau_{ads} = \frac{\rho_b Z}{v}\frac{q^*_{in}}{c_{in}} = \frac{(479(\text{kg/m}^3))(4.0\text{ m})}{0.33(\text{m/min})}$$
$$\times \left(\frac{201(\text{mg/g})}{2.0(\text{mg/L})}\right)\frac{1(\text{g/L})/(\text{kg/m}^3)}{1440(\text{min/d})} = 401\text{ d}$$

TABLE 8-5. Response of the MTZ to Various System and Operating Parameters

Parameter	Change in MTZ if Parameter is Doubled
Q	v_{MTZ} doubles; MTZ broadens (approximately doubles)
c_{in}	v_{MTZ} can increase by a factor related to the isotherm equation; for a Freundlich isotherm, the new v_{MTZ} is 2^{n-1} times the baseline v_{MTZ}; MTZ is predicted to broaden, but change is usually imperceptible
r_p	No change in v_{MTZ}; MTZ broadens
D_s	No change in v_{MTZ}; MTZ sharpens
k_L	No change in v_{MTZ}; MTZ sharpens, although effect is often negligible, since transport through boundary layer is not rate limiting

Similar calculations can be carried out the for the 10 mg/L influent, yielding results of $q_{in}^* = 395$ mg/L and $\tau_{ads} = 157$ d. Thus, as expected, $\tau_w \ll \tau_{ads}$ in both cases. If the MTZ were very thin, then complete breakthrough would occur after a time equal to $\tau_w + \tau_{ads}$, which is essentially the same as τ_{ads}. ∎

8.5 CHEMICAL REACTIONS IN FIXED BED ADSORPTION SYSTEMS

Adsorbed compounds often undergo chemical reactions, either directly with the solid to which they are bound or with other dissolved or adsorbed compounds. In general, solid surfaces might be thought of as likely locations for chemical reactions because adsorbates are more concentrated there than in bulk solution. Furthermore, electrons are relatively mobile in many solids, so that electron transfer (i.e., oxidation/reduction) reactions can sometimes take place between an oxidant at one location on the surface and a reductant at another location, without any direct contact between the primary reacting species. One important example of such a process is corrosion, in which an oxygen molecule adsorbed at one location on a surface might acquire electrons that are released from the pipe metal (e.g., iron or copper) at another location. The metal is thereby oxidized, and the pipe deteriorates.

The ability of metallic pipes to carry electrons vastly exceeds that of other solids; nevertheless, many nonmetallic solids provide a path for electron transfer that can dramatically increase the overall rate of redox reactions. For instance, the oxidation of manganous (Mn(II)) species by dissolved chlorine in water treatment is much more rapid when the Mn(II) species are adsorbed onto oxide surfaces than when they are in bulk solution (Knocke et al., 1991). In these cases, the oxidation products (Mn oxide particles) provide new adsorptive surfaces, so the reaction is auto-catalytic.

In addition to facilitating redox reactions, adsorption is often a key step in reactions leading to precipitation or dissolution of solids and in biodegradation reactions. The effect of such reactions on the overall system behavior can be evaluated by modifying the mass balance equation to consider the surface reaction. Consider, for example, an adsorbate that can be destroyed by a reaction either in solution or when it is adsorbed. Assuming that these reactions are first order with respect to the dissolved concentration and adsorption density, respectively, the reaction rate expressions are

$$r_V = -k_{1,\text{sol'n}} c \quad (8\text{-}64)$$

$$r_s = -k_{1,s} q \quad (8\text{-}65)$$

where r_s is the rate of the surface reaction, with units of mass of adsorbate per mass of adsorbent per unit time.

The set of equations needed to model the behavior of this substance in a plug flow, fixed bed system would be the same as shown in Table 8-3, except that the two mass balances would have to be modified to reflect the reactivity of the substance. The modified equations are as follows:

Solution phase mass balance

$$\frac{\partial c}{\partial t} = -\frac{v}{\varepsilon}\frac{\partial c}{\partial z} - k_L(c - c_{r_p})\frac{3}{r_p}\frac{1-\varepsilon}{\varepsilon} - k_{1,\text{sol'n}} c \quad (8\text{-}66)$$

Solid phase mass balance

$$\frac{\partial q}{\partial t} = D_s\left(\frac{\partial^2 q}{\partial r^2} - \frac{2}{r}\frac{\partial q}{\partial r}\right) - k_{1,s} q \quad (8\text{-}67)$$

A fixed bed adsorption process treating a nonreactive substance can never attain a steady state (except at complete breakthrough), because adsorbate steadily accumulates in the adsorbent. By contrast, adsorption processes treating reactive adsorbates are always predicted to reach a steady-state in which the advective transport of adsorbate out of the reactor is less than the advective input by the amount of decay (assuming that the reaction products desorb and are flushed out of the system). At the scale of individual particles, maintenance of the steady state requires that the adsorbate be transported through the boundary layer and into the solid at the same rate that it is destroyed on or in the solid. If these rates are sufficiently rapid, then excellent removal of the adsorbate from the solution passing through the reactor can be maintained, in theory, indefinitely.

The general scenario presented above describes the behavior of a solid catalyst. In the ideal case, such a material adsorbs a reactive substance, mediates its destruction, and then releases the products back to solution, so that the cycle can be repeated endlessly. Some adsorbents that operate in that way are used in environmental engineering, especially to foster redox reactions.

The catalysis of the oxidation of Mn(II) by metal oxide particles was alluded to previously. When drinking water that contains Mn(II) is passed through a sand filter, some portion of the metal adsorbs. The surface catalyzes the oxidation of the metal to Mn(III) or Mn(IV) species, which form insoluble oxides that tend to remain attached to the surface. Often, these metal oxides are even better adsorbents and catalysts for Mn(II) than the underlying sand. As a result, over the course of a few months to years, the sand in such systems becomes completely covered with the metal oxides, and the systems reach a pseudo-steady state in which most of the soluble Mn(II) is removed from water as it passes through the bed (Knocke et al., 1991; Merkle et al., 1997). (The solids in the system are not truly at steady state, since the metals accumulate continuously. However, the accumulation is usually insufficient to cause any problems inside the filter, and the solution phase does reach a steady state in terms of having stable influent and effluent compositions.)

Another important example of combined adsorption and reaction involves the biodegradation of adsorbed organics on GAC. Although activated carbon is an especially good adsorbent for hydrophobic solutes, it also binds mildly hydrophilic solutes reasonably well and can serve as a substrate onto which biofilms can attach. Therefore, over time, GAC columns treating either drinking water or wastewater can become efficient bioreactors, in which the GAC serves as a collector of molecules that can subsequently be consumed by the attached organisms. In such cases, the organisms can be viewed as carrying out a (bio) regeneration step. That is, by degrading adsorbed molecules and releasing the products to bulk solution, the organisms make adsorption sites available that had been previously occupied.

While bioregeneration is very attractive when it occurs in the idealized manner described here, in practice the benefits of adsorbent regeneration are mitigated by the blockage of surface sites and pore openings by the organisms. Nevertheless, reactors are sometimes designed to exploit the synergism between adsorption and biogrowth (Speitel et al., 1987; Speitel and DiGiano, 1987; Erlanson et al., 1997). And, even when a reactor is not designed with that synergism in mind, recognizing the potential for (and, in many cases, the inevitability of) biogrowth on adsorbents is sometimes essential for understanding the behavior of a system.

8.6 ESTIMATING LONG-TERM, FULL-SCALE PERFORMANCE OF FIXED BEDS FROM SHORT-TERM, BENCH-SCALE EXPERIMENTAL DATA

The equations derived in the preceding sections allow us to interpret experimental data, evaluate key system parameters, and predict the behavior of adsorption-based treatment systems. However, gathering such data from full-scale fixed bed systems is often impractical because of the long times required for equilibrium to be reached and for complete breakthrough to occur. The problem is especially severe for adsorption of trace contaminants onto high-capacity granular media such as GAC, in which case the time frame for collection of the necessary data can be months to years. To overcome this problem, a great deal of effort has been devoted to determining how the long-term behavior of full-scale systems can be predicted based on data from shorter-term experiments using much smaller systems. Fixed bed adsorption tests using reduced adsorbent particle size (r_p) and EBCT are commonly referred to as rapid small-scale column tests (RSSCTs).

Decreasing r_p and EBCT reduces the resistance imposed by the internal and external transport steps, respectively, in the overall adsorption process. For the data collected in these systems to be most useful, the reduction in these resistances should not alter their relative importance in the overall process. That is, if external mass transfer contributes 25% of the total resistance to adsorption in the large-scale system, values of r_p and EBCT should be chosen for the small-scale system that cause the external resistance to be 25% of the total in that system as well. When this goal is met, it is said that *similitude* has been achieved between the two systems.

In a fixed bed system, the effects of the various resistances to adsorption manifest themselves in the shape of the MTZ. Therefore, the goal of achieving similitude is sometimes expressed by reference to the MTZ: if we can choose conditions that cause the MTZ to have the same shape and to extend over the same fraction of the column in the two systems, then similitude has been achieved. In the water and wastewater treatment field, much of the work on identifying conditions that achieve similitude between large and small scale adsorption systems has been done by Crittenden, Summers, and their coworkers (e.g., Crittenden et al., 1986, 1987, 1991; Cummings and Summers, 1994; Summers et al., 1995), and their approach is summarized here.

Many issues that arise when one attempts to up- or down-scale a process can be addressed most easily by writing all the governing equations for the system in nondimensional terms. When that is done for a fixed bed adsorption system, the parameters z, t, and c are typically made nondimensional by dividing by Z, EBCT, and c_{in}, respectively.[6] In addition to these parameters, the nondimensionalized equations contain several coefficients that characterize the relative importance of advection, axial diffusion and dispersion, transport through the liquid boundary layer, pore diffusion, and surface diffusion in the reactor. The nondimensionalized equations describing different fixed bed adsorption reactors have the same general form with respect to the relationship among the dimensionless z, t, and c parameters, but they often have different values of the dimensionless coefficients.

If a small-scale system can be designed to have exactly the same dimensionless coefficients as a large-scale system, then the dimensionless form of the breakthrough curves (fractional breakthrough vs number of BVs treated) will be the same in the two systems, allowing us to predict the performance of the large scale system from relatively short tests using the smaller system. For instance, if the large system had an EBCT of 15 min, approximately 35,000 BVs of water could be treated in one year. If the EBCT in the small system was 0.2 min, the same number of BVs could be treated in a little over 4 days. Thus, if similitude could be achieved between the two systems, the fractional breakthrough after 1 year of operation of the full-scale system could be predicted from a 4-day test with the smaller system.

According to Crittenden et al. (1987), complete similitude can be achieved if six dimensionless coefficients are matched in the small- and large-scale systems. However,

---

[6] Note that the nondimensionalized time, t/EBCT, is the number of BVs treated.

if the void fractions (ε) and bed densities (ρ_b) in the small- and large-scale systems are identical, and if the isotherm equation and the value of D_s are the same for the particles with small and large r_p, then the requirements for similitude reduce to the following two criteria:

$$\frac{v_{S\text{-}S}}{v_{L\text{-}S}} = \frac{r_{p,L\text{-}S}}{r_{p,S\text{-}S}} \quad (8\text{-}68)$$

$$\frac{\text{EBCT}_{S\text{-}S}}{\text{EBCT}_{L\text{-}S}} = \frac{r_{p,S\text{-}S}^2}{r_{p,L\text{-}S}^2} \quad (8\text{-}69)$$

where S-S and L-S refer to the small- and large-scale systems, respectively.

Since EBCT equals Z/v, Equations 8-68 and 8-69 can be combined to specify the relationship between column length and particle size when similitude is achieved, as follows:

$$\frac{Z_{S\text{-}S}/v_{S\text{-}S}}{Z_{L\text{-}S}/v_{L\text{-}S}} = \frac{r_{p,S\text{-}S}^2}{r_{p,L\text{-}S}^2}$$

$$\frac{Z_{S\text{-}S}}{Z_{L\text{-}S}} = \frac{r_{p,S\text{-}S}^2}{r_{p,L\text{-}S}^2}\frac{v_{S\text{-}S}}{v_{L\text{-}S}} = \frac{r_{p,S\text{-}S}^2}{r_{p,L\text{-}S}^2}\frac{r_{p,L\text{-}S}}{r_{p,S\text{-}S}} = \frac{r_{p,S\text{-}S}}{r_{p,L\text{-}S}} \quad (8\text{-}70)$$

Because Equation 8-70 can be derived from Equations 8-68 and 8-69, the three equations are not independent. To achieve similitude, any two of the three must be satisfied; when that condition is met, the third equation will automatically be satisfied.

■ **EXAMPLE 8-9.** A fixed bed packed with GAC is being proposed to treat a water that has been contaminated with the pesticide dieldrin. The system under consideration would use GAC particles with a diameter of 0.6 mm, a column length of 3.5 m, and an EBCT of 10 min. A small-scale test using activated carbon with a diameter of 0.08 mm is to be conducted to predict the behavior of the full-scale system. What column length and EBCT should be used to achieve similitude between the small- and full-scale systems? How long should the small-scale test be run to simulate one year of operation at full scale?

Solution. According to Equation 8-70, the ratio of the lengths of the small- and large-scale systems should be the same as the ratio of the particle diameters. And, according to Equation 8-69, the ratio of the EBCT's should be the square of the ratio of the particle diameters. Thus

$$Z_{S\text{-}S} = \frac{r_{p,S\text{-}S}}{r_{p,L\text{-}S}} Z_{L\text{-}S} = \frac{0.08\,\text{mm}}{0.6\,\text{mm}} 3.5\,\text{m}$$

$$= 0.467\,\text{m} = 46.7\,\text{cm}$$

$$\text{EBCT}_{S\text{-}S} = \frac{r_{p,S\text{-}S}^2}{r_{p,L\text{-}S}^2}\text{EBCT}_{L\text{-}S} = \frac{(0.08\,\text{mm})^2}{(0.6\,\text{mm})^2} 10\,\text{min}$$

$$= 0.178\,\text{min} = 10.7\,\text{s}$$

To simulate 1 year of operation, the small-scale system should be operated until it has treated the same number of BVs of influent as the full-scale system would treat in 1 year. Because the EBCT of the small column is 1.78% that of the long column (0.178 min/10 min), the time required to simulate 1 year of operation would be 1.78% of a year, or 6.5 days. ■

As noted previously, the apparent value of D_s sometimes varies as a function of particle size, at least for adsorption onto activated carbon. Unfortunately, the conditions for similitude just described apply only to systems in which D_s is independent of r_p (i.e., the CD model, for which $m = 0$ in Equation 8-34). If $m \neq 0$, it is not possible to find parameter values for the small-scale system that satisfy all the requirements for similitude.

In such cases, Crittenden et al. (1987, 1991) have suggested that one should attempt to achieve close similitude between the large- and small-scale systems with respect to external mass transfer resistance (i.e., resistance to transport through the boundary layers around the particles) and internal resistance (resistance to transport inside the pores); because most beds operate as nearly ideal PFRs, they suggest that achieving similitude with respect to longer-range transport phenomena is less important.

The relative resistances associated with mass transfer through the boundary layer surrounding the particles and diffusion inside the particles can be matched in the two systems for any value of m if the following equation is satisfied:

$$\frac{\text{EBCT}_{S\text{-}S}}{\text{EBCT}_{L\text{-}S}} = \left(\frac{r_{p,S\text{-}S}}{r_{p,L\text{-}S}}\right)^{2-m} \quad (8\text{-}71)$$

Thus, for example, in a system that is characterized by PD ($m = 1$), similitude with respect to internal resistance can be achieved if the EBCT is proportional to the particle size. This contrasts with a CD system, in which the EBCT is proportional to the square of the particle size. The main consequence of Equation 8-71 is that, for a given ratio of particle sizes, the larger the value of m, the longer the EBCT that is required, and the longer the small-scale tests must be run to obtain data that can be used to predict the long-term behavior of the larger system. Because the plug flow approximation improves with increasing Reynolds number (Re), Crittenden et al. (1987, 1991) recommended that, in addition to using Equation 8-71, small-scale systems be designed so that Re ≥ 1 for flow around the particles.

Several researchers have compared the breakthrough curves for pilot- or full-scale adsorption columns with those predicted based on RSSCTs. The result of one such comparison is shown in Figure 8-18.

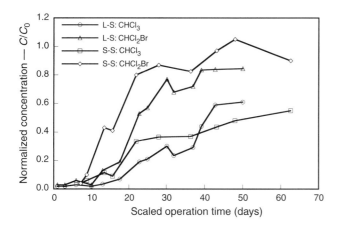

FIGURE 8-18. Comparison of predicted and observed breakthrough curves for two disinfection by-products formed during chlorination of drinking water–chloroform ($CHCl_3$) and bromodichloromethane ($CHCl_2Br$). The data from the small-scale (S-S) column were collected in slightly less than one week; they were then scaled to predict the behavior of the large-scale (L-S) column over a period of 7 weeks. *Source*: After Cummings and Summers (1994).

8.7 COMPETITIVE ADSORPTION IN COLUMN OPERATIONS: THE CHROMATOGRAPHIC EFFECT

Systems with Rapid Attainment of Adsorptive Equilibrium

When a solution containing multiple adsorbates is treated in a fixed bed adsorption system, the concentrations of adsorbates that are preferentially adsorbed decrease to a greater extent than the other adsorbates near the inlet. As a result, the solution becomes relatively enriched in weakly binding species as it moves through the column. The preference for one species over another generally reflects the relative adsorptive binding strengths, although kinetic factors (e.g., quicker movement into interior pores) could also contribute to the preference.

If the adsorbates bind to different groups of surface sites, or if they compete for the same group of sites, but the total adsorption density is insufficient to consume a substantial fraction of those sites, then they bind essentially independently. In such a case, the MTZ for each adsorbate forms and moves through the bed in an almost identical manner to what would occur in a single-adsorbate system. However, if the fraction of the surface sites occupied is large enough that competition between the adsorbates is significant, then the binding of each species will be less extensive than in a single-sorbate system.

Even if competition is significant, then the MTZ of the most preferred adsorbate will still develop and move downstream in a similar manner to its behavior in a single-adsorbate system, albeit at a higher velocity, because it will occupy a smaller fraction of the binding sites in any given layer of the bed. The MTZs of each of the less preferred species will also move through the column more rapidly than if it were the only adsorbate present. In the limiting case where the preferred species is strongly adsorbed and is present at a sufficiently large influent concentration that it occupies almost all the sites near the inlet, the less preferred species will adsorb negligibly until almost all of the preferred species has been removed from solution. In addition to this displacement downstream, the MTZ of the less preferred species might also change in other, sometimes dramatic, ways.

Consider a fixed bed adsorption system receiving an influent that contains comparable concentrations of a very strongly adsorbing species A and a weakly adsorbing species B. After the system has operated for some time, the adsorbent near the inlet will have equilibrated with the influent concentration of A (i.e., $q_A \approx q^*_{A,in}$), and a thin MTZ for species A will exist at some distance downstream. Because, for this limiting case scenario, we are assuming that species B competes negligibly for sites in the upstream section of the bed, the shape and location of the MTZ for species A are approximately the same as in a system that contains no B at all. The situation is shown schematically by the solid lines in Figures 8-19a and 8-19b, with the region where $q_A \approx q^*_{A,in}$ identified as zone I, and the MTZ identified as zone II. Downstream of the MTZ, c_A and q_A are both negligible (zones III and IV in the figure).

Because species B is a weak competitor for sites, and because we are assuming that $q^*_{A,in}$ is large enough that A occupies most of the available sites in the portion of the bed nearest the inlet, $q_B \approx 0$ in zone I. As c_A starts declining in zone II (the MTZ for species A), q_B starts increasing. We consider the relationship of q_B to $q^*_{B,in}$ shortly; for now, the only important point is that q_B increases. The MTZ for species A terminates slightly downstream, at which point species B is the only adsorbate remaining in solution. At that point, species B forms its own MTZ (in the region designated zone III in the figure), similar in shape (but not location or magnitude) to the one that would form if it had been the only adsorbate entering the bed. Downstream of the MTZ of species B, both c_A and c_B are negligible, and so q_A and q_B are as well (zone IV).

Now, consider the changes that occur in a small aliquot of influent as it passes through the reactor at the time depicted in Figure 8-19. When the solution first enters the column, neither A nor B adsorbs, since the adsorbent has already equilibrated with this solution (causing the adsorption densities to be $q_A \approx q^*_{A,in}$ and $q_B \approx 0$, respectively). Correspondingly, c_A and c_B remain at $c_{A,in}$ and $c_{B,in}$ throughout zone I.

Once the solution enters zone II, it encounters adsorbent that has previously been exposed to c_A values $<c_{A,in}$. As a

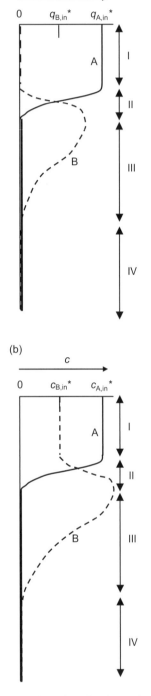

FIGURE 8-19. (a) Adsorption density and (b) concentration profiles of two adsorbable species in a continuous flow column adsorption system. Adsorbate A binds much more strongly than adsorbate B.

result, when the solution enters that zone, some A adsorbs, causing c_A to decline and q_A to increase. However, this is also a zone where some B has adsorbed (since the small amount of A reaching the zone previously did not occupy all the available sites). As q_A increases in this zone, some of the previously adsorbed B is displaced from the adsorbent sites. No B has been removed from the aliquot of influent up to this point, so the desorption of B causes c_B to increase above $c_{B,in}$. In essence, the adsorbent in this zone has collected B from earlier aliquots of influent and is adding it to the later-arriving aliquot.

Moving farther downstream, we reach the end of the MTZ for species A, where unoccupied sites are available to bind B. In fact, downstream of the MTZ of species A, the system behaves as if that location were the inlet of a bed that was receiving an influent containing species B as the only adsorbate. However, the "influent" at that point contains B at a concentration greater than $c_{B,in}$. Correspondingly, when this solution equilibrates with the local adsorbent, the adsorption density exceeds $q_{B,in}^*$.

Subsequent aliquots of influent cause the same scenario to be repeated continuously, so that the local, transiently high values of c_B and q_B grow ever larger, at locations that move steadily downstream. When this wave of ever-increasing concentrations of B reaches the outlet of the bed, the effluent concentration increases abruptly to a value that can be much larger than $c_{B,in}$.

The process of transient adsorption followed by desorption to generate a solution with a higher dissolved concentration of the desorbing species than is present in the feed is the basis for separating and concentrating species from a mixture by chromatography. Therefore, when this phenomenon occurs in a water treatment process, it is known as the *chromatographic effect*, and the high concentration of the adsorbate exiting the column is referred to as a *chromatographic peak*.

The highest value that c_B could attain anywhere in the column is the sum of $c_{B,in}$ and the maximum amount of B that can be displaced by adsorption of A. For instance, if the adsorbent has only one type of site, and molecules of A and B compete for that site-type, then the maximum concentration of B that can be displaced by adsorbing A molecules equals $c_{A,in}$. As a result, the maximum concentration of B that can exist anywhere in the column (or in the effluent), on a molar basis, is $c_{B,in} + c_{A,in}$. The upper limit on q_B would then be the value that is in equilibrium with the highest attainable value of c_B. Thus, for instance, if adsorption of B followed a Langmuir isotherm, and if the maximum attainable value of c_B were $\gg 1/K_B$, then q_B could reach q_{max} in the downstream portions of the bed, even if q_{max} were much greater than $q_{B,in}^*$.

If multiple adsorbates (A, B, C, ..., n) are present in the influent, the preceding scenario can develop at several points in a single bed. In a zone near the inlet, the adsorbent would be in equilibrium with the influent composition. The most strongly adsorbed species (A) would be selectively removed from solution in this region, and, at some distance downstream, essentially all of the A would be removed from solution. As a result, the remainder of the bed would contain only species B through n. Immediately downstream of where A disappears from the solution (i.e., downstream of the MTZ for species A), species B through n would all be present at concentrations greater than their concentrations in the influent to the bed.

Just downstream of the MTZ for species A, the most strongly adsorbing species still present in the solution (B) will be selectively removed, while those that bind less strongly will have a greater tendency to migrate deeper into the bed. Sorption of B will therefore induce a chromatographic effect on the other species present in this zone (C through n), so that the solution exiting the zone will contain those species at concentrations even greater than the concentrations at the inlet to the zone. The only qualitative difference between the sequence of events in this zone and the zone at the bed inlet is that the location of the second zone moves steadily downstream, as the upstream zone (where species A adsorbs) expands.

Following the above logic, in a fixed bed receiving an influent with n adsorbates, $n+1$ zones will develop, each containing one less adsorbate than the zone immediately upstream. In addition, the dissolved concentration of each adsorbate will increase at the boundaries between successive downstream zones, reaching a maximum concentration in the zone where it is the most strongly adsorbing species remaining in solution; at the end of that zone, the adsorbate undergoes a dramatic decrease to a concentration of essentially zero. All of the zones expand as the run proceeds, with the upstream zones pushing the downstream zones toward the outlet. This process leads to a characteristic concentration versus time profile in the effluent in which progressively more adsorbates appear stepwise (with the weakest-binding adsorbate appearing first). Each additional species that appears is at a concentration greater than its c_{in} value, except for the strongest adsorbate, which appears at c_{in}; with each appearance of an additional species, the concentrations of all the other species already present in the effluent decrease.

Such a sequence is shown for a generic system with four adsorbates in Figure 8-20. The adsorbates shown in the figure have progressively stronger affinity for the adsorbent in the order $D < C < B < A$. Initially, all the adsorbates are completely removed from solution, so the effluent contains none of them. Later, D appears, at the highest concentration it will ever have in the effluent. Subsequently, C appears, at the highest concentration *it* will ever have, and the concentration of D declines at the same time. The concentrations of C and D then remain constant until B appears in the effluent, again at the highest concentration that it will ever have, as the concentrations of C and D decline. This pattern then repeats itself when A breaks through the bed, and all four adsorbates are present in the effluent at their influent concentrations.

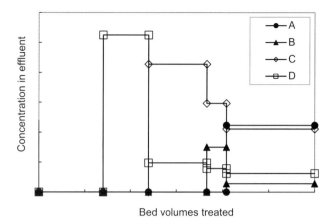

FIGURE 8-20. Generic effluent history obtained for a system in which four adsorbates are in the influent to an ion exchange column. The affinities of the adsorbates for the adsorbent increase in the order $D < C < B < A$. influent concentrations are those shown at the end of the run.

These trends are widely observed in full-scale treatment systems. However, quantitative predictions of the height and width of the various zones is complex, even in the limiting case of ideal plug flow through the column. The calculations for ion exchange reactions have been discussed by Clifford (1982),[7] and those for reactions in which adsorption is characterized by Freundlich isotherms and competition by the ideal-adsorbed solution theory (IAST) have been discussed by Crittenden et al. (1987); in both cases, the analysis is based on classic chromatography theory, adapted to the particular systems of interest. Computer programs that predict the effluent profiles for the two cases have been developed by the corresponding research groups (Tirupanangadu et al., 2002; CenCitt, 1999, respectively).

■ **EXAMPLE 8-10.** It is proposed to remove sulfate and nitrate from a groundwater source by ion exchange in a bed packed with resin that is pre-saturated with chloride. In the process, one charge equivalent of chloride (1 mol of Cl^-) is released for each charge equivalent of sulfate or nitrate that is adsorbed (0.5 mol of SO_4^{2-} or 1.0 mol of NO_3^-, respectively). The separation factors for sulfate and nitrate relative to chloride can be approximated as being constant throughout the bed, with values of 9.2 and 2.9, respectively. Some characteristics of the resin and the operation are as follows:

Apparent particle density of the resin ($\rho_{p,app}$): $1.10\,g/cm^3$
Packing density of the resin in the bed (ρ_b): $0.68\,g/cm^3$
Ion exchange capacity of the resin (q_{max}): $3.0\,meq/g$

---

[7] The calculations for ion exchange are based on the assumption that the separation factors are constant throughout the column. For heterovalent exchange, this assumption is not strictly valid, but it is usually a reasonable approximation.

362 ADSORPTION PROCESSES: REACTOR DESIGN AND ANALYSIS

Empty bed contact time (EBCT): 5 min

Influent sulfate concentration ($c_{SO_4^{2-},in}$):
  96 mg/L = 1.0 mmol/L = 2.0 meq/L

Influent nitrate concentration ($c_{NO_3^-,in}$):
  14 mg/L as N = 1.0 mmol/L = 1.0 meq/L

Influent chloride concentration ($c_{Cl^-,in}$):
  12 mg/L = 0.34 mmol/L = 0.34 meq/L

A computer program (EMCT 2.0; Tirupanangadu et al., 2002) predicts that three zones will be identifiable in the bed shortly after the run starts, and that, if the hydraulics in the bed approximate plug flow, the solution compositions in those zones will be as shown below. Assuming that the MTZs are of negligible thickness, determine the composition of the effluent from the bed as a function of time and of the number of BVs treated, until complete breakthrough.

	$c_{SO_4^{2-}}$ (meq/L)	$c_{NO_3^-}$ (meq/L)	c_{Cl^-} (meq/L)
Zone 1	2.00	1.00	0.34
Zone 2	0.00	2.93	0.41
Zone 3	0.00	0.00	3.34

Plot the number of equivalents of each ion adsorbed per liter of bed as a function of time and of the number of BVs treated, from the beginning of the run until complete breakthrough.

Solution. The solution composition in Zone 1 is identical to the influent composition; therefore, Zone 1 must be the zone nearest the inlet, where the resin has equilibrated with the influent. An aliquot of influent passes through this zone without any change in its composition. Then, essentially all the SO_4^{2-} is removed from that aliquot in the very thin MTZ that resides between Zones 1 and 2, so no SO_4^{2-} is present in the solution downstream of that MTZ.

When an aliquot of solution reaches the boundary between Zones 1 and 2, most of the SO_4^{2-} in that aliquot adsorbs, displacing some NO_3^- and Cl^- from the resin. Until this aliquot of solution arrived, the resin in that section of the bed had been exposed to solution containing only NO_3^- and Cl^-, and it had selectively adsorbed the NO_3^-. As a result, most of the ions that are displaced by the SO_4^{2-} will be NO_3^- ions. Correspondingly, at the boundary between Zones 1 and 2, the NO_3^- concentration increases from 1.00 to 2.93 meq/L, whereas the Cl^- concentration increases only from 0.34 to 0.41 meq/L.

After the SO_4^{2-} has been adsorbed from this aliquot of solution and the aliquot proceeds farther downstream, it passes through resin that has already equilibrated with solution having the same composition. As a result, no adsorption occurs, and the solution composition remains constant until the MTZ for NO_3^- is reached. At that point, essentially all the NO_3^- adsorbs, displacing Cl^- that had previously bound to the resin. Farther downstream, both the solution and the resin contain Cl^- as the only anion, so no more changes occur.

The composition of the resin phase in each zone can be determined based on the equations for competitive adsorption developed in Chapter 7. Specifically, Equation 7-92, repeated here as Equation 8-72, is directly applicable for computing the adsorption density of each ion in each zone of the bed. The results of those calculations are shown in the following table.

$$q_i = \frac{\alpha_{j/k} c_i}{\sum_{all\ j} \alpha_{j/k} c_j} q_{max} \qquad (8\text{-}72)$$

	$q_{SO_4^{2-}}$ (meq/g)	$q_{NO_3^-}$ (meq/g)	q_{Cl^-} (meq/g)
Zone 1	2.55	0.40	0.05
Zone 2	0.00	2.86	0.14
Zone 3	0.00	0.00	3.00

The rate at which Zone 1 expands to occupy progressively more of the bed can be computed using an approach very similar to that for computing the velocity of the MTZ of an adsorbate in a non-competitive system (Example 8-8). The preceding table indicates that 2.55 meq of SO_4^{2-} adsorbs onto each gram of resin in Zone 1. Since the influent contains 2.0 meq/L of SO_4^{2-}, and all of this SO_4^{2-} adsorbs, Zone 1 must expand to include an additional 2.0/2.55, or 0.78 g of resin for each liter of influent applied to the bed. The packing density of resin in the bed, ρ_b, is 0.68 g/cm³, or 680 g/L of bed, so Zone 1 expands at a rate given by

$$\begin{pmatrix} \text{Rate of expansion} \\ \text{of Zone 1} \end{pmatrix}$$

$$= \frac{0.78\text{ g of resin accrues to Zone 1 per L of influent}}{680\text{ g of resin per L of bed}}$$

$$= 1.15 \times 10^{-3}\text{ L of bed accrues to Zone 1/L of influent}$$

Inverting this result, we see that each 870 L of influent expands Zone 1 by 1.0 L of bed. Correspondingly, 870 BVs of influent would cause Zone 1 to occupy 1.0 BV of the bed; that is, the whole bed. Thus, we expect SO_4^{2-} to break through the bed after 870 BVs have been treated.

The movement of Zone 2 through the bed can be evaluated in a similar, but not identical way, because both of its boundaries (its beginning and end) move as the run proceeds. The easiest way to carry out the analysis is to note that, until the time when NO_3^- breaks through into the effluent, all of the NO_3^- supplied to the column is retained in Zones 1 and 2 (along with all the SO_4^{2-}). Based on the calculations just completed, Zone 1 expands at a rate of

0.78 g of resin per L of influent. Since $q_{NO_3^-}$ is 0.40 meq/g in Zone 1, that zone adsorbs (0.78)(0.40), or 0.312 meq of NO_3^- per L of influent. The influent contains 1.0 meq of NO_3^- per L, so 0.688 meq of NO_3^- per L must adsorb in Zone 2.

The adsorption density of NO_3^- in Zone 2 is 2.86 meq/g, so the rate at which resin mass accrues to Zone 2 can be computed as follows:

$$\begin{pmatrix} \text{Rate of expansion} \\ \text{of Zone 2} \end{pmatrix}$$

$$= \frac{0.688 \text{ meq NO}_3^- \text{ adsorbs in Zone 2 per L of influent}}{2.86 \text{ meq NO}_3^- \text{ adsorbs per gram of resin in Zone 2}}$$

$$= 0.24 \text{(gram of resin accrues to Zone 2)}/(\text{L of influent})$$

The total mass of resin associated with Zones 1 and 2 therefore grows at a rate of $0.78 + 0.24$, or 1.02 g of resin per L of influent. The corresponding volumetric growth rate of Zones 1 and 2 equals the product of the mass growth and the bed density; that is,

$$\begin{pmatrix} \text{Volumetric growth} \\ \text{of Zones 1} + 2 \end{pmatrix}$$

$$= \frac{1.02 \text{ g resin occupied per L of influent}}{680 \text{ g resin per L of bed}}$$

$$= 0.0015 \text{ L of bed occupied}/\text{L of influent}$$

Inverting the above result, we conclude that every 667 L of influent provides enough SO_4^{2-} and NO_3^- to cause the total volume associated with Zones 1 and 2 to increase by 1.0 L. Correspondingly, 667 BVs of influent would cause Zones 1 and 2 to occupy 1 BV of bed; that is, to occupy the whole bed. Thus, we conclude that Zone 2 will reach the effluent, and NO_3^- will break through the column, when 667 BVs of influent have been treated.

The results of these calculations are shown graphically in Figure 8-21. The conversion from BVs to time is based on the given EBCT of 5 min. NO_3^- breaks through after 55.6 h, and SO_4^{2-} does so after 72.5 h.

At the beginning of the treatment step, the resin is presaturated with Cl^-; that is, all the sites in the bed are occupied by Cl^-. The number of meq of chloride adsorbed per L of bed, designated here as m_{Cl^-}, can therefore be computed as

$$m_{Cl^-,\text{init}} = \rho_b q_{max}$$

$$= \left(0.68 \frac{\text{g}}{\text{cm}^3 \text{ of bed}}\right)(3.0 \text{ meq/g})(10^3 \text{ cm}^3/\text{L})$$

$$= 2040 \frac{\text{meq}}{\text{L of bed}}$$

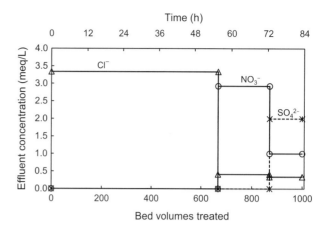

FIGURE 8-21. Effluent concentration profiles for the three ions in the example system.

Thereafter, m_{Cl^-} decreases by 3.0 meq for each liter of influent that passes through the bed (i.e., every 5 min), as Cl^- is released in exchange for the 2 meq of SO_4^{2-} and 1 meq of NO_3^- that adsorbs from each liter of influent. This process continues until the NO_3^- stops adsorbing; that is, until NO_3^- breakthrough.

During the period between breakthrough of NO_3^- and SO_4^{2-}, the influent contains 0.34 meq/L of Cl^-, and the effluent contains 0.41 meq/L. Thus, the amount of Cl^- adsorbed is declining at a rate of 0.07 meq per liter of influent. Given the EBCT of 5 min, the feed rate can be expressed as 0.2 L of influent per L of packed volume per minute. Therefore, the loss of Cl^- from the resin during this period is

Loss rate of Cl^- from bed

$$= (0.07 \text{ meq Cl}^-/\text{L of influent})$$

$$\times (0.2 \text{ L of influent}/[(\text{L of bed})(\text{min})])$$

$$= 0.014 \text{ meq Cl}^-/[(\text{L of bed})(\text{min})]$$

This loss continues until SO_4^{2-} breaks through, after which the influent and effluent both contain 0.34 meq/L Cl^-, and the (small) amount of adsorbed Cl^- remains constant.

The number of meq of NO_3^- adsorbed is initially zero. It then increases by 1.0 meq per liter of bed every 5 min, until NO_3^- breakthrough after 55.6 h. During the period when the NO_3^- chromatographic peak is exiting the bed, the effluent contains 2.93 meq/L NO_3^-, so the net loss of adsorbed NO_3^- is 1.93 meq/L of solution. This loss corresponds to (1.93 meq/L of influent) (0.2 L of influent per liter of bed per minute), or 0.386 meq per liter of bed per minute, until SO_4^{2-} breakthrough at $t = 72.5$ h. Thereafter, the amount of adsorbed NO_3^- remains constant.

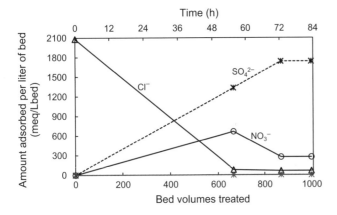

FIGURE 8-22. The distribution of adsorbed species as a function of the number of bed volumes treated in the example system.

The concentration of adsorbed sulfate increases by 2.0 meq/L every 5 min, from the beginning of the run until it breaks through into the effluent 72.5 h later. These trends are all shown in Figure 8-22. ∎

The preceding discussion and example make it clear that, in addition to reducing the time until the weaker adsorbate breaks through, competitive adsorption in fixed bed systems can generate a concentration of that constituent that is substantially greater than the influent concentration. As noted, the maximum value of that peak above the influent concentration is determined by the concentration(s) of the more strongly binding adsorbate(s). Thus, in the example system, the peak chloride concentration was 3.0 meq/L above its influent concentration, because that amount of chloride was released when all the sulfate and nitrate adsorbed. Correspondingly, the peak nitrate concentration was ~2.0 meq/L above its influent concentration, because that was the amount of nitrate released when all the sulfate adsorbed. (The actual peak was slightly less concentrated, because a small fraction of the SO_4^{2-} was exchanged for Cl^- rather than NO_3^-.)

While the consequences of the chromatographic effect might be serious under any circumstance, they can be especially severe when the weaker-binding adsorbate is a toxic compound that is present at only a trace concentration in the influent. For instance, consider a case like that in the example, but for an influent that also contains arsenate, which might bind as the monovalent ion $H_2AsO_4^-$. This ion typically adsorbs to strong base anion exchange resins with an affinity that is intermediate between the affinities of sulfate and nitrate. Making the same simplifying assumptions as in the example, if arsenate were present in the influent at a concentration of 1.0 μeq/L, it would form an extremely thin MTZ just downstream of the sulfate MTZ. In that case, virtually no arsenate would appear in the effluent until immediately before sulfate broke through. However, at least in theory, an arsenic spike of almost 2.0 meq/L (2000 times the influent concentration) could exit the bed at that instant.

In reality, the hydrodynamics in the bed (axial mixing), incomplete removal of both sulfate and arsenate prior to breakthrough, and the ability of arsenate to compete successfully for some resin sites even in the presence of sulfate would spread out and diminish the arsenate peak. Nevertheless, the qualitative point is clear: if adsorption is used to remove toxic compounds from a solution, a very large safety factor and/or backup removal systems must often be employed to assure that a large spike of that compound does not escape the bed and enter the effluent.

Competitive Adsorption in Systems That Do Not Reach Equilibrium Rapidly

Conceptually, modeling of competitive adsorption in fixed beds where adsorptive equilibrium is not attained instantaneously is a straightforward extension of the principles and equations developed previously. Usually, in such modeling, an assumption is made that the only interaction among the adsorbates is competition for the sites; that is, the parameters k_L and D_s are assumed to be the same for a given species in a given system regardless of the concentration of competing adsorbates. Then, mass balances are written for each constituent, for control volumes enclosing the solution, the solid, and the boundary layers around the adsorbent particles. The equations are all linked via the boundary conditions describing equilibrium at the solid/solution interface; the only difference between these equations and those for adsorption in single-adsorbate systems is that the isotherm for each adsorbate must take into account the presence of the competing adsorbates. If equilibration is not instantaneous, but local equilibrium is assumed to apply within the particle, then the expressions for local equilibrium must also reflect the competition.

Computer software is available for solving the equations that characterize competitive adsorption in fixed beds, and, if a system contains only a few well-defined and non-interacting adsorbates, such modeling is often quite successful. However, as noted in Chapter 7, a number of difficulties hamper our ability to model competitive adsorption in systems with complex mixtures of interacting adsorbates, and these difficulties are sometimes exacerbated in fixed bed systems. In particular, even when competitive adsorption can be characterized reasonably well in short-term batch tests, the extension of the results to fixed bed operations using the same adsorbent often leads to erroneous predictions.

For instance, in fixed bed GAC systems treating influents that contain trace organic compounds (e.g., pesticides or solvents) and higher concentrations of NOM, predictions based on batch tests often significantly under-estimate the competitive effects of the background compounds. That is, in practice, the trace compounds break through the beds earlier than predicted. In such systems, the trace compounds are often adsorbed more efficiently than most of the molecules that contribute to NOM, both because the trace compounds tend to be more hydrophobic and because they are smaller and therefore have access to more of the small internal pores.

The explanation for the poor correlation between batch and packed bed systems is thought to be related to the segregation of the MTZs of the two components, and the relatively slow reversibility of the sorption reaction. Specifically, in fixed bed operation, the MTZ of the NOM compounds precedes that of the trace, target compounds through the bed. In the upstream portion of the bed, the trace compounds and NOM compounds both adsorb simultaneously, with the trace compounds having a competitive advantage.[8]

Downstream of this region, the NOM compounds encounter clean adsorbent and adsorb extensively. However, in doing so, they might block many of the passageways to the smaller internal pore spaces. As a result, when the MTZ's of the trace compounds later move into the downstream portion of the bed, access to the pores is hindered or prevented altogether. Empirically, this effect is manifested as a substantial decrease in the apparent pore and surface diffusivity of the trace compound compared to the corresponding values on fresh adsorbent, or, if the diffusivity is lowered sufficiently, an apparent decline in the adsorptive capacity. Some efforts have been made to account for this so-called "preloading" effect, but they are inherently empirical due to the uncertain and inconsistent composition of the NOM.

In theory, this kinetic limitation on sorption of the smaller compounds should have no effect on the ultimate equilibrium. That is, eventually the smaller compounds would be expected to migrate into the pores and outcompete the natural compounds to the same extent that they would if they had arrived earlier. However, as a practical matter, the decreased effective diffusion rate can cause the trace compounds to break through the bed long before adsorptive equilibrium is reached. Thus, although the models developed in previous sections provide the essential framework for predicting and/or interpreting competitive adsorption in dynamic systems, one must be cautious to test whether the assumptions of the model apply as reliably in such systems as in single-adsorbate systems.

8.8 ADSORBENT REGENERATION

When the contaminant concentration in the effluent from an adsorption bed exceeds acceptable levels, the bed must be removed from service, and the adsorbent must be either regenerated or disposed of. Often, adsorption is uneconomical unless the adsorbent can be regenerated and reused.

One approach for regenerating spent adsorbent is to expose the solid to a solution containing a weakly adsorbing species at an overwhelming concentration. Using that approach, it is often possible to cause the weakly binding species to outcompete the more strongly binding adsorbates for the surface sites, even though the affinity of the weak adsorbate for the sites is relatively small.

If most of the sites are on the exterior surface of the adsorbent, or if transport into the interior of the adsorbent is rapid (as in ion exchange resins), then regeneration can be carried out rapidly *in situ*, without removing the adsorbent from the bed. Systems that can be regenerated in this manner typically target hydrophilic (usually ionic) adsorbates, such as metals or inorganic anions. In such cases, if the surface sites are weak acids or bases, the adsorbent might be regenerated by exposure to an acidic or basic solution, respectively. Strongly acidic or strongly basic adsorbents have negligible affinity for H^+ or OH^-, respectively, and hence must be regenerated with other ions, such as Na^+ or Cl^-. A typical reaction and conditions for each of these types of regeneration processes are shown in Table 8-6.

When an adsorbent that has been used to collect a multivalent contaminant is regenerated with a concentrated solution of a monovalent salt, the high ionic strength of the regenerant contributes to the regeneration efficiency through its effect on activity coefficients, above and beyond the effect of concentration on the separation factor, as described in Chapter 7. The reason is that increasing ionic strength decreases the activity coefficient of all dissolved ions, but the magnitude of the effect increases dramatically with increasing ionic charge. Therefore, even if the ionic strength were increased by an inert (nonsorbing) ion, it would diminish the activity of dissolved Ca^{2+}, for instance, more than the activity of Na^+. The decreased dissolved activity translates into a decreased tendency to adsorb, so the selectivity of Ca^{2+} over Na^+ diminishes with increasing ionic strength. Thus, when an adsorbent that has been used to collect Ca^{2+} from a low ionic strength solution like drinking water (e.g., in an ion exchange process) is exposed to a regenerant solution containing a high concentration of NaCl, regeneration is favored by two related, but separable mechanisms. The high concentration of Na^+ competes effectively with Ca^{2+} for the available binding sites, based strictly on mass action considerations, and the high ionic strength makes the Na^+ relatively more attractive to the resin than Ca^{2+}, independent of the concentrations.

The efficiency with which the adsorbate can be recovered from the solid during regeneration is usually very high (>95%) for ion exchange resins, but is often lower for oxide and other mineral adsorbents (80–90%). The adsorbate that is not recovered is often referred to as being irreversibly

---

[8] Keep in mind that NOM is a collection of a wide variety of organic compounds, with a spectrum of sizes, molecular weights, molecular charge, hydrophobicity, and other characteristics.

TABLE 8-6. Typical Conditions for Regenerating Adsorbents Used to Remove Ionic Adsorbates

Adsorbent	Regeneration Reaction	Typical Regenerant
Granular ferric hydroxide	$\equiv \text{FeO-Cu}^+ + 2\text{H}^+ \leftrightarrow \equiv \text{FeO-H}_2^+ + \text{Cu}^{2+}$	0.01 M HCl
Activated alumina	$\equiv \text{Al-F} + \text{OH}^- \leftrightarrow \equiv \text{AlO-H} + \text{F}^-$	0.01 M NaOH
Strong acid ion exchange resin	$\equiv \text{R-Ag} + \text{Na}^+ \leftrightarrow \equiv \text{R-Na} + \text{Ag}^+$	1.0 M NaCl
Strong base ion exchange resin	$\equiv \text{R}_2\text{-HAsO}_4 + 2\text{Cl}^- \leftrightarrow \equiv \text{R}_2\text{-Cl}_2 + \text{HAsO}_4^{2-}$	1.0 M NaCl

adsorbed, suggesting that it has bound to especially strong sites within the particle and therefore is not released as easily as adsorbate bound to average sites. Such a process would be expected to reduce the adsorption capacity of the adsorbent in future treatment cycles, and indeed such a reduction in capacity is often observed. However, in many cases, the accumulation of irreversibly bound adsorbate in the adsorbent particle has a surprisingly small effect on the adsorption capacity in subsequent adsorption cycles. In fact, in some cases, the cumulative amount of adsorbate remaining in the solid after many treatment cycles exceeds the total amount that adsorbed in the initial cycle, and yet the adsorbent continues to function effectively. In these cases, it might be that some transformation occurs that converts the adsorbate to another form, such as a precipitate, causing it to be retained but not to occupy a substantial number of surface sites (or, perhaps, to provide new sites).

An additional issue that might arise during regeneration is chemical transformation of the adsorbent itself, such as partial dissolution of the adsorbent. For example, partial dissolution of the solid leads to a steady decline in the adsorptive capacity of activated alumina, if the solid is regenerated by exposure to acidic or basic solutions.

As opposed to adsorbents that equilibrate rapidly with the solution phase and that are used to adsorb hydrophilic substances, adsorbents that are used to remove hydrophobic compounds and those that are microporous (and hence equilibrate slowly with solution) are not usually amenable to *in situ* regeneration. Both of these characteristics apply to GAC, so on-line regeneration of that adsorbent is impractical for at least two reasons. First, the hydrophobic adsorbates that are typically the target of GAC treatment processes are so strongly attracted to the solid, and so resistant to dissolution in water, that no reasonable modification of the aqueous phase composition can reverse the driving force for the adsorption reaction and cause desorption to be favorable. Second, even if the driving force could be reversed, the time required for the adsorbates to migrate from deep inside the pore structure to the bulk solution would keep the adsorbent out of service too long for the process to be economical.

Because of these problems, GAC is usually regenerated by removing it from the reactor and exposing it to extreme conditions similar to those used to activate it in the first place. Up to several percent of the GAC mass can be lost in this process. In addition, physical changes within the adsorbent particles, incomplete release of adsorbed compounds, and precipitation of inorganic compounds in the pores when the GAC is dried and heated can all diminish the adsorption capacity of the material that remains (Cannon et al., 1997). Nevertheless, such particles can often be reused several times, and ongoing research into improved methods for regeneration are likely to increase the adsorbent's effective lifetime.

8.9 DESIGN OPTIONS AND OPERATING STRATEGIES FOR FIXED BED REACTORS

Fixed bed adsorption reactors are used in a variety of configurations, each associated with a particular operating and regeneration strategy. A single bed might be used and completely regenerated as soon as the effluent concentration exceeds the acceptable level, or several beds might be operated in series or parallel. Although the decision about which design to use in a given application must include cost considerations, some simple mass balances can help identify operating strategies that are technically feasible and therefore worth further evaluation.

The Minimum Rate of Adsorbent Regeneration or Replacement

Consider a system in which, at intervals, some of the used adsorbent is removed and replaced by fresh or fully regenerated adsorbent. To minimize the amount of fresh adsorbent needed, the adsorbent being removed should contain the maximum possible adsorption density of contaminant, for the given influent composition; this maximum achievable adsorption density is q_{in}^*, the adsorbent density in equilibrium with the influent.[9]

A mass balance can be written on adsorbate in such a reactor from a time immediately before one regeneration step to a time immediately before another regeneration step, separated by an arbitrarily large number of adsorption/regeneration cycles. Assuming that the dissolved adsorbate

---

[9] A small portion of the adsorbent might equilibrate with a chromatographic peak of the adsorbate, but identifying and selectively regenerating this small amount of adsorbent is usually impractical.

concentrations and the adsorption densities at these two times are approximately equal, the amount of adsorbate released from the adsorbent in the intervening regeneration steps must equal the amount that adsorbed. Or, dividing both of these quantities by the time period of interest, the long-term average rates of adsorption and regeneration must be equal.

Define X as the long-term average rate (mass/time) at which spent adsorbent is either regenerated or removed from the bed and replaced by fresh adsorbent. Because the minimum value of X (X_{min}) corresponds to an operating procedure in which the adsorption density on the spent adsorbent is q_{in}^*, the mass balance is

$$\begin{pmatrix} \text{Rate at which} \\ \text{adsorbate is removed} \\ \text{from influent} \end{pmatrix} = \begin{pmatrix} \text{Rate of adsorbate} \\ \text{release from adsorbent} \\ \text{by regeneration} \end{pmatrix}$$

$$Q(c_{in} - c_{out,avg}) = X_{min}(q_{in}^* - q_{fresh}) \quad (8\text{-}73)$$

where $c_{out,avg}$ is the average adsorbate concentration in the effluent during the time of interest. Then, assuming that the adsorption density is zero on the fresh and/or regenerated adsorbent, the minimum regeneration rate can be computed as follows:

$$X_{min} = \frac{Q(c_{in} - c_{out,avg})}{q_{in}^*} \quad (8\text{-}74)$$

Interestingly, this computed value of X_{min} is independent of the number of packed beds in which the adsorbent is distributed, the details of the solid/liquid contacting arrangement (e.g., whether the beds are arranged for parallel or series flow), and whether the adsorbent addition and removal are continuous or intermittent.[10] Nevertheless, those design options do affect various aspects of system operation. In the following section, we explore the benefits and drawbacks of a few design options.

Design Options for Fixed Bed Adsorption Systems

Single Bed Designs The criteria for acceptable performance of an adsorption system might typically specify an instantaneous or short-term maximum effluent concentration that must never be exceeded, a long-term average concentration that cannot be exceeded, or both. In any of these cases, it is useful to think of the criteria in terms of the location of the MTZ when the system is regenerated. That is, the criteria can be thought of as limiting how much of the MTZ is allowed to leave the bed and enter the effluent stream before the adsorbent must be regenerated.

If the MTZ is extremely thin (i.e., if the breakthrough profile is very sharp) and easily monitored, we might choose a design that consists of a single, packed bed that is monitored just upstream of the effluent port. Then, the system could be operated until incipient breakthrough of the adsorbate, at which time the adsorbent could be regenerated or replaced. This situation is shown schematically in Figure 8-23a. In practice, a second bed would often be included in the design, both as a backup for upset conditions and so that flow could be treated continuously (even when one bed was being regenerated).

On the other hand, if the MTZ for the contaminant of interest is relatively long, and if the maximum acceptable effluent concentration is only a small fraction of the influent concentration, then use of a single bed for the sorption process would be impractical, even without consideration of upset conditions or transitory needs. The problem with such a system is that it would require that the media be regenerated or replaced at a time when a large fraction of the adsorbent's capacity remains unutilized, as demonstrated in

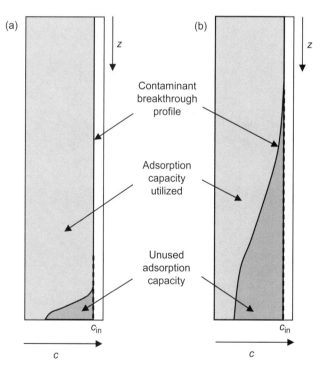

FIGURE 8-23. Comparison of concentration profiles and the utilization of adsorption capacity in two packed beds at the time when the contaminant concentration is the maximum allowable in the effluent. The bed on the left (a) has a much shorter MTZ than the one on the right (b), and so more of the adsorption capacity is utilized in bed (a).

---

[10] X_{min} can be also interpreted as the rate (mass of adsorbent per unit time) at which media that is in equilibrium with the influent is generated in (and therefore must be removed from) the bed. This rate is closely related to the rate at which the MTZ sweeps through the bed. Specifically, the volumetric rate at which portions of the bed reach equilibrium with the influent (volume of bed per unit time) can be expressed as the product $v_{MTZ}A_R$. Therefore, X_{min} and v_{MTZ} are related by $X_{min} = \rho_b v_{MTZ} A_R$.

Figure 8-23b. In such situations, it becomes attractive to decouple contaminant breakthrough from the timing of adsorbent regeneration and/or replacement. That is, it becomes attractive to operate the system in such a way that the effluent is always far from breakthrough, but the capacity of the media undergoing regeneration is nevertheless maximally utilized; that is, $q = q_{in}^*$. This goal can be achieved by treating the influent in a group of fixed beds in series, as described next.

Packed Adsorption Beds in Series: "Merry-Go-Round" Systems If adsorbent can be regenerated rapidly and on-site, as is typically the case for ion exchange systems and also many oxide-based adsorbents, treatment can often be efficiently carried out in a group of several packed beds in series. The decision about how many reactors to use, and how large each one should be, depends on a number of considerations. However, in all cases, the goal is to achieve an adsorption density of q_{in}^* throughout the first (most upstream) bed, and to regenerate that bed before the contaminant concentration in the effluent from the final (most downstream) bed exceeds the maximum acceptable value. Such an arrangement is shown schematically in Figure 8-24.

In the figure, Beds II, III, and IV are arranged in series and comprise the current treatment system. Bed II is near complete breakthrough, Bed III contains most of the MTZ, and Bed IV is polishing the flow, so that the effluent contains almost no contaminant. Bed I is offline and is being regenerated. A short time earlier, Bed I was in the most upstream position, followed by Bed II and Bed III, with Bed IV being out of service and undergoing regeneration. Shortly after the time shown, Bed II would be taken offline and regenerated, Beds III and IV would be moved upstream (so that the influent entered Bed III), and Bed I would be placed in the treatment sequence downstream of Bed IV. This arrangement is sometimes referred to as a merry-go-round system.

Because the minimum rate of adsorbent regeneration is independent of the number of beds in the system, a successful merry-go-round system could be designed with any number of beds >1. That is, we could choose to use only two beds, in which case the entire MTZ (or, at least, the portion of the MTZ with contaminant concentration greater than the maximum allowable concentration) would have to be contained in one bed while the second bed was being regenerated. Alternatively, we could use a much larger number of beds (say, 10) to accomplish the same treatment goal, in which case the MTZ could be spread among nine beds when one bed was being regenerated.

Generalizing this trend, for a system with n beds, the volume of adsorbent in each bed (V_{bed}) would have to be at least $V_{MTZ}/(n-1)$, where V_{MTZ} is the volume of bed occupied by the MTZ. The corresponding total volume of adsorbent in the entire system would be $V_{tot} = nV_{bed}$. Thus, the minimum total adsorbent requirement for a system designed as n fixed beds in series can be related to the volume occupied by the MTZ as follows:

$$V_{tot} = \frac{n}{n-1} V_{MTZ} \qquad (8\text{-}75)$$

$$M_{tot} = \frac{n}{n-1} \rho_p V_{MTZ} \qquad (8\text{-}76)$$

where M_{tot} is the total mass of adsorbent in the bed.

Equation 8-76 indicates that the adsorbent requirement decreases with increasing n, but that the marginal benefit of an additional bed diminishes as n increases. As n increases, the frequency of regeneration must also increase, because, with increasing n, each bed contains less adsorbent and hence becomes saturated (reaches q_{in}^*) faster. Remember, however, that the overall rate of media regeneration (mass regenerated per unit time) would be independent of n; that parameter is determined solely by the adsorbate loading rate (mass/time) and q_{in}^*. Thus, the trade-off to be considered when determining the number of beds is that as the number of beds increases, the total volume of media required decreases, but the capital cost of the mechanical parts of the system (beds, plumbing, etc.) increases. In the limit, systems can be designed with continuous removal and regeneration of small amounts of adsorbent, so that the $V_{ads,tot} \approx V_{MTZ}$. An example design for such a system is shown in Figure 8-25.

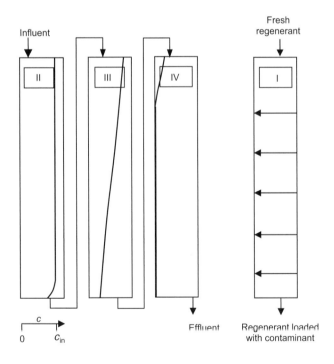

FIGURE 8-24. Schematic of a merry-go-round adsorption/regeneration process. Thick lines show the concentration profiles in the various reactors, with $c = 0$ on the left edge and $c = c_{in}$ on the right. Beds II, III, and IV are online and are operated in series to treat the influent, and Bed I is being regenerated offline.

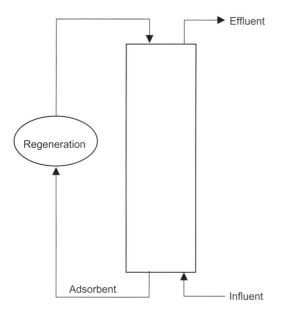

FIGURE 8-25. A packed bed adsorption system with continuous adsorbent regeneration. The system utilizes an upflow flow pattern. Adsorbent is continuously removed from the bottom of the bed, regenerated, and added to the top of the bed.

Packed Adsorption Beds in Parallel In some cases, it is not necessary to operate the system in a way that prevents any significant contaminant breakthrough, but only in a way that reduces the influent concentration substantially. Assume, for instance, that 75% removal of a target adsorbate is required on a steady basis. One strategy that accomplishes this goal is to construct a system that prevents any significant breakthrough and use it to treat only 75% of the total flow. The treated and untreated water could then be blended to yield an acceptable final product. However, if the influent concentration fluctuated significantly, such a design would transmit those fluctuations into the product water (via the untreated portion of the water).

An alternative design that accomplishes the same goal, but damps the fluctuations in influent and effluent concentrations, involves treating all the water in a group of packed beds in parallel. If the beds are operated such that the MTZ is at a different location in each bed, the adsorbent in each bed could be regenerated as (or somewhat after) that bed reached complete breakthrough, while the breakthrough from the other beds remained negligible. For instance, if 75% contaminant removal was required, a system with four beds could be installed. The beds would then be operated out of phase with one another, so that when the MTZ was near the top of one bed, it was at approximately 25%, 50%, and 75% of the total bed depth in the other three.

After a while, breakthrough would occur in the bed with the MTZ nearest the bottom. If the MTZs were sharp enough, that bed could be operated until essentially complete breakthrough, and then regenerated. In that way, even while that bed was experiencing complete breakthrough, the concentration in the blended effluent from all four beds would not exceed approximately 25% of the influent concentration. Transient increases and decreases in the influent concentration would cause all the MTZs to move through the beds at slightly faster or slower rates. However, they would also lead to corresponding fluctuations in the amounts adsorbed in the beds and so would not appear instantly as spikes in the effluent.

Numerous other operating strategies have been proposed for adsorption reactors, based on the particular needs being addressed (whether the contaminant is a trace or major component of the solution, how costly the adsorbent and valuable the adsorbate are, the extent of competition expected, etc.). In light of the diversity and ever growing number of adsorbents available, and the rapidly expanding list of compounds for which adsorptive treatment appears to be a good option, design of adsorption systems is likely to undergo substantial development in the coming years.

8.10 SUMMARY

Adsorption is carried out in a wide variety of reactors in water and wastewater treatment systems, including batch reactors, reactors with continuous flow of both solution and adsorbent, and those with flow of solution but not of the adsorbent. Any of these systems might be operated under conditions where the solid and solution throughout the reactor are continuously close to equilibrium with one another, or other conditions where only a portion or none of the system is close to equilibrium.

The behavior of reactors in which adsorption is occurring can be analyzed by writing mass balances describing the various components and compartments of the system. In the most general case, mass balances on the adsorbate in bulk solution, the boundary layers surrounding the adsorbent particles, and the particles themselves can be written and linked by the boundary conditions, since each of these control volumes shares a boundary with at least one of the others. At least in theory, when these mass balances are combined with known initial conditions and the values of various system-specific properties, the equations can be solved to yield the adsorbate concentration(s) throughout the system at all times.

As a practical matter, the equations are much easier to solve if certain simplifying assumptions are made. While the applicability of many of these assumptions is system-specific (e.g., well-mixed solution phase, rapid transport of the adsorbate throughout the adsorbent volume, negligible mass transfer resistance in the boundary layer), a few are applied almost universally to all adsorptive systems. The most important of these assumptions is that the adsorption reaction itself is never rate limiting. Put another way, the adsorbent surface and the solution immediately adjacent to it are assumed to attain equilibrium instantaneously.

Even with the assumptions noted above, the system of equations describing the temporal and spatial profiles of adsorbate in a system often have no closed-form solution. However, computer software is available for solving the equations numerically, and the results tend to follow a few, consistent patterns. In batch systems with slow approach to equilibrium, the soluble adsorbate concentration typically decreases rapidly initially, as the adsorbate binds to the exterior sites on the adsorbent. Thereafter, adsorption slows, being limited by the rate at which exterior surface sites become available due to the diffusive migration of adsorbate into the interior of the particle. The process continues, at an ever-diminishing rate, until the adsorption density throughout the particle is uniform and in equilibrium with the dissolved concentration that exists throughout bulk solution and also in the interior pores.

In fixed bed adsorption systems, the dissolved concentration of adsorbate usually develops a characteristic spatial profile: the adsorbate is present at its influent concentration near the inlet and for some distance downstream; the concentration is almost zero near the outlet and for some distance upstream of that location; and it changes rapidly as a function of distance in a short zone, called the MTZ, that connects these two limiting regions. The shape of the MTZ is controlled primarily by the resistance to adsorbate transport through the boundary layer and into the interior of the adsorbent particles, although the shape of the adsorption isotherm also plays a role. In general, the MTZ is predicted to be virtually a square wave if adsorptive equilibrium is reached rapidly, and it spreads out and acquires increasing curvature as mass transfer resistance (either internal or external) increases.

In the absence of reactions that destroy the adsorbate in the reactor, fixed beds never reach a steady state, because adsorbate accumulates continuously in the system. This accumulation is manifested as a steady migration of the MTZ from the influent to the effluent end of the bed. Although the shape of the MTZ depends on the kinetics of migration of adsorbate to the adsorbent surface and diffusion into the interior of the particle, the velocity of the MTZ through the column is equilibrium-controlled. When the MTZ passes through the effluent port, the contaminant is said to be breaking through the bed. Once complete breakthrough occurs, no more adsorbate is removed from water passing through the bed, and the adsorbent must be either regenerated or disposed of. On the other hand, if a chemical reaction that destroys the adsorbate occurs while it is in the bed or bound to the solid, then a steady state might be achieved in which the adsorbate exits the bed in the effluent stream at a steady rate that represents the difference between its loading rate and its rate of destruction. One example of such a process involves the biological degradation of adsorbed contaminants on the surface of activated carbon, in a process known as bioregeneration.

Microporous, granular adsorbents might take a very long time to reach equilibrium with an influent solution, because of the slow transport of adsorbate deep into the particles. To predict the performance of such systems, techniques have been developed to extrapolate data from short-term tests in small beds to much larger beds over much longer times of operation. These tests, widely referred to as RSSCTs, are based on similitude between the small and large systems. In cases where the apparent diffusivity of the adsorbate in the interior of the adsorbent particle is identical in the small- and large-scale systems, complete similitude can be achieved, and the results of small-scale tests can be used directly to predict the performance of large-scale systems. In other cases, the apparent diffusivity of the adsorbate on the small particles used in the RSSCTs differs from that on the larger particles used in the full-scale system, and similitude can be approximated, but not achieved completely. Although predictions based on RSSCTs are never perfect, they are often quite good, and they can be relied upon to provide valuable preliminary information for design.

Competition in adsorption beds diminishes the adsorption of all adsorbates, but especially that of weaker binding species. In systems that equilibrate rapidly, weakly binding adsorbates bind and are stored in the bed downstream of more strongly binding adsorbates. Then, when the stronger adsorbates migrate into the downstream region, the weaker ones are desorbed, causing their concentration in solution to exceed their influent concentrations transiently. When the so-called chromatographic peak of the weaker adsorbate reaches the effluent port, the concentration of that species can be (again, transiently) much higher than its influent concentration. If the species is toxic or harmful in other ways to consumers or downstream processes, the peak can have severe consequences.

In theory, competitive adsorption should have only a small effect on the strongest binding adsorbate in a system. However, if weakly binding adsorbates that bind downstream in the early part of an adsorption run do not desorb rapidly when the more strongly binding adsorbate arrives, the strong adsorbate might be precluded from adsorbing and hence might break through the bed much sooner than would be predicted based on competitive adsorption equilibrium.

Once the adsorbent in any system has reached equilibrium with the influent solution, the adsorbent is no longer capable of collecting adsorbate. Its usefulness can be recovered by a regeneration process, which might be carried out either *in situ* or after removing the adsorbent from the reactor, depending on the particular adsorbent in use. Adsorbents used to collect hydrophilic adsorbates such as metals or other ions can often be regenerated *in situ* by exposure to an aqueous solution with a different composition from the influent. On the other hand, adsorbents used to collect hydrophobic adsorbates usually require a more severe regeneration process, such as heating or exposure to a nonaqueous solvent, that is usually carried out in a separate reactor, often offsite.

The wide range of adsorbates that one might wish to remove from water, with similarly wide ranges of influent and target effluent concentrations, has led to a plethora of adsorption system designs. Assuming that the influent flow rate and composition are not subject to manipulation, the design choices for a fixed bed system include the identity and characteristics (e.g., size) of the adsorbent particles, the number of beds, bed dimensions, whether the beds are arranged in series or parallel, and possibly the composition of the regenerant. Given the regulatory trend requiring removal of ever more contaminants to lower levels, and the attractiveness of adsorption to achieve those goals, novel adsorbents and contacting schemes are likely to be the focus of much research in coming years.

REFERENCES

Cannon, F.S., Snoeyink, V.L., Lee, R.G., and Dagois, G. (1997) Effect of iron and sulfur on thermal regeneration of GAC. *JAWWA*, 89 (11), 111–122.

CenCITT (1999) *AdDesignS™: Adsorption Design Software for Windows*. The National Center for Clean Industrial Treatment Technologies, Michigan Tech. Univ., Houghton, MI.

Clifford, D.A. (1982) Multicomponent ion-exchange calculations for selected ion separations. *Indust. Eng. Chem. Fund.*, 21 (2), 141–153.

Crittenden, J.C., and Weber, W.J. (1978) Predictive model for design of fixed-bed adsorbers: parameter estimation and model development. *J. Environ. Eng. (ASCE)*, 104 (EE2), 185–197.

Crittenden J.C., Hutzler, N.J., Geyer, D.G., Oravitz, J.L., and Friedman, G. (1986) Transport of organic compounds with saturated groundwater flow: model development and parameter sensitivity. *Water Res.*, 22 (3), 271–284.

Crittenden, J.C., Berrigan, J.K., and Hand, D.W. (1986) Design of rapid fixed-bed adsorption tests for a constant diffusivity. *J. Water Poll. Control Fed.*, 58 (4), 312–319.

Crittenden, J.C., Berrigan, J.K., Hand, D.W., and Lykins, B.W., Jr., (1987) Design of rapid fixed-bed adsorption tests for nonconstant diffusivities. *J. Environ. Eng. (ASCE)*, 113 (2), 243–259.

Crittenden, J.C., Reddy, P.S., Arora, H., Trynoski, J., Hand, D.W., Perram, D.L., and Summers, R.S. (1991) Predicting GAC performance with rapid small-scale column tests. *JAWWA*, 83 (1), 77–87.

Cummings, L., and Summers, R.S. (1994) Using RSSCTs to predict field-scale GAC control of DBP formation. *JAWWA*, 86 (6), 88–97.

Erlanson, B.C., Dvorak, B.I., Speitel, G.E., Jr., and Lawler, D.F. (1997) Equilibrium model for biodegradation and adsorption of mixtures in GAC columns. *J. Environ. Eng. (ASCE)*, 123 (5), 469–478.

Hand, D.W., Crittenden, J.C., and Thacker, W.E. (1983) User-oriented batch reactor solutions to the homogeneous surface diffusion model. *J. Environ. Eng. (ASCE)*, 109 (1), 82–101.

Hand, D.W., Crittenden, J.C., and Thacker, W.E. (1984) Simplified models for design of fixed-bed adsorption systems. *J. Environ. Eng. (ASCE)*, 110 (2), 440–456.

Knocke, W.R., Occiano, S.C., and Hungate, R. (1991) Removal of soluble manganese by oxide-coated filter media: Sorption rate and removal mechanism issues. *JAWWA*, 83 (8), 64–69.

Merkle, P.B., Knocke, W.R., Gallagher, D.L., and Little, J.C. (1997) Dynamic model for soluble Mn^{2+} removal by oxide-coated filter media. *J. Environ. Eng. (ASCE)*, 123 (7), 650–658.

Najm, I. (1996) Mathematical modeling of PAC adsorption processes. *JAWWA*, 88 (10), 79–89.

Speitel, G.E., and DiGiano, F.A. (1987) The bioregeneration of GAC used to treat micropollutants. *JAWWA*, 79 (1), 64–73.

Speitel, G.E., Dovantzis, K., and DiGiano, F.A. (1987) Mathematical-modeling of bioregeneration in GAC columns. *J. Environ. Eng. (ASCE)*, 113 (1), 32–48.

Summers R.S., Hooper, S.M., Solarik G., Owen, D.M., and Hong, S.H. (1995) Bench-scale evaluation of GAC for NOM control. *JAWWA*, 87 (8), 69–80.

Tirupanangadu, M.S., Clifford D.A., and Guanhua, G. (2002) *EMCT Windows 2.0: A Visual Basic Application for Multicomponent Chromatography in Ion Exchange Columns*, Univ. of Houston, Dept. of Civil and Environmental Engineering, Houston, TX.

PROBLEMS

8-1. The following data were collected for uptake of trichlorophenol (TCP) on an activated carbon. The initial TCP concentration was 100 µg/L in all cases.

Carbon Dose (mg/L)	Final TCP (µg/L)
8	96
45	78
80	65
125	50
200	41
275	27
460	16
1100	5.2
3500	1.1
8000	0.4

(a) Determine the best-fit parameters for fitting the data with a Freundlich adsorption isotherm.

(b) An industrial plant must meet an effluent discharge limit of 1 g TCP/d. How much carbon would be used per day in a CFSTR in which adsorptive equilibrium is reached quickly, if process flow was 600,000 L/d of a waste containing 100 µg/L TCP?

(c) How much carbon would be required per day if this stream were treated in two sequential batch reactors, with half of the activated carbon added in each?

8-2. (a) An industry generates a waste stream containing 5.0 µmol/L of an organic contaminant and must meet an effluent limitation that allows a maximum of 1.0 mol/d of the contaminant to be discharged. Adsorption of the contaminant on PAC follows a Langmuir isotherm with $K_{ads} = 12.0$ L/mmol and $q_{max} = 1.8$ mmol/g. How much carbon would be used per liter of wastewater and per day in an equilibrium adsorption process if the process flow was 400,000 L/d?

(b) The plant is considering an expansion that would double its wastewater flow rate to 800,000 L/d. However, the discharge limitation will not change. Compute the increases in the required concentration and daily requirement for PAC.

(c) Determine the total daily requirement for activated carbon after the expansion if the treatment process is carried out in two steps: first, some activated carbon is added to the water and, after the system reaches equilibrium, the activated carbon is filtered out and a second, equal dose of activated carbon is added to achieve the treatment goal.

8-3. A pesticide manufacturer is using activated carbon to adsorb traces of the pesticide in the plant waste stream. The adsorption of the pesticide on the carbon can be characterized by the Freundlich isotherm $q = 14c^{0.3}$, where c is in mg/L and q is in mg of pesticide per gram of activated carbon.

Currently, to stay below the maximum allowing discharge concentration of 0.02 mg pesticide/L, the company is adding activated carbon at a dose of 200 mg per L of waste in a well-mixed tank. The carbon is then filtered before the water is discharged.

(a) Assuming that the adsorption reaction reaches equilibrium, estimate the pesticide concentration (mg/L) in the treated wastewater.

(b) An engineer claims that by placing the activated carbon in a column through which the water flowed, rather than adding it to the waste solution in a well-mixed tank, the same amount of activated carbon could treat at least 10 times as much water. Do you agree? Explain briefly.

8-4. A dissolved contaminant in a water supply is to be removed by adsorption onto activated carbon. The sorption reaction for this particular contaminant and activated carbon is characterized by the Langmuir adsorption isotherm, with $q_{max} = 77$ mg contaminant per gram activated carbon and $K_{ads} = 10$ L/mg contaminant.

The treatment process involves two steps: first, some activated carbon is added to the water and, after the system reaches equilibrium, the activated carbon is filtered out; then, a second, equal dose of activated carbon is added to the water. The mass of activated carbon added in each dose is identical.

(a) If the first dose reduces the concentration of dissolved contaminant from 1.0 to 0.3 mg/L, how much activated carbon is added per liter of water?

(b) Determine the concentration of contaminant remaining in solution after the second dose of activated carbon equilibrates and is filtered out of the water.

8-5. As a result of leakage from a storage tank, a groundwater aquifer has been contaminated with trichloroethylene (TCE). The concentration of TCE is 108 µg/L in an extended volume of the aquifer. The maximum contaminant level (MCL) established by the US EPA for this contaminant in drinking water is 5 µg/L. A water provider proposes to use activated carbon to treat the groundwater at a rate of 4.0 m³/min. The equilibrium adsorption isotherm for TCE on Calgon F400 activated carbon is $q = 0.034c^{0.65}$, where q is in grams sorbed per gram of activated carbon, and c is in mg/L. The treatment is to be achieved using PAC in a well mixed, steady-state reactor, followed by flocculation and settling. The detention time in the system is sufficient to assure that equilibrium is achieved prior to removal of the PAC.

(a) Calculate the daily requirement for PAC to achieve the treatment objective.

(b) Calculate the daily requirement for PAC if the treatment is achieved in a two stage contactor, with each stage having a residence time sufficient for equilibrium to be achieved, and the PAC from the first stage being removed from the suspension before the water enters the second stage. Equal amounts of fresh PAC are added to each of the two stages.

(c) Calculate the minimum possible PAC dose rate, based on a model treatment process in which the PAC reaches equilibrium with the influent solution. An idealized scenario in which such a situation would apply is counter-current flow of the PAC and water; that is, a system in which the PAC moves upstream (against the water flow) and exits at the location where the influent enters. Note that this scenario is different from that considered in part (b), in which the PAC is removed after each stage of treatment.

8-6. Determine whether PAC or GAC is a more efficient long-term method of activated carbon application to remove tetrachloroethene (commonly referred to as perchloroethylene, PCE) from a groundwater source, based on the carbon use rate (kg/d). The groundwater contains 0.200 mg/L PCE, and the maximum effluent concentration is 0.005 mg/L. The results of batch equilibrium tests can be fit by the Freundlich isotherm: $q = 150c^{0.51}$, where c is in mg/L and q is in mg/g. The flow rate is 1000 m³/day. Compare the carbon usage rates for the following two scenarios:

(a) PAC addition to a CFSTR operated at $c = 0.004$ mg/L.

(b) A GAC column operating as a PFR, that is 2.0 m long. The MTZ in the reactor has a length of 1.0 m, and the GAC is taken offline as soon as the front of the MTZ reaches the end of the column. Assume that the profile of adsorption density versus depth in the MTZ can be approximated as a symmetric, S-shaped curve.

8-7. The adsorption of TCE on a certain activated carbon is described by the following Freundlich isotherm:

$$q = 52c^{0.55}$$

where c is in mg/L and q is in mg/g. A groundwater containing 1.5 mg/L TCE is to be treated in an adsorption column at a flow rate of 12 m³/min. Estimate the daily consumption rate of activated carbon.

8-8. The following data are effluent concentrations from a test of a column packed with GAC and fed an influent containing 50 mg/L phenol. Both equilibrium and kinetics affect these results. You may assume that the effects of biodegradation, competition with NOM, and variations in water chemistry were negligible during the test. The effluent phenol concentration was essentially zero for the first 19 days of the test.

Time (Days)	Effluent Concentration (mg/L)
0.0	0.0
19.20	0.07
19.41	0.3
19.90	4.5
20.27	14.1
20.46	19.7
20.65	26.9
20.83	33.9
21.20	43.5
21.57	48.2
22.00	49.8
23.0	50.0

(a) Plot on a single figure both the above effluent data and the effluent profile expected if equilibrium is established instantaneously and the MTZ is negligibly thick, using c/c_{in} and time as axes.

(b) If breakthrough is defined as the time when $c = 70$ μg/L, what is the time to breakthrough for the column data?

(c) Find the fractional utilization of the activated carbon for this column. The fractional utilization is described as follows:

Fractional utilization

$$= \frac{\text{Mass of contaminant adsorbed at breakthrough}}{\text{Mass of contaminant adsorbed if all carbon reaches equilibrium with influent}}$$

(d) Discuss ways in which the fractional utilization of the adsorbent might be increased, without exceeding the breakthrough concentration.

8-9. Designers are considering using a fixed bed instead of a batch system to treat the water described in Problem 7-4. The bed has a diameter of 1.0 m and is packed with GAC to a depth of 3.5 m, and the bed density (ρ_b) is 790 kg/m³.

(a) What is the minimum rate (kg/d) at which the GAC must be disposed and/or regenerated? What conditions must be met for this minimum rate to be applicable? What treatment period (days) does this replacement rate correspond to?

(b) If the MTZ has a length of 1.5 m and has a symmetric "S" shape, and regeneration is required as soon as any concentration of Alachlor is seen in the effluent, what are the minimum GAC usage rate and the GAC replacement rate?

(c) Suggest an operating procedure for the regeneration step that would cause the GAC usage rate to approach that determined in part (a), even if the MTZ were 1.5 m, as in part (b)?

8-10. A fixed bed adsorber packed with GAC is being designed to treat groundwater that has been contaminated by TCE, at a concentration of 2 mg/L. Adsorption of TCE onto the GAC is characterized by the Freundlich isotherm: $q = 0.034c^{0.65}$, where q is in g TCE/g GAC and c is in mg TCE/L. The preliminary choices for the design parameters are as follows:

Flow rate	4 m^3/min
Maximum bed diameter	4 m
Bed depth	4 m
Superficial fluid velocity	10 m/h
GAC grain diameter	1 mm
Bed density, ρ_b	450 kg/m^3
Void fraction of bed, ε	0.4
Grain internal void fraction	0.65
Mass transfer zone (MTZ) length	2 m

(a) Calculate the EBCT, in minutes.

(b) What diameter of bed is needed? (*Note*: If the necessary diameter is greater than 4 m, you will need to use more than one bed, operated in parallel.)

(c) Calculate the residence time of a nonadsorbing tracer in the bed. Assume that diffusion of tracer into the internal pores of the GAC particles is negligible.

(d) Sketch a breakthrough curve (concentration versus time) for TCE, assuming that the feed contains 2 mg/L TCE and that the breakthrough profile is symmetric ("S"-shaped) and is centered around equilibrium plug flow behavior. Your plot should accurately show the times of (i) first breakthrough, (ii) $c_{out} = 0.5\ c_{in}$, and (iii) final breakthrough ($c_{out} = c_{in}$).

(e) Approximately, how many empty BVs of water can be treated before breakthrough occurs? Replot the breakthrough curve as concentration versus BVs fed (BV = empty BV).

(f) Assuming regeneration of the bed at the time of first breakthrough, estimate the average GAC usage rate. How much could the GAC usage rate be reduced if two identical columns with the above design were constructed and operated in series?

8-11. An activated carbon adsorption column has been operated to remove a particular contaminant from a solution in which it is the only adsorbable species. Adsorption in the system can be characterized by a Langmuir isotherm with $q_{max} = 18$ mg/g and $K_{ads} = 0.013$ L/mg. The system is operated until the contaminant, which is present in the influent at a concentration of 9 mg/L, breaks through the column completely. The column contains 150 kg of activated carbon.

(a) How much contaminant (in mg) is adsorbed in the column?

(b) The hydraulic residence time of fluid in the column is 10 min. After the contaminant has completely broken through the column as described above, the influent concentration suddenly increases to 18 mg/L. Would you expect the effluent concentration after 10 min (i.e., after the water with the higher influent concentration makes it through the column to the effluent end) to be substantially less than, approximately equal to, or substantially greater than 18 mg/L? Explain.

8-12. An adsorption column packed with activated carbon is operated until the effluent concentration is 4 mg/L, which is 50% of the influent concentration. For the contaminant being removed, adsorption on the activated carbon can be described by a Langmuir isotherm with $K_{ads} = 0.125$ L/mg and $q_{max} = 15$ mg of contaminant per gram of activated carbon. The column contains 700 kg of activated carbon. Based on your understanding of the shape of breakthrough curves, make a rough estimate of the total mass of contaminant that has adsorbed. Explain your reasoning.

8-13. An adsorption column is being used to treat a waste stream that is generated at a rate of 40 L/min for 12 h each day; no wastewater is generated during the remainder of the day. The following two modifications to the treatment process are being considered:

(a) Reducing the flow rate to 20 L/min but operating 24 h/d, thereby treating water at the same overall rate.

(b) Collecting the effluent during the 12-h treatment period and passing it through the column during the 12-h down time in the hopes of removing more contaminant.

Discuss the likely impacts on overall contaminant removal of each of these options. Consider both possibilities that adsorptive equilibrium is and is not reached quickly.

8-14. Two adsorbates, A and B, have been entering a column adsorption system for a long enough time that both have broken through completely. The breakthrough curves for both species are shown in

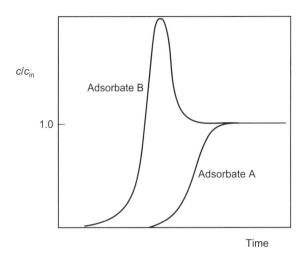

FIGURE 8-Pr14. Breakthrough curves for two adsorbates in a packed bed.

Figure 8-Pr14. The influent concentration of A then drops abruptly to zero, while that of B continues at a steady rate. In independent tests using only one adsorbate at a time, each adsorbs according to a Langmuir isotherm. Sketch the concentrations of A and B that you expect to see in the effluent for the subsequent time period, until a new steady state is reached. Explain your reasoning briefly.

8-15. Adsorption is one approach for removing NOM from drinking water. Because NOM is a mixture of diverse organic compounds, with correspondingly diverse adsorption properties, the dissolved concentration and adsorption density of NOM are frequently evaluated in terms of the organic carbon concentration. That is, c_{NOM} and q_{NOM} are frequently reported as mg DOC per liter of solution and per gram of adsorbent, respectively.

One way to evaluate the affinity of the NOM in a given raw water for an adsorbent is to perform a test in which various dilutions of the raw water are contacted with the adsorbent in batch systems. In each test, the suspension is mixed until equilibrium is presumed to have been attained, and the concentration and adsorption density of NOM, quantified in terms of DOC, are measured. The data can then be extrapolated to the adsorption density that would be in equilibrium with the DOC concentration in the raw water.

Alternatively, the original water can be diluted by various factors and passed through a column packed with the adsorbent. In that case, once the effluent DOC equals the influent value, the water is assumed to have equilibrated with the adsorbent in the column, and the equilibrium adsorption density can be computed based on the cumulative removal of DOC up to that point.

If tests such as those described above are carried out, the adsorption density in equilibrium with the full-strength raw water is typically higher in the column test than the extrapolated value in the batch test. Explain this result in terms of the heterogeneity of the molecules that make up the NOM.

8-16. Consider a well-mixed suspension of solids in an aqueous solution, in contact with an overlying gas phase. A particle-free solution flowing into the system contains reactant A, but no product P. Thus, particles are present in the initial suspension and are well mixed throughout the reactor, but no new particles are entering; however, particles do exit the reactor with the water. Solution exits the system at the same flow rate as it enters. In the liquid phase, the reaction $2A \leftrightarrow P$ is elementary in both directions, with forward and reverse rate constants k_2 and k_{-1}, respectively. These rate constants apply to the expression for the rate of destruction and formation of A; that is, to r_A. In addition, molecules of A can adsorb to sites on the suspended solids, with adsorption equilibrium described by the Langmuir isotherm.

The flux of A to the surface from bulk solution can be expressed as the product of a rate constant and the difference between the hypothetical, equilibrium adsorption density (based on the solution composition at that time) and the instantaneous adsorption density. During the time frame of interest, there is no significant diffusion of adsorbed A into the interior of the solids. Molecules of adsorbed A can be converted to P at a rate $r_{\sigma,A} = -k_{2,\sigma}\Gamma_A^2$. P is nonsorbable, so it enters solution as soon as it is formed at the particle surface.

The constants and variables that describe the system are summarized below.

Variables

c_A, c_P = instantaneous dissolved concentrations of A and P, respectively (mol/L);

Γ_A = instantaneous adsorption density of A (mol/m²);

M_A = instantaneous total mass of A in the system (mol);

t = time (s).

Constants

Q_L = liquid flow rate (L/s)

V_L = volume of liquid in system (L);

$c_{A,in}$ = influent concentration of dissolved A (mol/L);

$K_{mt,A}$ = mass transfer rate coefficient for adsorption of A (s^{-1});

$A_{S/L}^o$ = solid/liquid interfacial area in the reactor at time 0 (m²);

k_2, k_{-1} = forward and reverse rate constants for r_A for the reaction $2A(aq) \leftrightarrow P(aq)$; k_2 in $(mol/L)^{-1}s^{-1}$; k_{-1} in s^{-1};

$k_{2,\sigma}$ = rate constant for the reaction $2A(ads) \rightarrow P$ $m^2/(mol\ A\ s)$;

$K_{Lang,A}$ = Langmuir adsorption equilibrium constants for A on the particles (L/mol);

$\Gamma_{max,A}$ = maximum adsorption density of A (mol/m^2).

(a) Fill in the following table, indicating by a checkmark those processes that must be included in the specified mass balances.

Process	Affects Mass Balance on			
	c_A	c_P	Γ_A	M_A
Inflow of solution				
Outflow of solution				
Outflow of particles				
Adsorption/desorption				
Reaction in bulk liquid				
Reaction on particle surface				

(b) Write mass balances for dissolved A, dissolved P, adsorbed A, and total mass of A in the system at any future time t. The final expressions should be of the form:

$$\frac{dc_A}{dt} = \ldots;\quad \frac{dc_P}{dt} = \ldots;\quad \frac{d\Gamma_A}{dt} = \ldots;\quad \frac{dM_A}{dt} = \ldots$$

where the only variables to the right of the equal signs should be c_A, c_P, Γ_A, and t. Everything else should be constants. Use the notation shown above, and pay attention to units.

(Note: This problem sounds complicated at first, but it is straightforward if one recognizes that each process that can affect A and P corresponds to a separate term in the mass balance. By following the rules in Chapter 1 for what to include in a mass balance, and how to include it, you should find that the problem becomes very tractable.)

8-17. Compound X (MW 125) is released by an industry at a steady concentration of 500 mg/L in a process stream flowing at $2\ m^3/h$. This stream is mixed with a second stream, which has a flow rate of $3\ m^3/h$ and does not contain X. However, a competing adsorbate Y (MW 140) is present in the second stream at a concentration of 700 mg/L from 7 A.M. to 5 P.M. The industry wishes to treat the water to remove X and Y by adsorption onto PAC.

The adsorption of both X and Y onto the proposed adsorbent reaches equilibrium quickly in comparison with the contact time available. The binding of each adsorbate follows a Langmuir isotherm, with $q_{max} = 2$ mmol X or Y/g PAC, $K_{ads,X} = 200$ L/g X, and $K_{ads,Y} = 500$ L/g Y. The concentration of each contaminant must be reduced to ≤ 3 mg/L prior to discharge to the local sewage treatment plant. Your job is to identify which of several treatment options uses the least carbon to accomplish the goal.

A tank of volume $120\ m^3$ is available for use as the treatment system, yielding an overall residence time of 24 h. Three treatment options are being considered. Option 1 is a batch operation, in which the carbon would be added to the tank and the tank would be filled over the ensuing 24-h period, after which the water would be filtered to remove the carbon and discharged. Option 2 is to operate the tank as a CFSTR, with carbon added continuously at the rate necessary to meet the treatment objective. Option 3 is a more easily controlled, but more conservative version of Option 2: the carbon would be added at a steady rate determined to prevent the effluent concentration from ever exceeding the guideline. In this option, the effluent would be cleaner than required during most of the day.

(a) After a few days, the total concentrations of X and Y in the CFSTR versions of the treatment system will establish a stable daily pattern. That is, over any 24-h period, the total concentration of each pollutant in the reactor will vary in a consistent way, day after day. Determine this daily pattern for total X and total Y in the treatment system. (Hint: The equations needed to solve this part of the problem include mass balances on X and Y, equations describing equilibrium of X and Y with the PAC, and the effluent quality requirement.)

(b) Determine the carbon requirement for the batch treatment option.

(c) Because the system reaches equilibrium quickly, the equations used in part (b) are applicable at any instant in the CFSTR systems, although the numerical values to insert in those equations might not be the same. Using the results of parts (a) and (b), determine the daily dosing pattern and the total daily requirement of PAC for Option 2. Compare the total requirement with that for Options 1 and 3.

(d) Compute the efficiency of carbon usage in Options 1 and 2 by computing the fraction of the total available adsorption density that is used in the treatment process. Compare these values with the efficiency that would be obtained if the carbon came to equilibrium with the influent solution. How might one take better advantage of the available adsorption capacity of the carbon?

8-18. An industry produces two waste streams. One stream contains 0.5 mg/L m-chlorophenol and is generated at an average rate of 18 L/min, while the other flows at 50 L/min and contains no m-chlorophenol or other adsorbable contaminants. The discharge permit allows the industry to release 500 mg m-chlorophenol per day.

The plan is to treat the waste by adsorption onto PAC in a CFSTR, and then filter out the PAC before discharging the water. Sorption of m-chlorophenol onto the PAC that has been chosen conforms to the isotherm: $q = 59c^{0.22}$, where q is in mg/g and c is in mg/L. Assuming that equilibrium is reached in each case, compare the required PAC doses for scenarios in which the plant adds the adsorbent to the single waste containing the chlorophenol versus mixing the two streams before treatment. Discuss the result briefly.

8-19. As indicated in Problem 7.13, the following isotherms characterize adsorption of p-nitrophenol (PNP) and benzoic acid (BA) onto a particular PAC:

$$\text{PNP:} \quad q = 89c^{0.13} \quad \text{BA:} \quad q = 140c^{0.22}$$

where c is in mg/L, and q is in mg/g. (Note: if you solved Problem 7.13, some of the calculations from that problem can be applied directly to this one.)

(a) PAC is to be added to a waste stream flowing at a rate of 2×10^6 L/d to reduce the PNP concentration from 10 to 0.1 mg/L. How much PAC is required per liter of waste, and per day, if equilibrium is attained?

(b) What would the adsorption density and the daily GAC usage rate be if the waste described in part (a) were treated in a fixed bed packed with GAC, if the bed were operated until complete breakthrough? Assume that the isotherm equation is equally valid regardless of whether the activated carbon is present as GAC or PAC.

(c) If the apparent density of the GAC particles was 1.2 g/cm^3 and the packed bed had a void fraction of 0.38, how many BVs of feed could be treated before breakthrough, assuming that the breakthrough curve approximated a square wave?

(d) If the waste contained both 10 mg/L PNP and 15 mg/L BA, what would the adsorption density of each species be in a GAC column that was used until complete breakthrough of both contaminants, if competitive adsorption was consistent with the IAST? (Note: if you solved Problem 7.13, some of the calculations from that problem can be applied directly to this one.)

8-20. The key results of adsorption studies in columns are usually data for the effluent concentration (c_{out}) as a function of time. However, such results are often plotted on a normalized basis, as the ratio versus of the effluent concentration to the influent concentration (c_{out}/c_{in}) the number of BVs or pore volumes of water passed through the column.

In Figure 8-Pr20, the results of an adsorption study on a single compound are plotted in both of the ways described above, for a system with specified column geometry, column packing, Q_L, and influent composition. Sketch the expected results if the following changes were made in the system, while the other design and operational parameters remained constant. State any assumptions explicitly. Be as precise as you can be, indicating any points quantitatively if possible.

(a) The diameter of the column was halved. Assume that the kinetics of mass transfer from the solution to the surfaces of the adsorbent particles is not significantly altered.

(b) The length of the column was halved.

(c) The size of the GAC granules was halved. As in part (a), assume that the kinetics of mass transfer from the solution to the surfaces of the adsorbent particles is not significantly altered.

(d) The concentration of the influent was doubled. Assume that adsorption equilibrium is characterized by a Langmuir isotherm and the value of q in equilibrium with the original feed is 5% of q_{max}.

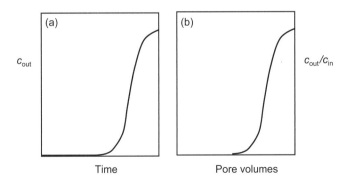

FIGURE 8-Pr20. Breakthrough curves for the same adsorbate in a packed bed, plotted in two different ways: (a) c_{out} versus time; (b) fractional breakthrough versus pore volumes treated.

(e) The concentration of the influent was doubled. Assume again that adsorption is characterized by a Langmuir isotherm, but in this case, the value of q in equilibrium with the original feed is 90% of q_{max}.

8-21. For a particular type of activated carbon, the Freundlich isotherm constants for the adsorption of toluene have been reported to be $k_f = 9920$ (μg/g) (L/μg)^n, and $n = 0.33$.

(a) Calculate a few points on this isotherm for liquid phase concentrations from 100 to 10,000 μg/L (i.e., 0.1 to 10 mg/L), and sketch figures of both the isotherm itself (i.e., with no transformation of the data) and a linearized version of the isotherm. Include the value of 2000 μg/L in the calculations, since that value will be used below.

(b) A continuous flow fixed bed adsorption system using GAC as the adsorbent has been designed to treat a solution containing 2 mg/L (i.e., 2000 μg/L) toluene. The superficial velocity is to be 175 m/d (approximately 0.2 cm/s or 7 m/h), and the proposed EBCT is 10 min. The anticipated bed density is 500 kg/m^3. Assuming complete utilization of the activated carbon in the bed, at what intervals do the operators have to replace the activated carbon (i.e., how long will the column remain in service?)? (Hint: carry out the calculations for a unit cross-sectional area of the bed.)

(c) Sketch a design that will allow complete utilization of the activated carbon in the bed described above, and provide a brief explanation of this design (and associated operation).

(d) After the bed is designed, it is discovered that the groundwater it treats also contains *ortho*-dichlorobenzene (*o*-DCB) at a concentration of 600 μg/L. The Freundlich isotherm parameters for adsorption of *o*-DCB on the GAC to be used (when *o*-DCB is the only compound present) are $k_f = 19,300$ (μg/g) (L/μg)^n and $n = 0.38$. Compare the adsorption densities and surface pressures of the two adsorbates once the activated carbon has equilibrated with the influent in systems in which each adsorbate is present alone. The specific area of the GAC is 950 m^2/g. Sketch qualitatively (i) the breakthrough curve for toluene when it is the only contaminant present, and (ii) the breakthrough curves (separate lines) for both toluene and *o*-DCB when both are present at the concentrations given.

8-22. Activated carbon is being used in a continuous flow adsorption column to remove phenol in an industrial waste stream. The influent contains 8 mg/L phenol and is to be treated at a flow rate of 1 m^3/min. When the system is operated to complete breakthrough, the breakthrough curve can be approximated as follows:

- Effluent concentration of zero for 45 days.
- A linear rise in effluent concentration from zero to c_{in} between days 46 and 55; that is, c_{out}/c_{in} equal to 0.1 on day 46, 0.2 on day 47, and so on. (This approximation is unrealistic, but it keeps the math simple.)

(a) Calculate the total mass of phenol adsorbed by the activated carbon.

(b) If the EBCT of the bed is 12 min and the apparent bed density is 460 kg/m^3, what is the adsorption density (q) in equilibrium with the influent liquid phase concentration?

(c) Assume that, in a new but identical bed packed with fresh GAC, the influent contains 5 mg/L of a second compound that adsorbs less strongly than phenol, in addition to 8 mg/L of phenol. Due to competition for adsorption sites, the adsorption density of phenol at complete breakthrough declines by 20%. Sketch the breakthrough curve for phenol in this situation, assuming that the mass transfer kinetics stay essentially the same as in the noncompetitive case.

(d) Assume that the second compound has a similar MTZ as phenol when it is the only compound present. Sketch the breakthrough curve for the second compound in the competitive situation; continue the curve until the time that phenol has completely broken through the bed.

9

PRECIPITATION AND DISSOLUTION PROCESSES

PAUL R. ANDERSON, M. M. BENJAMIN, AND D. F. LAWLER

9.1 Introduction
9.2 Fundamentals of precipitation processes
9.3 Precipitation dynamics: particle nucleation and growth
9.4 Modeling solution composition in precipitation reactions
9.5 Stoichiometric and equilibrium models for precipitation reactions
9.6 Solid dissolution reactions
9.7 Reactors for precipitation reactions
9.8 Summary
References
Problems

9.1 INTRODUCTION

This chapter describes precipitation and, to a lesser extent, dissolution reactions of importance in water and wastewater treatment processes; several examples of such reactions and processes are listed in Table 9-1. In many of these processes, the contaminant of concern is directly removed by conversion to a precipitate that can then be separated from the aqueous phase. Common target contaminants in such processes include cationic metals (Cu, Zn, Ni, Pb, Cr, and others) in industrial waste streams; Ca, Mg, Fe, and Mn in potable water sources; and phosphate (PO_4) in domestic wastewater or in the effluent from anaerobic digesters.

While the processes noted above are relatively common as approaches for removing contaminants from water, perhaps the most common precipitation process in water and wastewater treatment involves the formation of oxides of aluminum or iron when alum ($Al_2(SO_4)_3$) or ferric chloride ($FeCl_3$) is added to the water. This process, which is known as coagulation and is described in detail in Chapter 11, has long been used to facilitate agglomeration of colloids into larger flocs that can then be settled or filtered out of the suspension. More recently, the tendency of the oxides to adsorb dissolved contaminants (natural organic matter [NOM] in drinking water applications, and metals in industrial waste treatment) has been recognized as an important additional benefit of the process.

Situations in which the goal is to avoid precipitation of a solid or dissolve a solid that has already formed also arise occasionally. For example, precipitation of corrosion products or of $CaCO_3(s)$ can reduce the carrying capacity of water distribution pipes, as can precipitation of struvite ($MgNH_4PO_4 \cdot 6H_2O(s)$) in pipes carrying the products from anaerobic digestion of domestic sludge, and salts that precipitate in reverse osmosis (RO) processes can deposit on the membrane and dramatically reduce its permeability. In systems where such processes are likely to occur, the addition of complexing agents or pH adjustment is sometimes employed to interfere with or reverse the precipitation reaction.

Precipitation and dissolution also play critical roles in controlling the composition of natural systems via their effects on soil formation, weathering processes, and the transport of many constituents in lakes, streams, and the oceans. The principles presented in this chapter apply to all of these situations, although our focus is on engineered treatment processes.

Although precipitation is a common water and wastewater treatment process, comprehensive descriptions of precipitation (and especially precipitation kinetics) are rare in the traditional environmental engineering literature. One reason for this lack of attention is that precipitation is

---

Water Quality Engineering: Physical/Chemical Treatment Processes, First Edition. By Mark M. Benjamin, Desmond F. Lawler.
© 2013 John Wiley & Sons, Inc. Published 2013 by John Wiley & Sons, Inc.

TABLE 9-1. Precipitation Processes of Importance in Water and Wastewater Treatment

Constituent	Potential Problem	Treatment
Hardness (primarily Ca^{2+} and Mg^{2+})	Forms complexes with soap; scale formation in heated water systems (e.g., boilers)	Precipitation softening to generate $CaCO_3(s)$, $Mg(OH)_2(s)$
Fluoride	Fluorosis of teeth and bones	Precipitation of $CaF_2(s)$
Iron and Manganese	Discoloration of water and fixtures; forms deposits in distribution systems and on process equipment	Oxidation to precipitate $Fe(OH)_3(s)$ or $MnO_2(s)$; addition of complexing agents to prevent Fe precipitation in distribution system
Heavy metals (Cd, Cr(III), Cu, Ni, Pb, Zn)	Ecosystem or human health risk	Precipitation as hydroxide, carbonate, or sulfide solids
Phosphorus	Can facilitate eutrophication	Precipitation as various phosphate-containing solids, including $AlPO_4(s)$, $Ca_5(OH)(PO_4)_3(s)$, and $MgNH_4PO_4 \cdot 6\,H_2O(s)$

usually a relatively low-cost, efficient process for removing contaminants from solution. As a result, there has been little incentive to explore the underlying mechanisms and improve the process. Furthermore, in those cases where precipitation has been studied, the focus has usually been on the contaminant concentration remaining in solution, and relatively little attention has been paid to the characteristics of the solids or how those characteristics affect residual management.

Detailed studies of precipitation processes have been much more common in industries where greater emphasis is placed on the characteristics of the products. For example, the surface characteristics of particles used as catalysts or pigments or in data storage media are critical to the use of the product, and a substantial body of chemical engineering literature focuses on understanding and controlling these characteristics. That literature, however, emphasizes crystallization processes, which might not be the most appropriate model for the reactions that occur in water and wastewater treatment.

Sohnel and Garside (1992) distinguished between precipitation and crystallization, defining precipitation as a process that involves relatively insoluble materials that are present initially at a high degree of supersaturation. These conditions promote high nucleation rates and production of high concentrations (10^{11}–$10^{16}/cm^3$) of relatively small (0.1–1 μm) particles for which secondary processes such as aging, agglomeration, and coagulation can be significant. Because the value of the product depends on the physical characterestics of the particles, industrial crystallization processes focus or particle design and carefully control the supersaturation level and these secondary processes (Karpinski and Wey, 2002). By contrast, in most precipitation processes used for water and wastewater treatment, the supersaturation level is high initially, decreases rapidly when solids begin to precipitate, and then declines ever more slowly as the solution equilibrates with the solid.

Despite the relatively sparse attention to bulk and surface characteristics of precipitated solids in treatment operations, these characteristics often do play an important role in determining the success of the processes. For example, the size and density of precipitated particles are largely determined by the conditions during precipitation, and these characteristics control the effectiveness of subsequent solid/liquid separation. Also, as noted previously, adsorption is an important removal mechanism in many precipitation processes, and the surface properties of the precipitated solids are clearly important in these cases.

This chapter begins with an overview of precipitation reactions, in which the fundamental concepts and terminology are introduced. The subsequent section describes the initial steps of precipitation reactions, focusing on the thermodynamics and kinetics of the transition from soluble molecules to the smallest particles. The discussion in that section is generic, in that it applies to all particle formation processes without regard to the details of the solution chemistry. The next section brings traditional water chemistry concepts into the discussion, focusing on the dosages of reagents required to induce precipitation, the amounts of the solids that form, and the expected composition of the solution once equilibrium is reached. Both the reactions that form the solids directly and the ancillary acid/base and metal complexation reactions that can help determine whether and to what extent precipitation occurs are considered. The chapter concludes with brief sections on solid dissolution reactions and the types of reactors that are most commonly employed for precipitation processes in environmental engineering.

9.2 FUNDAMENTALS OF PRECIPITATION PROCESSES

Formation and Growth of Particles

Some of the early, comprehensive discussions of precipitation include those by Nielsen (1964) and Walton (1967). More recent works by Nylt et al. (1985), Randolph and Larson (1988) Schwartz and Myerson (2002) provide a chemical engineering perspective on crystallization processes in industry; Hochella and White (1990) and Stumm (1992) cover some of the same concepts, but their focus is on reactions at the mineral–water interface. Material presented in this section is compiled from all of these references.

When a water or wastewater is first dosed with reagents intended to induce a precipitation reaction, some of the solutes form *complexes* of just a few ions, which in turn come together to form slightly larger species referred to as *clusters*. Some clusters revert to their component units, but others grow large enough to be considered units of a separate, nonaqueous phase; such units are called *nuclei*. Like clusters, some nuclei undergo a reverse reaction and redissolve, but others grow into larger particles as additional solute molecules become integrated or they collide and agglomerate with other nuclei.

If nuclei form by reactions among dissolved species, without the participation of a pre-existing solid, the initial step in the precipitation process is known as *homogeneous* or *primary nucleation*. Often, however, suspended solids or colloidal particles that have a structure similar to a potential precipitate are present in the reactor and can act as seed particles on which nuclei of the new solid form. Such a scenario is referred to as *heterogeneous* or *secondary nucleation*.

Figure 9-1 offers a simplistic but useful picture of the surface of a growing particle once it has nucleated, illustrating the variety of sites where solute molecules (represented as cubes) can become incorporated. When a solute attaches, it can have from one to five of its sides in contact with the solid. It seems reasonable to assume that the stability of cube-surface interactions increases with increasing contact.

In this conceptual model, each building block that merges with the solid consists of a stoichiometric balance of the cations and anions. Alternatively, cation and anion components could arrive and integrate into the solid sequentially, or they could arrive and react at the surface before the integration step. Some evidence exists for each of these growth mechanisms (Chiang and Donohue, 1988), and it is possible that several of them proceed simultaneously, with the overall rate of growth of the solid being the sum of the rates of these parallel reactions.

Solute Transport, Surface Reactions, and Reversibility

To contribute to the growth of an existing solid, a solute that is initially in the bulk solution must go through a series of transport and reaction steps. As explained in Chapter 3, the overall reaction rate is limited by the step in the series that imposes the greatest resistance. The first step in the precipitation sequence is transport from bulk solution to the boundary layer surrounding a particle; as in gas transfer and adsorption, this step is generally assumed to impose little resistance on the overall process. In that case, the rate of solute arrival at the solid surface is limited by the rate of transport through the boundary layer surrounding the growing particle. This transport step, which is essentially identical to that in gas transfer and adsorption processes, is followed by adsorption of the solute to the surface.

Both the composition of the surface and the types of growth sites available influence how the adsorbed species subsequently react and integrate into the solid. If an adsorbed species has the same stoichiometry as the solid, it can be integrated directly to become part of the solid structure. On the other hand, if species reach the surface at a ratio that differs from their stoichiometric ratio in the solid, they must participate in one or more additional reactions before the integration step.

Constituents in solution other than those that form the solid can have dramatic effects on precipitation processes. For example, if other dissolved species combine with one of the precipitating ions, the activity of that ion and the driving force for precipitation are reduced. Precipitation can also be inhibited if dissolved species other than those that the solid comprises adsorb to the surface and block growth sites. In other cases, adsorbable species might increase growth rates, either by lowering the surface energy (Davey, 1979, 1982) or by creating more desirable surface nucleation sites (Zhang and Nancollas, 1990).

Although the preceding discussion focuses on processes leading to particle growth, dissolution reactions can proceed in parallel and compete with particle growth. As long as the solution is supersaturated, the forward (formation and growth) reaction rate exceeds the reverse rate, and net

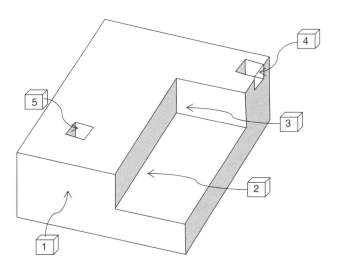

FIGURE 9-1. Particle growth occurs as solute migrates from solution and interacts with the surface. Numbers on the cubes refer to the number of sides of the solute cube in contact with the solid. Sites that generate one-sided contacts are often referred to as *terraces*, those that generate two-sided contacts as *steps* or *edges*, and those that have three or more sides of contact as *kinks*.

growth occurs. In a batch system, the chemical activity of the dissolved solutes decreases as the particles grow until the solution reaches equilibrium with the solid. At that point, the forward and reverse rates are equal, and net growth ceases. If the system is undersaturated with respect to the solid, the rate of the reverse reaction exceeds that of the forward reaction, so that the net reaction is dissolution.

Many precipitation processes in water and wastewater treatment begin at relatively high levels of supersaturation. Solids that initially form under these conditions often lack the repeating, long-range, three-dimensional pattern characteristic of crystals, because growth proceeds rapidly at many different surface sites, and new solute accumulates on the surface before the underlying layers are completely integrated. Because they lack long-range structure, these materials are known as *amorphous* solids. Amorphous solids tend to incorporate more water and to be less dense than their crystalline counterparts. According to Stumm (1992), solids that have a large unit cell are more likely to precipitate initially as amorphous materials. (The unit cell is the fundamental building block of a crystal; the complete crystal comprises an ordered array of many such cells.)

Amorphous solids form as an intermediate phase and are never the most stable solids that can exist in a system; given sufficient time, such solids always transform into more stable (crystalline) solids with long-range order. The rate of transformation depends strongly on environmental conditions such as pH, temperature, and the presence of surface-active agents in the system. The transformation might proceed entirely internally, or it could involve dissolution of the amorphous solid and precipitation of the more stable solid. If the rate of transformation is slow relative to the residence time of the particles, the system will not reach true equilibrium before the particles leave the reactor, but it might reach a virtually unchanging pseudo-equilibrium between the metastable (amorphous) solid and solution.

Fundamentals of Solid/Liquid Equilibrium

The Solubility Product   The key mathematical relationship needed to characterize precipitation and dissolution reactions is the equilibrium constant expression for a reaction between the solid and the chemical species that it comprises. Although such reactions can be written in a number of different forms, the most common representation has the solid (and, when appropriate, water) as the only reactant(s), and simple forms of the corresponding dissolved species as products. The equilibrium constant for a reaction written in this way is called the *solubility product* of the solid, K_{s0}. Several examples of such reactions and solubility products are shown in Table 9-2. In the table, a_i is the (dimensionless) chemical activity of species i.

TABLE 9-2.   Solubility Products of Various Solids of Interest[a]

Reaction	Solubility Product	Mineral	
$BaSO_4(s) \leftrightarrow Ba^{2+} + SO_4^{2-}$	$K_{s0} = \dfrac{a_{Ba^{2+}} a_{SO_4^{2-}}}{a_{BaSO_4(s)}} = 1.05 \times 10^{-10}$	Barite	(9-1)
$CaSO_4(s) \leftrightarrow Ca^{2+} + SO_4^{2-}$	$K_{s0} = \dfrac{a_{Ca^{2+}} a_{SO_4^{2-}}}{a_{CaSO_4(s)}} = 2.45 \times 10^{-5}$	Gypsum	(9-2)
$Al(OH)_3(s) \leftrightarrow Al^{3+} + 3\,OH^-$	$K_{s0} = \dfrac{a_{Al^{3+}} a_{OH^-}^3}{a_{Al(OH)_3(s)}} = 6.31 \times 10^{-32}$	(Amorphous)	(9-3)
$Fe(OH)_3(s) \leftrightarrow Fe^{3+} + 3\,OH^-$	$K_{s0} = \dfrac{a_{Fe^{3+}} a_{OH^-}^3}{a_{Fe(OH)_3(s)}} = 1.55 \times 10^{-39}$	(Amorphous)	(9-4)
$CaCO_3(s) \leftrightarrow Ca^{2+} + CO_3^{2-}$	$K_{s0} = \dfrac{a_{Ca^{2+}} a_{CO_3^{2-}}}{a_{CaCO_3(s)}} = 3.31 \times 10^{-9}$	Calcite	(9-5)
$Zn(OH)_2(s) \leftrightarrow Zn^{2+} + 2\,OH^-$	$K_{s0} = \dfrac{a_{Zn^{2+}} a_{OH^-}^2}{a_{Zn(OH)_2(s)}} = 2.98 \times 10^{-16}$	(Amorphous)	(9-6)
$ZnCO_3(H_2O)(s) \leftrightarrow Zn^{2+} + CO_3^{2-} + H_2O$	$K_{s0} = \dfrac{a_{Zn^{2+}} a_{CO_3^{2-}}}{a_{ZnCO_3(s)}} = 5.49 \times 10^{-11}$	Hydrated zinc carbonate	(9-7)
$ZnS(s) + H^+ \leftrightarrow Zn^{2+} + HS^-$	$K_{s0} = \dfrac{a_{Zn^{2+}} a_{HS^-}}{a_{ZnS(s)} a_{H^+}} = 3.55 \times 10^{-12}$	Sphalerite	(9-8)
$CdS(s) + H^+ \leftrightarrow Cd^{2+} + HS^-$	$K_{s0} = \dfrac{a_{Cd^{2+}} a_{HS^-}}{a_{CdS(s)} a_{H^+}} = 4.37 \times 10^{-15}$	Greenockite	(9-9)
$FeS(s) + H^+ \leftrightarrow Fe^{2+} + HS^-$	$K_{s0} = \dfrac{a_{Fe^{2+}} a_{HS^-}}{a_{FeS(s)} a_{H^+}} = 2.34 \times 10^{-3}$	Ferrous sulfide	(9-10)

[a] Equilibrium constants shown are from the database used by the computer program Visual Minteq (Gustafsson, 2011). These values were collated from numerous primary sources that are identified in the program documentation.

Although extensive compilations of data such as those shown in Table 9-2 have been published, it is important to appreciate the uncertainties associated with those data. The solubility products of solids that are frequently precipitated in treatment processes are often known to within an order of magnitude, but those for solids that are encountered less often might be much less well established. Furthermore, as noted above, the solid that first precipitates in a treatment process is often an amorphous solid that might have a solubility that differs by orders of magnitude from that of more crystalline solids with the identical stoichiometry. Without knowing which solid forms, it is impossible to select appropriate thermodynamic data to model the process. In such a case, if the treatment goal is to reduce the dissolved concentration of a target species below a specified concentration, it is probably best to assume that solubility is controlled by the amorphous solid (which yields the highest, and therefore most conservative estimate of solubility).

The Activity of Solid Phases and Solid Solutions Another complication that can arise when modeling precipitation processes is that the solids that form might be mixtures of two or more species. For example, a solid phase might be composed of 95% $Fe(OH)_3(s)$ and 5% $Al(OH)_3(s)$. If the $Al(OH)_3(s)$ is distributed randomly in the solid phase, the entire phase is referred to as a *solid solution*. Solid solutions can form when the molecular species that comprise the pure solid end-members have similar size and bonding characteristics.

By convention, the activity of any molecular species, i, in an ideal solid phase (one in which the activity coefficients of the constituents are all 1.0) equals the mole fraction, x_i, of that species in the phase. Thus, for example, the activity of the species $CaCO_3$ is 1.0 in a pure $CaCO_3(s)$ phase. Similarly, if the mixed solid described above were ideal, the activities of $Fe(OH)_3(s)$ and $Al(OH)_3(s)$ would be 0.95 and 0.05, respectively. If the solid solution were not ideal, then the activity of each constituent would be computed as the product of its mole fraction and an activity coefficient; that is, $a_i = \gamma_i x_i$. (Note that the activities of individual ions, such as Fe^{3+}, Al^{3+}, and OH^-, are undefined in the solid phase.)

As in aqueous solutions, the activity coefficient of a species in a solid is an indicator of how similarly that species behaves compared to its behavior in a well-defined reference environment (the standard state). For solids, the reference environment is the pure solid. When a molecule is in a solid solution, the bonding arrangement is likely to be more strained than in a pure solid, so the molecule is less "comfortable" than in its reference environment. As a result, its activity coefficient will be >1, meaning that the molecule has a greater tendency to escape (dissolve) from the solid solution than from a pure solid. If the mixed solid provides an environment that is too highly strained, then the species will not participate in the formation of a solid solution at all, but will instead either remain in solution or precipitate as a separate pure phase. For example, pure phases of $Fe(OH)_3(s)$ and $Al(OH)_3(s)$ might both precipitate in the same suspension, rather than form a single, solid solution. In this latter case, both the solids would have activities of 1.0, regardless of their relative concentrations in terms of moles of solid precipitated per liter of water.

The formation of solid solutions in water and wastewater treatment processes is poorly understood, and ignoring that possibility generally leads to conservative designs (i.e., designs that, if in error, over-predict the solute concentrations remaining in solution). Therefore, the usual assumption is that any solid that precipitates in a treatment process is a pure phase of a single species, with activity of 1.0. As a result, the term for the activity of the solid (i.e., the denominator in all the K_{s0} expressions in Table 9-2) is often left out of the solubility product expression. We adopt that convention here, except in cases where the discussion explicitly focuses on solids with nonunit activity.

Thermodynamics of Precipitation Reactions Like all equilibrium constants, the solubility product can be related to the Gibbs (free) energy of reaction, as follows:

$$K_{s0} = \exp\left(-\frac{\Delta \overline{G}_r^o}{RT}\right) \quad (9\text{-}11)$$

where $\Delta \overline{G}_r^o$ is the standard molar Gibbs energy of the reaction, R is the universal gas constant, and T is the absolute temperature. The standard Gibbs energy of reaction can be expressed in terms of the standard enthalpy and standard entropy of reaction as follows:

$$\Delta \overline{G}_r^o = \Delta \overline{H}_r^o - T \Delta \overline{S}_r^o \quad (9\text{-}12)$$

The standard Gibbs energy of reaction can be computed as the sum of the standard Gibbs energy of formation of the products minus that of the reactants:

$$\Delta \overline{G}_r^o = \sum_{\text{all } i} v_i \overline{G}_{f,i}^o \quad (9\text{-}13)$$

where $\overline{G}_{f,i}^o$ is the standard molar Gibbs energy of formation of species i, and v_i is the stoichiometric coefficient of i in the reaction of interest (positive for products, negative for reactants). Analogous expressions apply for the standard enthalpy and entropy of reaction. Values of these thermodynamic parameters for many species of interest in environmental engineering are available in a number of compilations (see, e.g., Stumm and Morgan (1996) or CRC (2008)).

In a solid material, ions are held in a rigid structure. Dissolved ions have much more mobility, but they are still in a somewhat structured environment, with each ion associated with a group of water molecules that are oriented by the local electrical field. These oriented water molecules, referred to as the *primary hydration sphere*, are nearly immobilized and move with the ion. A less well-defined, secondary group of water molecules near the solute (the *secondary hydration sphere*) is also oriented and has some tendency to move with the solute, but the association is weaker than for those in the primary group. The total number of water molecules in these groups depends on the ion, being higher for cations than anions and increasing with increasing ionic charge or decreasing ionic size. In dilute solutions, typical values are 3–5 for monovalent cations, ~10 for divalent cations, and 1–2 for monovalent anions (Bockris and Reddy, 1998).

The structured water plays an important role in the thermodynamics of precipitation. Consider, for example, the precipitation of the aluminum hydroxide solid *gibbsite* by the following reaction:

$$Al^{3+} + 3\,OH^- \leftrightarrow Al(OH)_3(s)$$

Thermodynamic data for this reaction under standard conditions are summarized in Table 9-3.

When the reaction proceeds under standard conditions, the three terms in Equation 9-12 can be calculated as follows:

$$\Delta \overline{G}_r^o = -1155 - (-489.4) - 3(-157.3)$$
$$= -193.7\,kJ/mol \quad (9\text{-}14a)$$

$$\Delta \overline{H}_r^o = -1293 - (-531.0) - 3(-230.0)$$
$$= -72.0\,kJ/mol \quad (9\text{-}14b)$$

$$T\Delta \overline{S}_r^o = 298[68.4 - (-308) - 3(-10.75)]\frac{1\,kJ}{10^3\,J}$$
$$= 121.8\,kJ/mol \quad (9\text{-}14c)$$

The standard enthalpy of reaction is negative, and the standard entropy is positive. According to reaction 9-12, both of these terms cause the standard Gibbs energy of reaction to decrease, and hence both make the reaction more thermodynamically favorable. The idea that precipitation could increase the entropy of the system might seem counterintuitive because the solid state appears to be more ordered than the aqueous phase. However, solvated aluminum ions are strongly coordinated with water. Disruption of this coordination reduces the restrictions on movement of the water molecules, causing the overall entropy change of the reaction to be positive.

The effect of temperature on the solubility product can be assessed by differentiating Equation 9-11 with respect to temperature:

$$\frac{d}{dT}\ln K_{eq} = \frac{d}{dT}\frac{\Delta \overline{G}_r^o}{RT} \quad (9\text{-}15)$$

Substituting $\Delta \overline{G}_r^o = \Delta \overline{H}_r^o - T\Delta \overline{S}_r^o$ and noting that $\Delta \overline{H}_r^o$ and $\Delta \overline{S}_r^o$ are approximately independent of temperature leads to the van't Hoff equation:

$$\ln \frac{K_{eq,T_2}}{K_{eq,T_1}} = \frac{\Delta \overline{H}_r^o}{R}\left(\frac{1}{T_1} - \frac{1}{T_2}\right) \quad (9\text{-}16)$$

Setting $T_1 = 298\,K$ (25°C), the solubility product at an arbitrary temperature T can then be determined from

$$K_{eq,T} = K_{eq,25°C}\exp\left[\frac{\Delta \overline{H}_r^o}{R}\left(\frac{1}{298\,K} - \frac{1}{T}\right)\right] \quad (9\text{-}17)$$

Because the change in enthalpy for a reaction can be positive or negative, the solubility product can increase or decrease with increasing temperature. For example, as temperature increases from 4°C to 30°C, the solubility product for the iron oxide mineral goethite decreases by more than two orders of magnitude, whereas the solubility product of gibbsite increases by more than one order of magnitude (Figure 9-2). The standard enthalpy of formation of calcite ($CaCO_3(s)$) is negative, so it becomes less soluble with increasing temperature. As a result, if water in which the solid is undersaturated is heated, K_{s0} can be exceeded, and precipitation can be initiated. This process sometimes occurs in hot water heaters and boilers, where precipitation of the solid on the heating coils can be problematic by interfering with the heat transfer process.

9.3 PRECIPITATION DYNAMICS: PARTICLE NUCLEATION AND GROWTH

As noted previously, much of the fundamental work on the dynamics of precipitation has been carried out using pure crystalline solids under conditions where the degree of supersaturation is closely controlled. Those types of conditions are rarely encountered in environmental engineering, but an understanding of the behavior of such systems and the

TABLE 9-3. Thermodynamic Data for Gibbsite and Its Components

Component	$\Delta \overline{G}_f^o$ (kJ/mol)	$\Delta \overline{H}_f^o$ (kJ/mol)	$\Delta \overline{S}_f^o$ (J/mol K)
Al^{3+}	−489.4	−531.0	−308
OH^-	−157.3	−230.0	−10.75
$Al(OH)_3(s)$	−1155	−1293	68.4

Source: From Stumm and Morgan (1996).

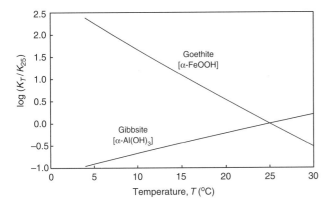

FIGURE 9-2. Relative change in the solubility products of goethite ($\Delta \overline{H}_r^\circ = -179.2\,\text{kJ/mol}$) and gibbsite ($\Delta \overline{H}_r^\circ = 72\,\text{kJ/mol}$) as a function of temperature. (The value of $\Delta \overline{H}_r^\circ$ for gibbsite used here equals $-\Delta \overline{H}_r^\circ$ used in the example in the text, because the graph is for dissolution of gibbsite, whereas the example is for precipitation.)

key concepts used to model them is nevertheless useful. This section provides an overview of the most widely used conceptual and mathematical models for nucleation and growth of ideal solids.

Thermodynamics of Nucleation

By definition, a solid first forms when an identifiable interface is established between the precipitating material and the surrounding solution; that is, when a molecular cluster is converted into a particle nucleus. The nucleus might then acquire more material and grow into a stable particle, or it might redissolve. Here, we analyze which of those two possibilities is more likely for nuclei of various sizes and solutions of various compositions. For convenience, we assume that the process is taking place at standard temperature and pressure.

To carry out the analysis, it is helpful to recall portions of the discussion of interfacial phenomena from the analysis of gas transfer (Chapter 5) and adsorption (Chapter 7). In Chapter 5, it was pointed out that, whereas molecules in the middle of a pure liquid experience essentially symmetric forces from their interactions with the surrounding molecules, those at a gas/liquid interface experience a net attraction toward the bulk liquid. A thermodynamic interpretation of this phenomenon is that the molecules at the surface are in a slightly higher (chemical) energy state than those in the interior. The consequence of this force imbalance is that, in the absence of other applied forces, liquid droplets acquire a spherical shape, thereby minimizing the interfacial area and the number of molecules in the higher energy state.

A similar situation applies to particles of a pure solid that are suspended in an aqueous solution; that is, molecules at the surface of the solid are subject to asymmetric forces and are therefore in a slightly stressed (higher energy) state compared to those in the interior. The "extra" chemical energy of the surface molecules is part of their chemical potential energy (their Gibbs free energy). In the case of solids, the internal forces orienting the molecules are stronger than those in a liquid, so that the particles do not necessarily acquire a spherical shape. Nevertheless, they do tend to acquire the shape that minimizes the interfacial area without placing excessive stress on the bonds within the solid (by distorting the crystal lattice).

Based on the preceding discussion, the Gibbs free energy of a whole particle can be expressed as the sum of the Gibbs free energy that the particle would have if all of its molecules were in the interior and an increment that accounts for the additional energy associated with the presence of the interface. Assuming that the number of molecules at the surface is proportional to the surface area, we can express this idea as follows:

$$\begin{pmatrix} \text{Molar Gibbs energy} \\ \text{of whole particle} \end{pmatrix} = \begin{pmatrix} \text{Molar Gibbs energy of} \\ \text{particle if all molecules} \\ \text{were in the interior} \end{pmatrix}$$

$$+ \begin{pmatrix} \text{Increment of Gibbs} \\ \text{energy per unit} \\ \text{area of surface} \end{pmatrix} \begin{pmatrix} \text{Surface area} \\ \text{per unit volume} \\ \text{of solid} \end{pmatrix} \begin{pmatrix} \text{Solid} \\ \text{volume} \\ \text{per mole} \end{pmatrix}$$

$$\overline{G}_p = \overline{G}_{p,\text{interior}} + \gamma_p \frac{A_p}{V_p} \overline{V}_p \qquad (9\text{-}18)$$

where γ_p is the interfacial tension (consistent with its definition in Equation 7-106), A_p is the surface area of the particle, V_p is the particle volume, and $\overline{V}_p$ is the particle's molar volume (i.e., volume per mole).

As a particle grows, its volume increases more rapidly than its surface area. As a result, the significance of the second term in Equation 9-18 diminishes, and for large particles, the molar Gibbs free energy of the solid is essentially identical to the molar Gibbs free energy of the interior molecules. If such a particle is a pure solid at standard temperature and pressure, its molar Gibbs energy is the standard molar Gibbs energy of the solid, $\overline{G}_p^\circ$. We can therefore substitute $\overline{G}_p^\circ$ for $\overline{G}_{p,\text{interior}}$ in Equation 9-18 to obtain the following expression for the molar Gibbs energy of a pure particle of any size:

$$\overline{G}_p = \overline{G}_p^\circ + \gamma_p \overline{V}_p \frac{A_p}{V_p} \qquad (9\text{-}19)$$

If the particle can be approximated as a sphere of diameter d_p, Equation 9-19 becomes

$$\overline{G}_p = \overline{G}_p^\circ + \gamma_p \overline{V}_p \frac{\pi d_p^2}{(1/6)\pi d_p^3} = \overline{G}_p^\circ + \frac{6\gamma_p \overline{V}_p}{d_p} \qquad (9\text{-}20)$$

Equations 9-19 and 9-20 are consistent with the idea that molecules at the surface of a pure phase have more Gibbs free energy than those in the interior, and that the average molar Gibbs energy of the entire solid (considering both the interior and surface molecules) declines as the particle grows, asymptotically approaching the standard molar Gibbs energy, $\overline{G}_p^o$. One way to interpret this result is that the standard state for a solid includes not only the specification of a standard temperature (typically, 25°C), pressure (1.0 atm), and composition (pure phase), but also a particle size (large enough that the extra free energy of the surface molecules is negligible in comparison to the total Gibbs energy of the solid). Thus, for a pure phase at standard temperature and pressure, the activity of the solid is 1.0 only in the limit of an infinitely large particle; for all particles of finite size, the activity is >1.0.

For any chemical species, the relationship between the molar Gibbs free energy and activity is

$$\overline{G}_i = \overline{G}_i^o + RT \ln a_i \qquad (9\text{-}21)$$

Applying Equation 9-21 to a spherical particle, we find

$$\overline{G}_p^o + \frac{6\gamma_p \overline{V}_p}{d_p} = \overline{G}_p^o + RT \ln a_p$$

$$a_p = \exp\left(\frac{6\gamma_p \overline{V}_p}{RT d_p}\right) \qquad (9\text{-}22)$$

Consistent with the preceding discussion, Equation 9-22 indicates that the activity of a solid is ~1.0 when the solid is present as large particles (large d_p), but that the activity increases as the particles shrink, so that very small particles might have activities substantially >1.0.

Now, consider an ideal solution containing concentrations c_M of a metal Me^{m+} and c_L of an anion L^{n-}, which can combine to form a solid, $Me_nL_m(s)$. Written in the conventional form used to define the solubility product (dissolution of the solid), the reaction is

$$Me_nL_m(s) \leftrightarrow n Me^{m+} + m L^{n-} \qquad (9\text{-}23)$$

Assume that the suspension contains relatively large particles of the solid, and that the product $a_M^n a_L^m$ (equal to $c_M^n c_L^m$ since the solution is ideal) is greater than K_{s0}, so the solution is supersaturated. Because the particles are large, the activity of the solid is 1.0, so

$$Q_{s0} = \frac{a_M^n a_L^m}{a_{M_nL_m(s)}} = \frac{c_M^n c_L^m}{1.0} > K_{s0} \qquad (9\text{-}24)$$

Under the circumstances, we expect additional solid to precipitate, causing the product $c_M^n c_L^m$ to decrease until it equals K_{s0}. However, we might reach a different conclusion if the particles were small enough that the extra energy of the surface molecules contributed significantly to $\overline{G}_p$. To explore that scenario, we again write the expression for the solubility quotient, but we express the activity of the solid as shown in Equation 9-22, rather than assigning it a value of 1.0:

$$Q_{s0} = \frac{a_M^n a_L^m}{a_{M_nL_m(s)}} = \frac{c_M^n c_L^m}{\exp\left((6\gamma_p \overline{V}_p)/(RT d_p)\right)} \qquad (9\text{-}25)$$

The molar free energy of the dissolution reaction can then be computed as follows:

$$\Delta \overline{G}_r = RT \ln \frac{Q_{s0}}{K_{s0}} = RT \ln \frac{c_M^n c_L^m \exp\left(-(6\gamma_p \overline{V}_p)/(RT d_p)\right)}{K_{s0}}$$

$$= RT \ln (SS) - \frac{6\gamma_p \overline{V}_p}{d_p} \qquad (9\text{-}26)$$

where SS is $c_M^n c_L^m / K_{s0}$, the degree of supersaturation of the solution with respect to precipitation of large particles.[1]

Equation 9-26 indicates the Gibbs free energy change per mole of $Me_nL_m(s)$ dissolving. The product of that quantity with the number of moles in a particle of diameter d_p yields the free energy of reaction for dissolution of one particle. Designating this quantity as $\overline{\overline{G}}_r$, we find

$$\frac{\text{Moles solid}}{\text{Particle}} = \frac{\text{Volume of solid/particle}}{\text{Volume of solid/mole}} = \frac{(1/6)\pi d_p^3}{\overline{V}_p}$$

$$\Delta \overline{\overline{G}}_r = \left(RT \ln (SS) - \frac{6\gamma_p \overline{V}_p}{d_p}\right)\left(\frac{(1/6)\pi d_p^3}{\overline{V}_p}\right)$$

$$= \frac{\pi d_p^3}{6\overline{V}_p} RT \ln(SS) - \pi d_p^2 \gamma_p \qquad (9\text{-}27)$$

Finally, since we are primarily interested in the precipitation reaction, we can write the free energy of that reaction (per particle) as the opposite of the expression in Equation 9-27:

$$\Delta \overline{\overline{G}}_{r,pptn} = -\frac{\pi d_p^3}{6\overline{V}_p} RT \ln(SS) + \pi d_p^2 \gamma_p \qquad (9\text{-}28)$$

---

[1] In some literature, SS is defined as $\left(c_M^n c_L^m / K_{s0}\right)^{1/(n+m)}$, which can be thought of as the degree of supersaturation for the precipitation reaction written as $(n/(n+m))Me + (m/(n+m))L \leftrightarrow Me_{n/(n+m)}L_{m/(n+m)}(s)$. Writing the reaction in this way defines one mole of solid as the mass of solid that contains a total of one mole of cations and anions, and hence facilitates comparison of SS values for solids with different stoichiometries.

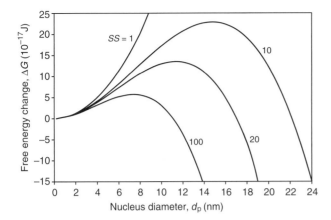

FIGURE 9-3. The free energy for formation of a single particle of diameter d_p from a solution with various degrees of supersaturation. Assumed properties of the precipitate were $\overline{V}_p = 21.1 \text{ cm}^3/\text{mol}$, $\gamma_p = 1.0 \text{ J/m}^2$, $T = 25°\text{C}$.

Equation 9-28 is plotted as a function of d_p for a representative solid at a few values of SS in Figure 9-3. For values of $SS \leq 1$, the free energy required to form a single particle from dissolved molecules is always positive or zero, so no precipitate is expected to form. The curves for $SS > 1$ all follow a pattern in which $\Delta\overline{\overline{G}}_{r,\text{pptn}}$ is positive and increasing with increasing particle size at low d_p, but then passes through a maximum, after which it decreases, eventually becoming negative. Because reactions proceed only when their Gibbs free energy change is <0, we infer that at a given value of SS larger than 1.0, small particles will have $\Delta\overline{\overline{G}}_{r,\text{pptn}} > 0$ and will not form, and, if such particles are added to the solution, they will dissolve. On the other hand, for the same value of SS, large particles are expected to form and be stable. In other words, a solution can simultaneously be supersaturated with respect to large particles (as indicated by the positive value of SS) and undersaturated with respect to small particles. Only particles of a single size (that causes $\Delta\overline{\overline{G}}_{r,\text{pptn}}$ to be 0) will be in equilibrium in such a solution and therefore will neither form spontaneously nor dissolve if they are already present. The plot indicates that the larger the degree of supersaturation of a solution (i.e., the larger the value of Q_{s0}/K_{s0}), the smaller the particles that are in equilibrium with it. Solving Equation 9-28 for $\Delta\overline{\overline{G}}_{r,\text{pptn}} = 0$ quantifies this relationship:

$$d_p\big|_{\text{at equilibrium}} = \frac{6\gamma_p \overline{V}_p}{RT \ln SS} \qquad (9\text{-}29)$$

While Equation 9-29 identifies the particle size that is in equilibrium with a given solution, it is of limited value, because it considers only the spontaneous appearance of fully formed, d_p-size particles from homogeneous solution, or their complete dissolution. A more relevant question is whether a particle of size d_p is likely to form from *incremental* growth of a slightly smaller particle, or to disappear due to incremental dissolution. That is, instead of $\Delta\overline{\overline{G}}_{r,\text{pptn}}$ for the reaction shown in Equation 9-23, we wish to explore the free energy change associated with the following reaction:

$$\text{Me}_n\text{L}_m(s)\big|_{d_p - d(d_p)} + dn \text{ Me}^{m+} + dm \text{ L}^{n-} \leftrightarrow \text{Me}_n\text{L}_m(s)\big|_{d_p} \qquad (9\text{-}30)$$

Figure 9-3 shows that, if SS is 20, more free energy is required to form a particle with a 6.1 nm diameter than to form one with a 6.0 nm diameter. Correspondingly, the free energy change accompanying conversion of a 6.0 to a 6.1 nm particle in such a solution is positive, so such particle growth would not be thermodynamically favorable. On the other hand, the opposite is true for conversion of a 16.0 to a 16.1 nm particle, so such growth would be favored in the given solution. Generalizing these results, we see that spontaneous dissolution is expected for particles smaller than a critical size ($d_{p,\text{crit}}$) that corresponds to the maximum in the $\Delta\overline{\overline{G}}_{r,\text{pptn}}$ versus d_p curve, and spontaneous growth is expected for particles larger than this critical size. Note that, for d_p slightly larger than $d_{p,\text{crit}}$, spontaneous growth is expected even though $\Delta\overline{\overline{G}}_{r,\text{pptn}}$ is positive. The positive value of $\Delta\overline{\overline{G}}_{r,\text{pptn}}$ indicates that the system would be more stable if the particle fully dissolved than if it remained in suspension. However, the fact that $d\Delta\overline{\overline{G}}_{r,\text{pptn}}/d(d_p)$ is negative indicates that the first, incremental steps in such a dissolution process would be thermodynamically unfavorable, so the reaction is more likely to proceed in the direction of growth than dissolution.

The value of $d_{p,\text{crit}}$ can be determined by setting $d\Delta\overline{\overline{G}}_{r,\text{pptn}}/d(d_p)$ to 0, yielding

$$d\Delta\frac{\overline{\overline{G}}_{r,\text{pptn}}}{d(d_p)} = \frac{\pi d_p^2}{\overline{V}_p} RT \ln(SS) - 4\pi d_p \gamma_p = 0$$

$$d_{p,\text{crit}} = \frac{4\overline{V}_p \gamma_p}{RT \ln(SS)} \qquad (9\text{-}31)$$

Comparison of Equations 9-29 and 9-31 indicates that $d_{p,\text{crit}}$ is always two-thirds as large as d_p of a particle that is in equilibrium with the solution. Substitution of Equation 9-31 into Equation 9-28 yields the following expression for $\Delta\overline{\overline{G}}_{d_{\text{crit}},\text{pptn}}$:

$$\Delta\overline{\overline{G}}_{d_{\text{crit}},\text{pptn}} = \frac{16\pi}{3} \frac{\overline{V}_p^2 \gamma_p^3}{(RT)^2} \frac{1}{(\ln SS)^2} \qquad (9\text{-}32)$$

Based on the preceding discussion, if a suspension initially contains particles with a range of sizes, the smaller particles will dissolve while the larger ones grow. As these reactions proceed, SS declines, causing $d_{p,\text{crit}}$ to increase and making ever larger particles unstable until, in theory, only a single, large particle remains in the suspension. This theoretical, ultimate condition is never reached in practice, but when a suspension is first prepared, one does observe that the larger particles grow while the smaller ones dissolve, a phenomenon referred to as *Ostwald ripening*.

Figure 9-3 suggests that the thermodynamic driving force always favors dissolution of the very small nuclei that must form before larger ones do. Under the circumstances, one might reasonably ask how new particles of finite size ever form from an initially particle-free solution. The answer is that, in a supersaturated solution, clusters of various sizes are constantly forming by random collisions among the solutes. Even though formation of larger clusters is not thermodynamically favorable at the macroscopic scale, a few clusters might collide with others, or with other solute molecules, and grow larger. If such collisions occur frequently compared to the rate of cluster dissolution, a cluster could grow larger than $d_{p,crit}$, after which further growth becomes thermodynamically favored. The process is analogous to what occurs in a chemical reaction between two solutes. In that case, the transition involves a collision between two separate molecules, converting their kinetic energy into chemical energy. If the energy input is large enough, the molecules reach an activated state (i.e., they form an activated complex), in which a rearrangement of the chemical bonds to generate the product molecule(s) becomes more favorable than returning to the original state. Essentially the same process occurs with nuclei, except that, in this case, the critical condition is associated with particle size rather than bond rearrangement. Based on this analogy, the rate of formation of stable particles, r_{nuc}, is commonly modeled by the Arrhenius equation:[2]

$$r_{nuc} = A \exp\left(-\Delta \overline{\overline{G}}_{d_{crit},pptn} \over k_B T\right) \quad (9\text{-}33)$$

where A is a constant that depends on the size of the unit cell in the solid, $\Delta \overline{\overline{G}}_{d_{crit},pptn}$ is the free energy required to form a particle of the critical diameter from the solutes, for the given SS, and k_B is the Boltzmann constant. Substituting Equation 9-32 into the above expression yields

$$r_{nuc} = A \exp\left\{-{16\pi\gamma_p^3 \overline{V}_p^2 \over 3(RT)^2 (\ln SS)^2} {1 \over k_B T}\right\} \quad (9\text{-}34)$$

For sparingly soluble salts, theoretical estimates of the pre-exponential factor in the rate expression are in the range

[2] In general, all combinations of clusters that generate units with identifiable solid/solution interfaces can be considered nucleation reactions. Here and in the remainder of the analysis, r_{nuc} is defined as the rate at which nuclei of sizes greater than or equal to d_{crit} are formed.

$10^{25} < A < 10^{30}$ nuclei/cm³ s (Randolph and Larson, 1988). However, experimentally determined values can be many orders of magnitude smaller.

The preceding discussion characterizes homogeneous precipitation, whereas most precipitation reactions in environmental systems are heterogeneous; that is, precipitation occurs on the surface of pre-existing solids. In such cases, the free energy required to initiate formation of a solid is less than in homogeneous precipitation, because, in addition to generating a new interface between the solution and the new precipitate, the process eliminates some of the high-energy interface between the solution and the pre-existing surface. (The new interface that is generated between the two solids typically has a lower surface energy than the solid/solution interface that existed.) This process reduces the activation energy for precipitation and allows spontaneous precipitation to begin once a smaller particle has formed (i.e., it lowers $d_{p,crit}$), but it does not change the general shape or qualitative interpretation of the $\Delta \overline{\overline{G}}_r$ versus d_p curve.

■ **EXAMPLE 9-1.** Taguchi et al. (1996) studied nucleation and growth kinetics of barium sulfate in a series of batch experiments and reported the following correlation between the nucleation rate and the saturation ratio:

$$\ln r_{nuc} = 28.7 - {99.5 \over (\ln SS)^2}$$

where r_{nuc} is in nuclei/m³ s. Estimate the pre-exponential factor in the Arrhenius equation and the apparent interfacial energy for barium sulfate precipitation, assuming that precipitation was homogeneous. Assume $T = 25°C$, and estimate the molar volume of the nuclei from the ratio of the molecular weight (233.4 g/mol) to the density of bulk BaSO₄(s) (4.50 g/cm³).

Solution. Taking the natural logarithm of both sides of Equation 9-34 yields

$$\ln r_{nuc} = \ln A - {16\pi\gamma_p^3 \overline{V}_p^2 \over 3(RT)^2 k_B T (\ln SS)^2}$$

Comparison of this equation with the correlation reported by Taguchi et al. shows that $\ln A = 28.7$, so

$$A = \exp(28.7) = 2.9 \times 10^{12} \text{nuclei/m}^3 \text{ s}$$
$$= 2.9 \times 10^6 \text{nuclei/cm}^3 \text{ s}$$

Similarly, the apparent surface energy can be estimated from a comparison of the arguments of the exponential terms in Equation 9-33 and the correlation of Taguchi et al.:

$${16\pi\gamma_p^3 \overline{V}_p^2 \over 3(RT)^2 k_B T} = 99.5$$

The molar volume of $BaSO_4(s)$ nuclei can be estimated as

$$\overline{V}_p = \frac{233.4\,\text{g/mol}}{4.50\,\text{g/cm}^3} = 51.9\,\text{cm}^3/\text{mol}$$

Rearranging the exponential argument in the expression for the nucleation rate, and solving for γ_p, we find

$$\gamma_p = \left\{ \frac{3(99.5)(RT)^2 k_B T}{16\pi \overline{V}_p^2} \right\}^{1/3}$$

$$= \left\{ \frac{3(99.5)[(8.314\,\text{J/mol K})(298\,\text{K})]^2 (1.380 \times 10^{-23}\,\text{J/K})(298\,\text{K})}{16\pi (51.9\,\text{cm}^3/\text{mol})^2 (10^{-6}\,\text{m}^3/\text{cm}^3)^2} \right\}^{1/3}$$

$$= 0.0382\,\text{J/m}^2 = 38.2\,\text{J/m}^2$$

The authors of the study noted that reported values of interfacial energy values are typically $>100\,\text{mJ/m}^2$ (sometimes much greater). They suggested that the smaller value inferred from their results could be because of heterogeneous nucleation proceeding simultaneously with homogeneous precipitation in their test systems. If the actual interfacial energy were in the typical range, the computed value of A would be many orders of magnitude larger and much closer to the range cited above (10^{25}–10^{30} nuclei/cm^3 s). ∎

Particle Growth and Size Distributions in Precipitation Reactors

The nucleation of stable particles is the first step in a precipitation process. However, in any reactor where precipitation is occurring, those nuclei rapidly grow into substantially larger particles. Because the gross rate of precipitation in a reactor is typically proportional to the total particle surface area available, the particle size distribution (PSD) in the reactor is a key parameter. We next describe how PSDs in such reactors are characterized and modeled. A more detailed discussion of methods for analyzing and reporting PSDs is provided in Chapter 11.

The raw data that are collected when analyzing PSDs are typically concentrations of particles within certain size (diameter) windows; the windows are often referred to as "bins." For instance, a PSD might include information such as that shown in Table 9-4.

The first three columns in the table are self-explanatory. The fourth column is the cumulative concentration of particles in the corresponding bin and all smaller bins. Thus, for instance, the entry for Bin 3 indicates that the suspension contains 7250 particles per milliliter that have diameters $<5.0\,\mu\text{m}$ (the upper limit of Bin 3). The final column in the table contains derived values, computed as the ratio of the particle concentration in a bin to the width of that bin. This value, known as the particle size distribution function, facilitates comparison of PSDs evaluated using instruments with different diameter cutoffs between the bins. The particle size distribution function (n) is a differential distribution, whereas the cumulative particle concentration (N) is a cumulative distribution. These two quantities are analogous to differential and cumulative distributions that we have encountered previously (e.g., the differential and cumulative residence time distributions, $E(t)$ and $F(t)$), and they bear the same relationship to one another; that is:

TABLE 9-4. Representations of the Particle Size Distribution in a Hypothetical Precipitation Reactor

Bin	Particle Diameter (d_p) Range (μm)	Particle Concentration in Bin, ΔN (#/mL)	Cumulative Particle Concentration, N (#/mL)	Particle Size Distribution Function, n (#/mL μm)
1	0.5–1.0	4500	4500	9000
2	1.0–2.0	2100	6600	2100
3	2.0–5.0	650	7250	217
4	5.0–10.0	76	7326	15.2
5	10.0–20.0	6	7332	0.6

$$n\{d_p\} = \frac{dN\{d_p\}}{dd_p} \quad (9\text{-}35)$$

$$N\{d_p\} = \int_0^{d_p} n\{d_p\}\,dd_p \quad (9\text{-}36)$$

The expected PSD function in a precipitation reactor for various scenarios of nucleation and growth can be modeled based on a mass and/or number balance on the particles in the system. For example, imagine that a precipitation reaction is proceeding in a CFSTR which receives an input stream that contains no particles. Both nucleation of new particles and growth of existing particles occur in the reactor, so a number balance on particles in bin i can be written as follows:

$$V\frac{d(\Delta N_i)}{dt} = Q(\Delta N_{\text{in},i} - \Delta N_i) + Vr_{G,\text{net},i} + Vr_{\text{nuc},i} \quad (9\text{-}37)$$

where $r_{G,\text{net},i}$ and $r_{\text{nuc},i}$ are the net rates of appearance of particles in the bin due to growth of existing particles and nucleation of new particles, respectively.[3] The subscript "net" is included in the term for particle growth, because that term includes both a positive contribution from growth of smaller particles (from bin $i-1$) and a negative

---
[3] Other mechanisms that convert particles from one size to another, such as particle agglomeration and particle fracturing, can also affect the particle size distribution and can be included in the analysis as well. Randolph and Larson (1988) describe how these kinds of processes can be included in the particle population balance.

contribution due to loss of particles from bin i when they grow into larger particles. $r_{\text{nuc},i}$ has no such subscript, because nucleation can cause new particles to appear, but cannot cause existing particles to disappear (i.e., dissolution of existing particles is ignored). Both $r_{G,\text{net},i}$ and $r_{\text{nuc},i}$ have dimensions of number of particles per unit volume per unit time.

The rate at which new material becomes incorporated into an existing particle is commonly assumed to be limited either by mass transfer across the liquid boundary layer surrounding the particle or by the chemical reaction incorporating the new material into the solid. In either case, the rate is expected to be proportional to the particle surface area. Therefore, for example, the rate of mass-transfer-limited particle growth in a reactor can be written as follows:

$$r_{G,m}(d_p) = k_{\text{mt}} A_{d_p} \Delta c \quad (9\text{-}38)$$

where $r_{G,m}(d_p)$ and A_{d_p} are, respectively, the mass-based growth rate (mass/time) and total area of d_p-size particles in the reactor; k_{mt} is the mass transfer coefficient for transport through the boundary layer; and Δc is the concentration difference of the precipitating substance across the boundary layer. The volumetric rate of growth of a single, d_p-size particle (r_{G,V_p}, volume/time) can be written as the product of the particle surface area and the rate at which its radius increases. Expressing the rate of radius increase as one-half the rate of increase in the diameter ($\dot{d}_p$, length/time), this relationship is

$$r_{G,V_p}(d_p) = \pi d_p^2 \frac{\dot{d}_p}{2} \quad (9\text{-}39)$$

The rate at which the mass of d_p-size particles in the reactor is increasing due to growth is the product of r_{G,V_p} with the particle density (ρ_p) and the number concentration of particles, so

$$r_{G,m}(d_p) = (\Delta N_{d_p}) \left[\rho_p \pi d_p^2 \frac{\dot{d}_p}{2} \right] \quad (9\text{-}40)$$

Similarly, the total surface area of d_p-size particles in the reactor is their number concentration times the area per particle:

$$A_{d_p} = (\Delta N_{d_p}) \left[\pi d_p^2 \right] \quad (9\text{-}41)$$

Substituting Equations 9-40 and 9-41 into Equation 9-38 yields

$$(\Delta N_{d_p}) \left[\rho_p \pi d_p^2 \frac{\dot{d}_p}{2} \right] = k_{\text{mt}} (\Delta N_{d_p}) \left[\pi d_p^2 \right] \Delta c$$

$$\dot{d}_p = 2 k_{\text{mt}} \Delta c / \rho_p \quad (9\text{-}42)$$

In general, the solute concentration at the particle surface is assumed to be the saturation concentration, so Δc equals $c_{\text{bulk}} - c_{\text{sat}}$ for all particles, regardless of their size. If k_{mt} is also independent of particle size, Equation 9-42 would imply that the diameters of all the particles in a given suspension grow at the same rate.

If growth is presumed to be limited by incorporation of material into the particles instead of by mass transfer, essentially the same result is obtained, except that $k_{\text{mt}} \Delta c$ is replaced by the reaction (i.e., incorporation) rate. For instance, if the reaction is first-order with a heterogeneous rate constant k_1 (length/time), the rate of diameter increase of all the particles in the suspension is

$$\dot{d}_p = \frac{2 k_1 c_{\text{bulk}}}{\rho_p} \quad (9\text{-}43)$$

Because, in truth, k_{mt} is thought to decrease with increasing particle diameter, it is possible that growth of some (smaller) particles will be controlled by chemical processes at their surface, while that of (larger) particles is controlled by mass transfer through the boundary layer. In such a case, not all particles in the reactor would grow at the same rate. Here, for simplicity, we ignore this possibility.

The preceding results can now be applied to derive a simple expression for the rate at which particles of size d_p appear or disappear in any well-mixed reactor due to the growth process. Consider particles in two adjacent bins, but now assume that the bins are separated by a differential size gap, $d(d_p)$. All the particles in each bin are approximated as having the diameter in the middle of the bin size range.[4] The time required for particles to grow from the size associated with the smaller to the larger bin is the ratio of the gap in particle diameters to the "velocity" of diameter increase, $\dot{d}_p$; that is

$$\begin{pmatrix} \text{Time for particle} \\ \text{growth from one} \\ \text{bin size to the next} \end{pmatrix} = \frac{\text{Diameter gap between bins}}{\text{Rate of particle diameter increase}}$$

$$= \frac{d(d_p)}{\dot{d}_p} \quad (9\text{-}44)$$

If all the particles in each bin are treated as though they have the same size (the average size in the bin) and they all grow at the same rate, the rate at which particles move from bin i to the next larger bin is simply the number of particles

---

[4] More sophisticated analyses (e.g., Hounslow et al. (1988); Kumar and Ramkrishna (1996)) consider the full distribution of particle sizes in each bin.

in bin i (ΔN_i) divided by the time given in Equation 9-44. Correspondingly, the net rate at which particles appear in bin i by particle growth is

$$r_{G,net,i} = \frac{\Delta N_{i-1}}{d(d_p)/\dot{d}_p} - \frac{\Delta N_i}{d(d_p)/\dot{d}_p}$$
$$= \dot{d}_p n_{i-1} - \dot{d}_p n_i$$
$$= -\dot{d}_p dn_i \qquad (9\text{-}45)$$

We can substitute this expression into Equation 9-37 to derive the steady-state PSD function in a CFSTR with a particle-free influent as follows. Because the reactor is operating at steady-state, the left side of Equation 9-37 is 0. In addition, since we have stipulated that the bins are differentially small, we can write the number concentration in a bin i, ΔN_i, as $n_i\, d(d_p)$. Making those substitutions, and replacing $r_{G,net,i}$ with the expression in Equation 9-45, we obtain

$$0 = -Qn_i d(d_p) - V\dot{d}_p\, dn_i + Vr_{nuc,i} \qquad (9\text{-}46)$$

Equation 9-46 can be applied separately to the smallest particles ($i=1$), which can form by nucleation but not by growth (since there are no smaller particles), and to all larger particles ($i \geq 2$), for which we assume net formation by growth overwhelms that by nucleation. Expressing the differential terms as finite differences, the equation for the smallest particles is

$$0 = -Qn_1(d_{p,1} - \cancel{d_{p,0}}) - V\dot{d}_p(n_1 - \cancel{n_0}) + Vr_{nuc,1}$$
$$= -n_1 d_{p,1} - \frac{V}{Q}\dot{d}_p n_1 + \frac{V}{Q}r_{nuc,1}$$
$$= -n_1(d_{p,1} - \tau\dot{d}_p) + \tau r_{nuc,1} \qquad (9\text{-}47)$$

$$n_1 = \frac{\tau r_{nuc}}{d_{p,1} + \tau\dot{d}_p} \qquad (9\text{-}48)$$

where τ is the mean hydraulic residence time, equal to V/Q, and $r_{nuc,1}$ has been simplified to r_{nuc} since we are assuming that all nucleation leads to the formation of size-1 particles. The denominator of Equation 9-48 is the sum of the diameter of the nuclei and the expected increase in the diameter of a nucleus if it stayed in the reactor for the mean hydraulic residence time. Since, on average, particles grow to be much larger than the nuclei while in the reactor, we assume that $d_{p,1} \ll \tau\dot{d}_p$, and the equation simplifies to

$$n_1 = \frac{r_{nuc}}{\dot{d}_p} \qquad (9\text{-}49)$$

For larger particles, we again start with Equation 9-46, but manipulate it as follows:

$$0 = -Qn_i d(d_p) - V\dot{d}_p\, dn_i \qquad (9\text{-}50)$$

$$\frac{dn_i}{n_i} = -\frac{1}{\tau\dot{d}_p}d(d_p) \qquad (9\text{-}51)$$

$$\ln\frac{n_j}{n_i} = -\frac{1}{\tau\dot{d}_p}(d_{p,j} - d_{p,i}) \qquad (9\text{-}52)$$

where i and j are any two arbitrary sizes (i.e., midpoints of size bins). Letting $i=1$ and substituting for n_1 from Equation 9-48 yields

$$\ln n_j = \ln n_1 - \frac{1}{\tau\dot{d}_p}(d_{p,j} - d_{p,1}) \qquad (9\text{-}53)$$

Again assuming that the particles of interest (and, as a practical matter, all particles measurable by a PSD analyzer) are much larger than the size of the nuclei (i.e., $d_{p,j} \gg d_{p,1}$), and substituting for n_1 from Equations 9-49 and 9-53 can be rewritten as

$$\ln n = \ln\frac{r_{nuc}}{\dot{d}_p} - \frac{1}{\tau\dot{d}_p}d_p \qquad (9\text{-}54a)$$

$$n = \frac{r_{nuc}}{\dot{d}_p}\exp\left(-\frac{d_p}{\tau\dot{d}_p}\right) \qquad (9\text{-}54b)$$

where the subscript j has been dropped since the equation applies to any particle size of interest.

Equation 9-54a indicates that, for precipitation in a CFSTR with a particle-free influent, a plot of $\ln n$ versus d_p should yield a straight line with slope equal to $-1/\tau\dot{d}_p$ and intercept $\ln(r_{nuc}/\dot{d}_p)$. Since τ would be known in advance, the slope of such a plot can be used to determine $\dot{d}_p$, and that value can be used in conjunction with the intercept to determine r_{nuc}.

■ **EXAMPLE 9-2.** A series of tests has been conducted in a laboratory-scale, well-mixed precipitation reactor with volume of 5 L, receiving feed at a rate of 0.3 L/min. Steady-state particle concentrations in the reactor (ΔN, #/L) were measured in several particle size intervals and were used to estimate the particle size distribution function, which could be described by the following equation: $\ln n = 15.42 - (0.03\,\mu m^{-1})d_p$, where the units of n are $(L\,\mu m)^{-1}$. (These units do not appear explicitly in the equation, but are necessary for its correct interpretation. Different units could be used for n, but then the first value on the right side [15.42] would change as well.) Estimate the particle growth rate and the nucleation rate in the system.

Solution. By writing the first term on the right in the equation for n as $\ln(\exp(15.42))$, we see that the particle size distribution function conforms to the form of Equation 9-54a, with $r_{nuc}/\dot{d}_p = \exp(15.42)$ and $1/(\tau \dot{d}_p) = 0.03\ \mu m^{-1}$. We can therefore find the growth rate and nucleation rates as follows:

$$\dot{d}_p = \frac{1}{1/(\tau \dot{d}_p)} \frac{1}{\tau} = \frac{1}{0.03/\mu m}\left(\frac{0.3\ L/min}{5\ L}\right) = 2\ \mu m/min$$

$$r_{nuc} = \frac{r_{nuc}}{\dot{d}_p}\dot{d}_p = [\exp(15.42)/L\ \mu m][2\ \mu m/min]$$

$$= 9.95 \times 10^6/L\ min \quad \blacksquare$$

The PSD function includes other useful information about the particles in the reactor. For example, the integral of this function from 0 to ∞ yields the particle concentration in the reactor. If the PSD function is given by Equation 9-54b, we find

$$N_p = \int_0^\infty \frac{r_{nuc}}{\dot{d}_p}\exp\left(-\frac{d_p}{\tau \dot{d}_p}\right)d(d_p)$$

$$= \frac{r_{nuc}}{\dot{d}_p}(-\tau \dot{d}_p)\left[\exp\left(-\frac{d_p}{\tau \dot{d}_p}\right)\right]_0^\infty = r_{nuc}\tau \quad (9\text{-}55)$$

Equation 9-55 confirms the intuitive expectation that, when precipitation is carried out in a CFSTR that receives a feed containing none of the particles of interest, the steady-state concentration of those particles is the product of the rate of new particle formation in the reactor and the mean hydraulic residence time.

In addition, by normalizing n to N (i.e., by taking the ratio n/N), we obtain a probability distribution function for the particle population, so that the expression $(n/N)\ d(d_p)$ indicates the likelihood that a particle in the system will have a diameter between d_p and $d_p + d(d_p)$. This probability distribution allows us to calculate mean values of the particle properties, per Equation 2.14. For example, the number-based, mean particle size is given by the ratio of the first moment of the diameter to the zeroth moment:[5]

$$d_{p,avg} = \frac{\int_0^\infty d_p[(r_{nuc}/\dot{d}_p)\exp(-(d_p/\tau \dot{d}_p))/N_p]d(d_p)}{\int_0^\infty [(r_{nuc}/\dot{d}_p)\exp(-(d_p/\tau \dot{d}_p))/N_p]d(d_p)} = \tau \dot{d}_p$$

$$(9\text{-}56)$$

---
[5] The integrals in the numerators of Equations 9-56–9-58 are of the general form $\int_0^\infty Ax^a \exp(-bx)dx$, where x is d_p, and A, a, and b are constants. Such integrals can be evaluated using integration by parts, the details of which can be found in any basic text on integral calculus.

Similarly, assuming spherical particles, the number-based, mean surface area and volume of the particles are

$$A_{p,avg} = \frac{\int_0^\infty \pi d_p^2[(r_{nuc}/\dot{d}_p)\exp(-(d_p/\tau \dot{d}_p))/N_p]d(d_p)}{\int_0^\infty [(r_{nuc}/\dot{d}_p)\exp(-(d_p/\tau \dot{d}_p))/N_p]d(d_p)}$$

$$= 2\pi(\tau \dot{d}_p)^2 = 2\pi d_{p,avg}^2 \quad (9\text{-}57)$$

$$V_{p,avg} = \frac{\int_0^\infty \pi \frac{d_p^3}{6}[(r_{nuc}/\dot{d}_p)\exp(-(d_p/\tau \dot{d}_p))/N_p]d(d_p)}{\int_0^\infty [(r_{nuc}/\dot{d}_p)\exp(-(d_p/\tau \dot{d}_p))/N_p]d(d_p)}$$

$$= \pi(\tau \dot{d}_p)^3 = \pi d_{p,avg}^3 \quad (9\text{-}58)$$

Multiplying this last expression by the particle density leads to an estimate of the average particle mass.

The preceding analysis was based on the assumption that the reactor influent contained no particles that could serve as seeds for growth of the target solid. In practice, particle-free raw waters are virtually never encountered. In many cases, particles present in the feed to a precipitation reactor provide good surfaces for nucleation of new solids that contain the target species, and in others, particles of the target solid itself are added to the feed in a process known as "seeding" the reactor. In either case, the precipitation reaction can proceed more easily (because it generates less new interfacial area than does nucleation of new particles), so contaminant removal is enhanced.

We can account for seeding in a CFSTR at steady state by modifying the number balance on i-size particles (Equation 9-49) as follows:

$$0 = Qn_i^{in}d(d_p) - Qn_i d(d_p) - V\dot{d}_p dn_i \quad (9\text{-}59)$$

where n_i^{in} is the particle size distribution function for i-size seed particles. Rearranging and solving for n_i, we obtain:

$$0 = -Q(n_i - n_i^{in})d(d_p) - V\dot{d}_p dn_i$$

$$0 = -(n_i - n_i^{in})d(d_p) - \tau \dot{d}_p dn_i$$

$$\frac{dn_i}{n_i - n_i^{in}} = -\frac{d(d_p)}{\tau \dot{d}_p} \quad (9\text{-}60)$$

As would be expected, Equation 9-60 simplifies to Equation 9-51 (and can then be integrated to yield Equations 9-54a and 9-54b) if no seed is added ($n_i^{in} = 0$).

In seeded systems, Equation 9-60 cannot be integrated analytically over the full range of i, because n_i^{in} is not a constant. Nevertheless, in many systems, the equation can be integrated over certain ranges. In particular, if $n_i^{in} = 0$ over a given size range, Equation 9-51 is valid over that range. Such a situation would apply, for example, if the seed particles all fell in a specified size range. In such a case, n_i^{in} would be finite (but not necessarily constant) over that range, and 0 outside of it. Equation 9-51 would then apply to particles that are both smaller and larger than the seed particles. Particles smaller than the smallest seed particles could be generated only by growth from nuclei, so the PSD for those particles would be identical to that for a system with no seed addition at all (as given by Equations 9-54a and b). For particles larger than the largest seed particles, Equation 9-51 would still apply, but Equations 9-54a and b would not since those particles would be generated by growth of both nuclei and seed. As noted, no analytical solution exists for the PSD function of particles in size range of the seed particles. However, the PSD function in that range can be determined by discretizing Equation 9-60 and solving for n_i as follows:

$$\frac{n_i - n_{i-1}}{n_i - n_i^{in}} = -\frac{\Delta d_p}{\tau \dot{d}_p} \quad (9\text{-}61)$$

$$n_i = \frac{n_{i-1} \tau \dot{d}_p + \Delta d_p n_i^{in}}{\tau \dot{d}_p + \Delta d_p} \quad (9\text{-}62)$$

The complete PSD function in a seeded CFSTR at steady state can therefore be developed by applying Equation 9-54b for particles up to the smallest seed particles, Equation 9-62 for particles in the size range of seed particles, and Equation 9-51 for larger particles. Alternatively, the numerical approximation (Equation 9-62) can be applied to all particles. In such a case, Equation 9-49 can be used to determine n_1, and Equation 9-62 can be used to compute n_i for all $i > 1$.

■ **EXAMPLE 9-3.** A CFSTR with $\tau = 15$ min is being used as a precipitation reactor and is achieving the desired treatment goal. Under current operating conditions, the nucleation rate is 3.0×10^7 particles/L min, and the particle growth rate is $1.0 \,\mu\text{m/min}$. However, the plant flow rate is scheduled to increase, which will cause τ to decrease, and you have been asked to assess whether seeding the reactor might allow the system to achieve the same precipitation efficiency (i.e., operate at the same SS) at the smaller τ. The particles have a density of $1.8 \,\text{g/cm}^3$.

(a) Determine the PSD function for the system as it is currently operating. What concentration of solids is being generated in the reactor?

(b) Seed particles are available in the 10–15 μm size range. Assuming the particles are uniformly distributed over that range, what would n^{in} be for those particles if $2.0 \,\text{g/L}$ of the particles were dosed into the reactor?

(c) What value of τ would cause the system with seed to operate at the same SS as the current, unseeded system?

Solution.

(a) The PSD function for the unseeded system can be developed directly from Equation 9-54b. Inserting the given values of r_{nuc}, $\dot{d}_p$, and τ yields

$$n = \frac{3.0 \times 10^7 / \text{L min}}{1.0 \,\mu\text{m/min}} \exp\left(-\frac{d_p}{(15 \text{ min})(1.0 \,\mu\text{m/min})}\right)$$

$$= (3.0 \times 10^7 / \text{L} \,\mu\text{m}) \exp\left(-\frac{d_p}{15 \,\mu\text{m}}\right)$$

The total number of particles in the reactor can be found from Equation 9-55:

$$N_{tot} = r_{nuc} \tau = (3.0 \times 10^7 / \text{L min})(15 \text{ min})$$

$$= 4.50 \times 10^8 / \text{L}$$

The average diameter and volume of particles are given by Equations 9-56 and 9-58, respectively:

$$d_{p,avg} = \tau \dot{d}_p = (15 \text{ min})(1.0 \,\mu\text{m/min}) = 15.0 \,\mu\text{m}$$

$$V_{p,avg} = \pi d_{p,avg}^3 = \pi (15.0 \,\mu\text{m})^3 (10^{-4} \,\text{cm}/\mu\text{m})^3$$

$$= 1.06 \times 10^{-8} \,\text{cm}^3$$

Finally, the mass-based particle concentration can be computed as

$$c_p = N_p \rho_p V_{p,avg}$$

$$= \left(\frac{4.5 \times 10^8}{\text{L}}\right)(1.8 \,\text{g/cm}^3)(1.06 \times 10^{-8} \,\text{cm}^3)$$

$$= 8.59 \,\text{g/L}$$

(b) For the seeded system, we are told that the particles are uniformly distributed over the size range from 10 to 15 μm, so their average diameter is 12.5 μm. For this relatively small size range, we can estimate the average volume of a particle as the volume of the average-sized particle:

$$V_{p,avg} = \pi \frac{d_{p,avg}^3}{6} = \pi \frac{(12.5 \,\mu\text{m})^3}{6} (10^{-4} \,\text{cm}/\mu\text{m})^3$$

$$= 1.02 \times 10^{-9} \,\text{cm}^3$$

(An exact calculation of $V_{p,avg}$ as the ratio of the first moment of V_p to the zeroth moment yields a value of $1.06 \times 10^{-9} \, cm^3$.) The average mass of a particle and the corresponding number concentration of particles and n^{in} in the 10–15-μm size range can then be computed as

$$m_{p,avg} = \rho_p V_{p,avg} = (1.8 \, g/cm^3)(1.02 \times 10^{-9} \, cm^3)$$
$$= 1.84 \times 10^{-9} \, g$$
$$N_{p,added} = \frac{c_{p,added}}{m_p} = \frac{2.0 \, g/L}{1.84 \times 10^{-9} \, g} = 1.09 \times 10^9/L$$
$$n^{in} = \frac{N_{p,added}}{\Delta d_p} = \frac{1.09 \times 10^9/L}{(15-10) \, \mu m} = 2.18 \times 10^8/(L \, \mu m)$$

(c) If the reactor with seed were exactly as efficient as the unseeded reactor, then SS would be the same in the two reactors, as would the concentration of precipitate formed. The total concentration of solid in the seeded reactor would therefore be the concentration in the unseeded reactor (8.59 g/L), plus the 2.0 g/L of seed added, for a total of 10.59 g/L.

The PSD function in the seeded reactor can be simulated by preparing a spreadsheet to model n in discrete particle-size ranges according to Equation 9-62. Values of d_p are first entered into one column. An expression for n of a small particle (smaller than any of the seed particles) can be written based on Equation 9-54b. The nucleation rate and particle growth rate in the seeded reactor will be identical to those in the unseeded reactor (because SS is the same), but τ in the seeded reactor is not yet known, so we write the expression for n of an arbitrarily small particle as

$$n_{d_1} = \frac{3.0 \times 10^7/L \, min}{1.0 \, \mu m/min} \exp\left(-\frac{d_1}{\tau_{seeded}(1.0 \, \mu m/min)}\right)$$

We can make a guess for τ_{seeded} and solve for d_1. We can then use Equation 9-62 to compute n for all larger particles (including the value of n^{in} from part (b) for particles in the size range of the seeded particles) and follow the same procedure as in part (a) to compute the concentration of solids in the reactor. We then iterate on τ_{seeded} until the computed concentration of solids in the seeded reactor is 10.59 g/L. The result is that the seeded reactor can perform as well with a mean hydraulic residence time of 5.8 min as the unseeded reactor does with $\tau = 15$ min. The PSD functions of the two reactors are compared in Figure 9-4. The effect of seeding is dramatic, not only for the size range of the added particles, but for all larger sizes as well. ∎

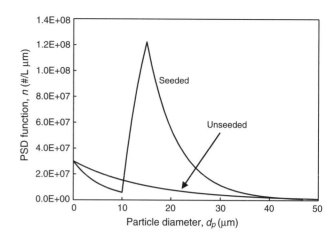

FIGURE 9-4. The PSD function for the unseeded and seeded reactors in the example system.

9.4 MODELING SOLUTION COMPOSITION IN PRECIPITATION REACTIONS

Models for the chemical changes that occur in a precipitation process can be divided into those that consider reaction kinetics and those based on the assumption that the reactions reach equilibrium. The equilibrium assumption is appropriate when the characteristic time for the precipitation reaction is short compared to the characteristic residence time in the system. Use of kinetic models for precipitation reactions is relatively rare in water and wastewater treatment; rather, the processes are typically designed based on the assumption that the reactions closely approach equilibrium.

Sometimes, a relatively simple analysis based on the reaction stoichiometry provides an adequate description of the process. This simplification is most appropriate in cases where the solution is highly supersaturated initially, and the residual concentrations remaining in solution are not crucial (e.g., if the goal is to precipitate a certain amount of metal oxide adsorbent, rather than to lower the concentration remaining in solution to a specified value). More often, though, a more sophisticated approach that retains the assumption of rapid equilibration is used; in these cases, it is often helpful to use chemical equilibrium software programs to analyze the system.

Determining the equilibrium solution composition in the presence of a solid precipitate is a central issue in water chemistry (see, e.g., textbooks by Pankow (1991), Morel and Hering (1993), Benjamin (2010), and Jensen (2003)). Some of the relevant computations for such systems are reviewed here. In addition to addressing the question of how much of the target species remains in solution (which is typically the focus in water chemistry texts), emphasis is also placed on computation of the amounts of reagents required to meet a treatment objective and the amount of solid sludge that is generated. In the following sections, we consider both stoichiometric and equilibrium models for precipitation, progressively accounting for greater complexity.

Quantitative Significance of the Solubility Product If the constituent ions of the solid behave ideally in solution (i.e., if they all have $\gamma_i = 1.0$), we can substitute the molar concentrations of those species for their activities.[6] For example, for a solution in equilibrium with pure $BaSO_4(s)$, the solubility product expression (Equation 9-1) can be simplified as follows:

$$K_{s0,BaSO_4(s)} = \frac{a_{Ba^{2+}} a_{SO_4^{2-}}}{a_{BaSO_4(s)}} = \frac{c_{Ba^{2+}} \gamma_{Ba^{2+}} c_{SO_4^{2-}} \gamma_{SO_4^{2-}}}{1.0}$$

$$= c_{Ba^{2+}} c_{SO_4^{2-}} = 1.05 \times 10^{-10} \quad (9\text{-}63)$$

Corresponding expressions could be written for all the other solubility products shown in Table 9-2. Note that Equation 9-63 applies only if $BaSO_4(s)$ is present, because the equation is based on the assumption that $a_{BaSO_4(s)} = 1.0$. If, in a given situation, the product $c_{Ba^{2+}} c_{SO_4^{2-}}$ is $<1.05 \times 10^{-10}$, then the solid would be undersaturated and would not form (or, if $BaSO_4(s)$ were in contact with the solution, it would dissolve until the product $c_{Ba^{2+}} c_{SO_4^{2-}}$ equaled K_{s0}).

■ **EXAMPLE 9-4.** The solubility product of $AgCl(s)$ (*cerargyrite*) is $10^{-9.75}$. Assuming that Ag^+ and Cl^- are the only forms in which Ag and Cl exist in solution, how much NaCl must be added to a solution initially containing 3 mg/L Ag^+ and no Cl^-, to reduce the equilibrium soluble Ag^+ concentration to 0.1 mg/L? Assume ideal behavior of both Ag^+ and Cl^-.

Solution. Since the ions behave ideally, we can make the same approximations for this system as those that led to Equation 9-63 for a system containing $BaSO_4(s)$, yielding

$$K_{s0} = c_{Ag^+} c_{Cl^-} = 10^{-9.75} = 1.78 \times 10^{-10}$$

The atomic weight of silver is 108, so 0.1 mg/L corresponds to

$$c_{Ag^+} = \frac{0.1\,\text{mg/L}}{108,000\,\text{mg/mol}} = 9.26 \times 10^{-7}\,\text{mol/L}$$

The concentration of Cl^- that is in equilibrium with this concentration of Ag^+ and with $AgCl(s)$ is

$$c_{Cl^-} = \frac{K_{s0}}{c_{Ag^+}} = \frac{1.78 \times 10^{-10}}{9.26 \times 10^{-7}} = 1.92 \times 10^{-4}\,\text{mol/L}$$

The mass of Ag^+ that precipitates is 2.9 mg/L, or 2.69×10^{-5} mol/L. The NaCl dose can be determined from a mass balance:

$$\begin{aligned}\text{Cl}^-\text{ dose required} &= \text{Cl precipitated} + \begin{array}{c}\text{Dissolved Cl}\\\text{at equilibrium}\end{array} - \begin{array}{c}\text{Cl present}\\\text{initially}\end{array}\\
&= 2.69 \times 10^{-5} + 1.92 \times 10^{-4} - 0\\
&= 2.19 \times 10^{-4}\,\text{mol/L}\end{aligned}$$

On a molar basis, the required NaCl dose is the same as the Cl^- dose, so

$$\begin{aligned}\text{NaCl dose required} &= (2.19 \times 10^{-4}\,\text{mol/L})(58,500\,\text{mg/mol})\\&= 12.8\,\text{mg/L}\end{aligned}$$
■

Accounting for Soluble Speciation of the Constituents of the Solid Although the solubility product establishes the relationship among a few dissolved species when the solution is equilibrated with a solid, the situation is often complicated by the reactions of those species with other solutes. For instance, dissolved metal ions can react with dissolved *ligands*[7] that are present in solution, such as OH^-, CO_3^{2-}, and $EDTA^{4-}$, and those same ligands might participate in acid/base reactions (to form HCO_3^-, H_2CO_3, and $H_x EDTA^{4-x}$).

Frequently, the parameter of interest in a treatment process is not the dissolved concentration of a particular species (e.g., Zn^{2+}), but rather the total dissolved concentration of an element in all species that contain it (e.g., Zn^{2+}, $ZnOH^+$, $Zn(CO_3)_2^{2-}$, $ZnEDTA^{2-}$). This interest has both a scientific and practical basis, especially for metals. First, although the different forms might behave differently in some important ways, they behave similarly in that they all are transported with the solution, whereas solid-phase Zn might settle or be filtered out of the solution; thus, if we are interested in the movement of Zn through a system, considering all the dissolved forms together is convenient and sensible. Second, measurement of the total concentration of dissolved metals is usually considerably easier than analysis of the individual

---

[6] The assumption that the solution is ideal simplifies the mathematical analysis of these systems considerably, and it is therefore made frequently in this chapter. For illustrative purposes, the effects of nonideality on solubility are shown in a few cases. In real systems, the assumption of ideal solute behavior is often not justified. In such cases, adjustments to account for the non-idealities can be made relatively easily using chemical equilibrium software.

[7] A ligand is a species that binds to a dissolved metal ion. The species formed from the combination of a metal ion and a ligand is called a *complex*, and, if the complex is very strong, it is referred to as a *chelate*. The complexes formed between most metal ions and EDTA are chelates. See any of the water chemistry texts cited earlier for a further discussion of these types of reactions.

species, so most regulations are written in terms of such total concentrations. In this chapter, we distinguish different groups of metal-containing species by referring to the total concentration (the sum of solid and soluble forms) as TOTM, and the total dissolved and solid-phase concentrations as TOTM(aq) and TOTM(s), respectively; corresponding definitions are used for the different groups of ligand-containing species.

The equilibrium composition of any aqueous system can be found by identifying all of the species that exist at equilibrium, and then writing an equal number of independent equations describing the constraints on them. Those equations include material balance expressions and appropriate equilibrium constant expressions (in some cases, a charge balance or proton condition is substituted for the material balance on H^+). When only a few species are present, and when the dominant reactions among solutes are acid/base reactions, the resulting family of equations can be solved analytically or graphically (using $\log c$ – pH diagrams). In more complicated systems, it is more practical to use computer programs that have been designed specifically to solve chemical equilibrium problems.

In an ideal solution, the concentration of a soluble metal–ligand complex with the formula ML_x is given by

$$c_{ML_x^{m-x}} = \beta_{ML_x^{m-x}} c_{M^{m+}} c_{L^{n-}}^x \qquad (9\text{-}64)$$

where $\beta_{ML_x^{m-x}}$ is the equilibrium constant for the following reaction for formation of the complex:

$$M^{m+} + xL^{n-} \leftrightarrow ML_x^{m-x} \qquad (9\text{-}65)$$

By convention, the equilibrium constant for $i = 1$ is usually written as $K_{L,1}$ rather than $\beta_{L,1}$.

If the solution contains several different ligands (e.g., OH^-, CO_3^{2-}, EDTA) that can form complexes with M^{m+}, a general expression for the total dissolved concentration of the metal, TOTM(aq), can be written as follows:[8]

$$\text{TOTM(aq)} = c_M \left(1 + \sum_{i=1}^{p} \beta_{M(L')_i} c_{L'}^i + \sum_{i=1}^{q} \beta_{M(L'')_i} c_{L''}^i + \ldots \right) \qquad (9\text{-}66)$$

In Equation 9-66, L' and L'' represent different ligands, and the product of c_M with the different terms in the parentheses accounts, respectively, for the contributions to TOTM(aq) of the free aquo ion, M^{m+}, all complexes of M with L', all complexes of M with L'', and so on.

---

[8] To reduce clutter, charge values are not shown in Equation 9-66 or subsequent equations involving the generic species M and L. However, charge values are shown when referring to real chemical species.

When a solid $M_nL_m(s)$ is present and in equilibrium with the solution, the dissolution reaction can be written as follows, and the corresponding solubility product expression can be manipulated to yield Equation 9-68 for the concentration of the free metal ion:

$$M_nL_m(s) \leftrightarrow nM^{m+} + mL^{n-}$$

$$K_{s0} = c_M^n c_L^m \qquad (9\text{-}67)$$

$$c_{M^{m+}} = \left(\frac{K_{s0}}{c_L^m} \right)^{1/n} \qquad (9\text{-}68)$$

Substituting Equation 9-68 into Equation 9-66 yields

$$\text{TOTM(aq)} = \left(\frac{K_{s0}}{c_L^m} \right)^{1/n}$$

$$\times \left(1 + \sum_{i=1}^{p} \beta_{M(L')_i} c_{L'}^i + \sum_{i=1}^{q} \beta_{M(L'')_i} c_{L''}^i + \ldots \right) \qquad (9\text{-}69)$$

Equation 9-69 provides a foundation for evaluating the total dissolved metal concentration in systems that reach equilibrium with solids. The first term on the right side is the concentration of the free metal, $c_{M^{m+}}$, in equilibrium with the solid; it establishes that $\text{TOTM(aq)} = c_{M^{m+}}$ when no complexing ligands are present ($c_{L'} = c_{L''} = \ldots = 0$). Whenever complexing ligands are present, $\text{TOTM(aq)} > c_{M^{m+}}$. In general, any number of complexing ligands could be present, and the precipitating ligand may or may not be one of these complexing ligands.

■ **EXAMPLE 9-5.** Copper, a common constituent in wastewaters from certain segments of the electronics industry, can be toxic to aquatic organisms. An industry is considering using precipitation of $Cu(OH)_2(s)$ as a pretreatment process to reduce the copper concentration in the water to <1.0 mg/L ($<10^{-4.81}$ M) before it is discharged to the local Publicly Owned Treatment Works (POTW). The untreated waste contains 10^{-3} M $CuCl_2$ (TOTCu $= 63.5$ mg/L) and no complexing ligands other than OH^-. Stability constants for formation of the Cu–OH complexes that are expected to be present at significant concentrations are given below, along with the solubility product of $Cu(OH)_2(s)$. Determine the total copper concentration that would remain in solution [TOTCu(aq)] as a function of pH, if the precipitation reaction reached equilibrium, for the range $4 <$ pH < 14. In what pH range is the process capable of lowering the dissolved copper concentration to the target level?

Reaction	Equilibrium Constant
$Cu^{2+} + OH^- \leftrightarrow CuOH^+$	$K_{OH,1} = 3.18 \times 10^6$
$Cu^{2+} + 2OH^- \leftrightarrow Cu(OH)_2^0$	$\beta_{OH,2} = 5.89 \times 10^{11}$
$Cu^{2+} + 3OH^- \leftrightarrow Cu(OH)_3^-$	$\beta_{OH,3} = 2.29 \times 10^{15}$
$Cu^{2+} + 4OH^- \leftrightarrow Cu(OH)_4^{2-}$	$\beta_{OH,4} = 1.86 \times 10^{16}$
$Cu(OH)_2(s) \leftrightarrow Cu^{2+} + 2OH^-$	$K_{s0} = c_{Cu^{2+}} c_{OH^-}^2 = 1.95 \times 10^{-19}$

Solution. The important species in this system are the solutes H^+, OH^-, Cl^-, Cu^{2+}, $CuOH^+$, $Cu(OH)_2^0$, $Cu(OH)_3^-$, $Cu(OH)_4^{2-}$, and the solid $Cu(OH)_2(s)$. Treating the concentrations of these nine species as unknowns, we need nine equations to determine the equilibrium composition of the system.

A material balance expression can be written for each family of compounds in the system. In this system, mass balances can be written on dissolved chloride and total (dissolved plus solid) copper in the equilibrium solution, as follows:

$$TOTCl(aq) = c_{Cl^-} = 2 \times 10^{-3}$$

$$TOTCu = c_{Cu^{2+}} + c_{CuOH^+} + c_{Cu(OH)_2^0} + c_{Cu(OH)_3^-}$$
$$+ c_{Cu(OH)_4^{2-}} + c_{Cu(OH)_2(s)} = 10^{-3}$$

In addition to the five equilibrium constants listed in the problem statement, the equilibrium constant for dissociation of water is applicable:

$$H_2O \leftrightarrow H^+ + OH^- \quad K_w = c_{H^+} c_{OH^-} = 10^{-14.0}$$

The materials balances and equilibrium expressions provide eight of the nine equations required to determine the system composition, leaving the ninth equation to specify the pH of the solution. Thus, for any given pH, we have nine equations to solve for the nine unknown concentrations. Methods for solving this collection of equations include analytical, graphical, and spreadsheet approaches, as well as chemical equilibrium computer programs. Complete descriptions of these methods can be found in the water chemistry texts listed previously.

A $\log c - pH$ diagram for this system is provided in Figure 9-5, showing the concentrations of all the dissolved, copper-containing species, as well as their sum (i.e., TOTCu (aq)), over the pH range of interest. TOTCu(aq) equals 10^{-3} M at pH < 6.16, indicating that no precipitation occurs in that pH range. (At pH's where no solid precipitates, the K_{s0} expression is not applicable, and the concentration of $Cu(OH)_2(s)$ is not relevant, so the mathematical system includes eight equations and eight unknowns.) As the pH increases above 6.16, TOTCu(aq) declines dramatically, reaching a minimum value of 1.48×10^{-7} M (9.4 μg/L)

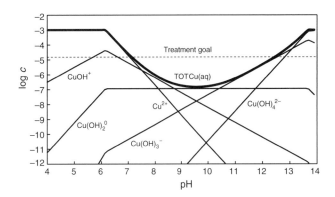

FIGURE 9-5. $\log c - pH$ diagram showing the major Cu species in a system with 10^{-3} M TOTCu and in which the only complexes of importance are $Cu(OH)_x$ species.

at pH 9.57. At higher pH, TOTCu(aq) increases, and, at pH > 13.6714 all of the copper is once again in solution. The speciation of TOTCu(aq) changes dramatically over the pH range shown, with Cu^{2+} dominating at low pH, followed sequentially by $CuOH^+$, $Cu(OH)_2^0$, and $Cu(OH)_3^-$ as the pH increases.

Based on these calculations, the industry could meet the treatment target by adjusting the pH to any value between 7.13 and 12.48, assuming that the system reached equilibrium and that the solids could be separated efficiently from the water after the precipitation reaction occurred. Although the minimum solubility and the pH at which it occurs vary widely among metals, the bowl-shaped pattern for TOTM as a function of pH is characteristic for most metals. ■

9.5 STOICHIOMETRIC AND EQUILIBRIUM MODELS FOR PRECIPITATION REACTIONS

We now consider the potential to precipitate a metal, M^{m+}, by addition of a ligand, L^{n-}, to form $M_nL_m(\underline{s})$, as shown in the following reaction (the reverse of reaction (9-67):

$$n M^{m+} + m L^{n-} \leftrightarrow M_nL_m(s) \quad (9\text{-}70)$$

The calculations for precipitation of AgCl(s) in Example 9-4 correspond to this scenario; here, we develop a more generic approach for solving such problems. In some processes, precipitation is initiated by the reverse sequence; that is, a ligand is already present in solution, and a metal is added to induce precipitation. Concepts presented in this section are useful for evaluating both kinds of processes.

We begin with the analysis of the simplest scenario—one in which M^{m+} and L^{n-} can participate only in the precipitation reaction. The dissolved concentration of each of these species after any amount of solids has precipitated can be expressed as the difference between its total concentration in the system and the amount in the solid. The molar concentrations of metal and ligand associated with the solid can be expressed as $nc_{Me_nL_m(s)}$ and $mc_{Me_nL_m(s)}$, respectively, so we can write:

$$\text{TOTM(aq)} = \text{TOTM} - nc_{M_nL_m(s)} \quad (9\text{-}71)$$

$$\text{TOTL(aq)} = \text{TOTL} - mc_{M_nL_m(s)} \quad (9\text{-}72)$$

These two expressions can then be combined to yield:

$$\text{TOTM(aq)} = \text{TOTM} - \frac{n}{m}\text{TOTL} + \frac{n}{m}\text{TOTL(aq)} \quad (9\text{-}73)$$

According to Equation 9-68, when the solid is present and the solutes behave ideally, the solubility product relates the concentrations of the free metal (M^{m+}) and free ligand (L^{n-}) via:

$$c_{M^{m+}} = \left(\frac{K_{s0}}{c_L^m}\right)^{1/n} \quad (9\text{-}68)$$

In the scenario under consideration (i.e., no reactions other than precipitation for either the metal or the ligand), we can substitute TOTM(aq) for c_M, and TOTL(aq) for c_L, so the equation becomes

$$\text{TOTL(aq)} = \left\{\frac{K_{s0}}{(\text{TOTM(aq)})^n}\right\}^{1/m} \quad (9\text{-}74)$$

Then, substituting Equation 9-74 into Equation 9-73, we obtain:

$$\text{TOTM(aq)} = \text{TOTM} - \frac{n}{m}(\text{TOTL}) + \frac{n}{m}\left\{\frac{K_{s0}}{(\text{TOTM(aq)})^n}\right\}^{1/m} \quad (9\text{-}75)$$

If the solutes do not behave ideally, essentially the same derivation applies, except that the solubility constant expression (Equation 9-68) must be written as $K_{s0} = (\gamma_M c_M)^n (\gamma_L c_L)^m$. In such a case, the analysis leads to the following expression instead of Equation 9-75:

$$\text{TOTM(aq)} = \text{TOTM} - \frac{n}{m}(\text{TOTL})$$
$$+ \frac{n}{m}\frac{1}{\gamma_L \gamma_M^{n/m}}\left\{\frac{K_{s0}}{(\text{TOTM(aq)})^n}\right\}^{1/m} \quad (9\text{-}76)$$

The difference represented by the first two terms on the right side of either Equation 9-75 or 9-76 indicates the Me^{m+} concentration that would remain in solution if every molecule of L in the system reacted with M to form the precipitate (i.e., if K_{s0} were 0). Such a limiting case scenario is referred to as *stoichiometric precipitation*. When stoichiometric precipitation is a reasonable model for the reaction, the concentration of TOTM(aq) decreases with increasing TOTL in proportion to the stoichiometric ratio, n/m. The last term on the right side of Equation 9-75 accounts for deviations from the stoichiometric precipitation model that arise because equilibrium precludes molecules of added ligand from reacting quantitatively with dissolved metal.

If m and n both equal 1, then Equation 9-75 can be solved for TOTM(aq) with the quadratic equation; the same is true of Equation 9-76, if the activity coefficients can be estimated independently. If either m or n does not equal 1, then the equations must be solved numerically.

Consider, for example, the precipitation of barite ($BaSO_4(s)$, $K_{s0} = 1.05 \times 10^{-10}$) by addition of Na_2SO_4 to a solution that initially contains 10^{-4} M $Ba(NO_3)_2$ (0.1 mmol/L, 13.7 mg/L Ba^{2+}) and no SO_4^{2-}. Because most of its salts are only slightly soluble, barium typically appears only in trace amounts in surface water and groundwaters. Higher concentrations can occur as the result of contamination from industrial uses in the manufacture of electronic components and metal alloys or from drilling muds used in gas and oil wells. Human health risks associated with acute exposure to barium include muscle weakness, upsets of the gastrointestinal system, and hypertension. To manage these risks, the USEPA has set a maximum contaminant level (MCL) of 1.0 mg/L (7.28×10^{-6} M) for barium in drinking water.

Ion exchange (described in Chapters 7 and 8) and reverse osmosis (a membrane-based process described in Chapter 15) are two common treatment technologies for removing barium from source waters. $BaSO_4(s)$ precipitation can play an important role in either of these technologies, but for different reasons. Precipitation of $BaSO_4(s)$ might be the treatment method of choice for removing Ba^{2+} from concentrated brine produced in an ion exchange process. On the other hand, precipitation of $BaSO_4(s)$ in reverse osmosis processes is undesirable, because the precipitate can reduce the rate at which water is purified by passage through the membrane.

If $BaSO_4(s)$ precipitates, we can apply Equation 9-75 with $n = m = 1$, yielding

$$\text{TOTBa(aq)} = 10^{-4} - \frac{1}{1}\text{TOTSO}_4$$
$$+ \frac{1}{1}\frac{1}{1.0(1.0)^{1/1}}\left\{\frac{1.05 \times 10^{-10}}{(\text{TOTBa(aq)})^1}\right\}^{1/1}$$
$$= 10^{-4} - \text{TOTSO}_4 + \frac{1.05 \times 10^{-10}}{\text{TOTBa(aq)}} \quad (9\text{-}77)$$

Equation 9-77 can be rearranged and solved using the quadratic equation. Results for various values of $TOTSO_4$ are shown in Figure 9-6, which reveals two regions of interest. In the first region, covering $TOTSO_4$ values from

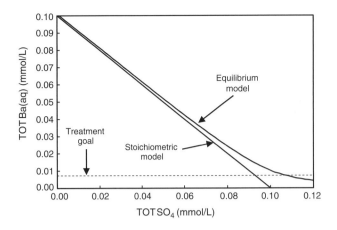

FIGURE 9-6. Total soluble barium concentration in a system with 0.1 mmol/L TOTBa and various TOTSO$_4$ concentrations.

0 to ~0.05 mmol/L, the stoichiometric model of precipitation appears to provide a good approximation for TOTBr (aq): each mole of added SO$_4^{2-}$ reacts almost quantitatively with one mole of Ba^{2+} to precipitate one mole of BaSO$_4$(s), so the plot is almost indistinguishable from a straight line with slope -1.

The behavior changes, however, as TOTSO$_4$ exceeds ~0.05 mmol/L. In this region, the linear relationship no longer applies, and the amount of TOTSO$_4$ required to precipitate a given amount of Ba^{2+} increases. In fact, because the curve becomes asymptotic to the horizontal axis, the figure shows that it is impossible to remove all the barium from solution no matter how large TOTSO$_4$ becomes.

Using the assumption of stoichiometric precipitation, we can estimate the value of TOTSO$_4$ required to lower TOTBa (aq) to the MCL of 1.0 mg/L (7.28×10^{-3} mmol/L) as the difference between the initial and final values of TOTBa(aq); this difference is 0.093 mmol/L. On the other hand, if the assumption were not applied, we could use Equation 9-77 to solve for the required reagent addition, yielding (with all values in mol/L):

$$\text{TOTSO}_4 = 10^{-4} + \frac{1.05 \times 10^{-10}}{\text{TOTBa(aq)}} - \text{TOTBa(aq)}$$

$$= 10^{-4} + \frac{1.05 \times 10^{-10}}{7.28 \times 10^{-6}} - 7.28 \times 10^{-6}$$

$$= 1.07 \times 10^{-4}$$

In one sense, the stoichiometric approach works reasonably well for estimating the TOTSO$_4$ that would be required to reach the MCL, underestimating the required dose by about 15%. In a full-scale system, TOTBa would undoubtedly vary over time, and an overdose of TOTSO$_4$ would likely be added anyway, to account for this variation and to provide a safety margin. Note also that the predicted amount of solids produced is the same using the two approaches since both approaches are based on the same amount of Ba^{2+} removal. This solids concentration is, to a close approximation, 0.1 mmol/L $\times$ 233.40 mg/mmol, or 23.34 mg/L.

Nevertheless, despite the small difference between the two computed TOTSO$_4$ doses, the effect of that dose on the final value of TOTBa(aq) is significant. Using Equation 9-77, we find that if the TOTSO$_4$ dose were based on the stoichiometric precipitation model, the residual TOTBa(aq) would be 2.0 mg/L, missing the MCL by 100%. Thus, even though the stoichiometric approximation might provide good estimates of the required reagent dose and the production of solids, the small error inherent in the approximation can cause a substantial error in the predicted solution composition.

The reason the stoichiometric approach works well for estimating the required sulfate dose in this case is that the treatment goal falls in the region of Figure 9-6, where the curve is approximately linear. Mathematically, this means that the last term on the right side of Equation 9-75 is negligible compared to the middle term. Such an outcome becomes less likely as the solubility of the solid under consideration increases.

■ **EXAMPLE 9-6.** Design calculations for a reverse osmosis system indicate that the calcium concentration in the reject water will reach 10^{-2} M TOTCa (401 mg/L). Addition of sulfuric acid (H$_2$SO$_4$) is a common pretreatment process for reverse osmosis systems, because it lowers the pH and thereby helps prevent precipitation of carbonate and hydroxide solids that can foul the membranes. However, such a pretreatment step increases the likelihood that gypsum (CaSO$_4$(s), $K_{s0} = 2.45 \times 10^{-5}$, MW = 136.15) will precipitate in the reject water.

Predict the amount of CaSO$_4$(s) that will form as a function of TOTSO$_4$, based on both a stoichiometric precipitation model and a model that takes chemical equilibrium into account. Assume that the solution is ideal and that the precipitation reaction is the only one in which the free metal and ligand ions participate.

Solution. In the stoichiometric approach, we assume that each mole of TOTSO$_4$ in the system reacts with one mole of TOTCa(aq) to precipitate one mole of gypsum. In that case, the total amount of CaSO$_4$(s) generated would equal TOTSO$_4$, up to a TOTSO$_4$ concentration equal to TOTCa, or 10^{-2} M. At that point, the total solids concentration would be 10^{-2} M $\times$ 136.15 g/mol, or 1.36 g/L. This result is shown by the broken line in Figure 9-7.

A more accurate, equilibrium-based analysis of the system can be carried out by substituting appropriate values into Equation 9-75 ($n = m = 1$ and K_{s0} as given in the problem

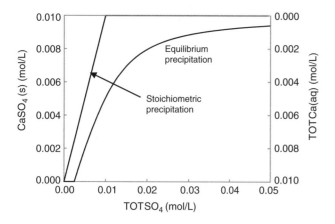

FIGURE 9-7. Precipitation of $CaSO_4(s)$ in a system with 0.01 M $TOTCa$ and various $TOTSO_4$, assuming ideal solute behavior.

statement) to compute $TOTCa(aq)$, and then using the following mass balance to compute $TOTCa(s)$:

$$TOTCa(s) = TOTCa - TOTCa(aq) \quad (9\text{-}78)$$

The results of the equilibrium calculations are shown by the solid line in Figure 9-7. These results differ from that using the stoichiometric model, in that $TOTSO_4$ must reach $\sim 2.45 \times 10^{-3}$ M before any precipitation occurs. They also differ from the results in the preceding example for precipitation of $BaSO_4(s)$ in that, once precipitation is initiated, it follows an approximately stoichiometric trend (a 1:1 molar ratio of $\Delta CaSO_4(s)$ to $\Delta TOTSO_4$) only until approximately one-half of the $TOTCa$ has precipitated; after that, substantially more than one mole of added $TOTSO_4$ is required to precipitate one mole of gypsum. Both of these deviations from the stoichiometric model reflect the higher solubility of $CaSO_4(s)$ compared to $BaSO_4(s)$. Substantial $TOTSO_4$ is needed to initiate precipitation because, at lower concentrations, the solubility product is not exceeded; similarly, the curvature in the plot of $TOTCa(s)$ versus $TOTSO_4$ occurs because, after approximately one-half of the $TOTCa$ has precipitated, $TOTCa(aq)$ and $TOTSO_4(aq)$ are of similar magnitude.

Thus, in this case, using the stoichiometric precipitation model overestimates the amount of precipitate formed. At $TOTSO_4 < 2.45 \times 10^{-3}$ M, the stoichiometric model predicts that solid forms, when in fact none would, and at a full stoichiometric concentration of 0.01 M, the model overestimates the amount of precipitate by approximately a factor of 2. ∎

As noted in the derivation of Equations 9-75 and 9-76, the equilibrium solubility of a solid can be affected by the formation of soluble complexes that include the constituents of the solid, and also by the ionic strength of the solution (primarily through its effect on activity coefficients). The effects of complex formation on solubility are system-specific and are considered in subsequent sections.

A few correlations between activity coefficients and ionic strength are shown in Table 9-5; more elaborate correlations that take into account the detailed ionic composition of the solution are also available (e.g., the specific ion interaction model of Pitzer [Langmuir, 1997; Benjamin, 2010]).

■ **EXAMPLE 9-7.**

(a) Recompute the concentration of $TOTCa(s)$ as a function of $TOTSO_4$, as in Example 9-6, but take nonideal behavior of Ca^{2+} and SO_4^{2-} into account. Consider solutions with ionic strengths of 0.03, 0.1, and 0.5 M (after precipitation). Use the Davies' equation to compute activity coefficients.

(b) Compute $TOTCa(s)$ once more for a solution with $I = 0.1$ M, but take into account the fact that Ca^{2+} and SO_4^{2-} can react to form a soluble complex, $CaSO_4(aq)$, with $K = 10^{2.36}$.

TABLE 9-5. Commonly Used Correlations Between Ionic Strength and Ion Activity Coefficients

Name	Equation	Comments and Applicable Conditions
Debye–Huckel limiting law	$\log \gamma = -Az^2 I^{1/2}$	$A = 1.82 \times 10^6 (\varepsilon T)^{-3/2}$, where ε is the dielectric constant of the medium, and I is the ionic strength[a] and z is the charge number of the ion. For water at 25°C, $A = 0.51$; applicable at $I < 0.005$ M
Extended Debye–Huckel	$\log \gamma = -Az^2 \dfrac{I^{1/2}}{1 + BaI^{1/2}}$	a is the ion size parameter, which has values between 3 and 9 (see water chemistry texts for listings); $B = 50.3(\varepsilon T)^{1/2}$; for water at 25°C, $B = 0.33$; applicable at $I < 0.1$ M
Davies	$\log \gamma = -Az^2 \left(\dfrac{I^{1/2}}{1 + I^{1/2}} - 0.3I \right)$	Applicable at $I < 0.5$ M

[a]Ionic strength is computed as $I = \sum_{\text{all } i} \frac{1}{2} c_i z_i^2$, where the summation is over all ions in solution. Although I is computed using c_i values in mol/L, it is frequently reported as a dimensionless value.
Source: From Benjamin (2010).

Solution.

(a) According to the Davies equation, the activity coefficients for Ca^{2+} and SO_4^{2-} are the same in a given solution, since both ions are divalent. For the solutions of interest, these activity coefficients are 0.52, 0.37, and 0.29 at ionic strengths of 0.03, 0.10, and 0.50 M, respectively. The concentration of precipitate formed at these ionic strengths, for the same range of $TOTSO_4$ as in Example 9-6, can be computed with the aid of Equation 9-76; the results are shown in Figure 9-8. Note that, although the calculations can be carried out even for an ionic strength of zero, the minimum ionic strength of a solution in equilibrium with $CaSO_4(s)$ is ~ 0.02 M, due to the Ca^{2+} and SO_4^{2-} in solution.

With increasing ionic strength of the solution, the activity coefficients of both Ca^{2+} and SO_4^{2-} decrease, so that larger concentrations of these ions are required to cause the solubility product (which is a product of activities, not concentrations) to be exceeded. For example, in the solution with $I = 0.5$ M, precipitation is not initiated until $TOTSO_4$ is ~ 0.029 M (as opposed to 0.002 M at $I = 0$); correspondingly, when $TOTSO_4$ equals 0.05 M, the prediction is that 37% of the TOTCa would precipitate if nonideal behavior is accounted for, but 94% would precipitate if the solutes behaved ideally. Once the solid forms, the increment of solid formed per unit increase in $TOTSO_4$ decreases with increasing ionic strength; that is, even after precipitation begins, the approximation that the reaction is stoichiometric becomes poorer as ionic strength increases.

(b) When the $CaSO_4(aq)$ complex is considered, the analysis of the solution composition at equilibrium becomes more difficult and is best solved using chemical equilibrium software. The results of the analysis are shown in Figure 9-8, which indicates that precipitation barely begins at the largest value of $TOTSO_4$ considered (0.05 M). The formation of the complex interferes with precipitation by "diverting" some of the calcium and sulfate in solution from free Ca^{2+} and SO_4^{2-} ions to a different species. ∎

■ **EXAMPLE 9-8.** The major components of an RO concentrate are summarized below.

(a) Using a chemical equilibrium model such as Visual Minteq, determine the saturation indices (i.e., SS values) of the solids calcite, aragonite, barite, gypsum, and magnesite in the original RO concentrate and after H_2SO_4 has been added to lower the pH to 6.0. How much H_2SO_4 must be added to the RO concentrate to lower the pH?

(b) What are the major species that make up TOTBa, TOTCa, and $TOTSO_4$ in the solution after pH adjustment? What are the activity coefficients for Ba^{2+}, Ca^{2+}, SO_4^{2-} in the solution?

Component	Value[a]	Component	Value
pH	7.78	$TOTSO_4$	2.92×10^{-2}
TOTCa	1.90×10^{-2}	TOTCl	3.10×10^{-2}
TOTMg	1.15×10^{-2}	$TOTCO_3$	1.10×10^{-2}
TOTNa	4.70×10^{-2}	TOTBa	1.46×10^{-5}

[a] All values in mol/L, except pH in pH units.

Solution.

(a) The solution can be modeled using Visual Minteq by inputting the total concentrations of the components listed in the table and specifying the known pH. When the program is run, the output indicates that the charge balance is satisfied to within 4.7%. Based on this result, we assume that all the major ionic species are accounted for, so that the calculated ionic strength of 0.120 M is accurate.

The H_2SO_4 addition required to lower the pH 6.0 can be determined in several different ways using Visual Minteq; perhaps the simplest is to carry out a numerical titration, adding various doses of H_2SO_4, until the target pH is reached. The result is that the required addition of H_2SO_4 is 0.33×10^{-2} M. The composition of the resulting solution can then be modeled using the same inputs as the initial solution, but changing the material balances on H^+ and SO_4^{2-} to account for the H_2SO_4 addition. (For those familiar with the program, these changes mean that TOTH

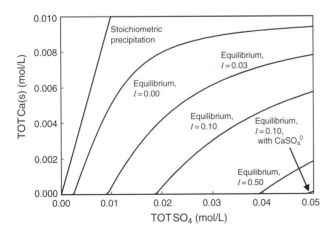

FIGURE 9-8. Effect of ionic strength on the precipitation of $CaSO_4(s)$ in a solution with 0.01 M TOTCa.

and TOTSO$_4$ change from 1.10 to 1.76 × 10^{-2} and from 2.92 to 3.25 × 10^{-2}, respectively.) The ionic strength of this solution is reported to be 0.121 M.

Using the computed ionic strengths in conjunction with the Davies equation to estimate activity coefficients, the program reports that the original solution is supersaturated with respect to the all the solids of interest (although the supersaturation of gypsum and magnesite is slight), whereas only barite and gypsum are supersaturated in the solution after pH adjustment. The results are summarized in the following table.

Mineral	SS pH 7.78	SS pH 6.0
Calcite	26.5	0.19
Aragonite	19	0.13
Barite	237	262
Gypsum	1.1	1.2
Magnesite	1.6	0.01

(b) Other relevant characteristics of the solution composition after the pH adjustment, including the major species that contribute to dissolved Ba, Ca, and SO$_4$, are summarized below.

$\gamma_{Ba^{2+}} = \gamma_{Ca^{2+}} = \gamma_{SO_4^{2-}} = 0.35$

TOTBa = 3.64 × 10^{-6}: 75% as Ba^{2+}, 23% as BaSO$_4$, 1% as BaHCO$_3^-$

TOTCa = 1.90 × 10^{-2}: 64% as Ca^{2+}, 33% as CaSO$_4$(aq), 1.7% as CaCl$^+$, 1.1% as CaHCO$_3^+$

TOTSO$_4$ = 3.25 × 10^{-2}: 62% as SO$_4^{2-}$, 21% as CaSO$_4$(aq), 11% as MgSO$_4$(aq), 5.5% as NaSO$_4^-$ ■

To summarize, a stoichiometric model of precipitation can sometimes provide useful estimates of the reagent dose required to remove a specified concentration of contaminant from solution or the amount of solids produced. However, this approach is appropriate only when one has confidence that essentially all the added reagent reacts with the target contaminant to form a solid. For certain solids (e.g., Fe(OH)$_3$), experience suggests that this approximation is virtually always applicable in systems of interest in environmental engineering. For others, one can assume that the approximation applies and then use the solubility constant expression to test that assumption for the maximum dose of added reagent; if it is applies for that case, it will also apply for all lower doses. High ionic strength of the solution (which reduces the activity coefficients of the reactants) and reactions of any of the constituents of the solid with other solutes can increase the solubility of the solid and thereby reduce the likelihood that the stoichiometric model will apply. If the stoichiometric model does not apply, the expected system composition after reagent addition (including the concentration of any solid that forms) can be predicted using chemical equilibrium software, assuming that the system reaches equilibrium within the time frame of the treatment process.

As has been noted, precipitation processes in environmental engineering are normally designed based primarily on past experience and, to the extent that they are analyzed within a theoretical framework, that framework is normally limited to equilibrium modeling. Nevertheless, in principle, chemical information about the extent of supersaturation can be linked to the kinetics-based analysis presented in the preceding section to improve the design process and explore many more treatment scenarios than can be tested empirically. The following example demonstrates how such an exploration might be carried.

■ EXAMPLE 9-9.

(a) Using the algorithm presented below, estimate the concentration of barite that will precipitate from the pH-adjusted RO concentrate considered in Example 9-8 in a steady-state CFSTR with values of τ ranging from 30 to 600 s. Assume that the barite has a density of 4.5 g/cm^3 and that the particle nucleation and growth rates can be modeled by the following expressions:[9]

$$r_{nuc} = 2.9 \times 10^{12} \frac{1}{m^3 \, s} \exp\left[-\frac{99.5}{(\ln SS)^2}\right]$$

$$\dot{d}_p = (1.0 \times 10^{-9} \, m/s)(SS)^{1.5}$$

(i) Select a reactor hydraulic residence time, τ.
(ii) Assume a value of supersaturation (SS) and calculate the corresponding solids concentration expected in the reactor.
(iii) Determine the concentrations of total dissolved Ba and SO$_4$ in the reactor (and therefore in the effluent) based on material balances.
(iv) Calculate the SS of barite in the effluent based on the results from step iii and the speciation in the pH 6 solution determined in Example 9-8. Assume that the distribution of TOTBa(aq) and TOTSO$_4$(aq) species remains as in that example as barite precipitates.
(v) Iterate on the assumed value of SS in step ii until it matches the value calculated in step iv (note that you can automate this process using software such as the *Solver* routine in Microsoft Excel).

---
[9] The given nucleation rate is based on data from Taguchi et al. (1996). Although the particle growth rate is similar to expressions that have been reported for growth of other insoluble precipitates, the exact expression is made up.

(vi) Choose another value of τ and repeat steps ii through v.

(b) Prepare a figure showing the particle concentration (mg/L) and effluent SS as a function of τ, and another figure showing particle diameter (μm) and the fraction of influent TOTBa precipitated over the same range of τ.

Solution.

(a) We can compute the concentration of solids (c_p) inside, and leaving, the CFSTR by

$$c_p = \rho_p N_p V_{p,avg}$$

The particle number concentration and the average particle volume can be expressed in terms of SS and τ using Equations 9-55 and 9-58, as follows:

$$N_p = r_{nuc}\tau = \left\{2.9 \times 10^{12}\frac{1}{m^3\,s}\exp\left[-\frac{99.5}{(\ln SS)^2}\right]\right\}\tau$$

$$V_{p,avg} = \pi d_{p,avg}^3 = \pi(\dot{d}_p\tau)^3 = \pi[(1\times 10^{-9}\,m/s)(SS)^{1.5}\tau]^3$$

$$= (3.14\times 10^{-27}\,m^3/s^3)(SS)^{4.5}\tau^3$$

Substituting these expressions and the given value for the density of barite into the equation for c_p, we obtain:

$$c_p = \rho_p N_p V_{p,avg}$$
$$= \{(4.5\,g/cm^3)(10^6\,cm^3/m^3)\}$$
$$\times \left\{2.9\times 10^{12}\frac{1}{m^3\,s}\tau\exp\left[-\frac{99.5}{(\ln SS)^2}\right]\right\}$$
$$\times \{(3.14\times 10^{-27}\,m^3/s^3)(SS)^{4.5}\tau^3\}$$
$$= (4.10\times 10^{-8}\,g/m^3\,s^4)\tau^4(SS)^{4.5}\exp\left[-\frac{99.5}{(\ln SS)^2}\right]$$

This expression can be used to compute the concentration of solids in the reactor (and the effluent) for any given values of τ and SS. Thus, we can choose a value of τ, guess the corresponding value of SS, and compute what c_p would be if that guess for SS were correct. We can then use the value of c_p to determine the concentrations of TOTBa and TOTSO$_4$ remaining in solution, and estimate the activities of Ba^{2+} and SO_4^{2-} by assuming that the speciation of TOTBa(aq) and TOTSO$_4$(aq) remains as determined in part (a). Finally, we can compute SS based on these estimated activities and compare it with the guessed value. If this value is different from the original guess for SS, we iterate on the guess until it is consistent with the calculated value. We then choose a different value of τ and repeat the process.

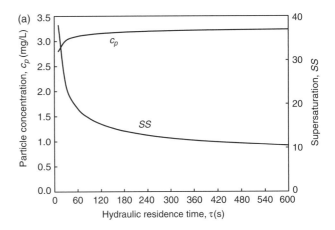

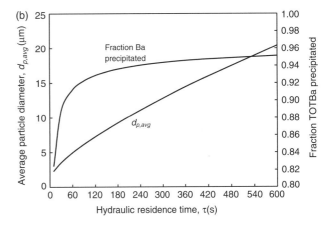

FIGURE 9-9. Effect of hydraulic residence time in the example CFSTR on (a) particle concentration and supersaturation and (b) average particle diameter and the fraction of Ba precipitated.

(b) The result of the calculations for a range of mean hydraulic residence times is shown in Figure 9-9a. The particle concentration and SS are both nonlinear functions of τ. Both parameters are relatively sensitive to changes in small values of τ, but approach asymptotic values with increasing τ. As a result, the incremental benefits of building a larger reactor grow smaller. Similarly, Figure 9-9b shows that the fraction of TOTBa precipitated increases with increasing τ, but incremental benefits diminish at residence times >2–3 min. On the other hand, the average particle diameter increases almost linearly with increasing τ, and the advantages of a large particle size for the subsequent solid/liquid separation step might provide a reason to design a reactor with a τ of several minutes. ∎

Precipitation of Hydroxide Solids

Metal Speciation and the Metal Hydroxide—pH Relationship The principal applications of precipitation in water and wastewater treatment involve cationic metals, with the goal being either to remove problematic metals (e.g., Pb^{2+}, Zn^{2+}, Cu^{2+}, Cd^{2+}, Ca^{2+}, Fe^{2+}) from solution or to precipitate metal hydroxides ($Al(OH)_3(s)$, $Fe(OH)_3(s)$) as coagulants or adsorbents. In this section, we consider factors that control the precipitation of metal hydroxides; the precipitation of other metal-containing solids (carbonates, phosphates, sulfides, etc.) is addressed in subsequent sections.

A key characteristic of metal ions in water is their tendency to form metal–ligand complexes. Some ligands (OH^-, CO_3^{2-}) are present in virtually all solutions of interest, while other ligands (natural organic matter, EDTA) are present only in waters from specific sources or that have been used in specific applications. If the concentrations of all ligands in the solution are known, the fraction of TOTM(aq) that is present as complexes can be computed with the aid of Equation 9-69; this fraction can vary from virtually 0 to almost 100%. As demonstrated in Example 9-7b, the formation of complexes reduces the tendency for precipitation to occur. If a metal is extensively and strongly complexed, removing it from solution by precipitation might be very difficult; indeed, ligands are sometimes added to water specifically to prevent precipitation of metals.

The formation of M–OH complexes (commonly referred to as hydrolysis of the metal) is possible in any aqueous solution, because of the ubiquity of OH^- ions in such solutions. Because c_{OH^-} is directly related to the solution pH, the relative concentrations of M^{m+} and all $M(OH)_x^{m-x}$ species are strongly pH-dependent, with the species with higher values of x accounting for a larger fraction of TOTM as pH increases.

The solubility of metal hydroxide solids in solutions where hydroxo-metal complexes are the only ones of importance is conveniently characterized on $\log c$ – pH plots. Such a plot for Zn speciation in a solution that is in equilibrium with amorphous $Zn(OH)_2(s)$ is shown in Figure 9-10; the stability constants used to prepare the plot are shown in Table 9-6.

Although the concentration of each dissolved Zn species changes monotonically with pH (with some species increasing and others decreasing), the sum of all dissolved species; that is, TOTZn(aq), has a characteristic bowl shape, passing

TABLE 9-6. Major Monomeric Zn–OH Complexes

Reaction	Log $^*\beta_i$
$Zn^{2+} + H_2O \leftrightarrow ZnOH^+ + H^+$	-8.997
$Zn^{2+} + 2H_2O \leftrightarrow Zn(OH)_2^0 + 2H^+$	-17.794
$Zn^{2+} + 3H_2O \leftrightarrow Zn(OH)_3^- + 3H^+$	-28.091
$Zn^{2+} + 4H_2O \leftrightarrow Zn(OH)_4^{2-} + 4H^+$	-40.488

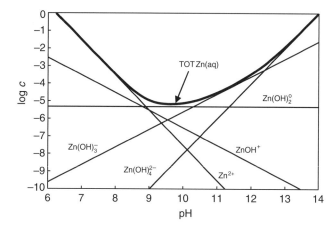

FIGURE 9-10. Zn speciation in an ideal solution that is equilibrated with amorphous $Zn(OH)_2(s)$.

through a minimum at some pH value. The pH at which the minimum in TOTM(aq) occurs, and the corresponding concentration, establish limits on how much treatment can be accomplished by precipitation of the metal hydroxide solid.

Situations are sometimes encountered in which the pH of a metal-containing solution is higher than that of the minimum solubility (i.e., higher than the pH at the bottom of the solubility "bowl"), so that precipitation can be induced by lowering the pH. However, the much more common situation is for the pH of the raw water to be below that of minimum solubility. In such cases, addition of hydrated lime ($Ca(OH)_2$) is often an inexpensive and effective way to raise the pH and induce precipitation.

Figure 9-10 specifies the concentrations of Zn species in solutions that are in equilibrium with amorphous $Zn(OH)_2(s)$. In most treatment applications, a solution with a known total Zn concentration (TOTZn) exists, in which case a more useful version of the diagram can be prepared showing the speciation both in the pH range where a solid forms and in ranges in which all the Zn is soluble. Such a diagram was presented previously for a system containing 10^{-3} M TOTCu (Figure 9-5); a diagram for a system containing 20 mg/L TOTZn (3.06×10^{-4} M) in a solution that also contains 0.05 M $NaNO_3$ (which contributes most of the ions affecting the ionic strength) is shown in Figure 9-11.

Acid–Base Requirements for Metal Hydroxide Precipitation The reactions shown in Table 9-6 indicate that Zn^{2+} and its hydrolyzed forms can interact with water in various ways that might either release or consume H^+ ions; that is, they might act as either acids or bases. In a solution in which the Zn species are the only weak acids and bases present, the change in the Zn speciation determines the acid or base requirement for the precipitation reaction. For example, consider a solution like the one characterized in Figure 9-11 (20 mg/L TOTZn, 0.05 M $NaNO_3$) that is initially at pH 7.0. All the Zn would be soluble in such a

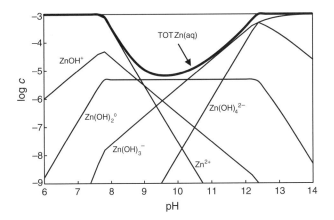

FIGURE 9-11. Log c-pH diagram for 3.06×10^{-4} M TOTZn in 0.05 M NaNO$_3$.

solution. If this solution were adjusted to pH 9.70 (at which the Zn solubility is minimized), and the system reached equilibrium, the vast majority of the Zn would precipitate. A summary of the distribution of TOTZn at the two pH values, and of the H$^+$ and OH$^-$ concentrations in the two solutions, is provided in Table 9-7. The calculations are based on the ionic strength of the original solution; that is, changes in the ionic strength due to the addition of base and the precipitation reaction have been ignored. Note that the product $c_{H^+}c_{OH^-}$ is larger than $10^{-14.0}$, because the equilibrium constant for the dissociation of water, K_w, requires that $a_{H^+}a_{OH^-} = 10^{-14.0}$, and in this solution, the activities of H$^+$ and OH$^-$ are less than their molar concentrations.

Comparison of the distribution of Zn in the two solutions makes it clear that the dominant conversion is from Zn^{2+} (which represents ~99% of TOTZn at pH 7.0) to Zn(OH)$_2$(s) (representing ~98% of TOTZn at pH 9.7). This conversion requires two OH$^-$ ions per Zn^{2+} ion undergoing the change, corresponding to an OH$^-$ requirement of ~2 × TOTZn, or 6.12×10^{-4} M. The changes in the concentrations of the hydrolyzed Zn species between the beginning and final conditions alter this value slightly, and these effects could

TABLE 9-7. Speciation in Solutions of 3.06×10^{-4} M (20 mg/L) TOTZn at Two pH Values

	Concentration, c (mol/L)		
Species	pH 7.0	pH 9.7	Δc
Zn^{2+}	3.04×10^{-4}	2.65×10^{-7}	-3.04×10^{-4}
ZnOH$^+$	1.67×10^{-6}	7.32×10^{-7}	-9.38×10^{-7}
Zn(OH)$_2^0$	2.18×10^{-8}	4.79×10^{-6}	$+4.77 \times 10^{-6}$
Zn(OH)$_3^-$	1.35×10^{-11}	1.48×10^{-6}	$+1.48 \times 10^{-6}$
Zn(OH)$_4^{2-}$	9.89×10^{-17}	5.44×10^{-9}	$+5.44 \times 10^{-9}$
TOTZn(aq)	3.06×10^{-4}	7.27×10^{-6}	-2.99×10^{-4}
Zn(OH)$_2$(s)	0.00	2.99×10^{-4}	$+2.99 \times 10^{-4}$
H$^+$	1.29×10^{-7}	2.57×10^{-10}	-1.29×10^{-7}
OH$^-$	1.29×10^{-7}	6.51×10^{-5}	$+6.50 \times 10^{-5}$

be taken into account in a more detailed accounting; for current purposes, though, the above estimate for the OH$^-$ consumption by Zn species is adequate.

Although the estimate of the OH$^-$ required to convert Zn^{2+} ions to Zn(OH)$_2$(s) is accurate, the total OH$^-$ requirement for the process is $>6.12 \times 10^{-4}$ M, because raising the pH from 7.0 to 9.7 requires that H$^+$ ions in the original solution be consumed, and that the concentration of OH$^-$ ions in the solution increase. As shown in the final two rows of the table, these changes require an additional 1.29×10^{-7} M OH$^-$ to react with >99% of the H$^+$ in the original solution, plus 6.50×10^{-5} M OH$^-$ to increase the concentration of OH$^-$ to its final value. The total OH$^-$ requirement to carry out the desired conversion can therefore be closely approximated as

$$[OH^-]_{required} \approx [OH^-]_{precipitated} + [OH^-]_{\substack{reacted \\ with\ H^+}}$$
$$+ [OH^-]_{\substack{in\ solution \\ at\ equilibrium}} - [OH^-]_{\substack{in\ initial \\ solution}}$$
$$= 2(3.06 \times 10^{-4}) + 1.29 \times 10^{-7}$$
$$+ 6.50 \times 10^{-5} - 1.29 \times 10^{-7}$$
$$= 6.77 \times 10^{-4}\ M$$

Since each mole of lime (Ca(OH)$_2$) releases two moles of OH$^-$, this OH$^-$ requirement could be met by adding 3.39×10^{-4} M Ca(OH)$_2$ to the pH 7.0 solution.

A plot of the change in pH and the amount of solid precipitated as a function of the OH$^-$ added shows that a stoichiometric precipitation model could describe the system quite well after the first 0.04 meq/L of OH$^-$ has been added (Figure 9-12). That is, the first 0.04 meq/L OH$^-$ added to the solution increases the pH to ~8.2 without precipitating any solid. Thereafter, each 0.02 meq/L OH$^-$ added causes

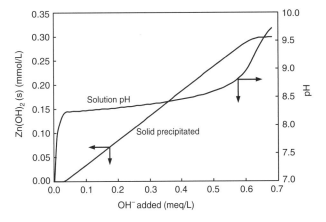

FIGURE 9-12. Response of a pH 7.0 solution containing 20 mg/L TOTZn (3.06×10^{-4} M) and 0.05 M NaNO$_3$ to addition of base. Although the example assumes the OH$^-$ is added as lime, the same plot would apply if it were added as caustic soda (NaOH) or any other strong base.

almost exactly 0.01 mmol/L of $Zn(OH)_2(s)$ to precipitate, until almost 99% of the TOTZn is present as the solid.

While the effect on the lime requirement of species other than those containing Zn is fairly small in this example system, that result is not necessarily a general one. Often, the pH of the solution to be treated is low enough that neutralization of H^+ requires a substantial lime dose, and in some cases, the presence of other weak acids in the solution has a similar effect.

■ **EXAMPLE 9-10.** Estimate the (hydrated) lime dosage necessary to increase the pH to 9.7 and thereby minimize TOTZn(aq), starting with the following two solution compositions. Assume that the $Zn(OH)_2(s)$ precipitation reaction reaches equilibrium in each case.

(a) 20 mg/L TOTZn, 0.05 M $NaNO_3$, pH 2.4.

(b) 20 mg/L TOTZn, 0.05 M $NaNO_3$, Alk = 60 mg/L as $CaCO_3$, pH 7.0. Assume that all of the alkalinity is contributed by the $H^+/H_2O/OH^-$ and carbonate acid/base groups, and ignore the formation of soluble complexes between Zn and CO_3 species and the possible precipitation of $ZnCO_3(s)$.

Solution.

(a) This solution contains the same TOTZn as the example system discussed previously, so the speciation of Zn at pH 9.7 will be identical to that in Table 9-7. Zn speciation in the initial solution (at pH 2.4) will be dominated even more strongly by Zn^{2+}, so the concentration of OH^- required to carry out the conversion of the Zn species (from essentially all Zn^{2+} to essentially all $Zn(OH)_2(s)$) will once again be 6.12×10^{-4} M. And, as before, the OH^- concentration will have to increase from essentially 0 (3.3×10^{-12} in this case) to 6.51×10^{-5} M; this OH^- requirement is therefore 6.51×10^{-5} M. However, in this case, the OH^- requirement for neutralizing the H^+ in the original solution will be much larger than before. The original solution contains 5.14×10^{-3} M H^+ (computed as the ratio $10^{-pH}/\gamma_{H^+}$, with $\gamma_{H^+} = 0.77$), and the solution at pH 9.7 contains 2.57×10^{-10} M, so 5.14×10^{-3} M OH^- is needed for the neutralization. The total OH^- requirement in this case is therefore

$$[OH^-]_{required} \approx [OH^-]_{precipitated} + [OH^-]_{\substack{reacted \\ with\ H^+}}$$
$$+ [OH^-]_{\substack{in\ solution \\ at\ equilibrium}} - [OH^-]_{\substack{in\ initial \\ solution}}$$
$$= 2(3.06 \times 10^{-4}) + 5.14 \times 10^{-3}$$
$$+ 6.51 \times 10^{-5} - 2.51 \times 10^{-12}$$
$$= 5.82 \times 10^{-3}\ M$$

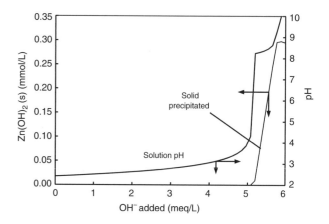

FIGURE 9-13. Response to addition of base of a solution identical to that characterized in Figure 9-12, except at an initial pH of 2.4.

The hydrated lime requirement would be one-half of this value, or 2.91×10^{-3} M (213 mg/L $Ca(OH)_2$), with almost 90% devoted to neutralizing the H^+. The pH change and amount of $Zn(OH)_2(s)$ precipitated at various points in the titration are shown in Figure 9-13.

Although Figures 9-12 and 9-13 look quite different from one another, that difference is largely due to the difference in scales. In fact, if the region from 5.15 to 5.85 meq/L OH^- added in Figure 9-13 were expanded, it would correspond exactly to the region from 0.0 to 0.7 meq/L OH^- added in Figure 9-13, because addition of 5.15 meq/L OH^- brings the latter system from pH 2.4 to pH 7.0. In other words, addition of 5.15 meq/L OH^- brings the more acidic solution exactly to the starting point of the less acidic one. Thus, like the solution that is initially at pH 7.0, the solution that is initially at pH 2.4 is characterized by stoichiometric precipitation of Zn $(OH)_2(s)$, but only after a more extensive "pretreatment" to bring the solution to the verge of precipitation.

(b) This solution is at the same pH as the solution that was considered earlier, but it differs in that it contains the weak acids H_2CO_3 and HCO_3^- in addition to H^+, OH^-, and Zn species. Because the initial pH is the same, the OH^- requirements for changing the Zn speciation and for changing the H^+ and OH^- concentrations are identical to those in the original example. However, in addition to those requirements, OH^- is required to change the speciation of the carbonate system.

The alkalinity of the initial solution is 60 mg/L as $CaCO_3$. The equivalent weight of $CaCO_3$ is 50, so, in units of equivalents per liter, the alkalinity is

$$\text{Alk} = \frac{60\,\text{mg/L as CaCO}_3}{50{,}000\,\text{mg CaCO}_3/\text{equiv}} = 1.2 \times 10^{-3}\,\text{eq/L}$$

The total concentration of carbonate species in the original solution can be determined from the known alkalinity, pH, and ionic strength (which is dominated by the NaNO$_3$ and affects the activity coefficients of all the ions). The exact calculations are tedious, though not difficult, and are most easily carried out using chemical equilibrium software. However, an excellent approximation for TOTCO$_3$ can be made by recognizing that, at pH less than $\sim$8.5, the concentration of bicarbonate ion is much greater than that of carbonate ion, and is typically much greater than that of H$^+$ and OH$^-$ as well. The problem statement specifies that that the alkalinity of the initial solution is entirely attributable to acid/base reactions of water and the carbonate group, so it can be expressed as

$$\text{Alk} = c_{\text{HCO}_3^-} + 2c_{\text{CO}_3^{2-}} + c_{\text{OH}^-} - c_{\text{H}^+} \quad (9\text{-}79)$$

where the alkalinity is in equivalents per liter, and the concentrations are in moles per liter. Applying the assumption that $c_{\text{HCO}_3^-}$ is much greater than the other terms on the right side (which we can test later to confirm that the calculations are accurate), we see that Alk is essentially equal to $c_{\text{HCO}_3^-}$, so $c_{\text{HCO}_3^-} \approx 1.2 \times 10^{-3}$ mol/L. The concentrations of the other carbonate species can then be computed from K_{a1} and K_{a2} for the carbonate group, in conjunction with activity coefficients based on the known ionic strength. For example, noting that the K_a expressions are relationships among chemical activities, and that the conventional standard state concentration for all solutes is 1.0 mol/L, we can compute the concentration of CO$_3^{2-}$ in the initial solution as follows:

$$K_{a2} = \frac{a_{\text{CO}_3^{2-}} a_{\text{H}^+}}{a_{\text{HCO}_3^-}} = \frac{\gamma_{\text{CO}_3^{2-}}(c_{\text{CO}_3^{2-}}/c_{\text{CO}_3^{2-},\text{standard state}})a_{\text{H}^+}}{\gamma_{\text{HCO}_3^-} c_{\text{HCO}_3^-}/c_{\text{HCO}_3^-,\text{standard state}}}$$

$$= \frac{\gamma_{\text{CO}_3^{2-}}}{\gamma_{\text{HCO}_3^-}} \frac{c_{\text{CO}_3^{2-}}}{c_{\text{HCO}_3^-}} a_{\text{H}^+}$$

$$c_{\text{CO}_3^{2-}} = K_{a2} \frac{\gamma_{\text{HCO}_3^-}}{\gamma_{\text{CO}_3^{2-}}} \frac{c_{\text{HCO}_3^-}}{a_{\text{H}^+}} \quad (9\text{-}80)$$

All the terms on the right side of Equation 9-80 are known or calculable (the activity coefficients can be estimated from an appropriate correlation with the ionic strength, and $a_{\text{H}^+} = 10^{-\text{pH}}$), so $c_{\text{CO}_3^{2-}}$ can be determined.[10] A similar calculation yields a value for $c_{\text{H}_2\text{CO}_3}$; the results are

[10] Note that, by definition, the pH equals the negative log of the activity of H$^+$, not the concentration of H$^+$.

TABLE 9-8. Carbonate Speciation in the Initial and Treated Solutions

Species	Concentration, c (mol/L)		
	pH 7.0	pH 9.7	Δc
H$_2$CO$_3$	2.09×10^{-4}	3.20×10^{-7}	-2.09×10^{-4}
HCO$_3^-$	1.20×10^{-3}	9.37×10^{-4}	-2.63×10^{-4}
CO$_3^{2-}$	1.21×10^{-6}	4.73×10^{-4}	$+4.72 \times 10^{-4}$
TOTCO$_3$(aq)	1.41×10^{-3}	1.41×10^{-3}	0.0
$\sum_{\text{CO}_3\,\text{species}}$ H$^+$	1.62×10^{-3}	9.38×10^{-4}	-6.82×10^{-4}

summarized in Table 9-8 and validate the assumption that $c_{\text{HCO}_3^-}$ is much greater than $c_{\text{CO}_3^{2-}}$, c_{OH^-}, or c_{H^+} (at pH 7, the latter two values are on the order of 10^{-7}). With this information, TOTCO$_3$(aq) can be computed, as can the total concentration of H$^+$ ions associated with the carbonate species $\left(\sum_{\text{CO}_3\,\text{species}} \text{H}^+ = 2c_{\text{H}_2\text{CO}_3} + c_{\text{HCO}_3^-}\right)$.

Since TOTCO$_3$(aq) remains constant when the pH is increased to 9.7, the K_a values can then be used in conjunction with the known TOTCO$_3$(aq) to compute the speciation at the higher pH. These results are also included in the table.

The calculations indicate that the carbonate species release 6.82×10^{-4} M H$^+$ when the pH is increased from 7.0 to 9.7. These H$^+$ ions must be neutralized for the desired pH change to be achieved, so they represent an additional demand for OH$^-$. Thus, in this case, the total OH$^-$ requirement equals that computed in the carbonate-free solution plus 6.82×10^{-4} M, for a total requirement of

$$[\text{OH}^-]\,\text{requirement} \approx 6.77 \times 10^{-4} + 6.82 \times 10^{-4}$$
$$= 1.36 \times 10^{-3}\,\text{M}$$

The net result is that approximately twice as much lime is required as in the absence of the carbonate species. The changes in this system over the whole range of OH$^-$ additions are shown in Figure 9-14.

As in the two cases analyzed previously, the precipitation reaction can be characterized quite well as proceeding according to a stoichiometric model after enough OH$^-$ has been added to bring the system to a critical condition. And, as in those previous cases, the critical condition is the pH of incipient precipitation, equal to 8.2. In the preceding examples, the OH$^-$ that had to be added to reach this pH was required almost entirely to neutralize free H$^+$ and to build up the free OH$^-$ concentration, whereas in the current scenario, a substantial amount of OH$^-$ is required to convert H$_2$CO$_3$ to HCO$_3^-$ (or, equivalently, to neutralize H$^+$ released by deprotonation of H$_2$CO$_3$). However, regardless, of the reaction(s) in which OH$^-$ participates, the key criterion for initiation of precipitation is that enough OH$^-$ be added to titrate the system to pH 8.2. ∎

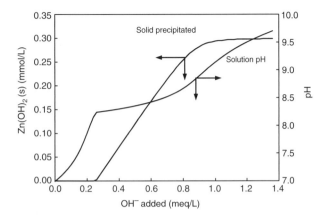

FIGURE 9-14. Response to addition of base of a solution identical to that characterized in Figure 9-12, except containing Alk equal to 60 mg/L as $CaCO_3$.

One other common application of hydroxide precipitation in water and wastewater treatment, but in which the emphasis is primarily on the amount of solids formed and not on the residual dissolved metal concentration, is formation of ferric or aluminum hydroxide in coagulation and/or adsorption processes. To form the solids in such processes, a salt of aluminum or iron is added to solution, and the metal precipitates. Because the precipitation reaction removes hydroxide ions from solution, it causes the pH to decline, and alkalinity must sometimes be added to assure that almost all of the metal precipitates and that the pH remains in an acceptable range for the precipitated solids to react with the contaminants in the water being treated.

■ **EXAMPLE 9-11.** A solution at pH 7.1 has an alkalinity of 24 mg/L as $CaCO_3$. The solution is to be dosed with 40 mg/L of alum ($Al_2(SO_4)_3 \cdot 14H_2O$) as part of a coagulation process. The process works best if the pH is ≥ 6.1. Assuming that the alkalinity is all from carbonate species and that essentially all the added Al ions precipitate as $Al(OH)_3(s)$, determine whether the final pH will be in the acceptable range and, if not, how much NaOH would have to be added to bring the pH to an acceptable value. Assume that the ionic strength of the solution is low enough that the solutes behave ideally (i.e., all activity coefficients are 1.0).

Solution. The MW of the alum is 594, so the alum dose is 40/594, or 0.067 mmol/L. The alkalinity of the original solution can be expressed in meq/L units as follows:

$$\text{Alk} = (24 \text{ mg as } CaCO_3/L)\left(\frac{1 \text{ meq}}{50 \text{ mg } CaCO_3}\right) = 0.48 \text{ meq/L}$$

The pH of the initial solution is 7.1. As explained in the preceding example, at that pH, CO_3^{2-}, H^+, and OH^- make negligible contributions to the alkalinity, so we can find the HCO_3^- concentration as follows:

$$\text{Alk} = c_{HCO_3^-} + 2c_{CO_3^{2-}} + c_{OH^-} - c_{H^+} \approx c_{HCO_3^-}$$

$$c_{HCO_3^-} = (0.48 \text{ meq/L})\left(\frac{1 \text{ mmol } HCO_3^-}{\text{meq}}\right) = 0.48 \text{ mmol/L}$$

The concentration of HCO_3^- can also be expressed as $\alpha_1(TOTCO_3)$, where α_1 is the conventional designation used in water chemistry to indicate the fraction of $TOTCO_3$ that is present as HCO_3^-. If the carbonate species behave ideally, α_1 is given by[11]

$$\alpha_1 = \frac{1}{((H^+)/K_{a1}) + 1 + (K_{a2}/(H^+))} \quad (9\text{-}81)$$

This equation indicates that α_1 equals 0.849 at pH 7.1, so

$$TOTCO_3 = \frac{c_{HCO_3^-}}{\alpha_1} = \frac{0.48 \text{ mmol/L}}{0.849} = 0.565 \text{ mmol/L}$$

The precipitation reaction consumes six equivalents of alkalinity per mole of alum added (three equivalents per mole of $Al(OH)_3(s)$ formed):

$$Al_2(SO_4)_3 + 6\,OH^- \rightarrow 2\,Al(OH)_3(s) + 3\,SO_4^{2-}$$

The final alkalinity of the solution can therefore be computed as

$$\text{Alk}_{\text{final}} = \text{Alk}_{\text{init}} - 6(\text{Alum dose})$$
$$= 0.48 \text{ meq/L}$$
$$\quad - 6 \text{ meq/mmol Alum}(0.067 \text{ mmol Alum/L})$$
$$= 0.078 \text{ meq/L}$$

The precipitation reaction consumes OH^-, so we expect the pH to decrease, meaning the final pH will be <7.1. We can therefore once again assume that Alk is approximately equal to the HCO_3^- concentration and, based on the K_a values, that CO_3^{2-} makes a negligible contribution to $TOTCO_3$. Then, noting that the reaction does not alter $TOTCO_3$, we can compute the H_2CO_3 concentration and the final pH as follows:

$$c_{H_2CO_3} = TOTCO_3 - c_{HCO_3^-} = (0.565 - 0.078) \text{ mmol/L}$$
$$= 0.487 \text{ mmol/L}$$
$$c_{H^+} = \frac{c_{H_2CO_3}}{c_{HCO_3^-}} K_{a1} = \frac{0.487 \text{ mmol/L}}{0.078 \text{ mmol/L}} 10^{-6.35}$$
$$= 2.79 \times 10^{-6} = 10^{-5.55}$$
$$\text{pH} = 5.55$$

---

[11] α values are, by definition, concentration ratios, whereas the expression on the right side of Equation 9-81 yields the ratio of the activity of HCO_3^- to the sum of the activities of H_2CO_3, HCO_3^-, and CO_3^{2-}. In nonideal solutions, a different expression for α_1 applies that includes the activity coefficients of the carbonate species.

The pH is below the target value. At the target pH of 6.1, α_1 is 0.36. If the pH is raised by adding NaOH, TOTCO$_3$ will remain the same, so the HCO$_3^-$ concentration will be 0.36 (0.565), or 0.20 mmol/L, and Alk will be 0.203 meq/L. Thus, the NaOH addition must increase Alk from 0.078 to 0.203 meq/L, a difference of 0.125 meq/L. Since each mole of NaOH contributes one equivalent of Alk, the required NaOH dose is 0.125 mmol/L. ∎

Precipitation of Carbonate Solids

Precipitation processes that form metal carbonate solids can offer several advantages over those that generate metal hydroxide solids. For example, if the carbonate solid is less soluble than its hydroxide counterpart at reasonable values of pH and TOTCO$_3$, the target metal concentration might be achieved with lower reagent additions and at lower cost. If the process is carried out at lower pH, post-process neutralization costs might be lower. Finally, because many carbonate solids have a higher density than do their hydroxide counterparts, settling and dewatering of carbonate solids can be more efficient.

Precipitative Softening One of the most common precipitation technologies in water treatment involves removing calcium (and sometimes magnesium and dissolved iron and manganese as well) by a process known as precipitative *water softening*. Although calcium is not associated with significant health risks, its removal can be desirable before industrial or domestic use of water for several reasons. For example, precipitated CaCO$_3$(s) can form scales that reduce heat transfer efficiency in industrial or domestic hot water systems. These precipitates can also clog distribution systems or water treatment membranes or foul meters and valves. In addition, both calcium and magnesium interact with nondetergent soaps, reducing the amount of suds formed and sometimes leaving a residual film on surfaces.

The divalent cations that contribute to these problems are collectively known as *hardness ions*. As a water quality parameter, total hardness is usually expressed as an equivalent concentration in units of mg/L as CaCO$_3$, meaning that the actual hardness in the solution (from all hardness ions) is equivalent to the hardness of a solution that contained the specified concentration of CaCO$_3$ and no other sources of hardness. This convention is consistent with the fact that the majority of the hardness in most fresh waters is contributed by Ca^{2+} ions, and it is also relatively convenient, because the molecular weight of CaCO$_3$ is 100.

■ **EXAMPLE 9-12.** A groundwater is reported to contain 225 mg/L total hardness, with contributions of 175 mg/L from Ca^{2+} ions and 50 mg/L from Mg^{2+} ions. What are the concentrations of Ca^{2+} and Mg^{2+} in meq/L, mmol/L, and mg/L?

Solution. The conversions are based on the fact that CaCO$_3$ has a MW of 100 and a charge of $+2$, so 50 mg of CaCO$_3$ is 1 meq. Ca and Mg have atomic weights of 40 and 24, respectively. The calculations are therefore as follows:

Ca^{2+}:
$$(175\,\text{mg/L as CaCO}_3)\left(\frac{1\,\text{meq}}{50\,\text{mg CaCO}_3}\right) = 3.50\,\text{meq/L Ca}^{2+}$$

$$(3.50\,\text{meq/L Ca}^{2+})\left(\frac{1\,\text{mmol Ca}^{2+}}{2\,\text{meq}}\right) = 1.75\,\text{mmol/L Ca}^{2+}$$

$$(1.75\,\text{mmol/L Ca}^{2+})(40\,\text{mg Ca}^{2+}/\text{mmol Ca}^{2+}) = 70\,\text{mg/L Ca}^{2+}$$

Mg^{2+}:
$$(50\,\text{mg/L as CaCO}_3)\left(\frac{1\,\text{meq}}{50\,\text{mg CaCO}_3}\right) = 1.00\,\text{meq/L Mg}^{2+}$$

$$(1.00\,\text{meq/L Ca}^{2+})\left(\frac{1\,\text{mmol Mg}^{2+}}{2\,\text{meq}}\right) = 0.50\,\text{mmol/L Mg}^{2+}$$

$$(0.50\,\text{mmol/L Mg}^{2+})(24\,\text{mg Mg}^{2+}/\text{mmol Mg}^{2+}) = 12\,\text{mg/L Mg}^{2+}$$
∎

Most common precipitation softening processes are variations of the so-called *lime-soda ash process*. Lime and soda ash are the common names for the chemicals CaO(s) and Na$_2$CO$_3$(s), respectively. Lime dissolves very slowly in water, so it is commonly converted to "hydrated lime" (Ca(OH)$_2$(s)) before being added in a treatment process; the hydration step is known as lime *slaking*.

$$\text{CaO(s)} + \text{H}_2\text{O} \leftrightarrow \text{Ca(OH)}_2 \quad (9\text{-}82)$$

After slaking, hydrated lime dissolves readily to release hydroxide:

$$\text{Ca(OH)}_2 \leftrightarrow \text{Ca}^{2+} + 2\,\text{OH}^- \quad (9\text{-}83)$$

The OH$^-$ ions that are released neutralize strong, and then progressively weaker, acid groups in solution. If enough lime is added to reach pH 10.3 or higher, most of the dissolved carbonic acid (H$_2$CO$_3$) and bicarbonate (HCO$_3^-$) are converted to carbonate (CO$_3^{2-}$) via the following reactions:

$$\text{H}_2\text{CO}_3 + \text{OH}^- \leftrightarrow \text{HCO}_3^- \quad (9\text{-}84)$$

$$\text{HCO}_3^- + \text{OH}^- \leftrightarrow \text{CO}_3^{2-} \quad (9\text{-}85)$$

The carbonate can then react with calcium in solution to produce CaCO$_3$(s):

$$\text{Ca}^{2+} + \text{CO}_3^{2-} \leftrightarrow \text{CaCO}_3(\text{s}) \quad (9\text{-}86)$$

If the solution does not contain enough TOTCO$_3$ to precipitate the desired amount of Ca^{2+}, soda ash (Na$_2$CO$_3$(s)) can be added.

Adding lime as part of a water softening process seems counter-intuitive, because it adds calcium to a solution which is being treated to reduce the calcium concentration. Lime, however, is an inexpensive and readily available source of alkalinity, and the additional calcium is relatively easy to precipitate subsequently as $CaCO_3(s)$. This approach results in additional solids formation and higher residuals management costs. An alternative approach is to use calcium-free sources of alkalinity such as NaOH (commonly called "caustic"), but these reagents are more expensive and often increase the overall cost of the process. Nevertheless, depending on the costs of sludge processing and disposal increase, this alternative approach might be attractive.

In the softening process, the reagents are added in a rapid-mix step to ensure uniform distribution. This mixing process is typically followed by a 30- to 60-min reaction time to promote particle aggregation and growth; the suspension is mixed during this step, but much less intensively than during the rapid-mix step. The softening step is typically followed by settling and filtration to separate the solids from the water. Before distribution of the treated water, a final pH adjustment is made to quench the reaction and adjust the water to the desired pH value.

The Stoichiometric Model of Precipitative Softening

Conventional descriptions of precipitation softening processes often suggest that the reagent requirements can be based on the stoichiometry of reactions 9-84 to 9-86, along with the acid/base balance of water itself. In such cases, the reagent doses can be calculated as follows.

(1) *Estimate dose of strong base* (OH^-) *required to deprotonate weak acids.* This dose is based on the assumptions that species of the carbonate group are the only weak acids in solution and that all of the added OH^- reacts with H_2CO_3 or HCO_3^- until those weak acids are completely converted to CO_3^{2-}. In this process, each mole of H_2CO_3 reacts with two moles of OH^-, and each mole of HCO_3^- reacts with one mole of OH^-. Therefore, the stoichiometric requirement for OH^- to deprotonate carbonate species is

$$[OH^-]_{CO_3} = 2c_{H_2CO_3, init} + c_{HCO_3^-, init} \quad (9\text{-}87)$$

(2) *Estimate dose of strong base needed to change the concentrations of* H^+ *and* OH^- *in the initial solution to those that apply at the final pH.* The pH must be at least 10.33 to cause most of the TOTCO$_3$ to be present as CO_3^{2-}, and softening is commonly carried out at pH values near that (the range 10.0–10.7 is common). At these pH values, c_{OH^-} is significant, and c_{H^+} is negligible. At the initial pH of the solution, chances are that both c_{OH^-} and c_{H^+} are negligible, so the OH^- dose required to adjust the H^+ and OH^- concentrations is approximately equal to the final value of c_{OH^-}. Assuming that OH^- can be treated as an ideal solute, this concentration is

$$[OH^-]_{pH} \approx c_{OH^-, fin} \approx 10^{pH_{fin} - 14} \quad (9\text{-}88)$$

where the first approximation is based on the idea that $c_{H^+, init}$, $c_{OH^-, init}$, and $c_{H^+, fin}$ are all negligible compared to $c_{OH^-, fin}$, and the second is based on the assumption that the activity coefficient of OH^- is ~1.0 after the lime addition. The OH^- requirement associated with this step is not considered explicitly in most discussions of stoichiometric softening, but rather is embedded in a safety factor that is added to account for unspecified OH^- requirements.

(3) *Determine the lime dose to provide the amount of* OH^- *computed in Steps 1 and 2.* This estimate is based on the assumption of complete dissociation of the hydrated lime:

$$\begin{aligned}[Ca(OH)_2]_{dose} &= 0.5\{[OH^-]_{CO_3} + [OH^-]_{pH}\} \\ &= c_{H_2CO_3, init} + 0.5 c_{HCO_3^-, init} \\ &\quad + 0.5 c_{OH^-, fin} \end{aligned} \quad (9\text{-}89)$$

(4) *Estimate the stoichiometric requirement for* TOTCO$_3$ *in the precipitation reaction.* Each mole of Ca^{2+} in the raw water and each mole of Ca^{2+} added with the lime can react with one mole of TOTCO$_3$ to form one mole of $CaCO_3(s)$. Assuming all the calcium in solution is in the form of Ca^{2+}, the stoichiometric TOTCO$_3$ requirement can be calculated as

$$\begin{aligned}[TOTCO_3]_{stoic} &= [TOTCa]_{\substack{after \\ lime}} \\ &= TOTCa_{init} + [Ca(OH)_2]_{dose} \\ &= c_{Ca^{2+}, init} + c_{H_2CO_3, init} \\ &\quad + 0.5 c_{HCO_3^-, init} + 0.5 c_{OH^-, fin}\end{aligned} \quad (9\text{-}90)$$

(5) *Determine whether the solution contains sufficient* TOTCO$_3$ *to meet the stoichiometric requirement and, if not, determine how much additional* TOTCO$_3$ *is needed.* If TOTCO$_3$ in the raw water is greater than the stoichiometric requirement computed in step 4, then no more carbonate need be added to the solution. However, if TOTCO$_3$ in the raw water is

less than [TOTCO$_3$]$_{\text{stoic}}$, then carbonate species will have to be added if one wishes to bring the final TOTCO$_3$ up to the stoichiometric value. In this latter case, soda ash (Na$_2$CO$_3$) is the most commonly used reagent for increasing TOTCO$_3$. Therefore, if [TOTCO$_3$]$_{\text{stoic}} \geq$ [TOTCO$_3$]$_{\text{init}}$, the required dose of soda ash is

$$[\text{Na}_2\text{CO}_3]_{\text{dose}} = [\text{TOTCO}_3]_{\text{stoic}} - [\text{TOTCO}_3]_{\text{init}}$$
$$= (c_{\text{Ca}^{2+},\text{init}} + c_{\text{H}_2\text{CO}_3,\text{init}} + 0.5 c_{\text{HCO}_3^-,\text{init}}$$
$$+ 0.5 c_{\text{OH}^-,\text{fin}}) - (c_{\text{H}_2\text{CO}_3,\text{init}} + c_{\text{HCO}_3^-,\text{init}})$$
$$= c_{\text{Ca}^{2+},\text{init}} - 0.5 c_{\text{HCO}_3^-,\text{init}} + 0.5 c_{\text{OH}^-,\text{fin}}$$
(9-91)

or, in sum:

$$[\text{Na}_2\text{CO}_3]_{\text{dose}} = \begin{cases} 0 & \text{if } [\text{TOTCO}_3]_{\text{stoic}} \leq [\text{TOTCO}_3]_{\text{init}} \\ c_{\text{Ca}^{2+}} - 0.5 c_{\text{HCO}_3^-} \\ + 0.5 c_{\text{OH}^-,\text{fin}} & \text{if } [\text{TOTCO}_3]_{\text{stoic}} > [\text{TOTCO}_3]_{\text{init}} \end{cases}$$
(9-92)

The procedure outlined above yields doses of lime and soda ash that cause TOTCO$_{3,\text{fin}}$ to be greater than or equal to TOTCa$_{\text{fin}}$ and most of the TOTCO$_3$ to be present as CO$_3^{2-}$. TOTCO$_{3,\text{fin}}$ and TOTCa$_{\text{fin}}$ will be equal in situations where soda ash is added. Because the precipitation reaction removes equal molar concentrations of TOTCO$_3$ and TOTCa, if TOTCO$_3$ = TOTCa before precipitation, that equality will also apply once the solid has precipitated and equilibrium has been reached. In such a case, the equilibrium concentration of TOTCa in an ideal solution can be estimated as follows:

$$K_{s0} = 3.31 \times 10^{-9} = (\text{Ca}^{2+})(\text{CO}_3^{2-})$$
$$\approx (\text{TOTCa})(\text{TOTCO}_3) = (\text{TOTCa})^2 \quad (9\text{-}93)$$
$$\text{TOTCa} = 5.75 \times 10^{-5} \quad (9\text{-}94)$$

This value of TOTCa corresponds to $\sim$2.3 mg/L. If TOTCO$_3$ > TOTCa after the various reagent doses, then the computed, equilibrium value of TOTCa would be even lower.

The steps summarized above reflect the classic stoichiometric approach to computing reagent doses for lime/soda ash softening. However, alternative approaches that generate less sludge are worth considering at locales where the stoichiometric approach leads to the conclusion that soda ash should be added along with the lime. For example, imagine that the computed doses are X mol/L lime and Y mol/L soda ash. If $X > Y$, we can (mentally) divide the lime dose into two portions of $X - Y$ and Y, and treat the overall dosing as though it involved addition of $X - Y$ mol/L of lime, plus a mixture containing Y mol/L of both lime and soda ash. This latter mixture is expected to react almost quantitatively as follows:

$$\text{Ca(OH)}_2 + \text{Na}_2\text{CO}_3 \rightarrow \text{CaCO}_3(s) + 2\text{Na}^+ + 2\text{OH}^-$$
(9-95)

Thus, the addition of Y mol/L of the two reagents has the same effect as adding $2Y$ mol/L NaOH plus Y mol/L CaCO$_3$(s). The solid produced by this reaction serves no useful role, and it contributes to the sludge burden requiring further processing and disposal. In essence, reaction 9.95 suggests that the point of adding these two reagents is to provide OH$^-$ to the solution, and the CaCO$_3$(s) is just an undesirable byproduct that appears because of the approach being used to generate the OH$^-$. If X were less than Y, essentially the same argument would apply, but the soda ash dose would be divided into two portions, and the overall (imagined) additions would be $Y - X$ mol/L of soda ash plus a mixture containing X mol/L of both reagents. In some cases, once the costs of reagents and sludge management are all taken into account, it might be more attractive to follow in practice a procedure suggested by the mental exercise above; that is, to add only enough lime or soda ash to reach a condition where TOTCa = TOTCO$_3$ (or, depending on the economics, even less than that), and then add NaOH to raise the pH to the target value, thereby injecting the desired amount of OH$^-$ into the solution without generating the CaCO$_3$(s) byproduct. Yet another approach for reducing the chemical costs involves adding the lime dose needed to reach the target pH, but adding less that the computed stoichiometric soda ash dose. After the sludge produced by this step is separated from the solution, additional carbonate can be injected by bubbling the solution with air. This process is discussed in greater detail shortly, in the section on recarbonation.

Numerous assumptions are embedded in the preceding calculations, almost all of which cause the estimate of TOTCa remaining in solution to be lower than the values that are actually obtained. Chief among these are the assumptions that TOTCO$_3$ is a good approximation for the concentration of CO$_3^{2-}$ after the lime addition; that the carbonate species are the only weak acids in the initial solution; that both Ca^{2+} and CO$_3^{2-}$ behave ideally; and that the solution reaches equilibrium with the solid. As a result, the actual value of TOTCa present after stoichiometric softening is invariably substantially higher than the value computed using Equation 9-94. On the other hand, the presence of calcium in finished drinking water is primarily an aesthetic problem, and concentrations up to $\sim$50 mg/L are usually not considered problematic. Therefore, these calculations often provide an acceptable guide for choosing reagent doses that will achieve satisfactory results.

EXAMPLE 9-13. The major ion composition of a hard water is shown below. The solution is at 25°C.

pH	8.30	K	14 mg/L
Ca	185 mg/L as $CaCO_3$	Alk	235 mg/L as $CaCO_3$
Mg	150 mg/L as $CaCO_3$	Cl	99 mg/L
Na	41 mg/L	SO_4	71 mg/L

(a) Carry out a charge balance on the solution.
(b) Estimate the concentrations of species in the carbonate acid/base group and of H^+ and OH^- in the raw water, assuming ideal solution behavior.
(c) Estimate the dosages of lime and, if needed, soda ash required for stoichiometric softening of this water at pH 10.7. Also estimate the concentration of $CaCO_3(s)$ produced.
(d) Estimate $TOTCO_3(aq)$ and the alkalinity in the softened solution. Based on those estimates and K_{s0} for calcite, estimate $TOTCa(aq)$ in the solution.

Solution.

(a) The concentrations of H^+ and OH^- can be estimated based on the pH, and the concentrations of the salt ions can be converted to meq/L units by dividing by their respective equivalent weights. Also, since the pH is 8.3, the HCO_3^- concentration in meq/L can be approximated as the Alk. The concentration of CO_3^{2-} can then be computed using K_{a2} for the carbonate system, as follows:

$$c_{CO_3^{2-},init} = \frac{c_{HCO_3} \cdot K_{a2}}{c_{H^+}} = \frac{(4.70 \times 10^{-3})(10^{-10.33})}{10^{-8.30}}$$
$$= 6.95 \times 10^{-5} \, mol/L$$

The calculations are summarized in Table 9-9.

The summation of positive charges in the final column is 8.84 meq/L, and that of negative charges is 9.06 meq/L. The charge balance therefore closes to within 2.4%, suggesting that all the major ions have been taken into account.

(b) The concentrations of all the species of interest except H_2CO_3 were computed in part (a). The H_2CO_3 concentration is given by

$$c_{H_2CO_3,init} = \frac{c_{HCO_3} \cdot c_{H^+}}{K_{a1}} = \frac{(4.70 \times 10^{-3})(10^{-8.30})}{10^{-6.35}}$$
$$= 5.27 \times 10^{-5}$$

(c) The final pH is specified to be 10.70, so the final concentration of OH^- after lime addition and precipitation is $10^{-3.30}$. Based on that value and the composition of the raw water as determined in parts a and b, the required dose of lime can calculated from Equation 9-89.

$$[Ca(OH)_2]_{dose} = c_{H_2CO_3,init} + 0.5c_{HCO_3^-,init} + 0.5c_{OH^-,fin}$$
$$= 5.27 \times 10^{-5} + 0.5(4.70 \times 10^{-3})$$
$$+ 0.5(10^{-3.30})$$
$$= 2.65 \times 10^{-3} \, mol/L$$

The stoichiometric requirement for $TOTCO_3$ equals $TOTCa$ in the solution after the lime addition. These values can be computed as

$$[TOTCO_3]_{stoic} = [TOTCa]_{\substack{after \\ lime}} = c_{Ca^{2+},init} + [Ca(OH)_2]_{dose}$$
$$= 1.85 \times 10^{-3} + 2.65 \times 10^{-3}$$
$$= 4.50 \times 10^{-3} \, mol/L$$

TABLE 9-9. Concentrations and Charges Associated with Ions in the Raw Water

Species	Concentration (mg/L)	MW	EW	Concentration (meq/L)	z_i (meq/mmol)	Concentration (mmol/L)	Charge (meq/L)
H^+					1	$10^{-5.30\,a}$	$+10^{-5.30}$
OH^-					1	$10^{-2.70\,a}$	$-10^{-2.70}$
Ca^{2+}	185 as $CaCO_3$		50	3.70	2	1.85	+3.70
Mg^{2+}	150 as $CaCO_3$		50	3.00	2	1.50	+3.00
Na^+	41 as Na	23	23	1.78	1	1.78	+1.78
K^+	14 as K	39	39	0.367	1	0.36	+0.36
Alk	235 as $CaCO_3$		50	4.70			
$HCO_3^{-\,b}$				4.70	1	4.70	−4.70
CO_3^{2-}						0.044	−0.09
Cl^-	99	35.5	35.5	2.79	1	2.79	−2.79
SO_4^{2-}	71	96	48	1.48	2	0.74	−1.48

[a] Based on H^+ concentration in moles per liter equal to 10^{-pH}, then multiplied by 10^3 to convert to mmol/L.
[b] HCO_3^- concentration assumed equal to Alk on meq/L basis.

TOTCO$_3$ in the raw water is

$$[\text{TOTCO}_3]_{\text{init}} = c_{\text{H}_2\text{CO}_3} + c_{\text{HCO}_3^-} + c_{\text{CO}_3^{2-}}$$
$$= 5.27 \times 10^{-5} + 4.70 \times 10^{-3} + 4.40 \times 10^{-5}$$
$$= 4.80 \times 10^{-3} \, \text{mol/L}$$

TOTCO$_3$ in the raw water is sufficient to meet the stoichiometric requirement, so no soda ash is needed.

Because stoichiometric softening is designed to precipitate the vast majority of the Ca in the system after all the reagents are added, the molar concentration of CaCO$_3$(s) can be approximated as TOTCa$_{\text{after}}^{\text{lime}}$. This quantity was computed above as 4.50×10^{-3} mol/L, or 450 mg/L. Note that only ∼41% of the precipitate derives from Ca that was in the raw water; the rest comes from Ca that was added with the lime.

(d) TOTCO$_3$(aq) in the softened solution is the difference between TOTCO$_3$ in the initial solution and the amount of CO$_3$ removed from solution by precipitation of CaCO$_3$(s):

$$\text{TOTCO}_3(\text{aq})_{\text{fin}} = \text{TOTCO}_3(\text{aq})_{\text{init}} - [\text{CaCO}_3(\text{s})]_{\text{fin}}$$
$$= 4.80 \times 10^{-3} - 4.50 \times 10^{-3}$$
$$= 3.0 \times 10^{-4}$$

Because the pH is known, the distribution of TOTCO$_3$(aq) in the softened water can be determined from the α values. At pH 10.7, α_1 and α_2 for the carbonate acid/base group are 0.30 and 0.70, respectively, so

$$(\text{HCO}_3^-) = \alpha_1(\text{TOTCO}_3) = 0.30(3.0 \times 10^{-4}) = 9.0 \times 10^{-5}$$
$$(\text{CO}_3^{2-}) = \alpha_2(\text{TOTCO}_3) = 0.70(3.0 \times 10^{-4}) = 2.1 \times 10^{-4}$$

The alkalinity can then be computed as follows:

$$\text{Alk} = (\text{HCO}_3^-) + 2(\text{CO}_3^{2-}) + (\text{OH}^-) - (\text{H}^+)$$
$$= 9.0 \times 10^{-5} + 2(2.1 \times 10^{-4}) + 10^{-3.30} - 10^{-10.70}$$
$$= 1.01 \times 10^{-3}$$

Finally, the equilibrium concentration of Ca^{2+} is given by

$$(\text{Ca}^{2+}) = \frac{K_{s0}}{(\text{CO}_3^{2-})} = \frac{3.31 \times 10^{-9}}{2.1 \times 10^{-4}} = 1.58 \times 10^{-5}$$

The assumption that most of the Ca is likely to precipitate is thus confirmed. ∎

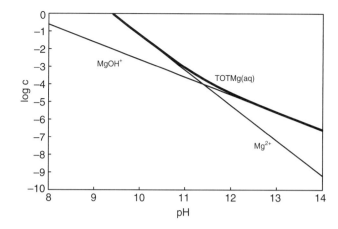

FIGURE 9-15. Log c − pH diagram for a solution in equilibrium with Mg(OH)$_2$(s).

Many source waters that contain high concentrations of calcium also contain substantial amounts of magnesium, and these two alkaline earth elements behave similarly in many ways. Therefore, magnesium is sometimes included as a target ion, along with calcium, in the softening process. MgCO$_3$(s) is considerably more soluble than CaCO$_3$(s) (K_{s0} for MgCO$_3$(s) is $10^{-7.46}$, or approximately one order of magnitude larger than K_{s0} for CaCO$_3$(s)), so the most common approach used to precipitate Mg^{2+} involves formation of Mg(OH)$_2$(s), with $K_{s0} = 10^{-9.21}$. The speciation of dissolved Mg in equilibrium with Mg(OH)$_2$(s) is shown in Figure 9-15.

When Mg^{2+} removal is desired in a softening process, the lime and soda ash dosages can be calculated based on a stoichiometric precipitation model, just as when Ca^{2+} is the only ion targeted for precipitation. The main difference in the calculations is that the dose of OH$^-$ must be sufficient to precipitate the Mg^{2+}, in addition to converting H$_2$CO$_3$ and HCO$_3^-$ to CO$_3^{2-}$ and building up the free OH$^-$ concentration. Because one mole of lime is required to provide the two moles of OH$^-$ needed for precipitation of one mole of Mg, the calculation of the required lime dose (Step 3 in the previous calculation) becomes

$$[\text{Ca(OH)}_2]_{\text{dose}} = 0.5\{[\text{OH}^-]_{\text{CO}_3} + [\text{OH}^-]_{\text{pH}} + [\text{OH}^-]_{\text{Mg}}\}$$
$$= c_{\text{H}_2\text{CO}_3,\text{init}} + 0.5 c_{\text{HCO}_3^-,\text{init}}$$
$$+ 0.5 c_{\text{OH}^-,\text{fin}} + c_{\text{Mg,init}} \quad (9\text{-}96)$$

This adjusted dose has ramifications for the remaining calculations, as follows:

$$[\text{TOTCO}_3]_{\text{stoic}} = [\text{TOTCa}]_{\text{after}}^{\text{lime}} = c_{\text{Ca}^{2+}} + [\text{Ca(OH)}_2]_{\text{dose}}$$
$$= c_{\text{Ca}^{2+},\text{init}} + c_{\text{H}_2\text{CO}_3,\text{init}} + 0.5 c_{\text{HCO}_3^-,\text{init}}$$
$$+ 0.5 c_{\text{OH}^-,\text{fin}} + c_{\text{Mg,init}} \quad (9\text{-}97)$$

$$[\text{Na}_2\text{CO}_3]_{\text{dose}} = \begin{cases} 0 & \text{if } [\text{TOTCO}_3]_{\text{stoic}} \leq [\text{TOTCO}_3]_{\text{init}} \\ c_{\text{Ca}^{2+},\text{init}} + c_{\text{Mg}^{2+},\text{init}} - 0.5 c_{\text{HCO}_3^-,\text{init}} + c_{\text{OH}^-,\text{fin}} & \\ & \text{if } [\text{TOTCO}_3]_{\text{stoic}} > [\text{TOTCO}_3]_{\text{init}} \end{cases}$$

(9-98)

Equation 9-98 indicates the soda ash dose required to cause TOTCO$_3$ to equal TOTCa after addition of all reagents. Adding that dose will assure that almost all the calcium in the system is precipitated as CaCO$_3$(s) at the final, high pH. However, as noted in the discussion of softening without Mg precipitation, an alternative approach is commonly used in which less than the stoichiometric dose of soda ash is added, and carbonate is provided to the solution after the solids produced at high pH have been removed.

Figure 9-15 indicates that a higher pH is generally required to achieve low TOTMg(aq) than is required just to precipitate CaCO$_3$(s); this factor can also can lead to substantial increases in the required reagent doses. In addition, particles and flocs of Mg(OH)$_2$(s) tend to settle much more slowly than those of CaCO$_3$(s), so precipitation of magnesium can significantly alter the requirements for downstream particle separation processes.

■ **EXAMPLE 9-14.** Repeat the calculations in Example 9-13, if the softening process is carried out at pH 11.4 to remove both Ca^{2+} and Mg^{2+}.

Solution. (a) Maintaining the assumption of ideal solution behavior, the concentration of OH$^-$ at pH 11.4 will be $10^{-2.86}$ mol/L. Then, applying Equations 9-96–9-98, we find

$$\begin{aligned}
[\text{Ca(OH)}_2]_{\text{dose}} &= c_{\text{H}_2\text{CO}_3,\text{init}} + 0.5 c_{\text{HCO}_3^-,\text{init}} + 0.5 c_{\text{OH}^-,\text{fin}} \\
&\quad + c_{\text{Mg,init}} \\
&= 5.27 \times 10^{-5} + 0.5(4.70 \times 10^{-3}) \\
&\quad + 0.5(2.51 \times 10^{-3}) + 1.50 \times 10^{-3} \\
&= 5.16 \times 10^{-3} \text{ mol/L} \\
[\text{TOTCO}_3]_{\text{stoic}} &= c_{\text{Ca}^{2+},\text{init}} + [\text{Ca(OH)}_2]_{\text{dose}} \\
&= 1.85 \times 10^{-3} + 5.16 \times 10^{-3} \\
&= 7.01 \times 10^{-3} \text{ mol/L}
\end{aligned}$$

In this case, the stoichiometric requirement for TOTCO$_3$ exceeds the amount available in the raw water, so some soda ash must be added, if stoichiometric softening is the goal. The required dose is:

$$\begin{aligned}
[\text{Na}_2\text{CO}_3]_{\text{dose}} &= [\text{TOTCO}_3]_{\text{stoic}} - [\text{TOTCO}_3]_{\text{init}} \\
&= 7.01 \times 10^{-3} - 4.80 \times 10^{-3} \\
&= 2.21 \times 10^{-3} \text{ mol/L}
\end{aligned}$$

Although the assumption that all the Mg initially present precipitates is less justified than the assumption of complete precipitation of the Ca in the system, we can use these assumptions to estimate the maximum concentration of each solid, and of total solids, that might form. The concentration of CaCO$_3$(s) that could form equals the total concentration of Ca in the system, which is the same as the stoichiometric value of TOTCO$_3$ computed above. Thus

$$\begin{aligned}
c_{\text{CaCO}_3(\text{s})} &= [\text{TOTCO}_3]_{\text{stoic}} = 7.01 \times 10^{-3} \text{ mol/L} = 701 \text{ mg/L} \\
c_{\text{Mg(OH)}_2(\text{s})} &= c_{\text{Mg}^{2+},\text{init}} \\
&= (1.50 \times 10^{-3} \text{ mol/L})(58{,}000 \text{ mg/mol}) \\
&= 87 \text{ mg/L}
\end{aligned}$$

The total concentration of solids precipitated is thus 788 mg/L, of which only ~35% is attributable to Ca and Mg ions in the raw water. ■

The Equilibrium Model of Precipitative Softening The approach of calculating reagent doses for softening based on a stoichiometric model was developed before high-speed computing and chemical equilibrium software became widely available. Although application of those doses usually leads to successful softening, the approach does not provide any insight into the sensitivity of the treated water composition to variations in the doses and therefore is not useful for optimizing the reagent doses when issues such as cost and solids production are important considerations. To determine whether nonstoichiometric softening can achieve the treatment goal, and to compare the effects of different dosing strategies in such systems, an equilibrium analysis is required. In this section, we use such an analysis to explore how alkalinity, pH, TOTCO$_3$(aq), TOTCa(aq), and TOTMg(aq) respond to the lime dose in a system with an initial composition identical to that given for the raw water in Examples 9-13 and 9-14.[12]

The first important insight that is obtained from this exercise is that most natural waters that are subjected to softening are supersaturated with respect to calcite even before any lime is added. That is, an equilibrium analysis based on the reported composition of the raw water indicates that CaCO$_3$(s) will precipitate from the unmodified solutions. The solutions are nevertheless stable in a practical sense; that is, solids do not readily precipitate if the solution composition is not altered. Possible causes of this stability include the substantial extent of supersaturation that is often needed to nucleate new solids, the presence of inhibitors in the water that adsorb to the surface of newly formed solids

---

[12] The calculations shown were carried out using the titration function and databases in the software package Mineql+® (Environmental Research Software, v.4.6, 2007); several other programs provide comparable capabilities.

and interfere with their growth, and inaccuracy of the assumption that all the TOTCa(aq) measured in the solution is in the form of Ca^{2+}. Whatever the cause of the stability, it is overcome when the extent of supersaturation is increased by lime addition, which leads to rapid precipitation of a large concentration of $CaCO_3(s)$ particles.

The raw water composition specified in Example 9-13, which is based on analysis of a real water supply, is significantly supersaturated with respect to $CaCO_3(s)$ ($K_{s0} = 3.31 \times 10^{-9}$), as shown by the following calculation:

$$Q_{s0} = \frac{a_{Ca^{2+}} a_{CO_3^{2-}}}{a_{CaCO_3(s)}} \approx c_{Ca^{2+}} c_{CO_3^{2-}}$$

$$= (1.85 \times 10^{-3})(4.40 \times 10^{-5}) = 8.14 \times 10^{-8} \quad (9\text{-}99)$$

Taking activity coefficients into account lowers the degree of supersaturation, but the result is still that the solution is supersaturated by more than an order of magnitude. Calculations using chemical equilibrium software indicate that, if this solution reached equilibrium, 5.39×10^{-4} mol/L calcite would precipitate, and the pH would drop from 8.30 to 7.46. In the following analysis, an arbitrary assumption has been made that no calcite precipitates until 0.5 mmol/L lime has been added.

The predicted changes in pH and the concentrations of various solutes and solids as the lime dose is increased from 0 to 4.0 mmol/L are shown in Figure 9-16. The simulation does not consider addition of soda ash. It will be useful to refer these figures throughout the following discussion.

Because of the assumption that no solids precipitate until 0.5 mmol/L lime has been added, each mmol/L of lime added up to that point increases TOTCa(aq) by 1 mmol/L and Alk by 2 meq/L, and also increases pH. The pH change alters the speciation of carbonate (not shown), but TOTCO$_3$(aq) remains constant. When the lime dose reaches 0.5 mmol/L, all the parameters undergo discontinuous changes, because solid is assumed to precipitate at that point, lowering TOTCa(aq), Alk, TOTCO$_3$(aq), and pH. Approximately 60% of TOTCa is predicted to precipitate, generating a pH drop of more than 1.7 units.

At higher lime doses, each mmol/L of lime causes TOTCa(aq), Alk, and TOTCO$_3$(aq) to decrease further, until almost all the calcium has been removed from solution, at a dose near 1.7 mmol/L. Because the pH is in the range 7.5–9.5 throughout this process, the dominant dissolved carbonate species is HCO_3^-, and the overall reaction conforms closely to the following:

$$Ca^{2+} + 2\,HCO_3^- + Ca(OH)_2 \rightarrow 2\,CaCO_3(s) + 2\,H_2O \quad (9\text{-}100)$$

When this reaction dominates, each mmol/L of lime added reduces TOTCa(aq) and TOTCO$_3$(aq) by 1 mmol/L

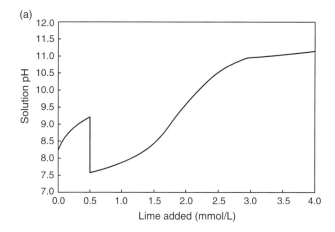

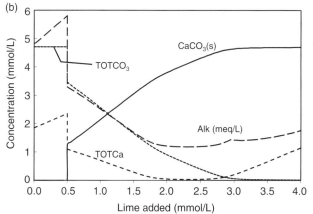

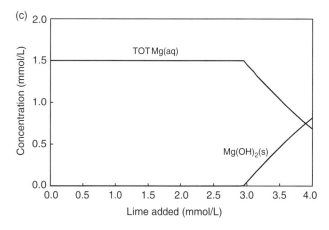

FIGURE 9-16. Response of the raw water from Example 9-13 to titration with lime, based on attainment of equilibrium after each lime dose. (a) pH; (b) Ca and CO$_3$ species; and (c) Mg species. At doses up to 0.5 mmol/L, the simulation assumes that reactions in solution equilibrate, but no solids precipitate; at higher doses, both reactions in solution and precipitation reactions are considered.

and Alk by 2 meq/L, while increasing the CaCO$_3$(s) concentration by 2 mmol/L; these changes are reflected in the slopes of the corresponding curves in Figure 9-16b.

When yet more lime is added, the Ca that enters with it is virtually the only Ca available. As long as HCO_3^- remains the dominant form of dissolved carbonate, this Ca reacts with the HCO_3^-, and the dominant reaction occurring is the following:

$$HCO_3^- + Ca(OH)_2 \rightarrow CaCO_3(s) + H_2O + OH^- \quad (9\text{-}101)$$

Because this reaction generates free OH^-, it causes the pH to increase. When the pH increases enough that a substantial fraction of the dissolved carbonate is present as CO_3^{2-}, the following reaction also becomes significant:

$$CO_3^{2-} + Ca(OH)_2 \rightarrow CaCO_3(s) + 2\,OH^- \quad (9\text{-}102)$$

These reactions, which dominate over the range of lime doses from $\sim$1.7 to 2.5 mmol/L, lower $TOTCO_3$(aq) and increase the pH, but they do not alter Alk or lower TOTCa(aq) significantly. At lime doses of $\sim$2.5–2.7 mmol/L, reactions 9-101 and (especially) 9-102 remain important, but the $TOTCO_3$(aq) has declined to the point where it is insufficient to cause precipitation of the added Ca, so a significant fraction of the lime simply dissolves and remains in solution, causing TOTCa(aq) and Alk to increase.

In the absence of dissolved Mg, most of any additional lime that was dosed would remain in solution, causing TOTCa(aq) and Alk to increase by 1 mmol/L and 2 meq/L, respectively, per mmol/L of lime added. However, at a lime dose of 2.7 mmol/L, $Mg(OH)_2$(s) begins to precipitate, and the dominant reaction becomes

$$Mg^{2+} + Ca(OH)_2 \rightarrow Mg(OH)_2(s) + Ca^{2+} \quad (9\text{-}103)$$

As long as this reaction dominates, each mmol/L of lime added reduces TOTMg(aq) by $\sim$1 mmol/L and increases TOTCa(aq) by 1 mmol/L, so it has essentially no effect on the total hardness remaining in solution. Alk also remains approximately constant in this region. If Mg removal were truly needed, and if one wished to keep the dissolved Ca concentration low in this high-pH solution, soda ash would have to be added in addition to the lime, so that the Ca^{2+} released in reaction 9-103 would precipitate. (Alternatively, as noted previously, NaOH could be added instead of lime once the lime dose reached $\sim$2.5 mol/L. Such an approach would raise the pH and precipitate $Mg(OH)_2$(s) without increasing the dissolved Ca concentration. Yet another approach is described shortly in the discussion of recarbonation.) At a lime dose of 4.0 mmol/L, the pH is predicted to be 11.2, with $\sim$64% of the Mg in the initial solution precipitated. Raising the pH further would remove more Mg from solution, but at a substantial cost in terms of reagent addition and sludge production (assuming that soda ash was added along with the lime to keep soluble Ca low).

TABLE 9-10. Comparison of Predicted System Composition Based on Stoichiometric and Equilibrium Models for Lime Softening[a]

	Model	
	Stoichiometric[b]	Equilibrium[b]
pH	10.70	10.93
TOTCa(aq)	1.58×10^{-5}	1.84×10^{-4}
TOTMg(aq)	1.50×10^{-3}	1.50×10^{-3}
$TOTCO_3$(aq)	3.00×10^{-4}	7.41×10^{-5}
Alk	8.91×10^{-4}	1.37×10^{-3}
$CaCO_3$(s)	4.50×10^{-3}	4.32×10^{-3}
$Mg(OH)_2$(s)	0	0

[a]Raw water as specified in Example 9-13; lime dose 2.65 mmol/L for both models.
[b]All values in mmol/L, except pH (dimensionless) and Alk (meq/L).

The results from Example 9-13, which characterized the expected behavior of the solution based on the assumption of stoichiometric precipitation, are compared with those based on the equilibrium analysis for the same lime dose in Table 9-10.

The table indicates that the prediction for the concentration of $CaCO_3$(s) formed based on the stoichiometric model is very close to that predicted based on an equilibrium analysis, and that the reagent doses computed using the traditional, stoichiometric approach are expected to achieve excellent softening; a comparison for the reagent dosages determined in Example 9-14 (that include precipitation of $Mg(OH)_2$(s)) would lead to the same conclusion. Thus, if one were interested solely in determining an acceptable set of dosages to soften a given water source, either approach could be used. However, the goal of a softening analysis is invariably more ambitious than that, requiring the analyst to compare a wide variety of alternative operating scenarios with respect to cost and other technical or practical considerations. The equilibrium analysis allows the response of the system to different dosing scenarios to be evaluated, whereas the stoichiometric analysis does not.

For example, for the water source considered in the preceding few examples, Figure 9-16 indicates that TOTCa(aq) can be reduced to <0.5 mmol/L (50 mg/L as $CaCO_3$) at a lime dose as low as 1.05 mmol/L, or about 40% of the stoichiometric dose. The figure also illustrates the dependence of sludge production on lime dose, and Figure 9-16c indicates the lime dose required to initiate $Mg(OH)_2$(s) precipitation. In addition, as has been noted, the equilibrium analysis also accounts for nonideal solute behavior and the contribution of soluble complexes to the total dissolved concentrations of Ca, Mg, and other species of interest. None of this information is available from a strictly stoichiometric analysis.

While an equilibrium analysis can provide accurate predictions for the solution composition once equilibrium is achieved, those predictions still suffer from the fact that the precipitation reactions often do not reach equilibrium in full-scale treatment plants. In addition, removal of the precipitated solids is invariably incomplete, so TOTCa in the water exiting the plant is always larger than the equilibrium value of TOTCa(aq), often by several fold. If $CaCO_3(s)$ is supersaturated in the water leaving the treatment plant, precipitation of the solid is expected in the distribution system. Historically, many systems have been operated with the express intent of promoting precipitation of small amounts of $CaCO_3(s)$ in the distribution system since the precipitate typically forms on pipe walls and is thought to interfere with corrosion of some pipe materials. On the other hand, if the solution is too supersaturated, enough $CaCO_3(s)$ could precipitate in the distribution system over time to significantly reduce the effective diameter of the pipes. To avoid such a situation, the treatment system can be designed to facilitate a closer approach to solid/liquid equilibrium. For example, solids recycle can be used to increase the solids concentration in the precipitation reactor and thereby accelerate the approach to equilibrium. The introduction of solids also reduces or eliminates the scenario shown at low lime doses in Figure 9-16, where $CaCO_3(s)$ does not precipitate even though the solution is supersaturated. Nason and Lawler (2009) reported that the benefits of seeding reactors used for precipitative softening include increases in the rates of precipitation and flocculation, leading to a dramatic reduction in the number of small particles formed compared to homogeneous nucleation and a corresponding shift in the particle mass distribution toward larger particle sizes.

In addition to the factors mentioned previously (failure to reach equilibrium and the presence of dissolved complexes of Ca), elevated values of TOTCa(aq) in the finished water can be caused by low temperatures. The standard enthalpy of reaction for precipitation of $CaCO_3(s)$, $\Delta \overline{H}_r^\circ$, is -12.5 kJ/mol. Therefore, according to the van't Hoff equation (Equation 9-16), the solubility of $CaCO_3(s)$ increases with decreasing temperature. For example, K_{s0} is 44% larger at 5°C than at 25°C.

The preceding discussion indicates that meeting the target value for hardness in a precipitative softening process (typically, around 2 meq/L, or 100 mg/L as $CaCO_3$ for potable water) is not particularly difficult, and that, with careful process control, substantially lower concentrations can be reached. However, in light of the unavoidable limitations of equilibrium modeling for predicting the overall performance of a softening process (those mentioned above, as well as temporal variability of the raw water quality), reagent doses in full-scale plants tend to be based on a combination of calculations and local experience. Also, as a practical matter, because pH is such a critical parameter and is more easily measured than either TOTCa or $TOTCO_3$, reagent dosages are often adjusted online to achieve a target pH, rather than based on stoichiometric calculations.

In some situations, especially for groundwaters with high hardness, it can make sense to treat only a portion of the influent to achieve very good hardness removal and then blend that treated solution with the remainder of the influent to meet the overall treatment objective. The blending not only dilutes the hardness in the untreated water but also uses the untreated water to partially neutralize the basicity of the softened water so that less acid is required to lower the pH to the acceptable range. This approach, known as *split-stream softening*, can save on both capital costs (because the treatment system can be designed for a smaller flow) and operating costs (because the reagent requirements are reduced). Split-stream softening is often a viable option for treating groundwaters with high hardness, but it is rarely appropriate for hard surface waters, where the entire flow would have to be treated to remove suspended particulate matter and perhaps natural organic matter.

Recarbonation of Softened Water One component of the reagent requirements for softening that has been referred to in the preceding discussion, but not quantified, arises from the need to lower the solution pH after the hardness ions have been removed. Although this neutralization can be accomplished by addition of a strong acid to the water, a common alternative method involves injecting $CO_2(g)$ into the water downstream of the precipitation step. This approach, known as *recarbonation*, has the advantages that it does not add more salt ions to the water (since hard waters often already contain substantial amounts of salt, it is desirable to avoid adding yet more) and it lowers the pH without lowering the alkalinity (which is already quite low after the precipitation step). It also provides a way to increase $TOTCO_3$ and precipitate some of the Ca remaining in solution after lime addition, and thereby serves as an alternative to addition of at least some soda ash. This point is apparent in Figure 9-17, which shows the changes in solution composition accompanying absorption of various amounts of CO_2 into the softened water characterized in Figure 9-16. For this analysis, it was assumed that lime was added at a dose of 4 mmol/L to the raw water, the resulting solution reached equilibrium, and then the precipitated solids ($CaCO_3(s)$ and $Mg(OH)_2(s)$) were all removed from the suspension before CO_2 was injected.

Doses of CO_2 up to ~ 1.0 mmol/L in this example system induce almost stoichiometric precipitation of $CaCO_3(s)$, because most of the added CO_2 combines with OH^- to form CO_3^{2-}, which then reacts with Ca^{2+}. The dominant reactions during this stage of recarbonation are as follows:

$$CO_2 + 2\,OH^- \rightarrow CO_3^{2-} + H_2O \qquad (9\text{-}104)$$

$$Ca^{2+} + CO_3^{2-} \rightarrow CaCO_3(s) \qquad (9\text{-}105)$$

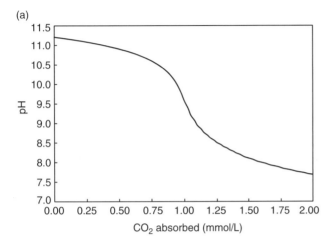

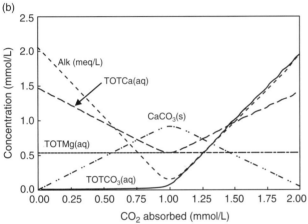

FIGURE 9-17. Changes in solution composition accompanying recarbonation of a softened water. For these calculations, the initial solution was assumed to have the composition corresponding to a lime dose of 4.0 mmol/L in Figure 9-16, but without any solids.

These reactions reduce the total dissolved Ca concentration from 1.47 to 0.54 mmol/L. The consumption of OH^- ions by the reactions causes the pH to decline, and the precipitation reaction reduces the alkalinity of the solution (by removing CO_3^{2-}). As the pH approaches pK_{a2} for the carbonate acid/base system (pH 10.33), a substantial portion of the dissolved CO_3^{2-} becomes protonated, causing $CaCO_3(s)$ to become undersaturated and start dissolving, according to the following reaction:

$$CaCO_3(s) + CO_2 + H_2O \rightarrow Ca^{2+} + 2HCO_3^- \quad (9\text{-}106)$$

If the newly precipitated $CaCO_3(s)$ (from the early parts of the recarbonation process) was not separated from the solution, this reaction would release Ca^{2+} back into solution and reverse the alkalinity loss that occurred previously. In fact, with additional CO_2 absorption, the reaction would continue until all the calcite that had formed at the beginning of the recarbonation process dissolved. For this example, that point would be reached after ~2.0 mmol/L CO_2 had entered solution, and the pH had dropped to 7.70.

In practice, CO_2 injection is limited to minimize dissolution of the newly formed $CaCO_3(s)$, and a second solid/liquid separation step is implemented at the condition near the peak in $CaCO_3(s)$ formation. Noting that Figure 9-17 characterizes a system in which no soda ash was added in the initial softening step, we see that a trade-off exists between adding soda ash in the first step and thereby reaching the desired, low total hardness concentration after a single stage of solid-/liquid separation at high pH, versus adding less or no soda ash in that step, but constructing two solid/liquid separation steps (one at high pH, and one after recarbonation). Based on equilibrium calculations, single-stage softening yields a softer water, but one with a higher Na^+ concentration. The hardness remaining in the water after two-stage softening is higher than that after single-stage softening, but is still in the acceptable range for most drinking water applications.

As with the softening process, recarbonation is generally carried out to a target final pH, rather than by controlling the CO_2 dose directly. Throughout the recarbonation process, all the TOTMg(aq) in the softened water remains in solution since the initial solution was saturated with $Mg(OH)_2(s)$ and the addition of CO_2 lowers the OH^- activity.

Figure 9-17 suggests that changes induced by recarbonation, like those in the softening process, could be described quite successfully using a stoichiometric precipitation model. However, as is true for the precipitation step, that approach does not provide the comprehensive understanding or insights that can be acquired by equilibrium-based modeling, nor does it provide the flexibility to easily consider alternative dosing scenarios. For this reason, and in light of the ready accessibility of chemical equilibrium software, designers and practitioners are gradually making greater use of the latter approach.

Precipitation of Other Metal Carbonates and Hydroxy-Carbonates Precipitation of solids containing carbonate ions is important in a number of contexts other than softening. One of the most pertinent of these is in water distribution systems in which the water comes into contact with lead. Lead is rarely present at concentrations of concern in the source waters used for water supply, but it can enter the water by corrosion of pipes or "service lines" (the small-diameter pipes between the water main and individual residences). In addition, brass and bronze meters and valves often contain lead, as did the solder that was commonly used in household plumbing until a few years ago. Lead can leach from any of these sources into the water, sometimes accumulating to problematic levels, especially in water that has been stagnant and in contact with the source of the lead for an hour or more (Edwards and Dudi, 2004).

The lead in the source materials is metallic; that is, it is in the form Pb(s), and it enters solution only after being oxidized to the divalent (Pb^{2+}) form.[13] The Pb^{2+} that is generated either forms solid phases on the pipe surface (for instance, the mineral litharge [PbO(s)] has frequently been found in the scales adjacent to metallic lead) or it can migrate into the water flowing through the pipe. The transport of Pb^{2+} is generally impeded by other corrosion products (i.e., scales) that have formed on the pipe surface, so the Pb^{2+} concentration increases in the solution immediately adjacent to the pipe. The Pb^{2+} can also react with OH^- and CO_3^{2-} ions in the solution to form both soluble complexes and, potentially, Pb-containing solids. This possibility is enhanced by fact that the electrochemical reduction of oxygen, which is an essential part of the corrosion reaction, generates OH^-, and therefore causes the pH near the pipe to be higher than in the bulk solution. For typical drinking water conditions, the lead solids that are most likely to precipitate are cerussite ($PbCO_3(s)$, $K_{s0} = 10^{-13.1}$) and hydrocerussite ($Pb_3(CO_3)_2(OH)_2(s)$, $K_{s0} = 10^{-45.5}$). Depending on the pH and alkalinity of the water, these solids are expected to place an upper limit on TOTPb(aq) of a few tens to a few hundreds of μg/L (Figure 9-18).

The concentration of soluble Pb in equilibrium with cerussite and hydrocerussite, while low, is still substantially greater than is considered acceptable in potable water. For example, the USEPA's Lead and Copper Rule (USEPA, 2007) specifies an "action level" for lead of 15 μg/L.[14] Fortunately, the slow diffusion of Pb species from the pipe walls (or from the precipitated solids) into the bulk solution and the subsequent dilution into the bulk flow usually prevent the soluble lead concentration in the bulk water from reaching the same levels at which it might exist near the walls. As a result, in most systems, the only locations where such high concentrations of soluble lead are found are the water near the source of the lead, and then only after periods of stagnation. However, extremely high lead concentrations can be found intermittently in tap water at some locations due to release of chunks of metallic lead (Lytle et al., 1993) or particulates containing oxidized lead (Edwards and Dudi, 2004; Triantafyllidou et al., 2007).

Corrosion "inhibitors" are sometimes added to the water to reduce lead solubility. Typically, these inhibitors contain orthophosphates and sometimes polyphosphates, which partially hydrolyze in the distribution system to form orthophosphates. In some cases, formulations containing zinc orthophosphate ($Zn_3(PO_4)_2(s)$) are used, but whether the zinc provides additional benefits is unclear. Orthophosphate-based inhibitors do not affect the electrochemical corrosion

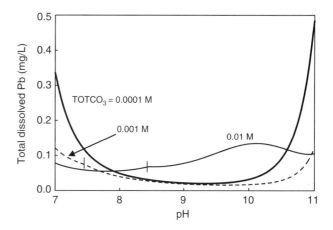

FIGURE 9-18. The total maximum soluble lead concentration in solutions equilibrated with carbonate-containing solids over a range of pH values and total dissolved carbonate concentrations. Ionic strength was fixed at $2 \times TOTCO_3$ for each simulation. The short vertical lines indicate the pH where the solid that is predicted to be present at equilibrium changes from cerussite (at lower pH) to hydrocerrusite (at higher pH); at 10^{-4} M $TOTCO_3$, hydrocerrusite is the solubility-controlling solid over the whole pH range shown. *Source*: After Schock and Lytle (2011).

process directly, but rather control lead solubility by promoting the precipitation of low-solubility, phosphate-containing solids near the pipe walls (Figure 9-19). Over time, the layer that contains these solids tends to become an increasingly effective barrier to diffusing species, thereby reducing the rate at which oxygen reaches the surface and decreasing the corrosion rate.

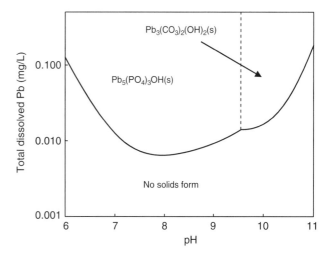

FIGURE 9-19. Predicted solubility of lead in a model drinking water containing 5×10^{-4} M $TOTCO_3$ and 0.5 mg/L $TOTPO_4$ (added as a corrosion inhibitor). No Pb-containing solid is expected in the region below the solid curve. In the upper regions, the solution is supersaturated with respect to the solid indicated, and on the lines separating different regions, the solution and/or solids on the two sides of the lines are in equilibrium. *Source*: After Schock and Lytle (2011).

---

[13] In some systems, the lead is oxidized to Pb(IV), which forms a highly insoluble oxide ($PbO_2(s)$). The Pb in this oxide might later be reduced to Pb(II) and enter solution.

[14] If a certain fraction (currently, 10%) of water samples at consumers' taps exceed the action level, the utility is required to take some measures to reduce the risk to public health. These measures might include notifying the public and/or altering the water composition to reduce its corrosivity.

Other Solids with pH-Dependent Metal and Ligand Speciation

For solids that contain cationic and anionic components that both exhibit pH-dependent changes in speciation, and in systems in which those components can react with other solutes as well as each other, prediction of the solution composition and the conditions under which solids are expected to precipitate is difficult without the aid of chemical equilibrium software. Even though these calculations would be daunting to carry out manually, the principles underlying the calculations are identical to those described in the preceding sections, and the data can often be readily interpreted once the calculations are complete.

Consider, for example, a wastewater containing $10^{-6}\,M$ TOTCd which is to be treated to reduce TOTCd(aq) to 2 μg/L ($1.8 \times 10^{-8}\,M$) or less. Like most heavy metals, Cd forms a fairly insoluble hydroxide solid, and also soluble complexes with OH^- ions. Assuming that $Cd(OH)_2$(am) forms when the pH of the solution is raised, the Cd solubility as a function of pH is as shown by the top curve in Figure 9-20a. In the hypothetical solution, $Cd(OH)_2$(am) is predicted to first form at pH 10.2, and then to continue precipitating as the pH is raised to 11.2. If the pH is raised even higher, the solid starts dissolving (due to the formation of $Cd(OH)_3^-$ and $Cd(OH)_4^{2-}$ complexes), and at pH 12.8 or higher, all the Cd in the system is once again dissolved. Even at the pH of minimum solubility, only ~70% of the Cd is predicted to precipitate. Thus, precipitation of $Cd(OH)_2$(s) is incapable of achieving the treatment objective.

Now, consider what happens if the solution contains various concentrations of sulfide species (H_2S(aq), HS^-, and S^{2-}, which we represent collectively as S(II)). Like OH^-, S^{2-} forms a solid with Cd^{2+} ("greenockite", with formula CdS(s)) and also forms soluble complexes ($CdHS^+$, $Cd(HS)_2^0$, etc.). However, the Cd—S bonds in both the solid and the complexes are considerably stronger than the corresponding Cd—OH bonds. The expected, equilibrium behavior of the Cd is shown for two, less-than-stoichiometric doses of TOTS(II) in Figure 9-20a. In each case, at pH slightly >2.0, CdS(s) precipitates until essentially all the sulfide is removed from solution. Thus, ~20% of the TOTCd precipitates when TOTS(II) is $2 \times 10^{-7}\,M$, and 50% precipitates when TOTS(II) is $5 \times 10^{-7}\,M$. The remaining Cd stays in solution until the solubility of $Cd(OH)_2$(s) is exceeded, at which point the curves in the systems with sulfide coincide with those in the sulfide-free case.[15] When the solubility of $Cd(OH)_2$(am) exceeds that of CdS(s) on the alkaline side of the minimum solubility, the sulfides once again precipitate.

The qualitative pattern of Cd solubility versus pH changes dramatically when the dose of TOTS(II) exceeds the

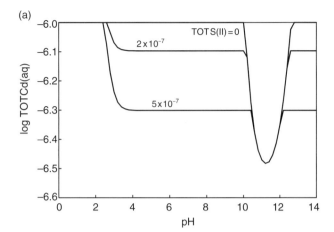

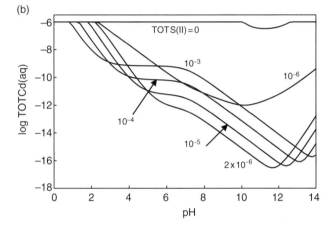

FIGURE 9-20. Solubility of Cd in a solution with $10^{-6}\,M$ TOTCd and a range of (a) low or (b) high concentrations of TOTS(II).

stoichiometric requirement, as shown in Figure 9-20b. In this case, the possibility of meeting the treatment target (if the system reaches equilibrium and the solids are efficiently separated from the solution) exists over a wide pH range. Furthermore, in these systems, CdS(s) is always less soluble than $Cd(OH)_2$(s), and the latter solid never forms.

When TOTCd = TOTS(II) = $10^{-6}\,M$, the Cd solubility curve is bowl-shaped, much as in systems where metal hydroxides precipitate. As in those systems, the bowl reflects the fact that the concentration of free aquo Cd^{2+} and cationic complexes (e.g., $CdOH^+$ and $CdHS^+$) decrease with increasing pH, whereas the concentrations of anionic complexes of Cd^{2+} (e.g., $Cd(OH)_3^-$ and $Cd(HS)_4^{2-}$) follow the opposite trend. At low pH, the former trend dominates, and at high pH, the latter one does.

When TOTS(II) is twice the stoichiometric dose (i.e., $2 \times 10^{-6}\,M$), even more CdS(s) precipitates at every pH, so TOTCd(aq) declines across the full pH range. However, when TOTS(II) is increased further, the trend reverses, and Cd becomes more soluble with increasing TOTS(II), at least up to pH 12. The explanation for this reversal is that, while

---

[15] When $Cd(OH)_2$(s) is less soluble than CdS(s), the CdS(s) dissolves, and $Cd(OH)_2$(s) is the only solid present. The details of this process are explained in most water chemistry texts.

the additional sulfide increases the driving force for CdS(s) precipitation, it does the same for formation of soluble Cd–HS complexes; the calculations tell us that, when TOTS(II) is greater than $\sim 2 \times 10^{-6}$, the effect on formation of Cd–HS complexes overwhelms the effect on CdS(s) formation, so solubility increases. The curves for TOTS(II) concentrations of 2×10^{-6} to 10^{-3} M also exhibit a shoulder that, in all cases, ends near pH 7. This shoulder reflects the influence of the acid/base chemistry of S(II) on the system, specifically the fact that pK_{a1} for H_2S is 7.0.

As noted, the calculations required to generate Figure 9-20 would be virtually unmanageable without the aid of a computer. However, no matter how complex the system, the governing equations are the mass balances and equilibrium relationships among the species, so that once the calculations have been carried out, the results are often readily interpretable.

Effects of Complexing Ligands on Metal Solubility

The discussion to this point has considered only OH^- and the anion that forms the solid as species that might form soluble complexes with metal cations. However, many other ligands might be present in the solution, and these ligands always increase the equilibrium solubility of the metal; indeed, if the affinity of such ligands for the metal is large enough, and they are present at sufficiently high concentration, they can prevent the solid from forming. In those cases, the metal might be precipitated by adding a different anion that combines with it to form a less soluble solid, adding a different metal that outcompetes the target metal for the complexing ligand, or adding a reagent (e.g., an oxidant) that reacts with and destroys the ligand.

As an example of the potential effects of strong complexing agents, consider again the solution containing 20 mg/L Zn (3.1×10^{-4} M) that was analyzed earlier in the chapter. As was shown in Figure 9-11, increasing the pH of the solution to 9.7 causes $Zn(OH)_2$(s) to be supersaturated, and, once equilibrium is attained, reduces TOTZn(aq) to ~ 0.45 mg/L (7×10^{-6} M). That calculation, however, assumes that the only soluble complexes of Zn that form in the solution are $Zn(OH)_x^{2-x}$ species. If a chelating agent that has a strong affinity for Zn^{2+}, such as EDTA, were present in the solution, it could also form complexes with Zn^{2+}, thereby lowering the activity of Zn^{2+} and reducing the level of supersaturation. In fact, the affinity of EDTA for Zn^{2+} is so great (log K for formation of $ZnEDTA^{2-}$ is 18.0) that essentially all the EDTA in solution binds with Zn^{2+} ions, at least up to pH ~ 11; if EDTA is present at equal or greater concentration than TOTZn, no $Zn(OH)_2$(s) will precipitate (Figure 9-21).

As noted above, approaches that might be pursued to precipitate zinc from such a solution include adding a precipitating ligand that can outcompete the EDTA for the Zn^{2+} (e.g., S(II)), adding a metal that can outcompete

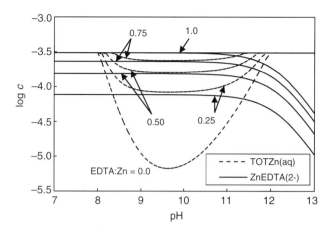

FIGURE 9-21. Effect of EDTA on Zn solubility. TOTZn $10^{-3.51}$ M (20 mg/L). Numbers indicate molar ratios of TOTEDTA to TOTZn.

the Zn^{2+} for EDTA, or destroying the EDTA in a chemical reaction. Although several other heavy metals have greater affinity than Zn^{2+} for EDTA, they are likely to be equally or more problematic as a constituent of the treated solution, so that option would probably be unattractive. Similarly, destruction of EDTA is extremely difficult, although it can be accomplished by some of the advanced oxidation processes described in Chapter 10. Thus, in this particular example, the only realistic option for precipitating the Zn would be to add a ligand that forms a less soluble solid than $Zn(OH)_2$(aq).

In some cases, complexing ligands or other reagents are added specifically to inhibit precipitation. For example, precipitation of calcite ($CaCO_3$(s)) and/or gypsum ($CaSO_4$(s)) is often of concern in reverse osmosis systems. Calcite formation can usually be controlled in such systems by lowering the solution pH, but control of gypsum precipitation requires addition of chemicals other than acids or bases; such chemicals are commonly referred to as anti-scalants. Anti-scalants, which are often proprietary mixtures of polyacrylates and polycarboxylates, or polyphosphates and polyphosphonates, are typically added at low dosages to inhibit scale formation or to ensure that any scale that does form is easily removed. The mechanisms by which anti-scalants disrupt the precipitation process are not completely understood, but they probably involve reactions with surface functional groups that interfere with the nucleation process (Shih et al., 2004; Shih et al., 2005; Rahardianto et al., 2006).

Precipitation Resulting from Redox Reactions

Some relatively common treatment processes in environmental engineering involve a combination of precipitation and oxidation/reduction (redox) reactions. For example, in the anoxic environments common to groundwaters, iron and manganese usually occur as Fe(II) and Mn(II), respectively.

In the presence of oxygen, chlorine, or other oxidizing agents, these reduced species can be oxidized to Fe(III) and Mn(IV), respectively, which are very insoluble and therefore tend to precipitate upon oxidation. These kinds of processes combine the redox reactions discussed in Chapter 10 with the metal hydroxide precipitation reactions presented in this chapter.

Treatment to remove soluble chromium from solution, which is of particular importance in wastewaters from the electroplating industry, is another example of a combined redox and precipitation process. In contrast with iron and manganese, however, the oxidized form, Cr(VI) (present primarily as $HCrO_4^-$ and CrO_4^{2-}), is much more soluble than the reduced form, Cr(III). Thus, upon reduction of Cr(VI) species, $Cr(OH)_3(s)$ tends to precipitate. Sulfur dioxide gas and sodium metabisulfite are commonly used as reducing agents in such processes. Sulfur dioxide poses greater health risks and requires more expensive metering equipment, but it is less expensive, so it is often used in larger systems (where the operators typically have more training). Chromium reduction with either reagent is more effective under acidic conditions, so the pH is typically lowered to between 2.0 and 3.0 with sulfuric acid before the reducing agent is added.

9.6 SOLID DISSOLUTION REACTIONS

In this section, we consider dissolution of solids; that is, the reverse process from that considered in the preceding portions of the chapter. Dissolution is important in numerous environmental engineering applications, as a desired outcome in some cases and a process to be minimized in others. A few examples include the release of soluble lead from lead carbonate and lead oxide solids in household plumbing; the potential recovery of aluminum from the $Al(OH)_3(s)$ that precipitates during coagulation with alum; the removal of mineral scales from reverse osmosis membranes; removal of solids containing transuranic elements from contaminated soils to prevent the release of those elements into groundwater; and the deterioration of concrete in tanks and concrete-lined pipes. Dissolution is also used widely in analytical procedures, for example, to determine the total concentration of metals in environmental samples.

Despite its widespread importance, dissolution has not been widely studied in environmental engineering, and no comprehensive understanding or modeling framework for characterizing the process (other than calculating the extent of dissolution expected at equilibrium) has been adopted. The understanding that does exist derives primarily from studies of geochemical weathering processes, in which rocks release solutes and are gradually transformed to soils and sediments. This work has been summarized and placed in the context of traditional water chemistry principles by Stumm (1992), and much of the following discussion is based on that work and the references therein.

In many dissolution reactions that occur in nature, solutes do not enter solution in the same stoichiometric ratios as exist in the solid. In such cases, the reaction is described as *incongruent dissolution*, and the products of the reaction include a new solid phase. For example, the mineral albite is converted to the clay montmorillonite via release of sodium and silica but not aluminum from the solid, via the following reaction:

$$2\,NaAlSi_3O_8(s) + 2\,H_2CO_3 + 4\,H_2O$$
$$\leftrightarrow Al_2Si_4O_{10}(OH)_2(s) + 2\,Na^+ + 2\,HCO_3^- + 2\,H_4SiO_4$$

The simpler solids that are commonly encountered in water and wastewater treatment processes typically undergo *congruent dissolution* and so are not converted into a different solid in the process.

Similar to solid precipitation, dissolution requires a sequence of transport and chemical reaction steps, including transport of reactants from the bulk solution to a point adjacent to the solid surface; adsorption of the reactants; one or more chemical reactions that detach constituents of the solid from the lattice and convert them to adsorbed species; desorption of the reaction products; and transport of the products to the bulk solution. The chemical reactions that convert species that are part of the solid lattice to adsorbed solutes often involve protonation, electron transfer (i.e., either oxidation or reduction of one of the atoms being released), and/or formation of a metal–ligand complex.

Adsorption and desorption reactions are assumed to reach equilibrium rapidly compared to the other steps in the sequence, so they are not considered to be the rate-limiting reactions for dissolution. Rather, the main rate limitation is usually imposed by transport steps for highly soluble solids and by chemical reaction steps for relatively insoluble solids. In the former case, the solute concentrations are assumed to be in equilibrium with the solid at the interface, and lower in bulk solution. The dissolution rate is controlled by the rate at which the solutes migrate through the liquid boundary layer and is proportional to the extent of solid-/solution disequilibrium (equal to $c_{surf} - c_{bulk}$) and the mass transfer coefficient (k_{mt}):

$$r_{diss} = k_{mt,i}(c_{i,surf} - c_{i,bulk}) \quad (9\text{-}107)$$

where i can be any of the reactants or products that exist in solution (e.g., a constituent released by the dissolving solid or a complexing agent that must bind to a metal ion in the solid to bring it into solution). Typically, the resistance to diffusion of larger solutes through the boundary layer is greater than that opposing diffusion of smaller solutes, so the dissolution rate is controlled by migration of the larger species.

In transport-controlled dissolution, the dissolution rate can be increased by increasing the intensity of mixing or by other modifications that reduce the resistance to transport of the rate-limiting reactant or product. However, in engineered water systems, transport-limited dissolution reactions typically proceed rapidly relative to the hydraulic residence time, so the solution composition can be calculated based on equilibrium with the solid, and the rate of dissolution is relatively unimportant. Thus, for example, if the temperature or the pH of water in a water distribution network decreases, calcite that is present on pipe walls dissolves rapidly enough to cause the water in the pipes to be in equilibrium with the calcite at the new temperature or pH shortly after the change occurs. As a result, most of the interest in solid dissolution rates focuses on relatively insoluble solids; in water and wastewater treatment, such solids include most metal oxides, hydroxides, carbonates, sulfides, and phosphates. Dissolution rates of these solids are controlled by reactions at the surface, most often the reactions that actually break the bonds in the lattice and convert the constituents to adsorbed species, which can then desorb and enter solution.

When dissolution is controlled by the rate of a surface reaction, the characteristic time for that reaction can be substantially longer than the hydraulic residence time in a treatment process. Under these conditions, the composition of the solution is approximately constant from the bulk solution up to the solid/solution interface, and intensifying the mixing of the fluid has almost no effect on dissolution. Rather, the dissolution rate can be altered only by changing the composition of the surface species; that is, the adsorbed reactants, intermediates, and products of the dissolution reaction. Such dissolution reactions are commonly treated separately depending on whether the key dissolving reagent is protons (*proton-promoted dissolution*) or a complexing ligand (*ligand-promoted dissolution*). Distinctions are also made based on whether the solid-phase cation (typically, a metal) is chemically reduced as part of the dissolution reaction (*reductive dissolution*). However, the reaction steps are fundamentally similar in all these cases, and they can all be understood in the context of a unified framework.

Dissolution by each of these mechanisms can proceed simultaneously at different types of surface sites (e.g., the site-types shown schematically in Figure 9-1). Typically, the reactions are presumed to proceed more rapidly at sites with fewer bonds to the bulk solid. Since the reactions at different types of sites proceed in parallel, the overall reaction rate is dominated by the site with the largest reaction rate.

Proton-promoted dissolution has been studied most extensively for metal oxides and is closely related to the state of protonation of surface oxide sites. As noted in Chapter 7, such sites can undergo reversible acid/base reactions which are commonly modeled as follows:

$$\equiv SOH \leftrightarrow \equiv SO^- + H^+ \qquad (9\text{-}108)$$

$$\equiv SOH + H^+ \leftrightarrow \equiv SOH_2^+ \qquad (9\text{-}109)$$

Protonation of a surface oxide weakens the bond between the oxide and the adjacent metal ion. If these bonds weaken sufficiently, the metal can break off of the lattice and be converted to an adsorbed ion, which can then desorb and enter the solution. A schematic representation of such a process is shown in Figure 9-22.

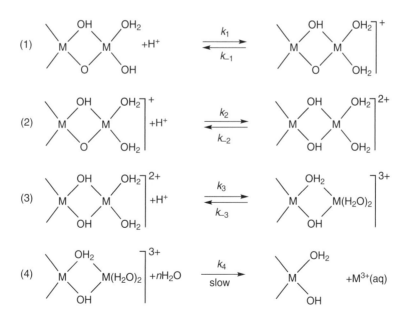

FIGURE 9-22. Schematic representation of the sequential protonation of oxide surface sites, leading to release of a structural metal ion, M^{3+}. *Source*: From Furrer and Stumm (1986), reprinted with permission.

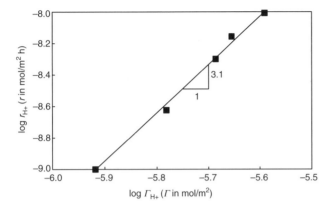

FIGURE 9-23. Dependence of the proton-promoted dissolution rate of an aluminum oxide (δ-Al_2O_3) on the adsorption density of

The reaction order of such processes with respect to the H^+ adsorption density (Γ_{H^+}) is commonly reported to be >1 and is often close to the charge on the central metal ion, suggesting that several oxide ions surrounding the central metal ion must be protonated before the metal breaks off of the lattice. For example, as shown in Figure 9-23, the proton-promoted dissolution rate of δ-Al_2O_3(s) is approximately third order with respect to Γ_{H^+}. (For such an analysis, Γ_{H^+} is defined to be 0 when the amount of H^+ bound to the surface is sufficient to cause the solid to have zero net surface charge.)

Lowering the pH of the solution increases the protonation of surface sites and hence facilitates the dissolution reaction. However, because the speciation of surface acid/base groups is relatively insensitive to changes in solution pH, the reaction order with respect to the activity of *dissolved* H^+ (i.e., $a_{H^+,\text{solution}}$, equal to 10^{-pH}) is typically much lower than the order with respect to Γ_{H^+}. For instance, Figure 9-23 indicates that the reaction order of δ-Al_2O_3 dissolution with respect to a_{H^+} is only ~ 0.33, compared to the reaction order of 3 with respect to Γ_{H^+}. Reaction orders with respect to a_{H^+} for proton-promoted dissolution of many oxides have been reported to be between 0 and 0.5.

In ligand-promoted dissolution of metal oxides, the Me—O bonds are broken because the metal is attracted more strongly to the ligand than to the structural oxide ion. As a result, the metal enters solution as a complex with the ligand, not as a hydrated, free metal ion. A schematic for the dissolution of a metal oxide promoted by a simple organic ligand (oxalate ion, with a formula of $HOOC$–COO^-) is shown in Figure 9-24. The oxalate adsorption reaction (Step 1) is presumed to reach equilibrium quickly. The metal-oxide bonds are then broken as a metal–oxalate complex forms and desorbs (Step 2). Finally, the oxide ions that had been in the second layer of the solid and are now at the surface become protonated to reproduce a surface that is essentially identical to the original one (Step 3).

In one study, the ligand-promoted dissolution rate of δ-Al_2O_3(s) was found to be directly proportional to the adsorption density of the ligand, Γ_L, for several organic ligands (Furrer and Stumm, 1986); that is, the order of the reaction with respect to Γ_L was 1.0. The data for oxalate-promoted dissolution are shown in Figure 9-25. As is true for proton-promoted dissolution, the reaction order with respect to Γ_L is typically higher than that with respect to the dissolved ligand concentration, c_L. For example, the two upper lines in Figure 9-25 indicate that a 50% increase in the oxalate-promoted dissolution rate was induced by a 50% increase in Γ_{Oxal}, but that these changes required a fivefold increase in c_{Oxal}.

Inorganic ligands can also promote dissolution; of special interest are OH^- ions, which can cause solids to dissolve rapidly at pH values where $Me(OH)_x^{n-x}$ complexes are significant dissolved species. Thus, contributions to the overall reaction rate can be made by surface sites that

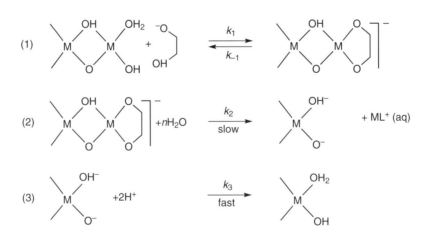

FIGURE 9-24. Schematic representation of the oxalate-promoted nonreductive dissolution of a metal oxide solid. *Source*: From Furrer and Stumm (1986); reprinted with permission.

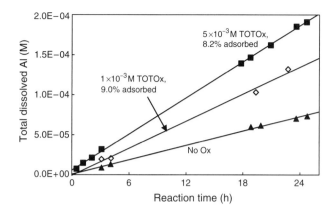

FIGURE 9-25. Effects of oxalate (Ox) concentration and adsorption on the dissolution rate of $\delta\text{-Al}_2\text{O}_3(s)$ at pH 3.5. *Source*: After Furrer and Stumm (1986).

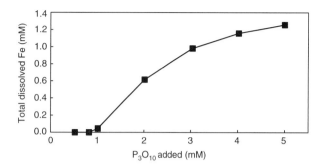

FIGURE 9-26. Equilibrium dissolved Fe in systems with various amounts of P_3O_{10} added. TOTFe $= 10^{-3}$ M, pH $= 5.0$. *Source*: After Lin and Benjamin (1990).

are protonated, deprotonated, and complexed with ligands; a baseline dissolution rate (k_{H_2O}) also exists under conditions where $\Gamma_{H^+} = \Gamma_{OH^-} = \Gamma_L = 0$. Stumm (1992) suggested that this rate can be treated as zero-order, so that the overall rate of nonreductive dissolution can be represented as follows:

$$r_{\substack{\text{overall,}\\\text{nonreductive}}} = k_{H^+}(\Gamma_{H^+})^a + k_{OH^-}(\Gamma_{OH^-})^b + k_L(\Gamma_L) + k_{H_2O}$$

(9-110)

where, in most cases, a and b are both >1.

Equation 9-110 describes the rate at which bonds between structural ions and the solid lattice are broken, which is a logical criterion for defining dissolution. However, the experimentally accessible parameter, and the one that is most often of practical interest, is the rate at which the constituents of the solid enter solution. In some cases, the dissolving ions remain adsorbed to the solid, so the solid "dissolves" (i.e., the lattice bonds break) without releasing its constituents to the solution. For example, tripolyphosphate ($P_3O_{10}^{5-}$) can cause ferrihydrite to dissolve by forming Fe–P_3O_{10} complexes. However, if TOTP$_3$O$_{10}$ is low compared to TOTFe, the complexes can adsorb, so that negligible Fe is present in solution at equilibrium. Such a scenario applies at TOTP$_3$O$_{10}$ < 1 mmol/L in Figure 9-26. Similar results have been reported by Szecsody et al. (1994) for dissolution of Fe oxides coated on sand when they were exposed to EDTA.

A related sequence of reaction steps is involved in reductive dissolution reactions. Perhaps, the most important of these reactions in environmental engineering are those that lead to dissolution of iron and manganese oxyhydroxides by reduction of trivalent (ferric) iron and tri- or tetravalent manganese to their divalent states (ferrous and manganous species, respectively). The Fe(II)- and Mn(II)-containing solid phases that might plausibly form in nature or water treatment systems are several orders of magnitude more soluble than the oxyhydroxides of the more oxidized metals, so the reduction reaction is usually accompanied by dissolution of the solid. In natural systems, these reactions occur in anaerobic water layers (e.g., the hypolimnion of a fresh or marine water body) and sediments, and they play a central role not only in the geochemical cycling of Fe and Mn, but also of trace elements that adsorb to the solids (metals and phosphorus) (Stumm and Morgan, 1996).

As is the case for Fe and Mn, the reduction of uranium and transuranic elements frequently increases their solubility dramatically; these reactions therefore play an important role in the movement of those elements away from sites where they are present naturally or have been released due to anthropogenic activity, and these reactions must be considered in the design of remediation strategies for those elements. Similarly, the reduction of Pb(IV) (which forms highly insoluble PbO$_2$(s)) to much more soluble Pb(II) species in drinking water distribution systems has been implicated as a possible explanation for the release of Pb into the water supply when the oxidation potential of the water declines because of a switch in disinfection practices (Rajasekharan et al., 2007).

The rates of reductive dissolution reactions depend strongly on the speciation of sites on the solid surface, so they bear many similarities to nonreductive ligand-promoted dissolution; the main difference is that, at some point in the sequence of reactions, a species that approaches or adsorbs to the surface donates one or more electrons to the solid. In some cases, the reaction requires, or at least is catalyzed by, absorption of light by the reductant (Waite et al., 1988). Once transferred, the electrons change the oxidation state of metal ions in the solid structure and thereby dramatically reduce their affinity for the surrounding ions.

Zinder et al. (1986) divided these reactions into two categories. In *proton-promoted reductive dissolution*, a reductant donates an electron to a solid in which the electron is relatively mobile. The reductant, which has been oxidized, plays no further role in the process and, presumably, returns to the bulk solution. Because of its mobility in the solid, the electron does not necessarily reduce a metal ion at the site where it entered the solid, but rather moves around until it encounters a surface metal ion that is "primed" to become reduced because of protonation of the adjacent oxide groups. The electron reduces that ion, causing the bonds holding the metal in the lattice to break and converting the metal into an adsorbed ion, which then desorbs and enters solution.

In *ligand-promoted reductive dissolution*, on the other hand, a reactant adsorbs to a site on the solid and donates one or more electrons directly to the surface metal ion at that site; because it both binds to a metal ion and donates electrons, the reactant is both a ligand and a reducing agent in this scenario. When the electrons are transferred, the metal ion is reduced, the ligand is oxidized, and again, both are converted to forms that are more likely to enter (or, in the case of the ligand, reenter) solution. As in nonreductive, ligand-promoted dissolution, the rate of the overall reaction is likely to be proportional to the adsorption density of the ligand/reductant. However, because reduction reactions are frequently accompanied by binding of H^+ to the reduced species, the reaction rate might depend on q_{H^+} in addition to q_L. Several other scenarios for reductive dissolution mechanisms and rate laws have been discussed by Stone and Morgan (1987), and reductive and nonreductive dissolution rates mediated by various ligands have been compared by Klewicki and Morgan (1999).

To summarize the key points in this section, solid dissolution is always a multistep process that includes mass transport, adsorption, and breakage of lattice bonds, and might include electron transfer and formation or dissociation of metal–ligand complexes as well. For relatively soluble solids, such as calcite, the overall dissolution rate can be controlled by transport of reactants or products through the liquid boundary layer adjacent to the solid. In most such cases, the characteristic time for the reaction is short compared to the hydraulic residence time in treatment processes, so an assumption that the solution is in equilibrium with the solid is usually justified. Characteristic times for dissolution of less soluble solids tend to be longer, so that they might not equilibrate with the solution in the time frame of typical water and wastewater treatment processes.

Any solid that is in contact with a solution that is undersaturated with respect to the solid is likely to react simultaneously at a variety of sites and via a variety of mechanisms. Sites that have fewer bonds to the bulk solid are likely to have higher intrinsic dissolution rate constants than those with more bonds. The contribution of a given type of site to the overall dissolution rate reflects the combined effect of its rate constant and its abundance on the surface.

Dissolution mechanisms that are significant in environmental engineering applications include some that are accelerated by high surface concentrations of protons, organic and inorganic ligands (including OH^-), and reducing agents. Relatively simple rate laws have been proposed for each of these mechanisms as a function of the surface concentration of a key species. However, even in those cases, the dependence of the rate on the more accessible variables of interest—the concentrations of the reactants in the bulk solution—can be complex and nonintuitive. The overall dissolution process can be complicated further by the role of light (which can catalyze reductive dissolution reactions), adsorption of the products of the reaction onto the remaining solid, or slow release of critical ligands via dissociation of pre-existing metal–ligand complexes. Although it has not been discussed explicitly, it should be clear that altering the water quality and modifying the speciation on a solid surface is a viable way not only for promoting, but also interfering with solid dissolution (e.g., by adding a ligand that adsorbs strongly but does not form strong soluble complexes with the constituents of the solid) (Stone and Morgan, 1984).

Because dissolution is rarely a critical issue in treatment processes, it has not been studied much in that context. However, an understanding of dissolution is relevant in at least some relatively uncontrolled environmental engineering applications, such as when complexing agents, acids, or bases are injected into the ground as part of a remediation process, or when changes in treatment processes alter the quality of water in the distribution system and thereby initiate dissolution of pipe scales. In those situations, the knowledge that has been acquired from studies of dissolution in natural environmental (i.e., geochemical) systems can be applied, but additional research that is applicable to the specific systems of interest will also be needed.

9.7 REACTORS FOR PRECIPITATION REACTIONS

As has been noted, the primary objective of many conventional precipitation processes in water and wastewater treatment is to remove dissolved contaminants from solution and achieve acceptable concentrations in the effluent. By far, the most common reactor type used for these processes is a CFSTR, in which the raw water is mixed rapidly with the precipitating agents and solids form almost instantly. The reagents are typically mixed in a highly turbulent regime for a short time (up to a minute or so), after which mixing continues at a reduced intensity for up to a few tens of minutes to provide time for additional precipitation and particle growth (by both accretion and agglomeration). Additional reagents might be added to facilitate particle collisions and growth of flocs in this second step, as described in Chapters 11 and 12.

As was also noted previously, the effects of reactor design and operation on process parameters other than the concentration of the target compound, such as the properties of the precipitated solids, have received relatively little attention in environmental engineering. Indeed, even the terminology used to describe these solids—residuals or sludge—suggests the low level of importance that has been ascribed to them historically. Recently, however, issues surrounding residuals management have gained prominence, because of the costs of processing and disposing of the solids (e.g., thickening, dewatering, and landfilling), the potential hazards that the solids pose, and the potential value of some of the solids in reuse or recycling scenarios (Tsang, 2004; Roth et al., 2008).

In response to these issues, design and operation strategies for precipitation processes are beginning to incorporate a life-cycle management approach that considers multiple objectives, including meeting the regulatory requirements for residuals management and minimizing reagent doses and energy use. An understanding of the fundamental steps in the precipitation process (primary and secondary nucleation, growth, agglomeration, and fracturing of particles) can help identify ways to improve reactor performance and achieve these multiple objectives.

Seeding the reactors with solids to provide surface nucleation sites can often lower the residual dissolved concentration of the target species, induce formation of larger, denser solids that settle more readily, and possibly also reduce reagent requirements. For example, the initial rate of Ca^{2+} precipitation (as $CaCO_3(s)$) in a water softening process was reported to increase from about 0.7 mg Ca^{2+}/L min in the absence of seed particles to around 2 mg Ca^{2+}/L min when 3 g/L of sludge from an operating precipitation softening plant was added (Chao and Westerhoff, 2002). In another study, seeding increased the efficiency of $CaCO_3(s)$ precipitation, but seed crystals from a water-softening facility were not as effective as pure calcite crystals, even though the former had a higher specific surface area (Mercer et al., 2005). It was suggested that organic matter, magnesium, and other impurities in the solids reduced the effectiveness of the seed particles from the operating facility.

Fluidized-bed reactors (FBRs) have also been investigated for various precipitation processes. Successful operation of such reactors requires a careful balance among the water quality parameters, the pellet concentration and size, and the water velocity through the reactor. For a given mass-based solids concentration, decreasing pellet diameter increases the amount of surface area available (on which the precipitation reaction can occur), but increases the risk of washout. By combining such a process with membranes for solid/liquid separation (discussed in Chapter 15), a process can be operated with small (high surface area) pellets and minimal risk of particle washout.

Guillard and Lewis (2001) noted several advantages of nickel carbonate precipitation in an FBR over that of conventional hydroxide precipitation in a CFSTR, including formation of a dense precipitate that was easily separated from solution, production of pellets that could be recovered and potentially reused, and the ability to implement the precipitation at lower pH, with lower chemical costs. Li et al. (2004, 2005) have also explored the use of FBRs for precipitation reactions, both for the removal of metal ions and for water softening.

Another FBR, the Crystalactor®, developed by the Dutch company DHV in the late 1970s, has been used commercially for water softening and for phosphate removal. In the phosphate removal process, calcium phosphate precipitates on the surface of sand particles that are added as a seed material. Carryover of small particles into the process effluent is sometimes a challenge, although that problem can often be overcome using conventional, granular media filtration to remove the particles (discussed in Chapter 14) (Eggers et al., 1991; Morgenroth, 2000).

Biological wastewater treatment processes can be designed and operated to enhance bacterial uptake of phosphate from the solution to help meet increasingly strict limitations on the total phosphorus concentration of the treatment plant effluent. The biosolids from such plants are usually stabilized in an anaerobic digester and then processed through a centrifuge, generating a solution with a total phosphorus concentration that can exceed 100 mg/L. By adding Mg^{2+} and other reagents to the phosphate-rich feed solution, a controlled precipitation process can be carried out that produces pellets of struvite ($MgNH_4PO_4 \cdot 6H_2O$), which can be used as a fertilizer. The process can be implemented in an FBR, where the feed solution and added reagents contact struvite particles that have precipitated previously and that serve as seed particles. Particles that grow large enough sink to the bottom of the bed, where they are removed.

In one such process designed to produce struvite from swine manure, the waste is sparged with air to purge CO_2, increase the pH, and provide mixing, and $MgCl_2$ is then added to induce struvite formation (Shepherd et al., 2007). In a related FBR process, known as the Phosnix® process, effluent from an anaerobic digester is fed to the base of the reactor, where NaOH and $MgCl_2$ or $Mg(OH)_2$ is added (Muench and Barr, 2001). The reactor contents are mixed and fluidized using air-sparging, and struvite particles form and grow in the bed until they become large enough to sink to the bottom of the reactor, from where they are periodically removed. In another approach (the REM-NUT® process), wastewater is treated by sequential cation and anion exchange to remove ammonium and phosphate, respectively (Liberti et al., 2001). Sodium chloride (NaCl) is then used to regenerate both ion exchange resins, the two regenerant solutions are mixed, and struvite is precipitated from the mixed regenerant by adding chemicals to achieve the desired Mg:N:P ratio and pH.

9.8 SUMMARY

Precipitation and, to a lesser extent, dissolution reactions are critical components of environmental engineering systems. These reactions are often intentionally promoted to achieve a treatment objective; in addition, they sometimes proceed in treatment or distribution systems in an uncontrolled manner. Precipitation is commonly used in water and wastewater treatment operations to remove hardness from drinking water or process water, iron and manganese from anoxic water supply sources, phosphate from domestic and agricultural wastewaters, and toxic metals from industrial wastewaters. The precipitated solids of greatest significance in these systems include $CaCO_3(s)$, $Fe(OH)_3(s)$, $MnO_2(s)$, hydroxyapatite, struvite, and a variety of metal hydroxides, oxides, sulfides, and carbonates. In addition, precipitation of $Fe(OH)_3(s)$ or $Al(OH)_3(s)$ is the central step in many coagulation processes (discussed in Chapter 11), and precipitation is a critical, albeit usually uncontrolled, phenomenon in a number of piping systems, where solids might form as a result of pipe corrosion or supersaturation of the solutions in the pipes.

Conceptual models of homogeneous precipitation consider the process to involve formation of soluble clusters of the precipitating species, which then grow into nuclei that have a (conceptually) well-defined interface with solution. Some nuclei dissolve, but others grow into stable particles. This process has been studied extensively in systems where the goal is to produce, to the extent possible, particles with uniform properties. However, in most environmental engineering processes, heterogeneous precipitation (i.e., precipitation onto existing particles that are much larger than the nuclei formed in homogeneous precipitation) dominates.

Similar to other reactions described in previous chapters, both thermodynamic and kinetic constraints control the extent to which precipitation and dissolution reactions occur in a given system. The thremodynamic constraint is quantified by the equilibrium constant for a reaction involving the solid; the most conventional reaction used in these cases is the solubility product, K_{s0}. If the activity quotient for the reaction, Q_{s0}, is greater than K_{s0}, precipitation of the solid is favored, and if Q_{s0} is less than K_{s0}, dissolution is favored. Although it is common to use K_{s0} values as though they apply to particles of any size, K_{s0} is the equilibrium constant only for the limiting case of an infinitely large particle, and smaller particles are more soluble than is suggested by this limiting case. The increase in solubility with decreasing particle size is usually negligible for particles that have grown to a few tens of nanometers, but it can be substantial for smaller particles, including nuclei in the early stages of growth. As a result, even if a solution is supersaturated with respect to formation of a large particle of a given solid, the formation of a small particle of the same solid might be thermodynamically unfavorable. The critical particle size required for precipitation to be thermodynamically favorable decreases with increasing supersaturation (i.e., increasing Q_{s0}) and can be predicted based on a theoretical analysis, at least for idealized conditions.

In cases where precipitation is thermodynamically favored, growth of existing particles occurs by a sequence of steps that includes transport of the precipitating species from bulk solution to the vicinity of the solid and then through the boundary layer at the fluid/solid interface. These transport steps are followed by adsorption onto the existing solid and a chemical reaction that incorporates the species into the solid. The resistances associated with transport through bulk solution and adsorption are usually considered negligible compared to those associated with the other steps, so either transport through the boundary layer or incorporation into the solid is assumed to be the rate-limiting step. In either case, the overall rate of precipitation is predicted to be proportional to the amount of solid surface area available and to the extent of disequilibrium between the bulk solution and the solid. If the disequilibrium is large, the extent of disequilibrium can often be approximated simply as the concentration of one of the precipitating species in the bulk solution. Combining a rate expression for precipitation with information about the hydrodynamics of a reactor can yield a prediction for the steady-state particle size distribution.

When a precipitation reaction proceeds, it alters the concentrations of the precipitating species directly, but it also alters the ionic strength of the solution and can have significant indirect effects on the concentrations of other solutes (e.g., if the reaction changes the solution pH). In early precipitation literature, these secondary effects were often ignored or accounted for only superficially, but more sophisticated analyzes are becoming prevalent, thanks largely to the availability of software that can model solution composition considering all the reactions simultaneously. This trend is evident in the approach to water softening. Early analyzes of the softening process used a stoichiometric model that, in essence, assumed that added reagents would react quantitatively with specific target species in the water. This model was used successfully for several decades to compute reagent doses that yielded softened water with acceptable quality. However, the approach was not useful for predicting the sensitivity of the final water quality to changes in the reagent doses, and hence could not be used to model alternative treatment scenarios or optimize the treatment process when the trade-off between costs and final water quality was considered. Chemical equilibrium modeling, while still idealized compared to the real process (because it ignores reaction kinetics and because it is limited by our incomplete knowledge about the solution composition, the crystal structure of the solids that form, and the equilibrium constants for many solids), provides a more complete perspective on the multiple and interacting reactions that occur in such systems, as well as the opportunity to explore a wider range of treatment alternatives. Similar statements apply to modeling of precipitation processes targeting removal of other species from solution.

The reactors in which controlled precipitation reactions are carried out are overwhelmingly designed as CFSTRs, both to ensure that the precipitating reagents are well-mixed into the water to be treated and to distribute the solids throughout the reactor and thereby provide widespread opportunities for the target species to be removed by heterogeneous precipitation. Fluidized bed reactors and sludge blanket reactors (essentially, fluidized beds with slowly settling solids and correspondingly low upflow velocities) are also fairly common.

The steps involved in solid dissolution are, in many ways, the reverse of the steps involved in precipitation. In particular, the rate-limiting step in dissolution processes is usually assumed to be either the breakage of surface bonds to convert a molecule of the solid to an adsorbed species, or transport of that species across the boundary layer surrounding the particle and into the bulk solution. Dissolution reactions are used much less frequently than precipitation reactions in water quality engineering, but they are important in specialized circumstances, such as the release of the solutes from the products of corrosion and the removal of scale from fouled membranes.

Even though precipitation and dissolution reactions have long been part of water and wastewater treatment systems, surprisingly little comprehensive analysis of these processes has been conducted. One reason for this lack of attention is that precipitation processes commonly used in water and wastewater treatment have provided cost-effective ways to achieve the primary treatment objectives. Those objectives, however, are changing, and interest is growing in meeting more stringent water quality standards while reducing energy requirements and cost. In addition, designers and operators are paying more attention to the residuals generated from precipitation processes, both in terms of management costs and potential resource recovery.

REFERENCES

Benjamin, M. (2010) *Water Chemistry*. Waveland Press, Long Grove, IL.

Bockris, J. O'M., and Reddy, A. K. M. (1998) *Modern Electrochemistry*. 2nd ed., Plenum Press, NY.

Chao, P. F., and Westerhoff, P. (2002) Assessment and optimization of chemical and physicochemical softening processes. *JAWWA*, 94 (3), 109–119.

Chiang, P., and Donohue, M. C. (1988) The effect of complex ions on crystal nucleation and growth. *J. Colloid Interface Sci.*, 126 (2), 579–591.

CRC (2008), In Lide, David R. (ed.), *Handbook of Chemistry and Physics*, 88th ed., National Institute of Standards and Technology, Washington, DC.

Davey, R. J. (1979) The control of crystal habit. In E. J. de Jong and S. J. Jancic, (eds.), *7th Symposium on Industrial Crystallization 78*. North Holland Publishing Company, Amsterdam, pp. 169–183.

Davey, R. J. (1982) Role of additives in precipitation processes. In S. J. Jancic and E. J. de Jong, (eds.), *8th Symposium on Industrial Crystallization 81*. North-Holland Publishing Company, Amsterdam, pp. 123–135.

Edwards, M., and Dudi, A. (2004) Role of chlorine and chloramine in corrosion of lead-bearing plumbing materials. *JAWWA*, 96 (10), 69–81.

Eggers, E., Dirkzwager A. H., and Van der Honing H. (1991) Full-scale experiences with phosphate crystallization in a crystalactor. *Water Sci. Technol.*, 23, 819–824.

Furrer, G., and Stumm, W. (1986) The coordination chemistry of weathering. I. Dissolution kinetics of δ-Al_2O_3 and BeO. *Geochim. et Cosmochim Acta*, 50, 1847–1860.

Guillard, D., and Lewis, A. E. (2001) Nickel carbonate precipitation in a fluidized-bed reactor. *Ind. Eng. Chem. Res.*, 40 (23), 5564–5569.

Gustafsson, J-P. (2011) *Visual Minteq*, version 3.0, updated Oct. 28, 2011. Accessed from http://www.lwr.kth.se/English/OurSoftware/vminteq/on Jan.12, 2012.

Hochella, M., Jr., and White, A. F. (1990) Mineral-water interface geochemistry: an overview. In Hochella, M. Jr. and White, A. F. (eds.), *Reviews in Mineralogy Volume 23: Mineral-Water Interface Geochemistry*, Mineralogical Society of America, Washington, D.C. pp. 1–16.

Hounslow, M. J., Ryall R. L., and Marshall V. R. (1988) A discretized population balance for nucleation, growth, and aggregation. *AIChE J.*, 34 (11), 1821–1832.

Jensen, J. N. (2003) *A Problem-Solving Approach to Aquatic Chemistry*. John Wiley & Sons, New York, NY.

Karpinski, P. H. and Wey, J. S. (2002) Precipitation processes. In Myerson, A. S. (ed.) *Handbook of Industrial Crystallization*, 2nd ed., Butterworth-Heinemann, Woburn, MA, pp. 141–160.

Klewicki, J. K., and Morgan, J. J. (1999) Dissolution of β-MnOOH particles by ligands: pyrophosphate, ethylenediaminetetraacetate, and citrate. *Geochim. Cosmochim. Acta*, 63, 3017–3024.

Kumar, S., and Ramkrishna, D. (1996) On the solution of population balance equations by discretization—I. A fixed pivot technique. *Chem. Eng. Sci.*, 51 (8), 1311–1332.

Langmuir, D. (1997) *Aqueous Environmental Geochemistry*. Prentice Hall, Upper Saddle River, NJ.

Li, C-W., Jian, J.-C., and Liao, J-C. (2004) Integrating membrane and fluidized bed pellet reactor for hardness removal. *JAWWA*, 96 (8), 151–158.

Li, C-W., Liang, Y. Mi., and Chen, Y. M. (2005) Combined ultrafiltration and suspended pellets for lead removal. *Separation and Purification Technol.*, 45, 213–219.

Liberti, L., Petruzzelli, D., and De Florio, L. (2001) REM NUT ion exchange plus struvite precipitation process. *Environ. Technol.*, 22, 1313–1324.

Lin, C-F., and Benjamin, M. M. (1990) Dissolution kinetics of minerals in the presence of sorbing and complexing ligands. *Environ. Sci. Technol.*, 24, 126–134.

Luo, B., Patterson, J. W., and Anderson, P. R. (1992) Kinetics of cadmium hydroxide precipitation. *Wat. Res.*, 26 (6), 745–751.

Lytle, D. A., Schock, M. R., Dues, N. R., and Clark, P. J. (1993) Investigating the preferential dissolution of lead from solder particulates. *JAWWA*, 85 (7), 104–110.

Mercer, K. L., Lin, Y. P., and Singer, P. C. (2005) Enhancing calcium carbonate precipitation by heterogeneous nucleation during chemical softening. *JAWWA*, 97 (12), 116–125.

Morel, F. M. M., and Hering, J. G. (1993) *Principles and Applications of Aquatic Chemistry*. Wiley-Interscience, NY.

Morgenroth, E. (2000) Opportunities for nutrient recovery in handling of animal residuals. In *Proceedings of the Water Environment Federation Animal Residuals Management 2000*, Kansas City, Missouri, Nov 12–14, 2000, pp. 531–543, Water Environment Federation, Alexandria, VA.

Muench, E. V., and Barr, K. (2001) Controlled struvite crystallization for removing phosphorus from anaerobic digester sidestreams. *Water Res.*, 35 (1), 151–159.

Nason, D. A., and Lawler, D. F. (2009) Particle size distribution dynamics during precipitative softening: declining solution composition. *Water Res.*, 43, 303–312.

Nielsen, A. E. (1964) *Kinetics of Precipitation*. Pergamon Press, Oxford.

Nyvlt, J., Söhnel, O., Matuchova, M., and Broul, M. (1985) *The Kinetics of Industrial Crystallization*. Elsevier, Amsterdam, The Netherlands.

Pankow, J. F. (1991) *Aquatic Chemistry Concepts*. Lewis Publishers, Chelsea, MI.

Rahardianto, A., Shih, W. Y. Lee, R. W., and Cohen, Y. (2006) Diagnostic characterization of gypsum scale formation and control in RO membrane desalination of brackish water. *J. Membr. Sci.*, 279, 655–668.

Rajasekharan, V. V., Clark, B. N., Boonsalee, S., and Switzer, J. A. (2007) Electrochemistry of free chlorine and monochloramine and its relevance to the presence of Pb in drinking water. *Environ. Sci. Technol.*, 41, 4252–4257.

Randolph, A. D., and Larson, M. A. (1988) *Theory of Particulate Processes. Analysis and Techniques of Continuous Crystallization*. 2nd ed., Academic Press, Inc., San Diego, CA.

Roth, D. K., Cornwell, D. A., Russell, J. S., Gross, M., Malmrose, P. E., and Wancho, L. (2008) Implementing residuals management: Cost implications for coagulation and softening plants. *JAWWA*, 100 (3), 81–93.

Schock, M. R., and Lytle, D. A. (2011) Internal corrosion and deposition control, Chapter 20. In Edzwald, J. K. (ed.), *Water Quality and Treatment*, 6th ed., McGraw Hill, New York, pp. 20.1–20.103.

Schwartz, A. M. and Myerson, A. S. (2002) Solutions and solution properties, Chapter 1. In Myerson, A. S. (ed.), *Handbook of Industrial Crystallization*, 2nd ed., Butterworth-Heinemann, Woburn, MA. pp. 1–32.

Shepherd, T. A., Burns, R. T., Moody, L. B., Raman, D. R., and Stalder, K. J. (2007) Development of an air sparged continuous flow reactor for struvite precipitation from two different liquid swine manure storage systems, International Symposium on Air Quality and Waste Management for Agriculture, Proceedings of the 16–19 September 2007 Conference (Broomfield, Colorado), ASABE Publication Number 701P0907cd.

Shih, W. Y., Albrecht, K., Glater, J., and Cohen, Y. (2004) A dual probe approach for evaluation of gypsum crystallization in response to antiscalant treatment. *Desalination*, 169, 213–221.

Shih, W. Y., Rahardianto, A., Lee, R. W., and Cohen, Y. (2005) Morphometric characterization of calcium sulfate dehydrate (gypsum) scale on reverse osmosis membranes. *J. Membr. Sci.*, 252, 253–263.

Sohnel, O., and Garside, J. (1992) *Precipitation. Basic Principles and Industrial Application*. Butterworth-Heinemann Ltd., Oxford.

Stone, A. T., and Morgan, J. J. (1984) Reduction and dissolution of manganese(III) and manganese(IV) oxides by organics. I. Reaction with hydroquinone. *Environ. Sci. Technol.*, 18, 450–456.

Stone, A. T., and Morgan, J. J. (1987) Reductive dissolution of metal oxides, Chapter 9. In Stumm, W. (ed.), *Aquatic Surface Chemistry*, Wiley-Interscience, New York, pp. 1–41.

Stumm, W. (1992) *Chemistry of the Solid-Water Interface*. Wiley-Interscience, New York.

Stumm, W., and Morgan, J. J. (1996) *Aquatic Chemistry*, 3rd ed., Wiley-Interscience, New York.

Szecsody, J. E., Zachara, J. M., and Bruckhart, P. L. (1994) Adsorption-dissolution reactions affecting the distribution and stability of Co-EDTA in iron oxide-coated sand. *Environ. Sci. Technol.*, 28, 1706–1716.

Taguchi, K., Garside, J., and Tavare, N. S. (1996) Nucleation and growth kinetics of barium sulphate in batch precipitation. *J. Crystal Growth*, 163, 318–328.

Triantafyllidou, S., Parks, J., and Edwards, M. (2007) Lead particles in potable water. *JAWWA*, 99 (6), 107–117.

Tsang, R. (2004) Introducing the water treatment plant residuals management committee. *JAWWA*, 96 (4), 50–52.

USEPA (2007) National Primary Drinking Water Regulations For Lead And Copper: Short-Term Regulatory Revisions And Clarifications. Federal Register (October 10, 2007), 72, 195, 57781-57820, http://www.epa.gov/fedrgstr/EPA-WATER/2007/October/Day-10/w19432.htm, Accessed Sep 29, 2010.

Waite, T. D., Wrigley, I. C., and Szymczak, R. (1988) Photoassisted dissolution of a colloidal manganese oxide in the presence of fulvic acid. *Environ. Sci. Technol.*, 22, 778–785.

Waite, T. D., Davis, J. A., Payne, T. E., Waychunas, G. E., and Xu, N. (1994) Uranium(VI) adsorption to ferrihydrite: application of a surface complexation model. *Geochimica Cosmochimica Acta*, 58, 5465–5478.

Walton, A. G. (1967) *The Formation and Properties of Precipitates*. Wiley Interscience, New York, NY.

Zhang, J., and Nancollas, G. H. (1990) Kink densities along a crystal surface step at low temperatures and under nonequilibrium conditions. *J. Crystal Growth*, 106 (2–3), 181–190.

Zinder, B., Furrer, G., and Stumm, W. (1986) The coordination chemistry of weathering. II. Dissolution of Fe(III) oxides. *Geochim. et Cosmochim. Acta*, 50, 1861–1869.

PROBLEMS

9-1. After passing through some old pipes containing lead, a solution contains 30 µg/L Pb^{2+}. The pH is 7.8, and the alkalinity of the solution is 275 mg/L as $CaCO_3$. If the system equilibrates rapidly and if the solutes behave ideally, will all the lead remain in solution or will some of it precipitate?

9-2. To minimize corrosion and the consequent release of lead and copper into the water supply, a utility has decided to increase the pH of their water to 9.3. The water is currently at pH 6.8 and contains 0.9 meq/L of alkalinity (same as 0.9 mmol/L). The water also contains 28 mg/L Ca^{2+}. They plan to achieve their goal by adding lime to the water. Assume ideal solution behavior.

(a) What is the total concentration of all carbonate species in the raw water?

(b) The lime dose required to raise the pH to the target value is 1.95×10^{-4} mol/L. Once the lime has been added, is $CaCO_3(s)$ likely to precipitate?

9-3. A municipal drinking water at pH 8.0 contains 3.22×10^{-6} mol/L $TOTPO_4$ (commonly designated as 0.1 mg/L $TOTPO_4$-P, meaning that the solution contains a total of 0.1 mg/L P, present in various phosphate species). After leaving the treatment plant, the water flows through galvanized pipes, which release Zn^{2+} when they corrode. What is the maximum concentration of Zn^{2+} that can be dissolved in the solution without causing $Zn_3(PO_4)_2(s)$ ($K_{s0} = 10^{-35.42}$) to precipitate? Assume ideal behavior of all solutes. The acid dissociation constants for H_3PO_4 are $K_{a1} = 10^{-2.15}$, $K_{a2} = 10^{-7.20}$, and $K_{a3} = 10^{-12.38}$.

9-4. A water source is at pH 7.4 and has alkalinity and Ca^{2+} concentrations of 1.1×10^{-3} equiv/L and 85 mg/L, respectively.

(a) Is $CaCO_3(s)$ undersaturated, saturated, or supersaturated in the solution? Assume the solutes behave ideally.

(b) What is the minimum (stoichiometric) dosage of lime (mol/L) needed to convert all the carbonate species to CO_3^{2-} (i.e., if the OH^- from the lime could be forced to react exclusively with carbonate species)?

(c) If the lime dosage computed in part (b) were added and the solution then reached equilibrium with $CaCO_3(s)$, how much solid would precipitate, in milligrams per liter? Assume that the approximation that all the carbonate species in the original solution are converted to CO_3^{2-} is acceptable both before and after the precipitation reaction occurs.

9-5. An industrial waste solution flowing at 150 L/min is at pH 2.0 and contains 10^{-2} mol/L $TOTZn^{2+}$. You wish to reduce this concentration to 1.0 mg/L by precipitating $Zn(OH)_2(s)$. Assuming the precipitation reaction proceeds to equilibrium and that the solution is ideal:

(a) How high must the pH be raised to achieve the treatment goal? Consider the reactions shown in Table 9-6.

(b) How much solid $Zn(OH)_2(s)$ will have to be landfilled per day?

9-6. The raw water supply for a community contains 18 mg/L total particulate matter. It is to be treated by addition of 60 mg alum $(Al_2(SO_4)_3 \cdot 14 H_2O)$ per liter of water treated. Because precipitation of $Al(OH)_3(s)$ generates acid (see Example 9-11), the pH decreases when the alum is added.

(a) For a total flow of 7500 m^3/day, compute the daily alum requirement, the concentration of solids in the water following alum addition, and the daily mass of sludge that must be disposed of, assuming the alum all precipitates as $Al(OH)_3(s)$.

(b) The water is initially at pH 7.5 and has [Alk] = 40 mg/L as $CaCO_3$. Essentially all the alkalinity is attributable to the carbonate acid/base system. It is desired to maintain a pH of at least 6.5 in the coagulated solution. Will chemical (base) addition be necessary when the alum is added? Assume ideal behavior of solutes.

(c) You should have found in part (b) that base will in fact have to be added to maintain the solution pH at 6.5 or higher. If the base that is added is NaOH, how much is required? Assume that no CO_2 is exchanged between the solution and the atmosphere during the alum addition and reaction.

(d) An alternative to addition of caustic for raising the pH is to bubble the solution with air after the alum addition and thereby strip out CO_2. Will it be possible to adjust the solution to pH 6.5 or higher using this approach? (Hint: recall what happens to alkalinity when CO_2 is added to or removed from a solution.)

9-7. Seeding a precipitative softening reactor with calcite can sometimes increase the efficiency of the process. Precipitation on the pre-existing solids requires transport of calcium to the particle surface (carbonate is also required, but calcium supply seems to limit the reaction), followed by the precipitation reaction, in which the ions are incorporated into the solid. A number of researchers (e.g., Chao and Westerhoff, 2002) have modeled the overall reaction as irreversible and second order with respect to the calcium concentration ($r_{\text{calcite}} = k_2(\text{Ca}^{2+})^2$).

In a system containing a bulk Ca^{2+} concentration of 2×10^{-3} mol/L, the overall reaction rate normalized to the calcite surface area is 4×10^{-6} mol/m² min. The diffusion coefficient for Ca^{2+} in the solution is 7.9×10^{-6} cm²/s, the diameter of the seed particles is 3 μm, and the water temperature is 20°C. For the experimental conditions, the mass transfer coefficient for Ca^{2+} transport through the liquid boundary layer around the particles can be estimated based on the following correlation for mass transfer during flow around a sphere:

$$k_{\text{mt}} = \frac{D_i}{d_p}\left[2 + 0.31\left(\frac{d_p}{\mu_L}\right)\left(\frac{D_i \rho_L}{\mu_L}\right)^{0.67}\left(\frac{\Delta\rho\,\mu_L g}{\rho_L\,\rho_L}\right)^{0.33}\right]$$

where $\Delta\rho = (\rho_p - \rho_L)$, and the other parameters have their usual meanings. The density of calcite is 2.71 g/cm³.

(a) What is the surface-normalized rate constant for the precipitation reaction, in L²/(m² min mol)?

(b) What is the mass transfer coefficient (m/s) for Ca^{2+} transport through the boundary layer in this system?

(c) What is the concentration (mol/L) of calcium at the particle/water interface? What do you infer about the resistance associated with mass transfer compared to that associated with the incorporation step? Briefly explain how you reached your conclusion.

9-8. The chemical composition of a groundwater supply for a community is summarized in the table below.

Constituent	Concentration[a]	Constituent	Concentration
Calcium	65	Alkalinity (as CaCO₃)	272
Magnesium	29		
Sodium	38	Chloride	35
Potassium	12	Nitrate (as N)	1.4
Iron(III)	0.36	Sulfate	36
		pH	7.5

[a] All values in mg/L, except pH.

(a) Carry out a charge balance on the reported composition.

(b) Estimate the lime (Ca(OH)_2) dose and, if necessary, soda ash (Na_2CO_3) dose for stoichiometric softening with a target pH = 10.9, assuming that CaCO_3(s) is the only solid that precipitates. Is the assumption valid? Assume ideal solution behavior.

(c) Estimate the concentration (g/L) of solids produced according to the stoichiometric precipitation model, if the reagent doses are those computed in part (b), and CaCO_3(s) is the only solid that precipitates.

(d) What is the predicted value of the total hardness remaining in solution, in mg/L as CaCO_3, for the conditions specified in part (c)?

(e) Suggest a dosing strategy based on the stoichiometric precipitation model that would reduce the total hardness to 50 mg/L as CaCO_3. (Hint: choose doses such that almost all the residual hardness is attributable to Mg.)

(f) Use a chemical equilibrium program to determine the equilibrium composition of the solution that would be generated by adding the reagent doses calculated in part (b). Allow the program to account for nonideal behavior of solutes, and to consider possible precipitation of both CaCO_3(s) and Mg(OH)_2(s). Compare the output for the solution pH and the concentration of solids with your answers to part (b).

9-9. The centrate from an anaerobic digestion process is at pH 7.0 and contains 78 mg/L TOTPO₄ as P (i.e., PO₄–P), 720 mg/L TOTNH₄ as N (NH₄–N), and 15.5 mg/L TOTMg. This solution is to be treated in a fluidized bed reactor to precipitate struvite ($pK_{s0} = 13.0$). The reactor is operated with a high recycle ratio (the ratio of the recycle flow rate to the feed flow rate), so the solution in the reactor can be assumed to be well-mixed. The reactor is to be operated at steady state with a struvite saturation index (SI, defined as log $[Q_{s0}/K_{s0}]$) of 1.0 in the well-mixed solution. Assume all solutes behave ideally.

(a) The desired SI can be achieved by adding MgCl_2, increasing the pH, or both. Prepare a figure showing the Mg^{2+} concentration in solution required to achieve the desired SI, over the range $7 \leq \text{pH} \leq 10$.

(b) The effluent from the precipitation reactor will be recycled to the biological treatment units. The target TOTPO₄-P concentration in this stream is 3.0 mg/L. Determine how much struvite (g/L) will be produced, and prepare a plot similar to that in part (a), but showing the required Mg^{2+} dose (mol/L, added as MgCl_2) as a function of pH, rather than the Mg^{2+} concentration in solution.

(c) Briefly discuss factors you would consider in deciding whether to add MgCl_2, NaOH, or both to accomplish the design goals.

9-10. Industrial waste solutions sometimes contain strong metal-complexing agents such as EDTA, which have been added to prevent metal precipitation in the process solutions. Although the formation of complexes is desirable in the production process, it can interfere with precipitation and removal of the metals during treatment of the waste streams. Consider a waste solution containing 325 µg/L TOTCu (5×10^{-6} mol/L). It is proposed to treat the solution by mixing with an alkaline solution that will increase the pH to 7.5 and cause the solution to contain 2×10^{-3} mol/L TOTCO$_3$, causing the mineral malachite (Cu$_2$(OH)$_2$CO$_3$(s), $K_{s0} = 10^{-33.47}$) to form. (Note: this problem can be solved manually with relatively few calculations, but it is also readily solved using computer equilibrium software.)

(a) If the treated solution reaches equilibrium, what would the concentrations of Cu^{2+} ion and total dissolved Cu (TOTCu(aq)) be? Assume that the only dissolved Cu species of significance are Cu^{2+} and CuOH$^+$, and that the solutes behave ideally. What fraction, if any, of the TOTCu originally in solution precipitates, and what is the concentration (mg/L) of malachite in the equilibrium system? The equilibrium constant for formation of CuOH$^+$ (K_1) is $10^{6.50}$.

(b) Repeat the calculations in part (a), assuming the solution also contains EDTA at a total molar concentration equal to that of copper. The pK_a's for H$_4$EDTA are 2.2, 3.1, 6.2, and 11.0. Based on these values, assume that HEDTA^{3-} is present at a much higher concentration than all other H$_x$EDTA^{x-4} ($x = 0, 2, 3,$ or 4) species. Also, assume that the dominant Cu–EDTA complex is CuEDTA^{2-}, with a formation constant of $10^{20.5}$.

Note: to solve this portion of the problem manually, you must make and test an assumption about the presence of the solid at equilibrium. If you assume that the solid is present, you can then solve for the EDTA^{4-} concentration that causes TOTEDTA to equal 5×10^{-6} mol/L, and test whether the concentration of CuEDTA^{4-} that is present under those conditions is consistent with the assumption. Alternatively, you can assume that no solid is present at equilibrium and solve for the speciation by solving the mass balances on TOTCu(aq) and TOTEDTA(aq) simultaneously, and then determining if the Cu^{2+} concentration is consistent with the assumption. If either assumption is made initially and then found to be wrong, the alternative assumption can be made to solve for the equilibrium speciation.

9-11. Waite et al. (1994) explored the speciation and adsorption of hexavalent uranium (U(VI)) under various environmental conditions. The "core" form of U(VI) is UO$_2^{2+}$, and this species can then be complexed with OH$^-$ or CO$_3^{2-}$ to form a variety of other soluble species, as summarized below.

Reaction	Log K
UO$_2^{2+}$ + OH$^-$ ↔ UO$_2$OH$^+$	8.8
UO$_2^{2+}$ + 2OH$^-$ ↔ UO$_2$(OH)$_2^0$	16.0
UO$_2^{2+}$ + 3OH$^-$ ↔ UO$_2$(OH)$_3^-$	22.0
UO$_2^{2+}$ + CO$_3^{2-}$ ↔ UO$_2$CO$_3^0$	9.7
UO$_2^{2+}$ + 2CO$_3^{2-}$ ↔ UO$_2$(CO$_3$)$_2^{2-}$	17.0
UO$_2^{2+}$ + 3CO$_3^{2-}$ ↔ UO$_2$(CO$_3$)$_3^{4-}$	21.63
β-UO$_2$(OH)$_2$(s) ↔ UO$_2^{2+}$ + 2OH$^-$	−23.07

(a) A solution at pH 8.0 contains 10^{-6} mol/L TOTU(VI) and no dissolved carbonate. Assuming the solutes behave ideally, do you expect β-UO$_2$(OH)$_2$(s) to precipitate? If so, how much (in moles of solid per liter of solution)?

(b) Repeat part (a), but assume the ionic strength of the solution is 0.15.

(c) Repeat part (b), but now assume that the solution is in equilibrium with the atmosphere, so the activity of H$_2$CO$_3^0$ is fixed at 10^{-5} mol/L.

9-12. A drinking water at pH 6.7 has alkalinity of 5.0×10^{-4} equiv/L. The water comes into contact with and equilibrates with malachite (Cu$_2$(OH)$_2$CO$_3$(s)) and cerrusite (PbCO$_3$(s)), both of which have formed on pipe surfaces due to corrosion. The solubility products of these two solids are $10^{-33.47}$ and $10^{-13.2}$, respectively.

(a) Assuming that the interactions between the solution and the solids lead to release of Cu^{2+} and Pb^{2+}, but do not release enough CO$_3^{2-}$ or OH$^-$ to alter the alkalinity or pH of the water, compute the concentrations of Cu^{2+} and Pb^{2+} that would be found in the water downstream of the corroded pipes. Assume ideal behavior.

(b) Would either of these concentrations cause the water to exceed the Action Level under the Lead and Copper Rule? If so, how much would the pH have to be increased to cause the water to be in compliance with the rule?

9-13. A softening process is being implemented in an effort to precipitate both CaCO$_3$(s) and Mg(OH)(s) from a feed solution that contains 190 mg/L Ca^{2+} and 45 mg/L Mg^{2+}. Addition of 215 mg/L lime [Ca(OH)$_2$] raises the pH from 7.0 to 11.3. At the final pH, the approximation (CO$_3^{2-}$) ≈ 0.9 TOTCO$_3$ is reasonable. The alkalinity of the raw water is 3.0×10^{-3} equiv/L. Assume ideal behavior.

(a) What is the alkalinity of the water after the lime addition, but before any precipitation has occurred?

(b) Estimate the concentration of Mg^{2+} that will remain in the water at equilibrium, assuming that the pH remains at 11.3 as the system equilibrates.

(c) Estimate the concentration of Ca^{2+} that will remain in the water at equilibrium, making the same assumption as in part (b).

9-14. An industrial wastewater at pH 6.50, contains 3.0 mg/L TOTNi, and has an alkalinity of 6.0×10^{-4} equiv/L from carbonate species. You wish to reduce the total soluble Ni concentration to 0.10 mg/L. Consider the following possible reactions of Ni, and assume the solutes all behave ideally. (If you use chemical equilibrium software to solve this problem, chances are that it will consider additional species that would be difficult to include in a manual solution. Therefore, the answers you get might be different using the two solution approaches.)

Reaction	Log K
$Ni^{2+} + OH^- \leftrightarrow NiOH^+$	4.103
$Ni^{2+} + 2OH^- \leftrightarrow Ni(OH)_2^0$	9.004
$Ni^{2+} + 3OH^- \leftrightarrow Ni(OH)_3^-$	12.009
$Ni(OH)_2(s) \leftrightarrow Ni^{2+} + 2OH^-$	-15.11
$NiCO_3(s) \leftrightarrow Ni^{2+} + CO_3^{2-}$	-11.20

(a) If precipitation of $NiCO_3(s)$ is kinetically inhibited, to what pH would the solution have to be raised to achieve the treatment goal by precipitation of $Ni(OH)_2(s)$?

(b) Estimate the dose of NaOH that would be required to reach the pH determined in part (a).

(c) If seed particles are added to the solution that facilitate the formation of $NiCO_3(s)$, how high would the pH have to be to reach the treatment goal? (Hint: assume that the precipitation occurs in the region where HCO_3^- is the only significant dissolved species of carbonate, and use this assumption to derive a relationship between (CO_3^{2-}) in the treated solution and pH. Then use that expression and K_{s0} to find (Ni^{2+}) as a function of pH, and find the pH where TOTNi(aq) meets the treatment criterion. Finally, test your initial assumption about the carbonate speciation, to make sure that it was valid.)

(d) How much NaOH would be required in the scenario described in part (c)?

9-15. An industry proposes to use local well water as a process water source. The major ion composition of the water summarized below. The solution is at 25°C and pH 8.00. Evaluate a precipitation softening process for this water by addressing the following:

Constituent	mg/L	Constituent	mg/L
Ca^{2+}	99	HCO_3^-	145
Mg^{2+}	42	SO_4^{2-}	94
Na^+	121	Cl^-	326
K^+	14	NO_3^-	23

(a) Test the charge balance condition for this analysis.

(b) Estimate the concentrations of species in the carbonate acid/base group and of H^+ and OH^- in the raw water. Assume ideal solution behavior.

(c) Use the stoichiometric softening approach to estimate the dosages of lime and soda ash (if required) to remove calcium and magnesium at a process pH of 10.7.

(d) Estimate the concentrations of $CaCO_3(s)$ and $Mg(OH)_2(s)$ produced.

(e) The process will include a recarbonation step that brings the solution to equilibrium with atmospheric CO_2 at pH 8.5 when the water enters their distribution system. Assume that, because of kinetic limitations, the precipitation reactions do not reach equilibrium during recarbonation. As a result, the effluent contains 3 mg/L TOTMg (aq) and 12 mg/L TOTCa(aq). Estimate the major ion composition of the finished water.

9-16. Luo et al. (1992) reported that the particle growth rate in $Cd(OH)_2(s)$ precipitation could be described by the following function:

$$\dot{d}_p = (10^{-1.79} \, \mu m/min)(SS)^{0.78}$$

You wish to remove Cd^{2+} from an industrial wastewater by precipitation of $Cd(OH)_2(s)$ ($pK_{s0} = 13.64$). The first step will be pH adjustment in a tank that operates as a CFSTR at steady state, with a hydraulic residence time of 60 min.

(a) If the pH in the tank is 11.0, the total dissolved Cd concentration is 0.5 mg/L, and the ionic strength is 0.01 mol/L, what will the number-average particle size be in the suspension? Assume that particles grow only by precipitation and not aggregation. Consider the formation of $CdOH^+$ and $Cd(OH)_2^0$ complexes ($K_1 = 10^{3.90}$ and $\beta_2 = 10^{7.71}$).

(b) What fraction of the particles in the effluent will be larger than 1 μm in diameter?

10

REDOX PROCESSES AND DISINFECTION

T. David Waite, D. F. Lawler, and M. M. Benjamin

10.1 Introduction
10.2 Basic principles and overview
10.3 Oxidative processes involving common oxidants
10.4 Advanced oxidation processes
10.5 Reductive processes
10.6 Electrochemical processes
10.7 Disinfection
10.8 Summary
References
Problems

10.1 INTRODUCTION

Redox processes are those involving the transfer of electrons from one species, which thereby becomes oxidized, to another, which becomes reduced. Redox reactions have dramatic effects on the physical, chemical, and biochemical properties of the reacting species. In water quality engineering, redox processes are often used to convert reactants into forms that are either less toxic or more easily removed from the aqueous stream of interest, and they are also the primary means used to disinfect waters.

Water and wastewater treatment processes that involve redox transformations are presented in this chapter. Basic principles of redox transformations are introduced first, followed by detailed consideration of particular oxidants and reductants of importance in water and wastewater treatment. Both conventional redox processes and *advanced oxidation processes* (AOPs) are described. In conventional redox processes, the oxidant or reductant is a (relatively) stable chemical species that reacts directly with the target contaminant. In contrast, in AOPs, redox-active chemicals are used to generate extremely reactive and therefore highly unstable intermediates (hydroxyl free radicals or other radical-like entities), which then react with the target species.[1] Many redox processes have both conventional and advanced components.

When oxidants are used to disinfect waters, they are likely to react with many different types of molecules in the cell walls, protoplasm, and interior structures, so the reactions do not have unique or even identifiable target molecules. Nevertheless, the reactions do involve oxidation and reduction, and the overall disinfection process is often modeled as a simple chemical reaction between the oxidant and microorganisims. Section 10.7 on disinfection describes the use of oxidants in such processes.

10.2 BASIC PRINCIPLES AND OVERVIEW

Applications of Redox Processes in Water and Wastewater Treatment

In this section, the variety of redox-based processes used in water and wastewater treatment and the basic chemistry underlying all redox reactions are introduced. Details of these applications are given subsequently in this chapter.

---

[1] Free radicals are species in which an electron orbital is occupied by a single electron. Since, for most molecules, the pairing of electrons in orbitals greatly stabilizes the molecules, free radicals have a strong tendency to undergo redox reactions, either oxidizing other chemicals so that the radical acquires an electron to pair with the existing one, or reducing other chemicals and losing the isolated electron in the process.

Water Quality Engineering: Physical/Chemical Treatment Processes, First Edition. By Mark M. Benjamin, Desmond F. Lawler.
© 2013 John Wiley & Sons, Inc. Published 2013 by John Wiley & Sons, Inc.

Oxidation

Control of Iron and Manganese
In natural aquatic systems, iron in the +3 oxidation state (Fe(III)) and manganese in the +3 or +4 oxidation state (Mn(III,IV)) form extremely insoluble oxide minerals that reside in the sediments of water bodies.[2] If the water becomes anoxic, as often happens in groundwater or in the lower layers of lakes and reservoirs, the metals can be reduced to much more soluble Fe(II) and Mn(II) species (*ferrous* and *manganous* species, respectively), causing their concentrations in solution to increase to levels up to hundreds or thousands of μg/L. If such water is used as a drinking water source, the redox state of the solution is likely to become more oxidizing, converting the metals back to insoluble Fe(III) (*ferric*) and Mn(III,IV) (*manganic*) species. These metals can then precipitate as reddish-orange or black deposits that can stain fixtures or clothing.

The most common approach for avoiding the problems associated with such a sequence is to induce the oxidation to occur under controlled conditions in a treatment plant and remove the solids that are produced before the water enters the distribution system. Ferrous iron is relatively easily oxidized by addition of weak oxidants such as oxygen or chlorine, but oxidation of manganous ions normally requires the use of either more powerful oxidants or a catalyst (often, an oxide mineral) in conjunction with the addition of oxygen or chlorine. Alternatively, the source water is sometimes aerated *in situ* to avoid or reverse the reduction of the Fe- and Mn-containing solids.

It should be noted that, even if ferrous iron is oxidized, the resulting product may be very fine colloids of iron oxides. These colloids can acquire a negative surface charge by adsorption of organic anions (NOM), which can prevent the solids from aggregating and allow them to pass through deep-bed filters in water treatment plants or membranes that are commonly used to remove particles before the analysis of water samples.

Destruction of Tastes and Odors
Various compounds exuded by algae and cyanobacteria, organic and inorganic sulfides released by the decay of organic matter, and organic compounds released from industrial sources can impart unpleasant tastes to drinking waters and odors to both drinking water and wastewaters. While adsorption onto activated carbon can remove many of these compounds, oxidative treatment has also been widespread. Chlorine, chlorine dioxide, ozone, and potassium permanganate, as well as AOPs, have all been used to induce such oxidation reactions.

Many chlorine-based oxidants are effective in mitigating odors from organics and reduced sulfur compounds. However, in some instances, chlorination can result in the formation of even more odorous entities; for example, chlorophenols typically have a stronger odor than their parent phenols. Ozone is the most effective oxidant for reducing odors associated with 2-methylisoborneol (MIB) and geosmin, which are produced by cyanobacteria and are the source of off-flavors and musty odors in drinking water; AOPs involving the combined use of ozone (O_3) and hydrogen peroxide (H_2O_2) are even more effective (Singer and Reckhow, 1999).

Color Removal
Color in drinking water results principally from the presence of natural organic matter (NOM). Oxidation using chemicals such as chlorine may reduce the color to some extent. However, addition of Fe- and Al-based coagulants, which precipitate and adsorb a substantial fraction of the color-forming compounds, is more effective than chemical transformation, particularly since reactions of the target compounds with chlorine, chlorine dioxide, and ozone are likely to generate troublesome disinfection by-products (DBPs), as described subsequently. Color in wastewater might arise from several sources. Textile dyes are particularly difficult to remove, but they can often be destroyed (or at least converted to colorless products) by AOPs.

Aid to Coagulation
As explained in more detail in Chapter 11, NOM in water can cause particles to resist aggregation, due primarily to the negative surface charge that results from adsorption of the organic matter to particle surfaces. Under some circumstances, oxidizing chemicals such as chlorine, potassium permanganate, and ozone can reduce this effect (Singer and Reckhow, 1999). The mechanisms underlying this benefit of oxidation are unclear, but could include oxidation of adsorbed organics to generate more polar molecules which desorb from the particles; alteration of the configuration of the adsorbed organics so that they more effectively bind coagulants such as Al(III) and Fe(III) at the particle surface; and oxidation of organics that form soluble metal–NOM complexes, causing them to release free metal ions which can then assist in particle coagulation. This process is effective in some water, but in others, oxidation does not significantly improve coagulation or does so only at the cost of leaving more NOM in the water (Edwards and Benjamin, 1992).

Oxidation of Synthetic Organic Chemicals
The concentration of synthetic organic chemicals in drinking water, treated wastewater, and industrial effluents may be reduced through the addition of oxidants. In the presence of particularly strong oxidants, the chemicals may be mineralized to CO_2. Alternatively, and more commonly, the

---

[2] Elements that participate in redox transformations are often represented by the atomic symbol followed by a roman numeral identifying its formal, assigned charge (*oxidation number* or *oxidation state*) in parentheses. Thus, for example, the designation Fe(III) indicates all species of iron in the +3 oxidation state, including the species Fe^{3+}, $FeOH^{2+}$, and $Fe(OH)_3(s)$ (and several others).

chemicals are transformed into less toxic and/or more biodegradable species. Advanced oxidation processes (AOPs) are particularly effective in degrading many organic contaminants. Because of the possibility that more toxic entities might form, the toxicity of the treated stream should be assessed before adopting a particular oxidation technology.

Destruction of Complexing Agents in Industrial Wastes
Various organic and inorganic compounds that are strong metal complexants are used in industrial processes and may cause problems if released to the environment. For example, large quantities of ethylenediamine tetra-acetic acid (EDTA) are used to remove calcium and/or magnesium-rich scale in boilers and pipelines, and cyanide is used in the leaching of gold. These agents can be destroyed by oxidation, sometimes in processes that require catalysis by transition metal ions such as Fe(III).

Reduction

Reductive processes, though much less common in water and waste treatment than oxidative processes, are used to remove certain contaminants or to minimize their impact. For example, sulfur dioxide and sodium bisulfite are commonly added to remove residual chlorine from waters, and zero-valent (metallic) iron barriers are proving to be an effective method of dehalogenating some groundwater contaminants such as trichloroethylene. Metals such as chromium and uranium, either in waste streams or subsurface environments, can be converted to less soluble states by reductive processes. Such processes are commonly carried out by addition of sulfite or ferrous iron.

Thermodynamic Aspects

A brief account of the thermodynamic principles underlying redox processes is presented here; additional details can be found in several texts (e.g., Morel and Hering (1993), Stumm and Morgan (1996); and Benjamin (2010)).

The defining feature of a redox reaction is the transfer of one or more electrons between two species, thereby increasing the oxidation number of the species that releases the electrons and decreasing that of the species that gains them. Free, hydrated electrons are extremely unstable in water, so a reaction that releases electrons never occurs in isolation; it always occurs in conjunction with another reaction that consumes the electron. Nevertheless, it is common to write reactions (sometimes called *half-reactions*) in which free electrons are shown as reactants or products. Consider first the simplest group of half-reactions, in which the species undergoing oxidation and reduction are the only two species in the reaction, other than the electron. Denoting the oxidized species as Ox and its conjugate reduced species as Red, we can write a generic such redox half-reaction and the associated mass law expression as follows:

$$\text{Ox} + n_e e^- \leftrightarrow \text{Red} \quad (10\text{-}1)$$

$$K = \frac{a_{\text{Red}}}{a_{\text{Ox}} a_{e^-}^{n_e}} \quad (10\text{-}2)$$

where a_i denotes the activity of species i, and K is the equilibrium constant for the reaction. Because free electrons are unstable in aqueous solution, evaluation of K using Equation 10-2 would be very difficult if we attempted to deduce the electron activity based on a measurement of the concentration of dissolved electrons. This problem is overcome by adopting the convention of assigning the electron activity a value of 1.0 in a system in which H^+ and $H_2(g)$ are in equilibrium and are both present at activities of 1.0. Based on this convention, the equilibrium constant for the following redox half-reaction becomes 1.0:

$$H^+ + e^- \leftrightarrow \tfrac{1}{2} H_2(g) \quad (10\text{-}3)$$

$$K = \frac{a_{H_2(g)}^{1/2}}{a_{H^+} a_{e^-}} = 1.0 \quad (10\text{-}4)$$

Knowing the value of K for the $H^+/H_2(g)$ couple, we can compute the activity of e^- in any other equilibrium system containing H^+ and $H_2(g)$. For example, in a pH 7.0 solution in equilibrium with a hydrogen partial pressure of 10^{-5} atm, the value of a_{e^-} would be

$$a_{e^-} = \frac{a_{H_2(g)}^{1/2}}{K a_{H^+}} = \frac{(10^{-5})^{0.5}}{(1.0)(10^{-7.0})} = 10^{+4.5} \quad (10\text{-}5)$$

Using the pX = $-\log$ X shorthand, we can also describe the activity of electrons in terms of pe. For example, for the conditions described in Equation 10-5, pe = $-\log (10^{4.5}) = -4.5$.

Once the value of a_{e^-} in a solution is established, the equilibrium constant of any redox half-reaction that is at equilibrium in the solution can be determined, in theory, by analysis of the activities of the species participating in that reaction. For example, assume that the solution described earlier (with pe = -4.5) also contains Fe^{2+} and Fe^{3+} in equilibrium with one another. These species can interconvert via the following redox half-reaction:

$$Fe^{3+} + e^- \leftrightarrow Fe^{2+} \quad (10\text{-}6)$$

By evaluating the activities of Fe^{2+} and Fe^{3+} in the solution, the equilibrium constant for this reaction could be quantified by inserting those values into the following expression:

$$K = \frac{a_{Fe^{2+}}}{a_{Fe^{3+}} a_{e^-}} = \frac{a_{Fe^{2+}}}{a_{Fe^{3+}}(10^{4.5})} \qquad (10\text{-}7)$$

For this particular reaction, the equilibrium constant K equals $10^{+13.03}$. This type of analysis has been conducted for many redox couples, and the equilibrium constants for these redox couples are tabulated in numerous collections. Some redox half-reactions of particular importance in treatment of water and wastewater and their associated equilibrium constants (log K values) are listed in Table 10-1.

As with other equilibrium constants, some conventional terminology has developed with regard to the method of representing redox reactions. Specifically, e° (the *standard electron activity*) is defined as the equilibrium constant for a redox half-reaction in which the reduced species releases one electron to form the corresponding oxidized species.

Equivalently, we can state that pe°; that is, $-\log a_{e^o}$, is the base-10 logarithm of the equilibrium constant for the redox half-reaction written as a reduction, normalized to a one-electron transfer. That is, pe° is log K for a reaction written as

$$\frac{1}{n_{e^-}} Ox + e^- \leftrightarrow \frac{1}{n_{e^-}} Red \qquad (10\text{-}8)$$

With this definition of pe°, Equation 10-2 can be manipulated to give

$$pe = pe^o - \frac{1}{n_{e^-}} \log \frac{a_{Red}}{a_{Ox}} \qquad (10\text{-}9)$$

Equation 10-9 shows that, for a half-reaction that can be written in the form of Reaction 10-8 (with only one species on each side of the reaction, other than e^-), the activities of the oxidized and reduced species are equal when the pe of the system equals pe°. The oxidized species will dominate

TABLE 10-1. Equilibrium Constants for Selected Redox Half-Reactions of Importance in Water and Wastewater Treatment

Reaction	Log K	pe°	pe° (W)	E_H^0 (mV)
$O_3(g) + 2H^+ + 2e^- \leftrightarrow O_2(g) + H_2O$	70.12	35.06	28.06	2069
$H_2O_2 + 2H^+ + 2e^- \leftrightarrow 2H_2O$	59.59	29.80	22.80	1758
$MnO_4^- + 4H^+ + 3e^- \leftrightarrow MnO_2(s) + 2H_2O$	86.22	28.74	19.41	1696
$2HOBr + 2H^+ + 2e^- \leftrightarrow Br_2(aq) + 2H_2O$	53.60	26.80	20.27	1581
$CrO_4^{2-} + 8H^+ + 3e^- \leftrightarrow Cr^{3+} + 4H_2O$	77.00	25.66	7.00	1514
$MnO_4^- + 8H^+ + 5e^- \leftrightarrow Mn^{2+} + 4H_2O$	127.82	25.56	14.36	1508
$Mn^{3+} + e^- \leftrightarrow Mn^{2+}$	25.51	25.51	25.51	1505
$ClO_2 + 4H^+ + 5e^- \leftrightarrow Cl^- + 2H_2O$	126.67	25.33	19.73	1495
$HOCl + H^+ + 2e^- \leftrightarrow Cl^- + H_2O$	50.20	25.10	21.60	1481
$BrO_3^- + 6H^+ + 6e^- \leftrightarrow Br^- + 3H_2O$	146.1	24.35	17.35	1437
$Cl_2(aq) + 2e^- \leftrightarrow 2Cl^-$	47.20	23.60	23.60	1392
$HOBr + H^+ + 2e^- \leftrightarrow Br^- + H_2O$	45.36	22.68	19.18	1338
$O_2(aq) + 4H^+ + 4e^- \leftrightarrow 2H_2O$	86.00	21.50	14.50	1268
$MnO_2(s) + 4H^+ + 2e^- \leftrightarrow Mn^{2+} + 2H_2O$	41.60	20.80	6.80	1227
$O_2(g) + 4H^+ + 4e^- \leftrightarrow 2H_2O$	83.12	20.78	13.78	1226
$SeO_4^{2-} + 3H^+ + 2e^- \leftrightarrow HSeO_3^- + H_2O$	36.31	18.15	4.15	1071
$ClO_2 + e^- \leftrightarrow ClO_2^-$	17.61	17.61	17.61	1039
$SO_4^{2-} + 2H^+ + 2e^- \leftrightarrow SO_3^- + H_2O$	27.16	13.58	6.58	801
$O_2(aq) + 2H^+ + 2e^- \leftrightarrow H_2O_2$	26.34	13.17	6.17	777
$Fe^{3+} + e^- \leftrightarrow Fe^{2+}$	13.03	13.03	13.03	769
$Cu_2O(s) + 2H^+ + 2e^- \leftrightarrow Cu(s) + H_2O$	15.80	7.90	0.90	466
$H_3AsO_4(aq) + 2H^+ + 2e^- \leftrightarrow H_3AsO_3(aq)$	18.98	9.49	2.49	560
$SO_4^{2-} + 9H^+ + 8e^- \leftrightarrow HS^- + 4H_2O$	33.68	4.21	-3.67	248
$2H^+ + 2e^- \leftrightarrow H_2(g)$	0.00	0.00	-7.00	0
$O_2(g) + H^+ + e^- \leftrightarrow H\dot{O}_2(aq)$	-2.20	-2.20	-9.20	-130
$Pb^{2+} + 2e^- \leftrightarrow Pb(s)$	-4.41	-2.20	-2.20	-130
$Ni^{2+} + 2e^- \leftrightarrow Ni(s)$	-8.47	-4.24	-4.24	-250
$Fe^{2+} + 2e^- \leftrightarrow Fe(s)$	-14.92	-7.46	-7.46	-440
$Zn^{2+} + 2e^- \leftrightarrow Zn(s)$	-25.83	-12.92	-12.92	-762

over the reduced species when the pe of the system is greater than pe°, and vice versa.[3] Recognition of this relationship also provides a means of ranking the relative tendencies for such redox reactions to proceed. Relationships like Equation 10-9 that express the pe as a function of pe° and the reaction quotient are known as the *Nernst equation*.

As can be seen from Table 10-1, many reduction half-reactions involve not only consumption of electrons, but also consumption of protons and production of water. In these cases, the simple relationship between pe and the dominant species noted in the preceding paragraph does not apply. A generic half-reaction for those reactions can be written as follows:

$$\text{Ox} + n_{H^+}H^+ + n_{e^-}e^- \leftrightarrow \text{Red} + n_w H_2O \quad (10\text{-}10)$$

(In some cases, water is the reduced species, in which case it would be treated as "Red" in the earlier reaction.) For this half-reaction, noting that the activity of water is 1.0, we can write

$$\text{pe} = \text{pe}^\circ - \frac{n_{H^+}}{n_{e^-}}\text{pH} - \frac{1}{n_{e^-}}\log \frac{a_{\text{Red}} a_w^{n_w}}{a_{\text{Ox}}} \quad (10\text{-}11)$$

Defining a new parameter, $\text{pe}^\circ_{\text{pH}} = \text{pe}^\circ - (n_{H^+}/n_{e^-})\text{pH}$, Equation 10-11 may be written as

$$\text{pe} = \text{pe}^\circ_{\text{pH}} - \frac{1}{n_{e^-}}\log \frac{a_{\text{Red}}}{a_{\text{Ox}}} \quad (10\text{-}12)$$

$\text{pe}^\circ_{\text{pH}}$ can be viewed as a conditional pe° that is specific to the given pH. By far the most common pH value used when reporting $\text{pe}^\circ_{\text{pH}}$ is 7.0; $\text{pe}^\circ_{7.0}$ is frequently represented as pe°(W), with the W signifying a neutral aqueous system. pe°(W) values are included for the half-reactions in Table 10-1, based on the identity:

$$\text{pe}^\circ(W) = \text{pe}^\circ - \frac{n_{H^+}}{n_{e^-}}(7.0) \quad (10\text{-}13)$$

Use of pe°(W) is convenient because it allows us to evaluate the relative stabilities of the oxidized and reduced species of a redox couple at pH 7.0, even if the reaction involves H$^+$ transfer, in the same way that we can for reactions that involve only the oxidized and reduced species. Thus, for example, if a reaction can be written in the form of Equation 10-10, then the oxidized species is expected to dominate over the reduced species at pH 7.0 if pe < pe°(W), whereas the reduced species is expected to dominate if pe > pe°(W). In other words, the higher the value of pe°(W), the more strongly oxidizing the redox couple is at pH 7.0 and the more stable the reduced species is compared to the oxidized species. Based on this idea, by comparing pe°(W) values of various redox couples, we can infer whether a redox reaction between those couples is thermodynamically favorable under conditions that are typical of natural water and many engineered water and wastewater treatment systems. For example, the HOCl/Cl$^-$ couple (pe°(W) = 21.60) is more strongly oxidizing than the MnO$_2$(s)/Mn^{2+} couple (pe°(W) = 6.80). As a result, we expect that hypochlorous acid (HOCl) (the oxidant of the more strongly oxidizing couple) would be able to oxidize Mn^{2+} (the reduced species in the less oxidizing couple), generating Cl$^-$ and MnO$_2$(s) as reaction products.

Because redox reactions often alter the acidity of the reactants (typically, oxidation generates species that are more acidic), the tendency for such reactions to proceed can be strongly pH dependent. However, for many redox-active elements, it is conventional to write redox reactions in terms of a particular species, regardless of whether that species is the dominant acid/base form of the element under the conditions of interest. For example, the oxidation of Fe(II) to Fe(III) is commonly written as the reverse of Reaction 10-6:

$$Fe^{2+} \leftrightarrow Fe^{3+} + e^- \quad (10\text{-}14)$$

This reaction does not consume or release H$^+$, so, at a given pe, the equilibrium ratio of Fe^{2+} to Fe^{3+} is independent of pH. Because pe° is log K for the one-electron reduction reaction, and Reaction 10-14 is a one-electron oxidation reaction, the equilibrium constant for the reaction is 10^{-pe°, or $10^{-13.03}$. And, since the reaction does not involve H$^+$ transfer, pe°(W) is the same as pe°; that is, 13.03.

At neutral pH, the dominant form of Fe(II) in clean water is Fe^{2+}, but almost all the Fe(III) is present as a solid ferric oxide, such as Fe(OH)$_3$(s). Reaction 10-14 can be rewritten in terms of these dominant species by combining that reaction with reactions for precipitation of Fe(OH)$_3$(s) and dissociation of water as follows:

$Fe^{2+} \leftrightarrow Fe^{3+} + e^-$	$\log K = -13.03$
$Fe^{3+} + 3OH^- \leftrightarrow Fe(OH)_3(s)$	$\log K = +45.20$
$3H_2O \leftrightarrow 3H^+ + 3OH^-$	$\log K = -42.00$
$Fe^{2+} + 3H_2O \leftrightarrow Fe(OH)_3(s) + 3H^+$	$\log K = -9.83$

$$(10\text{-}15)$$

In contrast to Reaction 10-14, Reaction 10-15 suggests that, under typical environmental conditions, three H$^+$ ions are released for every Fe atom oxidized, and that the reaction becomes increasingly favorable (in a thermodynamic sense) with increasing pH. Furthermore, because the equilibrium

---

[3] Note the similarity to monoprotic acid–base systems, where the activities of an acid and its conjugate base are equal when pH = pK_a, and the acid or base species dominates when the pH is less than or greater than pK_a.

constant for Reaction 10-15 is $10^{-9.83}$, pe° for the $Fe(OH)_3(s)/Fe^{2+}$ reaction is +9.83, and pe°(W) is 30.83. These values are substantially different from the corresponding values for Reaction 10-14. The implication is that, to assess the effect of pH on a redox reaction or the reaction's location in a redox sequence, it must be written in terms of the dominant acid/base species in each oxidation state.

The electron activity of a system can be converted to the *electrochemical potential* (E_H) via the expression

$$E_H = \frac{2.303RT}{\mathcal{F}} \text{pe} \quad (10\text{-}16)$$

where $\mathcal{F}$ (the Faraday constant) equals 96,485 coulombs/equiv of e^-. For a system at 25°C, the fraction in Equation 10-16 equals 59 mV, so $E_H = (59\,\text{mV})$ pe. The so-called standard redox potentials (E_H°) can be calculated from pe° values using this expression. Values of standard redox potentials are also shown in Table 10-1.

■ **EXAMPLE 10-1.** Determine the equilibrium constant for the oxidation of hydrogen bisulfide by hypochlorous acid.

Solution. Choosing the appropriate half-reactions from Table 10-1, normalizing to one electron in each, and reversing the bisulfide reaction to show it as an oxidation, we find the following:

$$\tfrac{1}{2}HOCl + e^- + \tfrac{1}{2}H^+ \leftrightarrow \tfrac{1}{2}Cl^- + \tfrac{1}{2}H_2O$$

$$K_{\text{Red'n}} = 10^{\text{pe}^\circ} = 10^{25.10}$$

$$\tfrac{1}{8}HS^- + \tfrac{1}{2}H_2O \leftrightarrow \tfrac{1}{8}SO_4^{2-} + e^- + \tfrac{9}{8}H^+$$

$$K_{\text{Oxid'n}} = 10^{-\text{pe}^\circ} = 10^{-4.21}$$

Summing the two half-reactions to obtain the full reaction, and multiplying the equilibrium constants to obtain the equilibrium constant for that reaction yields

$$\tfrac{1}{2}HOCl + \tfrac{1}{8}HS^- \leftrightarrow \tfrac{1}{2}Cl^- + \tfrac{1}{8}SO_4^{2-} + \tfrac{5}{8}H^+$$

$$K = K_{\text{Red'n}}K_{\text{Oxid'n}} = 10^{25.10}10^{-4.21} = 10^{20.89}$$

Finally, multiplying by eight (and therefore raising the equilibrium constant to the eighth power)

$$4HOCl + HS^- \leftrightarrow 4Cl^- + SO_4^{2-} + 5H^+$$

$$K = 10^{8(20.89)} = 10^{167.1}$$

For any solution normally encountered in environmental engineering systems, this reaction proceeds to completion; that is, bisulfide is completely oxidized to sulfate in the presence of sufficient chlorine. The reaction proceeds very rapidly, so that if a drinking water supply is taken from a source under reducing conditions (e.g., groundwater, hypolimnetic water), chlorine that is added reacts virtually instantaneously to oxidize any bisulfide (and many other inorganic species) in the water. These reactions can consume a significant amount of chlorine, as suggested by the four to one molar ratio in this example. ■

■ **EXAMPLE 10-2.** Although selenium is an essential dietary micronutrient, it can have adverse effects on aquatic life and on humans at concentrations that are still very low. The dominant oxidation states of selenium in natural water are +6 (selenate) and +4 (selenite). Compute the equilibrium ratio of TOTSe(VI) to TOTSe(IV) in a surface water in which the pe is controlled by the O_2/H_2O couple ($P_{O_2} = 0.21$ atm, pH = 7.5). Assume that the solutions behave ideally, so the activities of the solutes can be equated with their molar concentrations. At pH 7.5, the only species of Se(IV) that are significant are $HSeO_3^-$ and SeO_3^{2-}, and the only Se(VI) species of significance is SeO_4^{2-}. The concentrations of the two Se(IV) species are related by the following acid/base reaction:

$$HSeO_3^- \leftrightarrow H^+ + SeO_3^- \quad \log K = -8.4$$

Solution. The Se speciation can be determined using the Nernst equation, once the equilibrium value of pe is determined. In the aerobic surface water, the pe is controlled by the O_2/H_2O couple. Applying Equation 10-11 to that reaction, and noting that the reduced species formed in the reaction are two water molecules, we find

$$\text{pe} = \text{pe}^\circ - \frac{n_{H^+}}{n_{e^-}}\text{pH} - \frac{1}{n_{e^-}}\frac{a_{\text{Red}}}{a_{\text{Ox}}} = 20.78 - \tfrac{4}{4}\text{pH} - \tfrac{1}{4}\log\frac{a_{H_2O}^2}{a_{O_2(g)}}$$

The activity of water is 1.0 and that of gaseous O_2 is its partial pressure in atmospheres, so

$$\text{pe} = 20.78 - \tfrac{4}{4}(7.5) - \tfrac{1}{4}\log\tfrac{1.0}{0.21} = 13.11$$

By rearranging the Nernst equation and inserting the known value of pe along with the appropriate parameters for the Se redox reaction, we find

$$\log\frac{a_{\text{Red}}}{a_{\text{Ox}}} = n_{e^-}\left(\text{pe}^\circ - \text{pe} - \frac{n_{H^+}}{n_{e^-}}\text{pH}\right)$$

$$\log\frac{c_{HSeO_3^-}}{c_{SeO_4^{2-}}} = 2\left(18.15 - 13.11 - \tfrac{3}{2}(7.5)\right) = -12.42$$

$$\frac{c_{HSeO_3^-}}{c_{SeO_4^{2-}}} = 10^{-12.42}$$

This result indicates the ratio of individual Se species. Selenate ion is the only significant Se(VI) species present, so we can substitute TOTSe(VI) for the concentration of SeO_4^{2-}. However, SeO_3^{2-} contributes to TOTSe(IV), so we need to determine its concentration. According to the given acid/base reaction

$$\frac{c_{SeO_3^{2-}}}{c_{HSeO_3^-}} = \frac{10^{-8.40}}{c_{H^+}} = \frac{10^{-8.40}}{10^{-7.50}} = 10^{-0.90} = 0.13$$

$$TOTSe(IV) = 1.13 c_{HSeO_3^-}$$

Correspondingly

$$\frac{TOTSe(IV)}{TOTSe(VI)} = \frac{1.13 c_{HSeO_3^-}}{c_{SeO_4^{2-}}} = 1.13 \times 10^{-12.42}$$
$$= 4.30 \times 10^{-13}$$

Thus, selenium is expected to be present overwhelmingly in the oxidized state in aerobic surface waters. However, the speciation might be quite different in anaerobic environments (see Problem 10-1). ∎

Terminology for Oxidant Concentrations

In any engineering application, the point of adding an oxidant is to cause some of it to react and thereby be consumed. The concentration of oxidant that has been or might yet be consumed in a given system is referred to as the oxidant *demand*—the concentration that has been consumed as the demand that has been *exerted*, and the potential future consumption as the demand remaining. (A related, but slightly different definition is used when the disinfectant is ozone. That definition is given in the section dealing specifically with the use of ozone for disinfection.) The concentration that actually remains at any time is referred to as the oxidant *residual*. Thus, in the form of word equations

$$\binom{\text{Total oxidant}}{\text{demand}} = \binom{\text{Oxidant}}{\text{demand exerted}}$$
$$+ \binom{\text{Oxidant}}{\text{demand remaining}} \quad (10\text{-}17)$$

$$\binom{\text{Oxidant}}{\text{demand exerted}} = \binom{\text{Oxidant}}{\text{dose}} - \binom{\text{Oxidant}}{\text{residual}} \quad (10\text{-}18)$$

The use of these terms is especially common in discussions of disinfection, where the quantities are referred to as the *disinfectant demand* and the *disinfectant residual*.

Frequently, when an oxidant is dosed into a solution, some demand is exerted very quickly (the *instantaneous demand*), after which consumption continues at an ever-decreasing rate for at least hours and often days. In many cases, the instantaneous demand is caused by reactions between the oxidant and inorganic solutes such as Fe(II) and S(−II) species, as well as certain organic species. Subsequently, the slower demand is exerted primarily by dissolved or particulate organic matter. Oxidant decay during this latter phase can often be described by first-order kinetics (Haas and Kara, 1984):

$$c_{Ox}(t) = c_{Ox}(0)\exp(-k_1 t) \quad (10\text{-}19)$$

where k_1 is a first-order rate constant and $c_{Ox}(0)$ is the oxidant concentration after the instantaneous demand has been exerted.

Kinetics of Redox Reactions

While thermodynamic analysis provides insight into whether a proposed redox reaction can proceed, it provides no insight into the time scale of the transformation. In an oxic environment at pH 8, for example, thermodynamic considerations indicate that Mn^{2+} will transform to Mn(III) and/or Mn(IV) species, and Fe^{2+} will oxidize to the ferric state (Fe(III)). However, the rates of these processes are vastly different, with the characteristic reaction time for oxidation of Mn^{2+} on the order of years and that for Fe^{2+} oxidation on the order of a few minutes or less.

Because the time scale of redox processes is case-specific, little possibility exists for developing generic guidelines regarding the practicality of carrying out redox reactions in the time frames of interest. In view of this, we delay more detailed consideration of the kinetics of redox reactions to some specific example systems presented later in the chapter. The kinetics of important redox processes in environmental aquatic systems has been discussed by Stone and Morgan (1990) and Brezonik (1994).

10.3 OXIDATIVE PROCESSES INVOLVING COMMON OXIDANTS

In this section, reactions and applications involving the common oxidants—oxygen, chlorine, chloramines, chlorine dioxide, ozone, and potassium permanganate—are presented. These chemicals are widely used in the treatment of water and wastewater for the transformation of troublesome species into more readily removed or degraded forms and, in selected cases, for disinfection purposes. The reactions of these oxidants with organic and inorganic compounds are considered in this section, and reactions with organisms are considered in Section 10-6. AOPs involving free radical oxidants (particularly the hydroxyl radical) are considered in Section 10-3.

Oxygen

Oxygen is a key redox species in natural aquatic systems, and it also plays some role, albeit less prominent, in physical/chemical treatment systems. For example, dissolved oxygen (often provided by contacting the solution with air in a gas transfer process) is commonly used to oxidize ferrous to ferric iron in the treatment of drinking water (particularly groundwater) and thereby precipitate the iron. However, oxygen is too weak to serve as a direct oxidant for any organic compounds.

The residence times required for satisfactory iron and, potentially, manganese removal depend strongly on reactant concentrations and pH. The rate of oxidation of Fe(II) at a given pH is first order with respect to the concentrations of both total Fe(II) and O_2:

$$r_{\text{TOTFe(II)}} = -k c_{\text{TOTFe(II)}} c_{O_2(\text{aq})} \quad (10\text{-}20)$$

where TOTFe(II) represents the sum of all dissolved ferrous species (Fe^{2+}, $FeOH^+$, $Fe(OH)_2^o$, and so on). For a fixed concentration of dissolved oxygen (which might be maintained by equilibrating the water continuously with air or pure oxygen gas), the reaction is pseudo first order in $c_{\text{TOTFe(II)}}$, and the rate law may be written as

$$r_{\text{TOTFe(II)}} = -k_1 c_{\text{TOTFe(II)}} \quad (10\text{-}21)$$

where $k_1 = k c_{O_2(\text{aq})}$. The value of k or k_1 depends strongly on pH, as shown in Figures 10-1 and 10-2.

The pH-dependence of k_1 can be incorporated into the analysis by recognizing that each Fe(II) species undergoes the oxidation at a distinct rate. That is, Equations 10-20 and 10-21 can be written as

$$r_{\text{TOTFe(II)}} = -\left(k_{Fe^{2+}} c_{Fe^{2+}} c_{O_2(\text{aq})} + k_{FeOH^+} c_{FeOH^+} c_{O_2(\text{aq})} \right.$$
$$\left. + k_{Fe(OH)_2^o} c_{Fe(OH)_2^o} c_{O_2(\text{aq})} + \cdots \right) \quad (10\text{-}22)$$

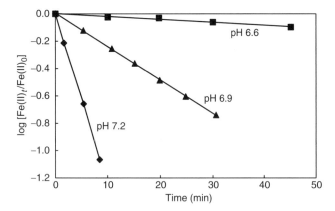

FIGURE 10-1. Removal of Fe(II) by oxygen in solutions of various pH. *Source*: From Stumm and Morgan (1996).

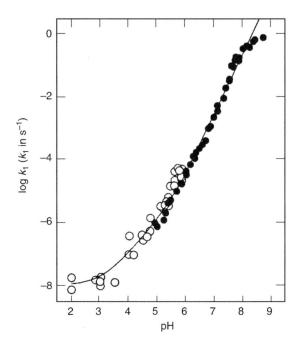

FIGURE 10-2. Effect of pH on the Fe(II) oxygenation rate in systems equilibrated with pure $O_2(g)$ ($P_{O_2} = 1$ atm). Open circles represent data by Singer and Stumm (1970); filled circles are from Millero et al. (1987). Solid line calculated using Equation 10-28 and values in Table 10-2. *Source*: After Stumm and Morgan (1996).

$$r_{\text{TOTFe(II)}} = -c_{O_2(\text{aq})} \sum_i k_i c_i \quad (10\text{-}23)$$

$$r_{\text{TOTFe(II)}} = -\sum_i k_{1,i} c_i \quad (10\text{-}24)$$

where the summations are over all Fe(II) species in solution and $k_{1,i} = k_i c_{O_2(\text{aq})}$. In conventional water chemistry notation, the fraction of the total TOTFe(II) concentration that is present as a species i is designated α_i, so Equation 10-23 can be rewritten as follows:

$$r_{\text{TOTFe(II)}} = -c_{O_2(\text{aq})} c_{\text{TOTFe(II)}} \sum_i k_i \frac{c_i}{c_{\text{TOTFe(II)}}}$$
$$= -c_{O_2(\text{aq})} c_{\text{TOTFe(II)}} \sum_i k_i \alpha_i \quad (10\text{-}25)$$

Comparing Equations 10-20 and 10-25, we find

$$k = \sum_i k_i \alpha_i \quad (10\text{-}26)$$

Thus, in Equation 10-20, k can be thought of as a weighted-average rate constant, based on the rate constants for the reactions of the individual ferrous species. Assuming ideal solution behavior (i.e., all activity coefficients equal to unity), each α_i value is expressible as an explicit function of the activity of H^+ (a_{H^+}, equal to 10^{-pH}). For Fe^{2+}, $FeOH^+$, and $Fe(OH)_2^o$, these expressions are

$$\alpha_{Fe^{2+}} = \frac{1}{1 + K_1 K_w/a_{H^+} + \beta_2 K_w^2/a_{H^+}^2} \quad (10\text{-}27a)$$

$$\alpha_{FeOH^+} = \frac{K_1 K_w/a_{H^+}}{1 + K_1 K_w/a_{H^+} + \beta_2 K_w^2/a_{H^+}^2} \quad (10\text{-}27b)$$

$$\alpha_{Fe(OH)_2^o} = \frac{\beta_2 K_w^2/a_{H^+}^2}{1 + K_1 K_w/a_{H^+} + \beta_2 K_w^2/a_{H^+}^2} \quad (10\text{-}27c)$$

where K_1 and β_2 are the equilibrium constants for formation of $FeOH^+$ and $Fe(OH)_2^o$ from Fe^{2+} and OH^-, and K_w is the dissociation constant for water. Assuming that these three species are the only ones that undergo oxidation at a significant rate, we obtain the following expression for k:

$$k = \frac{k_{Fe^{2+}} + k_{FeOH^+} K_1 K_w/a_{H^+} + k_{Fe(OH)_2^o}\beta_2 K_w^2/a_{H^+}^2}{1 + K_1 K_w/a_{H^+} + \beta_2 K_w^2/a_{H^+}^2} \quad (10\text{-}28)$$

The solid line in Figure 10-2 represents a fit of this expression to the data and yields the rate constants shown in Table 10-2.

TABLE 10-2. Rate Constants for Reaction of Fe(II) Species with O_2

Fe(II) Species, i	Second-Order Rate Constant, $k_i\,((mol/L)^{-1}\,s^{-1})$	Pseudo-First-Order Rate Constant, $k_{1,i}{}^a\,(s^{-1})$
Fe^{2+}	7.9×10^{-6}	1.0×10^{-8}
$FeOH^+$	25	3.2×10^{-2}
$Fe(OH)_2$	7.9×10^6	1.0×10^4

[a] For $c_{O_2} = 1.27 \times 10^{-3}$ mol/L $= 40.5$ mg/L, which is the value in equilibrium with pure $O_2(g)$ at $P = 1$ atm.
Source: From Wehrli (1990).

■ **EXAMPLE 10-3.** The influent to an activated sludge aeration tank contains 1.5 mg/L Fe(II). Diffusers supply sufficient oxygen to maintain a dissolved oxygen concentration of 2 mg/L. This tank behaves as a CFSTR with a detention time of 4 h and operates at a pH of 7.4.

(a) What is the expected effluent concentration of Fe(II)?
(b) Determine the relative contributions of the three Fe(II) species to TOTFe(II) and to the overall rate of oxidation. The hydrolysis reactions and constants for Fe(II) are as follows:

$$Fe^{+2} + OH^- \leftrightarrow FeOH^+ \quad K_1 = 10^{4.60}$$
$$Fe^{+2} + 2OH^- \leftrightarrow Fe(OH)_2^o \quad \beta_2 = 10^{7.51}$$

Solution.

(a) From Equation 10-28, we can calculate the overall, second-order rate coefficient as follows (the units are not shown explicitly, but are $[mol/L]^{-1}s^{-1}$ for all the k values):

$$k = \frac{7.9 \times 10^{-6} + (25)(10^{4.60})(10^{-14})/10^{-7.4} + (7.9 \times 10^6)(10^{7.51})(10^{-14})^2/(10^{-7.4})^2}{1 + (10^{4.60})(10^{-14})/10^{-7.4} + (10^{7.51})(10^{-14})^2/(10^{-7.4})^2} = 16.4\,(mol/L)^{-1}\,s^{-1}$$

The dissolved oxygen concentration in molar units is

$$c_{O_2(aq)} = (2.0\,mg/L)\frac{1\,mol}{32{,}000\,mg} = 6.25 \times 10^{-5}\,mol/L$$

Inserting these values into Equation 10-20, we find

$$r_{TOTFe(II)} = -k c_{TOTFe(II)} c_{O_2(aq)}$$
$$= -[16.4\,(mol/L)^{-1}\,s^{-1}]c_{TOTFe(II)}(6.25 \times 10^{-5}\,mol/L)$$
$$= -1.02 \times 10^{-3} c_{TOTFe(II)}$$

The result is the pseudo-first-order rate expression for oxidation of TOTFe(II) in the given system. We can insert this expression into the equation for the effluent concentration from a CFSTR (Equation 4-4) in which a first-order reaction is occurring to find

$$c_{Fe(II),out} = \frac{c_{Fe(II),in}}{1 + k_1 \tau} = \frac{1.5\,mg/L}{1 + (1.02 \times 10^{-3}\,s^{-1})(3600\,s/h)(4\,h)}$$
$$= 0.096\,mg/L$$

(b) The contributions of the three Fe(II) species to TOTFe(II) are given by the α values, as expressed in Equations 10-27a–c. The corresponding contributions to the overall oxidation rate are proportional to the values of the products $k_i\alpha_i$. These terms are quantified as follows:

Species	α_i	$k_i\alpha_i$	% of Total Rate
Fe^{2+}	0.99	7.9×10^{-6}	5×10^{-5}
$FeOH^+$	9.9×10^{-3}	0.25	1.5
$Fe(OH)_2^o$	2.0×10^{-6}	16.1	98.5

The values in the right column indicate the percentages of the overall oxidation rate that are attributable to the individual Fe(II) species. That is, they represent the ratios of $k_i\alpha_i/k$. Remarkably, even though $Fe(OH)_2^o$ accounts for only 0.0002% of the total Fe(II) in solution, it accounts for virtually all the oxidation that is occurring, due to the fact that $k_{Fe(OH)_2^o}$ is vastly larger than k_{FeOH^+} or $k_{Fe^{2+}}$. ■

The preceding discussion indicates that formation of hydroxyl complexes has a dramatic effect on the rate of Fe(II) oxidation by oxygen. Additionally, the presence of certain anions that form strong complexes with Fe(III) (such as phosphate) and the precipitation of ferric oxyhydroxide solids increase the Fe(II) oxidation rate. This rate enhancement may reflect a decrease in the importance of a reaction in which Fe(III) species are reduced back to Fe(II) species (Rose and Waite, 2002). In contrast, complexation of Fe(II) by Cl^-, SO_4^{2-}, and NOM results in a significant decrease in the oxidation rate. As a result, oxidation of Fe(II) in high ionic strength (briny) groundwater can take significantly longer than in pristine surface water, for otherwise equivalent conditions.

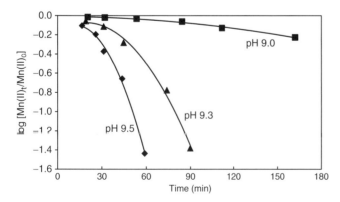

FIGURE 10-3. Removal of Mn(II) by oxygen in solutions of various pH. *Source*: From Stumm and Morgan (1996).

The oxidation of Mn(II) by oxygen occurs on time scales of interest in water treatment only at pH 9.5 and above, as shown in Figure 10-3. Thus, unlike the case for Fe(II), oxygenation is rarely used for the removal of Mn(II) from drinking water; rather, water containing excessive manganese is usually treated using more powerful oxidants such as chlorine, chlorine dioxide, or potassium permanganate, as discussed in the following section.

Unlike the case for ferrous iron, Mn(II) oxygenation does not appear to follow first-order kinetics with respect to Mn(II) at fixed pH. Rather, when the data from batch tests are analyzed based on a pseudo-first-order model, the apparent rate constant increases with time, as indicated by the increasing slopes with time in Figure 10-3. This trend has been attributed to autocatalysis as a result of Mn(II) adsorption to the Mn(IV) oxide solids formed in the oxygenation process, yielding a rate expression of the following form:

$$r_{Mn(II)} = -k_0 c_{Mn(II)} c_{O_2(aq)} - k c_{Mn(II)} c_{MnO_2(s)} c_{O_2(aq)} \quad (10\text{-}29)$$

Even with the autocatalysis, the oxidation of Mn(II) by oxygen is too slow in the pH range of interest to serve a useful role in most water treatment schemes. However, Mn(II) is also oxidized more rapidly heterogeneously (that is, on the solid) than in solution when other oxidants are employed, as shown in the subsequent sections. Heterogeneous oxidation of Mn(II) is often used in water treatment, where an oxidant is added to the water upstream of a granular media filter, allowing the oxidation to occur on the filter grains.

Chlorine

Chlorine (Cl_2) and its hydrolysis product, hypochlorous acid (HOCl), are used widely in both water and wastewater treatment, and for both disinfection and oxidative treatment of a wide variety of contaminants, including Fe(II), Mn(II), S(−II), and organic compounds.

Molecular chlorine is a gas at normal temperature and pressure. Although dissolution of the gas into an aqueous solution initially generates $Cl_2(aq)$, that species is unstable at near-neutral pH and disproportionates rapidly (with a characteristic reaction time on the order of 0.08 s) to form equal concentrations of hydrochloric acid (HCl) and hypochlorous acid (HOCl).[4] HCl is a strong acid that dissociates completely, whereas HOCl is a weak acid ($pK_a = 7.5$) that

---

[4] A molecular splitting process that involves oxidation and reduction of different atoms in the same molecule is referred to as *disproportionation*.

dissociates only partially. These reactions can be summarized as follows:

Reaction	Equilibrium Constant
$Cl_2(aq) \leftrightarrow Cl_2(g)$	$H_{pm} = 2.08 \times 10^5 \times \exp\left(-\dfrac{2818.5}{T}\right)$ atm/(mol/L) (10-30)
$Cl_2(aq) + H_2O \leftrightarrow H^+ + Cl^- + HOCl$	$K = 10^{-3.30}$ (10-31)
$HOCl \leftrightarrow H^+ + OCl^-$	$pK_a = 7.53$ (at 25°C) (10-32)

The disproportionation of $Cl_2(aq)$ (Reaction 10-31) can be written as the sum of two redox half-reactions in which the Cl_2 is oxidized and reduced, respectively. Those reactions, in turn, can be combined with each other and with the acid dissociation reaction (Equation 10-32) to generate redox half-reactions that relate the hypochlorous species (HOCl and OCl^-) to chloride (Cl^-). When written in the conventional way (as reductions), these four half-reactions are as follows:

$$HOCl + H^+ + e^- \leftrightarrow \tfrac{1}{2}Cl_2(aq) + H_2O \quad \log K = pe^\circ = 26.9$$
(10-33)

$$\tfrac{1}{2}Cl_2(aq) + e^- \leftrightarrow Cl^- \quad \log K = pe^\circ = 23.6$$
(10-34)

$$\tfrac{1}{2}HOCl + \tfrac{1}{2}H^+ + e^- \leftrightarrow \tfrac{1}{2}Cl^- + \tfrac{1}{2}H_2O \quad \log K = pe^\circ = 25.25$$
(10-35)

$$\tfrac{1}{2}OCl^- + H^+ + e^- \leftrightarrow \tfrac{1}{2}Cl^- + \tfrac{1}{2}H_2O \quad \log K = pe^\circ = 29.0$$
(10-36)

Both hypochlorous acid and hypochlorite anion are strong oxidants, with HOCl the more facile oxidant of the two. Since, in any given solution, HOCl and OCl^- are expected to be in continuous equilibrium with one another, the thermodynamic driving force is the same for the reaction of either of these species with a given reduced species. Thus, the greater oxidizing ability of HOCl over OCl^- must be due to kinetic factors, possibly related to the uncharged state of HOCl.

The equilibrium relationships among the various oxidation states and acid/base species of a chemical are often represented on *predominance area diagrams*, with the redox and acid/base status of the system indicated by the pe and pH, respectively. On such a diagram, the redox species that is present at the highest activity (i.e., the dominant redox

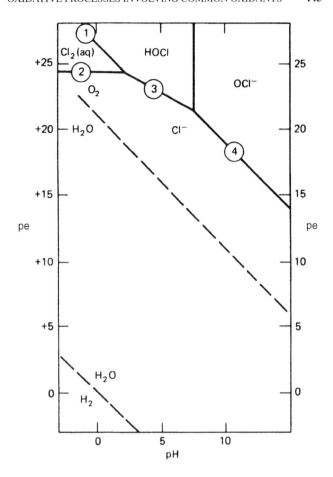

FIGURE 10-4. pe–pH diagram for the chlorine system $Cl_T = 0.04$ mol/L (1.42 g/L). Numbers ① through ④ refer to the equilibria described by Equations 10-33 to 10-36, respectively. *Source*: From Stumm and Morgan (1996); reprinted with permission.

species) is indicated in each region of pe-pH space. Such a diagram for Cl species in an aqueous system is shown in Figure 10-4. The region in which liquid water is stable is shown on the diagram by the broken lines (outside of that region, oxidation or reduction of water to $O_2(aq)$ or $H_2(aq)$, respectively, is thermodynamically favorable). Details of the preparation of such diagrams are available in most water chemistry texts.

Two key observations arise from this diagram. First, Cl^- is the dominant, stable form of chlorine in any aqueous solution; if a solution contains any $Cl_2(aq)$, HOCl, or OCl^-, the reduction of those species to Cl^- accompanied by oxidation of some of the water to $O_2(aq)$ is thermodynamically favorable. The only reason that these oxidized species of Cl can exist at significant concentrations in aqueous solutions is that their reactions with water tend to be very slow; for instance, at ambient temperatures, the half-life of concentrated sodium hypochlorite solutions ranges from a few months to a few years. Second, given that oxidized Cl species can be kinetically stable in solution, the most likely forms of such species are HOCl and OCl$^-$; Cl$_2$(aq) is a significant species only at pH values lower than those normally encountered in water and wastewater systems.

Hypochlorous acid and/or hypochlorite ion are usually dosed into solution in one of three ways: by dissolving Cl$_2$(g) into a side stream to form a concentrated solution of HOCl/OCl$^-$, and then mixing this side stream with the main flow; by adding a concentrated aqueous solution of sodium hypochlorite, NaOCl (i.e., bleach) into the water; or by adding solid or dissolved calcium hypochlorite, Ca(OCl)$_2$, into the solution. In the first case, a base might be needed to neutralize the acidity associated with the disproportionation reaction (Equation 10-31), and in the latter two cases, an acid might be needed to neutralize the basicity of the reagents. However, all of these approaches lead to the same species being present at equilibrium, and all are referred to as *chlorination* of the water.

The most common method of supplying chlorine to water and wastewater treatment plants has been as gaseous chlorine in pressurized tanks. However, chlorine gas is quite hazardous, so that issues of worker safety and public security (as well as the cost of employee training and safety equipment) have led some utilities to adopt hypochlorite over gaseous systems in recent years.

HOCl and OCl$^-$ (and, if it is significant, Cl$_2$(aq)) are collectively referred to as *free chlorine* or *free available chlorine* (FAC). Even though HOCl and OCl$^-$ are the dominant contributors to free chlorine in virtually all systems of interest in environmental engineering, the shorthand that is used most commonly to represent free chlorine is Cl$_2$, and that convention is adopted in this chapter. Correspondingly, concentrations of free chlorine (and of some other Cl-containing oxidants, as described in the following section) are frequently reported in units designated "mg/L as Cl$_2$". The concept is that the oxidizing capacity of the solution of interest (in electron equivalents per liter) is the same as would be present in a solution that contained the specified concentration of Cl$_2$. HOCl, OCl$^-$, and Cl$_2$ have two equivalents per mole, so their summation on the basis of mg/L as Cl$_2$ is the same as summing their concentrations on an equivalent per liter basis.

■ **EXAMPLE 10-4.** After disinfection, a wastewater contains free chlorine at a concentration of 2 mg/L as Cl$_2$ and is at pH 7. What free chlorine species are actually in the water and at what concentrations?

Solution. The statement that the free chlorine concentration is 2 mg/L as Cl$_2$ means that the actual total disinfectant concentration is the same as in a (hypothetical) solution that contained 2 mg/L Cl$_2$. Assuming that the concentration of Cl$_2$ (aq) is negligible, the total disinfectant concentration is the total hypochlorite concentration (i.e., the sum of HOCl and OCl$^-$). In meq/L, this concentration is

$$c_{TOTOCl} = c_{HOCl} + c_{OCl^-} = \left(\frac{2\,\text{mg/L Cl}_2}{70.9\,\text{mg/mmol}}\right)\left(\frac{2\,\text{meq Cl}_2}{\text{mmol Cl}_2}\right)$$
$$= 0.056\,\text{meq/L}$$

Both HOCl and OCl$^-$ can accept two electrons (i.e., they have 2 meq/mmol), so the total concentration can be expressed as a molar concentration as

$$c_{TOTOCl} = (0.056\,\text{meq/L})\left(\frac{1\,\text{mmol OCl}}{2\,\text{meq OCl}}\right) = 0.028\,\text{mmol/L}$$

We can then find the concentration of each species as follows:

$$c_{HOCl} = \alpha_0 c_{TOTOCl} = \frac{c_{HOCl}}{c_{HOCl} + c_{OCl^-}} c_{TOTOCl}$$
$$= \frac{c_{H^+}}{c_{H^+} + K_a} c_{TOTOCl}$$
$$= \frac{10^{-7}}{10^{-7} + 10^{-7.53}}(2.8 \times 10^{-5}\,\text{mol/L})$$
$$= 0.77(2.8 \times 10^{-5}\,\text{mol/L}) = 2.16 \times 10^{-5}\,\text{mol/L}$$
$$c_{OCl^-} = \alpha_1 c_{TOTOCl} = \frac{c_{OCl^-}}{c_{HOCl} + c_{OCl^-}} c_{TOTOCl}$$
$$= (1 - \alpha_0)c_{TOTOCl} = (1 - 0.77)(2.8 \times 10^{-5}\,\text{mol/L})$$
$$= 6.44 \times 10^{-6}\,\text{mol/L}$$

Using Equation 10-31, we find that $c_{Cl_2(aq)}$ is on the order of 10^{-13} mol/L, confirming the assumption that it contributes negligibly to TOTOCl. ■

Reactions of Free Chlorine with Inorganic Compounds

Reactions with Iron and Manganese Free chlorine is capable of oxidizing both ferrous iron (Fe(II)) and manganous manganese (Mn(II)) (see Table 10-1) and, in both cases, the rates increase significantly with increasing pH. However, whereas ferrous iron is oxidized rapidly by chlorine (characteristic reaction times of a few seconds) at neutral or slightly acidic pH (Figure 10-5), the oxidation of manganous ions is considerably slower, with characteristic reaction times in the order of 25 min or more even at pH 9.0 (Figure 10-6).

■ **EXAMPLE 10-5.** A groundwater being used as a water supply contains 1.5 mg/L TOTFe(II) and is at pH 8.4, so that most of the Fe(II) is in the form of Fe^{2+} (pK_{a1} for Fe^{2+} is 9.5). Write a balanced chemical reaction for the oxidation of Fe^{2+} by HOCl to generate Fe(III) in the form of Fe(OH)$_3$(s),

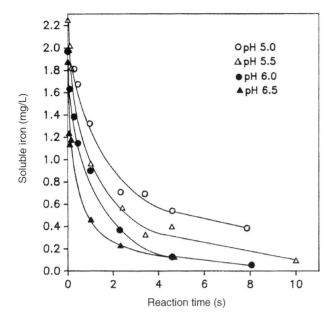

FIGURE 10-5. Kinetics of oxidation of ferrous iron by stoichiometric concentrations of free chlorine at various solution pHs and 25°C. *Source*: From Knocke et al. (1990); reprinted with permission. Copyright 1990 by American Water Works Association.

and find the dose of chlorine (as Cl_2) required for complete oxidation.

Solution. The oxidation of Fe(II) to Fe(III) is a one-electron transfer, and the reduction of HOCl to Cl^- is a two-electron transfer. Therefore, one molecule of HOCl oxidizes two molecules of iron. The balanced reaction is as follows:

$$2Fe^{+2} + HOCl + 5H_2O \rightarrow 2Fe(OH)_3(s) + Cl^- + 5H^+$$

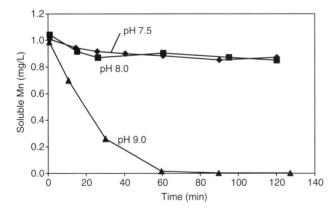

FIGURE 10-6. Kinetics of oxidation of Mn(II) by stoichiometric concentrations of free chlorine at 25°C. *Source*: From Knocke et al. (1990); reprinted with permission. Copyright 1990 by American Water Works Association.

The required dose of chlorine is found as follows:

$$(1.5\,\text{mg Fe/L})\left(\frac{1\,\text{mmol Fe}}{55.85\,\text{mg}}\right)\left(\frac{1\,\text{mmol HOCl}}{2\,\text{mmol Fe}}\right)$$
$$\times (2\,\text{meq HOCl/mmol HOCl})(1\,\text{meq}\,Cl_2/\text{meq HOCl})$$
$$\times (35.45\,\text{mg/meq}\,Cl_2) = 0.95\,\text{mg/L as}\,Cl_2 \quad\blacksquare$$

Reaction with Reduced Sulfur Compounds At times, removal of excess chlorine from solution is necessary, for example, before a disinfected water is discharged to a natural water body. This *dechlorination* process is normally accomplished by addition of reduced sulfur compounds such as dissolved sulfur dioxide (also known as sulfurous acid, H_2SO_3), sodium bisulfite ($NaHSO_3$), or sodium sulfite (Na_2SO_3) to the solution. These sulfite species constitute a diprotic acid/base system similar to the carbonate system, so that all can be considered to be in equilibrium with one another if any are present. They react quickly and completely with chlorine according to the following reactions:

$$SO_2(aq) + HOCl + H_2O \rightarrow SO_4^{2-} + Cl^- + 3H^+ \tag{10-37}$$

$$HSO_3^- + HOCl \rightarrow SO_4^{2-} + Cl^- + 2H^+ \tag{10-38}$$

$$SO_3^{2-} + HOCl \rightarrow SO_4^{2-} + Cl^- + H^+ \tag{10-39}$$

■ **EXAMPLE 10-6.** A wastewater treatment plant uses chlorine for disinfection, and then dechlorinates with sodium bisulfite before discharge to the receiving stream. If the free chlorine concentration is 2 mg/L as Cl_2, what dose of $NaHSO_3$ must be added to dechlorinate completely? If the water flow is $2\,\text{m}^3/\text{s}$, what flow rate of a 0.1 mol/L sodium bisulfite stock solution is required?

Solution. From Example 10-4, we know that 2 mg/L Cl_2 corresponds to a total hypochlorite concentration of 2.82×10^{-5} mol/L. According to Equation 10-38, 1 mole of HSO_3^- is required per mole of HOCl (or, more precisely, total hypochlorite), so

$$c_{NaHSO_3,\text{required}} = c_{SO_3^-,\text{required}} = 2.82 \times 10^{-5}\,\text{mol/L}$$

The molecular weight of $NaHSO_3$ is $[23+1+32+(3)(16)] = 104$, so

$$c_{NaHSO_3,\text{required}} = \left(2.82 \times 10^{-5}\,\text{mol/L}\right)(104\,\text{g/mol})$$
$$\times (1000\,\text{mg/g}) = 2.93\,\text{mol/L}$$

The required flow rate of the stock solution can be found as a mass balance at a junction, with two inputs (the main plant flow and the stock solution) and one output. Because the main flow has no bisulfite before the junction and the flow rate of the stock solution is expected to be insignificant in comparison to the main flow, we can write the mass balance by equating the mass/time of bisulfite into the junction (from the stock solution) with the mass/time of bisulfite out of the junction (assuming no reaction occurs there):

$$Q_{stock} c_{stock} = Q_{main} c_{out\ of\ junction}$$

$$Q_{stock} = \frac{Q_{main} c_{out\ of\ junction}}{c_{stock}}$$

$$= \frac{(2\,m^3/s)(2.82 \times 10^{-5}\,mol/L)}{0.1\,mol/L}$$

$$= 5.64 \times 10^{-4}\,m^3/s = 0.56\,L/s \qquad \blacksquare$$

Reactions with Bromide The presence of bromide ion (Br^-) in water undergoing oxidation is of concern, because it can lead to the formation of brominated organic compounds. For instance, brominated trihalomethane (THM) and haloacetic acid (HAA) species are formed during the chlorination of water containing Br^-. Amy et al. (1995) indicated that the mean Br^- concentration in randomly selected utility samples around the United States was approximately 62 μg/L. Generally, lakes exhibited lower Br^- levels than rivers and groundwater.

Hypochlorous acid oxidizes bromide to hypobromous acid (HOBr):

$$HOCl + Br^- \rightarrow HOBr + Cl^- \qquad (10\text{-}40)$$

A second-order rate constant of $2.95 \times 10^3\,M^{-1}\,s^{-1}$ has been reported for this reaction, which yields a characteristic reaction time for Br^- oxidation of approximately 11 s for a solution containing 30 μM HOCl (2 mg/L as Cl_2).

Hypobromous acid is a weaker oxidant than hypochlorous acid at neutral pH (Table 10-1) (which is why Reaction 10-40 proceeds to the right). However, hypobromous acid is also a slightly weaker acid than hypochlorous acid ($pK_a = 8.8$, compared to 7.53 for HOCl), so that at neutral or alkaline pH, the fraction of the total hypobromite that is protonated is larger than the corresponding fraction of the total hypochlorite; that is, (c_{HOBr}/TOTOBr) > (c_{HOCl}/ TOTOCl). Because the protonated forms of these species are more facile oxidants than the deprotonated (dissociated) forms, hypobromite species are usually more effective oxidants, overall, than hypochlorite species at $7 \leq pH \leq 9.5$.

Reactions with Organic Compounds The chlorine atom in HOCl or OCl^- carries a +1 charge and is a strong *electrophile* (an "electron-liking" entity), which accounts for the fact that these species are strong oxidants (i.e., the chlorine atoms HOCl and OCl^- "like" electrons enough that they have a strong tendency to acquire electrons and be converted to Cl^-). Such species react readily with electrons on *nucleophiles* (nucleus or positive-charge-liking entities), such as C–C, C–H, or S–H groups in organic compounds. Most reactions between HOCl or OCl^- and nucleophiles involve the transfer of electrons to the Cl species, generating Cl^-.

The specific organic functional groups that react with HOCl or OCl^- in complex environmental solutions are unclear, but they presumably include aromatic rings, aldehydes, ketones, alcohols, amino acids and sulfur compounds. Examples of two such redox reactions are shown in Figure 10-7.

In other cases, the Cl atom from the hypochlorite species becomes part of the organic molecule, either via substitution reactions (in which the Cl replaces a different atom in the molecule, typically H^+) or addition reactions (in which

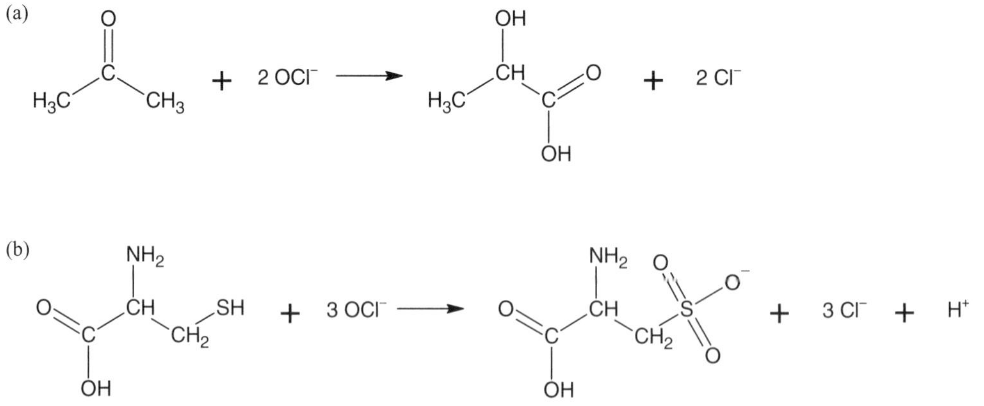

FIGURE 10-7. Oxidation of organic functional groups by OCl^-. (a) Oxidation of a ketone to an α-hydroxy carboxylic acid, and (b) oxidation of a sulfhydryl group to a sulfonyl group.

the Cl^+ and OH^- portions of the HOCl molecule bond separately to different locations on the organic molecule). The chlorinated organic compounds formed by such reactions are referred to collectively as disinfection by-products (DBPs). A great deal of attention has been focused on the formation of chlorinated DBPs because of their potential human and ecological health implications. However, it is important to recognize that, in most cases, the majority of the chlorine applied in the treatment of water is not incorporated into organic molecules but is reduced to chloride ion. For example, when drinking water is chlorinated, 90% or more of the OCl that reacts is typically reduced to Cl^-, while less than 10% becomes part of organic molecules. Given the importance of the formation of chlorinated DBPs, the remainder of this section focuses on substitution and addition reactions between HOCl and organic compounds.

Because most DBPs are thought to form by reactions with aromatic functional groups that have one or more attached hydroxyl groups, the reactions between chlorine and pure phenolic compounds are described first, followed by a discussion of the reactions of chlorine with the complex mix of organic compounds typically found in both drinking water sources and domestic wastewater.

The reactions of HOCl with phenolic compounds in drinking water have long been known to produce malodorous, poor tasting products. These reactions often proceed initially via substitution of Cl for H at the ortho (*o-*) and para (*p-*) positions on the aromatic ring (those immediately adjacent to and directly opposite to the OH group). This step (the first step shown in Figure 10-8) yields a mixture of 2- and 4-chlorophenol, which can then be further substituted to generate 2,4- and 2,6-dichlorophenols and 2,4,6-trichlorophenol (Larson and Weber, 1994). Further attack is possible, particularly at the *p*-position, leading eventually to ring cleavage and formation of stable products such as trichloroacetic acid, but it occurs slowly (Onodera et al., 1984).

The chlorine incorporation reactions can typically be described by second-order rate expressions of the form

$$r_{TOTCl_2} = -k_{obs} c_{TOTCl_2} c_{TOTPh} \quad (10\text{-}41)$$

where c_{TOTCl_2} and c_{TOTPh} are the total concentrations of free chlorine and phenolic compound (Ph), respectively. As can be seen from Figure 10-9, the observed rate constant exhibits a strong pH dependence. This pH dependence can be explained if HOCl and PhO^- are assumed to be the principal reactants, so that the rate law has the form

$$r_{TOTCl_2} = -k_2 c_{HOCl} c_{PhO^-} \quad (10\text{-}42)$$

Comparing Equations 10-41 and 10-42, we see that $k_{obs} = k_2 \alpha_{0,HOCl} \alpha_{1,Ph}$.

When the data are analyzed in this way, the second-order rate constant k_2 is reasonably constant between pH 6 and 12, suggesting that the assumption that the reactants are primarily HOCl and PhO^- is valid over this pH range. The rate of reaction is increased by substitutions on the phenolic ring that push electrons onto the ring and thereby make it a stronger nucleophile (so-called *activation* of the ring). Substituting groups that activate the ring include hydroxide (OH^-), alkyl (CH_3^-), aryl ($C_6H_5^-$), methoxy (OCH_3^-), and phenoxy ($OC_6H_5^-$) groups. Thus, the incorporation of OH^- into a benzene ring to form phenol can be viewed as an activating step that causes phenol to be more susceptible to Cl incorporation than benzene; unactivated aromatic rings such as benzene show little tendency to become chlorinated under water treatment conditions.

Although the reactions of free chlorine with a few well-characterized compounds, such as phenol, are important in specific systems, most molecules that form DBPs are larger, incompletely characterized organic compounds. The pool of organic compounds that is present in a natural water or a wastewater is typically quantified by the *total organic carbon* (TOC) concentration in the water, and, in natural aquatic systems, the compounds themselves are referred to collectively as *natural organic matter* (NOM). The compounds contributing to these collective measures are sometimes subcategorized based on various physical/chemical properties into *humic* and *nonhumic* fractions; *hydrophobic*, *transphilic*, and *hydrophilic* fractions; or biochemical categories such as saccharides, proteins, and so on. These separation and characterization techniques have been reviewed by Frimmel (1998).

Organic molecules that react with chlorine to form DBPs are commonly referred to as DBP *precursors*. Water subjected to chlorination tend to have an extremely diverse collection of such precursors, leading to the generation of an equally diverse set of DBPs. Numerous studies have been conducted to identify the NOM fractions that seem to be the "best" precursors. The general outcome is that all fractions can serve as precursors, with certain fractions being more likely than others to generate specific types of DBPs.

The formation of THMs during chlorination of drinking water was first recognized by Rook (1974) and Bellar et al. (1974). Since that time, several individual DBP species have been unambiguously identified. Most prominent among these are species with the basic structure of methane (CH_4) and acetic acid (CH_3COOH), but in which some of the hydrogen atoms have been replaced by Cl or, if the source water contains Br^-, by Br. However, methane and acetic acid molecules are not the precursors of the halogenated forms of those molecules; rather, the DBPs form by halogenation of functional groups on larger molecules, which then break down to release the one- and two-carbon DBPs that are detected.

Although some mono- and di-halogenated methane and acetic acid molecules are released by such reactions, in most cases, the tri-halogenated species are found in much higher

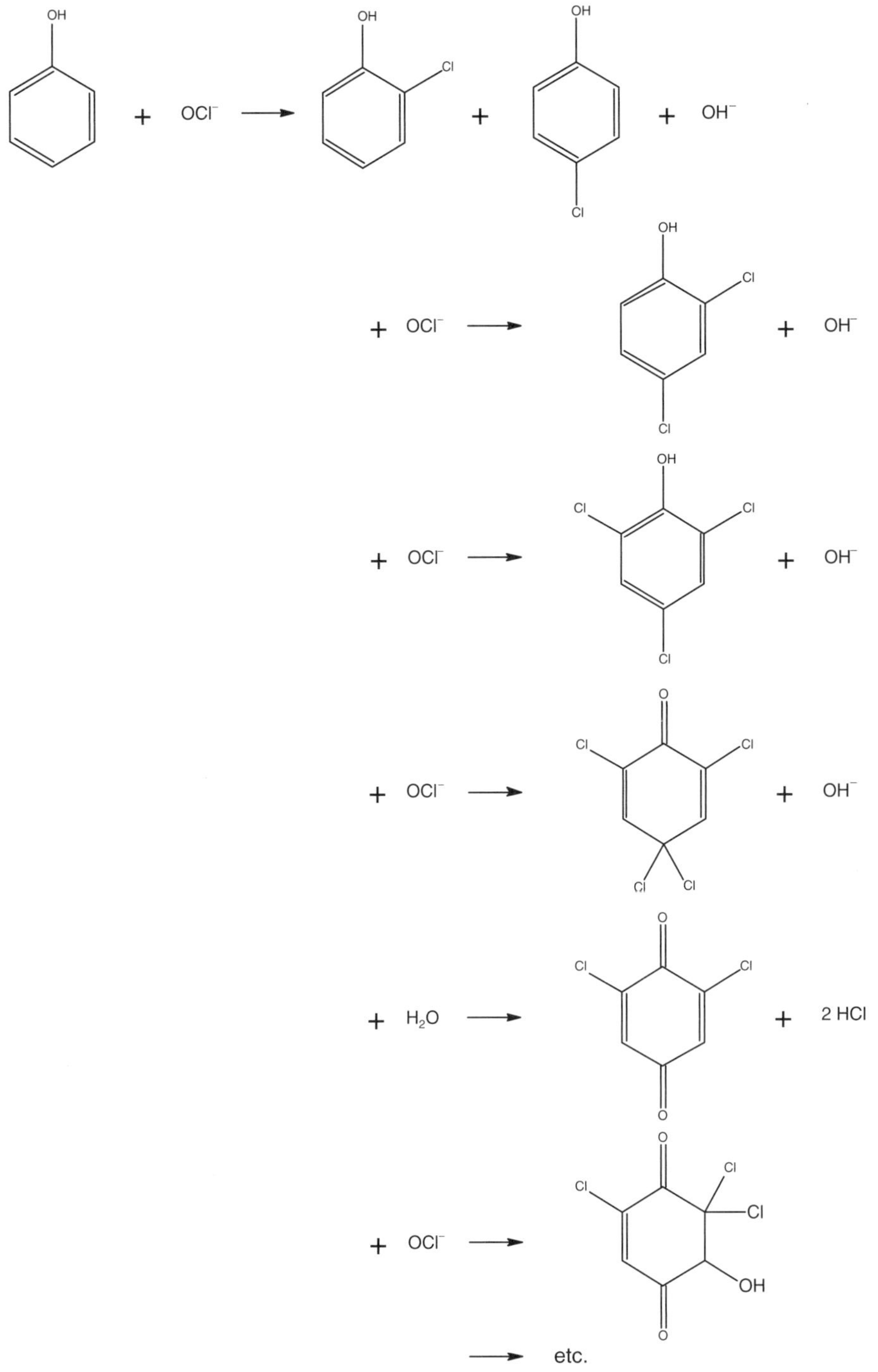

FIGURE 10-8. Pathway for reaction of hypochlorite with phenol.

concentrations. The implication is that the precursor molecules are relatively stable when one halogen attaches to a carbon atom, less stable when a second halogen attached, and still less stable once a third one attaches. The structures of chlorinated trihalomethanes (THMs) and haloacetic acids (HAAs) are illustrated in Figure 10-10. For the THMs, one, two, or all three of the chlorine atoms shown in the figure can be replaced by bromine, leading to four possible species. For the HAAs, bromine can substitute for any of the chlorine atoms, leading to nine possible species (two monohaloacetic acids, three dihaloacetic forms, and four trihaloacetic acid species). A key distinction between THMs and HAAs is that the former are volatile and therefore can easily be transferred from aqueous solutions into the gas phase, whereas the latter

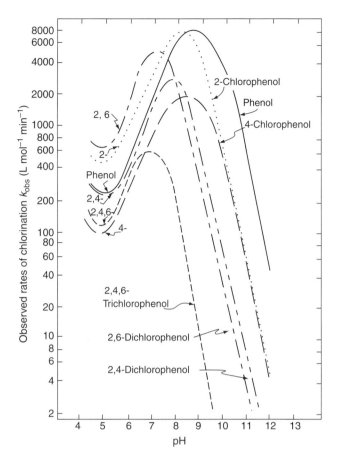

FIGURE 10-9. Variation in k_{obs} for chlorination of phenol and various chlorophenol intermediates. *Source*: From Stumm and Morgan (1996); reprinted with permission.

are essentially nonvolatile. Other specific DBPs that have been identified in most chlorinated drinking water, albeit often at very low levels, include chloral hydrate, haloacetonitriles, chloropicrin, and haloketones. Like THMs, these compounds are relatively volatile.

USEPA regulations establish maximum allowable levels (*maximum contaminant levels*, MCLs) of 80 and 60 μg/L for the sum of the four THMs (the *total trihalomethanes*, TTHMs) and the sum of five HAAs (HAA5), respectively, in drinking water.[5] The results of an early survey of the distribution of THMs in such water, made when the problems of THMs were first discovered, are shown in Figure 10-11. A later survey of drinking water in the United States (Arora et al., 1997) indicated that, unless existing treatment practices were modified, approximately 20% and 16% of the water treatment plants studied would exceed the MCLs for THMs and HAAs, respectively. In the intervening years, most plants have made changes in their operation (such as increasing the removal of NOM by enhanced coagulation, changing the location of first addition of disinfectant, or changing the primary or secondary disinfectant) to reduce the formation of THMs and HAAs.

The details of the reactions by which THMs and other DBPs are formed from complex precursors are not fully understood, but inferences have been drawn from studies of the chlorination of model compounds such as hydroxyl- and carboxyl-substituted, single aromatic rings such as resorcinol (1,3-dihydroxy benzene) and 3,5-dihydroxy benzoic acids. These compounds generate THMs via the so-called *haloform reaction*, which is initiated by the slight deprotonation of a carbon atom adjacent to a $C=O$ group (the α-carbon of a carbonyl compound). The deprotonated species exists in two resonance states, which together are referred to as an *enolate* anion. Enolate ions react with HOCl or HOBr by abstracting and bonding to the Cl^+ or Br^+ from the molecule; an example of this reaction sequence is shown in Figure 10-12.

Once the monochlorinated species is formed (second row in Figure 10-12), the identical sequence can repeat itself twice more to form the di-chloro (third row) and then tri-chloro species (fourth row). The tri-chlorinated form of the original molecule reacts with a hydroxyl group to generate a carboxylic acid and chloroform, as shown in bottom left of the figure. Formation of THMs tends to increase with increasing pH because of the role of the deprotonated form of the enolate ion and the reaction with hydroxyl ion that releases the THM from the chlorinated precursor.

Although the precursor of the enolate ion is shown in Figure 10-12 as a simple ketone, molecules that contain conjugated ketone and carboxylic acid groups (e.g., β-keto acids such as citric acid) or pairs of ketones are far more efficient precursors of THMs. Other functional groups that serve as good precursors of THMs have been discussed by Hanna et al. (1991) and Wu et al. (2000). While compounds as simple as those described earlier are unlikely to be present in natural water at significant concentrations, similar functional groups are expected to be present in NOM and may well represent the major active sites for attack by hypochlorous acid and hypochlorite ion.

The reactions leading to the formation of HAAs have been studied less extensively than those that form THMs. Presumably, the pathways for the formation of the two groups of DBPs have considerable overlap, with the species that are formed preferentially depending on the details of the chlorination process (solution pH, $Cl_2:DOC$ ratio) and the precursor structure. While most surveys indicate that HAAs are present in chlorinated water at molar concentrations slightly lower than those of THMs (Krasner et al., 1989), the relative proportions of these groups vary widely, depending on factors such as pH and bromide concentration

---

[5] The MCL for HAAs includes only five species (MCAA or monochloro-, DCAA or dichloro-, TCAA or trichloro-, MBAA or monobromo-, and DBAA or dibromo-acetic acid, referred to collectively as *HAA5*) because of limitations on the analysis of other HAAs at the time that the regulation was being developed. Targets of 50 μg/L for DCAA and 100 μg/L for TCAA have been set by the World Health Organization.

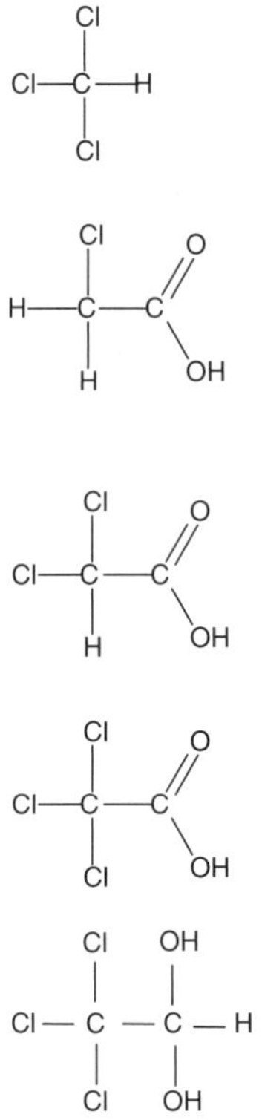

FIGURE 10-10. Common chlorinated disinfection by-products found in water treated with chlorine.

- Chloroform: $CHCl_3$ (example trihalomethane)
- Monochloroacetic acid (MCAA): $ClH_2C\text{-}COOH$ (example monohaloacetic acid)
- Dichloroacetic acid (DCAA): $Cl_2HC\text{-}COOH$ (example dihaloacetic acid)
- Trichloroacetic acid (TCAA): $Cl_3C\text{-}COOH$ (example trihaloacetic acid)
- Chloral hydrate (another low-molecular-weight by-product).

(Pourmoghaddas and Stevens, 1995). Most other DBPs that have been identified are found at concentrations that are lower than those of THMs and HAAs by an order of magnitude or more, and the pathways for the formation of those compounds have been explored less extensively.

As noted, bromine-containing DBPs are prevalent when water containing Br^- are chlorinated, due to the rapid oxidation of bromide to hypobromous acid, HOBr, and the subsequent participation of this species in halogenation reactions. In any realistic scenario, the HOCl dose is much greater than the Br^- concentration. As a result, when a solution containing Br^- is chlorinated, essentially all the Br^- is converted rapidly to OBr species, which then coexist with the OCl species. For example, complete oxidation of 120 μg/L Br^- (1.5×10^{-6} M) consumes only approximately 5% of a 2 mg/L Cl_2 (2.82×10^{-5} M) dose, so that both the hypobromite and hypochlorite oxidants are left in the solution. Both groups of oxidized halogens react with NOM, forming a mixture of halogenated DBPs, some containing only bromine (e.g., tribromomethane, $CHBr_3$, more commonly called bromoform), others only chlorine, and still others with both halogens. The substitution of HOBr into NOM is, in general, faster than that of HOCl, so that the ratio of bromine to chlorine in the DBPs is generally greater than the ratio of Br^- in the raw water to the Cl_2 dose. The extent of bromination of NOM is important because many brominated DBPs are thought to pose greater health risks than their chlorinated counterparts (WHO, 2000).

Because so many different halogenated DBPs are formed in any given system, and so many of those compounds have indeterminate structures, the cumulative concentration of all DBPs is sometimes reported in terms of the *total organic chlorine* (TOCl) or *total organic halogen* (TOX) concentration.[6] These quantities represent the total amount of Cl or halogen, respectively, in all the organic DBPs in the system, irrespective of the identity of the molecule to which they are attached. THMs and HAAs each usually represent 10–25% of the TOX, and it is rare for more than 50% of the TOX to be identified as specific compounds, even if sophisticated analytical protocols are used.

---

[6] The analytical test for TOCl or TOX involves collecting the halogenated organic molecules on an adsorbent, and then measuring the Cl or halogen concentration associated with those molecules. Because some DBPs might not be adsorbed in this process, the result is sometimes reported as the *adsorbable organic chlorine* (AOCl) or *adsorbable organic halogen* (AOX), respectively.

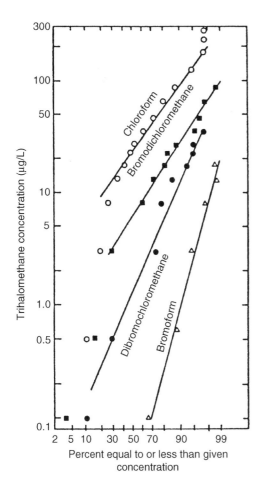

FIGURE 10-11. Distribution of concentrations of THMs in a survey of chlorinated drinking water in the United States. *Source*: From Symons et al. (1975); reprinted with permission.

In drinking water systems, chlorine is typically applied to water at dosages that provide a residual on the order of 0.2–1.0 mg/L as Cl_2 after the water has resided in the distribution system for hours to days. Laboratory studies of DBP formation reactions often include efforts to characterize either the concentration of DBPs expected to form under such typical chlorination scenarios, or the maximum concentrations expected based on the application of enough free chlorine to provide a Cl_2 residual of several milligrams per liter after a reaction period of several days. In the former case, the test is called a *simulated distribution system* (SDS) test, and in the latter, it is called a *formation potential* (FP) test. The results of formation potential tests are frequently reported in terms of the *trihalomethane formation potential* (THMFP), the *haloacetic acid formation potential* (HAAFP), *total organic halogen formation potential* (TOXFP), or similar parameters.

■ **EXAMPLE 10-7.** Korshin et al. (2004) reported the formation potential for several DBPs for the Jordan River (Utah). This sample of water had 4.8 mg/L dissolved organic carbon (DOC) and 233 μg/L bromide, and was dosed with 7.2 mg/L NaOCl *as Cl_2*. The resulting formation of DBPs was measured, and the values for the six compounds with the highest concentrations are reported in the following table. The first three compounds are trihalomethanes, while the latter three are haloacetic acids. Find the fraction of the DOC, the TOTOCl dose, and the total bromine in the system that is accounted for by these six compounds.

Compound	Concentration (μg/L)
$CHCl_3$	32.2
$CHCl_2Br$	47.5
$CHClBr_2$	41.7
$CH_2ClCOOH$	23.6
$CHCl_2COOH$	50.3
CCl_3COOH	29.4

Solution. In the following table, the given concentrations in μg/L for all compounds (column 2) are converted to molar units (column 4) by dividing by the molecular weights (column 3). Note that the atomic weight of Br (79.9) is considerably higher than that of Cl (35.45), so that the relative amounts of the three trihalomethane species on a molar basis are considerably different than on a mass basis.

Compound	Concentration (μg/L)	Molecular Weight	Concentration (μmol/L)	Carbon Content (μmol/L)	Chlorine Content (μmol/L)	Bromide Content (μmol/L)
$CHCl_3$	32.2	119.35	0.27	0.27	0.81	0
$CHCl_2Br$	47.5	163.8	0.29	0.29	0.58	0.29
$CHClBr_2$	41.7	208.25	0.20	0.20	0.20	0.40
$CH_2ClCOOH$	23.6	94.45	0.25	0.5	0.25	0
$CHCl_2COOH$	50.3	128.9	0.39	0.78	0.78	0
CCl_3COOH	29.4	163.35	0.18	0.36	0.54	0
			Total	2.40	3.16	0.69

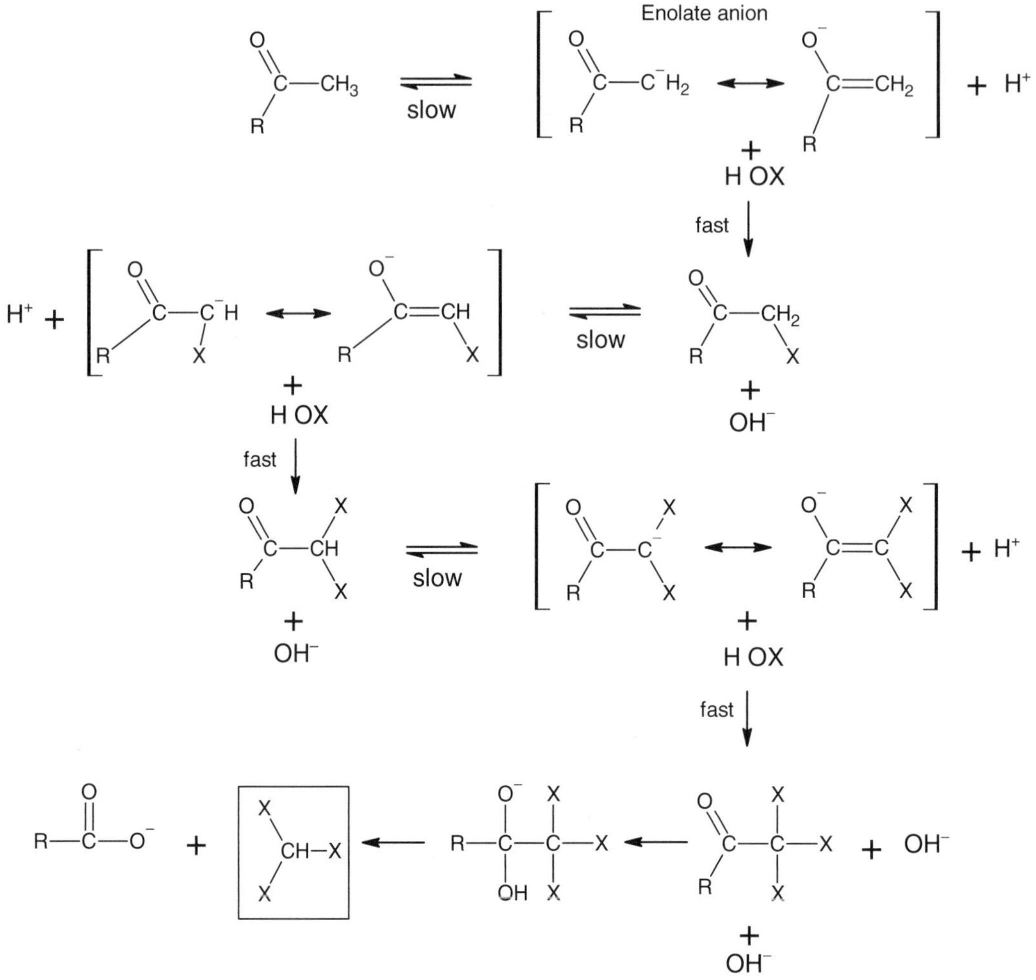

FIGURE 10-12. Pathway of the haloform reaction with methyl ketones. (X refers to any halogen.)
Source: From Morris and Baum (1978).

In columns 5–7, the molar concentration of each compound is multiplied by the number of C, Cl, and Br atoms in the molecule, respectively, to obtain the molar concentration of each atom accounted for by that compound. These values are summed at the bottom of each column to obtain the total molar concentrations of the three elements.

The DOC in the original water was

$$c_{DOC}(0) = 4.8\,\text{mg/L} = \frac{4.8\,\text{mg/L}}{12\,\text{mg/mmol}}(1000\,\mu\text{mol/mmol})$$

$$= 400\,\mu\text{mol/L}.$$

The fraction of carbon accounted for by the reported DBPs ($f_{C,DBP}$) is, therefore

$$f_{C,DBP} = \frac{c_{C,DBP}}{c_{C_o}} = \frac{2.40\,\mu\text{mol/L}}{400\,\mu\text{mol/L}} = 0.006, \text{ or } 0.6\%.$$

Noting that both NaOCl and have Cl_2 two equivalents per mole, the corresponding calculations for chlorine and bromine are as follows:

$$c_{Cl,added} = \left(7.2\frac{\text{mg}}{\text{L}}\text{ as }Cl_2\right)\left(\frac{1\,\text{mmol }Cl_2}{70.90\,\text{mg }Cl_2}\right)\left(\frac{2\,\text{meq}}{\text{mmol }Cl_2}\right)$$

$$\times \left(\frac{1\,\text{mmol NaOCl}}{2\,\text{meq}}\right)\left(\frac{1\,\text{mmol Cl}}{\text{mmol NaOCl}}\right)\left(\frac{1000\,\mu\text{mol}}{\text{mmol}}\right)$$

$$= 101\,\mu\text{mol/L Cl}$$

$$f_{Cl,DBP} = \frac{c_{Cl,DBP}}{c_{Cl,added}} = \frac{3.16\,\mu\text{mol/L}}{101\,\mu\text{mol/L}} = 0.032 \text{ or } 3.2\%.$$

$$c_{Br_o} = 233\,\mu\text{g/L} = \frac{233\,\mu\text{g/L}}{79.9\,\mu\text{g}/\mu\text{mol}} = 2.92\,\mu\text{mol/L}$$

$$f_{Br,DBP} = \frac{c_{Br,DBP}}{c_{Br_o}} = \frac{0.69\,\mu\text{mol/L}}{2.92\,\mu\text{mol/L}} = 0.237 \text{ or } 23.7\%.$$

Note that the high reactivity of the hypobromite species results in a large fraction of the bromine being incorporated into the DBPs, while only a small fraction of the added chlorine is incorporated. ■

The formation of both individual DBPs and the composite TOX often follows a pattern whereby some amount of the species is formed almost instantly (within seconds, before the first data point is collected), another portion forms relatively rapidly (over the course of several minutes to a few hours), and additional amounts form slowly but steadily for at least several days. These stages of DBP formation are often attributed to the reactions of different groups of precursors, although the overall process is more of a continuum than a sequence of distinct steps. An example of such a process is provided in Figure 10-13. That study was conducted under "THM formation potential conditions" (i.e., with a large Cl_2 residual present throughout the experiment). The authors defined the THM concentration that was generated during the first 3 h of reaction as the "rapid THMFP", and the concentration generated during 500 h of reaction as the "ultimate THMFP". A second-order model was found to describe the rate of THM formation during the second stage of the reaction:

$$r_{THM} = k[Cl_2][THMFP] = k[Cl_2]([THMFP]_{ult} - [THM])$$
(10-43)

where [THMFP] is the concentration of the (as yet unreacted) THM precursors remaining at time t, $[Cl_2]$ is the concentration of free chlorine at time t, and k is the second-order rate constant. It was suggested that the fast reacting THM precursors were resorcinol-like structures, while the slowly reacting precursors were other phenolic functional groups. Note that $[THMFP]_{ult}$ is the same thing as the final concentration of THMs formed. Thus, the form of the reaction expression shown on the right is similar to that of gas transfer reactions, in which the rate of reaction is proportional to the difference between the ultimate and current values.

Reactions of chlorine with other naturally occurring biopolymers are also of importance. For example, chlorination of chlorophyll, extracellular algal metabolites, and proteins can lead to high yields of THMs; the yield is defined as the amount of THM formed per unit of precursor, either in moles (for single known compounds) or grams of carbon (for mixtures, including natural waters). Similarly, the reaction of chlorine with lignin has historically been a key process in the bleaching of wood pulp and results in the production of a variety of oxidized and chlorinated species, including both high molecular weight "chloro-lignins" and smaller molecules (chlorinated aliphatic carboxylic acids, chloromuconic acids, and chlorinated phenolic species).

Chloramines

Formation of Chloramines In the presence of ammonia, free chlorine reacts rapidly and in a stepwise manner to form *mono-*, *di-*, and *tri-chloramine*, as shown. Trichloramine is also referred to as nitrogen trichloride.

$$NH_3 + HOCl \leftrightarrow NH_2Cl + H_2O \qquad (10\text{-}44)$$

$$NH_2Cl + HOCl \leftrightarrow NHCl_2 + H_2O \qquad (10\text{-}45)$$

$$NHCl_2 + HOCl \leftrightarrow NCl_3 + H_2O \qquad (10\text{-}46)$$

The chlorine atom in chloramines is in the +1 oxidation state (as it is in HOCl and OCl^-). Nevertheless, the properties of the hypochlorite species differ from those of the chloramines, so it is useful to distinguish between the two groups. To this end, chloramines are referred to as *combined* (as opposed to *free*) *chlorine*. The sum of the free and combined chlorine concentrations is referred to as *total chlorine*. Concentrations of combined and total chlorine, like those of free chlorine, are most commonly reported in units of milligrams per liter as Cl_2.

The ammonia that reacts to form chloramines may be present in water and wastewater as a result of decomposition of *N*-containing organic compounds, or it might be added to water either before or after dosing the water with chlorine, with the specific intent of forming chloramines as part of an overall disinfection strategy.

The dominant chloramine species that forms in a given solution depends on pH, the chlorine-to-ammonia ratio, the temperature, and the contact time. Trichloramine is not important under most conditions of interest due to either the preferential formation of the other forms and/or its instability. Within kinetic constraints (discussed below),

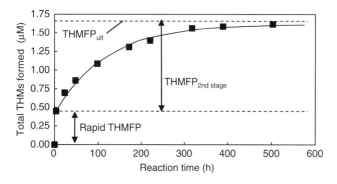

FIGURE 10-13. Formation of THM during chlorination of humic materials extracted from a Norwegian lake (pH 7, DOC 1.3 mg/L, phosphate buffer 5 mM, $[Cl_2]_o = 140\,\mu M$, $[THM]_{fin} = THMFP_{ult} = 1.62\,\mu M$, Rapid THMFP = 0.45 μM). Solid curve based on Equation 10-43, with $k = 0.017\,M^{-1}\,s^{-1}$. *Source*: After Gallard and von Gunten (2002).

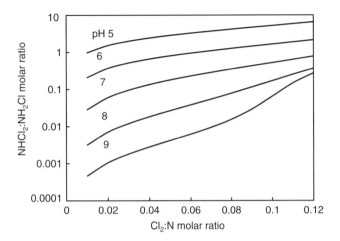

FIGURE 10-14. Effect of pH and Cl_2:N molar ratio on the equilibrium proportions of mono- and dichloramine (based on rate expressions given by Gray et al., 1978).

the relative concentrations of mono- and dichloramine are determined by the equilibrium of the following reaction:

$$H^+ + 2NH_2Cl \leftrightarrow NH_4^+ + NHCl_2 \qquad (10\text{-}47)$$

where $K_{eq} = \left(a_{NH_4^+} a_{NHCl_2}/a_{H^+} a_{NH_2Cl}^2\right) = 6.7 \times 10^5$ at 25°C (Gray et al., 1978). The initial rate of monochloramine formation when free chlorine is mixed with water containing ammonia is rapid ($t_{char} < 1$ min), with the result that monochloramine can accumulate to greater-than-equilibrium levels when the reagents are first mixed (Valentine and Jafvert, 1988). As pH decreases or the Cl_2:N ratio increases, the equilibrium ratio of dichloramine to monochloramine increases (Figure 10-14). (The chlorine-to-ammonia ratio is commonly designated as either Cl:N or Cl_2:N; in either case, the implication is that the chlorine dose is being expressed as Cl_2, and ammonia is being expressed as nitrogen [i.e., as NH_4–N].)[7]

Jafvert and Valentine (1992) presented a comprehensive model for the elementary reactions that form and decompose chloramines; reaction constants for some of the reactions were subsequently updated (Vikesland et al., 2001). The reaction scheme is summarized in Figure 10-15. Some key characteristics of the reaction sequence, particularly the formation and decay of the chloramines, are described next.

In the absence of other rapidly reacting Cl-consuming substances, the behavior of the chloramines is mostly dependent on the Cl_2:N molar ratio, but other solution characteristics (particularly pH and the presence of acid anions) also have noticeable effects. In general, when this ratio is less than one, virtually all of the free chlorine is consumed and monochloramine is the dominant species formed. The ratio of monochloramine to dichloramine decreases as the pH decreases from pH 8–6. Under these conditions, the species are relatively stable, decaying over a period of a few days.

However, when the Cl_2:N molar ratio is higher than approximately 1.5 (equivalent to Cl_2:N weight ratios of 7.6), the monochloramine and dichloramine that form decompose quite rapidly (with characteristic reaction times of a few minutes), so that any residual chlorine in the ultimate solution is present as free chlorine. The ammonia is oxidized, primarily to N_2, but some to NO_3^-. In the intermediate range ($1.0 < Cl_2$:N < 1.5), the decay of mono- and di-chloramine is much slower than at the higher ratios, but still faster than at the lower values. Some typical results from both experiments and model predictions for the rapid decay reactions (Cl_2:N > 1) are shown in Figure 10-16.

The reactions of free chlorine with ammonia are usually discussed in the context of the so-called *breakpoint chlorination curve*, an example of which is shown in Figure 10-17. In such a curve, the total residual chlorine is plotted as a function of the chlorine dose or the Cl_2:N ratio. In the absence of other species that consume chlorine rapidly, the chlorine residual is almost equal to the chlorine dose at low Cl_2:N ratios (region 1 in the figure). Under these conditions, almost all the chlorine residual is combined; that is, it is present as chloramines. As the Cl_2:N ratio increases, the chlorine residual also increases, but not quite stoichiometrically (region 2), because some of the added chlorine oxidizes N species rather than combines with them. As noted, several oxidation reactions might proceed, yielding N_2 or NO_3^- as products.

As the Cl_2:N ratio increases further, the oxidation of N-containing species becomes increasingly important, until it dominates the reactions that form chloramines. Under these conditions, not only the increment of chlorine that has just been added as Cl_2 or HOCl, but also the chlorine atoms in chloramines that had formed previously are reduced, and the nitrogen in the chloramines is oxidized. Since the chlorine that was in the chloramines contributed to the chlorine residual, this situation leads to the counterintuitive result that addition of free chlorine to the solution leads to a net decrease in the chlorine residual (region 3). In region 3, virtually all of the chlorine present is combined chlorine (i.e., chloramines) but a small amount is present as free chlorine; the relative amount of free chlorine increases (but is always small) as the Cl_2:N ratio increases.

---

[7] A Cl_2:N molar ratio of 1.0 yields a weight ratio of 71:14, or approximately 5. Most fundamental work is reported in terms of the molar ratio, but many reports in the environmental engineering literature use the weight ratio. Since the chemicals commonly used in treatment plants are Cl_2 gas and NH_3 gas, operators occasionally report weight ratios using these two gases, so that a molar ratio of 1.0 would give a weight ratio of 71:17, or 4.2. Weight ratios using Cl_2:NH_3 are never reported in the literature, so careful communication between engineers and treatment plant operators (who are accustomed to using different units from one another) is essential to ensure proper operation.

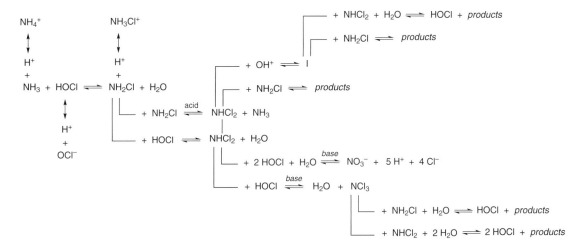

FIGURE 10-15. The formation and decomposition of chloramines. (I is an undetermined intermediate, and products include N_2, H_2O, Cl^-, H^+, NO_3^-, and other unidentified species.) *Source*: From Jafvert and Valentine (1992).

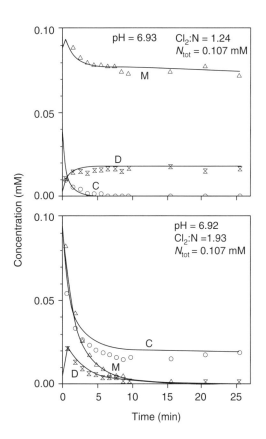

FIGURE 10-16. Sequential formation and decomposition of chloramine species. M refers to monochloramine, D to dichloramine, and C to free chlorine. The lower Cl_2:N molar ratio in the top figure allows the monochloramine to dominate in the time scale shown. The higher ratio in the bottom figure causes the rapid decay of the mono- and dichloramine, leaving the remaining chlorine present as free chlorine. *Source*: From Jafvert and Valentine (1992).

At a Cl_2:N molar ratio near 1.5 (corresponding to a Cl_2:N weight ratio of 7.6) the total chlorine residual passes through a minimum, known as the *breakpoint*. Once this point is passed (region 4), essentially all the ammonia has been oxidized, so any residual that is present is assumed to be free chlorine (though a small fraction of combined chlorine may remain). Stoichiometrically, the molar ratio of 1.5 is consistent with the oxidation of all the ammonia to molecular nitrogen (N_2).

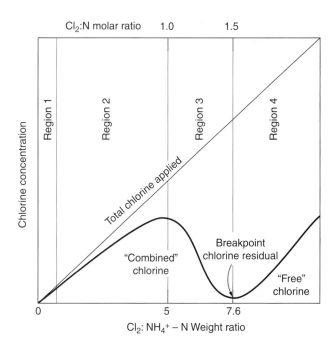

FIGURE 10-17. Concentration of generic chlorine species as a function of Cl_2:N weight ratio. *Source*: After White (1978).

■ **EXAMPLE 10-8.** In a certain municipal wastewater treatment plant, most of the ammonia originally present is nitrified (biologically oxidized to NO_3^-), but a concentration of 4.8 mg/L NH_3–N remains. What dose of Cl_2 is required to accomplish complete oxidation of the ammonia and leave a residual concentration of Cl_2 of 1.5 mg/L?

Solution. The dose required to accomplish the complete oxidation is estimated from the breakpoint curve as

$$(7.6 \text{ mg } Cl_2/\text{mg } NH_3\text{-N})(4.8 \text{ mg } NH_3\text{-N/L})$$
$$= 36.5 \text{ mg/L as } Cl_2.$$

Any chlorine added past the breakpoint remains in the water as free chlorine, so the required dose is approximately 38 mg/L as Cl_2. The answer is approximate because the value of 7.6 is inexact, and because it accounts only for the chlorine demand exerted by the ammonia. Reactions of chlorine with other reduced compounds in the wastewater would exert an additional demand and increase the required dose. Note that controlling the free residual concentration at doses beyond the breakpoint can be difficult because a small variation in the ammonia concentration can lead to a major swing (a factor of 7.6) in the chlorine dose required for oxidation. ■

Reactions of Chloramines with Inorganic Compounds
Chloramines are able to oxidize a range of reduced substances including ferrous iron, manganous ions and sulfidic species. The reaction between ferrous iron and monochloramine has been investigated in detail by Vikesland and Valentine (2002), who showed that the process is mediated by the formation of the free radical amidogen ($\dot{N}H_2$) and is catalyzed by the iron oxide solids that form.

■ **EXAMPLE 10-9.** The groundwater considered in Example 10-5 (containing 1.5 mg/L Fe(II)) is to be treated with monochloramine. Write a balanced chemical reaction for the oxidation of Fe^{2+} to $Fe(OH)_3(s)$ by monochloramine, and find the dose of monochloramine (as Cl_2) required for complete oxidation.

Solution. The oxidation of Fe(II) to Fe(III) (in this case, Fe^{2+} to $Fe(OH)_3(s)$) is a one-electron transfer, and the reduction of NH_2Cl to Cl^- is a two-electron transfer. Therefore, one molecule of NH_2Cl oxidizes two molecules of iron. The balanced equation is as follows:

$$2Fe^{2+} + NH_2Cl + 6H_2O \rightarrow 2Fe(OH)_3(s) + Cl^- + NH_3 + 5H^+$$

Because the concentrations of chloramines are expressed in the same units as for HOCl (i.e., both as Cl_2), the required dose is the same as in Example 10-5, or 0.95 mg/L as Cl_2. ■

Reaction of Chloramines with Organic Compounds. Although chloramines can react with organic matter to form some chlorinated DBPs, the extent of such reactions (i.e., the TOX formation) is much less than from an equivalent dose of free chlorine, with a few notable exceptions. In particular, formation of dichloroacetic acid and cyanogen chloride is often greater in chlorination than in chlorination. Little is known about the halogenated by-products of chloramination, except that they tend to be more hydrophilic and larger than the organic halides produced from free chlorine (Jensen et al., 1985). The decreased formation of THMs and most HAAs by chloramines compared to free chlorine has motivated many drinking water utilities to replace free chlorine with combined chlorine as the final disinfectant before the water is distributed to consumers.

Studies by Choi and Valentine (2002) and Mitch and Sedlak (2002) indicate that a probable human carcinogen, N-nitrosodimethylamine (NDMA), can be formed as a result of reactions between monochloramine and dimethylamine (DMA), which is a common component of human and animal waste and can remain in water even after secondary wastewater treatment. NDMA can also be formed by other reactions between organic and inorganic nitrogen species, including that between monochloramine and the DMA functional groups on anion exchange resins. It has been found in highly purified wastewater intended for recycle and in some treated drinking water and, in view of its likely high carcinogenicity, is recognized to be a particularly troublesome contaminant (Mitch et al., 2003).

Chlorine, Chloramines, and Lead Release in Water Distribution Systems. As noted previously, many drinking water utilities have replaced free chlorine with chloramines as the final disinfectant added during water treatment, because the chloramines generate much smaller quantities of THMs, HAAs, and most other chlorinated DBPs. While the switch in disinfectants is invariably successful in that regard, it has also been implicated as the cause of dramatic increase in the concentration of lead in tap water in a few locales. In the most intensely studied such system, lead levels in tap water in the Washington DC system increased immediately after the switch to combined chlorine was implemented, then decreased during a short period when it was reversed, and immediately increased when chloramines were used again (Edwards and Dudi, 2004). Subsequent testing confirmed that, although lead release from pure metallic lead pipe is not dramatically different when exposed to the two oxidants, lead release from brass fittings and, especially, from older pipes that had already accumulated lead-bearing scales is much greater when exposed to chloramines than free chlorine. Because the cognitive development of infants and children can be affected by very low exposure to lead (Fewtrell et al., 2004), and in view of an apparent link between the change in chloramines and instances of elevated blood lead levels in children (Miranda et al., 2007; Edwards et al., 2009), this redox-based water quality problem attracted a great deal of scientific and engineering research (as well as media and political attention). In Washington DC, the goal of controlling both DBP formation and lead release seems to have been achieved by using chloramine as the disinfectant and adding phosphate to immobilize lead.

The chemistry underlying these observations is complex and has not yet been fully characterized. However, the current understanding is that in any oxic system (even in the absence of a disinfectant), the surface of metallic lead (Pb(s)) is oxidized to Pb(II). For typical drinking water conditions, and in the absence of free chlorine, Pb(II) is likely to precipitate as cerrusite ($PbCO_3(s)$), hydrocerrusite ($Pb_3(OH)_2(CO_3)_2(s)$), or litharge ($PbO(s)$). While these solids are quite insoluble (see Figure 9-18), they can, under some circumstances, release enough lead to exceed the USEPA's action level of 15 μg/L. The rate and amount of dissolution can be significantly enhanced by the presence of NOM (Dryer and Korshin, 2007; Lin and Valentine 2008).

Theoretical calculations suggest that, under the oxidizing conditions generated by chloramine addition, most of the Pb(II) present should be oxidized to Pb(IV). In this oxidation state and at near-neutral pH, the lead is much less soluble than Pb(II) and precipitates as various $PbO_2(s)$ solids (Lytle and Schock, 2005). However, the oxidation of Pb(II) to Pb(IV) by chloramines is slow under drinking water conditions and does not proceed to a significant extent in time frames of interest.

In contrast, when free chlorine is present at residual concentrations commonly found in drinking water, the solution is considerably more oxidizing than when chloramines are present, and the Pb(II) is readily oxidized to Pb(IV). For this reason, water distribution systems that maintain a residual of free chlorine rarely have problems with excessive lead release, although $PbO_2(s)$ solids do accumulate in the corrosion scales. If these solids are exposed to unaltered NOM in the absence of free chlorine, they are prone to dissolution via the reduction of the Pb(IV) to Pb(II), but the dissolution does not occur when free chlorine is present. The difference between these two responses might be because the chlorine rapidly reoxidizes any Pb(II) that forms or because the chlorine reacts with the NOM to eliminate many of the NOM functional groups that react with the Pb(IV).

As noted above, when free chlorine is replaced with combined chlorine as the disinfectant, the oxidation potential usually remains above that necessary to stabilize Pb(IV), so switching disinfectants in otherwise pristine water should not induce reduction and dissolution of the lead. However, because chloramines do not react strongly with NOM and cannot rapidly reoxidize any Pb(II) that is generated by the reaction of NOM with Pb(IV), dissolution of $PbO_2(s)$ solids can and does occur when free chlorine is replaced with combined chlorine in a real distribution system. The release of lead into the water supply in such cases occurs not only because Pb(II) enters solution, but also because the bonds holding various lead-bearing solids in the scale can dissolve, leading to intermittent and unpredictable spikes in the total lead concentration in the water (Triantafyllidou et al., 2007). For reasons that are not clear at present, the switch to chloramines also seems to facilitate the release of particles of lead from solder under at least some circumstances (Edwards et al., 2009).

Chlorine Dioxide

Chlorine dioxide (ClO_2) is a relatively stable free radical ($O = \dot{C}l - O$) that is applied as an oxidant in a variety of industrial processes (e.g., cellulose bleaching) and as either an oxidant or a disinfectant in wastewater and drinking water treatment processes. Its use is attractive, in part, because it has a much lower tendency than free chlorine to produce chlorinated by-products, especially trihalomethanes and haloacetic acids. Chlorine dioxide is highly soluble in water and, once in solution, neither hydrolyzes nor dissociates, but remains as a neutral species. However, it is extremely volatile ($H_{pm} = 53.6$ atm/(mol/L)) and can be removed from dilute aqueous solutions with minimal aeration.

When ClO_2 oxidizes aqueous contaminants, it usually acquires a single electron and is thereby reduced to the chlorite anion (ClO_2^-):

$$ClO_2(aq) + e^- \leftrightarrow ClO_2^-$$
$$pe = 16.2, E_H^o = 0.954 \text{ V} \quad (10\text{-}48)$$

A major limitation of the use of chlorine dioxide, at least in drinking water applications, is that chlorite has a number of potential adverse health effects (based on studies with laboratory animals). In light of this, the USEPA has established an MCL of 1.0 mg/L for chlorite at representative locations in the distribution system, and a maximum concentration of ClO_2 (referred to in the regulation as a maximum residual disinfectant level, or MRDL) of 0.8 mg/L at the point of entry to the distribution system (USEPA, 1998a). A secondary concern is that, if present in tap water, chlorine dioxide can volatilize into the home or office environment and can lead to offensive chlorinous odors (Singer and Reckhow, 1999). Given that chlorine dioxide is rapidly consumed during water treatment and that up to 70% of the applied chlorine dioxide is reduced to chlorite, the practical upper limit for chlorine dioxide doses is approximately 1.5 mg/L unless chlorite is removed; such removal can be accomplished by the addition of a reductant (such as ferrous iron), which can reduce the chlorite to chloride according to the following half-reaction:

$$ClO_2^- + 4H^+ + 4e^- \rightarrow Cl^- + 2H_2O \qquad (10\text{-}49)$$

Another concern associated with residual chlorite in the distribution system is that it reacts with free chlorine to produce low levels of chlorine dioxide (which could cause the MRDL to be exceeded in the distribution system) and/or chlorate (ClO_3^-) by the following reactions:

$$HOCl + 2ClO_2^- \rightarrow 2ClO_2 + Cl^- + OH^- \qquad (10\text{-}50)$$

$$HOCl + ClO_2^- \rightarrow ClO_3^- + Cl^- + H^+ \qquad (10\text{-}51)$$

The presence of chlorate is generally considered undesirable, although no MCL has been established for it.

Generation of Chlorine Dioxide Chlorine dioxide gas is highly explosive at partial pressures exceeding 0.1 atm in air and thus cannot be compressed or stored commercially. For the same reason, it is unsafe to store concentrated aqueous solutions of ClO_2, because volatilization could release sufficient ClO_2 into the atmosphere to generate a hazard if the gas phase were closed or poorly ventilated. Therefore, for potable water treatment processes, aqueous stock solutions of between 0.1% and 0.5% ClO_2 by weight are commonly produced on-site, by one of several generation technologies (USEPA, 1999). The most common of these use sodium chlorite ($NaClO_2$) as the principal feedstock chemical, forming ClO_2 by oxidation of the chlorite ion with gaseous chlorine or hypochlorous acid or by disproportionation of the ion in a strong acid (typically, HCl):

$$2ClO_2^- + Cl_2(g) \leftrightarrow 2ClO_2 + 2Cl^- \qquad (10\text{-}52)$$

$$2ClO_2^- + HOCl \leftrightarrow 2ClO_2 + Cl^- + OH^- \qquad (10\text{-}53)$$

$$5ClO_2^- + 4H^+ \leftrightarrow 4ClO_2 + Cl^- + 2H_2O \qquad (10\text{-}54)$$

Careful control of the ClO_2 generation process is required to avoid the formation of chlorate (ClO_3^-); this control involves ensuring that the pH is kept low and that the proper stoichiometric ratios of the constituents are maintained. Although these processes are largely automated, they require knowledgeable operators; as a result, the use of ClO_2 has been limited primarily to larger treatment plants that have sufficient trained personnel. Emerging technologies for ClO_2 generation for water treatment include electrochemical processes, a process in which gaseous chlorine flows through chlorate embedded in an inert solid matrix, and a process similar to one currently used in the pulp and paper industry involving reduction of chlorate by acidified hydrogen peroxide.

Chlorine dioxide disproportionates in alkaline waters as follows:

$$2ClO_2 + 2OH^- \rightarrow ClO_2^- + ClO_3^- + H_2O \qquad (10\text{-}55)$$

Both ClO_2^- and ClO_3^- are undesirable end products, so that chlorine dioxide is an inappropriate choice for water that are high in pH or that are treated by precipitative softening.

Reactions of Chlorine Dioxide with Inorganic Compounds Chlorine dioxide reacts with a variety of inorganic species at rates that are typically first order with respect to each reactant (i.e., second order overall). Second-order rate constants derived by Hoigne and Bader (1994) for a number of species of importance in water and wastewater treatment are listed in Table 10-3. In addition to rate constants for reaction with both the acidic and basic forms of the various inorganic species (HA and A^-, respectively), values of k_{tot} (the overall rate constant for reaction with the sum of HA and A^-) extrapolated or interpolated to pH 8 are given for ease of comparison at a pH of relevance to water treatment. The rapid oxidation of Fe(II) and Mn(II) by ClO_2 (and also by ClO_2^-) has made it a popular choice for treatment of feed

TABLE 10-3. Second-Order Reaction-Rate Constants for Consumption of Chlorine Dioxide by Inorganic Compounds

Reactant	pK_a^a	k_{HA} (M^{-1} s^{-1})	k_{A^-} (M^{-1} s^{-1})	k_{tot} for pH 8 (M^{-1} s^{-1})
Water (H_2O/OH^-)	14		$\ll 0.05$	$\ll 0.05$
Hydrogen peroxide (H_2O_2/HO_2^-)	11.7	<0.1	$(1.3 \pm 0.2) \times 10^5$	30 ± 3
Bromide (Br^-)			$<10^{-4}$	<0.05
Iodide (I^-)			1400 ± 200	1400 ± 200
Ammonia (NH_4^+/NH_3)	9.24	$<10^{-6}$	$<10^{-6}$	$<10^{-6}$
Cyanide (HCN/CN^-)	9.21	$<10^{-5}$	8000 ± 2000	460 ± 100
Nitrite (HNO_2/NO_2^-)	3.15		113 ± 3	113
Iron(II) ($Fe^{2+}/FeOH^+$)	9.40	3000 ± 500		[b]
Manganese ($Mn^{2+}/MnOH^+$)	10.60		2×10^7	5×10^4
Ozone (O_3)		1370 ± 50		1400

[a] From database in Visual Minteq (Gustafsson, 2011).
[b] Precipitation of Fe(III) oxyhydroxides affects rate at pH 8.
Source: After Hoigne and Bader (1994).

water that contains these reduced species and also has a high DBP formation potential (which makes addition of free chlorine problematic until after substantial removal of NOM). However, the presence of organic matter can also limit the effectiveness of ClO_2 for oxidation of inorganic species such as Fe(II) and Mn(II), due to both competition between the organics and the metals for the oxidant and the formation of metal–organic complexes (Knocke et al., 1990; van Benschoten et al., 1992).

■ **EXAMPLE 10-10.** Write balanced chemical reactions for the oxidation of Fe^{2+} by both chlorine dioxide and chlorite. What dose of ClO_2 is required to oxidize the 1.5 mg/L Fe(II) in the groundwater considered in prior examples? Assume that 40% of the oxidation occurs by the one-electron transfer in which chlorine dioxide is reduced to chlorite, and that the other 60% occurs by the four-electron transfer that converts chlorite to chloride. What concentration of chlorite remains in solution, assuming no other demands exist for either oxidant.

Solution. The balanced reactions are as follows:

$$Fe^{2+} + ClO_2 + 3H_2O \rightarrow Fe(OH)_3(s) + ClO_2^- + 3H^+$$
$$4Fe^{2+} + ClO_2^- + 10H_2O \rightarrow 4Fe(OH)_3(s) + Cl^- + 8H^+$$

A total of 40% of the iron is oxidized according to the first equation, so the stoichiometric requirement for that reaction is

$$(0.4)(1.5 \text{ mg Fe/L})\left(\frac{1 \text{ mmol Fe}}{55.85 \text{ mg}}\right)(1 \text{ mmol } ClO_2/\text{mmol Fe})$$

$$= 0.011 \text{ mmol } ClO_2/L$$
$$= (0.011 \text{ mmol } ClO_2/L)(67.45 \text{ mg } ClO_2/\text{mmol } ClO_2)$$
$$= 0.73 \text{ mg } ClO_2/L$$

Because ClO_2 and ClO_2^- have identical molecular weights, the concentration of ClO_2^- formed as a product of the reaction is the same as the concentration of ClO_2 consumed; that is, 0.73 mg/L ClO_2^-. The other 60% of the Fe^{2+} is oxidized by a portion of this ClO_2^- according to the second reaction:

$$(0.6)(1.5 \text{ mg Fe/L})\left(\frac{1 \text{ mmol Fe}}{55.85 \text{ mg}}\right)\left(\frac{1 \text{ mmol } ClO_2^-}{4 \text{ mmol Fe}}\right)$$

$$= 0.0040 \text{ mmol } ClO_2^-/L$$
$$= (0.0040 \text{ mmol } ClO_2^-/L)$$
$$\quad \times (67.45 \text{ mg } ClO_2^-/\text{mmol } ClO_2)$$
$$= 0.27 \text{ mg } ClO_2^-/L$$

The remaining, unreacted chlorite is $(0.73 - 0.27)$ mg $ClO_2^-/L = 0.46$ mg ClO_2^-/L. The USEPA has set the MCL for chlorite in drinking water at 1 mg/L. If all of the oxidation of the Fe(II) were accomplished by the one-electron transfer reaction, the dose of ClO_2 and the residual ClO_2^- concentration would exceed this limit. ■

Reactions of Chlorine Dioxide with Organic Compounds
Chlorine dioxide undergoes a wide variety of redox reactions with organic matter to form oxidized organics and reduced Cl species (Singer and Reckhow, 1999). The dominant Cl-containing species generated is usually chlorite, accounting for 50–70% of the chlorine dioxide consumed. Like hypochlorite species, chlorine dioxide can initiate chlorine substitution reactions leading to the formation of chlorinated aliphatic and aromatic compounds; however, trihalomethanes have not been detected as reaction products, a major advantage of using chlorine dioxide as the first oxidant in a treatment train. Lalezary et al. (1986) compared the effectiveness of several oxidants for the oxidation of common taste and odor-causing compounds and found chlorine dioxide to be the most effective. However, removal of geosmin and 2-methylisoborneol (MIB) (the two compounds identified most often as sources of taste and odor in surface water supplies) by reaction with ClO_2 at concentrations that can be used in water treatment plants was only approximately 30%.

Chlorine dioxide reacts particularly selectively with phenols. Hoigne and Bader (1994) investigated the reaction of chlorine dioxide with a wide range of phenols (as well as a variety of other organic and inorganic species) and found that the reaction kinetics are first order in both the chlorine dioxide and phenol concentration, and second order overall. The effects of substituents on the reaction rate constants of the phenols are significant. Rate constants for the reaction of ClO_2 with selected phenols are shown in Figure 10-18. Since it is the phenoxide anion that reacts rapidly (up to 10^6 times as quickly as the nondissociated species), the apparent total reaction rate constant increases for most types of phenols by a factor of 10 per pH increment over most of the pH range of interest in water treatment.

Ozone

Ozone has been used for many years to treat drinking water, industrial wastewater, cooling water, swimming pools, and aquaria. It was first used as a potable water disinfectant in 1893 in Oudshoorn, The Netherlands, and in 1906, ozonation was installed in Nice, France, to treat domestic water supplies. The major goal of treatment has been to disinfect, though improvement of taste, reduction of color and odor, oxidation of iron and manganese, enhancement in biodegradability of organic compounds, and improvements in

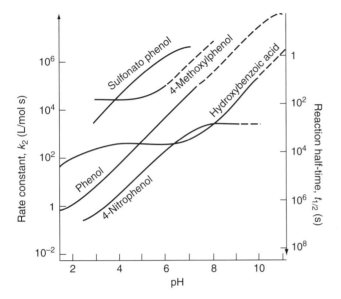

FIGURE 10-18. Reaction-rate constants and half-times of selected phenols in the presence of chlorine dioxide. Assumptions for $t_{1/2}$ computation: $c_{ClO_2} = 1\,\mu M$; $c_{phenol} \ll c_{ClO_2}$. *Source*: After Hoigne and Bader (1994).

flocculation and filtration performance have been important secondary goals. While ozone has been applied principally to drinking water (Langlais et al., 1991; USEPA, 1999), the application to wastewater is increasing thanks to advances in ozone generation and gas transfer technology (Metcalf and Eddy, 2003; Xu et al., 2002).

At room temperature, ozone exists as a colorless gas with a pungent odor that is readily detectable at concentrations as low as 0.02–0.05 ppm (by volume). Ozone is sparingly soluble in water. At 25°C, the Henry's constant (H_{pm}) is 107 atm/(mol/L), making ozone less volatile (i.e., more soluble) than oxygen, but significantly more volatile than chlorine. For example, the solubility of pure ozone at 1.0 atm pressure and 25°C is 450 mg/L. When ozone is used in water treatment, its gas-phase mole fraction is typically less than 12%; as a result, the mass transfer driving force is limited, with the result that dissolved ozone concentrations greater than about 2–3 mg/L are difficult to achieve.

Ozone is a powerful oxidant and can oxidize many substrates directly, as described shortly. Simultaneously, however, some ozone reacts with hydroxide ions to produce molecular oxygen and the hydroperoxy anion (HO_2^-, the conjugate base of hydrogen peroxide [H_2O_2]) (Equation 10-56). The latter product can react further with ozone to produce two free radicals: the uncharged hydroxyl radical ($\dot{O}H$), which can be thought of as a hydroxyl ion from which an electron has been abstracted, and the superoxide anion ($\dot{O}_2^-$), which can be thought of as an oxygen molecule to which an electron has bound (Equation 10-57). The net result of these reactions is the conversion of two molecules of ozone and one hydroxide ion into two molecules of oxygen and the two free radicals (Equation 10-58).

$$O_3 + OH^- \rightarrow HO_2^- + O_2 \qquad k = 70\,M^{-1}\,s^{-1} \quad (10\text{-}56)$$

$$O_3 + HO_2^- \rightarrow \dot{O}H + \dot{O}_2^- + O_2 \quad k = 2.8 \times 10^6\,M^{-1}\,s^{-1} \quad (10\text{-}57)$$

$$2O_3 + OH^- \rightarrow \dot{O}H + \dot{O}_2^- + 2O_2 \quad (10\text{-}58)$$

In this two-reaction sequence, the second reaction has a far higher rate constant than the first. As a result (as explained in Chapter 3), the intermediate HO_2^- does not build up to any appreciable concentration in any system, the overall rate of ozone decomposition (Equation 10-59) is dictated by the first reaction in the sequence, and the rate at which ozone decays via this pathway increases dramatically with increasing pH. That is,

$$r_{O_3,decomp} = -kc_{O_3}c_{OH^-} = -(70\,M^{-1}\,s^{-1})c_{O_3}c_{OH^-} \quad (10\text{-}59)$$

As noted previously, the existence of free radicals such as those generated by the earlier reaction sequence is the central feature of advanced oxidation processes (AOPs), which are addressed in a subsequent section of this chapter. For now, the main point is simply that ozone can react directly to oxidize substrates or it can decay and generate free radicals in the process. For ozone to be effective as a direct oxidant, the rate of the oxidation reaction must be faster than (or at least comparable to) the rate of ozone decomposition. In this section, we consider the generation of ozone and its direct reactions with substrates.

Ozone Generation Because ozone is an unstable molecule, it must be generated at the point of application. On-line production is typically accomplished using an ozone generator fed with either dried air or oxygen. In such generators, ozone is formed by an electrical (corona) discharge in an air gap between two electrodes separated by a dielectric (e.g., between glass tubes) (see Figure 10-19). The electrodes are connected to a high voltage generator that supplies a peak voltage in the 10–30 kV range. The amount of ozone produced depends on the frequency used, with low frequencies (50–60 Hz) being adequate for small applications (<250 kg/d), medium frequencies (<1000 Hz) suited to requirements of 1000 kg/d, and high frequencies (>1000 Hz) being used in sophisticated, high production units (Hoigne, 1998). If air is used, the most economical operation gives a product stream that contains about 2% ozone by volume. Enhancement of the amount of oxygen in the stream increases the economical yield of ozone; for example, use of pure oxygen enables the generation of a stream containing 8–12% ozone.

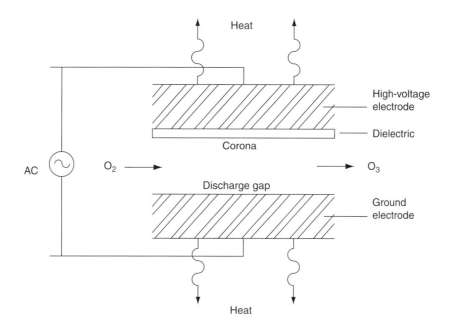

FIGURE 10-19. Schematic of a basic ozone generator. *Source*: USEPA (1999).

Efficient transfer of the ozone into the solution is critical and is normally achieved by diffusion of the ozone-laden gas into a counter-current flow of water via bubble diffuser contactors, injectors or turbine mixers (Langlais et al., 1991). Advantages and disadvantages of each method are summarized in Table 10-4. Any ozone that is not transferred to solution must be destroyed, as the off-gas concentrations are still well above fatal concentrations. For example, at 90% transfer efficiency, a 3% ozone feed stream will still contain 3000 ppm$_v$ of ozone in the off-gas (USEPA, 1999). The maximum acceptable exposure to gaseous ozone is 0.3 ppm$_v$ for 10 min. Ozone destruction is normally accomplished by converting the ozone to oxygen, either by exposure to high temperatures (>350°C) or with the aid of a catalyst at lower temperatures (Kameya and Urano, 2002).

■ **EXAMPLE 10-11.** In a laboratory-scale batch reactor, ozone is bubbled into a well-buffered solution at pH 6.0 and 25°C. The gas-phase ozone concentration is 0.025 atm, and a steady-state concentration of 9.5 mg/L O$_3$ is achieved in the liquid phase. No reactions other than the ozone decomposition occur, and the depth of the reactor is small enough that differences in the gas phase concentration with depth can be ignored. The Henry's constant for ozone, H_{pm}, is 107 atm/(mol/L). The problem is based on the work of Gurol and Singer (1982).

(a) What is the value of the gas transfer rate constant, $K_L a_L$?
(b) What will be the steady-state concentration of O$_3$ if the pH is raised to 8.0?

TABLE 10-4. Advantages and Disadvantages of Various Ozone Contactor Types

Contactor Type	Advantages	Disadvantages
Bubble diffuser	No moving parts Effective ozone transfer Low hydraulic headloss Operational simplicity	Deep contact basins (6–7 m) Vertical channeling of bubbles Maintenance of gaskets and piping
Injector dissolution	Injection and static mixing have no moving parts Very effective ozone transfer Contactor depth less than bubble diffusion	Additional headloss due to static mixers Turndown capability limited by injection system More complex operation and high cost
Turbine mixer	Enhanced ozone transfer due to high turbulence and small bubble size Contactor depth less than bubble diffusion Aspirating turbines can draw off-gas from other chambers for reuse Eliminates diffuser clogging concerns	Energy requirement Constant gas flow rate regardless of water flow or oxidant demand Maintenance requirements for turbine and motor

Source: After USEPA (1999).

Solution. (a) Because the composition of the gas phase in the reactor is approximately independent of depth, gas transfer is limited by the kinetics of interfacial transport, and the overall rate constant for gas transfer, k_I, can be equated with $K_L a_L$. The rate of ozone decay is characterized by Equation 10-59, so a mass balance on ozone in this batch reactor is as follows:

$$\begin{bmatrix} \text{Rate of change} \\ \text{of mass of ozone} \\ \text{in liquid phase} \\ \text{in reactor} \end{bmatrix} = \begin{bmatrix} \text{Rate of change} \\ \text{of mass of ozone} \\ \text{due to} \\ \text{gas transfer} \end{bmatrix}$$
$$+ \begin{bmatrix} \text{Rate of change} \\ \text{of mass of ozone} \\ \text{due to} \\ \text{ozone decomposition} \end{bmatrix}$$

$$0 = V K_L a_L (c_{O_3}{}^* - c_{L,O_3}) - V k c_{O_3} c_{OH^-}$$

We know the following values:

$$k = 70\,\text{M}^{-1}\,\text{s}^{-1}$$
$$c_{OH^-} = \frac{K_w}{[H^+]} = \frac{10^{-14}}{10^{-6}} = 10^{-8}\,\text{M}$$
$$c_{O_3} = (9.5\,\text{mg/L})\left(\frac{1\,\text{mol}}{48{,}000\,\text{mg}}\right) = 1.98 \times 10^{-4}\,\text{M}$$

We can find the concentration of ozone in equilibrium with the gas phase ($c_{O_3}{}^*$) from Henry's law:

$$c_{O_3}{}^* = \frac{P_{O_3(g)}}{H_{pm}} = \frac{0.025\,\text{atm}}{107\,\text{atm/M}} = 2.34 \times 10^{-4}\,\text{M}$$
$$= (2.34 \times 10^{-4}\,\text{M})(48{,}000\,\text{mg/mol}) = 11.2\,\text{mg/L}$$

The mass balance equation can therefore be rearranged to

$$K_L a_L = \frac{k c_{O_3} c_{OH^-}}{c_{O_3}{}^* - c_{O_3}}$$
$$= \frac{(70\,\text{M}^{-1}\,\text{s}^{-1})(1.98 \times 10^{-4}\,\text{M})(10^{-8}\,\text{M})}{2.34 \times 10^{-4}\,\text{M} - 1.98 \times 10^{-4}\,\text{M}}$$
$$= 3.85 \times 10^{-6}\,\text{s}^{-1}$$

(b) At pH 8, $c_{OH^-} = 10^{-6}$ M. The mass balance can be rearranged to find

$$c_{O_3} = \frac{K_L a_L\, c_{O_3}{}^*}{k c_{OH^-} + K_L a_L}$$
$$= \frac{(3.85 \times 10^{-6}\,\text{s}^{-1})(2.34 \times 10^{-4}\,\text{M})}{(70\,\text{M}^{-1}\,\text{s}^{-1})(10^{-6}\,\text{M}) + 3.85 \times 10^{-6}\,\text{s}^{-1}}$$
$$= 1.22 \times 10^{-5}\,\text{M} = 0.58\,\text{mg/L}$$

The example illustrates that the rapid kinetics of ozone decomposition limits the ozone concentration that can be achieved in the reactor, especially at pH greater than 7. ∎

Reaction of Ozone with Inorganic Compounds. Ozone reacts with a variety of inorganic compounds, most often via oxygen-atom transfer reactions such as the following:

$$Fe^{2+} + O_3 \rightarrow FeO^{2+} + O_2 \quad (10\text{-}60)$$
$$NO_2^- + O_3 \rightarrow NO_3^- + O_2 \quad (10\text{-}61)$$
$$Br^- + O_3 \rightarrow BrO^- + O_2 \quad (10\text{-}62)$$

Important aspects of several such reactions are summarized later based on the work of Hoigne (1998), and examples of second-order rate constants for related reactions are shown in Figure 10-20.

- *Hydrogen sulfide/bisulfide/sulfide* ($H_2S/HS^-/S^{2-}$). Hydrogen sulfide and its deprotonated analogs are generally oxidized by O_3 before O_3 is significantly consumed by any other solute. The reactions are of significance when groundwater containing sulfide are treated with ozone.
- *Sulfite species* (HSO_3^-/SO_3^{2-}). Sulfite species also react very rapidly with ozone. The reactions are particularly important in cloud chemistry, as they may control the oxidation of aqueous SO_2 species to sulfuric acid, a major cause of acid precipitation.

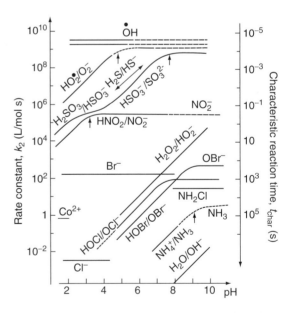

FIGURE 10-20. Second-order rate constants for direct reaction of ozone with inorganic solutes as a function of pH. Right scale, t_{char}, is the reaction time required to reduce the concentration of the solute to $1/e$ of its original value, for a constant $O_3(aq)$ concentration of 10 μM (approximately 0.5 mg/L) in a batch system. Vertical arrows indicate pK_a values of the indicated acid–base pairs. *Source*: After Hoigne (1998).

Iron(II), copper(I), manganese(II). Ozone oxidation of ferrous iron exhibits a surprisingly low rate constant (Loegager et al., 1992), whereas the oxidation of Cu(I) appears to proceed at almost diffusion controlled rates (Hoigne, 1998). Mn(II) oxidation by ozone is relatively rapid, with rate constants in the range from approximately 3×10^3 to $2 \times 10^4 \, M^{-1} s^{-1}$ (Reckhow et al., 1991); however, dissolved organic matter can often outcompete the Mn(II) for available ozone, so that residual Mn(II) remains in solution (Knocke et al., 1990).

Nitrite (NO_2^-). Nitrite is rapidly oxidized to nitrate (NO_3^-) by ozone. This reaction may be of use in removal of nitrite formed, for example, through incomplete microbiological denitrification.

Ammonia (NH_3). Ammonia reacts very slowly with ozone, and the protonated form (NH_4^+) reacts at a negligible rate, so that a 10-fold decrease in rate is observed for each decrease in pH unit at pH values below the pK_a (9.3).

Hypochlorous acid ($HOCl/OCl^-$). Ozone can react with free chlorine to generate chloride (77%) and chlorate (ClO_3^-) (23%). Hypochlorite ion reacts more rapidly than HOCl, but even for OCl^-, the characteristic reaction time for oxidation in the presence of $10 \, \mu M \, O_3$ is in the order of 1000 s. The characteristic times for reactions of ozone with other constituents typically present in potable water or wastewater are much shorter, so the reaction of ozone with hypochlorite species is almost never significant in these applications.

Chloramine (NH_2Cl). Ozone reacts with monochloramine at a significant rate to generate chloride and nitrate. This reaction is occasionally used to eliminate chloramines from swimming pool water to avoid the build up of eye irritants. The reaction is much less sensitive to pH than is the reaction of chloramines with free chlorine.

Bromide (Br^-). Ozone reacts with bromide to form hypobromite anion (OBr^-), with a characteristic time of ~1000 s in the presence of $0.5 \, mg/L \, O_3$. The OBr^- quickly establishes acid–base equilibrium with hypobromous acid, HOBr ($pK_a = 8.8$). HOBr may interact with NOM to form brominated organic by-products, including bromoform, brominated acetic acids, acetonitriles, bromopicrin, and (if ammonia is present) cyanogen bromide. However, von Gunten (2003a) noted that O_3 and HOBr react with similar functional groups and, given that the concentration of O_3 is always much higher than that of HOBr, concluded that the formation of bromo-organic compounds during ozonation is likely to be of minor importance.

While HOBr is stable to further attack by ozone, OBr^- is oxidized slowly, via BrO_2^-, to bromate (BrO_3^-). Hoigne (1998) reported that, in general, the level of bromate formation during drinking water ozonation is well within generally accepted quality criteria (the World Health Organization [WHO] provisional guideline is $25 \, \mu g/L$, and the proposed MCL of the USEPA and the European Union is $10 \, \mu g/L$). However, Song et al. (1997) suggested that bromate ion formation may be problematic in water containing more than $0.1 \, mg/L$ bromide. Bromate formation may be minimized either by pH depression (which promotes HOBr formation and increases O_3 stability) or by ammonia addition, which favors the formation of monobromamine:

$$HOBr + NH_3 \leftrightarrow NH_2Br + H_2O \qquad (10\text{-}63)$$

Iodide (I^-). Ozone reacts very rapidly with iodide to form hypoiodous acid (HOI), which can potentially react with NOM in a drinking water treatment system or be further oxidized to iodate (IO_3^-). Of particular concern is the possible formation of iodoform (CHI_3), which imparts unpleasant taste and odor to water. However, Bichsel and von Gunten (1999) concluded that the formation of significant concentrations of iodoorganic compounds in drinking water production is unlikely, because HOI transforms to iodate much more rapidly than it reacts with organic compounds.

Ozone decomposition. Although ozone undergoes autodecomposition to molecular oxygen, that process is slow in systems where direct reactions of ozone with target species are important (typically, at low pH). In contrast, the autodecomposition can be very important in systems in which ozone is used primarily as a vehicle for generating free radicals.

Direct Reaction of Ozone with Organic Compounds. Molecular ozone is a highly selective oxidant, with the rate of reaction strongly influenced by the electron density of the target reactant. A summary of the rates of reaction of ozone with various types of reactants is follows, based on the information provided by Hoigne (1998).

- *Alkanes, saturated alcohols, and chloroalkanes* do not react with molecular ozone at a significant rate.
- *Olefinic compounds* (i.e., compounds containing C–C double bonds) react with ozone within seconds. However, if the H atoms on the double bonded carbon atoms are replaced by chlorine, the double bond becomes deactivated, and no reaction occurs. Thus, common olefinic contaminants such as perchloroethylene and trichloroethylene cannot be oxidized by molecular ozone within a reasonable time.

- *Benzene and pyrene* react slowly with ozone, with a characteristic time of days for typical ozone concentrations. Substitution of the ring with methyl and/or methoxy groups activates it with respect to reactions with ozone, with the rate increasing by a factor of approximately seven per methyl group substituent.
- *Polycyclic aromatic hydrocarbons* typically react with molecular ozone within seconds.
- *Deprotonated phenols* (i.e., phenoxide anions) react with molecular ozone within seconds. These anions react approximately 10^6 times as fast as the corresponding protonated phenolic compounds. Thus, at pH values below the pK_a of the compound, the rate increases by a factor of 10 per pH unit increase; above the pK_a, oxidation often proceeds at nearly diffusion-controlled rates.
- *Carbohydrates* react very slowly with molecular ozone. However, they efficiently promote the chain reaction in which O_3 is transformed into hydroxyl radicals and are then oxidized by the secondary oxidants produced. These reactions are explained in the section on AOPs.
- *Amines* (RNH_2, etc.) *and amino acids* react quickly with molecular ozone when the amino group is not protonated. Thus, like phenols, their apparent rate constant increases 10-fold per pH unit when $pH < pK_a$.
- *N-containing heterocyclic compounds such as pyridine and atrazine* react very slowly with molecular ozone.
- *Natural organic matter* (NOM) possesses some functional groups that react quickly with molecular ozone, but the bulk of NOM appears to be relatively unreactive. Extensive reaction of NOM with ozone generates large amounts of oxalate ($HOOC-COO^-$), which does not react further.

According to Hoigne (1998), all primary reactions of ozone with dissolved organic compounds (M) can be formulated as bimolecular reactions with corresponding, second-order rate expressions:

$$O_3 + M \xrightarrow{k_M} M_{oxid} \qquad r_M = -k_M c_M c_{O_3(aq)} \qquad (10\text{-}64)$$

where k_M might include contributions from more than one reactive group in the molecule. Second-order rate constants for direct reactions of ozone with several organic solutes are given in Figure 10-21. The figure also shows the characteristic time for reaction with each solute. An extensive list of second-order rate constants for the reaction of ozone with specific organic contaminants has been compiled by von Gunten (2003a).

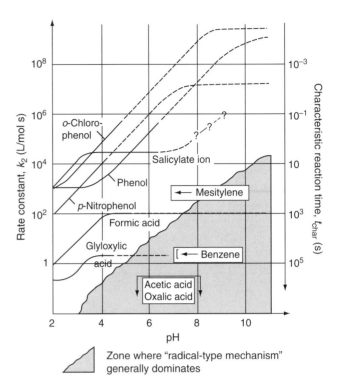

FIGURE 10-21. Examples of aqueous second-order rate constants, k_M, for direct reactions of ozone with dissociating organic solutes. The scale on the right shows the characteristic time for reaction, defined as the reaction time required to lower the concentration of the indicated solute to $1/e$ of its original value in a batch system, assuming a fixed dissolved ozone concentration of $10\,\mu M$ (approximately 0.5 mg/L, a concentration typical for conventional drinking water treatment). *Source*: After Hoigne (1998).

While many organic by-products are produced by ozonation of natural water (Richardson et al., 1999), most of these products are readily biodegradable. Therefore, drinking water treatment plants in which ozonation is used for disinfection and taste and odor control typically include a biological filtration step downstream of ozonation to remove the organics and avoid fouling in the distribution system.

Potassium Permanganate Permanganate ion (MnO_4^-, in which manganese is in the +VII oxidation state) is highly reactive in aqueous solution and can oxidize a wide range of inorganic and organic substances. In such reactions, permanganate is usually reduced to a mixture of Mn(III) and Mn(IV) species that precipitate as a black/brown mixture of manganese oxide and/or oxyhydroxide over most of the pH range of natural water. A redox half-reaction generating a Mn(IV) oxide is shown in Reaction 10-65.

$$MnO_4^- + 4H^+ + 3e^- \rightarrow MnO_2(s) + 2H_2O \qquad E_H^\circ = 1.68\,V \qquad (10\text{-}65)$$

Permanganate is used primarily (and relatively widely) in drinking water treatment for control of taste and odors, for removal of color, and for removal of dissolved iron and manganese. It may also be used to alleviate the formation of THMs and other DBPs by oxidizing precursors and thereby decreasing the demand for other disinfectants (USEPA, 1999). Although potassium permanganate has many potential uses as an oxidant, it is a poor disinfectant (USEPA, 1999).

Permanganate is also widely used for *in situ* destruction of some organic compounds, including chlorinated solvents and other nonaqueous phase liquids (NAPLs) at contaminated subsurface sites (Wickramanayake et al., 2000). Permanganate salts (e.g., $NaMnO_4$ and $KMnO_4$) are particularly attractive for subsurface remediation because the permanganate is more persistent than many other oxidants and can thus travel farther from the point of injection without breaking down. In addition, these compounds are relatively safe to handle in the field and do not generate large quantities of heat or gas on injection.

Generation of Permanganate Permanganate solutions are normally made on-site from potassium or sodium permanganate salts, which are supplied in dry form. A concentrated solution of 1–4% $KMnO_4$ is typically made from technical grade stock. While pure $KMnO_4$ is nonhygroscopic, the technical grades absorb some moisture and have a tendency to cake together.

Reactions of Permanganate with Ferrous and Manganous Species A major use of permanganate is for iron and manganese removal in drinking water treatment. In both cases, permanganate oxidizes the soluble divalent metals to their highly insoluble higher valence forms by reactions such as the following:

$$3Fe^{2+} + MnO_4^- + 7H_2O \rightarrow 3Fe(OH)_3(s) + MnO_2(s) + 5H^+ \quad (10\text{-}66)$$

$$3Mn^{2+} + 2MnO_4^- + 2H_2O \rightarrow 5MnO_2(s) + 4H^+ \quad (10\text{-}67)$$

Alkalinity is consumed in the earlier reactions in proportions of 1.67 equivalents per mole of ferrous iron oxidized and 1.33 equivalents per mole of manganous manganese oxidized. Expressing alkalinity in terms of the concentration of calcium carbonate (and recalling that 1 meq/L alkalinity corresponds to 50 mg/L as calcium carbonate), alkalinities of 1.49 and 1.21 mg/L as $CaCO_3$ are consumed per mg/L of Fe^{2+} and Mn^{2+} oxidized, respectively.

The stoichiometry of Reactions 10-66 and 10-67 indicates that the oxidation of one mole of iron or one mole of manganese requires 0.33 or 0.67 moles of permanganate, respectively (corresponding to 0.94 mg $KMnO_4$ per mg Fe^{2+} and 1.92 mg $KMnO_4$ per mg Mn^{2+}). In practice, the amounts of $KMnO_4$ required for the removal of the reduced species are generally less than suggested by the stoichiometry, a result that might reflect the removal of Fe^{2+} and Mn^{2+} ions by adsorption onto the oxyhydroxides formed. The presence of these solid products also appears to enhance the rate of oxidation of the soluble metal ions. Ferrous iron is oxidized very rapidly by potassium permanganate, though the presence of organic matter limits the effectiveness of the oxidant (Knocke et al., 1990).

The oxidation of Mn(II) by potassium permanganate has been modeled by a slightly modified version of Reaction 10-67 along with reactions for the adsorption of Mn(II) onto $MnO_2(s)$ and oxidation at the surface. These reactions and the proposed rate expressions for all of the reactions are as follows (van Benschoten et al., 1992):

$$3Mn^{2+} + 2MnO_4^- + 4OH^- \xrightarrow{k_1} 5MnO_2(s) + 2H_2O \quad (10\text{-}68)$$

$$Mn^{2+} + MnO_2(s) \xrightarrow{k_2} MnO_2 \equiv Mn^{2+} \quad (10\text{-}69)$$

$$3MnO_2 \equiv Mn^{2+} + 2MnO_4^- + 4OH^- \xrightarrow{k_3} 8MnO_2(s) + 2H_2O \quad (10\text{-}70)$$

$$r_{Mn^{2+}} = -3k_1 c_{Mn^{2+}} c_{MnO_4^-} c_{OH^-}^a - k_2 c_{Mn^{2+}} c_{MnO_2(s)} \quad (10\text{-}71)$$

$$r_{MnO_2(s)} = 5k_1 c_{Mn^{2+}} c_{MnO_4^-} c_{OH^-}^a - k_2 c_{Mn^{2+}} c_{MnO_2(s)} + 8k_3 c_{MnO_2 \equiv Mn^{2+}} c_{MnO_4^-} \quad (10\text{-}72)$$

$$r_{MnO_2 \equiv Mn} = k_2 c_{Mn^{2+}} c_{MnO_2(s)} - 3k_3 c_{MnO_2 \equiv Mn^{2+}} c_{MnO_4^-} \quad (10\text{-}73)$$

$$r_{MnO_4^-} = -2k_1 c_{Mn^{2+}} c_{MnO_4^-} c_{OH^-}^a - 2k_3 c_{MnO_2 \equiv Mn^{2+}} c_{MnO_4^-} \quad (10\text{-}74)$$

In these expressions, $MnO_2 \equiv Mn^{2+}$ represents an adsorbed Mn(II) ion on an $MnO_2(s)$ surface site. The expressions are based on an assumption that all the $MnO_2(s)$ is available to participate in such adsorption reactions; that is, surface layers of new precipitate do not block assess to adsorption sites in the underlying solid. Note that the rate of the solution-phase oxidation of Mn^{2+} (the first expression on the right in Equation 10-71) is assumed to have an ath-order dependence on the OH^- concentration, but the oxidation of adsorbed Mn(II) is assumed to be independent of the OH^- concentration, even though OH^- is consumed in the reaction.

A comparison of experimental data and model simulations at three different pH values is shown in Figure 10-22, and the reported, best-fit values for the rate constants, the exponent a, and the activation energies (which characterize the temperature dependence of k) are given in Table 10-5. A similar set of rate expressions was reported to apply for Mn(II) oxidation by ClO_2, so the parameter values for that oxidant are shown as well.

If the oxidation is carried out with an approximately stoichiometric dose of oxidant, the rate depends on the concentration of manganese to be oxidized. Thus, the results

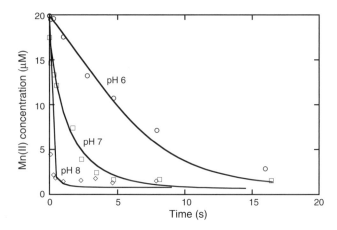

FIGURE 10-22. Model and experimental results for oxidation of Mn(II) by 105% of the stoichiometric requirement of potassium permanganate at 25°C and different pH values. *Source*: After van Benschoten et al. (1992).

shown in Figure 10-22 indicate a characteristic reaction time for the disappearance of Mn^{2+} in the order of 5 s for an initial Mn^{2+} concentration of 1 mg/L in a batch reactor. If the initial concentration were at the still problematic level of 50 μg/L, the characteristic reaction time would be in the order of 100 s, and a residence time of a few minutes would be needed to ensure greater than 95% oxidation in a PFR; the required residence time would be substantially greater in a CFSTR (see Chapter 4).

The presence of dissolved organic matter reduces the rate of Mn(II) oxidation significantly, either because of organic complexation of Mn(II) and/or because of the competitive demand for MnO_4^- by the organic matter. In a study by Carlson and Knocke (1999), a model that included competition between Mn(II) and organic matter for the oxidant yielded reasonable correspondence to experimental results.

■ **EXAMPLE 10-12.** Write a balanced chemical reaction for the oxidation of Fe^{2+} by permanganate, and find the dose of potassium permanganate required for complete oxidation of 1.5 mg/L Fe(II) (as in the groundwater considered in several previous examples). Assume that the permanganate is reduced to Mn(IV), forming manganese oxide solid.

Solution. The reduction of MnO_4^- to $MnO_2(s)$ is a three-electron transfer. Therefore, the balanced equation is as follows:

$$3Fe^{2+} + MnO_4^- + 7H_2O \rightarrow 3Fe(OH)_3(s) + MnO_2(s) + 5H^+$$

The required dose is then found to be

$$(1.5\ \text{mg Fe/L})\left(\frac{1\ \text{mmol Fe}}{55.85\ \text{mg}}\right)\left(\frac{1\ \text{mmol MnO}_4^-}{3\ \text{mmol Fe}}\right)$$

$$= 0.0090\ \text{mmol MnO}_4^-/\text{L}$$

$$= (0.0090\ \text{mmol MnO}_4^-/\text{L})$$

$$\times (158.04\ \text{mg KMnO}_4/\text{mmol MnO}_4^-)$$

$$= 1.41\ \text{mg KMnO}_4/\text{L} \qquad ■$$

■ **EXAMPLE 10-13.** A solution at pH 6.5 and 10°C contains 200 μg/L Mn(II), which is to be oxidized by the addition of permanganate. Assume that 110% of the required stoichiometric dose is to be added. Find the characteristic reaction time (t_char) for loss of Mn(II) from solution in a batch reactor that contains no $MnO_2(s)$, using the second definition of t_char given in Chapter 3 (i.e., the time to reach the ultimate concentration if the reaction continued at its initial rate). (Note that, as soon as the reaction begins, $MnO_2(s)$ will be generated, so the computed value of t_char is applicable only for the initial conditions.)

Solution. In the rate expression (Equation 10-71), the second term is zero under the circumstances described, because no $MnO_2(s)$ is present. We find the values for the first part of this expression as follows:

$$c_{Mn^{2+}} = \frac{200 \times 10^{-6}\ \text{g/L Mn}^{2+}}{54.94\ \text{g/mol}} = 3.64 \times 10^{-6}\ \text{M}$$

According to Equation 10-68, two moles of permanganate are required for every three moles of Mn(II) to be oxidized, and the specification of 110% of the stoichiometric dose yields

$$MnO_4^-\ \text{dose} = (1.1)\left(\tfrac{2}{3}\right)c_{Mn^{2+}} = (1.1)\left(\tfrac{2}{3}\right)(3.64 \times 10^{-6}\ \text{M})$$
$$= 2.67 \times 10^{-6}\ \text{M}$$

At pH 6.5, $c_{OH^-} = \dfrac{K_w}{[H^+]} = \dfrac{10^{-14}}{10^{-6.5}} = 10^{-7.5} = 3.16 \times 10^{-8}$ M.

TABLE 10-5. Model Parameter Estimates for Oxidation of Mn^{2+} by Potassium Permanganate and Chlorine Dioxide[a]

	Log k_1	Log k_2	Log k_3	$E_{Ar,1}$	$E_{Ar,2}$	$E_{Ar,3}$	a
KMnO$_4$	13.4	4.10	6.15	21.0	51.2	11.1	1.30
ClO$_2$	8.41	3.63	4.75	51.1	34.4	15.0	0.61

[a] k values are for $T = 25°C$, concentrations in mol/L, and time in s, E_{Ar} values are in kJ/mol.
Source: From van Benschoten et al. (1992).

According to Equation 3-87, the rate constant at 10°C can be computed from the given value at 25°C by

$$\ln \frac{k_{T_2}}{k_{T_1}} = \frac{E_{Ar}}{R}\left(\frac{1}{T_1} - \frac{1}{T_2}\right)$$

where T is the absolute temperature. Rearranging and substituting known values yields

$$k_{283\,K} = k_{298\,K} \exp\left[\frac{E_{Ar,1}}{R}\left(\frac{1}{T_1} - \frac{1}{T_2}\right)\right]$$

$$= 10^{13.4} \exp\left[\frac{21.0\,\text{kJ/mol}}{8.314 \times 10^{-3}\,\text{kJ/mol K}}\left(\frac{1}{298\,K} - \frac{1}{283\,K}\right)\right]$$

$$= 1.60 \times 10^{13}$$

Finally, we can find the characteristic reaction time for the oxidation of manganese. As noted in Table 10-5, k has units such that, when all concentrations are in molar units, r is in mol/L-s, so

$$t_{\text{char}} = \frac{c_{Mn^{2+},0}}{-r_{Mn^{2+},0}} = \frac{c_{Mn^{2+},0}}{k_1\left(c_{Mn^{2+},0}\right)^a \left(c_{MnO_4^-,0}\right)^b \left(c_{OH^-,0}\right)^c}$$

$$= \frac{3.64 \times 10^{-6}\,M}{\left(1.60 \times 10^{13}\right)\left(3.64 \times 10^{-6}\,M\right)\left(2.67 \times 10^{-6}\,M\right)\left(3.16 \times 10^{-8}\,M\right)^{1.12}}$$

$$= 132\,s$$

This calculated characteristic time is longer than the value based on the data in Figure 10-22 because the precipitation of $MnO_2(s)$ accelerates the removal of Mn(II) from solution, and this factor was not considered in the calculations here. ∎

Reactions of Permanganate with Organic Compounds. Potassium permanganate can oxidize a wide variety of organic compounds via pathways that include electron exchange, hydrogen atom abstraction, and direct donation of oxygen. The mechanism in any given reaction depends on the particular molecule and reaction conditions (pH, temperature, reactant concentrations, and solubility of target compound).

In drinking water plants, potassium permanganate is used at doses from 0.25 to 5 mg/L (or occasionally higher) for the removal of taste and odor-causing compounds. It is also used for DBP control by oxidizing organic precursors at the head of the treatment plant and thereby reducing their potential to react with disinfectants added subsequently. The effectiveness of this latter application remains in dispute, with some studies showing little effect of permanganate predosing and others showing significant reductions in THM production. No harmful by-products from oxidation of NOM by permanganate have been identified (USEPA, 1999).

As mentioned earlier, permanganate is also widely used for oxidative degradation of organic contaminants in subsurface environments. The ability of permanganate to degrade NAPLs such as trichloroethylene and tetrachloroethylene has been demonstrated both in controlled laboratory studies (Urynowicz and Siegrist, 2000) and in field trials (Clayton et al., 2000). Trace contaminants such as methyl *tert*-butyl ether (MTBE) are also oxidized by potassium permanganate, but at rates some 2–3 orders of magnitude lower than when hydroxyl radical-mediated AOPs are used (Damm et al., 2002). While the process has been shown to be effective in the degradation of a variety of organic contaminants, injection of permanganate into subsurface environments may oxidize, and thereby mobilize, trace elements such as chromium and selenium (Chambers et al., 2000). In addition, pore clogging as a result of $MnO_2(s)$ formation may be problematic. Despite these limitations, use of permanganate as a subsurface oxidant does appear to be a viable technology that is finding increasing use.

10.4 ADVANCED OXIDATION PROCESSES

Since the mid-1970s, a group of processes known as advanced oxidation processess (AOPs) have received intense scrutiny as a possible approach for oxidizing compounds that are resistant to conventional redox processes. The distinctive feature of AOPs is the generation and participation in the process of highly reactive oxidants such as hydroxyl radicals ($\dot{O}H$) or species that undergo many of the same reactions, such as the *ferryl* ion, $Fe(IV)O^{2+}$. As noted previously, *radicals* or *free radicals* are molecules containing an orbital occupied by a single (unpaired) electron.

In common parlance, the designation AOP is applied only to processes that have been developed in the past few decades and in which hydroxyl radicals or similar reactive species are the oxidants. Because ozonation has been used in water treatment for over a century, it is not usually considered an AOP. However, at alkaline pH, ozone can react with OH^- ions to generate $\dot{O}H$ radicals, so it shares many features with AOPs. Certain aspects of ozonation are therefore included in this section.

AOPs are often used in combination with other processes. For example, they may be used to oxidize nonbiodegradable compounds that remain after biological treatment of contaminated water, or to partially convert toxic, inhibitory, or refractory compounds to more biodegradable compounds before biological treatment.

The efficacy of an AOP depends primarily on the relative rates of generation and destruction of the $\dot{O}H$ (or other) radicals. The mechanism and rate of formation of these radicals are distinctive to the specific AOP in use, but once they are formed, most of the reactions they undergo are the same in all AOPs. (In some cases, the radicals undergo reactions with the reagents used to generate them in the first place, so the rates of those reactions also depend on the specific AOP being used.)

This section begins with a discussion of the reactions of $\dot{O}H$ radicals with target and nontarget compounds that are likely to be present in the water being treated. These reactions are generic AOP reactions, in which they occur regardless of which AOP is in use. Then, the reactions that generate the free radicals in specific AOPs and during ozonation, and any other reactions that are particularly significant in such systems, are discussed. The presentation relies heavily on the classic paper of Staehelin and Hoigné (1985), which elucidated many of the reaction pathways involved in the ozonation of organic compounds.

Reactions of OH Radicals with Inorganics

In practice, AOPs are used almost exclusively with the goal of oxidizing organics. Nevertheless, reactions of hydroxyl radicals with inorganic species are important in many AOPs. For instance, many Fe(II), As(III), and Cu(I) species can react with hydroxyl radicals to undergo one-electron oxidations. These transformations can alter the form and/or mobility of the element significantly, and they are sometimes used to aid in the removal of inorganic contaminants from water. For example, a variety of technologies for the removal of arsenic from drinking water are based on the hydroxyl radical-mediated oxidation of As(III) to As(IV) (Hug et al., 2001), which subsequently undergoes rapid oxidation to As(V). This oxidation is often carried out as a pretreatment step in arsenic removal processes, since As(V) is more readily removed than As(III) by adsorption onto oxide minerals, ion exchange, and several other treatment technologies. Generically, the one-electron transfer reactions that consume $\dot{O}H$ radicals while oxidizing inorganics can be expressed as follows:

$$\text{Inorg} + \dot{O}H \rightarrow \text{Inorg}^+ + OH^- \quad (10\text{-}75)$$

Some of the most important reactions between hydroxyl radicals and inorganic species involve weak bases such as HCO_3^- and $H_2PO_4^-$, in which case the $\dot{O}H$ can extract an electron (thereby becoming reduced to an OH^- ion) and convert the inorganic species to a free radical, for example,

$$HCO_3^- + \dot{O}H \leftrightarrow H\dot{C}O_3 + OH^- \leftrightarrow \dot{C}O_3^- + H_2O \quad (10\text{-}76)$$

$$H_2PO_4^- + \dot{O}H \leftrightarrow H_2\dot{P}O_4 + OH^- \leftrightarrow H\dot{P}O_4^- + H_2O \quad (10\text{-}77)$$

Hydroxyl radicals also react with sulfite (producing $\dot{S}O_3^-$) and Cl^- (producing $\dot{C}l$), but they apparently do not react with SO_4^{2-}, ClO_4^-, or NO_3^- (Buxton et al., 1988). The competition for hydroxyl radicals created by these inorganics, as well as the reactions between the inorganic radicals and target organic compounds, can significantly alter the course of oxidation, as described subsequently.

One other inorganic reaction that is relevant in any AOP is the combination of two $\dot{O}H$ radicals to generate hydrogen peroxide (H_2O_2) (Equation 10-78). Although H_2O_2 can oxidize some organics directly, it tends to be unreactive with the difficult-to-oxidize compounds that are the targets in AOP processes. Furthermore, H_2O_2 can react with other $\dot{O}H$ radicals via Reaction 10-79 to generate the even weaker oxidant O_2. Both Reactions 10-78 and 10-79 represent sinks for $\dot{O}H$ radicals; therefore, in most AOPs, they are considered undesirable pathways for the radicals to follow.

$$2\dot{O}H \rightarrow H_2O_2 \quad (10\text{-}78)$$

$$H_2O_2 + 2\dot{O}H \rightarrow O_2 + 2H_2O \quad (10\text{-}79)$$

Reactions of OH Radicals with Organics

Hydroxyl radicals react with organic compounds in a variety of ways. The primary reaction with saturated organic compounds (i.e., those with no double bonds) is hydrogen abstraction, as shown for methanol and *tert*-butyl alcohol as example compounds in Reactions 10-80 and 10-82, respectively. On the other hand, the primary reaction with unsaturated compounds is OH addition (e.g., Reaction 10-81, for a generic, substituted aromatic, which the reaction converts to an aliphatic ring).[8] These generalizations are, however, not universal; for example, the radicals react with some organics by electron transfer, without transfer of any atoms (e.g., Reaction 10-83, for oxalate).

$$CH_3OH + \dot{O}H \rightarrow \dot{C}H_2OH + H_2O \quad (10\text{-}80)$$

$$C_6H_5R + \dot{O}H \rightarrow H(OH)\dot{C}_6H_4R \quad (10\text{-}81)$$

$$(CH_3)_3C\text{-}OH + \dot{O}H \rightarrow (\dot{C}H_2)(CH_3)_2C\text{-}OH + H_2O \quad (10\text{-}82)$$

$$C_2O_4^{2-} + \dot{O}H \rightarrow \dot{C}_2O_4^- + OH^- \quad (10\text{-}83)$$

---

[8] The reaction shown here and the subsequent reactions for the substituted aromatic are based on a generic pathway described by Oturan and Pinson (1995). These authors reported that the aromatic initially reacts according to Reaction 10-81, but that the radical formed in Reaction 10-81 can then follow several different pathways, only one of which is presented here. The details of these pathways (including the location on the ring where oxidation occurs and the yields of the different reaction products) depend on the identity of the substituted substrate (i.e., what "R" is). Readers are referred to the original article for further details.

All of these reactions generate a carbon-centered radical by the transfer of one electron from a carbon atom in the organic molecule to the oxygen atom that had been part of the $\dot{O}H$ radical; the carbon is thereby oxidized (and is left with an unpaired electron), and the oxygen is reduced. The example reactions cited above for methanol, the substituted aromatic, and *tert*-butyl alcohol are shown as the first steps in the three reaction sequences displayed in Figure 10-23.

The organic radicals formed by the initial reactions with $\dot{O}H$ can undergo a variety of reactions with other solutes or with each other. One of the most important of these is the reaction with O_2 to form organoperoxy radicals (organic radicals with an O—O bond, represented generically as $R\dot{O}_2$). Such a reaction is shown as the first step in Reaction 10-84 and is also shown in each of the three pathways shown in Figure 10-23. With methanol or the substituted aromatic as the original substrate, these radicals can decompose spontaneously to generate a carbon-containing molecule that is more oxidized than the original reactant, plus a superoxide radical anion, $\dot{O}_2^-$ (second step in Reaction 10-84 and third step in Figure 10-23). However, the analogous reaction does not occur when *tert*-butyl alcohol is the substrate (Schuchmann and von Sonntag, 1979).

$$\dot{R} + O_2 \rightarrow R\dot{O}_2 \rightarrow R_{(2\,ox)} + \dot{O}_2^- + H^+ \quad (10\text{-}84)$$

The species shown as $R_{(2\,ox)}$ is the oxidized product of the reaction sequence and has two fewer electrons than the original reactant (R). At low pH, a significant portion of the $\dot{O}_2^-$ that is released can protonate to form its conjugate acid, $H\dot{O}_2$ ($pK_a = 4.8$):

$$\dot{O}_2^- + H^+ \leftrightarrow H\dot{O}_2 \quad (10\text{-}85)$$

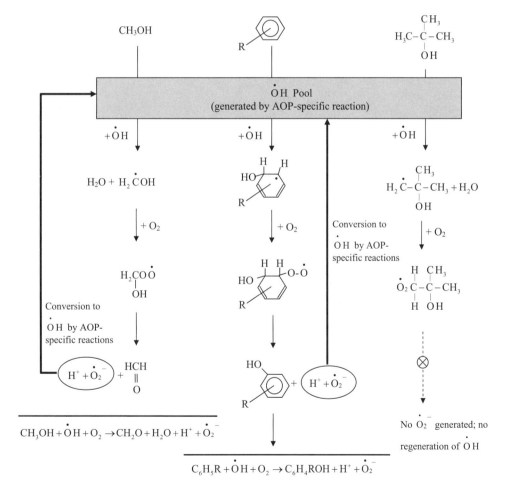

FIGURE 10-23. Schematic showing the reactions of methanol, a substituted benzene, and *tert*-butyl alcohol with hydroxyl radicals. Methanol and the substituted aromatic propagate the chain reaction, but *tert*-butyl alcohol terminates it.

The overall reactions from the initial reactants through this step for methanol and the aromatic as the target organics are

$$CH_3OH + \dot{O}H + O_2 \rightarrow CH_2O + H_2O + H^+ + \dot{O}_2^- \quad (10\text{-}86)$$

$$C_6H_5R + \dot{O}H + O_2 \rightarrow C_6H_4ROH + H^+ + \dot{O}_2^- \quad (10\text{-}87)$$

In many AOPs, the $\dot{O}_2^-$ and $H\dot{O}_2$ can undergo subsequent reactions that generate an $\dot{O}H$ radical, so that the whole cycle can be repeated. In such cases, the organic molecule that originally reacted with the $\dot{O}H$ is said to *propagate* the chain of oxidation reactions. This propagation step is shown schematically for methanol and benzene by the two thick arrows going back to the $\dot{O}H$ pool in Figure 10-23. The details of the propagation are different in different AOPs and are described subsequently. The reaction of $\dot{O}H$ radicals with *tert*-butyl alcohol does not generate $\dot{O}_2^-$, so no oxidative chain develops in that pathway.

The oxidized products of the reactions described earlier are often less toxic, more susceptible to biodegradation, or otherwise less problematic than the parent compound. As a result, even though AOPs can completely mineralize most organic contaminants to CO_2, H_2O, and inorganic ions such as chloride, such extensive conversions are not always necessary to achieve a satisfactory outcome. On the other hand, partial degradation may generate intermediates that are toxic to humans or inhibitory to biological treatment processes, in which case more complete oxidation is essential.

■ **EXAMPLE 10-14.** Relevant chemical reactions for the oxidation of atrazine ("At") by ozone, and the associated second-order rate constants, are given later. The first reaction is the direct oxidation of atrazine by ozone, the second and third describe the creation of hydroxyl radicals from ozone, and the last describes the oxidation of atrazine by the hydroxyl radical.

$O_3 + At \rightarrow$ Products $\qquad k_1 = 6\,M^{-1}\,s^{-1}$

$O_3 + OH^- \rightarrow HO_2^- + O_2 \qquad k_2 = 70\,M^{-1}\,s^{-1}$

$O_3 + HO_2^- \rightarrow \dot{O}H + \dot{O}_2^- + O_2 \quad k_3 = 2.8 \times 10^6\,M^{-1}\,s^{-1}$

$\dot{O}H + At \rightarrow$ Products $\qquad k_4 = 3 \times 10^9\,M^{-1}\,s^{-1}$

Determine the atrazine concentration as a function of time over a 10-min period in a batch reactor at pH 6.0 under the following circumstances:

(i) Initial atrazine concentration of 10 nM (2.16 μg/L) in the presence of 50 μM (2.4 mg/L) ozone, assuming oxidation occurs only by direct reaction with ozone;

(ii) The same conditions as in (i), but considering direct oxidation by ozone and indirect oxidation by reaction with hydroxyl radicals that are formed from ozone decomposition;

(iii) The same conditions as in (ii) except that the initial ozone concentration is 20 μM; and

(iv) The same conditions as in (ii) except that the initial atrazine concentration is 5 nM.

Solution. As shown in Chapter 3, the mass balance on any chemical i in a batch reactor has the form $dc_i/dt = \sum r_i$ where r_i is the rate of a reaction forming i, and the summation is overall reactions occurring in the reactor. Applying this idea to the four chemicals of interest in the given system (O_3, At, $\dot{O}H$, and HO_2^-), we obtain the rate expressions as shown; for case (i), where only the direct oxidation of atrazine by ozone is considered, the same equations apply, but the terms involving k_4 are omitted.

$$\frac{dc_{O_3}}{dt} = -k_1 c_{O_3} c_{At} - k_2 c_{O_3} c_{OH^-} - k_3 c_{O_3} c_{HO_2^-}$$

$$\frac{dc_{At}}{dt} = -k_1 c_{O_3} c_{At} - k_4 c_{\dot{O}H} c_{At}$$

$$\frac{dc_{\dot{O}H}}{dt} = k_3 c_{O_3} c_{HO_2^-} - k_4 c_{\dot{O}H} c_{At}$$

$$\frac{dc_{HO_2^-}}{dt} = k_2 c_{O_3} c_{OH^-} - k_3 c_{O_3} c_{HO_2^-}$$

For case (i), where only the direct oxidation of atrazine by ozone is considered, the equations are identical except that the terms involving k_4 are omitted. To predict the concentrations as a function of time, the equations are integrated simultaneously, utilizing the appropriate initial conditions. The initial concentrations of both HO_2^- and $\dot{O}H$ are zero in all cases, and that of OH^- is $10^{-8}\,M\,(=K_w/10^{-6})$, while those of O_3 and At are different for the various cases. Using software designed to solve these types of differential equations or a spreadsheet using small time steps, we obtain the results shown in Figure 10-24.

For the equivalent conditions, the degradation of atrazine by ozone alone (case i) is far slower than that by the combination of ozone and hydroxyl radical (case ii); in fact, as a result of the drastic difference in the rate constants k_1 and k_4, the hydroxyl radical concentration only has to reach $2 \times 10^{-17}\,M$ to degrade atrazine at a rate that matches the initial rate of degradation by ozone alone. The characteristic reaction times for atrazine degradation (i.e., the time to reach zero concentration if the initial rate of degradation continued indefinitely) are approximately 3200 s for case (i) and 270 s for case (ii).

The hydroxyl radical, $\dot{O}H$, is an intermediate in this set of reactions, created by the third reaction in the sequence and destroyed by the fourth. During much of the period of

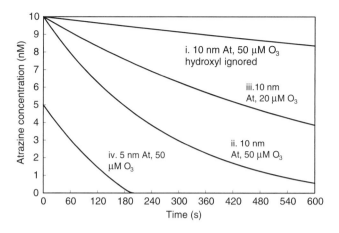

FIGURE 10-24. Atrazine degradation at different initial atrazine and ozone concentrations; pH 6.0.

interest, it is destroyed approximately as rapidly as it is formed, so that the concentration of ȮH reaches a pseudo steady state (i.e., $dc_{\dot{O}H}/dt \approx 0$) in which, according to the mass balance on ȮH, $k_3 c_{O_3} c_{HO_2^-} \approx k_4 c_{\dot{O}H} c_{At}$. Substituting this (approximate) equality into the mass balance on atrazine, the atrazine degradation rate can be seen to be (nearly) first order with respect to ozone concentration:

$$\frac{dc_{At}}{dt} \approx -k_1 c_{O_3} c_{At} - k_3 c_{O_3} c_{HO_2^-} = -(k_1 c_{At} + k_3 c_{HO_2^-}) c_{O_3}$$

As a result, reducing the initial ozone concentration to 20 μM (case iii) slows the initial atrazine degradation to approximately $\frac{2}{5}$ the rate at 50 μM (case ii), and the characteristic reaction time rises to approximately 670 s.

The atrazine reaction with hydroxyl radicals is so rapid that the overall degradation is controlled by the generation of these radicals, which is nearly independent of the atrazine concentration. As a result, for a given ozone concentration, atrazine decays at almost the same rate when the initial atrazine concentration is 5 nM (case iv) as when it is 10 nM (case ii). Correspondingly, the c_{At} versus t curves are nearly parallel for those two cases. The characteristic time for atrazine degradation for case (iv) is approximately 140 s, just slightly greater than half the value of 270 s for case (ii). ∎

∎ **EXAMPLE 10-15.** Compare the rate of atrazine degradation by ozone at pH 8.0 with that at pH 6.0 in solutions with an initial atrazine concentration of 10 nM and an initial ozone concentration of 20 μM (i.e., the same as case (iii) in Example 10-14).

Solution. The equations and solution methodology are identical to those of Example 10-14. The results are shown

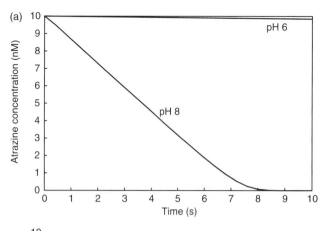

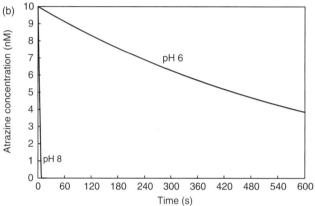

FIGURE 10-25. Degradation of atrazine at pH 6 and 8 in the presence of 20 μM ozone.

in Figure 10-25. The effect of pH is dramatic; only approximately 60% of the atrazine is oxidized in 10 min at pH 6, but essentially all of it is oxidized within 8 s at pH 8. The difference is attributable primarily to the more rapid production of hydroxyl radicals at the higher pH, according to the second and third reactions in the sequence shown in Example 10-14. ∎

The preceding discussion describes the primary pathway by which organics are oxidized and ȮH radicals are cycled in AOPs. However, a number of competing reactions divert the chemicals away from this path and thereby interfere with the treatment process. A few such diversionary reactions involving inorganic species were noted in the previous section, but others involve organics. For instance, although many hydroperoxy radicals undergo the second reaction shown in Equation 10-84, others cannot eject $\dot{O}_2^-$ radicals. In such cases, either the organic radicals generated by the initial reaction of the substrate with ȮH (i.e., Ṙ) or the hydroperoxy radicals (RȮ$_2$) might be destroyed by reactions with other constituents of the solution (e.g., redox-active metal ions like Fe^{3+} or Fe^{2+}, as shown in

Reactions 10-88 and 10-89) or by a dimerization reaction (e.g., Reaction 10-90).

$$\dot{R} + Fe^{3+} \rightarrow R^+ + Fe^{2+} \quad (10\text{-}88)$$

$$\dot{R} + Fe^{2+} \rightarrow R^- + Fe^{3+} \quad (10\text{-}89)$$

$$\dot{R} + \dot{R} \rightarrow R - R \quad (10\text{-}90)$$

(If $R\dot{O}_2$ is the reactant in the preceding reactions rather than $\dot{R}$, then one molecule of O_2 is released for each molecule of $R\dot{O}_2$ that reacts.) When an organic molecule follows one of these pathways, it is oxidized in the process, so a part of the treatment objective is achieved. However, the overall reaction sequence consumes hydroxyl radicals and does not replenish them. Such a sequence is shown for *tert*-butyl alcohol in Figure 10-23. Organics that react with hydroxyl radicals primarily in this way are said to *terminate* the oxidation chain reaction. Alkyl groups in organic molecules are important chain terminators in many systems.

Even if a hydroperoxy radical is capable of forming and propagating the oxidation chain (by ejecting an $\dot{O}_2^-$ radical), the organic (either $\dot{R}$ or $R\dot{O}_2$) might nevertheless be diverted onto an alternative path by a competing reaction. One such competing reaction that is of particular importance involves the carbonate radical, $\dot{CO}_3^-$ (which can form in AOPs via Reaction 10-76):

$$\dot{R} + \dot{CO}_3^- \rightarrow R^+ + CO_3^{2-} \quad (10\text{-}91)$$

In this case, one $\dot{OH}$ radical is consumed to generate each of the reactants ($\dot{R}$ and $\dot{CO}_3^-$), and no radicals are generated as products, so the reaction terminates the chain reaction. Molecules such as CO_3^{2-} that terminate the chain reaction by participating in a reaction that competes with a propagation step are called radical *scavengers*. Because of the efficiency of carbonate species as radical scavengers, AOPs are often ineffective in solutions at moderately alkaline pH that have high carbonate alkalinity. In addition to carbonate ions, molecules of NOM (which usually contain several carboxyl groups that behave similarly to CO_3^{2-} ions) are often important radical scavengers when natural water is treated using an AOP.

In Equation 10-91, the carbonate radical reacts with an organic molecule that has already been converted to a radical via a reaction with $\dot{OH}$. However, $\dot{CO}_3^-$ radicals can also react with unaltered molecules of substrate by reactions analogous to Equations 10-80, 10-81, or 10-83. When this occurs, $\dot{CO}_3^-$ plays the role that $\dot{OH}$ plays in those reactions, and the pathway propagates the chain. For example, the reaction analogous to Equation 10-80 would be

$$CH_3OH + \dot{CO}_3^- \rightarrow \dot{C}H_2OH + HCO_3^- \quad (10\text{-}92)$$

In Equation 10-92, the carbonate radical generates a molecule that can proceed through the chain reaction, so it is acting as a chain *promoter*, rather than a radical scavenger. In most systems, carbonate scavenges hydroxyl radicals at a far greater rate than it promotes the oxidation by reactions such as Reaction 10-92, so carbonate is generally considered to be a scavenger. On the other hand, phosphate radicals ($H_2\dot{P}O_4$, $H\dot{P}O_4^-$, and $\dot{P}O_4^{2-}$) can play the same roles as carbonate radicals, but their reactions as chain promoters tend to proceed much faster than their reactions as scavengers, so they are considered to be promoters of the oxidation chain.

Six reactions that can interfere with the desired oxidation reaction and regeneration of free radicals have been described earlier. These reactions are collected in Table 10-6, using methanol as the example target organic compound; the corresponding reaction pathways are shown schematically by the broken lines in Figure 10-26. The pathways are labeled G1–G6, with "G" designating that they are generic, in that they occur in all AOPs. Pathways G1–G3 involve only inorganic species and have the net effect of consuming free radicals without oxidizing any organics; those pathways are therefore particularly unattractive. Pathways G4–G6 do lead to oxidation of organics, but they nevertheless fail to regenerate free radicals. They are therefore less attractive than the chain propagating, cyclic pathway, labeled AOP-Cy and shown by bold, solid arrows. The reaction that initiates the chain by generating $\dot{OH}$ from starting materials is also shown and is labeled AOP-Init.

Note that pathways G5 and G6 have identical overall stoichiometries because the scavenger (HCO_3^-) that plays a key role in pathway G5 is not consumed. In pathway G5, bicarbonate remains in solution and acts as a catalyst for the

TABLE 10-6. Net Stoichiometries of Several Reaction Pathways Relevant in AOPs with Methanol as a Target Species

Pathway[a]	Reaction	
AOP-Cy[b]	$CH_3OH + \dot{OH} + O_2 \rightarrow$	
	$CH_2O + H_2O + H^+ + \dot{O}_2^-$	(10-86)
G1	$Inorg + \dot{OH} \rightarrow Inorg^+ + OH^-$	(10-75)
G2	$2\dot{OH} \rightarrow H_2O_2$	(10-78)
G3	$H_2O_2 + 2\dot{OH} \rightarrow 2H_2O + O_2$	(10-79)
G4	$2CH_3OH + 2\dot{OH} \rightarrow C_2H_6O_2 + 2H_2O$	(10-93)
G5, G6	$CH_3OH + 2\dot{OH} \rightarrow CH_2O + 2H_2O$	(10-94)

[a]As labeled in Figure 10-26.
[b]Excluding conversion of the superoxide radical back to a hydroxyl radical.

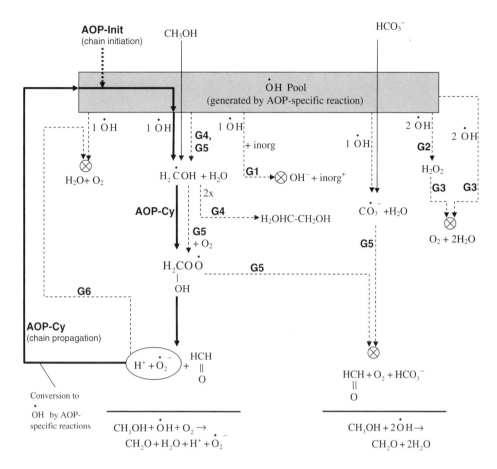

FIGURE 10-26. The oxidation of methanol by hydroxyl radicals in a chain promoting reaction (pathway AOP-Cy [bold lines, same as the two bold lines in Figure 10-23]) and by $C\dot{O}_3^-$ (right column), a scavenger that terminates the chain reaction. Other chain termination reactions include those with inorganics and dimerization. The symbol $\otimes$ indicates the endpoint of reaction sequences that compete with or terminate the chain reaction.

scavenging of $\dot{O}H$ radicals by diverting the radicals away from the chain-propagating path. That is, in the presence of substantial concentrations of carbonate, the rate at which $\dot{O}H$ radicals are destroyed via the net Reaction 10-94 is greatly increased.

■ **EXAMPLE 10-16.** Determine how the rates of atrazine degradation at pH 8.0 are influenced by the presence of carbonate alkalinity and/or NOM in comparison to the solution with neither of these constituents considered in Example 10-15. Assume that the direct reaction of ozone with NOM is slow enough that it can be ignored. All four solutions to be considered have an initial atrazine concentration of 10 nM and are dosed with 20 μM ozone. The differences among the solutions are summarized in the following table; solution 'i' was considered in Example 10-15.

Solution	c_{NOM} (mg/L as TOC)	Alkalinity (meq/L)
i	0.0	0.0
ii	0.0	0.1
iii	0.1	0.0
iv	0.1	0.1
v	1.0	0.0

The relevant reactions and rate constants not already shown in Example 10-14 are as follows:

$$\dot{O}H + NOM \rightarrow Products \quad k_5 = 2.5 \times 10^4 (mg/L)^{-1} s^{-1}$$

$$\dot{O}H + CO_3^{-2} \rightarrow C\dot{O}_3^- + OH^- \quad k_6 = 3.9 \times 10^8 M^{-1} s^{-1}$$

$$\dot{O}H + HCO_3^- \rightarrow C\dot{O}_3^- + H_2O \quad k_7 = 8.5 \times 10^6 M^{-1} s^{-1}$$

Solution. Both carbonate (and to a lesser extent bicarbonate) and NOM are hydroxyl radical scavengers that compete with atrazine for hydroxyl radicals. Therefore, both alkalinity and NOM are expected to reduce the rate of atrazine degradation.

The relationships among pH, alkalinity, and the concentrations of HCO_3^- and CO_3^{2-} are known from water chemistry. Based on these relationships, a solution at pH 8 with an alkalinity of 0.1 meq/L contains 0.1 mM HCO_3^- and 5 μM CO_3^{-2}. These concentrations are essentially constant throughout the time of reaction, because the total carbonate concentration is considerably higher than the ozone concentration, and the equilibration between HCO_3^- and CO_3^{-2} is extremely fast.

The volume-normalized mass balance expressions in the batch reactor for all constituents other than the hydroxyl radical are the same as in Example 10-14. Accounting for the reactions shown earlier, that expression becomes

$$\frac{dc_{\dot{O}H}}{dt} = k_3 c_{O_3} c_{HO_2^-} - k_4 c_{\dot{O}H} c_{At} - k_5 c_{\dot{O}H} c_{NOM}$$
$$- k_6 c_{\dot{O}H} c_{CO_3^{2-}} - k_7 c_{\dot{O}H} c_{HCO_3^-}$$

The competition for the hydroxyl radicals is evident in this equation; note that the analogous equation in the previous examples had only the first two terms on the right side, accounting for the generation of the radicals and their use by atrazine. Here, the radicals are consumed by atrazine, NOM, and both carbonate species. This competition causes a marked reduction in the rate of atrazine degradation, as shown in Figure 10-27. Both NOM alone and alkalinity alone substantially reduce the rate of atrazine degradation, though the effect of alkalinity is more dramatic at the chosen concentrations. When both of these hydroxyl radical scavengers are present, the rate of degradation of atrazine decreases even more. The concentrations used here for both of these constituents (alkalinity and NOM) are at the low end of those that would be encountered in potable water after it has been treated to remove most of the NOM, so that most natural waters would have even greater competition for the hydroxyl radicals than that illustrated in this example. ∎

Generation and Fate of OH Free Radicals in Ozonation and Some Specific AOPs

In the following sections, AOPs that are based on ultraviolet light (UV) plus H_2O_2, ozonation, light plus TiO_2, wet air oxidation, sonolysis, and Fenton's reagent (Fe(II) plus H_2O_2) are described. The focus of these sections is on the generation and destruction of $\dot{O}H$ radicals, with the understanding that once those radicals form, they can undergo all the generic reactions noted in the preceding sections. Thus,

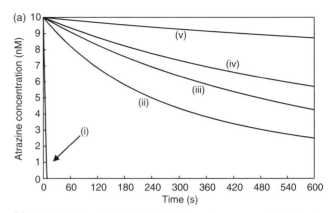

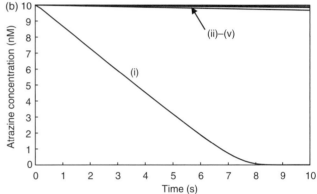

FIGURE 10-27. Effect of carbonate alkalinity and natural organic matter (NOM) on atrazine degradation by 20 μM ozone at pH 8. The labels next to the curves correspond to the solution conditions given in the table in the problem statement. Same data in both figures, with expanded scale on abscissa in (b). Note that the units on the abscissa are minutes in (a) and seconds in (b).

in terms of Figure 10-26, the focus in this section is entirely on reaction AOP-Init and pathway AOP-Cy.

UV/Hydrogen Peroxide Numerous successful full-scale wastewater plants use an AOP based on a combination of UV irradiation and H_2O_2 addition (Venkatadri and Peters, 1993). In this process, the initiation reaction involves the splitting of H_2O_2 when those molecules absorb energy from the UV light (shown below as "$h\nu$," where h is Planck's constant and ν is the frequency of light):

$$H_2O_2 + h\nu \rightarrow 2\dot{O}H \qquad (10\text{-}95)$$

Any light-induced molecular splitting is referred to as *photolysis*, so Equation 10-95 is an example of photolytic disproportionation.

H_2O_2 does not significantly absorb light with a wavelength (λ) above 300 nm, and even at lower wavelengths, the molar absorbance is low. Sunlight has very low intensity at such low wavelengths, so the process cannot utilize sunlight, and even with artificial irradiation, the fraction of incident

UV light that is absorbed is low unless the concentration of H_2O_2 is high. The photolysis of H_2O_2 has a high quantum yield,[9] so the light that is absorbed is utilized efficiently. The process is less effective for treating colored and turbid water, due to reduced penetration of UV light (Venkatadri and Peters, 1993).

Because HO_2^- (the conjugate base of H_2O_2) has a greater molar absorbance than H_2O_2, the irradiation is sometimes carried out at high pH, where formation of HO_2^- is favored ($pK_a = 11.7$):

$$H_2O_2 \leftrightarrow HO_2^- + H^+ \quad (10\text{-}96)$$

However, alkaline pH can sometimes reduce the overall process effectiveness, possibly due to the base-catalyzed decomposition of H_2O_2 (Venkatadri and Peters, 1993). The beneficial effect of high pH is likely to dominate at low H_2O_2 concentrations (i.e., when the fractional absorption of light is low), while the undesirable effect is more likely to dominate at high H_2O_2 concentrations.

In the UV/H_2O_2 AOP, the key reaction that closes the loop in the chain propagation reaction is the combination of two superoxide radical anions and two H^+ ions to regenerate an H_2O_2 molecule, as follows[10]:

$$2H^+ + 2\dot{O}_2^- \rightarrow 2H\dot{O}_2 \rightarrow H_2O_2 + O_2 \quad (10\text{-}97)$$

Thus, if both hydroxyl radicals produced by the photolysis of an H_2O_2 molecule follow the chain propagation pathway, then, using methanol as the example target species, the net stoichiometry of one cycle through that pathway would be

$$2CH_3OH + O_2 + h\nu \rightarrow 2CH_2O + 2H_2O \quad (10\text{-}98)$$

The key feature of this result is that the organic target is oxidized without consuming either free radicals or hydrogen peroxide; the only AOP-specific "reagent" that it consumes is UV light. The overall reaction sequence can be described as follows: photolysis of one H_2O_2 molecule generates two $\dot{O}H$ radicals; those radicals are consumed in the oxidation of the methanol, generating two superoxide radicals; then, the superoxide radicals recombine to generate a hydrogen peroxide molecule. Thus, in theory, if all the hydroxyl radicals followed pathway AOP-Cy, the process would continue until Reaction 10-98 reached equilibrium. That equilibrium lies so far to the right that the reaction would proceed until one of the reactants was essentially completely consumed. In reality, however, some $\dot{O}H$ radicals follow generic pathways G1–G4 and G6, so that some H_2O_2 is consumed. Pathway G5, which consumes $\dot{O}H$ radicals in most AOPs, does not do so in the UV/H_2O_2 AOP, thanks to the UV irradiation.

When the reactions that are relevant specifically to the UV/H_2O_2 AOP are added to Figure 10-26, we obtain Figure 10-28. When both the desired chain propagation and competing chain termination reactions are taken into consideration, an optimum H_2O_2 concentration can be identified. At concentrations less than this optimum, the production of free radicals is too slow to efficiently mediate the reaction with the target contaminants, and at higher concentrations, scavenging of hydroxyl radicals by H_2O_2 reduces the efficiency of the treatment process.

Ozone As noted in the introduction to this section, ozonation is usually not considered an AOP, because it has been in use for treatment of contaminated water for more than a century. And, indeed, the reactions involving direct oxidation of contaminants by ozone that were described earlier in this chapter bear few similarities to AOPs. On the other hand, reactions of ozone with OH^- ions do generate hydroxyl radicals, and these radicals behave as they do in AOPs. Thus, ozone can oxidize target contaminants via either of the two mechanisms, one analogous to conventional oxidation processes and the other analogous to AOPs (Figure 10-29). The relative importance of the two mechanisms depends on the chemical structure of the target contaminant and the operational conditions in the reactor (especially the solution pH).

Some characteristics of the direct reactions between ozone and reduced species have already been discussed. In this section, we first describe the generation of hydroxyl radicals in ozonation reactors and the reactions they undergo with ozone itself. We then consider the relative rates of the direct and indirect ($\dot{O}H$-mediated) oxidation reactions to gain an understanding of which mechanism is likely to be more important for particular target species and under particular conditions.

The formation of free radicals from ozone is usually initiated by its reaction with OH^- (Reaction 10-99) and hence is facilitated by alkaline pH. Therefore, the pH of solutions that are acidic or near neutral pH is usually increased before ozonation, if the objective is to carry out radical-based oxidation. Recall, however, that carbonate species are very effective at scavenging free radicals. As a result, ozonation is usually unattractive as an AOP-type

---

[9] Some light that is absorbed is subsequently emitted, without leading to a chemical transformation of the molecule. The quantum yield (ϕ) of a photochemical reaction at wavelength λ is the number of moles of the light-absorbing substance that react for each mole of photons (einstein) absorbed; that is: ϕ_λ = number of moles reacting/number of einsteins absorbed.

[10] In Reaction 10-97, two free radicals are consumed and none are generated. In general, H_2O_2 does not disproportionate to generate hydroxyl radicals, so the reaction appears to terminate the chain. However, in the UV/H_2O_2 AOP, the solution is irradiated with UV light, which does induce the disproportionation reaction. Therefore, in that AOP, generation of H_2O_2 propagates the chain rather than terminating it.

478 REDOX PROCESSES AND DISINFECTION

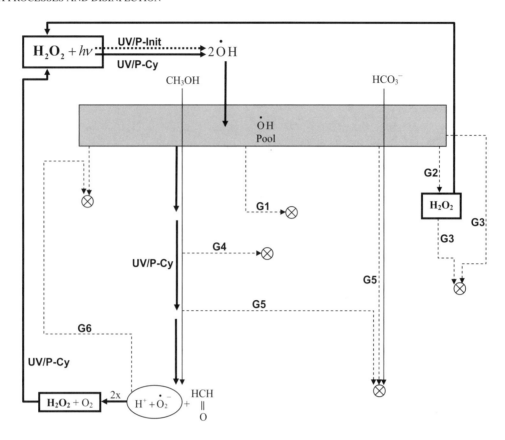

FIGURE 10-28. The UV/H_2O_2 AOP for oxidation of methanol. Reaction UV/P-Init is the initiation reaction generating $\dot{O}H$ radicals, and pathway UV/P-Cy is the chain propagation sequence. The reaction that regenerates H_2O_2 is shown in the box in the lower left. Pathways G1–G6 are shown in skeletal form to reduce clutter; details of those pathways are provided in Figure 10-26. As in Figure 10-26, ⊗ represents the endpoint of reactions that compete with or terminate the chain reaction.

process if the raw water contains even a moderate concentration of CO_3^{2-}. This constraint is sometimes expressed in terms of an upper limit on the alkalinity for a solution being considered for treatment by an AOP, but the real concern is the CO_3^{2-} concentration. Solutions with relatively high alkalinity are susceptible to AOP treatment if the vast majority of that alkalinity is present as HCO_3^- or other species, whereas those with lower alkalinity might be poor candidates for such treatment if a substantial fraction of the alkalinity is present as CO_3^{2-}.

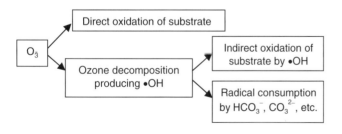

FIGURE 10-29. Pathways for oxidation of compounds by ozone.

The radical generation step in ozonation creates one superoxide anion and one hydroperoxy radical, which are in acid–base equilibrium via Reaction 10-85. The superoxide anions generated in the initiation step can subsequently react with ozone (Reaction 10-100) to produce ozonide radicals ($\dot{O}_3^-$), which then protonate and decompose to hydroxyl radicals (Reactions 10-101 and 10-102, respectively), releasing molecular oxygen in the process. The net effect of one instance of Reaction 10-99 and two instances each of Reactions 10-100 to 10-102 is shown in Reaction 10-103. These reactions constitute the most important $\dot{O}H$ generation (i.e., chain initiation) pathway in ozonation processes. Even though the net reaction consumes equal amounts of H^+ and OH^-, Reaction 10-99 is the rate-limiting step, so the rate increases substantially with increasing solution pH.

$$O_3 + OH^- \rightarrow 2\dot{O}_2^- + H^+ \quad (10\text{-}99)$$

$$O_3 + \dot{O}_2^- \rightarrow \dot{O}_3^- + O_2 \quad (10\text{-}100)$$

$$\dot{O}_3^- + H^+ \leftrightarrow H\dot{O}_3 \quad (10\text{-}101)$$

$$H\dot{O}_3 \rightarrow \dot{O}H + O_2 \quad (10\text{-}102)$$

$$3O_3 + OH^- + H^+ \rightarrow 2\dot{O}H + 4O_2 \quad (10\text{-}103)$$

Although ozone is the reagent that is generated and added to water to initiate the radical-based oxidation pathway described earlier, the ozonide species $\dot{O}_3^-$ and $H\dot{O}_3$ are the immediate precursors of the $\dot{O}H$ radicals and are convenient to use as the focal point for discussion of the initiation and propagation processes. For example, a second chain initiation pathway can be entered by a direct, one-electron transfer reaction from a solute M to molecular ozone:

$$O_3 + M \rightarrow \dot{O}_3^- + M^+ \quad (10\text{-}104)$$

These two chain initiation pathways are shown as Oz-Init1 and Oz-Init2, respectively, in Figure 10-30.

The chain propagation pathway for radical-mediated oxidation of methanol by ozone is shown by the bold solid lines in the figure (Oz-Cy). Like the initiation pathways, the propagation pathway generates an ozonide species ($H\dot{O}_3$) as the immediate precursor of the $\dot{O}H$ radical. Note that, unlike the case for the UV/H_2O_2 AOP, the chain propation sequence in ozonation consumes the central reagent that is added in the process (O_3). The net stoichiometry corresponding to one cycle through this pathway for the oxidation of methanol is

$$CH_3OH + O_3 \rightarrow CH_2O + O_2 + H_2O \quad (10\text{-}105)$$

Ozone itself may be a significant scavenger of hydroxyl radicals via Reaction 10-106.

$$\dot{O}H + O_3 \rightarrow H\dot{O}_2 + O_2 \quad (10\text{-}106)$$

However, scavenging by NOM and carbonate ion is far more important than by ozone in most water of interest; for this reason, Reaction 10-106 is not included in Figure 10-30.

As noted previously, many oxidizable substrates can react with either molecular ozone or $\dot{O}H$ radicals. Although the activation barrier for the direct oxidation by aqueous ozone is much larger than that for oxidation by hydroxyl radicals, the concentration of molecular ozone in ozonation processes is much larger than that of the radicals, so each oxidant dominates under some conditions. Because the production of hydroxyl radicals is facilitated by high pH, the hydroxyl

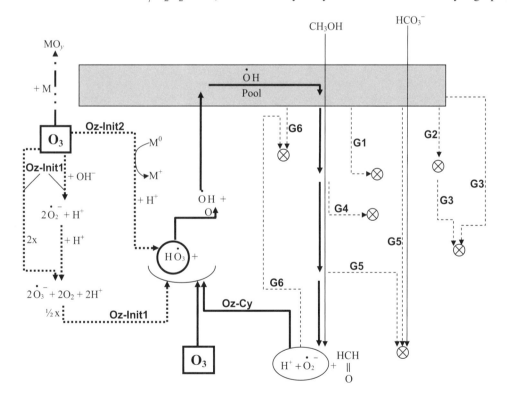

FIGURE 10-30. The radical-mediated ozonation of methanol. Two initiation pathways (Oz-Init1 and Oz-Init2) are shown by the bold dotted lines on the left. The chain propagation pathway (Oz-Cy) is shown by the bold solids lines and requires an input of one O_3 molecule per cycle. Pathways G1–G6 are shown in skeletal form, as in Figure 10-28. Also as in that figure, $\otimes$ represents the endpoint of reactions that compete with or terminate the chain reaction. Direct oxidation of a compound (not radical-mediated) is shown in the upper left.

radical-mediated oxidation pathways tend to dominate under those conditions (e.g., pH > 8), while direct oxidation with molecular ozone dominates under acidic conditions (Hoigne and Bader, 1976).

The variety and complexity of the reaction pathways makes prediction of the rates of substrate oxidation and ozone decomposition difficult other than in well-defined systems with a single or small number of solutes. Elovitz et al. (2000) suggested that the relative contributions of ozone and hydroxyl radicals to the degradation of a solute M can be formulated as follows. The overall rate of oxidation is

$$r_M = -k_{O_3} c_M c_{O_3} - k_{\dot{O}H} c_M c_{\dot{O}H} \quad (10\text{-}107)$$

where k_{O_3} and $k_{\dot{O}H}$ are the second-order rate constants for reaction of M with O_3 and $\dot{O}H$, respectively. Defining R_{ox} as the ratio of the concentrations of $\dot{O}H$ and O_3(aq); that is, $R_{ox} = c_{\dot{O}H}/c_{O_3}$, Equation 10-107 can be rewritten as

$$r_M = -(k_{O_3} + k_{\dot{O}H} R_{ox}) c_M c_{O_3} \quad (10\text{-}108)$$

The fraction of the oxidation of M that is mediated by $\dot{O}H$ radicals ($f_{\dot{O}H}$) can then be calculated as

$$f_{\dot{O}H} = \frac{k_{\dot{O}H} R_{ox}}{k_{O_3} + k_{\dot{O}H} R_{ox}} \quad (10\text{-}109)$$

Values of R_{ox}, $k_{\dot{O}H}$, and k_{O_3} must be determined experimentally, a task that is accomplished by measuring, throughout the process, the concentrations of ozone and of a probe molecule that is oxidized only by hydroxyl radicals. Once the reaction rate expression for the probe has been evaluated in separate, control experiments, the rate of reaction of M with both hydroxyl radicals and ozone can be determined by modeling the results obtained in such competition experiments. R_{ox} values in the range 10^{-9}–10^{-6} are typical for ozonation of natural water, with values $>10^{-7}$ observed either in the initial phase of an ozonation process or throughout an AOP in which hydroxyl radical production is enhanced (see section titled "O_3/UV and O_3/H_2O_2"). Values in the range from 10^{-9} to 10^{-7} are typical for the secondary (postinitial) stage of ozonation (von Gunten, 2003a,b).

Second-order rate constants for reaction of a variety of organic compounds with ozone and hydroxyl radicals are given in Table 10-7, and calculated $f_{\dot{O}H}$ values for a range of R_{ox} values are shown in Figure 10-31. Many pharmaceuticals and naturally occurring trace contaminants (such as the cyanobacterial toxin microcystin-LR) are readily degraded by direct reaction with ozone. On the other hand, many solvent organics and fuel additives are resistant to degradation by ozone and can only be degraded when the more strongly oxidizing hydroxyl radical is present at concentrations that can be achieved in an AOP, but not by ozonation alone.

TABLE 10-7. Rate Constants for the Oxidation of Selected Organic Compounds by Ozone and OH Radicals at Ambient Temperature[a]

Compound	k_{O_3} (M^{-1} s^{-1})	$t_{1/2}$[b]	$k_{\dot{O}H}$ (M^{-1} s^{-1})
Algal products			
Geosmin	<10	>1 h	8.2×10^9
2-Methylisoborneol	<10	>1 h	$\approx 3 \times 10^9$
Microcystin-LR	3.4×10^4	1 s	2.3×10^{10}
Pesticides			
Atrazine	6	96 min	3×10^9
Alachlor	3.8	151 min	7×10^9
Carbofuran	620	56 s	7×10^9
Dinoseb	1.5×10^5	0.23 s	4×10^9
Endrin	<0.02	>20 d	1×10^9
Methoxychlor	270	2 min	2×10^{10}
Solvents			
Chloroethene	1.4×10^4	2.5 s	1.2×10^{10}
cis-1,2-Dichloroethene	540	64 s	3.8×10^9
Trichloroethene	17	34 min	2.9×10^9
Tetrachloroethene	<0.1	>4 d	2×10^9
Chlorobenezene	0.75	13 h	5.6×10^9
Fuel additives			
Benzene	2	4.8 h	7.9×10^9
Toluene	14	41 min	5.1×10^9
o-Xylene	90	6.4 min	6.7×10^9
MTBE	0.14	2.8 d	1.9×10^9
t-BuOH	3×10^{-3}	133 d	6×10^8
Ethanol	0.37	26 h	1.9×10^9
Disinfection by-products			
Chloroform	≤0.1	≥100 h	5×10^7
Bromoform	≤0.2	≥50 h	1.3×10^8
Iodoform	<2	>5 h	7×10^9
Trichloroacetate	$<3 \times 10^{-5}$	>36 y	6×10^7
Pharmaceuticals			
Diclofenac[c]	$\sim 1 \times 10^6$	33 ms	7.5×10^9
Carbamazepine[c]	$\sim 3 \times 10^5$	0.1 s	8.8×10^9
Sulfamethoxazole[c]	$\sim 2.5 \times 10^6$	14 ms	5.5×10^9
17α-Ethinylestradiol[c]	$\sim 7 \times 10^9$	5 μs	9.8×10^9

[a]Data from von Gunten (2003a), except for $k_{\dot{O}H}$ for microcystin-LR, which is from Song et al. (2009).
[b]Estimated for 1 mg/L ozone.
[c]Rate constants are for the most reactive form of the pharmaceutical, which is typically the deprotonated form.

O_3/UV and O_3/H_2O_2 The production of hydroxyl radicals from ozone can be enhanced by irradiation with UV light, which photolyzes the ozone to produce singlet oxygen radicals (Ȯ) (Equation 10-110). These radicals combine with water to generate H_2O_2 (Equation 10-111), which, upon dissociation to hydroperoxide ion (HO_2^-), initiates ozone decomposition and formation of $H\dot{O}_2$ radicals (Equation 10-112). The $H\dot{O}_2$ radicals can then dissociate

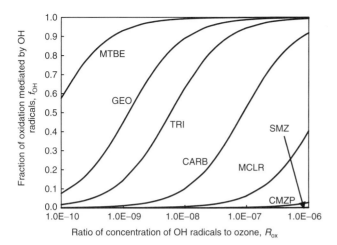

FIGURE 10-31. Relative importance of OH radicals versus ozone for oxidation of organic compounds under typical conditions for ozonation or AOP treatment of drinking water. (MTBE: methyl *tert*-butyl ether; GEO: geosmin; TRI: trichloroethene; CARB: carbofuran; MCLR: microcystin-LR; CMZP: carbamazepine; SMZ: sulfamethoxazole).

and generate hydroxyl radicals via Reactions 10-99 to 10-102, as described previously. Like H_2O_2, ozone does not significantly absorb light with a wavelength above 300 nm, so ozone decomposition and the concomitant generation of $\dot{O}H$ are not enhanced by sunlight.

$$O_3 + h\nu \rightarrow O_2 + \dot{O} \quad (10\text{-}110)$$

$$H_2O + \dot{O} \rightarrow H_2O_2 \quad (10\text{-}111)$$

$$O_3 + HO_2^- \rightarrow \dot{O}_3^- + H\dot{O}_2 \quad (10\text{-}112)$$

Hydroxyl radicals can also be formed by photolysis of H_2O_2 in this process, as occurs in the UV/H_2O_2 AOP (Reaction 10-95). Direct addition of hydrogen peroxide to ozone-containing solutions also generates hydroxyl radicals, via Reactions 10-112 and 10-99 to 10-102.

The combined use of ozone and hydrogen peroxide—known as the *peroxone* process—has found a niche in oxidizing difficult-to-treat organics such as taste and odor-causing compounds in drinking water, including geosmin and 2-MIB (USEPA, 1999 and references therein). Peroxone has also been shown to be effective in oxidizing halogenated compounds such as 1,1-dichloropropene, trichloroethylene, 1-chloro-pentane, and 1,2-dichloroethane (Masten and Hoigne, 1992). In addition, pharmaceuticals such as the lipid-lowering agent clofibric acid and the analgesics ibuprofen and diclofenac are effectively degraded by this combination of ozone and hydrogen peroxide (Zwiener and Frimmel, 2000). These pharmaceuticals have been found at low microgram per liter concentrations in sewage treatment plant effluents and source waters to drinking water treatment plants and are relatively resistant to attack by ozone alone.

The optimum peroxide and ozone doses appear to be relatively case specific, determined primarily by the reaction conditions and the reactivity of hydroxyl radicals with both the specific contaminant of concern and any radical scavengers present in the reaction matrix. For example, Glaze and Kang (1988) found the optimum peroxide:ozone ratio for TCE and PCE oxidation in groundwater to be 0.5 by weight; a similar ratio was found to be effective in a full-scale demonstration plant in Los Angeles, treating groundwater containing up to 447 mg/L TCE and 163 mg/L PCE (Karimi et al., 1997). Zwiener and Frimmel (2000) reported that more than 90% of the clofibric acid, ibuprofen, and diclofenac in river water samples (of DOC content 3.7 mg/L) was removed within 10 min by addition of 1.4 mg/L peroxide and 3.7 mg/L ozone (molar ratio of 2:1), and the removal of these compounds was increased to greater than 98% by increasing reagent addition to 1.8 mg/L peroxide and 5 mg/L ozone.

The scavenging capacity of the natural water (determined particularly by the DOC content and alkalinity of the water), the oxidant dose, and the reaction pH are all important determinants of overall effectiveness of the process, as demonstrated by Acero et al. (2000, 2001) in studies of the O_3- and O_3/H_2O_2-induced degradation of atrazine and MTBE. Before implementing such a treatment process, consideration must also be given to the range of organic degradation products that may be produced, as well as the ubiquitous formation of bromate that accompanies such ozonation reactions.

UV/Semiconductor When a semiconductor such as titanium dioxide (TiO_2) absorbs UV light, electrons gain energy and move from the valence band to the conduction band, leaving a positively charged "hole" in the valence band.[11] Holes (h^+_{VB}) and electrons (e^-_{CB}) can recombine or can react with molecules on the TiO_2 surface. Holes react with H_2O and OH^- to generate hydroxyl radicals, while electrons react with O_2 to form superoxide radicals ($\dot{O}_2^-$), as represented by the following reactions (Venkatadri and Peters, 1993):

$$TiO_2(s) + h\nu \rightarrow h^+_{VB} + e^-_{CB} \quad (10\text{-}113)$$

$$h^+_{VB} + e^-_{CB} \rightarrow heat \quad (10\text{-}114)$$

---

[11] In the band model of solids, the filled and vacant electron orbitals of isolated atoms form essentially continuous bands when these atoms are assembled into a lattice structure. The filled bonding orbitals form the valence band (VB) and the vacant antibonding orbitals form the conduction band (CB). These bands are separated by a forbidden region or band gap of energy E_g (usually given in units of electron volts [eV]). For some solids (e.g., Cu, Ag), the band gap is small ($E_g \ll k_B T$ where k_B is Boltzman's constant), enabling electrons to move from the VB to the CB with only a small energy of activation. (At 25°C, kT equals 2.57×10^{-2} eV.) However, other solids exhibit a large band gap and require the input of considerable energy to induce electrons to move into the CB. For example, E_g for TiO_2 lies in the range of 3.0–3.3 eV, which corresponds to a wavelength of 376–413 nm.

$$H_2O + h_{VB}^+ \rightarrow \dot{O}H + H^+ \quad (10\text{-}115)$$

$$OH^- + h_{VB}^+ \rightarrow \dot{O}H \quad (10\text{-}116)$$

$$O_2 + e_{CB}^- \rightarrow \dot{O}_2^- \quad (10\text{-}117)$$

In such systems, organic compounds can be directly oxidized by holes or reduced by electrons, but in most cases they are degraded by hydroxyl radicals. It is often assumed that degradation takes place on the $TiO_2(s)$ surface, although evidence for this is mixed, and it is also possible that degradation by $\dot{O}H$ occurs in solution (Sun and Pignatello, 1995). If hydroxyl radicals do react in solution, they will not diffuse very far from the $TiO_2(s)$ surface because of their high reactivity. Because the process is initiated by a reaction of the $TiO_2(s)$ with light, but the $TiO_2(s)$ is not consumed, the process is called *photocatalysis*.

$TiO_2(s)$ is stable, insoluble, nontoxic and inexpensive, so it is the most widely studied semiconductor in these processes (Venkatadri and Peters, 1993). Various $TiO_2(s)$ materials have different photocatalytic activities, depending on their bulk and surface properties (Lindner et al., 1997). However, photocatalytic degradation can also be mediated by several other semiconductors, including $ZnO(s)$, $Fe_2O_3(s)$, and $CdS(s)$.

$TiO_2(s)$ catalysts can be used in a suspension or immobilized on a support, eliminating the need for a subsequent separation step. Supports that have been used include plates, coils, tubes, glass beads, membranes, and quartz fibers. In most cases, immobilized $TiO_2(s)$ is less efficient than a suspension, due to the reduced number of active sites and mass transfer limitations.

Optimization of the catalyst concentration (when used in suspension) and solution pH is generally important. The optimum catalyst concentration depends mainly on the catalyst and reactor configuration. When the catalyst concentration is higher than the optimum value, the process is less efficient because of reduced light penetration. Although the optimum pH is normally specific for the compound being degraded, most compounds can be degraded over a wide range of pH values (Venkatadri and Peters, 1993; Lindner et al., 1997; Herrmann, 1999).

The fast recombination of holes and electrons is the main factor limiting the rate at which compounds are degraded by photocatalysis (Venkatadri and Peters, 1993). Various methods of improving the catalytic activity have been studied, including doping the catalyst with metal ions and adding alternative electron acceptors such as H_2O_2 or $S_2O_8^{2-}$. These methods normally increase the catalytic activity, but they inhibit the degradation of some compounds (Lindner et al., 1997). In addition to reducing the recombination of holes and electrons by reacting with electrons, H_2O_2 and $S_2O_8^{2-}$ provide additional routes for generating hydroxyl radicals (Reactions 10-118 through 10-120). Even if H_2O_2 is not doped into a catalyst, it might be generated during photocatalysis by the reduction of $H\dot{O}_2$ (Reaction 10-121) or the disproportionation (Reaction 10-97) of $H\dot{O}_2/\dot{O}_2^-$.

$$H_2O_2 + e_{CB}^- \rightarrow \dot{O}H + OH^- \quad (10\text{-}118)$$

$$S_2O_8^{2-} + e_{CB}^- \rightarrow S\dot{O}_4^- + SO_4^{2-} \quad (10\text{-}119)$$

$$S\dot{O}_4^- + H_2O \rightarrow SO_4^{2-} + H^+ + \dot{O}H \quad (10\text{-}120)$$

$$H\dot{O}_2 + e_{CB}^- \rightarrow HO_2^- \quad (10\text{-}121)$$

To have sufficient energy to initiate photocatalytic degradation using $TiO_2(s)$, light must have a wavelength below about 400 nm. About 5% of the energy of sunlight is in this range, and considerable effort has been devoted to the development of reactors for sunlight/$TiO_2(s)$ processes. Some reactor designs concentrate sunlight, while others do not. Although many laboratory and pilot-scale studies have been conducted exploring the use of $h\nu/TiO_2(s)$ to treat contaminated water and wastewater (Venkatadri and Peters, 1993), full-scale applications are still largely experimental (Herrmann, 1999).

Compounds that absorb visible light, such as dyes, can be degraded in light/semiconductor systems by an alternative mechanism known as photosensitization. This mechanism involves absorption of light to excite the dye molecule (to a form we designate as Dye*), which can then become oxidized by injecting an electron into the conduction band of the semiconductor:

$$Dye + h\nu \rightarrow Dye^* \quad (10\text{-}122)$$

$$Dye^* + TiO_2 \rightarrow Dye^{\bullet+} + e_{CB}^- \quad (10\text{-}123)$$

The dye radical cation (Dye$^{\bullet+}$) can then undergo further degradation. Recombination of the electron and the dye radical cation can be prevented if species such as oxygen are available to scavenge electrons (Reaction 10-117). Zhang et al. (1997) have demonstrated the photosensitized degradation of dyes using visible light and $TiO_2(s)$.

Wet Air Oxidation Wet air oxidation (WAO) uses O_2 as an oxidant at high temperatures and pressures, often exceeding 300°C and 10^5 kPa, respectively. Under these conditions, hydroxyl radicals are produced, and compounds can be oxidized by either oxygen or hydroxyl radicals. Either pure oxygen or air can be used as the O_2 supply (Lin and Ho, 1996). In a related process, known as wet peroxide oxidation, H_2O_2 is used as the oxidant instead of O_2. While WAO is considered a waste treatment technology today, it was first successfully commercialized in the 1930s and 1940s for the manufacture of artificial vanilla flavoring (vanillin). The WAO process was commercialized as the Zimmermann process, named after its developer, and the term "ZIMPRO process" remains a common synonym for WAO. Complete mineralization of the waste stream is difficult by WAO since some low molecular weight compounds (especially acetic and propionic acids and methanol) are resistant to oxidation. The nitrogen in organic nitrogen compounds is transformed into ammonia, which is also very stable under WAO conditions (Patria et al., 2004).

WAO is relatively expensive due to the high energy requirements. The addition of a catalyst can enable the use of lower temperatures and pressures and can also aid the removal of refractory compounds such as acetic acid and ammonia. Both dissolved and solid catalysts have been used, including metal salts (particularly salts of Fe and Cu), metal oxide powders, and metals immobilized on porous supports (Lin and Ho, 1996; Patria et al., 2004).

WAO is a thermal oxidation technology that is applied under wet conditions and, when compared to the dry thermal oxidation process of incineration, is relatively energy efficient, particularly if the oxidized effluent is cooled by heat exchange with the feed stream and the liquid effluent undergoes biological treatment. A variety of WAO processes have been patented and applied at full scale, including the noncatalytic VerTech process and the catalytic NS-LC, LOPROX, WPO, and ATHOS processes (Patria et al., 2004).

Sonolysis Ultrasonic irradiation of aqueous solutions that contain dissolved gas leads to the formation of cavitation bubbles. The collapse of these bubbles creates microenvironments with extremely high temperatures (4000–5000 K) and pressures (up to 10^6 kPa). Under these conditions, H_2O can dissociate into $\dot{O}H(g)$ and $\dot{H}(g)$ radicals, which can recombine, react with other species within the gas bubble, or diffuse out of the bubble into the bulk solution (Mason and Petrier, 2004). Nonvolatile molecules are degraded mainly by reaction with hydroxyl radicals in the bulk solution, while volatile molecules can also be degraded by reactions within the cavitation bubbles (Vinodgopal et al., 1998). Sonolysis has been shown to degrade a wide range of compounds including chlorinated aliphatics and aromatics (Hoffmann et al., 1996; Mason and Petrier, 2004), azo dyes, and NOM (Olson and Barbier, 1994). Considerable room exists for optimization of reaction conditions, particularly with regard to selection of pH and the frequency of the ultrasonic energy input (Hua and Hoffmann, 1997). As with all AOPs, attention must be given to the reaction products. However, in many instances, essentially complete mineralization is observed (Jiang et al., 2002).

Hydroxyl radicals, both within the gas bubbles and in the bulk solution, can recombine to form H_2O_2. The rate of ultrasonic degradation of target compounds can often be enhanced by the addition of Fe(II) species to the solution, which provide a secondary source of hydroxyl radicals through the reaction with H_2O_2 (Joseph et al., 2000). Combined use of photolysis and sonolysis has also been found to increase the rate of degradation of volatile organic compounds such as trichlororethane (TCA), TCE, and PCE beyond the additive effect of the oxidative technologies applied separately (Sato et al., 2001).

Both batch and flow sonication systems have been used at full scale, though higher intensity ultrasound is best applied in flowing systems. Either resonating tube reactors in which liquid passes through a pipe with ultrasonically vibrating walls or tubes containing resonating inserts are most commonly used. In the latter category, either tubuler or disc inserts can be used, with the application of a number of discs in series appearing particularly attractive for high power systems (Mason and Petrier, 2004).

Fenton-Based Systems

Dark Fenton Process In acidic solutions, hydrogen peroxide can oxidize Fe(II) to Fe(III) via a reaction that is thought to form hydroxyl radicals:

$$Fe(II) + H_2O_2 \rightarrow Fe(III) + \dot{O}H + OH^- \qquad (10\text{-}124)$$

Processes using these types of reactions were first described by Fenton (1894), and the reaction shown is often referred to as the Fenton reaction, but it is actually a specific example of a reaction known as the Haber–Weiss reaction (Haber and Weiss, 1934). Insight into the Fenton process has been provided by Barb et al. (1951) and Walling (1975).

In the Fenton process, a sequence of reactions is initiated by the oxidation of Fe(II). These reactions include the reduction of Fe(III) by peroxide to form the hydroperoxy radical (the forward direction of Equation 10-125), which can then reduce more Fe(III) (Equation 10-126, shown with the deprotonated hydroperoxy radical as the reactant). Hydroxyl radicals produced in Equation 10-124 can be scavenged by both Fe(II) (Equation 10-127) and H_2O_2 (Equation 10-128).

$$Fe(III) + H_2O_2 \leftrightarrow Fe(II) + H\dot{O}_2 + H^+ \qquad (10\text{-}125)$$

$$Fe(III) + \dot{O}_2^- \rightarrow Fe(II) + O_2 \qquad (10\text{-}126)$$

$$Fe(II) + \dot{O}H \rightarrow OH^- + Fe(III) \qquad (10\text{-}127)$$

$$H_2O_2 + \dot{O}H \rightarrow H\dot{O}_2 + H_2O \qquad (10\text{-}128)$$

The preceding discussion suggests that, in the Fenton process, iron can act as a catalyst for either the generation of $\dot{O}H$ radicals from H_2O_2 or the destruction of H_2O_2 without net formation of the radicals. The dominant reactions in these systems are still poorly understood, and considerable disagreement remains with regard to the exact intermediates that form, including whether or not hydroxyl radicals actually form (Kremer, 1999). Other intermediates that have been proposed include a hydrated Fe^{2+}–H_2O_2 complex and the ferryl ion (FeO^{2+}, with Fe in the +4 oxidation state) (Bossmann et al., 1998; Ensing et al., 2002). Empirically, Fenton-based systems effectively oxidize many organics that are difficult to oxidize in other ways, so it is apparent that the $\dot{O}H$ or other highly reactive species that are produced react more rapidly with the organics than with scavengers.

The reactions also indicate that, regardless of whether the process begins with Fe(II) or Fe(III) present, the other form of Fe will be generated, so that species in both oxidation states are present. The initial organic degradation rate is slower when the Fe is initially present in the +III oxidation state, because some of the Fe(III) must be reduced to Fe(II) before hydroxyl radicals (or ferryl species) are produced. However, the ultimate extent of degradation is independent of the initial iron oxidation state (Safarzadeh-Amiri et al., 1997). The reduction of Fe(III) (via Reactions 10-125 and 10-126) is generally much slower than the oxidation of Fe(II), so most of the iron in the system is expected to exist in the Fe(III) form once nearly steady-state conditions have been reached.

The Fenton process is most effective at a pH between 2 and 4, with an optimum pH ~3. If the solution is not sufficiently acidic, precipitation of ferric oxyhydroxide will inhibit this process by interfering with the regeneration of Fe(II) (Sun and Pignatello, 1993a). The presence of certain anions can also inhibit degradation by Fenton's reagent, either by scavenging hydroxyl radicals or by forming complexes with Fe(III) species.

A schematic for the reactions in a Fenton-based system containing methanol, similar to the ones developed previously for UV/H_2O_2 and ozonation, is presented in Figure 10-32. As described in the preceding discussion, the chain reaction can be initiated by addition of H_2O_2 and either Fe(II) or Fe(III), corresponding to pathways Fn-Init1 and Fn-Init2, respectively. The scavenging of superoxide radicals by reduction of Fe(III) (Reaction 10-126) is shown as pathway Fn-Term, and the chain propagating pathway is Fn-Cy. The net stoichiometry for oxidation of one molecule of methanol via one complete cycle of Fn-Cy is

$$CH_3OH + Fe^{2+} + \tfrac{1}{2}O_2 + \tfrac{1}{2}H_2O_2 \rightarrow CH_2O + Fe^{3+} + H_2O + OH^- \quad (10\text{-}129)$$

Note that one-half mole of H_2O_2 is consumed and one mole of Fe(II) is oxidized to Fe(III) for each mole of methanol that is oxidized via the chain-propagating pathway. The Fe(III) can subsequently be reduced back to Fe(II)

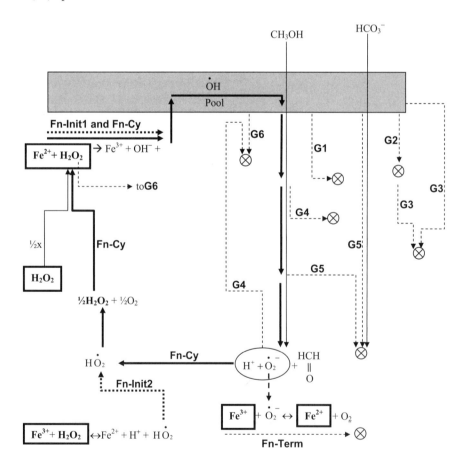

FIGURE 10-32. Reactions in a Fenton-based oxidation process, in a solution containing methanol. Chemicals shown in boxes are those that might be added as reagents in the process. Pathways G1–G6 are shown in skeletal form, as in Figure 10-28. Also as in that figure, ⊗ represents the endpoint of reactions that compete with or terminate the chain reaction.

(e.g., by Reaction 10-125), but at the cost of consuming more H_2O_2. For instance, if Reaction 10-125 occurs once for each instance of Reaction 10-129, and the $H\dot{O}_2$ then dissociates into H_2O_2 and O_2 via Reaction 10-97, the net reaction would be

$$CH_3OH + H_2O_2 \rightarrow CH_2O + 2H_2O \quad (10\text{-}130)$$

Light-Mediated Fenton Processes The degradation of compounds by the Fenton process can be strongly accelerated by irradiation with UV or a combination of UV and visible light, due to the photolysis of Fe(III) complexes such as $FeOH^{2+}$. These complexes undergo a photo-induced *ligand-to-metal charge transfer* (LMCT) reaction, producing Fe(II) and an oxidized ligand. For instance, when OH^- is the ligand, the reaction is

$$FeOH^{2+} + h\nu \rightarrow Fe^{2+} + \dot{O}H \quad (10\text{-}131)$$

The photoreduction of Fe(III) is faster than the equivalent reactions in the dark Fenton process. The production of $\dot{O}H$ (or another oxidant) by the reaction of Fe(II) with H_2O_2 is therefore enhanced in the photo-Fenton process, because additional $\dot{O}H$ is produced by Reaction 10-131 (Sun and Pignatello, 1993b). Photolysis of H_2O_2 is also possible, but this process would be expected to have only a minor effect due to the weak absorbance of light by H_2O_2.

Organic contaminants and their degradation intermediates may also form photoactive complexes with Fe(III) (Sun and Pignatello, 1993b). For instance, many ligands with a carboxyl group (RCO_2^-) form complexes with Fe(III) that also undergo a photo-induced LMCT reaction:

$$Fe(RCO_2)^{2+} + h\nu \rightarrow Fe^{2+} + \dot{R} + CO_2 \quad (10\text{-}132)$$

This reaction results in the reduction of Fe(III) and oxidation of the ligand, with evolution of CO_2 (Balzani and Carassiti, 1970). A wide variety of compounds including textile dyes are degraded by this type of light-assisted Fenton process (Xie et al., 2000).

An advantage of Fenton processes that involve light-absorbing complexes of ferric ions with an organic ligand is the possibility of extending the process to higher pH. The standard Fenton process becomes inoperable at pH above ~4.5 because of the tendency for ferric iron to hydrolyze and precipitate. The precipitated ferric species can then not be rereduced, and the process loses its catalytic capacity. Organic ligands that have a high affinity for ferric iron may prevent precipitation at the higher pH values and enable extension of the process to near neutral (or even higher pH) solutions. The organic ligand is likely to be consumed in the process, and its ability to outcompete OH^- ions for the Fe(III) will diminish as its concentration decreases.

An interesting extension of this ligand and light-mediated Fenton process involves the use of a strong ferric iron complexant such as oxalate, which forms a highly photoactive complex with Fe(III). The hydroxyl and/or organic radicals produced upon oxidation of photo-produced ferrous iron may then be used to degrade trace contaminants. This process has been named the *modified photo-Fenton process* and has been used to degrade a range of trace constituents. Sun and Pignatello (1993a) studied the degradation of pesticides by H_2O_2 and Fe(III) complexes at pH 6. They screened many ligands and found a number of Fe(III) complexes that had catalytic activity for the degradation of pesticides, as well as some Fe(III) complexes that were unreactive. The mechanism for pesticide degradation was analogous to that in the uncomplexed iron system—Fe(III) was reduced and then Fe(II) (typically present as an organic complex) reacted with H_2O_2, generating hydroxyl radicals (or some other oxidant such as a ferryl species). All active ligands were themselves degraded but, in many cases, the ligand degradation products also formed active Fe(III) complexes.

Heterogeneous Fenton Processes As noted in the preceding section, precipitation of iron oxides often makes the iron unreactive in Fenton processes. However, Gurol and Lin (1998) and Lin and Gurol (1998) reported that goethite and hydrogen peroxide could be effective in a heterogeneous Fenton process. Other iron oxides may also be effective in such processes, but most have been found to be less effective than goethite (Valentine and Wang, 1998). While the mechanism of the heterogeneous Fenton process is unclear, Lin and Gurol (1998) suggested that the degradation of hydrogen peroxide is catalyzed at ferric iron oxide surface sites ($\equiv Fe(III)-OH$); that is,

$$\equiv Fe(III)-OH + H_2O_2 \leftrightarrow \equiv Fe(III)-OH \cdot H_2O_2 \quad (10\text{-}133)$$

$$\equiv Fe(III)-OH \cdot H_2O_2 \rightarrow \equiv Fe(II) + H_2O + H\dot{O}_2 \quad (10\text{-}134)$$

$$\equiv Fe(II) + H_2O_2 \rightarrow \equiv Fe(III)-OH + \dot{O}H \quad (10\text{-}135)$$

$$\equiv Fe(III)-OH + H\dot{O}_2/\dot{O}_2^- \rightarrow \equiv Fe(II) + H_2O/OH^- + O_2 \quad (10\text{-}136)$$

The product of the first reaction in the sequence is an iron surface-peroxide complex that reduces the iron in the second step while creating the protonated superoxide radical. This process appears to have a number of advantages over the homogeneous process including the following:

- The iron oxide catalyst can be used over extended periods without requiring regeneration or replacement, and can be removed from the treated water by sedimentation or filtration.
- The rate of hydroxyl radical generation increases slightly with increasing pH in the range 5–9; in

contrast, in the homogeneous process, the reaction rate decreases sharply in this pH range, with a negligible rate of above about pH 4.5 if no complexing agents are used.

- The catalytic oxidation efficiency of goethite seems to be relatively insensitive to the inorganic carbonate concentration.
- The catalyst surface has a low affinity for bromide adsorption; hence the production of bromates is minimized.

Photolysis has been reported to enhance the rate of contaminant degradation by the heterogeneous Fenton process (He et al., 2002). In addition, heterogeneous Fenton and photo-Fenton processes using iron ions immobilized on perfluorinated membranes can be used to degrade organic compounds (Sabhi and Kiwi, 2001). These membranes are resistant to attack by hydroxyl radicals and, as such, are not degraded during treatment. These processes can be used to treat solutions at neutral pH with efficiencies reported to be close to those in acidic solutions. The membrane could be reused through many cycles. During its use, some Fe(III) might leach from the membrane, but the associated loss of activity may be acceptable (Sabhi and Kiwi, 2001).

Other support media for catalytic iron that have been used include zeolites (Pulgarin et al., 1995) and beads of cross-linked organic material, such as alginate gel beads (Fernandez et al., 2000).

Electrochemical Fenton Processes Two additional Fenton processes involve the use of *in situ* electrochemically produced reagents, differing mainly with respect to the form in which the iron enters the system. In cathodic Fenton processes, the iron is added as an Fe(II) or Fe(III) salt, whereas in anodic Fenton processes, the source of the iron is a sacrificial iron (Fe(0)) anode.

Cathodic Fenton Processes In cathodic processes, one or both of the reagents Fe(II) and H_2O_2 are produced *in situ*. The source of Fe(II) may be direct addition or reduction of Fe(III) at the cathode (Reaction 10-137). The source of H_2O_2 may be either direct addition or reduction of oxygen at the cathode (Reaction 10-138) (Ventura et al., 2002).

$$Fe(III) + e^- \rightarrow Fe(II) \quad (10\text{-}137)$$

$$O_2 + 2H^+ + 2e^- \rightarrow H_2O_2 \quad (10\text{-}138)$$

Simultaneous Fe(III) reduction and O_2 reduction can take place at the cathode at comparable rates. Since Fe(II) and H_2O_2 can be continuously produced at controlled rates, a more efficient and more complete degradation of the contaminant can be achieved than with classic Fenton systems, where all the reagents are typically added at the beginning of the reaction. This efficiency is achieved because competitive reactions, which consume the reagents without producing hydroxyl radicals, are less significant in the electrochemical process (Ventura et al., 2002).

In this process, the electrochemical cell is undivided and has an anode made from inert material such as platinum or platinized titanium. The cathode can be made from carbon-containing materials such as carbon felt (Oturan, 2000) or carbon-polytetrafluoroethylene (Boye et al., 2002) and is supplied with oxygen. Electro-Fenton systems may also be exposed to irradiation by UV/visible light, leading to photo-electro-Fenton processes (Boye et al., 2002).

Anodic Fenton Processes In the anodic Fenton process, an iron electrode is used as the anode and becomes the source of Fe(II). The process normally involves use of two cells, with a salt bridge providing electrical connectivity between the cells and a graphite electrode used as a cathode (Wang and Lemley, 2001). The process is normally operated at pH 2–3.

A modification of the electro-Fenton process, involving the use of a sacrificial iron anode, has been developed by Brillas and Casado (2002). This process makes use of an undivided electrochemical cell. Since the solution cannot be maintained in the acidic pH range, this process is characterized by the precipitation of iron oxyhydroxides.

Full-Scale Applications Environmental engineering applications of Fenton processes have been studied mainly at laboratory scale, with some pilot-scale studies (Venkatadri and Peters, 1993) and a few full-scale applications. Several full-scale plants in South Africa use Fenton's reagent to treat wastewater from the textile industry (Vandevivere et al., 1998). Commercial-scale Fenton installations exist in the United States for treatment of water contaminated with volatile organic compounds (VOCs) and semivolatile organic compounds (SVOCs) (USEPA, 1998b). Some of these processes include a commercial-scale photo-Fenton system that has been used for the treatment of contaminated groundwater and industrial wastewater (the Calgon Rayox® ENOX water treatment system) and several *in situ* processes in which hydrogen peroxide and a catalyst (a proprietary mixture of nonhazardous metallic salts) are injected into the subsurface environment via various techniques (Casey and Bergren, 1999; Greenberg et al., 1998).

10.5 REDUCTIVE PROCESSES

Sulfur-Based Systems

As indicated earlier, reduced sulfur species (mostly the S(IV) species $SO_2(aq)$, H_2SO_3, HSO_3^- and SO_3^{2-}) react rapidly with oxidants such as chlorine and are used in water treatment for dechlorination purposes. Additionally, sulfide species can be removed from solution by addition of ferric salts. Reduction of the Fe(III) to Fe(II) oxidizes some of the S(II) to S(VI) (i.e., SO_4^{2-}), and the Fe(II) that is generated can remove more S(II) by precipitation of the highly insoluble FeS(s). Reduced sulfur species are also effective in reducing Cr(VI) to the insoluble Cr(III) form, with sulfite (S(IV)) species commonly used to treat chromium-rich waste streams.

■ **EXAMPLE 10-17.** The treatment of wastewater that contain Cr(VI) as dichromate ($Cr_2O_7^{2-}$) begins by reducing the chromium to Cr(III). After the reduction of the Cr(VI) to Cr(III), the pH is raised to precipitate $Cr(OH)_3(s)$. The reduction is often carried out by adding sodium bisulfite ($NaHSO_3$) to the solution, in which the S(IV) is oxidized to sulfate (SO_4^{2-}). Such wastewater are produced from chrome-plating processes.

(a) Write a balanced chemical reaction showing the reduction of $Cr_2O_7^{2-}$ to Cr^{3+}, via oxidation of HSO_3^- to SO_4^{2-}. These species are the dominant ones in the various acid/base groups at pH 2.0 to 2.5, which is the pH range in which the reaction is normally carried out.

(b) On the basis of your answer to (a), comment on why this reaction is typically carried out at low pH.

(c) Find the dose of sodium bisulfite required for complete reduction of Cr(VI) species that are present at a concentration of 5 mg/L as TOTCr(VI).

Solution.

(a) Sodium bisulfite is a salt; that is, it splits into sodium ions (Na^+) and bisulfite ions (HSO_3^-) in water. The half-reaction for the oxidation of bisulfite is

$$HSO_3^- + H_2O \rightarrow SO_4^{2-} + 3H^+ + 2e^-$$

The half-reaction for the reduction of $Cr_2O_7^{2-}$ to Cr^{3+} is

$$Cr_2O_7^{2-} + 14H^+ + 6e^- \rightarrow 2Cr^{3+} + 7H_2O$$

To balance the electrons, we multiply the bisulfite half-reaction by three, and then add the two half-reactions to obtain the complete reaction:

$$Cr_2O_7^{2-} + 3HSO_3^- + 5H^+ \rightarrow 2Cr^{3+} + 3SO_4^{2-} + 4H_2O$$

(b) The reaction consumes protons ($5H^+$ on the left side of the reaction) and so is facilitated by a low pH, in that lowering the pH drives the equilibrium toward the right (i.e., more oxidation of S(IV) and reduction of Cr(VI)). This observation does not assure that the reaction will actually proceed more rapidly at lower pH. However, the kinetics of the reaction has been reported to be influenced by the speciation between H_2SO_3 and HSO_3^-, with the former leading to a faster reaction. The pK_{a1} value for H_2SO_3 is 1.8, so in this case, lowering the pH does increase the reaction rate.

(c) The required dose can be found from the stoichiometry of the balanced reaction:

$$(5\,mg\,Cr/L)\left(\frac{mmol\,Cr}{52.0\,mg\,Cr}\right)\left(\frac{3\,mmol\,HSO_3^-}{2\,mmol\,Cr}\right)$$
$$\times \left(104\,mg\,NaHSO_3/mmol\,HSO_3^-\right)$$
$$= 15\,mg/L\,NaHSO_3 \quad ■$$

Iron-Based Systems (Fe(II), Fe(s))

Ferrous iron is a powerful reductant and may be used to remove metals such as chromium from industrial waste streams by reduction. Under neutral to mildly alkaline conditions, the ferric iron that is generated is highly insoluble and precipitates as iron oxyhydroxide. The chromium is reduced to the trivalent state (Cr(III)), which has a strong tendency to adsorb onto and/or coprecipitate with the iron oxyhydroxide solids, for example,

$$CrO_4^{2-} + 3Fe^{2+} + 8H_2O \rightarrow 3Fe(OH)_3(s) + Cr(OH)_3(s) + 4H^+ \quad (10\text{-}139)$$

Similar reactions with Fe(II) would be expected to occur with elements such as uranium, which is soluble and highly mobile in the +VI oxidation state, but highly insoluble in the +IV oxidation state (readily forming $UO_2(s)$).

Zero-valent iron (Fe(s), sometimes shown as Fe° or designated as ZVI) is also a good reductant, releasing electrons that can reduce both organic and inorganic species. Fe(s) has been used successfully for the degradation of a wide range of contaminant organics in groundwater, including chlorinated and nitro-substituted entities (Tratnyek et al., 2003). Laboratory studies have confirmed the Fe(s) rapidly degrades atrazine, parathion (Gauch et al., 1999) and DDT (Sayles et al., 1997). In these latter cases, the organic is dechlorinated, as in the following generic reaction:

$$C_xH_yCl_z + zFe(s) \rightarrow C_xH_y^{z-} + zFe^{2+} + zCl^- \quad (10\text{-}140)$$

The reduction of chlorine and its simultaneous release from the organic molecule generally convert the organic to a form that is substantially more biodegradable.

Essentially all applications of this technology have involved creation of a permeable, reactive wall either by emplacement of iron metal in an excavated trench across a flow path or by injection of the iron into natural or engineered vertical fractures (Hocking et al., 2000). The possibility of using columns packed with Fe(s) also exists for *ex situ* treatment of wastewater streams. For example, Loraine et al. (2002) reported that packed-bed Fe(s) reactors could be a useful treatment option for pump-and-treat remediation of water contaminated with 1,2-dibromoethane (ethylene dibromide, EDB), a compound formerly used as an antiknock additive in leaded gasoline and aviation fuel. Similarly, Mantha et al. (2001) found that wastewater contaminated with nitrobenzene could be effectively treated using upflow, anaerobic columns containing Fe(s), though clogging problems associated with the oxidation (corrosion) of Fe metal and subsequent precipitation of iron precipitates occurred after some time.

Zhang et al. (1998) reported that the reactivity of nano-sized (1–100 nm) Fe(s) particles is higher than that of micro-sized particles, due in part to the increased surface area (average of $33.5\,m^2/g$ for the nano-sized particles compared to $0.9\,m^2/g$ for the commonly used micro-sized particles). These workers also reported a significantly higher (by up to 100 times) reactivity of nano-sized particles on a surface-area-normalized basis; however, more recent studies by Nurmi et al. (2005) revealed similar surface-area-normalized rate constants for nano and micro-sized Fe(s). The emplacement options for such particles are flexible since they may be pumped into aquifers where they will disperse along flowpaths, or they may be used to coat or impregnate other porous media (such as zeolites). However, concerns remain about such processes, because the ultimate environmental fate and effects of the nano-sized Fe(s) particles have not been extensively studied.

Catalyst metals such as palladium or platinum could also be placed on the Fe(s) surface. While expense would mitigate against large-scale use of such a bimetallic system, the presence of the catalyst lowers the activation energy for degradation substantially, thus broadening the application to a wider range of contaminants, reducing the formation of intermediates, and generally increasing the rate of degradation. Another advantage of such systems is that, if sulfide ions are present, the ferrous iron produced can act as a sink for those ions (through formation of FeS(s)), which otherwise tend to poison the Pd or Pt catalyst. It should also be noted that, in the presence of oxygen, Fe(s) may induce the oxidative degradation of contaminants as a result of the initiation of Fenton processes at or near the surface of the Fe(s) particles (Joo et al., 2005).

10.6 ELECTROCHEMICAL PROCESSES

As indicated earlier, electrochemical production of ferrous iron by oxidation of an iron anode and/or production of hydrogen peroxide at the cathode is an effective method for the continuous production of the reagents central to the operation of Fenton's reagent oxidation processes. Examples of the use of electrochemical methods for the direct transformation of contaminants are, however, relatively rare, due to both the energy costs associated with the process and the specialized equipment required.

Electrochemical oxidation of organic compounds by means other than Fenton-related processes generally occurs via hydroxyl radicals generated by water decomposition at the anode (Tahar and Savall, 1998); that is,

$$H_2O \rightarrow \dot{O}H + H^+ + e^- \qquad (10\text{-}141)$$

High oxygen overvoltage[12] anodes such as antimony-doped SnO_2-coated titanium anodes seem to be the most promising for this purpose (Panizza et al., 2000).

The possibility of direct reduction of contaminants at the cathode also exists. For example, Korshin and Jensen (2001) have shown that halogenated DBPs may be degraded by reductive means. Bromine-containing haloacetic acids were effectively dehalogenated at copper and gold electrodes. However, chlorine-containing haloacetic acids posed difficulties, because the monochloroacetate that formed could not be reduced directly.

10.7 DISINFECTION

Disinfection can be defined as the deliberate reduction of the number of viable pathogenic microorganisms in a system; disinfection of water and wastewater has relied heavily on the use of oxidants such as chlorine since the early 1900s. While classic waterborne diseases such as cholera and typhoid have been effectively controlled in developed countries by oxidant addition, a variety of organisms continue to cause problems, including viruses, certain bacteria (such as *Campylobacter* and *Mycobacteria*), and parasitic protozoans (particularly *Giardia* and *Cryptosporidium*) (Haas, 1999). Common oxidants added for disinfection purposes include molecular chlorine (Cl_2), sodium or calcium hypochlorite (NaOCl or $Ca(OCl)_2$), chlorine dioxide (ClO_2), and ozone (O_3). As discussed previously, ammonia is also often added either simultaneously or sequentially with chlorine to form chloramines to prolong the disinfecting action of chlorine (with monochloramine, NH_2Cl, being the primary disinfectant).

The addition of oxidants as disinfectants often leads to the concomitant production of disinfection by-products (DBPs)

---

[12] The overvoltage is a measure of the extent of disequilibrium of a redox reaction, expressed in terms of electrical potential. If a redox reaction occurs at an electrode surface, the electrode materials can have a significant, and sometimes controlling, effect on the reaction rate. The greater the resistance to the reaction, the greater is the disequilibrium (i.e., the greater the overvoltage) required for the reaction to proceed at a significant rate.

that pose some risk to human health. These compounds are normally formed through the reaction of the oxidant with natural organic matter (NOM) and are of particular concern in drinking water supplies. An extensive literature exists on the wide range of DBPs formed and their associated health impacts (see, e.g., WHO (2000); Craun et al. (2001); Bull et al. (2007)). Similar chlorinated products are generated from disinfection of wastewater and, depending upon the end disposal (or reuse) of the treated wastewater, may pose health or ecological risks.

Awareness of DBPs and the environmental effects of chlorine have led to a reduction in its use as a disinfectant, though alternative oxidants such as ozone are also recognized to produce a suite of troublesome compounds, particularly in water containing bromide. Ultraviolet (UV) radiation and microfiltration (MF) or ultrafiltration (UF) represent possible alternatives to oxidant addition. These processes do not provide any residual disinfecting capacity downstream of the treatment steps; this characteristic is attractive in wastewater treatment, where adding oxidants to a receiving stream is undesirable, but unattractive in drinking water treatment, where disinfecting capacity is usually desired in the distribution system. Secondary disinfectants must be added if a requirement exists for such residuals. Concerns about DBPs have focused attention on approaches for removing NOM in the early stages of water treatment and minimizing the load of organic matter generated within or transported to source water used for drinking purposes. Occasionally, DBPs are allowed to form but then are removed (by stripping, adsorption, or other treatment); generally, this approach is not favored because of the risk associated with many unknown or unidentified by-products.

Modeling Disinfection

Several empirical models have been developed to assist in the design and optimization of processes for the disinfection of water and wastewater. Chick (1908) investigated disinfection reaction rates by exposing bacteria to a constant disinfectant concentration in test tube reactors and enumerating survivors at successive time intervals. Chick considered the reaction between bacteria and the chemical disinfectant to be analogous to an elementary chemical reaction, with a reaction stoichiometry and rate expression as follows:

$$aX + nD \xrightarrow{k} P \text{ (inactive microbe)} \quad (10\text{-}142)$$

$$r_X = -k c_X^a c_D^n \quad (10\text{-}143)$$

where c_X represents the concentration of viable organisms (typically, a number concentration [e.g., number per mL]),

c_D is the concentration of disinfectant, and a and n are stoichiometric coefficients.

Assuming the disinfectant to be present in excess and a and n to be unity, Equation 10-143 can be simplified to the following expression, known as Chick's law:

$$r_X = -k^* c_X \quad (10\text{-}144)$$

where k^* is a pseudo-first-order reaction rate constant equal to kc_D. Chick's law states that the rate of inactivation of organisms is proportional to the number concentration remaining, for a given concentration of disinfectant. For a batch system in which the assumptions of Chick's law are met, and assuming that the disinfection is occurring much more rapidly than growth, a mass balance on organisms leads to the following familiar pseudo-first-order relationship:

$$V \frac{dc_X}{dt} = r_X V = -k^* c_X V$$

$$\int_{c_X(0)}^{c_X(t)} \frac{dc_X}{c_X} = -k^* \int_0^t dt$$

$$\ln \frac{c_X(t)}{c_X(0)} = -k^* t \quad (10\text{-}145\text{a})$$

$$\log \frac{c_X(t)}{c_X(0)} = -\frac{k^* t}{\ln(10)} = -\frac{k^* t}{2.303} = -k_{\text{Chick}} t \quad (10\text{-}145\text{b})$$

where $c_X(0)$ is the concentration of organisms at time zero and k_{Chick} equals $k^*/2.303$. This relationship is shown by the straight line in Figure 10-33. Such simple inactivation kinetics might be expected if the organisms present are genetically similar, if they are at a uniform stage of development, and if the inactivation occurs by a single interaction between the disinfectant and a sensitive site in the organism.

Watson (1908) accepted the approximation that $a = 1$ in Equation 10-143, but argued that the dependence of r_X on the disinfectant concentration should be shown explicitly; that is, he favored the following rate expression:

$$r_X = -k c_X c_D^n \quad (10\text{-}146)$$

The parameter n, referred to as the *coefficient of dilution*, can be viewed as the order of the reaction with respect to the disinfectant concentration and has been interpreted as the average number of disinfectant molecules that must "react" with the organism to inactivate it. If this expression is integrated, using similar assumptions as earlier (batch reactor, disinfection rate much greater than growth rate,

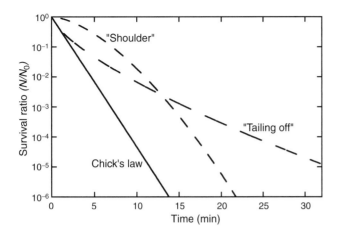

FIGURE 10-33. Generic patterns of microbial inactivation rate according to Chick's law and some common deviations from Chick's law are also shown. As indicated in Equation 10-145b, the slope of the line labeled "Chick's law" is $-k_{\text{Chick}}$. *Source*: Adapted from Haas (1999).

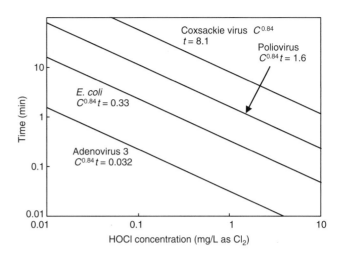

FIGURE 10-34. Time–concentration relationships for inactivation of various microorganisms by HOCl at 0–6°C. *Source*: After Berg (1964).

and constant disinfectant concentration), the so-called Chick–Watson expression is obtained:

$$\ln \frac{c_X(t)}{c_X(0)} = -k c_D^n t \quad \text{or} \quad \log \frac{c_X(t)}{c_X(0)} = -k_{\text{C-W}} c_D^n t \quad (10\text{-}147)$$

The expression on the right (based on $\log_{10}$) is more commonly used, and $k_{\text{C-W}}$ is referred to as the *coefficient of specific lethality*. By rearranging Equation 10-147, we see that $k_{\text{C-W}}$ and n can be determined from data indicating the exposure time required to achieve a given reduction in organism concentration at various disinfectant doses:

$$-n \log c_D + \log \frac{-\log\left(c_X(t)/c_X(0)\right)}{k_{\text{C-W}}} = \log t \quad (10\text{-}148)$$

If data are generated for a fixed degree of inactivation (fixed $c_X(t)/c_X(0)$), then the second term on the left side of Equation 10-148 is a constant, and the equation has the form of a linear equation. Thus, for a fixed degree of inactivation, a log–log plot of t versus c_D is expected to have a slope of $-n$ and an intercept of $\log([-\log(c_X(t)/c_X(0))]/k_{\text{C-W}})$; since $c_X(t)/c_X(0)$ is known, $k_{\text{C-W}}$ can be computed.

Because of the enormous range of microorganism concentrations that might be present in water of interest and the comparably large range of disinfection efficiencies desired, it is common to report such efficiencies as logarithmic values. For example, 99% inactivation of microorganisms (i.e., $1 - c_X(t)/c_X(0) = 0.99$) is referred to as "*2 logs of inactivation*" (because $\log(c_X(0)/c_X(t)) = 2$). Thus, Equation 10-147 can be rewritten as follows:

$$c_D^n t = -\frac{\log\left(c_X(t)/c_X(0)\right)}{k_{\text{C-W}}} = \frac{\log\left(c_X(0)/c_X(t)\right)}{k_{\text{C-W}}}$$

$$= \frac{\text{No. of logs of removal}}{k_{\text{C-W}}} \quad (10\text{-}149)$$

Equation 10-149 indicates that all combinations of c_D and t that yield the same value for the product $c_D^n t$ will accomplish the same degree of inactivation of an organism in a given water (i.e., for a given value of $k_{\text{C-W}}$). For instance, combinations of c_D and t that yield two logs of inactivation of various microorganisms are shown in Figure 10-34, for use of chlorine (i.e., HOCl) as the disinfectant.

The lower the required value of $c_D^n t$, the easier it is to achieve the specified level of inactivation. Thus, according to Equation 10-147, $k_{\text{C-W}}$ can be viewed as an indicator of either the relative potencies of different disinfectants toward a particular type of organism, or the relative susceptibility of different organisms to a given disinfectant: the larger the value of $k_{\text{C-W}}$, the greater is the disinfectant potency or the greater is the susceptibility of the organism.

In many cases, n is close to unity, in which case the product of concentration and time (commonly referred to as the CT *product*) needed to achieve a given degree of disinfection is fixed, independent of the absolute values of c_D and t (for a given temperature, pH, and so on, and maintaining the assumption that c_D is constant throughout the disinfection process). In the United States, regulations for disinfection of drinking water are based on the implicit assumption that the Chick–Watson law applies. Thus, the regulations specify the value of the CT product that must be met to achieve various levels of disinfection, where C is the disinfectant concentration in the effluent from the reactor (or set of reactors) under consideration, and T is a characteristic contact time, taken as the time when the value of the cumulative age distribution, $F(t)$, is 0.10 or 10%; this time is usually referred to as T_{10}. The CT values for a given degree of inactivation depend on the identities of the disinfectant and the target organism, the water temperature, and, in some cases, the solution pH. If the value of n in the

Chick–Watson law is thought to be different from unity for a given disinfectant–organism pair, the CT product required to achieve a given level of disinfection is different for different values of C. The use of the CT product was described in Chapter 4 (see Figure 4-10 and associated text).

As suggested by the curvilinear plots in Figure 10-33, common deviations from Chick–Watson kinetics include both "tailing off" and "shoulders." Shoulders often appear in inactivation curves for organisms that are incorporated into flocs; the flocs can make it difficult for the disinfectant to reach the organisms in the center of the floc. The flocs, which typically consist primarily of aluminum or iron oxide precipitates and/or microorganisms ("clumps") create a diffusional resistance for the disinfectant, decreasing the concentration of disinfectant to which microorganisms in the center of the floc are exposed. Shoulders in disinfection curves have also been interpreted as suggesting that more than one cell must be inactivated to achieve inactivation of a colony or plaque-forming unit, or that more than one target must be affected to kill a given cell (or that a given target must be affected several times) (Severin et al., 1984; Haas, 1980).

Tailing in inactivation curves might be due to (i) conversion of organisms to more resistant forms during inactivation (a process known as hardening), (ii) the existence of non-identical organisms or genetic variants of a single organism, with differing sensitivities to the disinfectant, (iii) protection of a subpopulation, or variations in received dose of disinfectant, or (iv) clumping of a subpopulation (Cerf, 1977). The hardening process and resultant tailing have been investigated particularly thoroughly, and an empirical rate law for this behavior has been proposed by Selleck et al. (1978) for the kinetics of wastewater disinfection with chlorine.

Hom (1972) proposed the following, modified expression for the inactivation rate to account for shoulders or tailing in disinfection curves:

$$r_X = -kmc_X t^{m-1} c_D^n \qquad (10\text{-}150)$$

where m is an empirical constant. Integration of this rate law for a batch system, provided c_D is constant, yields the Hom expression:

$$\int_{c_X(0)}^{c_X(t)} \frac{dc_X}{c_X} = -kmc_D^n \int_0^t t^{m-1} dt \qquad (10\text{-}151)$$

$$\log \frac{c_X(t)}{c_X(0)} = -k_{\text{Hom}} c_D^n t^m \qquad (10\text{-}152)$$

where k_{Hom} incorporates the translation from natural to base-10 logarithms. Equation 10-152 generates disinfection curves with shoulders or tailing if m is greater or less than one, respectively, and if $m = 1$, it simplifies to the Chick–Watson model (Equation 10-147).

The Chick, Chick–Watson, and Hom rate laws specify that the disinfectant concentration is approximately constant throughout the period of interest. If that assumption is not acceptable, the variable disinfectant concentration must be taken into account when the rate laws are integrated. For instance, in a batch reaction in which the disinfectant decays according to a first-order reaction, the residual concentration at any time is given by

$$c_D(t) = c_D(0)\exp(-k_D t) \qquad (10\text{-}153)$$

where k_D is the first-order rate constant and $c_D(0)$ is the disinfectant concentration after the instantaneous demand has been exerted. Equation 10-153 is identical to Equation 10-19, except for the substitution of D to represent a disinfectant, as opposed to "ox" to represent a generic oxidant. A pseudo-Chick–Watson model that accounts for a first-order rate of disinfectant demand during the contact time may be derived by combining Equations 10-146 and 10-153, yielding

$$r_X = -kc_X [c_D(0)]^n \exp(-k_D t n) \qquad (10\text{-}154)$$

Inserting Equation 10-154 into a mass balance on microorganisms in a batch system, integrating the resulting equation, and converting to a $\log_{10}$ basis, we obtain

$$\log \frac{c_X(t)}{c_X(0)} = -\frac{k'}{k_D n} \left(c_D^n(0) - c_D^n(t) \right) \qquad (10\text{-}155)$$

where, $k' = k/\ln(10)$.

Similarly, for disinfection curves with shoulders or tailing, a first-order disinfectant demand kinetic expression (Equation 10-19) can be combined with the Hom expression (Equation 10-150). If both m and n are greater than zero, such an approach yields the following closed form expression:

$$\log \frac{c_X(t)}{c_X(0)} = -\frac{k_{\text{Hom}} m (c_D(0))^n}{(nk_D)^m} \gamma(m, nk_D t) \qquad (10\text{-}156)$$

where γ is a standard mathematical function known as the incomplete gamma function (Haas and Joffe, 1994):

$$\gamma(\alpha, x) = \int_0^x e^{-z} z^{\alpha-1} dz \quad \alpha > 0, \; x \geq 0 \qquad (10\text{-}157)$$

If m is greater that approximately 0.3, Equation 10-156 can be approximated by the following equation (Haas and Joffe 1994; Gyurek and Finch, 1998):

$$\log \frac{c_X(t)}{c_X(0)} = -k_{\text{Hom}} m (c_D(0))^n t^m \eta \qquad (10\text{-}158)$$

where η is an "efficiency factor" that corrects the original Hom model (Equation 10-152) for disinfectant decay. η can be computed as

$$\eta = \left[\frac{1 - \exp(-nk_D t/m)}{nk_D t/m}\right]^m \tag{10-159}$$

A simpler variant of the Hom model incorporating disinfectant decay is the c_{avg}-Hom model, which is based on the concept of a geometric mean disinfectant residual given by

$$c_{D,avg} = \sqrt{c_D(0) \cdot c_D(t_f)} \tag{10-160}$$

where $c_D(0)$ and $c_D(t_f)$ are the disinfectant concentrations after the instantaneous demand has been exerted and at the end of the time period of interest, respectively. This approximation assumes that disinfectant decomposition follows first-order kinetics and that the final disinfectant residual is appreciably greater than zero. When this approximation is used, integration of the Hom rate law (Equation 10-150) with c_D^n replaced by $c_{D,avg}^n$ yields

$$\log \frac{c_X(t)}{c_X(0)} = -k_{Hom} c_{D,avg}^n t_f^m \tag{10-161}$$

EXAMPLE 10-18. In a batch test evaluating inactivation of *Giardia*, the free chlorine concentration declined from 3.30 to 2.04 mg/L as Cl_2 over the course of 1 h, and 2.8 logs of inactivation were achieved. Assuming the chlorine decay could be reasonably approximated as first order, develop the predicted curves for $\log[c_X/c_X(0)]$ versus time according to the following four models. Assume $n = 0.85$ and $m = 2.5$ in all cases. The rate constants have been chosen so that, in all cases, they correctly predict the conditions at the end of the test (i.e., 2.8 logs of inactivation after 1 h), for c_D expressed in milligrams per liter as Cl_2.

(i) The Chick–Watson model, with $k_{C-W} = 1.69 \times 10^{-2}$ (assuming the chlorine concentration remains at its initial value throughout the test).
(ii) The Hom model, with $k_{Hom} = 3.64 \times 10^{-5}$.
(iii) The Hom model using the "efficiency factor" approach to account for disinfectant decay, with $k_{Hom} = 4.45 \times 10^{-5}$.
(iv) The Hom model using the geometric mean disinfectant concentration to account for disinfectant decay, with $k_{Hom} = 4.46 \times 10^{-5}$.

Solution. The rate constant for chlorine decay can be evaluated from the information about the chlorine concentrations at time zero and remaining after 1 h of reaction. Substituting these values into the integrated mass balance for a substance undergoing first-order decay in a batch reactor, we find

$$\frac{c_D(t)}{c_D(0)} = \exp(-k_D t)$$

$$k_D = -\frac{1}{t} \ln \frac{c_D(t)}{c_D(0)} = -\frac{1}{60 \, min} \ln \frac{2.04 \, mg/L}{3.30 \, mg/L}$$

$$= 0.008 \, min^{-1}$$

(i) In Chick–Watson model, the disinfectant is assumed to remain constant at its initial value of 3.30 mg/L. Substituting the given information into Equation 10-147, we obtain

$$\log \frac{c_X(t)}{c_X(0)} = -k_{C-W} c_D^n t = -(1.69 \times 10^{-2} \, min^{-1})(3.30)t$$

(ii) The Hom model also assumes that the chlorine concentration is constant at its initial value, but it incorporates factors that cause the disinfection curve to have a tail or shoulder, depending on the value of m. Inserting the given values into Equation 10-152, we find

$$\log \frac{c_X(t)}{c_X(0)} = -k_{Hom} c_D^n t^m = -(3.64 \times 10^{-5})(3.30)t^{2.5}$$

(iii) When disinfectant decay is incorporated into the Hom model using the efficiency factor approach, Equations 10-158 and 10-159 can be used to describe the disinfection curve. Combining these two equations yields

$$\log \frac{c_X(t)}{c_X(0)} = -k_{Hom} m (c_D(0))^n t^m \left[\frac{1 - \exp(-nk_D t/m)}{(nk_D t/m)}\right]^m$$

$$= -(4.45 \times 10^{-5})(2.5)(3.30)^{0.85} t^{2.5}$$

$$\times \left[\frac{1 - \exp[-\{(0.85)(0.008 \, min^{-1})t/2.5\}]}{(0.85)(0.008 \, min^{-1})t/2.5}\right]^m$$

(iv) The geometric mean concentration of chlorine during the test is given by

$$c_{D,avg} = \sqrt{c_D(0) c_D(t_f)} = \sqrt{(3.30)(2.04)} = 2.60$$

Using this value as the constant chlorine concentration in Equation 10-161 yields

$$\log \frac{c_X(t)}{c_X(0)} = -k_{Hom} c_D^n t^m = -(4.46 \times 10^{-5})(2.60)^{0.85} t^{2.5}$$

All the four equations developed earlier yield the estimates for the predicted survival efficiency of *Giardia* as a function of time during the test, constrained in such a way

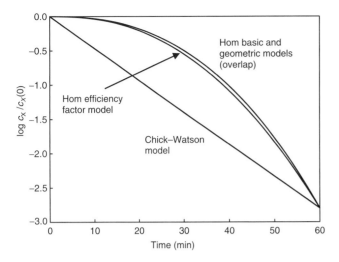

FIGURE 10-35. Comparison of predicted inactivation efficiency based on various models.

that they all predict identical survival efficiency (2.8 logs of inactivation) at the end of the test. The predictions are compared in Figure 10-35. The Chick–Watson model predicts that the logarithm of the survival efficiency declines linearly during the test, whereas all the curves based on the Hom model have shoulders, corresponding to an increasing inactivation rate as the test proceeds. The shapes of these latter curves are all similar, indicating that they would serve approximately equally well (or poorly) for modeling a real data set. Given the uncertainty and variability inherent in disinfection testing due to the range of resistances of individual organisms in the population, clumping of organisms, and other factors, it is unlikely that the different predictions by these version of the Hom model would be significant, so use of the simpler versions of the model (with the rate constant optimized for the selected version) is usually justified. ∎

Design and Operational Considerations

In general, the kinetics of microbial inactivation has been determined in batch systems. Real contactors, however, invariably have continuous flow, and they are likely to exhibit nonideal flow patterns as a result of nonuniform mixing and short-circuiting. If the residence time distribution of a reactor with continuous flow is known (e.g., from tracer experiments), then it is possible to develop reasonable estimates of inactivation efficiency in these systems (Lawler and Singer, 1993).

As discussed in Chapter 2, the residence time distribution functions $E(t)$ and $F(t)$ can be obtained from pulse or step tracer experiments. Using the $E(t)$ curve and the batch kinetic rate expression, the disinfection efficiency of a reactor with segregated flow can be predicted.[13] For such a condition, the survival ratio in a disinfection reactor can be described as follows:

$$\frac{c_{X,out}}{c_{X,in}} = \frac{\int_0^\infty c_X(t)E(t)\,dt}{c_{X,in}} \qquad (10\text{-}162)$$

Equation 10-162 can be used in conjunction with any desired expression for the kinetics of disinfection (i.e., any expression for obtaining values of $c_X(t)$).

■ **EXAMPLE 10-19.** Tracer tests for a disinfection reactor indicate that the RTD is consistent with that for three, equal-sized CFSTRs in series, with a total residence time of 1 h. If the water described in Example 10-18 is fed to this reactor and dosed with 2.8 mg/L TOTOCl as Cl_2, how many logs of *Giardia* inactivation are expected? Use an appropriate form of the Hom model and value of k_{Hom} to characterize inactivation kinetics.

Solution. In a CFSTR, the organisms are exposed to a constant disinfectant concentration, equal to that in the reactor and its effluent. Therefore, the basic Hom model is applicable to each (hypothetical) CFSTR in the sequence, using the disinfectant concentration in the reactor as c_D. Of the rate constants given in Example 10-18 for the Hom equation, the one based on the average value of c_D in the reactor is most applicable to the current situation.

The values of c_D in the three CFSTRs can be determined using Equation 4-30, which is applicable for first-order irreversible reactions in a series of CFSTRs. Written in terms of the variables of interest here, that equation is

$$c_{D,i} = c_{D,0}(1 + k_D \tau_i)^{-i}$$

where $c_{D,i}$ is the disinfectant concentration in the ith reactor, $c_{D,0}$ is the disinfectant dose to the first reactor, and τ_i is the hydraulic residence time in each reactor. Substituting known values, we find the disinfectant concentrations in the three reactors to be 2.84, 2.45, and 2.11 mg/L, respectively.

We next evaluate the inactivation efficiency in the first CFSTR in the sequence, using Equation 10-162 in conjunction with the basic Hom model. Rewriting Equation

---

[13] Segregated flow is almost certainly a good approximation for the microorganisms; that is, the individual organisms or clumps of organisms are likely to pass through the disinfection reactor without interacting significantly with one another. However, organisms might not behave in the same way as a soluble tracer, so the use of tracer data to infer the $E(t)$ function for the microorganisms might have some flaws. Segregated flow is not likely to be a good model for the disinfectant, but if the common model assumption that the disinfectant concentration is approximately constant throughout the period of disinfection is valid, then this concern is irrelevant.

10-161 to solve for the fractional survival of organisms as a function of time and disinfectant concentration in the reactor yields

$$\frac{c_X(t)}{c_X(0)} = \exp(-k_{\text{Hom}} c_D^n t^m) = \exp\left[-(4.46 \times 10^{-5}) c_D^{0.85} t^{2.5}\right]$$

This equation was derived for a batch reactor, but also applies to any segregated aliquot of water that is passing through the reactor with flow. We can therefore use it to compute the survival of organisms that have spent time t in a CFSTR by replacing $c_X(0)$ with $c_{X,\text{in}}$ and inserting the disinfectant concentration in the CFSTR for c_D. For instance, the fractional survival in an aliquot that spends 10 min in the first CFSTR before exiting is

$$\frac{c_X(10\,\text{min})}{c_{X,\text{in}}} = \exp\left[-(4.46 \times 10^{-5})(2.84)^{0.85}(10)^{2.5}\right] = 0.82$$

Substituting the expression for $c_X(t)/c_{X,\text{in}}$ applicable to the first CFSTR into Equation 10-162 along with the generic expression for $E(t)$ of a CFSTR yields

$$\frac{c_{X,1}}{c_{X,\text{in}}} = \int_0^\infty \frac{c_X(t)}{c_{X,\text{in}}} E(t)\,dt$$

$$= \int_0^\infty \{\exp[-k_{\text{Hom}}(c_D)^n (t)^m]\}\left\{\frac{1}{\tau_1}\exp\left(-\frac{t}{\tau_1}\right)\right\}dt$$

$$= \int_0^\infty \exp\left[-(4.46 \times 10^{-5})(2.84)^{0.85}(t)^{2.5}\right]\left[\frac{1}{20}\exp\left(-\frac{t}{20}\right)\right]dt$$

When this expression is integrated, it indicates that $c_{X,1}/c_{X,\text{in}}$ equals 0.548; that is, 54.8% of the organisms survive while passing through the first CFSTR. The corresponding values for the second and third reactors, using the same $E(t)$ function but different c_D values for the calculations, are 59.7% and 64.6%, respectively. The overall survival through the whole reactor is therefore

$$\frac{c_{X,\text{out}}}{c_{X,\text{in}}} = (0.548)(0.597)(0.646) = 0.211 = 21.1\%$$

Log(0.211) is -0.68, so the reactor is expected to achieve 0.68 logs of *Giardia* inactivation. ∎

Characteristic Performance of Specific Disinfectants

Chlorine Despite concerns related to both the ineffectiveness of chlorine for inactivation of certain organisms (particularly protozoans) and the formation of DBPs, chlorine remains the most widely used disinfectant for treatment of both potable water and wastewater. The first use of continuous chlorination for disinfection of a public water supply was in London in 1905 following an outbreak of typhoid fever. Chlorination of public water supplies became common practice in the United States following its adoption at the Bubbly Creek Filtration Plant in Chicago in 1908.

In drinking water treatment, chlorine may be added to an aqueous stream as the initial step in the treatment (*prechlorination*) or, more commonly, after other treatment processes (*postchlorination*). *Rechlorination* or *booster chlorination* may also be practiced where chlorine is added to treated water at one or more points in the distribution system. In wastewater treatment, disinfection only occurs after all biological treatment is completed.

The mechanisms by which organisms are inactivated by chlorine have been investigated by numerous workers (e.g., Shang and Blatchley, 2001) and appear to involve an increase in cell membrane permeability that leads to both a loss of cytoplasmic materials, such as proteins and nucleic acid, and an influx of chlorine, with resultant DNA damage. No gross disruption of cellular envelopes or significant change in zeta potential was observed from chlorination of bacterial suspensions at chlorine concentrations employed in water and wastewater disinfection (Jacangelo et al., 1991).

The CT values required to achieve various log inactivation values of *Giardia* and viruses for disinfection by free chlorine at selected temperatures and pH values are given in Table 10-8. The CT values depend on pH because HOCl is a far more powerful disinfectant than OCl$^-$, and the

TABLE 10-8. USEPA Disinfection Requirements by Free Chlorine

	Required CT (mg-min/L)[a]					
	Temperature (°C)			Temperature (°C)		
Log Inactivation	5	15	25	5	15	25
Giardia[b]	pH ≤ 6			pH 8		
1	39	19	10	72	41	20
2	77	39	19	144	81	41
3	116	58	29	216	122	61
Viruses	6 < pH < 9			pH 10		
2	4	2	1	30	15	7
3	6	3	1	44	22	11
4	8	4	2	60	30	15

[a]Units for concentrations in CT calculation are mg/L as Cl$_2$.
[b]CT values for disinfection of *Giardia* by chlorine depend on the chlorine concentration; the values shown are for 2 mg/L as Cl$_2$. The values for 1 mg/L as Cl$_2$ are ~10% less than those shown.

pK_a value of 7.53 for HOCl is in the pH range of interest; for example, for *Giardia*, the values at pH 8 (where OCl$^-$ is the dominant species) are much higher than at pH 6 (where HOCl is the dominant species). Also, because the value of n in the Chick–Watson law is not equal to 1.0 for chlorine, the required CT values are (slightly) dependent on the chlorine concentration.

Chloramines Chloramination of drinking water has been practiced in the United States for nearly 80 years for the purposes of both disinfection and taste and odor control. The major advantages of chloramination over chlorination are that it provides a more stable residual in water distribution systems and usually results in lower production of halogenated DBPs. However, because of its relatively weak ability to inactivate viruses and protozoans and its better penetration into pipe slimes (biofilms), chloramination is typically used as secondary rather than a primary disinfectant (USEPA, 1999).

Monochloramine is the preferred chloramine species for use in disinfecting drinking water not only because of efficacy, but also because it imparts less taste and odor than dichloramine and nitrogen trichloride do. Based on the rapid reaction of monochloramine with the four amino acids cysteine, cystine, methionine, and tryptophan in studies of *E. coli* inactivation, Jacangelo et al. (1991) deduced that inactivation by chloramines most likely involves inhibition of proteins or protein-mediated processes such as respiration.

Few studies of the mechanism of viral inactivation by chloramines have been carried out but, as for chlorine, the mechanism may depend on factors such as virus type and disinfectant concentration. For example, the site of attack on bacteriophage f2 seems to be the viral RNA (Olivieri et al., 1980), whereas poliovirus inactivation appears to involve the protein coat (Fujioka et al., 1983).

Experimental comparisons of the relative strengths of chloramines and chlorine as disinfectants have demonstrated consistently that chloramines are significantly weaker disinfectants. Some results from one recent comparison are shown in Table 10-9.

The bactericidal and viricidal efficiency of chloramine appears to be more temperature-sensitive than that of other disinfectants, so that chloramines are particularly inefficient at low temperatures (Wolfe et al., 1984). The effects of pH on chloramine disinfection efficacy are less clear (in part because pH also affects the physiological response of the organism) but, on balance, it appears that pH effects are relatively minor.

CT values for achieving various levels of *Giardia* cyst and virus inactivation with chloramines in the pH range 6–9 are shown in Table 10-10. The CT value required to achieve a one-log reduction of *Cryptosporidium* oocysts with monochloramine is in excess of 7200 mg-min/L

TABLE 10-9. CT Values for Microbial Inactivation at 5°C[a]

Disinfectant Agent	pH	E. coli (mg min/L)	Giardia lamblia (mg min/L)	Poliovirus 1 (mg min/L)
Free chlorine	6–7	0.034–0.05	32–46	1.1–2.5
Preformed chloramines	8–9	95–180	1470	768–3740
Chlorine dioxide	6–7	0.4–0.75	17	0.2–6.7
Ozone	6–7	0.02	1.3	0.1–0.2

[a]Table from Health Canada (2006), using data from Hoff (1986) for 99% inactivation of *E. coli* and Poliovirus 1, and from USEPA (1999) for 90% inactivation of *G. lamblia*.

(USEPA, 1999); it is unrealistic to achieve such values in drinking water treatment plants.

■ **EXAMPLE 10-20.** Water at pH 8.0 and 15°C is chlorinated as it is withdrawn from a lake and is then pumped to a drinking water treatment plant. The detention time in the pipe between the lake and the plant is 8 min, and the free chlorine concentration when the water reaches the plant is 2.0 mg/L as Cl_2. Sufficient ammonia to convert all of the free chlorine to monochloramine is mixed into the water in a CFSTR (commonly called a "rapid mix tank" in this context) just as the water enters the plant; a coagulant such as alum is added at the same time. The water then passes through flocculation and sedimentation tanks. (Coagulation, flocculation, and sedimentation are discussed in Chapters 11–13.)

A tracer test on the series of these three tanks indicates that the T_{10} value for that part of the treatment train is 81 min. Under current operation, the monochloramine concentration at the effluent of the sedimentation tank is

TABLE 10-10. CT Values for *Giardia* and Virus Inactivation Using Chloramines

	Required CT (mg-min/L)[a]		
	Temperature (°C)		
Log Inactivation	5	15	25
Giardia			
1	735	500	250
2	1470	1000	500
3	2200	1500	750
Viruses			
2	857	428	214
3	1423	712	356
4	1988	994	497

[a]Units for concentrations in CT calculation are mg/L as Cl_2.
Source: USEPA (1999).

1.6 mg/L as Cl_2. How many logs of inactivation of *Giardia* and viruses would this plant be credited with, according to US EPA regulations? (Note: At present, these regulations specify that, when the whole water treatment system is considered, a three-log inactivation of *Giardia* and a four-log inactivation of viruses must be achieved.)

Solution. The pipe is considered a plug flow reactor, so $T_{10,\text{pipe}} = \tau_{\text{pipe}} = 8$ min. Therefore

$$CT_{\text{pipe}} = (2.0\,\text{mg/L})(8\,\text{min}) = 16.0\,\text{mg min/L}$$

Because the disinfectant in the pipe is free chlorine, we use the values in Table 10-8 to determine the disinfection credit granted for this portion of the system. At pH 8.0 and 15°C, a CT value of 41 mg-min/L is required for one-log inactivation credit for *Giardia*, whereas a value of 4 mg min/L generates a four-log credit for virus disinfection. The value of 16.0 mg min/L is sufficient to accomplish all of the required virus disinfection. The relationship between CT and log inactivation is linear, so the credit for *Giardia* disinfection is

$$\log\text{ inactivation}_{Giardia,\text{pipe}} = \frac{16.0}{41} = 0.39$$

For the series of reactors, we find

$$CT_{\text{reactors}} = (1.6\,\text{mg/L})(81\,\text{min}) = 129.6\,\text{mg min/L}$$

For this section of the system, because the disinfectant in the reactors is chloramine, we use the values in Table 10-10 to compute the disinfection credit. Those values show that, at 15°C, 500 mg min/L is required for one-log inactivation of *Giardia*, so the credit for *Giardia* disinfection is

$$\log\text{ inactivation}_{Giardia,\text{reactors}} = \frac{129.6}{500} = 0.26$$

The total credit for inactivation of *Giardia* between the pump station and the effluent from the sedimentation tank is the sum of the two values above, or 0.65 logs, which is less than the EPA requirement. Although the value could be increased by using a higher dose of chlorine, the increased concentration of free chlorine in the pipe could lead to excessive DBP formation, and the increased concentration of chloramine in the finished water could generate off-tastes. Therefore, it is likely that at least some of the remaining required inactivation would have to be accomplished through physical removal of *Giardia* in particle removal processes (e.g., filtration); as noted, these processes are presented in subsequent chapters. ∎

Design and Operational Considerations. Important design and operational parameters for chloramination include the following:

- *Chlorine to Ammonia-Nitrogen* (Cl_2:NH_4–N) *Dosing Ratio*. Maintaining the Cl_2:NH_4–N mass ratio between 4.5:1 and 5:1 (i.e., near, but below, the stoichiometric ratio for complete conversion of the ammonia to monochloramine) enhances the formation of NH_2Cl, reduces the concentration of free ammonia (in comparison to lower doses of chlorine at the same ammonia concentration), and thereby reduces the likelihood of biologically mediated nitrification occurring in the distribution system.
- *pH*. pH in the range 7.5–9.0 is considered optimal.
- *Order of Addition*. Monochloramine can be formed by first adding ammonia and then chlorine, or vice versa. Ammonia is added first where formation of objectionable taste and odor compounds caused by reactions of chlorine with organic matter are a concern. However, in most drinking water systems, chlorine is added first, and the ammonia is added to "quench" the free chlorine residual at a point downstream. Typically, the point of ammonia addition is selected based on the trade-off between the benefits of increased disinfection (and, for regulatory compliance, increased CT) when free chlorine is present, and the drawbacks of increased DBP generation (USEPA, 1999). In wastewater treatment, chlorine is added after biological treatment and separation of the biomass from the treated water. Unless complete nitrification has been accomplished, ammonia is almost certain to be present in the treated wastewater, so that chloramines are formed and accomplish the required disinfection. In some plants, breakpoint chlorination is practiced to oxidize the ammonia; in those cases, the residual disinfectant is free chlorine.
- *Point of Application*. Disinfectant addition upstream of filters reduces biological growth on the filters, which has the salutary effect of keeping the filters clean and reducing the required backwashing frequency. However, it also reduces removal (i.e., biodegradation) of biologically degradable organic carbon (BDOC) compounds in the filters (USEPA, 1999).
- *Chloramine Residual*. The most appropriate residual concentration of chloramines (in a water distribution system) depends strongly on local conditions, but it is normally in the range 1.0–3.0 mg/L as Cl_2. In wastewater systems, the residual is quenched by the addition of sodium sulfite or other reducing agent.
- *Chemical and Biological Reactions That Consume Chloramines in Water Distribution Systems.* Chloramines can be consumed by reactions with pipe materials or reduced metals in pipe scales, and by biological

activity in water distribution systems. Ammonia used in the chloramination process can provide both energy and nutrient ammonia which may stimulate growth of autotrophic nitrifying bacteria in the distribution system; such growth could, in turn, cause increased nitrate levels in the finished water. An intermediate step in this reaction sequence generates small amounts of nitrite, which can accelerate decomposition of chloramines.

Chlorine Dioxide Chlorine dioxide is sometimes preferred over chlorine as an oxidant and disinfectant for water treatment because it produces fewer chlorinated organic by-products and virtually no chloroform. In addition, its disinfection efficiency is not strongly pH dependent, nor does it react with ammonia or oxidize bromide (Hoigne and Bader, 1994), and it is effective against chlorine-resistant *Cryptosporidium* oocysts (Clark et al., 2003).

Quantitative data were published as early as the 1940s demonstrating the efficacy of chlorine dioxide as a bactericide and viricide, but the disinfection mechanisms are not well understood. Noss et al. (1983) found that chlorine dioxide reacts readily with the amino acids cysteine, tryptophan, and tyrosine, but not with viral ribonucleic acid (RNA). On the basis of these results, they concluded that chlorine dioxide inactivated viruses by altering the viral capsid proteins. However, Alvarez and O'Brien (1982) showed that chlorine dioxide reacts with poliovirus RNA and impairs RNA synthesis. It has also been shown that chlorine dioxide reacts with free fatty acids and that it can disrupt the permeability of the outer membrane of bacteria (Aieta and Berg, 1986). It is thus unclear whether the primary mode of microbial inactivation by chlorine dioxide lies in attack on peripheral cell structures, nucleic acids, or both.

Although the disinfection efficiency is less sensitive to pH when chlorine dioxide is used than when chlorine is the disinfectant, studies on poliovirus 1, *E. coli*, *Cryptosporidium* and *Giardia* indicate that the degree of inactivation by chlorine dioxide increases slightly as pH increases. For example, one study showed a doubling in the level of inactivation of *Cryptosporidium* when the pH was increased from 6 to 8 for similar CT values (Liyanage et al., 1997).

In general, chlorine dioxide is approximately equivalent to chlorine as a bactericide. In comprehensive studies by Roberts et al. (1980) and Aieta et al. (1980) of the disinfection of secondary wastewater effluent by 2-, 5-, and 10-mg/L doses of chlorine and chlorine dioxide (as Cl_2), chlorine dioxide induced more rapid coliform inactivation than chlorine but, after 30 min of contact, chlorine was slightly more effective (Figure 10-36).

As is the case for chlorine and chloramines, chlorine dioxide is a more effective disinfectant at higher temperatures. As shown in Table 10-11, CT values expected to achieve any particular log inactivation of any of the regulated organisms decrease with increasing temperature. The CT values for *Cryptosporidium* included in the table

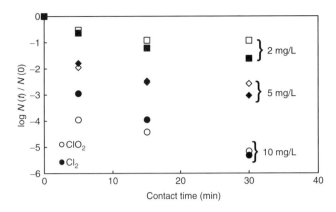

FIGURE 10-36. Comparison of germicidal efficiency of chlorine dioxide and chlorine. N is the concentration of total coliforms. *Source*: After Roberts et al. (1980).

(USEPA, 2003) are an indication that ClO_2 can be an effective disinfectant for this organism. However, the values are far higher than for *Giardia*, especially at the lower temperatures; the ratio of the required CT values for *Cryptosporidium* to *Giardia* is approximately 20 at 25°C but nearly 50 at 5°C. The temperature dependence of *Cryptosporidium* disinfection by ClO_2 was determined in bench-scale studies by Li et al. (2001).

The dose of chlorine dioxide that can be used in water treatment is limited by two constraints in US drinking water regulations—a maximum residual disinfectant level of 0.8 mg/L and the MCL of 1.0 mg/L for chlorite, a primary product of the use of ClO_2. At a ClO_2 concentration of 0.8 mg/L (or even a few tenths higher), adequate disinfection

TABLE 10-11. CT Values for Inactivation Using Chlorine Dioxide

	Required CT (mg-min/L)[a]		
	Temperature (°C)		
Log Inactivation	5	15	25
Giardia			
1	8.7	6.3	3.7
2	17	13	7.3
3	26	19	11
Viruses			
2	5.6	2.8	1.4
3	17.1	8.6	4.3
4	33.4	16.7	8.4
Cryptosporidium			
1	429	179	75
2	858	357	150
3	1286	536	226

[a]Units for concentrations in CT calculation are mg/L as ClO_2.
Source: USEPA (1991, 2003).

for *Cryptosporidium* can only be achieved with very long contact times, meaning that the use of chlorine dioxide as a sole disinfectant is impractical except where water temperatures are consistently high. However, it can contribute significantly to disinfection in cases where it is added primarily for another reason (e.g., oxidation of inorganic species) and another disinfectant is added downstream. In these instances, residual ClO_2 (and ClO_2^-) is sometimes removed by the addition of a reductant.

Ozone   Ozone is a powerful disinfectant that can typically achieve excellent disinfection with less contact time and at lower concentrations than free chlorine, chlorine dioxide, or monochloramine. However, because ozone autodecomposes, no residual is maintained in the distribution system. As a result, in the United States, Australia, and much of Europe, ozone disinfection is coupled with the addition of a secondary disinfectant such as chlorine, chloramines, or chlorine dioxide (USEPA, 1999).

Even very low concentrations of ozone are very effective against bacteria. For example, an ozone residual of 9 μg/L caused a four-log inactivation of *E. coli* in less than 1 min at a temperature of 12°C (Wuhrmann and Meyrath, 1955). Similar results have been reported for *Staphyloccus* sp. and *Pseudomonas fluorescens* inactivation; *Streptococcus faecalis* and *Mycobacterium tuberculosis* required contact times twice and six times as long, respectively, for the same reduction with the same dissolved ozone concentration (USEPA, 1999). Gram-positive cocci (*Staphylococcus*), the Gram-positive bacillae (*Bacillus*), and the Mycobacteria are more resistant to ozone than are Gram-negative bacteria such as *Escherichia coli*. Nevertheless, all bacterial strains are easily destroyed by relatively low levels of ozone (USEPA, 1999).

The mode of attack of ozone on bacteria appears to be oxidative degradation of the bacterial membrane via destruction of either the glycoproteins or glycolipids or certain amino acids such as tryptophan. In addition, ozone disrupts enzymatic activity of bacteria by acting on sulfhydral groups of certain enzymes. Ozone may also act on the nuclear material within the cell.

Viruses are typically more resistant to ozone than are actively growing bacteria, but less resistant than dormant forms of mycobacteria. Keller et al. (1974) reported that three-log removal of poliovirus 2 and coxsackie virus B3 was achieved in batch tests with a contact time of 5 min and ozone residuals of 0.8 and 1.7 mg/L, respectively; greater than five-log removal of coxsackie virus was achieved in a pilot plant with an ozone dosage of 1.45 mg/L, which provided a steady-state concentration of 0.28 mg/L.

The principle mode of virus inactivation appears to be destruction of the capsid, the protein and/or the membrane coat that surrounds the genomic nucleic acid (DNA or RNA) of a virus. This destruction liberates the RNA and disrupts

TABLE 10-12.   CT Values for Inactivation Using Ozone

Log Inactivation	Required CT (mg-min/L)[a]		
	Temperature (°C)		
	5	15	25
Giardia			
1	0.63	0.32	0.16
2	1.3	0.63	0.32
3	1.9	0.95	0.48
Viruses			
2	0.6	0.3	0.15
3	0.9	0.5	0.25
4	1.2	0.6	0.3
Cryptosporidium			
1	16	6.2	2.5
2	32	12	4.9
3	47	19	7.4

[a]Units for concentrations in CT calculation are mg/L as O_3.
Source: USEPA (1991, 2003).

attachment to the host. The "naked" RNA can then be inactivated by ozone at a much more rapid rate than RNA within the intact phage.

Protozoan cysts are much more resistant to ozone (and other disinfectants) than vegetative (actively growing) forms of bacteria and viruses. *G. lamblia* has sensitivity to ozone that is similar to the sporular forms of Mycobacteria, while other protozoa (such as *Naegleria*) are considerably more resistant. CT values required by the USEPA for *G. lamblia* inactivation are included in Table 10-12.

Cryptosporidium oocysts appear to be much more resistant to ozone than are many other protozoans (Peeters et al., 1989). Nevertheless, ozone is clearly one of the most effective disinfectants for *Cryptosporidium* (Gyurek et al., 1999). (It should be noted that studies of *Cryptosporidium* inactivation have often yielded ambiguous results, in part because of the difficulty, and different methods, of *Cryptosporidium* measurement.) Gyurek et al. (1999) and Li et al. (2001) investigated the inactivation of *Cryptosporidium* by ozone in phosphate-buffered solutions over the pH range 6–8 and at 1–37°C. The rate of inactivation was insensitive to pH in this range, but it decreased markedly with a decrease in temperature. *Cryptosporidium* inactivation kinetics was satisfactorily described by an *Incomplete gamma Hom* (IgH) expression (Equation 10-156), with the inactivation rate constants (k') adjusted for water temperature using the van't Hoff–Arrhenius relationship.

As was noted earlier, oxidant demand is computed slightly differently for ozone than for chlorine-based disinfectants. Although an assumption is commonly made that the disinfectant residual that is exerted over a given period of time is independent of the initial disinfectant dose, the assumption is often not valid, and the calculation of ozone demands takes that factor into account. To do so, the residual is analyzed after a given contact time for a range of doses. The results are then plotted, and the disinfectant demand is defined as the dose that is predicted (by extrapolation of the experimental values) to yield zero residual at the given contact time.

■ **EXAMPLE 10-21.** Determine the ozone demand at the three contact times investigated in the following data set. (Data adapted from Bonnelye and Richard, 1997)

Transferred O_3 (mg/L)	Residual O_3 at 2.5 min (mg/L)	Residual O_3 at 6 min (mg/L)	Residual O_3 at 10 min (mg/L)
1.5	1.1	0.7	0.5
1.3	0.8	0.4	0.3
0.9	0.4	0.2	0.1
0.6	0.2	0.1	Not analyzed

Solution. The data are plotted in Figure 10-37. The ozone demand is defined as the value of the x-intercept, which is 0.44, 0.52, and 0.66 mg/L for the 2.5-, 6-, and 10-min contact times, respectively. ■

CT values required for inactivation of *Giardia*, viruses, and *Cryptosporidium* by ozone are shown in Table 10-12 (USEPA, 1999, 2003). These requirements were largely based on the study by Li et al. (2001) described previously. Data obtained at pH 7.2 were assumed to apply for the pH range of 6–9.

An alternative to the approach used by the USEPA for assigning disinfection credit to specified CT values involves determination of the CT requirement for a given level of statistical confidence. Li et al. (2001) used the same data set

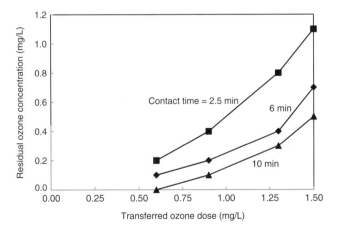

FIGURE 10-37. Example data for computing ozone demand.

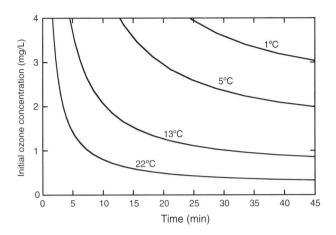

FIGURE 10-38. Ozone disinfection design guidelines for 90% confidence that 2.0 log units of inactivation of *Cryptosporidium parvum* will be achieved, based on IgH model for pH 6–8, for first-order ozone disappearance with a rate constant of 0.05 min^{-1}. *Source*: After Li et al. (2001).

as was used to develop Table 10-12 to develop the design curves for 2.0 log units of inactivation with 90% confidence shown in Figure 10-38. For these calculations, a rate constant for ozone disappearance of 0.05 min^{-1} was assumed. The dose required for 90% confidence that the desired inactivation will be achieved is approximately twice the dose given by the IgH model (which corresponds to two-log inactivation in the mean case). The authors noted that, to apply their suggested design methodology to natural waters, bench-scale tests would be necessary to validate model parameters for specific conditions, since high alkalinity, organic content, and other water quality parameters may influence ozone inactivation kinetics and ozone residual stability.

Rather than use the CT approach to characterize the expected extent of disinfection, von Gunten (2003b) modeled the disinfection of several microorganisms with ozone as a second-order reaction (first order with respect to both the ozone and microorganism concentrations, consistent with the Chick–Watson law with the exponent $n = 1$); rate constants from that study are shown in Table 10-13.

TABLE 10-13. Kinetics of the Inactivation of Microorganisms with Ozone at pH 7

Microorganism	k_{O_3} (M^{-1} s^{-1})	T (°C)
E. coli	1.04×10^5	20
B. subtilis spores	2.3×10^3	20
Rotavirus	6×10^4	20
G. lamblia cysts	2.3×10^4	25
Giardia muris cysts	1.2×10^4	25
C. parvum oocysts	6.7×10^2	20

Source: From von Gunten (2003b).

Disinfection of wastewaters using ozone has proven somewhat problematic because of the high competing demand for the oxidant exerted by dissolved organic matter. In early trials of wastewater disinfection, a desire to maintain significant dissolved ozone residuals led to the need for high (uneconomical) ozone dosages. Xu et al. (2002) investigated ozone disinfection of a variety of wastewater effluents and concluded that the CT approach was inappropriate, because it assumes, conservatively, that the only disinfectant concentration of importance is the residual concentration at the end of the treatment step. As a result, that approach predicts that no inactivation occurs if the residual is zero, whereas they observed 1–3 logs of inactivation of bacteria for typical ozone doses, even in the absence of ozone residual. An alternative model involving the simultaneous consumption of ozone by the microorganisms and the organic matrix was found to give much more satisfactory results. They concluded that transfer of ozone from the gas phase to the water was the critical step in fecal coliform inactivation with ozone, because of the rapid reaction between ozone and coliform bacteria. No difference in inactivation was found between systems with 2- and 10-min hydraulic retention times, for a given ozone dose transferred to the effluent. This result suggests that, for wastewater disinfection using ozone, ozone contactors should be designed to optimize mass transfer, with relatively less emphasis on contact time.

Ultraviolet Radiation Ultraviolet (UV) radiation is used increasingly in wastewater disinfection and, in conjunction with other disinfectants, is also finding increasing application in drinking water treatment. UV disinfection occurs as a result of photochemical damage to RNA and DNA within microorganisms (thus preventing reproduction) and, as such, is not strictly a redox process. However, a brief description of important aspects of the use of UV radiation in water and wastewater treatment is included here.

1. *Generation and Implementation of UV Radiation.* UV radiation occurs as electromagnetic waves with wavelengths between 100 and 400 nm, in the range between X rays and visible light. UV radiation is subclassified into smaller wavelength ranges as UV-A (315–400 nm), UV-B (280–315 nm), UV-C (200–280 nm), and vacuum UV (100–200 nm). In terms of germicidal effects, the optimum UV range is between 245 and 285 nm (USEPA, 1999); that is, primarily in the UV-C range.

   UV-emitting lamps used for disinfection are classified as either low pressure or medium pressure. (Medium-pressure lamps are sometimes also identified as high-pressure lamps.) Conventional low-pressure UV lamps are the most energy efficient source, with approximately 40% of the electrical input power

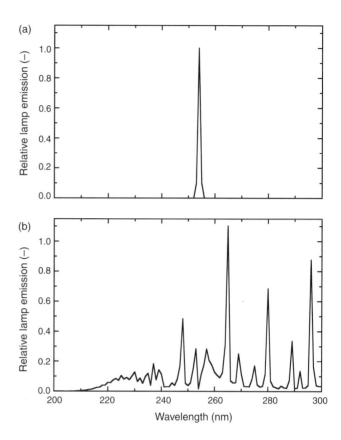

FIGURE 10-39. Spectral power distribution of UV lamps: (a) low pressure; (b) Medium pressure. *Source*: Data courtesy of Prof. Karl Linden of the University of Colorado.

converted into UV-C light output at wavelength 254 nm, which is close to the wavelength of the maximum germicidal effect (see Figure 10-39a). Medium pressure UV lamps produce more spectral lines and a continuum due to recombination radiation, as illustrated in Figure 10-39b. Medium-pressure lamps are less electrically efficient than low-pressure lamps, but they produce much greater UV-C light intensity and thus achieve a given UV light dosage in a much shorter irradiation time. In water and wastewater treatment applications, lamps are typically surrounded by quartz sheaths and are immersed in the flowing stream. The flow may be in a closed or open vessel and may be either parallel or perpendicular to the axes of the lamps. An alternative configuration involves flowing water through Teflon tubes (which are relatively transparent to UV radiation) surrounded by UV lamps (Haas, 1999).

The contact time required for UV disinfection is typically less than 1 min. Of greater importance than time is the proximity of the organism to the light source, which is usually assured by maintaining turbulent conditions and relatively small passageways between lamps in the reactor. Modeling of particle and fluid dynamics in the reactor can assist in optimizing design and operating conditions (Lyn et al., 1999).

The extent of destruction or inactivation of microorganisms by UV radiation is generally related to the UV dose, which is calculated as

$$D = It \qquad (10\text{-}163)$$

where D is the UV dose (mW s/cm^2, or mJ/cm^2), I is the (areal) intensity of the radiation (mW/cm^2), and t is the exposure time (s). For UV disinfection, the dose (sometimes referred to as the *fluence*) can be thought of as analogous to the CT product for chemical disinfectants.

2. *Disinfection Efficacy*. UV radiation is quite effective in inactivating bacteria and viruses, with the degree of inactivation proportional to both the microorganism concentration and the UV dose if the water is relatively free of other particulate matter; that is,

$$r_X = -kIc_X \qquad (10\text{-}164)$$

where k is the inactivation rate coefficient (cm^2/mW s). On integration in a batch system (where $r_X = -(dc_X/dt)$), Equation 10-164 yields

$$c_X(t) = c_X(0)\exp(-kIt) = c_X(0)\exp(-kD) \quad (10\text{-}165)$$

Such first order dependency has been observed by Snicer et al. (1996) in studies of MS-2 coliphage inactivation in groundwater (Figure 10-40). These investigators also compared the UV-susceptibility of MS-2 coliphage to that of hepatitis A virus, poliovirus, and rotavirus for 10 groundwater sources and found MS-2 to be approximately two to three times as resistant to UV disinfection as the three human pathogenic viruses. Bacteria have also been shown to be sensitive to UV radiation, with doses as low as 2–6 mW-s/cm^2 sufficient for one-log inactivation (USEPA, 1999).

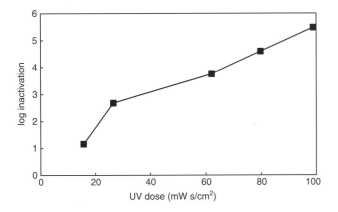

FIGURE 10-40. UV dose required for inactivation of MS-2 coliphage. *Source*: Adapted from Snicer et al. (1996).

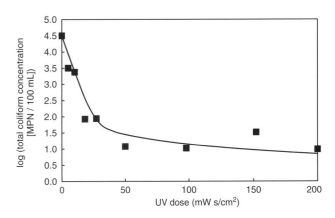

FIGURE 10-41. Typical response of coliform bacteria to UV light in wastewater secondary effluent. *Source*: From Loge et al. (2001).

The dose–response behavior of wastewater to UV disinfection follows similar first-order behavior at low UV doses but, as shown in an example data set in Figure 10-41, significant tailing (i.e., a reduced rate of inactivation) is observed at high doses. This tailing phenomenon (which is apparent in both UV and chlorination disinfection systems) has been attributed to the association of targeted organisms (commonly coliform bacteria) with particles in the bulk liquid medium (Loge et al., 2001).

Using recently developed techniques for the enumeration of particles associated with coliform bacteria, Emerick et al. (2000) demonstrated that a model of the following form provides a good description of the data:

$$c_x(t) = c_x(t)\exp(-kD) + \frac{c_P}{kD}(1 - \exp(-kD))$$
$$(10\text{-}166)$$

where c_P is the number concentration of particles larger than 10 μm (enumerated before the application of UV light) that contain one or more coliform bacteria, and the other parameters are as defined as in Equation 10-165. A major implication of the association of organisms with particles and flocs, and the protection of the organisms by those particles, is that a tertiary filtration step might be required in addition to UV disinfection to meet the discharge quality requirement at wastewater treatment plants (Loge et al., 2001).

Inactivation of protozoa by UV radiation has generally been thought to require much higher doses than inactivation of viruses and bacteria. However, recent studies by Linden et al. (2002) in which the infectivity of the target organism toward Mongolian gerbils was used as the measure of disinfection capability suggest that much lower doses may be adequate. For example, reduction of *G. lamblia* infectivity for gerbils after disinfection using collimated, nearly monochromatic (254 nm) UV radiation from a low-pressure mercury vapor lamp was very rapid, with greater than four-log inactivation attained at a dose of 1 mW s/cm^2. The implication is that earlier studies, which typically used some direct measure of cyst viability as the metric for quantifying disinfection efficiency, might have underestimated the amount of disinfection actually achieved, because many organisms that survived the irradiation might have nevertheless been rendered noninfective. In addition, the researchers found no evidence that *G. lamblia* cysts could "heal" (i.e., regain their infectivity, commonly referred to as *repairing* themselves) following doses of 16 and 40 mW s/cm^2 (which covers the normal range of UV doses used in water and wastewater disinfection).

High UV doses were also thought to be necessary for inactivation of *C. parvum* oocysts. However, a number of investigators (e.g., Shin et al., 2001) have shown that three-log reductions in oocyst infectivity could be achieved with a dose of only 3 mW s/cm^2, using infectivity in mice or *in vitro* cell culture as the metric for disinfection efficiency.

In proposed regulations, the USEPA has recommended the fluence values shown in Table 10-14 for the inactivation of the three regulated types of microorganisms. Unlike any of the chemical disinfectants, the dose required for *Cryptosporidium* inactivation is quite similar to that required for *Giardia*. Also, the values in the table make it clear that UV is relatively ineffective for virus removal.

3. *Residual Generation and Disinfection By-Product Formation.* UV radiation quickly dissipates in water, where it is either absorbed by or reflected off suspended and dissolved materials. As a result, a secondary disinfectant (such as chlorine or chloramine) is required in drinking water applications to maintain a residual throughout the distribution system.

The possibility of generating DBPs exists in a UV disinfection system, as any radiation not absorbed by the target organisms could excite dissolved constituents, leading (possibly) to the subsequent formation of reactive species such as superoxide radicals. Of particular interest is the effect of absorption of UV light by natural organic matter (NOM). However, the relatively low concentrations of NOM typically present in filtered water and the short contact time result in very little transformation of dissolved constituents. UV irradiation was also observed to have an insignificant effect on DBP formation by secondary disinfectants.

OH Free Radicals The role of OH radicals in disinfection processes is a subject of considerable debate. Consider, for example, the situation when water is ozonated, so that disinfection might occur via reactions of the organisms with either molecular ozone or hydroxyl radicals. The second-order rate constants for reaction of ozone with various organisms are shown in Table 10-13. Furthermore, for the secondary phase of ozonation (during which most of the disinfection will occur), the expected value of R_{ct} (the ratio of the concentrations of $\dot{O}H$ and ozone) is in the order of 10^{-8}. Based on that value and the data in Table 10-13, the rate constant for disinfection by $\dot{O}H$ would have to be very large for that disinfection pathway to be significant compared to disinfection by the ozone itself (von Gunten, 2003b). For example, to disinfect *C. parvum* oocysts with $\dot{O}H$ at a rate similar to that with molecular ozone, a rate constant in the order of $6.7 \times 10^{10} M^{-1} s^{-1}$ would be required. Such a value is approximately an order of magnitude higher than rate constants typically observed for reactions with $\dot{O}H$. Given that other organisms react with ozone even more rapidly than *C. parvum*, a significant role of hydroxyl radicals in disinfection seems unlikely. In AOPs, where R_{ct} values in the order of 10^{-6} can be expected, $\dot{O}H$ could theoretically play an important role. However, the nonspecificity of hydroxyl radical attack is likely to lower its effectiveness. For the other organisms shown in Table 10-13, even in AOPs, hydroxyl radicals are expected to have no effect on disinfection (von Gunten, 2003b).

10.8 SUMMARY

In this chapter, we have examined the application of electron transfer (i.e., redox) reactions to the transformation of species in water and wastewater. While an introduction to the theoretical aspects of redox reactions has been given, the major focus has been on the characteristics of redox reactions under conditions typically encountered in engineered treatment systems.

One widely used group of redox processes transforms inorganic contaminants from soluble to insoluble forms. Examples of these processes include the oxidation of soluble manganous and ferrous species to particulate manganic and ferric oxyhydroxides, respectively, during the treatment of potable water, and the reduction of hexavalent chromium species (dichromate [$Cr_2O_7^{2-}$] and chromate [CrO_4^{2-}]) to insoluble trivalent chromium [$Cr(OH)_3(s)$] in the treatment

TABLE 10-14. Fluence Values for Inactivation with UV Light

	Fluence Required (mW s/cm^2)		
Log Inactivation	*Giardia*	Viruses	*Cryptosporidium*
1	2.1	58	2.5
2	5.2	100	5.8
3	11	143	12
4	NA	186	NA

Source: USEPA (2003).

of electroplating wastes. Redox reactions of inorganic compounds can also be employed to form other, less troublesome, soluble molecules, with common examples including the oxidation of sulfide (S(II)) or sulfite (S(IV)) to sulfate (S(VI)) and the reduction of molecular chlorine to chloride.

The oxidative degradation of organic compounds via redox reactions is widespread and is likely to increase as the number of troublesome organic compounds recognized to be present in water increases. While oxidants such as chlorine and chlorine dioxide have some capability to degrade such organic compounds, stronger oxidants such as the hydroxyl free radical are of increasing importance because of their ability to degrade recalcitrant compounds. Such advanced oxidation processes (AOPs), as the use of free radicals is termed, can sometimes lead to complete mineralization of the organic substrate (i.e., the conversion of all the carbon in the substrate to carbonate species, in which carbon is in its highest oxidation state, +IV). More often, though, AOPs generate products that are incompletely oxidized; in such cases, biodegradation is sometimes used to complete the oxidation, especially if the products of the AOP reactions (referred to as intermediates in these cases, since they undergo subsequent reactions) are of health or ecological concern.

By far, the most widely adopted use of chemical oxidants in water and waste treatment is for disinfection. While the removal of pathogenic organisms from aqueous streams has provided enormous public health benefits, the possible generation of harmful disinfection by-products (DBPs) as a result of the addition of oxidants to organic-rich water continues to present challenges.

Chlorine (in the form of the hypochlorite species HOCl and OCl^-) continues to be the most widely used chemical oxidant and disinfectant, but its use has been curtailed in drinking water applications because its reactions with NOM in the source water generate halogenated DBPs. Chlorine readily oxidizes bromide to bromine, and bromine is often incorporated into the by-products along with chlorine. Many of these halogenated DBPs are known or suspected carcinogens, and the ones that are detected most commonly and at the highest concentrations (trihalomethanes and haloacetic acids) are regulated. The compounds that account for approximately one-half of the total organic halogen (TOX) in chlorinated drinking water have not yet been identified. Concern about DBPs has led to changes in water treatment practice, including efforts to remove the NOM and a shift to alternative disinfectants.

Other chlorine-based oxidants and disinfectants include chloramines and chlorine dioxide. Chloramines are formed *in situ* when chlorine reacts with ammonia that has been added to the water or is present naturally. When chloramines are the desired disinfectant, the molar ratio of chlorine to ammonia is maintained at less than one (i.e., a Cl_2:N mass ratio less than 5.1:1) to ensure that monochloramine (NH_2Cl) (rather than di- or tri-chloramines [$NHCl_2$ and NCl_3, respectively]) is the dominant reaction product. Chloramines are relatively weak oxidants/disinfectants, but they maintain a residual concentration in the water distribution system for longer than free chlorine does. Chlorine dioxide is a powerful disinfectant, but has the disadvantages that it must be created on-site and forms chlorite (a health concern) upon use.

Ozone is the most powerful direct oxidant that is widely used in water and wastewater treatment, and it plays a central role in many AOPs, in which even more powerful oxidants (hydroxyl radicals or similar species) are generated. The complex chemistry of AOPs requires careful control, but these processes are likely to be used increasingly as they are better understood. The most common AOP now in use relies on the combination of hydrogen peroxide and ozone to generate hydroxyl radicals. The disadvantages of ozone are that it leaves no residue in water and that it readily oxidizes bromide to bromate, which itself is a compound of health concern. UV light is the cleanest disinfection process, in which, as a physical process, it adds nothing to the water, although its reactions with compounds in the water might generate some new compounds of concern.

The challenges presented by redox treatment of water and waste are many, requiring development of innovative treatment processes that might use multiple oxidants and/or reductants in combination with other treatment technologies. Judging by the huge growth in recent years in research publications dealing with oxidation and reduction processes in water and wastewater treatment, we are certain to witness continuing developments in this area.

REFERENCES

Acero, J. L., Haderlein, S. H., Schmidt, T., Suter, M. J.-F., and von Gunten, U. (2001) MTBE oxidation by conventional ozonation and the combination ozone/hydrogen peroxide: efficiency of the process and bromate formation. *Environ. Sci. Technol.*, 35, 4252–4259.

Acero, J. L., Stemmler, K., and von Gunten, U. (2000) Degradation kinetics of atrazine and its degradation products with ozone and OH radicals: a predictive tool for drinking water treatment. *Environ. Sci. Technol.*, 34, 591–597.

Aieta, E. M., Berg, J. D., Roberts, P. V., and Cooper, R. C. (1980) Comparison of chlorine dioxide and chlorine in wastewater disinfection. *WPCF*, 52 (4), 810–822.

Aieta, E. and Berg, J. D. (1986) A review of chlorine dioxide in drinking water treatment. *JAWWA*, 78 (6), 62–72.

Alvarez, M. E. and O'Brien, R. T. (1982) Mechanisms of inactivation of poliovirus by chlorine dioxide and iodine. *Appl. Environ. Microbiol.*, 44 (5), 1064–1071.

Amy, G., Siddiqui, M., Zhai, W., DeBroux, J., and Odem, W. (1995) *Survey of Bromide in Drinking Water and Impacts on DBP Formation*. AWWA Research Foundation, Denver, CO.

Arora, H., LeChavellier, M. W., and Dixon, K. L. (1997) DBP occurrences survey. *JAWWA*, 89 (6), 80–93.

Balzani, V. and Carassiti, V. (1970) *Photochemistry of Coordination Compounds*. Academic Press, London.

Barb, W. G., Baxendale, J. H., George, P., and Hargrave, K. R. (1951) Reactions of ferrous and ferric ions with hydrogen peroxide. Part II. The ferric ion reaction. *Trans. Faraday Soc.*, 47, 591–616.

Bellar, T. L., Lichtenberg, J. J., and Kroner, R. C. (1974) The occurrence of organohalides in drinking water. *JAWWA*, 66 (12), 703–706.

Benjamin, M. M. (2010) *Water Chemistry*. Waveland Press, Long Grove, IL.

Berg, G. (1964) The virus hazard in water supplies. *J. N. Engl. Water Works Assoc.*, 78, 79–104.

Bichsel, Y. and von Gunten, U. (1999) Oxidation of iodide and hypoiodous acid in the disinfection of natural waters. *Environ. Sci. Technol.*, 33, 4040–4045.

Bonnelye, V. and Richard, Y. (1997) Changes in ozone demand of water during the treatment process. *Ozone Sci. Eng.*, 19 (4), 339–350.

Bossmann, S. H., Oliveros, E., Göb, S., Siegwart, S., Dahlen, E., Payawan, L., Straub, M., Wörner, M., and Braun, A. (1998) New evidence against hydroxyl radicals as reactive intermediates in the thermal and photochemically enhanced Fenton reactions. *J. Phys. Chem. A*, 102, 5542–5550.

Boye, B., Dieng, M. M., and Brillas, E. (2002) Degradation of herbicide 4-chlorophenoxyacetic acid by advanced electrochemical oxidation methods. *Environ. Sci. Technol.*, 36, 3030–3035.

Brezonik, P. L. (1994) *Chemical Kinetics and Process Dynamics in Aquatic Systems*. Lewis, Boca Raton, FL.

Brillas, E. and Casado, J. (2002) Aniline degradation by Electro-Fenton® and peroxi-coagulation processes using a flow reactor for wastewater treatment. *Chemosphere*, 47 (3), 241–246.

Bull, R. J., Reckhow, D. A., Rotello, V., Bull, O. M., and Kim, J. (2007) Use of toxicological and chemical models to prioritize DBP research. AwwaRF Report 91135F, 380 pp. IWA, London.

Buxton, G. V., Greenstock, C. L., Helman, W. P., and Ross, A. B. (1988) Critical review of rate constants for reactions of hydrated electrons, hydrogen atoms and hydroxyl radicals ($\cdot OH/\cdot O^-$) in aqueous solution. *J. Phys. Chem. Ref. Data*, 17 (2), 513–886.

Carlson, K. H. and Knocke, W. R. (1999) Modeling manganese oxidation with $KMnO_4$ for drinking water treatment. *J. Environ. Eng.*, 125, 892–896.

Casey, C. and Bergren, C. (1999) Chemical oxidation, natural attenuation drafted in Navy cleanup. *Pollution Engineering*, March 1999 Supplement.

Cerf, O. (1977) Tailing of survival curves of bacterial spores. *J. Appl. Bacteriol.*, 42, 1–19.

Chambers, J., Leavitt, A., Walti, C., Schreier, C. G., Melby, J., and Goldstein, L. (2000) In situ destruction of chlorinated solvents with $KMnO_4$ oxidizes chromium. In Wickramanayake, G. B., Gavaskar, A. R., and Chen, A. S. C. (eds.), *Chemical Oxidation and Reactive Barriers: Remediation of Chlorinated and Recalcitrant Compounds*. Battelle Press, Columbus, OH, pp. 49–55.

Chick, H. (1908) An investigation of the laws of disinfection. *J. Hygiene*, 8, 92–158.

Choi, J. and Valentine, R. L. (2002) Formation of *N*-nitrosodimethylamine (NDMA) from reaction of monochloramine: a new disinfection by-product. *Water Res.*, 36, 817–824.

Clark, R. M., Sivaganesan, M., Rice, E. W., and Chen, J. (2003) Development of a Ct equation for the inactivation of Cryptosporidium oocysts with chlorine dioxide. *Water Res.*, 37 (11), 2773–2783.

Clayton, W. S., Marvin, B. K., Pac, T., and Mott-Smith, E. (2000) A multisite field performance evaluation of in-situ chemical oxidation using permanganate. In Wickramanayake, G. B., Gavaskar, A. R., and Chen, A. S. C. (eds.), *Chemical Oxidation and Reactive Barriers: Remediation of Chlorinated and Recalcitrant Compounds*. Battelle Press, Columbus, OH, pp. 101–108.

Craun, G. F., Hauchman, F. S., and Robinson, D. E. (eds.) (2001) *Microbial Pathogens and Disinfection By-Products in Drinking Water: Health Effects and Management of Risks*. International Life Science Institute, Washington, DC.

Damm, J. H., Hardacre, C., Kalin, R. M., and Walsh, K. P. (2002) Kinetics of the oxidation of methyl tert-butyl ether (MTBE) by potassium permanganate. *Water Res.*, 36, 3638–3646.

Dryer, D. J. and Korshin, G. V. (2007) Investigation of the reduction of lead dioxide by natural organic matter. *Environ. Sci. Technol.*, 41, 5510–5514.

Edwards, M. and Benjamin, M. M. (1992) The effect of preozonation on coagulant-natural organic matter interactions. *JAWWA*, 84 (8), 63–72.

Edwards, M. and Dudi, A. (2004) Role of chlorine and chloramine in corrosion of lead-bearing plumbing materials. *JAWWA*, 96 (10), 69–81.

Edwards, M., Triantafyllidou, S., and Best, D. (2009) Elevated blood lead in young children due to lead-contaminated drinking water: Washington, DC, 2001–2004. *Environ. Sci. Technol.*, 43, 1618–1623.

Elovitz, M. S., von Gunten, U., and Kaiser, H.-P. (2000) Hydroxyl radical/ozone ratios during ozonation processes. II. The effect of temperature, pH, alkalinity and DOM properties. *Ozone Sci. Eng.*, 22, 123–150.

Emerick, R. W., Loge, F. J., Ginn, T. R., and Darby, J. L. (2000) Modeling the inactivation of particle-associated coliform bacteria. *Water Environ. Res.*, 72, 432–438.

Ensing, B., Buda, F., Blöchl, P. E., and Baerends, E. J. (2002). A Car-Parrinello study of the formation of oxidising intermediates from Fenton's reagent in aqueous solution. *Phys. Chem. Chem. Phys.*, 4, 3619–3627.

Fenton, H. J. H. (1894). Oxidation of tartaric acid in the presence of iron. *Chem. Soc. J. Lond.*, 65, 899–910.

Fernandez, J., Dhananjeyan, M. R., Kiwi, J., Senuna, Y., and Hilborn, J. (2000) Evidence for Fenton photo-assisted processes mediated by encapsulated Fe ions at biocompatible pH values. *J. Phys. Chem. B*, 104, 5298–5301.

Fewtrell, L. J., Pruss-Ustun, A., Landrigan, P., and Ayuso-Mateos, J. L. (2004) Estimating the global burden of disease of mild mental retardation and cardiovascular diseases from environmental lead exposure. *Environ. Res.*, 94 (2), 120–133.

Frimmel, F. H. (1998) Characterization of natural organic matter as major constituents in aquatic systems. *J. Contam. Hydrol.*, 35 (1–3), 201–216.

Fujioka, R. S., Tenno, K. M., and Loh, P. C. (1983) Mechanism of chloramine inactivation of poliovirus: a concern for regulators. In Jolley, R. L. (ed.), *Water Chlorination: Environmental Impacts and Health Affects*, Vol. 4. Ann Arbor Science, Inc., Ann Arbor, MI.

Gallard, H. and von Gunten, U. (2002) Chlorination of natural organic matter: kinetics of chlorination and of THM formation. *Water Res.*, 36, 65–74.

Gauch, A., Rima, J., Amine, C., and Martin-Bouyer, M. (1999). Rapid treatment of water contaminated with atrazine and parathion with zero-valent iron. *Chemosphere*, 39, 1309–1315.

Glaze, W. H., and Kang, J-W. (1988) Advanced oxidation processes for treating groundwater contaminated with TCE and PCE: laboratory studies. *JAWWA*, 88 (5), 57–63.

Gray, E., Margerum, D., and Huffman, R. (1978) Chloramine equilibria and the kinetics of disproportionation in aqueous solution. In Brinckman, F. E., and Bellama, J. M. (eds.), *Organometals and Metalloids, Occurrence and Fate in the Environment*, ACS Symposium Series No. 82. American Chemical Society, Washington, DC, pp. 264–277.

Greenberg, R. S., Andrews, T., Kakartla, P. K. C., and Watts, R. J. (1998) In-situ Fenton-like oxidation of volatile organics: laboratory, pilot, and full-scale demonstrations. *Remediation*, 8 (2), 29–42.

Gurol, M. D. and Lin, S. (1998) Continuous catalytic oxidation process. US Patent No. 5,755,977.

Gurol, M. D. and Singer, P. C. (1982) Kinetics of ozone decomposition: a dynamic approach. *Environ. Sci. Technol.*, 16 (7), 377–383.

Gustafsson, J. P. (2011) Visual MINTEQ, version 3.0, accessed 1/10/2012, KTH, Department of Land and Water Resources Engineering, Stockholm, Sweden.

Gyurek, L. L. and Finch, G. R. (1998) Modeling water treatment chemical disinfection kinetics. *J. Environ. Eng.*, 124, 783–793.

Gyurek, L. L., Li, H., Belosevic, M., and Finch, G. R. (1999) Ozone inactivation kinetics of cryptosporidium in phosphate buffer. *J. Environ. Eng.*, 125, 913–924.

Haas, C. N. (1980) A mechanistic kinetic model for chlorine disinfection. *Environ. Sci. Technol.*, 14, 339–340.

Haas, C. N. (1999) Disinfection, Ch. 14 in *Water Quality & Treatment: A Handbook of Community Water Supplies*. McGraw-Hill, Inc., New York.

Haas, C. N. and Joffe, J. (1994) Disinfection under dynamic conditions—modification of the Hom model for decay. *Environ. Sci. Technol.*, 28, 1367–1369.

Haas, C. N. and Kara, S. B. (1984) Kinetics of microbial inactivation by chlorine. II. Kinetics in the presence of chlorine demand. *Water Res.*, 18, 1451–1454.

Haber, F. and Weiss, J. J. (1934) The catalytic decomposition of hydrogen peroxide by iron salts. *Proc. R. Soc. (London), Ser. A*, 147, 332–345.

Hanna, J. V., Johnson, W. D., Quezada, R. A., Wilson, M. A., and Xiao-Qiao, L. (1991) Characterization of aqueous humic substances before and after chlorination. *Environ. Sci. Technol.*, 25, 1160–1164.

He, J., Tao, X., Ma, W., and Zhao, J. (2002) Heterogeneous photo-Fenton degradation of an azo dye in aqueous H_2O_2/iron oxide dispersions at neutral pHs. *Chem. Lett.*, 1, 86–87.

Health Canada (2006) *Guidelines for Canadian Drinking Water Quality: Guideline Technical Document (Total Coliforms)*. Ottawa, Ontario.

Herrmann, J. M. (1999) Heterogeneous photocatalysis: fundamentals and applications to the removal of various types of aqueous pollutants. *Catal. Today*, 53, 115–129.

Hocking, G., Wells, S., and Ospina, R. I. (2000). Deep reactive barriers for remediation of VOCs and heavy metals. In Wickramanayake, G. B., Gavaskar, A. R., and Chen, A. S. C. (eds.), *Chemical Oxidation and Reactive Barriers: Remediation of Chlorinated and Recalcitrant Compounds*. Battelle Press, Columbus, OH, pp. 307–314.

Hoff, J. C. (1986) *Inactivation of Microbial Agents by Chemical Disinfectants*. EPA/600/S2-86/067, USEPA, Cincinnati, OH.

Hoffmann, M. R., Hua, I., and Hochemer, R. (1996) Application of ultrasonic irradiation for the degradation of chemical contaminants in water. *Ultrasonic Sonochemistry*, 3, S163–S172.

Hoigne, J. (1998) Chemistry of aqueous ozone and transformation of pollutants by ozonation and advanced oxidation processes. In Hrubec, J. (ed.), *The Handbook of Environmental Chemistry. Vol. 5, Part C, Quality and Treatment of Drinking Water*. Springer-Verlag, Berlin, pp. 83–141.

Hoigné, J. and Bader, H. (1976) The role of hydroxyl radical reactions in ozonation processes in aqueous solutions. *Water Res.*, 10 (5), 377–386.

Hoigne, J. and Bader, H. (1994) Kinetics of reactions of chlorine dioxide (OClO) in water. I. Rate constants for inorganic and organic compounds. *Water Res.*, 28, 45–55.

Hom, L. W. (1972) Kinetics of chlorine disinfection in an ecosystem. *J. Sanit. Eng. Div. ASCE*, 98, 183–193.

Hua, I. and Hoffmann, M. R. (1997) Optimization of ultrasonic irradiation as an advanced oxidation technology. *Environ. Sci. Technol.*, 31, 2237–2243.

Hug, S. J., Canonica, L., Wegelin, M., Gechter, D., and von Gunten, U. (2001) Solar oxidation and removal of arsenic at circumneutral pH in iron containing waters. *Environ. Sci. Technol.*, 35, 2114–2121.

Jacangelo, J. G., Olivieri, V. P., and Kawata, K. (1991). Investigation of the mechanism of inactivation of Escherichia coli B. by monochloramine. *JAWWA*, 83, 80–87.

Jafvert, C. T. and Valentine, R. L. (1992) Reaction scheme for the chlorination of ammoniacal water. *Environ. Sci. Technol.*, 26, 577–586.

Jensen, J., Johnson, J., St. Aubin, J., and Christman, R. (1985) Effect of monochloramine on isolated fulvic acid. *Org. Geochem.*, 8, 71.

Jiang, Y., Petrier, C., and Waite, T. D. (2002) Kinetics and mechanisms of ultrasonic degradation of volatile chlorinated aromatics in aqueous solution. *Ultrasonics Sonochemistry*, 9, 163–168.

Joo, S-H., Feitz, A. J., Sedlak, D. L., and Waite, T. D. (2005). Quantification of the oxidizing capacity of nanoparticulate zero-valent iron. *Environ. Sci. Technol.*, 39, 1263–1268.

Joseph, J. M., Destaillates, H., Hung, H. M., and Hoffmann, M. R. (2000) The sonochemical degradation of azobenzene and related azo dyes: rate enhancements via Fenton's reactions. *J. Phys. Chem. A*, 104, 301–307.

Kameya, T. and Urano, K. (2002) Catalytic decomposition of ozone gas by a Pd impregnated MnO_2 catalyst. *J. Environ. Eng.*, 128, 286–292.

Karimi, A. A., Redman, J. A., Glaze, W. H., and Stolarik, G. F. (1997) Evaluation of an AOP for TCE and PCE removal. *JAWWA*, 89 (8), 41–53.

Keller, J. W., Morin, R. A., and Schaffernoth, T. J. (1974) Ozone disinfection pilot plant studies at Laconia, New Hampshire. *JAWWA*, 66 (12), 730.

Knocke, W. R., van Benschoten, J. E., Kearney, M., Soborski, A., and Reckhow, D. A. (1990) *Alternative Oxidants for the Removal of Soluble Iron and Manganese*. AWWA Research Foundation, Denver, CO.

Korshin, G. V., Benjamin, M. M., and Chang, H. S. (2004) *Modeling DBP Formation Kinetics: Mechanistic and Spectroscopic Approaches*. AWWA Research Foundation, Denver, CO.

Korshin, G. V. and Jensen, M. D. (2001) Electrochemical reduction of haloacetic acids and exploration of their removal by electrochemical treatment. *Electrochim. Acta*, 47 (5), 747–751.

Krasner, S. W., McGuire, M. J., Jacangelo, J. J., Patania, N. L., Reagan, K. M., and Aieta, E. M. (1989) The occurrence of disinfection by-products in US drinking waters. *JAWWA*, 81 (8), 41–53.

Kremer, M. L. (1999) Mechanism of the Fenton reaction. Evidence for a new intermediate. *Phys. Chem. Chem. Phys.*, 1, 3595–3605.

Lalezary, S., Pirbazari, M., and McGuire, M. J. (1986) Oxidation of five earthy-musty taste and odor compounds. *JAWWA*, 78 (3), 62–69.

Langlais, B., Reckhow, D. A., and Brink, D. R. (1991) *Ozone in Drinking Water Treatment: Application and Engineering*. AWWARF and Lewis Publishers, Boca Raton, FL.

Larson, R. A. and Weber, E. J. (1994) *Reaction Mechanisms in Environmental Organic Chemistry*. Lewis Publishers, CRC Press, Boca Raton, Florida.

Lawler, D. F. and Singer, P. C. (1993) Analyzing disinfection kinetics and reactor design: a conceptual approach versus the SWTR. *JAWWA*, 85 (11), 67–76.

Li, H., Gyürék, L. L., Finch, G. R., Smith, D. W., and Belosevic, M. (2001) Effect of temperature on ozone inactivation of cryptosporidium parvum in oxidant demand-free phosphate buffer. *J. Environ. Eng., ASCE*, 127 (5), 456–467.

Liang, L. and Singer, P. C. (2003) Factors influencing the formation and relative distribution of haloacetic acids and trihalomethanes in drinking water. *Environ. Sci. Technol.*, 37, 2920–2928.

Lin, S. and Gurol, M. D. (1998) Catalytic decomposition of hydrogen peroxide on iron oxide: Kinetics, mechanism, and implications. *Environ. Sci. Technol.*, 32, 1417–1423.

Lin, S. H. and Ho, S. J. (1996) Catalytic wet-air oxidation of high strength industrial wastewater. *Appl. Catal. B: Environ.*, 9, 133–147.

Lin, Y-P. and Valentine, R. L. (2008) The release of lead from the reduction of lead oxide (PbO_2) by natural organic matter. *Environ. Sci. Technol.*, 42, 760–765.

Linden, K. G., Shin, G. A., Faubert, G., Cairns, W., and Sobsey, M. D. (2002) UV disinfection of Giardia lamblia cysts in water. *Environ. Sci. Technol.*, 36, 2519–2522.

Lindner, M., Theurich, J., and Bahnemann, D. (1997) Photocatalytic degradation of organic compounds: accelerating the process efficiency. *Water Sci. Technol.*, 35 (4), 79–86.

Liyanage, L. R. J., Finch, G. R., and Belosevic, M. (1997) Effects of aqueous chlorine and oxychlorine compounds on Cryptosporidium parvum oocysts. *Environ. Sci. Technol.*, 31, 1992–1994.

Loegager, T., Holcman, J., Sehested, K., and Pedersen, T. (1992) Oxidation of ferrous ions by ozone in acidic solutions. *Inorg. Chem.*, 31 (17), 3523–3529.

Loge, F. J., Bourgeous, K., Emerick, R. W., and Darby, J. L. (2001) Variations in wastewater quality parameters influencing UV disinfection performance: Relative impact of filtration. *J. Environ. Eng.*, 127, 832–837.

Loraine, G. A. Burris, D. R., Li, L., and Schoolfield, J. (2002) Mass transfer effects on kinetics of dibromoethane reduction by zerovalent iron in packed-bed reactors. *J. Environ. Eng.*, 128, 85–93.

Lyn, D. A., Chiu, K., and Blatchley, E. R. (1999) Numerical modeling of flow and disinfection in UV disinfection channels. *J. Environ. Eng.*, 125, 17–26.

Lytle, D. A. and Schock, M. R. (2005) Formation of Pb(IV) oxides in chlorinated water. *JAWWA*, 97 (11), 102–114.

Mantha, R., Taylor, K. E., Biswas, N., and Bewtra, J. K. (2001) A continuous system for $Fe^°$ reduction of nitrobenzene in synthetic wastewater. *Environ. Sci. Technol.*, 35, 3231–3236.

Mason, T. J. and Petrier, C. (2004) Ultrasound processes. In Parsons, S. (ed.), *Advanced Oxidation Processes for Water and Wastewater Treatment*. IWA Publishing, London, pp. 185–208.

Masten, S. J. and Hoigne, J. (1992) Comparison of ozone and hydroxyl radical-induced oxidation of chlorinated hydrocarbons in water. *Ozone Sci. Eng.*, 14, 197–214.

Metcalf and Eddy (as revised by Tchobanoglous, G., Burton, F. L., and Stensel, H. D.) (2003) *Wastewater Engineering: Treatment and Reuse*, 4th edn. McGraw-Hill, New York.

Millero, F. J., Sotolonglo, S., and Izaguirre, M. (1987) The oxidation kinetics of Fe(II) in seawater. *Geochim. Cosmochim. Acta*, 51, 547–554.

Miranda, M. L., Kim, D., Hull, A. P., Paul, C. J., and Galiano, M. A. O. (2007) Changes in blood lead levels associated with use of chloramines in water treatment systems. *Environ. Health Perspect.*, 115 (2), 221–225.

Mitch, W. A. and Sedlak, D. L. (2002) Formation of N-nitrosodimethylamine (NDMA) from dimethylamine during chlorination. *Environ. Sci. Technol.*, 36, 588–595.

Mitch, W. A., Sharp, J. O., Trussell, R. R., Valentine, R. L., Alvarez-Cohen, L., and Sedlak, D. L. (2003). N-Nitrosodimethylamine as a drinking water contaminant: a review. *Environ. Eng. Sci.*, 20, 389–404.

Morel, F. M. M. and Hering, J. (1993) *Principles and Applications of Aquatic Chemistry*. Wiley-Interscience, New York.

Morris, J. C. and Baum, B. (1978) Precursors and mechanisms of haloform formation in the chlorination of water supplies. In Jolley R. L. et al. (eds.), *Water Chlorination: Environmental Impact and Health Effects*, Vol. 2, Ann Arbor Science Publishers, Inc., Ann Arbor, MI, pp. 29–48.

Noss, C. I., Dennis, W. H., and Olivieri, V. P. (1983) Reactivity of chlorine dioxide with nucleic acids and proteins. In Jolley, R. L., et al. (eds.), *Water Chlorination: Environmental Impacts and Health Effects*. Ann Arbor Science Publishers, Inc., Ann Arbor, MI.

Nurmi, J. T., Tratnyek, P. G., Sarathy, V., Baer, D. R., Amonette, J. E., Pecher, K., Wang, C. Linehan, J. C., Matson, D. W., Penn, R. L., and Driessen M. D. (2005) Characterization and properties of metallic iron nanoparticles: Spectroscopy, electrochemistry, and kinetics. *Environ. Sci. Technol.*, 39, 1221–1230.

Olivieri, V. P. et al. (1980) Reaction of chlorine and chloramines with nucleic acids under disinfection conditions. In Jolley, R. J., et al. (eds.), *Water Chlorination: Environmental Impact and Health Effects*, Vol. 3 Ann Arbor Science Publishers, Inc., Ann Arbor, MI.

Olson, T. M. and Barbier, P. (1994) Oxidation kinetics of natural organic matter by sonolysis and ozone. *Water Res.*, 28, 1383–1391.

Onodera, S., Yamada, K., Yamaji, Y., and Ishikura, S. (1984) Chemical changes of organic compounds in chlorinated water. IX. Formation of polychlorinated phenoxyphenols during the reaction of phenol with hypochlorite in dilute aqueous solution. *J. Chromatogr.*, 288, 91–100.

Oturan, M. A. (2000) An ecologically effective water treatment technique using electrochemically generated hydroxyl radicals for in situ destruction of organic pollutants: Application to herbicide 2,4-D. *J. Applied Electrochem.*, 30, 475–482.

Panizza, M., Bocca, C., and Cerisola, G. (2000) Electrochemical treatment of wastewater containing polyaromatic organic pollutants. *Water Res.*, 34, 2601–2605.

Patria, L., Maugans, C., Ellis, C., Belkhodja, M., Cretenot, D., Luck, F., and Copa, B. (2004) Wet air oxidation processes. In Parsons, S. (ed.), *Advanced Oxidation Processes for Water and Wastewater Treatment*. IWA Publishing, London, pp. 247–274.

Peeters, J. E., Mazas, E. A., Masschelein, W. J., Martinez de Maturana, I. V., and Debacker, E. (1989) Effect of disinfection of drinking water with ozone or chlorine dioxide on survival of Cryptosporidium parvum oocysts. *Appl. Environ. Microbiol.*, 55, 1519–1522.

Pourmoghaddas, H. and Stevens, A. A. (1995) Relationship between trihalomethanes and haloacetic acids with total organic halogen during chlorination. *Water Res.*, 29, 2059–2062.

Pulgarin, C., Peringer, P., Albers, P., and Kiwi, J. (1995) Effect of Fe-ZSM-5 zeolite on the photochemical and biochemical degradation of 4-nitrophenol. *J. Mol. Catal. A Chem.*, 95 (1), 61–74.

Roalson, S. R., Kweon, J. H., Lawler, D. F., and Speitel Jr., G. E. (2003) Enhanced softening: Effects of lime dose and chemical additions. *JAWWA*, 95 (11), 97–109.

Reckhow, D. A., Knocke, W. R., Kearns, M. J., and Parks, C. A. (1991) Oxidation of iron and manganese by ozone. *Ozone Sci. Eng.*, 13 (6), 675–695.

Richardson, S. D., Thurston Jr., A. D., Caughtran, T. V., Chen, P. H., Collette, T. W., Floyd, T. L., Schenck, K. M., Lykins Jr., B. W., Sun, G-R., and Majetich, G. (1999) Identification of new ozone disinfection byproducts in drinking water. *Environ. Sci. Technol.*, 33, 3368–3377.

Roberts, P. V., Aieta, E. M., Berg, J. D., and Chow, B. M. (1980) Chlorine dioxide for wastewater disinfection: a feasibility evaluation, EPA-600/281-092.

Rook, J. J. (1974) Formation of haloforms during chlorination of natural waters. *Water Treat. Exam.*, 23, 234–243.

Rose, A. L. and Waite, T. D. (2002) A kinetic model for Fe(II) oxidation in seawater in the absence and presence of natural organic matter. *Environ. Sci. Technol.*, 36, 433–444.

Sabhi, S. and Kiwi, J. (2001) Degradation of 2,4-dichlorophenol by immobilised iron catalysts. *Water Res.*, 35, 1994–2002.

Safarzadeh-Amiri, A., Bolton, J. R., and Cater, S. R. (1997) Ferrioxalate-mediated photodegradation of organic pollutants in contaminated water. *Water Res.*, 31, 787–798.

Sato, C., Hartenstein, D. S., and Motes, W. (2001) Photolysis of TCA, TCE, and PCE in flowthrough reactor system. *J. Environ. Eng. ASCE*, 127 (7), 620–629.

Sayles, G. D., You, G., Wang, M., and Kupferle, M. J. (1997) DDT, DDD and DDE dechlorination by zero-valent iron. *Environ. Sci. Technol.*, 31, 3448–3454.

Schuchmann, M. N. and von Sonntag, C. (1979) Hydroxyl radical induced oxidation of 2-methyl-2-propanol in oxygenated aqueous solution: A product and pulse radiolysis study. *J. Phys. Chem.*, 83, 780–784.

Selleck, R. E., Saunier, B. M., and Collins, H. F. (1978) Kinetics of bacterial deactivation with chlorine. *J. Environ. Eng. Div., ASCE*, 104, 1187–1212.

Severin, B. F., Suidan, M. T., and Engelbrecht, R. S. (1984) Series-event kinetic model for chemical disinfection. *J. Environ. Eng.*, 110, 430–439.

Shang, C. and Blatchley, E. R. (2001) Chlorination of pure bacterial cultures in aqueous solution. *Water Res.*, 35, 244–254.

Shin, G. A., Linden, K. G., Arrowwood, M. J., and Sobsey, M. D. (2001) Low-pressure UV inactivation and DNA repair potential of Cryptosporidium parvum oocysts. *Appl. Environ. Micobiol.*, 67, 3029–3032.

Singer, P. C. and Reckhow, D. A. (1999) Chemical oxidation, Ch. 12 in *Water Quality and Treatment. A Handbook of Community Water Supplies*. 5th edn., American Water Works Association, McGraw Hill.

Singer, P. C. and Stumm, W. (1970) Acid mine drainage: the rate determining step. *Science*, 167, 1121–1123.

Snicer, G. A., Malley, J. P., Margolin, A. B., and Hogan, S. P. (1996) Evaluation of ultraviolet technology in drinking water treatment. Presented at *AWWA Water Technology Conference*, Boston, MA.

Song, R., Westerhoff, P., Minear, R., and Amy, G. (1997) Bromate minimization during ozonation. *JAWWA*, 89 (6), 69–78.

Song, W., Xu, T., Cooper, W. J., Dionysiou, D. D., de la Cruz, A. A., and O'Shea, K. E. (2009) Radiolysis studies on the destruction of microcystin-LR in aqueous solution by hydroxyl radicals. *Environ. Sci. Technol.*, 43 (5), 1487–1492.

Staehelin, J. and Hoigné, J. (1985) Decomposition of ozone in water in the presence of organic solutes acting as promoters and inhibitors of radical chain reactions. *Environ. Sci. Technol.*, 19, 1206–1213.

Stone, A. T. and Morgan, J. J. (1990) Kinetics of chemical transformations in the environment. Ch. 1 in Stumm, W. (ed.), *Aquatic Chemical Kinetics*. Wiley-Interscience, New York, pp. 1–41.

Stumm, W. and Morgan, J. J. (1996) *Aquatic Chemistry: Chemical Equilibria and Rates in Natural Waters*, 3rd edn. Wiley-Interscience, New York.

Sun, Y. and Pignatello, J. J. (1993a) Activation of hydrogen peroxide by iron(III) chelates for abiotic degradation of herbicides and insecticides in water. *J. Agric. Food Chem.*, 41 (2), 308–312.

Sun, Y. and Pignatello, J. J. (1993b) Photochemical reactions involved in the total mineralization of 2,4-D by $Fe^{3+}/H_2O_2/UV$. *Environ. Sci. Technol.*, 27, 304–310.

Sun, Y. and Pignatello, J. J. (1995) Evidence for a surface dual hole-radical mechanism in the TiO_2 photocatalytic oxidation of 2,4-dichlorophenoxyacetic acid. *Environ. Sci. Technol.*, 29, 2065–2072.

Sung, W. and Morgan, J. J. (1980) Kinetics and product of ferrous iron oxygenation in aqueous systems. *Environ. Sci. Technol.*, 14 (5), 561–568.

Symons, J. M., Bellar, T. A., Carswell, J. K., Demarco, J., Kropp, K. L., Robeck, G. G., Seeger, D. R., Slocum, C. J., Smith, B. L., and Stevens, A. A. (1975) Natural organics reconnaissance survey for halogenated organics. *JAWWA*, 67 (11), 634–647.

Tahar, N. B. and Savall, A. (1998) Mechanistic aspects of phenol degradation by oxidation on a Ta/PbO_2 anode. *J. Electrochem. Soc.*, 145, 3427–3434.

Tratnyek, P. G., Scherer, M. M., Johnson, T. J., and Matheson, L. J. (2003) Permeable reactive barriers of iron and other zero-valent metals. In Tarr, M. A. (ed.), *Chemical Degradation Methods for Wastes and Pollutants: Environmental and Industrial Applications*. Marcel Dekker, New York, pp. 371–421.

Triantafyllidou, S., Parks, J., and Edwards, M. (2007) Lead particles in potable water. *JAWWA*, 99 (6), 107–117.

Urynowicz, M. A. and Siegrist, R. L. (2000) Chemical degradation of TCE DNAPL by permanganate. In Wickramanayake, G. B., Gavaskar, A. R., and Chen, A. S. C. (eds.), *Chemical Oxidation and Reactive Barriers: Remediation of Chlorinated and Recalcitrant Compounds*. Battelle Press, Columbus, OH, pp. 75–82.

USEPA (1991). Guidance Manual for Compliance with the Filtration and Disinfection Requirements for Public Water Systems Using Surface Water Sources. EPA 570391001.

USEPA (1998a) *Disinfectants and Disinfection By-products: Final Rule*. Federal Register 63, 241, 69478.

USEPA (1998b) *Handbook on Advanced Photochemical Oxidation Processes*. EPA/625/R-98/004, United States Environmental Protection Agency, December 1998.

USEPA (1999) *Alternative Disinfectants and Oxidants Guidance Manual*. United States Environmental Protection Agency Office of Water Report No. EPA 815-R-99-014.

USEPA (1999) *EPA Guidance Manual: Disinfection Profiling and Benchmarking*. EPA-815-R-99-013, Washington, DC.

USEPA (2003) *National Primary Drinking Water Regulations: Long Term 2 Enhanced Surface Water Treatment Rule; Proposed Rule*, Federal Register, 68, 154, 47640-47795.

Valentine, R. L. and Jafvert, C. T. (1988) General acid catalysis of monochloramine disproportionation. *Environ. Sci. Technol.*, 22, 691–696.

Valentine, R. L. and Wang, H. C. A. (1998) Iron oxide surface catalyzed oxidation of quinoline by hydrogen peroxide. *J. Environ. Eng.*, 124 (1), 31–38.

Van Benschoten, J. E., Lin, W., and Knocke, W. R. (1992) Kinetic modeling of manganese (II) oxidation by chlorine dioxide and potassium permanganate. *Environ. Sci. Technol.*, 26, 1327–1333.

Vandevivere, P. C., Bianchi, R., and Verstraete, W. (1998) Treatment and reuse of wastewater from the textile wet-processing industry: review of emerging technologies. *J. Chem. Technol. Biotechnol.*, 72, 289–302.

Venkatadri, R. and Peters, R. W. (1993) Chemical oxidation technologies: UV light/hydrogen peroxide, Fenton's reagent, and titanium dioxide-assisted photocatalysis. *Haz. Waste Haz. Mat.*, 10 (2), 107–149.

Ventura, A., Jacquet, G., Bermond, A., and Camel, V. (2002) Electrochemical generation of the Fenton's reagent: application to atrazine degradation. *Water Res.*, 36, 3517–3522.

Vikesland, P. J., Ozekin, K., and Valentine, R. L. (2001) Monochloramine decay in model and distribution system waters. *Water Res.*, 35, 1766–1776.

Vikesland, P. J. and Valentine, R. L. (2002) Modeling the kinetics of ferrous iron oxidation by monochloramine. *Environ. Sci. Technol.*, 36, 662–668.

Vinodgopal, K., Peller, J., Makogon, O., and Kamat, P. V. (1998) Ultrasonic mineralization of a reactive textile azo dye, Remazol Black B. *Water Res.*, 32, 3646–3650.

von Gunten, U. (2003a) Ozonation of drinking water: Part I. Oxidation kinetics and product formation. *Water Res.*, 37, 1443–1467.

von Gunten, U. (2003b) Ozonation of drinkingwater: Part II. Disinfection and by-product formation in presence of bromide, iodide or chlorine. *Water Res.*, 37, 1469–1487.

Walling, C. (1975) Fenton's reagent revisited. *Acc. Chem. Res.*, 8, 125–131.

Wang, Q. and Lemley, A. T. (2001) Kinetic model and optimization of 2,4-D degradation by anodic Fenton treatment. *Environ. Sci. Technol.*, 35, 4509–4514.

Watson, H. E. (1908) A note on the variation of the rate of disinfection with change in concentration of the disinfectant. *J. Hygiene*, 8, 536–542.

Wehrli, B. (1990) Redox reactions of metal ions at mineral surfaces. Ch. 11 in Stumm, W. (ed.), *Aquatic Chemical Kinetics*. Wiley-Interscience, New York, pp. 311–336.

White, G. C. (1978) *Handbook of Chlorination*. Litton Educational Publishing, New York.

WHO (World Health Organization) (2000). Environmental Health Criteria 216; Disinfectants and Disinfectant By-Products, available at www.who.int/ipcs/publications/ehc/ehc_216/en/, accessed Dec. 31, 2012.

Wickramanayake, G. B., Gavaskar, A. R., and Chen, A. S. C. (eds.) (2000) *Chemical Oxidation and Reactive Barriers: Remediation of Chlorinated and Recalcitrant Compounds*. Battelle Press, Columbus, OH.

Wolfe, R. L., Ward, N. R., and Olson, B. H. (1984) Inorganic chloramines as drinking water disinfectant: a review. *JAWWA*, 76 (5), 74–88.

Wu, W. W., Chadik, P. A., Davis, W. M., Delfino, J. J., and Powell, D. H. (2000) The effect of structural characteristics of humic substances on disinfection by-product formation in chlorination. In Barrett, S. E., Krasner, S. W., and Amy, G. L. (eds.), *Natural Organic Matter and Disinfection By-Products: Characterization and Control in Drinking Water*, ACS Symposium Series 761. American Chemical Society, Chapter 8, pp. 109–121.

Wuhrmann, K. and Meyrath, J. (1955) The bactericidal action of ozone solution. *Schweitz. J. Allgen. Pathol. Bakteriol.*, 18, 1060–1069 (in German).

Xie, Y., Chen, F., He, J., Zhao, J., and Wang, H. (2000) Photo-assisted degradation of dyes in the presence of Fe^{3+} and H_2O_2 under visible irradiation. *J. Photochem. Photobiol. A: Chem.*, 136, 235–240.

Xu, P., Janex, M-L., Savoye, P., Cockx, A., and Lazarova, V. (2002) Wastewater disinfection by ozone: main parameters for process design. *Water Res.*, 36, 1043–1055.

Zhang, W-X., Wang, C-B., and Lien, H-L. (1998) Treatment of chlorinated organic contaminants with nanoscale bimetallic particles. *Catal. Today*, 40, 387–395.

Zhang, F., Zhao, J., Zang, L., Shen, T., Hidaka, H., Pelizzetti, E., and Serpone, N. (1997) Photoassisted degradation of dye pollutants in aqueous TiO_2 dispersions under irradiation by visible light. *J. Mol. Catal. A: Chem.*, 120, 173–178.

Zwiener, C. and Frimmel, F. H. (2000) Oxidative treatment of pharmaceuticals in water. *Water Res.*, 34, 1881–1885.

PROBLEMS

10-1. Compare the equilibrium speciation of Se computed in Example 10-2 with that in anaerobic sediments in which the pe is controlled by the SO_4^{2-}/HS^- couple. Assume that $c_{SO_4^{2-}} = 10^{-3}$ mol/L, $c_{HS^-} = 10^{-5}$ mol/L, and pH = 8.5.

10-2. A utility plans to treat a water supply containing 0.8 mg/L Mn^{2+} by addition of ClO_2 to precipitate $MnO_2(s)$. If the reaction converts ClO_2 to ClO_2^-, how much $MnO_2(s)$ would be produced, and how much ClO_2 would be consumed, in one day of operation at a plant producing 40,000 m^3/d of water?

10-3. A metal plating facility has a waste stream at pH 6.5 that contains 6 mg/L of cyanide (expressed as CN^-). It is proposed to oxidize the cyanide to the innocuous cyanate ion (CNO^-) by addition of chlorine dioxide, which is reduced to chloride in the process.

(a) Write the balanced redox reaction and determine the dose of chlorine dioxide required to leave a residual chlorine dioxide concentration of 1 mg/L.

(b) As shown in Table 10-3, the oxidation rate of HCN is negligible in comparison with that of CN^-. If the dose of chlorine dioxide determined in part (a) is used, what detention time is required to oxidize 99% of the cyanide in a CFSTR, if the pH remains at 6.5? Assume instantaneous equilibrium of HCN/CN^-.

(c) If the pH is raised to 8.0, what detention time would be required in a CFSTR to meet the same requirements as in part (b)?

(d) For the same conditions as in parts (b) and (c), what detention times would be required in a PFR (or batch reactor)?

10-4. Equation 10-28 expresses the rate of oxidation of ferrous iron, Fe(II), as $r_{TotFe(II)} = -kc_{TotFe(II)}c_{O_2(aq)}$, where k is a complex function of pH, and Example 10-3 suggests that oxidation of $Fe(OH)_2^0$ accounts for almost all of the Fe(II) oxidation at pH 7.4. Plot k versus pH for the range $2 < pH < 12$, and determine the portion of this range over which oxidation of $Fe(OH)_2^0$ accounts for at least 90% of the overall Fe(II) oxidation.

10-5. A groundwater at 20°C and pH 6.5 to be used as a drinking water supply contains 1.0 mg/L of dissolved ferrous iron, Fe(II). The water is aerated to introduce oxygen for the oxidation of soluble Fe(II). The insoluble Fe(III) generated can then be removed by filtration. A contact basin after the aerator is provided to allow sufficient time for the oxidation reaction to occur. It is desired to lower the dissolved ferrous iron concentration to 0.3 mg/L, the secondary maximum contaminant level (SMCL) for iron in drinking water.

Assume that the aeration is sufficient to achieve equilibrium with the atmosphere ($p_{O_2} = 0.21$ atm) and that the dissolved oxygen concentration decreases negligibly in the contact basin.

(a) What detention time is required if the basin operates as a PFR? a CFSTR?

(b) What detention time would be required in each of the reactors in part (a) if the pH were raised to 7.5?

10-6. The kinetics of abiotic Fe(II) oxidation by O_2 in a batch reactor are characterized in Figure 10-Pr6. Extrapolation of the lines indicates that they pass through the following points:

Temperature (°C)	Time (min)	Fe(II) (μM)
5	24	1.0
15	178	1.0
25	240	2.08
30	240	20.6

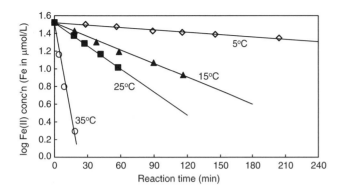

FIGURE 10-Pr6. Effect of temperature on the oxygenation kinetics of ferrous iron. Initial [Fe(II)] = 34.7 µmol/L, $P_{O_2} = 0.2$ atm; pH = 6.82; ionic strength adjusted to 0.11 mol/L with NaClO$_4$; alkalinity = 9 meq/L added as NaHCO$_3$. *Source*: After Sung and Morgan (1980).

(a) What is the order of the reaction in Fe(II)?

(b) Write an expression for $r_{\text{Fe(II)}}$ as a function of [Fe(II)] and an apparent rate constant, k_{obs}, where k_{obs} applies for specified conditions of pH, P_{O_2} and temperature, and estimate k_{obs} at each temperature for which data is presented.

(c) Test whether the dependence of the rate on temperature is consistent with the Arrhenius expression:

$$k = A \exp\left(-\frac{E_a}{RT}\right)$$

where k is the rate constant, A is a constant, and E_a is the activation energy for the reaction. If so, find the activation energy.

10-7. The pharmaceutical diclofenac can be oxidized either by direct reaction with ozone or by reaction with hydroxyl radicals, as indicated by the following reactions:

$$O_3 + \text{diclofenac} \xrightarrow{} \text{products} \quad k_{O_3} = 10^6 \, M^{-1} \, s^{-1}$$

$$\dot{O}H + \text{diclofenac} \xrightarrow{} \text{products} \quad k_{\dot{O}H} = 7.5 \times 10^9 \, M^{-1} \, s^{-1}$$

(a) In a water that contains no carbonate alkalinity, compare the rates of degradation of 10^{-8} M diclofenac under the following conditions: (i) pH 6.0, 20 µM O$_3$ dose; (ii) pH 8.0, 20 µM O$_3$ dose; (iii) pH 8.0, 2 µM O$_3$ dose.

(b) What effect would carbonate alkalinity have on the degradation rate?

(c) How much would the initial rate of diclofenac degradation differ from that calculated in part (a) if the solution contained 5 mg/L of solids to which the diclofenac could adsorb, and if diclofenac adsorption onto the solids were characterized by a Langmuir isotherm with K_{Lang} equal to 5×10^7 mol/mg and $q_{\max}$ equal to 10 µmol/g? Assume that adsorptive equilibrium is reached instantaneously and that only dissolved diclofenac can react with either ozone or hydroxyl radicals.

10-8. Molecular ozone is unstable in water and, at a given pH, decomposes by a first-order reaction. At pH 8.0, the half-life for autodecomposition of O$_3$ in a batch reactor is 2.0 min at 20°C. Ozone is bubbled into a buffered, pH 8.0 solution at 20°C in a completely mixed batch reactor. The overall gas transfer rate constant for the absorption of ozone by the solution, $K_L a_L$, is 5 h^{-1}, and the gas transfer can be assumed to be interfacially limited. The saturation concentration of O$_3$ in water at 20°C is 23 mg/L.

(a) Derive a generic expression for the concentration of dissolved ozone in the reactor at any time $\left(c^*_{O_3}(t)\right)$, in terms of the overall gas transfer rate constant, $K_L a_L$, the saturation concentration $\left(c^*_{O_3}\right)$; and the decomposition reaction rate constant (k_1). Assume that the initial concentration of dissolved ozone is zero.

(b) Find for the steady-state concentration of dissolved ozone in the reactor by applying the steady-state constraint directly to the mass balance on ozone. Show that this value is consistent with your expression from part (a) after the ozone has been bubbled into the solution for a long time.

10-9. When HOCl is added to water, it can enter into a number of competitive reactions, which ultimately determine the concentration and mix of THMs formed. Consider the following elementary reactions:

$$NH_3 + HOCl \xrightarrow{k_1} NH_2Cl + H_2O$$

$$NH_2Cl + HOCl \xrightarrow{k_2} NHCl_2 + H_2O$$

$$Br^- + HOCl \xrightarrow{k_3} HOBr + Cl^-$$

with $k_1 = 5 \times 10^6$ L-(mol s)$^{-1}$, $k_2 = 3.5 \times 10^2$ L-(mol s)$^{-1}$, and $k_3 = 2.95 \times 10^3$ L-(mol s)$^{-1}$

Assume that the reaction rate of NH$_4^+$ with HOCl, as well as the reaction rates of NH$_3$, NH$_2$Cl, and Br$^-$ with OCl$^-$ are all negligible compared to the rates of the earlier reactions.

A solution that is well buffered at pH 8.0 contains 0.5 mg L TOTNH$_3$–N and 0.5 mg/L Br$^-$. If the solution is dosed with 4 mg/L Cl_2, estimate the times at which (a) half of the $TOTNH_3$ and (b) half of the Br^- have reacted in a well-mixed batch system. Make sure you state and check any simplifying assumptions.

10-10. Why do we typically characterize a disinfection process by the amount of disinfectant remaining in the water after some time period (the "residual") rather than by the disinfectant dose applied.

10-11. A drinking water treatment plant has a target chlorine residual of 1.5 mg/L as Cl_2 of combined chlorine, which will be present primarily as NH_2Cl. The influent contains 0.5 mg/L Fe(II). How much chlorine and how much ammonia (as N) should be added to the water, assuming no other demands for chlorine are present?

10-12. If a small amount of the following items were added to a water supply, would the chlorine demand increase, decrease, or not change? Why? Assume that chlorine demand is quantified by adding enough chlorine to the sample to generate free chlorine residual, and then subtracting the residual from the dose.

Sodium bicarbonate: $NaHCO_3$; Sodium sulfide: Na_2S; Potassium permanganate: $KMnO_4$; Monochloramine: NH_2Cl; Alanine (an amino acid): CH_3–CH–NH_2–$COOH$; NH_3; NOM; NaCl; OCl^-.

10-13. On the descending leg of the breakpoint chlorination curve, an additional dose of 1 mg/L chlorine leads to a decrease in the chlorine residual. Why does the added chlorine not appear as a residual? Are chlorine atoms destroyed? Are they removed from the solution? If so, where do they go? If not, why are they not "counted" as part of the residual?

10-14. Figure 10-Pr14 shows a curve of chlorine residual versus chlorine dose for a groundwater sample that is influenced by discharge from a nearby septic tank. The major constituents that react with chlorine in the water are Fe(II) species, some organics that were not biologically degraded, and NH_3. The following treatment options are being considered:

(a) *Option 1*: Removal of 80% of the Fe(II) by exposure of the water to oxygen (air) and oxidation of the iron. This causes the iron to precipitate as $Fe(OH)_3(s)$.

(b) *Option 2*: Removal of 50% of the ammonia (e.g., by aeration or ion exchange).

(i) Sketch the curves of chlorine residual versus chlorine dose for each treatment

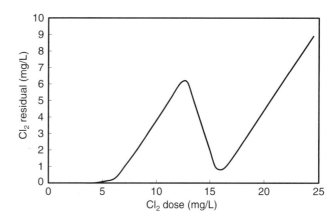

FIGURE 10-Pr14. A breakpoint chlorination curve for a groundwater contaminated by septic tank effluent.

option and write a sentence or two describing the rationale for your sketches.

(ii) Compare the expected disinfection efficiency and chloroform (DBP) formation in the three systems (no treatment, Option 1, Option 2) for a chlorine dose of 15 mg/L.

10-15. The domestic wastewater that has the breakpoint chlorination curve shown in Problem 14 contains 45 mg/L Cl^- before being dosed with chlorine. At the treatment plant, gaseous chlorine is used to prepare a concentrated solution of disinfectant, and the solution is then added to the water at the minimum dose needed to generate a total residual of 4.0 mg Cl_2/L.

(a) Does the chlorination process affect the Cl^- concentration in the water? If so, how much does the Cl^- concentration increase or decrease as a result of chlorination? If not, explain why not.

(b) Repeat part (a), if Cl_2 is added at the breakpoint dose.

(c) Five identical samples of the raw water are taken and are dosed with 5, 10, 15, 20, and 25 mg/L Cl_2 in the form of HOCl. In which of the samples is it likely that significant concentrations of THMs will be found, and why?

10-16. A treated wastewater contains 3.0 mg NH_3–N/L and 20 mg/L dissolved organic carbon, with average composition $C_5H_9O_4$. In principle, if exposed to sufficiently strong oxidizing conditions, the organic matter could get oxidized to CO_2 and H_2O.

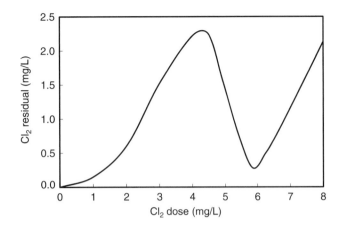

FIGURE 10-Pr17. Breakpoint chlorination curve for a natural water source.

However, when exposed to chlorine, only a portion of the potential oxidant demand is actually exerted. Assume (unrealistically) that 10% of the maximum potential demand is exerted instantaneously when the chlorine is first dosed into the solution, before any chlorine reacts with ammonia. Assume also that, when ammonia gets oxidized, the nitrogen is converted to N_2.

(a) What is the minimum dose of NaOCl (expressed as mg Cl_2/L) needed to generate a combined chlorine residual of 4.0 mg/L as Cl_2 immediately after dosing?

(b) What NaOCl dose is needed to generate a free chlorine residual of 1.0 mg/L as Cl_2?

10-17. Figure 10-Pr17 shows a breakpoint chlorination curve for a water supply source.

(a) Identify the chlorine dose or the range of doses where the following statements apply:
  (i) The free Cl residual is 0.5 mg/L.
  (ii) The combined Cl residual is 2.0 mg/L.
  (iii) The chlorine demand is 4.0 mg/L.
  (iv) When a small increment of chlorine is added, most of it undergoes no redox reaction.
  (v) When a small increment of chlorine is added, most of it reacts with N-containing compounds.
  (vi) When a small increment of chlorine is added, most of it is reduced to Cl^-.

(b) What is the chlorine demand when the chlorine dose is 7.0 mg/L?

(c) If the primary reason for disinfecting with chloramines is to reduce the formation of DBPs, in what order should the reagents be added; that is, should the ammonia be added first, followed by the chlorine, or should the chlorine be added first? Why?

10-18. A drinking water source is disinfected with calcium hypochlorite at a dose of 6.0 mg/L as Cl_2 when it first enters a treatment plant. The chlorine residual is measured at several points downstream, yielding the following results:

Location	Cl_2 Residual
In tank where Cl is dosed	3.8
Outlet of flocculation basin	2.0
Outlet of sedimentation basin	1.4

(a) If the flow rate through the plant is 1000 m^3/h, what is the daily requirement for $Ca(OCl)_2$, in kg?

(b) What is the chlorine demand between the dosing point and the outlet of the sedimentation basin?

(c) The TOX at the outlet from the sedimentation basis is 130 μg/L as Cl_2. Assuming that the Cl in TOX molecules is in the +1 oxidation state, what fraction of the chlorine demand does the TOX contribute?

(d) If the influent contained 16.0 mg/L Cl^-, what is the Cl^- concentration at the outlet of the sedimentation basin?

(e) The mean hydraulic residence time in the flocculation basin is 50 min, and that in the sedimentation basin is 2.5 h. For the computation of disinfection credits, these two treatment steps can be considered independent and additive. If the flocculation basin is completely mixed, and the sedimentation basis can be represented as a PFR, how much CT credit will the plant receive based on flow through the two tanks?

10-19. Various doses of ammonia are added to a treated drinking water that contains 2.0 mg/L of free chlorine. Fill in the following table to indicate the approximate concentrations of free chlorine residual, combined chlorine residual, free ammonia residual ($NH_3 + NH_4^+$), and residual nitrogen in the −III oxidation state, after the ammonia addition. Briefly explain your reasoning for each ammonia dose. All values in the table should be in mg/L units (as Cl_2 or as N, as appropriate).

NH$_3$–N Added	Free Residual Cl$_2$	Combined Cl$_2$	Free Residual NH$_3$	Residual N(–III)
0.6	–	–	–	–
0.4	–	–	–	–
0.3	–	–	–	–
0.1	–	–	–	–

10-20. Water at pH 8.0 is being disinfected with free chlorine. By what factor will the concentration of HOCl (the more active form of TOTOCl) increase if the pH is lowered to 7.0?

10-21. Discuss briefly the relative desirability of plug flow versus complete mixing for disinfection with free chlorine. Would your answer change at all if the disinfectant were combined chlorine?

10-22. Despite the fact that chlorine is used as a disinfectant, bacteria are present in seawater that contains more than 15,000 mg/L Cl$^-$. How is this possible?

10-23. Even though ozone is a very powerful disinfectant, it is never used as the sole disinfectant in water treatment plants in the United States. Why?

10-24. A wastewater contains 6 mg/L ammonia (NH$_3$) and is at pH 7.0 and 25°C. How much chlorine (Cl$_2$) must be added to attain a free chlorine residual of 1 mg/L? Compare the disinfecting power of this solution for poliovirus with one in which 12 mg/L is added to the raw water, based on the data in Table 10-9. Assume that, when ammonia is oxidized, half of the nitrogen is converted to N$_2$ and half is converted to NO$_3^-$.

10-25. When natural organic matter (NOM) is ozonated extensively, oxalic acid (H$_2$C$_2$O$_4$) and the species formed from its deprotonation (HC$_2$O$_4^-$ and C$_2$O$_4^{2-}$) represent a substantial fraction of the reaction products. A solution containing NOM with an average chemical formula of C$_6$H$_4$O$_3$ is dosed with 25 mg/L ozone and allowed to react for 30 min. An analyst determines that the ozone demand has been 15 mg/L. Assuming that the NOM that reacts all forms oxalate, determine the concentration of NOM (mg/L) that has reacted and the concentration of TOTOx that has formed (assuming all oxalate formed is present as the oxalate ion, C$_2$O$_4^{2-}$).

10-26. Table 10-8 shows CT values that are purported to achieve various levels of inactivation of *Giardia lamblia* cysts by free chlorine according to the USEPA's Surface Water Treatment Rule (SWTR). A footnote to the table indicates that these values depend on the free chlorine residual, and that the values shown are for a residual of 2 mg/L as Cl$_2$. The following table shows how the required CT product varies as a function of the chlorine residual for *Giardia* inactivation at pH 7.0 and 10°C.

Log Inactivation	Required CT Values (mg min/L)		
	@ 1 mg/L Cl$_2$	@ 2 mg/L Cl$_2$	@ 3 mg/L Cl$_2$
1	37	41	46
2	75	83	91
3	112	124	137

(a) Manipulate the Chick–Watson equation to develop a relationship showing how the product $c_D t$ is expected to vary as a function of c_D for a fixed number of "logs of inactivation" in a reactor with a uniform and constant chlorine concentration (c_D) and for a contact time t.

(b) The product $c_D t$, as those parameters are defined in the Chick–Watson equation, has a somewhat different meaning from the CT product as it is defined in the EPA's regulations. Nevertheless, when the regulations were developed, the Chick–Watson equation was used as a guideline for determining how the CT requirement should vary as a function of C. Revise the equation you developed in part (a) to obtain one that relates the ratio $c_D t$/logs of inactivation to c_D. Then use that equation in conjunction with the data given in the earlier table to estimate the Chick–Watson rate constant (k_{C-W}) and the coefficient of dilution (n) that the EPA used to determine the values shown in the table. That is, assuming the CT values in the table can be treated as values of the product $c_D t$ in the Chick–Watson equation, what values of k_{C-W} and n are implied?

10-27. A tracer study for a finished water reservoir at a drinking water treatment plant shows that the reservoir behaves as two equal-sized CFSTRs in series with a total residence time of 4 h. The chlorine dose is 3.0 mg/L as Cl$_2$, of which 0.5 mg/L is consumed in instantaneous reactions. Thereafter, the chlorine decays by a first-order reaction with a rate constant of 0.174 h^{-1}. The Chick–Watson inactivation rate constant for Giardia is 0.046 min^{-1}.

(a) What is the effluent chlorine concentration from each of the (two conceptual) reactors?

(b) Calculate the degree of Giardia inactivation using the tanks-in-series model, taking chlorine decay into consideration.

(c) Calculate the degree of Giardia inactivation based on an assumption of segregated flow by using the $E(t)$ curve for the reservoir, taking chlorine decay into consideration.

(d) Calculate the degree of inactivation in accordance with the SWTR using the T_{10} obtained from the tracer curve and the effluent chlorine concentration.

(e) Calculate the degree of inactivation if the reservoir behaved as a plug flow reactor with the same residence time, taking chlorine decay into consideration.

(f) Compare and discuss your findings.

10-28. Based on the results of a step input tracer test, the flow pattern in a disinfection reactor can be represented as a CFSTR with a 15-min hydraulic detention time and a plug flow reactor with a 5-min detention time, in series. Hypochlorous acid (HOCl) is being added at the reactor inlet, at a dose of 8 mg Cl_2/L.

When the influent is chlorinated in a separate, batch test, 1.5 mg/L Cl_2 is consumed before the first data point is analyzed, which is at $t = 30$ s. For the next 30 min, the Cl_2 concentration in solution decays approximately according to the rate expression $r_{Cl_2} = -0.02 - 0.15 c_{Cl_2}^{0.5}$, where r_{Cl_2} is in mg Cl_2/L-min and c_{Cl_2} is in mg Cl_2/L.

(a) Sketch the effluent tracer curve that you think led to the inferred reactor arrangement.

(b) Suggest a reason for the nearly complete consumption of added Cl_2 at low doses in the batch test.

(c) What is the value of T_{10} that would be used to compute the CT product for this system?

(d) What is the value C that would be used to compute the CT product for this system?

(e) The plant owners are considering installing a baffle that would convert the CFSTR portion of the system into two equal-sized CFSTRs in series. Would such a change increase, decrease, or have no effect on the values of C and T that would be used to compute the CT product? Explain your reasoning in a few sentences.

10-29. The data shown in Figure 10-34 were collected in solutions at 0–6°C and at a pH low enough that essentially all the TOTOCl was present as HOCl. If the reaction of poliovirus can be modeled as a conventional chemical reaction with activation energy of 8.2 kcal/mol K, what dose of TOTOCl must be applied to achieve 99% inactivation in 2 min in water at 20°C and pH 7.0? Assume that the solution exerts negligible chlorine demand during the test period and that HOCl is the only effective disinfectant species; that is, OCl^- contributes negligibly to disinfection. (Hint: Based on the results in Chapter 4, determine the order of the reaction with respect to poliovirus, and then compute a pseudo rate constant for the nth order disinfection reaction. This pseudo rate constant will incorporate the HOCl concentration. Then determine how the rate constant is expected to vary with temperature and pH, and use the result to answer the question.)

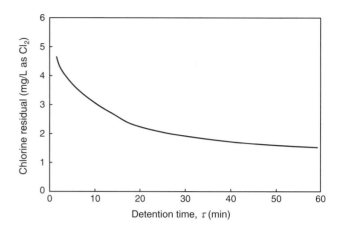

FIGURE 10-Pr30. Chlorine residual in the effluent of a CFSTR operated with a range of hydraulic detention times.

10-30. Figure 10-Pr30 shows the results of experiments in which the chlorine residual in a reactor was measured in a CFSTR with various hydraulic detention times. The initial dose was 5.3 mg/L as Cl_2.

(a) What is the chlorine demand exerted by the water if the residence time is 20 min?

(b) What CT credit would the plant receive if it operated the CFSTR with $\tau = 45$ min?

10-31. Assume that, at 25°C, the inactivation of *Giardia* (Gi) and *Cryptosporidium* (Cr) by ozone is described by the following equations:

$$\log(c_{Gi}/c_{Gi,0}) = -1.04 c_{O_3}^{0.12} t^{0.84}$$

$$\log(c_{Cr}/c_{Cr,0}) = -0.82 c_{O_3}^{0.23} t^{0.64}$$

(a) To assess the relative difficulty of inactivating these protozoa, plot $\log(c/c_0)$ for each organism as a function of the contact time for an ozone residual of 0.5 mg/L and for times up to 10 min.

(b) Also, for each organism, plot the time required to achieve two logs of inactivation as a function of the ozone residual, for a range of residuals from zero to 3.0 mg/L. How well do the results conform to the assumption made in the USEPA SWTR that CT should be constant to achieve a given degree of inactivation?

(c) Discuss the pros and cons of changing inactivation by changing either the disinfectant concentration or the contact time.

10-32. A generic reaction describing the formation of halogenated DBPs through the use of chlorine is

$$NOM + Cl_2 \rightarrow DBPs$$

Theoretically, one could limit the concentration of DPBs that reached a consumer by controlling any of the three "species" shown in this reaction. Briefly discuss how that might be done for each of the three species, and identify the advantages and disadvantages of each of these possible control methods.

10-33. A community that currently chlorinates its ammonia-free water supply to maintain a residual of 0.4 mg/L Cl_2 is considering adding ammonia to the water before chlorination.

(a) Why might they be considering this change?

(b) Should they alter the chlorine dose when they start the ammonia additions? If so, in what way? If not, why not?

(c) Should they choose the ammonia and chlorine doses to give a result less than, equal to, or beyond the breakpoint chlorination dose? Explain briefly.

10-34. Roalson et al. (2003) reported on the effects of softening on NOM removal and DBP formation. Some results for trihalomethane (THM) formation from the chlorination of raw water versus softened water in the study are given in Table 10-Pr34.

TABLE 10-Pr34.

Constituent	Trihalomethane Concentration (μg/L)	
	Raw Water	After Softening
Trichloromethane	64.7	33.1
Bromodichloromethane	45.8	37.0
Dibromochloromethane	27.5	31.6
Tribromomethane	2.4	4.9

(a) Find the percentage reduction in trihalomethane formation on a mass basis brought about by softening the water before chlorination.

(b) What percentage reduction in trihalomethane formation on a molar basis is brought about by softening?

(c) The bromine incorporation number is defined as the average number of bromine molecules incorporated into trihalomethanes, and is computed as the ratio of the total bromine content in the trihalomethanes to the total trihalomethane concentration, both expressed on a molar basis. Find the value of this parameter for both the raw and softened water, and comment on the change in the parameter brought about by softening.

10-35. Liang and Singer (2003) studied the formation of THMs and HAAs in raw and alum-coagulated water in batch reactors. Some results from the Manatee, NC water source are reported in Table 10-Pr35. The TOC values for the raw and coagulated waters were 8.1 and 3.9 mg/L, respectively.

(a) Compare and comment on the effect of coagulation on the formation of THMs and HAAs. Do the same for the effect of pH.

(b) According to the model expressed in Equation 10-43, if the chlorine concentration is approximately constant during the chlorination step, the reaction becomes pseudo first order, with a rate constant k' that incorporates the effect of

TABLE 10-Pr35.

	Total Trihalomethane Concentration (μM)				Total Haloacetic Acid Concentration (μM)			
	pH 6		pH 8		pH 6		pH 8	
Time (h)	Raw	Coagulated	Raw	Coagulated	Raw	Coagulated	Raw	Coagulated
1	0.89	0.27	1.43	0.47	1.25	0.27	1.21	0.28
2	1.01	0.32	1.66	0.49	1.41	0.31	1.40	0.32
4	1.19	0.37	1.90	0.61	1.66	0.38	1.61	0.35
8	1.37	0.44	2.30	0.76	1.81	0.43	1.82	0.43
24	1.66	0.50	2.84	1.04	2.20	0.58	1.99	0.57
72	2.10	0.79	3.41	1.31	2.65	0.77	2.10	0.64

chlorine ($k' = k[\text{Cl}_2]$). Estimate k' and the "ultimate" THM formation potential for the two data sets at pH 6. Repeat the process for the two HAA data sets at pH 6, assuming r_{HAA} is characterized by an equation analogous to Equation 10-43.

Note that Equation 10-43 was derived to characterize THM formation after any "instantaneous" THM formation has occurred. Since the data shown are all for reaction times of 1 h or longer, the changes in THM (or HAA) concentration in the data set meet that constraint. However, it would be incorrect to add a point (0, 0) to the data set or include such a point in the modeling analysis.

10-36. Regulations for disinfection by-products (DBPs) specify maximum contaminant levels (MCLs) for THMs and HAAs. The concentrations of these contaminants are analyzed at the outlet from the treatment plant and at various points in the water distribution system to assess whether the utility is meeting the regulation. In contrast, the regulations for copper and lead specify maximum values that are acceptable in the water coming out of consumers' taps. Sampling in consumers' homes is much more intrusive and inefficient than sampling at the treatment plant or in the distribution system. Why does the EPA insist that the tests for copper and lead be done on tap water, and not water in the larger distribution system?

PART III

REMOVAL OF PARTICLES FROM WATER

11

PARTICLE TREATMENT PROCESSES: COMMON ELEMENTS

11.1 Introduction
11.2 Particle stability
11.3 Chemicals commonly used for destabilization
11.4 Particle destabilization
11.5 Interactions of destabilizing chemicals with soluble materials
11.6 Mixing of chemicals into the water stream
11.7 Particle size distributions
11.8 Particle shape
11.9 Particle density
11.10 Fractal nature of flocs
11.11 Summary
References
Problems

The coagulation process must be considered as one of several interdependent components of a water purification facility, and coagulation efficiency must be related to subsequent treatment units (settling, filtration, sludge disposal). More attention should be directed to the physical and chemical properties of the flocs: density, shear strength, compressibility, filtrability. Trace quantities of impurities present in natural water (color, silica, proteins) can have dramatic effects on the physical and chemical properties of the resulting flocs, and thereby alter significantly their settling and filtering characteristics.

Stumm and O'Melia (1968)

11.1 INTRODUCTION

Many contaminants arrive at water and wastewater treatment plants as particles, others are converted into particles (by chemical precipitation or biological treatment) during treatment, and still others become attached to particles (by adsorption) in one of the treatment steps. Efficient removal of these particles is required for reasons of ecological and public health, as well as aesthetics, and is, therefore, a central task in most treatment plants. The remainder of this book is concerned primarily with particle removal processes.

Most treatment processes for particle removal fall into three categories. In gravity separation (settling or flotation), density differences between particles and water allow for their separation; the particles that are captured are removed in a relatively concentrated, low flow sludge stream, while most of the water flows out in a much cleaner effluent stream. In granular media filtration, the suspension (particles and water) is applied to the media, and particles are captured in the media by physical–chemical processes, while the water passes through; when the accumulation of particles in the media adversely affects the filter performance, the filter is temporarily taken out of service and the media is cleaned. In membrane filtration, most particles are removed by sieving—that is, the particles are larger than the pores of the membrane, so they are caught while the water passes through. Depending on the design, membranes can also remove soluble materials. In all three of these particle removal processes, the single influent flow containing some concentration of particles is separated into two effluent streams: the desired, cleaner stream with far fewer particles than the influent, and the inevitable dirtier stream that contains the captured particles. The engineering of these systems focuses on both the quality (particle concentration) and quantity of the effluent streams (fraction of influent water in each effluent). Details of gravity separation processes, granular media filtration, and membrane processes are presented in Chapters 13–15, respectively.

Often, particles in the influent suspension would not be effectively removed in these treatment processes without some pretreatment. For example, the particles in a suspension might settle too slowly for efficient removal in a typical

---

Water Quality Engineering: Physical/Chemical Treatment Processes, First Edition. By Mark M. Benjamin, Desmond F. Lawler.
© 2013 John Wiley & Sons, Inc. Published 2013 by John Wiley & Sons, Inc.

TABLE 11-1. Applications of Particle Processes

Application	Destabilization	Flocculation	Gravity Separations	Granular Media Filtration	Membrane Filtration
Drinking water (with surface water supply)	Always	Almost always	Almost always	Always one filtration process or the other	
Municipal wastewater	Sometimes	Rarely	Always	Often	Occasional
Industrial wastewater	Common	Common	Almost always	Rare	Rare
Industrial ultrapure process water	Rare	Rare	Rare	Common	Almost always

sedimentation tank. Bringing many small particles together to form a larger, more rapidly settling floc allows them to be removed more easily. However, the fact that the particles are not present as flocs in the influent suggests that some pretreatment process(es) might be needed to induce floc formation. These pretreatment processes typically include chemical alterations to the solution and/or mixing processes to bring about more particle–particle collisions.

The process of bringing particles together with the intention of creating larger particles is called *flocculation* and is the subject of Chapter 12. In this chapter, we focus on characteristics of particles that affect their behavior in water and on pretreatment processes that are designed to alter those characteristics and thereby enhance the performance of the subsequent particle treatment processes.

Particle removal processes are found in the process trains of the vast majority of water and wastewater treatment plants, as indicated in Table 11-1. Most modern drinking water plants using surface water sources have flocculation, gravity separation, and granular media filtration processes; increasingly, a membrane process is used in lieu of or in addition to granular media filtration. A schematic diagram of the most common design of drinking water treatment plants using surface water sources is shown in Figure 11-1. Flotation is sometimes used in lieu of sedimentation as the gravity separation process. When the raw water has a low particle concentration, the gravity separation step (and sometimes the flocculation process) can be eliminated.

Particle removal processes are also ubiquitous in municipal wastewater treatment plants, as suggested in the schematic diagram shown in Figure 11-2. Such plants always include primary sedimentation and almost always have secondary sedimentation (though some have flotation or, increasingly,

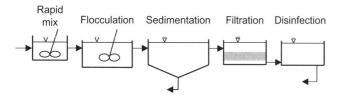

FIGURE 11-1. Schematic of a typical drinking water treatment plant using a surface water source.

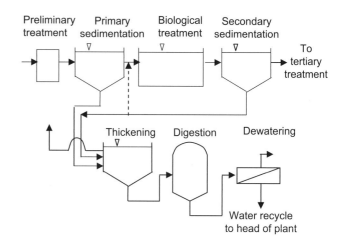

FIGURE 11-2. Schematic of a typical municipal wastewater treatment plant.

membrane filtration instead); some also have granular media filtration. These plants also have sludge concentrating processes, including gravity thickening and mechanical dewatering processes. In industrial settings, the production of process water for specific uses and the pretreatment of wastewaters before discharge to a municipal sewer system often require one or more of these particle removal processes since such plants vary widely depending on the contaminants in the wastewater, no typical schematic can be shown.

The success of most particle removal processes depends on the attachment of particles to other surfaces—either other particles in the suspension or a surface provided in the engineered system. O'Melia and Stumm (1967) introduced the concept that the removal of a particle in such processes can be viewed as consisting of two (more or less) independent steps: long-range transport of the particle to the vicinity of that other surface, and short-range transport to allow the particle to collide with and attach to that surface. Long-range transport is dominated by forces that are primarily physical in nature, while short-range transport and attachment rely on both physical and chemical forces. The distinctions between long and short range and between physical and chemical forces are often blurred, but the concept is still a useful one in understanding particle processes. Details of the long-range transport step differ among the various types of particle treatment processes and are considered in the following chapters. In this chapter, we focus on particle attachment and the short-range forces that allow (or do not allow) that attachment to occur; the forces involved in this step are essentially the same in all particle removal processes.

We begin this chapter by considering the primary reason why particles do *not* form flocs: the presence of an electrical charge on the particle surface. The origins of this surface charge are presented first, followed by the effects of that charge on the solution in the immediate vicinity of the particles. Next, we detail the short-range interactions as two similarly charged particles approach one another or as a charged particle approaches a similarly charged flat surface. In many cases, the electrical repulsion caused by the surface charges dominates such interactions, so that the particles cannot approach close enough to one another to form a floc, or the particle cannot reach and attach to the surface.

Overcoming this problem of particle *stability* (the inability to form flocs) is essential before most particle removal processes; this step is usually accomplished by adding certain chemicals to the suspension. The types of chemicals used for this *destabilization* are described in section "Chemicals Commonly Used for Destabilization," and the underlying physical–chemical mechanisms of particle destabilization are presented subsequently. These added chemicals might also interact with a variety of other soluble constituents already in the water, and that idea is also covered briefly.

Besides surface charge, other characteristics of particles, such as size, shape, and density, affect their behavior in water, and each of these is considered in the latter part of this chapter. Particle size distributions are changed by all of the particle treatment processes and also can be a principal determinant of the success of each process. Throughout Chapters 12–15, considerable emphasis is placed on understanding the effects of each treatment process on particle size distributions, and vice versa. Therefore, several common ways of presenting particle size distributions are introduced in this chapter.

Finally, many of the particles found in treatment systems are not individual particles, but rather are flocs—aggregates of a few or many particles sticking together. Such flocs often incorporate water as they are formed, so that the density, size, and shape of the flocs can be quite different from the corresponding characteristics of the original particles. The concepts of fractals and fractal dimensions can be useful to describe such flocs, and a brief section at the end of this chapter is devoted to this topic.

11.2 PARTICLE STABILITY

Particles in water often do not naturally attach to each other or other surfaces in a time frame that is useful for engineering processes. In such situations, we say that the particles are *stable*, because the suspension can stay essentially unchanged (with respect to particle size distribution) for a long time. Overcoming this stability; that is, destabilizing the particles, is usually essential for the successful removal of particles from water. In this section, we explore why particles are stable, and in the following sections, we describe the chemicals used for destabilization and the variety of ways in which particles can be destabilized.

The primary factor causing particles to be stable in water is their surface charge. If two particles in a suspension that have the same sign of surface charge approach one another, a repulsive interaction will alter their trajectories. As a result, particles that are heading toward a collision veer away from one another. If the repulsion is sufficiently strong, no collision occurs, and the attachment of one particle to another (or to any other surface) is prevented.

A second phenomenon tends to counteract the repulsion when the surfaces get quite close to one another. At such separations, attractive forces known as London-van der Waals forces become substantial and tend to bring (and keep) particles together. At sufficiently short separation distances, these attractive interactions are stronger than the electrical repulsion described above. Hence, if one particle approaches another particle or surface with sufficient momentum, it can overcome the repulsive interaction and get close enough to the other surface for the attractive forces to dominate. In that case, the particle attaches to the other particle or surface. For many suspensions, overcoming the repulsive interaction is a very rare event, so that the suspension is stable.

The distributions of ions and electrical potential around charged particles suspended in water were described in Chapter 7, where the effects of these distributions on adsorption reactions were described. These same distributions have profound implications for the interactions between pairs of suspended particles, or between a suspended particle and a grain of filter medium, so a brief review of that material is provided here. A key figure (Figure 7-43) from that discussion, showing the profile of electrical potential in the region around a charged particle, is reproduced in Figure 11-3.

As described in Chapter 7, the ionic distribution around a charged particle is commonly divided into two regions: a very thin region ($\sim$1 nm) next to the particle, in which ions are bound to the surface, and a much larger region (10–100 nm), where ions are in aqueous solution but at concentrations that differ from the bulk solution due to electrical interactions with the surface; this latter region is known as the diffuse layer. In the diffuse layer, ions with charges of the same sign as the particle (co-ions) are less concentrated than in the bulk solution, and ions with charges

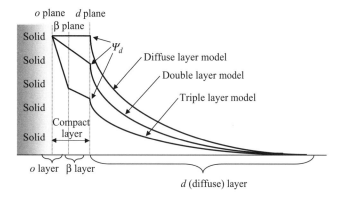

FIGURE 11-3. Conceptual models of the compact layer and diffuse layer of ions surrounding a charged particle surface. (Note that the scale of the compact layer and diffuse layers are somewhat misleading in this diagram; the compact layer is only one or a few molecular diameters thick (i.e., <1 nm generally), while the diffuse layer can stretch to 100 nm under common circumstances. Differences between the models are related to descriptions within the compact layer, not the diffuse layer.)

of the opposite sign (counterions) are more concentrated. The location near the particle at which the diffuse layer begins is idealized as a plane, referred to as the d-plane.

In the analysis of adsorption, the electrical potential profile in the region between the solid surface and the d-plane is critically important, because that is where many adsorbing ions reside. By contrast, when two particles approach one another, the electrical interactions between them become important at separation distances that are much larger than the thickness of these adsorption or compact layers. As a result, the electrical interactions are controlled largely by the profiles of electrical potential through the respective diffuse layers, and the distinctions among the models for the electrical profiles through the compact layers introduced in Chapter 7 are not of paramount importance. Therefore, when considering particle behavior in water, it is common to consider adsorbed species to be part of the solid and to define the interface between the particle and the solution to be at the beginning of the diffuse layer (i.e., at the d-plane). Correspondingly, when considering particle behavior, the value of ψ_d is commonly referred to as the surface potential.

Particle Charge

Particles in water usually carry a surface charge, which can originate in at least three different ways, as described briefly below. A more complete description of the origins of surface charge has been provided by Stumm and Morgan (1996).

Isomorphic Substitution When a solid forms in a solution, it often contains a small fraction of substituted species in the crystal lattice. The substitution occurs only if the species entering the solid lattice are approximately the same size as the species that they replace, so that the replacement can occur without stressing or deforming the lattice structure too severely. For this reason, the substitution is referred to as *isomorphic* (similar shape). If the charges on the two species of interest are not equal, the solid acquires a charge that is expressed at the surface. For example, if an aluminum ion (with a valence of $+3$) takes the place of a silicon ion (with a valence of $+4$), the solid acquires a net charge of -1.

Isomorphic substitution is common in clays, which are alumino-silicate solids composed of aluminum oxide (Al_2O_3) and silicon dioxide (SiO_2) layers in a repeating pattern. Different clays have different numbers and sequences of layers of each type. In clays, it is common for either ferrous iron (Fe^{2+}) or magnesium (Mg^{2+}) to be substituted for aluminum (Al^{3+}), and for aluminum to be substituted for silica (Si^{4+}); each of these isomorphic substitutions causes the clay particles to acquire a net negative charge.

Chemical Reactions at the Surface If the chemical species comprising the surface of a solid contain acid/base functional groups, the extent of ionization of these groups, and therefore the surface charge of the solid, depends on the pH of the water. Several common types of solids encountered in water and wastewater, including oxides and hydroxides, other inorganics, and biogenic organic solids have these characteristics.

Inorganic particles made up primarily of iron, aluminum, and silica oxides are common in natural waters, and these and several other hydrous metal oxides are often encountered in water and wastewater treatment as well. The surfaces of these particles interact with protons and hydroxide ions to form both positively and negatively charged surface species, commonly represented as $\equiv S - OH_2^+$, $\equiv S - OH$, and $\equiv S - O^-$, where $\equiv S$ refers to a site on the surface. Just as in solution chemistry, the dominant species shifts from the most protonated form (positively charged) at lower pH values to the least protonated (negatively charged) at higher pH, and the deprotonation reactions can be described by equilibrium constants (K_a). The change from one form to another occurs over a much wider pH range at the surface than for dissolved species, because the change in surface charge accompanying each reaction opposes additional reactions of the same type. For particles that interact in this way with protons and hydroxide ions, the surface charge is 0 at some particular pH. If no other ions contribute to the surface charge, this value is referred to as the pH_{zpc} (where zpc means the zero point of charge); the surface charge is positive at all pH values below the pH_{zpc} and negative at all pH values above the pH_{zpc}. If the surface charge is the result of interactions with other ions in addition to protons and hydroxide ions, the pH value at which it is 0 is called the isoelectric point (pH_{iep}).

Solids that are not oxides can also have acidic or basic surface species, causing their surface charge to depend on solution pH. For instance, surface CO_3 groups on $CaCO_3(s)$ particles can become protonated at low pH, causing the surface to be positively charged. Similar reactions can occur with the anionic surface groups of phosphates, sulfides, and other minerals.

Other ionizable functional groups also might be found on different kinds of particles. The surface of bacteria, for example, contains molecules with both carboxyl (—COOH) and amino (—NH_2) groups. Just as in solution, the carboxyl group is a stronger acid than the amino group (e.g., in ammonium ion, NH_4^+). As a result, at low pH, both types of groups tend to be protonated (—COOH and —NH_3^+), whereas at high pH, both tend to be deprotonated (—COO^- and —NH_2); in the intermediate pH range, most of the carboxyl groups are likely to be deprotonated, while most amino groups are protonated.

In municipal wastewater treatment, most of the particles (both those in the raw wastewater and those created in the biological treatment) are biogenic and therefore contain these carboxyl and amino groups. Because the surface concentration of the organic acids far exceeds that of the amino groups, these particles are generally negatively charged in the pH range of treatment (say, pH 6.5–8.5). Particles with these types of functional groups at the surface also have a pH_{zpc} and carry a positive charge at lower pH values and a negative charge at higher pH values; several examples are shown in Figure 11-4. In the figure, the surface charge is represented in terms of the electrophoretic mobility (EPM); some details about this measure of surface charge characteristics are given in the next section, but the important idea here is that the sign of EPM is the sign of the surface charge.

Adsorption on the Particle Surface Interactions between ions in solution and a surface can also generate surface charge. The ions can be either the same or different from those comprising the surface. For example, a suspension containing $CaCO_3(s)$ particles is bound to have both Ca^{2+} and CO_3^{2-} ions in solution. If the adsorption density of Ca^{2+} is greater than that of CO_3^{2-}, the imbalance of adsorbed ions will contribute a net positive charge to the surface. In this case, the imbalance might be described either in terms of unequal adsorption of the ions or *nonstoichiometric* dissolution of the solid. In other cases, surface complexes might form with ions that are different from those that form the solid, in which case the process is clearly one of adsorption. Examples include binding of phosphates on the surface of hydrous metal oxides and formation of surface complexes between calcium ions from solution and organic acid functional groups on the surface of a microorganism.

Perhaps, the most important example of adsorption that affects the surface charge of particles is that of large organic molecules, such as those identified as natural organic matter (NOM) in drinking water sources and microbial metabolites in wastewater. Such molecules are often strongly adsorbed onto both inorganic minerals and manufactured adsorbents (e.g., activated carbon and ion exchange resins). The functional groups that are part of these molecules are often charged in the pH range of interest, so that when the molecules adsorb, they contribute charge to the particle surface. It is now widely accepted that the adsorption of NOM onto particles in natural waters is almost always the principal determinant of their surface charge (O'Melia et al., 1999).

NOM is derived primarily from animal and plant decay and from the metabolism of plankton and aquatic bacteria. It enters freshwater systems (e.g., lakes, streams, groundwater) primarily via runoff and infiltration, but can also be produced directly in the aquatic system. The structure and chemical composition of NOM in water are complex and can vary substantially depending on its source, but in all cases it consists of both aliphatic units and aromatic rings that incorporate carboxyl, hydroxyl, amine, and carbonyl groups. Invariably, parts of these molecules are unionized, relatively nonpolar, and therefore hydrophobic, while other parts (those containing strongly polar or ionized functional groups) are hydrophilic. As a result, NOM molecules adsorb onto virtually all of the particles present in any suspension of interest, making the charge characteristics of the particles in any given water body much more similar than would be expected from their underlying chemical composition. This result is important for engineered systems, as we shall see.

The fact that the charge on particles in natural waters is dictated by NOM was first studied in seawater by Neihof and Loeb (1972). Subsequently, Hunter and Liss (1979) noted the similarity in surface charge characteristics of particles from estuaries and coastal waters from widely disparate sources. Numerous studies on freshwaters have indicated that the surface charge of particles is also

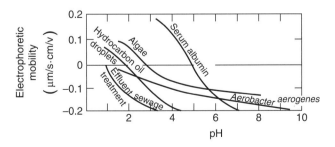

FIGURE 11-4. Effect of pH on surface charge for several types of particles. *Source*: Reprinted with permission from Stumm and Morgan (1996).

dictated by the adsorbed NOM. For example, Davis (1982) showed that the charge on pure γ-Al_2O_3 particles (50 mg/L) was changed from positive to negative in the pH range 6–9 when NOM (0.9 mg/L as dissolved organic carbon [DOC]) extracted from a natural water source was added to the suspension. Other examples of this phenomenon have been reported by Tipping and Higgins (1982), Ali et al. (1985), Edzwald (1993), Tiller and O'Melia (1993), and Chandrakanth and Amy (1996). Hence, in drinking water treatment of both surface waters and groundwaters, the surface charge on all the particles in the raw water is likely to be negative.

Characteristics of the Diffuse Layer

The interactions of particles as they approach one another, or as a particle approaches a flat plane, depend on the electrical potential profiles (and the closely related ionic charge distribution) in the solution adjacent to the two surfaces. In this section, we present two mathematical models that describe these profiles of charge and potential: the Gouy–Chapman model and the Debye–Hückel model. The Gouy–Chapman model is the more general of the two, but the Debye–Hückel is simpler and useful when the surface potential (i.e., the potential at the d-plane) is low.

Both of these models employ a mixture of the relevant variables that yields a variable (κ) with dimensions of inverse length. The value of κ can be found from the following equation:

$$\kappa = \left(\frac{e^2 \sum_{\text{all } i} n_{i,b} z_i^2}{\varepsilon_w k_B T}\right)^{1/2} \quad (11\text{-}1)$$

where e is elementary charge (1.6022×10^{-19} C), $n_{i,b}$ is concentration[1] of molecules of species i in the bulk solution (no. of molecules/m^3), z_i is charge number for ions of species i, k_B is Boltzmann's constant (1.3807×10^{-23} J/K), T is absolute temperature (K), ε_w is permittivity in water $= \varepsilon_0 \varepsilon_r$, where ε_0 is permittivity in vacuum $= 8.854 \times 10^{-12}$ C^2/J m, ε_r is relative permittivity of water[2] $= 78.5$ (at 298 K).

Equation 11-1 is valid in any consistent set of units; for the SI units given with the definitions of the variables, κ is in units of m^{-1}. Inserting all the constants, assuming the temperature is 25°C, recognizing that the ionic strength $I = \frac{1}{2} \sum c_{i,b} z_i^2$ (mol/L), and converting the length dimension to more appropriate units, Equation 11-1 simplifies to

$$\kappa \text{ (in units of nm}^{-1}\text{)} = 3.288\sqrt{I} \quad (11\text{-}2)$$

The Debye–Hückel model is a reasonable approximation when the surface potential is low, taken to mean that $\psi_d < k_B T / ze$ for a solution with an electrolyte made up of equi-charged anions and cations. For $z = 1$ and $T = 298$ K (or 25°C), $k_B T / ze$ equals 25.7 mV. When the surface potential is less than this value, the potential in solution drops approximately exponentially with distance, x, away from a charged planar surface, according to Equation 11-3:

$$\psi(x) = \psi_d \exp(-\kappa x) \quad (11\text{-}3)$$

Equation 11-3 suggests that κ^{-1} can be considered the characteristic length of the diffuse layer. In Chapter 3, we defined the inverse of a first order rate constant as a characteristic time because, in a batch reactor, it is the time required to reduce a concentration to $1/e$ of its original value. Here, κ^{-1} is defined as the characteristic length of the diffuse layer because, in this model, it is the distance required to reduce the potential to $1/e$ of its value at the surface. This characteristic length is often referred to as the Debye length or (somewhat inappropriately) the thickness of the diffuse layer.

In the Debye–Hückel model, the concentration of a species with charge number z varies with distance (x) away from a planar surface according to the following equation:

$$c_i(x) \cong c_{i,b}\left[1 - z_i \frac{e\psi(x)}{k_B T}\right] \quad (11\text{-}4)$$

Again, this equation is valid only for low values of ψ_d, but it holds for both co- and counterions; the product $z\psi$ is positive and negative for those two cases, respectively, so that the term in brackets is <1 for co-ions and >1 for counterions.

The Gouy–Chapman model for the variation of ionic concentration and potential with distance is general, so that the equations are not limited to low values of ψ_d. According to this model, the potential and ionic concentrations at a distance x from a planar surface can be determined for a symmetric electrolyte solution (one with the same charge number, z, on cations and anions) from the following equations:

$$\tanh\left(\frac{|z|\tilde{\psi}(x)}{4}\right) = \tanh\left(\frac{|z|\tilde{\psi}_d}{4}\right)\exp(-\kappa x) \quad (11\text{-}5)$$

$$c_i(x) = c_{i,b}\exp\left(\frac{z_i e\psi(x)}{k_B T}\right) = c_{i,b}\exp(-z_i\tilde{\psi}(x)) \quad (11\text{-}6)$$

where $\tilde{\psi} = e\psi/k_B T$ is a dimensionless potential, often referred to as the reduced potential.

---

[1] Note that $n_{i,b} = c_{i,b} N_A (1000 \text{ L/m}^3)$, where $c_{i,b}$ is the molar concentration of species i in the bulk solution, and N_A is Avogadro's Number (6.022×10^{23} molecules/mole).

[2] In older literature, the relative permittivity was called the dielectric constant.

The distribution of potential and charge adjacent to a surface according to these equations[3] is illustrated in Figure 11-5. For a given molar concentration of electrolyte, the difference in ionic strength between 1:1 and 2:2 electrolytes causes a substantial difference in the distributions of both potential and charge. Although the decay of potential appears to be exponential for both types of electrolytes in this figure, the results from the Gouy–Chapman model (Equation 11-5, used to create this figure) are slightly different from the Debye–Hückel (exponential decay) model (Equation 11-3). Careful inspection of the figure reveals that, for both conditions shown, the potential at $x = \kappa^{-1}$ or $\kappa x = 1$ is less than the value expected from the Debye–Hückel model ($25.7\exp(-1) = 9.45$ mV). The deviation is less for the monovalent case (9.27 mV) than the divalent case (8.82 mV).

The results illustrated are for the diffuse layer next to a flat plate, but they are also a good approximation for the layer surrounding a spherical particle, at distances much shorter than the particle radius. That is, the curvature of a large particle (e.g., a 1-mm diameter filter grain) has little effect on the behavior of a small particle (e.g., a 2-μm diameter) near its surface. Analytical solutions for the potential and charge as a function of the distance from a sphere are not available for the general case. When the Debye–Hückel approximation for low potentials is used, the potential decays away from a spherical surface as follows:

$$\psi(x) = \psi_d \frac{a_p}{r} \exp\left[-\kappa(r - a_p)\right] = \psi_d \frac{a_p}{a_p + x} \exp[-\kappa x] \quad (11\text{-}7)$$

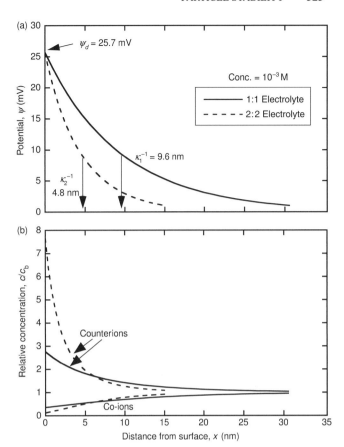

FIGURE 11-5. Diffuse layer potential (a) and ion concentrations (b) for a charged flat plate according to the Gouy–Chapman model.

In Equation 11-7, a_p is the radius of the spherical particle (including any adsorbed layers, i.e., to the d-plane) and r is the distance measured from the center of the particle, so that x is the distance measured from the d-plane. Comparing Equation 11-7 to Equation 11-3, we see from the pre-exponential terms that the potential decays more rapidly away from a sphere than a plane, reflecting the expanding volume away from the spherical surface.

Interaction of Charged Particles

In treatment processes designed to bring like-charged particles together to form flocs, or to bring a particle to a like-charged surface for capture, the interaction of the electrical fields adjacent to the two particles causes them to repel one another. It is convenient to characterize this repulsion in terms of the changes in various forms of energy in the system.

As the particles approach one another, the electrostatic repulsion increases. This repulsive interaction can be quantified as an increase in the electrical potential energy of the system. (Similarly, as explained subsequently, the attractive Van der Waals interactions between the particles can be

---

[3] Equation 11.5 is more complex than the Debye-Hückel model because $\tilde{\psi}$ is given as an implicit rather than explicit function of x. To solve, the equation can be rearranged to yield:

$$-\ln\left(\frac{\tanh(z\tilde{\psi}(x)/4)}{\tanh(z\tilde{\psi}_d/4)}\right) = \kappa x$$

One can choose a set of values for ψ between ψ_d and zero, convert these to values of $\tilde{\psi}$ (i.e., $\tilde{\psi}(x)$), and use this rearranged version of the equation to find the associated values of κx (and, for a given solution, x). Then Equation 11.6 is used to find the concentration of the counter-ions and co-ions.

Alternatively, it is possible to use the definition of the hyperbolic tangent to find an explicit equation for $\psi(x)$, with the following result:

$$\psi(x) = \frac{2k_B T}{ze} \ln\left(\frac{1 + \gamma_d \exp(-\kappa x)}{1 - \gamma_d \exp(-\kappa x)}\right)$$

where

$$\gamma_d = \frac{\exp(z\tilde{\psi}_d/2) - 1}{\exp(z\tilde{\psi}_d/2) + 1} = \frac{\exp(ze\psi_d/2k_B T) - 1}{\exp(ze\psi_d/2k_B T) + 1}$$

quantified as a *decrease* in the electrical potential energy of the system.) Since energy must be conserved, increases in potential energy are counterbalanced by an offsetting change in the particles' kinetic energy; that is, the particles slow down as they approach one another.

The potential energy associated with the interaction of the electrical double layers of the two particles, represented as V_R, is measured relative to a baseline condition in which those interactions are nonexistent (and V_R is defined to be 0). In theory, this baseline condition requires infinite separation of the particles; however, in practice, it can be defined as a separation that is large enough that, at every point along the line between the two particles, the electric potential of at least one of the particles is negligible.

The value of V_R at any separation distance can be computed as the work (energy) required to bring the particles from the baseline condition to that separation. Figure 11-6 illustrates the situation for two equal sized particles (with radius a_p); the separation distance between the particle surfaces on the line between the two centers is s, so that the separation distance between the two centers is $s + 2a_p$. In the lower part of the diagram, hypothetical profiles of ψ are shown for the two particles; for the conditions shown, the separation between the particles is large enough that either ψ_1 or ψ_2 is negligible at each point between them, so in this case, V_R would be 0.

The derivation of V_R as a function of separation distance is complex, but the result for symmetric electrolytes is relatively straightforward. For two identical spheres, the result is given by (Lyklema, 1978, as corrected by Stumm and Morgan, 1996)

$$V_{R,ss} = 64\pi \frac{n_b k_B T}{\kappa^2} \frac{a_p^2}{(s + 2a_p)} \left[\tanh\left(\frac{z\tilde{\psi}_d}{4}\right)\right]^2 \exp(-\kappa s) \tag{11-8}$$

where the subscript ss refers to sphere–sphere interactions, n_b is the number concentration of anions (or cations) in the bulk solution (e.g., no./m^3) and all other terms have been defined.[4] According to this equation, the repulsive energy diminishes with separation distance as a combination of an inverse function (second fraction on right, due to the particle curvature) and an exponential decay (last term, identical to the flat plate case). The repulsive interaction is a function of both the surface potential (ψ_d) and the solution conditions (κ). Since κ is proportional to $\sqrt{n_b}$, the term n_b/κ^2 is independent of ionic strength; therefore, the primary effect of the ionic strength is in the exponential term in this equation. On the other hand, the charge on the electrolyte (z) affects every term. Recall that κ is proportional to z, so that the first fraction shown decreases with increasing z value. Hence, in two solutions with the same ionic strength, but one made with monovalent electrolytes and the other with polyvalent electrolytes, the repulsive interaction of particles is smaller at any distance in the polyvalent solution. These effects are shown quantitatively after considering the energy associated with the (attractive) van der Waals interactions.

Van der Waals Attraction

The electron cloud around any molecule is in constant motion, deviating slightly from its most stable configuration, first in one direction, then in another. A force opposing these disturbances and attempting to drive the electron cloud to its lowest energy state always exists. However, the momentum of the electrons causes them to overshoot this lowest energy distribution, so the molecule is constantly shifting from one slightly unstable state to another. As a result, even in a highly symmetric, "nonpolar" molecule, the electron distribution is always slightly asymmetric. Such molecules are, therefore, actually slightly polar at any instant, and are nonpolar only on a time-averaged basis.

When two such molecules approach each other quite closely, the electrons in the outer orbit of each molecule affect the motion of the electrons in the other. If the

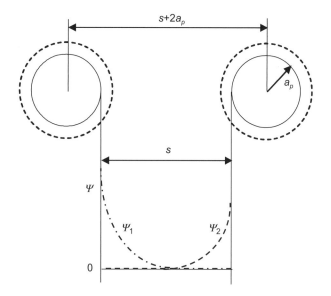

FIGURE 11-6. Schematic of the interaction of charged particles surrounded by diffuse layers. (Particles, including adsorbed layers within the d plane, have a radius a_p and are shown as solid circles; the surrounding diffuse layers have the characteristic thickness κ^{-1}, indicated by the dotted circles; and the separation distance between particle surfaces is s.)

---

[4] Recall that $n_b = c_b N_A (1000 \text{L/m}^3)$, where N_A is Avogadro's number, and that $k_B = R/N_A$, where R is the universal gas constant, so that $n_b k_B = c_b R (1000 \text{L/m}^3)$.

temporary polarities of the two molecules are the opposite of one another when they first interact, the interaction is favorable (attractive). For instance, if the electron cloud of molecule 1 is slightly more dense on the side facing molecule 2, and the electron cloud of molecule 2 is slightly less dense on the side facing molecule 1, then the initial interaction is attractive. An instant later, the electron cloud on molecule 1 might move to the opposite side. As it does so, the side facing molecule 2 becomes slightly positive, pulling the electron cloud of molecule 2 in that direction, and maintaining the attractive interaction. This process can repeat itself endlessly, so that a stable arrangement is established in which the electron clouds oscillate synchronously and the molecules constantly experience an attractive force toward one another. The force remains in effect until the molecules' kinetic energy causes them to separate.

If two particles approach each other very closely, the electron clouds of many of the molecules at the surface of each particle will be properly oriented to engage in attractive interactions of the sort described above with their counterparts on the other particle. Of course, many other molecules will not experience such attraction, including many that will experience a repulsion from the closest molecule on the other particle. However, if the initial orientations of two molecules are such that they are not attracted to one another, the electron cloud of each will continue to oscillate independently and randomly. Eventually, if the particles remain close to one another for many oscillations of the electron clouds, an attractive interaction between each pair of molecules on the opposing particles is almost certain to occur. Furthermore, because the attractive interactions tend to be stable and long-lived, while nonattractive and repulsive ones tend to be unstable and short-lived, the ultimate condition will be one in which the vast majority of the interactions are attractive. Even though each such interaction generates only a miniscule force, the combined effect of all of the attractive interactions can be quite strong.[5]

The interactions described above are operative only over very short distances and are known by several names, including van der Waals forces (after the person who first understood the molecular interactions), London forces (after the person who first understood the particle interactions), and dispersion forces.[6] These attractive forces depend on the *polarizability* of the particle (i.e., on the ease with which

molecules in the particle can be polarized by the movement of the electrons).

As with electrical repulsion, the van der Waals attraction is generally expressed in terms of energy. The material property that relates the attraction between particles to the separation distance is called the *Hamaker constant*, A_H, with units of energy (e.g., Joules). The Hamaker constant is related to the polarizability and has a range of values generally considered to be from 3×10^{-21} to 10^{-19} J, with a value of 10^{-20} J most commonly chosen. For two equi-sized, spherical particles, the energy associated with van der Waals attraction is expressed as follows:

$$V_{A,ss} = -\frac{A_H}{6}\left(\frac{2}{\bar{s}^2 - 4} + \frac{2}{\bar{s}^2} + \ln\frac{\bar{s}^2 - 4}{\bar{s}^2}\right) \quad (11\text{-}9)$$

where $\bar{s}$ is the dimensionless separation distance between the particle centers, given by

$$\bar{s} = \frac{s + 2a_p}{a_p} \quad (11\text{-}10)$$

When particles are touching each other, $s = 0$ and, therefore, $\bar{s} = 2$.

The magnitude of both the van der Waals attraction and the electrical repulsion for a typical situation of interest are illustrated in Figure 11-7. The total energy of interaction is the sum of these two contributions:

$$V_{Tot} = V_R + V_A \quad (11\text{-}11)$$

For the conditions illustrated in the figure, the value of κ^{-1}, the characteristic thickness of the diffuse layer, is ~ 17.5 nm. The particles exert essentially no influence on one another when far removed from each other. As they approach to within 50 nm of one another (three times the characteristic length of the diffuse layer), both energy terms become nonnegligible, with the repulsive term outweighing the attractive one for the conditions shown. In this example, the maximum (positive) energy of interaction occurs at a separation of ~ 16 nm; this maximum energy is called the energy barrier. The kinetic energy of the particles as they approach one another has to be sufficient to overcome this barrier if the particles are to form a floc.

Force is the negative derivative of energy with respect to distance, and it is instructive to consider the forces on these particles in addition to their interaction potential. At separation distances greater than that associated with the energy barrier, the slope of the V_{Tot} versus s plot is negative, meaning the force is positive and acting to keep the particles apart. At smaller separation distances, the force (i.e., negative derivative of the energy) is negative, meaning that the

---

[5] The process described can be thought of as analogous to what would occur if many small magnets were placed in a container that was then gently shaken. Before long, the magnets would form one large clump, all connected via favorable (positive to negative) interactions. This ultimate state is achieved not because positive-to-negative interactions occur more frequently than positive-to-positive or negative-to-negative, but because once a positive-to-negative interaction occurs, the magnets remain in that state, whereas they do not remain in the alternative orientations.

[6] An excellent description of these interactions, from the molecular origins to the integration over interacting bodies such as particles, is given by Hiemenz and Rajagopalan (1997).

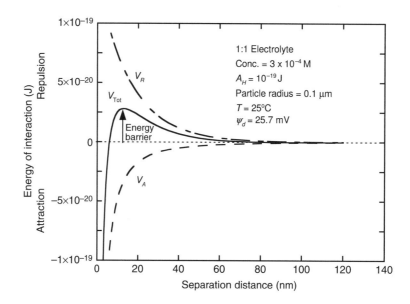

FIGURE 11-7. Energy of interaction as one particle approaches another. (Born repulsion, acting at extremely close distances, is not shown.)

particles are drawn together. Hence, getting over the energy barrier is the criterion for the particles to come together. In the example shown in the figure, the maximum negative slope occurs at a separation distance of ~24 nm, and calculations indicate the associated force on the particle is ~6×10^{-12} N. For comparison, the gravitational force on this small particle, even if its density is $2 \, g/cm^3$, is only in the range of 8×10^{-17} N. Thus, the energy barrier is substantial, and a suspension with the conditions indicated is likely to be quite stable. On the other hand, at close separation distances, the slope is very large and positive, meaning the particles are likely to stay together if they can overcome the energy barrier.

The discussion and figure have omitted another phenomenon that is important when particles are extremely close to one another, in the range of 1 nm. Under these circumstances, the particles' electron clouds interact to generate a strong repulsion known as Born repulsion. Inclusion of this repulsion in the analysis leads to a minimum in the energy curve at some extremely small separation distance. This point is known as the *primary minimum* of the energy curve, and it identifies the most stable distance of separation of the two particles.

Although not visible at the scale of Figure 11-7, the net interaction is slightly attractive at all separation distances greater than ~85 nm and passes through another minimum in that region. For the example system, this *secondary minimum* occurs at a separation of ~105 nm; the depth of this potential energy minimum is quite small for the conditions shown, so that only a very small amount of kinetic energy would be needed to move a particle away from this location. Therefore, it is unlikely that two particles would be held together at this separation distance. Other conditions can lead to a much deeper secondary minimum, which would hold particles together more strongly.

■ **EXAMPLE 11-1.** For the conditions shown in Figure 11-7, calculate the value of the repulsive, attractive, and total energy of interaction at a separation distance of 13 nm. Calculate the required relative velocity between two particles, each with a density of $1.5 \, g/cm^3$, to overcome this energy barrier, if they start at an infinite separation distance and are heading directly toward one another.

Solution. We choose to work in SI units because many of the constants have been given in those units. For the repulsive energy, we will use Equation 11-8, but some quantities in that equation must be calculated first, as follows.

Recognizing that, for a 1:1 electrolyte, the ionic strength is the same as the molar concentration, we find from Equation 11-2:

$$\kappa = 3.288\sqrt{I} = 3.288\sqrt{3 \times 10^{-4}} = 0.0569 \, nm^{-1}$$
$$= 5.69 \times 10^7 \, m^{-1}$$

The molecular concentration in the solution is the product of the molar concentration and Avogadro's number:

$$n_b = cN_A = (3 \times 10^{-4} \, mol/L)(6.022 \times 10^{23}/mol)(1000 \, L/m^3)$$
$$= 1.807 \times 10^{23}/m^3$$

For the conditions given, $\psi_d = 25.7$ mV and $T = 25°$C, so that $\tilde{\psi} = e\psi/k_BT = 1$. The repulsive interaction energy is then

$$V_{R,ss} = 64\pi \frac{n_b k_B T}{\kappa^2} \frac{a_p^2}{(s+2a_p)} \left[\tanh\left(\frac{z\tilde{\psi}_d}{4}\right)\right]^2 \exp(-\kappa s)$$

$$= 64\pi \frac{(1.807 \times 10^{23}/\text{m}^3)(1.38 \times 10^{-23} \text{kg m}^2/\text{s}^2 \text{ K})(298 \text{ K})}{(5.69 \times 10^7 \text{ m}^{-1})^2}$$

$$\times \frac{(1.0 \times 10^{-7} \text{ m})^2}{(1.3 \times 10^{-8} + 2.0 \times 10^{-7})\text{m}} \left[\tanh\left(\frac{1}{4}\right)\right]^2$$

$$\times \exp[(-0.0569 \text{ nm}^{-1})(13 \text{ nm})]$$

$$= 6.19 \times 10^{-20} \text{kg m}^2/\text{s}^2 = 6.19 \times 10^{-20} \text{ J}$$

For the attractive interaction, we begin by finding the dimensionless center-to-center separation distance:

$$\bar{s} = \frac{s + 2a_p}{a_p} = \frac{13 \text{ nm} + 2(100 \text{ nm})}{100 \text{ nm}} = 2.13$$

From Equation 11-9, the attractive energy of interaction is

$$V_{A,ss} = -\frac{A_H}{6}\left(\frac{2}{\bar{s}^2 - 4} + \frac{2}{\bar{s}^2} + \ln\frac{\bar{s}^2 - 4}{\bar{s}^2}\right)$$

$$= -\frac{10^{-19} \text{ J}}{6}\left(\frac{2}{2.13^2 - 4} + \frac{2}{2.13^2} + \ln\frac{2.13^2 - 4}{2.13^2}\right)$$

$$= -3.39 \times 10^{-20} \text{ J}$$

Finally, $V_{\text{Tot}} = V_R + V_A = 6.19 \times 10^{-20} \text{ J} - 3.39 \times 10^{-20} \text{ J} = 2.80 \times 10^{-20}$ J.

The positive sign indicates that the net interaction is repulsive. To overcome this energy barrier, a particle's kinetic energy would have to be greater than this value. The particle's mass is the product of its density (ρ_p) and volume (V_p):

$$m = \rho_p V_p = \rho_p \frac{4\pi a_p^3}{3} = (1500 \text{ kg/m}^3)\frac{4\pi(1.0 \times 10^{-7} \times \text{m})^3}{3}$$

$$= 6.28 \times 10^{-18} \text{ kg}$$

The kinetic energy ($\frac{1}{2}mv^2$, where v is the particle velocity) must exceed the net repulsive interaction energy, so that

$$\tfrac{1}{2}mv^2 \geq V_{\text{Tot}}$$

or

$$v \geq \sqrt{\frac{2V_{\text{Tot}}}{m}} = \sqrt{\frac{2(2.80 \times 10^{-20} \text{ J})}{6.28 \times 10^{-18} \text{ kg}}} = 0.0944 \text{ m/s} = 9.44 \text{ cm/s}$$

This velocity of approach between two particles is far higher than one would normally encounter in either natural waters or most treatment processes, so that a suspension with these characteristics would be quite stable. ∎

The effect of particle size on this picture of stability is shown in Figure 11-8. Under identical conditions of surface potential and electrolyte concentration, energy of interaction curves are shown for particles with radii of 0.1 and 1.0 μm. The differences in both the energy barrier and the secondary minimum are dramatic, suggesting that the secondary minimum might be able to capture the larger particles and hold them together, and also that the energy barrier that must be overcome by the larger particles to reach the primary minimum is significantly greater. Experience seems to indicate, however, that the differences in stability among different size particles are smaller than the figure suggests; whether this discrepancy is because of some error in the model, differences in the surface (i.e., d-plane) potential of different size particles, greater kinetic energy of the bigger particles,

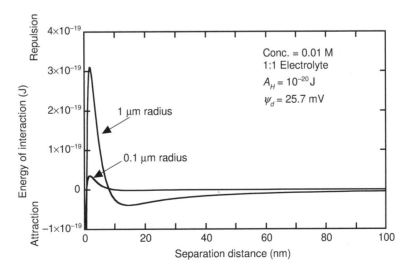

FIGURE 11-8. Effect of particle size on energy of interaction of two identical particles.

or perhaps other reasons is not known. The calculations in the figure do not include the retardation of van der Waals attraction (discussed subsequently), which would reduce the calculated depth of the secondary minimum since it reduces the attractive energy at greater separation distances. O'Melia and coworkers (Elimelech and O'Melia, 1990; O'Melia et al., 1997; and Hahn and O'Melia, 2004) have provided some evidence that capture in the secondary minimum might be important in filtration. Nevertheless, it is generally considered vital to reach the primary minimum to create long-lasting flocs and avoid their destruction (or detachment from filter grains) by hydraulic (shear) forces.

Interactions of a Particle and Flat Plate

Conceptually, the interactions of a charged particle (and its surrounding diffuse layer) with a charged flat plate (and its associated diffuse layer) are identical to those of two particles described above. However, the differences in geometry lead to differences in the equations describing the electrostatic repulsion and the van der Waals attraction. This case is important in at least two engineered systems of interest in water and wastewater treatment: membrane treatment, whether the membrane is a flat sheet or tubular, and granular media filtration, where the grains (in the range of 0.5 mm or more) are so much larger than the particles (0.5–50 μm) that they appear as a plane to a small particle.

According to Hogg et al. (1966), the potential energy associated with the interaction of the double layers of a flat plate and a particle is

$$V_{R,fp} = \pi \varepsilon_0 \varepsilon_r a_p \left\{ 2\psi_{d_f} \psi_{d_p} \ln\left[\frac{1+\exp(-\kappa s)}{1-\exp(-\kappa s)}\right] + \left(\psi_{d_f}^2 + \psi_{d_p}^2\right) \ln[1-\exp(-2\kappa s)] \right\} \quad (11\text{-}12)$$

where the subscript fp refers to the interactions of a flat plate (f) and a spherical particle (p), ψ_{d_f} and ψ_{d_p} refer to the surface potentials of the flat plate and the particle, respectively, and s is the separation distance between the plate and the particle (i.e., $s = x - a_p$, where x is the distance from the plane to the center of the particle). Note that this interaction energy can be either positive (repulsive, when the two potentials have the same sign) or negative (attractive, when the potential are of opposite sign).

The van der Waals attraction was given by the same investigators as follows:

$$V_{A,fs} = \frac{A_H a_p}{6s}\left(1 + \frac{14s}{\lambda}\right)^{-1} \quad (11\text{-}13)$$

The equation given is for the "retarded" van der Waals attraction, with the correction from the nonretarded form being the expression in the parentheses. Effects of the medium in transmitting the interacting oscillations of the dipoles are inadequately accounted for in the nonretarded van der Waals expressions. λ is the characteristic wavelength of the interaction and is generally thought to be ~ 100 nm (Anandarajah and Chen, 1995). At close separation distances, this retardation has little effect since it approaches the value of 1.

The energy of interaction between particles and a flat plate according to this model is shown for a few example systems in Figure 11-9. As before, the Born repulsion energy is omitted in the diagram, but it is very large at extremely close distances, so that a primary minimum in the energy curve is formed near the surface of the particles (i.e., at a separation distance <1 nm). The general shape of these curves is similar to those shown in Figures 11-7 and 11-8 for the interactions of two identical spherical particles. The

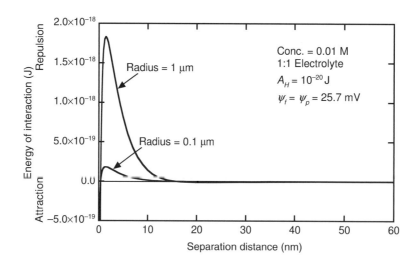

FIGURE 11-9. Effect of particle size on energy of interaction with a flat plate.

classic theory suggests that for a particle to be collected by the flat plate (e.g., for a particle to become attached to a media grain in a granular media filter), the energy barrier would have to be overcome so that the net interactive force would be positive and the particles would come together at the primary minimum.

Close inspection of the figure reveals that, like in particle–particle interactions, a secondary minimum (<0) of the energy curve occurs at much greater separation distances. Hahn and O'Melia (2004) have suggested that particles can be successfully captured in that secondary minimum in granular media filtration, and have presented both modeling results and experimental measurements that bear out this idea. Although the energy well in the secondary minimum is relatively small in comparison to that in the primary minimum, these investigators have argued that it is sufficient in comparison to the energy supplied by particle motion to prevent re-entrainment in many porous media applications.

Experimental Measurements Related to Charge and Potential

Measuring the surface charge or potential of suspended particles is of great interest when determining the stability of a suspension, or the degree to which added chemicals change that stability. Acid/base titrations of surface charge, electrophoresis, and measurement of streaming potential are the approaches used most commonly to characterize these parameters. Details of the theory behind these measurements are available elsewhere (Stumm and Morgan, 1996; Hiemenz and Rajagopalan, 1997; Hunter, 2001); only a brief synopsis is given here to explain how these measurements are used as indicators of particle stability.

Electrophoresis has been the most common of these measurement methods. When a suspension is placed in an electric field, charged particles in that suspension move toward the electrode of opposite charge. Each particle rapidly reaches a steady-state velocity, as the electrical force is balanced by fluid drag.[7] The *electrophoretic mobility* is the velocity (say, μm/s) divided by the electrical field strength (V/cm); if those units are used, as is common because it yields absolute values that are between 0 and ~5, the electrophoretic mobility is reported in mixed units of μm cm/V s. The sign of electrophoretic mobility corresponds with the sign of the surface charge; a negatively

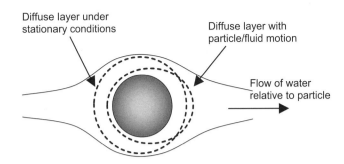

FIGURE 11-10. Conceptual diagram of shearing of diffuse layer by flow of water around particle.

charged particle moves toward the anode and is assigned a negative value of the electrophoretic mobility.[8]

When a particle moves during electrophoresis, the water motion around the particle alters the surrounding diffuse layer, as depicted in Figure 11-10. In fact, if it did not, the particle would not move, because the combination of the surface charge, compact layer, and diffuse layer is electrically neutral. Hence, what moves in the electric field is the particle and some of the surrounding ions, so that the electrophoretic mobility that is measured is a property not strictly of the particle, but of a composite of the particle plus some of the surrounding compact and diffuse layers. Much research has been done to determine the location of the shear plane relative to the particle surface, but the details are not critical for our understanding here. It is often assumed that the shear plane is at the border between the compact and diffuse layers (regardless of the model used for the compact layer).

The measured electrophoretic mobility is often converted to a second parameter related to the charge or potential: the *zeta potential* (ζ), which is the potential at the shear plane. While various models for the conversion exist (and depend on κ), the zeta potential is usually taken as the product of the electrophoretic mobility and a term that depends on the diffuse layer thickness and the dielectric properties and viscosity of the medium; details are provided by Hiemenz and Rajagopalan (1997) and Hunter (2001). If the shear plane is taken to be the same as the *d*-plane, then $\zeta = \psi_d$.

A third measure of the surface charge or potential characteristics is the *streaming current*. In a streaming current

---

[7] The given description is a simplified view of the reality. In addition to fluid drag on the particle surface, electrical work has to be done in the solution to move ions around the particle as it moves, and the associated force can substantially retard the motion of the particle. The relative importance of the fluid drag and this electrical force in determining the particle velocity depends on the dimensionless product κa. Details have been provided by Hunter (2001).

[8] The actual measurement is somewhat more complex than the simple description implies. Because glass (SiO_2) becomes negatively charged in contact with the solution and under the imposition of the electric field, cations from solution are in great excess near the glass surface. These cations migrate toward the cathode and their movement along the cell surface carries solution along with them. This process is known as electroosmosis. This movement of the water along the cell wall is necessarily counterbalanced in the closed cell by water flowing in the opposite direction in the center of the cell. With water flow in one direction at the walls and the opposite direction in the center of the cell (which can be either rectangular or cylindrical), the water velocity is zero at some layer in between. The velocity of the particles must be measured in that layer to be unaffected by the fluid flow.

detector, the particles are held (relatively) stationary, and fluid flows around the particles. In modern instrumentation, a solid cylinder (piston) is oscillated fairly slowly inside a vertical open cylinder of slightly larger diameter; the small annular space separating the two cylinders and a small reservoir below are filled with the suspension. Since the piston has a diameter much larger than the annular space, the fluid velocity induced as water escapes from or fills the annular space is far greater than the piston velocity, though that flow is still laminar. The particles are thought to adhere to the surfaces of both walls (Dentel and Kingery, 1989). The relatively high fluid velocity adjacent to these particles induces a shear plane, separating the diffuse layer from the particle. Since the diffuse layer around each particle contains a net excess of counterions, the moving charges constitute an electrical current. This current can be measured with electrodes at both ends of the annular space. Streaming current detectors have proven to be valuable on-line instruments to control the dosing of destabilizing chemicals, when the mechanism for destabilization involves surface charge neutralization (as explained subsequently).

11.3 CHEMICALS COMMONLY USED FOR DESTABILIZATION

Particle stability is undesirable for particle removal processes since those processes are successful only if particles can attach to one another or to a provided surface. Particles are destabilized by the addition of chemicals that change the characteristics of the solution, the surface charge, or the interactions between the particles. Before presenting the mechanisms of destabilization, it is useful to understand the properties and behavior of the chemicals that are used. In this section, the chemicals commonly used for destabilization are described.

Inorganic Species

The most common inorganic species used to promote particle destabilization are various formulations of aluminum and iron salts. These two metals have similar chemistries and, in many instances, either of them can be used quite effectively; in some cases, however, the differences in their chemistries can be important in determining why one formulation or metal might be more effective than another.

As noted in Chapter 9, aluminum can form several monomeric hydroxyl complexes (Al^{+3}, $Al(OH)^{+2}$, $Al(OH)_2^+$, $Al(OH)_3$, and $Al(OH)_4^-$). However, the aluminum species of greatest significance with respect to particle destabilization is usually the aluminum hydroxide precipitate, $Al(OH)_3$ (s). A log concentration diagram showing the solubility of as a function of pH is shown in Figure 11-11. The pH_{zpc} of $Al(OH)_3$ (s) is ~6.5, so the surface charge of

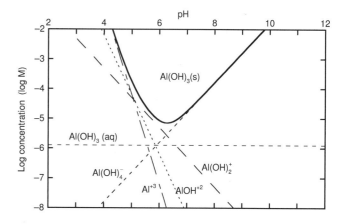

FIGURE 11-11. Aluminum speciation and solubility.

pure $Al(OH)_3$ (s) is positive at pH values below ~6.5 (the pH of minimum solubility) and negative at higher pH values.

Aluminum can also form complexes and/or precipitates with several common anions besides hydroxide, including phosphate (PO_4^{-3}), sulfate (SO_4^{-2}), and fluoride (F^-). Interactions between aluminum and organic matter are also critical in many applications; the interaction with NOM in drinking water treatment has been extensively studied, whereas the interactions with the organics in wastewater are less well known. The interactions of hydrous aluminum and iron oxide solids with NOM are explained in some detail in Chapter 7, and the effects of those interactions on particle charge and particle behavior are explained later in this chapter.

Iron has a similar chemistry to aluminum, but the solubility diagram for amorphous $Fe(OH)_3$ (s) (i.e., ferrihydrite) indicates that iron is generally less soluble throughout the entire pH range, and that the pH of minimum solubility is at a higher pH. These two properties often make iron a better (more versatile) choice than aluminum as the destabilizing chemical, especially when used at pH values greater than ~8.0.

The most common ways to add aluminum or iron are as the sulfate or chloride salts. Aluminum is most often added as a hydrated aluminum sulfate, commonly known as alum ($Al_2(SO_4)_3 \cdot xH_2O$), where x is typically in the range of 14–18, and iron is added as ferric sulfate ($Fe_2(SO_4)_3$) or ferric chloride ($FeCl_3$). In all of these cases, the addition and subsequent precipitation of the metal consumes three equivalents of alkalinity per mole of the metal (and therefore reduces the pH), as is clear from the following stoichiometry:

$$FeCl_3 + 3H_2O + 3HCO_3^- = Fe(OH)_3(s) + 3H_2CO_3 + 3Cl^-$$

$$Al_2(SO_4)_3 + 6H_2O + 6HCO_3^- = 2Al(OH)_3(s) + 6H_2CO_3 + 3SO_4^{2-}$$

When these metal salts are added to water or wastewater as destabilizing chemicals, the alkalinity and pH are reduced; that is, the metals are acids. As might be expected from the role of pH in determining speciation and solubility, the effectiveness of these chemicals is highly dependent on pH, so it must be carefully monitored and, in some cases, controlled. The original alkalinity of the water determines the expected change in pH. Low alkalinity waters are likely to require simultaneous addition of a base (e.g., $Ca(OH)_2$ or $NaHCO_3$) to neutralize the acid addition.

In well-oxygenated waters, iron can be added in the +II oxidation state (e.g., as ferrous sulfate, $FeSO_4$), allowing the dissolved oxygen to oxidize the iron as follows:

$$4Fe^{2+} + O_2 + 4H^+ \rightarrow 4Fe^{3+} + 2H_2O$$

As noted in Chapter 10, the rate of this oxidation reaction is relatively slow at low to neutral pH values but increases dramatically with increasing pH. Hence, the use of Fe(II) as a destabilizing chemical is limited to situations in which the pH, the dissolved oxygen content, and the alkalinity are all sufficiently high to facilitate the rapid oxidation of all the added Fe. If the oxidation is incomplete in the treatment plant but continues in a water distribution system, $Fe(OH)_3$ will form in the distribution system. The use of Fe(II) can be advantageous in some systems because the decrease in the oxygen concentration can help prevent the formation of bubbles in deep bed filters.

When either aluminum or iron salts are added to water, they can form some metal-hydroxo polynuclear complexes. Most evidence suggests that, in the case of aluminum, two dimeric species (molecules with two aluminum atoms) and one larger molecule (probably $Al_{13}O_4(OH)_{24}^{7+}$) form (Baes and Mesmer, 1976). Iron also forms polymers ($Fe_2(OH)_2^{+4}$, $Fe_3(OH)_4^{5+}$) but to a lesser extent than aluminum (Gray et al., 1995). These relatively large, highly charged species are surface-active, and hence adsorb rapidly to particles already in the water. However, the polymers are quite short-lived in typical coagulation (rapid-mix) processes used in water treatment. Their formation is a function of the pH and temperature of the water, the pH, and concentration of the stock solution, and the concentration and pH gradients induced in the suspension when the alum is added (and therefore of the mixing employed in the rapid-mix tank).

It is also possible to manufacture stable forms of polymeric alumino-hydroxyl species and to use these instead of alum as the destabilizing chemical. To prepare such chemicals, the pH of a stock solution of aluminum chloride (which is at low pH and therefore completely soluble) is raised very slowly by the addition of base. The slow addition leads to the formation of the polymeric forms (e.g., $Al_2(OH)_2^{4+}$, $Al_3(OH)_4^{5+}$, and $Al_{13}O_4(OH)_{24}^{7+}$) without the formation of the precipitate, $Al(OH)_3$ (s). The extent of polymerization is controlled by the final ratio of $[OH^-]_{added}/[Al]_T$. Although the species of interest are alumino-hydroxyl polymers, the product is referred to as polyaluminum chloride (PACl), in recognition of the starting material and presence of chloride in the final product. The identical process can be used with a starting material of aluminum sulfate (and the final product is then called polyaluminum sulfate); sulfate acts as a catalyst for $Al(OH)_3$ (s) precipitation, however, and so this product tends to be less stable in water. Although it is possible to manufacture preformed polymers of iron, the commercially available products are almost all polyaluminum chloride.

When PACl is added to water in a treatment scheme, the changed chemical conditions lead to a reequilibration of the speciation, and the precipitate $Al(OH)_3$ (s) is then formed. The use of preformed polyaluminum species, particularly, in removing humic substances, has been studied by Dempsey and coworkers (Dempsey, 1989; Dempsey et al., 1985). Other studies have investigated the relative effectiveness of the preformed polymers and the inorganic salts at different temperatures.

According to Van Benschoten and Edzwald (1990a, b), the polymeric species are unlikely to form when alum is used for destabilization, whereas the preformed polymers of polyaluminum chloride tend to be maintained in solution for periods longer than the typical detention times involved in engineered systems for water and wastewater. Hence, the operative species present in the water are considerably different depending on the source of aluminum. The $Al(OH)_3$ (s) precipitates that are formed from the two sources of aluminum are also different. Alum (or aluminum chloride) forms an amorphous precipitate of $Al(OH)_3$ (s) that incorporates a substantial amount of water, and so is gelatinous and fluffy. Polyaluminum chloride forms a more densely packed precipitate, with much less water incorporated.

Besides aluminum and iron, other inorganics can be added to water to aid or cause particle destabilization. Lime (CaO, or in its hydrated form, $Ca(OH)_2$) is added not primarily to destabilize suspended particles, but to precipitate calcium carbonate (in softening) or calcium phosphate (for phosphate removal). Nevertheless, these precipitates are effective at destabilizing existing particles in the water. Lime can also be added strictly to raise the pH of the water (because it is a base with two equivalents of alkalinity per mole of lime). For instance, lime is sometimes added as a base to counteract the acidic properties of aluminum or iron.

Organic Polymers

Several types of synthetic organic polymers are used for particle destabilization and to improve the strength of flocs that are created in the particle treatment processes. The most common types used in water and wastewater treatment are relatively low-molecular weight molecules that are positively charged in water: polydiallydimethyl ammonium chloride (often referred to as polyDADMAC) and epichlorohydrin dimethylamine (commonly referred to as epi-DMA). The

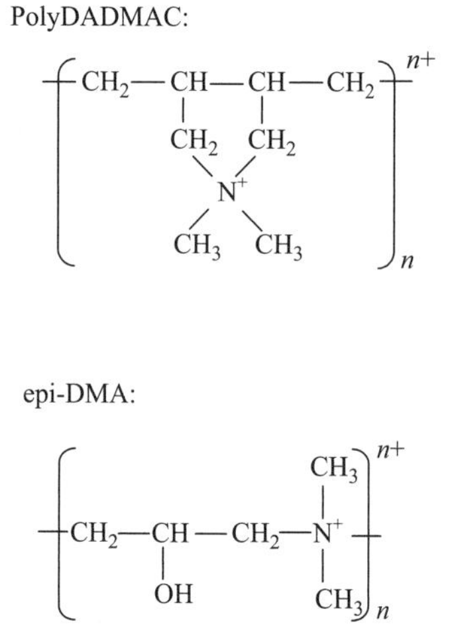

PolyDADMAC:

$$\left[-CH_2-CH-CH-CH_2- \atop {||\atop CH_2CH_2 \atop \diagdown\diagup \atop N^+ \atop \diagup\diagdown \atop CH_3CH_3} \right]_n^{n+}$$

epi-DMA:

$$\left[-CH_2-CH-CH_2-N^+- \atop {||\atop OHCH_3} \right]_n^{n+} \begin{array}{c} CH_3 \\ \end{array}$$

FIGURE 11-12. Chemical structures of the common low-molecular weight cationic polymers used for particle destabilization. The polymer consists of n of the monomers shown.

Nonionic: Acrylamide

$$\left[-CH_2-CH- \atop {|\atop C=O \atop |\atop NH_2} \right]_n$$

Cationic: Copolymer of acrylamide and quaternary ammonium monomers

$$\left[\left[-CH_2-CH- \atop {|\atop C=O\atop|\atop R\atop|\atop CH_3-N-CH_3\atop|\oplus\atop CH_3} \right]_x \left[-CH_2-CH- \atop {|\atop C=O\atop|\atop NH_2} \right]_y \right]_n$$

Anionic: Copolymer of acrylamide and acrylate

$$\left[\left[-CH_2-CH- \atop {|\atop C=O\atop|\atop O^\ominus} \right]_x \left[-CH_2-CH- \atop {|\atop C=O\atop|\atop NH_2} \right]_y \right]_n$$

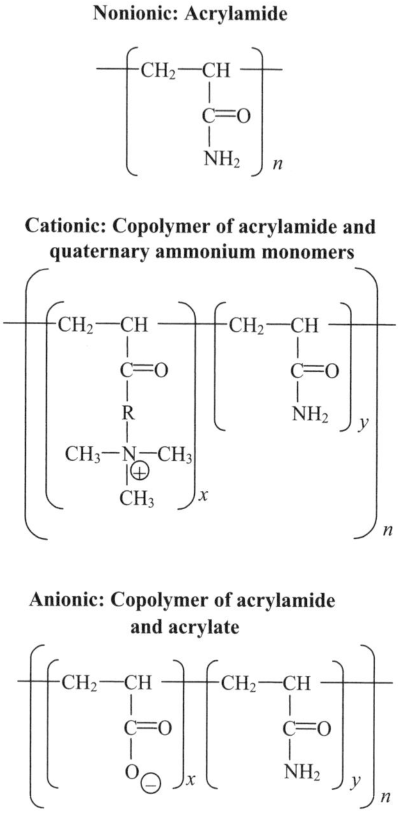

FIGURE 11-13. Chemical structures of the common high-molecular weight polymers used for particle destabilization. Nonionic polymers consist of n of the monomers shown. For both the cationic and anionic polymers, products with different numbers and ratios of x and y can be formulated. The copolymers shown are examples only; other groups can be used with the backbone of acrylamide to form these types of polymers. *Source*: Adapted from Mangravite (1983).

structures of the monomers that are linked together to form these two types of polymers are shown in Figure 11-12. In the case of the epi-DMA polyamines, the synthesis can lead to either linear or branched structures, and these different formulations could interact differently with the charged particle surfaces. In the figure, the chloride is not shown, as its departure from the molecule is what leaves the polymer positively charged. Both of these common polymers are available from several manufacturers; each manufacturer has many products with various molecular weights (n values on the figure) and, at each molecular weight, various values of the weight percent (i.e., dilution) of polymer in the formulation. Molecular weights for these polymers are usually in the range of 10^4–10^6. For these polymers, the charge density (charge per unit mass) is essentially independent of pH over a broad range; this fact is convenient for applications where the pH of the water can change rapidly and over a wide range (e.g., some industrial wastes), although the fact that the particle charge in the raw water might depend on pH can lead to an effect of pH on process performance.

Formulations with different molecular weights might have substantially different effectiveness, whereas the different active weight percents of polymer in the formulation primarily add convenience in matching the mixing requirements with each suspension and facility. As the molecular weight or the active weight percent increases, the viscosity of the solution increases; this property means that it might be difficult to disperse a stock solution with a high active weight percent of polymer uniformly into a suspension. If the application is in potable water use, the amount of free monomer in the product must be very low to avoid contaminating the drinking water; the polymers themselves are so surface-active that virtually all are removed in subsequent treatment.

A different class of polymers is based on acrylamide; as shown in Figure 11-13, the acrylamide alone leads to a nonionic polymer, but both cationic and anionic polymers can be made by copolymerization of the acrylamide with positive or negative organic functional groups. The polymer molecular weight can be varied by the number of the repeating units (n), and the charge density can be varied by the ratio of the number of acrylamide to either quaternary ammonium or acrylate monomers (y to x in the figure) (Mangravite et al., 1985). Again, many formulations are available, with different molecular weights (10^6–10^7), functional groups, and charge densities; these polymers are sold both in powder form and as emulsions, because aqueous solutions are quite unstable.

The essential feature of all of the organic polymers is that they are surface-active materials; that is, they adsorb to particle surfaces. The charge and size characteristics of these synthetic polymers then determine how they change the characteristics and interactions of the particles.

Some naturally occurring organic materials have also been shown to be effective destabilizing chemicals. The use of natural materials is particularly valuable in developing countries where the cost of synthetic organics or manufactured inorganic materials might be prohibitive. One natural material that has shown promising results is an extract made from *Moringa* seeds (Ndabigengesere and Narasiah, 1998), with the effective doses often being as low as with synthetic materials. A difficulty in using natural materials for drinking water applications is extracting the active ingredient, so that, when the chemical is used, the water is not contaminated with other organics that could promote biological growth or lead to other unwanted side effects.

11.4 PARTICLE DESTABILIZATION

The success of particle treatment processes depends on overcoming or eliminating the problem of stability induced by the repulsive energy barrier depicted in Figures 11-7 and 11-8. Particles can be destabilized in at least four different ways, each of which is discussed subsequently. These four destabilization mechanisms are (1) compression of the diffuse layer, (2) adsorption and charge neutralization, (3) enmeshment in a precipitate, and (4) adsorption and interparticle bridging.

Destabilization of a stable suspension is often induced by the addition of chemicals, commonly referred to as coagulants or flocculants. The degree of destabilization achieved is generally a nonlinear function of the concentration of the added chemical. The minimum concentration of the added chemical required for effective destabilization is referred to in some literature as the *critical coagulation concentration* (CCC). In the following sections, we explore how the various destabilization mechanisms work, including the effects of the concentration of the added chemical.

Some authors use the terms coagulant and flocculant interchangeably, while others distinguish between them. Even those authors who distinguish between the terms do not agree on the distinctions; some use coagulants to refer to chemicals that work by the precipitation mechanism and flocculants to refer to those that work by adsorption, others distinguish inorganic additives (coagulants) from organic polymers (flocculants), and still others distinguish chemicals that accomplish charge neutralization (coagulants) from those that work by interparticle bridging (flocculants). In addition, some of these chemicals are referred to by their use (*filter aids, dewatering aids*) when in fact the same chemicals might be used to accomplish destabilization for other particle removal processes. To avoid these terminology problems, we primarily use the term *destabilizing chemicals* to refer to all the chemicals of interest, although we sometimes refer to coagulants or flocculants, with the term flocculant reserved for high-molecular weight organic molecules that work by interparticle bridging.

Compression of the Diffuse Layer

If the ionic strength of a solution is increased, the availability of counterions to surround a charged particle increases, so that the volume or distance from the surface through which the diffuse layer extends is reduced. The effect of ionic strength on the composition and thickness of the diffuse layer was shown graphically in Figure 11-5 and is illustrated schematically in Figure 11-14. Higher ionic strength reduces the size of (compresses) the diffuse layer.

As shown in Figure 11-15, the reduction in the thickness of the diffuse layer with higher ionic strength tends to reduce the energy barrier that must be overcome to allow particles to collide; that is, it decreases the stability of the suspension. At some ionic strength, the net interactive energy is attractive at all separation distances. Since the ionic strength is a function of both the molar concentration and the valence of the dissolved electrolyte, increases in either of these parameters decrease the energy barrier, as indicated in the two parts of the figure.

Double layer compression is the basis of an empirical rule known as the Schulze-Hardy rule (1890s), which held that the ion concentration required to destabilize a suspension varies for mono-, di-, and trivalent compounds as $1 : 1/100 : 1/1000$. For this mechanism, the critical coagulation concentration can be quantified as the minimum salt concentration, which leads to a net interaction energy of zero ($V_{\text{Tot}} = 0$) at the energy barrier (where dV_{Tot}/ds is 0). The diffuse layer model predicts the CCC computed this way to depend on $1/z^6$, which translates to a ratio of $1 : 1/64 : 1/729$ in close agreement with the empirical Schulze-Hardy rule (Lyklema, 1978).

In full-scale engineered systems (especially for drinking water), compression of the double layer is not practical as a destabilization mechanism because the treated water would

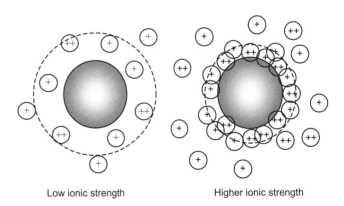

FIGURE 11-14. Schematic of the compression of the diffuse layer at high ionic strength. (Particle is assumed to be negatively charged; only the surrounding cations are shown for the sake of clarity. Characteristic thickness of the diffuse layer indicated by the dotted circles.)

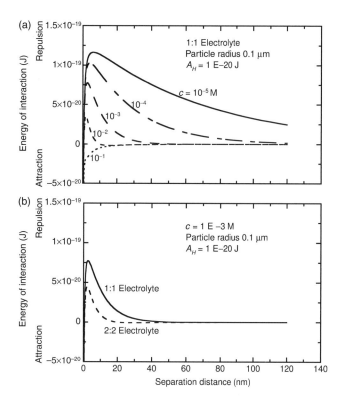

FIGURE 11-15. Effects of electrolyte concentration (a) and valence (b) on energy of interaction as particles approach one another.

Adsorption and Charge Neutralization

A second mechanism of destabilization is the neutralization of the charge on the particle by the adsorption of ions of the opposite charge. If just the right amount is adsorbed, the net charge on the particles becomes 0, the particles have no surrounding diffuse layer, electrostatic repulsion is nonexistent, and the problems that create particle stability are eliminated.

Adsorption of a charge neutralizing ion can occur because of specific chemical interactions between the destabilizing chemical and the surface of the particle, or because the molecule is hydrophobic and prefers to attach to a particle surface rather than stay in the bulk solution. If such adsorption occurs and the added chemical has a charge opposite to that of the particle, the net charge on the particle surface is reduced; in effect, the surface charge is being titrated by the addition of the destabilizing chemical. In the top portion of Figure 11-16, this effect is illustrated. The charge of the destabilizing chemical (positive in the illustration, opposite to the original surface charge) changes the net surface charge. At low doses of the chemical, the surface charge remains negative, but its magnitude is reduced; at some dose, the original surface charge has been neutralized and the net surface charge is 0. The driving force for this adsorption is not primarily electrostatic, so further addition of the destabilizing chemical beyond the dose needed for charge neutralization results in charge reversal, as illustrated.

have a high ionic strength (i.e., high salt content). However, it can be used in a situation in which the blending of two waste streams, one with high ionic strength and another with high particle content but low ionic strength, could lead to the effective flocculation of the particles without need for additional chemicals. Compression of the double layer has also frequently been employed in research on physical aspects of particle processes since this mechanism is the best understood mathematically. In that case, the chemicals that are added are inorganic electrolytes that do not change the pH, such as sodium chloride, sodium nitrate, or calcium chloride.

Diffuse layer compression is also important in estuaries where the salt content increases as the estuary approaches the ocean. Particles that are stable upstream become unstable, and the resulting coagulation of particles near the mouth leads to accumulations of sediment at or near the entrance to the ocean (Mehta et al., 1989). In addition, calculations indicate that, as the salt concentration increases, the depth of the secondary minimum in the total energy curve increases; this fact suggests that destabilization might occur to some degree by capture in the secondary minimum at concentrations less than that needed to make the interaction attractive at all distances.

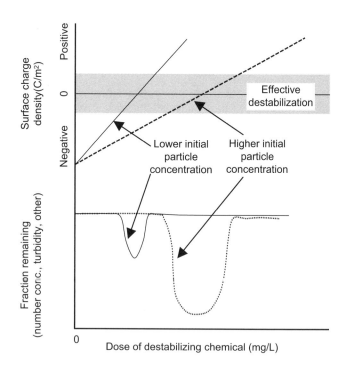

FIGURE 11-16. Particle destabilization by charge neutralization.

The bottom portion of Figure 11-16 shows the results of a procedure widely referred to as a *jar test*. In such a test, several samples of given suspension are subjected to various doses of the destabilizing chemical, rapidly mixed for a short period (15–90 s), slowly mixed for a longer period (15–45 min), and allowed to settle for some time period (20–60 min) or filtered with a coarse filter. Then, some measure of residual particle concentration (turbidity, suspended solids, or total particle number concentration) is taken. Jar testing is the standard testing procedure for determining the proper chemical dose (regardless of the mechanism of destabilization, and also irrespective of which particle removal process is being considered). A series of jars provides several points on one of the continuous curves shown. The most effective dose of the destabilizing chemical is that for which the chosen measure of particle content is lowest.

The (conceptual) results shown in Figure 11-16 indicate that destabilization is effective not only when the net particle surface charge is exactly 0, but also in a range of doses around that perfect dose. One explanation for this phenomenon is that the energy barrier has been lowered sufficiently that most particles in the suspension can overcome it. Another plausible reason has been offered by Gregory (1973) and Kasper (1971); their view, called the "patch model," holds that the destabilizing chemicals (generally relatively small organic polymers) adsorb in patches on the surface, forming an uneven charge distribution, with some areas still carrying the original surface charge and others (where extensive adsorption has occurred) being of opposite charge. A negative patch on one particle is attracted to the positive patch of another, resulting in effective destabilization even at doses below or above that required for exact charge neutralization.

The figure makes clear that, when so much destabilizing chemical has been added that the charge is reversed, the suspension is stable—no reduction in the chosen particle measure occurs. Overdosing of the chemical leads to a suspension that has the same problem as it did originally—charged, and therefore stable, particles. The term *restabilization* is used to describe this phenomenon. (Note that a suspension subjected to an overdose has never been destabilized, except perhaps quite briefly during the adsorption process, so the term is slightly misleading.) This phenomenon of overdosing and restabilization can sometimes lead to operational problems; if a suspension is stable (as indicated by poor removal in the subsequent treatment process), operators might assume that the current chemical addition represents an under-dosing and increase the dose. If, in reality, the current dose is an overdose, the correct decision is to reduce the amount of destabilizing chemical being added.

When particle destabilization is accomplished by adsorption and charge neutralization, the chemical dose required is correlated with the particle concentration; that is, there is a stoichiometric effect of particle concentration. This stoichiometry is actually between the dose required and the cumulative surface charge, which is usually assumed to be proportional to the total surface area concentration of the particles in the water. This stoichiometry is also illustrated conceptually in Figure 11-16 by the inclusion of curves for two suspensions with different particle concentrations but otherwise identical characteristics. The same chemical dose in the two suspensions leads to a lesser change in the surface charge of the higher concentration suspension because the dose is spread among a larger number of particles. Perhaps, not so obvious is that this spreading leads to a wider range of chemical doses that are effective; the same range of low (near 0) surface charge density leads to effective destabilization in both cases, but that range of charge densities is spread over a wider range of doses of destabilizing chemical for the higher particle concentration suspension. The stoichiometric effect can also lead to operational difficulties if the quality of the water to be treated (i.e., the particle concentration) is highly variable. In such a case, the dose of coagulant must be varied to match the variation in water quality. Use of on-line measurement of the streaming current, electrophoretic mobility, or zeta potential is particularly valuable to control the dose of destabilizing chemical in such situations.

Note also in Figure 11-16 that the effective destabilization zone is not only wider but also deeper for the higher concentration suspension. As shown subsequently in Chapter 12, higher concentrations of particles are more easily contacted with one another, resulting in a greater extent of flocculation in a fixed time.

Both inorganic chemicals (aluminum and iron) and charged organic polymers can be used for adsorption and charge neutralization. As has been noted, the surface charge on most particles encountered in environmental engineering is negative; hence, both the low-molecular weight cationic organic polymers (polyDADMAC and epiDMA) and the metal coagulants are commonly used for this purpose. Iron and aluminum are effective at pH values where their positive species dominate. Both of these inorganic coagulants also can form precipitates with soluble NOM at relatively low pH; this specific chemical interaction makes these inorganic coagulants particularly effective for particles whose surface charge is caused primarily by adsorption of NOM. At high doses of the metal salts, precipitation of the metal hydroxides occurs, as is described next.

Enmeshment in a Precipitate–Sweep Flocculation

A third mechanism of destabilization, enmeshment in a precipitate, occurs when an inorganic salt is added in such quantity that a precipitate is formed. Most commonly, the inorganic salt is $FeCl_3$ or alum, and the precipitate that forms is ferric hydroxide, $Fe(OH)_3(s)$, or aluminum

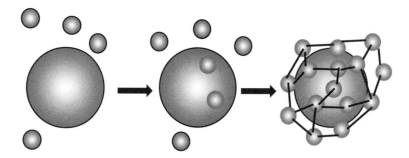

FIGURE 11-17. Schematic of the adsorption of metal ions (small circles) onto a particle (large circle) leading to the formation of a precipitate. The lines in the right part of the figure indicate the formation of the precipitate.

hydroxide, $Al(OH)_3(s)$. However, the same mechanism can occur when calcium carbonate, $CaCO_3(s)$, and (possibly) magnesium hydroxide, $Mg(OH)_2(s)$, are generated in a precipitative softening process, or when metal hydroxides are precipitated during treatment of industrial wastewaters.

In this destabilization mechanism, the cations that are the precursors of the precipitate first adsorb onto particles that are already in the water, just as occurs in the adsorption-and-charge-neutralization mechanism. However, in this mechanism, the surface catalyzes the formation of nuclei of the new solid, so that precipitation can proceed at a significant rate at lower coagulant doses than are required in the bulk solution. As a result, the original particles are caught up in the precipitate as it forms. This sequence of steps is shown schematically in Figure 11-17.

In most cases, the forming precipitates join with each other to form large flocs. One would expect that the hydroxide precipitates would be positively charged at normal pH values (because the zero point of charge is at alkaline pH for both $Al(OH)_3$ and $Fe(OH)_3$ solids). However, anions from solution (e.g., sulfate ions or NOM) adsorb onto the forming solids and neutralize the surface charge sufficiently to allow the flocs to continue to grow. This floc growth, combined with the fact that hydroxides incorporate a substantial amount of water into their formation, results in the rapid formation of large, fluffy flocs. If any of the original particles have not been caught up in the initiation of precipitation, they can be contacted quickly by these large flocs, leading to the notion of the particles being swept out of the suspension by the precipitate. As a result, this mechanism of destabilization is commonly referred to as *sweep flocculation* (or simply, sweep floc), even though most particles are probably destabilized by serving as sites of initiation of the precipitation reaction.

Because particles in the suspension act as nucleation sites, the rate of precipitation increases slightly with higher particle concentration. This leads to a mild inverse stoichiometry; that is, the more particles originally present, the less chemical necessary to accomplish effective destabilization.

Obviously, however, sufficient amounts of the metal salts must be added to achieve precipitation, and that often takes a factor of oversaturation of ~100, so this inverse stoichiometry is often minor in comparison to the chemical requirement for precipitation. In essence, the forming precipitate becomes the dominant species dictating the surface chemistry of all the particles in the suspension; as a result, this mechanism can be used for virtually any suspension as long as the pH is conducive to the precipitation.

Hence, aluminum and iron can effectively destabilize particles by two different mechanisms: adsorption and charge neutralization at low metal concentrations and sweep flocculation at higher concentrations. The pH of the solution and the nature and concentration of the particles in suspension influence these phenomena, so what constitutes "low" and "high" concentrations of the metals varies with the suspension. The solid line in Figure 11-18 illustrates that the two mechanisms are clearly differentiable and separate in some cases; as shown in greater detail subsequently, this situation generally applies at relatively low particle concentration and relatively low pH. If the pH is held constant, restabilization can only occur with increasing dose when adsorption (without precipitation) is the dominant

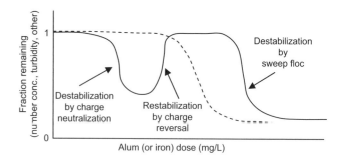

FIGURE 11-18. Destabilization by two different mechanisms by metals addition. (Solid line represents a given suspension at a pH, where the two mechanisms are separated; dotted line represents a more concentrated suspension at the same pH.)

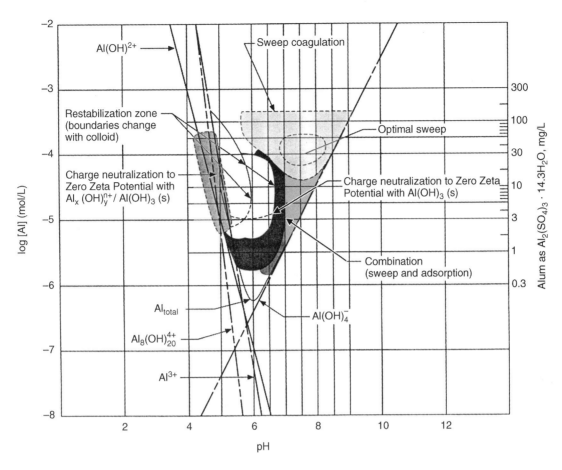

FIGURE 11-19. Alum destabilization diagram. *Source*: From Amirtharajah and Mills (1982), as redrawn in Edzwald et al. (1998).

mechanism of interaction between the metal ion and the particle surface. As noted above, restabilization occurs when the amount adsorbed is sufficient to reverse the original particle charge. Destabilization is accomplished at higher metal ion doses by enmeshment in the precipitate of the metal hydroxide.

If the particle concentration is relatively large, the charge neutralization and sweep flocculation regions can blend into one another, leading to results indicated by the dotted line in Figure 11-18. In this case, the destabilization can be accomplished by charge neutralization at doses below those required for sweep flocculation, but the dose where restabilization would begin is higher than that required for precipitation to be initiated. Hence, no restabilization occurs, and it is not possible to delineate where one mechanism of destabilization ends and the other begins.

Amirtharajah and coworkers summarized a large body of literature on the effectiveness of aluminum and iron salts as effective destabilizing chemicals and presented the results graphically. The original diagrams by Amirtharajah and Mills (1982) for aluminum and by Johnson and Amirtharajah (1983) for iron were subsequently updated, as shown in Figures 11-19 and 11-20, respectively. In each case, the authors reviewed many papers in which results of destabilization had been presented, and identified the effective zones of pH and coagulant dose for each mechanism on a log concentration versus pH solubility diagram for the hydroxides. These diagrams represent an excellent guideline for choosing a range of doses to investigate for a suspension whose characteristics are unknown.

Using the alum destabilization diagram in Figure 11-19, we can explore and compare the expected behavior of two suspensions, one at pH 6 and the other at pH 7.5, when they are dosed with alum. We assume that base is added simultaneously with the alum so that the pH is held constant. For the suspension at pH 6.0, we expect doses below $\sim$0.5 mg/L to have essentially no effect, doses in the range of 0.5–2 mg/L to be effective via the mechanism of adsorption and charge neutralization, doses in the range of 2–30 mg/L to be ineffective because the suspension has undergone charge reversal, and higher doses to cause precipitation of $Al(OH)_3$ (s) and destabilization by the sweep floc mechanism. As indicated on the diagram, the boundaries of these different regions vary with different suspensions, both

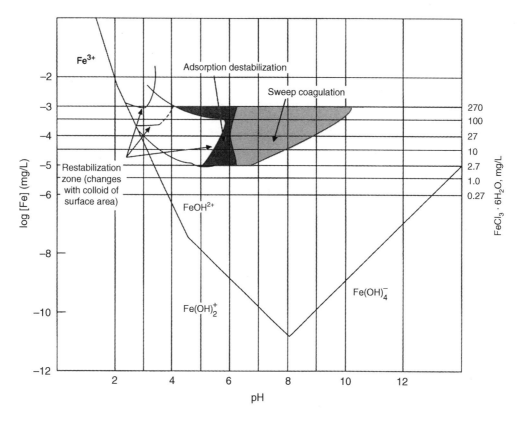

FIGURE 11-20. Iron destabilization diagram. *Source*: From Johnson and Amirtharajah (1983), as redrawn in Edzwald et al. (1998).

because of the particle concentration and because of the surface charge density.

For the suspension at pH 7.5, destabilization is effective at all doses higher than ∼3 mg/L. In the range 3–12 mg/L, destabilization is accomplished by a combination of charge neutralization and sweep flocculation; in this range, the destabilization might be initiated by charge neutralization, but the surface concentration of aluminum builds to the point where precipitation occurs. At higher doses, precipitation occurs rapidly and sweep flocculation is the mechanism of destabilization. Because alum is an acid, the pH would tend to decline as the alum dose is increased; hence, a real suspension that started at pH 7.5 might end up at pH 6.0 at higher doses (depending on the initial alkalinity), in which case the trend of destabilization with dose would be intermediate between those noted for the two constant pH suspensions.

Iron is less complex than aluminum in its behavior, primarily because it is less soluble. The distinction between charge neutralization and sweep floc is rarely clear at pH ≥ 6, because doses that are sufficient to cause charge neutralization are a few orders of magnitude higher than that needed to initiate precipitation. Hence, as indicated in Figure 11-20, the mechanism is thought to be primarily charge neutralization at lower pH values and enmeshment in a precipitate at higher pH values.

It is obvious by comparing Figures 11-19 and 11-20 that the region of insolubility of $Al(OH)_3$ (s) is much narrower than that for $Fe(OH)_3$ (s), which means that alum is effective as a coagulant in a smaller pH range than iron. For wastewater or drinking water applications in which the pH is higher than ∼8.0, iron is always the choice because of this difference in the solubilities of the metals.

Any other precipitate that is formed in solution as part of the overall treatment can also act as a coagulant. In the softening of drinking water, lime (CaO) is added to raise the pH and precipitate calcium carbonate ($CaCO_3$ (s)) and, at sufficiently high pH, magnesium hydroxide ($Mg(OH)_2$ (s)). Lime is also added to raise the pH to cause metal precipitation (e.g., $Cu(OH)_2$ (s), $Zn(OH)_2$ (s)) in many industrial wastewaters. Either lime or iron can be used to cause phosphate precipitation. All of these precipitates also act to capture other particles originally in the water, in the same manner as $Al(OH)_3$ (s) or $Fe(OH)_3$ (s).

An illustration of the effects of alum on particle destabilization is shown in Figure 11-21. In the experiments, aliquots of raw Mississippi River water (turbidity = 3.1 NTU) were adjusted to various pH values and either received no further chemical addition or were dosed with 50 mg/L of alum. All of the samples were subjected to identical jar test procedures (rapid mix for 60 s, slow mix for 30 min, and filtration

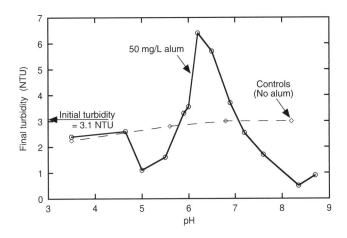

FIGURE 11-21. Particle destabilization of Mississippi river water with alum. *Source*: Adapted from Semmens and Field (1980).

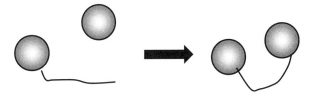

FIGURE 11-22. Schematic of interparticle bridging by a high-molecular weight polymer.

through a glass fiber filter). A small amount of turbidity removal occurred in the control samples (no alum addition), especially at the lower pH values where, presumably, the surface charge was lower. The alum samples show effective destabilization in two pH ranges, consistent with the general diagram shown in Figure 11-19. Destabilization was accomplished at pH values of 5.0 and 5.5 by charge neutralization and, at pH values above 7.0, by sweep flocculation.

The samples treated with alum in the range $5.8 < \text{pH} < 6.8$ all had final turbidity values considerably higher than the control samples. Apparently, in this pH range, $Al(OH)_3$ (s) was forming around the original particles, but the surface charge of the resulting particles or flocs was positive. The suspension, therefore, had a higher particle (mass) concentration than the original (or control) suspension, but the positive surface charge (due to the aluminum hydroxide speciation on the surface) caused the suspension to be stable. Such results are not unusual, especially with alum.

Adsorption and Interparticle Bridging

The final mechanism of destabilization is adsorption and interparticle bridging. The destabilizing chemicals used in this mechanism are usually the high-molecular weight (acrylamide-based) synthetic organic polymers described in the previous section. In this mechanism, it is thought that polymer molecules attach (adsorb) to the surfaces of two or more particles by some specific chemical interaction (not simply electrostatics) and form bridges between them. Figure 11-22 shows this idea conceptually. Large polymers are needed to accomplish destabilization by this mechanism, because the polymer bridge has to span two diffuse layers. Polymers are quite strong, however, and so once attached to two particles, the floc is unlikely to break apart. As a result, this mechanism of destabilization is particularly useful in situations in which there is substantial hydraulic shear or breakup is particularly undesirable, such as in dewatering applications and granular media filters.

Polymers with the same charge as the particle surface are sometimes effective in causing destabilization by this mechanism; that is, cationic, anionic, and nonionic polymers can each be used to bring about destabilization regardless of the charge on the surface. Thus, the attraction to the surface is not simply electrostatic. However, if the polymer used has the same charge as the surface (usually negative), then it is common that an inorganic polyvalent ion of the opposite charge (e.g., Ca^{+2}) has to be added simultaneously to make the polymer effective. This calcium might be acting as an indifferent electrolyte to compress the diffuse layer and allow the polymer to overcome the electrostatic repulsion and reach the particle surface, or it might form complexes with surface sites and/or ligand sites on anionic polymers, forming a particle-Ca-polyelectrolyte bridge that facilitates the attachment of the anionic polyelectrolyte to a negative surface.

Destabilization by bridging requires the same polymer molecule to attach to two different particles. As a result, the effectiveness depends somewhat independently on both the particle concentration and the polymer dose, and not just on their ratio as in the charge neutralization mechanism. When the polymer is introduced to the suspension and one end of a polymer molecule attaches to a particle, the other end sticks out into the surrounding fluid. In a low particle concentration suspension, the probability of that unattached end finding another particle is low, and so the molecule might bend around and attach to the same particle before finding another; in that case, no destabilization is achieved. This situation is depicted in Figure 11-23. In a high-concentration suspension, it is far more likely that the unattached end will find another particle before bending around the same particle. This particle concentration effect is another reason that

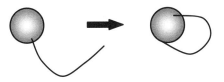

FIGURE 11-23. Ineffective destabilization at low particle concentration.

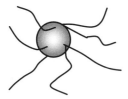

FIGURE 11-24. Overdosing of polymers for bridging mechanism.

this mechanism of destabilization is often used in sludge dewatering applications, where the particle concentration is very high. Similarly, it is rarely used in low concentration suspensions.

As with the charge neutralization mechanism, overdosing a suspension where destabilization by bridging is the goal can lead to restabilization, but for different reasons. If a high concentration of polymer is added to the suspension, one end of several polymer molecules can attach to each particle, taking up most of the adsorption sites; the unattached ends are then prevented from attaching to another particle, not only because most sites are already taken, but because the unattached ends are sterically hindered from making their way to the few untaken sites. A schematic diagram of a particle in an overdosed condition is shown in Figure 11-24; it is difficult to imagine the polymers of another, similarly overdosed, particle successfully finding their way to the surface of the one shown.

Hogg (1984) refined an earlier model of La Mer and coworkers (Smellie and La Mer, 1958; Healy and La Mer, 1962) to explain the effects of polymer dose. He assumed that the number of possible adsorption sites on a particle is proportional to the particle surface area, and that a polymer attached to a particle at one end was free to rotate so that the other end would tend to find an unoccupied site. He proposed that the effectiveness of a polymer dose was related to the probability that a polymer attached (at one end) to one particle would find an unoccupied site (for its other end) on another particle. (Note that, for this mechanism, effective destabilization involves two particles at once, and hence it is not possible to completely separate destabilization from the processes that lead to floc formation.) He found this probability by considering the inverse problem: that particles would *not* adhere to one another. This (inverse) situation would occur only if the sites on the two particles were both either completely filled by polymer, or completely devoid of polymer. If f_{occ} is the fractional coverage of the particles and m_i and m_j are the number of possible adsorption sites on particles i and j, respectively, he found $\alpha_{ij} = 1 - f^{m_i + m_j} - (1 - f_{occ})^{m_i + m_j}$.

For a suspension with particles that are all identical (same size and same number of possible adsorption sites), α_{ij} is close to 1.0 over a broad range of values of f_{occ}, meaning that destabilization is effective over a wide range of doses.

For two particles of different sizes, this model suggests that the range of doses (i.e., fractional surface coverage) that achieves effective destabilization broadens still further as the size ratio of the two particles (smaller to larger) involved in the interaction decreased. Since the desire in engineered systems is to find the *minimum* effective dose (surface coverage), overdosing to the point of restabilization is much less likely than in the case of charge neutralization. Unlike with charge reversal (where the surface charge can be measured), no independent measure is available to determine directly whether an ineffective dose represents an under-dose or overdose. Hence, empirical evidence directly related to the application (flocculation, filtration, or dewatering) is the only tool available to determine the optimal (i.e., minimum effective) dose in this mechanism.

The use of a high-molecular weight polymer does not necessarily mean that destabilization is accomplished by the bridging mechanism. Polymer charge often plays a significant role in destabilization, and the destabilization might well be accomplished by charge neutralization even for large polymers. What determines the mechanism is how the polymer is adsorbed to the surface. If it attaches at one end or forms large loops so that part of the molecule sticks out into the water and is available to attach to another particle, then the bridging mechanism is possible. But, it is also possible that the molecule will attach closely to the particle throughout the chain length. Walker and Grant (1996) have made the analogy that this latter phenomenon is like a zipper; after the first part of the molecule attaches to the surface, molecules along the chain keep attaching one by one, so that it closes onto the surface like a zipper.

11.5 INTERACTIONS OF DESTABILIZING CHEMICALS WITH SOLUBLE MATERIALS

It is now commonly accepted that the major source of surface charge of particles in most natural waters is adsorbed organic matter (O'Melia, 1989; Buffle et al., 1998). When aluminum is added to water containing both soluble NOM and particles with adsorbed NOM, the interactions between the aluminum species and NOM dictate the success or failure of destabilization. The aluminum destabilizes the particles by either charge neutralization or enmeshment in a precipitate of $Al(OH)_3$ (s); the NOM can also be removed to a significant extent either as an Al-NOM precipitate or by adsorption onto $Al(OH)_3$ (s). Generally, the ranges of pH and alum dose that achieve effective NOM removal overlap with the regions of effective particle removal. This overlap occurs because the charge on the particle is usually caused by the NOM, so the aluminum is interacting with the NOM in solution and on the particle in much the same way. However, the lower-pH, lower-alum-dose region where particle removal occurs by charge neutralization is generally where the aluminum-NOM precipitation occurs. At the higher-pH, higher-alum-dose region where sweep flocculation causes the particle removal, NOM removal occurs primarily by adsorption onto the high surface area of the precipitated amorphous $Al(OH)_3$ (s). Current US drinking water regulations require most utilities using surface water supplies to remove some NOM, and an acceptable method is by interaction with aluminum (or iron). To accomplish the desired NOM removal, the dose of alum is often increased beyond that required for effective particle destabilization; this process of achieving organics removal in addition to particle removal by the addition of higher doses of destabilizing chemical is known as *enhanced coagulation*. It can be accomplished by any of the destabilizing chemicals that lead to precipitation, with iron often being more effective than alum.

A dramatic example of the role of NOM in determining the effective dose for both particulate destabilization and NOM precipitation/adsorption was given by Edzwald (1993) and is reproduced in Figure 11-25. The Missouri river water had high particle concentration (turbidity of 670 NTU) but low dissolved organic carbon (DOC = 3 mg/L), whereas the Myrtle Beach water had a much lower particle concentration (turbidity = 30 NTU) and high DOC (20 mg/L). In jar tests, the dose of alum needed for effective destabilization required satisfying (titrating) the charge of the organic matter for both waters, which meant a low dose for Missouri river (despite the high particle concentration) and a high dose for Myrtle Beach. The removal of the DOC was in the solid form, either by adsorption onto precipitates of $Al(OH)_3$ (s) or by precipitation of an Al-NOM solid, although it is not clear from these results which of these phenomena occurred.

NOM also is removed by other precipitates that are formed in drinking water treatment. The interaction of NOM with iron is quite similar to that with aluminum, but is generally slightly more favorable; that is, iron is somewhat more effective for NOM removal in most waters. As the removal of NOM in drinking water treatment has become of greater importance, many water plants have changed from aluminum to iron as the primary coagulant to take advantage of this difference. NOM is also removed in softening, by both $CaCO_3$ (s) and $Mg(OH)_2$ (s) (Randkte et al., 1982; Liao and Randkte, 1986; Thompson et al., 1997; Roalson et al., 2003).

When aluminum or iron species are added to municipal wastewater during tertiary treatment, the metal is envisioned to form separate or mixed precipitates of metal phosphate and metal hydroxide. These precipitates help destabilize residual particles and adsorb some remaining organics. Hence, a tertiary treatment that includes aluminum or iron addition acts as a polishing step in enhancing the solids and organics removal in addition to accomplishing the main objective of phosphate precipitation.

As noted in Chapter 7, the metal precipitates that are formed through the additions of inorganics for particle destabilization are also effective adsorbents for other contaminants besides NOM. Arsenic is removed by aluminum or ferric hydroxide, heavy metals are adsorbed or coprecipitated, and silica can be removed in softening as a magnesium silicate. Although the primary reason to add these coagulants is typically to destabilize particles, the conditions (e.g., pH) and concentrations, and even the choice of coagulant, can be dictated by the desire to accomplish the removal of these other soluble contaminants.

Combinations of Additives

It is often helpful, or even necessary, to add more than one chemical to accomplish particle destabilization. The most obvious example is when pH needs to be controlled for effective destabilization, as is frequently the case for aluminum or iron additions. For NOM removal, a relatively low pH is often desirable, and so acid is sometimes added if the pH is not low enough after addition of alum or iron alone. On the other hand, the acidity of these coagulants sometimes must be offset by adding a base.

Since the surface charge of virtually all particles encountered in environmental engineering applications

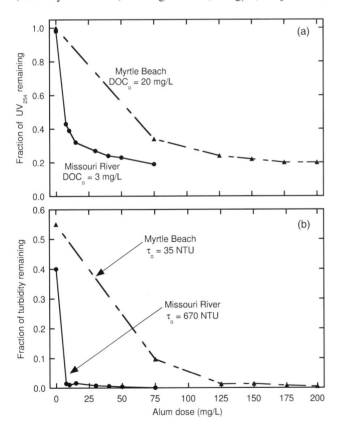

FIGURE 11-25. Effects of alum on removal of (a) organics and (b) turbidity. (All results at pH 7; UV_{254} taken as a measure of NOM content.) *Source*: Adapted from Edzwald (1993).

is a function of pH, the pH can be adjusted to accomplish some charge neutralization. It might well be cheaper to add an inorganic acid or base to reduce the surface charge somewhat before adding a charge neutralization polymer to complete the neutralization. Only rarely could the inorganic acid or base be used for the entire neutralization, because the pH$_{zpc}$ is usually outside the range that would be considered acceptable for the next treatment process or ultimate use, including discharge to the environment for wastewaters.

When the bridging mechanism is used and the charge of the polymer is the same as that of the particle, the same phenomena that tend to prevent particle–particle collisions will act to prevent polymer–particle collisions. To overcome this problem, the diffuse layer must be quite compact; hence, multivalent ions of opposite charge must be either present or added to accomplish destabilization. For example, if the particle charge is negative and an anionic polymer is used, calcium ions can be added (as, say, CaCl$_2$); in this case, it is common to add the inorganic species prior to the polymer.

Finally, bridging polymers are often added not so much for particle destabilization, but rather to enhance floc strength. In this situation, other chemicals (inorganic or organic) might be added first to accomplish the destabilization, after which the bridging polymers can be added to minimize floc breakup (in flocculation) or particle detachment (in deep bed filtration). Such polymer addition can also assist in dewatering processes.

11.6 MIXING OF CHEMICALS INTO THE WATER STREAM

To accomplish destabilization, the chemicals added must be effectively contacted with the particles in suspension. The two most common means of accomplishing that mixing are in a small CFSTR or within a pipe; in both cases, efficient mixing of the chemical with the suspension requires turbulent flow.

In turbulent flow, recognizable spatial flow patterns that persist for at least a short time are called eddies; within such eddies, a high degree of correlation exists between velocity at one point and that at another. Eddies vary in size from a maximum approaching the dimensions of the reactor to a minimum dependent on the ability of the fluid to transfer and dissipate energy. In turbulence, eddies with smaller and smaller length scales are formed as a result of the energy cascade; that is, energy is transferred from one packet of fluid to smaller ones until small enough eddies are formed that the energy is dissipated through viscous forces (Tritton, 1988). Kolmogorov's Universal Equilibrium Theory relates the net rate at which energy is transferred from large to small eddies to the overall energy dissipation rate. The key results of the theory (at least for our purposes) are two quantities,

known as the Kolmogorov microscales of length and time, computed as follows (Tennekes and Lumley, 1972):

$$\text{Length:} \quad \eta = \left(\frac{\nu^3}{\varepsilon}\right)^{1/4} \quad (11\text{-}14)$$

$$\text{Time:} \quad \tau_{\text{Kolmogorov}} = \left(\frac{\nu}{\varepsilon}\right)^{1/2} \quad (11\text{-}15)$$

where ν is the kinematic viscosity (cm^2/s; the ratio of the absolute viscosity, μ, to the fluid density, ρ_L) and ε is the energy dissipation per unit time per unit mass (cm^2/s^3). These scales represent the smallest scales of turbulent motion. Fluid motion in size ranges below that of the Kolmogorov length microscale is in some sense laminar since the viscous forces dominate the inertial forces.

In the environmental engineering literature, mixing intensity is usually described by the root mean square velocity gradient, $\overline{G}$, with dimensions of inverse time; a velocity gradient is the derivative of velocity over distance, which yields these units. For turbulent flows, this quantity is taken to be the inverse of the Kolmogorov timescale; that is,

$$\overline{G} = \frac{1}{\tau_{\text{Kolmogorov}}} = \left(\frac{\varepsilon}{\nu}\right)^{1/2} = \left(\frac{\varepsilon \rho_L}{\mu}\right)^{1/2} \quad (11\text{-}16)$$

The energy dissipation rate per unit mass, ε, is essentially a microscopic quantity that could vary over the space of a reactor and is difficult to measure without sophisticated instrumentation. Hence, it is common to estimate the average value of $\overline{G}$ in a reactor from macroscopic quantities, based on the assumption that the energy dissipated per unit time equals the power (i.e., energy/time) supplied:

$$\overline{G} = \left(\frac{\varepsilon \rho_L}{\mu}\right)^{1/2} = \left(\frac{P}{\mu V}\right)^{1/2} \quad (11\text{-}17)$$

where P is the power input (g cm^2/s^3) and V is the liquid volume in the reactor. Typical rapid mix units are designed as CFSTRs with $\overline{G}$ values in the range of 700–1000 s^{-1} (Amirtharajah and Tambo, 1991), so that the smallest characteristic time for mixing (i.e., $\tau_{\text{Kolmogorov}}$, the time for mixing within the smallest eddies) is on the order of 0.001 s.

Three other timescales are also of interest in considering the mixing requirements for adequate destabilization: the hydraulic detention time of the rapid mix unit (τ_{hyd}),[9] the time for mixing of the large-scale eddies from which the small eddies are formed (τ_{turb}), and the time for reaction of

---

[9] τ_{hyd} is V/Q; that is, it is the same quantity referred to throughout the book simply as τ. The subscript has been added here for consistency with the nomenclature being used for the other characteristic times of interest for mixing.

the destabilizing chemical with the particle surfaces (τ_{chem}). Of these, τ_{hyd} is a design parameter that could reasonably be set at values from <1 min up to many minutes. τ_{turb} is also controlled by design decisions, with an upper limit that might range from many minutes to as much as a few hours, if no mechanical mixing is provided, and a lower limit (less than, and perhaps much less than, a second) set by the practical constraints of how much mixing energy can be input to the water at acceptable cost; ways to quantify τ_{turb} are presented below. Finally, τ_{chem} reflects the inherent chemical kinetics of the chemical reactions that the added chemicals undergo (hydrolysis, precipitation, and/or adsorption) and, in contrast to τ_{hyd} and τ_{turb}, is beyond the control of the designers. The ultimate degree and uniformity of destabilization depend in a complex way on all of these timescales. However, it seems logical that good design would lead to τ_{hyd} being substantially greater than either τ_{chem} or τ_{turb}; in that case, the reagent would have time to mix throughout the reactor and to undergo whatever reactions it participates in before a significant fraction of either the reagent or the particles exited the reactor.

To ensure the most efficient use of the chemical, it would also seem desirable for τ_{turb} to be substantially smaller than τ_{chem}; in that case, the chemicals would be mixed into the smallest eddies uniformly before the reactions were completed. How essential this criterion is, however, depends on the nature of the coagulation reaction(s). For example, if large amounts of Al(OH)$_3$(s) solids precipitate in a small fraction of the flow (i.e., if $\tau_{chem} \ll \tau_{turb}$), and these solids are subsequently well distributed throughout the reactor, they are still likely to form an effective sweep floc (and would also effectively adsorb soluble components). In this case, it is likely that more aluminum salts would be required than if $\tau_{chem} > \tau_{turb}$, but at least a satisfactory result would be achieved.

On the other hand, polymeric coagulants are so surface active that, if they interact with particles in a small fraction of the flow before they are well distributed, they might overdose those particles and be ineffective as a destabilizing agent. Also, they are unlikely to desorb after the overdosed particles are mixed throughout the rest of the suspension, so that other particles would be underdosed. In this case, it would be important to distribute the chemicals as rapidly as possible, ideally such that $\tau_{turb} < \tau_{chem}$.

When the destabilizing chemical is first added to the water, it is likely to enter some or all of the large-scale eddies, which then mix with one another. If the reactor is a CFSTR (generally referred to in this context as the rapid mix tank) and if the chemical is added directly to the CFSTR (rather than into the influent pipe), most or all of the chemical is added to the single, large-scale eddy that passes the point of addition. This situation requires the greatest subsequent mixing of eddies, because the chemical starts in only a small fraction of the flow. If, on the other hand, the chemical is added to the influent pipe (and especially if it is added in a distributed manner throughout the cross-section of the flow), much of the distribution of the chemical throughout the flow is accomplished immediately; hence, this method of addition is considered preferable.

Most CFSTRs used as rapid mix tanks for destabilization are sized to provide a theoretical detention time (τ_{hyd}) in the range of 1–5 min, so it is useful to compare the mixing time of the large eddies to this value. In a CFSTR, a large mechanical mixer is generally used to create turbulent conditions, but in some cases, turbulence is induced hydraulically with over-and-under baffles. An impeller (such as a blade on a rotor) pumps a certain flow rate of water ($Q_{impeller}$) that depends on its rotational speed and geometry; the simplest way to estimate this flow is by the volume of water cut through by the impeller per unit time. A characteristic turnover time can then be calculated as the quotient of the reactor volume (V_{CFSTR}) and this impeller pumping rate; that is,

$$\tau_{turnover} = \frac{V_{CFSTR}}{Q_{impeller}} \quad (11\text{-}18)$$

$\tau_{turnover}$ provides an estimate of the characteristic time for large-scale mixing in the reactor; that is, it is an estimate of what was identified above as τ_{turb}. For full-scale rapid mix reactors built as CFSTRs, this characteristic time is in the range from 1 s to perhaps a minute (Oldshue and Trussell, 1991; David and Clark, 1991; Casson and Lawler, 1990). Thus, it appears that the criterion that τ_{hyd} be substantially greater than τ_{turb} would be met in most rapid mix tanks, with the exception of those with a combination of the shortest hydraulic residence time and the least intensive mixing (i.e., the smallest inputs of impeller energy). Clark et al. (1993) showed that, for a pilot scale (100 L) rapid mix reactor, $\tau_{turnover}$ was ~300 times $\tau_{Kolmogorov}$.

On the other hand, some evidence exists that $\tau_{Kolmogorov}$ is more important than $\tau_{turnover}$ in ensuring adequate mixing, especially if the destabilizing chemical enters the main flow stream in the pipe or channel just before the rapid mix tank rather than at a separate point within that tank (Amirtharajah and Trusler, 1986). That is, to ensure that every particle experiences nearly the same dose of destabilizing chemical, it is necessary to have a small value of $\tau_{Kolmogorov}$ since it is within the microscale that the ultimate interaction of the particles and chemical occurs.

The timescales of mixing, both within the microscale of turbulence and throughout the reactor, also need to be compared to the timescale for reaction of the destabilizing chemicals. For the addition of organic polymers for charge neutralization, the adsorption reaction itself is considered to be essentially instantaneous, and so the concern is only to transport the polymer into the same microscale eddy as the particle to be destabilized. When inorganic destabilizing agents (aluminum, iron, and others) are added, however, an additional concern is the reaction rates for hydrolysis and, in the case of sweep floc, precipitation. The stock solutions from which these chemicals are added into water are at low pH, and the speciation within those stock solutions depends on the concentration. When these solutions are diluted into the main body of water, the speciation is changed because both the pH and concentration are changed. The relative timescales of these reactions and of the mixing therefore become important in determining the resulting destabilization and the efficient use of the destabilizing chemicals.

The characteristic reaction times for the kinetics of hydrolysis of aluminum and iron and the formation of the hydroxide precipitates ($Al(OH)_3$ (s) and $Fe(OH)_3$ (s)) are not completely known. Dempsey et al. (1984) gave a synopsis of work on aluminum to that time, and noted the general agreement that hydrolysis among the monomeric species (Al^{+3}, $AlOH^{+2}$, $Al(OH)_2^+$, $Al(OH)_3$, and $Al(OH)_4^-$) was extremely fast, with characteristic times $<10^{-4}$ s. These reaction times are faster than any mixing times, and therefore these reactions are not of particular concern in this context; that is, they will achieve equilibrium. On the other hand, the formation of the dimer, $Al_2(OH)_2^{+4}$, is considerably slower, estimated to have a characteristic time of ~1 s at a total Al concentration of 10^{-3} M. According to Dempsey et al. (1984), "poor mixing (leading to concentration heterogeneities) . . . would result in more rapid dimerization and therefore faster polymerization and precipitation." Subsequently, Clark et al. (1993) presented experimental evidence that poor mixing led to increased formation of polymeric species. Klute and Amirtharajah (1991) indicated that effective charge neutralization using aluminum depended on the turbulence and on the concentration of the stock solution, with lower doses of total aluminum required when less concentrated stock solutions were used. On the other hand, precipitation reactions were complete in several seconds and were less sensitive to mixing and stock solution concentration.

Although CFSTRs are the most commonly used reactor designs for mixing destabilizing chemicals into water, that choice has some significant disadvantages. We know from the exit age distribution ($E(t)$ or $F(t)$) of a CFSTR that a substantial fraction of the fluid stays in the reactor for times much less than the theoretical detention time, while other parts of the fluid stay in the reactor for lengthy periods. Flocs are formed in this process, and the mixing can lead to breakup of flocs as well; hence, the detention time for which the particles are subjected to the intense mixing reflects a balance of the time needed for adequate mixing of the chemicals throughout the suspension and the time to avoid substantial breakup of flocs. These and other issues involved in mixing for particle destabilization in drinking water systems were reviewed by Amirtharajah et al. (1991).

Klute and Amirtharajah (1991) argued that turbulent pipe flow provided the best conditions for destabilization, as long as the destabilizing chemical was introduced at many points in a cross-section of the pipe. Turbulent pipe flow, which is essentially plug flow, avoids the problems of different packets of fluid being in the reactor for different times. The multipoint injection also reduces the disparity between $\tau_{turnover}$ and $\tau_{Kolmogorov}$. Klute and Amirtharajah presented evidence from Doell (1986) that the time for complete charge neutralization (measured by electrophoretic mobility) of silica particles by cationic polymers was inversely proportional to $\overline{G}$ or, in other words, directly proportional to $\tau_{Kolmogorov}$. They also interpreted the results in terms of the characteristic Kolmogorov length scale, and concluded that the limiting (best) case for charge neutralization by organic polymers would occur when $\eta_{Kolmogorov}/d_p = 1$, where d_p is the diameter of the particle. Lang (2002) noted that mixing of chemicals into water in a turbulent pipe flow section rather than a CFSTR would save operating costs, both because of reduced energy use and reduced chemical use.

While much research remains to be done on the interactions of chemicals, particles, and mixing, the emerging picture at this point is that a high degree of turbulence leads to more efficient chemical use for destabilization by adsorption and charge neutralization, whereas destabilization by precipitation is less sensitive to the mixing conditions.

11.7 PARTICLE SIZE DISTRIBUTIONS

Suspensions encountered in natural waters and in water and wastewater treatment plants have particles of many different sizes. Although precipitation under tightly controlled situations can lead to suspensions with almost uniform particles, virtually all suspensions of interest have broad and essentially continuous particle size distributions; that is, particles of every size between some lower and upper size limit exist. The lower limit is defined either by the limits of detection of the measurement device or by some arbitrary distinction on the basis of size between particulate and soluble matter. The upper limit is often determined by physical processes to which the suspension has been exposed; in any case, each sample of water has a maximum size particle.

Particle size (the abscissa in a graphical presentation of the size distribution) can be described and measured by length (diameter or radius), area (cross-sectional or surface), or volume. The amount of particulate matter in the water (the ordinate) can be described in terms of number, area, or volume. In addition, size distributions (like any distribution) can be expressed on a cumulative or differential basis, and can be normalized (or not). Finally, distributions can be expressed on a relative or absolute basis. As a result, a myriad of possibilities for expressing size distributions can

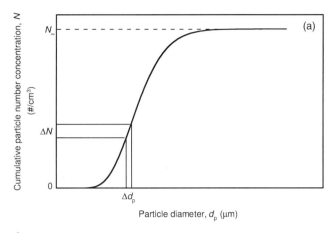
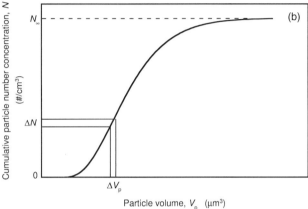

FIGURE 11-26. Cumulative particle size distributions on a number basis as a function of (a) particle diameter and (b) particle volume. (Values on ordinate represent the cumulative number concentration of particles with sizes less than or equal to the size on the abscissa.)

be used; fortunately, only a few are in common use. In this section, we start with generic distributions that help lead us to the commonly used ones.

Consider the cumulative number distributions shown in Figure 11-26. In Figure 11-26a, the size is expressed as particle diameter, whereas in Figure 11-26b, particle volume is chosen. In both parts, the ordinate expresses the cumulative number concentration of particles in the suspension from the lower size limit to the size under consideration. Note that the curves start at some finite size, as the concept of zero particle size has little or no meaning. If this distribution were measured by an instrument that responded to particle volume, points on the line shown in Figure 11-26b would be obtained as the raw data, and the values for diameter to create Figure 11-26a could be obtained by assuming the particles were spheres and using the relationship $d_p = (6V_p/\pi)^{1/3}$, where V_p and d_p are the particle volume and diameter, respectively. The particle diameter computed from this relationship is called the *equivalent spherical diameter*.

The cumulative distributions in Figure 11-26 are conceptually valuable but not particularly useful, for a few reasons. First, they tend to hide important aspects of the data that are more easily seen in differential distributions. Second, the ordinate is highly dependent on the lower limit of detection; two instruments with different lower limits of detection would produce two very different graphs for the same suspension when plotted this way, even if they performed identically in the size region measured by both. Using differential distributions overcomes these problems. Third, the use of an arithmetic scale for particle size (either diameter or volume) tends to compress the data for small sizes and spread out the data for large sizes, relative to their importance. For example, the differences in behavior and in concentration among 0.5, 1.0, and 2.0 μm (diameter) particles are far greater than the differences among 100.5, 101.0, and 102.0 μm particles, but the arithmetic scale suggests they have the same importance. Using a logarithmic scale for the abscissa overcomes this problem; on such a scale particles of 50, 100, and 200 μm are spread equivalently to the 0.5, 1.0, and 2.0 μm particles.

The slope at any point on Figure 11-26a or 11-26b, designated here as $n(d_p)$ and $n(V_p)$, respectively, is given by

$$n(d_p) = \frac{\Delta N}{\Delta d_p} = \frac{dN}{d(d_p)} \qquad (11\text{-}19)$$

$$n(V_p) = \frac{\Delta N}{\Delta V_p} = \frac{dN}{d(V_p)} \qquad (11\text{-}20)$$

where N is the cumulative number concentration at any size. Number concentrations are usually expressed as the number per cubic centimeter, and particle diameters are expressed in micrometers, so units associated with Equations 11-19 and 11-20 are $1/(\text{cm}^3\ \mu\text{m})$ and $1/(\text{cm}^3\ \mu\text{m}^3)$, respectively. A similar function can be defined on the basis of particle area if that is taken as the measure of size. Again, assuming the particles are spherical allows conversion among the different measures of size. Commonly, however, size is expressed as diameter (even if measurements are made on the basis of area or volume), and so the *particle size distribution function on the basis of diameter*, $n(d_p)$, is the common distribution encountered. We use this particle size distribution function frequently throughout the remainder of the book as the primary means of expressing the size distribution.

Values of $n(d_p)$ vary over many orders of magnitude for most distributions, and plots are usually made with a logarithmic scale for the ordinate. A typical plot for the particle size distribution function is shown in Figure 11-27; such a distribution might be typical for a flocculated suspension (i.e., an influent to a sedimentation tank, for either water or wastewater). The tangent lines shown on the figure are explained subsequently. Note that values below 0 are possible on such a figure; such a condition means that the slope

548 PARTICLE TREATMENT PROCESSES: COMMON ELEMENTS

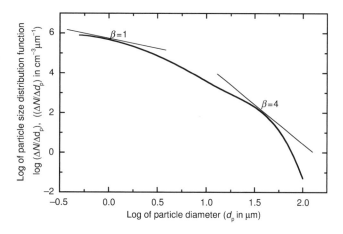

FIGURE 11-27. Typical particle size distribution function in water and wastewater.

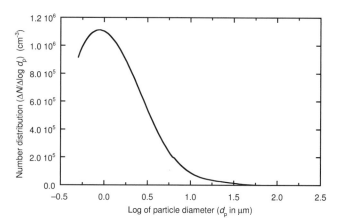

FIGURE 11-28. Typical number distribution. (Using same data as in Figure 11-27).

of the curve in Figure 11-26a is <1 (with associated units), so the logarithm of the slope is <0.

A second useful way to modify and present the information shown in Figure 11-26a is to divide ΔN for some small increment by the corresponding logarithmic increment in particle diameter, $\Delta (\log d_p)$, rather than the arithmetic size increment, Δd_p. The resulting, so-called *number distribution*, $\Delta N / \Delta \log d_p$, or, in the limit, $dN/d \log d_p$, can then be plotted as a function of the particle size as shown in Figure 11-28. The particular usefulness of this plot is that the area under the curve between any two values on the abscissa is the total number concentration (number/cm^3) of particles in that size range. This result follows directly from the definition of this distribution; that is,

$$\int_{\log d_{p1}}^{\log d_{p2}} \frac{dN}{d \log d_p} d \log d_p = \Delta N_{1-2} \quad (11\text{-}21)$$

Note that, if the entire measured size range is used in this integration, one obtains the total (measured) particle concentration, or N_∞. Besides this integration, a number distribution plot such as that shown in Figure 11-28 gives an immediate visual sense of the size range that contains most of the particles and the spread of the distribution.

For many applications of interest in environmental engineering, the number distribution might not be as important as the surface area or volume distribution. For example, adsorption is a surface phenomenon, so that one might prefer to know the total surface area concentration and the size range of particles that contains the most surface area for understanding adsorption onto a suspension. Similarly, since the suspended solids concentration is of paramount importance in sedimentation in wastewater plants, one might want to know how the volume concentration (which, when multiplied by particle density, yields the suspended solids

concentration) varies as a function of particle size. Assuming spherical particles, the number distribution value for each size increment i can be converted to the surface area and volume distributions by multiplying by the surface area and volume of a sphere, respectively[10]:

Surface area distribution: $\dfrac{\Delta S}{\Delta \log d_p} = \dfrac{\Delta N}{\Delta \log d_p} \pi d_p^2$ (11-22)

Volume distribution: $\dfrac{\Delta V}{\Delta \log d_p} = \dfrac{\Delta N}{\Delta \log d_p} \pi \dfrac{d_p^3}{6}$ (11-23)

Using these equations, the number distribution shown in Figure 11-28 can be converted to the surface and volume distributions shown in Figure 11-29. The form of these distributions means that, like the number distribution, the area under the curves between any two values of log diameter is the total surface area concentration or volume concentration of the suspension between those two values of the diameter. Note that the weighting by the surface area and volume of each particle size means that these distributions are shifted to larger sizes, relative to the number distribution. Virtually all of the particle volume concentration (Figure 11-29b) is located in the larger size range

---

[10] The concept of an area or volume distribution can be confusing, at first. A useful analogy is to consider a bunch of coins, say 7 pennies, 3 nickels, and 2 dimes. The number distribution has values of 7, 3, and 2 associated with the "sizes" 1, 5, and 10, respectively. The "value" distribution (analogous to the volume distribution) has values of 7, 15, and 20 associated with the sizes of 1, 5, and 10, respectively. In this example, the number distribution decreases with increasing size, but the value distribution increases with size. The surface and volume distributions are obtained in the same way as this example—multiplying the number concentration of each individual size by the surface area or volume associated with that size.

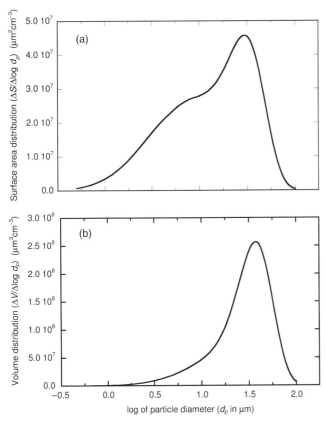

FIGURE 11-29. Calculated (a) surface area and (b) volume distributions. (Using same data as in Figure 11-27.)

($\log d_p > 0.8$, or $d_p > 6.3\,\mu$m), whereas most of the particle number concentration (Figure 11-28) is located in the range below this size. These differences can be important for different particle treatment processes. For example, as shown in subsequent chapters, flocculation can drastically reduce the concentration of small particles while having little effect on that of the large particles, whereas sedimentation (with no simultaneous flocculation) can drastically reduce the large particle concentration with little or no effect on the small particles. All such effects are more apparent on the (logarithmic) particle size distribution function graphs than on the (arithmetic) number and volume distribution graphs.

■ **EXAMPLE 11-2.** A particle counter analyzes particles in equal logarithmic increments of diameter, with $\Delta \log d_p = 0.1$, and d_p measured in micrometers. A 2-mL sample is measured in this particle counter, and in the size increment ("bin") that is centered at $\log d_p = 0.5$, the count is 1120 particles. Before measurement, this sample was created by diluting the original suspension 1:10 with particle-free dilution water. Find the values of $\log \Delta N/\Delta d_p$, $\Delta N/\Delta \log d_p$, and $\Delta V/\Delta \log d_p$ for this value of the diameter. Assume the particles are spherical.

Solution. We first find ΔN, the number concentration in the original sample in the measured size increment, as follows:

$$\Delta N = \left(\frac{\text{Number of particles counted}}{\text{Volume of sample measured}}\right)$$
$$\times \left(\frac{\text{Volume of measured sample}}{\text{Volume of original suspension}}\right)$$
$$= \frac{1120}{2\,\text{cm}^3}(10) = 5600/\text{cm}^3$$

For $\log d_p = 0.5$, $d_p = 10^{0.5} = 3.16\,\mu$m. The volume concentration is found as the product of the number concentration and the volume of this average particle[11]:

$$\Delta V = \Delta N\left(\pi\frac{d_p^3}{6}\right) = (5600\,\text{cm}^{-3})\left(\pi\frac{(3.16\,\mu\text{m})^3}{6}\right)$$
$$= 9.25 \times 10^4\,\mu\text{m}^3/\text{cm}^3$$

Because $\Delta \log d_p$ is given as 0.1, the values for the number and volume distributions for this size increment are

$$\frac{\Delta N}{\Delta \log d_p} = \frac{5600/\text{cm}^3}{0.1} = 5.60 \times 10^4\,\text{cm}^{-3}$$

$$\frac{\Delta V}{\Delta \log d_p} = \frac{9.25 \times 10^4\,\mu\text{m}^3/\text{cm}^3}{0.1} = 9.25 \times 10^5\,\mu\text{m}^3/\text{cm}^3$$

Next, we find the upper and lower limits of the size increment, and from that, the size increment itself. On a logarithmic basis, the limits are $\log d_p = 0.5 \pm 0.05$. The size increment therefore is

$$\Delta d_p = 10^{0.55} - 10^{0.45} = 3.548\,\mu\text{m} - 2.818\,\mu\text{m} = 0.730\,\mu\text{m}$$

Lastly, the value for the particle size distribution function for this size is

$$\log\frac{\Delta N}{\Delta d_p} = \log\frac{5600}{0.730} = 3.88 \qquad ■$$

Finally, it is useful to consider the relationships between the particle size distribution function on a number basis (Equation 11-19, and Figure 11-27) and the number, surface, and volume distributions. Recalling the mathematical identity that $d(\ln d_p) = d(d_p)/d_p$ and transforming to base 10 logarithms $\left(d(\log d_p) = d(d_p)/2.3 d_p\right)$ allows one

---

[11] In discretizing a continuous distribution, all particles in each increment are considered to be the same size, and so the "average" size is the only size in that increment. In reality, each increment contains a distribution of particle sizes and, if a greater amount of information were known about that distribution, a more accurate average size could be calculated.

to relate the number distribution and particle size distribution function as follows:

$$\frac{dN}{d\log d_p} = \frac{dN}{d\log d_p}\frac{d(d_p)}{d(d_p)} = \frac{d(d_p)}{d\log d_p}\frac{dN}{d(d_p)} = 2.3 d_p n(d_p) \quad (11\text{-}24)$$

Writing Equations 11-22 and 11-23 in differential form and substituting Equation 11-24 into them yields the following:

$$\frac{dS}{d\log d_p} = \frac{dN}{d\log d_p}\pi d_p^2 = 2.3\pi d_p^3 n(d_p) \quad (11\text{-}25)$$

$$\frac{dV}{d\log d_p} = \frac{dN}{d\log d_p}\pi \frac{d_p^3}{6} = \frac{2.3\pi}{6} d_p^4 n(d_p) \quad (11\text{-}26)$$

Consider a tangent to the line describing the logarithm of the particle size distribution function, or equivalently, consider a portion of the particle size distribution function that can be reasonably described by a straight line. In Figure 11-27, two such tangents are drawn, and the size region $0.5 < \log d_p < 1.2$ is nearly straight. Any of these lines can be described by an equation of the form:

$$\log(n(d_p)) = \log A - \beta \log(d_p) \quad (11\text{-}27)$$

where $\log A$ is the intercept (value of $\log(n(d_p))$ where $\log(d_p) = 0$) and β is the negative of the slope (written this way by convention since most such graphs have negative slopes and it is convenient to have positive values of β). After taking the antilogs, Equation 11-27 is transformed to

$$n(d_p) = \frac{dN}{d(d_p)} = A d_p^{-\beta} \quad (11\text{-}28)$$

If Equation 11-28 is then substituted into Equation 11-24, the result is as follows:

$$\frac{dN}{d\log d_p} = 2.3 A d_p^{1-\beta} \quad (11\text{-}29)$$

In this equation, if β has the value of one, the number distribution is independent of particle size; that is, the number distribution would be a horizontal line on a plot such as that shown in Figure 11-28. In that case, each equal logarithmic size increment ($\Delta \log d_p$) contains the same number concentration of particles. If β is <1, the exponent of d_p is positive, so that the number distribution increases with increasing particle diameter, and the number concentration of particles in equal logarithmic size increments increases as d_p increases. Similarly, if β is >1, the exponent of d_p is negative, and the number distribution decreases with increasing diameter. The implication is that the peak (or, at least, a local peak) of the number distribution occurs at the (log) diameter where the slope of the particle size distribution function is -1. This idea can be seen by comparing Figures 11-27 and 11-28; the log diameter value associated with the peak of the number distribution on Figure 11-28 is the same value where the tangent to the particle size distribution function is -1.

If Equation 11-28 is substituted into Equation 11-26 for the volume distribution, the result is

$$\frac{dV}{d\log d_p} = \frac{2.3\pi}{6} A d_p^{4-\beta} \quad (11\text{-}30)$$

In this equation, when $\beta = 4$, the volume distribution is independent of particle size (i.e., flat); when $\beta < 4$, the volume distribution is increasing; and when $\beta > 4$, the volume distribution is decreasing. Hence, the peak of the volume distribution occurs when the slope on the particle size distribution function graph is -4. Again, this relationship can be seen by comparing Figures 11-27 and 11-29b. The substitution for the surface area distribution is left to the reader; here, the critical value for β is three—separating the region where the surface area increases ($\beta < 3$) from that where it decreases ($\beta > 3$).

Experimental Measurements of Size Distributions

Size distributions can be measured in many ways. Historically, the only method was by direct observation with a microscope, but this method is tedious and limited to relatively large particles. A modern descendant of that method is to illuminate the particles in a small region of a suspension with a laser, photograph, or video with a magnifying lens, and use image analysis software to measure the particle size. Although not common, this method has been used to study the flocculation (i.e., growth in particle size) of particles (Chakraborti et al., 2000) and has the potential advantage of becoming an *in situ* measure of size distributions in certain reactors. Such an analysis responds to (i.e., measures) the cross-sectional area of particles or flocs.

The two most common techniques for measuring size distributions in water and wastewater use instruments that remove samples (continuously or in batch mode) from the suspension of interest and feed them through a small sensing zone. The size of the sensing zone and the velocity through that zone are arranged so that only one particle passes through at a time, at least if the suspension concentration is sufficiently low. As a particle passes through the sensing zone, an electrical signal related to the size of the particle is produced, and those signals are then sorted into size bins and counted. When the instrument is calibrated with spheres of different sizes, the upper and lower particle size limits for each bin are known, so that the counts in each bin can be converted to the particle size distribution. If only a few bins or channels are available, the data are often just reported as the number concentration for each particular size range; if many channels are available, the data are best converted to the distributions explained above.

In light blockage instruments, a laser light source shines in a direction perpendicular to the direction of flow in the sensing zone. As a particle passes through the light beam, the light is blocked from reaching the other side, and that light blockage is sensed by a photovoltaic cell located opposite to the beam. If the particle is opaque, the light that does not reach the photovoltaic cell is related to the cross-sectional area of the particle; each particle produces a voltage signal proportional to the particle area.

In electrical sensing zone instruments, the suspension is placed in an electrolyte solution and a constant current (or, in some instruments, voltage) is maintained between two electrodes on opposite sides of the sensing zone. As a particle passes through the small sensing zone and takes up a reasonable fraction of the sensing volume, the electrical resistance increases because the conductivity of the particle is negligible in comparison to the electrolyte solution. The resistance change causes a voltage peak (in constant current instruments) or current valley (in constant voltage instruments); these electrical changes are proportional to the particle volume.

Other methods of measuring the size distributions are possible, and some are relevant for particular circumstances. For example, microbalances have been used to study the sedimentation velocity distribution and, if the particle density is known, the data can be converted to a size distribution (assuming the Stokes settling velocity for the particles, explained in Chapter 13). Also, photon correlation spectroscopy measures the light scattered at several different angles at once by a suspension in a small region, and software is used to convert the signals from the many particles in the region into a size distribution. This method is limited to nearly monodisperse suspensions (i.e., suspensions with uniformly sized particles) and hence has limited value for typical suspensions encountered in treatment plants. It can, however, be very valuable for more theoretical studies with monodisperse suspensions.

11.8 PARTICLE SHAPE

Particles are often considered as spheres, especially in mathematical descriptions of behavior, because it is mathematically convenient. Measurements of particle size are also usually described in terms of an equivalent spherical diameter, regardless of the method of measurement.

In reality, of course, particles occur in an infinite variety of shapes. Some particles that are encountered in natural systems have a reasonably precise shape, because of the chemical or biological processes that form them; examples include many organisms and minerals. Other particles are fragments of solids that were once larger, but have been reduced in size due to erosion or other processes, perhaps creating quite irregular shapes. Finally, many particles of interest in engineered systems are precipitates formed during treatment or flocs created from smaller particles of irregular shape. Particles large enough to be seen by light microscopy can be evaluated for their shape, and images of smaller particles can be obtained using scanning electron microscopy or other techniques. When combined with image analysis software, measurements can be made to describe the shape in several mathematical ways.

When particle measurements are made in two dimensions (i.e., projected cross-sectional area), the shape is often characterized as a ratio of two lengthwise measurements. For example, a shape factor could be defined as the ratio of the longest axis (i.e., tip to tip distance) divided by the longest length of a line perpendicular to that axis. Another possibility is to incorporate an estimate of the perimeter from image analysis software as one of the measurements. Finally, the ratio of the projected area to the longest dimension (or some other specified length measurement) can be used as a shape factor. Each of these approaches yields a different result, but different particle shapes can be compared using any of these measurements.

The two most commonly used shape factors relate the particle surface area to its volume. For a spherical particle with diameter d_p, the surface area is πd_p^2 and the volume is $\pi d_p^3/6$, so the ratio of the area (A_p) to the volume (V_p) is $A_p/V_p = 6/d_p$. For nonspherical particles with a characteristic length d_p, the constant 6 is replaced by the shape factor ϕ; that is,

$$\phi = \frac{A_p d_p}{V_p} \qquad (11\text{-}31)$$

Spheres have a value of $\phi = 6$, but spheres have the minimum possible ratio of area to volume, so $\phi > 6$ for particles of any other shape. The other common measure is the sphericity, ψ, defined as the ratio of the area of a sphere to the area of the particle under consideration, when both have the same volume-equivalent diameter. This ratio is <1. These two descriptions of the shape are related, as $\psi = 6/\phi$. Often, one does not have the data to calculate these shape factors from direct measurements of the particle characteristics, so the parameters are inferred from the particle behavior or the effect of particles in specific systems. For example, as shown in Chapter 14, the head loss in a packed bed filter is influenced by the shape factor, so the shape factor can be estimated from studies of the head loss under various conditions.

TABLE 11-2. Densities of Water and Air at Various Temperatures

Temperature (°C)	Water Density (g/cm³)	Air Density (g/cm³)
0	1.001	1.29^{-3}
10	1.000	1.25^{-3}
20	0.998	1.21^{-3}
30	0.996	1.17^{-3}

11.9 PARTICLE DENSITY

The density of particles also varies widely among the types of particles frequently encountered in water and wastewater systems. As shown in Chapter 13, a major determinant of particle behavior in water is the difference between the particle density (ρ_p) and that of water (ρ_L). The density of water varies slightly with temperature but for most environmental temperatures is very nearly $1.0\,g/cm^3$ ($1000\,kg/m^3$). Various pure minerals have densities in the range of $4-5\,g/cm^3$, but particles, flocs, and organisms (including humans) that incorporate a large amount of water have densities that are barely greater than that of water. The density values (or range) for a variety of substances of interest in water and wastewater engineering are collected in Tables 11-2 and 11-3; air is included because of its role in flotation (see Chapter 13).

11.10 FRACTAL NATURE OF FLOCS

Various investigators have noted for many years that the flocs formed from primary particles (i.e., particles that are not expected to break into smaller units under any realistic conditions of interest) have characteristics that are different from the original particles. Even if the original particles were all uniform spheres (i.e., all had the same size and density), flocs formed from those particles differ from the original particles due to the water that is incorporated into the flocs they form. For mathematical simplicity,

TABLE 11-3. Densities of Various Particles of Interest in Water and Wastewater Treatment

Substance	Density (g/cm³)
Iron oxide	4.5
Sand	2.65
Clay	2.65
Calcium carbonate	2.4
Flocs of freshly precipitated hydroxides (Al, Fe, Mg)	1.01–1.05
Microorganisms (incl. activated sludge)	1.01–1.1

most flocculation modeling (as described in Chapter 12) has incorporated an assumption that particles coalesce (like oil droplets) into a sphere that conserves volume (i.e., the volume of the floc is the sum of the volumes of the two particles that form it). While such mathematical modeling has proven very fruitful, it clearly incorporates errors that could be important in predicting particle behavior in some circumstances.

To illustrate how some properties of flocs differ from those of the original particles, consider a floc made from four identical spheres, each with a diameter d_p, arranged in a pyramid shape; that is, three particles form one plane and the fourth sits on top, touching all three of the others. Plan and elevation views (i.e., in two dimensions) are shown in Figure 11-30. Although this hypothetical floc is quite densely packed (and certainly more densely packed than most real flocs), one can see that water might get trapped to some degree in the center region; although some water could flow through that region, most of the trapped water would move with the floc. Hence, the floc has some porosity that the original solid spheres did not, and the density of the overall floc (including the trapped water) would be less than

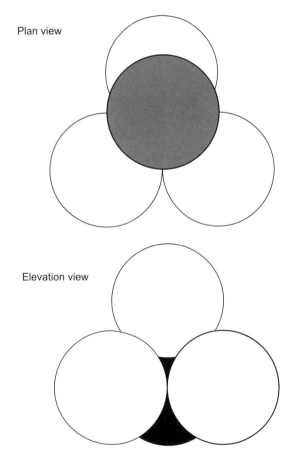

FIGURE 11-30. Hypothetical floc of four identical primary particles.

the density of the original particles (assuming they were denser than water).

Further, it is difficult to decide with much precision what the size of the floc is, regardless of whether that size is to be measured as a linear measure, an area, or a volume. For the hypothetical, four-particle floc, an image of the projected area taken as the plan view would show that the horizontal dimension yields the longest axis ($2d_p$), and the longest vertical (edge to edge) length is (after some geometrical considerations) $\{d_p(1+(\sqrt{3}/2))\}$. We could define the characteristic length (d_c) as either of these numbers, the average of the two, or by some other reasonable standard. If we then measured the perimeter of this projection of the floc, it would be some function of the characteristic length, but not necessarily to the first power. Similarly, the projected area in the plan view is something between three and four times the area of a single particle (A_p); in Figure 11-30, only the blackened area of the fourth particle is not blocked by the other three. The projected area in the elevation view is different and even less than in the plan view. If we chose a characteristic length that was greater than d_p (as any reasonable choice would dictate) the area could be expressed as a power function of the characteristic length, but that power would be less than the value of 2 expected for area as a function of size; that is,

$$\text{Area} \propto d_c^{D_2} \quad (11\text{-}32)$$

In Equation 11-32, D_2 is called the fractal dimension for area. The same logic can be applied to the volume of flocs; as long as the volume of the floc is considered to include some of the "trapped" water between the particles, then one would find that:

$$\text{Volume} \propto d_c^{D_3} \quad (11\text{-}33)$$

where D_3 is known as the fractal dimension for volume and has a value <3 (the normal value expected from Euclidian geometry and the assumption of coalescence). Closely related, the mass of solid matter in a floc is also proportional to $d_c^{D_3}$.

The above discussion makes clear that different methodologies exist for measuring the surface area, cross-sectional area, and volume of flocs, and also that one has choices about what is considered the characteristic length from which the other size properties are calculated. It is also clear that measurements of two different size properties of flocs are required to estimate the fractal dimensions of flocs; that is, one needs to have knowledge of, say, the length and floc volume to know what the fractal dimension is that relates them. For example, different investigators have chosen the longest dimension, the radius of gyration, and the circumference of a particle as the characteristic length. If, for a given floc, these quantities were directly proportional to one another, the areal (D_2) or volumetric (D_3) fractal dimension would be the same using any of the choices for the characteristic length (though the constants of proportionality would differ). However, if they were not linearly related, then the other fractal dimensions would depend on the choice. Hence, while the concept of fractal dimensions is relatively straightforward, the details of estimating and using fractal dimensions can be confusing. Some of the reports of fractal dimensions for flocs of interest in water and wastewater treatment are indicated below.

Li and Ganczarczyk (1989) calculated the fractal dimension (D_3) of particles based on earlier investigations of size or mass and settling velocity; as shown in Chapter 13, the settling velocity is expected to be proportional to the square of the characteristic length. The solids considered in these reports included activated sludge and trickling filter flocs, as well as several chemical flocs; values of D_3 were reported to be as low as 1.4 for activated sludge flocs and as high as 2.85 for ferric hydroxide flocs. Logan and coworkers (Logan and Kilps, 1995; Logan and Wilkinson, 1990; Jiang and Logan, 1991; Kilps et al., 1994) estimated the fractal dimensions of several types of flocs, using several different techniques for measurement of particle properties; the initial particles included monodisperse latex microspheres and various ocean particles. Estimates for D_3 varied from 1.39 to 1.92. Logan and Kilps (1995) also noted that the theory of fractal aggregates, according to Meakin (1988) suggests that, if $D_3 < 2$, then $D_2 = D_3$. Other investigators have thought it more reasonable to assume that $D_3 = D_2 + 1$ unless independent measurements of both the area and volume (as well as the normalizing characteristic length) are available. Lee et al. (2002) used a flocculation modeling approach to describe changes in measured particle size distributions during settling of marine sediments and reported values of D_3 between 2.6 and 3.0.

The concept of fractal dimensions is valuable in considering floc properties; it quantifies (even if roughly) the fact that flocs contain water and are irregularly shaped. As shown in Chapter 12, the fractal dimension has been considered by some investigators in modeling of flocculation. Even when it is not considered quantitatively, the idea of fractal dimensions is worth remembering in all considerations of particle and floc behavior in water and wastewater systems, as it represents a caution in interpreting particle measurements and models that treat (for obvious reasons of simplicity) flocs as nonporous spheres.

11.11 SUMMARY

This chapter serves as an introduction to the particle removal processes that are commonly used in water and wastewater treatment: flocculation, gravity separations (settling and flotation), and deep bed filtration. The success of these processes depends to a great extent on the tendency of particles to attach to each other or to other surfaces provided.

In many suspensions encountered in water and wastewater treatment, the particles tend to be stable; that is, they do not attach to one another or to other surfaces. This stability stems from the fact that their surfaces are charged. For most natural waters encountered in surface water treatment, the particle charge is primarily caused by the adsorption of NOM, and the dissociation of acidic groups of that NOM means that the surface charge is negative.

Charged surfaces attract ions of opposite charge from solution, leading to the existence of a diffuse layer of ions surrounding the particles. The charge and potential distribution in this diffuse layer as a function of distance away from the surface is well-described by the Gouy–Chapman model. When two similarly charged particles approach one another, the electrostatic repulsion of the particles (with their surrounding diffuse layers) alters the trajectories and often prevents the particles from attaching to one another.

Counterbalancing the effect of the electrical repulsion, to at least some extent, is van der Waals attraction; this attraction occurs because synchronously oscillating dipoles are induced in particles when they are very close to one another. This attraction is greater than the electrostatic repulsion when the particles are close to one another, but the repulsion is usually greater at larger separation distances. As a result, a critical separation distance exists; if the particles approach each other more closely than this critical distance, they are predicted to continue traveling toward one another to form a floc. The energy required to bring the particles from a very large separation distance to the critical distance is called the energy barrier. For two particles to get close enough to each other for the attractive force to dominate, they must have sufficient kinetic energy to overcome the energy barrier. For most suspensions of interest, only an occasional pair of particles has sufficient kinetic energy as they approach one another to overcome the energy barrier, making the suspension stable.

Four mechanisms to destabilize the particles; that is, to overcome this problem of stability, have been identified and utilized in environmental engineering processes, and all require the addition of chemicals. First, if salts are added to increase the ionic strength, the diffuse layer is compressed. This change diminishes the electrostatic repulsion between the particles while leaving the van der Waals attraction unchanged; if the repulsion is decreased sufficiently, the suspension is destabilized. This mechanism of destabilization is rarely used purposefully in engineering practice because increasing the ionic strength generally is undesirable. Second, molecules that are oppositely charged to the particle surface and that adsorb to that surface can be added to the suspension; when an appropriate dose of these molecules is added, the charge on the particle surfaces is titrated to (near) 0, so the repulsion is (nearly) eliminated and the suspension is destabilized. Both inorganic chemicals (most commonly metal salts) and synthetic organic compounds designed for this application are commonly used for this mechanism of destabilization. Control of the dose is critical in this mechanism since overdosing of the added chemical leads to a restabilization of the suspension. Third, inorganic precipitates (e.g., Al(OH)$_3$ (s), Fe(OH)$_3$ (s)) can be formed in a suspension via proper metal salt additions and pH control; the precipitates tend to form around particles originally in suspension so that the resulting sweep floc removes both the original particles and the precipitates. Lastly, high-molecular weight synthetic organics that adsorb to the particles in suspension and that are sufficiently long to bridge the diffuse layers of adjacent particles also destabilize suspensions. The second and third mechanisms, charge neutralization and sweep flocculation, are the most commonly used means of destabilization, although the bridging mechanism is used in processes with relatively high fluid shear rates (e.g., deep bed filtration) and/or high particle concentrations (sludge dewatering processes).

The dose of destabilizing chemical required usually depends on the charge on the particles, the number (or perhaps area) concentration of particles in the suspension, and the concentration of soluble constituents in the water (such as NOM) that can react with the destabilizing chemical. In highly idealized experimental systems, the amount of chemical required for effective destabilization might be calculable theoretically, but in practice, these requirements are virtually always determined experimentally.

All particle treatment processes alter the particle size distribution, and the initial size distribution can often dictate which treatment processes are likely to be effective. Most suspensions of interest in water and wastewater treatment have broad and essentially continuous size distributions, and particle behavior is strongly affected by size. Suspensions have an upper particle size limit, such that the probability of finding a particle larger than this size is remote. At the other extreme, definitions of a lower size limit are arbitrary—from molecules to macromolecules to particles, the size distribution is continuous. What is considered particulate material is usually operationally defined (e.g., what is caught on a particular filter membrane) or arbitrarily chosen as materials greater than a certain characteristic size (e.g., 100 nm). As a practical matter, the lower limit is often dictated by the ability to measure and distinguish particles of different sizes by some measuring method.

Although many possibilities exist for expressing and plotting size distributions, a few methods have come into common use. The size itself is most commonly expressed as a diameter, d_p since most particles are not spherical, this diameter is an equivalent spherical diameter based on whatever property is measured. For example, if a particle size measurement device responds to a projected area, the diameter is taken to be that of a sphere that would yield the same projected area. Size distributions are usually expressed in discrete form; that is, all particles within some small size range are considered to have the same (average) size. On a diameter basis, the size range of each discrete size increment might be expressed arithmetically (Δd_p) or logarithmically ($\Delta \log d_p$).

The number concentration (e.g., number per milliliter) of particles in a particular size range (ΔN) is usually normalized by the size range to create a value that is independent (or nearly so) of the size range; this normalization allows measurements taken by different instruments with different size ranges to be compared directly. Two normalized number concentrations in common use are $\Delta N/\Delta d_p$ (usually plotted logarithmically vs the log of the diameter and called the particle size distribution function) and $\Delta N/\Delta \log d_p$ (usually plotted arithmetically vs the log of the diameter and called the number distribution).

In many applications of interest, the number concentration is not as important as the surface area (ΔS) or volume (ΔV) concentration. These latter values are found for each size range as the product of the number concentration in that range and the surface area or volume of the average-sized particle in that range. The surface area and volume concentrations are also normalized when plotting the corresponding distributions (i.e., $\Delta S/\Delta \log d_p$ and $\Delta V/\Delta \log d_p$, both generally plotted arithmetically vs the log of the diameter and known as the surface area and volume distributions, respectively). These methods of plotting the size distributions allow one to see quickly the size ranges containing the largest concentration of particles, particle area, or particle volume.

Besides the particle surface charge and the size, other characteristics of particles that influence their behavior and therefore are important in particle removal processes include the density and shape. All contaminant removal processes take advantage of differences in properties between water and the contaminant; hence, particles with densities much different from water are easier to remove than those with a density nearly equal to water. Since a sphere is the most easily described particle shape and most models of particle processes begin by describing spheres, the shapes of other particles are described relative to spheres. The two most common descriptors for particle shape relate a particle's surface area to its volume in terms of the characteristic size; these descriptors are the shape factor, ϕ (which is 6 for spheres and has greater values for nonspheres) and the sphericity, ψ (which is 1 for spheres and has lower values for nonspheres).

Finally, as particles of various sizes and shapes form flocs that incorporate water, the relationships among the characteristic size, the surface or projected areas, and the volume (or mass) do not follow the geometric relationships of spheres. These relationships can be expressed in terms of fractal dimensions whereby, for example, the mass of a floc is not related to the third power of the diameter, but by some power <3. Fractal dimensions have been used by some investigators to describe the changing characteristics of particles through treatment processes.

All of the characteristics of particles noted above influence the particle behavior and therefore influence the effectiveness of the treatment processes that are described in the subsequent chapters. In turn, these treatment processes also often change the characteristics of the particles in suspension. These relationships between particle characteristics and particle removal processes are explored in the subsequent chapters.

REFERENCES

Ali, W., O'Melia, C. R., and Edzwald, J. K. (1985) Colloidal stability of particles in lakes: measurement and significance. *Wat. Sci. Technol.*, 17 (4/5), 701–712.

Amirtharajah, A., and Mills, K. M. (1982) Rapid-mix design for mechanisms of alum coagulation. *JAWWA*, 74 (4), 210–216.

Amirtharajah, A., and Tambo, N. (1991) Mixing in water treatment, Chapter 1 in Amirtharajah, A., Clark, M. M., and Trussell, R. R. (eds.), *Mixing in Coagulation and Flocculation*, American Water Works Association Research Foundation, Denver, CO.

Amirtharajah, A., and Trusler, S. L. (1986) Destabilization of particles by turbulent rapid mixing. *J. Environ. Eng. (ASCE)*, 112 (6), 1085–1108.

Amirtharajah, A., Clark, M. M., and Trussell, R. R. (eds.) (1991) *Mixing in Coagulation and Flocculation*. American Water Works Association Research Foundation, Denver, CO.

Anandarajah, A., and Chen, J. (1995) Single correction function for computing retarded van der Waals attraction. *J. Colloid Interface Sci.*, 176, 293–300.

Baes, C. F., and Mesmer, R. E. (1976) *The Hydrolysis of Cations*. Wiley-Interscience, New York.

Buffle, J., Wilkinson, K. J., Stoll, S., Filella, M., and Zhang, J. (1998) A generalized description of aquatic colloidal interactions: the three-colloidal component approach. *Environ. Sci. Technol.*, 32, 2887–2899.

Casson, L. W., and Lawler, D. F. (1990) Flocculation in turbulent flow: measurements and modeling of particle size distributions. *JAWWA*, 82 (8), 54–68.

Chakraborti, R. K., Atkinson, J. F., and Van Benschoten, J. E. (2000) Characterization of alum floc by image analysis. *Environ. Sci. Technol.*, 34, 3969–3976.

Chandrakanth, M. S., and Amy, G. L. (1996) Effects of ozone on the colloidal stability and aggregation of particles coated with natural organic matter. *Environ. Sci. Technol.*, 30 (2), 431–443.

Clark, M. M., Srivastava, R. M., and David, R. (1993) Mixing and aluminum precipitation. *Environ. Sci. Technol.*, 27, 2181–2189.

David, R., and Clark, M. M. (1991) Micromixing models and application to aluminum neutralization precipitation reactions, Chapter 5 in Amirtharajah, A., Clark, M. M., and Trussell, R. R.

(eds.), *Mixing in Coagulation and Flocculation*, American Water Works Association Research Foundation, Denver, CO.

Davis, J. A. (1982) Adsorption of natural dissolved organic matter at the oxide/water interface. *Geochim. Cosmochim. Acta*, 46, 2381–2393.

Dempsey, B. A. (1989) Reactions between fulvic acid and aluminum. Effects on the coagulation process. In *Aquatic Humic Substances: Influence on Fate and Treatment of Pollutants*, Advances in Chemistry Series #219, American Chemical Society, Washington, DC, pp. 409–24.

Dempsey, B. A., Ganho, R. M., and O'Melia, C. R. (1984) The coagulation of humic substances by means of aluminum salts. *JAWWA*, 76 (4), 141–150.

Dempsey, B. A., Sheu, H., Ahmed, T. M. T., and Mentink, J. (1985) Polyaluminum chloride and alum coagulation of clay-fulvic acid suspensions. *JAWWA*, 77 (3), 74–80.

Dentel, S. K., and Kingery, K. M. (1989) Theoretical principles of streaming current detection. *Water Sci. Technol.*, 21, 443–453.

Doell, B. (1986) *Die Kompensation der Oberflachenladung kolloidaler Silika-Suspensionen durch die Adsorption kationischer Polymere in turbulent durchstromten Rohrreaktoren*, Doctoral dissertation, Univ. of Karlsruhe, Germany.

Edzwald, J. K. (1993) Coagulation in drinking water treatment: particles, organics, and coagulants. *Water Sci. Technol.*, 27 (11), 21–35.

Edzwald, J. K., Bottero, J. Y., Ives, K. J., and Klute, R. (1998) Particle alteration and particle production processes, Ch. 4 in McEwen, J. B. (ed.), *Treatment Process Selection for Particle Removal*, American Water Works Association Research Foundation, Denver, CO.

Edzwald, J. K., Tobiason, J. E., and Kelley, M. B. (2005) Optimum coagulation conditions for *Cryptosporidium* removals. Proceedings, AWWA WQTC, Nov 2005, Quebec City, paper Tue ST2-1.

Elimelech, M., and O'Melia, C. R. (1990) Effect of particle size on collision efficiency in the deposition of Brownian particles with electrostatic energy barriers. *Langmuir*, 6, 1153–1163.

Gray, K. A., Yao, C., and O'Melia, C. R. (1995) Inorganic metal polymers: Preparation and characterization. *JAWWA*, 87 (4), 136–146.

Gregory, J. (1973) Rates of flocculation of latex particles by cationic polymers. *J. Colloid Interface Sci.*, 42 (2), 448–456.

Hahn, M., and O'Melia, C. R. (2004) Deposition and reentrainment of Brownian particles in porous media under unfavorable chemical conditions: some concepts and applications. *Environ. Sci. Technol.*, 38, 210–220.

Healy, T. W., and La Mer, V. K. (1962) The adsorption-flocculation reactions of a polymer with an aqueous colloidal dispersion. *J. Phys. Chem.*, 66, 1835–1838.

Hiemenz, P. C., and Rajagopalan, R. (1997) *Principles of Colloid and Surface Chemistry*. 3rd ed., Marcel Dekker, New York.

Hogg, R., Healy, T. W., and Furstenau, D. W. (1966) Mutual coagulation of colloidal suspensions. *Trans. Faraday Soc.*, 62, 1638–1651.

Hogg, R. (1984) Collision efficiency factors for polymer flocculation. *J. Colloid Interface Sci.*, 102 (1), 232–236.

Hunter, R. J. (2001) *Foundations of Colloid Science*. 2nd ed., Oxford University Press, Oxford, England.

Hunter, K. A., and Liss, P. S. (1979) The surface charge of suspended particles in estuarine and coastal waters. *Nature*, 282, 823–825.

Jiang, Q., and Logan, B. E. (1991) Fractal dimensions of aggregates determined from steady-state size distributions. *Environ. Sci. Technol.*, 25, 2031–2038.

Johnson, P. N., and Amirtharajah, A. (1983) Ferric chloride and alum as single and dual coagulants. *JAWWA*, 75 (5), 232–239.

Kasper, D. R. (1971) *Theoretical and Experimental Investigations of the Flocculation of Charged Particles in Aqueous Solutions by Polyelectrolytes of Opposite Charge*, Ph.D. dissertation, Cal. Inst. of Technol.

Kilps, J. R., Logan, B. E., and Alldredge, A. L. (1994) Fractal dimensions of marine snow aggregates determined from image analysis of *in situ* photographs. *Deep Sea Res.*, 41, 1159–1169.

Klute, R., and Amirtharajah, A. (1991) Particle destabilization and flocculation reactions in turbulent pipe flow, Chapter 6 in Amirtharajah, A., Clark, M. M., and Trussell, R. R. (eds.), *Mixing in Coagulation and Flocculation*, American Water Works Association Research Foundation, Denver, CO.

Lang, J. S. (2002) Optimize initial mixing with plug flow reactors. *Opflow*, American Water Works Assoc., 28 (12), 12–22.

Lawler, D. F., and Singer, P. C. (1984) Return flows from sludge treatment. *J. Water Poll. Control Fed.* 56 (2), 118–126.

Lee, D. G., Bonner, J. S., Garton, L. S., Ernest, N. S., and Autenrieth, R. L. (2002) Modeling coagulation kinetics incorporating fractal theories: comparison with observed data. *Water Res.*, 36, 1056–1066.

Li, D. H., and Ganczarczyk, J. (1989) Fractal geometry of particle aggregates generated in water and wastewater treatment processes. *Environ. Sci. Technol.*, 23, 1385–1389.

Liao, M. Y., and Randtke, S. J. (1986) Predicting the removal of soluble organic contaminants by lime-softening. *Water Res.*, 20, 27–35.

Logan, B. E., and Kilps, J. R. (1995) Fractal dimensions of aggregates formed in different fluid mechanical environments. *Water Res.*, 29, 443–453.

Logan, B. E., and Wilkinson, D. B. (1990) Fractal geometry of marine snow and other biological aggregates. *Limnol. Oceanography*, 35, 130–136.

Lyklema, J. (1978) Surface chemistry of colloids in connection with stability. In Ives, K. J. (ed.), *Scientific Basis of Flocculation*, NATO Advanced Study Institutes Series, pp. 3–36.

Mangravite, F. J. (1983) Synthesis and properties of polymers used in water treatment, in Proc. AWWA Seminar, Use of Organic Polyelectrolytes in Water Treatment, Amer. Water Works Assoc, Denver, CO pp. 1–16.

Mangravite, F. J., Jr., Leitz, C. R., and Galick, P. E. (1985) Organic polymeric flocculants: effect of charge density, molecular weight, and particle concentration. In Moudgil, B. M., and Somasundaran, P. (eds.), *Flocculation, Sedimentation and Consolidation*, Engineering Foundation, NY.

Meakin, P. (1988) Fractal aggregates. *Adv. Colloid Interface Sci.*, 28, 249–331.

Mehta, A. J., Hayter, E. J., Parker, W. R., Krone, R. B., and Teeter, A. M. (1989) Cohesive sediment transport. I. Process description. *J. Hydraulic Eng. (ASCE)*, 115, 1076–1093.

Ndabigengesere, A., and Narasiah, K. S. (1998) Quality of water treated by coagulation using *Moringa oleifera* seeds. *Water Res.*, 32, 781–791.

Neihof, R. A., and Loeb, G. I. (1972) The surface charge of particulate matter in seawater. *Limnol. Oceanography*, 17, 7–16.

Oldshue, J. Y., and Trussell, R. R. (1991) Design of impellers for mixing, Chapter 9 in Amirtharajah, A., Clark, M. M., and Trussell, R. R. (eds.), *Mixing in Coagulation and Flocculation*, American Water Works Association Research Foundation, Denver, CO.

O'Melia, C. R., (1989) Particle-particle interactions in aquatic systems. *Colloids Surfaces*, 39 (1–3), 255–271.

O'Melia, C. R., Becker, W. C., and Au, K. K. (1999) Removal of humic substances by coagulation. *Water Sci. Technol.*, 40 (9), 47–54.

O'Melia, C. R., Hahn, M. W., and Chen, C. (1997) Some effects of particle size in separation processes involving colloids. *Water Sci. Technol.*, 36 (4), 119–126.

O'Melia, C. R., and Stumm, W. (1967) Theory of water filtration. *JAWWA*, 59 (11), 1393–1412.

Posselt, H. S., Reidies, A. H., and Weber, W. J., Jr. (1968) Coagulation of hydrous manganese dioxide. *JAWWA*, 60 (1), 48–68.

Randkte, S. J., Thiel, C. E., Liao, M. Y., and Yamaya, C. N. (1982) Removing soluble organic contaminants by lime-softening. *JAWWA*, 74 (4), 192–202.

Roalson, S. R., Kweon, J. H., Lawler, D. F., and Speitel, G. E., Jr. (2003) Enhanced softening: effects of lime dose and chemical additions. *JAWWA*, 95 (11), 97–109.

Semmens, M. J., and Field, T. K. (1980) Coagulation: Experiences in organics removal. *JAWWA*, 72 (8), 476–483.

Sharp, E. L., Parsons, S. A., and Jefferson, B. (2004) The effects of changing NOM composition and characteristics on coagulation performance, optimisation, and control. *Water Sci. Technol: Water Supply*, 4 (4), 95–102.

Smellie, R. H., Jr., and La Mer, V. K. (1958) Flocculation, subsidence, and filtration of phosphate slimes. VI. A quantitative theory of filtration of flocculated suspensions. *J. Colloid Sci.*, 13, 589–599.

Stumm, W., and Morgan, J. J. (1996) *Aquatic Chemistry*. 3rd ed., Wiley Interscience, New York.

Stumm, W., and O'Melia, C. R. (1968) Stoichiometry of coagulation. *JAWWA*, 60 (5), 514–539.

Tennekes, H., and Lumley, J. L. (1972) *A First Course in Turbulence*. MIT Press, Cambridge, MA.

Thompson, J. D., White, M. C., Harrington, G. W., and Singer, P. C. (1997) Enhanced softening: factors influencing DBP precursor removal. *JAWWA*, 89 (6), 94–105.

Tiller, C. L., and O'Melia, C. R. (1993) Natural organic matter and colloidal stability: models and measurements. *Colloids Surfaces. A Physicochem. and Eng. Aspects*, 73, 89–102.

Tipping, E., and Higgins, D. C. (1982) The effect of adsorbed humic substances on the colloid stability of hematite particles. *Colloids Surfaces*, 5, 85–92.

Tritton, D. J. (1988) *Physical Fluid Dynamics*. 2nd ed., Oxford University Press.

Van Benschoten, J. E., and Edzwald, J. K. (1990a) Chemical aspects of coagulation using aluminum salts—I. Hydrolytic reactions of alum and polyaluminum chloride. *Water Res.*, 24, 1519–1526.

Van Benschoten, J. E., and Edzwald, J. K. (1990b) Chemical aspects of coagulation using aluminum salts—II. Coagulation of fulvic acid using alum and polyaluminum chloride. *Water Res.*, 24, 1527–1535.

Walker, H. W., and Grant, S. B. (1996) Factors influencing the flocculation of colloidal particles by a model anionic polyelectrolyte. *Colloids Surfaces A Physicohem. and Eng. Aspects*, 119, 229–239.

PROBLEMS

11-1. Prove that the two expressions for the (inverse of) the characteristic diffuse layer thickness (Equations 11-1 and 11-2) are equivalent.

11-2. (a) For a 1:1 electrolyte at a concentration of 2×10^{-3} M, compare the Debye–Huckel and the Gouy–Chapman models for the variation of ion concentration and potential as a function of distance from a flat plate at surface potential values of 33 and 53 mV. (The potential values are approximately half and double the value of 25.7 mV used to develop the figures shown in the chapter.) Assume the temperature is 25°C.

(b) Repeat the analysis for a 2:2 electrolyte (e.g., $CaSO_4$) for all of the same conditions as in part (a).

11-3. (a) Determine the energy of interaction (attractive, repulsive, and total) as a function of separation distance of two particles that have a radius of 0.5 μm, if the surface potential is 53 mV. The solution is a 1:1 electrolyte at a concentration of 2×10^{-3} M, the temperature is 25°C, and the particles' Hamaker constant is 10^{-20} J. Find the location and value of the energy barrier.

(b) Repeat the analysis for a 2:2 electrolyte for all of the same conditions as in part (a).

11-4. Sharp et al. (2004) fractionated the dissolved organic material in a natural water into four components and determined the charge density of each fraction by titrating solutions with poly-DADMAC, a cationic polymer; the (diluted) polymer solution had a charge density of 6.2 meq/g of solution. The results of the fractionation and titrations are shown in the following table.

558 PARTICLE TREATMENT PROCESSES: COMMON ELEMENTS

Component	Concentration (mg/L as C)	Charge Density (meq/g C)
Humic acids	3.5	6.8
Fulvic acids	2.4	4.2
Hydrophilic acids	2.0	0.06
Nonhydrophilic nonacids	1.0	<0.01 (nondetectable)

(a) The titrations were done using concentrated solutions of the different fractions, in each of which 250 mg of dissolved organic carbon (DOC) was added to 100 mL of water. Assuming that the density of the polymer solution is the same as water (1 g/mL), what was the dose of polymer (milligram polymer solution per liter of sample) required to neutralize the charge associated in the concentrated solution containing the humic acid fraction?

(b) Assuming the polymer is fed using the same diluted polymer solution described above, what dose of polymer (milligram polymer solution per liter of water) would be required to neutralize the charge associated with each fraction the DOC in the natural water? What would the doses be in terms of volume of polymer solution per volume of sample?

(c) The source water also has a solids concentration of 8 mg/L. For simplicity, assume that all particles are spherical with a diameter of 1.5 μm and a density of 1.5 g/cm^3. If the (negative) charge density on the particle surfaces is 0.2 C/m^2, what would be the additional required polymer dose, in mg/L, to neutralize these particles, if the same diluted polymer stock solution was used? If polymer were added to neutralize the charge on both the organics and the particles, what fraction of the polymer dose would be allocated to the dissolved organics?

(d) For the experiments reported, the poly-DADMAC was diluted to 0.1%w/w solution, but the product is sold in a 40%w/w solution. If the flow to the water treatment plant using this water as its source is 2 m^2/s, what is the daily requirement (in units of kg/d) for the product as sold?

11-5. A water treatment plant receives raw water with a pH of 7.2 and an alkalinity of 45 mg/L as CaCO$_3$. What will be the resulting pH of the water if they add 30 mg/L of alum (i.e., Al$_2$(SO$_4$)$_3$·16H$_2$O) and no other chemicals? Assume that the raw water has no other weak acid/base constituents besides the carbonate system.

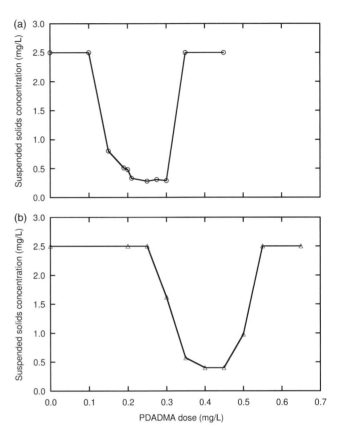

FIGURE 11-Pr6. Experimental results of jar tests at (a) pH 5.0 and (b) pH 7.0. *Source*: Reported by Posselt et al. (1968).

11-6. The results of jar tests are shown in Figure 11-Pr6, reproduced from Posselt et al. (1968). The jar tests were standard; that is, a small rapid mix period during which a destabilizing chemical was added, followed by a longer slow mixing period, then a settling period before a sample was taken. The solids were MnO$_2$ particles, and, as shown on the figure, the results were measured in terms of the suspended solids concentration. The destabilizing chemical was PDADMA; that is, polydiallydimethyl ammonium, a relatively low-molecular weight cationic polymer. Interpret (i.e., explain the causes of) both trends in these results—the effects of dose (same trend on both graphs) and the effects of pH (the difference between part (a) and part (b)).

11-7. Four jar tests were performed leading to the results shown below. All were performed using Min-u-sil 5 (almost pure SiO$_2$, with a density of 2.65 g/cm^3 and virtually all of the particles <5 μm in size) and Alum (Al$_2$(SO$_4$)$_3$·18 H$_2$O). The first three jar tests were performed at constant pH with varying alum dose; the last was performed at constant alum dose and varying pH. Since alum is an acid and would depress the pH, sodium bicarbonate (NaHCO$_3$) was provided as a buffer, and NaOH was added simultaneously with the alum to maintain the pH at the desired level. The results were obtained in a class laboratory by several different (inexperienced) people.

The jar tests were performed according to the following instructions:

(a) For each jar (different alum dose or pH condition), place 500 mL of water containing a known concentration of Min-u-sil 5 plus 2×10^{-3} M sodium bicarbonate buffer in a 600 mL beaker.

(b) Adjust the pH of each sample to the desired level by adding either strong acid or base.

(c) Maintain rapid mixing while simultaneously adding the desired alum dose and predetermined dose of NaOH.

(d) Place the beaker on the gang stirrer and mix for 20 min at 60 rpm.

(e) Remove the beaker and allow to sit quiescently for 20 min.

(f) With a wide-mouth pipet, slowly withdraw an aliquot of 25 mL of supernatant and transfer directly into a vial for the turbidity measurement.

(g) Record the final turbidity of the sample.

Draw appropriate graphs to analyze these results. Comment on the mechanisms of coagulation and the effects of concentration of suspended solids (Min-u-sil), concentration of alum, and pH. Be precise but complete in your explanations. You might find the graph from Amirtharajah (Figure 11-19) useful in interpreting these results. Note that the ordinate on the right side of that graph is labeled in mg/L as $Al_2(SO_4)_3 \cdot 14.3H_2O$, slightly different from the $Al_2(SO_4)_3 \cdot 18\ H_2O$ used in these experiments.

Experiment 1.
Min-u-sil conc. (initial): 100 mg/L
Turbidity of uncoagulated sample: 48 NTU
pH of all samples: 4.5

Alum Dose (mg/L as alum)	Turbidity (NTU)
0	43
1	42
2	43
3.5	43
6	43
10	41
20	44
35	43
60	43
100	43
200	3.8
350	0.84

Experiment 2.
Min-u-sil 5–100 mg/L
Turbidity of uncoagulated sample: 43 NTU
pH of all samples: 5.2

Alum Dose (mg/L as alum)	Turbidity (NTU)
0	32
1	24
2	12
3.5	12
6	12
10	10
20	41
35	41
60	43
100	43
200	2.4
350	1.5

Experiment 3.
Min-u-sil 5: 250 mg/L
pH = 5.2
Turbidity of uncoagulated sample: 160 NTU

Alum Dose (mg/L as alum)	Turbidity (NTU)
0	145
0.6	140
1	63
2	37
3.5	80
6	78
10	120
20	160
35	135
60	130
100	130
200	2.4
350	1.7

Experiment 4.
Min-u-sil 5: 100 mg/L
Turbidity of uncoagulated sample: 52 NTU
Alum dose for all samples: 5 mg/L

pH	Turbidity (NTU)
2.5	38
4.0	41
4.8	38
5.0	20
5.2	15
5.5	12
5.8	43
6.2	43
6.8	4.6

Experiment 4. (continued)

pH	Turbidity (NTU)
7.0	4.7
7.4	9.4
8.0	9.0
9.0	42
10.0	42
11.0	42

11-8. Sludge dewatering is aided by destabilization; that is, water can flow more rapidly out of the high concentration suspension known as sludge when the particles form larger flocs. One laboratory measure of the rate of dewatering is the capillary suction time (CST), a measure of the time for water to spread a certain distance through a particular paper filter in concentric circles from a central sludge source. Shorter values of CST indicate better dewaterability of the sludge.

Lawler and Singer (1984) reported on the dewaterability of anaerobically digested sludge. They used a relatively low-molecular weight cationic polymer to aid the dewatering by destabilizing the particles. The results from two experiments with the same sludge at a concentration of ∼3% solids (w/w %) are shown in the table below. The raw sludge was at a pH of 6.9, and the effects of adding polymer at various doses on the CST values at this pH are shown in the table. In a second experiment, the pH was adjusted to 6.0 before the addition of polymer, and the results from this test are also given in the table.

Interpret these results; that is, explain the trends with polymer addition and the effect of the pH adjustment.

Dose (mg/L)	Capillary Suction Times (s)	
	pH 6.9	pH 6.0
0	130	80
100	93	70
200	89	53
300	70	42
400	54	38
500	50	36

11-9. Given below is a table of values for the logarithms (base 10) of the diameter ($\log d_p$) and the particle size distribution function ($\log (\Delta N/\Delta d_p)$) for a hypothetical sample. ΔN represents the number of particles per cubic centimeter within a size increment Δd_p (with size measured in micrometers). For each pair shown, find the associated values for diameter (d_p) at the midpoint of $\Delta \log d_p$, the number concentration within each size range (ΔN), the number distribution ($\Delta N/\Delta \log d_p$), and the volume distribution ($\Delta V/\Delta \log d_p$). Plot (a) the particle size distribution function (given), (b) the number distribution, and (c) the volume distribution versus the log diameter values. For the number and volume distributions (parts (b) and (c)), use an arithmetic (not logarithmic) scale on the ordinate.

$\log d_p$	$\log (\Delta N/\Delta d_p)$
−0.3	4.90
−0.2	4.88
−0.1	4.82
0.0	4.73
0.1	4.61
0.2	4.47
0.3	4.30
0.4	4.11
0.5	3.90
0.6	3.66
0.7	3.40
0.8	3.12
0.9	2.82
1.0	2.49
1.1	2.15
1.2	1.79
1.3	1.40
1.4	1.00
1.5	0.58
1.6	0.14
1.7	−0.32

11-10. Edzwald et al. (2005) reported the following data from a set of experiments on a natural water to which *Cryptosporidium* oocysts had been spiked. The raw water had a turbidity of ∼2.2 NTU and a DOC concentration of 2.4 mg/L. All experiments were performed at 20°C. 20 mg/L of alum was added to all jars with a 30-s rapid mix step, followed by 20 min of flocculation (slow mixing), and then 60 min of sedimentation before sampling. The difference between the experiments was the pH. Interpret the result shown in the following table.

pH	Turbidity (NTU)	Particle Count (2–8 μm) (thousands/mL)	DOC (mg/L)	*Cryptosporidium* $\log_{10}$ Reduction
5.2	0.42	1.4	1.25	1.8
6.4	0.34	0.8	1.45	1.9
6.8	0.23	1.0	1.6	2.1
8.0	2.1	8.0	2.3	0.7

11-11. **(a)** Discuss the possible mechanisms involved in the destabilization of a suspension of mineral particles upon addition of alum. The suspension pH is initially 7.0, and the pH of the zero point of charge (PZC) of the particles is 6.2. Suggest three possible mechanisms for destabilization.

(b) If the pH of the PZC were 9.5, would the same mechanisms apply?

11-12. Olivine is a mixed iron-magnesium-silicate mineral, with a PZC of 11.5. The electrophoretic mobility of some olivine particles dispersed in deionized water is shown in Figure 11-Pr12 as a function of solution pH.

(a) Label the ordinate of the graph to indicate regions where the electrophoretic mobility is positive, zero, and negative.

(b) Indicate how you think the curve would change if the olivine were suspended in river water, and explain your reasoning.

11-13. A water containing 10 mg/L of DOC and 20 mg/L of clay particles is to be destabilized using a cationic polymer that has a charge density of 6 meq/g-active polymer. The polymer is supplied as a 50% by weight aqueous solution. Assume that the negative charge densities of the natural organic matter

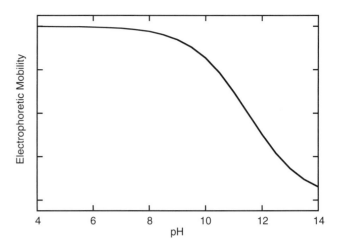

FIGURE 11-Pr12. Electrophoretic mobility of olivine.

and the clay particles are 10.0 and 1.0 meq/g, respectively.

(a) Calculate the required dose of the polymer solution (as purchased) to achieve destabilization by charge neutralization.

(b) What is the required dose if the DOC is reduced by 50%?

(c) What is the required dose if the clay is reduced by 50%?

12

FLOCCULATION

12.1 Introduction
12.2 Changes in particle size distributions by flocculation
12.3 Flocculation modeling
12.4 Collision frequency: Long-range force model
12.5 Collision efficiency: Short-range force model
12.6 Turbulence and turbulent flocculation
12.7 Floc breakup
12.8 Modeling of flocculation with fractal dimensions
12.9 Summary
　　Appendix 12A. Calculation equations for collision efficiency functions in the short-range force model
　　References
　　Problems

An important process in the treatment of water, sewage, and industrial wastes is the formation of suspended flocs which can be effectively removed from the liquid by settling or filtration. This process is known as flocculation or coagulation. Prior to 1920, the nature of the process was not well understood by sanitary engineers and was frequently confused with mixing—the purpose of which is to distribute the coagulating chemicals in the liquid being treated so as to promote solution of the chemical and completion of the chemical reactions. It has since been learned that flocculation is a physical process requiring gentle turbulence and time. Little progress has been made, however, in the scientific development of the principles involved and their application to design.

　　　　　　　　　　　　　　　　　Thomas R. Camp (1955)

12.1 INTRODUCTION

Particle removal is an essential part of most water and wastewater treatment systems and is usually accomplished in gravity separation (sedimentation or flotation) and/or filtration processes. Those removal processes require, or at least are facilitated by, the attachment of particles to one another or to some other surface. As a result, the processes would often be unsuccessful without chemical pretreatment to destabilize the particles, as described in Chapter 11. In many cases, a physical pretreatment process is also valuable as a way of inducing desired changes in the particle size distribution prior to particle removal. That pretreatment process, which typically involves providing detention time and mixing to induce particle collisions and attachment, is called *flocculation*. In flocculation, the particle size distribution changes from a large number of small particles to a smaller number of large particles (*flocs*). The term flocculation is used to refer to both the microscale process of particles coming together so that they can collide and grow, and the macroscale, engineered treatment process in which the microscopic steps occur.

Flocculation is used in various water treatment applications. For instance, in a standard treatment train to produce drinking water from surface water, a distinct flocculation step is included after particle destabilization and before particle removal (Letterman and Yiacoumi, 2011). However, further flocculation also occurs simultaneously in the particle removal processes that follow the engineered flocculation basin.

Flocculation is also an integral part of precipitation processes, such as those that are used to insolubilize metals in industrial wastewaters, phosphate in municipal wastewaters, and calcium (and sometimes magnesium) in water softening operations. In such processes, precipitates typically begin to form and flocculate in the rapid mix tank where chemical additions create the supersaturation conditions necessary for the solids to form. However, much of the floc formation, as well as some additional precipitation, occurs in a downstream tank designed specifically to facilitate that process. A similar sequence of reactors and reactions proceeds when precipitation is used as the mechanism for destabilizing particles that are already in the water (sweep flocculation). In the mathematical descriptions of flocculation that are presented in this chapter, the complication of solids being formed simultaneously is not

---

Water Quality Engineering: Physical/Chemical Treatment Processes, First Edition. By Mark M. Benjamin, Desmond F. Lawler.
© 2013 John Wiley & Sons, Inc. Published 2013 by John Wiley & Sons, Inc.

considered; the assumption is made that solid formation is complete prior to flocculation.

The term *coagulation* is often encountered in the literature in discussions of destabilization and flocculation. In this text, we avoid the use of that term because it has been used by different authors to mean different things. O'Melia (1972) used the term to refer to the combination of destabilization and flocculation, and that meaning has been widely used; these two pretreatment processes are often closely linked (e.g., for a suspension that is originally stable, flocculation would not occur without destabilization), so the use of a single term to refer them both together can be convenient. However, other authors use the term *coagulation* differently. Some use "flocculation" and "coagulation" to refer to different mechanisms of destabilization (and the related words flocculant and coagulant to refer to the chemicals responsible for that destabilization); in this terminology, coagulation usually refers to the addition of chemicals that cause the precipitation of new solids (metal hydroxides, calcium carbonate, or solids that incorporate contaminant ions such as NOM, phosphate, or arsenic). Others use coagulation to mean what we refer to as destabilization, and those authors generally use flocculation as we do here. Still others use the two terms flocculation and coagulation interchangeably, usually to mean the two processes of destabilization and flocculation together.

Because of the close link between the combination of destabilization and flocculation and the subsequent separation processes in treatment plants, the success of both destabilization and flocculation has often been judged by their effect on particle removal. Here, however, the process of flocculation is considered directly in terms of the formation of flocs by two-particle collisions and the consequent changes in the particle size distributions.

This chapter begins with a few examples of the changes in the particle size distribution accomplished by flocculation. We then give an overview of the processes that influence flocculation, by differentiating phenomena that influence long-range transport of particles from those that influence short-range transport. Then, the generic rate equation that describes the changes in the particle size distribution resulting from flocculation is developed, without regard to the specific mechanisms that cause or influence collisions. Next, the rates at which particles are brought into close proximity by long-range forces are developed. Finally, the effects of short-range forces are considered, some quantitatively. The result of these steps is a set of equations that, in theory, allow us to predict the changes in particle size distributions.

Throughout the development, several simplifying assumptions are made, and the latter part of this chapter describes some of the attempts that have been made to account properly for the complexity that has been ignored. Specifically, we address some issues associated with flocculation within turbulent flow conditions, the breakup of flocs, and the fractal nature of flocs. Linkages between engineering practice and the theoretical, mathematical approach to flocculation in this chapter are delineated at appropriate points throughout this chapter. Although the mathematical modeling is complex and still not completely successful in predicting what actually occurs in treatment plants, the qualitative insights that can be gained from the mathematical analysis are valuable both in designing new flocculation facilities and in troubleshooting existing facilities that experience poor performance.

12.2 CHANGES IN PARTICLE SIZE DISTRIBUTIONS BY FLOCCULATION

Flocculation has the objective of changing the size distribution to make particles more amenable to removal in subsequent processes. Before considering in detail how those changes are brought about, we consider a few examples from laboratory studies in which the size distribution was measured in detail. Similar changes in the size distribution occur in treatment plants, but most plant measurements provide less detailed information.

We first consider the changes in the size distribution when the mechanism of particle destabilization is charge neutralization. Particles of silica have a point of zero charge at approximately pH 2.5, so that flocculation at that pH is highly efficient. Results of batch experiments performed under those conditions are shown in Figure 12-1; the measurements were made with an instrument that responds to particle volume (which is conserved when particles join to form a floc) but are reported in terms of the equivalent spherical diameter. The dramatic and continuous loss of small particles (e.g., those <2 μm in diameter, or log d_p < 0.3) with increasing time is clear in these measurements, whereas the number concentration of larger particles (say, at 5 μm in diameter, or log d_p = 0.7) appears to be essentially constant. The apparent

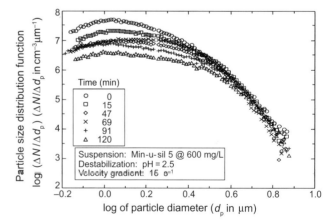

FIGURE 12-1. Changes in the size distribution in flocculation with charge neutralization as the destabilization mechanism. *Source*: Results from Broyan (1996).

lack of change in the number concentration of larger particles is explained by the counterbalancing of two phenomena. First, one expects the increase in the number concentration of larger particles to be much less than the decrease of small particles, because, for example, it takes 125 one-micrometer particles to yield the same particle volume as one 5-μm particle. The substantial loss in the number concentration of 1-μm particles is easily visible on the logarithmic scale of the ordinate, but one would expect to see a much smaller gain of the 5-μm particles. However, even that small gain was diminished (or even reversed) by sedimentation that removed some of the largest particles from the suspension in this small batch experiment.

Even the dramatic changes in the size distribution shown in Figure 12-1 might not be sufficient to facilitate the desired removal; that is, 5-μm particles are still small enough that they might not settle sufficiently rapidly in a subsequent sedimentation tank. Also, as shown later in this chapter, flocculation is strongly influenced by particle concentration, and the concentration of silica particles in the experiment reported in Figure 12-1 is unusually high. A suspension with a low concentration of predominantly small particles is better destabilized via the sweep floc mechanism; the new solids created by precipitation tend to be large and easier to remove. An example of sweep flocculation in batch experiments is shown in Figure 12-2, with the same results shown as both the volume distribution (Figure 12-2a) and the particle size distribution function (Figure 12-2b). The volume distribution changes dramatically, from the raw lake water, in which most of the measured particle volume is in particles less than 3 μm in diameter ($\log d_p < 0.5$), to the flocculated water (after 14 min), in which virtually all of the particle volume is in particles larger than 10 μm ($\log d_p > 1$). Recall that the area under the volume distribution between any two values of the abscissa represents the volume concentration (e.g., ppm_v) of particles in that size range. The particle size distribution function (Part B) shows the same results; this presentation allows one to see more completely the change in the size distribution—the greatest decrease (approximately one order of magnitude) occurred for the smallest particles measured, while all particles with $\log d_p > 0.75$ ($d_p > 5.6\,\mu m$) increased in number. The dramatic increase in the large particles in this experiment in comparison to the immeasurable increase in the previous one is due to the creation of the $Al(OH)_3$ (s) precipitate.

12.3 FLOCCULATION MODELING

The essence of flocculation is the collision of particles and their attachment to one another to form a floc. Forces that bring about long-range motion of particles (i.e., motion over distances approximately two or more orders of magnitude times the size of the largest particles under consideration) can bring particles close to one another, making collisions possible. The forces causing long-range motion include

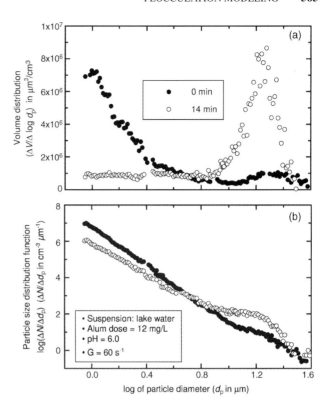

FIGURE 12-2. Changes in the particle size distribution brought about by sweep flocculation. (a) Volume distribution, and (b) particle size distribution function. *Source*: Results from Li (1996).

external forces on the fluid, which generate fluid motion; gravity, which causes vertical motion of the particles relative to the fluid; and thermal activity of molecules in solution, leading to Brownian motion of the particles. A simplified view of these collision mechanisms is shown in Figure 12-3. When particles get quite close to one another, other forces become significant; these short-range forces include, at the least, the van der Waals attraction and electric double layer repulsion (or, perhaps, attraction) explained in the previous chapter, as well as viscous forces relevant as the water separating the two particles moves out of the way.

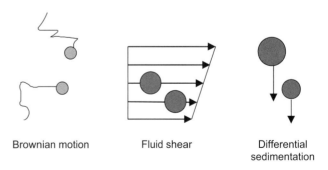

FIGURE 12-3. Particle paths leading to collisions by various mechanisms. (For fluid shear and differential sedimentation, the length of the arrows reflects the particle velocities.)

To understand the flocculation process, and especially the relevance of short-range and long-range forces, it is helpful to have a sense of the distances involved. Consider, for example, a monodisperse suspension (i.e., all particles having the same size) of particles with a diameter of 2 μm and a density of 1.1 g/cm³, at a total mass concentration of 4 mg/L. The mass concentration can be expressed as:

$$c = nm_p = n\pi \frac{d_p^3}{6} \rho_p \quad (12\text{-}1)$$

where c is the mass concentration (e.g., mg/L), n is the number concentration of particles, and m_p, d_p, and ρ_p are the mass, diameter, and density of the individual particles, respectively. Substituting the values into Equation 12-1:

$$4\,\text{mg/L} = n\pi \frac{(2 \times 10^{-4}\,\text{cm})^3}{6}\left(1.1 \times 10^3 \text{mg/cm}^3\right)\left(10^3 \text{cm}^3/\text{L}\right)$$
$$n = 0.87 \times 10^6/\text{cm}^3$$

Taking the inverse of the particle concentration ($1/n$) indicates that there is, on average, one particle in suspension per $1.15 \times 10^{-6}\,\text{cm}^3$ of water. This volume of water is contained in a cube with sides that are 105 μm. Thus, if one imagines adjacent cubes of water with a single 2-μm particle at the center of each (as shown in Figure 12-4), adjacent particles are a distance apart equal to the sides of the cube, or ∼50 diameters away. For a given total mass concentration (4 mg/L in this example) and a given particle density (1.1 g/cm³ here), the average particle–particle separation turns out to be the same, in terms of particle diameters, regardless of the particle size. That is, if the 4 mg/L suspension contained only particles of diameter 0.2 μm, the number concentration would be $0.87 \times 10^9\,\text{cm}^3$, and the separation between particles in a uniform distribution would be 10 μm. Thus, the average separation is ∼50 diameters for a 4 mg/L monodisperse suspension of 2-μm particles, 0.2-μm particles, or any other size particles.

Now consider the situation in a suspension containing 2 mg/L each of 2-μm and 0.2-μm diameter particles. On average, one 2-μm particle would be present in each 132-μm sided cube of water, and the nearest like-sized particle would

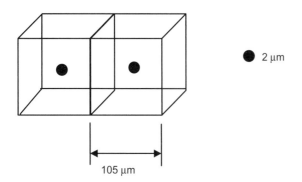

FIGURE 12-4. Particle separation distances. (Not to scale; particle would be nearly invisible if drawn to scale.)

be that distance away. However, that same cube would contain one thousand 0.2-μm particles, and the center-to-center distance between them would be only 13.2 μm. By extension, it would be fair to say that the distance from the 2-μm particle to the nearest 0.2 μm would average only 11.4 μm (the distance from a corner to the center of a 13.2 μm cube). Extending the analysis to consider a continuous size distribution in which the number concentration decreases with increasing size, the picture that emerges is that, for any individual particle of diameter d_p, many particles that are considerably smaller are likely to be nearby—within a distance of a few times d_p. On the other hand, the nearest particle of a considerably larger size is likely to be much farther away, typically at a distance of several tens times d_p. Of course, the numerical results depend on size distribution, but the qualitative results would be the same for most particle size distributions encountered in water and wastewater.

With this picture of interparticle distances in mind, it is instructive to compare those distances to the distances over which certain forces are significant. Consider first the viscous forces that come into play as a particle moves through water, and the water slides around it. The disturbance generated by this motion generally extends for a distance equal to a few tenths of the particle diameter. When two particles approach one another, the water being pushed out of the way by each particle also tends to carry the other particle out of the way. In addition, as the particles get quite close to one another, the cross-sectional area through which the water must flow to get out of the way is restricted, and so the resistive force increases. Thus, short-range viscous forces always oppose collisions, and they intensify as the particles get closer together.

Electrical double layer interactions usually extend over a distance that is somewhat shorter than the short-range viscous forces. Recall from Chapter 11 that double layer repulsion and the van der Waals attraction were significant only at separation distances in the tens of nanometers (10 nm = 0.01 μm).

Because of the different distances over which these short-range forces operate, even if electrical double layer repulsion is completely eliminated by chemical destabilization of the particles, the particles must invariably overcome an energy barrier at distances where they first feel each other's presence, before reaching a point where the van der Waals attraction dominates their interaction and they collide. Thus, two particles that are not initially moving in directions that would lead to a collision cannot be made to collide via the effects of short-range interactions. Rather, for flocculation to occur, two widely separated particles moving independently must first be moving in directions that would lead to a collision; even then, only a fraction of these impending collisions actually occur, because the first short-range forces that come into play cause the particles' paths to deviate from one another.

Rate Equation for Floc Formation

In light of the earlier discussion, flocculation can be viewed as a two-step, sequential process, analogous to a two-step elementary chemical reaction. In the first step, the particles are transported from locations where they have no influence on one another to locations where they do affect each other's movement, and in the second, they overcome the short-range barriers to collision and form a floc. Based on this analogy between flocculation and chemical reactions, and assuming that the formation of the floc is irreversible, the flocculation process might be represented as:

$$P_i + P_j \underset{\gamma_{b,ij}}{\overset{\beta_{ij}}{\longleftrightarrow}} P_i \equiv P_j \xrightarrow{\gamma_{f,ij}} P_k \qquad (12\text{-}2)$$

where P_i and P_j are two particles in the system (types "i" and "j"); $P_i\,P_j$ is a pair of particles that have approached each other closely enough that they affect each other's subsequent motion; P_k is a floc that is formed by the collision of P_i and P_j; and β_{ij}, $\gamma_{f,ij}$, and $\gamma_{b,ij}$ are rate constants for the respective reactions.

As noted, the rate constant β_{ij}, commonly called the collision frequency function, characterizes the transport of P_i and P_j particles to a separation distance where the short-range forces become significant, and $\gamma_{f,ij}$ and $\gamma_{b,ij}$ are, respectively, the forward and reverse rate constants for the two possible transitions from that condition; that is, either to form a floc or to separate again. The forces favoring floc formation (essentially the same ones that brought the particles close to one another plus van der Waals attraction) are incorporated into $\gamma_{f,ij}$, and those opposing floc formation are incorporated into $\gamma_{b,ij}$.

If one assumes that the concentration of particles in close proximity to one another ($P_i \equiv P_j$ pairs) is approximately at steady state, a balance on the number concentration of those particles (n_{ij}) in the system can be written as follows:

$$\frac{dn_{ij}}{dt} = 0 = \beta_{ij} n_i n_j - (\gamma_{b,ij} + \gamma_{f,ij}) n_{ij} \qquad (12\text{-}3)$$

$$n_{ij} = \frac{\beta_{ij} n_i n_j}{\gamma_{b,ij} + \gamma_{f,ij}} \qquad (12\text{-}4)$$

$$r_{k,i+j} = \gamma_{f,ij} n_{ij} = \frac{\gamma_{f,ij}}{\gamma_{b,ij} + \gamma_{f,ij}} \beta_{ij} n_i n_j \qquad (12\text{-}5a)$$

$$r_{k,i+j} = \alpha_{ij} \beta_{ij} n_i n_j \qquad (12\text{-}5b)$$

where $r_{k,i+j}$ is the rate of formation of flocs (P_k particles) via collisions between P_i and P_j particles, n_{ij} is the number concentration of $P_i\,P_j$ "proto-flocs," and $\alpha_{ij} \equiv \gamma_{f,ij}/(\gamma_{b,ij} + \gamma_{f,ij})$. (Note that $r_{k,i+j}$ is the rate of formation of type-k particles from specific type-i and type-j particles; the overall rate of formation of type-k particles in the system would have to consider their formation from collisions of other types of particles and their disappearance when they collide with other particles and become a part of larger flocs.)

Interpretation of the Rate Constant

In Equation 12-5b, the rate constant is the product of two parts, α and β. The meaning of β as the collision (i.e., close approach) frequency function has been noted. The conceptual meaning of α_{ij} can be seen as follows:

$$\alpha_{ij} \equiv \frac{\gamma_{f,ij}}{\gamma_{f,ij} + \gamma_{b,ij}} = \frac{\gamma_{f,ij} n_{ij}}{(\gamma_{f,ij} + \gamma_{b,ij})} = \frac{\gamma_{f,ij} n_{ij}}{\beta_{ij} n_i n_j}$$

$$\alpha_{ij} = \frac{\text{Rate of formation of } P_k \text{ flocs}}{\text{Rate of close approaches of } P_i \text{ and } P_j \text{ particles}}$$

(12-6)

Thus, α_{ij} can be defined as the fraction of close approaches of P_i and P_j particles that results in floc formation.

The analogy of flocculation to chemical reactions is entirely consistent with the physical description of flocculation provided previously. A close approach of two particles (as envisioned in the physical description of the process) is the same thing as the formation of the intermediate species $P_i \equiv P_j$ envisioned in the chemical description. Correspondingly, if we consider a close approach to be the critical prerequisite for a collision, α_{ij} can be interpreted in the physical description of the process as the fraction of the potential collisions (i.e., close approaches) that actually occur (i.e., that form a floc). Based on this understanding, α_{ij} is commonly called the *collision efficiency function*.

Flocculation modeling has progressed over time, with an increasing inclusion of phenomena that impact floc formation. For example, the distinction between forming "proto-flocs" and true flocs is relatively recent. Even with the increasing sophistication, however, the formation of flocs tends to be overestimated in the models because they consider an idealized view. As a result, the observed rate of flocculation is often less than the predicted rate. To account for this difference, an empirical correction factor, usually given the symbol α but here noted more explicitly as α_{emp}, is often used; that is,

$$\alpha_{\text{emp}} \equiv \frac{\text{Actual rate of floc formation}}{\text{Predicted rate of floc formation}} = \frac{\text{Actual } r_{k,\text{form}}}{\text{Predicted } r_{k,\text{form}}}.$$

The common shorthand description of α_{emp} is the *collision efficiency factor*, and we retain that terminology in this book. However, different modeling efforts have made different predictions of the rate of floc formation (i.e., different values of the denominator), and the values of α_{emp} vary depending on what phenomena are included in the mathematical predictions. Likewise, the verbal descriptions among authors vary, largely depending on the awareness of what has not been included in the predictions. Descriptions such as "the fraction of collisions that result in attachment," "the fraction of predicted collisions that really occur," and "the fraction of complete destabilization" have been used, but all of these are means of describing the above word fraction.

In early efforts to model flocculation, values of β for pairs of particles of various sizes were computed, considering the three long-range transport forces shown in Figure 12-3. Later modeling efforts have incorporated some of the short-range forces explicitly. Both of these approaches are explained in detail in following sections.

As it becomes possible to account for more forces explicitly, and to account more accurately for those that are known, fewer forces will need to be lumped into α_{emp}, a better understanding of the factors that control flocculation can be gained, and, presumably, better decisions can be made about how to design and operate flocculation systems. In that sense, the goal of flocculation modeling is to make $\alpha_{emp} = 1$, in which case all factors that influence successful flocculation will be accounted for explicitly.

Equation 12-5 is the fundamental equation used to model the formation of flocs from smaller particles. While this model is essentially identical to that for bimolecular chemical reactions, its application is inherently more complex than the applications of chemical reaction theory presented in Chapter 3, for several reasons. First, flocs of a given size can be formed by a very large number of combinations of smaller particles, even if only pairwise combinations are considered. For instance, the rate of formation of P_k-sized flocs must consider collisions between two particles that are both half as large as P_k, between one particle; that is, two-thirds as large as P_k and with another that is one-third as large, and so on. Thus, computation of the rate of formation of P_k-type particles must consider the contributions from many individual reactions occurring simultaneously.

Second, even when considering reactions between two particles of given sizes, several different reaction mechanisms or pathways can bring the particles close to one another. Three such pathways are noted earlier (fluid shear, sedimentation, Brownian motion). These pathways must be evaluated independently to understand and predict flocculation rates.

Third, unlike the situation in chemical reactions, virtually all products of a flocculation reaction (i.e., flocs) can themselves participate in subsequent reactions to form even larger flocs. And, finally, the factors controlling the rate constants for flocculation reactions depend not only on the chemical identities of the reactants and the system temperature but also on physical characteristics of the particles and the degree of mixing in the flocculation reactor.

Thus, while the conceptual aspects of two-particle collisions in flocculation can be treated by the formalism developed for chemical reactions in Chapter 3, the conversion of those concepts into useful results is more complex. A complete analysis of flocculation kinetics would require that α_{ij} and β_{ij} values be calculated for each possible pair of colliding particles in the system, considering all possible ways that they could be brought together, and that these values then be used in rate expressions describing the formation of flocs of all possible sizes. Ideally, the computation of α and β values would be based on fundamental principles of hydrodynamics and surface chemistry and would incorporate all possible phenomena that promote or impede collisions. The history (and ongoing research) of flocculation modeling can be viewed as a progression of attempts to understand and quantify more and more of the individual factors that control these two key variables.

In the following section, flocculation models are presented that mirror some of the key stages in the historical understanding of flocculation. First, the extension from a single type of particle (the formation of P_k particles given above) to an accounting of the changes of all sizes of particles is given. Next, a model that considers explicitly only the long-range forces that apply in laminar flow fields is presented; in essence, this model allows calculation of β_{ij} and lumps all other phenomena into the empirical correction factor, α_{emp}.

Lastly, complications associated with short-range forces, turbulent flow regimes, floc breakup, and the changing floc density caused by incorporation of water into flocs are discussed at various levels of sophistication, reflecting the current levels of understanding and modeling of these factors.

The Smoluchowski Equation

In 1917, Smoluchowski published what still remains the fundamental theory of flocculation. He wrote an equation for a "population balance" for flocculating particles in a suspension; that is, an equation describing what would happen to the number concentration (the population) of each size particle in a discrete distribution over time. This equation incorporated the following concepts:

- Particles of each size class can be formed by flocculation of smaller particles and lost by flocculating to become still larger particles,
- Only two-particle collisions need to be considered because the probability of three particles colliding at exactly the same instant is negligible in comparison to that of two particles colliding,
- In the formation of a floc, particle volume is conserved.

With these assumptions, the Smoluchowski equation can be expressed in words as follows. For a control volume in a batch system and considering only particles of a particular size (i.e., size class) k:

Rate of change with time of the number of particles of size k	=	Rate of formation of particles of size k by flocculation of smaller particles (the sum of whose volumes is size k)	−	Rate of loss of particles of size k by flocculation of size k particles with any size particles

Note that the form of the population balance is identical to that of a mass balance for a batch system, but here it is not a mass of a constituent that is of interest, but the number of particles of the particular size. Dividing by volume, and therefore considering concentrations rather than number, yields:

Rate of change with time of the number concentration of particles of size k = Rate of increase in the number concentration of particles of size k by flocculation of smaller particles (the sum of whose volumes is size k) − Rate of decrease in the number concentration of particles of size k by flocculation of size k particles with any size particles

Using the nomenclature developed above, this expression can be written for a single collision mechanism as:[1]

$$r_k = \frac{1}{2}\alpha_{\text{emp}} \sum_{\substack{\text{all } i \text{ and } j \\ \text{such that} \\ V_{p,i} + V_{p,j} = V_{p,k}}} \alpha_{ij}\beta_{ij}n_in_j - \alpha_{\text{emp}}n_k \sum_{\text{all } i} \alpha_{ik}\beta_{ik}n_i \quad (12\text{-}7)$$

where r_k is defined as the left side in the word equation immediately preceding (with dimensions $L^{-3}T^{-1}$); note that r_k is the net rate of formation of type-k particles from all possible combinations of type-i and type-j particles and the loss of those particles by flocculation with others, distinguishing it from $r_{k,i+j}$, which we defined as the rate of formation of size-k particles from a specific combination of type-i and type-j particles. i, j, and k are size categories of particles; n_i is the number concentration of particles of size i (L^{-3}); $V_{p,i}$ is the volume of particles of size i (L^3); β_{ij} is the collision frequency function for the long-range transport process under consideration, (L^3T^{-1}); α_{ij} is the dimensionless function accounting for short-range particle–particle interactions; α_{emp} is the empirical, dimensionless correction factor used to match experimental data to the equation.

The stipulation $V_{p,i} + V_{p,j} = V_{p,k}$ for the summation in the first term on the right in Equation 12-7 reflects the assumption that volume is conserved during collisions. Since i and j can take on all possible values (up to some maximum describing the largest particles expected in the suspension), the same collision is counted twice in the summation when $i \neq j$. For example, collisions between size 6 and size 9 particles would be counted when $i = 6$ and $j = 9$ and again when $j = 6$ and $i = 9$; the factor of $\frac{1}{2}$ preceding the summation corrects for this double counting. As explained by Benjamin (2011), the factor of $\frac{1}{2}$ is also needed when $i = j$ because the collision frequency function (as derived in Section 12-4) would double count these collisions. For the last term, a size k particle colliding with any size i particle (including another size k particle) will result in a loss of the size k particle, so that summation is taken over all i.

Because flocculation of size i and size j particles can proceed by several mechanisms, each of which operates (almost) independently, the overall rate of flocculation of such particles is the summation of the rates by each mechanism:

$$^{\text{tot}}r_i = {}^{\text{Sh}}r_i + {}^{\text{DS}}r_i + {}^{\text{Br}}r_i + \cdots \quad (12\text{-}8)$$

where the superscript on each term indicates whether it refers to flocculation brought about by fluid shear (Sh), differential sedimentation (DS), or Brownian motion (Br). In virtually all situations of interest, these are the only significant long-range transport mechanisms operating. However, novel treatment systems have been proposed in which, for instance, an electric or magnetic field might be applied to the system, leading to additional transport modes. The possibility that such alternative mechanisms of transport might be important is, therefore, indicated in the equation.

Each long-range transport mode is associated with a unique collision frequency function (${}^{\text{Sh}}\beta_{ij}$, ${}^{\text{DS}}\beta_{ij}$, and ${}^{\text{Br}}\beta_{ij}$, respectively), and a corresponding function accounting for forces relevant to short-range transport (${}^{\text{Sh}}\alpha_{ij}$, ${}^{\text{DS}}\alpha_{ij}$, and ${}^{\text{Br}}\alpha_{ij}$). Thus, for instance, the rate equation for flocculation by differential sedimentation would be written as follows:

$$^{\text{DS}}r_k = \frac{1}{2}\alpha_{\text{emp}} \sum_{\substack{\text{all } i \text{ and } j \\ \text{such that} \\ V_{p,i} + V_{p,j} = V_{p,k}}} {}^{\text{DS}}\alpha_{ij}{}^{\text{DS}}\beta_{ij}n_in_j - \alpha_{\text{emp}}n_k \sum_{\text{all } i} {}^{\text{DS}}\alpha_{ik}{}^{\text{DS}}\beta_{ik}n_i \quad (12\text{-}9)$$

Corresponding equations could be written for the flocculation rates by shear and Brownian motion, and the overall rate would be the summation of the three expressions, as shown in Equation 12-8.

At first glance, the dependence of α_{ij} on the long-range transport mechanism might seem odd. α_{ij} values account for the van der Waals attraction and for the viscous forces encountered as water escapes from the vicinity of the particles. The van der Waals attraction depends only on the separation of the particles and is independent of the collision mechanism, but the hydrodynamic interactions depend on the mechanism of interaction.

Particles, of course, do not know the mechanism that is bringing them together. Therefore, if a pair of particles was approaching one another by Brownian motion, and another

---

[1] Smoluchowski did not include the α terms at all, in the belief that all collisions were properly accounted for in the β term, so the equation he proposed was somewhat simpler than Equation 12-7. However, Equation 12-7 is a direct application of the number balance equation he proposed, updated in accordance with modern understanding.

pair was approaching one another in exactly the same way (along the same path and at the same relative velocity) by fluid shear, the α values for the two interactions would have to be identical. However, the velocity of approach at all separation distances is not the same between different mechanisms, and therefore the hydrodynamic interactions are not the same. Particles approaching one another by differential sedimentation are acted on by gravity and, in the absence of other particles, have velocities that are fixed by their physical properties (density, size) and are independent of their location. Particles approaching each other by fluid shear are generally assumed to be moving along parallel paths (as explained in Section 12-4), and their relative velocities depend on their location in the moving fluid and not on their physical properties. Hence, the short-range hydrodynamic interactions depend not only on the separation distance between the two particles but also on the fluid motion in the vicinity of the particles and the external forces acting on the individual particles. Thus, for a given pair of particles, each long-range transport mechanism is characterized by a different set of conditions describing the short-range interactions. For this reason, the α_{ij} values do effectively depend on the long-range transport mechanism bringing the particles together.

Accounting for the three mechanisms of flocculation together leads to the final formulation of the Smoluchowski equation as follows:

$$^{\text{Tot}}r_k = \frac{1}{2}\alpha_{\text{emp}} \sum_{\substack{\text{all } i \text{ and } j \\ \text{such that} \\ V_{p,i} + V_{p,j} \\ = V_{p,k}}} {}^{\text{Tot}}(\alpha_{ij}\beta_{ij})n_i n_j - \alpha_{\text{emp}} n_k \sum_{\text{all } i} {}^{\text{Tot}}(\alpha_{ik}\beta_{ik})n_i \quad (12\text{-}10a)$$

where

$$^{\text{Tot}}(\alpha_{ij}\beta_{ij}) = {}^{\text{Br}}\alpha_{ij}{}^{\text{Br}}\beta_{ij} + {}^{\text{Sh}}\alpha_{ij}{}^{\text{Sh}}\beta_{ij} + {}^{\text{DS}}\alpha_{ij}{}^{\text{DS}}\beta_{ij} \quad (12\text{-}10b)$$

Characteristic Reaction Times in Flocculation

In Chapter 3, one definition of the characteristic reaction time was given as the time required for the concentration to reach its ultimate value, if the initial rate of reaction was maintained. In flocculation, every particle size considered has its own rate of reaction, and that rate is initially negative for some particle sizes and positive for others. Flocculation reduces the number of small particles and increases the number concentration of large particles, but the distinction between small and large particles (or the distinction between sizes with initially increasing or decreasing particle number concentration) can vary according to the suspension.

Since the decrease in number concentration of small particles is usually the primary interest in flocculation, our focus with respect to determining a characteristic reaction time is on those particle sizes whose initial rate of reaction is negative. For such size particles, the ultimate concentration (if the rate continued unchanged) would be zero. Under those conditions, the characteristic reaction time is as follows:

$$t_{\text{char},k} = \frac{n_{k,o}}{-({}^{\text{Tot}}r_{k,o})}$$

$$= \frac{n_{k,o}}{-\frac{1}{2}\alpha_{\text{emp}} \sum_{\substack{\text{all } i \text{ and } j \\ \text{such that} \\ V_{p,i} + V_{p,j} \\ = V_{p,k}}} {}^{\text{Tot}}(\alpha_{ij}\beta_{ij})n_{i,o}n_{j,o} + n_{k,o}\alpha_{\text{emp}} \sum_{\text{all } i} {}^{\text{Tot}}(\alpha_{ik}\beta_{ik})n_{i,o}} \quad (12\text{-}11)$$

where the first equality is based on the second definition of t_{char} for a reactive chemical given in Chapter 3 (the time to reach the ultimate concentration if the reaction rate was to stay at the initial rate, or $t_{\text{char}} = c(0)/(-r(0))$ for a negative reaction rate). The second equality substitutes the rate expression for flocculation from Equation 12-10a. This equation appears complex, but it depends only on the initial conditions and hence can be calculated without great difficulty, once we know expressions for the various embedded parameters and functions.

As indicated by the subscript k in the notation, this characteristic time is for a particular size particle. We can simplify this equation considerably by choosing a size for which the first term on the right side of Equation 12-10 (expressing the increase in size-k particles) is negligible in comparison to the second term (the decrease in these particles); for that condition:

$$t_{\text{char},k*} \approx \frac{n_{k*,o}}{-({}^{\text{Tot}}r_{k*,o})} \approx \frac{1}{\alpha_{\text{emp}} \sum_{\text{all } i} {}^{\text{Tot}}(\alpha_{ik*}\beta_{ik*})n_{i,o}} \quad (12\text{-}12)$$

where the symbol k^* is used to denote the chosen size with the specification that the gain term for this size particle is negligible in comparison to the loss term. Finally, if the concentration of particles of each size in Equation 12-12 is expressed as the product of the fraction present in that size $(f_i = n_{i,o}/n_{\text{Tot},o})$ and the total number of particles originally present $(n_{\text{Tot},o} = \sum_{\text{all } i} n_{i,o})$, we can express the characteristic reaction time for this size particle as:

$$t_{\text{char},k*} \approx \frac{1}{\alpha_{\text{emp}} \sum_{\text{all } i} {}^{\text{Tot}}(\alpha_{ik}\beta_{ik})n_{i,o}}$$

$$= \frac{1}{n_{\text{Tot},o}\alpha_{\text{emp}} \sum_{\text{all } i} {}^{\text{Tot}}(\alpha_{ik}\beta_{ik})f_i}$$

$$= \frac{1}{\alpha_{\text{emp}}\gamma n_{\text{Tot},o}} \quad (12\text{-}13)$$

where γ is a constant that accounts for all of the collision frequency functions, the short-range collision efficiency term, and the relative particle size distribution. Hence, γ is a constant only for specific initial conditions. The final result is consistent with expectations for second-order reactions based on Chapter 3; that is, that the characteristic reaction time is the inverse of the product of the rate constant and the initial concentration. In this case, however, the rate "constant" $\alpha_{\text{emp}}\gamma$ is not truly constant, as f_i will change as flocculation proceeds.

Design Implications of the Smoluchowski Equation

Even without knowing any details of the collision frequency functions (β_{ij}) or the predicted collision efficiencies (α_{ij}), some design and operational ramifications of the Smoluchowski equation can be elucidated. Specifically, effects of particle concentration and chemical destabilization can be appreciated. The equation makes clear that flocculation is a second-order reaction. If two suspensions with identical relative size distributions, but one having a concentration 10 times the other ($n_{A,i} = 10 n_{B,i}$ for all i; e.g., suspended solids concentrations of 2 and 20 mg/L), are flocculated under otherwise identical conditions, the one with the higher concentration will flocculate 100 times as fast.

This concentration effect on the rate of flocculation has important ramifications for the design and operation of flocculation facilities. Recall from Chapter 4 that, as the order of the reaction increases above zero, the effects of the residence time distribution on the extent of reaction achieved increases. Assuming that the particles have residence times that are the same as the fluid, it might be valuable to design flocculation reactors to achieve a flow pattern that is nearly plug flow (the flow pattern that achieves the greatest amount of reaction for dissolved reactants undergoing second-order reactions). Plug flow is characterized by no mixing among fluid elements in the axial direction and perfect mixing in the transverse directions. Plug flow can be nearly achieved in turbulent pipe flow, but the level of turbulence required is likely to cause flocs to breakup (as explained subsequently), so this type of reactor is usually considered inappropriate for flocculation. On the other hand, mixing at a more gentle level promotes collisions between particles. Therefore, flocculators are generally built with gentle mixing in a reactor that has a residence time distribution equivalent to several CFSTRs in series, a design that approximates plug flow. As discussed in Chapter 2, such a residence time distribution can be obtained in various ways, but most involve baffling a large reactor to ensure that fluid (and, here, the particles in that fluid) cannot escape the reactor without spending considerable time getting from the influent to the effluent.

Second, the rate of flocculation can be increased, if the particle concentration in the flocculator is increased. The simplest method is to recycle sludge from a subsequent solid/liquid separation device (e.g., a sedimentation tank), but other methods can also be effective. For example, a reactor in which the suspension has a net upward flow will enable particles to stay in the tank longer than fluid (because their settling velocity will counterbalance the upward movement to some degree); the net effect is to have a particle residence time greater than the fluid residence time, and thereby have a volume concentration in the reactor greater than what is coming in or leaving. Other reactors have been developed that maintain a well-defined sludge (high particle concentration) layer that the suspension must pass through, yielding a zone of very high concentration. In all of these ways, the concentration in the flocculator is kept artificially high to increase the rate of flocculation and therefore the overall effectiveness of the process.

The particle concentration (and therefore the rate of flocculation) also can be increased by direct addition or precipitation of particles. The most common method is to use destabilization by the sweep floc mechanism, in which case the precipitation has dramatic effects on the particle size distribution by creating large precipitated particles. Destabilization by this method increases the particle volume concentration, often considerably. Without such precipitation, the flocculation of suspensions with low particle concentrations is very slow even with complete destabilization. As a result, the treatment of low turbidity waters for drinking water is usually accomplished by either a conventional process train (destabilization by sweep flocculation, followed by flocculation, sedimentation, and filtration) or by contact filtration (destabilization by charge neutralization followed by filtration only). The omission of flocculation and sedimentation in contact filtration plants is a recognition that these processes are not only unnecessary but ineffective for low turbidity suspensions—virtually no flocculation occurs in tanks with reasonable detention times, and as a result, virtually no sedimentation occurs either. The lack of flocculation is a direct result of the second-order particle concentration effect reflected in the Smoluchowski equation.

Another method to increase the particle concentration is to add some large dense solids such as silica or sand. If the original particles flocculate with these particles, the subsequent sedimentation process is highly effective because of the high density of the resulting flocs. This process is sometimes called *ballasted flocculation*.

In addition to the particle concentration effects, the effects of destabilization are also reflected in Equation 12-7. Chemical destabilization can affect α_{emp}, α_{ij}, or both. (The allocation of the effects between α_{emp} and α_{ij} is, to some degree, a matter of choice in the modeling.) Chemical destabilization is essential to the formation of flocs; without it, extremely few of the close approaches of two particles would result in the formation of a floc. The product of the two α terms can vary over several orders of magnitude depending on the degree of destabilization. The ramifications for design and operation are clear: if a flocculator is not performing well, the first place to look for possible explanations and for ways to improve performance is the chemical destabilization process.

■ **EXAMPLE 12-1.** Assuming that the rate product $\alpha_{\text{emp}}\gamma$ developed in Equation 12-13 remains constant throughout a flocculation process, compare the detention times required in a PFR and a CFSTR to achieve a 90% removal efficiency. Express the results (a) in terms of the product $\alpha_{\text{emp}}\gamma$ and the initial concentration of those particles and (b) in terms of the characteristic reaction time developed in Equation 12-13.

Solution. According to Chapter 4, the required detention time to reduce the reactant concentration from n_{in} to n_{out} via a second-order reaction in a PFR (using the flocculation symbols for the concentration and rate constant) is as follows:

$$\tau_{\text{pfr}} = \frac{1}{\alpha_{\text{emp}}\gamma n_{\text{in}}}\left[\frac{n_{\text{in}}}{n_{\text{out}}} - 1\right]$$

For 90% removal, $n_{\text{out}} = 0.1\, n_{\text{in}}$, so $n_{\text{in}}/n_{\text{out}} = 10$. Thus,

$$\tau_{\text{pfr}} = \frac{1}{\alpha_{\text{emp}}\gamma n_{k,\text{in}}}[10 - 1] = \frac{9}{\alpha_{\text{emp}}\gamma n_{k,\text{in}}} = 9t_{\text{char}}$$

For a CFSTR, the required detention time for a second-order reaction according to Chapter 4 is as follows:

$$\tau_{\text{cfstr}} = \frac{1}{\alpha_{\text{emp}}\gamma n_{\text{out}}}\left[\frac{n_{\text{in}}}{n_{\text{out}}} - 1\right]$$

Multiplying the numerator and denominator by n_{in} and making the same substitutions as above yields:

$$\tau_{\text{cfstr}} = \frac{1}{\alpha_{\text{emp}}\gamma n_{\text{out}}}\frac{n_{\text{in}}}{n_{\text{in}}}\left[\frac{n_{\text{in}}}{n_{\text{out}}} - 1\right] = \frac{1}{\alpha_{\text{emp}}\gamma n_{\text{in}}}\frac{n_{\text{in}}}{n_{\text{out}}}\left[\frac{n_{\text{in}}}{n_{\text{out}}} - 1\right]$$
$$= \frac{1}{\alpha_{\text{emp}}\gamma n_{\text{in}}}10[10-1] = \frac{90}{\alpha_{\text{emp}}\gamma n_{\text{in}}} = 90t_{\text{char}}$$

Taking the ratio of required detention times in the two types of ideal reactors yields:

$$\frac{\tau_{\text{cfstr}}}{\tau_{\text{pfr}}} = \frac{90t_{\text{char}}}{9t_{\text{char}}} = 10$$

Hence, to accomplish 90% removal, a CFSTR needs a detention time 10 times that of a PFR. More generally, for a fractional removal efficiency, η, the ratio $(\tau_{\text{cfstr}}/\tau_{\text{pfr}}) = 1/(1-\eta)$. Again, these results follow directly from those in Chapter 4, but here, for simplicity, we have made assumptions about the unimportance of the formation term in the Smoluchowski equation for the size under consideration and about the constancy of a rate expression that in fact does not stay constant. Despite the simplifications, the general conclusion that the flow pattern makes a difference consistent with the second-order reaction as explained in Chapter 4 is valid. ■

12.4 COLLISION FREQUENCY: LONG-RANGE FORCE MODEL

In the following discussion, the collision frequency function for collisions caused by fluid shear ($^{\text{Sh}}\beta_{ij}$,) in a laminar flow field is developed in some detail, based on the derivation shown by Friedlander (1977). The derivations of the corresponding functions for Brownian motion and differential sedimentation are then outlined and the result given, although they are not derived in as much detail.

Collisions by Fluid Shear

In any reactor except an ideal plug flow reactor, different elements of the fluid travel at different speeds. Thus, two particles traveling with the fluid but on different streamlines might be brought close to one another, allowing a collision. Flocculation via this mechanism is sometimes called *orthokinetic flocculation*.

Consider a laminar flow field subjected to shear; that is, a flow field in which the velocities in the horizontal (x) direction vary as a function of the vertical (y) position, as depicted in the left portion of Figure 12-5. Each particle (assumed spherical) is envisioned to flow on a straight path, with a velocity corresponding to that of the streamline that would pass through the center of the particle. Hence, particles that are at higher values of y travel faster than those at lower y values and can catch up to them after some time. If the particles are assumed to continue on rectilinear paths even as they approach one another, a collision will occur if the original vertical distance

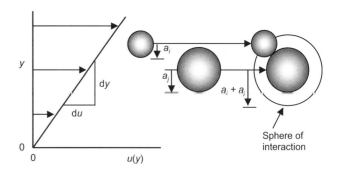

FIGURE 12-5. Conceptual view of collisions by fluid shear.

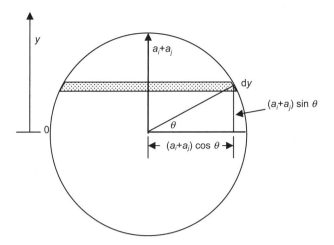

FIGURE 12-6. Sphere of interaction in two-particle collision by fluid shear.

between their center lines is less than or equal to the sum of their radii. This concept is illustrated in the right portion of Figure 12-5.

If the vertical coordinate is redefined so that its origin is at the center of particle j, then the center of an i-sized particle that has collided with particle j lies on the surface of a sphere centered at the origin and with radius $(a_i + a_j)$. Correspondingly, any i-sized particle whose center is not on the surface of that sphere has not collided with the j-sized particle. This sphere, called the sphere of interaction, is depicted in Figure 12-6, shown from a perspective that is rotated 90° from that in Figure 12-5 (i.e., what is left-to-right in Figure 12-5 is into the page in Figure 12-6). Thus, if the initial location and velocity of any i-sized particle are such that its center would eventually pass through the sphere of interaction between itself and a j-sized particle, the particles will collide.

Based on the earlier discussion, it is possible to develop an expression for the rate (#/time) at which particles of size i collide with one size j particle. As before, the coordinate system is Lagrangian, defined so that its origin is at (and moves with) the center of the size j particle (which is also the center of the sphere of interaction depicted in Figure 12-6. The rate at which size i particles approach the size j particle depends on the vertical coordinate of the size i particle: the larger the absolute value of y, the larger the rate of approach. If a differential strip of vertical dimension dy is defined on the surface of the sphere of interaction, the rate at which size i particles contact that strip (call this rate dW) is the product of the flux of particles toward the strip and the projected area of the strip perpendicular to the line of approach. The flux, in turn, is the product of the concentration of i-sized particles and the velocity of those particles relative to that of the single, size j particle under consideration. In words, the collision rate can be expressed as follows:

$$dW = \begin{bmatrix} \text{Concentration} \\ \text{of size } i \text{ particles} \\ \text{upstream of} \\ \text{and aiming for} \\ \text{the strip} \\ (1) \end{bmatrix} \begin{bmatrix} \text{Velocity of size} \\ i \text{ particles relative} \\ \text{to the size} \\ j \text{ particle under} \\ \text{consideration} \\ (2) \end{bmatrix} \begin{bmatrix} \text{Project} \\ \text{area of} \\ \text{the strip} \\ (3) \end{bmatrix}$$

In symbols, this expression can be developed as follows:

(1) Concentration $= n_i$, the same anywhere in the suspension
(2) Relative velocity $= y(du/dy) = yG$, where G is the velocity gradient. (Note that this expression accounts for the origin of the y-axis being at the center of the (moving) size j particle.)
(3) Projected area $=$ (length) (width) $= 2(a_i + a_j)(\cos\theta) \, dy$. As shown in Figure 12-6, the distance $(a_i + a_j)\cos\theta$ is the distance from the vertical centerline of the sphere to the edge at the angle θ. Choosing the center line-to-edge distance as the basis results in the multiplication by 2 to yield the required edge-to-edge length of this projected area but also dictates that the relevant range of θ is from 0 to $\pi/2$.

The vertical distance to the middle of the strip is $y = (a_i + a_j)\sin\theta$, so $dy = (a_i + a_j)\cos\theta \, d\theta$, and the projected area can be written as: $2(a_i + a_j)^2 (\cos^2\theta) \, d\theta$. Thus,

$$dW = n_i\{(a_i + a_j)\sin\theta(G)\}\{2(a_i + a_j)^2 \cos^2\theta \, d\theta\} \tag{12-14}$$

To find the overall rate at which i-sized particles collide with the j-sized particle, one needs to integrate dW over the entire sphere. The shaded area in Figure 12-6 represents a differential element in the top portion of the sphere; particles of size i whose centers are above the equator of the j particle and behind it in the direction of travel (as depicted in Figure 12-5) might catch up to the j particle and collide with it in that area. The j particle under consideration can also catch up to i-sized particles that are below the equator (and are therefore traveling slower) and in front of it in the direction of travel; although not shown, this condition is the equivalent of an i-sized particle below and to the right of the j particle in Figure 12-5. On the other hand, no i-sized particle that is below the equator of the j particle and trailing it will catch up to and collide with the j particle, nor will the j particle catch up to any i-sized particles that are above its equator and ahead of it. Hence, the collision rate in the lower, back and the upper, front parts of the sphere are zero, and these portions of the surface can be ignored in the integration. In terms of Figure 12-5, collisions can occur on the upper left (as shown) quadrant and the lower right quadrant, but cannot occur on the other two quadrants.

Since the system is symmetric, the rate of collisions on the lower, front portion of the sphere is identical to the rate on the upper back. Therefore, to find the overall rate of collisions, one can take the product of the three terms in Equation 12-14, integrate over the range of the angle θ from 0 to $\pi/2$, and multiply by 2 to account for both faces of the sphere that can collide with size i particles:

$$\oint_{\substack{\text{Entire}\\ \text{surface}}} dW = W_{\text{overall}} = 4n_i G(a_i + a_j)^3 \int_0^{\pi/2} \cos^2\theta \sin\theta\, d\theta \qquad (12\text{-}15)$$

Since $\sin\theta\, d\theta = -d(\cos\theta)$, the term inside the integral is equivalent to $-\cos^2\theta\, d(\cos\theta)$, and the integral equals $-(\frac{1}{3})\cos^3\theta\big|_0^{\pi/2}$. The value of this integral is $\frac{1}{3}$, so

$$W = \tfrac{4}{3} n_i G(a_i + a_j)^3 \qquad (12\text{-}16)$$

W is the rate of collisions with a single j-sized particle, so the cumulative rate of collisions of all j-sized particles in the suspension with i-sized particles is $n_j W$. The rate of collisions between i- and j-sized particles by shear is also defined in Equation 12-3 in terms of the collision frequency function as $^{Sh}\beta_{ij} n_i n_j$. Thus:

$$n_j W = \tfrac{4}{3} G(a_i + a_j)^3 n_i n_j = {}^{Sh}\beta_{ij} n_i n_j \qquad (12\text{-}17)$$

$$^{Sh}\beta_{ij} = \tfrac{4}{3} G(a_i + a_j)^3 = \tfrac{1}{6} G(d_i + d_j)^3 \qquad (12\text{-}18)$$

EXAMPLE 12-2. Find the values of the collision frequency function for collisions by fluid shear between 2-μm particles and 0.4-, 2-, and 30-μm particles, for a velocity gradient of 25 s^{-1}.

Solution. In cgs units, the diameters of 0.4, 2, and 30 μm are 4×10^{-5} cm, 2×10^{-4} cm, and 3×10^{-3} cm, respectively. Direct substitution into Equation 12-18 yields the desired results for the three types of two-particle collisions, as follows.

$$\begin{aligned}^{Sh}\beta(0.4, 2) &= \tfrac{1}{6} G(d_i + d_j)^3 \\ &= \tfrac{1}{6}(25\,\text{s}^{-1})\left[(4.0 \times 10^{-5}\,\text{cm}) + (2 \times 10^{-4}\,\text{cm})\right]^3 \\ &= 5.76 \times 10^{-11}\,\text{cm}^3/\text{s}\end{aligned}$$

$$\begin{aligned}^{Sh}\beta(2, 2) &= \tfrac{1}{6} G(d_i + d_j)^3 \\ &= \tfrac{1}{6}(25\,\text{s}^{-1})\left[(2.0 \times 10^{-4}\,\text{cm}) + (2 \times 10^{-4}\,\text{cm})\right]^3 \\ &= 2.67 \times 10^{-10}\,\text{cm}^3/\text{s}\end{aligned}$$

$$\begin{aligned}^{Sh}\beta(2, 30) &= \tfrac{1}{6} G(d_i + d_j)^3 \\ &= \tfrac{1}{6}(25\,\text{s}^{-1})\left[(2.0 \times 10^{-4}\,\text{cm}) + (3 \times 10^{-3}\,\text{cm})\right]^3 \\ &= 1.37 \times 10^{-7}\,\text{cm}^3/\text{s}\end{aligned}$$
∎

Calculations of the collision frequency function for fluid shear are shown in Figure 12-7 for three different values of

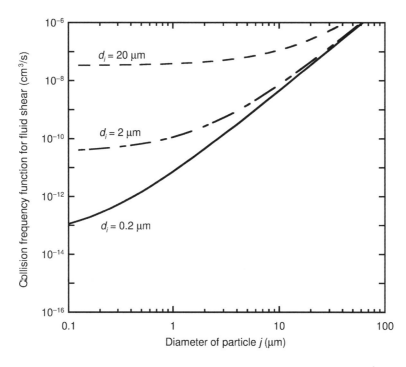

FIGURE 12-7. Collision frequency function for fluid shear ($G = 25\,\text{s}^{-1}$).

diameter for one particle (particle i in the figure) and a size range of three orders of magnitude for the second particle (from 0.1 to 100 μm). G values in the range of 15–50 s^{-1} are common in operating flocculators; for the calculations, the velocity gradient G was set at 25 s^{-1}. The relatively flat portion of the curve when $d_j < d_i$ and the fact that $^{Sh}\beta_{ij}$ values of all three curves are nearly equal at large d_j (where $d_j > d_i$ for all three values of d_i) indicate that the collision frequency function is strongly dependent on the size of the larger particle, but only slightly influenced by the smaller particle. The slope on this log–log graph approaches a value of 3.0 at high values of d_j, reflecting the exponent in Equation 12-18. The effect of the value of G, while not shown in the figure, is linear, as can be seen in Equation 12-18. Hence, doubling the velocity gradient to 50 s^{-1} or halving it to 12.5 s^{-1} would shift the lines up or down, respectively; the effect would be noticeable but not dramatic on this logarithmic graph.

Collisions by Differential Sedimentation

Collisions between particles can also occur because they settle at different velocities; the conceptual view is shown in Figure 12-8. The notation includes the settling velocity, v, for the two sizes of particles represented. The particle with the larger settling velocity (particle i) catches up to the smaller particle (particle j) as they both settle. The circle (i.e., oval in the view shown) at the bottom represents the capture cross-section; the center of that circle coincides with the center of the trajectory of the larger particle. If the trajectory of the center of the small (j) particle is anywhere

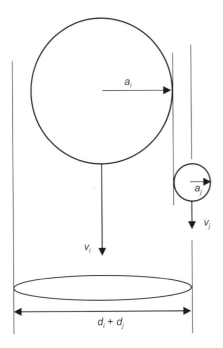

FIGURE 12-8. Conceptual diagram for deriving collision frequency function for differential sedimentation.

in that circle, the particles will collide. The small particle shown in Figure 12-8 has the *critical trajectory*, in that it will just barely collide with the big particle. For this critical trajectory, the smaller particle would attach to the larger at their equators when they collide, as indicated by the vertical tangent to both spheres.

The derivation of the collision frequency function for this mechanism is similar to that for the fluid shear mechanism; that is, the rate at which larger particles collide with a single, smaller one by falling onto it can be written as the product of the concentration of the larger particles, the relative settling velocities, and the projected area for potential collisions. The relative settling velocity is the difference in the settling velocities of the two particles, $|v_i - v_j|$. The projected area is a circle with radius $(a_i + a_j)$ or diameter $(d_i + d_j)$. The result is as follows:

$$^{DS}\beta_{ij} = |v_i - v_j|\left(\frac{\pi}{4}\right)(d_i + d_j)^2$$
$$= \frac{\pi g}{72\mu}(\rho_p - \rho_L)(d_i + d_j)^3|d_i - d_j| \quad (12\text{-}19)$$

where g is the gravitational constant (L/T^2); μ is the viscosity of the liquid (M/L/T); and ρ_p and ρ_L are the densities of the particles and liquid, respectively (M/L^3). The first expression is general, but the second assumes that both particles have the same density and settle according to Stokes' law.[2] This assumption is valid for most particles encountered in typical drinking water and wastewater plants. Note that, if the particles in the suspension had a distribution of densities, the collision frequency function would no longer be a function of the sizes of particles only. While accounting for these differences would not be difficult in the collision frequency function, it would complicate the overall modeling of flocculation considerably.

■ **EXAMPLE 12-3.** Find the values of the collision frequency function for collisions by differential sedimentation for 2-μm particles with 0.4, 2, and 30-μm particles, all with a density of 1.2 g/cm^3. Assume the temperature is 20°C, at which the density of water is 0.998 g/cm^3 and the viscosity is 1.002×10^{-2} g/cm s. Assume that all the particles settle according to Stokes' law. (The validity of this assumption is demonstrated in Chapter 13.)

Solution. Because we are assuming that Stokes' law applies, we can use the second form of Equation 12-19. Using more rounded values of the water density and viscosity would certainly be acceptable, but the greater detail is shown here for clarity. Again working in cgs units, the

---

[2] Stokes' law, which applies to particles with Reynolds numbers <0.1, is described in detail in Chapter 13. It states that $v = ((\rho_p - \rho_L)gd_p^2)/18\mu$. The second expression in Equation 12-19. is derived from the first by substituting this expression for v.

calculations proceed as follows:

$$^{DS}\beta_{ij} = \frac{\pi g}{72\mu}(\rho_p - \rho_L)(d_i + d_j)^3 |d_i - d_j|$$

$$^{DS}\beta_{0.4,2} = \frac{\pi(981\,\text{cm/s}^2)}{72(1.002 \times 10^{-2}\,\text{g/cm s})}\left((1.2 - 0.998)\text{g/cm}^3\right)\left((4.0 \times 10^{-5} + 2.0 \times 10^{-4})\text{cm}\right)^3 |(4.0 \times 10^{-5} - 2.0 \times 10^{-4})\text{cm}|$$

$$= 1.91 \times 10^{-12}\,\text{cm}^3/\text{s}$$

Substituting $d_j = 2 \times 10^{-4}$ cm or 3×10^{-3} cm into Equation 12-19 yields:

$$^{DS}\beta_{2,2} = 0 \text{ and } ^{DS}\beta_{2,30} = 7.92 \times 10^{-8}\,\text{cm}^3/\text{s}$$

The reason that $^{DS}\beta_{2,2} = 0$ is that the two particles have the same settling velocity, and hence cannot collide by differential sedimentation; the same result occurs for any two identical particles. ∎

The collision frequency function for differential sedimentation is plotted in Figure 12-9 in the same manner as for fluid shear in Figure 12-7. The collision frequency is very small when the particles are of almost the same size, and is zero when the two sizes are identical (indicated on the figure by the dip and discontinuity in each curve). If the particles are widely separated in size, the collision frequency is dominated by the larger particle. As is the case for fluid shear, this trend is indicated by the curves being almost flat when $d_i \gg d_j$ and being nearly identical when $d_j \gg d_i$.

The slope on this log–log plot approaches 4.0 when $d_j \gg d_i$, consistent with Equation 12-19.

The density can have a significant effect on the collision frequency, as the driving force for settling is the difference between the densities of particles and water. Particles in water and wastewater treatment plants have densities ranging from nearly the same as that of water to values approximating that of clay and silica (2.65 g/cm^3). If the particles had a density of 1.02 g/cm^3, all the lines in Figure 12-9 would be shifted down one order of magnitude, and if they had a density of 2.65 g/cm^3, the lines would be shifted upward slightly less than one order of magnitude.

■ **EXAMPLE 12-4.** Find the value of the collision frequency function for collisions by differential sedimentation between 2- and 30-μm particles with a density of 2.4 g/cm^3. Assume all of the same conditions as in the previous example, and compare the results with that example.

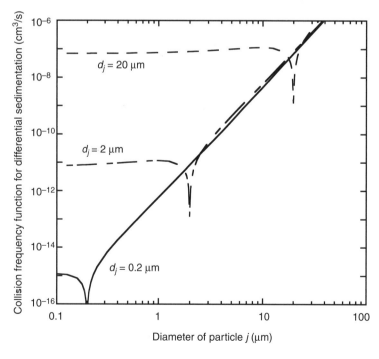

FIGURE 12-9. Collision frequency function for differential sedimentation (particle density = $\rho_p = 1.2$ g/cm^3).

Solution. The calculation is a direct application of Equation 12-19, as follows:

$$^{DS}\beta_{ij} = \frac{\pi g}{72\mu}(\rho_p - \rho_L)(d_i + d_j)^3|d_i - d_j|$$

$$^{DS}\beta_{2,30} = \frac{\pi(981\,\text{cm/s}^2)}{72(1.002 \times 10^{-2}\,\text{g/cm s})}((2.4 - 0.998)\text{g/cm}^3)((2.0 \times 10^{-4} + 3.0 \times 10^{-3})\text{cm})^3|(2.0 \times 10^{-4} - 3.0 \times 10^{-3})\text{cm}|$$

$$= 5.50 \times 10^{-7}\,\text{cm}^3/\text{s}$$

This value compares with $7.92 \times 10^{-8}\,\text{cm}^3/\text{s}$ for the lower density particles ($\rho_p = 1.2\,\text{g/cm}^3$) of the previous example. While the ratio of the particle densities between the two examples is two, the ratio of the values of $(\rho_p - \rho_L)$ is seven, leading to the sevenfold difference in the collision frequency functions. ∎

Collisions by Brownian Motion

Collisions between particles also occur by the Brownian motion of the particles; flocculation via this mechanism is sometimes called *perikinetic flocculation*. Brownian motion is caused by the random bombardment of molecules of the surrounding fluid on the particle surfaces. This phenomenon is identical to the causes of diffusion of soluble molecules that leads to a flux of material down a concentration gradient. Einstein (1905) was the first to combine the concepts of thermal motion of molecules in the liquid (expressed as a Maxwell–Boltzmann distribution) and the drag on particles (expressed by Stokes' law) to describe the Brownian motion as a diffusion coefficient. This Stokes–Einstein equation is

$$D_p = \frac{k_B T}{3\pi\mu d_p} \qquad (12\text{-}20)$$

The collision frequency function for Brownian motion is derived based on the assumption that the relative motion of particles i and j is the superposition of the motions of each. This assumption means that the diffusion coefficient describing the relative motion of the two colliding particles is the sum of their individual diffusion coefficients; that is,

$$D_{ij} = D_i + D_j = \frac{k_B T}{3\pi\mu}\left(\frac{1}{d_i} + \frac{1}{d_j}\right) \qquad (12\text{-}21)$$

The derivation of the collision frequency function, originally developed by Smoluchowski (1917), is not detailed here, but was shown by Friedlander (1977). As in the derivation of the collision frequency function for fluid shear or differential sedimentation, the analysis involves modeling the movement of i-sized particles to the surface of a single j-sized particle. In this case, the driving force is diffusion, and the rate of collisions is statistically based because the trajectories are random. A number balance can be written to compute the concentration profile of i-sized particles near a single j-sized particle, as a function of time $(n_i\{r,t\})$. In spherical coordinates, the equation is

$$\frac{\partial n_i}{\partial t} = D_{ij}\frac{1}{r^2}\frac{\partial}{\partial r}\left(r^2\frac{\partial n_i}{\partial r}\right) \qquad (12\text{-}22)$$

where r is the radial distance from the center of the j-sized particle.

This differential equation is first order in time and second order in space, and therefore requires one initial condition and two boundary conditions. One boundary condition is that the surface of the j particle is considered an infinite sink; that is, the concentration of size i particles is zero on the surface of the sphere of interaction at all times after time zero.[3] The second boundary condition describes the concentration of i-sized particles far from the size j particle as equal to the bulk concentration. The initial condition is that the concentration is the same everywhere (at the bulk concentration) at time zero.

The predicted rate of collisions due to Brownian motion can then be determined by computing the flux of i-sized particles to the surface of the j-sized particle according to Fick's law and multiplying the flux by the surface area of the j-sized particle. The resulting equation for the collision frequency function is

$$^{Br}\beta_{ij} = \frac{2k_B T}{3\mu}\left(\frac{1}{d_i} + \frac{1}{d_j}\right)(d_i + d_j) \qquad (12\text{-}23)$$

where k_B is Boltzmann's constant ($1.38 \times 10^{-16}\,\text{g cm}^2/\text{s}^2\,\text{K}$), T is absolute temperature (K), and the other variables have been defined.

The collision frequency function for Brownian motion is shown in Figure 12-10. While $^{Br}\beta_{ij}$ depends on the sizes of both particles, the effect of particle size is far less dramatic than in fluid shear and differential sedimentation. For each choice of d_i shown, $^{Br}\beta_{ij}$ varies by less than two orders of

---

[3] The point of this assumption is that once an i-sized particle reaches the surface, the two particles instantly become one larger particle; thus, the concentration of the smaller particle at the surface is zero, and back-diffusion from the surface is impossible.

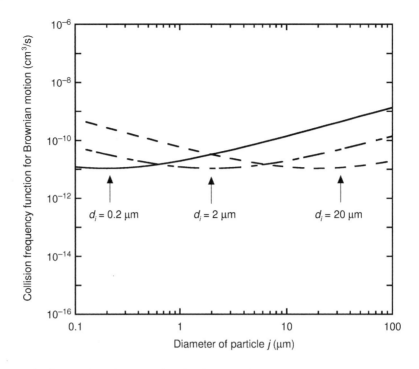

FIGURE 12-10. Collision frequency function for Brownian motion (temperature = 20°C = 293 K).

magnitude throughout the range of d_j. The function has a minimum when $d_i = d_j$ and is symmetric around that point; the value of that minimum is independent of particle size (and equal to $8k_BT/3\mu$). Temperature has some effect on collisions by this mechanism; lower temperature leads to lower collision frequency, both by the direct effect of T on the energy of the molecules in the suspension and by the fact that viscosity increases with decreasing temperature.

■ **EXAMPLE 12-5.** Find the values of the collision frequency function for collisions by Brownian motion for 2-μm particles with 0.4-, 2-, and 30-μm particles. Assume the temperature is 20°C, at which the viscosity of water is 1.002×10^{-2} g/cm s.

Solution. Direct substitution into Equation 12-23, using cgs units for all variables, yields the desired results for the collisions between 0.4- and 2-μm particles, as follows:

$$^{Br}\beta_{ij} = \frac{2k_BT}{3\mu}\left(\frac{1}{d_i} + \frac{1}{d_j}\right)(d_i + d_j)$$

$$^{Br}\beta_{0.4,2} = \frac{2(1.38 \times 10^{-16}\text{ g cm}^2/\text{s}^2\text{ K})293\text{ K}}{3(1.002 \times 10^{-2}\text{ g/cm s})}$$
$$\times \left(\frac{1}{4 \times 10^{-5}\text{ cm}} + \frac{1}{2 \times 10^{-4}\text{ cm}}\right)$$
$$\times (4 \times 10^{-5}\text{ cm} + 2 \times 10^{-4}\text{ cm})$$
$$= 1.94 \times 10^{-11}\text{ cm}^3/\text{s}$$

The other calculations are identical, but the units are omitted for brevity:

$$^{Br}\beta_{2,2} = \frac{2(1.38 \times 10^{-16})(293)}{3(1.002 \times 10^{-2})}\left(\frac{1}{2 \times 10^{-4}} + \frac{1}{2 \times 10^{-4}}\right)$$
$$\times (2 \times 10^{-4} + 2 \times 10^{-4})$$
$$= 1.08 \times 10^{-11}\text{ cm}^3/\text{s}$$

$$^{Br}\beta_{2,30} = \frac{2(1.38 \times 10^{-16})(293)}{3(1.002 \times 10^{-2})}\left(\frac{1}{2 \times 10^{-4}} + \frac{1}{3 \times 10^{-3}}\right)$$
$$\times (2 \times 10^{-4} + 3 \times 10^{-3})$$
$$= 4.59 \times 10^{-11}\text{ cm}^3/\text{s} \quad ■$$

The Total Collision Frequency Function

The preceding discussion describes three mechanisms that are capable of bringing particles from far apart to a point where they can collide. Since the mechanisms are essentially independent of one another, it is common to assume that the three collision frequency functions are additive. In the case under consideration, in which the influence of short-range forces is not included in an explicit way, the result is

$$^{tot}\beta_{ij} = {}^{Sh}\beta_{ij} + {}^{DS}\beta_{ij} + {}^{Br}\beta_{ij} \quad (12\text{-}24)$$

This assumption is identical to that made in writing Equation 12-8. Some investigators (e.g., Zeichner and

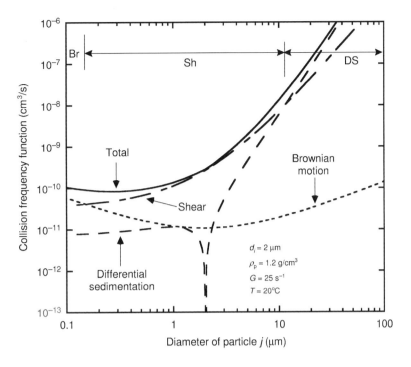

FIGURE 12-11. Total collision frequency function using the long-range force model.

Schowalter, 1977) have questioned the correctness of this assumption on the basis that the mechanisms are not completely independent, but to the accuracy of these collision frequency functions, it can be considered correct.

■ **EXAMPLE 12-6.** Find the total collision frequency function according to the long-range force model for collisions of 2 μm diameter particles with 0.4-, 2-, and 30-μm particles. The density of all particles is $1.2\,\text{g/cm}^3$, and the water temperature is 20°C (i.e., 293 K) where the density is $0.998\,\text{g/cm}^3$ and the viscosity is $1.002 \times 10^{-2}\,\text{g/cm s}$. For each pair of collisions, find the collision mechanism that contributes the most to the total collision frequency function value.

Solution. The collision frequency function values for each collision mechanism for these particles have been calculated in the previous examples. They are collected in the following table, and then summed for each collision pair.

According to the long-range force model, fluid shear contributes most to the overall collision frequency function for all three collision pairs under consideration. For the collisions between two 2-μm particles, shear is responsible for 96% of the total. Brownian motion is a significant contributor to the collisions between 0.4- and 2-μm particles, and differential sedimentation is substantial in the collisions of 2- and 30-μm particles. ■

A comparison of the magnitudes of the three terms in Equation 12-24 can indicate which transport process or processes are most significant in terms of providing possibilities for collisions (i.e., near approaches) between different particles. A graph illustrating the relative importance of the three mechanisms under conditions representative of water treatment plants is shown in Figure 12-11. Under conditions shown on the figure, it is clear that close approaches to 2-μm diameter particles are dominated by Brownian motion when the second particle (j) is quite small,

Diameters of Collision Pairs (μm)	Brownian Motion $^{Br}\beta_{ij}$ (cm^3/s)	Fluid Shear $^{Sh}\beta_{ij}$ (cm^3/s)	Differential Sedimentation $^{DS}\beta_{ij}$ (cm^3/s)	Total $^{Tot}\beta_{ij}$ (cm^3/s)	Dominant Collision Mechanism
0.4–2	1.94×10^{-11}	5.76×10^{-11}	1.91×10^{-12}	7.89×10^{-11}	Sh
2–2	1.08×10^{-11}	2.67×10^{-10}	0	2.78×10^{-10}	Sh
2–30	4.59×10^{-11}	1.37×10^{-7}	7.92×10^{-8}	2.16×10^{-7}	Sh

and dominated by differential sedimentation when the second particle is quite large. For all sizes in between, approaches by fluid shear are the most likely. The ranges where each mechanism dominates are a function of the choice of the nonvarying particle size (d_i) and all the parameters describing the particles, suspension, and operation. Generally, however, one can say that formation of proto-flocs between two small particles is dominated by Brownian motion, while collisions involving a very large particle and a much smaller particle are dominated by differential sedimentation, and all others by fluid shear.

Considering all possible combinations of particle sizes leads to a more general understanding of the relative importance of the three mechanisms in this model. In Figure 12-12, regions of dominance for each of the three mechanisms are shown for collisions between any two particle sizes. Only half of the region is used because the graph would be symmetric around the center line representing equi-sized particles; that is, the choice of i and j is arbitrary for collisions of any two sizes. The results in Figure 12-12 indicate that, when only long-range forces are considered, most collisions are predicted to occur by fluid shear. The regions are dependent on the specific conditions chosen, but the general conclusion that most collisions are predicted to occur by shear is general for any reasonable conditions. Increasing the velocity gradient or shear rate, G, increases the region of dominance by shear. Decreasing the temperature increases the fluid viscosity and decreases the regions dominated by Brownian motion and differential sedimentation.

Design Implications of the Long-Range Force Model

In the light of this analysis, it is not surprising that design and operation of flocculators has been based almost exclusively on fluid shear as the cause of collisions. Because the collision frequency is linearly dependent on G according to Equation 12-18, the highest mixing intensity that does not result in floc breakup has been thought best by many. Camp (1955) suggested that the dimensionless product $G\tau$ (where τ is the detention time) be used as the primary design factor for flocculation basins; the mathematical basis for the importance of this product relies on the assumption that all of the particles in the suspension are the same size. This assumption leads to simplified equations for the collision frequency (β) and the rate of flocculation that are suggested as an exercise at the end of this chapter. In a review of several existing facilities, Camp found that the product $G\tau$ ranged from ~23,000 to 210,000 and suggested that designers should use this range as a guideline. Even today, most flocculators are designed to meet this guideline.

According to the assumption made by Smoluchowski and accepted by most subsequent researchers, the total particle volume does not change during flocculation. This concentration can be represented as the dimensionless volume fraction, ϕ (i.e., ϕ = particle volume/suspension volume); Ives (1968) suggested that the effect of concentration on flocculation rate could be accounted for by including the floc volume fraction in the design parameter; that is, he suggested that the key parameter is the dimensionless product

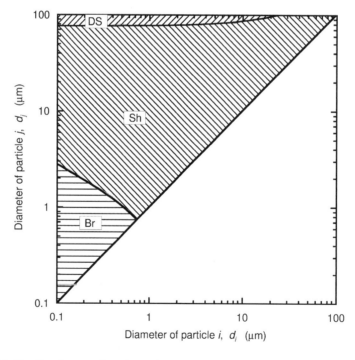

FIGURE 12-12. Dominant regions for each collision mechanism in the long-range force model. (Borders between regions vary somewhat with conditions.)

$\phi G\tau$. He argued that, in this way, the similarity in degree of flocculation achieved in sludge blanket clarifiers (where ϕ is high but G and τ are low) and in normal suspension flocculators (where ϕ is low but G and τ are high) could be explained. O'Melia (1972) extended the analysis to account for particle stability by including the collision efficiency factor α_{emp}; he suggested that all successful flocculation systems would have reasonably similar values of the dimensionless product $\alpha_{\text{emp}}\phi G\tau$.

The guidelines of Camp, Ives, and O'Melia are all applications of the idea that successful flocculators provide a detention time that is some multiple of a characteristic reaction time. This concept of design was suggested in Chapter 4, and used in Example 12-1. For example, the Camp guideline is equivalent to saying that the characteristic reaction time can be thought of as $1/G$ and that the design detention time should be 23,000–210,000 times that characteristic reaction time; such a definition of the characteristic reaction time would represent only a small fraction of the time required for the reaction to go to completion. The Ives and O'Melia guidelines include additional terms in the denominator of the characteristic reaction time (and are more consistent with the definition used throughout this book) and decrease the multiplier to obtain the design detention time.

As shown in Equation 12-13 (repeated here), the characteristic detention time for flocculation, for the cases in which the formation of the given size particles is negligible in comparison to the their loss by growth into larger particles, is as follows:

$$t_{\text{char}} = \frac{1}{n_{\text{Tot},o}\alpha_{\text{emp}} \sum_{\text{all }i}^{\text{Tot}}(\alpha_{ik}\beta_{ik})f_i} \quad (12\text{-}13)$$

The guidelines of the investigators cited earlier were based on the simplifications of considering monodisperse suspensions (thereby omitting the summation and the f_i terms in the denominator) and were developed for the long-range model (omitting the α_{ik} term). These investigators also considered shear as the only relevant collision mechanism and, therefore, could substitute G for $^{\text{Sh}}\beta$, since these two quantities are directly proportional. Finally, for a monodisperse suspension, ϕ is directly proportional to n_{Tot}. Hence, the final (O'Melia) guideline was the most complete in accounting for the various terms in the expression for t_{char}, whereas the earlier versions left some of these terms to be included in the (unknown) multiple of t_{char} that should be provided to obtain τ.

12.5 COLLISION EFFICIENCY: SHORT-RANGE FORCE MODEL

The collision frequency functions developed in the preceding section describe the movement of particles along paths that are unaffected by the existence of other particles. Such a view is realistic as long as the particles are far from one another, but it becomes unrealistic when the particles are in close proximity because it omits electrostatic effects (van der Waals attraction and electric double layer attraction or repulsion) and hydrodynamic effects (the flow of water around one particle affecting the motion of the other). These short-range phenomena are quantifiable as forces that are exerted on the two particles and that depend on the separation distance and the angle at which the two particles approach one another. Accounting for these short-range phenomena results in a modification of the collision frequencies developed above; we call this set of modifications the short-range force model.

The hydrodynamics associated with the movements of two closely spaced particles under low Reynolds number conditions has been of interest for a long time, and many investigators have reported results from calculations describing such forces. The key difference between these calculations and the classical calculations applicable to particles far from one another (the calculations leading to the β_{ij} functions) is that, when considering the motion of particles close to one another, the effect of each particle on the motion of the surrounding water is computed, and the effect of that water movement on the other particle is then taken into account. Accounting for this fluid motion around the particles, including the no-slip condition at the surface of each particle, reduces the predicted frequency of interparticle collisions; these effects are sometimes referred to as *hydrodynamic retardation*.

Interparticle forces due to van der Waals attraction and double layer repulsion can be incorporated into the same framework (Jeffrey and Onishi, 1984). The forces are computed as the negative gradient of the interparticle electrical potential ($F = -(\mathrm{d}\psi/\mathrm{d}r)$, where r is the distance between centers), as described in Chapter 11. The framework for these calculations is a procedure known as *trajectory analysis* and consists of evaluating the location of the smaller particle in a coordinate system that moves with the larger particle; that is, the analysis is based on a Lagrangian view, in which the origin of the coordinate system is always at the center of the larger particle. The use of trajectory analysis to predict particle collisions is illustrated next, for an example in which the long-range transport mechanism is differential sedimentation.

Consider the two-particle system shown in Figure 12-13, depicting the gravity settling of two spheres in proximity in a quiescent fluid (i.e., a fluid in which the only motion is induced by the particle motion). The larger particle (a_1) can be expected to settle faster than the smaller particle (with radius a_2). The plane of the two-dimensional polar coordinate system is set to be vertical and to pass through the centers of both particles.[4] The origin is chosen to be at

---

[4] A two-dimensional system is sufficient because spherical particles are symmetric and no forces are present that act in the third dimension.

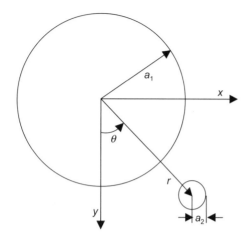

FIGURE 12-13. Two particles settling in close proximity: definition diagram for the effects of short-range forces on collision frequency.

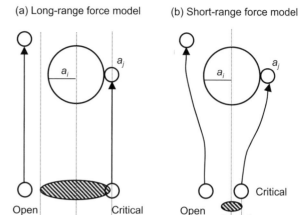

FIGURE 12-14. Trajectories of particles undergoing differential sedimentation according to the long-range and short-range models.

the center of the larger particle, and the angle θ is measured counterclockwise from a line directed vertically downward, as shown in the figure. The location of a second particle is defined by its center, with corresponding values of r (the center-to-center distance between the particles) and θ. The forces acting on particle 2 at any location, and therefore its acceleration toward or away from particle 1, can be computed based on the (induced) flow field at that location and the known properties of the system (in the case of gravity-induced motion, the densities of the particles and the fluid and the fluid viscosity).

The calculation is carried out by considering particle 2 at various initial locations, all far enough from particle 1 that interaction forces are negligible. For the current example, the initial velocities of both particles would be downward with magnitudes corresponding to the Stokes settling velocity. In successive short time steps, the forces and corresponding acceleration acting on particle 2 are evaluated, and its location is incremented; the process is repeated to determine the trajectory of particle 2 until it either collides with or passes by particle 1.

The calculations are carried out by computing the relative velocity v of the two particles (i.e., the difference in velocities of the two particles), resolved into r and θ directions. For each direction, the contributions of the long- and short-range factors are separated, that is:

$$\text{In } r \text{ direction}: v_r = v_{r,\text{gravity}} + v_{r,\text{interaction}} \qquad (12\text{-}25)$$

$$\text{In } \theta \text{ direction}: v_\theta = v_{\theta,\text{gravity}} + v_{\theta,\text{interaction}} \qquad (12\text{-}26)$$

In these equations, the interaction component of each velocity term is a function of the dimensionless separation distance, s, and the particle size ratio (λ, the ratio of smaller to larger particle radius).[5] The equation describing the relative motion of the particles is

$$\frac{dr(t)}{d\theta(t)} = r(t)\frac{v_r}{v_\theta} \qquad (12\text{-}27)$$

In trajectory analysis, this equation can be integrated numerically from any chosen initial condition to find the location of the second particle relative to the first as a function of time; that is, to find $r(t)$ and $\theta(t)$.

The principal value of trajectory analysis in the context of flocculation is that it can be used to calculate α_{ij} values for particles approaching one another by each of the three long-range transport mechanisms. The concept is illustrated in Figure 12-14. Trajectories of particles according to the long-range force approach are shown in Figure 12-14a, which is explained in earlier sections of this chapter (i.e., in which interactions are ignored), and trajectories of particles according to the short-range force approach are shown in Figure 12-14b, which accounts for the hydrodynamic effects and interparticle forces as two particles approach. Some possible trajectories of a small particle in relation to a large particle are depicted in a Lagrangian view (coordinate system moving with the large particle) in each case. Both particles are settling downward but, to the large particle with its faster settling velocity, it appears that the small particle is approaching from the bottom. Open trajectories (shown on the left side of the large particle for each model) are those that do not lead to a collision; the particles simply pass by

---

[5] The dimensionless separation distance between the centers is defined as the absolute separation divided by the average of the two particle radii; that is, $s \equiv r/((a_1 + a_2)/2) = 2r/(a_1 + a_2)$. When $s > 2$, the particles are separated, and when $s = 2$, contact occurs. The particle size ratio λ is defined as the ratio of the smaller radius to the larger one, and hence is always <1.

one another. The critical trajectory (shown on the right side for each model) is the one with the maximum horizontal separation distance (at an arbitrary large initial vertical separation) that leads to a collision.

In the long-range force model, the trajectories are rectilinear, so that the critical horizontal separation distance of the centers of the two particles when far removed from one another vertically is the sum of the radii of the two particles. Any size j particle whose center passes through the area defined by the circle with radius $a_i + a_j$ in Figure 12-14a will collide with the size i particle. In the short-range force model, the trajectories are curved and the minimum horizontal separation distance is much smaller, as indicated by X_c in Figure 12-14b. Any size j particle that passes through an area defined by a circle of radius X_c will collide with the size i particle; this area that leads to a collision or closed trajectory is known as the *capture cross-section*. In the trajectory analysis, the value of X_c is found by repeated trials with different values of the initial horizontal separation distance to find the largest distance that results in a collision. The collision efficiency function is then calculated as the ratio of the areas that lead to collisions in the curvilinear and rectilinear cases:

$$^{\text{DS}}\alpha_{ij} = \frac{X_c^2}{(a_i + a_j)^2} \quad (12\text{-}28)$$

The approach used to calculate the relative particle motion caused by fluid shear and the corresponding value of $^{\text{Sh}}\alpha_{ij}$ is nearly identical, although the details differ because the forces causing the motion are different. For transport by fluid shear, the direction of relative motion is not necessarily vertical but is in the direction of flow; Adler (1981a–d) is responsible for the corrections by fluid shear that are reported here. For Brownian motion, a different approach must be used to calculate the collision efficiency function in the short-range model because Brownian motion is statistically based. The analysis still accounts for the effects of the phenomena relevant at small separation distances, but the values are not derived from predicted trajectories of individual particles.

Equations 12-25, 12-26, and 12-27 make it clear why α_{ij} values depend on the long-range transport mechanism bringing the particles together. The components of velocity in the r and θ directions depend on the long-range transport mechanism, and these terms then determine the motion of the particle of interest relative to the other particle with which it might collide. As a practical matter, this result means that if two particles are brought to exactly the same relative positions and have the same velocities via shear in one instance, and via differential sedimentation in another, they would have different velocities at the next instant because of the differences in the long-range forces causing their motion, and therefore they would proceed along different paths to a collision.

The collision efficiency functions, α_{ij}, accounting for hydrodynamic effects and van der Waals forces (but not double layer interactions) for the three transport modes are shown in Figure 12-15. Details of the development of these figures are given by Han and Lawler (1991, 1992). Tables that allow calculation of α_{ij} for any combination of particle sizes are given in Appendix 12A. The choice to omit the double layer interactions stems from two considerations. First,

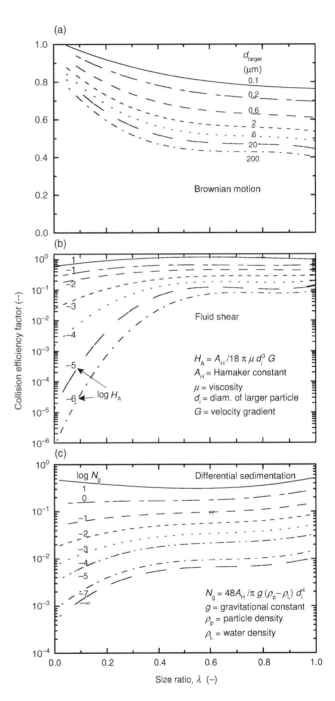

FIGURE 12-15. Collision efficiency functions in the short-range force model.

584 FLOCCULATION

if the destabilization is by charge neutralization and is complete, the double layer vanishes. Second, double layer interactions depend on solution and suspension conditions, making them less generalizable than van der Waals or hydrodynamic conditions. In essence, this choice means that double layer interactions are incorporated into α_{emp} for quantitative work.

For both fluid shear and differential sedimentation, all of the influencing variables besides the size ratio are incorporated into dimensionless numbers, commonly designated H_A in the case of fluid shear and N_g in the case of differential sedimentation. For each mechanism, these dimensionless parameters arise when the equations of motion are nondimensionalized. They both include the Hamaker constant (A), the size of the larger particle in the collision (d_l), and parameters associated with the particular collision mechanism, as follows:

$$H_A = \frac{A}{18\pi\mu d_l^3 G} \quad (12\text{-}29)$$

$$N_g = \frac{48A}{\pi(\rho_p - \rho_L)g d_l^4} \quad (12\text{-}30)$$

These dimensionless numbers are used in the calculation of the short-range collision efficiency functions, as indicated in Figure 12-15 and in Appendix 12A.

■ **EXAMPLE 12-7.** Estimate the collision efficiency functions according to the short-range force model for collisions of 2-μm diameter particles with 0.4-, 2-, and 30-μm particles by (a) Brownian motion, (b) fluid shear, and (c) differential sedimentation. Assume the Hamaker constant is 3×10^{-13} g cm^2/s^2, the density of the particles is 1.2 g/cm^3, and the temperature of the water is 20°C, at which the density is 0.998 g/cm^3 and the viscosity is 1.002×10^{-2} g/cm s. The velocity gradient is 25 s^{-1}.

Solution. We begin by calculating the λ values (ratio of smaller to larger diameter) for each of the three combinations of particle sizes, as follows:

$$\lambda_{0.4,2} = \frac{0.4\,\mu m}{2\,\mu m} = 0.2$$

$$\lambda_{2,2} = \frac{2\,\mu m}{2\,\mu m} = 1$$

$$\lambda_{2,30} = \frac{2\,\mu m}{30\,\mu m} = 0.06667$$

(a) *Brownian motion.* Before making exact calculations for these values, it is useful to get estimates from Figure 12-15a. On the graph, we can find or interpolate a line associated with the larger particle and then find the value of the collision efficiency at the appropriate λ values. For the cases at hand, we can estimate the following:

$^{Br}\alpha_{0.4,2} = 0.72 \quad ^{Br}\alpha_{2,2} = 0.54 \quad ^{Br}\alpha_{2,30} = 0.78$

For the larger particle being 2 μm, Table 12A-1 in Appendix 12A gives values for the coefficients (a, b, c, and d) to allow calculation of the $^{Br}\alpha$ values directly. For the collisions of the 2-μm particle with the 0.4- and 2-μm particles, the calculation is

$^{Br}\alpha_{i,j} = a + b\lambda + c\lambda^2 + d\lambda^3$

$^{Br}\alpha_{i,2} = 0.943 - 1.383\lambda + 1.725\lambda^2 - 0.748\lambda^3$

$^{Br}\alpha_{0.4,2} = 0.943 - 1.383(0.2) + 1.725(0.2)^2$
$\qquad - 0.748(0.2)^3 = 0.729$

$^{Br}\alpha_{2,2} = 0.943 - 1.383(1) + 1.725(1)^2$
$\qquad - 0.748(1)^3 = 0.537$

Table 12A-1 does not include values for the coefficients when the larger particle has a 30 μm diameter, but does show values for 20- and 60-μm particles. We can calculate $^{Br}\alpha$ values for both these values of the larger particle using $\lambda = 0.0667$ (for the collision of the 2- and 30-μm particles) and then interpolate between the two results.[6]

$^{Br}\alpha_{i,20} = 0.891 - 1.658\lambda + 2.221\lambda^2 - 1.009\lambda^3$

$^{Br}\alpha_{i,20,\lambda=0.06667} = 0.891 - 1.658(0.06667)$
$\qquad + 2.221(0.06667)^2 - 1.009(0.06667)^3$
$\qquad = 0.790$

$^{Br}\alpha_{i,60} = 0.871 - 1.739\lambda + 2.371\lambda^2 - 1.090\lambda^3$

$^{Br}\alpha_{i,60,\lambda=0.06667} = 0.891 - 1.739(0.06667)$
$\qquad + 2.371(0.06667)^2 - 1.090(0.06667)^3$
$\qquad = 0.765$

Finally, interpolating between these two values yields:

$^{Br}\alpha_{2,30} = 0.790 + \frac{30-20}{60-20}(0.765 - 0.790) = 0.784$

Note that, in all three cases, the estimates are very close to the calculated values; on this arithmetic graph, it is relatively easy to get good estimates of $^{Br}\alpha$ by eye.

(b) *Fluid shear.* The calculations begin with determination of the dimensionless parameter, H_A (and its logarithm), for each of the combinations, using

---

[6] Note that by using the λ value for the desired collision (in this case with a 30-μm particle as the larger) but using available values for the larger particle (here, 20 and 60 μm), the smaller particle is also being varied in this analysis. Alternatively, one could fix the smaller particle size (here, at 2 μm) rather than λ, and use the same two larger particles to find two α values, and then interpolate between them. The suggested approach of holding λ constant is simpler and at least as accurate.

Equation 12-29. For the three collision pairs, only two of the particles can be the larger, leading to two values of H_A:

For the 2-μm particle being the larger:

$$\log H_A = \log \frac{A}{18\pi\mu d_i^3 G}$$

$$= \log \frac{3 \times 10^{-13} \text{g cm}^2/\text{s}^2}{18\pi(1.002 \times 10^{-2} \text{g/cm s})(2 \times 10^{-4} \text{cm})^3 (25 \text{ s}^{-1})}$$

$$= \log(2.65 \times 10^{-3}) = -2.58$$

For the 30-μm particles:

$$\log H_A$$

$$= \log \frac{3 \times 10^{-13} \text{ g cm}^2/\text{s}^2}{18\pi(1.002 \times 10^{-2} \text{ g/cm s})(3 \times 10^{-3} \text{ cm})^3 (25 \text{ s}^{-1})}$$

$$= \log(7.84 \times 10^{-7}) = -6.11$$

With these values we can again estimate the collision efficiencies from Figure 12-15b, as follows:

$$^{Sh}\alpha_{0.4,2} = {}^{Sh}\alpha_{\lambda=0.2,\, \log H_A=-2.58} = 0.1$$

$$^{Sh}\alpha_{2,2} = {}^{Sh}\alpha_{\lambda=1,\, \log H_A=-2.58} = 0.35$$

$$^{Sh}\alpha_{2,30} = {}^{Sh}\alpha_{\lambda=0.06667,\, \log H_A=-6.11} = 1 \times 10^{-5}$$

Equations for calculating $^{Sh}\alpha$ values are provided in Appendix 12-A for specified values of $\log H_A$. We can use values of that parameter that surround the values of interest, calculate the $^{Sh}\alpha$ values at the appropriate λ values, and then interpolate between the two values obtained.

The equation for $^{Sh}\alpha$ is

$$^{Sh}\alpha = \frac{8}{(1+\lambda)^3} 10^{(a+b\lambda+c\lambda^2+d\lambda^3)}$$

For $\log H_A$ values of -3 and -2, we find

$$^{Sh}\alpha_{0.4,2,\, \log H_A=-3}$$

$$= \frac{8}{(1+\lambda)^3} 10^{(-2.523+5.550\lambda-6.098\lambda^2+2.553\lambda^3)}$$

$$= \frac{8}{(1+0.2)^3} 10^{(-2.523+5.550(0.2)-6.098(0.2)^2+2.553(0.2)^3)}$$

$$= 0.107$$

$$^{Sh}\alpha_{0.4,2,\, \log H_A=-2}$$

$$= \frac{8}{(1+\lambda)^3} 10^{(-1.704+3.116\lambda-2.881\lambda^2+1.121\lambda^3)}$$

$$= \frac{8}{(1+0.2)^3} 10^{(-1.704+3.116(0.2)-2.881(0.2)^2+1.121(0.2)^3)}$$

$$= 0.301$$

Finally, the desired value is found by interpolation of the above two values. Given that the parameter that separates the lines on the graph is $\log H_A$ and that the graph is logarithmic in α, it seems that the best approach is to use these two parameters in the interpolation. Hence, we find the log of the α values:

$$\log({}^{Sh}\alpha_{0.4,2,\, \log H_A=-3}) = \log(0.107) = -0.971$$

$$\log({}^{Sh}\alpha_{0.4,2,\, \log H_A=-2}) = \log(0.301) = -0.522$$

The interpolation then proceeds as follows:

$$\log({}^{Sh}\alpha_{0.4,2,\, \log H_A=-2.58}) = -0.971 + \frac{-3-(-2.58)}{-3-(-2)}$$
$$\times(-0.522-(-0.971))$$
$$= -0.782$$

and finally:

$$^{Sh}\alpha_{0.4,2,\, \log H_A=-2.58} = 10^{-0.782} = 0.165$$

Details of the calculations of the other two values are not shown, as they follow the same pattern as given here. The intermediate and final values are as follows:

$$^{Sh}\alpha_{2,2,\, \log H_A=-3} = 0.303$$

$$^{Sh}\alpha_{2,2,\, \log H_A=-2} = 0.449$$

$$^{Sh}\alpha_{2,2,\, \log H_A=-2.58} = 0.357$$

$$^{Sh}\alpha_{2,30,\, \log H_A=-7} = 4.62 \times 10^{-7}$$

$$4^{Sh}\alpha_{2,30,\, \log H_A=-6} = 1.15 \times 10^{-5}$$

$$^{Sh}\alpha_{2,30,\, \log H_A=-6.11} = 8.1 \times 10^{-6}$$

In this case, the original estimates based on reading the figure are not as accurate as in the Brownian motion case, but they still provide a quick and reasonable approximation.

(c) *Differential sedimentation.* The calculation of $^{DS}\alpha$ is very similar to that for fluid shear. We again begin by calculating the relevant parameter, in this case, $\log N_g$, for the two larger particles; this parameter is given by Equation 12-30.

$$\log N_g = \log \frac{48A}{\pi g(\rho_p - \rho_L)d_i^4}$$

$$\log N_{g,2}$$

$$= \log \frac{48(3 \times 10^{-13} \text{ g cm}^2/\text{s}^2)}{\pi(981 \text{ cm/s}^2)((1.2-0.998) \text{ g/cm}^3)(2 \times 10^{-4} \text{ cm})^4}$$

$$= \log(14.46) = 1.16$$

$$\log N_{g,30}$$

$$= \log \frac{48(3 \times 10^{-13} \text{g cm}^2/\text{s}^2)}{\pi(981 \text{ cm/s}^2)((1.2-0.998) \text{ g/cm}^3)(3 \times 10^{-3} \text{ cm})^4}$$

$$= \log(2.86 \times 10^{-4}) = -3.54$$

For the calculation with the 30-μm particle, we can proceed as with the fluid shear calculations: calculate $^{DS}\alpha$ values at $\log N_g = -4$ and -3, and make the interpolation on the logarithmic basis. Notice, however, that there are no calculation equations nor lines on Figure 12-15c for values of $\log N_g > 1.0$. From the trend of the values for lower $\log N_g$ values, it seems reasonable to assume that at a value of $\log N_g = 2$, no correction would be necessary; that is, that $^{DS}\alpha = 1$ for all values of λ. (Since it is unlikely that differential sedimentation will be a substantial mechanism for this situation of small particles and a relative low density, the assumption is not likely to be critical in any case.) We proceed with the calculations on that basis.

No collisions of identical particles occur by differential sedimentation, so only the values for the 0.4–2 and 2–30 μm particle collisions need be calculated. Since the calculations are performed identically to those for fluid shear, only the intermediate and final values are given.

$$^{DS}\alpha_{0.4,2,\,\log Ng=2} = 1$$
$$^{DS}\alpha_{0.4,2,\,\log Ng=1} = 0.380$$
$$^{DS}\alpha_{0.4,2,\,\log Ng=1.16} = 0.44$$
$$^{DS}\alpha_{2,30,\,\log Ng=-4} = 5.12 \times 10^{-3}$$
$$^{DS}\alpha_{2,30\,\log Ng=-3} = 0.0105$$
$$^{DS}\alpha_{2,30,\,\log Ng=-3.54} = 7.1 \times 10^{-3}$$

■ **EXAMPLE 12-8.** Find the total collision frequency function according to the short-range force model for collisions investigated in Example 12-7. For each pair of particles, find the long-range transport mechanism that contributes the most to the total collision frequency function value.

Solution. The values for the long-range collision frequency functions (β_{ij}) and the short-range collision efficiency functions (α_{ij}) for each collision mechanism for these particles have been calculated in the previous examples. The short-range collision frequency function is the product of these two values for each mechanism, and the total collision frequency function is the sum of these products for each collision pair. All of these values are collected in the following table for each colliding pair.

For collisions between 0.4- and 2-μm particles, the collision frequency is reduced by ~70% in the short-range force model in comparison to the long-range force model, and the mechanism with the largest contribution to the total switches from fluid shear to Brownian motion. For collisions between two 2-μm particles, the collision frequency is reduced by ~64% and shear remains the dominant mechanism. Predicted collisions between 2- and 30-μm particles are dramatically reduced (by 99.7%), and the dominant mechanism switches from fluid shear to differential sedimentation. Considering all of the α_{ij} values in the table, it is clear that the primary effect of the short-range forces is to dramatically reduce the collisions between large particles (say >10 μm diameter) and small particles (say $\lambda < 0.2$). ■

The results in Figure 12-15 and in the examples show that collision efficiency can be quite low (orders of magnitude smaller than one) for collisions by fluid shear and, to a lesser extent, by differential sedimentation, especially when the two particles are very different in size (low λ). As a result of the low collision efficiency, the overall predicted collision frequency (the product $\alpha_{ij}\beta_{ij}$) by these two mechanisms is considerably less than when the short-range forces are ignored (β_{ij} alone). On the other hand, the collision efficiency for collisions by Brownian motion are not nearly so low; the lowest value shown on Figure 12-15 is ~0.4; hence, for this mechanism, the differences between the long-range and short-range models is not great. An example of the total collision frequency function for the short-range force model, for reasonably typical conditions, is shown in Figure 12-16.

Collision Pairs and Functions	Brownian Motion (cm³/s)	Fluid Shear (cm³/s)	Differential Sedimentation (cm³/s)	Total (cm³/s)	Dominant Collision Mechanism
0.4–2 μm					
β_{ij}	1.94×10^{-11}	5.76×10^{-11}	1.91×10^{-12}	7.89×10^{-11}	Sh
α_{ij}	0.729	0.165	0.44	–	–
$\alpha_{ij}\beta_{ij}$	1.41×10^{-11}	9.50×10^{-12}	8.40×10^{-13}	2.44×10^{-11}	Br
2–2 μm					
β_{ij}	1.08×10^{-11}	2.67 E–10	0	2.78×10^{-10}	Sh
α_{ij}	0.537	0.357	–	–	–
$\alpha_{ij}\beta_{ij}$	5.80×10^{-12}	9.53×10^{-11}	0	1.01×10^{-10}	Sh
2–30 μm					
β_{ij}	4.59×10^{-11}	1.37×10^{-7}	7.92×10^{-8}	2.16×10^{-7}	Sh
α_{ij}	0.784	8.1×10^{-6}	7.1×10^{-3}	–	–
$\alpha_{ij}\beta_{ij}$	3.60×10^{-11}	1.11×10^{-12}	5.62×10^{-10}	5.99×10^{-10}	DS

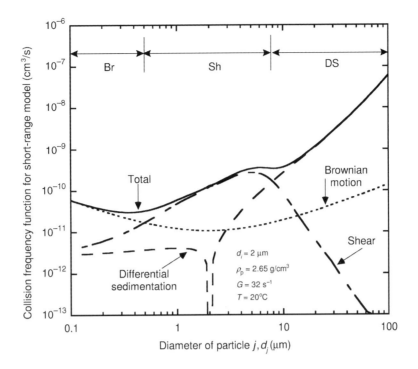

FIGURE 12-16. Total collision frequency function using the short-range force model. (Figure is specific to the conditions shown.)

A major difference in the short- and long-range models is that far fewer collisions are predicted to occur between particles of vastly different sizes in the short-range model; in this model, particles grow gradually through the size range rather than be scavenged by the largest particles in the suspension.

When the short-range forces are included in the prediction of collision frequency, not only is the rate of predicted collisions reduced but also the relative importance of the three mechanisms is altered for various combinations of the two sizes of particles. The dominant mechanism for predicted collisions, when short-range forces are included, is shown in Figure 12-17 for a particular set of conditions. Results from the long-range force model, delineated by the thick lines, are also shown for comparison; for the long-range model, Brownian motion dominates in the lower left corner (where both particles are small) and differential sedimentation dominates in the upper portion of the figure (where one particle is quite large). In the short-range model, Brownian motion is the primary cause of collisions for a small particle with any other size particle, fluid shear is only dominant when both particles are greater than ~ 1 μm and nearly the same size, and differential sedimentation is dominant when one particle is substantially greater in size than the other and both are greater than 1 μm. Fluid shear is predicted to be much less important as a cause of interparticle collisions in this short-range force model than in the long-range force model.

The difference in predictions between the long-range and the short-range force models for a flocculating suspension, as well as the ability of these models to fit experimental data, is shown in Figure 12-18. The model input for time zero, not shown, matched the experimental data for the measurable range but was extended to smaller sizes with a smooth curve, in recognition that some such particles existed and to allow for the predicted growth of particles into the measured range. Several observations can be made from the comparisons of measured and model-fit results. First, the amount of flocculation was not nearly as great in this experiment as in the higher concentration suspension (and longer time) shown in Figure 12-1, as expected. The loss of small particles was in the range of 0.4 log units, which translates to a 60% reduction in the number concentration. Second, the short-range force model gives a substantially better fit to the data than the long-range force model. The long-range model substantially overestimated both the reduction of small particles and the creation of large particles in comparison to the experimental results. The short-range model fits the experimental data quite well throughout the measured size range. These differences in predictions of the two models reflect the substantial reduction in collisions between large and small particles (i.e., at low λ values) brought about by the hydrodynamic corrections that are incorporated into the short-range force model. Finally, the best-fit lines for each model were determined by finding the value of α_{emp} that minimized the sum of squares of residuals between the model predictions and the experimental results; the resulting values are

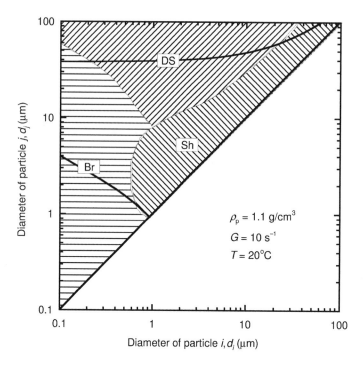

FIGURE 12-17. Dominant regions for each collision mechanism in the short-range force model. (The three shaded regions delineate the dominant regions for the short-range force model; the heavy lines delineate the boundaries in the long-range model, with Brownian motion dominant below the lower heavy line, differential sedimentation above the upper heavy line, and fluid shear in between.) (Comparison of the predictions from the long-range model in this figure with those in Figure 12-12 demonstrates the sensitivity of that model to the shear rate, G.)

indicated for each model on the figure. Because the long-range model predicts so many more collisions than the short-range model, the value of $\alpha_{\text{emp,LRM}}$ is smaller than $\alpha_{\text{emp,SRM}}$. Nevertheless, the single valued empirical collision efficiency function cannot properly account for the phenomena that are a function of the sizes of the colliding particles, and hence the long-range model cannot fit the data as well as the short-range model that more properly accounts for these effects.

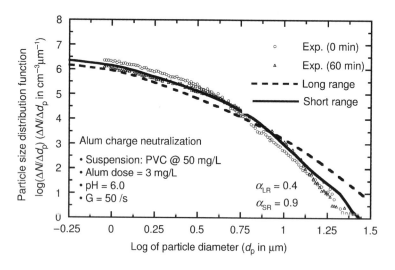

FIGURE 12-18. The fit of the long-range and short-range models to experimental data for a flocculating suspension. (The model input for time zero matched the experimental data but was extended to smaller sizes below the experimental limit of detection, as a smooth curve.) *Source*: From Li (1996).

Design Implications of the Short-Range Model

The short-range force model suggests that the designer of a flocculator does not have much control over the collisions that occur. Many of the collisions between particles occur by Brownian motion or differential sedimentation, and the parameters that control those collisions are generally not under the control of the designer or operator. The temperature of the water will influence the viscosity and density of the water, and increased temperature would increase the rate of collisions by all three mechanisms, but it is generally much too expensive to increase the temperature of water for this application, so that is not a realistic option.

The short-range force model suggests that the mixing intensity is not as important in causing collisions as is predicted by the long-range force model. The design implication is that sufficient mixing should be provided to keep particles in suspension; G values as low as $10\,s^{-1}$ are likely to be able to accomplish that goal in most suspensions. The proper range for G values is then thought to be between 10 and $50\,s^{-1}$, where the upper limit stems from a concern to avoid floc breakup. A substantial effort should be made by the designer to ensure a reasonably uniform mixing intensity throughout the flocculator to accomplish the goals of providing sufficient mixing to keep the particles in suspension but not enough to cause breakup.

12.6 TURBULENCE AND TURBULENT FLOCCULATION

Turbulence

The discussion of long-range transport via fluid shear in the preceding sections of this chapter implicitly assumed that the flow regime was laminar, by assuming that fluid motion throughout the system and at all times could be characterized by a single value of the shear rate (G). However, most full-scale flocculators are characterized by turbulent flow.

Turbulent flows have irregular fluctuations of velocity in all three directions. The instantaneous velocity in a given direction at any point can be described as the sum of a mean velocity and a fluctuating velocity at that point, as follows:

$$u_i(x_i, t) = U_i(x_i) + \tilde{u}_i(x_i, t) \quad (12\text{-}31)$$

where i is an index denoting each of the three perpendicular directions; x and t denote distance and time, respectively; u_i and U_i are the instantaneous and mean velocities in the direction i, respectively, at a given point, and $\tilde{u}_i$ is the fluctuation of the velocity at that point; that is, the difference between the mean and the instantaneous velocity.

The intensity of turbulence is related to the fluctuating component of the velocity. However, in steady flow at a given point, the mean value of the fluctuation is (by definition) zero, with positive and negative values canceling.

Therefore, the size of a typical fluctuation is quantified as the root mean square of the magnitude of the velocity fluctuation in the i direction, $\bar{u}_i$:

$$\bar{u}_i = \frac{1}{N_{\text{obs}}} \left(\sum_{\substack{t\,\text{or}\\x,y,z}} \tilde{u}_i^2 \right)^{1/2} \quad (12\text{-}32)$$

where the summation considers the value of $\tilde{u}_i$ at a single point at a statistically large number of times (N_{obs}), or at a single time over a large number of points in the system.

Recognizable spatial flow patterns that persist within turbulent flows for at least a short time are called eddies; within such flow patterns or eddies, a high degree of correlation exists between velocity at one point and that at another. Recall from Chapter 11 that the Kolmogorov theory relates the net rate at which energy is transferred from large to small eddies to the overall energy dissipation rate. This theory leads to the definition of the Kolmogorov microscales of length and time, computed as follows (Tennekes and Lumley, 1972):

$$\text{Length}: \eta_{\text{Kolmogorov}} = \left[\frac{v^3}{\varepsilon}\right]^{1/4} \quad (12\text{-}33)$$

$$\text{Time}: \tau_{\text{Kolmogorov}} = \left[\frac{v}{\varepsilon}\right]^{1/2} \quad (12\text{-}34)$$

where v is the kinematic viscosity (cm^2/s; the kinematic viscosity is the absolute viscosity, μ, divided by the fluid density, ρ) and ε is the energy dissipation per unit time per unit mass (cm^2/s^3). These scales represent the smallest scales of turbulent motion (TM).

Turbulent Flocculation

For pairs of particles that are smaller than the Kolmogorov length scale, the mechanism of flocculation is similar to that for flocculation in laminar flow fields, because the fluid motion is essentially laminar at that scale. For particles larger than this scale, the mechanism of flocculation can be viewed similarly to Brownian motion, except that the random motion is caused by turbulent eddies rather than molecular motion. The formalism for treating each of these limiting cases is well-developed, as described earlier in this chapter. The added complication applicable in turbulent flocculation compared to those limiting cases is the existence of a spectrum of values, rather than a single value, for a key variable. That is, whereas uniform shear throughout a system can be characterized by a single value of G, turbulent conditions are characterized by a spectrum of G (and velocity) values as the fluid energy cascades through eddies of decreasing size until the Kolmogorov microscale is reached. The spectrum over time for a given batch of fluid results in a spectrum over space at any instant in a reactor. As noted earlier, the velocity distribution near the particles affects both β and α values; therefore, turbulence is expected to affect both of these parameters. Attempts to model turbulent flocculation to date have focused on how to incorporate the turbulent energy spectrum into analysis of β and much of that work is summarized in the following paragraphs; little has been done with respect to its effect on α.

Camp and Stein (1943) recognized that the turbulence in flocculators required that the Smoluchowski approach for modeling collisions by shear be reconsidered. They accounted for the variation of the velocity gradient over space in the reactor by substituting the (root mean square) average gradient $(\overline{G})$ for the single valued G in the collision frequency function for laminar shear (Equation 12-18). As shown in Chapter 11, $(\overline{G})$ can be found as follows:

$$\overline{G} = \frac{1}{\tau_{\text{Kolmogorov}}} = \left(\frac{\varepsilon}{v}\right)^{1/2} = \left(\frac{P}{\mu V}\right)^{1/2} \quad (12\text{-}35)$$

where P is the power input, μ is the absolute viscosity of the water, and V is the volume of the reactor. This expression continues to be used widely to describe turbulent flocculation.

Saffman and Turner (1956) derived a collision frequency function for turbulent flocculation that was based on the dynamics of turbulent flow. Their interest was in the coagulation of water droplets in clouds, but Spielman (1978) later repeated the analysis for particles in water. These investigators focused on the collision of droplets (particles) that are much smaller than the Kolmogorov microscale and nearly the same size. They assumed the turbulence was isotropic (i.e., independent of direction) and that Taylor's (1935) estimate of the variance of the velocity fluctuation with distance was valid; that is,

$$\left[\text{Variance of } \frac{\partial \tilde{u}}{\partial x}\right] = \left(\frac{\partial \overline{u}}{\partial x}\right)^2 = \frac{\varepsilon}{15v} \quad (12\text{-}36)$$

With these assumptions, they found the collision frequency function for collisions brought about by turbulent motion (TM) to be

$$^{\text{TM}}\beta_{ij} = \left(\frac{\pi}{120}\right)^{1/2}\left(\frac{\varepsilon}{v}\right)^{1/2}(d_i + d_j)^3 \quad (12\text{-}37)$$

Levich (1962) expressed the effect of turbulence on particle motion as a diffusion coefficient and derived an expression for the collision frequency in the same way as that used for Brownian motion presented in Equations 12-20 to 12-23. The result for the collision frequency function for particles smaller than the Kolmogorov microscale is:

$$^{\text{TM}}\beta_{ij} = \left(\frac{3\pi}{2}\right)\chi(d_i + d_j)^3\left(\frac{\varepsilon}{v}\right)^{1/2} \quad (12\text{-}38)$$

where χ is an undefined (dimensionless) coefficient. Later investigators (Tambo and Watanabe, 1979; Delichatsios and Probstein, 1975) argued that the value of χ in Equation 12-38 should be taken as $(1/15)^{1/2}$ to be consistent with the Kolmogorov theory and Taylor's findings.

Argaman and Kaufman (1970) used a relationship from Corrsin (1962) to transform the turbulent energy spectrum into a diffusion coefficient for particles in a fluid. This approach was similar to that of Levich but these investigators considered particles larger than the microscale. Their analysis was limited to collisions between particles greatly different in size, stemming from the hypothesis that primary particles are removed principally by their collisions with large flocs composed of many such primary particles. For this condition (when the size of the small particle is negligible in comparison to the larger), they found the effective diffusion coefficient to be as follows:

$$D_{ij} = D_{\text{eff}} = K_s \overline{u}^2 a_F^2 \quad (12\text{-}39)$$

where a_F is the floc radius, and K_s is a (dimensional) coefficient of proportionality. Following the same procedure as was used for the collision frequency function for Brownian motion, the collision frequency function is found as:

$$^{\text{TM}}\beta_{ij} = {}^{\text{TM}}\beta_{1,F} = 4\pi(a_1 + a_F)K_s\overline{u}^2 a_F^2 \approx 4\pi K_s \overline{u}^2 a_F^2 \quad (12\text{-}40)$$

where the subscripts 1 and F refer to a primary particle and a floc. The assumption that the primary particle is much smaller than the floc leads to the final equation, but also limits its utility.

Delichatsios and Probstein (1975) developed a model for the flocculation of initially monodisperse suspensions in isotropic turbulent flows, considering particle sizes both larger and smaller than the Kolmogorov microscale. Their collision frequency function for heterodisperse collisions is

$$^{\text{TM}}\beta_{ij} = \left(\frac{\pi}{4}\right)(d_i + d_j)^2 u_r \quad (12\text{-}41)$$

where u_r is the relative velocity between particles during collision. This relative velocity, u_r, is taken as the root mean square turbulent velocity of two particles separated by a distance (d_a) that is equal to the average of the two particle diameters (i.e., $d_a = (d_i + d_j)/2$). In essence, their model defines the collision cross-section in the same way as was done for the collisions in a laminar velocity gradient and chooses a single value for the relative velocity of two colliding particles. They found different values of the relative velocity for three different scales of motion, as follows:

$$u_r = \left(\frac{1}{15}\right)^{1/2}\left(\frac{\varepsilon}{v}\right)^{1/2} d_a \quad \text{when } d_a < \eta \quad (12\text{-}42)$$

$$u_r = 1.37(\varepsilon d_a)^{1/3} \quad \text{when } \eta < d_a < L \quad (12\text{-}43)$$

$$u_r = (\varepsilon L)^{1/3} \quad \text{when } d_a = L \quad (12\text{-}44)$$

where L is the distance at which the correlation between fluid velocities effectively diminishes to zero (called the Eulerian macroscale of turbulence). These investigators presented the results of experiments performed in turbulent pipe flow and confirmed the model's dependence on the turbulence characteristics for particles smaller than the microscale. Attempts to check the model for sizes larger than the microscale were hampered by floc breakup.

In summary, four expressions for the collision frequency function in turbulent flow have been developed for the condition that the colliding particles are smaller than the length microscale. All show the same functionality with respect to the variables of interest, but differ in the numerical nondimensional coefficients. That is, all of these expressions have the form:

$$^{TM}\beta_{ij} = A(d_i + d_j)^3 \left(\frac{\varepsilon}{\upsilon}\right)^{1/2} \quad (12\text{-}45)$$

but the constant A has different values in each model, as follows:

Camp and Stein
$$A = \frac{1}{6} = 0.167 \quad (12\text{-}46)$$

Saffman and Turner
$$A = \left(\frac{\pi}{8}\right)^{1/2}\left(\frac{1}{15}\right)^{1/2} = 0.162 \quad (12\text{-}47)$$

Levich
$$A = \left(\frac{3\pi}{2}\right)\left(\frac{1}{15}\right)^{1/2} = 1.217 \quad (12\text{-}48)$$

Delichatsios and Probstein
$$A = \left(\frac{\pi}{8}\right)\left(\frac{1}{15}\right)^{1/2} = 0.101 \quad (12\text{-}49)$$

The differences demonstrate some uncertainty about the exact mechanism of collision in turbulent flocculation, but three of the four have quite similar values; additionally, they are quite similar to the collision frequency function for collisions by laminar fluid shear (Equation 12-18), in which the equivalent value of A is also $\frac{1}{6}$. Saffman and Turner's model is often considered most correct because of its sophisticated view of the collision mechanism. None of these mechanisms accounts for hydrodynamic retardation or other short-range forces in the collisions; that is, no attempts have been made to compute $^{TM}\alpha_{ij}$. Casson and Lawler (1990) assumed that Adler's (1981a) values for $^{Sh}\alpha_{ij}$ (developed for laminar shear fields) would be appropriate for turbulent collisions when the colliding particles were smaller than the microscale; their justification was that, below the microscale, the flow is essentially laminar.

Cleasby (1984) presented a review of studies on turbulent flocculation. He concluded that the expressions for collision frequencies for particles larger than the microscale were more valid than those derived for particles smaller than the microscale. At separation distances smaller than the microscale, he found that the relative velocities are proportional to $(\varepsilon\lambda_s)^{1/3}$ (where λ_s is the separation distance orthogonal to the velocity), while, at separation distances larger than the microscale, the relative velocities are proportional to $(\varepsilon/\upsilon)^{1/2}\lambda_s$, or $\overline{G}\lambda_s$. However, a later study (Hanson and Cleasby, 1990) concluded that the particles normally encountered in flocculation systems are smaller than the microscale.

Clark (1985, 1996) and Kramer and Clark (1997) noted that neither the original derivation of the collision frequency function for fluid shear by Smoluchowski nor the work of Camp and Stein properly accounted for the three-dimensional nature of turbulent flows. They argued that the three-dimensional strain rate of the fluid must be accounted for, but that by continuity, some components of the strain in any differential element of the fluid carry particles toward one another while others increase the separation of the particles; only those components that carry particles toward one another are relevant for (potential) collisions. They concluded that the G value of the Smoluchowski equation for collisions by fluid shear should be replaced by the maximum principal strain rate of the fluid.

The primary criticism that Kramer and Clark (1997) noted about the earlier work reviewed in this section is that the velocity gradient includes the energy that is dissipated in rotational flow, and that such flow does not contribute to interparticle collisions. Hence, even with a uniform velocity gradient, as in a Couette flow apparatus,[7] they believed that the velocity gradient was an overestimate of the maximum strain rate, and that therefore the rate of flocculation would be overestimated using the G value. They found that this problem was exacerbated in flows with a broad distribution of strain rates (or G values) that increase the rotational flow. Hence, they argued that a more proper estimate of $\overline{G}$ would account for the variance of the velocity gradient values throughout the fluid, as follows:

$$\overline{G} = \left(\frac{P}{\mu V} - \sigma^2_{G_L}\right)^{1/2} \quad (12\text{-}50)$$

where σ_{G_L} is the variance of the local values of the velocity gradient throughout the reactor. The implication is that the estimate of $\overline{G}$ by Equation 12-35 is too high, and that the error increases as the variation in the mixing intensity throughout the reactors increases. This result suggests that flocculation by fluid motion is more rapid if the mixing intensity is relatively uniform throughout the reactor.

---

[7] A Couette reactor is the annular space between two concentric cylinders; in most cases, one of the cylinders is stationary while the other rotates, but both could be rotated at different speeds. The no-slip condition means that, at each wall, the velocity relative to the wall is zero, and so (in laminar flow) a nearly uniform velocity gradient is set up across the annular space.

12.7 FLOC BREAKUP

Although the primary objective of flocculation is to increase the particle size, the mixing involved can subject flocs to shear fields that are strong enough to break them apart. While the concept that flocs are fragile is easy to understand, a thorough understanding of floc breakup has been elusive. The framework for considering breakup can be expressed by the following word equation:

Rate of change with time of the number concentration of particles of size k due to breakup	=	Rate of increase in the number concentration of particles of size k by breakup of a larger particle (floc) into fragments that include size k particles	−	Rate of decrease in the number concentration of particles of size k by breakup to smaller particles

The two terms on the right side can be added to the Smoluchowski equation (Equation 12-10) to form a total rate expression that includes both flocculation and breakup. Converting this word equation for breakup into a mathematical expression is more complex than the comparable terms for flocculation for three reasons:

- A criterion for determining when breakup occurs has to be set.
- The number and size of the products of the breakup of a floc are variable.
- Except in the case of an initially monodisperse suspension, a floc or particle of a given size can be constituted in a large number of different ways from primary particles, so that not all particles of a given size have the same susceptibility to breakup.

Given this complexity, simplifying assumptions have been necessary in the modeling of floc breakup. Most investigators have chosen to consider initially monodisperse suspensions; this assumption means, for example, that all 10-μm particles (flocs) contain the same number of original particles (of, say, 1 μm diameter). Without this assumption, one would have to differentiate various types of 10-μm particles, ranging from a single particle that was originally 10 μm and therefore could not break up to those that were made up of numerous smaller particles of various sizes. The result would be that many different breakup equations would have to be written for the same size particle, or a statistical description of how the concentration of each size class was constituted from various smaller particles would have to be supplied.

Even with an originally monodisperse suspension, a floc of, say, 30 original particles could break into a wide variety of smaller flocs. Breakup is usually assumed to result in two (and not more) fragments, with the idea that a floc breaks along a fault line with the weakest bonds. Two different approaches have been taken to describe this breakup. Early modelers assumed that the most likely breakup was an erosion of a single particle from the surface of the floc, so that the products were one primary particle and the remaining floc (of 29 original particles in this example). More recently, a statistical distribution of possible breakup products has been assumed; in our example of a 30-particle floc, some probability has been assigned to every possible two-product outcome. Some investigators have allowed for more than two fragments, and have done so either by assuming a constant number of fragments (usually three or four and usually all of the same size) or by allowing a distribution of both the number of fragments and their sizes (but maintaining a number balance of the particles).

Finally, a criterion for when breakup occurs has to be set. Kramer and Clark (1999) provided an excellent summary of different approaches taken by different investigators. One approach is to assume that flocs have a limiting strength and break when that strength is exceeded. This approach requires a description of the flow field so that the stresses to which flocs are subjected can be described; Adler and Mills (1979) took this approach and found the strain rate in each direction in a flow field to set a breakup criterion. A second approach has been to use a maximum strain rate directly as a criterion for breakup; this approach is somewhat more empirical (in that no direct description of the floc strength is required). Another empirical approach has been to focus on the particle size directly rather than on the fluid mechanics. In this approach, one can allow for an increasing probability of breakup with larger flocs, including setting both lower and upper thresholds of floc size for breakup, with no breakup below the lower threshold and a probability of one for breakup of particles at the upper threshold. Some of the studies incorporating these approaches are described in the following paragraphs.

Argaman and Kaufman (1970) considered floc breakup to occur by surface erosion; that is, primary particles would be pulled off the surface of a floc one by one. They considered a suspension as having only two types of particles—primary particles (or the original small particles) and flocs (envisioned as substantial aggregates made of many primary particles). Their work was done before particle counting could be done easily, and hence they had only turbidity measurements available as indicators of the particle size distribution and concentration. They considered the rate of creation of primary particles from floc breakup to be proportional to the surface area of flocs and the shear rate ($\overline{G}$), and concluded that, in a single CFSTR with hydraulic residence time τ, the steady-state equation for the fraction of primary particles remaining could be expressed as:

$$\frac{n_{\text{out}}}{n_{\text{in}}} = \frac{1 + K_B \overline{G}^2 \tau}{1 + K_A K_s \phi \overline{G} \tau} \qquad (12\text{-}51)$$

where n_{out} is the number concentration of primary particles in the effluent, n_{in} is the number concentration of primary particles in the influent, K_A is an aggregation constant (dimensionless), K_s is the same coefficient as in Equation 12-39, and K_B is a breakup constant (with dimensions of time). The authors arrived at this equation only after making several assumptions, including that the average floc size was inversely proportional to $\overline{G}$; it is this assumption that leads to the $\overline{G}^2$ term in the numerator of Equation 12-51. The values of K_A and K_B were found from fitting experimental results. Unfortunately, turbidity measurements are not sensitive enough to judge the validity of the model.

Parker et al. (1972) extended the analysis of Argaman and Kaufman and noted different modes of possible floc breakup for different conditions of turbulence. They proposed that a suspension had a maximum stable floc size that was proportional to $\overline{G}^{-s}$, where s is a constant exponent (>1); this result means that the maximum floc size decreases with increasing $\overline{G}$ value. Andreu-Villegas and Letterman (1976) also investigated the effects of velocity gradient in flocculation using this framework. Sonntag and Russel (1987) used a similar framework, substituting the turbulent strain rate for the velocity gradient. Some investigators (e.g., Tambo and Hozumi, 1979) who have assumed that a maximum floc size exists have found that the value of s is less than one, meaning that the maximum size can increase with the velocity gradient, but less than proportionally.

Hanson and Cleasby (1990) subjected flocs formed with 45 min of slow mixing to the same conditions used during the rapid mixing phase. A stake and stator reactor, which had a relatively uniform mixing intensity, caused very little breakup of the alum/kaolin flocs at 20°C, but a substantial amount at 5°C. A turbine impeller, which generated regions of intense turbulence near the impeller but much lower intensities elsewhere, caused significant breakup at both temperatures. These results support the idea that, both to promote flocculation and to minimize floc breakup, the mixing intensity should be reasonably uniform throughout the reactor.

Pandya and Spielman (1982) presented a semitheoretical/semiempirical study of floc breakup using an elegant system for inserting preformed flocs into a well-defined shear field undergoing extensional flow. They photographed two modes of breakup: random periodic splitting of flocs and quasi-continuous erosion of fine particles from the outer surfaces of flocs. Floc splitting was modeled with a probability distribution for the breakup of a parent floc into two or more daughter fragments. For erosion, probability distributions were also used for the characteristic size of the erosion product, ensuring that the total particle volume concentration in the suspension was conserved. These terms yielded a population balance equation that not only had terms for the increase and decrease of the concentration of each particle size by flocculation (such as Equation 12-7) but also included terms for an increase and decrease of concentration of each size by both methods of breakup (erosion and splitting). As in all breakup models, the terms for floc breakup could only be quantified by fitting to experimental results, but their framework was insightful into the mechanisms of floc breakup.

Kramer and Clark (1999) developed a model that accounted for much of the insight of the previous investigators. They made the common assumption of an initially monodisperse suspension, so that the composition of every floc class was understood as a set number of original particles. They assumed that flocs broke into two fragments, the sizes of which could vary among all of the possibilities that yielded the total number of original particles in the floc. Based on the mechanisms of failure described in the solid mechanics literature, they concluded that floc failure (breakup) would occur from stresses acting normal to the floc (rather than from shear stresses). These stresses depend on the maximum strain-rate tensor, which can be determined for a known flow field. Their final model for breakup included three empirical parameters: a breakup constant and exponents for the effects of viscosity and floc size on breakup. Adjusting these parameters allowed predictions of size distributions that accounted for both flocculation and breakup, ranging from almost no net flocculation (due to high breakup) to an accumulation of a large fraction of the original particles in the largest size class considered (due to very little breakup).

With the assumption that breakup is undesirable, the implication of the breakup modeling done by these various investigators, and particularly that by Kramer and Clark, is that flocculation reactors should be designed to have relatively uniform mixing intensities throughout the reactor. Kramer and Clark describe this mixing intensity in terms of the maximum strain rate, but that is closely related to the traditional value. If the mixing intensity varies considerably across the reactor, substantial breakup will occur in the portions of the reactor with the highest intensities, thereby acting against the flocculation desired. Many traditionally designed flocculators with paddles rotating near stationary baffles have locations of intense mixing intensity and others with minimal mixing.

Ducoste and Clark (1998) and Ducoste (2002) addressed the question of scale-up between laboratory reactors and full-scale reactors with respect to turbulent flocculation and breakup. They hypothesized that the mechanism of breakup was different for flocs larger and smaller than the Kolmogorov microscale, and that the regions of the reactor with these two different scales were considerably different for full-scale versus pilot and laboratory scale reactors. Their experimental work had indicated that the maximum stable floc size decreased with an increase in tank size (among laboratory scale reactors) even when the turbulence levels (indicated by the same values of the microscale) were the same. Their model proposed that flocs larger than the Kolmogorov microscale in the impeller region of a reactor broke up by fracture based on the local energy dissipation rate, whereas flocs smaller than the microscale broke up by erosion according to a critical velocity. With these two criteria and the recognition that the impeller zone is a smaller fraction of the total volume as reactor size increases, they successfully modeled the experimentally measured changes in size distributions in several reactors of different sizes and turbulence levels. Their results suggest that capturing all of the phenomena of floc growth and breakup in a mathematical model is quite complex, but apparently achievable. As with many of the results noted in this section, a ramification of these results is that minimizing the variation in the energy dissipation rate (degree of turbulent mixing) in different parts of a reactor is likely to lead to better flocculation.

Yukselen and Gregory (2004) reported experiments on floc breakage and the possibility of regrowing flocs after breakage. They formed flocs of clay particles under various chemical conditions in a standard jar test apparatus, then subjected the flocs to intense shear for a short period, and then returned to a normal shear rate to induce flocculation. Using an optical monitor, they measured a "flocculation index" that is correlated with floc size and found that both the initial breakup during the high shear ($G = 520\,s^{-1}$) and the regrowth of flocs were strongly influenced by the mechanism of destabilization (sweep floc vs adsorption and charge neutralization). Generally, flocs formed by alum (or polyaluminum chloride) via precipitation of $Al(OH)_3$ (s) were much less able to reflocculate than those formed by polyelectrolytes via adsorption and charge neutralization. The reasons for these results are not entirely clear, but the results suggest in yet another way that avoiding regions of high shear rates in flocculators or subsequent particle processing prior to removal should be avoided if possible.

12.8 MODELING OF FLOCCULATION WITH FRACTAL DIMENSIONS

As noted in Chapter 11, flocs of particles are often characterized by fractal dimensions that are less than their Euclidian counterparts; that is, the projected area, particle volume, and mass are functions of the characteristic length raised to powers less than 2, 3, and 3, respectively. This phenomenon occurs because flocs do not coalesce (for example, as oil droplets would) when they are formed

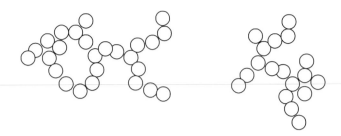

FIGURE 12-19. Schematic representation of the fractal nature of flocs.

from smaller particles, but rather have void spaces that are filled with water. Several estimates of fractal dimensions of flocs encountered in water and wastewater treatment were noted in Chapter 11. A schematic two-dimensional representation of flocs composed of monodisperse original particles is shown in Figure 12-19; the picture makes it clear that a floc is not a coalesced sphere of the original particles but is spread out in a relatively nonstructured way. While the schematic shown is simply a sketch, mathematical modeling of floc structure in a specified flow field has been done by various investigators, beginning with Vold (1963).

In the modeling with the Smoluchowski equation described earlier, the fractal nature of flocs was not accounted for with either the long-range or short-range models. In those models, flocs were assumed to coalesce into spheres with a total volume equal to the volume of the flocculating particles. The coalescence assumption is mathematically convenient but clearly incorrect. Models that incorporate the fractal dimensions of flocs avoid the error introduced by the coalescence assumption, but most have incorporated other mathematical simplifications to make the modeling tractable.

As with the modeling that incorporates floc breakup, most models that consider the fractal aspects of flocs are based on the assumption that the initial suspension consists of monodisperse spheres. With that assumption, all particles that are some larger size consist of a set number of the original particles and contain a set amount of water (which depends on the fractal dimension). Jiang and Logan (1991) proposed a model of flocculation that incorporated the concepts of fractal dimension. Their model is closely related to the long-range force model noted earlier; in essence they rewrote the collision frequency functions for the mechanisms of Brownian motion, fluid shear, and differential sedimentation using the idea that a floc diameter was proportional not to the solid volume raised to 1/3 but rather to the inverse of the fractal dimension. Jiang and Logan's model was quite general, allowing for nonspherical particles and flocs that had a different packing density than the smaller flocs from which they were created.

A somewhat simpler version of this model was proposed by Lee et al. (2000, 2002), and that model is explained here.

They assumed that all flocs have the same fractal dimension and that, when two flocs collide and attach, both the *solid* volume and the floc volume are conserved, so that the new floc has the same fractal dimension as the colliding flocs. This assumption means that, if the fractal dimension were three, the model would be the same as the long-range force model. They referred to this model as the coalesced fractal sphere model, since it maintains most of the assumptions of the coalesced sphere model. With these assumptions, the relationship between the number of original particles in a floc (N), the floc diameter (d_f), the diameter of the original particles (d_p), and the (volumetric) fractal dimension (D_3) is as follows:

$$N = \left(\frac{d_f}{d_p}\right)^{D_3} \quad (12\text{-}52)$$

The definition of the floc diameter (d_f) and floc volume (V_f) is closely related and maintain the Euclidian relationship, $V_f = (\pi/6)d_f^3$; that is, the floc volume is considered to be the sphere that encases all of the solids in the floc, so that the diameter is the longest linear dimension of the floc. From these relationships, one can calculate several other properties of the fractal flocs, and these are shown in Table 12-1. In the equations, the subscripts p and f refer to the original particle and flocs, respectively, V_p is the volume of the original type of particle, and V_s is the *solid* volume of flocs of made from the original particles.

The collision frequency functions in this model are shown in Table 12-2 for the three collision mechanisms; these can be used directly in the long-range force model in lieu of the collision frequency functions presented earlier in this chapter. In these equations, the volumes (V_i and V_j) are the solid volumes. For differential sedimentation, the settling velocity takes on a somewhat different functionality when the fractal dimension is less than 2 than when it is larger, because the surface area that experiences drag is different. As noted in Chapter 11, many believe that the theory of fractal aggregates leads to the conclusion that, when $D_3 < 2$, then $D_2 = D_3$; this theory was followed in determining the settling velocities. Therefore, the collision frequency function (which includes the differences in the settling velocities of the two particles that can collide) also has two forms depending on whether the volumetric fractal dimension is greater than or less than 2.

Lee et al. (2000) ran a number of simulations with this model and reported that, if D_3 is considered a fitting parameter, it has much the same effect as the inverse of α_{emp} in the long-range force model—that is, a lower value of D_3 leads to more rapid flocculation since a floc with the same *solid* volume has a larger *aggregate* volume and therefore sweeps out more solution (and particles in that solution) in the same time.

In addition to the differences in geometric considerations noted earlier, the model of Jiang and Logan (1991) also allowed for a varied description of the settling velocities for different size particles, since the drag on the particles can change functionality at higher Reynolds number (as explained in Chapter 13). Jiang and Logan also found values for the steady-state slopes of particle size distribution functions when fresh small particles are continuously supplied to the system, following earlier work by Hunt (1982) on the coalesced-sphere, long-range force model. Both Lee et al. and Jiang and Logan showed that flocculation is predicted to

TABLE 12-1. Properties of Fractal Flocs

Floc Property	Equation	Equation Number
Solid volume	$V_s = NV_p = \frac{\pi}{6}d_p^{3-D_3}d_f^{D_3}$	(12-53)
Solid mass	$m_f = \rho_p V_s = \frac{\pi}{6}\rho_p d_p^{3-D_3}d_f^{D_3}$	(12-54)
Floc diameter	$d_f = \left(\frac{6}{\pi}\right)^{1/3}(V_p)^{((1/3)-(1/D_3))}(V_s)^{1/D_3}$	(12-55)

TABLE 12-2. Collision Frequency Functions for Fractal Flocs

Collision Mechanism	Equation	Equation Number
Fluid shear	$_{\text{Fractal}}^{\text{Sh}}\beta_{ij} = \frac{1}{\pi}G(V_p)^{1-(3/D_3)}\left(V_i^{1/D_3} + V_j^{1/D_3}\right)^3$	(12-56)
Differential sedimentation $2 \leq D_3 \leq 3$	$_{\text{Fractal}}^{\text{DS}}\beta_{ij} = \frac{g}{12\mu}\left(\frac{\pi}{6}\right)^{-(1/3)}(\rho_p - \rho_L)(V_p)^{((1/3)-(1/D_3))}\left(V_i^{1/D_3} + V_j^{1/D_3}\right)^2\left|V_i^{(D_3-1)/D_3} - V_j^{(D_3-1)/D_3}\right|$	(12-57a)
Differential sedimentation $D_3 \leq 2$	$_{\text{Fractal}}^{\text{DS}}\beta_{ij} = \frac{g}{12\mu}\left(\frac{\pi}{6}\right)^{-(1/3)}(\rho_p - \rho_L)(V_p)^{((4/3)-(3/D_3))}\left(V_i^{1/D_3} + V_j^{1/D_3}\right)^2\left|V_i^{1/D_3} - V_j^{1/D_3}\right|$	(12-57b)
Brownian motion	$_{\text{Fractal}}^{\text{Br}}\beta_{ij} = \frac{2k_BT}{3\mu}\left(\frac{1}{V_i^{1/D_3}} + \frac{1}{V_j^{1/D_3}}\right)\left(V_i^{1/D_3} + V_j^{1/D_3}\right)$	(12-58)

be far faster when the fractal size is accounted for than when it is not. These investigators did not explicitly account for the possibility that water could flow *through* a fractal (porous) floc, but their use of the long-range (rather than the short-range) collision frequency functions is equivalent to that assumption.

■ **EXAMPLE 12-9.** In Example 12-2, we evaluated the collision frequency function for the long-range force model for collisions between 2 and 30-μm (diameter) particles by fluid shear at a G value of 25 s^{-1}. The sizes represent coalesced spheres, so that the fractal diameters would be greater than these values. Re-evaluate this collision frequency function for collisions between these two particles using the fractal collision model, assuming that the original (nonfractal) particles were 1 μm in diameter and that the fractal dimension is 2.1.

Solution. We begin by finding the volumes of the baseline (nonfractal) particle and the two flocs that collide.

$$V_p = V(1) = \frac{\pi}{6}d_p^3 = \frac{\pi}{6}(1 \times 10^{-4} \text{ cm})^3 = 5.236 \times 10^{-13} \text{ cm}^3$$

$$V_i = V(2) = \frac{\pi}{6}d_p^3 = \frac{\pi}{6}(2 \times 10^{-4} \text{ cm})^3 = 4.189 \times 10^{-12} \text{ cm}^3$$

$$V_j = V(30) = \frac{\pi}{6}d_p^3 = \frac{\pi}{6}(3 \times 10^{-3} \text{ cm})^3 = 1.414 \times 10^{-8} \text{ cm}^3$$

We now use the collision frequency function for fractals colliding due to fluid shear (Equation 12-56) to find the desired value:

$$\text{Sh}_{\text{Fractal}}\beta_{ij} = \frac{1}{\pi}G(V_p)^{1-(3/D_3)}\left(V_i^{1/D_3} + V_j^{1/D_3}\right)^3$$

$$= \frac{1}{\pi}(25 \text{ s}^{-1})(5.236 \times 10^{-13})^{1-(3/2.1)}$$

$$\times \left[(4.189 \times 10^{-12})^{1/2.1} + (1.414 \times 10^{-8})^{1/2.1}\right]^3 \text{cm}^3$$

$$= 9.49 \times 10^{-6} \text{cm}^3/\text{s}$$

The resulting value for the collision frequency is almost 70 times that found in Example 12-2 for the coalesced-sphere, long-range model. The example is somewhat extreme, given that a 30 μm floc is composed of 27,000 (= (30 μm/1 μm)^3) original particles, but emphasizes that the volume of a fractal floc is much greater than its stated (nonfractal) size. ■

As noted in Chapter 11, many investigators have reported measurements of flocs in terms of fractal dimensions. Some have considered the idea that the fractal dimension is determined (or at least influenced) by the degree of particle destabilization (expressed in terms of α_{emp}) and by the collision mechanism. By any collision mechanism, a higher value of α_{emp} is thought to lead to a lower fractal dimension;

that is, if virtually any collision is expected to be successful, a highly porous floc can be created. By contrast, if the surface chemistry and double layer are such that most collisions predicted by the long-range model are not successful in leading to a floc, then only those collisions which lead rapidly to many points of contact between the colliding (fractal) flocs will be successful; such flocs will be denser and therefore have a higher fractal dimension. The mechanism of destabilization also influences the fractal dimension, with sweep flocculation leading to flocs of lower fractal dimensions than those formed after adsorption and charge neutralization.

Similarly, it is generally thought that collisions by Brownian motion lead to a lower fractal dimension than collisions by fluid shear. In particular, if the fluid shear (or mixing intensity) is great enough to cause some floc breakup, only flocs that are held together by multiple particle–particle contact points will be strong enough to avoid breakup. In this view, floc breakup has the advantage of causing denser and stronger flocs to be created, and such flocs are more robust in subsequent removal processes (Serra et al., 1997). Jiang and Logan (1991) found from their dimensional analysis that flocs created by fluid shear would have a fractal dimension greater than 2.4, whereas flocs created by differential sedimentation would have lower fractal dimensions, perhaps as low as 1.6.

12.9 SUMMARY

In this chapter, the theory and practice of flocculation have been presented. Flocculation changes the size distribution of a suspension without removing any particle mass, by combining originally disparate particles into flocs. Hence, particles tend to shift from the smaller size ranges in a suspension to larger sizes; the objective of flocculation is to make the particles easier to remove from suspension in subsequent treatment processes, or to otherwise improve the operation of those processes.

Particles move in water by moving with the fluid, by settling, and by Brownian motion and, therefore, they approach one another by these same mechanisms. If two particles are suitably destabilized, their approach to one another by any of these mechanisms can lead to collision and attachment—the formation of a floc from the two original particles. Equations to describe the collision frequency for each of these mechanisms, when short-range forces are ignored, have been known for many years. They are presented in this chapter as Equation 12-18 for particles moving in a fluid shear field (or equivalently as Equation 12-45 for particles in turbulent flow), Equation 12-19 for differential sedimentation, and Equation 12-23 for Brownian motion. Flocculation within turbulent conditions is quite complex, so that our understanding of this phenomenon is weaker than under laminar conditions; nevertheless, virtually all full-scale flocculators have turbulent flow.

When particles come near each other, some phenomena that are ignored in the long-range force model influence the particle motion and, therefore, impact the formation of flocs. These phenomena include van der Waals attraction between particles, electrostatic repulsion of particles, and the effects of water flow in the near vicinity of each particle. In the modeling reported in this chapter, a means of calculating corrections to the collision frequency functions to account for the van der Waals attraction and the hydrodynamic effects are presented; this model is called the short-range force model herein. Although it is possible within the same framework to account for electrostatic repulsion, that possibility has not been included here, under the assumption that successful particle destabilization would reduce that repulsion to near zero. Accounting for these short-range effects reduces the predicted collisions by all three mechanisms. However, the corrections for collisions by Brownian motion are relatively minor, whereas the corrections for collisions between particles of vastly different sizes by both fluid motion and differential sedimentation are quite substantial. These changes from the long-range to the short-range model result in predictions of a more gradual growth of particles through the distribution, as collisions between particles of nearly the same size take on a greater importance than those of widely variant sizes.

Flocs can be fragile; that is, flocs can be broken into fragments to recreate smaller particles. If floc breakup is excessive, it can negate the effects of flocculation, and hence it is usually desirable to minimize breakup in full-scale reactors. Recent developments suggest that breakup is related to the maximum strain rate imposed on the fluid, so that designs of flocculators that avoid great variations in strain rate are likely to minimize floc breakup.

Traditional modeling of flocculation assumes that flocs behave as coalesced spheres—that is, the assumption is that the particle volume is conserved when two particles form a floc. In fact, however, flocs incorporate space (water) between the particles when they form, so that they take on a fractal dimension that is less than the expected Euclidean value for area or volume. Fractal flocs can also allow water to flow through them to some degree, thereby reducing the need for the hydrodynamic corrections built into the short-range force model.

On a quantitative basis, it is difficult to use the modeling described here to design a flocculation reactor for a specific application. However, several general ramifications of the models can be applied to the design and operation of all flocculation reactors. Flocculation is a second-order reaction with respect to particle number concentration, as the Smoluchowski equation makes clear. That suggests that the residence time distribution of flocculation reactors should be made as close to plug flow as possible. On the other hand, mixing is important to keep particles in suspension and cause collisions by fluid motion. Balancing the need for mixing and the benefits of a plug-flow-like residence time distribution has led to many flocculators being built as a series of a few CFSTRs. The concentration effect also means that flocculation is improved (i.e., occurs to a greater extent) if solids are recycled (internally or externally), so that the solids residence time is longer than the water residence time, and the concentration in the reactor is higher than that of the influent. This concentration effect of flocculation is the reason why destabilization via the sweep floc mechanism is often used—the creation of new particles by precipitation not only destabilizes the existing particles but also increases the size and number of particles in the system, all effects that increase the rate of flocculation.

Besides the theoretical residence time and the residence time distribution, the mixing intensity (usually expressed by the velocity gradient, G, in flocculators) is the primary variable under the control of the designer of flocculation reactors. Typical designs use G values in the range of $10-50\,s^{-1}$. Primary concerns of designers should be to ensure that the mixing intensity is both sufficient to keep the particles in suspension and distributed reasonably evenly throughout the reactor (or subreactor) to avoid floc breakup. Doing so allows flocculation by all of the collision mechanisms, even though many of the parameters that influence the rate of collisions by the various mechanisms are not susceptible to the control of the designer or operator.

APPENDIX 12A. CALCULATION EQUATIONS FOR COLLISION EFFICIENCY FUNCTIONS IN THE SHORT-RANGE FORCE MODEL

Calculation equations for collision efficiency functions for Brownian motion, fluid shear, and differential sedimentation are shown in Tables 12A-1, 12A-2, and 12A-3, respectively.

TABLE 12A-1. Collision Efficiency for Brownian Motion

Diameter of Larger Particle (μm)	a	b	c	d
0.1	1.025	−0.626	0.516	−0.152
0.2	1.007	−0.860	0.870	−0.322
0.6	0.976	−1.155	1.342	−0.554
1.0	0.962	−1.263	1.522	−0.645
2	0.943	−1.383	1.725	−0.748
6	0.916	−1.533	1.991	−0.887
10	0.905	−1.587	2.087	−0.936
20	0.891	−1.658	2.221	−1.009
60	0.871	−1.739	2.371	−1.090
200	0.863	−1.775	2.439	−1.125

$^{Br}\alpha_{ij} = a + b\lambda + c\lambda^2 + d\lambda^3$.
λ = size ratio ($0 < \lambda \leq 1$).

TABLE 12A-2. Collision Efficiency for Fluid Shear

log H_A	a	b	c	d
1	−1.128	2.498	−2.042	0.671
0	−1.228	2.498	−2.042	0.671
−1	−1.482	3.189	−3.468	1.581
−2	−1.704	3.116	−2.881	1.121
−3	−2.523	5.550	−6.098	2.553
−4	−3.723	10.039	−12.569	5.557
−5	−5.775	18.267	−24.344	10.992
−6	−7.037	20.829	−25.589	10.755
−7	−8.773	25.663	−30.703	12.555
−8	−9.733	30.663	−35.703	14.555

$$^{Sh}\alpha_{ij} = \frac{8}{(1+\lambda)^3} 10^{a+b\lambda+c\lambda^2+d\lambda^3}$$

$$H_A = \frac{A}{18\pi\mu d_l^3 G}$$

λ = size ratio ($0 < \lambda \leq 1$); d_1 = diameter of larger particle; A = Hamaker constant; μ = absolute viscosity; G = velocity gradient.

TABLE 12A-3. Collision Efficiency for Differential Sedimentation

log N_g	a	b	c	d
1	−0.3184	−0.548	0.0865	0.501
0	−0.840	0.423	−1.069	0.930
−1	−1.320	1.318	−2.170	1.361
−2	−1.757	2.137	−3.229	1.794
−3	−2.152	2.880	−4.232	2.230
−4	−2.505	3.547	−5.186	2.668
−5	−2.815	4.137	−6.088	3.108
−6	−3.084	4.652	−6.940	3.551
−7	−3.310	5.090	−7.742	3.996
−∞	−3.928	6.423	−9.449	4.614

$$^{DS}\alpha_{ij} = 10^{a+b\lambda+c\lambda^2+d\lambda^3}$$

$$N_g = \frac{48A}{\pi g(\rho_p - \rho_L)d_1^4}$$

λ = size ratio ($0 < \lambda \leq 1$); d_1 = diameter of larger particle; A = Hamaker constant; ρ_p, ρ_L = density of particle and fluid, respectively; g = gravitational constant.

REFERENCES

Adler, P. M. (1981a) Heterocoagulation in shear flow. *J. Colloid Interface Sci.*, 83 (1), 106–115.

Adler, P. M. (1981b) Interaction of unequal spheres I. Hydrodynamic interaction: colloidal forces. *J. Colloid Interface Sci.*, 84 (2), 461–474.

Adler, P. M. (1981c) Interaction of unequal spheres II. Conducting spheres. *J. Colloid Interface Sci.*, 84 (2), 475–488.

Adler, P. M. (1981d) Interaction of unequal spheres III. Experimental. *J. Colloid Interface Sci.*, 84 (2), 489–496.

Adler, P. M., and Mills, P. M. (1979) Motion and rupture of a porous sphere in a linear flow field. *J. Rheology*, 23 (1), 25–37.

Andreu-Villegas, R., and Letterman, R. D. (1976) Optimizing flocculator power input. *J. Environ. Eng. Div. (ASCE)*, 102 (EE2), 251–263.

Argaman, Y., and Kaufman, W. J. (1970) Turbulence and flocculation. *J. San. Eng. Div. (ASCE)*, 96 (SA2), 223–241.

Benjamin, M. M. (2011) Clarification of a common misunderstanding of collision frequencies in the Smoluchowski equation. *J. Environ. Eng.*, 137 (4), 297–300.

Broyan, J. L. (1996) Flocculation of SiO_2 suspensions in a couette flow reactor. M.S. thesis. Department of Civil Engineering, University of Texas at Austin.

Camp, T. R. (1955) Flocculation and flocculation basins. *ASCE Trans.*, 120, 1–16. (Also available in *Civil Engineering Classics: Outstanding Papers of Thomas R. Camp*, ASCE, New York, 1973).

Camp, T. R., and Stein, P. C. (1943) Velocity gradients and internal work in fluid motion. *J. Boston Soc. Civil Eng.*, 30, 219–237.

Casson, L. W., and Lawler, D. F. (1990) Flocculation in turbulent flow: measurements and modeling of particle size distributions. *JAWWA*, 82 (8), 54–68.

Cleasby, J. L. (1984) Is velocity gradient a valid turbulent flocculation parameter? *J.. Environ. Eng. (ASCE)*, 110 (5), 875–897.

Clark, M. M. (1985) Critique of Camp and Stein's RMS velocity gradient. *J. Environ. Eng. (ASCE)*, 111 (6), 741–754.

Clark, M. M. (1996) *Transport Modeling for Environmental Engineers and Scientists*. Wiley-Interscience, New York.

Corrsin, S. (1962) Theories of turbulent diffusion. *Mechanique de la Turbulence*. Edition Du Centre National de la Recherche Scientifique, Paris (in French).

Delichatsios, M. A., and Probstein, R. F. (1975) Coagulation in turbulent flow: theory and experiment. *J. Colloid Interface Sci.*, 51 (3), 394–405.

Ducoste, J. (2002) A two-scale PBM for modeling turbulent flocculation in water treatment processes. *Chem. Eng. Sci.*, 57, 2157–2168.

Ducoste, J. J., and Clark, M. M. (1998) The influence of tank size and impeller geometry on turbulent flocculation: I. Experimental. *Environ. Eng. Sci.*, 15 (3), 215–225.

Einstein, A. (1905) Über die von der molekularkinetischen Theorie der Wärme geforderte Bewegung von in ruhenden Flüssigkeiten suspendierten Teilchen. *Ann. Phys.*, 17, 549–560 (in German). Available in English as: On the movement of small particles suspended in a stationary liquid demanded by the molecular-kinetic theory of heat, in Einstein, A. (1926) *Investigations on the Theory of the Brownian Movement* (R. Fürth, ed.), republished by Dover Publications (1956).

Friedlander, S. K. (1977) *Smoke, Dust and Haze*. Wiley-Interscience, New York.

Han, M., and Lawler, D. F. (1991) Interactions of two settling spheres: settling rates and collision efficiency. *J. Hyd. Eng. (ASCE)*, 117 (10), 1269–1289.

Han, M. Y., and Lawler, D. F. (1992) The (relative) insignificance of G in flocculation. *JAWWA*, 84 (10), 79–91.

Hanson, A. T., and Cleasby, J. L. (1990) The effects of temperature on turbulent flocculation: fluid dynamics and chemistry. *JAWWA*, 82 (11), 56–73.

Hunt, J. R. (1982) Particle dynamics in seawater: Implications for predicting the fate of discharged particles. *Environ. Sci. Technol.*, 16, 303–309.

Ives, K. J. (1968) Theory of operation of sludge blanket clarifiers. *Proc. Inst. Civil Engineers (UK)*, 39, 243.

Jeffrey, D. J., and Onishi, Y. (1984) Calculation of the resistance and mobility functions for two unequal rigid spheres in low Reynolds number flow. *J. Fluid Mech.*, 139, 261–290.

Jiang, Q., and Logan, B. E. (1991) Fractal dimensions of aggregates determined from steady-state size distributions. *Environ. Sci. Technol.*, 25, 2031–2038.

Kramer, T. A., and Clark, M. M. (1997) Influence of strain-rate on coagulation kinetics. *J. Environ. Eng. (ASCE)*, 123, 444–452.

Kramer, T. A., and Clark, M. M. (1999) Incorporation of aggregate breakup in the simulation of orthokinetic coagulation. *J. Colloid Interface Sci.*, 216, 116–126.

Lee, D. G., Bonner, J. S., Garton, L. S., Ernest, A. N. S., and Autenrieth, R. L. (2000) Modeling coagulation kinetics incorporating fractal theories: a fractal rectilinear approach. *Water Res.*, 34, 1987–2000.

Lee, D. G., Bonner, J. S., Garton, L. S., Ernest, A. N. S., and Autenrieth, R. L. (2002) Modeling coagulation kinetics incorporating fractal theories: comparison with observed data. *Water Res.*, 36, 1056–1066.

Letterman, R. D. and Yiacoumi S. (2011) Coagulation and flocculation. Chapter 8 in Edzwald, J. E. (ed.), *Water Qualilty and Treatment: A Handbook on Drinking Water*, McGraw-Hill, New York, pp. 8.1–8.81.

Levich, V. G. (1962) *Physical Hydrodynamics*. Prentice-Hall, Englewood Cliffs, N.J.

Li, J. (1996) Rectilinear vs. curvilinear models of flocculation: experimental tests. Ph.D. dissertation. Department of Civil Engineering, University of Texas.

O'Melia, C. R. (1972) Coagulation and flocculation. In Weber, W. J. Jr., (ed.), *Physicochemical Processes for Water Quality Control*, Wiley-Interscience, New York.

Pandya, J. D., and Spielman, L. A. (1982) Floc breakage in agitated suspensions: theory and data processing strategy. *J. Colloid Interface Sci.*, 90 (2), 517–531.

Parker, D. S., Kaufman, W. J., and Jenkins, D. (1972) Floc breakup in turbulent flocculation processes. *J. San. Eng. Div. (ASCE)*, 98 (SA1), 79–99.

Saffman, P. G., and Turner, J. S. (1956) On the collision of drops in turbulent clouds. *J. Fluid Mech.*, 1 (16), 16–30.

Serra, T., Colomer, J., and Casamitjana, X. (1997) Aggregation and breakup of particles in a shear flow. *J. Colloid Interface Sci.*, 187, 46–473.

Smoluchowski, M. (1917) Versuch einer mathematischen theorie der koagulations-kinetik kolloider losungen. *Z. Physik. Chem.*, 92 (2), 129–168 (in German).

Sonntag, R. C., and Russel, W. B. (1987) Elastic properties of flocculated networks. *J. Colloid Interface Sci.*, 116 (2), 485–489.

Spielman, L. A. (1978) Hydrodynamic aspects of flocculation. In Ives, K. J. (ed.), *The Scientific Basis of Flocculation*, Sijthoff and Noordhoff, Alphen aan den Rijn, The Netherlands, pp. 63–88.

Tambo, N., and Hozumi, H. (1979) Physical characteristics of flocs II. Strength of floc. *Water Res.*, 13, 421–428.

Tambo, N., and Watanabe, Y. (1979) Physical aspect of flocculation process I. Fundamental treatise. *Water Res.*, 13, 429–439.

Taylor, G. I. (1935) Statistical theory of turbulence, Parts I – IV. *Proc. Royal Soc. A*, 151, 421–478.

Tennekes, H., and Lumley, J. L. (1972) *A First Course in Turbulence*, MIT Press, Cambridge, MA.

Vold, M. J. (1963) Computer simulation of floc formation in a colloidal suspension. *J. Colloid Sci.*, 18, 684–695.

Yukselen, M. A., and Gregory, J. (2004) The reversibility of floc breakage. *Int. J. Miner. Process.*, 73, 251–259.

Zeichner, G. R., and Schowalter, W. R. (1977) Use of trajectory analysis to study stability of colloidal dispersions in flow fields. *AIChE J.*, 23, 243–254.

PROBLEMS

12-1. In most of the examples in this chapter, collisions between 2-μm particles and 0.4-, 2-, and 30-μm particles were considered. Choose another "primary" particle (say, 5 μm) and either the same (0.4 and 30 μm) or different particles as the "secondary" particles, and redo the calculations in the examples that are summarized in the table shown in the solution to Example 12-8.

If possible, choose different secondary particles from your colleagues, and then combine all the results to see the trends for the whole range of likely particles in suspensions.

12-2. Set up a spreadsheet to calculate values of the collision frequency function according to the long-range force model for all three collision mechanisms and their total for any suspension of particles, with one particle size being held constant and the other varying at equal logarithmic intervals from 0.1 to 100 μm diameter. With minimal changes, the user should be able to specify:

(a) Any particle size to be the one held constant.

(b) Particle density.

(c) Water temperature, and therefore viscosity and density.

(d) Velocity gradient.

For a default, use $\Delta \log d_p$ as 0.1, yielding 31 sizes in the range $-1.0 \leq \log d_p \leq 2.0$. Choose the particle that is closest to 1.5 μm as the standard size, a particle density of 1.4 g/cm^3, a velocity gradient of 20 s^{-1}, and the viscosity and water density used throughout this chapter.

12-3. Calculate the collision efficiency function values according to the short-range force model for all three collision mechanisms for the same standard particle used in Problem 12-2 with both the smallest (0.1 μm) and the largest (100 μm) particle in the range. Make a small table that shows both the long-range and the short-range collision frequency values for each of the mechanisms and the total.

12-4. Traditionally, the kinetics of flocculation has been simplified by assuming that all of the particles in a suspension at any instant are the same size (i.e., $V_{p,i} = V_{p,j}$). This assumption is implemented by recognizing that when two identical i particles collide to form a new k particle, the total number of particles in the system declines by one. Then, the mass and volume of all the particles in the system are assumed to instantly be redistributed, such that the suspension is once again monodisperse, but still with one fewer particle than before the collision (and hence with each particle being slightly larger than before the collision). The particles are then once again all referred to as type i, although the size of the particles is now slightly greater than an instant earlier. Note that, under this assumption, differential sedimentation is not an operative mechanism.

(a) Using this assumption, show that the Smoluchowski equation simplifies to:
$$^{\text{tot}}r_{i+k} = -\tfrac{1}{2}\alpha_{\text{emp}}\alpha_{ii}\beta_{ii}n^2$$

where $^{\text{tot}}r_{i+k}$ is the sum of the rates of formation of i and k particles, and hence the overall rate of formation of all particles in the system, and $n = n_i$, the total number concentration of particles in suspension at any instant. Note that, despite the absence of any k particles in the system at any instant, the rate of formation of k particles is nonzero, and this rate must be included in the calculation of the overall rate of particle formation.

(b) Assuming that Brownian motion is the only mechanism for collision, show that the simplification derived in part (a) for a monodisperse suspension leads to the following rate of flocculation:
$$^{\text{Br}}r_n = -\alpha_{\text{emp}}{}^{\text{Br}}\alpha_{ii}\frac{4k_B T}{3\mu}n^2$$

(c) Assuming that fluid shear is the dominant mechanism for collision, show that the simplification derived in part (a) for a monodisperse suspension leads to the following rate of flocculation:
$$^{\text{Sh}}r_n = -\tfrac{2}{3}\alpha_{\text{emp}}{}^{\text{Sh}}\alpha_{ii}Gd_i^3 n^2$$

(d) Define ϕ as the volume of particles per volume of suspension, and show that
$$\phi = n_{t=0}\frac{\pi d_{t=0}^3}{6} = n_{\text{any }t}\frac{\pi d_{\text{any }t}^3}{6} = n\frac{\pi d^3}{6}$$

Substitute the above equation into the expression for $^{\text{Sh}}r_n$ found in part (c) to find:
$$^{\text{Sh}}r_n = -\frac{4}{\pi}\alpha\phi Gn$$

where α is the product $\alpha_{\text{emp}}{}^{\text{Sh}}\alpha_{ii}$. This expression is commonly cited as the rate of flocculation.

12-5. Consider a completely destabilized suspension with only two types of particles, with $d_p = 1.5$ μm and $d_p = 20$ μm. For such a suspension, no more of the smaller particles can be created by flocculation, so the flocculation rate expression for the small size has only has the loss term.

(a) Under these circumstances, write the rate equation for the smaller size particle.

(b) Find an expression for the characteristic reaction time for flocculation of these particles in terms of the collision frequency function, the collision efficiency function, and the concentration of the larger size particles.

(c) To simplify further, assume that the long-range force model is valid, and that collisions by fluid shear are so dominant that other mechanisms can be ignored. Develop an expression for the characteristic reaction time and evaluate that expression under the following circumstances:

- The number concentration of the 20-μm particles is $10^5/\text{cm}^3$.
- The number concentration of the 1.5-μm particles is $10^6/\text{cm}^3$.
- The velocity gradient is $20\,\text{s}^{-1}$.
- $\alpha_{\text{emp}} = 1$.

(d) Repeat part (c), except now assume that collisions by differential sedimentation are so dominant that other mechanisms can be ignored. Assume that the density of the particles is $1.4\,\text{g/cm}^3$ and that the water temperature is 20°C.

(e) For the collisions described in part (d), estimate the collision efficiency function according to the short-range force model from the appropriate graph displayed in this chapter, and then revise your estimate of the characteristic reaction time accordingly. Assume that the Hamaker constant is $10^{-13}\,\text{g}\,\text{cm}^2/\text{s}^2$.

(f) Flocculators are typically built with a detention time in the range of 20–30 min. Are your answers to parts (c), (d), and (e) reasonable in comparison to this standard design practice? Comment briefly.

12-6. The quality of raw water for drinking water supplies varies considerably among various sources. For example, Mississippi River water might contain 250 mg/L of clay or mineral particles with a density of 2.5 g/cm^3 and a diameter of ~1 μm. Alternatively, Quabbin Reservoir water (the supply for the City of Boston) might contain 0.55 mg/L of particles that are mostly algae with a density of 1.05 g/cm^3 and a diameter of 4 μm.

(a) Calculate the particle volume concentration (dimensionless and as ppm$_{vol}$) and the particle number concentration (#/mL) for each water source.

(b) Suppose that the particles in each of the two waters are to be destabilized with a cationic polymer such that the collision efficiency factor, α_{emp}, is 0.5. A flocculation tank with a 30-min detention time and a mean velocity gradient (G) of 30 s^{-1} is to be utilized. Assuming that the tank operates as an ideal plug flow reactor, calculate the fractional reduction in particle number achieved by Brownian motion and fluid shear flocculation mechanisms (independently) for each water. Assume a water temperature of 20°C. (Note that equations for monodisperse suspensions are given in Problem 12-4.)

(c) For the case of the Mississippi River water, what reduction in particle number is achieved by flocculation by fluid shear if the tank is modeled as a single CFSTR? as 3 equal volume CFSTRs in series (with the same total volume)?

(d) Suppose that a decision is made to treat the Quabbin Reservoir by adding 50 mg/L of alum $(Al_2(SO_4)_3 \cdot 14.3H_2O)$ at pH 7. Assume that all of the alum precipitates as $Al(OH)_3(s)$ and that the floc particles have a density of 1.05 g/cm^3. What is the new total particle volume concentration? If the Al flocs have a diameter of 20 μm, what is the initial collision rate between algae and Al floc? How does this compare to the collision rate without alum addition? (Use the fluid shear collision mechanism for the interaction of particles of two different diameters ($\beta(i,j)$). Assume that α is 0.5).

13

GRAVITY SEPARATIONS

13.1 Introduction
13.2 Engineered systems for gravity separations
13.3 Sedimentation of individual particles
13.4 Batch sedimentation: Type I
13.5 Batch sedimentation: Type II
13.6 Continuous flow ideal settling
13.7 Effects of nonideal flow on sedimentation reactor performance
13.8 Thickening
13.9 Flotation
13.10 Summary
References
Problems

A discussion of the subject from a theoretical standpoint ... may lead to a better understanding of it, to the collection of better data, and to improvements in design. The processes which take place in sedimentation are extremely complex; to discuss them at once in their entirety seems hopeless. First, conditions much simpler than those which actually exist must be assumed, and from these simple assumptions the more complex conditions can be approached.

A. Hazen, *On Sedimentation* (1904)

13.1 INTRODUCTION

Solid/liquid separation lies at the heart of many treatment systems, because many pollutants are ultimately removed from water in the solid phase. Sedimentation (settling) and, to a lesser extent, flotation, are very commonly used solid/liquid separation processes that rely on gravity as the driving force. These processes have one influent (a suspension with some particle concentration, c_{in}) and two effluents, a relatively clean stream (i.e., one with a particle concentration, c_{out}, less than the influent, generally called simply the effluent) and a relatively dirty stream (one with a particle concentration, c_u, higher than the influent, generally called the sludge or underflow). When handling relatively dilute streams, the sludge flow (Q_u) is small in comparison with the influent flow, so that the effluent flow (Q_{out}) is nearly as large as the influent flow (Q_{in}). In conventional sedimentation systems, as illustrated in Figure 13-1, the effluent comes out the top of the reactor, and the sludge is removed from the bottom; flotation systems and some sedimentation systems have more complex arrangements. While we might hope to minimize the solids concentration in the effluent and simultaneously maximize the concentration in the sludge, real designs invariably must emphasize the characteristics of one of those streams at the expense of the other.

Gravity separation appears deceptively simple: a suspension left to sit quiescently for a long period usually leads to a great degree of solid/liquid separation, and one might expect similar separation when the suspension is induced to flow slowly from an inlet to an outlet. However, modern treatment systems often require a high continuous throughput, making for a more complicated process. As with all particle treatment processes, the effectiveness of gravity solid/liquid separation devices depends on physical and chemical characteristics of the particles and water, as well as on the reactor design and operation.

Traditionally, four types of sedimentation (or, identically, flotation) have been defined, based on the concentration of the particles and their flocculent nature. When the particle concentration is low and the particles are not very "sticky" (i.e., do not flocculate with one another), each particle settles with little or no influence from other particles in the suspension. This phenomenon is called Type I sedimentation. When the particle concentration is low and the particles do have a strong tendency to flocculate, flocculation occurs simultaneously with sedimentation, influencing the degree of sedimentation achieved. This type of sedimentation is

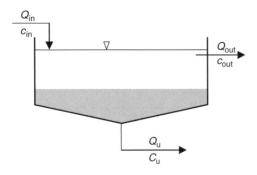

FIGURE 13-1. Schematic of a continuous flow sedimentation tank.

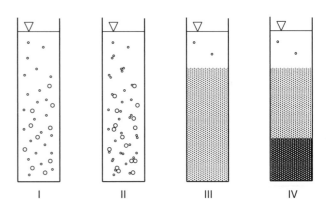

FIGURE 13-2. Schematic of various types of sedimentation.

called Type II sedimentation. When the particle concentration is high, particle settling is influenced by the upward flow of water caused by the displacement at lower levels; remarkably, above some concentration, a well-defined interface forms between a highly concentrated sludge zone and a much cleaner liquid zone above it. In batch systems, the interface gradually falls to lower depths. This phenomenon, often referred to as zone settling, constitutes Type III settling; an exact understanding of the causes for the formation of this well-defined interface is still lacking. (Water also flows upward in Types I and II sedimentation, but the particle volume concentration is sufficiently low that the hindrance created by that flow can be ignored.) Type III settling can occur whether the particles form flocs with one another or not, but it occurs in flocculent systems at lower concentrations than in nonflocculent systems. Finally, for some suspensions at high concentrations, the rate of subsidence of the well-defined solid/liquid interface is determined by the collapse of the floc structure below. This phenomenon occurs when the particles (flocs) incorporate a considerable amount of water; the weight of overlying solids can exceed the support capacity of the flocs at the bottom, so these flocs collapse and water is squeezed out. This phenomenon is known as Type IV sedimentation, or compression.

A schematic diagram that illustrates the distinctions among the different types of sedimentation is shown in Figure 13-2. The diagram is a caricature in the sense that it overstates the differences among the various types. The distinction between settling of low concentration suspensions (Types I and II) and high concentration suspensions (Types III and IV) is usually clear—the existence of the well-defined interface that characterizes Types III and IV sedimentation is, by definition, easy to see. However, the distinctions between Types I and II sedimentation, and between Types III and IV, are usually blurry, in part because the degree of "flocculent nature" is difficult to quantify, and in part because the word "flocculent" has related but slightly different meanings for Type II and

Type IV sedimentation. In both cases, particles are being brought into contact with one another. For Type II sedimentation, the term is used identically as it is in Chapter 12 to characterize the tendency of particles to stick to one another. For Type IV sedimentation, particles are forced together by the squeezing out of water between them in the concentrated suspension.

In engineered systems, the primary concern in Types I and II sedimentation is usually the effluent quality; that is, achieving a low value of c_{out}. As a result, the analysis of Types I and II sedimentation focuses on individual particles or flocs, with the goal being to predict how (and how many) particles get into the effluent stream. In Types III and IV sedimentation, the concern can be either to maximize the rate of removal of the particle mass (the product of the underflow rate and the underflow concentration) or to maximize the solids concentration of the underflow while maintaining an acceptable effluent concentration. Hence, the focus in Types III and IV sedimentation is on the behavior of particles in a highly concentrated environment and on the interaction of water flow with those particles.

The sedimentation categories defined earlier can also be used with respect to flotation, although only two (Types II and III) of the four categories apply well in most applications in environmental engineering. For Type I flotation, the particles of concern would have to be less dense than water, and this condition is rarely met. Generally, the particles of interest are naturally denser than water and are made less dense than water by the attachment of air bubbles to their surfaces; for this attachment process to occur, the surfaces must be sticky. Conceptually, the process of attaching a bubble to a particle is identical to attaching two particles together, so all of the concerns about particle–particle interactions that were the subject of the previous two chapters come into play for flotation. However, the chemical conditions required for particle–bubble destabilization are not necessarily the same as for particle–particle destabilization. Often, the bubbles are formed on the surface of the particles; the water is made to be supersaturated with respect to air (i.e., nitrogen and oxygen), and the particle surfaces act as nucleation sites for the creation of bubbles.

In addition to the particle concentration and flocculent nature, a third characteristic of suspensions that is used to characterize settling behavior is the heterodispersity (i.e., variation in the properties) of the particles in suspension. Heterodispersity can include several characteristics of the particles (size, surface charge, shape, and density), but primarily, we consider the size heterodispersity, or the size distribution. For Type I settling, heterodispersity dictates the sensitivity of the expected particle removal efficiency to the primary design variable, as shown subsequently. For Type II sedimentation, heterodispersity influences the rate of flocculation (as shown in Chapter 12) and, therefore, sedimentation. Finally, for Type III sedimentation, increasing heterodispersity increases the particle concentration required to form the characteristic well-defined solid/liquid interface.

The chapter begins with a brief presentation of some of the common ways in which gravity separation systems are used in water and wastewater treatment practice. Then the analysis starts with a discussion of the simplest sedimentation system—a batch system with a single particle in an infinite medium. Gradually, complexity is added, first by considering heterodisperse suspensions of both Type I and Type II in batch systems. The behavior of both Type I and II suspensions in continuous flow systems is then delineated for several types of ideal flow configurations. The effect of nonideal flow in the most common types of sedimentation systems is then considered. Sections on thickening (Types III and IV) and flotation conclude this chapter.

13.2 ENGINEERED SYSTEMS FOR GRAVITY SEPARATIONS

Gravity separation systems are the most commonly used process in water and wastewater treatment. Typical municipal wastewater plants have both primary and secondary settling tanks, with the primary tanks designed to remove most of the solids in the influent, and the secondary units designed to remove the microorganisms that are created in the biological treatment that occurs after primary sedimentation. Most wastewater treatment plants also have a separate sludge thickener. If the contaminants in an industrial wastewater are mostly dissolved materials, the treatment plant would not need a primary sedimentation tank but would require solid-/liquid separation after either biological treatment or chemical precipitation. Drinking water treatment plants for surface waters generally have sedimentation tanks after flocculation, although plants that treat waters with low particle concentrations and that do not use sweep flocculation are sometimes designed without a gravity separation unit. Flotation is used less commonly than sedimentation, but it is used increasingly after biological treatment in wastewater or when the particles in a water supply are not dense (e.g., when the dominant particles are algal cells).

Sedimentation tanks are built in a variety of designs, but the two basic types are rectangular and circular. Rectangular tanks have two advantages. In large plants, where several units are necessary, rectangular tanks can use common wall construction between adjacent units, saving construction costs and utilizing the space efficiently. They can also be easily designed with an inlet arrangement that spreads the flow evenly over the entire influent wall of the tank. The even distribution of the flow is valuable, as shown in Section 13.7. An example of rectangular sedimentation tanks at a large wastewater treatment plant is shown in Figure 13-3.

Despite the cited advantages of rectangular tanks, circular tanks are more common, primarily because the sludge withdrawal system can be much simpler than in rectangular reactors. A slowly rotating arm that scrapes sludge toward a pit near the center of the bottom of the tank has far fewer moving parts than the systems that are needed for rectangular tanks. The primary sedimentation tanks in wastewater treatment plants also usually have a scum collection system that operates at the water surface and is driven by the same rotating arms. The advantage of the circular sludge collection system can be incorporated into square or rectangular tanks, as long as the length is an integer multiple of the width; arms that extend and retract are built into the system to clean the corners of the tank. An aerial view of a wastewater treatment plant that includes several circular sedimentation tanks is shown in Figure 13-4.

In circular tanks with primarily horizontal flow, the water comes into the center of the tank and flows radially outward toward the perimeter, where it is collected. As a result, the velocity of the water decreases as it flows from the small inlet zone toward the tank perimeter. Design of the inlet zone to spread the flow horizontally and vertically is quite important to prevent short-circuiting. The outlet structures in these tanks are virtually always overflow weirs, so that the flow has a vertical (upward) component as it approaches the outlet. Some circular tanks, especially those in which both flocculation and sedimentation are expected to occur, are built with an inverted cone for the inlet, so that water has to flow down underneath the outside edge of the cone and then flow back up on the outside of the cone to reach the weirs. This design aids the spreading of the flow and helps avoid short-circuiting, but it causes the flow to the outlet zone to be primarily vertical. As we shall see subsequently, such upflow reactors are theoretically not as efficient as horizontal flow reactors; nevertheless, the advantages of better flow distribution might overcome this theoretical disadvantage of upflow.

Similarly, in both rectangular and circular sedimentation tanks, designing the outlet zone is so that the flow of water

FIGURE 13-3. Rectangular settling tank at Tampa, Florida Wastewater Treatment Plant. *Source*: Image captured March 2011 from www.tampagov.net/dept_wastewater/information_resources/advanced_wastewater_treatment_plant/Virtual_Tour/Step_1_-_preliminary_treatment.asp.

through the tank is approximately uniform is essential. In circular tanks with horizontal flow, the outlets are usually overflow weirs on the perimeter of the tank. In circular tanks with an upward flow, weirs are also added radially, like spokes on a wheel. In rectangular reactors, the effluent collection is at the opposite wall from the influent, and might consist of overflow weirs, submerged weirs, or a perforated effluent wall.

An interesting modification of conventional settling tank design is to fill a tank with modules that have multiple inclined surfaces. Particles fall onto these surfaces, slide down the incline, and then drop to the bottom. Such settling reactors are called tube or plate settlers (depending on how close the vertical spacers are to each other); an example is shown in Figure 13-5. Tube or plate settlers provide a much greater collection area within the same floor area and, as we

FIGURE 13-4. Wastewater treatment plant with several circular sedimentation tanks. (The four largest circular tanks in the picture are trickling filters, but the six circular tanks in the foreground are sedimentation tanks.) *Source*: Image from http://denr.sd.gov/des/sw/MunicipalWastewaterTreatmentPlantAnatomy.aspx, March, 2011.

FIGURE 13-5. Example of plate settlers. No water in this tank for the picture; water would flow up from the bottom between the plates and then horizontally into the effluent launders (collection channels) that are visible in the picture. *Source*: Image taken March 2011 from www.norfolk.gov/Utilities/produce/LlamellaPlates.jpg.

shall see, improve the collection efficiency. Because of the presence of so much equipment in the tank, the design of the inlet, outlet, and sludge collection mechanisms is more complex than in conventional systems.

13.3 SEDIMENTATION OF INDIVIDUAL PARTICLES

Stokes' Law

The basis of much of our understanding of gravity separations is Stokes' law. Stokes (1851) considered the behavior of a single spherical particle in an infinite fluid medium; the settling of such a particle under the influence of gravity is the interest here. As illustrated in Figure 13-6, three forces act on the particle: gravity, buoyancy, and drag. When these three forces balance (i.e., when there is no particle acceleration), the relative velocity of the particle with respect to the fluid is called the terminal settling velocity, v. The gravity force, F_g, acts downward (taken as positive here) and is found from a direct application of Newton's second law:

$$F_g = m_p g = \rho_p V_p g = \rho_p \left(\tfrac{1}{6}\pi d_p^3\right)g \quad (13\text{-}1)$$

where the subscript p refers to the particle, m is mass, ρ is the density, V is volume, d is diameter, and g is the gravitational constant.

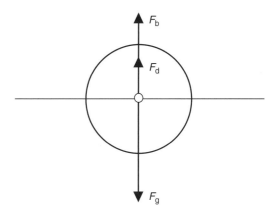

FIGURE 13-6. Force balance on particle settling in infinite flow field.

The buoyancy force, F_b, acts upward and is similar to the gravitational force, reflecting the displacement of fluid of density ρ_L by the particle:

$$F_b = -m_L g = -\rho_L V_p g = -\rho_L \left(\tfrac{1}{6}\pi d_p^3\right)g \quad (13\text{-}2)$$

where the subscript 'L' refers to the liquid and all other symbols have been identified.

The movement of the particle causes water to flow around it. The mathematical determination of the drag force created by that flow is nontrivial; an elegant derivation is given in Bird et al. (2006) and is summarized here. Consideration is restricted to "creeping flow" conditions; that is, those for which the inertial terms in the Navier–Stokes equations[1] can be neglected. This criterion translates to situations in which no turbulent eddies form downstream of the particle, corresponding to Reynolds numbers with respect to the particle less than approximately 0.1. The particle Reynolds number can be calculated as follows:

$$\text{Re} = \frac{\rho_L v d_p}{\mu} \quad (13\text{-}3)$$

where μ is the viscosity of water.

The problem under consideration (settling of a particle with a velocity v with respect to the otherwise stationary fluid) is identical to the flow of fluid at velocity v (far removed from the surface) around a stationary particle. Because of the symmetry around the sphere, the solution

---
[1] The Navier–Stokes equations are a set of differential equations that relate the acceleration of a fluid to other properties of the system, such as the pressure gradient and the fluid viscosity. The equations are mathematically complex and have analytical solutions only for certain, highly idealized limiting cases. They can, however, be solved numerically for more realistic cases, and provide essential information for the behavior of the fluid. A description of these equations can be found in any fluid mechanics text.

is most easily found using spherical coordinates, with distance r from the center of the particle and angle θ from the vertical. The three boundary conditions are zero velocity in both the r and θ directions at the surface of the particle, and a velocity approaching v as the distance from the center of the particle approaches infinity. The result is that the fluid velocity components in the r and θ directions are as follows:

$$v_r = v\left[1 - \frac{3}{2}\left(\frac{r_p}{r}\right) + \frac{1}{2}\left(\frac{r_p}{r}\right)^3\right]\cos\theta \quad (13\text{-}4)$$

$$v_\theta = v\left[1 - \frac{3}{4}\left(\frac{r_p}{r}\right) - \frac{1}{4}\left(\frac{r_p}{r}\right)^3\right]\sin\theta \quad (13\text{-}5)$$

Here, r_p is radius of the particle. The drag force on the particle can be found as the volumetric integral of the energy dissipated per unit area, considering concentric spheres in the fluid from $r = r_p$ to $r = \infty$.[2] The result, often known as Stokes' law, is as follows:

$$F_d = -6\pi\mu v r_p = -3\pi\mu v d_p \quad (13\text{-}6)$$

where the negative sign indicates that this force is in the opposite direction of the velocity.

The condition that the three forces on the particle balance so that the particle is at its terminal settling velocity leads to the following:

$$F_g + F_b + F_d = 0 \quad (13\text{-}7)$$

or, after substitution of Equations 13-1, 13-2, and 13-6 and simplification

$$v = \frac{(\rho_p - \rho_L)g d_p^2}{18\mu} \quad (13\text{-}8)$$

This expression is also often known as Stokes' law; in this case more properly referred to as Stokes' law for settling, and the velocity described by Equation 13-8 is usually called the Stokes' velocity. Because most particles are denser than water, the velocity is usually downward and settling is being described; however, nothing in the derivation imposes such a constraint, so Stokes' velocity is equally valid in describing flotation of particles whose density is less than that of the surrounding liquid.

The limitations imposed (or assumptions made) in reaching Equation 13-8 include a spherical particle, low Reynolds number (neglecting inertial terms), and an infinite fluid

---

[2] The same result can be obtained by integrating, over the entire particle surface, the component in the direction of flow of both the pressure distribution normal to the surface and the shear stress tangential to the surface. See Bird et al. (2006) for details.

(infinite distance to propagate the flow disturbance induced by the particle).[3] These conditions are rarely met for the solid/liquid separations required in environmental engineering practice, but the deviations are often sufficiently minor that Stokes' law can be applied. Furthermore, even in conditions where it cannot be applied, Stokes' law provides a baseline for comparison with other results.

■ **EXAMPLE 13-1.** Calculate the Stokes' settling velocity in water at 20°C of a spherical particle with a diameter of 10 μm and a density of 1.5 g/cm³. At that temperature, the density of water is 0.998 g/cm³ and the absolute viscosity is 1.002×10^{-2} g/cm s. Check that the assumptions of Stokes' law are met.

Solution. Application of Stokes' law (Equation 13-8) yields the following:

$$v = \frac{(\rho_p - \rho_L)g d_p^2}{18\mu}$$

$$= \frac{\{1.5(\text{g/cm}^3) - 0.998(\text{g/cm}^3)\}(981(\text{cm}^3/\text{s}^2))(10^{-3}\text{cm})^2}{18(1.002 \times 10^{-2}(\text{g/cm s}))}$$

$$= 2.73 \times 10^{-3}(\text{cm/s})$$

To check that Stokes' law applies, we have to confirm that the Reynolds number is less than 0.1. From Equation 13-3

$$\text{Re} = \frac{\rho_L v d_p}{\mu}$$

$$= \frac{(0.998\,\text{g/cm}^3)(2.73 \times 10^{-3}\,\text{cm/s})(10^{-3}\,\text{cm})}{(1.002 \times 10^{-2}\,\text{g/cm s})}$$

$$= 2.72 \times 10^{-4}$$

The Reynolds number is well within the criterion for the applicability of Stokes' law. ■

---

[3] An additional simplification is the assumption that the force balance is zero; that is, the calculation yields the terminal velocity and does not describe the approach to that velocity. If, instead of zero, the net force is set to the product of mass $\{(\pi/6)\rho_p d_p^3\}$ and acceleration (dv/dt), and the expressions for the three forces are summed, one can find, after several steps of calculus and algebra, the following expression for the vertical velocity (assumed to start at zero): $v(t) = \left[\{(\rho_p - \rho_L)g d_p^2\}/18\mu\right]\left[1 - \exp\left(-18\mu/\rho_p d_p^2 t\right)\right]$. The exponential term rapidly approaches zero, leaving the Stokes' law expression for the velocity. Any particle for which Stokes' law applies will attain 95% of its terminal velocity in approximately a millisecond or less, so an assumption that the particle reaches its terminal velocity essentially instantly is almost always reasonable. As in several other applications in this book, one can think of the inverse of the fraction in the exponential term as a characteristic time in the achievement of a steady settling velocity.

Suspensions that are encountered in water and wastewater treatment are heterodisperse, as explained in Chapter 11. For sedimentation, the most important distribution is the settling velocity distribution, because particles with higher settling velocities are more easily removed. This distribution can be determined empirically in sedimentation experiments, as explained subsequently. Alternatively, one can estimate the settling velocity distribution from a measured particle size distribution if the particle density and shape are known (and assumed to be uniform for all the particles) by finding the settling velocity of each size class of particles.

In sedimentation, the primary interest is usually the removal of mass, rather than the reduction in particle numbers or particle surface area. Hence, the settling velocity distribution is generally defined in terms of the mass fraction of the suspension with settling velocities in different increments. The most common way of describing the distribution is on a relative and cumulative basis; that is, the fraction of the mass of the suspension that has settling velocity less than or equal to each successive settling velocity. The conversion of a measured particle size distribution to this cumulative settling velocity distribution, symbolized as $f(v)$, is illustrated in the following example.

■ **EXAMPLE 13-2.** The particle size distribution of a wastewater sample is shown in the first three columns of Table 13-1. Assume that the temperature is 20°C, at which the viscosity is 1.002×10^{-2} g/cm s and the density is 0.998 g/cm^3, and that all of the particles in the suspension are spherical and have a density of 1.5 g/cm^3.

(a) Find the Stokes' settling velocity of each size class considered in units of m/h. Check whether the criterion for using Stokes' law is met for each size class.

(b) For each size class, find the (cumulative) volume fraction of particles with settling velocity equal to or less than the settling velocity for that class. Plot these cumulative volume fractions versus the associated settling velocities; this plot is the cumulative settling velocity distribution of the suspension, $f(v)$.

TABLE 13-1. Particle Size Distributions in Batch Settling

$\log d_p$ (d_p in μm)[a]	d_p (μm)[a]	Particle Size Distribution Function $\log \Delta N/\Delta d_p$ ($\Delta N/\Delta d_p$ in cm^{-3} μm^{-1})	Volume Distribution $\Delta V/\Delta \log d_p$ (μm^3/cm^3)	Cumulative Volume Fraction (–)	Settling Velocity (m/h)	Reynolds Number (–)
−0.3	0.50	5.60	3.04E+04	1.53E−05	2.46E−04	3.43E−08
−0.2	0.63	5.57	7.10E+04	5.12E−05	3.91E−04	6.84E−08
−0.1	0.79	5.51	1.56E+05	1.30E−04	6.19E−04	1.37E−07
0.0	1.00	5.44	3.30E+05	2.97E−04	9.81E−04	2.73E−07
0.1	1.26	5.35	6.75E+05	6.38E−04	1.55E−03	5.44E−07
0.2	1.58	5.24	1.33E+06	1.31E−03	2.46E−03	1.08E−06
0.3	2.00	5.13	2.56E+06	2.60E−03	3.91E−03	2.16E−06
0.4	2.51	4.99	4.74E+06	5.00E−03	6.19E−03	4.32E−06
0.5	3.16	4.85	8.50E+06	9.30E−03	9.81E−03	8.62E−06
0.6	3.98	4.69	1.47E+07	1.67E−02	1.55E−02	1.72E−05
0.7	5.01	4.51	2.46E+07	0.029	2.46E−02	3.43E−05
0.8	6.31	4.31	3.94E+07	0.049	3.91E−02	6.84E−05
0.9	7.94	4.10	6.07E+07	0.080	6.19E−02	1.37E−04
1.0	10.00	3.87	8.93E+07	0.125	9.82E−02	2.72E−04
1.1	12.59	3.62	1.25E+08	0.188	0.155	5.44E−04
1.2	15.85	3.34	1.67E+08	0.272	0.246	1.08E−03
1.3	19.95	3.04	2.10E+08	0.378	0.391	2.16E−03
1.4	25.12	2.71	2.49E+08	0.504	0.619	4.32E−03
1.5	31.62	2.36	2.76E+08	0.644	0.981	8.62E−03
1.6	39.81	1.90	2.42E+08	0.766	1.55	0.017
1.7	50.12	1.40	1.92E+08	0.863	2.46	0.034
1.8	63.10	0.85	1.36E+08	0.932	3.91	0.068
1.9	79.43	0.25	8.63E+07	0.976	6.19	0.137
2.0	100.00	−0.40	4.81E+07	1.000	9.82	0.272

[a]The values in the first two columns correspond to the middle of the logarithmic interval detected by the particle size analyzer. For example, the row with a value of $\log d_p = 0.0$ in the first column gives information for a size interval of $-0.05 < \log d_p < 0.05$, corresponding to $0.89 < d_p < 1.12$ μm.

610 GRAVITY SEPARATIONS

Solution.

(a) Each size class shown in the table has a logarithmic interval of diameter ($\Delta \log d_p$) = 0.1, and the diameter in the center of that logarithmic interval is shown in the second column. The calculations for the Stokes' settling velocity are identical to that shown in Example 13-1. The conversion to the units of m/h for a 10-μm particle (the particle size considered in Example 13-1) is as follows:

$$v = 2.73 \times 10^{-3}\,\text{cm/s}$$
$$= \left(2.73 \times 10^{-3}\,\text{cm/s}\right)\left(\frac{1\,\text{m}}{100\,\text{cm}}\right)\left(\frac{3600\,\text{s}}{1\,\text{h}}\right)$$
$$= 9.82 \times 10^{-2}\,\text{m/h}$$

The Re values in the last column reveal that the largest two sizes (slightly) violate the criterion for the application of Stokes' law. In Section titled "Inertial Effects on Sedimentation", we consider how to make a better estimate of the settling velocity under such conditions. The corrections for these upper two size classes are minor.

(b) The given particle size distribution function $\log(\Delta N/\Delta d_p)$ is converted to the volume distribution $(\Delta V/\Delta \log d_p)$ as described in Chapter 11 (See Example 11-2). The total volume concentration is found as the area under the curve of the volume distribution; that is,

$$\text{Total volume concentration} = \int (dV/d\log d_p)\,d(\log d_p),$$

which can be approximated as $\sum\left((\Delta V/\Delta \log d_p)\Delta \log d_p\right)$. Since $\Delta \log d_p$ is constant (0.1) in this case, it can be taken out of the summation, and the total volume concentration is just the sum of the values in the fourth column, multiplied by 0.1. For the distribution shown, the total volume concentration is $198 \times 10^6\,\mu\text{m}^3/\text{cm}^3$, or 198 parts per million by volume (ppm$_v$). For a particle density of 1.5 g/cm^3, the suspended solids concentration is

$$c_{ss} = \left(198 \times 10^6\,\mu\text{m}_p^3/\text{cm}_L^3\right)\left(1.5\left(\text{g/cm}_L^3\right)\right)\left(10^3\,\text{mg/g}\right)$$
$$\times \left(\frac{\text{cm}_p^3}{10^{12}\,\mu\text{m}_p^3}\right)\left(10^3\,\text{cm}_L^3/\text{L}\right) = 297\,\text{mg/L}$$

This size distribution might be representative of the influent to a primary sedimentation tank in a wastewater plant, or the influent to a sedimentation tank following chemical precipitation in drinking water or industrial wastewater treatment.

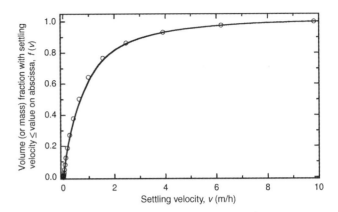

FIGURE 13-7. Cumulative settling velocity distribution for given particle size distribution.

The cumulative volume (or mass) fraction is the ratio of the running sum of the volume concentration (from the lowest particle size measured) to the total volume concentration. The cumulative settling velocity distribution can be plotted as this cumulative volume fraction versus the settling velocity for each size class (i.e., velocity class) considered, as shown in Figure 13-7. The first several size classes account for only a very small fraction of the total mass, and are therefore crowded together in the lower left corner of this graph. The utility of this distribution is demonstrated throughout the chapter. ■

Inertial Effects on Sedimentation

In the development of Stokes' law as shown earlier, the frictional forces opposing particle motion were considered, but the inertial effects of fluid flow around the particles were neglected. The total drag force on the particle is a combination of the drag due to friction and that due to inertial effects. The source and magnitude of these inertial effects are considered here.

Consider a stationary particle with water flowing around it, and imagine that the velocity of water is gradually increased from zero to some high value. When the fluid is stationary, the drag force is zero because the friction between the fluid and the particle surface is zero and because there is no kinetic contribution to the pressure on the sphere. (The stagnant pressure variation with the depth of water leads to the buoyancy force.) The situation changes as flow begins. For very slow velocity, both the kinetic contribution to the pressure distribution on the surface of the sphere and the friction drag increase linearly with the velocity, with the friction drag being responsible for two-thirds of the total drag. Both of these effects are accounted for in the drag force expressed in Equation 13-6. As the velocity increases to create a Reynolds number greater than approximately 0.1, however, the dynamic pressure of the flow ($0.5 \rho v^2$) begins to create significant pressure variations across the surface of the sphere, creating substantially more pressure drag. Additionally, at a Reynolds number somewhere between 1 and 20 (17, according to Jenson, 1959), the flow begins to separate from the rear surface of the sphere, forming a low-pressure wake behind the sphere. As the velocity continues to increase, the point on the circumference at which the flow separates from the sphere's surface moves forward, increasing the size of the low-pressure region on the rear of the sphere. Both the increased pressure variation and the flow separation are associated with the inertia of the flowing fluid and lead to greater pressure drag with higher flow velocity. As the velocity increases, the pressure drag becomes larger than the frictional drag. For example, at Re = 1000, pressure drag represents approximately 95% of the total drag.

The Stokes analysis, therefore, underestimates the drag at Reynolds numbers greater than 0.1. A general expression for the drag force, valid at all values of Re, is

$$F_d = \pm \tfrac{1}{2} C_d A_p \rho_L v^2 = \pm \tfrac{1}{2} C_d \pi r_p^2 \rho_L v^2 = \pm \tfrac{1}{8} C_d \pi d_p^2 \rho_L v^2 \quad (13\text{-}9)$$

where C_d is the (dimensionless) drag coefficient and A_p is the projected area of the particle (facing the flow). For the low Reynolds numbers at which Stokes' law for settling applies, the combination of Equations 13-6 and 13-9 yields

$$C_d = \frac{24}{\text{Re}} \quad (13\text{-}10)$$

For higher Reynolds numbers, several correlations between C_d and Re have been developed from experimental data.[4] Following the approach of Haider and Levenspiel (1989), Brown and Lawler (2003) found the best correlation to be as follows:

$$C_d = \frac{24}{\text{Re}}\left(1 + 0.150\,\text{Re}^{0.681}\right) + \frac{0.407}{1 + (8710/\text{Re})} \quad (13\text{-}11)$$

Equation 13-11 agrees with Equation 13-10 at low Reynolds numbers, but provides a much greater value of C_d at higher Reynolds numbers.

When Equation 13-9 is substituted into the force balance (Equation 13-7), the resulting expression for the settling velocity is

$$v = \left\{\frac{4}{3}\left(\frac{\rho_p - \rho_L}{\rho_L}\right)\frac{g d_p}{C_d}\right\}^{1/2} \quad (13\text{-}12)$$

For particles with Re < 0.1,[5] the settling velocity can be found from Stokes' law (Equation 13-8). For particles with higher Re number, an iterative (trial and error) approach can be used to find the settling velocity that simultaneously satisfies Equations 13-3, 13-11, and 13-12. One can begin the iterative process by estimating the velocity from Stokes' law (Equation 13-8).

The iterative approach can be eliminated if a direct correlation for settling velocity, rather than the drag coefficient, is utilized. To develop this correlation in a general way, the particle size and velocity are cast in dimensionless terms as follows:

$$\tilde{d} = \left(\frac{3}{4}C_d \text{Re}^2\right)^{1/3} = d_p\left[\frac{g\rho_L(\rho_p - \rho_L)}{\mu^2}\right]^{1/3} \quad (13\text{-}13)$$

$$\tilde{v} = \left(\frac{3\text{Re}}{4 C_d}\right)^{1/3} = v\left[\frac{\rho_L^2}{g\mu(\rho_p - \rho_L)}\right]^{1/3} \quad (13\text{-}14)$$

For 0.1 < Re < 4000 (essentially all conditions of interest in water treatment applications have Re < 4000), Brown and Lawler (2003) found the best correlation to be as follows:

$$\tilde{v} = \frac{\tilde{d}^2\left(22.5 + \tilde{d}^{2.046}\right)}{0.0258\tilde{d}^{4.046} + 2.81\tilde{d}^{3.046} + 18\tilde{d}^{2.046} + 405} \quad (13\text{-}15)$$

To calculate the settling velocity (v), one calculates the dimensionless diameter ($\tilde{d}$) from Equation 13-13, then the dimensionless velocity ($\tilde{v}$) from Equation 13-15, and finally the velocity from a rearrangement of Equation 13-14. Using the dimensionless quantities, the criterion for the applicability of Stokes' law (Re < 0.1, where C_d is given by Equation 13-10) translates approximately to $\tilde{d} < 1$. Hence, one can determine directly from Equation 13-13 whether Stokes' law is applicable.

In practice, most particles of interest in water and wastewater treatment have a sufficiently small settling velocity that Stokes' law applies, but some exceptions are common as

---

[4] The correlation that has been used most widely in environmental engineering was reported by Fair et al. (1968), as follows: $C_d = (24/\text{Re}) + (3/\sqrt{\text{Re}}) + 0.34$ for Re ≥ 0.3. The correlation by Brown and Lawler yields a better fit to the data and hence is recommended.

[5] Some authors use a more conservative cutoff value for the Reynolds numbers, such as 0.05 or even 0.01, while others use more liberal values such as 0.3 or 1.0. The reader can calculate the error using the complete expression for values between 0.01 and 1.

well. The sand media in deep bed filtration (whose settling velocity has to be known for calculations and decisions about backwashing velocity) and particles captured in grit chambers typically have Reynolds numbers greater than 0.1, and therefore have settling velocities less than the Stokes' velocity. Large particles or flocs of relatively dense inorganic precipitates (e.g., calcium carbonate precipitated in softening) can also have Reynolds numbers that are too large to use Stokes' law directly. Also, although the preceding discussion focuses on the behavior of spherical particles with known properties, *any* nonflocculating particle reaches a terminal settling velocity quickly in water. For many such particles, the settling velocity, v, can only be determined empirically; nevertheless, it is common to assume that particles are spheres with the settling velocity calculated accordingly.

■ **EXAMPLE 13-3.** Calculate the settling velocity at 20°C of a sand grain that is considered to be spherical with a diameter of 1 mm. (Such particles are used for filter media.) The density of sand is 2.65 g/cm^3.

Solution. The properties of water at this temperature were given in Example 13-1. It is wise to begin by seeing if Stokes' law applies. The diameter of interest here is 100 times that of the one considered in Example 13-1. If the particles had equal densities, the Stokes' velocity of the particle of interest would be 10^4 times that of the particle considered in the previous example, and the associated Reynolds number would be 10^6 times as big. The different densities of the particles would have only a minor effect on these ratios, so this mental calculation shows that Re would be far greater than 0.1 in the current case, and therefore, that Stokes' law would not apply.

To find the settling velocity, we first find the dimensionless diameter from Equation 13-13:

$$\tilde{d} = d_p \left[\frac{g\rho_L(\rho_p - \rho_L)}{\mu^2} \right]^{1/3}$$

$$= 0.1\,\text{cm} \left[\frac{(981\,\text{cm/s}^3)(0.998\,\text{g/cm}^3)(2.65 - 0.998)\text{g/cm}^3}{(1.002 \times 10^{-2}\,\text{g/cm s})^2} \right]^{1/3}$$

$$= 25.3$$

Note that $\tilde{d} > 1$, confirming that Stokes' law does not apply. We then use Equation 13-15 to find the dimensionless velocity:

$$\tilde{v} = \frac{\tilde{d}^2 \left(22.5 + \tilde{d}^{2.046} \right)}{0.0258\tilde{d}^{4.046} + 2.81\tilde{d}^{3.046} + 18\tilde{d}^{2.046} + 405}$$

$$= \frac{(25.3)^2 \left(22.5 + 25.3^{2.046} \right)}{0.0258(25.3)^{4.046} + 2.81(25.3)^{3.046} + 18(25.3)^{2.046} + 405}$$

$$= 6.20$$

Finally, the settling velocity is found after inverting Equation 13-14:

$$v = \tilde{v} \left[\frac{g\mu(\rho_p - \rho_L)}{\rho_L^2} \right]^{1/3}$$

$$= 6.20 \left[\frac{(981\,\text{cm/s}^2)(1.002 \times 10^{-2}\,\text{g/cm s})((2.65 - 0.998)\text{g/cm}^3)}{(0.998\,\text{g/cm}^3)^2} \right]^{1/3}$$

$$= 15.7\,\text{cm/s} \qquad ■$$

13.4 BATCH SEDIMENTATION: TYPE I

A typical Type I suspension of interest in environmental engineering includes particles with a virtually continuous distribution of v values. We next consider the settling properties of such a suspension and, in the process, establish an approach for determining the settling velocity distribution of particles in any Type I suspension. In several cases, we consider a batch sedimentation experiment in which, at time zero, a suspension is placed in a vessel such that the particles are dispersed uniformly. We assume that the fluid in that suspension is completely stagnant, and that the suspension is sufficiently dilute that adjacent particles do not influence one another. In essence, this means that each particle behaves as if it were the only particle present. In the sections that follow, we consider suspensions of increasing complexity.

Monodisperse Suspension

Consider first a suspension in which all of the particles are identical and therefore have the same settling velocity. If we could measure the concentration of particles at a depth z over time, what would we see? As particles originally present at depth z settle to greater depths, they would (for a while) be replaced by an equal number of particles falling from a layer above. The process would continue until particles that started at the top of the suspension had settled past the depth of the measurement layer, after which the concentration would be zero. The time at which the step change in concentration would occur is the quotient of the depth and the settling velocity. The result can be stated succinctly as follows:

$$c(z,t) = \begin{cases} c_0 & \text{for } t \leq \frac{z}{v} \text{ or } z \leq vt \\ 0 & \text{for } t > \frac{z}{v} \text{ or } z < vt \end{cases} \quad (13\text{-}16\text{a})$$

where z is the distance from the top of the water column. The result can be written even more succinctly in terms of the

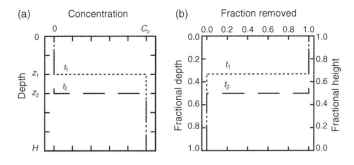

FIGURE 13-8. Batch sedimentation profiles for a monodisperse suspension.

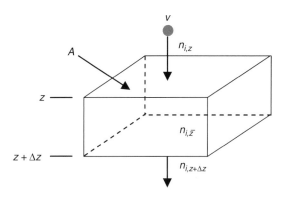

FIGURE 13-9. Control volume for considering settling.

Heaviside function,[6] as follows:

$$c(z,t) = c_0 H_{z=vt}(t) \qquad (13\text{-}16b)$$

The result is illustrated by the profiles in Figure 13-8. In Figure 13-8a, the concentration is shown as a function of depth; in Figure 13-8b, the fraction removed is shown as a function of the fractional depth (measured from the top) or fractional height (measured from the bottom). As indicated in Figure 13-8a, the particle concentration is zero from the top of the reactor ($z = 0$) to the labeled depths (z_1 and z_2 at t_1 and t_2, respectively), and is the initial concentration, c_o, from those depths to the bottom. In reality, this picture only considers the active settling zone; the accumulation of solids in a sludge layer is ignored (or, more properly, is considered to occur at greater depths). If the settling velocity of the particles is v, then $z_1 = vt_1$ and $z_2 = vt_2$. Figure 13-8b is a mirror image of Figure 13-8a: the fraction removed is one from the top of the reactor to the labeled depths, and is zero from those depths to the bottom.

Although the results for this monodisperse system can be obtained directly from conceptual arguments, it is useful to consider a more fundamental mathematical approach. For this analysis, we consider how the number of particles changes in a small differential element in the reactor, as shown in Figure 13-9.

No water flows into or out of this volume, and we assume that particle transport by diffusion (or Brownian motion) is negligible. Also, in Type I sedimentation, no creation or destruction of particles occurs by reaction (precipitation or flocculation). However, particles can enter the control volume by sedimentation from above and be lost by sedimentation to lower depths, as indicated

by the arrows in the figure. In essence, sedimentation represents another way besides advection, diffusion, or dispersion (as discussed in Chapter 1) for mass to penetrate the control volume. Hence the particle number balance can be written as follows:

$$\begin{bmatrix} \text{Rate of change of} \\ \text{number of particles} \\ \text{stored in control volume} \end{bmatrix} = \begin{bmatrix} \text{Rate of which} \\ \text{particles settle into} \\ \text{the control volume} \\ \text{from above} \end{bmatrix}$$

$$- \begin{bmatrix} \text{Rate of which} \\ \text{particles settle out of} \\ \text{the control volume} \\ \text{to below} \end{bmatrix}$$

The number of particles in the control volume is the product of the number concentration, $n_{\bar{z}}$, and the volume of the element ($A\Delta z$). The concentration of particles inside the element is denoted with the subscript $\bar{z}$, whereas the concentrations at the boundaries are denoted by the subscripts z (at the top) and $z + \Delta x$ at the bottom. The rate at which particles settle in or out is the product of the flux (itself the product of settling velocity and concentration at the appropriate height) and the area. Putting these concepts into symbols yields

$$\frac{d(A\Delta z)n_{i,\bar{z}}}{dt} = v_i n_{i,z} A - v_i n_{i,z+\Delta z} A \qquad (13\text{-}17)$$

where n_i is the number concentration of particles of size i (the only size under consideration here, but in subsequent situations we will consider multiple sizes), and v_i is the settling velocity of those particles. Recognizing that A is constant and therefore can be cancelled in all terms, and dividing both sides by Δz, yields the following:

$$\frac{dn_{i,\bar{z}}}{dt} = \frac{v_i n_{i,z} - v_i n_{i,z+\Delta z}}{\Delta z} \qquad (13\text{-}18)$$

---

[6] Recall from Chapter 2 that the Heaviside function, $H_k(t)$, is a step function with a value of zero for $t < k$ and a value of 1.0 at $t \geq k$.

In this form, the equation is directly interpretable as follows: the rate at which the concentration of particles changes in a differential element at any depth is the difference between the terms characterizing the rate at which particles settle into that element from above and the rate at which they settle out to the regions below. This form of the equation is useful in numerical analysis.

Allowing the length increment in Equation 13-18 to shrink to infinitesimal dimensions yields the following partial differential equation:

$$\frac{\partial n_i}{\partial t} = -v_i \frac{\partial n_i}{\partial z} \quad (13\text{-}19)$$

This partial differential equation is first order in both time and space (depth), and so requires an initial condition and a boundary condition. If we replace n by c, this equation becomes identical to that shown in Chapter 2 for the step input of conservative tracer to a plug flow reactor. The initial and boundary conditions in the two systems are also essentially the same—in each case, the concentration is the same everywhere at time zero and is constant at the boundary at all $t > 0$. Hence, as in the plug flow response to a step tracer input, the solution is the initial concentration times the Heaviside function, with the step occurring at the depth z equal to the product of v and t, as shown in Equation 13-16b. In the tracer test, the original clean water is replaced by water containing tracer; here, regions containing particles are replaced by regions that are particle-free. There, the velocity at which the interface moved was the fluid flow velocity; here, the velocity is the settling velocity.

Bimodal Suspension

Consider next a bimodal Type I suspension; that is, a suspension with two different types of particles (A and B), each with a different settling velocity. Assume that 25% of the particles are of Type A, with a settling velocity (v_A), and the remaining 75% of the particles (the Type B particles) have a settling velocity (v_B) that is twice as large (i.e., $v_B = 2v_A$). Again imagine that this suspension is completely mixed at time zero, and then is allowed to settle under quiescent conditions.

At any depth z at early times (specifically, $t < z/v_B$), both Type A and Type B particles are detected at their initial concentrations. At time $t = z/v_B$, however, the last B-type particle (one that started at the top of the suspension) passes the observation point, and the particles detected are all A particles, whose concentration is only 25% of the total original concentration. Similarly, at time $t = z/v_A$, the last Type A particle passes, and the particle concentration at z drops to zero. The profile for each type of particle is the same as for the monodisperse suspension (a Heaviside function),

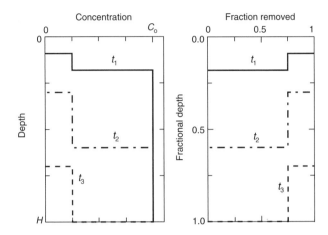

FIGURE 13-10. Batch sedimentation profiles for bimodal suspension.

and the total concentration is the sum of the two. The profiles for concentration and fractional removal with depth, therefore, proceed as illustrated in Figure 13-10. At times t_1 and t_2, the concentration is zero near the top, the same as the original concentration at the bottom, and changes from 0 to 0.25 c_o and then to c_o at depths related to the settling velocities of the two types of particles. By t_3, however, all of the larger particles have made it to the sludge layer, so only the smaller particles remain anywhere in the suspension.

These results for concentration at a given time, t, can be expressed mathematically as follows:

$$c(z,t) = \begin{cases} 0 & \text{for } z < v_A t \\ c_{A,0} & \text{for } v_A t \leq z \leq v_B t \\ c_{A,0} + c_{B,0} & \text{for } z \geq v_B t \end{cases} \quad (13\text{-}20a)$$

or, more generally and succinctly,

$$c(z,t) = \sum_{\text{all } i} c_i \big|_{v_i < z/t} = \sum_{\text{all } i} c_{0,i} H_{v_i t}(z) \quad (13\text{-}20b)$$

Note that the equations and explanations are written here in terms of number concentration, but they could just as easily be written in terms of volume or mass concentration. Also, the inclusion in Equation 13-20 of a separate term for each type (size, settling velocity) of particle in the system stems from the fact that, in a formal derivation, a separate Equation 13-19 must be written and solved for each type.

Heterodisperse Suspension

Real suspensions encountered in water and wastewater treatment often have particles with diameters ranging (at least) from 0.1 μm to 100 μm; this three order of magnitude range of sizes translates to a range spanning six orders of magnitude in settling velocities, assuming uniform density of particles (and all with settling velocities that yield low Reynolds Numbers). In essence, this range means that some particles will settle infinitesimal distances while others settle the several meters (3–6 m) of depth provided in typical settling tanks. As noted elsewhere, it is reasonable to treat this distribution of sizes (or settling velocities) as essentially a continuous function.

The extension of the earlier analysis to a realistic suspension that has a wide variety of particles is straightforward. The relevant equation has already been derived (Equation 13-20b). At any given depth (z) and time (t), the suspension would be devoid of particles with settling velocity greater than (z/t) and would contain all particles with lesser settling velocities at their original concentrations, consistent with the analysis of the simpler suspensions given earlier.

■ **EXAMPLE 13-4.** The suspension described in Example 13-2 is allowed to settle in a quiescent batch reactor. Find the particle size distribution at a depth of 25 cm after 15 min. Plot the particle size distribution function and the volume distribution at time zero and at the indicated time and depth. Also find the mass (or volume) fraction removed from the suspension at this time and depth.

Solution. At $t = 15$ min, all particles with

$$v > (z/t) = 25\,\text{cm}/15\,\text{min} = 0.25\,\text{m}/0.25\,\text{h} = 1\,\text{m/h}$$

will have settled more than 25 cm. As a result, all particles with $v > 1$ m/h that started above the observation point will have settled past it and will therefore be present at zero concentration. On the other hand, all particles with lower settling velocities will be present at their original concentrations. Reference to Table 13-1 indicates that particles with $\log d_p \leq 1.6$ have settling velocities greater than 1.0 m/h and particles with $\log d_p \leq 1.5$ have settling velocities less than this value. Hence the particle size distribution function will be identical to the initial distribution up to $\log d_p = 1.5$, and will have a value of zero at higher diameters. The results of both ways of plotting the size distribution are shown in Figure 13-11. Table 13-1 also indicates that the cumulative volume fraction of particles with $\log d_p \leq 1.5$ is 0.644, so this fraction remains in suspension at the indicated time and depth, and the other 35.6% is removed. ■

The profiles (i.e., the change as a function of the depth) of the concentration and the fraction remaining from a batch test on a real suspension with a nearly continuous settling velocity distribution might look like those shown in Figure 13-12. At the bottom ($z = H$), the fractional removal at a given time (t) would equal the fraction of particles with settling velocity greater than (H/t); we designate this

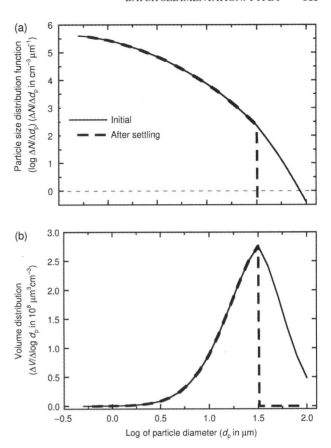

FIGURE 13-11. Changes in particle size distribution at a particular time and depth during Type I batch settling.

fractional removal as η^*. If all the particles in the suspension have a density greater than that of water, all would have a finite settling velocity, and therefore the predicted concentration at $z = 0$ would be zero at all $t > 0$. In practice, however, most suspensions contain some particles with such low settling velocities that they remain very near the surface for long times, as indicated in Figure 13-12. In addition, of course, it is not possible to meet the ideal conditions of these hypothetical experiments, so some finite concentration of particles is always present at the top of the suspension.

Three trends reflected in Figure 13-12 are worth noting. First, at a given time (any one profile shown), concentration increases (or fractional removal decreases) with increasing depth, because fewer particles have settling velocity greater than (z/t) as z increases. Second, as time increases, the concentration at a given depth decreases (or fractional removal increases), because more particles have settling velocities greater than (z/t) as t increases. Third, the concentration (or fraction removed) is constant at equal values of (z/t); for example, the fraction removed is the same at a depth of 2 m after 10 min of settling as it was at 1 m after 5 min. This trend means that a vertical slice in either

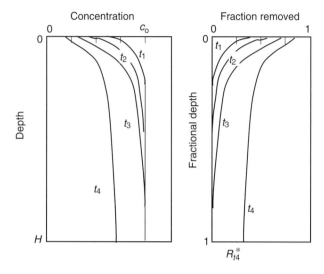

FIGURE 13-12. Batch sedimentation profiles for a heterogeneous suspension.

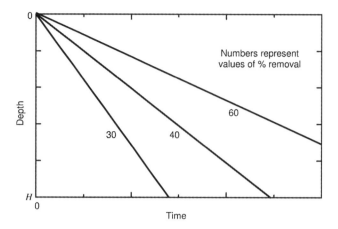

FIGURE 13-13. Isopleths of constant removal in Batch Type I sedimentation.

part of Figure 13-12 cuts through the profiles at equal values of (z/t). This idea is expressed mathematically as follows:

$$c(z_j, t_j) = c(z_i, t_i) \quad \text{if} \quad \frac{z_j}{t_j} = \frac{z_i}{t_i} \tag{13-21}$$

Hence, if the concentration at a particular time and location ($c(z_j, t_j)$) is known, Equation 13-21 can be used to find other times and locations where the same concentration will exist. The same idea is illustrated in terms of removal efficiency in Figure 13-13. Any straight line starting at the origin has a constant value of (z/t), and so the value of c (and hence η) is also constant along such a line.[7] Correspondingly, for Type I settling of a given suspension, η depends only on (z/t). That is, for a given suspension, η is the same for a given value of (z/t) regardless of the height of the column or the duration of the test. This important result has dramatic implications for the design of continuous flow sedimentation reactors, which is discussed in Section 13.6.

The preceding discussion demonstrates that the profiles of concentration and fractional removal for quiescent, Type I settling can be found in different ways. Most directly, measuring the particle concentrations in samples taken at several depths and times in a batch sedimentation experiment leads directly to the curves shown in Figure 13-12. However, the concentrations measured at any single depth at various times or at any single time at various depths also yield sufficient information, since Equation 13-21 allows us to extend the data to other times and depths.

In any of these types of tests (assuming the conditions are truly quiescent), one is measuring a property of the suspension: the cumulative settling velocity distribution. The fraction removed (on number, volume, mass, or other basis) at depth z and time t is the fraction of the suspension (on the same basis) with settling velocity, v, greater than (z/t). Similarly, the fraction remaining at that time and depth is the fraction with settling velocity less than or equal to that value of v. Throughout the remainder of this chapter, the symbol $f(v)$ is used to denote the cumulative settling velocity distribution—the fraction of the suspension with a settling velocity less than or equal to v.

A few examples of $f(v)$ functions are shown in Figure 13-14a. As with any relative cumulative distribution, this distribution must increase monotonically from zero to one; the shape of the distribution is not otherwise constrained. The same distributions are shown in Figure 13-14b in the differential form and denoted as $e(v)$; that is, the settling velocity distributions shown in (b) are the derivatives of those in (a) and represent the probability density functions for the settling velocities of the three suspensions.[8] In this figure, suspension I is monodisperse, whereas suspensions II and III have particles with a wide variety of settling velocities. Suspension II consists of a large fraction with low settling velocities and a small fraction with very high settling velocities relative to the monodisperse suspension's settling velocity. Suspension III has a small fraction with low settling velocities, and a larger fraction with relatively high settling velocities. One can imagine that the role of flocculation before sedimentation is to change a suspension with a distribution like that of suspension II to one like that of suspension III. For any single suspension (except a monodisperse suspension), the distributions would be different depending on whether particle number, particle volume, or

---

[7] Curves along which some parameter of interest is constant are referred to as isopleths. Thus, Figure 13-13 shows three isopleths of constant removal efficiency, and demonstrates that such isopleths are linear for Type I settling.

[8] The settling velocity distribution, $e(v)$, and the cumulative settling velocity distribution, $f(v)$, are related to each other in exactly the same way as the residence time distribution, $E(t)$, and the cumulative age distribution, $F(t)$, introduced in Chapter 2. The total area under the curves (with the limits of the independent variable going from zero to infinity) of both $e(v)$ and $E(t)$ is one. And, the cumulative distributions, $f(v)$ and $F(t)$, are defined as the running sum of the density functions in each case, so their ranges are from zero to one.

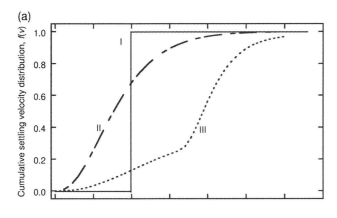

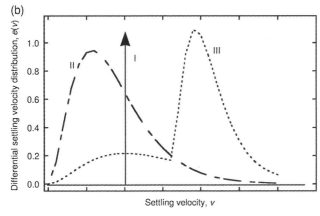

FIGURE 13-14. Various cumulative and differential settling velocity distributions.

mass is chosen as the basis of measurement. Most commonly, the interest in sedimentation is mass removal, so the mass (or, identically if all particles have the same density, particle volume) is usually used as the basis. The utility of these distributions is shown subsequently in the analysis of removal in continuous flow reactors.

■ **EXAMPLE 13-5.** Batch test with single port for Type I settling.

A batch sedimentation test on a Type I suspension is carried out in a column with a single port located 1 m below the water surface. Samples are taken from that port at various times and analyzed for the suspended solids concentration. The resulting data are shown in the first two columns of Table 13-2.

Determine (a) the expected vertical profile of concentration in the column, assuming a very great total column depth, at times of 20, 50, and 120 min, (b) the fractional removal at the same three times as a function of fractional depth, assuming a total depth of 3 m, and (c) the cumulative settling velocity distribution for the suspension.

Solution.

(a) The fraction remaining (Column 3 in the table) at each time is found by dividing the concentration at each time by the initial concentration, 220 mg/L. The fraction removed (Column 4) is, of course, one minus the fraction remaining.

To find the expected profile at the specified times, the relationship shown in Equation 13-21 is used. The experimental results (first two columns of Table 13-2) form a set of values of $c(z_i, t_i)$ with z_i

TABLE 13-2. Data and Analysis for Batch Sedimentation with Single Port

| Time (min) | Water Quality (at a Depth of 1 m) | | | Depths (m) at which Stated Concentrations are Found | | | Settling Velocity (m/h) |
	Concentration (mg/L)	Fraction Remaining (–)	Fraction Removed (–)	20 min	50 min	120 min	
(1)	(2)	(3)	(4)	(5)	(6)	(7)	(8)
0	220	1.00	0.00				
10	208	0.95	0.05	2.00	5.00	12.0	6.0
15	189	0.86	0.14	1.33	3.33	8.0	4.0
20	172	0.78	0.22	1.00	2.50	6.0	3.0
30	147	0.67	0.33	0.67	1.67	4.0	2.0
40	128	0.58	0.42	0.50	1.25	3.0	1.5
50	113	0.51	0.49	0.40	1.00	2.4	1.2
60	102	0.46	0.54	0.33	0.83	2.0	1.0
80	84	0.38	0.62	0.25	0.63	1.5	0.75
100	72	0.33	0.67	0.20	0.50	1.2	0.60
120	63	0.29	0.71	0.17	0.42	1.0	0.50
150	53	0.24	0.76	0.13	0.33	0.80	0.40
180	46	0.21	0.79	0.11	0.28	0.67	0.33
240	36	0.16	0.84	0.08	0.21	0.50	0.25
300	29	0.13	0.87	0.07	0.17	0.40	0.20

constant at 1 m. Each specified time (20, 50, and 120 min) is a value of t_j, and the equation is used to find the depth, z_j, at which the concentration shown in Column 2 will occur. For example, the experimental results indicate that $c(z_i, t_i)$ at $z_i = 1$ m and $t_i = 60$ min is 102 mg/L. (The row with these results is in bold print, since it is used throughout this example to illustrate the calculations.) That same concentration (and therefore fraction remaining and fraction removed) will be present after 20 min at a depth z_j (from the top) calculated as follows:

$$z_j = \frac{z_i}{t_i} t_i = \frac{1\,\text{m}}{60\,\text{min}} 20\,\text{min} = 0.33\,\text{m}$$

Using the same equation for the 50 and 120 min times yields the depths of 0.83 m and 2.0 m for this same concentration; these depths are shown in Columns 5, 6, and 7 in the table. The depths at which all of the measured concentrations (Column 2) will be found for the three chosen settling times are shown in the profiles columns (Columns 5, 6, and 7) and also shown graphically in Figure 13-15a. By interpolation of the raw data, this method could be used to find the depth at which any specific concentration value (not just those found in the experiment) would be located at any time.

(b) The fractional removal at the 1 m depth of the port is given in Column 4. The fractional depth at which this same fractional removal will be obtained in a 3-m deep column can be computed by dividing the depth values in the profiles columns (Columns 5–7) by 3 m (not shown). For example, we found in part (a) that a concentration of 102 mg/L (or a fractional removal of 0.54) is expected at a depth of 0.33 m after 20 min of settling; this point translates to a

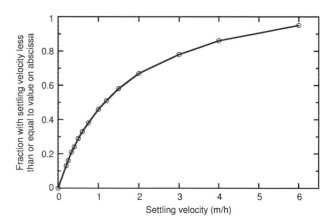

FIGURE 13-16. Cumulative settling velocity distribution.

fractional depth of 0.11 (i.e., 0.33 m/3 m) in the 20-min profile. Values >1.0 for the fractional depth (corresponding to values >3.0 m in the profiles columns) mean that the associated concentration is no longer found anywhere in the suspension (above the sludge layer). The results are shown in Figure 13-15b.

(c) The cumulative settling velocity distribution is a direct interpretation of the data. The fraction that remains at any depth, z, and time, t, is the fraction that has a settling velocity $\leq (z/t)$. The values of the settling velocity in the last column are obtained from the raw data as the depth of 1 m divided by the times given in the first column (along with the conversion from minutes to hours). The associated data in Column 3 of Table 13-2 are then plotted versus the values in the last column, as shown in Figure 13-16; this plot is the cumulative settling velocity distribution, $f(v)$, of the suspension—the fraction with settling velocity less than or equal to the value on the abscissa. ∎

13.5 BATCH SEDIMENTATION: TYPE II

As described in the introduction, sedimentation with simultaneous flocculation is known as Type II sedimentation. The flocs that are created in this process settle differently than the "parent" particles—generally faster as the particles grow in size. Particles that are small (and therefore have low settling velocities) are removed from the size distribution at a particular location primarily by flocculation (to create larger particles) and only to a small degree by sedimentation. Large particles are still removed primarily by sedimentation as in Type I settling, but flocculation influences the population in a given large size class by creating flocs of that size from

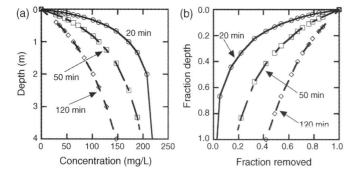

FIGURE 13-15. Concentration profiles and fractional removals expected in batch reactor. (In part (b), reactor is assumed to be 3 m deep.)

smaller particles. These ideas are in sharp contrast to Type I sedimentation, in which the size distribution at any depth is changed only by elimination of particles whose settling velocity exceeds the ratio of depth to time since settling began.

In engineered systems of interest in water and wastewater treatment, virtually all sedimentation of dilute suspensions is Type II (flocculent) sedimentation. In drinking water treatment, sedimentation immediately follows flocculation, and flocs continue to form and grow in the sedimentation process. The addition of chemicals before flocculation ensures the flocculent nature of the particles. In typical municipal wastewater treatment plants, sedimentation occurs both before and after biological treatment, and both suspensions are naturally flocculent, though (especially for primary treatment) they can be made more so by the addition of chemical flocculants upstream of the sedimentation reactor. Precipitation processes, whether undertaken as a means of destabilizing particles originally in the water or as a treatment process to remove metals or other soluble contaminants, generate particles that are naturally flocculent. Therefore, all of the considerations regarding particle destabilization and surface chemistry discussed in earlier chapters are relevant to flocculent sedimentation.

Conceptually, Type II sedimentation is easy to understand at the microscopic level. Two particles collide and form a single floc that settles faster than either of the original particles. Mathematically, this process can be formulated as a combination of the Smoluchowski equation to describe flocculation and the Stokes equation to describe sedimentation. Macroscopically, the process is nearly identical to Type I sedimentation, although the fact that the size distribution can be changed by both flocculation and sedimentation means that, in practice, Type II sedimentation is more effective than Type I (for initially equivalent suspensions) and, in analysis, some simplifications that apply to Type I sedimentation do not apply to Type II. In this section, we present both the mathematical analysis at the microscopic level and some of the macroscopic manifestations of Type II sedimentation.

Mathematical Analysis

Consider a particle number balance on a Type II suspension in a differential volume element such as the one shown previously in Figure 13-9. As in the analogous Type I system, there is no flow (advection) into or out of this element, and we assume that net concentration changes by diffusion of particles (i.e., Brownian motion) across the element boundaries are negligible. Also as in the Type I system, particles can enter the top (at z) and exit from the bottom (at $z + \Delta z$) by sedimentation, and thereby change the concentration in the control volume. However, unlike in Type I suspensions, the concentration of a particular size particle in the element can also be changed by flocculation. For a given size particle, the concentration can increase by flocculation of smaller particles to form a floc of that size, and it can decrease by collision and attachment of that size particle with any other particle to form a still bigger one. Therefore, the resulting volume-normalized number balance for particles of a particular size (k) at a particular location (z) includes the gain and loss terms for flocculation (developed in Chapter 12) and the net loss by sedimentation (see Equation 13-19), as follows:

$$\frac{\partial n_{k,z}}{\partial t} = \left[\frac{1}{2} \alpha_{\text{emp}} \sum_{i+j=k} \alpha_{ij} \beta_{ij} n_{i,z} n_{j,z} \right] - \left[\alpha_{\text{emp}} n_{k,z} \sum_{\text{All } i} \alpha_{ik} \beta_{ik} n_{i,z} \right]$$
$$- \left[v_k \frac{\partial n_k}{\partial z} \right] \quad (13\text{-}22)$$

Equation 13-22 is not solvable analytically and, therefore, must be solved numerically. To do so, it is useful to discretize the space domain (z). A number balance on the differential element surrounding the depth $\bar{z}$ (as in Figure 13-9) can be written in words and symbols as follows:

$$\begin{bmatrix} \text{Rate of change} \\ \text{of number} \\ \text{concentration} \\ \text{of } k\text{-particles} \\ \text{in a} \\ \text{differential} \\ \text{layer at depth } \bar{z} \end{bmatrix} = \begin{bmatrix} \text{Rate of increase} \\ \text{in concentration} \\ \text{of } k\text{-particles} \\ \text{at depth } \bar{z} \text{ by} \\ \text{flocculation of} \\ i \text{ and } j \text{ particles} \\ \text{to form} \\ k\text{-particles} \end{bmatrix} - \begin{bmatrix} \text{Rate of decrease} \\ \text{in concentration} \\ \text{of } k\text{-particles} \\ \text{at depth } \bar{z} \text{ by} \\ \text{flocculation of} \\ k\text{-particles} \\ \text{with any} \\ \text{particles} \end{bmatrix}$$
$$+ \begin{bmatrix} \text{Rate of increase} \\ \text{in concentration} \\ \text{of } k\text{-particles} \\ \text{at depth } \bar{z} \text{ by} \\ \text{sedimentation} \\ \text{from above} \end{bmatrix} - \begin{bmatrix} \text{Rate of decrease} \\ \text{in concentration} \\ \text{of } k\text{-particles} \\ \text{at depth } \bar{z} \text{ by} \\ \text{sedimentation} \\ \text{to} \\ \text{lower depths} \end{bmatrix}$$

$$\frac{dn_{k,\bar{z}}}{dt} = \left[\frac{1}{2} \alpha_{\text{emp}} \sum_{i+j=k} \alpha_{ij} \beta_{ij} n_{i,\bar{z}} n_{j,\bar{z}} \right] - \left[\alpha_{\text{emp}} n_{k,\bar{z}} \sum_{\text{All } i} \alpha_{ik} \beta_{ik} n_{i,\bar{z}} \right]$$
$$+ \left[v_k \frac{n_{k,z}}{\Delta z} \right] - \left[v_k \frac{n_{k,z+\Delta z}}{\Delta z} \right] \quad (13\text{-}23)$$

A substantial computer program is required to integrate Equation 13-23 numerically to predict the changes in the particle size distribution as a function of time and depth in a batch reactor. However, some useful insights can be gleaned directly from the equation, especially when compared to Type I sedimentation. Consider a destabilized suspension ($\alpha_{\text{emp}} = 1$) with a broad particle size distribution and all particles having a density greater than that of water. For a differential element near the top of the column, large particles originally in the suspension will settle to lower depths rather quickly, as in Type I suspensions. After that, large particles might appear in that control volume, but only if they are created by flocculation in that or higher elements. Flocculation does not create high number concentrations of such particles (because it takes many smaller particles to create such a floc), so the number concentration is expected to remain low.

For small particles, the settling velocity is quite low, so the gain or loss by sedimentation will be low (and nearly balanced, because the concentration gradient of small particles will not be great, also as for Type I suspensions). However, the number concentration of small particles in most suspensions is (originally) high, so flocculation might be substantial. As in flocculation, we expect a loss of the very smallest sizes in suspension (and specifically in the differential element under consideration) because the loss of these small particles (the second term) is likely to outweigh the gain by flocculation of still smaller ones (the first term). Hence, over time, the size distribution is likely to show a loss of the largest particles (dominated by sedimentation), a loss of the smallest particles (dominated by the loss term of flocculation), and the least change for particles in the intermediate size range (where gain by flocculation is significant in comparison to, and perhaps even greater than, loss by sedimentation).

A Type I suspension with the same original particle size distribution would have lost almost no small particles, and would have lost all of the largest particles in the same differential element. When summed on a mass or particle volume basis over the entire distribution, the loss will necessarily be greater in Type II than in Type I, because flocculation conserves mass or volume while it creates particles that settle faster. On average, the particles in any differential element will have faster settling velocities for the Type II suspension. Therefore, the sedimentation flux (in and out) is increased by the effects of flocculation.

Next consider how this picture changes as we move from the top of the column to lower depths. The same general logic describes each differential element. But, as in Type I, sedimentation creates an overall mass concentration gradient with increasing concentrations from the top to the bottom by moving bigger particles down faster than smaller particles. As the number concentration decreases in any layer, the rate of flocculation also decreases, because flocculation is a second order reaction in particle number and depends on the heterodispersity of the suspension. As depth increases, the number concentration of intermediate and large particles increases, so flocculation is likely to be faster at lower depths than higher depths. In an ideal batch reactor (no vertical water flow at all), it is conceivable that the number concentration of small particles (in some defined increment of size)

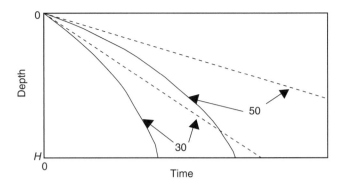

FIGURE 13-17. Isopleths of constant removal for Type I and Type II batch sedimentation. (Straight broken lines for Type I; curved solid lines for Type II. Numbers represent percent removal.)

would be less at some point lower in the water column than higher up. Nevertheless, the overall mass concentration must increase with depth, as flocculation cannot create mass, and sedimentation can only carry mass downward.

Experimental Analysis

When flocculation occurs simultaneously with sedimentation, the fractional removal is not constant at a given value of the ratio (z/t) as it is for sedimentation alone (i.e., for Type I suspensions). To understand why, consider the behavior of a particular particle during flocculent sedimentation. When that particle becomes incorporated into a floc with other particles, its settling velocity increases, and hence it reaches a greater depth at some particular time than it would without flocculation. Flocculation continues to occur as time goes on, so the original particle is incorporated into larger and larger flocs, and its settling velocity continually increases; that is, the particle accelerates. This phenomenon occurs for every particle in the suspension, so the overall movement of particles (over the entire size range of the suspension) is faster in Type II sedimentation than Type I, and therefore the total concentration of particles (on a number, volume, or mass basis) at any depth and time is less than it would be under nonflocculent conditions.[9] The continual acceleration as long as flocculation continues means that the difference between flocculent and nonflocculent conditions (for the same initial suspension) grows continuously with increasing time or depth. This concept is summarized in Figure 13-17, which shows how isopleths of constant overall removal efficiency in a batch reactor curve downward in Type II sedimentation but are straight for Type I suspensions.

The rate of flocculation is strongly dependent (i.e., second order) on the number concentration of particles present. Hence, the extent to which flocculation affects

---

[9] Note that this statement does not preclude the possibility that the concentration of a particular size particle might be greater at a particular time and depth in Type II than in Type I.

sedimentation is dependent on particle concentration. Two suspensions with identical initial *relative* size distributions but with different concentrations (absolute size distributions) will flocculate at different rates and therefore settle at different rates, with the initially higher concentration suspension losing mass more quickly. In other words, the extent of curvature of the lines for Type II sedimentation is dependent on particle concentration, with greater curvature for higher initial concentration. This fact leads to the consequence that, in drinking water treatment, high quality raw waters (i.e., waters with low initial particle concentrations) are sometimes treated without sedimentation. In such cases, after destabilization by flocculent addition, the water is either put immediately into the filtration process (*contact filtration*) or passed through a flocculation step and then to filtration (*direct filtration*). The reason for omitting sedimentation in these cases is not so much that it is unnecessary but that it accomplishes very little—a direct consequence of the concentration effect in flocculent sedimentation.

For Type I suspensions, the cumulative settling velocity distribution, $f(v)$, is a property of the (original) suspension, and does not change. For Type II suspensions, this property can still be defined, but it does change with time (and the ability to determine its value depends to some degree on the depth of the column used for the test). The conceptual results shown in Figure 13-17 for a Type II suspension can be used virtually identically as for a Type I suspension to calculate a settling velocity distribution. Here, v is defined as the *average* settling velocity, determined as the total distance of fall (distance to a port from the top of the water column) divided by the time since the beginning of the test. The value of $f(v)$ for that time is the fraction of the suspension remaining at that time and port. This idea is illustrated in Figure 13-18a, which shows the (curved) isopleths for two values of the fraction remaining (0.25 and 0.60).

Consider the isopleth for the fraction remaining equal to 0.25. Each point on that curve designates a time (t_p) and depth (z_p) at which 25% of the initial particle mass concentration remains in the suspension. That is, each point represents $f(v) = 0.25$, where v is an average velocity calculated as z_p/t_p (the negative slopes of the broken lines in the figure) and designated as $v_{t,f}$. Thus, after 60 min, the depth on that line is 0.42 m, so $v_{60,0.25} = (0.42\,\text{m}/60\,\text{min}) = 0.007\,\text{m}/\text{min}$. After 90 min, $v_{90,0.25} = (0.90\,\text{m}/90\,\text{min}) = 0.010\,\text{m}/\text{min}$. The fact that $v_{90,0.25} > v_{60,0.25}$ indicates that the particles representing the lower 25% of the mass distribution aggregated and therefore accelerated during the period from 60 to 90 min. Correspondingly, the fact that the isopleth is always concave downward indicates that the particles accelerated during the entire time they were settling. In contrast, for a Type I suspension, isopleths are linear, indicating that the $v_{t,0.25}$ would be the same at all values of t, and that the particles do not flocculate and accelerate.

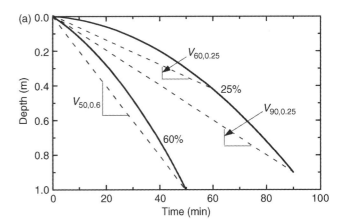
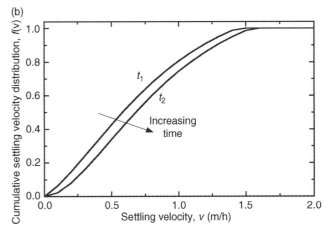

FIGURE 13-18. Experimental analysis of settling data for Type II suspensions. (In part (a), the solid lines represent isopleths for 25 and 60% of the suspension remaining.)

The same information is illustrated in a different format in Figure 13-18b. This presentation of the results emphasizes that the cumulative settling velocity distribution shifts to the right as the duration of settling increases; that is, for a given value of $f(v)$, v gets larger as settling proceeds. And, for a given value of v, $f(v)$ decreases over time; that is, the fraction of the suspension with a settling velocity less than a particular v decreases with time.

For Type II suspensions, the experimental apparatus can impose limits on the availability of data. For example, for the system characterized in Figure 13-18a, one cannot find the average settling velocity corresponding to any $f(v) > 0.6$ after 50 min in this 1 m column. That is, the part of the cumulative settling velocity distribution associated with high settling velocities cannot be found; the only way to overcome this problem is to use a deeper column for the test. For this reason, it is commonly recommended that batch-settling tests are carried out in columns that are as deep as the full-scale settling reactor is expected to be. Fortunately, as we will see subsequently, the exact shape of this (upper) part of the cumulative settling velocity distribution is not important beyond a certain v value for many reactors, and that recommendation can be somewhat relaxed.

13.6 CONTINUOUS FLOW IDEAL SETTLING

In modern water and wastewater treatment systems, sedimentation is usually accomplished in continuous flow reactors. Particle removal occurs in these reactors because the pathway that a particle takes in the reactor is different from the pathway that water takes. Particles tend to fall to the bottom and are removed in a concentrated stream, and the relatively clean water is removed from a different location (end wall or the top, depending on the design). In this section, we consider the removal that occurs in a variety of continuous flow sedimentation tanks; in the different types of reactors, water flows in different ways and so the removal that is accomplished is different. We begin by giving a conceptual framework relevant to all the types of reactors, and then, within that framework, consider idealized versions of the common designs of sedimentation facilities for dilute suspensions.

The overall performance of a sedimentation reactor is measured as the removal efficiency—the fraction of the influent particle concentration (usually measured in terms of the particle mass concentration) that does *not* exit the reactor in the (relatively clean) effluent stream. That overall performance is dependent on the characteristics of both the suspension and the reactor. To the extent possible, it is convenient to separate the effects of the suspension being treated from the effects of the reactor design on that overall performance.

Separating Influences of Suspension and Reactor

The key characteristic of a suspension with respect to its settling properties is its cumulative settling velocity distribution, $f(v)$. As noted in the preceding section, $f(v)$ is an unchanging characteristic of Type I suspensions, but it changes as particles flocculate in Type II suspensions. The changes in $f(v)$ depend on the availability of other particles, which, in continuous flow reactors, depends on both location and the flow pattern. As a result, it is difficult to predict the steady state $f(v)$ that will develop in such reactors.

The characteristics of the reactor that affect the overall performance are described by another function. In any sedimentation reactor, particles with higher settling velocities fall to the bottom more quickly than those with lower settling velocities, so particles with higher settling velocities tend to be removed to a greater extent (at least by sedimentation alone). In different types of continuous flow sedimentation reactors, the removal of particles with a particular settling velocity, v, can be different because the flow patterns are different. We characterize this influence of the flow pattern on particle removal by defining a function that describes the expected fractional removal efficiency of particles that have settling velocity, v, in a particular reactor. We call this function the *reactor settling potential function*, and represent it as $F(v)$. This function exists for a sedimentation reactor independent of whether particles with any particular settling velocity (or, for that matter, any particles at all) are present in the reactor. That is, as emphasized by its name, this function describes the *potential* for the reactor to remove particles with a particular settling velocity. In the following sections, we assess various types of sedimentation reactors for their ability to remove particles with different settling velocities; that is, we find $F(v)$ for various reactor types. First, however, we develop a generic approach for designing and/or analyzing sedimentation reactors, based on independent determination of $f(v)$ and $F(v)$.

The overall removal efficiency in a continuous flow sedimentation reactor can be found by summing, over all increments of settling velocity, the product of the fractional removal of particles in each increment and the fraction of the suspension that is in that increment; that is

$$\begin{bmatrix} \text{Overall} \\ \text{fractional} \\ \text{removal} \\ \text{of particles} \\ \text{in the reactor} \end{bmatrix} = \sum_{\substack{\text{all increments} \\ \text{of } \Delta v}} \begin{bmatrix} \text{Fractional removal} \\ \text{of particles that settle} \\ \text{at an average velocity} \\ \text{between } v \text{ and } v + \Delta v \\ \text{while in the reactor} \end{bmatrix}$$

$$\times \begin{bmatrix} \text{Fraction} \\ \text{of the influent} \\ \text{suspension that settles} \\ \text{at an average velocity} \\ \text{between } v \text{ and } v + \Delta v \\ \text{while in the reactor} \end{bmatrix} \quad (13\text{-}24)$$

We have just defined the fractional removal of particles with a settling velocity, v, as the reactor settling potential function, $F(v)$; hence, the first term on the right side of Equation 13-24 can be taken as an average value of this function over the interval between v and $v + \Delta v$, written as $\overline{F(v)}$. The second term is $(f(v + \Delta v) - f(v))$, or $\Delta f(v)$. Thus, the word equation can be converted to symbols as follows:

$$\eta_{\text{tot}} = \sum_{\text{all } \Delta v} \overline{F(v)} \Delta f(v) \quad (13\text{-}25)$$

The subscript "tot" is included in the removal efficiency to emphasize that the summation is over the entire suspension. For numerical work in which $f(v)$ is only known at certain increments, this formulation (Equation 13-25) is quite useful. However, we can also write $\Delta f(v)$ as $(\Delta f(v)/\Delta v)\Delta v$; that is,

$$\eta_{\text{tot}} = \sum_{\text{all } \Delta v} \overline{F(v)} \frac{\Delta f(v)}{\Delta v} \Delta v \quad (13\text{-}26)$$

In the limit as the increments go to zero, the first term on the right becomes just $F(v)$ and the second becomes $(df/dv)dv$, or simply df. The term (df/dv) is the slope of the cumulative settling velocity distribution; that is, it is the (differential) settling velocity distribution, given the symbol $e(v)$, and having dimensions of inverse velocity. For differentially small increments of settling velocity, therefore, we can rewrite Equation 13-25 in any of the following ways:

$$\eta_{\text{tot}} = \int_{v=0}^{v=\infty} F(v)\frac{df(v)}{dv}dv = \int_{v=0}^{v=\infty} F(v)e(v)dv \quad (13\text{-}27a)$$

$$\eta_{\text{tot}} = \int_{f(v=0)}^{f(v=\infty)} F(v)df = \int_{0}^{1} F(v)df \quad (13\text{-}27b)$$

The conversion of Equation 13-27a to Equation 13-27b shows how the limits of integration change when the integration is considered over increments of f rather than increments of v.

Equation 13-27 has been written assuming that none of the particles has a settling velocity less than zero. For a suspension that includes some particles that float, the lower limit of integration of Equation 13-27b would have to be modified to account for such particles. For convenience, in most of the remainder of this chapter (except when considering flotation explicitly), we assume that $v > 0$ for all particles and use the limits shown in Equation 13-27.

Interpreting the Reactor Settling Potential Function

The results in Equations 13-26 and 13-27 are completely general for any sedimentation reactor and combine properties of the reactor expressed in $F(v)$ with properties of the suspension expressed in $f(v)$. Before quantifying $F(v)$ for various types of reactors, it is useful to consider some likely general properties of this function. No reactor will remove particles that do not settle at all (those with $v = 0$), so we expect $F(0) = 0$ in all sedimentation reactors. Further, it seems likely (although not formally necessary) that, for any sedimentation reactor, particles with a settling velocity greater than some critical value (v^*) will all be captured, so that $F(v > v^*) = 1$. Hence, in the analysis for each type of sedimentation tank, we can divide the analysis into two parts: find the critical settling velocity above which all particles are removed, and then find the fractional removal (if any) of particles with settling velocities less than this critical value. Rewriting Equation 13-27 to show these two components explicitly yields the following:

$$\eta_{\text{tot}} = \int_{0}^{1} F(v)df = \int_{0}^{f(v^*)} F(v)df + \int_{f(v^*)}^{1} F(v)df$$

$$= \int_{0}^{f(v^*)} F(v)df + \int_{f(v^*)}^{1} df = \int_{0}^{f(v^*)} F(v)df + (1 - f(v^*))$$

(13-28)

An interesting consequence of this analysis is that, for $v > v^*$, the shape of the cumulative settling velocity distribution, $f(v)$, is inconsequential in terms of the overall removal achieved. Hence, when determining the settling velocity distribution, the part that characterizes particles with settling velocities less than v^* is of greater importance.

We now consider a graphical interpretation of the integrations shown in Equations 13-27 and 13-28. An arbitrary reactor settling potential function $F(v)$ is shown in Figure 13-19a. Subsequently, we will see the kind of reactor and flow pattern

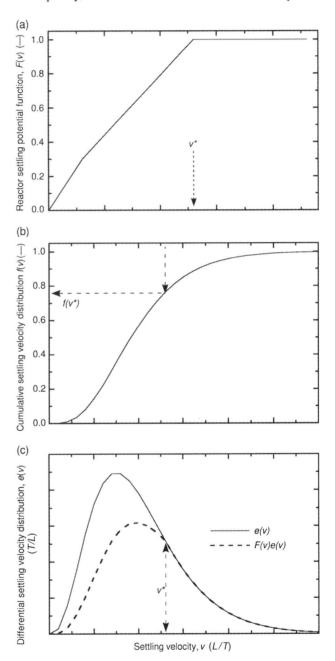

FIGURE 13-19. Graphical representation of the overall fractional removal as $\int_0^\infty F(v)(df/dv)dv$.

that leads to this particular shape for $F(v)$, but all that matters now is that it has the expected characteristics—it increases from zero to one over a range of settling velocities. The minimum value of v for which $F(v)=1$ is labeled as v^* on the figure.

In Figure 13-19b, a cumulative settling velocity distribution, $f(v)$, is shown. This distribution also has arbitrary but expected characteristics—a range of settling velocities and a monotonic increase (as a *cumulative* distribution) as it goes from 0 to 1. These two functions are entirely independent of one another; that is, the reactor described by $F(v)$ in Figure 13-19a could receive any suspension, and the influent suspension described by $f(v)$ in Figure 13-19b could be put into any reactor. However, when this suspension is put into this reactor, the removal efficiency attained is described by Equations 13-27 and 13-28 which link the two functions. We first note $f(v^*)$—the value on the cumulative settling velocity distribution associated with the critical settling velocity in the reactor, or, in other words, the value of v for which all $F(v > v^*) = 1$.

To evaluate Equation 13-27a, we need to characterize the differential settling velocity distribution, $e(v)$. Recall that this function is found as the derivative of $f(v)$ and that its significance is that the fraction of the mass of the (influent) suspension with settling velocities between v and $v + dv$ is $e(v)dv$. Hence, the total area under the curve of $e(v)dv$, shown in Figure 13-19c, is one (the whole suspension). As shown in Figure 13-19a, each value of the (average) settling velocity, v, is associated with a specific fractional removal in the reactor, $F(v)$. When each differential increment of the differential settling velocity distribution, $e(v)dv$, is multiplied by $F(v)$, we obtain the fractional removal of the part of the suspension with that increment of settling velocity. The lower (broken) curve in Figure 13-19c represents values of this product, where $F(v)$ is from Figure 13-19a and $e(v)dv$ is from the upper, solid curve in Figure 13-19c; that is, the lower curve is a plot of $F(v)\,e(v)dv$. The area under this curve is thus η_{out}, and the process of preparing this curve and computing the area is the graphical equivalent of carrying out the integration shown in Equation 13-27a. Note that the two curves in Figure 13-19c are contiguous for all $v > v^*$, since all particles with such settling velocities are removed.

Finally, since each value of v has associated values of both $F(v)$ and $f(v)$, these values can be plotted against one another as shown in Figure 13-20. For both axes, the scale is from 0 to 1, and so the area of the total graph is 1.0 and represents the entire possible removal space. A differential area under this curve for some interval df is $F(v)df$, so the area under the entire curve is $\int_0^1 F(v)df$. According to Equations 13-27b and 13-28, this integral equals the overall fractional removal, η_{tot}. Thus, this plot provides an alternative graphical approach for evaluating η_{tot} from the $F(v)$ and $f(v)$ functions. Note that $F(v)$ (which is formally $F(f(v))$ on this curve) reaches a value of 1.0 at $f(v^*)$.

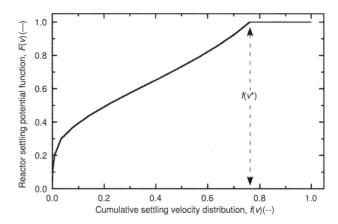

FIGURE 13-20. Graphical representation of overall fractional removal as $\int_0^1 F(v)df$.

The preceding analysis represents a generic approach to finding the expected removal efficiency in any continuous flow sedimentation reactor. We next consider all of the common types of continuous flow sedimentation reactors to find their reactor settling potential functions, $F(v)$, and illustrate how these reactors are expected to perform.

Upflow Reactor

We first consider an ideal upflow reactor, a schematic diagram of which is shown in Figure 13-21. This reactor is an idealization of an upflow clarifier, a common design for sedimentation tanks in water and wastewater treatment plants. In our idealization, the reactor has vertical flow from the bottom to the top, and the flow is completely uniform across the horizontal (surface) area; that is, the flow is perfect plug flow in the upward direction. The influent comes in through a perforated bottom and water flows out through multiple submerged weirs at the top.[10] The vertical water velocity is the flow rate divided by the surface area; this term, (Q_{out}/surface area), is called the *overflow rate*, the *surface overflow rate*, or *surface loading rate* for this and all sedimentation reactors; we designate this term as v_{OR}.

Each particle's vertical velocity is the difference between the (upward) water velocity and the particle's (downward) settling velocity. All particles with settling velocity less than the vertical water velocity would rise (although more slowly than the water), flow out the top with the water, and not be captured at all. On the contrary, particles with settling velocities greater than the vertical water velocity (overflow rate)

---

[10] Real upflow sedimentation tanks usually contain some baffling that imparts some horizontal motion to the water or restricts the area of flow to less than the full surface area at the point of entrance near the bottom of the reactor.

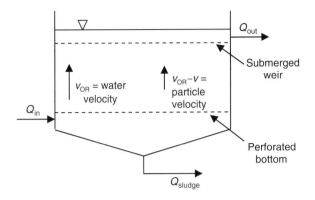

FIGURE 13-21. Schematic of ideal upflow sedimentation reactor.

would have a net downward velocity, so they would never rise in the reactor and would be captured completely. As a result, the critical settling velocity, v^*, is the same as the overflow rate, v_{OR}. Converting these ideas into symbols, the reactor settling potential function for the ideal upflow reactor is

$$F(v) \text{ vertical reactor} = \begin{cases} 0 & \text{for } v < v^* \\ 1 & \text{for } v \geq v^* \end{cases} \quad (13\text{-}29\text{a})$$

or

$$F(v) \text{ vertical reactor} = H_{v^*}(v) \quad (13\text{-}29\text{b})$$

where $v^* = v_{OR}$.

The $F(v)$ curve is shown in Figure 13-22 for this ideal vertical flow reactor. Note once again that the function being plotted is a characteristic of the reactor (in this case, an ideal, vertical-flow PFR), and has nothing to do with the characteristics of the suspension that is to be treated in the reactor.

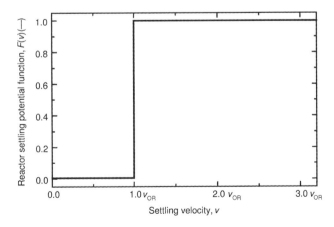

FIGURE 13-22. Reactor settling potential function for an ideal upflow reactor.

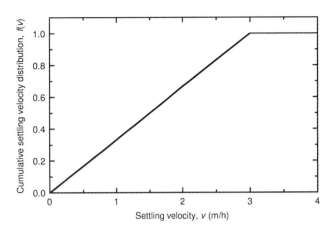

FIGURE 13-23. Cumulative settling velocity of suspension in Example 13-6.

■ **EXAMPLE 13-6.** The cumulative settling velocity distribution, $f(v)$, of a certain Type I suspension is found to be reasonably approximated as shown in Figure 13-23. Estimate the removal efficiency expected for this suspension in an ideal upflow settling tank with an overflow rate of 1.2 m/h.

Solution. The cumulative settling velocity distribution shown in the graph can be described mathematically as follows:

$$f(v) = \begin{cases} \frac{1}{3}v & \text{for } v > 3 \text{ m/h} \\ 1 & \text{for } v \geq 3 \text{ m/h} \end{cases}$$

For this reactor, $v^* = v_{OR} = 1.2$ m/h, and for this suspension, $f(v^*) = \left(\frac{1}{3}\right)(1.2) = 0.4$. Using Equations 13-28 and 13-29, the removal efficiency in the vertical flow reactor can be found as follows:

$$\eta_{tot} = \int_0^1 F(v) df = \int_0^{f(v_{OR})} F(v) df + (1 - f(v_{OR}))$$

$$= 1 + (1 - f(v_{OR})) = 1 - 0.40 = 0.60 \quad ■$$

For Type I suspensions, $f(v)$ is an unchanging property of the influent suspension and, for an upflow reactor, $F(v)$ depends solely on v_{OR}; the removal efficiency for a given influent in an upflow reactor, therefore, depends only on the overflow rate. Consider, for example, two upflow reactors with the same cross-sectional area receiving the same suspension at the same flow rate, with one being quite shallow and the other much deeper. Both reactors have the same overflow rate (v_{OR}), but the deeper one has a much longer hydraulic detention time. In both reactors, particles with settling velocities less than the overflow rate rise, and those with settling velocity greater than the overflow rate fall. Particles with $v < v_{OR}$ stay in both reactors longer than the fluid (and for different amounts of time in the two reactors). Nevertheless, all such particles eventually leave in the effluent of both reactors, so the removal efficiency is the same.

Next, consider the behavior of a flocculent (Type II) suspension in an ideal upflow reactor. For flocculent suspensions, particles that come into the reactor with a settling velocity less than the overflow rate might flocculate to form particles with a settling velocity greater than the overflow rate. Such particles would begin to rise in the reactor but, when their settling velocity exceeded the overflow rate, they would fall to the bottom and be caught. Since particles have a net velocity that equals the difference between the water rise velocity and the particle's settling velocity, some particles stay in the reactor for times much longer than the hydraulic detention time. As a result, the time available for flocculation can be substantial, and flocculation leads to the removal of some particles that have initial settling velocities less than the overflow rate.

For Type II suspensions, therefore, the overflow rate does not completely dictate the removal that occurs in ideal upflow reactors. Consider again the deep and shallow reactors with the same overflow rate receiving the same suspension. The shallow reactor has a short detention time and therefore allows little time for flocculation to influence the settling velocity distribution (by the change in the particle size distribution). The deep reactor, with its longer liquid detention time (and even longer particle detention times, for particles with initial $v < v_{OR}$), allows more flocculation, and therefore more sedimentation to occur. For Type II suspensions, the overflow rate would still be the primary design or operational variable, but the detention time (depth) would also play a role in the removal that would be accomplished.

The flocculation that can occur in an upflow reactor is somewhat limited by the availability of particles, however. Particles that enter the reactor with settling velocity greater than the overflow rate immediately leave the suspension; hence, these larger particles are not available to flocculate with smaller particles. In the short-range force model for flocculation, it is thought that collisions between particles of widely variant sizes are not common, so this immediate loss of part of the particle size distribution might not be a severe detriment to the amount of flocculation that would occur. Nevertheless, the fact that particles with $v > v_{OR}$ never enter the settling zone and therefore are not available for flocculation is likely to reduce the amount of flocculation, and therefore reduce the sedimentation, in comparison to other reactors (presented below) in which this immediate separation does not occur.

■ **EXAMPLE 13-7.** The influent to an upflow clarifier with an overflow rate of 1 m/h is as described in Example 13-2. Assume the suspension is nonflocculent. Find the percent removal on a mass (or volume) basis and plot the particle size distribution of the influent and effluent, both as the particle size distribution function and the volume distribution.

Solution. The effluent will contain no particles with a settling velocity greater than 1 m/h, and will contain the same concentration as the influent of particles with lower settling velocities. This situation is identical to that described in Example 13-4 in a batch experiment at a depth and time corresponding to a settling velocity of 1 m/h. Therefore, the graphs are identical to those presented in Example 13-4, and the removal efficiency, as calculated in that example, is 35.6%. ■

Ideal Horizontal Flow Reactors

Rectangular Reactor Figure 13-24 is a schematic of a rectangular, continuous-flow settling tank, with horizontal flow. In this idealized tank, the inlet and outlet zones are separated from the settling zone by walls with multiple ports; each port imposes a small, equal head loss, so the flow is distributed uniformly both vertically and horizontally. The sludge zone is considered to be isolated from the flow in the main portion of the tank also, so that once a particle enters the sludge zone, it is irreversibly removed. The ideal settling zone is treated as a PFR with horizontal flow; any mixing in the vertical direction would mix dirtier water from deeper in the reactor with cleaner water higher in the reactor and so is undesirable. This ideal settling zone has length L, height H, and width W, yielding a total volume V.

Our task is to determine the reactor settling potential function, $F(v)$, for this ideal horizontal reactor. We do so by considering the fractional removal efficiency of Type I particles with settling velocity v in the reactor. Because the flow is plug flow, a Lagrangian view of a column of water (suspension) as it flows from the influent to the effluent is identical to a batch test, with each horizontal location corresponding to a particular time. When this column of the suspension reaches the effluent wall, the suspension would have the same vertical profile of concentrations as would be present in the batch test after a time equal to the detention time, τ, in the reactor with flow.

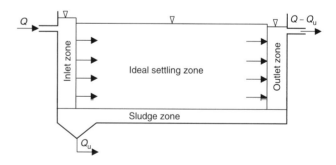

FIGURE 13-24. Ideal rectangular settling tank with horizontal flow.

Consider an aliquot of the effluent taken at some particular depth, z, as it crosses the outlet zone wall. As shown previously for batch sedimentation of Type I particles, that aliquot will contain the same concentration as the influent for particles with settling velocity less than (z/τ) and no particles with settling velocity greater than that. Hence, the removal efficiency of particles with settling velocity v can be equated to the fraction of the (effluent) flow from which these particles have been completely removed. For example, if in a given reactor, Type I particles with a settling velocity of 0.5 m/h would be removed from 70% of the flow (i.e., from 70% of the height of the water column), then $F(0.5\,\text{m/h})$ would equal 0.70.

Every particle in this ideal tank has a horizontal velocity (v_{hor}) equal to the fluid velocity, and a vertical velocity equal to its settling velocity. The overall velocity is the vector sum of these two components; the direction of that vector sum yields the direction of the trajectory of the particle. The trajectories of two particular hypothetical Type I particles in such a reactor are shown in Figure 13-25. First, consider a particle that enters the reactor at the very top of the water column and just settles to the bottom as it reaches the effluent end. The settling velocity of this particle is the critical settling velocity, v^*; all particles with higher settling velocities will be caught completely because they will have steeper trajectories and therefore will reach the sludge zone and not the effluent wall, irrespective of the depth at which they enter the settling zone. The figure makes clear that v^* is the settling velocity of a particle that settles the entire depth H in the detention time, τ. Using the definition of the detention time ($\tau = (V/Q)$), we can find that this critical settling velocity is related to the design and operation of the ideal horizontal flow reactor as follows:

$$v^* = \frac{H}{\tau} = \frac{HQ}{V} = \frac{HQ}{HLW} = \frac{Q}{LW} = \frac{\text{Flow rate}}{\text{Surface area}} = v_{\text{OR}} \quad (13\text{-}30)$$

Thus, for this ideal horizontal flow reactor, the critical velocity is the overflow rate, just as it was for the vertical flow reactor.

We have implicitly assumed in this analysis that the flow rate does not change across the reactor, referring only to Q rather than Q_{in} or Q_{out}. Most properly, Q_{out} should be specified, but the sludge flow is often so small as to be unimportant for this analysis. This assumption that the sludge flow is small is conservative and, in some cases, is not very accurate. For example, in secondary sedimentation tanks after biological treatment in wastewater treatment plants, the flow of sludge out the bottom can be a substantial percentage (often 25% or so) of the total flow. This situation leads to a net downward flow that aids sedimentation.

The trajectories of two identical particles with settling velocity $v_1 < v_{\text{OR}}$ are also shown in Figure 13-25: one that just reaches the bottom at the effluent end of the reactor, and the other that comes into the reactor at the very top. While in the reactor, each of these particles settles a vertical distance h_1 equal to $v_1\tau$. Any particle with settling velocity v_1 that enters the reactor at a point within a distance h_1 of the bottom will be removed (lost to the sludge layer below). Conversely, a particle with that settling velocity that enters at a distance larger than h_1 above the bottom will appear in the effluent. Since the flow is uniformly distributed throughout the depth at the influent, the fraction of the flow that enters within a distance h_1 of the bottom is (h_1/H). Further, since $\tau = (h_1/v_1) = (H/v_{\text{OR}})$, the ratio (h_1/H) equals (v_1/v_{OR}). Generalizing this result, the fractional removal efficiency in the reactor of particles with any settling velocity $v < v_{\text{OR}}$ equals (v/v_{OR}). Combining the result with the one presented above for particles with $v \geq v_{\text{OR}}$, the complete $F(v)$ function for an ideal horizontal PFR can be represented as follows:

$$F(v) = \begin{cases} 1 & \text{for } v \geq v_{\text{OR}} \\ \dfrac{v}{v_{\text{OR}}} & \text{for } v < v_{\text{OR}} \end{cases} \quad (13\text{-}31)$$

A plot of Equation 13-31 is shown in Figure 13-26.

It is instructive to compare Equation 13-31 and Figure 13-26 for ideal horizontal flow reactors with Equation 13-29a and

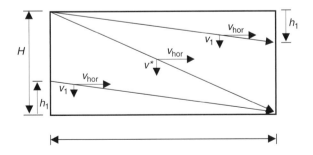

FIGURE 13-25. Critical trajectory and removal of particles with $v < v_{\text{OR}}$ in an ideal horizontal flow reactor. (Only ideal settling zone from Figure 13-24 is shown.)

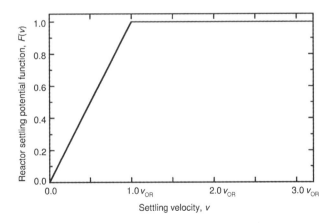

FIGURE 13-26. Reactor settling potential function for an ideal horizontal plug flow reactor.

Figure 13-22, which apply to ideal upflow reactors. Both types of reactors remove all particles with a settling velocity greater than the overflow rate. Upflow reactors accomplish no removal of particles with lower settling velocities, whereas horizontal flow reactors remove smaller particles at an efficiency that equals the ratio of their settling velocity to the overflow rate. Hence, ideal horizontal flow reactors are intrinsically better than ideal upflow reactors, at least with respect to removal efficiency. Recall, however, that particles with $v < v_{OR}$ can stay in ideal upflow reactors for times greater than the hydraulic detention time, τ, whereas all particles that are not caught at the bottom of an ideal horizontal flow reactor exit at the detention time. For Type II suspensions, this extra time for flocculation in upflow reactors diminishes the advantages of horizontal flow reactors.

The result shown in Equation 13-31 for particles with $v < v_{OR}$ can also be derived by considering a hypothetical particle with velocity v_1 that enters at the very top of the reactor. Like the one considered earlier, this particle settles a distance $h_1 = v_1 \tau_{PFR}$ as it travels across the reactor, so it leaves in the effluent at a distance h_1 below the top of the water column. Water that enters the reactor within a distance h_1 of the top of the basin, and therefore leaves above this height, does not contain any particles with settling velocity $> v_1$. Since the effluent flow is uniformly distributed from top to bottom, the fractional removal efficiency of particles with any particular settling velocity $v < v_{OR}$ is (h/H), and $F(v)$ equals (h/H) (or (v/v_{OR})) for these particles. This result is, of course, the same one we obtained earlier.

The preceding discussion indicates that the settling potential function, $F(v)$, of a continuous flow reactor can be thought of in a variety of equivalent ways:

- $F(v)$ is the expected fractional removal of particles that have an average settling velocity v while in the reactor,
- $F(v)$ is the fraction of the effluent flow that is devoid of particles with average settling velocity $\geq v$, and
- $F(v)$ is the fraction of the influent flow that loses all of its particles with average settling velocity $\geq v$ to the sludge layer.

These various descriptions of the reactor settling potential function are equally valid for all types of reactors. For example, in an ideal upflow reactor, all of the effluent flow is devoid of particles with $v \geq v_{OR}$, and none of that flow is devoid of particles with lesser settling velocities.

■ **EXAMPLE 13-8.** Find the overall removal efficiency in an ideal horizontal flow settling tank for the same suspension considered in Example 13-6, and at the same overflow rate of 1.2 m/h.

Solution. The mathematical description of the cumulative settling velocity distribution of the suspension was given in Example 13-6. As in that example, we use Equation 13-28 to solve for the removal efficiency, but here, we use the settling potential function for an ideal horizontal flow reactor (Equation 13-31) as follows:

$$\eta_{\text{out,ideal horizontal}} = \int_0^1 F(v)df = \int_0^{f(v_{OR})} F(v)df + (1 - f(v_{OR}))$$

$$= \int_0^{f(v_{OR})} \frac{v}{v_{OR}} df + (1 - f(v_{OR}))$$

For this suspension

$$f(v_{OR}) = \left(\frac{1}{3}\right)(1.2) = 0.40$$

$$1 - f(v_{OR}) = 1 - 0.40 = 0.60$$

In the region of $v < v_{OR}$, $(df(v)/dv) = \left(\frac{1}{3}\right)$, so that $df = \left(\frac{1}{3}\right)dv$, and the integral term in the removal equation becomes

$$\int_0^{f(v_{OR})} \frac{v}{v_{OR}} df = \int_0^{v_{OR}} \frac{v}{v_{OR}} \frac{1}{3} dv = \int_0^{1.2} \frac{v}{1.2} \frac{1}{3} dv$$

$$= \left.\frac{v^2}{(2)(1.2)(3)}\right|_0^{1.2} = \frac{1.44}{7.2} = 0.20$$

The overall removal, therefore, is the sum of the two components:

$$\eta_{\text{out}} = 0.20 + 0.60 = 0.80$$

Note that the removal in the ideal horizontal flow reactor (80%) is substantially greater than the removal in the ideal upflow reactor (60%). ■

■ **EXAMPLE 13-9.** Find the expected removal efficiency in an ideal horizontal flow reactor with an overflow rate of 1.1 m/h for the Type I suspension whose settling velocity distribution was found in Example 13-5. Assume that the suspension contains no particles with $v > 8$ m/h.

Solution. In this case, since $f(v)$ is not given as an explicit mathematical function, we carry out the required integration numerically. The settling velocity distribution found in Example 13-5 is given in the first two columns of the spreadsheet shown in Table 13-3. A few values from the earlier spreadsheet have been omitted because the intervals of $f(v)$ were too small to be of consequence, and a value of v when $f(v) = 1$ has been assumed; the result is insensitive to that assumption. In addition to the values obtained

TABLE 13-3. Calculation of Removal Efficiency in Ideal Horizontal Flow Reactor with $v_{OR} = 1.1$ m/h

Settling Velocity, v (m/h)	Cumulative Settling Velocity Distribution $f(v)$ (–)	Interval of f Δf (–)	Removal Efficiency at Settling Velocity v $F(v)$ (–)	Removal Efficiency for Interval $\overline{F(v)}$ (–)	Removal in Interval $\overline{F(v)}\Delta f$ (–)
0	0		0		
0.2	0.13	0.13	0.182	0.091	0.012
0.33	0.21	0.08	0.300	0.241	0.019
0.5	0.29	0.08	0.455	0.377	0.030
0.75	0.38	0.09	0.682	0.568	0.051
1	0.46	0.08	0.909	0.795	0.064
1.1	0.484	0.024	1	0.954	0.023
1.5	0.58	0.096	1	1	0.096
2	0.67	0.09	1	1	0.09
3	0.78	0.11	1	1	0.11
4	0.86	0.08	1	1	0.08
6	0.95	0.09	1	1	0.09
8	1.0	0.05	1	1	0.05
				Sum:	0.715

directly from Example 13-5, a value of $f(v)$ is interpolated at $v^* = v_{OR} = 1.1$ m/h from the adjacent values of $f(v)$; that is,

$$f(1.1) = f(1.0) + \left(\frac{1.1 - 1.0}{1.5 - 1.0}\right)(f(1.5) - f(1.0))$$
$$= 0.46 + (0.2)(0.58 - 0.46) = 0.484$$

The calculation of the removal efficiency is an application of Equation 13-26. The third column ($\Delta f(v)$) is found as the difference between adjacent values in the second column, and each value is shown in the row associated with the endpoint of the interval. The fourth column is the value of $F(v)$ associated with the value of v in the first column. For this horizontal flow reactor, $F(v)$ is calculated as the minimum of (v/v_{OR}) or one, in accordance with Equation 13-31. The fifth column is the average value of $F(v)$ in the interval $\overline{F(v)}_i = (F(v_{i-1}) + F(v_i))/2$ and is shown at the endpoint of the interval. Finally, the last column shows the product of the values in Columns 3 and 5, $\left(\overline{F(v)}_i \Delta f_i\right)$, for each interval. The overall removal is the sum of the values in this column, in accordance with Equation 13-26, and is shown at the bottom of the column; that is, $\eta_{out} = 0.715$, or 71.5%.

The spreadsheet is shown for all values of v, but it could be cut off at the value of v_{OR} and yield the same result by using a discretized form of Equation 13-28, as follows:

$$\eta_{tot} = \sum_{\text{all } i \text{ with } v \leq v_{OR}} \left(\overline{F(v)}\Delta f_i\right)_i + (1 - f(v_{OR}))$$
$$= [0.012 + 0.019 + 0.030 + 0.051 + 0.064 + 0.023] + (1 - 0.484)$$
$$= 0.715 \qquad \blacksquare$$

Circular Reactor Consider next a circular, continuous flow sedimentation tank, with the same idealities as the rectangular tank considered earlier; that is, an effluent wall that collects liquid from all layers equally, an influent distribution that is also evenly distributed radially and vertically, and a flow pattern that is perfect horizontal plug flow and does not allow any vertical mixing. In circular tanks, the influent comes in at the center of the tank and flows radially outward toward the effluent structure at the outside edge. As shown in Figure 13-27, the influent is

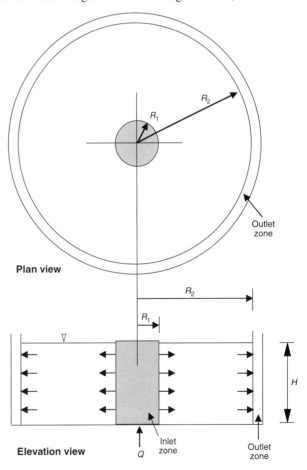

FIGURE 13-27. Circular continuous flow sedimentation tank.

630 GRAVITY SEPARATIONS

TABLE 13-4. Particle Removal in Ideal Horizontal Flow Reactor: Type I

$\log d_p$ (d_p in μm)	Influent Particle Size Distribution Function $(\log \Delta N/\Delta d_p)_{in}$ ($\Delta N/\Delta d_p$ in cm^{-3} μm^{-1})	Influent Volume Distribution $(\Delta V/\Delta \log d_p)_{in}$ (μm^3/cm^3)	Settling Velocity v (m/h)	Removal Efficiency $F(v)$ (–)	Effluent Volume Distribution $(\Delta V/\Delta \log d_p)_{out}$ (μm^3/cm^3)	Effluent Particle Size Distribution Function $(\log \Delta N/\Delta d_p)_{out}$ ($\Delta N/\Delta d_p$ in cm^{-3} μm^{-1})
−0.3	5.60	3.04E+04	2.46E−04	2.46E−04	3.03E+04	5.60
−0.2	5.57	7.10E+04	3.91E−04	3.91E−04	7.09E+04	5.57
−0.1	5.51	1.56E+05	6.19E−04	6.19E−04	1.56E+05	5.51
0.0	5.44	3.30E+05	9.81E−04	9.81E−04	3.30E+05	5.44
0.1	5.35	6.75E+05	1.55E−03	1.55E−03	6.74E+05	5.35
0.2	5.24	1.33E+06	2.46E−03	2.46E−03	1.33E+06	5.24
0.3	5.13	2.56E+06	3.91E−03	3.91E−03	2.55E+06	5.12
0.4	4.99	4.74E+06	6.19E−03	6.19E−03	4.71E+06	4.99
0.5	4.85	8.50E+06	9.81E−03	9.81E−03	8.42E+06	4.84
0.6	4.69	1.47E+07	1.55E−02	1.55E−02	1.45E+07	4.68
0.7	4.51	2.46E+07	2.46E−02	2.46E−02	2.40E+07	4.50
0.8	4.31	3.94E+07	3.91E−02	3.91E−02	3.79E+07	4.30
0.9	4.10	6.07E+07	6.19E−02	6.19E−02	5.69E+07	4.07
1.0	3.87	8.93E+07	9.82E−02	9.82E−02	8.05E+07	3.82
1.1	3.62	1.25E+08	0.155	0.155	1.06E+08	3.54
1.2	3.34	1.67E+08	0.246	0.246	1.26E+08	3.22
1.3	3.04	2.10E+08	0.391	0.391	1.28E+08	2.83
1.4	2.71	2.49E+08	0.619	0.619	9.50E+07	2.30
1.5	2.36	2.76E+08	0.981	0.981	5.27E+06	0.64
1.6	1.90	2.42E+08	1.55	1.0	0	–
1.7	1.40	1.92E+08	2.46	1.0	0	–
1.8	0.85	1.36E+08	3.91	1.0	0	–
1.9	0.25	8.63E+07	6.19	1.0	0	–
2.0	−0.40	4.81E+07	9.82	1.0	0	–

brought into the sedimentation zone at a radial distance R_1 from the center point and travels to the outside edge at a distance R_2 from the center. The analysis is set up in cylindrical coordinates to match the shape of the tank; r is the distance from the center point, ϕ is the angle from an arbitrary reference radius line, and z is the depth. The horizontal distance of interest varies from R_1 to R_2 as water flows from the influent structure to the effluent wall; the depth varies from zero at the top to the full depth, H; and the angle ϕ varies from 0 to 2π.

Because we are assuming ideal plug flow, the travel time for every aliquot of fluid is the theoretical detention time, $\tau = (V/Q) = \left(\pi(R_1^2 - R_1^2)H/Q\right)$. The horizontal velocity slows along the pathway from R_1 to R_2, but the vertical velocity of a particle is always its settling velocity. Hence, the trajectory of each individual particle is curved, but every particle falls the same distance in this circular reactor as it would in a rectangular reactor with the same value of τ or in a time t equal to τ in a batch reactor. Hence, the expected vertical profile of concentrations at R_2 in the continuous flow reactor is the same as in the batch reactor at time τ. We conclude, therefore, that the reactor settling potential function, $F(v)$, is the same for circular tanks as for rectangular tanks (Figure 13-26 and Equation 13-31).

■ **EXAMPLE 13-10.** The influent to an ideal horizontal flow reactor with an overflow rate of 1 m/h is as described in Example 13-2. Assume the suspension is nonflocculent. Plot the particle size distribution of the influent and effluent, both as the particle size distribution function and the volume distribution, and find the percent removal on a mass (or volume) basis.

Solution. The calculations are summarized in Table 13-4. The first four columns are repeated from Table 13-1. Each size class has an associated settling velocity, v_i, and the removal efficiency of particles in that size class is described by $F(v_i)$, given by Equation 13-31. For size classes with $v_i < v_{OR}$, $F(v) = (v_i/v_{OR})$; in this case, $v_{OR} = 1$ m/h, so the (dimensionless) numerical values for $F(v)$ are the same as the (dimensional) v_i values for those classes. For size classes with $v_i > v_{OR}$, $F(v) = 1$. The removal efficiencies range from infinitesimally small for the smallest particles to 100% for the largest five size

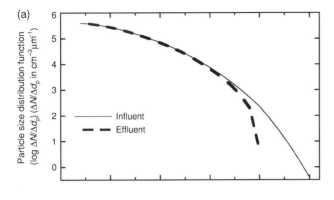

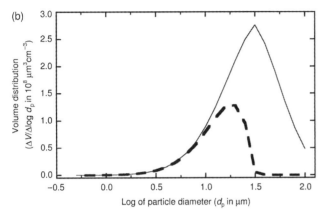

FIGURE 13-28. Changes in particle size distributions in ideal horizontal flow reactor.

classes. Only five size classes have more than 10% and less than 100% removal.

For each size class i, the effluent concentration is the product of the fraction remaining, $(1 - F(v_i))$, and the influent concentration. So, the effluent volume distribution values are found as follows:

$$\left(\frac{\Delta V_i}{\Delta \log d_p}\right)_{\text{out}} = (1 - F(v_i))\left(\frac{\Delta V_i}{\Delta \log d_p}\right)_{\text{in}}$$

These values, shown in the sixth column of Table 13-4, can then be converted to the particle size distribution function values, $(\log(\Delta N_i/\Delta d_{p,i}))$, using the techniques shown in Chapter 11.[11] The results are shown in the final column in the table. Plots of the influent and effluent particle size distribution functions and volume distributions are shown in Figure 13-28.

The total particle volume concentrations for the influent and effluent are found by summing all of the values of $(\Delta V_i/\Delta \log d_p)$ for each stream and multiplying by the logarithmic size increment (0.1 from Example 13-2).

---

[11] Alternatively, one can calculate these values directly. An example for the value at $\log d_p = 1.1$ is as follows: $[\log(\Delta N/\Delta d_p)]_{\text{out}} = \log\left[\left(10^{\log(\Delta N/\Delta d_p)}\right)_{\text{in}}(1 - F(v))\right] = \log[(10^{3.62})(1 - 0.155)] = 3.54$.

Doing so yields a total particle volume concentration of $198 \times 10^6\ \mu\text{m}^3/\text{cm}^3$ for the influent (as reported in Example 13-2) and $69.2 \times 10^6\ \mu\text{m}^3/\text{cm}^3$ for the effluent. The fractional removal of particle volume, therefore, is $1 - (69.2/198) = 0.65$. This process is identical to finding the fraction of the influent suspension represented by each size class, $\Delta f_i = (\Delta V_i/\sum_{\text{all }i}\Delta V_i)$ (where the $\Delta \log d_p$ is omitted because it cancels), and then finding the overall removal as $\eta_{\text{out}} = \sum_{\text{all }i}F(v_i)\Delta f_i$. Note that the removal efficiency of the horizontal flow reactor (65%) is considerably greater than that of the upflow reactor (36%, reported in Example 13-7). The result of 65% removal is typical of primary sedimentation tanks in municipal wastewater treatment plants. ∎

Type I versus Type II Suspensions in Horizontal Flow Reactors The same considerations about Type I and Type II suspensions that were noted for the ideal upflow reactor also apply to the horizontal reactors. That is, the removal efficiency for a given suspension depends only on the overflow rate for Type I suspensions, but on both the overflow rate and detention time (or depth) for Type II suspensions.

Again, for Type I suspensions, $f(v)$ is an unchanging characteristic of the influent suspension, and $F(v)$ depends only on the ratio of v and v_{OR}, so that the overall removal efficiency for a given suspension depends on v_{OR} only. Hence, a deep reactor and a shallow reactor, both with the same surface area and both receiving the same influent flow (and hence having the same overflow rate), will have the same removal efficiency. One might think that the deeper reactor would perform better because of the greater time available for settling, or worse because of the greater distance that particles must fall to hit bottom, but these two factors exactly offset one another.

∎ **EXAMPLE 13-11.** For the Type I suspension whose cumulative settling velocity distribution was found in Example 13-5, plot the removal efficiency as a function of overflow rate between 0.25 and 6 m/h.

Solution. The spreadsheet reflected in Table 13-3 can be evaluated at different overflow rates, and the removal efficiency determined at each. The values of $F(v)$ (the fourth column in that spreadsheet) depend on v_{OR} according to Equation 13-31. At an overflow rate of 4 m/h, for example, only the last three entries in that column would be one, and the values above that would be calculated as (v/v_{OR}). Hence, above the last three rows, all of the values in the last three columns of the table would be lower than that determined for the overflow rate of 1.1 m/h shown in the table, and therefore the removal efficiency (the sum at the bottom right of the table) would be lower. The resulting data for several

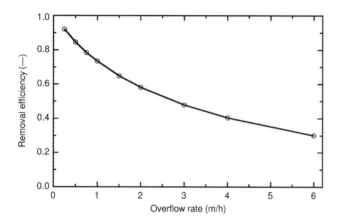

FIGURE 13-29. Effect of overflow rate on removal efficiency for a Type I suspension in a horizontal flow reactor.

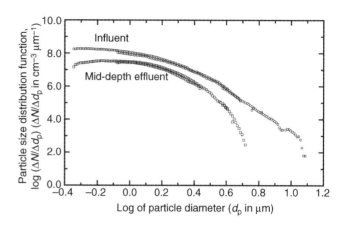

FIGURE 13-30. Changes in the particle size distribution during continuous flow flocculent (Type II) sedimentation.

overflow rates are shown in Figure 13-29. Note that increasing the overflow rate (i.e., increasing the flow rate or decreasing the surface area) decreases the removal efficiency. So, for example, primary sedimentation basins in wastewater treatment plants (without prior flow equalization basins) might achieve substantially reduced removal efficiencies during peak flow periods in comparison to average conditions. ∎

For Type II suspensions, $f(v)$ changes with time in a batch reactor due to the flocculation of the suspension, and therefore it changes with detention time in a continuous flow reactor. The effects of the simultaneous flocculation and sedimentation in continuous flow sedimentation can be illustrated with some experimental results that were obtained using a four-compartment reactor, with the compartments separated by perforated walls to prevent short-circuiting. A small compartment distributed the flow evenly into the first compartment. For each experiment, the suspension was made from a supply of dry particles and was destabilized as it flowed into the distribution compartment, so that no flocculation occurred before the sedimentation unit; that is, under all conditions, the influent (relative) size distribution (and therefore the settling velocity distribution) was the same. After steady-state conditions were achieved, samples were removed from each compartment at different depths and the size distribution was measured.

An example illustrating the change in the size distribution between the influent to the system and the effluent from mid-depth in the fourth compartment is shown in Figure 13-30. For the larger particles ($\log d_p > 0.3$, or $d_p > 5\,\mu m$), the removal (difference between influent and effluent) increased with increasing size, as expected from the increasing settling velocities, and the largest particles were removed completely. For the smaller particles ($\log d_p < 0.1$, or $d_p < 1.3\,\mu m$), the removal increased with decreasing size, a characteristic of flocculation (as smaller particles grow into larger particles). Particles in the intermediate size range were removed least efficiently, because the gain by flocculation of smaller particles nearly balanced the losses by flocculation into larger particles and by sedimentation. In this example, the net change is a loss of particles of all sizes due to this combination of flocculation and sedimentation. The trends shown in the figure occur in most systems encountered in environmental engineering.

Further results from the same research illustrate how variables that are important in flocculation influence the amount of sedimentation that can occur. For example, results of experiments using different influent suspension concentrations (200 and 600 mg/L) and different chemical conditions—at the pH of the zero point of charge (ZPC) and at a pH 1.5 units away from the ZPC—are shown in Figure 13-31.

For the two suspensions with the same initial concentration, removal efficiency (i.e., $\eta_{tot} = 1 - \text{fraction remaining}$) was greater at the ZPC than at the non-ZPC pH. For the two

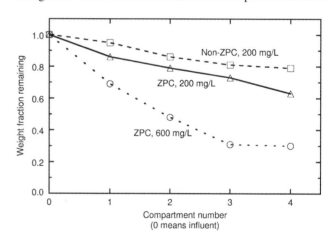

FIGURE 13-31. Effects of particle destabilization and particle concentration on removal efficiency in Type II, continuous flow sedimentation. Results from mid-depth ports in each compartment under steady-state conditions. *Source*: Wirojanagud (1983).

suspensions with the same chemical conditions, removal was substantially higher for the suspension with the higher initial concentration. In Type I sedimentation, the three lines would lie on top of one another; the dramatic effect of flocculation is illustrated by the differences in these lines. By the last compartment, approximately 80% of the particle mass remained in suspension in the system where flocculation was minimal (non-ZPC, low concentration) but only approximately 30% remained in suspension in the system where flocculation was promoted (ZPC, high concentration). In real systems, the effects of higher concentration are often taken advantage of by recycling sludge in precipitation processes (e.g., softening in water treatment, or metals precipitation in industrial situations), an approach that seeds the reaction and improves sedimentation.

Correspondence of Batch and Continuous Flow Reactors for Ideal Horizontal Flow For both rectangular and circular ideal PFRs with horizontal flow, the suspension is equivalent to a moving batch, with different horizontal locations in the continuous flow reactors corresponding to different times in a batch reactor. We can take advantage of this correspondence to simplify the analysis of removal in these reactors.

The flow rate through a differential area of the effluent wall is the product of that area and the horizontal velocity. The total effluent flow, Q_{out}, is the integral of this horizontal velocity (or volumetric water flux) over the area of the outlet wall; that is,

For rectangular reactors:

$$Q_{out} = \int_0^H \int_0^W v_h(L,w,z)\,dw\,dz = v_h WH \quad (13\text{-}32)$$

For circular reactors:

$$Q_{out} = \int_0^H \int_0^{2\pi} v_h(R_2,\phi,z) R_2\,d\phi\,dz = v_{h,R_2}(2\pi R_2)H \quad (13\text{-}33)$$

The integral expressions in the middle of these two equations are completely general, whereas the last form on the right side in both cases is for the ideal case where the flow is uniform across the entire wall.

The concentration in the effluent, c_{out}, is the weighted average of the concentrations across the effluent wall, with the weighting factor being the horizontal (effluent) flow rate just described; that is,

For rectangular reactors:

$$c_{out} = \frac{\int_0^H \int_0^W c(w,z) v_h(w,z)\,dw\,dz}{\int_0^H \int_0^W v_h(w,z)\,dw\,dz} = \frac{v_h WH}{} = \frac{1}{H}\int_0^H c(z)\,dz$$
(13-34)

For circular reactors:

$$c_{out} = \frac{\int_0^{2\pi}\int_0^H c(z,\phi,R_2) v_h(z,\phi,R_2)\,dz\,R_2\,d\phi}{\int_0^{2\pi}\int_0^H v_h(z,\phi,R_2)\,dz\,R_2\,d\phi} = \frac{1}{H}\int_0^H c(z)\,dz$$
(13-35)

Again, the first equality in each case is completely general, allowing for nonuniform distribution of the flow across the effluent wall, while the second applies to the ideal case of uniform flow across the wall. With ideal flow, the concentration varies only with depth, not with location along the width (w) or perimeter (ϕ) of the effluent wall, so the concentration in both cases is expressed only as $c(z)$. Note that the concentration under consideration can be any measure of particle concentration; suspended solids concentration, total volume concentration, and total number concentration are the most commonly used measures (but, of course, would give different results for the same suspension).

The overall removal efficiency, η_{out}, for any reactor is $1 - (c_{out}/c_{in})$. Therefore, for the ideal horizontal flow sedimentation reactor, the overall removal efficiency can be expressed in any of the following ways:

$$\eta_{out} = 1 - \frac{c_{out}}{c_{in}} = 1 - \frac{1}{c_{in}H}\int_0^H c(z)\,dz = \frac{1}{H}\int_0^H \eta(z)\,dz = \int_0^1 \eta(z')\,dz'$$
(13-36)

where z is depth, z' is the normalized (fractional) depth, $\eta(z)$ is the removal efficiency as a function of depth, and $\eta(z')$ is the removal efficiency as a function of fractional depth.

The expressions in Equations 13-34, 13-35, and 13-36 can be evaluated with the aid of a plot showing the profiles for concentration and fractional removal efficiency at the effluent wall of the reactor, such as shown in Figure 13-32. According to Equations 13-34 and 13-35, the particle concentration in the mixed effluent of the ideal horizontal flow reactors (rectangular or circular) is the area below and to the left of the profile shown in Figure 13-32a, divided by the total depth, H. Similarly, according to Equation 13-36, the fractional removal efficiency is the area above and to the left of the profile shown in Figure 13-32b. Hence, the figure provides an immediate visual picture of the performance of a continuous flow reactor.

The analysis of ideal horizontal plug flow reactors performed by the numerical, analytical, or graphical integration of the concentration (or removal) at the effluent wall must, of course, lead to the same result as that found from a numerical or analytical evaluation of the integral, $\int_0^1 F(v)\,df$. The correspondence is exact, because the same batch experiments are used to find the $f(v)$ values (corresponding, in the case of Type II suspensions, to a specific time and depth), and

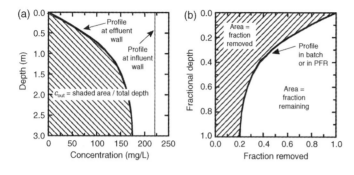

FIGURE 13-32. Using batch sedimentation results to predict behavior in ideal horizontal flow continuous flow settlers.

the integration of $F(v)df$ essentially reconstructs the area above the profile in Figure 13-32.

The conditions at the effluent wall of a PFR with a hydraulic residence time, τ_{PFR}, are identical to those in a batch reactor after time $t = \tau_{PFR}$. Therefore, the profiles of concentration and removal at the effluent wall of the PFR can be predicted and/or reproduced exactly in a batch settling test, if a column of the same height as the PFR is used and the batch settling time equals τ_{PFR}. Using this approach, batch tests can be used to explore the maximum particle removal efficiency that could be achieved for various possible designs of full-scale continuous flow reactors (maximum, because the full-scale reactor would likely be less ideal than the batch system). For flocculent suspensions (Type II), the batch data must be obtained at the times and over the depths (with multiple sampling ports) that are under consideration for the full-scale reactor. For nonflocculent suspensions (Type I), however, we have already established that the concentration in a batch reactor is a function only of the ratio of depth and time, not of each independently, so that the data required to produce plots such as Figure 13-32 for Type I suspensions can be obtained from data at any single depth over time, or any single time over depth.

To predict the expected profile at the effluent end of an ideal, plug flow, horizontal tank requires the integration of Equation 13-23, the number balance equation considering both flocculation and sedimentation. Recall that, in a Lagrangian view of a plug flow reactor, this equation is the same as for a batch system. Hence, numerical integration of Equation 13-23 for all particle sizes simultaneously would predict the required vertical profile. Programs that carry out such calculations exist (e.g., Lawler et al., 1980; Ramaley et al., 1981; Valioulis and List, 1984) and are likely to be used more frequently in the future, especially if they can be improved to account for realistic reactor configurations and less-than-ideal flow characteristics. In the meanwhile, most designs of full-scale reactors are based either on experimental results (i.e., batch tests at full depth, with sampling at several depths over time) or simply on experience with similar suspensions.

Tube Settlers

While horizontal and vertical flow reactors with a single floor for particle collection are the most common sedimentation reactors in use, tube settlers are increasingly used. A schematic of a simple tube settler is shown in Figure 13-33. We turn our attention now to the development of the reactor settling potential function for these reactors.

Before delving into the mathematical analysis of tube settlers, it is worthwhile thinking about how we expect their geometry to affect particle removal. We focus on the two features of tube settlers that distinguish them from horizontal PFRs: the short distance that particles need to fall before striking a surface, at which point they are considered to be permanently removed, and the upward velocity component of the water flow. Each of these features can be evaluated by considering a limiting case, as follows.

If N false floors are evenly spaced in a horizontal PFR (so that, including the real floor, there are $N+1$ surfaces onto which particles can fall), then the system becomes equivalent to a set of $N+1$ parallel ideal PFRs, each with overflow rate $v_{OR,original}/(N+1)$. Decreasing the overflow rate increases removal efficiency for a given suspension (by decreasing the critical settling velocity), so this modification would appear to be a useful way to improve system performance without increasing the footprint of the reactor. In fact, such reactors with several horizontal (or near horizontal) floors have been built, but the difficulties associated with sludge collection from such systems made the design impractical.

The other limiting case consists of adding an upward component to the bulk water flow without decreasing H (or increasing the available floor area of the reactor) at all. In the limit, this design becomes the ideal vertical reactor described previously. In this limiting case, particles with settling velocities less than the upward water velocity are not removed at all, and the overall removal efficiency of the reactor is less than that in a horizontal PFR of the same dimensions.

In a tube settler, particles have a shorter distance to fall than in a horizontal PFR with the same overall dimensions (or, equivalently, the available horizontal surface area is greater than in the PFR), and the suspension has an upward velocity component. The former factor increases the removal efficiency of particles, and the latter decreases it, so we might reasonably expect that the net effect of converting a horizontal PFR to a tube settler could be either beneficial or detrimental, depending on which of these factors is more important. Of course, we can also expect

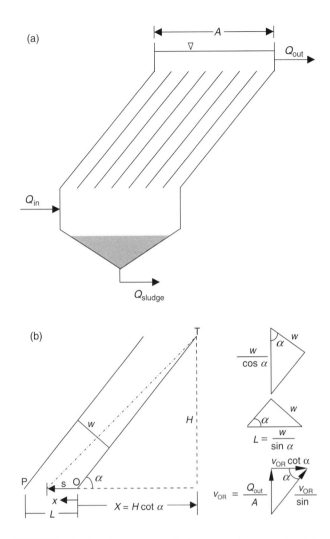

FIGURE 13-33. Schematic and definition diagrams for tube settlers.

that is, ($H \cos \alpha / w$) represents the number (N) of vertical "stages" between the influent and the effluent. (If one draws a vertical line starting at the point O in the left portion of Figure 13-33b until hitting the top wall of the tube, then draws a horizontal line until hitting the lower wall again, and continues to draw vertical and horizontal lines in this way until reaching the top, there will be N steps.) If a cross-section of the tank has plates or tubes throughout the depth H (as would occur in a large tank), N is the number of plates or sections in a vertical plane.

The horizontal distance between tubes (and therefore the size of the opening in the horizontal dimension at both the top and bottom) is $L = (w/\sin \alpha)$, as indicated in the middle figure on the right. The bottom small figure is the velocity vector. If the flow is ideal plug flow through each tube, the water velocity parallel to the tube is $(Q/(A \sin \alpha))$. The vertical component of the water velocity (with positive up) is simply (Q/A), defined here as v_{OR} (i.e., the overflow rate). The horizontal component of the water velocity is $v_{OR} \cot \alpha$.

Consider next the behavior of particles. Particles are assumed to be removed if they hit the right (or bottom) wall of the tube. Particles have a horizontal velocity that is the same as the water, but the vertical velocity of a particle is the difference between the (upward) vertical water velocity and the (downward) settling velocity of that particle in quiescent conditions; that is, $v_{OR} - v$. (Note that v_{OR} is defined as positive up and v is defined as positive down, following our intuition.) Since the vertical velocity of the particle is less than that of water, particles stay in the reactor longer than water; throughout that time, they have the same horizontal velocity as the water, so that they tend to move toward the right wall as they rise. Of course, if $v > v_{OR}$, then the particle does not rise into the tube at all, and the removal of such particles is complete.

The removal of particles that do enter the tube is more interesting. The analytical approach is to follow backwards the trajectory of a particle with settling velocity, v, from the point T (at the top right of the tube in Figure 13-33b) and determine the position (s, a particular value of the variable x) where it entered the tube. Any particle with the same settling velocity that entered to the right of s (i.e., at $x < s$) is captured, whereas any particle with that settling velocity that entered to the left of s (i.e., at $x > s$) is not removed. Therefore, assuming the particles are uniformly distributed across the entrance, the ratio (s/L) is the fractional removal of particles with that settling velocity; that is, $F(v) = (s/L)$.

If the value of s is greater than or equal to L for a given value of v, the implication is that *all* the particles with velocity v that enter the tube fall fast enough to hit the wall before they reach the outlet. The smallest value of v that meets this criterion is the critical velocity, v^*. In Figure 13-33b, v^* corresponds to a trajectory from point

that realistic designs are those in which the beneficial effects outweigh the detrimental ones.

The active settling zone in a tube settler is a set of inclined tubes or plates. The simplest construction of such a reactor is depicted in Figure 13-33a; in such a reactor with an elevation view shaped as a parallelogram, the flow enters at the bottom left and travels up and to the right. Sets of tubes can also be constructed to fit (or retro-fit) into a rectangular reactor. A single tube is sketched in Figure 13-33b. The design variables, in addition to the overall surface area (A) of the tank, include the depth (H) of the tank, the angle of incline (α) of each tube, and the spacing (w) between the two sides of each tube.

On the right portion of Figure 13-33b are three small figures that aid in understanding the geometry and flow in a tubular reactor. The top diagram indicates that the vertical separation distance between tubes is ($w/\cos \alpha$). Since the total vertical distance is H, the ratio of these two quantities;

P to point T, and to $(s/L) = 1.0$. Thus, we can express the reactor settling potential function in terms of s and L as follows:

$$F(v) = \begin{cases} s/L & \text{for } s < L, \text{ i.e., } v < v^* \\ 1.0 & \text{for } s \geq L, \text{ i.e., } v \geq v^* \end{cases} \quad (13\text{-}37)$$

Although the above expressions are correct, the fact that $F(v)$ equals s/L when $v < v^*$ is not particularly useful to us, since we do not yet know how to compute s for a given v. The challenge, then, is to relate s to v. To do so, we note that the time, τ_p, for a particle with settling velocity, v, to travel the vertical distance H is

$$\tau_p = \frac{H}{v_{OR} - v} \quad (13\text{-}38)$$

The horizontal distance traveled is the product of the horizontal velocity, v_h, and the travel time; for a particle that strikes point T, the distance is the sum of the horizontal distance taken up by the tube (X) and s; that is,

$$s + X = v_h \tau_p = (v_{OR} \cot \alpha)\left(\frac{H}{v_{OR} - v}\right) \quad (13\text{-}39)$$

As can be seen in Figure 13-33, $X = H \cot \alpha$. Substituting this expression into Equation 13-39 and rearranging yields

$$s = (v_{OR} \cot \alpha)\left(\frac{H}{v_{OR} - v}\right) - H \cot \alpha$$
$$= H \cot \alpha \left(\frac{v_{OR}}{v_{OR} - v} - 1\right)$$
$$= H \cot \alpha \left(\frac{v}{v_{OR} - v}\right) \quad (13\text{-}40)$$

The fractional removal of particles with settling velocity $v < v^*$ can be found, therefore, as

$$\begin{bmatrix} \text{Fractional removal} \\ \text{of particles with} \\ v < v^* \end{bmatrix} = F(v)_{v<v^*} = \frac{s}{L} = \frac{H \cot \alpha (v/(v_{OR} - v))}{w/\sin \alpha}$$
$$= \frac{H}{w} \cos \alpha \left(\frac{v}{v_{OR} - v}\right)$$
$$= N\left(\frac{v}{v_{OR} - v}\right) \quad (13\text{-}41)$$

As noted earlier, the critical velocity, v^*, corresponds to $s = L$. Substituting $s/L = 1$ into Equation 13-41 and solving for the corresponding value of v $(=v^*)$ yields

$$v^* = \frac{v_{OR}}{(H \cos \alpha / w) + 1} = \frac{v_{OR}}{N + 1} \quad (13\text{-}42)$$

Thus, the complete $F(v)$ function for tube settlers can be written in terms of v as

$$F(v) = \begin{cases} N\left(\dfrac{v}{v_{OR} - v}\right) & \text{if } v < v^* \\ 1 & \text{if } v \geq v^* \end{cases} \quad (13\text{-}43\text{a})$$

where

$$v^* = \frac{v_{OR}}{(H \cos \alpha / w) + 1} = \frac{v_{OR}}{N + 1} = \frac{Q}{A(N + 1)} \quad (13\text{-}43\text{b})$$

and

$$N = \frac{H \cos \alpha}{w} \quad (13\text{-}43\text{c})$$

Equation 13-43 indicates that the reactor settling potential function, $F(v)$, for a tube settler depends on only two composite parameters: N, which incorporates the effects of the design variables H, w, and α; and the overflow rate, which, for a given flow rate, depends only on the surface area of the tank. The effects of N and v_{OR} on $F(v)$ are shown in Figure 13-34; $F(v)$ for an ideal horizontal reactor is included for reference. For a given v_{OR}, the four curves become progressively steeper as N increases and reach the terminal value of $F(v) = 1$ at progressively lower values of v (i.e., v^* decreases as N increases, as shown in Equation 13-43b).

For a given suspension, higher values of $F(v)$ correspond to improved removal of particles, so the plot indicates that increasing N improves performance. This result is not surprising, because increasing N corresponds to more opportunities for particles to reach a plate. In the absence of other considerations (cost, minimum practical spacing distance between plates), adding more stages would always improve performance.

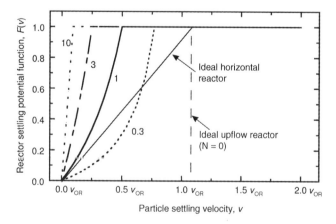

FIGURE 13-34. Settling potential function for tube settlers at various values of N.

The figure also indicates that when $N < 1$, the performance of the reactor would be worse than that of an ideal PFR with horizontal flow for some slowly settling particles but better than the ideal PFR with vertical flow. Physically, values of N between 0 and 1 correspond to reactors in which the particle collection plates are far removed from one another; this condition might also be thought of as an approximation for reactors in which the water enters with an even distribution through an entrance wall, but leaves by flowing over weirs at the top of the water surface. Note that there is some penalty associated with adding the upflow velocity component to the flow. The case where $N = 0$ corresponds to the vertical flow reactor described previously. The reasonable range of N values for design is $3 \leq N \leq 10$, so in practice, the conversion of a horizontal PFR to a tube settling system should always improve settler performance, albeit at a cost in terms of both capital and operating expenses.[12]

The graph makes clear the advantages of tube settlers over more conventional settling tank designs. Particles with settling velocities much smaller than the overflow rate can be completely removed in tube settlers. Hence, it is possible to use a much larger overflow rate (i.e., smaller area) to accomplish the same overall removal as in an upflow or horizontal flow tank. Tube settlers are often used in industrial settings where the waste treatment is accomplished within closed spaces and the value of space (area) is high. Another example where space is costly is in extremely cold climates where the treatment has to be enclosed to avoid freezing. Finally, tubes have often been used to retrofit existing settling tanks that have exceeded their original design flow; converting an existing settling tank (either horizontal flow or upflow) into a tube settler accommodates the greater flow in the same area and improves removal efficiency.

As was shown for reactor types considered previously, the preceding analysis of tube settlers is valid for either Type I or Type II suspensions, but the differences between the two deserve consideration. For Type I, the cumulative settling velocity distribution, $f(v)$, is a characteristic of the influent that is unchanged by the reactor, and hence its determination in a batch experiment is sufficient. For Type II, however, $f(v)$ is influenced by the flocculation that can occur. As in an upflow reactor, the largest particles never enter the settling zone of a tube settler and are not available to flocculate with the smaller particles. In addition, the overflow rates ($v_{OR} = (Q/A)$) of tube settlers are typically much higher than those of either upflow or horizontal flow reactors, so that the theoretical detention times ($\tau = (Q/A)$) are often relatively short; flocculation is also limited by these relatively short detention times. However, the particle detention time is greater than the hydraulic detention time (especially for particles with a settling velocity just less than v^*), so that some particles are available for flocculation for longer times than the hydraulic detention time. No batch test can truly emulate this set of changing conditions; a reasonable compromise might be to obtain $f(v)$ data in a batch experiment done at the depth of the full-scale reactor and for a time equivalent to the hydraulic detention time. Such a test would overestimate the effects of the biggest particles in suspension, but underestimate the effects of the increased particle detention time, in comparison with the continuous flow tube settler. These effects cannot be expected to offset each other exactly.

■ **EXAMPLE 13-12.** For the same suspension described in Example 13-6, find the expected removal efficiency in a tube settler with overflow rate $v_{OR} = 2.4$ m/h and $N = 3$. (Note that the overflow rate is doubled in comparison with the previous example in recognition that tube settlers should accomplish good removal in less space.)

Solution. The expected removal efficiency is found from Equation 13-28 using the reactor settling potential function for tube settlers described in Equation 13-43 with the value of $N = 3$. That is,

$$\eta_{out} = \int_0^{f(v^*)} F(v) df + (1 - f(v^*))$$

where $F(v) = N(v/(v_{OR} - v))$ for $v < v^*$.

Using Equation 13-43b, we find v^* as follows:

$$v^* = \frac{v_{OR}}{N+1} = \frac{v_{OR}}{3+1} = \frac{v_{OR}}{4} = 0.25 v_{OR}$$

From Example 13-6, we know that, in this range of v, $f(v) = \left(\frac{1}{3}\right)v$, so that

$$f(v^*) = \frac{0.25 v_{OR}}{3} = \frac{(0.25)(2.4)}{3} = 0.20,$$

and therefore $1 - f(v^*) = 1 - 0.20 = 0.80$.

We also know from Example 13-6 that, for $v < v^*$, $df = \left(\frac{1}{3}\right)dv$, so that

$$\int_0^{f(v^*)} F(v)df = \int_0^{f(0.25 v_{OR})} N\left(\frac{v}{v_{OR} - v}\right) df = \int_0^{0.25 v_{OR}} 3\left(\frac{v}{v_{OR} - v}\right)\frac{1}{3} dv$$

$$= \int_0^{0.6} \left(\frac{v}{2.4 - v}\right) dv = (2.4 - v)\Big|_0^{0.6} - 2.4 \ln(2.4 - v)\Big|_0^{0.6}$$

$$= 1.8 - 2.4 - 2.4 \ln\left(\frac{1.8}{2.4}\right) = -0.60 + 0.69 = 0.09$$

---

[12] The results for $N = 3$ and $N = 10$ in the figure appear to be reasonably straight lines. A lower approximation for the equation of the straight line can be obtained by assuming that $v \ll v_{OR}$ in Equation 13-43, yielding $F(v) = N(v/v_{OR})$ for $v < v^*$. An upper approximation is found as a straight line from the origin to $F(v) = 1$ at $v = v^* = v_{OR}/(N+1)$, yielding $F(v) = (N+1)(v/v_{OR})$. As N increases, the difference between these two approximations decreases. The linear approximations suggest that a tube reactor behaves like an ideal horizontal reactor with an overflow rate between $1/N$ and $1/(N+1)$ times the actual overflow rate.

Therefore, the overall removal in the reactor is

$$\eta_{out} = (1 - f(v^*)) + \int_0^{f(v^*)} F(v)df = 0.80 + 0.90 = 0.89.$$

Note that, as could be expected from the graph of $F(v)$ in Figure 13-34, the removal efficiency is greater in this reactor than in either the upflow or horizontal flow reactors, even though the overflow rate is twice the value used in the earlier examples for those reactors. The extra plates available for particle collection reduce v^* to a value considerably less than v_{OR} and thereby improve the removal efficiency. ■

■ **EXAMPLE 13-13.** Find the expected removal efficiency in a tube settler with an overflow rate of 4 m/h and $N = 5$ for the Type I suspension whose settling velocity distribution was found in Example 13-5.

Solution. The analysis follows the same sequence as shown in Example 13-9 for the ideal horizontal reactor, and is shown in Table 13-5. The only difference is in the values of for this reactor; these values are calculated according to Equation 13-43 and are shown in the fourth column of the table. For the conditions described

$$v^* = \frac{v_{OR}}{N+1} = \frac{4\,\text{m/h}}{5+1} = 0.667\,\text{m/h}$$

and so $F(v) = 1$ for all $v > 0.667$ m/h. The value of $f(v^*)$ is found by interpolating between $f(0.5) = 0.29$ and $f(0.75) = 0.38$ to be $f(0.667) = 0.35$. The spreadsheet is truncated past this point, and the removal is found as

$$\eta_{out} = (1 - f(v^*)) + \sum_{\text{all } i \text{ with } v \leq v^*} \left(\overline{F(v)}df\right)_i$$

$$= (1 - 0.35) + 0.114 = 0.794, \text{ or } 79.4\% \quad ■$$

Summary of Sedimentation in Ideal Flow Reactors

In the preceding sections, we have explored the expected behavior, under ideal (plug flow) conditions, of three types of continuous flow sedimentation reactors: upflow, horizontal flow, and tube settlers. For each of these reactor types, we have derived an equation for the reactor settling potential function, $F(v)$, that describes the expected fractional removal of particles with an average settling velocity, v, while in the reactor. In all three cases, complete removal is expected for particles with an average settling velocity greater than some critical value, v^*; that is, $F(v) = 1$ for all $v > v^*$. For upflow and horizontal flow reactors, the critical settling velocity is the overflow rate, defined as the flow rate divided by the (horizontal) surface area, $v^* = v_{OR} = Q/A$. The great value of tube settlers is that the critical settling velocity can be much smaller than the overflow rate; that is, $v^* = v_{OR}/(N+1)$, where N accounts for several geometric details of the design (the height of the reactor, the spacing between tubes, and the angle of the tubes relative to horizontal) and $N + 1$ can be thought of as the number of surfaces ("floors") that one crosses on a vertical line through the reactor. For the same overflow rate in the three types of reactors, the removal efficiency will be greatest for a tube settler (assuming that $N \geq 1$), followed by the horizontal flow reactor, and then the upflow reactor.

The actual removal achieved in a reactor depends on the settling velocity distribution of the suspension that is treated, which can be characterized by the cumulative settling velocity distribution, $f(v)$. This distribution is an unchanging characteristic of the suspension for Type I suspensions, because the average velocity of any particle is the same as its instantaneous or initial value. For Type II suspensions, flocculation changes the settling velocity distribution of the suspension as time proceeds, and therefore the average settling velocity for particles changes. Hence, for Type II suspensions in continuous flow systems, $f(v)$ is not entirely independent of the reactor. This complication causes little difficulty in the case of the ideal horizontal flow reactor, since such a reactor behaves as a moving batch reactor. For these reactors, the experimental determination of $f(v)$ carried out in a batch reactor at the same depth as the full-scale continuous flow reactor, and for a time equivalent to the detention time, yields the proper distribution. For the other ideal continuous flow reactors, no batch experiment can

TABLE 13-5. Calculation of Removal Efficiency in Tube Settler

Settling Velocity, v (m/h)	Cumulative Settling Velocity Distribution $f(v)$ (–)	Interval of f Δf (–)	Removal Efficiency at v $F(v)$ (–)	Removal Efficiency for Interval $\overline{F(v)}$ (–)	Removal in Interval $\overline{F(v)}\Delta f$ (–)
0	0		0		
0.2	0.13	0.13	0.263	0.132	0.017
0.33	0.21	0.08	0.450	0.356	0.029
0.5	0.29	0.08	0.714	0.582	0.047
0.667	0.35	0.06	1	0.857	0.051
				Sum	0.144

exactly emulate the flocculation conditions, since particle detention times are different from hydraulic detention times, and some of the largest particles are removed from the suspension before it enters the settling (or reaction) zone.

13.7 EFFECTS OF NONIDEAL FLOW ON SEDIMENTATION REACTOR PERFORMANCE

To this point, we have considered only highly idealized flow patterns in the reactors. In reality, the flow patterns in sedimentation and flotation reactors are considerably more complicated. Except for tube settlers, these reactors are generally large, open tanks. They often have some baffling to spread out the influent and direct the flow relatively uniformly away from that point. Additional baffles are sometimes installed to prevent short-circuiting. Considerable attention is paid to the design of the effluent collection system to spread that collection over a broad area of the tank, almost always involving the use of weirs that impose an upward velocity to the water. Nevertheless, most tanks have a relatively large fraction of the volume where the water flows unimpeded, and this condition leads to substantial deviations from the plug flow envisioned throughout the previous analysis.

The incoming flow itself causes mixing in the reactor as the energy of the entering water is dissipated. Most sedimentation facilities are outdoors and exposed to the wind. A relatively small wind across the top water surface can have a substantial effect on the flow pattern of a sedimentation reactor, both by imparting energy to the water and by creating a small elevation difference between the upwind and downwind sides of a reactor. Especially if the effluent is captured through overflow weirs (as opposed to submerged weirs), this small elevation imbalance can lead to large differences in the fraction of the flow that leaves different points in the effluent capture system.

Another common cause of nonideal (i.e., nonplug) flow in sedimentation reactors is density differences between the influent suspension and the water already in the tank. These density differences might be caused by temperature differences, or just by the concentration of particles in the suspension. In this section, we consider the impacts of various types of nonideal flow on the expected performance of sedimentation facilities.

Nonideal, Tiered Flow

We begin by considering a special type of nonideality: segregated, horizontal flow in which different elevations in the reactor have different horizontal velocities. We refer to this situation as tiered flow. As an example, imagine a rectangular reactor in which only 10% of the flow goes through the top half of the reactor, and the remaining 90% of the flow goes through the bottom half, as shown in

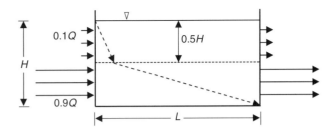

FIGURE 13-35. Example of tiered flow in horizontal flow, rectangular reactor. (Trajectory of a particular particle shown by dotted line explained in text.)

Figure 13-35. As in the previous examples, we carry out the analysis by determining $F(v)$ for the reactor.

Consider what happens to a Type I particle with $v = (H/\tau)$ that starts at the top of the reactor; recall that this settling velocity was the critical one (i.e., v^*) in an ideal horizontal flow sedimentation reactor. Under ideal plug flow conditions (same flow at all levels), the horizontal velocity (of both water and any particle) is the length L of the reactor divided by the hydraulic detention time, τ; that is, $v_{h,ideal} = (L/\tau) = (Q/HW)$. In those conditions, the critical particle would reach the half-depth at the half-length ($0.5L$). In the reactor with the specified bifurcated flow, on the other hand, the horizontal velocity in the top half is only 0.2 of the ideal flow velocity $((0.1Q/0.5HW) = 0.2v_{h,ideal})$, while the vertical (settling) velocity remains the same at $v = (H/\tau)$. So, when the particle falls $0.5H$ (in time 0.5τ), it will only have traveled $0.2*0.5L = 0.1L$; that is when it reaches half-depth, it will only have traveled one-tenth of the length of the reactor.

The particle then falls into the bottom half of the reactor, where its horizontal velocity is $(9/5)\,v_{h,ideal}$. Therefore, in the time it takes for the particle to reach bottom (0.5τ), it will travel horizontally $(9/5)*0.5L = 0.9L$ in the lower half of the reactor. The total horizontal distance it travels will be $0.1L$ in the top and $0.9L$ in the bottom, or $1.0L$; that is, the particle will just reach the bottom as it is about to exit the effluent end of the reactor. In other words, the same particle is the critical particle in the tiered system as in the ideal system, though the trajectories in the two cases are different (linear in the ideal case, and a dogleg pattern as shown in Figure 13-35 in the tiered one); that is, $v^*_{tiered} = v^*_{ideal}$. This result is completely general; that is, it is obtained for any fraction of the flow going into any horizontal fraction of the reactor.

The next question is whether the fractional removal of particles with settling velocities less than the overflow rate remains the same when the flow is bifurcated in this manner. To answer this question, imagine the same reactor examined earlier, and consider the trajectory of particles with a settling velocity exactly one-half the overflow rate ($v = v_{OR}/2$). Half of these particles would be removed in the ideal basin (the one that came in below the half-depth); that is,

$F(v_{OR}/2)_{ideal} = 0.5$. The question is, what fraction would be removed in the reactor with tiered flow? The question can be reformulated by identifying the height (depth) at which a particle with $v = v_{OR}/2$ would have to enter to be caught (reach the bottom) just as it is about to exit, and then determining what fraction of the flow comes in at or below that height.

As noted earlier, the horizontal velocity in the bottom half of the reactor is $(9/5)\,v_{h,ideal}$. Assuming for now that the particle in question enters in the lower half of the reactor, it would stay in the reactor for $(5/9)\,\tau$ and would fall a distance of $(5/9)\tau(v_{OR}/2)$, or $5/18\,H$. That is, a particle with $v = v_{OR}/2$ that strikes the bottom of the reactor just as it is about to exit would have to enter the reactor at a height equal to $5/18\,H$ above the bottom. This result confirms that the particle did indeed enter the reactor in the bottom half.

A height of $5/18\,H$ corresponds to $5/9$ of the height of the fast-moving, bottom layer of water. The portion of the reactor at or below that point is handling 50% of the total flow (i.e., $5/9*90\%$), so 50% of these particles would be caught; that is, $F(v_{OR}/2)_{tiered} = 0.5$, identical to the result for the ideal case. Again, this result is completely general, and so we conclude that the removal for Type I sedimentation in any tank with horizontally segregated nonideal flow is identical to that in an ideal case; that is, $F(v)$ is described properly by Equation 13-31.

Nonideal, Channeled Flow

Now consider another bifurcated flow reactor, but this time with the flow split side-by-side by a vertical plane along the direction of the flow; that is, standing at the influent end of a horizontal reactor, an observer might see that the left half gets $0.1Q$ and the right half gets $0.9Q$, as shown in Figure 13-36a. We call this as channeled flow. This reactor would have the same residence time distribution as is the case considered previously. Again, consider the particles that start at the top and whose settling velocity is equal to the average overflow rate; the trajectories of these particles are indicated in Figure 13-36b. On the left (slow) side, these particles reach the bottom only $0.2L$ along the reactor. On the right (fast) side, they do not reach the bottom and therefore are not captured. In the ideal (nonbifurcated) reactor, 100% of these particles would be removed. Clearly, the overall removal efficiency is not the same in a channeled flow reactor as in the ideal case.

This reactor behaves as two ideal settling tanks in parallel. If the overall tank is considered to have an overflow rate of v_{OR} ($=Q/A$), then the left side has an overflow rate of $0.2v_{OR}$ ($=0.1Q/0.5A$) and the right side has an overflow rate of $1.8\,v_{OR}$. Each of the two tanks would have the reactor settling function of an ideal tank with its respective overflow rate, but when put in terms of the overall v_{OR}, the two functions would be different. Designating the left side with the subscript one and the right side with subscript two, the two reactor settling potential functions are as follows:

$$F_1(v) = \begin{cases} \dfrac{v}{0.2v_{OR}} & \text{for } v < 0.2v_{OR} \\ 1 & \text{for } v \geq 0.2v_{OR} \end{cases} \quad (13\text{-}44\text{a})$$

$$F_2(v) = \begin{cases} \dfrac{v}{1.8v_{OR}} & \text{for } v < 1.8v_{OR} \\ 1 & \text{for } v \geq 1.8v_{OR} \end{cases} \quad (13\text{-}44\text{b})$$

These functions are illustrated in Figure 13-37, where the settling velocities are expressed relative to the overflow rate, v_{OR}.

The overall $F(v)$ is the weighted sum of the two components, with the weighting factor being the fraction of the flow through each half; that is, $F(v) = 0.1\,F_1(v) + 0.9\,F_2(v)$,

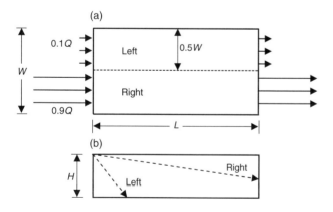

FIGURE 13-36. Example of channeled nonideal flow in horizontal flow, rectangular reactor: (a) plan view; (b) elevation view. (Left and right in plan view imagines that the observer is standing behind the flow; trajectories indicated in elevation view are explained in the text.)

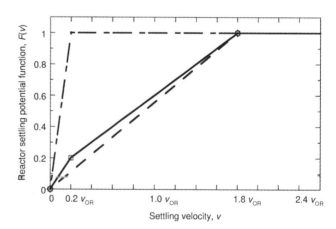

FIGURE 13-37. Settling potential function for a nonideal channeled flow reactor. (One-tenth of the flow in one-half side of the reactor, and the rest in the other half.)

so that the overall settling potential distribution is described as

$$F(v) = \begin{cases} \dfrac{v}{v_{OR}} & 0 < v < 0.1 v_{OR} \\ 0.1 + 0.5(v/v_{OR}) & 0.2 v_{OR} < v < 1.8 v_{OR} \\ 1 & v > 1.8 v_{OR} \end{cases} \quad (13\text{-}45)$$

The removal efficiency for Type I sedimentation can be predicted in this case, because the flow path for every parcel of fluid (and therefore particle) is known. The example given defines a specific (and unrealistic) nonideal flow pattern. Nevertheless, the implications can be generalized. Flow that is not evenly distributed across the horizontal dimension of a rectangular reactor (or radially in a circular reactor), even though it is distributed evenly in the vertical direction, results in the ideal (but small) removal of particles with a low settling velocity, and a less than ideal removal of more rapidly settling particles. In the example, particles with a settling velocity equal to the overflow rate are only 60% removed ($F(v_{OR}) = 0.6$), instead of the 100% removal expected in an ideal tank. Also, for this example, particles have to have a settling velocity $>1.8 v_{OR}$ to be completely removed; that is, $v^* = 1.8 v_{OR}$. Clearly, failure to distribute influent and to collect effluent evenly across the reactor can be costly in terms of removal efficiency, so influent and effluent structures that do distribute the flow evenly across the tank are of great value.

■ **EXAMPLE 13-14.** For the same suspension described in Example 13-6, find the expected removal efficiency in the nonideal, channeled flow reactor described earlier, with an overflow rate v_{OR} equal to 1.2 m/h.

Solution. The expected removal efficiency is found from Equation 13-28 using the reactor settling potential function for the specific nonideal flow condition described in Equation 13-45. That is,

$$\eta_{tot} = (1 - f(v^*)) + \int_0^{f(v^*)} F(v) df$$

where

$$f(v) = \begin{cases} \dfrac{v}{v_{OR}} & 0 < v < 0.2 v_{OR} \\ 0.1 + 0.5(v/v_{OR}) & 0.2 v_{OR} < v < 1.8 v_{OR} \\ 1 & v > 1.8 v_{OR} \end{cases}$$

According to the suspension's settling velocity distribution given in Example 13-6,

$$f(v^*) = f(1.8 v_{OR}) = f((1.8)(1.2 \text{ m/h})) = f(2.16 \text{ m/h})$$
$$= \frac{2.16}{3} = 0.72$$

so that $1 - f(v^*) = 0.28$.

The integration has to be carried out in two separate regions because of the break in $F(v)$ at $v = 0.2 v_{OR} = (0.2)(1.2 \text{ m/h}) = 0.24 \text{ m/h}$; at this value of v, $F(v) = 0.2$. Recall from Example 13-6 that, everywhere in the region of interest ($0 < v < 1.8 v_{OR}$ or $0 < v < 2.16 \text{ m/h}$), $f(v) = (v/3)$ and $df = \left(\frac{1}{3}\right) dv$. The integration is

$$\int_0^{f(v^*)} F(v) df = \int_0^{f(0.2 v_{OR})} \frac{v}{v_{OR}} df + \int_{f(0.2 v_{OR})}^{f(1.8 v_{OR})} \left(0.1 + 0.5 \frac{v}{v_{OR}}\right) df$$

$$= \int_0^{0.2 v_{OR}} \frac{v}{v_{OR}} \frac{1}{3} dv + \int_{0.2 v_{OR}}^{1.8 v_{OR}} \left(0.1 + 0.5 \frac{v}{v_{OR}}\right) \frac{1}{3} dv$$

$$= \int_0^{0.24} \frac{v}{1.2} \frac{1}{3} dv + \int_{0.24}^{2.16} \frac{0.1}{3} dv + \int_{0.24}^{2.16} \left(0.5 \frac{v}{1.3}\right) \frac{1}{3} dv$$

$$= \left.\frac{v^2}{(2)(1.2)(3)}\right|_0^{0.24} + \left.0.033 v\right|_{0.24}^{2.16} + \left.\frac{0.5 v^2}{(2)(1.2)(3)}\right|_{0.24}^{2.16}$$

$$= 0.008 + 0.064 + 0.320 = 0.392 \approx 0.39$$

Adding the two portions of the result, the final answer is

$$\eta_{out} = (1 - f(v^*)) + \int_0^{f(v^*)} f(v) df = 0.28 + 0.39 = 0.67$$

This result is less than the removal achieved in the ideal horizontal flow reactor at the same overflow rate ($\eta_{out} = 0.80$), as expected, but that it is higher than that achieved in the ideal upflow reactor ($\eta_{out} = 0.60$). ■

The two cases of nonideal flow discussed earlier (tiered and channeled flow) are quite specialized; both have every parcel of fluid that enters behaving as if in a plug flow reactor. In terms of flow, both can be represented as two reactors in parallel. In contrast, when considering particles, the first case (tiered flow) represents two reactors in series, since particles fall from one into the other, whereas the second (channeled flow) is two reactors in parallel. Relative to the ideal plug flow case, tiered flow yields the same particle removal efficiency for Type I sedimentation, but channeled flow yields a lesser percent removal—the better removal of the half of the reactor with slow flow does not make up for the worse removal of the fast half. Tiered flow uses all of the surface area equally, just as in the ideal case, whereas the channeled flow does not. These results mean that, in real reactors, building influent and effluent structures that spread the flow equally across the surface area is considerably more important than spreading the flow equally throughout the depth.

The quantitative prediction of results for Type II sedimentation, for both the ideal and nonideal flow reactors considered earlier, requires a substantial computer program. Qualitatively, however, we can imagine the effects. Both of the nonideal reactors have a cumulative exit age distribution ($F(t)$, not to be confused with $F(v)$), with two steps—from 0 to 0.9 at $t = 0.56\tau$ (i.e., $(0.5V/0.9Q)$) and from 0.9 to 1.0 at $t = 5\tau$. Recall from Chapter 4 that, for any reaction with order greater than zero, a plug flow reactor achieves the best results; in comparison with the plug flow case, the detriment to flocculation (a second-order reaction) caused by half the flow having a detention time of 0.56τ cannot be compensated completely by the benefit of the other half having a detention time of 5τ. Less flocculation (and therefore less sedimentation) will occur in either one of these reactors than in the ideal plug flow case. Type II sedimentation will, of course, still lead to greater removal than Type I in identical reactors. Nevertheless, we would expect that the concepts described earlier—that different nonidealities in horizontal (tiered) or vertical (channeled) segregation lead to different effects on sedimentation—still hold true.

Quantitative predictions for a particular influent particle size distribution in the various types of reactors confirm these ideas, as shown by the volume distribution in Figure 13-38.

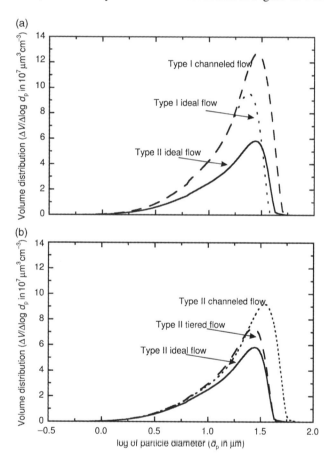

FIGURE 13-38. Effects of flocculation (Type II vs. Type I) and flow patterns on effluent volume distributions in horizontal flow sedimentation reactors. (Particle density $= 1.4\,\text{g/cm}^3$; overflow rate $= 1\,\text{m/h}$, approximately equivalent to the settling velocity of particles with size $\log d_p = 1.6$.)

The influent particle size distribution is not shown, as the peak would be approximately three times as high as the highest value of the effluent size distributions in the figure and would obscure the differences in the results. The effect of flocculation that occurs within the sedimentation reactor is shown in Figure 13-38a as the difference between the Type I (nonflocculent) and Type II (flocculent) ideal cases; as expected, the total volume concentration (area under the curve shown) is less for the Type II case. In other words, the removal efficiency is greater for the Type II case. The fact that large particles are created in flocculation is evident by the higher concentration of the largest particles in the Type II case than in the Type I case. The loss of small particles by flocculation is obscured in this view, since the volume distribution is dominated by the larger particles. Also apparent in Figure 13-38a is the lesser removal in the nonideal, channeled flow reactor in comparison with the ideal plug flow reactor; the tiered flow reactor is identical in Type I as the ideal case.

For Type II sedimentation, both types of nonideality in the flow pattern result in worse removal than in the ideal, plug flow reactor, as shown in Figure 13-38b. However, as one might expect as an extension of the Type I results, the tiered flow reactor gives a size distribution (and therefore a removal efficiency) that is intermediary between the ideal flow and the channeled flow cases. Hence, the conclusion reached earlier in relation to the Type I results holds true for flocculent sedimentation as well; that is, for sedimentation in reactors with horizontal flow, distributing the influent uniformly in the horizontal plane is more important than distributing it uniformly throughout the depth of the reactor. Creating influent and effluent structures that distribute the flow across the area of the reactor is vital to achieve the best results.

Mixed Flow

Thus far, we have restricted our analysis of nonideal flow to cases where the flow is strictly horizontal and segregated—no mixing of one parcel of fluid with another. The analysis of the ideal vertical flow reactors and tube settlers with $N < 1$ suggests that vertical flow reduces the fractional removal of particles with a settling velocity less than the critical velocity. The examples of nonideal cases with strictly horizontal flow suggest that flow nonidealities yield quite different results depending on whether the nonideality is in a vertical or horizontal direction. But in all these cases, the effluent concentration can be predicted if the settling velocity distribution of particles in the influent and the pathway of every fluid parcel are known. The flow segregation leads to the result that the trajectory of any entering particle (or any particle at any point in the reactor and at any time) is completely determined.

The analysis of sedimentation within mixed flow; that is, flow for which different parcels of fluid mix with one another, is not so straightforward. Consider a simple case of two small elements of an overall batch reactor that are adjacent in the vertical dimension and that have an exchange flow between them. These two elements, besides having a flux of particles caused by gravity at any layer, have a flux of particles associated with the exchange flow. Particles can move downward from the top element to the bottom element at a velocity that is the sum of the exchange velocity (assuming for the moment that it is constant across the interface between the two elements) and the settling velocity; particles move upward at a velocity that is the difference of these two velocities. Hazen (1904) considered an extreme case of such exchange velocities in which the particle concentration was maintained uniform throughout the vertical dimension (i.e., the reactor was essentially well-mixed), but particles were allowed to settle at the bottom of the reactor without resuspension.

In less severe cases, a concentration gradient exists in the water column (with higher concentrations at deeper depths), but the particle concentration profile is not as simple to predict as in the case with no vertical motion. An exact mass balance could be written and solved, if the exchange flows in a whole series of vertical elements were known exactly and reasonable assumptions were made about the concentration profiles within each element. When vertical exchange flows combine with Type I settling, some particles will be found at higher locations than they would in the absence of mixing. Numerical solutions of flow patterns in reactors, generated by computational fluid mechanics programs, could be used to determine the exchange flows and therefore the concentration profile at any point in a steady-state reactor. Several such models have been presented, including those by Schamber and Larock (1981), Adams and Rodi (1990), and Zhou and McCorquodale (1992). Unfortunately, the results are difficult to generalize, because the solutions are generated for specific geometries and conditions. Such models are quite useful in detailed design, allowing one to test the effects of inlet and outlet designs, baffles, and overall geometry.

A simpler approach to describing vertical exchange flows is through a turbulence model. In such a model, a dispersion (or turbulent eddy) coefficient describes the exchange across an interface, so that the net flux of particles of a given size between different elements conforms to Fick's laws. If we consider a continuous flow reactor with uniform plug flow in the horizontal direction (x) but allow turbulent mixing (as well as particle settling) in the vertical direction (z), the number balance for particles of size i in a differential element at steady state yields

$$v_x \frac{\partial n_i}{\partial x} = \varepsilon_z \frac{\partial^2 n_i}{\partial z^2} - v_i \frac{\partial n_i}{\partial z} \quad (13\text{-}46)$$

where the horizontal velocity, v_x, is assumed constant with depth for simplification, ε_z is the dispersion coefficient for water (and therefore particles, since they are assumed to move with the flow) across a horizontal interface, n_i is the number concentration of particles of size i, and v_i is the settling velocity. This formulation for vertical mixing was first described by Dobbins (1944) for the batch case and was then extended by Camp (1946) to describe settling in continuous flow tanks, as is done here. Camp also assumed that no materials were scoured from the bottom, just as for the ideal tank, meaning that once a particle is removed from the settling zone, it is lost forever. He envisioned that the source of the turbulence (vertical mixing) was the no-slip condition at the bottom of the reactor.

With these assumptions, Camp was able to find a solution for the expected removal of particles of any settling velocity in a rectangular sedimentation tank. He found that two dimensionless numbers uniquely determined the removal efficiency for any particle: the ratio of the particle's settling velocity to the overflow rate (v/v_{OR}), and the quantity ($vH/2\varepsilon_z$), where H is the full depth of the tank. The solution is mathematically cumbersome, involving a series formulation in which coefficients are found as the successive roots of a transcendental equation. Camp presented a graphical solution (nomograph) that would allow its use in a relatively straightforward manner. The resulting solution for $F(v)$ showed a gradual bending away from the theoretical line for the ideal horizontal reactor as v/v_{OR} increased, with the degree of curvature determined by the value of ε_z. El Baroudi (1969) presented a methodology for determining ε_z from tracer results, thereby making Camp's results more useful.

Ostendorf and Botkin (1987) took the view that the turbulence was caused by the inlet conditions. They first considered the limiting case of "strong diffusion" in which a uniform concentration (of a given size of particle) was present throughout the reactor, but that settling and removal was accomplished at the bottom. This is the same case that Camp considered. In this case, dispersion exactly offsets settling everywhere except at the bottom. For this condition, they found

$$F(v)_{\text{strong}} = 1 - \exp\left(-\frac{v}{v_{OR}}\right) \quad (13\text{-}47\text{a})$$

Under less turbulent conditions, a concentration gradient would exist throughout the vertical dimension of the reactor, and the result for the settling potential function was as follows:

$$f(v)_{\text{mild}} = 1 - \lambda \operatorname{ierfc}\left(\frac{(v/v_{OR}) - 1}{2\lambda}\right) \quad (13\text{-}47\text{b})$$

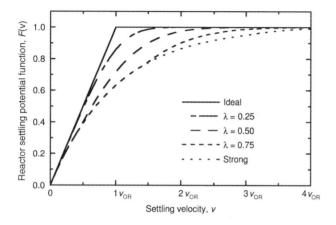

FIGURE 13-39. Reactor settling potential function for horizontal flow reactor with vertical turbulence. *Source*: Calculated from Ostendorf and Botkin model.

where $\lambda = \left(\sqrt{\varepsilon/v_{OR}H}\right)$ (a nondimensionalized dispersivity). The function ierfc is the integrated complementary error function, equal to the integral of the complementary error function from the value of its argument to infinity.[13] It is mathematically possible for Equation 13-47b to yield a value less than Equation 13-47a, but that condition cannot occur physically, so one must also specify that $F(v)$ is the maximum of the values obtained from the two equations.

Results calculated from this model are shown in Figure 13-39. The family of curves indicates that the removal is quite constrained in the region between the limiting cases of the ideal reactor and strong diffusion. Values of the parameter λ are more likely to be in the range of 0.25–0.5 than higher values. While it might be difficult to estimate the value of λ (or, equivalently, ε_z) for a given reactor, the results shown give a good indication of the expected reduction in removal efficiency caused by vertical mixing in a sedimentation reactor.

■ **EXAMPLE 13-15.** Find the expected removal efficiency in a horizontal flow reactor for the Type I suspension whose settling velocity distribution was found in Example 13-5 if the reactor has strong vertical turbulence and an overflow rate of 1.1 m/h.

Solution. We again follow the same type of spreadsheet used in Example 13-9, except that the values for $F(v)$ are found from Equation 13-47a. Here, as shown in Table 13-6, 60% removal is expected, as compared to 71.5% under ideal conditions (from Example 13-9). ■

Summary of Nonideal Flow Effects

Nonideal flow is a substantial problem in sedimentation facilities, especially in Type II sedimentation, which is the predominant type of sedimentation in environmental engineering applications. Any type of nonideality (i.e., any deviations from plug flow conditions) reduces the amount of flocculation that will occur and therefore reduces the removal by sedimentation (because fewer large particles will be created). For stratified flow in which the velocity is different at different depths, the effects on pure sedimentation (Type I and the sedimentation aspects of Type II) are nil, as the two effects of horizontal velocity and vertical settling distance offset each other exactly; this finding is in accordance with the concept that overflow rate is the determinant of effluent quality for Type I sedimentation. However, nonideality in the horizontal dimension (unequal flow to different parts of the effluent structure across the tank) or by mixing in the vertical dimension reduces the removal that can be achieved, with a more substantial effect on the larger particles that sedimentation facilities are designed to remove. Because of the adverse effects of nonideal flow on both Type I and Type II sedimentation, designers and operators of sedimentation facilities attempt to create and maintain structures for the inlet and outlet that spread the flow evenly (with minimal velocity variation) across the reactor and thereby create nearly plug flow conditions.

13.8 THICKENING

As noted in the introduction of this chapter, the sedimentation (and flotation) behavior of suspensions changes dramatically at high solids concentrations. As illustrated in Figure 13-40, a suspension with a sufficiently high suspended solids concentration exhibits a well-defined interface between the settling solids and the liquid above when settling in a batch reactor. This phenomenon is known as zone settling or hindered settling. In zone settling, it appears that all of the particles in a given layer settle at the same rate; certainly, the particles in the top layer are settling uniformly, since such uniform settling is what forms the well-defined solid/liquid interface. The reason for this phenomenon being occured is not completely understood, but the term "hindered" settling reflects

---

[13] Common spreadsheet programs do not include a built-in function for ierfc. Instead, ierfc(x) is calculated by the following formula: ierfc(x) = $(\exp(-x^2)/\sqrt{\pi}) - x\,\text{erfc}(x)$. However, if x is negative (i.e., when $v < v_{OR}$), this expression might not be directly usable, because some spreadsheets do not include an expression for a negative argument of erfc. This complication can be overcome by taking advantage of the following properties of the error function:

$$\text{erf}(-x) = -\text{erf}(x)$$
$$\text{erfc}(x) = 1 - \text{erf}(x)$$

These two expressions combine to yield erfc($-x$) = $1 + \text{erf}(x)$. Hence, for negative values of x: ierfc(x) = $(\exp(-x^2)/\sqrt{\pi}) - x(1 + \text{erf}(|x|))$.

TABLE 13-6. Calculation of Removal Efficiency in a Horizontal Flow Reactor with Vertical Turbulence

Settling Velocity, v (m/h)	Cumulative Settling Velocity Distribution $f(v)$ (–)	Interval of f Δf (–)	Removal Efficiency at v $F(v)$ (–)	Removal Efficiency for Interval $\overline{F(v)}$ (–)	Removal in Interval $\overline{F(v)}\Delta f$ (–)
0	0		0		
0.2	0.13	0.13	0.166	0.083	0.011
0.33	0.21	0.08	0.259	0.213	0.017
0.5	0.29	0.08	0.365	0.312	0.026
0.75	0.38	0.09	0.494	0.430	0.039
1	0.46	0.08	0.597	0.546	0.044
1.5	0.58	0.12	0.744	0.671	0.080
2	0.67	0.09	0.838	0.791	0.071
3	0.78	0.11	0.935	0.886	0.097
4	0.86	0.08	0.974	0.954	0.076
6	0.95	0.09	0.996	0.985	0.089
8	1.0	0.05	0.999	0.998	0.050
				Sum	0.600

the idea that the behavior of individual particles is hindered by the existence of many near neighbors. Apparently, the hydrodynamics of water escaping through the settling suspension in such concentrated systems prevents particles that would have widely variant individual settling velocities from passing each other easily.

The minimum solids concentration at which this phenomenon occurs depends on several characteristics of the particles, including their size and density (i.e., individual settling velocities), the breadth of the settling velocity and particle size distributions, the ability of the particles to form flocs, and the amount of water that is incorporated into the flocs. Suspensions that are flocculent and have narrow distributions of particles with (individual) low settling velocities tend to exhibit zone settling at lower concentrations than those that are nonflocculent and have a broad distribution of particles with higher settling velocities. Essentially all sludges[14] that are generated in water and wastewater treatment exhibit zone settling. This class of suspensions includes those from alum- or iron-based coagulation or calcium carbonate precipitation in water treatment, metals precipitation in industrial wastewater treatment, and both primary and secondary settling in wastewater treatment.

Batch Thickening

When a suspension undergoes zone settling in a batch reactor, the solid/liquid interface falls over time; that is, the depth from the water surface to the solid/liquid interface increases with time, as shown in Figure 13-41. The settling rate (the slope at any point on the curve) is generally constant for some time, but later the motion decreases and eventually stops. The initial constant rate of fall is the settling velocity (v_z) associated with the initial concentration throughout the suspension (and therefore associated with the concentration of the layer at the interface). At the bottom of the reactor, the settling causes the concentration to increase; concentrations higher than the initial concentration propagate upward from the bottom. Eventually, these successively higher concentrations reach the interface, and the increased concentration at the interface causes the fall of the interface to slow down. The behavior in the latter stages of this process can be complex and is considered subsequently; we focus first only on the initial constant rate of fall.

If we carry out another experiment on a suspension with the same type of particles (density, relative size distribution)

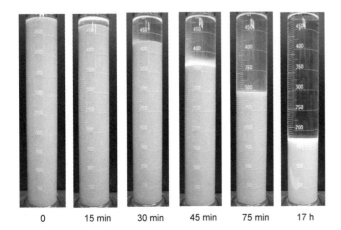

FIGURE 13-40. Example of thickening: time lapse of the sedimentation of sludge from a softening water treatment plant.

[14] The term "sludge" has fallen into disfavor among many in the field but is used here as the most general term available to describe the concentrated suspensions that are formed in water and wastewater treatment plants. Terms such as "biosolids" or "solid residuals" are commonly used in wastewater and water treatment, respectively. "Sludge" refers to the suspension; that is, both the solids and the water those solids are in.

646 GRAVITY SEPARATIONS

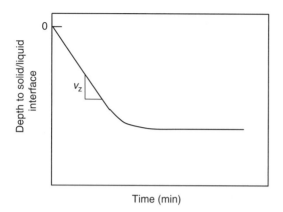

FIGURE 13-41. Fall of the solid/liquid interface in zone settling.

but at a higher initial concentration, the initial slope would be lower and the ultimate height of the sludge layer would be higher, as indicated in Figure 13-42; that is, the zone settling velocity decreases with increasing concentration.

Several investigators have developed models based on theoretical considerations to describe the relationship between the zone settling velocity and the (initial) solids concentration. A few of these relationships for an idealized situation in which the suspension consists of uniform particles are summarized in Equations 13-48 through 13-50. In these equations, the solids concentration, ϕ, is expressed as the solids volume fraction (volume of solids per volume of suspension), and the velocity is calculated relative to the Stokes velocity (v_s) of the individual particles.

Brinkman (1947):

$$\frac{v_z}{v_s} = 1 + \frac{3\phi}{4}\left(1 - \sqrt{\frac{8}{\phi} - 3}\right) \tag{13-48}$$

Richardson and Zaki (1954):

$$\frac{v_z}{v_s} = (1 - \phi)^{4.65} \tag{13-49}$$

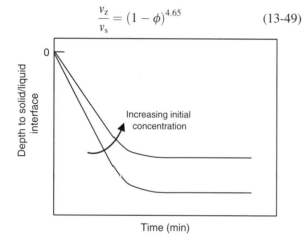

FIGURE 13-42. Effect of initial solids concentration on the fall of interface.

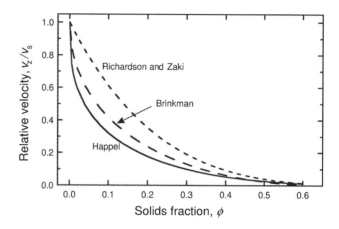

FIGURE 13-43. Mathematical models for the effect of solids concentration on settling velocity.

Happel (1958):

$$\frac{v_z}{v_s} = \frac{3 - (9/2)\phi^{1/3} + (9/2)\phi^{5/3} - 3\phi^2}{3 + 2\phi^{5/3}} \tag{13-50}$$

These relationships are plotted in Figure 13-43. Although they differ quantitatively, all indicate a substantial decrease in the settling rate with increasing solids concentrations. At solids volume fractions in the range of $0.1 < \phi < 0.4$ (a reasonable range for treatment plant sludges), the settling velocities are in the range from approximately 10–60% of the Stokes velocity. Unfortunately, these models are of little practical value in treatment plant analysis and design, because real sludges do not meet the assumptions underlying the models' formulations. Also, it is difficult to determine the solids fraction of a flocculent sludge, because water is incorporated into the flocs of solids in most cases. Nevertheless, these models are useful in as much as they provide mathematical representations of the general behavior expected.

For the heterodisperse concentrated suspensions encountered in water and wastewater, the relationship between the settling velocity and concentration is more often described with empirical models. Some investigators (e.g., Michaels and Bolger, 1962; Dick and Ewing, 1967; Scott, 1968; Knocke, 1986) have used the Richardson and Zaki model (Equation 13-49) as a template, describing the relationship as follows:

$$v_i = v_o\left(1 - \frac{c_i}{\rho_f}\right)^{4.65} \tag{13-51}$$

where ρ_f is the average floc density. Heterodisperse suspensions do not have a unique, limiting case, settling velocity at infinite dilution (i.e., a settling velocity that corresponds to the Stokes velocity for monodisperse suspensions). Therefore, the value of v_o in Equation 13-51 has no true physical meaning but is used as a reference velocity in fitting data to this model. For the model described in Equation 13-51, values of v_o and ρ_f are found from the intercept and slope of a plot of $v_i^{(1/4.65)}$ versus c_i.

Based on experiments with activated sludge, Vesilind (1968) proposed another empirical relationship:

$$v_i = v_o \exp(-kc_i) \qquad (13\text{-}52)$$

Keinath (1985) used this relationship in modeling the behavior of continuous flow thickeners for activated sludge. The parameters of this model (v_o and k) are found from a semi-log plot of settling velocity as a function of solids concentration.

Solids Flux

The analysis of thickening is carried out using a quantity known as the *solids flux*—the mass of solids that passes through a unit area per unit time—as the key parameter. In batch systems, the solids flux[15] is the product of the zone settling velocity (length/time) and the solids concentration (in mass per volume); that is, $G_i = v_i c_i$. Each experiment like those depicted in Figures 13-41 and 13-42 leads to a single value of v_i (from the straight line portion of the curve) associated with the initial solids concentration, c_i. Because the settling velocity varies with solids concentration, the batch flux is also a function of concentration; an example of that relationship is shown in Figure 13-44. The solids flux tends toward zero at high concentration, because the velocity approaches zero and also tends toward zero at low concentration, because the concentration approaches zero; the curve, therefore, goes through a maximum at some intermediate concentration. The solids flux curve is shown as a broken line at low concentrations in Figure 13-44, because, at low concentrations, the phenomenon of zone settling does not exist; that is, the particles settle individually and not *en masse* as they do in zone settling. At some high (but finite) concentration, gravity alone would be incapable of thickening the suspension any further, and the velocity would be zero; however, in practice, performing experiments at such high concentrations is difficult, so experimental results are usually like the line in the figure—tending toward zero, but not at zero at the upper end of the concentration range studied.

The solids flux curve characterizes the ability of the suspension to pass solids downward (or upward in flotation) and allows analysis of what can (and cannot) happen during settling. Consider a suspension whose solids flux curve is described in Figure 13-45 and which has an initial concentration, c_o, uniformly distributed throughout the depth of a batch reactor. As noted, layers of higher concentration must

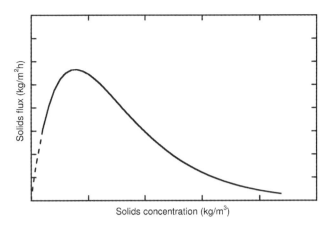

FIGURE 13-44. Solids flux curve.

form at the bottom of the reactor, since the total mass of solids in the reactor stays constant and the water above the interface has essentially zero concentration. The flux curve is useful in determining the behavior of the suspension below the interface, assuming that the curve represents a characteristic of the suspension that is valid throughout the suspension (and not just the interface). In the following analysis, we use the solids flux curve to determine how the profile of solids concentration (concentration versus depth) changes with time in a batch reactor.

Imagine a layer of some thickness forming at the bottom of the reactor (shortly after the beginning of the batch experiment) with the concentration c_1, as shown in Figure 13-45. The concentration above is assumed to still be c_0 so that only these two concentrations exist in the reactor. According to Figure 13-45, a layer with concentration c_1 would have a higher flux rate than the initial suspension, meaning that it would be able to pass solids downward faster than it received them from the suspension above (at c_0). That is, at the interface between the bottom of the layer at c_0 and

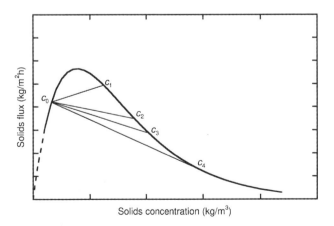

FIGURE 13-45. Analysis of possible concentrations in a thickening column.

---

[15] Throughout the rest of the book, fluxes are described by the symbol J, but here we use the symbol G for consistency with the literature about thickening.

the top of the layer at c_1, solids cannot be supplied fast enough from the layer at c_0 to sustain a layer at concentration c_1; therefore, this concentration layer at c_1 would not be stable and cannot exist anywhere in the column.

Next, imagine a layer forming at the bottom of the reactor with the concentration c_2, also shown in Figure 13-45. This layer does not have the problem noted for the imaginary layer with concentration c_1, and hence it appears that this layer would propagate upward. In that case, the boundary between the bottom of the layer at c_0 and the top of the layer at c_2 would move upward, so that the thickness of the layer at c_2 would increase. Again, we assume (for the moment) that no concentration other than c_0 and c_2 exists in the suspension. As a result, the concentration changes abruptly from c_0 to c_2 at some depth. With these assumptions, we can write a mass balance for the solids in a thin section that contains the boundary between the two layers and a small distance on either side of that boundary, as depicted by the shaded region in Figure 13-46. The dotted line represents the boundary between the two concentrations, with c_0 above this line and c_2 below it. This entire control volume is below the solid/liquid interface so that the concentration c_0 extends above the top boundary of this section.

The mass balance for solids in this control volume is

$$\begin{bmatrix} \text{Rate of change} \\ \text{of mass of solids} \\ \text{in the control volume} \end{bmatrix} = \begin{bmatrix} \text{Rate of input} \\ \text{of mass of solids} \\ \text{by settling from} \\ \text{above the control volume} \end{bmatrix}$$
$$- \begin{bmatrix} \text{Rate of output} \\ \text{of mass of solids} \\ \text{by settling to} \\ \text{below the control volume} \end{bmatrix}$$

$$\frac{d(c_0 z_0 + c_2 z_2)A}{dt} = c_0 v_0 A - c_2 v_2 A \quad (13\text{-}53)$$

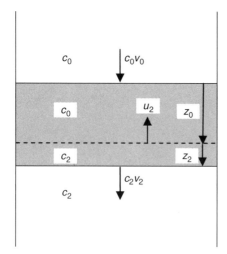

FIGURE 13-46. Schematic of a control volume containing the boundary between layers at concentrations c_0 and c_2.

where A is the horizontal cross-sectional area of the column. Dividing by A and expressing the cv terms on the right side as solids fluxes leads to

$$\frac{d(c_0 z_0 + c_2 z_2)}{dt} = G_0 - G_2 \quad (13\text{-}54)$$

Since G_0 is greater than G_2, the mass of solids will increase in the control volume, and the boundary between the two layers will rise; that is, z_0 will decrease and z_2 will increase at the same rate. The rate of this upward propagation of the boundary is $-(dz_0/dt) = (dz_2/dt) = u_2$.

The assumption that c_0 and c_2 are the only concentrations present means that they are constant, so the left side of Equation 13-54 becomes

$$\frac{d(c_0 z_0 + c_2 z_2)}{dt} = c_0 \frac{dz_0}{dt} + c_2 \frac{dz_2}{dt} = (c_0 - c_2)u_2 \quad (13\text{-}55)$$

Substituting into Equation 13-54 yields

$$(c_2 - c_0)u_2 = G_0 - G_2 \quad (13\text{-}56)$$

Rearranging, the velocity of upward propagation of the layer with concentration c_2 is

$$u_2 = \frac{G_0 - G_2}{c_2 - c_0} = -\frac{G_2 - G_0}{c_2 - c_0} \quad (13\text{-}57)$$

This velocity of upward propagation is the negative of the slope of the line connecting the two points on the flux curve associated with c_o and c_2; this line is shown in Figure 13-45.[16] If we then consider the possibility of another layer at slightly higher concentration (say c_3), we can see that the upward propagation of a layer with concentration c_3 would be faster than that of c_2. This result means that the layer of concentration c_2 would not really exist, because it would be overtaken by layers of higher concentration. The result also confirms that no concentration between c_0 and c_2 would exist, as assumed. The layer at c_3 would also not exist, because it too would be overtaken by a layer with yet higher concentration with a higher rate of propagation.

Extension of this reasoning leads to the conclusion that the lowest concentration layer that could exist and propagate upward from the bottom is the one designated c_4 on Figure 13-45, corresponding to the tangent to the flux curve

---

[16] Note that the same analysis could be carried out for the concentration c_1 considered earlier. In that case, the velocity of propagation would be $u_1 = -(G_1 - G_0)/(c_1 - c_0)$. Since both the numerator and denominator in this case are positive, the interface between these two layers is propagated downward, meaning (as stated earlier) that the layer at concentration c_1 could not be sustained. The layer at c_0 would propagate downward and would eventually extend to the bottom of the reactor, restoring the initial condition.

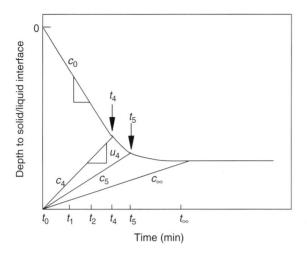

FIGURE 13-47. Interaction of the fall of the solid/liquid interface and the rise of higher concentration layers.

from c_0. This concentration gives the maximum velocity of propagation, namely $u_4 = (G_4 - G_0)/(c_4 - c_0)$; hence, if it forms at the bottom of the reactor immediately after initiation of the settling test, its upper boundary will propagate upward at this rate. This layer overtakes the initial concentration, c_0, when it breaks the surface of the solid/liquid interface, and at that time, the interface velocity slows down. This situation is depicted in Figure 13-47 on a batch settling curve; for all $t < t_4$, the concentration at (i.e., just below) the solid/liquid interface is the initial concentration, c_0, but at t_4, the concentration suddenly jumps to the concentration c_4 that has propagated upward from the bottom at the velocity, u_4.

Layers with higher concentrations than c_4 will also form and propagate upward at slower rates. Kynch (1952) hypothesized that all possible concentrations up to the highest that is ultimately achieved would form immediately at the bottom of the reactor and propagate upward at the rates reflective of their concentrations. The concentration profiles that would develop after various times for the suspension considered in Figures 13-45 and 13-47 are shown in Figure 13-48, based on the Kynch assumption. At time zero, the concentration is uniform throughout the reactor at c_0. As time progresses (t_1 through t_3), the depth (from the top) of the interface increases and the concentration just below the interface is still c_0; the thickness of the layer with concentration c_0 decreases, as the height measured from the bottom to the layer at concentration c_4 increases. At greater depths, the ultimate concentration (c_∞) forms at the bottom immediately after t_0, and the concentration decreases gradually from that concentration at the bottom to c_4 for all times less than t_4. At times greater than t_4, the concentration is greater than c_4 at all depths below the interface. At time infinity (not shown), the concentration would be uniform at c_∞ throughout the depth. These profiles stem directly from the flux characteristics detailed earlier.

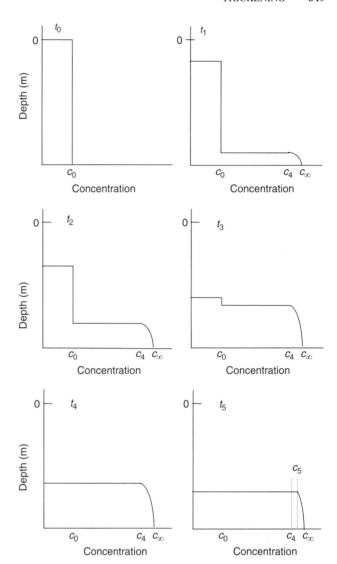

FIGURE 13-48. Concentration profiles during batch thickening for an incompressible suspension.

If the Kynch assumption were valid, one could determine the complete flux curve from a single batch test by measuring the concentration at the interface over time. However, the assumption is not valid for the suspensions normally encountered in environmental engineering applications, because the suspensions are compressible. In such cases, the weight of the solids in the upper portion of the sludge can compress (squeeze more water from) the layers below. Over time, this weight increases, so the solids concentration at the bottom increases steadily as settling proceeds, and the maximum concentration does not form at the bottom of the suspension until all settling ceases.

Most of the earlier analysis is valid for compressible suspensions. However, the achievement of the final interface height in a single batch experiment is likely to be a long, slow process. And, when that height is achieved, the concentration from the interface to the bottom of the reactor will not be uniform, but rather will increase gradually with depth from the interface to the bottom of the reactor.

For compressible suspensions at high concentrations, the settling velocity depends not only on the solids concentration, but also on the compressibility. Hence, the flux curve is influenced by the depth of the suspension during the batch settling test. Karl and Wells (1999) developed an elegant mathematical model of thickening that includes the compressibility of the sludges, and many of the references within that paper also contain theoretical analyses of sludge compression. However, most investigators have taken a less theoretical approach to obtain the flux curve that is relevant for design and operation. Dick and Ewing (1967), using activated sludge, showed the effect of the initial height of the suspension on the settling velocity and flux curves for compressible suspensions. Because knowledge of the batch flux curve is essential for proper design of continuous flow thickeners (as shown earlier), those authors recommended determining the settling characteristics of the suspension using batch reactors at the full expected depth of the sludge layer in continuous flow reactors. Vesilind (1968) considered various factors that would influence the batch settling characteristics of real suspensions. He also used activated sludge and found that the diameter of the batch test vessel could have a dramatic influence on the settling velocity, with wall effects causing a slowing of the settling in most cases. The effects were obvious even in settling vessels with diameters greater than 20 cm. Importantly, though, he found that the wall effects could be essentially eliminated if very slow stirring (e.g., 1 rpm) with narrow bars was incorporated into the test apparatus.

Continuous Flow Thickening

Continuous flow gravity thickeners, as indicated in the introduction of this chapter, have one influent flow and two effluent flows—one effluent (at the top) has a low solids concentration and the other (at the bottom) has a much higher concentration. In fact, the purpose of the thickener is to make that bottom concentration high and the flow that contains those solids quite low in comparison with the influent conditions. A schematic of a continuous flow gravity thickener is shown in Figure 13-49. Continuous flow thickening can also be accomplished in other types of devices, including flotation thickeners (considered in Section "Sludge Thickening by Dissolved Air Flotation") and belt thickeners in which the water drains through a porous belt while the solids are retained (and which are not considered in this book). The analysis presented in this section is largely based on the work of Dick (1972, 1989) and Vesilind (1979); an analysis that extends the work of Kynch to consider simultaneous clarification (Types I and II settling) and thickening in continuous flow thickeners under a variety

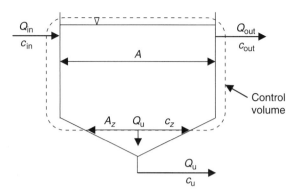

FIGURE 13-49. Schematic diagram of a continuous flow gravity thickener.

of loading conditions was presented by Lev et al. (1986). A more extensive treatise on the treatment of wastewater sludges, an update of the Vesilind (1979) work, is in Sanin et al. (2011).

In continuous flow thickening, solids move toward the bottom of the reactor not only by settling but also by the net downward flow of suspension induced by the sludge withdrawal at the bottom. In the main part of the reactor where the cross-sectional area is A, this withdrawal adds an average downward velocity, $u = (Q_u/A)$, where Q_u is the underflow rate. The solids handling capacity (i.e., flux through a horizontal plane) at any concentration in a continuous flow thickener is therefore increased, relative to that in a batch reactor, by this net downward velocity. The settling of solids relative to the fluid is the same as in the batch case, so that the flux in a continuous flow reactor is the sum of the batch flux (G_i) and the underflow flux (the product of the underflow velocity, u, and the solids concentration).

Thickeners have slanted bottoms with sludge rakes to carry the suspension toward the underflow outlet, which has a small area. This slanted bottom area is included in Figure 13-49. In the bottom portion of the thickener, the cross-sectional area varies with height; we designate the area at depth z as A_z.

In the analysis that follows, we consider mass balances for three different control volumes, all at steady state. These three control volumes are (i) the entire reactor, (ii) just the top portion of the reactor (with the bottom of the control volume below the solid/liquid interface,[17] assuming there is one, but above the slanted bottom portion), and (iii) the top portion of the reactor and some of the slanted portion

---

[17] The term "solid/liquid interface" is a shorthand way of referring to the well-defined interface between the settling solids and water, and should not be confused with the same term as used in adsorption and other surface reactions. For surface reactions, the term is precise in referring to a boundary between two phases; in thickening, the boundary is between two solids concentrations—one nearly zero and the other quite high.

This last control volume is indicated in Figure 13-49; the other two are similar but have lower and higher positions for the bottom of the control volume, respectively.

A mass balance on the solids in the entire continuous flow thickener at steady state (assuming no reaction that creates or destroys solids) yields

$$Q_{in}c_{in} = Q_{out}c_{out} + Q_u c_u \quad (13\text{-}58)$$

In a thickener, c_{out} is expected to be quite low (since the concentration is very low above the well-defined solid/liquid interface), and so it is common, though not necessary to assume that all of the solids go out in the underflow; that is,

$$Q_{in}c_{in} = Q_u c_u \quad (13\text{-}59)$$

Dividing by the cross-sectional area of the main part of the reactor expresses this mass balance in terms of solids flux:

$$\frac{Q_{in}c_{in}}{A} = \frac{Q_u c_u}{A} \quad (13\text{-}60\text{a})$$

or, after substitutions

$$G_{applied} = u c_u \quad (13\text{-}60\text{b})$$

For the second mass balance, we consider the top portion of the thickener only; that is, we choose the bottom of the control volume to be above the slanted portion but below the (assumed) solid/liquid interface. If the concentration at this height is c_i, the downward flux of solids through that bottom boundary of the control volume is $v_i c_i + u c_i$ (the sum of the batch flux and net downward flux caused by the underflow). The mass balance (after dividing by the area, A, and again assuming the effluent solids that escape out the top are negligible) is as follows:

$$\frac{Q_{in}c_{in}}{A} = G_{applied} = v_i c_i + u c_i \quad (13\text{-}61)$$

The two mass balances (Equations 13-60b and 13-61) are considered in relation to a batch flux curve in Figure 13-50. (We show subsequently why the conditions depicted are considered an underloaded thickener.) According to Equation 13-60b, a line that connects $G_{applied}$ on the ordinate to c_u on the abscissa has the slope of $-u$ as indicated; that is, all points on that line, for concentrations c_i between zero and c_u, are described by the expression, $G = G_{applied} - u c_i$. We call this line the *operating line*. According to Equation 13-61, however, $G_{applied} - u c_i = v_i c_i$. Only one point on the operating line satisfies that equation—where that line crosses the batch flux curve. Therefore, only one concentration can exist in the top portion of the reactor under the

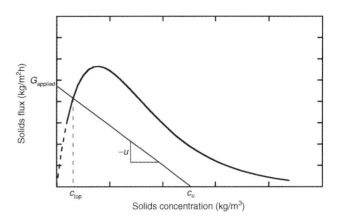

FIGURE 13-50. Analysis of an underloaded continuous flow thickener.

assumed circumstances of applied flux, underflow concentration, and steady state; that concentration is indicated as c_{top} in Figure 13-50.

Finally, a third steady-state mass balance is carried out on the portion of the thickener above some height z within the slanted portion at the bottom and yields the following:

$$Q_{in}c_{in} = Q_{out}c_{out} + Q_u c_z + G_{b,c_z} A_z \quad (13\text{-}62)$$

where c_z is the concentration at depth z, and G_{b,c_z} is the batch flux at the concentration c_z. As before, we assume that $Q_{out}c_{out}$ is negligible, divide by the (full) area of the thickener, and use the same definitions for $G_{applied}$ and u. With those substitutions, we can rewrite Equation 13-62 as follows:

$$G_{applied} = u c_z + G_{b,c_z}\left(\frac{A_z}{A}\right) \quad (13\text{-}63)$$

It is useful to multiply this equation by A/A_z to obtain the following:

$$G_{applied}\left(\frac{A}{A_z}\right) = u\left(\frac{A}{A_z}\right) c_z + G_{b,c_z} \quad (13\text{-}64)$$

In words, the left side of Equation 13-64 describes the flux of solids that arrive at the depth z from the top; because $A > A_z$, this applied solids flux is greater in the bottom portion of the thickener than in the upper portion. As before, the applied flux is processed downward by a combination of the downward suspension flow and the flux of solids relative to the suspension, and this combination is expressed by the right side of the equation. The first term on the right describes the increasing velocity of the suspension with decreasing area in the bottom of the thickener, and the second expresses the flux relative to the suspension flow (i.e., the batch flux). The situation is depicted graphically in

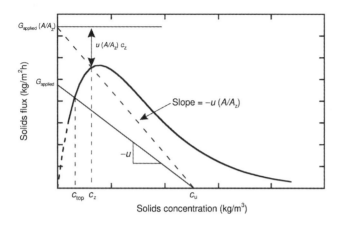

FIGURE 13-51. Analysis in the bottom, slanted portion of an underloaded thickener.

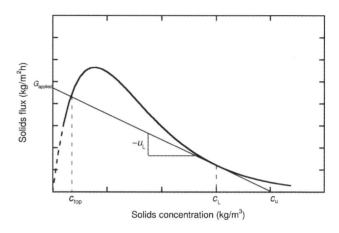

FIGURE 13-52. Analysis of a critically loaded continuous flow thickener.

Figure 13-51; as in the upper portion of the thickener, only a single point on the graph satisfies Equation 13-64; this point is at the intersection of the batch flux curve and the negatively sloped line from the intercept on the ordinate at $G_{\text{applied}}(A/A_z)$ to the intercept on the abscissa at c_u. The concentration at this point is c_z. Because A_z diminishes continuously to nearly zero at the withdrawal point at the bottom of the thickener, the intercept on the ordinate continuously increases, the (negative) slope of the line to c_u continuously increases (becoming vertical in the limit as A_z goes to zero), and c_z continuously increases to c_u in the bottom of the reactor.

It is useful to recall at this point the analysis of batch thickening. As indicated in Figure 13-45 and Equation 13-57, a layer with a concentration higher than the initial concentration could be propagated upward in the batch thickener at a velocity equal to that of the (negative) slope of a line connecting the two points on the batch flux curve for the two concentrations. Transferring that idea to the analysis of the (underloaded) continuous flow thickener, the imposed underflow velocity (u) is greater than the slope of any such line connecting the batch fluxes associated with c_{top} and a concentration less than or equal to c_u. Therefore, no layers at any concentration can be propagated upward (and expanded to any thickness) in the continuous flow reactor under the conditions we have considered, and the concentration rises continuously in the bottom slanted portion of the reactor. At any value of z and corresponding A_z, the applied flux is met by a combination of the underflow flux and the batch (or suspension) flux, but the batch flux never limits the ability to pass the solids fast enough to the layer (and concentration) below. In fact, the applied flux could be increased and the system would still be able to transport all the solids to the bottom and remove them in the underflow (hence, the designation of this operating state as "underloaded").

The operation of the thickener represented in Figure 13-50 could be changed by reducing the underflow rate, Q_u (and correspondingly, u). This change reduces the magnitude of the (negative) slope of the operating line and increases the value of c_u. Such a change represents an improvement in the operation, because a high value of c_u is a primary objective in thickening. The limit of such improvement is represented by the line in Figure 13-52; this operating line is drawn from the value of G_{applied} on the ordinate as a tangent to the batch flux curve. The slope of this limiting operating line is designated u_L (L for limiting). The point of tangency represents a second condition (besides c_{top}) at which the two mass balances (Equations 13-60b and 13-61) are satisfied; the concentration at this point on the curve is designated as c_L. This concentration is stable in the operation of a thickener; that is, a layer (of any thickness) at this concentration would be self-sustaining in the upper part of the thickener, because the velocity at which the suspension could propagate the concentration c_L upward in a batch reactor (in a suspension with c_{init} equal to c_{top}) is exactly matched by the velocity at which the underflow is pulling the suspension down (i.e., the underflow velocity). In the upper part of the thickener, layers at both concentrations, c_{top} and c_L, could exist; whether both existed and at what height the concentration changed from one to the other would depend on the history of the solids loading before the (current) steady-state condition. We call this condition "critically loaded", because, as shown subsequently, this condition represents the maximum G_{applied} for which all of the solids can be captured and the desired c_u can be obtained.

For the bottom, slanted section of the thickener, the analysis is identical to that given earlier for the underloaded thickener. The concentration decreases from c_u at the bottom to c_L at the top of the slanted portion. The location of the solid/liquid interface (and, if it exists, the transition from c_L to c_{top}) occurs in the full (cylindrical) section of the thickener.

Finally, it is useful to consider what would happen if an operator tried to achieve an underflow concentration higher

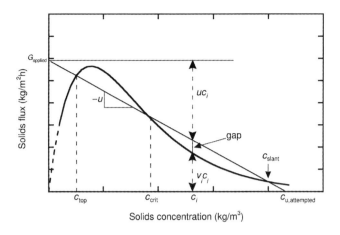

FIGURE 13-53. Analysis of an overloaded continuous flow thickener.

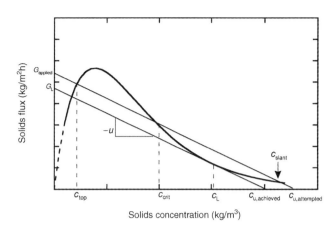

FIGURE 13-54. Analysis of the achievable solids handling in an overloaded thickener.

than that described earlier as the limiting condition. This situation is depicted in Figure 13-53 by the (attempted) operating line with $G_{applied}$ on the ordinate and $c_{u,attempted}$ on the abscissa. In this case, the operating line crosses the batch flux curve not only at c_{top} but also at the concentration labeled c_{crit} (for critical). According to our previous analysis for the top portion of the thickener, both of these concentrations could exist at steady state. Further, our analysis for the bottom portion would suggest that the concentration labeled c_{slant} would exist at the top of the bottom portion of the reactor (if $c_{u,attempted}$ were achieved at the bottom). Therefore, all three of these concentrations could conceivably exist at steady state in the upper part of the thickener.

We consider next what would happen if a concentration between c_{crit} and c_{slant} existed in the upper part of the reactor; such a case is indicated on Figure 13-53 with the concentration c_i. As shown in the figure, the sum of the underflow flux (uc_i) and the batch flux (v_ic_i) for that concentration is less than the applied flux (by the amount labeled "gap"); that is, in a layer at this concentration, solids could not pass downward at sufficient speed to match the applied flux. A layer at this concentration could, therefore, increase in thickness. The same idea can be seen by considering that the slope of a line on the batch flux curve between c_{crit} and c_i is greater than u; that means that, in a batch reactor, a concentration layer at c_i would propagate upward from the bottom of the reactor at a velocity greater than u. Because, in the continuous flow reactor under consideration, the underflow velocity is less than that propagation velocity, the layer at c_i would propagate upward. In that case, the mass balance (Equation 13-61) is not met; more solids are being applied to the thickener than can be successfully thickened to the desired underflow concentration. Note that, in both the underloaded and critically loaded conditions discussed earlier, the operating line is always below the batch flux curve for all concentrations greater than c_{top}. That means that, if one posits any concentration (c_i) greater than c_{top} in

the upper part of the thickener, the sum $v_ic_i + uc_i$ is always greater than $G_{applied}$, so that that concentration cannot be sustained.

Returning to the overloaded case, the slope of the operating line is determined by the underflow rate, and the maximum solids loading that can be thickened successfully at that underflow rate is a tangent to the under side of the flux curve. Therefore, as shown in Figure 13-54, the actual underflow concentration achieved ($c_{u,achieved}$) is less than $c_{u,attempted}$, and the areal loading of solids successfully processed (G_L) to that concentration is less than $G_{applied}$. The excess solids loading is the difference between the applied and limiting fluxes; this amount is equivalent to the maximum gap depicted in Figure 13-53. These excess solids will eventually go out the top of the tank. A mass balance (i.e., Equation 13-58) combined with the water balance ($Q_{in} = Q_{out} + Q_u$) yields

$$c_{out} = \frac{Q_{in}c_{in} - Q_uc_u}{Q_{out}} = \frac{(G_{applied} - G_L)A}{Q_{in} - Q_u} \quad (13\text{-}65)$$

The overload can be handled operationally without causing this effluent concentration to rise by increasing the underflow rate (steepening the operating line) and being satisfied with the consequent lower underflow concentration.

In all of the figures shown in this section, specific values of the solids concentration are not given, because the values that can be achieved vary considerably for the different types of suspensions encountered in environmental engineering. Suspensions of activated sludge from secondary wastewater treatment and alum sludges from water treatment might achieve underflow concentrations in the range of $20\,kg/m^3$. These suspensions incorporate a substantial amount of water into the flocs and the particles are barely denser than water. On the other hand, thickening a calcium carbonate suspension from water softening might achieve an underflow concentration of $80\,kg/m^3$; calcium carbonate particles are relatively dense (~2.4 g/cm³) and do not incorporate much water in the flocs. Often, sludge concentrations are reported not in mass/volume units but as percent solids; this occurs because it is simpler to measure a mass of the thickened suspension than a volume, so the percent solids represents a measure of mass of solids per mass of suspension. Nevertheless, the figures are shown here in mass/volume units as it is more fundamental to thickening theory.

■ **EXAMPLE 13-16.** The data shown in the first two columns in Table 13-7 were obtained in a series of batch thickening tests on activated sludge.

(a) A continuous flow thickener is to be designed to treat this sludge. The influent flow rate is 30 m³/h and the sludge is to be thickened from 6.5 g/L to 17.5 g/L. What diameter circular thickener is required?
(b) Identify the limiting concentration.
(c) If the solids loading increases by 50% after the thickener designed in part (a) exists, what is the maximum underflow concentration that can be achieved at steady state without losing excessive solids in the effluent flow?

Solution. We begin by calculating the batch solids flux, and converting units as appropriate for the full-scale situation. The results are shown in the third column of the table. To illustrate, the first row value is calculated as follows:

$$G_i = c_i v_i = (1.8\,\text{g/L})(5.0\,\text{cm/min})(1000\,\text{L/m}^3)(\text{m}/100\,\text{cm})$$
$$\times (60\,\text{min/h})(\text{kg}/1000\,\text{g}) = 5.4\,\text{kg/m}^2\,\text{h}$$

Note that concentrations in g/L are the same value in kg/m³.

TABLE 13-7. Activated Sludge Thickening Characteristics

Initial Concentration (g/L)	Settling Velocity (cm/min)	Batch Solids Flux (kg/m² h)
1.8	5.0	5.40
2.0	4.5	5.40
2.7	3.6	5.83
3.5	2.9	6.09
4.0	1.6	3.84
5.3	0.93	2.96
6.5	1.02	3.98
7.8	0.43	2.01
9.5	0.21	1.20
10.3	0.22	1.36
12.5	0.14	1.05
14.0	0.058	0.487
17.0	0.032	0.326
18.0	0.033	0.356

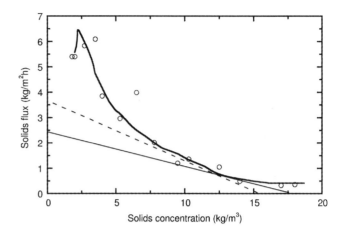

FIGURE 13-55. Batch solids flux curve for activated sludge in Example 13-16.

(a) The batch solids flux data are plotted as a function of the solids concentration in Figure 13-55. The data are somewhat scattered, but a reasonable curve is drawn through the points. A tangent is drawn from the target underflow solids concentration of 17.5 g/L (or 17.5 kg/m³) on the axis to the underneath side of the solids flux curve, and extended to the intercept on the ordinate, as shown by the solid line in Figure 13-55. That intercept is found to be 2.4 kg/m² h.

The influent solids loading rate is $Q_{in}c_{in}$, or $(30\,\text{m}^3/\text{h})(6.5\,\text{kg/m}^3) = 195\,\text{kg/h}$. The required area can then be found as the ratio of the limiting applied flux and this solids loading rate:

$$A = \frac{\text{Loading rate}}{\text{Limiting flux}} = \frac{195\,\text{kg/h}}{2.4\,\text{kg/m}^2\,\text{h}} = 81.3\,\text{m}^2$$

The required diameter is found from this area:

$$d = \sqrt{\frac{4A}{\pi}} = \sqrt{\frac{(4)(81.3\,\text{m}^2)}{\pi}} = 10.2\,\text{m}$$

(b) The limiting solids concentration is the value of the concentration at the point of tangency on the flux curve. This value is read directly as 14 kg/m³.

(c) When the solids loading increases by 50%, the applied flux will increase by 50%:

$$G_{\text{applied,new}} = 1.5 G_{\text{applied,original}} = (1.5)(2.4\,\text{kg/m}^2\,\text{h})$$
$$= 3.6\,\text{kg/m}^2\,\text{h}$$

Using this value on the ordinate of the graph, a new tangent line (dashed line in the figure) is drawn to the flux curve, and the value of the concentration where that line meets the abscissa is the maximum achievable steady-state underflow concentration. That value is read from the figure as 15.2 kg/m³ or 15.2 g/L. ∎

Design of Continuous Flow Gravity Thickeners

The design process for continuous flow thickeners can also be illustrated with a figure similar to those used for the operational analysis, assuming the flux curve for the suspension is known (or can be estimated). On Figure 13-56, several tangent lines to the batch flux curve are drawn from different values of c_u on the abscissa, and each line yields a different intercept on the ordinate. These intercepts represent the maximum (or limiting) values of the applied flux (i.e., G_L) for each corresponding value of the underflow concentration; we designate this limiting flux as G_L. The family of lines indicates that a higher desired underflow concentration yields a lower value of the limiting applied flux. At the time of design, one is likely to know the maximum solids loading that might be applied to the thickener; that is, $(Q_{in}c_{in})_{max}$. For each value of the desired underflow concentration, one can find the required area from the corresponding value of G_L, as follows:

$$A \frac{(Q_{in}c_{in})_{max}}{G_L} \quad (13\text{-}66)$$

As the desired c_u increases for a given suspension, G_L decreases, and therefore the required area increases. Hence, the critical design variable for thickeners is the cross-sectional area, with higher values of area being able to achieve higher values of c_u (and consequently lower values of Q_u). Hence, design represents an economic trade-off between the costs of increased area of a thickener or increased costs of further processing of a greater underflow in downstream sludge handling processes.

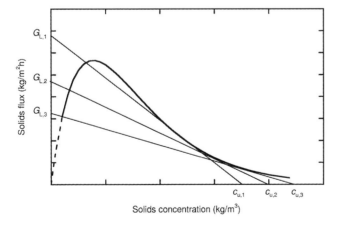

FIGURE 13-56. Analysis for design of continuous flow thickeners.

13.9 FLOTATION

Like sedimentation, flotation of particles suspended in water is driven by density differences, with flotation occurring when the density of a particle is less than that of the solution. Few particles encountered in environmental engineering have such low densities, but particles with a density just greater than that of water can be induced to float if air bubbles (one or more) become attached to them. Therefore, in environmental engineering systems, flotation almost always involves the creation of air bubbles in a suspension of particles, the attachment of the bubbles to the particles, and the subsequent rise of the particles to the top of the system, where they are removed. Although the bubbles are sometimes injected into the system by blowing air into the water through diffusers with very small pores, a much more common approach for generating bubbles is to cause the solution to become so supersaturated with air that gas bubbles form spontaneously as the system approaches gas/liquid equilibrium. Such processes are referred to as *dissolved air flotation* (DAF). In such a process, a portion of the flow is saturated with air at a high pressure, causing large amounts of air (i.e., its constituents) to dissolve. This flow is then mixed with the remaining flow under normal atmospheric conditions (i.e., $P_{tot} \approx 1$ atm). The reduction in pressure in the former portion reduces the solubility of the air, and bubbles form directly in the suspension as the new equilibrium condition is approached. Some of the bubbles form on the particle surfaces, and others form in solution and subsequently attach to the particles. If the density of the resulting particle–bubble floc is less than that of the solution, the floc will rise.

■ **EXAMPLE 13-17.** Determine (a) the settling velocity of a 15-μm diameter particle with a density of 1.1 g/cm³ suspended in water at 15°C and (b) the rise velocity of that particle after a 30-μm diameter bubble has attached to it. At 15°C, water has a density of 1.0 g/cm³ and a viscosity of 1.139×10^{-2} g/cm s, and moist air has a density of 1.226×10^{-3} g/cm³ (Giacomo, 1982).

Solution.

(a) The settling velocity is determined from Stokes' law (Equation 13-8).

$$v_s = \frac{(\rho_p - \rho)gd_p^2}{18\mu}$$

$$= \frac{((1.1 - 1.0)\text{g/cm}^3)(981\,\text{cm/s}^2)(15 \times 10^{-4}\,\text{cm})^2}{18(1.139 \times 10^{-2}\,\text{g/cm s})}$$

$$= 1.08 \times 10^{-3}\,\text{cm/s}$$

$$= (1.08 \times 10^{-3}\,\text{cm/s})(3600\,\text{s/h})(1\,\text{m}/100\,\text{cm})$$

$$= 0.0388\,\text{m/h}$$

(b) When a bubble (subscript bb) is attached to a particle (subscript p), the total volume of the particle–bubble floc (subscript pbb) is conserved, as follows:

$$V_{pbb} = V_p + V_{bb} = \frac{\pi}{6}\left(d_p^3 + d_{bb}^3\right)$$

$$= \frac{\pi}{6}\left((15^3 + 30^3)\,\mu m^3\right) = 1.59 \times 10^4\,\mu m^3$$

$$= 1.59 \times 10^{-8}\,cm^3$$

The equivalent spherical diameter of the particle–bubble floc is

$$d_{pbb} = \left(\frac{6V_{pbb}}{\pi}\right)^{1/3} = \left((15^3 + 30^3)\,\mu m^3\right)^{1/3} = 31.2\,m$$

The particle diameter is half that of the bubble, but its volume is only $1/8$ that of the bubble, so the equivalent spherical diameter of the particle–bubble floc is only $(9/8)^{1/3}$ times the bubble diameter.

The density of the particle–bubble floc is its mass divided by its volume, or

$$\rho_{pbb} = \frac{mass_p + mass_{bb}}{V_{pbb}} = \frac{\rho_p V_p + \rho_{bb} V_{bb}}{V_{pbb}}$$

$$= \frac{(1.1\,g/cm^3)(\pi/6)(15\,\mu m)^3 + (1.226 \times 10^{-3}\,g/cm^3)(\pi/6)(30\,\mu m)^3}{1.59 \times 10^4\,\mu m^3}$$

$$= 0.123\,g/cm^3$$

With the density of the floc being less than that of water, it will move upward. The rise velocity of the particle–bubble floc can be determined using Stokes' law, under the assumption that the floc can be treated as a spherical particle. Recognizing that the rise velocity is the opposite of the settling velocity, we find

$$v_{rise} = -v_{settle} = \frac{(\rho - \rho_{pbb})gd_{pbb}^2}{18\mu}$$

$$= \frac{[(1.0 - 0.123)(g/cm^3)](981\,cm/s^2)(31.2 \times 10^{-4}\,cm)^2}{18(1.139 \times 10^{-2}\,g/cm\,s)}$$

$$= 4.08 \times 10^{-2}\,cm/s = (4.08 \times 10^{-2}\,cm/s)(3600\,s/h)$$

$$\times \left(\frac{1\,m}{100\,cm}\right) = 1.47\,m/h$$

Note that this rise velocity is considerably greater than the settling velocity of the original particle, so that, in this case, flotation can achieve solid/liquid separation more rapidly than settling. Interestingly, this result is obtained even though the particle chosen has a higher density than those usually considered for flotation. ∎

The formation of bubbles in a supersaturated solution is a familiar phenomenon. For example, the pressure in an unopened bottle of a carbonated beverage is considerably greater than atmospheric, and therefore much more gas (in this case, carbon dioxide) dissolves in the drink than would dissolve at atmospheric pressure. This fact is a direct consequence of Henry's law, as discussed in Chapter 5. After the bottle is opened, the solution is at atmospheric pressure, and some of the dissolved gas must be released to reach equilibrium with the new, lower pressure. This gas is released by a combination of passive diffusion across the gas/liquid interface and formation of bubbles within the liquid, which then rise and enter the overlying gas phase.

The criterion for bubble formation within a bulk liquid is that the sum of the equilibrium partial pressures of all the dissolved gases exceed the static pressure at the location where the bubble forms. For example, if a solution is in equilibrium with air at a pressure of 3 atm, then the dissolved oxygen and nitrogen in the solution will have equilibrium partial pressures of approximately 0.63 and 2.37 atmospheres, respectively (corresponding to 0.21^*3 atm and 0.79^*3 atm, respectively); the sum of the partial pressures is therefore 3 atm. When such a solution is exposed to atmospheric pressure, bubbles will be able to form in the bulk liquid, and both oxygen and nitrogen will evolve from the solution until a new equilibrium state is achieved. The bubbles that form will have a composition that is in equilibrium with the solution and the ambient pressure at that location. Thus, if this solution were at a depth of greater than 20 m of water, the total pressure on the water would exceed 3 atm, and no bubbles would form.

Similarly, if the dissolved oxygen and nitrogen in a solution had equilibrium partial pressures of 0.3 atm and 0.6 atm, respectively, and if the solution were in contact with the normal atmosphere, then the solution would be supersaturated with oxygen and undersaturated with nitrogen. In this case, oxygen would exit the solution, and nitrogen would enter it. However, the oxygen would escape from the solution strictly by passive diffusion across the air/water interface; no bubbles would form, because the total partial pressure of all the gases in the solution (0.9 atm) would be insufficient to overcome the local hydrostatic pressure opposing the formation of such bubbles.

In general, it is easier for bubbles to form at solid/water interfaces than in bulk solution. (This is why bubbles in carbonated beverages appear to form preferentially on the walls of the container, rather than in the middle of the liquid.) Therefore, in a flotation system, many bubbles form on particles that are suspended in the water; others might form in the bulk solution and intercept suspended particles as the bubbles rise. If the density of the particle–bubble combination is less than that of the solution, the particle will be carried toward the water surface.

Particles with a high density can be floated by the attachment of a sufficient volume of bubbles, but such particles (unless quite small) settle well and hence are not good candidates for flotation. On the other hand, particles (or flocs) with a density just slightly greater than water do not settle well, and, once attached to bubbles, often rise at a much higher velocity than they settle. Therefore, flotation is considered as an attractive particle separation process only when the particles of interest have a density not much greater than that of water. A second criterion is that the particles need to be floatable with a reasonable input of dissolved air. Two types of suspensions that are commonly encountered in environmental engineering meet these criteria and therefore are often subjected to dissolved air flotation for solid/liquid separation:

- Low concentration suspensions (Types I and II) in drinking water, when the natural particles (or flocs after coagulation) are relatively low in density and relatively few in number. The most common application of flotation for this type of suspension has been for water sources with consistently low turbidity and with algae as the major particle type. Typically, these conditions apply to lake water but not to river water, which often contains denser particles and is more susceptible to turbidity spikes after rain storms. Nevertheless, flotation has been used extensively (particularly in Scandinavia, England, Japan, and South Africa) for typical alum- or iron-coagulated river or lake sources.

- High concentration suspensions (Types III and IV) for thickening of low density particles. The most common application in environmental engineering is for biological sludges—especially the thickening of activated sludge, sometimes after sedimentation. This application is also useful for metal hydroxide sludges and food processing wastes. High concentration suspensions are susceptible to flotation, because they do not require a bubble for every particle; since they settle or rise *en masse*, the well-defined solid/liquid interface that is characteristic of thickening will form and allow the particles to rise when the density of the sludge-bubble mixture is less than that of water.

Details of how flotation is done for these two types of applications are described later. Many aspects of the two applications are essentially identical, and those common elements are explained first.

Flotation Sytems Overview

A schematic diagram of a typical flotation system for a dilute suspension is shown in Figure 13-57; systems designed for concentrated suspensions differ in ways described subsequently, but also have the following components.

- *Saturator*, a high pressure (3.5–6 atm) tank to which both air and water are supplied, and in which, at steady state, virtually all of the injected air dissolves into the

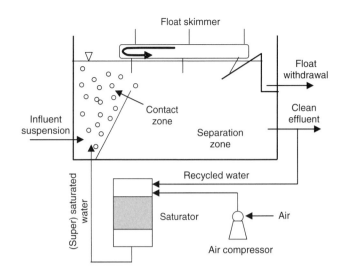

FIGURE 13-57. Schematic of a flotation system. (This diagram reflects a rectangular unit designed for low concentration suspensions. Details of the contact and separation zones are somewhat different in flotation units built with different geometries or for high concentration suspensions.)

water. The water supply to the saturator is typically recycled clean water, either from the flotation tank effluent or a cleaner water source from elsewhere in the water or wastewater plant. This water source must have a relatively low particle concentration to avoid clogging either the packing (if a packed bed saturator is used) or the nozzles at the entrance to the air/suspension contact chamber.

- *Air/suspension contact chamber*, in which a flow of water that has been saturated with air at a high pressure is introduced to the suspension to be treated at atmospheric pressure. As already noted, the reduced pressure causes bubbles to form in the water, and these bubbles attach to the particles of the suspension.

- *Separation zone*, in which the particle–bubble flocs float to the top of the liquid.

- *Particle removal system*, including a skimmer that moves the particles on the top (typically called the *float*) into a collection area. In high concentration systems (and occasionally in low concentration systems), a bottom skimmer (sludge scraper) is also provided to remove particles that settle to the bottom. In drinking water applications in which discharge to a sewer is allowable, the float can be withdrawn hydraulically by raising the water level and discharging the sludge over a weir; this system is used in lieu of a skimmer.

Since the saturator and the physical–chemical process leading to the creation of bubbles are essentially identical for both the low concentration and the high concentration systems, they are considered first in what follows. The contact and separation zones are somewhat different in the two applications, and they are considered subsequently.

Saturator

Saturators are gas absorption devices that run at relatively high pressure (3.5–6 atm, or approximately 350–600 kPa) to allow large amounts of air to dissolve into the water. Several types of saturators are available, with differences in how the air is introduced and contacted with water, but the basic principles are the same for all. In large plants, the most common type of saturator is a packed bed, similar to those described for stripping and absorption in Chapter 6, except for the pressure and the absence of an exiting gas stream. In smaller plants, unpacked towers are often used. Packed beds are generally more efficient but can require more careful operation (and perhaps more maintenance) due to the potential buildup of particles.

Air consists primarily of nitrogen and oxygen; in the design and operation of saturators, it is common to ignore the other, minor constituents. Further, because air is taken in from the atmosphere (i.e., at atmospheric pressure) and ultimately released in the flotation unit at atmospheric pressure, it is often acceptable and convenient to consider air as a single species. Nevertheless, it is useful to begin by considering the behavior of nitrogen and oxygen separately.

Figure 13-58a shows the equilibrium relationship between the gas and liquid phases for both nitrogen and oxygen according to Henry's law, with the gas-phase concentrations given in atmospheres and the liquid-phase concentrations in mg/L. In these units, the Henry's constants (H_{pc}) at 25°C for nitrogen and oxygen are 5.57×10^{-2} and 2.47×10^{-2} atm L/mg, respectively. At a total pressure of 1 atm (consisting of 0.79 atm N_2 and 0.21 atm O_2), the equilibrium liquid-phase concentrations are 14.2 and 8.5 mg/L, respectively. The correspondence between the gas and liquid concentrations for this condition is indicated by the dotted lines in the figure. Adding the liquid-phase concentrations for the two species, the result is a total dissolved air concentration of 22.7 mg/L.

Considering air as a single component thus leads to a Henry's constant (at 25°C and for the particular makeup of air stated earlier) of $(1\,\text{atm}/22.7\,\text{mg/L}) = 4.41 \times 10^{-2}$ (atm L/mg), as illustrated by the heavy line for normal air in Figure 13-58b. More generally (for other temperatures and air compositions), the Henry's constant for air can be found as

$$H_{\text{air}} = \frac{H_{N_2} H_{O_2}}{y_{O_2} H_{N_2} + y_{N_2} H_{O_2}} = \frac{H_{N_2} H_{O_2}}{y_{O_2} H_{N_2} + (1 - y_{O_2}) H_{O_2}} \tag{13-67}$$

where H_{O_2} and H_{N_2} are the Henry's constants for oxygen and nitrogen, respectively, and y_i is the mole fraction of

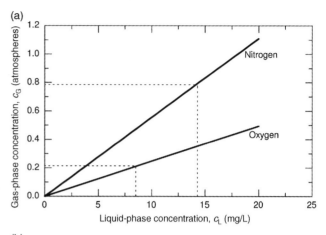

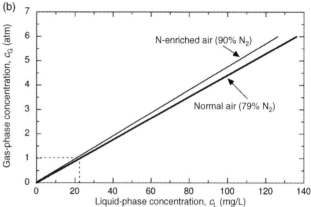

FIGURE 13-58. Henry's law relationship for air.

constituent i in the air. The expression on the far right emphasizes that air is considered to consist of only nitrogen and oxygen.

Figure 13-58b also illustrates that the elevated pressure of a saturator allows the dissolution of far more air into the water than can happen at normal atmospheric pressure. Since Henry's law is linear, a pressure in the saturator of 5 atm leads to an equilibrium value in the liquid that is five times the equilibrium value at 1 atm. The actual concentrations in both the gas and liquid phases in a saturator, however, are not necessarily those shown at equilibrium for normal air in Figure 13-58b, for two reasons. First, saturators are gas transfer devices and are not typically designed to achieve equilibrium; like the packed towers for gas transfer discussed in Chapter 6, equilibrium between the two phases is approached, but not necessarily achieved, as the liquid travels through the packing. The efficiency of a saturator is defined as the fractional elimination of the extent of disequilibrium; that is,

$$\eta_s = \frac{c_{L,s,\text{out}} - c_{L,s,\text{in}}}{c_{L,s}^* - c_{L,s,\text{in}}} \tag{13-68}$$

where the subscript s refers to the saturator (to differentiate it from other parts of a flotation system, subsequently), *in* and *out* refer to the saturator influent and effluent, respectively, c_L is the liquid-phase concentration of dissolved air, and $c_{L,s}^*$ is the liquid-phase concentration of air that would exist at equilibrium with the gas phase in the reactor. The equilibrium value, $c_{L,s}^*$, would be that associated with the inlet gas phase to the saturator.

The second reason that the liquid effluent from the saturator is not equilibrated with a hypothetical, pressurized air phase with the same composition as normal air is that oxygen is more soluble in water that nitrogen (i.e., its Henry's constant is lower). As a result, in the saturator, oxygen is preferentially dissolved (and thereby preferentially removed from the gas) compared to nitrogen. The liquid in the saturator approaches equilibrium with this N-enriched gas, not pure air. Therefore, even if gas/liquid equilibrium were attained, the equilibrium condition would not be the one indicated for normal air in Figure 13-58b. The fact that saturators are operated with the minimum possible air injection rate (so that there is no gaseous exit stream, and the cost associated with pressurizing the air is minimized) maximizes the difference between the composition of the gas inside the saturator and normal air. The lighter line in Figure 13-58b is the equilibrium line assuming that the air in the saturator is 90% N_2 and 10% O_2; the two lines shown represent the extremes that might be found in real systems.

We can analyze the gas transfer processes occurring at steady state in the saturator and the flotation tank by writing a set of mass balances on nitrogen and oxygen in various control volumes.[18] Throughout the analysis, we assume that the saturator is operated with no gaseous effluent stream. The first mass balance is written for a control volume (CV) that includes the air compressor, the saturator, and the liquid phase in the flotation tank. This CV has liquid entering in the influent suspension and exiting via the clean water and float withdrawal streams, and it has gas entering via the compressor and exiting by the formation of bubbles in the flotation tank. (Because the CV is defined to exclude the bubbles, bubble formation takes mass out of the CV.) The mass balances on N_2 and O_2 in this CV have the following form:

$$Q_L c_{L,in,i} + Q_{G,in} c_{G,in,i} = Q_L c_{L,out,i} + Q_{G,out} c_{G,out,i} \quad (13\text{-}69)$$

where i is either N_2 or O_2, and G,out is understood to refer to the bubbles. The two liquid effluent streams (float and clean effluent in Figure 13-57) are assumed to have the same composition, so they are combined and treated as a single stream with flow rate (Q_L) equal to the liquid influent flow rate. On the other hand, the gas flow rates in and out of the CV might differ, so $Q_{G,out}$ is not constrained to equal $Q_{G,in}$.

Assuming that gas/liquid equilibrium is achieved in the flotation tank, we can apply Henry's law to each species in that tank; that is,

$$c_{L,out,i} = H_{cc,i} c_{G,out,i} \quad (13\text{-}70)$$

We also know that the total pressure inside an average bubble equals the hydrostatic at mid-depth in the flotation tank. Assuming that the depth of the tank is small enough that its effect on pressure can be ignored, and applying Equation 5-19 to convert from partial pressures to mass-based concentrations in the gas phase, the total pressure can be related to the gas-phase concentrations by[19]

$$P_{tot} = \sum_i P_i = RT \sum_i \frac{c_{G,out,i}}{MW_i} \quad (13\text{-}71)$$

Presuming that the compositions and flow rates of the two influent streams are known, the five equations identified earlier (two with the form of Equation 13-69 [one each for N_2 and O_2], two with the form of Equation 13-70 and Equation 13-71) contain five unknowns (c_{L,out,N_2}, c_{L,out,O_2}, c_{G,out,N_2}, c_{G,out,O_2}, and $Q_{G,out}$). They can therefore be solved to find the compositions and flow rates of the effluent streams.

These five equations rely on two assumptions: gas/liquid equilibrium is attained in the flotation tank, and the system is at steady state. Note, however, that no assumption is made about the conditions in the saturator. As a result, the results are independent of how the saturator is operated (e.g., its pressure or efficiency) and of the composition of the pressurized solution exiting it.

In general, the solution to the five equations is not obvious and must be determined by calculation. However, one limiting case has an interesting and simple outcome that is worth noting. If the influent solution is in equilibrium with the atmosphere (both atmospheric N_2 and O_2), then the equations are satisfied if neither the gas nor the liquid undergoes any net change in composition or flow rate as it moves through the system. That is, in this limiting case, no net gas transfer occurs in the system, and the effect of pressurizing and then depressurizing the water and air is simply to convert

---

[18] In the literature, the mass balances are often written on air (i.e., on the combination of nitrogen and oxygen, considered as a single entity). However, in that case, different Henry's constants for air must be computed for the (initially) unknown compositions of the gas phases in the saturator and in the bubbles in the flotation tank. Under the circumstances, considering the two species independently, each with a known, constant Henry's constant, seems more straightforward.

[19] Considering i to include only N_2 and O_2 ignores the contribution of water vapor to the total pressure. At 25°C, the vapor pressure of water is 0.03 atm. This vapor pressure contributes to the total pressure, so, in an air bubble that has equilibrated with an aqueous solution at 25°C, the sum of the partial pressures of N_2 and O_2 is not P_{tot}, but only $P_{tot} - 0.03$ atm. This correction is minor for the current discussion and is therefore ignored.

EXAMPLE 13-18. Flotation is to be used to thicken the sludge from a biological treatment process operating at 25°C. At this temperature, H_{cc} values are 63.8 L_G/L_L for N_2 and 32.3 L_G/L_L for O_2. Thanks to the intense aeration in the aeration tank, the liquid phase of the sludge is equilibrated with atmospheric N_2. However, because biological activity consumes O_2 so rapidly, the sludge contains only 1 mg/L dissolved O_2. Air is supplied to the saturator at a rate corresponding to 0.04 L per liter of water fed to the flotation system.

Assuming that gas/liquid equilibrium is achieved in the flotation tank, that no oxygen is consumed in that tank, and that the system is at steady state, what volume of bubbles is generated per liter of water fed to the system, and what are the compositions of the liquid and bubbles in the tank?

Solution. Dividing through Equation 13-69 by Q_L, we can apply it to N_2 and O_2 to write

$$c_{L,in,N_2} + \frac{Q_{G,in}}{Q_L}c_{G,in,N_2} = c_{L,out,N_2} + \frac{Q_{G,out}}{Q_L}c_{G,out,N_2}$$

$$c_{L,in,O_2} + \frac{Q_{G,in}}{Q_L}c_{G,in,O_2} = c_{L,out,O_2} + \frac{Q_{G,out}}{Q_L}c_{G,out,O_2}$$

Because the solution in the flotation tank is equilibrated with the bubbles, the following Henry's law relationships apply in the tank:

$$c_{L,out,N_2} = \frac{c_{G,out,N_2}}{H_{cc,N_2}}$$

$$c_{L,out,O_2} = \frac{c_{G,out,O_2}}{H_{cc,O_2}}$$

Also, ignoring the effect of depth on the pressure inside the bubbles, we can write Equation 13-71 as

$$P_{tot} = 1.0\,\text{atm} = RT\left(\frac{c_{G,out,N_2}}{MW_{N_2}} + \frac{c_{G,out,O_2}}{MW_{O_2}}\right)$$

The gas-phase concentrations of N_2 and O_2 in the atmosphere can be written in mass-based units based on their partial pressures, as follows:

$$c_{G,in,N_2} = P_{N_2}\frac{MW_{N_2}}{RT} = (0.79\,\text{atm})\frac{28\,\text{g/mol}}{(0.082\,\text{atm}\,L_G/\text{mol K})(298\,\text{K})}(1000\,\text{mg/g}) = 905\,\text{mg}/L_G$$

$$c_{G,in,O_2} = P_{O_2}\frac{MW_{O_2}}{RT} = (0.21\,\text{atm})\frac{32\,\text{g/mol}}{(0.082\,\text{atm}\,L_G/\text{mol K})(298\,\text{K})}(1000\,\text{mg/g}) = 275\,\text{mg}/L_G$$

The dissolved N_2 concentration in the influent can then be computed based on the fact (assumption) that the solution is in equilibrium with atmospheric N_2:

$$c_{L,in,N_2} = \frac{c_{G,in,N_2}}{H_{cc,N_2}} = \frac{905\,\text{mg}/L_G}{63.8\,L_G/L_L} = 14.19\,\text{mg}/L_L$$

Values of c_{L,in,O_2}, $(Q_{G,in}/Q_L)$, H_{cc,N_2}, H_{cc,O_2}, and T are given in the problem statement, and MW_{N_2} and MW_{O_2} are known. With this information, the five equations characterizing the system can be solved, yielding the following results:

$$\frac{Q_{G,out}}{Q_L} = 0.0354\,(L_G/L_L)$$

$c_{L,out,N_2} = 15.48\,\text{mg/L}$ $c_{G,out,N_2} = 988\,\text{mg/L}$

$c_{L,out,O_2} = 5.60\,\text{mg/L}$ $c_{G,out,O_2} = 181\,\text{mg/L}$

Converting the N_2 and O_2 concentrations in the bubbles to partial pressures, we find that $P_{G,out,N_2} = 0.862\,\text{atm}$, and $P_{G,out,O_2} = 0.138\,\text{atm}$; that is, the gas in the saturator (and bubbles formed in the contact zone) are enriched in N_2 and depleted in O_2 compared to the atmosphere (i.e., compared to the gas that was injected into the saturator). Also, by comparing $(Q_{G,out}/Q_L)$ with $(Q_{G,in}/Q_L)$, we see that gas flow rate exiting the system as bubbles is approximately 12% less than the air flow injected into the system. The decrease in gas flow rate and the change in gas-phase composition are closely related. Most of the decrease in gas flow is due to the fact that the influent solution was undersaturated with O_2, so that a net transfer of O_2 from the gas phase into solution occurred as the two phases equilibrated. In addition, when oxygen exited the gas bubbles, the partial pressure of N_2 in the bubbles increased slightly (because its mole fraction increased, while the total pressure in the flotation tank remained the same as in the feed). As a result, a driving force was developed for transfer of N_2 out of the bubbles. The loss of both N_2 and O_2 from the gas phase accounts for the overall reduction in Q_G from the compressor to the flotation tank. Furthermore, a larger fraction of O_2 in the air had to transfer into solution to reach equilibrium than was the case for N_2, so the net change in composition of the bubble represented an enrichment in N_2. ∎

Bubble Formation

The preceding analysis yields the information about the gas flow rate (i.e., the volumetric bubble formation rate) in the flotation tank. However, more important parameters are likely to be the number concentration of bubbles and their volume fraction in the contact zone, where they are generated. The number concentration of bubbles formed in the flotation unit is of critical importance, especially in dilute suspension applications where efficient operation generally requires that the bubble concentration greatly exceed the particle concentration. For the thickening of concentrated suspensions, where the suspension tends to behave *en masse* in forming a solid/liquid interface, the bubble concentration need not greatly exceed the particle concentration, but the concentration of bubbles required is still much greater than for dilute suspensions.

Two equivalent approaches can be taken to determine the volumetric bubble formation rate in the contact zone of the flotation unit, one of which focuses on the gas phase and the other on the liquid phase. For both approaches, we consider only the limiting case where gas/liquid equilibrium is achieved within this zone. In both cases, we seek the volume of bubbles per volume of liquid and the number concentration of bubbles in that zone. We define the volume of bubbles per volume of liquid as a volumetric fraction:

$$\phi_{G,CZ} = \frac{V_{G,CZ}}{V_{L,CZ}}$$

The maximum number concentration and volume fraction of bubbles in the contact zone of the tank can be evaluated based on a mass balance around the zone where the water from the saturator mixes with the influent. In the limiting case where gas/liquid equilibrium is achieved within this zone, the generation of gas proceeds at the same rate as gas eventually exits the flotation tank, $Q_{G,out}$. Recognizing that each volume can be expressed as the product of the flow rate and detention time in a flowing system, the volume fraction of bubbles in the contact zone would be

$$\phi_{G,CZ} = \frac{V_{G,CZ}}{V_{L,CZ}} = \frac{Q_{G,out}\tau_{bb}}{V_{L,CZ}} = \frac{Q_{G,out}\tau_{bb}}{Q_{L,in}\tau} \quad (13\text{-}72)$$

where τ_{bb} and τ are the detention times in the contact zone of bubbles and water,[20] respectively. In flotation systems,

---

[20] τ is defined here as the ratio of the volume of the contact zone and the flow rate of the influent suspension (i.e., the water to be treated). As shown in Figure 13-56, the contact zone also receives water from the saturator, so a different definition of detention time that included that second flow rate would be possible. However, it is conventional in recycle systems to define τ as done here, and the mathematics of mass balances is carried out to account for the additional flow in terms of this definition.

τ_{bb} is necessarily less than τ, since both the liquid (with particles) and the bubbles are traveling upward and the bubbles have a rise velocity relative to water; correlations of the rise velocity with bubble diameter are given in Chapter 5.

The number of bubbles formed per unit time is found as the volumetric flow rate of the exiting gas divided by the volume of an (average) bubble:

$$\begin{bmatrix}\text{Number of bubbles}\\\text{formed per unit time}\end{bmatrix} = \dot{N}_{bb} = \frac{Q_{G,out}}{V_{bb}} = \frac{Q_{G,out}}{\pi d_{bb}^3/6} = \frac{6Q_{G,out}}{\pi d_{bb}^3} \quad (13\text{-}73)$$

As shown subsequently, the number concentration of bubbles in the contact zone is of interest in the modeling of particle removal. As with the volumetric fraction, this value depends on the detention time of bubbles in that zone:

$$n_{bb} = \frac{\dot{N}_{bb}\tau_{bb}}{V_{L,cz}} = \frac{\dot{N}_{bb}\tau_{bb}}{Q_{L,in}\tau} \quad (13\text{-}74)$$

As a practical matter, both ϕ_{bb} and n_{bb} are difficult to evaluate with much precision in real systems, for several reasons. First, the volume of the contact zone is not entirely distinct from the volume of the separation zone and, therefore, is not known well. Second, the detention time of the bubbles can be estimated from the depth and rise velocity, but the size of the bubbles (and therefore the rise velocity of a free bubble) is not known precisely. Third, many bubbles attach to particles (the whole point of flotation systems!) and, therefore, do not rise with nearly as high a velocity as a free, or unattached, bubble; estimates of τ_{bb} based on rise velocities of bubbles will be low. And lastly, the volumetric gas flow rate, the starting point of all the calculations, is based on an equilibrium calculation that represents an ideal. Despite these difficulties, the equations provide a useful guide for estimating ϕ_{bb} and n_{bb}, and are particularly valuable in investigating the changes that might be brought about by different designs or operating strategies.

The second approach to determining the amount of bubble formation is to consider the changes in the dissolved air in the liquid phase. When the water stream from the saturator enters the flotation unit, the pressure suddenly drops from that of the saturator to atmospheric pressure (again ignoring the depth effects on the pressure). As a result, the driving force for gas transfer favors driving the dissolved air out of solution and into the gaseous phase. The maximum possible rate of bubble formation, expressed as mass of air per unit time, is the product of the saturator flow and the difference in the saturator effluent concentration (which is the same thing as the contact

662   GRAVITY SEPARATIONS

zone inlet concentration) and the equilibrium concentration in the flotation unit (typically at 1 atm); that is,

$$\begin{bmatrix} \text{Mass/time of} \\ \text{bubbles created} \end{bmatrix}_{\max} = \dot{M}_{bb} = Q_{L,s}(c_{L,s,out} - c_{L,cz}) \quad (13\text{-}75)$$

The liquid flow to the saturator is often (though not always) recycled clean water from the flotation tank, in which case $Q_{L,s}$ can be expressed as $rQ_{L,in}$, where r is the recycle ratio. With equilibrium assumed at both the saturator effluent and in the contact zone, the liquid-phase concentrations can be written in terms of the pressure $c_L = (P_G/H_{air})$, where the appropriate H_{air} reflects the nitrogen and oxygen composition of the air in the saturator (and bubbles). Making these substitutions and assuming again that the pressure in the contact zone is 1 atm, Equation 13-75 can be written as

$$\dot{M}_{bb} = \frac{rQ_{L,in}}{H_{air}}(P_s - P_{cz}) = \frac{rQ_{L,in}}{H_{air}}(P_s - 1) \quad (13\text{-}76)$$

The volumetric formation rate of bubbles is the mass rate divided by the density of air. On a molar basis, the density of gas is calculated from the ideal gas law:

$$\frac{n_{mol}}{V_G} = \frac{P}{RT} \quad (13\text{-}77)$$

Converting to a mass basis (g/L) requires multiplying by the molecular weight; for any composition of air, we find

$$\rho_{air} = \frac{MW_{air}P_{CZ}}{RT} = \frac{(y_{O_2}MW_{O_2} + y_{N_2}MW_{N_2})P_{CZ}}{RT} \quad (13\text{-}78)$$

Substituting known quantities, the calculation becomes

$$\rho_{air} = \frac{(y_{O_2}(32\,\text{g/mol}) + y_{N_2}(28\,\text{g/mol}))(1\,\text{atm})}{(0.082\,\text{atm}\,L_G/\text{mol K})T} \quad (13\text{-}79)$$

Finally, we can find the volumetric rate of formation of bubbles:

$$Q_{G,bb} = Q_{G,out} = \frac{\dot{M}_{bb}}{\rho_{air}} \quad (13\text{-}80)$$

Substituting from Equations 13-76 and 13-79 results in

$$Q_{G,out} = \frac{M_{bb}}{\rho_{air}} = \frac{rQ_{L,in}(P_s - 1)(0.082\,\text{atm-}L_G/\text{mol-K})T}{H_{air}(y_{O_2}(32\pi\,\text{g/mol}) + y_{N_2}(28\,\text{g/mol}))(1\,\text{atm})} \quad (13\text{-}81)$$

This analysis based on the liquid phase has assumed that equilibrium is achieved in the saturator. If the efficiency of the saturator is accounted for, and if the efficiency is the same for both nitrogen and oxygen (as suggested by Haarhoff and Steinbach, 1996), the equation is modified by multiplying the saturator pressure (P_s) by the efficiency (η_s).

With this alternative means of calculating $Q_{G,out}$, Equations 13-73 and 13-74 can be used to find the rate of bubble formation and the bubble concentration.

Typically, the water stream from the saturator is released through a nozzle against an impinger, and the design of the nozzle and impinger affects both the rate of formation and size of the bubbles (Rykaart and Haarhoff, 1995). The bubbles formed in DAF typically have diameters ranging between 10 and 100 μm, with the vast majority <50 μm and the average in the range 30–40 μm (Han et al., 2002a; Rees et al.,1980; Fukushi et al., 1995). Han et al. (2002b) showed that the bubble size was independent of the saturator pressure above 3.5 atm, but was dramatically larger at lower pressures.

To complete the analysis, we assess the conditions in the saturator and in the water exiting that unit and entering the contact zone of the flotation tank. To do so, we write one more mass balance, this time choosing the saturator as the control volume. The analysis is closely analogous to the one conducted on the whole system and expressed mathematically in Equations 13-69 through 13-71. Once again, we have one liquid and one gaseous influent stream of known composition, and one effluent liquid stream. However, in this case, there is no effluent gas stream, and the liquid and gas do not necessarily reach equilibrium. The absence of an effluent gas stream is handled simply by setting $Q_{G,out}$ to zero in Equation 13-69, and the absence of gas/liquid equilibrium is accounted for by using Equation 13-68, with an assumed value of η_s in place of Equation 13-70. Using this approach, the total pressure required in the saturator (to achieve the goal of complete dissolution of the influent gas for a given $Q_{L,s}$) or the required $Q_{L,s}$ (to achieve the goal for a given P_{tot} in the saturator) can be determined.

■ **EXAMPLE 13-19.** The system described in Example 13-18 is operated with an influent flow rate, Q_L, of 50 L/s and a recycle ratio, r, of 70%. The saturator operates without any gaseous effluent and with an efficiency (η_s) of 85% for both N_2 and O_2. The nozzle and impinger used in the system generate bubbles with an average diameter of 30 μm. Assume that the ratio of bubble detention time to the liquid detention time in the contact zone is 0.3.

(a) Determine the compositions of the gas phase in the saturator and the liquid exiting the saturator. What is the total pressure in the saturator?

(b) What is the rate of bubble formation in the contact zone?

(c) What are the volume fraction of gas and the number concentration of bubbles in the contact zone?

Solution.

(a) We know the composition and flow rate of each stream entering the saturator, and we know the flow rate of the only exiting stream, so we can find its composition via mass balances on N_2 and O_2. Noting that the composition of the liquid entering the saturator is the same as that exiting the flotation tank (designated $c_{L,f,out}$ in the equations), the calculations for N_2 are

$$rQ_L c_{L,f,out,N_2} + Q_{G,s,in} c_{G,s,in,N_2} = rQ_L c_{L,s,out,N_2}$$

$$(0.70)(50\,L_L/s)(15.48\,mg/L_L)$$
$$+(0.04\,L_G/L_L)(50\,L_L/s)(905\,mg/L_L)$$
$$= (0.70)(50\,L_L/s)c_{L,s,out,N_2}$$

$$c_{L,s,out,N_2} = 67.2\,mg/L_L$$

The analogous calculation for O_2 indicates that $c_{L,s,out,O_2} = 21.3\,mg/L_L$. Knowing the composition of the liquid exiting the saturator, we can use a rearranged version of Equation 13-68 to find the concentration that would be in equilibrium with the gas in the saturator, and then the composition of the gas. For instance, again carrying out the calculations for N_2 (and again noting that the concentration entering the saturator is the concentration exiting the flotation tank):

$$c^*_{L,s,out,N_2} = \frac{c_{L,s,out,N_2} - c_{L,f,out,N_2}}{\eta_{s,N_2}} + c_{L,f,out,N_2}$$

$$= \frac{67.2\,mg/L_L - 15.48\,mg/L_L}{0.85} + 15.48\,mg/L_L$$

$$= 76.3\,mg/L_L$$

$$c_{G,s,N_2} = H_{N_2} c^*_{L,s,out,N_2} = (63.8\,L_L/L_G)(76.3\,mg/L_L)$$

$$= 4870\,mg/L_G$$

$$P_{s,N_2} = \frac{RT c_{G,s,N_2}}{MW_{N_2}}$$

$$= \frac{(0.082\,atm\,L_G/mol\,K)(298\,K)(4870\,mg/L_G)}{28,000\,mg/mol}$$

$$= 4.25\,atm$$

The analogous calculations for O_2 yield $c^*_{L,s,out,O_2} = 24.1\,mg/L_L$ and $P_{s,O_2} = 0.59\,atm$. The total pressure in the saturator is therefore 4.25 atm $+ 0.59\,atm$, or 4.84 atm. The gas is 87.7% N_2, due to the preferential dissolution of O_2.

(b) The rate of bubble formation can be determined from Equation 13-73:

$$\dot{N}_{bb} = \frac{6Q_{G,out}}{\pi d^3_{bb}} = \frac{6(Q_{G,out}/Q_{L,in})Q_{L,in}}{\pi d^3_{bb}}$$

$$= \frac{6(0.0354\,L_G/L_L)(50\,L_L/s)(1000\,cm^3/L_L)}{\pi (30\,\mu m)^3 (cm^3/10^{12}\,\mu m^3)}$$

$$= 1.25 \times 10^{11}/s$$

(c) The volume fraction of gas in the water is found from Equation 13-72:

$$\phi_{G,CZ} = \frac{Q_{G,out}\tau_{bb}}{Q_{L,in}\tau} = (0.0354\,L_G/L_L)(0.3)$$

$$= 0.011\,L_G/L_L, \text{ or approximately 1\%.}$$

The number concentration of bubbles is calculated from Equation 13-74:

$$n_{bb} = \frac{\dot{N}_{bb}\tau_{bb}}{Q_{L,in}\tau} = \frac{(1.25 \times 10^{11}/s)(0.3)}{50\,L_L/s} = 7.5 \times 10^8/L_L$$

∎

Flotation for Low Concentration Suspensions

We next turn our attention to the two common applications of flotation, and focus first on the low concentration suspensions. The primary objectives in designing and operating flotation units for dilute suspensions are (1) to ensure a high frequency of encounters (i.e., potential collisions) between the bubbles and the particles in suspension; (2) to ensure that those encounters actually do lead to attachment of the bubbles to the particles; and (3) to allow sufficient time and induce an appropriate flow pattern so that the bubble–particle flocs rise to the surface and are efficiently separated from the clean water below. The first two of these objectives relate to the performance of the contact zone, while the last objective relates to the performance of the separation zone. Two models of bubble–particle collisions have been proposed and are presented briefly here; modeling of the separation zone is presented subsequently.

Contact Zone Modeling

Calculations like those in Example 13-17 for the rise velocity of bubble–particle flocs show that, if a bubble with diameter d_{bb} attaches to a particle with diameter d_p, the rise velocity for the particle–bubble floc is almost independent of d_p if $d_p/d_{bb} < 0.5$. That is, for particles in that size range, the volume, density, and rise velocity are dominated by the properties of the bubble. When flotation is used under such conditions, a single bubble might collect several particles or small flocs as it rises toward the surface. On the other hand, for initial particle sizes that are significantly larger than the bubbles, the reverse is true: several bubbles must attach to a single particle (floc) to accomplish flotation.

The two models that have been proposed for the interactions between bubbles and particles in the contact zone of a flotation unit are referred to as the *flocculation model* and the *filtration model*. Both models assume that particles attach only to bubbles, not to other particles. A key difference between these models is in their assumptions about the d_p/d_{bb} ratio. The flocculation model assumes that this ratio is relatively large and that the particle–bubble composites consist of one particle and from one to many bubbles; in contrast, the filtration model assumes that this ratio is small, and that the composites consist of one bubble and from one to many particles. Thus, rather than representing opposing views of the processes occurring in the contact zone, these two models are complementary; both situations can arise in water and wastewater treatment processes, depending on the prior history (treatment) of the particles.

Flocculation Model Several Japanese coworkers (Tambo and Fukushi, 1985; Fukushi et al., 1995, 1998; Matsui et al., 1998) have modeled bubble–particle interactions based on a heterogeneous flocculation model; that is, flocculation between bubbles and particles. They assumed that collisions are caused (only) by turbulent fluid motion, and hence they use the collision frequency function for turbulent motion according to Levich (Equation 12-48); that is,

$$\beta_{bb,i} = \left(\frac{3\pi}{2}\right)\left(\frac{1}{15}\right)^{1/2}\left(\frac{\varepsilon}{v}\right)^{1/2}(d_{bb} + d_i)^3 \quad (12\text{-}48)$$

where the subscript i refers to a particle size class, v is the kinematic viscosity (cm^2/s) (equivalent to the absolute viscosity divided by the water density), and ε is the energy dissipation per unit time per unit mass (cm^2/s^3). All bubbles are assumed to be the same size, and no short-range forces are considered. Because this model is designed for systems in which the particles are substantially larger than the bubbles, the collision frequency function is essentially a function of the particle size only (i.e., $(d_{bb} + d_i)^3 \approx d_i^3$). $\beta_{bb,i}$ is assumed to remain constant for a given size particle regardless of how many bubbles area attached to it.

The assumption that particle–particle flocculation does not occur means that the total number of size-i particles remains constant while a particle is in the contact zone. However, size-i particles can successively accumulate bubbles in that zone, from zero up to some maximum number ($j_{max,i}$) of attached bubbles. The rate of this change can be characterized by a kinetic expression for the number concentration of size-i particles with any particular number (j) of bubbles attached. The concentration of size-i particles with j bubbles attached ($n_{i,j}$) can increase (by the attachment of another bubble to an i-sized particle with $j-1$ bubbles) or decrease (by the attachment of another bubble to form a unit with $j+1$ bubbles). The corresponding kinetic expression is

$$r_{i,j} = \alpha_{i,j-1}\beta_{bb,i}n_{bb}n_{i,j-1} - \alpha_{i,j}\beta_{bb,i}n_{bb}n_{i,j} \quad (13\text{-}82)$$

where it is understood that only the second term is applicable for the bare particle ($j = 0$).

The maximum number of bubbles that can attach to a particle is related to the cross-sectional areas of both the particle and the bubbles, and the collision efficiency of the first bubble ($\alpha_{i,0}$) with a bare particle:

$$j_{max,i} = \alpha_{i,0}\left(\frac{d_i}{d_{bb}}\right)^2 \quad (13\text{-}83)$$

The idea that the maximum number of bubbles that can attach to a particle depends on the collision efficiency of the first bubble might seem counterintuitive, but a plausible argument is as follows. Like solid particles, bubbles in a solution acquire a surface charge. Since all bubbles in the contact zone have the same composition, they acquire the same surface charge and repel one another. Therefore, to achieve good particle–bubble flocculation, the particles must have a surface charge of the opposite sign, or at least near neutral. As more bubbles attach to a particle, the charge on the particle–bubble composite approaches that of the bubbles alone, and it becomes increasingly difficult to add yet another bubble. The larger the repulsion, the smaller the value of $\alpha_{i,0}$, and therefore the smaller the value of $j_{max,i}$. This concept is accounted for in the model through the collision efficiency term, as follows:

$$\alpha_{i,j} = \alpha_{i,0}\left(1 - \frac{j}{j_{max,i}}\right) \quad (13\text{-}84)$$

Finally, the rate of change of the number concentration of free bubbles remaining in solution can be equated to the rate at which bubbles are lost in all of the particle–bubble collisions, considering all particle sizes; that is,

$$r_{bb} = -\sum_{\text{all } i}\beta_{bb,i}n_{bb}\left[\sum_{j=1}^{m_i}\alpha_{i,j}n_{i,j}\right] \quad (13\text{-}85)$$

Because the total number of particles with a certain (original) size stays constant, the equations are quite tractable and can be solved analytically by Laplace transform to yield $n_{i,j}$ at any time; that is, the number concentration of each size particle with each possible number of attached bubbles. In a continuous flow system, the time over which these rate equations are integrated is the rise time of a bubble through the contact zone, and the resulting distribution can then be used as input to a model for the performance of the separation zone.

Fukushi et al. (1995) used this model to simulate interactions in the contact zone of a flotation system receiving an input suspension that was well flocculated with alum, forming aluminum hydroxide flocs with sizes ranging from 100 μm to approximately 600 μm. Such particles or flocs can assimilate large number of bubbles and be made to float with a substantial rise velocity. The model results suggested that, during the residence time of a typical bubble, several bubbles can attach to each particle (floc). These results emphasize the need to supply sufficient bubbles to attach several bubbles to each particle and the value of flocculation pretreatment to reduce the number and increase the size of influent particles.

Filtration Model Edzwald and coworkers (Malley and Edzwald, 1991; Edzwald, 1995; Haarhoff and Edzwald, 2004) have evolved a model for collisions between particles and bubbles that is conceptually related to a filtration model. The model for filtration, known as the single spherical collector model, is explained more fully in Chapter 14; only the broad concepts are outlined here.

The basic idea of the model, as it applies to flotation, is that bubbles serve as "collectors" of small particles as the bubbles rise through the water. The model treats the interaction between particles of size i and bubbles as an elementary second-order reaction:

$$r_i = -k_2 n_i n_{bb} \quad (13\text{-}86)$$

The rate constant k_2 is envisioned to be the product of three terms:

- The volume of suspension swept out by a single bubble per unit time.
- The efficiency with which individual bubbles collect particles that are initially in the path that the bubbles take as they rise. This efficiency, referred to as the *single bubble removal efficiency* and symbolized as η, is defined as the ratio of the rate at which particles are predicted to collide with a bubble to the rate at which particles and bubbles approach one another.
- A collision efficiency factor, defined as the fraction of predicted collisions between bubbles and particles that results in attachment.

The first of these terms, the volume of suspension swept out by a single bubble per unit time, is the rise rate of a bubble, v_{bb}, multiplied by the cross-sectional area of a bubble, $A_{bb} = (\pi d_{bb}^2/4)$. The last term is given the symbol α_{pbb} and is defined just as in flocculation. In this model, it is generally taken to be an empirical factor that accounts for all otherwise unaccounted for phenomena as particles and bubbles approach one another. In the hydrodynamic modeling incorporated into bubble–particle interactions, the streamlines of flow around a rising bubble are accounted for, but the short-range forces of van der Waals attraction and electrostatic attraction are not. Han et al. (2001) have used trajectory analysis to calculate theoretical values of α_{pbb} that account for the effects of particle and bubble size.

The middle term in the calculation—the single bubble removal efficiency—is in many ways the heart of this model. Much as in flocculation, trajectory analyses can be carried out to determine the rate at which particles are expected to collide with a bubble as both move in suspension, normalized by the concentrations of bubbles and particles. Each bubble is assumed to rise vertically at its (Stokes) rise velocity, whereas the particles can move by Brownian motion, fluid motion, and settling. Particle–bubble collisions can occur through each of these types of particle motion, and the resulting single bubble removal efficiencies can then be found. The results are shown below.

For collisions caused by Brownian motion of the particle,

$$\eta_{BM} = 0.9 \left(\frac{kT}{\mu d_p d_{bb} v_{bb}} \right)^{2/3} = 6.18 \left(\frac{kT}{g \rho_L d_p} \right)^{2/3} \left(\frac{1}{d_{bb}^2} \right) \quad (13\text{-}87)$$

where the latter term incorporates the bubble rise velocity as calculated from Stokes' law (Equation 13-8) and the approximation that $(\rho_L - \rho_{bb}) \approx \rho_L$. According to this equation, the rate of collisions by Brownian motion decreases as either the particle size or bubble size decreases.

For collisions caused by particle motion due to fluid motion (called *interception* of a particle by a collector in the filtration model, and hence the subscript I),

$$\eta_I = \left(\frac{d_p}{d_{bb}} + 1 \right)^2 - \frac{3}{2}\left(\frac{d_p}{d_{bb}} + 1 \right) + \frac{1}{2}\left(\frac{d_p}{d_{bb}} + 1 \right)^{-1} \quad (13\text{-}88)$$

For collisions caused by sedimentation, accounting for both the rising bubble and the falling particle,

$$\eta_S = \frac{(\rho_p - \rho_L) g d_p^2}{18 \mu v_{bb}} = \frac{\rho_p - \rho_L}{\rho_L - \rho_{bb}} \left(\frac{d_p}{d_{bb}} \right)^2 \quad (13\text{-}89)$$

Finally, as in flocculation modeling, the different mechanisms are assumed to act independently, so that the total single bubble removal efficiency is found as the sum of the efficiencies associated with the three mechanisms:

$$\eta_T = \eta_{BM} + \eta_I + \eta_S \quad (13\text{-}90)$$

A fourth mechanism leading to collisions, accounting for the inertia of particles and bubbles, has also been proposed, but it is generally much less significant than the mechanisms described earlier, and hence is ignored here. The single bubble removal efficiency for sedimentation is also generally considerably less than that of interception and probably could be ignored as well. The results of calculations using

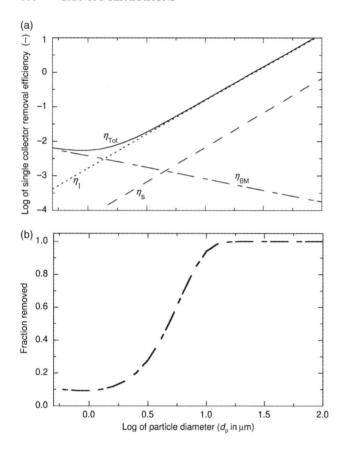

FIGURE 13-59. Results of the filtration model of the contact zone. (Conditions: $\phi_{bb} = 3 \times 10^{-3}\,L_G/L_L$, $d_{bb} = 30\,\mu m$, $T = 293\,K$, $\tau_{CZ} = 4\,min$, $\rho_p = 1.06\,g/cm^3$, $\alpha_{pbb} = 1$.)

Equations 13-88–13-90 are shown in Figure 13-59 for reasonable values of all of the variables. The results in this figure make it clear that particles with a diameter around 1 μm (log $d_p = 0$) are the most difficult to collect, with smaller particles being captured more efficiently by Brownian motion and larger particles by interception.

Putting all these terms together results in the following equation for the rate constant of Equation 13-86:

$$k_2 = \alpha_{pbb}\eta_T v_{bb}\frac{\pi d_{bb}^2}{4} \qquad (13\text{-}91)$$

In using this model, it is common to use the volume concentration of bubbles instead of the number concentration; with the bubbles considered uniform spheres, the volume concentration is the product of the number concentration and $\pi(d_{bb}^3/6)$. This quantity is considered a constant in the model. Also, bubbles are envisioned to be able to collide with and collect multiple particles. Incorporating the idea of a constant bubble volume concentration turns the rate equation into a pseudo-first-order equation. Furthermore, if the flow is considered plug flow in the contact zone (so that $(dn_i/dt) = r_i$ in a differential control volume along a vertical pathway in that zone), the rate equation can be integrated directly to yield the ratio of the effluent to the influent concentration of particles of size i as follows:

$$\frac{n_{i,out}}{n_{i,in}} = \exp\left(-\frac{3}{2}\frac{\alpha_{pbb}\eta_T \phi_{bb} v_{bb}\tau_{cz}}{d_{bb}}\right) \qquad (13\text{-}92)$$

The total concentration of particles that remain unattached to a bubble can then be found by summing over all the particle sizes:

$$n_{T,out} = \sum_{all\ i} n_{i,in}\exp\left(-\frac{3}{2}\frac{\alpha_{pbb}\eta_{T,i}\phi_{bb} v_{bb}\tau_{cz}}{d_{bb}}\right) \qquad (13\text{-}93)$$

Example results from calculations using this model of particle–bubble interactions in the contact zone are presented in Figure 13-59. The comparison of the three mechanisms of contact in Figure 13-59a suggests that, for all sizes greater than approximately 1 μm (log $d_p = 0.0$), interception is the dominant mechanism; particles of smaller sizes are most likely to collide with bubbles by Brownian motion, but the collision frequency is low. The fractional removal of particles (i.e., the fraction that collides with and attaches to bubbles in the contact zone) is shown in Figure 13-59b and reflects the same idea: virtually all particles greater than 10 μm (log $d_p = 1.0$) collide with bubbles, and very few particles smaller than 1 μm (log $d_p = 0.0$) are collected by bubbles. The collection efficiency dramatically rises in the narrow particle size range in between. While the results depend on the specific conditions indicated on the figure, the model is relatively insensitive over normal ranges of most of the parameters, so these results can be taken as reasonably general.

The interpretation of these results is clear—the effective use of flotation for dilute suspensions requires pretreatment by flocculation to dramatically reduce the number concentration of very small particles, but flocculation need not produce very large flocs. For this reason, Edzwald and coworkers (Edzwald et al., 1992, 1999; Valade et al., 1996) have argued that flocculation with metal coagulants (alum or iron using the sweep floc mechanism) can be done with quite short detention times (5–10 min) when preceding flotation units rather than the 20–60 min normally provided before sedimentation.

Both models for the contact zone—the flocculation model proposed by Tambo and coworkers and the filtration model proposed by Edzwald and coworkers—depend on favorable short-range interactions between particles and bubbles. Both models use the α_{pbb} term to describe these interactions. Han and coworkers have focused on this aspect of flotation through both mathematical modeling (Han et al., 2001; Han, 2001) and experimental investigations of the surface charge characteristics of bubbles (Han and Dockko, 1999). Their results indicate that particle–bubble interactions can be very favorable, as one might expect, if the pretreatment results in the particles having a positive charge and the bubbles a negative charge. Their measurements suggest that bubbles have a ζ-potential of approximately -25 mV at near-neutral pH. Fukushi et al. (1995) suggested that α_{pbb} has a maximum value in the range of 0.3–0.4.

Comparison of the Contact Zone Models The two models for the collisions of bubbles and particles in the contact zone have both been reported by their originators to be successful in modeling flotation behavior. Despite apparent conceptual differences, the two models share many important characteristics. Both models take a second-order reaction approach, and both focus entirely on particle–bubble collisions and ignore any further particle–particle flocculation that could occur in the contact zone.

Both models also essentially agree that the bubble–particle interactions are primarily caused by the relative motion brought about by the fluid motion; however, the fluid motion envisioned in the flocculation model is turbulence induced by the bubbles, and that in the filtration model is considered laminar around the rising bubble. The flocculation model could easily be modified to account for the other mechanisms of contact (Brownian motion, sedimentation); the filtration model ostensibly includes these mechanisms but, in most cases, they are computed to be negligible in comparison to interception.

As noted earlier, a substantial difference that underlies the bases of the two models is the relative sizes of the bubbles and the particles. In the flocculation model, it is generally assumed that the particles (flocs, really) are large in comparison to the bubbles; hence, several bubbles are envisioned to be able to collide with a single particle. This view leads to an accounting based primarily on the number of (unattached) bubbles. In the filtration model, it is generally assumed that the particles are small in relation to bubbles; hence, several particles can be attached to a single bubble, and the primary accounting is of unattached particles. These different views lead to different expectations for pretreatment: the flocculation model suggests that extensive flocculation of particles to create large flocs is valuable, whereas the filtration model suggests that more moderate flocculation (to reduce the number concentration of very small particles but not create very large flocs) is sufficient. In drinking water applications where flotation is used after sweep flocculation (precipitation of aluminum or ferric hydroxide), recent designs use a short flocculation period to make relatively small flocs. In that case, the filtration model is most relevant. In wastewater applications where flotation is used in lieu of sedimentation for the separation of biological flocs after activated sludge treatment, the flocs are quite large, and so the flocculation model is most relevant.

Separation Zone Modeling

As in sedimentation, the most common approach to designing the solid/liquid separators used in flotation has been based on hydraulic loading (the equivalent of overflow rate, as the flow rate divided by the surface area of the separation zone of the tank). In flotation, the effluent flow exits at or near the bottom of the separation zone, either at the end of that zone (in horizontal tanks) or around the perimeter (in circular tanks). An increasingly common design for dilute suspensions is to build a false floor with many holes to use the available area uniformly and avoid areas of high velocity that could impede the solid/liquid separation. An extension of that idea is to stack a flotation unit above a deep bed filter; this design is also becoming more common because of the reduced footprint requirements and its intrinsic simplicity. The hydraulic loading rates (flow rate per area) in these units are generally set by the filtration rate, since it is usually lower than the common values for flotation units.

Theoretically, particles with a rise rate (calculated as in Example 13-17) greater than the hydraulic loading will all be caught. In practice, it has been found that excellent separation is achieved at even higher hydraulic loading rates. Odegaard (2001) reported using flotation with hydraulic loading rates in the range of 5–15 m/h in lieu of primary sedimentation in wastewater treatment plants, after adding alum for phosphate precipitation. With proper pretreatment (to achieve particle flocs in the range of 10 μm or larger and favorable bubble–particle interactions), hydraulic loading rates of 15 m/h are now common for algal-laden lake waters (Haarhoff and Edzwald, 2004), and rates in the range of 20–40 m/h are used occasionally. Calculations of rise velocities indicate that bubbles must be in the range of 80–100 μm to achieve such high rise velocities. While bubbles in that size range have been reported, many bubbles are smaller than that size range, and therefore have smaller rise velocities. Thus, it appears that the design hydraulic loading rates can be higher than the rise rates of many of the individual bubbles in the system and still achieve excellent removal. Reported hydraulic loading rates are usually based on the entire area of the flotation unit, and not just the separation zone area; using the smaller area of the separation zone would result in still higher values.

The success of flotation even when the hydraulic loading rate is higher than the bubble rise rate apparently occurs because of a density effect that aids the solid/liquid separation. Lundh et al. (2001) made simultaneous measurements of the air (bubble) content (using a turbidimeter, since bubbles deflect light) and fluid velocities (using an acoustic Doppler velocimeter). They found evidence of flow segregation—the bubble laden part of the water contained most of the particles and separated from the denser bubble-free water. The less dense part of the water floated on top of the denser water, effectively reducing the overflow rate, since only a fraction of the flow (containing most of the particles) participated in the effective zone for flotation. In addition, the portion with the high bubble concentrations exhibited nearly perfect horizontal flow near the top surface of the water, minimizing the distance that particles had to move upward to be collected. Thus, they found that, in addition to being essential for forming particle–bubble flocs, the bubbles helped establish and maintain a flow pattern that enhanced particle removal. With increasing hydraulic loading rates, bubble concentrations had to be increased to maintain this effect.

Sludge Thickening by Dissolved Air Flotation

The second major application of DAF is for the thickening of sludge. The most common use of this application is to thicken biosolids from the activated sludge process, but it has also been used for metal hydroxide precipitates, combined primary and secondary sludge at wastewater treatment plants, and other concentrated suspensions. This application grew out of the mining industry, where it has been used extensively to concentrate and, in some cases, separate ores. The fundamental principles are the same in this application as in the flotation of dilute suspensions, but the differences of particle concentration bring differences in operation as well. The major differences from the flotation of dilute suspension can be summarized as follows:

- Contact between bubbles and particles, a major issue in dilute suspensions, is not an issue in thickening, because the particle concentration is so high. As a result, compressed air is introduced directly into the suspension influent in the entry pipe or in the central distribution core in a circular flotation unit, and no separate contact zone is required.
- A major design and operational variable for flotation thickeners is the air-to-solids (A/S) ratio: the mass of air supplied per unit time divided by the mass of solids (particles) supplied per unit time. This same parameter could be used for dilute suspensions, where the requirement is to supply sufficient bubbles to meet the particle demand, but only in thickening is it considered a major operational variable. The rise rate of the solid–liquid interface in batch tests, the concentration of thickened sludge achieved in continuous flow thickeners, and the capture efficiency of the solids all increase dramatically with increasing A/S ratio at low levels of this parameter, but these effects are asymptotic at some higher A/S ratio (Wood and Dick, 1973; Gulas et al., 1978).
- The top of the float rises above the water level, and the distance above the water level can be controlled by the designer. Water, therefore, drains out of the portion of the float that is above the water level. This drainage leads to substantially higher solids concentrations than can be achieved in conventional gravity thickeners.

The air-to-solids (mass) ratio used in flotation thickeners for biosolids from activated sludge plants typically varies in the range of 0.002–0.05. As noted earlier, increasing this ratio over some range increases the float solids concentration, but this improvement levels off at higher A/S values. The value of the A/S ratio at which the leveling off occurs depends on sludge characteristics such as particle size, density, and surface charge, but these relationships have not been delineated. Gehr and Henry (1978) defined a *thickening parameter* for batch flotation tests that accounted for both the A/S ratio and the volume of particles in the test. They claimed that it was a more robust parameter to describe flotation thickening, but it has not been generally adopted.

In designing flotation thickeners, one can perform batch tests using a laboratory-scale apparatus that allows variation of the pressure of the saturator. Wood and Dick (1973) described an apparatus that allowed for the simultaneous introduction of the suspension and the pressurized saturated stream, as in a full-scale unit. They found that this approach reduced some of the drawbacks of previous-batch testing devices in which the saturated air stream entered at the bottom of a stationary batch of the suspension. Nevertheless, they concluded that, as in gravity thickening, batch tests should be carried out at depths that are as close to full-scale as possible, and with column diameters that are greater than approximately 20 cm to avoid wall effects. The relatively large scale of the testing apparatus required to obtain useful design data (meaning that a large volume of sludge is required) has resulted in few reports of laboratory-scale data. Rather, design of flotation units is typically based either on pilot-scale units or, when the application is similar to existing situations, experience.

Based on the dissertation of Wood (1970), both Dick (1972) and Vesilind (1979) suggested that the design and operation of flotation thickeners be based on flux theory, identically as gravity thickeners, except that flux curves should be developed at various A/S ratios. In this way, one could determine the area required to achieve a desired float concentration (not considering drainage) for a given mass application rate as a function of the A/S ratio. Drainage would then be accounted for separately, and would increase the float concentration that could be achieved. As in gravity thickening, the required area would be set by the capacity of the sludge to transmit solids. Theoretically, this approach appears to be correct, but no reports in the literature of a design carried out this way are available. Wood (1970) included a batch flux plot obtained in a laboratory flotation unit but noted the difficulty of obtaining data at solids concentrations that exceed the limiting solids concentration. This difficulty stems in part from the high saturation pressures required to maintain a constant A/S ratio without excessive dilution of the suspension.

As in gravity thickening, polymers are often used to increase the solids flux that can be achieved in flotation. Gehr and Henry (1978) and Sugahara and Oku (1993) reported that cationic polymer addition enhanced the flotation performance by increasing the float solids concentration achieved at otherwise identical conditions. The mechanism of destabilization that led to larger particle sizes and better flotation performance were apparently adsorption and interparticle bridging.

Despite the apparent lack of a coherent theory for the design and operation of flotation thickeners, they are built and operated successfully in practice. This success can be attributed to a combination of conservative designs based on prior experience, an inherent robustness of the process, and the availability of several operating variables and strategies (control of the A/S ratio, addition of polymers, control of float depth above the water, speed of sludge withdrawal mechanisms) that can improve the ability of a flotation thickener to handle a given solids loading rate.

13.10 SUMMARY

Gravity separations by sedimentation and, to a much lesser extent, flotation are responsible for the vast majority of the ultimate separation of contaminants from water and wastewater. In many cases, contaminants that are not in particulate form in the influent to a treatment system are converted to particles (chemically or biologically) before the gravity separation system. Hence, most water and wastewater treatment systems contain one or more gravity separation tanks. Gravity separation is not universal, however; when a water or wastewater has a low particle concentration, gravity separation is relatively ineffective and unnecessary. In such situations, solid/liquid separation is accomplished by deep bed filtration or membrane filtration, as explained in subsequent chapters.

Gravity separation is primarily driven by Stokes' law; more than 150 years ago, Stokes found that, for a single particle in laminar conditions, the settling velocity (or rise velocity in flotation) was proportional to the difference in density between the particle and surrounding fluid and to the square of the size of the particle, and inversely proportional to the viscosity of the fluid. For the particles of primary interest in most water and wastewater applications, laminar conditions do apply. However, virtually, all systems of interest have sufficient particle concentration that particle interactions add complexity to Stokes' findings.

Most particles in gravity separation systems in water and wastewater treatment have a tendency to flocculate; that is, they form flocs as particle collisions occur during sedimentation or flotation. This fact aids solid/liquid separation, as the resulting flocs have higher settling velocities than the original particles. In general, sedimentation would be far less successful were it not for the flocculation that occurs before and during the process, and flotation would be completely unsuccessful were it not for the flocculation of air bubbles with particles. Flocculation complicates the mathematical analysis of gravity separation systems, however, so it is not uncommon for quantitative analysis to focus on nonflocculent systems, with the benefits of flocculation being considered qualitatively or semiquantitatively. Nevertheless, progress is continually made on the quantitative analysis of flocculent suspensions.

For sedimentation of dilute suspensions in continuous flow systems, the mathematical analysis presented in this chapter considered the interaction of the cumulative settling velocity distribution ($f(v)$) and the reactor settling potential function ($F(v)$, a characteristic of the reactor design). Theoretical expressions for $F(v)$ for all of the common ideal reactors were presented, and some characteristics of nonideal flow systems were also investigated. Common reactors include horizontal flow reactors (both rectangular and circular), vertical flow reactors, and mixed horizontal and vertical flow reactors (including tube settlers); the reactor settling potential functions for these reactors were presented, and they explain the major characteristics of sedimentation in continuous flow systems. In all cases, two velocities proved important: the overflow rate (v_{OR}, the effluent flow rate divided by the horizontal surface area of the reactor) and the critical settling velocity (v^*, the minimum settling velocity for which the removal efficiency is 100%). For the simplest ideal cases, these two characteristics are the same, but more complex (and more efficient) reactors like tube settlers are characterized by v^* values less than v_{OR}.

The cumulative settling velocity, $f(v)$; that is used in the analysis is defined as the fraction of the influent suspension that has an average settling velocity $\leq v$ while in the reactor. For nonflocculent (Type I) suspensions, the settling velocity of particles is unchanged while the particles are in the reactor, and so the average settling velocity for any particle is the same as its settling velocity in the influent. For such suspensions, $f(v)$ can be determined experimentally from batch settling tests, or calculated from the influent particle size distribution. Therefore, the removal efficiency can be predicted for Type I suspensions in many types of reactors, because both $f(v)$ and $F(v)$ can be determined or are known.

For Type II suspensions, the average settling velocity of particles changes while particles are in a reactor because flocs are formed, and the rate of formation of flocs depends on the concentration, size distribution, and surface chemistry of the particles and the flow pattern within the reactor. The concentration and size distribution are influenced by the water flow and reactor itself, and so $f(v)$ is influenced by the reactor size and flow pattern. For such suspensions, $f(v)$ can be easily determined in a batch test only in the case of ideal horizontal plug flow, since time in a batch test is equivalent to location in a plug flow system operating at steady state. The ability to predict behavior of flocculent suspensions in gravity separation tanks is, therefore, not as great as for nonflocculent suspensions. Theoretically, a thorough analysis is possible, based on a combination of Smoluchowski's equation for flocculation, Stokes' equation for sedimentation, and a computational fluid dynamics solution for the fluid flow.

Suspensions with high particle concentrations undergo thickening under the influence of gravity. In thickening, the flow of water through the solids is hindered by the large number of particles in close proximity, and so the macroscopic behavior of suspensions undergoing thickening is considerably different from that of dilute suspensions. Specifically, particles appear to settle *en masse* with little or no movement of one particle past another. The analysis of thickening, both for batch and continuous flow systems, is based on the solids flux through any horizontal plane.

While sedimentation is far more common, flotation is also used frequently to separate particles from water and can be far more effective than sedimentation when the particle densities are close to that of water. Flotation is used both for relatively dilute suspensions and for concentrated suspensions. For both applications, some water (usually a recirculation stream) is saturated with air at relatively high pressure so that, when that water is released into the lower pressure zone of a reactor, bubbles form as the solution approaches equilibrium between the liquid and gas phases. For dilute suspensions, achieving excellent contact between bubbles and particles has been the subject of substantial mathematical modeling. Once a bubble is attached to the particle, the rise rate is generally quite high, so that removal is not difficult. For concentrated suspensions, contact between particles and bubbles is essentially unavoidable, and the issue is simply to provide sufficient bubble volume to float the particles and to remove the subsequent sludge layer (the "float"). In drinking water treatment, the most common application of flotation is when the source water has a relatively low particle concentration at all times and when many of those particles (at least during some period of the year) are algae. In wastewater treatment, flotation has been used primarily for thickening of biological sludges or chemical sludges from the precipitation of metals as hydroxides.

REFERENCES

Adams, E., and Rodi, W. (1990) Modeling flow and mixing in sedimentation tanks. *J. Hydraul. Eng.*, 116 (7), 895–913.

Bird, R. B., Stewart, W. E., and Lightfoot, E. N. (2006) *Transport Phenomena*, 2nd edn., John Wiley & Sons, New York.

Brinkman, H. C. (1947) A calculation of the viscous force exerted by a flowing fluid on a dense swarm of particles. *App. Sci. Res. A*, 1, 27–34.

Brown, P. P., and Lawler, D. F. (2003) Sphere drag and settling velocity revisited. *J. Environ. Eng. (ASCE)*, 129 (3), 222–231.

Camp, T. R. (1946) Sedimentation and the design of settling tanks. *Trans. ASCE*, 111 (3), 895–936.

Dick, R. I. (1972) Sludge treatment, Ch. 12. In Weber, W. J. Jr., (ed.), *Physicochemical Processes for Water Quality Control*. Wiley Interscience, New York.

Dick, R. I. (1989) Fundamental accounting of the performance of thickeners. *Proceedings, first annual meeting of the American Filtration Society*, Ocean City, Maryland, March, 1988, pp. 299–304.

Dick, R. I., and Ewing, B. B. (1967) Evaluation of activated sludge thickening theories. *J. Sanit. Eng. Div. (ASCE)*, 83 (SA4), 9–30.

Dobbins, W. E. (1944) Effect of turbulence on sedimentation. *Trans. ASCE*, 109 (1), 629–656.

Edzwald, J. K. (1995) Principles and applications of dissolved air flotation. *Water Sci. Technol.*, 31 (3–4), 1–23.

Edzwald, J. K., Tobiason, J. E., Amato, T., and Maggi, L. J. (1999) Integrating high rate dissolved air flotation technology into plant design. *JAWWA*, 91 (12), 41–53.

Edzwald, J. K., Walsh, J. P., Kaminski, G. S., and Dunn, H. J. (1992) Flocculation and air requirements for dissolved air flotation. *JAWWA*, 84 (3), 92–100.

El Baroudi, H. M. (1969) Characterization of settling tanks by eddy diffusion. *J. Environ. Eng., ASCE*, 95 (SA3) 527–544.

Fair, G. M., Geyer, J. C., and Okun, D. A. (1968) *Water and Wastewater Engineering*, Vol. 2, *Water Purification and Wastewater Treatment and Disposal*, Wiley, New York.

Fukushi, K., Matsui, Y., and Tambo, N. (1998) Dissolved air flotation: Experiments and kinetic analysis. *J. Water Supply Res. Technol.–Aqua* 47, 76–86.

Fukushi, K., Tambo, N., and Matsui, Y. (1995) A kinetic model for dissolved air flotation in water and wastewater treatment. *Water Sci. Technol.*, 31 (3–4), 37–47.

Gehr, R., and Henry, J. G. (1978) Measuring and predicting flotation performance. *J. Water Pollut. Control Fed.*, 50 (2), 203–215.

Giacomo, P. (1982) Equation for the determination of the density of moist air. *Metrologia*, 18 (1), 33–40.

Gulas, V., Benefield, L., and Randall, C., (1978) Factors affecting the design of dissolved air flotation systems. *J. Water Pollut. Control Fed.*, 50, 1835–1840.

Haarhoff, J., and Edzwald, J. K. (2004) Dissolved air flotation modeling: insights and shortcomings. *J. Water Supply: Res. Technol.—Aqua*, 53 (3), 127–150.

Haarhoff, J., and Steinbach, S. (1996) A model for the prediction of the air composition in pressure saturators. *Water Res.*, 30, 3074–3082.

Haider, A., and Levenspiel, O. (1989) Drag coefficient and terminal velocity of spherical and nonspherical particles. *Powder Technol.*, 58, 63–70.

Han, M. Y. (2001) Modeling of DAF: the effect of particle and bubble characteristics. *J. Water Supply: Res. Technol.—Aqua*, 51, 27–34.

Han, M. Y., and Dockko, S. (1999) Zeta potential measurement of bubbles in DAF process and its effect on the removal efficiency. *Water Supply*, 17 (3/4), 177–182.

Han, M. Y., Kim, W. T., and Dockko, S. (2001) Collision efficiency factor of bubble and particle (α_{bp}) in DAF: theory and experimental verification. *Water Sci. Technol.*, 43 (8), 139–144.

Han, M. Y., Park, Y. H., and Yu, T. J. (2002a) Development of new method of measuring bubble size. *Water Sci. Technol.: Water Supply*, 2 (2), 77–83.

Han, M. Y., Park, Y. H., Lee, J., and Shim, J. S. (2002b) Effect of pressure on bubble size in dissolved air flotation. *Water Sci. Technol: Water Supply*, 2, 41–46.

Happel, J. (1958) Viscous flow in multiparticle systems: Slow motion of fluids relative to beds of spherical particles. *AIChE J.*, 4 (2), 197–201.

Hazen, A. (1904) On sedimentation. *Trans. ASCE*, 53 (2), 45–71.

Jenson, V. G. (1959) Viscous flow round a sphere at low Reynolds number (≤ 40). *Proc. R. Soc., Ser. A*, 249 (1258), 346–366.

Karl, J. R., and Wells, S. A. (1999) Numerical model of sedimentation/thickening with inertial effects. *J. Environ. Eng. (ASCE)*, 125 (9), 792–806.

Keinath, T. M. (1985) Operational dynamics and control of secondary clarifiers. *J. Water Pollut. Control Fed.*, 57 (7), 770–776.

Knocke, W. R. (1986) Effects of floc volume variations on activated sludge thickening characteristics. *J. Water Pollut. Control Fed.*, 58 (7), 784–791.

Kynch, G. J. (1952) A theory of sedimentation. *Trans. Faraday Soc.*, 48, 166–176.

Lawler, D. F., O'Melia, C. R., and Tobiason, J. E. (1980) Integral water treatment plant design: from particle size to plant performance. Ch. 16. In Kavanaugh, M., and Leckie, J. O. (eds.), *Particulates in Water, Advances in Chemistry Series*, 189, American Chemical Society, 353–388.

Lev, O., Rubin, E., and Sheintuch, M. (1986) Steady state analysis of a continuous clarifier-thickener system. *AIChE J.*, 32 (9), 1516–1525.

Lundh, M., Jönsson, L., and Dahlquist, J. (2001) The flow structure in the separation zone of a DAF pilot plant and the relation with bubble concentration. *Water Sci. Technol.*, 43 (8), 185–194.

Malley, J. P., Jr., and Edzwald, J. K. (1991) Concepts for dissolved air flotation treatment of drinking waters. *J. Water Supply: Res. Technol.—Aqua*, 40, 7–17.

Matsui, Y., Kukushi, K., and Tambo, N. (1998) Modeling, simulation and operational parameters of dissolved air flotation. *J. Water Supply Res. Technol.–Aqua* 47, 9–20.

Michaels, A. S., and Bolger, J. C. (1962) Settling rates and sediment volumes of flocculated kaolin suspensions. *Ind. Eng. Chem. Fund.*, 1 (1), 24–33.

Odegaard, H. (2001) The use of dissolved air flotation in municipal wastewater treatment. *Water Sci. Technol.*, 43 (8), 75–81.

Ostendorf, D. W., and Botkin, B. C. (1987) Sediment diffusion in primary shallow rectangular clarifiers. *J. Environ. Eng. (ASCE)*, 113 (3), 595–611.

Ramaley, B. L., Lawler, D. F., Wright, W. C., and O'Melia, C. R. (1981) Integral analysis of water plant performance. *J. Environ. Eng. Div. (ASCE)*, 107 (EE3), 547.

Rees, A. J., Rodman, D. J., and Zabel, T. F. (1980) Dissolved air flotation for solid-liquid separation. *J. Separation Process Technol.*, 1 (3), 19–23.

Richardson, J. F., and Zaki, W. N. (1954) The sedimentation of a suspension of uniform spheres under conditions of viscous flow. *Chem Eng. Sci.*, 3 (2), 65–73.

Rykaart, E. M., and Haarhoff, J. (1995) Behaviour of air injection nozzles in dissolved air flotation. *Water Sci. Technol.*, 31 (3–4), 25–35.

Sanin, F. D., Clarkson, W. W., and Vesilind, P. A. (2011) *Sludge Engineering: The Treatment and Disposal of Wastewater Sludges*, DEStech Publications, Lancaster, PA.

Schamber, D. R., and Larock, B. E. (1981) Numerical analysis of flow in sedimentation basins. *J. Hydraul. Div. ASCE*, 107 (5), 575–591.

Scott, K. J. (1968) Thickening of calcium carbonate slurries. *Ind. Eng. Chem. Fund.*, 7 (3), 484–490.

Stokes, C. G. (1851) On the effect of internal friction on the motion of pendulums. *Trans. Camb. Phil. Soc.*, 9 (2), 8–106. (republished in 1980 in Stokes, C. G., *Mathematical and Physical Papers, Volume 1.* Cambridge University Press, Cambridge, England.).

Sugahara, M., and Oku, S. (1993) Parameters influencing sludge thickening by dissolved air flotation. *Water Sci. Technol.*, 28 (1), 87–90.

Tambo, N., and Fukushi, K. (1985). A kinetic study of dissolved air flotation. *JAWWA*, 610, 2–11.

Valade, M. T., Edzwald, J. K., Tobiason, J. E., Dahlquist, J., Hedberg, T., and Amato, T. (1996) Particle removal by flotation and filtration: pretreatment effects. *JAWWA*, 88 (12), 35–47.

Valioulis, I. A., and List, E. J. (1984) Numerical simulation of a sedimentation basin. 1. Model development. *Environ. Sci. Technol.*, 18, 242–247.

Vesilind, P. A. (1968) Discussion of 'evaluation of activated sludge thickening theories' by Dick and Ewing. *J. San. Eng. Div. (ASCE)*, 94 (SA1), 185–191.

Vesilind, P. A. (1979) *Treatment and Disposal of Wastewater Sludges* (revised edn.), Ann Arbor Science, Ann Arbor, MI.

Wirojanagud, W. 1983. *Particle Size Distributions in Flocculent Sedimentation*, PhD Dissertation, University of Texas at Austin,

Wood, R. F. (1970) *The Effect of Sludge Characteristics upon the Flotation of Bulked Activated Sludge*, PhD Dissertation, University of Illinois, Urbana-Champaign, IL.

Wood, R. F., and Dick, R. I. (1973). Factors influencing batch flotation tests. *J. Water Pollut. Control Fed.*, 45 (2), 304–315.

Zhou, S., and McCorquodale, J. (1992) Modeling of rectangular settling tanks. *J. Hydraul. Eng.*, 118 (10), 1391–1405.

PROBLEMS

13-1. In Example 13-2, the Reynolds number for the settling of the two largest particles violated the criterion for the applicability of Stokes' law. Find the correct settling velocities for these two particles, and comment on the extent of error if the criterion were that Re had to be less than 0.3 (as some researchers have suggested).

13-2. Several different types of particles can be encountered in a wastewater treatment plant. The influent can contain pieces of glass, sand, and small inorganic and organic particles; the aeration basin might contain large activated sludge flocs. These five types of particles could have the characteristics noted in Table 13-Pr2. Determine the settling velocity of each of these types of particles, and note where each of these particles is likely to be removed in a standard secondary wastewater treatment plant.

TABLE 13-Pr2. Typical Particles in a Wastewater Treatment Plant

Designation	Description	Diameter	Density (g/cm³)
A	Glass	10 mm	2.65
B	Sand	0.6 mm	2.65
C	Inorganic	40 μm	1.8
D	Organic	20 μm	1.2
E	Activated sludge	250 μm	1.03

13-3. **(a)** The derivation of the Stokes settling velocity given in the chapter incorporates the assumption that the sum of the forces is zero; that is, it assumes that the terminal settling velocity has been reached. For a particle that begins at rest, the velocity increases until it reaches the terminal velocity. Derive the expression for the velocity as a function of time for a spherical particle that begins at rest and for which the Reynolds number is sufficiently low that the drag force is described by the Stokes expression.

(b) For a 10 μm spherical particle with a density of 1.5 g/cm³ (i.e., the particle described in Example 13-1), calculate the time taken to reach 95% of the terminal velocity.

13-4. A histogram describing the settling velocities of nonflocculating (Type I) particles in a water supply is shown in Figure 13-Pr4. The number under each bar indicates the value of v_s at the end of the range; for example, the first bar represents particles with settling velocities between 0 and 0.2 cm/min. The distribution is bimodal; that is, it has two distinct peaks. If a smooth curve is drawn through the data, the areas under the curve are such that approximately 50% of the suspended solids concentration is included in each group.

(a) Sketch a plot of the cumulative settling velocity distribution (fraction of particles with settling velocity (v_s) less than v vs. v) for this system. Determine the expected removal in a continuous flow ideal horizontal sedimentation tank if the overflow rate is 2.2 cm/min.

(b) If the water passes through an ideal horizontal flow sedimentation basin with an overflow rate of 2.2 cm/min, sketch a curve, with the same axes as in the plot shown, that describes the settling velocity distribution of the particles in the effluent. Confirm the result from (a).

(c) For the settling tank described in (a) and (b), would overall removal efficiency be improved if it turned out that only particles with settling velocity greater than 2.2 cm/min could flocculate? Explain.

13-5. An industry treats its wastewater in a sedimentation basin with dimensions 25 m (L) × 10 m (W) × 3 m (H); this basin can be considered to have ideal, horizontal plug flow. The particles in the water are primarily nonflocculating, with a settling velocity distribution as shown in Figure 13-Pr5. The industry operates at

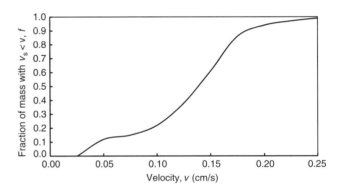

FIGURE 13-Pr5. Cumulative settling velocity distribution on a mass basis.

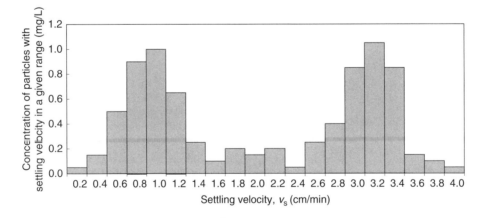

FIGURE 13-Pr4. Settling velocity distribution.

half-strength during the night, leading to a two-step wastewater production cycle. Every day, they produce wastewater at a rate of $18\,\text{m}^3/\text{min}$ between 6 AM and 10 PM, and only half that rate from 10 PM until 6 AM. Except for the flow rate, the wastewater characteristics are the same during the two parts of the day.

(a) At the current time, wastewater is treated at the same rate it is produced; that is, at two different rates in each 24-h period. Find the removal efficiency at each flow rate and the overall removal efficiency, on a mass basis, for each day. Ignore any nonsteady behavior at the transition between the flow rates.

(b) The industry is considering installing an equalization basin that will allow it to treat the same flow rate at all times during the day. What will be the expected removal efficiency in the sedimentation basin if it treats this equalized flow each day? Will the total daily mass of solids in the effluent from the settling basin be less under this operating strategy than the current situation?

13-6. A batch settling test on a Type II suspension has yielded the following data for removal of particulate mass as a function of time and depth.

Time (min)	Depth (m)			
	0.6	1.2	1.8	2.4
	Percentage Removal of Particles (Based on Mass)			
0	0	0	0	0
10	21	18	13	12
20	31	24	22	19
30	50	34	30	28
45	60	51	44	41
60	78	58	54	52
90	85	82	78	69

(a) Determine the expected removal in a continuous flow sedimentation tank that has ideal horizontal plug flow, if the depth of the settling zone is 1.8 m and the detention time is 45 min.

(b) Determine the expected removal in a continuous flow sedimentation tank that has ideal horizontal plug flow, if the depth of the settling zone is 2.4 m and the detention time is 60 min. (Note that the systems described in (a) and (b) have identical overflow rates.)

(c) If neither of the reactors described in (a) and (b) yielded sufficient removal efficiency, which of the following ways of increasing the detention time in the reactor would more likely lead to better removal: (i) increasing the surface area, or (ii) increasing the depth.

13-7. For the same suspension as described in Example 13-2 and utilized in Example 13-4, find the expected profile of the solids concentration as a function of depth after 4 h in a batch settling test, assuming the suspension was uniformly mixed at time zero and quiescent thereafter. Assume the total depth of the reactor is 3 m. Express the results also as the fractional removal as a function of fractional depth of the reactor.

13-8. Example 13-2 gives a particle size distribution that is utilized in Example 13-10 as the influent to a continuous flow horizontal sedimentation basin. For each of the reactors described below, use the same influent distribution, consider the particles to be nonflocculent, and calculate and plot the expected effluent particle size distribution, expressed both as the volume distribution and the particle size distribution function. In addition, find the overall percent removal on both the number and volume basis.

(a) An upflow clarifier, with the same overflow rate as given in Example 13-10.

(b) A tube settler with $N=4$ and an overflow rate three times that given in Example 13-10.

(c) A nonideal flow reactor that has 30% of the flow going through one side (half) of the reactor, and the other 70% going through the other half of the reactor, assume each side has ideal horizontal plug flow.

(d) A nonideal horizontal flow reactor with a value of $\lambda = 0.30$.

13-9. The cumulative settling velocity distributions of three different suspensions are characterized as follows.

$$f_1(v) = \begin{cases} \dfrac{v}{v+1} & \text{for } v \leq 2\,\text{m/h} \\ (1/3)v & \text{for } 1 < v < 3\,\text{m/h} \\ 1 & \text{for } v \leq 3\,\text{m/h} \end{cases}$$

$$f_2(v) = \begin{cases} (1/3)v & \text{for } v < 3\,\text{m/h} \\ 1 & \text{for } v \geq 3\,\text{m/h} \end{cases}$$

$$f_3(v) = \begin{cases} (1/6)v^2 & \text{for } v \leq 2\,\text{m/h} \\ (1/3)v & \text{for } 2 < v < 3\,\text{m/h} \\ 1 & \text{for } v \geq 3\,\text{m/h} \end{cases}$$

These three suspensions exhibit the identical settling characteristics for particles with $v > 2\,\text{m/h}$, but they have different distributions for particles with lower settling velocities. The distribution of the second suspension (i.e., $f_2(v)$) was used in several examples in this chapter to find the expected removal efficiency in different types of reactors. For the ideal upflow reactor and the ideal, horizontal flow reactor, the removal efficiency was found for an overflow rate of 1.2 m/h.

(a) Plot these three distributions on a single graph, and briefly comment on the differences among these distributions.

(b) Find the expected removal efficiency for suspension 1 in an ideal upflow reactor with an overflow rate of 1.2 m/h.

(c) Find the expected removal efficiency for suspension 3 in an ideal upflow reactor with an overflow rate of 1.2 m/h.

(d) Find the expected removal efficiency for suspension 1 in an ideal, horizontal flow reactor with an overflow rate of 1.2 m/h.

(e) Find the expected removal efficiency for suspension 3 in an ideal, horizontal flow reactor with an overflow rate of 1.2 m/h.

(f) Show, in a well-designed table, the expected removal efficiencies for all three suspensions in both types of ideal reactors, when the overflow rate is 1.2 m/h. Comment on how the shapes of the settling velocity distributions affect the removals achieved in each type of reactor.

13-10. For the same three suspensions described in Problem 13–9, investigate the effects of the overflow rate on the expected removal efficiency in both an upflow reactor and an ideal, horizontal flow reactor. That is, find and plot the expected removal efficiencies as a function of the overflow rate in the range of 0–2 m/h. Comment briefly on the effects of the shapes of the distribution on the results for the two different reactor types.

13-11. Figure 13-Pr11 shows the settling velocity distribution of noninteracting particles in a water supply source. The water is currently being treated in a horizontal flow sedimentation basin, which is assumed to behave ideally.

(a) In a test of the basin's performance, uniform "tracer" particles that are known to have a terminal settling velocity of 0.025 cm/s are injected into the influent. Analysis of the effluent shows that it contains only approximately 20% of the tracer particles that were injected, and the rest have been removed by settling. Estimate the overflow rate of the basin.

(b) The flow rate of the sedimentation basin is 0.5 m³/s, and it is 4 m deep. Estimate its surface area and the hydraulic residence time.

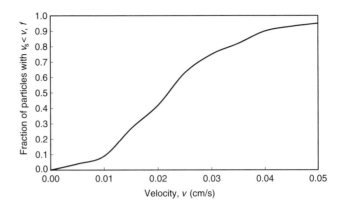

FIGURE 13-Pr11. Cumulative settling velocity distribution on a mass basis.

(c) The surface area estimated based on the tracer test turns out to be 25% less than the geometric surface area. Suggest possible reasons for the discrepancy, other than bad data.

(d) Imagine that you could take a snapshot of an aliquot of raw water that entered the basin right in the middle of the water column (i.e., halfway between the top and bottom), and that you could then keep track of all the particles that were in that snapshot as the water flowed through the basin. What fraction of the particles that were in the snapshot would hit the bottom of the basin and be removed as the water passed through the system?

13-12. Consider a few different suspensions, each containing (spherical) particles with diameters of 2, 6, 20, 60, and 200 μm. In suspension A, the particles have a density of 2.5 g/cm³ and the water temperature is 25°C. In suspension B, the particles have a density of 1.05 g/cm³ and the water temperature is again 25°C. In suspension C, the particles have a density of 2.5 g/cm³ and the water temperature is 5°C. (At 25°C, water has a density of 0.997 g/cm³ and a viscosity of 0.0089 g/cm s; at 5°C, the density is 1.000 g/cm³ and the viscosity is 0.01518 g/cm s.)

(a) Calculate the settling velocities of all of the particles in each suspension, accounting for inertial effects if necessary.

(b) If an ideal horizontal flow sedimentation tank at a water treatment plant is operated to remove 20 μm particles of density 2.5 g/cm³ completely, what overflow rates can be used in the summer (25°C) and winter (5°C)?

13-13. The following data are from a laboratory settling analysis performed on a dilute suspension of discrete particles. The sampling port was located at a depth of 1 m in a 2 m deep column.

Settling time (min)	6	12	24	48	96	192
Weight fraction remaining	0.56	0.48	0.37	0.19	0.05	0.02

What overflow rate is required to remove 85% of the suspended solids in a continuous flow ideal horizontal sedimentation tank?

13-14. (a) Calculate the settling velocity of 12 μm spherical floc particles with a density of 1.15 g/cm³ in summer (20°C) and winter (4°C).

(b) Calculate the rise velocity of the same floc particles after the attachment of a single air bubble of 40 μm diameter for the same two temperatures. The density of air is 1.27 kg/m³ at 4°C and 1.19 kg/m³ at 20°C.

(c) Comment on the results of the calculations in (a) and (b) with respect to sedimentation and DAF tank designs.

13-15. Determine the number of bubbles that must attach to the following particles to reduce the density to less than water so they will float. Perform all calculations for 20°C and use a mean bubble diameter of 40 μm.

- $d_p = 8$ μm with $\rho_p = 1050$ kg/m³ (algae)
- $d_p = 8$ μm with $\rho_p = 2650$ kg/m³ (clay mineral)
- $d_p = 80$ μm with $\rho_p = 1050$ kg/m³ (large, light floc)
- $d_p = 80$ μm with $\rho_p = 2650$ kg/m³ (large, dense floc)

14

GRANULAR MEDIA FILTRATION

14.1 Introduction
14.2 A typical filter run
14.3 General mathematical description of particle removal: Iwasaki's model
14.4 Clean bed removal
14.5 Predicted clean bed removal in standard water and wastewater treatment filters
14.6 Head loss in a clean filter bed
14.7 Filtration dynamics: Experimental findings of changes with time
14.8 Models of filtration dynamics
14.9 Filter cleaning
14.10 Summary
Problems

It is within the realm of possibilities that in due time a team of highly trained and properly equipped scientific men may be able to analyze the water characteristics, the filtering media characteristics and determine with mathematical exactness the results which may be obtained under a given set of conditions or for any number of different sets of conditions. It is not so easy for the practical minded engineer to anticipate with confidence that these scientific developments can be applied on a practical scale so that the many complications due to the fickleness of mother nature may be neutralized or provided for and a filtered water produced with greater comparative economy than with less refined methods.

Stanley (1937)

14.1 INTRODUCTION

With only minor modifications (e.g., to recognize the substantial entry of women into the profession), Stanley's remark in 1937 describing the status of filtration knowledge could well have been written today. The ability to model the behavior of granular media filters has grown tremendously in the intervening years, but our understanding of the process is still far from complete, and the use of mathematical models in design and operation is still limited. In this chapter, a mathematical framework for granular media filtration is presented and insights garnered from this framework are explained. Much of the understanding of filtration has come from a myriad of experimental studies and from the operation of full-scale filters, and insights from these sources are included and integrated with the mathematical perspective.

At first glance, granular media filtration appears simple: a suspension is passed through a layer of sand (or other media), particles are caught on the media, and water that is much cleaner than the influent appears in the effluent. Occasionally, the media need to be cleaned, either because the water coming out is no longer sufficiently clean or because the media are so clogged with the captured particles that water is not flowing through easily enough. Although the presentation below introduces a good deal of complexity, it is useful to remember this simple picture.

Filtration is used almost universally in drinking water treatment if the source is surface water. In this application, filtration is (virtually always) the final particle removal process, so that superb removal efficiency is desired, for both aesthetic and public health reasons. Granular media filtration is also used increasingly as a tertiary treatment for wastewater, to remove particles and reduce the biochemical oxygen demand (since most particles at this point in a treatment plant are microorganisms or fragments of microorganisms from secondary treatment).

The focus of this chapter is on rapid rate,[1] depth (or granular media) filtration for the removal of particles. Such filtration is the most common type encountered in drinking water and wastewater treatment, although other types of filtration also play some role. "Slow" filtration, the historic predecessor of the current "rapid" filtration, is still used in developing countries as well as in rural areas in the developed world, because it requires less maintenance and fewer operational decisions to produce acceptable water quality. Slow filtration is quite similar in many ways to rapid filtration, but most of the removal occurs in a biologically active layer that grows up above the media. Straining (the capture of particles because they do not fit through the pore spaces) and biological activity are important removal processes in slow depth filtration, whereas these phenomena are much less important in rapid rate filters. The terms "slow" and "rapid" are arbitrarily defined and have a fuzzy border between them.

On the other hand, membrane filtration, in which water passes through a thin membrane while particles do not, differs from granular media filtration in several ways. In membrane filtration, the pore sizes of the membranes are usually smaller than the particles to be collected, and therefore, most removal occurs by straining at the surface. In contrast, straining is usually unimportant in rapid, granular media filtration. Membrane filtration has been used often in the production of ultraclean water for industrial purposes and food preparation, and its application in drinking water and wastewater treatment is expanding rapidly. Membrane filtration is covered in Chapter 15.

Filtration can also be used for highly concentrated slurries; that is, as a sludge treatment process. In such cases, the particles themselves form a tightly packed matrix or "cake," and this cake becomes more important than the supporting mesh as a filter medium. Sludge filtration is not considered in this book.

The term filtration is also used for some processes in water and wastewater treatment that are not used primarily for particle removal. For example, trickling filters in wastewater treatment remove soluble organic compounds by sorption into a biofilm (a biological slime layer) where they are oxidized. In addition, filters whose primary purpose is to remove particles are increasingly used to accomplish additional treatment objectives. For example, the chemical oxidation of reduced inorganic species, such as Fe^{+2} and Mn^{+2}, is sometimes accomplished through adsorption and surface reaction in deep bed filters that also remove particles (Hargette and Knocke, 2001). Similarly, iron-oxide-coated sand has been used to adsorb natural organic matter (Chang et al., 1997) or arsenic (Petrusevski et al., 2002). Granular activated carbon is now frequently used as a filter media, allowing adsorption of various soluble contaminants to occur in the same treatment unit as particle removal. Biological filtration, in which a biofilm is builtup on the filter media, can be used in drinking water treatment to further oxidize natural organic matter after partial oxidation by ozone. These processes are characterized as filtration, because they comprise granular media packed in a bed through which water flows; the surface area of the media provides sites for desired reactions to occur. While such applications are important, they are not considered in this chapter; our focus is only on the primary use of filtration: particle removal.

The insignificance of straining as a particle removal mechanism in rapid, granular media filtration might seem surprising since straining is the primary, if not the only, meaning of filtration as the term is used colloquially. The fact that straining cannot be important in this context becomes clear if we consider the three length scales of interest in granular media filters: the particle size range (generally 0.1–100 μm), the media size range (0.4–2.0 mm), and the bed depth range (0.4–2.0 m). Except for a combination of the largest particles in a suspension and the smallest media grains (a combination that is rarely encountered, as it would constitute poor design), these scales are quite different from one another. To express these differences in a more familiar scale, the ratios of particle diameter to media grain to depth of bed in a typical filter are similar to the size ratios of a marble (or smaller ball bearing) to a basketball to an 80-story building. As a result, a media grain appears almost as a flat surface to a particle, and the pores between grains are huge openings through which most particles could easily pass, based on the size alone; thus, the removal of the particles must be attributed predominantly to some process(es) other than straining. Similarly, these ratios indicate that the diameter of a grain can be considered as a differential length in comparison to the bed depth. Remembering these length scales provides a useful perspective throughout this chapter.

As Stanley articulated in the quotation opening this chapter, granular media filtration is a complex subject. This complexity stems from four sources, as summarized in Table 14-1. First, two (rather than the usual one) dependent variables are of primary interest: the effluent quality and the head loss[2] through the filter. Second, the process is

---

[1] The rate of filtration, or the filtration velocity, refers to the flow rate per unit cross-sectional area. The distinction between "rapid" and "slow" filtration is based on this velocity, with rapid filters usually having values of 3–30 m/h, and slow filters having values <2 m/h.

[2] Head loss is a measure of the (potential) energy lost as water flows through the filter and is explained in detail subsequently. As particles accumulate in the filter, they make it more difficult for water to flow through. This increasing "difficulty" is measured by the rise in head loss.

TABLE 14-1. Variables and Modes of Operation in Granular Media Filtration

Category	Property of Interest	Parameter
Dependent variables	Effluent quality	Turbidity
		Particle size distribution
		Suspended solids concentration
	Loss of energy	Head loss
Independent variables	Time or cumulative volume processed (since previous backwash)	Time, volume per unit area
	Characteristics of operation	Filtration velocity (flow rate per cross-sectional area)
	Characteristics of media	Size and size distribution
		Density
		Shape
		Porosity
		Depth
		Surface chemistry
	Characteristics of particles in suspension	Size distribution
		Concentration
		Density
		Shape
		Surface charge/surface chemistry
	Characteristics of solution	Temperature
		Density
		Viscosity
		Chemical composition
Operational modes	Filtration	
	Backwash	

never at steady state: effluent quality changes with time (often improving for a while at the beginning of a treatment cycle and later deteriorating), and the head loss rises continuously. Third, several independent variables affect the process, including characteristics of the filter operation, the media, the particles in suspension, and the solution. Finally, the need to clean the media relatively frequently indicates that the filter operates in two modes: filtration and cleaning. The cleaning, or backwashing, means that filtration has some aspects that are similar to a batch system (the water treated during a filter "run"—the time from one backwash to another—forms a batch) and other aspects that are indicative of a continuous flow system.

A Typical Filter

A schematic diagram of a granular media filter for a drinking water application is shown in Figure 14-1; wastewater filters are identical except they do not have valve E (explained subsequently). The need for two modes of operation—filtration and backwashing—means that the physical system is somewhat more complex than many other treatment processes. The sand (or other media) is supported by gravel that rests on the perforated filter floor above the underdrain system. This figure is drawn for a typical gravity-driven filter, but *pressure filters* are sometimes used, especially in small plants. In such filters, the filter media is in an enclosed pressurized vessel, with the pressure supplied by a pump. The details of how water enters and leaves the vessel are slightly different than those shown.

During the filtration cycle, water flows in through valve A, down through the sand, and out through valve B. In this period, the water level is below the top of the backwash

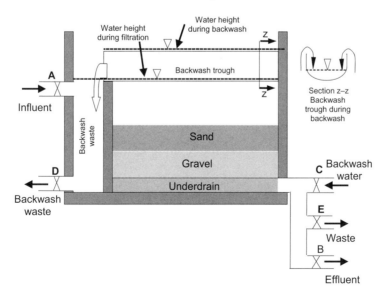

FIGURE 14-1. Schematic diagram of a gravity-driven granular media filter.

water trough, and valves C, D, and E are closed. As explained subsequently, most modern filters are operated at a constant filtration velocity, which is accomplished by including a rate controller at valve B.

During the backwash period, water flows in through valve C, up through the sand, over the edge into the trough (see section z–z in the figure), out of the end of the trough, and through valve D to waste. In this period, the water level is just over the top of the trough, and valves A, B, and E are closed. The water used for backwashing (that enters through valve C) is previously saved filter effluent (i.e., the cleanest water available in the plant). The (upward) water flow during backwashing is substantially higher than the (downward) flow during filtration; the high water flow leads to fluidization of the sand media, so that the bed is *expanded* (the top of the sand is higher during backwashing).

After backwash, the water that exits the filter in the first few minutes of the next filter run often contains a substantial (and unacceptable) particle concentration; in a drinking water plant, valve E is open (and B is closed) during this short period so that this water does not enter the water supply. In a wastewater plant, this brief period of poorer filtration is not a significant problem, and valve E does not exist. The water that is labeled as going "to waste" (from valve D during backwash and valve E in the first few minutes of filtration) is usually saved, treated to some extent (typically by sedimentation), and then recycled to the head of the treatment plant for retreatment. In this way, only the water in the sludge from the sedimentation of the backwash water is actually lost. With this recycling, the *recovery* of water (i.e., the fraction of influent water that becomes product water) is quite high (>99%) in granular media filtration, even though the fraction of the filtered water that is used for backwashing (2–4%) can be substantial.

Brief transition periods are needed between the two modes of operation. At the end of a filter run, valve A is closed a few minutes before valve B, so that the water drains from the filter to (approximately) the top of the media. Then, valve D is opened before the backwashing begins (by opening valve C) to ensure that the filter does not overflow. Similarly, at the end of the backwashing, valve C is closed to shut off the backwash supply before valve D is shut to allow water to drain from the gullet above valve D. Only after valve D is closed are valves B and then A opened.

The schematic in Figure 14-1 does not include a few aspects of backwashing equipment that are common in granular media filters. Many filters include a *surface wash* system to break up accumulations of particles at the very top of the sand (or other) media. Surface wash is accomplished by rotating arms that send jets of water down toward the top of the media to break up small cakes that might have builtup. Also, many modern filters include *air scour* systems that inject air at the same time as the backwash water to create greater turbulence and thereby facilitate release of collected particles from the media grains. Details about backwashing are explained in a section near the end of this chapter.

Although Figure 14-1 indicates the granular media as sand (only), many filters have two different types of media. In this case, media with the larger size is on top and media with the smaller size is on bottom (and supported by the gravel). To achieve this distribution, the top media must have a lower settling velocity than the bottom media, and therefore must have a much lower density (since it is of larger size) than the media on the bottom. The most common combination for such *dual media filters* consists of anthracite coal on top and sand below.

When a filter is being backwashed, the water that would otherwise be directed to this filter must be directed to others. Hence, a few (or several) filters are always built in parallel so that the total plant flow can be processed even when a filter is unavailable due to backwashing. If, for example, five filters are installed and one is taken out of service for backwashing, the flow rate to each of the other four increases by 25%; this hydraulic surge can cause a loss of particles that were previously captured in the other filters.

14.2 A TYPICAL FILTER RUN

The patterns followed by the two dependent variables—effluent quality and head loss (or pressure development)—during a typical filter run are shown as conceptual graphs in Figure 14-2. For effluent quality, the results are shown generically as effluent concentration; in a specific application, effluent quality is measured as turbidity, suspended solid concentration or total particle number concentration larger than some minimum size.

The abscissa in Figure 14-2 is labeled as either time or cumulative volume of water put through the filter since the previous backwash. Filters are often designed and operated to achieve a (nearly) constant flow rate, in which case the cumulative volume throughput is linearly related to time. The results depicted in Figure 14-2 assume such a constant rate of filtration. Filters can also be designed as declining rate filters in which the flow rate declines as the friction in the filter increases due to the accumulation of particles removed from the suspension. In such filters, the head loss through the filter is nearly constant with time (contrary to what is shown in this figure), and the cumulative throughput must be calculated incrementally to account for the changes in flow rate.

This figure depicts a time period that would typically last for 1–3 days for drinking water applications and approximately half these values for wastewater filters. This figure also ignores the very rapid changes in effluent quality that

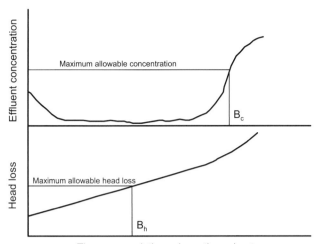

FIGURE 14-2. Typical filter performance in a single run. (The points B_c and B_h represent the times or throughput volumes at which backwashing is necessary because the maximum allowable effluent concentration or maximum allowable head loss is exceeded, respectively. A normal filter run would be terminated when the first of these points was reached, but the graph is extended here for illustrative purposes.)

often occur in the first few minutes of a run, as backwash water is replaced in the filter by fresh influent. The filter performance during these few minutes has been studied in some detail and is described subsequently. At the beginning of a filter run, the bed of media is clean, a certain effluent concentration (or removal efficiency) is achieved, and the head loss is at some specific, nonzero value. Well-tested mathematical models exist to describe both removal efficiency and head loss under these initial clean bed conditions, at least for idealized systems.

As time proceeds, the effluent concentration first decreases (i.e., the water quality improves) for a while (10 minutes to a few hours), then is nearly steady for a considerable period, and finally begins to increase steadily. The improvement in effluent quality in the early period is known as *ripening*. The intermediate period is often referred to and thought of as a steady-state period. During this period, events at the microscopic level (improved removal of some sizes or types of particles, less removal of other particles, breakoff of previously captured particles, recapture of particles that are broken off) tend to balance each other so that the effluent quality, at least as measured by composite parameters such as turbidity or suspended solid concentration, remains reasonably constant. However, as suggested in this figure, some fluctuations in effluent quality are apparent. As noted above, this period typically lasts for many hours. At still longer times, the effluent quality clearly deteriorates and breakthrough occurs. At such a point, the desired effluent quality is not achieved and the filter must be cleaned

(backwashed) again. During backwashing, the filter is out of service and clean water (i.e., previously filtered water) is being used for the backwashing. Long filter runs are desirable, because they reduce the fraction of time that a filter is out of service and the fraction of product water that must be sacrificed for backwashing.

Particles accumulate in the filter throughout the entire run. This accumulation causes the water to flow through more circuitous and constricted pathways and the particles encounter more surface area as time proceeds. The increased path length and resistance to flow lead to the monotonic increase in head loss indicated in this figure. It is common in filtration (and other water applications) to express differences in the available energy per unit volume of fluid (ΔE_V) between two points as the head loss (h_L, with dimensions of length) between these two points

$$h_L = \frac{\Delta E_V}{\rho_L g} \quad (14\text{-}1)$$

where ρ_L is the density of water, and g is the gravitational constant. The head loss represents the energy expended per unit volume of water to pass the water through the filter.[3] Head loss often rises nearly linearly with time, but sometimes exhibits a stronger rise, especially later in a run, as indicated in this figure. The available head is limited either by physical constraints (in gravity flow applications) or by economic considerations (in pressurized filters), so backwashing becomes necessary when the head loss exceeds a certain value.

The preceding discussion indicates that backwashing is necessitated by either of two events: deterioration of effluent quality to unacceptable levels or excessive head loss. Which condition is met first (and therefore dictates that backwashing should occur) depends on several design and operational variables. Traditional designs in drinking water treatment cause the head loss criterion for backwashing to be met considerably earlier than the effluent concentration criterion, as suggested in this figure. Experience with a particular filter treating a particular water often results in an operating rule to backwash filters after a certain operational time before either one of the critical conditions is met.

An alternative abscissa for Figure 14-2 is the volume throughput per unit area of the filter. This normalization simplifies comparison among filters of different dimensions. Normalizing the cumulative volume throughput by area

---

[3] Any change in energy, ΔE, of water, can be expressed as an equivalent change in the elevation of the water (h_L) by $\Delta E = mgh_L$. If this equation is normalized to a unit volume of water, the left side has units of energy per unit volume, which is the same as force per unit area or pressure; the right side becomes $\rho_L g h_L$. Thus, a change in available energy can be expressed as an equivalent change in elevation, or head, as shown in Equation 14-1. According to the Bernoulli equation, applicable to a steady flow of an incompressible fluid (as in a filter), the available energy is the sum of three terms, representing pressure above a chosen datum (usually atmospheric), kinetic energy, and potential energy above a defined datum.

leads to a net dimension of length (e.g., m^3/m^2 or just m); this length can be thought of as the (imaginary) height of a water column that started above the filter and that has passed through the filter in the time since the last backwash. The value of the cumulative volume throughput per unit area at the time of backwash is called the *(gross) normalized filter run production* ("gross" because it ignores the water used for backwashing that would be subtracted from this value to find the net production). This value has also been called the *unit filter run volume;* this term is not used herein because of the potential confusion in using the term "volume" for a parameter that actually has dimensions of length. Traditional (older) drinking water filters achieve run times of $\sim$40–50 h at a filtration velocity (flow rate divided by area) of 5 m/h, or a normalized filter run production of 200–250 m (i.e., m^3/m^2). Modern designs often set a minimum normalized production of 400 m per run as a design goal (Kawamura, 1999).

This description of a typical filter run demonstrates that the design and operation of filters constitute a balance among several competing requirements and desires:

(1) Effluent should be of sufficient quality in the early part of a run that little or no water needs to be wasted or recycled before meeting the effluent guidelines.
(2) Effluent should meet water quality guidelines over a long time period (or, more precisely, until a high normalized production is achieved).
(3) Head loss should increase slowly enough that the filter can operate until a high normalized production is achieved.
(4) Requirements for backwashing flow and volume must be reasonable.

Unfortunately, many changes in design and operation that help achieve one of these goals make it more difficult to achieve another. For example, increasing the size of the media grains in a filter reduces the rate at which head loss increases, but also leads to higher effluent particle concentration and greater backwash flow requirements. The problem of higher effluent concentration might be overcome with a deeper bed of media, at some cost in head loss. Such tradeoffs must be considered in the context of the application. In drinking water filtration, protection of the public health is of paramount importance. In this case, choices for design and operation that would lead to extended filter runs but increase the risk of poor effluent quality for some period are usually considered unacceptable. It is for this reason that, as noted above, drinking water filters are almost always designed so that the maximum head loss criterion for backwashing is met before the effluent quality criterion.

On the other hand, the choice might be different for wastewater filters. Although such filters are designed to improve effluent quality, the consequences of slightly worse quality are not as great in most wastewater applications, and designers are more willing to sacrifice effluent quality for longer runs or less head loss. Therefore, designs of wastewater filters often reflect a closer balance between the times elapsed before the two backwashing criteria are reached. In both drinking water and wastewater filtration, a long-term trend in design and operation has been to increase the media size, increase the velocity, and, to a lesser extent, increase the bed depth.

Faced with the complexity of filtration, one temptation is to treat the filter as a "black box," assuming that understanding the details of what occurs in this black box is beyond reason. In this case, design and operation involve, primarily, emulating what has worked in similar situations before. This approach has some merits; the wealth of experience in filtration has great value and must be accounted for in any new design. Nevertheless, an understanding based on fundamental concepts, even if incomplete, can aid the interpretation of this experience and more efficiently lead to improvements in design and operation. Such an approach is taken in this chapter.

In the following sections, models for the removal of particles in a clean filter are considered first; the similarities between the removal mechanisms in filtration and particle growth mechanisms in flocculation are emphasized. Two commonly used and relatively simple models are presented in some detail, followed by some complications that have been considered by various investigators. Second, some analysis of the ramifications of these models for design is given. Third, a view of the head loss in a clean filter bed caused by the flow of water through the porous media is presented. Fourth, the dynamics of filtration is considered, with a discussion of the effects of the design and operational variables on the filter behavior, followed by presentation of a few models for the ripening of filters. The associated models for the increase in head loss are also presented. Finally, an explanation of filter backwashing is presented. The viewpoint throughout this chapter is primarily microscopic; that is, at the scale of particles; macroscopic models and experimental results of various investigators are interpreted from this particle viewpoint.

The emphasis within this chapter is on the physical aspects of filtration; however, it is essential to recognize that the chemical aspects (i.e., particle destabilization) are equally important. Indeed, if a conventionally designed filter is not removing particles adequately, the cause is far more likely to be chemical than physical. The omission of these considerations here stems from the belief that the chemical aspects of filtration must be addressed upstream of the point where particles enter a filter. Destabilization of particles; that is, the chemical aspects of filtration, is considered in Chapter 11.

14.3 GENERAL MATHEMATICAL DESCRIPTION OF PARTICLE REMOVAL: IWASAKI'S MODEL

Consider a small increment of filter bed length (Δz) as depicted in Figure 14-3. The suspension passes through the randomly packed media with a net reduction in particle concentration, as particles are caught on the media. If interactions between suspended particles are negligible and the bed is considered as uniform (i.e., with a fixed number of media grains per unit length), a probabilistic argument is useful for deriving one of the fundamental equations describing filtration. Define λ as the probability of particle capture per unit length; this indicates that, in a short increment of length (Δz), the probability of capture is $\lambda \Delta z$ and the probability of passage is $(1 - \lambda \Delta z)$. If N_z and $N_{z+\Delta z}$ represent the number concentrations of particles at locations z and $z + \Delta z$, respectively, then

$$N_{z+\Delta z} = (1 - \lambda \Delta z) N_z \qquad (14\text{-}2)$$

Rearranging and defining ΔN

$$\frac{N_{z+\Delta z} - N_z}{\Delta z} = \frac{\Delta N}{\Delta z} = -\lambda N_z \qquad (14\text{-}3)$$

Taking the limits as Δz approaches 0 yields

$$\frac{\partial N}{\partial z} = -\lambda N \qquad (14\text{-}4)$$

where the partial derivative is used to reflect the fact that, in real filters, the conditions change over both depth and time. This equation, which might be considered as the fundamental equation describing filtration, was first formulated based on empirical observations by Iwasaki (1937). A more sophisticated probabilistic approach was published by Hsiung and Cleasby (1968). The parameter of the equation, λ, is known as the filter coefficient, with dimensions of inverse length (i.e., probability *per unit length*).

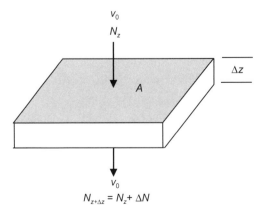

FIGURE 14-3. Differential element of a filter.

Unfortunately, Equation 14-4, in the form shown, does not accomplish the central goal of filtration modeling, namely, to relate the probability of capture to the characteristics of the filter media, operation, or suspension. The approaches to this problem are explained below. A second problem is that the bed is not really homogeneous down to infinitesimal dimensions. At a length (in both horizontal and vertical dimensions) on the order of the size of a single filter grain or less, the specific geometry must be accounted for in some way, and the probability of capture is different at different points in this geometry. As a result, to model filtration from first principles, we must describe the behavior of the fluid and the particles at a length scale comparable to the size of a single grain and then expand to the scale of the whole filter.

Following the approach used by Iwasaki, we next write a number balance (i.e., identical in concept to a mass balance, but using the number of particles as the variable of interest) on the unit element. Because particles that are removed from suspension must be captured on the media, this balance can be expressed as follows:

$$\begin{bmatrix} \text{Rate of loss} \\ \text{of particles} \\ \text{from suspension} \end{bmatrix} = \begin{bmatrix} \text{Rate of gain} \\ \text{of particles} \\ \text{deposited} \\ \text{on media} \end{bmatrix} \qquad (14\text{-}5)$$

We consider an aliquot of water as it travels the incremental distance, Δz, in time increment, Δt. The left side of Equation 14-5 is the difference between the rates at which particles enter and leave the differential element[4]; that is,

$$\begin{bmatrix} \text{Rate of loss} \\ \text{of particles} \\ \text{from suspension} \end{bmatrix} = Av_0 N_z - Av_0(N_z + \Delta N) = Av_0 \Delta N \qquad (14\text{-}6)$$

where A is the cross-sectional area of the volume element under consideration, and v_0 is the superficial velocity (the velocity if no media were in the bed, equal to the flow rate divided by the cross-sectional area), as illustrated in Figure 14-3.

To quantify the right side of Equation 14-5, we define the specific deposit, σ, as the number of particles deposited per unit volume of media. The right side is then the product of $\Delta \sigma / \Delta t$ and the volume of media in the differential element. And, the volume of media in the small increment

---

[4] The terms in Equation 14-6 are the product of the area and the flux (itself the product of velocity and number concentration). In truth, the area available for flow is εA, where ε is the porosity (defined in the subsequent paragraph), and the average velocity through the pore space is v_0/ε. The porosity terms cancel, leading to the equation shown.

of the bed is $A\Delta z(1-\varepsilon)$, where ε is the bed porosity. Hence,

$$\begin{bmatrix} \text{Rate of gain} \\ \text{of particles} \\ \text{deposited on media} \end{bmatrix} = \frac{\Delta\sigma}{\Delta t} A\Delta z(1-\varepsilon) \quad (14\text{-}7)$$

Combining Equations 14-5–14-7 and rearranging yields

$$-\frac{v_0}{1-\varepsilon}\frac{\Delta N}{\Delta z} = \frac{\Delta\sigma}{\Delta t} \quad (14\text{-}8)$$

Iwasaki (1937) also realized that the capture of particles altered the probability of further capture, and found that removal was improved (i.e., ripening occurred) by the buildup of captured solids. To express this phenomenon mathematically, he proposed that the filter coefficient increased linearly with the specific deposit

$$\lambda = \lambda_0 + a\sigma \quad (14\text{-}9)$$

where λ_0 is the initial filter coefficient and a is a ripening coefficient, with dimensions of length squared. Many other models for the time-dependent behavior of filters have been proposed since Iwasaki's original model, but this is still often considered adequate if a filter run is terminated before particle breakthrough becomes important. Although Iwasaki's original work was expressed in particle number, many investigators have expressed his equations in other dimensions; for example, the particle concentration is expressed in units of mass per volume or volume (of particles) per volume (of suspension), with consequent changes in the units of the specific deposit.

14.4 CLEAN BED REMOVAL

Consider again a thin layer of a filter bed, as shown in Figure 14-3. A number balance around a small volume element for the number of particles in *suspension* can be expressed as follows:

$$\begin{bmatrix} \text{Rate of change} \\ \text{of number of} \\ \text{particles in} \\ \text{suspension in the} \\ \text{volume element} \end{bmatrix} = \begin{bmatrix} \text{Rate at which} \\ \text{particles enter} \\ \text{the volume} \\ \text{element} \end{bmatrix}$$

$$- \begin{bmatrix} \text{Rate at which} \\ \text{particles leave} \\ \text{the volume} \\ \text{element} \end{bmatrix} - \begin{bmatrix} \text{Rate of removal} \\ \text{of particles from} \\ \text{suspension by} \\ \text{attachment of} \\ \text{filter grains} \end{bmatrix}$$

The above equation can be expressed in symbols as

$$\frac{\Delta(NA\varepsilon\Delta z)}{\Delta t} = v_0 A N_z - v_0 A(N_z + \Delta N) - A\varepsilon\Delta z r_p$$

$$= -v_0 A \Delta N - A\varepsilon\Delta z r_p \quad (14\text{-}10)$$

The term, $-r_p$, is the rate at which particles are removed from suspension by the pseudo-reaction of attachment to filter grains in the control volume, and the rest of the variables are as described previously. The dimensions of r_p are number per volume per time or simply 1/(volume×time) since particle number is dimensionless.

As a first approximation, it is reasonable to consider that the filter is at quasi-steady state; that is, that two successive aliquots of water (representing a small Δt) approaching the same layer of the filter experience virtually the same conditions in this layer, even though the change in N over the depth of this layer can be substantial. Over longer periods of time, the layer might change substantially, but this is not of interest here. In other words, the two terms on the right side of Equation 14-10 are approximately equal (and opposite) to one another, but the magnitude of each is (or can be) much larger than the left side. With such an assumption, the left side of Equation 14-10 is considered 0, and the equation can be rearranged to yield

$$r_p \Delta V_L = -v_0 A \Delta N \quad (14\text{-}11)$$

where ΔV_L is the volume of liquid in the differential control volume, equivalent to $A\varepsilon\Delta z$. To proceed further in developing an equation for particle removal in a filter, we must choose a conceptual model for the geometry of the media grains and the spaces between them and the relationship between the differential size and the whole filter bed. When combined with an analysis that predicts the motion of suspended particles in the differential element, the geometric model can be used to calculate the removal efficiency. Several geometric models and various levels of complexity for the forces considered in the analysis of particle motion have been considered by different investigators. Two of the most common (and closely related) models are presented below.

The Single Spherical Isolated Collector Model

The simplest geometric model for filtration is that of the single media grain (referred to as a collector) unaffected by its neighbors. This model was originally developed to describe air filtration and was first applied to water filtration by Yao (1968) and Yao et al. (1971). In the model, the flow of particles toward the collector is considered; the specific flow pattern is chosen subsequently but is not important at this stage. The removal in a differential layer of the filter (i.e., in the control volume described above) is considered to be the product of the removal by one collector and the number of collectors in the layer; that is

$$r_p \Delta V_L = \begin{bmatrix} \text{Rate of removal of particles} \\ \text{due to one collector} \end{bmatrix} \begin{bmatrix} \text{Number of} \\ \text{collectors in layer} \end{bmatrix} \quad (14\text{-}12)$$

To find the removal by one collector, the same approach as used for flocculation and sedimentation is followed; that is, we consider long-range transport separately from other factors (acting at very close separation distances) that affect particle removal. Also as in flocculation and sedimentation, long-range transport is assumed to occur via advection (i.e., transport with the flowing water), sedimentation, and Brownian motion. In modeling long-range transport, particles are considered to be approaching the collector if they pass through the top of an imaginary cylinder that extends above the single grain for a distance that is long relative to the grain diameter, d_c, as shown in Figure 14-4. The distance is chosen far enough above the grain that the streamlines are parallel.

The rate of approach of particles to a single collector (number/time) is the product of the number concentration, the approach or superficial velocity, and the projected area of the collector; that is,

$$\begin{bmatrix} \text{Rate at which} \\ \text{particles approach} \\ \text{collector} \end{bmatrix} = N v_0 \frac{\pi d_c^2}{4} \quad (14\text{-}13)$$

Based on an analysis that considers the three modes of long-range transport noted above, some of the approaching particles are predicted to be transported to (collide with) the

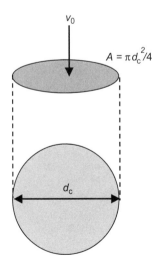

FIGURE 14-4. Conceptual diagram for approach of particles toward collector. (The dotted lines representing the distance above the filter grain should be much greater than d_c.)

surface of the filter grain. The ratio of the predicted rates of collision and approach is considered as the single collector transport efficiency, η; that is,

$$\eta \equiv \frac{\text{Rate at which particles are predicted to collide with collector, considering only long-range transport}}{\text{Rate at which particles approach collector}} \quad (14\text{-}14)$$

The details of the analysis of η for each mode of transport are provided below; as of now, the conceptual definition (the ratio in Equation 14-14) is sufficient.

As in flocculation, the fraction of the approaching particles that actually collides with and attaches to the media grain is less than η, due to short-range forces that (mostly) interfere with collisions (O'Melia and Stumm, 1967). We account for these forces by introducing an attachment efficiency factor, α, defined (again, as in flocculation) as the ratio of the actual to predicted rate of collisions. Thus, α "corrects" the predictions so that they match experimental observations, implicitly accounting both for known short-range forces (hydrodynamic forces, van der Waals attraction, and interactions between the electrical double layers) and for any other, as yet unidentified, interactions between the particle and the collector grain

$$\alpha \equiv \frac{\text{Rate at which particles collide with and attach to the collector}}{\text{Rate at which particles are predicted to collide with collector, considering only long-range transport}} \quad (14\text{-}15)$$

Because particles that attach to the collector are removed from the suspension, the removal due to one collector can be expressed as the product of α, η, and the rate at which particles approach a single collector from afar as given in Equation 14-13. Thus,

$$\begin{bmatrix} \text{Removal} \\ \text{due to one} \\ \text{collector} \end{bmatrix} = \alpha \eta \frac{N v_0 \pi d_c^2}{4} \quad (14\text{-}16)$$

Finally, we can consider the differential layer in Figure 14-3 to be made of an assemblage of collectors, and the number of these collectors can be determined as follows:

$$\begin{bmatrix} \text{Number of} \\ \text{collectors} \\ \text{in layer of} \\ \text{depth } \Delta z \end{bmatrix} = \frac{\text{Total volume of collector media}}{\text{Volume of one collector}}$$

$$= \frac{A \Delta z (1 - \varepsilon)}{\pi d_c^3 / 6} \quad (14\text{-}17)$$

Combining Equations 14-11, 14-12, 14-16, and 14-17 yields

$$-v_0 A \Delta N = \alpha \eta \left[\frac{N v_0 \pi d_c^2}{4} \right] \left[\frac{6 A \Delta z (1 - \varepsilon)}{\pi d_c^3} \right] \quad (14\text{-}18)$$

Shrinking the differential element to infinitesimal dimensions and simplifying Equation 14-18 gives

$$\frac{dN}{N} = -\frac{3}{2} \frac{(1 - \varepsilon) \alpha \eta}{d_c} dz \quad (14\text{-}19)$$

Integrating across the total length of the bed (Z), with an influent concentration of N_{in} and an effluent concentration of N_{out} yields

$$\int_{N_{in}}^{N_{out}} \frac{dN}{N} = -\frac{3}{2} \int_0^Z \frac{(1 - \varepsilon) \alpha \eta}{d_c} dz \quad (14\text{-}20)$$

Assuming that all characteristics of the bed and suspended particles are the same throughout the filter, all the terms on the right side except dz can be taken out of the integral, and the following results are obtained:

$$\ln \frac{N_{out}}{N_{in}} = -\frac{3}{2} \frac{(1 - \varepsilon) \alpha \eta}{d_c} Z \quad (14\text{-}21)$$

$$N_{out} = N_{in} \exp \left[-\frac{3}{2} \frac{(1 - \varepsilon) \alpha \eta}{d_c} Z \right] \quad (14\text{-}22)$$

Either the differential (Equation 14-19) or integral (Equations 14-21 or 14-22) form can be considered as the fundamental equation describing particle removal in filters according to the single spherical collector model. To use either of these equations for predictions of filter performance, however, we need to determine how the single collector transport efficiency, η, is related to the filter design and operational characteristics, and this is shown in the next section.

Throughout this book, we have defined the removal efficiency of various processes as ($1 - $ (effluent concentration/influent concentration)). Here, we represent this term by, η_{filter}, so that

$$\eta_{filter} = 1 - \frac{N_{out}}{N_{in}} \quad (14\text{-}23)$$

Note that η_{filter} refers to the removal in an entire filter bed, whereas η refers to the removal by a single collector in this filter.

Removal Mechanisms and Transport Efficiencies for a Single Isolated Collector

A conceptual view of the three mechanisms of long-range transport of particles to a media grain (collector) that are

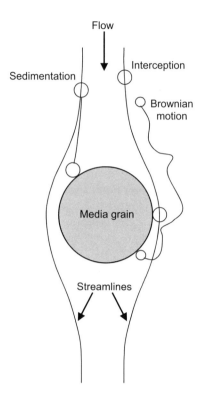

FIGURE 14-5. Particle removal mechanisms in the single isolated collector model. (The ratio of particle sizes to media grain size is shown much larger than in real systems for the sake of clarity in this figure.)

included in the calculation of η is shown in Figure 14-5. As described above, these mechanisms are interception, sedimentation, and Brownian motion.

Because this model originally considered the collector to exist in an infinite medium, the flow pattern used in the analysis was that around a single sphere with no neighbors. Assuming that the velocity is sufficiently slow that inertial terms can be ignored, the Navier–Stokes equations can be applied to describe such flow. In the following paragraphs, the expressions derived for η for each attachment mechanism are presented, based on these assumptions. Subsequent analyses have considered more complex (and more realistic) flow patterns, but the simpler analysis is adequate for the current purposes.

Particle–Collector Collisions by Interception For particles moving with the fluid, the center of the particle is considered to flow along a streamline. The contact between the particle and the collector occurs if the center of the particle comes within one particle radius of the surface of the collector; that is, the collector "intercepts" the particle. As a result, this mechanism of removal is known as interception. A critical trajectory can be defined as the one in which the particle is just captured at the equator of the collector; this condition is depicted in Figure 14-6. Thus, if we ignore the

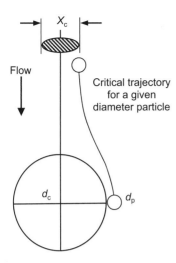

FIGURE 14-6. Conceptual diagram of the critical trajectory for capture by interception.

short-range forces in the force balance, the critical trajectory for interception is the one for which the streamline is at a horizontal distance $d_p/2$ from the equator of the media grain. Since the particle is assumed to move with the fluid, the critical trajectory can be identified by solving the equations of motion for the fluid, going backwards in time and using the particle capture at the equator as the initial condition. In this analysis, the flow is restricted to laminar conditions.

At a distance far above the collector, the streamlines are parallel (and vertical) and unaffected by the collector. The distance from the vertical centerline of the collector (projected up) to the critical trajectory at such a location above the collector is the radius of a circle that defines the capture cross-section; that is, particles whose centers pass through this circle are intercepted. This circle, with diameter X_c, is depicted in Figure 14-6. The single collector transport efficiency is the ratio of the flux of particles through the capture cross-section to the flux that enters the full cylinder shown in Figure 14-4. Assuming that both the flow and the particles in the flow are uniformly distributed at this location above the collector, the single collector transport efficiency is calculated as the ratio of the area of this critical circle to the projected area of the collector; that is, $\eta_I = \left(\frac{\pi}{4}X_c^2 / \frac{\pi}{4}d_c^2\right) = (X_c/d_c)^2$. The diameter of this critical circle, X_c, is proportional to the particle size, so that this single collector transport efficiency can be shown to be as follows[5] (Yao, 1968; Tien, 1989):

$$\eta_I = \frac{3}{2}\left(\frac{d_p}{d_c}\right)^2 \quad (14\text{-}24)$$

---

[5] The equation shown is an approximation for the case that $d_p \ll d_c$. Yao (1968) indicated that the full equation should be $\eta_I = \frac{1}{2}(d_p/d_c)^2(3 - (d_p/(d_c + d_p)))$ but that the approximation is virtually always valid.

■ **EXAMPLE 14-1.** Find the single collector transport efficiency by interception of 2-μm particles in a filter with a media size of 0.6 mm.

Solution. The problem is a direct application of Equation 14-24. Converting both diameters into centimeters, we find

$$\eta_I = \frac{3}{2}\left(\frac{2 \times 10^{-4}\,\text{cm}}{6 \times 10^{-2}\,\text{cm}}\right)^2 = 1.67 \times 10^{-5}$$

This result indicates that only approximately one of every 60,000 2-μm particles that approach a single collector is captured by interception. While this probability is quite small, remember that an aliquot of water (and the particles in it) will pass approximately a thousand collectors as it moves through the depth of a full-scale filter. Even then, this result suggests that collection of 2-μm particles by interception is not very efficient. ■

Particle–Collector Collisions by Sedimentation Because of the force of gravity, particles can bend away from the streamlines and contact the collector by sedimentation. The critical trajectory for this mechanism is found by integration of the equations of motion for the particle including the force of gravity on the particles. Hence, moving along the streamline (and capture by interception) is a limiting case for these equations of motion that occurs when the particle density is the same as that of water. When the critical trajectory is found (the one that captures the particle at the collector equator), the single collector removal efficiency is found in the same way as for interception alone; this situation is essentially identical to that depicted in Figure 14-6, except that the particle trajectory bends away from the fluid streamline so that the value of X_c is larger and therefore the single collector removal efficiency is larger. This single collector removal efficiency includes the combined effects of interception and sedimentation; we represent this quantity as η_{I+g}, the single collector removal efficiency for the two phenomena together.

When the effects of gravity are included, the equations of motion cannot be solved analytically, but they can be solved by numerical integration. Yao (1968) calculated values of η_{I+g} for several conditions and also had calculated results for interception alone (η_I) under the same conditions. Assuming that the effects of sedimentation were additive to interception, he calculated the single collector transport efficiency for sedimentation (gravity) alone (η_g) by subtracting the two

$$\eta_g = \eta_{I+g} - \eta_I \quad (14\text{-}25)$$

He found that these results for η_g over a wide range of conditions were well approximated as the ratio of the Stokes settling velocity and the approach velocity; that is,

$$\eta_g = \frac{v_s}{v_0} = \frac{(\rho_p - \rho_L)g d_p^2}{18\mu v_0} \qquad (14\text{-}26)$$

where v_s is the Stokes velocity of the particle, ρ_p and ρ_L are the densities of the particle and fluid, respectively, g is the gravitational constant, and μ is the absolute viscosity of the fluid.

Historically, filtration research and practice have emphasized the removal of particles $>\sim 1\,\mu\text{m}$, in part because of the greater ability to measure them and in part because of interest in bacteria and protozoa. Sedimentation is the dominant mechanism for removal of these particles in a filter.

■ **EXAMPLE 14-2.** Find the single collector transport efficiency by sedimentation of 2-μm particles with a density of 1.08 g/cm³ in a filter at 25°C with a superficial filtration velocity of 5 m/h (= 0.14 cm/s).

Solution. At 25°C, the viscosity of water is 8.91×10^{-3} g/cm s and the density is 0.997 g/cm³. The problem is a direct application of Equation 14-26, as follows:

$$\eta_g = \frac{(\rho_p - \rho_L)g d_p^2}{18\mu v_0}$$

$$= \frac{[(1.08 - 0.997)\,\text{g/cm}^3](981\,\text{cm/s}^2)(2 \times 10^{-4}\,\text{cm})^2}{18(8.91 \times 10^{-3}\,\text{g/cm s})(0.14\,\text{cm/s})}$$

$$= 1.45 \times 10^{-4}$$

Note that the transport efficiency by sedimentation (for these specific conditions) is approximately an order of magnitude higher than that by interception found in Example 14-1. For a given filter and given suspension, the ratio η_g/N_I is the same for all particle sizes since the expressions for both mechanisms have a dependence on d_p^2. ■

Particle–Collector Collisions by Brownian Motion A trajectory analysis cannot be performed to model transport by Brownian motion because the particle motion is not deterministic but probabilistic. Brownian motion is caused by the unequal distribution around the particle surface of collisions between the particle and molecules in solution at any given instant. The approach to finding the single collector transport efficiency for this mechanism is taken from Levich (1962), who solved the convective/diffusion mass balance equation for the particle concentration in the flow field surrounding a single collector

$$\frac{\partial N}{\partial t} = -\mathbf{v}\cdot\nabla N + D_p \nabla^2 N \qquad (14\text{-}27)$$

where t is time, $\mathbf{v}$ is the local velocity vector (variant with location and different from v_0), and D_p is the diffusion coefficient describing the Brownian motion of particles. The form of the equation is identical to that for a nonreactive substance in a reactor with advection and dispersion (Equation 1-80 in Chapter 1). The collector is treated as an infinite sink, so that the concentration vanishes at the surface. This sink creates a concentration gradient from higher values in solution to lower values next to the surface, leading to the diffusive particle flux toward the surface. Just as was done in Chapter 2 for a reactor whose flow pattern is plug flow with dispersion, the problem is nondimensionalized in terms of the Peclet number (Pe), expressing the ratio of the advective transport to the diffusive transport, with the characteristic length taken as the diameter of the collector

$$\text{Pe} = \frac{v_0 d_c}{D_p} \qquad (14\text{-}28)$$

The diffusion coefficient for the particle is estimated from the Stokes–Einstein equation (Equation 12-20), as was done for the Brownian motion of particles in flocculation

$$D_p = \frac{k_B T}{3\pi \mu d_p} \qquad (14\text{-}29)$$

where k_B is Boltzmann's constant and T is the absolute temperature. The solution, when expressed in terms of the single collector transport efficiency, was found by Yao (1968) to be as follows:

$$\eta_{Br} = 4.04\,\text{Pe}^{-2/3} = 0.905\left(\frac{k_B T}{\mu d_c d_p v_0}\right)^{2/3} \qquad (14\text{-}30)$$

■ **EXAMPLE 14-3.** Find the single collector transport efficiency by Brownian motion for a 2-μm diameter particle at 25°C in a filter with a filtration velocity of 5 m/h and a media size of 0.6 mm.

Solution. The solution is a direct application of Equation 14-30. The conversions of several parameters into CGS units are shown in the previous examples.

$$\eta_{Br} = 0.905\left(\frac{k_B T}{\mu d_c d_p v_0}\right)^{2/3}$$

$$= 0.905\left(\frac{(1.38 \times 10^{-16}\,\text{g cm}^2/\text{s}^2\,\text{K})298\,\text{K}}{(8.91 \times 10^{-3}\,\text{g/cm s})(0.06\,\text{cm})(2 \times 10^{-4}\,\text{cm})(0.14\,\text{cm/s})}\right)^{2/3}$$

$$= 1.78 \times 10^{-4}$$

This value is quite similar to that for sedimentation; recall, however, that with increasing size, the single collector transport efficiency by Brownian motion decreases, whereas it increases for sedimentation. ■

Overall Particle–Collector Collision Efficiency The particle transport by the three transport mechanisms is assumed to be additive, so that the total single collector transport efficiency is found as

$$\eta = \eta_I + \eta_g + \eta_{Br} \quad (14\text{-}31)$$

This equation assumes that a particle that is predicted to be removed by interception or sedimentation will not be simultaneously predicted to be removed by Brownian motion. Yao (1968) performed a numerical analysis considering the three mechanisms individually and simultaneously and concluded that this summation of the three separate collision efficiencies was reasonable.

The use of the model to predict removal in a filter involves finding values for the three single collector removal efficiencies (Equations 14-24, 14-26, and 14-30) for the specified operating conditions for each type of particle (size, density) to be considered. Equation 14-31 can be used to find the overall single collector removal efficiency for each particle type, and Equation 14-22 can then be used to determine the behavior within the full filter. Since α accounts for all phenomena that are not accounted for explicitly in the derivation of the single collector removal efficiencies, its value cannot be predicted or determined analytically. It is taken as an empirical fitting parameter for this model. Yao et al. (1971) published experimental results that were in reasonably good agreement with the model; of particular importance was the agreement of the trends of removal with respect to particle size, as explained subsequently.

■ **EXAMPLE 14-4.** Find the expected removal efficiency for the particle and filter characteristics specified in Examples 14-1 to 14-3. Additional specifications are that the media depth is 60 cm, the porosity is 0.40, the particles are perfectly destabilized, and all other short-range forces can be ignored; that is, $\alpha = 1$.

Solution. Using Equation 14-31 and the answers from the previous examples, we find

$$\eta = \eta_I + \eta_g + \eta_{Br} = 1.67 \times 10^{-5} + 1.45 \times 10^{-4} \\ + 1.78 \times 10^{-4} = 3.40 \times 10^{-4}$$

We can then find the expected fraction remaining from Equation 14-22, slightly rearranged as

$$\frac{N_{out}}{N_{in}} = \exp\left[-\frac{3}{2}\frac{(1-\varepsilon)\alpha\eta}{d_c}Z\right]$$
$$= \exp\left[-\frac{3}{2}\frac{(1-0.40)(1)(3.40\times 10^{-4})}{0.06\,\text{cm}}60\,\text{cm}\right] = 0.74$$

Finally, the fraction removed is $1 - (N_{out}/N_{in}) = 0.26$ or 26% removal. Such a removal level is not sufficient for most

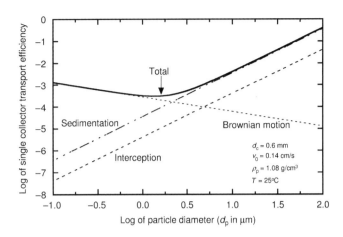

FIGURE 14-7. Single collector transport efficiencies according to the isolated spherical collector model. (Conditions: $d_c = 0.6$ mm; $v_0 = 0.14$ cm/s (~ 5 m/h); $\rho_p = 1.08$ g/cm^3; $T = 25°$C at which $\rho_L = 0.997$ g/cm^3 and $\mu = 8.91 \times 10^{-3}$ g/cm s.)

applications. However, the removal improves with ripening, and this size particle happens to be one with nearly the minimum collection efficiency. Therefore, the expected removal efficiency during the majority of the filter run and for most particle sizes is expected to be substantially $>26\%$. ■

Some aspects of the model (or similar models discussed below) are particularly useful for explaining the effects of the many variables that influence the removal of particles in filtration. The single collector transport efficiency for each mechanism and the total for all mechanisms are shown in Figure 14-7 for specified conditions that are in the range typically used in water treatment practice. One important result shown is that some critical size has a minimum transport efficiency. As the particle size decreases below this critical size, removal efficiency increases because of increased Brownian motion; as particle size increases above this critical size, removal efficiency increases by interception and sedimentation.

The critical size depends on the characteristics of collectors, particles, and operation, but within the range of these variables found in water and wastewater treatment, the critical size does not vary greatly. The lines for interception and sedimentation on this log–log graph have a slope of 2, reflecting the exponent in Equations 14-24 and 14-26; the line for Brownian motion has a slope of $-2/3$, reflecting the exponent in Equation 14-30. Recall that the sedimentation collision efficiency function represents the increment of removal over that expected from interception alone. As noted above, for almost all conditions encountered in water and wastewater treatment in standard high-rate filtration, sedimentation adds appreciably to interception as a removal mechanism. This fact is illustrated in Figure 14-5, which shows that sedimentation enhances particle transport toward the collector (relative to a particle that follows the streamlines).

690 GRANULAR MEDIA FILTRATION

TABLE 14-2. Changes in Particle Size Distributions in Filtration

Log of Particle Diameter log(d_p)	Particle Diameter d_p (μm)	Influent Log of Particle Size Distribution Function log$(\Delta N/\Delta d_p)_{in}$	Total Single Collector Removal Efficiency η	Fraction Remaining $N_{i,out}/N_{i,in}$	Effluent Log of Particle Size Distribution Function log$(\Delta N/\Delta d_p)_{out}$
−0.3	0.50	5.60	4.57E−04	0.663	5.42
−0.2	0.63	5.57	3.99E−04	0.698	5.41
−0.1	0.79	5.51	3.54E−04	0.727	5.37
0.0	1.00	5.44	3.22E−04	0.748	5.31
0.1	1.26	5.35	3.06E−04	0.759	5.23
0.2	1.58	5.24	3.09E−04	0.757	5.12
0.3	2.00	5.13	3.39E−04	0.737	4.99
0.4	2.51	4.99	4.08E−04	0.693	4.83
0.5	3.16	4.85	5.35E−04	0.618	4.64
0.6	3.98	4.68	7.53E−04	0.508	4.39
0.7	5.01	4.51	1.11E−03	0.368	4.07
0.8	6.31	4.31	1.69E−03	0.218	3.65
0.9	7.94	4.10	2.62E−03	0.094	3.07
1.0	10.00	3.86	4.10E−03	0.025	2.26
1.1	12.59	3.60	6.46E−03	0.003	1.08
1.2	15.85	3.32	1.02E−02	0.0	0.0
1.3	19.95	3.01	1.61E−02	0.0	
1.4	25.12	2.66	2.55E−02	0.0	
1.5	31.62	2.27	4.05E−02	0.0	
1.6	39.81	1.75	6.41E−02	0.0	
1.7	50.12	1.13			
1.8	63.10	0.29			

■ **EXAMPLE 14-5.** Determine the effluent particle size distribution from the filter described in the previous examples, if the influent distribution to the filter is as described in the first three columns of Table 14-2. The characteristics of the particles, water, and filter have been given in the above examples.

Solution. The calculation of η for each size follows the methods shown in Examples 14-1 to 14-4. The calculations are summarized in Table 14-2. Note that the value of η for the 2-μm particle agrees with the value shown in Example 14-4. All others are calculated in an identical manner. The effluent particle size distribution function value can be calculated from the influent value as follows:

$$\log\left(\frac{\Delta N}{\Delta d_p}\right)_{i,out} = \log\left[\left(\frac{\Delta N}{\Delta d_p}\right)_{i,in} \frac{N_{i,out}}{N_{i,in}}\right]$$

$$= \log\left[\left(10^{\log(\Delta N/\Delta d_p)_{i,in}}\right) \frac{N_{i,out}}{N_{i,in}}\right]$$

The influent and effluent particle size distribution functions are plotted in Figure 14-8. The removal of the larger particles is essentially complete, whereas the removal of the smaller particles is not as great. The overall removal efficiency on a number basis is 40%, whereas it is 97% on a volume basis. (These results for the removal efficiency can be calculated by converting the particle size distribution functions into the number and volume distributions for the influent and effluent.) ■

Single Spherical Collector in Packed Medium Model

Rajagopalan and Tien (1976), Rajagopalan et al. (1982), and Tien (1989) have presented a closely related but more sophisticated model than that of Yao et al. (1971), one which accounts for van der Waals attraction, the influence

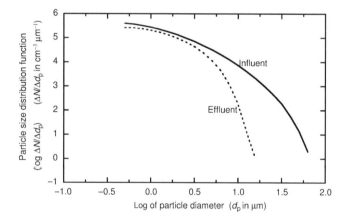

FIGURE 14-8. Effect of filtration on the particle size distribution function. (All filtration conditions given in previous examples; isolated collector model used for calculations.)

of neighboring collectors on the flow pattern, and hydrodynamic retardation. The influence of neighboring collectors was assessed with the use of Happel's sphere-in-cell model, originally developed as a means of describing the head loss as fluid passed through a packed medium. In this model, each spherical grain is considered to be surrounded by an imaginary spherical envelope, with the size of the cell (sphere and the surrounding envelope) determined to match the porosity of the overall medium (Happel, 1958). That is, if the radius of the particle (media grain, in the case of filtration) is a (equivalent to $d_c/2$), the radius of the cell, b, is chosen so that $(a^3/b^3) = 1 - \varepsilon$. The boundary condition on the cell, in applying the Navier–Stokes equation of motion, is that there is no shear at the outer cell boundary. In using this model for filtration modeling, the starting point of the trajectory analysis is at the edge of the fluid envelope rather than at an infinite distance above the collector.

Hydrodynamic retardation results from the fact that the approaching particle influences the flow pattern around the collector, especially when the separation distance is small. The forces and torques imposed on the particle were first described quantitatively by Goren and O'Neil (1971); the concepts are identical to those described in Chapter 12 for two approaching particles in flocculation.

The model of Rajagopalan and Tien cannot be solved analytically, so numerous trajectory analyses were performed numerically. In these trajectory analyses, particles were considered non-Brownian; that is, Brownian motion was ignored because the random motion could not be accounted for directly. The effects of Brownian motion are considered separately, as shown below. The trajectory results were correlated in terms of three dimensionless terms: a gravity number (N_G), a London van der Waals attraction (N_{Lo}) term, and the particle–collector size ratio (N_R). Sedimentation and interception were not separated in this model. Although Rajagopalan and Tien reported their results in terms of the unit bed element approach[6] (see Tien, 1989), the results are translated here and presented in terms of the single collector transport efficiency, for ease of comparison with the equations presented above. The results are as follows.

For combined interception and sedimentation

$$\eta_{I+g} = A_s N_{Lo}^{0.125} N_R^{1.875} + 0.00338 A_s N_G^{1.2} N_R^{-0.4} \quad (14\text{-}32)$$

where

$$A_s = \frac{2(1-p^5)}{2 - 3p + 3p^5 - 2p^6} \quad (14\text{-}33)$$

$$p = \frac{a}{b} = (1-\varepsilon)^{1/3} \quad (14\text{-}34)$$

$$N_{Lo} = \frac{4A_H}{9\pi\mu d_p^2 v_0} \quad (14\text{-}35)$$

$$N_R = \frac{d_p}{d_c} \quad (14\text{-}36)$$

$$N_G = \frac{(\rho_p - \rho_L)g d_p^2}{18\mu v_0}, \text{ and} \quad (14\text{-}37)$$

A_H is Hamaker's constant.

The A_s term is a parameter of the Happel sphere-in-cell model, so the influence of this choice of the flow description is explicit in this filtration model.

■ **EXAMPLE 14-6.** Find the single collector transport efficiency by the combination of interception and sedimentation according to the packed medium model of Rajagopalan and Tien for the 2-μm particle in the filter described in the earlier examples. Consider the Hamaker constant for this particle in water to be 3×10^{-13} g cm²/s².

Solution. We note that the expression for N_G in Equation 14-37 is the same as the expression for η_g in the single collector model, and its value under these conditions was found in Example 14-2. From that example, $N_G = 1.45 \times 10^{-4}$.

From Equation 14-36,

$$N_R = \frac{d_p}{d_c} = \frac{2 \times 10^{-4} \text{ cm}}{6 \times 10^{-2} \text{ cm}} = 0.0033$$

---

[6] In the unit bed element approach, the entire filter is assumed to consist of a stack of horizontal unit bed elements in series, each with a thickness on the order of a grain diameter (although the exact thickness depends on the porosity). The removal efficiency of a single unit bed element is calculated for a chosen porous media model and chosen geometry for the pores or collectors within the element. The overall removal efficiency of the entire bed is then calculated as: $E = 1 - \sum_{i=1}^{M}(1 - e_i)$, where e_i is the efficiency of a single unit bed element and M is the number of bed elements in the entire filter. This efficiency, e_i, can be related to the filter coefficient and the length of a unit bed element. With various coworkers, Tien has developed a number of models in this framework, with different choices for the porous media flow model and for the geometry. Reported here are the results for the single spherical collector model (for which e_i is related to η) and the Happel flow model. Fundamentally, the unit bed element approach assumes that the Iwasaki formulation (Equation 14-4) is valid at lengths less than that of the unit bed element and not above; the Yao et al. model essentially assumes the opposite.

From Equation 14-35,

$$N_{Lo} = \frac{4A_H}{9\pi\mu d_p^2 v_0}$$

$$= \frac{4(3 \times 10^{-13}\,\mathrm{g\,cm^2/s^2})}{9\pi(8.91 \times 10^{-3}\,\mathrm{g\,cm^2/s^2})(2 \times 10^{-4}\,\mathrm{cm})^2(0.14\,\mathrm{cm/s})}$$

$$= 8.51 \times 10^{-4}$$

From Equation 14-34,

$$p = (1-\varepsilon)^{1/3} = (0.6)^{1/3} = 0.843.$$

From Equation 14-33,

$$A_s = \frac{2(1-p^5)}{2 - 3p + 3p^5 - 2p^6}$$

$$= \frac{2(1 - 0.843^5)}{2 - 3(0.843) + 3(0.843)^5 - 2(0.843)^6} = 38.0$$

Finally, from Equation 14-32, we find

$$\eta_{I+g} = A_s N_{Lo}^{0.125} N_R^{1.875} + 0.00338 A_s N_G^{1.2} N_R^{-0.4}$$

$$= (38)(8.51 \times 10^{-4})^{0.125}(0.0033)^{1.875}$$

$$+ 0.00338(38)(1.45 \times 10^{-4})^{1.2}(0.0033)^{-0.4}$$

$$= 3.81 \times 10^{-4} \qquad \blacksquare$$

The single collector efficiency due to Brownian motion in the Rajagopalan and Tien model is identical to that in the simple, isolated collector model, except for a modification to account for the different assumptions about the velocity distribution around the collectors.[7] The resulting expression for η_{Br} (Cookson, 1970) is

$$\eta_{Br} = 4A_s^{1/3}\mathrm{Pe}^{-2/3} = 0.9 A_s^{1/3}\left(\frac{k_B T}{\mu d_c d_p v_0}\right)^{2/3} \qquad (14\text{-}38)$$

As in the Yao model, the single collector removal efficiencies for different mechanisms are considered as additive ($\eta = \eta_{I+g} + \eta_{Br}$), and the overall filter performance can be predicted by applying Equation 14-22. Again, α cannot be predicted and is used as an empirical fitting parameter ($0 < \alpha < 1$), but the value in this model should be higher than that in the Yao model because more of the short-range phenomena that interfere with collisions have been accounted for explicitly in the formulation of the model.

■ **EXAMPLE 14-7.** Find the expected overall removal efficiency, according to the packed media model, of the 2-μm particle in the filter described in the previous examples.

Solution. We begin by finding the single collector removal efficiency by Brownian motion. As noted above, the expression in Equation 14-38 is similar to that in the single collector model except for the flow model correction.

$$\eta_{Br} = 0.9 A_s^{1/3}\left(\frac{k_B T}{\mu d_c d_p v_0}\right)^{2/3}$$

$$= 0.9(38)^{1/3}\left(\frac{(1.38 \times 10^{-16}\,\mathrm{g\,cm^2/s^2\,K})298\,\mathrm{K}}{(8.91 \times 10^{-3}\,\mathrm{g/cm\,s})(0.06\,\mathrm{cm})(2 \times 10^{-4}\,\mathrm{cm})(0.14\,\mathrm{cm/s})}\right)^{2/3}$$

$$= 5.94 \times 10^{-4}$$

Summing the different individual removal efficiencies leads to

$$\eta = \eta_{Br} + \eta_{I+g} = (5.94 \times 10^{-4}) + (3.81 \times 10^{-4}) = 9.75 \times 10^{-4}$$

Finally, the overall fraction remaining is found from Equation 14-22

$$\frac{N_{out}}{N_{in}} = \exp\left[-\frac{3}{2}\frac{(1-\varepsilon)\alpha\eta}{d_c}Z\right]$$

$$= \exp\left[-\frac{3}{2}\frac{(1-0.40)(1)(9.75 \times 10^{-4})}{0.06\,\mathrm{cm}}60\,\mathrm{cm}\right] = 0.416$$

and the removal, η_{filter}, is $1 - (N_{out}/N_{in}) = 1 - 0.416 = 0.584$.

The predicted removal efficiency is somewhat higher in the packed medium model than in the original single collector model; much of this difference stems from the more sophisticated flow model used. ■

Updated Packed Medium Model

As indicated above, the trajectory analysis that underlies the Rajagopalan and Tien model considered interception and sedimentation but not Brownian motion; transport by Brownian motion was considered separately (and identically to how it was treated by earlier investigators). Tufenkji and Elimelech (2004) included particle diffusion (Brownian motion) within the context of the flow regime, so that all three long-range transport mechanisms were considered simultaneously. They also considered the role of van der Waals attraction and hydrodynamic retardation on the single collector removal by Brownian motion; these phenomena had been ignored by Cookson (1970) and therefore by

---

[7] Yao et al. (1971) also reported this Cookson–Happel modification to η_{Br} and recommended its use. It is reported in this chapter as associated with the packed medium model of Rajagopalan and Tien rather than with the isolated collector model of Yao et al. primarily because the separation is more consistent with other aspects of the two models, but also to demonstrate in the examples and figures the effect of this Happel flow parameter.

Rajagopalan and Tien. Using an approach similar to that of Rajagopalan and Tien (i.e., solving the equations of motion including Brownian diffusion over various conditions, and then expressing the results by a regression equation), they found the following equation for η_{Br}:

$$\eta_{Br} = 2.4 A_s^{1/3} \text{Pe}^{-0.715} N_R^{-0.081} N_{vdW}^{0.052} \quad (14\text{-}39)$$

where
$$N_{vdW} = \frac{A_H}{k_B T} \quad (14\text{-}40)$$

N_{vdW} is the van der Waals number, and all other variables are as defined earlier. This equation yields a lower collection efficiency by Brownian motion than that found from Equation 14-38, but substantially higher efficiency than without the flow correction (Equation 14-30); these differences also indicate that the equation proposed by Tufenkji and Elimelech fits the experimental results reported by Yao et al. (1971) better than either of the other equations.

Tufenkji and Elimelech (2004) also presented revised equations for collection by interception and sedimentation. The results are quite similar, but not identical, to those reported by Rajagopalan and Tien, because these recent investigators were more explicit in accounting for the temperature effects on van der Waals attraction and hydrodynamic retardation. Just as earlier investigators (in both filtration and flocculation), they assumed that the surface charge on the particles was zero, so that electrostatic interactions could be ignored. Their definitions of variables were identical to those of Rajagopalan and Tien, except for the addition of the van der Waals number and a difference in the numerical constant included in their *attraction number* in comparison to Rajagopalan and Tien's London number

$$N_A = \frac{A_H}{3\pi \mu d_p^2 v_0} \quad (14\text{-}41)$$

The results of their numerical solutions could be well described by the following regression equations for the single collector removal efficiencies by interception and sedimentation:

$$\eta_I = 0.55 A_s N_R^{1.675} N_A^{0.125} \quad (14\text{-}42)$$

$$\eta_g = 0.22 N_R^{-0.24} N_G^{1.11} N_{vdW}^{0.053} \quad (14\text{-}43)$$

Tufenkji and Elimelech (2004) compared the experimental results from several investigators to the three models explained above (Yao et al., including the Cookson correction for η_{Br}; Rajagopalan and Tien; and their own model) and showed that their model gave the overall best fit to these data.

Other Advanced Models

The Yao model presented above is highly simplified; it can be considered analogous to the rectilinear model for collisions in flocculation. Since the development of the model by Yao (1968), several investigators have improved the mathematical formulation by reducing the number of simplifications incorporated into the model. The Rajagopalan and Tien model is representative of a class of more complex models that account more realistically for the flow field, the geometry of the bed and forces acting on the particles. Some other proposed modifications to Yao's model are described briefly below.

First, modifications to the equation for the flow field that account for neighboring collectors have been considered. The Yao model considered the Stokes flow equation, which was developed for the single particle (collector) in an infinite medium. Besides the Happel model used by Rajagopalan and Tien, the model of flow in porous media by Brinkman (1949) has often been used. Even these models represent simplifications of the true flow field; designed to account for the head loss through the porous media but not the exact flow pattern, they average some characteristics of the filter. For example, they are two-dimensional representations that ignore contact points between media grains.

Second, the geometry of a real filter bed is obviously considerably more complicated than that of a single sphere. Several geometries have been considered by different investigators, including constricted tube models (Payatakes et al., 1973a,b) and assemblages of spheres in known geometries. After choosing a geometry, the flow equations and forces on particles can be assimilated into the trajectory analysis, and various mechanisms for collection can be studied. Tien, with various coworkers, has investigated this area of filtration modeling extensively; this work is discussed in detail in Tien (1989). Cushing and Lawler (1998) considered a densely packed cubic array of spheres as the geometry so as to determine the effects of contact points between filter grains on the flow pattern and particle trajectories; after correcting for the porosity of a real filter (much greater than in dense packing), they expressed the results for non-Brownian transport efficiency in a form similar to that of Rajagopalan and Tien but with different coefficients and exponents, as follows:

$$\eta_{I+g} = 0.029 N_{Lo}^{0.012} N_R^{0.023} + 0.48 N_G^{1.8} N_R^{-0.38} \quad (14\text{-}44)$$

When used instead of Equation 14-32, this model exhibits greater removal of small particles and a less dramatic effect of particle size on removal efficiency with trends that are consistent with some reported experimental results.

Third, additional forces have been considered in the analysis of particle motion in the vicinity of collectors. Identically to the short-range model of particle collisions in flocculation, these forces include van der Waals attraction, electrostatic repulsion, and the hydrodynamic interactions as a particle approaches the collector. The hydrodynamic effects (often referred to as hydrodynamic retardation, since they act to hinder collisions) account for the increased viscous forces as the water between the approaching particle and the collector moves out of the way. All these effects are accounted for in the trajectory analysis in the same manner as in flocculation. Including these effects in the transport equations increases the degree to which the model incorporates the true motion of particles in a filter and decreases the effect of the empirical factor α. The Rajagopalan and Tien model does account for these forces to some degree, but other methods are possible.

All the models presented above implicitly assume that, for particles to be collected, they must fall into the primary minimum of the energy curve describing the electrostatic interactions between the particle and collector.[8] However, by using a Monte Carlo simulation approach in combination with the equations of motion and the interaction energies at every location, Hahn and O'Melia (2004) provided convincing evidence that particles in the Brownian size range could also be collected by coming to rest at the location of the secondary minimum of this curve. They indicated that these results could be well modeled by calculating a theoretical collision efficiency factor. They proposed that small particles, similar to molecules, have a distribution of velocities that can be adequately described by the Maxwell distribution, f_M (developed for molecules)

$$f_M(v_p) = 4\pi \left(\frac{m_p}{2\pi k_B T}\right)^{3/2} v_p^2 \exp\left(-\frac{\frac{1}{2}m_p v_p^2}{k_B T}\right) \quad (14\text{-}45)$$

where m_p and v_p are the mass and velocity of particles, respectively, and f_M is the probability that a given particle will have a velocity between v_p and $v_p + dv_p$, divided by dv_p.

In Equation 14-45, the argument of the exponential is the ratio of kinetic energy to thermal energy. If a particle has an amount of kinetic energy that exceeds the depth of the secondary minimum, it will not be captured; similarly, if a particle is captured in the secondary minimum and, by collisions with other particles, it acquires more kinetic energy than the depth of the secondary minimum, it will escape. Therefore, this energy is considered as the activation energy for escape from the secondary minimum. Because $f_M(v_p)$ is a probability distribution function, the integral of $f_M(v_p)dv_p$ over any interval in v_p equals the fraction of particles that have velocities falling in this interval, and its integral over all values of v_p is one. Therefore, Hahn and O'Melia could define a theoretical collision efficiency factor for capture in the secondary minimum (α) as follows:

$$\alpha = 1 - \begin{pmatrix} \text{Fraction of particles with} \\ \text{kinetic energy greater than} \\ \text{activation energy} \end{pmatrix}$$
$$= 1 - \begin{pmatrix} \text{Fraction of particles} \\ \text{with } v_p > v_{p,\text{act}} \end{pmatrix} = 1 - \int_{v_{p,\text{act}}}^{\infty} f_M(v_p)$$
$$(14\text{-}46)$$

where $v_{p,\text{act}}$ is the velocity required for a particle to have kinetic energy equal to the activation energy for escape from the secondary minimum.

As explained in Chapter 11, the depth of the secondary minimum (here considered as the activation energy) is determined by the characteristics of the solution, surface potential on the particle, and Hamaker constant. Under carefully controlled conditions, these values are all known. Hahn and O'Melia showed excellent agreement of this model with the previously reported experimental results.

14.5 PREDICTED CLEAN BED REMOVAL IN STANDARD WATER AND WASTEWATER TREATMENT FILTERS

It is instructive to explore the predictions of the models for clean bed particle removal in what might be considered a standard design and the effects of various design and operating variables on these predictions. Consider first a deep bed filter for drinking water, and also that earlier treatment includes alum flocculation and sedimentation. The particles remaining will have a low density (assuming sweep floc conditions) and a broad size distribution; here, we assume that the density is 1.08 g/cm^3 and that particles might range from 0.1 to 100 μm. The removal efficiency for a clean filter can be predicted from the Yao et al. model using Equations 14-22, 14-24, 14-26, 14-30, and 14-31; the predicted removal efficiency is shown in Figure 14-9 for conditions specified in this figure.

The results for conditions such that the attachment efficiency (α) is 1 are shown by the solid line in Figure 14-9. The strong effect of particle size is apparent, with essentially complete removal of all particles greater than 10 μm ($\log d_p > 1.0$), minimum removal ($\sim$25%) of particles 1.3 μm in diameter ($\log d_p = 0.1$), and good but not excellent removal of the smallest particles. Such removal would be considered inadequate for drinking water purposes: many particles would pass through the filter and the resulting particle number concentration or turbidity would be too high. Because the results shown characterize the filter performance right at the beginning of the filter run, the filter could still operate successfully, but only after it ripens;

---

[8] Recall from Chapter 11 that, when two like-charged particles approach one another, the combination of van der Waals attraction, electrical repulsion, and Born repulsion leads to two local maxima in the net attractive force: one at a shorter separation distance, where the net attraction is very strong (the primary minimum in interaction energy) and one at a greater separation where the interaction is weaker, but still attractive (the secondary minimum).

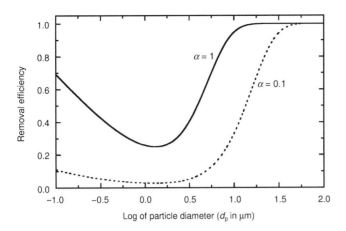

FIGURE 14-9. Effects of attachment efficiency on overall filter removal efficiency. (Same conditions as in Figure 14-7; also $Z = 60$ cm; $\varepsilon = 0.40$.)

that is, after removal increases during the early part of the run. Without ripening, filters for drinking water applications would need to be designed and operated more stringently; that is, with greater media depth, smaller media, lower velocity, or some combination of these three. Fortunately, substantial ripening does occur and removal is quickly improved over that predicted in this example.

As noted previously, control of the surface chemistry to destabilize the particles is an essential part of filtration. The attachment efficiency factor, α, must account for all short-range phenomena and various factors not accounted for explicitly in the formulation of the model; these factors include van der Waals attraction (in some of the modeling), electrostatic interactions, effects of interparticle bridging or enmeshment in a floc, deviations from random packing, nonspherical media, incomplete description of the flow field, and perhaps other imperfections in the model assumptions. Of these factors, the destabilization effects (charge neutralization, enmeshment in a precipitate, or interparticle bridging) are the most variable among suspensions and the most controllable by the operator. It is not uncommon, therefore, for α to be referred to as a measure of the destabilization achieved.

The dramatic effect of the attachment efficiency is demonstrated in Figure 14-9 by comparison of predicted results for complete destabilization ($\alpha = 1$) and poor destabilization ($\alpha = 0.1$). (α values can be orders of magnitude <0.1 so this latter condition represents a case where the destabilization, although not perfect, is within the range that might be achieved in operating treatment plants.) Except for the very largest particles (of which one would expect very few at this stage of treatment), removal is dramatically less with $\alpha = 0.1$ than with $\alpha = 1$. For example, 10μm ($\log d_p = 1.0$) particles, which are almost completely removed (98%) in the perfect destabilization case, have only a 31% removal efficiency under the less than ideal destabilization conditions considered.

A primary conclusion in the original Yao, et al. (1971) paper was that, in filters of standard design treating typical suspensions, opportunities for collisions between particles in suspension and media grains (i.e., for long-range transport to bring particles into the proximity of media grains) are plentiful, so that excellent removal should be achieved as long as the proper chemical (destabilization) conditions are maintained.[9] As suggested by the model results in Figure 14-9, poor removal is expected when destabilization is poor. Several laboratory studies have confirmed these effects of destabilization (e.g., Habibian and O'Melia, 1975) and thereby confirm both the importance of proper surface chemistry and the unimportance of straining in deep bed filtration (because removal by straining would be insensitive to surface chemistry). A study of operating full-scale water treatment facilities (Cleasby et al., 1992) also concluded that improper destabilization chemistry was the primary cause of poor particle removal (or detachment of captured particles) in deep bed filters. If a filter is achieving less than satisfactory removal, the first factor to investigate is the destabilization chemistry.

A comparison of the three models presented above for the typical conditions is shown in Figure 14-10. The predictions for small particles (i.e., those in the range for which Brownian motion dominates) are quite different. The difference in this range between the models of Yao and Rajagopalan and Tien stems entirely from the use of the Happel flow model in Rajagopalan and Tien's work and the isolated

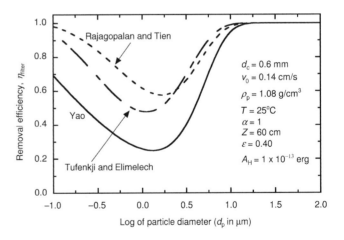

FIGURE 14-10. Comparison of the three most common filtration models. (All conditions same as in Figure 14-9; $\alpha = 1$.)

---

[9] This conclusion is not so obvious from Figure 14-9, Yao et al. drew this conclusion based in part on their recommendation to use the Cookson–Happel equation for η_{Br}. As shown subsequently, this equation predicts greater removal efficiency for small particles than that calculated by Equation 14-30 and shown in Figure 14-9.

collector flow model in Yao's work; as noted above, both models used the same expression for the capture by Brownian motion except for the flow model. The Rajagopalan and Tien model predicts the best removal for small particle sizes and the Tufenkji and Elimelech model predicts the best removal for the larger sizes, but the basic conclusions given above for the Yao et al. model still hold true; that is, ripening is necessary to achieve excellent removal, both large and small particles are captured better than some intermediate size in the range of 1 μm, and excellent destabilization is necessary for good particle removal.

Design Tradeoffs

Because designers have many variables under their control, it is valuable to consider what these models suggest about the tradeoffs among the principal design variables: the size of the media (d_c), the depth of the media (Z), and the filtration velocity (v_0). In considering these tradeoffs, the Yao et al. model is easiest to work with because of its mathematical simplicity and the complete separation of the different mechanisms; in the discussion below, the algebraic result is shown from the Yao et al. model, but corresponding numerical results from all three models are shown.

The predicted effects from tradeoffs among filtration velocity, filter depth, and media size are illustrated in Figure 14-11 for the three models. The baseline or standard condition is the same as shown in Figure 14-10. As noted above, the primary interest in filtration has traditionally been the removal of particles >~1 μm, and these particles (at least according to the Yao et al. model) are primarily removed by the sedimentation mechanism within rapid rate filters. The alternative designs shown in Figure 14-11 were chosen in such a way that the removal by sedimentation according to this model would be the same for all four conditions. If the removal were accomplished by sedimentation alone (i.e., $\eta = \eta_g$), the removal in the full depth of the filter would be computed by inserting the single collector removal efficiency for sedimentation, η_g (Equation 14-26), into Equation 14-22, yielding

$$\ln \frac{N_{out}}{N_{in}} = -\frac{3}{2} \frac{(1-\varepsilon)\alpha(\rho_p - \rho_L)g d_p^2}{18\mu v_0} \frac{Z}{d_c} \quad (14\text{-}47)$$

In such a case, two filters (A and B) with the same porosity would have the same predicted removal efficiency for the same suspension if the following condition was met:

$$\frac{Z_A}{v_{oA} d_{cA}} = \frac{Z_B}{v_{oB} d_{cB}} \quad (14\text{-}48)$$

According to this equation, if a designer wants to change from one media size to a larger size (from d_{cA} to d_{cB}) and

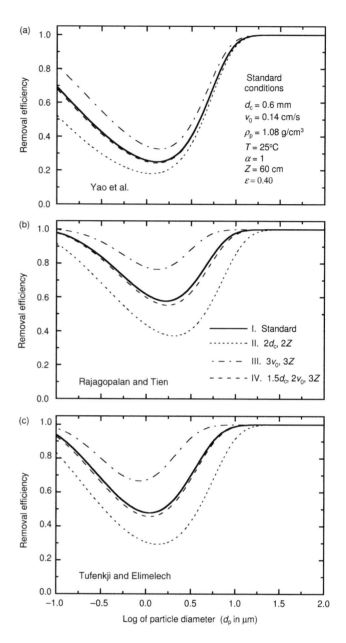

FIGURE 14-11. Effects of design tradeoffs among the major variables: depth, velocity, and media size according to several models. (Standard case as in Figure 14-10.)

maintain the same removal efficiency for large particles, the depth of media should be increased proportionally (such that $Z_B/Z_A = d_{cB}/d_{cA}$). Similarly, an increase in velocity should be offset with a proportional increase in media depth or a proportional decrease in media size.

The three conditions shown in Figure 14-11 besides the standard condition (Case I) all satisfy the criterion of Equation 14-48 relative to the standard condition; that is, if Case I is considered as filter A, the three other cases are different options for filter B that satisfy Equation 14-48. Consider first the results in Figure 14-11a (the Yao et al.

model). For particles larger than the size corresponding to the minimum removal efficiency, the predicted performance of the filter is quite similar for all four cases. This result is expected, because sedimentation is the dominant mechanism for these particles, and the conditions have been chosen to make removal by this mechanism identical in all four cases. Even so, the results are not identical in this range because collisions induced by interception and Brownian motion, while relatively unimportant, are not negligible.

We consider next what would be required for removal of particles by interception to be identical in two filters. A combination of Equations 14-24 and 14-22 suggests that the following equation must be satisfied to meet this condition according to the Yao et al. model

$$\frac{Z_A}{d_{cA}^3} = \frac{Z_B}{d_{cB}^3} \quad (14\text{-}49)$$

Since interception relies on strictly geometric factors, the velocity of water flow through the column is predicted to have no effect on particle removal efficiency by this mechanism. Equation 14-49 indicates that the diameter of the collector grains is very important in determining the frequency of particle–collector collisions by interception. For instance, if we were to double the media size, the equation predicts that the filter column would have to lengthened by a factor of eight in order to maintain equivalent interception-induced collision rates.

In practice, the media size used in filters has gradually increased over the past 50 years, but designers have included minimal changes in the depth of the bed (certainly not more than a proportional increase); the fact that these filters have operated satisfactorily (i.e., with little change in effluent quality) suggests that interception (at least as described in this model) is not the dominant mechanism of removal.

For smaller particles (i.e., for sizes below the diameter of minimum removal efficiency), Brownian motion is the dominant transport mechanism, and the removal efficiencies under the four conditions shown in Figure 14-11a are substantially different. For removal by Brownian motion to be identical for different designs (A and B), the following condition (found from Equations 14-22 and either 14-30 or 14-38) must be met according to the Yao et al. model

$$\frac{Z_A}{v_{0A}^{2/3} d_{cA}^{5/3}} = \frac{Z_B}{v_{0B}^{2/3} d_{cB}^{5/3}} \quad (14\text{-}50)$$

This condition is very nearly met in the comparison of the baseline case with Case IV; the difference in the ratio expressed in Equation 14-50 between these two cases is <5%. As a consequence, the predictions for the removal of small particles are almost identical in the two cases. The other conditions (Cases II and III), however, yield different results. For Case II, changing the length proportionally to the change in media size is less than what is required to keep the removal of small particles the same, and so the removal efficiency is less than in the standard case for these particles. For Case III, changing the length proportionally to the change in filtration velocity is more than what is required to keep the removal of small particles the same, and so the removal efficiency is greater than in the standard case for these particles.

The predicted removal efficiencies according to the other two models are shown in Figure 14-11b,c for the same four cases. The differences among these models for the standard conditions are explained along with Figure 14-10. The tradeoffs among depth, velocity, and media size according to the Tufenkji and Elimelech model are nearly identical to those of the Rajagopalan and Tien model, in terms of changes from their respective standard conditions, but both of these models predict a greater change in removal efficiency among the different conditions than the Yao et al. model. Nevertheless, the predictions of all three models suggest that condition III (triple the standard velocity and filter depth) leads to better removal, whereas condition II (double the media size and depth) leads to worse removal than the standard condition. These results suggest that, in terms of the resulting water quality, increasing the media size is a more risky alternative than increasing the velocity. For small particles, the Tufenkji and Elimelech model is intermediate between the other models for all conditions; recall that the only difference between the Rajagopalan and Tien model and the Yao et al. model for the small particles is the inclusion of the Happel flow parameter.

Another difference among the models is the effect of various design changes on the diameter associated with the least removal efficiency; for all the models, this point reflects the size below which removal by Brownian motion becomes the most significant. In this regard, the Yao et al. model is quite insensitive to the changes shown, although this value decreases relative to the standard condition for the increased media size and increases for the increased velocity. The other two models have a more sensitive response to these design and operational changes, and suggest the opposite effect: the condition that leads to the smallest diameter associated with the lowest removal efficiency is the one with increased velocity, and that diameter is increased for the condition with increased diameter. Across all the results shown for all three models, this diameter associated with the least removal efficiency varies from ~0.8 to 2.2 μm (log diameters of −0.1 and 0.35, respectively).

These conclusions about the differences among the three models depend to some extent on the specific conditions being considered. For instance, if a much higher particle density is chosen (e.g., $> 2.0\,g/cm^3$), the right portions of the curves in the Rajagopalan and Tien and the Tufenkji and Elimelech models fall much closer to one another and, in that way, become much more similar to the predictions of the Yao et al. model. Additional investigation of the tradeoffs among various design and operating parameters is instructive and can be carried out by developing spreadsheets to compute various transport efficiencies under different conditions.

14.6 HEAD LOSS IN A CLEAN FILTER BED

Several models have been developed to describe the loss of energy as water passes through the porous media of a filter. For laminar flow (low Reynolds number) conditions, the most commonly used is the Carman–Kozeny equation, named after the original investigators responsible for its development (Kozeny, 1927; Carman, 1937).

The driving force for water flow through a length Δz of the media equals the product of the difference in available energy per unit volume (ΔE_V) over this length and the cross-sectional area of water. For the system under consideration (illustrated in Figure 14-12), the cross-sectional area of the water is assumed to be the product of the porosity and the total area; that is, the area porosity is assumed to be equivalent to the volume porosity, a reasonable assumption for random packing. Equation 14-1 indicates that ΔE_V is $\rho_L g h_L$, so

$$F_{\text{water flow}} = (\Delta E_V)A\varepsilon = \rho_L g h_L A\varepsilon \quad (14\text{-}51)$$

The media grains exert a frictional resisting force on the water at the grain/water interface, which generates fluid shear throughout the aqueous phase. The magnitude of this force is the product of the stress on the media (τ) and the total surface area of the grains in the section of interest. Expressing the surface area over the full length of the bed (Z) as the product of the surface area per unit volume of bed (S) and the volume of the bed in this length, we obtain

$$F_{\text{resisting flow}} = \tau SAZ \quad (14\text{-}52)$$

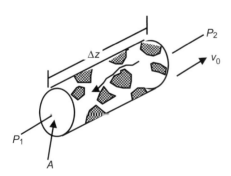

FIGURE 14-12. Definition sketch for pressure drop in flow through porous media.

A number of substitutions can be made into Equation 14-52 to make it more useful in this situation. First, a dimensional analysis[10] is carried out, assuming that the stress is a function of the fluid viscosity (μ), fluid density (ρ_L), hydraulic radius (r_H), and average interstitial velocity in the pores (v_{pore}). The hydraulic radius is a characteristic length that is related to the geometry of the flow path and that serves as an indicator of the geometric contribution to the resistance to flow. For well-defined flow paths (e.g., pipes or open channels), r_H is normally defined as the ratio of the wetted area to the wetted perimeter. Unfortunately, a wetted perimeter cannot be easily defined for the complex flow path in the system of interest here. Under the circumstances, one reasonable definition of r_H might be the ratio of the wetted volume to the wetted area. However, this definition implicitly assumes that, as water moves through the bed, its flow is resisted by the average surface area per unit volume. This assumption would be valid if the water followed a linear path from the entrance to the exit. In truth, though, the water takes a tortuous path, and its sideways motion causes it to encounter additional surface area without encountering additional volume. To account for this fact, we define r_H as the ratio of the wetted volume to the product of the wetted area and the tortuosity, where the tortuosity is the ratio of the true distance traveled by the water (Z_e) to the length of the bed (Z); that is

$$r_H = \frac{\varepsilon}{S(Z_e/Z)} = \frac{\varepsilon}{S} \frac{Z}{Z_e} \quad (14\text{-}53)$$

The dimensional analysis yields the following relationship:

$$\frac{\tau}{\rho_L v_{\text{pore}}^2} = k_a \frac{\mu}{\rho_L v_{\text{pore}} r_H} \quad (14\text{-}54)$$

where k_a is a constant of proportionality. The average interstitial velocity, v_{pore}, is related to the superficial velocity, v_0, as follows:

$$v_{\text{pore}} = \frac{v_0}{\varepsilon} \frac{Z_e}{Z} \quad (14\text{-}55)$$

The first fraction on the right side of Equation 14-55 equals the average fluid velocity in the z direction inside the bed, and the second fraction accounts for the fact that the actual path is not just in this direction, but is tortuous.

---

[10] In a dimensional analysis, a dependent variable of interest (z) is assumed to be related to several independent variables ($s, t, u, \ldots$) by an equation of the form: $z = s^a t^b u^c \ldots$. In many cases, the need for dimensional consistency in this relationship establishes what the values of the exponents (a, b, and c, in this example) must be. It is common to then combine the variables into dimensionless groups and express the final relationship in terms of these groups, as is done in Equation 14-54.

Finally, the surface area per unit volume of bed, S, can be related to the porosity and the surface area per unit volume of media (S_0) as follows:

$$S = S_0(1 - \varepsilon) \qquad (14\text{-}56)$$

Substituting Equations 14-53 to 14-56 into Equation 14-52 yields

$$F_{\text{resisting flow}} = k_a \mu v_0 S_0^2 A \frac{Z_e^2}{Z} \frac{(1-\varepsilon)^2}{\varepsilon^2} \qquad (14\text{-}57)$$

If the water flow is steady, then the driving force for flow must equal the force resisting flow, so Equation 14-57 can be equated with Equation 14-51 to yield an expression for the superficial velocity v_0 as a function of the physical properties of the system and the pressure differential across the bed

$$\begin{aligned}
v_0 &= \frac{1}{k_a} \frac{\varepsilon^3}{(1-\varepsilon)^2} \frac{1}{S_0^2} \frac{\rho_L g h_L}{\mu} \frac{Z}{Z_e^2} \\
&= \frac{1}{k_a(Z_e/Z)^2} \frac{\varepsilon^3}{(1-\varepsilon)^2} \frac{1}{S_0^2} \frac{\rho_L g h_L}{\mu} \frac{1}{Z} \\
&= \frac{1}{k} \frac{\varepsilon^3}{(1-\varepsilon)^2} \frac{1}{S_0^2} \frac{\rho_L g h_L}{\mu} \frac{1}{Z} \qquad (14\text{-}58)
\end{aligned}$$

where $k \equiv k_a(Z_e/Z)^2$.

The tortuosity, Z_e/Z, and the constant of proportionality, k_a, are both functions of the geometry. They can be derived from fundamental principles for well-defined geometries, but they can only be estimated for random packing (as is common in filters and other environmental systems). Typically, these two terms are lumped into a single empirical constant, k, as shown in Equation 14-58. Solving this equation for the head loss per unit length of the filter yields the Carman–Kozeny equation

$$\frac{h_L}{Z} = k \frac{\mu}{\rho_L g} \frac{(1-\varepsilon)^2}{\varepsilon^3} S_0^2 v_0 \qquad (14\text{-}59)$$

k is usually found to be between 4.5 and 6 (and most commonly assumed to be 5) for random packed media.

The Carman–Kozeny equation is considered to be a fundamental equation describing the head loss in a clean filter bed as a function of properties of the fluid, properties of the media, and design and operating variables. The head loss is directly proportional to the length (or depth) of the filter and also directly proportional to the superficial velocity (not the velocity squared, as is typically the case for turbulent flow). For spherical media, the specific surface area, S_0, is the ratio of the surface area of a sphere (πd_c^2) to the volume of a sphere ($\pi d_c^3/6$), yielding $S_0 = 6/d_c$. For nonspherical media, the constant 6 is replaced by a shape factor, ϕ, with a value greater than 6 (because a sphere has the minimum surface area per unit volume of any shape). For common media in use in filters (sand, charcoal, and activated carbon), a value of ϕ of 7.5 is common.[11]

■ **EXAMPLE 14-8.** Find the head loss according to the Carman–Kozeny equation for the filter described in earlier examples. Assume that the value of the constant, k, is 5 and that the sand grains have a shape factor of 7.5.

Solution. The characteristics of the filter are: media depth = 60 cm, media size = 0.6 mm, porosity = 0.40, temperature of water = 25°C at which the density = 0.997 g/cm^3 and the viscosity = 8.91 × 10^{-3} g/cm s. The solution is a direct application of Equation 14-59.

$$\begin{aligned}
h_L &= k \frac{\mu}{\rho_L g} \frac{(1-\varepsilon)^2}{\varepsilon^3} S_0^2 v_0 Z \\
&= 5 \frac{8.91 \times 10^{-3} \text{ g/cm s}}{(0.997 \text{ g/cm}^3)(981 \text{ cm/s}^2)} \\
&\quad \times \left(\frac{(1-0.40)^2}{0.40^3}\right)\left(\frac{7.5}{0.06 \text{ cm}}\right)^2 (0.14 \text{ cm/s})(60 \text{ cm}) \\
&= 33.6 \text{ cm}
\end{aligned}$$ ■

Several other equations to describe the head loss in filters have also been used. These equations are similar but not identical to the Carman–Kozeny equation with respect to the functionality of certain variables and the numerical results under certain conditions. Tien (1989) tabulated many of these equations, and Trussell and Chang (1999) summarized how the more common equations were derived. The Carman–Kozeny equation assumes creeping flow around the media grains (i.e., the inertial terms in the Navier–Stokes equation are ignored); while this assumption has been reasonable for traditional filter designs, modern filters often have much higher velocity and the assumption can be questionable. Ergun (1952) proposed an equation that combined two terms—the first with the same functionality as the Carman–Kozeny equation and the second to account for inertial effects, as follows:

$$h_L = k_1 \frac{\mu}{\rho_L g} \frac{(1-\varepsilon)^2}{\varepsilon^3} S_0^2 v_0 Z + k_2 \frac{1}{g} \frac{(1-\varepsilon)}{\varepsilon^3} S_0 v_0^2 Z \qquad (14\text{-}60)$$

---

[11] Some authors write this equation slightly differently, by taking the factor of 6^2 out of the S_0 term and introducing the sphericity (Ψ) into this term. Sphericity is the ratio of the surface area of a sphere to the surface area of the grain, when both have the same volume-equivalent diameter, and therefore is less than one. It is equivalent to $6/\phi$. Multiplying the common value of 5 for k by 6^2 leads to a leading coefficient value of 180.

The values of k_1 and k_2, based on Ergun's original work, are considered to be 4.17 and 0.29, respectively. For spheres (where $S_0 = 6/d_c$), the factors of 6 can be incorporated into the coefficients, yielding values of 150 and 1.75, respectively, and S_0 is replaced by $1/d_c$, as Ergun originally reported. The second term contributes a significant fraction of the overall head loss only for filters with quite high velocities and large media. The first term in Equation 14-60 has a lower coefficient but otherwise identical functionality as the Carman–Kozeny equation, and therefore this equation yields a lower predicted head loss when the second term is negligible. Later investigators have generally reported values that make the Carman–Kozeny and Ergun equations agree at low velocities; for example, MacDonald et al. (1979) reported k_1 and k_2 to be 5 and 0.3 (when the equation is written as in Equation 14-60). Trussell and Chang (1999) proposed that different coefficients are required for different types of media and suggested a range of values for some commonly used media. Given that not all media grains are identical in size and shape, they also noted the difficulty in choosing a single equivalent diameter to use in any of these equations; the usual choice is the "effective size."[12]

14.7 FILTRATION DYNAMICS: EXPERIMENTAL FINDINGS OF CHANGES WITH TIME

The behavior of filters throughout a run is quite dynamic, as indicated above. Some of the experimental studies that elucidate this behavior are indicated here, but a complete compilation of the literature is not included. In the subsequent section, the mathematical modeling of filtration dynamics is given. Periodically, reviews of the filtration literature have been written; examples include those by Ives (1971, 1980), Herzig et al. (1970), O'Melia (1985), and Amirtharajah (1988). Tien (1989) provided a comprehensive view of theoretical work on filtration, and many experimental studies are also cited throughout that book. The chapters in other books (Cleasby, 1972; Cleasby and Logsdon, 1999; MWH, 2005; and Tobiason et al., 2011) also summarize a large body of literature on filtration.

Filters exhibit several different phenomena over the course of a filter run. Immediately after backwashing, the effluent quality can undergo dramatic short-term changes as the residual backwash water is replaced with fresh influent. Ripening (improved removal with time), breakoff (deteriorating removal associated with the detachment of previously retained particles), and breakthrough (deteriorating removal with time caused either by breakoff or lack of capture) can all follow. Different parts of a filter bed might well be at different stages of this process at the same time; for example, the top of a bed might be experiencing breakthrough while the bottom is still ripening. Similarly, different particle sizes within a suspension might be at different stages of this process at the same time and location within the bed. These phenomena are influenced by all the major independent variables as well. Meanwhile, the head loss inexorably rises and is influenced by all the major variables and the phenomena influencing the capture of particles.

Immediately After Backwashing

Amirtharajah and Wetstein (1980) presented a comprehensive study of filter behavior in the brief period immediately after backwashing. They showed experimental evidence that the effluent concentration of particles is low for some short period, rises for another short period, begins to fall, rises a second time, and then falls continuously for an extended period. They envisioned three remnants of backwash water (water in the underdrain system, in the media itself, and above the media) to be responsible for this complicated behavior. The lag period before the first rise was caused by clean backwash water in the underdrain system below the filter. The first rise and peak was caused by backwash water that was within the media when backwashing was ended. They found that this water could be dirtier than the backwash water above the media and attributed this observation to collisions among the media grains as the bed contracted from its fluidized state during backwashing, allowing particles still attached to the media to become dislodged. The second rise and peak were caused by the mixing of new influent with the remnant of backwash water above the media at the end of backwashing, combined with the relatively poor removal of particles by clean media. A long period of improving quality followed the second peak; this period is the ripening period that has been noted throughout this chapter. In laboratory experiments in which the bed is completely clean and filled with clean water prior to initiation, only one of these peaks (the equivalent of the second) is noticeable.

Amburgey et al. (2003, 2004) proposed that several steps in the above sequence could be eliminated, or at least greatly reduced in severity, by extending the backwash period at a slower flow rate than the full backwash rate so as to replace all the water in the filter with clean backwash water before ending the backwash. These investigators also found that, in some cases, the effluent quality immediately after the restart of the filter could be improved if this start were delayed by a period of time—even up to 1 h. The benefits of such a change in operating procedure would have to be weighed against potential disadvantages brought on by the increased hydraulic load on the other filters in service. Other investigators

---

[12] Filter media are generally subjected to a sieving test, in which the percent by weight that passes through each sieve opening size is recorded. The effective size, d_e, is the value of d_{10}, the sieve opening through which 10% of the media (by weight) passes. A second characteristic, the uniformity coefficient, is also defined from this test as the ratio d_{60}/d_{10}; values in the range of 1.3–1.5 are common.

(e.g., Colton et al., 1996) have studied the effects of a "slow start" to a filter run (running the filter at a lower filtration velocity until ripening is complete).

The fact that relatively poor effluent quality is often obtained at the beginning of a filter run was recognized early in the twentieth century. To deal with this phenomenon, most filters in drinking water treatment plants are equipped with the ability to waste (send back to the head of the plant) the water that initially comes through; this ability is reflected in Figure 14-1 by valve E. As noted earlier, wastewater filters generally do not have this capability. Even in drinking water systems, some filters have been designed without the possibility of wasting, with the thought that the small amount of breakthrough at that stage of a filter run was insignificant when averaged over an entire filter run. The better use of destabilizing chemicals in water treatment, sometimes added to the backwash water near the end of the backwash period (Harris, 1970), minimizes the period of poor removal, and this practice contributed to the idea that filters could be designed without the capability of wasting. However, with the current concern in drinking water applications about *Giardia* and *Cryptosporidium* (pathogenic organisms which, because of their rather impenetrable cysts, are very poorly disinfected by conventional means and therefore must be physically removed for protection of the public health), interest in the possibility of breakthrough in this early stage is quite high. Some evidence suggests that a substantial fraction of all the particles that are not captured during a filter run can break through in the initial few minutes.

Ripening

After the initial flushing of water and particles in the bed prior to beginning a new filter run, particle removal generally improves for a significant time period, because collected particles act as additional media in helping to capture particles in suspension. The reasons are both physical and chemical. Physically, retained particles protrude into the void space; the flow pattern, and therefore the trajectory of suspended particles, is determined more by the media grains than by the already-captured particle. The protrusion of a captured particle into the flow field considerably increases the opportunity for capture, so that ripening occurs. Chemically, proper destabilization ensures that the particle–particle interactions will be favorable for attachment, perhaps more so than the particle–media interactions. Numerous studies have shown the effects of favorable particle–particle surface chemistry on ripening, perhaps none more directly than those from Habibian and O'Melia (1975) shown in Figure 14-13. The results of jar tests, which characterize particle–particle interactions only, indicated that the optimum dose of polymer for the particular suspension and polymer investigated was 0.07 mg/L (Figure 14-13a), and that both higher and lower doses resulted in a very stable suspension. When this optimum dose was used during filtration, the filter ripened quickly and gave excellent effluent quality (Figure 14-13b, Filter 4), whereas all other doses gave poor results. Because the filter receiving the optimally destabilized suspension was the only one with substantial removal, it was also the only one that had a substantial increase in head loss (Figure 14-13c).

In a creative study allowing independent variation of the particle–media and particle–particle chemical interactions, Elimelech and O'Melia (1990) showed that effective ripening required both particle–media and particle–particle interactions to be favorable. As noted above, good ripening is essential for excellent removal in most common filter designs. Ripening can have dramatic effects on filtration efficiency over the first 10–40 min of a typical filter run (e.g., Cleasby and Baumann, 1962; Tobiason et al., 1993), but can also continue for extended periods of several hours (e.g., Iwasaki, 1937; Clark et al., 1992).

Breakthrough

The concept that filters have a finite capacity for removal is probably obvious and certainly well documented. As the deposit grows, the interstitial velocity rises, with the result that it is both more difficult for additional particles to attach and more likely that a previously retained particle will experience shear forces that exceed the attachment force. Either a lessening of attachment or some degree of detachment will lead to a reduction in particle capture; that is, to breakthrough.

Evidence of breakthrough has been reported for many years (e.g., Eliassen, 1941; Adin and Rebhun, 1987; Ginn et al., 1992; Lewis and Manz, 1991; Moran et al., 1993a; Hahn et al., 2004). As noted subsequently, macroscopic models of filtration have included terms to describe breakthrough in a few different ways, reflecting different ideas of why breakthrough occurs.

One of the more perplexing questions associated with breakthrough has been whether it is caused primarily by the lack of capture (i.e., particles passing directly through the filter without being caught) or by detachment of previously retained particles. The question has been perplexing because effluent measurements alone, in a normal filter run, cannot differentiate between the different causes of breakthrough. Increasing evidence from experiments designed to overcome this problem suggests that detachment is the primary cause of breakthrough. The early evidence of the importance of detachment came from experiments using radio-labeled particles for part of a filter run and monitoring the possible migration of captured particles after switching the influent to nonlabeled particles (Stanley, 1955); overall, these experiments were somewhat inconclusive, but some clearly demonstrated detachment. Logsdon et al. (1981) used a similar concept, spiking the influent with *Giardia* cysts for a period and then terminating the spiking and monitoring the effluent

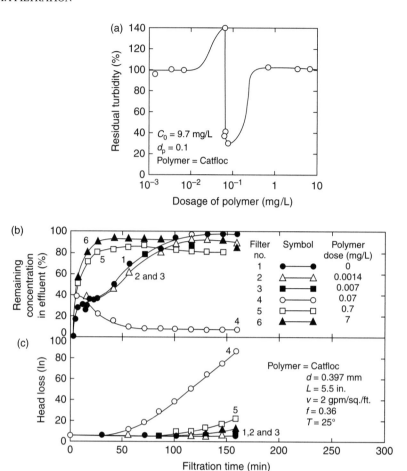

FIGURE 14-13. Comparison of jar test results (a) with filter performance (b and c), precoated media. *Source*: From Habibian and O'Melia (1975).

to see if cysts were detected after the spiking was stopped; *Giardia* cysts were found well after the spiking, providing unambiguous evidence of detachment.

Payatakes et al. (1981) used a two-dimensional apparatus equivalent to using rods (of very short length) as the filter media to simulate the grains of a real filter; observations made through a video camera provided direct evidence of the re-entrainment of captured particles and flocs. Ives (1989) observed grains and pores in three-dimensional filter beds through a fiber optic endoscope. He found that when a considerable amount of deposit had buildup on the top of a media grain, some particle detachment took place, apparently the result of instabilities caused by arriving particles. An analogy was made to snow deposition on mountainsides, with occasional occurrences of small avalanches. Moran et al. (1993b) performed experiments in which a long period of the buildup of deposit from an influent at one particle concentration was followed by a brief period with an influent at much lower particle concentration; after the switch to the low concentration, the suspension concentration at various depths in the bed was higher than the (reduced) influent concentration, again providing incontrovertible evidence of detachment of particles and/or flocs.

An example of both ripening and breakthrough, and how they occur throughout the size range, is shown in Figure 14-14. The results are from laboratory experiments using the effluent from sedimentation at a drinking water softening plant at the conditions specified in this figure. In Figure 14-14, the (log) removal is indicated for each size by the vertical separation between the influent curve and the curves at Port G at the two times shown. In Figure 14–14a, ripening is indicated by the greater removal at 66 min than at 17 min for particles throughout the size range investigated. Note the general increase in removal with increasing size throughout the measured size range at both times; this effect is conceptually consistent with, but much less dramatic than, that predicted by the models for clean bed removal discussed above. In Figure 14-14b, the results at 2682 min (~46 h) indicate breakthrough for most of the size range—the removal efficiency has decreased in comparison to that at 66 min for all but the smallest sizes. For the very smallest sizes shown, better removal was still apparent at the long time period in comparison to that at 66 min.

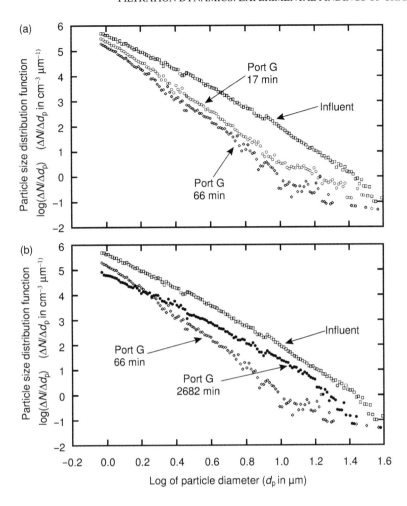

FIGURE 14-14. Filter ripening (a) and breakthrough (b) throughout a size distribution. (Filter conditions: Port G at a depth of 746 mm, filtration velocity = 6.5 m/h; sand media = 1.85 mm; influent suspension was effluent from a sedimentation tank at a water softening plant, so most particles were $CaCO_3$.)

A major cause of detachment of previously captured particles in water and wastewater treatment plants is an increase in the filtration velocity. Such changes in the filtration velocity occur regularly when one filter in a bank of filters operating in parallel is taken out of service for backwashing. To keep a constant overall flow rate in the plant, the other filters take the flow from the missing one and thereby experience a hydraulic surge; especially if this change occurs rapidly, this surge causes detachment of previously retained particles (Kawamura, 2000).

Head Loss

The total head in a gravity filter is the sum of the elevation head (h_{el}) and the pressure head (h_p); the elevation head, by definition, decreases linearly (at a slope of 1) with depth throughout the filter at all times. In a filter with no flow, the pressure head increases linearly with depth (also at a slope of 1), so the total head is constant through the filter, as shown in Figure 14-15a. This figure includes only the media and water above the media; that is, the gravel support and underdrain system below the media are ignored.

In a clean bed with flow (Figure 14-15b), the pressure head increases at a slope of 1 from the water level to the top of the media, as in the no-flow case. If the media bed has uniform characteristics, the change in pressure head through the bed is also linear, but the slope is less than that above the bed, because energy is expended to overcome the friction between the water and the grains. (The slope of the pressure head line in Figure 14-15b is visually steeper than in Figure 14-15a, but since the dependent variable of head is plotted on the horizontal axis, the derivative of pressure head with respect to depth is less than 1.) If the Carman–Kozeny expression holds, the slope of this portion of the line for pressure head can be determined from Equation 14-59. The total head (h_{tot}) is, therefore, constant from the water level to

704 GRANULAR MEDIA FILTRATION

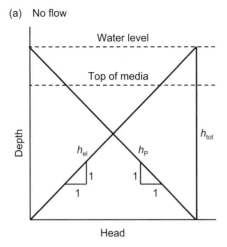

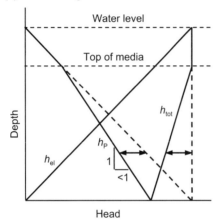

FIGURE 14-15. Total head and head loss in granular media filters under conditions of no flow (a) and flow through a clean bed (b).

the top of the media, and then decreases linearly with depth to the bottom of the filter. The head loss at any depth is the difference between the total head at the top of the filter and the total head at that depth. Since the elevation head is independent of the flow and the total head is a constant at static (no-flow) conditions, the head loss at any depth is also equal to the difference between the pressure heads under static and flow conditions at this depth; this fact is indicated in Figure 14-15b by equal lengths of lines (with double-headed arrows) at the same depth.

Figure 14-16 represents a conceptual picture of the head loss development throughout a filter run. Because the loss of total head from the top of the media to a given location equals the loss of pressure head at this location relative to the no-flow condition, only the pressure head is shown in Figure 14-16a. The static (no-flow) head is again represented by a line with 1:1 slope in this figure, and the pressure head with flow through the clean media (time zero of the filter run, indicated as t_0 in this figure) is again indicated by a straight line with lesser increase in head with increasing depth.

As particles are captured and the deposit grows during a filter run, the pressure head is no longer a linear function of depth; for example, conditions that might be found early in a filter run are represented by the line labeled t_1 in Figure 14-16a. More particles are captured near the top of the media than at lower depths, both because the concentration is higher (and removal with length is proportional to the concentration for any size, as shown by Equations 14-4 or 14-19), and because, after a short time, the already-captured particles help to capture the later ones. As a result, the friction losses are much higher per unit depth near the top of the media than at lower depths, and the pressure head increases much less than in the static case, and can even decrease. Early in a filter run, few particles reach the bottom part of the filter, and so the rate of rise of the pressure head is nearly the same as in the clean bed (t_0) case. That is, most of the *additional* head loss caused by the captured particles (i.e., the difference between the head loss at time zero and the head loss at t_1) occurs in the top portion of the bed.

As time proceeds, ripening tends to exacerbate this situation, so even more head loss occurs at the top

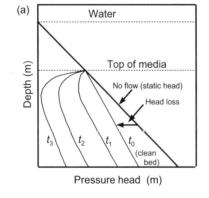 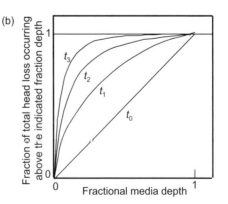

FIGURE 14-16. Development of head loss over time and depth in a monomedia filter. (a) Total head, and (b) fractional head loss with depth.

(times t_2 and t_3 in Figure 14-16a). At still later times (not shown), breakoff (detachment) of some particles or flocs occurs near the top of the filter, and these particles are recaptured lower in the filter, so the head loss begins to be spread out more over the filter depth. Figure 14-16b illustrates these ideas on a fractional basis.

Although not shown, we can imagine a fourth time after the start of the filter run in Figure 14-16a, at which the head would penetrate the vertical axis—that is, a negative pressure head could occur. In such a situation, the pressure would be less than atmospheric pressure. It is essential to avoid this situation in real filters because, as explained in Chapter 13, it would allow dissolved gases (oxygen, nitrogen) that had equilibrated with the water at atmospheric pressure to come out of solution. These gases could form bubbles that would interfere with the operation of the filter.

Captured particles increase the head loss both because they provide additional surface and therefore increase the frictional resistance to water flow and because they reduce the available pore volume for the flow. These phenomena have somewhat independent effects on the increase in head loss. Many experiments have indicated that the capture of the same volume of particles of different sizes causes different amounts of incremental head loss: smaller particles generate greater head loss, presumably because of their greater surface area. Also, capturing the same mass (and same surface area) of particles on media of different sizes results in less additional head loss for the larger media. As in many areas of filtration, the effects are often difficult to separate because the pattern of removal with filter depth (or the morphology of the deposit) is also changed with the change in any design variable. A few of the models that account for these changes in head loss are described subsequently. First, however, we consider the effects of design and operational variables on filter behavior in a conceptual way.

Filtration Dynamics: Effects of Design and Operational Variables

Earlier, we investigated the effects of design and operational variables on clean bed removal using the clean bed models. Now, we consider the effects of these variables on filter behavior throughout a filtration run; prior to investigating mathematical models that have been proposed to describe these effects, we use a conceptual approach. Of interest are both the primary design and operational variables (media size, bed depth, and filtration velocity) and the influent particle concentration. We investigate how the progression of effluent quality and head loss through a filter run is changed by an increase in one or more of these variables. Typical values for these variables are shown in Table 14-3. In the past 30 years, designers have extended the range of each of these variables beyond those that had been used for many years prior, although many filters are still designed and

TABLE 14-3. Typical Range of Major Design Variables for Rapid Rate Filtration

Design Variable	Traditional Range	Modern Range
Media size[a] (mm)	0.4–0.9	0.45–2.0
Depth of media (m)	0.3–1.0	0.5–2.5
Filtration velocity (m/h)	5–10	5–30

[a]Effective size.

operated within the range labeled as traditional. As noted previously, designers have tended to increase the depth of the filter as they have increased either the media size or filtration velocity, although often less than the proportionately. The analysis below helps explain the effects of these design variables and provides some justification for the newer design approaches.

Bed Depth In Figure 14-17, conceptual graphs for the effluent concentration and head loss are shown for a standard filter and for a second filter with a bed deeper than the standard. All other conditions are identical between the two filters. No specific values are given on either axis, but the standard filter can be considered as a traditional drinking water filter. Such filters are on the conservative end of modern designs, with a bed depth on the order of 40 cm of 0.5 mm sand, and a filtration velocity of 5 m/h. In such a standard filter, the head loss criterion for backwashing would be met well before the effluent concentration criterion, assuming adequate particle destabilization, and this trend is indicated in this figure.

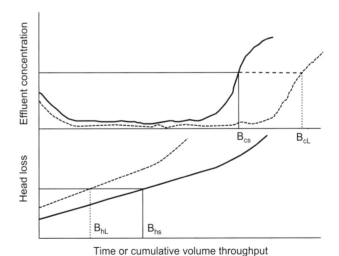

FIGURE 14-17. Effects of increased bed depth on filter performance. (Solid lines and subscript S refer to the "standard" filter, and dotted lines and subscript L refer to the filter with increased depth. Times or throughput volumes when backwashing would be necessary are indicated, using the terminology of Figure 14-1.)

Consider first the effect of the bed depth on effluent quality. Under clean bed conditions, the effluent concentration would be lower for the deeper filter than for the standard case, as indicated by Equations 14-4, 14-21, or 14-22. Note that a heterodisperse suspension will change dramatically through a filter, so it is too simplistic to think that the change in overall concentration (e.g., turbidity, suspended solids, or total particle number) can be calculated directly from Equation 14-22, but it is certainly true that the concentration would be reduced. As the run proceeds, the greater depth of the new filter will continue to produce cleaner water, and hence the effluent concentration is always below that of the standard filter. It is reasonable to think that the capacity of a filter to hold particles is roughly proportional to its depth, so the time or cumulative throughput when substantial breakthrough occurs can be expected to increase approximately proportionally to the increase in depth.

According to the Carman–Kozeny equation (Equation 14-59), the initial head loss is directly proportional to the depth, so the initial head loss is greater for the deeper filter. The deeper filter also captures more particles than the standard filter at every instant so the head loss can be expected to rise more rapidly. However, because the standard filter achieves excellent removal, the extra depth of the deeper filter receives water with very few particles to capture, and the increase in the slope of the head loss curve is quite slight.

If the head loss criterion to backwash is the same value for the two filters, then the time to backwash or the gross production of the deeper filter would be reduced compared with the standard filter. Since the primary difference in the two head loss curves is the difference at the start of the run, the amount of the reduction in time or production depends on the fraction of the overall head loss that is accounted for by the initial head loss. The advantage of extra depth (assuming that the head loss criterion dictates the end of the run in both cases) is the better removal achieved, but it comes at the expense of a shorter run.

Media Size A modern trend in filter design in water treatment plants is to increase the media size. Consider again the changes in effluent quality and head loss between the same standard filter described above and one with larger media, but all other characteristics are the same. A conceptual graph describing reasonable expectations is shown in Figure 14-18.

At the start of the filter run, according to Equation 14-22, an increase in the media size results in an increase in the effluent concentration; depending on the model used and the dominant mechanism of collection, the effects can vary in the range from d_c^{-1} to d_c^{-3}. Since interception (which leads to the -3 exponent) is rarely dominant, the expected effect corresponds to -1 exponent for the particles that can be easily detected and that contribute most of the volume and mass in most filters. In the early stages of the run, ripening

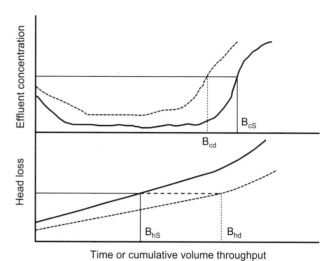

FIGURE 14-18. Effects of increased media size on filter performance. (Subscript d refers to filter with increased media size; all other symbols defined in Figure 14-17.)

can be expected to occur somewhat more slowly with the larger media, given that fewer particles are captured to act as additional collectors. The reduced removal efficiency can be expected to continue through the central part of the run, as less surface and fewer unit bed elements (to use Rajagapolan and Tien's term) are available for collection. It is perhaps not intuitively obvious that breakthrough will happen earlier in the filter with the larger media, but experimental results demonstrate this effect. Although the pores are larger, so that the velocity gradient surrounding the media grains is less severe (and this effect does mitigate the breakthrough), the dominant effect is the reduced collector surface area, leading to reduced capacity to capture and hold particles in the filter.

Increasing media size provides a benefit in terms of head loss. The initial head loss is less in the filter with larger grains, since head loss is inversely proportional to the square of the media size according to Equation 14-59. The head loss also rises more slowly, in small part because of the reduced collection of particles, but mostly because the larger pores mean less restriction of the flow; in other words, the same reasons that the head loss is less at the start of the run are maintained throughout the run.

Considered in light of the criteria for backwashing, these results are quite useful. Increasing the media size reduces the time or cumulative volume throughput when the effluent concentration criterion will be met but increases the time for the head loss criterion to be met. If the standard filter meets the head loss criterion first, then an increase in media size will bring the times to meet the two criteria closer together. These ideas were confirmed experimentally by O'Melia (1974); a conceptual graph demonstrating this idea is shown in Figure 14-19. The locations of the two lines are dependent on the filter conditions, the suspension being filtered, and the

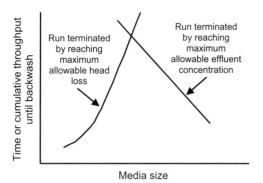

FIGURE 14-19. Effects of media size on filter run length. *Source*: Adapted from O'Melia (1974).

chosen values for the maximum allowable effluent concentration and maximum head loss allowed before backwashing is considered necessary.

The most efficient design (i.e., the one that maximizes the production of each filter run) results when the two criteria are met simultaneously. Most filters, however, meet one of the criteria to trigger backwashing sooner than the other. As noted above, conventional filters in drinking water plants have been designed to meet the head loss criterion substantially sooner than the water quality criterion, but recent trends have been to increase the media size and the depth. The result is a filter system that is closer to the water quality criterion (but still meets it) when the head loss criterion is exceeded. The simultaneous trend to monitor the effluent quality of every filter in a bank of filters also helps ensure that the effluent guidelines are not exceeded. Wastewater filters have been designed with larger media for many years, and tend to operate closer to the optimum condition.

Many filters are dual media filters or even trimedia filters. Such filters have a coarser layer of media on top and a finer layer on bottom. These filters combine the advantages of slower head loss development of the large media and the higher removal efficiency of the small media, and spread out the development of the head loss over a larger portion of the bed. Hence, if the total media depth were the same as for the two filters whose behavior is described in Figure 14-18, the trends for effluent concentration and head loss for the dual media filter would be intermediary between the two lines shown. For the media to remain properly aligned after backwashing, the settling velocity of the larger media must be less than the settling velocity of the fine media; this goal is accomplished by using anthracite coal (density $\approx 1.5\,\text{g/cm}^3$) or activated carbon (density $\approx 1.2\,\text{g/cm}^3$) for the coarse media, and sand (density $\approx 2.65\,\text{g/cm}^3$) for the fine media.

Filtration Velocity The final of the primary design and operational variables that we explore is the filtration velocity. We again consider the changes in trends for the effluent concentration and head loss between a standard filter and a modified one, this time one with an increased flow rate. Conceptual graphs comparing the performance of these filters are shown in Figure 14-20, with separate graphs showing the performance as a function of time and cumulative throughput. The separation is necessary because the relationship between the two variables is different for the two filters.

At the start of the filter run, the effluent concentration is higher for the filter with the higher superficial velocity, consistent with predictions from the single collector removal efficiencies by sedimentation (Equation 14-26) and Brownian motion (Equation 14-30) (or similar effects in the packed medium model, Equations 14-32 and 14-38). The higher velocity makes it less likely that a particle will settle

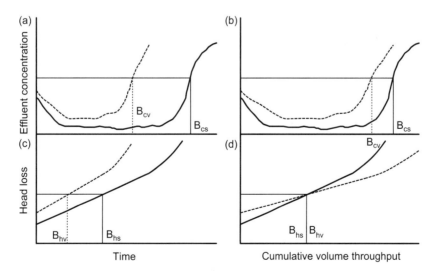

FIGURE 14-20. Effects of increased filtration velocity on performance. (Subscript v refers to filter with increased filtration velocity; all other symbols defined in Figure 14-13.)

onto any particular collector or will move by Brownian motion to the collector surface as it flows by a media grain, because the time in the vicinity of each grain is shorter. These effects continue through the ripening period and the long (nearly) steady-state period of a filter run, so the effluent concentration is always higher in the filter with the higher v_0 at comparable times (Figure 14-20a). In terms of the cumulative throughput, the number of particles that enter the filter per unit time is proportional to the filtration velocity, but the capture is slightly less than proportional. On average, particles penetrate the bed deeper before being captured, so the development of the deposit over time and depth is different in the two filters. In the filter with the higher velocity, higher shear forces lead to breakoff of captured particles (the primary cause of breakthrough) after fewer particles have been captured (although this effect is mitigated somewhat by the differences in deposit morphology); generally, the capacity of the filter to hold particles is likely to decrease some. These ideas indicate that both the time (Figure 14-20a) and, to a much lesser extent, the cumulative volume throughput (Figure 14-20b) are reduced before the effluent concentration criterion for backwashing is reached.

The head loss is initially higher in the filter with the larger v_0, as it is directly proportional to velocity according to the Carman–Kozeny equation (Equation 14-59) and nearly so in the other equations for head loss in the clean bed. As time proceeds (Figure 14-20c), more particles are captured per unit time in the new filter than the standard one, so the head loss rises more rapidly. However, on the basis of cumulative throughput, the results appear different (Figure 14-20d). The slightly lower removal per unit volume treated would reduce the slope of the head loss versus cumulative throughput slightly. But a more important effect that also reduces the slope in this figure is the spreading out of the capture through the filter bed. The deposit in the top of the filter would contain fewer particles and also likely be more compact (since greater shear would prevent loose particle accumulation), and these effects lead to less head loss per unit of deposit per increment of velocity, and therefore a slower rise in head loss on the basis of volume throughput. This trend is illustrated in Figure 14-20d and indicates that the lines for the rise in head loss could eventually cross. Whether the lines cross before or after the head loss backwash criterion is met depends on details of the two designs and the suspension involved. In this figure, the two lines are shown to cross just when the head loss criterion for backwashing is met, as a means of emphasizing that the value of D_h, expressed as cumulative volume throughput, could increase or decrease with an increase in filtration velocity.

For the case in which the standard filter is the conventional one described at the start of this section, it is likely that the effect of increasing the filtration velocity would be to increase the filter throughput. This analysis again assumes that the cause of ending the filter run for the standard filter is the head loss criterion. Such a change also brings the throughput required to meet both criteria more closely together—that is, it increases the throughput to meet the head loss criterion and decreases the throughput to meet the effluent concentration criterion. Relative to traditional designs, therefore, an increased filtration velocity is likely a more efficient design. If such were the case, the capacity of an existing plant could be increased without capital expenditure; for a new plant, the flow could be treated using fewer filters because the volume of water treated per unit area could be increased.

Depth, Media Size, and Velocity In the discussion of design tradeoffs for clean bed removal efficiency earlier in the chapter, it was suggested that approximately equal removal, at least for larger particles, would be achieved in two filters that had the same value for the ratio $Z/v_0 d_c$. It is useful now to consider the behavior of two such filters in which one has the traditional design and the other has greater values of all three variables. As the run progresses, it is reasonable to think that the effluent concentration of the modified filter will progress similarly to the standard filter, when the comparison is based on cumulative throughput. Ripening, the long central portion of the filter run, and breakthrough are all expected to be reasonably similar for the two conditions, as the increased filter depth makes up for the reduced removal per unit length caused by the increased media size and velocity. Moran et al. (1993a) demonstrated these effects.

The comparison is more complex for the initial (clean bed) head loss in the two filters. According to the Carman–Kozeny equation (Equation 14-59), the relative head loss between two filters can be calculated as follows:

$$\frac{h_{L,2}}{h_{L,1}} = \left(\frac{v_2}{v_1}\right)\left(\frac{Z_2}{Z_1}\right)\left(\frac{d_1}{d_2}\right)^2 \quad (14\text{-}61)$$

If we then add the constraint that the two filters are designed to achieve nearly the same removal by having the same ratio of $Z/v_0 d_c$ (i.e., that $(Z_2/Z_1) = (v_2 d_2 / v_1 d_1)$), we find that Equation 14-61 simplifies as follows:

$$\frac{h_{L,2}}{h_{L,1}} = \left(\frac{v_2}{v_1}\right)^2 \left(\frac{d_1}{d_2}\right) \quad (14\text{-}62)$$

This equation indicates that the initial head loss could go up or down relative to the standard condition, depending on the relative increases of velocity and media size (and the associated increase in depth); if both are increased by the same amount (say, 50%), then the initial head loss will rise by that same amount. As the run progresses, both the increase in velocity and the increase in media size are expected to reduce the rate of buildup of head loss as a function of cumulative throughput, and the effect of added length on the slope of the head loss curve would be negligible; these effects are shown individually in Figures 14-17, 14-18, and 14-20. Altogether, the head loss is expected to rise much less dramatically in the modified filter than in the standard one. Therefore, if head loss is the controlling criterion for backwashing, the cumulative throughput per unit area should increase with these changes in design. These expectations are consistent with modern design practice. In recent years, filters have been designed with increases in all three variables relative to traditional practice, and the throughput has been increased as a result. Generally, designers have not increased the depth of the media as much as Equation 14-48 would suggest, but the overall effects have still been to increase the throughput before backwashing is necessitated.

Influent Concentration The effects of influent concentration on the effluent concentration and head loss development are relatively straightforward to ascertain. It is clear from Equation 14-22 that the effluent concentration at the start of a filter run is proportional to the influent concentration. Deep bed filters are not absolute barriers, and some particles are likely to pass through at all times. If the relative particle size distributions of two suspensions are the same, the suspension with lower concentration will have a lower effluent concentration at the start of a filter run. As the run progresses, ripening will be slower for the suspension with the lower concentration, but it is likely that the effluent concentration for the suspension will always remain lower, and breakthrough will occur later. Head loss will initially be the same for the two suspensions, but the rise in head loss will be slower for the suspension with lower concentration since fewer particles are captured per unit volume of throughput (or per time). Hence, the filter run will be longer.

This analysis points out the value of a solid/liquid separation process preceding filtration—typically sedimentation or flotation. Any improvement in these processes that reduces their effluent concentration translates directly to an improved filter effluent quality and also increases the production of each filter run. These processes are only unnecessary if the raw water is of quite high quality; in this case, these solid/liquid separation processes are relatively ineffective and it becomes more cost effective to filter the water directly, with or without flocculation. Flocculation is generally helpful in improving filter performance even in the absence of a sedimentation step; the increased particle size leads to a more gradual rise in the head loss during the run.

Summary of Effects of Independent Variables The preceding discussion of the effects of several design and operational variables on filtration performance, and especially on the time or cumulative volume until one or the other of the criteria for initiating the backwash cycle, is summarized in Table 14-4. The table emphasizes the effects of

TABLE 14-4. Summary of Effects of Independent Variables on Length of Filter Run

Independent Variable		Head Loss		Effluent Quality	
		Time	Cumulative Volume	Time	Cumulative Volume
Depth	↑	↓[a]	↓	↑	↑
Media size	↑	↑	↑	↓	↓
Velocity	↑	↓	↔	↓↓	↓
Influent concentration	↑	↓	↓	↓	↓

[a] A down arrow in the head loss or effluent quality columns indicates that an increase in the independent variable (left column) leads to a decrease in the time (or cumulative volume) when the criterion for backwashing because of excessive head loss or poor effluent quality is met.

increases in each of the independent variables on the two backwash criteria because modern designers have tended to increase these variables in comparison to earlier values. Each of these effects was discussed individually above. Except in the case of changes in the filtration velocity, the effects on either time or cumulative volume throughput are identical since they are proportional to one another.

In the earlier discussion, the fact that increases in media size had opposite effects on reaching the two backwashing criteria was noted. The table makes clear that depth has the same type of tradeoff (but opposite in direction). These effects of depth have led to two consequences as designs have changed in the last 30 years. First, as noted above, designers have often *not* increased the depth proportionately to increases in media size or velocity; apparently, they have thought that the effluent quality with lesser increases in depth would still be adequate and that the water quality benefits associated with greater depth would not justify the increased cost of more frequent backwashing.

Second, the opposite effects of depth and media size show the advantages of dual media filters. These filters, with the larger media on top and smaller media on bottom, have the advantage of large media in slowing head loss development and the advantage of small media in ensuring high-quality effluent. By providing designers with multiple variables (depth and size of two different media), dual media filters can be more finely tuned than monomedia filters to achieve acceptable combinations of effluent quality and normalized production for each filter run.

14.8 MODELS OF FILTRATION DYNAMICS

The ability to predict or model mathematically the dynamic behavior of filters is imperfect, but some models have been proposed that make reasonable predictions under some circumstances and, in any case, give valuable insight into

TABLE 14-5. Selected Macroscopic Models for Effluent Quality and Head Loss Development[a]

Effluent Quality		Head Loss	
Expression for λ/λ_0	References	Expression for $(\partial h_L/\partial z)/(\partial h_L/\partial z)_0$	References
$\left(1+\dfrac{b\sigma}{\varepsilon}\right)^{n_1}\left(1-\dfrac{\sigma}{\varepsilon}\right)^{n_2}$	Mackrle et al. (1965)	$\left(1+\dfrac{d\sigma}{\varepsilon}\right)$	Mints (1966)
$1+b\sigma-\dfrac{a\sigma^2}{\varepsilon-\sigma}$	Ives (1960); Ives and Sholji (1965)	$1+d\sigma$	Mehter et al. (1970)
$\left(1+\dfrac{b\sigma}{\varepsilon}\right)^{n_1}\left(1-\dfrac{\sigma}{\varepsilon}\right)^{n_2}\left(1-\dfrac{\sigma}{\sigma_{\text{ult}}}\right)^{n_3}$	Ives (1969)	$\left(1+\dfrac{d\sigma}{\varepsilon}\right)^{m_1}\left(1-\dfrac{\sigma}{\varepsilon}\right)^{m_2}$	Ives (1969)

[a] In these expressions, the specific deposit is expressed as volume of deposit per volume of bed. Adjustable parameters are a, b, d, n_i, m_i, and σ_{ult}.

filter behavior. The focus below is on the microscopic models; that is, those that attempt to describe the changes in removal efficiency and head loss over time on the basis of changes on the microscopic (media grain and particle) level. Nevertheless, a few of the macroscopic models; that is, those that describe changes on the bed level, are noted first.

Macroscopic Models

Numerous macroscopic filtration models for both particle removal and head loss development have been proposed over the years. The key equations from a few of these models are summarized in Table 14-5, adapted from Tien (1989); only the models for effluent quality that show the possibility of both ripening and breakthrough are shown. The concepts behind each model are not discussed in detail here. For removal, the models express the change in the filter coefficient primarily as a function of the specific deposit (σ); in the Ives (1969) formulation, the model assumes that there is a maximum or ultimate value of the specific deposit (σ_{ult}). For head loss development, the models are somewhat simpler, but reflect a variety of thoughts on how the increasing deposit (or reduction in porosity) changes the head loss.

Ripening Model of O'Melia and Ali

O'Melia and Ali (1978) extended the work of Yao et al. (1971) for clean bed removal to include filter ripening, and also extended the Carman–Kozeny equation (Equation 14-59) for head loss in a ripening filter. They considered only the possibility of improved removal caused by the captured particles acting as additional collectors; breakoff or detachment of particles or flocs was not considered. The geometry considered was the same single spherical collector used by Yao et al., and no change was considered in either the porosity of the bed or the diameter of the collector as particles were captured. The conceptual underpinnings are summarized as follows:

(1) At any instant, the "collector" is considered to be the original collector (media grain) and the associated retained particles,

(2) Only a fraction (β) of retained particles can collect additional particles, and each captured particle can collect at most one additional particle,

(3) Only a fraction (α_p) of predicted collisions between particles in suspension and a retained particle is successful; this fraction is not necessarily the same as the corresponding ratio (α) for collisions between particles and a clean collector (media grain),

(4) Only a fraction (β') of particles that are captured creates additional head loss, and

(5) Additional head loss occurs because of the change in the ratio of media surface area to media volume; this ratio is in the Carman–Kozeny equation (with the media in this model considered as the original media and the already-captured particles).

The single collector removal efficiency of a filter grain and its associated retained particles is then described as

$$\gamma_r = \frac{\begin{pmatrix}\text{Rate at which particles}\\\text{collide with}\\\text{and are collected by}\\\text{a filter grain}\end{pmatrix}+N_r\begin{pmatrix}\text{Rate at which particles}\\\text{collide with}\\\text{and are collected by}\\\text{a single retained particle}\end{pmatrix}}{\text{Rate at which particles approach the filter grain}}$$

where N_r is the number of particles that are retained on a filter grain and that can act as collectors. Translating this conceptual definition into an equation gives:

$$\gamma_r = \alpha\eta + N_r\alpha_p\eta_p\left(\frac{d_p}{d_c}\right)^2 \qquad (14\text{-}63)$$

where η_p is the transport or collision efficiency between a particle in suspension and a retained particle. γ_r is normalized to the filter grain whereas η_p is normalized to a retained particle, so the factor $(d_p/d_c)^2$ accounts for this difference.

O'Melia and Ali also did not directly account for the changes in particle trajectories (and therefore particle capture). Rather, an accounting of the retained particles that act as additional collectors was made as follows:

$$\begin{bmatrix} \text{Rate} \\ \text{of change} \\ \text{of } N_r \end{bmatrix} = \begin{bmatrix} \text{Fraction of} \\ \text{retained particles} \\ \text{that act as} \\ \text{additional} \\ \text{collectors} \end{bmatrix}$$

$$\times \frac{\begin{bmatrix} \text{Rate at which particles} \\ \text{attach to original collector} \end{bmatrix}}{\begin{bmatrix} \text{Rate at which particles} \\ \text{approach original collector} \end{bmatrix}} \begin{bmatrix} \text{Rate at which} \\ \text{particles approach} \\ \text{original collector} \end{bmatrix}$$

or

$$\frac{\partial N_r}{\partial t} = \beta[\eta\alpha]\left[v_0 N \frac{\pi}{4} d_c^2\right] \qquad (14\text{-}64)$$

A number balance on the particles in suspension in a thin layer (Δz) of the filter is identical to that shown above in Equation 14-10, and the resulting equation is

$$\frac{\partial}{\partial t} A\varepsilon \Delta z N = v_0 A N - v_0 A \left(N + \frac{\partial N}{\partial z}\Delta z\right) - \frac{(3/2)\gamma_r A v_0 N (1-\varepsilon)\Delta z}{d_c} \qquad (14\text{-}65)$$

The form of the last term stems from Equation 14-18, but it is revised to account for the definition of the single collector removal efficiency during ripening. Dividing by the common term $A\Delta z$, assuming that the porosity is constant, and rearranging yields

$$\varepsilon \frac{\partial N}{\partial t} + v_0 \frac{\partial N}{\partial z} + \frac{(3/2)\gamma_r v_0 N (1-\varepsilon)}{d_c} = 0 \qquad (14\text{-}66)$$

The change in filtration efficiency with time and depth is then found by solving Equations 14-63, 14-64, and 14-66 simultaneously using a numerical approach. The approach involves some simplifying assumptions to develop a recursive equation that allows calculation of γ_r at one time step based on its value in the previous time step and the capture that occurs during this time step. The original paper describing this model (O'Melia and Ali, 1978) did not include the factor of ε in the first term of Equation 14-66, an error apparently carried through by several investigators.

To model the development of head loss during filtration, O'Melia and Ali began with the Carman–Kozeny equation (Equation 14-59) but expanded the terms for the specific surface area to include captured particles that contribute to head loss. In this view, the increase in head loss is entirely due to the increased surface area that the water contacts as it flows through the filter. In this model, the S_0 term in the Carman–Kozeny equation is defined as the ratio of the surface area of collectors and these particles that contribute to head loss to the volume of collectors and particles; assuming both collectors and particles are spherical, we can write

$$S_0^2 = \left(\frac{A_c + \beta' A_p}{V_c + V_p}\right)^2 = \left(\frac{N_c \pi d_c^2 + \beta' N_p \pi d_p^2}{(N_c \pi d_c^3/6) + (N_p \pi d_p^3/6)}\right)^2$$

$$= \frac{36}{d_c^2}\left(\frac{1 + \beta'\left(\frac{N_p}{N_c}\right)(d_p/d_c)^2}{1 + \left(\frac{N_p}{N_c}\right)(d_p/d_c)^3}\right)^2 \qquad (14\text{-}67)$$

Here, subscripts c and p refer to original collectors and retained particles, respectively, and N, A, and V refer to the total number, surface area, and volume, respectively.

This model has a number of simplifying assumptions that can be questioned, but it has provided a framework for considering the improved removal during ripening and the increase in head loss throughout the run. It contains a strong empirical component, with five empirical parameters (α, α_p, η_p, β, and β'). However, as explained by Darby et al. (1992), the three parameters α_p, η_p, and β always appear as a product in the solution of the model, and this product can be considered as the fraction of particles approaching a retained particle that act to collect additional particles when captured. In this view, the model has three parameters that must be fit to experimental data.

As an example of the use of the model, a figure presented by Darby et al. (1992) is shown in Figure 14-21. The sensitivity of the model to various parameters (as well as the process of calibrating the model to the experimental results, represented as circles) is indicated by showing both the results for the best-fit parameter values (solid lines) and values that are 50% higher and lower than the best values for the three parameters. The experiments were performed with 2-μm latex spheres at their zero point of charge (optimal destabilization), consistent with the fact that an α value of 1.0 gave the best fit of the model to the experimental data. Using values for either the attachment efficiency factor (α) or the ripening factor ($\alpha_p\eta_p\beta$) that are lower than the best-fit values decreases the predicted removal (higher fraction remaining, Figure 14-21a) throughout the time period, as expected. The head loss (Figure 14-21b) is plotted as the increase in the hydraulic gradient (obtained by subtracting the initial head loss from each value and then dividing by the filter depth).

The two central problems of the model are suggested by careful inspection of these results. First, as strictly a ripening model, the model shows the effluent quality to improve always, and does not include either the possibility of an apparent steady state or reduced removal (breakthrough) after some period. Second, the form of the head loss model always causes a concave upward head loss increase, whereas the experimental results shown suggest an approximately

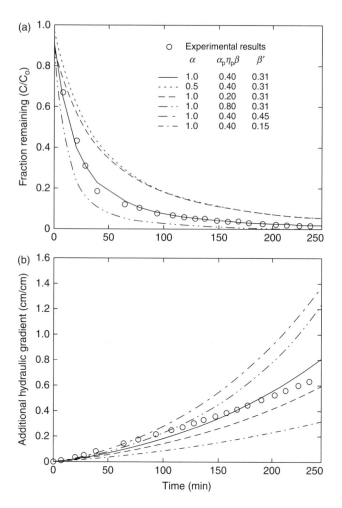

FIGURE 14-21. Modeling filter dynamics for (a) ripening and (b) head loss. *Source*: From Darby et al. (1992).

linear increase in head loss. Generally, experimental results indicate a (nearly) linear increase in head loss for an extended period, followed by a period of a concave upward increase.

Numerous efforts to improve the O'Melia and Ali model have been undertaken. While largely successful in improving the fit to a given set of experimental results, these efforts also involve the generation of a greater number of (empirical) fitting parameters, leaving the model less likely to be useful for predictive purposes; that is, for predicting filter performance under different circumstances than used for the calibration. Nevertheless, these efforts are valuable in understanding the shortcomings of the original model and represent progress toward the development of better models. One line of suggested improvements has been to limit the amount of removal that can occur. Tare and Venkobachar (1985) assumed that the amount of surface area of retained particles was limited and that removal efficiency would decrease as this limit was approached. Vigneswaran and Chang (1986), based on earlier work by Adin and Rebhun (1977), assumed the removal rate would decrease as the hydraulic gradient increased. They also accounted for the change in the porosity within the filter caused by the retained particles, and included the concept that the deposit itself would be porous. Subsequently, Vigneswaran and Tulachan (1988) proposed that both the direct deposition on media and the amount of total deposit were limited. They suggested that the rate of direct deposition decreased as the available surface area decreased. The rate of deposition onto previously retained particles was thought to decrease as a minimum allowable bed porosity (or, alternatively, a maximum deposit) was approached. This model allows a complete filter cycle (initial removal, ripening, a period of nearly steady-state good removal, and breakthrough) to be modeled, at the expense of additional fitting parameters beyond those in the original model.

Tobiason and Vigneswaran (1994) used this same approach to model the particle removal in their experiments. To model the head loss, they considered a few different approaches. First, they modified the original O'Melia and Ali head loss equation (Equation 14-67 substituted into Equation 14-59) to include both the change in porosity within the filter and the porosity of the deposit, consistent with the changes made in the particle removal side of the model. This change, however, still led to a concave upward plot for the increase in head loss under all circumstances. They found that a model in which the increase in head loss is assumed to be directly proportional to the increase in surface area of the deposited particles was most satisfactory in describing their results. Several investigators (e.g., Moran, et al., 1993a; Veerapeneni and Wiesner, 1993) have found a similar dependence on surface area.

To summarize, efforts to model changes in both the removal efficiency and head loss over the course of a filter run, beginning with the work of O'Melia and Ali (1978) and continuing through several iterations and improvements to this original model, suggest a primary dependence on the surface area of captured particles. However, these models do not account for breakthrough or detachment; as a result, their value for design and operation is limited in situations where filter run time is limited by breakthrough.

Ripening Models of Tien and Coworkers

As noted above, Tien and various coworkers have developed a series of clean bed filtration models involving different choices for the geometry of the unit bed element, including an isolated sphere model and constricted tube models. Within these frameworks, they have also investigated the improvement in removal efficiency and the increase in head loss caused by deposited particles. Payatakes et al. (1974) assumed a smooth deposit morphology and investigated both the capillary and Brinkman flow models to predict the increase in head loss, but found that both seriously underestimated the head loss development. Tien et al. (1979) predicted both the removal efficiency and head loss during ripening by considering a spherical cell model and a smooth deposit for the early stages of ripening and a constricted tube model with deposition for later stages. The fit to experimental data was fair, although a lack of sufficient data from carefully controlled experiments prevented a full test of the model. Tien (1989) summarized several attempts to describe mathematically the development of head loss with various deposit morphologies; he concluded that the complexities of deposit morphology and the stochastic nature of deposition would make complete, simultaneous modeling of both head loss increases and ripening very difficult.

Mackie et al. Ripening Model

Mackie et al. (1987) developed another model to describe the dynamics of deep bed filtration that accounts for both the improvement and subsequent deterioration of removal efficiency. As in all models that account for ripening, they accepted the idea that retained particles act as additional collectors for subsequent particles. They adopted both the Happel sphere-in-cell model to characterize the geometry and flow and the Tien unit bed element approach, noting that others had sometimes made an erroneous translation from the efficiency of a unit bed element to the filter coefficient. The two key differentiating elements of the model are as follows. First, they considered both microscopic and macroscopic effects of deposits on the flow field. In this context, macroscopic means on the order of the media grain size, and microscopic means on the order of the size of particles in suspension. On the macroscopic level, they suggested that, after a significant amount of deposit accumulated on the originally spherical collector, the shape would be changed to have a "snow-cap" or dome shape on top of the sphere; the Happel flow equations for this deformed shape were calculated, with some simplifying assumptions. In the numerical modeling, the shape (and porosity) was updated at reasonable intervals but not after every new particle was deposited. On the microscopic level, they calculated the effect that an individual collected particle would have on the average probability of capturing an additional particle. Because effects of every single deposited particle are not accounted for explicitly, an empirical fitting parameter (a fraction of predicted collisions that result in attachment) is included.

The second distinguishing element of the model is the idea that reductions in removal efficiency result primarily from increases in interstitial velocities as the deposit grows. The authors suggest that particles cannot be collected if the tangential velocity exceeds a critical value at a certain location, namely, where the center of a particle in suspension would have to be if the particle's edge was just touching the grain or deposit. This critical velocity is also taken as a fitting parameter, although physically reasonable values less than the average interstitial velocity should be chosen. An important ramification of this approach was that the effects of particle size could be explicitly accounted for; the center of large particles would be at a greater distance from the deposit or grain surface at their critical trajectory, so they would exceed the critical tangential velocity at lower amounts of deposit than small particles. The result is that the model predicts that the breakthrough of large particles occurs much earlier than that for small particles, a result that is consistent with experimental results. A major benefit of the Mackie approach is this explicit accounting for different particle sizes, which was not incorporated into the original O'Melia and Ali model (although several of the investigators noted above did include a technique for accounting for different sizes) nor into the ripening models of Tien and coworkers.

Mackie et al. (1987) also showed how the model fit the results from a few experiments. The model parameters were fit to one set of results and these values were then used to predict the results of other experiments. While the predictions were imperfect, they were quite reasonable for describing the effects of particle size, filter depth, grain size, and velocity. Unfortunately, the model has not been applied to other suspensions or experiments by other investigators, so its general utility is uncertain. The model focuses on removal only and not on head loss, although an extension to predict the head loss would be reasonably straightforward within the context of the calculations already performed in the model.

Use and Value of Dynamic Models

The various models for the dynamic behavior of filters presented above reflect an effort to understand and predict this behavior; the quest described by Stanley in the quote at the start of this chapter is ongoing. If one believes that the purpose of modeling is to be able to predict *a priori* the full behavior of filters, then the modeling efforts to date have been a failure. However, if the purpose of modeling is viewed as providing insight into filter behavior that might be useful for design and operation, then the modeling has been quite successful, even if incomplete.

The dynamic models give the designer further insight into the tradeoffs among the design variables than the clean bed models described above. The O'Melia and Ali model, for example, suggests that ripening is primarily a function of the surface area of the captured particles; that is, a function of $N_c d_p^2$, as shown in Equation 14-67. Combining this idea with the normalizations for different velocity and media size suggested in Equation 14-48 gives a basis for comparing experiments with different conditions over the entire filter runs. For example, Moran et al. (1993a) and Kau and Lawler (1995) compared removals in experiments with different media sizes as a function of the cumulative surface area captured at different depths. When compared at the same depth and time, removals appeared very different between

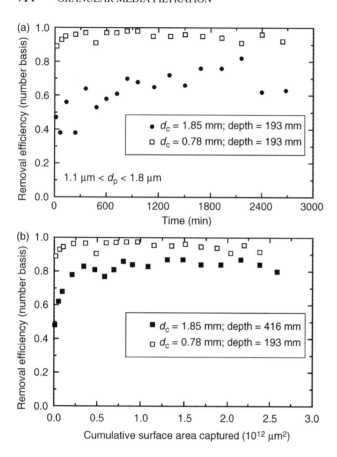

FIGURE 14-22. Filtration results with different media sizes: (a) compared over time at the same depth and (b) normalized for the captured particles and compared at different depths. *Source*: Adapted from Kau and Lawler (1995).

two experiments with two different media sizes, as shown in Figure 14-22a. However, when compared at a nearly constant ratio of depth to media diameter, the results were far more similar, as shown in Figure 14-22b. For the ratio of depth to diameter to be identical in these two experiments, the samples from the experiment with $d_c = 1.85$ mm should have been taken at a depth of 457 mm; if such a port had been available, the results in (b) would have been closer together. Also, recall from Figure 14-11 that the Rajagopalan and Tien model predicts that the removal at the lower depth and larger media would be not quite as extensive as that for the smaller media and shorter bed depth. These results not only confirm some of the concepts incorporated into the models, but also allow designers to make rational decisions about the trade-offs of different filter designs. While the models are insufficient to design filters without pilot testing, they might reduce the amount of pilot testing necessary by allowing translation of results from one condition to another. Nevertheless, Kau and Lawler (1995) also showed that, when breakthrough was significant, the normalizations were not as effective.

The dynamic modeling of head loss also provides insight that can be useful in design and operation. Several investigators (e.g., Veerapeneni and Wiesner, 1993; Tobiason et al., 1993; Darby and Lawler, 1990) have compared experiments at the same surface area captured, a normalization suggested by Equation 14-67. This normalization accounts well for the effects of different particle sizes at the same filtration conditions. When comparing experiments performed under different conditions, the same removal (at least for large particles) is expected at depths that are scaled proportionally to changes in media size or velocity (Equation 14-48); using the surface area captured to predict the associated rise in head loss at these different depths, different experiments can give dramatically different head losses. For example, experiments with different media sizes show far less rise in the head loss in experiments with large media, because the change in porosity caused by the same surface area of captured particles is far less for the larger media. Such effects are more understandable with the aid of the models.

14.9 FILTER CLEANING

Throughout this chapter, we have referred to the backwashing of the filter—the operation of cleaning the filter when either the head loss or effluent concentration has exceeded set criteria. The principles of backwashing are outlined here. Extensive work by Amirtharajah (e.g., Amirtharajah, 1978, 1984; Hewitt and Amirtharajah, 1984) and Cleasby (e.g., Cleasby et al., 1977; Cleasby and Fan, 1981) forms the basis for this understanding of backwashing. Many of the design and operational details of backwashing, as for all aspects of filtration, are described by Cleasby and Logsdon (1999) and by Tobiason et al. (2011).

Backwashing refers to the passing of water upwards through a filter to obtain adequate removal of the captured particles; however, backwashing alone is not sufficient unless the particles are only loosely attached to the filter media. In most water and wastewater applications, the particles are attached reasonably strongly and backwashing is supplemented by a surface wash system and/or air scour; each of these is described below.

Backwashing (without air scour) generally requires that the packed media be fluidized; that is, that the filter grains no longer be supported by each other but by the force of the rising fluid. At an upflow superficial velocity lower than that required for fluidization, the bed does not move, and the head loss through the filter media can be calculated using either the Carman–Kozeny (Equation 14-59) or Ergun (Equation 14-60) equation as appropriate. As the velocity increases from zero to the point where fluidization occurs, the head loss increases linearly at low values and then more steeply (as the second term of the Ergun equation becomes significant).

At velocities above the minimum required for fluidization, the bed expands; that is, the media grains separate from one another, so the porosity of the bed increases. Under these conditions, the net downward force (gravity minus buoyancy) on the media is balanced by the upward drag force exerted by the moving fluid, so the bed remains stationary. (Individual grains do move upward or downward transiently, as the drag force temporarily exceeds the net gravitational force, or vice versa, in the turbulent flow.) If the upward velocity is then increased, the bed expands further, reaching a new steady-state condition in which the increased drag due to the higher velocity is compensated by the decreased drag due to the increased porosity; in this case, the net drag is the same as before (exactly balancing the net gravitational force). This process is limited, however; at some velocity, the porosity of the bed is large enough that increasing it still further has only a small effect on the drag. Further increase in the velocity leads to a net upward force on the grains, and they are washed out of the filter. For obvious reasons, backwashing velocities are never allowed to approach this value, and the bed expansion is usually small. A conceptual curve describing the relationships between the head loss and upflow superficial velocity in the normal operating range (i.e., at velocities below those leading to washout) is shown in Figure 14-23.

Of interest is the minimum superficial velocity required for fluidization (v_{mf}); that is, the point of intersection of the increasing head loss below fluidization and the constant head loss under fluidized conditions. To determine this value, we find the head loss at the point where fluidization would just begin; at this point, the depth and porosity of the media are the same as during filtration, but the weight of the media is (just barely) supported by the upflowing liquid. As in the case of particles settling through a quiescent water column (see the derivation of Stokes' law for settling in Chapter 13), the buoyancy force partially counterbalances the gravitational force of the particles. The net weight of the media (gravity minus buoyancy) is, therefore, the product of the difference in media and water densities, the volume of media, and the gravitational constant; that is,

$$\begin{bmatrix} \text{Net weight of media} \\ \text{(gravity minus buoyancy)} \end{bmatrix} = (\rho_p - \rho_L)AZ(1-\varepsilon)g \quad (14\text{-}68)$$

This force must be counterbalanced by the upflow of the water to achieve fluidization. The force of the upflowing water can be found as the product of the area and the head loss through the media; that is,

$$\begin{bmatrix} \text{Force on the media} \\ \text{from upflow} \end{bmatrix} = A\rho_L g h_L \quad (14\text{-}69)$$

where h_L is the head loss through the media. Equating the two expressions above and canceling common terms yields the head loss needed to accomplish fluidization, $h_{L,f}$

$$h_{L,f} = \frac{\rho_p - \rho_L}{\rho_L} Z(1-\varepsilon) \quad (14\text{-}70)$$

This head loss is shown by the horizontal line in Figure 14-23. When this head loss is equated with that described by the Ergun equation (Equation 14-60), a quadratic equation is found for v_{mf}, the minimum superficial velocity required to begin fluidization during backwashing.

$$k_2 \frac{1}{g} \frac{(1-\varepsilon)}{\varepsilon^3} S_0 v_{inf}^2 Z + k_1 \frac{\mu}{\rho_L g} \frac{(1-\varepsilon)^2}{\varepsilon^3} S_0^2 v_{inf} Z \\ - \frac{(\rho_p - \rho_L)}{\rho_L} Z(1-\varepsilon) = 0 \quad (14\text{-}71)$$

After canceling the common terms, the equation simplifies to yield

$$k_2 \frac{1}{g} \frac{1}{\varepsilon^3} S_0 v_{mf}^2 + k_1 \frac{\mu}{\rho_L g} \frac{(1-\varepsilon)}{\varepsilon^3} S_0^2 v_{mf} - \frac{\rho_p - \rho_L}{\rho_L} = 0 \quad (14\text{-}72)$$

The resulting expression for v_{mf} is a complex function of the porosity of the bed, the diameter (or specific surface area) and density of the grains, and the temperature (as it influences the viscosity and density of water), but it is easily evaluated numerically for a specific situation. The media size is generally not uniform, so the size recommended for use in calculation is the 90th percentile size (d_{90}, for which 90% of the particles by weight are smaller) to ensure fluidization of the entire bed (Tobiason et al., 2011). Because of the uncertainty in the coefficients for the Ergun equation, experiments are likely to be required to determine exact values for a given filter.

During backwash, it is usually desirable to use a velocity greater than the minimum to achieve fluidization. Doing so

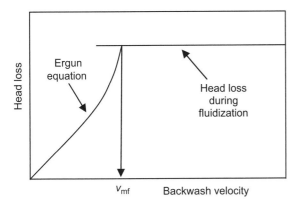

FIGURE 14-23. Effects of backwash velocity on head loss: determination of superficial backwash velocity required to cause fluidization.

expands the bed; that is, the porosity increases as the velocity rises. Because the filter walls contain the media, the expansion occurs only vertically. Equating the mass of the media under normal filtration (unexpanded) conditions to that when the bed is expanded leads to the following relationship between the depth of media and the porosity under unexpanded (no subscripts) and expanded (subscript ex) conditions

$$\frac{Z_{ex}}{Z} = \frac{1-\varepsilon}{1-\varepsilon_{ex}} \qquad (14\text{-}73)$$

The relationships between porosity and velocity have been determined empirically, and several such relationships have been proposed. One common approach has been to use the same relationships found to describe the declining settling velocity of solids under increasingly concentrated conditions; that is, zone (or Type III) settling. Most of these relationships use the settling velocity of an individual particle in an infinite medium (Stokes velocity, if the Reynolds number is sufficiently low) as the basis. The Richardson and Zaki (1954) equation is perhaps the most common and can be written as follows:

$$v_{ex} = v_s \varepsilon_{ex}^{4.65} \qquad (14\text{-}74)$$

where v_{ex} is the required superficial backwash velocity and v_s is the settling velocity of the media grains. Despite some disagreement on the degree of expansion (if any) that should be achieved for excellent backwashing, most filters that do not have air scour systems (explained subsequently) are backwashed with an expansion of 10–20% (i.e., $Z_{ex}/Z = 1.1 - 1.2$). In any case, it appears that the degree of expansion does not play a substantial role in backwashing.

Surface Wash

By itself, backwashing is often insufficient to clean the media. A particular problem is the top few centimeters of the filter bed, which are often so clogged with collected particles that the filter grains are held together by the collected particles. That is, a cake has formed in the top part of the filter and can sometimes rise *en masse* with backwashing alone. To break up this mat of filter grains and collected particles, a high-velocity *downward* water flow is often injected before the start of the vertically upward flow of the backwash. In such a surface wash system, a set of nozzles just above the top of the filter bed directs a high-velocity flow stream at an angle into the top of the filter bed before (for 1–3 min) and during backwashing. The nozzles can be fixed in a set of pipes that run across the surface of the filter or can rotate to cover the entire bed (except the corners). They are generally located about 10 cm above the bed during normal operation and might become submerged below the top of the media during the expansion caused by backwashing. The surface wash is stopped at least a few minutes before the end of backwashing. The overall flow rate is only a small fraction of the backwash flow, but the small nozzles impart a high velocity that causes the media clumps to collide with one another, thereby loosening the captured particles. Since surface wash systems necessarily only affect the top several centimeters of the bed, they are most effective for monomedia filters with relatively small media size—systems that capture most of the particles in this top portion of the bed.

Air Scour

A second way to induce rapid movement of the media grains relative to one another during backwashing, and therefore induce the collisions and abrasion thought essential for effective cleaning of the filter, is by providing air bubbles that travel up through the filter media. Air is supplied in a separate system from the backwash flow, but is introduced through nozzles at the bottom of the media. The goal in air scour is to have the bubbles form, collapse, and reform a number of times as they rise; this condition has been called collapse-pulsing. Amirtharajah showed both experimentally (Hewitt and Amirtharajah, 1984) and theoretically (Amirtharajah, 1984) that backwashing is most effective in systems with air scour if the water velocity is considerably below that required for fluidization of the bed. The smaller pores in an unfluidized bed increase the difficulty of the air travel, and this condition helps to form large bubbles from the coalescence of smaller ones; in turn, the weight of the media in the unfluidized bed helps to cause rapid deformation (which Amirtharajah called "collapse") of the bubbles. Amirtharajah recommended that the water velocity be in the range from 20 to 40% of v_{mf}. The theoretical development consisted of three components: (i) an analysis based on principles of soil mechanics for the conditions of soil collapse, (ii) a relationship between air pressure and the rate of bubble formation, accounting for the surface tension of bubbles and frictional losses as they are formed, and (iii) the effects of water flow on the stress required to create the collapse condition. The resulting equation is complex, but ultimately relates the properties of the media, the bed (depth of media and depth of overlying water), superficial air velocity, and backwash superficial velocity (as a fraction of the fluidization velocity) for optimal backwashing (i.e., to accomplish the collapse-pulsing condition). The final equation is

$$P_a - \frac{4\gamma}{k_1 d_{60}} - k_2 v_a^2 = \rho_L g Z \left\{ 1 + \frac{H}{Z} + \left(\frac{\rho_p - \rho_L}{\rho_L}\right)(1-\varepsilon_0) \right.$$
$$\left. \times \left[\tan^2\left(45 - \frac{\phi}{2}\right) + \left(1 - \tan^2\left(45 - \frac{\phi}{2}\right)\right) \frac{v_b}{v_{inf}}\right]\right\}$$
$$(14\text{-}75)$$

where P_a is the air pressure (pascals = kg/s² m), γ the surface tension of water (kg/s²) (7.28 × 10⁻² kg/s² for water at 20°C), k_1 the empirical dimensionless constant, the fraction of the effective size that characterizes the pore size (−), d_{60} the effective diameter of the media grains, the size that 60% of the mass of media is less than (m), k_2 an empirical constant, a friction coefficient relating pressure drop to air velocity (kg/m³), v_a the superficial air velocity, air flow rate per unit (horizontal) area (m/s), ρ_p, ρ_L the density of media grains and water, respectively (kg/m³), g the gravitational constant (9.81 m/s²), Z the depth of media (m), H the depth of water above media (during backwashing) (m), ϕ the angle of repose of media when submerged (degrees) (i.e., the maximum angle from the horizontal that the media can sit without slumping), ε_0 the porosity of media (with no fluidization) (−), and v_b the superficial water velocity during backwashing (m/s).

Only three of the parameters are empirical, and Amirtharajah gave values for all three for sand. The first, k_1, is narrowly constrained between 7 and 10% (the range of the average pore size as a percentage of the effective size in a packed bed), and Amirtharajah used one set of data to find the value 0.082 (i.e., 8.2%, nearly in the middle of the expected range). This value is not likely to change dramatically for any media, although the porosity for anthracite is greater than for sand, so the value could be slightly greater. The second, k_2, was found to have the value of 6.32×10^5 kg/m³ for sand, and this value might well change considerably for other media. The third empirical parameter is ϕ, the angle of repose for the material, and this was found experimentally to be 33.2° for sand. On the right side of the equation, every term inside the brackets is written in dimensionless terms, so that any consistent units can be used for each of these terms, but overall, the equation has dimensions of pressure.

The important result of this equation is that, for a given filter, the backwash water velocity (and therefore the fractional fluidization, v_b/v_{mf}) is inversely related to the square of the superficial air velocity (v_a^2). Within the range of fractional fluidization values where collapse-pulsing can occur ($\sim 0.2 < (v_b/v_{mf}) < 0.4$), the required air velocity can be determined. This relationship is demonstrated in the example below, adapted from the Amirtharajah (1984) article. Subsequently, Amirtharajah (1993) reported empirical results for other media (anthracite, granular activated carbon, dual media) in the form of the inverse square relationship of backwash velocity to the air velocity, but they did not evaluate the theoretical constants. These empirical results are at least qualitatively consistent with the ideas expressed in Equation 14-75, in that the coefficients of the equation change in the expected direction, relative to sand, for the lower density and higher porosity associated with anthracite and activated carbon beds.

■ **EXAMPLE 14-9.** A sand filter has the following characteristics: media size (d_{60}) = 0.62 mm; media depth = 0.76 m; media porosity = 0.40; depth of water above media during backwashing = 0.963 m; sand density = 2.52 g/cm³ = 2520 kg/m³. At an air pressure of 26.8 kPa (= 2.68×10^4 kg/m s²), determine the required superficial air velocities to accomplish the "collapse-pulsing" condition at several values of backwash water velocity (expressed as a fraction of the minimum fluidization velocity). Use the values of empirical constants reported by Amirtharajah.

Solution. The equation to describe collapse-pulsing (Equation 14-75) is

$$P_a - \frac{4\gamma}{k_1 d_{60}} - k_2 v_a^2 = \rho_L g Z \left\{ 1 + \frac{H}{Z} + \left(\frac{\rho_p - \rho_L}{\rho_L}\right)(1 - \varepsilon_0) \right.$$
$$\left. \times \left[\tan^2\left(45 - \frac{\phi}{2}\right) + \left(1 - \tan^2\left(45 - \frac{\phi}{2}\right)\right) \frac{v_b}{v_{inf}} \right] \right\}$$

Substituting the values in metric (SI) units for the described filter, with each term having units of kg/m s²

$$2.68 \times 10^4 - \frac{4(7.28 \times 10^{-2})}{0.082(6.2 \times 10^{-4})} - (6.32 \times 10^5) v_a^2$$
$$= (1000)(9.81)(0.76)\left\{ 1 + \frac{0.963}{0.76} + \left(\frac{2520 - 1000}{1000}\right)(1 - 0.40) \right.$$
$$\left. \times \left[\tan^2\left(45 - \frac{33.2}{2}\right) + \left(1 - \tan^2\left(45 - \frac{33.2}{2}\right)\right) \frac{v_b}{v_{mf}} \right] \right\}$$

$$2.107 \times 10^4 - (6.32 \times 10^5) v_a^2 = 7456$$
$$\times \left\{ 2.267 + 0.912 \left[0.2924 + (1 - 0.2924) \frac{v_b}{v_{mf}} \right] \right\}$$

$$v_a^2 = 3.407 \times 10^{-3} - 7.633 \times 10^{-3} \frac{v_b}{v_{mf}}$$

The following values are then found by substitution

Fractional Fluidization v_b/v_{mf}	(Air Velocity)² (m/s)²	Air Velocity (m/h)
0.20	1.89E−03	156
0.25	1.50E−03	139
0.30	1.11E−03	120
0.35	7.35E−04	98
0.40	3.54E−04	68

■

14.10 SUMMARY

Deep bed filtration is very commonly used as a means to remove particles from water and to achieve very low concentrations of particles in the treated water. With common designs for drinking water and wastewater (postsecondary treatment) applications, excellent filtrate quality can be attained for long times and high values of volume throughput per unit area, as long as the particles have been adequately destabilized. If inadequate effluent quality is found, the most likely cause is poor destabilization.

Filter performance is described by two dependent variables—the effluent concentration and the head loss through the filter. Either of these performance characteristics can limit the filter run and require that the filter media be cleaned (backwashed). Traditional designs for filters have generally been conservative, in the sense that they have caused filters to require backwashing because of excessive head loss at a time much earlier than the effluent quality would demand. Current designs use some combination of higher velocity and larger media, combined with deeper beds, to increase the production of each filter run, and to bring the times when backwashing would be required by the two criteria closer together. These changes have been made possible by better understanding and utilization of particle destabilization, and the increasingly common practice of monitoring the effluent quality of every filter in a bank of filters in parallel, among other factors.

Filter effluent quality is very difficult to predict, although many mathematical models have been proposed. These models yield important insights into the behavior of particles in filters and are quite useful in understanding filtration behavior, but experimental results are still required for quantitative work with filters. The models are far better at predicting the changes that will ensue if some design and operational variables are changed from a known condition than predicting results without calibration. In this sense, the models are quite useful to limit the amount of pilot-scale testing required if a new filter is to be built or if changes to an existing filter are contemplated, but they do not eliminate the need for such pilot testing. Useful starting points for using experimental results from one condition to predict what will happen in another are Equations 14-48 for effluent quality and 14-61 or 14-62 for head loss development.

Cleaning the filter media at the end of every filter run is accomplished by backwashing; that is, pumping water up through the filter bed; to be effective for most waters, the backwashing is accompanied by either a surface wash or air scour (or sometimes both). During backwashing of any one filter, the other filters in a bank of filters in parallel generally take on additional flow; this increase in the flow (or filtration velocity) can cause detachment of particles in these filters, especially if done abruptly.

Our understanding of granular media filtration, both in theoretical and practical terms, has increased continually over its long history, but this trend needs to continue, so that someday the pessimism of Stanley in the quote at the start of this chapter will have been ultimately proven unfounded.

REFERENCES

Adin, A., and Rebhun, M. (1977) A model to predict concentration and head-loss profiles in filtration. *JAWWA*, 69 (8), 444–453.

Adin, A., and Rebhun, M. (1987) Deep-bed filtration: accumulation-detachment model parameters. *Chem. Eng. Sci.*, 42, 1213–1219.

Amburgey, J. E., Amirtharajah, A., Brouckaert, B. M., and Spivey, N. C. (2003) An enhanced backwashing technique for improved filter ripening. *JAWWA*, 95 (12), 81–94.

Amburgey, J. E., Amirtharajah, A., Brouckaert, B. M., and Spivey, N. C. (2004) Effect of washwater chemistry and delayed start on filter ripening. *JAWWA*, 96 (1), 97–110.

Amirtharajah, A. (1978) Optimum backwashing of sand filters. *J. Environ. Eng. Div. (ASCE)*, 104 (EE5), 917–932.

Amirtharajah, A. (1984) Fundamentals and theory of air scour. *J. Environ. Eng. Div. (ASCE)*, 110 (3), 573–590.

Amirtharajah, A. (1988) Some theoretical and conceptual views of filtration. *JAWWA*, 80 (12), 36–46.

Amirtharajah, A. (1993) Optimum backwashing of filters with air scour: a review. *Water Sci. Technol.*, 27 (10), 195–211.

Amirtharajah, A., and Wetstein, D. P. (1980) Initial degradation of effluent quality during filtration. *JAWWA*, 72 (9), 518–524.

Brinkman, H. C. (1949) Calculations on the flow of heterogeneous mixtures through porous media. *Appl. Sci. Res.*, A1, 333–346.

Carman, P. (1937) Fluid flow through granular beds. *Trans., Inst. Chem. Eng.*, 15, 150–166.

Chang, Y., Li, C. H., and Benjamin, M. M. (1997) Iron oxide-coated media for NOM sorption and particulate filtration. *JAWWA*, 89 (5), 100–113.

Clark, S. C., Lawler, D. F., and Cushing, R. S. (1992) Contact filtration: particle size and ripening. *JAWWA*, 84 (12), 61–71.

Cleasby, J. L. (1972) Filtration, Chapter 4 In Weber, W. W. (ed.), *Physicochemical Processes for Water and Wastewater*, John Wiley & Sons, New York.

Cleasby, J. L., Arboleda, A., Burns, D. E., Prendiville, P. W., and Savage, E. S. (1977) Backwashing of granular filters. *JAWWA*, 69 (2), 115–126.

Cleasby, J. L., and Baumann, E. R. (1962) Selection of sand filtration rates. *JAWWA*, 54 (5), 579–602.

Cleasby, J. L., and Fan, K. S. (1981) Predicting fluidization and expansion of filter media. *J. Environ. Eng. Div. (ASCE)*, 107 (EE3), 455–471.

Cleasby, J. L., and Logsdon, G. S. (1999) Granular bed and precoat filtration, Chapter 8 In *Water Quality and Treatment*, 5th ed., American Water Works Association, McGraw Hill, New York.

Cleasby, J. L., Sindt, G. L., Watson, D. A., and Baumann, E. R. (1992) *Design and Operation Guidelines for Optimization of the High-Rate Filtration Process: Plant Demonstration Studies*, AWWA Research Foundation Report, Denver, CO.

Colton, J. F., Hillis, P., and Fitzpatrick, S. B. (1996) Filter backwash and start-up strategies for enhanced particulate removal. *Water Res.*, 30, 2502–2507.

Cookson, J. T. (1970) Removal of submicron particles in packed beds. *Environ. Sci. Technol.*, 4, 124–134.

Cushing, R. S., and Lawler, D. F. (1998) Depth filtration: fundamental investigation through three-dimensional trajectory analysis. *Environ. Sci. Technol.*, 32, 3793–3801.

Darby, J. L., and Lawler, D. F. (1990) Ripening in depth filtration: effect of particle size on removal and head loss. *Environ. Sci. Technol.*, 24 (7), 1069–1079.

Darby, J. L., Attanasio, R. E., and Lawler, D. F. (1992) Filtration of heterodisperse suspensions: modeling of particle removal and head loss. *Water Res.*, 26, 711–726.

Eliassen, R. (1941) Clogging of rapid sand filters. *JAWWA*, 33 (5), 926–942.

Elimelech, M., and O'Melia, C. R. (1990) Kinetics of deposition of colloidal particles in porous media. *Environ. Sci. Technol.*, 24, 1528–1536.

Ergun, S. (1952) Fluid flow through packed columns. *Chem. Eng. Prog.*, 48, 89–94.

Ginn, T. M., Jr., Amirtharajah, A., and Karr, P. R. (1992) Effects of particle detachment in granular media filtration. *JAWWA*, 84 (2), 66–76.

Goren, S. L., and O'Neil, M. E. (1971) Hydrodynamic resistance to a particle of a dilute suspension when in the neighborhood of a large obstacle. *Chem. Eng. Sci.*, 26 (3), 325–338.

Habibian, M. T., and O'Melia, C. R. (1975) Particles, polymers, and performance in filtration. *J. Environ. Eng. Div. (ASCE)*, 101, EE4, 567–583.

Hahn, M. W., and O'Melia, C. R. (2004) Deposition and reentrainment of Brownian particles in porous media under unfavorable chemical conditions: some concepts and applications. *Environ. Sci. Technol.*, 38, 210–220.

Hahn, M., Abadzic, D., and O'Melia. C. R. (2004) Aquasols: on the role of secondary minima. *Environ. Sci. Technol.*, 38, 5915–5924.

Happel, J. (1958) Viscous flow in multiparticle systems: slow motion of fluids relative to beds of spherical particles. *AIChE J.*, 4 (2), 197–201.

Hargette, A. C., and Knocke, W. R. (2001) Assessment of fate of manganese in oxide-coated filtration systems. *J. Environ. Eng. (ASCE)*, 127 (12), 1132–1138.

Harris, W. L. (1970) High rate filter efficiency. *JAWWA*, 62 (8), 515.

Herzig, J. P., Leclerc, D. M., and LeGoff, P. (1970) Flow of suspensions through porous media—application to deep bed filtration. *Indust. Eng. Chem.*, 62 (5), 8–35.

Hewitt, S. R., and Amirtharajah, A. (1984) Air dynamics through filter media during air scour. *J. Environ. Eng. (ASCE)*, 110 (3), 591–606.

Hsiung, K. Y., and Cleasby, J. L. (1968) Prediction of filter performance. *J. San. Eng. Div. (ASCE)*, 94 (SA6), 1043–1069.

Ives, K. J. (1960) Rational design of filters. Proceedings, Institution of Civil Engineers, London, England, 16, 189–193.

Ives, K. J. (1969) Theory of filtration, Special Subject No. 7, *International Water Supply Congress and Exhibition*, Vienna.

Ives, K. J. (1980) Deep bed filtration: theory and practice. *Filtration Separation*, 17, 157–166.

Ives, K. J. (1971) Filtration of water and waste water. *CRC Crit. Rev. Environ. Control*, 2 (2), 293–335.

Ives, K. J. (1989) Filtration studied with endoscopes. *Water Res.*, 23 (7), 861–866.

Ives, K. J., and Sholji, I. (1965) Research on variables affecting filtration. *J. San. Eng. Div. (ASCE)*, 91 (SA4), 1–18.

Iwasaki, T. (1937) Some notes on filtration. *JAWWA*, 29 (10), 1591–1602.

Kau, S. M., and Lawler, D. F. (1995) Dynamics of deep-bed filtration: velocity, depth, and media. *J. Environ. Eng., ASCE*, 121 (12), 850–859.

Kawamura, S. (1999) Design and operation of high-rate filters. *JAWWA*, 91 (12), 77–90.

Kawamura, S. (2000) *Integrated Design and Operation of Water Treatment Facilities*. 2nd ed. John Wiley & Sons, New York.

Kozeny, J. (1927) Über kapillare Leitung des Wasser im Boden. *Sitzungsber. Akad. Wiss. Wien*, 136, 271–306.

Levich, V. G. (1962) *Physicochemical Hydrodynamics*. Prentice-Hall, Englewood Cliffs, NJ.

Lewis, C. M., and Manz, D. H. (1991) Light scatter particle counting: improving filtered water quality. *J. Environ. Eng., ASCE*, 117 (2), 209–223.

Logsdon, G. S., Symons, J. M., Hoye, R. L.Jr., and Arozarena, M. M. (1981) Alternative filtration methods for removal of *Giardia* cysts and cyst models. *JAWWA*, 73 (2), 111–118.

MacDonald, I. F., El-Sayed, M. S., Mow, K., and Dullien, F. A. L. (1979) Flow through porous media—the Ergun equation revisited. *Ind. Eng. Chem. Fundamen.* 18 (3), 199–208.

Mackie, R. I., Horner, R. M. W., and Jarvis, R. J. (1987) Dynamic modeling of deep-bed filtration. *AIChE J.*, 33 (11), 1761–1775.

Mackrle, V., Draka, O., and Svec, J., (1965) Hydrodynamics of the Disposal of Low Level Liquid Radioactive Wastes in Soil, *International Atomic Energy Agency*, Report No. 98, Vienna.

Mehter, A. A., Turina, R. M., and Tien, C. (1970) *Filtration in Deep Beds of Granular Activated Carbon*, Research Report No. 70-3, FWPCA Grant No. 17020 OZO, Syracuse University.

Mints, D. M. (1966) *Modern Theory of Filtration*, Special Report No. 10, International Water Supply Congress, Barcelona.

Moran, D. C., Moran, M. C., Cushing, R. S., and Lawler, D. F. (1993a) Particle behavior in deep-bed filtration: part 1-ripening and breakthrough. *JAWWA*, 85 (12), 69–81.

Moran, M. C., Moran, D. C., Cushing, R. S., and Lawler, D. F. (1993b) Particle behavior in deep-bed filtration: part 2-particle detachment. *JAWWA*, 85 (12), 82–93.

MWH (Montgomery Watson Harza) (2005) Granular filtration. Chapter 11 In *Water Treatment Principles and Design*, 2nd ed., John Wiley and Sons, New York.

O'Melia, C. R. (1974) *The Role of Polyelectrolytes in Filtration Processes*, Report No. EPA-670/2-74-032, U.S. Environmental Protection Agency, Cincinnati, OH.

O'Melia, C. R., and Stumm, W. (1967) Theory of water filtration. *JAWWA*, 59 (11), 1393–1412.

O'Melia, C. R., and Ali, W. (1978) The role of retained particles in deep bed filtration. *Progr. Water Technol.*, 10 (5/6), 167–182.

O'Melia, C. R. (1985) Particles, pretreatment, and performance in water filtration *J. Environ. Eng. Div., ASCE*, 111 (6), 874–890.

Payatakes, A. C., Park, H. Y., and Petrie, J. (1981) A visual study of particle deposition and reentrainment during depth filtration of hydrosols with a polyelectrolyte. *Chem. Eng. Sci.*, 36, 1319–1335.

Payatakes, A. C., Rajagopalan, R., and Tien, C. (1974) Application of porous media models to the study of deep bed filtration. *Can. J. Chem. Eng.*, 52, 722–731.

Payatakes, A. C., Tien, C., and Turian, R. M. (1973a) A new model for granular porous media: Part I. Model formulation. *AIChE J.*, 19, 58–67.

Payatakes, A. C., Tien, C., and Turian, R. M. (1973b) A new model for granular porous media: Part II. Numerical solution of steady state incompressible Newtonian flow through periodically constricted tubes. *AIChE J.*, 19, 58–67.

Petrusevski, B., Sharma, S. K., Kruis, F., Omeruglu, P., and Schippers, J. C. (2002) Family filter with iron-coated sand: solution for arsenic removal in rural areas. *Water Sci. Technology: Water Supply*, 2 (5–6), 127–133.

Rajagopalan, R., and Tien, C. (1976) Trajectory analysis of deepbed filtration with the sphere-in-cell porous media model. *AIChE J.*, 22 (3), 523–533.

Rajagopalan, R., Tien, C., Pfeffer, R., and Tardos, G. (1982) Letter to the editor. *AIChE J.*, 28 (5), 871.

Richardson, J. F., and Zaki, W. N. (1954) Sedimentation and fluidisation: part I. *Trans. Inst. Chem. Eng.*, 32, 35–53.

Stanley, D. R. (1955) Sand filtration studied with radiotracers. *J. San. Eng. Div., ASCE*, 81 (1), 592–596.

Stanley, W. E. (1937) Discussion of T. Iwasaki's Some notes on filtration. *JAWWA*, 29 (10), 1591–1602.

Tare, V., and Venkobachar, C. (1985) New conceptual formulation for predicting filter performance. *Env. Sci. Technol.*, 19, 497–499.

Tien, C. (1989) *Granular Filtration of Aerosols and Hydrosols*. Butterworth Publishers, Stoneham, MA.

Tien, C., Turian, R. M., and Pendse, H. (1979) Simulation of the dynamic behavior of deep bed filters. *AIChE J.*, 25 (3), 385–395.

Tobiason, J. E., Cleasby, J. L., Logsdon, G. S., and O'Melia, C. R. (2011) Granular media filtration, Chapter 10 In *Water Quality and Treatment*, 6th ed., American Water Works Association, McGraw Hill, New York.

Tobiason, J. E., Johnson, G. S., Westerhoff, P. K., and Vigneswaran, B. (1993) Particle size and chemical effects on contact filtration performance. *J. Environ. Eng., ASCE*, 119, 520–539.

Tobiason, J. E., and Vigneswaran, B. (1994) Evaluation of a modified model for deep bed filtration. *Water Res.*, 28 (2), 335–342.

Trussell, R. R., and Chang, M. (1999) Review of flow through porous media as applied to head loss in water filters. *J. Environ. Eng.*, ASCE, 125 (11), 998–1006.

Tufenkji, N., and Elimelech, M. (2004) Correlation equation for predicting single-collector efficiency in physicochemical filtration in saturated porous media. *Env. Sci. Technol.*, 38 (2), 529–536.

Veerapeneni, S., and Wiesner, M. R. (1993) Role of suspension polydispersivity in granular media filtration. *J. Environ. Eng., ASCE*, 119 (1), 172–190.

Vigneswaran, S., and Chang, J. S. (1986) Mathematical modelling of the entire cycle of deep bed filtration. *Water Air Soil Pollut.*, 29, 155–164.

Vigneswaran, S., and Tulachan, R. K. (1988) Mathematical modeling of transient behavior of deep bed filtration. *Water Res.*, 22 (9), 1093–1100.

Yao, K. M., Habibian, M. T., and O'Melia, C. R. (1971) Water and wastewater filtration: concepts and applications. *Env. Sci. Technol.*, 5 (11), 1105–1112.

Yao, K. M. (1968) *Influence of Suspended Particle Size on the Transport Aspect of Water Filtration*, PhD Dissertation, Department of Environmental Sciences and Engineering, University of North Carolina, Chapel Hill, N.C.

PROBLEMS

14-1. If a picture is worth 1,000 words, an experience might be worth 10,000. Go to an operating wastewater or drinking water treatment plant and investigate a filter under the guidance of a knowledgeable person (operator or engineer associated with the plant). Witness the operation of a backwash. Go down below the filter to see the piping and valving. Some things to look and ask for include:

(a) The number of valves and their function to allow for normal operation, backwashing (including, perhaps, surface wash or air scour), and, in some cases in drinking water plants, the possibility of wasting filtered water rather than accepting it.

(b) The channels for influent water flow during normal operation and effluent backwash flow, including troughs that set the height of water above the filter during backwashing.

(c) Monitoring and recording equipment for head loss and effluent quality, and monitoring records (if any) for the head loss and effluent concentration during a single filter run. Determine what measure(s) the plant uses for effluent quality.

(d) The operating rule for the decision when to backwash and the usual amount of time between successive backwashes.

(e) Determine some details about backwashing methodology; for example, how the flow is (gradually) reduced to the filter when beginning a backwash cycle and the timing of air scour or surface wash relative to the initiation of the main backwash flow.

(f) Values for the primary design and operating characteristics: media size(s) and depth(s), filtration velocity under average and peak conditions, (surface) area of the filter, and backwash flow or velocity.

(g) The rate controller (assuming the filter you see is a constant flow rate filter).

14-2. The influent to a particular filter is a monodisperse suspension of 2-μm particles with a density of 1.05 g/cm^3. Assume that the Hamaker constant for the particles is 10^{-20} J. The filter has the following characteristics: depth of media = 75 cm, grain diameter = 0.6 mm, porosity = 0.4, filter plan area = 25 m^2, influent flow = 150 m^3/h, temperature = 25°C.

 (a) Assuming $\alpha = 1$, what is the expected removal efficiency in this filter according to the Tufenkji and Elimelech model?

 (b) Plot $N/N_{influent}$ as a function of depth in this filter.

 (c) Calculate the head loss across this filter bed using the Carman–Kozeny equation.

14-3. If the suspended solid concentration of the particles that enter the filter described in Problem 2 is 1.5 mg/L, and if we assume that all particles are captured in the filter, what fraction of the original pore space is occupied by the removed particles after a 24-h period?

14-4. Perform the mass balance described above Equation 14-73 to derive that equation, which describes the relationship between porosity and media depth under normal and expanded conditions.

14-5. All the details required to generate the values shown in Figures 14-7, 14-9, 14-10, and 14-11 are given in those figures.

 (a) Use a spreadsheet to determine these values.

 (b) Systematically change various parameters associated with the filter design and operation to see how the single collector removal efficiencies and overall filter removals are changed by the changes in the parameters you choose. Make appropriate graphs to document these changes and comment briefly.

 (c) Add a line to Figure 14-10 for the array-of-spheres model expressed in Equation 14-44; use the same equation for Brownian motion as the Rajagopalan and Tien model (Equation 14-38).

14-6. Make a photocopy of Figure 14-7, and indicate how the lines for each mechanism would shift with increases in all the filtration variables: filtration velocity, media depth, media size, water temperature, and particle density. Also indicate how each increase would affect the point of minimum collection efficiency (both horizontally and vertically).

14-7. Two filters, one labeled "traditional" and the other "modern," are described below. These values are not necessarily realistic; some parameters have been varied between the two in a rather arbitrary manner just to demonstrate the effect of this parameter (e.g., media density). For simplicity, both filters are given as monomedia filters, although many filters (both traditional and modern) are constructed as dual media filters. Assume that the media grains are all identical in size.

The particle size distribution function for the influent to these two filters is given in the following table. Assume that the particle density of this suspension is 1.1 g/cm^3 and that the Hamaker constant for these particles is 10^{-13} erg.

The temperature of the water is 25°C, at which the water density is 0.997 g/cm^3 and the viscosity is 8.91×10^{-3} g/cm s. The value of the Boltzmann constant is 1.38×10^{-16} g cm^2/s^2 K.

Filter Characteristics

Parameter	Symbol	Units	Traditional Filter	Modern Filter
Media size (diameter)	d_c	mm	0.5	1.8
Media depth	Z	cm	35	160
Filtration velocity	v	m/h	5	18
Shape factor	ϕ	—	6	7.5
Media density	ρ_c	g/cm^3	2.65 (sand)	1.5 (anthracite)
Bed porosity	ε	—	0.40	0.48

Particle Size Distribution[a]

$\log d_p$	$\log(\Delta N/\Delta d_p)$	$\log d_p$	$\log(\Delta N/\Delta d_p)$
−0.3	4.90	0.8	3.12
−0.2	4.88	0.9	2.82
−0.1	4.82	1.0	2.49
0.0	4.73	1.1	2.15
0.1	4.61	1.2	1.79
0.2	4.47	1.3	1.40
0.3	4.30	1.4	1.00
0.4	4.11	1.5	0.58
0.5	3.90	1.6	0.14
0.6	3.66	1.7	−0.32
0.7	3.40		

[a] d_p in μm, ΔN in cm^{-3}.

 (a) For all the sizes given in the size distribution (i.e., for every $\log d_p$ value between −0.3 and +1.7 at 0.1 increments), find the total single collector removal efficiency and the removal efficiency in the whole filter according to the Yao et al. model, assuming $\alpha = 1.0$. Use this information to find the log of the particle size distribution function log($\Delta N/\Delta d_p$), the number distribution ($\Delta N/\Delta \log d_p$), and the volume distribution ($\Delta V/\Delta \log d_p$) of the filter effluent. Also determine the overall removal efficiency (% removal) on the basis of (i) particle number concentration and (ii) particle volume (suspended solid) concentration.

(b) Repeat (a) using the Rajagapolan and Tien model.

(c) Repeat (a) using the Tufenkji and Elimelech model.

(d) Determine the head loss expected at the start of a filter run, according to the Carman–Kozeny and Ergun equations. For the Ergun equation, use the values $k_1 = 5$ and $k_2 = 0.3$ (i.e., so that the first term of the Ergun equation is identical to the Carman–Kozeny equation).

(e) Using the Ergun equation for the head loss, determine the following values associated with backwashing: (i) the minimum velocity required to accomplish fluidization (v_{mf}), (ii) the velocity required to accomplish 10% expansion of the bed (i.e., $Z_e/Z = 1.10$), and (iii) the water velocity required if air scouring is used and 30% of v_{mf} is to be the design criterion.

14-8. A wastewater treatment plant has an average flow of 50,000 m^3/d, but the peak flow entering the plant is considerably higher. The plant has had only secondary treatment until now, but filters are to be installed to achieve a higher effluent quality. Assume that the peak flow that reaches the end of the plant is 100,000 m^3/d. The designer wants the filtration velocity never to exceed 18 m/h. The maximum velocity can occur if one filter is out of service for backwashing at the same time that the peak flow is experienced.

(a) What is the minimum filter area that must be in service during the critical period?

(b) If filters are to be built square and to have a maximum size of 8.0 m on a side, how many filters have to be built?

(c) Determine the percent increase in the flow (or filtration velocity) to the remaining filters when one is taken out of service for backwashing.

(d) When one filter is taken out of service for backwashing, the flow to the other filters that remain in service increases. Show the relationship between the percent of this increase and the number of filters in a plant, and discuss the tradeoffs in design that must be considered in light of this relationship.

14-9. Figure 14-19 is a conceptual diagram showing how the filter run time (or volume throughput) varies with the media size, based on the time to reach the two termination criteria—excessive head loss or excessive effluent concentration. The locations of the two lines on such a figure depend on all the other design and operational variables as well as on the values set for the two criteria. On separate graphs for the variation in each parameter described below, show how the two lines would move on such a figure; that is, each figure will have four lines, two for each criteria with one line for the "standard" condition and the other for the changed condition specified. In some cases, it is possible that one of the lines will not move. If possible, show not only whether you expect the shift to be to the left or right but also whether it would change the slope or shape of the curve. If both lines shift, indicate whether you expect the optimum media size (i.e., the one that gives maximum production) to shift to larger or smaller sizes.

(a) Reduce the maximum allowable head loss.

(b) Reduce the maximum allowable effluent concentration.

(c) Reduce the depth of the media.

(d) Reduce the filtration velocity (choose volume throughput as the ordinate).

(e) Reduce the influent concentration.

(f) Change the influent size distribution so that it had more small particles but fewer large particles. This could occur, for example, by eliminating flocculation and switching from "direct filtration" (filtration after particle destabilization and flocculation, but no gravity separation) to "contact filtration" (filtration after particle destabilization only). Assume that the effluent concentration is measured in suspended solid concentration, which is dominated by large particles.

(g) Same as (f) but assume that the effluent concentration is measured by turbidity or total particle number concentration, which are dominated by small particles.

14-10. A laboratory-scale granular media filter was run with a constant flow rate of a clay suspension. A separate feed pump for alum was used to control the alum dose; which was varied as indicated in the following table.

Time Period (min)	Alum Dose (mg/L as alum)
0–90	2.5
90–121	0
121–180	2.5
180–300	1.0
300–320	3.0

TABLE 14-Pr10. Effluent Quality as a Function of Time

Time (min)	Turbidity (NTU)	Time (min)	Turbidity (NTU)	Time (min)	Turbidity (NTU)	Time (min)	Turbidity (NTU)
1	43	70	2.0	124	2.5	210	38
9	7	80	2.0	130	5.0	220	39
20	5	90	2.0	140	5.0	260	45
30	4.5	95	32	160	4.5	300	46
40	3.5	100	36	180	4.0	305	4.0
50	4.0	110	39	200	21	310	3.0
60	2.0	120	38	205	30	320	2.5

The turbidity of the filter effluent was monitored frequently, and the results shown in Table 14-Pr10 were obtained. Plot the data appropriately, and explain these results.

14-11. To achieve removal of some soluble contaminants, many water utilities now use more coagulant (metal salts) than they did when their plants were designed. In some cases, these plants were designed as "direct filtration" plants; that is, the rapid mix basin, where coagulant is added, was followed by a flocculation basin, but the water was then applied directly to a granular media filter. However, when they increase the coagulant dose, they might find it necessary to add sedimentation or flotation before filtration. Why would the increase in coagulant dose cause them to make such a decision?

14-12. (a) For an aqueous suspension at 15°C and particles of density $\rho_p = 1.15$ g/cm^3, develop curves of log η versus d_p and identify the particle size of minimum filtration efficiency according to the Yao et al. model for granular media filtration. Consider four possible operating conditions: $v_0 = 0.2$ or 0.5 cm/s; grain diameters $d = 0.4$ or 1.5 mm.

(b) What is the removal efficiency of the particles identified as having the minimum removal efficiency for the $v_0 = 0.2$ cm/s, $d_c = 0.4$ mm case in an ideal horizontal flow sedimentation basin with an overflow rate of 40 m^3/m^2 d?

(c) Compare the result in (b) with the removal efficiency of the same particles in a clean filter of the same dimensions and with a filtration velocity equal to the overflow rate of the sedimentation basin; that is, the removal that would be obtained if the sedimentation basin were filled with sand and converted into a downflow filter. Assume that the packed depth of the filter is 1.5 m, that its porosity is 0.38, and that the incoming particles are fully destabilized. Comment on the comparison.

14-13. A pilot-scale filter with a cross-sectional area of 0.25 m^2, a depth of 1.5 m, and a porosity of 0.38 is being operated at a flow rate of 60 L/min. The water temperature is 25°C at which the viscosity is 0.9×10^{-2} g/(cm s) and the density is 1.00 g/cm^3. Although the influent particle distribution is heterodisperse, the filter was designed so that the single collector removal efficiencies according to the Yao model for 0.1-μm-diameter particles by sedimentation and interception are $\eta_g = 1.0 \times 10^{-6}$ and $\eta_I = 6.0 \times 10^{-8}$.

The volume distribution of the influent particles is shown in Figure 14-Pr13. In the graph, V is in μm^3/cm^3 and d_p is in μm. There are no particles in the influent with log $d_p > 2.5$ or log $d_p < -3$. The function can be represented by three line segments:

- For the size region $-3 < \log d_p < -2$:

  $dV/d(\log d_p) = (3 \times 10^6 + 10^6 \log d_p)$ μm^3/cm^3.

- For the size region $-2 < \log d_p < 1$:

  $dV/d(\log d_p) = 10^6$ μm^3/cm^3.

- For the size region $1 < \log d_p < 2.5$:

  $dV/d(\log d_p) = (1.667 \times 10^6 - 0.667$

  $\times 10^6 \log d_p)$ μm^3/cm^3.

(a) Compute the diameter of the collector grains and the density of the incoming particles.

(b) Plot log η versus log d_p.

(c) Calculate and plot the volume distribution $(dV/d\log d_p)$ for the effluent particles. Show the influent distribution on the same graph.

(d) Compute the total mass concentration (mg/L) of particles in the influent and the effluent.

14-14. The analysis of the particle size distribution in the influent and effluent of a sand filter suggests that a significant fraction of the incoming particles with diameters around 1 μm pass through and appear in the effluent, yet most of the incoming particles with diameters around 0.1 μm are removed.

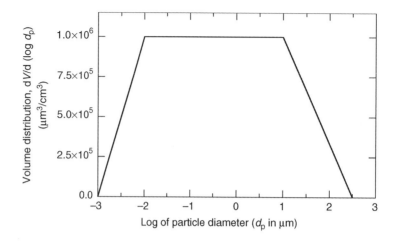

FIGURE 14-Pr13. Volume distribution of the filter influent.

(a) How is it possible for a filter to remove small particles and yet not remove larger ones?

(b) How does density affect particle removal in the system described? Specifically, would denser particles be removed more or less efficiently than less dense particles, if both particles were 0.1 μm in diameter? Answer the same question for particles that are 10 μm in diameter.

14-15. Figure 14-Pr15 shows several curves describing the hydrodynamic head at various depths in a filter as a function of time.

(a) Why do most of the curves have two distinct portions?

(b) The filter run must be terminated and the backwashing cycle begun when one of three criteria is met: the head loss across the bed is greater than 3.5 m, significant turbidity breakthrough occurs, or air bubbles start forming in the filter bed. Assuming that the influent water has equilibrated with air at 1.0 atm, estimate when each of these criteria would be violated.

14-16. A sand filter being used in a water treatment facility consistently reaches its water quality criterion for backwashing before the criterion for head loss is exceeded. Why is this mode of operating undesirable? What are some remedies that could be investigated to improve this situation without a major redesign of the facilities?

14-17. A graduate student is performing a series of tests on a laboratory-scale filter under various destabilization conditions for the particles in the influent; all experiments are run with identical hydraulic conditions in the same filter. The runs are short and only designed to emulate the early period in a full-scale filter at a water plant. Under most conditions, the results are what might be considered typical of a filter run during the ripening period: mediocre removal at first, with better effluent quality as time proceeds. However, in some experiments,

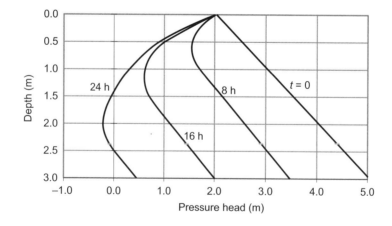

FIGURE 14-Pr15. Pressure head profiles at various times.

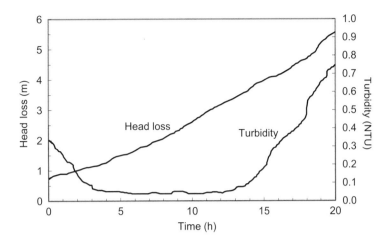

FIGURE 14-Pr18. Head loss and effluent quality in an operating filter.

the results are nearly the opposite: excellent (virtually 100%) removal for several minutes followed by an increasing effluent particle concentration until virtually no removal occurs. Explain these results.

14-18. The head loss and effluent during a typical filter run at a drinking water treatment plant are shown in Figure 14-Pr18. The maximum head loss at which the filter can be operated efficiently is 2.5 m, and the turbidity criterion is 0.2 NTU.

(a) Approximately how long should the operator run the filter before backwashing it?

(b) Explain briefly why the head loss increases at an approximately steady rate over time.

(c) Describe the likely effects of the following changes on the amount of water that can be processed in a filter run. Assume in all cases that the changes are fairly small, so that whatever limits the run length in the baseline (initial) condition still limits it after the change has been implemented.

  (i) Adding a small amount of polymeric coagulant (and thereby improving the destabilization) as the water enters the filter bed.

  (ii) Adding a small amount of sand to the bed to make the column deeper.

  (iii) Replacing the sand in the bed with slightly larger grained material with the same depth. (Assume the porosity remains the same, but recognize that the pore size will increase.).

(d) On a copy of the plot, sketch new curves showing the expected head loss and turbidity if option (i) (coagulant addition) is implemented.

(e) On another copy of the plot, sketch new curves showing the expected head loss and turbidity if option (ii) (sand addition) is implemented. Assume that the additional sand increases the bed depth by 15%.

14-19. Assuming that the fluid velocity through a filter is slow, so that there is no wake downstream of a sand grain, how is it possible for a small particle suspended in the influent to become attached to the *bottom* of a sand grain?

14-20. When sand filters are used to treat drinking water, they are almost always designed and operated so that the head loss criterion for backwashing is reached before turbidity breakthrough. On the other hand, when filters are used to remove particles from treated sewage, operating with either the turbidity breakthrough or head loss limitation occurring first is considered acceptable. Why the difference?

14-21. It has been suggested that all the mechanisms that cause particles to be removed in filter columns (interception, Brownian motion, and sedimentation) are also applicable in coagulation basins. Yet, filters are much more efficient at removing particles than coagulation/flocculation/settling basins.

(a) If the movements of particles in filter beds are so similar to those in coagulation basins, why are filters generally much more efficient at removing particles?

(b) Sweep coagulation is used to alter the particle size distribution in coagulation processes upstream of sedimentation basins, but generally not upstream of filtration processes. Explain this difference in the pretreatment processes.

14-22. Figure 14-Pr22 shows the changes in the profile of the *pressure* head in a granular media filter that is operated with a constant *total* head (and therefore declining flow as the run proceeds). Sketch the total head versus depth in the filter at the five times shown. (Time $t=0$ is just before the filter run

726 GRANULAR MEDIA FILTRATION

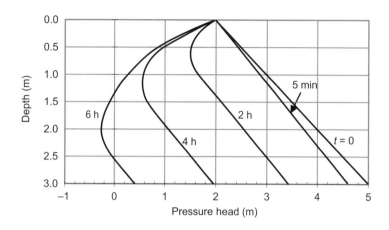

FIGURE 14-Pr22. Pressure head profiles during filter operation.

starts.) Define the datum for total head as the bottom of the filter media.

14-23. Several curves of pressure versus depth through a sand filter are shown in Figure 14-Pr23, each corresponding to a different time during the filter run. The hydraulic (pressure) head at the top of the sand (indicated in the figure as depth $= 0$ m) is 4.0 m at all times. The filter bed is 2.0 m deep and must be backwashed when particles start breaking through into the effluent in significant amounts, or when the head loss exceeds 2.2 m. The filter is operated in constant flow mode.

(a) Why are the curves near the bottom of the filter approximately parallel to one another regardless of how long the run has lasted?

(b) What is the pressure head at the bottom of the filter under no-flow conditions?

(c) Estimate the time when you think each of the criteria for backwashing the media is first met, and explain how you made this determination.

(d) One option available to the system operator is to add 0.5 m of sand to the column. If this were done, what do you think the head loss through the bed would be 2 h after the beginning of a filter cycle?

(e) If the criteria for head loss and turbidity breakthrough remain the same, do you think that adding the sand would increase, decrease, or not affect the length of a filter cycle?

14-24. A plot characterizing head loss in a filter run is shown in Figure 14-Pr24. The total depth of sand in the filter is 3 m.

(a) A plot of head loss versus time for this filter is approximately linear, even though the head loss versus depth curve is highly nonlinear. Explain briefly why the head loss increases at an approximately steady rate over time.

(b) Explain in a sentence or two why the top portion of the head loss versus depth curve is approximately linear at any given time, with approximately the same slope from one time to the next.

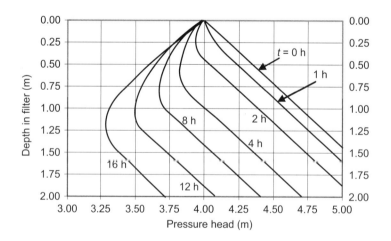

FIGURE 14-Pr23. Head profiles over time in an operating filter.

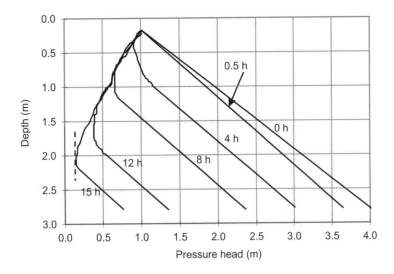

FIGURE 14-Pr24. Pressure head profiles over time in an operating filter.

(c) For the data at $t = 8$ h, at what depth in the column is the total head largest? At what depth is the pressure largest? At what depth is the head loss largest?

(d) Still considering the data at $t = 8$ h, what fraction of the head loss can be attributed to particles trapped in the sand, as opposed to the sand grains themselves?

(e) What is the local gradient of head loss versus depth (in units of m/m) at the point indicated by the dashed line in the figure; that is, at a point where a tangent to the curve is vertical. (Hint: keep in mind that the figure shows pressure versus depth, and head loss is a measure of the decrease in total head, not just the pressure head.)

14-25. If we could obtain images of a very small particle in the vicinity of a sand grain in the middle of a filter, we might see the following. As the particle moves downward with the water, it first moves away from the sand grain, then abruptly changes directions and moves toward the grain. It then continues moving toward the grain, but at a progressively slower velocity. Then, rather suddenly, it accelerates toward the grain and attaches to it. Describe the factors that you think cause these various motions.

14-26. The profile of hydrodynamic pressure in a rapid sand filter 6 h after backwashing is shown in Figure 14-Pr26. The head loss through the filter at the beginning of the filter run was 0.3 m. Based on this information, sketch curves of head loss across the whole filter and turbidity of the effluent from the beginning of the run until $t = 36$ h.

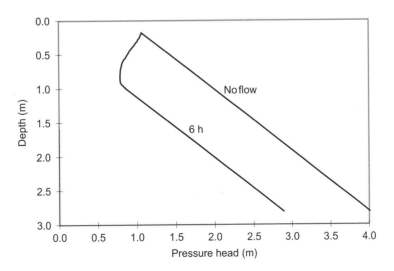

FIGURE 14-Pr26. Pressure head profiles with no flow and after 6 h of operation.

PART IV

MEMBRANE-BASED WATER AND WASTEWATER TREATMENT

15

MEMBRANE PROCESSES

MARK BENJAMIN, DESMOND LAWLER, AND MARK WIESNER

15.1 Introduction
15.2 Overview of membrane system operation
15.3 Membranes, modules, and the mechanics of membrane treatment
15.4 Parameters used to describe membrane systems
15.5 Overview of pressure-driven membrane systems
15.6 Quantifying driving forces in membrane systems
15.7 Quantitative modeling of pressure-driven membrane systems
15.8 Modeling transport of water and contaminants from bulk solution to the surface of pressure-driven membranes
15.9 Effects of crossflow on permeation and fouling
15.10 Electrodialysis
15.11 Modeling dense membrane systems using irreversible thermodynamics
15.12 Summary
References
Problems

15.1 INTRODUCTION

A membrane is a thin layer of semipermeable material that is capable of separating substances based on their physical or chemical properties. The variety of membrane systems in use and the range of materials that they can separate from aqueous solutions are broad. In fact, one membrane process or another has been proposed as a replacement for most of the separation processes encountered in water and wastewater engineering. Such processes generate at least two outlet streams—one that is cleaner than the feed and another containing the constituents that have been separated from it. The efficiency of the overall process is typically evaluated in the context of two issues: how well the membrane separates the constituents from one another and how rapidly the influent can be processed per unit of membrane used. The ideal situation is to achieve very efficient separation at a high rate of treatment; but in practice, one goal is usually compromised to achieve the desired performance with respect to the other.

The applications of membranes in environmental engineering have not been limited to treatment of aqueous streams. For instance, membranes have been used to purify organic solvents, separate constituents of gases from each other, introduce gases to liquid streams, dewater sludges, and strip volatile compounds from solutions. Membranes are also of growing importance in environmental analytical procedures (e.g., molecular weight fractionations, determination of contaminant partitioning between phases, and as components of species-selective probes). However, in this chapter, we focus on the use of membranes to purify aqueous streams. An excellent, more comprehensive treatise on membrane systems has been provided by Strathmann et al. (2006), and the one focusing strictly on microfiltration (MF) and ultrafiltration (UF) has been prepared by Zeman and Zydney (1996).

This chapter begins with a description of the hardware of membrane systems—the membranes themselves and the modules in which they are utilized—along with a broad overview of how such systems are operated and the terminology used to describe the systems. Following these sections, an overview is presented of the most common membrane systems used in environmental engineering applications—those driven by an imposed transmembrane pressure (TMP) differential. We then turn our attention to the quantitative modeling of transport in membrane systems,

---

Water Quality Engineering: Physical/Chemical Treatment Processes, First Edition. By Mark M. Benjamin, Desmond F. Lawler.
© 2013 John Wiley & Sons, Inc. Published 2013 by John Wiley & Sons, Inc.

beginning with consideration of the forces responsible for transport through the membranes and then applying these ideas to pressure-driven systems, with separate sections dealing with transport up to and through the membrane. The final sections consider membrane processes driven by a differential in electrical potential (rather than pressure), and a more advanced theoretical analysis of transport in pressure-driven systems. Throughout, an effort is made to relate the theoretical analysis of membrane processes based on mathematical models with the practical aspects of membrane processes as they are used in environmental engineering applications.

15.2 OVERVIEW OF MEMBRANE SYSTEM OPERATION

The key features of a membrane system for treating an aqueous solution are highlighted in Figure 15-1. At the beginning of a treatment cycle, the feed enters the system and contacts one side of the membrane. At this point, a driving force for migration of various constituents across the membrane either exists already (if the solutions on opposite sides of the membrane have different compositions) or is imposed (e.g., by applying a pressure differential across the membrane). The overall driving force might be physical, chemical, electrical, or some combination of these. The different types of driving force are discussed in detail later in this chapter; for now, the critical points are simply that a driving force for change in the system exists, and in response to this force, some constituents begin to migrate through the membrane.

The membrane impedes the transport of all the constituents that reach it, but it does so selectively, interfering with the transport of some constituents more than others. Several mechanisms can contribute to this selectivity. Perhaps, the

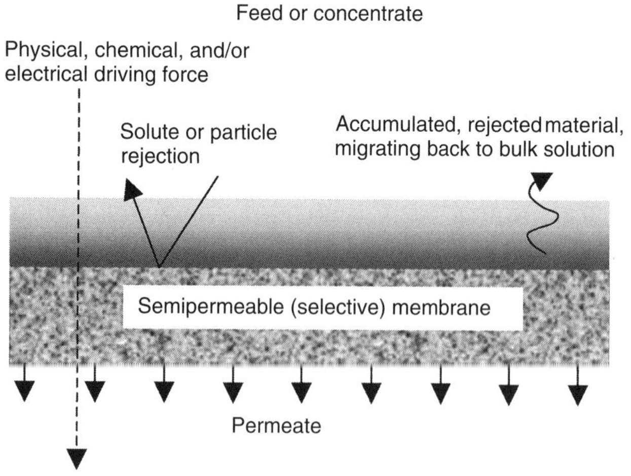

FIGURE 15-1. Schematic representation of the key processes occurring in a membrane treatment system.

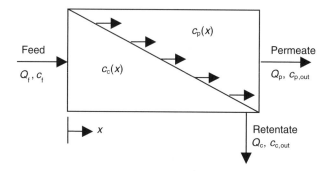

FIGURE 15-2. Flows into and out of a contaminant-rejecting membrane system. The subscript c on the retentate refers to its common designation as the *concentrate*. Both c_c and c_p might vary with x. $c_{c,out}$ is the concentration of the exiting rejectate stream. Permeate is generated all along the membrane, so $c_{p,out}$ represents a blend of the fluid permeating at all values of x.

simplest is mechanical sieving—particles and molecules that are larger than the openings in the membrane are completely rejected, whereas those that are smaller have the potential to enter and pass through the membrane. However, several other physical and chemical factors also play a role in determining the extent to which a particular material passes through the membrane. Depending on the membrane structure and operating conditions, a system might be used primarily to remove suspended particles, colloids, macromolecules, or small molecules from the feed. Throughout this chapter, we refer to specific subgroups of these constituents when appropriate, and we refer to the complete collection of nonaqueous constituents in the feed generically as "contaminants."

Most full-scale membrane processes operate with continuous or semicontinuous flow. The membrane and the principal flow streams into and out of such a system are often represented by a simple schematic, such as the one shown in Figure 15-2. The flow streams are referred to as the *influent* or *feed*, the *retentate*, and the *permeate*.[1] In most systems of interest, water passes through the membrane, and other constituents in the feed are selectively rejected. As a result, the contaminants on the feed side of the membrane are more concentrated than in the inflow, and the retentate is referred to as the *concentrate*. In systems where salt becomes concentrated on one side of the membrane (either by selective rejection or selective passage), the solution containing the concentrated salt is often referred to as the *brine*.

---

[1] An extensive list of definitions for terms related to membrane systems has been adopted by the International Union of Pure and Applied Chemistry (IUPAC). Unless otherwise noted, the terminology in this chapter is consistent with these conventions. The complete list has been published by Koros et al. (1996). An updated list is maintained on the website of the North American Membrane Society, at http://www.che.utexas.edu/nams/NAMSHP.html.

The species that are rejected by a membrane accumulate near the membrane surface and can either aid or impede separation of the contaminants that approach the membrane subsequently. If this layer of accumulated material increases the resistance to transport of contaminants more than that of water, it tends to improve the rejection efficiency. On the other hand, the increased concentration of rejected species increases the driving force for migration of those species through the membrane, so any increase in rejection is selflimiting. The accumulation of rejected species on or inside the membrane also invariably increases the resistance to transport of constituents that are intended to pass through the membrane, and in particular to transport of water; this process is referred to as *membrane fouling*. In membrane systems used for water and wastewater treatment, common foulants include colloidal organic detritus (e.g., fragments of cells and/or other biological structures); macromolecules such as natural organic matter (NOM); inorganic colloids such as clays, $CaCO_3(s)$, and Fe or Al oxides; solutes that might adsorb to the membrane such as silica; and microorganisms and their secretions. In many cases, the contaminants originally attach to the membrane individually, but subsequently coagulate or precipitate to form a composite that is a more severe foulant than the individual constituents.

The accumulation of rejected species near the membrane surface, and therefore fouling, can sometimes be reduced by providing a steady flow of water parallel to the surface with a sufficient velocity to scour the surface (Figure 15-3a and 15-3b). This flow is referred to both as *crossflow* (because it is perpendicular to the flow through the membrane) and as *tangential flow* (because it is parallel to the membrane). In other cases, motion is induced in the membranes themselves, either by vibration or rotation or by bubbling air through the adjacent solution. In addition, in some systems, water is intermittently pulsed from the permeate to the feed side in an attempt to dislodge any contaminants that accumulate in the pores or next to the membrane (Figure 15-3c); this process is referred to as *back-flushing* or *backwashing*. *Fast-flushing*, in which a burst of high velocity water is passed over the membrane surface (as an alternative or in addition to continuous crossflow), was commonly applied in the past, but is used rarely at present.

Fluid cleaning typically restores some, but not all, of the membrane permeability that is lost due to fouling. Any deterioration in membrane performance that is not reversed by flushing is referred to as (hydraulically) *irreversible fouling*. Hydraulically irreversible fouling can often be largely overcome by contacting the membrane with chemical reagents, such as acids, bases, detergents, or oxidants. In some systems, a mild chemical cleaning is carried out once to several times per week by adding chemicals to the solution used to backwash the membrane, and perhaps altering the backwashing protocol. At longer intervals, all systems require

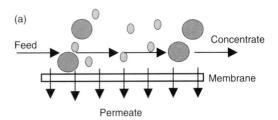

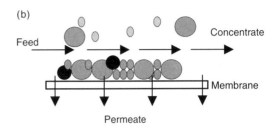

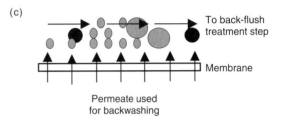

FIGURE 15-3. Approaches for minimizing buildup of contaminants on membranes. Crossflow in a system with (a) a clean membrane and (b) a membrane on which contaminants have deposited. (c) Back-flushing.

more aggressive chemical cleanings; such cleanings are sometimes referred to as *cleaning-in-place* (CIP). In most cases, for a system to be economical, the need for a CIP must arise not more frequently than approximately once per month, and preferably only once every several months.

Membrane fouling is extremely complex, for several reasons. First, different species have different effects on the passage of water, and these effects are not necessarily a simple function of the concentration or mass buildup of the rejected species. Second, rejected species do not simply accumulate near the membrane—they are also transported back into the bulk solution by various processes that depend on the species' properties and the local hydrodynamic regime. Finally, rejected constituents may undergo physical/chemical transformations near the surface, such as precipitation or coagulation. Understanding, modeling, and interfering with the fouling process constitute a major thrust of research in membrane technology. A few approaches for minimizing fouling in practice, as well as for modeling it theoretically, are presented in greater detail later in this chapter.

15.3 MEMBRANES, MODULES, AND THE MECHANICS OF MEMBRANE TREATMENT

Membrane Structure, Composition, and Interactions with Water

Membranes are fabricated from a variety of starting materials, and they are implemented in various physical forms. In this section, we describe a few physical/chemical features of the membranes used most commonly in water-focused environmental engineering applications and relate those features to membrane performance.

Transport through membranes occurs through small openings in the membrane material. Although the sizes of these openings range widely (sometimes even within a single membrane), two extreme cases can be considered. At one extreme are membranes in which fluid moves predominantly by advection through distinct pores. Most of the contaminant rejection achieved by these *porous membranes* occurs at the membrane surface, although some contaminants might enter the pores and get trapped before they exit. Porous membranes typically remove particles and most colloids, and those with the smallest pores can remove large molecules as well. At the other extreme are so-called *dense* or *nonporous membranes*, in which distinct pores are not discernible even when the membranes are scrutinized at high magnification.[2] These membranes are used to achieve efficient removal of small molecules from solution. Dense membranes are commonly modeled as a homogeneous, nonaqueous phase into which molecules dissolve on the feed side and from which they elute on the permeate side. In such models, selectivity is attributed to different solubilities and rates of diffusion of different molecules in the membrane phase: more soluble molecules and those with higher diffusivities selectively permeate, while those that are less soluble or have lower diffusivities are selectively rejected.

Important distinctions also exist among macroscopic features of membranes. For example, *symmetric membranes* have an approximately uniform composition throughout their entire thickness, so that slices through different layers of the membrane parallel to its surface appear essentially identical. Images of a symmetric membrane are shown in Figure 15-4a. In contrast, *asymmetric membranes* combine a thin membrane skin that is responsible for contaminant rejection with a much thicker layer of material that provides mechanical support (Figure 15-4b); the support layer plays almost no role in the separation of the constituents and has negligible hydraulic resistance. Because the rejecting skin of

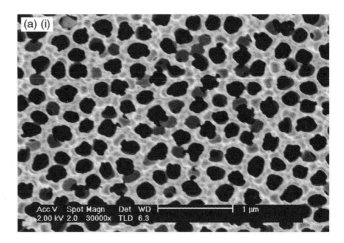

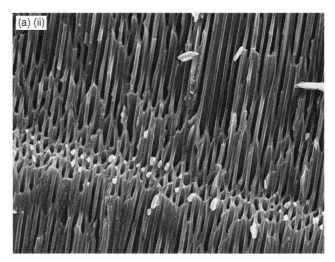

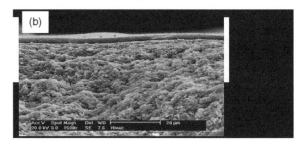

FIGURE 15-4. Scanning electron micrographs of membrane surfaces. (a) (i) Top and (ii) cross-sectional views of a symmetric membrane. (b) An asymmetric membrane. The skin near the top of the photograph is responsible for contaminant rejection; the thicker portion of the membrane below the skin is the more porous support structure. *Source:* (a) Z. Hendren, (b) M. Wiesner.

---

[2] One interpretation of this observation is that the spaces through which the permeating species move are generated transiently to accommodate these species, and that the spaces do not exist in the absence of the permeate, much as the open space occupied by a gas bubble is continuously generated (and the space that the bubble formerly occupied disappears) as the bubble rises through a column of water (Wijmans and Baker, 1995).

an asymmetric membrane is thin (typically $\leq 1\,\mu m$), the energy required to force water through it is relatively small. The support layer is relatively thick (tens to hundreds of micrometers), but it has only a small hydraulic resistance. As a result, asymmetric membranes combine high rejection of contaminants (imparted by the characteristics of the skin)

TABLE 15-1. Characteristics of Various Organic Membranes

Membrane Material	Hydrophobicity/ Hydrophilicity	Chlorine Tolerance	pH Tolerance	Temperature Range (°C)	Typical Applications[a]
Cellulosic (cellulose acetate, cellulose triacetate, cellulose nitrate, cellulose esters)	Hydrophilic	High	4–8.5	50	UF/NF/RO
Polysulfones/polyethersulfones	Hydrophobic/moderately hydrophilic	Moderate	1–12	0–80	UF, substrate for RO thin film composites
Polyvinylidene fluoride (PVDF)	Hydrophobic	High	2–10	130–150	MF/UF
Polypropylene	Moderately hydrophobic	Low	0.5–14	50	MF
Polyacrylonitriles (PAN)	Hydrophobic	Moderate	2–10	60	UF
Polytetrafluoroethylene (PTFE)	Hydrophobic	High	1–12	260	MF
Polyamides	Hydrophilic	Low	2–8	45	NF/RO

[a] MF, UF, NF, and RO refer to microfiltration, ultrafiltration, nanofiltration, and reverse osmosis, respectively, which are the major categories of pressure-driven membrane processes. The size of the contaminants removed by these processes decreases in the order listed. These processes are discussed in subsequent sections of the chapter.

with good mechanical strength (imparted by the support) and low resistance to permeate flow (due to the minimum thickness of the skin). Most dense membranes are asymmetric, whereas porous membranes might be either symmetric or asymmetric.

The chemical characteristics of membranes—both their bulk composition and their surface chemistry—also play an important role in determining their performance. Although membranes are fabricated from various materials, organic polymers are by far the most common starting materials for the membranes used in water and wastewater treatment systems. Ceramic membranes are used in specialized applications, where their excellent mechanical strength and their tolerance to oxidants and extremes of pH and temperature might be required. Historically, ceramic membranes have tended to be substantially more expensive than polymeric membranes, due to the relatively low membrane area that can be packed into a given volume as well as the costs of production. However, their cost has declined in recent years, and they are now competitive with polymeric membranes in some water and wastewater applications.

Membranes made of materials that have relatively unfavorable interactions with water molecules (e.g., polytetrafluoroethylene (Teflon®)) are referred to as *hydrophobic membranes*. Dissolving water into a phase of such materials is difficult, so they are rarely used to prepare dense membranes; however, they are used fairly frequently for producing porous membranes or as support materials in asymmetric membranes. Hydrophobic solutes (such as many organic solutes) tend to accumulate on or in hydrophobic membranes, leading to membrane fouling.

Hydrophilic membranes, on the other hand, have highly polar or ionized functional groups on their surfaces and therefore interact favorably with water molecules, thereby facilitating water transport. However, these same functional groups make the membranes more vulnerable than their hydrophobic counterparts to attack by oxidants and to biodegradation. Hydrophilic membranes are, in most cases, less likely than hydrophobic membranes to become fouled via adsorption of organic matter. Unfortunately, fouling by NOM can be severe on both hydrophobic and hydrophilic membranes, since molecules of NOM often have both hydrophobic and hydrophilic components. Some characteristics of membranes made of various organic polymers are summarized in Table 15-1.

Many hydrophilic membranes acquire a pH-dependent surface charge via ionization of weakly acidic or basic functional groups in the membrane structure (primarily carboxyl and amino groups, in the case of polymeric membranes). The pH at which the positive and negative charges exactly balance, so that the membrane is uncharged, is called the *isoelectric point*; at pHs below the isoelectric point, the net charge is positive, and at pHs above it, the net charge is negative (Figure 15-5). In general, greater surface charge correlates with greater hydrophilicity. For membranes used to treat aqueous solutions, the isoelectric point is invariably at

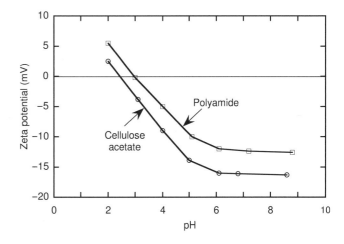

FIGURE 15-5. Effect of pH on surface charge (expressed in terms of zeta potential) of two RO membranes. *Source:* Adapted from Childress and Deshmukh (1998).

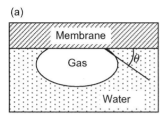 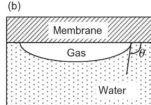

FIGURE 15-6. Measurement of contact angle. The membrane pictured in (a) is more hydrophilic membrane than that in (b).

TABLE 15-3. Typical Transmembrane Pressures and Recoveries for Pressure Driven Membrane Process

Membrane Process	Transmembrane Pressure, TMP (kPa)	System Recovery (%)[a]
Microfiltration	10–100	90–99+
Ultrafiltration	50–300	85–95+
Nanofiltration	200–1500	75–90+
Reverse osmosis	500–8000	60–90

[a] Defined as the ratio of permeate flow rate to feed flow rate.

pH < 4, so the surface charge is negative at the pH of natural waters. The magnitude of this charge can be reduced by adsorption of Ca^{2+} or other divalent cations to the membrane or increased by adsorption of NOM (Elimelech and Childress, 1996). In general, because membranes and NOM molecules both carry a negative charge, higher calcium concentration and lower pH values tend to exacerbate fouling by NOM (by reducing the electrical repulsion between the NOM molecules and the membrane).

The hydrophobicity or hydrophilicity of a membrane can be quantified through measurements of the contact angle, θ, defined as the angle, measured in the liquid, that is formed at the intersection of vapor, solid, and liquid phases. The contact angle can be measured by trapping a bubble of gas beneath a submerged membrane (Figure 15-6). The bubble tends to flatten out beneath the membrane, but this tendency is opposed by the attraction of the membrane for water. The greater the membrane/water affinity (i.e., the greater the hydrophilicity of the membrane), the smaller is the contact angle. Geens et al. (2004) characterized membranes with contact angles near to 35°, 75°, and 115° as very hydrophilic, moderately hydrophobic, and hydrophobic, respectively.

Driving Forces for Membrane Processes

Although several different forces might help drive the transport of water and contaminants in membrane systems, a single driving force frequently dominates over the others. The identity of this dominant driving force is often used to group and/or name membrane processes, as indicated in Table 15-2.

TABLE 15-2. Important Driving Forces in Various Membrane Processes

Main Driving Force	Examples of Membrane Processes
Pressure gradient	Microfiltration, ultrafiltration, nanofiltration, reverse osmosis, piezodialysis
Concentration gradient	Dialysis
Gradient in electrical potential	Electrodialysis, electroösmosis
Temperature gradient	Thermoösmosis

To date, environmental engineering applications of membranes have been dominated by the first four pressure-driven processes listed in this table, along with dialysis and electrodialysis; piezodialysis, electroösmosis, and thermoösmosis have been applied only for occasional laboratory separations.

Pressure-Driven Processes Pressure-driven membrane processes are generally classified as *microfiltration* (MF), *ultrafiltration* (UF), *nanofiltration* (NF), or *reverse osmosis* (RO), based on the types of materials they reject and the mechanisms by which transport and rejection occur. RO is sometimes referred to as *hyperfiltration*. The progression from MF to UF, NF, and finally RO corresponds to a steady decrease in the maximum size of material that can permeate the membranes. The pressure differential between the bulk concentrate and permeate (the TMP) increases, and the fraction of the feed that can be economically recovered decreases, in the same sequence (Table 15-3).

■ **EXAMPLE 15-1.** What height would a column of water have to be to exert a pressure equal to 15 or 4500 kPa (typical TMPs for MF and RO, respectively)?

Solution. From fluid mechanics, we know that the relationship between the height of a column of water and the pressure at its base is $P = \rho_L g h$. Therefore, the heights of columns of water that would have the specified pressures are

$$h_{MF} = \frac{P}{\rho_L g} = \frac{(15{,}000\,\text{Pa})(1.0(\text{kg/m s}^2)/\text{Pa})}{(1000\,\text{kg/m}^3)(9.81\,\text{m/s}^2)} = 1.53\,\text{m}$$

$$h_{RO} = \frac{P}{\rho_L g} = \frac{(4.5 \times 10^6\,\text{Pa})(1.0(\text{kg/m s}^2)/\text{Pa})}{(1000\,\text{kg/m}^3)(9.81\,\text{m/s}^2)} = 459\,\text{m}$$ ■

MF membranes are porous, with characteristic pore diameters on the order of 0.01–2 μm. These membranes are capable of removing particles and some colloidal matter from the feed, but virtually no solutes. They are usually rated based on the nominal diameter of the pores and thus the size of contaminants that the membrane can be expected to reject.

However, many MF membranes are manufactured in a way that generates a wide range of effective pore sizes, so some particles larger than the nominal pore size might appear in the permeate, and some particles smaller than that size might be removed.

At the other extreme, RO membranes are dense membranes that reject colloidal and particulate matter completely, so all the permeating species can be considered to be of molecular dimensions. The membranes are usually rated based on the rejection of a standard salt such as NaCl or the fractional reduction in conductivity from the feed to the permeate when a standard ionic solution is treated. Typical values for solute rejection by RO membranes range from 90% to over 99%.

Similar to MF membranes, UF membranes are porous, and they do not reject salt ions or other small molecules. However, UF membranes typically do reject some macromolecules. NF membranes have been modeled using both the porous and dense membrane paradigms. They are usually considered to reject particles and colloids completely, and many can efficiently (>90%) reject large organic molecules that generate color and serve as precursors for disinfection by-products (DBPs) in natural waters. Some can also reject inorganic molecules as small as the divalent ions that cause hardness and, to a lesser extent, monovalent ions (usually with an efficiency of <70%).

While it is easy to distinguish between RO and MF membranes, the distinctions between RO and NF, NF and UF, and UF and MF membranes are less clear. The IUPAC has attempted to clarify the matter by defining MF as a process that rejects particles and macromolecules larger than 0.1 μm, UF as a process that rejects species that are larger than "about 2 nm," and NF as a process that rejects particles and solutes that are smaller than about 2 nm; other workers have placed the cutoff between UF and NF at slightly larger sizes, but always <10 nm. The IUPAC definition of RO refers to the selective transport of water against an osmotic gradient (defined subsequently in this chapter); however, this criterion is also met by many NF and some UF systems, although the osmotic gradient in those systems is generally much smaller than in RO systems. In general, these processes are probably best viewed as operationally defined segments along a continuum.

Figure 15-7 provides a comparison of the size of materials often encountered in aqueous streams with the characteristic sizes of contaminants removed by various membrane processes. This figure suggests that most conventional separations in environmental engineering could be achieved by pressure-driven membrane processes. Many such applications have been explored. For example, MF and UF have been used for solid–liquid separation instead of conventional packed bed filters, sedimentation basins, and dissolved air flotation systems. MF is also now widely used to separate biosolids from solution in biological waste treatment processes, in which case the reactor is referred to as a *membrane bioreactor* (MBR). MF and UF membranes are effective barriers to passage of protozoa, bacteria, and, in the case of UF, viruses, so they can be useful for disinfecting potable

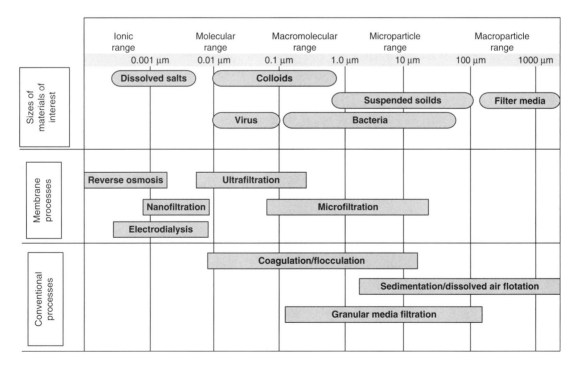

FIGURE 15-7. Typical sizes of contaminants that might be present in water, and of the materials that various treatment process can effectively remove from water.

water or wastewater. These processes have also been widely used for separating metal precipitates and removing emulsified oils, paints, and dyes from industrial wastewaters. Both RO and NF have been used to recover dissolved metals in industrial processes and as alternatives to activated carbon adsorption for the removal of pesticides from drinking water.

Processes Driven Primarily by Concentration Differences or Electrical Forces In some (usually small-scale) membrane systems for water purification, the main species that migrate through the membrane are solutes rather than water; in such processes, the separation of solutes from water is referred to as *dialysis*. Dialysis is driven by a difference across the membrane in the electrochemical activity of the solute. The dominant factor controlling this activity is often the solute's concentration; hence, the process is commonly described as being driven by a concentration difference.

If the solute is ionic, a difference in electrical potential across the membrane can enhance the transport caused by a concentration gradient or even drive ions against an unfavorable concentration gradient. When an electrical field is used to drive ionic transport through a membrane, the process is called *electrodialysis* (ED). The membranes used for ED are essentially sheets of ion exchange resin and are called *ion exchange membranes*. ED is sometimes applied to desalt waters of low to moderate salinity and also for recovery of metals from industrial processes. ED systems are described in greater detail later in this chapter.

It is also possible to create a concentration gradient and a driving force for contaminant transport by temporarily converting the contaminant into another species within the membrane in a process known as *facilitated transport*. In this process, the component of interest reacts within the membrane to form a new species (e.g., a metal might combine with a ligand to form a complex), which then diffuses across the membrane and decomposes to reform the original species at the other side. The process relies on the use of materials and operating conditions that cause the complex to be soluble and stable in the membrane environment, but much less so in solution. For example, a difference in the pH of the bulk solutions might cause the complex to form on the feed side of the membrane and dissociate on the product side. The release of the species of interest on the product side maintains the concentration gradient across the membrane and induces more transport. This process has been proposed to recover metals from dilute solutions such as metal finishing wastes.

Configuration and Hydraulics of Membrane Systems

For economic reasons, it is desirable to pack large quantities of membrane area into a limited volume. To accomplish this, numerous discrete subunits of membrane material, called membrane *elements*, are often packaged together in units

FIGURE 15-8. A membrane array. Each module could contain from a few to several thousand membrane elements. *Source:* Yujung Chang.

referred to as membrane *modules*. In RO and NF systems, the modules are commonly called *pressure vessels*. Multiple modules that are connected by piping to form an independent processing unit are referred to as *arrays*, one or more of which are combined to form the full-scale system. A membrane array is shown in Figure 15-8.

Configuration of Membrane Elements In some cases, membrane modules contain stacks of flat membrane sheets, with alternating sheets facing opposite directions. Thus, the permeate side of each membrane faces the permeate side of another membrane, and the feed side of one faces the feed side of another (Figure 15-9). This arrangement is called a *plate-and-frame* system. Such systems are operated by injecting feed into the spaces between the feed sides of adjacent membrane sheets. Water then permeates through both sheets and into the space between two permeate sides from where it is collected. Thin sheets of screening known as spacers are placed between each pair of membranes, so that influent can flow easily into the feed side of each membrane, and permeate can flow easily out to the collection pipe. Dialysis systems often use plate-and-frame modules.

Spiral-wound membranes represent a variation of the plate-and-frame idea. In this arrangement, pairs of membrane sheets are glued together along three edges, with the permeate sides facing one another. A spacer is placed inside the sack formed by the two membranes, and the open ends of several sacks are attached to slits running along the length of a collection tube. Spacers are also attached to the outside each sack, and the ensemble is wound around the central collection tube, as shown in Figure 15-10. Feed enters the unit axially between adjacent sacks, flows along a path that is kept open by the feed-side spacers, and can pass through

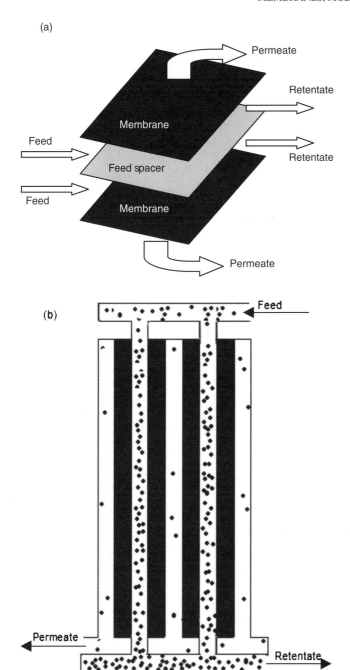

FIGURE 15-9. Schematic representation of one layer of a plate-and-frame membrane system: (a) the arrangement of membrane and spacers; (b) changes in contaminant concentrations as water flows through the element.

the membrane on either side of the feed-side flow path. After crossing the membrane, the permeate follows a spiral path inside a sack until it reaches the collection tube. Multiple tubes are sometimes contained within a single pressure vessel, with the concentrate from one tube serving as the feed for the next.

Membranes are also produced in cylindrical shapes at various scales. In these instances, flow may be introduced under pressure to the inside of the cylinder, so that permeation occurs outward through the membrane walls (inside-out), or the feed may be introduced outside the membrane, with permeation occurring inward (outside-in). Outside-in flow has the advantage that relatively large particles can be present in the feed and yet not clog the flow path. If the diameter of the cylinder is greater than approximately 1 mm, the element is called a *tubular* membrane. Tubular membranes typically utilize inside-out flow. If the diameter of the cylinder is <1 mm, the cylinder is referred to as a *hollow fiber*. Hollow fiber membranes may have either inside-out or outside-in flow, but outside-in is more common. Finally, if the cylinder diameter is extremely small (~0.1 μm, close to the diameter of a strand of hair), it is referred to as a *capillary fiber* membrane. Capillary fibers are always outside-in membranes. In all of these cases, tens (tubular) to thousands (hollow fiber or capillary) of individual membrane elements are assembled in parallel to form a membrane module.

In many hollow fiber or capillary membrane systems, the fibers are "potted" into a resin that seals off each end of the module. The fibers penetrate through the resin, and a separate port in the side of the module allows water to enter or leave (Figure 15-11). In crossflow filtration, both ends of each fiber are open, whereas in dead-end filtration, the downstream end is plugged by the resin. In units with inside-out flow, pressurized feed enters the module through the open ends of the fibers, and permeate passes through the fiber walls into the "shell side" of the module, from where it exits through the side port; in units with outside-in flow, the flow directions are reversed.

In other systems, outside-in hollow fiber membranes are immersed into the feed solution in an unpressurized basin, and a permeate-side vacuum is applied to induce transport across the membrane (Figure 15-12). These systems are limited to low-pressure (MF, UF) applications because the maximum available pressure difference is the sum of atmospheric pressure and the hydrostatic pressure associated with the water depth.

The surface area of membrane available in a module divided by the volume of the module is referred to as the *packing density*. Typical ranges for the packing densities of different membrane geometries are shown in Table 15-4.

As the packing density increases, less space is available within the module to allow rejected materials, such as particles, to circulate freely without obstructing the flow. If flow through the module is obstructed, large pressure drops can occur, resulting in increased operational costs and potential damage to the module. To avoid such situations, greater degrees of pretreatment for particle removal are required for membrane modules with higher packing

(a) Spiral-wound module

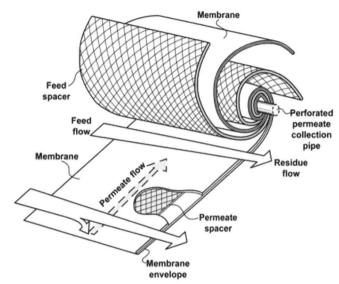

(b) Spiral-wound module cross-section

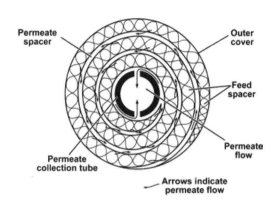

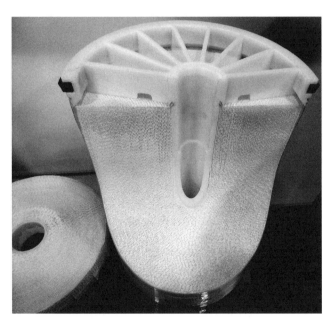

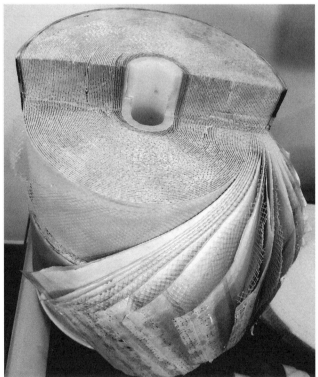

FIGURE 15-10. Spiral-wound membrane modules. (a) Schematic representation (http://www.netl.doe.gov/technologies/coalpower/ewr/co2/pubs/5312-43085%20MTR%20spiral-wound%20polymeric%20membrane.pdf; accessed May 1, 2012); (b) a full-scale module, cut open after testing, to show the membrane leaves, the spacers, and the central post where the permeate collects. *Source:* Yujung Chang.

densities. Typically, membrane systems that are designed to remove dissolved materials (NF, RO) utilize modules with higher packing densities (spiral-wound or capillary fiber) and stricter pretreatment requirements. Membranes designed primarily for particle removal (MF, UF) use modules with lower packing densities (hollow fiber or tubular), in which the flow path between membrane surfaces is sufficient to avoid clogging of the module on the feed side.

FIGURE 15-11. Hollow fiber membrane modules. (a) End views of a full-scale module, showing the fibers potted in an epoxy resin; (i) cross-section of the whole module; (ii) closeup showing individual hollow fibers. (b) Side view of a laboratory-scale module, showing the outside of the fibers and the opening on the shell side where feed enters for outside-in flow, and permeate exits for inside-out flow. *Source:* (a) Pierre Kwan, (b) Joel Migdal.

Small systems, often designed to be mobile and therefore useful as test rigs or for emergency applications, sometimes use hollow fiber or tubular NF or RO membranes, thereby minimizing the need for pretreatment. Such systems are

FIGURE 15-12. An immersible membrane system. Units like the one shown, with thousands of hollow fiber membranes, are submerged in a tank, and a vacuum is applied to the fiber interiors to draw water from the tank through the membranes. *Source:* From www.gewater.com/products/equipment/mf_uf_mbr/zeeweed_500.jsp, Accessed April 23, 2012.

typically operated with a lower fraction of the feed water recovered as clean permeate than larger systems. Adequate crossflow is required in these systems to reduce fouling associated with accumulation of rejected species near the membrane surface.

Configuration of Membrane Arrays The most common flow patterns in membrane modules are referred to as dead-end, feed-and-bleed, and cascade (or staged) filtration. In *dead-end filtration*, all the fluid entering the unit is filtered through the membrane and exits as permeate. In such systems, it is common to arrange several membrane arrays in parallel to treat the feed flow. Periodically, an array is taken out of service, and the accumulated materials are removed by a

TABLE 15-4. Approximate Packing Densities for a Number of Membrane Geometries

Membrane Geometry	Approximate Packing Density (m^2/m^3)
Capillary	5000–8000
Spiral-wound	700–2000
Hollow fiber	1000–2000
Flat (plate-and-frame)	200–500
Tubular	100–300

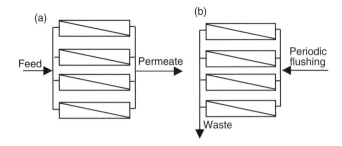

FIGURE 15-13. A single array of four modules, configured for dead-end filtration. (a) Normal operation and (b) back-flushing.

hydraulic flush, chemical cleaning, or both. One array in such a system is shown schematically in Figure 15-13.

In a *feed-and-bleed* configuration, most of the water entering the membrane modules exits as permeate, but a continuous stream of concentrate is also removed ("bled") from the system (Figure 15-14). This latter stream is either discharged to waste or processed to separate the concentrated contaminants from the water, so that the water can be sent back to the membrane unit for recovery. Both dead-end and feed-and-bleed systems can be operated with or without crossflow. If crossflow is used in these types of systems, then only a small portion of the flow entering a membrane element is removed as permeate during any single pass through the module. However, the concentrate is recycled through the element several times, so that ultimately the majority (or, in the case of dead-end systems, all) of the feed exits as permeate.

Dead-end and feed-and-bleed configurations are most typically encountered in applications where porous membranes are used and the target contaminants are particles. In such systems, pretreatment is usually limited, and periodic back-flushing is applied to remove materials that have been rejected by the membranes and accumulated in the module. Typical times between back-flushing events are on the order of several minutes to 1–2 h.

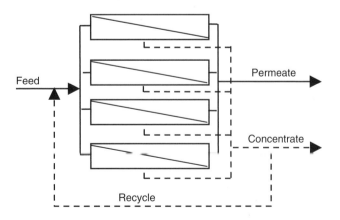

FIGURE 15-14. A feed-and-bleed membrane system with crossflow.

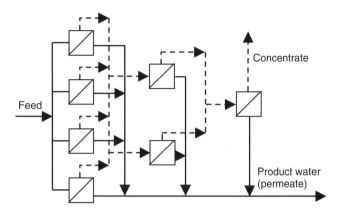

FIGURE 15-15. Staged (or cascade) configuration of membrane arrays.

In high-pressure (RO and NF) systems, arrays of membranes are usually arranged in a *cascade* or *staged* configuration in which the concentrate from one stage serves as the feed for a subsequent stage (Figure 15-15). This configuration is used to increase the overall production of clean water, while exposing fewer membranes to the worst quality of water. Arrays are periodically removed from service for chemical cleaning. The number of stages feasible for an operating plant is rarely greater than three.

15.4 PARAMETERS USED TO DESCRIBE MEMBRANE SYSTEMS

Location, Concentration, and Pressure

Concentrations, pressures, and other parameters are of interest at several different locations in membrane systems, including the bulk concentrate and permeate solutions, the interior of the membrane, and immediately adjacent to each side of the two membrane/solution interfaces. Differences in the parameters between various locations are also of interest. In this chapter, the subscripts c, p, and m are used to describe locations in the concentrate, permeate, and the membrane, respectively; in the absence of an additional identifier, these subscripts refer to locations away from the interface (i.e., in the bulk solution or the interior of the membrane). Locations at the concentrate/membrane and membrane/permeate interfaces are designated by c/m and m/p, respectively. Thus, for example, $c_{i,m_{m/p}}$ refers to the concentration of species i in the membrane at the membrane/permeate interface.

Parameter differences between two locations are designated by an upper-case delta (Δ) in combination with the same indicators of location. For example, $\Delta_{m_m} P$ is used to designate the pressure difference across the membrane, with both pressures measured inside the membrane; $\Delta_{m_{c/p}} P$ also refers to the pressure difference across the membrane, but with values measured immediately adjacent to the membrane, in the concentrate and permeate solutions. In addition, int is used to designate a difference across an interface (e.g., $\Delta_{int_{c/m}}P$ refers to the pressure difference across the concentrate/membrane interface). Although they are not necessarily equal, $\Delta_{m_m}P$, $\Delta_{m_{c/p}}P$, and $\Delta_{c/p}P$ are all referred to in the literature as the *transmembrane pressure* (TMP); the latter definition (TMP = $\Delta_{c/p}P$) is most common, and this definition is adopted here. To be consistent with the convention used most often in the membrane literature, the difference, ΔX, in any parameter X between two locations is computed as the value of X at the location closer to the concentrate minus its value at the location closer to the permeate.

Transport of Fluid and Solutes

Recovery If a membrane system is used to produce a clean permeate stream from a feed of lower quality, the *recovery* (or *relative recovery*), r, of the system is defined as the fraction of the feed water (f) that becomes product water; that is[3]

$$r = \frac{Q_p}{Q_f} = 1 - \frac{Q_c}{Q_f} \qquad (15\text{-}1)$$

Flux The rates at which water and contaminants are transported to and/or through a membrane are typically normalized to the surface area of the membrane (A_m) and expressed as *fluxes*. In general, fluxes are reported on a mass or molar basis (mass or moles of constituent per unit membrane area per unit time). However, the transport of water or bulk solution is also commonly reported as a volumetric flux (volume of water or solution per unit membrane area per time). As in other systems, fluxes in membrane systems are usually designated by the letter J (J_w for flux of water, J_i for flux of a contaminant i, and J_V for flux of the total solution). In this chapter, the designation J_i always refers to the molar flux of i; J_i values can be converted into mass fluxes by multiplication with the molar mass of i.

The volumetric flux is the ratio of a volumetric flow rate toward a surface to the area of that surface; it can therefore be interpreted as the fluid velocity perpendicular to the specified surface. In this chapter, volumetric fluxes are identified by adding a caret (^) over the J. For all practical purposes, the volumetric flux of the overall solution ($\hat{J}_V$) equals the volumetric flux of water alone ($\hat{J}_w$) in all membrane systems used to treat aqueous streams; however, when deriving theoretical relationships, it is sometimes important to distinguish between these quantities. $\hat{J}_V$ is commonly referred to as the *permeation rate* or *permeation velocity*, v_w:

$$\hat{J}_V \approx \hat{J}_w = \frac{Q_p}{A_m} = v_w \qquad (15\text{-}2)$$

Note that, although v_w is referred to as the permeation velocity, it actually refers to the nominal velocity approaching (i.e., outside) the membrane; the velocity in the pores of a porous membrane would be v_w/ε_m, where ε_m is the membrane porosity.

A nominal velocity can also be computed to describe the movement of a contaminant i up to or through a membrane. Specifically, we can interpret the ratio of the molar flux of i to its molar concentration (c_i) as its nominal velocity; that is,

$$v_i = \frac{J_i}{c_i} \qquad (15\text{-}3)$$

When Equation 15-3 is applied to contaminants as they approach the membrane, v_i corresponds to the average, net velocity of the contaminants perpendicular to the membrane. Thus, v_i accounts both for the movement of contaminants with the bulk water approaching the membrane and also for any transport that they undergo independent of the water (e.g., by diffusion). The same statement applies to the contaminants in the pores of porous membranes, once a correction is made for the membrane porosity. The nominal velocity of contaminants (or water) through dense membranes can also be computed using Equation 15-3, but only in conjunction with an estimate for the concentration of the constituent in the membrane phase.

■ **EXAMPLE 15-2.** A test system containing a single spiral-wound element with $30\,m^2$ of membrane area receives a constant feed flow of $5\,m^3/h$ and is operated at a permeate flux of $25\,L/m^2\,h$. Calculate: (a) the recovery for this module; (b) the average velocity of the water toward the membrane; and (c) the recovery for a pressure vessel with three such elements in series, assuming that the recovery is identical in each element.

Solution.

(a) The permeate flow from the module and the recovery for the element (r_{el}) can be computed as follows:

$$Q_p = \hat{J}_V A_m = (25\,L/m^2\,h)30\,m^2 = 750\,L/h$$

$$r_{el} = \frac{Q_p}{Q_f} = \frac{750\,L/h}{(5\,m^3/h)(1000\,L/m^3)} = 0.15 = 15\%$$

---

[3] The IUPAC convention defines this quantity as the relative recovery of water and assigns it the symbol $\eta_{\eta,w}$. In the environmental engineering literature, the parameter is conventionally given the symbol r, and we adopt that convention in this chapter.

(b) The velocity of water toward the membrane equals the volumetric flux. Converting the given flux into more conventional units of velocity, we find

$$\hat{J}_V = 25 \, \text{L/m}^2 \, \text{h} \left(\frac{1 \, \text{m}^3}{1000 \, \text{L}}\right) \left(100 \, \frac{\text{cm}}{\text{m}}\right)$$
$$= 2.5 \, \text{cm/h}$$

The velocity is very small, making it clear why large amounts of surface area are required to produce a substantial overall flow.

(c) The second and third elements following in series will each also recover 15% of the flow entering them, so

$$Q_{p,2} = 0.15 Q_{\text{in},2} = 0.15 \left[(1.0 - 0.15) Q_{\text{in},1}\right]$$
$$= 0.13 Q_{\text{in},1}$$
$$Q_{p,3} = 0.15 Q_{\text{in},3} = 0.15 \left[(1.0 - 0.15) Q_{\text{in},2}\right]$$
$$= 0.15 \left[(1.0 - 0.15)^2 Q_{\text{in},1}\right] = 0.11 Q_{\text{in},1}$$

or, in general, for n elements:

$$Q_{p,n} = r_{\text{el}}(1.0 - r_{\text{el}})^{n-1} Q_{\text{in},1}$$

For the three-element pressure vessel in this example, the overall recovery is $0.15 + 0.13 + 0.11$, or 0.39; that is, 39%. ∎

Specific Flux and Permeance In pressure-driven membrane processes, the flux per unit pressure differential between the two solutions adjacent to the membrane is referred to as the *specific flux* or *permeance* and is commonly given the symbol k_w. As we will see shortly, the "effective" pressure differential that provides the net driving force for permeation is often different from the hydrostatic pressure differential. Anticipating that distinction, we represent the net driving force here as ΔP_{eff}, so the permeance is computed as

$$k_w = \frac{\hat{J}_V}{\Delta_{m_{c/p}} P_{\text{eff}}} \quad (15\text{-}4)$$

If a membrane is clean and clean water is the feed, the effective pressure differential across the membrane equals the hydrostatic pressure differential. In addition, since the membrane imposes the only resistance to permeation in such a system, that pressure differential equals the differential across any region that includes the membrane, so $\Delta_{c/p} P_{\text{eff}} = \Delta_{c/p} P = \Delta_{m_m} P = \Delta_{m_{c/p}} P$. The permeance under these limiting conditions is of special significance in membrane modeling and is designated here as k_w^*:

$$k_w^* = \left(\frac{\hat{J}_V}{\Delta_{c/p} P_{\text{eff}}} = \frac{\hat{J}_V}{\Delta_{c/p} P} = \frac{\hat{J}_V}{\Delta_{m_m} P} = \frac{\hat{J}_V}{\Delta_{m_{c/p}} P}\right) \text{ clean membrane,} \atop \text{clean water feed}$$
(15-5)

Under most circumstances of interest, the permeance of membranes for solutions of interest is very close to that for pure water, $f_{\text{spiral} \atop \text{wound}} = 6.23 \text{Re}^{-0.3}$. We will make this approximation throughout the chapter and refer to k_w^* simply as the permeance of the solution.

If we model flow through a porous membrane as fully developed flow through isolated capillaries, we can interpret k_w^* in terms of fundamental parameters without any empirical terms. For example, applying the Hagen–Poiseuille equation[4] to this situation, the permeate flux through a membrane with pores having a radius of r_{pore} is expected to be

$$\hat{J}_V = \varepsilon_m \frac{r_{\text{pore}}^2 \Delta_{m_{c/p}} P_{\text{eff}}}{8 \mu \xi \delta_m} = \frac{A_{\text{pore}}}{A_m} \frac{r_{\text{pore}}^2 \Delta_{m_{c/p}} P_{\text{eff}}}{8 \mu \xi \delta_m}$$
$$= \frac{N_{\text{pores}}}{A_m} \frac{\pi r_{\text{pore}}^4 \Delta_{m_{c/p}} P_{\text{eff}}}{8 \mu \xi \delta_m} \quad (15\text{-}6)$$

where ε_m is the membrane porosity, equal to the ratio of the open pore area (A_{pore}) to the gross area of the membrane surface (A_m); ξ is the pore tortuosity factor; δ_m is the membrane thickness; and N_{pores}/A_m is the number of pores per unit area of membrane. The final equality in Equation 15-6 is based on expressing the open pore area as the product of the number of pores and the area per pore (πr_{pore}^2). Comparing Equations 15-4 and 15-6, we obtain

$$k_w^* = \frac{N_{\text{pores}}}{A_m} \frac{\pi r_{\text{pore}}^4}{8 \mu \xi \delta_m} \quad (15\text{-}7)$$

Resistance The resistance that a membrane system imposes on the transport of a species can be quantified as the ratio of the driving force required to generate the flux to the flux itself.

$$\text{Resistance} = \frac{\text{Driving force for flux}}{\text{Flux}} \quad (15\text{-}8)$$

This calculation is analogous to that for resistance in electrical systems, in which the voltage (V) is the driving force for the current (I), and the resistance ($\mathcal{R}$) can be computed as $\mathcal{R} = V/I$. Although this idea can be applied

---

[4] The Hagen–Poiseuille equation is a classic equation in fluid mechanics describing the relationship between flow rate and frictional head loss under laminar flow conditions. Its derivation can be found in any introductory fluid mechanics text. Flow that conforms to this equation is often referred to as "Poiseuille flow."

to any species and any membrane system, it is most often applied to characterize transport of water or the bulk solution in systems in which the driving force is a pressure differential. For such an application, the generic relationship shown in Equation 15-8 can be written as follows:

$$\overline{\overline{\mathscr{R}}} \equiv \frac{\Delta P_{\text{eff}}}{\hat{J}_V} \qquad (15\text{-}9)$$

where $\overline{\overline{\mathscr{R}}}$ is the resistance of any specified region across which the effective pressure drop is ΔP_{eff}.

The resistance to permeation, as defined in Equation 15-9, is a composite parameter that depends on the properties of membrane, fluid, and any accumulated foulants. For example, the resistance to permeation through the membrane increases in proportion to the viscosity (μ) of the fluid and the thickness of the membrane (δ_m). Frequently, the dependence on viscosity is separated from the other factors, so that the computed resistance reflects only the properties of the membrane and foulants. Here, we represent the resistance based on this alternative definition as $\mathscr{R}$:

$$\mathscr{R} \equiv \frac{\Delta P_{\text{eff}}}{\mu \hat{J}_V} \qquad (15\text{-}10)$$

When defined as in Equation 15-9, membrane resistance has dimensions of $ML^{-2}T^{-1}$ (e.g., kg/m² s), whereas when defined as in Equation 15-10, the dimensions are inverse length, L^{-1} (e.g., m^{-1}). In this chapter, the term "resistance" refers to the expression in Equation 15-10 unless otherwise indicated.

Based on Equation 15-10, the total resistance to transport of solution between the bulk concentrate and permeate ($\mathscr{R}_{\text{tot}}$, including the contributions of both the membrane and the foulants) can be written as

$$\mathscr{R}_{\text{tot}} = \frac{\Delta_{c/p} P_{\text{eff}}}{\mu \hat{J}_V} \qquad (15\text{-}11)$$

However, if a system is tested with a clean membrane and with clean water as the feed, then all the resistance can be attributed to the membrane itself, and $\mathscr{R}_{\text{tot}}$ indicates the resistance of the membrane alone, $\mathscr{R}_m$:

$$\mathscr{R}_m = \frac{\Delta_{m_{c/p}} P_{\text{eff}}}{\mu \hat{J}_V} \bigg|_{\substack{\text{clean membrane} \\ \text{and clean water feed}}}$$
$$= \frac{\Delta_{c/p} P_{\text{eff}}}{\mu \hat{J}_V} \bigg|_{\substack{\text{clean membrane} \\ \text{and clean water feed}}} \qquad (15\text{-}12)$$

Comparison of Equations 15-5 and 15-12 shows that $\mathscr{R}_m = 1/\mu k_w^*$. Typical values for membrane resistance in MF, UF, NF, and RO systems are listed in Table 15-5. Approaches for quantifying the contributions of foulants

TABLE 15-5. Approximate Values of Membrane Resistance and Transmembrane Pressures (TMPs) for Pressure-Driven Membrane Processes

Process	Typical Volumetric Flux, $\hat{J}_V$ (L/m² h)	Typical Membrane Resistance, $\mathscr{R}_m$ (m^{-1})
Microfiltration	50–300	$1 \times 10^{11} - 1 \times 10^{12}$
Ultrafiltration	30–150	$1 \times 10^{12} - 1 \times 10^{13}$
Nanofiltration	20–50	$1 \times 10^{13} - 1 \times 10^{14}$
Reverse osmosis	5–40	$5 \times 10^{13} - 1 \times 10^{15}$

and other factors to the overall resistance are described subsequently.

Permeability Two parameters that are closely related to resistance and that are sometimes also used to characterize pressure-driven membranes are the *permeability coefficient*, k_V, and the *intrinsic permeability*, $k_{V,\text{intr}}$. The intrinsic permeability (with dimensions of L²) can be defined as the inverse of the ratio of membrane resistance to membrane thickness:

$$k_{V,\text{intr}} \equiv \left(\frac{\mathscr{R}_m}{\delta_m}\right)^{-1} = \frac{\mu \hat{J}_V}{\Delta_{m_{c/p}} P / \delta_m} \qquad (15\text{-}13)$$

The pressure difference in Equation 15-13 is between the solutions adjacent to the membrane, so that any resistance associated with transport across the membrane/solution interfaces is included for the calculation, along with that in the interior of the membrane.

The permeability coefficient is defined as the corresponding ratio when Equation 15-9 (rather than Equation 15-10) is used to define the resistance. This coefficient equals the volumetric flux divided by the TMP gradient and has dimensions of L³T/M. It can also be computed as the ratio of the intrinsic permeability to the viscosity[5]:

$$k_V \equiv \frac{\hat{J}_V}{\Delta_{m_{c/p}} P / \delta_m} = \frac{k_{V,\text{intr}}}{\mu} \qquad (15\text{-}14)$$

By rearranging Equation 15-14, we can obtain the following expressions for the volumetric flux:

$$\hat{J}_V = k_{V,\text{intr}} \frac{\Delta_{m_{c/p}} P / \delta_m}{\mu} = k_V \frac{\Delta_{m_{c/p}} P}{\delta_m} \qquad (15\text{-}15)$$

Equation 15-15 is a close analog to Darcy's law for flow of a solution through porous media (such as granular-media filters), emphasizing the fundamental similarity of membrane filtration to this process, even though the two processes operate at very different length scales.

---

[5] The permeability of a membrane is frequently represented in the literature as k_m. Here, we use the subscript V to emphasize that the permeability describes the volumetric flow of solution through the membrane.

Permeabilities are also sometimes reported for individual species based on an equation analogous to Equation 15-14, but using the molar or mass flux instead of the volumetric flux:

$$k_i \equiv \frac{J_i}{\Delta_{m_{c/p}}P/\delta_m} \qquad (15\text{-}16)$$

Note that, by comparing Equations 15-12 and 15-13 with Equation 15-6, we see that the Hagen–Poiseuille equation provides a theoretical approach for predicting $k_{V,\text{intr}}$ and $\mathcal{R}_m$ of a porous membrane from purely geometric considerations, as long as the flow through the pores is fully developed and laminar.

$$k_{V,\text{intr}} \equiv \frac{A_{\text{pore}} r_{\text{pore}}^2}{8 A_m \xi} \qquad (15\text{-}17)$$

$$\mathcal{R}_m = \frac{8 A_m \xi \delta_m}{A_{\text{pore}} r_{\text{pore}}^2} \qquad (15\text{-}18)$$

■ **EXAMPLE 15-3.** A symmetric membrane has a thickness of 75 μm, pores with a nominal size of 0.04 μm and a tortuosity of 1.5, and a ratio of pore area to membrane area of 0.60. Assuming that the flow can be treated as Poiseuille flow, predict:

(a) Intrinsic permeability of the membrane
(b) Membrane resistance
(c) Flux through the membrane, in L/m² h, if it is used in a system that is fed pure water at 25°C ($\mu = 1.002 \times 10^{-3}$ kg/m s), and the TMP is 20 kPa. (Note that, because the test fluid is pure water, all the pressure drop between the concentrate and permeate occurs inside the membrane.)
(d) Flow rate through the membrane, if it is used in a laboratory-scale apparatus that holds circular membrane discs with a diameter of 6.5 cm, and the system is operated as described in part (c).

Solution.

(a) The intrinsic permeability can be found from Equation 15-17

$$k_{V,\text{intr}} = \frac{A_{\text{pore}}}{A_m} \frac{r_{\text{pore}}^2}{8\xi} = 0.60 \frac{(4 \times 10^{-8}\,\text{m})^2}{8(1.5)}$$
$$= 8.0 \times 10^{-17}\,\text{m}^2$$

(b) The membrane resistance is computed using Equation 15-18.

$$\mathcal{R}_m = 8 \frac{A_m}{A_{\text{pore}}} \frac{\xi \delta_m}{r_{\text{pore}}^2} = 8 \frac{1}{0.60} \frac{(1.5)(7.50 \times 10^{-5}\,\text{m})}{(4 \times 10^{-8}\,\text{m})^2}$$
$$= 9.38 \times 10^{11}\,\text{m}^{-1}$$

(c) Because we are assuming that $\Delta_{m_{c/p}}P = \Delta_{c/p}P = $ TMP, the flux can be found from Equation 15-6. Noting that $20\,\text{kPa} = 20,000\,\text{N/m}^2 = 2.0 \times 10^4\,\text{kg/m s}^2$:

$$\hat{J}_V = \frac{A_{\text{pore}}}{A_m} \frac{r_{\text{pore}}^2}{8\mu\xi\delta_m} \Delta_{m_{c/p}}P$$
$$= (2.13 \times 10^{-5}\,\text{m}^3/\text{m}^2\,\text{s})(3600\,\text{s/h})(1000\,\text{L/m}^3)$$
$$= 76.6\,\text{L/m}^2\,\text{h}$$
$$= 0.60 \frac{(4 \times 10^{-8}\,\text{m})^2}{8(1.002 \times 10^{-3}\,\text{kg/m s})(1.5)(7.5 \times 10^{-5}\,\text{m})}$$
$$\times (2.0 \times 10^4\,\text{kg/m s}^2) = 2.13 \times 10^{-5}\,\text{m}^3/\text{m}^2\,\text{s}$$

(d) The expected flow rate is the product of the flux and membrane area.

$$Q = \hat{J}_V A_m = \hat{J}_V \left(\pi \frac{d_m^2}{4}\right)$$
$$= (76.6\,\text{L/m}^2\,\text{h})\left(\pi \frac{(6.5 \times 10^{-2}\,\text{m})^2}{4}\right) = 0.254\,\text{L/h}$$
$$= (0.254\,\text{L/h})(1000\,\text{mL/L})(1\,\text{h}/60\,\text{min})$$
$$= 4.24\,\text{mL/min} \qquad ■$$

Effect of Temperature on Water Permeation According to Equation 15-10, permeate flux increases with decreasing viscosity (μ) of the permeating water, so increased permeation rates can be achieved at higher temperatures. It is often useful to specify the permeation rate at a reference temperature (e.g., 20°C) for purposes of comparison. If data for permeate flux are available at one temperature, a first approximation for the flux at a different temperature can be obtained by multiplying the flux at the baseline temperature by the ratio of the viscosities at the two temperatures; that is,

$$\hat{J}_{V,T_2} = \hat{J}_{V,T_1} \frac{\mu_{T_1}}{\mu_{T_2}} \qquad (15\text{-}19)$$

However, a change in temperature can generate various effects that go beyond that of viscosity. Therefore, the overall effect of temperature on permeate flux is sometimes described using an approach that is more flexible than Equation 15-19, such as with an Arrhenius-type equation similar to the following equation:

$$\hat{J}_{V,T} = \hat{J}_{V,T_{\text{ref}}} \exp\left[s\left(\frac{1}{T_{\text{ref}}} - \frac{1}{T}\right)\right] \qquad (15\text{-}20)$$

where $\hat{J}_{V,T}$ is the permeate flux at (absolute) temperature T, $\hat{J}_{V,T_{\text{ref}}}$ is the flux at the reference temperature, and s is an empirical constant that must be evaluated for each membrane.

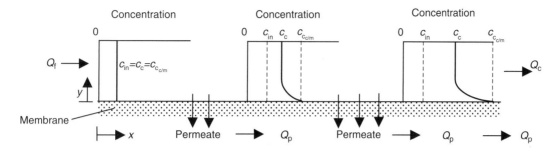

FIGURE 15-16. Concentration profiles near the membrane wall at various locations in a system with crossflow. The left-most profile applies at the inlet, while those in the center and on the right apply at locations progressively downstream.

■ **EXAMPLE 15-4.** Sharma et al. (2003) reported that the following equation describes the temperature dependence of flux through an NF membrane at a given $\Delta_{m_{c/p}}P$, using a reference temperature of 298 K (25°C)

$$\hat{J}_{V,T} = \hat{J}_{V,298} \exp\left(2617\left[\frac{1}{298} - \frac{1}{T}\right]\right)$$

Water temperatures at a facility using this type of membrane for potable water treatment range from 3°C in the winter to 21°C in the summer; the corresponding viscosities are 1.62×10^{-3} and 9.8×10^{-4} Pa s, respectively. Estimate the ratio of wintertime to summertime flux, if $\Delta_{m_{c/p}}P$ remains constant. Also, determine the same ratio if the only effect of temperature were on viscosity.

Solution. The ratio of wintertime to summertime flow can be calculated directly using the given equation. The result is

$$\frac{\hat{J}_{V,3°C}}{\hat{J}_{V,21°C}} = \frac{\hat{J}_{V,25°C}\exp[2617((1/298) - 1/(273+3))]}{\hat{J}_{V,25°C}\exp[2617((1/298) - 1/(273+21))]}$$
$$= 0.56$$

Thus, the permeate flux when the water is coldest is slightly more than half of its value when the water is warmest.

If the change in flux were strictly a response to the increase in viscosity, the ratio of winter to summer fluxes would conform to Equation 15-19. Rearranging that equation and inserting the given values gives

$$\frac{\hat{J}_{V,3°C}}{\hat{J}_{V,21°C}} = \frac{\mu_{21°C}}{\mu_{3°C}} = \frac{9.8 \times 10^{-4}\text{ Pa s}}{1.62 \times 10^{-3}\text{ Pa s}} = 0.60$$

The predicted decline in the flux considering only the effect of viscosity (40%) is thus within 10% of the observed decline (44%). ■

Membrane Selectivity: Rejection and Separation

The parameters defined in the preceding section are used primarily to characterize the permeation of bulk solution through membranes. We next define the parameters that are used to characterize the efficiency with which membranes separate water from other constituents of the feed. Many of these parameters involve ratios of concentrations at various points in membrane systems. To provide a context for the definitions, some of the expected trends in the contaminant concentration as a function of location in a system with crossflow are shown in Figure 15-16.

In this figure, distance from the inlet is designated by x, and distance away from the membrane is designated by y. The figure shows typical contaminant concentration profiles at three x values, corresponding to the inlet, the middle, and the exit of a membrane element. At the inlet, the concentration is uniform in the y direction and is equal to the feed concentration. As the feed moves downstream, its bulk concentration (c_c) increases, due to the selective loss of clean water from the feed flow channel to the permeate. In addition, a nonuniform concentration profile develops in the y direction, with the highest concentration near the membrane, where the contaminants are rejected. Both because c_c is increasing as the water flows and because water is being removed from the suspension that is migrating downstream near the membrane, $c_{c_{c/m}}$ increases with increasing x, but not necessarily at the same rate that c_c increases. The average concentration at a given value of x is designated as $c_{c,\text{avg}}$. All the fluid at that value of x has a concentration between $c_{c_{c/m}}$ and c_c, so $c_c < c_{c,\text{avg}} < c_{c_{c/m}}$.

The concentration in the permeate (c_p) is not represented in the figure. In theory, the profile of c_p should be qualitatively similar to that of c_c, albeit with much lower values for the concentrations. However, because the concentrations on the permeate side are all small, variations in c_p with both x and y are generally ignored.

Polarization The ratio of the concentration at the m/c interface to that in the adjacent bulk solution (c_c) is sometimes called the *polarization factor* (PF) or the *polarization modulus* (PM) of i:

$$\text{PF}_i \equiv \frac{c_{i,c_{c/m}}}{c_{i,c}} \quad (15\text{-}21)$$

Note that PF_i is a local variable whose value is likely to change from the system entrance to the exit.

Rejection The *rejection* of a species i is given the symbol $\hat{R}_i$ and is defined as 1.0 minus the ratio of the concentrations of i on the permeate and feed sides of the membrane. Like PF_i, $\hat{R}_i$ is a local parameter, whose value must be computed based on the conditions at a particular location (x) in a membrane system. Values of $\hat{R}_i$ can be computed using the bulk, interfacial, or average concentration on the feed side. The IUPAC recommends referring to the value computed using c_c as the *apparent rejection factor*, and that computed using $c_{c_{c/m}}$ as the *intrinsic rejection factor*:

$$\hat{R}_{i,\text{app}} \equiv 1 - \frac{c_{i,p}}{c_{i,c}} \tag{15-22}$$

$$\hat{R}_{i,\text{intr}} \equiv 1 - \frac{c_{i,p}}{c_{i,c_{c/m}}} \tag{15-23}$$

The bulk concentration at a specific location in a membrane system is rarely measurable, and measurement of the concentration right at the membrane surface is even more problematic. As a result, characterization of $\hat{R}_{i,\text{app}}$ or $\hat{R}_{i,\text{intr}}$ is always based on a combination of empirical data, assumptions, and model calculations.

Although the local rejection factor is important for fundamental modeling of membrane performance, the primary interest in design calculations is more often the *overall rejection factor*, R_i, which characterizes the rejection based on the feed and permeate of the whole system[6]:

$$R_i \equiv 1 - \frac{c_{i,p,\text{out}}}{c_{i,f}} \tag{15-24}$$

The overall rejection R_i can be related to the concentration in the concentrate solution at the system outlet ($c_{c,\text{out}}$) by writing a mass balance around the whole membrane system and assuming that the system is at steady state, as follows:

$$\begin{pmatrix} \text{Rate of change of} \\ \text{mass of contaminant} \\ \text{stored in system} \end{pmatrix} = \begin{pmatrix} \text{Rate at which} \\ \text{contaminant enters} \\ \text{in feed} \end{pmatrix}$$
$$- \begin{pmatrix} \text{Rate at which} \\ \text{contaminant leaves} \\ \text{in permeate} \end{pmatrix} - \begin{pmatrix} \text{Rate at which} \\ \text{contaminant leaves} \\ \text{in concentrate} \end{pmatrix}$$

---

[6] R is essentially the same parameter that we have referred to throughout this book (when considering contaminant removal by other processes) as the removal efficiency, η. We use R instead of η_i for consistency with most published work in environmental engineering. The IUPAC does not define a term comparable to R.

$$0 = Q_f c_f - Q_p c_{p,\text{out}} - Q_c c_{c,\text{out}} \tag{15-25}$$

$$c_{c,\text{out}} = \frac{Q_f c_f - Q_p c_{p,\text{out}}}{Q_c} \tag{15-26}$$

Substituting for Q_p and Q_c from Equation 15-1 and for $c_{p,\text{out}}$ from Equation 15-24 (and dropping the designation i), we obtain

$$c_{c,\text{out}} = \frac{Q_f c_f - r Q_f (1-R) c_f}{(1-r) Q_f} = c_f \frac{1 - r(1-R)}{1-r} \tag{15-27}$$

The value of R computed using Equation 15-24 characterizes the overall contaminant rejection efficiency in terms of concentrations. However, because the flow rates of the feed and permeate streams are not equal, R does not indicate the rejection efficiency with respect to contaminant mass. To characterize that efficiency, we define the *overall mass rejection coefficient*, R_{mass}, as the fraction of the contaminant mass in the influent that is rejected by the membrane. R_{mass} can be computed as

$$R_{\text{mass}} = \frac{Q_c C_{c,\text{out}}}{Q_f C_f} = 1 - \frac{Q_p C_{p,\text{out}}}{Q_f C_f} = 1 - r\frac{C_{p,\text{out}}}{C_f} = 1 - r(1-R) \tag{15-28}$$

■ **EXAMPLE 15-5.** An RO system operates at an overall recovery, r, of 80%. The overall rejection efficiency for Na^+ in this system is 98%. If the Na^+ concentration in the feed is 150 mg/L, what is its concentration in the concentrate exiting the system, and what is the overall mass rejection coefficient?

Solution. Substituting the given values into Equation 15-27, we calculate

$$c_{c,\text{out}} = c_f \frac{1 - r(1-R)}{1-r}$$
$$= (150 \text{ mg/L}) \frac{1 - 0.80(1 - 0.98)}{1 - 0.80} = 738 \text{ mg/L}$$

The overall mass rejection coefficient is found by directly applying Equation 15-28 as

$$R_{\text{mass}} = 1 - r(1-R) = 1 - 0.80(1 - 0.98) = 0.984$$

Thus, 80% of the water but only 1.6% of the Na^+ exits in the product, whereas 20% of the water and 98.4% of the Na^+ are in the concentrate or waste stream. ■

Challenge Tests and MWCO Tests evaluating the rejection of a specific target contaminant that has been added to the feed of a membrane system are commonly referred to as *challenge tests*. If contaminants with a range of sizes are added in a challenge test, the results can, in theory, be interpreted in terms of the membrane's pore size distribution. However, this interpretation is severely limited by the rapid formation of a layer of rejected material that may itself remove contaminants, regardless of whether they would be removed by the clean membrane. Contaminant removal inside membrane pores (as opposed to on the membrane surface) also confounds the interpretation of the results of such tests (Sharma et al., 2003).

The *molecular weight cutoff* (MWCO) for a membrane is the smallest size of molecule for which a specified percentage (typically 90%) is rejected (ASTM, 2001). Rejection as a function of molecular weight for UF and NF membranes is sometimes determined by filtering a range of compounds with similar structures but with different molecular weights, such as a series of dextrans or proteins. The shape, charge, and flexibility of the molecules used in the challenge test affect their ability to pass through the membrane and therefore can play an important role in determining the measured MWCO. The operating conditions used in the test can also affect passage of the molecules, so the evaluation must be performed using a well-defined protocol, which usually includes the use of low contaminant concentrations, low fluxes, and low TMPs. Once MWCOs are determined for a series of membranes, the membranes can be used to estimate the molecular weight distribution of an unknown population of molecules, such as NOM, in a water sample.

To eliminate the effects of molecular charge on the test, evaluations of the MWCO are usually carried out using neutral species. However, RO and NF membranes are often employed primarily to remove salt ions from solution, in which case MWCOs are determined based on the ability of the membrane to reject ions of different valences and/or sizes. Salt passage is typically evaluated using single salts to avoid interactions that occur in mixtures. Most frequently, $MgSO_4$ or some other divalent salt is used to test nanofilters, and either $MgSO_4$ or $NaCl$ is used to test RO membranes.

Separation The *separation coefficient*, $S_{C,A/B}$, of a membrane for solute A over solute B refers to the relative ability of the two species to pass through the membrane. The separation coefficient is defined in terms of the concentration-based rejection efficiency, as follows:

$$S_{C,A/B} = \frac{1-R_A}{1-R_B} = \frac{(c_A/c_B)_p}{(c_A/c_B)_c} \quad (15\text{-}29)$$

The larger the separation coefficient, the more easily species A passes through the membrane relative to species B, or, equivalently, the more efficiently the membrane rejects species B compared to species A. $S_{C,A/B}$ is a local parameter; the value of this parameter evaluated at the exit from the system is referred to as the *separation factor*, $S_{F,A/B}$

$$S_{F,A/B} = S_{C,A/B}\big|_{\text{exit}} = \frac{(c_A/c_B)_{p,\text{out}}}{(c_A/c_B)_{c,\text{out}}} \quad (15\text{-}30)$$

15.5 OVERVIEW OF PRESSURE-DRIVEN MEMBRANE SYSTEMS

Similarities and Differences Among MF, UF, NF, and RO

Pressure-driven membrane systems (i.e., MF, UF, NF, and RO) account for the vast majority of membrane systems used in environmental engineering. In all these systems, a gradient in the hydrostatic pressure drives the transport of both water and contaminants from the bulk concentrate to the bulk permeate. Contaminants that are rejected by the membrane accumulate near the membrane on the feed side, from where they can migrate back to the bulk concentrate. The accumulation of contaminants opposes fluid transport, and the resistance that the water encounters as it passes by the contaminants and through the membrane is overcome by dissipation of the TMP. At least in principle, in systems with continuous flows of feed, concentrate, and permeate, this scenario can lead a steady state in which contaminants are transported toward and away from the membrane at equal rates. Models for the performance of all pressure-based membrane processes must account for these phenomena and therefore have a great deal in common.

By the same token, significant distinctions exist among MF, UF, NF, and RO processes, and these differences must also be reflected in the model concepts or assumptions. One important distinction is based on the size of the rejected contaminants, and in particular, whether they are considered to be solutes or particles. This distinction is manifested in the mechanism by which the contaminants affect the hydraulic resistance: rejected particles affect resistance via the drag force that they exert on the water passing around them, whereas rejected solutes do so by their effect on the osmotic pressure of the solution (discussed shortly). A second important distinction is based on whether the membranes are considered porous or dense: transport through porous membranes is modeled as advection, whereas that through dense membranes is modeled as diffusion. As has been noted, MF and UF membranes are porous, and RO membranes are dense, but NF membranes are intermediate between the two limiting cases and have been modeled in both ways.

The two distinctions noted above overlap, in that the contaminants rejected by porous membranes are generally viewed as particles, whereas those rejected by dense membranes are viewed as solutes (because most particles in the feed to such systems are removed in pretreatment steps). However, the overlap is not perfect, at least in part because the distinctions between porous and dense membranes, and also between large solutes and small particles, are ambiguous.

Operation and Trends in the Performance of Pressure-Based Membrane Systems

If a pressure differential is applied between the solutions on the two sides of a clean membrane, water and some (ideally, small) portion of the contaminants in the feed solution begin to permeate through the membrane. At least for a short while thereafter, the dominant resistance to permeation is provided by the membrane itself. As a result, for a given feed composition, we expect the permeation rate to be proportional to the net driving force, equal to the TMP less any countervailing force generated by the difference in the solution compositions on the two sides of the membrane. However, if the membrane is accomplishing the intended objective, this situation cannot last: rejected contaminants accumulate near the surface, increasing the overall resistance to permeation, and causing the specific flux ($\hat{J}_V/\text{TMP}$) to decline; that is, the membrane begins to foul. The decline in specific flux might be manifested as either a decrease in flux (if the TMP is held constant) or an increase in the required TMP (if the TMP is adjusted to maintain a constant flux); both constant-flux and constant-TMP (usually referred to as constant-pressure) systems are used commercially, but constant-flux operation is more common.

Typically, the specific flux declines for a while and then reaches one of two limiting conditions: either the operation stabilizes at a near-steady-state condition (stable flux and TMP), or the fouling becomes so severe that the operation must be halted to clean the membrane. These possibilities are illustrated schematically in Figure 15-17. In systems that reach steady state, the rate at which foulants are carried to

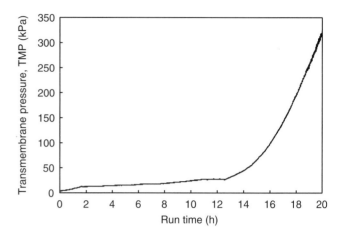

FIGURE 15-18. The effect of fouling by natural organic matter on TMP during constant-flux filtration. Fouling can increase dramatically over a short period. Data are for filtration of Lake Washington water (2.4 mg/L DOC) through a polyethersulfone MF membrane with nominal pore size of 0.1 μm. $\hat{J}_V = 97 \, \text{L/m}^2 \, \text{h}$, no back-flushing.

the membrane by advection is balanced by the sum of the rates at which they pass through the membrane and at which they return to the bulk concentrate by diffusion and other processes that are explained subsequently; as a result, the environment near the membrane is unchanging. Severe fouling, on the other hand, occurs when conditions favor continuous accumulation of foulants near the membrane. In systems operated with constant TMP, this latter situation ultimately causes the flux to become too small for continued operation to be economical; in constant-flux systems, severe fouling leads to a condition in which the target flux cannot be achieved at any reasonable (and perhaps not at any) TMP. The transition from acceptable performance to severe fouling can be gradual or rapid; an example of the latter situation is shown in Figure 15-18 for fouling of an MF membrane by NOM. Most commercial systems are operated under conditions where severe fouling would occur if operation were continued indefinitely, but that scenario is avoided by back-flushing while the extent of fouling is still acceptably small.

Although the preceding discussion focuses on TMP and flux as the primary parameters affecting fouling, a similar fouling pattern has been observed in response to several other parameters, including recovery, filtration cycle time, and various characteristics of the feed. That is, fouling is relatively mild over a certain range of values of the parameter of interest, but then can increase dramatically when some critical value is exceeded. The transition is usually attributed to a change in the physicochemical nature of the deposited foulants, such as conversion of dissolved or adsorbed macromolecules into a three-dimensional gel near the membrane surface (Chang and Benjamin, 1996), collapse of a "loose" cake of rejected particles to a more compact one (Harmant and Aimar,

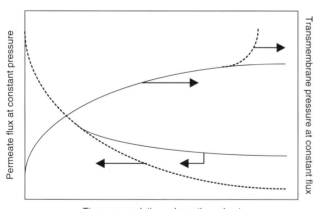

FIGURE 15-17. Membrane fouling as indicated by either an increase in TMP or a decrease in permeate flux. Solid lines characterize a system that eventually reaches steady state, and broken lines characterize a system that fails, either by the flux becoming unacceptably small or the TMP becoming unacceptably large.

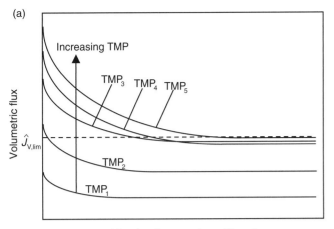

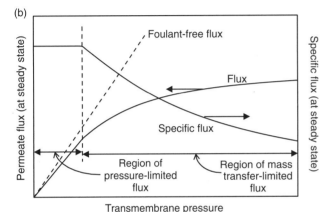

FIGURE 15-19. A typical response of permeate flux through a porous membrane to changes in the transmembrane pressure. (a) Changes in flux over time at five different TMPs. The steady-state flux has an upper limit that cannot be exceeded, regardless of how large the TMP becomes. (b) Dependence of the steady-state flux and specific flux on TMP.

applied. The fact that the flux remains constant when the driving force increases under these conditions indicates that the overall hydraulic resistance must be increasing in proportion to the driving force. The increase in resistance might be caused by accumulation of foulant near the membrane, compression of the membrane, and/or changes in the structure of the accumulated material.

The steady-state fluxes from Figure 15-19a are plotted as a function of TMP in Figure 15-19b. At low pressures, the steady-state flux is proportional to the TMP, so the specific flux is constant; in this range, the flux is said to be *pressure-limited*. At higher TMPs, the increase in steady-state flux is less than proportional to the increase in TMP, and at still higher values, the limiting flux is reached. In this range, the specific flux declines with increasing TMP, and the flux is said to be *mass-transfer-limited*.

The preceding discussion focuses on fouling during a single filtration cycle. Such cycles typically last from several minutes to a few hours in systems with porous membranes, after which the membranes are back-flushed. Figure 15-20 illustrates the performance of an MF membrane that was back-flushed with water every 20 min in laboratory testing. Each back-flush event, lasting for a period of several seconds, reversed the majority of the fouling that occurred during the preceding cycle, but not all of it. As a result, the cumulative amount of hydraulically irreversible fouling steadily increased. Such a trend is commonly observed in full-scale systems, though at quite different time frames: significant reversible fouling can occur over a period of minutes to hours, but irreversible fouling typically builds up much more slowly than shown in the figure, over days to

1996; Tarabara et al., 2004), or increased penetration of foulants into the cake layer adjacent to the membrane or into the membrane itself (Zhang and Song, 2000; Chellam and Jacangelo, 1998). The exact value of the critical parameter is feed- and system-specific and therefore must be determined based on laboratory and pilot studies.

If a given membrane system and feed are tested in constant-pressure mode over a range of TMPs, the system response might be similar to that shown in Figure 15-19a. The performance characteristics at each TMP are similar to those shown for the constant-pressure systems in Figure 15-17. In addition, however, while the initial flux increases with increasing TMP, the steady-state flux does so only up to some limiting value ($\hat{J}_{V,\lim}$); steady-state fluxes larger than $\hat{J}_{V,\lim}$ cannot be achieved no matter how large a TMP is

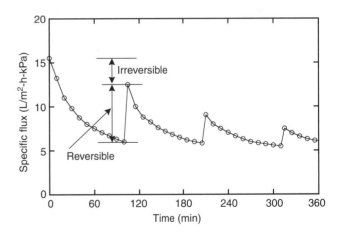

FIGURE 15-20. A common fouling pattern in porous membranes. The steep increases in specific flux correspond to times when the membrane was back-flushed. The arrows indicate the amount of fouling that can and cannot be eliminated by the back-flushing (hydraulically reversible and irreversible fouling, respectively).

weeks. The trends shown in the figure occur whether the membrane is operated at constant pressure or constant flux.

The decreased effectiveness of back-flushing over time is usually attributed to incomplete removal of foulants and a shift in the nature of the foulants remaining on the membrane. For example, fouling of MF and UF membranes by organic matter adsorbed inside the membrane pores may steadily increase, so that after several cycles, it dominates over fouling by particulate materials (which are relatively easily dislodged by back-flushing). Also, particulate foulants may consolidate, interact with other foulants, or migrate into membrane pores. As noted previously, at some point (typically after a few weeks to a few months), it becomes necessary to remove the membranes from service for more extensive chemical cleaning.

Back-flushing is usually impractical in NF and RO systems, due to the very low performance of the membranes. Therefore, in NF and RO systems, operational conditions are chosen such that fouling occurs much more gradually than in MF or UF. Nevertheless, fouling does develop over time, and eventually the permeate flux or TMP reaches a critical value at which it is necessary to take the membranes out of service for chemical cleaning.

The reagents used for chemical cleaning of membranes depend on both the nature of the foulant and the composition of the membranes. Precipitated inorganic foulants can often be dissolved with acids and/or reducing agents (e.g., citric acid can be used as a chemical reductant to dissolve iron oxide precipitates). Alkaline solutions, sometimes amended with enzymes, surfactants, or oxidants, are preferred when the foulants are organic. Heating the cleaning solution usually improves both the kinetics and the overall efficiency of the cleaning. The specific reagents used are often proprietary mixtures; the composition of those mixtures and the details of the cleaning sequences are usually recommended by membrane manufacturers and tend to be determined by experience and trial-and-error rather than by theory.

15.6 QUANTIFYING DRIVING FORCES IN MEMBRANE SYSTEMS

Energy and Driving Force in Membrane Systems

Any force, F, can be interpreted as a spatial gradient in "available" or "free" energy, E. The force is in the direction of decreasing energy so, for example, for a gradient in the x direction:

$$F_{\substack{\text{in}+x \\ \text{direction}}} = -\frac{dE}{dx} \quad (15\text{-}31)$$

Applying this idea to a chemical species in a membrane system, if the species moves a short distance, the decrease in the total available energy in the system per unit distance moved can be interpreted as the force favoring this movement; that is, as the driving force for the movement. If we designate the distance perpendicular to the membrane by y, with y assigned a value of zero at the concentrate/membrane interface and increasing into the membrane, the force on the fluid in the direction of permeation is

$$F_{\substack{\text{concentrate} \\ \text{to permeate}}} = F_{\substack{\text{in}+y \\ \text{direction}}} = -\frac{dE}{dy} \quad (15\text{-}32)$$

In membrane systems, the overall energy gradient driving transport might be associated with gradients in static pressure, electrical potential, and/or the system composition. However, in many systems, the gradient in one type of energy dominates all others and is the only one that needs to be considered. For instance, in MF and UF systems, solutes are usually not rejected by the membranes to an appreciable extent (so the gradients in solute concentration are negligible throughout the system), and no electrical gradient is applied. Therefore, the transmembrane gradient in pressure is the only significant factor affecting the decline in available energy of water and solutes as they move from the concentrate to the permeate.

In other cases, two or more types of energy contribute significantly to the gradient. In NF and RO systems, for example, the only externally imposed driving force is the pressure differential across the membrane. However, the selective rejection of solutes can lead to a significant difference in the composition of the solutions on opposite sides of the membrane, causing a concentration-based driving force to develop as well; both the imposed and induced driving forces must be accounted for to understand the system performance.

The second law of thermodynamics indicates whether, and in what direction, a system will undergo a spontaneous change. Although the law is usually expressed in terms of total universal entropy, under certain circumstances it can be formulated in more intuitive terms related to the available energy in the system. In particular, if the temperature and pressure of a system are constant and equal to the temperature and pressure in the local surroundings, the second law indicates that spontaneous change will always occur in the direction that causes a decline in a function known as the *Gibbs (free) energy* (G) of the system. Furthermore, if the material in the system is an incompressible fluid, the constraint of constant pressure can be relaxed, and spontaneous change will occur in the direction that leads to a decline in the Gibbs energy even if the pressure changes. This latter situation characterizes virtually all membrane systems used for water treatment.

The Gibbs energy includes contributions from many sources (chemical bonds within and among molecules in the system, the pressure applied to the system, gravitational potential energy, electrical potential energy, etc.). However, our only interest is in *changes* in the Gibbs energy, and the only contributors to G that are likely to change when water and other constituents pass through a membrane are related to the chemical composition of the solution, the pressure, and (in some cases) the electrical potential.

The Gibbs energy per mole of a species i (either water or a solute, in systems of interest here) is called either the *molar Gibbs energy* of i ($\overline{G}_i$) or the *chemical potential* of i (μ_i). $\overline{G}_i$ can be interpreted conceptually as the increment in total Gibbs energy of the system per mole of i added, when a differential number of moles of i (dn_i) is added and the temperature, pressure, electrical potential, and the number of moles of all species other than i are constant, that is

$$\mu_i = \overline{G}_i \equiv \left.\frac{\partial G}{\partial n_i}\right|_{P,T,\psi,n_{j\neq i}} \quad (15\text{-}33)$$

$\overline{G}_i$ can be evaluated for any combination of P, T, and ψ, and for any system composition. When evaluated under *standard* conditions (typically $P = 1.0$ atm, $T = 25°C$, $\psi = 0$ V, and a system composition such that the chemical activity of i, a_i, is 1.0), it is referred to as the standard molar Gibbs energy of i, designated as $\overline{G}_i^o$. Values of $\overline{G}_i^o$ are tabulated for water and most common solutes of interest. Designating standard conditions for P, T, and ψ by a superscript ^{o}, the value of $\overline{G}_i$ under any other (i.e., nonstandard) conditions is given by

$$\overline{G}_i = \overline{G}_i^o + \int_{P^o}^{P} \overline{V}_i dP + \int_{T^o}^{T} \overline{S}_i dT + z_i \mathcal{F} \int_{\psi^o}^{\psi} d\psi + RT \ln a_i \quad (15\text{-}34)$$

where $\overline{V}_i$ is the molar volume of i; $\overline{S}_i$ is the molar entropy of i; z_i is the charge on i (including sign); $\mathcal{F}$ is the Faraday constant; and R and T are the universal gas constant and the absolute temperature, respectively.

For species in a liquid phase, $\overline{V}_i$ is essentially independent of P and so can be taken out of the first integral on the right side of Equation 15-34. In addition, we will focus on systems at constant temperature, in which the second integral equals zero. Then, making the conventional choices of 1.0 atm and 0 V for P^o and ψ^o, respectively, the equation can be rearranged and rewritten as

$$\overline{G}_i = \overline{G}_i^o + RT \ln a_i + \overline{V}_i(P - 1.0 \text{ atm}) + z_i \mathcal{F} \psi \quad (15\text{-}35)$$

In the remainder of this chapter, we refer to the sum of the first two terms on the right side of Equation 15-35 as $\overline{G}_{\text{chem},i}$, the third term on the right side as $\overline{G}_{\text{mech},i}$, and the final term on the right as $\overline{G}_{\text{elec},i}$. These definitions are summarized in Table 15-6.

Finally, we can use Equation 15-35 to compute the difference in $\overline{G}_i$ between two locations that differ with respect to

TABLE 15-6. Important Types of Molar Gibbs Energy in Membrane Systems

Type of Energy	Expression for $\overline{G}_i$	
Chemical (concentration-based)	$\overline{G}_{\text{chem},i} = \overline{G}_{\text{chem},i}^o + RT \ln a_i$	(15-36)
Mechanical (pressure-based)	$\overline{G}_{\text{mech},i} = \overline{V}_i(P - 1.0 \text{ atm})$	(15-37)
Electrical	$\overline{G}_{\text{elec},i} = z_i \mathcal{F} \psi$	(15-38)
Total available energy[a]	$\overline{G}_{\text{tot},i} = \overline{G}_{\text{chem},i} + \overline{G}_{\text{mech},i} + \overline{G}_{\text{elec},i}$	(15-39)

[a] In a typical membrane system; in some systems, other types of energy would have to be included.

composition, pressure, and/or electrical potential as

$$\Delta_{1-2}\overline{G}_i = RT \ln \frac{a_{i,2}}{a_{i,1}} + \overline{V}_i(P_2 - P_1) + z_i \mathcal{F}(\psi_2 - \psi_1) \quad (15\text{-}40)$$

■ **EXAMPLE 15-6.** The activity of water is 0.99 on the concentrate side of an RO membrane and 1.0 on the permeate side, and the absolute pressures on the two sides are 1600 and 100 kPa, respectively. These values apply throughout the bulk solutions and up to the membrane surface. The system is at 25°C.

(a) Compute the transmembrane differences in (i) mechanical, (ii) chemical, and (iii) total molar Gibbs energy of water.
(b) If the membrane is 1.5 μm thick, what is the average driving force for transport of water through the membrane, expressed as newtons per mole of water transferred.

Solution.

(a) Following the sign convention established previously, we define $\Delta_{m_{c/p}}\overline{G}_w$ as the difference $\overline{G}_{w,c} - \overline{G}_{w,p}$, where w represents water.

(i) The transmembrane difference in $\overline{G}_{\text{mech},w}$ can be computed by applying Equation 15-37 to the two sides of the membrane. To use that equation, we need to know the molar volume of water, which can be calculated from the density of water, as follows:

$$\overline{V}_w = \frac{MW_w}{\rho_w} = \frac{18 \text{ g/mol}}{10^6 \text{ g/m}^3} = 1.80 \times 10^{-5} \text{ m}^3/\text{mol}$$

Then, noting that $1\,\text{kPa} = 1000\,\text{Pa} = 1000\,\text{N/m}^2 = 1000\,\text{J/m}^3$, we compute

$$\Delta_{m_{c/p}}\overline{G}_{\text{mech,w}} = \overline{V}_w \Delta_{m_{c/p}} P = \overline{V}_w(P_c - P_p)$$
$$= (1.80 \times 10^{-5}\,\text{m}^3/\text{mol})(1600\,\text{kPa} - 100\,\text{kPa})$$
$$\times (10^3\,\text{J/m}^3\,\text{kPa}) = +27.0\,\text{J/mol}$$

(ii) The transmembrane difference in chemical energy per mole of water can be computed by a similar process, except that Equation 15-36 is used instead of Equation 15-37. Noting that the standard molar free energy of water, $\overline{G}_w^o$, is the same in the two solutions, the transmembrane difference in chemical energy is

$$\Delta_{m_{c/p}}\overline{G}_{\text{chem,w}} = \left(\overline{G}_{w,c}^o + RT \ln a_{w,c}\right) - \left(\overline{G}_{w,p}^o + RT \ln a_{w,p}\right)$$
$$= RT \ln \frac{a_{w,c}}{a_{w,p}}$$
$$= (8.314\,\text{J/mol K})(298\,\text{K}) \ln \frac{0.99}{1.0}$$
$$= -24.9\,\text{J/mol}$$

(iii) The transmembrane difference in the total available energy, $\Delta_{m_{c/p}}\overline{G}_{\text{tot,w}}$, is the sum of the differences in mechanical and chemical energy, or $+2.1\,\text{J/mol}$. The sign of this value indicates that the available energy is greater on the concentrate side than the permeate side of the membrane, as it must be for permeation to occur.

(b) Normalizing the force and energy terms in Equation 15-32 to one mole of water yields:

$$\overline{F}_w = -\frac{d\overline{G}_w}{dy}$$

The average value of $\overline{F}_w$ across the membrane can be determined by replacing the differentials in this equation with finite differences across the membrane ($\Delta_{m_{c/p}}$). However, because of the convention we have adopted for the meaning of $\Delta_{m_{c/p}}$, an increase in $\overline{G}_w$ or y across the membrane corresponds to a negative value of $\Delta_{m_{c/p}}\overline{G}_w$ or $\Delta_{m_{c/p}} y$, respectively. Also, because we are assuming that the concentrations adjacent to the membrane are the same as those in the bulk, $\Delta_{m_{c/p}}\overline{G}_w = \Delta_{c/p}\overline{G}_w$. We can therefore rewrite the equation for $\overline{F}_w$ across the membrane as

$$\overline{F}_{w,\text{avg}} = -\frac{-\Delta_{m_{c/p}}\overline{G}_w}{-\Delta_{m_{c/p}}y} = -\frac{-\Delta_{c/p}\overline{G}_w}{y_{p_{m/p}} - y_{c_{c/m}}} = +\frac{\Delta_{c/p}\overline{G}_w}{\delta_m - 0}$$
$$= \frac{2.1\,\text{J/mol}}{1.5 \times 10^{-6}\,\text{m}} = 1.4 \times 10^6\,\text{J/mol m}$$
$$= 1.4 \times 10^6\,\text{N/mol}$$

The net force is positive (i.e., favorable) for movement of water from the concentrate to the permeate side of the membrane. ■

The Osmotic Pressure

When solutes enter an aqueous solution, they decrease the activity of the water (a_w) and therefore decrease $\overline{G}_{\text{chem,w}}$. Equation 15-36 indicates that the change in $\overline{G}_{\text{chem,w}}$ relative to pure water is given by

$$\underbrace{\overline{G}_{\text{chem,w}}}_{\text{actual solution}} - \underbrace{\overline{G}_{\text{chem,w}}}_{\text{pure water}} = \left(\overline{G}_w^o + RT \ln a_w\right)$$
$$- \underbrace{\left(\overline{G}_w^o + \cancel{RT \ln a_w}\right)}_{\text{pure water}} = RT \ln a_w \quad (15\text{-}41)$$

where the $\ln a_w$ term for pure water in the second expression is zero, because the activity of pure water is 1.0.

It is common to express the change in $\overline{G}_{\text{chem,w}}$ caused by the entry of solutes into the solution in terms of an equivalent change in pressure; that is, as the change in pressure which, if applied to the solution, would have an equivalent effect on $\overline{G}_{\text{tot,w}}$. Assuming that the solution has the same molar volume as pure water, this pressure change can be determined by equating the expression in Equation 15-41 with that for the change in $\overline{G}_{\text{mech,w}}$ induced by a pressure change, ΔP, according to Equation 15-37

$$\begin{pmatrix}\text{Actual effect of}\\ \text{solutes on } \overline{G}_{\text{chem,w}}\end{pmatrix} = RT \ln a_w$$
$$= \begin{pmatrix}\text{Hypothetical effect of a}\\ \text{pressure change on } \overline{G}_{\text{mech,w}}\end{pmatrix} = \overline{V}_w \Delta P_{\text{equiv}}$$

$$\Delta P_{\text{equiv}} = \frac{RT}{\overline{V}_w} \ln a_w \quad (15\text{-}42)$$

$$-\Delta P_{\text{equiv}} = -\frac{RT}{\overline{V}_w} \ln a_w \equiv \Pi \quad (15\text{-}43)$$

where $-\Delta P_{\text{equiv}}$ is positive and equals the *reduction* in pressure that would have the same effect on $\overline{G}_{\text{tot,w}}$ as the solutes do. This quantity is known as the *osmotic pressure* and, as indicated in Equation 15-43, is designated as Π; the idea of the osmotic pressure is shown schematically in Figure 15-21.

The solutions of interest when treating water in membrane systems are almost invariably dilute (e.g., even in seawater, the mole fraction of H_2O is ~0.98). As a result, a_w is close to 1.0 in both the concentrate and permeate solutions, and $\ln a_w \approx a_w - 1$. Furthermore, assuming that the activity coefficient of water is 1.0, we can equate a_w with

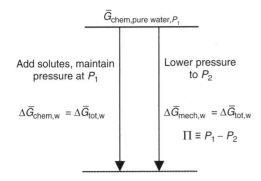

FIGURE 15-21. The underlying concept defining the osmotic pressure. The change indicated on the left is the actual change of interest in the solution; the one on the right is the imagined change that has the same effect on $\overline{G}_{tot,w}$.

the mole fraction of water. That is, $a_w = c_w/(c_w + \sum_i c_i)$, where the summation is over all solutes in the solution. Substituting these approximations into Equation 15-43, we obtain

$$\Pi = -\frac{RT}{\overline{V}_w}\left(\frac{c_w}{c_w + \sum_i c_i} - 1\right) = \frac{RT}{\overline{V}_w}\frac{\sum_i c_i}{c_w + \sum_i c_i} \quad (15\text{-}44)$$

Finally, we note that the sum $c_{w,1} + \sum_i c_{i,1}$ is the total molar concentration of all species (water and solutes) in solution. Since molar concentration (mol/L) is the inverse of molar volume (L/mol), this sum equals $\overline{V}^{-1}$, and, since we have already assumed that $\overline{V} \approx \overline{V}_w$, Equation 15-44 simplifies to

$$\Pi = \frac{RT}{\overline{V}_w}\frac{\sum_i c_i}{(\overline{V}_w)^{-1}} = RT\sum_i c_i \quad (15\text{-}45)$$

Equation 15-45 is an approximate version of Equation 15-43, and both equations are known as the *van't Hoff equation*. Equation 15-45 yields an acceptably close value for Π in the vast majority of systems of interest. If a solution is sufficiently concentrated that Equation 15-45 does not apply, Π can be estimated by the following equation (Carnahan and Starling, 1969):

$$\Pi = k_B T n \frac{1 + \phi + \phi^2 - \phi^3}{(1-\phi)^3} = RTc\frac{1 + \phi + \phi^2 - \phi^3}{(1-\phi)^3}$$
$$= \Pi_{\text{van't Hoff}} \Phi \quad (15\text{-}46)$$

where k_B is the Boltzmann constant; n, c, and ϕ are, respectively, the number concentration, the molar concentration, and the volume fraction of all solute molecules in the solution, and Φ equals the fraction shown in the middle equalities. The equality between the second and third expressions is based on the fact that the Boltzmann constant (k_B) equals the universal gas constant (R) divided by Avogadro's number. The correction factor for nonideality is such that, in highly concentrated solutions, the actual osmotic pressure is greater than would be predicted by Equation 15-45. However, the correction is small for volume fractions less than several percent.

When membranes reject solutes from a feed solution, the permeate has a lower solute content, and therefore a lower osmotic pressure, than the concentrate. The transmembrane difference in osmotic pressures ($\Delta_{m_{c/p}}\Pi$) establishes a driving force (i.e., a gradient in $\overline{G}_{chem,w}$) for clean water to migrate from the permeate back into the concentrate, which is opposite to the direction that we want. This driving force leads to a classical manifestation of osmotic pressure—the migration of water across a salt-rejecting membrane from the less salty to the more salty side, if the mechanical pressure difference across the membrane is less than the difference in osmotic pressure.

This phenomenon is illustrated in Figure 15-22, in which an ideal membrane that rejects all solutes separates a column of a solution containing solutes (on Side A) from a column of pure water (on Side B). Initially, the heights of the columns are equal, so, assuming the solutions have approximately equal densities, the mechanical pressure at any height h is uniform across the membrane. However, the presence of the solutes causes the activity, and therefore the molar Gibbs energy, of the water on Side A to be lower than on Side B, so a driving force exists for water to migrate into Side A. Water moves in this direction, increasing the mechanical pressure

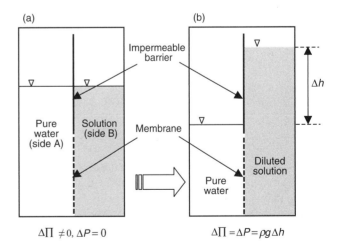

FIGURE 15-22. Water transport across a membrane driven by a difference in the osmotic pressure. The initial solutions are assumed to have nearly identical densities. (a) The initial condition, in which the mechanical pressure is equal on the two sides of the membrane. (b) The equilibrium condition, in which the difference in mechanical pressure across the membrane equals the difference in osmotic pressures of the two solutions.

on side A, and decreasing it on side B until the mechanical pressure difference driving water toward side B exactly balances the osmotic pressure difference driving it toward side A; that is, until $\overline{G}_{tot,w}$ is identical on the two sides. Applying this result to RO systems, we infer that a mechanical pressure at least equal to $\Delta_{m_c/p} \Pi$ must be applied across the membrane to counteract this composition-driven transport of water. That is, a mechanical pressure equal to $\Delta_{m_c/p} \Pi$ must be applied to the concentrate just to avoid reverse permeation, and an even greater pressure must be applied if clean water is to be produced.

The pressure exerted on an object can be interpreted as the object's mechanical potential energy per unit volume. Therefore, the osmotic pressure of a solution can be interpreted as the amount by which the presence of solutes lowers the potential energy per unit volume of water, compared with pure water. Correspondingly, at least this amount of energy per unit volume must be expended to convert the solution into pure water (i.e., to remove all the solutes from the solution) by any process, including RO. Thus, even if an RO process were 100% efficient (which, of course, no process is), an energy input per unit volume equal to $\Delta_{c/p} \Pi$ would be required to operate the system.

■ **EXAMPLE 15-7.** The major ion composition of seawater is shown below in both mass- and mole-based concentration units.

Ion	MW	Concentration g/L	Concentration mol/L
Na^+	23	10.77	0.47
Mg^{2+}	24	1.29	0.05
Ca^{2+}	40	0.41	0.01
K^+	39	0.40	0.01
Cl^-	35.5	19.35	0.55
SO_4^{2-}	96	2.71	0.03
Total		34.93	1.12

(a) What is the osmotic pressure of seawater at 25°C?

(b) What is the osmotic pressure on the concentrate side of a reverse osmosis system that is fed seawater and that is operating at steady state with 40% recovery and 99.5% rejection of salts? The permeate is recirculated at a rate sufficient that the concentrate can be treated as well mixed (i.e., no gradient from inlet to outlet).

(c) What is the minimum theoretical power requirement to operate the system in part b to produce $1 \, m^3/h$ of water? (Note: To increase the pressure by ΔP in a fluid with flow rate Q, the power received by the fluid [$\mathscr{P}_{fluid}$] must be $Q\Delta P$.)

(d) How much power would be required to operate the system in part (c) if the feed pump is 75% efficient at converting electrical power into mechanical power, and if the operating TMP is twice the value of Π computed in part (b)?

Solution.

(a) The osmotic pressure of seawater can be computed to within a few percent using Equation 15-45:

$$\Pi \approx RT \sum c_i$$
$$= (0.0821 \, L \, atm/mol \, K)(298 \, K)(1.12 \, mol/L)$$
$$= 27.4 \, atm = 27.4 \, atm \left(\frac{101.3 \, kPa}{atm}\right) = 2776 \, kPa$$

(b) The system is operating with 40% recovery and 99.5% rejection, so the ion concentration in the concentrate solution exiting the system can be calculated using Equation 15-27:

$$c_{c,out} = c_f \frac{1 - r(1-R)}{1-r}$$
$$= (1.12 \, mol/L) \frac{1 - 0.4(1 - 0.995)}{1 - 0.40}$$
$$= 1.86 \, mol/L$$

Because the solution on the concentrate side of the membrane is well mixed, c_c is the concentration throughout the system on that side of the membrane. The osmotic pressure of this solution is

$$\Pi = (0.0821 \, L \, atm/mol \, K)(298 \, K)(1.86 \, mol/L)$$
$$\times \left(\frac{101.3 \, kPa}{atm}\right) = 4617 \, kPa$$

(c) The power requirement is $Q * TMP$, and the minimum TMP is $\Delta \Pi$. Since the permeate is assumed to contain negligible salt, $\Delta \Pi$ is just the osmotic pressure of the concentrate, as computed in part (b). The minimum power requirement is therefore

$$\mathscr{P} = Q(\Delta \Pi) = (1 \, m^3/h)(4617 \, kPa)\left(\frac{1000 \, N/m^2}{1 \, kPa}\right)$$
$$\times \left(\frac{1 \, h}{3600 \, s}\right)\left(\frac{1 \, kW}{1000 \, N \, m/s}\right) = 1.28 \, kW$$

(d) Based on the problem statement, the TMP is twice the value of Π from (b), so TMP = 9234 kPa. The power required to pressurize seawater to 9234 kPa and treat it by RO at a rate of 1 m³/h, using a pump with 75% efficiency, is

$$\mathscr{P}_{\text{pump}} = \frac{Q(\text{TMP})}{\eta_{\text{pump}}} = \frac{(1\,\text{m}^3/\text{h})(9234\,\text{kPa})}{0.75}\left(\frac{1000\,\text{N/m}^2}{1\,\text{kPa}}\right)$$
$$\times\left(\frac{1\,\text{h}}{3600\,\text{s}}\right)\left(\frac{1\,\text{kW}}{1000\,\text{N m/s}}\right) = 3.42\,\text{kW} \quad\blacksquare$$

Historically, the osmotic pressure has been used exclusively to characterize the effect of solutes on the molar Gibbs energy of water. However, the idea of an equivalent pressure can also be instructive when applied directly to solutes, and in particular to differences in solute activities between two solutions. For instance, in membrane systems, it is useful to compute the pressure that would have to be applied across the membrane to generate the same difference in $G_{\text{tot},i}$ as is actually present due to a difference in the activity of i between the two bulk solutions. Based on Equations 15-36 and 15-37, we can compute this equivalent pressure differential ($\Delta_{\text{c/p}}P_{i,\text{equiv}}$) as

$$\overline{V}_i\Delta_{\text{c/p}}P_{i,\text{equiv}} = \left(G_i^\circ + RT\ln a_i\right)_\text{c} - \left(G_i^\circ + RT\ln a_i\right)_\text{p}$$

$$\Delta_{\text{c/p}}P_{i,\text{equiv}} = \frac{RT}{\overline{V}_i}\ln\frac{a_{i,\text{c}}}{a_{i,\text{p}}} \quad (15\text{-}47)$$

If a solute is selectively rejected, $a_{i,\text{c}}$ is greater than $a_{i,\text{p}}$, so $\Delta_{\text{c/p}}P_{i,\text{equiv}}$ is positive. However, if i is water (which is selectively permitted to permeate, causing $a_{\text{w,c}}$ to be less than $a_{\text{w,p}}$), $\Delta_{\text{c/p}}P_{i,\text{equiv}}$ is negative (and equals $-\Pi$).

■ **EXAMPLE 15-8.** Compute the TMP differential that would generate a gradient in $\overline{G}_{\text{mech, Na}^+}$ that is equivalent to the actual gradient generated by the transmembrane difference in Na⁺ concentrations in the preceding example system. Compare the result with the actual TMP gradient in part (d) of that example.

Solution. The equivalent pressure differential can be computed using Equation 15-47. Since the Na⁺ rejection is 99.5%, the ratio c_p/c_f is 100/0.5 or 200. However, the concentrate solution is more concentrated than the feed; we saw in Example 15-7b that c_c/c_f was 1.86/1.12, so the ratio c_c/c_f is

$$\frac{c_\text{c}}{c_\text{p}} = \frac{c_\text{c}}{c_\text{f}}\frac{c_\text{f}}{c_\text{p}} = \frac{1.86}{1.12}\frac{1}{0.005} = 332$$

Assuming (unrealistically) that the activity coefficient for Na⁺ is the same in the concentrate and permeate solutions, the activity ratio is the same as the concentration ratio, so

$$\Delta_{\text{c/p}}P_{i,\text{equiv}} = \frac{RT}{\overline{V}_i}\ln\frac{a_{i,\text{c}}}{a_{i,\text{p}}}$$
$$= \frac{(8.314\,\text{J/mol K})(298\,\text{K})(1\,\text{kPa L/J})}{1\,\text{L}/55.6\,\text{mol}}\ln(332)$$
$$= 800{,}000\,\text{kPa} = (800{,}000\,\text{kPa})\left(\frac{1\,\text{atm}}{101.3\,\text{kPa}}\right)$$
$$= 7894\,\text{atm}$$

The equivalent pressure far exceeds the applied hydrostatic pressure, confirming that the chemical driving force for Na⁺ transport dominates over the mechanical driving force. ■

Relative Magnitudes of Different Driving Forces for Transport in Membrane Systems

To gain an appreciation for the magnitudes of the various terms contributing to the driving force for transport of water and solutes in different membrane systems, consider two hypothetical, but reasonable scenarios: 2% rejection of a macromolecule such as NOM in a UF system with $\Delta_{\text{c/p}}P = 15\,\text{kPa}$, and 99.5% rejection of a small molecule, such as Na⁺, in an RO system with $\Delta_{\text{c/p}}P = 5000\,\text{kPa}$. To calculate $\overline{G}_i$ of the different species, we make the following assumptions:

- The NOM molecules are all identical, with a molar volume five times that of water
- The molar volume of Na⁺ is the same as that of water,
- The solutions are ideal and at 25°C
- The electrical potential is constant throughout the system ($\Delta_{\text{c/p}}\psi = 0$)
- The total solute concentration (used to compute the osmotic pressure) is 10^{-2} mol/L in the bulk concentrate solution in both systems. (Note that this concentration accounts for all solutes in each solution, not just the target species.)

Using the equations in Table 15-6, we can calculate the various contributions to $\Delta_{\text{c/p}}\overline{G}_{\text{tot},i}$ (the change in $\overline{G}_{\text{tot},i}$ between the bulk concentrate and bulk permeate solutions) for each species in each system; the results are shown in Table 15-7. $\Delta_{\text{c/p}}\overline{G}_{\text{tot},i}$ is defined as the value of $\overline{G}_{\text{tot},i}$ on the concentrate side of the membrane minus that on the permeate side, so a positive value of $\Delta_{\text{c/p}}\overline{G}_{\text{tot},i}$ indicates that the $\overline{G}_{\text{tot},i}$ declines in the direction of permeation and therefore favors transport from the concentrate to the permeate.

The value of $\Delta_{\text{c/p}}\overline{G}_{\text{mech},i}$ is the product of the molar volume of the species being considered and the pressure

TABLE 15-7. Contributions to the Transmembrane Energy Difference in Typical UF and RO Systems

	UF System		RO System	
	NOM	H$_2$O	Na$^+$	H$_2$O
$\Delta_{c/p}\overline{G}_{mech,i}{}^a$	1.35	0.27	90.0	90.0
$\Delta_{c/p}\overline{G}_{chem,i}$	50.1	0.0	13,127	−0.44
$\Delta_{c/p}\overline{G}_{elec,i}$	0.0	0.0	0.0	0.0
$\Delta_{c/p}\overline{G}_{tot,i}$	51.4	0.27	13,217	88.2

^{a}All values in J/mol. Assumed conditions described in the text.

difference between the two solutions. It is therefore larger for the NOM molecules than for water in the UF system, and identical for water and Na$^+$ in the RO system; it is also larger for all species in the RO than the UF system, because of the larger $\Delta_{c/p}P$. More importantly, the absolute value of $\Delta_{c/p}\overline{G}_{chem,i}$ is much larger than that of $\Delta_{c/p}\overline{G}_{mech,i}$ for the solute in both systems, whereas the reverse is true for water.

As a result, in both systems, the mechanical energy difference is the primary driving force for transport of water, whereas the chemical energy difference is the primary driving force for the solute.

Note also that the absolute values of $\Delta_{c/p}\overline{G}_{chem,i}$ and $\Delta_{c/p}\overline{G}_{tot,i}$ are much larger for both solutes than for water. This difference arises because $\Delta_{c/p}\overline{G}_{chem,i}$ depends on the ratio of the chemical activities of the species of interest on the two sides of the membrane:

$$\Delta_{c/p}\overline{G}_{chem,i} = RT \ln \frac{a_{i,c}}{a_{i,p}} \quad (15\text{-}48)$$

This ratio typically ranges from 1.0 to a minimum of approximately 0.97 for water, but can range from 1.0 to 10^4 or more for solutes. Therefore, not only does $\Delta_{c/p}\overline{G}_{chem,i}$ have opposite signs for solutes and water (it favors transport of solutes into the permeate, but of water back into the concentrate), but its magnitude is invariably far greater for solutes than for water.

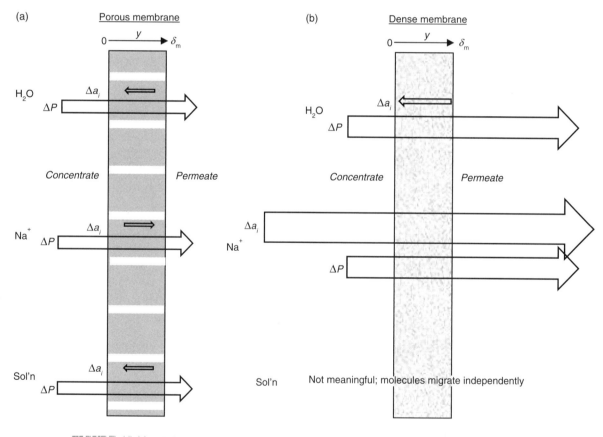

FIGURE 15-23. Schematic illustrating the relative magnitudes of the mechanical and chemical driving forces for water and a solute in typical (a) porous and (b) dense membrane systems. The driving forces can be thought of as applying to the whole solution in porous membranes (as a weighted average of the forces on the molecules in solution), but such a representation has no meaning in dense membranes, because no aqueous phase exists inside dense membranes. Note that these driving forces are for overall transport from the concentrate to the permeate; as we will see, they do not necessarily apply at every point along that path.

The observations for this example system can be generalized to most real systems. In particular, in pressure-driven membrane systems that reject solutes to any extent at all, the overall driving force for transport of solutes is dominated by the chemical energy term ($\Delta_{c/p}\overline{G}_{chem,i}$), but the overall driving force for transport of water is dominated by the mechanical energy term ($\Delta_{c/p}\overline{G}_{mech,w}$). In MF and UF systems, $\Delta_{c/p}\overline{G}_{chem,w}$ is negligible compared with $\Delta_{c/p}\overline{G}_{mech,w}$; in RO and, to a lesser extent, NF systems, $\Delta_{c/p}\overline{G}_{chem,w}$ can be significant, but it is nevertheless always smaller in magnitude than $\Delta_{c/p}\overline{G}_{mech,w}$, since that is the only way that water permeation in the desired direction can occur. These results are illustrated in Figure 15-23.

The above generalizations characterize the overall change in molar Gibbs energy between the bulk concentrate and the bulk permeate; they therefore allow us to identify the relative magnitudes of different components of the driving force for transport of water and solutes between those two reservoirs. Note, however, that molar Gibbs energy of a species can be interconverted among its various forms under certain circumstances. As we will see, such inter-conversions do occur in some membrane systems so that, for instance, the major driving force for water transport is converted from a gradient in $\Delta\overline{G}_{mech,w}$ into a gradient in $\Delta\overline{G}_{chem,w}$ over most of the region of interest in systems with dense membranes. Nevertheless, the preceding discussion is useful both for identifying the major driving forces at the gross system level, and also to provide a general understanding of the relationship between energy changes and driving forces in membrane systems.

15.7 QUANTITATIVE MODELING OF PRESSURE-DRIVEN MEMBRANE SYSTEMS

The modeling of pressure-driven membrane systems shares several general characteristics with the modeling of gas transfer and adsorption systems. All these processes involve the transfer of species from an aqueous solution into either a different phase or a second aqueous solution with a different composition. And, in each case, we need to model transport from a bulk aqueous phase into and across one or more boundaries and through one or more interfacial regions. In addition, in each process, the bulk liquid phase (or, in the case of membranes, the concentrate) might be spatially uniform, or it might vary from the inlet to the outlet of the system.

In modeling of pressure-driven membrane systems, as in the other processes, we typically assume that transport through the bulk solutions and equilibration across phase boundaries are both rapid; we also assume that the permeate is uniform from the membrane surface to the bulk solution, so no transport modeling is applied to that region. The potential rate-limiting steps are therefore transport through the concentrate-side liquid boundary layer and through the membrane itself. The analysis consists of developing separate models for transport through each of these regions and then linking these models by the constraints that (i) at steady-state, the net flux of each species must be identical through both regions; (ii) if the membrane is modeled as a nonaqueous phase, equilibrium obtains at both membrane/solution interfaces; and (iii) if the bulk concentrate composition changes as a function of location, the solution entering any segment of the system is identical to the solution leaving the segment immediately upstream. Each of these steps has analogs in the approaches used in earlier chapters to model gas transfer and adsorption systems.

The similarities noted above notwithstanding, membrane systems are more complex than gas transfer or adsorption systems, primarily because transport through the liquid boundary layer in membrane systems occurs by both advection and diffusion, whereas in gas transfer and adsorption systems, transport is exclusively diffusive. As a result, membrane modeling must take into account not only concentration gradients, but also the pressure gradient, as potential driving forces. In addition, water and contaminants are transported at different rates through the boundary layer adjacent to membranes, and the transport of each can affect the other. For example, if the contaminants are particles, the resistance they impose on the water migrating past them increases as their concentration increases.

In the following sections, we first present two widely used models for transport through the membrane interior, and then relate the analysis to models for transport through the concentrate-side boundary layer for systems in which the bulk concentrate composition is uniform. Finally, we consider how the models can be modified to accommodate possible changes in the bulk concentrate composition from the inlet to the outlet of the system.

Conceptual Models for Transport in Pressure-Driven Membranes

As has been noted, transport through membranes is widely modeled according to one of two paradigms. In the *pore-flow* model, well-defined pores are assumed to penetrate through the membrane, so that the aqueous phase is continuous from the concentrate to the permeate. In this model, transport of water through the pores is modeled as strictly advective. Often, contaminants that enter a pore in an aliquot of solution are assumed to move at the same velocity as the water, so the contaminant concentration remains constant as the aliquot moves through the pore. In this scenario, referred to as *strongly coupled transport* of the water and contaminants, the only location where the constituents behave independently is at the mouth of the pores, where some species are selectively rejected due to steric, chemical, and/or electrical factors.

At least two factors can help decouple the movement of contaminants from that of the water and invalidate the assumption of strongly coupled transport: diffusion and hindered transport. The contaminant concentration in the permeate is, of course, lower than that in the concentrate, and this difference generates a diffusive driving force for contaminant transport from the concentrate to the permeate that adds to the advective driving force. If diffusive transport is significant, the flux of contaminants into the permeate is higher, and rejection is lower, than would otherwise be the case. This trend is at least partially mitigated by *hindered transport*, which refers to the effect of the nonuniform velocity profile in the pores on particle motion. Because the flow inside membrane pores is laminar, the fluid velocity varies with the distance from the pore centerline. Correspondingly, because the contaminants are of finite size, each contaminant is exposed to a range of fluid velocities. Detailed analyses of the effects of this nonuniform velocity field indicate that it impedes contaminant transport compared with the case of uniform flow at the average velocity.

In contrast to the pore-flow model, the *solution-diffusion* model envisions the membrane as a continuous, nonporous phase that completely separates the concentrate and permeate solutions; that is, the aqueous phase is not continuous from the concentrate to the permeate. Transport between the two solutions is assumed to occur via dissolution of constituents into the membrane, migration of those species across the membrane as discrete molecules (not as a bulk fluid), and transfer out of the membrane phase and into the permeate solution. Correspondingly, in the solution-diffusion model, selective rejection is attributed to differences in the solubility and/or diffusivity of different constituents in the membrane phase. In contrast to the pore-flow model, the dissolution of each constituent into the membrane and its migration across the membrane is assumed to be minimally affected by the movement of other constituents; that is, the model is based on *uncoupled transport*. Therefore, in this case, analysis of transport through the membrane focuses on the behavior of individual species, not on an aliquot of solution.

The conceptual bases for both the pore-flow and solution-diffusion models were first suggested in the mid-1800s, but virtually all quantitative modeling of membrane processes was based on the pore-flow model through the middle of the last century (Wijmans and Baker, 1995). However, the paradigm began to shift in the 1960s, when the mathematics of the solution-diffusion model was developed and refined by Kraus et al. (1964), Lonsdale et al. (1965), and Thau et al. (1966). Since that time, the pore-flow model has been used primarily to characterize microfiltration (MF) and ultrafiltration (UF), and the solution-diffusion model has been used to model reverse osmosis (RO); both approaches have been used to model nanofiltration (NF).

These two models for transport across the concentrate/membrane interface and through the membrane itself are developed next. The objective is to derive equations that relate the fluxes of the various constituents, and therefore the rejection of contaminants, to the relevant driving forces. In addition, the analyses lead to predictions for the profiles of various parameters (e.g., concentrations, pressure, and molar Gibbs energy) throughout the system.

The Pore-Flow Model

Changes in Solution Composition at the Concentrate/Membrane Interface According to the limiting case of the pore-flow model, in which transport of contaminants is tightly coupled to that of water within the pore, contaminant rejection is determined entirely by processes at the concentrate/membrane interface. The mechanism(s) responsible for the rejection might be steric (the water enters the pores, but the contaminants are too large to do so), electrical (the contaminants are charged and therefore repelled from the membrane, whereas the water is not), or chemical (the chemical properties of the contaminants and membrane are such that they repel one another). We consider first the simple case of uncharged, spherical contaminants that are rejected strictly due to steric constraints.

Clearly, if the spheres are larger than the membrane pores, they will be rejected completely and will not appear in the permeate. Contaminants that are smaller than the pores can also be rejected, if they are not appropriately aligned with the pore opening when they approach the membrane. For example, a sphere of radius r_i will be rejected if its center is within a distance r_i of the edge of the pore (Figure 15-24). As a result, inside the pore, the particle concentration will be

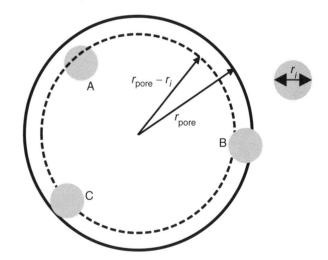

FIGURE 15-24. Rejection of spherical contaminants by cylindrical pores. Particle A can enter the pore, but particle B cannot. Particle C is as close to the wall as a particle of this size can be and still enter the pore. By convention, a particle's location is defined by its center, so the annular space is considered to have a particle concentration of zero.

finite within a core zone defined by a cylinder that is concentric with the pore and has radius $r_{\text{pore}} - r_i$, and will be zero in an annulus between this core zone and the perimeter of the pore. Assuming that the particle concentration in the solution that enters the core zone is the same as in the bulk feed, the fraction of the contaminants that enters and passes through the pore (the *sieve coefficient* for *i*-sized contaminants) will equal the fraction of the total flow that enters the core zone. This idea can be expressed mathematically for particles with $r_i < r_{\text{pore}}$ as follows:

$$\phi_i = 1 - \hat{R}_i = \frac{\text{Rate at which particles enter pore}}{\text{Rate at which particles approach pore}} = \frac{\text{Rate at which particles enter core}}{\text{Rate at which particles approach pore}}$$

$$= \frac{Q_{\text{core}} \phi_{\text{f}}}{Q_{\text{pore}} \phi_{\text{f}}} = \frac{\int_0^{r_{\text{pore}} - r_i} v(r) 2\pi r \, dr}{\int_0^{r_{\text{pore}}} v(r) 2\pi r \, dr} = \frac{\int_0^{r_{\text{pore}} - r_i} v(r) r \, dr}{\int_0^{r_{\text{pore}}} v(r) r \, dr}$$

(15-49)

where ϕ_i is the sieve coefficient for *i*-sized contaminants, $\hat{R}_i$ is the intrinsic rejection efficiency of *i*-sized contaminants,[7] and $v(r)$ is the velocity of water entering the pore at a distance r from the center of the pore.

If water entered the pore at a uniform velocity across the pore's cross-section, then $v(r)$ would be constant (independent of r), and the expected rejection efficiency would equal the fraction of the cross-sectional area from which the particles are excluded (the annulus). However, it is more common to assume that water enters the pore with a fully developed laminar velocity profile, in which case the velocity is greater near the center than near the perimeter, causing the rejection efficiency to be lower than for uniform flow. The predicted particle rejection efficiency for either of these velocity profiles can be derived by integrating Equation 15-49, yielding the following results for contaminants smaller than the pore diameter (Ferry, 1936):

$$\hat{R}_{i,\text{fl}} = 1 - (1 - \lambda_i)^2 \quad (15\text{-}50\text{a})$$

$$\hat{R}_{i,\text{pa}} = 1 - \left[2(1 - \lambda_i)^2 - (1 - \lambda_i)^4\right] \quad (15\text{-}50\text{b})$$

In these equations, λ_i is the nondimensionalized contaminant diameter, equal to d_i/d_{pore}, or, equivalently, r_i/r_{pore}, and fl and pa refer to the assumed velocity profiles: fl for a flat (uniform) profile, and pa for a parabolic profile, characteristic of Poiseuille flow. As has been noted, R_i is 1.0 for all contaminants larger than the pore diameter (i.e., for $\lambda_I \geq 1$).

[7] Recall that the *intrinsic rejection*, $\hat{R}_i$, is the ratio of the contaminant concentration in the permeate to that in the concentrate immediately adjacent to the membrane. The *rejection*, R_i, is the corresponding ratio based on the concentration in the bulk permeate.

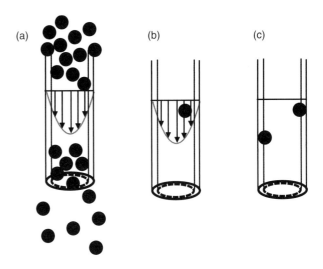

FIGURE 15-25. Phenomena that can decouple transport of contaminants from that of water as they pass through pores. (a) Contaminants are carried by the water through the center (core) portion of the pore, but they are excluded from the area near the perimeter. Clean water flows through the perimeter area, so when the two portions of the flow mix in the permeate, the contaminant concentration is less than in the core, setting up a concentration gradient that induces diffusion from the concentrate to the permeate; (b) a contaminant is exposed to a range of far-field fluid velocities, the net effect of which is to cause it to be transported less rapidly than if it were exposed only to the velocity through its center; (c) contaminants might adsorb to pore walls.

In more sophisticated analyses, some decoupling of water and contaminant transport inside the pore is included by accounting for diffusion and hindered transport, and possibly for adsorption as well (Figure 15-25). In that case, contaminant rejection is estimated by solving a modified form of the convective-diffusion equation (Deen, 1987)

$$\langle J_{i,\text{m}} \rangle = \overline{k}_{\text{D}} D_i \frac{d \langle c_i \rangle}{dz} + \overline{k}_{\text{C}} \langle v_{\text{w,m}} \rangle \langle c_i \rangle \quad (15\text{-}51)$$

where $J_{i,\text{m}}$, D_i, and c_i are the flux, diffusivity, and concentration of the contaminant, respectively, $v_{\text{w,m}}$ is the fluid velocity inside the pore, z is the distance from the pore entrance, and the angle brackets signify that the parameter values are averaged across the pore's cross-section. $\overline{k}_{\text{D}}$ and $\overline{k}_{\text{C}}$ are factors that account for the hindrance generated by the nonuniform velocity profile inside the pore, which causes the contaminant to move more slowly in the direction of water flow than it would in a uniform velocity field. Numerical expressions for $\overline{k}_{\text{D}}$ and $\overline{k}_{\text{C}}$ have been developed for several particle and pore geometries (Dechadilok and

Deen, 2006).[8] Note that $J_{i,m}$ and $v_{w,m}$ refer to values inside the pores, not in the fluid approaching the membrane; these parameters can be converted into those that apply just outside the membrane by multiplying by the membrane porosity.

Equation 15-51 can be solved once boundary conditions are specified, with the most common boundary conditions reflecting an assumption that the particle concentration just inside the pore is in equilibrium with that in the bulk solution just outside. This equilibrium is quantified by a partition coefficient, Φ, that accounts for the rejection of particles at both ends of the pore by steric, electrical, and/or chemical interactions:

$$\Phi = \frac{\langle c_{m_{c/m}} \rangle}{c_c} = \frac{\langle c_{m_{m/p}} \rangle}{c_p} \qquad (15\text{-}52)$$

For a spherical contaminant that is rejected solely by steric interactions, $\Phi = (1-\lambda)^2$. If the bulk permeate consists solely of water and contaminants that have passed through the pores, the solution to Equation 15-51 using the boundary conditions shown in Equation 15-52 is (Dechadilok and Deen, 2006):

$$\langle J_m \rangle = \frac{\Phi \overline{k}_C \langle v_{w,m} \rangle c_c}{1 - (1 - \Phi \overline{k}_C)\exp(-\text{Pe})} \qquad (15\text{-}53)$$

where Pe is the pore Peclet number, equal to $\overline{k}_C \langle v_{w,m} \rangle L_{\text{pore}}/\overline{k}_D D$, and L_{pore} is the length of the pore. The Peclet number is an indication of the ratio of convective to diffusive flux inside the pore.

The concentration of contaminant entering the permeate can be computed as the ratio of the mass flux of contaminant to the volumetric flux of solution:

$$c_p = \frac{\text{Mass rate at which contaminant enters permeate (e.g., mol/min)}}{\text{Volumetric rate at which solution enters permeate (e.g., L/min)}} = \frac{\langle J_m \rangle A_m}{\langle \hat{J}_{V,m} \rangle A_m}$$

$$= \frac{\langle J_m \rangle}{\langle v_{w,m} \rangle} \qquad (15\text{-}54)$$

---

[8] $\overline{k}_D$ and $\overline{k}_C$ also account for the fact that Equation 15-51 is written in terms of average parameter values across the whole pore cross-section, rather than across the core zone, which is the only region where the contaminants can actually reside. This choice can cause $\overline{k}_D$ and $\overline{k}_C$ to have values >1, which, at first glance, seems to imply that the flux is enhanced rather than hindered. However, if the equation were written to consider only the conditions in the core zone, those parameters would always have values <1, confirming the idea that the non-uniform velocity profile always impedes contaminant transport.

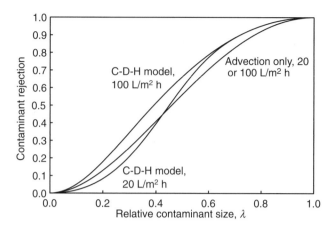

FIGURE 15-26. Predicted rejection of spherical contaminants by cylindrical pores in the absence of electrical or chemical interactions between the contaminants and the membrane. The advection-only model corresponds to Equations 15-50b and yields the same predicted rejection for any flux. The C-D-H model corresponds to Equation 15-55, using $D = 6.7 \times 10^{-11}$ m^2/s.

The rejection efficiency for spheres according to the convective-diffusion-hindered (C-D-H) transport model can therefore be computed as follows:

$$\hat{R}_{i,\text{C-D-H}} = 1 - \frac{c_p}{c_c} = 1 - \frac{\langle J_m \rangle / \langle v_{w,m} \rangle}{c_c}$$

$$= 1 - \frac{\Phi \overline{k}_C}{1 - (1 - \Phi \overline{k}_C)\exp(-\text{Pe})} \qquad (15\text{-}55)$$

The relationships in Equations 15-50 and 15-55 are plotted in Figure 15-26 for volumetric fluxes (inside the pore) of 20 and 100 L/m^2 h. The predicted rejection efficiency using Equation 15-50 depends only on λ and so is the same at the two fluxes. Adding diffusion as a potential transport mechanism decreases rejection, whereas adding hindrance increases it. Diffusion tends to have a larger effect at small λ and low volumetric flux, so under these conditions, the convection-diffusion-hindrance (C-D-H) model predicts less rejection than the convection-only model, whereas at higher λ and volumetric flux, the opposite is true. Both these trends are apparent in the figure. When combined with a contaminant size distribution in the influent, the relationships shown in the figure can be used to predict the corresponding size distribution in the permeate, as is demonstrated for a simple system with only advective transport in the following example.

■ **EXAMPLE 15-9.** Predict the overall percentage removal of particle number and mass (volume) if the effluent of the ideal horizontal settling tank analyzed in Example 13-10 is fed to an MF system with uniform pores that are 1.0 μm in diameter. Assume a parabolic velocity profile at the entrance to the pores, and that unhindered convection is the only significant contaminant transport mechanism inside the pores.

Solution. The calculations in Example 13-10 yielded values for the particle size distribution (PSD) function of the effluent from the settling tank, expressed as discrete values applicable to various increments of $\log d$. Thus, for example, the value of the PSD function listed for $\log d = -0.3$ actually characterizes all the particles with nominal diameters in the range $-0.35 < \log d < -0.25$. To compute the number and volume concentrations of particles in that size increment, we carry out the following calculations:

$$\log\left(\frac{\Delta N}{\Delta d}\right)_f = 5.60; \quad \frac{\Delta N_f}{\Delta d} = 10^{5.60} = 3.98 \times 10^5 \text{ cm}^{-3} \text{ μm}^{-1}$$

Range of d: $10^{-0.35} < d < 10^{-0.25}$;

$$\Delta d = \left(10^{-0.25} - 10^{-0.35}\right) \text{μm} = 0.116 \text{ μm}$$

$$\Delta N_f = \left(\frac{\Delta N_f}{\Delta d}\right)\Delta d = \left(3.98 \times 10^5 \text{ cm}^{-3} \text{ μm}^{-1}\right)(0.116 \text{ μm})$$

$$= 4.60 \times 10^4 \text{ cm}^{-3}$$

$$\Delta V_f = \Delta N_f \left(\frac{\pi d^3}{6}\right) = \left(4.60 \times 10^4 \text{ cm}^{-3}\right)\frac{\pi \left(10^{-0.3} \text{ μm}\right)^3}{6}$$

$$= 3040 \text{ μm}^3/\text{cm}^3$$

Similar calculations can be carried out for each size range. However, the membrane being used to treat the effluent has 1.0-μm pores ($\log d = 0.0$), so it rejects all particles larger than that size. Since that size falls in the middle of a size range in the reported PSD function ($-0.05 < \log d < 0.05$), we can estimate the number concentration of particles in the range $-0.05 < \log d < 0.0$ by assuming that the value of the PSDF for the full range ($-0.05 < \log d < 0.05$) also applies to the lower half of the range ($-0.05 < \log d < 0.0$). The results are shown in Table 15-8.

For each particle size range i, λ_i is calculated as

$$\lambda_i = \frac{d_i}{d_{\text{pore}}} = \frac{\left(10^{(\log d_i)_{\text{mean}}}\right) \text{μm}}{1 \text{ μm}}$$

where $(\log d_i)_{\text{mean}}$ is the mean value of $\log d$ in the range. The removal efficiency for each group of particles with $d < 1$ μm can then be calculated with Equation 15-50b, and the particle concentrations in the permeate on both a number and volume basis can be calculated for each particle size as the product of $(1 - R_i)$ and either ΔN_f or ΔV_f, respectively. Particles larger than 1 μm are all rejected by the membrane, so for those particles, the rejection efficiency is 1. The results of these calculations are shown in Table 15-9.

$$\hat{R}_{\text{Tot Num}} = 1 - \frac{\Delta N_p}{\Delta N_f} = 1 - \frac{38,900}{686,000} = 0.943 = 94.3\%$$

$$\hat{R}_{\text{Tot Vol}} = 1 - \frac{\Delta V_p}{\Delta V_f} = 1 - \frac{4500}{6.94 \times 10^7} = 0.99994 = 99.994\%$$

These results indicate that essentially all the particle mass is removed but, on a number basis, more than 5% of the

TABLE 15-8. PSD of Feed to MF System (Particles with $d \leq 1$ μm only)

Range of $\log d_i$ (d_i in μm)	Mean $\log d_i$ (d_i in μm)	PSDF $\log(\Delta N/\Delta d_i)_f$ ($\Delta N/\Delta d_i$ in cm^{-3} μm^{-1})	Number Concentration ΔN_f (cm^{-3})	Volume Concentration ΔV_f (μm^3 cm^{-3})
−0.35 to −0.25	−0.30	5.60	46,000	3,040
−0.25 to −0.15	−0.20	5.57	54,100	7,120
−0.15 to −0.05	−0.10	5.51	59,300	15,600
−0.05 to 0.0	−0.025	5.44	30,000	13,200
>0.0	N/A	N/A	497,000	6.93 × 10^7
		SUM	686,000	6.94 × 10^7

TABLE 15-9. Removal of Particles of Various Sizes by the Membrane

Range of $\log d_i$ (d_i in μm)	Mean $\log d_i$ (d_i in μm)	Sieve Coefficient λ	Removal Efficiency (%)	Number Concentration Remaining, ΔN_p (cm^{-3})	Volume Concentration Remaining, ΔV_p (μm^3 cm^{-3})
−0.35 to −0.25	−0.30	0.501	56.4	20,062	1,322
−0.25 to −0.15	−0.20	0.631	74.6	13,732	1,806
−0.15 to −0.05	−0.10	0.794	91.7	4,912	1,289
−0.05 to 0.0	−0.025	0.944	99.7	187	82
>0.0	N/A	N/A	100.0	0	0
			Sum	38,900	4,500

particles remain. In reality, membranes with a nominal pore diameter of 1 μm have a pore size distribution that would include some pores greater than that value (and some less), and the removal would not be as great as indicated by these calculations. ■

Permeation of Solution As noted previously, in porous membranes, the driving force for transport of water through the membrane is almost entirely associated with the pressure gradient. We reached this conclusion based on the idea that the activity of water changes negligibly across the membrane, along with the assumption that the solution was not exposed to an external electrical field. Therefore, normalizing Equation 15-32 to one mole of water and expanding the expression for total molar Gibbs energy, we can write the molar driving force for permeation of water through a porous membrane as

$$\overline{F}_{\text{perm,w,m}} = -\frac{d\overline{G}_{\text{tot,w,m}}}{dy} = -\overline{V}_w \frac{dP_m}{dy} - RT\frac{d\ln a_{w,m}}{dy} - z_w \mathcal{F}\frac{d\psi_m}{dy} \quad (15\text{-}56)$$

where y is defined as before, with a value of 0 at the concentrate/membrane interface and increasing into the membrane. Assuming tight coupling between the water and the contaminants inside the pore, the same equation applies to transport of the whole solution (s), an assumption that we incorporate by rewriting the expression as

$$\overline{F}_{\text{perm,s,m}} = -\overline{V}_s \frac{dP_m}{dy} \quad (15\text{-}57)$$

The solution velocity inside the pore is constant, so the driving force for permeation must be balanced by an equal and opposite force resisting the flow, attributable to friction between the solution and the membrane. Based on classical fluid dynamics (if the pores are envisioned to be isolated channels) or on flow through porous media (if they are envisioned to be an interconnected network, similar to the pores in packed beds), we postulate that the resisting force per mole, $\overline{F}_{\text{resist,s,m}}$, is proportional to the solution velocity in the pore. We define $\overline{F}_{\text{resist,s,m}}$ to be positive in the $-y$ direction and therefore opposite to the direction of permeation. Noting that the fluid velocity in the pore equals the fluid velocity approaching the membrane divided by the membrane surface porosity (ε_m), we can write

$$\overline{F}_{\text{resist,s,m}} = \hat{k}_{s,m} v_{s,m} = \frac{\hat{k}_{s,m}}{\varepsilon_m} \hat{J}_V \quad (15\text{-}58)$$

where $\hat{k}_{s,m}$ is the proportionality factor between the velocity in the pore and the resisting force.

Equating the force resisting flow with the permeation force, rearranging to separate the differential terms, and integrating, we obtain an equation for $\hat{J}_V$:

$$\frac{\hat{k}_{s,m}}{\varepsilon_m}\hat{J}_V = -\overline{V}_s \frac{dP_m}{dy} \quad (15\text{-}59)$$

$$\hat{J}_V \int_0^{\delta_m} dy = -\frac{\varepsilon_m \overline{V}_s}{\hat{k}_{s,m}} \int_{P_m@y=0}^{P_m@\delta_m} dP_m \quad (15\text{-}60)$$

$$\hat{J}_V = -\frac{\varepsilon_m \overline{V}_s}{\hat{k}_{s,m}} \frac{(-\Delta_{m_m}P)}{\delta_m} = K_V \Delta_{m_m}P \quad (15\text{-}61)$$

where $K_V = \varepsilon_m \overline{V}_s / \hat{k}_{s,m}\delta_m$. In a steady-state system, $\hat{J}_V$, ε_m, $\overline{V}_s$, and $\hat{k}_{s,m}$ are all constant. Equation 15-59 indicates that, in such a system, the pressure gradient through the membrane must be constant, As a result, Equation 15-61 indicates that the volumetric flux through the membrane is proportional to the TMP difference, measured inside the membrane.

Relating Permeation to Contaminant Rejection In most situations where the pore-flow model is applied, the contaminants are particles. In such cases, the pressure is assumed to remain almost constant as water moves from just outside to just inside the pore. Furthermore, inside the pore, the effect of the particles on the flow is generally ignored. Therefore, for a given (porous) membrane that rejects only particles, the permeation rate can be related to the pressure drop between the two solutions immediately adjacent to the membrane ($\Delta_{m_c/p}P$), independent of the nature or concentration of particles in the feed and of the rejection efficiency.

While the preceding statements apply when the contaminants are particles, the pore-flow model is also sometimes applied to rejection of macromolecules by porous membranes, in which case the contaminants can be considered soluble. In such a case, rejection of the solutes at the pore mouth leads to an abrupt decrease in the osmotic pressure, which in turn generates a driving force for water to move from inside the pore back to the concentrate solution. For net water transport toward the permeate to occur, this osmotic driving force must be counteracted by a pressure-based driving force of at least the same magnitude. Thus, the pressure just inside the pore, the driving force for solution permeation through the pore, and the permeation rate all decrease with increasing contaminant rejection (i.e., increasing $\Delta_{\text{int}_{c/m}}\Pi$). Depending on the details of the velocity profile inside the pore, either no change or a similar small decline in the osmotic pressure might occur as the permeate exits the pore.

Based on the preceding discussion, the pressure drop between the solutions adjacent to the membrane ($\Delta_{m_c/p}P$) can be related to the pressure difference across the interior of the membrane ($\Delta_{m_m}P$) as follows:

$$\Delta_{m_{c/p}}P = \Delta_{int_{c/m}}P + \Delta_{m_m}P + \Delta_{int_{m/p}}P$$
$$= \Delta_{int_{c/m}}\Pi + \Delta_{m_m}P + \Delta_{int_{m/p}}\Pi$$
$$= \Delta_{m_{c/p}}\Pi + \Delta_{m_m}P \quad (15\text{-}62)$$

where $\Delta_{m_{c/p}}\Pi$ can be substituted for $\Delta_{int_{c/m}}\Pi + \Delta_{int_{m/p}}\Pi$ because the solution does not change composition while inside the pore (at least for the limiting-case model of perfectly coupled transport). Equation 15-62 can be substituted into Equation 15-61 to yield

$$\hat{J}_V = K_V\left(\Delta_{m_{c/p}}P - \Delta_{m_{c/p}}\Pi\right) = K_V\Delta_{m_{c/p}}P_{eff} \quad (15\text{-}63)$$

where the effective pressure differential, ΔP_{eff}, is defined as the difference between the hydrostatic pressure differential (ΔP) and the osmotic pressure differential ($\Delta \Pi$). Equation 15-64 indicates that the flux of solution through the membrane is directly proportional to the effective pressure difference between the two solutions adjacent to the membrane.

K_V is the proportionality factor relating the volumetric flux to the effective pressure difference across the membrane. This proportionality factor was defined earlier (Equation 15-4) as the permeance of the membrane, k_w, for passage of the solution of interest; we also made the assumption that the permeance for the solution could be approximated as that for pure water (k_w^*), so Equation 15-63 can be rewritten as

$$\hat{J}_V = k_w\Delta_{m_{c/p}}P_{eff} \approx k_w^*\Delta_{m_{c/p}}P_{eff} \quad (15\text{-}64)$$

A semiempirical modification is sometimes applied to the osmotic pressure term in Equation 15-64, so that the equation is written as follows:

$$\hat{J}_V = k_w\left(\Delta_{m_{c/p}}P - \sigma\Delta_{m_{c/p}}\Pi\right) \quad (15\text{-}65)$$

where σ has a value between 0 and 1 and is known as the *reflection coefficient*. By inspection, Equation 15-65 becomes identical to Equation 15-64 if $\sigma = 1$. A mechanistic interpretation can be ascribed to the reflection coefficient within the context of models that account for incomplete coupling between transport of water and solutes, but for current purposes, σ can be considered as an empirical factor used to fit the model predictions to experimental data.

The most important practical implication of the pore-flow model with tight coupling is that, for a given membrane and a given feed solution, the rejection efficiency by clean membranes is predicted to be independent of the water permeation rate or the TMP. If the TMP is altered, the fluxes of the contaminant and water are predicted to increase by equal percentages (though not necessarily the same percentage as the TMP increase), so that the rejection remains the same. As noted previously, if the constraint of tight coupling is relaxed and contaminant diffusion and/or hindered transport inside the pore is significant, then rejection will depend on the water permeation rate and TMP; however, these dependences are generally weak for the membrane operating conditions typically used in water and wastewater treatment.

■ **EXAMPLE 15-10.** An industrial wastewater at 15°C contains 2.8 mmol/L of dissolved organic molecules and is being treated by a "tight UF" membrane. When the concentrate solution is well mixed and the TMP is 75 kPa, the volumetric flux is 40 L/m² h, the recovery (r) is 90%, and the rejection (R) is 60%. What rejection and volumetric flux are expected according to the limiting-case (i.e., tightly coupled transport) pore-flow model if the TMP is increased to 100 kPa and the bleed rate is adjusted to maintain the same recovery?

Solution. For tightly coupled transport, the pore-flow model indicates that rejection is independent of the TMP or the solution flux. Rejection will therefore remain at 60% when the TMP is increased.

The flux at the higher TMP will be related to the new conditions in the system by Equation 15-64. Assuming that the permeance of the membrane does not change significantly when the TMP is increased, we can write

$$k_{w,100\,kPa} \approx k_{w,75\,kPa} = \left(\frac{\hat{J}_V}{\Delta_{m_m}P_{eff}}\right)_{75\,kPa}$$

To find ΔP_{eff} under the baseline (75 kPa) operating conditions, we need to know c_c in that system, which we can compute from Equation 15-27

$$c_c = c_f\frac{1 - r(1 - R)}{1 - r}$$
$$= (2.8 \times 10^{-3}\,\text{mol/L})\frac{1 - 0.9[1 - 0.60]}{1 - 0.9} = 0.0179\,\text{mol/L}$$

The rejection is 60%, so the concentration in the permeate is $(1 - r)c_f$. The concentration change between the concentrate and permeate and the corresponding difference in osmotic pressure are therefore

$$\Delta_{c/m}c = c_c - (1 - r)c_f = (0.0179 - [0.40]0.0028)\,\text{mol/L}$$
$$= 0.0168\,\text{mol/L}$$
$$\Delta_{c/m}\Pi = RT\Delta_{c/m}c$$
$$= (0.0821\,\text{L atm/mol K})(288\,\text{K})(0.0168\,\text{mol/L})$$
$$\times (101.3\,\text{kPa/atm}) = 40.2\,\text{kPa}$$

We can now calculate k_w as

$$k_w = \frac{\hat{J}_V}{\Delta_{m_{c/p}} P - \Delta_{m_{c/p}} \Pi} = \frac{40\,\text{L/m}^2\,\text{h}}{(75 - 40.2)\,\text{kPa}}$$
$$= 1.15\,(\text{L/m}^2\,\text{h})/\text{kPa}$$

As noted above, the rejection does not change when the TMP is increased. As a result, if the system is operated with the same recovery, c_c will still be 0.0179 mol/L, and the osmotic pressure change across the membrane will remain the same as well. The expected flux can therefore be computed as

$$\hat{J}_V = k_w \left(\Delta_{m_{c/p}} P - \Delta_{m_{c/p}} \Pi \right)$$
$$= 1.15\left((\text{L/m}^2\,\text{h})/\text{kPa}\right)[(100 - 40.2)\,\text{kPa}]$$
$$= 68.8\,\text{L/m}^2\,\text{h}$$

The result indicates that a 33% increase in TMP leads to an increase of ~72% in the flux, with no effect on the contaminant rejection efficiency. Remember, however, that this calculation is based on the (unrealistic) assumption that the increase in TMP will have no effect on the osmotic pressure and the hydrodynamic pressure at the concentrate/membrane interface; the changes in those parameters in response to a change in TMP are considered shortly. ∎

The Solution–Diffusion Model

In the solution-diffusion model, contrary to the pore-flow model, different species are assumed to move through the membrane independently; that is, the model is one of *uncoupled transport*. Mathematically, the implication of this assumption is that the driving force for each species depends on the gradient in the total molar Gibbs energy of that species alone, not on the gradients in the molar Gibbs energies of other species. Thus, contrary to the pore-flow model, the transport of each species must be analyzed individually. The analysis requires that we characterize the key parameters—pressure and concentration—that determine $\overline{G}_{tot,i}$ of each species at each location inside the membrane.

The Transmembrane Pressure Profile According to the Solution–Diffusion Model The pressure inside a membrane cannot be measured directly, so in the solution–diffusion model, as in the pore-flow model, it is evaluated based on a theoretical analysis. The pressure at a point in any phase equals one-third of the sum of the three normal (orthogonal) stresses on a differential control volume centered at that point; that is, $P \equiv (\sigma_{xx} + \sigma_{yy} + \sigma_{zz})/3$. If the phase is a liquid, then all the nonorthogonal stresses on the control volume (σ_{xy}, σ_{xz}, σ_{yz}, etc.) are zero, and if it is stagnant, a force balance leads to the conclusion that the pressure must be constant throughout the phase. In the solution-diffusion model, the membrane is treated as though it were such a (nonaqueous) stagnant liquid, and this uniform pressure profile is assumed to apply.[9]

In the case of a flat-sheet membrane, pressure is applied from the concentrate side, and an equal resisting force is supplied by the underlying support layer to hold the membrane in place. In this case, the constant pressure throughout the membrane is equated with the applied, high pressure, and an abrupt decrease in pressure is assumed to occur at the membrane/support layer interface (i.e., the membrane/permeate interface) (Paul and Ebra-Lima, 1970). The migration of permeating species through the membrane is assumed to have no effect on this pressure profile. The pressure inside a hollow fiber or tubular membrane is also assumed to be uniform, although its value is not necessarily either $P_{int_{c/m}}$ or P_p (Paul, 1972).

Concentration Changes at the Membrane/Solution Interfaces The solution-diffusion model assumes that equilibrium is reached rapidly at the membrane/solution interfaces, so that $\overline{G}_{tot,i}$ of each species i is identical on the two sides of each interface. Applying this assumption across the concentrate/membrane interface (where no pressure change occurs), and assuming that the molar volume of i is the same in the membrane and aqueous phases yields

$$\overline{G}^o_{i,m_{c/m}} + RT \ln a_{i,m_{c/m}} + \cancel{\overline{V}_i P_{m_{c/m}}} = \overline{G}^o_{i,c_{c/m}} + RT \ln a_{i,c_{c/m}} + \cancel{\overline{V}_i P_{c_{c/m}}}$$
(15-66)

$$\ln \frac{a_{i,m_{c/m}}}{a_{i,c_{c/m}}} = \frac{\overline{G}^o_{i,c_{c/m}} - \overline{G}^o_{i,m_{c/m}}}{RT} = \frac{\Delta_{int_{c/m}} \overline{G}^o_i}{RT} = \ln K_{eq,i,aq/m}$$
(15-67)

$$\frac{a_{i,m_{c/m}}}{a_{i,c_{c/m}}} = K_{eq,i,aq/m}$$
(15-68)

where $K_{eq,i,aq/m}$ is the equilibrium constant for the dissolution of i from the aqueous solution into the membrane phase, and the final equality in Equation 15-67 follows directly from the thermodynamic definition of the equilibrium constant.[10]

The activity of i in any phase is $\gamma_i c_i / c_{i,\text{std.state}}$, so

$$K_{eq,i,aq/m} = \frac{a_{i,m_{c/m}}}{a_{i,c_{c/m}}} = \frac{\gamma_{i,m} c_{i,int_m}/c_{i,m,\text{std.state}}}{\gamma_{i,aq} c_{i,int_{aq}}/c_{i,aq,\text{std.state}}}$$
(15-69)

---

[9] Clearly, the analogy between the membrane and a liquid is imperfect—if the membrane were truly a liquid, it would be unable to resist the pressure differential and would flow into the porous support layer. Nevertheless, the analogy is often invoked.

[10] In general, for a chemical transformation, $-\Delta \overline{G}^o_r / RT = \ln K_{eq}$, where $\overline{G}^o_r$ is defined as $-\Delta \overline{G}^o$ of the products minus that of the reactants. In this case, the "product" is i in the membrane, and the "reactant" is i in the aqueous solution, so $-\Delta \overline{G}^o_r$ corresponds to $+\Delta_{int_{c/m}} \overline{G}^o_i$.

The conventional standard state concentration of water in an aqueous solution is 55.6 mol/L (i.e., a mole fraction of 1.0), and that for solutes is 1.0 mol/L, but conventions for standard state concentrations in the membrane phase are less well established. One convenient option is to use the same standard state concentrations as are used for the aqueous solution; that is, $c_{i,\text{m,std.state}} = c_{i,\text{aq,std.state}}$. (Note, however, that in the membrane, the concentrations are in units of moles per liter of membrane, not moles per liter of water). If, in addition, we assume that all species behave ideally in both phases (i.e., all γ_i values are 1.0), the right side of Equation 15-69 simplifies to the ratio of concentrations on the membrane and solution sides of the interface. As in other systems, this equilibrium ratio of concentrations in two phases at equal pressure is referred to as a distribution or partition coefficient. Designating this coefficient as $K_{D,\text{aq/m}}$, we can write, for the concentrate/ membrane interface[11]

$$K_{D,i,\text{aq/m}} \equiv \frac{c_{i,\text{m}_{c/m}}}{c_{i,\text{c}_{c/m}}} \quad (15\text{-}70)$$

In principle, $K_{D,i,\text{aq/m}}$ can be determined experimentally by equilibrating the membrane with a bulk solution of known composition and analyzing the amount of i that enters the membrane phase. In practice, however, such experiments are difficult to carry out with the extremely thin membranes that are in common use. Since water is a nearly pure phase in the aqueous solution but relatively dilute in the membrane phase, $K_{D,\text{w,aq/m}}$ is bound to be <1. And, since contaminants are selectively rejected by the membranes, their distribution coefficients are expected to be less than that of water and therefore also <1.

A similar analysis can be applied across the membrane/permeate interface. However, at that location, the pressure change is nonzero, so

$$\overline{G}^o_{i,\text{m}_{m/p}} + RT \ln a_{i,\text{m}_{m/p}} + \overline{V}_i P_{\text{m}_{m/p}} = \overline{G}^o_{i,\text{p}_{m/p}} + RT \ln a_{i,\text{p}_{m/p}} + \overline{V}_i P_{\text{p}_{m/p}} \quad (15\text{-}71)$$

$$\ln \frac{a_{i,\text{m}_{m/p}}}{a_{i,\text{p}_{m/p}}} = \frac{\overline{G}^o_{i,\text{p}_{m/p}} - \overline{G}^o_{i,\text{m}_{m/p}}}{RT} + \frac{\overline{V}_i (P_{\text{p}_{m/p}} - P_{\text{m}_{m/p}})}{RT}$$

$$= -\frac{\Delta_{\text{int}_{m/p}} \overline{G}^o_i}{RT} + \frac{\overline{V}_i (-\Delta_{\text{int}_{m/p}} P)}{RT}$$

$$= \ln K_{\text{eq},i,\text{aq/m}} - \frac{\overline{V}_i \Delta_{\text{int}_{m/p}} P}{RT} \quad (15\text{-}72)$$

$$\frac{a_{i,\text{m}_{m/p}}}{a_{i,\text{p}_{m/p}}} = K_{\text{eq},i,\text{aq/m}} \exp\left(-\frac{\overline{V}_i \Delta_{\text{int}_{m/p}} P}{RT}\right) \quad (15\text{-}73)$$

[11] For the conventions used here, $K_{D,i,\text{aq/m}}$ equals $K_{\text{eq,m/aq}}$. Note, however, that this equality is based on the choice of standard states. If different standard state conditions are chosen, the same value would apply for $K_{D,i,\text{aq/m}}$, but not for $K_{\text{eq},i,\text{aq/m}}$.

Making the same choices as before for the standard state concentrations, and again assuming that $\gamma_i = 1.0$, we can rewrite the activity ratio as a concentration ratio

$$\frac{c_{i,\text{m}_{m/p}}}{c_{i,\text{p}_{m/p}}} = K_{D,i,\text{aq/m}} \exp\left(-\frac{\overline{V}_i \Delta_{\text{int}_{m/p}} P}{RT}\right) \quad (15\text{-}74)$$

Equations 15-73 and 15-74 indicate that, even though equilibrium applies across the membrane/permeate interface, the ratio of chemical activities across that interface does not correspond to the equilibrium constant, nor does the ratio of concentrations correspond to the distribution coefficient. Rather, the ratios are less than the corresponding K values by an amount that depends on the pressure drop across the interface (which, according to the model, is also the pressure drop between the solutions adjacent to the membrane).

Permeation Through the Membrane and the Concentration Profile Across the Membrane As has been noted, in the solution-diffusion model, the pressure inside the membrane is assumed to be constant. Therefore, assuming that the electrical potential (ψ) is also constant, any gradient in $\overline{G}_{\text{tot},i}$ inside the membrane must be associated with a gradient in chemical activity. Equating the gradient in $\overline{G}_{\text{tot},i}$ with the driving force for permeation per mole of i, we obtain

$$\overline{F}_{\text{perm},i,\text{m}} = -\frac{d\overline{G}_{\text{tot},i,\text{m}}}{dy} = -\overline{V}_i \frac{dP_m}{dy} - RT \frac{d \ln a_{i,\text{m}}}{dy} \quad (15\text{-}75)$$

Assuming that $\gamma_i = 1.0$

$$\overline{F}_{\text{perm},i,\text{m}} = -RT \frac{d \ln(c_{i,\text{m}}/c_{i,\text{m,std.state}})}{dy} \quad (15\text{-}76)$$

We next assume that the only force resisting transport of i is a friction-like interaction between molecules of i and the membrane, and that the magnitude of the resisting force is proportional to the average, nominal velocity of the molecules in the direction of permeation, that is

$$\overline{F}_{\text{resist},i,\text{m}} = \hat{k}_{i,\text{m}} v_{i,\text{m}} = \hat{k}_{i,\text{m}} \frac{J_i}{c_{i,\text{m}}} \quad (15\text{-}77)$$

where $c_{i,\text{m}}$ is the molar concentration of i in the membrane phase, and the second equality is based on Equation 15-3. Solving for J_i and substituting the expression for $\overline{F}_{\text{perm},i,\text{m}}$ from Equation 15-75 for $\overline{F}_{\text{resist},i,\text{m}}$, we obtain

$$J_i = \frac{c_{i,\text{m}}}{\hat{k}_{i,\text{m}}} \overline{F}_{\text{resist},i,\text{m}} = -\frac{c_{i,\text{m}}}{\hat{k}_{i,\text{m}}} \left(RT \frac{d \ln(c_{i,\text{m}}/c_{i,\text{m,std.state}})}{dy}\right) \quad (15\text{-}78)$$

Because $d \ln x$ equals $(1/x)dx$, the logarithmic differential in Equation 15-78 can be written as

$$d \ln \frac{c_{i,m}}{c_{i,m,\text{std.state}}} = \frac{c_{i,m,\text{std.state}}}{c_{i,m}} d \frac{c_{i,m}}{c_{i,m,\text{std.state}}} = \frac{1}{c_{i,m}} dc_{i,m} \tag{15-79}$$

Substituting this expression into Equation 15-78 yields

$$J_i = -\frac{c_{i,m}}{\hat{k}_{i,m}}\left(RT\frac{(1/c_{i,m})dc_{i,m}}{dy}\right) = -\frac{RT}{\hat{k}_{i,m}}\frac{dc_{i,m}}{dy}$$

$$= -D_{i,m}\frac{dc_{i,m}}{dy} \tag{15-80}$$

where $D_{i,m}$, equal to $RT/\hat{k}_{i,m}$, is the diffusion coefficient for i in the membrane phase. Because the flux of a given species through the membrane, J_i, must be the same at all locations in the membrane at steady state, Equation 15-80 indicates that $dc_{i,m}/dy$ is constant; that is, that the concentration profile through the membrane is linear and equal to $-\Delta_{m_m} c_i/\delta_m$. Rearranging and integrating Equation 15-80, we obtain

$$J_i \delta_m = D_{i,m}\Delta_{m_m} c_i \tag{15-81}$$

$$J_i = \frac{D_{i,m}}{\delta_m}\Delta_{m_m} c_i = k_{\text{mt},i,m}\Delta_{m_m} c_i \tag{15-82}$$

where $k_{\text{mt},i,m}$ is the mass transfer coefficient for transport of i through the membrane (analogous to the ratio D/δ that appears in the equations for transport of volatile species in the stagnant-film model for gas transfer (Equations 5-36 and 5-41)). As in Chapter 5, the mass transfer coefficient has dimensions of length per time and can be thought of as a characteristic velocity for diffusive transport, in this case diffusive transport of i through the membrane.

■ **EXAMPLE 15-11.** An RO system is operated at 25°C with a TMP of 3000 kPa. The concentrate solution is well mixed and contains 0.2 mol/L CaCl$_2$, and the permeate contains 0.01 mol/L CaCl$_2$; no other solutes are present. The distribution coefficients for water and the salt ions are 0.03 and 0.02, respectively. Assuming that all $\gamma_i = 1.0$ and that the molar volumes of the salt ions are identical to that of water, describe the profiles of concentration and pressure for all species in the system.

Solution. According to the solution-diffusion model, the pressure remains at 3000 kPa from the concentrate side of the membrane to the membrane/permeate interface, where it undergoes a step decline to atmospheric pressure. The total concentration of salt ions in the concentrate solution is 0.6 mol/L (0.2 mol/L Ca^{2+} and 0.4 mol/L Cl$^-$), so the water concentration is $55.56 - 0.6$ or 54.96 mol/L. A similar calculation indicates that the permeate contains 0.01 mol/L Ca^{2+}, 0.02 mol/L Cl$^-$, and 55.53 mol/L H$_2$O. The Ca^{2+} concentrations in the membrane at the two interfaces can then be computed using Equations 15-70 and 15-74

$$c_{\text{Ca}^{2+},m_{c/m}} = c_{\text{Ca}^{2+},c_{/m}}K_{D,\text{Ca}^{2+},\text{aq}/m} = (0.2\,\text{mol/L})0.02 = 0.004\,\text{mol/L}$$

$$c_{\text{Ca}^{2+},m_{m/p}} = c_{\text{Ca}^{2+},p_{m/p}}K_{D,\text{Ca}^{2+},\text{aq}/m}\exp\left(-\frac{\overline{V}_{\text{Ca}^{2+}}\Delta_{\text{int}_{m/p}}P}{RT}\right)$$

$$= (0.01\,\text{mol/L})(0.02)$$

$$\times \exp\left(-\frac{(0.018\,\text{L/mol})(3000\,\text{kPa})}{(0.0821\,\text{L atm/mol K})(298\,\text{K})(101.3\,\text{kPa/atm})}\right)$$

$$= 1.96 \times 10^{-4}\,\text{mol/L}$$

The corresponding calculations indicate that the Cl$^-$ concentration inside the membrane declines from 0.008 mol/L on the concentrate side to 3.91×10^{-4} mol/L on the permeate side, while the H$_2$O concentration decreases from 1.649 to 1.630 mol/L. The concentration profiles within the membrane are linear in all cases, and the concentrations then undergo a step increase to the values in the permeate solution. (Remember that, inside the membrane, the concentrations are in moles per liter of the membrane phase.) ■

In Equation 15-82, the term $\Delta_{m_m} c_i$ represents the change in the concentration of i across the membrane, with both concentrations measured in the membrane phase. To express the flux as a function of more accessible parameters, we can substitute for $\Delta_{m_m} c_i$ from Equations 15-70 and 15-74, as follows:

$$J_i = k_{\text{mt},i,m}\left(c_{i,m_{c/m}} - c_{i,m_{m/p}}\right)$$

$$= k_{\text{mt},i,m}\left(K_{D,i,\text{aq}/m}c_{i,c_{/m}} - K_{D,i,\text{aq}/m}c_{i,p_{m/p}}\exp\left(-\frac{\overline{V}_i\Delta_{\text{int}_{m/p}}P}{RT}\right)\right) \tag{15-83}$$

For realistic operating conditions, the argument of the exponential in Equation 15-83 is always close to zero; for example, at 20°C and for the molar volume of water, the argument equals -0.007, -0.022, and -0.037 for $\Delta_{\text{int}_{m/p}}P$ values of 10, 30, and 50 atm, respectively. Under the circumstances, $\exp\left(-\overline{V}_i\Delta_{\text{int}_{m/p}}P/RT\right)$ can be approximated as $1 - \overline{V}_i\Delta_{\text{int}_{m/p}}P/RT$, so we can write

$$J_i = k_{\text{mt},i,m}\left(K_{D,i,\text{aq}/m}c_{i,c_{/m}} - K_{D,i,\text{aq}/m}c_{i,p_{m/p}}\left(1 - \frac{\overline{V}_i\Delta_{\text{int}_{m/p}}P}{RT}\right)\right)$$

$$= K_{D,i,\text{aq}/m}k_{\text{mt},i,m}\left(c_{i,c_{/m}} - c_{i,p_{m/p}} + c_{i,p_{m/p}}\left(\frac{\overline{V}_i\Delta_{\text{int}_{m/p}}P}{RT}\right)\right) \tag{15-84}$$

$$= K_{D,i,\text{aq}/m}k_{\text{mt},i,m}\left(\Delta_{m_{c/p}} c_i + c_{i,p_{m/p}}\left(\frac{\overline{V}_i\Delta_{\text{int}_{m/p}}P}{RT}\right)\right) \tag{15-85}$$

The first term inside the brackets in Equation 15-85 is the concentration difference between the solutions adjacent to the two sides of the membrane, and the second term is typically a few percent of the concentration on the permeate side. For contaminants, represented here as a species A, the first term is typically far greater than the second, so the second can be ignored. In this case,

$$J_A = K_{D,A,aq/m} k_{mt,A,m} \Delta_{m_{c/p}} c_A$$
$$= k_{mt,A,aq/m} \Delta_{m_{c/p}} c_A \quad (15\text{-}86)$$

where $k_{mt,A,aq/m}$ equals $K_{D,A,aq/m} k_{mt,A,m}$ and is a mass transfer coefficient for transport of A through the membrane, based on the aqueous-phase concentrations adjacent to the membrane.

For water, the second term in the summation in Equation 15-85 is often comparable in magnitude to the first, so it cannot be ignored. Nevertheless, the equation can be simplified by noting that the concentration of water in the permeate solution is close to that of pure water, so $c_{w,p,int_{m/p}} \overline{V}_w \approx 1.0$. The equation can therefore be manipulated as follows:

$$J_w = K_{D,w,aq/m} k_{mt,w,m} \left(\Delta c_w + \frac{\Delta_{int_{m/p}} P}{RT} \right)$$
$$= \frac{K_{D,w,aq/m} k_{mt,w,m}}{RT} \left(RT \Delta_{m_{c/p}} c_w + \Delta_{int_{m/p}} P \right) \quad (15\text{-}87)$$
$$= \frac{K_{D,w,aq/m} k_{mt,w,m}}{RT} \left(-\Delta_{m_{c/p}} \Pi + \Delta_{int_{m/p}} P \right) \quad (15\text{-}88)$$

Because the pressure is constant from the solution at the concentrate/membrane interface to the membrane side of the membrane/permeate interface, $\Delta_{m_{c/p}} P = \Delta_{m_{c/p}} P$, and we can write

$$J_w = \frac{K_{D,w,aq/m} k_{mt,w,m}}{RT} \left(\Delta_{m_{c/p}} P - \Delta_{m_{c/p}} \Pi \right) \quad (15\text{-}89)$$

Finally, converting the molar flux of water into a volumetric flux, and assuming that the volumetric flux of solution is essentially identical to that of the water alone, we find

$$\hat{J}_V \approx J_w \overline{V}_w = \frac{\overline{V}_w K_{D,w,aq/m} k_{mt,w,m}}{RT} \left(\Delta_{m_{c/p}} P - \Delta_{m_{c/p}} \Pi \right)$$
$$= k_w \left(\Delta_{m_{c/p}} P - \Delta_{m_{c/p}} \Pi \right) \quad (15\text{-}90)$$

where k_w equals $(\overline{V}_w K_{D,w,aq/m} k_{mt,A,m})/RT$ and, by comparison with Equation 15-4, is seen to be the permeance of the membrane for passage of the given solution. Again assuming that this permeance equals k_w^*, the permeance of the membrane for pure water, we can write

$$\hat{J}_V = k_w^* \left(\Delta_{m_{c/p}} P - \Delta_{m_{c/p}} \Pi \right) \quad (15\text{-}91)$$

As we did in Equation 15-54, we can equate the concentration of contaminant in the permeate with the ratio of the mass flux of contaminant to the volumetric flux of solution. In this case, the equality yields

$$c_{A,p_{m/p}} = \frac{\text{Mass rate at which contaminant enters permeate (e.g., mol/min)}}{\text{Volumetric rate at which solution enters permeate (e.g., L/min)}} = \frac{J_A A_m}{\hat{J}_V A_m} = \frac{J_A}{\hat{J}_V}$$
$$(15\text{-}92)$$

Substituting for J_A and $\hat{J}_V$ from Equations 15-86 and 15-91, respectively, and solving for $c_{A,p_{m/p}}$, we find

$$c_{A,p_{m/p}} = \frac{k_{mt,A,aq/m} \Delta_{m_{c/p}} c_A}{k_w^* \left(\Delta_{m_{c/p}} P - \Delta_{m_{c/p}} \Pi \right)} = \frac{k_{mt,A,aq/m} \left(c_{A,c_{/m}} - c_{A,p_{m/p}} \right)}{k_w^* \left(\Delta_{m_{c/p}} P - \Delta_{m_{c/p}} \Pi \right)}$$

$$c_{A,p_{m/p}} \left(1 + \frac{k_{mt,A,aq/m}}{k_w^* \left(\Delta_{m_{c/p}} P - \Delta_{m_{c/p}} \Pi \right)} \right) = \frac{k_{mt,A,aq/m} c_{A,c_{/m}}}{k_w^* \left(\Delta_{m_{c/p}} P - \Delta_{m_{c/p}} \Pi \right)}$$
$$(15\text{-}93)$$

$$c_{A,p_{m/p}} = \frac{k_{mt,A,aq/m} c_{A,c_{/m}}}{k_w^* \left(\Delta_{m_{c/p}} P - \Delta_{m_{c/p}} \Pi \right) + k_{mt,A,aq/m}} \quad (15\text{-}94)$$

Equations 15-91 and 15-94 represent a comprehensive solution to the equations describing transport of water and contaminants across the membrane according to the solution-diffusion model. To utilize the equations, values are required for the clean-water permeance, k_w^*, and the mass transfer coefficient for transport through the membrane for each contaminant in the system (i.e., $k_{mt,A,aq/m}$ for each contaminant), both of which can be evaluated experimentally. The other parameters in the equations relate to conditions at the membrane/solution interfaces. The pressure and solution composition at the m/p interface are typically assumed to equal those in the bulk permeate, but those at the c/m interface can only be determined by relating the equations derived in this section to a model for transport through, and the pressure in, the liquid boundary layer on the concentrate side of the membrane. However, if transport through the membrane is the rate-limiting step for transport of all the species, then the concentrations and pressure at the c/m interface can be approximated as those in the bulk concentrate, and Equation 15-94 becomes

$$c_{A,p} = \frac{k_{mt,A,aq/m} c_{A,c}}{k_w^* \left(\Delta_{c/p} P - \Delta_{c/p} \Pi \right) + k_{mt,A,aq/m}} \quad (15\text{-}95)$$

In a typical application, $\Delta_{c/p}P$ and the concentration of each contaminant in the concentrate ($c_{i,c}$) would be known. One equation similar to Equation 15-95 could be written for each of the n contaminants, yielding a set of n equations in $n+1$ unknowns (the n unknown $c_{i,p}$ values and $\Delta_{c/p}\Pi$). The van't Hoff equation provides an additional relationship among these unknowns, so the equations can be solved to determine all the $c_{i,p}$. Once $\Delta_{c/p}\Pi$ has been determined, the water flux can be computed using Equation 15-91.

Finally, we can solve for the intrinsic rejection efficiency, $\hat{R}_{A,\text{int}}$ (as defined in Equation 15-23), by substituting $\hat{J}_V$ for the denominator in Equation 15-93 (based on Equation 15-91) and rearranging the equation, as follows:

$$\frac{c_{A,p_{m/p}}}{c_{A,c_{c/m}}} = \frac{k_{\text{mt},A,\text{aq}/m}}{\hat{J}_V}\left(1 - \frac{c_{A,p_{m/p}}}{c_{A,c_{c/m}}}\right) \quad (15\text{-}96)$$

$$1 - \hat{R}_A = \frac{k_{\text{mt},A,\text{aq}/m}}{\hat{J}_V}\hat{R}_A \quad (15\text{-}97)$$

$$\hat{R}_A = \left(1 + \frac{k_{\text{mt},A,\text{aq}/m}}{\hat{J}_V}\right)^{-1} = \frac{\hat{J}_V}{\hat{J}_V + k_{\text{mt},A,\text{aq}/m}} \quad (15\text{-}98)$$

The first equality in Equation 15-98 indicates that the intrinsic rejection depends on the ratio of the characteristic velocity for diffusive transport of the contaminant through the membrane ($k_{\text{mt},A,\text{aq}/m}$) to the characteristic velocity for transport of bulk solution ($\hat{J}_V$). It also indicates that increasing the TMP, and therefore the volumetric flux, increases the rejection of any contaminant. This result is obtained because the increased pressure has a far greater effect on the flux of water than on the flux of contaminants.

■ **EXAMPLE 15-12.** The system described in Example 15-11 has a volumetric flux of $30\,\text{L/m}^2\,\text{h}$.

(a) What is the mass transfer coefficient for transport of CaCl$_2$ through the membrane? (In a complex solution containing multiple salts, Ca^{2+} and Cl^- might have different mass transfer coefficients, but in a solution with a single salt, the anions and cations must have equal coefficients to maintain electrical neutrality.)

(b) Estimate the clean-water permeance of the membrane.

(c) What volumetric flux and salt rejection efficiency would be expected if the TMP were increased to 4000 kPa, assuming the CaCl$_2$ concentration in the concentrate remained at 0.2 mol/L?

Solution.

(a) The mass transfer coefficient can be computed by rearranging Equation 15-98 and substituting the given values. The salt rejection is 95% (from 0.2 to 0.01 mol/L), so

$$k_{\text{mt},\text{Ca}^{2+},\text{aq}/m} = \hat{J}_V\left(\frac{1}{\hat{R}_{\text{Ca}^{2+}}} - 1\right) = (30\,\text{L/m}^2\,\text{h})\left(\frac{1}{0.95} - 1\right)$$
$$= 1.58\,\text{L/m}^2\,\text{h}$$
$$= (1.58\,\text{L/m}^2\,\text{h})\left(\frac{1\,\text{m}^3}{1000\,\text{L}}\right)(100\,\text{cm/m})\left(\frac{1\,\text{h}}{3600\,\text{s}}\right)$$
$$= 4.39 \times 10^{-5}\,\text{cm/s}$$

(b) The clean-water permeance can be computed from Equation 15-95, once the osmotic pressure difference across the membrane is determined. The total contaminant concentration on the concentrate side of the membrane is 0.6 mol/L (0.2 mol/L Ca^{2+} + 0.4 mol/L Cl$^-$), and that on the permeate side is 0.03 mol/L. The osmotic pressure difference across the membrane is therefore

$$\Delta_{c/p}\Pi = RT\Delta_{c/p}\left(\sum_i c_i\right) = (0.0821\,\text{L atm/mol K})(298\,\text{K})$$
$$\times (0.57\,\text{mol/L})(101.3\,\text{kPa/atm}) = 1413\,\text{kPa}$$

Then, solving Equation 15-95 for k_w^* and substituting the known values, we find

$$k_w^* = \frac{k_{\text{mt},\text{Ca}^{2+},\text{aq}/m}\left((c_{\text{Ca}^{2+},c}/c_{\text{Ca}^{2+},p}) - 1\right)}{\Delta_{c/p}P - \Delta_{c/p}\Pi}$$
$$= \frac{(1.58\,\text{L/m}^2\,\text{h})((0.2\,\text{mol/L}/0.01\,\text{mol/L}) - 1)}{3000\,\text{kPa} - 1413\,\text{kPa}}$$
$$= 1.89 \times 10^{-2}\,\text{L/m}^2\,\text{h kPa}$$

(c) When the applied pressure is increased, both the flux and the rejection change, but the mass transfer coefficient and the permeance remain the same (assuming that the permeance for the concentrate equals that for clean water). Therefore, Equation 15-95 contains $c_{\text{Ca}^{2+},p}$ and $\Delta_{c/p}\Pi$ as the only unknowns. The van't Hoff equation (i.e., the relationship between $\Delta_{c/p}\Pi$ and $\Delta_{c/p}(\sum_i c_i)$ shown in (b)) provides a second independent equation relating these parameters, so we can solve for the unknowns. These values can then be substituted into Equation 15-91 to determine the volumetric flux. The results are

$$c_{\text{Ca}^{2+},p} = 6.33 \times 10^{-3}\,\text{mol/L} \quad \hat{R}_{\text{Ca}^{2+}} = 0.968$$
$$\hat{J}_V = 48.4\,\text{L/m}^2\,\text{h} \quad \Delta_{c/p}\Pi = 1440\,\text{kPa}$$

The concentration and flux of Cl$^-$ would be twice the corresponding values for Ca^{2+}, and the rejection would be the same for the two ions. Consistent with the prior discussion, the flux increases and the concentrations of Ca^{2+} and Cl^- in the permeate decrease when the TMP increases, so the rejection increases. ∎

Comparison of Predicted Fluxes and Transmembrane Parameter Profiles in the Pore-Flow and Solution–Diffusion Models

As is clear from a comparison between Equations 15-91 and 15-64, the pore-flow and solution-diffusion models make essentially identical predictions for the flux of liquid through a membrane—both predict that the flux is directly proportional to the effective pressure differential (hydrostatic pressure minus osmotic pressure) between the solutions immediately adjacent to the membrane. On the other hand, the models lead to different predictions for the flux and rejection of contaminants. The solution-diffusion model predicts that, for a given $\Delta_{m_{c/p}} c_A$, contaminant flux is independent of $\Delta_{m_{c/p}} P$ (Equation 15-86). When this result is combined with the strong dependence of water flux on $\Delta_{m_{c/p}} P$, contaminant rejection is predicted to increase with increasing $\Delta_{m_{c/p}} P$. In contrast, in the limiting case of the pore-flow model, contaminant rejection is predicted to be independent of $\Delta_{m_{c/p}} P$; increasing $\Delta_{m_{c/p}} P$ increases the fluxes of both contaminants and water by equal percentages, so that rejection remains constant. While more sophisticated versions of the pore-flow model do predict a dependence of rejection on $\Delta_{m_{c/p}} P$, the predicted effect of an increase in this parameter is generally smaller than (and in some cases, in the opposite direction from) the effect predicted by the solution-diffusion model. The different predictions regarding the effect of the applied pressure on rejection represent the key practical distinction between the two models, and the sensitivity of rejection by dense membranes to $\Delta_{m_{c/p}} P$ provides some of the strongest evidence supporting the conceptual basis of the solution-diffusion model for these membranes.

The derivations in the preceding sections also lead to predictions for the transmembrane profiles of several important parameters according to the two models. Consider first the profiles inside porous membranes for the limiting case of perfectly coupled transport. The tight coupling in such systems causes the contaminants to move through the membrane at the same velocity as the water, so that the solution composition is identical at all locations inside the pores (except that the contaminant concentration is zero in the region right next to the membrane, from which it is sterically excluded). Correspondingly, the contaminant concentration and activity, the water concentration and activity, and the osmotic pressure are all constant throughout this region, although they might undergo discontinuous changes from just outside to just inside the pores at the two interfaces. On the other hand, the hydrostatic pressure declines linearly inside the membrane from the concentrate to the permeate side. The linear profile of P, along with an unchanging solution composition, leads to linear profiles for $\overline{G}_{\text{tot,w}}$ and $\overline{G}_{\text{tot},i}$. All these profiles are shown in Figure 15-27a

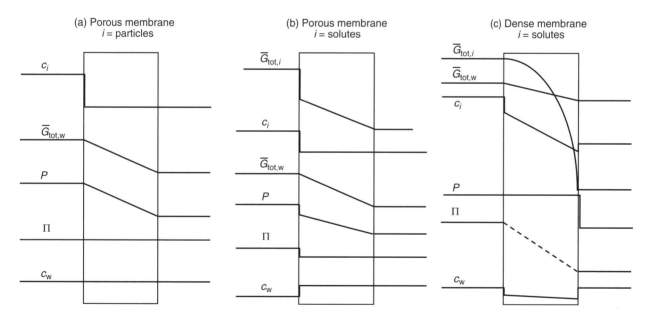

FIGURE 15-27. Profiles of several parameters across membranes. (a) Porous membranes and particulate contaminants; (b) porous membranes and dissolved contaminants; (c) dense membranes and dissolved contaminants. For porous membranes, the values of c_i, $\overline{G}_{\text{tot},i}$, $\overline{G}_{\text{tot,w}}$, and Π vary across the pore cross-section, because contaminants are excluded from the zone immediately adjacent to the membrane surface; for these parameters, the profiles characterize the average value across the pore cross-section. Although the trend shown for each parameter is correct, the profiles are not drawn to scale.

for particulate contaminants and in Figure 15-27b for soluble contaminants. (If the contaminants are particles, they are not part of the aqueous phase, and neither their chemical activity nor their molar energy is related to their concentration. Therefore, no profiles are shown in Figure 15-27a for a_i or $\overline{G}_{\text{tot},i}$.)

In contrast to the situation inside the pores, water and contaminants are not tightly coupled at the pore mouth, where the contaminant concentration decreases abruptly. If the contaminants being rejected are particles, then the composition of the solution phase, the osmotic pressure, and the hydrostatic pressure are all continuous across the mouth, as shown in Figure 15-27a. On the other hand, if the rejected contaminants are solutes, the solution becomes depleted in these species and correspondingly enriched in water across the interface, leading to the discontinuities in c_w, c_i, and Π shown in Figure 15-27b. As has been noted, the hydrostatic pressure must decrease by an amount at least equal to the decrease in osmotic pressure to maintain net permeation of water. If the osmotic pressure change is the only significant factor causing the pressure change across the c/m interface, then $\overline{G}_{\text{chem},w}$ increases by the same amount as $\overline{G}_{\text{mech},w}$ decreases, so $\overline{G}_{\text{tot},w}$ remains constant. Solutes also loose pressure-based Gibbs energy at the pore mouth; but, contrary to water, their concentration decreases, so they loose chemical Gibbs energy as well, and $\overline{G}_{\text{tot},i}$ decreases. Taking all these considerations into account, the pore-flow model suggests that the concentrations of all rejected contaminants undergo a discontinuous change at the interface and do not change thereafter, and the driving force for migration of both water and contaminants once they enter the membrane is a linear gradient in the hydrostatic pressure.

The corresponding predictions of the solution-diffusion model, in which transport of contaminants (all of which are assumed to be present as solutes in the concentrate) is uncoupled from that of water, are shown in Figure 15-27c. In this model, the hydrostatic pressure is constant between the concentrate solution at the concentrate/membrane interface and the membrane side of the membrane/permeate interface, and the driving force for transport of any species inside the membrane is associated with its concentration gradient. The concentration gradient is predicted to be linear for both water and contaminants, but the fractional change in the concentration of water from the concentrate to the permeate side of the membrane is small (virtually always <5%), whereas that of contaminants is typically much larger.

In this model, the concentrations of all permeating species undergo step changes at both aqueous solution/membrane interfaces, in such a way that equilibrium is maintained across the interfaces for each species. The osmotic pressure inside a dense membrane is undefined, since the concept is based on the bulk phase being water. Nevertheless, the osmotic pressures of the bulk concentrate and permeate solutions are well defined, and since the concentration of water in the membrane is approximately constant while that of contaminant declines linearly, it seems reasonable to represent a pseudo-osmotic pressure as declining linearly from the concentrate to the permeate side of the membrane. All these considerations are reflected in Figure 15-27c.

Summary of Transport Through Membranes

In the preceding sections, the transport of water and contaminants through pressure-driven membranes has been described and modeled for various scenarios. The key conclusions of the analysis include the following:

- The transport of contaminants might be uncoupled, weakly coupled, or strongly coupled to that of water. Strong coupling applies to most contaminants in porous membrane systems and is accounted for in the limiting-case scenario of the pore-flow model; uncoupled transport applies in most dense membrane systems and is modeled using the solution-diffusion model. Weak coupling is an intermediate case that requires more complex modeling.

- In any pressure-based membrane system, the primary driving force for transport of water from the concentrate solution to the permeate solution is the hydrostatic pressure applied to the concentrate. In porous membranes, this driving force is steadily expended as solution flows through the pores and the pressure steadily declines. In dense membranes, the pressure-based driving force between the two bulk aqueous solutions is converted into a concentration-based driving force inside the membrane. This conversion occurs because of the step change in pressure at the membrane/permeate interface along with the assumed rapid achievement of equilibrium at both membrane/solution interfaces. As in porous membranes, the driving force for transport is steadily expended as species migrate through the membrane, but in this case, the decrease in $G_{\text{tot},i}$ is caused by the steady decline in concentration while the hydrostatic pressure remains constant.

- In the pore-flow model with strong coupling, rejection of contaminants occurs only at the pore mouth. If the constraint of strong coupling is relaxed, diffusive transport tends to decrease rejection, while hindrance to transport caused by the nonuniform velocity profile in the pore tends to increase rejection. In the solution-diffusion model, rejection reflects the different tendencies of contaminants and water to dissolve into the membrane and diffuse through it.

- In the limiting case of the pore-flow model with tight coupling, the primary driving force for contaminant transport through the pore is the gradient in hydrostatic pressure, and the contaminant concentration is constant throughout the core zone of the pore (and zero in the annulus just inside the pore perimeter). Even though the overall difference in $\overline{G}_{tot,i}$ for contaminants between the concentrate and permeate is dominated by $\overline{G}_{chem,i}$, all the decline in $\overline{G}_{chem,i}$ occurs at the pore mouth, so it does not contribute to the driving force for transport after the contaminant enters the pore. If diffusive transport through the pore is considered, then both $\overline{G}_{chem,i}$ and $\overline{G}_{mech,i}$ decline as contaminants move through the pore. According to the solution-diffusion model, the dominant driving force for transport of solutes through the membrane is the gradient in their concentration, as it is for transport of water.

- The limiting case of the pore-flow model with tight coupling predicts that an increase in the TMP differential will lead to identical fractional increases in the fluxes of water and all contaminants, so the rejection efficiency for contaminants remains the same. If diffusion and hindered transport are included in this analysis, some dependence of rejection on the TMP difference is expected, with diffusion decreasing rejection and hindrance increasing it. The solution-diffusion model predicts that an increase in the TMP differential will lead to a much larger fractional increase in flux of water than flux of contaminants, and hence will lead to an increase in contaminant rejection efficiency.

15.8 MODELING TRANSPORT OF WATER AND CONTAMINANTS FROM BULK SOLUTION TO THE SURFACE OF PRESSURE-DRIVEN MEMBRANES

Overview

In the region between the bulk concentrate and the membrane surface, water and contaminants are somewhat coupled (they are advected toward the membrane together), but they also behave independently, in that contaminants are selectively rejected and diffuse back toward the bulk concentrate. Models for transport through this region are developed next; these models are then related to the models for transport through membranes, yielding composite models that allow overall permeation and rejection to be calculated as a function of the known conditions in the bulk concentrate, the system hydrodynamics, and the properties of the membrane. The limiting case of *frontal filtration*; that is, filtration in which the solution velocity near the membrane is strictly perpendicular to the membrane, is analyzed first. Although true frontal filtration is rarely encountered outside the laboratory, it is a reasonable approximation for dead-end membrane systems, in which the water usually has only a small component of flow parallel to the membrane. In addition, the analysis establishes a baseline to which systems with crossflow can subsequently be compared.

Transport through the boundary layer adjacent to membranes is independent of the nature of the membrane (porous vs dense) and also, to a large extent, the nature of the contaminant (particulate vs soluble). For both types of contaminants, net transport represents a balance between advection toward the membrane and diffusion away from it.[12] Therefore, a model for contaminant transport in this region is developed first, irrespective of the contaminant identity. The effect of the contaminant on water transport does depend on whether the contaminant is particulate or soluble, and this effect is explored subsequently.

Physicochemical State of Contaminants Near the Membrane

The rejected contaminants that accumulate near a membrane are commonly envisioned as residing in one or two layers. One layer, which is presumed to be present in any membrane system, contains relatively dispersed contaminants that are capable of diffusing back into the bulk solution. The concentration in this layer increases as the membrane is approached. Formation of this layer is referred to as *concentration polarization* (CP), and the layer itself is referred to as either a *concentration boundary layer* or a *concentration polarization layer*. The increased contaminant concentration adjacent to the membrane increases the osmotic pressure if the contaminant is dissolved, and increases the frictional resistance to water flow if the contaminant is particulate. It also increases the tendency for adsorption onto the membrane and the likelihood that a compact layer will form (as described shortly). All these processes can interfere with water flux. In addition, the increased concentrations increase the driving force for solute transport through the membrane, decreasing the rejection efficiency.

If the contaminant concentration in the CP layer reaches some critical value, a second layer can form between the CP layer and the membrane. This layer is usually envisioned to have a fixed concentration throughout and is referred to here as a *compact layer*. Three kinds of compact layers are often distinguished, based on their constituents:

- If the layer comprises discrete particles that were present in the influent, it is called a *cake* layer
- If the layer comprises colloids (such as macromolecules) that have coagulated near the membrane surface, it is called a *gel* layer
- If the layer comprises molecules that entered the system in solution and have precipitated as a well-defined solid, it is called a *scale*.

---

[12] In some literature, the random movement of molecules induced by their bombardment on all sides by water molecules is referred to as diffusion, and the analogous movement of particles is referred to as *Brownian motion*. The equations governing these processes are identical, and no distinction is made between them here.

774 MEMBRANE PROCESSES

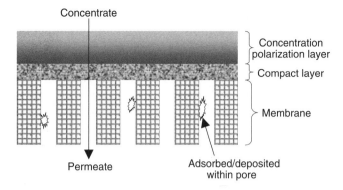

FIGURE 15-28. Fouling mechanisms in pressure-driven membrane processes.

Compact layers are usually presumed to remain immobile during a treatment cycle (i.e., until the membrane is back-flushed or chemically cleaned). When present, they can contribute substantially to the resistance of the system.

A schematic showing the two layers of rejected contaminants and also the possible accumulation of contaminants inside membrane pores is provided in Figure 15-28. Both layers are considered relevant in both frontal and crossflow filtration systems, but the relative importance of the layers can differ in the two types of systems. In frontal filtration, the likelihood that a compact layer (especially a cake or gel layer) will form is high, and the contaminant concentration in that layer is likely to be far greater than in the CP layer; the compact layer is also likely to account for most of the fouling. In crossflow systems, compact layers are somewhat less likely to form, so the CP layer can play a larger role in membrane fouling; however, even in these systems, the contribution of the CP layer to fouling is often small. In porous membrane systems, partial or complete blockage of pore openings can also contribute significantly to fouling.

Transport Through the Boundary Layer and Concentration Polarization in Frontal Filtration

We begin the analysis by considering a system similar to that shown schematically in Figure 15-29a, in which influent flows steadily into a mixed reservoir in contact with a membrane. Although the bulk liquid in the reservoir is mixed, the mixing does not extend all the way to the membrane surface, where a CP layer forms. Water containing a contaminant concentration less than that in the feed permeates through the membrane at a volumetric flux $\hat{J}_V$, corresponding to a permeation velocity v_w, and concentrate with a contaminant concentration larger than that in the feed is withdrawn from the reservoir. For simplicity, we assume that the system is at steady state and that it contains only water and a group of identical contaminants (molecules or monodisperse particles) which the membrane rejects with

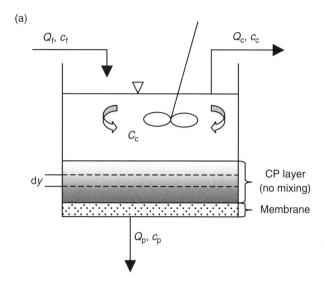

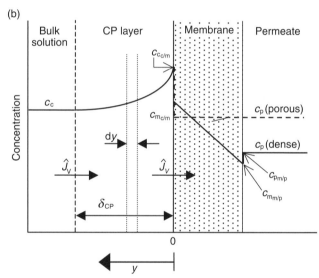

FIGURE 15-29. (a) Schematic of a frontal filtration system that could be operated at steady state. (b) Concentration profiles in the system in part (a). The predicted contaminant profile varies slightly among different models. Note: the thickness of the CP layer is greatly exaggerated compared with the size of the bulk solution, for typical applications.

some efficiency that is known as a function of volumetric flux, $\hat{J}_V$.

By definition, the CP layer spans the solution region in which a concentration gradient exists, from either the surface of the membrane or the surface of the compact layer to the location where the concentration is essentially that in the bulk concentrate (c_c). We assume, for now, that no compact layer forms; that is, the CP layer extends all the way to the membrane surface. A mass balance on contaminant in a control volume of width dy between two planes in the CP layer can be expressed as

$$\begin{bmatrix} \text{Rate of change of mass} \\ \text{of contaminant stored} \\ \text{in the control volume} \end{bmatrix} = \begin{bmatrix} \text{Net rate at which} \\ \text{contaminant enters} \\ \text{control volume at } y + dy \end{bmatrix}$$
$$- \begin{bmatrix} \text{Net rate at which} \\ \text{contaminant leaves} \\ \text{control volume at } y \end{bmatrix}$$

Since we are assuming steady-state, the storage term is zero, so the two terms on the right side equal one another. Dividing each of these terms by the membrane area, we conclude that the contaminant flux at y must equal the flux at $y + dy$, or, more generally, the flux must be the same across a plane at any value of y. Writing this flux as the sum of advective and diffusive components, we have:

$$v_w c + D \frac{dc}{dy} = J_{CP} = \text{constant} \quad (15\text{-}99)$$

where the plus sign preceding the diffusive term reflects the fact that flux in the $+y$ direction is $-D\,dc/dy$, and we wish to quantify the flux in the $-y$ direction (toward the membrane).

A second mass balance can be written around a control volume spanning the region from a plane at any value of y in the CP layer to a parallel plane on the permeate side of the membrane. Again applying the steady-state constraint, it is clear that the flux of contaminant across the plane in the permeate solution must equal J_{CP}. However, because the permeate is assumed to have a uniform composition, the contaminant flux on the permeate side of membrane is strictly advective and is given by $v_w c_p$. Thus, combining the result of this mass balance with Equation 15-99, we obtain:

$$v_w c + D \frac{dc}{dy} = J_{CP} = J_p = v_w c_p \quad (15\text{-}100)$$

$$v_w(c - c_p) + D \frac{dc}{dy} = 0 \quad (15\text{-}101)$$

$$-\frac{v_w}{D} dy = \frac{dc}{c - c_p} \quad (15\text{-}102)$$

Equation 15-102 can be integrated in conjunction with the boundary condition that the concentration is c_c at the outer edge of the CP layer:

$$c = c_c \quad \text{at } y = \delta_{CP} \quad (15\text{-}103)$$

The result of the integration is as follows:

$$c(y) - c_p = (c_c - c_p) \exp\left(\frac{v_w}{D}(\delta_{CP} - y)\right) \quad (15\text{-}104)$$

The quantity $c - c_p$ is the difference between the concentration at some location y and the concentration in the permeate. This difference is sometimes referred to as the 'reduced concentration' of the contaminant, $\hat{c}$. Using this notation, Equation 15-104 and the corresponding expression for the contaminant concentration at the concentrate/membrane boundary are

$$\hat{c}(y) = \hat{c}_c \exp\left(\frac{v_w}{D}(\delta_{CP} - y)\right) \quad (15\text{-}105)$$

$$\hat{c}(0) = \hat{c}_{c_{c/m}} = \hat{c}_c \exp\left(\frac{v_w}{D} \delta_{CP}\right) \quad (15\text{-}106)$$

Equation 15-105 indicates that the reduced contaminant concentration changes exponentially between its boundary values of $\hat{c}_{c_{c/m}}$ at $y = 0$ and $\hat{c}_c$ at $y = \delta_{CP}$. In many cases, the contaminant rejection is very efficient, so c_p is much less than c_c, and therefore necessarily much less than $\hat{c}_{c_{c/m}}$. In that case, the absolute concentration is essentially identical to the reduced concentration, and Equations 15-105 and 15-106 simplify to

$$c(y) = c_c \exp\left(\frac{v_w}{D}(\delta_{CP} - y)\right) \quad (15\text{-}107)$$

$$c_{c_{c/m}} = c_c \exp\left(\frac{v_w}{D} \delta_{CP}\right) = c_c \exp\left(\frac{v_w}{D/\delta_{CP}}\right) \quad (15\text{-}108)$$

For mathematical simplicity, we make the assumption that $c_p \ll c_c$ in the remainder of the analysis.

To use any of the preceding equations in a practical calculation, the thickness of the concentration boundary layer (δ_{CP}) must be known. Logically, we might expect δ_{CP} to depend primarily on the intensity of mixing of the bulk solution, which has a direct effect on the fluid dynamics near the membrane. Approaches for estimating δ_{CP} based on this idea are available in the literature for specific system geometries (Clark, 1996). Alternatively, estimates are sometimes made for the composite parameter D/δ_{CP}, which appears in the denominator of the exponential argument in Equation 15-108. This parameter combination has units of velocity and can be interpreted as the mass transfer coefficient, $k_{\text{mt,CP}}$, for transport of contaminant through the CP layer from the membrane to the bulk solution:

$$k_{\text{mt,CP}} = \frac{D}{\delta_{CP}} \quad (15\text{-}109)$$

The diffusion coefficient depends on the identity of the contaminant, and, for a given physical system (geometry, mechanical mixing), the thickness of the CP layer tends to increase as the diffusion coefficient increases. However, the change in δ_{CP} is generally less dramatic than the change in D, so $k_{\text{mt,CP}}$ values tend to increase with increasing D. As a result, $k_{\text{mt,CP}}$ values are larger for molecules than for colloids or particles in the same system.

Substituting Equation 15-109 into Equations 15-107 and 15-108 yields

$$c(y) = c_c \exp\left(\frac{1 - (y/\delta_{CP})}{k_{mt,CP}} v_w\right) \quad (15\text{-}110)$$

$$c_{c_{c/m}} = c_c \exp\left(\frac{v_w}{k_{mt,CP}}\right) \quad (15\text{-}111)$$

Taking logarithms of both sides of Equation 15-111, solving for v_w, and substituting $\hat{J}_V$ for v_w, we obtain the following, frequently cited form of the equation:

$$\hat{J}_V = k_{mt,CP} \ln \frac{c_{c_{c/m}}}{c_c} \quad (15\text{-}112)$$

Also, recalling the definition of the polarization factor, PF, as $c_{c_{c/m}}/c_c$, we can write:

$$\text{PF} = \frac{c_{c_{c/m}}}{c_c} = \exp\left(\frac{\hat{J}_V}{k_{mt,CP}}\right) \quad (15\text{-}113)$$

The conceptual model presented above and the resulting equations (Equations 15-110 to 15-113) are known as the *film model* for transport to the membrane.

The replacement of D/δ_{CP} with $k_{mt,CP}$ in the governing equations of the film model makes those expressions more generic. In essence, it suggests that the flux of contaminant away from the membrane can be described by a characteristic velocity ($k_{mt,CP}$), without specifying the mechanism generating that velocity. In systems where transport is solely by diffusion and the film is stagnant, that characteristic velocity equals D/δ_{CP}; in other cases (such as systems with crossflow, which are discussed subsequently), transport away from the membrane might be caused primarily by other mechanisms, in which case $k_{mt,CP}$ would be a characteristic velocity associated with those mechanisms, rather than diffusion. The assumption that is implicit in writing Equations 15-110 through 15-113 in terms of $k_{mt,CP}$ is that, even though a real system might differ substantially from the model scenario in which back-transport is by diffusion only, the shape of the concentration profile in the CP layer is the same as in that scenario.

A discussion of specific correlations that are used to estimate $k_{mt,CP}$ in different systems is postponed until the section that addresses crossflow filtration. For now, the critical point is simply that approaches do exist for estimating $k_{mt,CP}$ based on known system parameters. Once a value of $k_{mt,CP}$ is estimated from such correlations, a characteristic thickness of the CP layer can be computed using Equation 15-109, and the concentration profile in that layer (including the value of $c_{c_{c/m}}$) can be computed using Equation 15-110.

Returning to our model system, the concentration in the bulk concentrate at steady state can be related to that in the feed by Equation 15-27. Substituting this expression (with the assumption that $R \approx 1$) and the equality $v_w = \hat{J}_V$ into Equation 15-111, we obtain

$$c_{c_{c/m}} = \frac{c_f}{1-r} \exp\left(\frac{\hat{J}_V}{k_{mt,CP}}\right) \quad (15\text{-}114)$$

The recovery, r, can be expressed as $r = Q_p/Q_f = \hat{J}_V A_m/Q_f$, so:

$$c_{c_{c/m}} = \frac{c_f}{1 - (\hat{J}_V A_m/Q_f)} \exp\left(\frac{\hat{J}_V}{k_{mt,CP}}\right) \quad (15\text{-}115)$$

In a hypothetical application of Equation 15-115, we might be interested in a system in which an influent with a known flow rate and composition is being fed to a membrane with a known area and which rejects the contaminant(s) of interest almost completely. If we estimated $k_{mt,CP}$ for any incoming contaminant, we could use Equation 15-115 to solve for $c_{c_{c/m}}$ of that constituent as a function of the permeate flux, $\hat{J}_V$, and then determine the concentration profile for the contaminant through the CP layer via Equation 15-110. The same process could then be repeated to estimate the profiles of other contaminants, so that the complete contaminant profile in the CP layer would be known for any selected permeation rate.

The derivation for systems in which R is substantially <1.0 is very similar and leads to the following final result:

$$c_{c_{c/m}} = c_f \frac{1-r(1-R)}{1-r} \exp\left(\frac{\hat{J}_V}{k_{mt,CP}}\right) - c_p \left[\exp\left(\frac{\hat{J}_V}{k_{mt,CP}}\right) - 1\right] \quad (15\text{-}116\text{a})$$

$$= c_p + \left(c_f \frac{1-r(1-R)}{1-r} - c_p\right) \exp\left(\frac{\hat{J}_V}{k_{mt,CP}}\right) \quad (15\text{-}116\text{b})$$

■ **EXAMPLE 15-13.** Water containing 8 mg/L of suspended solids is being treated by frontal filtration through an MF membrane at a flux of 80 L/m² h, at a recovery of 85% and with virtually complete particle rejection. The mass transfer coefficient for a particular type of particle that accounts for 10% of the total solid load is estimated to be 4.8×10^{-6} m/s, and the diffusion coefficient of that type of particle is estimated to be 3×10^{-7} cm²/s, based on the Stokes–Einstein equation.

(a) Compute the concentration of this type of particle in the bulk concentrate.

(b) Estimate the thickness of the CP layer for the particles of interest.

(c) Compute the concentration profile of the particles through the CP layer; what is $c_{c_{c/m}}$?

(d) What is the polarization factor for the particles?

Solution.

(a) Noting that the feed concentration of the particles of interest is 10% of 8 mg/L, or 0.8 mg/L, we can use Equation 15-27 to find their concentration in the bulk concentrate:

$$c_c = c_f \frac{1 - r(1-R)}{1-r} = (0.8 \text{ mg/L}) \frac{1 - 0.85(1-1)}{1 - 0.85}$$
$$= 5.3 \text{ mg/L}$$

(b) The thickness of the CP layer can be computed by direct substitution into Equation 15-109:

$$\delta_{CP} = \frac{D}{k_{mt,CP}} = \frac{3.0 \times 10^{-7} \text{ cm}^2/\text{s}}{(4.8 \times 10^{-6} \text{ m/s})(100 \text{ cm/m})}$$
$$= 6.25 \times 10^{-4} \text{ cm} = 6.25 \text{ μm}$$

(c) The concentration of the particles throughout the CP layer can be found using Equation 15-110, yielding the profile shown in Figure 15-30. The concentration at the concentrate/membrane interface is then found from Equation 15-114 as

$$c_{c_{c/m}} = \frac{c_f}{1-r} \exp\left(\frac{\hat{J}_V}{k_{mt,CP}}\right)$$
$$= \frac{0.8 \text{ mg/L}}{1 - 0.85} \exp\left(\frac{[80 \text{ L/m}^2 \text{ h}][1 \text{ m}^3/1000 \text{ L}]}{[4.8 \times 10^{-6} \text{ m/s}][3600 \text{ s/h}]}\right)$$
$$= 547 \text{ mg/L}$$

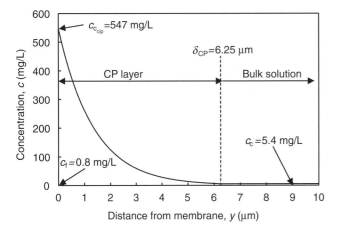

FIGURE 15-30. Concentration profile through the CP layer for the example system.

(d) Finally, based on the definition of the polarization factor, we find

$$\text{PF} = \frac{c_{c_{c/m}}}{c_c} = \frac{547 \text{ mg/L}}{5.3 \text{ mg/L}} = 102$$

Thus, the concentration of this type of particle is predicted to increase from 0.8 mg/L in the feed to 5.3 mg/L in the bulk concentrate simply due to selective removal of water via the permeate flow, and then to increase exponentially from 5.3 to 547 mg/L over the 6.25-μm thickness of the CP layer. The concentration at any point in the CP layer can be calculated from Equation 15-110; the resulting concentration profile is shown in Figure 15-30. Note that the analysis does not impose a condition of a smooth transition from the bulk concentration to that at the edge of the CP layer, so the slope of the c versus y plot is discontinuous at this location (according to the model), even though the concentration itself is continuous. Note also that the analysis implicitly assumes that D is unaffected by concentration and therefore is the same throughout the CP layer. ∎

■ **EXAMPLE 15-14.** Compute the PF for a molecular solute with $D = 10^{-5}$ cm^2/s in a system operated with the same flux as in Example 15-13, if $k_{mt,CP}$ for the solute is 9.6×10^{-5} m/s (i.e., 20 times as large as that for the particles analyzed in that example).

Solution. The PF can again be computed as in Example 15-13, yielding:

$$\text{PF} = \exp\left(\frac{\hat{J}_V}{k_{mt,CP}}\right)$$
$$= \exp\left(\frac{[80 \text{ L/m}^2 \text{ h}][1 \text{ m}^3/1000 \text{ L}]}{[9.6 \times 10^{-5} \text{ m/s}][3600 \text{ s/h}]}\right) = 1.26$$

The increase in contaminant concentration from the bulk to the membrane is only 26%, compared with >10,000% for the particles in the preceding example. This difference, which can be attributed to the much greater ease with which the molecules can migrate back to the bulk solution, has important implications for system operation, as will be shown. ∎

Relating Permeation to TMP in Systems with Frontal Filtration

The film model is useful for relating the permeate flux to contaminant concentrations in the CP layer. However, the relationship of most interest is likely to be between the flux and the applied pressure, since the pressure is the primary controllable variable in most systems. Assuming that the pressure on the permeate side of the membrane is atmospheric, the applied pressure can be equated with the TMP or $\Delta_{c/p}P$. We have already developed an equation relating the steady-state flux to the changes in hydrostatic and osmotic pressure across the membrane, and have noted that the same equation is obtained regardless of whether the membrane is viewed as porous or dense (Equations 15-64 and 15-89, respectively). That equation is linked to the conditions in the CP layer in two ways: first, the osmotic pressure in solution at the concentrate/membrane interface (in Equation 15-89) must be consistent with the concentration profile in the CP layer (computed using Equation 15-111 or 15-112), and second, the flux of water through the membrane must be the same as that through the CP layer. In addition, the sum of the pressure changes through the CP layer and the membrane must yield the TMP; that is, TMP = $\Delta_{c/p}P$ = $\Delta_{CP}P + \Delta_{m_c/p}P$. In this section, we derive a relationship between the flux and the contaminant concentration profile in the CP layer and show how that expression can be combined with other available information to determine the volumetric flux, $\hat{J}_V$, that is generated by any given, applied TMP.

The Coupling Force Exerted by Rejected Contaminants on the Permeate Flow At steady state, the CP layer is characterized by net movement of water from the bulk solution toward the membrane, but slower movement of the contaminants (or, in the case of complete rejection and steady state, no net movement). As a result, in aggregate, the water and contaminants must move past one another. In doing so, the species interact in a way that impedes the migration of both water to the membrane and contaminants back to the bulk concentrate. If the contaminants are particles, the resistance to water flow is the drag force exerted on the water, whereas if they are solutes, the resistance can be attributed to electrostatic interactions of the different molecules as they slide past one another.

Flow of water through the CP layer can be analyzed using essentially the same arguments as for flow through membranes, with the only change being that the resistance is exerted by the rejected contaminants rather than the membrane material. The force per mole driving the water through the CP layer ($\overline{F}_{perm,w}$, defined as positive toward the membrane (i.e., in the $-y$ direction)) can be equated with the negative gradient of its total molar Gibbs energy (i.e., $-d\overline{G}_{tot,w}/d(-y)$, or $+d\overline{G}_{tot,w}/dy$)[13]

$$\overline{F}_{w,perm} = \frac{d\overline{G}_{w,tot}}{dy} = \overline{V}_w \frac{dP}{dy} + RT \frac{d \ln a_w}{dy} \quad (15\text{-}117)$$

$$-\overline{V}_w \frac{dP}{dy} - \overline{V}_w \frac{d\Pi}{dy} = \overline{V}_w \left(\frac{dP}{dy} - \frac{d\Pi}{dy} \right) \quad (15\text{-}118)$$

Note that P decreases and Π increases as the membrane is approached, so both terms in parentheses in Equation 15-118 contribute positively to the force driving water toward the membrane.

The fact that the water moves at a constant velocity indicates that it experiences no net force,[14] so the force resisting the flow must be equal and opposite to $\overline{F}_{w,perm}$. Designating this force as $\overline{F}_{w,resist}$, defined as positive toward the bulk concentrate, we obtain

$$\overline{F}_{w,resist} = \overline{F}_{w,perm} = \overline{V}_w \left(\frac{dP}{dy} - \frac{d\Pi}{dy} \right) \quad (15\text{-}119)$$

Effect of Rejected Particles on Flux If the contaminants are particles, the solution has a constant composition throughout the CP layer. In this case, $d\Pi/dy$ is zero, Equation 15-119 simplifies to

$$\overline{F}_{w,resist} = \overline{V}_w \frac{dP}{dy} \quad (15\text{-}120)$$

The Reynolds number for flow in the CP layer is invariably very small, so if the particle concentration is small enough that the flow around each particle is unaffected by the presence of the other particles, the drag force exerted by a single particle (represented here as $\overline{F}^*_{w,resist}$) is given by Stokes' law (Equation 13-6). In terms of the parameter and sign conventions being used in this derivation, Stokes' law is

$$\overline{F}^*_{w,resist} = 3\pi\mu d_p v_w \quad (15\text{-}121)$$

Because $\overline{F}^*_{w,resist}$ is the drag force exerted by a single particle, the drag force experienced by water as it passes through a layer where the number concentration of particles (#/L) is n can be expressed as either $n\overline{F}^*_{w,resist}$ (force per liter of water) or $n\overline{V}_w\overline{F}^*_{w,resist}$ (force per mole of water). Substituting this latter expression for $\overline{F}^*_{w,resist}$ into Equation 15-120, and then substituting for $\overline{F}^*_{w,resist}$ based on Equation 15-121, we obtain:

$$\overline{F}^*_{w,resist} n\overline{V}_w = \overline{V}_w \frac{dP}{dy} \quad (15\text{-}122)$$

$$\frac{dP}{dy} = 3\pi\mu d_p v_w n \quad (15\text{-}123)$$

According to the film model, the variation in n through the CP layer is as shown in Equation 15-107. Substituting

---

[13] As in the analysis of flow through membranes, the constituent being modeled is the whole solution (s) if the contaminants are particles, and just water (w) if the contaminants are solutes. Assuming that $\hat{J}_V \approx \hat{J}_w$, the distinction between these two scenarios has no effect on the equation for transport, and the derivation here applies to both.

[14] If the contaminants are particles and their concentration in the CP layer increases to the point where they occupy a significant fraction of the volume, the water must accelerate as it passes through the layer, and a force would be required to induce that acceleration. However, the velocities of interest are so small that the required force would in all cases be negligible compared to the drag force associated with flow past the particles' surfaces.

that expression for n, replacing v_w by $\hat{J}_V$, and integrating, we find

$$\frac{dP}{dy} = 3\pi\mu d_p \hat{J}_V n_c \exp\left(\frac{\hat{J}_V}{D}(\delta_{CP} - y)\right) \quad (15\text{-}124)$$

$$\int_{P(0)}^{P(\delta_{CP})} dP = 3\pi\mu d_p \hat{J}_V n_c \int_0^{\delta_{CP}} \exp\left(\frac{\hat{J}_V}{D}(\delta_{CP} - y)\right) dy$$

$$\Delta_{CP}P = -3\pi\mu d_p \hat{J}_V n_c \frac{D}{\hat{J}_V}\left[\exp\left(\frac{\hat{J}_V}{D}(\delta_{CP} - y)\right)\right]_{y=0}^{y=\delta_{CP}}$$
$$(15\text{-}125)$$

$$= 3\pi\mu d_p n_c D\left(\exp\left(\frac{\hat{J}_V}{D}\delta_{CP}\right) - 1\right) \quad (15\text{-}126a)$$

$$= 3\pi\mu d_p n_c D\left(\exp\left(\frac{\hat{J}_V}{k_{mt,CP}}\right) - 1\right) \quad (15\text{-}126b)$$

Earlier, we derived an expression for the pressure change across the membrane (Equation 15-64). Noting that, if the contaminants are particles, the osmotic pressure remains constant across the membrane ($\Delta_{m_{c/p}}\Pi = 0$), we can rearrange that expression as follows:

$$\hat{J}_V = k_w^* \Delta_{m_{c/p}} P_{eff} = k_w^*\left(\Delta_{m_{c/p}}P - \Delta_{m_{c/p}}\Pi\right) = k_w^* \Delta_{m_{c/p}}P$$

$$\Delta_{m_{c/p}}P = \frac{\hat{J}_V}{k_w^*} \quad (15\text{-}127)$$

By combining Equations 15-126b and 15-127, we obtain a relationship relating the pressure loss between the bulk concentrate and permeate to the permeate flux as follows:

$$\Delta_{c/p}P = \text{TMP} = \Delta_{CP}P + \Delta_{m_{c/p}}P$$
$$= 3\pi\mu_L d_p D n_c \left(\exp\left(\frac{\hat{J}_V}{k_{mt,CP}}\right) - 1\right) + \frac{\hat{J}_V}{k_w^*}$$
$$(15\text{-}128)$$

For a given membrane, feed composition, and mixing intensity, all the terms on the right side of Equation 15-128 other than $\hat{J}_V$ are known or can be estimated from independent measurements or correlations. Therefore, the TMP required to achieve a given permeate flux can be computed directly by inserting the flux into the right side of the equation. Alternatively, for a given TMP, the equation can be solved numerically to determine the flux that will be achieved, or, if experimental data are available for flux at various TMPs, the equation can be used to identify the best-fit value of one of the model parameters, such as $k_{mt,CP}$ or δ_{CP}.

Equation 15-128 relates the permeate flux directly to the pressure loss across the CP layer and the membrane itself. In essence, the result acknowledges that the TMP is the driving force for transport between the bulk concentrate and the permeate, and that any loss of pressure as the solution migrates through the CP layer diminishes the driving force available to push the solution through the membrane; therefore, the greater the loss of pressure in the CP layer, the lower the permeation rate is. The equation was derived based on the assumption that the particles behave independently. If the particle concentration in the CP layer becomes sufficiently large that this assumption fails, a more complex relationship between concentration and the pressure gradient (e.g., the Happel (1958) cell model) can be utilized. In addition, at contaminant concentrations corresponding to volume fractions of a few percent or more, the effective diffusivity might be lower and the effective viscosity higher than the values calculated or observed in dilute solutions. Nevertheless, in each of these cases, the general approach to the analysis remains as shown above.

■ **EXAMPLE 15-15.** A membrane has a flux of 80 L/m^2 h when fed clean water in a frontal filtration system when the TMP is 20 kPa and the temperature is 20°C. Determine the TMP required to generate fluxes of 0–100 L/m^2 h, and the corresponding polarization factors, if the feed to the membrane contains 20-nm particles that the membrane rejects with 96% efficiency. The particles have a density of 1.1 g/cm^3, and their concentration in the feed is 3 mg/L, corresponding to a number concentration of 1.63×10^{13}/mL. The mass transfer coefficient in the CP layer, k_{mt}, is estimated to be 2×10^{-6} m/s, and the Brownian diffusion coefficient is estimated as 2.15×10^{-7} cm^2/s based on the Stokes–Einstein equation. How much does the presence of the particles increase the TMP required to maintain a flux of 80 L/m^2 h?

Solution. The relationship between TMP and flux is given by Equation 15-128. In that equation, the first term on the right side is the pressure loss through the CP layer and the second term is the pressure loss through the membrane. The permeance of the membrane is assumed to be the same for a feed containing contaminant as it is for a feed of clean water, so we can find k_w^* from the results of the test with clean water:

$$k_w^* = \frac{\hat{J}_V}{\Delta_{m_{c/p}}P} = \frac{80\,\text{L/m}^2\,\text{h}}{20\,\text{kPa}} = 4.0\,(\text{L/m}^2\,\text{h})/\text{kPa}$$

Values of μ, d_p, D_{Br}, and n_c are all known; converted into SI units, these values are 10^{-3} kg/m s, 2×10^{-8} m, 2.15×10^{-11} m^2/s, and 1.63×10^{19}/m^3, respectively. The TMP required to maintain the flux at 80 L/m^2 h can therefore be computed as follows:

$$\begin{aligned}
\mathrm{TMP} &= 3\pi\mu d_\mathrm{p} D_\mathrm{Br} n_\mathrm{c}\left(\exp\left(\frac{\hat{J}_\mathrm{V}}{k_{\mathrm{mt,CP}}}\right) - 1\right) + \frac{\hat{J}_\mathrm{V}}{k_\mathrm{w}^*}\\
&= 3\pi\left(10^{-3}\,\mathrm{kg/m\,s}\right)\left(2\times 10^{-8}\,\mathrm{m}\right)\left(2.15\times 10^{-11}\,\mathrm{m^2/s}\right)\left(\frac{1.63\times 10^{19}}{\mathrm{m^3}}\right)\\
&\quad \times \left(\exp\left(\frac{(80\,\mathrm{L/m^2\,h})((1\,\mathrm{m^3}/1000\,\mathrm{L})/(3600\,\mathrm{s/h}))}{2.0\times 10^{-6}\,\mathrm{m/s}}\right) - 1\right)\left[\frac{1\,\mathrm{kPa}}{1000\,\mathrm{N/m^2}}\right]\left[\frac{1\,\mathrm{N}}{\mathrm{kg\,m/s^2}}\right] + \frac{80\,\mathrm{L/m^2\,h}}{4.0(\mathrm{L/m^2\,h})/\mathrm{kPa}}\\
&= 24.4\,\mathrm{kPa}
\end{aligned}$$

For the given flux, the CP layer has only a small effect on the TMP. The results obtained using the same approach for other fluxes (i.e., using the same equation and changing only the value of $\hat{J}_\mathrm{V}$) are shown in Figure 15-31. The polarization is modest and the effect of the CP layer on the TMP is minimal until the flux reaches a critical range ($\sim 70\,\mathrm{L/m^2\,h}$), after which both the PF and the required TMP increase dramatically. As we will see shortly, other factors come into play that can prevent the flux and PF from reaching the highest values shown in the graph. ■

Effect of Rejected Solutes on Flux As is true when the contaminants are particles, the equations derived previously for transport through the membrane still apply when the contaminants are solutes, but the presence of the CP layer changes the flux, because the conditions in the concentrate solution adjacent to the membrane change. If the contaminants are all solutes, the water experiences no drag force as it moves through the CP layer, so the pressure loss in that layer is negligible, and the pressure loss across the membrane, $\Delta_{\mathrm{m_{c/p}}} P$, can be equated with the entire pressure loss from the bulk concentrate to the bulk permeate (i.e., the TMP).[15] Note that, in such a case, water is transported from the bulk concentrate to the membrane *solely* by the increase in osmotic pressure induced by the rejection of the solutes at the membrane surface. Substituting TMP for in $\Delta_{\mathrm{m_{c/p}}}P$ into Equation 15-64 yields

$$\begin{aligned}
\hat{J}_\mathrm{V} &= k_\mathrm{w}^*\left(\mathrm{TMP} - \Delta_{\mathrm{m_{c/p}}}\Pi\right)\\
&= k_\mathrm{w}^*\left(\mathrm{TMP} - \Pi_{\mathrm{c_{c/m}}} + \Pi_\mathrm{p}\right) \qquad (15\text{-}129\mathrm{a})\\
&\approx k_\mathrm{w}^*\left(\mathrm{TMP} - \Pi_{\mathrm{c_{c/m}}}\right) \qquad (15\text{-}129\mathrm{b})
\end{aligned}$$

where the approximation in Equation 15-129b applies if rejection is efficient, so the osmotic pressure of the permeate is much less than that of the concentrate solution adjacent to the membrane. Concentration polarization increases $\Pi_{\mathrm{c_{c/m}}}$, and Equation 15-129 indicates that such a change decreases

[15] Note that, because the hydrostatic pressure is constant from the bulk concentrate to the membrane surface, the *only* driving force transporting water toward the membrane in these systems is the increase in osmotic pressure generated by concentration polarization of the solutes.

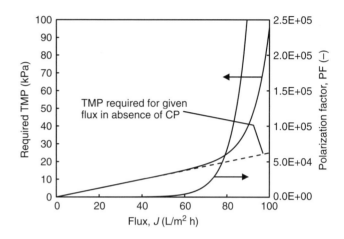

FIGURE 15-31. The TMP required to achieve various fluxes, and the corresponding polarization factor, in the example system.

the flux, even though the water experiences no loss in hydrostatic pressure as it travels through the CP layer. This result reflects the fact that the molar Gibbs energy of the water declines as it passes through that layer, and the energy available to drive water across the membrane is therefore less than in the absence of polarization.

According to the film model, the flux through the CP layer for nearly complete rejection is given by:

$$\hat{J}_\mathrm{V} = k_{\mathrm{mt,CP}} \ln \frac{c_{\mathrm{c_{c/m}}}}{c_\mathrm{c}} \qquad (15\text{-}112)$$

Equating the flux through the CP layer with that through the membrane and applying the approximate form of the van't Hoff equation, we obtain

$$k_\mathrm{w}^*\left(\mathrm{TMP} - \Pi_{\mathrm{c_{c/m}}}\right) = k_{\mathrm{mt,CP}} \ln \frac{c_{\mathrm{c_{c/m}}}}{c_\mathrm{c}} \qquad (15\text{-}130\mathrm{a})$$

$$k_\mathrm{w}^*\left(\mathrm{TMP} - RT c_{\mathrm{c_{c/m}}}\right) = k_{\mathrm{mt,CP}} \ln \frac{c_{\mathrm{c_{c/m}}}}{c_\mathrm{c}} \qquad (15\text{-}130\mathrm{b})$$

$$\mathrm{TMP} = \frac{k_{\mathrm{mt,CP}}}{k_\mathrm{w}^*}\ln\frac{c_{\mathrm{c_{c/m}}}}{c_\mathrm{c}} + RT c_{\mathrm{c_{c/m}}} \qquad (15\text{-}131)$$

For a typical scenario, the feed concentration is known, so c_c for any desired recovery can be computed from a mass balance (Equation 15-27). k_w^* and $k_{\mathrm{mt,CP}}$ can be determined from independent experiments or estimated from hydrodynamic correlations. Then, a desired flux can be chosen, Equation 15-112 can be solved to determine the corresponding concentration adjacent to the membrane ($c_{c_{c/m}}$), and Equation 15-131 can be solved to find the TMP required to generate that flux. Alternatively, if the TMP is known, Equation 15-131 can be solved to determine $c_{c_{c/m}}$, and that value can be inserted into Equation 15-109 to find the flux.

If the rejection is substantially <1.0, then Equation 15-106 must be used instead of Equation 15-112 to relate flux to the concentration profile in the CP layer, and the subsequent algebra is more complicated than shown here. However, the general analysis is the same and again leads to expressions that can be solved to determine the overall system performance as a function of known or controllable parameters.

■ **EXAMPLE 15-16.** The first stage of a one-pass, spiral-wound RO system is to be operated with 25°C seawater as feed, 20% recovery, and a TMP of 4000 kPa. When the system is operated with this TMP but with pure water as feed, the flux is 50 L/m² h. For the anticipated system geometry and operational conditions, the mass transfer coefficient in the CP layer is anticipated to be 0.005 cm/s. Estimate the volumetric flux, based on the bulk concentration in the middle of the module. Assume that the salts in the feed are completely rejected.

Solution. The total solute concentration in seawater was computed in Example 15-7 as 1.12 mol/L. Assuming that the flux is approximately uniform along the length of the membrane, 10% of the water in the feed will permeate through the membrane by the time the feed reaches the middle of the module, so the concentration in the bulk solution at that location will be (1.12 mol/L)/0.9 or 1.24 mol/L. The permeance of the membrane (k_w^*) can be computed based on the clean-water flux

$$k_w^* = \left.\frac{\hat{J}_V}{\text{TMP}}\right|_{\substack{\text{clean water}\\\text{feed}}} = \frac{50\,\text{L/m}^2\,\text{h}}{4000\,\text{kPa}}$$
$$= 1.25 \times 10^{-2}\,\text{L}/(\text{m}^2\,\text{h kPa})$$
$$= \left(1.25 \times 10^{-2}\,\text{L}/(\text{m}^2\,\text{h kPa})\right)(1000\,\text{cm}^3/\text{L})$$
$$\times \left(\frac{1\,\text{m}^2}{10^4\,\text{cm}^2}\right)\left(\frac{1\,\text{h}}{3600\,\text{s}}\right)$$
$$= 3.47 \times 10^{-7}\,\text{cm/s kPa}$$

Then, substituting known values into Equation 15-131, we find

$$\text{TMP} = \frac{k_{\text{mt,CP}}}{k_w^*}\ln\frac{c_{c_{c/m}}}{c_c} + RTc_{c_{c/m}}$$

$$4000\,\text{kPa} = \left(\frac{5.0 \times 10^{-3}\,\text{cm/s}}{3.47 \times 10^{-7}\,\text{cm/s kPa}}\right)\ln\frac{c_{c_{c/m}}}{1.24\,\text{mol/L}}$$
$$+ (0.082\,\text{L atm/mol K})(298\,\text{K})(101.3\,\text{kPa/atm})c_{c_{c/m}}$$

A numerical solution indicates that this equation is satisfied when $c_{c_{c/m}} = 1.31$ mol/L. Inserting that result into Equation 15-131, we can find the flux

$$\hat{J}_V = k_{\text{mt,CP}}\ln\frac{c_{c_{c/m}}}{c_c} = (0.005\,\text{cm/s})\ln\frac{1.31\,\text{mol/L}}{1.24\,\text{mol/L}}$$
$$= 2.75 \times 10^{-4}\,\text{cm/s}$$
$$= (2.75 \times 10^{-4}\,\text{cm/s})\left(\frac{\text{L}}{1000\,\text{cm}^3}\right)(10^4\,\text{cm}^2/\text{m}^2)(3600\,\text{s/h})$$
$$= 9.88\,\text{L/m}^2\,\text{h}$$

The salts decrease the flux by more than 80% compared with that for clean water. The decline can be attributed to the presence of a high osmotic pressure at the concentrate/membrane interface and the corresponding decrease in the effective driving force for water transport across the membrane. Most of the increase in osmotic pressure is due to the salt concentration in the bulk concentrate; concentration polarization contributes a small amount, increasing the concentration adjacent to the membrane by 5.6% over c_c. ■

Expressing the Effects of the CP Layer on Permeation in Terms of Resistance The effect of the CP layer on permeation of water is sometimes expressed in terms of the hydraulic resistance imparted by the layer. This resistance can be computed based on Equation 15-9, with the recognition that ΔP_{eff} is $\Delta P - \Delta\Pi$; that is,

$$\mathcal{R}_{\text{CP}} = \frac{\Delta_{\text{CP}}P - \Delta_{\text{CP}}\Pi}{\mu\hat{J}_V} \quad (15\text{-}132)$$

Calculations of the hydraulic resistance imposed by different sublayers of the CP layer (by applying Equation 15-132 over short distances δy within the layer) suggest that the majority of the resistance is contributed by a small region that resides nearest the membrane surface and contains the highest contaminant concentration.

According to the *resistances-in-series* model, the overall resistance to permeation can be computed as the sum of the resistance of the CP layer and the resistance across the membrane. Noting that $\hat{J}_V$ is the same throughout, this summation yields

$$\mathcal{R}_{\text{tot}} = \mathcal{R}_{\text{CP}} + \mathcal{R}_{m_{c/p}} = \frac{\Delta_{\text{CP}}P - \Delta_{\text{CP}}\Pi}{\mu\hat{J}_V} + \frac{\Delta_{m_{c/p}}P - \Delta_{m_{c/p}}\Pi}{\mu\hat{J}_V}$$
$$(15\text{-}133a)$$

$$= \frac{\Delta_{c/p}P - \Delta_{c/p}\Pi}{\mu\hat{J}_V} = \frac{\text{TMP} - \Delta_{c/p}\Pi}{\mu\hat{J}_V} \quad (15\text{-}133b)$$

The resistance attributable to the membrane (the final term in Equation 15-133a) is, to a good approximation, independent of the composition of the solution and the operational conditions and is therefore frequently equated to the resistance for permeation of pure water under standard conditions. In contrast, the resistance due to the CP layer is strongly dependent on both the solution composition (through its effect on the osmotic pressure) and the operational conditions (through their effect on the extent of concentration polarization). The resistance of the CP layer is usually, but not always, substantially less than that of the membrane.

Comparing the Effects of Particles and Solutes on Flux Through the CP Layer Comparing the $\hat{J}_V$-TMP relationships applicable for rejection of solutes and particles, it is clear that osmotic pressure plays the same role in solute-rejecting systems that hydrostatic pressure does in particle-rejecting systems. For example, if a membrane rejects only particles, P declines through the CP layer, while Π remains constant, whereas if it rejects only solutes (because the feed contains no particles), Π increases while P remains constant. (Recall that, because Π is effectively a negative pressure (vacuum) applied to the water, an increase in Π has the same effect on the water as a decrease in P.) Both types of change cause the molar Gibbs energy of water to decrease as the membrane is approached. The driving force for water movement through the layer is the gradient in $\overline{G}_w$, which is proportional to the gradient in P if the contaminants are particles and the gradient in Π if they are solutes.

In most systems, the rejected contaminants can be identified unambiguously as particles or solutes, so the appropriate equations for modeling the $\hat{J}_V$ − TMP relationship are clear. However, some rejected materials fall in the middle of the particle–solute spectrum and can generate significant gradients in both hydrostatic and osmotic pressure across the CP layer; that is, they can affect the movement of water via both physical (frictional) and chemical (osmotic) interactions. Furthermore, a given membrane might reject some contaminants that are best represented as solutes and others that are best represented as particles. In such cases, according to Equation 15-119, the net force on the water is the summation of the forces associated with the two types of gradients; that is, the effects of solutes and particles on the overall resisting force are additive.

Despite the understanding that both rejected particles and rejected solutes affect water movement through the CP layer, the current state-of-the-art is such that only a single type of contaminant is considered in any single analysis. In addition, that contaminant is almost invariably treated strictly as a particle or a solute (generating, respectively, either a decline in hydrostatic pressure or an increase in osmotic pressure across the CP layer, but not both). If the particle/solute nature of the contaminant is ambiguous, researchers typically make an arbitrary decision to treat it as one of the limiting cases and compute the corresponding $\Delta_{CP}P$ or $\Delta_{CP}\Pi$. After crossing the CP layer, the water proceeds through the membrane, losing more energy in the process.

Normally, we can quantify the amounts and types of available energy that the water contains in the two bulk solutions unambiguously, and the relationship of the overall energy loss to the flux is the key practical outcome of the preceding analyses. For such an assessment, the question of whether the energy loss experienced by the water as it passes through the CP layer is entirely physical, entirely chemical, or some combination of the two, is unimportant. Nevertheless, this issue remains a topic of some theoretical interest, both to facilitate a more fundamental understanding of the process and to develop more accurate predictive models of membrane performance.

The Formation of Cakes, Gels, or Scales, and the Limiting Flux

Formation of a Compact Layer and the Definition of c_{lim} In the film model, the concentration of contaminants in the CP layer is unconstrained; that is, the concentration at the membrane surface could grow indefinitely without violating any of the model equations. Practically, however, an upper limit must exist for $c_{c/m}$, imposed by the maximum packing density if the contaminants are particles, by the possibility that the contaminants will coagulate to form a gel-like phase if they are macromolecules, and by the possibility of precipitation of a well-defined solid if they are smaller solutes. In any of these cases, the key feature of the system is that the CP layer extends from the bulk solution to some location near the membrane, but that a *compact layer* with an approximately constant contaminant concentration exists between this location and the membrane itself. In this section, we consider the behavior of steady-state systems in which such a layer exists.

Equation 15-112 expresses the predicted relationship between permeation flux and the concentrations at the edges of the CP layer, based on the film model (and assuming that $c_p \ll c_c$). Although the highest concentration in the CP layer is identified in that equation as $c_{c/m}$, the same general analysis applies to the CP layer if it terminates at a boundary with a compact layer slightly away from the membrane. The only substantive difference is that the concentration at the interior boundary of the CP layer depends on the system operating parameters in the absence of a compact layer, whereas it equals whatever concentration induces the formation of a compact layer when such a layer is present. Referring to this concentration (the maximum contaminant concentration that can exist in the CP layer) as c_{lim}, we can rewrite Equations 15-112 and 15-115 for a steady-state frontal filtration system containing a compact layer as follows:

$$\hat{J}_V = k_{mt,CP} \ln \frac{c_{lim}}{c_c} \qquad (15\text{-}134)$$

$$c_{lim} = \frac{c_f}{1 - (\hat{J}_V A_m / Q_f)} \exp\left(\frac{\hat{J}_V}{k_{mt,CP}}\right) \qquad (15\text{-}135)$$

The Limiting Flux and the Film-Gel Model Consider how Equation 15-135 applies to a system receiving a given feed flow and in which a compact layer is present, but with adjustable TMP. In such a system, c_f, A_m, and Q_f would all be known, and the mass transfer coefficient, $k_{mt,CP}$ (or, equivalently, δ_{CP}) could be estimated based on the system hydrodynamics. In addition, although c_{lim} might not be known, it is a fixed value for the given system. As a result, only one value of $\hat{J}_V$ would satisfy the equation. That is, if a compact layer forms in a frontal filtration system with a given influent flow rate, composition, and hydrodynamic regime, the steady-state volumetric flux is fully determined. Perhaps surprisingly, since none of the parameters noted above is likely to change when the TMP is changed, this result indicates that the flux in such systems is independent of the TMP.[16] Combining this result with the preceding analysis for systems in which compact layers do not form, we conclude that, in a frontal filtration system with a given feed composition and geometry, the flux will increase with increasing TMP until a compact layer forms and will remain essentially constant no matter how much the TMP is increased thereafter. Equations 15-115 and 15-135 are therefore consistent with the qualitative observations noted previously (Figure 15-19). The value of the maximum flux is referred to as the *limiting flux*, $\hat{J}_{V,lim}$. When the film model for conditions in the CP layer is combined with the idea of a limiting concentration and flux in the presence of a compact layer, it is referred to as the *film-gel* model.

The constancy of $\hat{J}_V$ in the presence of a compact layer, regardless of the TMP, implies that c_c must also be independent of TMP in such systems (based on Equation 15-134). This fact, combined with a constant c_{lim}, leads to the conclusion that the contaminant concentration profile and the pressure loss across the CP layer are also independent of TMP. It is logical to ask, then: if essentially everything about the CP layer remains constant when the TMP is increased, where is the additional applied pressure being dissipated?

We can answer this question by considering how a system that is initially at steady state and that contains a compact layer might respond to a step increase in TMP. At least in the short term, such a step change would undoubtedly cause the flux of both water and contaminant toward the membrane to increase. During this transient, the permeate flux would increase (i.e., the additional water approaching the membrane would pass through it), but the additional contaminants that approach the membrane would be rejected (recall that we are assuming $R = 1$). The preceding analysis indicates that the rejected contaminants do not accumulate in the CP layer; therefore, they must accumulate in the compact layer, causing this layer to thicken. As the layer thickens, it imposes a greater resistance on flow, so the pressure loss across it increases, and the flux through it declines. Eventually, the increased pressure loss through the compact layer exactly compensates the increase in the applied TMP, so the flux returns to its original value. Thus, once the compact layer achieves its new steady-state thickness, the pressure loss through both the CP layer and the membrane are exactly what they were before the increase in TMP, the pressure loss through the compact layer has increased by exactly the same amount as the TMP increased, and the flux through all three layers is exactly what it was previously.

The Hydraulic Resistance of the Compact Layer According to the resistances-in-series model, the resistance of the compact layer can be added to that of the CP layer and the membrane to obtain the overall resistance to permeation. Since, in the presence of a compact layer, the flux is the limiting flux, we can write

$$\mathscr{R}_{tot} = \mathscr{R}_{CP} + \mathscr{R}_{compact} + \mathscr{R}_{m_{c/m}}$$
$$= \frac{\Delta_{CP} P - \Delta_{CP} \Pi}{\mu \hat{J}_{V,lim}} + \frac{\Delta_{compact} P}{\mu \hat{J}_{V,lim}} + \frac{\Delta_{m_{c/m}} P - \Delta_{m_{c/m}} \Pi}{\mu \hat{J}_{V,lim}}$$
(15-136a)

$$= \frac{\Delta_{c/p} P - \Delta_{c/p} \Pi}{\mu \hat{J}_{V,lim}} = \frac{TMP - \Delta_{c/p} \Pi}{\mu \hat{J}_{V,lim}} \quad (15\text{-}136b)$$

Note that, because the solution composition is constant throughout the compact layer, $\Delta_{compact} \Pi$ is zero. Other than the replacement of $\hat{J}_V$ by $\hat{J}_{V,lim}$, Equation 15-136 has the same form as Equation 15-133, which applies in the absence of a compact layer. If no compact layer is present, an increase in TMP leads to a less-than proportionate increase in $\hat{J}_V$, so $\mathscr{R}_{tot}$ increases (and the specific flux decreases); if a compact layer is present, an increase in TMP leads to no increase in $\hat{J}_V$, and $\mathscr{R}_{tot}$ increases (and specific flux decreases) even more rapidly. As noted previously, in most frontal filtration membrane systems where compact layers form, the resistance contributed by the compact layer overwhelms that contributed by the CP layer and the membrane.

The resistance $\mathscr{R}_{compact}$ is sometimes represented as the product of the *specific resistance* (resistance per unit thickness), $\hat{\mathscr{R}}_{compact}$, and the layer's thickness, $\delta_{compact}$. If the layer is incompressible and is composed of uniform particles, an expression for flow through porous media, such as the Carman–Kozeny equation (Equation 14-59), might be used to predict the specific resistance

$$\hat{\mathscr{R}}_{compact} = \frac{180(1 - \varepsilon_{compact})^2}{d_p^2 \varepsilon_{compact}^3} \quad (15\text{-}137)$$

where $\varepsilon_{compact}$ is the porosity of the compact layer, and d_p is the diameter of the particles that form it. The value of

---

[16] This statement assumes that the cake is incompressible. The effect of compressibility is discussed subsequently.

$\varepsilon_{\text{compact}}$ in such systems would typically be in the range 0.3–0.4.

If, on the other hand, the compact layer is a gel composed of coagulated macromolecules, it is likely to be compressible, and in the compressed state it might have porosities as low as 0.1 or even smaller. In that case, a limiting flux is still predicted to exist, but the increase in $\hat{\mathcal{R}}_{\text{compact}}$ when the TMP is increased is attributable to the combined effects of increases in the thickness and specific resistance of the compact layer. Finally, if the compact layer is a precipitated scale, its porosity could be anywhere in a wide range, but might be close to zero.

Based on these considerations, formation of cake layers from particles present in the feed stream (as might occur in MF and UF processes) is undesirable but is usually tolerated. Formation of gel layers (in UF and NF processes) is considered more problematic, so pretreatment and operational procedures are often designed to minimize this possibility. The worst-case scenario is formation of tight scales (in NF or RO processes), which can cause such severe fouling that it is considered absolutely unacceptable, and all necessary measures are taken to prevent such an occurrence.

Estimating c_{lim} and k_{mt} in Systems with Compact Layers and the Flux Paradox The fact that c must be c_{lim} at the inner edge of the CP layer when a compact layer is present allows $k_{\text{mt,CP}}$ and c_{lim} to be evaluated experimentally, as follows. Assume that experiments are conducted in a setup similar to that shown in Figure 15-29, using a range of feed concentrations but at the same mixing intensity (so that $k_{\text{mt,CP}}$ remains approximately constant). For each feed concentration, the system can be operated at a large enough TMP that a compact layer forms, and the limiting flux can then be determined. According to Equation 15-134, the behavior of such systems is characterized by the following relationship:

$$\hat{J}_{V,\text{lim}} = k_{\text{mt,CP}} \ln \frac{c_{\text{lim}}}{c_c} = k_{\text{mt,CP}} \ln c_{\text{lim}} - k_{\text{mt,CP}} \ln c_c \quad (15\text{-}138)$$

Equation 15-138 indicates that $k_{\text{mt,CP}}$ and c_{lim} can be estimated from the slope and x-intercept of a plot of $\hat{J}_{V,\text{lim}}$ versus $\ln c_c$, as shown by the straight line in Figure 15-32. Test systems used for such experiments utilize large enough feed concentrations and feed reservoirs that c_c can generally be approximated as c_f, so Equation 15-138 and Figure 15-32 are usually shown with c_f replacing c_c as the independent variable. Note that, as c_f and c_c decrease, higher fluxes and TMPs are required to reach c_{lim}; therefore, the TMP increases from right to left in the figure. In the limit where c_c equals c_{lim}, the flux is predicted to be zero, because any finite flux would cause the compact layer to grow indefinitely.

If the compact layer is compressible, then c_{lim} is larger and $\hat{J}_{V,\text{lim}}$ is smaller at high TMP than they would be if it

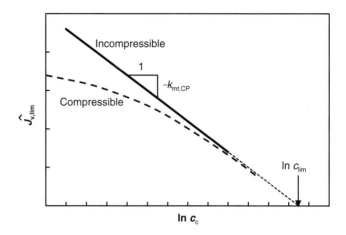

FIGURE 15-32. A plot that can be used to estimate $k_{\text{mt,CP}}$ and c_{lim}.

were incompressible. The effect of compressibility is to cause the experimental curve to fall below the expected straight line at higher TMPs (lower c_c) and to deviate increasingly from linearity as c_c decreases, as shown by the curved line in the figure.

When experiments needed to prepare a plot like that in Figure 15-32 are carried out with model contaminants, the limiting concentration is often found to be in the expected range for close packing (volume fractions on the order of 0.6 for rigid particles, and up to 0.9 for deformable particles). In contrast, the fluxes in such experiments and the corresponding, inferred values of $k_{\text{mt,CP}}$ are often orders of magnitude larger than those expected based on an independent estimate of δ_{CP} and the value of D computed using the Stokes–Einstein equation. The discrepancy is particularly severe for particles in the size range from one to several micrometers and has been referred to as "the flux paradox" (Green and Belfort, 1980).

Attempts to explain the flux paradox have focused on the possibility that fluid motion in the CP layer increases the transport of particles from that layer back to the bulk solution, causing the particles to have an effective diffusivity much larger than their Brownian diffusivity. Because fluid motion near the membrane is a central feature of crossflow filtration, we defer further discussion of the flux paradox to the section on crossflow.

Concentration Polarization and Precipitative Fouling As has been noted, precipitation of an inorganic scale next to the membrane can cause severe fouling of RO and NF membranes. Scale-forming species that are commonly found in the feed to such systems include calcium, magnesium, aluminum, and iron (primarily ferrous), any of which might precipitate as a hydroxide, carbonate, or sulfate solid, and also silica, which can form a solid directly or as a metal silicate. As described in Chapter 9, the condition necessary for a cation, Ct^{x+}, and an anion, An^{y-}, to precipitate as a solid $Ct_yAn_x(s)$ is that the solubility quotient, Q_{s0}, exceeds the solubility product, K_{s0}

$$Q_{s0} = a_{Ct}^y a_{An}^x = \left(\gamma_{Ct} \frac{c_{Ct}}{c_{Ct,std.state}}\right)^y \left(\gamma_{An} \frac{c_{An}}{c_{An,std.state}}\right)^x > K_{s0}$$
(15-139)

where a_i is the activity of species i, and $c_{i,std.state}$ is the concentration of i in the standard state. The approximation that activity coefficients are close to 1.0 is generally not reasonable for the conditions in the CP layer when precipitation is imminent, so values of γ_i must be estimated using correlations such as the Davies equation.[17]

Applying Equation 15-27 to compute the concentration of each rejected species in the bulk solution in a frontal filtration system (and dropping the subscript out), we have

$$c_c = c_f \frac{1 - r(1 - R)}{1 - r}$$
(15-27)

In addition, for each ion, the concentration at the membrane surface equals the product of the concentration in the bulk concentrate and the polarization factor, PF:

$$c_{c_{c/m}} = c_c(PF)$$
(15-140)

By combining the preceding equations with an expression for the PF (e.g., Equation 15-113), we can compute the ion activity product at the membrane surface in the absence of precipitation. If, for the given combination of feed concentrations, rejection, and recovery, $Q_{s0,c_{c/m}} > K_{s0}$, then scale formation is thermodynamically possible, and the system operation might have to be modified to prevent precipitation.

Strategies for avoiding precipitative scaling can focus on reducing the concentration of either the free anion or the free cation. For example, acid might be added to reduce the concentrations of bases such as hydroxide and carbonate; feed water might be pretreated by lime softening, precipitation, or ion exchange to remove scale-forming cations; or anti-scaling agents such as hexametaphosphate might be added to form complexes with metals and thereby impede precipitation or to poison crystal growth. Alternatively, the system can be operated at lower recovery, thereby concentrating the solutes less in the concentrate solution.

■ **EXAMPLE 15-17.** Spiral-wound NF membranes are being used to soften a groundwater that contains 4.8×10^{-3} M Ca^{2+}, 9.0×10^{-4} M SO_4^{2-}, and 7.8×10^{-3} M Cl^- as its major constituents. The system achieves 93% overall rejection of all

---
[17] The Davies equation and other approaches for estimating activity coefficients are discussed briefly in Chapter 9 and extensively in most water chemistry textbooks.

these ions in a process that approximates frontal filtration. Recirculation of concentrate to the inlet is sufficiently rapid that the concentrate can be considered to have uniform concentration along the entire length of the element. Neglecting the effects of other electrolytes, estimate the maximum recovery at which a system can operate without pretreatment and without any risk of scaling, if the polarization factor for all three ions is 1.4. Assume that gypsum ($CaSO_4 \cdot 2H_2O$) ($K_{s0} = 2.5 \times 10^{-5}$) controls calcium sulfate solubility and that activity coefficients can be estimated using the Davies equation.

Solution. The highest concentrations of Ca^{2+} and SO_4^{2-} in the system will be those at the membrane surface. At this location, the concentrations of all three ions can be expressed as $(PF)c_c$. In addition, the concentrations in the bulk concentrate can be expressed in terms of the feed concentration, the recovery, and the rejection via Equation 15-27. The concentrations at the surface can therefore be related to those in the feed by

$$c_{c_{c/m}} = (PF)\left(\frac{1-r(1-R)}{1-r}\right)c_f = (1.4)\left(\frac{1-r(1-0.93)}{1-r}\right)c_f$$

$$= \frac{1.4 - 0.098r}{1-r}c_f$$

When the concentrations of Ca^{2+} and SO_4^{2-} are just sufficient to precipitate gypsum, the solubility quotient will equal the solubility product for gypsum. Therefore, we want to ensure that Q_{s0} does not exceed the solubility product:

$$a_{Ca^{2+},c/m}a_{SO_4^{2-},c/m} = \left(\gamma_{Ca^{2+}}\frac{c_{Ca^{2+}}}{c_{Ca^{2+},std.}}\right)\left(\gamma_{SO_4^{2-}}\frac{c_{SO_4^{2-}}}{c_{SO_4^{2-},std.}}\right) < 2.5 \times 10^{-5}$$

By convention, $c_{Ca^{2+},std.} = c_{SO_4^{2-},std.} = 1.0\,\text{mol/L}$. According to the Davies equation, the activity coefficient of an ion depends only on the ionic strength of the solution and the absolute value of the valence of the ion of interest, so $\gamma_{Ca^{2+}} = \gamma_{SO_4^{2-}}$. Therefore, representing the activity coefficient of both these species as γ_{di}, we can write the criterion for preventing precipitation at the surface as follows:

$$\left[\gamma_{di}\left(\frac{1.4 - 0.098r}{1-r}\right)\left(\frac{c_{Ca^{2+},f}}{1.0\,\text{mol/L}}\right)\right]$$
$$\times \left[\gamma_{di}\left(\frac{1.4 - 0.098r}{1-r}\right)\left(\frac{c_{SO_4^{2-},f}}{1.0\,\text{mol/L}}\right)\right] < 2.5 \times 10^{-5}$$

Grouping the terms and substituting the given values for the feed concentrations, we find

$$\left(\frac{1.4 - 0.098r}{1-r}\right)^2 \gamma_{di}^2 (4.8 \times 10^{-3})(9.0 \times 10^{-4}) < 2.5 \times 10^{-5}$$

$$\left(\frac{1.4 - 0.098r}{1-r}\right)^2 \gamma_{di}^2 \leq 5.79$$

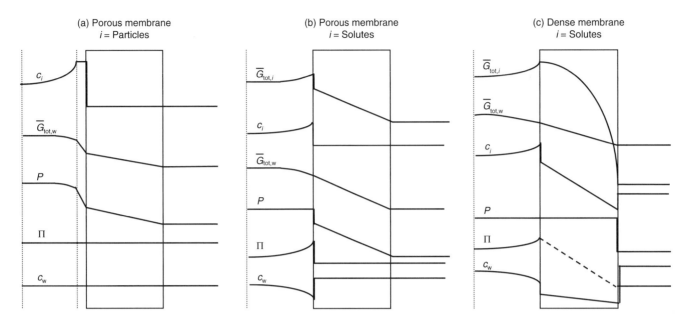

FIGURE 15-33. Profiles of various parameters of interest from the bulk concentrate to the bulk permeate, with a CP layer present. (a) Rejection of particles by a porous membrane, assuming a compact layer is present; (b) rejection of solutes by a porous membrane, assuming a compact layer is absent; (c) rejection of solutes by a dense membrane, in the absence of a compact layer. The broken lines on the concentrate side of the membrane indicate the edges of the CP layer and the compact layers.

The Davies equation, which relates γ_{di} to the ionic strength (I), is

$$\log \gamma_i = -0.51 z_i^2 \left(\frac{I^{1/2}}{1 + I^{1/2}} - 0.3 I \right)$$

Utilizing the expression derived earlier for $c_{c_c/m}$, the ionic strength at the concentrate/membrane interface can be computed as

$$I = 0.5 \sum c_i z_i^2$$
$$= 0.5 \left[c_{\text{Ca}^{2+},c/m}(2)^2 + c_{\text{SO}_4^{2-},c/m}(-2)^2 + c_{\text{Cl}^-,c/m}(-1)^2 \right]$$
$$= \frac{1.4 - 0.098r}{1-r} \left(2c_{\text{Ca}^{2+},f} + 2c_{\text{SO}_4^{2-},f} + 0.5 c_{\text{Cl}^-,f} \right)$$
$$= \frac{1.4 - 0.098r}{1-r} (0.0153)$$

The preceding equation for I, the Davies equation relating I to γ_{di}, and the equation expressing the criterion for preventing precipitation provide three equations in three unknowns (r, γ_{di}, and I). These equations can be solved simultaneously to determine the minimum value of r at which precipitation might occur; the result is $r = 0.81$. At this recovery, the concentrations of Ca^{2+} and SO_4^{2-} at the wall are 3.37×10^{-2} M and 6.32×10^{-3} M, respectively, or approximately seven times the feed concentration. ∎

Relating Parameter Profiles in the CP Layer with Those in the Membrane

The preceding sections establish several characteristics of the physical/chemical environment in the CP layer. By combining information about the CP layer with that derived previously about the conditions at the membrane/solution interfaces and inside the membrane, we can develop the profiles of key parameters from the bulk concentrate to the bulk permeate, as shown in Figure 15-33. The profile in part (a) is based on an assumption that a compact layer is present; if no such layer is present, the conditions shown in the CP layer would apply up to the membrane surface.

Consistent with the preceding discussion, the concentration of contaminant increases steadily from the bulk concentrate to either the limiting concentration at the edge of the compact layer (as in Figure 15-33a) or to some value less than c_{lim} at the membrane surface (as in Figures 15-33b,c). In systems where the contaminant is a solute (Figures 15-33b,c), the osmotic pressure increases in parallel with the contaminant concentration, and the hydrostatic pressure is constant from the bulk concentrate to the membrane surface. The combination of a uniform hydrostatic pressure and an increasing osmotic pressure causes $\overline{G}_{\text{tot},w}$ to decrease steadily from the bulk concentrate toward the membrane, drawing water in that direction.

In contrast to the case for water, the solute concentration gradient in the CP layer favors transport from the membrane toward the bulk concentrate. Since there is no pressure gradient to counteract this driving force, the gradient that develops in $\overline{G}_{\text{tot},i}$ suggests that net transport of solute should be toward the bulk solution, yet we know that the solute actually moves in the opposite direction. This apparent anomaly can be resolved by recognizing that $\overline{G}_{\text{tot},i}$ accounts for the available energy of i in isolation, but does not account for any coupling with other components of the system. In any functioning system, solute transport is partially coupled to transport of water, so that transport of i responds not only to the gradient in $\overline{G}_{\text{tot},i}$, but also to the gradient in $\overline{G}_{\text{tot,w}}$. The fact that the net transport of i is toward the membrane indicates that the latter factor is dominant.

Nonsteady-State Fouling During Frontal or Dead-End Filtration

Essentially all the discussion and analysis to this point has focused on steady-state systems. However, in reality, membrane systems are rarely operated under steady-state conditions. In this section, we consider some of the models that have been developed to describe the changes in the relationship between flux and TMP that occur as contaminants accumulate in frontal filtration membrane systems. Such a situation would arise in a system like that shown in Figure 15-29 if the rate at which contaminants exited in the permeate and concentrate streams ($Q_p c_p + Q_c c_c$) was less than the rate at which they entered in the feed ($Q_f c_f$). In that scenario, the contaminants might accumulate in the reservoir overlying the membrane, in the CP and compact layers adjacent to the membrane, and/or in the membrane pores; presumably, the resistance to permeation would steadily increase as well. The nonsteady-state scenario that has been most commonly analyzed is for the limiting case of constant-pressure, dead-end filtration with particulate contaminants, porous membranes, and complete capture of the incoming solids in a cake layer or within the membrane pores (i.e., negligible accumulation of particles in the CP layer, and negligible contribution of that layer to the total resistance to flow).

The starting point for most efforts to characterize the gradual development of fouling is work by Hermia (1982), who derived equations that he referred to as "blocking laws" to predict the decline in permeate flow during constant-pressure, dead-end filtration for four scenarios. In the simplest scenario, the resistance to flow is envisioned to reside entirely at the surface of the membrane. Rejected particles are envisioned to block access to some of the open ("active") membrane area, so that the reduction in the permeate flow rate is proportional to the area that is covered by the particles. Also, the amount of open membrane area that is blocked is assumed to be directly proportional to the volume of water filtered, implying that a given mass of foulants blocks the same amount of open area regardless of how much area has already been blocked.[18] This scenario is referred to as *complete blocking*.

In the second scenario, the resistance imparted by the foulants is again assumed to reside entirely at the surface of the membrane, and permeate flux is again envisioned to decline in proportion to the decline in the total open membrane area. However, in this case, some of the particles that reach the membrane are assumed to cover area that has already been covered by previously rejected particles. The flow pattern approaching the membrane is assumed to remain constant throughout the process. In such a case, if, say, 10% of the original pore area has been blocked, newly arriving particles are expected to cover only 90% as much active area as they would if they were approaching a clean membrane. Hermia referred to this scenario as *intermediate blocking*.

The third scenario attributes most of the resistance imparted by the foulants not to accumulation of particles on the surface of the membrane, but to attachment of the particles to the interior surfaces of the pores. The particles are envisioned to coat the pores and interfere with flux by steadily narrowing the effective pore diameter. This scenario is referred to as *standard blocking* of the pores.

Finally, Hermia considered that the particles might form a steadily growing cake layer on the membrane surface, thereby increasing the resistance to permeation without any specific interactions with the membrane pores. This scenario was referred to as *cake filtration*.

Hermia analyzed each of these scenarios to obtain expressions for the expected decline of the permeation flow rate over time during constant-pressure filtration; that is, he developed expressions for dQ_p/dt. For example, for the complete blocking model, the derivation begins by defining a term that we refer to as the "active flux," $\hat{J}_{\text{V,active}}$. This quantity is defined as the ratio of the permeate flow rate to the area of the membrane that is actually open (as opposed to the conventional definition of flux, in which the flow is normalized to the surface area of the whole membrane. According to the complete blocking model, $\hat{J}_{\text{V,active}}$ remains constant as a filtration run proceeds, because the flow through the area that is still active (i.e., open) is unaffected by the rejected particles; all that changes during the run is that the total amount of active area declines. Thus, we can write, at any time during the run

$$\hat{J}_{\text{V,active}} = \frac{Q_p}{A_{\text{active}}} \quad (15\text{-}141)$$

Based on the underlying concept of the complete blocking model, the active area remaining declines steadily with the

---

[18] Hermia suggested that every rejected particle blocked an area equal to its projected cross-section. However, that constraint is overly restrictive; the analysis applies as long as a given amount of rejected solids blocks a fixed amount of area, regardless of the relationship to the cross-sectional area of the rejected particles.

cumulative volume filtered (V_{cum}), that is

$$A_{\text{active},t} = A_{\text{active},0} - \sigma V_{\text{cum}} \quad (15\text{-}142)$$

where σ is the active area blocked per unit volume of permeate. Multiplying by $\hat{J}_{V,\text{active}}$, noting that, at any time, $\hat{J}_{V,\text{active}} A_{\text{active}} = Q_p$, and differentiating with respect to time, we obtain

$$Q_p = Q_{p,0} - \hat{J}_{V,\text{active}} \sigma V_{\text{cum}} \quad (15\text{-}143)$$

$$\frac{dQ_p}{dt} = 0 - \hat{J}_{V,\text{active}} \sigma \frac{dV_{\text{cum}}}{dt} = -\hat{J}_{V,\text{active}} \sigma Q_p \quad (15\text{-}144)$$

Rearranging and integrating:

$$\int_{Q_{p,0}}^{Q_{p,t}} \frac{dQ_p}{Q_p} = -\hat{J}_{V,\text{active}} \sigma \int_0^t dt \quad (15\text{-}145)$$

$$\ln \frac{Q_{p,t}}{Q_{p,0}} = -\hat{J}_{V,\text{active}} \sigma t \quad (15\text{-}146)$$

$$Q_{p,t} = Q_{p,0} \exp(-\hat{J}_{V,\text{active}} \sigma t) \quad (15\text{-}147)$$

Hermia chose to report the same result in a slightly different format by defining the product $\hat{J}_{V,\text{active}} \sigma$ in Equation 15-143 as K_b, and then inverting the equation to obtain

$$\frac{1}{Q_p} = \frac{1}{Q_{p,0} - K_b V_{\text{cum}}} \quad (15\text{-}148)$$

He then differentiated Equation 15-148 with respect to V_{cum} to obtain

$$\frac{d}{dV_{\text{cum}}} \frac{1}{dV_{\text{cum}}/dt} = \frac{d}{dV_{\text{cum}}} \frac{1}{Q_{p,0} - K_b V_{\text{cum}}} \quad (15\text{-}149)$$

$$\frac{d^2 t}{dV_{\text{cum}}^2} = K_b \left(\frac{1}{Q_{p,0} - K_b V_{\text{cum}}} \right)^2 = \frac{K_b}{Q_p^2} \quad (15\text{-}150)$$

$$\frac{d^2 t}{dV_{\text{cum}}^2} = K_b \left(\frac{dt}{dV_{\text{cum}}} \right)^2 \quad (15\text{-}151)$$

The advantage of expressing the result as in Equation 15-151 is that the other fouling mechanisms modeled by Hermia lead to expressions with similar forms. The derivation of the cake filtration model is presented next, because it is widely used and has relevance for one of the standard tests for characterizing membranes. The final results from the analyses of the intermediate blocking and standard blocking models are presented as well, but the derivations are not shown; those derivations were presented by Hermia (1982).

The derivation of the cake filtration model begins with a rearrangement of Equation 15-136 for a system that rejects particles only (so that $\Delta_{c/p} \Pi = 0$)

$$\hat{J}_V = \frac{\Delta_{c/p} P}{\mu \left(\mathscr{R}_{m_{c/p}} + \mathscr{R}_{\text{CP}} + \mathscr{R}_{\text{cake}} \right)} \quad (15\text{-}152)$$

Writing the flux as $(dV_{\text{cum}}/dt)/(A_m)$, assuming that $\mathscr{R}_{\text{CP}}$ is negligible compared with $\mathscr{R}_{m_{c/p}}$ and $\mathscr{R}_{\text{cake}}$, multiplying by A_m, and inverting both sides yields

$$\frac{dt}{dV_{\text{cum}}} = \frac{\mu \mathscr{R}_{m_{c/p}}}{A_m \Delta_{c/p} P} + \frac{\mu \mathscr{R}_{\text{cake}}}{A_m \Delta_{c/p} P} = \frac{\mu \mathscr{R}_{m_{c/p}}}{A_m \Delta_{c/p} P} + \frac{\mu \hat{\mathscr{R}}_{\text{cake}} \delta_{\text{cake}}}{A_m \Delta_{c/p} P} \quad (15\text{-}153)$$

Assuming that all the particles fed to the membrane are captured, we can write the following mass balance:

Mass of particles fed up to time t = Mass of particles deposited in the cake

The mass concentration of particles, in both the feed and the cake, can be expressed as the product of their volume fraction (ϕ_f and ϕ_{cake}, respectively) and the particle density (ρ_p), so the mass balance can be written as

$$V_{\text{cum}} \phi_f \rho_p = A_m \delta_{\text{cake}} \phi_{\text{cake}} \rho_p \quad (15\text{-}154)$$

$$\delta_{\text{cake}} = \frac{V_{\text{cum}} \phi_f}{A_m \phi_{\text{cake}}} \quad (15\text{-}155)$$

Substitution of Equation 15-155 into Equation 15-153 yields

$$\frac{dt}{dV_{\text{cum}}} = \frac{\mu \mathscr{R}_{m_{c/p}}}{A_m \Delta_{c/p} P} + \frac{\mu \hat{\mathscr{R}}_{\text{cake}} V_{\text{cum}} \phi_f}{A_m^2 \Delta_{c/p} P \phi_{\text{cake}}} \quad (15\text{-}156)$$

$$dt = \left(\frac{\mu \mathscr{R}_{m_{c/p}}}{A_m \Delta_{c/p} P} + \frac{\mu \hat{\mathscr{R}}_{\text{cake}} V_{\text{cum}} \phi_f}{A_m^2 \Delta_{c/p} P \phi_{\text{cake}}} \right) dV_{\text{cum}} \quad (15\text{-}157)$$

Assuming that the properties of the cake ($\hat{\mathscr{R}}_{\text{cake}}$ and ϕ_{cake}) are constant, we can integrate Equation 15-157 from time zero to time t, and then divide by V_{cum} to obtain

$$\frac{t}{V_{\text{cum}}} = \frac{\mu \mathscr{R}_{m_{c/p}}}{A_m \Delta_{c/p} P} + \frac{\mu \hat{\mathscr{R}}_{\text{cake}} \phi_f}{2 A_m^2 \Delta_{c/p} P \phi_{\text{cake}}} V_{\text{cum}} \quad (15\text{-}158)$$

Equation 15-158 suggests that a plot of t/V_{cum} versus V_{cum} should yield a straight line, from which both $\mathscr{R}_{m_{c/p}}$ and $\hat{\mathscr{R}}_{\text{cake}}$ can be determined (from the intercept and slope, respectively), assuming that the other parameters are known. Finally, differentiating Equation 15-156 with respect to V_{cum}

$$\frac{d^2 t}{dV_{\text{cum}}^2} = \frac{\mu \hat{\mathscr{R}}_{\text{cake}} \phi_f}{2 A_m^2 \Delta_{c/p} P \phi_{\text{cake}}} = K_c \quad (15\text{-}159)$$

TABLE 15-10. Hermia's Models for Nonsteady-State Fouling in Frontal Filtration

Model	d^2t/dV^2_{cum}	Q_p
Complete blocking	$K_b \left(\dfrac{dt}{dV_{cum}}\right)^2$	$Q_{p,t} = Q_{p,0} \exp(-K_b t)$
Intermediate blocking	$K_i \left(\dfrac{dt}{dV_{cum}}\right)^1$	$Q_{p,t} = Q_{p,0} \exp(-K_i V_{cum})$
Standard blocking	$K_s \left(\dfrac{dt}{dV_{cum}}\right)^{3/2}$	$Q_{p,t} = \dfrac{Q_{p,0}}{(1 + 0.5 K_s Q_{p,0} t)^2}$
Cake filtration	$K_c \left(\dfrac{dt}{dV_{cum}}\right)^0 = K_c$	$Q_{p,t} = \dfrac{Q_{p,0}}{(1 + 2 K_c Q_{p,0}^2 t)^{1/2}}$

Like the result for the complete blocking model (Equation 15-151), Equation 15-159 can be considered to have the following form, albeit with the trivial result that $n = 0$:

$$\frac{d^2 t}{dV^2_{cum}} = K_c \left(\frac{dt}{dV_{cum}}\right)^n \qquad (15\text{-}160)$$

As shown in Table 15-10, the results for the intermediate blocking and standard blocking models can be expressed in the same form, with each fouling scenario having a different value of n. Thus, if one can evaluate n from experimental data, one can determine if the system behavior is consistent with any of the model scenarios.

To evaluate n for a given system, the cumulative filtrate volume, V_{cum}, is measured over time. The reciprocals of the first and second derivatives of this function are then approximated numerically, and their logarithms are plotted against one another. The slope that best approximates a fit to the data provides the value of n.

■ **EXAMPLE 15-18.** A laboratory UF test on a suspension is performed in a cell with a membrane area of 15 cm² at a TMP of 20 kPa, and the permeate volume is measured every 10 s; a subset of the data is shown in the first two columns of the following table. The high frequency of the raw data collection allows good estimates to be made of the instantaneous flow rate $(Q_p = (dV)/(dt))$, and from these data, of the instantaneous values of $(dV)/(dt)$, $(dt)/(dV)$, and $(d^2t)/(dV^2)$. These estimates are shown in the final three columns of the following table. (Note that the values in these final columns are approximate, instantaneous values that apply at the time indicated, based on data collected at 10-s intervals around the time indicated in the first column; they are not derived directly from the data in the second column, which shows cumulative values since the beginning of the test.)

(a) Convert the data into values of flux (in L/m² h) and plot the results as a function of time.
(b) Determine whether the data fit one of the Hermia models for fouling.

Experimental Data For Flux Decline in Ultrafiltration

Time (min)	V_{cum} (cm³)	dV/dt (cm³/min)	dt/dV (min/cm³)	d^2t/dV^2 (min/cm⁶)
0	0	1.60	0.63	0.012
5	7.5	1.40	0.71	0.016
10	13.9	1.20	0.83	0.020
15	19.5	1.04	0.96	0.025
20	24.4	0.90	1.11	0.033
25	28.7	0.79	1.27	0.046
30	32.2	0.70	1.43	0.060
35	35.5	0.60	1.67	0.077
40	38.4	0.52	1.92	0.101
45	40.8	0.45	2.22	0.140
50	42.8	0.39	2.56	0.185
55	44.6	0.34	2.94	0.241
60	46.1	0.30	3.33	0.320

Solution. Relevant calculations for both parts of the problem are summarized in the following table.

Time (min)	Flow (mL/min)	Flux (L/(m² h))	ln (dt/dV)	ln (d²t/dV²)
0	1.60	64.0	−0.47	−4.42
5	1.40	56.0	−0.34	−4.14
10	1.20	48.0	−0.18	−3.91
15	1.04	41.6	−0.04	−3.69
20	0.90	36.0	0.11	−3.41
25	0.79	31.6	0.24	−3.08
30	0.70	28.0	0.36	−2.81
35	0.60	24.0	0.51	−2.56
40	0.52	20.8	0.65	−2.29
45	0.45	18.0	0.80	−1.97
50	0.39	15.6	0.94	−1.69
55	0.34	13.6	1.08	−1.42
60	0.30	12.0	1.20	−1.14

(a) The permeate flux is the flow rate (given in the third column in the data table) divided by the membrane area. For example, the first value is

$$\hat{J}_V = \frac{Q_p}{A} = \left(\frac{1.60 \text{ cm}^3/\text{min}}{15 \text{ cm}^2}\right)\left(\frac{1 \text{ L}}{1000 \text{ cm}^3}\right)(10^4 \text{ cm}^2/\text{m}^2)$$
$$\times \left(\frac{60 \text{ min}}{1 \text{ h}}\right) = 64 \text{ L/m}^2 \text{ h}$$

The flux throughout the test is given in the third column of the table and is plotted in Figure 15-34.

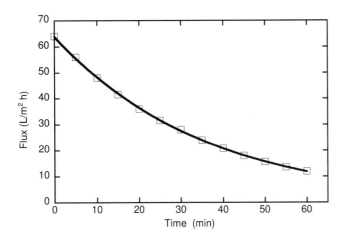

FIGURE 15-34. Flux decline in the example system.

(b) The fit of the data to the Hermia models can be tested by plotting values of $\ln((d^2t)/(dV^2))$ versus $\ln((dt)/(dV))$ (or, identically, $\log_{10}$ values of each). The values are shown in the previous table and are plotted in Figure 15-35. The data fall on a straight line with a slope of approximately 2.0 (regression analysis yields a value of 1.97). This result indicates that the complete blocking model is appropriate for these data. Another indication of the appropriateness of that model (not shown) is that the flow (or flux) data exhibit an exponential decline (i.e., a plot of $\ln Q$ versus t is a straight line). ■

Hermia's model equations are based on the assumption that a single fouling mechanism dominates in a given system. Ho and Zydney (2000) argued that a more like scenario, at least for the systems they studied involving MF of proteins, consists of a gradual transition from pore blockage by the

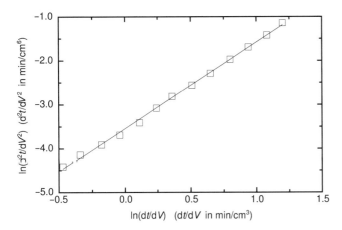

FIGURE 15-35. Hermia analysis of the data from the example system.

TABLE 15-11. Predicted Nonsteady-State TMP Buildup in Constant-Flux Filtration[a]

Compressibility of Deposited Foulant	Deposit Compressibility Equation	Nonsteady-State Expression for $\Delta_{c/p} P$ (i.e., TMP)
Incompressible	$\alpha_{cake} = k_0$	$\dfrac{d^2(\Delta_{c/p}P)}{dV_{cum}^2} = 0 \left(\dfrac{d\Delta_{c/p}P}{dV_{cum}}\right)^n = 0$
Linear compressibility	$\alpha_{cake} = \alpha_{cake}^o \times (1 + k_1 \Delta_{c/p} P)$	$\dfrac{d^2(\Delta_{c/p}P)}{dV_{cum}^2} = K_1 \left(\dfrac{d\Delta_{c/p}P}{dV_{cum}}\right)^{3/2}$
Power law compressibility	$\alpha_{cake} = k_{n_1} (\Delta_{c/p} P)^{n_2}$	

[a] α_{cake} and α_{cake}^o are the specific resistance of the cake at any TMP and at the limit of zero TMP, respectively; k_o is a composite of various measurable parameters; and k_1, k_n, n_1, n_2, and K_n are empirical fitting parameters (K_1 is a specific combination of k_1 and various measurable parameters).

Source: After Chellam and Xu (2006).

contaminants that reach the bare membrane surface to cake formation as the surface becomes covered and multiple layers of foulants accumulate; they presented a model that fit their data reasonably well for that entire process.

Hermia's equations apply exclusively to constant-pressure filtration, but a similar analysis can be carried out for constant-flux filtration. The analysis for incompressible cakes was presented by Hlavacek and Bouchet (1993) and Suarez and Veza (2000); Chellam and Xu (2006) later extended that analysis to compressible cakes. In an effort to emphasize the parallelism between these systems and the constant-pressure systems considered by Hermia, the equations have been expressed as relationships with the following general form:

$$\frac{d^2(\Delta_{c/p}P)}{dV_{cum}^2} = K \left(\frac{d\Delta_{c/p}P}{dV_{cum}}\right)^n \quad (15\text{-}161)$$

However, not all the cases of interest can be fit to such an equation. The key results of the analysis are summarized in Table 15-11; the derivations and some experimental verification are provided by Chellam and Xu (2006) and the references cited therein.

Empirical Measures of Fouling: The MFI and SDI Tests A commonly reported empirical measure of the contribution of cake or gel layers to the overall hydraulic resistance in membrane systems is based on Hermia's cake filtration equation. This measure, known as the *modified fouling index* (MFI), is computed by measuring the volumetric flux at the beginning and end (and, in some cases, at intermediate times) of a period of constant-pressure, dead-end filtration. Filtrate volume is recorded every 30 s over a 15-min period. The data are then fitted to Equation 15-158, and the MFI is defined as the fraction (assumed to be constant) in the second term on the right side of the equation. The left side of the equation (i.e., the ratio t/V_{cum}) is the inverse of the average permeate flow rate, $\overline{Q}_p$, since the beginning of the test, and the first term on the right is the inverse of the initial flow rate (i.e., the flow rate when V_{cum} is negligible). Therefore, since the MFI is specifically for a test duration of 15 min, Equation 15-158 can be rewritten in terms of the MFI as follows:

$$\frac{1}{\overline{Q}_{p,15\,\text{min}}} = \frac{1}{Q_{p,0}} + (\text{MFI})V_{\text{cum},15\,\text{min}} \quad (15\text{-}162)$$

A similar parameter, known as the *silt density index* (SDI), is also commonly used to evaluate the potential for a given feed stream to foul membranes (especially RO membranes). The experimental setup for determining the SDI is the same as that for the MFI test. The fractional decline in the flux from the initial value is expressed as a percentage and divided by the time interval (Δt) to obtain the SDI, as follows:

$$\text{SDI} = \frac{\left(1 - (\hat{J}_{V,t}/\hat{J}_{V,0})\right)100\%}{\Delta t} \quad (15\text{-}163)$$

The SDI can thus be thought of as a standardized way to express the rate of flux decline for the particular feed solution of interest.

The ASTM standard for this test (ASTM, 2007) specifies use of a membrane with 0.45-μm-diameter pores in an unstirred, dead-end, batch filtration cell under a constant pressure of 207 kPa.[19] The standard conditions for the test include a 47-mm diameter filter and a fixed volume of 500 mL for each measurement; if a membrane of different size is used, the volume is adjusted proportionally to the membrane area. With the volume of water and the area of the membrane specified, the only variation when measuring the flux is the time required for filtration of the given volume of water at the beginning of the test ($t_{\text{filt},0}$) and after time Δt has elapsed ($t_{\text{filt},t}$). Hence, the SDI is usually calculated as follows:

$$\text{SDI} = \frac{\left(1 - (t_{\text{filt},0}/t_{\text{filt},t})\right)100\%}{\Delta t} \quad (15\text{-}164)$$

where Δt is in minutes and $t_{\text{filt},0}$ and $t_{\text{filt},t}$ are in identical units, usually seconds. According to Equation 15-164, SDI has formal units of min^{-1}; by convention, however, it is always reported as a dimensionless number.

In the standard test, Δt is set at 15 min. If no fouling of the membrane occurred during this period, the time to filter 500 mL would not change during the test (i.e., $t_{\text{filt},t}$ would

equal $t_{\text{filt},0}$) and the value of the SDI would be zero. Generally, some fouling does occur, so that $t_{\text{filt},t} > t_{\text{filt},0}$ and the SDI is greater than zero. If the membrane became completely clogged during the test, $t_{\text{filt},t}$ would be infinite, and the SDI would be $(100/15) = 6.67$; however, such a test is not considered valid. Rather, in such a case, a shorter Δt is used (10 min, or, if necessary, 5 min), and higher values of SDI can, therefore, be obtained. In performing the test, analysts typically obtain the data for $t_{\text{filt},t}$ after each of these periods (values of Δt equal to 5, 10, and 15 min), and then use the data for the longest Δt that gives a value of $\left(1 - (t_{\text{filt},0})/(t_{\text{filt},t})\right) < 0.75$ (i.e., the longest t for which the flux after Δt is at least 25% of the flux through the clean membrane). The SDI should be reported with Δt indicated (e.g., SDI$_{10}$), but that is rarely done.

Because the test membrane is specified to have 0.45-μm pores, the SDI test evaluates (roughly) the likelihood that particles or relatively large colloids in the feed will foul a membrane, but it does not address the possibility that smaller colloids or solutes will do so. Nevertheless, historically, the SDI test has been used to investigate the potential for a given feed to foul all types of pressure-driven membrane systems, including RO systems. In general, SDI values in excess of ~3.5 are considered indicative of a high potential for fouling of outside-in hollow fiber MF and UF membranes and spiral-wound NF or RO membranes. While the SDI can be a useful qualitative indicator of the need for pretreatment, it is an inadequate indicator of membrane performance for design or control purposes, because the conditions used in the test do not resemble those during membrane operation. SDI values less than ~2 do not correlate well with membrane fouling observed in practice.

■ **EXAMPLE 15-19.** In a standard SDI test, the time to filter 500 mL at time 0 was 31 s. After 5, 10, and 15 min, the times to filter this volume were 40, 53, and 70 s, respectively. Find the value of the SDI.

Solution. The SDI should be calculated using the longest time for which the flux is at least 25% of the initial flux; that is, the longest time for which $\left(1 - (t_{\text{filt},0}/t_{\text{filt},t})\right) \leq 0.75$. Therefore, we begin by testing the value at 15 min against this criterion:

$$\left(1 - (t_{\text{filt},0}/t_{\text{filt},15})\right) = (1 - (31\,\text{s}/70\,\text{s})) = 0.56$$

Since the criterion is met, we can use the data from the 15-min time, and the data from the earlier times can be ignored. The SDI is found as follows:

$$\text{SDI}_{15} = \frac{\left(1 - (t_{\text{filt},0}/t_{\text{filt},t})\right)100\%}{\Delta t} = \frac{(1 - (31\,\text{s}/70\,\text{s}))100\%}{15\,\text{min}} = 3.8 \quad ■$$

---

[19] Membranes with nominal pore size of 0.45 μm are widely available. Depending on the manufacturer, all the pore openings might actually be close to that diameter, or they might be widely distributed around that value.

Summary: Modeling Membrane Performance in Frontal Filtration

Our analysis of frontal filtration is now complete. Although the analysis was restricted to idealized scenarios (e.g., a feed containing either a single solute or a monodisperse particle population, and uniform pores if the membrane is porous), the results provide a useful, quantitative framework for predicting the effects of several parameters on system performance and for understanding the behavior of more realistic systems.

The key steps were to derive separate expressions for transport of water and contaminants through the membrane and through the CP layer (and the compact layer, if it is present), and then to link those expressions via various physical/chemical constraints or assumptions (e.g., equal flux of solution throughout the system, TMP equal to the sum of the pressure losses through different regions, continuity of concentrations in the aqueous phase, and equilibrium across phase boundaries). For transport across the solution/membrane boundaries and through the membrane, we considered two models—the pore-flow and solution-diffusion models—that correspond to limiting cases of strong and negligible coupling, respectively, between water and the contaminants. The key practical difference between these models is that, at least for the limiting case of perfect coupling between contaminants and water, the pore-flow model predicts that rejection efficiency is independent of the TMP differential ($\Delta_{m_{c/p}} P$) and water flux, whereas the solution-diffusion model predicts that rejection increases when those parameters increase.

The analysis of transport across the CP layer showed that the relationship between flux and the concentration profile has the same form for particles and solutes, but the different types of contaminants exert their effects on water flux via two different mechanisms. Particles exert a drag force on water passing around them and cause a loss of pressure that reduces the driving force for solution transport across the membrane. Accumulation of solutes in the CP layer, on the other hand, causes the chemical activity of water to decline as it approaches the membrane, thereby reducing the driving force for water transport via the effect on $\overline{G}_{chem,w}$. Regardless of the mechanism, increasing polarization leads to increasing resistance and decreased water flux.

The maximum contaminant concentration achievable in the CP layer is limited by the physical/chemical properties of the contaminant (e.g., the shape, surface charge, and compressibility of particles, the tendency for macromolecules to coagulate into a gel, and the tendency for molecular solutes to precipitate as solids). When the resistances of the CP layer, the compact layer (if present), and the membrane are all taken into account, the permeate flux at low TMP is predicted to be small, and the CP layer is predicted to extend to the membrane surface. At the low end of this range of TMPs, the CP layer contributes negligibly to the overall resistance, so the total resistance to permeation is approximately equal to the (constant) resistance imposed by the membrane alone, and $\hat{J}_V$ increases in direct proportion to the TMP. As the TMP and flux increase, the CP layer becomes thicker and the concentration of rejected species near the wall increases. At some point, the resistance of this layer becomes significant compared with that of the membrane, and the overall resistance increases significantly with increasing TMP. As a result, an increase in TMP still leads to an increase in flux, but the increase is less than proportionate (i.e., the specific flux, $\hat{J}_V/\text{TMP}$, decreases). The exact TMP at which this transition occurs and the relationship between $\hat{J}_V$ and TMP in this transitional region depend on the properties of the rejected species and on the operational details of the system. The system is said to be mass transport-limited under these conditions.

If the TMP is increased further, the limiting concentration of rejected species is reached near the membrane surface, and a compact layer of uniform composition is assumed to form. From this point onwards, increasing the TMP leads to an increase in the thickness of the compact layer (and, if the layer is compressible, its density), and a corresponding increase in the hydraulic resistance. Under these conditions, the CP layer has a fixed thickness and concentration profile, so its contribution to the overall resistance is fixed, independent of the TMP. When the TMP is increased further, the thickness (and, possibly, the density) of the compact layer increases just enough to balance the increased driving force for permeation, so the permeation flux remains constant, independent of TMP; this maximum flux is known as the limiting flux.

Nonsteady-state models have been developed to predict the performance of frontal filtration systems, and these models can, in principle, be used to distinguish among different mechanisms of fouling, based on the pattern of flux decline over time. They also form the basis for commonly used tests to characterize the fouling tendency of a given influent to membrane systems.

15.9 EFFECTS OF CROSSFLOW ON PERMEATION AND FOULING

General Considerations in Modeling Fluid Flow and Particle Transport in Crossflow Filtration

In crossflow filtration systems, the fluid has significant velocity components in both the axial and transverse directions (parallel and perpendicular to the membrane, respectively), and both velocity components vary as a function of location within the system. Based on the conventions used in fluid dynamics, the axial coordinate is designated as x and increases from the inlet to the outlet of the membrane element, and the axial velocity is designated as u. If the feed channel

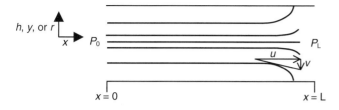

FIGURE 15-36. System definition diagram for flow through a channel bounded by membranes (either above and below the channel or forming a cylindrical channel).

can be modeled as a thin slit (e.g., the region between two flat sheets in a plate-and-frame system, or between adjacent layers of a spiral-wound membrane), the distance in the transverse direction is represented by y or h and might be measured either from the membrane surface or the centerline of the channel; if the channel is cylindrical (hollow fiber or tubular membranes with inside-out flow), transverse distances are almost always measured from the centerline and represented by r. In either case, transverse velocities are represented by v. These coordinate definitions are shown for a cylindrical channel in Figure 15-36.

In virtually all membrane systems with crossflow, the axial velocity (u) declines steadily from the entrance to the exit (increasing x) as water is lost from the feed channel by permeation. The axial velocity also decreases as the membrane surface is approached, due to the shear force that the membrane exerts on the fluid; this factor is important only in the laminar boundary layer if the bulk flow is turbulent, but throughout the whole channel if the flow is laminar. The velocity toward the membrane (i.e., the transverse velocity, v) varies as a function of the axial location because the TMP declines with increasing x, due to friction-induced pressure loss in the feed channel (from P_0 to P_L in the figure); in systems with cylindrical feed channels, v also varies in the radial direction due to the increasing flow cross-section as the membrane is approached.

The equations describing flow of pure water in such systems, though complicated, can be solved. However, including the interactions of colloids and particles with the fluid and with one another in such an analysis is extremely difficult, largely because of our incomplete understanding of the physical/chemical state of concentrated suspensions, in which the particles and water molecules experience significant short-range forces due to electrostatic and van der Waals interactions. As a result, several simplifications are frequently applied in modeling the fluid dynamics in crossflow systems. First, the axial velocity is modeled without considering the contaminants at all; that is, the axial velocity profile is modeled as that for flow of pure water. The axial pressure loss, which depends primarily on the axial velocity profile, is typically also computed based on flow of pure water. In addition, the effects of permeation on the axial velocity profile and of the abrupt transition in flow conditions at the entrance to the membrane element are frequently ignored.

Whereas the presence of contaminants is generally ignored in modeling axial flow in systems with crossflow, the accumulation of contaminants in both a CP layer and (possibly) a compact layer must be taken into account when modeling flow in the transverse direction. Typically, the transverse velocity at a given location and time is assumed to correspond to the steady-state velocity that would be found for frontal filtration at that location, considering the instantaneous, local values of the bulk concentration, the TMP, the contaminant concentration profile through the CP layer, and the compact layer thickness. Depending on the sophistication of the model, variations in any of these parameters as a function of axial location and/or time might be included.

Both axial and transverse fluid flow are central to the modeling of contaminant transport in crossflow filtration systems. The transverse fluid flow is important for the obvious reason that that flow carries the contaminants toward the membrane. The role of the axial flow profile, especially very close to the membrane, is more subtle. Although the flow near the membrane (or near the compact layer, if such a layer is present) accounts for only a small portion of the total axial flow rate, water at that location has the highest concentration of contaminants, so it might account for a significant fraction of the contaminant transport from the inlet to the exit of the element. In addition, the fluid shear rate in this region (du/dy or du/dr, commonly represented as γ_{Sh}) often controls the transport rate of colloids and particles away from the membrane, via mechanisms that are described shortly.

In this section, models that have been proposed to describe both fluid flow and contaminant transport in crossflow filtration are presented, utilizing the prior analysis of frontal filtration as the foundation. The analysis emphasizes steady-state operation, even though steady state is almost never achieved in practice; such models have dominated the literature, because steady-state systems are easier to study experimentally and to model mathematically.

Fluid Flow in Crossflow Filtration

Typically, in systems with crossflow, anywhere from <10% to ~30% of the fluid entering a membrane element passes through the membrane and exits as permeate, and the remainder exits as concentrate. The concentrate may then be recycled to the module entrance for additional passes at the membrane. In such systems, expressions for flow within impervious channels are often reasonably accurate for describing the axial flow.

The flow inside hollow fiber membranes is almost always laminar (indicated by Reynolds number <2100, for flow with a circular cross-section), and the velocity profile within each fiber can be approximated as parabolic, as shown in Figure 15-37. In such cases, the axial velocity of the fluid, u,

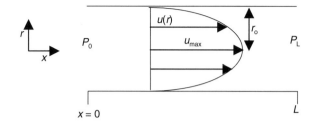

FIGURE 15-37. The velocity profile in an impervious, cylindrical tube with laminar flow (Poiseuille flow).

is a function of the radial distance, r, from the center of the tube is

$$u(r) = u_{max}\left[1 - \left(\frac{r}{r_o}\right)^2\right] = 2\bar{u}\left[1 - \left(\frac{r}{r_o}\right)^2\right] \quad (15\text{-}165)$$

where r_o is the radius of the tube, u_{max} is the maximum velocity (which occurs at the center of the tube [$r=0$]), and $\bar{u}$ is the average velocity, defined as Q/A; as indicated, $\bar{u} = u_{max}/2$. u_{max} can be expressed as a function of other system parameters as follows:

$$u_{max} = \frac{(P_0 - P_L)r_o^2}{4\mu_L L} \quad (15\text{-}166)$$

where P_0 and P_L are the pressures at the inlet ($x=0$) and exit ($x=L$) of the element, respectively.

If the flow channel can be represented as the space between two parallel sheets, the flow is again laminar in most circumstances of interest. Defining the height of the opening as $2h_o$, the axial velocity is

$$u(h) = u_{max}\left[1 - \left(\frac{h}{h_o}\right)^2\right] = \frac{3}{2}\bar{u}\left[1 - \left(\frac{h}{h_o}\right)^2\right] \quad (15\text{-}167)$$

where h is the distance in the transverse direction measured from the centerline, $\bar{u}$ is $\tfrac{2}{3}u_{max}$, and u_{max} can be found as

$$u_{max} = \frac{(P_o - P_L)h_o^2}{2\mu_L L} \quad (15\text{-}168)$$

The fact that the axial distance, x, does not appear in Equation 15-165 (for cylindrical membranes) or Equation 15-167 (for spiral-wound membranes) implies that the axial velocity, u, at a given distance from the centerline is constant throughout the membrane length. We know, however, that fluid is permeating through the membrane, so the average velocity of the fluid must decrease with distance x. If we assume that the permeation velocity, v_w, does not change with distance along the membrane, then the concentrate flow rate decreases linearly with x. In that case, designating the cross-sectional area available for flow as

A_{cross} and the membrane surface area between the inlet and location x as $A_{m,x}$, the average crossflow velocity at any axial distance from the inlet can be computed as follows:

$$Q_x = Q_{in} - \hat{J}_V A_{m,x} \quad (15\text{-}169)$$

Substituting an expression for $A_{m,x}$ for cylindrical membranes with diameter d_o, we can develop an expression for the average velocity at location x ($\bar{u}_x$), as follows:

$$Q_x = Q_{in} - \hat{J}_V(\pi d_o x) \quad (15\text{-}170)$$

$$\bar{u}_x = \frac{Q_x}{A_{cross}} = \frac{Q_{in} - \hat{J}_V(\pi d_o x)}{A_{cross}} = \bar{u}_{in} - \frac{\hat{J}_V(\pi d_o x)}{\pi(d_o^2/4)}$$

$$= \bar{u}_{in} - \frac{4\hat{J}_V x}{d_o} \quad (15\text{-}171)$$

The corresponding expressions for flow rate and average velocity in membrane systems with rectangular channels and permeate flow through two membranes (one on each side of the channel) are

$$Q_x = Q_{in} - \hat{J}_V A_{m,x} = Q_{in} - \hat{J}_V(2Wx) \quad (15\text{-}172)$$

$$\bar{u}_x = \frac{Q_x}{A_{cross}} = \frac{Q_{in} - 2\hat{J}_V Wx}{2Wh} = \bar{u}_{in} - \frac{\hat{J}_V x}{h} \quad (15\text{-}173)$$

where W is the width of the channel (assumed sufficiently large that it does not affect the flow distribution). Even when the reduction in axial flow rate due to permeation is taken into account, the axial flow profile is usually assumed to be the one that would apply for the given, local axial flow rate in an impermeable channel. A review of the hydrodynamics of flow through porous channels has been provided by Chellam et al. (1995).

The shear rate at any point in a system with crossflow can be found by differentiating the axial velocity at that point with respect to distance in the transverse direction. When this differentiation is carried out for an impermeable channel with laminar flow, the resulting expressions for the shear rate at the channel wall are

Rectangular cross-section

$$\gamma_{Sh,wall} = \frac{3Q}{2Wh_o^2} = \frac{3\bar{u}}{h_o} \quad (15\text{-}174)$$

Circular cross-section

$$\gamma_{Sh,wall} = \frac{4Q}{\pi r_o^3} = \frac{4\bar{u}}{r_o} = \frac{8\bar{u}}{d_o} \quad (15\text{-}175)$$

The shear rate decreases with distance from the wall. Furthermore, permeation flow induces some slip at the wall, which reduces the shear rate compared with that in an impermeable channel. As a result, the actual shear rate in the CP layer of a system with crossflow varies with location and is always less than that given by Equations 15-174 and 15-175. Nevertheless, because the CP layer is very thin, the shear rate is almost universally assumed to be constant and equal to the value calculated with Equations 15-174 or 15-175 throughout the layer.

■ **EXAMPLE 15-20.** Compute the average axial fluid velocities at the midpoint and outlet of a hollow fiber membrane system with inside-out flow. Also, estimate the shear rate at the membrane surface at the axial midpoint of the fibers. The fibers have an internal diameter of 0.06 cm and a length of 1.2 m. The crossflow velocity at the system inlet is 90 cm/s, the permeate flux is 80 L/m² h, and $T = 20°C$.

Solution. The desired velocities can be computed by direct substitution into Equation 15-171:

$$\bar{u}_{mid} = \bar{u}_{in} - \frac{4\hat{J}_V x_{mid}}{d_o} = 90\,\text{cm/s}$$

$$- \frac{4(80\,\text{L/m}^2\,\text{h})(0.6\,\text{m})(1\,\text{m}/100\,\text{cm})(1000\,\text{cm}^3/1\,\text{L})(1\,\text{h}/3600\,\text{s})}{0.06\,\text{cm}}$$

$$= 81.1\,\text{cm/s}$$

The crossflow velocity decreases by approximately 10% from the inlet to the midpoint. Because the permeation rate is assumed to be constant along the membrane length, the crossflow velocity would decline by the same amount (8.9 cm/s) between the midpoint and the outlet, so $\bar{u}_L$ at the outlet would be $81.1 - 8.9$ or 72.2, cm/s. Note that the axial velocities are at least three orders of magnitude larger than typical transverse (permeation) velocities.

The shear rate at the membrane surface at the midpoint of the fiber can be estimated using Equation 15-175, if the flow is laminar. The Reynolds number for flow through the fiber is

$$\text{Re} = \frac{\rho_L d_o \bar{u}}{\mu_L} = \frac{(1\,\text{g/cm}^3)(0.06\,\text{cm})(81.1\,\text{cm/s})}{0.01\,\text{g/cm s}} = 487$$

Since Re < 2100, the flow is laminar, and Equation 15-175 applies:

$$\gamma_{Sh,wall} = \frac{8\bar{u}}{d_o} = \frac{8(81.1\,\text{cm/s})}{0.06\,\text{cm}} = 10,800\,\text{s}^{-1}\quad ■$$

The Pressure Profile on the Concentrate Side of Crossflow Membrane Systems

The axial pressure drop and power requirements are critical operational parameters affecting the economics of membrane systems. In fact, the cost associated with pumping water at relatively high velocities through small fibers has been the major impetus for a trend to decrease or eliminate crossflow in many membrane systems in recent years. To a good approximation, the headloss across the element (i.e., from the entry to the exit), Δh_L, can be calculated from the Darcy–Weisbach equation using the average velocity at the axial midpoint of the membrane ($\bar{u}_{mid}$)

$$h_L = f\frac{L}{d_o}\frac{\bar{u}_{mid}^2}{2g} \qquad (15\text{-}176)$$

where f is the friction factor, and the other parameters are as defined previously. The value of f depends on the Reynolds number, Re, and can be approximated from the following correlations for flow through smooth pipes:

$$f_{\text{cylindrical}} = \begin{cases} \dfrac{64}{\text{Re}} & \text{Re} < 2100\,(\text{laminar flow}) \\ \dfrac{0.316}{\text{Re}^{0.25}} & 3000 < \text{Re} < 10^5 \end{cases} \qquad (15\text{-}177)$$

For Reynolds numbers between 2100 and 3000, the flow is transitional between laminar and turbulent, and the value of f does not conform to a simple mathematical representation. Flow in hollow fiber membranes is typically laminar, and that in tubular membranes is typically in the transitional or turbulent regimes.

The spacers separating the membrane sheets in spiral-wound membrane modules force concentrate to accelerate and pass through narrow gaps between the spacer and the membrane repeatedly as it moves through the module. As a result, frictional pressure losses through such modules are much larger than those through hollow fiber membranes, even if the nominal spacing between the membrane sheets in the spiral-wound module is similar to the diameter of the fibers. Schock and Miquel (1987) suggested the following empirical expression for the friction factor for flow through spiral-wound modules with spacers:

$$f_{\substack{\text{spiral} \\ \text{wound}}} = 6.23\,\text{Re}^{-0.3} \qquad (15\text{-}178)$$

The frictional pressure loss across the element is given by[20]

$$\Delta P_{L,\text{fric}} = \rho_L g h_L = f\frac{L}{d_o}\left(\frac{1}{2}\rho_L \bar{u}_{mid}^2\right) \qquad (15\text{-}179)$$

---

[20] The subscript *fric* is shown on ΔP in Equation 15-179, because the expression accounts only for pressure losses due to friction. If the module is not horizontal, the overall pressure change across the element will also reflect changes in elevation.

As an approximation, Equations 15-176–15-179 can be extended to cylindrical membranes with outside-in flow by assuming that the water flows through spaces in the module that can be represented by equivalent tubes. The hydraulic diameter of the hypothetical tubes, d_h, can be calculated as four times the hydraulic radius (r_h), which is defined as the ratio of the void volume in the module to the wetted (membrane) surface area[21]:

$$d_h = 4r_h = \frac{4(\text{Void volume})}{\text{Membrane surface area}} \quad (15\text{-}180)$$

The hydraulic diameter is proportional to the reciprocal of the packing density of the membrane module. Thus, membranes with high packing densities are characterized by small hydraulic diameters for flow within the module.

When applied to outside-in flow through cylindrical (hollow fiber or tubular) membranes packed in a cylindrical module, Equation 15-180 becomes

$$d_h = \frac{4(\text{Module volume} - \text{Membrane volume})}{\text{Membrane surface area}}$$

$$= \frac{4\left(\pi d_{\text{mod}}^2/4\right)L - N\left[\pi d_{\text{out}}^2/4\right)L\right]}{N\pi d_{\text{out}} L} \quad (15\text{-}181)$$

$$= \frac{d_{\text{mod}}^2 - N d_{\text{out}}^2}{N d_{\text{out}}}$$

where d_{mod} is the diameter of a module, d_{out} is the outside diameter of a membrane element, and N is the number of elements in a module.

■ **EXAMPLE 15-21.** Calculate the frictional pressure loss and the headloss across the membrane elements in three systems with crossflow that all have the following properties and operating parameters:

- Module (and element) length: 100 cm
- Module diameter: 7.5 cm
- Average velocity at inlet: 50 cm/s
- Permeate flux: 80 L/m² h

The feed solution is at 20°C, at which the viscosity and density are 0.01 g/cm s and 1.0 g/cm³, respectively. Consider the following membrane modules.

(a) A module packed with hollow fibers with 0.05-cm inner diameters, at a packing density of 2×10^3 m²/m³, operated with inside-out flow;

(b) A module containing tubular membranes with 1.6-cm inside diameters, at a packing density of 10^2 m²/m³, operated with inside-out flow;

(c) The same packing density as in part (a), but operated with outside-in flow using membranes with 0.075-cm outside diameters.

Solution. The headloss and frictional pressure loss across any of the elements can be calculated using Equations 15-176 and 15-179, respectively. To carry out these calculations, it is convenient to first express the flux in different units, as follows:

$$\hat{J}_V = v_w = 80\,\text{L/m}^2\,\text{h}\,(1000\,\text{cm}^3/\text{L})\left(\frac{1\,\text{m}^2}{10^4\,\text{cm}^2}\right)\left(\frac{1\,\text{h}}{3600\,\text{s}}\right)$$

$$= 2.22 \times 10^{-3}\,\text{cm/s}$$

To use Equation 15-176, we need to know value of the friction factor, f. This value depends on the Reynolds number, which differs in the different systems.

(a) In system a, the average crossflow velocity at the midpoint of the membrane, $\bar{u}_{\text{mid}}$, can be calculated from Equation 15-171:

$$\bar{u}_{\text{mid}} = \bar{u}_{\text{in}} - \frac{2\hat{J}_V L}{d_o}$$

$$= 50\,\text{cm/s} - \frac{2(2.22 \times 10^{-3}\,\text{cm/s})(100\,\text{cm})}{0.05\,\text{cm}}$$

$$= 41.1\,\text{cm/s}$$

For a cylindrical channel, the hydraulic diameter equals the geometric diameter, so the Reynolds number can be found as

$$\text{Re} = \frac{\rho_L d_h \bar{u}_{\text{mid}}}{\mu_L}$$

$$= \frac{(1\,\text{g/cm}^3)(0.05\,\text{cm})(41.1\,\text{cm/s})}{0.01\,\text{g/cm s}} = 206$$

Since the flow path is cylindrical and the Reynolds number is <2100, the flow is laminar. The friction factor is therefore given by

$$f = \frac{64}{\text{Re}} = \frac{64}{206} = 0.311$$

Inserting values into Equations 15-176 and 15-179, we find:

$$h_L = f\frac{L}{d_o}\frac{\bar{u}_{\text{mid}}^2}{2g} = (0.311)\left(\frac{100\,\text{cm}}{0.05\,\text{cm}}\right)\frac{(41.1\,\text{cm/s})^2}{2(981\,\text{cm/s}^2)} = 536\,\text{cm}$$

$$\Delta P_{L,\text{fric}} = \Delta h_L \rho_L g = (536\,\text{cm})(1\,\text{g/cm}^3)(981\,\text{cm/s}^2)$$

$$= 5.26 \times 10^5\,\text{dyne/cm}^2$$

$$= 5.26 \times 10^5\,\text{dyne/cm}^2\left(\frac{1\,\text{kPa}}{10^4\,\text{dyne/cm}^2}\right) = 52.6\,\text{kPa}$$

---

[21] The hydraulic radius of a channel, r_h, is defined as the ratio of the void volume to the wetted surface area, so $d_h = 4r_h$. By these definitions, the hydraulic diameter of a filled cylinder is the same as its geometric diameter, but the hydraulic radius is one-half the geometric radius.

(b) The procedure is essentially the same for the module with tubular membranes, except that in this case, the flow is turbulent, so the expression for f is different. The calculations are as follows:

$$\bar{u}_{mid} = \bar{u}_{in} - \frac{2\hat{J}_{V,avg}L}{d_o}$$

$$= 50\,cm/s - \frac{2(2.22 \times 10^{-3}\,cm/s)(100\,cm)}{1.6\,cm}$$

$$= 49.7\,cm/s$$

$$Re = \frac{\rho_L d_h \bar{u}_{mid}}{\mu_L} = \frac{(1\,g/cm^3)(1.6\,cm)(49.7\,cm/s)}{0.01\,g/cm\,s} = 7956$$

$$f = \frac{0.316}{Re^{0.25}} = \frac{0.316}{(7956)^{0.25}} = 0.0334$$

$$h_L = f\frac{L}{d_o}\frac{\bar{u}_{mid}^2}{2g} = (0.0334)\left(\frac{100\,cm}{1.6\,cm}\right)\frac{(49.7\,cm/s)^2}{2(981\,cm/s^2)}$$

$$= 2.63\,cm$$

$$\Delta P_{L,fric} = \Delta h_L \rho_L g = (2.63\,cm)(1\,g/cm^3)(981\,cm/s^2)$$

$$= 2.58 \times 10^3\,dyne/cm^2$$

$$= 2.58 \times 10^3\,dyne/cm^2 \left(\frac{1\,kPa}{10^4\,dyne/cm^2}\right)$$

$$= 0.26\,kPa$$

The headloss is only ~0.5% of that in the hollow fiber system in part (a), while the system produces 5% as much permeate (based on the ratio of the packing densities).

(c) In this case, the flow is outside-in, so the hydraulic diameter, d_h, differs from the geometric diameter of the fibers (d_o). We can calculate d_h using Equation 15-181, if we first determine the number of fibers in a module, N, as follows:

$$V_{mod} = \pi\left(\frac{d_{mod}}{2}\right)^2 L_{mod} = \pi\left(\frac{7.5\,cm}{2}\right)^2 (100\,cm)$$

$$= 4418\,cm^3 = 4.42 \times 10^{-3}\,m^3$$

$$A_{m,tot} = (V_{mod})(Packing\,density)$$

$$= (4.42 \times 10^{-3}\,m^3)(2 \times 10^3\,m^2/m^3) = 8.84\,m^2$$

$$A_{m,one\,fiber} = \pi d_{out} L = \pi(0.075\,cm)(100\,cm)$$

$$= 23.6\,cm^2 = 2.36 \times 10^{-3}\,m^2$$

$$N = \frac{A_{m,tot}}{A_{m,one\,fiber}} = \frac{8.84\,m^2/module}{2.36 \times 10^{-3}\,m^2/fiber} = 3750\,fibers/module$$

Then, substituting values into Equation 15-181:

$$d_h = \frac{d_{mod}^2 - Nd_{out}^2}{Nd_{out}} = \frac{(7.5\,cm)^2 - 3750(0.075\,cm)^2}{3750(0.075\,cm)} = 0.125\,cm$$

The calculation of the average velocity inside the module follows the derivation of Equation 15-171, but differs because of the different flow geometries. The flow rate entering the feed ("shell") side of the module is the product of the fluid velocity (given as 50 cm/s) and the area available to the flow. This area equals the difference between the cross-sectional area of the module and the area occupied by the fibers. Thus,

$$A_{available,mod} = A_{xs,mod} - A_{xs,fibers} = A_{xs,mod} - NA_{xs,one\,fiber}$$

$$= \frac{\pi(7.5\,cm)^2}{4} - 3750\frac{\pi(0.075\,cm)^2}{4} = 27.6\,cm^2$$

$$Q_{in} = \bar{u}_{in} A_{available} = (50\,cm/s)(27.6\,cm^2) = 1381\,cm^3/s$$

$$Q_{out} = Q_{in} - \hat{J}_V A_{m,tot}$$

$$= 1381\,cm^3/s - (2.22 \times 10^{-3}\,cm/s)(8.84 \times 10^4\,cm^2)$$

$$= 1184\,cm^3/s$$

$$\bar{u}_{out} = \frac{Q_{out}}{A_{available}} = \frac{1184\,cm^3/s}{27.6\,cm^2} = 42.9\,cm/s$$

$$\bar{u}_{mid} = \frac{\bar{u}_{in} + \bar{u}_{out}}{2} = \frac{(50 + 42.9)\,cm/s}{2} = 46.4\,cm/s$$

Now, following the same procedure as in parts (a) and (b), we find

$$Re = \frac{\rho_L d_h \bar{u}_{mid}}{\mu_L} = \frac{(1\,g/cm^3)(0.125\,cm)(46.4\,cm/s)}{0.01\,g/cm\,s}$$

$$= 581\,(laminar)$$

$$f = \frac{64}{Re} = \frac{64}{581} = 0.110$$

$$h_L = f\frac{L}{d_h}\frac{\bar{u}_{mid}^2}{2g} = (0.110)\left(\frac{100\,cm}{0.125\,cm}\right)\frac{(46.4\,cm/s)^2}{2(981\,cm/s^2)}$$

$$= 97.0\,cm$$

$$\Delta P_{L,fric} = \Delta h_L \rho_L g = (97.0\,cm)(1\,g/cm^3)(981\,cm/s^2)$$

$$= 9.51 \times 10^4\,dyne/cm^2$$

$$= 9.51 \times 10^4\,dyne/cm^2\left(\frac{1\,kPa}{10^4\,dyne/cm^2}\right)$$

$$= 9.51\,kPa$$

The pressure loss across the module in this case is intermediate between that in the systems analyzed in parts (a) and (b). The key physical differences among the three systems that lead to this result are their hydraulic diameters (0.05, 1.20, and 0.125 cm in parts (a)–(c), respectively). The hydraulic diameter appears directly in the equation for pressure loss (Equation 15-179), with a decrease in d_h leading to an increase in $\Delta P_{L,fric}$. The hydraulic diameter also has an indirect effect on frictional pressure loss, since it enters into the calculation of the Reynolds number and thereby helps determine the flow regime and friction factor, f. Specifically, f increases with decreasing Re, and is much more sensitive to Re in laminar than turbulent flow. The fact that the systems in parts (a) and (c) had laminar flow, and that Re was greater in part (c) than part (a) also contributes to the significant difference in pressure loss across the three systems. ∎

Contaminant Transport Mechanisms in Crossflow Filtration

As in frontal filtration, the extent to which rejected contaminants accumulate near the surface of membranes in systems with crossflow reflects a balance between processes that carry the contaminants toward the membrane and those that carry them away, either back to the bulk solution or parallel to the membrane and out of the system. Therefore, the analysis of contaminant transport in crossflow systems follows essentially the same steps as in a frontal filtration system, but the presence of crossflow introduces some new terms into the mass balance equation and alters the values of others. Davis (1992) has reviewed these issues for both colloidal and particulate contaminants.

In frontal filtration, contaminant transport is dominated by advective transport toward the membrane and diffusion away from it. As a result, the mass balance on contaminants (Equation 15-99) is essentially identical to the one-dimensional advective-diffusion equation derived in Chapter 1 (Equation 1-34) for a nonreactive constituent. Equation 15-99 is repeated in a rearranged form here, with the subscript on the diffusivity modified to emphasize the fact that it is typically understood to account only for Brownian motion.

$$0 = v\frac{dc}{dy} + D_{Br}\frac{d^2c}{dy^2} \qquad (15\text{-}182)$$

The control volume used to derive Equation 15-99 was a differential layer of thickness dy; the boundaries of the control volume in the plane parallel to the membrane were not specified, because the system composition and pressure were assumed to be uniform in that plane. In contrast, in a system with crossflow, variations are expected in the system conditions in the axial (x) direction, leading to both advective and diffusive transport in that direction. In almost all analyses, the net axial transport due to diffusion is assumed to be negligible compared with that induced by the crossflow, so the mass balance equation includes only one term (the advective term) to account for axial transport. Adding this term, and anticipating that other factors might contribute to diffusivity and that other transport terms might be relevant, we can write a generic mass balance equation for contaminants in steady-state, crossflow filtration as follows:

$$0 = \frac{\partial uc}{\partial x} + \frac{\partial vc}{\partial y} + \frac{\partial}{\partial y}\left(D_{tot}\frac{\partial c}{\partial y}\right) + \text{other transport terms} \qquad (15\text{-}183)$$

The equation is written with partial differentials, because c can vary in both the x and y directions. In addition, the velocities and diffusivity are placed inside the differential operators in recognition of the fact that they, like concentration, might vary with location in the system.

Concentrations of particles and colloids in crossflow systems are often expressed in terms of the volume fraction (ϕ) of the suspension that they occupy, in which case Equation 15-183 becomes:

$$0 = \frac{\partial u\phi}{\partial x} + \frac{\partial v\phi}{\partial y} + \frac{\partial}{\partial y}\left(D_{tot}\frac{\partial \phi}{\partial y}\right) + \text{other transport terms} \qquad (15\text{-}184)$$

Brownian and Shear-Induced Diffusion Brownian motion of contaminant species is generated primarily by interactions of those species with molecules in the solution phase. If the contaminants are sufficiently concentrated, collisions among those species might also contribute significantly to their motion; by the same token, at large contaminant concentrations, steric interferences might reduce the Brownian motion. These phenomena—both induced motion and interference with such motion—can affect contaminant species in any solution or suspension, whether it is mixed or stagnant. However, if the fluid is mixed mechanically, the motion induced by the mixing invariably overwhelms diffusive transport, so diffusion is important only in fluids that are nearly stagnant.

In a system with crossflow, a shear flow exists near the membrane, meaning that adjacent layers of water flow past one another parallel to the membrane. If particles are being rejected by the membrane, a concentration gradient exists due to concentration polarization. This situation is shown in Figure 15-38. The shear flow causes particles in each layer to collide with those in adjacent layers as the layers slide by one another. Each collision causes both particles to change direction slightly, with the particle closer to the membrane

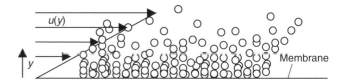

FIGURE 15-38. Schematic demonstration of shear-induced diffusion. Because of the concentration gradient, the particles are knocked upward more often than they are knocked downward, leading to net transport away from the membrane.

being pushed still closer, and the particle farther from the membrane being pushed even farther away. Because of the concentration gradient, any particle is more likely to collide with one closer to the membrane than with one farther away, so the net transport arising from all the collisions is away from the membrane (i.e., from higher to lower particle concentration). Furthermore, given the mechanism inducing this transport, the net rate of transport is expected to be proportional to both the particle concentration gradient and the shear rate. The direction of net motion and the dependence on the concentration gradient make this phenomenon similar to Brownian diffusion, and it has therefore been named *shear-induced diffusion* (Zydney and Colton, 1986; Davis and Leighton, 1987).[22] Note, however, Brownian and shear-induced diffusion differ in important ways: shear-induced diffusion increases with increasing shear rate and increasing particle size, whereas Brownian motion is insensitive to shear rate and decreases with increasing particle size.

Shear-induced diffusion is usually treated as additive with Brownian diffusion:

$$D_{tot} = D_{Br} + D_{Sh} \quad (15\text{-}185)$$

The Brownian diffusion coefficient, D_{Br}, for contaminants of (equivalent) diameter d_p can be calculated from the Stokes–Einstein equation[23]:

$$D_{Br} = \frac{k_B T}{3\pi\mu d_p} \quad (15\text{-}186)$$

where k_B is the Boltzmann constant and T is the absolute temperature.

A number of expressions have been proposed for estimating D_{Sh}. For example, Zydney and Colton (1986) recommended using the following equations to compute D_{Sh} for rigid, spherical particles, based on data reported by Eckstein et al. (1977) from studies of concentrated suspensions subjected to uniform shear:

$$D_{Sh} = \begin{cases} 0.0375 d_p^2 \gamma_{Sh} \phi & \phi < 0.2 \\ 0.0075 d_p^2 \gamma_{Sh} & \phi \geq 0.2 \end{cases} \quad (15\text{-}187)$$

The two parts of Equation 15-187 imply that D_{Sh} increases with increasing particle concentration up to

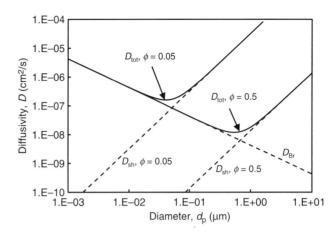

FIGURE 15-39. Contributions of Brownian motion and shear to D_{tot} for hard spheres. $\gamma_{Sh,wall} = 3750\, s^{-1}$ and $T = 20°C$ were used in all calculations. D_{Sh} computed using Equation 15-188.

$\phi = 0.2$, above which it remains constant.[24] Leighton and Acrivos (1987) reported that the following modification to the equation improved the fit to experimental data, for ϕ up to 0.5:

$$D_{Sh} = \frac{\tilde{D}_{Sh}(\phi)}{4} d_p^2 \gamma_{Sh} \quad (15\text{-}188)$$

where $\tilde{D}_{Sh}(\phi)$ is a dimensionless function that can be computed as

$$\tilde{D}_{Sh}(\phi) = 0.33\phi^2(1 + 0.5\exp(8.8\phi)) \quad (15\text{-}189)$$

Wiesner et al. (1989) pointed out that, because D_{Br} decreases with increasing particle size, whereas D_{Sh} follows the opposite trend, D_{tot} passes through a minimum at a specific particle size, as shown in Figure 15-39. The lower value of ϕ (0.05) shown in Figure 15-39 is in the range that might be expected in concentrated suspensions of highly hydrated solids, such as microorganisms, precipitated coagulants such as $Al(OH)_3(s)$ or $Fe(OH)_3(s)$, and many other solids that are likely to be found in water and wastewater treatment applications. Thus, for the conditions that apply near the membrane in typical UF and MF systems with crossflow, the minimum in D_{tot} is expected to occur for particles with diameters of a few tenths of a micrometer. This result suggests that shear-induced diffusion is the dominant back-transport mechanism for inorganic solids

---

[22] Shear-induced diffusion is also sometimes referred to as *lateral migration*, because the particle motion is "lateral" (i.e., sideways) compared with the fluid motion that induces the motion. Note, however, that whereas the phrase "shear-induced diffusion" refers to a specific transport mechanism, "lateral migration" is used generically to refer to any (or all) of several transport mechanisms that cause particles to migrate away from the membrane.

[23] This expression was also used to describe Brownian motion of particles in Chapter 12.

[24] The applicability of the Eckstein et al. study to membrane filtration of colloids is open to question, since the particles used in that study were two to three orders of magnitude larger, and the shear rates two to three orders of magnitude smaller, than those typically encountered in membrane systems. Also, Zydney and Colton's equation appears to overestimate the values reported by Eckstein et al. by 20–50%. Nevertheless, this equation has been widely cited and used by subsequent modelers.

and bacteria (i.e., particle with d_p larger than $\sim 1\,\mu m$), whereas Brownian diffusion dominates for macromolecules and viruses ($d_p < \sim 0.04\,\mu m$).

Advective contaminant flux toward the membrane equals $v_w c_i$ and is insensitive to contaminant size. Therefore, if back-transport is dominated by diffusive processes, the only factor that is expected to differentiate the behavior of small and large contaminants is their different values of D_{tot}. As a result, contaminants in the size range corresponding to the lowest values of D_{tot} are expected to accumulate preferentially in the thick suspensions and/or compact layers adjacent to the membrane surface and to have a disproportionate effect on fouling.

The preceding discussion implicitly assumes that the behavior of individual particles in a polydisperse suspension can be modeled based on their behavior in monodisperse suspensions. This assumption is unrealistic, but, unfortunately, more realistic theories for particle transport and membrane performance in polydisperse suspensions are very complex and are largely undeveloped. Qualitatively, it has been suggested that the presence of larger particles in the layers of rejected material near a membrane surface might be beneficial, because they can help produce a more porous cake with less hydraulic resistance. Larger particles may also enhance the transport of smaller particles out of the cake layer by sweeping or scouring them from the membrane surface. Based on these observations and modeling results such as those shown in Figure 15-39, Wiesner et al. (1989) noted that increasing the crossflow velocity with the intention of scouring materials from the membrane surface might sometimes have the counter-intuitive and counter-intended effect of decreasing permeate flux. Such an outcome could arise if the higher crossflow velocity decreased the fraction of larger particles in the cake due to the larger effect of shear-induced diffusion on those particles than on smaller ones. Because of these and similar complexities, selection of the crossflow velocity for a particular application remains largely empirical.

■ **EXAMPLE 15-22.** Estimate the total diffusion coefficient (D_{tot}) at 20°C of particles that are 0.05, 0.5, and 5 μm in diameter, in the system described in Example 15-20, in a region of a concentration polarization layer where the particle volume fraction is 0.05.

Solution. The total diffusion coefficient is the sum of the coefficients for Brownian and shear-induced diffusion, as per Equation 15-185. We can estimate D_{Sh} using Equation 15-188 as

$$D_{Sh} = \frac{\tilde{D}_{Sh}(\phi)}{4} d_p^2 \gamma$$

The dimensionless diffusivity function for use in Equation 15-188 can be computed from Equation 15-189:

$$\tilde{D}_{Sh} = 0.33\phi^2(1 + 0.5\exp(8.8\phi))$$
$$= 0.33(0.05)^2(1 + 0.5\exp[8.8(0.05)]) = 1.47 \times 10^{-3}$$

We approximate γ_{Sh} throughout the CP layer as $\gamma_{Sh,wall}$, which was computed to be 10,800/s in Example 15-20. We can then calculate the shear-induced diffusivity for 0.5-μm particles as

$$D_{Sh} = \frac{\tilde{D}_{Sh}(\phi)}{4} d_p^2 \gamma_{Sh} = \frac{1.47 \times 10^{-3}}{4}(0.5 \times 10^{-4}\,cm)^2(10,800/s)$$
$$= 9.92 \times 10^{-9}\,cm^2/s$$

The Brownian diffusion coefficient of these 0.5-μm particles can be obtained from the Stokes–Einstein equation (Equation 15-186):

$$D_{Br} = \frac{k_B T}{3\pi\mu d_p} = \frac{(1.38 \times 10^{-16}\,g\,cm^2/s^2\,K)293\,K}{3\pi(0.01\,g/cm\,s)(0.5 \times 10^{-4}\,cm)}$$
$$= 8.58 \times 10^{-9}\,cm^2/s$$

D_{tot} for these particles is then

$$D_{tot} = D_{Sh} + D_{Br}$$
$$= 9.92 \times 10^{-9}\,cm^2/s + 8.58 \times 10^{-9}\,cm^2/s$$
$$= 1.85 \times 10^{-8}\,cm^2/s$$

The values for the 0.05- and 5-μm particles can be calculated similarly, and the results are summarized in the following table.

Diameter (μm)	D_{Sh} (cm^2/s)	D_{Br} (cm^2/s)	D_{tot} (cm^2/s)
0.05	9.92×10^{-11}	8.58×10^{-8}	8.59×10^{-8}
0.5	9.92×10^{-9}	8.58×10^{-9}	1.85×10^{-8}
5.0	9.92×10^{-7}	8.58×10^{-10}	9.93×10^{-7}

D_{tot} is dominated by D_{Sh} for the largest particles and by D_{Br} for the smallest, and both terms make comparable contributions to D_{tot} for the 0.5-μm particles. For all these particles, D_{tot} is approximately 1.5–3 orders of magnitude smaller than the diffusivity of solutes (typically, on the order of $10^{-5}\,cm^2/s$). ■

Deterministic Transport Mechanisms in Crossflow Filtration In addition to diffusion, the motion of an isolated particle in a membrane system is likely to be affected by several deterministic forces, including a drag force, F_{drag}, due to fluid flow; the gravitational force, F_{grav}; an inertial lift force, F_{lift}, due to differential fluid velocities across different portions of the particle's surface; London-van der Waals attractive forces between the particle and other particles or the membrane, F_{vdW}; and electrical forces, F_{elec}, due to the interaction of the particle's surface charge with that of other particles, with the membrane, or with an externally imposed

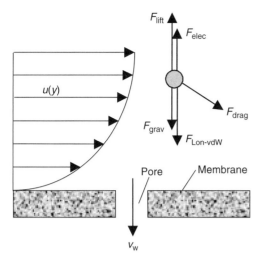

FIGURE 15-40. Some deterministic forces on particles in a membrane system with crossflow and downward permeation. The electrical and gravitational forces might be in other directions, depending on the charge on the particle and membrane and the orientation of the module, respectively.

electrical field. Some of these forces are illustrated schematically in Figure 15-40.

With the exception of the inertial lift force, all the forces shown in the schematic are potentially relevant both in systems with and without crossflow, and all of them have been discussed in advanced analyses of membrane systems (Cohen and Probstein, 1986; Wiesner et al., 1989). However, the gravitational force is often small for particles in membrane systems,[25] and the London-van der Waals and electrical forces are highly system-specific. Therefore, the only deterministic particle transport mechanism that is considered quantitatively in this section is inertial lift.

In crossflow filtration, inertial lift occurs in the laminar boundary layer adjacent to the membrane as a consequence of the transverse gradient in the axial fluid velocity (i.e., the same characteristic that causes shear-induced diffusion); this velocity gradient causes the pressure in the higher velocity region (farther from the membrane) to be lower than that in the lower velocity region, which in turn causes particles to experience a lift force perpendicular to and away from the membrane (Porter, 1972). Inertial lift does not occur in

---

[25] The gravitational force depends on the density of the particle and approaches zero as the particle approaches neutral buoyancy. Even for particles that are not neutrally buoyant, the gravitational force is zero if the membranes are oriented vertically (e.g., as hollow fibers arrayed in vertical modules). For other designs (e.g., hollow fibers in horizontal modules), the net gravitational force is positive in half of the flow (the flow moving toward a lower membrane surface) and negative in the other half (the flow moving toward an upper surface). Therefore, the gravitational force tends to be significant only in plate-and-frame membrane systems with horizontal membranes. Such systems are often used in laboratory experiments, but are much less common in full-scale applications.

turbulent flow. Because of the complex hydrodynamics in the region near the membrane, the lift velocity for a given particle varies significantly as a function of distance from the membrane, even if the shear rate is constant. Belfort and coworkers (Altena and Belfort, 1984; Drew et al., 1991) have proposed the following approximation for the lift velocity:

$$v_{\text{lift}} = \frac{b \rho_L d_p^3 \gamma_{\text{Sh,wall}}^2}{128 \mu} \quad (15\text{-}190)$$

where b is a value that varies with distance from the membrane and depends on the Reynolds number characterizing flow through the membrane channel. The maximum lift velocity occurs a small distance from the membrane surface, where b has its maximum value. This value, $b_{\max}$, has been estimated as follows (Drew et al., 1991):

$$b_{\max} = \begin{cases} 1.3 & \text{for Re} \ll 1, \text{cylindrical membrane} \\ 1.6 & \text{for Re} \ll 1, \text{slit membrane} \\ 0.577 & \text{for } 1 < \text{Re} < 2100, \text{all membranes} \end{cases}$$

Equation 15-190 indicates that the lift velocity increases with the cube of the particle size and the square of the shear rate. Thus, similar to shear-induced diffusion, vertical lift is most significant for large particles in systems with high flow rates and/or narrow channels.

It was noted previously that the more rapidly contaminants diffuse, the less likely they are to accumulate near the membrane surface. A similar statement applies to contaminants that are subject to inertial lift. However, since the inertial lift velocity is deterministic, we can state with certainty that particle deposition will not occur if the maximum inertial lift velocity, $v_{\text{lift,max}}$, is greater than the permeation velocity, v_w, because particles that experience such a lift velocity will not be transported toward the membrane past the location of $b_{\max}$. Thus, in a system with a given crossflow velocity, a critical particle size exists above which particles are not expected to reach the membrane surface (or the surface of the cake or gel layer, if such a layer forms). Particles larger than this critical size are therefore not expected to contribute significantly to fouling. Particles that are smaller than this size experience an inertial lift velocity less than the permeation velocity and can reach the inner boundary of the CP layer and become part of the surface deposit, but their net velocity toward that boundary will decrease as their size (and their lift velocity) increases.

■ **EXAMPLE 15-23.** Estimate the size of the largest particle that will be able to overcome inertial lift and deposit on the membrane in the system described in Example 15-20.

Solution. The size of the largest particle that will be able to overcome inertial lift and deposit on the membrane can be calculated by equating the permeation velocity with the maximum lift velocity; that is, $v_w = v_{lift}$. The permeation velocity is the same as the volumetric flux, and the maximum lift velocity is given by Equation 15-190, so this equality can be written as

$$\hat{J}_V = \frac{b_{max}\rho_L d_p^3 \gamma_{Sh,wall}^2}{128\mu}$$

The value of b_{max} depends on the channel Reynolds number, which we computed in Example 15-20 to be 487. Because the Reynolds number is >1, the appropriate value of b_{max} is 0.577. The shear rate at the membrane wall was also determined in Example 15-20, as 10,800/s. Substituting these values into the equality between the flux and the lift velocity, we find:

$$80 \, L/m^2 \, h \left(\frac{1 \, m^3}{1000 \, L}\right)(100 \, cm/m)\left(\frac{1 \, h}{3600 \, s}\right)$$
$$= \frac{0.577(1.0 \, g/cm^3)d_p^3(10,800/s)^2}{128(0.01 \, g/cm \, s)} 2.22 \times 10^{-3} \, cm/s$$
$$= \left(5.26 \times 10^7 \frac{1}{cm^2 \, s}\right)d_p^3$$

$d_p = 3.48 \times 10^{-4} \, cm = 3.48 \, \mu m$

Thus, particles larger than 3.48 μm in diameter are not expected to enter the CP layer and hence will not contribute to fouling. ∎

Relative Importance of Different Back-Transport Mechanisms The relative importance of Brownian diffusion, shear-induced diffusion, and inertial lift as back-transport processes in systems with crossflow depends strongly on the contaminant size, as shown for typical operating conditions in Figure 15-41. In this figure, the contributions of the different transport mechanisms are shown as mass transfer coefficients, which can be thought of as characteristic velocities for transport from the membrane back to the bulk concentrate. Remember that, at least in the case of Brownian and shear-induced diffusion, these velocities are useful indicators of the magnitudes of the individual terms, but they do not represent real or uniform velocities that apply throughout the CP layer. A range of typical permeation velocities (i.e., fluxes) in environmental engineering membrane applications is also shown in the figure. If the characteristic velocity for back-transport exceeds the permeation flux, the chance that the particles will be transported to and deposited on the membrane surface is small. Contaminants in the size range from a few tens of nanometers to a few micrometers have back-transport velocities that are comparable to or smaller than typical permeation velocities, so those contaminants are likely to reach the membrane. On the other hand, particles that are larger than 10 μm (i.e., in the

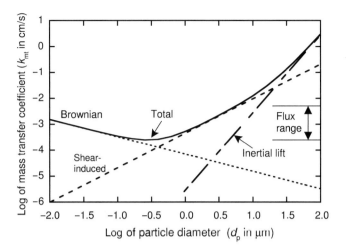

FIGURE 15-41. Particle mass transport away from a membrane surface in a crossflow system, calculated for a hypothetical inside-out hollow fiber. Conditions: $\gamma_{Sh,wall} = 2500 \, s^{-1}$, $L_{el} = 100 \, cm$, $c_b = 60 \, mg/L$, $c_{wall} = 20,000 \, mg/L$, D_{Br} from Stokes–Einstein equation. Flux range indicated spans the entire spectrum normally encountered in environmental engineering membrane systems.

size range where inertial lift is an important back-transport mechanism) are not expected to reach the membrane in most environmental engineering membrane systems with crossflow.

Modeling Contaminant Transport and Flux in Crossflow Filtration Systems

Overview If shear-induced diffusion and inertial lift are included explicitly in the steady-state mass balance on contaminant in a crossflow system, and if the other transport terms (e.g., those due to electrical or gravitational forces) are unimportant, Equation 15-183 becomes

$$0 = -\frac{\partial uc}{\partial x} + \frac{\partial vc}{\partial y} + \frac{\partial}{\partial y}\frac{(D_{Br} + D_{Sh})\partial c}{\partial y} - \frac{\partial v_{lift}c}{\partial y} \quad (15\text{-}191)$$

where the signs are based on u being positive from the inlet to the exit of the element, v being positive in the direction of permeation, y being positive away from the membrane, and v_{lift} being positive in the direction of lift (away from the membrane).

In theory, we could solve Equation 15-191 if we could quantify the three parameters related to back-transport (D_{Br}, D_{Sh}, v_{lift}), write complete expressions to characterize u and v, and specify appropriate boundary conditions (one in x and two in y). As has been noted in the preceding sections, our understanding of the physical/chemical conditions in crossflow systems (especially those that develop very near the membrane) is incomplete. Therefore, simplified and approximate expressions are used in conjunction with

Equation 15-191 to derive predictions for the concentration profiles and to relate those profiles to the applied pressure. In the following sections, some of the key model simplifications and the resulting predictions for flux as a function of TMP, feed composition, and membrane characteristics are presented. While many of the model results have been tested in laboratory-scale experiments, they have rarely been used to design or describe full-scale systems. Nevertheless, in the discussion, an effort is made to relate the models to full-scale membrane systems. As with several other processes described in this text, the models can provide insights for design and operation of such systems, even though a direct quantitative application of the models to design is not yet possible. The development and application of a state-of-the-art model for an RO system have been well described by Hoek et al. (2008).

Assumptions Commonly Used in Modeling Crossflow Filtration Several features that are common to most models of crossflow filtration are shown in Figure 15-42. The curved lines in the figure represent the edges of the CP layers and correspond to concentration isopleths along which $c \approx c_c$. The size of the CP layer is greatly exaggerated in the figure; typically, the maximum thickness of this layer would be no more than a few percent (and often <1%) of the channel cross-section. The growth of the CP layer from negligible thickness at the entrance to a limiting thickness downstream represents a key difference between crossflow and frontal filtration. For typical crossflow velocities and system geometries, the CP layer is much thinner than the velocity boundary layer, and it develops over a longer distance. Therefore, whereas the velocity boundary layer is usually assumed to be fully established near the entrance to the membrane element, the CP layer

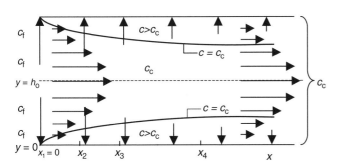

FIGURE 15-42. Variations in several parameters in crossflow filtration. Axial and transverse velocities are indicated by the thin and thick arrows, respectively, with the length of the arrows indicating the relative magnitude within each type. The edge of the concentration boundary layer is represented by the curved lines. The figure characterizes a rectangular flow channel bounded by permeable membranes, but essentially the same figure would apply for a cylindrical channel.

can expand for a significant distance downstream and might still be expanding when the concentrate exits the element (Kruelen et al. 1993).

The horizontal arrows at the inlet and outlet of the element represent the axial velocity vectors (u). Because the CP layer is thin compared with the velocity boundary layer, the shear rate is assumed to be constant throughout the CP layer (i.e., the axial velocity is assumed to increase linearly with distance from the membrane surface, which we now designate $y = 0$ (the same location identified as the "wall" previously)):

$$u = \gamma_{\text{Sh}, y=0} y \quad \text{for } y < \delta_{\text{CP}} \quad (15\text{-}192)$$

In addition, it is usually assumed that the permeation flow through the membrane segment being analyzed is small compared with the axial flow; that is, that the recovery in the segment is small (say, less than a few percent). If this assumption is unreasonable to apply over the full length of the membrane element, the segment length considered in any single analysis can be reduced until the assumption is valid; the segments can then be linked mathematically to model the whole element. The assumption that the recovery is small allows us to treat the bulk concentration, c_c, as constant throughout the segment. Combining this assumption with an assumption that the flow is fully developed within a short distance of the entrance to the segment allows us to approximate the axial velocity profile throughout the segment as fully developed flow through a channel with impermeable walls.

In most modeling, the hydrostatic pressure in the bulk concentrate is also assumed to be constant in the segment being analyzed; such an assumption is justified if the pressure loss from the inlet to the outlet of the segment is small compared with the TMP. This condition is met over the full length of the element in many practical applications of MF and UF membranes, but often not for NF or RO membranes. In cases where the condition is not met, the length of the segment being analyzed can be reduced until the assumption is acceptable.

The Mechanics of Crossflow Filtration Modeling To predict the performance of a particular crossflow filtration system, the preceding assumptions are combined with other assumptions and with operational information about the system to model the behavior in a segment at the entrance to a membrane element. The conditions at the terminus of that segment are then computed and used as the inlet conditions for the next segment. The process can be visualized as shown in Figure 15-43, in which a membrane element is represented as a sequence of three segments. Water flows through the element from left to right, and permeate is removed through the membrane in each segment. The crossflow is treated as constant within a

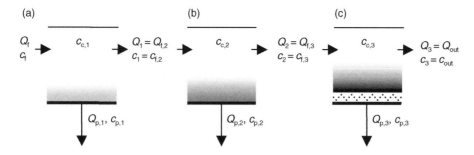

FIGURE 15-43. A schematic representation of a model for crossflow filtration. The stippled band in the final segment represents a compact layer of constant composition.

segment, and then to undergo a step decrease when the water exits, so that the flow to downstream segments is adjusted for the water lost to permeation upstream. Correspondingly, c_c is assumed to be constant within a segment, but to undergo a step increase between segments. The thickness of the CP layer also increases in the downstream direction, as does the concentration at the membrane wall ($c_{c_{c/m}}$) until it reaches $c_{\lim}$, at which point a compact layer forms; from that point downstream, $c_{c_{c/m}}$ and the concentration throughout the compact layer remain at $c_{\lim}$. (In the figure, a compact layer is envisioned to form in the third segment.)

The schematics for the first two segments and the third segment in Figure 15-43 are essentially identical to those presented earlier for frontal filtration systems in the absence and presence of a compact layer, respectively. Much of the prior analysis of frontal filtration therefore applies to each of these segments independently. However, the changing thickness of the CP layer along the membrane length and the possible movement of material in the CP layer from one segment to the next distinguish the overall behavior of a system with crossflow from a single, frontal filtration system.

Modeling the Thickness of the CP Layer and $k_{mt,CP}$ in Crossflow Filtration In modeling crossflow systems, the key consequence of the growth of the CP layer with increasing axial distance from the entrance to the element is that it decreases the mass transfer coefficient in the CP layer (recall that, if the CP layer is stagnant, $k_{mt,CP}$ can be interpreted as D_{tot}/δ_{CP}). Therefore, a good deal of effort has been devoted to estimating $k_{mt,CP}$ as a function of x. The relationship is commonly presented as a correlation among three dimensionless numbers—the Sherwood number (Sh), the Reynolds number (Re), and the Schmidt number (Sc)—in the following form:

$$\text{Sh}(x) = A\text{Re}^a \text{Sc}^b \left(\frac{d_h}{x}\right)^c \quad (15\text{-}193)$$

where A is a constant that depends on the channel geometry, and

$$\text{Sh} = \frac{k_{mt} d_h}{D_{tot}} \quad (15\text{-}194)$$

$$\text{Re} = \frac{d_h \bar{u} \rho_L}{\mu} \quad (15\text{-}195)$$

$$\text{Sc} = \frac{\mu}{D_{tot} \rho_L} \quad (15\text{-}196)$$

The preceding equations apply regardless of whether D_{tot} is dominated by Brownian or shear-induced diffusion, or whether both mechanisms contribute significantly to it. However, D_{Sh} depends on the particle concentration (as per Equations 15-187 and 15-189), so if D_{Sh} contributes significantly to D_{tot}, D_{tot} must be evaluated for a specific concentration. Because a large value of D_{Sh} is usually required to fit the experimental data, D_{Sh} is almost always assigned the value corresponding to $c_{c_{c/m}}$; that is, the largest value expected anywhere in the CP layer.

The form of Equation 15-193 used most often to model crossflow filtration systems is based on an analogy with an equation derived originally by Lévêque for heat transfer from solid surfaces to fluids flowing past them, and subsequently applied to mass transfer in analogous systems (Probstein, 1994). This approach was first proposed by Blatt et al. (1970), who suggested that, for most membrane systems of interest, the exponents in Equation 15-193 are $a = b = c = 1/3$, and the value of A is 1.24 for a rectangular slit and 1.09 for a circular cross-section. Substituting these values and the definitions of Sh, Re, and Sc into Equation 15-193 and rearranging yields explicit expressions for $k_{mt,CP}(x)$. A few such expressions for systems with various geometries are given in Table 15-12.

Frequently, rather than applying a different value of $k_{mt,CP}$ at each axial location being modeled, a single, average value of that parameter is assumed to be valid over the whole

TABLE 15-12. Various Forms of the Lévêque-Based Correlation for k_{mt} in Crossflow Membrane Systems

	$\overline{k}_{mt,CP}$		$\overline{k}_{mt,CP}$	
General	$A\text{Re}^{1/3}\text{Sc}^{1/3}\left(\dfrac{d_h}{x}\right)^{1/3}\dfrac{D_{tot}}{d_h} = A\left(\dfrac{D_{tot}^2 \bar{u}}{d_h x}\right)^{1/3} \overline{k}_{mt,CP}$	(15-198)	$\dfrac{3}{2}A\text{Re}^{1/3}\text{Sc}^{1/3}\left(\dfrac{d_h}{L_{el}}\right)^{1/3}\dfrac{D_{tot}}{d_h} = \dfrac{3}{2}A\left(\dfrac{D_{tot}^2 \bar{u}}{d_h L_{el}}\right)^{1/3}$	(15-199)
Slit ($A = 1.24$)	$0.78\left(\dfrac{D_{tot}^2 \bar{u}}{h_o x}\right)^{1/3} = 0.544\left(\dfrac{D_{tot}^2 \gamma_{\text{Sh},y=0}}{x}\right)^{1/3}$	(15-200)	$1.18\left(\dfrac{D_{tot}^2 \bar{u}}{h_o L_{el}}\right)^{1/3} = 0.816\left(\dfrac{D_{tot}^2 \gamma_{\text{Sh},y=0}}{L_{el}}\right)^{1/3}$	(15-201)
Circular ($A = 1.09$)	$1.09\left(\dfrac{D_{tot}^2 \bar{u}}{d_o x}\right)^{1/3} = 0.544\left(\dfrac{D_{tot}^2 \gamma_{\text{Sh},y=0}}{x}\right)^{1/3}$	(15-202)	$1.63\left(\dfrac{D_{tot}^2 \bar{u}}{d_o L_{el}}\right)^{1/3} = 0.816\left(\dfrac{D_{tot}^2 \gamma_{\text{Sh},y=0}}{L_{el}}\right)^{1/3}$	(15-203)

Source: After Blatt et al. (1970).

length of the membrane element. This average value can be computed from the local values as follows:[26]

$$\overline{k}_{mt,CP} = \frac{\int_0^{L_{el}} k_{mt,CP} dx}{\int_0^L dx} \quad (15\text{-}197)$$

Expressions for both $k_{mt,CP}$ versus x and $\overline{k}_{mt,CP}$ are shown in Table 15-12.

Systems with a Ubiquitous Compact Layer: Applying the Film-Gel Model in Systems with Crossflow Filtration The Leveque equation for heat transfer assumes that the temperature of the solid surface is uniform, and the analogous assumption underlying the equations in Table 15-12 is that the concentration at $y = 0$ is constant over the axial span being modeled. This constraint is most closely approximated in systems in which a compact layer forms close to the entrance to the element and is present along virtually its entire length, in which case $c_{y=0} = c_{\lim}$ at all x. The analysis of such systems usually assumes that back-transport is dominated by diffusive processes. In that case, the term in Equation 15-191 involving v_{lift} is negligible, and the equation becomes:[27]

$$0 = -\frac{\partial uc}{\partial x} + \frac{\partial vc}{\partial y} + D_{tot}\frac{\partial^2 c}{\partial y^2} \quad (15\text{-}204)$$

Approaches for estimating D_{tot} have been described, and, as noted previously, the axial velocity field (u as a function of y) is modeled using the equations for fully developed, impermeable flow. Thus, Equation 15-204 can be solved by identifying a relationship between v and other system parameters and by writing appropriate boundary conditions. The boundary conditions establish that the contaminant concentration is uniform at the entrance to the membrane element and characterize the conditions at the inner and outer edges of the CP layer. If we assume that the change in c_c along the element is small, these boundary conditions are

$$c = c_f \quad \text{at } x = 0 \quad (15\text{-}205)$$
$$c = c_{\lim} \quad \text{at } y = 0 \quad (15\text{-}206)$$
$$c = c_c \quad \text{at } y = \delta_{CP} \quad (15\text{-}207)$$

An equation relating v to other system parameters can be written by making an assumption about how a pseudo-steady-state concentration of contaminant is maintained in the segment. The simplest such assumption is that the net advective transport of contaminant toward the membrane by the permeating flow is balanced completely by diffusion back to the bulk concentrate, not by downstream transport within the CP layer. This assumption, which applies exactly in the case of frontal filtration and was quantified by Equation 15-101 in our analysis of those systems, can be expressed mathematically

---

[26] This computation involves taking the ratio of the first (spatial) moment of the flux to the zeroth moment, consistent with the general approach for computing average values presented in Chapter 2.

[27] In addition to the assumption that back-transport occurs primarily by diffusion, the conversion of Equation 15-191 to Equation 15-204 assumes that D_{tot} is independent of c. Some models take into account the decrease in the diffusion coefficient with increasing contaminant concentration due to interactions among the contaminants in concentrated suspensions, but that effect (which is different from shear-induced diffusion and occurs even in the absence of shear) is ignored here.

as follows.

$$v(c - c_p) = -D_{tot}\frac{dc}{dy} \quad (15\text{-}208)$$

When this equation is utilized in conjunction with the others developed previously, the model for the contaminant profile in a crossflow system becomes identical to the film-gel model for frontal filtration, except that δ_{CP}, $k_{mt,CP}$, and $\hat{J}_{V,lim}$ vary with axial location. The resulting equations for the contaminant concentration profile and flux as a function of x, and for the average flux over the whole element, are as follows (Blatt et al., 1970; Porter, 1972):

$$c(x,y) = c_c \exp\left(\frac{v_w(x)}{D_{tot}}[\delta_{CP}(x) - y]\right) \quad (15\text{-}209)$$

$$\hat{J}_{V,lim}(x) = \frac{D_{tot}}{\delta_{CP}(x)}\ln\frac{c_{lim}}{c_c} = k_{mt,CP}(x)\ln\frac{c_{lim}}{c_c} \quad (15\text{-}210)$$

$$\hat{J}_{V,lim,avg} = \frac{\int_0^{L_{el}} \hat{J}_{V,lim}(x)dx}{\int_0^{L_{el}} dx} = \overline{k}_{mt,CP}\ln\frac{c_{lim}}{c_c} \quad (15\text{-}211)$$

Note that $\hat{J}_{V,lim,avg}$ is a length-averaged value of *limiting* fluxes. That is, because a compact layer is assumed to be present throughout, the flux at each location is a limiting flux, but the value of that limiting flux nevertheless varies along the element length, due to the gradual thickening of the CP layer.

As noted previously, if the change in c_c is large enough from the entrance to the exit of the element that assigning it a constant, average value seems inappropriate, Equation 15-211 can be applied to sequential, short segments of the element, and the value of c_c applied can be different in each segment (as indicated in Figure 15-43). In this case, the value of $\overline{k}_{mt,CP}$ in a segment between x_1 and x_2 should not be computed as though a new CP layer started to form at the beginning of each segment; rather, it can be computed based on the expressions for $k_{mt,CP}(x)$ in Table 15-12 and the following, more general version of Equation 15-197:

$$\overline{k}_{mt,CP,x_1\text{-}x_2} = \frac{\int_{x_1}^{x_2} k_{mt,CP} dx}{\int_{x_1}^{x_2} dx} \quad (15\text{-}212)$$

■ **EXAMPLE 15-24.** For a system with the following properties and operating parameters, compute the expected profiles of k_{mt} and $\hat{J}_V$ as a function of x in the element, and also the average values of those parameters across the whole element, according to the film-gel model. Assume that a compact layer is present throughout the whole element. The system uses crossflow with recycle, and the value given as c_f is the concentration entering the element, after the raw feed has mixed with the recycled concentrate. The recovery during a single pass is small enough that c_c can be approximated by this value of c_f throughout the element. Note that $y = 0$ is defined to be the inner boundary of the CP layer (i.e., the location where the concentration first reaches c_{lim}), and therefore is a distance $\delta_{compact}$ away from the membrane surface).

$c_f = 60\,\text{mg/L}$; $\quad c_{lim} = 20{,}000\,\text{mg/L}$; $\quad \gamma_{y=0} = 2500/\text{s}$;
$D_{tot} = 1 \times 10^{-6}\,\text{cm}^2/\text{s}$; $\quad L_{el} = 75\,\text{cm}$

Solution. We can use the final expressions in Equations 15-202 and 15-210, respectively, to estimate the local mass transfer coefficient and limiting permeation flux at various locations in the element. For example, at $x = 1\,\text{cm}$, the calculations are

$$k_{mt}|_{1\,\text{cm}} = 0.544\left(\frac{D_{tot}^2 \gamma_{Sh,y=0}}{x}\right)^{1/3}$$

$$= 0.544\left(\frac{[1.0 \times 10^{-6}\,\text{cm}^2/\text{s}]^2[2500/\text{s}]}{1\,\text{cm}}\right)^{1/3}$$

$$= 7.38 \times 10^{-4}\,\text{cm/s}$$

$$\hat{J}_{V,lim}(1\,\text{cm}) = k_{mt,CP}|_{1\,\text{cm}}\ln\frac{c_{lim}}{c_c} = (7.38 \times 10^{-4}\,\text{cm/s})$$

$$\times \ln\frac{20{,}000\,\text{mg/L}}{60\,\text{mg/L}} = 4.29 \times 10^{-3}\,\text{cm/s}$$

$$= 4.29 \times 10^{-3}\,\text{cm/s}\left(\frac{1\,\text{L}}{1000\,\text{cm}^3}\right)(10^4\,\text{cm}^2/\text{m}^2)(3600\,\text{s/h})$$

$$= 154.4\,\text{L/m}^2\,\text{h}$$

Similar calculations can be carried out for different values of x, yielding the results shown in Figure 15-44. Both the local flux and the local mass transfer coefficient decrease by more than 75% along the membrane length.

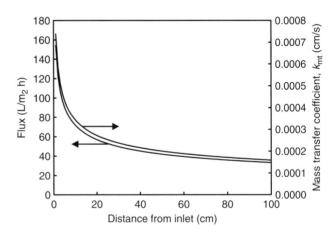

FIGURE 15-44. Changes in the permeate flux and mass transfer coefficient through the CP layer as water moves through the membrane fiber in the example system.

Equations 15-203 and 15-211 can be used to compute the average values, yielding

$$\overline{k}_{mt,CP} = 0.816 \left(\frac{D_{tot}^2 \gamma_{Sh,y=0}}{L_{el}} \right)^{1/3}$$

$$= 0.816 \left(\frac{[1.0 \times 10^{-6} \, cm^2/s]^2 [2500/s]}{75 \, cm} \right)^{1/3}$$

$$= 2.63 \times 10^{-4} \, cm/s$$

$$\hat{J}_{V,lim,avg} = \overline{k}_{mt,CP} \ln \frac{c_{lim}}{c_c} = (2.63 \times 10^{-4} \, cm/s) \ln \frac{20{,}000 \, mg/L}{60 \, mg/L}$$

$$= 1.53 \times 10^{-3} \, cm/s$$

$$= 1.53 \times 10^{-3} \, cm/s \left(\frac{1 \, L}{1000 \, cm^3} \right) (10^4 \, cm^2/m^2)(3600 \, s/h)$$

$$= 54.9 \, L/m^2 h \qquad \blacksquare$$

Effect of Crossflow Velocity on Flux. If contaminant back-transport is dominated by Brownian diffusion, D_{tot} is insensitive to the crossflow velocity. In that case, the combination of Equation 15-211 with the equations in Table 15-12 leads to the prediction that flux varies with the one-third power of $\gamma_{y=0}$ or $\overline{u}$. On the other hand, if contaminant back-transport is dominated by shear-induced diffusion, the limiting flux is directly proportional to $\gamma_{Sh,y=0}$ or $\overline{u}$. Thus, not surprisingly, the advantages of crossflow are considerably greater if shear-induced diffusion is significant than if back-transport is dominated by Brownian motion alone.

Modified Versions of the File and Film-Gel Models for Systems with Crossflow Two alternatives to the film and film-gel models have been used fairly widely to describe crossflow systems, based on different choices for one of the boundary conditions and for the equation establishing a pseudo-steady state in the CP layer. Although these modifications can be applied regardless of whether a compact layer is present, the mathematics are simpler for cases where such a layer is present throughout the system, and the models have been used primarily to describe such systems.

In the film and film-gel models, the CP layer is envisioned to have an identifiable, abrupt edge at $y = \delta_{CP}$, where $c = c_c$ and the slope of the concentration profile is discontinuous. Because of this feature, the thickness of the CP layer and the mass transfer coefficient in the layer are important model parameters that must be determined independently, such as by analogy to the Lévêque analysis of heat transfer. The abrupt transition from the CP layer to the bulk concentrate is unrealistic, and a number of researchers (e.g., Probstein et al., 1978; Trettin and Doshi, 1980) have argued that it might be preferable to represent the CP layer as extending indefinitely in the y direction, with the contaminant concentration approaching c_c asymptotically. In that case, the boundary condition becomes

$$c = c_c \quad \text{at } y = \infty \qquad (15\text{-}213)$$

The different concentration profiles expected using Equation 15-207 versus Equation 15-213 as a boundary condition for systems with compact layers throughout are shown in Figure 15-45. At each value of x, the concentration decays from c_{lim} at $y = 0$ to c_c at some distance from the edge of the compact layer. Because the contaminant profile in the CP layer is affected by the crossflow in addition to the permeation flow and diffusion, the profile is not necessarily exponential

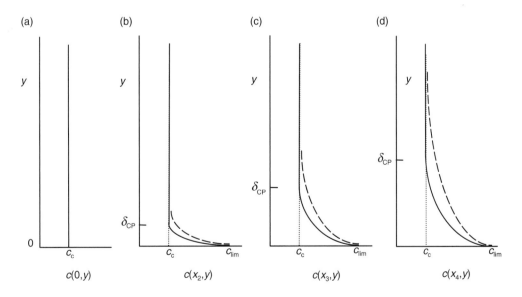

FIGURE 15-45. Particle concentration profiles in crossflow filtration in systems where a compact layer is present at all axial locations. The four figures represent axial locations progressively farther downstream. Solid curves and broken curves correspond (qualitatively) to Equations 15-207 and 15-213, respectively.

(as it is in frontal filtration), but it is usually determined to be only slightly different from exponential. When Equation 15-207 is used, the decline in c ceases abruptly at $y = \delta_{CP}$ and $c = c_c$ at all $y > \delta_{CP}$, whereas when Equation 15-213 is used, the decline is steady and c_c is approached asymptotically at large values of y. In either case, the thickness of the CP layer increases with increasing x.

Trettin and Doshi (1980) reported that, for systems in which a compact layer is present throughout, predictions of $\hat{J}_{V,\text{lim,avg}}$ using Equation 15-213 as the boundary condition approached those based on the film-gel model when the feed concentration (which was assumed to equal c_c) was >0.25 c_{lim} and the permeation rate was small. However, if the feed concentration was substantially less than c_{lim} (as is typically the case), the model predictions differed substantially, with the predictions that used Equation 15-213 fitting experimental data better than those based on the film-gel model (Equation 15-211). If c_c/c_{lim} was less than approximately 0.1, the limiting flux was fit much better by the following equation than by Equation 15-211:[28]

$$\hat{J}_{V,\text{lim,avg}} = \left(\frac{3}{2}\right)^{2/3} \left(\frac{D_{\text{tot}}^2 \gamma_{\text{Sh},y=0}}{L_{\text{el}}}\right)^{1/3} \left(\frac{c_{\text{lim}}}{c_c}\right)^{1/3}$$

$$= 1.31 \left(\frac{D_{\text{tot}}^2 \gamma_{\text{Sh},y=0}}{L_{\text{el}}}\right)^{1/3} \left(\frac{c_{\text{lim}}}{c_c}\right)^{1/3} \quad (15\text{-}214)$$

■ **EXAMPLE 15-25.** For the same system and assumptions as in Example 15-24, compute the average flux according to the Trettin–Doshi model.

Solution. Substituting the given values into Equation 15-214, assuming $c_c = c_f$, we find:

$$\hat{J}_{V,\text{lim,avg}} = \left(\frac{3}{2}\right)^{2/3} \left(\frac{D_{\text{tot}}^2 \gamma_{y=0}}{L_{\text{el}}}\right)^{1/3} \left(\frac{c_{\text{lim}}}{c_c}\right)^{1/3}$$

$$= \left(\frac{3}{2}\right)^{2/3} \left(\frac{[1.0 \times 10^{-6}\,\text{cm}^2/\text{s}]^2 [2500/\text{s}]}{75\,\text{cm}}\right)^{1/3}$$

$$\times \left(\frac{20{,}000\,\text{mg/L}}{60\,\text{mg/L}}\right)^{1/3} = 2.92 \times 10^{-3}\,\text{cm/s}$$

$$= (2.92 \times 10^{-3}\,\text{cm/s})\left(\frac{1\,\text{L}}{1000\,\text{cm}^3}\right)(10^4\,\text{cm}^2/\text{m}^2)(3600\,\text{s/h})$$

$$= 105.2\,\text{L/m}^2\,\text{h}$$

For this example system, in which $c_f/c_{\text{lim}} < 0.01$, the predicted average flux is almost twice that computed using the film-gel model. ■

An alternative modification to the film and film-gel models is based on the idea that material in the CP layer can migrate downstream and exit the membrane element, and that this movement might be a significant factor in maintaining steady-state conditions in the element. This possibility can be incorporated into the analysis by writing a mass balance on contaminant in a control volume that includes the entire CP layer between the entrance to the element and some arbitrary location downstream (x_1). Such a control volume for a membrane with a rectangular flow channel is shown schematically in Figure 15-46. The volume has a length x_1, a width W_{el} (into the page), and a height y^* such that, at the exit of the element, y^* is between the edges of the CP and the velocity boundary layers; as a result, at $x = x_1$, $c(y^*) \approx c_c$ and $u(y^*) = \gamma_{\text{Sh},y=0}\,y^*$.

At steady state, a mass balance on contaminant in this control volume can be written as follows:

$$0 = \int_0^{y^*} c(0,y)u(y)W_{\text{el}}\,dy - \int_0^{y^*} c(x_1,y)u(y)W_{\text{el}}\,dy$$

$$+ \int_0^{x_1} c_f v(x)W_{\text{el}}\,dx - \int_0^{x_1} c_p v(x)W_{\text{el}}\,dx \quad (15\text{-}215)$$

where the first term on the right side accounts for contaminant entering the control volume via axial advection at the inlet (mass/time); the second term accounts for contaminant exiting with the axial flow at x_1; the third term accounts for contaminant entering via advection from the bulk solution (i.e., across the $y = y^*$ boundary) between the inlet and x_1; and the final term accounts for contaminant exiting by permeation through the membrane. Diffusive transport out of the

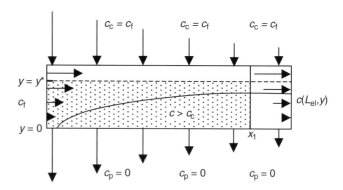

FIGURE 15-46. The control volume (stippled area) for a mass balance equating contaminant transport into and out of the CP layer from $x = 0$ to x_1. The height of the control volume is such that it contains the CP layer but is still within the velocity boundary layer.

---

[28] Trettin and Doshi (1980) derived Equation 15-214 using a mathematical technique known as the integral method, which requires an assumption for the functional form of the c versus y profile. The equation is exact for the assumed form of the profile, but since that form has not been confirmed, the equation should be viewed as approximate. In addition to Equation 15-214, Trettin and Doshi presented a more complex equation for computing $\hat{J}_{V,\text{lim,avg}}$ over the full range of c_{lim}/c_f ratios, but Equation 15-214 (which applies if $c_f/c_{\text{lim}} \ll 1$) is much more widely cited.

CP layer is not included in the mass balance because the boundary of the control volume at y^* is specified to be outside the CP layer, at a location where the concentration gradient is negligible.

The first and second terms in Equation 15-215 can be combined based on the assumption that the velocity profiles are identical at the entrance and x_1, and the third and fourth terms can be combined as well. Then, dividing by W_{el}, noting that $c(0, y) = c_f = c_c$, and rearranging, we can write

$$\int_0^{x_1} (c_c - c_p)v(x)dx = \int_0^{y^*} [c(x_1, y) - c_c]u(y)dy \quad (15\text{-}216)$$

Because $c(x_1, y) \approx c_c$ at all $y > y^*$, the term in brackets on the right side of Equation 15-216 is zero when $y > y^*$, so the upper limit of integration can be changed to ∞ without changing the value of the integral. Making this change, assuming that c_p is approximately constant axially, and dropping the nonessential arguments of v, c, and u, we obtain

$$(c_c - c_p)\int_0^{x_1} v\,dx = \int_0^\infty u[c(x_1) - c_c]dy \quad (15\text{-}217)$$

Equation 15-217, sometimes referred to as the *flux conservation* equation, is a mathematical statement of the requirement that, at steady state, all the contaminant that enters the specified control volume must ultimately be carried out of the element, either in the downstream flow (either as part of the concentrate or the more slowly flowing CP layer) or the permeate.

Using Equation 15-217 as the basis for steady state and the boundary condition that $c = c_c$ at $y = \infty$ (Equation 15-213), Song (1998) obtained the following equations for the concentration profile and the local and average permeate fluxes in a system with a compact layer throughout:

$$c(y) - c_c = (c_{\lim} - c_c)\exp\left(-\frac{\hat{J}_{V,\lim}(x)}{D_{tot}}y\right) \quad (15\text{-}218)$$

$$\hat{J}_{V,\lim}(x) = \left(\frac{2}{3}\right)^{1/3}\left(\frac{D_{tot}^2 \gamma_{Sh,wall}}{x}\right)^{1/3}\left(\frac{c_{\lim}}{c_c} - 1\right)^{1/3} \quad (15\text{-}219)$$

$$\hat{J}_{V,\lim,avg} = \frac{\int_0^{L_{el}} \hat{J}_{V,\lim}(x)dx}{\int_0^{L_{el}} dx}$$

$$= \left(\frac{3}{2}\right)^{2/3}\left(\frac{D_{tot}^2 \gamma_{Sh,y=0}}{L_{el}}\right)^{1/3}\left(\frac{c_{\lim}}{c_c} - 1\right)^{1/3}$$

$$= 1.31\left(\frac{D_{tot}^2 \gamma_{Sh,y=0}}{L_{el}}\right)^{1/3}\left(\frac{c_{\lim}}{c_c} - 1\right)^{1/3} \quad (15\text{-}220)$$

This equation is identical to the result obtained by Trettin and Doshi for systems in which $c_c \ll c_{\lim}$ (Equation 15-214), except for the subtraction of 1 from $c_{\lim}/c_c$ in Equation 15-220. If $c_c \ll c_{\lim}$, this subtraction has a negligible effect on the calculation, so the two models are essentially identical if the feed concentration is much less than the limiting concentration. If the feed concentration is close to the limiting concentration, the Trettin–Doshi model approximates the film-gel model (Equation 15-211), and the subtraction of 1 from $c_{\lim}/c_c$ has a significant effect on the calculation of $\hat{J}_{V,\lim,avg}$ using the Song model, so the results of Song's model diverge from those of the other models.

■ **EXAMPLE 15-26.** For the same system and assumptions as in Example 15-24, compute the average flux according to the Song model.

Solution. The calculations involve direct substitution into Equation 15-220

$$\hat{J}_{V,\lim,avg} = \left(\frac{3}{2}\right)^{2/3}\left(\frac{D_{tot}^2 \gamma_{Sh,y=0}}{L_{el}}\right)^{1/3}\left(\frac{c_{\lim}}{c_c} - 1\right)^{1/3}$$

$$= \left(\frac{3}{2}\right)^{2/3}\left(\frac{[1.0 \times 10^{-6}\,\text{cm}^2/\text{s}]^2[2500/\text{s}]}{75\,\text{cm}}\right)^{1/3}$$

$$\times \left(\frac{20{,}000\,\text{mg/L}}{60\,\text{mg/L}} - 1\right)^{1/3} = 2.92 \times 10^{-3}\,\lim\,\text{cm/s}$$

$$= 2.92 \times 10^{-3}\,\text{cm/s}\left(\frac{1\,\text{L}}{1000\,\text{cm}^3}\right)(10^4\,\text{cm}^2/\text{m}^2)(3600\,\text{s/h})$$

$$= 105.2\,\text{L/m}^2\,\text{h}$$

The result is virtually identical to that for the Trettin–Doshi model, consistent with the fact that $c_f \ll c_{\lim}$. ■

Relating Flux to TMP in Crossflow Systems with Ubiquitous Compact Layers. In frontal filtration systems containing a compact layer, $\hat{J}_V$ is independent of the TMP. For the same reason, in systems with crossflow, the flux at any location in the element (and therefore the average flux) is insensitive to the TMP if a compact layer is present throughout the axial length. That is, the same (limiting) average flux will be achieved at any TMP, as long as the TMP is large enough to induce formation of a compact layer. Under these conditions, the TMP is dissipated almost entirely in the compact layer and the membrane, and the only effect of increasing the TMP is predicted to be an increase the thickness of the compact layer everywhere in the element.

Systems with no Compact Layer Adapting the Film Model to Systems with Crossflow. The analysis of a membrane system with crossflow and with no compact layer is more complex than that presented above because, in the absence of a compact layer, the contaminant concentration at the inner boundary of the CP layer ($c_{c_{c/m}}$) is both unknown and changing as a function of x. In addition to increasing the mathematical complexity, the variation in $c_{c_{c/m}}$ makes the correlations in Table 15-12 less appropriate for estimating $k_{mt,CP}$. Therefore, the simplifying assumption is frequently made that $c_{c_{c/m}}$ is constant (i.e., independent of x) at some average, initially unspecified value, so that the equations in Table 15-12 can still be used to calculate $k_{mt,CP}(x)$ and $\overline{k}_{mt,CP}$.

When the simplification of a constant $c_{c_{c/m}}$ is combined with appropriate boundary conditions and an equation establishing how steady state is maintained, the mass balance on contaminant (Equation 15-204) can be solved to determine the concentration profile in the CP layer and the flux at any value of x. In fact, if Equation 15-207 ($c = c_c$ at $y = \delta_{CP}$) is used as the boundary condition at the outer edge of the CP layer and $\overline{k}_{mt,CP}$ is used as the value of $k_{mt,CP}$ at all x, the system of equations becomes identical to those for a frontal filtration system in the absence of a compact layer. If the rejection is complete, the solution to those equations is essentially identical to that shown in Equation 15-112, with $\overline{k}_{mt,CP}$ replacing $k_{mt,CP}$

$$\hat{J}_{V,avg} = \overline{k}_{mt,CP} \ln \frac{c_{c_{c/m}}}{c_c} \quad (15\text{-}221)$$

Relating Flux to TMP in Systems with Crossflow and no Compact Layer. In the absence of a compact layer, the flux depends on the TMP. If we make the same assumptions and apply the same boundary conditions that led to Equation 15-221, the predicted $\hat{J}_{V,avg}$ versus TMP relationship for a system with crossflow is identical to that for a frontal filtration system, with average values over the element length in the crossflow system replacing the corresponding uniform values in frontal filtration. For frontal filtration systems, we derived different relationships between TMP and $\hat{J}_V$ depending on whether the contaminants were particles or solutes (Equations 15-128 and 15-131, respectively). A similar situation applies in crossflow filtration, yielding

Particles:

$$\text{TMP} = 3\pi\mu_L d_p D_{Br} n_c \left(\exp\left(\frac{\hat{J}_{V,avg}}{\overline{k}_{mt,CP}} \right) - 1 \right) + \frac{\hat{J}_{V,avg}}{k_w^*} \quad (15\text{-}222)$$

Solutes:

$$k_w^* \left(\text{TMP} - \overline{\Pi}_{c_{c/m}} \right) = \overline{k}_{mt,CP} \ln \frac{\overline{c}_{c_{c/m}}}{c_c} \quad (15\text{-}223)$$

The TMP-$\hat{J}_{V,avg}$ relationship shown in Equation 15-222 is based on the idealization that the headloss for flow past the particles can be computed based on Stokes flow past an isolated particle; as in frontal filtration systems, if idealization does not apply, a more complex relationship can be postulated.

If the contaminants are particles, all the parameters in Equation 15-222 other than $\hat{J}_{V,avg}$ and TMP are likely to be known or calculable, so that the TMP required to achieve a given average flux can be computed directly. If the contaminants are solutes, Equation 15-118 can be combined with Equation 15-221 to yield

$$\hat{J}_{V,avg} = k_w^* \left(\text{TMP} - \overline{\Pi}_{c_{c/m}} \right) = \overline{k}_{mt,CP} \ln \frac{\overline{c}_{c_{c/m}}}{c_c} \quad (15\text{-}224)$$

For a given TMP, the equality between the middle and final expressions in Equation 15-224 can be solved iteratively with an equation relating $\overline{\Pi}_{c_{c/m}}$ to $\overline{c}_{c_{c/m}}$ to determine the unique values of those parameters that satisfy the equations; the average flux can then be determined using either of the expressions in Equation 15-224.

Because $\overline{\Pi}_{c_{c/m}}$ increases and $\overline{k}_{mt,CP}$ decreases with increasing element length, $\hat{J}_{V,avg}$ decreases as the element length increases, regardless of whether the contaminants are particles or solutes. Thus, as we would expect intuitively, the highest fluxes achievable with a given TMP occur for very short elements (because the CP layer remains relatively thin throughout the element). In the design process, the advantages of increasing average flux with decreasing element length must be traded off against the accompanying decrease in membrane area per module.

While the use of Equations 15-222–15-224 is attractive because of their relative simplicity and similarity to the equations for the flux in frontal filtration systems, they suffer from the very assumptions that make them so simple. Elimelech and co-workers (Song and Elimelech, 1995; Elimelech and Bhattacharjee, 1998) were particularly critical of the use of correlations from nonpermeable systems to estimate $\overline{k}_{mt,CP}$ (the equations in Table 15-12) and also of the assumed constancy of $c_{c_{c/m}}$ (or, equivalently, the use of an average value of $c_{c_{c/m}}$ to characterize the whole membrane element). They pointed out that, by using the boundary condition that c approached c_c at $y = \infty$ (Equation 15-213), the need to estimate $\delta_{CP}(x)$ and $\overline{k}_{mt,CP}$ could be eliminated. Using this boundary condition, along with the flux conservation equation (Equation 15-217) to define how steady state is maintained and the assumption that the contaminant is completely rejected, they derived the following relationship between axial location and other parameters in the system, applicable to dissolved contaminants in the absence of a compact layer:[29]

$$x = \alpha \frac{\Delta_{CP}\Pi_{ideal}}{\Delta_{CP}\Pi} \left\{ \frac{\text{TMP}_{eff} - 3\Delta_{CP}\Pi}{(\text{TMP}_{eff} + \Delta_{CP}\Pi)^3} - \frac{1}{(\text{TMP}_{eff})^2} \right\} \quad (15\text{-}225)$$

---

[29] Equation 15-225 appears substantially different from any equation presented by Elimelech and Bhattacharjee (1998), but it is equivalent to their result.

where

- $\alpha = (\gamma_{\text{Sh},y=0} D_{\text{Br}}^2 (\mu_L \mathcal{R}_m)^3)/(6\Pi_{\text{c,ideal}})$, a parameter defined for convenience that can be evaluated based on the known information about the system inputs and properties.
- $\Delta_{\text{CP}}\Pi_{\text{ideal}} = RT(c_c - c_{c_{c/m}})$, the osmotic pressure difference across the CP layer based on ideal solution behavior (i.e., the van't Hoff equation (Equation 15-131)).
- $\text{TMP}_{\text{eff}} = \text{TMP} - \Delta_{c/p}\Pi$, the effective pressure difference (i.e., the net driving force) between the bulk concentrate and the permeate.

In Equation 15-225, $\Delta_{\text{CP}}\Pi$ is the actual osmotic pressure differential across the CP layer and differs from $\Delta_{\text{CP}}\Pi_{\text{ideal}}$ if the contaminant becomes sufficiently concentrated near the membrane that the osmotic pressure at that location cannot be approximated by the van't Hoff expression. As noted in the discussion of Equation 15-46, this situation is likely to arise only if $\phi_{c_{c/m}}$ is larger than several percent, a scenario that rarely occurs except in systems where macromolecules are the target contaminants. Although Equation 15-225 was derived for dissolved contaminants, Elimelech and Bhattacharjee (1998) suggested that it could also be used to model systems with particulate contaminants whose back-transport was dominated by Brownian motion. In essence, their approach characterizes the molar Gibbs energy lost by water passing through a concentrated suspension of small particles as an equivalent change in the osmotic pressure of the water, and then uses this value of $\Delta_{\text{CP}}\Pi$ in the equation.

Typically, all the terms on the right side of Equation 15-225 other than $\Delta_{\text{CP}}\Pi$ and $E_{\text{CP}}\Pi_{\text{ideal}}$ would be known, and those unknown values can be determined for any given value of $c_{c_{c/m}}$. Therefore, one can choose any value of $c_{c_{c/m}}$ and evaluate the right side of Equation 15-225 to determine the value of x at which this value of $c_{c_{c/m}}$ would be found. By repeating this procedure for several values of $c_{c_{c/m}}$, the complete $c_{c_{c/m}}$ versus x profile can be established. This profile can then be used to determine the $\Delta_{m_{c/p}}\Pi$ versus x profile. Finally, if the contaminants are solutes and no compact layer forms, the pressure at the membrane surface is the same as that in the bulk, so $\Delta_{m_{c/p}}P = \text{TMP}$, and the flux at any x can be expressed as

$$\hat{J}_V(x) = \frac{\text{TMP} - \Delta_{m_{c/p}}\Pi(x)}{\mu_L \mathcal{R}_{m_{c/p}}} \qquad (15\text{-}226)$$

Equation 15-226 can be used to determine the permeation flux at the axial location, x, where that value of $c_{c_{c/m}}(x)$ is found. The average flux over the whole membrane can then be determined by an integration analogous to that in Equation 15-197.

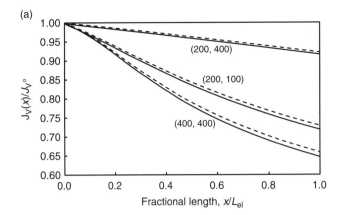

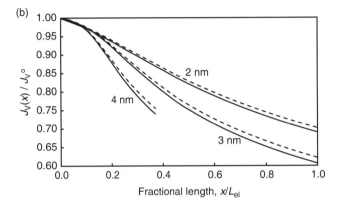

FIGURE 15-47. Predicted axial variation in permeate flux in systems with no compact layer. In both figures, $L_{\text{el}} = 0.5$ m, $\phi_f = 10^{-3}$; $\mu_L \mathcal{R}_{m_{c/p}} = 2 \times 10^{10}$ Pa s/m. In both figures, $L_{\text{el}} = 0.5$ m, $\phi_f = 10^{-3}$; $\mu_L \mathcal{R}_{m_{c/p}} = 2 \times 10^{10}$ Pa s/m. In(a), $d_p = 8$ nm, and values shown in parentheses are TMP in kPa and shear rate in s^{-1}; in (b), $\gamma_{\text{Sh,wall}} = 400$ s^{-1}, and values shown are the particle diameter. In both parts, solid curves are exact numerical solutions to the governing equations, and broken curves are approximations that are easier to calculate. *Source:* After Elimelech and Bhattacharjee (1998).

Elimelech and Bhattacharjee (1998) presented a graphical approach for carrying out these calculations, and an analytical approach that yielded explicit expressions for $\hat{J}_V(x)$ and $\hat{J}_{V,\text{avg}}$ for systems in which $c_{c_{c/m}}$ does not change significantly over the axial distance being analyzed. The predicted effects of TMP, c_f, the shear rate, and the size of the contaminant species on the permeation flux as a function of axial distance according to this model are shown in Figure 15-47. As with the models presented previously, the local permeate flux declines dramatically as water moves from the inlet to the outlet of the element.

Systems in which a Compact Layer is Present along only a Portion of the Membrane Element

If a compact layer forms in the system, it will first appear at the value of x where $c_{c_{c/m}}$ first equals c_{lim}. This value of x, which we will refer to as x_{crit}, can be computed by using c_{lim} to calculate $\Delta_{CP}\Pi$ and $\Delta_{CP}\Pi_{ideal}$ in Equation 15-225. If $x_{crit} > L_{el}$, then a compact layer does not form anywhere in the system. At the other extreme, if $x_{crit} \ll L_{el}$, then a compact layer will be present along almost the entire axial length, and the flux can be computed using one of the equations derived previously for such a system (e.g., Equation 15-220).

In the intermediate case where x_{crit} is less than L_{el}, but is still a significant fraction of L_{el}, a compact layer is expected to be present between x_{crit} and L_{el}, but not between the inlet and x_{crit}. In that case, the average flux from the inlet to x_{crit} can be determined using the procedure for a system with no compact layer, replacing L_{el} by x_{crit}. The flux in the portion of the membrane where the compact layer is present can then be computed based on a modification of one of the equations developed for situations in which the compact layer is present over the whole membrane length, such as Equation 15-220. That equation cannot be applied directly, because it assumes that the CP layer has zero thickness at the location where the compact layer first appears, whereas in the current scenario, the CP layer is already partially developed at x_{crit}. Assuming that the growth of the CP layer is unaffected by the formation of the compact layer (other than by the location of $y = 0$ moving away from the membrane surface), the average flux in the region from x_{crit} to L_{el} can be computed by determining $\hat{J}_{V,lim,avg}$ from 0 to L_{el} according to Equation 15-220, and then subtracting the value from 0 to x_{crit}; that is,

$$\hat{J}_{V,lim,avg}\Big|_{x_{crit}}^{L_{el}} = \hat{J}_{V,lim,avg}\Big|_{0}^{L_{el}} - \hat{J}_{V,lim,avg}\Big|_{0}^{x_{crit}} \quad (15\text{-}227)$$

where each term on the right side is computed using Equation 15-220.

EXAMPLE 15-27.

(a) In several previous examples, the flux through a membrane element was calculated assuming that a compact layer was present along the entire element length. For the same system conditions, determine if and where in the system a compact layer first forms, using the Elimelech and Bhattacharjee model for the system in which the TMP is 700 kPa, $\mathscr{R}_{m_{c/p}}$ is 2×10^{13}/m, and the water temperature is $25°C$. The contaminant has an MW of 1000 and is rejected virtually completely by the membrane. The osmotic pressure throughout the system is generated primarily by the target contaminant and can be calculated using the correlation of Carnahan and Starling (Equation 15-46). Assume that the volume fraction of the contaminant is 4.5×10^{-4} in the feed and 0.15 in a solution with the limiting concentration.

(b) What is the flux at the location computed in (a)?

Solution.

(a) The location at which $c_{c_{c/m}}$ first equals c_{lim} can be computed using Equation 15-225. However, to use that equation, we need to compute the osmotic pressure at various locations in the system. The osmotic pressure in the permeate is zero, since we are assuming that the contaminant is completely rejected by the membrane. Given the MW of 1000, we find:

$$c_f = c_c = \frac{60 \text{ mg/L}}{10^6 \text{ mg/mol}} = 6 \times 10^{-5} \text{ mol/L};$$

$$c_{lim} = \frac{20{,}000 \text{ mg/L}}{10^6 \text{ mg/mol}} = 2 \times 10^{-2} \text{ mol/L}$$

The correction factor (Φ) that accounts for non-ideality in the Carnahan–Starling equation can be evaluated in the concentrate (i.e., feed) solution and at the location where the compact layer forms as

$$\Phi_c = \frac{1 + \phi_c + \phi_c^2 - \phi_c^3}{(1 - \phi_c)^3} = 1.002;$$

$$\Phi_{lim} = \frac{1 + \phi_{lim} + \phi_{lim}^2 - \phi_{lim}^3}{(1 - \phi_{lim})^3} = 1.904$$

Accordingly, the osmotic pressures in the bulk concentrate and in the solution containing the contaminant at c_{lim} are

$$\Pi_c = RTc_c\Phi_c = (8.314 \text{ kPa L/mol K})(298 \text{ K}) \times (6 \times 10^{-5} \text{ mol/L})(1.002) = 0.15 \text{ kPa}$$

$$\Pi_{lim} = RTc_{lim}\Phi_{lim} = (8.314 \text{ kPa L/mol K})(298 \text{ K}) \times (2 \times 10^{-2} \text{ mol/L})(1.904) = 94.33 \text{ kPa}$$

The corresponding values of Π in ideal solutions (i.e., according to the van't Hoff equation) are computed as in the preceding calculations, except without the final term (Φ) in the product. The resulting values for Π_{ideal}, $\Delta_{CP}\Pi$, and $\Delta_{CP}\Pi_{ideal}$ are

$$\Pi_{c,ideal} = RTc_c = 0.15 \text{ kPa}$$

$$\Pi_{lim,ideal} = RTc_{lim} = 49.55 \text{ kPa}$$

$$\Delta_{CP}\Pi = \Pi_c - \Pi_{lim} = (0.15 - 94.33) \text{ kPa} = -94.18 \text{ kPa}$$

$$\Delta_{CP}\Pi_{ideal} = \Pi_{c,ideal} - \Pi_{lim,ideal} = (0.15 - 49.55) \text{ kPa}$$
$$= -49.40 \text{ kPa}$$

The other values needed to use Equation 15-225 can be computed as follows:

$$\alpha = \frac{\gamma_{\text{Sh,wall}} D_{\text{Br}}^2 \left(\mu_L \mathcal{R}_{m_{c/p}}\right)^3}{6\Pi_{c,\text{ideal}}}$$

$$= \frac{(2500/\text{s})\left([1.0 \times 10^{-6}\,\text{cm}^2/\text{s}][10^{-4}\,\text{m}^2/\text{cm}^2]\right)^2 \left([1.00 \times 10^{-6}\,\text{kPa s}][2 \times 10^{13}/\text{m}]\right)^3}{6(0.15\,\text{kPa})}$$

$$= 2.24 \times 10^5\,\text{m kPa}^2$$

$$\text{TMP}_{\text{eff}} = \text{TMP} - \Delta_{\text{tot}}\Pi = \text{TMP} - (\Pi_c - \Pi_p) = [700 - (0.15 - 0)]\,\text{kPa} = 699.85\,\text{kPa}$$

Because $\Delta_{\text{CP}}\Pi$, and $\Delta_{\text{CP}}\Pi_{\text{ideal}}$ have been calculated based on $c_{c_{c/m}} = c_{\text{lim}}$, the value of x determined by inserting these values into Equation 15-225 will be x_{crit}. This value is

$$x_{\text{crit}} = \alpha \frac{\Delta_{\text{CP}}\Pi_{\text{ideal}}}{\Delta_{\text{CP}}\Pi} \left\{ \frac{\text{TMP}_{\text{eff}} - 3\Delta_{\text{CP}}\Pi}{(\text{TMP}_{\text{eff}} + \Delta_{\text{CP}}\Pi)^3} - \frac{1}{(\text{TMP}_{\text{eff}})^2} \right\}$$

$$= (2.24 \times 10^5\,\text{m kPa}^2)\left(\frac{-49.40\,\text{kPa}}{-94.18\,\text{kPa}}\right)$$

$$= \left\{ \frac{699.85\,\text{kPa} - 3(-94.18\,\text{kPa})}{(699.85\,\text{kPa} - 94.18\,\text{kPa})^3} - \frac{1}{(699.85\,\text{kPa})^2} \right\}$$

$$= 0.28\,\text{m} = 28\,\text{cm}$$

(b) The flux at $x = 0.28$ m is given by Equation 15-226 as

$$\hat{J}_V(x) = \frac{\text{TMP} - \Delta_{m_{c/p}}\Pi(x)}{\mu_L \mathcal{R}_m} = \frac{\text{TMP} - (\Pi_{\text{lim}} - \Pi_p)}{\mu_L \mathcal{R}_m}$$

$$= \frac{700\,\text{kPa} - (94.34 - 0)\,\text{kPa}}{(1.00 \times 10^{-6}\,\text{kPa s})(2 \times 10^{13}/\text{m})}(100\,\text{cm/m})$$

$$= 3.02 \times 10^{-3}\,\text{cm/s}$$

$$= 3.02 \times 10^{-3}\,\text{cm/s}\left(\frac{1\,\text{L}}{1000\,\text{cm}^3}\right)(10^4\,\text{cm}^2/\text{m}^2)(3600\,\text{s/h})$$

$$= 109.0\,\text{L/m}^2\,\text{h} \qquad \blacksquare$$

Modeling Crossflow Filtration of Particles Subject to Significant Inertial Lift For particles with nominal diameters of several micrometers or more, inertial lift can become the dominant back-transport mechanism, so diffusive transport term in Equation 15-191 can be ignored. If we also assume that the concentration in the bulk concentrate is constant and evaluate the equation at $y = 0$, we obtain

$$0 = \frac{\partial vc}{\partial y} - \frac{\partial v_{\text{lift}} c}{\partial y} \quad \text{at } y = 0 \qquad (15\text{-}228)$$

By inspection, this equation is satisfied when $v = v_{\text{lift}}$. Thus, steady state is achieved when the advective flux of particles toward the membrane at the surface of the cake layer exactly balances their flux away from the membrane at that location due to inertial lift.

Davis (1992) quantified the relative hydraulic resistances associated with the cake layer and the membrane in such systems in terms of a dimensionless parameter β, equal to the ratio of the resistance of a cake filling the entire channel to that of the membrane alone. Thus, for a rectangular channel with membranes on both sides and a half-height h_o, Davis defined β by

$$\beta = \frac{h_o \hat{\mathcal{R}}_{\text{cake}}}{\mathcal{R}_m} \qquad (15\text{-}229)$$

where $\hat{\mathcal{R}}_{\text{cake}}$ is the specific resistance of the cake (i.e., the resistance per unit thickness). Making this substitution, the combined resistance of the cake and membrane can be written as follows:

$$\mathcal{R}_{\text{cake}} = \frac{\beta \mathcal{R}_m \delta_{\text{cake}}}{h_o} \qquad (15\text{-}230)$$

$$\mathcal{R}_{\text{tot}} = \mathcal{R}_m + \mathcal{R}_{\text{cake}} = \mathcal{R}_m\left(1 + \frac{\beta \delta_{\text{cake}}}{h_o}\right) \qquad (15\text{-}231)$$

Davis distinguished two limiting cases of Equation 15-231, corresponding to systems in which the resistance associated with the cake layer was either much greater or much less than that of the membrane. If the cake resistance dominates ($\beta \gg 1$), a thin cake layer develops and the shear rate at $y = 0$ is almost the same as in the absence of a cake. However, the permeation velocity is significantly less than through the bare membrane due to the resistance imposed by the cake. In this case, as when Brownian or shear-induced diffusion is the dominant back-transport mechanism, a limiting flux is reached that cannot be increased by increasing the TMP; increasing the TMP simply increases the thickness of the high-resistance cake layer proportionately.

In contrast, if the membrane resistance dominates ($\beta \ll 1$), a thick cake layer develops that significantly reduces the cross-section available for crossflow, increases the axial velocity, and therefore increases the shear rate at $y = 0$. However, in contrast

TABLE 15-13. Expressions for Flux and Cake Layer Thickness Under Various Limiting Conditions, for Systems with Back-Transport Dominated by Inertial Lift

	$\hat{J}_V$		δ_{cake}	
$\beta \gg 1$	v^o_{lift}	(15-232)	$\dfrac{\mathcal{R}_m}{\hat{\mathcal{R}}_{cake}}\left(\dfrac{\hat{J}^o_V}{v^o_{lift}} - 1\right)$	(15-233)
$\beta \ll 1$	$\hat{J}^o_V \left(= \dfrac{TMP}{\mu_L \mathcal{R}_m}\right)$	(15-234)	$h_o\left[1 - \left(\dfrac{v^o_{lift}}{\hat{J}^o_V}\right)^{1/4}\right]$ (rectangular)	(15-235)
			$r_o\left[1 - \left(\dfrac{v^o_{lift}}{\hat{J}^o_V}\right)^{1/5}\right]$ (circular)	(15-236)

to the case where $\beta \gg 1$, the cake layer imposes negligible resistance on the permeate flow. In this case, the permeation rate is proportional to the TMP, as would be the case in the absence of a cake layer. Regardless of the location of the dominant resistance, the flux in systems where inertial lift dominates back-transport is approximately independent of axial location, because the resistance to permeation arises from a mechanism that is unrelated to the development of the CP layer.

Equations for the predicted flux and cake layer thickness for the two limiting cases are shown in Table 15-13. Other values of β represent intermediate conditions, in which the cake contributes different proportions of the overall hydraulic resistance. Davis quantified both the limiting cases and intermediate cases in terms of the two dimensionless ratios that are used as the axes in Figure 15-48. The abscissa is the ratio of the flux through the bare membrane at the given TMP ($\hat{J}^o_V$, given by Equation 15-12) to the inertial lift at the membrane surface for the given particle size and crossflow velocity in the absence of a cake (v^o_{lift}, given by Equation 15-190). The ordinate is the ratio of the actual flux ($\hat{J}_V$) to the flux through the bare membrane at the same TMP ($\hat{J}^o_V$). The impact of many plausible changes in system configuration or operational conditions can be explored using Figure 15-48. For instance, increasing the TMP while maintaining a fixed crossflow velocity shifts the system from right to left along a single curve, and increasing the crossflow velocity while holding TMP fixed shifts it horizontally from left to right.

Summary of Equations for Modeling Crossflow Filtration The key assumptions and equations presented in the preceding sections for crossflow filtration systems operating at steady state are summarized in Table 15-14.

Nonsteady-State Fouling Patterns in Crossflow Filtration The expressions derived by Hermia (Table 15-10) for the nonsteady-state trend in flux as contaminants accumulate on a membrane cannot be used for systems with crossflow, because with crossflow, the quantity of material deposited is not necessarily directly related to the volume of permeate produced. A few researchers have attempted to develop analogous expressions for systems with crossflow by adjusting the net particle deposition rate to account for the effect of the crossflow. The approaches that have been used have assumed that particles could be lost from a deposited cake layer at a rate that was proportional to the cake thickness (Wiesner et al., 1992) or that the effective permeation velocity for particle transport to the membrane could be expressed as the difference between the actual permeation velocity and a characteristic velocity for particle back-transport (Kilduff et al., 2002). This characteristic back-transport velocity is conceptually related to the back-transport mechanisms

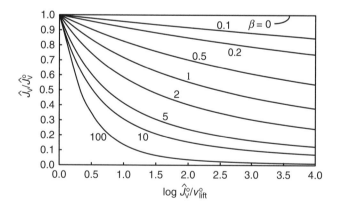

FIGURE 15-48. Predicted permeate flux in cylindrical membranes where back-transport is dominated by inertial lift. $\hat{J}^o_V$ is the flux through the clean membrane for the given TMP (computed according to Equation 15-12), and v^o_{lift} is the lift velocity for the given particle size and the shear rate at the wall in the absence of a cake (computed according to Equation 15-190). *Source:* After Davis (1992)

TABLE 15-14. Key Equations and Assumptions Used to Model Steady-State Crossflow Filtration

Mass balance on contaminant in CP layer

$$0 = -\frac{\partial uc}{\partial x} + \frac{\partial vc}{\partial y} + \frac{\partial (D_{Br} + D_{Sh})\partial c}{\partial y} - \frac{\partial v_{lift}c}{\partial y} \quad (15\text{-}191)$$

Coefficients and other parameter values appearing in Equation 15-191

$$u = \gamma_{Sh,y=0}\, y \quad (15\text{-}192)$$

v evaluated based on system geometry and simplifying assumptions

$$D_{Br} = \frac{k_B T}{3\pi\mu_L d_p} \quad (15\text{-}186)$$

$$D_{Sh} = \tilde{D}_{Sh}(\phi) a_p^2 \gamma_{Sh,y=0} = \frac{\tilde{D}_{Sh}(\phi)}{4} d_p^2 \gamma_{Sh,y=0} \quad (15\text{-}188)$$

$$v_{lift} = \frac{b\rho_L d_p^3 \gamma_{Sh,y=0}^2}{128\mu_L} \quad (15\text{-}190)$$

Assumed dimensions of the CP layer

CP layer is much thinner than velocity boundary layer ($\delta_{CP} \ll \delta_{vel}$)

$\delta_{CP} = 0$ at $x = 0$, and grows steadily with x thereafter; details of the $\delta_{CP}(x)$ profile are model-dependent

Boundary conditions[a,b]

At $x = 0$: $\quad c = c_f \quad (15\text{-}205)$

At $y = 0$: $\quad c = \begin{cases} c_c \exp(v/k_{mt,CP}) \\ \quad \text{(no compact layer)} \quad (15\text{-}211) \\ c_{lim} \quad \text{(compact layer present)} \quad (15\text{-}206) \end{cases}$

At $y = \delta_{CP}$: $\quad c = c_c \quad (15\text{-}207)$

or

At $y = \infty$: $\quad c = c_c \quad (15\text{-}213)$

Assumption regarding how steady state is maintained in CP layer

Back-transport dominated by diffusion:

At $y = 0$: $\quad v(c - c_p) = -D_{tot}\dfrac{dc}{dy} \quad (15\text{-}208)$

or

At any x_1: $\quad (c_c - c_p)\displaystyle\int_0^{x_1} v\,dx = \int_0^\infty u[c(x_1) - c_c]dy$

$\quad (15\text{-}217)$

Back-transport dominated by inertial lift:

At $y = 0$: $\quad 0 = \dfrac{\partial vc}{\partial dy} - \dfrac{\partial v_{lift}c}{\partial dy} \quad (15\text{-}228)$

(continued)

TABLE 15-14. *(Continued)*

Link between v and TMP in the absence of a compact layer

Particles: $\quad \text{TMP} = 3\pi\mu_L d_p D_{Br} n_c \left(\exp\left(\dfrac{\hat{J}_{V,avg}}{\overline{k}_{mt,CP}}\right) - 1\right) + \dfrac{\hat{J}_{V,avg}}{k_w^*}$

$\quad (15\text{-}222)$

Solutes: $\quad k_w^*\left(\text{TMP} - \overline{\Pi}_{c_{c/m}}\right) = \overline{k}_{mt,CP} \ln \dfrac{\overline{c}_{c_{c/m}}}{c_c} \quad (15\text{-}224)$

[a] The boundary condition at $y = 0$ is Equation 15-111 in the absence of the compact layer and Equation 15-206 if a compact layer is present.

[b] The condition at $y = 0$ can be either Equation 15-207 or 15-213, depending on the assumed thickness of the CP layer.

discussed in the preceding sections, but it is presumed to account for all back-transport mechanisms collectively and is evaluated purely empirically.

In at least one study (Hong et al., 1997), the results suggested that loss of material from the compact layer was negligible during the approach to steady state. That is, while the compact layer was growing (a period that lasted up to a few hours), essentially all the contaminant transported to the membrane entered the compact layer and remained there. The growth of the layer could therefore be modeled based on a mass balance with no term for particle loss from the cake, and the corresponding pressure loss could be modeled based on equations for flow through porous media.

These models for the transient stage of membrane operation are all based on idealized assumptions about the transport of contaminant to and (in some cases) away from the membrane, and also about the effect of the contaminant on flux and pressure loss once it arrives at the surface. None of these models is in widespread use. Many systems are back-flushed sufficiently frequently that they operate at nonsteady state much of the time, so improved understanding and modeling of this stage represents a pressing need and challenge for future research. In the absence of modeling that is sufficiently accurate to influence operational strategies, those strategies (e.g., the frequency of backwashing) are largely empirical.

Summary of Modeling Approaches and Results for Crossflow Filtration Crossflow is applied in membrane systems to reduce the accumulation of rejected material near the membrane and thereby achieve higher permeation fluxes at a given TMP. The relevant tradeoff is the energy requirement to induce the crossflow versus the improvement in flux. In recent years, the use of high-velocity crossflow in MF and UF systems for treatment of water and wastewater has declined, due to the energy costs and the development of effective systems with outside-in flow. Crossflow is still used extensively in NF and RO systems.

The approach used to analyze systems with crossflow is similar to that for frontal filtration systems: a mass balance on the species targeted for rejection is combined with a model for fluid flow, and the two are solved simultaneously with appropriate boundary conditions to predict the contaminant profile in the CP layer. That profile is then related to the volumetric flux, with the pressure profile in the system playing a central role if no compact layer forms. As in frontal filtration, the extent of polarization is much greater for macromolecules and particles (in MF and UF systems) than for small solutes (in RO systems). However, the equations are more complex when crossflow is included, because the axial component of fluid velocity complicates the flow pattern and causes the CP layer to be nonuniform, growing in the direction of bulk flow. Although mathematical models for crossflow filtration have grown more sophisticated over time, several simplifying assumptions are still required in all the models to make them tractable.

As in frontal filtration, concentration polarization in systems with crossflow can lead to a steep concentration gradient across the CP layer, from c_c at its outer edge to either $c_{c_c/m}$ or c_{lim} at its inner edge ($y = 0$). Because both the thickness of the CP layer and the value of c_c grow with distance from the inlet, $c_{y=0}$ increases rapidly with distance downstream of the inlet. Since the CP layer is negligibly thick at the inlet, no compact layer can be present at that location, but such a layer might form at any point downstream, or not at all, depending on the operating conditions. Within the portion of the membrane element lacking a compact layer, the concentration profiles and flux are sensitive to the TMP, but in the portion where a compact layer is present, the dependence of flux on TMP disappears. Increasing either the feed concentration or the TMP, or decreasing the crossflow velocity, increases the likelihood that a compact layer will form and, if one does form, causes the location where it first forms to move upstream (closer to the inlet). Regardless of whether a compact layer forms, the flux is predicted to decline, often by a significant fraction, with axial distance (x) because of the steady growth of the CP layer. Because the flux varies with axial location, the average flux across the whole membrane must be determined by an integration along the entire membrane length.

The details of the predicted concentration profile through the CP layer and adjacent to the membrane depend on several model assumptions, in addition to the known features of the system. Among these assumptions are the nature of the boundary conditions (e.g., whether the CP layer is treated as having an abrupt versus an indistinct boundary with the bulk solution), how steady state is achieved in the compact layer, and which back-transport mechanism dominates. Many of the modeling assumptions are linked to the size of the targeted species, because different mechanisms of back-transport dominate for different sizes of contaminants. Back-transport of molecular and macromolecular contaminants is dominated by diffusion, whereas shear-induced diffusion dominates back-transport for colloids and small particles (~0.1 to a few micrometers), and inertial lift dominates for larger particles.

Regardless of the size of the contaminant or the dominant back-transport mechanism, the overall, effective pressure differential that drives transport of water (TMP $- \Delta_{c/p}\Pi$) must be dissipated between the bulk concentrate and the bulk permeate. Portions of the available energy associated with this differential are utilized to drive the permeation flow from the bulk concentrate to the surface of the compact layer (if one exists), through the compact layer, and then through the membrane. Once water reaches the compact layer or the membrane surface, the presence or absence of crossflow becomes irrelevant, and the transport and energy loss phenomena are identical to those in frontal filtration systems.

For species whose back-transport is dominated by shear-induced diffusion or inertial lift, typical operating conditions cause a compact layer (usually considered to be a cake) to form over almost the entire membrane. When back-transport is dominated by shear-induced diffusion, the cake is usually considered to be thin and therefore to have negligible effect on the crossflow velocity profile, but if inertial lift is important, the possibility that the cake will significantly reduce the flow cross-section must be considered.

In most systems, the majority of the TMP is thought to be dissipated as water passes through the compact layer; the resistance of the membrane is usually substantially less than that of the compact layer, and the resistance of the CP layer is almost always negligible by comparison.

Although crossflow systems operate under nonsteady-state conditions most of the time, available models that account for such conditions are in their infancy by comparison with steady-state models, and new research is required to address this situation.

15.10 ELECTRODIALYSIS

Electrodialysis (ED) is a membrane process with two essential differences from the systems described in the earlier parts of the chapter: it uses a gradient in electrical potential, rather than a pressure gradient, as the driving force for separation, and it generates clean water by selective permeation of ionic contaminants, rather than water, through the membrane. These systems use *ion exchange membranes*, which are essentially sheets of ion exchange resin that have fixed (immobile) charges of one sign incorporated into their structure. For example, cation exchange membranes contain a high concentration of fixed anionic groups. The principle of electroneutrality requires that these fixed negative charges be balanced (at some reasonable length scale) by positive ions (cations), but these cations are in solution in the interstices of the membrane and are free to move under the influence of an applied electrical field. In ED, cations enter and pass through the membrane in a continuous stream under the influence of such a field, whereas the anions remain fixed in place.

The ED process has several potential uses. In some cases, the primary product is not the relatively clean water from which ions have been separated, but the relatively concentrated stream; examples include the production of sea salt from sea water, the conversion of neutral salts into separate solutions of acids and bases, and the recovery of acids from industrial wastewaters. Nevertheless, the primary use of ED that is discussed here is the generation of a clean-water stream from one with a much higher ionic content. Although ED has been used for the production of drinking water from seawater, this degree of desalination is now generally accomplished more economically by RO. ED is more useful for treating brackish water—water with an ionic content (mostly Na^+ and Cl^-) too high for direct use as potable water, but much lower than sea water. It has also been used for softening (removal of Ca^{2+} and Mg^{2+}) or for removal of specific ions (e.g., nitrate, sulfate, or arsenate) from relatively low ionic strength waters. An interesting possibility is to use ED to increase the recovery in RO treatment; the influent to the ED process would be the concentrate from RO, and the clean-water effluent from ED would be recycled to the RO input or mixed directly with RO effluent for drinking water. ED might also be useful for the treatment of wastewaters with high ionic content that are not good candidates for treatment by precipitation or other, simpler methodologies. In some cases, both the production of a relatively clean stream and the recovery and reuse of a concentrated stream (e.g., containing acids or bases for cleaning operations) might be possible. Many of the applications of ED in water and wastewater treatment have been reviewed by Schoeman and Thompson (1996), Strathmann (2004), and Tanaka (2007).

ED is accomplished in units with tens or even hundreds of pairs of alternating cation and anion exchange membranes that all lie between a cathode and an anode; a schematic diagram of an ED unit is shown in Figure 15-49. The space between two adjacent membranes is called a *cell*, and the complete collection of membranes is referred to as a *stack*. The thickness of the membranes is in the range of 0.1–0.6 mm, and the spacing between them is in the range of 0.3–2 mm. In this figure, the membranes are labeled with the sign of their fixed charge; that

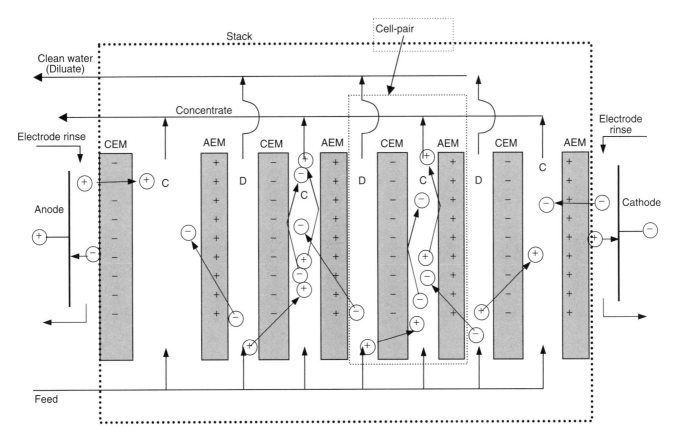

FIGURE 15-49. Schematic of electrodialysis system. Real systems have tens to hundreds of cell-pairs in a stack; CEM and AEM refer to cation and anion exchange membranes, respectively; positive and negative symbols within membranes refer to the fixed charge; positive and negative symbols that are circled refer to cations and anions in solution, respectively; D and C refer to diluate and concentrate, respectively. *Source:* Adapted from Strathmann (2004).

is, cation exchange membranes are shown with negative signs, and anion exchange membranes with positive signs.

The feed enters the cells at one edge of the stack (the bottom in the figure) and, as the water travels the pathways between the membranes, cations are attracted to the cathode and anions to the anode. Cations can pass through cation exchange membranes easily but, at least in the ideal case, cannot pass through anion exchange membranes, and the opposite is true for anions. As a result of both the movement and blockage of ions, some cells lose ions of both types of charge while the adjacent cells collect ions of both signs. The water with the increased concentration of ions is called the *concentrate*, and the water with the decreased ion concentration is called the (or, sometimes, *dialysate*). The repeating unit in a stack is a *cell-pair*, consisting of one of each type of membrane and one of each type of cell (concentrate and diluate). The solutions in the outer compartments that contain the electrodes are isolated from the feed, clean water, and concentrate, for reasons that are explained below. Typically, the potential drop between the electrodes is controlled at a preselected value intended to achieve the desired water quality in the exiting diluate (i.e., product) stream. The potential drop is commonly in the range of 0.5–1.5 V/cell-pair and, since the number of cell-pairs in a stack is often in the range of 50–100, the total voltage across the whole ED unit can be quite high.

If the system shown in Figure 15-49 were completely successful, each concentrate cell would contain all the cations that entered that cell in the feed as well as those that started in the adjacent cell to the left and passed through the cation exchange membrane. The same cell would also contain all the anions that entered it in the feed, along with those that started in the adjacent cell to the right and passed through the anion exchange membrane. Assuming that all the cells received the same feed flow rate, the maximum recovery of clean water would be 50% (the water in all of the diluate cells), and the concentration of all of the ions in the concentrate would be twice that in the raw water. To increase the recovery and the ionic concentration of the concentrate, successive units in series could be fed the concentrate from the previous unit or some of the concentrate effluent could be recycled to the concentrate influent.

Electrodialysis reversal (EDR) is essentially the same process; the difference is that the polarization of the anode and cathode is reversed a few times per hour, so that the concentrate cells under one condition become the diluate cells in the opposite condition. This reversal prevents charged macromolecules and charged particles (that would be attracted to the membrane surface but be unable to pass through) from accumulating and fouling the membrane. For a short period after each reversal, all water from both types of cells is either wasted or recycled to the influent to prevent contamination of the product water; the idea is that the reduced recovery in the short term is more than compensated by the long-term reduction in fouling. In the following analysis, no distinction is made between ED and EDR, since the two processes differ only in their piping, valving, and electrical controls.

A simplified diagram that shows a one-cell ED system and that emphasizes the reactions at the cathode and anode is shown in Figure 15-50. For typical operating conditions, reduction of water generates hydrogen gas and hydroxyl ions at the cathode, thereby raising the pH. At the anode, water is oxidized, generating oxygen gas and protons, and lowering the pH. In some applications, other products can be formed at the electrodes; for example, chlorine gas can form at the anode when ED is used for desalination of brackish or

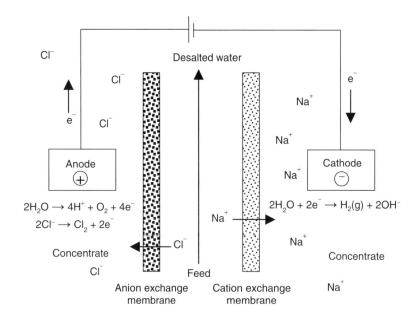

FIGURE 15-50. A single-cell electrodialysis unit showing the reactions at the electrodes.

seawater (where Cl⁻ concentrations are high). The products of these reactions are undesirable in the final product water or even in the concentrate. For this reason, the water in those cells is usually isolated and replenished continuously, as shown in Figure 15-49.

Transport in Systems with a Gradient in Electrical Potential

We begin the analysis of ED systems by considering the transport of mass and charge in solutions that are subject to a gradient in electrical potential, ψ. The electrical current through such solutions is carried by migrating ions, so transport of mass and charge is linked. Specifically, the current density carried by a particular ion (the current carried by the ion per unit area, $\hat{I}_i$) is related to the mass flux of that ion (moles transported per unit area per unit time, J_i) as follows:

$$\hat{I}_i = z_i \mathcal{F} J_i \tag{15-237}$$

where z_i is the ionic charge and $\mathcal{F}$ is Faraday's constant. Faraday's constant is the absolute value of the charge per mole of electrons; that is, it is $N_A|e|$, where N_A is Avogadro's number and e is the charge on an electron (given in coulombs, C, as -1.602×10^{-19} C). One mole of charge is commonly referred to as one equivalent (equiv), so Faraday's constant is $\mathcal{F} = 96{,}485$ C/equiv. Using this expression for $\mathcal{F}$, considering z_i to have units of equivalents of charge per mole of i, and expressing J_i in mol/m² s, Equation 15-237 yields the current density in A/m².

The overall current density in the solution, $\hat{I}$, is the sum of the current densities associated with the different ions:

$$\hat{I} = \sum_{\text{all } i} \hat{I}_i = \mathcal{F} \sum_{\text{all } i} z_i J_i \tag{15-238}$$

Both current density and flux have directionality as well as magnitude; that is, they are vectors. For an ED system, we define the x and y directions, respectively, as perpendicular to and parallel to the membranes, with positive x pointing from the anode to the cathode and positive y pointing from the fluid inlet to the outlet. With these definitions, the electrical gradient drives cations in the positive x direction and anions in the negative x direction. As a result, z_i and J_i are both positive for cations and both negative for anions, so both cations and anions make positive contributions to the current density.

As is true in other systems we have considered previously, mass transport can occur in ED systems by either advection or diffusion. However, if an electrical gradient is present, charged species also move in response to this gradient by *electromigration*. Thus, the total molar flux of species i at any location in an ED system is

$$J_i = J_{i,\text{adv}} + J_{i,\text{D}} + J_{i,\psi} \tag{15-239}$$

where the terms on the right side are the fluxes due to advection, diffusion, and electromigration, respectively. In the following sections, the equations used to model advection and diffusion are reviewed briefly and are related to the corresponding current densities via Equation 15-237. The electromigration flux is then explored in greater detail, since it is what distinguishes ED systems from those that we have considered previously.

Transport due to Advection As explained in Chapter 2, the advective flux of species i in the x direction is given by the product of its concentration and the x component of the fluid velocity as

$$J_{i,\text{adv}} = c_i v_x \tag{15-240}$$

The corresponding current density carried by species i and by all species in solution are

$$\hat{I}_{i,\text{adv}} = z_i \mathcal{F} c_i v_x \tag{15-241}$$

$$\hat{I}_{\text{adv}} = \sum_{\text{all } i} \hat{I}_{i,\text{adv}} = \mathcal{F} v_x \sum_{\text{all } i} z_i c_i \tag{15-242}$$

In an ED system, the current is strictly in the x direction, but the advective flow is either zero (inside ideal membranes) or in the y direction (in the concentrate and diluate cells); that is, $v_x = 0$ everywhere. As a result, in such systems, advection is important for transporting ions from the inlet to the outlet, but it makes no contribution to carrying the current. More fundamentally, with rare (microscopic) exceptions, the summation shown in Equation 15-242 is zero according to the principle of electroneutrality.[30] Hence, while advection can be important for mass flux, it rarely contributes to the electrical current.

Transport due to Diffusion The diffusive flux of species i in the x direction and the corresponding current density are given by[31]

$$J_{i,\text{D}} = -D_i \frac{dc_i}{dx} \tag{15-243}$$

$$\hat{I}_{i,\text{D}} = z_i \mathcal{F} J_{i,\text{D}} = -z_i \mathcal{F} D_i \frac{dc_i}{dx} \tag{15-244}$$

---

[30] As explained by Newman and Thomas-Alyea (2004), electroneutrality is not really a fundamental law of nature, but occurs as a consequence of the Poisson relationship, which suggests that it would take a large electrical force to maintain a charge separation. In the absence of such a force, electroneutrality occurs. Although it is possible to operate ED systems with such charge separation (and this is sometimes done), we limit our discussion to systems where electroneutrality occurs everywhere.

[31] Formally, one would write this equation in terms of the activity rather than concentration. However, it is common practice in the ED field to incorporate the correction for nonideal solute behavior into the diffusivity.

The overall current density associated with diffusion of all species is therefore

$$\hat{I}_D = \mathcal{F} \sum_{\text{all } i} z_i J_{i,D} = -\mathcal{F} \sum_{\text{all } i} z_i D_i \frac{dc_i}{dx} \quad (15\text{-}245)$$

Equation 15-245 applies to diffusion in a membrane as well as in bulk solution. However, diffusion coefficient values are typically much lower in ion exchange membranes than in aqueous solutions, so to generate the same diffusive flux in a membrane as in solution, a much larger gradient in concentration (or activity) would be needed in the membrane.

Transport in Solution due to an Electric Field   Ohm's law states that current and potential are proportional to each another. One expression of Ohm's law is that the current equals the product of the *electrical conductance* and the voltage; that is, the constant of proportionality is the conductance (which is the reciprocal of the electrical resistance). The SI unit for conductance is Siemens (S).[32]

The conductance of a solution depends on both its ionic content and the system geometry. By normalizing the conductance to eliminate its dependence on the system geometry, we can obtain a parameter that isolates the role of the ionic composition in determining the ability of the solution to carry current. This normalization, which is accomplished by multiplying the conductance by the distance the current travels and dividing by the cross-sectional area through which the current travels, yields a parameter referred to as the *conductivity* (κ), with units such as S/m or μS/cm. This normalization also allows writing Ohm's law in terms of the current density and the potential gradient, with κ as the constant of proportionality, as follows:

$$\hat{I}_\psi = -\kappa \frac{d\psi}{dx} \quad (15\text{-}246)$$

The negative sign in Equation 15-246 is required because positive current is defined as flowing from the positive electrode (anode) to the negative electrode (cathode), which is the direction of decreasing electrical potential.

The solution conductivity can be expressed in several ways, thereby leading to several ways of expressing the total current density carried by all ions in solution ($\hat{I}_\psi$) and the current density carried by an individual ion ($\hat{I}_{i,\psi}$) as follows:

$$\kappa = \mathcal{F} \sum_{\text{all } i} u_i z_i c_i = \sum_{\text{all } i} \lambda_i z_i c_i = \frac{\mathcal{F}^2}{RT} \sum_{\text{all } i} D_i z_i^2 c_i \quad (15\text{-}247)$$

$$\hat{I}_\psi = -\kappa \frac{d\psi}{dx} = -\mathcal{F} \frac{d\psi}{dx} \sum_{\text{all } i} u_i z_i c_i = -\frac{d\psi}{dx} \sum_{\text{all } i} \lambda_i z_i c_i$$
$$= -\frac{\mathcal{F}^2}{RT} \frac{d\psi}{dx} \sum_{\text{all } i} D_i z_i^2 c_i \quad (15\text{-}248)$$

$$\hat{I}_{i,\psi} = -\mathcal{F} u_i z_i c_i \frac{d\psi}{dx} = -\lambda_i z_i c_i \frac{d\psi}{dx} = -\frac{\mathcal{F}^2}{RT} D_i z_i^2 c_i \frac{d\psi}{dx} \quad (15\text{-}249)$$

Equation 15-247 indicates that the solution conductivity is a conservative property; that is, the conductivity is found as the sum of contributions from all the ions, and (ideally) each ion's contribution is proportional to its concentration, The three terms u_i (ionic mobility), λ_i (specific equivalent ionic conductance), and D_i (diffusion coefficient) are related parameters that all describe how easily molecules of an ionic species i move through an aqueous solution under the influence of various driving forces; these parameters and the relationships among them are described next.

The *ionic mobility* (u_i) is the electrically induced velocity of an ion through an aqueous solution normalized (divided) by the field strength, with resulting units such as m^2 s^{-1} V^{-1}; this parameter expresses the same concept for ions as the electrophoretic mobility does for particles (as explained in Chapter 11). The current density associated with ion i, $\hat{I}_{i,\psi}$, can be thought of as the flux of charge attributable to movement of those ions, and, like all fluxes, can be interpreted as the product of a velocity and a concentration. In that context (and recalling that $\mathcal{F} = eN_A$), the expression $-\mathcal{F} u_i z_i c_i (d\psi)/(dx)$ can be thought of as the product of the ion velocity (itself the product of the ion velocity per unit field strength [u_i] and the field strength [$(d\psi)/(dx)$]) and the concentration of charges (found as the product of the number concentration of ions [$N_A c_i$] and the charge per ion [ez_i]).

The most direct measure of the contribution of each ionic species to the solution conductivity is the *specific equivalent ionic conductance*, λ_i. Values of λ_i have been found for all common ions and are tabulated in the *Handbook of Chemistry and Physics* (CRC, 2009) and many textbooks. After the value of λ for any single ion is established, λ for a second ion can be determined by measuring the conductance of a solution containing those two ions as the only solutes and applying Equation 15-250. That process can then be repeated for various solutions, each containing only one ion with an unknown λ, to determine λ, of all ions of interest.

$$\kappa_j = \sum_{\text{all } i} \lambda_{i,j} z_i c_{i,j} \quad (15\text{-}250)$$

The product of the molar concentration (c_i) and the (absolute value of) ionic charge($|z_i|$) yields concentration units of equiv/L, and the units of λ are conventionally reported in S cm^2/equiv. To determine the first value, a second independent measure is needed; that measure is the fraction of the overall current carried by each ion in a binary system.

---

[32] Siemens were formerly called mhos, where mho is the reverse spelling of ohm, consistent with the notion that conductance is the reciprocal of resistance. For solutions, one Siemen (1 S = 1 A/V) represents a very large conductance, so values of solution conductance are commonly reported in micro Siemens (μS).

Experimental techniques have been developed to determine that value (Robinson and Stokes, 1959); but nowadays, the tabulated values for λ_i are used virtually universally.

Both the specific equivalent ionic conductance and the ionic mobility (u_i) provide ways to quantify the ability of a particular type of ion to carry current. These two parameters are related by

$$\lambda_i = u_i \mathcal{F} \tag{15-251}$$

Formally, both λ_i and u_i have the same sign as the charge on the ion, z_i; hence, the products $u_i z_i$ and $\lambda_i z_i$ in Equations 15-248 and 15-249 are always positive. The same result can be obtained if the negative signs are omitted when reporting values of λ_i and u_i for anions, and if the products $u_i z_i$ and $\lambda_i z_i$ are written as $u_i|z_i|$ and $\lambda_i|z_i|$, respectively. This convention is often used in the literature.

The movement of an ion in response to an electrical potential gradient is closely related to its movement in response to a gradient in concentration or chemical activity (i.e., its diffusivity). In both cases, the driving force is a gradient in electrochemical activity, and the resisting force is primarily provided by the water molecules past which the solute moves. This relationship can be expressed quantitatively as follows:

$$D_i = \frac{RT}{z_i \mathcal{F}} u_i = \frac{RT}{z_i \mathcal{F}^2} \lambda_i \tag{15-252}$$

Substitution of these relationships for either u_i or λ_i into Equations 15-248 and 15-249 yields the final expressions in those equations. These final expressions are the ones used most commonly in the analysis of ED systems because, as shown below, both diffusion and electromigration are important in some regions of interest, and the same ion characteristic (D_i) is used in the description of both processes.

The values of λ_i and D_i for a few ions are shown in Table 15-15. These values make it clear that, in an aqueous solution, protons (H^+) and hydroxide (OH^-) ions migrate through solution much more readily than other ions[33] and hence have the ability to carry much more current. Because the interest in ED is usually to remove other cations and anions, the solutions to be treated are maintained at near-neutral pH, where the concentrations of both H^+ and OH^- are low and their contributions to the overall current density are low. The data in Table 15-15 indicate that cations generally have a lower specific conductance than anions; as a result, cation transport usually limits the overall transport rate in ED systems. As noted, the data are given for "infinite dilution" or zero concentration. In solutions con-

TABLE 15-15. Specific Equivalent Ionic Conductance and Diffusivity of Selected Ions[8]

Cation	$+\lambda_0$ (S cm^2/equiv)	D_0 (cm^2/s)	Anion	$-\lambda_0$ (S cm^2/equiv)	D_0 (cm^2/s)
H^+	349.8	9.312×10^{-5}	OH^-	198.0	5.260×10^{-5}
Na^+	50.11	1.334×10^{-5}	Cl^-	76.34	2.032×10^{-5}
K^+	73.52	1.957×10^{-5}	Br^-	78.3	2.084×10^{-5}
Ca^{2+}	59.50	0.7920×10^{-5}	HCO_3^-	41.5	1.105×10^{-5}
Mg^{2+}	53.06	0.7063×10^{-5}	NO_3^-	71.44	1.902×10^{-5}
			SO_4^{2-}	80	1.065×10^{-5}

[a]Newman and Thomas-Alyea (2004); the subscript "0" on λ and D indicates that the values shown for the specific equivalent ionic conductance and diffusivity are based on extrapolation to an infinitely dilute (i.e., ideal) solution, in which the activity coefficients of all ions are 1.0.

taining a single salt, the conductance (or diffusivity) decreases somewhat with increased concentration, because the increased concentration leads to an increase in the ionic strength and a decrease in the ion activity coefficients; values of κ for various salt concentrations in single-salt solutions are tabulated in the *Handbook for Chemistry and Physics* (CRC, 2009). Semiempirical, semitheoretical equations to describe this effect have been formulated and are reviewed in several textbooks (Bockris and Reddy, 1998; Horvath, 1985; Kortum, 1965; Wright, 2007; Walker, 2010).

Invoking Equation 15-237 again, the molar flux of a species due to electromigration can be computed as follows:

$$J_{i,\psi} = \frac{\hat{I}_{i,\psi}}{z_i \mathcal{F}} \tag{15-253}$$

$$= -u_i c_i \frac{d\psi}{dx} = -\frac{\lambda_i}{\mathcal{F}} c_i \frac{d\psi}{dx} = -\frac{\mathcal{F}}{RT} D_i z_i c_i \frac{d\psi}{dx} \tag{15-254}$$

Overall Transport and Current Densities in Systems with a Gradient in Electrical Potential Using the final expression in Equation 15-254 to express the electrically induced flux, we can now rewrite Equation 15-239 for transport in the x direction as follows:

$$J_{i,x} = v_x c_i - D_i \frac{dc_i}{dx} - \frac{\mathcal{F}}{RT} D_i z_i c_i \frac{d\psi}{dx} \tag{15-255}$$

This expression (or the equivalent using any of the expressions for in Equation 15-254) is known as the extended Nernst–Planck equation; when the advective term is omitted, the equation is known simply as the Nernst–Planck equation. The corresponding equation for current density (assuming that the current is only in the x direction, but ignoring the subscript x on these terms) is

$$\hat{I}_i = \hat{I}_{i,\text{adv}} + \hat{I}_{i,D} + \hat{I}_{i,\psi} = z_i \mathcal{F} v_x c_i - D_i z_i \mathcal{F} \frac{dc_i}{dx} - \frac{\mathcal{F}^2}{RT} D_i z_i^2 c_i \frac{d\psi}{dx} \tag{15-256}$$

As noted previously, advection does not carry any current in ED systems. Therefore, in ED systems, the advective

---

[33] The higher mobility of H^+ (or, more correctly, H_3O^+) and OH^- reflects the fact that, in addition to migration of intact ions, these species can, in effect, move through solution by the transfer of protons from one water molecule to the next. This mode of transport is not available to any other solute.

terms in Equation 15-256 can be ignored, and the equation simplifies to the Nernst–Planck equation

$$\text{If } \hat{I}_{\text{adv}} = 0: \quad \hat{I}_i = -D_i z_i \mathcal{F} \frac{dc_i}{dx} - \frac{\mathcal{F}^2}{RT} D_i z_i^2 c_i \frac{d\psi}{dx} \quad (15\text{-}257)$$

In addition, as is explained below, the ion concentration is uniform in some regions in ED systems. In these regions, diffusion is negligible, so Equation 15-256 can be simplified even more as

$$\text{If } \hat{I}_D = \hat{I}_{\text{adv}} = 0: \quad \hat{I}_i = -\frac{\mathcal{F}^2}{RT} D_i z_i^2 c_i \frac{d\psi}{dx} \quad (15\text{-}258)$$

Applying Equation 15-258 to the collection of all ions in solution yields

$$\text{If } \hat{I}_D = \hat{I}_{\text{adv}} = 0: \quad \hat{I} = -\mathcal{F} \sum_{\text{all } i} z_i J_i = -\frac{\mathcal{F}^2}{RT} \frac{d\psi}{dx} \sum_{\text{all } i} D_i z_i^2 c_i \quad (15\text{-}259)$$

Equation 15-259 indicates that, in regions where transport occurs only by electromigration, the current density is directly proportional to the potential gradient. The potential drop from one point to another (from x_1 to x_2) in such a region can be computed by integration of Equation 15-259, yielding

$$\Delta \psi_{x_1 - x_2} = \frac{RT}{\mathcal{F}^2 \left(\sum_{\text{all } i} D_i z_i^2 c_i \right)} \hat{I}(x_1 - x_2) \quad (15\text{-}260)$$

The usefulness of these simplified forms of Equation 15-257 is demonstrated in the following section.

Modeling Electrodialysis Systems

Overview ED systems are, of course, three dimensional. Most often, the flow is vertical (and upward) so that particles or bubbles are likely to be carried through the system rather than accumulate on the membranes. As water travels from the influent to the effluent (the y direction), ions move in the x direction due to both electromigration and diffusion (which is induced near the membranes by the combination of electromigration and ion rejection). As a result of these transport mechanisms and selective transport of particular ions across the membranes, the concentration of ions in the diluate cells decreases steadily in the direction of water flow, whereas that in the concentrate cells increases.

In this section, we build a mathematical model of ED systems, using the equations derived previously to evaluate ion transport in the x and y directions. The analysis considers an idealized ED system at steady state. The idealizations include plug flow of the bulk solutions in both the diluate and concentrate cells, stagnant boundary layers next to the membranes, and 100% perm-selective membranes (i.e., the cation exchange membrane rejects anions completely and the anion exchange membrane rejects cations completely). In addition, we assume that the solution contains only a single salt (NaCl in the example) that behaves ideally, and that the salt concentration and solution pH are such that the contributions of H^+ and OH^- to the current density are negligible.

We begin by considering transport in the x direction in a single cell-pair, and then combine the result with an analysis of advection in the y direction to develop a two-dimensional model for that cell-pair. We then extend the analysis to consider an arbitrary number of cell-pairs in parallel, as in a real ED system. At any given (x, y) value, we expect the concentration to be the same throughout the depth of the system (i.e., for all z values); therefore, the two-dimensional, multiple-cell-pair model is sufficient to describe the true three-dimensional system.

Transport in the x Direction in a Single Cell-Pair To model salt separation in a single cell-pair, we must understand the transport of both cations and anions through eight regions:

- Anion exchange membrane
- Boundary layer on the diluate side of the anion exchange membrane
- Bulk diluate solution
- Boundary layer on the diluate side of the cation exchange membrane
- Cation exchange membrane
- Boundary layer on the concentrate side of the cation exchange membrane
- Bulk concentrate solution
- Boundary layer on the concentrate side of the anion exchange membrane.

At any specific value of y, the ion gradients and potential drop in each region of the system adjust to satisfy two requirements: that the overall potential drop (per cell-pair) equals the imposed value and that the electrical current is the same in all these regions. In each region, electroneutrality must be maintained (at some macromolecular length scale), so cations and anions do not act independently of one another. The presentation is based in part on a general description of electrochemical systems by Bard and Faulkner (2001).

To aid in the analysis, it is useful to introduce the concept of the *transport number*, t_i, defined as the fraction of the total current carried by species i:

$$t_i = \hat{I}_i / \hat{I} \quad (15\text{-}261)$$

Because the membranes are assumed to completely reject the co-ion (Na^+ in the anion exchange membrane and Cl^- in the cation exchange membrane), all of the current within each membrane is carried by the counter-ions. The transport number of the mobile ion in each membrane is therefore unity, whereas that of the rejected ion is zero. Using an overbar to denote a quantity in the membrane, the membrane transport numbers for the NaCl system are

$$\bar{t}_{Na^+,\text{CEM}} = 1 \quad \bar{t}_{Cl^-,\text{CEM}} = 0 \quad \bar{t}_{Cl^-,\text{AEM}} = 1 \quad \bar{t}_{Na^+,\text{AEM}} = 0$$

In both the bulk concentrate and bulk diluate, the solutions are uniform in the x direction and hence have no concentration gradient. Therefore, the current is carried entirely by electromigration. Writing $\hat{I}_{i,\psi}$ in terms of specific equivalent ionic conductance, and adopting the convention in which λ_i is written as a positive value for both cations and anions, the transport number for Na^+ in either the bulk diluate or bulk concentrate can be evaluated as follows:

$$t_{Na} = \frac{\hat{I}_{Na}}{\hat{I}_{tot}} = \frac{\hat{I}_{Na}}{\hat{I}_{Na} + \hat{I}_{Cl}} =$$

$$= \frac{\lambda_{Na^+}|z_{Na^+}|c_{Na^+}(d\psi/dx)_{bulk}}{\lambda_{Na^+}|z_{Na^+}|c_{Na^+}(d\psi/dx)_{bulk} + \lambda_{Cl^-}|z_{Cl^-}|c_{Cl^-}(d\psi/dx)_{bulk}}$$

$$= \frac{\lambda_{Na}}{\lambda_{Na} + \lambda_{Cl}} = \frac{50.11}{50.11 + 76.34} = 0.396 \quad (15\text{-}262)$$

The corresponding calculation for Cl^- (or the knowledge that $t_{Na^+} + t_{Cl^-} = 1$ in this single-salt solution) yields the result $t_{Cl^-} = 0.604$. The simplification of the full equation takes advantage of the principle of electroneutrality for this binary salt ($|z_{Na^+}|c_{Na^+} = |z_{Cl^-}|c_{Cl^-}$); this fact applies for any binary salt, so generally

$$t_{cat,bulk} = \frac{\lambda_{cat}}{\lambda_{cat} + \lambda_{an}} \quad \text{and} \quad t_{an,bulk} = \frac{\lambda_{an}}{\lambda_{cat} + \lambda_{an}} \quad (15\text{-}263)$$

As indicated, the same calculation applies to both the bulk diluate and bulk concentrate, despite the fact that the concentrations, c_i, and potential gradients, $(d\psi)/(dx)$, are different in those two solutions. For simplicity in the following example, we round these values to $t_{Na^+} = 0.4$ and $t_{Cl^-} = 0.6$.

Consider a horizontal plane in the idealized ED system containing a single cell-pair, and assume that, at the particular y location, the current density is $100 \, A/m^2$. A schematic diagram indicating the contributions to this current density due to transport of Na^+ and Cl^- in the various regions of the system is shown in Figure 15-51; all cell-pairs in a full system would be the same as the one

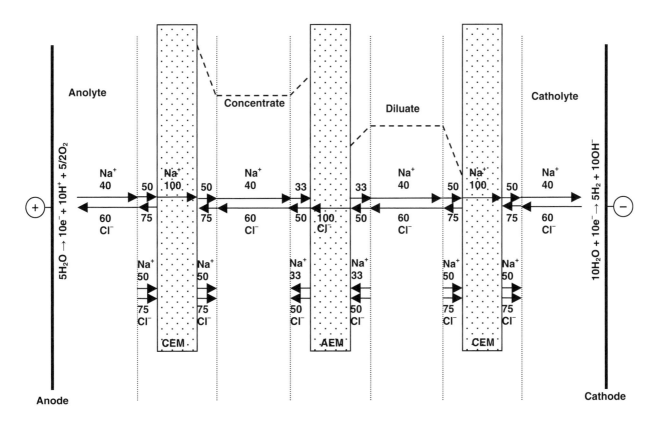

FIGURE 15-51. Current density carried by specific ions on a particular plane in a single-cell-pair ED system with sodium chloride as the only solute. The upper pairs of arrows and numbers indicate the current density (in A/m^2) caused by electromigration for each ion, while the lower pairs indicate the current density caused by diffusion. The concentration profiles in the concentrate and diluate are sketched in the top part of the diagram. *Source:* Adapted from Bard and Faulkner (2001).

shown.[34] For this solution, the transport numbers for both membranes and for the bulk solutions have already been calculated, and so the current densities for those regions are simply the product of the transport numbers and total current density. In each membrane, these values are 100 A/m^2 for the selected ion (i.e., the one that passes through the membrane) and zero for the rejected (or excluded) ion. In both the diluate and concentrate bulk solutions, the values are 40 and 60 A/m^2 for sodium and chloride, respectively.

In the boundary layers adjacent to each membrane, current is carried by both electromigration and diffusion. For the selected ion (sel) in any boundary layer, electromigration and diffusion transport ions in the same direction and both contribute to the total current density. For the rejected ion (rej), electromigration in one direction exactly offsets diffusion in the other, because no net transport of that ion occurs (at steady state) in the boundary layer. We can apply Equation 15-257 to the rejected ion to obtain

$$\hat{I}_{\text{rej}} = 0 = -D_{\text{rej}} z_{\text{rej}} \mathcal{F} \frac{dc_{\text{rej}}}{dx} - \frac{\mathcal{F}^2}{RT} D_{\text{rej}} z_{\text{rej}}^2 c_{\text{rej}} \frac{d\psi}{dx} \quad (15\text{-}264\text{a})$$

or after rearrangement:

$$\frac{dc_{\text{rej}}}{dx} = -\frac{\mathcal{F}}{RT} z_{\text{rej}} c_{\text{rej}} \frac{d\psi}{dx} \quad (15\text{-}264\text{b})$$

Electroneutrality holds within the boundary layer and, for the binary salt, is expressed as

$$z_{\text{rej}} c_{\text{rej}} = -z_{\text{sel}} c_{\text{sel}} \quad (15\text{-}265)$$

Taking the derivative yields

$$z_{\text{rej}} \frac{dc_{\text{rej}}}{dx} = -z_{\text{sel}} \frac{dc_{\text{sel}}}{dx} \quad (15\text{-}266)$$

Equation 15-266 indicates that the concentration gradient of the cation and anion will be identical for a symmetric electrolyte, but not for asymmetric electrolytes. Substituting both Equations 15-265 and 15-266 into Equation 15-264b yields the following equation in terms of the selected ion concentration:

$$\frac{dc_{\text{sel}}}{dx} = -\frac{\mathcal{F}}{RT} z_{\text{rej}} c_{\text{sel}} \frac{d\psi}{dx} \quad (15\text{-}267)$$

The selected ion carries the entire current in the boundary layer, and so applying Equation 15-257 to that ion yields

$$\hat{I}_{\text{sel}} = \hat{I} = -D_{\text{sel}} z_{\text{sel}} \mathcal{F} \frac{dc_{\text{sel}}}{dx} - \frac{\mathcal{F}^2}{RT} D_{\text{sel}} z_{\text{sel}}^2 c_{\text{sel}} \frac{d\psi}{dx} \quad (15\text{-}268)$$

Substituting $(dc_{\text{sel}})/(dx)$ from Equation 15-267 yields

$$\hat{I} = D_{\text{sel}} \frac{\mathcal{F}^2}{RT} z_{\text{sel}} z_{\text{rej}} c_{\text{sel}} \frac{d\psi}{dx} - \frac{\mathcal{F}^2}{RT} D_{\text{sel}} z_{\text{sel}}^2 c_{\text{sel}} \frac{d\psi}{dx} \quad (15\text{-}269\text{a})$$

$$\hat{I} = \left(D_{\text{sel}} \frac{\mathcal{F}^2}{RT} z_{\text{sel}} c_{\text{sel}} \frac{d\psi}{dx} \right) (z_{\text{rej}} - z_{\text{sel}}) \quad (15\text{-}269\text{b})$$

The first term on the right side in Equation 15-269a is the current density in the boundary layer attributable to diffusion of the selected ion, and the second is the current density attributable to its electromigration. The fraction of the total current carried by the selected ion and attributable to each of these mechanisms is therefore

$$\hat{I}_{D,\text{sel}}/\hat{I} = \frac{z_{\text{rej}}}{z_{\text{rej}} - z_{\text{sel}}} \quad (15\text{-}270\text{a})$$

$$\hat{I}_{\psi,\text{sel}}/\hat{I} = -\frac{z_{\text{sel}}}{z_{\text{rej}} - z_{\text{sel}}} = \frac{z_{\text{sel}}}{z_{\text{sel}} - z_{\text{rej}}} \quad (15\text{-}270\text{b})$$

Thus, for example, for a symmetric electrolyte (such as NaCl), diffusion and electromigration of the selected ion each carries one-half of the total current through each boundary layer. For CaCl$_2$, diffusion carries one-third of the current near the cation exchange membrane and electromigration two-thirds, but these ratios are reversed near the anion exchange membrane.

Furthermore, by comparing the first term on the right side in Equation 15-264 with that in Equation 15-268, we see that the current densities attributable to diffusion for the two ions in all of the boundary layers are in the ratio of their diffusion coefficients; that is,

$$\left(\frac{\hat{I}_{D,\text{rej}}}{\hat{I}_{D,\text{sel}}} \right)_{\text{Boundary layer}} = \frac{-D_{\text{rej}} z_{\text{rej}} \mathcal{F} (dc_{\text{rej}}/dx)}{-D_{\text{sel}} z_{\text{sel}} \mathcal{F} (dc_{\text{sel}}/dx)}$$

$$= \frac{D_{\text{rej}} z_{\text{sel}} \mathcal{F} (dc_{\text{sel}}/dx)}{-D_{\text{sel}} z_{\text{sel}} \mathcal{F} (dc_{\text{sel}}/dx)} = -\frac{D_{\text{rej}}}{D_{\text{sel}}}$$

$$(15\text{-}271)$$

The negative sign stems from the fact that the diffusive contributions to the current from the two ions oppose one another. For systems with only one cation and one anion, all the boundary layers are characterized by the following relationship:

$$-\left(\frac{\hat{I}_{D,\text{an}}}{\hat{I}_{D,\text{cat}}} \right)_{\text{Boundary layer}} = \frac{D_{\text{an}}}{D_{\text{cat}}} = \frac{\lambda_{\text{an}}}{\lambda_{\text{cat}}} \quad (15\text{-}272)$$

where the subscripts refer to the anion and cation. For NaCl, as depicted in Figure 15-51, diffusion of Cl$^-$ contributes 1.5 times as much current density as does diffusion of Na$^+$ through each boundary layer:

$$-\left(\frac{\hat{I}_{D,\text{Cl}}}{\hat{I}_{D,\text{Na}}} \right)_{\text{Boundary layer}} = \frac{D_{\text{Cl}}}{D_{\text{Na}}} = \frac{\lambda_{\text{Cl}}}{\lambda_{\text{Na}}} = \frac{0.6}{0.4} = 1.5$$

$$(15\text{-}273)$$

---

[34] The conditions in the vicinity of the electrodes (i.e., those experienced by the anolyte and catholyte) are more complex than indicated here because of the production of protons and hydroxyl ions, but the concern here is only with the example cell-pair in the middle of the figure.

Applying Equations 15-270a, 15-270b, and 15-273 to the situation depicted in Figure 15-51 results in the apportioning of the current densities to the two ions and the two mechanisms in every boundary layer as shown.

The concentration profiles in the boundary layers are also shown schematically in Figure 15-51. Note that, in each cell, the concentration gradients are greater in the boundary layers adjacent to the cation exchange membrane than those adjacent to the anion exchange membrane. This situation also arises from the combinations of Equations 15-270a and 15-207b and 15-273. For the 1:1 electrolyte depicted in the figure, diffusion of the selected ion must carry half the current (50 A/m^2) in every boundary layer, and that takes a lesser gradient for chloride (with its higher diffusion coefficient) than for sodium.

The interest in ED, of course, is not to pass current through a solution for its own sake, but to transport ions from the diluate to the concentrate. The current densities shown in Figure 15-51 can all be translated into molar fluxes via Equation 15-237. Since $|z| = 1$ for both Na$^+$ and Cl$^-$ (the only ions in this solution), these translations are the same for both ions. For example, for the current density $\hat{I} = 100$ A/m^2 in Figure 15-51 (and recalling that 1 A = 1 C/s)

$$J_i = \frac{100\,\text{C/m}^2\,\text{s}}{(1\,\text{eq/mol})(96{,}485\,\text{C/eq})} = 1.04 \times 10^{-3}\,\text{mol/m}^2\,\text{s}$$
$$= 1.04\,\text{mmol/m}^2\,\text{s}$$

Both ions are transported out of the diluate (sodium to the right and chloride to the left in the figure) and into the concentrate (sodium from the left and chloride from the right) at this flux.

Relating Ion Fluxes, Electrical Current Density, and the Electrical Potential Difference The preceding section identifies the contributions of different ions to a fixed total current density at various points in a horizontal plane of an ED system at steady state. We next consider how large the electrical potential difference across a cell must be to establish and maintain such a current density, and how the potential and concentrations change across different portions of the cell.

In each of the bulk solutions and within each of the membranes of an ED system, the concentrations are uniform at a particular y value, and the current is associated strictly with electromigration. As a result, the potential drop across each of these regions can be computed using Equation 15-260 (with different diffusion coefficients and concentrations in the membrane from those in solution). As noted previously, this equation indicates that the potential gradient is linear across the region of interest, with a slope that is directly proportional to the current.

The concentration–current–potential relationships in the boundary layers are more complicated, because both electromigration and diffusion are operative. We develop this relationship next, for an idealized, binary-salt, steady-state system with known current density and known solution composition in the bulk solutions of both the diluate and concentrate. It is convenient to define the length dimension x to be zero at the membrane surface and to increase away from the membrane, regardless of the direction of current flow. The maximum value of x is the boundary layer thickness, δ; that is, the interface between the boundary layer and the bulk solution is at $x = \delta$.

As noted previously, the current density of the selected ion equals the full current density in each boundary layer (Equation 15-268), and the fraction of that current density attributable to diffusion is $(z_{\text{rej}})/(z_{\text{rej}} - z_{\text{sel}})$ (Equation 15-270a). Combining these ideas, we can express the current density of the selected ion attributable to diffusion as

$$\frac{z_{\text{rej}}}{z_{\text{rej}} - z_{\text{sel}}}\hat{I} = -D_{\text{sel}} z_{\text{sel}} \mathcal{F}\frac{dc_{\text{sel}}}{dx} \qquad (15\text{-}274)$$

Rearranging and integrating yields an expression for the concentration profile of that ion

$$\int_{c(x_1)}^{c(x_2)} dc_{\text{sel}} = -\frac{z_{\text{rej}}}{z_{\text{sel}}(z_{\text{rej}} - z_{\text{sel}})}\frac{\hat{I}}{D_{\text{sel}}\mathcal{F}}\int_{x_1}^{x_2} dx$$

$$c_{\text{sel}}(x_2) = c_{\text{sel}}(x_1) - \frac{z_{\text{rej}}}{z_{\text{sel}}(z_{\text{rej}} - z_{\text{sel}})}\frac{\hat{I}}{D_{\text{sel}}\mathcal{F}}(x_2 - x_1)$$

$$(15\text{-}275)$$

Equation 15-275 describes the difference in concentration of the selected ion between any two points in any of the boundary layers in the system (i.e., on either side of either type of ion exchange membrane). Equations 15-274 and 15-275 indicate that the selected ion concentration profile is linear through the boundary layer. The profile of the rejected ion can be calculated by combining the electroneutrality equation (Equation 15-265) with Equation 15-275, and is therefore also linear. Our primary interest is in the change in concentrations between the bulk solution and the membrane. However, Equation 15-275 is awkward to use directly for that calculation because so many of the terms in the equation have signs that depend on whether we are analyzing the boundary layer adjacent to the cation or anion exchange membrane, and whether this layer is in the concentrate or diluate cell. This complication can be largely overcome by eliminating some of the signs and separating the analysis of cations from that of anions. The results are shown in Table 15-16. In the equations, δ is the thickness of the boundary layer (without directionality), x is the distance from the membrane to the location of interest, $\hat{I}$ is always taken as positive, and the − and + signs apply to the diluate and concentrate solutions, respectively.

Once the concentration profiles of both ions are established, the profile of electrical potential can be derived by

TABLE 15-16. Model of Idealized System: Boundary Layer Concentrations, Potentials, and Fluxes

Near Anion Exchange Membrane[a,b,c]	Near Cation Exchange Membrane[a,b,c]

$$c_{\text{cat}}(x) = c_{\text{cat,bulk}}\left[1 \mp \frac{\hat{I}\delta(1-x/\delta)}{z'\mathcal{F}D_{\text{an}}c_{\text{cat,bulk}}}\right] \quad (15\text{-}277)$$

$$c_{\text{cat}}(x) = c_{\text{cat,bulk}}\left[1 \mp \frac{\hat{I}\delta(1-x/\delta)}{z''\mathcal{F}D_{\text{cat}}c_{\text{cat,bulk}}}\right] \quad (15\text{-}279)$$

$$c_{\text{cat,M}} = c_{\text{cat,bulk}}\left[1 \mp \frac{\hat{I}\delta}{z'\mathcal{F}D_{\text{an}}c_{\text{cat,bulk}}}\right] \quad (15\text{-}278)$$

$$c_{\text{cat,M}} = c_{\text{cat,bulk}}\left[1 \mp \frac{\hat{I}\delta}{z''\mathcal{F}D_{\text{cat}}c_{\text{cat,bulk}}}\right] \quad (15\text{-}280)$$

$$c_{\text{an}} = \frac{z_{\text{cat}}}{|z_{\text{an}}|}c_{\text{cat}} \quad (15\text{-}281)$$

$$c_{\text{an}} = \frac{z_{\text{cat}}}{|z_{\text{an}}|}c_{\text{cat}} \quad (15\text{-}281)$$

$$\psi(x) - \psi_M = -\frac{RT}{z_{\text{cat}}\mathcal{F}}\ln\left[\frac{1 \mp \dfrac{\hat{I}\delta(1-x/\delta)}{z'\mathcal{F}D_{\text{an}}c_{\text{cat,bulk}}}}{1 \mp \dfrac{\hat{I}\delta}{z'\mathcal{F}D_{\text{an}}c_{\text{cat,bulk}}}}\right] \quad (15\text{-}282)$$

$$\psi(x) - \psi_M = \frac{RT}{|z_{\text{an}}|\mathcal{F}}\ln\left[\frac{1 \mp \dfrac{\hat{I}\delta(1-x/\delta)}{z''\mathcal{F}D_{\text{cat}}c_{\text{cat,bulk}}}}{1 \mp \dfrac{\hat{I}\delta}{z''\mathcal{F}D_{\text{cat}}c_{\text{cat,bulk}}}}\right] \quad (15\text{-}283)$$

$$\Delta\psi_{\text{BL}} = -\frac{RT}{z_{\text{cat}}\mathcal{F}}\ln\left(1 \mp \frac{\hat{I}\delta}{z'\mathcal{F}D_{\text{an}}c_{\text{cat,bulk}}}\right) \quad (15\text{-}284)$$

$$\Delta\psi_{\text{BL}} = \frac{RT}{|z_{\text{an}}|\mathcal{F}}\ln\left(1 \mp \frac{\hat{I}\delta}{z''\mathcal{F}D_{\text{cat}}c_{\text{cat,bulk}}}\right) \quad (15\text{-}285)$$

$$J_{\text{cat}} = 0 \quad (15\text{-}286)$$

$$J_{\text{cat}} = \frac{\hat{I}}{z_{\text{cat}}\mathcal{F}} \quad (15\text{-}288)$$

$$J_{\text{an}} = -\frac{\hat{I}}{|z_{\text{an}}|\mathcal{F}} \quad (15\text{-}287)$$

$$J_{\text{an}} = 0 \quad (15\text{-}289)$$

[a] In all the equations with the $\mp$ symbol, the negative sign is for the diluate and the positive for the concentrate. $\hat{I}$ is always taken as positive; absolute value signs are omitted to avoid clutter.

[b] $z' \equiv z_{\text{cat}} + |z_{\text{an}}|$; $\quad z'' \equiv z_{\text{cat}} + \dfrac{z_{\text{cat}}^2}{z_{\text{an}}}$; $\quad \Delta\Psi_{\text{BL}} \equiv \Psi_M - \Psi(\delta)$.

[c] The absolute value signs could be omitted in the expressions for z' and z'', with consequent sign changes from $+$ to $-$ in each of those equations. The form shown makes it easier to see what terms are additive, but in numerical modeling, it is likely to be simpler to use the true value and opposite signs in the appropriate terms.

any of the relationships between ψ and c_{sel} or c_{rej}. For example, multiplying both sides of Equation 15-264b by dx, rearranging and integrating, we find

$$\int_{c(\delta)}^{c(x)} \frac{dc_{\text{rej}}}{c_{\text{rej}}} = -\frac{\mathcal{F}}{RT}z_{\text{rej}}\int_{\psi(\delta)}^{\psi(x)} d\psi \quad (15\text{-}276)$$

$$\psi(x) = \psi(\delta) - \frac{RT}{z_{\text{rej}}\mathcal{F}}\ln\frac{c_{\text{rej}}(x)}{c_{\text{rej}}(\delta)}$$

The result indicates that ψ is a logarithmic function of c_{rej} (and therefore c_{sel}) in the boundary layers. Again converting this expression to put it in terms of the cations and anions, the potential profile across each boundary layer can be calculated, as also shown in Table 15-16.

■ **EXAMPLE 15-28.** An ED system has the following characteristics:

- The influent is a neutral pH solution of NaCl at a concentration of 0.1 M and a temperature of 25°C, and the same influent is applied at the same flow rate to both the diluate and concentrate cells;

- The distance between adjacent membranes is 1 mm, the boundary layers adjacent to both membranes are 50 μm thick, and both membranes are 250 μm thick;
- The cation exchange membrane has a fixed charge density of 3900 eq/m^3 and the diffusivity of Na$^+$ in the membrane is 1.75×10^{-10} m^2/s;
- The anion exchange membrane has a fixed charge density of 1100 eq/m^3 and the diffusivity of Cl$^-$ in the membrane is 9.68×10^{-11} m^2/s.[35]

The system is operated at a potential drop that yields a current density at the entrance layer of 200 A/m^2. Determine the concentration and potential drop profiles across a cell-pair at the entrance where the concentrations in the bulk solutions of the diluate and the concentrate are the same.

Solution. Since the pH is neutral, the current densities carried by protons and hydroxide ions are assumed to be negligible and are, therefore, ignored. In the cation exchange

[35] Characteristics of the membranes are adapted from Pourcelly et al. (1996) and from Amang et al. (2003).

membrane (cation exchange membrane), the potential gradient is constant throughout the membrane thickness (because the concentration is constant) and can be found from a rearrangement of Equation 15-249, recognizing that only the Na$^+$ ions are involved in the current flow

$$\left(\frac{d\psi}{dx}\right)_{CEM} = -\frac{RT}{\mathcal{F}^2 z_{Na^+}{}^2 D_{CEM,Na^+} c_{CEM}} \hat{I}$$

$$= -\frac{(8.314\,J/mol\,K)(298\,K)(200\,A/m^2)(1C/A\,s)}{(96,485\,C/eq)^2(1\,eq/mol)^2(1.75 \times 10^{-10}\,m^2/s)(3900\,eq/m^3)}$$

$$= -78.0\,V/m$$

$$\Delta\psi_{CEM} = \left(\frac{d\psi}{dx}\right)_{CEM} d_{CEM} = (-78.0\,V/m)(250\,\mu m)((10^{-6}\,m)/(\mu m))$$

$$= -0.0195\,V = -19.5\,mV$$

The calculation of the potential drop through the anion exchange membrane (AEM) is similar, with chloride being the moving ion; the results are that $((d\psi)/(dx))_{AEM} = -500.3\,V/m$ and $\Delta\psi_{AEM} = -0.1250\,V = -125.0\,mV$. The much higher potential drop in this anion exchange membrane is caused by the lower charge density and lower diffusivity in comparison to the cation exchange membrane.

The concentrations of both bulk solutions (i.e., the diluate and concentrate) are the same at the entrance to the ED system, and therefore the potential gradients are also the same. This potential gradient can be calculated from Equation 15-259, and the potential drops can then be calculated as the product of the gradient with the width of the bulk solutions, d_{bulk}. This width is the distance between membranes (1 mm or 1000 μm) minus twice the boundary layer thickness, or 900 μm.

The chloride concentration at the membrane surface is the same, since both ions carry a single charge and electroneutrality prevails.

The diffusion of sodium also controls the concentration gradient on the diluate side of the cation exchange membrane, but here electromigration and diffusion are in the same direction. Because the bulk concentration in the diluate is the same as in the concentrate in this case (at the entrance to the system), the calculation is identical except for a negative sign before the fraction term; therefore

$$c_{Na^+,CEM,dil} = (0.1\,mol/L)(1 - 0.388) = 0.0612\,mol/L$$

Again, electroneutrality requires that $c_{Cl^-,CEM,dil} = 0.0612\,mol/L$.

The calculations at the surface of the anion membrane are similar, except that it is the diffusion of chloride that controls the process and enters into the equation. The results are

$$c_{Cl^-,AEM,con} = c_{Na^+,AEM,con} = 0.1255\,mol/L$$

and $c_{Cl^-,AEM,dil} = c_{Na^+,AEM,dil} = 0.0745\,mol/L$.

The potential drop is a nonlinear function of distance through each boundary layer, so a complete profile would require a calculation at several points. However, our interest here is only in determining the overall potential drop in each boundary layer. As an example, we use Equation 15-285 for the concentrate side of

$$\left(\frac{d\psi}{dx}\right)_{bulk} = -\frac{RT}{\mathcal{F}^2 \sum_{all\,i} z_i^2 D_i c_i} \hat{I} = -\frac{RT}{\mathcal{F}^2 z^2 c_{bulk} \sum_{all\,i} D_i} \hat{I}$$

$$= -\frac{(8.314\,J/mol\,K)(298\,K)(200\,A/m^2)(1\,C/A\,s)}{(96,485\,C/mol)^2(0.1\,mol/L)(1000\,L/m^3)(1.334 \times 10^{-9}) + (2.032 \times 10^{-9}\,m^2/s)} = -158.2\,\frac{V}{m}$$

$$\Delta\psi_{bulk} = \left(\frac{d\psi}{dx}\right)_{bulk} d_{bulk} = (-158.2\,V/m)(900\,\mu m)(10^{-6}\,m/\mu m) = -0.1424\,V = -142.4\,mV$$

In the boundary layer on the concentrate side of the cation exchange membrane, we can find the concentration at the membrane surface ($x = 0$) from Equation 15-280.

$$c_{Na^+,CEM,conc} = c_{Na^+,bulk}\left[1 + \frac{\hat{I}\delta}{(z_{Na^+} + (z_{Na^+}{}^2/|z_{Cl^-}|))\mathcal{F}D_{Na^+}c_{Na^+,bulk}}\right]$$

$$= 0.1\,mol/L\left[1 + \frac{(200\,A/m^2)(50 \times 10^{-6}\,m)(1\,C/A\,s)}{(2\,equiv/mol)(96,485\,C/equiv)(1.334 \times 10^{-9}\,m^2/s)(0.1\,mol/L)(1000\,L/m^3)}\right]$$

$$= (0.1\,mol/L)(1 + 0.388) = 0.1388\,mol/L$$

the CEM membrane.[36]

$$\Delta\psi_{BL,CEM,conc} = \frac{RT}{|z_{an}|\mathcal{F}}\ln\left[1 + \frac{\hat{I}\delta}{(z_{cat} + (z_{cat}^2)/(|z_{an}|))\mathcal{F}D_{cat}c_{cat,bulk}}\right]$$

$$= \frac{(8.314\,\text{J/mol K})298\,\text{K}}{(1\,\text{eq/mol})(96,485\,\text{C/eq})}\ln\left[1 + \frac{(200(\text{A/m}^2))(5\times 10^{-5}\,\text{m})}{(2\,\text{eq/mol})(96,485\,\text{C/eq})(1.334\times 10^{-9}\,\text{m}^2/\text{s})(100\,\text{mol/m}^3)}\right]$$

$$= (0.0256\,\text{V})\ln[1.388] = 0.0084\,\text{V} = 8.4\,\text{mV}$$

The loss of potential through each part of the cell-pair is summarized in the following table. The total potential drop is the sum of all of the values in this table, or 463.8 mV (i.e., almost one-half volt per cell-pair). The potential drop through the entire system would be this value multiplied by the number of cell-pairs, plus whatever potential drop occurs in the electrode cells at both ends of the system. The concentration and potential profiles through the cell-pair are shown in Figure 15-52.

	Diluate				Concentrate		
AEM	BL_{AEM}	Bulk	BL_{CEM}	CEM	BL_{CEM}	Bulk	BL_{AEM}
$\Delta\psi$ (mV) −125.1	−7.6	−142.4	−12.6	−19.5	−8.4	−142.4	−5.8

■

Analysis of the Two-Dimensional ED System

In ED systems, the solution to be treated travels at a constant flow rate (in the y direction) and (ideally) by plug flow through the diluate channels, while ions are continuously subjected to a driving force (in the x direction) that transports them to the boundary of this solution and into another phase (the ion exchange membrane). The contaminants then pass through the membrane before entering the concentrate solution, which transports them out of the system.

Although the primary driving force for contaminant transport and the geometry of the system are unique to ED, the general process bears many similarities to other multiphase, contaminant removal systems with plug flow of solution that we have considered previously, including gas transfer columns, fixed- or moving-bed adsorptive contactors, and granular-media filtration. To model all these systems at steady state, we wrote mass balances at two scales: one at the macroscopic scale of the reactor (to characterize gross contaminant removal between the inlet and a point downstream in the reactor or at its exit), and the other over a differentially thin layer in the reactor (to characterize transport out of the feed solution at that location). We follow a similar procedure here to model two-dimensional steady-state ED systems.

Macroscopic Mass Balance on ED Reactor The macroscopic mass balance for any particular ion with a control volume from the entrance to a particular value of y in the ED system is

$$Q_{conc}c_{i,conc,in} + Q_{dil}c_{i,dil,in} = Q_{conc}c_{i,conc}(y) + Q_{dil}c_{i,dil}(y)$$
(15-290)

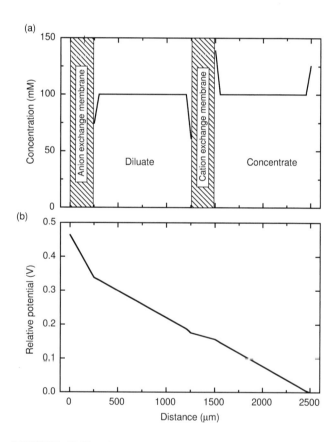

FIGURE 15-52. Concentration and potential profiles through one cell-pair for Example 15-28.

---

[36] To reduce clutter, the identities of 1 V = 1 J/C and 1 C = 1 A s are not shown in the calculation.

Assuming that no water flows through the membrane, Equation 15-290 can be rearranged to find:

$$c_{i,\text{conc}}(y) = \frac{Q_{\text{conc}}c_{i,\text{conc,in}} + Q_{\text{dil}}c_{i,\text{dil,in}} - Q_{\text{dil}}c_{i,\text{dil}}(y)}{Q_{\text{conc}}} \quad (15\text{-}291)$$

If, as is commonly true, both the concentrations and the flow rates entering the diluate and concentrate cells are equal, Equation 15-291 can be rewritten as follows:

$$c_{i,\text{conc}}(y) = 2c_{i,\text{in}} - c_{i,\text{dil}}(y) \quad (15\text{-}292)$$

Using Equation 15-291 or (15-292) as appropriate, we can compute the concentration of each ion in the concentrate solution at any given value of y, if the concentration in the diluate solution is known.

Microscopic Mass Balance on a Differential Element in an ED Reactor We next evaluate the conditions in a control volume that is differentially thick in the direction of bulk flow (y direction), considering various parts of the system (the bulk solutions, the boundary layers, and the membranes) separately. We begin by writing a mass balance on the cation in a control volume that comprises the fluid in a dy-thick layer of the bulk diluate, as shown schematically in Figure 15-53.

Cations enter and leave this control volume by advection, and they also leave by transport into the boundary layer adjacent to the cation exchange membrane. At steady state, the flux of cations into the boundary layer (in the x direction) equals its flux into and through the membrane, so we can equate the flux into the membrane with the net flux (in $-$ out) of the bulk solution. Taking these factors into account, along with the assumption of steady state, we find

$$\frac{\partial c_{\text{cat}}}{\partial t}dV = Q\left[c_{\text{cat}}(y) - \left\{c_{\text{cat}}(y) + \frac{dc_{\text{cat}}(y)}{dy}dy\right\}\right] - J_{\text{cat,CEM}}dA \quad (15\text{-}293)$$

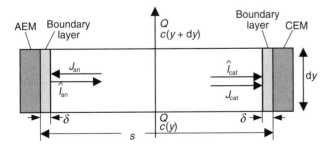

FIGURE 15-53. Control volume for a mass balance on ion i in the bulk diluate. The layer is assumed to be well-mixed, so the concentrations exiting the layer are the same as those inside it.

$$0 = -Qdc_{\text{cat}} - J_{\text{cat,CEM}}dA \quad (15\text{-}294)$$

where dV and dA are, respectively, the volume of the control volume and the area of the cation exchange membrane bordering it. Writing Q and dA in terms of the fluid velocity (v_y), the width of the membrane in the z direction (W_m, into the page in Figure 15-53), and the width of the plug flow (i.e., bulk) portion of the cell ($W_{\text{bulk}} = s - 2\delta$, the distance between the membranes minus the portion of the cell occupied by the boundary layers) yields

$$0 = -vW_mW_{\text{bulk}}dc_{\text{cat}} = -J_{\text{cat,CEM}}W_m dy \quad (15\text{-}295)$$

In the differential element, the gradient of concentration in the direction of flow (the y direction) is found by rearranging Equation 15-295 to find

$$\frac{dc_{\text{cat}}}{dy} = -\frac{J_{\text{cat,CEM}}}{vW_{\text{bulk}}} \quad (15\text{-}296)$$

Theoretically, we could separate the variables and integrate from the inlet ($y = 0$) to the outlet ($y = L_{\text{cell}}$) to obtain the concentration at the outlet end. However, the mass flux is related to the current (which varies along the cell length) and the current is related to the (uniform) potential drop through a cell, so such an integration is not possible analytically. Rather, a numerical integration of the differential through the cell is necessary. To perform this integration, the relationships among potential, current, and flux in the various parts of a cell-pair must be known.

The preceding example exploring the conditions at the entrance ($y = 0$) illustrates most of the calculations necessary in the analysis of full (two-dimensional) systems, but that example was simplified by the specification of the current. In real systems, the potential drop per cell-pair is known (and is constant with y) and the current density at a given y (or Δy increment) must be determined; this situation is the opposite of that considered in the example. Since the potential drop in the boundary layers is a nonlinear function of the distance in the boundary layer and the adjacent bulk concentration, the current density for the particular bulk concentrations and voltage drop per cell-pair must be found by trial-and-error. As y increases along the length of an ED cell, the concentration of the diluate decreases (increasing the fraction of the cell-pair potential drop through that portion of the cell) and the concentration of the concentrate increases (reducing the relative potential drop through that layer). In response to the diminishing concentration in the diluate, the current density decreases with increasing y. Nevertheless, all the necessary equations to solve for the removal achieved in an ED system (or the required size of an ED system for a desired removal) have been presented, as illustrated in the following example.

EXAMPLE 15-29. An ED system is operated at a liquid loading rate (flow per unit area) that leads to a velocity in both bulk solutions of 0.1 m/s, generating 30-μm-thick boundary layers adjacent to each membrane; the membranes are 0.5 mm thick and are separated by 1 mm. The length of the membranes is 2 m and the potential drop is set at 0.8 V/cell-pair. The influent solution and membrane characteristics (diffusivities and ion capacities) are identical to those in Example 15-28. Determine the current density and the concentrations of the diluate and concentrate along the cell length.

Solution. At the influent, the concentration in both the bulk and diluate are known (and equal). A trial-and-error solution to find the current density that yields the specified potential drop for those bulk concentrations is performed first, using all the equations illustrated in Example 15-28. An initial guess for the current density at $y = 0$ is made and the associated potential drop across one cell is calculated as in Example 15-28. The (guessed) current density is then adjusted up or down repeatedly until the value that yields the stated potential drop is reached; for this first case of the influent concentration in both bulk solutions, the current density is found to be 263.6 A/m^2. The flux through the (ideal) cationic membrane $(J_{cat,CEM})$ is calculated from this current density using Equation 15-237 (with the result being 2.732×10^{-3} mol/m^2 s), and the concentration gradient $(dc_{cat})/(dy)$ is then calculated from Equation 15-297:

$$\frac{dc_{cat,dil}}{dy} = -\frac{J_{p,CEM}}{vW_{PF}} = -\frac{2.732 \times 10^{-3} \,\text{mol/m}^2\,\text{s}}{(0.1\,\text{m/s})\big((1000 - 2(30)) \times 10^{-6}\,\text{m}\big)}$$
$$= -29.06 \,\text{mol/m}^4 = -29.06 \,\text{mmol/L m}$$

For numerical integration, we can write

$$c_{cat,dil}(y + \Delta y) = c_{cat,dil}(y) + \left(\frac{dc_{cat,dil}}{dy}\right)_y \Delta y \quad (15\text{-}297)$$

The results for the concentrations of both the concentrate and diluate, obtained by numerical integration of Equation 15-297 and the use of Equation 15-292, are shown in Figure 15-54a, with a nearly linear drop in the diluate concentration throughout the cell length (and identical rise in the concentrate). In truth, the values of $(dc_{cat,dil})/(dy)$ diminish continuously (albeit slightly) throughout the length; the plot of the current density as a function of length shown in Figure 15-54b has considerably more curvature. ■

Ramifications for Design of Electrodialysis Systems
Although the mathematical model that has been presented

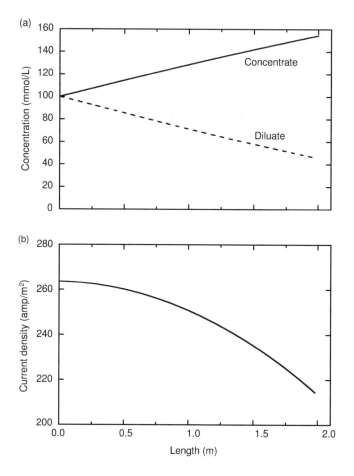

FIGURE 15-54. Electrodialysis performance along the cell length: (a) concentration; (b) current density.

incorporates many simplifications, several ramifications for the design and operation of ED systems can be seen. The influence of hydraulic characteristics of an ED system is most obvious in the thickness of the boundary layers, which is determined by the turbulence in the cells. Designers generally include spacers in the system expressly to promote turbulence within the cells to reduce this boundary layer thickness. ED systems are also generally operated with relatively high superficial velocities in the cells (in the range of 3–15 cm/s), which also promotes turbulence. Diminishing the boundary layer thickness reduces the distance through which diffusion controls the process, and therefore reduces the potential drop in the diluate boundary layers; in turn, the reduced potential drop increases the current density (and separation rate) for a given applied voltage. Increasing the velocity also decreases the detention time in a cell (or requires longer cells) and increases the headloss. The issue of superficial velocity therefore ultimately represents a tradeoff between capital and operating costs (for a given desired effluent quality).

The model also suggests that a larger boundary layer in the concentrate could be helpful, because the concentration in the boundary layers is greater than that in the bulk, and therefore the potential drop would be decreased with a larger boundary layer. A larger boundary layer in the concentrate cell could be accomplished by using a lower velocity in the concentrate; theoretically, that could be accomplished either with a larger cross-sectional area or a lower flow rate in the concentrate cell in comparison to the diluate cell. However, some ED systems are EDR systems, in which case it would not be possible to make the geometry of the concentrate cells different from the diluate cells. Further, using different flow rates to create different velocities in adjacent cells could result in pressure differences between adjacent cells, and since the ion exchange membranes are not impermeable, an advective flow from the diluate into the concentrate would be created. Therefore, the apparent advantage of a larger boundary layer in the concentrate cells is illusory.

However, a separate means of increasing the concentration in the concentrate (and thereby reducing the voltage drop in that cell) is available and is often used. Some of the concentrate is recycled from the effluent back to the influent of the concentrate cells, as shown in Figure 15-55. In this way, the influent concentration of the concentrate is greater than that of the diluate, and the potential drop through the concentrate is less than in the nonrecycle case throughout the unit. The flow rates in both cells are the same (to avoid the pressure difference noted above), so greater than 50% of the influent goes through the diluate cells, thereby increasing the recovery (r) in comparison to the no recycle case. For example, if a recovery of 80% is desired, 80% of the influent goes to the diluate cells, while the remaining 20% is joined by three times that amount ($(2r-1) = 2(0.8) - 1 = 0.6$) that has been recycled from the concentrate effluent to yield equal overall flow rates in both cell types.

The Limiting Current. The only electrical control in an ED system is the value of the applied voltage. As noted previously, ED systems generally operate with potential drops in the range of 0.5–1.5 V/cell-pair. Increasing the voltage per cell-pair directly increases the current density, and therefore increases the ion separation rate and the removal efficiency in a given unit. However, the voltage cannot be increased

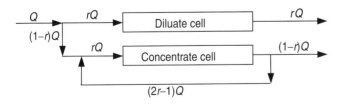

FIGURE 15-55. Recycling of concentrate effluent to increase recovery and reduce the voltage drop in a cell-pair.

indefinitely without difficulty. As indicated in Equations 15-277 and 15-279, the concentration in the diluate boundary layer diminishes linearly from the bulk concentration to both membrane surfaces. The steepness of the concentration gradient increases with increasing applied voltage (or, more precisely, with increasing current density) until, at some current density, the calculated concentration is zero at one of the membrane surfaces; this current density is known as the *limiting* current. In most cases, the limiting current density calculated based on conditions adjacent to the cation exchange membrane is lower than that based on conditions adjacent to the anion exchange membrane, because the diffusivity of cations is generally less that that of anions. Assuming that that is true, the limiting current can be calculated from Equation 15-280 as

$$\hat{I}_{\lim} = \frac{c_{\text{cat,dil,bulk}}\left(z_{\text{cat}} + (z_{\text{cat}}^2/|z_{\text{an}}|)\right)\mathcal{F}D_{\text{cat}}}{\delta} \quad (15\text{-}298)$$

For the membrane system and influent described in Example 15-29, the maximum voltage that can be applied without exceeding the limiting current at the system outlet (i.e., $y = 2\,\text{m}$) can be obtained by trial-and-error and is approximately 1.16 V/cell-pair.

In the physical system, the concentration at the membrane surface cannot really reach zero; rather, the fact that such a concentration is predicted reflects a failure of one or more of the assumptions used in the model development. In this case, two assumptions of the model as presented above might not be met. First, as explained in the following section (see Equation 15-300 and associated discussion), another source of potential drop than those considered is the junction potential between the solutions on either side of each membrane. Under most circumstances, this potential drop is negligible; however, when the concentration near the surface of the membrane in the diluate gets quite low, this potential drop becomes significant. Since the total available potential is fixed in any system, an increased loss of potential near the surface of the membrane leaves less potential drop available to drive electromigration, and so the concentration at the membrane surface does not decrease as much as the model predicts.

The second assumption that might fail as the limiting current is approached is that essentially all of the current is carried by the target ions (Na^+ and Cl^- in the example system) and a negligible portion is carried by H^+, OH^-, or other ions in the solution. In reality, as the concentration of the target ions in the boundary layer approaches zero, the diffusive and electromigration currents both become small—the former because the concentration gradient is limited and the latter because the concentration itself is limited. Under these conditions, the electrical resistance of the solution is high, so a large driving force is required to maintain even a small current; mathematically, this result is shown in Equations 15-284 and 15-285, which indicate that

$\Delta\psi$ approaches infinity as c_{bulk} approaches zero. At some point, the current carried by the target ions becomes sufficiently small that a non-negligible fraction of the current is carried by other ions. If the pH of the solution is near neutral, however, the concentration of the target ions near the surface would need to be quite low for the transport number of either hydroxide or hydronium to rise above 0.05. As with the first case, an increased expenditure of potential near the membrane surface means that less is available in other parts of the system, so the overall transport is limited. In addition, any current carried by nontarget ions provides no benefit in terms of the removal of target ions from the diluate solution.

According to Equation 15-298, when the diluate concentration is high, the limiting current is large, so the actual current density can be high without approaching the limiting value. A high current is desirable, since the ion separation rate is directly proportional to the current, and a greater separation will occur per unit length (in the y direction) of the system. On the contrary, when the diluate concentration is low, the current must be low to remain below $\hat{I}_{lim}$. To take advantage of the high-concentration region near the inlet of the system while avoiding problems near the outlet, ED systems are often designed to have several units in series, with sequentially lower applied voltages per cell-pair.

The effect of the bulk concentration of the diluate indicated in Equation 15-298 helps explain the limitation of ED as a treatment process. Excellent removal efficiency means that c_{bulk} becomes very low, but a low concentration limits the current that can be carried, and the current dictates the separation that can be achieved. Thus, very low diluate (i.e., effluent) concentrations cannot be achieved in ED systems with reasonable residence times and applied voltages; hence, ED is best used when the desired effluent quality can contain an ionic concentration of a few millimolar. Drinking water regulations in the United States require that the total dissolved solids be <500 mg/L, and reducing salt concentrations to below this limit with ED is certainly possible. However, certain industrial applications (pharmaceutical and semi-conductor manufacturing, for example) require ultrapure water with far lower ion concentrations, so ED is inappropriate for those applications. A related technology called electrodeionization (EDI, which is essentially a hybrid between ED and ion exchange technologies) can be used to achieve product waters with much lower ionic concentrations.

Complications of Real Systems

The preceding presentation illustrates the essential principles of ED and suggests the critical design and operational features of ED systems. Real systems are, of course, more complicated. Here, we consider briefly some complexities of real systems that were ignored in the preceding presentation.

Water Flow Through Ion Exchange Membranes Several mechanisms can carry water through the ion exchange membranes. Ion exchange membranes are not impermeable, and so a pressure difference between adjacent cells can cause water to move from the higher to the lower pressure cell, carrying both cations and anions with it. In real systems, a small excess pressure is often imposed on the diluate side, so that any such movement is from the diluate to the concentrate, rather than *vice versa*. This choice reflects the decision that a small loss of water from the diluate has less serious consequences than contamination of the diluate by leakage from the concentrate.

Many ions, especially cations, are strongly hydrated, and the waters of hydration are carried along with the ion through the membrane; this phenomenon leads to a loss of water from the diluate and a gain of water in the concentrate. This flux of water also carries other water molecules with it by the friction (viscosity) of water, a phenomenon known as electroosmosis. Also, the activity of water is higher in a diluate cell than in an adjacent concentrate cell, leading to an osmotic driving force for water to pass from the diluate to the concentrate.

Nonideal Behavior of Membranes Membranes are never ideal; that is, they are not able to accomplish a complete separation of cations and anions. Ion exchange membranes are semiporous, so a low concentration of ions of the same charge as the fixed charge (i.e., those that theoretically should be rejected) can migrate through the small pores. The diffusive and electromigration driving forces for such ions are both from the concentrate to the diluate. This transport is enhanced by the high concentration at the boundary between the membrane and the concentrate. A second source of nonideality is the passage of protons or hydroxyl ions through the membrane. Because the specific equivalent ionic conductance of these ions is much higher than that of other ions, they can contribute to the current even at relatively low concentrations, and thereby reduce the current carried by the target ions.

Multicomponent Systems The preceding analysis has been limited to single-salt systems. Real waters subjected to ED treatment contain several different anions (e.g., chloride, sulfate, bicarbonate, carbonate, and others) and cations (e.g., sodium, potassium, calcium, magnesium, and others) at significant concentrations. For such complex systems, the transport equations for ED systems cannot be solved analytically. Nevertheless, the principles elucidated for the single-salt case can be generalized to more complicated solutions. For example, the transport numbers in the bulk solution can be calculated from the diffusivity values and concentrations of the ions. Ion exchange membranes (like ion exchange resins, as explained in Chapter 7) exhibit selectivity factors (a measure of the preference for one ion over another) and this selectivity determines the transport numbers in the membrane. Ultimately, the principles of diffusion, electromigration, and electroneutrality that dictate the behavior of all of the ions are the same as in the simple systems.

Overlimiting Current According to Ohm's law, the current density achieved in an ED system is proportional to the applied potential, and this relationship is followed until the current approaches the limiting value. If the applied potential is increased a small amount beyond this value, the current remains (essentially) at the same limiting current value. However, at some higher value of applied potential, the current density increases dramatically. In this region, the membranes still generally perform well in separating salts. Considerable attention has been devoted to understanding this phenomenon; although a few theories have been elucidated, none is universally accepted. Some early explanations, such as leakage of co-ions through the membrane and high amounts of electro-osmosis, have been found lacking, but other theories such as electro-convection induced by the electrical double layer immediately adjacent to the membrane (where electroneutrality does not occur) and hydrodynamic convection in the boundary layer have merit; these ideas are explored in more advanced textbooks on ED (Strathmann, 2004; Tanaka, 2007).

Additional Sources of Potential Drop The potential drop associated with the transport of ions through all eight regions of a cell-pair was delineated above. Two other sources of potential drop also are present in ED systems, though both are generally small in full-scale operating systems. The first is the potential drop at the electrodes themselves and in the electrode rinse cells. To maintain the disequilibrium in the system that drives ions through the boundary layers and the membranes throughout the system, the potential associated with the electrode reactions (see Figure 15-50), an *overpotential* that causes the disequilibrium and the potential associated with transport in the electrode rinse cells must be continuously invested. Typically, the total of these three requirements is in the range of 1.5–5 V, increasing with the applied potential.

The disparity in charge density near the surface of the membranes (high charge density on the membrane surface and much lower in the adjacent solutions) leads to a potential drop across the membrane that can be reasonably estimated by the junction potential between two solutions, using the conditions in the solution phase at the borders with the membrane on the concentrate and diluate side. The junction potential between two adjacent solutions is calculated as (Bard and Faulkner, 2001)

$$\psi_j = \psi_{sol\,1} - \psi_{sol\,2} - \frac{RT}{\mathcal{F}} \sum_i \int_{sol\,1}^{sol\,2} \frac{t_i}{z_i} d\ln a_i \quad (15\text{-}299)$$

The direction from solution 1 (sol 1) to solution 2 (sol 2) is in the direction of the current; for the cation exchange membrane, sol 1 is the diluate and sol 2 is the concentrate, whereas the reverse is true for the anion exchange membrane. If the membrane is ideal, only the selected ion is relevant, as its transport number is unity. If we further assume an ideal solution of a binary salt, the junction potential between the concentrate and diluate on opposite sides of the membrane can be calculated as

$$\psi_{j,mem} = -\frac{RT}{\mathcal{F}} \frac{1}{|z_{sel}|} \ln \frac{c_{conc,m}}{c_{dil,m}} \quad (15\text{-}300)$$

where the subscript "sel" designates the selected ion that travels through the membrane, the subscripts on the solution concentrations designate the locations as the concentrate and diluate sides of the membrane, and the absolute value for the charge of the selected ion allows always writing the logarithm term as the ratio of concentrate to diluate.

■ **EXAMPLE 15-30** Find the junction potentials across the cation and anion exchange membranes associated with the conditions of Example 15-28.

Solution. The solution is a simple plug-in to Equation 15-300 using values for the ion concentrations at the membrane surfaces. Adjacent to the cation exchange membrane, the concentrations (of both sodium and chloride) were found to be 0.1388 mol/L and 0.0612 mol/L on the concentrate and diluate sides, respectively. The selected ion is sodium, so $|z_{sel}| = z_{Na} = 1$, and the potential drop is

$$\psi_{j,CEM} = -\frac{RT}{z_{Na}\mathcal{F}} \ln \frac{c_{conc,m}}{c_{dil,m}} = -\frac{(8.314\,J/mol\,K)298\,K}{(1\,eq/mol)(96,485\,C/eq)}$$
$$\times \ln \frac{0.1388\,mol/L}{0.0612\,mol/L} = -0.0210\,V = -21.0\,mV$$

Similarly, the concentrations adjacent to the anion exchange membrane were found to be 0.1255 mol/L and 0.0745 mol/L, and the associated potential drop across the anion exchange membrane is

$$\psi_{j,AEM} = -\frac{RT}{|z_{Cl}|\mathcal{F}} \ln \frac{c_{conc,m}}{c_{dil,m}} = -\frac{(8.314\,J/mol\,K)298\,K}{(1\,eq/mol)(96,485\,C/eq)}$$
$$\times \ln \frac{0.1255\,mol/L}{0.0745\,mol/L} = -0.0134\,V = -13.4\,mV$$

The net potential drop caused by the two junction potentials is the sum of the two calculated values or −34.4 mV per cell-pair. This example is for the entrance conditions where the bulk diluate and concentrate solutions are at the same concentration; as removal occurs through the ED system, the ratio of the concentrations would increase, and so the potential drops (and the net) would increase. ■

Summary

ED is a proven technology for removing ions from water. Although some of its earlier uses are now more often accomplished by RO, ED is a viable, and perhaps the best available, technology in some niches. Currently, its most common use for the production of drinking water is in the treatment of mildly brackish waters—those, for example with TDS values <2500 mg/L—to produce drinking water with <500 mg/L TDS. In addition, the treatment of RO concentrate by ED and subsequent RO treatment of the diluate offers the possibility to increase water recovery, while reducing the tendency of the solution to foul the RO membranes. ED also can be used to treat a variety of industrial wastes with high ionic content, either alone or as a pretreatment for waters with other problematic constituents. Although the principles of ED have been presented above primarily in terms of simple salt systems, ED can be applied to complex waters with a variety of ionic compounds. Further development of highly selective membranes could lead to targeted uses such as the removal of fluoride or arsenic ions, although such membranes are not available now.

15.11 MODELING DENSE MEMBRANE SYSTEMS USING IRREVERSIBLE THERMODYNAMICS

In the homogeneous solution-diffusion model developed in preceding sections for transport through dense membranes, transport of water and all solutes through the membrane was assumed to be uncoupled; that is, each constituent that could permeate the membrane was assumed to respond only to the diffusional driving force associated with its concentration gradient, regardless of the driving force on the other constituents. This model is sometimes referred to as a two-parameter model, because in a prototype system containing only water and a single solute, two parameters are needed to characterize transport through the membrane: one each describing the response of water and the solute to the driving forces on them ($\Delta_{m_{c/p}} P$ and $\Delta_{m_{c/p}} \Pi$). However, in some situations, it has been found necessary to account for interactions among the migrating species. These interactions are often attributed to a resistive force that the faster moving species encounter as they pass molecules of the slower moving species, slowing down the former and accelerating the latter. If only one solute can permeate the membrane (in addition to water), then one additional parameter is needed to characterize the solute-water interactions, and the corresponding models are sometimes referred to as three-parameter models. (In either model, more parameters would be required if more solutes were considered.)

The most widely used mathematical model for simulating these interactions is based on a generalized theory for the rates of physical/chemical processes known as irreversible thermodynamics (or, more formally, the thermodynamics of irreversible processes). Whereas classical thermodynamics focuses on the ultimate, equilibrium state that a closed system is expected to attain and the energy changes that occur between the initial state and that final state, irreversible thermodynamics focuses on the rates at which processes occur when systems are not at equilibrium. It is particularly useful for characterizing systems that are poised at nonequilibrium, but steady-state, conditions (Denbigh, 1981).

The underlying premise of irreversible thermodynamics is that the rate of any physical/chemical process can be related to the extent of disequilibrium of all the processes that can occur in the system. Applied to membranes, the implication is that the flux of each species across the membrane depends not only on the extent of disequilibrium of that species considered in isolation, $\Delta \mu_{i,\text{tot}}$ (as is assumed in the two-parameter models), but also on the extent of disequilibrium of all the species that can cross the membrane. Correspondingly, the rate at which the overall process approaches equilibrium is indicated by the rate of dissipation of available energy in the whole system, considering all the species present. The rate of energy dissipation associated with transport of species i across a membrane can be expressed as $J_i \Delta_{m_{c/p}} \mu_{i,\text{tot}} A_m$. It is common to normalize this rate of energy dissipation to the membrane area, as $J_i \Delta_{m_{c/p}} \mu_{i,\text{tot}}$, with dimensions corresponding to energy per unit area per unit time. The rate of dissipation of available energy when all permeating species are considered can therefore be computed as

$$\Phi = \sum_{\text{all } i} J_i \Delta_{m_{c/p}} \mu_{i,\text{tot}} \quad (15\text{-}301)$$

where Φ is known as the *energy dissipation function*, with dimensions corresponding to energy per unit membrane area per time.[37]

Although Equation 15-301 is written as the sum of $J_i \Delta_{m_{c/p}} \mu_{i,\text{tot}}$ terms, it is possible to obtain the value of the energy dissipation function in a wide variety of ways, each of which conforms to the following, generic equation:

$$\Phi = \sum_{\text{all } i} (\text{Flow of } i)(\text{Driving force for the flow of } i)$$

$$(15\text{-}302)$$

The terms "flow" and "force" are used in Equation 15-302 in a colloquial, rather than a formal, sense, consistent with their usage in the general irreversible thermodynamics literature (e.g., the driving force is not restricted to be an energy

---

[37] In the fundamental development of irreversible thermodynamics, rates of irreversible processes are characterized in terms of the rate of entropy production, dS/dt. However, in an isothermal process, like a membrane system, $dS/dt = (-dE_{\text{tot}}/dt)/T$, where T is the absolute temperature, so the rate of entropy production is directly proportional to the rate of dissipation of available energy.

gradient). For instance, in the formulation shown in Equation 15-301, the "flow" is a flux and the "driving force" is $\Delta_{m_{c/p}}\mu_{i,\text{tot}}$. Regardless of the parameters chosen for the flow and force terms, their product must have the dimensions of energy per unit area per time to yield the dissipation function as the result. Therefore, once a certain parameter is chosen as the force, the parameter that must be chosen for the flow is determined, and vice versa. Forces and flows that meet this criterion are referred to as *conjugates* of one another.

To explore some of the implications of Equation 15-301, consider a three-component system containing water and solutes A and B. We can write Equation 15-301 for the system as follows:

$$\Phi = J_w \Delta_{m_{c/p}}\mu_{w,\text{tot}} + J_A \Delta_{m_{c/p}}\mu_{A,\text{tot}} + J_B \Delta_{m_{c/p}}\mu_{B,\text{tot}}$$
(15-303)

According to irreversible thermodynamics theory, if a process is not too far from equilibrium, not only does its rate depend on the driving forces for all the nonequilibrated processes in the system, but the dependence is linear. Based on this principle, in a membrane system that contains m permeating species and that is not too far from equilibrium, we can express the flux of species i across the membrane as follows:[38]

$$J_i = L_{i1}F_1 + L_{i2}F_2 + \cdots + L_{im}F_m = \sum_{k=1}^{m} L_{ik}F_k \quad (15\text{-}304)$$

where F_k is the force driving the transport of species k across the membrane, and L_{ik} is the coefficient describing the effect of F_k on the flux of i. The driving force F_k might be quantified as $\Delta_{m_{c/p}}\mu_{k,\text{tot}}$, as suggested above, or in other ways that are described shortly. Equations like to (15-304), which describes the rate of a process in terms of a weighted, linear summation of forces, are referred to as *phenomenological equations*, and the L_{ik} values are called *phenomenological coefficients*. Phenomenological coefficients that relate the transport of one species to the driving force for another (i.e., all L_{ik} other than those for which $i = k$) are called *coupling coefficients*. If all these coupling coefficients are zero, the model reduces to the uncoupled model presented earlier in this chapter. It is common to separate the terms in the summation in Equation 15-304 into those that refer to a single species and those that couple different species, as follows:

$$J_i = L_{ii}F_i + \sum_{\text{all } k \neq i} L_{ik}F_k \quad (15\text{-}305)$$

---

[38] Although the constraint that the system be "not too far from equilibrium" is not quantified, available data suggests that all practical applications of membrane technology in water treatment meet this constraint.

Three important relationships applicable to the phenomenological coefficients are worth noting. First, all L_{ii} must be greater than zero; this requirement establishes that, in the absence of coupling (i.e., if all $L_{ik} = 0$ for $i \neq k$), the flux of i will be in the direction of F_i. Second, the phenomenological coefficients are symmetric, indicating that, in all cases, $L_{ik} = L_{ki}$. This equality, known as the Onsager reciprocal relationship, derives from an assumption that the processes occurring in the system are reversible at the microscopic scale, even though when summed, they lead to a macroscopically irreversible change. This relationship is central to the theory of irreversible thermodynamics and is discussed in more detailed descriptions of this topic. Finally, the requirement that macroscopic change in the system always be in the direction that increases entropy (decreases available energy) leads to the conclusion that, in all cases, $L_{ii}L_{kk} \geq L_{ik}L_{ki} = L_{ik}^2$. One implication of this relationship is that, if L_{ii} is zero, then L_{ik} must be zero for all k; that is, if a species does not pass through a membrane in response to a gradient in its own available energy, then it cannot be transported through the membrane via coupled transport.

The starting point for the analysis of coupled transport through the membrane is Equation 15-40 applied across the membrane, as follows:

$$\Delta_{m_{c/p}}\overline{G}_i = \overline{V}_i \Delta_{m_{c/p}} P + RT\Delta_{m_{c/p}} \ln a_i + z_i \mathcal{F}\Delta_{m_{c/p}}\psi \quad (15\text{-}40)$$

It is convenient to eliminate the logarithmic function from this equation, which can be accomplished by noting that, for a differential change in a variable x, $d \ln x$ equals dx/x. Thus, for a finite change in x, $\Delta \ln x$ can be approximated by $\Delta x/x^*$, where x^* is a value somewhere in the interval covered by Δx. For the equality to be exact, x^* must be defined by

$$x^* \equiv \frac{\Delta x}{\Delta \ln x} \quad (15\text{-}306)$$

If Δx is small compared with the absolute values of x in the interval, then x^* is almost half-way between the two end values; if not, x^* is closer to the lower of the two values. The value of x^* is therefore a kind of weighted mean value (weighted toward the lower end of the interval) and is known as the *logarithmic mean* of x, x_{lm}. Utilizing this definition, Equation 15-56 can be rewritten as follows:

$$\Delta_{m_{c/p}}\overline{G}_i = \overline{V}_i \Delta_{m_{c/p}} P + RT \frac{\Delta_{m_{c/p}} a_i}{a_{i,\text{lm}}} + z_i \mathcal{F}\Delta_{m_{c/p}}\psi \quad (15\text{-}307)$$

Assuming that the solutes behave ideally ($a_i = c_i$) and that electrically driven transport is negligible, and applying the approximation that $RTc_i = \Pi_i$ (Equation 15-131), we can write for the two solutes in our hypothetical three-component system:

$$\Delta_{m_{c/p}} G_A = \overline{V}_A \Delta_{m_{c/p}} P + \frac{\Delta_{m_{c/p}} \Pi_A}{c_{A,lm}} \quad (15\text{-}308a)$$

$$\Delta_{m_{c/p}} \mu_{B,tot} = \overline{V}_B \Delta_{m_{c/p}} P + \frac{\Delta_{m_{c/p}} \Pi_B}{c_{B,lm}} \quad (15\text{-}308b)$$

Equation 15-307 also applies to the water, but in this case, it is more useful to apply the van't Hoff equation (in the form $\Delta \Pi = -RT \Delta \ln a_w$) to Equation 15-56, yielding

$$\Delta_{m_{c/p}} \overline{G}_w = \overline{V}_w \Delta_{m_{c/p}} P - \overline{V}_w \Delta_{m_{c/p}} \Pi = \overline{V}_w \left(\Delta_{m_{c/p}} P - \Delta_{m_{c/p}} \Pi \right)$$
(15-309)

Substituting Equations 15-308 and 15-309 into Equation 15-303 and rearranging, we find

$$\Phi = J_w \overline{V}_w \left(\Delta_{m_{c/p}} P - \Delta_{m_{c/p}} \Pi \right) + J_A \left(\overline{V}_A \Delta_{m_{c/p}} P + \frac{\Delta_{m_{c/p}} \Pi_A}{c_{A,lm}} \right)$$
$$+ J_B \left(\overline{V}_B \Delta_{m_{c/p}} P + \frac{\Delta_{m_{c/p}} \Pi_B}{c_{B,lm}} \right) \quad (15\text{-}310a)$$

$$= \left(J_w \overline{V}_w + J_A \overline{V}_A + J_B \overline{V}_B \right) \Delta_{m_{c/p}} P$$
$$+ \left(\frac{J_A}{c_{A,lm}} \Delta_{m_{c/p}} \Pi_A + \frac{J_B}{c_{B,lm}} \Delta_{m_{c/p}} \Pi_B - J_w \overline{V}_w \Delta_{m_{c/p}} \Pi \right)$$
(15-310b)

Noting that $\Pi = \Pi_A + \Pi_B$, we can rewrite this equation as

$$\Phi = \left(J_w \overline{V}_w + J_A \overline{V}_A + J_B \overline{V}_B \right) \Delta_{m_{c/p}} P$$
$$+ \left(\frac{J_A}{c_{A,lm}} - J_w \overline{V}_w \right) \Delta_{m_{c/p}} \Pi_A$$
$$+ \left(\frac{J_B}{c_{B,lm}} - J_w \overline{V}_w \right) \Delta_{m_{c/p}} \Pi_B \quad (15\text{-}311a)$$

$$= \hat{J}_V \Delta_{m_{c/p}} P + \hat{J}_{A/w} \Delta_{m_{c/p}} \Pi_A + \hat{J}_{B/w} \Delta_{m_{c/p}} \Pi_B \quad (15\text{-}311b)$$

where $\hat{J}_V$, $\hat{J}_{A/w}$, and $\hat{J}_{B/w}$ are volumetric fluxes representing, respectively, the three quantities in parentheses in Equation 15-311a.

Equations 15-303 and 15-311 provide two expressions for the dissipation function in the same system. In the former equation, the driving forces are transmembrane differences in available energy of individual species ($\Delta_{m_{c/p}} \mu_{i,tot}$), and the flows are fluxes of those species, whereas in the latter, the driving forces are transmembrane differences in mechanical and osmotic pressure ($\Delta_{m_{c/p}} P$ and $\Delta_{m_{c/p}} \Pi$), and the flows are more complicated expressions whose meaning is explained shortly. The terms on the right side of these two equations can be manipulated in other ways to yield yet other formulations, with each resulting equation having a unique set of driving forces and conjugate flows. No matter what formulation is used, however, the number of forces and flows is always the same, and always equal to the number of species that can permeate the membrane.

Next, consider the meaning of the various fluxes in Equation 15-311b. The flux in the first term, $\hat{J}_V$, is the summation of three other terms, each of which is the product of the molar flux and the molar volume of one of the species in the system. Thus, each term in the summation represents the volumetric flux of one species, and the sum of the three terms equals the total volumetric flux across the membrane. Because a volumetric flux can also be interpreted as a velocity, $\hat{J}_V$ also represents the average velocity of the solution up to and through the membrane.

Deciphering the solution up to and through of $\hat{J}_{A/w}$ and $\hat{J}_{B/w}$ is not quite as straightforward. Like $\hat{J}_V$, these quantities have the dimensions of a volumetric flux or velocity. Consider $\hat{J}_{A/w}$, which is defined as the difference $(J_A)/(c_{A,lm}) - J_w \overline{V}_w$. Recalling that, for any substance i, $J_i = c_i v_i$, we can interpret $J_A/c_{A,lm}$ as the net or apparent velocity of molecules of A through the membrane at the location where the concentration of A is $c_{A,lm}$. The molecules closer to the concentrate side of the membrane have a lower velocity, and those closer to the permeate have a higher velocity; given that $c_{lm,A}$ is a weighted mean concentration in the membrane, $J_A/c_{A,lm}$ can be thought of as an overall characteristic velocity of A during its transport through the membrane.

The second term in the expression that defines $\hat{J}_{A/w}$ is the product $J_w \overline{V}_w$, which, as noted previously, can be interpreted as either the volumetric flux or the velocity of water through the membrane. (Formally, the velocity of water changes with location in the membrane, just as the velocity of the solutes does. However, because the concentration of water is almost constant throughout the membrane, the change in its velocity is insignificant.) Therefore, the difference $(J_A)/(c_{A,lm}) - J_w \overline{V}_w$, or $J_{A/w,V}$, equals the velocity of A relative to that of water; a similar statement applies to $J_{B/w,V}$.

The overall energy dissipation function, as expressed in Equation 15-311, is thus seen to be a summation of three terms, each of which is a product of a flow and a conjugate force. One term accounts for energy dissipation associated with the flow of the overall solution, which is driven by a pressure difference; this conjugate linkage makes sense, since the applied pressure affects all the constituents of the solution indiscriminately. The other terms account for energy dissipation associated with the flow of the constituents relative to one another. The dissipation equation includes one term for the relative velocities of A and water and another for the relative velocities of B and water; the relative velocities of A and B are accounted for implicitly by the difference between these latter two terms. These (relative) flows are driven by the concentration gradients of the solutes and can be quantified in terms of the osmotic pressures contributed by those solutes. Again, these conjugate linkages make sense, since the concentration gradients affect the movement of solutes relative to water.

While the absolute value of the dissipation function under various operating conditions is of interest for some applications, the main value of Equation 15-311 for our purposes is to identify an acceptable combination of conjugate pairs of forces and flows for insertion into Equation 15-305.[39] Using the conjugate pairs identified in Equation 15-311, we can write a version of Equation 15-305 for each constituent, as follows:

$$\hat{J}_V = L_{VV}\Delta_{m_{c/p}}P + L_{VA}\Delta_{m_{c/p}}\Pi_A + L_{VB}\Delta_{m_{c/p}}\Pi_B \quad (15\text{-}312)$$

$$\hat{J}_{A/w} = L_{AV}\Delta_{m_{c/p}}P + L_{AA}\Delta_{m_{c/p}}\Pi_A + L_{AB}\Delta_{m_{c/p}}\Pi_B \quad (15\text{-}313)$$

$$\hat{J}_{B/w} = L_{BV}\Delta_{m_{c/p}}P + L_{BA}\Delta_{m_{c/p}}\Pi_A + L_{BB}\Delta_{m_{c/p}}\Pi_B \quad (15\text{-}314)$$

Equations 15-312–15-314 comprise a mathematical model for describing and/or predicting fluxes in a three-component membrane system. The equations have nine coefficients, three of which can be eliminated by applying the Onsager reciprocal relationship ($L_{ik} = L_{ki}$); the remaining six can be evaluated experimentally, and the significance of several of these can be interpreted by inspection. For instance, according to Equation 15-312, in a system with no osmotic gradient ($\Delta_{m_{c/p}}\Pi_A = \Delta_{m_{c/p}}\Pi_B = 0$), the volumetric flux of solution is given by $L_{VV}\Delta_{m_{c/p}}P$. Thus, we can evaluate the phenomenological coefficient L_{VV} by measuring the volumetric flux in an experiment in which a TMP is applied to a system with identical solutions on the two sides of the membrane. Note that, by comparison of this result with Equation 15-15, we can relate L_{VV} to the permeability or intrinsic permeability of the membrane to the given solution as

$$L_{VV} = \frac{k_{V,\text{intr}}}{\delta_m \mu} = \frac{k_V}{\delta_m} \quad (15\text{-}315)$$

The other phenomenological coefficients can be evaluated in similar experiments, but using different solutions on the two sides of the membrane, and measuring the fluxes of all the constituents.

The volumetric flux of solution is usually a parameter of interest, so Equation 15-312 is useful as presented. On the other hand, Equations 15-313 and 15-314 describe the volumetric fluxes of species A and B relative to the flux of water, which is not the most useful way of characterizing the movement of those solutes; a better way would be as absolute fluxes (i.e., not relative to water) and on a mass or molar (not volumetric) basis. The equations can be transformed to yield approximate expressions that are more directly useful as follows. Based on the definitions of $\hat{J}_V$ and $\hat{J}_{A/w}$, we can write

$$(\hat{J}_V + \hat{J}_{A/w})c_{A,\text{lm}} = \left[\left(\cancel{J_w\overline{V}_w} + J_A\overline{V}_A + J_B\overline{V}_B\right) + \left(\frac{J_A}{c_{A,\text{lm}}} - \cancel{J_w\overline{V}_w}\right)\right]c_{A,\text{lm}}$$

$$= J_A\left(\overline{V}_A c_{A,\text{lm}} + \frac{J_B}{J_A}\overline{V}_B c_{A,\text{lm}} + 1\right)$$

(15-316)

The product $\overline{V}_A c_{A,\text{lm}}$ is the volume fraction of solute A in a solution that contains A at a molar concentration of $c_{A,\text{lm}}$; it is therefore much less than 1.0. The second term in parentheses is complicated to describe in words, but if we assume that $\overline{V}_B$ is close to $\overline{V}_A$, we see that this term equals the ratio of the fluxes of B and A times the volume fraction of A, and therefore is also likely to be much less than 1.0. As a result, we can write an expression for the molar flux of A, and by extension, a similar one for the molar flux of B, as follows:

$$(\hat{J}_V + \hat{J}_{A/w})c_{A,\text{lm}} \approx J_A \quad (15\text{-}317)$$

$$(\hat{J}_V + \hat{J}_{B/w})c_{B,\text{lm}} \approx J_B \quad (15\text{-}318)$$

Substitution of the expressions for $\hat{J}_V$, $\hat{J}_{A/w}$, and $\hat{J}_{B/w}$ from Equations 15-312 to 15-314 into Equations 15-317 and 15-318 yields the desired (approximate) expressions for the mass or molar fluxes of A and B.

Although the phenomenological coefficients that appear in Equations 15-312 to 15-314 can be evaluated experimentally using the procedures described, the fact that those coefficients can depend strongly on the composition of the feed solution makes that approach problematic. It has been suggested that certain combinations of the coefficients are likely to vary much less than the individual coefficients, so it is helpful to recast the equations in terms of the combined coefficients. The most useful of these coefficients are referred to as the *reflection coefficient*, σ_i, and the *solute permeability coefficient*, ω_{ij}. The definitions of these parameters and the phenomenological equations they lead to are as follows:

$$\sigma_i \equiv -\frac{L_{iV}}{L_{VV}} \quad (15\text{-}319)$$

$$\omega_{ij} \equiv \left(L_{ij} - \sigma_i\sigma_j L_{VV}\right)c_{i,\text{lm}} \quad (15\text{-}320)$$

$$\hat{J}_V = L_{VV}\left(\Delta_{m_{c/p}}P - \sum_{i=1}^{m}\sigma_i\Delta_{m_{c/p}}\Pi_i\right) \quad (15\text{-}321)$$

$$J_i = c_{i,\text{lm}}(1-\sigma_i)\hat{J}_V + \sum_{j=1}^{m}\omega_{ij}\Delta_{m_{c/p}}\Pi_j \quad (15\text{-}322)$$

---

[39] Note that, like the dissipation function, the phenomenological equations can be written using a wide variety of forces and flows. In all cases, the number of independent phenomenological equations equals the number of species that can permeate through the membrane.

To understand the conceptual meaning of σ_i and ω_{ij}, it is convenient to consider a system containing only a single solute, A. In that case, Equations 15-321 and 15-322 become

$$\hat{J}_V = L_{VV}\left(\Delta_{m_{c/p}}P - \sigma_A \Delta_{m_{c/p}}\Pi\right) \quad (15\text{-}323)$$

$$J_A = c_{A,lm}(1 - \sigma_A)\hat{J}_V + \omega_{AA}\Delta_{m_{c/p}}\Pi \quad (15\text{-}324)$$

where the subscript has been dropped from $\Delta_{m_{c/p}}\Pi_A$, since $\Pi_A = \Pi_{tot} = \Pi$. Two limiting values of σ_A are of interest. If $\sigma_A = 0$ then, according to Equation 15-319, L_{AV} must be zero. This condition implies that, when the system is subjected to a pressure-based driving force, the permeating species exert no force on one another; that is, neither species holds back nor pulls forward the other. For this to occur, the two species must travel through the membrane at the same speed, implying that the membrane is absolutely nonselective. In such a situation, according to Equation 15-323, the flux of solution through the membrane is directly proportional to the TMP. Note that, if $\sigma_i = 0$, ω_{iV} must be zero as well.

On the other hand, $\sigma_A = 1.0$ implies that the membrane is perfectly selective; that is, it completely rejects the solute. In this case, the solute exerts maximum effect on the water (holding it back by lowering the activity of water to the maximum extent possible on the concentrate side of the membrane), and the flux of solution is proportional to the difference between the mechanical and osmotic pressures. At intermediate values of σ_A, the resistance that the solute imposes on the flux of solution is intermediate between these extremes. Thus, we can conclude that σ_A is a direct indicator of the efficiency with which the membrane rejects solute i.

The easiest way to interpret ω_{AA} is to consider a hypothetical system in which $\Delta_{m_{c/p}}P$ and $\Delta_{m_{c/p}}\Pi$ are nonzero, but the net volume flux across the membrane ($\hat{J}_V$) is zero, because the volumetric fluxes of water and solute balance one another. Note that, in this case, the water must be flowing from the permeate to the concentrate side of the membrane; that is, in the direction of normal osmosis, not RO. In such a system, according to Equation 15-324, ω_{AA} is the molar flux of solute per unit osmotic pressure difference. Under these conditions, Equation 15-323 indicates that σ_A can be computed as the ratio $\Delta_{m_{c/p}}P/\Delta_{m_{c/p}}\Pi$. Thus, experiments conducted under conditions of no net permeation can yield values for both σ_A and ω_{AA}.

The preceding analysis has some important ramifications regarding the effect of $\Delta_{m_{c/p}}P$ on the rejection of A. First, at low TMPs, $\Delta_{m_{c/p}}P$ will be less than $\sigma_A \Delta_m \Pi$, so the flux of water computed using Equation 15-323 will be negative; that is, water will migrate from the permeate to the concentrate. The flux of solute computed using Equation 15-324 is always positive, so the computed rejection under these conditions will be negative. Even if a TMP equal to $\sigma_A \Delta_{m_{c/p}}\Pi$ is applied, migration of water will cease, but migration of solute will continue (in response to the combined driving forces of the mechanical pressure and the concentration gradient). As a result, the solute concentration in the permeate will gradually increase. Only when a pressure greater than $\sigma_A \Delta_{m_{c/p}}\Pi$ is applied will water begin to migrate toward the permeate, and only at some greater TMP will net rejection of solute occur. The minimum TMP required to achieve net rejection depends on σ_A. If the membrane is perfectly selective ($\sigma_A = 1$), then any pressure greater than the osmotic pressure will generate clean permeate. As σ_A decreases, greater pressures are required to produce a permeate that is cleaner than the feed or concentrate; in the limit of σ_A equal to zero, no amount of pressure will achieve that goal, since a reflection coefficient of zero implies that the membrane is completely nonselective for water over the solute.

Equations 15-323 and 15-324 indicate that, as long as $\sigma_i < 1$, $\hat{J}_V$ and J_i both increase with increasing TMP. However, $\hat{J}_V$ increases more rapidly than J_i, causing rejection to increase with increasing TMP or solution flux. The quantitative expression describing this relationship is

$$R_A = \frac{\sigma_A - \sigma_A \exp(-a\hat{J}_V)}{1 - \sigma_A \exp(-a\hat{J}_V)} \quad (15\text{-}325)$$

where $a = (1 - \sigma_A)/(k_{A,m}) = (1 - \sigma_A)(D_{A,m}/\delta_m)$.

Based on Equation 15-325, R_A approaches a limiting value equal to σ_A at very large values of $\hat{J}_V$; that is, σ_A represents the maximum possible rejection of A that can be achieved by the membrane.

15.12 SUMMARY

This chapter has provided an introduction to the numerous applications of membranes in water and wastewater treatment processes and to the conceptual and mathematical approaches that have been developed for understanding those processes. Interim summaries have been provided at several points in this chapter, to which the reader is referred for more details of specific aspects of membrane operation and modeling; only an abbreviated restatement of some of the salient points from those summaries is provided here.

Membrane processes have developed rapidly in the past few decades, and the field continues to evolve rapidly. In environmental engineering, the vast majority of membrane systems utilize an applied pressure to drive solution through the membrane from the feed (concentrate) side to the permeate side, as the membrane selectively rejects dissolved and/or particulate contaminants. The applied pressure increases and the size of the contaminants that can be rejected decreases in the sequence from microfiltration (MF) to ultrafiltration (UF) to nanofiltration (NF) to reverse osmosis (RO). In some applications, the primary driving force is a gradient in chemical activity (dialysis) or electrical potential (ED or EDR), and solutes (rather than water) are the dominant species transported across the membrane.

In pressure-driven systems, membranes are commonly represented as either being porous (MF and UF) and characterized by coupled transport of water and contaminants, or dense (RO) and characterized by uncoupled transport; NF membranes have been modeled using both paradigms. Both water and contaminants are envisioned to enter and be transported through porous membranes primarily by advection, with contaminant rejection occurring primarily at the entrances to the pores. In contrast, entry into and transport through dense membranes is envisioned to occur via dissolution and diffusion, respectively, so rejection occurs both at the interface and in the membrane interior. In the limiting-case scenarios of perfectly coupled transport in porous membranes, contaminant rejection is predicted to be independent of the transmembrane pressure (TMP) and permeation flux, whereas it is predicted to increase with increases in those parameters in dense membranes.

Overall transport in and performance of membranes depends not only on processes occurring at the surface of and inside the membrane itself, but also in the liquid layer adjacent to it on the concentrate side (the concentration polarization (CP) layer). Rejected contaminants accumulate in this layer and migrate by diffusion back toward the bulk concentrate; in addition, if they become sufficiently concentrated, they can form a compact layer (a gel, cake, or scale) immediately adjacent to the membrane. Regardless of whether or not a compact layer forms, the accumulated contaminant imposes a hydraulic resistance opposing solution flow. The deterioration in membrane performance associated with this resistance and any resistance generated by contaminant that has accumulated inside the membrane is referred to as membrane fouling. Such fouling can be mitigated by crossflow and can be partially reversed by intermittent back-flushing; over time, however, hydraulically irreversible fouling increases to the point where chemical cleaning is required. Cleaning agents used in such cases might include acids, bases, oxidants, reductants, and/or detergents.

Mathematical modeling of membrane performance requires simultaneous solution of equations for transport both parallel and perpendicular to the membrane in the solution phases, rejection of contaminant at the membrane/solution interfaces, and transport through the membranes themselves. These equations have been solved with various levels of sophistication and employing various assumptions about the nature of the contaminants (in particular, soluble vs particulate), the membranes (porous vs dense), and the system geometry. Interactions among the contaminants in the CP layer, and the corresponding fouling they induce, are complex and have proven particularly difficult to model.

The future of membranes in environmental engineering applications is very promising. Because of their ability to provide a virtually absolute barrier to contaminants, and because they can be designed to exclude contaminants of almost any specified size, membranes provide the most reliable and versatile approach for contaminant removal that has ever been developed. Production of membranes from new materials and/or via new techniques and development of new approaches for preventing or overcoming fouling will likely expand the use of this technology to the point where it is the dominant tool for water quality modification in the coming generation.

REFERENCES

ASTM (2007) *SDI D 4189-07: Standard Test Method for Silt Density Index (SDI) of Water*. ASTM International, West Conshohocken, PA. DOI: 10.1520/D4189-07, www.astm.org

ASTM (2001) *E 1343-90 (2001): Standard Test Method for Molecular Weight Cutoff Evaluation of Flat Sheet Ultrafiltration Membranes*. ASTM International, West Conshohocken, PA. DOI: 10.1520/E1343-90R01, www.astm.org

Altena, F. W., and Belfort, G. (1984) Lateral migration of spherical particles in porous flow channels: application to membrane filtration. *Chem. Eng. Sci.*, 49 (2), 343–355.

Bard, A. J., and Faulkner, L. R. (2001) *Electrochemical Methods*, 2nd edn., John Wiley & Sons, New York.

Blatt, W. F., Dravid, A., Michaels, A. S., and Nelsen, L. (1970) Solute polarization and cake formation in membrane ultrafiltration: causes, consequences, and control techniques. In Flinn, J. E. (ed.), *Membrane Science and Technology*, Plenum Press, New York.

Bockris, J. O. M., and Reddy, A. K. N. (1998) *Modern Electrochemistry I: Ionics*. Plenum Press, New York.

Carnahan, N. F., and Starling, K. E. (1969) Equation of state for nonattracting rigid spheres. *J. Chem. Phys.*, 51, 635–636.

Chang, Y-J., and Benjamin, M. M. (1996) Iron oxide adsorption and UF to remove NOM and control fouling. *JAWWA*, 88 (12), 74–88.

Chellam, S., and Xu, W. (2006) Blocking laws analysis of dead-end constant flux microfiltration of compressible cakes. *J. Colloid Interface Sci.*, 301, 248–257.

Chellam, S., and Jacangelo, J. G. (1998) Existence of critical recovery and impacts of operational mode on potable water microfiltration. *J. Environ. Eng. (ASCE)*, 124, 1211–1219.

Chellam, S., Wiesner, M. R., and Dawson, C. (1995) Laminar flow in porous ducts. *Rev. Chem. Eng.*, 11 (1), 53–99.

Childress, A. E., and Deshmukh, S. S. (1998). Effect of humic substances and anionic surfactants on the surface charge and performance of reverse osmosis membranes. *Desalination*, 118, 167–174.

Clark, M. M. (1996) *Transport Modeling for Environmental Engineers and Scientists*. Wiley-Interscience, New York, NY.

Cohen, R. D., and Probstein, R. F. (1986) Colloidal fouling of reverse osmosis membranes. *J. Colloid Interface Sci.*, 114 (1), 194–207.

CRC (2009) In Vanysek, P. (ed.), *Handbook of Chemistry and Physics*, CRC Press/Taylor and Francis, Cleveland, OH.

Davis, R. H. (1992) Modeling of fouling of crossflow microfiltration membranes. *Separ. Purif. Meth.*, 21 (2), 75–126.

Davis, R. H., and Leighton, D. T. (1987) Shear-induced transport of a particle layer along a porous wall. *Chem. Eng. Sci.*, 42 (2), 275–281.

Dechadilok, P., and Deen, W. M. (2006) Hindrance factors for diffusion and convection in pores. *Ind. Eng. Chem. Rev.*, 45, 6953–6959.

Deen, W. M. (1987) Hindered transport of large molecules in liquid-filled pores. *AIChE J.*, 33, 1409–1425.

Denbigh, K. (1981) *The Principles of Chemical Equilibrium: With Applications in Chemistry and Chemical Engineering*. Cambridge University Press, New York.

Drew, D. A., Schonberg, J. A., and Belfort, G. (1991) Lateral inertial migration of a small sphere in fast laminar flow through a membrane duct. *Chem. Eng. Sci.*, 46, 3219–3224.

Eckstein, E. C., Bailey, D. G., and Shapiro, A. H. (1977) Self-diffusion of particles in shear-flow of a suspension. *J. Fluid Mech.*, 79, 191–208.

Elimelech, M., and Bhattacharjee, S. (1998) A novel approach for modeling concentration polarization in crossflow membrane filtration based on the equivalence of osmotic pressure model and filtration theory. *J. Membr. Sci.*, 145, 223–241.

Elimelech, M., and Childress, A. E. (1996) *Water Treatment Technology Program Report No. 10*, U. S. Dept. of the Interior, Bureau of Reclamation, Washington, D.C.

Ferry, J. D. (1936) Statistical evaluation of sieve constants in ultrafiltration. *J. Gen. Physiol.*, 20, 95–104.

Geens, J., van der Bruggen, B., and Vandacasteele, C. (2004) Characterization of the solvent stability of polymeric nanofiltration membranes by measurement of contact angles and swelling. *Chem. Eng. Sci.*, 59 (5), 1161–1164.

Green, G., and Belfort, G. (1980) Fouling of ultrafiltration membranes: lateral migration and particle trajectory model. *Desalination*, 35, 129–147.

Happel, J. (1958) Viscous flow in multiparticle systems: slow motion of fluids relative to beds of spherical particles. *AIChE J.*, 4 (2), 197–201.

Harmant, P., and Aimar, P. (1996) Coagulation of colloids retained by porous wall. *AIChE J.*, 42 (12), 3523.

Hermia, J. (1982) Constant pressure blocking filtration laws—application to power-law non-Newtonian fluids. *Trans. Inst. Chem. Eng.*, 60, 183–187.

Hlavacek, M., and Bouchet, F. (1993) Constant flowrate blocking laws and an example of their application to dead-end microfiltration of protein of solutions. *J. Membr. Sci.*, 82, 285–295.

Ho, C. C., and Zydney, A. L. (2000) A combined pore blockage and cake filtration model for protein fouling during microfiltration. *J. Colloid Interface Sci.*, 232, 389–399.

Hoek, E. M. V., Allred, J., Knoell, T., and Jeong, B.-H. (2008) Modeling the effects of fouling on full-scale reverse osmosis processes. *J. Membr. Sci.*, 314, 33–49.

Hong, S., Faibish, R. S., and Elimelech, M. (1997) Kinetics of permeate flux decline in crossflow membrane filtration of colloidal suspensions. *J. Colloid Interface Sci.*, 196, 267–277.

Horvath, A. L. (1985) *Handbook of Aqueous Electrolyte Solutions: Physical Properties, Estimation, and Correlation Methods*. Ellis Horwood Limited, Halsted Press, John Wiley & Sons, New York.

Kilduff, J. E., Mattaraj, S., Sensibaugh, J., Pieracci, J. P., Yuan, Y., and Belfort, G. (2002) Modeling flux decline during nanofiltration of NOM with poly(arylsulfone) membranes modified using UV-assisted graft polymerization. *Environ. Eng. Sci.*, 19 (6), 477–495.

Koros, W. J., Ma, Y. H., and Shimidz, T. (1996) Terminology for membrane processes (IUPAC Recommendations). *J. Membr. Sci.*, 120 (2), 149–159.

Kortum, G. (1965) *Treatise on Electrochemistry*. Elsevier Pub. Co., Amsterdam, New York.

Kraus, K. A., Raridon, R. J., and Baldwin, W. H. (1964). Properties of organic-water mixtures. I. Activity coefficients of sodium chloride, potassium chloride, and barium nitrate in saturated water mixtures of glycol, glycerol, and their acetates. Model solution for hyperfiltration membranes. *J. Am. Chem. Soc.*, 86, 2571–2576.

Kruelen, H., Smolders, C.A., Versteeg, G. F., and van Swaaij, W. P. M. (1993) Microporous hollow fibre membrane modules as gas–liquid contactors. Part 1: physical mass transfer processes. *J. Membr. Sci.*, 78, 197–216.

Leighton, D., and Acrivos, A. (1987) The shear-induced migration of particles in concentrated suspensions. *J. Fluid Mech.*, 181, 415–439.

Lonsdale, H. K., Merten, U., and Riley, R. L. (1965) Transport properties of cellulose acetate osmotic membranes. *J. Appl. Polym. Sci.*, 9, 1341–1362.

Newman, J. S., and Thomas-Alyea, K. E. (2004) *Electrochemical Systems*. John Wiley & Sons, Hoboken, NJ.

Paul, D. R., and Ebra-Lima, O. M. (1970). Pressure-induced diffusion of organic liquids through highly swollen polymer membranes. *J. Appl. Polym. Sci.*, 14, 2201.

Paul, D. R. (1972) The role of membrane pressure in reverse osmosis. *J. Appl. Polym. Sci.*, 16, 771.

Porter, M.C. (1972) Ultrafiltration of colloidal suspensions. *Adv. Separation Tech.*, 68, 21–30.

Probstein, R. F., Shen, J. S., and Leung, W. F. (1978) Ultrafiltration of macromolecular solutions at high polarization in laminar channel flow. *Desalination*, 24, 1/2/3, 1–16.

Probstein, R. F. (1994) *Physicochemical Hydrodynamics: An Introduction*. 2nd edn., Wiley Interscience, New York.

Robinson, R. A., and Stokes, R. H. (1959) *Electrolyte Solutions—The Measurement and Interpretation of Conductance, Chemical Potential and Diffusion in Solutions of Simple Electrolytes*. Butterworths Scientific Publications, London.

Sharma, R. R., Agrawal, R., and Chellam, S. (2003) Temperature effects on sieving characteristics of polymeric nanofiltration membranes: pore size distributions and solute rejection. *J. Membr. Sci.*, 223 (1–2), 69–87.

Schock, G., and Miquel, A. (1987) Mass transfer and pressure loss in spiral wound modules. *Desalination*, 64, 339.

Schoeman, J. J., and Thompson, M. A. (1996) *Water Treatment Membrane Processes*. AWWA, Denver, CO.

Song, L. (1998) A new model for the calculation of the limiting flux in ultrafiltration. *J. Membr. Sci.*, 144, 173–185.

Song, L., and Elimelech, M. (1995) Theory of concentration polarization in crossflow filtration. *J. Chem. Soc. Faraday Trans.*, 91, 3389–3398.

Strathmann, H. (2004) *Ion-Exchange Membrane Separation Processes*. Membrane Science and Technology Series, Elsevier, Boston.

Strathmann, H., Giorno, L., and Drioli, E. (2006) *An Introduction to Membrane Science and Technology*. Institute on Membrane Technology, CNR-ITM, Consiglio Nazionale Delle Ricerche, Rome.

Suarez, J. A., and Veza, J. M. (2000) Dead-end microfiltration as advanced treatment for wastewater. *Desalination*, 127, 47–58.

Tanaka, Y. (2007) *Ion Exchange Membranes: Fundamentals and Applications*. Membrane Science and Technology Series, Elsevier, Boston.

Tarabara, V. V., Koyuncu, I., and Wiesner, M. (2004) Effect of hydrodynamics and solution chemistry on permeate flux in crossflow filtration: direct experimental observation of filter cake cross sections. *J. Membr. Sci.*, 241, 65–78.

Thau, G., Bloch, R., and Kedem, O. (1966) Water transport in porous and non-porous membranes. *Desalination*, 1, 129–138.

Trettin, D. R., and Doshi, M. R. (1980) Limiting flux in ultrafiltration of macromolecular solutions. *Chem. Eng. Commun.*, 4, 507–522.

Walker, W. S. (2010) *Improving Recovery in Reverse Osmosis Desalination of Inland Brackish Groundwaters via Electrodialysis*, Dissertation, University of Texas at Austin, Austin, TX.

Wiesner, M. R., Veerapaneni, S., and Brejchova, D. (1992) Improvements in membrane microfiltration using coagulation pretreatment. In Klute, R., and Hahn, H. H. (eds.), *Chemical Water and Wastewater Treatment II*, Springer-Verlag, Berlin, Heidelberg.

Wiesner, M.R., Clark, M.M., and Mallevialle, J. (1989) Membrane filtration of coagulated suspensions. *J. Environ. Eng. (ASCE)*, 115 (1), 20–40.

Wijmans, J. G., and Baker, R. W. (1995) The solution-diffusion model: a review. *J. Membr. Sci.*, 107, 1–21.

Xu, W., and Chellem, S. (2005) Initial stages of bacterial fevling during deal-end filtration. *Environ. Sci. Technol.*, 39, 6470–6476.

Wright, M. R. (2007) *An Introduction to Aqueous Electrolyte Solutions*. John Wiley & Sons, Chichester.

Zeman, L. J., and Zydney, A. L. (1996) *Microfiltration and Ultrafiltration: Principles and Applications*. Marcel Dekker, New York.

Zhang, M., and Song, L. (2000) Mechanisms and parameters affecting flux decline in cross-flow microfiltration and ultrafiltration of colloids. *Environ. Sci. Technol.*, 34, 3767–3773.

Zydney, A. L. and Colton, C. K. (1986). A concentration polarization models for the filtrate flux in cross-flow microfiltration of particulate suspensions. *Chem. Eng. Commun.*, 47, 1–21.

PROBLEMS

15-1. Calculate the feed flow and the total number of modules (pressure vessels) required to produce a treated flow of $150\,m^3/h$ in the system described in Example 15-2.

15-2. RO membranes are to be operated at an average permeate flux of $30\,L/m^2\,h$, using spiral-wound membrane elements with a packing density of $1500\,m^2/m^3$. Each element is 10 cm in diameter and 95 cm long. How many membrane elements would be required to produce a continuous flow of $20{,}000\,m^3/d$?

15-3. What fraction of the fluid entering the fiber permeates through the membrane in a single pass in the system described in Example 15-20?

15-4. A typical municipal water usage rate in the US is $350\,L/person\,d$. A dead-end MF membrane system is being designed for a city with a population of 1 million inhabitants. The proposed system will have a recovery of 92% and is designed to have a flux of $75\,L/m^2\,h$ when meeting 120% of the average demand rate with 10% of the system is out of service for cleaning.

   (a) Determine the required membrane area.

   (b) What is the velocity of water toward the membrane, in cm/min, under baseline operating conditions (average demand, and all membrane modules active)?

   (c) Compare your answers to parts (a) and (b) to the corresponding values for treating the baseline feed flow rate in a conventional granular-media filter with an overflow rate of $300\,L/m^2\,min$.

   (d) What volume of concentrate is generated daily under baseline operating conditions?

15-5. Why is it common to backwash MF and UF membranes, but not RO membranes?

15-6. A two-stage membrane array contains four modules in the first stage and two in the second, with the concentrate from the first stage serving as the feed to the second. The feed has a conductivity of $1800\,\mu S/cm$, and each stage is operated with 35% recovery. The salt rejection, as indicated by the reduction in conductivity between the feed and the permeate, is 90% and 82% in the two stages, respectively. The permeates from the two stages are blended before distribution to the community.

(a) What is the overall recovery of the system?

(b) What is the conductivity of the concentrate from the second stage?

(c) What is the overall salt rejection, based on the concentrations in the feed and the blended permeate? Assume that conductivity is a conservative parameter.

15-7. A tubular MF system treating an industrial wastewater is operated with an influent flow rate of 120 L/min, a flux of 45 L/m² h, and a recovery of 95%. Each membrane module contains 3.0 m² of membrane. The feed contains 6.2 mg/L of suspended solids, and the overall solid rejection efficiency is 99.5%.

(a) How many modules are required?

(b) What are the concentrations of suspended solids in the permeate and concentrate?

(c) What mass of solids is sent to the solid settling pond daily?

15-8. Speculate as to why are MF and UF membrane elements typically cylindrical, as opposed to the more common spiral-wound geometry of NF and RO elements.

15-9. A nanofilter being used to soften a water supply is operated at 20°C, with a pressure differential between the bulk solutions of 900 kPa and a recovery of 60%. The concentration of major ions in the feed solution and the rejection efficiency of each ion are shown below. The concentrations of all species are shown in mg/L, except HCO_3^- is in mg/L as $CaCO_3$, meaning that the true concentration of HCO_3^- in meq/L is the same as the number of meq/L that would be associated with the given number of mg/L of $CaCO_3$.

	c_f	R		c_f	R
Na^+	65	0.71	Cl^-	176	0.79
Ca^{2+}	110	0.970	HCO_3^-	145	0.980
Mg^{2+}	21	0.975	SO_4^{2-}	110	0.985

(a) What are the concentrations of the ions in the concentrate and permeate solutions? Report all concentrations in the same units as in the influent table.

(b) What are the osmotic pressure difference and the effective TMP across the membrane?

(c) Determine $\Delta_{m_c/p}\overline{G}_{chem}$, $\Delta_{m_c/p}\overline{G}_{mech}$, and $\Delta_{m_c/p}\overline{G}_{tot}$ for Ca^{2+} and for water. Include signs. Assume that the molar volume of Ca^{2+} is the same as that of water, and use the Davies equation to correct for activity coefficients.

15-10. Viruses and bacteria with nominal diameters of 60 nm and 0.70 μm, respectively, are present in a solution at 15°C that is being fed to a microfilter. The membrane has cylindrical pores that are 1.0 μm in diameter and 25 μm long. The porosity of the membrane is estimated to be 80%. The system is being operated with a permeate flux of 100 L/m² h.

(a) What is the Reynolds number for flow through the pores?

(b) What fraction of each group of organisms is expected to be rejected by the microfilter at the beginning of a filter run, if the flow regime is assumed to be fully developed at the pore entrance and transport is strictly by convection? What about if the filter had 0.2-μm pores?

(c) The diffusive and convective hindrance factors for particles with radii that are 30% of the pore radius have been estimated as 0.415 and 1.32, respectively. Considering these factors and the possibility that particles can move through the pores by diffusion as well as convection, make a new estimate for the initial rejection efficiency of the viruses in the membrane with 0.2-μm pores. Estimate the virus's diffusivity from the Stokes–Einstein equation. Do you think advection or diffusion is more responsible for contaminant transport?

(d) Assume that a 10-μm-thick layer of the bacteria has accumulated on the filter with 1.0-μm pores. Treating the bacteria and virus particles as spheres and assuming that the viruses are perfectly destabilized, what fractional removal of the viruses is expected in the bacterial layer according to the Yao et al. granular-media filtration model developed in Chapter 14? Assume that the bacterial layer has a porosity of 0.38.

15-11. Explain why, according to the solution-diffusion model, the ratio of contaminant concentrations in the dissolved and membrane phases equals K_d at the c/m interface, but not at the m/p interface, even though equilibrium is assumed to apply at both locations.

15-12. Surface water containing NOM at a concentration of 4.5 mg/L TOC is treated by ultrafiltration, at a flux of 75 L/m² h. The filter rejects only 3% of the TOC, which forms a gel on the membrane surface. During each backwashing cycle, 80% of the organic matter that has accumulated since the last backwashing is removed; the remaining 20% remains on the membrane until a chemical cleaning is carried out.

If the organics are 45% C by mass and the gel 20% organics by mass (most of the mass of the gel is water) and has a density of ~1.0 g/cm^3, how thick would the gel be after 1 month (30 days) of operation? (Note: gels become harder to remove that longer the time between backwashing steps. The answer to this question should give you some understanding of why backwashing is carried out so frequently.)

15-13. $\Delta_{c/p}\overline{G}_{tot,i}$ for a solute such as a protein is dominated by the difference in concentrations between the concentrate and permeate solutions. Yet, in a separation process using a porous membrane, transport of the protein through the membrane is thought to be mostly advective, not diffusive. Explain this apparent anomaly.

15-14. A membrane has a pore size distribution that can be described as three distinct groups. One group of pores has diameters clustered around 0.20 μm, another has diameters clustered around 0.10 μm, and the third has diameters around 0.05 μm. Each group accounts for approximately one-third of the total open area on the membrane. What fraction of the volumetric flow passes through each group of pores? Estimate the overall rejection efficiency for particles that are 0.04 μm in diameter, if Ferry's model applies and the flow is fully developed as the water enters the pores.

15-15. The mass transfer coefficient for contaminants in the CP layer has dimensions of velocity and can be thought of as a characteristic velocity for transport away from the membrane. For a frontal filtration membrane system (i.e., one with no crossflow), would you expect $k_{mt,CP}$ for a 1-μm particle to be bigger, smaller, or about the same as for a Na$^+$ ion? Explain your reasoning.

15-16. What is the main driving force causing water to move from the bulk concentrate to the c/m interface in a steady-state system with a porous membrane and particulate contaminants? What about in a system with a dense membrane and soluble contaminants? What are the main forces resisting transport of water between those two locations in each type of system?

15-17. The data shown in Figure 15-Pr17 were collected from a batch test in which seawater was filtered through an RO membrane in a stirred cell with dead-end flow. The TMP was constant at 7000 kPa, and the feed was at 20°C. The plot shows the volumetric flux as a function of the "concentration factor" defined as the apparent concentration of salts in the feed reservoir normalized to that at the beginning of the test. (The concentration

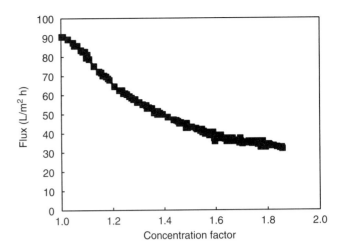

FIGURE 15-Pr17. Flux decline during a batch test of a reverse osmosis system operated at fixed TMP.

is referred to as "apparent" because it is possible that some salts precipitated in the cell and therefore were not dissolved in the concentrated feed solution.)

When deionized (DI) water was filtered through the clean membrane, the flux was stable at 140 L/m^2 h. The decline in flux when seawater was filtered can be attributed to a combination of membrane fouling and the increased osmotic pressure as clean water is removed in the permeate flow and the solution in the feed reservoir becomes progressively more concentrated.

(a) Compare the initial flux with seawater as feed to the expected flux, based on an assumption that the difference between the fluxes of seawater and DI water is entirely due to the difference in osmotic pressures.

(b) What is the resistance of the membrane to permeation of seawater?

(c) What is the resistance attributable to fouling when 40% of the original solution in the feed reservoir has permeated through the membrane, assuming nearly complete rejection of all solutes and no precipitation of salts in the feed reservoir?

15-18. The only solute in the feed to an RO system has an MW of 95 and is present at a concentration of 20,000 mg/L. The system operates at 20°C with 60% recovery and 90% rejection.

(a) What is the minimum TMP required to generate water flow from the concentrate to the permeate?

(b) What is the activity of water in the permeate?

(c) If the actual TMP is 50% greater than the minimum computed in part (a), and the solution/membrane distribution coefficient for water is

0.03, what is the concentration of water inside the membrane at the m/p interface?

15-19. The flux of water across an RO membrane at 25°C is 35 litars per square meter per hour (LMH) when the TMP is 7000 kPa, and the solute concentrations in the concentrate and permeate are 0.28 M and 4×10^{-3} M, respectively.

(a) What is the flux of solute under these conditions?

(b) What would the instantaneous fluxes of water and solute be if the pressure were suddenly reduced to 5000 kPa, while all other system parameters remained the same? Your answers can be approximate if you state and explain any simplifying assumptions that you make.

15-20. A brackish water is to be treated by RO using a membrane that has a permeance, determined from tests with clean water, of 0.040 L/m² h kPa. The feed flow rate is 2.2 m³/min, its temperature is 15°C, and its major ion composition is as shown below.

Constituent	Concentration[a]
pH	8.0
Conductivity	1253
Calcium (Ca)	28
Chloride (Cl)	260
Magnesium (Mg)	11.9
Nitrate (NO$_3$)	11.8
Silicon (Si)	8.0
Sodium (Na)	245
Sulfate (SO$_4$)	145
Total inorganic carbon	27

[a] Values in mg/L, except for pH (pH units) and conductivity (μS/cm)

(a) The system is operated with a recovery of 75% and an average rejection of 94%, which can be assumed to apply to all solutes. If the polarization factor is 1.08, what pressure must be applied to the solution to achieve a flux of 30 L/m² h, if the membrane permeance under these conditions is 90% of that when it is fed clean water? (Note that this calculation only accounts for the pressure required to force water through a clean membrane. The actual required TMP could be considerably greater due to membrane fouling.)

(b) Because of fouling, the actual average TMP is four times the value computed in part (a). If the headloss between the feed and the exiting concentrate is 2350 kPa, what is the power required to pressurize the influent, if the pump is 72% efficient?

(c) What fraction of the energy applied to the feed remains in the concentrate as it exits the RO pressure vessels?

15-21. Xu and Chellam (2005) carried out batch filtration tests at constant pressure to determine the fouling properties of bacterial suspensions and reported the following results for one test. Their experimental system used a laboratory-scale filter with 4.2 cm² of accessible membrane area. Determine whether the results are consistent with any of the "blocking" models proposed by Hermia.

Time (s)	Flux (L/m² s)	Time (s)	Flux (L/m² s)
0	2.20E−01	1650	5.28E−02
150	1.69E−01	1800	4.84E−02
300	1.41E−01	1950	4.62E−02
450	1.19E−01	2100	4.40E−02
600	1.03E−01	2250	4.18E−02
750	9.02E−02	2400	3.96E−02
900	7.70E−02	2550	3.74E−02
1050	7.04E−02	2700	3.52E−02
1200	6.60E−02	2850	3.41E−02
1350	5.94E−02	3000	3.30E−02
1500	5.72E−02		

15-22. Estimate the headloss for crossflow through 0.8- and 1.2-mm diameter hollow fibers. The hollow fibers are 1.8 m long, and the average crossflow velocity is 0.9 m/s. Water temperature is 10°C.

15-23. (a) Compute the headloss, in m, when water at 20°C passes through a 25-nm pore that is 3 μm long, at a flux of 100 L/m² h. The membrane porosity is 0.9. Assume that the flow is fully developed and laminar throughout the pore. What would the pressure loss be across such a pore, in kPa?

(b) Compare the result in part (a) with the headloss for the same flux through a cake layer comprising 1-μm particles, if the layer is 30 μm thick, and headloss is characterized by the Carmen–Kozeny equation. Assume that $k = 5$ in the Carmen–Kozeny equation and that the layer porosity is 0.4.

15-24. The conditions at the location where a compact layer is first expected to form in a particular crossflow membrane system were determined in Example 15-27. Determine and plot the profiles of $c_{c_{c/m}}$ versus x and $\hat{J}_V$ versus x from the inlet to this location. The volume fraction of the contaminant equals $7.5 \times 10^{-6} c$, where c is in mg/L.

15-25. Describe some key similarities and key differences between Brownian diffusion and shear-induced diffusion. Why is Brownian diffusion independent of the shear rate?

15-26. Describe some key similarities and differences between transport across a gas/liquid interface, transport across the pore mouth in the pore-flow model, and transport across the concentrate/membrane interface in the solution-diffusion model.

15-27. Hoek et al. (2008) describe performance of a pilot-scale, spiral-wound RO array that consisted of six modules (pressure vessels) in a 3:2:1 cascade array. That is, the influent was split and distributed into three modules; the concentrate exiting these modules served as the feed to the next stage, which consisted of two modules in parallel; and the concentrate from the second stage was the feed to the third stage, which consisted of a single module.

Each module contained seven membrane elements in series, and each element comprised 31 membrane sheets (leaves), separated by spacers and wrapped such that the gap between leaves was approximately 0.65 mm. Feed could enter each element between any two membrane sheets or between the outside sheet and the wall, so the element can be envisioned to have 31 separate influent channels. The leaves were 0.91 m long (in the axial direction) and 0.71 m wide (in the direction of the spiral). The spacers occupied 20% of the volume between the leaves (i.e., the spacer porosity was 0.8). The hydraulic diameter was estimated to be 1.0 mm, based on geometry and fitting to experimental data. Tests were conducted using a constant feed flow rate of 11.4 L/s at 20°C.

(a) Determine the flow rates entering and leaving each module in the various stages, for average recoveries in the three stages of 65%, 45%, and 25%, respectively.

(b) Find the average axial velocity in the modules in each stage. Ignore the curvature of the spiral-wound module, so that the cross-section of the concentrate flow channel can be treated as a single, long, thin rectangle.

(c) Estimate the axial pressure loss through each module, and through the whole array, at the beginning of the test. (As the test proceeded, foulant accumulated in the channels and reduced the effective cross-section, causing an increase in pressure loss.) Compare your result with the pressure losses through typical hollow fiber modules for MF or UF.

15-28. In some membrane-based treatment processes, particles are added to adsorb potential foulants and thereby prevent the foulants from reaching the membrane surface. In some cases, the adsorbents appear to be most effective when they form a cake layer directly on the membrane, whereas in others it is preferable to keep the particles in the bulk suspension. The particles of interest fall in the size range from a few nanometers to a few micrometers.

For a hollow fiber membrane with a 0.8-mm diameter, operated at a constant flux of 60 L/m^2 h and a temperature of 25°C, determine the crossflow velocities, if any, that would cause the adsorbent particles to be deposited and to remain in place, for particles with nominal diameters of 20 nm, 1 μm, and 10 μm. Consider the range of crossflow velocities from 2 to 300 cm/s. The density of the particles is 1.7 g/cm^3. Assume that the solids volume fraction adjacent to the membrane is 11%. What back-transport mechanism is dominant for each type of particle?)

INDEX

Acid/base reactions:
    metal hydroxide precipitation, 404–409
    particle treatment systems, charge and potential measurements, 531–532
    volatile compounds, gas transfer, 199–202
Acrylamide polymers, particle destabilization chemicals, 534–535
Activated carbon:
    adsorption process:
        basic principles, 257
        competitive adsorption, ideal adsorption model, 305–306
        fixed bed adsorption systems, chemical reactions, 357
        slow adsorption equilibrium systems, 343
Activated complex, molecular collisions, 85–87
Activity coefficient:
    effect on metal solubility 398-401
    gas/liquid equilibrium:
        Henry's law and, 163–164
        reference states, 198–199
        Raoult's law and, 164, 168
Additives, particle destabilization, soluble materials interactions, 543–544
Adsorbate:
    defined, 259
    ion adsorption, electrical potential, interfacial region, 315–318
Adsorbent:
    defined, 259
    regeneration, 365–366
        minimum rates, 366–369
Adsorbent particles, slow adsorption equilibrium systems:
    pore *vs.* surface diffusion, 341–343
    transport-limited adsorption rates, batch reactors, 345–350
Adsorption. *See also* Competitive adsorption
    defined, 259

fixed bed reactors:
    chemical reactions in, 356–357
    design options and strategies, 366–369
    mass transfer zone movement, 354–356
    minimum rate of adsorbent regeneration/replacement, 366–367
    performance evaluation, 357–359
    rapidly equilibrating systems, 333–340
        equation summary, 340
        mass balance on, 335–336
        plug flow reactors, 336–340
        qualitative analysis, 333–335
    packed adsorption beds in series, 368–369
    slow equilibrium systems, transport-limited adsorption rates, 350–354
isotherms, adsorption equilibrium, 269–296
    bidentate adsorption, 281–282
    competitive adsorption:
        Freundlich isotherms, 294–296
        ion exchange equilibrium, 284–294
        site-binding paradigm, 283–284
    ideal adsorbed solution model, 302–306
    distribution or partition coefficient, 282–283
    site-binding paradigm, 270–281
molecular-scale models, 266–268
    electrically induced partitioning, ion adsorption, 268
    phase transfer reaction, 267–268
    surface complexation reactions, 267
particle destabilization:
    charge neutralization, 536–537
    interparticle bridging, 541–542
particle stability, 522
particle surface charge, 523–524

Water Quality Engineering: Physical/Chemical Treatment Processes, First Edition. By Mark M. Benjamin, Desmond F. Lawler.
© 2013 John Wiley & Sons, Inc. Published 2013 by John Wiley & Sons, Inc.

Adsorption (*Continued*)
  Polanyi adsorption model and isotherm, 306–314
    basic principles, 306–313
    linear, Langmuir and Freundlich isotherms, 313–314
  rapidly equilibrating systems, 328–340
    batch systems, 328–331
    competitive adsorption, 359–364
    continuous flow reactors, 331–332
    fixed bed reactors, 333–340
      equation summary, 340
      mass balance on, 335–336
      plug flow reactors, 336–340
      qualitative analysis, 333–335
      sequential batch reactors, 332–333
  rate processes overview, 327–328
  slowly equilibrating systems, 340–354
    competitive adsorption, 364–365
    pore *vs.* surface diffusion, porous adsorbent particles, 341–343
    transport-limited adsorption rates:
      batch reactors, 343–350
      fixed-bed systems, 350–354
  solid dissolution reactions, 422–426
  surface charge interactions, 314–320
  surface interactions and reactions, modeling of, 314
    charged interfaces and electrostatic ion sorption, 314–319
    pore condensation and surface precipitation, 319–320
  surface pressure modeling, 297–306
    basic principles, 296–297
    competitive adsorption and, 302–306
    surface tension/isotherm data, computation using, 297–302
  terminology and mechanisms, 259–262
  water and wastewater treatment, 262–266
    activated carbon applications, 262–264
    cationic metal sorption, iron and aluminum oxides, 265–266
    metal hydroxide solids, 266
    natural organic matter sorption, drinking water coagulation, 264–265
Adsorption density, defined, 259
Adsorption equilibrium, isotherms, 269–296
  bidentate adsorption, 281–282
  competitive adsorption:
    Freundlich isotherms, 294–296
    ion exchange equilibrium, 284–294
    site-binding paradigm, 283–284
  distribution or partition coefficient, 282–283
  site-binding paradigm, 270–281
Adsorption isotherms, defined, 260
Advanced oxidation processes (AOPs):
  free radicals, ozone generation, 462
  hydroxyl radicals:
    Fenton-based systems, 483–486
      anodic processes, 486
      electrochemical processes, 486
      full-scale applications, 486
      heterogeneous processes, 485–486
      light-mediated processes, 485
    inorganic compound reactions, 470
    organic compound reactions, 470–476
    ozonation, 476–483
      ozone, 477–480
      peroxone, 481
      singlet oxygen radicals, 480–481
      sonolysis, 483
      ultraviolet radiation/hydrogen peroxide, 476–477
      ultraviolet radiation/semiconductor, 481–482
      wet air oxidation, 482–483
  synthetic organic compound degradation, 437
  taste and odor reduction, 436
Advection:
  counter-current gas transfer, operating line mass balance, 227–229
  crossflow filtration systems, contaminant transport, 800
  defined, 4
  electrodialysis transport, 819
  mass balance equations:
    arbitrary flow systems, 20–21
    unidirectional flow systems, 10–11
  one-dimensional advective diffusion equation, 19–20
Advective flux:
  gas transfer rate, well-mixed liquid phase systems, 217–218
  mass balance equations, unidirectional flow systems, 10–11
Aeration rate:
  gas transfer, batch reactors, 214–216
  mass balance equations, 18
Air flow rate, gas transfer, steady state continuous flow reactors, 222–225
Air-in-water reactors, gas transfer systems, 159–162
Air scour, granular-media filtration:
  filter cleaning, 716–717
Air/suspension contact chamber, gravity separations, flotation, 657
Alcohols, redox reactions, ozone, 465–466
Alkanes, redox reactions, ozone, 465–466
Alum, particle destabilization, sweep flocculation, 537–541
Aluminum hydroxides:
  adsorption:
    cationic sorption process, 265–266
    surface charge interactions, 314–320
  adsorption process and, basic principles, 257
  particle destabilization, 532–533
    precipitate-sweep flocculation, 538–541
  precipitation, 408–409
Amines, redox reactions, ozone, 466
Ammonia, redox reactions:
  chloramine formation, 455–456
  ozone interaction, 465
Amorphous solids, precipitation and, 382
Anodic electrochemical processes, 488
Anodic Fenton process, hydroxyl radical oxidation, 486
Apparent rejection factor, membrane system selectivity, 748
Arbitrary flow system, concentration gradient, mass balance equations, 20–23
  advection term, 20–21
  diffusion/dispersion terms, 21–23

overall mass balance, 23
storage and reaction terms, 23
Arrays, membrane systems, 738
configuration, 741–742
Arrhenius equation:
gas transfer rate constants, temperature effects, 195
precipitation, particle growth and formation thermodynamics, 388–389
Asymmetric membranes, structure and composition, 734–736
Atrazine:
advanced oxidation process, hydroxyl radicals, 472–476
redox reactions, ozone, 466
Attraction number, granular-media filtration, packed medium model, 693
Autocatalytic reactions, molecular collisions, 86–87
Autoretardants, molecular collisions, 87
Axial velocity, crossflow filtration systems, 793–795

Back-flushing:
membrane systems, 733
pressure-driven membrane systems, performance evaluation, 74
Back-transport mechanisms, crossflow filtration systems:
contaminant transport, 802
contaminant transport modeling, inertial lift and, 813–816
Backwashing:
granular-media filtration:
filter cleaning, 714–717
filtration run process, 681–682
time-related effects, 700–701
membrane systems, overview, 733
Ballasted flocculation, 571–572
Batch reactors:
gravity separations, ideal horizontal flow, 633–634
irreversible reaction mass balance, 89–90
rapid adsorption equilibrium, 328–331
sequential batch reactors, 332–333
slow adsorption equilibrium systems, transported-limited adsorption rates, 343–350
well-mixed liquid phase systems, gas transfer in, 213–216
Batch sedimentation, gravity separations:
type I batch sedimentation, 612–618
bimodal suspension, 614
heterodisperse suspension, 614–618
monodisperse suspension, 612–614
type II batch sedimentation, 618–621
experimental analysis, 620–621
mathematical analysis, 619–620
Batch thickening, gravity separations, 645–647
Bed depth, granular-media filtration design, 705–706
Bed volume (BV), defined 332
Benzene, redox reactions, ozone, 466
Bidentate adsorption, isotherms, 281–283
Bimolecular reactions:
kinetics, 83
redox reactions, ozone, 466
Biochemical oxygen demand, advective inflow/outflow, 18

Biological treatment, gas-in-liquid systems, overall gas transfer rate coefficient, 243–246
Boltzmann constant:
membrane systems, osmotic pressure, 755–757
particle stability, 524–525
Boltzmann distribution, molecular collisions, 86–87
Booster chlorination, chlorine disinfectant performance, 494–495
Born repulsion, particle stability:
flat plate-charged particle interaction, 530–531
Van der Waals attractions, 527–530
Boundary layer:
bulk water/contaminant transport modeling, pressure-driven membrane systems, 774–777
electrodialysis transport kinetics, 825–828
Breakpoint chlorination, redox reactions, chloramine formation, 456–459
Breakthrough curve:
granular-media filtration, time-related effects, 701–703
rapid adsorption equilibrium systems, fixed bed reactors, 334–340
transport-limited adsorption rates, fixed bed reactors, 350–354
Breakup, flocculation modeling, 592–594
Bridging mechanism, particle destabilization, adsorption and, 541–542
Brine solution, membrane systems, overview, 732–733
Bromide, redox reactions, 448
organic compound interactions, 452–455
ozone interaction, 465
Brownian motion:
crossflow filtration systems, contaminant transport, 798–800
flocculation collision mechanisms, 577–578
collision efficiency equations, 597–598
short-range force model, 584–589
particle-collector collisions, granular-media filtration, 688
particle collisions, gravity separations, 665–667
Bubble formation:
gas-in-liquid systems:
gas transfer rate coefficients and, 188–192
well-mixed liquid phase systems, 208–213
gravity separations, flotation effects, 656–657, 661–663
Bulk density, 336
Bulk water/contaminant transport modeling, pressure-driven membrane systems, 773–792
boundary layer and concentration polarization, 774–777
compact layer:
flux paradox, 784
formation, 782
hydraulic resistance, 783–784
concentration polarization and precipitative fouling, 784–786
CP layer:
parameter profiles, 786–787
permeation effects and resistance, 781–782
flux effects:
CP layer, particle/solute comparisons, 782
rejected particles, 778–780
rejected solutes, 780–781
frontal filtration systems:
nonsteady-state fouling, 787–791

Bulk water/contaminant transport modeling (*Continued*)
    performance evaluation, 792
        permeation/TMP relationship, 777–782
        limiting flux, film-gel model, 783
        nonsteady-state fouling, frontal and dead-end filtration, 787–791
            modified fouling and silt density indices, 790–791
        rejected contaminant, coupling force on permeate flow, 778

Cadmium, precipitation, pH-dependent metal-ligand speciation, 420–421
Cake filtration:
    crossflow filtration systems, contaminant transport modeling, layer thickness and inertial lift, 813–816
    nonsteady-state fouling, frontal and dead-end filtration systems, 787–791
Calcium-carbonate water softeners:
    mass balance equations, 5–7
    particle destabilization chemicals, 533
        preciptate-sweep flocculation, 540–541
    precipitative softening, 409–418
Capillary fiber membrane systems, 739, 741
Capture cross-section, flocculation collision efficiency, short-range force model, 583–589
Carbohydrates, redox reactions, ozone, 466
Carbonate ions, advanced oxidation process, hydrogen peroxide addition, 479–480
Carbonate solids:
    metal hydroxide precipitation, speciation, 405–409
    precipitation, 409–419
        equilibrium model, 414–419
        lime softening, stoichiometric/equilibrium model comparison, 416–419
        metal carbonates and hydroxycarbonates, 418–419
        pH requirements, 414
        precipitative softening, 409–410
        recarbonation, softened water, 417–419
        stoichiometric model, 410–413
Carbon dioxide hydration, reaction kinetics, 83
Carman–Kozeny equation, granular-media filtration:
    bed depth assessment, 706
    clean bed removal, 699–700
    ripening models, 710–714
    velocity parameters, 707–708
Cascade air stripping systems, 241
Catalysts, molecular collisions, 86–87
Cathodic electrochemical processes, 488
Cathodic Fenton processes, hydroxyl radical oxidation, 486
Cation-exchange membranes:
    electrodialysis:
        single-cell transport, 822–825
        two-dimensional electrodialysis system, 829–832
        water flow, 832
    electrodialysis and, 816–818
Cationic sorption, iron and aluminum oxides, 265–266
Cell-pair unit, electrodialysis, 818
    transport kinetics, 822–825
Cerussite, lead precipitation, 418–419

Challenge tests, membrane system selectivity, 748–749
Channeled flow effects, nonideal flow reactors, gravity separations, 640–642
Characteristic flow time:
    non-steady state reactor performance, continuous flow stirred tank reactors, 143–144
    steady state performance evaluation, continuous flow stirred tank reactors, 123
Characteristic reaction times, reversible reaction kinetics, 103–104
Characteristics reaction times, irreversible reaction analysis, 97–99
Charged interfacial structure, ion adsorption, electrostatic contribution, 314–319
Charge neutralization, particle destabilization, adsorption and, 536–537
Charge-potential relationships, ion adsorption, adsorbate profile and electrical potential, 317–318
Chelating solids, precipitation, speciation, 395–396
Chemical adsorption (chemisorption), defined, 259
Chemical disinfection, 488–500
    series-based continuous flow stirred tank reactors, steady state performance evaluation, 133–134
Chemical potential, membrane systems, driving force and, 753–754
Chemical reactions:
    fixed bed adsorption systems, 356–357
    gas/liquid equilibrium, Henry's law and, 162
    mass balance equations, unidirectional flow systems, 15–16
    molecular collisions, 84–87
    particle charge, 522–523
    sequential reaction kinetics, steady state equilibrium and, 112–114
Chemical structure, gas/liquid equilibrium and, 167–168
Chick's law, disinfection processes, 489–490
Chick–Watson law:
    disinfection processes, 490–492
    continuous flow stirred tank reactors in series, steady state performance evaluation, 133–134
Chloramines:
    disinfectant performance, 495–497
    redox reactions, 455–459
        formation, 455–456
        inorganic compounds, 458
        organic compounds, 458
        ozone interaction, 465
        water distribution systems, chlorine/lead release and, 458–459
Chlorine:
    disinfectant performance, 494–495
    natural organic matter reactions, 25
    reaction kinetics, 87–88
    redox reactions, 444–455
        bromide reactions, 448
        chloramine formation, 458–459
        inorganic compounds, free chlorine reactions, 446–455
        iron and manganese reactions, 446–447

organic compound reactions, 448–455
sulfur compound reduction, 447–448
Chlorine dioxide:
  disinfectant performance, 497–498
  redox reactions, 459–461
    generation, 460
    inorganic compounds, 460–461
    organic compounds, 461
    potassium permanganate, 468–469
Chlorine-nitrogen molar ratio:
  chloramine disinfectants, 496
  redox reactions, chloramine formation, 456
Chlorine-to-ammonia ratio:
  chloramine disinfectants, 496
  redox reactions, chloramine formation, 455–456
Chloroalkanes, redox reactions, ozone, 465–466
Chlorophenol intermediates, redox reactions, chlorine-organic compound interactions, 449–451
Chromatographic effect, competitive adsorption, column reactors, 359–365
Chromium compounds:
  precipitation, redox reactions, 422
  reductive processes, iron-based systems, 487–488
Circular reactors, gravity separations, 629–631
Clean bed removal, granular-media filtration, 683–694
  head loss, 698–699
  packed medium single spherical collector, 690–692
  removal mechanisms and transport efficiencies, 686–690
  single spherical isolated collector model, 683–686
  water/wastewater treatment systems, 694–698
  Yao model, 693–694
Cleaning-in-place (CIP), membrane systems, 733
Closed boundary reactors:
  dispersion modeling, 54–55
  residence time distributions, mean hydraulic retention time, 37–42
Coagulation. See Flocculation; Destabilization
Coefficient of dilution, disinfection processes, 489–490
Coefficient of specific lethality, disinfection processes, 490
Coefficient of variation, residence time distributions, probability statistics, 41–42
Co-ions, ion adsorption, adsorbate profile and electrical potential, 316–318
Collision efficiency factor:
  flocculation modeling, 567–568
  short-range force model equations, 597–598
Collision efficiency function, flocculation modeling, 567
Collision mechanisms:
  flocculation:
    fractal dimensions, 595–596
    long-range force model, 572–581
      Brownian motion, 577–578
      design implications, 580–581
      differential sedimentation, 575–577
      fluid shear collisions, 572–575
      total collision frequency, 578–580
    short-range force model, 581–589

calculation equations, 597–598
design implications, 589
flocculation modeling, basic parameters, 565–566
Color removal, oxidation, 436
Column contactors, gas transfer:
  operating line mass balance, 226–229
  pressure loss and liquid holdup, 233–235
  tower mass balance design equation, 229–233
Column height, gas transfer column contactors, design constraints, 236–241
Column size, gas transfer column contactors design and, 240
Combined chlorine, redox reactions, formation, 455
Compact layer:
  bulk water/contaminant transport modeling:
    flux paradox, 784
    formation, 782
    hydraulic resistance, 783–784
  crossflow filtration systems, contaminant transport modeling, 805–816
    TMP-flux relations, 809–816
  ion adsorption, adsorbate profile and electrical potential, 317–318
  pressure-driven membrane systems, bulk water/contaminant transport modeling, physicochemical properties, 773–774
Competitive adsorption:
  column reactors, chromatographic effect, 359–365
  ion adsorption, equilibrium constants, electrostatic contribution, 318–319
  isotherms, adsorption equilibrium:
    Freundlich isotherms, 294–296
    ion exchange equilibrium, 284–294
    site-binding paradigm, 283–284
      Freundlich isotherms, 294–296
  Langmuir isotherm, 283–284
  rapid adsorption equilibrium attainment, 359–364
  slow adsorption equilibrium, 364–365
  surface pressure models, 297–306
    ideal adsorbed solution theory, 302–306
Complete blocking scenario, nonsteady-state fouling, frontal and dead-end filtration systems, 787–791
Complexing ligands, precipitation, metal solubility effects, 421
Complex structure formation, precipitation, 381
  solid speciation, 395–396
Computational fluid dynamics (CFD), continuous flow reactors, basic principles, 30
Concentrate:
  crossflow filtration systems, pressure profile, 795–798
  electrodialysis, 818
  membrane systems, overview, 732–733
  pressure-driven membrane systems, pore-flow model, concentrate/membrane interface, 760–764
Concentration equalization, continuous flow reactors, 66–69
Concentration gradient:
  irreversible reactions, CFSTR/PFR comparisons, 126–129
  mass balance:
    arbitrary flow directions, 20–23

Concentration gradient (*Continued*)
　　advection term, 20–21
　　diffusion/dispersion terms, 21–23
　　overall mass balance, 23
　　storage and reaction terms, 23
　　unidirectional flow system and, 7–20
　　　advection term, 10–11
　　　chemical reaction term, 15–16
　　　differential one-dimensional mass balance, 18–20
　　　diffusion/dispersion terms, 11–15
　　　overall mass balance terms, 16–18
　　　storage term, 8–9
Concentration polarization (CP) layer:
　　crossflow filtration systems, contaminant transport layer, 803–816
　　frontal filtration systems:
　　　particle/solute effects, flux comparisons, 782
　　　permeation and TMP, 777–782
　　　resistance, permeation effects, 781–782
　　pressure-driven membrane systems:
　　　bulk water/contaminant transport modeling:
　　　　boundary layer transport, 774–777
　　　　defined, 773–774
　　　　membrane/parameter profile comparisons, 786–787
　　　　precipitative fouling, 784–786
Concurrent flow/concentration equalization, continuous flow reactors, 69–70
Conductivity, electrodialysis transport, 820–821
Consecutive reactions. *See* Sequential reactions
Constant capacitance model, ion adsorption, adsorbate profile and electrical potential, 317–318
Constant diffusivity, slow adsorption equilibrium systems, 343
Contact angles, hydrophilic membrane systems, 736
Contact zone modeling, gravity separations, flotation, 663–667
Continuity equation, three-dimensional mass balance, 24–25
Continuous flow reactors:
　　differential equations, Laplace transforms, 71–73
　　equalization, 62–70
　　　concentration equalization, 66–69
　　　concurrent flow and concentration equalization, 69–70
　　　flow equalization, 63–66
　　gravity separations, 622–639
　　　horizontal flow reactors, 626–634
　　　　batch and continuous flow reactors, 633–634
　　　　circular reactors, 629–631
　　　　rectangular reactors, 626–629
　　　　sedimentation, 638–639
　　　　type I *vs.* type II suspension, 631–633
　　　ideal horizontal flow, 633–634
　　　reactor settling potential function, 623–624
　　　suspension/reactor separation, 622–623
　　　thickening effects, 650–655
　　　tube settlers, 634–639
　　　upflow reactor, 624–626
　　hydraulic characteristics overview, 29–30
　　ideal reactors, 42–48
　　　continuous flow stirred tank reactors, 42

　　　pulse input, 45–47
　　　step input, 47–48
　　nonideal reactors, 48–62
　　　hydraulic behavior indices, 62
　　　residence time distributions, 50–62
　　　　continuous flow stirred tank reactors in series, 55–57
　　　　nonequivalent continuous flow stirred tank reactors in series, 62
　　　　plug flow reactors in parallel and series, 58–61
　　　　plug flow reactors with dispersion, 50–55
　　　　short-circuiting and dead space modeling, 57–58
　　　tracer output, 48–50
　　　convolution integral, 48–50
　　performance evaluation:
　　　nonideal flow reactors, 135–141
　　　　remaining fraction:
　　　　　dispersion model, 140–141
　　　　　exit age distribution, 136–140
　　　　steady state performance, 141
　　　non-steady-state conditions, 141–146
　　　　continuous flow stirred tank reactors, 142–144
　　　　conversion extent, 144–146
　　　　plug flow reactor conversion, 141–142
　　　steady state reaction:
　　　　continuous flow stirred tank reactor, 121–123
　　　　continuous flow stirred tank reactor in series, 130–134
　　　　irreversible reactions, 126–129
　　　　multiple ideal reactors, 130–135
　　　　parallel CFSTRS and PFRs, 135
　　　　plug flow reactor, 123–126
　　　　　in series, 130
　　　　rate expressions, 135
　　　　reversible reactions, 129–130
　　　　single ideal reactor, 121–130
　　plug flow reactors, 42–45
　　　fixed frame pulse input, 43–44
　　　moving frame pulse input, 44–45
　　rapid adsorption equilibrium, water and adsorbent flow mechanics, 331–332
　　residence time distributions, 30–42
　　　mean hydraulic detention time, 37–42
　　　probability distribution statistics, 37–42
　　　pulse input response, 33–35
　　　step input response, 35–37
　　　tracers, 30–33
　　steady state, gas transfer, 220–225
　　　continuous flow stirred tank reactors, 220–225
　　　plug flow reactors, 220
Continuous flow stirred tank reactors (CFSTRs). *See also* Series-based continuous flow stirred tank reactors
　　concentration equalization, 66–69
　　disinfection processes, 493–494
　　equalization, 63–70
　　gas transfer, steady state systems, 220–225
　　　overall gas transfer rate coefficient, 244–246
　　hydraulic characteristics, 42

non-steady-state conditions:
    nonideal flow reactor performance, 144–146
    performance evaluation, 142–144
    particle destabilization chemicals, water stream mixing, 544–546
    particle size distribution and precipitation in, 389–394
    precipitation reactions, 426–427
    pulse input, 45–47
        Laplace transform, 71–73
    rapid adsorption equilibrium, batch reactors, 330–331
    in series:
        nonequivalent CFSTRs, 62
        residence time distributions, 55–57
    short circuit and dead space modeling, 57–58
    slow adsorption equilibrium systems, transported-limited adsorption rates, 349–350
    steady state performance, 121–123
        irreversible reactions, 126–129
        nth order reactions, 126
        plug flow reactor comparisons, 126–129
        reversible reactions, 129–130
    step input, 47–49
Control volume (CV):
    mass balance equations, 3–4, 6–7
        arbitrary flow systems:
            advection, 20–21
            overall mass balance, 23
            storage and reaction terms, 23
            three-dimensional mass balance, 24–25
        unidirectional flow systems, 7–20
            advection, 10–11
            chemical reactions, 15–16
            diffusion and dispersion, 11–15
        one-dimensional mass balance, unidirectional flow systems, 18–20
Convective-diffusion equation, pressure-driven membrane systems, pore-flow model, concentrate/membrane interface, 761–764
Convective-diffusion-hindered (C-D-H) transport model, pressure-driven membrane systems, 762–764
Convolution integral, nonideal reactors, tracer input and output, 48–50
Copper, redox reactions, ozone interaction, 465
Corrosion inhibitors, lead precipitation, 419
Counter-current gas transfer. *See also* Column contactors
    design summary, 241–243
    operating line mass balance, 226–229
Counter-ions, ion adsorption, adsorbate profile and electrical potential, 316–318
Coupled transport and reaction, gas transfer, 183–187
Coupling coefficients, dense membrane systems modeling, 835–838
Coupling force, frontal filtration systems, permeation and TMP, 778
Critical coagulation concentration (CCC), particle destabilization process, 535
Critical size, particle-collector collisions, granular-media filtration, 689–690

Crossflow, defined, 733
Crossflow air stripping systems, 241
Crossflow filtration systems, 792–816
    compact layer flux/TMP interactions, 808
    compact layer with partial membrane element, 811–813
    concentrate pressure profile, 795–798
    contaminant transport mechanisms, 798–802
        back-transport mechanisms, 802
        Brownian and shear-induced diffusion, 798–800
        deterministic transport mechanisms, 800–802
    contaminant transport modeling, 802–816
    CP layer thickness, 804–805
    file and film-gel models, 807–808
    film-gel model, 805–807
    film model without compact layer, 808–809
    fluid flow:
        modeling parameters, 792–793
        processes, 793–795
    flux modeling, 802–816
        TMP, no compact layer, 809–811
    inertial particle lift, 813–814
    modeling assumptions, 803
    modeling equations, 814–816
    modeling mechanics, 803–804
    nonsteady-state fouling patterns, 814–816
    particle transport modeling, 792–793
    summary of modeling approaches, 815–816
Cross-sectional area, gas transfer column contactors, design constraints, 236–241
CT product:
    chlorine dioxide disinfectant performance, 497–498
    chlorine disinfectant performance, 494–495
    disinfection processes, 490–492
    ozone disinfectant performance, 498–500
Cummins and Westrick approximation, column contactor gas towers, design criteria and, 236–240
Cumulative age distribution $(F(t))$:
    continuous flow reactors, 31
        residence time distributions, pulse input response, 34–35
        step input response, 35–37
    continuous flow stirred tank reactors in series, residence time distributions, 56–57
    hydraulic behavior indices, 62
    nonideal reactors:
        non-steady-state conditions, performance evaluation, 145–146
        residence time distributions, 50–62
        tracer input and output, 49–50
    plug flow reactors, 42–45
        parallel and series-based models, 59–61
Cumulative probability function, residence time distribution, continuous flow reactors, 31
Cumulative site occupation function, Freundlich isotherm, 276–281
Cumulative surface site distribution function, adsorption equilibrium isotherms, 270–271

Current density:
   electrodialysis, ion fluxes and, 825–828
   electrodialysis transport kinetics, 819–822
      electrical potential gradient, 821–822

Darcy's law, membrane systems, permeability, 745–746
Darcy–Weisbach equation, crossflow filtration systems, pressure profile, 795–798
Dark Fenton process, hydroxyl radicals, 483–485
Davies equation:
   gas transfer, activity coefficient, 199–202
   precipitative fouling, pressure-driven membrane systems, 785–786
Dead-end filtration:
   membrane arrays, 741–742
   nonsteady-state fouling, 787–791
Dead space:
   continuous flow reactors, residence time distributions, mean hydraulic retention time, 37–42
   nonideal reactors, residence time distributions, 57–58
Debye–Huckel equation:
   gas transfer, activity coefficient, 199–202
   particle stability, 524–525
Dechlorination, redox reactions, sulfur reduction, 447–448
Decomposition process, redox reactions, ozone, 465
Degrees of freedom:
   gas transfer column contactors, design feasibility criteria, 236–240
   well-mixed liquid phase systems, gas transfer, design constraints, 226
Dense membrane systems:
   irreversible thermodynamics modeling, 834–838
   structure and composition, 734–736
Design constraints and choices:
   disinfection processes, 493–494
      chloramine disinfectants, 496–497
   fixed bed reactors, 366–369
   flocculation modeling:
      long-rang force model, 580–581
      short-range model, 589
      Smoluchowski equation, 571–572
   gas transfer column contactors:
      column size and, 240
      feasibility criteria, 236–240
      geometric parameters, 236–241
      influent stream and treatment objectives, 236
      pressure loss and liquid buildup, 235
      tower mass balance equation, 229–233
   granular-media filtration:
      bed depth, 705–706
      clean bed removal, 696–698
      filtration dynamics, 705–709
      influent concentration, 709
      media size, 706–707
      velocity parameters, 707–708
   gravity separations, continuous flow thickening, 655
   two-dimensional electrodialysis system, 830–832
   well-mixed liquid phase systems, gas transfer, 226
Desorption:
   defined, 259
   solid dissolution reactions, 422–426
Destabilization, particle treatment process, 535–542
   adsorption:
      charge neutralization, 536–537
      interparticle bridging, 541–542
   chemicals used in, 532–535
      inorganic compounds, 532–533
      organic polymers, 533–535
      soluble materials interaction, 542–544
         additive combinations, 543–544
      water stream mixing, 544–546
   diffuse layer compression, 535–536
   precipitate-sweep flocculation enmeshment, 537–541
Detention times:
   irreversible reactions:
      CFSTR/PFR comparisons, 126–129
      continuous flow stirred tank reactors, steady state performance:
         first-order rate constant, 123
         pulse input, 46–47
      plug flow reactors, steady state performance, 124–125
      reversible reactions, reactor performance evaluation, 129–130
   continuous flow stirred tank reactors in series, steady state performance evaluation, 131–134
Deterministic transport mechanisms, crossflow filtration systems, contaminant transport, 800–801
Dialysis. *See also* Electrodialysis
   membrane systems, electrical forces and concentration differences, 738
Differential equations, Laplace transform solutions, 70–73
Differential reaction rate analysis, irreversible reactions, 96–97
Differential sedimentation, flocculation collision mechanisms, 575–577
   efficiency equations, 598
   short-range force model, 585–589
Differential site occupation function, Freundlich isotherm, 277–281
Differential surface site distribution function, adsorption equilibrium isotherms, 270–271
Diffuse layer structures:
   ion adsorption, adsorbate profile and electrical potential, 317–318
   particle destabilization process, compression, 535–536
   particle stability, 524–525
Diffuser systems, gas transfer systems, 159–162
Diffusion:
   defined, 4
   electrodialysis transport, 819–820
   mass balance equations:
      arbitrary flow systems, 21–23
      unidirectional flow systems, 11–15
   one-dimensional advective diffusion equation, 19–20
Diffusion coefficient, unidirectional flow systems, 11–15
Diffusion-controlled reactions, molecular collision frequency, 84

Dimensionless parameters:
  crossflow filtration systems, contaminant transport layer modeling, 804–816
  gas-in-liquid systems, mass transfer correlations, 189–192
  Henry's constants, 165–167
  mass balance equations, 6–7
  mechanistic gas transfer models, limiting-case scenarios, 175
  plug flow reactors, residence time distribution, 51–55
Dirac delta function:
  continuous flow reactors, tracer studies, residence time distributions, 32–33
  plug flow reactors:
    cumulative and exit age distribution, 42–45
    Eulerian view, 43–44
Disequilibrium rate, column contactor gas tower design, 231–233
Disinfectant demand, defined, 441
Disinfectant residual, defined, 441
Disinfection byproducts (DBPs):
  chemical reactions, unidirectional flow systems, 16
  chloramine disinfectants, 495–497
  chlorine disinfectant performance, 494–495
  production of, 488
  reaction kinetics, 87–88
  redox reactions:
    chloramine reaction, 458
    chlorine-organic compound interactions, 449–455
  ultraviolet radiation disinfection, 502
Disinfection processes:
  chloramine disinfectant performance, 495–497
  chlorine dioxide disinfectant performance, 497–498
  chlorine disinfectant performance, 494–496
  design and operational issues, 493–494
  empirical modeling, 489–493
  hydroxyl radicals, 502
  ozone disinfectant performance, 498–500
  ultraviolet radiation, 500–502
Dispersion:
  defined, 4–5
  mass balance equations:
    arbitrary flow systems, 21–23
    unidirectional flow systems, 11–15
  plug flow reactors, residence time distribution, 50–55
Dispersion coefficient, unidirectional flow systems, 12–15
Dissolution:
  gas/liquid equilibrium, 167–168
    Henry's law and, 162
  solid dissolution reactions, 422–426
Dissolved air flotation, gravity separations, 655
  sludge thickening, 668–669
Dissolved oxygen (DO):
  gas/liquid equilibrium, 168–169
  redox reactions, oxygen-iron species, 443–444
  well-mixed liquid phase systems, gas and water transfer, 210–213
Distribution coefficient:
  adsorption equilibrium, 269–270
  gas/liquid equilibrium, Henry's law and, 163–164
Divergence theorem, three-dimensional mass balance, 24–25

Donnan equilibrium:
  ion adsorption, electrically induced partitioning, 268
  ion exchange reactions, 292–294
Donnan exclusion, Ion exchange adsorption, 294
d-plane, particle stability and, 521–522
Drinking water coagulation:
  adsorption process, natural organic matter, 264–265
  particle treatment applications, 519–521
Driving forces, membrane systems, 736–738
  concentration differences and electrical forces, 738
  energy characteristics, 752–754
  osmotic pressure, 754–757
  pressure-driven processes, 736–738
  quantification, 752–759
  transport magnitudes, 757–759
Dual media filters, granular-media filtration, 680
Dual-spike methodology, gas transfer, steady state systems, overall gas transfer rate coefficient, 244–246

Eckert curves, column contactor gas towers, design criteria and, 236–240
Eddy characteristics, particle destabilization chemicals, 544–546
Effective mass transfer coefficient, gas transfer, coupled transport and reaction, 184–187
Effective residence time, plug flow reactors, residence time distributions, 53–55
Efficiency factor:
  disinfection processes, 491–492
  particle-collector collisions, granular-media filtration, 688
Electrical conductance, electrodialysis transport, 820–821
Electrical double layer (EDL), ion adsorption, adsorbate profile and electrical potential, 317–318
Electrical forces, membrane systems, 738
Electrical potential:
  electrodialysis transport kinetics, 819–822
    advection transport, 819
    current densities, 820–828
    diffusion transport, 819–820
    ion fluxes and current density, 825–828
    solution in electric field, 820
  ion adsorption:
    adsorbate profile, interfacial region, 315–318
    charged interfacial structure, 314–319
    surface binding effects, 314–315
  particle stability, 522
Electrochemical potential, redox processes, thermodynamics, 440–441
Electrochemical processes:
  Fenton processes, hydroxyl radical oxidation, 486
  redox reactions, 488
Electrodeionization (EDI), two-dimensional electrodialysis system, 832
Electrodialysis (ED), 816–834
  basic principles and applications, 816–819
  defined, 738
  ion-exchange membranes, water flow, 832

Electrodialysis (ED) (*Continued*)
   modeling parameters, 822
      ion fluxes, current density and electrical potential difference, 825–828
         single cell-pair transport, 822–825
      multicomponent systems, 832–833
      nonideal membrane behavior, 832
      overlimiting current, 833
      potential drop sources, 833
      real system complications, 832–833
   schematic, 817
   single-cell unit, 818
   transport systems, electrical potential gradient, 819–822
      advective transport, 819
      current densities, 820–828
      diffusive transport, 819–820
      two-dimensional systems, 828–832
         design issues, 830–831
         limiting current, 831–832
         macroscopic mass balance, 828–829
         microscopic mass balance, 829–830
Electrodialysis reversal (EDR), defined, 818
Electromigration, electrodialysis transport, 819–822
Electroosmosis, electrodialysis and, 832
Electrophoretic mobility (EPM), particle surface charge, 523
   measurements, 531–532
Electrostatic contribution:
   charged particle interaction, 525–526
   ion adsorption:
      charged interfacial structure, 314–319
      equilibrium constants, competitive adsorption, 318–319
Elementary reaction:
   defined, 82
   kinetics, 84–87
Empty bed contact time (EBCT):
   fixed bed reactor performance, 357–359
   rapid adsorption equilibrium, fixed bed reactors, 338–340
Energy costs, gas transfer column contactors, design criteria and, 239–240
Energy dissipation function, dense membrane systems modeling, 834–838
Energy parameters, membrane systems, driving force and, 752–754
Enhancement factor, gas transfer, coupled transport and reaction, 184–187
Enmeshment, particle destabilization, precipitate-sweep flocculation, 537–541
Enthalpy, precipitation reactions, 384
Enzyme-limited reactions, sequential reaction kinetics, 113–114
Epichlorohydrindimethylamine, particle destabilization chemicals, 533–535
Equalization:
   continuous flow reactors, 62–70
      concentration equalization, 66–69
      concurrent flow and concentration equalization, 69–70
      flow equalization, 63–66
   mass balance equations, 9
      unidirectional flow systems, storage term, mass balance equations, 9
      wastewater production example, 64–66
Equilibrium constants:
   adsorption models, 268–269
   gas/liquid equilibrium, Henry's law and, 162–164
   Henry's constants, 197-199
   ion adsorption, competitive adsorption, 318–319
   redox reactions, thermodynamics, 438–441
Equilibrium models:
   precipitation, 397–422
      carbonate solids precipitation, 409–419
      complexing ligands, metal solubility, 421
      hydroxide solids precipitation, 404–409
      ionic strength and activity coefficients, 400
      metal and hydroxy carbonates, 418–419
      pH-dependent metal and ligand speciation, 420–421
      precipitative softening, 409–417
      redox reactions, 421–422
      soft water recarbonation, 417–418
   precipitative carbonate softening, 414–417
Equivalent spherical diameter, particle size distribution, particle treatment process, 547–551
Error function, 14–15
Ethylenediamine tetra-acetic acid (EDTA):
   oxidation, 437
   precipitation:
      complexing ligands, 421
      hydroxide solids, 404–405
      soluble speciation, 395–396
Eulerian view, plug flow reactors:
   cumulative and exit age distribution, 43–45
   residence time distribution, dispersion, 50–55
   steady state performance evaluation, 124–125
Exit age distribution *(E(t))*:
   concentration equalization, 66–69
   continuous flow reactors, 30–31
      pulse input response, 33–35
      step input response, 35–37
   continuous flow stirred tank reactors in series, residence time distributions, 56–57
   gas transfer, steady state systems, overall gas transfer rate coefficient, 244–246
   hydraulic behavior indices, 62
   nonideal reactors:
      residence time distributions, 50–62
         plug flow reactors, 52–55
      tracer input and output, 49–50
   plug flow reactors, 42–45
      reactors in parallel and series, 59–61
   continuous flow stirred tank reactors in series, steady state performance evaluation, 133–134
Extent of disequilibrium, reversible reaction kinetics, 105–107

Facilitated transport, membrane processes, 738
Faraday constant:
   electrodialysis transport kinetics, 819–822
   redox processes, thermodynamics, 440–441

Fast-flushing, membrane systems, overview, 733
Feasibility criteria, gas transfer column contactors, design constraints and, 236–240
Feed-and-bleed membrane configuration, 742
Fenton-based systems, advanced oxidation processes, hydroxyl radicals, 483–486
  anodic processes, 486
  electrochemical processes, 486
  full-scale applications, 486
  heterogeneous processes, 485–486
  light-mediated processes, 485
Ferric hydroxides:
  adsorption process and, basic principles, 257
  precipitation, 408–409
Ferrihydrite, cationic sorption, 265–266
Ferryl ion, advanced oxidation process, 469
Fick's law:
  diffusion, unidirectional flow systems, 12–15
  mechanistic gas transfer models:
    fluid packets, 173–174
    limiting-case scenarios, 174–175
  pore diffusion, slow adsorption equilibrium systems, 341–343
Film-gel model:
  bulk water/contaminant transport modeling, compact layer modeling, 783
  crossflow filtration systems, contaminant transport modeling, 805–816
Film models:
  crossflow filtration systems, contaminant transport modeling, 809–816
  pressure-driven membrane systems, bulk water/contaminant transport modeling, 776–777
Filter cleaning, granular-media filtration, 714–717
  air scour, 716–717
  surface wash, 716
Filtration systems. *See also* Membrane systems
  granular-media filtration:
    backwashing, 699–700
    breakthrough, 701–703
    clean bed removal, 683–694
      head loss, 698–699
      packed medium single spherical collector, 690–692
      removal mechanisms and transport efficiencies, 686–691
      water/wastewater treatment systems, 694–698
      Yao model, 693–694
    design/operational criteria, 705–709
      bed depth, 705–706
      influent concentration, 709
      media size, 706–707
      velocity parameters, 707–708
    dynamics models, 709–714
    filter cleaning, 714–717
      air scour, 716–717
      surface wash, 716
    filter components, 679–680
    filter run, 680–682
    head loss, 703–705
    macroscopic models, 710
    particle removal mathematics, Iwasaki model, 682–683
    ripening, 701
      Mackie model, 713
      O'Melia–Ali model, 710–712
      Tien model, 712–713
    time-related dynamics, 699–709
  gravity separations, flotation, 665
First-order irreversible reactions:
  CFSTR/PFR comparisons, 126–129
  continuous flow stirred reactors, 122–123
  series-based continuous flow stirred tank reactors, steady state performance evaluation, 131–134
Fixed bed reactors:
  chemical reactions, 356–357
  design options, 366–369
  mass transfer zone movement, 354–356
  performance evaluation, 357–359
  rapid adsorption equilibrium systems, 333–340
    equation summary, 340
    mass balance on, 335–336
    plug flow reactors, 336–340
    qualitative analysis, 333–335
  slow adsorption equilibrium systems, transported-limited adsorption rates, 350–354
Fixed-film gas transfer model, fluid packets, gas/solution interface, 171–174
Fixed frame of reference. *See* Eulerian view
Flat plate, charged particle interactions, 530–531
Flocculation:
  characteristic reaction times, 570–571
  collision mechanisms:
    long-range force model, 572–581
      Brownian motion, 577–578
      design implications, 580–581
      differential sedimentation, 575–577
      fluid shear collisions, 572–575
      total collision frequency, 578–580
    short-range force model, 581–589
      calculation equations, 597–598
      design implications, 589
  defined, 520
  floc breakup, 592–594
  formation rate equation, 567
  fractal dimensions modeling, 594–596
  modeling, 565–572
  particle destabilization, precipitate-sweep flocculation, 537–541
  particle size distribution changes, 564–565
  particle treatment systems, 521
    fractal characteristics, 552–553
  rate constant interpretation, 567–568
  Smoluchowski equation, 568–570
    design issues, 571–572
  turbulence, 589
  turbulent flocculation, 589–591
Flocculation model, gravity separations, flotation, 664–665

Flotation, gravity separations, 655–669
　bubble formation, 661–663
　contact zone modeling, 663–667
　　filtration model, 665–667
　　flocculation model, 664–665
　　model comparisons, 667
　low concentration suspensions, 663
Flow equalization, continuous flow reactors, 63–66
Flow streams:
　crossflow filtration systems, 792–795
　membrane systems, 732–733
Fluence values, ultraviolet radiation disinfection, 502
Fluid dynamics, gas transfer models, 170–171
Fluidized-bed reactors (FBRs), precipitation reactions, 427
Fluid packets, gas transfer, mass balance, gas/solution interface, 171–174
Fluid shear, flocculation collision mechanisms, 572–575
　efficiency equations, 598
　short-range force model, 584–589
Fluid transport:
　crossflow filtration systems:
　　modeling parameters, 792–793
　　processes, 793–795
　membrane systems, 743–747
　　flux velocity, 743–744
　　permeability, 745–746
　　recovery, 743
　　resistance, 744–745
　　specific flux and permeance, 744
　　temperature effects, water permeation, 746–747
Flux conservation equation, crossflow filtration systems, contaminant transport modeling, 809–816
Flux properties:
　crossflow filtration systems, contaminant transport modeling:
　　inertial lift and, 813–816
　　TMP-flux relations, 809–816
　frontal filtration systems:
　　particle/solute effects, CP layer comparisons, 782
　　rejected particles, flux effects, 778–780
　　rejected solutes, flux effects, 780–782
　mechanistic gas transfer models, limiting-case scenarios, 174–176
　membrane systems, fluid/solute transport, 743–744
　pressure-driven membrane systems:
　　compact layer modeling and flux paradox, 784
　　performance evaluation, 750–752
　　pore flow and solution-diffusion models, 771–772
　unidirectional flow systems, 12–15
Force, particle stability, Van der Waals attraction, 527–530
Formation potential (FP) test, redox reactions, organic compound interactions, 453–455
Fractal dimensions:
　flocculants, particle treatment systems, 552–553
　flocculation modeling, 594–596
Free chlorine/free available chlorine, redox reactions, 446–455
Free radicals. See also Hydroxyl radicals
　advanced oxidation processes:
　　basic principles, 469–470
　　ozone interaction, 462
Freundlich isotherm:
　adsorption equilibrium:
　　multisite isotherms, 281
　　partition coefficient, 282–283
　　Polanyi isotherm, 313–314
　　surface pressure adsorption model, 298–302
　competitive adsorption:
　　column reactors, chromatographic effect, 361–364
　　ideal adsorption model, 302–306
　　site-binding models, 294–296
　mathematical properties, 260–261
　rapid adsorption equilibrium, batch reactors, 328–331
　semicontinuous site surface distribution, 276–281
　slow adsorption equilibrium systems, transported-limited adsorption rates, batch reactors, 346–350
Frictional pressure loss, crossflow filtration systems, pressure profile, 795–798
Frontal filtration:
　bulk water/contaminant transport modeling, boundary layer and concentration polarization transport, 774–777
　nonsteady-state fouling, 787–791
　performance evaluation, 792
　permeation-TMP relation, 777–782
Froude number, liquid-in-gas systems, mass transfer coefficients, 193–195
Full-scale systems, gas transfer:
　rate equations overview, 207
　spatial variations, solution and gas concentrations, 226–241
　　design equation applications, 236
　　　column size sensitivity and parameter value uncertainty, 240
　　　influent stream, treatment objectives and, 236–240
　　non-packed column systems, 240–241
　　operating line mass balance, 226–229
　　pressure loss and liquid holdup, 233–235
　　tower design equations, 229–233
　volatile species acid/base reactions, transfer rate, 199–202
　well-mixed liquid phase systems case study, 207–226
　　batch liquid reactors, 213–216
　　flow design constraints and choices, 226
　　limiting cases, general kinetic expression, 216–220
　　　macroscopic (advective) limitation, 217–218
　　　microscopic (interfacial) limitation, 218–219
　　　overall gas transfer rate, 219–220
　　overall gas transfer rate expression, 207–213
　　continuous liquid flow stirred tank reactors in series, 225
　　steady state continuous liquid flow reactors, 220–225
　　　continuous flow stirred tank reactors, 220–225
　　　plug flow reactors, 220

Gas absorption, defined, 155
Gas flow rate, gas-in-liquid systems, hydrodynamic effects, 192
Gas-in-liquid systems:
　gas transfer:
　　overall gas transfer rate coefficient, 243–246
　　rate coefficient estimation, 188–192

system-specific parameters, 187–196
gas transfer, well-mixed liquid phase systems, 208–213
overview of, 159–162
Gas/liquid equilibrium:
gas transfer, 155–157
Henry's law, 162–169
c_L, c_G, and Henry's law constant, 164–167
partition coefficients, equilibrium constants and formal definitions, 162–164
structure and temperature, 167–169
volatilization and dissolution, 162
Gas-phase mass transfer coefficient, mechanistic gas transfer models, 176–179
Gas phase resistance:
counter-current gas transfer, 227–229
gas transfer rate coefficient:
liquid phase combined with, 179–181
liquid phase compared with, 181–183
Gas stripping:
adsorption comparison with, 258
counter-current gas transfer, operating line mass balance, 227–229
defined, 155
design summary for, 241–243
gas transfer column contactors, design criteria, 239–240
non-packed column designs, 240–241
pressure loss and liquid holdup, 233–235
well-mixed liquid phase systems, gas transfer, macroscopic (advective) limitation, 217–218
Gas transfer:
adsorption comparison with, 258
Gibbs adsorption isotherm, 298–302
engineered transfer systems, 159–162
environmental engineering applications, 155
gas and liquid phase changes, 170
gas/liquid equilibrium, 155–157
Henry's law, 162–169
c_L, c_G, and Henry's law constant, 164–167
partition coefficients, equilibrium constants and formal definitions, 162–164
volatilization and dissolution, 162
Henry's constants, concentration and activity coefficient conventions, 197–199
physicochemical environment conventions, 198–199
real gas transfer systems, volatile activity coefficient, 199
reference states, activity coefficients based on, 198–199
standard state concentrations, 197
hydrodynamic and operating conditions, 187–196
gas-in-liquid systems, 188–192
liquid-in-gas systems, 192–195
solution effects, 195–196
temperature effects, 195
transfer rate coefficient estimates, 188–195
transfer rate constants, effects on, 195–196
mass balance incorporation, 157–158
mechanistic models, 170–179
fluid dynamics and interfacial mass transport, 170–171

fluid packet interfacial transfer, 171–174
limiting case residence times, 174–175
mass balance, volatile gas/solution interface, 171–179
packet age and residence time distribution, 175
transfer coefficient, 175–179
molecular diffusion, 196–197
overall transfer rate coefficient, 179–187
combined gas-liquid phase resistance, 179–181
coupled transport and reaction, 183–187
gas-phase/liquid-phase resistance comparisons, 181–183
rate equations overview, 207
spatial variations, solution and gas concentrations, 226–241
design equation applications, 236
column size sensitivity and parameter value uncertainty, 240
influent stream, treatment objectives and, 236–240
non-packed column systems, 240–241
operating line mass balance, 226–229
pressure loss and liquid holdup, 233–235
tower design equations, 229–233
transport and reaction kinetics, 157
volatile species acid/base reactions, transfer rate, 199–202
well-mixed liquid phase systems case study, 207–226
batch liquid reactors, 213–216
flow design constraints and choices, 226
limiting cases, general kinetic expression, 216–220
macroscopic (advective) limitation, 217–218
microscopic (interfacial) limitation, 218–219
overall gas transfer rate, 219–220
overall gas transfer rate expression, 207–213
continuous liquid flow stirred tank reactors in series, 225
steady state continuous liquid flow reactors, 220–225
continuous flow stirred tank reactors, 220–225
plug flow reactors, 220
Gas transfer capacity factor:
column contactor gas tower design, 231–233
constraints in, 241–243
design feasibility and, 237–240
pressure loss and liquid buildup, 233–235
counter-current gas transfer, mass balance, 229
steady state continuous flow reactors, 222–225
Gas transfer coefficient, mechanistic gas transfer, 175–179
Gas transfer rate coefficients:
estimation methods, 188–195
gas-in-liquid systems, 188–192
Gas transfer rate constants, temperature effects, 195
Gas volume, gas/liquid equilibrium, 164–167
Geometric volume, continuous flow reactors, residence time distributions, mean hydraulic retention time, 37–42
Geosmin, redox reactions, chlorine dioxide interaction, 461
Gibbs adsorption isotherm, surface pressure adsorption computation, 298–302
Gibbs free energy:
ion adsorption, surface binding effects, 315
membrane systems, driving force and energy, 752–754
Polanyi adsorption model, 308–314
precipitation, particle growth and formation thermodynamics, 385–389

Gibbs free energy (*Continued*)
  precipitation reactions, 383–384
  sequential reaction kinetics, thermodynamics, 112
  surface pressure adsorption models, 297–306
Gouy–Chapman model, particle stability, 524–525
Granular activated carbon:
  activated carbon preparation, adsorption, water and wastewater treatment, 262–264
  adsorbent regeneration, 365–366
Granular-media filtration:
  backwashing, 699–700
  breakthrough, 701–703
  clean bed removal, 683–694
    head loss, 698–699
    packed medium model update, 692–693
    packed medium single spherical collector, 690–692
    removal mechanisms and transport efficiencies, 686–692
    single spherical isolated collector model, 683–686
    water/wastewater treatment systems, 694–698
    Yao model, 693–694
  design/operational criteria, 705–709
    bed depth, 705–706
    influent concentration, 709
    media size, 706–707
    velocity parameters, 707–708
  dynamics models, 709–714
  filter cleaning, 714–717
    air scour, 716–717
    surface wash, 716
  filter components, 679–680
  filter run, 680–682
  head loss, 703–705
  macroscopic models, 710
  particle removal mathematics, Iwasaki model, 682–683
  ripening, 701
    Mackie model, 713
    O'Melia–Ali model, 710–712
    Tien model, 712–713
  time-related dynamics, 700–709
Gravity separations:
  continuous flow ideal settling, 622–639
    ideal horizontal flow reactors, 626–634
      batch and continuous flow reactors, 633–634
        circular reactors, 629–631
        rectangular reactors, 626–629
        sedimentation, 638–639
        type I *vs.* type II suspension, 631–633
      reactor settling potential function, 623–624
      tube settlers, 634–639
      upflow reactor, 624–626
  engineered systems, 605–607
  flotation, 655–669
    bubble formation, 661–663
    contact zone modeling, 663–667
      filtration model, 665–667
      flocculation model, 664–665
      model comparisons, 667
    low concentration suspensions, 663
    saturator, 658–660
    sludge thickening, dissolved air flotation, 668–669
    systems overview, 657–658
  particle sedimentation, 607–612
    inertial effects, 610–612
    Stokes' law, 607–610
  sedimentation reactors, nonideal flow effects, 639–644
    channel flow, 640–642
    mixed flow, 642–644
    tiered flow, 639–640
  thickening, 644–655
    batch thickening, 645–647
    continuous flow thickening, 650–655
      design constraints, 655
    solids flux, 647–650
  type I batch sedimentation, 612–618
    bimodal suspension, 614
    heterodisperse suspension, 614–618
    monodisperse suspension, 612–614
  type II batch sedimentation, 618–621
    experimental analysis, 620–621
    mathematical analysis, 619–620

Haber–Weiss reaction, hydrogen peroxide oxidation, 483–486
Hagen–Poiseuille equation, membrane systems, permeate flux, 744
Half-reactions, redox thermodynamics, 437–441
Half-times, irreversible reaction rate analysis, 91–93
Haloacetic acid (HAA), redox reactions, 448
  chloramine reaction, 458
  chlorine-organic compound interactions, 450–455
Haloacetic acid formation potential (HAAFP), redox reactions, 453–455
Hamaker constant, particle stability, Van der Waals attractions, 527–530
Hardness ions, precipitative carbonate softening, 409–410
  equilibrium model, 417–419
  stoichiometric model, 411–413
Head loss, granular-media filtration:
  clean bed removal, 698–700
  macroscopic models, 710
  time-related effects, 703–705
Heaviside function:
  continuous flow reactors, tracer studies, residence time distributions, 32–33
  type I batch sedimentation, monodisperse suspension, 612–614
Height of a (gas) transfer unit (HTU), column contactor gas tower design, 232–233
  constraints in, 242–243
Henry's constant:
  common units, 165
  concentration conventions and activity coefficients, 197–199
  dimensionless constants, liquid and gas phase resistance, 181–183
  environmentally important gases, 165

gas/liquid equilibrium:
    defined, 163–164
    parameters effecting, 168–169
    ideal gas law, 164–167
Henry's law:
    gas/liquid equilibrium, 156–157, 162–169
        acid/base reactions, 199–202
        c_L, c_G, and Henry's law constant, 164–167
        partition coefficients, equilibrium constants and formal definitions, 162–164
        structure and temperature, 167–169
        volatilization and dissolution, 162
    gravity separations, flotation effects, 656–657
    linear isotherms, 260
    ozone generation, 464
    reference states, activity coefficient and, 198–199
Hermia model, nonsteady-state fouling, frontal and dead-end filtration systems, 787–791
Heterodisperse suspension, type I batch sedimentation, 614–618
Heterogeneous Fenton processes, hydroxyl radical oxidation, 485–486
Heterogeneous nucleation, particle growth and formation, precipitation and, 381
Heterovalent ion exchange, competitive adsorption, Langmuir isotherm, 286, 288–294
Hexavalent chromium, reaction kinetics, 82
Hindered settling, gravity separations, 644–655
Hollow fiber membrane systems, 739, 741
    crossflow filtration systems, fluid flow and, 793–795
Homogeneous nucleation, particle growth and formation, precipitation and, 381
Homogeneous surface diffusion model (HSDM), slow adsorption equilibrium systems, 343
Homovalent ion exchange, competitive adsorption, Langmuir isotherm, 286–288
Hom rate law, disinfection processes, 491–492
Horizontal flow reactors, gravity separations, 626–634
    batch and continuous flow reactors, 633–634
    circular reactors, 629–631
    rectangular reactors, 626–629
    type I vs. type II suspension, 631–633
Humic fractions, redox reactions, chlorine-organic compound interactions, 449–455
Hydration sphere, precipitation reactions, 384
Hydraulic behavior indices:
    conversion extent and, 146–147
    nonideal reactors, 62
Hydraulics, membrane system configuration and, 738–742
Hydrocerussite, lead precipitation, 418–419
Hydrodynamic conditions, gas transfer effects, 187–196
Hydrodynamic retardation, flocculation collision efficiency, short-range force model, 581–589
Hydrofluoric acid, reversible reaction kinetics, 100
Hydrogen bonding, gas/liquid equilibrium, 167–168

Hydrogen peroxide, advanced oxidation process:
    ozone interaction, 476–480
    ozonide/ultraviolet interaction, 480–481
Hydrogen sulfide/bisulfide/sulfide, redox reactions, ozone interaction, 464
Hydrophilicity:
    gas/liquid equilibrium, 167–168
    membrane systems, 735–736
    redox reactions, chlorine-organic compound interactions, 449–455
Hydrophobic effect:
    gas/liquid equilibrium, 168–169
    redox reactions, chlorine-organic compound interactions, 449–455
Hydrophobic membranes, 735–736
Hydroxide solids, precipitation, 404–409
Hydroxycarbonates, precipitation, 418–419
Hydroxyl radicals:
    advanced oxidation processes:
        Fenton-based systems, 483–486
            anodic processes, 486
            electrochemical processes, 486
            full-scale applications, 486
            heterogeneous processes, 485–486
            light-mediated processes, 485
        inorganic compound reactions, 470
        organic compound reactions, 470–476
        ozonation, 476–483
            ozone, 477–480
            peroxone, 481
            singlet oxygen radicals, 480–481
            sonolysis, 483
            ultraviolet radiation/hydrogen peroxide, 476–477
            ultraviolet radiation/semiconductor, 481–482
            wet air oxidation, 482–483
    disinfection performance, 502
    electrochemical processes, 488
Hyperfiltration. See Reverse osmosis (RO) membranes
Hypobromous acid, redox reactions, 448
    organic compound interactions, 452–455
    ozone interaction, 465
Hypochlorite anion, redox reactions, 443–455
    organic compounds, 448–455
    ozone interaction, 465
Hypochlorous acid (HOCl), redox reactions, 444–455
    bromide, 448
    organic compounds, 448–455
    ozone interaction, 465

Ideal adsorbed solution theory (IAST), competitive adsorption:
    column reactors, chromatographic effect, 361–364
    surface pressure models, 302–306
Ideal continuous flow reactors, 42–48
    continuous flow stirred tank reactors, 42
        pulse input, 45–47
        step input, 47–48
    defined, 29–30

Ideal continuous flow reactors (*Continued*)
   gravity separations, 622–639
      horizontal flow reactors, 626–634
         batch and continuous flow reactors, 633–634
         circular reactors, 629–631
         rectangular reactors, 626–629
         sedimentation, 638–639
         type I *vs.* type II suspension, 631–633
      ideal horizontal flow, 633–634
      reactor settling potential function, 623–624
      suspension/reactor separation, 622–623
      tube settlers, 634–639
      upflow reactor, 624–626
   plug flow reactors, 42–45
      fixed frame pulse input, 43–44
      moving frame pulse input, 44–45
   steady state reaction performance:
      continuous flow stirred tank reactor, 121–123
         first-order irreversible reactions, 122–123
         non-first-order irreversible reactions, 123
      irreversible reactions, CSFSTRs *vs.* PFRs, 126–129
      multiple ideal reactors, 130–135
      parallel CFSTRS and PFRs, 135
      plug flow reactor, 123–126
         Eulerian view, 124–125
         irreversible nth-order reaction, 125–126
         Lagrangian view, 125
      rate expression derivation, 135
      reversible reactions, 129–130
      continuous flow stirred tank reactors in series, 126–129
      plug flow reactors in series, 126
      single ideal reactor, 121–130
Ideal gas law, gas/liquid equilibrium, Henry's law constant and, 164–167
Incomplete gamma Hom (IgH) expression, ozone disinfectant performance, 498–500
Industrial wastes, oxidation, 437
Inertia, gravity separation, particle sedimentation, 610–612
Inertial lift, crossflow filtration systems, particle filtration and, 813–816
Influent stream:
   gas transfer column contactors, design constraints and, 236
   granular-media filtration design criteria, 709
   membrane systems, overview, 732–733
Injected gas systems, well-mixed liquid phase systems, gas transfer, 212–213
Inner sphere complexes, ion adsorption, adsorbate profile and electrical potential, 316–318
Inorganic compounds:
   particle charge, surface interactions, 522–523
   particle destabilization, 532–533
      charge neutralization, 537
   redox reactions:
      chloramine reaction, 458
      chlorine dioxide, 460–461
      free chlorine, 446–455

      ozone interaction, 464–465
      potassium permanganate, 467–469
Instantaneous demand, defined, 441
Integral method, irreversible reaction rate analysis, 89–90
Interception, particle-collector collisions, granular-media filtration, 686–687
Interfacial transport:
   adsorption equilibrium and, 267–268
      surface charge interactions, 314–319
   column contactor gas tower design, 230–233
   counter-current gas transfer, mass balance, 229
   ion adsorption, adsorbate profile and electrical potential, 315–318
   mechanistic gas transfer models:
      fluid dynamics and mass transport, 170–171
      fluid packets, gas/solution interface, 171–174
      limiting-case scenarios, 174–175
      volatile species, gas/solution interface, 171–179
   surface pressure adsorption models, surface tension and, 297–306
   well-mixed liquid phase systems, gas transfer, microscopic (interfacial) limitation, 218–219
Intermediate blocking scenario, nonsteady-state fouling, frontal and dead-end filtration systems, 787–791
Intermittent treatment procedures, activated carbon preparation, adsorption, water and wastewater treatment, 262–264
Interparticle bridging:
   flocculation collision efficiency, short-range force model, 581–589
   particle destabilization, adsorption and, 541–542
Intrinsic binding constant, ion adsorption, surface binding effects, 315
Intrinsic permeability, membrane systems, fluid/solute transport, 745–746
Intrinsic rejection factor:
   membrane system selectivity, 748
   pressure-driven membrane systems, pore-flow model, concentrate/membrane interface, 761–764
Iodide, redox reactions, ozone interaction, 465
Ion activity coefficients, precipitation, stoichiometric and equilibrium models, 400
Ion exchange adsorption:
   adsorbent regeneration, 365–366
   basic properties, 258
   competitive adsorption:
      Donnan equilibrium, 292–294
      Langmuir isotherm, 284–289
      nomenclature and conventions, 289–294
   electrically induced partitioning, 268
   electrodialysis, water flow and, 832
   surface charge interactions, 314–319
      adsorbate profile, interfacial region electrical potential, 315–318
      electrical potential effects, 314–315
      electrostatic contribution, equilibrium constants, 318–319
   surface complexation, site-binding models, 267
Ion exchange membranes, electrodialysis and, 738

Ion fluxes, electrodialysis, 825–828
Ionic mobility, electrodialysis transport, 820–821
Iron compounds:
    ion adsorption, adsorbate profile and electrical potential, 315–318
    oxidation, dark Fenton process, 483–486
    particle destabilization, 532–533
    precipitation, redox reactions, 421–422
    redox reactions:
        chlorine, 446–447
        oxidation, 436
        oxygen species, 442–444
        ozone interaction, 465
        potassium permanganate, 467–469
        reductive processes, 486–488
Iron hydroxides:
    adsorption:
        cationic sorption process, 265–266
        surface charge interactions, 314–320
    particle destabilization, precipitate-sweep flocculation, 537–541
    solid dissolution, 425–426
Irreversible fouling, defined, 733
Irreversible reaction kinetics, 88–99
    assumption of irreversibility, 105–107
    batch reactor mass balance, 89
    comparative CFSTR/PFR analysis, 126–129
    differential reaction rate analysis, 96–97
    half-time reactions, 91–93
    integral reaction rate analysis, 89–91
    multiple reactive species concentrations, 93–96
    nonpower-law rates, 97
    plug flow nth order reactions, steady state performance, 125–126
    reaction times, 97–99
    steady state performance evaluation:
        CFSTRS/PFR comparisons, 126–129
        continuous flow stirred tank reactors, 122–123
Irreversible thermodynamics, dense membrane systems modeling, 834–838
Isoelectric point,
    membrane systems, 735–736
    particle charge, 522-524
Isomorphic substitution, particle charge, 522
Isotherms. *See also specific isotherms*, e.g., Freundlich isotherm
    adsorption equilibrium, 269–296
        bidentate adsorption, 281–282
        competitive adsorption:
            Freundlich isotherms, 294–296
            ion exchange equilibrium, 284–294
            site-binding paradigm, 283–284
        distribution or partition coefficient, 282–283
        site-binding models, 270–281
            Freundlich isotherm, 276–281
            multisite Langmuir isotherms, 274–276
            non-Langmuir behavior, 273–274
            single-site Langmuir isotherm, 271–273
            surface site distribution, 270–271
        surface pressure computation, 297–302
    defined, 260
    Gibbs adsorption isotherm, 298–302
    linear isotherm, 260
    Polanyi adsorption model, 306–314
    Polanyi isotherm, 313–314
    rapid adsorption equilibrium, batch reactors, 328–331
    slow adsorption equilibrium systems, pore/surface diffusion, 342–343
Iwasaki model, granular-media filtration, 682–683

Jar test:
    granular-media filtration, breakthrough effects, 702–703
    particle destabilization, charge neutralization, 537

Kolmogorov's universal equilibrium theory:
    flocculation mechanisms, turbulence, 589
    particle destabilization chemicals, water stream mixing, 544–546

Lagrangian view, plug flow reactors:
    cumulative and exit age distribution, 43–45
    residence time distribution, dispersion, 50–55
    steady state performance evaluation, 125
Langmuir isotherm:
    adsorption equilibrium:
        bidentate adsorption, 282–283
        competitive adsorption, 283–294
            heterovalent absorption, 288–294
            homovalent adsorption, 286–288
        multisite isotherms, 274–276, 281
        non-Langmuir behavior, 273–274
        partition coefficient, 282–283
        Polanyi isotherm, 313–314
        single-site isotherm, 271–273
        surface pressure adsorption model, 298–302
    competitive adsorption, column reactors, chromatographic effect, 360–364
    mathematical properties, 260–261
    rapid adsorption equilibrium, batch reactors, 328–331
Laplace transforms:
    concentration equalization, 67–69
    differential equation solutions, 70–73
    nonideal reactors, tracer input and output, 49–50
    non-steady state reactor performance, continuous flow stirred tank reactors, 142–144
    three-dimensional mass balance, 24–25
Lead:
    precipitation of, 418–419
    redox reactions, chloramine interaction, 458–459
    solid dissolution, 425–426
Lévêque correlation, crossflow filtration systems, contaminant transport modeling, 805–816
Ligand-metal interactions. *See* Metal-ligand interactions
    light-mediated Fenton processes, hydroxyl radical oxidation, 485
Ligand-promoted dissolution, 423–426

Light blockage instruments, particle size distribution, particle treatment process, 551
Light-mediated Fenton processes, hydrogen peroxide oxidation, 485
Lime-soda ash process:
  particle destabilization chemicals, 533
    precipitate-sweep flocculation, 540–541
    precipitative carbonate softening, 409–410
      equilibrium model, 416–417
      stoichiometric model, 410–413
Limiting cases:
  crossflow filtration systems, contaminant transport modeling, 813–816
  mechanistic gas transfer models, 174–175
  reversible reaction kinetics, 103–107
    nearly complete reactions, 106–107
    very rapid/very slow equilibrium approaches, 104–105
  well-mixed liquid phase systems, gas transfer:
    kinetic expression, 216–220
    macroscopic (advective) limitation, 217–218
    microscopic (interfacial) limitation, 218–219
    rate limitations summary, 219–220
    steady state continuous flow reactors, 222–225
Limiting current, two-dimensional electrodialysis system, 831–832
Limiting flux, bulk water/contaminant transport modeling, compact layer modeling, 783
Liquid flow rate, gas transfer, steady state continuous flow reactors, 222–225
Liquid holdup, column contactor gas towers, 233–235
Liquid-in-gas systems:
  gas transfer:
    rate coefficient estimation, 192–195
    system-specific parameters, 187–196
  overview of, 159–162
Liquid-phase gas transfer coefficient, mechanistic gas transfer models, 176–179
Liquid phase mass balance:
  counter-current gas transfer, 227–229
  gas transfer, volatile compounds, acid/base reactions, 200–202
  well-mixed liquid phase systems, gas and water transfer, 210–213
Liquid-phase mass transfer coefficient, mechanistic gas transfer models, 176–179
Liquid phase resistance, gas transfer rate coefficient:
  gas phase combined with, 179–181
  gas phase compared with, 179–181
Loading rates, gas transfer column contactors, design constraints, 236–241
Logarithmic mean, dense membrane systems modeling, 835–838
Logs of inactivation, disinfection processes, 490–492
London forces:
  crossflow filtration systems, deterministic contaminant transport, 800–801
  granular-media filtration, packed medium model, 693
  particle stability, 527–530
Long-range force model, flocculation collision mechanisms, 572–581
  Brownian motion, 577–578
  design implications, 580–581
  differential sedimentation, 575–577
  fluid shear collisions, 572–575
  short-range force model vs., 587–588
  total collision frequency, 578–580

Macropore properties, activated carbon, 262–264
Macroscopic (advective) limitation:
  counter-current gas transfer, mass balance, 229
  gas transfer rate, well-mixed liquid phase systems, 217–218
Macroscopic mass balance, two-dimensional electrodialysis system, 828–829
Macroscopic models, granular-media filtration design criteria, 710
Macroscopic timescales, particle destabilization chemicals, water stream mixing, 544–546
Magnesium, precipitative softening:
  equilibrium model, 415–417
  ion concentrations and charges, 413–414
Manganese compounds:
  precipitation, 421–422
  redox reactions:
    chlorine, 446–447
    oxidation, 436
    oxygen species, 444
    ozone, 465
    potassium permanganate, 466–469
  solid dissolution, 425–426
Mass balance:
  adsorption/regeneration cycles, 366–369
  arbitrary flow directions, 20–23
    advection term, 20–21
    diffusion/dispersion terms, 21–23
    overall mass balance, 23
    storage and reaction terms, 23
  basic principles, 3–6
  defined, 3
  flow equalization, 63–66
  gas transfer processes, 157–158
    batch reactors, 213–216
    operating line, 226–229
    steady state continuous flow reactors, 221–225
    tower design equations, 229–233
    volatile species, gas/solution interface, 171–179
    well-mixed liquid phase systems, 211–213
  gas transfer systems, 207
  irreversible reactions, batch reactors, 89–90
  non-steady state performance evaluation, continuous flow reactors, 141–146
  plug flow reactors:
    Eulerian view, 43–44
    Lagrangian view, 44–45
    steady state performance evaluation, 124–125
  rapid adsorption equilibrium:
    batch reactors, 328–331

continuous flow water/adsorbent systems, 331–332
   fixed bed reactors, 335–336
slow adsorption equilibrium systems, transported-limited adsorption rates, batch reactors, 345–350
steady state performance evaluation:
   continuous flow stirred tank reactors, 121–123
   reversible reactions, 129–130
summary of equations, 25–26
three-dimensional mass balance, differential form, 24–25
unidirectional flow, 7–20
   advection term, 10–11
   chemical reaction term, 15–16
   differential one-dimensional mass balance, 18–20
   diffusion/dispersion terms, 11–15
   overall mass balance terms, 16–18
   simplified balances, 17–18
   storage term, 8–9
Mass transfer-limited flux, pressure-driven membrane systems, performance evaluation, 751–752
Mass transfer zone (MTZ):
   competitive adsorption:
      column reactors, chromatographic effect, 359–364
      slow adsorption equilibrium, 365
   fixed bed adsorbers, 354–356
      performance evaluation, 357–359
      single-bed design, 367–369
   rapid adsorption equilibrium systems, fixed bed reactors, 334–340
   transport-limited adsorption, fixed bed reactors, 350–354
Mass transport:
   gas-in-liquid systems, dimensionless parameters, 189–192
   gas transfer models, 170–171
   liquid-in-gas systems, 192–195
Maximum mixedness models, nonideal reactors, 61
Maxwell–Boltzman distribution, flocculation collision, Brownian motion, 577–578
Mean hydraulic retention time, continuous flow reactors, residence time distributions, 37–42
Mechanistic gas transfer models, 170–179
   fluid dynamics and interfacial mass transport, 170–171
   fluid packet interfacial transfer, 171–174
   limiting case residence times, 174–175
   mass balance, volatile gas/solution interface, 171–179
   packet age and residence time distribution, 175
   transfer coefficient, 175–179
Media size, granular-media filtration design, 706–707
Membrane elements, configuration and hydraulics, 738–742
Membrane fouling:
   crossflow filtration systems, nonsteady-state patterns, 814–816
   defined, 733
   pressure-driven membrane systems:
      concentration polarization and, 784–786
      nonsteady-state fouling, 787–791
      performance evaluation, 750–752
Membrane modules, membrane system configuration and, 738–742

Membrane systems:
   basic properties and applications, 731–732
   chemical composition, 735–736
   configuration, 738–742
      array configuration, 741–742
      element configuration, 738–741
   crossflow filtration, 792–816
      compact layer flux/TMP interactions, 808
      concentrate pressure profile, 795–798
      contaminant transport mechanisms, 798–802
         back-transport mechanisms, 802
         Brownian and shear-induced diffusion, 798–800
         deterministic transport mechanisms, 800–802
      contaminant transport modeling, 802–816
      CP layer thickness, 804–805
      film and film-gel models, 807–808
      film-gel model, 805–807
      film model without compact layer, 808–809
      fluid flow:
         modeling parameters, 792–793
         processes, 793–795
      flux modeling, 802–816
         TMP, no compact layer, 809–811
      inertial particle lift, 813–814
      modeling assumptions, 803
      modeling equations, 814–816
      nonsteady-state fouling patterns, 814–816
      particle transport modeling, 792–793
      summary of modeling approaches, 815–816
   dense membrane systems, irreversible thermodynamics, 834–838
   driving forces, 736–738
      concentration differences and electrical forces, 738
      energy characteristics, 752–754
      osmotic pressure, 754–757
      pressure-driven processes, 736–738
      quantification, 752–759
      transport magnitudes, 757–759
   electrodialysis, 816–834
      basic principles and applications, 816–819
      ion-exchange membranes, water flow, 832
      modeling parameters, 822
         ion fluxes, current density and electrical potential difference, 825–828
         single cell-pair transport, 822–825
      multicomponent systems, 832–833
      nonideal membrane behavior, 832
      overlimiting current, 833
      potential drop sources, 833
      real system complications, 832–833
      transport systems, electrical potential gradient, 819–822
         advective transport, 819
         current densities, 820–828
         diffusion transport, 819–820
         solution in electric field, 820
         two-dimensional systems, 828–832

Membrane systems (*Continued*)
　　design issues, 830–831
　　　limiting current, 831–832
　　　macroscopic mass balance, 828–829
　　　microscopic mass balance, 829–830
　　fluid/solute transport parameters, 743–747
　　　flux velocity, 743–744
　　　permeability, 745–746
　　　recovery, 743
　　　resistance, 744–745
　　　specific flux and permeance, 744
　　　temperature effects, water permeation, 746–747
　　location, concentration, and pressure parameters, 742–743
　　organic membrane characteristics, 735–736
　　pressure-driven systems:
　　　bulk solutions, water/contaminant transport modeling, 773–792
　　　　boundary layer and concentration polarization, 774–777
　　　　compact layer:
　　　　　flux paradox, 784
　　　　　formation, 782
　　　　　hydraulic resistance, 783–784
　　　　concentration polarization and precipitative fouling, 784–786
　　　　contaminant physicochemistry, 773–774
　　　　CP layer:
　　　　　parameter profiles, 786–787
　　　　　permeation effects and resistance, 781–782
　　　　flux effects:
　　　　　CP layer, particle/solute comparisons, 782
　　　　　rejected particles, 778–780
　　　　　rejected solutes, 780–781
　　　　frontal filtration systems:
　　　　　nonsteady-state fouling, 787–791
　　　　　performance evaluation, 792
　　　　　permeation/TMP relationship, 777–782
　　　　limiting flux, film-gel model, 783
　　　　nonsteady-state fouling, frontal and dead-end filtration, 787–791
　　　　　modified fouling and silt density indices, 790–791
　　　　rejected contaminant, coupling force on permeate flow, 778
　　performance evaluation, 750–752
　　quantitative modeling, 759–773
　　　pore-flow model, 760–766
　　　　contaminant rejection, solution permeation, 764–766
　　　　predicted flux/transmembrane parameter comparisons, 771–773
　　　　solution composition, concentration/membrane interface, 760–764
　　　　solution permeation, 764
　　　predicted flux/transmembrane parameter comparisons, 771–773
　　　solution-diffusion model, 766–771
　　　　membrane permeation and concentration profile, 767–771
　　　　predicted flux/transmembrane parameter comparisons, 771–773
　　　　transmembrane pressure profile, 766–767
　　　transport conceptual models, 759–760
　　　transport summary, 772–773
　　schematic representation, 732
　　selectivity parameters, 747–749
　　　challenge tests and molecular weight cutoff, 748–749
　　　rejection, 748
　　　separation coefficient, 749
　　structure, composition and water interactions, 734–736
　　system operations, 732–733
"Merry-go-round" adsorption systems, design options, 368–369
Metal carbonates, precipitation, 418–419
Metal hydroxides:
　　adsorption reactors, 266
　　particle destabilization, precipitate-sweep flocculation, 538–541
　　precipitation, speciation and pH relationship, 404–409
Metal-ligand interactions, precipitation:
　　carbonate solids precipitation, 409–419
　　complexing ligands, metal solubility, 421
　　hydroxide solids precipitation, 404–409
　　ionic strength and activity coefficients, 400
　　metal and hydroxy carbonates, 418–419
　　pH-dependent metal and ligand speciation, 420–421
　　precipitative softening, 409–417
　　redox reactions, 421–422
　　soft water recarbonization, 417–418
　　stoichiometric and equilibrium models, 397–422
Metal oxides:
　　adsorption process:
　　　cationic sorption process, 265–266
　　　natural organic matter, drinking water coagulation, 264–265
Methanol, advanced oxidation process, hydroxyl radicals, 471–476
2-Methylisoborneol (MIB), redox reactions, chlorine dioxide interaction, 461
Microfiltration (MF) membranes:
　　characteristics, 736–737
　　classification, 736
　　driving force and energy in, 752–754
　　packing density, 740–742
　　pore-flow model, 760
　　pressure-driven systems, overview, 749–752
　　transmembrane energy difference, 758–759
Microorganism inactivation kinetics, ozone disinfectant performance, 499–500
Micropore properties, activated carbon preparation, adsorption, water and wastewater treatment, 262–264

Microscopic (interfacial) limitation:
  counter-current gas transfer, mass balance, 229
  well-mixed liquid phase systems, gas transfer, 218–219
Microscopic mass balance, two-dimensional electrodialysis system, 829–830
Microscopic timescales, particle destabilization chemicals, water stream mixing, 544–546
Mixed flow effects, nonideal flow reactors, gravity separations, 642–644
Mixed ion exchange system, competitive adsorption, Langmuir isotherm, 286–294
Mixing patterns:
  particle destabilization, chemicals in water stream, 544–546
  plug flow reactors, early vs. late mixing, 58–61
  reactor performance evaluation, CFSTR/PFR comparisons, 128–129
Modified fouling index (MFI), nonsteady-state fouling, frontal and dead-end filtration systems, 790–791
Molar adsorption potential, Polanyi adsorption model, 306–314
Molar Gibbs energy:
  membrane systems:
    driving force and, 753–757
    transmembrane energy difference, 758–759
  pressure-driven membrane systems:
    pore-flow model, solution permeation, 764
    solution-diffusion model, 766–771
Molecular collisions:
  energetics, 84–87
  frequency, 84
Molecular-scale adsorption models, 266–268
  electrically induced partitioning, ion adsorption, 268
  interface-adsorption equilibrium, 267–268
  phase transfer reaction, 267–268
  surface complexation reaction, 267
Molecular weight cutoff (MWCO), membrane system selectivity, 749
Monochloramine:
  disinfectant performance, 495–497
  irreversible reaction rate analysis, 90–91
Monodisperse suspension, type I batch sedimentation, 612–614
Morrill index, hydraulic behavior, 62
Moving frame of reference. See Lagrangian view
Multicomponent electrodialysis systems, 832–833
Multiple ideal reactors, steady state performance evaluation, 130–135
Multiple reactive species, reaction rate analysis, 93–96
Multisite isotherms, adsorption equilibrium, Langmuir/Freundlich comparisons, 281

Nanofiltration (NF) membranes:
  cascade/staged configuration, 742
  characteristics, 737
  classification, 736
  conceptual models, 760
  driving force and energy, 752–754
  molecular weight cutoff, 749

  packing density, 740–742
  pressure-driven systems, overview, 749–752
  pressure vessels, 738
  transmembrane energy difference, 758–759
Natural organic matter (NOM):
  adsorption process:
    basic principles, 257
    competitive adsorption, ideal adsorption model, 306
    drinking water coagulation, 264–265
    particle charge, 523–524
    water and wastewater treatment systems, activated carbon, 263–264
  advanced oxidation process, hydrogen peroxide addition, 479–480
  competitive adsorption, site-binding models, Freundlich isotherm, 295–296
  disinfection byproducts, 489
  membrane systems:
    structure and composition, 735–736
    transmembrane energy difference, 758–759
  particle destabilization:
    charge neutralization, 537
    chemicals used for, 532–535
    soluble materials interactions, 542–544
  pressure-driven membrane systems, performance evaluation, 750–752
  reaction kinetics, 87–88
  redox reactions:
    chlorine-organic compound interactions, 449–455
    oxidation, 436
    ozone, 466
  slow adsorption equilibrium systems, 343
    competitive adsorption, 365
Nernst equation, redox processes, thermodynamics, 439–441
Nernst–Planck equation, electrodialysis transport kinetics, 821–822
Nitrites, redox reactions, ozone interaction, 465
N-nitrosodimethylamine (NDMA), redox reactions, chloramine reaction, 458
Nonelementary reactions:
  kinetics of, 87–88
  temperature dependence, 114–115
Nonequivalent series-based continuous flow stirred tank reactors, 62
Non-first-order irreversible reactions, continuous flow stirred tank reactors, 123
Nonhumic fractions, redox reactions, chlorine-organic compound interactions, 449–455
Nonideal flow reactors:
  defined, 29–30
  gravity separations, sedimentation effects, 639–644
    channel flow, 640–642
    mixed flow, 642–644
    tiered flow, 639–640
  non-steady state conditions, performance under, 144–146
  performance evaluation, 135–141

Nonideal flow reactors (*Continued*)
Nonideal membrane behavior, electrodialysis, 832
Nonideal reactors, hydraulic characteristics, 48–62
    hydraulic behavior indices, 62
    residence time distributions, 50–62
        continuous flow stirred tank reactors in series, 55–57
        nonequivalent continuous flow stirred tank reactors in series, 62
        plug flow reactors in parallel and series, 58–61
        plug flow reactors with dispersion, 50–55
        short-circuiting and dead space modeling, 57–58
Nonporous membranes, structure and composition, 734–736
Nonpower-law rates, irreversible reaction analysis, 97
Nonspecific adsorption, defined, 259
Non-steady-state conditions:
    continuous flow reactor performance, 141–146
        continuous flow stirred tank reactors, 142–144
        conversion extent, 144–146
        plug flow reactor conversion, 141–142
    membrane fouling:
        crossflow filtration systems, 814–816
        frontal and dead-end filtration systems, 787–791
    nonideal flow reactor performance, 144–146
    slow adsorption equilibrium systems, transported-limited adsorption rates, batch reactors, 346–350
Normalized filter run production, granular-media filtration, 682
nth-order reactions:
    continuous flow stirred tank reactors, steady state performance, 123–124
    plug flow reactors, steady state performance evaluation, 125–126
Nucleation, particle growth and formation, precipitation and, 381, 384–394
    thermodynamics, 385–389
Number distribution, particle size distribution, particle treatment process, 548–551
Number of (gas) transfer units (NTU), column contactor gas tower design, 232–233
    constraints in, 242–243

Odor reduction, oxidation, 436
Off-gas sampling, gas transfer, steady state systems, overall gas transfer rate coefficient, 245–246
Ohm's law, electrodialysis transport, 820–821
Olefinic compounds, redox reactions, ozone, 465–466
Onda correlations:
    gas transfer column contactors:
        design constraints, 236–241
        design feasibility criteria, 236–240
    liquid-in-gas systems, mass transfer coefficients, 193–195
One-dimensional mass balance, differential form, unidirectional flow systems, 18–20
Open-boundary reactors, residence time distributions, mean hydraulic retention time, 38–42
Operating lines:
    counter-current gas transfer, mass balance, 226–229
    gravity separations, continuous flow thickening, 651–655

Organic compounds. *See also* Synthetic organic compounds (SOCs)
    advanced oxidation process, hydroxyl radicals, 470–476
    particle destabilization chemicals, 533–535
    redox reactions:
        chloramine, 458
        chlorine, 448–455
        chlorine dioxide, 461
        ozone, 465–466
        potassium permanganate, 469
Orthokinetic flocculation, collision mechanisms, 572–575
Orthophosphate inhibitors, lead precipitation, 419
Osmotic pressure, membrane systems, 754–757
Ostwald ripening, precipitation, particle growth and formation thermodynamics, 388–389
Outer sphere complexes, ion adsorption, adsorbate profile and electrical potential, 316–318
Outwardly directed unit normal, mass balance equations, arbitrary flow systems, 21
Overall gas transfer rate coefficient (K_L):
    gas and liquid phase resistance:
        combined resistance, 179–181
        comparisons of, 181–183
    gas-in-liquid systems, biological treatment applications, 243–246
    well-mixed liquid phase systems, gas transfer, 211–213
        batch reactors, 214–216
        limiting cases, kinetic expression, 217–220
        macroscopic (advective) limitation, 217–218
        microscopic (interfacial) limitation, 218–219
        rate limitation summary, 219–220
Overall mass balance equation, arbitrary flow systems, 23
Overall mass rejection coefficient, membrane system selectivity, 748
Overall rejection factor, membrane system selectivity, 748
Overflow rate,
    horizontal flow reactor, gravity separations, 626–631
    nonideal flow reactors, gravity separations, 638–644
    tube settlers, gravity separations, 634–638
    upflow reactor, gravity separations, 624–626
Overlimiting current, electrodialysis systems, 833
Oxidant concentrations. *See also specific oxidants*, e.g., Oxygen
    redox processes, terminology, 441
Oxidation, redox processes, 436–437
Oxygen, redox processes, 441–444
Oxygen values, gas-in-liquid systems, mass transfer coefficients, 189–192
Ozone:
    advanced oxidation processes, hydroxyl radicals, 476–483
        ozone, 477–480
        peroxone, 481
        singlet oxygen radicals, 480–481
        sonolysis, 483
        ultraviolet radiation/hydrogen peroxide, 476–477
        ultraviolet radiation/semiconductor, 481–482
        wet air oxidation, 482–483
    contactor types, generator systems, 463
    disinfectant performance, 498–500

redox reactions, 461–466
  generation, 462–464
  inorganic compounds, 465–466
  organic compounds, 464–465
 taste and odor reduction, 436
Ozonide radicals, advanced oxidation process:
 hydrogen peroxide addition, 478–480
 ultraviolet radiation and, 480–481

Packed bed systems. *See also* Column contactors; Counter-current gas transfer
 adsorption:
  activated carbon preparation, 263–264
  series-based "merry-go-round" systems, 368–369
Packed medium model, granular-media filtration, 690–694
Packet age, mechanistic gas transfer, residence time distributions and, 175
Packing density:
 membrane system configuration, 739–742
 rapid adsorption equilibrium, fixed bed reactors, 336
Packing media, gas transfer column contactors, design feasibility and, 237–240
Palladium catalysts, reductive processes, iron-based systems, 488
Parallel-based continuous stirred-tank reactors, steady state reaction performance, 135
Parallel packed bed adsorption systems, design options, 369
Parallel plug flow reactors:
 segregated flow and early *vs.* late mixing, 58–61
 steady state performance, 135
Partial surface pressure, competitive adsorption, ideal adsorption model, 302–306
Particle-collector collisions, granular-media filtration, 686–690
 Brownian motion, 688
 efficiency, 689–690
 interception, 686–687
 sedimentation, 687–688
Particle density, particle treatment process, 552
Particle formation and growth, precipitation, 380–381
 dynamics, 384–394
 nucleation mechanics, 384–394
 size distribution, 389–394
 thermodynamics, 383–389
Particle sedimentation
 crossflow membrane filtration systems, inertial lift and, 813–816
 gravity separation, 607–612
  inertial effects, 610–612
  Stokes' law, 607–610
 particle-collector collisions, granular-media filtration, 687–688
Particle shape, particle treatment process, 551–552
Particle size distribution:
 flocculation, 564–565
 granular-media filtration, 689–690
 gravity separation, batch setting, 609, 614–616
 gravity separation, continuous flow settling, 630–632
 particle stability, Van der Waals attractions, 529–530

 particle treatment process, 546–551
 precipitation reactors, 389–394
 pressure-driven membrane systems, pore-flow models, 763–764
Particle stability. *See also* Destabilization
 particle treatment process, 521–532
  charge and potential measurements, 531–532
  charged particle interaction, 525–526
  diffuse layer characteristics, 524–525
  particle charge, 522–524
   isomorphic substitution, 522
   surface adsorption, 523–524
   surface chemical reactions, 522–523
  particle-flat plate interactions, 530–531
  Van der Waals attraction, 526–530
Particle transport, crossflow filtration systems, 792–795
Particle treatment process:
 applications, 519–521
 destabilization chemicals, 532–535
  inorganic compounds, 532–533
  organic polymers, 533–535
  soluble materials interaction, 542–544
   additive combinations, 543–544
   water stream mixing, 544–546
 destabilization process, 535–542
  adsorption:
   charge neutralization, 536–537
   interparticle bridging, 541–542
  diffuse layer compression, 535–536
  precipitate-sweep flocculation enmeshment, 537–541
 flocs fractal characteristics, 552–553
 gravity separations, flotation, 657–658
 particle density, 552
 particle shape, 551–552
 particle size distributions, 546–551
  measurements, 550–551
 particle stability, 521–532
  charge and potential measurements, 531–532
  charged particle interaction, 525–526
  diffuse layer characteristics, 524–525
  particle charge, 522–524
   isomorphic substitution, 522
   surface adsorption, 523–524
   surface chemical reactions, 522–523
  particle-flat plate interactions, 530–531
  Van der Waals attraction, 526–530
Partition coefficients:
 adsorption equilibrium, 269–270
  isotherm properties, 282–283
 gas/liquid equilibrium, Henry's law and, 162–164
Peclet number:
 plug flow reactors, residence time distribution, 51–55
 pressure-driven membrane systems, pore-flow model, concentrate/membrane interface, 762–764
Penetration gas transfer model, residence time distribution, 171–174

Perikinetic flocculation, collision mechanisms, 577–578
Permeability, membrane systems, fluid/solute transport, 745–746
Permeability coefficient, membrane systems, fluid/solute transport, 745–746
Permeance, membrane systems, fluid/solute transport, 744
Permeation rate/velocity:
  frontal filtration, CP layer resistance, permeation effects, 781–782
  membrane systems:
    fluid/solute transport, 743–744
    temperature effects, water permeation, 746–747
  pressure-driven membrane systems:
    pore-flow models, solution permeation, 764
    solution-diffusion model, 767–771
    TMP and, frontal filtration systems, 777–782
pH-adsorption edge, iron and aluminum oxides, 265–266
Phase transfer adsorption:
  equilibrium constants, 269
  ionic adsorbates, 319–320
  solid/solution interface, 267–268
Phenolic compounds, redox reactions:
  chlorine dioxide interaction, 461
  chlorine-organic compound interactions, 449–451
  ozone, 466
Phenomenological equations and coefficients, dense membrane systems modeling, 835–838
pH levels:
  chloramine disinfectants, 496
  particle charge, surface interactions, 522–523
  particle destabilization, precipitate-sweep flocculation, 538–541
  precipitation:
    metal hydroxides, 404–409
    metal-ligand speciation, 420–421
  redox reactions, chloramine formation, 456
Photocatalysis, advanced oxidation process, ultraviolet radiation, 482
Photolysis, advanced oxidation process, hydrogen peroxide addition, 476–480
Physical adsorption, defined, 259
Physicochemical environment:
  Henry's constants, standard state conditions, 198
  real gas transfer conditions, 198–199
Plate-and-frame membrane systems, 738–739
Platinum catalysts, reductive processes, iron-based systems, 488
Plug flow reactors. *See also* Parallel plug flow reactors; Series plug flow reactors
  aqueous phase concentration, limiting cases, 219–220
  gas transfer, steady state systems, 220
    aqueous-phase reactions, 222–223
  gravity separations, horizontal flow, 626–634
  hydraulic characteristics, 42–45
  mass balance equations, diffusion and dispersion, 15, 19
  non-steady state performance:
    conversion extent, 141–142
    nonideal reactors, 145–146
  one-dimensional mass balance equations, 19–20
  parallel and series models, 58–61

  pulse input:
    Eulerian view, 43–44
    Lagrangian view, 44–45
  rapid adsorption equilibrium, 336–340
  residence time distribution, dispersion, 50–55
  slow adsorption equilibrium systems:
    fixed bed systems, 350–354
    transport-limited adsorption rates, 349–350
  steady state performance:
    irreversible reactions, 126–129
      CFSTR comparisons, 126–129
    reversible reactions, 129–130
  steady state performance evaluation, 123–126
    Eulerian view, 124–125
    irreversible nth-order reactions, 125–126
    Lagrangian view, 125
Poiseuille flow, pressure-driven membrane systems, pore-flow model, concentrate/membrane interface, 761–764
Polanyi adsorption model, phase transfer, 267–268
Polanyi adsorption model and isotherm, 306–314
  basic principles, 306–313
  linear, Langmuir and Freundlich isotherms, 313–314
Polarizability, particle stability, Van der Waals attractions, 527–530
Polarization factor/modulus (PF/PM), membrane system selectivity, 747–748
Polyaluminum chloride (PACl), particle destabilization chemicals, 533
Polycyclic aromatic hydrocarbons, redox reactions, ozone, 466
Polydiallyldimethyl ammonium chloride, particle destabilization chemicals, 533–535
Polymer compounds, particle destabilization chemicals, 533–535
  adsorption and interparticle bridging, 542
Population balance, flocculation modeling, Smoluchowski equation, 568–570
Pore condensation, phase transfer adsorption, ionic adsorbates, 319–320
Pore diffusion:
  adsorption rate, 328
  slow adsorption equilibrium systems, adsorbent particles, 341–343
Pore-flow model, pressure-driven membrane systems, 760–766
  contaminant rejection, solution permeation, 764–766
  flux predictions, 771–772
  predicted flux/transmembrane parameter comparisons, 771–773
  solution composition, concentration/membrane interface, 760–764
  solution permeation, 764
  transport parameters, 759–760
Pore/surface diffusion model (PSDM), slow adsorption equilibrium systems, 343
Porous membranes:
  pressure-driven membrane systems, performance evaluation, 751–752
  structure and composition, 734–736
Postchlorination, chlorine disinfectant performance, 494–495

Potential drop profiles:
    electrodialysis systems, 833
    electrodialysis transport kinetics, 825–828
Potential energy:
    charged particle interaction, 526
        flat plate, 530–531
    Polanyi adsorption model, 306–314
Powdered activated carbon (PAC):
    activated carbon preparation, adsorption, water and wastewater treatment, 262–264
    slow adsorption equilibrium systems, transport-limited adsorption rates, batch reactors, 343, 349–350
Power law, nonelementary reactions, 88
Prechlorination, chlorine disinfectant performance, 494–495
Precipitate-sweep flocculation, particle destabilization, 537–541
Precipitation:
    basic principles, 380–384
    particle formation and growth, 380–381
        dynamics, 384–394
        size distribution, 389–394
        thermodynamics, 383–389
    reactor design criteria, 426–427
    solid constituents, soluble speciation, 395–396
    solid/liquid equilibrium, 382–384
    solubility product quantitation, 395
    solute transport, surface reactions and reversibility, 381–382
    solution composition modeling, 394–397
    stoichiometric and equilibrium models, 397–422
        carbonate solids precipitation, 409–419
        complexing ligands, metal solubility, 421
        hydroxide solids precipitation, 404–409
        ionic strength and activity coefficients, 400
        metal and hydroxy carbonates, 418–419
        pH-dependent metal and ligand speciation, 420–421
        precipitative softening, 409–417
        redox reactions, 421–422
        soft water recarbonization, 417–418
Precipitative fouling, pressure-driven membrane systems, concentration polarization and, 784–786
Precipitative softening, carbonate solid precipitation, 409–410
Precursors, disinfection byproducts, redox reactions, chlorine-organic compound interactions, 449–455
Predominance area diagrams, redox reactions, chlorine species, 443–455
Pressure differentials, pressure-driven membrane systems, pore-flow model, permeation and contaminant rejection, 765
Pressure-driven membranes, classification, 736
Pressure-driven membrane systems:
    bulk solutions, water/contaminant transport modeling, 773–792
        boundary layer and concentration polarization, 774–777
        compact layer:
            flux paradox, 784
            formation, 782
            hydraulic resistance, 783–784
            concentration polarization and precipitative fouling, 784–786
        CP layer:
            parameter profiles, 786–787
            permeation effects and resistance, 781–782
        flux effects:
            CP layer, particle/solute comparisons, 782
            rejected particles, 778–780
            rejected solutes, 780–781
        frontal filtration systems:
            nonsteady-state fouling, 787–791
            performance evaluation, 792
            permeation/TMP relationship, 777–782
        limiting flux, film-gel model, 783
        nonsteady-state fouling, frontal and dead-end filtration, 787–791
            modified fouling and silt density indices, 790–791
        rejected contaminant, coupling force on permeate flow, 778
    overview, 749–752
    performance evaluation, 750–752
    quantitative modeling, 759–773
        pore-flow model, 760–766
            contaminant rejection, solution permeation, 764–766
            predicted flux/transmembrane parameter comparisons, 771–773
            solution composition, concentrate/membrane interface, 760–764
            solution permeation, 764
        predicted flux/transmembrane parameter comparisons, 771–773
        solution-diffusion model, 766–771
            membrane permeation and concentration profile, 767–771
            predicted flux/transmembrane parameter comparisons, 771–773
            transmembrane pressure profile, 766–767
        transport conceptual models, 759–760
    transport summary, 772–773
Pressure filters, granular-media filtration, 679–680
Pressure gradient:
    column contactor gas towers:
        design criteria and, 239–240
        measurements, 233–235
    crossflow filtration systems, concentrate profile, 795–798
    gas transfer, well-mixed liquid phase systems, 208–213
    membrane systems, parameters, 742–743
Pressure-limited flux, pressure-driven membrane systems, performance evaluation, 751–752
Pressure vessels, membrane systems, 738
Primary minimum (energy curve), particle stability, Van der Waals attractions, 528–530
Primary nucleation, particle growth and formation, precipitation and, 381
Probability density function (PDF), residence time distribution, continuous flow reactors, 31
Probability distribution statistics, continuous flow reactors, residence time distributions, 37–42

Proportional diffusivity, slow adsorption equilibrium systems, 343
Proton-promoted dissolution, 423–426
Proton-promoted reductive dissolution, solid dissolution, 425–426
Pseudo-first-order rate constant:
    disinfection processes, 488–489
    well-mixed liquid phase systems, gas transfer, 217–220
Pseudo-Henry's constant, gas transfer, acid/base reactions, 200–202
Pulse input:
    continuous flow reactors, residence time distributions:
        response analysis, 33–35
        tracer studies, 32–33, 36–37
    continuous flow stirred tank reactors, 45–47
        Laplace transform, 71–73
    plug flow reactors:
        Eulerian view, 43–44
        Lagrangian view, 44–45
Pyrene, redox reactions, ozone, 466
Pyridines, redox reactions, ozone, 466
Pyrolysis, activated carbon preparation, adsorption, water and wastewater treatment, 262–264

Quantitative structure-activity relationships (QSARs), gas/liquid equilibrium, 168–169

Raoult's law:
    gas/liquid equilibrium, 156–157, 164
        volatility parameters, 168–169
    reference states, activity coefficient and, 198–199
Rapid adsorption equilibrium systems, 328–340
    batch reactors, 328–331
    competitive adsorption, 359–364
    continuous flow reactors, 331–332
    fixed bed reactors, 333–340
        equation summary, 340
        mass balance on, 335–336
        plug flow reactors, 336–340
        qualitative analysis, 333–335
    sequential batch reactors, 332–333
Rapid small-scale column tests (RSSCTs), fixed bed reactor performance, 357–359, 370–371
Rate constants. *See also* Reaction kinetics
    advanced oxidation process, hydroxyl radical/ozone oxidation, 480
    floc formulation, 567
    redox reactions, oxygen-iron species, 443–444
Rate-controlling process, sequential reaction kinetics, 108–111
Rate equation, floc formation, 567
Rate expressions, ideal continuous-flow reactors, steady state reaction performance, 135
Rate of encounters, molecular collisions, 84
Reaction kinetics:
    elementary reactions, 84–87
        molecular collisions:
            energetics, 84–87
            frequency, 84
        flocculation modeling, 570–571

gas transfer processes, 157
irreversible reactions, 88–99
    batch reactor mass balance, 89
    differential reaction rate analysis, 96–97
    half-time, 91–93
    integral reaction rate analysis, 89–91
    nonpower-law rates, 97
    reaction times, 97–99
mass balance equation, 81–82
nonelementary reactions, 87–88, 114–115
reversible reactions, 99–107
    disequilibrium, 102–103
    limiting case rate expressions, 104–107
    nearly complete reactions, 106–107
    reaction quotients, irreversibility assumption, 105–106
    times and limiting cases characteristics, 103–104
    unidirectional/bidirectional reactions, 102–103
    very slow/very rapid equilibrium, 104–105
sequential reactions, 107–114
    rate-controlling step, 108–111
    steady state and chemical equilibrium, 112–114
    thermodynamics, 111–112
temperature dependence, nonelementary reactions, 114–115
terminology, 82–83
Reaction mechanism/pathway, defined, 82
Reaction quotients, reversible reaction kinetics, 105–107
Reaction rate constants, defined, 83
Reaction term, mass balance equations, arbitrary flow systems, 23
Reactor systems, precipitation reactions, 426–427
Reagent dose estimation:
    carbonate solid precipitation, stoichiometric model, 410–413
    precipitation, stoichiometric and equilibrium models, 402
Recarbonation, softened water, carbonate precipitation, 417–418
Rechlorination, chlorine disinfectant performance, 494–495
Recovery, membrane systems, fluid/solute transport, 743
Rectangular reactors, gravity separations, 626–628
Redox reactions. *See also* Advanced oxidation processes
    basic principles and overview, 435–441
    chloramines, 455–459
        formation, 455–456
        inorganic compounds, 458
        organic compounds, 458
        water distribution systems, chlorine/lead release and, 458–459
    chlorine dioxide, 459–461
        generation, 460
        inorganic compounds, 460–461
        organic compounds, 461
    chlorine species, 444–455
        bromide reactions, 448
        inorganic compounds, free chlorine reactions, 446–453
        iron and manganese reactions, 446–447
        organic compound reactions, 448–455
    kinetics, 441
    oxidant concentration terminology, 441
    oxygen species, 441–444
    ozone, 461–466

generation, 462–464
  inorganic compounds, 465–466
  organic compounds, 464–465
 potassium permanganate, 466–469
  ferrous/manganous species reactions, 467–469
  generation, 467
  organic compounds, 469
 precipitation, 421–422
 reduction, 437
 reductive processes:
  iron-based systems, 487–488
  sulfur-based systems, 486–487
 thermodynamics, 437–441
 water/wastewater treatment, 435
Reductive dissolution, 423–426
Reductive processes, redox reactions, 437
 iron-based systems, 487–488
 sulfur-based systems, 486–487
Reference states, gas transfer, activity coefficient and, 198–199
Reflection coefficient:
 dense membrane systems modeling, 837–838
 pressure-driven membrane systems, pore-flow model, permeation and contaminant rejection, 765
Regeneration analysis, adsorbent regeneration, 365–366
Rejection:
 membrane system selectivity, 748
 pressure-driven membrane systems, pore-flow model, concentrate/membrane interface, 761–764
  permeation and contaminant rejection, 764–765
Repulsion
 charged particle interaction, 525–526
 particle stability, Van der Waals attractions, 527–530
Residence time distributions (RTDs):
 continuous flow reactors, 30–42
  mean hydraulic detention time, 37–42
  plug flow reactors, 42–45
  probability distribution statistics, 37–42
  pulse input response, 33–35
  step input response, 35–37
  tracer studies, 31–33
 defined, 29–30
 disinfection processes, 493–494
 gas transfer, steady state systems, overall gas transfer rate coefficient, 244–246
 mechanistic gas transfer:
  fluid packets, gas/solution interface, 171–174
  packet age and, 175
 nonideal reactors, 50–62
  continuous flow stirred tank reactors in series, 55–57
  nonequivalent continuous flow stirred tank reactors in series, 62
  short-circuiting and dead space modeling, 57–58
 plug flow reactors:
  with dispersion, 50–55
  in parallel and series, 58–61
 series-based continuous flow stirred tank reactors, steady state performance evaluation, 132–134

slow adsorption equilibrium systems, transport-limited adsorption rates, batch reactors, 349–350
Residuals concentration, chloramine disinfectant performance, 496–497
Resistance, membrane systems:
 compact layer modeling, 783–784
 fluid/solute transport, 744–745
 frontal filtration, CP layer resistance, permeation effects, 781–782
Resistances-in-series model, CP layer resistance, permeation effects, 781–782
Restabilization, particle destabilization, charge neutralization, 537
Retentate, membrane systems, 732–733
Reverse osmosis (RO) membranes:
 cascade/staged configuration, 742
 characteristics, 737
 classification, 736
 driving force and energy, 752–754
 molecular weight cutoff, 749
 packing density, 740–742
 pH levels and, 735–736
 pressure-driven systems, overview, 749–752
 pressure vessels, 738
 solution-diffusion model, 760
 transmembrane energy difference, 757–759
Reverse rate constant, reversible reactions, reactor performance evaluation, 129–130
Reversible reaction kinetics, 99–107
 disequilibrium, 102–103
 limiting case rate expressions, 104–107
 nearly complete reactions, 106–107
 reaction quotients, irreversibility assumption, 105–106
 steady state performance evaluation, 129–130
 times and limiting cases characteristics, 103–104
 unidirectional/bidirectional reactions, 102–103
 very slow/very rapid equilibrium, 104–105
Reynolds number:
 crossflow filtration systems:
  contaminant transport layer modeling, 804–816
  pressure profile, 795–798
 gravity separation, inertial effects, 610–612
 liquid-in-gas systems, mass transfer coefficients, 193–195
Ripening, granular-media filtration:
 filtration run process, 681
 Mackie model, 713
 O'Melia–Ali model, 710–712
 Tien model, 712–713
 time-related effects, 701
Rise velocity, gas-in-liquid systems, 188–192

Salt effects, gas/liquid equilibrium, 169
Saturators, gravity separations, flotation, 657–660
Scavenger compounds, advanced oxidation process, hydrogen peroxide addition, 479–480
Schmidt number, crossflow filtration systems, contaminant transport modeling, 804–816

Schulze–Hardy rule, particle destabilization process, diffuse layer compression, 535–536
Secondary minimum (energy curve), particle stability, Van der Waals attractions, 528–530
Secondary nucleation, particle growth and formation, precipitation and, 381
Second-order decay reaction:
　CFSTR/PFR comparisons, 126–129
　　series continuous flow stirred tank reactors, steady state performance evaluation, 132–134
Second-order reaction rate constants, redox reactions:
　chlorine dioxide, 460–461
　ozone, 466
Segregated flow systems, plug flow reactors in parallel and series, 58–61
Selectivity parameters, membrane systems, 747–749
　challenge tests and molecular weight cutoff, 748–749
　polarization, 747–748
　rejection, 748
　separation coefficient, 749
Semicontinous site surfaces, Freundlich isotherm, 276–281
Sensing zone measurements, particle size distribution, particle treatment process, 550–551
Separation coefficient, membrane systems, 749
Separation factor:
　competitive adsorption, ion exchange reactions, 289–294
　membrane systems, 749
Separation zone, gravity separations:
　flotation, 657
　modeling, 667–668
Sequential reaction kinetics, 107–114
　rapid adsorption equilibrium, sequential batch reactors, 332–333
　rate-controlling step, 108–111
　steady state and chemical equilibrium, 112–114
　thermodynamics, 111–112
Sequential reactions, defined, 87–88
Series-based continuous flow stirred tank reactors:
　gas transfer reactions, 225
　nonequivalent CFSTRs, 62
　steady state performance evaluation, 130–134
Series-based packed bed adsorption systems, design options, 368–369
Series-based plug flow reactors:
　segregated flow and early vs. late mixing, 58–61
　steady state performance evaluation, 130
Series reactions. See Sequential reactions
Settling potential functions, gravity separations, ideal continuous flow reactors, 623–624
Shape factor, particle shape, particle treatment process, 551–552
Shear-induced diffusion, crossflow filtration systems, contaminant transport, 798–800
Sherwood number, crossflow filtration systems, contaminant transport layer modeling, 804–816
Short-circuiting models, nonideal reactors, residence time distributions, 57–58
Short-range force model, flocculation collision efficiency, 581–589
　calculation equations, 597–598
　design implications, 589
Sieve coefficient, pressure-driven membrane systems, pore-flow model, concentrate/membrane interface, 761–764
Silt density index (SDI), nonsteady-state fouling, frontal and dead-end filtration systems, 791
Similitude, fixed bed adsorbers, performance evaluation, 357–359
Simulated distribution system (SDS) test, redox reactions, organic compound interactions, 453–455
Single bubble removal efficiency, gravity separations, flotation, 665
Single-site adsorption equilibrium:
　Langmuir isotherm, 271–273
　rapid adsorption, batch reactors vs., 330–331
Single sphere model, granular-media filtration:
　isolated collector model, 683–686
　　particle-collector collisions, 686–691
　　　Brownian motion, 688
　　　efficiency, 689–690
　　　interception, 686–687
　　　sedimentation, 687–688
　packed medium model, 690–694
Site-binding models, adsorption equilibrium:
　competitive adsorption, Freundlich isotherms, 294–296
　electrically induced partitioning, 268
　equilibrium constants, 268–269
　Freundlich isotherms, 294–296
　ion exchange equilibrium, 284–294
　isotherms, 270–281
　　Freundlich isotherm, 276–281
　　multisite Langmuir isotherm, 274–276
　　single-site Langmuir isotherm, 271–273
　　surface site distribution functions, 270–271
　site-binding paradigm, 283–284
　surface complexation, 267
Slaking, precipitative carbonate softening, 409–410
Slow adsorption equilibrium systems, 340–354
　competitive adsorption, 364–365
　pore vs. surface diffusion, porous adsorbent particles, 341–343
　transport-limited adsorption rates:
　　batch reactors, 343–350
　　fixed-bed systems, 350–354
Sludge thickening
　dissolved air flotation, 668–669
　gravity separations, 645-655
Smoluchowski equation, flocculation modeling, 568–570
　design issues, 571–572
Solid/liquid equilibrium, precipitation and, 382–383
Solid phase activity, precipitation, 383–384
Solids:
　dissolution reactions, 422–426
　precipitation, 383–384
　speciation, 395–396
Solids flux, gravity separations, 647–654
Solute permeability coefficient, dense membrane systems modeling, 837–838

Solute transport:
  membrane systems, 743–747
    flux velocity, 743–744
    frontal filtration systems, rejected solutes, flux effects, 780–782
    permeability, 745–746
    recovery, 743
    resistance, 744–745
    specific flux and permeance, 744
    temperature effects, water permeation, 746–747
  precipitation and, 381–382
Solution composition modeling, precipitation, 394–397
Solution-diffusion model, pressure-driven membrane systems, 766–771
  flux predictions, 771–772
  membrane permeation and concentration profile, 767–771
  predicted flux/transmembrane parameter comparisons, 771–773
  transmembrane pressure profile, 766–767
  transport parameters, 760
Sonolysis, advanced oxidation process, 483
Sorbate, defined, 259
Sorbent, defined, 259
Sorption, defined, 259
Spatial flow patterns, particle destabilization chemicals, 544–546
Spatially uniform composition, batch reactors, aqueous phase concentration, 219–220
Spatial variations, solution and gas concentrations, gas transfer systems, 226–241
  design equation applications, 236
    column size sensitivity and parameter value uncertainty, 240
    influent stream, treatment objectives and, 236–240
    non-packed column systems, 240–241
    operating line mass balance, 226–229
    pressure loss and liquid holdup, 233–235
    tower design equations, 229–233
Specific adsorption, defined, 259
Specific equivalent ionic conductance, electrodialysis transport, 820–821
Specific flux, membrane systems, fluid/solute transport, 744
Specific power input, gas-in-liquid systems, hydrodynamic effects, 191–192
Specific resistance, membrane systems, compact layer modeling, 783–784
Specific surface area (SSA):
  adsorption equilibrium constants, 269
    surface pressure adsorption model, 298–302
  competitive adsorption, ideal adsorption model, 302–306
  defined, 259
Spiral-wound membrane system, 738–740
Split-stream softening, carbonate precipitation, 417
Stagnant-film gas transfer model, fluid packets, gas/solution interface, 171–174
Standard blocking, nonsteady-state fouling, frontal and dead-end filtration systems, 787–791
Standard cubic meters per hour (SCM/h), gas transfer, well-mixed liquid phase systems, 208–213

Standard deviation, residence time distributions, probability statistics, 41–42
Standard electron activity, redox thermodynamics, 438–441
Standard state conditions, Henry's constants and, 198–199
Steady state conditions:
  gas transfer, continuous liquid flow reactors, 220–225
    continuous flow stirred tank reactors, 220–225
    plug flow reactors, 220
  ideal continuous flow reactor performance:
    continuous flow stirred tank reactor, 121–123
      first-order irreversible reactions, 122–123
      non-first-order irreversible reactions, 123
    irreversible reactions, CSFSTRs vs. PFRs, 126–129
    multiple ideal reactors, 130–135
    parallel CFSTRS and PFRs, 135
    plug flow reactor, 123–126
      Eulerian view, 124–125
      irreversible nth-order reaction, 125–126
      Lagrangian view, 125
    rate expression derivation, 135
    reversible reactions, 129–130
    series continuous flow stirred tank reactors, 126–129
    series plug flow reactors, 126
    single ideal reactor, 121–130
  nonideal continuous flow reactor performance, 135–141
    fraction remaining measurements:
      dispersion model, 140–141
      exit age distribution, 136–140
      residence time distribution, 141
  sequential reaction kinetics, 112–114
  unidirectional flow systems, storage term, mass balance equations, 8–9
Step input:
  continuous flow reactors, tracer studies, residence time distributions, 32–33, 35–37
  continuous flow stirred tank reactors, 47–49
  non-steady state reactor performance, 141–142
Stern layer ion adsorption, adsorbate profile and electrical potential, 317–318
Stoichiometric reaction:
  advanced oxidation process, hydroxyl radicals, organic compounds, 474–476
  particle destabilization, charge neutralization, 537
  precipitation, 397–422
    carbonate solids precipitation, 409–419
    complexing ligands, metal solubility, 421
    hydroxide solids precipitation, 404–409
    ionic strength and activity coefficients, 400
    metal and hydroxy carbonates, 418–419
    pH-dependent metal and ligand speciation, 420–421
    precipitative softening, 409–417
    redox reactions, 421–422
    soft water recarbonation, 417–418
  precipitative softening, 410–413
Stokes–Einstein equation:
  flocculation collision, Brownian motion, 577–578
  molecular collision frequency, 84

Stokes flow, granular-media filtration, 693–694
Stokes' law, gravity separation, particle sedimentation, 607–610
Stokes velocity, gravity separations, batch thickening, 646–647
Storage term, mass balance equations:
　arbitrary flow systems, 23
　unidirectional flow systems, 7–10
Streaming current, particle surface charge, measurements, 531–532
Strongly-coupled transport, pressure-driven membrane systems, conceptual model, 759–760
Substrate-limited reactions, sequential reaction kinetics, 113–114
Sulfides, precipitation, pH-dependent metal-ligand speciation, 420–421
Sulfites, redox reactions, ozone interaction, 464
Sulfur compounds, redox reactions:
　chlorine, 447–448
　reductive processes, 486–487
Superficial velocity, rapid adsorption equilibrium, fixed bed reactors, 336
Supersaturation:
　precipitation, stoichiometric and equilibrium models, 402–422
　precipitation and, 382
Surface aerators, gas transfer systems, 159–162
Surface area distribution, particle size distribution, particle treatment process, 548–551
Surface charge interactions:
　adsorption models, 314–320
　　adsorbate profile, interfacial region electrical potential, 315–318
　　electrical potential effects, 314–315
　　electrostatic contribution, equilibrium constants, 318–319
　particle treatment systems:
　　adsorption, 523–524
　　basic principles, 520–521
　　charge and potential measurements, 531–532
　　chemical reactions, 522–523
　　diffuse layer structures, 524–525
　　isomorphic substitution, 522
　　particle stability, 521–524
　solid dissolution reactions, 423–426
Surface complexation:
　adsorption equilibrium, 267
　ion adsorption, adsorbate profile and electrical potential, 316–318
Surface diffusion:
　adsorption rate, 328
　precipitation and, 381–382
　slow adsorption equilibrium systems, adsorbent particles, 341–343
Surface ligands, adsorption equilibrium, 267
Surface precipitation, ionic adsorbates, 319–320
Surface pressure adsorption models, 297–306
　basic principles, 296–297
　competitive adsorption and, 302–306
　surface tension/isotherm data, computation using, 297–302
Surface renewal model, residence time distribution, 171–174

Surface site distribution functions, adsorption equilibrium isotherms, 270–271
Surface tension, surface pressure adsorption models, 297–306
　pressure computation, 297–302
Surface wash systems, granular-media filtration, 680
　filter cleaning, 716
Surface Water Treatment Rule, series continuous flow stirred tank reactors, steady state performance evaluation, 134
Sweep flocculation, particle destabilization, 538–541
Symmetric membranes, structure and composition, 734–736
Synthetic organic compounds (SOCs):
　competitive adsorption, site-binding models, Freundlich isotherm, 295–296
　oxidation, 436–437
　particle destabilization chemicals, 533–535

Tangential flow, defined, 733
Taste destruction, oxidation, 436
Temperature dependence:
　gas/liquid equilibrium, 167–168
　gas transfer, well-mixed liquid phase systems, 208–213
　gas transfer rate constants, 195
　membrane systems, water permeation, 746–747
　nonelementary reaction kinetics, 114–115
Thermodynamics:
　dense membrane systems modeling, 834–838
　membrane systems, driving force and energy, 752–754
　precipitation, 383–384
　　particle growth and formation, 385–389
　redox reactions, 437–441
　sequential reaction kinetics, 111–112
Thickening, gravity separations, 644–655
　batch thickening, 645–647
　continuous flow thickening, 650–655
　　design constraints, 655
　flotation thickening, 668–669
　solids flux, 647–650
Three-dimensional mass balance, differential form, 24–25
Tiered flow effects, nonideal flow reactors, gravity separations, 639–640
Titanium dioxide, advanced oxidation process, ultraviolet radiation, 481–482
Total chlorine, redox reactions, formation, 455
Total collision frequency, flocculation collision, 578–580
Total dissolved and solid-phase concentrations (TOTM(aq)/TOTM(s)), precipitation:
　hydroxide solids, 404–409
　speciation, 396
　stoichiometric and equilibrium models, 398–422
Total organic carbon (TOC), redox reactions, chlorine-organic compound interactions, 449–455
Total organic chlorine (TOCl), redox reactions, organic compound interactions, 452–455
Total organic halogen (TOX), redox reactions, organic compound interactions, 452–455
Total organic halogen formation potential (TOXFP), redox reactions, 453–455

Tracer studies:
  continuous flow reactors:
    hydraulics analysis, 29–30
    residence time distributions, 31–33
      pulse input response, 33–35
  continuous flow stirred tank reactors, 49–50
  nonideal reactors, 48–50
  plug flow reactors, residence time distribution, 51–55
  reactor characterization, 40–42
Trajectory analysis, flocculation collision efficiency, short-range force model, 581–589
Transient response:
  non-steady state reactor performance, 141–146
  reactor performance evaluation, CFSTR/PFR comparisons, 128–129
Transition state theory, molecular collisions, 85–87
Transmembrane energy difference, driving force magnitude, 757–759
Transmembrane gradient, membrane systems, driving force and energy, 752–754
Transmembrane pressure (TMP):
  crossflow filtration systems:
    contaminant transport modeling, 809–816
    fluid flow and, 793–795
  frontal filtration systems, permeation and, 777–782
  overview of membrane systems, 731–732
  pressure-driven membrane systems:
    compact layer modeling, limiting flux and film-gel model, 783
    flux predictions, 771–772
    performance evaluation, 750–752
    pore-flow model, permeation and contaminant rejection, 765–766
    recovery and, 736
    solution-diffusion model, 766
Transphilic fractions, redox reactions, chlorine-organic compound interactions, 449–455
Transport kinetics:
  electrodialysis, electrical potential gradient, 819–822
    advective transport, 819
    current densities, 820–828
    diffusive transport, 819–820
    solution in electric field, 820
  gas transfer processes, 157
  membrane systems, driving force magnitude, 757–759
  pressure-driven membrane systems:
    conceptual models, 759–760
    summary of, 772–773
  solid dissolution reactions, 422–426
  transport-limited adsorption rates:
    batch reactors, 343–350
    fixed bed reactors, 350–354
Transport number, electrodialysis, single cell-pair, 822–825
Trichloroethylene (TCE), gas transfer column contactors, design criteria and, 236–240
Trihalomethane (THM), redox reactions, 448
  chloramine reaction, 458
  chlorine-organic compound interactions, 449–455
Trihalomethane formation potential (THMFP), redox reactions, 453–455
Triple-layer model (TLM), ion adsorption, adsorbate profile and electrical potential, 315–318
Tube settlers, gravity separations, 634–639
Tubular membrane systems, 739
Turbulence, flocculation mechanisms, 589
Turbulent flocculation, basic principles, 589–591
Two-dimensional electrodialysis system, 828–832
  design issues, 830–831
  limiting current, 831–832
  macroscopic mass balance, 828–829
  microscopic mass balance, 829–830
Two-film gas transfer model, fluid packets, gas/solution interface, 171–174

Ultrafiltration (UF) membranes:
  characteristics, 737
  classification, 736
  driving force and energy in, 752–754
  molecular weight cutoff, 749
  packing density, 740–742
  pore-flow model, 760
  pressure-driven systems, overview, 749–752
  research and applications, 731–732
  transmembrane energy difference, 757–759
Ultraviolet radiation:
  advanced oxidation process:
    hydrogen peroxide addition, 476–480
    ozonation, hydroxyl radicals, 476–480
    ozonide radical/hydrogen peroxide, 480–481
  disinfectant performance, 500–502
    efficacy evaluation, 501–502
    generation and implementation, 500–501
Uncoupled transport, pressure-driven membrane systems:
  modeling of, 760
  solution-diffusion model, 766–771
Unidirectional flow system, mass balance, concentration gradient and, 7–20
  advection term, 10–11
  chemical reaction term, 15–16
  differential one-dimensional mass balance, 18–20
  diffusion/dispersion terms, 11–15
  overall mass balance terms, 16–18
  storage term, 8–9
Unit filter run, granular-media filtration, 682
Upflow reactor, gravity separations, 624–626
Uranium/transuranic elements, solid dissolution, 425–426

Van der Waals attraction:
  crossflow filtration systems, deterministic contaminant transport, 800–801
  flocculation collision efficiency, short-range force model, 581–589
  granular-media filtration, packed medium model, 693
  particle stability, 526–530
    flat plate-charged particle interaction, 530–531

Van't Hoff equation:
    dense membrane systems modeling, 836–838
    membrane systems, osmotic pressure, 755–757
Vapor pressure, gas/liquid equilibrium, 168–169
Variance, residence time distributions, mean hydraulic retention time, 41–42
Velocity parameters:
    granular-media filtration design, 707–708
        air scour, 716–717
    membrane systems, fluid/solute transport, 743–744
Velocity vector, arbitrary flow systems, advection, 20–21
Viscosity measurements, particle destabilization chemicals, water stream mixing, 544–546
Volatile organic compounds (VOCs):
    gas transfer:
        acid/base reactions, 199–202
        activity coefficient, 199
        batch reactors, 214–216
        diffusivities, 177–179
        gas/liquid equilibrium, 155–157
        macroscopic (advective) limitation, 217–218
        mass balance, gas/solution interface, 171–179
    Henry's constants, 165–167
    removal systems, mass balance equations, 6–7
Volatilization, gas/liquid equilibrium, Henry's law and, 162
Volume distribution, particle size distribution, particle treatment process, 548–551
Volume-normalized mass balances, defined, 6
Volumetric adsorption potential, Polanyi adsorption model, 307–314
Volumetric flux ratio:
    membrane systems:
        fluid/solute transport, 743–744
        permeability, 745–746
    pressure-driven membrane systems, pore-flow model, concentrate/membrane interface, 762–764
        permeation and contaminant rejection, 765

Water distribution systems:
    chloramine disinfectant performance, 496–497
    redox reactions, chloramine interaction, 458–459
Water softening:
    carbonate solid precipitation, 409–410
        recarbonation, 418
        stoichiometric model, 410–413
    mass balance equations, 5–7
Weber number, liquid-in-gas systems, mass transfer coefficients, 193–195
Well-mixed liquid phase systems:
    adsorption process, activated carbon preparation, 262–264
    gas transfer, 207–226
        batch liquid reactors, 213–216
        flow design constraints and choices, 226
        limiting cases, general kinetic expression, 216–220
            macroscopic (advective) limitation, 217–218
            microscopic (interfacial) limitation, 218–219
            overall gas transfer rate, 219–220
        overall gas transfer rate expression, 207–213
        series continuous liquid flow stirred tank reactors, 225
        steady state continuous liquid flow reactors, 220–225
            continuous flow stirred tank reactors, 220–225
            plug flow reactors, 220
    particle destabilization chemicals, 533
    rapid adsorption equilibrium, 330–331
    slow adsorption equilibrium systems, transport-limited adsorption rates, batch reactors, 346–350
Wet air oxidation, advanced oxidation process, 482–483

Yao model, granular-media filtration, 693–694

Zero-valent iron (ZVI), redox reactions, reductive processes, 487–488
Zeta potential, particle surface charge, measurements, 531–532
Zinc orthophosphate inhibitors:
    lead precipitation, 419
    precipitation, complexing ligands, 421
Zone settling, gravity separations, 644–655